COXETER BIALGEBRAS

The goal of this monograph is to develop Hopf theory in the setting of a real reflection arrangement. The central notion is that of a Coxeter bialgebra which generalizes the classical notion of a connected graded Hopf algebra. The authors also introduce the more structured notion of a Coxeter bimonoid and connect the two notions via a family of functors called Fock functors. These generalize similar functors connecting Hopf monoids in the category of Joyal species and connected graded Hopf algebras.

The building blocks of the theory are geometric objects associated to a reflection arrangement such as faces, flats, lunes, and their orbits under the action of the Coxeter group. A generalized notion of zeta and Möbius function play a fundamental role in all aspects of the theory, including exp-log correspondences and results such as the Poincaré–Birkhoff–Witt theorem. The Tits algebra and its invariant subalgebra also play key roles.

This monograph opens a new chapter in Coxeter theory as well as in Hopf theory, connecting the two. It also relates fruitfully to many other areas of mathematics such as discrete geometry, semigroup theory, associative algebras, algebraic Lie theory, operads, and category theory. It is carefully written, with effective use of tables, diagrams, pictures, and summaries. It will be of interest to students and researchers alike.

Encyclopedia of Mathematics and Its Applications

This series is devoted to significant topics or themes that have wide application in mathematics or mathematical science and for which a detailed development of the abstract theory is less important than a thorough and concrete exploration of the implications and applications.

Books in the **Encyclopedia of Mathematics and Its Applications** cover their subjects comprehensively. Less important results may be summarized as exercises at the ends of chapters. For technicalities, readers can be referred to the bibliography, which is expected to be comprehensive. As a result, volumes are encyclopedic references or manageable guides to major subjects.

ENCYCLOPEDIA OF MATHEMATICS AND ITS APPLICATIONS

All the titles listed below can be obtained from good booksellers or from Cambridge University Press. For a complete series listing visit
www.cambridge.org/mathematics.

136 A. Kristály, V. D. Rădulescu and C. Varga *Variational Principles in Mathematical Physics, Geometry, and Economics*
137 J. Berstel and C. Reutenauer *Noncommutative Rational Series with Applications*
138 B. Courcelle and J. Engelfriet *Graph Structure and Monadic Second-Order Logic*
139 M. Fiedler *Matrices and Graphs in Geometry*
140 N. Vakil *Real Analysis through Modern Infinitesimals*
141 R. B. Paris *Hadamard Expansions and Hyperasymptotic Evaluation*
142 Y. Crama and P. L. Hammer *Boolean Functions*
143 A. Arapostathis, V. S. Borkar and M. K. Ghosh *Ergodic Control of Diffusion Processes*
144 N. Caspard, B. Leclerc and B. Monjardet *Finite Ordered Sets*
145 D. Z. Arov and H. Dym *Bitangential Direct and Inverse Problems for Systems of Integral and Differential Equations*
146 G. Dassios *Ellipsoidal Harmonics*
147 L. W. Beineke and R. J. Wilson (eds.) with O. R. Oellermann *Topics in Structural Graph Theory*
148 L. Berlyand, A. G. Kolpakov and A. Novikov *Introduction to the Network Approximation Method for Materials Modeling*
149 M. Baake and U. Grimm *Aperiodic Order I: A Mathematical Invitation*
150 J. Borwein *et al.* *Lattice Sums Then and Now*
151 R. Schneider *Convex Bodies: The Brunn–Minkowski Theory (Second Edition)*
152 G. Da Prato and J. Zabczyk *Stochastic Equations in Infinite Dimensions (Second Edition)*
153 D. Hofmann, G. J. Seal and W. Tholen (eds.) *Monoidal Topology*
154 M. Cabrera García and Á. Rodríguez Palacios *Non-Associative Normed Algebras I: The Vidav–Palmer and Gelfand–Naimark Theorems*
155 C. F. Dunkl and Y. Xu *Orthogonal Polynomials of Several Variables (Second Edition)*
156 L. W. Beineke and R. J. Wilson (eds.) with B. Toft *Topics in Chromatic Graph Theory*
157 T. Mora *Solving Polynomial Equation Systems III: Algebraic Solving*
158 T. Mora *Solving Polynomial Equation Systems IV: Buchberger Theory and Beyond*
159 V. Berthé and M. Rigo (eds.) *Combinatorics, Words and Symbolic Dynamics*
160 B. Rubin *Introduction to Radon Transforms: With Elements of Fractional Calculus and Harmonic Analysis*
161 M. Ghergu and S. D. Taliaferro *Isolated Singularities in Partial Differential Inequalities*
162 G. Molica Bisci, V. D. Radulescu and R. Servadei *Variational Methods for Nonlocal Fractional Problems*
163 S. Wagon *The Banach–Tarski Paradox (Second Edition)*
164 K. Broughan *Equivalents of the Riemann Hypothesis I: Arithmetic Equivalents*
165 K. Broughan *Equivalents of the Riemann Hypothesis II: Analytic Equivalents*
166 M. Baake and U. Grimm (eds.) *Aperiodic Order II: Crystallography and Almost Periodicity*
167 M. Cabrera García and Á. Rodríguez Palacios *Non-Associative Normed Algebras II: Representation Theory and the Zel'manov Approach*
168 A. Yu. Khrennikov, S. V. Kozyrev and W. A. Zúñiga-Galindo *Ultrametric Pseudodifferential Equations and Applications*
169 S. R. Finch *Mathematical Constants II*
170 J. Krajíček *Proof Complexity*
171 D. Bulacu, S. Caenepeel, F. Panaite and F. Van Oystaeyen *Quasi-Hopf Algebras*
172 P. McMullen *Geometric Regular Polytopes*
173 M. Aguiar and S. Mahajan *Bimonoids for Hyperplane Arrangements*
174 M. Barski and J. Zabczyk *Mathematics of the Bond Market: A Lévy Processes Approach*
175 T. R. Bielecki, J. Jakubowski and M. Niewęgłowski *Structured Dependence between Stochastic Processes*
176 A. A. Borovkov, V. V. Ulyanov and Mikhail Zhitlukhin *Asymptotic Analysis of Random Walks: Light-Tailed Distributions*
177 Y.-K. Chan *Foundations of Constructive Probability Theory*
178 L. W. Beineke, M. C. Golumbic and R. J. Wilson (eds.) *Topics in Algorithmic Graph Theory*
179 H.-L. Gau and P. Y. Wu *Numerical Ranges of Hilbert Space Operators*
180 P. A. Martin *Time-Domain Scattering*
181 M. D. de la Iglesia *Orthogonal Polynomials in the Spectral Analysis of Markov Processes*
182 A. E. Brouwer and H. Van Maldeghem *Strongly Regular Graphs*
183 D. Z. Arov and O. J. Staffans *Linear State/Signal Systems*
184 A. A. Borovkov *Compound Renewal Processes*
185 D. Bridges, H. Ishihara, M. Rathjen and H. Schwichtenberg (eds.) *Handbook of Constructive Mathematics*

ENCYCLOPEDIA OF MATHEMATICS AND ITS APPLICATIONS

Coxeter Bialgebras

MARCELO AGUIAR
Cornell University, Ithaca

SWAPNEEL MAHAJAN
Indian Institute of Technology, Mumbai

CAMBRIDGE
UNIVERSITY PRESS

CAMBRIDGE
UNIVERSITY PRESS

University Printing House, Cambridge CB2 8BS, United Kingdom

One Liberty Plaza, 20th Floor, New York, NY 10006, USA

477 Williamstown Road, Port Melbourne, VIC 3207, Australia

314–321, 3rd Floor, Plot 3, Splendor Forum, Jasola District Centre,
New Delhi – 110025, India

103 Penang Road, #05–06/07, Visioncrest Commercial, Singapore 238467

Cambridge University Press is part of the University of Cambridge.

It furthers the University's mission by disseminating knowledge in the pursuit of education, learning, and research at the highest international levels of excellence.

www.cambridge.org
Information on this title: www.cambridge.org/9781009243773
DOI: 10.1017/9781009243759

First published 2023

Printed in the United Kingdom by TJ Books Limited, Padstow Cornwall

A catalogue record for this publication is available from the British Library.

ISBN 978-1-009-24377-3 Hardback

Contents

Preface xi

Introduction 1

Preliminaries 17

Chapter 1. Coxeter groups and reflection arrangements 19
- 1.1. Face-types, flat-types, biface-types 20
- 1.2. Lune-types and bilune-types 29
- 1.3. Stabilizer subgroups of a Coxeter group 34
- 1.4. Gallery distance, gate property, length 35
- 1.5. Modules over the Coxeter group 37
- 1.6. Descents, global descents, shuffles 38
- 1.7. Groupoid of biface-types 40
- 1.8. Face-type enumeration 48
- 1.9. Enumeration in the Coxeter group algebra 60
- 1.10. Invariant flat-, lune-, bilune-incidence algebras 63
- 1.11. Noncommutative Zaslavsky formulas and q-analogues 75
- 1.12. Invariant Birkhoff, Tits, Janus algebras 77
- 1.13. Invariant Lie and Zie elements 84
- 1.14. Coxeter–Birkhoff, Coxeter–Tits, Coxeter–Janus algebras 85
- 1.15. Coordinate arrangement 87
- 1.16. Braid arrangement 90
- 1.17. Notes 95

Part I. Coxeter species 99

Chapter 2. Coxeter species and Coxeter bimonoids 101
- 2.1. Support and type morphisms 104
- 2.2. Induced and local morphisms 109
- 2.3. Coxeter species 110
- 2.4. Coxeter monoids and Coxeter comonoids 116
- 2.5. Coxeter bimonoids 122
- 2.6. (Co)commutative Coxeter (co)monoids 127
- 2.7. Species vs Coxeter species 135
- 2.8. Signature functors on Coxeter species 137
- 2.9. Coxeter (co, bi)monoids as functor categories 139

2.10. Bimonads on Coxeter species 147
2.11. Opposite transformation 153
2.12. Joyal species and Coxeter species 154
2.13. Notes 156

Chapter 3. Basic theory of Coxeter bimonoids 159
3.1. Cauchy powers of a Coxeter species 159
3.2. Primitive and decomposable filtrations 160
3.3. Universal constructions of Coxeter bimonoids 164
3.4. (Co)abelianizations of Coxeter (co)monoids 174
3.5. Hadamard product. Convolution Coxeter monoid 176
3.6. Internal hom for Coxeter comonoids 180
3.7. Universal measuring Coxeter comonoid 183
3.8. Exp-log between (co)derivations and (co)monoid morphisms 185
3.9. Logarithm of the identity map 190
3.10. Exp-log between primitive and group-like series 193
3.11. Norm transformation on Coxeter bimonoids 200
3.12. Characteristic operations by faces 203
3.13. Algebra modules and Coxeter bimonoids 205
3.14. Antipodes of Coxeter bimonoids 210
3.15. Notes 213

Chapter 4. Examples of Coxeter bimonoids 216
4.1. Coxeter species characteristic of chambers 216
4.2. Exponential Coxeter bimonoid 218
4.3. Coxeter bimonoid of chambers 222
4.4. Coxeter bimonoid of flats 225
4.5. Coxeter bimonoid of faces 228
4.6. Coxeter bimonoid of top-nested faces 233
4.7. Coxeter bimonoid of pairs of chambers 235
4.8. Coxeter Lie species and Coxeter Zie species 237
4.9. Decoration functor on Coxeter species 238
4.10. Notes 244

Chapter 5. Coxeter operads 246
5.1. Coxeter dispecies 247
5.2. Coxeter operads 248
5.3. Hadamard product 252
5.4. Coxeter operad monoids 252
5.5. Coxeter commutative, associative, Lie operads 255
5.6. Notes 260

Chapter 6. Coxeter Lie monoids 261
6.1. Coxeter Lie monoids and Coxeter Lie comonoids 262
6.2. (Co)commutator (co)bracket 264
6.3. Primitive and indecomposable parts of Coxeter bimonoids 265
6.4. (Co)free Coxeter Lie (co)monoids 266

6.5. Universal (co)enveloping Coxeter (co)monoids 269
6.6. Joyal Lie monoids and Coxeter Lie monoids 271
6.7. Notes 272

Chapter 7. Structure theory of Coxeter bimonoids 273
7.1. Borel–Hopf for (co)commutative Coxeter bimonoids 274
7.2. Leray–Samelson for bicommutative Coxeter bimonoids 275
7.3. Rigidity of Coxeter q-bimonoids for q not a root of unity 277
7.4. Loday–Ronco for Coxeter 0-bimonoids 279
7.5. Hoffman–Newman–Radford 280
7.6. Poincaré–Birkhoff–Witt 284
7.7. Cartier–Milnor–Moore 287
7.8. Coxeter bimonoids for a rank-one arrangement 288
7.9. Notes 297

Part II. Coxeter spaces 299

Chapter 8. Coxeter spaces and Coxeter bialgebras 301
8.1. Support and type morphisms 304
8.2. Induced and local morphisms 311
8.3. Opposite and adjacent support-type morphisms 314
8.4. Coxeter spaces 317
8.5. Coxeter algebras and Coxeter coalgebras 324
8.6. Coxeter bialgebras 331
8.7. Coxeter q-bialgebras 341
8.8. (Co)commutative Coxeter (co)algebras 347
8.9. Coxeter bialgebras and face-type enumeration 355
8.10. Signature functor on Coxeter spaces 356
8.11. Coxeter (co, bi)algebras as functor categories 358
8.12. Bimonads on Coxeter spaces 365
8.13. Opposite transformation 372
8.14. Graded vector spaces and Coxeter spaces 373
8.15. Notes 379

Chapter 9. Basic theory of Coxeter bialgebras 382
9.1. Cauchy powers of a Coxeter space 383
9.2. Primitive and decomposable filtrations 385
9.3. Universal constructions of Coxeter bialgebras 388
9.4. (Co)abelianizations of Coxeter (co)algebras 399
9.5. Hadamard product. Convolution Coxeter algebra 403
9.6. Internal hom for Coxeter coalgebras 406
9.7. Universal measuring Coxeter coalgebra 409
9.8. Exp-log between (co)derivations and (co)algebra morphisms 412
9.9. Logarithm of the identity map 418
9.10. Exp-log between primitive and group-like series 421
9.11. Norm transformation on Coxeter bialgebras 430
9.12. Bialgebra axiom on the primitive part 437

9.13. Characteristic operations by face-types 439
9.14. Antipodes of Coxeter bialgebras 445
9.15. Notes 451

Chapter 10. Examples of Coxeter bialgebras 456
10.1. Coxeter space characteristic of chambers 457
10.2. Coxeter bialgebra of polynomials 459
10.3. Tensor, shuffle, symmetric, exterior Coxeter bialgebras 471
10.4. Coxeter bialgebras of face-types 483
10.5. Coxeter bialgebra of flat-types 505
10.6. Coxeter bialgebra of symmetries 510
10.7. Notes 515

Chapter 11. Coxeter operad algebras 522
11.1. Nested flats, nested top-lunes, nested face-types 523
11.2. Coxeter dispecies 529
11.3. Coxeter operads 531
11.4. Coxeter spaces as a module over Coxeter dispecies 537
11.5. Coxeter operad algebras 542
11.6. Coxeter commutative, associative, Lie operads revisited 547
11.7. Signature functor on Coxeter dispecies 554
11.8. May operads and Coxeter operads 557
11.9. Notes 560

Chapter 12. Coxeter Lie algebras 561
12.1. Coxeter Lie algebras and Coxeter Lie coalgebras 562
12.2. (Co)commutator (co)bracket 567
12.3. Primitive and indecomposable parts of Coxeter bialgebras 568
12.4. (Co)free Coxeter Lie (co)algebras 571
12.5. Universal (co)enveloping Coxeter (co)algebras 576
12.6. Graded Lie algebras and Coxeter Lie algebras 582
12.7. Notes 585

Chapter 13. Structure theory of Coxeter bialgebras 586
13.1. Borel–Hopf for (co)commutative Coxeter bialgebras 588
13.2. Leray–Samelson for bicommutative Coxeter bialgebras 595
13.3. Rigidity of Coxeter q-bialgebras for q not a root of unity 600
13.4. Loday–Ronco for Coxeter 0-bialgebras 606
13.5. Hoffman–Newman–Radford 607
13.6. Poincaré–Birkhoff–Witt 614
13.7. Cartier–Milnor–Moore 620
13.8. Coxeter bialgebras for a rank-one arrangement 623
13.9. Notes 638

Part III. Fock functors 641

Chapter 14. Fock functors 643
14.1. Full Fock functors 644

14.2. Deformed full Fock functors 651
14.3. Full Fock functors and signature functors 653
14.4. Bosonic and fermionic Fock functors 656
14.5. Full Fock functors from bosonic Fock functors 661
14.6. Fock functors and commutativity 662
14.7. Norm transformation between Fock functors 671
14.8. Anyonic Fock functor 674
14.9. Fock functors and Coxeter operads 676
14.10. Decorated Fock functors 682
14.11. Notes 684

Chapter 15. Coxeter bimonoids and Coxeter bialgebras 687
15.1. Coxeter bimonoids to Coxeter bialgebras 689
15.2. Commutativity under Fock functors 694
15.3. Norm map between Fock spaces 695
15.4. (Co)freeness under Fock functors 697
15.5. Nonnegativity of the Boolean transform 701
15.6. Coxeter operad monoids to Coxeter operad algebras 702
15.7. Coxeter Lie monoids to Coxeter Lie algebras 705
15.8. Primitive and indecomposable parts under Fock functors 708
15.9. Universal enveloping constructions under Fock functors 710
15.10. Notes 712

Chapter 16. Adjoints of Fock functors 714
16.1. Trivialization functor 717
16.2. Hadamard product with chambers 721
16.3. Adjoints of bosonic Fock functors 722
16.4. Lifts of adjunctions involving bosonic Fock functors 725
16.5. Bosonic Fock functors and abelianization 739
16.6. Adjoints preserving commutativity 740
16.7. Adjoints of fermionic Fock functors 745
16.8. Adjoints of full Fock functors 745
16.9. Coxeter operad algebras to Coxeter operad monoids 750
16.10. Coxeter Lie algebras to Coxeter Lie monoids 753
16.11. (Co)freeness under Fock functor adjoints 756
16.12. Notes 757

Chapter 17. Structure theory under Fock functors 759
17.1. Series of Coxeter species and Coxeter spaces 759
17.2. Actions of the invariant bilune-incidence algebra 762
17.3. Internal hom for Coxeter species and Coxeter spaces 766
17.4. Norm maps on Coxeter species and Coxeter spaces 769
17.5. Transformation between bosonic Fock functor adjoints 772
17.6. Antipodes under Fock functors 776
17.7. Structure theorems under Fock functors 778
17.8. Lie theory under Fock functors 782
17.9. Notes 785

Chapter 18. Examples of Fock spaces 786
18.1. Coxeter space characteristic of chambers as a Fock space 787
18.2. Coxeter bialgebra of polynomials as a Fock space 788
18.3. Tensor Coxeter bialgebra as a Fock space 794
18.4. Coxeter bialgebra of face-types as a bosonic Fock space 798
18.5. Coxeter bialgebra of flat-types as a bosonic Fock space 804
18.6. Coxeter bialgebra of symmetries as a bosonic Fock space 806
18.7. Fock functors for a rank-one arrangement 807
18.8. Notes 815

Appendix A. Category theory 816
A.1. Bimonads and bilax functors 816
A.2. Semidirect products 821
A.3. Notes 824

References 826

List of Notations 842

List of Tables 852

List of Figures 855

List of Summaries 857

Author Index 858

Subject Index 863

Preface

Goal

Species and bimonoids for linear hyperplane arrangements were introduced and studied in our previous monograph [**11**]. It marks the beginning of Hopf and Lie theory for arrangements. Faces of the arrangement play a critical role in this theory. The Tits product on faces, for instance, enters into the formulation of the bimonoid axiom.

The present work aims to introduce and study Coxeter species and Coxeter spaces. These are defined in the special setting of linear reflection arrangements. We formulate Coxeter species using faces; they carry more structure than species, the additional structure comes from the action of the Coxeter group of the reflection arrangement on faces. We formulate Coxeter spaces using face-types; they carry less structure than Coxeter species, making their study considerably harder. Similarly, we define and study Coxeter bimonoids and Coxeter bialgebras.

The classical picture is of Joyal species and graded vector spaces, and of Joyal bimonoids and graded bialgebras. This picture connects to the one above via braid arrangements (which are a family of linear reflection arrangements). A summary is provided in Table I below.

TABLE I. Coxeter bimonoids and Joyal bimonoids.

Starting data	Objects of interest	
Hyperplane arrangement	Species Bimonoids	
Reflection arrangement	Coxeter species Coxeter bimonoids	Coxeter spaces Coxeter bialgebras
Braid arrangement (classical)	Joyal species Joyal bimonoids	Graded vector spaces Graded bialgebras

The relevance of braid arrangements to Hopf theory was brought out and promoted in all our earlier works [**6**], [**7**], [**8**]. Along with combinatorial notions such as shuffles and integer compositions, we emphasized related geometric notions such as the Tits product, gate property, distance functions, lunes, and so on, thus making connections to Coxeter theory more transparent, and paving the way for [**11**] and the present work.

The theory of operads for linear hyperplane arrangements was also initiated in our earlier monographs [**10**, Section 15.9], [**11**, Chapter 4]. In the present work, we introduce the notion of a Coxeter operad for a linear reflection arrangement. Coxeter operads relate to classical operads (which we call May operads) via braid arrangements. See the summary in Table II below.

TABLE II. Coxeter operads and May operads.

Starting data	Objects of interest
Hyperplane arrangement	Operads
Reflection arrangement	Coxeter operads
Braid arrangement (classical)	May operads

Historical note

The notion of a Coxeter bialgebra matured out of [**6**], where Coxeter theory and Hopf theory were linked together for the first time. In a similar vein, the notion of a Coxeter bimonoid matured out of [**7**], where the theory of Joyal bimonoids was laid out in detail. We formulated Coxeter bimonoids, Coxeter bialgebras and Fock functors linking them in the year 2011, and their basic theory became clear to us by the time [**8**] was published in 2013. It necessarily involved Coxeter Lie monoids and Coxeter Lie algebras which can be viewed as objects of algebraic Lie theory. There was also the intertwining story of Coxeter operads.

Writing all this up in a coherent manner presented us with many expository problems. We first needed to write [**10**] which could then serve as a general reference for hyperplane arrangements. This book was published in 2017 after a lot of administrative delays. Next, we felt that it would be easier to explain the theory of species, operads, Lie monoids and bimonoids for hyperplane arrangements which was less structured but worked in a more general setting and had direct connections to the abstract theory of semigroups. This led to the monograph [**11**] published in 2020. While working on this monograph, we also carefully looked at the early history of Hopf algebras; interested readers can see the end-of-chapter notes in [**11**]. Our nomenclature for structure theorems in Hopf theory is based on this study.

The above developments were also instrumental in our discovery of the category of lunes, of noncommutative zeta and Möbius functions, of the q-Janus algebra, and of the groupoid of biface-types.

Synopsis

We now briefly summarize the contents of this monograph.

Objects. For a linear reflection arrangement, we introduce the category of Coxeter species and notions of Coxeter monoid, Coxeter comonoid, Coxeter bimonoid, Coxeter Lie monoid therein. For a scalar q, we introduce Coxeter q-bimonoids, with $q = 1$ recovering Coxeter bimonoids. For $q = -1$, we use the term signed Coxeter bimonoid. These generalize corresponding notions in Joyal species [**7**, Chapters 8 and 9] from braid arrangements to any linear reflection arrangement. Similarly, we introduce the category of Coxeter spaces and notions of Coxeter algebra, Coxeter coalgebra, (signed) Coxeter bialgebra, Coxeter q-bialgebra, Coxeter Lie algebra therein. These generalize familiar notions in graded vector spaces, namely, graded algebra, graded coalgebra, (signed) graded bialgebra, graded q-bialgebra, graded Lie algebra.

For a linear reflection arrangement, we introduce the notion of a Coxeter operad generalizing the classical notion of a May operad pertaining to braid arrangements. We define the Coxeter commutative operad, Coxeter associative operad, Coxeter Lie operad. Modules over these Coxeter operads in the category of Coxeter species are commutative Coxeter monoids, Coxeter monoids, Coxeter Lie monoids, respectively, and in the category of Coxeter spaces are commutative Coxeter algebras, Coxeter algebras, Coxeter Lie algebras, respectively. We also consider their dual Coxeter cooperads whose comodules yield Coxeter comonoids, and so on. Coxeter (co)operads give rise to (co)monads on Coxeter species and on Coxeter spaces; thus, Coxeter monoids and Coxeter algebras can also be viewed as algebras over a monad, and so on.

Theory. We formulate Coxeter bimonoids in terms of faces and Coxeter bialgebras in terms of face-types; their theories proceed largely in parallel with each other. (There are alternative formulations of Coxeter bimonoids in terms of flats and lunes, and of Coxeter bialgebras in terms of top-nested faces, top-lunes.) We discuss primitive and decomposable filtrations, universal constructions, (co)abelianization constructions, Hadamard product, enrichment over Coxeter comonoids and Coxeter coalgebras, the antipode and Takeuchi formula, exp-log correspondences, q-norm map with connections to the q-logarithm, characteristic operations. We present structure results such as the Loday–Ronco theorem, its q-generalization for q not a root of unity, the Leray–Samelson theorem, the Borel–Hopf theorem, the Hoffman–Newman–Radford rigidity theorems, and the Poincaré–Birkhoff–Witt and Cartier–Milnor–Moore theorems. Noncommutative zeta and Möbius functions, their q-analogues and representation theory of the (invariant) Tits algebra and (invariant) q-Janus algebra play a prominent role.

Examples. Examples of Coxeter bimonoids include the exponential Coxeter bimonoid and Coxeter bimonoids of chambers, flats, faces, top-nested faces, top-lunes and pairs of chambers. These generalize the exponential Joyal bimonoid and Joyal bimonoids of linear orders, set partitions, set compositions, linear set compositions, linear set partitions, and pairs of linear orders [**7**, Chapter 12].

Examples of Coxeter bialgebras include the Coxeter bialgebra of polynomials, the tensor and shuffle Coxeter bialgebras and Coxeter bialgebras of flat-types, face-types, and symmetries. These generalize the Hopf algebra of polynomials, the tensor and shuffle Hopf algebras, and Hopf algebras of symmetric functions, noncommutative symmetric functions, quasisymmetric functions and permutations. Interestingly, several identities related to face-type enumeration follow from the study of these Coxeter bialgebras.

Interconnections. The category of Coxeter species is richer than that of Coxeter spaces (just as the category of Joyal species is in relation to that of graded vector spaces). Similarities and differences between the two categories can be conceptually understood via a family of functors called Fock functors which can be pictured as

$$\{\text{Coxeter species}\} \xrightarrow{\text{Fock functors}} \{\text{Coxeter spaces}\}.$$

Fock functors carry Coxeter bimonoids to Coxeter bialgebras. For instance, they carry the exponential Coxeter bimonoid to the Coxeter bialgebra of polynomials. Adjoints of Fock functors go in the other direction starting with

$$\{\text{Coxeter spaces}\} \xrightarrow{\text{Fock functor adjoints}} \{\text{Coxeter species}\}.$$

Such adjoints allow us to exhibit Coxeter bialgebras as a reflective and coreflective subcategory of Coxeter bimonoids. They provide a theoretical tool to establish results on Coxeter bialgebras using corresponding results on Coxeter bimonoids. We provide many illustrations of this nature.

Prerequisites

We assume basic familiarity with posets, monoids, groups, group representations and (co)invariants, associative algebras (such as monoid algebras, group algebras, incidence algebras), modules over algebras. Other than that, prerequisites for reading this book pertain to three main areas: category theory, Hopf and Lie theory, reflection arrangements and Coxeter groups. They are elaborated below.

Category theory. We assume familiarity with category theory at the level of [**25**], [**226**], [**307**], [**309**]. Concepts such as functors, natural transformations, groupoids, functor categories, equivalences between categories, presentation of categories, adjunctions and universal properties are used without explanation. Commutative diagrams form an essential component of our exposition. Bimonads and bilax functors, and semidirect products are briefly recalled in Appendix A. Some other concepts which appear in this book are $\Bbbk$-linear categories, enrichment over monoidal categories, double categories.

Hopf and Lie theory. This book presents Hopf and Lie theory first for Coxeter species and then for Coxeter spaces. The theory for Coxeter species (for reflection arrangements) proceeds in analogy with the theory for species (for hyperplane arrangements) developed in detail in [**11**]. To appreciate the role of the Tits monoid and related geometric ideas in everything that we do, some basic familiarity with [**11**] is highly desirable. Having said this, our discussion on Coxeter species, though compact, is fairly self-contained, with occasional references to [**11**].

Coxeter bimonoids generalize Joyal bimonoids [**7**], [**8**], while Coxeter bialgebras generalize connected graded Hopf algebras [**345**]. Since we develop everything from first principles, familiarity with either Joyal bimonoids or Hopf algebras is not strictly required, though some exposure to these notions is useful for motivational purposes.

Reflection arrangements and Coxeter groups. We assume familiarity with reflection arrangements at the level of [**10**, Chapters 1, 5, 6]. Some specialized topics that we require include noncommutative zeta and Möbius functions, representation theory of the (invariant) Tits algebra, and identities related to face-type enumeration [**10**, Chapters 9, 15, 16]. To keep the book self-contained, this material (along with some new results) is reviewed here in an introductory chapter. The introductory chapter of [**11**] on hyperplane arrangements is also a valuable reference.

Readership

This monograph would be of interest to students and researchers working in the areas of hyperplane arrangements, Coxeter groups, discrete geometry, semigroup theory, associative algebras, Joyal species, May operads, Hopf algebras, algebraic Lie theory, category theory.

Organization

The book is organized into three parts:

- Part I introduces Coxeter species and Coxeter bimonoids and develops their basic structure theory,
- Part II does the same for Coxeter spaces and Coxeter bialgebras,
- Part III develops the theory of Fock functors which connects Coxeter species and Coxeter spaces.

Parts I and II are independent of each other. Objects in Part I are more structured than objects in Part II making them easier to study. Nonetheless, in terms of presentation, we treat both of them equally. Part III connects the two. The organization of Part I is similar to that of [**11**], and does for Coxeter species what [**11**] did for species. The two expositions supplement each other and enable the reader to benefit from both. Parts I and III may also be compared to [**7**, Parts II and III], and do for Coxeter species and Coxeter spaces what [**7**] did for Joyal species and graded vector spaces.

The chart of interdependence of chapters is shown in Figure I. Chapter 1 is introductory in nature. The dotted horizontal lines bring out the parallel between Parts I and II. Part III is to be largely read in order. The thick horizontal arrow indicates that it depends on both Parts I and II. Chapters 4, 10, 18 are not shown in the chart; they consist of examples and can be read in parallel with the theory developed in other chapters.

Tables, pictures, summaries form an essential component of our exposition. Examples and exercises (of varying levels of difficulty and with generous hints) are interspersed throughout the book. A Notes section at the end of

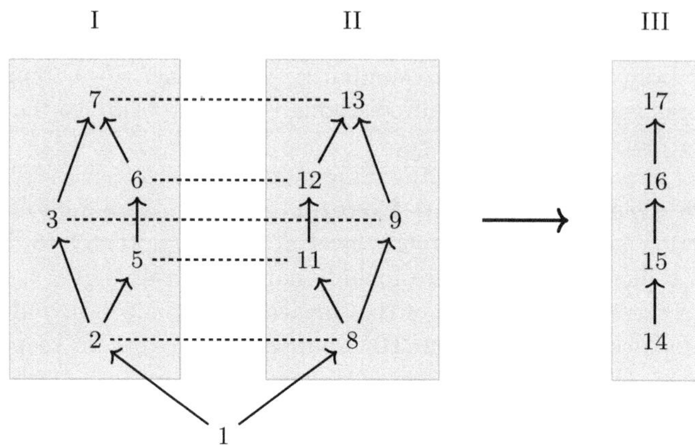

FIGURE I. Chart of interdependence of chapters.

each chapter connects to the classical literature with explanations and references. A list of notations, a list of tables, a list of figures, a list of summaries, an author index, a subject index are provided at the end of the book.

Teaching

The book is suitable for a two-semester sequence at the graduate level. It can also be used selectively for courses of shorter duration or theme-based seminars. Here are some guidelines; they can also be used for self-study.

For a systematic approach:

- The introductory Chapter 1, supplemented by [**10**], can be used to teach a course on Coxeter groups. It contains many new ideas and perspectives not to be found in standard books on Coxeter theory.
- Alternatively, one can focus entirely on Part I on Coxeter species and supplement it with material on species in [**11**]. Numerous subplans are possible here; see in particular those outlined on [**11**, Preface, page xvi]. We elaborate slightly. First develop basic concepts from Chapter 2 in parallel with some examples from Chapter 4. Then decide on which topics to cover from Chapter 3 and move ahead from there.
- More ambitiously, one can focus entirely on Part II on Coxeter spaces, and follow the same plan as above. Namely: First develop basic concepts from Chapter 8 in parallel with some examples from Chapter 10. Then decide on which topics to cover from Chapter 9 and move ahead.
- Alternatively, one can focus on Part III on Fock functors. The way to go would be to first develop basic familiarity with the main objects in the first two parts, namely, Coxeter bimonoids and Coxeter bialgebras, and then study how they interact with each other via Fock functors.
- Chapters 5 and 11, supplemented by [**11**, Chapter 4] can be used to teach a course on Coxeter operads.

For an exploratory approach:

- One can focus on a specific aspect of Hopf theory such as the Hadamard product or exp-log correspondences for both Coxeter species and Coxeter spaces and see how they compare.
- A course can be built around a rich example such as the Coxeter bialgebra of face-types, using it to grasp different aspects of the theory.
- One can also focus on a specific topic from Chapter 1 and explore how it connects to Coxeter bialgebras. Possible topics are shuffles, modules over the Coxeter group, face-type enumeration, zeta and Möbius functions, invariant Tits and invariant q-Janus algebras.
- One can approach everything through the lens of category theory. For instance, one can focus on bimonads and bilax functors, and see how the main objects in the book fit into this framework. A possible subtopic here could be adjunctions and universal constructions. Similarly, one can study the different finite categories (groupoids) built out of geometric objects such as faces, flats, face-types, and so on.
- One can also look at "visual" aspects of Coxeter bialgebras. We explain many concepts, calculations, arguments with pictures (in the groupoid of biface-types) involving vertices, edges, triangles, and arrows. The reader can come up with alternative pictures by changing parameter values, and also visualize one dimension higher with tetrahedra.

The book can also be used as an additional reference while teaching standard courses on category theory, Hopf algebras, Lie algebras, Joyal species, and May operads.

New developments

The extension of Hopf theory to hyperplane arrangements done in [**11**] brought forth many new ideas; see [**11**, Preface, pages xvii–xix]. The special setting of reflection arrangements pursued in the present book substantiates these ideas as well as adds several new features. They are listed below.

Reflection arrangements.

- emergence of the groupoid of biface-types with face-types as objects and biface-types as morphisms,
- role of the invariant Tits and invariant q-Janus algebras and connection of their structure constants to face-type enumeration,
- emergence of identities in the group algebra of the Coxeter group related to shuffles.

Coxeter species and Coxeter spaces.

- distinction between species for a hyperplane arrangement, and Coxeter species for a reflection arrangement,
- emphasis on "higher operations" involving faces and face-types as opposed to "binary operations" involving vertices and vertex-types,
- set-theoretic nature of the bimonoid axiom for Coxeter bimonoids as opposed to linear nature of the bialgebra axiom for Coxeter bialgebras;

moreover, the former involves only support morphisms, while the latter mixes support and type morphisms,

- connection between the isomorphism problem for the general q-norm map on Coxeter q-bimonoids (or on Coxeter q-bialgebras) and the q-logarithm for q not a root of unity,
- connection between representation theory of algebras and Hopf theory that relates Coxeter–Birkhoff algebra modules to bicommutative Coxeter bimonoids, Coxeter–Tits algebra modules to (co)commutative Coxeter bimonoids, Coxeter–Janus algebra modules to Coxeter bimonoids,

- construction of the decoration functor on Coxeter species starting with a module over the Coxeter group,
- embedding the category of Coxeter group modules inside the categories of Coxeter species and Coxeter spaces,

- clarity on how field characteristic issues enter into the Coxeter story via invariant noncommutative zeta and Möbius functions,
- deducing results for Coxeter species from those for species, thus, avoiding field characteristic issues wherever possible,

- distinction between the o-signature functor and u-signature functor on Coxeter species constructed from signature spaces over and under flats,
- seamless passage between unsigned and signed worlds of both Coxeter species and Coxeter spaces via the u-signature functor,

- derivation of several enumeration identities in Coxeter theory involving face-types and shuffles from existence of specific Coxeter bialgebras.

Coxeter operads.

- distinction between operads for a hyperplane arrangement, and Coxeter operads for a reflection arrangement,
- emergence of the monoidal category of Coxeter dispecies under the substitution product leading to the notion of Coxeter operads,
- interpreting Coxeter species and Coxeter spaces as module categories over Coxeter dispecies leading to notions of Coxeter operad monoids and Coxeter operad algebras,
- introduction of the Coxeter commutative, associative, Lie operads involving a rich interplay of flats, top-lunes, face-types (which is specific to the Coxeter context).

Fock functors.

- $\Bbbk$-linearity of the categories of Coxeter (co, bi)monoids and Coxeter (co, bi)algebras, and also of Fock functors and their adjoints (made possible by the absence of the traditional tensor product of vector spaces),
- interactions of the bosonic Fock functor with (primitive and group-like) series of Coxeter bimonoids and of Coxeter bialgebras and their exp-log correspondences,

- interactions of Fock functors with the general q-norm map on Coxeter q-bimonoids and on Coxeter q-bialgebras,
- viewing Coxeter bialgebras as a reflective and coreflective subcategory of Coxeter bimonoids via adjoints of bosonic Fock functors, and deducing results for Coxeter bialgebras from those for Coxeter bimonoids,
- proving nonnegativity of the Boolean transform of the Hilbert type series of Coxeter bimonoids using Fock functors.

Perspectives and future work

This monograph along with the previous one [**11**] opens a vast new area of research. Needless to say, Coxeter bimonoids, Coxeter bialgebras, Fock functors, Coxeter operads are fundamental objects worthy of further study. Two important perspectives on this work are summarized below.

(1) It gives geometric insights into the theory of connected graded Hopf algebras, with combinatorics of integer compositions interpreted as geometry of face-types. This upgrades definitions (for instance, by employing higher operations), improves known results (for instance, by systematically employing zeta and Möbius series), and points to new results (for instance, on connected graded q-bialgebras). These are indicated in the end-of-chapter notes and we plan to explain this in detail in a separate work. Similar comments apply to the theory of May operads.

(2) It brings a Hopf as well as an operadic perspective to the theory of hyperplane arrangements and reflection arrangements (and more generally to the theory of finite semigroups). Along with giving new proofs of known results, it has led to several new concepts such as the category of lunes, Lie and Zie elements, and noncommutative zeta and Möbius functions which we first presented in [**10**]. We plan to explain further ideas such as operadic homology for arrangements in future work.

Acknowledgements

We warmly thank our mentors Ken Brown and Steve Chase for their advice and encouragement over the years. We continue to find inspiration in the work of André Joyal, Jacques Tits, and Gian-Carlo Rota.

It is a pleasure to thank Cambridge University Press for their continued support of our work. Special thanks to Roger Astley, Anna Scriven, and Clare Dennison.

Aguiar supported by Simons Foundation grant 586087.

Introduction

We now describe the contents of the monograph in more detail, along with pointers to important results. For organizational purposes, the book has been divided into three parts; they are preceded by an introductory chapter on reflection arrangements.

Reflection arrangements. (Chapter 1.) A linear hyperplane arrangement $\mathcal{A}$ is a finite collection of hyperplanes in a real vector space passing through the origin. It has many associated geometric objects such as faces, chambers, flats, bifaces, lunes, bilunes, and so on (Table 1.1). Some illustrative pictures are shown below.

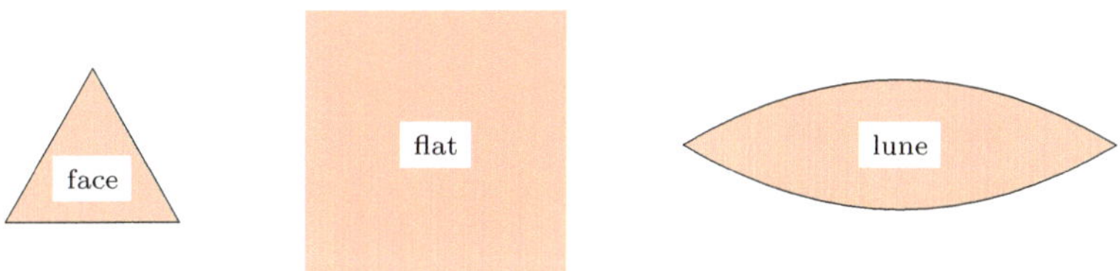

Unlike faces and lunes, flats do not have a boundary; in fact they are subspaces, so the middle picture is to be interpreted as extending indefinitely to the plane of the paper.

The set of faces, flats, bifaces carry monoid structures called the Tits monoid, Birkhoff monoid, Janus monoid, respectively. Linearizing them over a field $\Bbbk$ yields the Tits algebra, Birkhoff algebra, Janus algebra, respectively. We also construct certain incidence algebras called the flat-incidence algebra, lune-incidence algebra, bilune-incidence algebra. They contain various interesting kinds of zeta and Möbius functions.

We say $\mathcal{A}$ is a *linear reflection arrangement* if it is preserved by reflection in any of its hyperplanes (assuming an inner product on the ambient vector space). The group generated by these reflections is finite and is called the *Coxeter group* of $\mathcal{A}$. We denote it by W. By taking W-orbits of faces, flats, bifaces, lunes, and so on, we obtain notions of face-types, flat-types, biface-types, lune-types, and so on (Table 1.2). We introduce the groupoid of biface-types whose objects are face-types and morphisms are biface-types (Section 1.7). A wealth of enumeration identities related to face-types and shuffles are given in Sections 1.8 and 1.9. Similarly, we have the W-invariant subalgebras of all algebras mentioned above. We call them the invariant Tits algebra, invariant lune-incidence algebra, and so on (Sections 1.10 and 1.12).

The braid arrangement, which is an important example of a reflection arrangement, is reviewed in Section 1.16.

Parts I and II

We give a combined introduction to Coxeter species and Coxeter spaces. They are formulated in terms of faces and face-types, respectively; Table III summarizes the main notations. Hopf theories of these two objects proceed largely in parallel with each other. We bring out these similarities, often by putting concepts next to each other in tabular form. At the same time and more importantly, we also point out subtle differences between the two.

TABLE III. Coxeter species and Coxeter spaces.

Faces			Face-types		
Coxeter species	p, q	β, w	Coxeter spaces	V, W	σ
Coxeter monoids	a, b	μ_A^F	Coxeter algebras	A, B	μ_Z^T
Coxeter comonoids	c, d	Δ_A^F	Coxeter coalgebras	C, D	Δ_Z^T
Coxeter bimonoids	h, k	μ_A^F, Δ_A^F	Coxeter bialgebras	H, K	μ_Z^T, Δ_Z^T

Coxeter species and Coxeter spaces. (Chapters 2 and 8.) Fix a reflection arrangement $\mathcal{A}$ with Coxeter group W.

Coxeter species. A *Coxeter species* p is a family of vector spaces $\mathsf{p}[F]$, one for each face F, together with linear maps

$$\beta_{G,F} : \mathsf{p}[F] \to \mathsf{p}[G] \quad \text{and} \quad w : \mathsf{p}[F] \to \mathsf{p}[wF],$$

called support and type morphisms, the former whenever F and G have the same support, and the latter for each F and $w \in W$, subject to suitable compatibility axioms, see (2.12a) and (2.12b). A *Coxeter monoid* a is a Coxeter species together with linear maps

$$\mu_A^F : \mathsf{a}[F] \to \mathsf{a}[A],$$

one for each $A \leq F$, subject to naturality, associativity, unitality axioms, see (2.28a) and (2.28b). A *Coxeter comonoid* c is defined dually using linear maps

$$\Delta_A^F : \mathsf{c}[A] \to \mathsf{a}[F]$$

for $A \leq F$, subject to naturality, coassociativity, counitality axioms, see (2.36a) and (2.36b). We refer to μ as the product and to Δ as the coproduct. A *Coxeter bimonoid* h is a Coxeter species which carries a Coxeter monoid and a Coxeter comonoid structure, and satisfies the bimonoid axiom, see (2.39). It involves the Tits product on faces. The data in a Coxeter bimonoid is shown in Summary 2.1.

The above notions can also be formulated in terms of flats and lunes, see Summary 2.2. Further, they have commutative analogues obtained using the (co)commutativity axiom, see (2.49) and (2.57) for faces and (2.52) and (2.58) for flats and lunes. We also define a *Coxeter q-bimonoid* for a scalar

q by replacing the bimonoid axiom by the q-bimonoid axiom, see (2.42) and (2.45). The q-distance function on faces plays a starring role here. For $q = -1$, we use the term signed Coxeter bimonoid.

Coxeter spaces. A *Coxeter space* V is a family of vector spaces $\mathrm{V}[T]$, one for each face-type T, together with linear maps

$$\sigma : \mathrm{V}[T] \to \mathrm{V}[U],$$

one for each biface-type σ, subject to suitable compatibility axioms (involving composition of biface-types), see (8.28). A *Coxeter algebra* is a Coxeter space together with linear maps

$$\mu_Z^T : \mathrm{V}[T] \to \mathrm{V}[Z],$$

one for each $Z \leq T$, subject to naturality, associativity, unitality axioms, see (8.41). There is a dual notion of a *Coxeter coalgebra*, and a self-dual notion of a *Coxeter bialgebra*. The bialgebra axiom involves a sum and is not set-theoretic, see (8.64). It involves biprojection of biface-types which again has the Tits product at its core. The data in a Coxeter bialgebra is shown in Summary 8.3.

The above notions can also be formulated in terms of top-nested faces starting with support and type morphisms

$$\beta_{G,D,F,C} : \mathrm{V}[F, C] \to \mathrm{V}[G, D] \quad \text{and} \quad w : \mathrm{V}[F, C] \to \mathrm{V}[wF, wC],$$

and so on, see Summary 8.1. Another possibility is to formulate them in terms of top-lunes, see Summary 8.2. Further, they have commutative analogues obtained using the (co)commutativity axiom, see (8.79) and (8.88) for top-nested faces, (8.82), (8.83) and (8.90) for top-lunes, (8.86) and (8.91) for face-types. We also define a *Coxeter q-bialgebra* for a scalar q by replacing the bialgebra axiom by the q-bialgebra axiom, see (8.66), (8.69), (8.73). For $q = -1$, we use the term signed Coxeter bialgebra.

Coxeter species vs Coxeter spaces. Coxeter species carry more structure than Coxeter spaces. To appreciate this point, suppose $u, v \in W$ are such that $u(F) = v(F) = G$ and $u(C) = D$ and $v(C) = E$ as shown in Figure II.

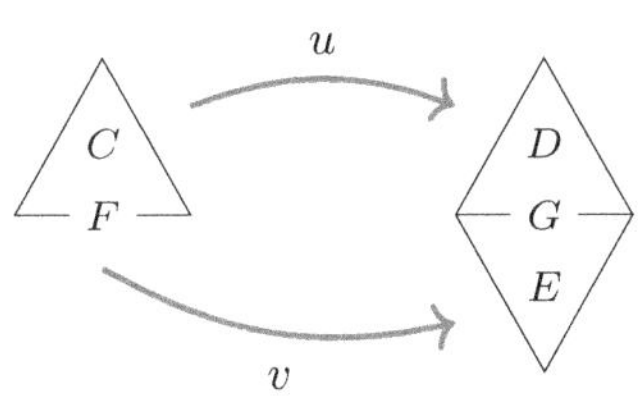

FIGURE II. Coxeter species vs Coxeter spaces.

For a Coxeter species p, we have $u, v : \mathsf{p}[F] \to \mathsf{p}[G]$, while for a Coxeter space V, we have $u : \mathrm{V}[F, C] \to \mathrm{V}[G, D]$ and $v : \mathrm{V}[F, C] \to \mathrm{V}[G, E]$. The key observation is: In the first case, u and v link the same two objects, while in the second case, they link different objects.

Duality functor. Duality was mentioned in the above discussion. More formally: There is a duality functor on the category of Coxeter species which interchanges Coxeter monoids and Coxeter comonoids, and preserves Coxeter q-bimonoids (Sections 2.3.8 and 2.5.9). There is a similar duality functor on the category of Coxeter spaces which interchanges Coxeter algebras and Coxeter coalgebras, and preserves Coxeter q-bialgebras (Sections 8.4.10 and 8.7.8).

Signature functors. There are ways to connect the unsigned and signed worlds. For Coxeter species, we have the u-signature functor and o-signature functor, see Section 2.8. They are related to signature spaces under and over flats, respectively. For Coxeter spaces, we have the u-signature functor, see Section 8.10. The additional structure present in a Coxeter species allows for two signature functors as opposed to only one for Coxeter spaces.

Functor categories. A salient feature of the categories of Coxeter monoids, Coxeter comonoids, Coxeter bimonoids is that they are functor categories just like the category of Coxeter species (Section 2.9). The same holds for the categories of Coxeter (co, bi)algebras (Section 8.11). This is possible due to the absence of the traditional tensor product of vector spaces in the construction of these categories.

Bimonads and bimonad bialgebras. The preceding notions can be put into a categorical framework using monads, comonads, bimonads (Appendix A.1). There exists a bimonad $(\mathcal{T}, \mathcal{T}^\vee, \lambda)$ on the category of Coxeter species whose bialgebras are Coxeter bimonoids. Similarly, there exists a bimonad $(\overline{\mathcal{T}}, \overline{\mathcal{T}}^\vee, \overline{\lambda})$ on the category of Coxeter spaces whose bialgebras are Coxeter bialgebras. These results along with their companions are stated in Propositions 2.71 and 8.115. For commutative aspects, see Propositions 2.78 and 8.121. We mention in particular the bimonads $(\mathcal{S}, \mathcal{S}^\vee, \lambda)$ and $(\overline{\mathcal{S}}, \overline{\mathcal{S}}^\vee, \overline{\lambda})$.

Examples. (Chapters 4 and 10.) We now turn to examples.

Coxeter bimonoids. The *exponential Coxeter bimonoid* E is defined by setting $\mathsf{E}[A] := \Bbbk$ for all faces A, and all structure maps

$$\beta_{G,F} : \mathsf{E}[F] \to \mathsf{E}[G], \qquad w : \mathsf{E}[F] \to \mathsf{E}[wF],$$
$$\mu_A^F : \mathsf{E}[F] \to \mathsf{E}[A], \qquad \Delta_A^F : \mathsf{E}[A] \to \mathsf{E}[F]$$

to be identities. The bimonoid axiom holds trivially. More generally, for a W-module M, we have the *decorated exponential Coxeter bimonoid* E_M whose components are all M; structure maps involving β, μ, Δ are identities, while those involving w employ the action of w on M. See Tables 4.1 and 4.2 for closely related objects.

More examples of Coxeter bimonoids based on chambers, faces, flats, top-nested faces, top-lunes, pairs of chambers are given in Chapter 4.

Coxeter bialgebras. The *Coxeter bialgebra of polynomials* $\Bbbk[x]$ is defined by setting $\Bbbk[x][T] := \Bbbk$ for all face-types T. The linear maps σ are all identities. For clarity, we write x^T for the basis element $1 \in \Bbbk[x][T]$. The product and coproduct are

$$\mu_Z^T : \mathbb{k}[x][T] \to \mathbb{k}[x][Z] \qquad\qquad \Delta_Z^T : \mathbb{k}[x][Z] \to \mathbb{k}[x][T]$$
$$x^T \mapsto x^Z \qquad\qquad x^Z \mapsto d_{T/Z}\, x^T,$$

where $d_{T/Z}$ is the number of faces of type T greater than some fixed face of type Z. (For the braid arrangement, $d_{T/Z}$ is a product of multinomial coefficients.) More generally, we have the Coxeter q-bialgebra of polynomials $\mathbb{k}_q[x]$ obtained by replacing $d_{T/Z}$ by a suitable face-type enumeration polynomial $d_{T/Z}(q)$ (which generalizes the Poincaré polynomial of W). The verification of the q-bialgebra axiom for $\mathbb{k}_q[x]$ is nontrivial and equivalent to the gate identity (1.59). As companions to $\mathbb{k}[x]$, we define the *divided power Coxeter bialgebra* $\mathbb{k}\{x\}$ and *signed Coxeter bialgebra of dual numbers* $\mathbb{k}[dx]$, see Table 10.1.

More generally, for a W-module M, we define the *tensor Coxeter q-bialgebra* $\overline{\mathcal{T}_q}(M)$, *shuffle Coxeter q-bialgebra* $\overline{\mathcal{T}_q}^{\vee}(M)$, and *symmetric Coxeter bialgebras* $\bar{\mathcal{S}}(M)$ and $\bar{\mathcal{S}}^{\vee}(M)$, *exterior signed Coxeter bialgebras* $\bar{\mathcal{E}}(M)$ and $\bar{\mathcal{E}}^{\vee}(M)$, see Table 10.3. The W-action on M by shuffles enters into (co)product formulas. The q-bialgebra axioms for $\overline{\mathcal{T}_q}(M)$ and $\overline{\mathcal{T}_q}^{\vee}(M)$ are equivalent to identities (1.69) in the group algebra of W.

More examples of Coxeter bialgebras based on face-types, flat-types, Coxeter symmetries are given in Chapter 10.

Face-type enumeration. (Sections 1.8 and 1.9.) As noted above, (co)products of Coxeter bialgebras are quite intricate and tied to ideas from face-type enumeration. A systematic summary of such connections is given in Table IV. Identities in the left column are used to prove results in the right column. Interestingly, the logical flow can be reversed. That is, one can first prove results in the right column via the technology of universal constructions or Fock functors, and then use them to deduce identities in the left column.

TABLE IV. Coxeter bialgebras and face-type enumeration.

Face-type enumeration	Relevance to Coxeter bialgebras
polynomials (1.38), (1.41), (1.52)	Observation 8.99 (Section 8.9)
multiplicative property (1.32)	Section 10.2.4, Lemma 10.12, Section 10.2.17
gate identity (1.59)	Lemma 9.89, Section 10.2.4
gate identity (1.57), identity (1.40)	Section 10.4.4
gate identity (1.58)	Lemma 10.79
alternating sign identities (1.63)	Lemma 9.130, antipode formulas (10.30), (10.116), (10.117)
multiplicative property (1.66), identities (1.69), (1.70), (1.71)	Section 10.3.6
alternating sign identity (1.72)	antipode formulas (10.57) and (10.58)

The above phenomena are not seen for Coxeter bimonoids where the bimonoid axiom is set-theoretic and easier to handle. This simplicity happens because of the extra structure present in faces in comparison to face-types which keeps enumeration polynomials hidden from view.

Shuffles. (Section 1.6.) For any face-type T, we have the notion of T-shuffles. These are Coxeter symmetries obtained by taking Tits product of a reference chamber C_r with faces of type T.

Shuffles become visible in the study of Coxeter bialgebras when we focus on the S-component, where S is the maximum face-type. The starting point is that $\mathrm{V}[S]$ is a W-module for any Coxeter space V (Exercise 8.38). The action of T-shuffles on $\mathrm{H}[S]$ is present in the q-bialgebra axiom of H. We also already noted that shuffles appear in (co)product formulas of the tensor and shuffle Coxeter q-bialgebras; this is because they are "freely generated" by their S-components. See Table V for precise pointers.

TABLE V. Coxeter bialgebras and shuffles.

T-shuffles, T-deshuffles	(1.21), (1.22), (1.23)
(co)projection of biface-types	Exercise 1.35
q-bialgebra axiom	Exercises 8.73 and 8.76
tensor and shuffle Coxeter q-bialgebras	formulas (10.37) and (10.39)

Hopf theory for reflection arrangements. (Chapters 3 and 9.) We develop the basic theory of Coxeter bimonoids and Coxeter bialgebras discussing primitive and decomposable filtrations, universal constructions, abelianization and coabelianization, Hadamard product and enrichment, exp-log correspondences, norm transformation, characteristic operations, antipode. Pointers to important formulas and results are summarized in Table VI. This can be compared with the theory for bimonoids for hyperplane arrangements developed in [**11**, Part II]. Some specific points are highlighted below.

Species vs Coxeter species. The bimonoid and (co)commutativity axioms for a Coxeter species do not involve type morphisms. The same is true for maps involved in exp-log correspondences. As a consequence, results on Coxeter species can be formally deduced from corresponding results on species proved in [**11**], see Remark 2.52. In contrast, for Coxeter spaces, both support and type morphisms intervene in the bialgebra and (co)commutativity axioms. This explains why in Table I in the preface, in the last column, there is no analogue of species and bimonoids (involving only support morphisms).

The connection between species and Coxeter species can be made precise using semidirect products, see (2.67) and Lemma 2.70. This is analogous to the connection between vector spaces and W-modules. There is more that one can do with W-modules than with vector spaces. The same is true for Coxeter species. For instance, type morphisms in a Coxeter species allow us to connect it to Coxeter spaces via Fock functors.

TABLE VI. Hopf theory for Coxeter species and Coxeter spaces.

Hopf theory	Coxeter species	Coxeter spaces
Cauchy powers	Formula (3.1)	Formula (9.1)
com. Cauchy powers	Formula (3.4)	Formulas (9.4), (9.5)
primitive and decomposable filtrations	Table 3.1	Table 9.1
	(3.9) and (3.12)	(9.9) and (9.12)
primitive and indecomposable part functors	Lemma 3.5	Lemma 9.8
universal constructions and adjunctions	Table 3.2	Table 9.2
	Summary 3.1	Summary 9.1
(co)abelianization adjunctions	Summary 3.2	Summary 9.2
Hadamard product	Table 3.3	Table 9.3
	Formula (3.53)	Formula (9.59)
enrichment via internal hom	Section 3.6.3	Section 9.6.3
enrichment via universal measuring	Section 3.7.4	Section 9.7.4
exp-log correspondences ((co)derivation version)	Table 3.4	Table 9.4
	Theorem 3.37	Theorem 9.56
	Theorem 3.43	Theorem 9.61
exp-log correspondences (series version)	Theorem 3.56	Theorem 9.76
	Theorem 3.59	Theorem 9.80
	Lemma 3.66	Lemma 9.85
pq map	Proposition 3.44	Proposition 9.65
logarithm of identity	Proposition 3.46	Proposition 9.67
	Proposition 3.50	Proposition 9.72
general q-norm map	Theorem 3.79	Theorem 9.110
characteristic operation	Formula (3.149)	Formula (9.155)
two-sided characteristic operation	Formula (3.157)	Formula (9.164)
antipode	Formula (3.159)	Formula (9.171)
rank one (toy example)	Section 7.8	Section 13.8

Universal constructions. There is a complete parallel between universal constructions for Coxeter species and Coxeter spaces. These constructions follow from the existence of the bimonad $(\mathcal{T}, \mathcal{T}^\vee, \lambda)$ on Coxeter species and bimonad $(\overline{\mathcal{T}}, \overline{\mathcal{T}}^\vee, \overline{\lambda})$ on Coxeter spaces. Thus, $\mathcal{T}(\mathsf{p})$ is the free Coxeter monoid on the Coxeter species p, while $\overline{\mathcal{T}}(\mathrm{V})$ is the free Coxeter algebra on the Coxeter space V. Explicitly,

$$\mathcal{T}(\mathsf{p})[A] = \bigoplus_{F:\, A \leq F} \mathsf{p}[F] \quad \text{and} \quad \overline{\mathcal{T}}(\mathrm{V})[Z] = \bigoplus_{T:\, Z \leq T} \mathrm{V}[T],$$

with product components μ_A^F and μ_Z^T given by inclusion. Cofree objects are constructed dually using $\mathcal{T}^\vee$ and $\overline{\mathcal{T}}^\vee$. Going one step further, these free and cofree objects are used to build universal Coxeter bimonoids and Coxeter bialgebras by using the mixed distributive laws λ and $\overline{\lambda}$.

For commutative aspects, we need to bring in the monads $\mathcal{S}$, $\overline{\mathcal{S}}$ and comonads $\mathcal{S}^\vee$, $\overline{\mathcal{S}}^\vee$. Thus, $\mathcal{S}(\mathsf{p})$ is the free commutative Coxeter monoid on the Coxeter species p, and $\overline{\mathcal{S}}(\mathrm{V})$ is the free commutative Coxeter algebra on the Coxeter space V, and so on.

Hadamard product. Coxeter bimonoids possess certain features not to be seen for Coxeter bialgebras. The key distinction is that the bimonoid axiom is set-theoretic, while the bialgebra axiom is not. As a result, Hadamard product preserves Coxeter bimonoids, but it does not preserve Coxeter bialgebras in general. This yields the *decoration functor* on Coxeter species obtained by taking Hadamard product with E_M (Section 4.9) which has no analogue for Coxeter spaces. Similarly, we have the biconvolution Coxeter bimonoid built out of two Coxeter bimonoids, but no analogous construction for Coxeter bialgebras. See Remarks 9.35 and 9.38. See also Theorem 7.12 which is specific to Coxeter bimonoids.

Exp-log correspondences. Interestingly, exp-log correspondence (say between primitive and group-like series) is not unique in general; it depends on the choice of an invariant noncommutative zeta and Möbius function. The same remark applies to logarithm of the identity map. There are two situations where we do have uniqueness, namely, the bicommutative setting and the q-setting for q not a root of unity. In these situations, the logarithm of the identity map induces an isomorphism from the indecomposable part to the primitive part, and its inverse is the $\mathfrak{pq}$ map. The latter is always defined (that is, with no dependence on any noncommutative zeta or Möbius function).

There is no difference between the free Coxeter algebra and free commutative Coxeter algebra on one generator. In contrast, the free Coxeter monoid and free commutative Coxeter monoid on one generator differ. In our notations,

$$\mathcal{T}(\mathsf{x}) = \mathsf{\Gamma} \neq \mathsf{E} = \mathcal{S}(\mathsf{x}) \quad \text{and} \quad \overline{\mathcal{T}}(\mathrm{X}) = \Bbbk[x] = \overline{\mathcal{S}}(\mathrm{X}).$$

This gives rise to two notions of exponential series for Coxeter monoids, but only one notion for Coxeter algebras.

Characteristic operations. Coxeter bimonoids admit characteristic operations by faces, while Coxeter bialgebras admit characteristic operations by face-types. To say this precisely, recall that the Tits algebra has a basis indexed by faces, while the invariant Tits algebra has a basis indexed by face-types. For a cocommutative Coxeter bimonoid h and cocommutative Coxeter bialgebra H,

$$F \cdot x := \mu_O^F \Delta_O^F(x) \quad \text{and} \quad T \cdot x := \mu_\emptyset^T \Delta_\emptyset^T(x),$$

respectively, define a left action of the Tits algebra on $\mathsf{h}[O]$ and of the invariant Tits algebra on $\mathrm{H}[\emptyset]$. Here O is the smallest face, and $\emptyset$ is the smallest face-type.

Moreover, for h as above, the Tits algebra action on $\mathsf{h}[O]$ combined with the W-action on $\mathsf{h}[O]$ via type morphisms yields an action of the Coxeter–Tits algebra on $\mathsf{h}[O]$. (The Coxeter–Tits algebra is the semidirect product of the action of the Coxeter group on the Tits algebra.) In fact, this leads to an equivalence between categories of cocommutative Coxeter bimonoids and left Coxeter–Tits algebra modules (Proposition 3.86). This and other similar results are summarized in Proposition 3.82.

q not a root of unity. We prove many results for Coxeter q-bimonoids and Coxeter q-bialgebras under the assumption that q is not a root of unity. This assumption while convenient to state is stronger than what is required. Since our theory is local to a reflection arrangement, the results remain valid provided we stay away from certain specific roots of unity, namely, the zeroes of the determinants of the Varchenko matrices in Theorem 1.56 and its proof (which ensures that the two-sided q-zeta and q-Möbius functions exist). For instance, in rank one, $q \neq \pm 1$ suffices, see [**11**, Remark 13.78] in this regard.

Coxeter operads. (Chapters 5 and 11.) A Coxeter dispecies $\mathbf{p}$ can be formulated using flats or top-lunes or face-types as a family of vector spaces

- $\mathbf{p}[\mathrm{X}, \mathrm{Y}]$, one for each pair of flats $\mathrm{X} \leq \mathrm{Y}$, or
- $\mathbf{p}[\mathrm{L}, \mathrm{M}]$, one for each pair of top-lunes $\mathrm{L} \leq \mathrm{M}$, or
- $\mathbf{p}[T, U]$, one for each pair of face-types $T \leq U$.

In each of these three setups, the components are connected by linear maps subject to suitable compatibilities. This richness of setups is specific to the Coxeter context and not to be seen for dispecies considered in [**11**, Chapter 4].

The category of Coxeter dispecies is a monoidal category under the substitution product, see formulas (11.10), (11.11), (11.15). Monoids wrt this product are Coxeter operads. The substitution map of a Coxeter operad is summarized in Table 11.1. (It can be compared to multiplication of upper triangular matrices.)

Coxeter operad monoids and Coxeter operad algebras. The category of Coxeter species is a left module category over the category of Coxeter dispecies via (5.9). (This structure can be compared to how an upper triangular matrix acts on a column vector.) Thus, for any Coxeter operad $\mathbf{a}$, we have the category of left $\mathbf{a}$-modules in Coxeter species. We call it the category of Coxeter $\mathbf{a}$-monoids.

Similarly, the category of Coxeter spaces is a left module category over the category of Coxeter dispecies via any of the equivalent formulas (11.25), (11.28), (11.29). Thus, for any Coxeter operad **a**, we have the category of left **a**-modules in Coxeter spaces. We call it the category of Coxeter **a**-algebras. The structure map of a Coxeter **a**-algebra is summarized in Table 11.2.

Coxeter commutative, associative, Lie operads. The three fundamental examples of Coxeter operads are the Coxeter commutative, associative, Lie operads which we denote **Com**, **As**, **Lie**, respectively. We develop them in all three setups: flats, top-lunes, face-types. Each setup has its advantage and appeal. A Coxeter **As**-monoid is the same as a Coxeter monoid, while a Coxeter **As**-algebra is the same as a Coxeter algebra. Such results along with connections to monad algebras are summarized in Tables 5.1 and 11.4.

In contrast to **Com** and **As**, it is considerably harder to formulate **Lie** and establish its presentation as a binary quadratic structure. The core work for this was done in [**10**, Sections 10.6 and 14.5], [**11**, Example 4.12].

Lie theory for reflection arrangements. (Chapters 6 and 12.) Along with Coxeter (co, bi)monoids, we also have the notion of a Coxeter Lie monoid. Similarly, we have the notion of a Coxeter Lie algebra. These are Coxeter **Lie**-monoids and Coxeter **Lie**-algebras, respectively, associated to the Coxeter operad **Lie**.

TABLE VII. Lie theory for Coxeter species and Coxeter spaces.

Lie theory	Coxeter species	Coxeter spaces
(co)commutator (co)bracket	Formula (6.4)	Formula (12.10)
	Formula (6.7)	Formula (12.15)
primitive and indecomposable parts	Proposition 6.1	Proposition 12.3
	Proposition 6.2	Proposition 12.4
(co)free Lie construction	Theorem 6.3	Theorem 12.7
	Theorem 6.5	Theorem 12.12
universal (co)enveloping construction	Theorem 6.8	Theorem 12.21
	Theorem 6.11	Theorem 12.26

Pointers to important concepts and results are summarized in Table VII. This can be compared with the theory for Lie monoids for hyperplane arrangements developed in [**11**, Chapter 16].

Binary operations vs higher operations. The product maps μ_A^F and μ_Z^T in a Coxeter monoid and Coxeter algebra are "higher operations" arising from elements of the Coxeter associative operad **As**. More traditionally, using the presentation of **As**, one can consider "binary operations", that is, μ_A^F and μ_Z^T only when $A \lessdot F$ and $Z \lessdot T$ subject to the covering associativity axiom. The same can be done for **Com** and **Lie**. (The Lie case is much harder as mentioned earlier.) This approach is summarized in Table VIII, see also Section 8.5.9. Also compare with [**11**, Table 16.1].

TABLE VIII. Relations in Coxeter operad presentations.

Coxeter operad	relations	Coxeter species	Coxeter spaces
As	covering associativity axiom	third diagram in (2.31)	second diagram in (8.43)
Com	covering commutativity axiom	Exercise 2.36	Lemma 8.97
Lie	antisymmetry and Jacobi identity	(6.1b) and (6.1c)	(12.3b) and (12.3c)

Structure results. (Chapters 7 and 13.) Structure results for Coxeter bimonoids and Coxeter bialgebras are summarized in Table IX; see also the summaries provided in Tables 7.1 and 13.1. The Leray–Samelson theorem, for instance, yields an equivalence between categories of Coxeter species and bicommutative Coxeter bimonoids, and between categories of Coxeter spaces and bicommutative Coxeter bialgebras. The Poincaré–Birkhoff–Witt and Cartier–Milnor–Moore theorems make precise the connection between Hopf and Lie theory. These structure results can be compared with structure results for bimonoids for hyperplane arrangements in [**11**, Part III].

TABLE IX. Comparison of structure theorems.

Theorem	Coxeter species	Coxeter spaces
Borel–Hopf	Theorems 7.1 and 7.2	Theorems 13.1 and 13.8
Leray–Samelson	Theorems 7.4 and 7.5	Theorems 13.14 and 13.17
Loday–Ronco	Theorems 7.10 and 7.11	Theorems 13.35 and 13.37
q-generalization for q not a root of unity	Theorems 7.8 and 7.9	Theorems 13.25 and 13.30
Hoffman–Newman–Radford	Theorems 7.14 and 7.15	Theorems 13.39 and 13.40
	Theorems 7.17 and 7.18	Theorems 13.44 and 13.46
Poincaré–Birkhoff–Witt	Theorems 7.20 and 7.23	Theorems 13.53 and 13.59
Cartier–Milnor–Moore	Theorems 7.26 and 7.27	Theorems 13.65 and 13.68

Zeta and Möbius functions in their various avatars play a starring role in these theorems. They intervene in the definitions of the involved maps; a compact summary is given in Table X. We use the terms Borel–Hopf isomorphisms, Leray–Samelson isomorphisms, and so on for these maps. Also of relevance for the structure theorems is the representation theory of the (invariant) Tits algebra and (invariant) q-Janus algebra for q not a root of unity; idempotents in these algebras allow us to get finer information about the involved isomorphisms.

TABLE X. Structure theorems via zeta and Möbius functions.

Theorem	zeta, Möbius	Coxeter species	Coxeter spaces
Borel–Hopf	$\boldsymbol{\zeta}, \boldsymbol{\mu}$	(7.1), (7.2)	(13.1), (13.2)
Leray–Samelson	ζ, μ	(7.5), (7.6)	(13.12), (13.13)
Loday–Ronco	$\boldsymbol{\zeta}_0, \boldsymbol{\mu}_0$	(7.14), (7.15)	(13.26), (13.27)
q-generalization for q not a root of unity	ζ_q, μ_q	(7.10), (7.11)	(13.17), (13.18)
Hoffman–Newman–Radford	ζ, μ	(7.16), (7.17)	(13.30), (13.31)
	ζ_q, μ_q	(7.22), (7.23)	(13.36), (13.37)
Poincaré–Birkhoff–Witt	$\boldsymbol{\zeta}, \boldsymbol{\mu}$	(7.28), (7.29)	(13.47), (13.49)

Joyal species and graded vector spaces. Classical Hopf–Lie theory. For the braid arrangement, faces and face-types correspond to set compositions and integer compositions, respectively, see Table 1.9. This provides the link between Coxeter bimonoids and Joyal bimonoids, and between Coxeter bialgebras and graded bialgebras. For more details, see Sections 2.12 and 8.14. For comparison with important classical examples such as the Joyal bimonoid of linear orders, see Table 4.3 and with the tensor Hopf algebra and Hopf algebra of quasisymmetric functions, see Tables 10.8, 10.9, 10.10. For links between Coxeter Lie monoids and Joyal Lie monoids, and between Coxeter Lie algebras and graded Lie algebras, see Sections 6.6 and 12.6.

Part III

Fock functors connect the world of Coxeter species to the world of Coxeter spaces. They provide a conceptual understanding of the similarities and differences between the two which were summarized in Tables VI, VII, IX, X. Since Coxeter species are more structured than Coxeter spaces, Fock functors, as a general rule, involve loss of information.

Fock functors. (Chapter 14.) Fock functors come in various flavors. We have full Fock functors $\mathcal{K}$ and $\mathcal{K}^\vee$, bosonic Fock functors $\overline{\mathcal{K}}$ and $\overline{\mathcal{K}}^\vee$, fermionic Fock functors $\overline{\mathcal{K}}_{-1}$ and $\overline{\mathcal{K}}^\vee_{-1}$, and so on. A summary is provided in Table 14.1.

Full Fock functors. Let us start with $\mathcal{K}$ and $\mathcal{K}^\vee$. As functors from Coxeter species to Coxeter spaces, the two are the same. So let us focus on $\mathcal{K}$. For a Coxeter species p, the Coxeter space $\mathcal{K}(\mathsf{p})$ on the (F, C)-component is

$$\mathcal{K}(\mathsf{p})[F, C] := \mathsf{p}[F],$$

see (14.2). This can also be formulated in terms of face-types, see (14.14). Support and type morphisms of $\mathcal{K}(\mathsf{p})$ are induced from those of p. Since p

is a Coxeter species, the subgroup W_F of W which fixes F acts on $\mathsf{p}[F]$. A key point to appreciate is that information about these actions is lost in the passage from p to $\mathcal{K}(\mathsf{p})$.

Bosonic and fermionic Fock functors. Let us now look at the bosonic Fock functors $\overline{\mathcal{K}}$ and $\overline{\mathcal{K}}^\vee$. They are defined by

$$\overline{\mathcal{K}}(\mathsf{p})[F,C] := \mathsf{p}[F]_{W_F} \quad \text{and} \quad \overline{\mathcal{K}}^\vee(\mathsf{p})[F,C] := \mathsf{p}[F]^{W_F},$$

where we take W_F-coinvariants and W_F-invariants, respectively, see (14.27) and (14.30). The situation for fermionic Fock functors $\overline{\mathcal{K}}_{-1}$ and $\overline{\mathcal{K}}^\vee_{-1}$ is similar, we again take coinvariants and invariants of $\mathsf{p}[F]$ but with the action twisted by the sign representation of W_F.

Bilax properties. We have the bimonad $(\mathcal{T}, \mathcal{T}^\vee, \lambda)$ on the category of Coxeter species and bimonad $(\overline{\mathcal{T}}, \overline{\mathcal{T}}^\vee, \overline{\lambda})$ on the category of Coxeter spaces. The critical observation is that Fock functors are bilax wrt these bimonads. See Theorems 14.2 and 14.4 for full Fock functors, Theorems 14.14 and 14.15 for bosonic Fock functors, Theorem 14.16 for fermionic Fock functors. The bosonic Fock functors are also bilax wrt bimonads $(\mathcal{S}, \mathcal{S}^\vee, \lambda)$ and $(\overline{\mathcal{S}}, \overline{\mathcal{S}}^\vee, \overline{\lambda})$; this is part of Theorem 14.30.

Interestingly, the bilax structures of $\mathcal{K}$ and $\mathcal{K}^\vee$ can be deformed by a scalar q resulting in bilax functors $\mathcal{K}_q$ and $\mathcal{K}^\vee_q$, see Theorems 14.7 and 14.8. We refer to these as the q-full Fock functors or also as deformed full Fock functors. For $q = -1$, we use the term signed full Fock functors. Bosonic Fock functors link to full Fock functors, while fermionic Fock functors link to signed full Fock functors.

As suggested by the notation, the bilax structures of $\mathcal{K}_q$ and $\mathcal{K}^\vee_q$ are conjugates of each other wrt duality. The same remark applies to bosonic and to fermionic Fock functors.

Interrelationships among Fock functors. There are many relationships among Fock functors. See for instance Proposition 14.12 involving the o-signature functor which links parameter values q and $-q$ of the deformed full Fock functors, (14.34) which links bosonic and fermionic Fock functors, and Proposition 14.22 involving the Coxeter q-bimonoid of chambers which links the q-full Fock functors and bosonic Fock functors. We also have the q-norm transformation κ_q between $\mathcal{K}_q$ and $\mathcal{K}^\vee_q$ which is a morphism of bilax functors (Proposition 14.40). On a Coxeter species p, on the (F, C)-component, it is given by action of $\sum_{w \in W_F} q^{l(w)} w$. This map is an isomorphism when q is not a root of unity (Proposition 14.48).

Decorated Fock functors. Fock functors can be decorated by using a W-module M. We think of M as the space of decorations. Decorated Fock functors are summarized in Table 14.2. When $M = \Bbbk$ with trivial W-action, we recover usual Fock functors.

Coxeter bimonoids to Coxeter bialgebras. (Chapter 15.) Since Fock functors are bilax, they send Coxeter bimonoids to Coxeter bialgebras (Theorem 15.1). We use the term Fock space for values of Fock functors. Their products and coproducts are described by formulas (15.2) and (15.3). Relationships among Fock functors imply relationships among corresponding Fock spaces, see for instance Theorem 15.2, Proposition 15.8, Proposition 15.16. Coxeter bimonoids are much easier to construct than Coxeter bialgebras; thus, Fock functors provide a simple recipe to construct Coxeter bialgebras and morphisms among them.

The commutativity properties of Fock functors are summarized in Theorems 15.11 and 15.12. For instance, the full Fock functor $\mathcal{K}$ preserves cocommutativity, dually, $\mathcal{K}^{\vee}$ preserves commutativity, while bosonic Fock functors preserve both. Interactions of Fock functors with universal constructions are treated in Section 15.4. See in particular Propositions 15.24, 15.25, 15.29 which say that bosonic Fock functors preserve (co)freeness. As an application we show that full Fock spaces are (co)free as Coxeter (co)algebras (Theorem 15.31). A further application related to dimensions of the components of a Coxeter bimonoid is given in Proposition 15.33.

Coxeter Lie monoids to Coxeter Lie algebras. (Chapter 15.) The full Fock functor $\mathcal{K}^{\vee}$ and bosonic Fock functors $\overline{\mathcal{K}}^{\vee}$ and $\overline{\mathcal{K}}$ send Coxeter Lie monoids to Coxeter Lie algebras; formulas in terms of Lie brackets are given in (15.32) and (15.33). The reason why we have $\mathcal{K}^{\vee}$ and not $\mathcal{K}$ is linked to the fact that $\mathcal{K}^{\vee}$ preserves commutativity. Dually, $\mathcal{K}$, $\overline{\mathcal{K}}$, $\overline{\mathcal{K}}^{\vee}$ send Coxeter Lie comonoids to Coxeter Lie coalgebras, see formulas (15.34) and (15.35). Now we have $\mathcal{K}$ and not $\mathcal{K}^{\vee}$ because $\mathcal{K}$ preserves cocommutativity.

The primitive part functors go from Coxeter bimonoids to Coxeter Lie monoids, and from Coxeter bialgebras to Coxeter Lie algebras. Similarly, the indecomposable part functors go from Coxeter bimonoids to Coxeter Lie comonoids, and from Coxeter bialgebras to Coxeter Lie coalgebras. The universal (co)enveloping functors go in the other direction. Interactions of Fock functors with primitive part functors are summarized in Propositions 15.46 and 15.47, and with indecomposable part functors in Propositions 15.48 and 15.49. Similarly, interactions of Fock functors with universal (co)enveloping functors are discussed in Section 15.9.

Adjoints of Fock functors. (Chapter 16.) We introduce the trivialization functor $\mathsf{t}_{(-)}$ from Coxeter spaces to Coxeter species; it is the left and right adjoint of bosonic Fock functors (Propositions 16.12 and 16.13). We then lift these adjunctions to Coxeter (co, bi)algebras and Coxeter (co, bi)monoids, see Summaries 16.1 and 16.2. Lax and colax structures on the trivialization functor play a central role here. See the adjunctions in Propositions 16.20 and 16.29. Also see the adjunctions in Propositions 16.35 and 16.48 which involve the functor $\overline{\mathcal{Z}}^{\vee}$ and those in Propositions 16.42 and 16.57 which involve the functor $\overline{\mathcal{Z}}$. Both $\overline{\mathcal{Z}}^{\vee}$ and $\overline{\mathcal{Z}}$ are built out of $\mathsf{t}_{(-)}$ using the adjoint lifting theorem. Moreover, they have commutative analogues which we denote $\overline{\mathcal{Z}}^{\vee}_{ab}$ and $\overline{\mathcal{Z}}^{coab}$; related adjunctions are given in Propositions 16.68, 16.70, 16.73.

Bosonic Fock functors commute with (co)abelianization. A simple way to deduce this is to observe that bosonic Fock functor adjoints commute with inclusion functors (Exercise 16.65).

For adjoints of full Fock functors, see Summary 16.3. The starting adjunctions are in Propositions 16.81 and 16.83; they involve the functor $\mathcal{L}$ obtained by taking Hadamard product of $\mathsf{t}_{(-)}$ with the Coxeter species of chambers. The lifts of these adjunctions are in Propositions 16.84 and 16.86 and in Propositions 16.87 and 16.89.

For adjoints involving Lie objects, see Summary 16.5.

Reflective and coreflective subcategories. (Chapter 16.) The trivialization functor $\mathsf{t}_{(-)}$ is a section of $\overline{\mathcal{K}}$ and $\overline{\mathcal{K}}^{\vee}$ (Propositions 16.14 and 16.15). Thus, every Coxeter space is a bosonic Fock space. More formally, Coxeter spaces is a reflective and coreflective subcategory of Coxeter species via $\mathsf{t}_{(-)}$ (Proposition 16.17).

There are similar results for Coxeter (co)algebras and Coxeter (co)monoids involving $\mathsf{t}_{(-)}$, $\overline{\mathcal{Z}}$, $\overline{\mathcal{Z}}^{\vee}$ (Propositions 16.62 and 16.63). Further, for a scalar q, Coxeter q-bialgebras is a reflective and coreflective subcategory of Coxeter q-bimonoids via $\overline{\mathcal{Z}}$ and $\overline{\mathcal{Z}}^{\vee}$ (Theorem 16.64). Thus, every Coxeter q-bialgebra H arises as the value of $\overline{\mathcal{K}}^{\vee}$ on the Coxeter q-bimonoid $\overline{\mathcal{Z}}^{\vee}(\mathrm{H})$, and as the value of $\overline{\mathcal{K}}$ on $\overline{\mathcal{Z}}(\mathrm{H})$ (Theorems 16.51 and 16.58).

There are similar results in the Lie case: Every Coxeter Lie algebra is the value of bosonic Fock functors on some Coxeter Lie monoid (Propositions 16.100 and 16.107).

Hopf theory for reflection arrangements under Fock functors. (Chapter 17.) We now see some further instances of how Fock functors connect the theory of Coxeter species to the theory of Coxeter spaces.

Exp-log correspondences between primitive and group-like series of a Coxeter bimonoid h identify with those of the Coxeter bialgebra $\overline{\mathcal{K}}^{\vee}(\mathsf{h})$ (Theorem 17.9). The starting point is to identify series of a Coxeter species p with those of the Coxeter space $\overline{\mathcal{K}}^{\vee}(\mathsf{p})$ (Lemma 17.1), and then identify actions of the invariant lune-incidence algebra on series of a Coxeter monoid a and the Coxeter algebra $\overline{\mathcal{K}}^{\vee}(\mathsf{a})$ (Proposition 17.3). Similar results comparing actions of the invariant bilune-incidence algebra on Coxeter species maps and Coxeter space maps are given in Lemma 17.11, Exercise 17.15, Lemma 17.18. For connections related to internal hom and universal measuring for Coxeter species and Coxeter spaces, see Lemmas 17.21, 17.24, 17.28.

The general q-norm map for Coxeter species can be related to the general q-norm map for Coxeter spaces using Fock functors (Lemma 17.31). They can also be related using bosonic Fock functor adjoints as in diagram (17.37), or in a mixed manner as in diagrams (17.30) and (17.31). There is also a result linking the isomorphism problem for the two general q-norm maps when q is not a root of unity (Theorem 17.41).

Bosonic Fock functors preserve antipodes (Proposition 17.44), but full Fock functors do not in general. The first fact can be used to derive the antipode formula of a (co)free Coxeter bialgebra from the one for a (co)free Coxeter bimonoid (Exercise 17.46).

It is also possible to prove structure theorems for Coxeter spaces (such as Leray–Samelson) using corresponding theorems for Coxeter species; this is elaborated in Sections 17.7 and 17.8.

Examples. (Chapter 18.) Applying Fock functors to the exponential Coxeter bimonoid E yields the Coxeter bialgebra of polynomials $\Bbbk[x]$ and its relatives. For instance,

$$\mathcal{K}(\mathsf{E}) = \overline{\mathcal{K}}(\mathsf{E}) = \Bbbk[x], \quad \mathcal{K}^{\vee}(\mathsf{E}) = \overline{\mathcal{K}}^{\vee}(\mathsf{E}) = \Bbbk\{x\}, \quad \overline{\mathcal{K}}_{-1}(\mathsf{E}) = \overline{\mathcal{K}}^{\vee}_{-1}(\mathsf{E}) = \Bbbk[dx].$$

See Lemmas 18.3 and 18.4. This implies the gate identity (1.59) as explained in Section 18.2.3; see also Table IV. The Coxeter bialgebra of polynomials also arises from the Coxeter bimonoid of chambers (Lemma 18.6). More generally, for a W-module M, the decorated exponential Coxeter bimonoid E_M yields the tensor Coxeter bialgebra $\overline{\mathcal{T}}(M)$ and its relatives (Lemmas 18.19 and 18.21). These Coxeter bialgebras can also be obtained by applying decorated Fock functors to E.

These and other examples involving faces and face-types, flats and flat-types, pairs of chambers and Coxeter symmetries are listed in Table 18.1. Once a Coxeter bialgebra is interpreted as the Fock space of a Coxeter bimonoid, properties of the latter can be used to deduce properties of the former. See for instance in Section 18.2.5 how freeness properties of E imply those of $\Bbbk[x]$. Similarly, Observation 18.20 shows how commutativity properties of E_M imply those of $\overline{\mathcal{T}}(M)$.

Fock functors and Coxeter operads. (Sections 14.9, 15.6, 16.9.) For any Coxeter operad $\mathbf{a}$, we understand how certain Fock functors carry Coxeter $\mathbf{a}$-monoids to Coxeter $\mathbf{a}$-algebras (Theorems 15.35 and 15.38). These follow from their lax properties wrt monads on Coxeter species and Coxeter spaces induced by the action of $\mathbf{a}$ (Theorems 14.52, 14.53, 14.54). When $\mathbf{a}$ is the Coxeter commutative, associative, Lie operads, we recover some of our earlier results. For adjoints of Fock functors in the generality of Coxeter operads, see Summary 16.4 as well as Proposition 16.94 and Table 16.1.

Preliminaries

CHAPTER 1

Coxeter groups and reflection arrangements

In this introductory chapter, we review Coxeter groups and reflection arrangements. We set up basic terminology and notations and put down all important results that will be required in this book. We discuss:

- faces, chambers, flats, bifaces, lunes, bilunes,
- face-types, flat-types, biface-types, lune-types, bilune-types,
- support and type maps, base and case maps,
- groupoid of biface-types,
- Tits monoid, (invariant) Tits algebra, Coxeter–Tits monoid, Janus monoid, (invariant) q-Janus algebra, Coxeter–Janus monoid,
- gallery distance, gate property, length of a Coxeter symmetry,
- descents and global descents, shuffles,
- face-type enumeration, gate identities, alternating sign identities,
- enumerative identities in the Coxeter group algebra,
- (invariant) lune-incidence algebra, (invariant) bilune-incidence algebra, zeta and Möbius functions, Zaslavsky formulas,
- braid arrangement, Chu–Vandermonde identity, zeta and Möbius series.

Notions such as faces, flats, lunes, Tits monoid, lune-incidence algebra (which make no reference to the Coxeter group) can be defined in the generality of hyperplane arrangements. However, since our main interest here is reflection arrangements, we may not always bother with these finer distinctions.

TABLE 1.1. Objects associated to a reflection arrangement $\mathcal{A}$.

$\Sigma[\mathcal{A}]$	faces	A, B, F, G, H, K	
$\Gamma[\mathcal{A}]$	chambers	C, D, E	
$\Pi[\mathcal{A}]$	flats	X, Y, Z, W	s(F), b(L), c(L)
J[$\mathcal{A}$]	bifaces	(F, F') with $\mathrm{s}(F) = \mathrm{s}(F')$	
	nested faces	(H, G) with $H \leq G$	
$\Lambda[\mathcal{A}]$	lunes	L, M, N, P, Q	s(H, G)
	local bifaces	(A, F, F')	
	bilunes	(L, L′)	s(A, F, F')

The main objects associated to a reflection arrangement $\mathcal{A}$ along with their notations are summarized in Table 1.1. In the first column, $\Sigma[\mathcal{A}]$ denotes

TABLE 1.2. Object-types associated to a reflection arrangement $\mathcal{A}$.

$\overline{\Sigma}[\mathcal{A}]$	face-types	Z, T, U, V, Y	$\mathrm{t}(F)$
$\overline{\Pi}[\mathcal{A}]$	flat-types	λ, μ, ρ	$\mathrm{t}(\mathrm{X})$ or $\mathrm{s}(T)$
$\overline{\mathrm{J}}[\mathcal{A}]$	biface-types	(T, σ, T')	$\mathrm{t}(F, F')$
	nested face-types	(T, U) with $T \leq U$	$\mathrm{t}(H, G)$
$\overline{\Lambda}[\mathcal{A}]$	lune-types		$\mathrm{t}(\mathrm{L})$ or $\mathrm{s}(T, U)$
	local biface-types	(Z, T, σ, T')	$\mathrm{t}(A, F, F')$
	bilune-types		$\mathrm{t}(\mathrm{L}, \mathrm{L}')$ or $\mathrm{s}(Z, T, \sigma, T')$

the set of all faces of $\mathcal{A}$, and so on. Note: Some of the sets have not been given any notations. The last column shows alternative ways to denote some of the objects. Thus, a flat X can also be the support $\mathrm{s}(F)$ of a face F, or the base $\mathrm{b}(\mathrm{L})$ or case $\mathrm{c}(\mathrm{L})$ of a lune L, and so on.

The Coxeter group W of $\mathcal{A}$ acts on the objects in Table 1.1. The W-orbits of these actions are shown in Table 1.2. The type map t indicates the orbit to which an object belongs. We call the face-type $\mathrm{t}(F)$ as the type of the face F, and so on. Warning: Despite similarity in notation, $\mathrm{t}(F, F')$ and $\mathrm{t}(H, G)$ stand for different things since (F, F') is a biface while (H, G) is a nested face, so their W-orbits are quite different.

Standard notations related to posets, categories, vector spaces that we use in this chapter and elsewhere in the book are given in the list of notations at the end of the book. The main references for this chapter are [**10**] and [**11**, Chapter 1]. Readers with basic familiarity with reflection arrangements can skip ahead to Parts I and II and return to this chapter when necessary.

1.1. Face-types, flat-types, biface-types

We begin with the most basic geometric objects associated to a reflection arrangement, namely, faces, face-types, flats, flat-types, bifaces, biface-types. See Tables 1.1 and 1.2.

1.1.1. Coxeter groups and reflection arrangements. A *linear hyperplane arrangement* $\mathcal{A}$ is a finite set of hyperplanes passing through the origin (codimension-one subspaces) in a fixed real vector space. The latter is the *ambient space* of $\mathcal{A}$. We assume that it is equipped with an inner product. We say $\mathcal{A}$ is a *linear reflection arrangement* if reflection in any hyperplane of $\mathcal{A}$ preserves $\mathcal{A}$. The group generated by these reflections is finite and is called the *Coxeter group* of $\mathcal{A}$. We denote it by W. We refer to an element of W as a Coxeter symmetry, and denote it by letters u, v, w. We write e for the identity element of W. A linear reflection arrangement is necessarily simplicial.

Convention 1.1. Unless stated otherwise, all arrangements in this book are assumed to be linear. Also, we take all W-actions to be left actions.

TABLE 1.3. Examples of reflection arrangements.

reflection arrangement	hyperplanes
coordinate arrangement	$x_i = 0$ for $1 \leq i \leq n$
braid arrangement or type A	$x_i = x_j$ for $1 \leq i, j \leq n$
arrangement of type B	$x_i = \pm x_j$, $x_i = 0$ for $1 \leq i, j \leq n$
arrangement of type D	$x_i = \pm x_j$ for $1 \leq i, j \leq n$

Some important examples of reflection arrangements in $\mathbb{R}^n$ are shown in Table 1.3. The coordinate and braid arrangements are discussed further in Sections 1.15 and 1.16. Additional information can be found in [**10**, Chapter 6]. We also mention the *dihedral arrangement* in $\mathbb{R}^2$ consisting of d equally spaced lines passing through the origin for any $d \geq 2$.

1.1.2. Pointed reflection arrangements. A *pointed reflection arrangement* is a pair $(\mathcal{A}, C_r)$, where $\mathcal{A}$ is a reflection arrangement and C_r is a chamber of $\mathcal{A}$. (The notion of chambers is elaborated below.) We use the letter $\boldsymbol{\alpha}$ to denote a pointed reflection arrangement. We refer to C_r as the *reference chamber*.

1.1.3. Faces and chambers. Tits monoid. Let $\mathcal{A}$ be an arrangement. The hyperplanes of $\mathcal{A}$ dissect the ambient space into *faces*. Let $\Sigma[\mathcal{A}]$ denote the set of faces. It is a graded poset under inclusion. The minimum element is called the *central face*. (It is the intersection of all hyperplanes.) A rank-one face is called a *vertex*, a rank-two face an *edge*, a corank-one face a *panel*. A maximal face is called a *chamber*. Let $\Gamma[\mathcal{A}]$ denote the set of chambers. We denote faces by letters A, B, F, G, H, K and chambers by letters C, D, E. The central face is denoted O. The intersection of two faces is a face, so meets exist in $\Sigma[\mathcal{A}]$. We denote the meet of F and G by $F \wedge G$. The rank of $\mathcal{A}$, denoted $\mathrm{rk}(\mathcal{A})$, is by definition the rank of the graded poset $\Sigma[\mathcal{A}]$.

A convenient way to visualize $\mathcal{A}$ is as a regular cell complex. It is obtained by taking quotient of $\mathcal{A}$ by the central face and then intersecting it with the unit sphere. The cell complex of an arrangement of six hyperplanes in $\mathbb{R}^3$ is shown below. Faces of $\mathcal{A}$ are precisely the vertices, edges, triangles, quadrilaterals, and so on.

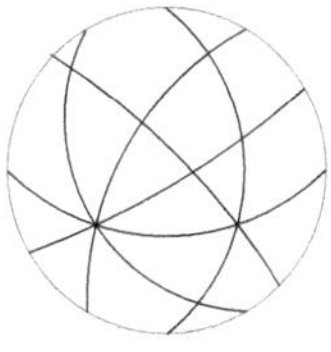

For a reflection arrangement $\mathcal{A}$, the regular cell complex is called the *Coxeter complex* of $\mathcal{A}$. It is a simplicial complex.

The poset of faces $\Sigma[\mathcal{A}]$ carries a monoid structure, with the central face as the identity element. We call it the *Tits monoid*, see [**10**, Section 1.4] and [**11**, Section 1.1.2]. For faces F and G, we denote their Tits product by FG. For F a face and C a chamber, FC is a chamber, thus, the set of chambers

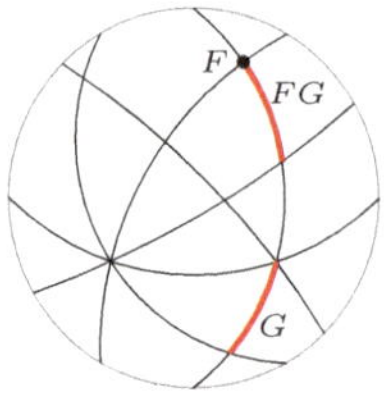

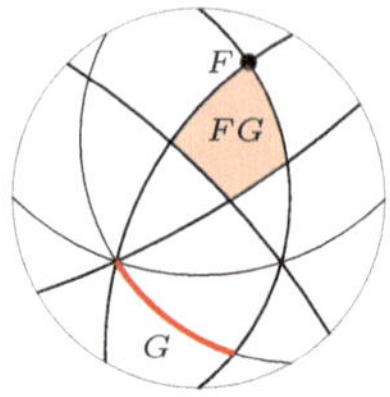

FIGURE 1.1. Tits product of faces.

$\Gamma[\mathcal{A}]$ is a left $\Sigma[\mathcal{A}]$-set. The Tits monoid is an example of a *left regular band*, that is, $FGF = FG$ for any F and G. In particular, $FF = F$ for any F. Some illustrations of the Tits product are shown in Figure 1.1. We also say that FG is the projection of G on F.

For a reflection arrangement $\mathcal{A}$, the Coxeter group W acts on the set of faces $\Sigma[\mathcal{A}]$. Moreover, the action preserves the Tits product, that is, $w(FG) = w(F)w(G)$ for faces F and G, and $w \in W$. The action restricts to the set of chambers $\Gamma[\mathcal{A}]$. Moreover, this action is simply transitive. That is, given C and D, there is a unique $w \in W$ such that $w(C) = D$. It follows that for a pointed reflection arrangement $(\mathcal{A}, C_r)$,

$$\Gamma[\mathcal{A}] \cong W, \qquad wC_r \leftrightarrow w, \quad C_r \leftrightarrow e,$$

as left W-sets.

1.1.4. Flats. Birkhoff monoid. Let $\mathcal{A}$ be an arrangement. A *flat* of $\mathcal{A}$ is a subspace obtained by intersecting some subset of hyperplanes in $\mathcal{A}$. Let $\Pi[\mathcal{A}]$ denote the set of flats. It is a graded lattice under inclusion. We denote flats by letters X, Y, Z, W. We denote the meet of X and Y by $\mathrm{X} \wedge \mathrm{Y}$, join by $\mathrm{X} \vee \mathrm{Y}$, minimum flat by $\bot$, maximum flat by $\top$. We view $\Pi[\mathcal{A}]$ as a monoid with product given by join, with $\bot$ as the identity element. We call it the *Birkhoff monoid*. It is a commutative left regular band.

Alternatively, a flat is an equivalence class in the set of faces under the relation

$$A \sim B \iff AB = A \text{ and } BA = B.$$

We denote the class of A by $\mathrm{s}(A)$ and call it the *support* of A. This defines an order-preserving map

$$\mathrm{s} : \Sigma[\mathcal{A}] \twoheadrightarrow \Pi[\mathcal{A}], \qquad A \mapsto \mathrm{s}(A). \tag{1.1}$$

In fact, it is a morphism from the Tits monoid to the Birkhoff monoid.

For a reflection arrangement $\mathcal{A}$, the Coxeter group W acts on the set of flats $\Pi[\mathcal{A}]$. Moreover, the action preserves the Birkhoff product (join), that is, $w(\mathrm{X} \vee \mathrm{Y}) = w(\mathrm{X}) \vee w(\mathrm{Y})$ for flats X and Y, and $w \in W$.

1.1.5. Bifaces. Janus monoid. Let $\mathcal{A}$ be an arrangement. A *biface* of $\mathcal{A}$ is a pair (F, F') of faces such that $\mathrm{s}(F) = \mathrm{s}(F')$. See Figure 1.2; the shading indicates that F and F' have the same support. Let $\mathrm{J}[\mathcal{A}]$ denote the set of bifaces. It carries a monoid structure defined by $(F, F')(G, G') := (FG, G'F')$

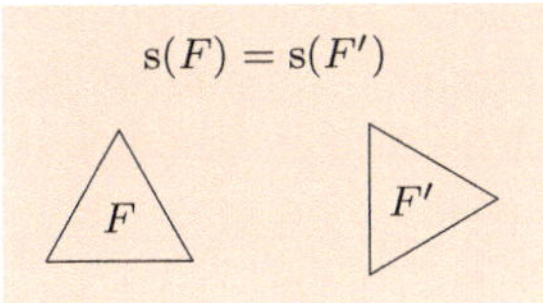

FIGURE 1.2. Visualizing a biface.

with identity element (O, O). We call it the *Janus monoid*. It is a *regular band*, that is, an idempotent monoid in which the axiom $xyxzx = xyzx$ holds.

Observe: There is a commutative diagram of monoids

$$\begin{array}{ccccccc} \mathrm{J}[\mathcal{A}] & \twoheadrightarrow & \Sigma[\mathcal{A}]^{\mathrm{op}} & \qquad & (F, F') & \longmapsto & F' \\ \downarrow & & \downarrow \scriptstyle{\mathrm{s}} & & \downarrow & & \downarrow \\ \Sigma[\mathcal{A}] & \underset{\mathrm{s}}{\twoheadrightarrow} & \Pi[\mathcal{A}] & & F & \longmapsto & \mathrm{s}(F) = \mathrm{s}(F'). \end{array} \tag{1.2}$$

For a reflection arrangement $\mathcal{A}$, the Coxeter group W acts on the set of bifaces $\mathrm{J}[\mathcal{A}]$ via $w(F, F') := (wF, wF')$. It follows that $w((F, F')(G, G')) = w(F, F')w(G, G')$. Moreover, (1.2) is a diagram of W-actions on monoids.

Remark 1.2. Let G and K be faces of the same support and $F \leq G$. Then, in general, there does *not* exist a face $H \leq K$ which has the same support as F. See the illustration below.

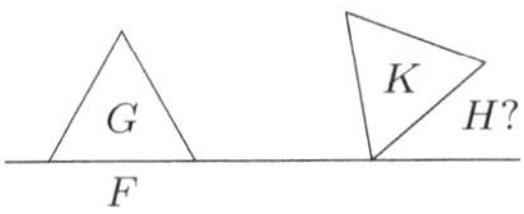

However, if H does exist, then it is unique. This is because distinct subfaces of a face have distinct supports [**10**, Proposition 1.17].

1.1.6. Arrangements over and under flats. Let $\mathcal{A}$ be a reflection arrangement. For a flat X, the arrangement over X, denoted $\mathcal{A}_{\mathrm{X}}$, consists of hyperplanes of $\mathcal{A}$ which contain X. In fact, $\mathcal{A}_{\mathrm{X}}$ is a reflection arrangement.

For a face F, the *star* of F consists of all faces K greater than F, while the *top-star* of F consists of all chambers C greater than F. For clarity, we may denote them by K/F and C/F. For a face F, let $\mathcal{A}_F := \mathcal{A}_{\mathrm{s}(F)}$, and identify the set of faces and chambers of $\mathcal{A}_F$ with the star and top-star of F, respectively. Flats of $\mathcal{A}_F$ are flats of $\mathcal{A}$ which contain F. For clarity, we may denote them by X/F.

For a flat X, the arrangement under X, denoted $\mathcal{A}^{\mathrm{X}}$, has ambient space X and consists of hyerplanes obtained by intersecting hyperplanes of $\mathcal{A}$ with X. However, $\mathcal{A}^{\mathrm{X}}$ is not a reflection arrangement in general. This leads to the notion of good reflection arrangements discussed in [**10**, Section 5.7].

1.1.7. Coxeter system. Let $(\mathcal{A}, C_r)$ be a pointed reflection arrangement. Let S be the set of reflections in the walls of C_r. (The walls of C_r are precisely the supports of the panels of C_r.) Fact: S generates the Coxeter group W of $\mathcal{A}$. The pair (W, S) is called a *Coxeter system*.

1.1.8. Face-types. Let $\boldsymbol{\alpha} = (\mathcal{A}, C_r)$ be a pointed reflection arrangement. A *face-type* of $\mathcal{A}$ is an orbit of the W-action on faces. The *type* of a face F, denoted $\mathrm{t}(F)$, is the orbit to which it belongs. Let $\overline{\Sigma}[\mathcal{A}]$ denote the set of face-types. The canonical map

$$\mathrm{t} : \Sigma[\mathcal{A}] \twoheadrightarrow \overline{\Sigma}[\mathcal{A}], \qquad F \mapsto \mathrm{t}(F) \tag{1.3}$$

is called the *type map*. Face-types identify with faces of C_r. In other words, each face of C_r has a distinct type, and all face-types arise in this manner. Face-types also identify with subsets of S. We denote them by letters Z, T, U, V. They form a Boolean poset under inclusion. The type of the central face is $\emptyset$, while the type of any chamber is S. We use the term *vertex-type* to refer to the type of a vertex, *edge-type* to refer to the type of an edge, and so on. There is only one chamber-type, namely, S. For the type of the central face, we use the term *central face-type*.

Example 1.3 (Ranks two and three). We now provide several illustrations of face-types in reflection arrangements of ranks two and three. They serve as toy models for understanding concepts, results, and calculations involving face-types.

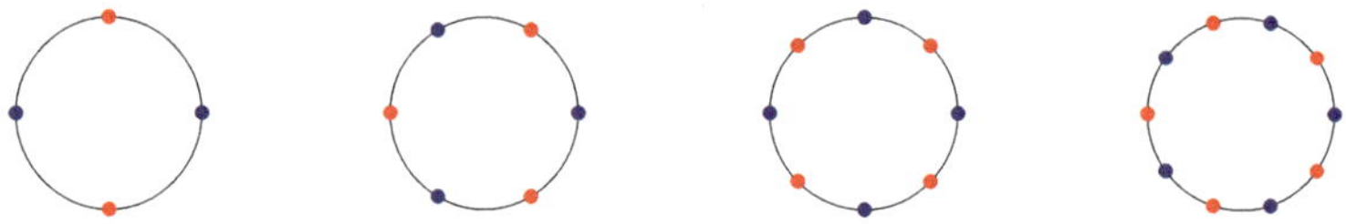

$d = 2$ $d = 3$ $d = 4$ $d = 5$

FIGURE 1.3. Rank-two dihedral arrangements.

Figure 1.3 shows the dihedral arrangements of d lines for $d = 2, 3, 4, 5$. Face-types are indicated by colors. Vertex-types are red and blue, and the chamber-type is red-blue. For d odd, red and blue vertex-types are opposite, while for d even, they are self-opposite. Opposition is formally introduced in Section 1.1.11.

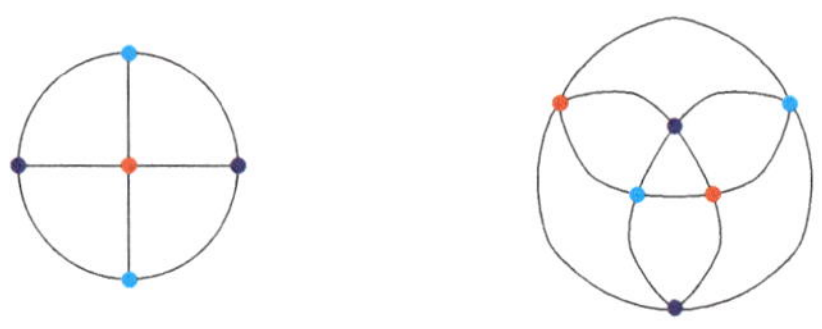

FIGURE 1.4. Rank-three coordinate arrangement.

In Figure 1.4, the left picture is a front view of the rank-three coordinate arrangement with 4 triangles visible out of 8. The right picture is a flattening of the left picture in which 7 triangles are visible, with one triangle on the back side. Vertex-types are red, blue, cyan, edge-types are red-blue, red-cyan, blue-cyan, and the chamber-type is red-blue-cyan.

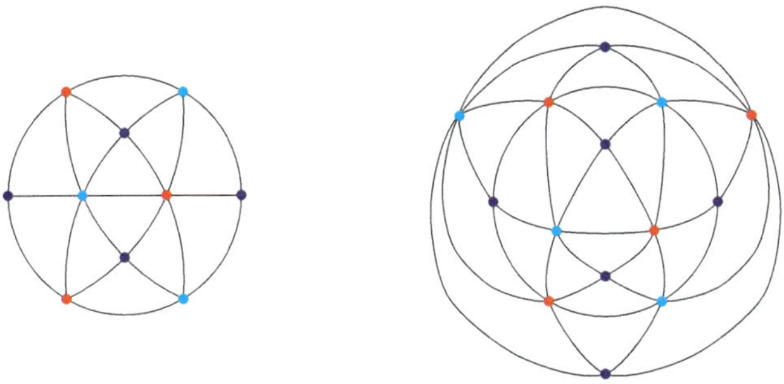

FIGURE 1.5. Rank-three braid arrangement (Type A).

In Figure 1.5, the left picture is a front view of the rank-three braid arrangement. It has 24 triangles of which 12 are visible. The right picture is a flattening of the left picture with 23 triangles visible. The color coding is done as in Figure 1.4. Red and cyan vertex types are opposite, while the blue vertex-type is self-opposite.

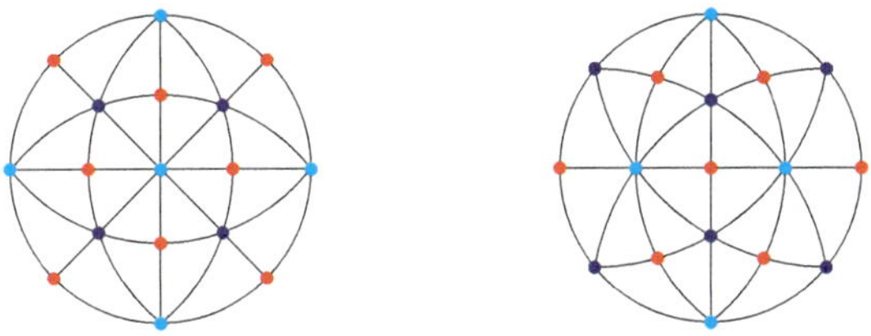

FIGURE 1.6. Rank-three arrangement of type B.

Figure 1.6 shows two different front views of the rank-three arrangement of type B. It has 48 triangles of which 24 are visible in either picture. In contrast to type A, all vertex-types are now self-opposite. □

Fact 1.4. Let A and B be faces of the same support. If F and G are faces greater than A and $w(F) = G$ for some $w \in W$, then $w(BF) = BG$. If F and G are faces greater than A and $\mathrm{t}(F) = \mathrm{t}(G)$, then $\mathrm{t}(BF) = \mathrm{t}(BG)$.

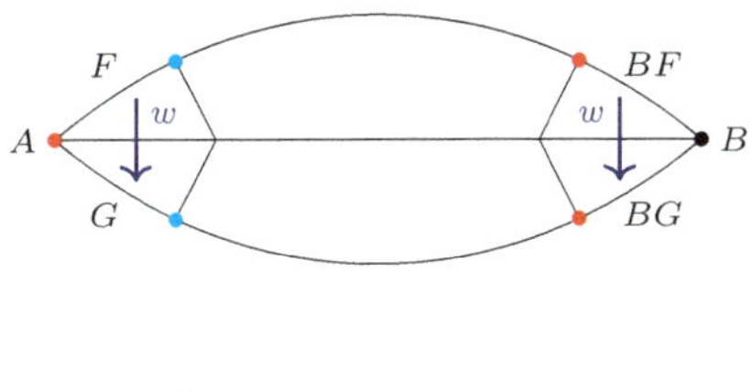

PROOF. See [**10**, Lemma 5.2]. □

Observation 1.5. For any face-type T, the set of chambers $\Gamma[\mathcal{A}]$ is the disjoint union of top-stars of faces of type T. See the illustration below for the dihedral arrangement of 3 lines.

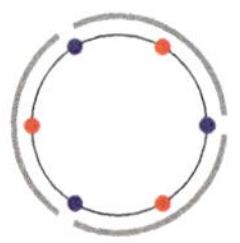

top-stars for the red vertex-type

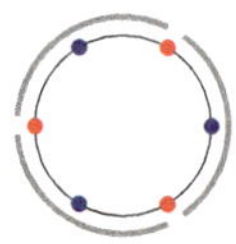

top-stars for the blue vertex-type

More generally, for face-types $U \geq T$, the set of faces of type U can be written as a disjoint union indexed by faces of type T. Setting $U := S$ recovers the previous situation.

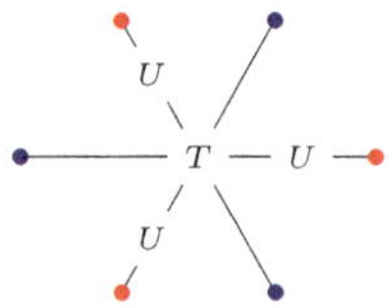

An illustration for T a vertex-type and U an edge-type is shown above. The three edges of type U go together indexed by their common vertex of type T.

Going over and under flats. Let $\mathcal{A}$ be a reflection arrangement. Then so is $\mathcal{A}_{\mathrm{X}}$. Its face-types can be understood as follows. For $\boldsymbol{\alpha} = (\mathcal{A}, C_r)$ and $T \leq S$, consider the pointed reflection arrangement

$$\boldsymbol{\alpha}_T := (\mathcal{A}_F, C_r/F),$$

where F is the face of C_r of type T. Face-types of $\boldsymbol{\alpha}_T$ identify with subsets of S which contain T. We denote a typical face-type by U/T, with $T \leq U \leq S$.

In contrast, $\mathcal{A}^{\mathrm{X}}$ may not be a reflection arrangement, and even when it is, one usually loses face-type information present in X during this passage. For example, the outer circle in the left picture in Figure 1.5 is a reflection arrangement, namely, the dihedral arrangement of 3 lines. However, the intricate color pattern on the circle is not to be seen for the dihedral arrangement.

1.1.9. Flat-types. For a reflection arrangement $\mathcal{A}$, a *flat-type* of $\mathcal{A}$ is an orbit of the W-action on flats. The *type* of a flat X, denoted $\mathrm{t}(\mathrm{X})$, is the orbit to which it belongs. Let $\overline{\Pi}[\mathcal{A}]$ denote the set of flat-types. For face-types T and U, we say $T \sim U$ if there exist faces F and G of the same support which have types T and U, respectively. Equivalence classes of this relation correspond to flat-types. We write $\mathrm{s}(T)$ for the class of T and call it the *support* of T.

Example 1.6 (Rank three). Consider the rank-three braid arrangement in Figure 1.5. A picture of its rank-two flat is shown in Figure 1.7. Observe: s(blue-cyan) = s(red-blue) = s(red-cyan). Thus, all three edge-types have the same support.

In contrast, for the type B arrangement in Figure 1.6, s(blue-cyan) = s(red-blue) and it is different from s(red-cyan). In further contrast, for the coordinate arrangement in Figure 1.4, s(blue-cyan), s(red-blue), s(red-cyan) are all different. □

FIGURE 1.7. Rank-two flat in the rank-three braid arrangement.

For any face, the type of its support equals the support of its type. We call this the *support-type* of that face. Thus, we have the commutative diagram

$$
\begin{array}{ccc} \Sigma[\mathcal{A}] & \xrightarrow{\ \mathrm{t}\ } & \overline{\Sigma}[\mathcal{A}] \\ {\scriptstyle \mathrm{s}}\downarrow & & \downarrow{\scriptstyle \mathrm{s}} \\ \Pi[\mathcal{A}] & \xrightarrow[\ \mathrm{t}\ ]{} & \overline{\Pi}[\mathcal{A}] \end{array} \qquad\qquad \begin{array}{ccc} \text{face} & \xmapsto{\ \mathrm{t}\ } & \text{face-type} \\ {\scriptstyle \mathrm{s}}\downarrow & & \downarrow{\scriptstyle \mathrm{s}} \\ \text{flat} & \xmapsto[\ \mathrm{t}\ ]{} & \text{flat-type}. \end{array} \tag{1.4}
$$

We denote flat-types by letters λ, μ, ρ. There is a partial order on flat-types: $\lambda \leq \mu$ if there are faces F and G with $F \leq G$ such that the support-types of F and G are λ and μ, respectively. Equivalently, instead of faces we can use face-types or flats. The poset of flat-types has unique minimum and maximum elements which we denote λ_{min} and λ_{max}, respectively.

For a face-type T, a flat-type of $\boldsymbol{\alpha}_T$ is generally denoted λ/T (with the minimum and maximum flat-types denoted $(\lambda/T)_{\mathrm{min}}$ and $(\lambda/T)_{\mathrm{max}}$).

Remark 1.7. Note very carefully the difference between $\mathrm{s}(U/T) = \mathrm{s}(V/T)$ and $\mathrm{s}(U) = \mathrm{s}(V)$. The former is stronger than the latter in general. For example, in the rank-three braid arrangement in Figure 1.5, s(red-blue) = s(cyan-blue) but s(red-blue/blue) $\neq$ s(cyan-blue/blue). This is the same as [**10**, Remark 16.36]. For related comments, see Section 10.5.1.

Definition 1.8. Fix a flat-type λ. Suppose T is a face-type of support λ, and X is a flat of type λ. Let $|\lambda|$ denote the number of faces of type T and support X. It only depends on λ and not on the particular choice of T and X. Observe: $|\lambda_{\mathrm{min}}| = 1$ and $|\lambda_{\mathrm{max}}| = |W|$. In general, $|\lambda|$ divides $|W|$. Since $\mathrm{s}(T)$ is a flat-type, it makes sense to write $|\mathrm{s}(T)|$.

While dealing with flat-types of $\boldsymbol{\alpha}_Z$, we write $|\lambda/Z|$ and $|\mathrm{s}(T/Z)|$. The latter is the number of faces of type T/Z on any flat of type $\mathrm{s}(T/Z)$.

Example 1.9 (Rank three). Let us look at some rank-two flats in rank-three reflection arrangements. In Figure 1.7, the flat shown supports six edges, with each edge-type occurring twice. Thus $|\lambda| = 2$, where λ is the type of this flat. In Figure 1.6, in the first picture, for the outer circle-type, $|\lambda| = 8$, while in the second picture, for the outer circle-type, $|\lambda| = 4$. In Figure 1.4, for the support of any edge-type, $|\lambda| = 4$. □

Definition 1.10 (Invariant family of scalars). The following families of scalars specify equivalent pieces of data.

- (ξ_F) indexed by faces such that $\xi_F = \xi_G$ when $\mathrm{s}(\mathrm{t}(F)) = \mathrm{s}(\mathrm{t}(G))$.
- (ξ_{X}) indexed by flats such that $\xi_{\mathrm{X}} = \xi_{\mathrm{Y}}$ when $\mathrm{t}(\mathrm{X}) = \mathrm{t}(\mathrm{Y})$.
- (ξ_T) indexed by face-types such that $\xi_T = \xi_U$ when $\mathrm{s}(T) = \mathrm{s}(U)$.
- (ξ_λ) indexed by flat-types.

We denote any of them by ξ, and refer to ξ as an *invariant family of scalars*.

Observe: For any face-type Z, an invariant family of scalars ξ of $\boldsymbol{\alpha}$ induces an invariant family of scalars of $\boldsymbol{\alpha}_Z$, again denoted ξ, by setting $\xi_{T/Z} := \xi_T$.

1.1.10. Biface-types. For a reflection arrangement $\mathcal{A}$, a *biface-type* of $\mathcal{A}$ is an orbit of the W-action on bifaces. The *type* of a biface (F', F), denoted $\mathrm{t}(F', F)$, is the orbit to which it belongs. Let $\overline{\mathrm{J}}[\mathcal{A}]$ denote the set of biface-types. We use letters σ, τ, ρ to denote biface-types. For clarity, we also write σ as a triple (T', σ, T), where T and T' are types of F and F' with $\mathrm{t}(F', F) = \sigma$. In fact, later in Section 1.7, we will see that biface-types can be interpreted as morphisms in a category whose objects are face-types.

Note very carefully: A biface-type carries finer information than a pair of face-types (T', T) with $\mathrm{s}(T) = \mathrm{s}(T')$. That is, $\mathrm{t}(F', F) \neq (\mathrm{t}(F'), \mathrm{t}(F))$. For instance, in the rank-one arrangement with chambers C and $\overline{C}$, the bifaces (C, C) and $(C, \overline{C})$ represent different biface-types, but have the same underlying pair of face-types, namely, (S, S).

1.1.11. Opposition map. Every face F has an *opposite face* $\overline{F}$. The opposition map on faces

$$\Sigma[\mathcal{A}] \to \Sigma[\mathcal{A}], \qquad F \mapsto \overline{F}$$

commutes with the action of W. In other words, $w\overline{F} = \overline{wF}$. For faces $A \leq F$, the opposite of F in the star of A is $A\overline{F}$. We can think of it as the opposite of F local to A. See the illustrations in Figure 1.8. When $A = O$, this recovers the usual notion of opposite.

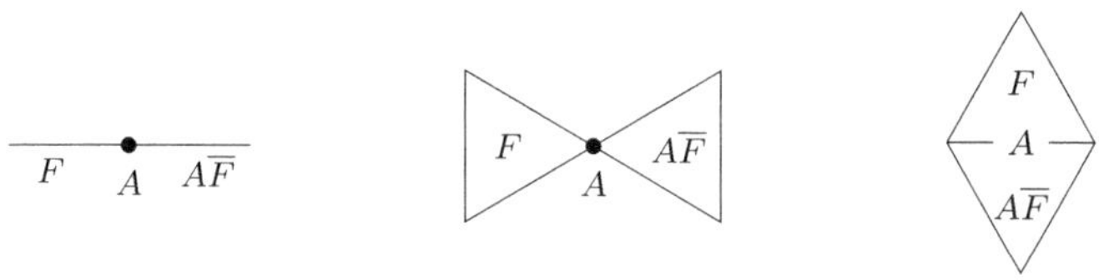

FIGURE 1.8. Opposite faces in stars.

The opposition map on faces induces a map on the set of face-types

$$\overline{\Sigma}[\mathcal{A}] \to \overline{\Sigma}[\mathcal{A}], \qquad T \mapsto \overline{T}.$$

That is, if F has type T, then $\overline{F}$ has type $\overline{T}$. More generally, for face-types $Z \leq T$, pick a face A of type Z, a face F of type T which is greater than A, and let $Z\overline{T}$ denote the type of $A\overline{F}$. It is clear that the face-type $Z\overline{T}$ does not depend on the particular choice of A and F.

Fact 1.11. Let C and D be opposite chambers in the star of A, that is, $D = A\overline{C}$ and $C = A\overline{D}$. Then, for any $w \in W$, $w(C)$ and $w(D)$ are opposites in the star of $w(A)$.

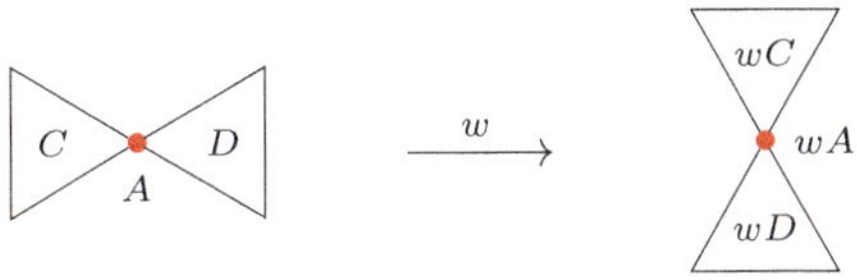

PROOF. See [**10**, Lemma 5.4]. □

Fact 1.12. Let F be a face and C a chamber. Define $u \in W$ by $u(FC) = C$. Then $u(\overline{F}C) = \overline{F_0}C$, where F_0 is the face of C of the same type as F.

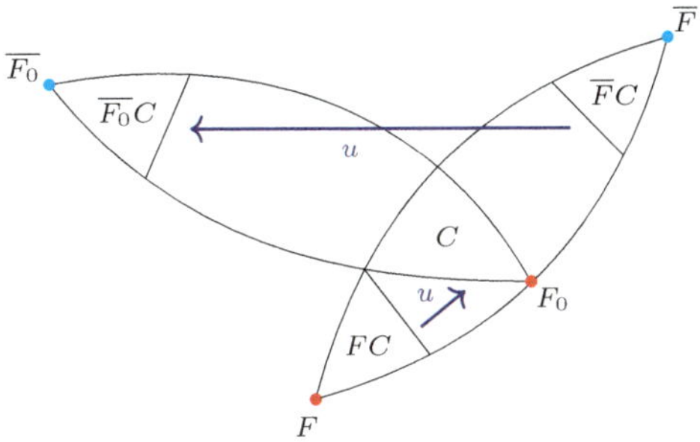

PROOF. See [**10**, Lemma 5.5]. □

1.2. Lune-types and bilune-types

We now review nested faces, lunes, nested face-types, lune-types. We also discuss local bifaces, bilunes, local biface-types, bilune-types. See Tables 1.1 and 1.2. For supplementary information on lunes, see [**10**, Chapters 3 and 4].

1.2.1. Nested faces and lunes. Let $\mathcal{A}$ be an arrangement. A *nested face* of $\mathcal{A}$ is a pair of faces (A, F) such that $A \leq F$. A *lune* of $\mathcal{A}$ is an equivalence class in the set of nested faces under the relation

$$(A, F) \sim (B, G) \iff AB = A,\ BA = B,\ AG = F,\ BF = G.$$

We write $\mathrm{s}(A, F)$ for the class of (A, F) and call it the *support* of (A, F). Lunes of the form $\mathrm{s}(A, A)$ identify with the flat $\mathrm{s}(A)$, while lunes of the form $\mathrm{s}(O, F)$ identify with the face F. In this way, every face and flat is a lune. A *top-nested face* is a pair of faces (A, C) such that $A \leq C$ and C is a chamber. The support of a top-nested face is a *top-lune*. Let $\Lambda[\mathcal{A}]$ denote the set of lunes, and $\widehat{\Lambda}[\mathcal{A}]$ denote the set of top-lunes. We denote lunes by letters L, M, N, P, Q.

To each nested face (A, F), we associate the subset of faces

$$\{H \mid AH = F \text{ and } \mathrm{s}(H) = \mathrm{s}(F)\}.$$

(For a top-nested face, this is a subset of chambers.) This subset only depends on the lune $\mathrm{s}(A, F)$. Moreover, distinct lunes yield distinct subsets. These subsets can thus be used to visualize lunes. Two illustrations of top-lunes are shown below. They are not fully visible; some part is on the back side.

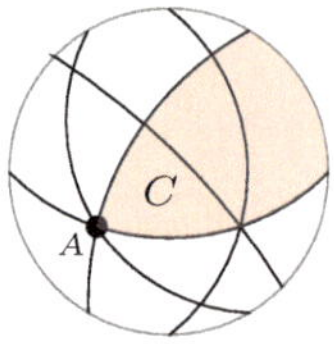

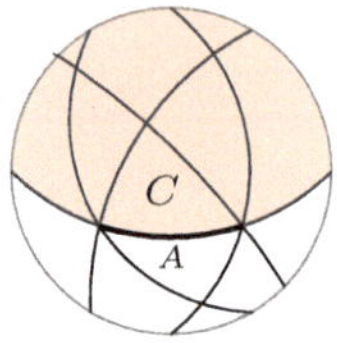

The first picture explains the origin of the term "lune".

For the lune $s(A, F)$, the flats $s(A)$ and $s(F)$ only depend on the lune, and are, respectively, called the *base* and *case* of that lune. We denote the base and case of L by $b(L)$ and $c(L)$, respectively. A lune L is a face iff $b(L) = \bot$, and a flat iff $b(L) = c(L)$. The *slack* of L is defined by $sk(L) := rk(c(L)) - rk(b(L))$, the difference between the ranks of the case and base of L. Lunes of slack 0 are precisely flats. The slack of a face is its rank in the poset of faces.

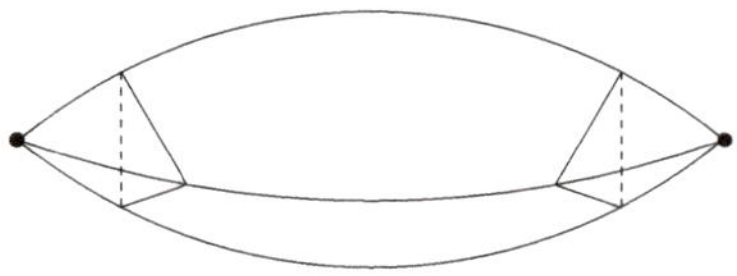

FIGURE 1.9. Lune in a rank-four arrangement.

Figure 1.9 shows a top-lune of slack three in a rank-four arrangement. The two black vertices are opposite to each other, and together with the two marked tetrahedra define top-nested faces whose support is this top-lune.

Every lune L has an *opposite* $\overline{L}$ defined as follows. For $L = s(A, F)$, let $\overline{L} = s(A, A\overline{F})$. When L is a face, $\overline{L}$ is indeed its opposite face. Also, a lune equals its opposite iff it is a flat.

Lunes of $\mathcal{A}$ with base X correspond to faces of $\mathcal{A}_X$, with opposite lunes corresponding to opposite faces. More specifically, lunes of $\mathcal{A}$ with base X and case Y correspond to chambers of $\mathcal{A}_X^Y$, with opposite lunes corresponding to opposite chambers.

1.2.2. Category of lunes. For an arrangement $\mathcal{A}$, define the *category of lunes* as follows. Objects are flats and morphisms are lunes. More precisely, a morphism from Y to X is a lune L with base X and case Y. Identity morphisms are flats. Composition of morphisms is defined by $s(F, G) \circ s(G, H) = s(F, H)$.

For a reflection arrangement $\mathcal{A}$, the Coxeter group W acts on the category of lunes. In particular, for $w \in W$ and composable lunes L and M, $w(L \circ M) = wL \circ wM$.

1.2.3. Action of the Birkhoff monoid on lunes. For an arrangement $\mathcal{A}$, the Tits monoid $\Sigma[\mathcal{A}]$ acts on the set of lunes $\Lambda[\mathcal{A}]$ via

$$A \cdot s(F, G) = s(AF, AG).$$

Further, this induces an action of the Birkhoff monoid $\Pi[\mathcal{A}]$ on the set of lunes $\Lambda[\mathcal{A}]$. For a flat X and a lune L, we write $\mathrm{X} \cdot \mathrm{L}$ for the action of X on L.

In fact, the Birkhoff monoid acts on the category of lunes. In particular, for a flat X and composable lunes L and M,

$$\mathrm{X} \cdot (\mathrm{L} \circ \mathrm{M}) = (\mathrm{X} \cdot \mathrm{L}) \circ (\mathrm{X} \cdot \mathrm{M}). \tag{1.5}$$

1.2.4. Partial orders on top-lunes. The inclusion partial order on top-lunes is as follows. For top-lunes L and M, we say $\mathrm{L} \leq \mathrm{M}$ if there exist faces $A \leq F$ and $A \leq C$ such that $\mathrm{L} = \mathrm{s}(A, C)$ and $\mathrm{M} = \mathrm{s}(F, FC)$.

Fact 1.13. For top-lunes $\mathrm{L} \leq \mathrm{N}$, and a flat Y such that $\mathrm{b(L)} \leq \mathrm{Y} \leq \mathrm{b(N)}$, there exists a unique top-lune M with base Y such that $\mathrm{L} \leq \mathrm{M} \leq \mathrm{N}$.

PROOF. See [**10**, Lemma 4.6]. □

We now define another partial order on top-lunes, denoted $\leq'$. For top-lunes L and M, we say $\mathrm{L} \leq' \mathrm{M}$ if there exists a lune N such that $\mathrm{L} = \mathrm{N} \circ \mathrm{M}$. (The case of N must be the base of M, so N is in general not a top-lune.) In this situation, the lune N is unique, and we denote it by L/M, thus, $\mathrm{L} = (\mathrm{L}/\mathrm{M}) \circ \mathrm{M}$. Observe: For $\mathrm{L} \leq' \mathrm{M} \leq' \mathrm{N}$, we have $(\mathrm{L}/\mathrm{M}) \circ (\mathrm{M}/\mathrm{N}) = \mathrm{L}/\mathrm{N}$.

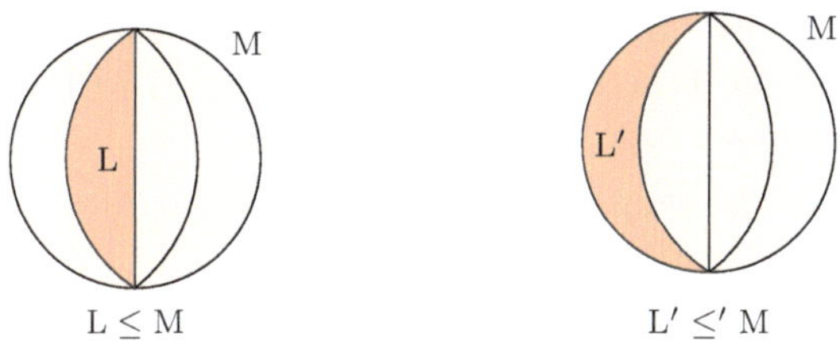

FIGURE 1.10. Partial orders on top-lunes.

A comparison of the two partial orders on top-lunes is shown in Figure 1.10. In both pictures, M is the visible hemisphere, and L and L′ are portions between two adjacent longitudes.

Exercise 1.14. Let $\mathrm{L} \leq' \mathrm{M}$ and $\mathrm{L} = \mathrm{s}(A, C)$. Show: There exists a unique face F such that $A \leq F \leq C$ and $\mathrm{M} = \mathrm{s}(F, C)$. Deduce: $\mathrm{L} \leq' \mathrm{M}$ implies $\mathrm{L} \leq \mathrm{M}$.

Exercise 1.15. Show: $\mathrm{L} \lessdot' \mathrm{M}$ iff there exists a lune N of slack one such that $\mathrm{L} = \mathrm{N} \circ \mathrm{M}$.

Exercise 1.16. Show: For any $\mathrm{L} \leq \mathrm{M}$, there exists L′ such that $\mathrm{b(L')} = \mathrm{b(L)}$ and $\mathrm{L'} \leq' \mathrm{M}$. (See Figure 1.10.)

Exercise 1.17. Show: If $\mathrm{L} \leq' \mathrm{N}$ and $\mathrm{L} \leq \mathrm{M} \leq \mathrm{N}$, then $\mathrm{M} \leq' \mathrm{N}$, however $\mathrm{L} \leq' \mathrm{M}$ does not hold in general. See the illustration below.

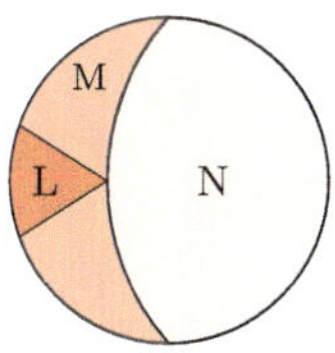

Here, N is the visible hemisphere, M is the portion between two adjacent longitudes, and L is a chamber.

1.2.5. Local bifaces and bilunes. Let $\mathcal{A}$ be an arrangement. A *local biface* of $\mathcal{A}$ is a triple (A, F, F') such that $F, F' \geq A$ and $\mathrm{s}(F) = \mathrm{s}(F')$. A *bilune* of $\mathcal{A}$ is an equivalence class in the set of local bifaces under the relation

$$(A, F, F') \sim (B, G, G') \iff AB = A,\ BA = B,\\ AG = F,\ BF = G,\ AG' = F',\ BF' = G'.$$

We write $\mathrm{s}(A, F, F')$ for the class of (A, F, F') and call it the *support* of (A, F, F'). It yields the two lunes $\mathrm{s}(A, F)$ and $\mathrm{s}(A, F')$. Alternatively, a bilune is a pair of lunes $(\mathrm{L}, \mathrm{L}')$ such that $\mathrm{b}(\mathrm{L}) = \mathrm{b}(\mathrm{L}')$ and $\mathrm{c}(\mathrm{L}) = \mathrm{c}(\mathrm{L}')$. We refer to these two flats as the base and case of $(\mathrm{L}, \mathrm{L}')$ and denote them by $\mathrm{b}(\mathrm{L}, \mathrm{L}')$ and $\mathrm{c}(\mathrm{L}, \mathrm{L}')$, respectively.

Bilunes of the form $\mathrm{s}(O, F, F')$ identify with the biface (F, F'). A bilune $(\mathrm{L}, \mathrm{L}')$ is a biface iff $\mathrm{b}(\mathrm{L}, \mathrm{L}') = \perp$.

1.2.6. Category of bilunes. For an arrangement $\mathcal{A}$, define the *category of bilunes* as follows. Its objects are flats and morphisms are bilunes. More precisely, a morphism from Y to X is a bilune $(\mathrm{L}, \mathrm{L}')$ with base X and case Y. Composition is done coordinatewise, that is,

$$(\mathrm{L}, \mathrm{L}') \circ (\mathrm{M}, \mathrm{M}') := (\mathrm{L} \circ \mathrm{M}, \mathrm{L}' \circ \mathrm{M}').$$

Identity morphisms are bilunes of the form (X, X) for some flat X.

1.2.7. Nested face-types and lune-types. For a reflection arrangement $\mathcal{A}$, a *nested face-type* of $\mathcal{A}$ is a pair (T, U) of face-types such that $T \leq U$. Alternatively, a nested face-type is an orbit of the W-action on nested faces. The *type* of a nested face (F, G), denoted $\mathrm{t}(F, G)$, is the orbit to which it belongs, namely, $(\mathrm{t}(F), \mathrm{t}(G))$.

For nested face-types (T, U) and (T', U'), we say $(T, U) \sim (T', U')$ if there exist nested faces (F, G) and (F', G') of the same support and of types (T, U) and (T', U'), respectively. A *lune-type* is an equivalence class under this relation on nested face-types. We write $\mathrm{s}(T, U)$ for the class of (T, U) and call it the *support* of (T, U). Alternatively, a lune-type is an orbit of the W-action on lunes. The *type* of a lune L, denoted $\mathrm{t}(\mathrm{L})$, is the orbit to which it belongs. We have the commutative diagram

$$\begin{array}{ccc} \text{nested face} & \overset{\mathrm{t}}{\longmapsto} & \text{nested face-type} \\ {\scriptstyle \mathrm{s}}\big\downarrow & & \big\downarrow{\scriptstyle \mathrm{s}} \\ \text{lune} & \underset{\mathrm{t}}{\longmapsto} & \text{lune-type.} \end{array} \tag{1.6}$$

This extends diagram (1.4): we view a face F as the nested face (F, F), and the flat $\mathrm{s}(F)$ as the lune $\mathrm{s}(F, F)$.

We also have the commutative diagram

$$
\begin{array}{ccc}
\text{top-nested face} & \overset{\mathrm{t}}{\longmapsto} & \text{top-nested face-type} \\
{\scriptstyle \mathrm{s}}\big\downarrow & & \big\downarrow{\scriptstyle \mathrm{s}} \\
\text{top-lune} & \underset{\mathrm{t}}{\longmapsto} & \text{top-lune-type}
\end{array}
\tag{1.7}
$$

which sits inside diagram (1.6). Top-nested face-types are in bijection with face-types. Moreover, for face-types T and T', $\mathrm{s}(T,S) = \mathrm{s}(T',S)$ iff $\mathrm{s}(T) = \mathrm{s}(T')$; hence, top-lune-types are in bijection with flat-types.

Let $\overline{\Lambda}[\mathcal{A}]$ denote the set of lune-types.

1.2.8. Local biface-types and bilune-types. For a reflection arrangement $\mathcal{A}$, a *local biface-type* of $\mathcal{A}$ is an orbit of the W-action on local bifaces. The *type* of a local biface (A, F', F), denoted $\mathrm{t}(A, F', F)$, is the orbit to which it belongs. For clarity, we write a local biface-type σ as a tuple (Z, T', σ, T), where Z, T, T' are types of A, F, F' with $\mathrm{t}(A, F', F) = \sigma$. In the language of Section 1.7.6, a local biface-type (Z, T', σ, T) is a morphism $\sigma : T \to T'$ local to Z.

As for biface-types, a local biface-type also carries finer information than a triple of face-types (Z, T', T) with $T, T' \geq Z$ and $\mathrm{s}(T/Z) = \mathrm{s}(T'/Z)$. That is, $\mathrm{t}(A, F', F) \neq (\mathrm{t}(A), \mathrm{t}(F'), \mathrm{t}(F))$.

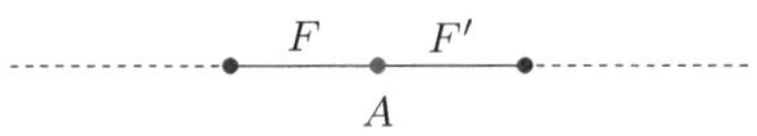

The picture shows part of a rank-two flat in a reflection arrangement. The local bifaces (A, F, F) and (A, F', F) represent different local biface-types, but have the same underlying triple of face-types.

We say two local biface-types are equivalent if there exist two local bifaces of the same support whose types are the given local biface-types. A *bilune-type* is an equivalence class under this relation on local biface-types. We write $\mathrm{s}(Z, T', \sigma, T)$ for the class of (Z, T', σ, T) and call it the *support* of (Z, T', σ, T). Alternatively, a bilune-type is an orbit of the W-action on bilunes. The *type* of a bilune $(\mathrm{L}', \mathrm{L})$, denoted $\mathrm{t}(\mathrm{L}', \mathrm{L})$, is the orbit to which it belongs. We have the commutative diagram

$$
\begin{array}{ccc}
\text{local biface} & \overset{\mathrm{t}}{\longmapsto} & \text{local biface-type} \\
{\scriptstyle \mathrm{s}}\big\downarrow & & \big\downarrow{\scriptstyle \mathrm{s}} \\
\text{bilune} & \underset{\mathrm{t}}{\longmapsto} & \text{bilune-type.}
\end{array}
\tag{1.8}
$$

This extends diagram (1.6): we view a nested face (A, F) as the local biface (A, F, F), and a lune L as the bilune (L, L).

Diagrams (1.4), (1.6), (1.8) can be seen as instances of a general diagram, see [**10**, Section 5.2.2, Diagram (5.2)].

1.3. Stabilizer subgroups of a Coxeter group

Let $\mathcal{A}$ be a reflection arrangement with Coxeter group W. We now look at stabilizer subgroups of W arising from the W-action on faces, flats, top-lunes, and interrelationships between these subgroups.

1.3.1. Stabilizer subgroups. A semidirect product. We say a Coxeter symmetry fixes a flat X *pointwise* if it fixes every point of X, with the latter viewed as a subspace of the ambient space.

For a face F, put

$$W_F := \{w \in W \mid w(F) = F\}.$$

This is the subgroup of W which leaves F invariant. Fact: Elements of W_F fix the flat $\mathrm{s}(F)$ pointwise. Thus $W_F = W_G$ whenever $\mathrm{s}(F) = \mathrm{s}(G)$. The cardinality of W_F is the number of chambers greater than F.

For a flat X, put

$$\widehat{W}_{\mathrm{X}} := \{w \in W \mid w(\mathrm{X}) = \mathrm{X}\}.$$

Let W_{X} be the subgroup of $\widehat{W}_{\mathrm{X}}$ which fixes X pointwise. It is a normal subgroup of $\widehat{W}_{\mathrm{X}}$. It is a Coxeter group in its own right; the corresponding reflection arrangement is $\mathcal{A}_{\mathrm{X}}$. So the cardinality of W_{X} is the number of chambers in $\mathcal{A}_{\mathrm{X}}$, or equivalently, the number of top-lunes with base X. Note: $W_{\mathrm{X}} = W_F$ whenever $\mathrm{s}(F) = \mathrm{X}$. The number $|\lambda|$ in Definition 1.8 with $\lambda = \mathrm{t}(\mathrm{X})$ is a numerical measure of the difference between the groups W_{X} and $\widehat{W}_{\mathrm{X}}$. This claim is made precise below.

For a top-lune L, let

$$W_{\mathrm{L}} := \{w \in W \mid w(\mathrm{L}) = \mathrm{L}\}.$$

For a top-lune L with base X, there is an isomorphism of groups

$$\widehat{W}_{\mathrm{X}} \xrightarrow{\cong} W_{\mathrm{X}} \rtimes W_{\mathrm{L}}, \tag{1.9}$$

the latter being a semidirect product group, see [**10**, Proposition 5.8]. The group W_{L} has cardinality $|\lambda|$, where λ is the type of the base of L.

1.3.2. Conjugacy classes. Observe: W_F and W_G are conjugate subgroups if F and G have the same type. For a face-type T, let W_T denote the conjugacy class of subgroups W_F with F of type T. We have a canonical representative for W_T, namely, W_{F_0}, where F_0 is the face of C_r of type T. From now on, we always view W_T in this manner as a subgroup of W.

Similarly, $\widehat{W}_{\mathrm{X}}$ and $\widehat{W}_{\mathrm{Y}}$ are conjugate subgroups if X and Y have the same type. For a flat-type λ, let W_λ denote the conjugacy class of subgroups $\widehat{W}_{\mathrm{X}}$ with X of type λ. In contrast to face-types, there is no canonical representative for W_λ, but still we allow ourselves to treat it as a subgroup (in which case it is assumed that some particular representative has been chosen).

Similarly, W_{L} and W_{M} are conjugate subgroups if L and M have the same type. For a top-lune-type ρ, let W_ρ denote the conjugacy class of subgroups W_{L} with L of type ρ. By picking a particular representative, we view W_ρ as a subgroup of W.

Exercise 1.18. Let L and L′ be top-lunes, and F and F' be faces supported by the bases of L and L′, respectively, such that $\mathrm{t}(F) = \mathrm{t}(F')$. Show: There exists a unique $u \in W$ such that $u(\mathrm{L}) = \mathrm{L}'$ and $u(F) = F'$. When L and L′ have the same base X, the element u belongs to $\widehat{W}_{\mathrm{X}}$, and when $F = F'$, it belongs to W_F.

1.4. Gallery distance, gate property, length

We review notions of gallery distance, gate property, q-distance function on faces, length of a Coxeter symmetry (with emphasis on longest elements) and weak left Bruhat order.

1.4.1. Gallery distance. Let $\mathcal{A}$ be an arrangement. For chambers C and D, let $\mathrm{dist}(C, D)$ denote the number of hyperplanes which separate C and D. This is the *gallery distance* between C and D. For chambers C, D, E, we write $C \text{--} D \text{--} E$ if $\mathrm{dist}(C, E) = \mathrm{dist}(C, D) + \mathrm{dist}(D, E)$. We represent this schematically as follows.

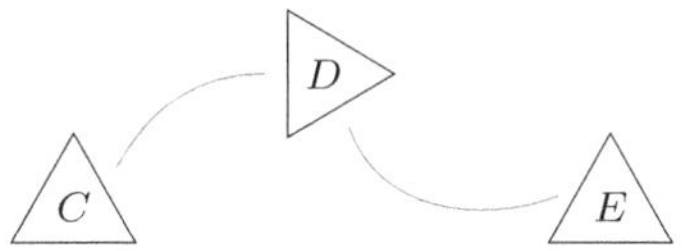

This is a minimal gallery from C to E passing through D. (Formally, a minimal gallery from C to D is a sequence of distinct chambers starting at C and ending at D such that adjacent chambers share a panel and supports of these panels are all distinct.)

More generally, for faces F and G, let $\mathrm{dist}(F, G)$ denote the number of hyperplanes which separate F and G. We have $\mathrm{dist}(F, G) = \mathrm{dist}(FG, GF)$, and $\mathrm{dist}(F, G) = 0$ iff $FG = GF$. If F and G have the same support, and $F \leq H$, then $\mathrm{dist}(F, G) = \mathrm{dist}(H, GH)$. For faces F, G, H of the same support, we write $F \text{--} G \text{--} H$ if $\mathrm{dist}(F, H) = \mathrm{dist}(F, G) + \mathrm{dist}(G, H)$. This is a minimal gallery from F to H passing through G.

Recall: Lunes of $\mathcal{A}$ with base X correspond to faces of $\mathcal{A}_{\mathrm{X}}$. For lunes L and M with base X, define $\mathrm{dist}(\mathrm{L}, \mathrm{M})$ to be the distance between faces L and M in the arrangement $\mathcal{A}_{\mathrm{X}}$.

1.4.2. Gate property. The *gate property* says that for chambers C and D, and H a face of D, we have a minimal gallery $D \text{--} HC \text{--} C$. More generally, for faces F, H, K with $H \leq K$, we have $KF \text{--} HFK \text{--} FK$. As a special case, when F and K have the same support with $H \leq K$, we have $K \text{--} HF \text{--} F$. This minimal gallery is illustrated in Figure 1.11.

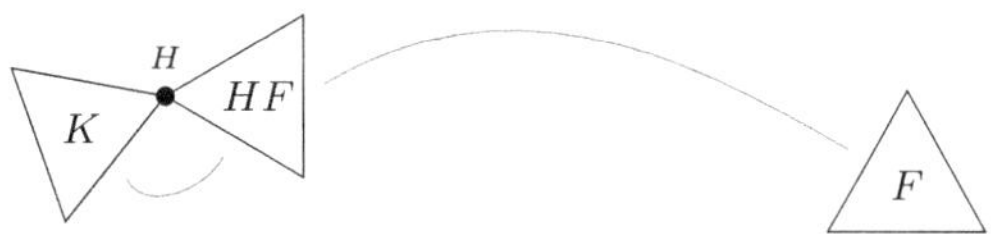

FIGURE 1.11. Gate property in terms of a minimal gallery.

1.4.3. q-distance function on faces. For a scalar q, define the *q-distance function* v_q on pairs of faces by

$$(v_q)_{F,G} := q^{\mathrm{dist}(F,G)}. \tag{1.10}$$

For $q = 1$, note that $v_1 \equiv 1$. For $q = -1$, we employ the term *signed distance function.*

1.4.4. W-invariant distance functions. The properties of the q-distance function v_q can be used to define an abstract notion of distance function v, see [**11**, Section 1.4.5] for a precise definition. For a reflection arrangement, we say v is W-invariant if $v_{F,G} = v_{wF,wG}$ for $w \in W$. Such a v is necessarily symmetric, that is, $v_{F,G} = v_{G,F}$, see [**10**, Section 8.3.3]. The basic example of a W-invariant distance function is indeed v_q.

1.4.5. Length of a Coxeter symmetry. Let (W, S) be a Coxeter system. The *length* of an element $w \in W$, denoted $l(w)$, is the smallest integer such that w can be expressed as a word in the generating set S whose length is that integer. Equivalently,

$$l(w) = \mathrm{dist}(C_r, wC_r), \tag{1.11}$$

where C_r is the reference chamber. (That the two definitions match requires an argument.) Since gallery distance is W-invariant,

$$\mathrm{dist}(uC_r, vC_r) = l(u^{-1}v)$$

for $u, v \in W$. For chambers uC_r and vC_r, define

$$d(uC_r, vC_r) := u^{-1}v. \tag{1.12}$$

This is the *W-valued gallery distance* between uC_r and vC_r. It is invariant under the diagonal action of W. Further,

$$d(E, C) = d(E, D)d(D, C) \tag{1.13}$$

for any chambers C, D, E.

Observe: $l(w) = l(w^{-1})$ and $d(C, D) = d(D, C)^{-1}$.

1.4.6. Longest elements and opposition. Let $w_\emptyset$ denote the longest element in W. It is the Coxeter symmetry which sends C_r to $\overline{C_r}$, and thus, $l(w_\emptyset) = \mathrm{dist}(C_r, \overline{C_r})$. The inverse of $w_\emptyset$ is itself. More generally, for a face-type T, let w_T denote the longest element in W_T. It is the Coxeter symmetry which sends C_r to $A\overline{C_r}$, where A is the face of C_r of type T, and thus, $l(w_T) = \mathrm{dist}(C_r, A\overline{C_r})$.

For a face-type T, define $\mathrm{dist}(T, \overline{T}) := \mathrm{dist}(F, \overline{F})$, where F is any face of type T. Alternatively,

$$\mathrm{dist}(T, \overline{T}) = l(w_\emptyset) - l(w_T). \tag{1.14}$$

This can be deduced from the gate property. In particular, $\mathrm{dist}(S, \overline{S}) = l(w_\emptyset)$. More generally, for face-types $Z \leq T$, define $\mathrm{dist}(T, Z\overline{T})$ by applying the above definition to $\boldsymbol{\alpha}_Z$. Alternatively, $\mathrm{dist}(T, Z\overline{T}) = l(w_Z) - l(w_T)$. In particular, $\mathrm{dist}(S, Z\overline{S}) = l(w_Z)$.

Observe: For face-types $Z \leq T$,

$$\operatorname{dist}(T, \overline{T}) = \operatorname{dist}(Z, \overline{Z}) + \operatorname{dist}(T, Z\overline{T}). \tag{1.15}$$

Exercise 1.19. Deduce (1.14) as a special case of (1.15).

Example 1.20 (Rank two). Consider the dihedral arrangement of d lines. Then

$$\operatorname{dist}(\emptyset, \overline{\emptyset}) = 0, \quad \operatorname{dist}(T, \overline{T}) = d - 1, \quad \operatorname{dist}(S, \overline{S}) = d,$$

with T a vertex-type. Further, for Z a vertex-type and T an edge-type, (1.15) takes the form $d = (d-1) + 1$.

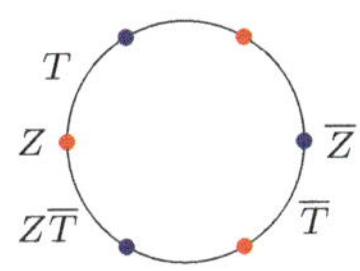

An illustration for $d = 3$ is shown above. □

1.4.7. Bruhat order. For $u, v \in W$, we say $u \leq v$ in the *weak left Bruhat order* on W if there exist chambers C, D, E such that $d(C, D) = u^{-1}$, $d(C, E) = v^{-1}$ and $C \,\text{--}\, D \,\text{--}\, E$. Alternatively,

$$u \leq v \text{ in } W \iff C_r \,\text{--}\, u^{-1}C_r \,\text{--}\, v^{-1}C_r. \tag{1.16}$$

In terms of the length function, $u \leq v$ iff $l(v) = l(vu^{-1}) + l(u)$.

More generally, for a face-type T, we have the weak left Bruhat order on the subgroup W_T which we denote $\leq_T$.

1.5. Modules over the Coxeter group

We set up terminology and notation for modules over a Coxeter group W. They play an important role in the theory developed in this book. They are implicit in Coxeter species and Coxeter spaces via functors (2.20), (2.21), and (8.30), and explicit in the decoration functor on Coxeter species (Section 4.9), the tensor and shuffle Coxeter bialgebras (Section 10.3), and the decorated Fock functors (Section 14.10).

1.5.1. Modules over the Coxeter group. Let $\Bbbk$ be a fixed field. By a left W-module M, we mean a left module M over $\Bbbk W$, the group algebra of W. It is the same as a $\Bbbk$-linear representation of W. We write $w \cdot m$ or $w(m)$ for the action of $w \in W$ on $m \in M$. The latter is consistent with the notation $w(F)$ that we use for the action of W on faces.

1.5.2. Trivial and sign representations. The *trivial representation* of W is the one-dimensional module $\Bbbk$ on which $w \in W$ acts by identity. The *sign representation* of W is the one-dimensional module $\Bbbk$ on which $w \in W$ acts by scalar multiplication by $(-1)^{l(w)}$. Recall that $l(w)$ depends on the reference chamber C_r; however, the parity of $l(w)$ is independent of this choice, see [**10**, Exercise 5.6]. For instance, a reflection always has odd length.

1.5.3. Operations on modules. Let M be a left W-module. Let M_- denote the left W-module obtained by tensoring M with the sign representation of W. Similarly, let M^* denote the dual left W-module with action given by $(w \cdot f)(m) := f(w^{-1} \cdot m)$ for $w \in W$, $f \in M^*$, $m \in M$. Observe:

$$(M_-)_- = M \quad \text{and} \quad (M^*)^* = M,$$

the latter when M is finite-dimensional.

1.5.4. Invariants and coinvariants. For a left W-module M, we write M_W and M^W for W-coinvariants and W-invariants of M, respectively. Since W_T is a subgroup of W, the notations M_{W_T} and M^{W_T} make sense for any face-type T. For general definitions of group (co)invariants, see for instance [**7**, Section 2.5.1].

Exercise 1.21. Let F be the face of C_r of type T. Let G be a face of type U of the same support as F. Define $w \in W$ by $w(GC_r) = C_r$. Check:

$$W_T \to W_U, \qquad u \mapsto wuw^{-1}$$

is a group isomorphism. Moreover, it is length-preserving. For a left W-module M, the map

$$M \to M, \qquad m \mapsto w \cdot m$$

is an isomorphism of modules, with the domain regarded as a W_T-module and codomain as a W_U-module. In other words, $w \cdot (u \cdot m) = wuw^{-1} \cdot (w \cdot m)$ for $m \in M$ and $u \in W_T$.

1.6. Descents, global descents, shuffles

We review notions of descents, global descents, shuffles. Descents and global descents come in two flavors: They are defined for pairs of chambers and for Coxeter symmetries. The two situations are related by the W-valued gallery distance. Moreover, these ideas are closely connected to the notion of shuffles, particularly, to longest shuffles.

We will see later that shuffles are implicit in the face-type formulation of Coxeter bialgebras (Section 8.7.3) and explicit in definitions of the tensor and shuffle Coxeter bialgebras (Section 10.3). See also Table V in the main introduction.

1.6.1. Descents and global descents of pairs of chambers. Let $\mathcal{A}$ be a simplicial arrangement. For chambers C and D, define $\mathrm{des}(C, D)$ to be the smallest face H of D such that $HC = D$. In other words,

$$HC = D \iff \mathrm{des}(C, D) \leq H \leq D. \tag{1.17}$$

We say $\mathrm{des}(C, D)$ is the *descent* of D wrt C. Similarly, for chambers C and D, define

$$\mathrm{gdes}(C, D) := \overline{C} \wedge D, \tag{1.18}$$

the meet of $\overline{C}$ and D. We say $\mathrm{gdes}(C, D)$ is the *global descent* of D wrt C. It is a face of D.

1.6.2. Descents and global descents of Coxeter symmetries. Let $\boldsymbol{\alpha} = (\mathcal{A}, C_r)$ be a pointed reflection arrangement. For $w \in W$, define $\mathrm{des}(w)$ to be the type of the face $\mathrm{des}(C_r, wC_r)$, with the latter defined by (1.17). We say $\mathrm{des}(w)$ is the *descent* of w. Similarly, for $w \in W$, define $\mathrm{gdes}(w)$ to be the type of the face $\mathrm{gdes}(C_r, wC_r)$, with the latter defined by (1.18). We say $\mathrm{gdes}(w)$ is the *global descent* of w. We may also apply these notions to $\boldsymbol{\alpha}_Z$ for a face-type Z. In this situation, for clarity, we use notations $\mathrm{des}_Z(w)$ and $\mathrm{gdes}_Z(w)$ for $w \in W_Z$.

The descent and global descent maps on pairs of chambers and on Coxeter symmetries fit into commutative diagrams

$$\begin{array}{ccc} \Gamma[\mathcal{A}] \times \Gamma[\mathcal{A}] & \xrightarrow{\mathrm{des}} & \Sigma[\mathcal{A}] \\ {\scriptstyle d}\downarrow & & \downarrow{\scriptstyle \mathrm{t}} \\ W & \xrightarrow[\mathrm{des}]{} & \overline{\Sigma}[\mathcal{A}] \end{array} \qquad \begin{array}{ccc} \Gamma[\mathcal{A}] \times \Gamma[\mathcal{A}] & \xrightarrow{\mathrm{gdes}} & \Sigma[\mathcal{A}] \\ {\scriptstyle d}\downarrow & & \downarrow{\scriptstyle \mathrm{t}} \\ W & \xrightarrow[\mathrm{gdes}]{} & \overline{\Sigma}[\mathcal{A}], \end{array} \tag{1.19}$$

where d is the W-valued gallery distance in (1.12).

1.6.3. Descents of pairs of faces. Now let F and G be any faces in a simplicial arrangement such that $GF = G$. Then FG and $F\overline{G}$ have the same support as G. So they are all chambers in the arrangement under the flat $\mathrm{s}(G)$. Define

$$\mathrm{des}(F, G) := \mathrm{des}(FG, G) \quad \text{and} \quad \overline{\mathrm{des}}(F, G) := \mathrm{des}(F\overline{G}, G), \tag{1.20}$$

where each rhs employs the definition of $\mathrm{des}(C, D)$. Both are faces of G. When F and G are both chambers, the two coincide and agree with the previous definition.

1.6.4. Shuffles. Fix a pointed reflection arrangement $\boldsymbol{\alpha} = (\mathcal{A}, C_r)$. For a face-type T, define the set of T-shuffles by

$$\mathrm{Sh}_T := \{\zeta \in W \mid \mathrm{des}(\zeta) \leq T\}. \tag{1.21}$$

As special cases, $\mathrm{Sh}_\emptyset = \{e\}$ and $\mathrm{Sh}_S = W$. A chamber obtained by projecting C_r on a face of type T is called a T-gate wrt C_r. Since $\zeta = d(C_r, \zeta C_r)$, it follows from (1.17) and the first diagram in (1.19) that all T-shuffles are of the form $d(C_r, D)$, as D varies over all T-gates. In other words,

$$\mathrm{Sh}_T = \{\zeta \in W \mid \zeta(C_r) \text{ is a } T\text{-gate}\}. \tag{1.22}$$

The two extreme choices of T-gates are C_r and $F\overline{C_r}$, with F the face of $\overline{C_r}$ of type T. These yield T-shuffles of shortest and longest lengths, respectively. Let us denote the longest T-shuffle by $\theta(T)$.

A "dual" way to think about T-shuffles is as follows. For a face-type T, define the set of T-deshuffles Sh^*_T by

$$\zeta \in \mathrm{Sh}_T \iff \zeta^{-1} \in \mathrm{Sh}^*_T\,.$$

We have

$$\mathrm{Sh}^*_T = \{\zeta^{-1} \in W \mid F(\zeta^{-1} C_r) = C_r\}, \tag{1.23}$$

with F the face of C_r of type T. In other words, T-deshuffles correspond to chambers in the T-based top-lune at C_r. See the illustrations in Figure 1.12 with T the red vertex-type.

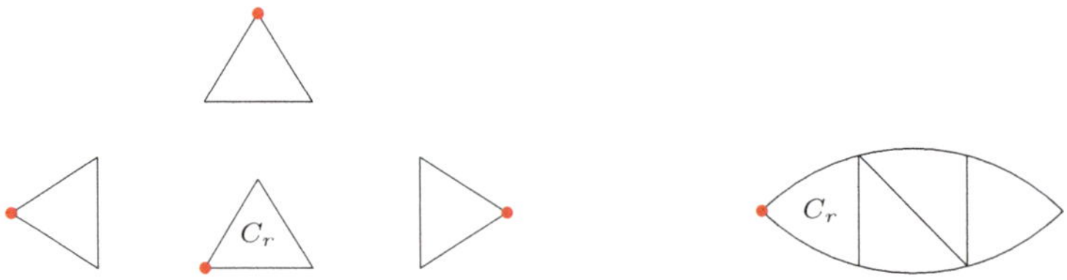

set of T-shuffles Sh_T set of T-deshuffles Sh^*_T

FIGURE 1.12. Two dual ways to picture shuffles.

We may also apply the above notions to $\boldsymbol{\alpha}_Z$ for a face-type Z. For $T \geq Z$, we write $\mathrm{Sh}_{T/Z}$ for the set of T/Z-shuffles in $\boldsymbol{\alpha}_Z$. Thus, $\mathrm{Sh}_{S/Z} = W_Z$. In general, $\mathrm{Sh}_{T/Z}$ is a subset of W_Z and hence of W. Important identities in the group algebra of W involving shuffles are given in Section 1.9.

1.6.5. Adjunctions. The maps des and gdes are left and right adjoints, respectively, to θ (with all three viewed as order-preserving maps between W and $\overline{\Sigma}[\mathcal{A}]$). That is, for a face-type T and $w \in W$,

$$\begin{aligned} \mathrm{des}(w) \leq T &\iff w \leq \theta(T), \\ T \leq \mathrm{gdes}(w) &\iff \theta(T) \leq w. \end{aligned} \tag{1.24}$$

There are similar adjunctions for des and gdes of pairs of chambers, see [**6**, Proposition 5.2.5].

1.7. Groupoid of biface-types

We introduce the groupoid of biface-types of a reflection arrangement. Its objects are face-types and morphisms are biface-types. Its connected components are indexed by flat-types. We define notions of biprojection, induced morphisms, local and opposite morphisms, Coxeter symmetry of a morphism, length of a morphism. We also reformulate the gate property in this language. The example of the coordinate arrangement is treated in Section 1.15.2 and of the braid arrangement in Section 8.14.1. A different approach to the groupoid of biface-types is given later in Section 8.1, see in particular Section 8.1.5. For similar ideas, see also the categories discussed in Section 2.1.

The groupoid of biface-types has connections to face-type enumeration (Section 1.8), to the invariant bilune-incidence algebra (Section 1.10), to the invariant q-Janus algebra (Section 1.12).

Notation 1.22. Let C_W denote the one-object category of the Coxeter group W.

1.7.1. Groupoid of biface-types. Let $\boldsymbol{\alpha} = (\mathcal{A}, C_r)$ be a pointed reflection arrangement. Define the category $\boldsymbol{\alpha}\text{-}\overline{\mathsf{Cox}}$ as follows. Objects are face-types and morphisms are biface-types. More precisely, a morphism $\sigma : T \to T'$ is a biface-type σ such that $\mathrm{t}(F) = T$ and $\mathrm{t}(F') = T'$ for any biface (F', F) which represents σ. Composition of morphisms is as follows. If (G, F) represents $\sigma : \mathrm{t}(F) \to \mathrm{t}(G)$ and (H, G) represents $\tau : \mathrm{t}(G) \to \mathrm{t}(H)$, then (H, F) represents $\tau\sigma : \mathrm{t}(F) \to \mathrm{t}(H)$. Thus,

$$\mathrm{t}(H, G)\,\mathrm{t}(G, F) = \mathrm{t}(H, F).$$

The identity morphism at T, denoted id_T, equals $\mathrm{t}(F, F)$ with F any face of type T. The category $\boldsymbol{\alpha}$-$\overline{\mathtt{Cox}}$ is a groupoid: if $\sigma = \mathrm{t}(F', F)$, then $\sigma^{-1} = \mathrm{t}(F, F')$. We call it the *groupoid of biface-types*.

Morphisms in $\boldsymbol{\alpha}$-$\overline{\mathtt{Cox}}$ and their composition can be visualized as in Figure 1.13. Also see the related Figure 1.2. The shading emphasizes that the involved face-types have the same support. In later pictures of morphisms, we will omit the shading.

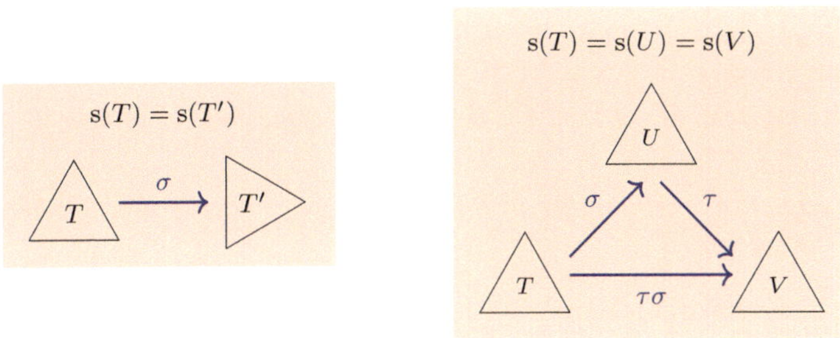

FIGURE 1.13. Morphisms in the groupoid of biface-types.

Example 1.23 (Ranks one and two). For a rank-one arrangement with face-types $\emptyset \lessdot S$, the groupoid of biface-types is

with $\sigma^2 = \mathrm{id}_S$. It has two connected components. Explicitly, $\mathrm{id}_\emptyset = \mathrm{t}(O, O)$, $\mathrm{id}_S = \mathrm{t}(C, C) = \mathrm{t}(\overline{C}, \overline{C})$ and $\sigma = \mathrm{t}(C, \overline{C}) = \mathrm{t}(\overline{C}, C)$, where C and $\overline{C}$ are the two chambers.

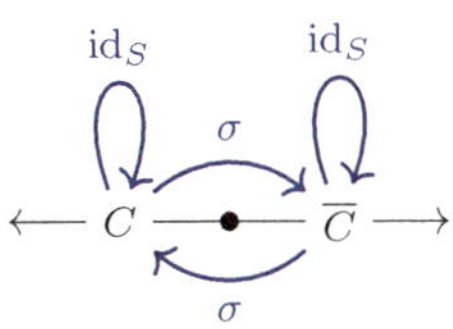

The picture above shows the two ways to visualize the morphism id_S and the two ways to visualize $\sigma : S \to S$ in the rank-one arrangement.

For the rank-two dihedral arrangement of d lines with face-types $\emptyset \lessdot T, U \lessdot S$, the groupoid of biface-types has 4 objects. For d odd, it has 3 connected components, while for d even, it has 4 connected components. Those involving vertex-types T

and U are

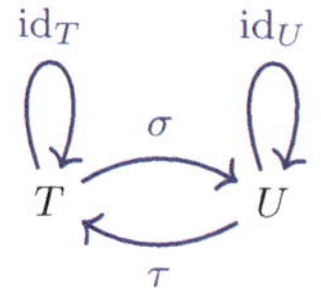

d odd

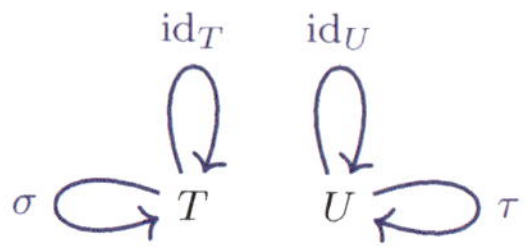

d even

with $\tau\sigma = \mathrm{id}_T$, $\sigma\tau = \mathrm{id}_U$ for d odd, and $\sigma^2 = \mathrm{id}_T$, $\tau^2 = \mathrm{id}_U$ for d even. The dependence on the parity of d arises because the opposite of a vertex of type T (or of type U) has either type T or type U depending on parity of d. See Figure 1.3. The remaining two components are as follows. The central face-type $\emptyset$ and its identity morphism are one component. The chamber-type S and its $2d$ morphisms are another component. □

Lemma 1.24. *Let $\sigma : T \to T'$ be a morphism. Then:*

(1) *For any face F of type T, there is a unique face F' of type T' such that $\mathrm{t}(F', F) = \sigma$.*
(2) *For any face F' of type T', there is a unique face F of type T such that $\mathrm{t}(F', F) = \sigma$.*

As a consequence, the number of bifaces of type σ equals the number of faces of type T or of type T'.

Proof. The two items imply each other by using σ^{-1}. Let us prove the first item. It has two parts, namely, existence of F' and uniqueness of F'. Existence: Start with any $\mathrm{t}(H', H) = \sigma$, and take any $w \in W$ such that $wH = F$, and put $F' = wH'$. Then $\mathrm{t}(F', F) = \sigma$. Uniqueness: Suppose $\mathrm{t}(F', F) = \mathrm{t}(F'', F) = \sigma$. Then there exists a $w \in W$ such that $w(F) = F$ and $w(F') = F''$. Since $w(F) = F$, it fixes $\mathrm{s}(F)$ pointwise, so $w(F') = F'$. Thus, $F' = F''$. □

Example 1.25 (Rank three). In the rank-three braid arrangement, let σ be the morphism from red-cyan to red-blue shown in the left picture below.

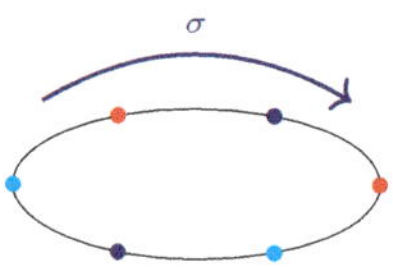

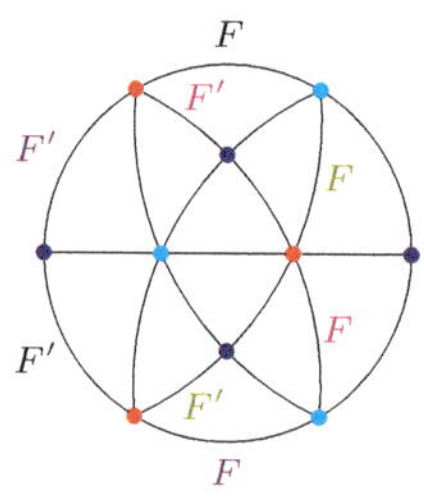

Four bifaces (F', F) which represent σ are shown in the right picture, distinguished from one another by color codes. Picking any F of type red-cyan determines F' uniquely, and similarly, picking any F' of type red-blue determines F uniquely. The two are supported on the same flat separated by a red-blue edge. There are eight more bifaces (F', F) which represent σ; in all of these either F or F' or both are on the backside. □

1.7.2. Connected components. In a groupoid, two objects lie in the same connected component iff there is a morphism from one object to the other. For related discussion including the skeleton of a groupoid, see [**7**, Section A.3.3].

Exercise 1.26. Show: There is a morphism from T to T' in $\boldsymbol{\alpha}\text{-}\overline{\mathtt{Cox}}$ iff T and T' have the same support (and in particular the same rank). In this case, there are $|\lambda|$ morphisms from T to T', where $\lambda = \mathrm{s}(T) = \mathrm{s}(T')$ and $|\lambda|$ is as in Definition 1.8. (Use Lemma 1.24.)

As a consequence:

Lemma 1.27. *The connected components of* $\boldsymbol{\alpha}\text{-}\overline{\mathtt{Cox}}$ *are indexed by flat-types. Instead of flat-types, one may also use top-lune-types, see* (1.7).

Example 1.28 (Rank three). The three pictures below show the connected components of the groupoid of biface-types indexed by rank-two flat-types in the braid arrangement and the arrangement of type B of rank three.

The number on each loop indicates the number of morphisms from that object to itself. The number in the middle of a bidirectional arrow indicates that there are those many morphisms in either direction. The numbers 2, 8, 4 agree with those for $|\lambda|$ in Example 1.9.

We leave it as an exercise to work out the composition rule for these morphisms, and in particular, to determine the groups of order 8 and 4 consisting of loops on objects. □

Exercise 1.29 (Minimum, maximum flat-types). Check: For the minimum flat-type, the connected component consists of the central face-type $\emptyset$ and its identity morphism. For the maximum flat-type, the connected component consists of the chamber-type S and its $|W|$ morphisms. More precisely,

$$\boldsymbol{\alpha}\text{-}\overline{\mathtt{Cox}}(S,S) \xrightarrow{\cong} W, \qquad \mathrm{t}(D,C) \mapsto d(D,C) \tag{1.25}$$

is a group isomorphism, with d the W-valued gallery distance in (1.12). (Use (1.13).) Following Notation 1.22, we have an inclusion functor

$$\mathsf{C}_W \hookrightarrow \boldsymbol{\alpha}\text{-}\overline{\mathtt{Cox}} \tag{1.26}$$

which identifies C_W with the connected component of S.

The skeleton of the groupoid $\boldsymbol{\alpha}\text{-}\overline{\mathtt{Cox}}$ amounts to picking, for each flat-type λ, a face-type T of support λ and equipping it with the group $\boldsymbol{\alpha}\text{-}\overline{\mathtt{Cox}}(T,T)$. This is the same as the skeletal category in Exercise 8.11.

1.7.3. Coxeter symmetry of a biface-type. We build on Lemma 1.24. For a biface-type σ, there are two canonical bifaces of type σ, namely, those whose either first or second coordinate is a face of C_r. Related ideas are presented below.

Lemma 1.30. *For a biface-type σ, the following two ways to define the element $w_\sigma \in W$ are equivalent.*

(1) $w_\sigma(F'C_r) = C_r$, *where* $\mathrm{t}(F', F) = \sigma$ *with* $F \leq C_r$.
(2) $w_\sigma(C_r) = GC_r$, *where* $\mathrm{t}(G', G) = \sigma$ *with* $G' \leq C_r$.

See the illustration in Figure 1.14. Note: $w_\sigma(F) = G$ and $w_\sigma(F') = G'$.

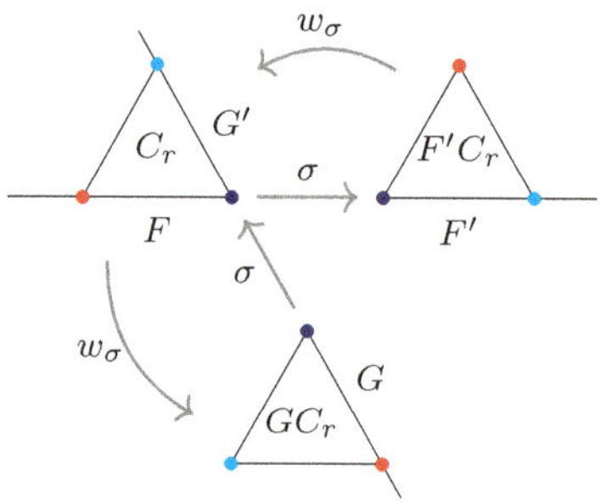

FIGURE 1.14. Coxeter symmetry w_σ of a biface-type σ.

PROOF. Let w_σ be defined as in item (1). Firstly, $w_\sigma(F') = G'$ since w_σ is type-preserving. Further, since $\mathrm{t}(F', F) = \mathrm{t}(G', G)$, Lemma 1.24 implies $w_\sigma(F) = G$. Now $C_r = F(F'C_r)$. Hence, $w_\sigma(C_r) = w_\sigma(F)w_\sigma(F'C_r) = GC_r$. This is how w_σ is defined in item (2). □

We refer to w_σ as the Coxeter symmetry of σ. In the language of Section 8.1.2, w_σ is the Coxeter symmetry which appears in either of the two normal forms of σ, see (8.2).

Exercise 1.31. Show: For σ and τ composable morphisms, $w_\tau w_\sigma = w_{\tau\sigma}$ and $w_{\mathrm{id}} = \mathrm{id}$. A helpful picture is shown below.

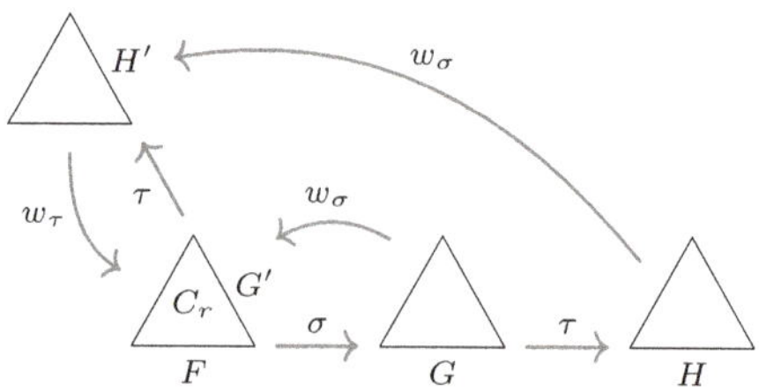

This defines a functor (groupoid morphism)

$$\alpha\text{-}\overline{\mathsf{Cox}} \to \mathsf{C}_W \tag{1.27}$$

which on objects sends T to the unique object of C_W, and on morphisms sends σ to w_σ. As a formal consequence, $(w_\sigma)^{-1} = w_{\sigma^{-1}}$.

Exercise 1.32. Check: For $\sigma : S \to S$, $w_\sigma = d(D, C)$, where C and D are any chambers with $\mathrm{t}(D, C) = \sigma$. In other words, the group isomorphism (1.25) sends σ to w_σ. Thus, the inclusion functor (1.26) is a section of (1.27).

1.7.4. Face-types biprojected by a biface-type. Let $\sigma : U \to U'$ be a morphism in $\boldsymbol{\alpha}\text{-}\overline{\mathtt{Cox}}$. Let $T \leq U$ and $T' \leq U'$. We say σ

- *projects* T on T' if $F'F = G'$,
- *coprojects* T' on T if $FF' = G$,
- *biprojects* T and T' if $FF' = G$ and $F'F = G'$,

where (G', G) is any biface which represents σ, and F is the face of G of type T, and F' is the face of G' of type T'. This can be visualized as in Figure 1.15.

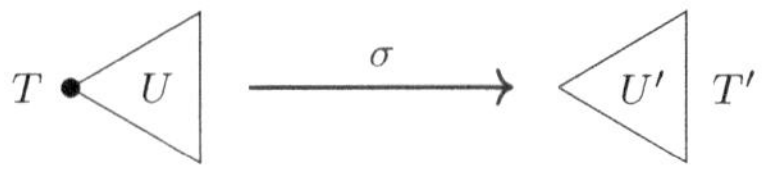

σ biprojects T and T'

The special cases $T' = U'$ and $T = U$ are, respectively, highlighted below.

σ coprojects U' on T $\qquad$ σ projects U on T'

FIGURE 1.15. Face-types biprojected by a biface-type.

Example 1.33 (Rank three). In the rank-three arrangement of type B, consider the following three examples of σ, with U and U' as edge-types and T and T' as vertex-types. Since σ is between edge-types, all the action is happening on a rank-two flat.

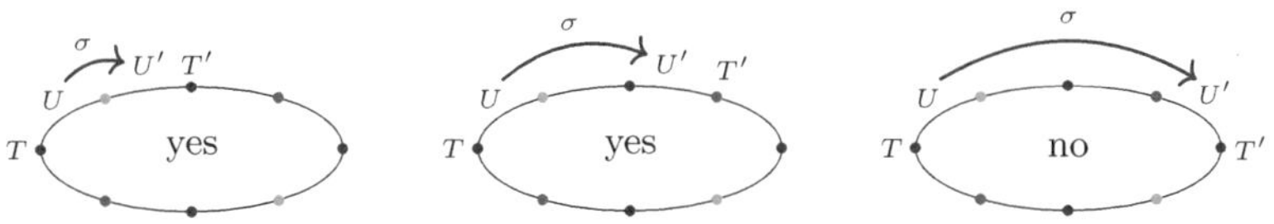

In the first two cases, σ biprojects T and T', while in the last one, it does not. □

Observation 1.34. Let $\sigma : U \to U'$ with $T \leq U$ and $T' \leq U'$. Then the following are equivalent.

- σ biprojects T and T'.
- σ projects T on T' and coprojects T' on T.
- σ projects T on T' and coprojects U' on T.
- σ projects U on T' and coprojects T' on T.

Also,

- σ projects T on T' iff σ^{-1} coprojects T on T',
- σ coprojects T' on T iff σ^{-1} projects T' on T,
- σ biprojects T and T' iff σ^{-1} biprojects T' and T.

Exercise 1.35 (Shuffles). Recall shuffles from Section 1.6.4. Check:

- $\sigma : S \to S$ projects S on T iff w_σ is a T-deshuffle.
- $\sigma : S \to S$ coprojects S on T iff w_σ is a T-shuffle.

The two statements imply each other by taking inverses. (Use Exercise 1.32.)

Exercise 1.36. Let F and F' be faces of C_r of types T and T', respectively. Check: The following pieces of data are equivalent.

(1) A morphism σ which biprojects T and T'.
(2) A face G' of type T' such that $FG' \leq C_r$.
(3) A face G of type T such that $F'G \leq C_r$.

The bifaces $(G'F, FG')$ and $(F'G, GF')$ represent σ.

1.7.5. Induced morphisms. Let $\sigma : T \to T'$ be a morphism. Write $\sigma = \mathrm{t}(H', H)$. Let $F \geq H$ and $F' \geq H'$ be such that $F = HF'$ and $F' = H'F$. This induces a morphism $\mathrm{t}(F', F)$ which we write as $\sigma : U \to U'$ (where U and U' are the types of F and F'). Note very carefully: The same letter σ is used for the given morphism and its induced morphism. We represent them schematically as

$$\begin{array}{ccc} U & \xrightarrow{\sigma} & U' \\ | & & | \\ T & \xrightarrow[\sigma]{} & T'. \end{array} \tag{1.28}$$

Also see the illustrations in Figure 1.16. In the first picture, T and T' are types of two vertices opposite to each other.

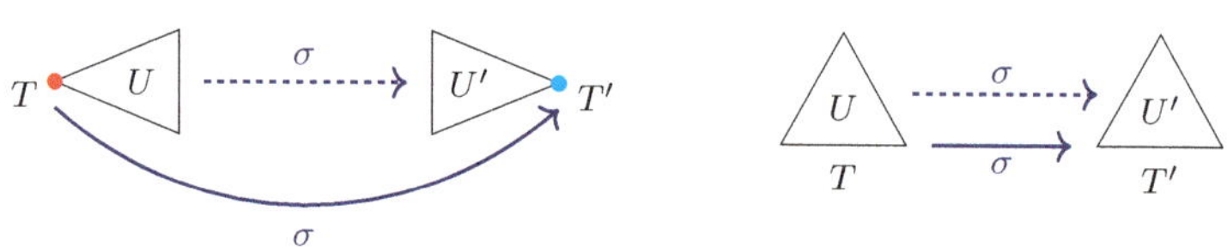

FIGURE 1.16. Induced morphisms in the groupoid of biface-types.

We write $U' = \sigma_*(U)$ and $U = \sigma^*(U')$. See also the explanation after (8.12). The w_σ defined in Lemma 1.30 for an induced σ matches the w_σ for the original σ, see Exercise 8.20.

Exercise 1.37. Show: The nested face-types (T, U) and (T', U') have the same support iff there exists a morphism $\sigma : T \to T'$ such that $\sigma_*(U) = U'$, or equivalently, $U = \sigma^*(U')$. In other words, $\mathrm{s}(T, U) = \mathrm{s}(T', U')$ iff there exists a diagram of the form (1.28).

Exercise 1.38. Give an example of $T \leq U$ and $\sigma : T \to T$ for which $\sigma_*(U) \neq U$.

1.7.6. Local morphisms. A morphism $\sigma : T \to T'$ is *local* to a face-type Z if $Z \leq T$ and $Z \leq T'$ and σ can be represented by a biface (F', F) such that F and F' share a common face A of type Z. Let $\boldsymbol{\alpha}_Z\text{-}\overline{\texttt{Cox}}$ denote the subcategory of $\boldsymbol{\alpha}\text{-}\overline{\texttt{Cox}}$ whose objects are face-types greater than Z and morphisms are those local to Z. See also the discussion in Section 8.2.2.

1.7.7. Opposite morphisms. For a face-type T, let $\beta_{\overline{T},T}$ denote the morphism from T to $\overline{T}$ represented by the biface $(\overline{F}, F)$ with F of type T. Observe: $\beta_{T,\overline{T}}\beta_{\overline{T},T} = \mathrm{id}$.

More generally, for any $Z \leq T$, let $\beta_{Z\overline{T},T}$ denote the morphism from T to $Z\overline{T}$ represented by $(A\overline{F}, F)$ with A of type Z, F of type T, and $A \leq F$. It is illustrated in Figure 1.17. In the first and third pictures, T covers Z.

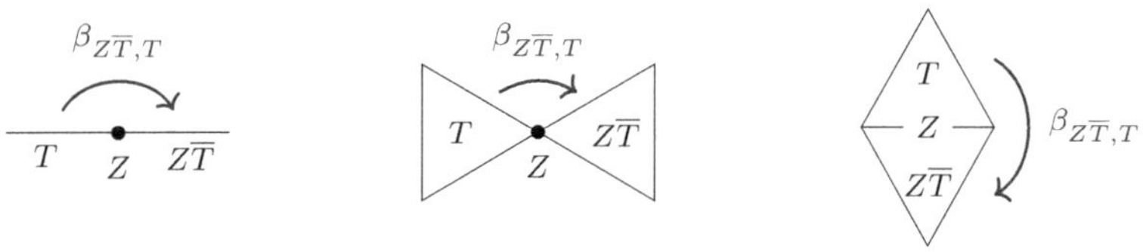

FIGURE 1.17. Opposite morphisms in the groupoid of biface-types.

The morphism $\beta_{Z\overline{T},T}$ is local to Z. When $Z = \emptyset$, we recover $\beta_{\overline{T},T}$. An alternative viewpoint on these morphisms is given in (8.17) and (8.19).

Observation 1.39. The Coxeter symmetry w_σ of the morphism $\sigma := \beta_{\overline{S},S}$ is $w_\emptyset$, the longest element in W. More generally, for a face-type Z, the Coxeter symmetry of $\sigma := \beta_{Z\overline{S},S}$ is w_Z.

1.7.8. Length of a biface-type. For a biface-type σ, define its *length* by

$$l(\sigma) := \mathrm{dist}(F, F'), \tag{1.29}$$

where $\mathrm{t}(F', F) = \sigma$. Observe: $l(\sigma) = l(\sigma^{-1})$. Also, $l(\sigma) = 0$ iff $\sigma = \mathrm{id}_T$ for some face-type T.

1.7.9. Gate property. For a morphism $\rho : U \to V$ and $T \leq V$, there exists a unique $\sigma : U \to V'$ which projects U on T and a unique $\tau : V' \to V$ local to T such that $\rho = \tau\sigma$ and $l(\rho) = l(\tau) + l(\sigma)$. This is the *gate property* for the groupoid of biface-types. We say that $\tau\sigma$ is the factorization of the morphism ρ wrt the face-type T. It is illustrated in Figure 1.18. It reformulates the usual gate property which is in terms of a minimal gallery. Compare Figure 1.18 with Figure 1.11. The translation is $\rho = \mathrm{t}(K, F)$, $\tau = \mathrm{t}(K, HF)$, $\sigma = \mathrm{t}(HF, F)$ and $T = \mathrm{t}(H)$.

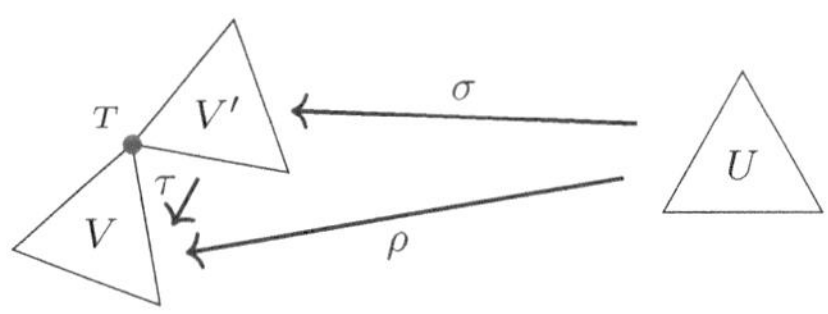

FIGURE 1.18. Gate property in terms of biface-types.

Note: There is a "dual" way of formulating the gate property by replacing morphisms σ, τ, ρ by their inverses. It can be pictured as in Figure 1.18 by reversing the direction of the arrows.

Exercise 1.40. Let $\sigma : T \to T'$ be a morphism and $\sigma : U \to U'$ a morphism induced from it as in (1.28). Let V be any face-type. Check: There is a unique bijection between morphisms τ starting from U which biproject T and V, and morphisms ρ starting from U' which biproject T' and V, with the property that $\rho\sigma\tau^{-1}$ is local to V. See the left picture below.

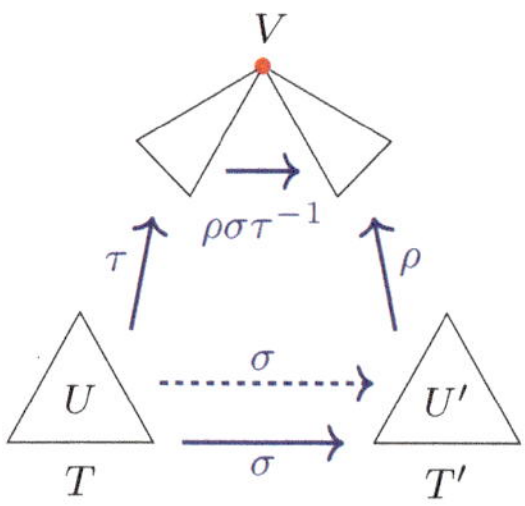

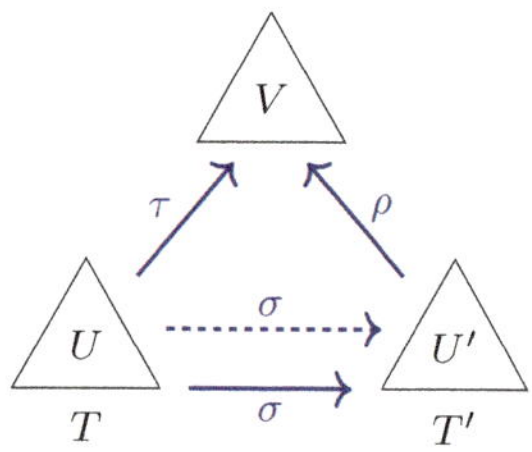

As a special case: There is a unique bijection between morphisms $\tau : U \to V$ which coproject V on T, and morphisms $\rho : U' \to V$ which coproject V on T' defined by $\tau = \rho\sigma$. See the right picture above.

Exercise 1.41. Deduce: For face-types T and V, there is a bijection between morphisms τ which biproject T and V, and morphisms ρ which biproject $\overline{T}$ and V defined as follows. If $\mathrm{t}(G, F) = \tau$, then $\mathrm{t}(B\overline{G}, \overline{A}F) = \rho$, where A and B are faces of F and G of types T and V, respectively. See the picture below.

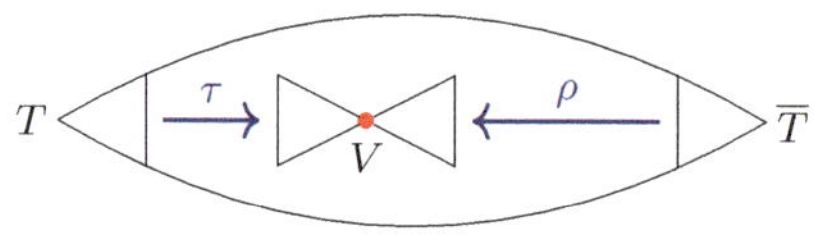

(Apply Exercise 1.40 to $\sigma = \beta_{\overline{T},T}$. In this situation, we also have $l(\sigma) = l(\tau) + l(\rho\sigma\tau^{-1}) + l(\rho)$. See [**10**, Proposition 1.64], [**11**, (1.20)] in this regard.)

1.8. Face-type enumeration

Consider the polynomial ring $\mathbb{Z}[q]$ in the variable q. We define several polynomials in q related to face-type enumeration in a reflection arrangement and bring out their interconnections, see Table 1.4. The polynomial $d_T(q)$ enumerates faces of type T. It specializes to the Poincaré polynomial of the arrangement when $T := S$. More generally, $a^{TUT'}(q)$ enumerates faces of type T' in a lune of type $\mathrm{s}(T, U)$. More refined counting is done by $a^{TUT'U'}(q)$ where (T, U) and (T', U') play a symmetric role. Each of these enumerations can also be done local to a face-type Z; we indicate that by the subscript Z.

Enumeration polynomials have intimate connections to the groupoid of biface-types (Section 1.7). They are also related to zeta and Möbius functions (Section 1.10) and to structure constants of the invariant Tits algebra and invariant q-Janus algebra (Section 1.12). We record several identities involving these polynomials under the broad umbrella of gate identities and alternating

TABLE 1.4. Face-type enumeration polynomials.

Polynomials in q		Interrelationships
$a^{TUT'U'}(q)$	$a_Z^{TUT'U'}(q)$	$a^{TUT'U'}(q) = a^{T'U'TU}(q)$
$a^{TUT'}(q)$	$a_Z^{TUT'}(q)$	$a^{TUT'}(q) = \sum_{U'} a^{TUT'U'}(q)$
$a^{TT'}(q)$	$a_Z^{TT'}(q)$	$a^{TT'}(q) = a^{T'T}(q) = \sum_{U,U'} a^{TUT'U'}(q)$
$a^{TT'}(q,\xi)$	$a_Z^{TT'}(q,\xi)$	$a^{TT'}(q, \xi \equiv 1) = a^{TT'}(q)$
$d_T(q)$	$d_{T/Z}(q)$	$d_T(q) = a^{SST}(q) = a^{TSS}(q) = a^{ST}(q)$
$c^{TT'}(q)$	$c_Z^{TT'}(q)$	$c^{TT'}(q) = c^{T'T}(q) = a^{TTT'}(q)$

sign identities. These can also be viewed as numerical identities by specializing q to a scalar in any field $\Bbbk$. Supplementary information on some of this material can be found in [**10**, Sections 16.4, 16.5, 16.10].

The above identities can be proved directly as in [**10**]. More conceptual proofs using ideas developed in this monograph are given later. We elaborate briefly. Face-type enumeration is intimately connected to the study of Coxeter bialgebras in Part II, see Section 8.9. The gate identities follow from the existence of the Coxeter q-bialgebras of polynomials and of face-types (Sections 10.2.4 and 10.4.4) which can be established using either universal constructions or Fock functors (Sections 18.2.3 and 18.4.2). Similarly, the alternating sign identities follow from antipode considerations. See also Table IV in the main introduction.

In this section, we fix a pointed reflection arrangement $\boldsymbol{\alpha} = (\mathcal{A}, C_r)$ with Coxeter group W.

1.8.1. Counting of face-types. For a face-type T in $\boldsymbol{\alpha}$, define d_T to be the number of faces of type T. In particular, $d_\emptyset = 1$ and d_S is the number of chambers which is the same as $|W|$. More generally, for face-types $T \leq U$, define $d_{U/T}$ to be the number of faces of type U/T in $\boldsymbol{\alpha}_T$.

For a face-type T, define the polynomial

$$d_T(q) := \sum_{F:\, \mathrm{t}(F)=T} q^{\mathrm{dist}(C_r, FC_r)} = \sum_{\zeta \in \mathrm{Sh}_T} q^{l(\zeta)}. \tag{1.30}$$

The first sum is over all faces F of type T. The second sum is over all T-shuffles (1.22), and $l(\zeta)$ is the length of ζ wrt C_r as in (1.11). We refer to $d_T(q)$ as the *face-type enumerator*. Note: $d_T(1) = d_T$. Alternatively, using the dual viewpoint of T-deshuffles given by (1.23),

$$d_T(q) = \sum_{C:\, FC=C_r} q^{\mathrm{dist}(C_r, C)} = \sum_{\zeta \in \mathrm{Sh}_T} q^{l(\zeta^{-1})}, \tag{1.31}$$

where F is the face of C_r of type T. Thus, we sum over chambers C in the T-based top-lune at C_r, and each summand contributes a power of q.

More generally, for face-types $T \le U$, define the polynomial $d_{U/T}(q)$ by applying (1.30) to the face-type U/T of $\boldsymbol{\alpha}_T$.

Lemma 1.42. *For face-types $T \le U \le V$,*

$$d_{V/T}(q) = d_{V/U}(q)\, d_{U/T}(q). \tag{1.32}$$

PROOF. We can prove this using either definition (1.30) or (1.31). Let us work with the latter. An illustration of the identity $d_U(q) = d_{U/T}(q) d_T(q)$ is shown in the picture below, with T a vertex-type and U an edge-type.

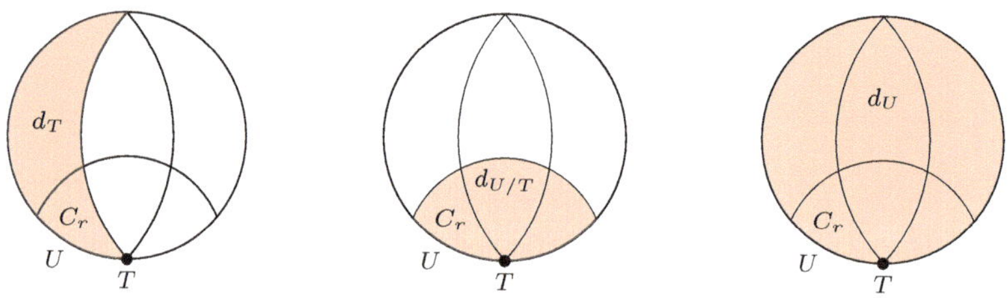

The number of chambers in the shaded regions are d_T, $d_{U/T}$, d_U, respectively. More generally, q-counting them using their distance from C_r yields the polynomials $d_T(q)$, $d_{U/T}(q)$, $d_U(q)$, respectively. The gate property ensures that powers of q match properly. □

We refer to (1.32) as the *multiplicative property* of face-type enumerators. It is part of a class of identities called gate identities, see Exercise 1.63.

Exercise 1.43. Show:

(1) If $\mathrm{s}(T, U) = \mathrm{s}(T', U')$, then $d_{U/T}(q) = d_{U'/T'}(q)$.
(2) If $\mathrm{s}(T) = \mathrm{s}(T')$, then $d_T(q) = d_{T'}(q)$.
(3) If $\mathrm{s}(T) = \mathrm{s}(T')$, then $d_T(-1) = (-1)^{\mathrm{dist}(F,F')} d_{T'}(-1)$, where F and F' are faces of the same support of types T and T', respectively.

(For item (1), use Fact 1.4. For item (2), use the multiplicative property (1.32) and item (1). Warning: item (3) is different from item (2) specialized to $q = -1$.)

1.8.2. Poincaré polynomial. Let us specialize the face-type enumerator to the maximum face-type S. We refer to $d_S(q)$ as the *Poincaré polynomial* of W. Explicitly,

$$d_S(q) = \sum_{C} q^{\mathrm{dist}(C_r, C)} = \sum_{w \in W} q^{l(w)}. \tag{1.33}$$

The first sum is over all chambers C, while the second sum is over all elements of W. Fact: This polynomial factorizes as

$$d_S(q) = \prod_{i=1}^{n} (1 + q + \cdots + q^{e_i}) \tag{1.34}$$

for certain positive integers $e_1, \dots, e_n$, with $n := \mathrm{rk}(S)$. These integers are called the *exponents* of the Coxeter system (W, S). The number 1 is always an exponent whenever $\operatorname{rk}(\boldsymbol{\alpha}) \geq 1$. As a consequence,

$$d_S(-1) = \sum_{w \in W} (-1)^{l(w)} = \begin{cases} 1 & \text{if } \boldsymbol{\alpha} \text{ has rank zero,} \\ 0 & \text{otherwise.} \end{cases} \tag{1.35}$$

This also follows from Exercise 1.43, item (3) (assuming $\operatorname{char}(\Bbbk) \neq 2$).

More generally, for a face-type T, the polynomial $d_{S/T}(q)$ is the Poincaré polynomial of W_T. It is given by

$$d_{S/T}(q) = \sum_{w \in W_T} q^{l(w)}. \tag{1.36}$$

As a special case of the multiplicative property (1.32),

$$d_S(q) = d_{S/T}(q) d_T(q). \tag{1.37}$$

Thus, any face-type enumerator $d_T(q)$ can be expressed as the quotient of two Poincaré polynomials.

1.8.3. Counting of face-types in lunes.

1.8.3. Counting of face-types in lunes. We now turn to more general considerations.

For face-types $T \leq U$ and T' in $\boldsymbol{\alpha}$, define $a^{TUT'}$ to be the number of faces F of type T' such that $HF = G$, where (H, G) is any fixed nested face of type (T, U). This can be interpreted as counting faces of type T' in the interior of the lune $\mathrm{s}(H, G)$. More generally, for face-types $Z \leq T \leq U$ and $Z \leq T'$, by applying the above definition to $\boldsymbol{\alpha}_Z$, we obtain the numbers $a_Z^{TUT'}$.

For face-types $T \leq U$ and T', define the polynomial

$$a^{TUT'}(q) := \sum_F q^{\operatorname{dist}(H,F)}, \tag{1.38}$$

with the sum over all faces F of type T' such that $HF = G$, where (H, G) is any fixed nested face of type (T, U). Note: $a^{TUT'}(1) = a^{TUT'}$. More generally, by applying (1.38) to $\boldsymbol{\alpha}_Z$, we obtain polynomials $a_Z^{TUT'}(q)$.

Exercise 1.44. Show: If $\mathrm{s}(T, U) = \mathrm{s}(T', U')$, then $a^{TUV} = a^{T'U'V}$ for any face-type V. However, under the same hypothesis, $a^{TUV}(q) \neq a^{T'U'V}(q)$ in general.

Special case. The face-type enumerators (1.30) are special cases of the polynomials (1.38) as follows. For a face-type T',

$$d_{T'}(q) = a^{SST'}(q) = a^{T'SS}(q). \tag{1.39}$$

In particular, $a^{SSS}(q)$ is the Poincaré polynomial of W.

Example 1.45 (Ranks one and two). For a rank-one arrangement with face-types $\emptyset \lessdot S$,

$$a^{\emptyset\emptyset\emptyset}(q) = a^{\emptyset SS}(q) = a^{SS\emptyset}(q) = 1, \ a^{\emptyset S\emptyset}(q) = a^{\emptyset\emptyset S}(q) = 0, \ a^{SSS}(q) = 1 + q.$$

In particular, $d_\emptyset(q) = 1$ and $d_S(q) = 1 + q$.

For the dihedral arrangement of d lines with face-types $\emptyset \lessdot T, U \lessdot S$,

$$a^{TTS}(q) = a^{UUS}(q) = 0,$$
$$a^{TSS}(q) = a^{USS}(q) = a^{SST}(q) = a^{SSU}(q) = 1 + q + \cdots + q^{d-1},$$
$$a^{SSS}(q) = (1 + q)(1 + q + \cdots + q^{d-1}),$$

$$a^{TTT}(q) = a^{UUU}(q) = \begin{cases} 1 & \text{if } d \text{ is odd,} \\ 1 + q^{d-1} & \text{if } d \text{ is even,} \end{cases}$$

$$a^{TTU}(q) = a^{UUT}(q) = \begin{cases} q^{d-1} & \text{if } d \text{ is odd,} \\ 0 & \text{if } d \text{ is even,} \end{cases}$$

$$a^{TSU}(q) = a^{UST}(q) = \begin{cases} 1 + q^2 + q^4 + \cdots + q^{d-3} & \text{if } d \text{ is odd,} \\ 1 + q^2 + q^4 + \cdots + q^{d-2} & \text{if } d \text{ is even,} \end{cases}$$

$$a^{TST}(q) = a^{USU}(q) = \begin{cases} q(1 + q^2 + q^4 + \cdots + q^{d-3}) & \text{if } d \text{ is odd,} \\ q(1 + q^2 + q^4 + \cdots + q^{d-4}) & \text{if } d \text{ is even and } d \geq 4, \\ 0 & \text{if } d = 2, \end{cases}$$

$$a^{\emptyset VY}(q) = a^{VY\emptyset}(q) = \begin{cases} 1 & \text{if } V = Y, \\ 0 & \text{otherwise,} \end{cases}$$

where V and Y are any of the face-types $\emptyset, T, U, S$. In particular,

$$d_\emptyset(q) = 1, \quad d_T(q) = d_U(q) = 1 + q + \cdots + q^{d-1}, \quad d_S(q) = (1+q)(1+q+\cdots+q^{d-1}),$$

the last one being the Poincaré polynomial. □

Exercise 1.46 (Rank three). The polynomials $a^{TUT'}(q)$ in some rank three arrangements are shown below, with T = red, U = red-blue, and T' varying over vertex-types.

	Figure 1.4	Figure 1.5	Figure 1.6
T' = red	0	q	0
T' = blue	1	1	$1 + q^5$
T' = cyan	0	0	q^2

Check these entries. Work out similar tables for other choices of T and U.

Exercise 1.47. Check: For face-types Z, V, Y,

$$a_Z^{ZVY}(q) = a_Z^{VYZ}(q) = \begin{cases} 1 & \text{if } V = Y, \\ 0 & \text{otherwise.} \end{cases} \tag{1.40}$$

1.8.4. Pairs of nested face-types. More refined counting in lunes.

For nested face-types (T, U) and (T', U'), define $a^{TUT'U'}$ to be the number of nested faces (F', G') of type (T', U') such that $FF' = G$ and $F'F = G'$ for a fixed nested face (F, G) of type (T, U). This is a count associated to the lune $\mathrm{s}(F, G)$. There is an inbuilt symmetry, namely, $a^{TUT'U'} = a^{T'U'TU}$. The latter is the number of nested faces (F, G) of type (T, U) such that $FF' = G$ and $F'F = G'$ for a fixed nested face (F', G') of type (T', U'). This is a count associated to the lune $\mathrm{s}(F', G')$.

A typical way to visualize a configuration of nested faces which contributes to $a^{TUT'U'}$ is shown below. Note: $\mathrm{s}(U) = \mathrm{s}(U')$.

Compare with the first picture in Figure 1.15. This connection is made precise in Section 1.8.11.

With notation as above, define the polynomial

$$a^{TUT'U'}(q) := \sum q^{\mathrm{dist}(G,G')}, \tag{1.41}$$

with the sum either over (F, G) or over (F', G') depending on the viewpoint. Note: $a^{TUT'U'}(1) = a^{TUT'U'}$. Symmetry continues to hold, that is,

$$a^{TUT'U'}(q) = a^{T'U'TU}(q). \tag{1.42}$$

The above polynomial can also be expressed as

$$a^{TUT'U'}(q) = \sum q^{l(w)}, \tag{1.43}$$

with the sum over those $w \in W$ such that

$$F_0'(wF_0) = G_0', \ (wF_0)F_0' = wG_0, \ (wG_0)C = wC,$$

where F_0, G_0, F_0', G_0' are faces of C_r of types T, U, T', U', respectively. More generally, by applying (1.41) to $\boldsymbol{\alpha}_Z$, we obtain polynomials $a_Z^{TUT'U'}(q)$.

Special case. The polynomials (1.38) are present in (1.41) as follows.

$$\sum_{U'} a^{TUT'U'}(q) = a^{TUT'}(q), \qquad \sum_{U} a^{TUT'U'}(q) = a^{T'U'T}(q). \tag{1.44a}$$

$$a^{TUT'T'}(q) = \begin{cases} a^{TUT'}(q) & \text{if } \mathrm{s}(T') = \mathrm{s}(U), \\ 0 & \text{otherwise.} \end{cases} \tag{1.44b}$$

$$a^{TST'S}(q) = a^{TST'}(q). \tag{1.44c}$$

Exercise 1.48 (Dual way of counting). We now give a dual approach to (1.38). Use either identity in (1.44a) to show: For face-types $T \leq U$ and T',

$$a^{TUT'}(q) = \sum_{H'} q^{\mathrm{dist}(H',F')}. \tag{1.45}$$

The sum is over all faces H' of type T such that $F'H' \leq E$ and $H'F'$ has type U. Here E is any fixed chamber and F' is its face of type T'. This is a count associated to the top-lune $\mathrm{s}(F', E)$. It is a top-lune of type $\mathrm{s}(T', S)$.

Exercise 1.49. Show: For face-types T and T',

$$a^{T'T'T}(q) = \sum_{\substack{U:\, U \geq T \\ \mathrm{s}(U) = \mathrm{s}(T')}} a^{TUT'}(q). \tag{1.46}$$

(Use (1.42), (1.44a), (1.44b).)

1.8.5. Pairs of face-types. For face-types T and T', define

$$a^{TT'}(q) := \sum_{U,U'} a^{TUT'U'}(q). \tag{1.47}$$

The sum is over both U and U'. Thus, $a^{TT'}(q) = a^{T'T}(q)$. Also note from (1.44a) that

$$a^{TT'}(q) = \sum_{U} a^{TUT'}(q) = \sum_{U'} a^{T'U'T}(q).$$

More generally, by applying (1.47) to $\boldsymbol{\alpha}_Z$, we obtain polynomials $a_Z^{TT'}(q)$.

Observe:

$$a^{ST}(q) = a^{SST}(q) = d_T(q).$$

In particular, $a^{SS}(q)$ is the Poincaré polynomial of W.

Example 1.50 (Ranks one and two). For a rank-one arrangement with face-types $\emptyset \lessdot S$,

$$a^{\emptyset\emptyset}(q) = a^{\emptyset S}(q) = a^{S\emptyset}(q) = 1, \quad a^{SS}(q) = 1 + q.$$

For the dihedral arrangement of d lines with face-types $\emptyset \lessdot T, U \lessdot S$,

$$\begin{gathered} a^{\emptyset\emptyset}(q) = a^{\emptyset T}(q) = a^{T\emptyset}(q) = a^{\emptyset U}(q) = a^{U\emptyset}(q) = a^{\emptyset S}(q) = a^{S\emptyset}(q) = 1, \\ a^{TS}(q) = a^{US}(q) = a^{ST}(q) = a^{SU}(q) = 1 + q + \cdots + q^{d-1}, \\ a^{SS}(q) = (1+q)(1 + q + \cdots + q^{d-1}), \\ a^{TT}(q) = a^{UU}(q) = \begin{cases} 1 + q(1 + q^2 + q^4 + \cdots + q^{d-3}) & \text{if } d \text{ is odd}, \\ 1 + q(1 + q^2 + q^4 + \cdots + q^{d-2}) & \text{if } d \text{ is even}, \end{cases} \\ a^{TU}(q) = a^{UT}(q) = \begin{cases} 1 + q^2 + q^4 + \cdots + q^{d-1} & \text{if } d \text{ is odd}, \\ 1 + q^2 + q^4 + \cdots + q^{d-2} & \text{if } d \text{ is even}. \end{cases} \end{gathered}$$

These can be derived using the calculations in Example 1.45. □

Fact 1.51. For face-types T and T', the number of faces of type T' in a T-based top-lune equals the number of faces of type T in a T'-based top-lune. The special case $T' = S$ yields: The number of faces of type T equals the number of chambers in a T-based top-lune. See the bijection between (1.22) and (1.23).

PROOF. This expresses the symmetry $a^{TT'}(1) = a^{T'T}(1)$. The fact also follows from Exercise 1.36. See (1.64c) below in this regard. □

Exercise 1.52. Recall descents from Section 1.6. Check: For face-types T and T',

$$a^{TT'}(q) = \sum_{\substack{w \in W: \\ \operatorname{des}(w) \leq T \\ \operatorname{des}(w^{-1}) \leq T'}} q^{l(w)}. \tag{1.48}$$

Equivalently, the sum is over those $w \in W$ such that $(wF_0)C_r = wC_r$ and $F_0'(wC_r) = C_r$, where F_0 and F_0' are faces of C_r of types T and T', respectively.

1.8.6. Pairs of face-types and an invariant family of scalars.

We now generalize further. Let (ξ_T) be an invariant family of scalars indexed by face-types as in Definition 1.10. Define the polynomial

$$a^{TT'}(q, \xi) := \sum_{U, U'} a^{TUT'U'}(q)\, \xi_U = \sum_{U, U'} a^{TUT'U'}(q)\, \xi_{U'}. \tag{1.49}$$

When $\xi \equiv 1$, we recover the polynomials $a^{TT'}(q)$.

Exercise 1.53. Prove the second equality in (1.49). (Use: $\mathrm{s}(U) = \mathrm{s}(U')$ implies $\xi_U = \xi_{U'}$.)

Example 1.54 (Rank one). For a rank-one arrangement with face-types $\emptyset \lessdot S$,

$$a^{\emptyset\emptyset}(q,\xi) = \xi_\emptyset, \quad a^{\emptyset S}(q,\xi) = a^{S\emptyset}(q,\xi) = \xi_S, \quad a^{SS}(q,\xi) = (1+q)\,\xi_S.$$

This generalizes what is written for rank one in Example 1.50. The calculation for the dihedral arrangement generalizes similarly. □

It follows from (1.44a) that

$$a^{TT'}(q,\xi) = \sum_U a^{TUT'}(q)\,\xi_U = \sum_{U'} a^{T'U'T}(q)\,\xi_{U'}. \tag{1.50}$$

Symmetry continues to hold, that is,

$$a^{TT'}(q,\xi) = a^{T'T}(q,\xi). \tag{1.51}$$

More generally, we have polynomials $a_Z^{TT'}(q,\xi)$ obtained by applying (1.49) to $\boldsymbol{\alpha}_Z$ with the invariant family ξ restricted to face-types greater than Z.

Exercise 1.55. Show: The number $a^{TT'}(1,\xi)$ only depends on $\mathrm{s}(T)$ and $\mathrm{s}(T')$. That is, $a^{TT'}(1,\xi) = a^{UU'}(1,\xi)$ if $\mathrm{s}(T) = \mathrm{s}(U)$ and $\mathrm{s}(T') = \mathrm{s}(U')$.

1.8.7. Pairs of face-types of the same support. Varchenko matrices. For face-types T and T' of the same support, define the polynomial

$$c^{TT'}(q) := \sum_{\substack{G:\,\mathrm{t}(G)=T' \\ \mathrm{s}(G)=\mathrm{s}(F)}} q^{\mathrm{dist}(F,G)}, \tag{1.52}$$

with F a fixed face of type T. The two extreme cases are $c^{\emptyset\emptyset}(q) = 1$ and $c^{SS}(q) = d_S(q)$. Also, $c^{TT'}(1) = |\mathrm{s}(T)| = |\mathrm{s}(T')|$, with $|\lambda|$ as in Definition 1.8. More generally, by applying (1.52) to $\boldsymbol{\alpha}_Z$, we obtain polynomials $c_Z^{TT'}(q)$.

Special case. The polynomials (1.52) are special cases of (1.41) as follows. For $\mathrm{s}(T) = \mathrm{s}(T')$,

$$c^{TT'}(q) = a^{TTT'T'}(q) = a^{TTT'}(q) = a^{T'T'T}(q). \tag{1.53}$$

It follows that

$$c^{TT'}(q) = c^{T'T}(q). \tag{1.54}$$

For a flat-type λ, define the matrix A_λ indexed by face-types of support λ with (T,T') entry equal to $c^{TT'}(q)$. We call A_λ the *Varchenko matrix* of the flat-type λ. Let us now specialize q to a scalar in a field $\Bbbk$, and view A_λ as a matrix with entries in $\Bbbk$.

Theorem 1.56. *For q not a root of unity, for a flat-type λ, the Varchenko matrix A_λ is invertible.*

PROOF. This can be deduced from [**11**, Theorem 1.10] which pertains to the invertibility of Varchenko matrices A_{X} indexed by flats X. One way to proceed is as follows. Arrange A_{X} in block form so that faces of the same type are in the same block. If there is a linear relation say between the rows of A_λ, then it induces a relation between the rows of A_{X} with each coefficient occurring $|\lambda|$ times. An alternative way to proceed is given later in Exercise 1.99. □

The two extreme cases are $A_{\lambda_{\min}} = [1]$ and $A_{\lambda_{\max}} = [d_S(q)]$. Both matrices have size 1. The factorization (1.34) shows that $A_{\lambda_{\max}}$ is indeed invertible when q is not a root of unity.

Example 1.57 (Ranks one and two). For a rank-one arrangement with face-types $\emptyset \lessdot S$,

$$c^{\emptyset\emptyset}(q) = 1 \quad \text{and} \quad c^{SS}(q) = 1 + q.$$

For the dihedral arrangement of d lines with face-types $\emptyset \lessdot T, U \lessdot S$,

$$c^{\emptyset\emptyset}(q) = 1, \qquad c^{SS}(q) = (1+q)(1+q+\cdots+q^{d-1}),$$
$$c^{TT}(q) = c^{UU}(q) = 1 \text{ if } d \text{ is odd}, \qquad c^{TU}(q) = c^{UT}(q) = q^{d-1} \text{ if } d \text{ is odd},$$
$$c^{TT}(q) = c^{UU}(q) = 1 + q^{d-1} \text{ if } d \text{ is even}.$$

One may explicitly check Theorem 1.56 in these cases. The Varchenko matrices are either of size 1 or 2. □

Example 1.58 (Rank-three). Consider the rank-two flat of the rank-three braid arrangement in Figure 1.7. Let λ denote its type. The corresponding Varchenko matrix A_λ is

	red-blue	red-cyan	blue-cyan
red-blue	$1+q$	q^2+q^3	q^4+q^5
red-cyan	q^2+q^3	$1+q^5$	q^2+q^3
blue-cyan	q^4+q^5	q^2+q^3	$1+q$

Its determinant is $(1-q)^3(1+q)^4(1+q^2)^2(1+q+q^2+q^3+q^4)$. All its zeroes are roots of unity, so Theorem 1.56 holds.

For the rank-three arrangement of type B in Figure 1.6, for the flat-types represented by the two outer circles, the Varchenko matrices are

	red-blue	blue-cyan
red-blue	$1+q+q^7+q^8$	$q^2+q^3+q^5+q^6$
blue-cyan	$q^2+q^3+q^5+q^6$	$1+q^3+q^5+q^8$

and

	red-cyan
red-cyan	$1+q+q^3+2q^4+q^5+q^7+q^8$

One may check as before that Theorem 1.56 holds for these two matrices. Note very carefully: The entry in the latter matrix is *not* the Poincaré polynomial of the dihedral arrangement of 4 lines. □

The example of the coordinate arrangement is done in Exercise 1.122.

1.8.8. Constant terms of enumeration polynomials.

The constant term of a polynomial in q is obtained by setting $q = 0$. For the various enumeration polynomials, we have the following.

Exercise 1.59. Check:

$$d_T(0) = 1, \quad a^{TT'}(0,\xi) = \xi_{T\cup T'}, \quad c^{TT'}(0) = \begin{cases} 1 & \text{if } T = T', \\ 0 & \text{otherwise}, \end{cases} \tag{1.55}$$

$$a^{TUT'}(0) = \begin{cases} 1 & \text{if } T\cup T' = U, \\ 0 & \text{otherwise}, \end{cases} \quad a^{TUT'U'}(0) = \begin{cases} 1 & \text{if } T\cup T' = U = U', \\ 0 & \text{otherwise}. \end{cases} \tag{1.56}$$

1.8.9. Gate identities. We now recall an important identity of polynomials proved in [**10**, Lemma 16.26].

For face-types $T \leq V \leq Y$ and T',

$$a^{VYT'}(q) = \sum_{U:\, U \geq T} a_T^{VYU}(q)\, a^{TUT'}(q). \tag{1.57}$$

The proof is encapsulated in Figure 1.19. It makes use of the gate property. We call (1.57) the *gate identity* for face-type enumeration.

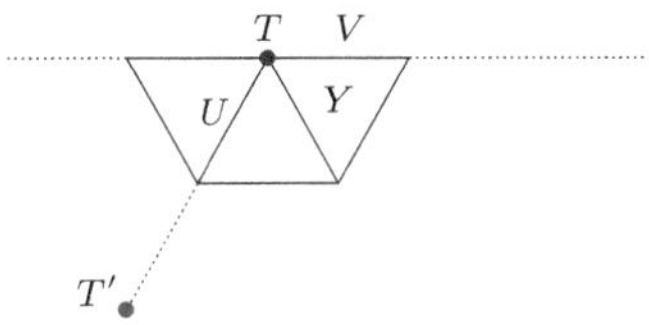

FIGURE 1.19. Gate identity for face-type enumeration.

Example 1.60 (Rank three). Let us see how the gate identity (1.57) works in specific examples. Consider the rank-three arrangement of type B in Figure 1.6. Let T = red, V = red-cyan, Y = red-blue-cyan. For T' = blue, we get

$$\begin{array}{ccccccccc} 1+q^2+q^3+q^5 & = & 0 & + & 0 & + & 1(1+q^5) & + & (1+q)q^2, \\ & & \text{red} & & \text{red-cyan} & & \text{red-blue} & & \text{red-blue-cyan} \end{array}$$

while for T' = red, we get

$$\begin{array}{ccccccccc} q+q^2+q^4+q^5 & = & 0 & + & 0 & + & 0 & + & (1+q)(q+q^4). \\ & & \text{red} & & \text{red-cyan} & & \text{red-blue} & & \text{red-blue-cyan} \end{array}$$

The face-type U is indicated below each term in the rhs. □

Exercise 1.61. Show: In the gate identity (1.57), the term $U := T$ contributes to the rhs iff $V = Y$ and $\mathrm{s}(T) \geq \mathrm{s}(T')$. In this case, the contribution equals $a^{TTT'}(q)$. (Use (1.40).)

Exercise 1.62. Derive the following consequences of the gate identity.

- For an invariant family of scalars ξ, for face-types $T \leq V$ and T',

$$a^{VT'}(q, \xi) = \sum_{U:\, U \geq T} a_T^{VU}(q, \xi)\, a^{TUT'}(q). \tag{1.58}$$

  (Multiply (1.57) by ξ_Y and then sum over all $Y \geq V$.)
- For face-types T and T',

$$d_{T'}(q) = \sum_{U:\, U \geq T} d_{U/T}(q)\, a^{TUT'}(q). \tag{1.59}$$

  (Set $V = Y := S$ in (1.57) and use (1.39), or set $V := S$ in (1.58).)
- For face-types $T \leq V$ and T' with $\mathrm{s}(V) = \mathrm{s}(T')$,

$$c^{VT'}(q) = \sum_{\substack{U:\, U \geq T \\ \mathrm{s}(U/T) = \mathrm{s}(V/T)}} c_T^{VU}(q)\, a^{TUT'}(q), \tag{1.60}$$

$$|\mathrm{s}(V)| = |\mathrm{s}(V/T)| \sum_{\substack{U:\, U \geq T \\ \mathrm{s}(U/T)=\mathrm{s}(V/T)}} a^{TUT'}, \tag{1.61}$$

with notation as in Definition 1.8. (For the first identity, set $V = Y$ in (1.57) and use (1.53). The second identity is the special case $q = 1$.)

We refer to all of the above also as the gate identities.

Exercise 1.63. Set $Y = T' := S$ in (1.57) or $T' := S$ and $\xi \equiv 1$ in (1.58) to deduce the identity $d_V(q) = d_{V/T}(q)d_T(q)$ for face-types $T \leq V$. The more general multiplicative property (1.32) also follows.

Exercise 1.64. A double-nested face is a triple (F, G, H) of faces with $F \leq G \leq H$, and a double-nested face-type is a triple (T, U, V) of face-types with $T \leq U \leq V$.

For double-nested face-types (T, U, V) and (T', U', V'), define the polynomial $a^{TUVT'U'V'}(q)$ by summing $q^{\mathrm{dist}(G,G')}$ over all (F', G', H') of type (T', U', V') such that $FF' = G$, $F'F = G'$, $FH' = H$, $F'H = H'$, where (F, G, H) is a fixed double-nested face of type (T, U, V).

Show: For face-types $T \leq Y$ and $T' \leq V'$,

$$a^{T'V'Y}(q) = \sum_{\substack{U,V: \\ T \leq U \leq V}} a_T^{UVY}(q)\, a^{TUVT'V'}(q), \tag{1.62}$$

where $a^{TUVT'V'}(q)$ is obtained by summing $a^{TUVT'U'V'}(q)$ over U'. See the illustration below.

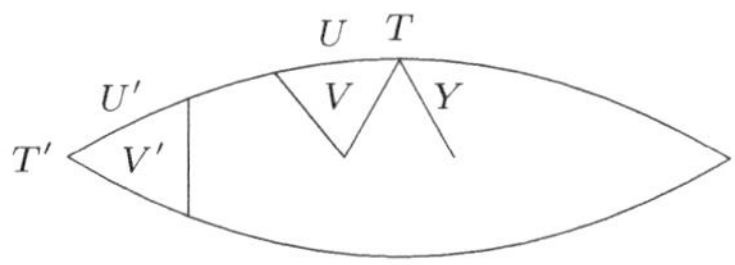

We refer to (1.62) also as a gate identity.

1.8.10. Alternating sign identities.

We now record some identities which involve computations of alternating sums. We refer to these collectively as the *alternating sign identities*.

We have

$$\sum_T (-1)^{\mathrm{rk}(T)} d_T(q) = (-1)^{\mathrm{rk}(S)} q^{l(w_\emptyset)}. \tag{1.63a}$$

The sum is over all face-types T.

For face-types $T \leq U$,

$$\sum_{T'} (-1)^{\mathrm{rk}(T')} a^{TUT'}(q) = (-1)^{\mathrm{rk}(U)} q^{\mathrm{dist}(T,\overline{T})}. \tag{1.63b}$$

The sum is over all face-types T'.

For face-types U and V,

$$\sum_{T:\, T \leq V} (-1)^{\mathrm{rk}(T)} a^{TVU}(q) = \begin{cases} (-1)^{\mathrm{rk}(V)} q^{\mathrm{dist}(U,\overline{U})} & \text{if } V \geq \overline{U}, \\ 0 & \text{otherwise.} \end{cases} \tag{1.63c}$$

Exercise 1.65. Check the alternating sign identities (1.63) in ranks one and two by using Example 1.45.

1.8.11. Groupoid of biface-types. Recall from Section 1.7 the groupoid of biface-types. Face-type enumeration polynomials can be expressed using morphisms in this category as follows.

Exercise 1.66. Check:

$$a_Z^{TUT'U'}(q) = \sum_{\substack{\sigma\colon U \to U' \\ \text{local to } Z \\ \text{biprojects } T \text{ and } T'}} q^{l(\sigma)}. \tag{1.64a}$$

If $\mathrm{s}(U/Z) = \mathrm{s}(T'/Z)$, then

$$a_Z^{TUT'}(q) = \sum_{\substack{\sigma\colon U \to T' \\ \text{local to } Z \\ \text{coprojects } T' \text{ on } T}} q^{l(\sigma)}. \tag{1.64b}$$

In particular, the number of terms that appear in these sums are $a_Z^{TUT'U'}$ and $a_Z^{TUT'}$, respectively. Setting $Z = \emptyset$, we see that $a^{TUT'U'}(q)$ and $a^{TUT'}(q)$ are given by the above sums with the condition "local to Z" removed. (To deduce (1.64b) from (1.64a), apply (1.44b) to $\boldsymbol{\alpha}_Z$.)

Similarly,

$$a_Z^{TT'}(q) = \sum_{\substack{\sigma \text{ local to } Z \\ \text{biprojects } T \text{ and } T'}} q^{l(\sigma)}, \tag{1.64c}$$

and for $\mathrm{s}(T/Z) = \mathrm{s}(T'/Z)$,

$$c_Z^{TT'}(q) = \sum_{\substack{\sigma\colon T \to T' \\ \text{local to } Z}} q^{l(\sigma)}. \tag{1.64d}$$

The more general polynomials $a_Z^{TT'}(q, \xi)$ can be written similarly by replacing $q^{l(\sigma)}$ by $q^{l(\sigma)}\xi_\sigma$ with $\xi_\sigma := \xi_U = \xi_{U'}$ for $\sigma : U \to U'$.

Exercise 1.67. Let $\sigma : T \to T'$ be a morphism and $\sigma : U \to U'$ a morphism induced from it as in (1.28). Show: $d_{U/T} = d_{U'/T'}$, and more generally, $d_{U/T}(q) = d_{U'/T'}(q)$ for a scalar q. (Combine Exercise 1.37 with Exercise 1.43, item (1).)

Exercise 1.68. Let $\sigma : T \to T'$ be a morphism local to Z. Show: $d_{T/Z} = d_{T'/Z}$, and more generally, $d_{T/Z}(q) = d_{T'/Z}(q)$ for a scalar q. Also

$$d_{T/Z}(-1) = (-1)^{l(\sigma)} d_{T'/Z}(-1).$$

(Combine Exercise 1.26 with Exercise 1.43, items (2), (3) and apply to $\boldsymbol{\alpha}_Z$.)

Exercise 1.69. Let $\sigma : T \to T'$ be a morphism local to Z and $\sigma : V \to V'$ a morphism induced from it. Show: $a_Z^{TVU} = a_Z^{T'V'U}$ for any face-type U greater than Z. (Combine Exercise 1.37 with Exercise 1.44 and apply to $\boldsymbol{\alpha}_Z$.)

Exercise 1.70. Consider the diagram of induced morphisms

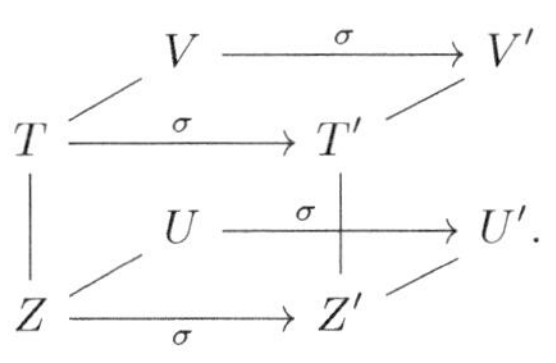

Show: $a_Z^{TVU}(q) = a_{Z'}^{T'V'U'}(q)$. For any invariant family of scalars ξ, $a_Z^{TU}(q,\xi) = a_{Z'}^{T'U'}(q,\xi)$. (Multiply the previous identity on either side by ξ_V and $\xi_{V'}$, and sum over V and V'.)

1.9. Enumeration in the Coxeter group algebra

We now generalize some of the numerical identities in Section 1.8 by viewing them in the group algebra of W. There are two equivalent ways to write each identity; we pass from one to the other by replacing each element of W by its inverse. We recover the original identity by replacing each element by the scalar 1.

These group algebra identities can also be proved conceptually using ideas of this monograph via the tensor and shuffle Coxeter q-bialgebras whose existence can be established using either universal constructions or Fock functors, see Section 10.3.6 and Section 18.3. These generalize similar comments made at the end of the introduction to Section 1.8.

1.9.1. Augmentation map. Fix a field $\Bbbk$. Define the *augmentation map*

$$\Bbbk W \to \Bbbk, \qquad w \mapsto 1 \tag{1.65}$$

extended by linearity. It is an algebra morphism.

1.9.2. Gate identities. Recall shuffles from Section 1.6.4.

Exercise 1.71. For face-types $T \le U \le V$, there is a bijection

$$\mathrm{Sh}_{U/T} \times \mathrm{Sh}_{V/U} \xrightarrow{\cong} \mathrm{Sh}_{V/T}, \qquad (\sigma,\tau) \mapsto \sigma\tau,$$

where $\sigma\tau$ is the product in W, and moreover $l(\sigma\tau) = l(\sigma) + l(\tau)$. This bijection can be equivalently written as

$$\mathrm{Sh}^*_{V/U} \times \mathrm{Sh}^*_{U/T} \xrightarrow{\cong} \mathrm{Sh}^*_{V/T}, \qquad (\tau^{-1},\sigma^{-1}) \mapsto \tau^{-1}\sigma^{-1}.$$

Establish these bijections independently by employing (1.22) and (1.23), respectively, and using the gate property. We refer to these bijections as the *multiplicative property* of shuffles.

Deduce: For face-types $T \le U \le V$,

$$\begin{gathered}\sum_{\rho\in \mathrm{Sh}_{V/T}} q^{l(\rho)}\rho = \Big(\sum_{\sigma\in \mathrm{Sh}_{U/T}} q^{l(\sigma)}\sigma\Big)\Big(\sum_{\tau\in \mathrm{Sh}_{V/U}} q^{l(\tau)}\tau\Big),\\ \sum_{\rho\in \mathrm{Sh}_{V/T}} q^{l(\rho)}\rho^{-1} = \Big(\sum_{\tau\in \mathrm{Sh}_{V/U}} q^{l(\tau)}\tau^{-1}\Big)\Big(\sum_{\sigma\in \mathrm{Sh}_{U/T}} q^{l(\sigma)}\sigma^{-1}\Big).\end{gathered} \tag{1.66}$$

Applying the augmentation map to (1.66) recovers the multiplicative property (1.32) of face-type enumerators.

Exercise 1.72. We now highlight an important special case of Exercise 1.71 arising from $\mathrm{Sh}_{S/T} = W_T$. For a face-type T, there are length-preserving bijections

$$\begin{array}{ll} \mathrm{Sh}_T \times W_T \xrightarrow{\cong} W & \qquad W_T \times \mathrm{Sh}^*_T \xrightarrow{\cong} W \\ (v,u) \longmapsto vu, & \qquad (u^{-1},v^{-1}) \longmapsto u^{-1}v^{-1}. \end{array} \tag{1.67}$$

Deduce: For a face-type T,

$$\begin{aligned}\sum_{w\in W} q^{l(w)} w &= \Big(\sum_{v\in \mathrm{Sh}_T} q^{l(v)} v\Big)\Big(\sum_{u\in W_T} q^{l(u)} u\Big),\\ \sum_{w\in W} q^{l(w)} w^{-1} &= \Big(\sum_{u\in W_T} q^{l(u)} u^{-1}\Big)\Big(\sum_{v\in \mathrm{Sh}_T} q^{l(v)} v^{-1}\Big).\end{aligned} \tag{1.68}$$

These are special cases of (1.66). Applying the augmentation map recovers the multiplicative property (1.37).

Exercise 1.73. For face-types $Z \leq T$ and $Z \leq T'$,

$$\begin{aligned}\sum_{\tau\in \mathrm{Sh}_{T/Z}} q^{l(\tau)} \tau &= \sum_{U,\,U'} \Big(\sum_{\sigma\in \mathrm{Sh}_{U'/T'}} q^{l(\sigma)} \sigma\Big)\Big(\sum_{\substack{\rho:\, U\to U'\\ \text{local to } Z\\ \text{biprojects } T \text{ and } T'}} q^{l(\rho)} w_\rho\Big),\\ \sum_{\tau\in \mathrm{Sh}_{T'/Z}} q^{l(\tau)} \tau^{-1} &= \sum_{U,\,U'} \Big(\sum_{\substack{\rho:\, U\to U'\\ \text{local to } Z\\ \text{biprojects } T \text{ and } T'}} q^{l(\rho)} w_\rho\Big)\Big(\sum_{\sigma\in \mathrm{Sh}_{U/T}} q^{l(\sigma)} \sigma^{-1}\Big),\end{aligned} \tag{1.69}$$

with w_ρ as in Lemma 1.30. Establish both identities independently by employing (1.22) and (1.23), respectively, and using the gate property. An illustration for the second identity for $Z = \emptyset$ is shown below.

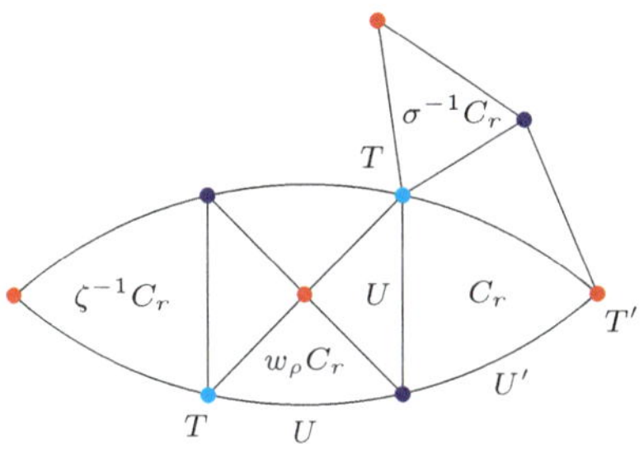

Also see [**10**, Lemma 3.39] in this regard.

Applying the augmentation map to (1.69) recovers the gate identity (1.59).

Exercise 1.74. Let $\sigma : Z \to Z'$ be a morphism in $\boldsymbol{\alpha}$-$\overline{\texttt{Cox}}$, and $\sigma : T \to T'$ be a morphism induced from it with $Z \leq T$ and $Z' \leq T'$ as in (1.28). Show:

$$\begin{aligned}w_\sigma\Big(\sum_{\zeta\in \mathrm{Sh}_{T/Z}} q^{l(\zeta)} \zeta\Big) &= \Big(\sum_{\zeta'\in \mathrm{Sh}_{T'/Z'}} q^{l(\zeta')} \zeta'\Big) w_\sigma,\\ w_\sigma\Big(\sum_{\zeta\in \mathrm{Sh}_{T/Z}} q^{l(\zeta)} \zeta^{-1}\Big) &= \Big(\sum_{\zeta'\in \mathrm{Sh}_{T'/Z'}} q^{l(\zeta')} (\zeta')^{-1}\Big) w_\sigma,\end{aligned} \tag{1.70}$$

with w_σ as in Lemma 1.30. An illustration for the second identity is shown below with Z the red vertex-type, Z' the blue vertex-type, T the red and blue edge-type, T' the blue and green edge-type. The biface consisting of the right blue vertex and left red vertex (which are opposite to each other) represents σ.

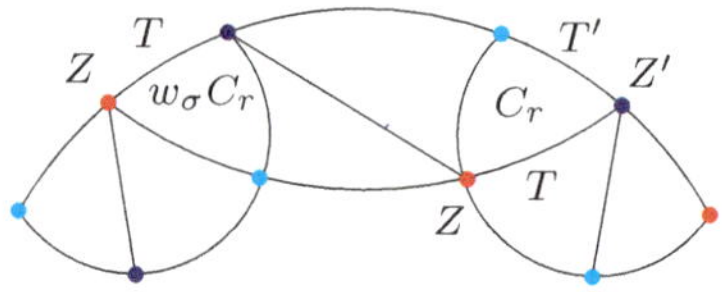

Applying the augmentation map to (1.70) recovers the result of Exercise 1.67.

Exercise 1.75. Show: For a morphism $\sigma : T \to T'$ in $\boldsymbol{\alpha}$-$\overline{\mathsf{Cox}}$ local to Z,

$$\begin{aligned}\Big(\sum_{\zeta' \in \mathrm{Sh}_{T'/Z}} \zeta'\Big) w_\sigma &= \sum_{\zeta \in \mathrm{Sh}_{T/Z}} \zeta, \\ w_\sigma \Big(\sum_{\zeta \in \mathrm{Sh}_{T/Z}} \zeta^{-1}\Big) &= \sum_{\zeta' \in \mathrm{Sh}_{T'/Z}} (\zeta')^{-1},\end{aligned} \tag{1.71}$$

with w_σ as in Lemma 1.30. Interestingly, applying the augmentation map recovers only a part of Exercise 1.68. Identity (1.71) does not work with powers of q in general. It does work, however, for $q = -1$ with $(-1)^{l(\zeta)}$, $(-1)^{l(\zeta')}$, $(-1)^{l(w_\sigma)}$ inserted appropriately.

1.9.3. Alternating sign identity.

We have

$$\begin{aligned}\sum_T (-1)^{\mathrm{rk}(T)} \sum_{\zeta \in \mathrm{Sh}_T} q^{l(\zeta)} \zeta &= (-1)^{\mathrm{rk}(S)} q^{l(w_\emptyset)} w_\emptyset, \\ \sum_T (-1)^{\mathrm{rk}(T)} \sum_{\zeta \in \mathrm{Sh}_T} q^{l(\zeta)} \zeta^{-1} &= (-1)^{\mathrm{rk}(S)} q^{l(w_\emptyset)} w_\emptyset.\end{aligned} \tag{1.72}$$

The first sum in either identity is over all face-types T, while the second sum is over all T-shuffles. Applying the augmentation map to (1.72) recovers the alternating sign identity (1.63a).

1.9.4. Invertible elements in the Coxeter group algebra.

Fact 1.76. For q not a root of unity, the element $\sum_{w \in W} q^{l(w)} w$ is invertible in the group algebra of W.

Proof. See [**10**, Lemma 8.28]. It is a consequence of the Varchenko factorization theorem on distance functions [**10**, Formula (8.41)]. □

A generalization is given below.

Exercise 1.77. Show: For q not a root of unity, for a face-type T, the elements $\sum_{\zeta \in \mathrm{Sh}_T} q^{l(\zeta)} \zeta$ and $\sum_{\zeta \in \mathrm{Sh}_T} q^{l(\zeta)} \zeta^{-1}$ are invertible in the group algebra of W. (Use Fact 1.76 and identities (1.68).)

Apply the augmentation map to deduce: For a face-type T, the scalar $d_T(q)$ is invertible for q not a root of unity. (Alternatively and more directly, use (1.34) and (1.37), following the same logical thread.)

1.10. Invariant flat-, lune-, bilune-incidence algebras

Fix a reflection arrangement $\mathcal{A}$ with Coxeter group W. We discuss the flat-incidence algebra, lune-incidence algebra, bilune-incidence algebra, and zeta and Möbius functions therein. We also discuss their W-invariant subalgebras, and algebra morphisms linking them. See Tables 1.5 and 1.6. Zeta and Möbius functions are characterized by the lune-additivity and noncommutative Weisner formulas. Face-type enumeration polynomials enter the story when we express these formulas in terms of face-types. The groupoid of biface-types enters naturally into the study of the invariant bilune-incidence algebra. Supplementary information on some of this material can be found in [**10**, Sections 15.1, 15.2, 15.3, 16.7], [**11**, Sections 1.5 and 1.6]. A general reference for incidence algebras of posets and categories is [**10**, Appendix C].

All our algebras are defined over a fixed field $\Bbbk$.

TABLE 1.5. Incidence algebras.

Incidence algebras		zeta, Möbius functions
$\mathrm{I}_{\mathsf{flat}}[\mathcal{A}]^W$	invariant flat-incidence algebra	ζ, μ (unique) ζ_q, μ_q (unique)
$\mathrm{I}_{\mathsf{face}}[\mathcal{A}]^W$	invariant face-incidence algebra	
$\mathrm{I}_{\mathsf{lune}}[\mathcal{A}]^W$	invariant lune-incidence algebra	$\boldsymbol{\zeta}$, $\boldsymbol{\mu}$ (not unique) $\boldsymbol{\zeta}_q$, $\boldsymbol{\mu}_q$ (unique)
$\mathrm{I}_{\mathsf{biface}}[\mathcal{A}]^W$	invariant biface-incidence algebra	
$\mathrm{I}_{\mathsf{bilune}}[\mathcal{A}]^W$	invariant bilune-incidence algebra	ζ_q, μ_q (unique)

TABLE 1.6. Algebra morphisms between incidence algebras.

Algebra morphisms	Values on zeta, Möbius functions
$\mathrm{bc} : \mathrm{I}_{\mathsf{lune}}[\mathcal{A}]^W \to \mathrm{I}_{\mathsf{flat}}[\mathcal{A}]^W$	$\mathrm{bc}(\boldsymbol{\zeta}) = \zeta$, $\mathrm{bc}(\boldsymbol{\mu}) = \mu$ $\mathrm{bc}(\boldsymbol{\zeta}_q) = \zeta_q$, $\mathrm{bc}(\boldsymbol{\mu}_q) = \mu_q$
$\mathrm{p}, \mathrm{q} : \mathrm{I}_{\mathsf{bilune}}[\mathcal{A}]^W \to \mathrm{I}_{\mathsf{lune}}[\mathcal{A}]^W$	$\mathrm{p}(\zeta_q) = \mathrm{q}(\zeta_q) = \boldsymbol{\zeta}_q$ $\mathrm{p}(\mu_q) = \mathrm{q}(\mu_q) = \boldsymbol{\mu}_q$

We will see later that zeta and Möbius functions play a fundamental role in the theory of Coxeter bimonoids and Coxeter bialgebras, including exp-log correspondences (Chapters 3 and 9) and results such as the Poincaré–Birkhoff–Witt theorem (Chapters 7 and 13).

1.10.1. Invariant flat-incidence algebra. A *nested flat* is a pair of flats (X, Y) such that $\mathrm{X} \leq \mathrm{Y}$. The *flat-incidence algebra*, denoted $\mathrm{I}_{\mathsf{flat}}[\mathcal{A}]$, is the incidence algebra of the poset of flats. It consists of functions s on nested flats, with the product of s and t given by

$$(st)(\mathrm{X},\mathrm{Z}) = \sum_{\mathrm{Y}:\, \mathrm{X}\leq\mathrm{Y}\leq\mathrm{Z}} s(\mathrm{X},\mathrm{Y})t(\mathrm{Y},\mathrm{Z}). \tag{1.73}$$

We denote its W-invariant subalgebra by $\mathrm{I}_{\mathrm{flat}}[\mathcal{A}]^W$ and call it the *invariant flat-incidence algebra*. It consists of functions s on nested flats which are invariant, that is, $s(\mathrm{X},\mathrm{Z}) = s(w\mathrm{X}, w\mathrm{Z})$ for $w \in W$.

1.10.2. Zeta and Möbius functions. The *zeta function* ζ is the element of the flat-incidence algebra defined by $\zeta(\mathrm{X},\mathrm{Y}) = 1$ for all flats $\mathrm{X} \leq \mathrm{Y}$. The *Möbius function* μ is the element of the flat-incidence algebra such that $\mu(\mathrm{X},\mathrm{X}) = 1$ for all X and the *Weisner formula* holds, that is,

$$\sum_{\mathrm{W}:\, \mathrm{W}\geq\mathrm{X},\, \mathrm{W}\vee\mathrm{Y}=\mathrm{Z}} \mu(\mathrm{X},\mathrm{W}) = 0 \tag{1.74}$$

for all flats $\mathrm{X} < \mathrm{Y} \leq \mathrm{Z}$.

The zeta and Möbius functions are inverses of each other in the flat-incidence algebra. In fact, they belong to the invariant flat-incidence algebra.

1.10.3. Invariant lune-incidence algebra. Let us now look at faces. The *face-incidence algebra*, denoted $\mathrm{I}_{\mathrm{face}}[\mathcal{A}]$, is the incidence algebra of the poset of faces. It consists of functions s on nested faces, with the product of s and t given by

$$(st)(F,H) = \sum_{G:\, F\leq G\leq H} s(F,G)t(G,H). \tag{1.75}$$

We denote its W-invariant subalgebra by $\mathrm{I}_{\mathrm{face}}[\mathcal{A}]^W$. It consists of functions s on nested faces which are invariant, that is, $s(H,G) = s(wH, wG)$ for $w \in W$. In this case, the expression $s(T,U)$ makes sense. In fact, $\mathrm{I}_{\mathrm{face}}[\mathcal{A}]^W$ identifies with the incidence algebra of the poset of face-types. It consists of functions s on nested face-types, with the product of s and t given by

$$(st)(T,V) = \sum_{U:\, T\leq U\leq V} s(T,U)t(U,V). \tag{1.76}$$

The *lune-incidence algebra*, denoted $\mathrm{I}_{\mathrm{lune}}[\mathcal{A}]$, is the subalgebra of the face-incidence algebra consisting of functions s such that $s(A,F) = s(B,G)$ whenever (A,F) and (B,G) have the same support. It is the same as the incidence algebra of the category of lunes. It has a basis indexed by lunes. We denote its W-invariant subalgebra by $\mathrm{I}_{\mathrm{lune}}[\mathcal{A}]^W$ and call it the *invariant lune-incidence algebra*. The latter is also a subalgebra of the incidence algebra of the poset of face-types consisting of those s such that $s(T,U) = s(T',U')$ whenever (T,U) and (T',U') have the same support. Thus, it has a basis indexed by lune-types.

1.10.4. Base-case map. The *base-case map* is the algebra morphism

$$\mathrm{bc} : \mathrm{I}_{\mathrm{lune}}[\mathcal{A}] \to \mathrm{I}_{\mathrm{flat}}[\mathcal{A}] \tag{1.77}$$

defined by

$$\mathrm{bc}(t)(\mathrm{X},\mathrm{Y}) := \sum_{F:\, F \geq A,\, \mathrm{s}(F)=\mathrm{Y}} t(A,F),$$

where A is a fixed face of support X.

Example 1.78 (Base-case). Let Y be the rank-two flat below and $\mathrm{X} = \mathrm{s}(A)$.

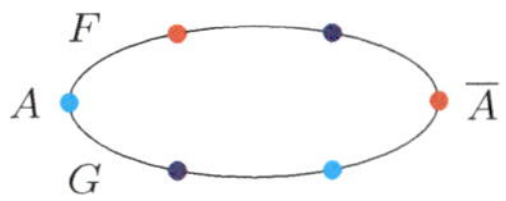

Then $\mathrm{bc}(t)(\mathrm{X},\mathrm{Y}) = t(A,F) + t(A,G)$. This can also be interpreted as the sum of values of t on the two semicircles $\mathrm{s}(A,F)$ and $\mathrm{s}(A,G)$ which split Y. □

The base-case map is a map of left W-modules. So it induces a map from the invariant lune-incidence algebra to the invariant flat-incidence algebra

$$\mathrm{bc} : \mathrm{I}_{\mathrm{lune}}[\mathcal{A}]^W \to \mathrm{I}_{\mathrm{flat}}[\mathcal{A}]^W. \tag{1.78}$$

Exercise 1.79. Show: The base-case map (1.77) is surjective. If $\mathrm{char}(\Bbbk) \nmid |W|$, then the induced map (1.78) is also surjective.

1.10.5. Noncommutative zeta and Möbius functions. A *noncommutative zeta function* $\boldsymbol{\zeta}$ is an element of the lune-incidence algebra such that $\boldsymbol{\zeta}(A,A) = 1$ for all A and the *lune-additivity formula* holds, that is,

$$\boldsymbol{\zeta}(H,G) = \sum_{\substack{F:\, F \geq A,\, HF=G \\ \mathrm{s}(F)=\mathrm{s}(G)}} \boldsymbol{\zeta}(A,F) \tag{1.79}$$

for all faces $A \leq H \leq G$. As a particular case, the *flat-additivity formula* says that for any flat $\mathrm{X} \geq \mathrm{s}(A)$,

$$\sum_{F:\, F \geq A,\, \mathrm{s}(F)=\mathrm{X}} \boldsymbol{\zeta}(A,F) = 1. \tag{1.80}$$

A *noncommutative Möbius function* $\boldsymbol{\mu}$ is an element of the lune-incidence algebra such that $\boldsymbol{\mu}(A,A) = 1$ for all A and the *noncommutative Weisner formula* holds, that is,

$$\sum_{F:\, F \geq A,\, HF=G} \boldsymbol{\mu}(A,F) = 0 \tag{1.81}$$

for all faces $A < H \leq G$.

Noncommutative zeta and Möbius functions are inverse to each other in the lune-incidence algebra, see [**10**, Theorem 15.28], [**11**, Theorem 1.14]. Moreover, the dimension of the affine space of all noncommutative zeta or Möbius functions is equal to the number of faces minus the number of flats.

Fact 1.80. For noncommutative zeta and Möbius functions $\boldsymbol{\zeta}, \boldsymbol{\mu} \in \mathrm{I}_{\mathrm{lune}}[\mathcal{A}]$,

$$\mathrm{bc}(\boldsymbol{\zeta}) = \zeta \quad \text{and} \quad \mathrm{bc}(\boldsymbol{\mu}) = \mu,$$

where ζ and μ are the zeta and Möbius functions in $\mathrm{I}_{\mathrm{flat}}[\mathcal{A}]$.

PROOF. See [**10**, Lemmas 15.17 and 15.23] and [**11**, Lemma 1.17]. □

A noncommutative zeta function $\boldsymbol{\zeta}$ is *invariant* if $\boldsymbol{\zeta}(H, G) = \boldsymbol{\zeta}(wH, wG)$ for $w \in W$. Similarly, a noncommutative Möbius function $\boldsymbol{\mu}$ is *invariant* if $\boldsymbol{\mu}(H, G) = \boldsymbol{\mu}(wH, wG)$ for $w \in W$. In this situation, the expressions $\boldsymbol{\zeta}(T, U)$ and $\boldsymbol{\mu}(T, U)$ make sense.

Fact 1.81. Let $\mathcal{A}$ be a reflection arrangement with Coxeter group W. For $\mathrm{rk}(\mathcal{A}) \geq 1$, the following are equivalent.

- An invariant noncommutative zeta function of $\mathcal{A}$ exists.
- An invariant noncommutative Möbius function of $\mathcal{A}$ exists.
- $\mathrm{char}(\mathbb{k}) \nmid |W|$.

Moreover, if $\mathrm{char}(\mathbb{k}) \nmid |W|$, then the dimension of the affine space of all invariant noncommutative zeta or Möbius functions is equal to the number of face-types minus the number of flat-types.

PROOF. See [**10**, Lemma 16.42 and Theorem 16.48, items (1), (2)]. □

Recall the face-type enumeration scalars a_Z^{TVU} defined in Section 1.8.3. For an invariant noncommutative zeta function $\boldsymbol{\zeta}$, we have $\boldsymbol{\zeta}(Z, Z) = 1$ for all Z and the lune-additivity formula (1.79) takes the form

$$\boldsymbol{\zeta}(T, V) = \sum_{\substack{U:\, U \geq Z \\ \mathrm{s}(U/Z) = \mathrm{s}(V/Z)}} a_Z^{TVU}\, \boldsymbol{\zeta}(Z, U) \tag{1.82}$$

for all face-types $Z \leq T \leq V$. The flat-additivity formula (1.80) becomes

$$\sum_{\substack{U:\, U \geq Z \\ \mathrm{s}(U/Z) = \mathrm{s}(V/Z)}} a_Z^{VVU}\, \boldsymbol{\zeta}(Z, U) = |\mathrm{s}(V/Z)| \sum_{\substack{U:\, U \geq Z \\ \mathrm{s}(U/Z) = \mathrm{s}(V/Z)}} \boldsymbol{\zeta}(Z, U) = 1 \tag{1.83}$$

for all face-types $Z \leq V$, with $|\mathrm{s}(V/Z)|$ as in Definition 1.8.

Example 1.82 (Lune-additivity, flat-additivity). Consider a rank-two flat in the rank-three braid arrangement as in Figure 1.7. It is redrawn below with two semicircles marked in bold. Both are unions of three edges.

In the left picture, lune-additivity for the marked semicircle, that is, (1.82) for $Z = \emptyset$, $T =$ cyan, $V =$ cyan-red says

$$\boldsymbol{\zeta}(\text{cyan}, \text{cyan-red}) = \boldsymbol{\zeta}(\emptyset, \text{cyan-red}) + 2\,\boldsymbol{\zeta}(\emptyset, \text{blue-red}).$$

The coefficient 2 indicates that there are two edges of type blue-red on that semicircle, that is, $a^{TVU} = 2$ for $U =$ blue-red. Similarly, in the right picture, for the marked semicircle, lune-additivity says

$$\boldsymbol{\zeta}(\text{blue}, \text{blue-red}) = \boldsymbol{\zeta}(\emptyset, \text{blue-red}) + \boldsymbol{\zeta}(\emptyset, \text{cyan-red}) + \boldsymbol{\zeta}(\emptyset, \text{blue-cyan}).$$

Flat-additivity (1.83) with $Z = \emptyset$ for the above flat says

$$2\big(\boldsymbol{\zeta}(\emptyset, \text{blue-cyan}) + \boldsymbol{\zeta}(\emptyset, \text{blue-red}) + \boldsymbol{\zeta}(\emptyset, \text{cyan-red})\big) = 1.$$

Note: $|\lambda| = 2$ for the flat-type λ under discussion; this explains the 2 in front. □

Exercise 1.83. Use Fact 1.80 or put $V = S$ in (1.83) to deduce the formula

$$\boldsymbol{\zeta}(Z, S) = \frac{1}{d_{S/Z}} = \frac{d_Z}{d_S} = \frac{1}{|W_Z|}. \tag{1.84}$$

However, other values of $\boldsymbol{\zeta}$ are not uniquely determined in general.

For an invariant noncommutative Möbius function $\boldsymbol{\mu}$, we have $\boldsymbol{\mu}(Z, Z) = 1$ for all Z and the noncommutative Weisner formula (1.81) takes the form

$$\sum_{U:\, U \geq Z} a_Z^{TVU}\, \boldsymbol{\mu}(Z, U) = 0 \tag{1.85}$$

for all face-types $Z < T \leq V$.

Example 1.84 (Noncommutative Weisner). Consider the following top-lune in the rank-three braid arrangement.

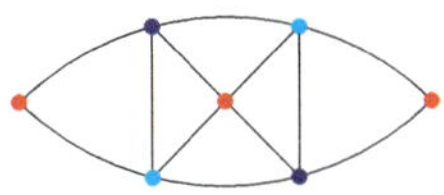

Noncommutative Weisner in this case, that is, (1.85) for $Z = \emptyset$, $T = $ red, $V =$ blue-cyan-red says

$$\boldsymbol{\mu}(\emptyset, \text{red}) + 2\,\boldsymbol{\mu}(\emptyset, \text{blue-cyan}) + 2\,\boldsymbol{\mu}(\emptyset, \text{blue-red}) + 2\,\boldsymbol{\mu}(\emptyset, \text{cyan-red}) \\ + 6\,\boldsymbol{\mu}(\emptyset, \text{blue-cyan-red}) = 0.$$

The coefficients indicate how many times a given face-type appears in the interior of the top-lune. □

Exercise 1.85. Use Fact 1.80 to deduce the formula

$$\boldsymbol{\mu}(Z, S) = \frac{\mu(\mathrm{X}, \top)}{|W_{\mathrm{X}}|}, \tag{1.86}$$

where X is any flat such that $\mathrm{t}(\mathrm{X}) = \mathrm{s}(Z)$. However, other values of $\boldsymbol{\mu}$ are not uniquely determined in general.

Example 1.86 (Rank one). Let $\mathrm{char}(\Bbbk) \neq 2$. For a rank-one arrangement with face-types $\emptyset \lessdot S$, there is a unique invariant $\boldsymbol{\zeta}$ and $\boldsymbol{\mu}$, namely,

$$\boldsymbol{\zeta}(\emptyset, \emptyset) = 1, \quad \boldsymbol{\zeta}(S, S) = 1, \quad \boldsymbol{\zeta}(\emptyset, S) = 1/2,$$
$$\boldsymbol{\mu}(\emptyset, \emptyset) = 1, \quad \boldsymbol{\mu}(S, S) = 1, \quad \boldsymbol{\mu}(\emptyset, S) = -1/2.$$

The values on $(\emptyset, S)$ are consistent with formulas (1.84) and (1.86). □

Exercise 1.87. Show: For any invariant noncommutative zeta function $\boldsymbol{\zeta}$, for face-types T and T',

$$|\mathrm{s}(T')| \sum_{\substack{U:\, U \geq T \\ \mathrm{s}(U) = \mathrm{s}(T')}} \boldsymbol{\zeta}(T, U) = a^{T'T'T}, \tag{1.87}$$

with $|\mathrm{s}(T')|$ as in Definition 1.8. The rhs counts the number of faces of type T on a flat of type $\mathrm{s}(T')$. (Use gate identity (1.61), flat-additivity formula (1.83), identity (1.46) for $q = 1$.) When $\mathrm{s}(T) = \mathrm{s}(T')$, (1.87) reduces to $|\mathrm{s}(T')| = a^{T'T'T} = |\mathrm{s}(T)|$.

Exercise 1.88. Show: For any invariant noncommutative Möbius function $\boldsymbol{\mu}$, for face-types $T \leq V$ and T',

$$\sum_{U:\, U \geq T'} a^{TVU} \boldsymbol{\mu}(T', U) = \sum_{\substack{Y:\, V \geq Y \geq T \\ \mathrm{s}(Y) = \mathrm{s}(T')}} a^{TYT'} \boldsymbol{\mu}(Y, V). \tag{1.88}$$

The rhs is a simplification of the lhs. (Use gate identity (1.62) and noncommutative Weisner formula (1.85).) An equivalent form of (1.88) is given later in (1.140).

Exercise 1.89. Show: For any invariant noncommutative Möbius function $\boldsymbol{\mu}$, for a face-type T,

$$\sum_{U:\, U \geq T} \boldsymbol{\mu}(T, U)\, d_U = \begin{cases} d_S & \text{if } T = S, \\ 0 & \text{otherwise.} \end{cases} \tag{1.89}$$

(Either write $d_U = d_T d_{U/T}$, pull d_T outside the sum, and apply (1.85) or, alternatively, specialize (1.88). In either case, use (1.39).)

1.10.6. Noncommutative q-zeta and q-Möbius functions.

Let q be a scalar which is not a root of unity. Recall from [**11**, Section 1.5.9] the *noncommutative q-zeta function* $\boldsymbol{\zeta}_q$ and *noncommutative q-Möbius function* $\boldsymbol{\mu}_q$. These are mutually inverse elements of the invariant lune-incidence algebra. The element $\boldsymbol{\zeta}_q$ is such that $\boldsymbol{\zeta}_q(A, A) = 1$ for all A, and the *q-lune-additivity formula* holds, that is,

$$\boldsymbol{\zeta}_q(H, G) = \sum_{\substack{F:\, F \geq A,\, HF = G \\ \mathrm{s}(F) = \mathrm{s}(G)}} \boldsymbol{\zeta}_q(A, F)\, q^{\mathrm{dist}(F,G)} \tag{1.90}$$

for all faces $A \leq H \leq G$. As a particular case, the *q-flat-additivity formula* says

$$\sum_{\substack{F:\, F \geq A \\ \mathrm{s}(F) = \mathrm{s}(G)}} \boldsymbol{\zeta}_q(A, F)\, q^{\mathrm{dist}(F,G)} = 1 \tag{1.91}$$

for all faces $A \leq G$. The element $\boldsymbol{\mu}_q$ is such that $\boldsymbol{\mu}_q(A, A) = 1$ for all A, and the *noncommutative q-Weisner formula* holds, that is,

$$\sum_{F:\, F \geq A,\, HF = G} \boldsymbol{\mu}_q(A, F)\, q^{\mathrm{dist}(F,G)} = 0 \tag{1.92}$$

for all faces $A < H \leq G$.

Now recall the face-type enumeration polynomials $a_Z^{TVU}(q)$ obtained by applying (1.38) to $\boldsymbol{\alpha}_Z$, and their specializations $c_Z^{TT'}(q)$ obtained by applying (1.52) to $\boldsymbol{\alpha}_Z$. We now express $\boldsymbol{\zeta}_q$ and $\boldsymbol{\mu}_q$ in terms of face-types. We have $\boldsymbol{\zeta}_q(Z, Z) = 1$ for all Z and the q-lune-additivity formula (1.90) takes the form

$$\boldsymbol{\zeta}_q(T, V) = \sum_{\substack{U:\, U \geq Z \\ \mathrm{s}(U/Z) = \mathrm{s}(V/Z)}} a_Z^{TVU}(q)\, \boldsymbol{\zeta}_q(Z, U) \tag{1.93}$$

for all face-types $Z \leq T \leq V$. The q-flat-additivity formula (1.91) becomes

$$\sum_{\substack{U:\,U\geq Z\\ \mathrm{s}(U/Z)=\mathrm{s}(V/Z)}} a_Z^{VVU}(q)\,\zeta_q(Z,U) = \sum_{\substack{U:\,U\geq Z\\ \mathrm{s}(U/Z)=\mathrm{s}(V/Z)}} c_Z^{VU}(q)\,\zeta_q(Z,U) = 1 \tag{1.94}$$

for all face-types $Z \leq V$.

Exercise 1.90. Put $V = S$ in (1.94) to deduce the formula

$$\zeta_q(Z,S) = \frac{1}{d_{S/Z}(q)} = \frac{d_Z(q)}{d_S(q)}. \tag{1.95}$$

Similarly, we have $\mu_q(Z,Z) = 1$ for all Z and the noncommutative q-Weisner formula (1.92) takes the form

$$\sum_{U:\,U\geq Z} a_Z^{TVU}(q)\,\mu_q(Z,U) = 0 \tag{1.96}$$

for all face-types $Z < T \leq V$.

Example 1.91 (Ranks one and two). Let q be a scalar which is not a root of unity. For a rank-one arrangement with face-types $\emptyset \lessdot S$,

$$\zeta_q(\emptyset,S) = \frac{1}{1+q} \quad \text{and} \quad \mu_q(\emptyset,S) = \frac{-1}{1+q}.$$

For the dihedral arrangement of d lines, for face-types $\emptyset \lessdot T \lessdot S$,

$$\zeta_q(\emptyset,T) = \frac{1}{1+q^{d-1}}, \quad \zeta_q(T,S) = \frac{1}{1+q}, \quad \zeta_q(\emptyset,S) = \frac{1-q}{(1-q^d)(1+q)},$$
$$\mu_q(\emptyset,T) = \frac{-1}{1+q^{d-1}}, \quad \mu_q(T,S) = \frac{-1}{1+q}, \quad \mu_q(\emptyset,S) = \frac{1-q^{d-1}}{(1-q^d)(1+q^{d-1})}.$$

Using the calculations in Example 1.45, one may check that the above ζ_q and μ_q indeed satisfy the q-lune-additivity formula (1.93) and the noncommutative q-Weisner formula (1.96). □

Exercise 1.92. Check: The q-lune-additivity formula (1.93) is a tautology when $Z = T \leq V$. (Use identity (1.40).)

Exercise 1.93. Show: For q not a root of unity, for face-types T and T',

$$\sum_{\substack{U:\,U\geq T\\ \mathrm{s}(U)=\mathrm{s}(T')}} \zeta_q(T,U)\,c^{UT'}(q) = a^{T'T'T}(q). \tag{1.97}$$

(Use gate identity (1.60), q-flat-additivity formula (1.94), identity (1.46).) This is the q-analogue of identity (1.87). When $\mathrm{s}(T) = \mathrm{s}(T')$, (1.97) reduces to $c^{TT'}(q) = a^{T'T'T}(q) = c^{T'T}(q)$ which we saw in (1.53) and (1.54).

Exercise 1.94. Show: For q not a root of unity, for face-types $T \leq V$ and T',

$$\sum_{U:\,U\geq T'} a^{TVU}(q)\,\mu_q(T',U) = \sum_{\substack{Y:\,V\geq Y\geq T\\ \mathrm{s}(Y)=\mathrm{s}(T')}} a^{TYT'}(q)\,\mu_q(Y,V). \tag{1.98}$$

This is the q-analogue of identity (1.88). (Use gate identity (1.62) and noncommutative q-Weisner formula (1.96).)

Exercise 1.95. Show: For q not a root of unity, for a face-type T,

$$\sum_{U:\,U\geq T} \boldsymbol{\mu}_q(T,U)\, d_U(q) = \begin{cases} d_S(q) & \text{if } T=S, \\ 0 & \text{otherwise.} \end{cases} \tag{1.99}$$

This is the q-analogue of formula (1.89).

1.10.7. q-zeta and q-Möbius functions.

For q not a root of unity, define the *q-zeta function* ζ_q and *q-Möbius function* μ_q by

$$\zeta_q := \mathrm{bc}(\boldsymbol{\zeta}_q) \quad \text{and} \quad \mu_q := \mathrm{bc}(\boldsymbol{\mu}_q), \tag{1.100}$$

where bc is the base-case map (1.78). These are mutually inverse elements of the invariant flat-incidence algebra.

1.10.8. Projective lune-incidence algebra.

For any $s \in \mathrm{I}_{\mathrm{lune}}[\mathcal{A}]$, define $\overline{s} \in \mathrm{I}_{\mathrm{lune}}[\mathcal{A}]$ by $\overline{s}(A,F) := s(A, A\overline{F})$. We say s is projective if $s = \overline{s}$. The map $s \mapsto \overline{s}$ is an involution of the lune-incidence algebra. Hence, projective functions in the lune-incidence algebra form a subalgebra. We call it the *projective lune-incidence algebra*.

If s is invariant, then so is $\overline{s}$. In this situation, it makes sense to write $\overline{s}(Z,T) = s(Z, Z\overline{T})$. Thus, the involution $s \mapsto \overline{s}$ restricts to the invariant lune-incidence algebra yielding the *projective invariant lune-incidence algebra*.

If $\boldsymbol{\zeta}$ and $\boldsymbol{\mu}$ are (invariant) noncommutative zeta and Möbius functions, then so are $\overline{\boldsymbol{\zeta}}$ and $\overline{\boldsymbol{\mu}}$. For q not a root of unity, the noncommutative q-zeta and q-Möbius functions $\boldsymbol{\zeta}_q$ and $\boldsymbol{\mu}_q$ belong to the projective invariant lune-incidence algebra.

1.10.9. Invariant bilune-incidence algebra.

The *biface-incidence algebra*, denoted $\mathrm{I}_{\mathrm{biface}}[\mathcal{A}]$, consists of functions s on local bifaces, with the product of s and t defined by

$$(st)(A,G,G') = \sum_{\substack{(F,F')\colon \mathrm{s}(F)=\mathrm{s}(F') \\ A\leq F\leq G \\ A\leq F'\leq G'}} s(A,F,F')t(F',F'G,G'). \tag{1.101}$$

We denote its W-invariant subalgebra by $\mathrm{I}_{\mathrm{biface}}[\mathcal{A}]^W$ and call it the *invariant biface-incidence algebra*. It has a basis of local biface-types. In this case, the expression $s(Z,T,\sigma,T')$ makes sense. The product can be expressed using the groupoid of biface-types as

$$(st)(Z,U,\rho,U') = \sum s(Z,T,\sigma,T')t(T',U'',\tau,U'). \tag{1.102}$$

The sum is over all morphisms $\sigma : T' \to T$ local to Z (with $T \leq U$ and $T' \leq U'$) and $\tau : U' \to U''$ local to T' with $U'' = \sigma^*(U)$ such that $\rho = \sigma\tau$. In this composite, we use the induced morphism $\sigma : U'' \to U$. In other words, we sum over diagrams of the form

$$\begin{array}{ccccc} & & \overset{\rho}{\frown} & & \\ U & \xleftarrow{\sigma} & U'' & \xleftarrow{\tau} & U' \\ & \diagdown & & \diagdown \quad \diagup & \\ & T & \xleftarrow{\sigma} & T' & \\ & & \diagdown \; \diagup & & \\ & & Z. & & \end{array} \tag{1.103}$$

Two illustrations of diagram (1.103) for $Z = \emptyset$ are shown in Figure 1.20.

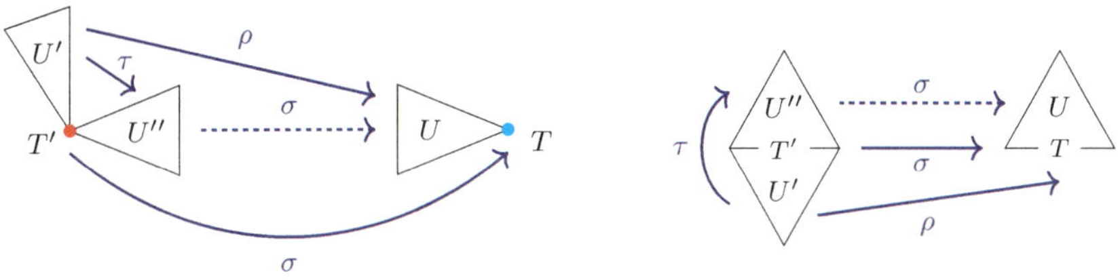

FIGURE 1.20. Product in the invariant biface-incidence algebra.

Observation 1.96. Consider diagrams of the form (1.103), with $Z = \emptyset$ and the morphism $\rho : U' \to U$ given. There is one such diagram for each face-type $T' \leq U'$ for which the face H in Remark 1.2 exists for $\mathrm{t}(K, G) = \rho$ and F the face of G of type T'. In this situation: $\mathrm{t}(H, F) = \sigma$ and, in particular, $T = \mathrm{t}(H)$. This determines the induced σ, and then τ. Also, $\rho = \sigma\tau$ equals the factorization of ρ yielded by the "dual" gate property for the face-type T'. Note: The extreme choices $T' := \emptyset$ and $T' := U'$ always work.

The *bilune-incidence algebra*, denoted $\mathrm{I}_{\mathrm{bilune}}[\mathcal{A}]$, is the subalgebra of the biface-incidence algebra consisting of functions s such that $s(A, F, F') = s(B, G, G')$ whenever (A, F, F') and (B, G, G') have the same support. It is the same as the incidence algebra of the category of bilunes. It has a basis indexed by bilunes. We denote its W-invariant subalgebra by $\mathrm{I}_{\mathrm{bilune}}[\mathcal{A}]^W$ and call it the *invariant bilune-incidence algebra*. It has a basis of bilune-types. It is also a subalgebra of the invariant biface-incidence algebra consisting of functions which have the same value on local biface-types of the same support.

Remark 1.97. There is another associative product on functions on local bifaces, with $t(F', F'G, G')$ replaced by $t(F, G, FG')$ in (1.101). The two products match on the bilune-incidence algebra. Similarly, in the alternative to (1.102), we sum over diagrams (1.103), but with the parallelogram on the right and triangle on the left. The two products match on the invariant bilune-incidence algebra.

1.10.10. Relating lune- and bilune-incidence algebras. There are algebra morphisms

$$\mathrm{p}, \mathrm{q} : \mathrm{I}_{\mathrm{bilune}}[\mathcal{A}] \to \mathrm{I}_{\mathrm{lune}}[\mathcal{A}] \quad \text{and} \quad \mathrm{i} : \mathrm{I}_{\mathrm{lune}}[\mathcal{A}] \to \mathrm{I}_{\mathrm{bilune}}[\mathcal{A}]$$

defined by

$$
\begin{aligned}
\mathrm{p}(s)(A,F) &:= \sum_{\substack{F'\colon F'\geq A\\ \mathrm{s}(F)=\mathrm{s}(F')}} s(A,F,F'),\\
\mathrm{q}(s)(A,F') &:= \sum_{\substack{F\colon F\geq A\\ \mathrm{s}(F)=\mathrm{s}(F')}} s(A,F,F'),\\
\mathrm{i}(s)(A,F,F') &:= \begin{cases} s(A,F) & \text{if } F=F',\\ 0 & \text{otherwise.}\end{cases}
\end{aligned}
\tag{1.104}
$$

These are maps of left W-modules and induce algebra morphisms

$$\mathrm{p},\mathrm{q}: \mathrm{I}_{\mathrm{bilune}}[\mathcal{A}]^W \to \mathrm{I}_{\mathrm{lune}}[\mathcal{A}]^W \quad \text{and} \quad \mathrm{i}: \mathrm{I}_{\mathrm{lune}}[\mathcal{A}]^W \to \mathrm{I}_{\mathrm{bilune}}[\mathcal{A}]^W.$$

In terms of face-types,

$$
\begin{aligned}
\mathrm{p}(s)(\mathrm{Z},\mathrm{T}) &= \sum_{\substack{\mathrm{T}'\colon\\ \mathrm{T}'\geq \mathrm{Z}}} \sum_{\substack{\sigma\colon \mathrm{T}'\to\mathrm{T}\\ \text{local to Z}}} s(\mathrm{Z},\mathrm{T},\sigma,\mathrm{T}'),\\
\mathrm{q}(s)(\mathrm{Z},\mathrm{T}') &= \sum_{\substack{\mathrm{T}\colon\\ \mathrm{T}\geq \mathrm{Z}}} \sum_{\substack{\sigma\colon \mathrm{T}'\to\mathrm{T}\\ \text{local to Z}}} s(\mathrm{Z},\mathrm{T},\sigma,\mathrm{T}'),\\
\mathrm{i}(s)(\mathrm{Z},\mathrm{T},\sigma,\mathrm{T}') &= \begin{cases} s(\mathrm{Z},\mathrm{T}) & \text{if } \sigma=\mathrm{id},\\ 0 & \text{otherwise.}\end{cases}
\end{aligned}
\tag{1.105}
$$

The map i is a section to both p and q. As a consequence, i is injective and p and q are surjective.

1.10.11. Two-sided q-zeta and q-Möbius functions. Let q be a scalar which is not a root of unity. Recall from [**11**, Section 1.6.6] the *two-sided q-zeta function* ζ_q and *two-sided q-Möbius function* μ_q. These are mutually inverse elements of the invariant bilune-incidence algebra. The element ζ_q is such that $\zeta_q(A,A,A) = 1$ for all A, and the *two-sided q-lune-additivity formula* holds, that is,

$$\sum_{\substack{F\colon F\geq A,\, HF=G\\ \mathrm{s}(F)=\mathrm{s}(G)}} \zeta_q(A,F,G')\, q^{\mathrm{dist}(F,G)} = \begin{cases} \zeta_q(H,G,G') & \text{if } G'\geq H,\\ 0 & \text{otherwise}\end{cases} \tag{1.106}$$

for all faces $A \leq H \leq G$ and $A \leq G'$ with $\mathrm{s}(G) = \mathrm{s}(G')$. As a particular case, the *two-sided q-flat-additivity formula* says

$$\sum_{\substack{F\colon F\geq A\\ \mathrm{s}(F)=\mathrm{s}(G)=\mathrm{s}(G')}} \zeta_q(A,G',F)\, q^{\mathrm{dist}(F,G)} = \begin{cases} 1 & \text{if } G=G',\\ 0 & \text{if } G\neq G'\end{cases} \tag{1.107}$$

for all faces $A \leq G, G'$ with $\mathrm{s}(G) = \mathrm{s}(G')$. The element μ_q is such that $\mu_q(A,A,A) = 1$ for all A, and the *two-sided q-Weisner formula* holds, that is,

$$\sum_{\substack{(F,F')\colon F,F'\geq A,\, \mathrm{s}(F)=\mathrm{s}(F') \\ HF=G,\, F'H=G'}} \boldsymbol{\mu}_q(A,F,F')\, q^{\mathrm{dist}(F,G)} = 0 \tag{1.108}$$

for all faces $A < H \leq G$ and $A < G'$ with $\mathrm{s}(G) = \mathrm{s}(G')$.

In terms of face-types: We have $\boldsymbol{\zeta}_q(Z, Z, \mathrm{id}, Z) = 1$ for all Z and the two-sided q-lune-additivity formula (1.106) takes the form

$$\sum_{\substack{\sigma\colon V'\to T \text{ local to } Z \\ \tau\colon T\to V \text{ local to } Z \\ \rho=\tau\sigma \\ \tau \text{ projects } T \text{ on } U}} \boldsymbol{\zeta}_q(Z,T,\sigma,V')\, q^{l(\tau)} = \begin{cases} \boldsymbol{\zeta}_q(U,V,\rho,V') & \text{if } \rho \text{ is local to } U, \\ 0 & \text{otherwise} \end{cases} \tag{1.109}$$

for all face-types $Z \leq U \leq V$ and any morphism $\rho : V' \to V$ local to Z. An illustration with $Z = \emptyset$ is shown in Figure 1.21. The two red morphisms labeled ρ indicate the two alternatives in (1.109).

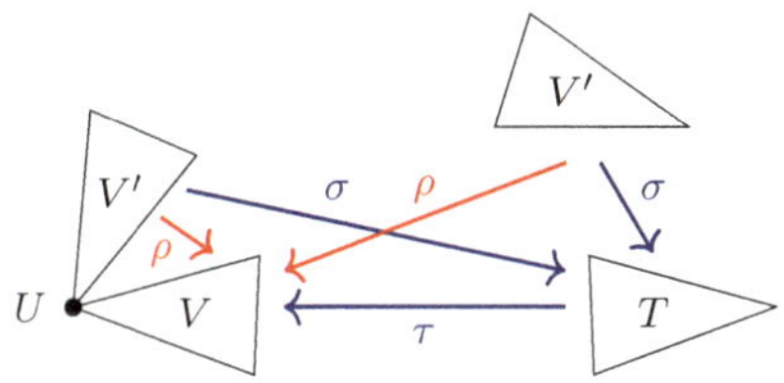

FIGURE 1.21. Two-sided q-lune-additivity formula.

The two-sided q-flat-additivity formula (1.107) becomes

$$\sum_{\substack{\sigma\colon V'\to T \text{ local to } Z \\ \tau\colon T\to V \text{ local to } Z \\ \rho=\tau\sigma}} \boldsymbol{\zeta}_q(Z,T,\sigma,V')\, q^{l(\tau)} = \begin{cases} 1 & \text{if } \rho = \mathrm{id}, \\ 0 & \text{otherwise} \end{cases} \tag{1.110}$$

for any morphism $\rho : V' \to V$ local to Z. This formula is the special case $U = V$ of (1.109).

We have $\boldsymbol{\mu}_q(Z, Z, \mathrm{id}, Z) = 1$ for all Z and the two-sided q-Weisner formula (1.108) takes the form

$$\sum_{\substack{T'\colon T'\leq V' \\ \sigma\colon T'\to T \text{ local to } Z \\ \tau\colon \sigma_*(V')\to V \text{ local to } Z \\ \rho=\tau\sigma \\ \tau \text{ biprojects } T \text{ and } U}} \boldsymbol{\mu}_q(Z,T,\sigma,T')\, q^{l(\tau)} = 0 \tag{1.111}$$

for all face-types $Z < U \leq V$ and any morphism $\rho : V' \to V$ local to Z. Note very carefully: In $\rho = \tau\sigma$, we use $\sigma : V' \to \sigma_*(V')$ induced from $\sigma : T' \to T$. An illustration with $Z = \emptyset$ is shown in Figure 1.22.

Note: There is also a “dual” way of formulating (1.109), (1.110), (1.111) by replacing morphisms σ, τ, ρ by their inverses.

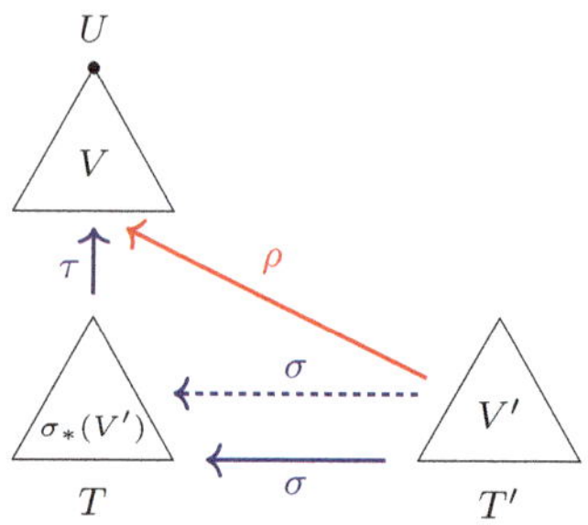

FIGURE 1.22. Two-sided q-Weisner formula.

Example 1.98 (Rank one). Let q be a scalar which is not a root of unity. For a rank-one arrangement with face-types $\emptyset \lessdot S$,

$$\zeta_q(\emptyset, S, \mathrm{id}, S) = \frac{1}{1-q^2}, \quad \zeta_q(\emptyset, S, \sigma, S) = \frac{-q}{1-q^2},$$
$$\mu_q(\emptyset, S, \mathrm{id}, S) = \frac{-1}{1-q^2}, \quad \mu_q(\emptyset, S, \sigma, S) = \frac{q}{1-q^2},$$

with the morphism σ as in the first paragraph in Example 1.23. □

Exercise 1.99 (Inverse of the Varchenko matrix). Recall from Section 1.8.7: The Varchenko matrix of a flat-type λ/Z is defined by $A_{\lambda/Z} := (c_Z^{TT'}(q))$. Its entries are the polynomials (1.64d), with $\mathrm{s}(T/Z) = \mathrm{s}(T'/Z) = \lambda/Z$. A way to prove Theorem 1.56 is given below.

Let q be a scalar which is not a root of unity. For $\mathrm{s}(T/Z) = \mathrm{s}(T'/Z)$, define

$$b_Z^{TT'}(q) := \sum_{\substack{\sigma: T \to T' \\ \text{local to } Z}} \zeta_q(Z, T', \sigma, T). \tag{1.112}$$

Check: For any flat-type λ/Z, the matrices $(c_Z^{TT'}(q))$ and $(b_Z^{TT'}(q))$ are inverses of each other, with T and T' subject to the condition $\mathrm{s}(T/Z) = \mathrm{s}(T'/Z) = \lambda/Z$. (Use the two-sided q-flat-additivity formula (1.110).)

Specializing to the maximum flat-type in $\boldsymbol{\alpha}_Z$, deduce: For any face-type Z,

$$\sum_{\substack{\sigma: S \to S \\ \text{local to } Z}} \zeta_q(Z, S, \sigma, S) = \frac{1}{d_{S/Z}(q)}, \tag{1.113}$$

the reciprocal of the Poincaré polynomial of $\boldsymbol{\alpha}_Z$.

For more general considerations, see Lemma 9.101 and Exercise 9.103.

Fact 1.100. We have $\mathrm{p}(\zeta_q) = \mathrm{q}(\zeta_q) = \boldsymbol{\zeta}_q$ and $\mathrm{p}(\mu_q) = \mathrm{q}(\mu_q) = \boldsymbol{\mu}_q$. This explains why we see the same values in (1.95) and (1.113).

PROOF. See [**11**, Lemma 1.37]. □

Exercise 1.101 (Two-sided to one-sided). Check: Summing (1.109) over ρ with varying V' but fixed V yields (1.93), while summing (1.111) similarly yields (1.96).

1.10.12. 0-zeta and 0-Möbius functions. We now specialize to $q = 0$. The *noncommutative* 0-*zeta function* $\boldsymbol{\zeta}_0$ and *noncommutative* 0-*Möbius function* $\boldsymbol{\mu}_0$ have explicit formulas given by

$$\boldsymbol{\zeta}_0(A, F) = 1 \quad \text{and} \quad \boldsymbol{\mu}_0(A, F) = (-1)^{\mathrm{rk}(F/A)}. \tag{1.114}$$

Equivalently,

$$\boldsymbol{\zeta}_0(Z, T) = 1 \quad \text{and} \quad \boldsymbol{\mu}_0(Z, T) = (-1)^{\mathrm{rk}(T/Z)}. \tag{1.115}$$

Applying the base-case map, we see that the 0-*zeta function* ζ_0 and 0-*Möbius function* μ_0 are

$$\zeta_0(\mathrm{X}, \mathrm{Y}) = c_{\mathrm{X}}^{\mathrm{Y}} \quad \text{and} \quad \mu_0(\mathrm{X}, \mathrm{Y}) = (-1)^{\mathrm{rk}(\mathrm{Y}/\mathrm{X})}\, c_{\mathrm{X}}^{\mathrm{Y}}, \tag{1.116}$$

where $c_{\mathrm{X}}^{\mathrm{Y}}$ is the number of chambers in $\mathcal{A}_{\mathrm{X}}^{\mathrm{Y}}$, the arrangement over X and under Y.

Similarly, the *two-sided* 0-*zeta function* ζ_0 and *two-sided* 0-*Möbius function* μ_0 have explicit formulas given by

$$\begin{aligned} \zeta_0(A, F, F') &= \begin{cases} 1 & \text{if } F = F', \\ 0 & \text{otherwise,} \end{cases} \\ \mu_0(A, F, F') &= \begin{cases} (-1)^{\mathrm{rk}(F/A)} & \text{if } F = F', \\ 0 & \text{otherwise.} \end{cases} \end{aligned} \tag{1.117}$$

Equivalently,

$$\begin{aligned} \zeta_0(Z, T, \sigma, T') &= \begin{cases} 1 & \text{if } T = T' \text{ and } \sigma = \mathrm{id}, \\ 0 & \text{otherwise,} \end{cases} \\ \mu_0(Z, T, \sigma, T') &= \begin{cases} (-1)^{\mathrm{rk}(T/Z)} & \text{if } T = T' \text{ and } \sigma = \mathrm{id}, \\ 0 & \text{otherwise.} \end{cases} \end{aligned} \tag{1.118}$$

1.11. Noncommutative Zaslavsky formulas and q-analogues

We briefly discuss the noncommutative Zaslavsky formulas and their q-analogues. They are formulated in terms of zeta and Möbius functions. They have connections to antipodes of Coxeter bimonoids and Coxeter bialgebras (Sections 3.14 and 9.14).

1.11.1. Noncommutative Zaslavsky formulas. Let $\mathcal{A}$ be any arrangement. For mutually inverse noncommutative zeta and Möbius functions $\boldsymbol{\zeta}$ and $\boldsymbol{\mu}$, for faces $A \leq G$,

$$\boldsymbol{\zeta}(A, A\overline{G}) = \sum_{F:\, A \leq F \leq G} (-1)^{\mathrm{rk}(G/F)} \boldsymbol{\zeta}(F, G), \tag{1.119a}$$

$$\boldsymbol{\mu}(A, A\overline{G}) = (-1)^{\mathrm{rk}(G/A)} \sum_{F:\, A \leq F \leq G} \boldsymbol{\mu}(A, F), \tag{1.119b}$$

$$\sum_{F:\, A \leq F \leq G} (-1)^{\mathrm{rk}(G/F)} \boldsymbol{\zeta}(A, A\overline{F})\, \boldsymbol{\mu}(F, G) = 1, \tag{1.119c}$$

$$\sum_{F:\, A \le F \le G} (-1)^{\mathrm{rk}(G/F)} \boldsymbol{\zeta}(A, F)\, \boldsymbol{\mu}(F, F\overline{G}) = 1. \tag{1.119d}$$

We refer to either (1.119c) or (1.119d) as the *noncommutative Zaslavsky formula*. For more details, see [**11**, Section 1.8].

Exercise 1.102. Deduce the *Zaslavsky formula*: For any arrangement $\mathcal{A}$,

$$\sum_{\mathrm{X}} |\mu(\mathrm{X}, \top)| = c(\mathcal{A}), \tag{1.120}$$

where $c(\mathcal{A})$ is the number of chambers of $\mathcal{A}$. (Use Fact 1.80.)

Now let $\mathcal{A}$ be a reflection arrangement. Identities (1.119) can be reformulated in terms of face-types by replacing faces A, F, G by face-types Z, T, U, respectively. In other words: For mutually inverse invariant noncommutative zeta and Möbius functions $\boldsymbol{\zeta}$ and $\boldsymbol{\mu}$, for face-types $Z \le U$,

$$\boldsymbol{\zeta}(Z, Z\overline{U}) = \sum_{T:\, Z \le T \le U} (-1)^{\mathrm{rk}(U/T)} \boldsymbol{\zeta}(T, U), \tag{1.121a}$$

$$\boldsymbol{\mu}(Z, Z\overline{U}) = (-1)^{\mathrm{rk}(U/Z)} \sum_{T:\, Z \le T \le U} \boldsymbol{\mu}(Z, T), \tag{1.121b}$$

$$\sum_{T:\, Z \le T \le U} (-1)^{\mathrm{rk}(U/T)} \boldsymbol{\zeta}(Z, Z\overline{T})\, \boldsymbol{\mu}(T, U) = 1, \tag{1.121c}$$

$$\sum_{T:\, Z \le T \le U} (-1)^{\mathrm{rk}(U/T)} \boldsymbol{\zeta}(Z, T)\, \boldsymbol{\mu}(T, T\overline{U}) = 1. \tag{1.121d}$$

1.11.2. Noncommutative q-Zaslavsky formulas.

Let $\mathcal{A}$ be any arrangement. For q not a root of unity, for faces $A \le G$,

$$\boldsymbol{\zeta}_q(A, A\overline{G})\, q^{\mathrm{dist}(G, A\overline{G})} = \sum_{F:\, A \le F \le G} (-1)^{\mathrm{rk}(G/F)} \boldsymbol{\zeta}_q(F, G), \tag{1.122a}$$

$$\boldsymbol{\mu}_q(A, A\overline{G}) = (-1)^{\mathrm{rk}(G/A)} \sum_{F:\, A \le F \le G} \boldsymbol{\mu}_q(A, F)\, q^{\mathrm{dist}(F, A\overline{F})}, \tag{1.122b}$$

$$\sum_{F:\, A \le F \le G} (-1)^{\mathrm{rk}(G/F)} \boldsymbol{\zeta}_q(A, A\overline{F})\, \boldsymbol{\mu}_q(F, G)\, q^{\mathrm{dist}(F, A\overline{F})} = 1, \tag{1.122c}$$

$$\sum_{F:\, A \le F \le G} (-1)^{\mathrm{rk}(G/F)} \boldsymbol{\zeta}_q(A, F)\, \boldsymbol{\mu}_q(F, F\overline{G})\, q^{\mathrm{dist}(F, A\overline{F})} = 1. \tag{1.122d}$$

We refer to either (1.122c) or (1.122d) as the *noncommutative q-Zaslavsky formula*.

Exercise 1.103. Deduce the *q-Zaslavsky formula*: For any arrangement $\mathcal{A}$,

$$\sum_{\mathrm{X}} (-1)^{\mathrm{rk}(\mathcal{A}_{\mathrm{X}})} q^{g(\mathrm{X})}\, \zeta_q(\bot, \mathrm{X})\, \mu_q(\mathrm{X}, \top) = c(\mathcal{A}), \tag{1.123}$$

where $c(\mathcal{A})$ is the number of chambers of $\mathcal{A}$, and $g(\mathrm{X}) := \mathrm{dist}(F, \overline{F})$ for any face F of support X.

Now let $\mathcal{A}$ be a reflection arrangement. Identities (1.122) can be reformulated in terms of face-types as follows. For q not a root of unity, for face-types $Z \le U$,

$$\boldsymbol{\zeta}_q(Z, Z\overline{U})\, q^{\mathrm{dist}(U,Z\overline{U})} = \sum_{T:\, Z \le T \le U} (-1)^{\mathrm{rk}(U/T)} \boldsymbol{\zeta}_q(T, U), \tag{1.124a}$$

$$\boldsymbol{\mu}_q(Z, Z\overline{U}) = (-1)^{\mathrm{rk}(U/Z)} \sum_{T:\, Z \le T \le U} \boldsymbol{\mu}_q(Z, T)\, q^{\mathrm{dist}(T,Z\overline{T})}, \tag{1.124b}$$

$$\sum_{T:\, Z \le T \le U} (-1)^{\mathrm{rk}(U/T)} \boldsymbol{\zeta}_q(Z, Z\overline{T})\, \boldsymbol{\mu}_q(T, U)\, q^{\mathrm{dist}(T,Z\overline{T})} = 1, \tag{1.124c}$$

$$\sum_{T:\, Z \le T \le U} (-1)^{\mathrm{rk}(U/T)} \boldsymbol{\zeta}_q(Z, T)\, \boldsymbol{\mu}_q(T, T\overline{U})\, q^{\mathrm{dist}(T,Z\overline{T})} = 1. \tag{1.124d}$$

1.11.3. Solomon coefficients. Let $\mathcal{A}$ be any arrangement. Let $\boldsymbol{\mu}$ be a noncommutative Möbius function. For a local biface (A, G, H), define

$$\mathrm{sln}_A^{G,H} := \sum_{F:\, F \ge A,\, FH = G} \boldsymbol{\mu}(A, F). \tag{1.125}$$

We refer to (1.125) as a *Solomon coefficient*. Observe: For $A \le H$,

$$\mathrm{sln}_A^{H,H} = (-1)^{\mathrm{rk}(H/A)} \boldsymbol{\mu}(A, A\overline{H}) \quad \text{and} \quad \mathrm{sln}_A^{A\overline{H},H} = \boldsymbol{\mu}(A, A\overline{H}).$$

We used (1.119b).

Now let $\mathcal{A}$ be a reflection arrangement. Let $\boldsymbol{\mu}$ be an invariant noncommutative Möbius function. For a local biface-type (Z, U, α, T), define

$$\mathrm{sln}_Z^{\alpha} := \sum_{\substack{T':\, Z \le T' \le U \\ \alpha \text{ projects } T \text{ on } T'}} \boldsymbol{\mu}(Z, T'), \tag{1.126}$$

with terminology as in Section 1.7.4. Observe: For $Z \le T$,

$$\mathrm{sln}_Z^{\mathrm{id}_T} = (-1)^{\mathrm{rk}(T/Z)} \boldsymbol{\mu}(Z, Z\overline{T}) \quad \text{and} \quad \mathrm{sln}_Z^{\beta_{Z\overline{T},T}} = \boldsymbol{\mu}(Z, Z\overline{T}).$$

We used (1.121b).

Exercise 1.104. Check: For face-types $U, V \ge Z$ with $\mathrm{s}(U/Z) = \mathrm{s}(V/Z)$,

$$\sum_{T:\, Z \le T \le V} \boldsymbol{\mu}(Z, T)\, a_Z^{TVU} = \sum_{\substack{\alpha:\, U \to V \\ \text{local to } Z}} \mathrm{sln}_Z^{\alpha}. \tag{1.127}$$

(Use (1.64b) for $q = 1$.)

1.12. Invariant Birkhoff, Tits, Janus algebras

Fix a reflection arrangement $\mathcal{A}$ with Coxeter group W. We discuss the Birkhoff algebra, Tits algebra, Janus algebra, and their W-invariant subalgebras with emphasis on families of orthogonal idempotents, thus highlighting their connection to zeta and Möbius functions. See Table 1.7. Face-type enumeration polynomials enter the story as structure constants of the invariant Tits algebra and invariant q-Janus algebra. For q not a root of unity, the invariant q-Janus algebra is the incidence algebra of the groupoid of biface-types. For supplementary information on some of this material, see [**10**, Sections 9.1, 9.3, 15.5, 16.1, 16.2, 16.8] and [**11**, Section 1.9].

All our algebras are over a fixed field $\Bbbk$.

TABLE 1.7. Birkhoff, Tits, Janus algebras.

$\sqcap[\mathcal{A}]$ $\mathtt{H}_X$, $\mathtt{Q}_X$	Birkhoff algebra	$\sqcap[\mathcal{A}]^W$ $\mathtt{H}_\lambda$, $\mathtt{Q}_\lambda$	invariant Birkhoff algebra
$\Sigma[\mathcal{A}]$ $\mathtt{H}_F$, $\mathtt{Q}_F$, $\mathtt{E}_X$	Tits algebra	$\Sigma[\mathcal{A}]^W$ $\mathtt{H}_T$, $\mathtt{Q}_T$, $\mathtt{E}_\lambda$	invariant Tits algebra
$\mathsf{J}[\mathcal{A}]$ $\mathtt{H}_{(F',F)}$	Janus algebra	$\mathsf{J}[\mathcal{A}]^W$ $\mathtt{H}_{(T',\sigma,T)}$	invariant Janus algebra
$\mathsf{J}_q[\mathcal{A}]$ $\mathtt{H}_{(F',F)}$, $\mathtt{Q}_{(F',F)}$	q-Janus algebra for q not a root of unity	$\mathsf{J}_q[\mathcal{A}]^W$ $\mathtt{H}_{(T',\sigma,T)}$, $\mathtt{Q}_{(T',\sigma,T)}$	invariant q-Janus algebra for q not a root of unity

We will see later that the invariant Tits algebra and invariant q-Janus algebra give rise to characteristic operations on Coxeter bialgebras (Section 9.13) which can be used to study structure theorems such as the Borel–Hopf theorem (Sections 13.1 and 13.3).

1.12.1. Invariant Birkhoff algebra. The linearization of the Birkhoff monoid is the *Birkhoff algebra*. We denote it by $\sqcap[\mathcal{A}]$, and use the letter $\mathtt{H}$ for its canonical basis. Thus,

$$\mathtt{H}_X \cdot \mathtt{H}_Y := \mathtt{H}_{X \vee Y}. \tag{1.128}$$

Define the $\mathtt{Q}$-basis of $\sqcap[\mathcal{A}]$ by either of the two equivalent formulas

$$\mathtt{H}_X = \sum_{Y:\, Y \geq X} \mathtt{Q}_Y, \qquad \mathtt{Q}_X = \sum_{Y:\, Y \geq X} \mu(X, Y)\, \mathtt{H}_Y, \tag{1.129}$$

where μ is the Möbius function in the invariant flat-incidence algebra. Fact: The Birkhoff algebra is a split-semisimple commutative algebra (of dimension equal to the number of flats) and $\mathtt{Q}$ is its basis of primitive idempotents.

Let $\operatorname{char}(\Bbbk) \nmid |W|$. The Coxeter group W acts on $\sqcap[\mathcal{A}]$ via $w \cdot \mathtt{H}_X := \mathtt{H}_{w(X)}$. Let $\sqcap[\mathcal{A}]^W$ denote the subalgebra consisting of W-invariant elements. We call it the *invariant Birkhoff algebra*. It has a basis consisting of flat-types: For each flat-type λ, set

$$\mathtt{H}_\lambda := |\lambda| \sum_{X:\, t(X) = \lambda} \mathtt{H}_X \quad \text{and} \quad \mathtt{Q}_\lambda := \sum_{X:\, t(X) = \lambda} \mathtt{Q}_X, \tag{1.130}$$

with $|\lambda|$ as in Definition 1.8. As λ varies over all flat-types, these define $\mathtt{H}$- and $\mathtt{Q}$-bases of the invariant Birkhoff algebra. The invariant Birkhoff algebra is a split-semisimple commutative algebra (of dimension equal to the number of flat-types) and $\mathtt{Q}$ is its basis of primitive idempotents. We have

$$\mathtt{H}_\mu = \sum_{\lambda:\, \mu \leq \lambda} r_{\lambda\mu} \mathtt{Q}_\lambda, \tag{1.131}$$

where $r_{\lambda\mu}$ is $|\mu|$ times the number of flats of type μ contained in a given flat of type λ.

1.12.2. Invariant Tits algebra. The linearization of the Tits monoid is the *Tits algebra*. We denote it by $\Sigma[\mathcal{A}]$, and use the letter $\mathtt{H}$ for its canonical basis. Thus,

$$\mathtt{H}_F \cdot \mathtt{H}_G := \mathtt{H}_{FG}. \tag{1.132}$$

Let $\mathrm{char}(\Bbbk) \nmid |W|$. Now fix an invariant noncommutative zeta function $\boldsymbol{\zeta}$ and its inverse Möbius function $\boldsymbol{\mu}$ in the invariant lune-incidence algebra. Define a $\mathtt{Q}$-basis of $\Sigma[\mathcal{A}]$ by either of the two equivalent formulas

$$\begin{aligned} \mathtt{H}_F &= \sum_{G:\, F \leq G} \boldsymbol{\zeta}(F, G)\, \mathtt{Q}_G, \\ \mathtt{Q}_F &= \sum_{G:\, F \leq G} \boldsymbol{\mu}(F, G)\, \mathtt{H}_G. \end{aligned} \tag{1.133}$$

For each flat X, put

$$\mathtt{E}_{\mathrm{X}} := \sum_{F:\, \mathrm{s}(F) = \mathrm{X}} \boldsymbol{\zeta}(O, F)\, \mathtt{Q}_F. \tag{1.134}$$

We call $\mathtt{E}_{\mathrm{X}}$ an *eulerian idempotent* and $\mathtt{E}_\perp$ a *first eulerian idempotent*. Fact: The $\mathtt{E}_{\mathrm{X}}$, as X varies over all flats, is a complete system of primitive orthogonal idempotents of $\Sigma[\mathcal{A}]$. We refer to $\mathtt{E}$ as an *invariant eulerian family*.

The Coxeter group W acts on $\Sigma[\mathcal{A}]$ via $w \cdot \mathtt{H}_F := \mathtt{H}_{w(F)}$. Let $\Sigma[\mathcal{A}]^W$ denote the subalgebra consisting of W-invariant elements. We call it the *invariant Tits algebra*. It has a basis consisting of face-types: For each face-type T, set

$$\mathtt{H}_T := \sum_{F:\, \mathrm{t}(F) = T} \mathtt{H}_F \quad \text{and} \quad \mathtt{Q}_T := \sum_{F:\, \mathrm{t}(F) = T} \mathtt{Q}_F. \tag{1.135}$$

As T varies over all face-types, these define $\mathtt{H}$- and $\mathtt{Q}$-bases of the invariant Tits algebra. Using Observation 1.5,

$$\begin{aligned} \mathtt{H}_T &= \sum_{U:\, T \leq U} \boldsymbol{\zeta}(T, U)\, \mathtt{Q}_U, \\ \mathtt{Q}_T &= \sum_{U:\, T \leq U} \boldsymbol{\mu}(T, U)\, \mathtt{H}_U. \end{aligned} \tag{1.136}$$

For each flat-type λ, set

$$\mathtt{E}_\lambda := \sum_{\mathrm{X}:\, \mathrm{t}(\mathrm{X}) = \lambda} \mathtt{E}_{\mathrm{X}}. \tag{1.137}$$

We call this an *invariant eulerian idempotent*. We deduce from (1.134) that

$$\mathtt{E}_\lambda = \sum_{T:\,\mathrm{s}(T)=\lambda} \zeta(\emptyset, T)\, \mathtt{Q}_T. \tag{1.138}$$

In particular, for the minimum and maximum flat-types,

$$\mathtt{E}_{\lambda_{\min}} = \mathtt{Q}_\emptyset \quad \text{and} \quad \mathtt{E}_{\lambda_{\max}} = \tfrac{1}{|W|}\mathtt{Q}_S = \tfrac{1}{|W|}\mathtt{H}_S.$$

The $\mathtt{E}_\lambda$, as λ varies over all flat-types, yield a complete system of primitive orthogonal idempotents of $\Sigma[\mathcal{A}]^W$.

Example 1.105 (Rank one). Let $\mathrm{char}(\Bbbk) \neq 2$. The invariant Tits algebra in rank one is 2-dimensional with basis elements $\mathtt{H}_\emptyset$ and $\mathtt{H}_S$ and product rule $\mathtt{H}_S \cdot \mathtt{H}_S = 2\,\mathtt{H}_S$ with $\mathtt{H}_\emptyset$ as the unit element. The $\mathtt{Q}$-basis elements are $\mathtt{Q}_\emptyset = \mathtt{H}_\emptyset - \frac{1}{2}\mathtt{H}_S$ and $\mathtt{Q}_S = \mathtt{H}_S$. The invariant eulerian idempotents are

$$\mathtt{E}_{\lambda_{\min}} = \mathtt{H}_\emptyset - \tfrac{1}{2}\mathtt{H}_S \quad \text{and} \quad \mathtt{E}_{\lambda_{\max}} = \tfrac{1}{2}\mathtt{H}_S.$$

These idempotents are mutually orthogonal and add up to the unit element. □

Exercise 1.106 (Structure constants). Show: The $\mathtt{H}$-basis elements of the invariant Tits algebra multiply as follows. For face-types T and T',

$$\mathtt{H}_T \cdot \mathtt{H}_{T'} = \sum_{U:\,U \geq T} a^{TUT'}\, \mathtt{H}_U, \tag{1.139}$$

where $a^{TUT'}$ are the numbers defined in Section 1.8.3. These are structure constants of $\Sigma[\mathcal{A}]^W$ in the $\mathtt{H}$-basis. Moreover, for face-types T and T',

$$\mathtt{H}_T \cdot \mathtt{Q}_{T'} = \sum_{\substack{Y:\,Y \geq T \\ \mathrm{s}(Y)=\mathrm{s}(T')}} a^{TYT'}\, \mathtt{Q}_Y. \tag{1.140}$$

This is equivalent to identity (1.88). (Express $\mathtt{Q}$ in terms of $\mathtt{H}$, multiply out the lhs, and compare the coefficient of $\mathtt{H}_V$ on both sides for each $V \geq T$.)

1.12.3. Invariant q-Janus algebra.

The linearization of the Janus monoid is the *Janus algebra*. We denote it by $\mathsf{J}[\mathcal{A}]$, and use the letter $\mathtt{H}$ for its canonical basis. More generally, for a scalar q, we have the *q-Janus algebra* $\mathsf{J}_q[\mathcal{A}]$ with product defined by

$$\mathtt{H}_{(F',F)} \cdot \mathtt{H}_{(G',G)} := q^{\mathrm{dist}(F,G')}\, \mathtt{H}_{(F'G',GF)}. \tag{1.141}$$

The Coxeter group W acts on $\mathsf{J}[\mathcal{A}]$ via $w \cdot \mathtt{H}_{(F',F)} := \mathtt{H}_{(wF',wF)}$. Let $\mathsf{J}[\mathcal{A}]^W$ denote the subalgebra consisting of W-invariant elements. We call it the *invariant Janus algebra*. More generally, for a scalar q, we have the *invariant q-Janus algebra* $\mathsf{J}_q[\mathcal{A}]^W$. It has a basis consisting of biface-types: For a biface-type σ, let $\mathtt{H}_\sigma$ denote the sum of $\mathtt{H}_{(F',F)}$ over all bifaces (F', F) of type σ. For clarity, we may write $\mathtt{H}_{(T',\sigma,T)}$ instead of $\mathtt{H}_\sigma$. As σ varies, these define the $\mathtt{H}$-basis of $\mathsf{J}_q[\mathcal{A}]^W$.

Observation 1.107. For a scalar q, the q-Janus algebra $\mathsf{J}_q[\mathcal{A}]$ is isomorphic to its opposite via the map $\mathtt{H}_{(F',F)} \mapsto \mathtt{H}_{(F,F')}$, and the invariant q-Janus algebra $\mathsf{J}_q[\mathcal{A}]^W$ is isomorphic to its opposite via the map $\mathtt{H}_\sigma \to \mathtt{H}_{\sigma^{-1}}$.

Now suppose q is not a root of unity. One can then define the $\mathtt{Q}$-basis and $\mathtt{Q}^d$-basis of $\mathtt{J}_q[\mathcal{A}]$, see [**10**, Section 9.9.8], [**11**, Section 1.9.4]. The product in the $\mathtt{Q}$-basis is

$$\mathtt{Q}_{(F',F)} \cdot \mathtt{Q}_{(G',G)} = \begin{cases} \mathtt{Q}_{(F',G)} & \text{if } F = G', \\ 0 & \text{otherwise.} \end{cases} \tag{1.142}$$

The same formula holds with $\mathtt{Q}^d$ instead of $\mathtt{Q}$. The action of the Coxeter group W on the $\mathtt{Q}$-basis is given by $w \cdot \mathtt{Q}_{(F',F)} = \mathtt{Q}_{(wF',wF)}$. For a biface-type σ, let $\mathtt{Q}_\sigma$ denote the sum of $\mathtt{Q}_{(F',F)}$ over all bifaces (F',F) of type σ. For clarity, we may write $\mathtt{Q}_{(T',\sigma,T)}$ instead of $\mathtt{Q}_\sigma$. As σ varies, these define the $\mathtt{Q}$-basis of $\mathtt{J}_q[\mathcal{A}]^W$. Using formula (1.142) and Lemma 1.24, we derive that

$$\mathtt{Q}_\sigma \cdot \mathtt{Q}_\tau = \begin{cases} \mathtt{Q}_{\sigma\tau} & \text{if } \sigma \text{ and } \tau \text{ are composable,} \\ 0 & \text{otherwise.} \end{cases} \tag{1.143}$$

Similarly, we have elements $\mathtt{Q}^d_\sigma$ and the above formula with $\mathtt{Q}^d$ instead of $\mathtt{Q}$. We deduce:

Theorem 1.108. *For q not a root of unity, the invariant q-Janus algebra $\mathtt{J}_q[\mathcal{A}]^W$ is the incidence algebra of the groupoid of biface-types. The elements $\mathtt{Q}_{(T,\mathrm{id},T)}$, as T varies over face-types, are mutually orthogonal idempotents which add up to the unit element. The same statement holds for $\mathtt{Q}^d_{(T,\mathrm{id},T)}$.*

Example 1.109 (Rank one). For a scalar q, the invariant q-Janus algebra in rank one is 3-dimensional with basis elements $\mathtt{H}_{(\emptyset,\mathrm{id},\emptyset)}$, $\mathtt{H}_{(S,\mathrm{id},S)}$, $\mathtt{H}_{(S,\sigma,S)}$ and product rule

$$\begin{aligned} \mathtt{H}_{(S,\mathrm{id},S)} \cdot \mathtt{H}_{(S,\mathrm{id},S)} = \mathtt{H}_{(S,\sigma,S)} \cdot \mathtt{H}_{(S,\sigma,S)} &= \mathtt{H}_{(S,\mathrm{id},S)} + q\,\mathtt{H}_{(S,\sigma,S)}, \\ \mathtt{H}_{(S,\mathrm{id},S)} \cdot \mathtt{H}_{(S,\sigma,S)} = \mathtt{H}_{(S,\sigma,S)} \cdot \mathtt{H}_{(S,\mathrm{id},S)} &= q\,\mathtt{H}_{(S,\mathrm{id},S)} + \mathtt{H}_{(S,\sigma,S)}, \end{aligned}$$

with $\mathtt{H}_{(\emptyset,\mathrm{id},\emptyset)}$ as the unit element.

In rank one, for $q \neq \pm 1$, the $\mathtt{Q}$-basis and $\mathtt{Q}^d$-basis are identical and are given by

$$\begin{aligned} \mathtt{Q}_{(S,\mathrm{id},S)} = \mathtt{Q}^d_{(S,\mathrm{id},S)} &= \tfrac{1}{1-q^2}\big(\mathtt{H}_{(S,\mathrm{id},S)} - q\,\mathtt{H}_{(S,\sigma,S)}\big), \\ \mathtt{Q}_{(S,\sigma,S)} = \mathtt{Q}^d_{(S,\sigma,S)} &= \tfrac{1}{1-q^2}\big(\mathtt{H}_{(S,\sigma,S)} - q\,\mathtt{H}_{(S,\mathrm{id},S)}\big), \\ \mathtt{Q}_{(\emptyset,\mathrm{id},\emptyset)} = \mathtt{Q}^d_{(\emptyset,\mathrm{id},\emptyset)} &= \mathtt{H}_{(\emptyset,\mathrm{id},\emptyset)} - \tfrac{1}{1-q^2}\big(\mathtt{H}_{(S,\mathrm{id},S)} - q\,\mathtt{H}_{(S,\sigma,S)}\big). \end{aligned}$$

The coefficients are closely related to the formulas for ζ_q and μ_q in Example 1.98. Also observe: $\mathtt{Q}_{(\emptyset,\mathrm{id},\emptyset)}$ and $\mathtt{Q}_{(S,\mathrm{id},S)}$ are orthogonal idempotents which add up to the unit element. □

In general,

$$\mathtt{Q}_{(\emptyset,\mathrm{id},\emptyset)} = \mathtt{Q}^d_{(\emptyset,\mathrm{id},\emptyset)} = \sum_{\sigma:\, T' \to T} \mu_q(\emptyset, T, \sigma, T')\,\mathtt{H}_{(T,\sigma,T')}. \tag{1.144}$$

Also,

$$\begin{aligned} \mathtt{Q}_{(T,\mathrm{id},T)} &= \sum_{\sigma:\, T' \to T} \zeta_q(\emptyset, T, \sigma, T')\,\mathtt{H}_{(T,\sigma,T')} + \text{higher terms}, \\ \mathtt{Q}^d_{(T',\mathrm{id},T')} &= \sum_{\sigma:\, T' \to T} \zeta_q(\emptyset, T, \sigma, T')\,\mathtt{H}_{(T,\sigma,T')} + \text{higher terms}, \end{aligned} \tag{1.145}$$

where higher terms are those involving face-types of rank strictly greater than that of T and T'. One can also write a more precise formula along the lines of [**11**, Formula (1.128)].

Exercise 1.110 (Structure constants). Let us see how the $\mathtt{H}$-basis elements multiply in $\mathsf{J}_q[\mathcal{A}]^W$. For biface-types σ and τ, write

$$\mathtt{H}_\sigma \cdot \mathtt{H}_\tau = \sum_\rho a^{\sigma\rho\tau}(q)\, \mathtt{H}_\rho \tag{1.146}$$

for suitable polynomials $a^{\sigma\rho\tau}(q)$. These are structure constants of $\mathsf{J}_q[\mathcal{A}]^W$ in the $\mathtt{H}$-basis. We shorten the scalar $a^{\sigma\rho\tau}(1)$ to $a^{\sigma\rho\tau}$.

Show: For biface-types σ, τ, ρ, the scalar $a^{\sigma\rho\tau}$ is either 0 or 1. More generally, $a^{\sigma\rho\tau}(q)$ is either 0 or a power of q. More precisely, for $\sigma : T \to T'$ and $\tau : U \to U'$,

$$a^{\sigma\rho\tau}(q) = \begin{cases} q^{l(\sigma^{-1}\rho\tau^{-1})} & \text{if } \sigma^{-1}\rho\tau^{-1} \text{ biprojects } U' \text{ and } T, \\ 0 & \text{otherwise.} \end{cases} \tag{1.147}$$

In the first alternative, ρ goes from a face-type greater than U to a face-type greater than T'. The term biprojects is as in Section 1.7.4. An illustration is shown in Figure 1.23 with T, T', U, U' as edge-types. In formula (1.147), in the composite $\sigma^{-1}\rho\tau^{-1}$, the morphisms σ and τ are induced from the given σ and τ. In the picture, these are shown as dotted arrows linking triangle-types.

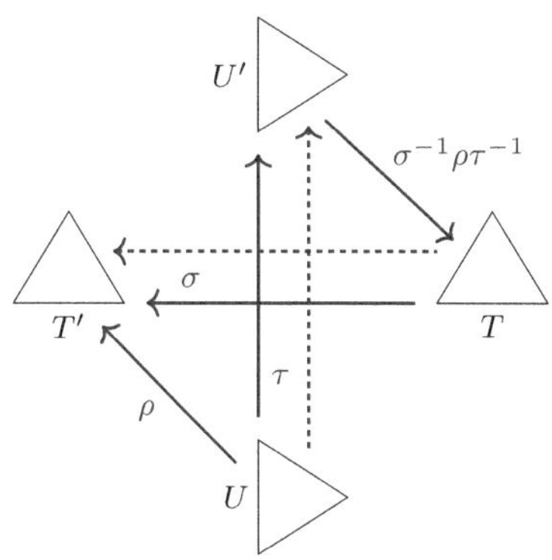

FIGURE 1.23. Structure constants of the invariant q-Janus algebra.

For face-types T and T' and biface-type ρ,

$$a^{\mathrm{id}_{T'}\,\rho\,\mathrm{id}_T}(q) = \begin{cases} q^{l(\rho)} & \text{if } \rho \text{ biprojects } T \text{ and } T', \\ 0 & \text{otherwise.} \end{cases} \tag{1.148}$$

This is the special case of formula (1.147) with $\sigma := \mathrm{id}_{T'}$ and $\tau := \mathrm{id}_T$.

We now connect the polynomials $a^{TUT'U'}(q)$ in Section 1.8.4 to structure constants of $\mathsf{J}_q[\mathcal{A}]^W$. Check: For a scalar q,

$$a^{TUT'U'}(q) = \sum_{\rho:\, U \to U'} a^{\mathrm{id}_{T'}\,\rho\,\mathrm{id}_T}(q). \tag{1.149}$$

(Use formula (1.64a) with $Z = \emptyset$ and formula (1.148).)

1.12.4. Invariant support map. The linearization of the map (1.1) yields an algebra morphism from the Tits algebra to the Birkhoff algebra. We call it the *support map*. The kernel of the support map is the radical of the Tits algebra. Thus, the Tits algebra is elementary, that is, the quotient by its radical is a split-semisimple commutative algebra. By linearizing (1.2), we obtain the commutative diagram of algebras

$$\begin{array}{ccc} \mathsf{J}[\mathcal{A}] & \twoheadrightarrow & \Sigma[\mathcal{A}]^{\mathrm{op}} \\ \downarrow & & \downarrow \mathrm{s} \\ \Sigma[\mathcal{A}] & \xrightarrow[\mathrm{s}]{} & \Pi[\mathcal{A}] \end{array} \qquad \begin{array}{ccc} \mathtt{H}_{(F,F')} & \longmapsto & \mathtt{H}_{F'} \\ \downarrow & & \downarrow \\ \mathtt{H}_F & \longmapsto & \mathtt{H}_{\mathrm{s}(F)} = \mathtt{H}_{\mathrm{s}(F')}. \end{array} \tag{1.150}$$

The Janus algebra is also elementary; its radical is the kernel of the composite map $\mathsf{J}[\mathcal{A}] \to \Pi[\mathcal{A}]$.

Let $\mathrm{char}(\Bbbk) \nmid |W|$. The support map restricts to a linear map

$$\mathrm{s} : \Sigma[\mathcal{A}]^W \twoheadrightarrow \Pi[\mathcal{A}]^W.$$

We call this the *invariant support map*. The kernel of the invariant support map is the radical of the invariant Tits algebra. Thus, the invariant Tits algebra is also elementary. (For a general result, see [**10**, Lemma D.46].) By taking W-invariants in (1.150), we obtain the commutative diagram of algebras

$$\begin{array}{ccc} \mathsf{J}[\mathcal{A}]^W & \twoheadrightarrow & (\Sigma[\mathcal{A}]^{\mathrm{op}})^W \\ \downarrow & & \downarrow \mathrm{s} \\ \Sigma[\mathcal{A}]^W & \xrightarrow[\mathrm{s}]{} & \Pi[\mathcal{A}]^W \end{array} \qquad \begin{array}{ccc} \mathtt{H}_{(T,\sigma,T')} & \longmapsto & \mathtt{H}_{T'} \\ \downarrow & & \downarrow \\ \mathtt{H}_T & \longmapsto & \mathtt{H}_{\mathrm{s}(T)} = \mathtt{H}_{\mathrm{s}(T')}. \end{array} \tag{1.151}$$

The invariant Janus algebra is also elementary; its radical is the kernel of the composite map $\mathsf{J}[\mathcal{A}]^W \to \Pi[\mathcal{A}]^W$.

Exercise 1.111. Check:

$$\mathrm{s}(\mathtt{E}_{\mathrm{X}}) = \mathtt{Q}_{\mathrm{X}}, \quad \mathrm{s}(\mathtt{Q}_F) = \mathtt{Q}_{\mathrm{s}(F)}, \quad \mathrm{s}(\mathtt{E}_\lambda) = \mathtt{Q}_\lambda, \quad \mathrm{s}(\mathtt{Q}_T) = |\mathrm{s}(T)|\, \mathtt{Q}_{\mathrm{s}(T)}.$$

1.12.5. Solomon descent algebra. Recall shuffles from Section 1.6.4. The *Solomon descent algebra* is the subalgebra of $\Bbbk W$ linearly spanned by the elements $\sum_{\zeta \in \mathrm{Sh}_T} \zeta$, as T varies over face-types. The map

$$\Sigma[\mathcal{A}]^W \hookrightarrow \Bbbk W, \qquad \mathtt{H}_T \mapsto \sum_{\zeta \in \mathrm{Sh}_T} \zeta^{-1} \tag{1.152}$$

is an injective algebra morphism. It follows that the invariant Tits algebra is isomorphic to the opposite of the Solomon descent algebra. For more details, see [**10**, Section 16.3].

Similarly, for a scalar q, the map

$$\mathsf{J}_q[\mathcal{A}]^W \to \Bbbk W, \qquad \mathtt{H}_{(T,\sigma,T')} \mapsto \sum_{\zeta \in \mathrm{Sh}_{T'}} q^{l(\zeta)} w_\sigma \zeta^{-1} \tag{1.153}$$

is an algebra morphism, with w_σ as in Lemma 1.30. For $q = 1$, it factors through $\mathsf{J}[\mathcal{A}]^W \twoheadrightarrow \Sigma[\mathcal{A}]^W$ to recover (1.152). This follows from (1.71).

Exercise 1.112. Show: For a scalar q, there is an algebra morphism

$$\mathsf{J}_q[\mathcal{A}]^W \to \Bbbk W, \qquad \mathtt{H}_{(T,\sigma,T')} \mapsto \sum_{\zeta \in \mathrm{Sh}_T} q^{l(\zeta)} \zeta w_\sigma. \tag{1.154}$$

(Both algebras are isomorphic to their opposites. For the former, see Observation 1.107. For the latter, linearize the map $w \mapsto w^{-1}$. Now conjugate (1.153) by these isomorphisms to obtain (1.154).)

A conceptual understanding of these algebra morphisms using ideas developed in this monograph is given later in Exercise 10.52.

Example 1.113 (Rank one). In rank one, (1.152) specializes to the isomorphism

$$\Sigma[\mathcal{A}]^W \xrightarrow{\cong} \Bbbk W, \qquad \mathtt{H}_\emptyset \mapsto e,\ \mathtt{H}_S \mapsto e + w,\ \mathtt{H}_S - \mathtt{H}_\emptyset \mapsto w,$$

while (1.153) and (1.154) both specialize to

$$\mathsf{J}_q[\mathcal{A}]^W \to \Bbbk W, \qquad \mathtt{H}_{(\emptyset,\mathrm{id},\emptyset)} \mapsto e,\ \mathtt{H}_{(S,\mathrm{id},S)} \mapsto e + q\,w,\ \mathtt{H}_{(S,\sigma,S)} \mapsto q\,e + w.$$

That these are algebra morphisms can be checked directly. □

1.12.6. Arrangement over a face-type. We may apply the preceding discussion to $\boldsymbol{\alpha}_Z$. Thus, for each face-type T greater than Z, put

$$\mathtt{H}_{T/Z} := \sum_{F:\, F \geq A,\, \mathrm{t}(F/A) = T/Z} \mathtt{H}_{F/A}, \tag{1.155}$$

where A is the face of C_r of type Z. This defines the H-basis of the invariant Tits algebra of $\boldsymbol{\alpha}_Z$. Observe:

$$\mathtt{H}_{T/Z} \cdot \mathtt{H}_{T'/Z} := \sum_{U:\, U \geq T} a_Z^{TUT'} \mathtt{H}_{U/Z}. \tag{1.156}$$

Similarly, we have Q-basis elements $\mathtt{Q}_{T/Z}$ and invariant eulerian idempotents $\mathtt{E}_{\lambda/Z}$. For the invariant Birkhoff algebra of $\boldsymbol{\alpha}_Z$, we have idempotents $\mathtt{Q}_{\lambda/Z}$. Similar remarks apply to the invariant q-Janus algebra of $\boldsymbol{\alpha}_Z$.

1.13. Invariant Lie and Zie elements

We introduced Lie and Zie elements of an arrangement in [**10**, Chapter 10], [**11**, Section 1.12]. The space of Lie elements $\mathsf{Lie}[\mathcal{A}]$ is a certain subspace of $\Gamma[\mathcal{A}]$ which is the linearization of the set of chambers. In a similar vein, the space of Zie elements $\mathsf{Zie}[\mathcal{A}]$ is a certain subspace of $\Sigma[\mathcal{A}]$. These two subspaces are defined by linear systems of equations involving top-lunes and lunes, respectively. We do not write them here; they are given later in Section 4.8 in the context of Coxeter species, see (4.95) and (4.98).

For a reflection arrangement $\mathcal{A}$, we also have the subspace $\mathsf{Lie}[\mathcal{A}]^W$ of W-invariant Lie elements, and similarly the subspace $\mathsf{Zie}[\mathcal{A}]^W$ of W-invariant Zie elements. See [**10**, Section 16.6].

For an arrangement $\mathcal{A}$, a Zie element is *special* if its coefficient of the central face is 1. A family of special Zie elements, one for each arrangement $\mathcal{A}_{\mathrm{X}}$, as X varies over all flats, is equivalent to a noncommutative Möbius function. Similarly, for a reflection arrangement $\mathcal{A}$, a family of invariant special Zie elements indexed by flats is equivalent to an invariant noncommutative Möbius function. See [**10**, Theorems 15.44 and 16.48, items (7), (8)], [**11**,

Lemma 1.13]. A typical example of an invariant special Zie element is the $\mathtt{Q}$-basis element $\mathtt{Q}_\emptyset$ as in (1.136).

1.14. Coxeter–Birkhoff, Coxeter–Tits, Coxeter–Janus algebras

The action of the Coxeter group W on the Birkhoff, Tits, Janus monoids (algebras) allows us to construct their semidirect products. These are the Coxeter–Birkhoff, Coxeter–Tits, Coxeter–Janus monoids (algebras). We will see later that the categories of modules over these algebras are related to the categories of Coxeter bimonoids and their commutative counterparts (Section 3.13). General information on semidirect product monoids is given in Appendix A.2.3.

1.14.1. Coxeter–Tits monoid. The Coxeter group W acts on the Tits monoid $\Sigma[\mathcal{A}]$. So we can form their semidirect product

$$\Sigma[\mathcal{A}] \rtimes W = \{(F, w) \mid F \in \Sigma[\mathcal{A}],\ w \in W\}.$$

This is a monoid with product defined by

$$(G, v) \cdot (F, u) := (Gv(F), vu). \tag{1.157}$$

The unit element is (O, e). We refer to $\Sigma[\mathcal{A}] \rtimes W$ as the *Coxeter–Tits monoid* and abbreviate it to $\mathrm{W}\Sigma[\mathcal{A}]$.

The discussion above applies to any group acting on any monoid. For instance, W also acts on the opposite of the Tits monoid. Set $\mathrm{W}\Sigma[\mathcal{A}]^{\mathrm{op}} := \Sigma[\mathcal{A}]^{\mathrm{op}} \rtimes W$. Its product is

$$(G, v) \cdot (F, u) := (v(F)G, vu). \tag{1.158}$$

This object is isomorphic to the opposite monoid of $\mathrm{W}\Sigma[\mathcal{A}]$:

Exercise 1.114. Check: The map

$$(\mathrm{W}\Sigma[\mathcal{A}])^{\mathrm{op}} \to \mathrm{W}\Sigma[\mathcal{A}]^{\mathrm{op}}, \qquad (F, w) \mapsto (w^{-1}(F), w^{-1})$$

is an isomorphism of monoids.

1.14.2. Coxeter–Birkhoff monoid. Similarly, the action of the Coxeter group W on the Birkhoff monoid $\Pi[\mathcal{A}]$ yields the semidirect product

$$\Pi[\mathcal{A}] \rtimes W = \{(\mathrm{X}, w) \mid \mathrm{X} \in \Pi[\mathcal{A}],\ w \in W\}.$$

This is a monoid under

$$(\mathrm{Y}, v) \cdot (\mathrm{X}, u) := (\mathrm{Y} \vee v(\mathrm{X}), vu). \tag{1.159}$$

The unit element is $(\bot, e)$. We refer to $\Pi[\mathcal{A}] \rtimes W$ as the *Coxeter–Birkhoff monoid* and abbreviate it to $\mathrm{W}\Pi[\mathcal{A}]$.

Exercise 1.115. Check: The map

$$(\mathrm{W}\Pi[\mathcal{A}])^{\mathrm{op}} \to \mathrm{W}\Pi[\mathcal{A}], \qquad (\mathrm{X}, w) \mapsto (w^{-1}(\mathrm{X}), w^{-1})$$

is an isomorphism of monoids.

1.14.3. Coxeter–Janus monoid. Finally, the action of W on the Janus monoid $\mathrm{J}[\mathcal{A}]$ yields

$$\mathrm{J}[\mathcal{A}] \rtimes W = \{(F, F', w) \mid (F, F') \in \mathrm{J}[\mathcal{A}],\ w \in W\}.$$

This is a monoid under

$$(G, G', v) \cdot (F, F', u) := (Gv(F), v(F')G', vu). \tag{1.160}$$

The unit element is (O, O, e). We refer to $\mathrm{J}[\mathcal{A}] \rtimes W$ as the *Coxeter–Janus monoid* and abbreviate it to $\mathrm{WJ}[\mathcal{A}]$.

Exercise 1.116. Check: The map

$$\mathrm{WJ}[\mathcal{A}] \to (\mathrm{WJ}[\mathcal{A}])^{\mathrm{op}}, \qquad (F, F', w) \mapsto (w^{-1}(F'), w^{-1}(F), w^{-1})$$

is an isomorphism of monoids.

1.14.4. Linearization of monoids. The linearization of the above monoids yields algebras. They are summarized in Table 1.8.

TABLE 1.8. Monoids and their linearized algebras.

Monoid	Linearized algebra
Coxeter–Tits monoid $\mathrm{W\Sigma}[\mathcal{A}]$	Coxeter–Tits algebra $\mathsf{W\Sigma}[\mathcal{A}]$
$\mathrm{W\Sigma}[\mathcal{A}]^{\mathrm{op}}$	$\mathsf{W\Sigma}[\mathcal{A}]^{\mathrm{op}}$
Coxeter–Birkhoff monoid $\mathrm{W\Pi}[\mathcal{A}]$	Coxeter–Birkhoff algebra $\mathsf{W\Pi}[\mathcal{A}]$
Coxeter–Janus monoid $\mathrm{WJ}[\mathcal{A}]$	Coxeter–Janus algebra $\mathsf{WJ}[\mathcal{A}]$

The Coxeter–Janus algebra also has a q-analogue, namely, the *Coxeter-q-Janus algebra* $\mathsf{WJ}_q[\mathcal{A}]$. It arises from the action of W on the q-Janus algebra.

1.14.5. Modules. Since each one of the monoids in Table 1.8 is a semi-direct product, their modules can be described in terms of an action of the Coxeter group and an action of the Tits or Birkhoff or Janus monoid with a compatibility condition. Let us make this explicit.

A left $\mathsf{W\Sigma}[\mathcal{A}]$-module is the same as a vector space M which is a left W-module and a left $\Sigma[\mathcal{A}]$-module, with the two actions compatible via

$$w(\mathtt{H}_F \cdot m) = w(\mathtt{H}_F) \cdot w(m) \tag{1.161}$$

for $m \in M$, $w \in W$, $F \in \Sigma[\mathcal{A}]$. Here we have written $\mathtt{H}_F \cdot m$ for the action of $\Sigma[\mathcal{A}]$ and $w(m)$ for the action of W.

Similarly, a left $\mathsf{W\Sigma}[\mathcal{A}]^{\mathrm{op}}$-module is the same as a vector space M which is a left W-module and a right $\Sigma[\mathcal{A}]$-module, with the two actions compatible via

$$w(m \cdot \mathtt{H}_F) = w(m) \cdot w(\mathtt{H}_F) \tag{1.162}$$

for $m \in M$, $w \in W$, $F \in \Sigma[\mathcal{A}]$. In view of Exercise 1.114, it is also the same as a right $\mathsf{W\Sigma}[\mathcal{A}]$-module.

A left $\mathsf{W\Pi}[\mathcal{A}]$-module is the same as a vector space M which is a $\mathsf{\Pi}[\mathcal{A}]$-module and a left W-module, with the two actions compatible via

$$w(\mathtt{H}_{\mathrm{X}} \cdot m) = w(\mathtt{H}_{\mathrm{X}}) \cdot w(m) \tag{1.163}$$

for $m \in M$, $w \in W$, $\mathrm{X} \in \Pi[\mathcal{A}]$. (We do not need to specify left or right for a $\mathsf{\Pi}[\mathcal{A}]$-module since $\mathsf{\Pi}[\mathcal{A}]$ is commutative.) A right $\mathsf{W\Pi}[\mathcal{A}]$-module admits a similar description with W acting on the right. We do not need to consider this explicitly in view of Exercise 1.115.

A left $\mathsf{WJ}[\mathcal{A}]$-module is the same as a vector space M which is a left W-module and a left $\mathsf{J}[\mathcal{A}]$-module, with the two actions compatible via

$$w(\mathtt{H}_{(F,F')} \cdot m) = w(\mathtt{H}_{(F,F')}) \cdot w(m) \tag{1.164}$$

for $m \in M$, $w \in W$, $(F, F') \in \mathrm{J}[\mathcal{A}]$. We can also consider right $\mathsf{WJ}[\mathcal{A}]$-modules, but it is not required in view of Exercise 1.116.

More generally, a left $\mathsf{WJ}_q[\mathcal{A}]$-module combines a left W-module and a left $\mathsf{J}_q[\mathcal{A}]$-module with compatibility (1.164).

1.15. Coordinate arrangement

The coordinate arrangement is an example of a reflection arrangement, see Table 1.3. It is discussed in detail in [**10**, Section 6.1]. We now specialize some concepts from preceding sections to this arrangement. The rank-three coordinate arrangement is shown in Figure 1.4. The first picture in Figure 1.3 is the rank-two coordinate arrangement.

1.15.1. Coordinate arrangement. The coordinate arrangement of rank n, denoted $\mathcal{C}^n$, is defined by the hyperplanes $x_i = 0$ for $1 \leq i \leq n$. The ambient space is $\mathbb{R}^n$. The Coxeter group is $\mathbb{Z}_2^n$, the product of n copies of $\mathbb{Z}_2$. The generator of the i-th copy of $\mathbb{Z}_2$ acts on $\mathbb{R}^n$ by changing the i-th coordinate to its negative.

Faces of $\mathcal{C}^n$ are n-tuples in which each entry is one of 0, $+$, $-$. The partial order on faces for $n = 3$ is illustrated below.

$$000 \leq 0-0 \leq +-0 \leq +-+.$$

In other words, to go up in the partial order, we replace 0 by either $+$ or $-$. Chambers are n-tuples with no zero entries. By convention, the reference chamber is the one with all entries $+$. The Tits product of faces F and G is as follows. The i-th entry of FG equals the i-th entry of F if it is nonzero, else it equals the i-th entry of G. For example, for $n = 3$,

$$(0-+)(-0-) = --+, \quad (-00)(+0+) = -0+, \quad (0+0)(---) = -+-.$$

Two faces have the same support iff their sets of zero entries match.

Face-types are subsets T of $[n]$, with partial order given by inclusion. For a face F, its type $\mathrm{t}(F)$ is the subset consisting of the nonzero entries of F. The rank of a face-type T, denoted $\mathrm{rk}(T)$, is the cardinality of T. Two face-types have the same support iff they are equal.

Exercise 1.117. Check: For any coordinate arrangement, the face H in Remark 1.2 always exists.

Exercise 1.118 (Shuffles). Recall shuffles from Section 1.6.4. Check: For any face-type T of $\mathcal{C}^n$, the sets of T-shuffles and T-deshuffles coincide, and consist of those $\zeta \in \mathbb{Z}_2^n$ which act by identity on elements of $[n] \setminus T$.

1.15.2. Groupoid of biface-types.

For the coordinate arrangement $\mathcal{C}^n$, we now make explicit the groupoid of biface-types from Section 1.7. Objects are face-types which are the same as subsets $T \subseteq [n]$. There are no morphisms from T to U for $T \neq U$. The automorphism group at T is $\mathbb{Z}_2^T$, a product of copies of $\mathbb{Z}_2$, one for each element of T. We say $\sigma : T \to T$ preserves some elements of T and flips the rest (depending on which element of $\mathbb{Z}_2$ is chosen for each element of T). A biface (F, F') has type σ iff the flips of σ occur in those entries where F and F' differ (that is, if one is $+$, then the other is $-$).

Exercise 1.119. Check: The concept of biprojection specializes as follows.

- Let $\sigma : U \to U$ with $T, T' \leq U$. Then σ biprojects T and T' iff $T \cup T' = U$ and σ preserves all elements of $U \setminus (T \cap T')$. Note very carefully: There is no condition on what σ does on elements of $T \cap T'$.
- Let $\sigma : U \to U$ with $T \leq U$. Then σ projects U on T iff σ coprojects U on T iff σ preserves all elements of $U \setminus T$.

Similarly, the concepts of induced morphisms, local morphisms, and so on specialize as follows.

- For $\sigma : T \to T$ and $T \leq U$, the induced morphism $\sigma : U \to U$ is the one which flips the same elements as the original σ; in particular, it preserves all elements of $U \setminus T$.
- A morphism $\sigma : T \to T$ is local to Z if $Z \leq T$ and σ preserves all elements of Z (with no condition on what σ does on elements of $T \setminus Z$).
- For a face-type T, the morphism $\beta_{\overline{T},T}$ flips all elements of T. Note: $\overline{T} = T$. More generally, the morphism $\beta_{Z\overline{T},T}$ preserves all elements of Z and flips all elements of $T \setminus Z$.
- For $\sigma : T \to T$, its length $l(\sigma)$ is the number of elements of T that σ flips.
- For a morphism $\sigma \in \mathbb{Z}_2^T$, the Coxeter symmetry $w_\sigma \in \mathbb{Z}_2^n$ is the same as σ on elements of T, and identity on elements of $[n] \setminus T$.
- For $\rho : U \to U$ and $T \leq U$, the gate property says $\rho = \tau\sigma$, where σ flips only elements of T, and τ flips only elements of $U \setminus T$, both as specified by ρ.

Exercise 1.120. Verify directly Exercise 1.40 for the arrangement $\mathcal{C}^n$. (Here $T = T'$ and $U = U' = V \cup T$. Morphisms τ and ρ, which correspond to each other under the bijection, both preserve all elements of $U \setminus (V \cap T)$, and differ on those elements of $V \cap T$ which σ flips. Exercise 1.41 is the special case where τ and ρ differ on all elements of $V \cap T$.)

1.15.3. Face-type enumeration.

We now specialize some of Section 1.8 to $\mathcal{C}^n$. Let q be a scalar. For a face-type T,

$$d_T(q) = (1+q)^{\mathrm{rk}(T)}.$$

In particular, the Poincaré polynomial (1.34) is $d_S(q) = (1+q)^n$. Similarly,

$$a^{TUT'U'}(q) = \begin{cases} (1+q)^{\mathrm{rk}(T\cap T')} & \text{if } T \cup T' = U = U', \\ 0 & \text{otherwise,} \end{cases}$$

$$a^{TUT'}(q) = \begin{cases} (1+q)^{\mathrm{rk}(T\cap T')} & \text{if } T \cup T' = U, \\ 0 & \text{otherwise,} \end{cases}$$

$$a^{TT'}(q,\xi) = (1+q)^{\mathrm{rk}(T\cap T')}\xi_{T\cup T'}.$$

More generally, these polynomials can also be defined local to a face-type. For example, $d_{U/T}(q) = (1+q)^{\mathrm{rk}(U/T)}$, and so on.

Exercise 1.121. Verify directly the gate identity (1.57) for the arrangement $\mathcal{C}^n$. (When $Y = V \cup T'$, both sides are powers of $1+q$, with only the summand for $U = T \cup T'$ contributing to the rhs.)

Exercise 1.122. Observe: $c^{TT}(q) = (1+q)^{\mathrm{rk}(T)}$. Deduce Theorem 1.56 for $\mathcal{C}^n$. (For any flat-type λ, the Varchenko matrix A_λ has size 1 with entry $(1+q)^{\mathrm{rk}(\lambda)}$.)

1.15.4. Zeta and Möbius functions.

We now specialize some of Section 1.10 to $\mathcal{C}^n$. Let $\mathrm{char}(\Bbbk) \neq 2$. For any coordinate arrangement, there is a unique invariant $\boldsymbol{\zeta}$ and $\boldsymbol{\mu}$, namely,

$$\boldsymbol{\zeta}(T,U) = \left(\frac{1}{2}\right)^{\mathrm{rk}(U/T)} \quad \text{and} \quad \boldsymbol{\mu}(T,U) = \left(\frac{-1}{2}\right)^{\mathrm{rk}(U/T)}.$$

Since the support map from face-types to flat-types is a bijection, uniqueness follows from Fact 1.81. One can also use formula (1.76) to check that the above $\boldsymbol{\zeta}$ and $\boldsymbol{\mu}$ are indeed inverses. In rank one, the above formulas reduce to those in Example 1.86.

For q not a root of unity,

$$\boldsymbol{\zeta}_q(T,U) = \left(\frac{1}{1+q}\right)^{\mathrm{rk}(U/T)} \quad \text{and} \quad \boldsymbol{\mu}_q(T,U) = \left(\frac{-1}{1+q}\right)^{\mathrm{rk}(U/T)}.$$

This generalizes the rank-one calculation in Example 1.91. One can check directly that the above $\boldsymbol{\zeta}_q$ and $\boldsymbol{\mu}_q$ are inverses, and they satisfy formulas (1.93) and (1.96).

Similarly, for q not a root of unity,

$$\zeta_q(Z,T,\sigma,T) = \frac{(-q)^{l(\sigma)}}{(1-q^2)^{\mathrm{rk}(T/Z)}} \quad \text{and} \quad \mu_q(Z,T,\sigma,T) = \frac{(-1)^{\mathrm{rk}(T/Z)-l(\sigma)}q^{l(\sigma)}}{(1-q^2)^{\mathrm{rk}(T/Z)}}.$$

This generalizes the rank-one calculation in Example 1.98.

Exercise 1.123. Check: For the arrangement $\mathcal{C}^n$, the product (1.102) specializes to

$$(st)(Z,U,\rho,U) = \sum_{T:\, Z \le T \le U} s(Z,T,\sigma,T)t(T,U,\tau,U), \tag{1.165}$$

with σ and τ determined by ρ as follows. The flips of σ on elements of $T \setminus Z$ are exactly those of ρ. Similarly, the flips of τ on elements of $U \setminus T$ are exactly those of ρ. See Observation 1.96 and Exercise 1.117 in this regard.

Check directly: The above ζ_q and μ_q are inverses under the product (1.165) and satisfy formulas (1.109) and (1.111).

1.16. Braid arrangement

The braid arrangement is an important example of a reflection arrangement, see Table 1.3. We now specialize some concepts from preceding sections to this arrangement. Table 1.9 shows the rich combinatorics involved. For additional entries, see [**10**, Table 6.2]. The rank-three braid arrangement is shown in Figure 1.5; for a more detailed picture, see [**10**, page 144]. Additional information can be found in [**10**, Sections 6.3, 6.4, 6.5, 6.6].

Apart from standard material, we also make a connection to the Chu–Vandermonde identity, and introduce zeta and Möbius series.

TABLE 1.9. Geometry and combinatorics.

Object		Object-type	
face	set composition	face-type	integer composition
chamber	linear order	chamber-type	integer composition with all parts 1
central face	set composition with one block	central face-type	integer composition with one part
flat	set partition	flat-type	integer partition
biface	pair of set compositions with the same blocks	biface-type	pair of integer compositions with a size-preserving bijection between their parts
lune	set partition with a composition of each block	lune-type	integer partition with a composition of each part

1.16.1. Braid arrangement. Fix a nonempty finite set I. The *braid arrangement* on I, denoted $\mathcal{B}^I$, consists of the hyperplanes $x_a = x_b$ in $\mathbb{R}^I$, as a and b vary over elements of I with $a \neq b$. It is a reflection arrangement, the Coxeter group is the group of all bijections from I to itself. We set $\beta^n := (\mathcal{B}^{[n]}, 1|\dots|n)$ with $[n] := \{1, \dots, n\}$. This is a pointed reflection arrangement with reference chamber $C_r = 1|\dots|n$. (Chambers correspond to linear orders.) We call β^n the braid arrangement on n letters. Note: Its Coxeter group is the symmetric group S_n on n letters.

1.16.2. Faces and flats. Faces of $\mathcal{B}^I$ identify with compositions of I. Recall: A *composition* of I is a finite sequence $(I_1, \dots, I_k)$ of disjoint nonempty subsets of I whose union is I. The subsets I_i are the *blocks* of the composition. We write $F \vDash I$ to indicate that $F = (I_1, \dots, I_k)$ is a composition of I. The partial order on compositions is refinement. For example, for $I = \{a, b, c, d, e\}$,

$$abcde \leq ac|bde \leq c|a|be|d \leq c|a|e|b|d.$$

A chamber corresponds to a composition of I with singleton blocks which is the same as a linear order on I. The opposite of a composition is obtained by reversing the order of its blocks, that is, the opposite of $F = (I_1, \dots, I_k)$ is

$\overline{F} = (I_k, \ldots, I_1)$. The *Tits product* of $F = (S_1, \ldots, S_p)$ and $G = (T_1, \ldots, T_q)$ is the composition

$$FG := (A_{11}, \ldots, A_{1q}, \ldots, A_{p1}, \ldots, A_{pq})\widehat{\ }, \tag{1.166}$$

where $A_{ij} = S_i \cap T_j$ and the hat indicates that empty intersections are removed. For example,

$$(ac|bde)(abe|cd) = a|c|be|d \quad \text{and} \quad (abe|cd)(ac|bde) = a|be|c|d.$$

Flats of $\mathcal{B}^I$ identify with partitions of I. Recall: A *partition* X of I is a collection X of disjoint nonempty subsets of I whose union is I. These subsets are the *blocks* of the partition. We write $\mathrm{X} \vdash I$ to indicate that X is a partition of I. The partial order on partitions is refinement. For example, for $I = \{a, b, c, d, e\}$,

$$\{abcde\} \leq \{ac, bde\} \leq \{a, be, c, d\} \leq \{a, b, c, d, e\}.$$

The *support* of a composition F of I is the partition $\mathrm{s}(F)$ of I obtained by forgetting the order among the blocks.

1.16.3. Face-types and flat-types. Face-types and flat-types of β^n identify with compositions and partitions of n. Recall: Let $n \geq 1$. A *composition* $\alpha = (a_1, \ldots, a_k)$ of n is a finite sequence of positive integers such that $a_1 + \cdots + a_k = n$. The opposite of $\alpha = (a_1, \ldots, a_k)$ is given by $\overline{\alpha} = (a_k, \ldots, a_1)$. A *partition* $\lambda = (l_1, \ldots, l_k)$ of n is a finite sequence of positive integers such that $l_1 \geq l_2 \geq \cdots \geq l_k$ and $l_1 + \cdots + l_k = n$. We write $\alpha \vDash n$ and $\lambda \vdash n$ to indicate that α is a composition of n, and λ is a partition of n. The numbers a_i and l_i are the *parts* of α and λ. The partial order on compositions is given by refinement. For example,

$$(5) \leq (2, 3) \leq (1, 1, 2, 1) \leq (1, 1, 1, 1, 1).$$

For partitions, $\lambda \leq \mu$ if μ is obtained by refining each part in λ, and rearranging the parts in descending order. For example,

$$(5) \leq (3, 2) \leq (2, 1, 1, 1) \leq (1, 1, 1, 1, 1).$$

The support of a composition of n is the partition of n obtained by reordering the parts in decreasing order.

Exercise 1.124. Verify the combinatorial descriptions of bifaces, biface-types, lunes, lune-types given in Table 1.9. Biface-types are developed in more detail in Section 8.14.

For set compositions $F \leq G$, let $(G/F)_i$ denote the set composition consisting of those contiguous blocks of G which refine the i-th block of F. Similarly, for integer compositions $\alpha \leq \beta$, we define integer compositions $(\beta/\alpha)_i$.

1.16.4. Poincaré polynomial. We now specialize some of Section 1.8 to β^n. Let q be a scalar. Let the i-th *q-number* be

$$(i)_q := 1 + q + \cdots + q^{i-1}.$$

For a positive integer n, define the *q-factorial* of n by

$$(n)_q! := (n)_q (n-1)_q \ldots (1)_q.$$

More generally, for a composition $(a_1, \dots, a_k)$ of n, define the *q-multinomial coefficient* by

$$\binom{n}{a_1 \dots a_k}_q = \frac{(n)_q!}{(a_1)_q! \dots (a_k)_q!}.$$

The composition $(1, \dots, 1)$ recovers the q-factorial. When the composition has two parts, we use the term *q-binomial coefficient*.

For the braid arrangement $\boldsymbol{\beta}^n$, the Poincaré polynomial (1.34) is

$$d_S(q) = (n)_q!. \tag{1.167}$$

More generally, for a composition T of n, $d_T(q)$ is the corresponding q-multinomial coefficient. The result of Exercise 1.43, item (2) says that compositions with the same underlying partition have the same q-multinomial coefficient.

Exercise 1.125. Check: The polynomials $a^{TUT'U'}(q)$ in (1.41) for the braid arrangement $\boldsymbol{\beta}^n$ specialize as follows. For a scalar q, for compositions α, β, α', β' of n with $\alpha \leq \beta$, $\alpha' \leq \beta'$, α with k parts, α' with l parts,

$$a^{\alpha\beta\alpha'\beta'}(q) = \sum_c q^{\sum_{i<i',\, j>j'} c_{ij} c_{i'j'}}, \tag{1.168}$$

where $c = (c_{ij})$ runs over all $k \times l$ matrices of nonnegative integers whose row sums yield α, column sums yield α',

$$\beta = (c_{11}, \dots, c_{1l}, \dots, c_{k1}, \dots, c_{kl})\widehat{\ } \quad \text{and} \quad \beta' = (c_{11}, \dots, c_{k1}, \dots, c_{1l}, \dots, c_{kl})\widehat{\ },$$

where hats indicate that zero entries are removed. The power of q in the above formula can be interpreted as the "distance" between the matrix c and its transpose.

The polynomials $a^{TUT'}(q)$ in (1.38) can be described similarly by using (1.44a). Thus, we define $a^{\alpha\beta\alpha'}(q)$ as above by dropping the condition on β'. For $q = 1$, these are structure constants of the Solomon descent algebra of compositions, see (1.139) and Section 1.12.5.

Exercise 1.126. Check: The gate identity (1.59) for $\boldsymbol{\beta}^n$ specializes as follows. For a scalar q, for compositions $(a_1, \dots, a_k)$ and $(b_1, \dots, b_l)$ of n,

$$\binom{n}{b_1 \dots b_l}_q = \sum_c \binom{a_1}{c_{11} \dots c_{1l}}_q \cdots \binom{a_k}{c_{k1} \dots c_{kl}}_q q^{\sum_{i<i',\, j>j'} c_{ij} c_{i'j'}}, \tag{1.169}$$

where $c = (c_{ij})$ runs over all $k \times l$ matrices of nonnegative integers whose row sums are $a_1, \dots, a_k$ and column sums are $b_1, \dots, b_l$. When both compositions have two parts, we get the *q-Chu–Vandermonde identity*

$$\binom{m+n}{k}_q = \sum_{i+j=k} \binom{m}{i}_q \binom{n}{j}_q q^{(m-i)j}. \tag{1.170}$$

This arises from the special case of (1.59) when T and T' are vertex-types.

1.16.5. Zeta and Möbius series. We now turn to content related to Section 1.10 for $\boldsymbol{\beta}^n$. This discussion makes use of shuffles and quasishuffles of integer compositions; for definitions, see for instance [**6**, Definition 3.2.2]. For example, for compositions $(2, 1, 5)$ and $(3, 2, 4)$,

$$(3, 2, 2, 1, 5, 4) \text{ is a shuffle and } (2+3, 2, 1, 5+4) \text{ is a quasishuffle.}$$

In a similar manner, one can (quasi)shuffle three or more compositions. Every shuffle is a quasishuffle, but not conversely.

A *zeta series* is a family of scalars $(\boldsymbol{\zeta}_\gamma)$, as γ varies over compositions of positive integers, such that $\boldsymbol{\zeta}_{(n)} = 1$ for all n and

$$\boldsymbol{\zeta}_\alpha \boldsymbol{\zeta}_\beta = \sum_{\gamma \text{ a shuffle of } \alpha \text{ and } \beta} \boldsymbol{\zeta}_\gamma \tag{1.171}$$

for any compositions α and β.

A *Möbius series* is a family of scalars $(\boldsymbol{\mu}_\gamma)$, as γ varies over compositions of positive integers, such that $\boldsymbol{\mu}_{(n)} = 1$ for all n and

$$\sum_{\gamma \text{ a quasishuffle of } \alpha \text{ and } \beta} \boldsymbol{\mu}_\gamma = 0 \tag{1.172}$$

for any compositions α and β.

Exercise 1.127. Check: For a zeta series $\boldsymbol{\zeta}$,

$$\boldsymbol{\zeta}_{\alpha_1} \cdots \boldsymbol{\zeta}_{\alpha_k} = \sum_{\gamma \text{ a shuffle of } \alpha_1, \dots, \alpha_k} \boldsymbol{\zeta}_\gamma \tag{1.173}$$

for any compositions $\alpha_1, \dots, \alpha_k$. We call (1.173) the *lune-additivity formula* (series version). We recover (1.171) when $k = 2$. As another special case, when all compositions have one part,

$$\sum_{\gamma \text{ a shuffle of } (a_1), \dots, (a_k)} \boldsymbol{\zeta}_\gamma = 1 \tag{1.174}$$

for any positive integers $a_1, \dots, a_k$. We call (1.174) the *flat-additivity formula* (series version).

For a Möbius series $\boldsymbol{\mu}$,

$$\sum_{\gamma \text{ a quasishuffle of } \alpha_1, \dots, \alpha_k} \boldsymbol{\mu}_\gamma = 0 \tag{1.175}$$

for any compositions $\alpha_1, \dots, \alpha_k$ with $k \geq 2$. We call (1.175) the *noncommutative Weisner formula* (series version). We recover (1.172) when $k = 2$.

Zeta and Möbius series yield invariant noncommutative zeta and Möbius functions, respectively, one for each braid arrangement, by setting

$$\boldsymbol{\zeta}(\alpha, \beta) := \prod_i \boldsymbol{\zeta}_{(\beta/\alpha)_i} \quad \text{and} \quad \boldsymbol{\mu}(\alpha, \beta) := \prod_i \boldsymbol{\mu}_{(\beta/\alpha)_i}. \tag{1.176}$$

This follows by relating (1.173) to either (1.79) or (1.82) and (1.175) to either (1.81) or (1.85). Note: We can recover the series from the functions by restricting to pairs (α, β) where α has only one part. Also, the inverse relationship between invariant noncommutative zeta and Möbius functions sets up a bijection between zeta and Möbius series.

Example 1.128. Write $\alpha = (a_1, \dots, a_k)$. The standard example of mutually inverse zeta and Möbius series is

$$\boldsymbol{\zeta}_\alpha = \frac{1}{k!} \quad \text{and} \quad \boldsymbol{\mu}_\alpha = \frac{(-1)^{k-1}}{k}. \tag{1.177}$$

This arises from the exponential and logarithmic power series; for the precise connection, see [**11**, Example 9.140].

Another interesting example is

$$\zeta_\alpha = \prod_{i=1}^{k} \frac{a_i}{a_1 + \cdots + a_i} \quad \text{and} \quad \boldsymbol{\mu}_\alpha = \frac{(-1)^{k-1} a_k}{a_1 + \cdots + a_k}. \tag{1.178}$$

We leave it as an exercise to check that these are indeed mutually inverse zeta and Möbius series. We refer to this Möbius series as the *Dynkin series*. □

A *Zie series* is a sequence $(\mathtt{Q}_{(n)})$, where $\mathtt{Q}_{(n)}$ is an invariant special Zie element in the braid arrangement $\boldsymbol{\beta}^n$. There is a bijection between Zie series and Möbius series specified by

$$\mathtt{Q}_{(n)} = \sum_{\alpha \text{ a composition of } n} \boldsymbol{\mu}_\alpha \, \mathtt{H}_\alpha. \tag{1.179}$$

Thus, zeta series, Möbius series, Zie series are equivalent pieces of data. A Zie series can be extended to a $\mathtt{Q}$-basis by

$$\begin{aligned} \mathtt{H}_\alpha &= \sum_{\beta:\, \alpha \leq \beta} \zeta(\alpha, \beta)\, \mathtt{Q}_\beta, \\ \mathtt{Q}_\alpha &= \sum_{\beta:\, \alpha \leq \beta} \boldsymbol{\mu}(\alpha, \beta)\, \mathtt{H}_\beta. \end{aligned} \tag{1.180}$$

These formulas specialize (1.136). The elements

$$\mathtt{E}_\lambda = \sum_{\alpha:\, \mathrm{s}(\alpha) = \lambda} \zeta_\alpha \, \mathtt{Q}_\alpha, \tag{1.181}$$

as λ varies over partitions of n, are the *Krob–Leclerc–Thibon idempotents*. They specialize (1.138). Example (1.177) yields the *Garsia–Reutenauer idempotents*.

1.16.6. q-zeta and q-Möbius series. For γ a quasishuffle of $(a_1, \ldots, a_k)$ and $(b_1, \ldots, b_l)$, define the *length* of γ by

$$l(\gamma) := \sum_{b_j \text{ appears strictly before } a_i \text{ in } \gamma} a_i b_j. \tag{1.182}$$

This definition extends to quasishuffles of more than two compositions. It also applies to shuffles (by viewing them as quasishuffles).

For q not a root of unity, we have the *q-zeta series* $\boldsymbol{\zeta}_q$ and *q-Möbius series* $\boldsymbol{\mu}_q$. They consist of families of scalars $(\boldsymbol{\zeta}_{q,\gamma})$ and $(\boldsymbol{\mu}_{q,\gamma})$ indexed by compositions γ of positive integers satisfying the *q-lune-additivity formula* and *noncommutative q-Weisner formula* (series version). These formulas are defined by inserting q power the length of the (quasi)shuffle in (1.173) and (1.175). Similarly, we also have the *q-flat-additivity formula*. In contrast to zeta and Möbius series, the q-zeta and q-Möbius series are unique. Moreover, they are projective. They yield the noncommutative q-zeta and q-Möbius functions, respectively, for each braid arrangement.

In a similar manner, for q not a root of unity, we have the *two-sided q-zeta series* ζ_q and *two-sided q-Möbius series* μ_q. Instead of compositions, they are indexed by biface-types in the braid arrangement which are described in Table 1.9.

1.17. Notes

Basic references for this chapter are [**10**] and [**11**, Chapter 1]. For material specific to reflection arrangements, see [**10**, Chapters 5, 6, 16]. For a similar selection of topics, see also [**6**, Chapters 1 and 2] and [**7**, Chapter 10]. This chapter, though introductory in nature, contains many new ideas and results. Specific pointers to our previous monographs [**6**], [**7**], [**10**], [**11**] along with other important historical sources are mentioned below.

1.17.1. Coxeter groups. Coxeter groups originated in work of Coxeter [**91**]; see also the paper by Witt [**362**]. The foundations of Coxeter theory (directed to the study of buildings) were laid down by Tits [**353**].

1.17.2. Birkhoff, Tits, Janus. (Sections 1.1, 1.12, 1.14.) The Birkhoff, Tits, Janus monoids are fundamental objects in the theory of hyperplane arrangements. This terminology was introduced in [**10**], [**11**] and is continued in this book.

Birkhoff monoid. The Birkhoff monoid is an instance of the construction of an idempotent commutative monoid from a semilattice which is explained in Birkhoff's book [**45**, Section I.5], see also [**10**, Example E.2]. Posets of set partitions and integer partitions (flats and flat-types of the braid arrangement) also appeared in his work [**43**, Section 18], [**45**, Section I.8, Examples 9 and 10]. The split-semisimplicity of the Birkhoff algebra originated in work of Solomon [**331**, Theorem 1] and Greene [**158**, Section 1].

For the invariant Birkhoff algebra, see for instance [**6**, Lemmas 2.6.5 and 5.7.1], [**10**, Section 16.1]. The partial order on flat-types in Section 1.1.9 is considered in [**6**, Section 2.6], [**10**, Section 5.2.3] and agrees with the one in [**315**, Section 5, Remark 5.1]. The numbers $|\lambda|$ in Definition 1.8 are defined in [**6**, Definition 5.7.4], [**10**, Section 5.5.1].

Tits monoid. The Tits monoid arose in independent work of Tits [**353**, Section 2.30], [**354**, Proposition 1] and Bland [**48**, Section 5, page 62]. The gate property also originated in work of Tits [**353**, Section 3.19.6], see also Ziegler's thesis [**366**, page 130]. It has been highlighted in all our monographs [**6**, Section 1.2.4], [**7**, Section 10.5], [**10**, Section 1.10 and Appendix A.2], [**11**, Section 1.4]. The (opposite) Tits algebra of the braid arrangement appeared in work of Epstein, Glaser, Stora [**119**, Section 1], [**120**, Section 4.1], see also [**118**, Definition 2.1]. The Tits algebra was studied by Bidigare, Hanlon, Rockmore [**40**], [**41**], Brown and Diaconis [**64**], [**65**], [**66**], Schocker [**321**], Saliola [**316**], and later in great detail in our monograph [**10**].

The invariant Tits algebra is isomorphic to the opposite of the Solomon descent algebra (Section 1.12.5). This is a result of Bidigare [**40**, Theorem 3.8.1], see also [**64**, Section 9.7], [**7**, Section 10.8], [**10**, Theorem 16.8]. The Solomon descent algebra was first defined by Solomon [**333**, Theorem 1]. It was studied further by Bergeron, Bergeron, Howlett, Taylor [**35**], Saliola [**315**], and in more detail in [**10**, Chapter 16]. The Solomon descent algebra of compositions (which arises from the braid arrangement) was studied by Garsia and Reutenauer [**144**]. Some earlier references are [**123**, Theorem 7], [**152**, Corollary 12], [**274**].

Janus monoid. The Janus monoid, Janus algebra and more generally the q-Janus algebra were introduced in [**10**, Sections 1.5 and 9.9], [**11**, Section 1.9]. The invariant q-Janus algebra was not explicitly mentioned in [**10**]. In particular, Exercise 1.110, identity (1.149), Theorem 1.108, and algebra morphisms (1.153), (1.154) are new.

Coxeter analogues. The Coxeter–Birkhoff, Coxeter–Tits, Coxeter–Janus monoids in Section 1.14 were introduced in [**10**, Section 5.6].

1.17.3. Lunes. (Section 1.2.) Lunes hold a special place in ancient mathematical history. Hippocrates of Chios famously achieved the first calculation of the exact area of a curved figure: the planar region bounded by a 180-degree circular arc and a 90-degree circular arc, the lune of Hippocrates. Lunes were used by Bhaskaracharya, under the name "vapraka", for computing the surface area of a sphere [**38**, Goladhyaya, Bhuvanakosha, Verses 58–61]. Leonardo da Vinci often employed lunulae as ornamental designs and in attempts at the quadrature of the circle.

The basic theory of lunes for hyperplane arrangements (including their connection to nested faces) has been developed and promoted in all our monographs [**6**, Sections 2.3, 2.4, 2.5], [**7**, Sections 10.9, 12.5, 12.7], [**10**, Chapters 3, 4, 7, 15], [**11**, Sections 1.2, 1.5, 1.6, 7.7]. This includes the fundamental construction of the category of lunes [**10**, Chapter 4] and the consideration of local bifaces and bilunes [**11**, Section 1.6]. The partial order $\leq$ on top-lunes was defined in [**10**, Section 4.1]; the other partial order $\leq'$ was not explicitly considered in [**10**]. Diagrams (1.4) and (1.6) involving flat-types and lune-types are considered in [**10**, Diagrams (5.3) and (5.4)], however diagram (1.8) involving bilune-types is new.

Some other references where lunes have appeared are [**357**, Section 7], [**41**, Section 4], [**42**, Section 6.1], [**246**, Lemma 5 and Figure 1.9]. For top-nested faces, see [**366**, Construction 5.5.2], [**21**, Sections 5 and 7], [**47**, Definition 2.5.5], [**285**, page 175], [**121**, Chapter 5], [**97**, Sections 2.1 and 3.6].

1.17.4. Descents, global descents, shuffles. (Section 1.6.)

Descents. The notion of descent of a chamber D wrt another chamber C, denoted $\mathrm{des}(C, D)$, originated in work of Brown [**64**, Section 9.3]; observation (1.17) was given in [**64**, Proposition 4]. These ideas were developed further in [**246**, Section 1.6.2], [**6**, Chapter 5], [**7**, Section 10.7], [**10**, Chapter 7]. The related notion of descent $\mathrm{des}(w)$ of a Coxeter group element $w \in W$ is well-known, see for instance [**6**, Section 5.3.1], [**46**, Definition (1.20)], [**94**, Definition 4.7.1].

The more general notion of descent involving faces, denoted $\mathrm{des}(F, G)$, was developed by us in an unpublished manuscript written at the same time as [**6**], and appeared briefly in [**10**, Section 7.1.1], see also [**7**, last paragraph in Section 10.7]. It subsumes for instance the notion of quasi-descent which appeared in [**109**].

Global descents. The notion of global descent of a chamber D wrt another chamber C, denoted $\mathrm{gdes}(C, D)$, first appeared in [**6**, Section 5.2.6]. The related notion of global descent $\mathrm{gdes}(w)$ of a Coxeter group element $w \in W$ was given in [**6**, Section 5.3.2]; the adjunctions (1.24) were given in [**6**, Proposition 5.3.5]. This material built on the special case of global descents of permutations which had appeared earlier in [**12**, Section 2.4]. See also [**7**, Sections 10.7.1–10.7.3].

Shuffles. The geometric approach to shuffles was initiated in our monographs [**6**, Section 5.3.5], [**7**, Section 10.6], [**10**, Sections 6.5.4 and 16.3.1]. Classical shuffles (for an integer composition with two parts) appeared in early work of Eilenberg and Mac Lane [**117**, Section 5], [**236**, pages 10–11], [**237**, Chapter VIII, Theorem 8.8].

1.17.5. Groupoid of biface-types. (Section 1.7.) Consideration of this groupoid and all results related to it are new.

1.17.6. Face-type enumeration. (Section 1.8.) A basic reference for the face-type enumerators $d_T(q)$ is [**10**, Section 16.4], see also [**46**, Section 7.1]. The factorization (1.34) of the Poincaré polynomial $d_S(q)$ was proved by Chevalley [**87**] for Weyl groups and by Solomon [**330**, Corollary 2.3] for Coxeter groups, see also [**195**,

Theorem on page 73], [**46**, Theorem 7.1.5]. The Poincaré polynomial for hyperplane arrangements is considered in [**335**], [**337**, Section 6.1]. The connection of the numbers $a^{TUT'}$ (structure constants of the invariant Tits algebra) with lunes was brought forth in our monographs [**6**, Lemma 2.6.3], [**10**, Lemmas 16.19 and 16.20]. The polynomials $a^{TT'}(q)$ in (1.47), (1.48) were defined in [**10**, Section 16.10.2], [**7**, Equation (10.138)], however, their generalizations $a^{TT'}(q, \xi)$ in (1.49) (using an invariant family of scalars) are new. The numbers $a^{TT'}$ had appeared earlier somewhat implicitly in [**6**, Lemma 2.6.2]. The polynomials $c^{TT'}(q)$ in (1.52), the Varchenko matrix A_λ, Theorem 1.56 and Exercise 1.99 were not explicitly considered in [**10**].

For connection of face-type enumeration to Coxeter bialgebras, see Notes 8.15.4.

Gate identities and alternating sign identities. The gate identity (1.57) for $q = 1$ appeared in work of Bergeron, Bergeron, Howlett, Taylor [**35**, Proposition 2.6]. Its connection to the gate property along with the q-generalization was explained in [**10**, Lemmas 16.21 and 16.26]. Identity (1.62) was proved in [**10**, Lemma 16.56]. The alternating sign identities in Section 1.8.10 are related to the descent, lune, Witt identities [**10**, Sections 7.1, 7.2, 7.3], [**11**, Section 1.7]. Formula (1.63a) is due to Solomon [**330**, Formula (7)], see also [**10**, Formula (16.19)], [**46**, Formula (7.3)], [**195**, Proposition on page 21], [**340**, Corollary 1.29]; the case $q = 1$ goes back to Witt [**362**, Satz 3, Formula (8)].

q-numbers. (Section 1.16.4.) For information related to q-numbers, see for instance the books [**209**, Section IV.2], [**17**, Chapter 10], [**204**], [**18**, Chapter 7], [**13**, Section 1.6], [**338**, Section 1.7], [**122**]. The q-binomial coefficients go back at least to Gauss [**145**]. For Exercise 1.125: The formula for $a^{\alpha\beta\alpha'}(q)$ for $q = 1$ was given in [**10**, Lemma 16.58]; earlier sources are [**144**, Proposition 1.1], [**22**], [**49**, Section 4]. The formula for $a^{\alpha\beta\alpha'\beta'}(q)$ was not explicitly written in [**10**]. For Exercise 1.126: An early reference to the q-Chu–Vandermonde identity (1.170) is by Jensen [**198**, Formulas (9) and (10)], see also [**156**, Formula (2.13)]. The original sources for $q = 1$ are the book by Chu [**88**] and the paper by Vandermonde [**356**, Second formula on page 492]. A generalization in the case $q = 1$ is given by Carlitz [**73**, Formula (10)] and Tauber [**349**, Formula (8)], see also [**308**, Chapter 1, Problem 17]. For a further generalization to the q-multinomial identity (1.169) for $q = 1$, see for instance [**31**, Theorem 2].

Enumeration identities in the Coxeter group algebra. (Section 1.9.) The bijection (1.67) is well-known; see for instance [**10**, Lemma 16.6], [**2**, Proposition 2.20 and Remark 2.22]. We do not know of any reference for the more general result in Exercise 1.71. Identities (1.69), (1.70), (1.71) also look new.

1.17.7. Incidence algebras. Zeta and Möbius functions. (Sections 1.10 and 1.11.) Classical Möbius functions of posets originated in work of Möbius [**268**] and their early theory was developed by Weisner [**359**], Ward [**358**], Hall [**171**] and Rota [**313**]. The lune-incidence algebra and noncommutative zeta and Möbius functions therein were introduced in our monograph [**10**, Chapter 15]. A salient feature of a noncommutative zeta function is that it is not identically 1. In fact, the flat-additivity formula (1.80) is "interpretable" as a stationary distribution of a certain random walk on faces of support X, see [**10**, Chapter 12]. The q-zeta and q-Möbius functions, along with their two-sided versions were introduced in our next monograph [**11**, Sections 1.5 and 1.6]. Formulas (1.82) and (1.85) were obtained in [**10**, Lemma 16.35], however, their q-analogues (1.93) and (1.96) are new. Similarly, formulas (1.109), (1.110), (1.111) are new. Formulas (1.84) and (1.86) for $\boldsymbol{\zeta}(Z, S)$

and $\boldsymbol{\mu}(Z,S)$ were not explicitly written in [**10**]. Formula (1.89) was given in [**10**, Lemma 16.39], however, its q-analogue (1.99) is new. Formula (1.140) was given in [**10**, Exercise 16.47], however, the equivalent identity (1.88) appears here for the first time.

Zaslavsky formulas. Formula (1.120) is independently due to Zaslavsky [**364**, Theorem A] and Las Vergnas [**221**, Proposition 8.1], [**222**, Theorem 3.1]. The noncommutative Zaslavsky formulas (1.119) were first established in our monograph [**11**, Section 1.8]; the Witt identity plays a key role in the proof. Their formulation in terms of face-types (1.121) is new. The noncommutative q-Zaslavsky formulas, for q not a root of unity, are also new.

Solomon coefficients. The Solomon coefficients (1.125) were defined in our monograph [**11**, Section 1.8.3]. Their formulation in terms of face-types (1.126) is new. Identity (1.127) is also new. For the braid arrangement, the Solomon coefficients were discussed in [**11**, Section 17.7.4]. They go back to Solomon [**332**, Formula (2.3)] and Bialynicki-Birula, Mielnik, Plebański [**39**, Formula (3.7)], [**264**, Formula (10.12)]; some recent references are [**284**, Proposition 7.5], [**109**, Theorem 4.8 and Corollary 5.11], [**110**, Theorems 2.1 and 7.2], with the connection to the braid arrangement manifest in [**10**, Formula (12.51)]. For more historical information, see [**11**, Notes to Chapter 17, page 712].

Zeta and Möbius series. (Section 1.16.5.) Zeta and Möbius series were implicitly considered in [**10**, Theorem 16.48 and Section 16.11.6]. For α a composition, $\boldsymbol{\zeta}_\alpha = \mathbf{u}^\alpha$, where $\mathbf{u}$ is an invariant homogeneous section in the sense of [**10**, Section 16.8.1]. For Example (1.177), see [**10**, Formulas (16.56), (16.76), (16.77)]. Example (1.178) of the Dynkin series is highlighted in [**10**, Exercise 16.63]. The Krob–Leclerc–Thibon idempotents were first defined in [**215**, Section 3.4]. They generalized the Garsia–Reutenauer idempotents which had appeared earlier in [**144**, Theorems 3.1 and 3.2], see also [**215**, Section 3.3], [**10**, Formula (16.78)]. However, [**144**] and [**215**] make no contact with zeta or Möbius functions. The formula for the zeta series (inverse to the Dynkin series) in (1.178) appeared implicitly in [**108**, Theorem 4.1], [**250**, Theorem 6], [**111**, Theorem 2.1].

Consideration of q-zeta and q-Möbius series appears to be new. More details will be given in future work.

Zie series. In the language of [**149**, Section 5.2], invariant special Zie elements for the braid arrangement are Lie idempotents that live in the Solomon descent algebra. For more references on Lie idempotents, see [**215**, Section 3.1], [**305**, Section 8.4]. Zie series are equivalent to Lie idempotent series considered in [**111**, Section 2].

Special Zie elements (not necessarily invariant) inside the Tits algebra of the braid arrangement were first considered by Schocker [**321**, pages 562–563]. The terminology of Zie elements was introduced later in [**10**, Chapter 10].

1.17.8. General references. Some useful books that deal with Coxeter groups are by Abramenko and Brown [**2**, Chapters 1, 2, 3], Björner and Brenti [**46**], Borovik and Borovik [**57**], Borovik, Gelfand, White [**58**, Chapters 5 and 7], Bourbaki [**61**], Carter [**76**], Contou-Carrère [**90**, Chapters 8 and 9], Coxeter [**92**, Chapter XI], Davis [**94**], Garrett [**142**], Grove and Benson [**161**], Hiller [**184**], Humphreys [**195**], Kane [**205**], McMullen and Schulte [**256**, Chapter 3], Petersen [**293**, Chapter 11], Ronan [**310**], Thomas [**352**, Part I], Weiss [**360**].

Part I

Coxeter species

CHAPTER 2

Coxeter species and Coxeter bimonoids

We begin the study of Coxeter species, and monoids, comonoids, bimonoids therein. Let $\mathcal{A}$ be a fixed reflection arrangement with Coxeter group W. Faces, flats, lunes of $\mathcal{A}$ play a central role in our study. A useful summary is provided in Tables 2.1 and 2.2. Standard notations related to categories are given in the list of notations at the end of the book.

Coxeter species. We introduce the notion of a Coxeter $\mathcal{A}$-species. It can be formulated in terms of either faces or flats of $\mathcal{A}$. In terms of faces: A Coxeter $\mathcal{A}$-species p attaches a vector space $\mathsf{p}[F]$ to every face F of $\mathcal{A}$ along with linear maps

$$\beta_{G,F} : \mathsf{p}[F] \to \mathsf{p}[G] \quad \text{and} \quad w : \mathsf{p}[F] \to \mathsf{p}[wF],$$

the former whenever $\mathrm{s}(F) = \mathrm{s}(G)$, and the latter for each F and $w \in W$. These are support and type morphisms, respectively. They are subject to certain conditions, which we call support-support, support-type, type-type, support-unitality, type-unitality axioms.

In terms of flats: A Coxeter $\mathcal{A}$-species p attaches a vector space $\mathsf{p}[\mathrm{X}]$ to every flat X of $\mathcal{A}$ along with linear maps

$$w : \mathsf{p}[\mathrm{X}] \to \mathsf{p}[w\mathrm{X}]$$

called type morphisms satisfying type-type and type-unitality axioms. The difference with the previous formulation is the absence of support morphisms and axioms involving them. This happens because faces are replaced by flats. (Recall that the two are related by the support map.)

Coxeter monoids and comonoids. Next, we introduce the notion of a Coxeter $\mathcal{A}$-monoid. It is a Coxeter $\mathcal{A}$-species a equipped with linear maps

$$\mu_A^F : \mathsf{a}[F] \to \mathsf{a}[A],$$

one for each $F \geq A$. These are product components of a. They are subject to support-naturality, type-naturality, associativity, unitality axioms. Dually, a Coxeter $\mathcal{A}$-comonoid is a Coxeter $\mathcal{A}$-species c equipped with linear maps

$$\Delta_A^F : \mathsf{c}[A] \to \mathsf{c}[F],$$

one for each $F \geq A$, called coproduct components of c, subject to support-naturality, type-naturality, coassociativity, counitality axioms. Alternatively, one may only use those (co)product components for which $A \lessdot F$ along with the covering (co)associativity axiom.

TABLE 2.1. Axioms for Coxeter species and (co, bi)monoids.

Notion		Axioms
Coxeter species	faces	support-support, support-type, type-type, support-unitality, type-unitality axioms
	flats	type-type, type-unitality axioms
Coxeter (co)monoid	faces	support-naturality, type-naturality, (co)associativity, (co)unitality axioms
	flats lunes	type-naturality, (co)associativity, (co)unitality axioms
Coxeter bimonoid	faces	bimonoid axiom (pentagon)
	flats lunes	bimonoid axiom (square)
Coxeter q-bimonoid	faces	q-bimonoid axiom (pentagon)
	flats lunes	q-bimonoid axiom (square)
(co)com. Coxeter (co)monoid	faces	(co)commutativity axiom
	flats	type-naturality, (co)associativity, (co)unitality axioms
bicom. Coxeter bimonoid	flats	bicommutative bimonoid axiom (square)

In terms of flats, a Coxeter $\mathcal{A}$-monoid is a Coxeter $\mathcal{A}$-species a equipped with linear maps

$$\mu_{\mathrm{L}} : \mathsf{a}[\mathrm{X}] \to \mathsf{a}[\mathrm{Z}],$$

indexed by lunes L with base Z and case X, subject to type-naturality, associativity, unitality axioms. Note here the absence of the support-naturality axiom. Dually, a Coxeter $\mathcal{A}$-comonoid c is defined using linear maps

$$\Delta_{\mathrm{L}} : \mathsf{c}[\mathrm{Z}] \to \mathsf{c}[\mathrm{X}]$$

indexed by lunes L with base Z and case X. Alternatively, one may only use lunes of slack 1 along with the covering (co)associativity axiom.

Coxeter bimonoids. By combining notions of Coxeter $\mathcal{A}$-monoid and Coxeter $\mathcal{A}$-comonoid, we obtain the self-dual notion of a Coxeter $\mathcal{A}$-bimonoid. The product and coproduct satisfy a compatibility condition called the bimonoid axiom. It involves the Tits product on faces. It can also be phrased in terms of lunes and the action of the Birkhoff monoid on it. Interestingly, for

TABLE 2.2. Alternative axioms for Coxeter (co)monoids.

Notion	Alternative axioms
Coxeter (co)monoid	covering (co)associativity axiom
(co)commutative Coxeter (co)monoid	opposition (co)commutativity axiom
	covering (co)commutativity axiom

a scalar q, one can deform the bimonoid axiom using the q-distance function. This yields the more general notion of a Coxeter $\mathcal{A}$-q-bimonoid. For $q = -1$, we use the term signed Coxeter $\mathcal{A}$-bimonoid. We also pay attention to the case $q = 0$.

Examples of Coxeter bimonoids are given in Chapter 4, and these could be read in parallel with the present chapter. The toy example of rank one is developed in Section 7.8.

(Co)commutative Coxeter (co)monoids. There is a notion of commutativity for a Coxeter $\mathcal{A}$-monoid, and cocommutativity for a Coxeter $\mathcal{A}$-comonoid. A Coxeter $\mathcal{A}$-bimonoid may be commutative, cocommutative, both or neither. In terms of flats, a commutative Coxeter $\mathcal{A}$-monoid can be described using linear maps

$$\mu_{\mathrm{Z}}^{\mathrm{X}} : \mathsf{a}[\mathrm{X}] \to \mathsf{a}[\mathrm{Z}],$$

one for each $\mathrm{X} \geq \mathrm{Z}$. Interestingly, this formulation makes no direct reference to Coxeter $\mathcal{A}$-monoids. The same remarks apply to cocommutative Coxeter $\mathcal{A}$-comonoids.

There is also a notion of signed commutativity for Coxeter $\mathcal{A}$-monoids and signed cocommutativity for Coxeter $\mathcal{A}$-comonoids.

Signature functors. The categories of Coxeter bimonoids and signed Coxeter bimonoids are isomorphic. The same holds for the categories of commutative Coxeter monoids and signed commutative Coxeter monoids, and so on. We construct two explicit isomorphisms called the u-signature functor and o-signature functor by appropriately twisting support and type morphisms of the underlying Coxeter species.

Species. For any hyperplane arrangement $\mathcal{A}$, one can define the notion of $\mathcal{A}$-species, followed by notions of $\mathcal{A}$-monoid, $\mathcal{A}$-comonoid, $\mathcal{A}$-bimonoid. They were studied in depth in our previous monograph [**11**]. By imposing on them additional structure of type morphisms (which make sense only when $\mathcal{A}$ is a reflection arrangement), one obtains the notion of Coxeter $\mathcal{A}$-species, and so on. This connection can be used to establish results on Coxeter $\mathcal{A}$-species using corresponding results on $\mathcal{A}$-species proved in [**11**].

Bimonads on Coxeter species. Monads, comonads, bimonads on a category are reviewed in Appendix A.1. They provide the categorical framework for Coxeter monoids, comonoids, bimonoids. More precisely: We construct a bimonad on Coxeter $\mathcal{A}$-species which we denote $(\mathcal{T}, \mathcal{T}^\vee, \lambda)$. A $\mathcal{T}$-algebra is the same as a Coxeter $\mathcal{A}$-monoid, a $\mathcal{T}^\vee$-coalgebra is the same as a Coxeter $\mathcal{A}$-comonoid, and a $(\mathcal{T}, \mathcal{T}^\vee, \lambda)$-bialgebra is the same as a Coxeter $\mathcal{A}$-bimonoid. Commutative aspects of the theory are handled in a similar manner.

Joyal bimonoids. Connections to classical Hopf theory. The notion of a Coxeter $\mathcal{A}$-bimonoid, when specialized to braid arrangements $\mathcal{B}^I$, relates to the classical notion of a Joyal bimonoid. More precisely, there is a functor from connected Joyal species to Coxeter $\mathcal{B}^I$-species which induces functors from connected Joyal (co, bi)monoids to Coxeter $\mathcal{B}^I$-(co, bi)monoids. Note: "connected" is equivalent to "positive".

Convention 2.1. Whenever convenient, we drop $\mathcal{A}$ from the notation, and write Coxeter species instead of Coxeter $\mathcal{A}$-species, Coxeter monoid instead of Coxeter $\mathcal{A}$-monoid, and so on.

2.1. Support and type morphisms

Starting with a reflection arrangement $\mathcal{A}$, we construct a finite category $\mathcal{A}$-Cox. It is a groupoid. Its objects are faces of $\mathcal{A}$. Its morphisms are generated by two kinds of morphisms called support and type morphisms. The former are between faces of the same support, the latter are between faces of the same type. They are subject to 5 relations called support-support, support-type, type-type, support-unitality, type-unitality relations. By analyzing these relations, we obtain two normal forms for morphisms in $\mathcal{A}$-Cox. We also construct a category $\mathcal{A}$-Cox' equivalent to $\mathcal{A}$-Cox whose objects are flats of $\mathcal{A}$, and whose morphisms are called type morphisms. See Table 2.3.

TABLE 2.3. Base categories for Coxeter species.

Groupoid	Objects	Morphisms
$\mathcal{A}$-Cox	faces	support and type morphisms
$\mathcal{A}$-Cox'	flats	type morphisms

2.1.1. Category with faces as objects. Fix a reflection arrangement $\mathcal{A}$. Let W denote its Coxeter group. We write e for the identity element of W. By convention, W acts on the set of faces $\Sigma[\mathcal{A}]$ on the left.

Define the category $\mathcal{A}$-Cox by a presentation. An object is a face of $\mathcal{A}$. There are two different kinds of generating morphisms. Whenever F and G have the same support, there is a unique *support morphism*

$$\beta_{G,F} : F \to G,$$

and whenever F and F' have the same type, there are *type morphisms*

$$w : F \to F',$$

one for each $w \in W$ satisfying $w(F) = F'$. These generating morphisms are subject to the 5 relations

$$
\begin{array}{ccc}
& G & \\
{\scriptstyle \beta_{G,F}} \nearrow & & \searrow {\scriptstyle \beta_{H,G}} \\
F & \xrightarrow[\beta_{H,F}]{} & H
\end{array}
\qquad
\begin{array}{ccc}
F & \xrightarrow{\beta_{G,F}} & G \\
{\scriptstyle w} \downarrow & & \downarrow {\scriptstyle w} \\
F' & \xrightarrow[\beta_{G',F'}]{} & G'
\end{array}
\qquad
\begin{array}{ccc}
& F' & \\
{\scriptstyle u} \nearrow & & \searrow {\scriptstyle v} \\
F & \xrightarrow[vu]{} & F''
\end{array}
\tag{2.1a}
$$

$$
(F \xrightarrow{\beta_{F,F}} F) = \mathrm{id} \qquad\qquad (F \xrightarrow{e} F) = \mathrm{id}\,. \tag{2.1b}
$$

In (2.1a),

- in the first diagram, $\mathrm{s}(F) = \mathrm{s}(G) = \mathrm{s}(H)$,
- in the second diagram, $\mathrm{s}(F) = \mathrm{s}(G)$, $\mathrm{s}(F') = \mathrm{s}(G')$, $w(F) = F'$, $w(G) = G'$, and
- in the third diagram, $u(F) = F'$ and $v(F') = F''$.

In (2.1b), F is any face. Note: Morphisms labeled by β are between faces of the same support, and those labeled by u, v, w are between faces of the same type. (Whenever convenient, we shorten $\beta_{G,F}$ to β.)

We refer to diagrams (2.1a) as the *support-support relation*, *support-type relation*, *type-type relation*, respectively, and to conditions (2.1b) as the *support-unitality relation* and *type-unitality relation*, respectively.

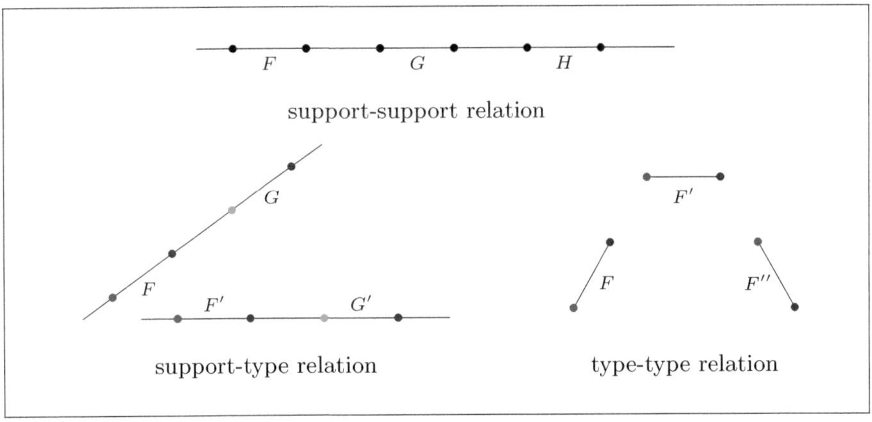

FIGURE 2.1. Support-support, support-type, type-type relations.

See the pictures in Figure 2.1. In relations involving type, vertices of the same type are shown with the same color.

Lemma 2.2. *The category $\mathcal{A}$-*Cox *is a groupoid. The support morphisms*

$$\beta_{G,F} : F \to G \quad \text{and} \quad \beta_{F,G} : G \to F$$

are inverses, while the type morphisms

$$w : F \to F' \quad \text{and} \quad w^{-1} : F' \to F$$

are inverses.

PROOF. It suffices to establish the above claims about support and type morphisms. For the first, we use support-support and support-unitality relations, while for the second, we use type-type and type-unitality relations. □

2.1.2. Normal forms for morphisms. We refer to an arbitrary morphism in $\mathcal{A}$-Cox as a *support-type morphism*. It has two normal forms:

Lemma 2.3. *Every morphism in $\mathcal{A}$-Cox can be uniquely expressed as*

- *a type morphism composed with a support morphism, and*
- *a support morphism composed with a type morphism.*

PROOF. We explain the first statement; the second is similar. Let α be a morphism in $\mathcal{A}$-Cox. If it is a support morphism, then we write

$$(F \xrightarrow{\beta_{G,F}} G) = (F \xrightarrow{e} F \xrightarrow{\beta_{G,F}} G),$$

if it is a type morphism, then we write

$$(F \xrightarrow{w} F') = (F \xrightarrow{w} F' \xrightarrow{\beta_{F',F'}} F'),$$

and if it is an identity morphism, then we write

$$(F \xrightarrow{\mathrm{id}} F) = (F \xrightarrow{e} F \xrightarrow{\beta_{F,F}} F).$$

These hold by support-unitality and type-unitality relations.

In general, we have $\alpha = (F \to F_1 \to \cdots \to F_n \to G)$, where each arrow is either a support or type morphism. Using a support-support (type-type) relation, two adjacent support (type) morphisms can be shortened to a single support (type) morphism, while the order of adjacent support and type morphisms can be switched using a support-type relation. So α can be reduced to the form $F \xrightarrow{w} F' \xrightarrow{\beta} G$. Moreover, this form is unique: w is determined by the formula $w = u_k \dots u_1$, where $u_1, \dots, u_k$ are type morphisms which appear in α in that order. This further forces $F' = w(F)$. □

The two normal forms in Lemma 2.3 can be compared as follows. For a morphism σ in $\mathcal{A}$-Cox, in normal form,

$$\sigma = \beta w \iff \sigma = w\beta. \tag{2.2}$$

This follows from the support-type relation. Note very carefully: The same $w \in W$ occurs on both sides.

Similarly, for composable morphisms σ and τ in $\mathcal{A}$-Cox, in normal form,

$$\begin{aligned} \sigma = \beta u \text{ and } \tau = \beta v &\implies \tau\sigma = \beta v u, \\ \sigma = u\beta \text{ and } \tau = v\beta &\implies \tau\sigma = v u \beta. \end{aligned} \tag{2.3}$$

These normal forms provide two complementary ways to think about $\mathcal{A}$-Cox. The first one is as follows.

Proposition 2.4. *Objects of $\mathcal{A}$-Cox are faces of $\mathcal{A}$. A morphism $F \to G$ is an element $w \in W$ such that $w(F)$ and G have the same support. A morphism u composed with a morphism v is the morphism vu (product in the group W).*

An illustration is shown below.

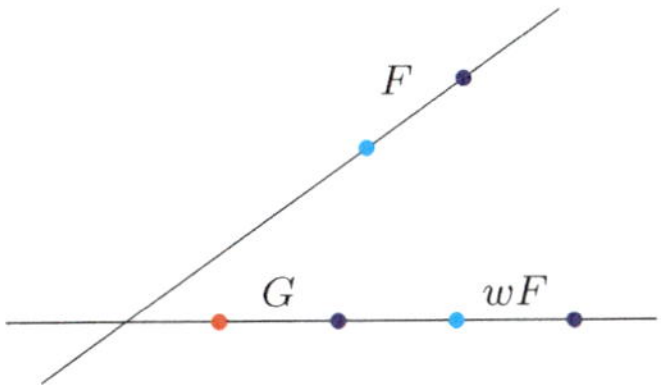

A morphism can be written as

$$\begin{array}{ccc} & F' & \\ {}^{w}\nearrow & & \searrow^{\beta} \\ F & & G, \end{array}$$

and composition as

$$\begin{array}{ccccccc} & & & F'' & & & \\ & & {}^{v}\nearrow & & \searrow^{\beta} & & \\ & F' & & & & G' & \\ {}^{u}\nearrow & & \searrow^{\beta} & & {}^{v}\nearrow & & \searrow^{\beta} \\ F & & & G & & & H. \end{array}$$

The complementary way to think about $\mathcal{A}$-Cox is as follows.

Proposition 2.5. *Objects of $\mathcal{A}$-Cox are faces of $\mathcal{A}$. A morphism $F \to G$ is an element $w \in W$ such that F and $w^{-1}(G)$ have the same support. A morphism u composed with a morphism v is the morphism vu (product in the group W).*

A morphism can be written as

$$\begin{array}{ccc} & G' & \\ {}^{\beta}\nearrow & & \searrow^{w} \\ F & & G, \end{array}$$

and composition as

$$\begin{array}{ccccccc} & & & H'' & & & \\ & & {}^{\beta}\nearrow & & \searrow^{u} & & \\ & G' & & & & H' & \\ {}^{\beta}\nearrow & & \searrow^{u} & & {}^{\beta}\nearrow & & \searrow^{v} \\ F & & & G & & & H. \end{array}$$

Compare and contrast with previous pictures.

Observation 2.6. For a morphism σ in $\mathcal{A}$-Cox, in normal form,

$$\sigma = \beta w = w\beta \iff \sigma^{-1} = \beta w^{-1} = w^{-1}\beta. \tag{2.4}$$

In long notation, for $\sigma : F \to G$, the normal forms are

$$\sigma = \beta_{G,wF}\, w = w\, \beta_{w^{-1}G,F} \quad \text{and} \quad \sigma^{-1} = \beta_{F,w^{-1}G}\, w^{-1} = w^{-1}\, \beta_{wF,G}.$$

This follows from Lemma 2.2.

Exercise 2.7. Show: There is a morphism from F to G in $\mathcal{A}$-Cox iff F and G have the same support-type. In particular, existence of a morphism from F to G implies that F and G have the same rank.

2.1.3. Category with flats as objects. Define the category $\mathcal{A}$-Cox' as follows. Objects are flats of $\mathcal{A}$. Morphisms are

$$w : \mathrm{X} \to \mathrm{X}'$$

with $w \in W$ such that $w(\mathrm{X}) = \mathrm{X}'$. We refer to these as *type morphisms.* A morphism u composed with a morphism v is the morphism vu (product in the group W). In other words,

$$\mathcal{A}\text{-}\mathsf{Cox}' = \mathsf{el}_W(\Pi[\mathcal{A}]), \tag{2.5}$$

the groupoid of elements for the action of W on the set of flats $\Pi[\mathcal{A}]$. For the general definition of the groupoid of elements, see Appendix A.2.3.

We now relate the category with faces as objects to the category with flats as objects. Define a functor

$$\mathcal{A}\text{-}\mathsf{Cox} \to \mathcal{A}\text{-}\mathsf{Cox}' \tag{2.6}$$

which on objects sends F to its support $\mathrm{s}(F)$, and on morphisms sends $\beta : F \to G$ to $\mathrm{id} : \mathrm{s}(F) \to \mathrm{s}(G)$ and $w : F \to F'$ to $w : \mathrm{s}(F) \to \mathrm{s}(F')$.

Proposition 2.8. *The functor* (2.6) *is an equivalence of categories.*

PROOF. We describe a functor in the other direction. For each flat X, pick a face sec(X) with support X. On objects, send X to sec(X), and on morphisms, send $w : \mathrm{X} \to \mathrm{X}'$ to $\beta w : \mathrm{sec}(\mathrm{X}) \to \mathrm{sec}(\mathrm{X}')$. That this assignment respects composition of morphisms follows from Proposition 2.4. It is straightforward to check that this functor yields an equivalence. □

2.1.4. Category with flat-types as objects. In a groupoid C, any full subcategory which includes at least one object from each connected component of C is equivalent to C. In this case, the inclusion functor is full and faithful and essentially surjective on objects, and hence is an equivalence [**238**, Section IV.4].

A full subcategory of $\mathcal{A}$-Cox of this kind appeared in the proof of Proposition 2.8 consisting of faces of the form sec(X). It is isomorphic to $\mathcal{A}$-Cox'. Continuing further, we can go to a skeleton of the groupoid by picking exactly one object from each connected component of $\mathcal{A}$-Cox'. The connected components are indexed by flat-types. This yields the following.

Proposition 2.9 (Skeletal category). *The category $\mathcal{A}$-Cox is equivalent to the skeletal category obtained by taking disjoint union of the one-object categories associated to the subgroups W_λ of W, one for each flat-type λ.*

2.1.5. Connected components. We highlight two connected components of the category $\mathcal{A}$-Cox, and its equivalent category $\mathcal{A}$-Cox'.

Exercise 2.10 (Minimum, maximum flat-types). Check: The subcategories consisting of

- the central face O and morphisms from O to itself in $\mathcal{A}$-Cox, and
- the minimum flat $\perp$ and morphisms from $\perp$ to itself in $\mathcal{A}$-Cox'

are connected components and they correspond to each other under the equivalence in Proposition 2.8. Moreover, they are the same as C_W, the group W viewed as a one-object category.

Similarly, the subcategories consisting of

- chambers and morphisms between them in $\mathcal{A}$-$\mathtt{Cox}$, and
- the maximum flat $\top$ and morphisms from $\top$ to itself in $\mathcal{A}$-$\mathtt{Cox}'$

are connected components which correspond to each other. The latter is the same as C_W.

In terms of Proposition 2.9, the above connected components correspond to the minimum flat-type $\lambda_{\min} = \mathrm{s}(\emptyset)$ and maximum flat-type $\lambda_{\max} = \mathrm{s}(S)$, respectively.

2.2. Induced and local morphisms

We define the notion of induced morphisms in the category $\mathcal{A}$-$\mathtt{Cox}$. We also define classes of morphisms in $\mathcal{A}$-$\mathtt{Cox}$ local to a face.

2.2.1. Induced morphisms. Let $\sigma : A \to B$ be a morphism in $\mathcal{A}$-$\mathtt{Cox}$. For a face G greater than B, define $\sigma^*(G)$ to be the face greater than A inductively as follows.

- For $\sigma = \mathrm{id}$, put $\sigma^*(G) = G$.
- For $\sigma = \beta$, put $\sigma^*(G) = AG$, and for $\sigma = w$, put $\sigma^*(G) = w^{-1}(G)$.
- For composable morphisms σ and τ, put $(\tau\sigma)^*(G) = \sigma^*(\tau^*(G))$.

One may check that it is well-defined. Further, there is an induced morphism $\sigma^*(G) \to G$ which we continue to denote σ. It is id for id, β for β, w for w, and respects compositions, that is, for $\sigma : A \to B$, $\tau : B \to K$ and $K \le G$,

$$((\tau\sigma)^*(G) \xrightarrow{\tau\sigma} G) = (\sigma^*\tau^*(G) \xrightarrow{\sigma} \tau^*(G) \xrightarrow{\tau} G).$$

Similarly, given $\sigma : A \to B$, for a face F greater than A, one defines the face $\sigma_*(F)$ greater than B and an induced morphism $\sigma : F \to \sigma_*(F)$.

Induced morphisms can be visualized as follows.

$$\begin{array}{ccc} F & \xrightarrow{\sigma} & \sigma_*(F) \\ | & & | \\ A & \xrightarrow[\sigma]{} & B \end{array} \qquad\qquad \begin{array}{ccc} \sigma^*(G) & \xrightarrow{\sigma} & G \\ | & & | \\ A & \xrightarrow[\sigma]{} & B \end{array} \tag{2.7}$$

Exercise 2.11. Check: For $\sigma : A \to B$ and $A \le F$, $B \le G$,

$$G = \sigma_*(F) \iff F = \sigma^*(G).$$

Moreover, in this situation, $(F \xrightarrow{\sigma} \sigma_*(F)) = (\sigma^*(G) \xrightarrow{\sigma} G)$.

Exercise 2.12. Check: If $\sigma : A \to B$ in normal form is $\sigma = \beta w = w\beta$, then any induced $\sigma : F \to G$ (with $F = \sigma^*(G)$ and $G = \sigma_*(F)$) also has normal form $\sigma = \beta w = w\beta$ (with the same w).

Exercise 2.13. Check: The following compatibility holds.

$$\begin{array}{ccc} G & \xrightarrow{\sigma} & \sigma_*(G) \\ | & & | \\ F & \xrightarrow{\sigma} & \sigma_*(F) \\ | & & | \\ A & \xrightarrow[\sigma]{} & B \end{array} \qquad\qquad \begin{array}{ccc} \sigma^*(H) & \xrightarrow{\sigma} & H \\ | & & | \\ \sigma^*(G) & \xrightarrow{\sigma} & G \\ | & & | \\ A & \xrightarrow[\sigma]{} & B \end{array} \tag{2.8}$$

In the left diagram, we start with $\sigma : A \to B$ and $A \leq F \leq G$, while in the right diagram, we start with $\sigma : A \to B$ and $B \leq G \leq H$.

2.2.2. Morphisms local to a face. A morphism $\sigma : F \to G$ in $\mathcal{A}$-$\mathtt{Cox}$ is *local* to a face A if $A \leq F, G$ and σ is obtained by composing support and type morphisms involving only faces greater than A. Equivalently, σ is local to A if $A \leq F, G$ and in normal form $\sigma = \beta w = w\beta$ with $w(A) = A$.

For each face A, we have the subcategory $\mathcal{A}_A$-$\mathtt{Cox}$ of $\mathcal{A}$-$\mathtt{Cox}$ whose objects are faces greater than A and morphisms are those local to A. As implied by the notation, this subcategory is an instance of $\mathcal{A}$-$\mathtt{Cox}$ with $\mathcal{A}$ replaced by $\mathcal{A}_A$.

Lemma 2.14. *A morphism $\sigma : A \to A'$ in $\mathcal{A}$-$\mathtt{Cox}$ induces an isomorphism*

$$\mathcal{A}_A\text{-}\mathtt{Cox} \to \mathcal{A}_{A'}\text{-}\mathtt{Cox}, \qquad F \mapsto \sigma_*(F), \quad \tau \mapsto \sigma\tau\sigma^{-1} \tag{2.9}$$

of categories. Further, it preserves support and type morphisms.

PROOF. Observe: σ_* is a bijection on objects with inverse σ^*, see Exercise 2.11. We now check that conjugation by σ sends morphisms local to A to morphisms local to A'. It suffices to do this when σ is a support morphism and when σ is a type morphism.

- σ is a support morphism. The conjugate of $\beta : F \to G$ (local to A) is $\beta : \sigma_*(F) \to \sigma_*(G)$ (local to B) by the support-support relation. The conjugate of $w : F \to F'$ (local to A) is $w : \sigma_*(F) \to \sigma_*(F')$ (local to B) by the support-type relation.
- σ is a type morphism, say u. The conjugate of $\beta : F \to G$ (local to A) is $\beta : \sigma_*(F) \to \sigma_*(G)$ (local to B) by the support-type relation. The conjugate of $w : F \to F'$ (local to A) is $uwu^{-1} : \sigma_*(F) \to \sigma_*(F')$ (local to B) by the type-type relation.

It is also clear that conjugation by σ respects composition of morphisms. □

Exercise 2.15. Deduce: A morphism $\sigma : A \to A'$ in $\mathcal{A}$-$\mathtt{Cox}$ induces a group isomorphism

$$W_A \to W_{A'}, \qquad u \mapsto wuw^{-1},$$

where $\sigma = \beta w = w\beta$ in normal form.

2.3. Coxeter species

We introduce Coxeter species. Let $\Bbbk$ be a field. This field will remain fixed throughout the book. Let Vec denote the category of vector spaces over $\Bbbk$. For a reflection arrangement $\mathcal{A}$, the category of Coxeter $\mathcal{A}$-species is the category of functors from $\mathcal{A}$-$\mathtt{Cox}$ to Vec, where $\mathcal{A}$-$\mathtt{Cox}$ is the category introduced in Section 2.1. One may also employ instead the equivalent category $\mathcal{A}$-$\mathtt{Cox}'$. Thus, a Coxeter species can be formulated using either faces or flats, see the first entry in Table 2.1. It makes sense to talk of Coxeter subspecies, Coxeter quotient species, the dual of a Coxeter species. One can also consider Coxeter set-species which take values in the category of sets (instead of vector spaces).

2.3.1. Coxeter species in terms of faces. Recall the category $\mathcal{A}$-Cox from Section 2.1. A *Coxeter $\mathcal{A}$-species* is a functor

$$\mathsf{p} : \mathcal{A}\text{-}\mathsf{Cox} \to \mathsf{Vec}.$$

A map of Coxeter $\mathcal{A}$-species $\mathsf{p} \to \mathsf{q}$ is a natural transformation. This defines the category of Coxeter $\mathcal{A}$-species which we denote $\mathcal{A}$-CSp. It is a functor category, and we also write

$$\mathcal{A}\text{-}\mathsf{CSp} := [\mathcal{A}\text{-}\mathsf{Cox}, \mathsf{Vec}].$$

For Coxeter $\mathcal{A}$-species p and q, let $\mathcal{A}$-$\mathsf{CSp}(\mathsf{p}, \mathsf{q})$ denote the space of all maps from p to q.

The value of a Coxeter $\mathcal{A}$-species p on an object F is denoted $\mathsf{p}[F]$. This is the F-component of p. Thus, a Coxeter $\mathcal{A}$-species p consists of a family of vector spaces $\mathsf{p}[F]$, one for each face F, equipped with linear maps

$$\sigma : \mathsf{p}[F] \to \mathsf{p}[G],$$

one for each morphism $\sigma : F \to G$ in $\mathcal{A}$-Cox, such that the diagrams

$$\begin{array}{ccc} & \mathsf{p}[G] & \\ {\scriptstyle\sigma}\nearrow & & \searrow{\scriptstyle\tau} \\ \mathsf{p}[F] & \xrightarrow[\tau\sigma]{} & \mathsf{p}[H] \end{array} \qquad\qquad (\mathsf{p}[F] \xrightarrow{\mathrm{id}} \mathsf{p}[F]) = \mathrm{id} \tag{2.10}$$

commute, the former for all $\sigma : F \to G$ and $\tau : G \to H$, and the latter for every F. Similarly, a map of Coxeter $\mathcal{A}$-species $f : \mathsf{p} \to \mathsf{q}$ consists of a family of linear maps

$$f_F : \mathsf{p}[F] \to \mathsf{q}[F],$$

one for each face F, such that the diagram

$$\begin{array}{ccc} \mathsf{p}[F] & \xrightarrow{f_F} & \mathsf{q}[F] \\ {\scriptstyle\sigma}\big\downarrow & & \big\downarrow{\scriptstyle\sigma} \\ \mathsf{p}[G] & \xrightarrow[f_G]{} & \mathsf{q}[G] \end{array} \tag{2.11}$$

commutes for all $\sigma : F \to G$. The map f_F is the F-component of f.

Using the presentation of $\mathcal{A}$-Cox, we obtain the following description. A Coxeter $\mathcal{A}$-species p consists of a family of vector spaces $\mathsf{p}[F]$, one for each face F. Whenever F and G have the same support, there is a linear map

$$\beta_{G,F} : \mathsf{p}[F] \to \mathsf{p}[G],$$

and whenever F and F' have the same type, there are linear maps

$$w : \mathsf{p}[F] \to \mathsf{p}[F'],$$

one for each $w \in W$ satisfying $w(F) = F'$. These are support and type morphisms of p, respectively. We also refer to them as structure maps of p.

(To emphasize dependence on p, we may sometimes write $\mathsf{p}[w]$ instead of w, and so on.) The following diagrams must commute.

$$
\text{(2.12a)} \qquad
\begin{array}{ccc}
 & \mathsf{p}[G] & \\
{\scriptstyle \beta_{G,F}}\nearrow & & \searrow{\scriptstyle \beta_{H,G}} \\
\mathsf{p}[F] & \xrightarrow[\beta_{H,F}]{} & \mathsf{p}[H]
\end{array}
\qquad
\begin{array}{ccc}
\mathsf{p}[F] & \xrightarrow{\beta_{G,F}} & \mathsf{p}[G] \\
{\scriptstyle w}\big\downarrow & & \big\downarrow{\scriptstyle w} \\
\mathsf{p}[F'] & \xrightarrow[\beta_{G',F'}]{} & \mathsf{p}[G']
\end{array}
\qquad
\begin{array}{ccc}
 & \mathsf{p}[F'] & \\
{\scriptstyle u}\nearrow & & \searrow{\scriptstyle v} \\
\mathsf{p}[F] & \xrightarrow[vu]{} & \mathsf{p}[F'']
\end{array}
$$

$$
\text{(2.12b)} \qquad (\mathsf{p}[F] \xrightarrow{\beta_{F,F}} \mathsf{p}[F]) = \mathrm{id} \qquad\qquad (\mathsf{p}[F] \xrightarrow{e} \mathsf{p}[F]) = \mathrm{id}
$$

The conventions are as in (2.1a)–(2.1b). We refer to diagrams (2.12a) as the *support-support axiom*, *support-type axiom*, *type-type axiom*, respectively. See the pictures in Figure 2.1. Similarly, we refer to conditions (2.12b) as the *support-unitality axiom* and *type-unitality axiom*, respectively.

A map of Coxeter $\mathcal{A}$-species $f : \mathsf{p} \to \mathsf{q}$ consists of a family of linear maps

$$f_F : \mathsf{p}[F] \to \mathsf{q}[F],$$

one for each face F, such that the diagrams

$$
\text{(2.13)} \qquad
\begin{array}{ccc}
\mathsf{p}[F] & \xrightarrow{f_F} & \mathsf{q}[F] \\
{\scriptstyle \beta_{G,F}}\big\downarrow & & \big\downarrow{\scriptstyle \beta_{G,F}} \\
\mathsf{p}[G] & \xrightarrow[f_G]{} & \mathsf{q}[G]
\end{array}
\qquad\qquad
\begin{array}{ccc}
\mathsf{p}[F] & \xrightarrow{f_F} & \mathsf{q}[F] \\
{\scriptstyle w}\big\downarrow & & \big\downarrow{\scriptstyle w} \\
\mathsf{p}[F'] & \xrightarrow[f_{F'}]{} & \mathsf{q}[F']
\end{array}
$$

commute, the former whenever $\mathrm{s}(F) = \mathrm{s}(G)$ and latter whenever $w(F) = F'$.

A Coxeter $\mathcal{A}$-species p is *finite-dimensional* if its F-component $\mathsf{p}[F]$ has finite dimension for all faces F.

An important instance of the support-type axiom is given below.

Lemma 2.16. *Let F be a face and C a chamber both greater than A. Let F_0 be the face of C of the same type as F. Then for a Coxeter $\mathcal{A}$-species p, the diagram*

$$
\text{(2.14)} \qquad
\begin{array}{ccc}
\mathsf{p}[F] & \xrightarrow{\beta_{A\overline{F},F}} & \mathsf{p}[A\overline{F}] \\
{\scriptstyle w(FC)=C}\big\downarrow & & \big\downarrow{\scriptstyle w(A\overline{F}C)=A\overline{F_0}C} \\
\mathsf{p}[F_0] & \xrightarrow[\beta_{A\overline{F_0},F_0}]{} & \mathsf{p}[A\overline{F_0}]
\end{array}
$$

commutes.

Proof. The point to note is that the Coxeter symmetry w, which sends FC to C, sends $A\overline{F}C$ to $A\overline{F_0}C$. This is the content of Fact 1.12 applied to the star of A. □

Note: The direction of the vertical arrows in diagram (2.14) can be reversed by replacing w by w^{-1}. Similarly, the direction of the horizontal arrows can be reversed.

2.3.2. Coxeter species in terms of flats. Proposition 2.8 yields the following alternative description of a Coxeter species. It is in terms of flats instead of faces.

Proposition 2.17. *A Coxeter $\mathcal{A}$-species* p *is a family of vector spaces* $\mathsf{p}[\mathrm{X}]$, *one for each flat* X, *equipped with linear maps*

$$w : \mathsf{p}[\mathrm{X}] \to \mathsf{p}[\mathrm{X}']$$

whenever $w(\mathrm{X}) = \mathrm{X}'$, *such that the diagrams*

$$\begin{array}{ccc} & \mathsf{p}[\mathrm{X}'] & \\ {}^{u}\nearrow & & \searrow^{v} \\ \mathsf{p}[\mathrm{X}] & \xrightarrow[vu]{} & \mathsf{p}[\mathrm{X}''] \end{array} \qquad (\mathsf{p}[\mathrm{X}] \xrightarrow{e} \mathsf{p}[\mathrm{X}]) = \mathrm{id} \tag{2.15}$$

commute, the former whenever $u(\mathrm{X}) = \mathrm{X}'$ *and* $v(\mathrm{X}') = \mathrm{X}''$, *and the latter for every* X.

A map of Coxeter $\mathcal{A}$-species $f : \mathsf{p} \to \mathsf{q}$ *is a family of linear maps*

$$f_{\mathrm{X}} : \mathsf{p}[\mathrm{X}] \to \mathsf{q}[\mathrm{X}],$$

one for each flat X, *such that the diagram*

$$\begin{array}{ccc} \mathsf{p}[\mathrm{X}] & \xrightarrow{f_{\mathrm{X}}} & \mathsf{q}[\mathrm{X}] \\ {\scriptstyle w}\downarrow & & \downarrow{\scriptstyle w} \\ \mathsf{p}[\mathrm{X}'] & \xrightarrow[f_{\mathrm{X}'}]{} & \mathsf{q}[\mathrm{X}'] \end{array} \tag{2.16}$$

commutes whenever $w(\mathrm{X}) = \mathrm{X}'$.

We refer to the linear maps w as type morphisms of p, and to (2.15) as the *type-type axiom* and *type-unitality axiom*, respectively.

2.3.3. Relating faces and flats. Let us elaborate on the connection between the two viewpoints on Coxeter $\mathcal{A}$-species. Starting with a Coxeter $\mathcal{A}$-species p as in Proposition 2.17, set $\mathsf{p}[F] := \mathsf{p}[\mathrm{X}]$ for all F with support X, and $\beta_{G,F} = \mathrm{id}$ whenever F and G have the same support. This yields a Coxeter $\mathcal{A}$-species in the first sense.

Conversely, given a Coxeter $\mathcal{A}$-species in the first sense, set $\mathsf{p}[\mathrm{X}] := \mathsf{p}[H]$ where H is an arbitrary but fixed face of support X, and we get a Coxeter $\mathcal{A}$-species as in Proposition 2.17. The map $w : \mathsf{p}[\mathrm{X}] \to \mathsf{p}[w\mathrm{X}]$ is defined as the composite

$$\mathsf{p}[\mathrm{X}] = \mathsf{p}[H] \xrightarrow{w} \mathsf{p}[wH] \xrightarrow{\beta_{G,wH}} \mathsf{p}[G] = \mathsf{p}[w\mathrm{X}],$$

where H and G are the chosen faces for flats X and $w\mathrm{X}$, respectively. Note: For a face F with support X, we have inverse isomorphisms

$$\beta_{F,\mathrm{X}} : \mathsf{p}[\mathrm{X}] \to \mathsf{p}[F] \quad \text{and} \quad \beta_{\mathrm{X},F} : \mathsf{p}[F] \to \mathsf{p}[\mathrm{X}]. \tag{2.17}$$

Moreover, the diagrams

$$\begin{array}{ccc} \mathsf{p}[F] & \xrightarrow{\beta_{G,F}} & \mathsf{p}[G] \\ & {\scriptstyle \beta_{\mathrm{X},F}} \searrow \quad \swarrow {\scriptstyle \beta_{\mathrm{X},G}} & \\ & \mathsf{p}[\mathrm{X}] & \end{array} \qquad \begin{array}{ccc} \mathsf{p}[F] & \xrightarrow{w} & \mathsf{p}[wF] \\ {\scriptstyle \beta_{\mathrm{X},F}} \downarrow & & \downarrow {\scriptstyle \beta_{w\mathrm{X},wF}} \\ \mathsf{p}[\mathrm{X}] & \xrightarrow[w]{} & \mathsf{p}[w\mathrm{X}] \end{array} \tag{2.18}$$

commute. These two diagrams can be combined into one as follows. For a morphism $\sigma : F \to G$ in $\mathcal{A}$-Cox with normal form $\sigma = \beta w = w\beta$, the diagram

$$\begin{array}{ccc} \mathsf{p}[F] & \xrightarrow{\sigma} & \mathsf{p}[G] \\ {\scriptstyle \beta_{\mathrm{X},F}} \downarrow & & \downarrow {\scriptstyle \beta_{w\mathrm{X},G}} \\ \mathsf{p}[\mathrm{X}] & \xrightarrow[w]{} & \mathsf{p}[w\mathrm{X}] \end{array} \tag{2.19}$$

commutes.

Remark 2.18. In certain situations, one wants to use both the face and flat formulations of a Coxeter species at the same time. In such cases, we make use of maps (2.17) and diagrams (2.18) and (2.19).

Exercise 2.19 (W-modules). Check: For a Coxeter $\mathcal{A}$-species p,

- in terms of faces, the components $\mathsf{p}[O]$ and $\mathsf{p}[C]$ for any chamber C are left W-modules, and
- in terms of flats, the components $\mathsf{p}[\bot]$ and $\mathsf{p}[\top]$ are left W-modules.

This yields functors

$$\mathcal{A}\text{-}\mathsf{CSp} \to W\text{-}\mathsf{Mod}, \qquad \mathsf{p} \mapsto \mathsf{p}[\bot], \tag{2.20}$$

$$\mathcal{A}\text{-}\mathsf{CSp} \to W\text{-}\mathsf{Mod}, \qquad \mathsf{p} \mapsto \mathsf{p}[\top]. \tag{2.21}$$

(See Exercise 2.10.) This viewpoint is further developed in Proposition 2.20 below.

2.3.4. Coxeter species in terms of flat-types. We may use Proposition 2.9 to give yet another formulation of Coxeter species as follows.

Proposition 2.20. *A Coxeter $\mathcal{A}$-species p is a family of W_λ-modules, one for each flat-type λ. A map of Coxeter $\mathcal{A}$-species is a family of W_λ-module maps, one for each flat-type λ.*

2.3.5. Product and coproduct. The *zero Coxeter species* is the Coxeter $\mathcal{A}$-species with all components zero. It is both the initial and terminal object in the category of Coxeter $\mathcal{A}$-species.

For Coxeter $\mathcal{A}$-species p and q, define their *direct sum* $\mathsf{p}+\mathsf{q}$ by either of the two equivalent formulas

$$\begin{aligned} (\mathsf{p}+\mathsf{q})[F] &:= \mathsf{p}[F] \oplus \mathsf{q}[F], \\ (\mathsf{p}+\mathsf{q})[\mathrm{X}] &:= \mathsf{p}[\mathrm{X}] \oplus \mathsf{q}[\mathrm{X}]. \end{aligned} \tag{2.22}$$

The structure maps of $\mathsf{p}+\mathsf{q}$ are induced from those of p and q. Direct sum is both the product and coproduct in the category of Coxeter $\mathcal{A}$-species.

2.3.6. Coxeter subspecies and Coxeter quotient species. A $\Bbbk$-*linear category* is a category in which each hom-set carries a vector space structure over $\Bbbk$ and composition of morphisms is $\Bbbk$-bilinear. A functor between such categories is $\Bbbk$-*linear* if maps on hom-sets are $\Bbbk$-linear [**7**, Section 1.1.4].

The category of Coxeter species is $\Bbbk$-linear: If $f, g : \mathsf{p} \to \mathsf{q}$ are maps of Coxeter species, then so is $af + bg$ for any $a, b \in \Bbbk$ with

$$(af + bg)_F := af_F + bg_F, \qquad (af + bg)_{\mathrm{X}} := af_{\mathrm{X}} + bg_{\mathrm{X}}. \tag{2.23}$$

This can be seen from (2.13) and (2.16). We can also talk of Coxeter subspecies and Coxeter quotient species, and kernel, cokernel, image, coimage of a map f. We employ the standard notations $\ker(f)$, $\operatorname{coker}(f)$, $\operatorname{im}(f)$, $\operatorname{coim}(f)$. One can also talk of injectivity and surjectivity of f. These notions amount to working componentwise and employing corresponding notions for vector spaces on each component. The parallel discussion for species is in [**11**, Section 2.6].

2.3.7. (Co)invariants wrt automorphism group of type morphisms. Let p be a Coxeter $\mathcal{A}$-species. Then, for each face A, the component $\mathsf{p}[A]$ is a left W_A-module. This follows from the type-type and type-unitality axioms. We write

$$\mathsf{p}[A]^{W_A} \quad \text{and} \quad \mathsf{p}[A]_{W_A}$$

for the spaces of W_A-invariants and W_A-coinvariants, respectively. Further, for a morphism $\sigma : A \to B$, we have induced commutative diagrams

$$\begin{array}{ccc} \mathsf{p}[A]^{W_A} & \dashrightarrow & \mathsf{p}[B]^{W_B} \\ \downarrow & & \downarrow \\ \mathsf{p}[A] & \xrightarrow[\sigma]{} & \mathsf{p}[B] \end{array} \qquad\qquad \begin{array}{ccc} \mathsf{p}[A] & \xrightarrow{\sigma} & \mathsf{p}[B] \\ \downarrow & & \downarrow \\ \mathsf{p}[A]_{W_A} & \dashrightarrow & \mathsf{p}[B]_{W_B}. \end{array} \tag{2.24}$$

This is because the conjugate of a type morphism is a type morphism by Lemma 2.14.

Similarly, let $f : \mathsf{p} \to \mathsf{q}$ be a map of Coxeter $\mathcal{A}$-species. Then by the second diagram in (2.13), the component $f_A : \mathsf{p}[A] \to \mathsf{q}[A]$ is a map of left W_A-modules. This induces commutative diagrams

$$\begin{array}{ccc} \mathsf{p}[A]^{W_A} & \dashrightarrow & \mathsf{q}[A]^{W_A} \\ \downarrow & & \downarrow \\ \mathsf{p}[A] & \xrightarrow[f_A]{} & \mathsf{q}[A] \end{array} \qquad\qquad \begin{array}{ccc} \mathsf{p}[A] & \xrightarrow{f_A} & \mathsf{q}[A] \\ \downarrow & & \downarrow \\ \mathsf{p}[A]_{W_A} & \dashrightarrow & \mathsf{q}[A]_{W_A}. \end{array} \tag{2.25}$$

2.3.8. Duality functor on Coxeter species. For p a Coxeter $\mathcal{A}$-species, define its dual Coxeter $\mathcal{A}$-species p^* as follows. For a face F, let

$$\mathsf{p}^*[F] := \mathsf{p}[F]^*,$$

where $\mathsf{p}[F]^*$ denotes the dual of the vector space $\mathsf{p}[F]$. For a morphism $\sigma : A \to B$ in $\mathcal{A}$-Cox, we have an invertible linear map $\sigma : \mathsf{p}[A] \to \mathsf{p}[B]$. Dualize it and take its inverse to get a linear map $\mathsf{p}^*[A] \to \mathsf{p}^*[B]$, and define this to be the structure map of p^* corresponding to σ.

Moreover, a map $\mathsf{p} \to \mathsf{q}$ of Coxeter $\mathcal{A}$-species induces a map $\mathsf{q}^* \to \mathsf{p}^*$. This defines a functor

$$(-)^* : \mathcal{A}\text{-}\mathtt{CSp} \to \mathcal{A}\text{-}\mathtt{CSp}^{\mathrm{op}}. \tag{2.26}$$

This is the *duality functor* on Coxeter species.

Remark 2.21. The above construction works for any groupoid in place of $\mathcal{A}$-$\mathtt{Cox}$. For the groupoid $\mathcal{A}$-$\mathtt{Hyp}$ (defined in Section 2.7), it yields the duality functor on $\mathcal{A}$-species discussed in [**11**, Section 2.9]. Other examples are given later in Section 5.2.7 and Section 8.4.10.

2.3.9. Coxeter set-species. Let Set denote the category of sets. By replacing the codomain Vec by Set in the definition of Coxeter species, one obtains the notion of Coxeter set-species. In other words, a *Coxeter $\mathcal{A}$-set-species* is a functor

$$\mathrm{p} : \mathcal{A}\text{-}\mathtt{Cox} \to \mathsf{Set}.$$

A map of Coxeter $\mathcal{A}$-set-species $\mathrm{p} \to \mathrm{q}$ is a natural transformation. We denote the category of Coxeter $\mathcal{A}$-set-species by

$$\mathcal{A}\text{-}\mathtt{CSetSp} = [\mathcal{A}\text{-}\mathtt{Cox}, \mathsf{Set}].$$

Consider the functor

$$\Bbbk(-) : \mathsf{Set} \longrightarrow \mathsf{Vec},$$

which sends a set to the vector space with basis the given set. Composing a Coxeter set-species p with this functor yields a Coxeter species, which we denote $\Bbbk\mathrm{p}$. Thus, we have

$$\Bbbk(-) : \mathcal{A}\text{-}\mathtt{CSetSp} \longrightarrow \mathcal{A}\text{-}\mathtt{CSp}. \tag{2.27}$$

We call this the *linearization functor*.

2.3.10. Coxeter b-species and Coxeter lb-species. The notion of Coxeter species can be generalized by weakening the support-support axiom as follows. For a Coxeter lb-species, we only require the support-support axiom in (2.12a) whenever $F \text{--} G \text{--} H$. For a Coxeter b-species, we require in addition that each support morphism $\beta_{G,F}$ is a linear isomorphism. Observe: A Coxeter b-species is a Coxeter species if in addition $\beta_{G,F}^{-1} = \beta_{F,G}$.

A parallel more detailed discussion for $\mathcal{A}$-species is given in [**11**, Section 3.8]. Prefixes b and lb stand for braiding and lax braiding, respectively. No prefix (which is our usual setting) means the braiding is a symmetry.

2.4. Coxeter monoids and Coxeter comonoids

In this section and the next, we introduce Coxeter monoids, Coxeter comonoids, Coxeter bimonoids. Moreover, for a scalar q, we define Coxeter q-bimonoids, with $q = 1$ recovering Coxeter bimonoids. For $q = -1$, we use the term signed Coxeter bimonoids. We also pay attention to the case $q = 0$. A summary of the resulting categories along with their notations is given in Table 2.4. Here $\mathcal{A}$ denotes a fixed reflection arrangement. Many times, particularly in later chapters, we will drop $\mathcal{A}$ from the notation, and write $\mathsf{Mon}(\mathsf{CSp})$ and so on. This is consistent with Convention 2.1.

TABLE 2.4. Categories of Coxeter monoids, comonoids, bimonoids.

Category	Description
$\mathsf{Mon}(\mathcal{A}\text{-}\mathsf{CSp})$	Coxeter $\mathcal{A}$-monoids
$\mathsf{Comon}(\mathcal{A}\text{-}\mathsf{CSp})$	Coxeter $\mathcal{A}$-comonoids
$\mathsf{Bimon}(\mathcal{A}\text{-}\mathsf{CSp})$	Coxeter $\mathcal{A}$-bimonoids
$q\text{-}\mathsf{Bimon}(\mathcal{A}\text{-}\mathsf{CSp})$	Coxeter $\mathcal{A}$-q-bimonoids
$(-1)\text{-}\mathsf{Bimon}(\mathcal{A}\text{-}\mathsf{CSp})$	signed Coxeter $\mathcal{A}$-bimonoids
$0\text{-}\mathsf{Bimon}(\mathcal{A}\text{-}\mathsf{CSp})$	Coxeter $\mathcal{A}$-0-bimonoids

Coxeter (co, bi)monoids are structured objects in the category of Coxeter species. We formulate them in terms of faces, and also in terms of flats and lunes. They involve product and coproduct maps subject to a bunch of axioms, see the second, third, fourth entries in Table 2.1, and the first entry in Table 2.2.

For Coxeter monoids a and b, let $\mathsf{Mon}(\mathcal{A}\text{-}\mathsf{CSp})(\mathsf{a}, \mathsf{b})$ denote the space of all Coxeter monoid morphisms from a to b. Similar notations are employed for the remaining categories in Table 2.4.

2.4.1. Coxeter monoids in terms of faces. Recall Coxeter $\mathcal{A}$-species from Section 2.3.1. In the present context, we use the letter a instead of p. A *Coxeter $\mathcal{A}$-monoid* is a Coxeter $\mathcal{A}$-species a equipped with linear maps

$$\mu_A^F : \mathsf{a}[F] \to \mathsf{a}[A],$$

one for each $A \leq F$, such that the diagrams

$$\text{(2.28a)} \qquad \begin{array}{ccc} \mathsf{a}[F] & \xrightarrow{\beta_{BF,F}} & \mathsf{a}[BF] \\ {\scriptstyle \mu_A^F}\downarrow & & \downarrow{\scriptstyle \mu_B^{BF}} \\ \mathsf{a}[A] & \xrightarrow[\beta_{B,A}]{} & \mathsf{a}[B] \end{array} \qquad \begin{array}{ccc} \mathsf{a}[F] & \xrightarrow{w} & \mathsf{a}[F'] \\ {\scriptstyle \mu_A^F}\downarrow & & \downarrow{\scriptstyle \mu_{A'}^{F'}} \\ \mathsf{a}[A] & \xrightarrow[w]{} & \mathsf{a}[A'] \end{array}$$

$$\text{(2.28b)} \qquad \begin{array}{ccc} & \mathsf{a}[F] & \\ {\scriptstyle \mu_F^G}\nearrow & & \searrow{\scriptstyle \mu_A^F} \\ \mathsf{a}[G] & \xrightarrow[\mu_A^G]{} & \mathsf{a}[A] \end{array} \qquad (\mathsf{a}[A] \xrightarrow{\mu_A^A} \mathsf{a}[A]) = \mathrm{id}$$

commute.

- In the first diagram in (2.28a), $A \leq F$ and $\mathrm{s}(A) = \mathrm{s}(B)$, and in the second, $A \leq F$ and $w(A) = A'$, $w(F) = F'$.
- In the first diagram in (2.28b), $A \leq F \leq G$, and in the second, A is any face.

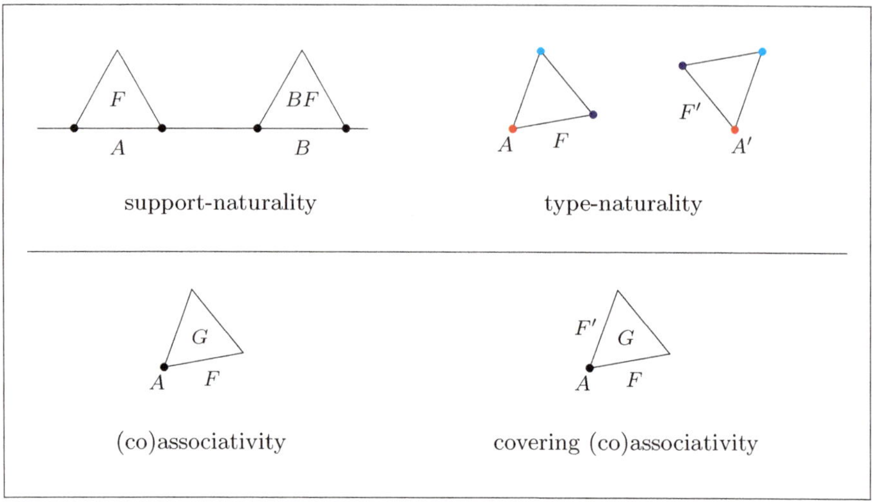

FIGURE 2.2. Naturality and (co)associativity axioms.

We denote a Coxeter $\mathcal{A}$-monoid by a pair (a, μ), or simply by a with μ understood. The maps μ_A^F are product components of a. We refer to diagrams (2.28a) as the *support-naturality axiom* and *type-naturality axiom*, respectively, and to diagrams (2.28b) as the *associativity axiom* and *unitality axiom*, respectively. The first two pictures in Figure 2.2 illustrate naturality; the third picture illustrates associativity. In the picture for support-naturality, the horizontal line indicates that A and B have the same support. Similarly, in the picture for type-naturality, the colors on the vertices indicate that A and A' have the same type, and F and F' have the same type.

A morphism $f : \mathsf{a} \to \mathsf{b}$ of Coxeter $\mathcal{A}$-monoids is a map of Coxeter $\mathcal{A}$-species such that for each $A \leq F$, the diagram

(2.29)
$$\begin{array}{ccc} \mathsf{a}[F] & \xrightarrow{f_F} & \mathsf{b}[F] \\ {\scriptstyle \mu_A^F}\downarrow & & \downarrow{\scriptstyle \mu_A^F} \\ \mathsf{a}[A] & \xrightarrow[f_A]{} & \mathsf{b}[A] \end{array}$$

commutes.

Observation 2.22. Support-naturality and type-naturality axioms for a Coxeter $\mathcal{A}$-monoid a can be combined into one *naturality axiom* as follows. For a morphism $\sigma : A \to B$ in $\mathcal{A}$-Cox, the diagram

(2.30)
$$\begin{array}{ccc} \mathsf{a}[F] & \xrightarrow{\sigma} & \mathsf{a}[G] \\ {\scriptstyle \mu_A^F}\downarrow & & \downarrow{\scriptstyle \mu_B^G} \\ \mathsf{a}[A] & \xrightarrow[\sigma]{} & \mathsf{a}[B] \end{array}$$

commutes for $G = \sigma_*(F)$, or equivalently, $F = \sigma^*(G)$, with notation as in (2.7).

- Specializing σ to β and w, respectively, in (2.30) yields (2.28a).

- Pasting diagrams in (2.28a) for β and w given by $\sigma = \beta w = w\beta$ yields (2.30).

2.4.2. Covering associativity axiom in terms of faces. We now discuss a slightly different formulation of a Coxeter monoid which involves fewer product components, namely, only those μ_A^F where F covers A.

Proposition 2.23. *A Coxeter $\mathcal{A}$-monoid is a Coxeter $\mathcal{A}$-species* a *equipped with linear maps*

$$\mu_A^F : \mathsf{a}[F] \to \mathsf{a}[A]$$

whenever F covers A, such that the diagrams

$$(2.31)\qquad \begin{array}{ccc} \mathsf{a}[F] & \xrightarrow{\beta_{BF,F}} & \mathsf{a}[BF] \\ {\scriptstyle \mu_A^F}\downarrow & & \downarrow{\scriptstyle \mu_B^{BF}} \\ \mathsf{a}[A] & \xrightarrow[\beta_{B,A}]{} & \mathsf{a}[B] \end{array} \qquad \begin{array}{ccc} \mathsf{a}[F] & \xrightarrow{w} & \mathsf{a}[F'] \\ {\scriptstyle \mu_A^F}\downarrow & & \downarrow{\scriptstyle \mu_{A'}^{F'}} \\ \mathsf{a}[A] & \xrightarrow[w]{} & \mathsf{a}[A'] \end{array} \qquad \begin{array}{ccc} \mathsf{a}[G] & \xrightarrow{\mu_{F'}^G} & \mathsf{a}[F'] \\ {\scriptstyle \mu_F^G}\downarrow & & \downarrow{\scriptstyle \mu_A^{F'}} \\ \mathsf{a}[F] & \xrightarrow[\mu_A^F]{} & \mathsf{a}[A] \end{array}$$

commute, where arrows labeled by μ go from a face to a face it covers. Explicitly, in the first diagram, F and BF cover A and B, respectively, in the second diagram, F and F' cover A and A', respectively, and in the last diagram, G covers F and F', and they in turn cover A.

A morphism $f : \mathsf{a} \to \mathsf{b}$ of Coxeter $\mathcal{A}$-monoids is a map of Coxeter $\mathcal{A}$-species such that diagram (2.29) *commutes whenever F covers A.*

PROOF. Recall that the poset of faces is strongly connected [**11**, Lemma 1.4, item (1)]. Hence, by using [**11**, Proposition 1.5], we replace the associativity axiom by the third diagram above. This is the critical change. The unitality axiom drops out. The naturality axioms stay as they are. □

The first two diagrams in (2.31) are instances of the support-naturality and type-naturality axioms (2.28a). They can be combined into one using Observation 2.22. They are all squares. We refer to the third diagram in (2.31) as the *covering associativity axiom.* It is a square in contrast to the associativity axiom, the first diagram in (2.28b), which is a triangle. Also contrast the last two pictures in Figure 2.2.

2.4.3. Coxeter monoids in terms of flats and lunes. We now formulate Coxeter monoids in terms of flats and lunes by building on the description of Coxeter species in Proposition 2.17. It makes use of the composite $\mathrm{L} \circ \mathrm{M}$ of lunes L and M in the category of lunes (Section 1.2.2).

Proposition 2.24. *A Coxeter $\mathcal{A}$-monoid is a Coxeter $\mathcal{A}$-species* a *equipped with linear maps*

$$\mu_{\mathrm{L}} : \mathsf{a}[\mathrm{X}] \to \mathsf{a}[\mathrm{Z}],$$

indexed by lunes L *with base* Z *and case* X, *such that the diagrams*

$$(2.32)\qquad \begin{array}{ccc} \mathsf{a}[\mathrm{X}] & \xrightarrow{w} & \mathsf{a}[\mathrm{X}'] \\ {\scriptstyle \mu_{\mathrm{L}}}\downarrow & & \downarrow{\scriptstyle \mu_{\mathrm{L}'}} \\ \mathsf{a}[\mathrm{Y}] & \xrightarrow[w]{} & \mathsf{a}[\mathrm{Y}'] \end{array} \qquad \begin{array}{ccccc} & & \mathsf{a}[\mathrm{Y}] & & \\ & {\scriptstyle \mu_{\mathrm{M}}}\nearrow & & \searrow{\scriptstyle \mu_{\mathrm{L}}} & \\ \mathsf{a}[\mathrm{X}] & & \xrightarrow[\mu_{\mathrm{L}\circ\mathrm{M}}]{} & & \mathsf{a}[\mathrm{Z}] \end{array} \qquad (\mathsf{a}[\mathrm{Z}] \xrightarrow{\mu_{\mathrm{Z}}} \mathsf{a}[\mathrm{Z}]) = \mathrm{id}$$

commute. In the first diagram, L *has base* Y *and case* X*, and* $w(\mathrm{X}) = \mathrm{X}'$, $w(\mathrm{Y}) = \mathrm{Y}'$, $w(\mathrm{L}) = \mathrm{L}'$*. In the second diagram,* L *has base* Z *and case* Y*, while* M *has base* Y *and case* X*. In the third diagram,* Z *is any flat.*

A morphism of Coxeter $\mathcal{A}$*-monoids* $f : \mathsf{a} \to \mathsf{b}$ *is a map of Coxeter* $\mathcal{A}$*-species such that the diagram*

$$\begin{array}{ccc} \mathsf{a}[\mathrm{X}] & \xrightarrow{f_{\mathrm{X}}} & \mathsf{b}[\mathrm{X}] \\ {\scriptstyle \mu_{\mathrm{L}}}\downarrow & & \downarrow{\scriptstyle \mu_{\mathrm{L}}} \\ \mathsf{a}[\mathrm{Z}] & \xrightarrow[f_{\mathrm{Z}}]{} & \mathsf{b}[\mathrm{Z}] \end{array} \tag{2.33}$$

commutes for all lunes L *with base* Z *and case* X*.*

PROOF. Product components μ^F_A and μ_{L} in the two formulations of a Coxeter $\mathcal{A}$-monoid a are linked by the diagram

$$\begin{array}{ccc} \mathsf{a}[F] & \xrightarrow{\beta_{\mathrm{X},F}} & \mathsf{a}[\mathrm{X}] \\ {\scriptstyle \mu^F_A}\downarrow & & \downarrow{\scriptstyle \mu_{\mathrm{L}}} \\ \mathsf{a}[A] & \xrightarrow[\beta_{\mathrm{Z},A}]{} & \mathsf{a}[\mathrm{Z}], \end{array} \tag{2.34}$$

where $\mathrm{X} = \mathrm{s}(F)$, $\mathrm{Z} = \mathrm{s}(A)$ and $\mathrm{L} = \mathrm{s}(A, F)$. The vertical maps define each other. If we start with the μ^F_A, then the μ_{L} are well-defined due to the support-naturality axiom. Conversely, if we start with the μ_{L}, then the μ^F_A satisfy the support-naturality axiom. Also, there is a clear correspondence between the remaining axioms. The checks make use of diagrams (2.18). □

We refer to diagrams (2.32) as the *type-naturality axiom*, *associativity axiom*, *unitality axiom*, respectively.

2.4.4. Covering associativity axiom in terms of flats and lunes. An alternative formulation using fewer product components, namely, only those μ_{L} with L of slack 1 is given below.

Proposition 2.25. *A Coxeter* $\mathcal{A}$*-monoid is a Coxeter* $\mathcal{A}$*-species* a *equipped with linear maps*

$$\mu_{\mathrm{L}} : \mathsf{a}[\mathrm{X}] \to \mathsf{a}[\mathrm{Z}],$$

indexed by lunes L *of slack* 1 *with base* Z *and case* X*, such that the diagrams*

$$\begin{array}{ccc} \mathsf{a}[\mathrm{X}] & \xrightarrow{w} & \mathsf{a}[\mathrm{X}'] \\ {\scriptstyle \mu_{\mathrm{L}}}\downarrow & & \downarrow{\scriptstyle \mu_{\mathrm{L}'}} \\ \mathsf{a}[\mathrm{Y}] & \xrightarrow[w]{} & \mathsf{a}[\mathrm{Y}'] \end{array} \qquad \begin{array}{ccc} \mathsf{a}[\mathrm{X}] & \xrightarrow{\mu_{\mathrm{M}'}} & \mathsf{a}[\mathrm{Y}'] \\ {\scriptstyle \mu_{\mathrm{M}}}\downarrow & & \downarrow{\scriptstyle \mu_{\mathrm{L}'}} \\ \mathsf{a}[\mathrm{Y}] & \xrightarrow[\mu_{\mathrm{L}}]{} & \mathsf{a}[\mathrm{Z}] \end{array} \tag{2.35}$$

commute. In the first diagram, L *and* L′ *have slack* 1*, while in the second diagram,* L, M, L′, M′ *have slack* 1 *and* $\mathrm{L} \circ \mathrm{M} = \mathrm{L}' \circ \mathrm{M}'$*. (Note: The latter has slack* 2*.)*

A morphism $f : \mathsf{a} \to \mathsf{b}$ *of Coxeter* $\mathcal{A}$*-monoids is a map of Coxeter* $\mathcal{A}$*-species such that diagram* (2.33) *commutes whenever* L *has slack* 1*.*

PROOF. This follows by reformulating the description in Proposition 2.23 in terms of flats and lunes (as in the previous proof). A cover relation of faces translates to a lune of slack 1.

Alternatively, one can deduce this result from Proposition 2.24 by using the presentation of the category of lunes [**10**, Proposition 4.42]. We recall that the latter is obtained by using that the poset of faces is strongly connected. □

The first diagram in (2.35) is an instance of the type-naturality axiom. We refer to the second diagram in (2.35) as the *covering associativity axiom*. It is a square instead of a triangle.

2.4.5. Coxeter comonoids in terms of faces. We now turn to the dual notion. A *Coxeter $\mathcal{A}$-comonoid* is a Coxeter $\mathcal{A}$-species c equipped with linear maps

$$\Delta_A^F : \mathsf{c}[A] \to \mathsf{c}[F],$$

one for each $A \leq F$, such that the diagrams

(2.36a)
$$\begin{array}{ccc} \mathsf{c}[F] & \xrightarrow{\beta_{BF,F}} & \mathsf{c}[BF] \\ {\scriptstyle \Delta_A^F}\uparrow & & \uparrow{\scriptstyle \Delta_B^{BF}} \\ \mathsf{c}[A] & \xrightarrow[\beta_{B,A}]{} & \mathsf{c}[B] \end{array} \qquad \begin{array}{ccc} \mathsf{c}[F] & \xrightarrow{w} & \mathsf{c}[F'] \\ {\scriptstyle \Delta_A^F}\uparrow & & \uparrow{\scriptstyle \Delta_{A'}^{F'}} \\ \mathsf{c}[A] & \xrightarrow[w]{} & \mathsf{c}[A'] \end{array}$$

(2.36b)
$$\begin{array}{ccccc} & & \mathsf{c}[F] & & \\ & {\scriptstyle \Delta_A^F}\nearrow & & \searrow{\scriptstyle \Delta_F^G} & \\ \mathsf{c}[A] & & \xrightarrow[\Delta_A^G]{} & & \mathsf{c}[G] \end{array} \qquad (\mathsf{c}[A] \xrightarrow{\Delta_A^A} \mathsf{c}[A]) = \mathrm{id}$$

commute. (The conventions are as in the Coxeter monoid case.)

We denote a Coxeter $\mathcal{A}$-comonoid by a pair (c, Δ), or simply by c with Δ understood. The maps Δ_A^F are coproduct components of c. We refer to diagrams (2.36a) as the *support-naturality axiom* and *type-naturality axiom*, respectively, and to diagrams (2.36b) as the *coassociativity axiom* and *counitality axiom*, respectively.

A morphism $f : \mathsf{c} \to \mathsf{d}$ of Coxeter $\mathcal{A}$-comonoids is a map of Coxeter $\mathcal{A}$-species such that for each $A \leq F$, the diagram

(2.37)
$$\begin{array}{ccc} \mathsf{c}[F] & \xrightarrow{f_F} & \mathsf{d}[F] \\ {\scriptstyle \Delta_A^F}\uparrow & & \uparrow{\scriptstyle \Delta_A^F} \\ \mathsf{c}[A] & \xrightarrow[f_A]{} & \mathsf{d}[A] \end{array}$$

commutes.

2.4.6. Covering coassociativity axiom in terms of faces. The dual of Proposition 2.23 is stated below.

Proposition 2.26. *A Coxeter $\mathcal{A}$-comonoid is a Coxeter $\mathcal{A}$-species* c *equipped with linear maps*

$$\Delta_A^F : \mathsf{c}[A] \to \mathsf{c}[F]$$

whenever F covers A, such that the diagrams

$$
\begin{array}{ccc}
\mathsf{c}[F] & \xrightarrow{\beta_{BF,F}} & \mathsf{c}[BF] \\
{\scriptstyle \Delta_A^F}\uparrow & & \uparrow{\scriptstyle \Delta_B^{BF}} \\
\mathsf{c}[A] & \xrightarrow[\beta_{B,A}]{} & \mathsf{c}[B]
\end{array}
\qquad
\begin{array}{ccc}
\mathsf{c}[F] & \xrightarrow{w} & \mathsf{c}[F'] \\
{\scriptstyle \Delta_A^F}\uparrow & & \uparrow{\scriptstyle \Delta_{A'}^{F'}} \\
\mathsf{c}[A] & \xrightarrow[w]{} & \mathsf{c}[A']
\end{array}
\qquad
\begin{array}{ccc}
\mathsf{c}[F'] & \xrightarrow{\Delta_{F'}^G} & \mathsf{c}[G] \\
{\scriptstyle \Delta_A^{F'}}\uparrow & & \uparrow{\scriptstyle \Delta_F^G} \\
\mathsf{c}[A] & \xrightarrow[\Delta_A^F]{} & \mathsf{c}[F]
\end{array}
\tag{2.38}
$$

commute. (The conventions are as in the Coxeter monoid case.)

A morphism $f : \mathsf{c} \to \mathsf{d}$ of Coxeter $\mathcal{A}$-comonoids is a map of Coxeter $\mathcal{A}$-species such that diagram (2.37) *commutes whenever F covers A.*

The first two diagrams in (2.38) are instances of the support-naturality and type-naturality axioms (2.36a). We refer to the third diagram in (2.38) as the *covering coassociativity axiom*. It is a square in contrast to the coassociativity axiom, the first diagram in (2.36b), which is a triangle.

2.4.7. Coxeter comonoids in terms of flats and lunes. One may also formulate Coxeter comonoids in terms of flats and lunes along the lines of Propositions 2.24 and 2.25 using linear maps

$$\Delta_{\mathrm{L}} : \mathsf{c}[\mathrm{Z}] \to \mathsf{c}[\mathrm{X}]$$

(with lunes L of slack 1 for the latter result).

2.5. Coxeter bimonoids

We build on the discussion in Section 2.4, combining notions of a Coxeter monoid and a Coxeter comonoid to obtain the notion of a Coxeter bimonoid.

2.5.1. Coxeter bimonoids in terms of faces. A *Coxeter $\mathcal{A}$-bimonoid* is a triple $(\mathsf{h}, \mu, \Delta)$, where h is a Coxeter $\mathcal{A}$-species, (h, μ) is a Coxeter $\mathcal{A}$-monoid, (h, Δ) is a Coxeter $\mathcal{A}$-comonoid, and for faces $A \leq F$ and $A \leq G$, the diagram

$$
\begin{array}{ccccc}
\mathsf{h}[F] & \xrightarrow{\mu_A^F} & \mathsf{h}[A] & \xrightarrow{\Delta_A^G} & \mathsf{h}[G] \\
{\scriptstyle \Delta_F^{FG}}\downarrow & & & & \uparrow{\scriptstyle \mu_G^{GF}} \\
\mathsf{h}[FG] & & \xrightarrow[\beta_{GF,FG}]{} & & \mathsf{h}[GF]
\end{array}
\tag{2.39}
$$

commutes.

We refer to diagram (2.39) as the *bimonoid axiom*. See the first picture in Figure 2.3. As a shorthand, we write the bimonoid axiom as $\Delta\mu = \mu\beta\Delta$. Note very carefully that it does not involve type morphisms.

A morphism of Coxeter $\mathcal{A}$-bimonoids is a map of the underlying Coxeter $\mathcal{A}$-species which is a morphism of the underlying Coxeter $\mathcal{A}$-monoids and Coxeter $\mathcal{A}$-comonoids.

The data present in a Coxeter $\mathcal{A}$-bimonoid is shown in Summary 2.1.

Lemma 2.27. *For defining a Coxeter $\mathcal{A}$-bimonoid, it suffices to require the bimonoid axiom* (2.39) *only for faces $A \lessdot F$ and $A \lessdot G$.*

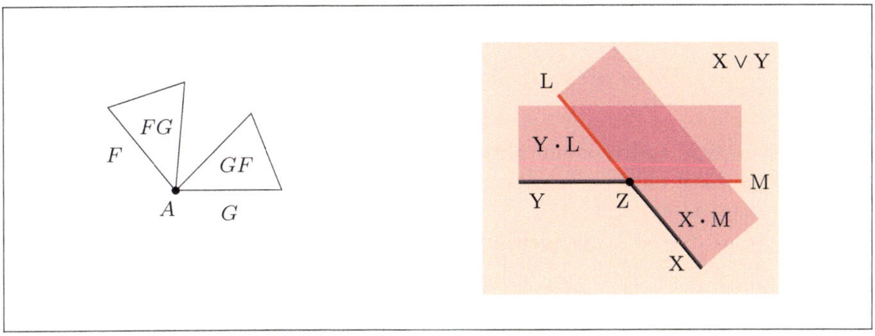

FIGURE 2.3. Bimonoid axiom.

A Coxeter $\mathcal{A}$-bimonoid h consists of linear maps

$$\beta_{G,F} : \mathsf{h}[F] \to \mathsf{h}[G], \qquad w : \mathsf{h}[F] \to \mathsf{h}[wF],$$

$$\mu_A^F : \mathsf{h}[F] \to \mathsf{h}[A], \qquad \Delta_A^F : \mathsf{h}[A] \to \mathsf{h}[F],$$

which satisfy

- support-support, support-type, type-type, support-unitality, type-unitality axioms (2.12a) + (2.12b),
- support-naturality and type-naturality axioms (2.28a) + (2.36a),
- (co)associativity and (co)unitality axioms (2.28b) + (2.36b),
- bimonoid axiom (2.39).

There are a total of 14 axioms.

SUMMARY 2.1. Coxeter bimonoid in terms of faces.

This relates to Propositions 2.23 and 2.26. However, note very carefully: $A \lessdot F$ and $A \lessdot G$ does not imply $F \lessdot FG$ or $G \lessdot GF$ in general.

PROOF. We repeatedly use

(2.40)

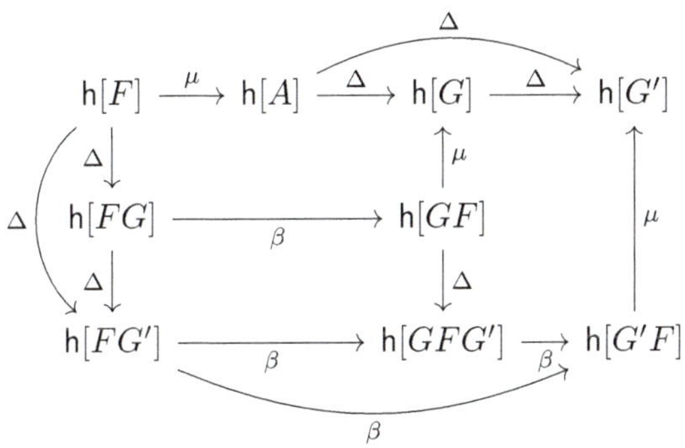

and its dual diagram. The subscripts and superscripts on μ, Δ, β have been omitted for convenience. □

A Coxeter $\mathcal{A}$-bimonoid h consists of linear maps

$$\mu_{\mathrm{L}} : \mathsf{h}[\mathrm{X}] \to \mathsf{h}[\mathrm{Z}], \quad w : \mathsf{h}[\mathrm{X}] \to \mathsf{h}[w\mathrm{X}], \quad \Delta_{\mathrm{L}} : \mathsf{h}[\mathrm{Z}] \to \mathsf{h}[\mathrm{X}],$$

which satisfy

- type-type and type-unitality axioms (2.15),
- type-naturality, (co)associativity, (co)unitality axioms (2.32) + its dual,
- bimonoid axiom (2.41).

There are a total of 9 axioms.

SUMMARY 2.2. Coxeter bimonoid in terms of flats and lunes.

2.5.2. Coxeter bimonoids in terms of flats and lunes. We now formulate Coxeter bimonoids in terms of flats and lunes by building on the description of Coxeter species in Proposition 2.17, Coxeter monoids in Proposition 2.24, and the dual description for Coxeter comonoids.

Proposition 2.28. *A Coxeter $\mathcal{A}$-bimonoid is the same as a triple $(\mathsf{h}, \mu, \Delta)$, where h is a Coxeter $\mathcal{A}$-species, (h, μ) is a Coxeter $\mathcal{A}$-monoid, (h, Δ) is a Coxeter $\mathcal{A}$-comonoid, and such that for flats $\mathrm{Z} \leq \mathrm{X}$, $\mathrm{Z} \leq \mathrm{Y}$, lune L with base Z and case X, and lune M with base Z and case Y, the diagram*

$$\begin{array}{ccc} \mathsf{h}[\mathrm{X}] & \xrightarrow{\mu_{\mathrm{L}}} & \mathsf{h}[\mathrm{Z}] \\ {\scriptstyle \Delta_{\mathrm{X}\cdot\mathrm{M}}}\big\downarrow & & \big\downarrow{\scriptstyle \Delta_{\mathrm{M}}} \\ \mathsf{h}[\mathrm{X} \vee \mathrm{Y}] & \xrightarrow[\mu_{\mathrm{Y}\cdot\mathrm{L}}]{} & \mathsf{h}[\mathrm{Y}] \end{array} \tag{2.41}$$

commutes.

PROOF. We translate between diagrams (2.39) and (2.41) using diagrams (2.18), (2.34) and its dual. □

We refer to diagram (2.41) as the *bimonoid axiom*. It is a square in contrast to (2.39) which is a pentagon. Also compare the two pictures in Figure 2.3.

The data present in a Coxeter $\mathcal{A}$-bimonoid is shown in Summary 2.2. Compare and contrast with Summary 2.1.

Exercise 2.29. Use Lemma 2.27 to deduce: For defining a Coxeter $\mathcal{A}$-bimonoid, it suffices to require the bimonoid axiom (2.41) only for lunes L and M both of slack 1.

2.5.3. Coxeter q-bimonoids in terms of faces. We now turn to deformed Coxeter bimonoids. They are obtained by deforming the bimonoid axiom (2.39) using the q-distance function.

For a scalar q, a *Coxeter $\mathcal{A}$-q-bimonoid* is a triple $(\mathsf{h}, \mu, \Delta)$, where h is a Coxeter $\mathcal{A}$-species, (h, μ) is a Coxeter $\mathcal{A}$-monoid, (h, Δ) is a Coxeter $\mathcal{A}$-comonoid, and for faces $A \leq F$ and $A \leq G$, the diagram

$$\begin{array}{ccccc} \mathsf{h}[F] & \xrightarrow{\mu_A^F} & \mathsf{h}[A] & \xrightarrow{\Delta_A^G} & \mathsf{h}[G] \\ {\scriptstyle \Delta_F^{FG}}\big\downarrow & & & & \big\uparrow{\scriptstyle \mu_G^{GF}} \\ \mathsf{h}[FG] & & \xrightarrow[(\beta_q)_{GF,FG}]{} & & \mathsf{h}[GF] \end{array} \tag{2.42}$$

commutes, where

$$(\beta_q)_{GF,FG} := q^{\operatorname{dist}(GF,FG)} \beta_{GF,FG}. \tag{2.43}$$

We refer to diagram (2.42) as the *q-bimonoid axiom*. As a shorthand, we write it as $\Delta\mu = \mu\beta_q\Delta$.

When $q = 1$, we recover the notion of Coxeter $\mathcal{A}$-bimonoid. For $q = -1$, we get a different notion which we call a *signed Coxeter $\mathcal{A}$-bimonoid.* Similarly, for the (-1)-bimonoid axiom, we employ the term *signed bimonoid axiom.*

A morphism of Coxeter $\mathcal{A}$-q-bimonoids is a map of the underlying Coxeter $\mathcal{A}$-species which is a morphism of the underlying Coxeter $\mathcal{A}$-monoids and Coxeter $\mathcal{A}$-comonoids. Note: q plays no role here.

Exercise 2.30. Show: For defining a Coxeter $\mathcal{A}$-q-bimonoid, it suffices to require the q-bimonoid axiom (2.42) only for faces $A \lessdot F$ and $A \lessdot G$. This generalizes Lemma 2.27. (Repeatedly use diagram (2.40) and its dual with β_q instead of β. The bottom-horizontal composite is again β_q by the gate property.

An illustration is shown above with $A = O$.)

Example 2.31 (Rank zero). Let $\mathcal{A}$ be a rank-zero reflection arrangement. The category of Coxeter $\mathcal{A}$-species is isomorphic to the category of vector spaces. Explicitly, a Coxeter $\mathcal{A}$-species p corresponds to the vector space $\mathsf{p}[O]$. (Recall: O is the only face in $\mathcal{A}$.) In fact, in a similar manner, the categories of Coxeter $\mathcal{A}$-monoids, Coxeter $\mathcal{A}$-comonoids, Coxeter $\mathcal{A}$-q-bimonoids are all isomorphic to the category of vector spaces. The scalar q plays no role. □

2.5.4. Coxeter 0-bimonoids in terms of faces. Let us now specialize to $q = 0$. A Coxeter $\mathcal{A}$-0-bimonoid is a triple $(\mathsf{h}, \mu, \Delta)$, where h is a Coxeter $\mathcal{A}$-species, (h, μ) is a Coxeter $\mathcal{A}$-monoid, (h, Δ) is a Coxeter $\mathcal{A}$-comonoid such that for faces $A \leq F$ and $A \leq G$, the first diagram below commutes if $FG = GF$, and the second diagram below commutes if $FG \neq GF$.

$$\begin{array}{ccc} \mathsf{h}[F] & \xrightarrow{\mu_A^F} & \mathsf{h}[A] \\ {\scriptstyle \Delta_F^{FG}}\big\downarrow & & \big\downarrow{\scriptstyle \Delta_A^G} \\ \mathsf{h}[FG] = \mathsf{h}[GF] & \xrightarrow[\mu_G^{GF}]{} & \mathsf{h}[G] \end{array} \qquad \begin{array}{ccccc} & & \mathsf{h}[A] & & \\ & {\scriptstyle \mu_A^F}\nearrow & & \searrow{\scriptstyle \Delta_A^G} & \\ \mathsf{h}[F] & & \xrightarrow[0]{} & & \mathsf{h}[G] \end{array} \tag{2.44}$$

We refer to diagrams (2.44) as the 0-*bimonoid axiom*. It is the case $q = 0$ of the q-bimonoid axiom.

2.5.5. Coxeter q-bimonoids in terms of flats and lunes. For a scalar q, one can formulate a Coxeter $\mathcal{A}$-q-bimonoid in terms of flats and lunes. The description is as in Proposition 2.28 with the bimonoid axiom (2.41) replaced by the diagram

$$\begin{array}{ccc} \mathsf{h}[\mathrm{X}] & \xrightarrow{\mu_{\mathrm{L}}} & \mathsf{h}[\mathrm{Z}] \\ {\scriptstyle q^{\operatorname{dist}(\mathrm{L},\mathrm{M})}\Delta_{\mathrm{X}\cdot\mathrm{M}}}\downarrow & & \downarrow{\scriptstyle \Delta_{\mathrm{M}}} \\ \mathsf{h}[\mathrm{X}\vee\mathrm{Y}] & \xrightarrow[\mu_{\mathrm{Y}\cdot\mathrm{L}}]{} & \mathsf{h}[\mathrm{Y}], \end{array} \tag{2.45}$$

where $\operatorname{dist}(\mathrm{L},\mathrm{M})$ is the distance between lunes L and M, with the lunes viewed as faces in the arrangement over Z.

The proof is similar; we now translate between diagrams (2.42) and (2.45). We refer to diagram (2.45) as the *q-bimonoid axiom*.

The data present in a Coxeter $\mathcal{A}$-q-bimonoid (in terms of faces, and in terms of flats and lunes) can be shown as in Summaries 2.1 and 2.2 with the bimonoid axiom replaced by the q-bimonoid axiom.

2.5.6. Coxeter 0-bimonoids in terms of flats and lunes. For $q = 0$, the q-bimonoid axiom (2.45) simplifies as follows.

For a lune L with base Z and case X, and lune M with base Z and case Y, the first diagram below commutes if $\mathrm{L}\circ(\mathrm{X}\cdot\mathrm{M}) = \mathrm{M}\circ(\mathrm{Y}\cdot\mathrm{L})$, and the second diagram below commutes if not.

$$\begin{array}{ccc} \mathsf{h}[\mathrm{X}] & \xrightarrow{\mu_{\mathrm{L}}} & \mathsf{h}[\mathrm{Z}] \\ {\scriptstyle \Delta_{\mathrm{X}\cdot\mathrm{M}}}\downarrow & & \downarrow{\scriptstyle \Delta_{\mathrm{M}}} \\ \mathsf{h}[\mathrm{X}\vee\mathrm{Y}] & \xrightarrow[\mu_{\mathrm{Y}\cdot\mathrm{L}}]{} & \mathsf{h}[\mathrm{Y}] \end{array} \qquad \begin{array}{ccccc} & & \mathsf{h}[\mathrm{Z}] & & \\ & {\scriptstyle \mu_{\mathrm{L}}}\nearrow & & \searrow{\scriptstyle \Delta_{\mathrm{M}}} & \\ \mathsf{h}[\mathrm{X}] & & \xrightarrow[0]{} & & \mathsf{h}[\mathrm{Y}] \end{array} \tag{2.46}$$

We refer to diagrams (2.46) as the 0-*bimonoid axiom*. Compare with (2.44).

2.5.7. Coxeter υ-bimonoids. Let υ be any W-invariant distance function on $\mathcal{A}$ as in Section 1.4.4. For a Coxeter $\mathcal{A}$-species p, let β_υ denote the family of linear maps

$$(\beta_\upsilon)_{G,F} : \mathsf{p}[F] \to \mathsf{p}[G], \qquad (\beta_\upsilon)_{G,F} := \upsilon_{F,G}\beta_{G,F} \tag{2.47}$$

whenever $\mathrm{s}(F) = \mathrm{s}(G)$. It does not matter whether we write $\upsilon_{F,G}$ or $\upsilon_{G,F}$ since υ is symmetric.

A *Coxeter $\mathcal{A}$-υ-bimonoid* is a triple (h,μ,Δ), where h is a Coxeter $\mathcal{A}$-species, (h,μ) is a Coxeter $\mathcal{A}$-monoid, (h,Δ) is a Coxeter $\mathcal{A}$-comonoid, and for faces $A \le F$ and $A \le G$, the diagram

$$\begin{array}{ccccc} \mathsf{h}[F] & \xrightarrow{\mu_A^F} & \mathsf{h}[A] & \xrightarrow{\Delta_A^G} & \mathsf{h}[G] \\ {\scriptstyle \Delta_F^{FG}}\downarrow & & & & \uparrow{\scriptstyle \mu_G^{GF}} \\ \mathsf{h}[FG] & & \xrightarrow[(\beta_\upsilon)_{GF,FG}]{} & & \mathsf{h}[GF] \end{array} \tag{2.48}$$

commutes. We refer to diagram (2.48) as the *v-bimonoid axiom*. It may also be formulated in terms of flats and lunes.

By letting v be the q-distance function, one recovers the notion of Coxeter $\mathcal{A}$-q-bimonoid.

2.5.8. Coxeter submonoids and Coxeter quotient monoids. A Coxeter submonoid of a Coxeter monoid a is a Coxeter subspecies of a to which product components of a restrict. Similarly, one defines a Coxeter quotient monoid of a. We employ similar terminology for Coxeter comonoids and Coxeter q-bimonoids.

The zero Coxeter monoid is the Coxeter monoid with all components zero. Its product components are all necessarily zero. Similarly, we have the zero Coxeter comonoid and zero Coxeter q-bimonoid.

Exercise 2.32. Check: Direct sum (2.22) preserves Coxeter monoids, Coxeter comonoids, and Coxeter q-bimonoids. Explicitly, if h and k are Coxeter q-bimonoids, then so is $\mathsf{h} + \mathsf{k}$ with product and coproduct induced from those of h and k.

Exercise 2.33. Check: If $f, g : \mathsf{a} \to \mathsf{b}$ are morphisms of Coxeter monoids, then so is $af + bg$ as defined in (2.23). (Use (2.29) or (2.33).) The same holds for morphisms of Coxeter comonoids and Coxeter q-bimonoids. So these are all $\Bbbk$-linear categories. For related ideas, see Section 2.9.

2.5.9. Duality. Recall from Section 2.3.8 the duality functor on Coxeter species. The dual of a Coxeter monoid is a Coxeter comonoid, and vice versa obtained by dualizing (co)product components. The dual of a Coxeter bimonoid is a Coxeter bimonoid. More generally, for a scalar q, the dual of a Coxeter q-bimonoid is a Coxeter q-bimonoid.

A Coxeter q-bimonoid h is *self-dual* if $\mathsf{h} \cong \mathsf{h}^*$. Similarly, a morphism of Coxeter q-bimonoids $f : \mathsf{h} \to \mathsf{h}^*$ is *self-dual* if $f = f^*$. (It is implicit here that h is finite-dimensional.)

2.6. (Co)commutative Coxeter (co)monoids

There is a notion of commutativity for Coxeter monoids and cocommutativity for Coxeter comonoids. A Coxeter bimonoid may be commutative, cocommutative, both, or neither. There is also a notion of signed commutativity for Coxeter monoids, signed cocommutativity for Coxeter comonoids, and signed (co)commutativity for signed Coxeter bimonoids. A summary of the resulting categories along with their notations is given in Tables 2.5 and 2.6. Here $\mathcal{A}$ denotes a fixed reflection arrangement.

We define commutativity using the commutativity axiom. We also give variants called the opposition commutativity axiom and covering commutativity axiom. We formulate them in terms of faces, and also in terms of flats and lunes. See the last two entries in Table 2.1 and the second entry in Table 2.2.

2.6.1. Commutative Coxeter monoids in terms of faces. Recall Coxeter $\mathcal{A}$-monoids from Section 2.4.1. A Coxeter $\mathcal{A}$-monoid a is *commutative* if

TABLE 2.5. Categories of (co)commutative Coxeter (co)monoids.

Category	Description
$\mathsf{Mon}^{\mathrm{co}}(\mathcal{A}\text{-}\mathsf{CSp})$	com. Coxeter $\mathcal{A}$-monoids
${}^{\mathrm{co}}\mathsf{Comon}(\mathcal{A}\text{-}\mathsf{CSp})$	cocom. Coxeter $\mathcal{A}$-comonoids
$\mathsf{Bimon}^{\mathrm{co}}(\mathcal{A}\text{-}\mathsf{CSp})$	com. Coxeter $\mathcal{A}$-bimonoids
${}^{\mathrm{co}}\mathsf{Bimon}(\mathcal{A}\text{-}\mathsf{CSp})$	cocom. Coxeter $\mathcal{A}$-bimonoids
${}^{\mathrm{co}}\mathsf{Bimon}^{\mathrm{co}}(\mathcal{A}\text{-}\mathsf{CSp})$	bicom. Coxeter $\mathcal{A}$-bimonoids

TABLE 2.6. Categories of signed (co)commutative Coxeter (co)monoids.

Category	Description
$(-1)\text{-}\mathsf{Mon}^{\mathrm{co}}(\mathcal{A}\text{-}\mathsf{CSp})$	signed com. Coxeter $\mathcal{A}$-monoids
$(-1)\text{-}{}^{\mathrm{co}}\mathsf{Comon}(\mathcal{A}\text{-}\mathsf{CSp})$	signed cocom. Coxeter $\mathcal{A}$-comonoids
$(-1)\text{-}\mathsf{Bimon}^{\mathrm{co}}(\mathcal{A}\text{-}\mathsf{CSp})$	signed com. signed Coxeter $\mathcal{A}$-bimonoids
$(-1)\text{-}{}^{\mathrm{co}}\mathsf{Bimon}(\mathcal{A}\text{-}\mathsf{CSp})$	signed cocom. signed Coxeter $\mathcal{A}$-bimonoids
$(-1)\text{-}{}^{\mathrm{co}}\mathsf{Bimon}^{\mathrm{co}}(\mathcal{A}\text{-}\mathsf{CSp})$	signed bicom. signed Coxeter $\mathcal{A}$-bimonoids

the diagram

$$
\begin{array}{ccc}
\mathsf{a}[F] & \xrightarrow{\beta_{G,F}} & \mathsf{a}[G] \\
{\scriptstyle \mu_A^F} \searrow & & \swarrow {\scriptstyle \mu_A^G} \\
& \mathsf{a}[A] &
\end{array}
\tag{2.49}
$$

commutes whenever $A \leq F$, $A \leq G$, $\mathrm{s}(F) = \mathrm{s}(G)$. We refer to diagram (2.49) as the *commutativity axiom*. As a shorthand, we write it as $\mu\beta = \mu$. Note very carefully that it does not involve type morphisms.

Exercise 2.34. Show: A Coxeter $\mathcal{A}$-monoid a is commutative iff the diagram

$$
\begin{array}{ccc}
\mathsf{a}[F] & \xrightarrow{\beta_{G,F}} & \mathsf{a}[G] \\
{\scriptstyle \mu_A^F} \downarrow & & \downarrow {\scriptstyle \mu_B^G} \\
\mathsf{a}[A] & \xrightarrow[\beta_{B,A}]{} & \mathsf{a}[B]
\end{array}
\tag{2.50}
$$

commutes whenever $\mathrm{s}(A) = \mathrm{s}(B)$, $\mathrm{s}(F) = \mathrm{s}(G)$, $A \leq F$, $B \leq G$. (For the forward direction, combine the commutativity, support-naturality and support-support axioms. The backward direction is clear.)

2.6.2. Covering commutativity axiom in terms of faces.

Exercise 2.35. Show: A Coxeter $\mathcal{A}$-monoid a is commutative iff the diagram

$$
\begin{array}{ccc}
\mathsf{a}[F] & \xrightarrow{\beta_{A\overline{F},F}} & \mathsf{a}[A\overline{F}] \\
{\scriptstyle \mu_A^F} \searrow & & \swarrow {\scriptstyle \mu_A^{A\overline{F}}} \\
& \mathsf{a}[A] &
\end{array}
\tag{2.51}
$$

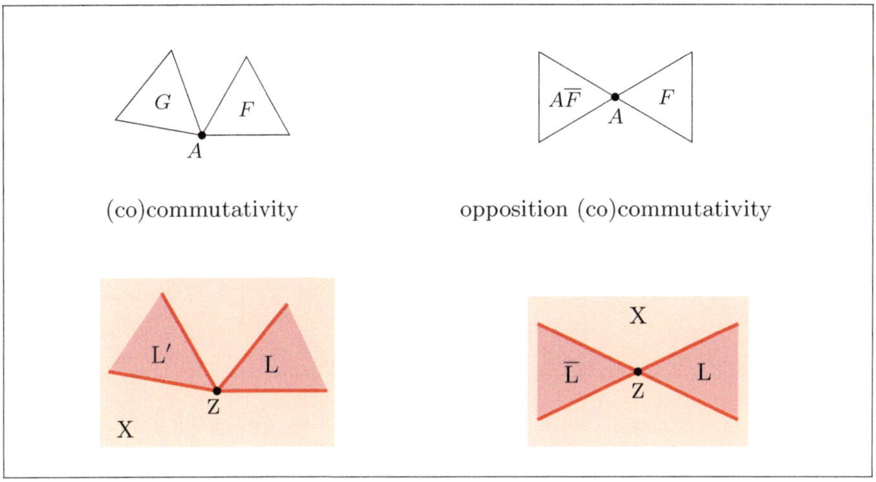

FIGURE 2.4. (Co)commutativity axiom for Coxeter (co)monoids.

commutes whenever $A \leq F$. We refer to diagram (2.51) as the *opposition commutativity axiom*. For a comparison with the commutativity axiom, see the first two pictures in Figure 2.4.

A stronger result is given below.

Exercise 2.36. Show: For a a Coxeter $\mathcal{A}$-monoid, the following are equivalent.

- a is commutative.
- Commutativity axiom (2.49) holds whenever $A \lessdot F$ and $A \lessdot G$.
- Opposition commutativity axiom (2.51) holds whenever $A \lessdot F$.

In these particular cases, the two axioms in fact coincide. We call this the *covering commutativity axiom*. See the first two pictures in Figure 2.5.

For Coxeter $\mathcal{A}$-monoids as formulated in Proposition 2.23 (with product components involving cover relations), one can define commutativity using the covering commutativity axiom.

2.6.3. Commutative Coxeter monoids in terms of flats.

Commutativity can also be formulated in terms of flats and lunes by using the description of Coxeter monoids in Proposition 2.24.

Lemma 2.37. *A Coxeter $\mathcal{A}$-monoid* a *is commutative iff*

$$(\mathsf{a}[\mathrm{X}] \xrightarrow{\mu_{\mathrm{L}}} \mathsf{a}[\mathrm{Z}]) = (\mathsf{a}[\mathrm{X}] \xrightarrow{\mu_{\mathrm{L}'}} \mathsf{a}[\mathrm{Z}]) \tag{2.52}$$

whenever $\mathrm{Z} \leq \mathrm{X}$ *and lunes* L *and* L' *both have base* Z *and case* X.

PROOF. We translate between diagrams (2.49) and (2.52) using diagrams (2.18) and (2.34). □

We refer to (2.52) as the *commutativity axiom*.

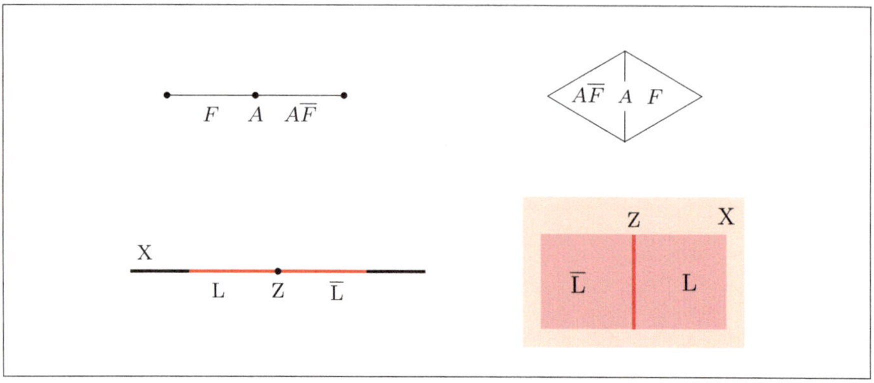

FIGURE 2.5. Covering (co)commutativity axiom for Coxeter (co)monoids.

Proposition 2.38. *A commutative Coxeter $\mathcal{A}$-monoid is a Coxeter $\mathcal{A}$-species* a *equipped with linear maps*

$$\mu_{\mathrm{Z}}^{\mathrm{X}} : \mathsf{a}[\mathrm{X}] \to \mathsf{a}[\mathrm{Z}],$$

one for each $\mathrm{Z} \leq \mathrm{X}$*, such that the diagrams*

$$\begin{array}{ccc} \mathsf{a}[\mathrm{X}] & \xrightarrow{w} & \mathsf{a}[\mathrm{X}'] \\ {\scriptstyle \mu_{\mathrm{Y}}^{\mathrm{X}}}\downarrow & & \downarrow{\scriptstyle \mu_{\mathrm{Y}'}^{\mathrm{X}'}} \\ \mathsf{a}[\mathrm{Y}] & \xrightarrow[w]{} & \mathsf{a}[\mathrm{Y}'] \end{array} \qquad \begin{array}{ccc} & \mathsf{a}[\mathrm{Y}] & \\ {\scriptstyle \mu_{\mathrm{Y}}^{\mathrm{X}}}\nearrow & & \searrow{\scriptstyle \mu_{\mathrm{Z}}^{\mathrm{Y}}} \\ \mathsf{a}[\mathrm{X}] & \xrightarrow[\mu_{\mathrm{Z}}^{\mathrm{X}}]{} & \mathsf{a}[\mathrm{Z}] \end{array} \qquad (\mathsf{a}[\mathrm{Z}] \xrightarrow{\mu_{\mathrm{Z}}^{\mathrm{Z}}} \mathsf{a}[\mathrm{Z}]) = \mathrm{id} \tag{2.53}$$

commute, the first for all $\mathrm{Y} \leq \mathrm{X}$ *and* $w \in W$ *with* $w(\mathrm{Y}) = \mathrm{Y}'$ *and* $w(\mathrm{X}) = \mathrm{X}'$*, the second for every* $\mathrm{Z} \leq \mathrm{Y} \leq \mathrm{X}$*, and the third for every* Z.

A morphism of commutative Coxeter $\mathcal{A}$-monoids $f : \mathsf{a} \to \mathsf{b}$ *is a map of Coxeter $\mathcal{A}$-species such that the diagram*

$$\begin{array}{ccc} \mathsf{a}[\mathrm{X}] & \xrightarrow{f_{\mathrm{X}}} & \mathsf{b}[\mathrm{X}] \\ {\scriptstyle \mu_{\mathrm{Z}}^{\mathrm{X}}}\downarrow & & \downarrow{\scriptstyle \mu_{\mathrm{Z}}^{\mathrm{X}}} \\ \mathsf{a}[\mathrm{Z}] & \xrightarrow[f_{\mathrm{Z}}]{} & \mathsf{b}[\mathrm{Z}] \end{array} \tag{2.54}$$

commutes whenever $\mathrm{Z} \leq \mathrm{X}$.

PROOF. This follows from Proposition 2.24 and Lemma 2.37. Diagrams (2.32) and (2.33) reduce to diagrams (2.53) and (2.54), respectively, with μ_{L} reducing to $\mu_{\mathrm{Y}}^{\mathrm{X}}$ for any lune L with base Y and case X. □

We refer to diagrams (2.53) as the *type-naturality axiom*, *associativity axiom*, *unitality axiom*, respectively.

2.6.4. Covering commutativity axiom in terms of flats and lunes.

Exercise 2.39. Show: A Coxeter $\mathcal{A}$-monoid a is commutative iff

$$(\mathsf{a}[\mathrm{X}] \xrightarrow{\mu_{\mathrm{L}}} \mathsf{a}[\mathrm{Z}]) = (\mathsf{a}[\mathrm{X}] \xrightarrow{\mu_{\overline{\mathrm{L}}}} \mathsf{a}[\mathrm{Z}]) \tag{2.55}$$

whenever $\mathrm{Z} \leq \mathrm{X}$ and L has base Z and case X. We refer to (2.55) as the *opposition commutativity axiom*. (Use Exercise 2.35.) For a comparison with the commutativity axiom (2.52), see the last two pictures in Figure 2.4.

Exercise 2.40. Show: For a a Coxeter $\mathcal{A}$-monoid, the following are equivalent.

- a is commutative.
- Commutativity axiom (2.52) holds for lunes L and L′ of slack 1.
- Opposition commutativity axiom (2.55) holds for lunes L of slack 1.

In these particular cases, the two axioms coincide. We call this the *covering commutativity axiom*. (Use Exercise 2.36.) See the last two pictures in Figure 2.5.

For Coxeter $\mathcal{A}$-monoids as formulated in Proposition 2.25, one can define commutativity using the covering commutativity axiom.

Proposition 2.41. *A commutative Coxeter $\mathcal{A}$-monoid is a Coxeter $\mathcal{A}$-species* a *equipped with linear maps*

$$\mu_{\mathrm{Z}}^{\mathrm{X}} : \mathsf{a}[\mathrm{X}] \to \mathsf{a}[\mathrm{Z}]$$

whenever X *covers* Z, *such that the diagrams*

$$\begin{array}{ccc} \mathsf{a}[\mathrm{X}] & \xrightarrow{w} & \mathsf{a}[\mathrm{X}'] \\ {\scriptstyle \mu_{\mathrm{Y}}^{\mathrm{X}}}\downarrow & & \downarrow{\scriptstyle \mu_{\mathrm{Y}'}^{\mathrm{X}'}} \\ \mathsf{a}[\mathrm{Y}] & \xrightarrow[w]{} & \mathsf{a}[\mathrm{Y}'] \end{array} \qquad \begin{array}{ccc} \mathsf{a}[\mathrm{X}] & \xrightarrow{\mu_{\mathrm{Y}'}^{\mathrm{X}}} & \mathsf{a}[\mathrm{Y}'] \\ {\scriptstyle \mu_{\mathrm{Y}}^{\mathrm{X}}}\downarrow & & \downarrow{\scriptstyle \mu_{\mathrm{Z}}^{\mathrm{Y}'}} \\ \mathsf{a}[\mathrm{Y}] & \xrightarrow[\mu_{\mathrm{Z}}^{\mathrm{Y}}]{} & \mathsf{a}[\mathrm{Z}] \end{array} \tag{2.56}$$

commute, where arrows labeled by μ go from a flat to a flat it covers. Explicitly, in the first diagram, X *and* X′ *cover* Y *and* Y′, *respectively, and in the second diagram,* X *covers both* Y *and* Y′, *and they in turn cover* Z.

A morphism of commutative Coxeter $\mathcal{A}$-monoids $f : \mathsf{a} \to \mathsf{b}$ is a map of Coxeter $\mathcal{A}$-species such that diagram (2.54) *commutes whenever* X *covers* Z.

PROOF. This follows from Proposition 2.25 and Exercise 2.40. Diagrams (2.35) reduce to diagrams (2.56). Alternatively, one can deduce this result from Proposition 2.38 by using that the poset of flats is strongly connected [**11**, Lemma 1.4, item (2)]. □

We refer to the second diagram in (2.56) as the *covering associativity axiom*.

2.6.5. Cocommutative Coxeter comonoids in terms of faces. We now briefly consider the dual notion. Recall Coxeter $\mathcal{A}$-comonoids from Section 2.4.5. A Coxeter $\mathcal{A}$-comonoid c is *cocommutative* if the diagram

$$\begin{array}{ccc} \mathsf{c}[F] & \xrightarrow{\beta_{G,F}} & \mathsf{c}[G] \\ {\scriptstyle \Delta_A^F}\nwarrow & & \nearrow{\scriptstyle \Delta_A^G} \\ & \mathsf{c}[A] & \end{array} \tag{2.57}$$

commutes whenever $A \leq F$, $A \leq G$, $\mathrm{s}(F) = \mathrm{s}(G)$. We refer to diagram (2.57) as the *cocommutativity axiom*. As a shorthand, we write it as $\beta\Delta = \Delta$. It does not involve type morphisms.

The duals to Exercises 2.34, 2.35, 2.36 hold. Thus, we have the *opposition cocommutativity axiom* and *covering cocommutativity axiom.*

2.6.6. Cocommutative Coxeter comonoids in terms of flats. The dual of Lemma 2.37 says:

Lemma 2.42. *A Coxeter $\mathcal{A}$-comonoid* c *is cocommutative iff*

$$(\mathsf{c}[\mathrm{Z}] \xrightarrow{\Delta_{\mathrm{L}}} \mathsf{c}[\mathrm{X}]) = (\mathsf{c}[\mathrm{Z}] \xrightarrow{\Delta_{\mathrm{L}'}} \mathsf{c}[\mathrm{X}]) \tag{2.58}$$

whenever $\mathrm{Z} \leq \mathrm{X}$ *and lunes* L *and* L' *both have base* Z *and case* X.

We refer to (2.58) as the *cocommutativity axiom.* The duals to Exercises 2.39 and 2.40 also hold.

As a consequence, cocommutative Coxeter comonoids can be formulated purely in terms of flats along the lines of Propositions 2.38 and 2.41 using linear maps

$$\Delta_{\mathrm{Z}}^{\mathrm{X}} : \mathsf{c}[\mathrm{Z}] \to \mathsf{c}[\mathrm{X}],$$

(with covering generators for the latter result).

2.6.7. (Co, bi)commutative Coxeter bimonoids. The notion of commutativity also applies to Coxeter bimonoids. We say a Coxeter $\mathcal{A}$-bimonoid is

- *commutative* if its underlying Coxeter $\mathcal{A}$-monoid is commutative,
- *cocommutative* if its underlying Coxeter $\mathcal{A}$-comonoid is cocommutative,
- *bicommutative* if it is both commutative and cocommutative.

Bicommutative Coxeter bimonoids can be nicely reformulated in terms of the maps $\mu_{\mathrm{Z}}^{\mathrm{X}}$ and $\Delta_{\mathrm{Z}}^{\mathrm{Y}}$ by building on Proposition 2.38 and its dual.

Proposition 2.43. *A bicommutative Coxeter $\mathcal{A}$-bimonoid is the same as a triple* $(\mathsf{h}, \mu, \Delta)$, *where* h *is a Coxeter $\mathcal{A}$-species,* (h, μ) *is a commutative Coxeter $\mathcal{A}$-monoid,* (h, Δ) *is a cocommutative Coxeter $\mathcal{A}$-comonoid, and such that for flats* $\mathrm{Z} \leq \mathrm{X}$ *and* $\mathrm{Z} \leq \mathrm{Y}$, *the diagram*

$$\begin{array}{ccc} \mathsf{h}[\mathrm{X}] & \xrightarrow{\mu_{\mathrm{Z}}^{\mathrm{X}}} & \mathsf{h}[\mathrm{Z}] \\ {\scriptstyle \Delta_{\mathrm{X}}^{\mathrm{X}\vee\mathrm{Y}}}\downarrow & & \downarrow{\scriptstyle \Delta_{\mathrm{Z}}^{\mathrm{Y}}} \\ \mathsf{h}[\mathrm{X}\vee\mathrm{Y}] & \xrightarrow[\mu_{\mathrm{Y}}^{\mathrm{X}\vee\mathrm{Y}}]{} & \mathsf{h}[\mathrm{Y}] \end{array} \tag{2.59}$$

commutes.

PROOF. This follows from Proposition 2.28 and Lemma 2.37 and its dual. Diagram (2.41) reduces to diagram (2.59). □

We refer to diagram (2.59) as the *bicommutative bimonoid axiom.* As a shorthand, we write it as $\Delta\mu = \mu\Delta$.

The data present in a bicommutative Coxeter $\mathcal{A}$-bimonoid is shown in Summary 2.3. Compare and contrast with Summary 2.2.

> A bicommutative Coxeter $\mathcal{A}$-bimonoid h consists of linear maps
> $$\mu_{\mathrm{Z}}^{\mathrm{X}} : \mathsf{h}[\mathrm{X}] \to \mathsf{h}[\mathrm{Z}], \quad w : \mathsf{h}[\mathrm{X}] \to \mathsf{h}[w\mathrm{X}], \quad \Delta_{\mathrm{Z}}^{\mathrm{X}} : \mathsf{h}[\mathrm{Z}] \to \mathsf{h}[\mathrm{X}],$$
> which satisfy
> - type-type and type-unitality axioms (2.15),
> - type-naturality, (co)associativity, (co)unitality axioms (2.53) + its dual,
> - bicommutative bimonoid axiom (2.59).
>
> There are a total of 9 axioms.

SUMMARY 2.3. Bicommutative Coxeter bimonoid in terms of flats.

Exercise 2.44. Show: For a bicommutative Coxeter $\mathcal{A}$-bimonoid, it suffices to require the bicommutative bimonoid axiom (2.59) only for flats $\mathrm{Z} \lessdot \mathrm{X}$ and $\mathrm{Z} \lessdot \mathrm{Y}$. Note very carefully: $\mathrm{X} \lessdot \mathrm{X} \vee \mathrm{Y}$ or $\mathrm{Y} \lessdot \mathrm{X} \vee \mathrm{Y}$ may not necessarily hold in this situation. (Repeatedly use

$$\begin{array}{ccccc} \mathsf{h}[\mathrm{X}] & \xrightarrow{\Delta} & \mathsf{h}[\mathrm{X} \vee \mathrm{Y}] & \xrightarrow{\Delta} & \mathsf{h}[\mathrm{X} \vee \mathrm{Y}'] \\ {\scriptstyle \mu}\downarrow & & \downarrow{\scriptstyle \mu} & & \downarrow{\scriptstyle \mu} \\ \mathsf{h}[\mathrm{Z}] & \xrightarrow[\Delta]{} & \mathsf{h}[\mathrm{Y}] & \xrightarrow[\Delta]{} & \mathsf{h}[\mathrm{Y}'] \end{array}$$

and its dual diagram.) Compare with Lemma 2.27.

Observation 2.45. For a commutative (cocommutative) Coxeter $\mathcal{A}$-bimonoid, the bimonoid axiom (2.41) reduces to the left (right) diagram below.

(2.60)
$$\begin{array}{ccc} \mathsf{h}[\mathrm{X}] & \xrightarrow{\mu_{\mathrm{Z}}^{\mathrm{X}}} & \mathsf{h}[\mathrm{Z}] \\ {\scriptstyle \Delta_{\mathrm{X}\cdot\mathrm{M}}}\downarrow & & \downarrow{\scriptstyle \Delta_{\mathrm{M}}} \\ \mathsf{h}[\mathrm{X} \vee \mathrm{Y}] & \xrightarrow[\mu_{\mathrm{Y}}^{\mathrm{X}\vee\mathrm{Y}}]{} & \mathsf{h}[\mathrm{Y}] \end{array} \qquad \begin{array}{ccc} \mathsf{h}[\mathrm{X}] & \xrightarrow{\mu_{\mathrm{L}}} & \mathsf{h}[\mathrm{Z}] \\ {\scriptstyle \Delta_{\mathrm{X}}^{\mathrm{X}\vee\mathrm{Y}}}\downarrow & & \downarrow{\scriptstyle \Delta_{\mathrm{Z}}^{\mathrm{Y}}} \\ \mathsf{h}[\mathrm{X} \vee \mathrm{Y}] & \xrightarrow[\mu_{\mathrm{Y}\cdot\mathrm{L}}]{} & \mathsf{h}[\mathrm{Y}] \end{array}$$

These are the *commutative bimonoid axiom* and *cocommutative bimonoid axiom*, respectively.

Exercise 2.46. Write down the data present in a (co)commutative Coxeter $\mathcal{A}$-bimonoid in terms of flats and lunes along the lines of Summaries 2.2 and 2.3.

2.6.8. Signed (co)commutativity in terms of faces. We now turn to signed aspects. A Coxeter $\mathcal{A}$-monoid a is *signed commutative* if the diagram

(2.61)
$$\begin{array}{ccccc} \mathsf{a}[F] & & \xrightarrow{(\beta_{-1})_{G,F}} & & \mathsf{a}[G] \\ & {\scriptstyle \mu_A^F}\searrow & & \swarrow{\scriptstyle \mu_A^G} & \\ & & \mathsf{a}[A] & & \end{array}$$

commutes whenever $A \leq F$, $A \leq G$, $\mathrm{s}(F) = \mathrm{s}(G)$. We refer to diagram (2.61) as the *signed commutativity axiom*. The notation $(\beta_{-1})_{G,F}$ is as in (2.43) with $q = -1$.

Dually, a Coxeter $\mathcal{A}$-comonoid c is *signed cocommutative* if the diagram

(2.62)
$$\begin{array}{ccc} \mathsf{c}[F] & \xrightarrow{(\beta_{-1})_{G,F}} & \mathsf{c}[G] \\ {}_{\Delta_A^F}\nwarrow & & \nearrow_{\Delta_A^G} \\ & \mathsf{c}[A] & \end{array}$$

commutes whenever $A \leq F$, $A \leq G$, $\mathrm{s}(F) = \mathrm{s}(G)$. We refer to diagram (2.62) as the *signed cocommutativity axiom*.

We say a signed Coxeter $\mathcal{A}$-bimonoid is *signed (co)commutative* if its underlying Coxeter $\mathcal{A}$-(co)monoid is signed (co)commutative. Similarly, a signed Coxeter $\mathcal{A}$-bimonoid is *signed bicommutative* if it is both signed commutative and signed cocommutative.

2.6.9. Signed (co)commutativity in terms of flats and lunes. Signed (co)commutativity can also be formulated in terms of flats and lunes by using the description of Coxeter (co)monoids in Proposition 2.24 and its dual.

Exercise 2.47. Check: A Coxeter $\mathcal{A}$-monoid a is signed commutative iff

(2.63)
$$(\mathsf{a}[\mathrm{X}] \xrightarrow{\mu_{\mathrm{L}}} \mathsf{a}[\mathrm{Z}]) = (-1)^{\mathrm{dist}(\mathrm{L},\mathrm{L}')}(\mathsf{a}[\mathrm{X}] \xrightarrow{\mu_{\mathrm{L}'}} \mathsf{a}[\mathrm{Z}])$$

whenever $\mathrm{Z} \leq \mathrm{X}$ and lunes L and L' both have base Z and case X.

Dually, a Coxeter $\mathcal{A}$-comonoid c is signed cocommutative iff

(2.64)
$$(\mathsf{c}[\mathrm{Z}] \xrightarrow{\Delta_{\mathrm{L}}} \mathsf{c}[\mathrm{X}]) = (-1)^{\mathrm{dist}(\mathrm{L},\mathrm{L}')}(\mathsf{c}[\mathrm{Z}] \xrightarrow{\Delta_{\mathrm{L}'}} \mathsf{c}[\mathrm{X}])$$

whenever $\mathrm{Z} \leq \mathrm{X}$ and lunes L and L' both have base Z and case X.

These are signed analogues of (2.52) and (2.58).

As a consequence, one can formulate signed commutative Coxeter monoids in terms of flats along the lines of Proposition 2.38. To handle signs properly, one needs the following definition.

Definition 2.48. For flats $\mathrm{Z} \leq \mathrm{X}$, define $\mathbf{E}^-[\mathrm{Z},\mathrm{X}]$ to be the one-dimensional vector space spanned by symbols $\mathtt{H}_{\mathrm{L}}$, with L a lune of base Z and case X, subject to relations

$$\mathtt{H}_{\mathrm{L}} = (-1)^{\mathrm{dist}(\mathrm{L},\mathrm{L}')}\mathtt{H}_{\mathrm{L}'}.$$

We write $\mathtt{H}_{[\mathrm{L}]}$ for the class of $\mathtt{H}_{\mathrm{L}}$. Any such element is a basis of $\mathbf{E}^-[\mathrm{Z},\mathrm{X}]$. Alternatively, one may employ symbols $\mathtt{H}_{F/A}$ with $\mathrm{s}(F/A) = \mathrm{X}/\mathrm{Z}$, and relations

$$\mathtt{H}_{F/A} = (-1)^{\mathrm{dist}(F',A'F)}\mathtt{H}_{F'/A'}.$$

We write $\mathtt{H}_{[F/A]}$ for the class of $\mathtt{H}_{F/A}$. The symbols $\mathtt{H}_{\mathrm{L}}$ and $\mathtt{H}_{F/A}$ are related by $\mathrm{L} = \mathrm{s}(A,F)$.

We refer to $\mathbf{E}^-[\mathrm{Z},\mathrm{X}]$ as the *signature space* over Z and under X. Important special cases are $\mathbf{E}^-[\bot,\mathrm{X}]$ and $\mathbf{E}^-[\mathrm{X},\top]$ which we refer to as signature spaces under X and over X, respectively.

A signed commutative Coxeter $\mathcal{A}$-monoid a consists of linear maps

$$\mu_{\mathrm{Z}}^{\mathrm{X}} : \mathbf{E}^-[\mathrm{Z},\mathrm{X}] \otimes \mathsf{a}[\mathrm{X}] \to \mathsf{a}[\mathrm{Z}]$$

whenever $\mathrm{Z} \leq \mathrm{X}$, subject to appropriate type-naturality, associativity, unitality axioms. Similar formulations can be given for signed cocommutative

Coxeter comonoids and signed bicommutative signed Coxeter bimonoids. In the context of species (as opposed to Coxeter species), they were given in [**11**, Propositions 2.36 and 2.37].

2.6.10. Coxeter double (co)monoids and Eckmann–Hilton. Define the category of Coxeter double monoids as follows. A *Coxeter $\mathcal{A}$-double monoid* is a triple (a, μ, μ'), where a is a Coxeter $\mathcal{A}$-species, (a, μ) and (a, μ') are Coxeter $\mathcal{A}$-monoids, and for faces $A \le F$ and $A \le G$, the diagram

$$\begin{array}{ccccc} \mathsf{a}[F] & \xrightarrow{\mu_A^F} & \mathsf{a}[A] & \xleftarrow{(\mu')_A^G} & \mathsf{a}[G] \\ {\scriptstyle (\mu')_F^{FG}}\uparrow & & & & \uparrow{\scriptstyle \mu_G^{GF}} \\ \mathsf{a}[FG] & & \xrightarrow[\beta_{GF,FG}]{} & & \mathsf{a}[GF] \end{array} \tag{2.65}$$

commutes. We refer to diagram (2.65) as the *double monoid axiom*. A morphism of Coxeter $\mathcal{A}$-double monoids is a map of the underlying Coxeter $\mathcal{A}$-species which is a morphism of Coxeter $\mathcal{A}$-monoids for both μ and μ'.

The category of Coxeter double comonoids is defined dually.

Exercise 2.49. Show: The category of Coxeter double (co)monoids is equivalent to the category of (co)commutative Coxeter (co)monoids. This fact is an avatar of the Eckmann–Hilton argument.

2.7. Species vs Coxeter species

The theory of $\mathcal{A}$-species was laid out in [**11**]. An $\mathcal{A}$-species is defined for a fixed hyperplane arrangement $\mathcal{A}$, and, in particular, makes sense when $\mathcal{A}$ is a reflection arrangement. We recall this notion and compare it with the notion of Coxeter $\mathcal{A}$-species. We also compare monoids, comonoids, bimonoids in $\mathcal{A}$-species with Coxeter monoids, Coxeter comonoids, Coxeter bimonoids.

2.7.1. Species. Let $\mathcal{A}$ be a hyperplane arrangement. Define the category $\mathcal{A}$-$\mathtt{Hyp}$ by a presentation. Objects are faces of $\mathcal{A}$. Morphisms are generated by

$$\beta_{G,F} : F \to G$$

whenever $\mathrm{s}(F) = \mathrm{s}(G)$, subject to the relations

$$\begin{array}{ccc} & G & \\ {\scriptstyle \beta_{G,F}}\nearrow & & \searrow{\scriptstyle \beta_{H,G}} \\ F & \xrightarrow[\beta_{H,F}]{} & H \end{array} \qquad\qquad (F \xrightarrow{\beta_{F,F}} F) = \mathrm{id},$$

the former whenever $\mathrm{s}(F) = \mathrm{s}(G) = \mathrm{s}(H)$, and the latter for any F.

Observe: $\mathcal{A}$-$\mathtt{Hyp}$ is a disjoint union of indiscrete categories. In particular, it is a groupoid. The inverse of $\beta_{G,F} : F \to G$ is $\beta_{F,G} : G \to F$.

An $\mathcal{A}$-*species* is a functor

$$\mathsf{p} : \mathcal{A}\text{-}\mathsf{Hyp} \to \mathsf{Vec}.$$

A map of $\mathcal{A}$-species $\mathsf{p} \to \mathsf{q}$ is a natural transformation. This defines the category of $\mathcal{A}$-species which we denote $\mathcal{A}\text{-}\mathsf{Sp}$. It is a functor category, and we also write

$$\mathcal{A}\text{-}\mathsf{Sp} = [\mathcal{A}\text{-}\mathtt{Hyp}, \mathsf{Vec}].$$

The value of an $\mathcal{A}$-species p on an object F is denoted $\mathsf{p}[F]$. This is the F-component of p. Using the presentation of $\mathcal{A}\text{-}\mathtt{Hyp}$, we obtain:

An $\mathcal{A}$-species p consists of a family of vector spaces $\mathsf{p}[F]$, one for each face F of $\mathcal{A}$, together with linear maps

$$\beta_{G,F} : \mathsf{p}[F] \to \mathsf{p}[G]$$

whenever $\mathrm{s}(F) = \mathrm{s}(G)$, such that the diagrams

$$\begin{array}{ccc} & \mathsf{p}[G] & \\ {\scriptstyle \beta_{G,F}} \nearrow & & \searrow {\scriptstyle \beta_{H,G}} \\ \mathsf{p}[F] & \xrightarrow[\beta_{H,F}]{} & \mathsf{p}[H] \end{array} \qquad (\mathsf{p}[F] \xrightarrow{\beta_{F,F}} \mathsf{p}[F]) = \mathrm{id} \tag{2.66}$$

commute, the former whenever $\mathrm{s}(F) = \mathrm{s}(G) = \mathrm{s}(H)$, and the latter for any F.

For more details, see [**11**, Section 2.1].

2.7.2. Semidirect product groupoid. Let $\mathcal{A}$ be a reflection arrangement with Coxeter group W. The action of W on the set of faces of $\mathcal{A}$ has the property that if two faces F and G have the same support, then the faces $w(F)$ and $w(G)$ have the same support for any $w \in W$. It follows that there is an action of W on the category $\mathcal{A}\text{-}\mathtt{Hyp}$ defined as follows. For $w \in W$, the functor $\mathcal{F}_w : \mathcal{A}\text{-}\mathtt{Hyp} \to \mathcal{A}\text{-}\mathtt{Hyp}$ is given by

$$\mathcal{F}_w(F) = w(F) \quad \text{and} \quad \mathcal{F}_w(\beta_{G,F}) = \beta_{w(G),w(F)}.$$

The resulting semidirect product groupoid is $\mathcal{A}\text{-}\mathtt{Cox}$; we write this as

$$\mathcal{A}\text{-}\mathtt{Cox} = \mathcal{A}\text{-}\mathtt{Hyp} \rtimes W. \tag{2.67}$$

Background information on semidirect products is given in Appendix A.2.3. The support-type relation in (2.1a) is an instance of (A.22).

The semidirect product (2.67) makes sense with W replaced by any of its subgroups. In particular, choosing the trivial subgroup of W recovers $\mathcal{A}\text{-}\mathtt{Hyp}$.

2.7.3. Species vs Coxeter species. Since $\mathcal{A}\text{-}\mathtt{Hyp}$ is a subcategory of $\mathcal{A}\text{-}\mathtt{Cox}$, every Coxeter species is also a species. Conversely, one can view a Coxeter species as a species equipped with additional structure of type morphisms.

Observation 2.50. A Coxeter $\mathcal{A}$-species is an $\mathcal{A}$-species p equipped with linear maps

$$w : \mathsf{p}[F] \to \mathsf{p}[F'],$$

such that the diagrams

$$\begin{array}{ccc} & \mathsf{p}[F'] & \\ {\scriptstyle u} \nearrow & & \searrow {\scriptstyle v} \\ \mathsf{p}[F] & \xrightarrow[vu]{} & \mathsf{p}[F''] \end{array} \qquad \begin{array}{ccc} \mathsf{p}[F] & \xrightarrow{\beta_{G,F}} & \mathsf{p}[G] \\ {\scriptstyle w}\downarrow & & \downarrow{\scriptstyle w} \\ \mathsf{p}[F'] & \xrightarrow[\beta_{G',F'}]{} & \mathsf{p}[G'] \end{array} \qquad (\mathsf{p}[F] \xrightarrow{e} \mathsf{p}[F]) = \mathrm{id} \tag{2.68}$$

commute.

The above are precisely the three axioms which involve type morphisms, namely, type-type, support-type, type-unitality.

In this regard, compare Proposition 2.17 with the description of species in terms of flats [**11**, Proposition 2.5].

2.7.4. Bimonoids vs Coxeter bimonoids.

Monoids, comonoids, bimonoids in $\mathcal{A}$-species, along with their commutative and signed counterparts, are defined in [**11**, Sections 2.2–2.5]. They are defined in the same manner as Coxeter monoids, Coxeter comonoids, Coxeter bimonoids with the difference that type morphisms and axioms involving them are absent. In particular, every Coxeter (co, bi)monoid is a (co, bi)monoid in species. Conversely, one can view a Coxeter (co, bi)monoid as a (co, bi)monoid equipped with additional structure of type morphisms. Let us say this precisely.

Observation 2.51. A Coxeter $\mathcal{A}$-monoid is an $\mathcal{A}$-monoid a where in addition a carries a Coxeter $\mathcal{A}$-species structure such that the second diagram in (2.28a) commutes. A morphism of Coxeter $\mathcal{A}$-monoids is a morphism of the underlying $\mathcal{A}$-monoids which is also a map of the underlying Coxeter $\mathcal{A}$-species.

Dually, a Coxeter $\mathcal{A}$-comonoid is an $\mathcal{A}$-comonoid c where in addition c carries a Coxeter $\mathcal{A}$-species structure such that the second diagram in (2.36a) commutes. A morphism of Coxeter $\mathcal{A}$-comonoids is a morphism of the underlying $\mathcal{A}$-comonoids which is also a map of the underlying Coxeter $\mathcal{A}$-species.

A Coxeter $\mathcal{A}$-bimonoid is an $\mathcal{A}$-bimonoid h where in addition h carries a Coxeter $\mathcal{A}$-species structure such that the second diagrams in (2.28a) and (2.36a) commute.

A Coxeter $\mathcal{A}$-monoid is commutative if its underlying $\mathcal{A}$-monoid is commutative. Dually, a Coxeter $\mathcal{A}$-comonoid is cocommutative if its underlying $\mathcal{A}$-comonoid is cocommutative. A Coxeter $\mathcal{A}$-bimonoid is (co, bi)commutative if the underlying $\mathcal{A}$-bimonoid is (co, bi)commutative.

Remark 2.52. Many results on Coxeter $\mathcal{A}$-bimonoids can be formally deduced from corresponding results on $\mathcal{A}$-bimonoids. For some illustrative examples, see Remark 3.3, Remark 3.8, Proposition 3.44, Proposition 3.49, Theorem 3.79, Proposition 6.1, Proposition 6.4, Theorem 6.7, Theorem 7.1, Theorem 7.4, Theorem 7.7, Theorem 7.14, Theorem 7.20, Theorem 7.26. Note very carefully that some results need the extra hypothesis $\mathrm{char}(\Bbbk) \nmid |W|$.

2.8. Signature functors on Coxeter species

The unsigned and signed worlds can be connected by two simple constructions called the u-signature functor and o-signature functor. The idea is to twist support morphisms by the signed distance function and type morphisms by the length function. The letters u and o stand for under and over; this terminology will become clear in Section 4.2.6 where we relate these functors to signature spaces under and over flats.

2.8.1. Signature functor (under version).

For a Coxeter $\mathcal{A}$-species p, define the Coxeter $\mathcal{A}$-species p_{-} with F-component $\mathsf{p}[F]$, and support and type morphisms

$$(2.69)\qquad \begin{array}{ccc} \mathsf{p}_-[F] & \xrightarrow{\beta_{G,F}} & \mathsf{p}_-[G] \\ \| & & \| \\ \mathsf{p}[F] & \xrightarrow[(-1)^{\mathrm{dist}(F,G)}\beta_{G,F}]{} & \mathsf{p}[G] \end{array} \qquad \begin{array}{ccc} \mathsf{p}_-[F] & \xrightarrow{w} & \mathsf{p}_-[wF] \\ \| & & \| \\ \mathsf{p}[F] & \xrightarrow[w]{} & \mathsf{p}[wF]. \end{array}$$

Further, a map $f : \mathsf{p} \to \mathsf{q}$ of Coxeter $\mathcal{A}$-species induces a map $f : \mathsf{p}_- \to \mathsf{q}_-$ of Coxeter $\mathcal{A}$-species. Thus, we have a functor

$$(2.70)\qquad (-)_- : \mathcal{A}\text{-}\mathsf{CSp} \to \mathcal{A}\text{-}\mathsf{CSp}, \qquad \mathsf{p} \mapsto \mathsf{p}_-.$$

This is the *u-signature functor*. It is an isomorphism of categories (with its inverse being itself). It is also its own left and right adjoint.

Observation 2.53. If a is a Coxeter $\mathcal{A}$-monoid, then so is a_- with product components μ_A^F the same as those of a. Further, if a is commutative, then a_- is signed commutative, and vice versa. Dually, if c is a Coxeter $\mathcal{A}$-comonoid, then so is c_-. Further, if c is cocommutative, then c_- is signed cocommutative, and vice versa. Combining the two situations, if h is a Coxeter $\mathcal{A}$-q-bimonoid, then h_- is a Coxeter $\mathcal{A}$-$(-q)$-bimonoid. In particular, if h is a Coxeter $\mathcal{A}$-bimonoid, then h_- is a signed Coxeter $\mathcal{A}$-bimonoid, and vice versa.

This leads to the following.

Lemma 2.54. *For a scalar q, there is an isomorphism of categories*

$$(2.71)\qquad (-)_- : q\text{-}\mathsf{Bimon}(\mathcal{A}\text{-}\mathsf{CSp}) \to (-q)\text{-}\mathsf{Bimon}(\mathcal{A}\text{-}\mathsf{CSp}), \qquad \mathsf{h} \mapsto \mathsf{h}_-$$

whose left and right adjoint is again $(-)_-$. In particular, the categories of Coxeter $\mathcal{A}$-bimonoids and signed Coxeter $\mathcal{A}$-bimonoids are isomorphic.

Similarly, the categories of (co)commutative Coxeter $\mathcal{A}$-(co)monoids and signed (co)commutative Coxeter $\mathcal{A}$-(co)monoids are isomorphic.

2.8.2. Signature functor (over version). For a Coxeter $\mathcal{A}$-species p, define the Coxeter $\mathcal{A}$-species p^- with F-component $\mathsf{p}[F]$, and support and type morphisms

$$(2.72)\qquad \begin{array}{ccc} \mathsf{p}^-[F] & \xrightarrow{\beta_{G,F}} & \mathsf{p}^-[G] \\ \| & & \| \\ \mathsf{p}[F] & \xrightarrow[(-1)^{\mathrm{dist}(F,G)}\beta_{G,F}]{} & \mathsf{p}[G] \end{array} \qquad \begin{array}{ccc} \mathsf{p}^-[F] & \xrightarrow{w} & \mathsf{p}^-[wF] \\ \| & & \| \\ \mathsf{p}[F] & \xrightarrow[(-1)^{l(w)}w]{} & \mathsf{p}[wF]. \end{array}$$

A map $f : \mathsf{p} \to \mathsf{q}$ of Coxeter $\mathcal{A}$-species induces a map $f : \mathsf{p}^- \to \mathsf{q}^-$ of Coxeter $\mathcal{A}$-species, and yields the functor

$$(2.73)\qquad (-)^- : \mathcal{A}\text{-}\mathsf{CSp} \to \mathcal{A}\text{-}\mathsf{CSp}, \qquad \mathsf{p} \mapsto \mathsf{p}^-.$$

This is the *o-signature functor*. It satisfies all properties of the u-signature functor mentioned above; we do not repeat them separately.

2.8.3. Composing under and over. The u-signature and o-signature functors commute. Thus, we also have the Coxeter species $\mathsf{p}_-^- := (\mathsf{p}_-)^- = (\mathsf{p}^-)_-$. Its support and type morphisms are

$$\begin{array}{ccc} \mathsf{p}_-^-[F] & \overset{\beta_{G,F}}{\dashrightarrow} & \mathsf{p}_-^-[G] \\ \| & & \| \\ \mathsf{p}[F] & \underset{\beta_{G,F}}{\longrightarrow} & \mathsf{p}[G] \end{array} \qquad \begin{array}{ccc} \mathsf{p}_-^-[F] & \overset{w}{\dashrightarrow} & \mathsf{p}_-^-[wF] \\ \| & & \| \\ \mathsf{p}[F] & \underset{(-1)^{l(w)}w}{\longrightarrow} & \mathsf{p}[wF]. \end{array} \tag{2.74}$$

Exercise 2.55. Show: The Coxeter species p, p_-, p^-, p_-^- are not isomorphic in general. See Exercise 4.8 for an important special case.

2.9. Coxeter (co, bi)monoids as functor categories

For a reflection arrangement $\mathcal{A}$, recall that the category of Coxeter species $\mathcal{A}$-CSp is a functor category, namely, it is the category of functors from $\mathcal{A}$-$\mathtt{Cox}$ to Vec. We now show that the categories of Coxeter monoids, Coxeter comonoids, Coxeter bimonoids and their commutative counterparts are also functor categories. In particular, they are all $\Bbbk$-linear categories (as already noted in Exercise 2.33). The notations that we employ for the base categories are summarized in Table 2.7. Similar to $\mathcal{A}$-$\mathtt{Cox}$, these base categories are all defined by presentations. The objects are either faces or flats, while the relations encode the various axioms listed in Table 2.1.

TABLE 2.7. Coxeter (co, bi)monoids as functor categories.

C	$[\mathsf{C}, \mathsf{Vec}]$	C	$[\mathsf{C}, \mathsf{Vec}]$
$\mathcal{A}$-$\mathtt{Cox}^{\mathrm{d}}$	Coxeter $\mathcal{A}$-monoids	$\mathcal{A}$-$\mathtt{Cox}_{\mathrm{c}}^{\mathrm{d}}$	Coxeter $\mathcal{A}$-bimonoids
$\mathcal{A}$-$\mathtt{Cox}_{\mathrm{c}}$	Coxeter $\mathcal{A}$-comonoids	$\mathcal{A}$-$\mathtt{Cox}_{\mathrm{c}}^{\mathrm{e}}$	com. Cox. $\mathcal{A}$-bimonoids
$\mathcal{A}$-$\mathtt{Cox}^{\mathrm{e}}$	com. Cox. $\mathcal{A}$-monoids	$\mathcal{A}$-$\mathtt{Cox}_{\mathrm{r}}^{\mathrm{d}}$	cocom. Cox. $\mathcal{A}$-bimonoids
$\mathcal{A}$-$\mathtt{Cox}_{\mathrm{r}}$	cocom. Cox. $\mathcal{A}$-comonoids	$\mathcal{A}$-$\mathtt{Cox}_{\mathrm{r}}^{\mathrm{e}}$	bicom. Cox. $\mathcal{A}$-bimonoids

Similar comments apply to $\mathcal{A}$-species. The analogue of Table 2.7 for (co, bi)monoids in $\mathcal{A}$-species is [**11**, Table 2.3].

2.9.1. Coxeter (co)monoids in terms of faces. Define the category $\mathcal{A}$-$\mathtt{Cox}_{\mathrm{c}}$ by a presentation. Objects are faces of $\mathcal{A}$. Morphisms are generated by

$$F \xrightarrow{\Delta} G, \quad F \xrightarrow{w} F', \quad A \xrightarrow{\beta} B,$$

subject to the 9 relations

$$\begin{array}{ccc} & G & \\ {}^{\beta}\nearrow & & \searrow^{\beta} \\ F & \xrightarrow[\beta]{} & H \end{array} \qquad \begin{array}{ccc} & F' & \\ {}^{u}\nearrow & & \searrow^{v} \\ F & \xrightarrow[vu]{} & F'' \end{array} \qquad \begin{array}{ccc} & G & \\ {}^{\Delta}\nearrow & & \searrow^{\Delta} \\ F & \xrightarrow[\Delta]{} & H \end{array} \tag{2.75a}$$

(2.75b)
$$\begin{array}{ccc} F & \xrightarrow{\beta} & G \\ {\scriptstyle w}\big\downarrow & & \big\downarrow{\scriptstyle w} \\ F' & \xrightarrow[\beta]{} & G' \end{array} \qquad \begin{array}{ccc} AG & \xrightarrow{\beta} & G \\ {\scriptstyle \Delta}\big\uparrow & & \big\uparrow{\scriptstyle \Delta} \\ A & \xrightarrow[\beta]{} & B \end{array} \qquad \begin{array}{ccc} F & \xrightarrow{w} & F' \\ {\scriptstyle \Delta}\big\uparrow & & \big\uparrow{\scriptstyle \Delta} \\ A & \xrightarrow[w]{} & A' \end{array}$$

(2.75c)
$$(F \xrightarrow{\Delta} F) = \mathrm{id} \qquad (F \xrightarrow{e} F) = \mathrm{id} \qquad (F \xrightarrow{\beta} F) = \mathrm{id}\,.$$

By convention, Δ goes from a smaller face to a bigger face, β is between faces of the same support, and w goes from a face F to $w(F)$.

We can also give a presentation for $\mathcal{A}\text{-}\mathsf{Cox}_{\mathsf{c}}$ by using a smaller set of generators. Namely, we allow $\Delta : F \to G$ only when G covers F, and modify the third relation in (2.75a) to

(2.76)
$$\begin{array}{ccc} F' & \xrightarrow{\Delta} & G \\ {\scriptstyle \Delta}\big\uparrow & & \big\uparrow{\scriptstyle \Delta} \\ A & \xrightarrow[\Delta]{} & F \end{array}$$

whenever G covers F and F', and they in turn cover A. It is also understood that in the second relation in (2.75b), G covers B and hence AG covers A, in the third relation in (2.75b), F covers A and hence F' covers A', and the first relation in (2.75c) is removed.

Exercise 2.56. Show: The category $\mathcal{A}\text{-}\mathsf{Cox}_{\mathsf{c}}$ is generated by the category $\mathcal{A}\text{-}\mathsf{Cox}$ and morphisms $\Delta : F \to G$ subject to the relations

(2.77)
$$\begin{array}{ccc} \sigma^*(G) & \xrightarrow{\sigma} & G \\ {\scriptstyle \Delta}\big\uparrow & & \big\uparrow{\scriptstyle \Delta} \\ A & \xrightarrow[\sigma]{} & B \end{array} \qquad \begin{array}{ccc} & G & \\ {\scriptstyle \Delta}\nearrow & & \searrow{\scriptstyle \Delta} \\ F & \xrightarrow[\Delta]{} & H \end{array} \qquad (F \xrightarrow{\Delta} F) = \mathrm{id}\,.$$

In the first relation, $\sigma : A \to B$ is any morphism in $\mathcal{A}\text{-}\mathsf{Cox}$, with the remaining notation as in (2.7).

Alternatively, $\mathcal{A}\text{-}\mathsf{Cox}_{\mathsf{c}}$ is generated by $\mathcal{A}\text{-}\mathsf{Cox}$ and $\Delta : F \to G$ whenever G covers F, subject to relation (2.76) and the first relation in (2.77) with G covering B.

Let $\mathcal{A}\text{-}\mathsf{Cox}^{\mathsf{d}}$ denote the category obtained from $\mathcal{A}\text{-}\mathsf{Cox}_{\mathsf{c}}$ by reversing arrows labeled Δ, and relabeling them μ. In other words, morphisms of $\mathcal{A}\text{-}\mathsf{Cox}^{\mathsf{d}}$ are generated by

$$G \xrightarrow{\mu} F, \quad F \xrightarrow{w} F', \quad A \xrightarrow{\beta} B.$$

The relations are written down in a similar manner. For the smaller set of generators, we allow $\mu : G \to F$ only when G covers F.

Proposition 2.57. *The category of Coxeter $\mathcal{A}$-comonoids is equivalent to the functor category $[\mathcal{A}\text{-}\mathsf{Cox}_{\mathsf{c}}, \mathsf{Vec}]$. Dually, the category of Coxeter $\mathcal{A}$-monoids is equivalent to $[\mathcal{A}\text{-}\mathsf{Cox}^{\mathsf{d}}, \mathsf{Vec}]$.*

PROOF. We explain the first statement. Using the presentation of $\mathcal{A}\text{-}\mathsf{Cox}_{\mathsf{c}}$, we see: A functor $\mathsf{p} : \mathcal{A}\text{-}\mathsf{Cox}_{\mathsf{c}} \to \mathsf{Vec}$ consists of a family of vector spaces $\mathsf{p}[F]$, one for each face F, equipped with linear maps

$$\mathsf{p}[F] \xrightarrow{\Delta} \mathsf{p}[G], \quad \mathsf{p}[F] \xrightarrow{w} \mathsf{p}[F'], \quad \mathsf{p}[A] \xrightarrow{\beta} \mathsf{p}[B],$$

such that the diagrams

$$
\begin{array}{ccc}
\begin{array}{ccc} & \mathsf{p}[G] & \\ {\scriptstyle\beta}\nearrow & & \searrow{\scriptstyle\beta} \\ \mathsf{p}[F] & \xrightarrow[\beta]{} & \mathsf{p}[H] \end{array}
&
\begin{array}{ccc} & \mathsf{p}[F'] & \\ {\scriptstyle u}\nearrow & & \searrow{\scriptstyle v} \\ \mathsf{p}[F] & \xrightarrow[vu]{} & \mathsf{p}[F''] \end{array}
&
\begin{array}{ccc} & \mathsf{p}[G] & \\ {\scriptstyle\Delta}\nearrow & & \searrow{\scriptstyle\Delta} \\ \mathsf{p}[F] & \xrightarrow[\Delta]{} & \mathsf{p}[H] \end{array}
\end{array}
$$

$$
\begin{array}{ccc}
\begin{array}{ccc} \mathsf{p}[F] & \xrightarrow{\beta} & \mathsf{p}[G] \\ {\scriptstyle w}\downarrow & & \downarrow{\scriptstyle w} \\ \mathsf{p}[F'] & \xrightarrow[\beta]{} & \mathsf{p}[G'] \end{array}
&
\begin{array}{ccc} \mathsf{p}[AG] & \xrightarrow{\beta} & \mathsf{p}[G] \\ {\scriptstyle\Delta}\uparrow & & \uparrow{\scriptstyle\Delta} \\ \mathsf{p}[A] & \xrightarrow[\beta]{} & \mathsf{p}[B] \end{array}
&
\begin{array}{ccc} \mathsf{p}[F] & \xrightarrow{w} & \mathsf{p}[F'] \\ {\scriptstyle\Delta}\uparrow & & \uparrow{\scriptstyle\Delta} \\ \mathsf{p}[A] & \xrightarrow[w]{} & \mathsf{p}[A'] \end{array}
\end{array}
$$

$$(\mathsf{p}[F] \xrightarrow{\Delta} \mathsf{p}[F]) = \mathrm{id} \qquad (\mathsf{p}[F] \xrightarrow{e} \mathsf{p}[F]) = \mathrm{id} \qquad (\mathsf{p}[F] \xrightarrow{\beta} \mathsf{p}[F]) = \mathrm{id}$$

commute. The data involving β and w says that p is a Coxeter $\mathcal{A}$-species, while that involving Δ turns it into a Coxeter $\mathcal{A}$-comonoid. If we work instead with the smaller presentation of $\mathcal{A}\text{-}\mathsf{Cox}_{\mathsf{c}}$ with covering generators for Δ, then we obtain the description of a Coxeter $\mathcal{A}$-comonoid in Proposition 2.26. □

Remark 2.58. The relations used to present categories $\mathcal{A}\text{-}\mathsf{Cox}_{\mathsf{c}}$ and $\mathcal{A}\text{-}\mathsf{Cox}^{\mathsf{d}}$ can be named using the nomenclature in Table 2.1. For instance, diagrams (2.75a) are the *support-support relation*, *type-type relation*, *coassociativity relation*, respectively. The same remark applies to categories that appear in the discussion below.

2.9.2. Coxeter (co)monoids in terms of flats and lunes. Define the category $\mathcal{A}\text{-}\mathsf{Cox}'_{\mathsf{c}}$ by a presentation. Objects are flats of $\mathcal{A}$. Morphisms are generated by

$$\mathrm{Y} \xrightarrow{\Delta_{\mathrm{L}}} \mathrm{X}, \qquad \mathrm{X} \xrightarrow{w} \mathrm{X}',$$

with L a lune of base Y and case X, and $\mathrm{X}' = w\mathrm{X}$, subject to the relations

$$
\begin{array}{ccc}
\begin{array}{ccc} & \mathrm{Y} & \\ {\scriptstyle\Delta_{\mathrm{L}}}\nearrow & & \searrow{\scriptstyle\Delta_{\mathrm{M}}} \\ \mathrm{Z} & \xrightarrow[\Delta_{\mathrm{L}\circ\mathrm{M}}]{} & \mathrm{X} \end{array}
&
\begin{array}{ccc} \mathrm{X} & \xrightarrow{w} & \mathrm{X}' \\ {\scriptstyle\Delta_{\mathrm{L}}}\uparrow & & \uparrow{\scriptstyle\Delta_{w\mathrm{L}}} \\ \mathrm{Y} & \xrightarrow[w]{} & \mathrm{Y}' \end{array}
&
\begin{array}{ccc} & \mathrm{X}' & \\ {\scriptstyle u}\nearrow & & \searrow{\scriptstyle v} \\ \mathrm{X} & \xrightarrow[vu]{} & \mathrm{X}'' \end{array}
\end{array}
\tag{2.78a}
$$

$$(\mathrm{X} \xrightarrow{\Delta_{\mathrm{X}}} \mathrm{X}) = \mathrm{id} \qquad (\mathrm{X} \xrightarrow{e} \mathrm{X}) = \mathrm{id}\,. \tag{2.78b}$$

Exercise 2.59. Give a smaller presentation for $\mathcal{A}\text{-}\mathsf{Cox}'_{\mathsf{c}}$ by using generators $\mathrm{Y} \xrightarrow{\Delta_{\mathrm{L}}} \mathrm{X}$ and $\mathrm{X} \xrightarrow{w} \mathrm{X}'$, the former only when L has slack 1.

Exercise 2.60. Show: The categories $\mathcal{A}\text{-}\mathsf{Cox}_{\mathsf{c}}$ and $\mathcal{A}\text{-}\mathsf{Cox}'_{\mathsf{c}}$ are equivalent. More precisely, the functor $\mathcal{A}\text{-}\mathsf{Cox}_{\mathsf{c}} \to \mathcal{A}\text{-}\mathsf{Cox}'_{\mathsf{c}}$ which on objects sends F to $\mathrm{s}(F)$, and on morphisms sends

$$
\begin{aligned}
(F \xrightarrow{\Delta} G) &\longmapsto (\mathrm{s}(F) \xrightarrow{\Delta_{\mathrm{s}(F,G)}} \mathrm{s}(G)) \\
(F \xrightarrow{w} F') &\longmapsto (\mathrm{s}(F) \xrightarrow{w} \mathrm{s}(F')) \\
(A \xrightarrow{\beta} B) &\longmapsto (\mathrm{s}(A) \xrightarrow{\mathrm{id}} \mathrm{s}(B))
\end{aligned}
$$

is an equivalence. Hence, the category of Coxeter $\mathcal{A}$-comonoids is equivalent to the functor category $[\mathcal{A}\text{-}\mathsf{Cox}'_{\mathsf{c}}, \mathsf{Vec}]$. This corresponds to the description of Coxeter $\mathcal{A}$-comonoids in terms of flats and lunes mentioned in Section 2.4.7.

Let $(\mathcal{A}\text{-}\mathsf{Cox}^{\mathsf{d}})'$ denote the category obtained from $\mathcal{A}\text{-}\mathsf{Cox}'_{\mathsf{c}}$ by reversing arrows labeled Δ_{L}, and relabeling them μ_{L}. It is equivalent to the category $\mathcal{A}\text{-}\mathsf{Cox}^{\mathsf{d}}$. Hence the category of Coxeter $\mathcal{A}$-monoids is equivalent to the functor category $[(\mathcal{A}\text{-}\mathsf{Cox}^{\mathsf{d}})', \mathsf{Vec}]$. This corresponds to the description of Coxeter $\mathcal{A}$-monoids in terms of flats and lunes in Propositions 2.24 and 2.25.

2.9.3. Coxeter bimonoids in terms of faces.

Define the category $\mathcal{A}\text{-}\mathsf{Cox}^{\mathsf{d}}_{\mathsf{c}}$ by a presentation. Objects are faces of $\mathcal{A}$. Morphisms are generated by

$$F \xrightarrow{\Delta} G, \quad G \xrightarrow{\mu} F, \quad F \xrightarrow{w} F', \quad A \xrightarrow{\beta} B,$$

subject to the 14 relations

(2.79a)
$$\begin{array}{ccc} & G & \\ {}^{\mu}\nearrow & & \searrow^{\mu} \\ H & \xrightarrow[\mu]{} & F \end{array} \qquad \begin{array}{ccc} & G & \\ {}^{\beta}\nearrow & & \searrow^{\beta} \\ F & \xrightarrow[\beta]{} & H \end{array} \qquad \begin{array}{ccc} & F' & \\ {}^{u}\nearrow & & \searrow^{v} \\ F & \xrightarrow[vu]{} & F'' \end{array} \qquad \begin{array}{ccc} & G & \\ {}^{\Delta}\nearrow & & \searrow^{\Delta} \\ F & \xrightarrow[\Delta]{} & H \end{array}$$

(2.79b)
$$\begin{array}{ccc} G & \xrightarrow{w} & G' \\ {\scriptstyle\mu}\downarrow & & \downarrow{\scriptstyle\mu} \\ A & \xrightarrow[w]{} & A' \end{array} \qquad \begin{array}{ccc} F & \xrightarrow{\beta} & G \\ {\scriptstyle w}\downarrow & & \downarrow{\scriptstyle w} \\ F' & \xrightarrow[\beta]{} & G' \end{array} \qquad \begin{array}{ccc} F & \xrightarrow{w} & F' \\ {\scriptstyle\Delta}\uparrow & & \uparrow{\scriptstyle\Delta} \\ A & \xrightarrow[w]{} & A' \end{array}$$

(2.79c)
$$\begin{array}{ccc} F & \xrightarrow{\beta} & BF \\ {\scriptstyle\mu}\downarrow & & \downarrow{\scriptstyle\mu} \\ A & \xrightarrow[\beta]{} & B \end{array} \qquad \begin{array}{ccccc} FG & & \xrightarrow{\beta} & & GF \\ {\scriptstyle\Delta}\uparrow & & & & \downarrow{\scriptstyle\mu} \\ F & & & & G \\ & {}_{\mu}\searrow & & \nearrow_{\Delta} & \\ & & A & & \end{array} \qquad \begin{array}{ccc} AG & \xrightarrow{\beta} & G \\ {\scriptstyle\Delta}\uparrow & & \uparrow{\scriptstyle\Delta} \\ A & \xrightarrow[\beta]{} & B \end{array}$$

(2.79d) $(F \xrightarrow{\mu} F) = \mathrm{id} \quad (F \xrightarrow{\beta} F) = \mathrm{id} \quad (F \xrightarrow{\Delta} F) = \mathrm{id} \quad (F \xrightarrow{e} F) = \mathrm{id}.$

By convention, Δ goes from a smaller face to a bigger face, μ from a bigger face to a smaller face, β is between faces of the same support, and w goes from a face F to $w(F)$.

Exercise 2.61. Give a smaller presentation for $\mathcal{A}\text{-}\mathsf{Cox}^{\mathsf{d}}_{\mathsf{c}}$ by using generators $\beta : A \to B$, $w : F \to F'$, $\Delta : F \to G$, $\mu : G \to F$ with the latter two only when G covers F.

Proposition 2.62. *The category of Coxeter $\mathcal{A}$-bimonoids is equivalent to the functor category $[\mathcal{A}\text{-}\mathsf{Cox}^{\mathsf{d}}_{\mathsf{c}}, \mathsf{Vec}]$.*

PROOF. This is similar to the proof of Proposition 2.57. The 14 relations listed above translate to the 14 axioms listed in Summary 2.1. The middle diagram in (2.79c) is the *bimonoid relation*. It translates to the bimonoid axiom (2.39). ☐

2.9.4. Coxeter bimonoids in terms of flats and lunes. Define the category $(\mathcal{A}\text{-}\mathsf{Cox}^{\mathsf{d}}_{\mathsf{c}})'$ by a presentation. Objects are flats of $\mathcal{A}$. Morphisms are generated by

$$\mathrm{Y} \xrightarrow{\Delta_{\mathrm{L}}} \mathrm{X}, \quad \mathrm{X} \xrightarrow{w} \mathrm{X}', \quad \mathrm{X} \xrightarrow{\mu_{\mathrm{L}}} \mathrm{Y},$$

with L a lune of base Y and case X, and $\mathrm{X}' = w\mathrm{X}$, subject to the 9 relations

(2.80a)
$$\begin{array}{ccc} & \mathrm{Y} & \\ {\scriptstyle\Delta_{\mathrm{L}}}\nearrow & & \searrow{\scriptstyle\Delta_{\mathrm{M}}} \\ \mathrm{Z} & \xrightarrow[\Delta_{\mathrm{L}\circ\mathrm{M}}]{} & \mathrm{X} \end{array} \qquad \begin{array}{ccc} & \mathrm{X}' & \\ {\scriptstyle u}\nearrow & & \searrow{\scriptstyle v} \\ \mathrm{X} & \xrightarrow[vu]{} & \mathrm{X}'' \end{array} \qquad \begin{array}{ccc} & \mathrm{Y} & \\ {\scriptstyle\mu_{\mathrm{M}}}\nearrow & & \searrow{\scriptstyle\mu_{\mathrm{L}}} \\ \mathrm{X} & \xrightarrow[\mu_{\mathrm{L}\circ\mathrm{M}}]{} & \mathrm{Z} \end{array}$$

(2.80b)
$$\begin{array}{ccc} \mathrm{X} & \xrightarrow{w} & \mathrm{X}' \\ {\scriptstyle\Delta_{\mathrm{L}}}\uparrow & & \uparrow{\scriptstyle\Delta_{w\mathrm{L}}} \\ \mathrm{Y} & \xrightarrow[w]{} & \mathrm{Y}' \end{array} \qquad \begin{array}{ccccc} & & \mathrm{X}\vee\mathrm{Y} & & \\ & {\scriptstyle\Delta_{\mathrm{X}\cdot\mathrm{M}}}\nearrow & & \searrow{\scriptstyle\mu_{\mathrm{Y}\cdot\mathrm{L}}} & \\ \mathrm{X} & & & & \mathrm{Y} \\ & {\scriptstyle\mu_{\mathrm{L}}}\searrow & & \nearrow{\scriptstyle\Delta_{\mathrm{M}}} & \\ & & \mathrm{Z} & & \end{array} \qquad \begin{array}{ccc} \mathrm{X} & \xrightarrow{w} & \mathrm{X}' \\ {\scriptstyle\mu_{\mathrm{L}}}\downarrow & & \downarrow{\scriptstyle\mu_{w\mathrm{L}}} \\ \mathrm{Y} & \xrightarrow[w]{} & \mathrm{Y}' \end{array}$$

(2.80c)
$$(\mathrm{X} \xrightarrow{\Delta_{\mathrm{X}}} \mathrm{X}) = \mathrm{id} \qquad (\mathrm{X} \xrightarrow{e} \mathrm{X}) = \mathrm{id} \qquad (\mathrm{X} \xrightarrow{\mu_{\mathrm{X}}} \mathrm{X}) = \mathrm{id}\,.$$

We build on the result of Exercise 2.60 as follows.

Exercise 2.63. Show: The categories $\mathcal{A}\text{-}\mathsf{Cox}^{\mathsf{d}}_{\mathsf{c}}$ and $(\mathcal{A}\text{-}\mathsf{Cox}^{\mathsf{d}}_{\mathsf{c}})'$ are equivalent. Deduce: The category of Coxeter $\mathcal{A}$-bimonoids is equivalent to the functor category $[(\mathcal{A}\text{-}\mathsf{Cox}^{\mathsf{d}}_{\mathsf{c}})', \mathsf{Vec}]$. This corresponds to the description of Coxeter $\mathcal{A}$-bimonoids in terms of flats and lunes in Proposition 2.28. The 9 relations listed above translate to the 9 axioms listed in Summary 2.2. The middle diagram in (2.80b) is the *bimonoid relation*. It translates to the bimonoid axiom (2.41).

2.9.5. (Co)commutative Coxeter (co)monoids. Define the category $\mathcal{A}\text{-}\mathsf{Cox}_{\mathsf{r}}$ by a presentation. Objects are flats of $\mathcal{A}$. Morphisms are generated by

$$\mathrm{Y} \xrightarrow{\Delta} \mathrm{X}, \qquad \mathrm{X} \xrightarrow{w} \mathrm{X}',$$

subject to the 5 relations

(2.81a)
$$\begin{array}{ccc} & \mathrm{Y} & \\ {\scriptstyle\Delta}\nearrow & & \searrow{\scriptstyle\Delta} \\ \mathrm{Z} & \xrightarrow[\Delta]{} & \mathrm{X} \end{array} \qquad \begin{array}{ccc} \mathrm{X} & \xrightarrow{w} & \mathrm{X}' \\ {\scriptstyle\Delta}\uparrow & & \uparrow{\scriptstyle\Delta} \\ \mathrm{Y} & \xrightarrow[w]{} & \mathrm{Y}' \end{array} \qquad \begin{array}{ccc} & \mathrm{X}' & \\ {\scriptstyle u}\nearrow & & \searrow{\scriptstyle v} \\ \mathrm{X} & \xrightarrow[vu]{} & \mathrm{X}'' \end{array}$$

(2.81b)
$$(\mathrm{X} \xrightarrow{\Delta} \mathrm{X}) = \mathrm{id} \qquad (\mathrm{X} \xrightarrow{e} \mathrm{X}) = \mathrm{id}\,.$$

By convention, Δ goes from a smaller flat to a bigger flat, and w goes from a flat X to $w(\mathrm{X})$.

The category $\mathcal{A}\text{-}\mathsf{Cox}_{\mathsf{r}}$ is equivalent to the category $\mathcal{A}\text{-}\mathsf{Cox}_{\mathsf{c}}$ subject to the *cocommutativity relation*

(2.82)
$$\begin{array}{ccc} F & \xrightarrow{\beta} & G. \\ {\scriptstyle\Delta}\nwarrow & & \nearrow{\scriptstyle\Delta} \\ & A & \end{array}$$

This can be unified with the middle relation in (2.75b) by using the more general relation

$$\begin{array}{ccc} F & \xrightarrow{\beta} & G \\ {\scriptstyle\Delta}\uparrow & & \uparrow{\scriptstyle\Delta} \\ A & \xrightarrow[\beta]{} & B \end{array} \tag{2.83}$$

whenever $\mathrm{s}(A) = \mathrm{s}(B)$, $\mathrm{s}(F) = \mathrm{s}(G)$, $F \geq A$, $G \geq B$. In other words, $\mathcal{A}\text{-}\mathtt{Cox}_{\mathtt{r}}$ is equivalent to the category defined exactly like $\mathcal{A}\text{-}\mathtt{Cox}_{\mathtt{c}}$ but with the middle relation in (2.75b) replaced by (2.83).

Let $\mathcal{A}\text{-}\mathtt{Cox}^{\mathtt{e}}$ denote the category obtained from $\mathcal{A}\text{-}\mathtt{Cox}_{\mathtt{r}}$ by reversing arrows labeled Δ, and relabeling them μ. The category $\mathcal{A}\text{-}\mathtt{Cox}^{\mathtt{e}}$ is equivalent to the category $\mathcal{A}\text{-}\mathtt{Cox}^{\mathtt{d}}$ subject to the *commutativity relation*

$$\begin{array}{ccc} F & \xrightarrow{\beta} & G \\ & {\scriptstyle\mu}\searrow \quad \swarrow{\scriptstyle\mu} & \\ & A. & \end{array} \tag{2.84}$$

This can be replaced by the more general relation

$$\begin{array}{ccc} F & \xrightarrow{\beta} & G \\ {\scriptstyle\mu}\downarrow & & \downarrow{\scriptstyle\mu} \\ A & \xrightarrow[\beta]{} & B \end{array} \tag{2.85}$$

whenever $\mathrm{s}(A) = \mathrm{s}(B)$, $\mathrm{s}(F) = \mathrm{s}(G)$, $F \geq A$, $G \geq B$.

Exercise 2.64. Give smaller presentations for the categories $\mathcal{A}\text{-}\mathtt{Cox}_{\mathtt{r}}$ and $\mathcal{A}\text{-}\mathtt{Cox}^{\mathtt{e}}$ by restricting the generators μ and Δ to go between flats whose ranks differ by 1.

Proposition 2.65. *The category of cocommutative Coxeter $\mathcal{A}$-comonoids is equivalent to the functor category $[\mathcal{A}\text{-}\mathtt{Cox}_{\mathtt{r}}, \mathsf{Vec}]$. Dually, the category of commutative Coxeter $\mathcal{A}$-monoids is equivalent to $[\mathcal{A}\text{-}\mathtt{Cox}^{\mathtt{e}}, \mathsf{Vec}]$.*

The presentation for $\mathcal{A}\text{-}\mathtt{Cox}^{\mathtt{e}}$ yields the description of a commutative Coxeter $\mathcal{A}$-monoid in Proposition 2.38. The smaller presentation with covering generators corresponds to the description in Proposition 2.41. The relation (2.84) translates to the commutativity axiom (2.49), while (2.85) translates to (2.50). Similar comments apply to $\mathcal{A}\text{-}\mathtt{Cox}_{\mathtt{r}}$.

Exercise 2.66. Check: There is a functor $\mathcal{A}\text{-}\mathtt{Cox}'_{\mathtt{c}} \to \mathcal{A}\text{-}\mathtt{Cox}_{\mathtt{r}}$ which sends

$$(\mathrm{Y} \xrightarrow{\Delta_{\mathrm{L}}} \mathrm{X}) \longmapsto (\mathrm{Y} \xrightarrow{\Delta} \mathrm{X}) \quad \text{and} \quad (\mathrm{X} \xrightarrow{w} \mathrm{X}') \longmapsto (\mathrm{X} \xrightarrow{w} \mathrm{X}').$$

This induces a functor $[\mathcal{A}\text{-}\mathtt{Cox}_{\mathtt{r}}, \mathsf{Vec}] \to [\mathcal{A}\text{-}\mathtt{Cox}'_{\mathtt{c}}, \mathsf{Vec}]$. It is the same as the inclusion of the category of cocommutative $\mathcal{A}$-comonoids into the category of $\mathcal{A}$-comonoids. Dual statements apply to $\mathcal{A}$-monoids.

2.9.6. (Co)commutative Coxeter bimonoids. Define $\mathcal{A}\text{-}\mathsf{Cox}^{\mathsf{d}}_{\mathsf{r}}$ to be the category $\mathcal{A}\text{-}\mathsf{Cox}^{\mathsf{d}}_{\mathsf{c}}$ subject to the cocommutativity relation (2.82). Equivalently, it is the category defined exactly like $\mathcal{A}\text{-}\mathsf{Cox}^{\mathsf{d}}_{\mathsf{c}}$ but with the last relation in (2.79c) replaced by (2.83).

Similarly, define $\mathcal{A}\text{-}\mathsf{Cox}^{\mathsf{e}}_{\mathsf{c}}$ to be the category $\mathcal{A}\text{-}\mathsf{Cox}^{\mathsf{d}}_{\mathsf{c}}$ subject to the commutativity relation (2.84). Equivalently, it is the category defined exactly like $\mathcal{A}\text{-}\mathsf{Cox}^{\mathsf{d}}_{\mathsf{c}}$ but with the first relation in (2.79c) replaced by (2.85).

Proposition 2.67. *The category of cocommutative Coxeter $\mathcal{A}$-bimonoids is equivalent to the functor category $[\mathcal{A}\text{-}\mathsf{Cox}^{\mathsf{d}}_{\mathsf{r}}, \mathsf{Vec}]$. Dually, the category of commutative Coxeter $\mathcal{A}$-bimonoids is equivalent to $[\mathcal{A}\text{-}\mathsf{Cox}^{\mathsf{e}}_{\mathsf{c}}, \mathsf{Vec}]$.*

2.9.7. Bicommutative Coxeter bimonoids. Define the category $\mathcal{A}\text{-}\mathsf{Cox}^{\mathsf{e}}_{\mathsf{r}}$ by a presentation. Objects are flats of $\mathcal{A}$. Morphisms are generated by

$$\mathrm{Y} \xrightarrow{\Delta} \mathrm{X}, \quad \mathrm{X} \xrightarrow{w} \mathrm{X}', \quad \mathrm{X} \xrightarrow{\mu} \mathrm{Y},$$

subject to the 9 relations

(2.86a)

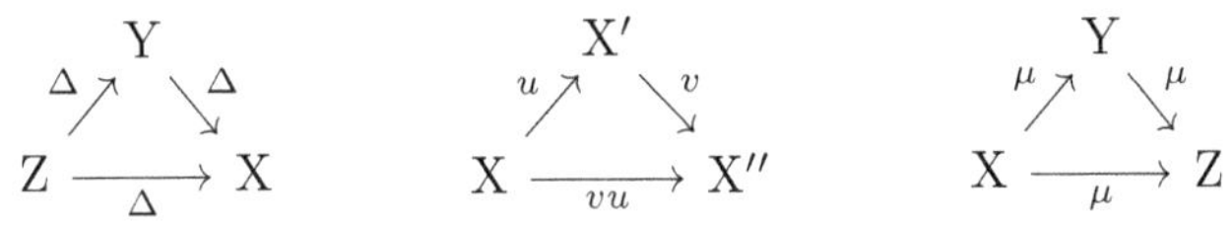

(2.86b)

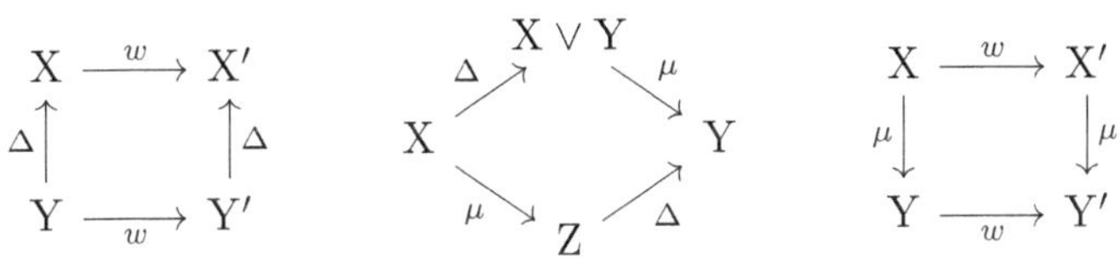

(2.86c) $\qquad (\mathrm{X} \xrightarrow{\Delta} \mathrm{X}) = \mathrm{id} \qquad (\mathrm{X} \xrightarrow{e} \mathrm{X}) = \mathrm{id} \qquad (\mathrm{X} \xrightarrow{\mu} \mathrm{X}) = \mathrm{id}\,.$

By convention, Δ goes from a smaller flat to a bigger flat, μ goes from a bigger flat to a smaller flat, and $w \in W$ goes from a flat X to $w(\mathrm{X})$.

The category $\mathcal{A}\text{-}\mathsf{Cox}^{\mathsf{e}}_{\mathsf{r}}$ is equivalent to the category $\mathcal{A}\text{-}\mathsf{Cox}^{\mathsf{d}}_{\mathsf{c}}$ subject to the extra relations (2.82) and (2.84). Equivalently, it is the category defined exactly like $\mathcal{A}\text{-}\mathsf{Cox}^{\mathsf{d}}_{\mathsf{c}}$ with relations (2.85) and (2.83) replacing the first and third relations, respectively, in (2.79c).

Exercise 2.68. Give a smaller presentation for $\mathcal{A}\text{-}\mathsf{Cox}^{\mathsf{e}}_{\mathsf{r}}$ by using generators $\mathrm{X} \xrightarrow{w} \mathrm{X}'$, $\mathrm{Y} \xrightarrow{\Delta} \mathrm{X}$, $\mathrm{X} \xrightarrow{\mu} \mathrm{Y}$ with the latter two only when X covers Y.

Proposition 2.69. *The category of bicommutative Coxeter $\mathcal{A}$-bimonoids is equivalent to the functor category $[\mathcal{A}\text{-}\mathsf{Cox}^{\mathsf{e}}_{\mathsf{r}}, \mathsf{Vec}]$.*

The 9 relations listed above translate to the 9 axioms listed in Summary 2.3. The middle diagram in (2.86b) is the *bicommutative bimonoid relation.* It translates to the bicommutative bimonoid axiom (2.59).

2.9.8. Normal forms for morphisms. Similar to the situation for Coxeter species (Section 2.1.2), morphisms in the base categories in Table 2.7 have nice normal forms. They are summarized in Table 2.8. By convention, we place the type morphism w at the beginning of the normal form. This is possible since w commutes with Δ, β, μ.

TABLE 2.8. Normal forms for morphisms.

Category	Normal form	Category	Normal form
$\mathcal{A}\text{-}\mathtt{Cox}^{\mathtt{d}}$	$F \xrightarrow{w} F' \xrightarrow{\beta} G' \xrightarrow{\mu} G$	$\mathcal{A}\text{-}\mathtt{Cox}^{\mathtt{d}}_{\mathtt{c}}$	$F \xrightarrow{w} F' \xrightarrow{\Delta} F'' \xrightarrow{\beta} G'' \xrightarrow{\mu} G$
$\mathcal{A}\text{-}\mathtt{Cox}_{\mathtt{c}}$	$F \xrightarrow{w} F' \xrightarrow{\Delta} G' \xrightarrow{\beta} G$	$\mathcal{A}\text{-}\mathtt{Cox}^{\mathtt{e}}_{\mathtt{c}}$	$F \xrightarrow{w} F' \xrightarrow{\Delta} H \xrightarrow{\beta} GH \xrightarrow{\mu} G$
$\mathcal{A}\text{-}\mathtt{Cox}^{\mathtt{e}}$	$\mathrm{X} \xrightarrow{w} \mathrm{X}' \xrightarrow{\mu} \mathrm{Y}$	$\mathcal{A}\text{-}\mathtt{Cox}^{\mathtt{d}}_{\mathtt{r}}$	$F \xrightarrow{w} F' \xrightarrow{\Delta} F'H \xrightarrow{\beta} H \xrightarrow{\mu} G$
$\mathcal{A}\text{-}\mathtt{Cox}_{\mathtt{r}}$	$\mathrm{X} \xrightarrow{w} \mathrm{X}' \xrightarrow{\Delta} \mathrm{Y}$	$\mathcal{A}\text{-}\mathtt{Cox}^{\mathtt{e}}_{\mathtt{r}}$	$\mathrm{X} \xrightarrow{w} \mathrm{X}' \xrightarrow{\Delta} \mathrm{Y}' \xrightarrow{\mu} \mathrm{Y}$

In words,

- a morphism in $\mathcal{A}\text{-}\mathtt{Cox}^{\mathtt{d}}$ amounts to choosing a G' and a w such that $G' \geq G$ and $\mathrm{s}(G') = \mathrm{s}(w(F))$,
- a morphism in $\mathcal{A}\text{-}\mathtt{Cox}_{\mathtt{c}}$ amounts to choosing a G' and a w such that $G' \geq w(F)$ and $\mathrm{s}(G') = \mathrm{s}(G)$.

Similar descriptions can be given for the remaining categories. For instance,

- a morphism in $\mathcal{A}\text{-}\mathtt{Cox}^{\mathtt{d}}_{\mathtt{r}}$ amounts to choosing a H and a w such that $H \geq G$ and $\mathrm{s}(H) \geq \mathrm{s}(w(F))$,
- a morphism in $\mathcal{A}\text{-}\mathtt{Cox}^{\mathtt{e}}_{\mathtt{c}}$ amounts to choosing a H and a w such that $\mathrm{s}(H) \geq \mathrm{s}(G)$ and $H \geq w(F)$,
- a morphism in $\mathcal{A}\text{-}\mathtt{Cox}^{\mathtt{e}}_{\mathtt{r}}$ amounts to choosing a Y' and a w such that Y' is greater than both $w(\mathrm{X})$ and Y.

The analogous table for $\mathcal{A}$-species is [**11**, Table 2.4]. The key difference among the two is the absence of type morphisms in the latter. This is made more precise below.

2.9.9. Semidirect product groupoids. Recall from Section 2.7.2 that W acts on $\mathcal{A}\text{-}\mathtt{Hyp}$, and $\mathcal{A}\text{-}\mathtt{Cox}$ is the resulting semidirect product groupoid. In a similar manner, W acts on all categories in [**11**, Table 2.3], and the resulting semidirect product groupoids are the categories in Table 2.7. In other words:

Lemma 2.70. *We have*

$$\begin{aligned} \mathcal{A}\text{-}\mathtt{Cox}^{\mathtt{d}} &= \mathcal{A}\text{-}\mathtt{Hyp}^{\mathtt{d}} \rtimes W, & \mathcal{A}\text{-}\mathtt{Cox}^{\mathtt{d}}_{\mathtt{c}} &= \mathcal{A}\text{-}\mathtt{Hyp}^{\mathtt{d}}_{\mathtt{c}} \rtimes W, \\ \mathcal{A}\text{-}\mathtt{Cox}_{\mathtt{c}} &= \mathcal{A}\text{-}\mathtt{Hyp}_{\mathtt{c}} \rtimes W, & \mathcal{A}\text{-}\mathtt{Cox}^{\mathtt{e}}_{\mathtt{c}} &= \mathcal{A}\text{-}\mathtt{Hyp}^{\mathtt{e}}_{\mathtt{c}} \rtimes W, \\ \mathcal{A}\text{-}\mathtt{Cox}^{\mathtt{e}} &= \mathcal{A}\text{-}\mathtt{Hyp}^{\mathtt{e}} \rtimes W, & \mathcal{A}\text{-}\mathtt{Cox}^{\mathtt{d}}_{\mathtt{r}} &= \mathcal{A}\text{-}\mathtt{Hyp}^{\mathtt{d}}_{\mathtt{r}} \rtimes W, \\ \mathcal{A}\text{-}\mathtt{Cox}_{\mathtt{r}} &= \mathcal{A}\text{-}\mathtt{Hyp}_{\mathtt{r}} \rtimes W, & \mathcal{A}\text{-}\mathtt{Cox}^{\mathtt{e}}_{\mathtt{r}} &= \mathcal{A}\text{-}\mathtt{Hyp}^{\mathtt{e}}_{\mathtt{r}} \rtimes W. \end{aligned}$$

Let us focus on one of the entries. The categories $\mathcal{A}\text{-}\mathtt{Cox}_{\mathtt{c}}$ and $\mathcal{A}\text{-}\mathtt{Hyp}_{\mathtt{c}}$ have the same objects, while a morphism in $\mathcal{A}\text{-}\mathtt{Cox}_{\mathtt{c}}$ consists of an element of W and a morphism in $\mathcal{A}\text{-}\mathtt{Hyp}_{\mathtt{c}}$. A normal form for a morphism in $\mathcal{A}\text{-}\mathtt{Hyp}_{\mathtt{c}}$ is

$F' \xrightarrow{\Delta} G' \xrightarrow{\beta} G$, see [11, (2.69)]. By tagging on an element of W, we obtain a normal form for a morphism in $\mathcal{A}$-$\mathsf{Cox_c}$. This is precisely what is given in Table 2.8.

2.10. Bimonads on Coxeter species

This section assumes basic familiarity with monads and their algebras, comonads and their coalgebras, and bimonads and their bialgebras. These notions are reviewed in Appendix A.1.

We define a monad $\mathcal{T}$ on Coxeter species whose algebras are Coxeter monoids, a comonad $\mathcal{T}^\vee$ whose coalgebras are Coxeter comonoids, and a bimonad $(\mathcal{T}, \mathcal{T}^\vee, \lambda)$ whose bialgebras are Coxeter bimonoids. These also have commutative, signed commutative and q-analogues. These constructions are analogous to the ones on $\mathcal{A}$-species; the latter are discussed in detail in [11, Chapter 3], so we will be brief here.

2.10.1. Coxeter bimonoids in terms of faces. Let $\mathcal{A}$ be a reflection arrangement. The basic result is stated below.

Proposition 2.71. *The following are equivalences of categories.*

$$\begin{aligned} \mathcal{T}\text{-}\mathit{algebras} &\cong \mathit{Coxeter}\ \mathcal{A}\text{-}\mathit{monoids} \\ \mathcal{T}^\vee\text{-}\mathit{coalgebras} &\cong \mathit{Coxeter}\ \mathcal{A}\text{-}\mathit{comonoids} \\ (\mathcal{T}, \mathcal{T}^\vee, \lambda)\text{-}\mathit{bialgebras} &\cong \mathit{Coxeter}\ \mathcal{A}\text{-}\mathit{bimonoids} \\ (\mathcal{T}, \mathcal{T}^\vee, \lambda_q)\text{-}\mathit{bialgebras} &\cong \mathit{Coxeter}\ \mathcal{A}\text{-}q\text{-}\mathit{bimonoids} \end{aligned}$$

Let us first construct the functor $\mathcal{T} : \mathcal{A}\text{-}\mathsf{CSp} \to \mathcal{A}\text{-}\mathsf{CSp}$. For a Coxeter $\mathcal{A}$-species p, define the Coxeter $\mathcal{A}$-species $\mathcal{T}(\mathsf{p})$ on the A-component by

$$\mathcal{T}(\mathsf{p})[A] := \bigoplus_{F:\, A \leq F} \mathsf{p}[F]. \tag{2.87}$$

Support and type morphisms of $\mathcal{T}(\mathsf{p})$ are defined from those of p as follows.

- Suppose $\mathrm{s}(A) = \mathrm{s}(B)$. Then, for every F greater than A, there is a face $G := BF$ greater than B with $\mathrm{s}(G) = \mathrm{s}(F)$. The sum of the linear maps $\beta_{G,F} : \mathsf{p}[F] \to \mathsf{p}[G]$ yields the linear map $\beta_{B,A} : \mathcal{T}(\mathsf{p})[A] \to \mathcal{T}(\mathsf{p})[B]$.
- Suppose $w \in W$ and $w(A) = A'$. Then, for every F greater than A, there is a face $F' := w(F)$ greater than A'. The sum of the linear maps $w : \mathsf{p}[F] \to \mathsf{p}[F']$ yields the linear map $w : \mathcal{T}(\mathsf{p})[A] \to \mathcal{T}(\mathsf{p})[A']$.

More abstractly, a morphism $\sigma : A \to B$ induces a morphism $\sigma : F \to G$, where $G = \sigma_*(F)$ as in (2.7). The sum of the linear maps $\sigma : \mathsf{p}[F] \to \mathsf{p}[G]$ yields the linear map

$$\sigma : \mathcal{T}(\mathsf{p})[A] \to \mathcal{T}(\mathsf{p})[B].$$

This turns $\mathcal{T}(\mathsf{p})$ into a Coxeter $\mathcal{A}$-species. Further, a map $f : \mathsf{p} \to \mathsf{q}$ of Coxeter $\mathcal{A}$-species yields a map $\mathcal{T}(f) : \mathcal{T}(\mathsf{p}) \to \mathcal{T}(\mathsf{q})$ of Coxeter $\mathcal{A}$-species whose A-component $\mathcal{T}(f)_A$ is the sum of the linear maps $f_F : \mathsf{p}[F] \to \mathsf{q}[F]$ over all $F \geq A$. Thus, $\mathcal{T}$ is a functor. Note: It is $\Bbbk$-linear.

We now specify the monad structure on $\mathcal{T}$. For that, first observe:

$$\mathcal{T}\mathcal{T}(\mathsf{p})[A] = \bigoplus_{(F,G):\, A \leq F \leq G} \mathsf{p}[G].$$

The sum is over both F and G. The structure map

$$\mathcal{T}\mathcal{T} \to \mathcal{T}, \qquad \bigoplus_{(F,G):\, A \leq F \leq G} \mathsf{p}[G] \to \bigoplus_{G:\, A \leq G} \mathsf{p}[G]$$

evaluated on p on the A-component sends each summand in the lhs identically to the matching summand in the rhs, while

$$\mathrm{id} \to \mathcal{T}, \qquad \mathsf{p}[A] \to \bigoplus_{F:\, A \leq F} \mathsf{p}[F]$$

evaluated on p on the A-component is given by inclusion.

One can check that $\mathcal{T}$-algebras are the same as Coxeter $\mathcal{A}$-monoids (Section 2.4.1). For a $\mathcal{T}$-algebra a, the product component μ_A^F of the corresponding Coxeter $\mathcal{A}$-monoid is the F-component of $\mathcal{T}(\mathsf{a})[A] \to \mathsf{a}[A]$. For more details, see the analogous result for monoids in species in [**11**, Proposition 3.2].

The comonad $\mathcal{T}^\vee$ on Coxeter $\mathcal{A}$-species is constructed dually. As a functor, $\mathcal{T}^\vee = \mathcal{T}$. The comonad structure on $\mathcal{T}^\vee$ is as follows. The map

$$\mathcal{T}^\vee \to \mathcal{T}^\vee\mathcal{T}^\vee, \qquad \bigoplus_{G:\, A \leq G} \mathsf{p}[G] \to \bigoplus_{(F,G):\, A \leq F \leq G} \mathsf{p}[G]$$

evaluated on p on the A-component sends each summand in the lhs identically to all matching summands in the rhs, while

$$\mathcal{T}^\vee \to \mathrm{id}, \qquad \bigoplus_{F:\, A \leq F} \mathsf{p}[F] \to \mathsf{p}[A]$$

evaluated on p on the A-component sends $\mathsf{p}[A]$ to itself, and all other summands to zero. Coalgebras over $\mathcal{T}^\vee$ are the same as Coxeter $\mathcal{A}$-comonoids (Section 2.4.5).

Now define a mixed distributive law between $\mathcal{T}$ and $\mathcal{T}^\vee$ as follows. The map

$$\lambda : \mathcal{T}\mathcal{T}^\vee \to \mathcal{T}^\vee\mathcal{T}, \qquad \bigoplus_{(F,G):\, A \leq F \leq G} \mathsf{p}[G] \to \bigoplus_{(F',G'):\, A \leq F' \leq G'} \mathsf{p}[G'] \tag{2.88}$$

evaluated on p on the A-component has matrix-components

$$\big(\lambda_{F,G,F',G'} : \mathsf{p}[G] \to \mathsf{p}[G']\big) := \begin{cases} \beta_{G',G} & \text{if } FF' = G \text{ and } F'F = G', \\ 0 & \text{otherwise.} \end{cases}$$

Note very carefully that it does not involve type morphisms.

Remark 2.72 (Matrix-component). The term *matrix-component* (used above and elsewhere in the book) refers to a component of a linear map from a direct sum to a direct sum, starting from a particular summand and ending in a particular summand. See [**11**, Appendix A.4] for more explanation.

Exercise 2.73. Check: (2.88) is a natural transformation. In particular, evaluated on p, it is a map of Coxeter $\mathcal{A}$-species, that is, diagrams (2.13) commute. (Use support-support and support-type axioms.)

Exercise 2.74. Check: For faces G, F', G' all greater than A,

$$\big(\lambda_{A,G,F',G'} : \mathsf{p}[G] \to \mathsf{p}[G']\big) = \begin{cases} \mathrm{id} & \text{if } F' = G' = G, \\ 0 & \text{otherwise.} \end{cases}$$

By symmetry, there is a similar statement for $\lambda_{F,G,A,G'}$.

Theorem 2.75. *The triple $(\mathcal{T}, \mathcal{T}^\vee, \lambda)$ is a bimonad on Coxeter $\mathcal{A}$-species. Its bialgebras are the same as Coxeter $\mathcal{A}$-bimonoids.*

PROOF. This can be deduced from the corresponding result for species [**11**, Theorem 3.4 and Proposition 3.5]. This is because type morphisms do not play any role in the calculations. For convenience, we briefly recall the checking of the left diagram in (A.7a). Evaluated on p on the A-component, it is as follows.

$$(2.89)\qquad \begin{array}{ccccc} \displaystyle\bigoplus_{A \le F \le G \le H} \mathsf{p}[H] & \xrightarrow{\ \lambda\ } & \displaystyle\bigoplus_{A \le F' \le G' \le H'} \mathsf{p}[H'] & \xrightarrow{\ \lambda\ } & \displaystyle\bigoplus_{A \le F' \le G'' \le H''} \mathsf{p}[H''] \\ \uparrow & & & & \uparrow \\ \displaystyle\bigoplus_{A \le F \le H} \mathsf{p}[H] & & \xrightarrow[\lambda]{\qquad\qquad\qquad} & & \displaystyle\bigoplus_{A \le G'' \le H''} \mathsf{p}[H''] \end{array}$$

The sums are over faces other than A (which is fixed). Fix $F \le H$ and $F' \le G'' \le H''$, and consider the matrix-component $\mathsf{p}[H] \to \mathsf{p}[H'']$ following both directions in the diagram.

Going across and up, the matrix-component is nonzero and equals $\beta_{H'',H}$ iff conditions in the rhs below hold.

$$FF' \le H, \quad F'FG'' = F'H, \quad G''F = H'' \iff FG'' = H, \quad G''F = H''.$$

Going up and across, the matrix-component is nonzero and equals $\beta_{H'',H}$ iff conditions in the lhs above hold. (In this case, $G = FF'$, $G' = F'F$, $H' = F'H$ are forced.) For details on the equivalence of conditions in the lhs and rhs above, see the proof of [**11**, Theorem 3.4].

The diagrams in (A.7b) can be checked using Exercise 2.74. □

More generally, for a scalar q, one can define the mixed distributive law

$$(2.90)\qquad \lambda_q : \mathcal{T}\mathcal{T}^\vee \to \mathcal{T}^\vee\mathcal{T}, \qquad (\lambda_q)_{F,G,F',G'} := q^{\mathrm{dist}(G,G')} \lambda_{F,G,F',G'}.$$

It is a deformation of λ which one recovers by setting $q = 1$. In fact, for any W-invariant distance function υ, one can define

$$(2.91)\qquad \lambda_\upsilon : \mathcal{T}\mathcal{T}^\vee \to \mathcal{T}^\vee\mathcal{T}, \qquad (\lambda_\upsilon)_{F,G,F',G'} := \upsilon_{G,G'} \lambda_{F,G,F',G'}.$$

One recovers the previous case by letting υ be the q-distance function.

Exercise 2.76. Check: (2.90) and more generally (2.91) is a natural transformation. Note very carefully how W-invariance of the distance function υ gets used in the latter check.

Theorem 2.77. *For a scalar q, the triple $(\mathcal{T}, \mathcal{T}^\vee, \lambda_q)$ is a bimonad on Coxeter $\mathcal{A}$-species whose bialgebras are the same as Coxeter $\mathcal{A}$-q-bimonoids. More generally, for any W-invariant distance function υ, $(\mathcal{T}, \mathcal{T}^\vee, \lambda_\upsilon)$ is a bimonad whose bialgebras are the same as Coxeter $\mathcal{A}$-υ-bimonoids.*

PROOF. We rerun the previous proof, with λ replaced by λ_q. The additional step is to check that powers of q match along the two directions in diagram (2.89). This uses the gate property. More generally, for a distance function υ, recall that the gate property is among its defining axioms [**11**, (1.30)], see also [**10**, Section E.2]. Also Exercise 2.74 works in this generality. □

2.10.2. Coxeter bimonoids in terms of flats and lunes. We now recast the above construction in terms of flats and lunes. For a Coxeter $\mathcal{A}$-species p, the Coxeter $\mathcal{A}$-species $\mathcal{T}(\mathsf{p})$ on the Z-component is

$$\mathcal{T}(\mathsf{p})[\mathrm{Z}] = \bigoplus_{\mathrm{L}:\, \mathrm{b(L)=Z}} \mathsf{p}[\mathrm{c(L)}]. \tag{2.92}$$

The sum is over all lunes L with base Z, while the summands evaluate p at the case of L. For $w \in W$, the type morphism

$$w : \mathcal{T}(\mathsf{p})[\mathrm{Z}] \to \mathcal{T}(\mathsf{p})[w\mathrm{Z}]$$

is the sum of the linear maps $w : \mathsf{p}[\mathrm{c(L)}] \to \mathsf{p}[\mathrm{c}(w\mathrm{L})]$ over all lunes L with base Z.

The monad structure on $\mathcal{T}$ is as follows. The map

$$\mathcal{T}\mathcal{T} \to \mathcal{T}, \qquad \bigoplus_{\substack{(\mathrm{L,M}):\, \mathrm{b(L)=Z} \\ \mathrm{c(L)=b(M)}}} \mathsf{p}[\mathrm{c(M)}] \to \bigoplus_{\mathrm{N}:\, \mathrm{b(N)=Z}} \mathsf{p}[\mathrm{c(N)}]$$

evaluated on p on the Z-component sends the summand indexed by (L, M) to the summand indexed by $\mathrm{N} := \mathrm{L} \circ \mathrm{M}$ via the identity map $\mathsf{p}[\mathrm{c(M)}] \to \mathsf{p}[\mathrm{c(L \circ M)}]$, while

$$\mathrm{id} \to \mathcal{T}, \qquad \mathsf{p}[\mathrm{Z}] \to \bigoplus_{\mathrm{L}:\, \mathrm{b(L)=Z}} \mathsf{p}[\mathrm{c(L)}]$$

evaluated on p on the Z-component is given by inclusion into the summand $\mathrm{L} := \mathrm{Z}$. Algebras over $\mathcal{T}$ are the same as Coxeter $\mathcal{A}$-monoids (as formulated in Proposition 2.24).

The comonad $\mathcal{T}^\vee$ is constructed dually. Its coalgebras are the same as Coxeter $\mathcal{A}$-comonoids (as mentioned in Section 2.4.7).

The mixed distributive law

$$\lambda : \mathcal{T}\mathcal{T}^\vee \to \mathcal{T}^\vee\mathcal{T}, \qquad \bigoplus_{\substack{(\mathrm{L,M}):\, \mathrm{b(L)=Z} \\ \mathrm{c(L)=b(M)}}} \mathsf{p}[\mathrm{c(M)}] \to \bigoplus_{\substack{(\mathrm{L',M'}):\, \mathrm{b(L')=Z} \\ \mathrm{c(L')=b(M')}}} \mathsf{p}[\mathrm{c(M')}] \tag{2.93}$$

evaluated on p on the Z-component has matrix-components

$$\lambda_{\mathrm{L,M,L',M'}} := \begin{cases} \mathsf{p}[\mathrm{c(M)}] \xrightarrow{\mathrm{id}} \mathsf{p}[\mathrm{c(M')}] & \text{if } \mathrm{b(M)} \cdot \mathrm{L'} = \mathrm{M} \text{ and } \mathrm{b(M')} \cdot \mathrm{L} = \mathrm{M'}, \\ 0 & \text{otherwise.} \end{cases}$$

Bialgebras over the bimonad $(\mathcal{T}, \mathcal{T}^\vee, \lambda)$ are the same as Coxeter $\mathcal{A}$-bimonoids (as formulated in Proposition 2.28).

More generally, for any W-invariant distance function υ, by suitably deforming (2.93), we obtain the bimonad $(\mathcal{T}, \mathcal{T}^\vee, \lambda_\upsilon)$ whose bialgebras are Coxeter $\mathcal{A}$-υ-bimonoids.

2.10.3. Bicommutative Coxeter bimonoids. We now turn to commutative aspects. The basic result is stated below.

Proposition 2.78. *The following are equivalences of categories.*

$$\begin{aligned} \mathcal{S}\text{-algebras} &\cong \text{com. Coxeter } \mathcal{A}\text{-monoids} \\ \mathcal{S}^\vee\text{-coalgebras} &\cong \text{cocom. Coxeter } \mathcal{A}\text{-comonoids} \\ (\mathcal{S}, \mathcal{T}^\vee, \lambda)\text{-bialgebras} &\cong \text{com. Coxeter } \mathcal{A}\text{-bimonoids} \\ (\mathcal{T}, \mathcal{S}^\vee, \lambda)\text{-bialgebras} &\cong \text{cocom. Coxeter } \mathcal{A}\text{-bimonoids} \\ (\mathcal{S}, \mathcal{S}^\vee, \lambda)\text{-bialgebras} &\cong \text{bicom. Coxeter } \mathcal{A}\text{-bimonoids} \end{aligned}$$

We now construct the monad $\mathcal{S} : \mathcal{A}\text{-}\mathsf{CSp} \to \mathcal{A}\text{-}\mathsf{CSp}$. We use the formulation of Coxeter $\mathcal{A}$-species in Proposition 2.17. For a Coxeter $\mathcal{A}$-species p, define the Coxeter $\mathcal{A}$-species $\mathcal{S}(\mathsf{p})$ on the Z-component by

$$\mathcal{S}(\mathsf{p})[\mathrm{Z}] := \bigoplus_{\mathrm{X}:\, \mathrm{Z} \leq \mathrm{X}} \mathsf{p}[\mathrm{X}]. \tag{2.94}$$

Type morphisms of $\mathcal{S}(\mathsf{p})$ are defined as follows. Suppose $w \in W$ and $w(\mathrm{Z}) = \mathrm{Z'}$. Then, for every $\mathrm{X} \geq \mathrm{Z}$, there is a corresponding flat $\mathrm{X'} := w(\mathrm{X})$ greater than $\mathrm{Z'}$. The sum of the linear maps $w : \mathsf{p}[\mathrm{X}] \to \mathsf{p}[\mathrm{X'}]$ yields the linear map

$$w : \mathcal{S}(\mathsf{p})[\mathrm{Z}] \to \mathcal{S}(\mathsf{p})[\mathrm{Z'}].$$

The monad structure on $\mathcal{S}$ is defined in a manner similar to $\mathcal{T}$ (with faces replaced by flats). One can check that $\mathcal{S}$-algebras are the same as commutative Coxeter $\mathcal{A}$-monoids (as formulated in Proposition 2.38).

The comonad $\mathcal{S}^\vee$ on Coxeter $\mathcal{A}$-species is constructed dually, and its coalgebras are the same as cocommutative Coxeter $\mathcal{A}$-comonoids. As a functor, $\mathcal{S}^\vee = \mathcal{S}$. The comonad structure on $\mathcal{S}^\vee$ is defined in a manner similar to $\mathcal{T}^\vee$.

Define the *abelianization*

$$\pi : \mathcal{T} \twoheadrightarrow \mathcal{S} \tag{2.95}$$

as follows. Evaluated on a Coxeter $\mathcal{A}$-species p, on the A-component, on the F-summand, it is given by $\beta_{\mathrm{s}(F),F}$, with the latter as in (2.17). Formally, (id, π) is a lax functor of monads from $\mathcal{S}$ to $\mathcal{T}$ in the sense of (A.2).

Dually, define the *coabelianization*

$$\pi^\vee : \mathcal{S}^\vee \hookrightarrow \mathcal{T}^\vee \tag{2.96}$$

as follows. Evaluated on a Coxeter $\mathcal{A}$-species p, on the A-component, on the X-summand, it is given by $\sum_F \beta_{F,\mathrm{X}}$, where F runs over faces greater than A which have support X. Formally, $(\mathrm{id}, \pi^\vee)$ is a colax functor of comonads from $\mathcal{S}^\vee$ to $\mathcal{T}^\vee$ in the sense of (A.5).

The mixed distributive law between $\mathcal{T}$ and $\mathcal{T}^\vee$, via (co)abelianization, induces three more mixed distributive laws: one between $\mathcal{S}$ and $\mathcal{T}^\vee$, one between $\mathcal{T}$ and $\mathcal{S}^\vee$, one between $\mathcal{S}$ and $\mathcal{S}^\vee$. We denote all of them by λ. See the first diagram in (2.100) below. The law between $\mathcal{S}$ and $\mathcal{S}^\vee$, namely,

$$\lambda : \mathcal{S}\mathcal{S}^\vee \to \mathcal{S}^\vee\mathcal{S}, \qquad \bigoplus_{(\mathrm{X},\mathrm{Y}) :\, \mathrm{Z} \leq \mathrm{X} \leq \mathrm{Y}} \mathsf{p}[\mathrm{Y}] \to \bigoplus_{(\mathrm{X}',\mathrm{Y}') :\, \mathrm{Z} \leq \mathrm{X}' \leq \mathrm{Y}'} \mathsf{p}[\mathrm{Y}'] \tag{2.97}$$

evaluated on p on the Z-component has matrix-components given by the identity map when $\mathrm{Y} = \mathrm{X} \vee \mathrm{X}' = \mathrm{Y}'$, and 0 otherwise.

Exercise 2.79. Check: (2.97) is a natural transformation. In particular, evaluated on p, it is a map of Coxeter $\mathcal{A}$-species, that is, diagram (2.16) commutes.

The last three entries in Proposition 2.78 may also now be verified.

2.10.4. Signed bicommutative signed Coxeter bimonoids. Recall that (co)commutative Coxeter (co)monoids have signed analogues. As expected, there is a monad $\mathcal{E}$ whose algebras are signed commutative Coxeter monoids, and a comonad $\mathcal{E}^\vee$ whose coalgebras are signed cocommutative Coxeter comonoids. Explicitly: For a Coxeter $\mathcal{A}$-species p,

$$\mathcal{E}(\mathsf{p})[\mathrm{Z}] := \bigoplus_{\mathrm{X} :\, \mathrm{Z} \leq \mathrm{X}} \mathbf{E}^-[\mathrm{Z}, \mathrm{X}] \otimes \mathsf{p}[\mathrm{X}], \tag{2.98}$$

with $\mathbf{E}^-[\mathrm{Z}, \mathrm{X}]$ the signature space over Z and under X as in Definition 2.48. The complete result is stated below.

Proposition 2.80. *The following are equivalences of categories.*

$$\begin{aligned}
\mathcal{E}\text{-algebras} &\cong \textit{signed com. Coxeter } \mathcal{A}\textit{-monoids} \\
\mathcal{E}^\vee\text{-coalgebras} &\cong \textit{signed cocom. Coxeter } \mathcal{A}\textit{-comonoids} \\
(\mathcal{E}, \mathcal{T}^\vee, \lambda_{-1})\text{-bialgebras} &\cong \textit{signed com. signed Coxeter } \mathcal{A}\textit{-bimonoids} \\
(\mathcal{T}, \mathcal{E}^\vee, \lambda_{-1})\text{-bialgebras} &\cong \textit{signed cocom. signed Coxeter } \mathcal{A}\textit{-bimonoids} \\
(\mathcal{E}, \mathcal{E}^\vee, \lambda_{-1})\text{-bialgebras} &\cong \textit{signed bicom. signed Coxeter } \mathcal{A}\textit{-bimonoids}
\end{aligned}$$

There are surjective and injective natural transformations

$$\pi_{-1} : \mathcal{T} \to \mathcal{E} \qquad \text{and} \qquad \pi^\vee_{-1} : \mathcal{E}^\vee \to \mathcal{T}^\vee. \tag{2.99}$$

These are the *signed abelianization* and *signed coabelianization*, respectively. Evaluated on a Coxeter $\mathcal{A}$-species p, on the A-component, on the F-summand, the former is given by $\mathtt{H}_{[F/A]} \otimes \beta_{\mathrm{s}(F),F}$. The latter is defined dually.

The mixed distributive law λ_{-1} between $\mathcal{T}$ and $\mathcal{T}^\vee$ induces three more mixed distributive laws via signed (co)abelianization. See the second diagram in (2.100) below. This leads to the last three entries in Proposition 2.80.

2.10.5. Bimonad summary. We have the following commutative diagrams of bilax functors between bimonads.

$$\begin{array}{ccc} (\mathcal{S},\mathcal{S}^\vee,\lambda) & \longrightarrow & (\mathcal{T},\mathcal{S}^\vee,\lambda) \\ \downarrow & & \downarrow \\ (\mathcal{S},\mathcal{T}^\vee,\lambda) & \longrightarrow & (\mathcal{T},\mathcal{T}^\vee,\lambda) \end{array} \qquad \begin{array}{ccc} (\mathcal{E},\mathcal{E}^\vee,\lambda_{-1}) & \longrightarrow & (\mathcal{T},\mathcal{E}^\vee,\lambda_{-1}) \\ \downarrow & & \downarrow \\ (\mathcal{E},\mathcal{T}^\vee,\lambda_{-1}) & \longrightarrow & (\mathcal{T},\mathcal{T}^\vee,\lambda_{-1}) \end{array} \tag{2.100}$$

The functors are bilax in the sense of (A.8) with $\mathcal{F} :=$ id and the (co)lax structure given by (signed) (co)abelianization.

Exercise 2.81. Recall the u-signature functor (2.70). Check: For a Coxeter species p, we have natural identifications

$$\mathcal{T}(\mathsf{p}_-)_- = \mathcal{T}(\mathsf{p}), \qquad \mathcal{S}(\mathsf{p}_-)_- = \mathcal{E}(\mathsf{p}), \qquad \mathcal{E}(\mathsf{p}_-)_- = \mathcal{S}(\mathsf{p})$$

as Coxeter species. The same holds for the o-signature functor (2.73).

We mention that duality on Coxeter species is a bilax functor between $(\mathcal{T},\mathcal{T}^\vee,\lambda_q)$ and $(\mathcal{T}^\vee,\mathcal{T},\lambda_q)$, with the latter bimonad on the opposite category. This explains why duality preserves Coxeter q-bimonoids.

2.11. Opposite transformation

We introduce the opposite transformation for Coxeter species. It links a face to its locally opposite face. This construction yields op and cop constructions for Coxeter species. A parallel more detailed discussion for species is given in [**11**, Sections 2.10 and 3.4].

2.11.1. Opposite transformation. Recall from (2.87) the functor $\mathcal{T}$ on Coxeter species. There is a natural transformation $\tau : \mathcal{T} \to \mathcal{T}$. On a Coxeter species p, the A-component of $\tau_{\mathsf{p}} : \mathcal{T}(\mathsf{p}) \to \mathcal{T}(\mathsf{p})$ is defined by summing the linear maps

$$\beta_{A\overline{F},F} : \mathsf{p}[F] \to \mathsf{p}[A\overline{F}]$$

over all $F \geq A$. We call τ the *opposite transformation*. Since $\mathcal{T} = \mathcal{T}^\vee$ as functors, we can also write $\tau : \mathcal{T}^\vee \to \mathcal{T}^\vee$.

Moreover, (id, τ) is a lax functor of monads from $\mathcal{T}$ to $\mathcal{T}$, and a colax functor of comonads from $\mathcal{T}^\vee$ to $\mathcal{T}^\vee$. Going further, the functors $(\mathrm{id}, \tau, \mathrm{id})$ and $(\mathrm{id}, \mathrm{id}, \tau)$ both from $(\mathcal{T},\mathcal{T}^\vee,\lambda)$ to itself are bilax.

2.11.2. Deformed opposite transformation. More generally, for a scalar q, define $\tau_q : \mathcal{T} \to \mathcal{T}$ by employing the linear maps

$$(\beta_q)_{A\overline{F},F} := q^{\mathrm{dist}(F,A\overline{F})} \beta_{A\overline{F},F}.$$

For $q = 1$, we recover the opposite transformation. For $q = -1$, we use the term *signed opposite transformation*.

Proposition 2.82. *Let $q \neq 0$. The functors* $(\mathrm{id}, \tau_{q^{-1}}, \mathrm{id})$ *and* $(\mathrm{id}, \mathrm{id}, \tau_{q^{-1}})$ *both from*

$$(\mathcal{T},\mathcal{T}^\vee,\lambda_q) \to (\mathcal{T},\mathcal{T}^\vee,\lambda_{q^{-1}})$$

are bilax.

We leave this as an exercise.

2.11.3. Op and cop constructions. Fix $q \neq 0$. A Coxeter monoid (a, μ) has an *opposite Coxeter monoid* $\mathsf{a}^{\mathrm{op}} := (\mathsf{a}, \mu\beta_q)$ with the same underlying Coxeter species and with product $\mu\beta_q$ defined by

$$(\mu\beta_q)^F_A : \mathsf{a}[F] \xrightarrow{(\beta_q)_{A\overline{F},F}} \mathsf{a}[A\overline{F}] \xrightarrow{\mu^{A\overline{F}}_A} \mathsf{a}[A].$$

We write ${}^{\mathrm{op}}\mathsf{a} := (\mathsf{a}, \mu\beta_{q^{-1}})$. Dually, a Coxeter comonoid (c, Δ) has an *opposite Coxeter comonoid* ${}^{\mathrm{cop}}\mathsf{c} := (\mathsf{c}, \beta_q\Delta)$ with

$$(\beta_q\Delta)^F_A : \mathsf{c}[A] \xrightarrow{\Delta^{A\overline{F}}_A} \mathsf{c}[A\overline{F}] \xrightarrow{(\beta_q)_{F,A\overline{F}}} \mathsf{c}[F].$$

We write $\mathsf{c}^{\mathrm{cop}} := (\mathsf{c}, \beta_{q^{-1}}\Delta)$.

Proposition 2.83. *For $q \neq 0$, for $\mathsf{h} = (\mathsf{h}, \mu, \Delta)$ a Coxeter q-bimonoid,*

$$\mathsf{h}^{\mathrm{cop}} := (\mathsf{h}, \mu, \beta_{q^{-1}}\Delta) \quad \textit{and} \quad {}^{\mathrm{op}}\mathsf{h} := (\mathsf{h}, \mu\beta_{q^{-1}}, \Delta)$$

are Coxeter q^{-1}-bimonoids, and

$$\mathsf{h}^{\mathrm{op,cop}} := (\mathsf{h}, \mu\beta_q, \beta_{q^{-1}}\Delta) \quad \textit{and} \quad {}^{\mathrm{op,cop}}\mathsf{h} := (\mathsf{h}, \mu\beta_{q^{-1}}, \beta_q\Delta)$$

are Coxeter q-bimonoids. In fact,

$$\mathsf{h}^{\mathrm{op,cop}} = {}^{\mathrm{op}}(\mathsf{h}^{\mathrm{cop}}) \quad \textit{and} \quad {}^{\mathrm{op,cop}}\mathsf{h} = ({}^{\mathrm{op}}\mathsf{h})^{\mathrm{cop}}.$$

This can be either checked directly or formally deduced from Proposition 2.82.

The above considerations carry over to the more general setting of Coxeter b-species (Section 2.3.10). Here instead of $\beta_{q^{-1}}$ we employ the inverse of β_q.

2.12. Joyal species and Coxeter species

The category of Joyal species and monoids, comonoids, bimonoids therein were reviewed in [**11**, Section 2.16]. We explain how Joyal species give rise to Coxeter $\mathcal{B}^J$-species, where $\mathcal{B}^J$ is the braid arrangement on the set J. Moreover, this construction carries Joyal (co, bi)monoids to Coxeter $\mathcal{B}^J$-(co, bi)monoids. This builds on ideas in [**11**, Sections 2.16.6 and 2.16.7].

2.12.1. Coxeter $\mathcal{B}^J$-species. Fix a finite set J. We specialize the category $\mathcal{A}$-$\mathtt{Cox}$ and the notion of Coxeter $\mathcal{A}$-species to $\mathcal{A} := \mathcal{B}^J$, and make them explicit by means of the combinatorial descriptions in Section 1.16.

The category $\mathcal{B}^J$-$\mathtt{Cox}$ is as follows. Objects are compositions of J. For compositions F and G of J, a morphism $F \to G$ is a bijection $\sigma : J \to J$ such that $\sigma(F)$ and G have the same support. Morphisms are composed by composing the defining bijections. Any morphism σ can be uniquely expressed as a composite of a support morphism and a type morphism. These go between compositions of the same support and of the same type, respectively, and are denoted

$$\beta_{G,F} : F \to G \quad \text{and} \quad w : F \to wF.$$

In a support morphism, the defining bijection $\sigma = \mathrm{id}$, while in a type morphism, $\sigma = w$.

A Coxeter $\mathcal{B}^J$-species p consists of vector spaces $\mathsf{p}[F]$, one for each composition F of J, equipped with linear maps $\sigma : \mathsf{p}[F] \to \mathsf{p}[G]$, one for each bijection σ such that $\sigma(F)$ and G have the same support, which respect composition of bijections as in diagrams (2.10). Equivalently, p can be formulated using support and type morphisms satisfying diagrams (2.12a) and (2.12b).

2.12.2. Joyal species to Coxeter $\mathcal{B}^J$-species. Let p be a Joyal species. For a finite set J, we construct a Coxeter $\mathcal{B}^J$-species p^J as follows. For a composition $F = (S_1, \dots, S_k)$ of J, set

$$\mathsf{p}^J[F] := \mathsf{p}[S_1] \otimes \cdots \otimes \mathsf{p}[S_k]. \tag{2.101}$$

Support and type morphisms are as follows.

- Whenever F and G have the same support, define the linear map
$$\beta_{G,F} : \mathsf{p}^J[F] \to \mathsf{p}^J[G]$$
by permuting the tensor factors.
- Whenever $wF = F'$, define the linear map
$$w : \mathsf{p}^J[F] \to \mathsf{p}^J[F']$$
by taking tensor product of linear maps $\sigma_i : \mathsf{p}[S_i] \to \mathsf{p}[S'_i]$ for $1 \leq i \leq k$. Here $F = (S_1, \dots, S_k)$ and $F' = (S'_1, \dots, S'_k)$ with w inducing bijections $\sigma_i : S_i \to S'_i$ for all i.

It is straightforward to verify axioms (2.12a) and (2.12b).

Example 2.84. Let $J = \{a, b, c, d, e, f, g\}$ and $F = bde|af|cg$. Then
$$\mathsf{p}^J[bde|af|cg] = \mathsf{p}[bde] \otimes \mathsf{p}[af] \otimes \mathsf{p}[cg].$$
Consider the face $G = af|cg|bde$ which has the same support as F. The map

$$\begin{array}{ccc} \mathsf{p}^J[bde|af|cg] & \overset{\beta_{G,F}}{\dashrightarrow} & \mathsf{p}^J[af|cg|bde] \\ \| & & \| \\ \mathsf{p}[bde] \otimes \mathsf{p}[af] \otimes \mathsf{p}[cg] & \longrightarrow & \mathsf{p}[af] \otimes \mathsf{p}[cg] \otimes \mathsf{p}[bde] \end{array}$$

permutes the three tensor factors.

Consider the bijection w which sends a, b, c, d, e, f, g to c, g, f, a, e, b, d, respectively. Then $wF = F' = gae|cb|fd$, and

$$\begin{array}{ccc} \mathsf{p}^J[bde|af|cg] & \overset{w}{\dashrightarrow} & \mathsf{p}^J[gae|cb|fd] \\ \| & & \| \\ \mathsf{p}[bde] \otimes \mathsf{p}[af] \otimes \mathsf{p}[cg] & \underset{\sigma_1 \otimes \sigma_2 \otimes \sigma_3}{\longrightarrow} & \mathsf{p}[gae] \otimes \mathsf{p}[cb] \otimes \mathsf{p}[fd], \end{array}$$

where σ_1 sends b, d, e to g, a, e, respectively, σ_2 sends a, f to c, b, respectively, σ_3 sends c, g to f, d, respectively. $\square$

2.12.3. Joyal (co, bi)monoids to Coxeter $\mathcal{B}^J$-(co, bi)monoids. Now let a be a Joyal monoid. Then, for a subset S of J and composition F of S, by associativity, there is an iterated product map $\mathsf{a}^S[F] \to \mathsf{a}[S]$. For $A \leq F$, both compositions of J, define
$$\mu_A^F : \mathsf{a}^J[F] \to \mathsf{a}^J[A]$$
by tensoring the maps $\mathsf{a}^{S_i}[F^i] \to \mathsf{a}[S_i]$, where $A = (S_1, \dots, S_k)$, and F^i is the part of F which refines S_i. This turns a^J into a Coxeter $\mathcal{B}^J$-monoid.

Example 2.85. For $J = \{a, b, c, d, e, f, g\}$, $F = bd|e|af|cg$, $A = bde|afcg$,

$$\begin{array}{ccc} \mathsf{a}^J[bd|e|af|cg] & \xrightarrow{\mu_A^F} & \mathsf{a}^J[bde|afcg] \\ \| & & \| \\ \mathsf{a}[bd] \otimes \mathsf{a}[e] \otimes \mathsf{a}[af] \otimes \mathsf{a}[cg] & \xrightarrow[\mu^{bd,e} \otimes \mu^{af,cg}]{} & \mathsf{a}[bde] \otimes \mathsf{a}[afcg], \end{array}$$

where $\mu^{bd,e}$ and $\mu^{af,cg}$ are product components of the Joyal monoid a. □

In a similar manner, a Joyal comonoid yields a Coxeter $\mathcal{B}^J$-comonoid. This defines functors from the category of Joyal (co)monoids to the category of Coxeter $\mathcal{B}^J$-(co)monoids.

These functors preserve (signed) (co)commutativity. Further, they also induce a functor from the category of Joyal q-bimonoids to the category of Coxeter $\mathcal{B}^J$-q-bimonoids for a scalar q.

2.13. Notes

Coxeter species and Coxeter bimonoids for reflection arrangements are presented here for the first time. We formulated them in the year 2010 when we were in the process of finalizing [**7**]. They generalize classical notions of Joyal species and Joyal bimonoids [**7**, Part II] as explained in Section 2.12. A detailed theory of species and bimonoids in the more general but less structured setting of hyperplane arrangements was developed in our previous monograph [**11**]. The connection between species and Coxeter species is elaborated in Section 2.7.

Warning. In [**7**], Joyal species are called species, and Joyal bimonoids are called bimonoids in species. The terms “Joyal species” and “Joyal bimonoids” were first used in [**11**, Section 2.16]. We continue this usage in the present book. The term “species” is now reserved for species for hyperplane arrangements.

2.13.1. Joyal bimonoids. Joyal species, as we know them today, were introduced and studied by Joyal [**202**]. Notable earlier attempts to define “species of structures” were by Ehresmann [**116**, Chapter II] and Bourbaki [**59**], [**60**, Section IV.1]. Related ideas are present in Benabou’s thesis [**32**] and in the books by Hasse and Michler [**177**, Chapter X], Bucur and Deleanu [**71**, Section 4.4], Arbib and Manes [**19**, Chapter 6], Adámek [**3**, Definition on page 18].

Joyal set-species are treated in the books by Bergeron, Labelle, Leroux [**36**] and Méndez [**259**]. The study of algebraic structures (monoids, comonoids, bimonoids, Hopf monoids, Lie monoids) in Joyal species was initiated by Joyal [**202**, Section 7.1], [**203**, Section 4.2]. It was carried further in our monographs [**7**, Parts II and III], [**8**]; iterated products and higher compatibilities were emphasized in [**8**, Section 10]. Classical analogues of (co)product components μ_A^F and Δ_A^F of Coxeter (co)monoids are the higher (co)product maps of Joyal (co)monoids [**8**, (30) and (32)]. (The idea of indexing iterated products by elements of suitable posets was already present in [**6**, Sections 6.3–6.8].)

A Joyal species p can also be formulated as a sequence $\mathsf{p}[0], \mathsf{p}[1], \ldots,$ where each $\mathsf{p}[n]$ is a S_n-module, with S_n the symmetric group on n letters [**7**, page 237]. This is the classical analogue of Proposition 2.20; in more detail, for each n, for the braid arrangement on n letters, $\mathsf{p}[\perp]$ corresponds to the S_n-module $\mathsf{p}[n]$ of the Joyal species, while the other $\mathsf{p}[\mathrm{X}]$ correspond to tensor products of the S_j-modules $\mathsf{p}[j]$ for $1 \leq j \leq n$. Similarly, (2.20) corresponds to the functor which sends a Joyal species p to the S_n-module $\mathsf{p}[n]$, while (2.21) corresponds to the functor which sends a Joyal species p to the S_n-module $\mathsf{p}[1]^{\otimes n}$.

The Cauchy product of Joyal species in the language of S_n-modules was treated by Kelly [**210**, Formula (2.2)]. (This paper, though published in 2005, dates back to 1971.) Monoids wrt the Cauchy product were considered by Barratt [**30**, Definition 3] under the name "twisted algebras". Twisted algebras and twisted Hopf algebras were studied by Stover [**341**, Sections 3 and 4]. Some later references prior to the publication of [**7**] are [**131**, Section 1.2.10], [**231**], [**290**], [**291**], [**292**]; a recent book reference is [**84**, Section 9.4].

2.13.2. Coxeter bimonoids for reflection arrangements.

This chapter can be viewed as the Coxeter analogue of [**11**, Chapter 2]. The description of monoids in species in terms of flats and lunes is briefly mentioned in [**11**, Remark 4.31]. The discussion here in Section 2.4.3 (in the Coxeter setting) is more elaborate. Also, bimonoid axioms (2.41) and (2.60) appear here for the first time. Section 2.10 on bimonads can be viewed as the Coxeter analogue of [**11**, Chapter 3]. The formulation in Section 2.10.2 in terms of flats and lunes is new.

Signature functors. The o-signature functor on species was discussed in detail in [**11**, Section 8.10]. There it was defined as Hadamard product with the signed exponential species E^- as in (4.17) (with E^- defined in terms of signature spaces over flats). The description of support and type morphisms as in (2.72) lends new perspective. The u-signature functor on species (without any specific name) was considered in [**11**, Section 2.5.4]. There the definition was as in (2.69); the description as in (4.17) (with E_- defined in terms of signature spaces under flats) was not given. A comparison of (2.69) and (2.72) shows that the underlying species of p_- and p^- are identical; this fact is equivalent to the isomorphism discussed in [**11**, Section 8.10.7].

Eckmann–Hilton argument. Exercise 2.49 is the Coxeter species analogue of [**11**, Proposition 2.27]. The same result for Coxeter spaces is given later in Exercise 8.98. The classical Eckmann–Hilton argument originated in work of Eckmann and Hilton [**112**, Theorem 4.17], [**185**, Proposition 14.13 and Theorem 14.14]. More references are given on [**11**, pages 132–133].

2.13.3. Double categories in Hopf theory.

A double category is defined for instance in the books by Brown, Higgins, Sivera [**69**, Section 6.1], Grandis [**157**, Chapter 3], Johnson and Yau [**199**, Section 12.3]; the concept goes back to Ehresmann [**115**, Section II.4, page 389], [**116**, Section I.3.A, Définition 30]. A vacant double groupoid is defined by Mackenzie [**241**, Section 2, Definition 2.11]; it has an associated diagonal groupoid [**241**, Section 2, pages 200–201]. This construction is also given by Fiedorowicz and Loday [**125**, Section 2]; they work with categories instead of groupoids and use the term "star condition" for the vacant condition.

The category $\mathcal{A}$-$\mathtt{Cox}$ is an example of a diagonal groupoid associated to a vacant double groupoid. We have two groupoids at hand. There is $\mathcal{A}$-$\mathtt{Hyp}$ with its morphisms β, and there is $\mathsf{el}_W(\Sigma[\mathcal{A}])$, the groupoid of elements for the action of W on the set of faces $\Sigma[\mathcal{A}]$. There is a double groupoid with $\mathcal{A}$-$\mathtt{Hyp}$ as its horizontal groupoid, $\mathsf{el}_W(\Sigma[\mathcal{A}])$ as its vertical groupoid, and the 2-cells (squares) being the support-type relations in (2.1a). This double groupoid is vacant, so it has an associated diagonal groupoid. Propositions 2.4 and 2.5 essentially discuss this construction. Lemma 2.3 says that the diagonal groupoid is generated by the horizontal and vertical groupoids. Similar remarks apply to the category $\mathcal{A}$-$\overline{\mathtt{Cox}}$ that is defined in Chapter 8.

Diagrams (2.7) (pertaining to induced morphisms) can be viewed as 2-cells of a vacant double category whose horizontal category is $\mathcal{A}$-$\mathtt{Cox}$ and vertical category is the one associated to the poset of faces. Diagrams (2.8) illustrate vertical composition of 2-cells. These 2-cells are used to formulate the naturality axiom (2.30) of a Coxeter monoid (or of a Coxeter comonoid by reversing the convention on the vertical arrows). We can also consider the smaller double category whose horizontal category is $\mathcal{A}$-$\mathtt{Hyp}$ instead of $\mathcal{A}$-$\mathtt{Cox}$. Its 2-cells can be used to formulate the naturality axiom of an $\mathcal{A}$-monoid or $\mathcal{A}$-comonoid in species [**11**, First diagrams in (2.8) and (2.10)]; the 2-cell for a $\mathcal{A}$-comonoid is shown in [**11**, Second diagram in (2.71a)]. Similar remarks apply to diagrams (1.28) and (8.11) (pertaining to induced morphisms in $\boldsymbol{\alpha}$-$\overline{\mathtt{Cox}}$ and $\mathcal{A}$-$\overline{\mathtt{Cox}}$).

We also mention that there is a vacant double category whose horizontal and vertical categories coincide and are the category of lunes. Further, the endomorphism monoid of the minimum flat in its associated diagonal category is the Janus monoid. Relevant results are [**10**, Lemma 4.22 and Proposition 4.31]. Moreover, an $\mathcal{A}$-bimonoid can be interpreted as a Mackey functor (as defined by Hartl, Pirashvili, Vespa [**176**, Section 1.3]) on this double category. For this, the bimonoid axiom (2.41) in terms of flats and lunes is more convenient. This improves upon the comments made on [**11**, bottom of page 133].

The relevance of double categories to Hopf theory and operads with more details on the above examples will be explained separately.

CHAPTER 3

Basic theory of Coxeter bimonoids

In Chapter 2, we introduced Coxeter species, and then developed notions of Coxeter monoids, Coxeter comonoids, Coxeter bimonoids. More generally, for a scalar q, we defined Coxeter q-bimonoids. The case $q = 1$ recovered Coxeter bimonoids; $q = -1$ yielded signed Coxeter bimonoids. We also saw commutative Coxeter monoids and cocommutative Coxeter comonoids. We now proceed with the basic theory of these objects. It is the Coxeter analogue of the theory of bimonoids in species developed in [**11**, Part II]. The extra part is to understand the role of type morphisms (connected to the Coxeter group W); this is mostly straightforward. One point which merits attention is that we need to assume $\mathrm{char}(\Bbbk) \nmid |W|$ whenever we use invariant noncommutative zeta and Möbius functions. To avoid repetition, we do not go over the Coxeter analogue of every result in [**11**, Part II].

We discuss primitive and decomposable filtrations, universal constructions such as the free Coxeter monoid, (co)abelianization of a Coxeter (co)monoid, Hadamard product of Coxeter species, enrichment of Coxeter (co)monoids over Coxeter comonoids, exp-log correspondences, q-norm maps with connections to the q-logarithm, characteristic operations by faces and Coxeter–Tits algebra modules, antipode of a Coxeter bimonoid.

Examples illustrating the above theory are provided in Chapter 4 and they could be read in parallel with the present chapter. The toy example of rank one is developed in Section 7.8.

3.1. Cauchy powers of a Coxeter species

We introduce Cauchy powers and commutative Cauchy powers of a Coxeter species. Coproduct components of a Coxeter comonoid can be encapsulated as maps from the Coxeter comonoid to its Cauchy powers. Dually, product components of a Coxeter monoid can be encapsulated as maps to the Coxeter monoid from its Cauchy powers.

3.1.1. Cauchy powers of a Coxeter species.

Let p be a Coxeter species. For $k \geq 1$, define the Coxeter species p^k by

$$\mathsf{p}^k[A] := \bigoplus_{\substack{F:\, F \geq A \\ \mathrm{rk}(F/A) = k-1}} \mathsf{p}[F]. \tag{3.1}$$

The structure maps of p^k are as follows. A morphism $\sigma : A \to B$ induces a morphism $\sigma : F \to G$, where $G = \sigma_*(F)$ as in (2.7). The sum of the linear maps $\sigma : \mathsf{p}[F] \to \mathsf{p}[G]$ yields the linear map

$$\sigma : \mathsf{p}^k[A] \to \mathsf{p}^k[B].$$

We call p^k the k-th *Cauchy power* of the Coxeter species p. When $k = 1$, there is only one summand in (3.1) for $F = A$, and so $\mathsf{p}^1 = \mathsf{p}$.

For a Coxeter comonoid c, for $k \geq 1$, define

$$\Delta^{k-1} : \mathsf{c} \to \mathsf{c}^k, \qquad \Delta_A^{k-1} := (\Delta_A^F)_{\mathrm{rk}(F/A)=k-1}. \tag{3.2}$$

Note: $\Delta^0 = \mathrm{id}$ since $\Delta_A^A = \mathrm{id}$ by the counitality axiom in (2.36b).

Dually, for a Coxeter monoid a, for $k \geq 1$, define

$$\mu^{k-1} : \mathsf{a}^k \to \mathsf{a}, \qquad \mu_A^{k-1} := (\mu_A^F)_{\mathrm{rk}(F/A)=k-1}. \tag{3.3}$$

Note: $\mu^0 = \mathrm{id}$ since $\mu_A^A = \mathrm{id}$ by the unitality axiom in (2.28b).

Exercise 3.1. Check: Δ^{k-1} and μ^{k-1} are maps of Coxeter species. (Use naturality axiom (2.30) and its dual.)

3.1.2. Commutative Cauchy powers of a Coxeter species. We now turn to commutative aspects for which we employ the formulations in Propositions 2.17 and 2.38.

Let p be a Coxeter species. For $k \geq 1$, define the Coxeter species $\mathsf{p}^{\overline{k}}$ by

$$\mathsf{p}^{\overline{k}}[\mathrm{Z}] := \bigoplus_{\substack{\mathrm{X}:\, \mathrm{X} \geq \mathrm{Z} \\ \mathrm{rk}(\mathrm{X}/\mathrm{Z}) = k-1}} \mathsf{p}[\mathrm{X}] \tag{3.4}$$

with type morphisms as follows. For $w : \mathrm{Z} \to w\mathrm{Z}$, the sum of the linear maps $w : \mathsf{p}[\mathrm{X}] \to \mathsf{p}[w\mathrm{X}]$ over all $\mathrm{X} \geq \mathrm{Z}$ with $\mathrm{rk}(\mathrm{X}/\mathrm{Z}) = k - 1$ yields the linear map

$$w : \mathsf{p}^{\overline{k}}[\mathrm{Z}] \to \mathsf{p}^{\overline{k}}[w\mathrm{Z}].$$

We call $\mathsf{p}^{\overline{k}}$ the k-th *commutative Cauchy power* of the Coxeter species p.

For a cocommutative Coxeter comonoid c, for $k \geq 1$, define

$$\Delta^{\overline{k-1}} : \mathsf{c} \to \mathsf{c}^{\overline{k}}, \qquad \Delta_{\mathrm{Z}}^{\overline{k-1}} := (\Delta_{\mathrm{Z}}^{\mathrm{X}})_{\mathrm{rk}(\mathrm{X}/\mathrm{Z})=k-1}. \tag{3.5}$$

Dually, for a commutative Coxeter monoid a, for $k \geq 1$, define

$$\mu^{\overline{k-1}} : \mathsf{a}^{\overline{k}} \to \mathsf{a}, \qquad \mu_{\mathrm{Z}}^{\overline{k-1}} := (\mu_{\mathrm{Z}}^{\mathrm{X}})_{\mathrm{rk}(\mathrm{X}/\mathrm{Z})=k-1}. \tag{3.6}$$

Note: $\Delta^{\overline{0}} = \mathrm{id}$ and $\mu^{\overline{0}} = \mathrm{id}$ since $\Delta_{\mathrm{Z}}^{\mathrm{Z}} = \mathrm{id}$ and $\mu_{\mathrm{Z}}^{\mathrm{Z}} = \mathrm{id}$.

3.2. Primitive and decomposable filtrations

Any Coxeter comonoid carries a filtration called the primitive filtration. It arises by taking kernels of coproduct components from the Coxeter comonoid to its Cauchy powers. The first term is the primitive part of the Coxeter comonoid. Dually, any Coxeter monoid carries a filtration called the decomposable filtration. It arises by taking images of product components to the Coxeter monoid from its Cauchy powers. The first term is the decomposable part of the Coxeter monoid. Moding it out yields its indecomposable part. See Table 3.1.

TABLE 3.1. Primitive and decomposable filtrations.

Coxeter comonoid c	primitive part $\mathcal{P}(\mathsf{c})$ primitive filtration $\mathcal{P}_1(\mathsf{c}) \subseteq \mathcal{P}_2(\mathsf{c}) \subseteq \cdots \subseteq \mathsf{c}$
Coxeter monoid a	decomposable part $\mathcal{D}(\mathsf{a})$ decomposable filtration $\mathsf{a} \supseteq \mathcal{D}_1(\mathsf{a}) \supseteq \mathcal{D}_2(\mathsf{a}) \supseteq \dots$ indecomposable part $\mathcal{Q}(\mathsf{a})$

We formulate trivial Coxeter comonoids (monoids) and relate them to the primitive (indecomposable) part functor. We also introduce the related notions of coderivations and derivations.

A parallel more detailed discussion of these ideas for species is given in [**11**, Chapter 5].

3.2.1. Primitive filtration of a Coxeter comonoid. Let c be a Coxeter comonoid. Define the Coxeter subspecies $\mathcal{P}(\mathsf{c})$ by

$$\mathcal{P}(\mathsf{c})[A] := \bigcap_{F:\, F>A} \ker(\Delta_A^F : \mathsf{c}[A] \to \mathsf{c}[F]). \tag{3.7}$$

We refer to $\mathcal{P}(\mathsf{c})$ as the *primitive part* of c. This construction is functorial in c and yields the *primitive part functor* from the category of Coxeter comonoids to the category of Coxeter species.

More generally, for $k \geq 1$, define the Coxeter subspecies $\mathcal{P}_k(\mathsf{c})$ by

$$\mathcal{P}_k(\mathsf{c})[A] := \bigcap_{\substack{F:\, F\geq A \\ \mathrm{rk}(F/A)\geq k}} \ker(\Delta_A^F : \mathsf{c}[A] \to \mathsf{c}[F]). \tag{3.8}$$

We refer to $\mathcal{P}_k(\mathsf{c})$ as the k-th term of the *primitive filtration* of c. The first term is the primitive part of c, that is, $\mathcal{P}_1(\mathsf{c}) = \mathcal{P}(\mathsf{c})$. Equivalently, in terms of (3.2), $\mathcal{P}_k(\mathsf{c}) = \ker \Delta^k$. Observe:

$$\mathcal{P}_1(\mathsf{c}) \subseteq \mathcal{P}_2(\mathsf{c}) \subseteq \cdots \subseteq \mathsf{c} \quad \text{with} \quad \bigcup_{k\geq 1} \mathcal{P}_k(\mathsf{c}) = \mathsf{c}. \tag{3.9}$$

More precisely, $\mathcal{P}_k(\mathsf{c}) = \mathsf{c}$ as soon as k exceeds the rank of the arrangement.

Proposition 3.2. *A morphism $f : \mathsf{c} \to \mathsf{d}$ of Coxeter comonoids is injective iff the restriction $f : \mathcal{P}(\mathsf{c}) \to \mathcal{P}(\mathsf{d})$ is injective.*

PROOF. Forward implication is clear. For backward implication: Suppose f is injective on the primitive part. We do a backward induction on the partial order on faces. For any chamber C, the component f_C is injective since $\mathsf{c}[C]$ belongs to the primitive part. For a given face F, by induction hypothesis, f_G is injective for all $G > F$. We want to show f_F is injective. Suppose $f_F(z) = 0$. Then, for any $G > F$, by (2.37), $f_G \Delta_F^G(z) = \Delta_F^G f_F(z) = 0$, and since f_G is injective, $\Delta_F^G(z) = 0$. So z is primitive, and hence $z = 0$. Thus f_F is injective which completes the induction step. $\square$

Remark 3.3. One can also deduce the above result as a formal consequence of the corresponding result for comonoids in species [**11**, Proposition 5.18] since the primitive part only depends on the coproduct and not on type morphisms.

3.2.2. Decomposable filtration of a Coxeter monoid.

Let a be a Coxeter monoid. Define the Coxeter subspecies $\mathcal{D}(\mathsf{a})$ by

$$\mathcal{D}(\mathsf{a})[A] := \sum_{F:\, F>A} \operatorname{im}(\mu_A^F : \mathsf{a}[F] \to \mathsf{a}[A]). \tag{3.10}$$

We refer to $\mathcal{D}(\mathsf{a})$ as the *decomposable part* of a.

More generally, for $k \geq 1$, define the Coxeter subspecies $\mathcal{D}_k(\mathsf{a})$ by

$$\mathcal{D}_k(\mathsf{a})[A] := \sum_{\substack{F:\, F\geq A \\ \operatorname{rk}(F/A)\geq k}} \operatorname{im}(\mu_A^F : \mathsf{a}[F] \to \mathsf{a}[A]). \tag{3.11}$$

This is the k-th term of the *decomposable filtration* of a. The first term is the decomposable part of a, that is, $\mathcal{D}_1(\mathsf{a}) = \mathcal{D}(\mathsf{a})$. Equivalently, in terms of (3.3), $\mathcal{D}_k(\mathsf{a}) = \operatorname{im} \mu^k$. Observe:

$$\mathsf{a} \supseteq \mathcal{D}_1(\mathsf{a}) \supseteq \mathcal{D}_2(\mathsf{a}) \supseteq \dots \quad \text{with} \quad \bigcap_{k\geq 1} \mathcal{D}_k(\mathsf{a}) = 0. \tag{3.12}$$

More precisely, $\mathcal{D}_k(\mathsf{a}) = 0$ as soon as k exceeds the rank of the arrangement.

3.2.3. Indecomposable part of a Coxeter monoid.

Let a be a Coxeter monoid. The *indecomposable part* of a is the Coxeter quotient species

$$\mathcal{Q}(\mathsf{a}) := \mathsf{a}/\mathcal{D}(\mathsf{a}). \tag{3.13}$$

Equivalently, $\mathcal{Q}(\mathsf{a}) = \operatorname{coker} \mu^1$. This construction is functorial in a and yields the *indecomposable part functor* from the category of Coxeter monoids to the category of Coxeter species.

Proposition 3.4. *A morphism $f : \mathsf{a} \to \mathsf{b}$ of Coxeter monoids is surjective iff the induced map $f : \mathcal{Q}(\mathsf{a}) \to \mathcal{Q}(\mathsf{b})$ is surjective.*

This result is the dual of Proposition 3.2 in view of (3.14) below and can be formally deduced from it.

3.2.4. Duality.

For c a Coxeter comonoid, $\mathcal{P}_k(\mathsf{c})$ and $\mathcal{D}_k(\mathsf{c}^*)$ are orthogonal complements of each other under the canonical pairing between c and c^*. Dually, for a a Coxeter monoid, $\mathcal{D}_k(\mathsf{a})$ and $\mathcal{P}_k(\mathsf{a}^*)$ are orthogonal complements of each other under the canonical pairing between a and a^*.

The case $k = 1$ can also be formulated as follows. For a Coxeter comonoid c and Coxeter monoid a,

$$\mathcal{P}(\mathsf{c})^* \cong \mathcal{Q}(\mathsf{c}^*) \quad \text{and} \quad \mathcal{Q}(\mathsf{a})^* \cong \mathcal{P}(\mathsf{a}^*). \tag{3.14}$$

3.2.5. Commutative aspects. The primitive filtration of a cocommutative Coxeter comonoid c, and dually, the decomposable filtration of a commutative Coxeter monoid a can be phrased in terms of flats (instead of faces) as follows.

$$\mathcal{P}_k(\mathsf{c})[\mathrm{Z}] = \bigcap_{\substack{\mathrm{X}:\, \mathrm{X} \geq \mathrm{Z} \\ \mathrm{rk}(\mathrm{X}/\mathrm{Z}) \geq k}} \ker(\Delta_{\mathrm{Z}}^{\mathrm{X}} : \mathsf{c}[\mathrm{Z}] \to \mathsf{c}[\mathrm{X}]), \tag{3.15}$$

$$\mathcal{D}_k(\mathsf{a})[\mathrm{Z}] = \sum_{\substack{\mathrm{X}:\, \mathrm{X} \geq \mathrm{Z} \\ \mathrm{rk}(\mathrm{X}/\mathrm{Z}) \geq k}} \mathrm{im}(\mu_{\mathrm{Z}}^{\mathrm{X}} : \mathsf{a}[\mathrm{X}] \to \mathsf{a}[\mathrm{Z}]). \tag{3.16}$$

Equivalently, in terms of (3.5) and (3.6), $\mathcal{P}_k(\mathsf{c}) = \ker \Delta^{\overline{k}}$ and $\mathcal{D}_k(\mathsf{a}) = \mathrm{im}\, \mu^{\overline{k}}$.

3.2.6. Trivial Coxeter (co)monoids. Every Coxeter species can be turned into a Coxeter monoid and a Coxeter comonoid in a trivial way by setting

$$\mu_A^F := \begin{cases} 0 & \text{if } F > A, \\ \mathrm{id} & \text{if } F = A, \end{cases} \qquad \Delta_A^F := \begin{cases} 0 & \text{if } F > A, \\ \mathrm{id} & \text{if } F = A. \end{cases} \tag{3.17}$$

We refer to this as a *trivial Coxeter monoid* and *trivial Coxeter comonoid*, respectively. This defines functors

$$\mathit{trv} : \mathsf{CSp} \to \mathsf{Mon}(\mathsf{CSp}) \quad \text{and} \quad \mathit{trv} : \mathsf{CSp} \to \mathsf{Comon}(\mathsf{CSp}). \tag{3.18}$$

These are the *trivial Coxeter monoid functor* and *trivial Coxeter comonoid functor*, respectively. In fact, they map into the subcategories of commutative Coxeter monoids and cocommutative Coxeter comonoids, respectively.

Lemma 3.5. *The trivial Coxeter comonoid functor is left adjoint to the primitive part functor. Dually, the trivial Coxeter monoid functor is right adjoint to the indecomposable part functor.*

This is displayed below.

$$\mathsf{CSp} \underset{\mathcal{P}}{\overset{\mathit{trv}}{\rightleftarrows}} \mathsf{Comon}(\mathsf{CSp}) \qquad \mathsf{Mon}(\mathsf{CSp}) \underset{\mathit{trv}}{\overset{\mathcal{Q}}{\rightleftarrows}} \mathsf{CSp} \tag{3.19}$$

Explicitly:

- For c a trivial Coxeter comonoid, a morphism of Coxeter comonoids $\mathsf{c} \to \mathsf{d}$ is the same as a map of Coxeter species $\mathsf{c} \to \mathcal{P}(\mathsf{d})$.
- For a a trivial Coxeter monoid, a morphism of Coxeter monoids $\mathsf{b} \to \mathsf{a}$ is the same as a map of Coxeter species $\mathcal{Q}(\mathsf{b}) \to \mathsf{a}$.

3.2.7. Derivations and coderivations. We now define (co, bi)derivations.

- For a Coxeter species p and Coxeter comonoid c, a map of Coxeter species $f : \mathsf{p} \to \mathsf{c}$ is a *coderivation* if it maps into $\mathcal{P}(\mathsf{c})$.
- For a Coxeter monoid a and Coxeter species p, a map of Coxeter species $f : \mathsf{a} \to \mathsf{p}$ is a *derivation* if it factors through $\mathcal{Q}(\mathsf{a})$.
- For a Coxeter monoid a and Coxeter comonoid c, a map of Coxeter species $f : \mathsf{a} \to \mathsf{c}$ is a *biderivation* if it maps into $\mathcal{P}(\mathsf{c})$ and factors through $\mathcal{Q}(\mathsf{a})$.

The above definitions are illustrated below.

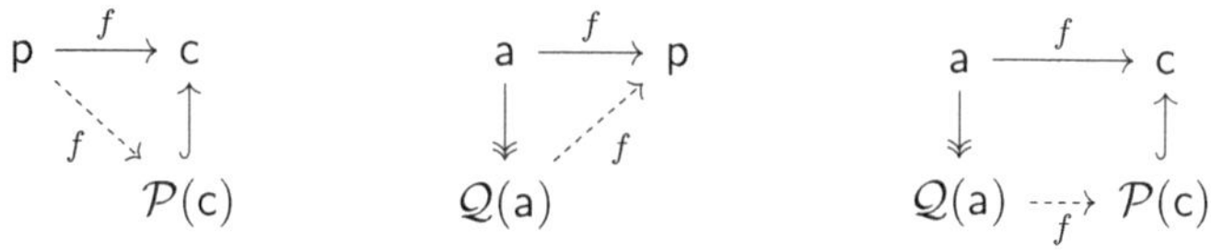

Note: A map is a biderivation iff it is both a derivation and a coderivation.

Observation 3.6. We have:

- A map f is a coderivation iff $\Delta_A^F f_A = 0$ for $F > A$. Thus, a map of Coxeter species $f : \mathsf{p} \to \mathsf{c}$ is a coderivation iff f is a morphism of Coxeter comonoids (with trivial coproduct on p).
- A map f is a derivation iff $f_A \mu_A^F = 0$ for $F > A$. Thus, a map of Coxeter species $f : \mathsf{a} \to \mathsf{p}$ is a derivation iff f is a morphism of Coxeter monoids (with trivial product on p).

This can also be seen as a consequence of adjunctions (3.19).

3.3. Universal constructions of Coxeter bimonoids

We construct the free Coxeter monoid $\mathcal{T}(\mathsf{p})$ and cofree Coxeter comonoid $\mathcal{T}^\vee(\mathsf{p})$ on a Coxeter species p. In addition, for a scalar q, we construct the free Coxeter q-bimonoid $\mathcal{T}_q(\mathsf{c})$ on a Coxeter comonoid c, and the cofree Coxeter q-bimonoid $\mathcal{T}_q^\vee(\mathsf{a})$ on a Coxeter monoid a. Every Coxeter species can be viewed as a Coxeter (co)monoid whose nontrivial (co)product components are zero. Thus, to every Coxeter species p, we have Coxeter q-bimonoids $\mathcal{T}_q(\mathsf{p})$ and $\mathcal{T}_q^\vee(\mathsf{p})$. These universal objects along with the terminology for their products and coproducts are summarized in Table 3.2.

TABLE 3.2. Universal Coxeter bimonoids.

Starting data	Coxeter q-bimonoid	Product	Coproduct
Coxeter comonoid c	$\mathcal{T}_q(\mathsf{c})$	concatenation	q-dequasishuffle
Coxeter monoid a	$\mathcal{T}_q^\vee(\mathsf{a})$	q-quasishuffle	deconcatenation
Coxeter species p	$\mathcal{T}_q(\mathsf{p})$	concatenation	q-deshuffle
Coxeter species p	$\mathcal{T}_q^\vee(\mathsf{p})$	q-shuffle	deconcatenation
Coxeter species p	$\mathcal{T}_0(\mathsf{p}) = \mathcal{T}_0^\vee(\mathsf{p})$ for $q = 0$	concatenation = 0-shuffle	deconcatenation = 0-deshuffle

We also discuss the free commutative Coxeter monoid $\mathcal{S}(\mathsf{p})$ and cofree cocommutative Coxeter comonoid $\mathcal{S}^\vee(\mathsf{p})$ on a Coxeter species p, and so on. These have signed analogues which we denote $\mathcal{E}(\mathsf{p})$ and $\mathcal{E}^\vee(\mathsf{p})$, and so on.

Free and cofree objects are left and right adjoints, respectively, to forgetful functors which we denote *frg*. The main adjunctions are listed in Summary 3.1. Some ways to approach them are listed below.

Coxeter species $\underset{\mathit{frg}}{\overset{\mathcal{T}}{\rightleftarrows}}$ Coxeter monoid	Coxeter comonoid $\underset{\mathcal{T}^\vee}{\overset{\mathit{frg}}{\rightleftarrows}}$ Coxeter species
Coxeter comonoid $\underset{\mathit{frg}}{\overset{\mathcal{T}_q}{\rightleftarrows}}$ Coxeter q-bimonoid	Coxeter q-bimonoid $\underset{\mathcal{T}_q^\vee}{\overset{\mathit{frg}}{\rightleftarrows}}$ Coxeter monoid
Coxeter species $\underset{\mathcal{P}}{\overset{\mathcal{T}_q}{\rightleftarrows}}$ Coxeter q-bimonoid	Coxeter q-bimonoid $\underset{\mathcal{T}_q^\vee}{\overset{\mathcal{Q}}{\rightleftarrows}}$ Coxeter species
Coxeter species $\underset{\mathit{frg}}{\overset{\mathcal{S}}{\rightleftarrows}}$ commutative Coxeter monoid	cocommutative Coxeter comonoid $\underset{\mathcal{S}^\vee}{\overset{\mathit{frg}}{\rightleftarrows}}$ Coxeter species
Coxeter comonoid $\underset{\mathit{frg}}{\overset{\mathcal{S}}{\rightleftarrows}}$ commutative Coxeter bimonoid	cocommutative Coxeter bimonoid $\underset{\mathcal{S}^\vee}{\overset{\mathit{frg}}{\rightleftarrows}}$ Coxeter monoid
Coxeter species $\underset{\mathcal{P}}{\overset{\mathcal{S}}{\rightleftarrows}}$ bicommutative Coxeter bimonoid	bicommutative Coxeter bimonoid $\underset{\mathcal{S}^\vee}{\overset{\mathcal{Q}}{\rightleftarrows}}$ Coxeter species

SUMMARY 3.1. Adjunctions involving Coxeter (co, bi)monoids.

- The adjunctions can be established directly using our explicit constructions of the universal objects. We will not get into this.
- The existence of universal objects can be seen as a formal consequence of the existence of certain bimonads on Coxeter species which we constructed in Section 2.10. This approach was explained in detail in the context of species in [**11**, Chapter 6], so we will not repeat it here.
- One can construct the universal objects by taking corresponding universal objects in species and understanding how type morphisms act on face-components (which remain the same as before). We will explain this briefly.

3.3.1. Free Coxeter monoid on a Coxeter species. For a Coxeter species p, recall from (2.87) the Coxeter species $\mathcal{T}(\mathsf{p})$ defined by

$$\mathcal{T}(\mathsf{p})[A] := \bigoplus_{F:\, A \leq F} \mathsf{p}[F]. \tag{3.20}$$

Observe:

$$\mathcal{T}(\mathsf{p}) = \mathsf{p} + \mathsf{p}^2 + \mathsf{p}^3 + \dots, \tag{3.21}$$

the sum of Cauchy powers of p defined in (3.1). It carries a Coxeter monoid structure. The product component

$$\mu_A^F : \mathcal{T}(\mathsf{p})[F] \to \mathcal{T}(\mathsf{p})[A]$$

is the inclusion map which sends the summand $\mathsf{p}[H]$ (with $H \geq F$) to itself. We call μ the *concatenation product*. This yields a functor

$$\mathcal{T} : \mathsf{CSp} \to \mathsf{Mon}(\mathsf{CSp}).$$

It is left adjoint to the forgetful functor. The unit of the adjunction is the inclusion $\mathsf{p} \hookrightarrow \mathcal{T}(\mathsf{p})$ arising from (3.21). Thus, $\mathcal{T}(\mathsf{p})$ is the *free Coxeter monoid* on the Coxeter species p. Its universal property is stated below.

Theorem 3.7. *Let* a *be a Coxeter monoid,* p *a Coxeter species,* $f : \mathsf{p} \to \mathsf{a}$ *a map of Coxeter species. Then there exists a unique morphism of Coxeter monoids* $\hat{f} : \mathcal{T}(\mathsf{p}) \to \mathsf{a}$ *such that the diagram*

$$\begin{array}{ccc} \mathcal{T}(\mathsf{p}) & \overset{\hat{f}}{\dashrightarrow} & \mathsf{a} \\ \uparrow & \nearrow_{f} & \\ \mathsf{p} & & \end{array}$$

commutes.

Explicitly, the map $\hat{f}$, on the A-component, on the F-summand, is

$$\mathsf{p}[F] \xrightarrow{f_F} \mathsf{a}[F] \xrightarrow{\mu_A^F} \mathsf{a}[A]. \tag{3.22}$$

Remark 3.8 (Species vs Coxeter species). The adjunction between $\mathcal{T}$ and *frg*, or equivalently, the universal property in Theorem 3.7 can be deduced from the corresponding result for species [**11**, Theorems 6.1 and 6.2]. In more detail: Let a be a Coxeter monoid and p a Coxeter species. Let $f : \mathsf{p} \to \mathsf{a}$ be a map of species and $\hat{f} : \mathcal{T}(\mathsf{p}) \to \mathsf{a}$ be the corresponding map of monoids given by formula (3.22) as in [**11**, Theorem 6.2]. Then f is a map of Coxeter species iff $\hat{f}$ is a map of Coxeter species. The diagram for checking the latter is shown below.

$$\begin{array}{ccccc} \mathsf{p}[F] & \xrightarrow{f_F} & \mathsf{a}[F] & \xrightarrow{\mu_A^F} & \mathsf{a}[A] \\ {\scriptstyle w}\downarrow & & {\scriptstyle w}\downarrow & & \downarrow{\scriptstyle w} \\ \mathsf{p}[wF] & \xrightarrow[f_{wF}]{} & \mathsf{a}[wF] & \xrightarrow[\mu_{wA}^{wF}]{} & \mathsf{a}[wA] \end{array}$$

The right square commutes since a is a Coxeter monoid. The left square commutes iff f is a map of Coxeter species.

3.3.2. Free Coxeter q-bimonoid on a Coxeter comonoid.

Let q be a scalar. For a Coxeter comonoid c, the free Coxeter monoid $\mathcal{T}(\mathsf{c})$ carries a Coxeter q-bimonoid structure. To indicate dependence on q, we write $\mathcal{T}_q(\mathsf{c})$. Its coproduct component Δ_A^G is defined by

$$\begin{array}{ccccc} \mathcal{T}_q(\mathsf{c})[A] & & \overset{\Delta_A^G}{\dashrightarrow} & & \mathcal{T}_q(\mathsf{c})[G] \\ \uparrow & & & & \uparrow \\ \mathsf{c}[H] & \xrightarrow[\Delta_H^{HG}]{} & \mathsf{c}[HG] & \xrightarrow[(\beta_q)_{GH,HG}]{} & \mathsf{c}[GH] \end{array} \tag{3.23}$$

for each $A \leq H$, with β_q as in (2.43). We call Δ the *q-dequasishuffle coproduct*. This yields a functor

$$\mathcal{T}_q : \mathsf{Comon}(\mathsf{CSp}) \to q\text{-}\mathsf{Bimon}(\mathsf{CSp}). \tag{3.24}$$

It is left adjoint to the forgetful functor. Thus, $\mathcal{T}_q(\mathsf{c})$ is the *free Coxeter q-bimonoid* on the Coxeter comonoid c. Its universal property is stated below.

Theorem 3.9. *Let q be a scalar. Let h be a Coxeter q-bimonoid, c a Coxeter comonoid, $f : \mathsf{c} \to \mathsf{h}$ a morphism of Coxeter comonoids. Then there exists a unique morphism of Coxeter q-bimonoids $\hat{f} : \mathcal{T}_q(\mathsf{c}) \to \mathsf{h}$ such that the diagram*

$$\begin{array}{ccc} \mathcal{T}_q(\mathsf{c}) & \overset{\hat{f}}{\dashrightarrow} & \mathsf{h} \\ \uparrow & \nearrow_{f} & \\ \mathsf{c} & & \end{array}$$

commutes.

Explicitly, the map $\hat{f}$ is given as before by formula (3.22).

Remark 3.10. The adjunction between $\mathcal{T}_q$ and *frg*, or equivalently, the universal property in Theorem 3.9 can be deduced from the corresponding result for species [**11**, Theorems 6.5 and 6.6]. This builds on Remark 3.8.

3.3.3. Left adjoint of the primitive part functor. Precomposing (3.24) with the trivial Coxeter comonoid functor (3.18) yields the functor

$$\mathcal{T}_q : \mathsf{CSp} \to q\text{-}\mathsf{Bimon}(\mathsf{CSp}). \tag{3.25}$$

(We continue to denote it by $\mathcal{T}_q$.) For a Coxeter species p, the product and coproduct of $\mathcal{T}_q(\mathsf{p})$ are

$$\begin{array}{ccc} \mathcal{T}_q(\mathsf{p})[F] & \overset{\mu_A^F}{\dashrightarrow} & \mathcal{T}_q(\mathsf{p})[A] \\ \uparrow & & \uparrow \\ \mathsf{p}[H] & \xrightarrow[\mathrm{id}]{} & \mathsf{p}[H] \end{array} \qquad \begin{array}{ccc} \mathcal{T}_q(\mathsf{p})[A] & \overset{\Delta_A^G}{\dashrightarrow} & \mathcal{T}_q(\mathsf{p})[G] \\ \uparrow & & \uparrow \\ \mathsf{p}[H] & \xrightarrow[(\beta_q)_{GH,HG}]{} & \begin{cases} \mathsf{p}[GH] & \text{if } HG = H, \\ 0 & \text{otherwise.} \end{cases} \end{array} \tag{3.26}$$

The product is concatenation as before, but note how the coproduct has simplified from (3.23). We call it the *q-deshuffle coproduct*.

Using the first adjunction in (3.19), we deduce that the functor $\mathcal{T}_q$ above is left adjoint to the primitive part functor (from Coxeter q-bimonoids to Coxeter species). This fact is phrased below as a universal property.

Theorem 3.11. *Let q be a scalar. Let h be a Coxeter q-bimonoid, p a Coxeter species, $f : \mathsf{p} \to \mathcal{P}(\mathsf{h})$ a map of Coxeter species. Then there exists a unique morphism of Coxeter q-bimonoids $\hat{f} : \mathcal{T}_q(\mathsf{p}) \to \mathsf{h}$ such that the diagram*

$$\begin{array}{ccc} \mathcal{T}_q(\mathsf{p}) & \overset{\hat{f}}{\dashrightarrow} & \mathsf{h} \\ \uparrow & & \uparrow \\ \mathsf{p} & \xrightarrow[f]{} & \mathcal{P}(\mathsf{h}) \end{array}$$

commutes.

Explicitly, the map $\hat{f}$, on the A-component, on the F-summand, is

$$\mathsf{p}[F] \xrightarrow{f_F} \mathcal{P}(\mathsf{h})[F] \hookrightarrow \mathsf{h}[F] \xrightarrow{\mu_A^F} \mathsf{h}[A]. \tag{3.27}$$

3.3.4. Cofree Coxeter comonoid on a Coxeter species. Dually, one can construct the *cofree Coxeter comonoid* on a Coxeter species p. We denote it by $\mathcal{T}^\vee(\mathsf{p})$. As a Coxeter species, it equals $\mathcal{T}(\mathsf{p})$ and is given by (3.20). The coproduct component

$$\Delta_A^F : \mathcal{T}^\vee(\mathsf{p})[A] \to \mathcal{T}^\vee(\mathsf{p})[F]$$

is the projection which sends the summand $\mathsf{p}[H]$ (with $H \geq A$) to itself if $H \geq F$, and to 0 otherwise. We call Δ the *deconcatenation coproduct*. This yields a functor

$$\mathcal{T}^\vee : \mathsf{CSp} \to \mathsf{Comon(CSp)}$$

which is right adjoint to the forgetful functor. The counit of the adjunction is the projection $\mathcal{T}^\vee(\mathsf{p}) \twoheadrightarrow \mathsf{p}$ arising from (3.21). The universal property of $\mathcal{T}^\vee(\mathsf{p})$ can be phrased as follows.

Theorem 3.12. *Let c be a Coxeter comonoid, p a Coxeter species, $f : \mathsf{c} \to \mathsf{p}$ a map of Coxeter species. Then there exists a unique morphism of Coxeter comonoids $\hat{f} : \mathsf{c} \to \mathcal{T}^\vee(\mathsf{p})$ such that the diagram*

$$\begin{array}{ccc} \mathsf{c} & \overset{\hat{f}}{\dashrightarrow} & \mathcal{T}^\vee(\mathsf{p}) \\ & {\scriptstyle f}\searrow & \big\downarrow \\ & & \mathsf{p} \end{array}$$

commutes.

Explicitly, the map $\hat{f}$, on the A-component, into the F-summand, is

$$\mathsf{c}[A] \xrightarrow{\Delta_A^F} \mathsf{c}[F] \xrightarrow{f_F} \mathsf{p}[F]. \tag{3.28}$$

Remark 3.13. The adjunction between $\mathcal{T}^\vee$ and *frg*, or equivalently, the universal property in Theorem 3.12 can be deduced from the corresponding result for species [**11**, Theorems 6.9 and 6.10] as in Remark 3.8.

3.3.5. Cofree Coxeter q-bimonoid on a Coxeter monoid. Let q be a scalar. For a Coxeter monoid a, the cofree Coxeter comonoid $\mathcal{T}^\vee(\mathsf{a})$ carries a Coxeter q-bimonoid structure. To indicate dependence on q, we write $\mathcal{T}_q^\vee(\mathsf{a})$. Its product component μ_A^G is defined by

$$\begin{array}{ccccc} \mathcal{T}_q^\vee(\mathsf{a})[G] & & \overset{\mu_A^G}{\dashrightarrow} & & \mathcal{T}_q^\vee(\mathsf{a})[A] \\ \big\uparrow & & & & \big\uparrow \\ \mathsf{a}[K] & \xrightarrow[((\beta_q)_{HG,GH})]{} & \displaystyle\bigoplus_{\substack{H:\, A \leq H \\ GH = K}} \mathsf{a}[HG] & \xrightarrow[(\mu_H^{HG})]{} & \displaystyle\bigoplus_{\substack{H:\, A \leq H \\ GH = K}} \mathsf{a}[H] \end{array} \tag{3.29}$$

for each $G \le K$. For the second map, the matrix-component is μ_H^{HG} on matching indices, and zero otherwise. We call μ the *q-quasishuffle product.* This yields a functor

$$\mathcal{T}_q^\vee : \mathsf{Mon}(\mathsf{CSp}) \to q\text{-}\mathsf{Bimon}(\mathsf{CSp}). \tag{3.30}$$

It is right adjoint to the forgetful functor. Thus, $\mathcal{T}_q^\vee(\mathsf{a})$ is the *cofree Coxeter q-bimonoid* on the Coxeter monoid a. Its universal property is stated below.

Theorem 3.14. *Let q be a scalar. Let h be a Coxeter q-bimonoid, a a Coxeter monoid, $f : \mathsf{h} \to \mathsf{a}$ a morphism of Coxeter monoids. Then there exists a unique morphism of Coxeter q-bimonoids $\hat{f} : \mathsf{h} \to \mathcal{T}_q^\vee(\mathsf{a})$ such that the diagram*

$$\begin{array}{ccc} \mathsf{h} & \overset{\hat{f}}{\dashrightarrow} & \mathcal{T}_q^\vee(\mathsf{a}) \\ & {\scriptstyle f}\searrow & \downarrow \\ & & \mathsf{a} \end{array}$$

commutes.

Explicitly, the map $\hat{f}$ is given as before by formula (3.28).

Remark 3.15. The adjunction between $\mathcal{T}_q^\vee$ and *frg*, or equivalently, the universal property in Theorem 3.14 can be deduced from the corresponding result for species [**11**, Theorems 6.12 and 6.13]. This builds on Remark 3.13.

3.3.6. Right adjoint of the indecomposable part functor.

Precomposing (3.30) with the trivial Coxeter monoid functor (3.18) yields the functor

$$\mathcal{T}_q^\vee : \mathsf{CSp} \to q\text{-}\mathsf{Bimon}(\mathsf{CSp}). \tag{3.31}$$

(We continue to denote it by $\mathcal{T}_q^\vee$.) For a Coxeter species p, the product and coproduct of $\mathcal{T}_q^\vee(\mathsf{p})$ are

$$\begin{array}{ccc} \mathcal{T}_q^\vee(\mathsf{p})[G] & \overset{\mu_A^G}{\dashrightarrow} & \mathcal{T}_q^\vee(\mathsf{p})[A] \\ \uparrow & & \uparrow \\ \mathsf{p}[K] & \xrightarrow[((\beta_q)_{HG,GH})]{} & \displaystyle\bigoplus_{\substack{H:\, GH=K \\ HG=H \\ A\le H}} \mathsf{p}[H] \end{array} \qquad \begin{array}{ccc} \mathcal{T}_q^\vee(\mathsf{p})[A] & \overset{\Delta_A^F}{\dashrightarrow} & \mathcal{T}_q^\vee(\mathsf{p})[F] \\ \uparrow & & \uparrow \\ \mathsf{p}[H] & \longrightarrow & \begin{cases} \mathsf{p}[H] & \text{if } H \ge F, \\ 0 & \text{otherwise.} \end{cases} \end{array} \tag{3.32}$$

The coproduct is deconcatenation as before, but the product has simplified from (3.29). We call it the *q-shuffle product.*

Using the second adjunction in (3.19), we deduce that the functor $\mathcal{T}_q^\vee$ above is right adjoint to the indecomposable part functor (from Coxeter q-bimonoids to Coxeter species). This fact is phrased below as a universal property.

Theorem 3.16. *Let q be a scalar. Let h be a Coxeter q-bimonoid, p a Coxeter species, $f : \mathcal{Q}(\mathsf{h}) \to \mathsf{p}$ a map of Coxeter species. Then there exists a unique* *morphism of Coxeter q-bimonoids* $\hat{f} : \mathsf{h} \to \mathcal{T}_q^\vee(\mathsf{p})$ *such that the diagram*

$$\begin{array}{ccc} \mathsf{h} & \overset{\hat{f}}{\dashrightarrow} & \mathcal{T}_q^\vee(\mathsf{p}) \\ \downarrow & & \downarrow \\ \mathcal{Q}(\mathsf{h}) & \xrightarrow[f]{} & \mathsf{p} \end{array}$$

commutes.

Explicitly, the map $\hat{f}$, on the A-component, into the F-summand, is

$$\mathsf{h}[A] \xrightarrow{\Delta_A^F} \mathsf{h}[F] \twoheadrightarrow \mathcal{Q}(\mathsf{h})[F] \xrightarrow{f_F} \mathsf{p}[F]. \tag{3.33}$$

Exercise 3.17 (Conjugation by signature functors). Recall the o-signature functor (2.73). Check: For a scalar q, for a Coxeter comonoid c and Coxeter monoid a, there are isomorphisms

$$\mathcal{T}_q(\mathsf{c}^-) = \mathcal{T}_{-q}(\mathsf{c})^- \quad \text{and} \quad \mathcal{T}_q^\vee(\mathsf{a}^-) = \mathcal{T}_{-q}^\vee(\mathsf{a})^-$$

of Coxeter q-bimonoids. In particular, for a Coxeter species p,

$$\mathcal{T}_q(\mathsf{p}^-) = \mathcal{T}_{-q}(\mathsf{p})^- \quad \text{and} \quad \mathcal{T}_q^\vee(\mathsf{p}^-) = \mathcal{T}_{-q}^\vee(\mathsf{p})^-$$

as Coxeter q-bimonoids. The same holds for the u-signature functor (2.70).

3.3.7. Free and cofree Coxeter 0-bimonoid. Let us specialize to $q = 0$. For a Coxeter species p, the product and coproduct of $\mathcal{T}_0(\mathsf{p})$ in (3.26) coincide with those of $\mathcal{T}_0^\vee(\mathsf{p})$ in (3.32). They are given by concatenation (0-shuffle) and deconcatenation (0-deshuffle), respectively. They are rewritten below.

$$\begin{array}{ccc} \mathcal{T}_0(\mathsf{p})[F] & \overset{\mu_A^F}{\dashrightarrow} & \mathcal{T}_0(\mathsf{p})[A] \\ \uparrow & & \uparrow \\ \mathsf{p}[H] & \xrightarrow[\text{id}]{} & \mathsf{p}[H] \end{array} \qquad \begin{array}{ccc} \mathcal{T}_0(\mathsf{p})[A] & \overset{\Delta_A^F}{\dashrightarrow} & \mathcal{T}_0(\mathsf{p})[F] \\ \uparrow & & \uparrow \\ \mathsf{p}[H] & \longrightarrow & \begin{cases} \mathsf{p}[H] & \text{if } H \geq F, \\ 0 & \text{otherwise.} \end{cases} \end{array} \tag{3.34}$$

This is the sense in which concatenation and deconcatenation are compatible.

3.3.8. Free commutative Coxeter monoid. For a Coxeter species p, recall from (2.94) the Coxeter species $\mathcal{S}(\mathsf{p})$ defined by

$$\mathcal{S}(\mathsf{p})[\mathrm{Z}] := \bigoplus_{\mathrm{X}:\, \mathrm{Z} \leq \mathrm{X}} \mathsf{p}[\mathrm{X}]. \tag{3.35}$$

Observe:

$$\mathcal{S}(\mathsf{p}) = \mathsf{p} + \mathsf{p}^{\overline{2}} + \mathsf{p}^{\overline{3}} + \dots, \tag{3.36}$$

the sum of commutative Cauchy powers of p defined in (3.4). It carries a commutative Coxeter monoid structure. The product component

$$\mu_{\mathrm{Z}}^{\mathrm{X}} : \mathcal{S}(\mathsf{p})[\mathrm{X}] \to \mathcal{S}(\mathsf{p})[\mathrm{Z}]$$

is the inclusion map which sends the summand $\mathsf{p}[\mathrm{Y}]$ (with $\mathrm{Y} \geq \mathrm{X}$) to itself. Equivalently, $\mathcal{S}(\mathsf{p})$ is the Coxeter quotient monoid of $\mathcal{T}(\mathsf{p})$ under the surjection $\mathcal{T}(\mathsf{p}) \twoheadrightarrow \mathcal{S}(\mathsf{p})$ in (2.95). This yields a functor

$$\mathcal{S} : \mathsf{CSp} \to \mathsf{Mon}^{\mathsf{co}}(\mathsf{CSp}).$$

It is left adjoint to the forgetful functor. The unit of the adjunction is the inclusion $\mathsf{p} \hookrightarrow \mathcal{S}(\mathsf{p})$ arising from (3.36). Thus, $\mathcal{S}(\mathsf{p})$ is the *free commutative Coxeter monoid* on the Coxeter species p. Its universal property is stated below. It is the commutative analogue of Theorem 3.7.

Theorem 3.18. *Let* a *be a commutative Coxeter monoid,* p *a Coxeter species,* $f : \mathsf{p} \to \mathsf{a}$ *a map of Coxeter species. Then there exists a unique morphism of Coxeter monoids* $\hat{f} : \mathcal{S}(\mathsf{p}) \to \mathsf{a}$ *such that the diagram*

$$\begin{array}{ccc} \mathcal{S}(\mathsf{p}) & \overset{\hat{f}}{\dashrightarrow} & \mathsf{a} \\ \uparrow & \nearrow_{f} & \\ \mathsf{p} & & \end{array}$$

commutes.

Explicitly, the map $\hat{f}$, on the Z-component, on the X-summand, is

$$\mathsf{p}[\mathrm{X}] \xrightarrow{f_{\mathrm{X}}} \mathsf{a}[\mathrm{X}] \xrightarrow{\mu_{\mathrm{Z}}^{\mathrm{X}}} \mathsf{a}[\mathrm{Z}]. \tag{3.37}$$

3.3.9. Free commutative Coxeter bimonoid on a Coxeter comonoid. For c a Coxeter comonoid, $\mathcal{S}(\mathsf{c})$ carries a Coxeter bimonoid structure as a quotient of $\mathcal{T}(\mathsf{c})$. This is the free commutative Coxeter bimonoid on c. Further, if c is cocommutative, then $\mathcal{S}(\mathsf{c})$ is bicommutative. Formally, we have functors

$$\mathcal{S} : \mathsf{Comon}(\mathsf{CSp}) \to \mathsf{Bimon}^{\mathsf{co}}(\mathsf{CSp}) \quad \text{and} \quad \mathcal{S} : {}^{\mathsf{co}}\mathsf{Comon}(\mathsf{CSp}) \to {}^{\mathsf{co}}\mathsf{Bimon}^{\mathsf{co}}(\mathsf{CSp})$$

which are left adjoints to the respective forgetful functors. By precomposing the latter with the trivial Coxeter comonoid functor, we obtain the functor

$$\mathcal{S} : \mathsf{CSp} \to {}^{\mathsf{co}}\mathsf{Bimon}^{\mathsf{co}}(\mathsf{CSp}). \tag{3.38}$$

It is left adjoint to the primitive part functor (from bicommutative Coxeter bimonoids to Coxeter species). This fact is phrased below as a universal property. It is the commutative analogue of Theorem 3.11.

Theorem 3.19. *Let* h *be a bicommutative Coxeter bimonoid,* p *a Coxeter species,* $f : \mathsf{p} \to \mathcal{P}(\mathsf{h})$ *a map of Coxeter species. Then there exists a unique morphism of Coxeter bimonoids* $\hat{f} : \mathcal{S}(\mathsf{p}) \to \mathsf{h}$ *such that the diagram*

$$\begin{array}{ccc} \mathcal{S}(\mathsf{p}) & \overset{\hat{f}}{\dashrightarrow} & \mathsf{h} \\ \uparrow & & \uparrow \\ \mathsf{p} & \underset{f}{\longrightarrow} & \mathcal{P}(\mathsf{h}) \end{array}$$

commutes.

Explicitly, the map $\hat{f}$, on the Z-component, on the X-summand, is

$$\mathsf{p}[\mathrm{X}] \xrightarrow{f_{\mathrm{X}}} \mathcal{P}(\mathsf{h})[\mathrm{X}] \hookrightarrow \mathsf{h}[\mathrm{X}] \xrightarrow{\mu_{\mathrm{Z}}^{\mathrm{X}}} \mathsf{h}[\mathrm{Z}]. \tag{3.39}$$

3.3.10. Cofree cocommutative Coxeter comonoid. Dually, one can construct the *cofree cocommutative Coxeter comonoid* on a Coxeter species p. It is denoted $\mathcal{S}^\vee(\mathsf{p})$. As a Coxeter species, it equals $\mathcal{S}(\mathsf{p})$ and is given by (3.35). The coproduct component

$$\Delta_{\mathrm{Z}}^{\mathrm{X}} : \mathcal{S}^\vee(\mathsf{p})[\mathrm{Z}] \to \mathcal{S}^\vee(\mathsf{p})[\mathrm{X}]$$

is the projection which sends the summand $\mathsf{p}[\mathrm{Y}]$ (with $\mathrm{Y} \geq \mathrm{Z}$) to itself if $\mathrm{Y} \geq \mathrm{X}$, and to 0 otherwise. Equivalently, $\mathcal{S}^\vee(\mathsf{p})$ is the Coxeter subcomonoid of $\mathcal{T}^\vee(\mathsf{p})$ under the injection $\mathcal{S}^\vee(\mathsf{p}) \hookrightarrow \mathcal{T}^\vee(\mathsf{p})$ in (2.96). This yields a functor

$$\mathcal{S}^\vee : \mathsf{CSp} \to {}^{\mathsf{co}}\mathsf{Comon}(\mathsf{CSp}).$$

It is right adjoint to the forgetful functor. The counit of the adjunction is the projection $\mathcal{S}^\vee(\mathsf{p}) \twoheadrightarrow \mathsf{p}$ arising from (3.36). The universal property of $\mathcal{S}^\vee(\mathsf{p})$ can be phrased as follows. It is the commutative analogue of Theorem 3.12.

Theorem 3.20. *Let c be a cocommutative Coxeter comonoid, p a Coxeter species, $f : \mathsf{c} \to \mathsf{p}$ a map of Coxeter species. Then there exists a unique morphism of Coxeter comonoids $\hat{f} : \mathsf{c} \to \mathcal{S}^\vee(\mathsf{p})$ such that the diagram*

$$\begin{array}{ccc} \mathsf{c} & \overset{\hat{f}}{\dashrightarrow} & \mathcal{S}^\vee(\mathsf{p}) \\ & {\scriptstyle f}\searrow & \downarrow \\ & & \mathsf{p} \end{array}$$

commutes.

Explicitly, the map $\hat{f}$, on the Z-component, into the X-summand, is

$$\mathsf{c}[\mathrm{Z}] \xrightarrow{\Delta_{\mathrm{Z}}^{\mathrm{X}}} \mathsf{c}[\mathrm{X}] \xrightarrow{f_{\mathrm{X}}} \mathsf{p}[\mathrm{X}]. \tag{3.40}$$

3.3.11. Cofree cocommutative Coxeter bimonoid on a Coxeter monoid. Dually, for a a Coxeter monoid, $\mathcal{S}^\vee(\mathsf{a})$ carries a Coxeter bimonoid structure. It sits inside $\mathcal{T}^\vee(\mathsf{a})$. This is the cofree cocommutative Coxeter bimonoid on a. Further, if a is commutative, then $\mathcal{S}^\vee(\mathsf{a})$ is bicommutative. Formally, we have functors

$$\mathcal{S}^\vee : \mathsf{Mon}(\mathsf{CSp}) \to {}^{\mathsf{co}}\mathsf{Bimon}(\mathsf{CSp}) \quad \text{and} \quad \mathcal{S}^\vee : \mathsf{Mon}^{\mathsf{co}}(\mathsf{CSp}) \to {}^{\mathsf{co}}\mathsf{Bimon}^{\mathsf{co}}(\mathsf{CSp})$$

which are right adjoints to the respective forgetful functors. By precomposing the latter with the trivial Coxeter monoid functor, we obtain the functor

$$\mathcal{S}^\vee : \mathsf{CSp} \to {}^{\mathsf{co}}\mathsf{Bimon}^{\mathsf{co}}(\mathsf{CSp}). \tag{3.41}$$

It is right adjoint to the indecomposable part functor (from bicommutative Coxeter bimonoids to Coxeter species). This fact is phrased below as a universal property. It is the commutative analogue of Theorem 3.16.

Theorem 3.21. *Let* h *be a bicommutative Coxeter bimonoid,* p *a Coxeter species,* $f : \mathcal{Q}(\mathsf{h}) \to \mathsf{p}$ *a map of Coxeter species. Then there exists a unique morphism of Coxeter bimonoids* $\hat{f} : \mathsf{h} \to \mathcal{S}^\vee(\mathsf{p})$ *such that the diagram*

$$\begin{array}{ccc} \mathsf{h} & \overset{\hat{f}}{\dashrightarrow} & \mathcal{S}^\vee(\mathsf{p}) \\ \downarrow & & \downarrow \\ \mathcal{Q}(\mathsf{h}) & \underset{f}{\longrightarrow} & \mathsf{p} \end{array}$$

commutes.

Explicitly, the map $\hat{f}$, on the Z-component, into the X-summand, is

$$\mathsf{h}[\mathrm{Z}] \xrightarrow{\Delta_{\mathrm{Z}}^{\mathrm{X}}} \mathsf{h}[\mathrm{X}] \twoheadrightarrow \mathcal{Q}(\mathsf{h})[\mathrm{X}] \xrightarrow{f_{\mathrm{X}}} \mathsf{p}[\mathrm{X}]. \tag{3.42}$$

Exercise 3.22. We have defined two functors, namely, $\mathcal{S}$ and $\mathcal{S}^\vee$ from Coxeter species to bicommutative Coxeter bimonoids. In fact, check that $\mathcal{S} = \mathcal{S}^\vee$.

3.3.12. Signed analogue. For a Coxeter species p, we have the *free signed commutative Coxeter monoid* $\mathcal{E}(\mathsf{p})$ and *cofree signed cocommutative Coxeter comonoid* $\mathcal{E}^\vee(\mathsf{p})$. They can be defined as

$$\mathcal{E}(\mathsf{p}) := \mathcal{S}(\mathsf{p}_-)_- \quad \text{and} \quad \mathcal{E}^\vee(\mathsf{p}) := \mathcal{S}^\vee(\mathsf{p}_-)_- \tag{3.43}$$

by conjugating $\mathcal{S}$ and $\mathcal{S}^\vee$ by the u-signature functor (2.70). As Coxeter species, both are equal to (2.98) (as mentioned in Exercise 2.81). The first is a Coxeter quotient monoid of $\mathcal{T}_{-1}(\mathsf{p}) = \mathcal{T}(\mathsf{p}_-)_-$, while the second is a Coxeter subcomonoid of $\mathcal{T}_{-1}^\vee(\mathsf{p}) = \mathcal{T}^\vee(\mathsf{p}_-)_-$. See Exercise 3.17 in this regard.

For a Coxeter comonoid c, we have the free signed commutative signed Coxeter bimonoid $\mathcal{E}(\mathsf{c})$, and for a Coxeter monoid a, we have the cofree signed cocommutative signed Coxeter bimonoid $\mathcal{E}^\vee(\mathsf{a})$. They can be defined as

$$\mathcal{E}(\mathsf{c}) := \mathcal{S}(\mathsf{c}_-)_- \quad \text{and} \quad \mathcal{E}^\vee(\mathsf{a}) := \mathcal{S}^\vee(\mathsf{a}_-)_-. \tag{3.44}$$

The first is a signed Coxeter quotient bimonoid of $\mathcal{T}_{-1}(\mathsf{c})$, while the second is a signed Coxeter subbimonoid of $\mathcal{T}_{-1}^\vee(\mathsf{a})$.

The above discussion can also be carried out by using the o-signature functor (2.73) in place of the u-signature functor.

3.3.13. Alternative description of Coxeter bimonoids. For a Coxeter monoid (a, μ), putting together product components μ_A^F yields a map of Coxeter species $\mu : \mathcal{T}(\mathsf{a}) \to \mathsf{a}$. Dually, for a Coxeter comonoid (c, Δ), we have a map of Coxeter species $\Delta : \mathsf{c} \to \mathcal{T}^\vee(\mathsf{c})$.

Exercise 3.23. Let q be a scalar. Let h be a Coxeter species equipped with maps μ and Δ such that (h, μ) is a Coxeter monoid and (h, Δ) is a Coxeter comonoid. Show: The following are equivalent.

(1) $(\mathsf{h}, \mu, \Delta)$ is a Coxeter q-bimonoid.
(2) $\mu : \mathcal{T}_q(\mathsf{h}) \to \mathsf{h}$ is a morphism of Coxeter comonoids.
(3) $\Delta : \mathsf{h} \to \mathcal{T}_q^\vee(\mathsf{h})$ is a morphism of Coxeter monoids.

The coproduct of $\mathcal{T}_q(\mathsf{h})$ is given by q-dequasishuffle (3.23) with h viewed as a Coxeter comonoid, while the product of $\mathcal{T}_q^\vee(\mathsf{h})$ is given by q-quasishuffle (3.29) with h viewed as a Coxeter monoid.

3.4. (Co)abelianizations of Coxeter (co)monoids

We discuss the (co)abelianization construction which turns a Coxeter (co)monoid into a (co)commutative Coxeter (co)monoid. It also applies to Coxeter bimonoids. The universality of these constructions is shown in Summary 3.2. We also have signed analogues.

Coxeter monoid	$\underset{inc}{\overset{(-)_{ab}}{\rightleftarrows}}$	com. Coxeter monoid	cocom. Coxeter comonoid	$\underset{(-)^{coab}}{\overset{inc}{\rightleftarrows}}$	Coxeter comonoid
Coxeter bimonoid	$\underset{inc}{\overset{(-)_{ab}}{\rightleftarrows}}$	com. Coxeter bimonoid	cocom. Coxeter bimonoid	$\underset{(-)^{coab}}{\overset{inc}{\rightleftarrows}}$	Coxeter bimonoid
cocom. Coxeter bimonoid	$\underset{inc}{\overset{(-)_{ab}}{\rightleftarrows}}$	bicom. Coxeter bimonoid	bicom. Coxeter bimonoid	$\underset{(-)^{coab}}{\overset{inc}{\rightleftarrows}}$	com. Coxeter bimonoid

SUMMARY 3.2. Adjunctions involving (co)abelianization.

A parallel more detailed discussion of these ideas for species is given in [**11**, Section 2.7].

3.4.1. Abelianization of a Coxeter monoid. Every Coxeter monoid a has a largest commutative Coxeter quotient monoid a_{ab} called the *abelianization* of a. The quotient map $\pi_{\mathsf{a}} : \mathsf{a} \twoheadrightarrow \mathsf{a}_{ab}$ is called the *abelianization map*. Explicitly, the A-component of its kernel is spanned by the elements

$$\mu_A^F(x) - \mu_A^G \beta_{G,F}(x) \ \in \mathsf{a}[A], \tag{3.45}$$

as x, F, G vary (with F and G both greater than A and of the same support).

A morphism of Coxeter monoids f from a Coxeter monoid a to a commutative Coxeter monoid b factors through π_{a} yielding a commutative diagram of Coxeter monoids

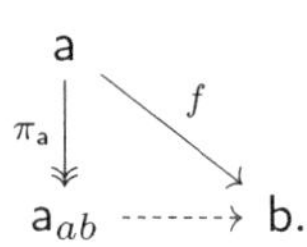

Thus, abelianization is left adjoint to the inclusion functor:

$$\mathsf{Mon}(\mathsf{CSp}) \underset{inc}{\overset{(-)_{ab}}{\rightleftarrows}} \mathsf{Mon}^{\mathsf{co}}(\mathsf{CSp}). \tag{3.46}$$

Hence, commutative Coxeter monoids is a reflective subcategory of Coxeter monoids via the inclusion functor (since this is full, faithful, and has a left adjoint). Further, (3.46) induces an adjunction between categories of Coxeter bimonoids and commutative Coxeter bimonoids, and between cocommutative Coxeter bimonoids and bicommutative Coxeter bimonoids.

3.4.2. Coabelianization of a Coxeter comonoid. Dually, every Coxeter comonoid c has a largest cocommutative Coxeter subcomonoid c^{coab} called the *coabelianization* of c. The inclusion map $\pi_{\mathsf{c}}^{\vee} : \mathsf{c}^{coab} \hookrightarrow \mathsf{c}$ is called the *coabelianization map*. Explicitly, $\mathsf{c}^{coab}[A]$ consists of those $x \in \mathsf{c}[A]$ for which

$$\beta_{G,F}\Delta_A^F(x) = \Delta_A^G(x) \tag{3.47}$$

for all faces F and G both greater than A and of the same support.

The image of a morphism of Coxeter comonoids f from a cocommutative Coxeter comonoid d to a Coxeter comonoid c lies inside c^{coab} yielding a commutative diagram

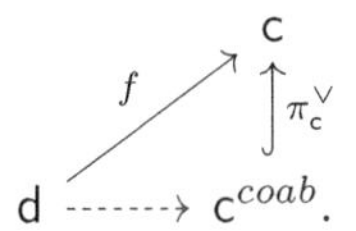

Thus, coabelianization is right adjoint to the inclusion functor:

$$ {}^{\mathrm{co}}\mathsf{Comon}(\mathsf{CSp}) \underset{(-)^{coab}}{\overset{inc}{\rightleftarrows}} \mathsf{Comon}(\mathsf{CSp}). \tag{3.48}$$

Hence, cocommutative Coxeter comonoids is a coreflective subcategory of Coxeter comonoids via the inclusion functor (since this is full, faithful, and has a right adjoint). Further, (3.48) induces an adjunction between categories of Coxeter bimonoids and cocommutative Coxeter bimonoids, and between commutative Coxeter bimonoids and bicommutative Coxeter bimonoids.

3.4.3. Signed analogue. By replacing $\beta_{G,F}$ by $(-1)^{\mathrm{dist}(F,G)}\beta_{G,F}$, the above discussion can be carried out in the signed setting. In this context, we use terms *signed abelianization* and *signed coabelianization*, and notations a_{-ab} and c^{-coab}. Formally,

$$(\mathsf{a}_-)_{ab} = (\mathsf{a}_{-ab})_- \quad \text{and} \quad (\mathsf{c}_-)^{coab} = (\mathsf{c}^{-coab})_-,$$

with $(-)_-$ being the u-signature functor (2.70). The same holds for the o-signature functor (2.73).

3.4.4. Universal constructions. For a Coxeter comonoid c and Coxeter monoid a,

$$\mathcal{S}(\mathsf{c}) = \mathcal{T}(\mathsf{c})_{ab} \quad \text{and} \quad \mathcal{S}^{\vee}(\mathsf{a}) = \mathcal{T}^{\vee}(\mathsf{a})^{coab} \tag{3.49}$$

as Coxeter bimonoids. In particular, this applies when c is a trivial Coxeter comonoid, and when a is a trivial Coxeter monoid. In this case, we write: For a Coxeter species p,

$$\mathcal{S}(\mathsf{p}) = \mathcal{T}(\mathsf{p})_{ab} \quad \text{and} \quad \mathcal{S}^{\vee}(\mathsf{p}) = \mathcal{T}^{\vee}(\mathsf{p})^{coab} \tag{3.50}$$

as bicommutative Coxeter bimonoids. The abelianization map $\mathcal{T}(\mathsf{p}) \twoheadrightarrow \mathcal{S}(\mathsf{p})$ and coabelianization map $\mathcal{S}^{\vee}(\mathsf{p}) \hookrightarrow \mathcal{T}^{\vee}(\mathsf{p})$ are the canonical surjection and injection, respectively.

Similarly, for a Coxeter comonoid c and Coxeter monoid a,

$$\mathcal{E}(\mathsf{c}) = \mathcal{T}_{-1}(\mathsf{c})_{-ab} \quad \text{and} \quad \mathcal{E}^{\vee}(\mathsf{a}) = \mathcal{T}_{-1}^{\vee}(\mathsf{a})^{-coab} \tag{3.51}$$

as signed Coxeter bimonoids. In particular, for a Coxeter species p,

$$\mathcal{E}(\mathsf{p}) = \mathcal{T}_{-1}(\mathsf{p})_{-ab} \quad \text{and} \quad \mathcal{E}^{\vee}(\mathsf{p}) = \mathcal{T}^{\vee}_{-1}(\mathsf{p})^{-coab} \tag{3.52}$$

as signed bicommutative signed Coxeter bimonoids. The signed abelianization map $\mathcal{T}_{-1}(\mathsf{p}) \twoheadrightarrow \mathcal{E}(\mathsf{p})$ and signed coabelianization map $\mathcal{E}^{\vee}(\mathsf{p}) \hookrightarrow \mathcal{T}^{\vee}_{-1}(\mathsf{p})$ are the canonical surjection and injection, respectively.

3.5. Hadamard product. Convolution Coxeter monoid

We introduce the Hadamard product of Coxeter species. It defines a symmetric monoidal structure on the category of Coxeter species, with the exponential Coxeter species as the unit object. Moreover, this structure extends to Coxeter monoids, comonoids, bimonoids. We show that the Hadamard product of (co)free Coxeter (co)monoids is again (co)free. We then construct the internal hom for the Hadamard product of Coxeter species, and discuss related constructions of the convolution Coxeter monoid, coconvolution Coxeter comonoid, biconvolution Coxeter bimonoid. These ideas are continued in Sections 3.6 and 3.7. A part summary is provided in Table 3.3.

TABLE 3.3. Hadamard product of Coxeter species.

$\mathsf{p} \times \mathsf{q}$	Hadamard product of Coxeter species p and q
$\mathrm{hom}^{\times}(\mathsf{p}, \mathsf{q})$	internal hom of p and q
$\mathsf{c} \times \mathsf{d}$	Hadamard product of Coxeter comonoids c and d
	copower of d by c
$\mathcal{C}(\mathsf{c}, \mathsf{d})$	enrichment of Coxeter comonoids over itself internal hom of c and d
	power of d by c
$\overline{\mathcal{C}}(\mathsf{a}, \mathsf{b})$	enrichment of Coxeter monoids over Coxeter comonoids universal measuring Coxeter comonoid
$\mathsf{c} \triangleright \mathsf{a}$	copower of a by c
$\mathrm{hom}^{\times}(\mathsf{c}, \mathsf{a})$	convolution Coxeter monoid
	power of a by c

The Hadamard product for species was studied in [**11**, Chapter 8]. Since the details are similar, we will be brief here.

3.5.1. Exponential Coxeter species. Let E be the Coxeter species defined by $\mathsf{E}[F] = \Bbbk$ for all faces F, with support and type morphisms all identities. This is the *exponential Coxeter species*. Further, it carries a Coxeter bimonoid structure with product and coproduct components also defined to be identities. This is the *exponential Coxeter bimonoid*. It is self-dual and bicommutative. This object is studied in detail in Section 4.2.

3.5.2. Hadamard product. For Coxeter species p and q, define the Coxeter species $\mathsf{p} \times \mathsf{q}$ by

$$(\mathsf{p} \times \mathsf{q})[F] := \mathsf{p}[F] \otimes \mathsf{q}[F]. \tag{3.53}$$

For a support morphism $\beta_{G,F} : F \to G$ and type morphism $w : F \to F'$, there are induced linear maps

$$\begin{gathered} \mathsf{p}[F] \otimes \mathsf{q}[F] \xrightarrow{\beta_{G,F} \otimes \beta_{G,F}} \mathsf{p}[G] \otimes \mathsf{q}[G] \\ \mathsf{p}[F] \otimes \mathsf{q}[F] \xrightarrow{w \otimes w} \mathsf{p}[F'] \otimes \mathsf{q}[F'], \end{gathered} \tag{3.54}$$

which turn $\mathsf{p} \times \mathsf{q}$ into a Coxeter species. In terms of flats,

$$(\mathsf{p} \times \mathsf{q})[\mathrm{X}] = \mathsf{p}[\mathrm{X}] \otimes \mathsf{q}[\mathrm{X}] \tag{3.55}$$

with $\mathsf{p}[\mathrm{X}] \otimes \mathsf{q}[\mathrm{X}] \xrightarrow{w \otimes w} \mathsf{p}[\mathrm{X}'] \otimes \mathsf{q}[\mathrm{X}']$ for the type morphism $w : \mathrm{X} \to \mathrm{X}'$. This is the *Hadamard product* of p and q. It defines a monoidal structure on the category of Coxeter species CSp. The unit object is the exponential Coxeter species E. Interchanging the order of the tensor factors in (3.53) yields a braiding $\mathsf{p} \times \mathsf{q} \to \mathsf{q} \times \mathsf{p}$ (which is in fact a symmetry). Thus, CSp is a symmetric monoidal category under the Hadamard product.

Observation 3.24. Hadamard product preserves Coxeter monoids and comonoids. More precisely: If a and b are Coxeter monoids, then so is $\mathsf{a} \times \mathsf{b}$ with product components in the left diagram below. If c and d are Coxeter comonoids, then so is $\mathsf{c} \times \mathsf{d}$ with coproduct components in the right diagram below.

$$\begin{array}{ccc} (\mathsf{a} \times \mathsf{b})[F] & \overset{\mu_A^F}{\dashrightarrow} & (\mathsf{a} \times \mathsf{b})[A] \\ \| & & \| \\ \mathsf{a}[F] \otimes \mathsf{b}[F] & \xrightarrow[\mu_A^F \otimes \mu_A^F]{} & \mathsf{a}[A] \otimes \mathsf{b}[A] \end{array} \qquad \begin{array}{ccc} (\mathsf{c} \times \mathsf{d})[A] & \overset{\Delta_A^F}{\dashrightarrow} & (\mathsf{c} \times \mathsf{d})[F] \\ \| & & \| \\ \mathsf{c}[A] \otimes \mathsf{d}[A] & \xrightarrow[\Delta_A^F \otimes \Delta_A^F]{} & \mathsf{c}[F] \otimes \mathsf{d}[F] \end{array} \tag{3.56}$$

This can also be formulated in terms of flats and lunes. Hadamard product of (co)commutative Coxeter (co)monoids is again (co)commutative. Hadamard product also preserves Coxeter bimonoids. In fact: For scalars p and q, if h is a Coxeter p-bimonoid and k is a Coxeter q-bimonoid, then $\mathsf{h} \times \mathsf{k}$ is a Coxeter pq-bimonoid.

Thus, the Hadamard product defines a symmetric monoidal structure on $\mathsf{Mon}(\mathsf{CSp})$, $\mathsf{Comon}(\mathsf{CSp})$, $\mathsf{Bimon}(\mathsf{CSp})$, and their commutative counterparts. The unit object is E in all cases.

3.5.3. Bilax structure. We refer to the functor $\mathsf{CSp} \times \mathsf{CSp} \to \mathsf{CSp}$ which sends (p, q) to $\mathsf{p} \times \mathsf{q}$ as the *Hadamard functor*. Its properties can be understood more formally as follows.

Theorem 3.25. *For scalars p and q, the Hadamard functor*

$$(\times, \varphi, \psi) : (\mathcal{T}, \mathcal{T}^{\vee}, \lambda_p) \times (\mathcal{T}, \mathcal{T}^{\vee}, \lambda_q) \to (\mathcal{T}, \mathcal{T}^{\vee}, \lambda_{pq})$$

is bilax, with φ and ψ given by inclusion and projection. Thus, for a Coxeter q-bimonoid k, the functor

$$(-) \times \mathsf{k} : (\mathcal{T}, \mathcal{T}^{\vee}, \lambda_p) \to (\mathcal{T}, \mathcal{T}^{\vee}, \lambda_{pq})$$

is bilax.

PROOF. This can be deduced from the corresponding result for species [**11**, Theorems 8.4 and 8.6]. The additional observation is that components of φ and ψ are maps of Coxeter species. $\square$

3.5.4. Hadamard product of (co)free Coxeter (co)monoids. For Coxeter species p and q, define the Coxeter species $\mathsf{p} \star \mathsf{q}$ by

$$(\mathsf{p} \star \mathsf{q})[A] := \bigoplus_{F \wedge G = A} \mathsf{p}[F] \otimes \mathsf{q}[G]. \tag{3.57}$$

The sum is over all pairs (F, G) of faces whose meet is A. Support and type morphisms of $\mathsf{p} \star \mathsf{q}$ are induced from those of p and q.

Proposition 3.26. *For Coxeter species p and q, there is a natural isomorphism of Coxeter 0-bimonoids*

$$\mathcal{T}_0(\mathsf{p} \star \mathsf{q}) \cong \mathcal{T}_0(\mathsf{p}) \times \mathcal{T}_0(\mathsf{q}).$$

As a consequence, Hadamard product of (co)free Coxeter (co)monoids is again (co)free.

In the above identification, the A-component of both sides is a direct sum of $\mathsf{p}[F] \otimes \mathsf{q}[G]$, as F, G vary over faces greater than A.

PROOF. This can be deduced from the corresponding result for species [**11**, Proposition 15.2]. The additional observation is that the identification is a map of Coxeter species. $\square$

Similarly, for Coxeter species p and q, define the Coxeter species $\mathsf{p} \diamond \mathsf{q}$ by

$$(\mathsf{p} \diamond \mathsf{q})[\mathrm{Z}] := \bigoplus_{\mathrm{X} \wedge \mathrm{Y} = \mathrm{Z}} \mathsf{p}[\mathrm{X}] \otimes \mathsf{q}[\mathrm{Y}]. \tag{3.58}$$

The sum is over all pairs (X, Y) of flats whose meet is Z. Type morphisms of $\mathsf{p} \diamond \mathsf{q}$ are induced from those of p and q.

Proposition 3.27. *For Coxeter species p and q, there is a natural isomorphism of bicommutative Coxeter bimonoids*

$$\mathcal{S}(\mathsf{p} \diamond \mathsf{q}) \cong \mathcal{S}(\mathsf{p}) \times \mathcal{S}(\mathsf{q}).$$

As a consequence, Hadamard product of (co)free (co)commutative Coxeter (co)monoids is again (co)free (co)commutative.

PROOF. This can be deduced from the corresponding result for species [**11**, Proposition 15.4]. $\square$

3.5.5. Internal hom. For Coxeter species p and q, let $\mathrm{hom}^{\times}(\mathsf{p}, \mathsf{q})$ denote the Coxeter species defined by

$$\mathrm{hom}^{\times}(\mathsf{p}, \mathsf{q})[A] := \mathrm{Hom}_{\Bbbk}(\mathsf{p}[A], \mathsf{q}[A]). \tag{3.59}$$

The rhs is the space of all linear maps from $\mathsf{p}[A]$ to $\mathsf{q}[A]$. For a support morphism $\beta_{B,A} : A \to B$ and type morphism $w : A \to A'$, a linear map

$f : \mathsf{p}[A] \to \mathsf{q}[A]$ induces linear maps

$$\begin{array}{ccc} \mathsf{p}[B] & \dashrightarrow & \mathsf{q}[B] \\ {\scriptstyle \beta_{B,A}^{-1}}\downarrow & & \uparrow{\scriptstyle \beta_{B,A}} \\ \mathsf{p}[A] & \xrightarrow[f]{} & \mathsf{q}[A] \end{array} \qquad \begin{array}{ccc} \mathsf{p}[A'] & \dashrightarrow & \mathsf{q}[A'] \\ {\scriptstyle w^{-1}}\downarrow & & \uparrow{\scriptstyle w} \\ \mathsf{p}[A] & \xrightarrow[f]{} & \mathsf{q}[A]. \end{array} \tag{3.60}$$

With these structure maps, $\hom^{\times}(\mathsf{p},\mathsf{q})$ is a Coxeter species. In terms of flats: $\hom^{\times}(\mathsf{p},\mathsf{q})[\mathrm{Z}]$ is the space of all linear maps from $\mathsf{p}[\mathrm{Z}]$ to $\mathsf{q}[\mathrm{Z}]$. For $w : \mathrm{Z} \to \mathrm{Z}'$, a linear map $f : \mathsf{p}[\mathrm{Z}] \to \mathsf{q}[\mathrm{Z}]$ induces

$$\mathsf{p}[\mathrm{Z}'] \xrightarrow{w^{-1}} \mathsf{p}[\mathrm{Z}] \xrightarrow{f} \mathsf{q}[\mathrm{Z}] \xrightarrow{w} \mathsf{q}[\mathrm{Z}'].$$

This construction is the internal hom for the Hadamard product on the category of Coxeter species. That is: For Coxeter species p, m, n, there is a natural bijection

$$\mathsf{CSp}(\mathsf{p} \times \mathsf{m}, \mathsf{n}) \xrightarrow{\cong} \mathsf{CSp}\big(\mathsf{p}, \hom^{\times}(\mathsf{m},\mathsf{n})\big). \tag{3.61}$$

In particular, for Coxeter species m and n,

$$\mathsf{CSp}(\mathsf{m}, \mathsf{n}) \cong \mathsf{CSp}\big(\mathsf{E}, \hom^{\times}(\mathsf{m},\mathsf{n})\big). \tag{3.62}$$

For a Coxeter species p, set

$$\mathrm{end}^{\times}(\mathsf{p}) := \hom^{\times}(\mathsf{p},\mathsf{p}). \tag{3.63}$$

3.5.6. Convolution Coxeter monoid. For c a Coxeter comonoid and a a Coxeter monoid, $\hom^{\times}(\mathsf{c},\mathsf{a})$ carries a Coxeter monoid structure with product components

$$\mu_A^F : \hom^{\times}(\mathsf{c},\mathsf{a})[F] \to \hom^{\times}(\mathsf{c},\mathsf{a})[A], \qquad f \mapsto \mu_A^F f \Delta_A^F. \tag{3.64}$$

We refer to $\hom^{\times}(\mathsf{c},\mathsf{a})$ as the *convolution Coxeter monoid*. In terms of flats and lunes,

$$\mu_{\mathrm{L}} : \hom^{\times}(\mathsf{c},\mathsf{a})[\mathrm{X}] \to \hom^{\times}(\mathsf{c},\mathsf{a})[\mathrm{Z}], \qquad f \mapsto \mu_{\mathrm{L}} f \Delta_{\mathrm{L}} \tag{3.65}$$

with $\mathrm{X} = \mathrm{c}(\mathrm{L})$ and $\mathrm{Z} = \mathrm{b}(\mathrm{L})$.

Dually, for a a Coxeter monoid and c a Coxeter comonoid, $\hom^{\times}(\mathsf{a},\mathsf{c})$ carries a Coxeter comonoid structure with coproduct components

$$\Delta_A^F : \hom^{\times}(\mathsf{a},\mathsf{c})[A] \to \hom^{\times}(\mathsf{a},\mathsf{c})[F], \qquad f \mapsto \Delta_A^F f \mu_A^F. \tag{3.66}$$

We refer to $\hom^{\times}(\mathsf{a},\mathsf{c})$ as the *coconvolution Coxeter comonoid*. It can also be formulated in terms of flats and lunes.

For Coxeter bimonoids h and k, by combining the preceding constructions, we obtain Coxeter monoid and comonoid structures on $\hom^{\times}(\mathsf{h},\mathsf{k})$ which turn it into a Coxeter bimonoid. We call it the *biconvolution Coxeter bimonoid*. In particular, for a Coxeter bimonoid h, we have the biconvolution Coxeter bimonoid $\mathrm{end}^{\times}(\mathsf{h})$.

More generally, for scalars p and q, if h is a Coxeter p-bimonoid and k is a Coxeter q-bimonoid, then $\hom^{\times}(\mathsf{h},\mathsf{k})$ is a Coxeter pq-bimonoid.

Special case. For a Coxeter comonoid c and Coxeter monoid a,

$$\mathsf{c}^* \cong \mathrm{hom}^\times(\mathsf{c}, \mathsf{E}) \quad \text{and} \quad \mathsf{a} \cong \mathrm{hom}^\times(\mathsf{E}, \mathsf{a}) \tag{3.67}$$

as Coxeter monoids, and

$$\mathsf{a}^* \cong \mathrm{hom}^\times(\mathsf{a}, \mathsf{E}) \quad \text{and} \quad \mathsf{c} \cong \mathrm{hom}^\times(\mathsf{E}, \mathsf{c}) \tag{3.68}$$

as Coxeter comonoids.

3.6. Internal hom for Coxeter comonoids

We construct the internal hom for Coxeter comonoids wrt the Hadamard product. We write $\mathcal{C}(\mathsf{c}, \mathsf{d})$ for internal hom of c and d. For a cocommutative Coxeter comonoid c and Coxeter bimonoid k, the Coxeter comonoid $\mathcal{C}(\mathsf{c}, \mathsf{k})$ carries a Coxeter bimonoid structure. In addition, if c carries a Coxeter bimonoid structure, then $\mathcal{C}(\mathsf{c}, \mathsf{k})$ can be realized as a Coxeter subbimonoid of the biconvolution Coxeter bimonoid $\mathrm{hom}^\times(\mathsf{c}, \mathsf{k})$.

The existence of internal hom implies that the category of Coxeter comonoids is enriched over itself, and it possesses powers and copowers. See Table 3.3. For background on enrichment and the notion of powers and copowers, see for instance [**11**, Appendix B].

The discussion in this section may be supplemented with the parallel more detailed discussion for species in [**11**, Sections 8.4 and 8.5].

3.6.1. Internal hom for Coxeter comonoids. For Coxeter comonoids c and d, we construct a Coxeter comonoid $\mathcal{C}(\mathsf{c}, \mathsf{d})$ as follows. An element of $\mathcal{C}(\mathsf{c}, \mathsf{d})[A]$ is a family of linear maps

$$f_{F/A} : \mathsf{c}[F] \to \mathsf{d}[F],$$

one for each $A \leq F$, such that the condition $f_{G/A}\Delta^G_F = \Delta^G_F f_{F/A}$ holds for each $A \leq F \leq G$. We refer to $f_{A/A}$ as the base term, and to $f_{F/A}$ for $F > A$ as the higher terms. Now let $\mathrm{s}(A) = \mathrm{s}(B)$, $A' = w(A)$, and $(f_{F/A})$ be an element of $\mathcal{C}(\mathsf{c}, \mathsf{d})[A]$. For each $F \geq A$, consider the composite maps

$$g_{G/B} : \mathsf{c}[G] \xrightarrow{\beta^{-1}_{G,F}} \mathsf{c}[F] \xrightarrow{f_{F/A}} \mathsf{d}[F] \xrightarrow{\beta_{G,F}} \mathsf{d}[G]$$

$$h_{F'/A'} : \mathsf{c}[F'] \xrightarrow{w^{-1}} \mathsf{c}[F] \xrightarrow{f_{F/A}} \mathsf{d}[F] \xrightarrow{w} \mathsf{d}[F']$$

with $G := BF$ and $F' := w(F)$. The linear maps

$$\begin{aligned} \beta_{B,A} : \mathcal{C}(\mathsf{c}, \mathsf{d})[A] &\to \mathcal{C}(\mathsf{c}, \mathsf{d})[B] & \qquad w : \mathcal{C}(\mathsf{c}, \mathsf{d})[A] &\to \mathcal{C}(\mathsf{c}, \mathsf{d})[A'] \\ (f_{F/A}) &\mapsto (g_{G/B}) & (f_{F/A}) &\mapsto (h_{F'/A'}) \end{aligned} \tag{3.69}$$

turn $\mathcal{C}(\mathsf{c}, \mathsf{d})$ into a Coxeter species. The coproduct component

$$\Delta^G_A : \mathcal{C}(\mathsf{c}, \mathsf{d})[A] \to \mathcal{C}(\mathsf{c}, \mathsf{d})[G], \qquad (f_{H/A}) \mapsto (f_{H/G}), \tag{3.70}$$

is defined by restricting families to faces greater than G, that is, $f_{H/G} := f_{H/A}$ for $H \geq G$. This completes the construction of the Coxeter comonoid $\mathcal{C}(\mathsf{c}, \mathsf{d})$. In fact, this is the internal hom for the Hadamard product of Coxeter comonoids. That is: For Coxeter comonoids c, d, m, there is a natural bijection

$$\mathsf{Comon}(\mathsf{CSp})(\mathsf{m} \times \mathsf{c}, \mathsf{d}) \xrightarrow{\cong} \mathsf{Comon}(\mathsf{CSp})\big(\mathsf{m}, \mathcal{C}(\mathsf{c}, \mathsf{d})\big). \tag{3.71}$$

In particular, for Coxeter comonoids c and d,

$$\mathsf{Comon}(\mathsf{CSp})(\mathsf{c},\mathsf{d}) \cong \mathsf{Comon}(\mathsf{CSp})\big(\mathsf{E}, \mathcal{C}(\mathsf{c},\mathsf{d})\big). \tag{3.72}$$

Exercise 3.28 (Primitive part). Show: For Coxeter comonoids c and d,

$$\mathcal{P}(\mathcal{C}(\mathsf{c},\mathsf{d})) = \mathrm{hom}^{\times}(\mathsf{c}, \mathcal{P}(\mathsf{d})). \tag{3.73}$$

(Higher terms of any family in the primitive part of $\mathcal{C}(\mathsf{c},\mathsf{d})$ must necessarily be zero.)

Remark 3.29. Recall from Proposition 2.57 that Coxeter comonoids is a functor category. There is a general result about existence of internal hom in a functor category, see for instance [**11**, Theorem B.14]. Its proof spells out the general construction. When specialized to Coxeter comonoids, it yields the above explicit description of the internal hom.

3.6.2. Coxeter bimonoid structure.

We build on the preceding construction. For c a cocommutative Coxeter comonoid and k a Coxeter bimonoid, $\mathcal{C}(\mathsf{c},\mathsf{k})$ carries a Coxeter bimonoid structure with product components

$$\mu_A^F : \mathcal{C}(\mathsf{c},\mathsf{k})[F] \to \mathcal{C}(\mathsf{c},\mathsf{k})[A], \qquad f \mapsto \tilde{f}, \tag{3.74}$$

where for $K \geq A$, the map $\tilde{f}_{K/A}$ is defined as the composite

$$\mathsf{c}[K] \xrightarrow{\Delta_K^{KF}} \mathsf{c}[KF] \xrightarrow{\beta_{FK,KF}} \mathsf{c}[FK] \xrightarrow{f_{FK/F}} \mathsf{k}[FK] \xrightarrow{\beta_{KF,FK}} \mathsf{k}[KF] \xrightarrow{\mu_K^{KF}} \mathsf{k}[K].$$

For $K \geq F$, we have $\tilde{f}_{K/A} = f_{K/F}$; thus $\tilde{f}$ extends f. One checks that $\tilde{f}$ indeed belongs to $\mathcal{C}(\mathsf{c},\mathsf{k})[A]$; see [**11**, top diagram on page 352] for details. The remaining checks are indicated below.

- <u>Naturality axiom for μ wrt both support and type morphisms.</u>

$$\begin{array}{ccccccccccc}
\mathsf{c}[K] & \xrightarrow{\Delta} & \mathsf{c}[KF] & \xrightarrow{\beta} & \mathsf{c}[FK] & \xrightarrow{f_{FK/F}} & \mathsf{k}[FK] & \xrightarrow{\beta} & \mathsf{k}[KF] & \xrightarrow{\mu} & \mathsf{k}[K] \\
{\scriptstyle \beta/w^{-1}}\uparrow & & {\scriptstyle \beta/w^{-1}}\uparrow & & {\scriptstyle \beta/w^{-1}}\uparrow & & \downarrow{\scriptstyle \beta/w} & & \downarrow{\scriptstyle \beta/w} & & \downarrow{\scriptstyle \beta/w} \\
\mathsf{c}[K'] & \xrightarrow[\Delta]{} & \mathsf{c}[K'F'] & \xrightarrow[\beta]{} & \mathsf{c}[F'K'] & & \mathsf{k}[F'K'] & \xrightarrow[\beta]{} & \mathsf{k}[K'F'] & \xrightarrow[\mu]{} & \mathsf{k}[K']
\end{array}$$

  Product components involved are μ_A^F and $\mu_{A'}^{F'}$. Also $K \geq A$, $K' \geq A'$. In the support case, $\mathrm{s}(A) = \mathrm{s}(A')$ with $F' = A'F$, $K' = A'K$, while in the type case, $w(A) = A'$, $w(F) = F'$, $w(K) = K'$. An illustration is provided below.

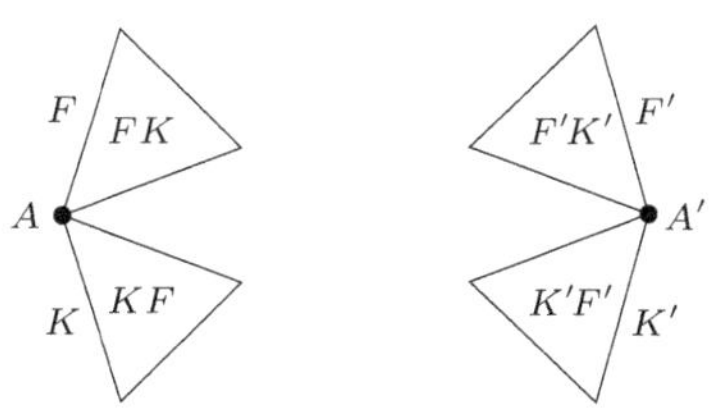

- Associativity axiom for μ.

$$
\begin{array}{ccccccccccc}
\mathsf{c}[K] & \xrightarrow{\Delta} & \mathsf{c}[KG] & \xrightarrow{\beta} & \mathsf{c}[GK] & \xrightarrow{f_{GK/G}} & \mathsf{k}[GK] & \xrightarrow{\beta} & \mathsf{k}[KG] & \xrightarrow{\mu} & \mathsf{k}[K] \\
\Delta\big\downarrow & \overset{\Delta}{\nearrow} & & \overset{\beta}{\searrow} & \big\uparrow\beta & & \beta\big\downarrow & \overset{\beta}{\nearrow} & & \overset{\mu}{\searrow} & \big\uparrow\mu \\
\mathsf{c}[KF] & \xrightarrow[\beta]{} & \mathsf{c}[FK] & \xrightarrow[\Delta]{} & \mathsf{c}[FKG] & & \mathsf{k}[FKG] & \xrightarrow[\mu]{} & \mathsf{k}[FK] & \xrightarrow[\beta]{} & \mathsf{k}[KF]
\end{array}
$$

Here $A \leq F \leq G$ and $A \leq K$. An illustration is shown below with $A = O$.

- Bimonoid axiom.

$$
\begin{array}{ccccccccccc}
\mathsf{c}[K] & \xrightarrow{\Delta} & \mathsf{c}[KF] & \xrightarrow{\beta} & \mathsf{c}[FK] & \xrightarrow{f_{FK/F}} & \mathsf{k}[FK] & \xrightarrow{\beta} & \mathsf{k}[KF] & \xrightarrow{\mu} & \mathsf{k}[K] \\
\Delta\big\downarrow & & & & & & & & & & \big\uparrow\mu \\
\mathsf{c}[KF] & \xrightarrow[\beta]{} & \mathsf{c}[GFK] & \xrightarrow[\beta]{} & \mathsf{c}[FK] & \xrightarrow[f_{FK/F}]{} & \mathsf{k}[FK] & \xrightarrow[\beta]{} & \mathsf{k}[GFK] & \xrightarrow[\beta]{} & \mathsf{k}[KF]
\end{array}
$$

Here $A \leq F$ and $A \leq G \leq K$. An illustration is shown below with $A = O$.

If k is commutative, then so is $\mathcal{C}(\mathsf{c}, \mathsf{k})$.

Exercise 3.30. Let h and k be Coxeter bimonoids, with h cocommutative. Check: The map

$$\mathcal{C}(\mathsf{h}, \mathsf{k}) \hookrightarrow \mathrm{hom}^{\times}(\mathsf{h}, \mathsf{k}), \qquad (f_{F/A}) \mapsto f_{A/A} \tag{3.75}$$

on the A-component, is an injective morphism of Coxeter bimonoids. In addition, if k is (co, bi)commutative, then so is $\mathcal{C}(\mathsf{h}, \mathsf{k})$.

Deduce: For h a cocommutative Coxeter bimonoid, $\mathcal{C}(\mathsf{h}, \mathsf{h})$ is a cocommutative Coxeter subbimonoid of $\mathrm{end}^{\times}(\mathsf{h})$.

Exercise 3.31 (Deformation). Let q be a scalar. Use the gate property to check: For c a cocommutative Coxeter comonoid and k a Coxeter q-bimonoid, the Coxeter comonoid $\mathcal{C}(\mathsf{c}, \mathsf{k})$ carries a Coxeter q-bimonoid structure with product component μ^F_A defined as in (3.74) with the composite map multiplied by $q^{\mathrm{dist}(F,K)}$. We denote it by $\mathcal{C}(\mathsf{c}, \mathsf{k})_q$. Setting $q = 1$ recovers $\mathcal{C}(\mathsf{c}, \mathsf{k})$.

Check: For h a cocommutative Coxeter bimonoid and k a Coxeter q-bimonoid, the map (3.75) is an injective morphism of Coxeter q-bimonoids.

3.6.3. Enrichment. The category of Coxeter comonoids is a symmetric monoidal category under the Hadamard product. Moreover, we saw that the internal hom exists. Hence, this category is enriched over itself with hom-objects $\mathcal{C}(\mathsf{c},\mathsf{d})$. That is, there are morphisms of Coxeter comonoids

$$\mathcal{C}(\mathsf{n},\mathsf{p}) \times \mathcal{C}(\mathsf{m},\mathsf{n}) \to \mathcal{C}(\mathsf{m},\mathsf{p}) \qquad \text{and} \qquad \mathsf{E} \to \mathcal{C}(\mathsf{c},\mathsf{c}),$$

the former for any Coxeter comonoids m, n, p, and the latter for any Coxeter comonoid c. The former is defined by composing families, and the latter by using families consisting of identity maps. The identification (3.72) says that the underlying category of this enriched category is indeed the category of Coxeter comonoids.

This enriched category possesses powers and copowers. For Coxeter comonoids c, d, m, the copower of c by m is $\mathsf{m} \times \mathsf{c}$, while the power of d by m is $\mathcal{C}(\mathsf{m},\mathsf{d})$. That is:

$$\mathcal{C}(\mathsf{m} \times \mathsf{c},\mathsf{d}) \cong \mathcal{C}(\mathsf{m},\mathcal{C}(\mathsf{c},\mathsf{d})) \cong \mathcal{C}(\mathsf{c},\mathcal{C}(\mathsf{m},\mathsf{d})). \tag{3.76}$$

Observe: The first isomorphism implies (3.71) by taking Coxeter comonoid morphisms from E and using (3.72).

An abstract setup for the above discussion is given for instance in [**11**, Example B.10].

3.7. Universal measuring Coxeter comonoid

This section proceeds in parallel with Section 3.6. We construct the universal measuring Coxeter comonoid $\overline{\mathcal{C}}(\mathsf{a},\mathsf{b})$ from one Coxeter monoid a to another Coxeter monoid b. For a commutative Coxeter monoid a and Coxeter bimonoid h, the Coxeter comonoid $\overline{\mathcal{C}}(\mathsf{h},\mathsf{a})$ carries a Coxeter bimonoid structure. In addition, if a carries a Coxeter bimonoid structure, then $\overline{\mathcal{C}}(\mathsf{h},\mathsf{a})$ can be realized as a Coxeter subbimonoid of the biconvolution Coxeter bimonoid $\mathrm{hom}^{\times}(\mathsf{h},\mathsf{a})$.

The universal measuring Coxeter comonoid can be used to enrich the category of Coxeter monoids over the category of Coxeter comonoids. This enriched category possesses powers and copowers. See Table 3.3.

3.7.1. Universal measuring Coxeter comonoid. For Coxeter monoids a and b, we construct a Coxeter comonoid $\overline{\mathcal{C}}(\mathsf{a},\mathsf{b})$ as follows. An element of $\overline{\mathcal{C}}(\mathsf{a},\mathsf{b})[A]$ is a family of linear maps

$$f_{F/A} : \mathsf{a}[F] \to \mathsf{b}[F],$$

one for each $A \leq F$, such that the condition $\mu_F^G f_{G/A} = f_{F/A} \mu_F^G$ holds for each $A \leq F \leq G$. Support and type morphisms and coproduct components are defined in a manner similar to $\mathcal{C}(\mathsf{c},\mathsf{d})$, see (3.69) and (3.70). We call $\overline{\mathcal{C}}(\mathsf{a},\mathsf{b})$ the *universal measuring Coxeter comonoid*. It satisfies the following universal property. For a Coxeter comonoid c and Coxeter monoids a and b, there is a natural bijection

$$\mathsf{Mon}(\mathsf{CSp})\big(\mathsf{a},\mathrm{hom}^{\times}(\mathsf{c},\mathsf{b})\big) \xrightarrow{\cong} \mathsf{Comon}(\mathsf{CSp})\big(\mathsf{c},\overline{\mathcal{C}}(\mathsf{a},\mathsf{b})\big), \tag{3.77}$$

where $\mathrm{hom}^{\times}(\mathsf{c},\mathsf{b})$ is the convolution Coxeter monoid. In particular, for Coxeter monoids a and b,

$$\mathsf{Mon}(\mathsf{CSp})(\mathsf{a},\mathsf{b}) \cong \mathsf{Comon}(\mathsf{CSp})\big(\mathsf{E},\overline{\mathcal{C}}(\mathsf{a},\mathsf{b})\big). \tag{3.78}$$

Exercise 3.32 (Primitive part). Show: For Coxeter monoids a and b,

$$\mathcal{P}(\overline{\mathcal{C}}(\mathsf{a},\mathsf{b})) = \mathrm{hom}^{\times}(\mathcal{Q}(\mathsf{a}),\mathsf{b}). \tag{3.79}$$

Compare with Exercise 3.28.

3.7.2. Coxeter bimonoid structure. We build on the preceding construction. For h a Coxeter bimonoid and a a commutative Coxeter monoid, $\overline{\mathcal{C}}(\mathsf{h},\mathsf{a})$ carries a Coxeter bimonoid structure with product components

$$\mu_A^F : \overline{\mathcal{C}}(\mathsf{h},\mathsf{a})[F] \to \overline{\mathcal{C}}(\mathsf{h},\mathsf{a})[A], \qquad f \mapsto \tilde{f}, \tag{3.80}$$

where for $K \geq A$, the map $\tilde{f}_{K/A}$ is defined as the composite

$$\mathsf{h}[K] \xrightarrow{\Delta_K^{KF}} \mathsf{h}[KF] \xrightarrow{\beta_{FK,KF}} \mathsf{h}[FK] \xrightarrow{f_{FK/F}} \mathsf{a}[FK] \xrightarrow{\beta_{KF,FK}} \mathsf{a}[KF] \xrightarrow{\mu_K^{KF}} \mathsf{a}[K].$$

The checks are similar to the one for $\mathcal{C}(\mathsf{c},\mathsf{k})$ in Section 3.6.2 and amount to reversing appropriate arrows. If h is cocommutative, then $\overline{\mathcal{C}}(\mathsf{h},\mathsf{a})$ is commutative.

Exercise 3.33. Let h and k be Coxeter bimonoids, with k commutative. Check: The map

$$\overline{\mathcal{C}}(\mathsf{h},\mathsf{k}) \hookrightarrow \mathrm{hom}^{\times}(\mathsf{h},\mathsf{k}), \qquad (f_{F/A}) \mapsto f_{A/A} \tag{3.81}$$

on the A-component, is an injective morphism of Coxeter bimonoids. In addition, if h is commutative, then $\overline{\mathcal{C}}(\mathsf{h},\mathsf{k})$ is cocommutative, and if h is cocommutative, then $\overline{\mathcal{C}}(\mathsf{h},\mathsf{k})$ is commutative.

Deduce: For h a commutative Coxeter bimonoid, $\overline{\mathcal{C}}(\mathsf{h},\mathsf{h})$ is a cocommutative Coxeter subbimonoid of $\mathrm{end}^{\times}(\mathsf{h})$.

Compare with Exercise 3.30.

Exercise 3.34 (Deformation). Let q be a scalar. Check: For h a Coxeter q-bimonoid and a a commutative Coxeter monoid, the Coxeter comonoid $\overline{\mathcal{C}}(\mathsf{h},\mathsf{a})$ carries a Coxeter q-bimonoid structure with product component μ_A^F defined as in (3.80) with the composite map multiplied by $q^{\mathrm{dist}(F,K)}$. We denote it by $\overline{\mathcal{C}}(\mathsf{h},\mathsf{a})_q$.

Check: For h a Coxeter q-bimonoid and k a commutative Coxeter bimonoid, the map (3.81) is an injective morphism of Coxeter q-bimonoids.

Compare with Exercise 3.31.

3.7.3. A quotient of the free Coxeter monoid. For a Coxeter comonoid c and Coxeter monoid a, define $\mathsf{c} \triangleright \mathsf{a}$ to be the following Coxeter quotient monoid of $\mathcal{T}(\mathsf{c} \times \mathsf{a})$. Its A-component is the quotient of $\mathcal{T}(\mathsf{c} \times \mathsf{a})[A]$ by the linear span of elements of the form

$$x \otimes \mu_F^G(y) - \Delta_F^G(x) \otimes y, \tag{3.82}$$

for any $x \in \mathsf{c}[F]$ and $y \in \mathsf{a}[G]$ with $A \leq F \leq G$. It satisfies the following universal property. For a Coxeter comonoid c and Coxeter monoids a and b, there is a natural bijection

$$\mathsf{Mon}(\mathsf{CSp})(\mathsf{c} \triangleright \mathsf{a},\mathsf{b}) \xrightarrow{\cong} \mathsf{Comon}(\mathsf{CSp})\big(\mathsf{c},\overline{\mathcal{C}}(\mathsf{a},\mathsf{b})\big). \tag{3.83}$$

3.7.4. Enrichment. The category of Coxeter monoids is enriched over the symmetric monoidal category of Coxeter comonoids wrt the Hadamard product with hom-objects $\overline{\mathcal{C}}(\mathsf{a}, \mathsf{b})$. That is, there are morphisms of Coxeter comonoids

$$\overline{\mathcal{C}}(\mathsf{n}, \mathsf{p}) \times \overline{\mathcal{C}}(\mathsf{m}, \mathsf{n}) \to \overline{\mathcal{C}}(\mathsf{m}, \mathsf{p}) \qquad \text{and} \qquad \mathsf{E} \to \overline{\mathcal{C}}(\mathsf{a}, \mathsf{a}),$$

the former for any Coxeter monoids m, n, p, and the latter for any Coxeter monoid a. The former is defined by composing families, and the latter by using families consisting of identity maps. The identification (3.78) says that the underlying category of this enriched category is indeed the category of Coxeter monoids.

This enriched category possesses powers and copowers. For a Coxeter comonoid c and Coxeter monoids a and b, the copower of a by c is $\mathsf{c} \triangleright \mathsf{a}$, while the power of b by c is the convolution Coxeter monoid $\mathrm{hom}^{\times}(\mathsf{c}, \mathsf{b})$. That is:

$$\overline{\mathcal{C}}(\mathsf{c} \triangleright \mathsf{a}, \mathsf{b}) \cong \mathcal{C}(\mathsf{c}, \overline{\mathcal{C}}(\mathsf{a}, \mathsf{b})) \cong \overline{\mathcal{C}}(\mathsf{a}, \mathrm{hom}^{\times}(\mathsf{c}, \mathsf{b})). \tag{3.84}$$

Observe: The second isomorphism implies (3.77) by taking Coxeter comonoid morphisms from E and using (3.72) and (3.78). In a similar manner, the first isomorphism implies (3.83).

3.8. Exp-log between (co)derivations and (co)monoid morphisms

Recall the invariant lune-incidence algebra, invariant flat-incidence algebra and invariant bilune-incidence algebra from Section 1.10. We define actions of these algebras on the space of all maps from a (cocommutative) Coxeter comonoid c to a (commutative) Coxeter monoid a. We then use these to formulate various exp-log correspondences by employing zeta and Möbius functions. See Table 3.4. They provide bijections between spaces of (co, bi)derivations and spaces of Coxeter (co, bi)monoid morphisms. These ideas are continued in Sections 3.9 and 3.10.

TABLE 3.4. Exp-log on Coxeter species.

algebra action	exp-log and zeta-Möbius
$\mathrm{I}_{\mathrm{lune}}[\mathcal{A}]^W$ on $\mathsf{CSp}(\mathsf{c}, \mathsf{a})$	$\mathbf{exp}(f) = \boldsymbol{\zeta} \circ f$, $\mathbf{log}(f) = \boldsymbol{\mu} \circ f$
$\mathrm{I}_{\mathrm{flat}}[\mathcal{A}]^W$ on $\mathsf{CSp}(\mathsf{c}, \mathsf{a})$	$\exp(f) = \zeta \circ f$, $\log(f) = \mu \circ f$
$\mathrm{I}_{\mathrm{bilune}}[\mathcal{A}]^W$ on $\mathsf{CSp}(\mathsf{c}, \mathsf{a})$	$\mathbf{exp}_q(f) = \zeta_q \circ f$, $\mathbf{log}_q(f) = \mu_q \circ f$
$\mathrm{I}_{\mathrm{lune}}[\mathcal{A}]^W$ on $\mathscr{S}(\mathsf{a})$	$\mathbf{exp}(v) = \boldsymbol{\zeta} \circ v$, $\mathbf{log}(v) = \boldsymbol{\mu} \circ v$ $\mathbf{exp}_q(v) = \boldsymbol{\zeta}_q \circ v$, $\mathbf{log}_q(v) = \boldsymbol{\mu}_q \circ v$

Exp-log correspondences for species were studied in [**11**, Chapter 9]. Since the details are similar, we will be brief here.

3.8.1. Action of the invariant lune-incidence algebra. Let c be a Coxeter comonoid and a a Coxeter monoid. For $s \in \mathrm{I}_{\mathrm{lune}}[\mathcal{A}]^W$ and $f : \mathsf{c} \to \mathsf{a}$ a map of Coxeter species, define $s \circ f : \mathsf{c} \to \mathsf{a}$ by

$$(s \circ f)_A := \sum_{F:\, F \geq A} s(A, F)\, \mu_A^F f_F \Delta_A^F. \tag{3.85}$$

This is a map of Coxeter species, that is, diagram (2.11) commutes. To see this, fix $\sigma : A \to A'$, and for any $F \geq A$, consider the commutative diagram

$$\begin{array}{ccccccc}
\mathsf{c}[A] & \xrightarrow{\Delta_A^F} & \mathsf{c}[F] & \xrightarrow{f_F} & \mathsf{a}[F] & \xrightarrow{\mu_A^F} & \mathsf{a}[A] \\
\sigma\downarrow & & \sigma\downarrow & & \downarrow\sigma & & \downarrow\sigma \\
\mathsf{c}[A'] & \xrightarrow[\Delta_{A'}^{F'}]{} & \mathsf{c}[F'] & \xrightarrow[f_{F'}]{} & \mathsf{a}[F'] & \xrightarrow[\mu_{A'}^{F'}]{} & \mathsf{a}[A'].
\end{array}$$

The middle square is an instance of (2.11), while the side squares are instances of the naturality axiom (2.30) and its dual. Now multiply the above diagram by $s(A, F) = s(A', F')$ and sum over all $F \geq A$.

Exercise 3.35. Check: The assignment $(s, f) \mapsto s \circ f$ defines a left action of $\mathrm{I}_{\mathrm{lune}}[\mathcal{A}]^W$ on $\mathsf{CSp}(\mathsf{c}, \mathsf{a})$. That is, it is linear in s and f, and for $s, t \in \mathrm{I}_{\mathrm{lune}}[\mathcal{A}]^W$,

$$s \circ (t \circ f) = st \circ f \quad \text{and} \quad \delta \circ f = f,$$

with δ the unit element of $\mathrm{I}_{\mathrm{lune}}[\mathcal{A}]^W$. Thus, $\mathsf{CSp}(\mathsf{c}, \mathsf{a})$ is a left module over $\mathrm{I}_{\mathrm{lune}}[\mathcal{A}]^W$.

Exercise 3.36 (Action in terms of flats and lunes). Let c be a Coxeter comonoid and a a Coxeter monoid as formulated in Proposition 2.24 and its dual. Check: The action (3.85) can be rewritten as

$$(s \circ f)_{\mathrm{Z}} = \sum_{\mathrm{L}:\, \mathrm{b}(\mathrm{L}) = \mathrm{Z}} s(\mathrm{L})\, \mu_{\mathrm{L}} f_{\mathrm{c}(\mathrm{L})} \Delta_{\mathrm{L}}. \tag{3.86}$$

Also check directly that this is a map of Coxeter species, that is, diagram (2.16) commutes, and that it defines an action.

3.8.2. Exp-log on Coxeter species maps. Let $\mathrm{char}(\Bbbk) \nmid |W|$. Fix an invariant noncommutative zeta function $\boldsymbol{\zeta}$ and Möbius function $\boldsymbol{\mu}$ which are inverses of each other in $\mathrm{I}_{\mathrm{lune}}[\mathcal{A}]^W$. For a map of Coxeter species $f : \mathsf{c} \to \mathsf{a}$, we say $\boldsymbol{\zeta} \circ f$ is an *exponential* of f and $\boldsymbol{\mu} \circ f$ is a *logarithm* of f. We also use notations $\mathbf{exp}(f)$ and $\mathbf{log}(f)$. Using formula (3.85),

$$\mathbf{exp}(f)_A = \sum_{F:\, F \geq A} \boldsymbol{\zeta}(A, F)\, \mu_A^F f_F \Delta_A^F, \tag{3.87a}$$

$$\mathbf{log}(f)_A = \sum_{F:\, F \geq A} \boldsymbol{\mu}(A, F)\, \mu_A^F f_F \Delta_A^F. \tag{3.87b}$$

Since $\boldsymbol{\zeta}$ and $\boldsymbol{\mu}$ are inverses, we obtain inverse bijections

$$\mathsf{CSp}(\mathsf{c}, \mathsf{a}) \underset{\mathbf{log}}{\overset{\mathbf{exp}}{\rightleftarrows}} \mathsf{CSp}(\mathsf{c}, \mathsf{a}). \tag{3.88}$$

We refer to (3.88) as an *exp-log correspondence*. It depends on $\boldsymbol{\zeta}$ and $\boldsymbol{\mu}$.

3.8.3. Exp-log between (co)derivations and (co)monoid morphisms. Under additional hypotheses, the correspondence (3.88) restricts as follows.

Theorem 3.37. *Let* $\mathrm{char}(\Bbbk) \nmid |W|$. *For a cocommutative Coxeter comonoid* c *and Coxeter bimonoid* k, *we have inverse bijections*

$$\mathsf{CSp}(\mathsf{c}, \mathcal{P}(\mathsf{k})) \underset{\mathbf{log}}{\overset{\mathbf{exp}}{\rightleftarrows}} \mathsf{Comon}(\mathsf{CSp})(\mathsf{c}, \mathsf{k}). \tag{3.89}$$

Dually, for a Coxeter bimonoid h *and commutative Coxeter monoid* a, *we have inverse bijections*

$$\mathsf{CSp}(\mathcal{Q}(\mathsf{h}), \mathsf{a}) \underset{\mathbf{log}}{\overset{\mathbf{exp}}{\rightleftarrows}} \mathsf{Mon}(\mathsf{CSp})(\mathsf{h}, \mathsf{a}). \tag{3.90}$$

The first statement says: For $f : \mathsf{c} \to \mathsf{k}$ a coderivation, $\mathbf{exp}(f) : \mathsf{c} \to \mathsf{k}$ is a Coxeter comonoid morphism, and for $g : \mathsf{c} \to \mathsf{k}$ a Coxeter comonoid morphism, $\mathbf{log}(g) : \mathsf{c} \to \mathsf{k}$ is a coderivation.

The second statement says: For $f : \mathsf{h} \to \mathsf{a}$ a derivation, $\mathbf{exp}(f) : \mathsf{h} \to \mathsf{a}$ is a Coxeter monoid morphism, and for $g : \mathsf{h} \to \mathsf{a}$ a Coxeter monoid morphism, $\mathbf{log}(g) : \mathsf{h} \to \mathsf{a}$ is a derivation.

PROOF. This can be deduced from the corresponding result for species [**11**, Theorems 9.11 and 9.12]. This is because type morphisms do not play any role in the calculation. The two directions use the lune-additivity formula (1.79) and noncommutative Weisner formula (1.81), respectively. □

3.8.4. Action of the invariant flat-incidence algebra. Let c be a cocommutative Coxeter comonoid and a a commutative Coxeter monoid. For $s \in \mathrm{I}_{\mathrm{flat}}[\mathcal{A}]^W$ and $f : \mathsf{c} \to \mathsf{a}$ a map of Coxeter species,

$$(s \circ f)_{\mathrm{Z}} := \sum_{\mathrm{X}:\, \mathrm{X} \geq \mathrm{Z}} s(\mathrm{Z}, \mathrm{X})\, \mu_{\mathrm{Z}}^{\mathrm{X}} f_{\mathrm{X}} \Delta_{\mathrm{Z}}^{\mathrm{X}} \tag{3.91}$$

defines a left action of $\mathrm{I}_{\mathrm{flat}}[\mathcal{A}]^W$ on $\mathsf{CSp}(\mathsf{c}, \mathsf{a})$.

Under the commutativity hypothesis, when $\mathrm{char}(\Bbbk) \nmid |W|$, the action (3.85), or equivalently, (3.86) of $\mathrm{I}_{\mathrm{lune}}[\mathcal{A}]^W$ factors through the base-case map (1.78) to yield an action of $\mathrm{I}_{\mathrm{flat}}[\mathcal{A}]^W$ on $\mathsf{CSp}(\mathsf{c}, \mathsf{a})$ which coincides with (3.91).

3.8.5. Commutative exp-log. Let ζ and μ be the zeta function and Möbius function in $\mathrm{I}_{\mathrm{flat}}[\mathcal{A}]^W$. For a map of Coxeter species $f : \mathsf{c} \to \mathsf{a}$ from a cocommutative Coxeter comonoid c to a commutative Coxeter monoid a, we say $\exp(f) := \zeta \circ f$ and $\log(f) := \mu \circ f$ are the *exponential* and *logarithm* of f, respectively. Using formula (3.91),

$$\exp(f)_{\mathrm{Z}} = \sum_{\mathrm{X}:\, \mathrm{X} \geq \mathrm{Z}} \mu_{\mathrm{Z}}^{\mathrm{X}} f_{\mathrm{X}} \Delta_{\mathrm{Z}}^{\mathrm{X}}, \tag{3.92a}$$

$$\log(f)_{\mathrm{Z}} = \sum_{\mathrm{X}:\, \mathrm{X} \geq \mathrm{Z}} \mu(\mathrm{Z}, \mathrm{X})\, \mu_{\mathrm{Z}}^{\mathrm{X}} f_{\mathrm{X}} \Delta_{\mathrm{Z}}^{\mathrm{X}}. \tag{3.92b}$$

(Recall: $\zeta \equiv 1$.) Since ζ and μ are inverses, we obtain inverse bijections

$$\mathsf{CSp}(\mathsf{c},\mathsf{a}) \underset{\log}{\overset{\exp}{\rightleftarrows}} \mathsf{CSp}(\mathsf{c},\mathsf{a}). \tag{3.93}$$

This is the *commutative exp-log correspondence*. When $\mathrm{char}(\Bbbk) \nmid |W|$, (3.92a) and (3.92b) are induced from (3.87a) and (3.87b) in view of Fact 1.80, and all exp-log correspondences (3.88) reduce to (3.93).

Theorem 3.38. *For a cocommutative Coxeter comonoid* c *and commutative Coxeter bimonoid* k, *we have inverse bijections*

$$\mathsf{CSp}\big(\mathsf{c},\mathcal{P}(\mathsf{k})\big) \underset{\log}{\overset{\exp}{\rightleftarrows}} \mathsf{Comon}(\mathsf{CSp})(\mathsf{c},\mathsf{k}). \tag{3.94}$$

For a cocommutative Coxeter bimonoid h *and commutative Coxeter monoid* a, *we have inverse bijections*

$$\mathsf{CSp}\big(\mathcal{Q}(\mathsf{h}),\mathsf{a}\big) \underset{\log}{\overset{\exp}{\rightleftarrows}} \mathsf{Mon}(\mathsf{CSp})(\mathsf{h},\mathsf{a}). \tag{3.95}$$

For a cocommutative Coxeter bimonoid h *and commutative Coxeter bimonoid* k, *we have inverse bijections*

$$\mathsf{CSp}\big(\mathcal{Q}(\mathsf{h}),\mathcal{P}(\mathsf{k})\big) \underset{\log}{\overset{\exp}{\rightleftarrows}} \mathsf{Bimon}(\mathsf{CSp})(\mathsf{h},\mathsf{k}). \tag{3.96}$$

The first two statements can be made explicit as after Theorem 3.37. The third statement says: For $f : \mathsf{h} \to \mathsf{k}$ a biderivation, $\exp(f) : \mathsf{h} \to \mathsf{k}$ is a Coxeter bimonoid morphism, and for $g : \mathsf{h} \to \mathsf{k}$ a Coxeter bimonoid morphism, $\log(g) : \mathsf{h} \to \mathsf{k}$ is a biderivation.

PROOF. This can be deduced from the corresponding result for species [**11**, Theorems 9.40, 9.41, 9.43 and Remark 9.42]. It makes use of the Weisner formula (1.74). When $\mathrm{char}(\Bbbk) \nmid |W|$, the first two statements also follow from Theorem 3.37 via the base-case map. The third statement follows by combining the first two. □

3.8.6. Action of the invariant bilune-incidence algebra. Let c be a Coxeter comonoid and a a Coxeter monoid. For $s \in \mathrm{I}_{\mathrm{bilune}}[\mathcal{A}]^W$ and $f : \mathsf{c} \to \mathsf{a}$ a map of Coxeter species, define $s \circ f : \mathsf{c} \to \mathsf{a}$ by

$$\begin{aligned}(s \circ f)_A &:= \sum_{\substack{F,F' \geq A \\ \mathrm{s}(F)=\mathrm{s}(F')}} s(A,F,F')\, \mu_A^F \beta_{F,F'} f_{F'} \Delta_A^{F'} \\ &= \sum_{\substack{F,F' \geq A \\ \mathrm{s}(F)=\mathrm{s}(F')}} s(A,F,F')\, \mu_A^F f_F \beta_{F,F'} \Delta_A^{F'}.\end{aligned} \tag{3.97}$$

The sums are over both F and F'.

Exercise 3.39. Check: $s \circ f$ is a map of Coxeter species. Moreover, the assignment $(s,f) \mapsto s \circ f$ defines a left action of $\mathrm{I}_{\mathrm{bilune}}[\mathcal{A}]^W$ on $\mathsf{CSp}(\mathsf{c},\mathsf{a})$.

Exercise 3.40 (Action in terms of flats and lunes). This is a continuation of Exercise 3.36. Check: The action (3.97) can be rewritten as

$$(s \circ f)_{\mathrm{Z}} = \sum_{(\mathrm{L},\mathrm{L}')\colon \mathrm{b}(\mathrm{L},\mathrm{L}')=\mathrm{Z}} s(\mathrm{L},\mathrm{L}')\, \mu_{\mathrm{L}} f_{\mathrm{c}(\mathrm{L},\mathrm{L}')} \Delta_{\mathrm{L}'}. \tag{3.98}$$

The sum is over all bilunes (L,L') with base Z. Also check directly that this is a map of Coxeter species, and that it defines an action.

Exercise 3.41 (Comparison of actions). Recall algebra morphisms p, q, i from (1.104). They can be used to relate the actions (3.85) and (3.97). Check:

- The action (3.97) restricts via the map i to the action (3.85).
- When c is cocommutative, the action (3.97) factors through the map p to induce the action (3.85).
- When a is commutative, the action (3.97) factors through the map q to induce the action (3.85).
- When c is cocommutative and a is commutative, the action factors further through the base-case map (1.78) to induce the action (3.91) (assuming $\mathrm{char}(\Bbbk) \nmid |W|$).

3.8.7. q-exp-log. Let q be a scalar which is not a root of unity. Recall the two-sided q-zeta function $\boldsymbol{\zeta}_q$ and two-sided q-Möbius function $\boldsymbol{\mu}_q$. These are elements of the invariant bilune-incidence algebra. For a map of Coxeter species $f : \mathsf{c} \to \mathsf{a}$, we say $\mathbf{exp}_q(f) := \boldsymbol{\zeta}_q \circ f$ and $\mathbf{log}_q(f) := \boldsymbol{\mu}_q \circ f$ are the *q-exponential* and *q-logarithm* of f, respectively. Using formula (3.97),

$$\mathbf{exp}_q(f)_A = \sum_{\substack{F,F' \geq A \\ \mathrm{s}(F)=\mathrm{s}(F')}} \zeta_q(A,F,F')\, \mu_A^F \beta_{F,F'} f_{F'} \Delta_A^{F'}, \tag{3.99a}$$

$$\mathbf{log}_q(f)_A = \sum_{\substack{F,F' \geq A \\ \mathrm{s}(F)=\mathrm{s}(F')}} \boldsymbol{\mu}_q(A,F,F')\, \mu_A^F \beta_{F,F'} f_{F'} \Delta_A^{F'}. \tag{3.99b}$$

Since $\boldsymbol{\zeta}_q$ and $\boldsymbol{\mu}_q$ are inverses, we obtain inverse bijections

$$\mathsf{CSp}(\mathsf{c},\mathsf{a}) \underset{\mathbf{log}_q}{\overset{\mathbf{exp}_q}{\rightleftarrows}} \mathsf{CSp}(\mathsf{c},\mathsf{a}). \tag{3.100}$$

This is the *q-exp-log correspondence.*

Observation 3.42. If either c is cocommutative or a is commutative, then (3.99a) and (3.99b) specialize to

$$\mathbf{exp}_q(f)_A = \sum_{F\colon F \geq A} \zeta_q(A,F)\, \mu_A^F f_F \Delta_A^F, \tag{3.101a}$$

$$\mathbf{log}_q(f)_A = \sum_{F\colon F \geq A} \boldsymbol{\mu}_q(A,F)\, \mu_A^F f_F \Delta_A^F. \tag{3.101b}$$

Further, if c is cocommutative and a is commutative, then

$$\mathbf{exp}_q(f)_{\mathrm{Z}} = \sum_{\mathrm{X}\colon \mathrm{X} \geq \mathrm{Z}} \zeta_q(\mathrm{Z},\mathrm{X})\, \mu_{\mathrm{Z}}^{\mathrm{X}} f_{\mathrm{X}} \Delta_{\mathrm{Z}}^{\mathrm{X}}, \tag{3.101c}$$

$$\mathbf{log}_q(f)_{\mathrm{Z}} = \sum_{\mathrm{X}\colon \mathrm{X} \geq \mathrm{Z}} \mu_q(\mathrm{Z},\mathrm{X})\, \mu_{\mathrm{Z}}^{\mathrm{X}} f_{\mathrm{X}} \Delta_{\mathrm{Z}}^{\mathrm{X}}. \tag{3.101d}$$

This follows from Exercise 3.41 and Fact 1.100 and (1.100).

Theorem 3.43. *Suppose q is not a root of unity. Then: for a Coxeter comonoid c and Coxeter q-bimonoid k, we have inverse bijections*

$$\mathsf{CSp}(\mathsf{c}, \mathcal{P}(\mathsf{k})) \underset{\mathbf{log}_q}{\overset{\mathbf{exp}_q}{\rightleftarrows}} \mathsf{Comon}(\mathsf{CSp})(\mathsf{c}, \mathsf{k}), \tag{3.102}$$

for a Coxeter q-bimonoid h and Coxeter monoid a, we have inverse bijections

$$\mathsf{CSp}(\mathcal{Q}(\mathsf{h}), \mathsf{a}) \underset{\mathbf{log}_q}{\overset{\mathbf{exp}_q}{\rightleftarrows}} \mathsf{Mon}(\mathsf{CSp})(\mathsf{h}, \mathsf{a}), \tag{3.103}$$

for Coxeter q-bimonoids h and k, we have inverse bijections

$$\mathsf{CSp}(\mathcal{Q}(\mathsf{h}), \mathcal{P}(\mathsf{k})) \underset{\mathbf{log}_q}{\overset{\mathbf{exp}_q}{\rightleftarrows}} q\text{-}\mathsf{Bimon}(\mathsf{CSp})(\mathsf{h}, \mathsf{k}). \tag{3.104}$$

PROOF. This can be deduced from the corresponding result for species [**11**, Theorems 9.78, 9.79, 9.81]. This is because type morphisms do not play any role in the calculation. The two directions use the two-sided q-lune-additivity formula (1.106) and two-sided q-Weisner formula (1.108), respectively. ☐

3.8.8. 0-exp-log. The case $q = 0$ is of interest. Using formulas (1.117) for ζ_0 and μ_0, formulas (3.99a) and (3.99b) specialize to

$$\mathbf{exp}_0(f)_A = \sum_{F:\, F \geq A} \mu_A^F f_F \Delta_A^F, \tag{3.105a}$$

$$\mathbf{log}_0(f)_A = \sum_{F:\, F \geq A} (-1)^{\mathrm{rk}(F/A)}\, \mu_A^F f_F \Delta_A^F. \tag{3.105b}$$

For $q = 0$, we refer to (3.100) as the 0-*exp-log correspondence*.

3.9. Logarithm of the identity map

For any Coxeter bimonoid, there is a natural map from its primitive part to its indecomposable part. It is closely related to the logarithm of the identity map of a Coxeter bimonoid. These considerations extend to the q-setting.

3.9.1. Primitive part to indecomposable part. For a scalar q, for a Coxeter q-bimonoid h, we have the canonical map

$$\mathrm{pq}_{\mathsf{h}} : \mathcal{P}(\mathsf{h}) \hookrightarrow \mathsf{h} \twoheadrightarrow \mathcal{Q}(\mathsf{h}). \tag{3.106}$$

This map is natural in h, that is, for a morphism $f : \mathsf{h} \to \mathsf{k}$ of Coxeter q-bimonoids, the diagram

$$\begin{array}{ccc} \mathcal{P}(\mathsf{h}) & \xrightarrow{f} & \mathcal{P}(\mathsf{k}) \\ {\scriptstyle \mathrm{pq}_{\mathsf{h}}}\downarrow & & \downarrow{\scriptstyle \mathrm{pq}_{\mathsf{k}}} \\ \mathcal{Q}(\mathsf{h}) & \underset{f}{\longrightarrow} & \mathcal{Q}(\mathsf{k}) \end{array} \tag{3.107}$$

commutes. This defines a natural transformation pq from $\mathcal{P}$ to $\mathcal{Q}$ (viewing both as functors from Coxeter q-bimonoids to Coxeter species).

Proposition 3.44. *Let* h *be a Coxeter bimonoid. Then:*

(1) h *is cocommutative iff the map* $\mathrm{pq}_{\mathsf{h}} : \mathcal{P}(\mathsf{h}) \to \mathcal{Q}(\mathsf{h})$ *is surjective.*
(2) h *is commutative iff the map* $\mathrm{pq}_{\mathsf{h}} : \mathcal{P}(\mathsf{h}) \to \mathcal{Q}(\mathsf{h})$ *is injective.*
(3) h *is bicommutative iff the map* $\mathrm{pq}_{\mathsf{h}} : \mathcal{P}(\mathsf{h}) \to \mathcal{Q}(\mathsf{h})$ *is bijective.*

PROOF. This is a formal consequence of the corresponding result for species [**11**, Proposition 5.56]. □

Proposition 3.45. *For a Coxeter q-bimonoid* h *for q not a root of unity, the map* $\mathrm{pq}_{\mathsf{h}} : \mathcal{P}(\mathsf{h}) \to \mathcal{Q}(\mathsf{h})$ *is bijective.*

PROOF. This is a formal consequence of the corresponding result for species [**11**, Proposition 5.58]. □

An improvement of this result is given in Proposition 3.50 below.

3.9.2. Logarithm of the identity map. Let $\mathrm{char}(\Bbbk) \nmid |W|$. Fix an invariant noncommutative Möbius function $\boldsymbol{\mu}$. For a Coxeter q-bimonoid h, we have the operator $\mathbf{log}(\mathrm{id}) : \mathsf{h} \to \mathsf{h}$. This is a *logarithm of the identity map* on h. By formula (3.87b),

$$\mathbf{log}(\mathrm{id})_A = \sum_{F:\, F \geq A} \boldsymbol{\mu}(A, F)\, \mu_A^F \Delta_A^F. \tag{3.108}$$

To emphasize dependence on h, we may also write $\mathbf{log}(\mathrm{id}_{\mathsf{h}})$.

Proposition 3.46. *Let* $\mathrm{char}(\Bbbk) \nmid |W|$. *Let* h *be a Coxeter bimonoid. For* h *cocommutative,* $\mathbf{log}(\mathrm{id})$ *is an idempotent operator on* h *with image* $\mathcal{P}(\mathsf{h})$. *For* h *commutative,* $\mathbf{log}(\mathrm{id})$ *is an idempotent operator on* h *with coimage* $\mathcal{Q}(\mathsf{h})$.

PROOF. We explain the first statement. Suppose h is cocommutative. That $\mathbf{log}(\mathrm{id})$ maps to $\mathcal{P}(\mathsf{h})$ follows from (3.89). That it is identity on $\mathcal{P}(\mathsf{h})$ follows from formula (3.108). □

The two situations when h is either cocommutative or commutative are, respectively, illustrated below.

(3.109)

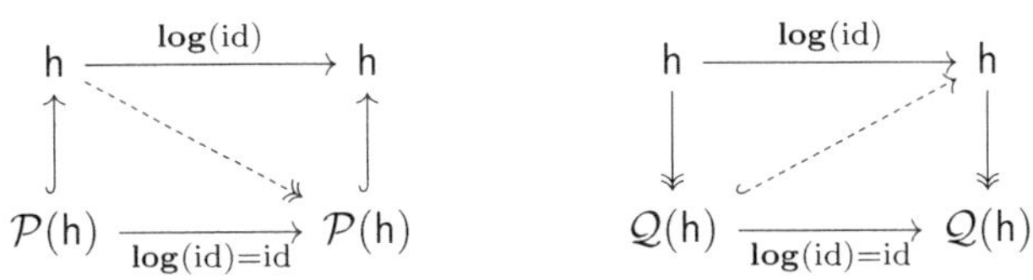

Exercise 3.47. Check: For h cocommutative, the first diagram below commutes. For h commutative, the second diagram below commutes.

(3.110)

The horizontal maps are the canonical maps.

Let us now turn to commutative aspects. By formula (3.92b), the *logarithm of the identity map* on a bicommutative Coxeter bimonoid h is

$$\log(\mathrm{id})_{\mathrm{Z}} = \sum_{\mathrm{X}\colon \mathrm{X}\geq \mathrm{Z}} \mu(\mathrm{Z},\mathrm{X})\, \mu_{\mathrm{Z}}^{\mathrm{X}} \Delta_{\mathrm{Z}}^{\mathrm{X}}. \tag{3.111}$$

In this case, (3.108) reduces to the above formula via the base-case map.

Proposition 3.48. *For* h *a bicommutative Coxeter bimonoid,* log(id) *is an idempotent operator on* h *with image* $\mathcal{P}(\mathsf{h})$ *and coimage* $\mathcal{Q}(\mathsf{h})$ *yielding the commutative diagram of Coxeter species*

$$\begin{array}{ccc} \mathsf{h} & \xrightarrow{\log(\mathrm{id})} & \mathsf{h} \\ \downarrow & & \uparrow \\ \mathcal{Q}(\mathsf{h}) & \xrightarrow[\cong]{} & \mathcal{P}(\mathsf{h}). \end{array} \tag{3.112}$$

In particular, $\mathcal{P}(\mathsf{h})$ *and* $\mathcal{Q}(\mathsf{h})$ *are isomorphic as Coxeter species. Moreover, the inverse of* $\mathcal{Q}(\mathsf{h}) \to \mathcal{P}(\mathsf{h})$ *is the map* pq_{h} *in* (3.106).

PROOF. One may employ (3.96). When $\mathrm{char}(\Bbbk) \nmid |W|$, this can also be deduced from Proposition 3.46. □

Proposition 3.49. *For bicommutative Coxeter bimonoids* h *and* k, *the bijection* (3.96) *preserves isomorphisms.*

PROOF. This is a formal consequence of the corresponding result for species [**11**, Proposition 9.55]. □

Let q be a scalar which is not a root of unity. Let h be a Coxeter p-bimonoid for a scalar p possibly different from q. By formula (3.99b), the *q-logarithm of the identity map* on h is

$$\mathbf{log}_q(\mathrm{id})_A = \sum_{\substack{F,F'\geq A \\ \mathrm{s}(F)=\mathrm{s}(F')}} \mu_q(A,F,F')\, \mu_A^F \beta_{F,F'} \Delta_A^{F'}. \tag{3.113}$$

In particular, the 0-*logarithm of the identity map* on h is

$$\mathbf{log}_0(\mathrm{id})_A = \sum_{F\colon F\geq A} (-1)^{\mathrm{rk}(F/A)}\, \mu_A^F \Delta_A^F. \tag{3.114}$$

This is a special case of (3.105b).

Proposition 3.50. *For* h *a Coxeter q-bimonoid for q not a root of unity,* $\mathbf{log}_q(\mathrm{id})$ *is an idempotent operator on* h *with image* $\mathcal{P}(\mathsf{h})$ *and coimage* $\mathcal{Q}(\mathsf{h})$ *yielding the commutative diagram of Coxeter species*

$$\begin{array}{ccc} \mathsf{h} & \xrightarrow{\mathbf{log}_q(\mathrm{id})} & \mathsf{h} \\ \downarrow & & \uparrow \\ \mathcal{Q}(\mathsf{h}) & \xrightarrow[\cong]{} & \mathcal{P}(\mathsf{h}). \end{array} \tag{3.115}$$

In particular, $\mathcal{P}(\mathsf{h})$ *and* $\mathcal{Q}(\mathsf{h})$ *are isomorphic as Coxeter species. Moreover, the inverse of* $\mathcal{Q}(\mathsf{h}) \to \mathcal{P}(\mathsf{h})$ *is the map* pq_{h} *in* (3.106).

PROOF. One may employ (3.104). □

Proposition 3.45 follows from the above result.

Proposition 3.51. *For q not a root of unity, for Coxeter q-bimonoids* h *and* k, *the bijection* (3.104) *preserves isomorphisms.*

PROOF. This is a formal consequence of the corresponding result for species [**11**, Proposition 9.91]. □

A basic example of this result is given by Proposition 3.50. Namely, for q not a root of unity, for a Coxeter q-bimonoid h, the isomorphism of Coxeter q-bimonoids $\mathrm{id} : \mathsf{h} \to \mathsf{h}$ corresponds to the isomorphism of Coxeter species $\mathbf{log}_q(\mathrm{id}) : \mathcal{Q}(\mathsf{h}) \to \mathcal{P}(\mathsf{h})$ under the bijection (3.104).

3.10. Exp-log between primitive and group-like series

We define the space of series of a Coxeter species. We also introduce primitive and group-like series of a Coxeter comonoid. The invariant lune-incidence algebra acts on the space of series of a Coxeter monoid. This yields exp-log correspondences, one for each invariant noncommutative zeta or Möbius function. Each such correspondence restricts to a bijection between primitive series and group-like series of any Coxeter bimonoid. This story has a commutative as well as a q-analogue for q not a root of unity. Moreover, exp-log correspondences in this section relate to those in Section 3.8 via the convolution Coxeter monoid, internal hom for Coxeter comonoids, universal measuring Coxeter comonoid. We also consider linear and exponential series of a Coxeter monoid; for a Coxeter bimonoid, a linear series is primitive while an exponential series is group-like. There is also a stronger notion of equi-exponential series.

3.10.1. Series of a Coxeter species. Let p be a Coxeter species. A *series* of p is a family of elements $v_F \in \mathsf{p}[F]$, one for each face F, such that

$$\beta_{G,F}(v_F) = v_G \quad \text{and} \quad w(v_F) = v_{wF},$$

the former whenever $\mathrm{s}(F) = \mathrm{s}(G)$ and the latter for all $w \in W$. In terms of flats: A series of p is a family of elements $v_{\mathrm{X}} \in \mathsf{p}[\mathrm{X}]$, one for each flat X, such that

$$w(v_{\mathrm{X}}) = v_{w\mathrm{X}}$$

for all $w \in W$. Let $\mathscr{S}(\mathsf{p})$ denote the space of series of p. This defines a functor $\mathscr{S}$ from the category of Coxeter species to the category of vector spaces.

Example 3.52 (Internal hom). For Coxeter species p and q, using (3.60):

$$\mathscr{S}(\hom^{\times}(\mathsf{p},\mathsf{q})) = \mathsf{CSp}(\mathsf{p},\mathsf{q}). \tag{3.116}$$

As special cases, for a Coxeter species p,

$$\mathscr{S}(\mathsf{p}) = \mathsf{CSp}(\mathsf{E},\mathsf{p}) \quad \text{and} \quad \mathscr{S}(\mathsf{p}^*) = \mathsf{CSp}(\mathsf{p},\mathsf{E}). \tag{3.117}$$

Interestingly, (3.116) can be formally deduced from the first statement in (3.117) by using (3.62). □

Remark 3.53. Note very carefully: A series of a Coxeter species is different from a series of its underlying species. The latter was defined in [**11**, Section 9.5.1] and only involves the condition $\beta_{G,F}(v_F) = v_G$.

3.10.2. Primitive and group-like series of a Coxeter comonoid.

Let c be a Coxeter comonoid. We define below two subspaces of $\mathscr{S}(\mathsf{c})$, namely, $\mathscr{P}(\mathsf{c})$ and $\mathscr{G}(\mathsf{c})$.

- A series v of c is *primitive* if $\Delta_A^F(v_A) = 0$ for all $F > A$. Let $\mathscr{P}(\mathsf{c})$ denote the space of primitive series of c. Note: $\mathscr{P}(\mathsf{c}) = \mathscr{S}(\mathcal{P}(\mathsf{c}))$.
- A series v of c is *group-like* if $\Delta_A^F(v_A) = v_F$ for all $F \geq A$. Let $\mathscr{G}(\mathsf{c})$ denote the space of group-like series of c.

In terms of flats and lunes: A series v of c is primitive if $\Delta_{\mathrm{L}}(v_{\mathrm{b(L)}}) = 0$ for all lunes L of slack at least 1, and group-like if $\Delta_{\mathrm{L}}(v_{\mathrm{b(L)}}) = v_{\mathrm{c(L)}}$ for all lunes L. For c a cocommutative Coxeter comonoid, observe that a series v of c is

- primitive if $\Delta_{\mathrm{X}}^{\mathrm{Y}}(v_{\mathrm{X}}) = 0$ for all $\mathrm{Y} > \mathrm{X}$,
- group-like if $\Delta_{\mathrm{X}}^{\mathrm{Y}}(v_{\mathrm{X}}) = v_{\mathrm{Y}}$ for all $\mathrm{Y} \geq \mathrm{X}$.

This definition is entirely in terms of flats (with no dependence on lunes).

Example 3.54 (Enrichment). Recall from Section 3.6.1 the Coxeter comonoid $\mathcal{C}(\mathsf{c},\mathsf{d})$ associated to Coxeter comonoids c and d. We have

$$\begin{aligned} \mathscr{P}(\mathcal{C}(\mathsf{c},\mathsf{d})) &= \mathsf{CSp}(\mathsf{c},\mathcal{P}(\mathsf{d})), \\ \mathscr{G}(\mathcal{C}(\mathsf{c},\mathsf{d})) &= \mathsf{Comon}(\mathsf{CSp})(\mathsf{c},\mathsf{d}). \end{aligned} \tag{3.118}$$

As a special case, for a Coxeter comonoid c,

$$\mathscr{P}(\mathsf{c}) = \mathsf{CSp}(\mathsf{E},\mathcal{P}(\mathsf{c})) \quad \text{and} \quad \mathscr{G}(\mathsf{c}) = \mathsf{Comon}(\mathsf{CSp})(\mathsf{E},\mathsf{c}). \tag{3.119}$$

Interestingly, these two statements can be used to formally deduce (3.118) by using (3.62), (3.72), (3.73).

Recall from Section 3.7.1 the Coxeter comonoid $\overline{\mathcal{C}}(\mathsf{a},\mathsf{b})$ associated to Coxeter monoids a and b. We have

$$\begin{aligned} \mathscr{P}(\overline{\mathcal{C}}(\mathsf{a},\mathsf{b})) &= \mathsf{CSp}(\mathcal{Q}(\mathsf{a}),\mathsf{b}), \\ \mathscr{G}(\overline{\mathcal{C}}(\mathsf{a},\mathsf{b})) &= \mathsf{Mon}(\mathsf{CSp})(\mathsf{a},\mathsf{b}). \end{aligned} \tag{3.120}$$

As a special case, for a Coxeter monoid a,

$$\mathscr{P}(\mathsf{a}^*) = \mathsf{CSp}(\mathcal{Q}(\mathsf{a}),\mathsf{E}) \quad \text{and} \quad \mathscr{G}(\mathsf{a}^*) = \mathsf{Mon}(\mathsf{CSp})(\mathsf{a},\mathsf{E}). \tag{3.121}$$

This is how primitive series link to (co)derivations and group-like series to Coxeter (co)monoid morphisms. □

3.10.3. Action of the invariant lune-incidence algebra.

Let a be a Coxeter monoid. For $s \in \mathrm{I}_{\mathrm{lune}}[\mathcal{A}]^W$ and $v \in \mathscr{S}(\mathsf{a})$, define $s \circ v \in \mathscr{S}(\mathsf{a})$ by

$$(s \circ v)_A := \sum_{F:\, F \geq A} s(A,F)\, \mu_A^F(v_F). \tag{3.122}$$

Exercise 3.55. Check: $s \circ v$ is a series of a, and the assignment $(s, v) \mapsto s \circ v$ defines a left action of $\mathrm{I}_{\mathrm{lune}}[\mathcal{A}]^W$ on $\mathscr{S}(\mathsf{a})$.

In terms of flats and lunes: The action (3.122) can be rewritten as

$$(s \circ v)_{\mathrm{Z}} = \sum_{\mathrm{L}:\, \mathrm{b(L)}=\mathrm{Z}} s(\mathrm{L})\, \mu_{\mathrm{L}}(v_{\mathrm{c(L)}}). \tag{3.123}$$

3.10.4. Exp-log on series of a Coxeter monoid. Let $\mathrm{char}(\Bbbk) \nmid |W|$. Fix an invariant noncommutative zeta function $\boldsymbol{\zeta}$ and Möbius function $\boldsymbol{\mu}$ which are inverses of each other in $\mathrm{I}_{\mathrm{lune}}[\mathcal{A}]^W$. For $v \in \mathscr{S}(\mathsf{a})$, define

$$\mathbf{exp}(v)_A := \sum_{F:\, F \geq A} \boldsymbol{\zeta}(A,F)\, \mu_A^F(v_F), \tag{3.124a}$$

$$\mathbf{log}(v)_A := \sum_{F:\, F \geq A} \boldsymbol{\mu}(A,F)\, \mu_A^F(v_F). \tag{3.124b}$$

Since $\boldsymbol{\zeta}$ and $\boldsymbol{\mu}$ are inverses, we obtain inverse bijections

$$\mathscr{S}(\mathsf{a}) \underset{\mathbf{log}}{\overset{\mathbf{exp}}{\rightleftarrows}} \mathscr{S}(\mathsf{a}). \tag{3.125}$$

We refer to (3.125) as an *exp-log correspondence.*

3.10.5. Exp-log between primitive and group-like series. When a carries a Coxeter bimonoid structure, (3.125) restricts as follows.

Theorem 3.56. *Let* $\mathrm{char}(\Bbbk) \nmid |W|$. *For a Coxeter bimonoid* h, *we have inverse bijections*

$$\mathscr{P}(\mathsf{h}) \underset{\mathbf{log}}{\overset{\mathbf{exp}}{\rightleftarrows}} \mathscr{G}(\mathsf{h}). \tag{3.126}$$

These bijections depend on $\boldsymbol{\zeta}$ and $\boldsymbol{\mu}$.

PROOF. This can be deduced from the corresponding result for species [**11**, Theorem 9.117]. This is because type morphisms do not play any role in the calculation. The two directions use the lune-additivity formula (1.79) and noncommutative Weisner formula (1.81), respectively. □

Exercise 3.57 (Comparison of actions). Let us now compare the two actions (3.85) and (3.122) of the invariant lune-incidence algebra. In fact, either one can be seen as a special case of the other. Check:

(1) For c a Coxeter comonoid and a a Coxeter monoid, the action (3.122) on the space of series of the convolution Coxeter monoid $\mathrm{hom}^{\times}(\mathsf{c}, \mathsf{a})$ coincides with the action (3.85) via the identification (3.116) with $\mathsf{p} := \mathsf{c}$ and $\mathsf{q} := \mathsf{a}$. (Use product formula (3.64).) In a similar manner, (3.123) relates to (3.86) using formula (3.65).

(2) For a a Coxeter monoid, the action (3.85) for $\mathsf{c} := \mathsf{E}$ coincides with the action (3.122) via the identification (3.117).

Deduce: The exp-log correspondences (3.88) and (3.125) imply each other.

Exercise 3.58 (Comparison of exp-logs). The exp-log correspondences (3.89) and (3.126) imply each other as follows.

(1) Apply (3.126) to the Coxeter bimonoid $\mathcal{C}(\mathsf{c}, \mathsf{k})$ from Section 3.6.2 to obtain (3.89) via the identification (3.118).

(2) Apply (3.89) with $\mathsf{c} := \mathsf{E}$ to obtain (3.126) via the identification (3.119).

Similarly, apply (3.126) to the Coxeter bimonoid $\overline{\mathcal{C}}(\mathsf{h}, \mathsf{a})$ from Section 3.7.2 to obtain (3.90) via the identification (3.120).

3.10.6. q-exp-log. Let q be a scalar which is not a root of unity. Recall the noncommutative q-zeta function $\boldsymbol{\zeta}_q$ and noncommutative q-Möbius function $\boldsymbol{\mu}_q$. For $v \in \mathscr{S}(\mathsf{a})$, define

$$\mathbf{exp}_q(v)_A := \sum_{F:\, F \geq A} \boldsymbol{\zeta}_q(A, F)\, \mu_A^F(v_F), \tag{3.127a}$$

$$\mathbf{log}_q(v)_A := \sum_{F:\, F \geq A} \boldsymbol{\mu}_q(A, F)\, \mu_A^F(v_F). \tag{3.127b}$$

Since $\boldsymbol{\zeta}_q$ and $\boldsymbol{\mu}_q$ are inverses, we obtain inverse bijections

$$\mathscr{S}(\mathsf{a}) \underset{\mathbf{log}_q}{\overset{\mathbf{exp}_q}{\rightleftarrows}} \mathscr{S}(\mathsf{a}). \tag{3.128}$$

We refer to (3.128) as the *q-exp-log correspondence*.

Theorem 3.59. *For a Coxeter q-bimonoid* h *for q not a root of unity, we have inverse bijections*

$$\mathscr{P}(\mathsf{h}) \underset{\mathbf{log}_q}{\overset{\mathbf{exp}_q}{\rightleftarrows}} \mathscr{G}(\mathsf{h}). \tag{3.129}$$

PROOF. This can be deduced from the corresponding result for species [**11**, Theorem 9.122]. The calculations are similar to those in Theorem 3.56. We now employ the q-lune-additivity formula (1.90) and noncommutative q-Weisner formula (1.92). □

Exercise 3.60 (Comparison of q-exp-logs). Let us now consider the analogue of Exercise 3.58 for q not a root of unity.

(1) Apply (3.129) to the Coxeter q-bimonoid $\mathcal{C}(\mathsf{c}, \mathsf{k})_q$ from Exercise 3.31 to obtain (3.102) (for c cocommutative) via the identification (3.118).
(2) Apply (3.102) with $\mathsf{c} := \mathsf{E}$ to obtain (3.129) via the identification (3.119).

Similarly, apply (3.129) to the Coxeter q-bimonoid $\overline{\mathcal{C}}(\mathsf{h}, \mathsf{a})_q$ from Exercise 3.34 to obtain (3.103) (for a commutative) via the identification (3.120).

3.10.7. Linear series of a Coxeter species. Let p be a Coxeter species. A series x of p is *linear* if $x_F = 0$ whenever F is not a chamber. In terms of flats: A series x of p is linear if $x_{\mathrm{Y}} = 0$ whenever $\mathrm{Y} \neq \top$. Observe: A linear series of p is the same as an element $x_\top \in \mathsf{p}[\top]^W$. Let $\mathscr{X}(\mathsf{p})$ denote the space of linear series of p. We mention that

$$\mathscr{X}(\mathsf{p}) = \mathsf{CSp}(\mathsf{x}, \mathsf{p}), \tag{3.130}$$

where x is the Coxeter species defined later in Section 4.1.

Observe: For a Coxeter comonoid c, any linear series of c is primitive, that is, $\mathscr{X}(\mathsf{c}) \subseteq \mathscr{P}(\mathsf{c})$.

Example 3.61. We build on Example 3.52. For Coxeter species p and q,

$$\mathscr{X}(\hom^\times(\mathsf{p}, \mathsf{q})) = W\text{-}\mathsf{Mod}(\mathsf{p}[\top], \mathsf{q}[\top]).$$

In words, a linear series of $\hom^\times(\mathsf{p}, \mathsf{q})$ is the same as a map $f : \mathsf{p} \to \mathsf{q}$ of Coxeter species such that $f_{\mathrm{X}} = 0$ if $\mathrm{X} \neq \top$. This is the same as a map $f_\top : \mathsf{p}[\top] \to \mathsf{q}[\top]$ of W-modules. See also Exercise 4.2. □

3.10.8. Exponential series of a Coxeter monoid. Let a be a Coxeter monoid. A series e of a is *exponential* if for face-types $T \geq Z$ and any face A of type Z,

$$\sum_{\substack{F:\, F \geq A \\ \mathrm{t}(F/A)=T/Z}} \mu_A^F(e_F) = d_{T/Z}\, e_A, \tag{3.131}$$

where $d_{T/Z}$ is the number of faces of type T/Z in $\boldsymbol{\alpha}_Z$ (Section 1.8.1). If $\mathrm{char}(\Bbbk) \nmid |W|$, then we can divide equation (3.131) by $d_{T/Z}$, and interpret e_A as the average value of the summands $\mu_A^F(e_F)$. Let $\mathscr{E}(\mathsf{a})$ denote the space of exponential series of a.

A series e of a is *equi-exponential* if $\mu_A^F(e_F) = e_A$ for all $F \geq A$. Let $\mathscr{E}^o(\mathsf{a})$ denote the space of equi-exponential series of a. Observe:

$$\mathscr{E}^o(\mathsf{a}) = \mathsf{Mon}(\mathsf{CSp})(\mathsf{E}, \mathsf{a}). \tag{3.132}$$

Any equi-exponential series is exponential (since there are $d_{T/Z}$ summands in the lhs of (3.131) and each one of them evaluates to e_A). Thus, $\mathscr{E}^o(\mathsf{a}) \subseteq \mathscr{E}(\mathsf{a})$. An interpretation similar to (3.132) for $\mathscr{E}(\mathsf{a})$ is given later in Section 4.3.9.

Exercise 3.62. Another way to formulate exponential series using a subset of the defining conditions is given below. Let a be a Coxeter monoid. Show:

(1) Let $\mathrm{char}(\Bbbk) \nmid |W|$. A series e of a is exponential iff for any face A,

$$\sum_{C:\, C \geq A} \mu_A^C(e_C) = |W_A|\, e_A.$$

(Recall: $|W_A| = d_{S/Z}$ where $\mathrm{t}(A) = Z$. Now use Observation 1.5 and the multiplicative property (1.37) for $q = 1$ applied to $\boldsymbol{\alpha}_Z$.)

(2) A series e of a is equi-exponential iff $\mu_A^C(e_C) = e_A$ for all $C \geq A$.

Exercise 3.63. Let $\mathrm{char}(\Bbbk) \nmid |W|$. Show: For a commutative Coxeter monoid a, notions of exponential and equi-exponential series coincide, that is, $\mathscr{E}(\mathsf{a}) = \mathscr{E}^o(\mathsf{a})$.

Exercise 3.64. Show: For a a free Coxeter monoid, $\mathscr{E}^o(\mathsf{a}) = 0$ (assuming $\mathrm{rk}(\mathcal{A}) \geq 1$), that is, a free Coxeter monoid does not have any nonzero equi-exponential series.

Exercise 3.65 (Exponential series in terms of flats and lunes). Let $\mathrm{char}(\Bbbk) \nmid |W|$. Let a be a Coxeter monoid. Check: A series e of a is exponential if for face-types $T \geq Z$ and any flat X with $\mathrm{t}(\mathrm{X}) = \mathrm{s}(Z)$,

$$\sum_{\substack{\mathrm{L}:\, \mathrm{b}(\mathrm{L})=\mathrm{X} \\ \mathrm{t}(\mathrm{L})=\mathrm{s}(Z,T)}} \mu_{\mathrm{L}}(e_{\mathrm{c}(\mathrm{L})}) = d_{T/Z}\, e_{\mathrm{X}}. \tag{3.133}$$

This condition translates to (3.131) via $\mathrm{L} = \mathrm{s}(A, F)$. Equivalently, a series e of a is exponential if for any flat X,

$$\sum_{\substack{\mathrm{L}:\, \mathrm{b}(\mathrm{L})=\mathrm{X} \\ \mathrm{c}(\mathrm{L})=\top}} \mu_{\mathrm{L}}(e_{\top}) = |W_{\mathrm{X}}|\, e_{\mathrm{X}}.$$

The sum is over all top-lunes with base X.

A series e of a is equi-exponential if $\mu_{\mathrm{L}}(e_{\mathrm{c}(\mathrm{L})}) = e_{\mathrm{b}(\mathrm{L})}$ for all lunes L, or equivalently, if $\mu_{\mathrm{L}}(e_{\top}) = e_{\mathrm{b}(\mathrm{L})}$ for all top-lunes L.

For a commutative Coxeter monoid a, a series e of a is exponential if $\mu_{\mathrm{X}}^{\mathrm{Y}}(e_{\mathrm{Y}}) = e_{\mathrm{X}}$ for all $\mathrm{Y} \geq \mathrm{X}$.

More generally, for a scalar q, define a q-exponential series of a Coxeter monoid a by replacing $d_{T/Z}$ by $d_{T/Z}(q)$ in condition (3.131). Let $\mathscr{E}_q(\mathsf{a})$ denote the space of q-exponential series of a.

We have a q-analogue of Exercise 3.62, item (1) with $|W_A|$ replaced by the Poincaré polynomial of $\mathcal{A}_A$.

3.10.9. Exp-log between linear and exponential series.

Lemma 3.66. *Let* $\operatorname{char}(\Bbbk) \nmid |W|$. *For a Coxeter monoid* a, *any exp-log correspondence* (3.125) *restricts to inverse bijections*

$$\mathscr{X}(\mathsf{a}) \underset{\textbf{log}}{\overset{\textbf{exp}}{\rightleftarrows}} \mathscr{E}(\mathsf{a}). \tag{3.134}$$

Moreover, these bijections are independent of $\boldsymbol{\zeta}$ *and* $\boldsymbol{\mu}$.

Explicitly, for x a linear series and e an exponential series,

$$\textbf{exp}(x)_A = \frac{1}{|W_A|} \sum_{C:\, C \geq A} \mu_A^C(x_C) \quad \text{and} \quad \textbf{log}(e)_A = \begin{cases} e_C & \text{if } A = C, \\ 0 & \text{otherwise.} \end{cases} \tag{3.135}$$

We leave these calculations as an exercise. A key point to note is that since $\boldsymbol{\zeta}$ is invariant, $\boldsymbol{\zeta}(A, C)$ equals $\frac{1}{|W_A|}$ independent of C. Also note: $e_C = x_C$ for all C when x and e correspond to each other under (3.134). Formulas (3.135) can also be written in terms of flats and lunes.

Corollary 3.67. *Let* $\operatorname{char}(\Bbbk) \nmid |W|$. *For a Coxeter monoid* a, *elements* $z \in \mathsf{a}[\top]^W$ *parametrize exponential series* e *of* a *via*

$$e_{\mathrm{X}} = \frac{1}{|W_{\mathrm{X}}|} \sum_{\substack{\mathrm{L}:\, \mathrm{b(L)=X} \\ \mathrm{c(L)}=\top}} \mu_{\mathrm{L}}(z).$$

The sum is over all top-lunes with base X.

Exercise 3.68. Let $\operatorname{char}(\Bbbk) \nmid |W|$. Show: For x a linear series of a, the following are equivalent.

(1) $\textbf{exp}(x)$ is an equi-exponential series of a.
(2) $\mu_A^C(x_C) = \mu_A^D(x_D)$ whenever C and D are adjacent chambers with A as their common panel.
(3) $\mu_A^C(x_C) = \mu_A^D(x_D)$ for all $A \leq C, D$.

These conditions capture equi-exponential series among exponential series.

Exercise 3.69. Let $\operatorname{char}(\Bbbk) \nmid |W|$. Check: For a Coxeter monoid a,

$$\mathscr{X}(\mathsf{a}) = \mathscr{X}(\mathsf{a}_{ab}) \quad \text{and} \quad \mathscr{E}(\mathsf{a}) = \mathscr{E}(\mathsf{a}_{ab})$$

induced from the abelianization map $\pi_{\mathsf{a}} : \mathsf{a} \twoheadrightarrow \mathsf{a}_{ab}$. Thus, the correspondence (3.134) only depends on the abelianization of a. This result is reformulated later in Exercise 4.16.

As a special case: For a Coxeter species p,

$$\mathscr{X}(\mathcal{T}(\mathsf{p})) = \mathscr{X}(\mathcal{S}(\mathsf{p})) = \mathscr{X}(\mathsf{p}) \quad \text{and} \quad \mathscr{E}(\mathcal{T}(\mathsf{p})) = \mathscr{E}(\mathcal{S}(\mathsf{p}))$$

induced from the abelianization map $\mathcal{T}(\mathsf{p}) \twoheadrightarrow \mathcal{S}(\mathsf{p})$. Note: $\mathscr{E}^o(\mathcal{T}(\mathsf{p})) = 0$ assuming $\operatorname{rk}(\mathcal{A}) \geq 1$, while $\mathscr{E}^o(\mathcal{S}(\mathsf{p})) = \mathscr{E}(\mathcal{S}(\mathsf{p}))$.

For q not a root of unity, we have a q-analogue of Lemma 3.66 which gives inverse bijections

$$\mathscr{X}(\mathsf{a}) \underset{\mathbf{log}_q}{\overset{\mathbf{exp}_q}{\rightleftarrows}} \mathscr{E}_q(\mathsf{a}) \tag{3.136}$$

by restricting the q-exp-log correspondence (3.128). In formula (3.135), we replace $|W_A|$ by the Poincaré polynomial of $\mathcal{A}_A$.

Example 3.70 (Convolution Coxeter monoid). A linear series of $\mathrm{hom}^\times(\mathsf{c}, \mathsf{a})$ is the same as a map $f_\mathrm{T} : \mathsf{c}[\mathrm{T}] \to \mathsf{a}[\mathrm{T}]$ of W-modules as in Example 3.61. By using (3.134), we deduce: An exponential series of $\mathrm{hom}^\times(\mathsf{c}, \mathsf{a})$ is a family of linear maps $g_\mathrm{X} : \mathsf{c}[\mathrm{X}] \to \mathsf{a}[\mathrm{X}]$, one for each flat X, such that g_T is a map of W-modules and for each X, $g_\mathrm{X} = \frac{1}{|W_\mathrm{X}|} \sum_\mathrm{L} \mu_\mathrm{L} g_\mathrm{T} \Delta_\mathrm{L}$ with the sum over all top-lunes L with base X. See Exercise 3.57 in this regard. There is a similar statement for q-exponential series. □

Lemma 3.71. *Let* $\mathrm{char}(\Bbbk) \nmid |W|$. *For a Coxeter bimonoid* h, *any exponential series of* h *is group-like, that is,* $\mathscr{E}(\mathsf{h}) \subseteq \mathscr{G}(\mathsf{h})$.

PROOF. We can proceed directly. Alternatively: By (3.134), $\mathbf{log}(e)$ is a linear series. Hence, it is primitive, which by (3.126) implies that e is group-like. □

To summarize:

Proposition 3.72. *For* $\mathrm{char}(\Bbbk) \nmid |W|$, *for a Coxeter bimonoid* h, *the diagram*

$$\begin{array}{ccccc} \mathscr{X}(\mathsf{h}) & \hookrightarrow & \mathscr{P}(\mathsf{h}) & \hookrightarrow & \mathscr{S}(\mathsf{h}) \\ \updownarrow & & \updownarrow & & \updownarrow \\ \mathscr{E}(\mathsf{h}) & \hookrightarrow & \mathscr{G}(\mathsf{h}) & \hookrightarrow & \mathscr{S}(\mathsf{h}) \end{array} \tag{3.137}$$

commutes. The vertical identifications are the exp-log correspondences in (3.125), (3.126), (3.134).

For q not a root of unity, we have q-analogues of Lemma 3.71 and Proposition 3.72 for a Coxeter q-bimonoid h with q-exponential series contained in group-like series, that is, $\mathscr{E}_q(\mathsf{h}) \subseteq \mathscr{G}(\mathsf{h})$.

Exercise 3.73. Check: Linear and exponential series of the Coxeter bimonoid $\mathcal{C}(\mathsf{c}, \mathsf{k})$ are in bijection with those of $\mathrm{hom}^\times(\mathsf{c}, \mathsf{k})$ induced by the map (3.75) with $\mathsf{h} := \mathsf{c}$. A similar statement applies to the Coxeter bimonoid $\overline{\mathcal{C}}(\mathsf{h}, \mathsf{a})$ and the map (3.81).

3.10.10. Action of the invariant flat-incidence algebra.

For a commutative Coxeter monoid a, the invariant flat-incidence algebra acts on $\mathscr{S}(\mathsf{a})$ by

$$(s \circ v)_\mathrm{Z} := \sum_{\mathrm{X}:\, \mathrm{X} \geq \mathrm{Z}} s(\mathrm{Z}, \mathrm{X})\, \mu_\mathrm{Z}^\mathrm{X}(v_\mathrm{X}). \tag{3.138}$$

Equivalently, when a is commutative, the action (3.122), or equivalently, (3.123) factors through the base-case map (1.78) to yield the above action (assuming $\mathrm{char}(\Bbbk) \nmid |W|$).

3.10.11. Commutative exp-log. The exponential and logarithm of a series v are defined by the action of the zeta function ζ and Möbius function μ in the invariant flat-incidence algebra. In other words,

$$\exp(v)_{\mathrm{Z}} := \sum_{\mathrm{X}\colon \mathrm{X}\geq \mathrm{Z}} \mu_{\mathrm{Z}}^{\mathrm{X}}(v_{\mathrm{X}}), \tag{3.139a}$$

$$\log(v)_{\mathrm{Z}} := \sum_{\mathrm{X}\colon \mathrm{X}\geq \mathrm{Z}} \mu(\mathrm{Z},\mathrm{X})\, \mu_{\mathrm{Z}}^{\mathrm{X}}(v_{\mathrm{X}}). \tag{3.139b}$$

(Recall: $\zeta \equiv 1$.) Since ζ and μ are inverses, we obtain inverse bijections

$$\mathscr{S}(\mathsf{a}) \underset{\log}{\overset{\exp}{\rightleftarrows}} \mathscr{S}(\mathsf{a}). \tag{3.140}$$

This is the *commutative exp-log correspondence*. When $\mathrm{char}(\Bbbk) \nmid |W|$, (3.139a) and (3.139b) are induced from (3.124a) and (3.124b) in view of Fact 1.80, and all exp-log correspondences (3.125) reduce to (3.140).

Theorem 3.74. *For a bicommutative Coxeter bimonoid* h, *we have inverse bijections*

$$\mathscr{P}(\mathsf{h}) \underset{\log}{\overset{\exp}{\rightleftarrows}} \mathscr{G}(\mathsf{h}). \tag{3.141}$$

PROOF. This can be deduced from the corresponding result for species [**11**, Theorem 9.124]. It makes use of the Weisner formula (1.74). When $\mathrm{char}(\Bbbk) \nmid |W|$, it also follows from Theorem 3.56 via the base-case map. □

For a commutative Coxeter monoid a, the correspondence (3.140) restricts to inverse bijections

$$\mathscr{X}(\mathsf{a}) \underset{\log}{\overset{\exp}{\rightleftarrows}} \mathscr{E}(\mathsf{a}). \tag{3.142}$$

Explicitly: A linear series x of a is the same as an element $x_{\top} \in \mathsf{a}[\top]^W$. The corresponding exponential series e of a is $e_{\mathrm{X}} = \mu_{\mathrm{X}}^{\top}(x_{\top})$. Note: $e_{\top} = x_{\top}$.

3.11. Norm transformation on Coxeter bimonoids

For a scalar q, the q-norm transformation κ_q is a natural transformation between functors $\mathcal{T}_q$ and $\mathcal{T}_q^{\vee}$ from the category of Coxeter species to the category of Coxeter q-bimonoids. Thus, for every Coxeter species p, we have the q-norm map $(\kappa_q)_{\mathsf{p}}$ from $\mathcal{T}_q(\mathsf{p})$ to $\mathcal{T}_q^{\vee}(\mathsf{p})$ which is a morphism of Coxeter q-bimonoids. It arises from freeness of $\mathcal{T}_q$ and cofreeness of $\mathcal{T}_q^{\vee}$. It is an isomorphism when q is not a root of unity. For $q = 0$, κ_0 is in fact the identity. It is of interest to understand the image of κ_q when q is a root of unity. The cases $q = \pm 1$ yield functors $\mathcal{S}$ and $\mathcal{E}$, respectively.

More generally, we consider arbitrary Coxeter q-bimonoid morphisms of the form $\mathcal{T}_q(\mathsf{c}) \to \mathcal{T}_q^{\vee}(\mathsf{a})$ for c a Coxeter comonoid and a a Coxeter monoid. For q not a root of unity, we give a necessary and sufficient condition for any such map to be an isomorphism. Interestingly, it involves the q-logarithm.

A parallel more detailed discussion for species is given in [**11**, Section 6.9].

3.11.1. q-norm map. Let q be a scalar. View $\mathcal{T}_q$ and $\mathcal{T}_q^\vee$ as functors from the category of Coxeter species to the category of Coxeter q-bimonoids as in (3.25) and (3.31). Define the *q-norm transformation*

$$\kappa_q : \mathcal{T}_q \to \mathcal{T}_q^\vee \tag{3.143}$$

as follows. For a Coxeter species p, the map

$$(\kappa_q)_{\mathsf{p}} : \mathcal{T}_q(\mathsf{p}) \to \mathcal{T}_q^\vee(\mathsf{p}), \tag{3.144}$$

on the A-component, on the F-summand and into the F'-summand, is

$$\mathsf{p}[F] \to \mathsf{p}[F'], \qquad v \mapsto \begin{cases} q^{\operatorname{dist}(F,F')}\beta_{F',F}(v) & \text{if } \mathrm{s}(F') = \mathrm{s}(F), \\ 0 & \text{otherwise.} \end{cases}$$

We call (3.144) the *q-norm map*. It is a morphism of Coxeter q-bimonoids. It preserves Cauchy powers of p, and further, it is identity on the first Cauchy power $\mathsf{p}^1 = \mathsf{p}$.

Exercise 3.75. Check the universality of the q-norm map as stated below.

- In Theorem 3.11, for $\mathsf{h} := \mathcal{T}_q^\vee(\mathsf{p})$ and $f :=$ id, the map $\hat{f}$ equals (3.144).
- In Theorem 3.16, for $\mathsf{h} := \mathcal{T}_q(\mathsf{p})$ and $f :=$ id, the map $\hat{f}$ equals (3.144).

3.11.2. Norm map and signed norm map. Consider the cases $q = \pm 1$. We refer to κ_1 as the norm transformation and to κ_{-1} as the signed norm transformation. They factor as follows.

$$\begin{array}{ccc} \mathcal{T} & \xrightarrow{\kappa_1} & \mathcal{T}^\vee \\ \downarrow & & \uparrow \\ \mathcal{S} & \xrightarrow[\cong]{} & \mathcal{S}^\vee \end{array} \qquad\qquad \begin{array}{ccc} \mathcal{T}_{-1} & \xrightarrow{\kappa_{-1}} & \mathcal{T}_{-1}^\vee \\ \downarrow & & \uparrow \\ \mathcal{E} & \xrightarrow[\cong]{} & \mathcal{E}^\vee \end{array} \tag{3.145}$$

The two diagrams are conjugates of each other wrt the u-signature functor (2.70). Let us focus on the first diagram. Functors $\mathcal{S}$ and $\mathcal{S}^\vee$ are as in (3.38) and (3.41), respectively. In fact, $\mathcal{S} = \mathcal{S}^\vee$ by Exercise 3.22. The vertical maps are abelianization and coabelianization, see (3.50).

3.11.3. General q-norm map. For a scalar q, for a Coxeter comonoid c and Coxeter monoid a, there is a natural bijection

$$\mathsf{CSp}(\mathsf{c}, \mathsf{a}) \xrightarrow{\cong} q\text{-}\mathsf{Bimon}(\mathsf{CSp})(\mathcal{T}_q(\mathsf{c}), \mathcal{T}_q^\vee(\mathsf{a})), \qquad f \mapsto \widehat{f}, \tag{3.146}$$

where $\widehat{f}$, on the A-component, on the F-summand and into the G-summand, is

$$\begin{array}{ccccccc} & & & \mathsf{c}[GF] & & & \\ & & \overset{(\beta_q)_{GF,FG}}{\nearrow} & & \overset{f_{GF}}{\searrow} & & \\ \mathsf{c}[F] & \xrightarrow{\Delta_F^{FG}} \mathsf{c}[FG] & & & & \mathsf{a}[GF] & \xrightarrow{\mu_G^{GF}} \mathsf{a}[G], \\ & & \underset{f_{FG}}{\searrow} & & \underset{(\beta_q)_{GF,FG}}{\nearrow} & & \\ & & & \mathsf{a}[FG] & & & \end{array}$$

with β_q as in (2.43). The diamond commutes, so the two ways to go from $\mathsf{c}[F]$ to $\mathsf{a}[G]$ match. We call $\widehat{f}$ the general q-norm map associated to f.

Special case. Let p be a Coxeter species. View it as a trivial Coxeter comonoid and trivial Coxeter monoid, and apply (3.146) to $f := \mathrm{id}$. Then $\widehat{f} : \mathcal{T}_q(\mathsf{p}) \to \mathcal{T}_q^\vee(\mathsf{p})$ coincides with the q-norm map (3.144).

Remark 3.76. The bijection (3.146) can be obtained by restricting the corresponding bijection for species [**11**, (6.77)]. More precisely: Let c be a Coxeter comonoid and a a Coxeter monoid. Let $f : \mathsf{c} \to \mathsf{a}$ be a map of species and $\widehat{f}$ its image under the bijection in [**11**, (6.77)]. Then f is a map of Coxeter species iff $\widehat{f}$ is a map of Coxeter species. Compare with Remark 3.8.

3.11.4. General norm map. Let us now look at the case $q = 1$. For a cocommutative Coxeter comonoid c and commutative Coxeter monoid a, there is a natural bijection

$$\mathsf{CSp}(\mathsf{c},\mathsf{a}) \xrightarrow{\cong} {}^{\mathsf{co}}\mathsf{Bimon}^{\mathsf{co}}(\mathsf{CSp})(\mathcal{S}(\mathsf{c}),\mathcal{S}^\vee(\mathsf{a})), \qquad f \mapsto \widehat{f}, \tag{3.147}$$

where $\widehat{f}$, on the Z-component, on the X-summand and into the Y-summand, is

$$\mathsf{c}[\mathrm{X}] \xrightarrow{\Delta_{\mathrm{X}}^{\mathrm{X}\vee\mathrm{Y}}} \mathsf{c}[\mathrm{X}\vee\mathrm{Y}] \xrightarrow{f_{\mathrm{X}\vee\mathrm{Y}}} \mathsf{a}[\mathrm{X}\vee\mathrm{Y}] \xrightarrow{\mu_{\mathrm{Y}}^{\mathrm{X}\vee\mathrm{Y}}} \mathsf{a}[\mathrm{Y}].$$

This $\widehat{f}$ is induced from the $\widehat{f}$ in (3.146) via (co)abelianization. We continue to refer to $\widehat{f} : \mathcal{S}(\mathsf{c}) \to \mathcal{S}^\vee(\mathsf{a})$ as the general norm map associated to f.

Theorem 3.77. *For c a cocommutative Coxeter comonoid, a a commutative Coxeter monoid, let f and $\widehat{f}$ be as in (3.147). Then*

$$\widehat{f} : \mathcal{S}(\mathsf{c}) \to \mathcal{S}^\vee(\mathsf{a}) \text{ is an isomorphism of Coxeter bimonoids}$$

iff

$$\log(f) : \mathsf{c} \to \mathsf{a} \text{ is an isomorphism of Coxeter species.}$$

In addition, if either c or a is trivial, then $\widehat{f}$ is an isomorphism of Coxeter bimonoids iff f is an isomorphism of Coxeter species.

PROOF. This is a formal consequence of the corresponding result for species [**11**, Exercise 9.60]. A relevant result is Proposition 3.49. For a more constructive approach, see Theorem 7.16. □

3.11.5. q-norm map for q not a root of unity. For $q = 0$, observe that $\kappa_0 = \mathrm{id}$ which is the canonical identification $\mathcal{T}_0 = \mathcal{T}_0^\vee$. More generally, for q not a root of unity (which includes $q = 0$), the q-norm map is an isomorphism:

Lemma 3.78. *Suppose q is not a root of unity. Then the q-norm transformation κ_q is an isomorphism. More precisely, for a Coxeter species p, the q-norm map (3.144) is an isomorphism of Coxeter q-bimonoids, which moreover induces an isomorphism of Coxeter species $\mathsf{p}^k \to \mathsf{p}^k$ for each $k \geq 1$.*

Explicitly, the inverse of (3.144), namely,

$$\mathcal{T}_q^\vee(\mathsf{p}) \to \mathcal{T}_q(\mathsf{p}), \tag{3.148}$$

on the A-component, on the F'-summand and into the F-summand, is

$$\mathsf{p}[F'] \to \mathsf{p}[F], \qquad v \mapsto \begin{cases} \zeta_q(A,F,F')\beta_{F,F'}(v) & \text{if } \mathrm{s}(F') = \mathrm{s}(F), \\ 0 & \text{otherwise.} \end{cases}$$

PROOF. This can be checked using the two-sided q-flat-additivity formula (1.107). See [**11**, Exercise 6.78] in this regard. □

Let us now go to the general q-norm map.

Theorem 3.79. *Suppose q is not a root of unity. For c a Coxeter comonoid, a a Coxeter monoid, let f and $\widehat{f}$ be as in (3.146). Then*

$$\widehat{f} : \mathcal{T}_q(\mathsf{c}) \to \mathcal{T}_q^\vee(\mathsf{a}) \text{ is an isomorphism of Coxeter } q\text{-bimonoids}$$
$$\text{iff}$$
$$\mathbf{log}_q(f) : \mathsf{c} \to \mathsf{a} \text{ is an isomorphism of Coxeter species.}$$

In addition, if either c or a is trivial, then $\widehat{f}$ is an isomorphism of Coxeter q-bimonoids iff f is an isomorphism of Coxeter species.

PROOF. This is a formal consequence of the corresponding result for species [**11**, Exercise 9.97]. A relevant result is Proposition 3.51. For the last part, note that $\mathbf{log}_q(f) = f$ if either c or a is trivial. For a more constructive approach, see Theorem 7.19. □

We recover Lemma 3.78 from Theorem 3.79 by setting $\mathsf{c} = \mathsf{a} = \mathsf{p}$ with trivial Coxeter (co)monoid structures and $f = \mathrm{id}$.

3.12. Characteristic operations by faces

In this section, we make use of the Coxeter bimonoid of faces Σ defined later in Section 4.5. Each component $\Sigma[A]$ is an algebra, namely, the Tits algebra of $\mathcal{A}_A$. We also use the Coxeter bimonoid of flats Π from Section 4.4. Each component $\Pi[\mathrm{Z}]$ is an algebra, namely, the Birkhoff algebra of $\mathcal{A}_\mathrm{Z}$.

3.12.1. Characteristic operations. Let h be a Coxeter bimonoid. For a face A, for $z \in \Sigma[A]$ and $h \in \mathsf{h}[A]$, define an element $z \cdot h \in \mathsf{h}[A]$ by

$$z \cdot h := \sum_{F:\, F \geq A} a^{F/A}\, \mu_A^F \Delta_A^F(h) \quad \text{with} \quad z = \sum_{F:\, F \geq A} a^{F/A}\, \mathtt{H}_{F/A}. \tag{3.149}$$

In particular,

$$\mathtt{H}_{F/A} \cdot h := \mu_A^F \Delta_A^F(h). \tag{3.150}$$

We refer to this as a *characteristic operation*.

For a Coxeter bimonoid h, define a morphism of Coxeter bimonoids

$$\Psi : \Sigma \to \mathrm{end}^\times(\mathsf{h}) \tag{3.151}$$

with A-component

$$\Psi_A : \Sigma[A] \to \mathrm{end}^\times(\mathsf{h})[A], \qquad \Psi_A(\mathtt{H}_{F/A}) := \mu_A^F \Delta_A^F : \mathsf{h}[A] \to \mathsf{h}[A].$$

Equivalently, by (3.149),

$$\Psi_A(z)(h) = z \cdot h$$

for all $z \in \Sigma[A]$ and $h \in \mathsf{h}[A]$.

Exercise 3.80. Check: The map (3.151) is indeed a morphism of Coxeter bimonoids. Moreover, for h cocommutative, it factors as in the left diagram below, while for h commutative, it factors as in the right diagram below.

$$\begin{array}{ccc} \Sigma & \longrightarrow & \mathrm{end}^{\times}(\mathsf{h}) \\ & \searrow \quad \nearrow & \\ & \mathcal{C}(\mathsf{h},\mathsf{h}) & \end{array} \qquad \begin{array}{ccc} \Sigma & \longrightarrow & \mathrm{end}^{\times}(\mathsf{h}) \\ & \searrow \quad \nearrow & \\ & \overline{\mathcal{C}}(\mathsf{h},\mathsf{h}) & \end{array} \tag{3.152}$$

The inclusion maps are as in (3.75) and (3.81) with $\mathsf{h} = \mathsf{k}$.

For related ideas, see Exercise 4.20.

3.12.2. Commutative characteristic operations.

Let h be a bicommutative Coxeter bimonoid. For a flat Z, for $z \in \Pi[\mathrm{Z}]$ and $h \in \mathsf{h}[\mathrm{Z}]$, define an element $z \cdot h \in \mathsf{h}[\mathrm{Z}]$ by

$$z \cdot h := \sum_{\mathrm{X}:\, \mathrm{X} \geq \mathrm{Z}} a^{\mathrm{X/Z}}\, \mu_{\mathrm{Z}}^{\mathrm{X}} \Delta_{\mathrm{Z}}^{\mathrm{X}}(h) \quad \text{with} \quad z = \sum_{\mathrm{X}:\, \mathrm{X} \geq \mathrm{Z}} a^{\mathrm{X/Z}}\, \mathtt{H}_{\mathrm{X/Z}}. \tag{3.153}$$

In particular,

$$\mathtt{H}_{\mathrm{X/Z}} \cdot h := \mu_{\mathrm{Z}}^{\mathrm{X}} \Delta_{\mathrm{Z}}^{\mathrm{X}}(h). \tag{3.154}$$

We refer to this as a *commutative characteristic operation.*

Define a morphism of Coxeter bimonoids

$$\Psi : \Pi \to \mathrm{end}^{\times}(\mathsf{h}) \tag{3.155}$$

with Z-component

$$\Psi_{\mathrm{Z}} : \Pi[\mathrm{Z}] \to \mathrm{end}^{\times}(\mathsf{h})[\mathrm{Z}], \qquad \Psi_{\mathrm{Z}}(\mathtt{H}_{\mathrm{X/Z}}) := \mu_{\mathrm{Z}}^{\mathrm{X}} \Delta_{\mathrm{Z}}^{\mathrm{X}} : \mathsf{h}[\mathrm{Z}] \to \mathsf{h}[\mathrm{Z}].$$

Equivalently by (3.153),

$$\Psi_{\mathrm{Z}}(z)(h) = z \cdot h$$

for all $z \in \Pi[\mathrm{Z}]$ and $h \in \mathsf{h}[\mathrm{Z}]$.

Exercise 3.81. Check: For a bicommutative Coxeter bimonoid h, we have the commutative diagram of Coxeter bimonoids

$$\begin{array}{ccc} \Sigma & & \\ \downarrow & \searrow & \\ \Pi & \longrightarrow & \mathrm{end}^{\times}(\mathsf{h}). \end{array} \tag{3.156}$$

The vertical map is the support map (4.76), while the oblique and horizontal maps are (3.151) and (3.155), respectively. In this situation, the two diagrams in (3.152) can be combined into one (with Π in place of Σ).

3.12.3. Two-sided characteristic operations.

For a scalar q, and a face A, let $\mathsf{J}_q[A]$ denote the q-Janus algebra of $\mathcal{A}_A$.

Let h be a Coxeter q-bimonoid. For a face A, for $z \in \mathsf{J}_q[A]$ and $h \in \mathsf{h}[A]$, define an element $z \cdot h \in \mathsf{h}[A]$ by

$$z \cdot h := \sum_{\substack{(F,F'):\, F,F' \geq A \\ \mathrm{s}(F)=\mathrm{s}(F')}} a^{F/A,F'/A}\, \mu_A^F \beta_{F,F'} \Delta_A^{F'}(h) \tag{3.157}$$

with

$$z = \sum_{\substack{(F,F')\colon F,F' \geq A \\ \mathrm{s}(F)=\mathrm{s}(F')}} a^{F/A,F'/A}\, \mathtt{H}_{(F/A,F'/A)}.$$

In particular,

$$\mathtt{H}_{(F/A,F'/A)} \cdot h := \mu_A^F \beta_{F,F'} \Delta_A^{F'}(h). \tag{3.158}$$

We refer to this as a *two-sided characteristic operation*. Note very carefully: β and not β_q is employed in this definition.

3.13. Algebra modules and Coxeter bimonoids

Let $\mathcal{A}$ be a reflection arrangement with Coxeter group W. Recall from Section 1.14 the Coxeter–Tits algebra $\mathsf{W\Sigma}[\mathcal{A}]$, Coxeter–Birkhoff algebra $\mathsf{W\Pi}[\mathcal{A}]$ and Coxeter–Janus algebra $\mathsf{WJ}[\mathcal{A}]$. The latter also has a q-analogue, namely, the Coxeter-q-Janus algebra $\mathsf{WJ}_q[\mathcal{A}]$. We now relate the categories of modules over these algebras to Coxeter bimonoids and their commutative counterparts.

Proposition 3.82. *There are equivalences of categories:*

$$\begin{aligned}
\textit{(left) } \mathsf{WJ}[\mathcal{A}]\textit{-modules} &\cong \textit{Coxeter } \mathcal{A}\textit{-bimonoids} \\
\textit{(left) } \mathsf{WJ}_q[\mathcal{A}]\textit{-modules} &\cong \textit{Coxeter } \mathcal{A}\textit{-}q\textit{-bimonoids} \\
\textit{left } \mathsf{W\Sigma}[\mathcal{A}]\textit{-modules} &\cong \textit{cocommutative Coxeter } \mathcal{A}\textit{-bimonoids} \\
\textit{right } \mathsf{W\Sigma}[\mathcal{A}]\textit{-modules} &\cong \textit{commutative Coxeter } \mathcal{A}\textit{-bimonoids} \\
\textit{(left) } \mathsf{W\Pi}[\mathcal{A}]\textit{-modules} &\cong \textit{bicommutative Coxeter } \mathcal{A}\textit{-bimonoids.}
\end{aligned}$$

These equivalences will be established in the course of the discussion below. They build on the categorical equivalences for $\mathcal{A}$-species established in [**11**, Chapter 11] which we recall below.

Proposition 3.83. *There are equivalences of categories:*

$$\begin{aligned}
\textit{(left) } \mathsf{J}[\mathcal{A}]\textit{-modules} &\cong \mathcal{A}\textit{-bimonoids} \\
\textit{(left) } \mathsf{J}_q[\mathcal{A}]\textit{-modules} &\cong \mathcal{A}\textit{-}q\textit{-bimonoids} \\
\textit{left } \Sigma[\mathcal{A}]\textit{-modules} &\cong \textit{cocommutative } \mathcal{A}\textit{-bimonoids} \\
\textit{right } \Sigma[\mathcal{A}]\textit{-modules} &\cong \textit{commutative } \mathcal{A}\textit{-bimonoids} \\
\Pi[\mathcal{A}]\textit{-modules} &\cong \textit{bicommutative } \mathcal{A}\textit{-bimonoids.}
\end{aligned}$$

We give two approaches, the first one proceeds directly, while the second one is more formal and uses Karoubi envelopes.

3.13.1. Modules over the Coxeter–Janus algebra.

Proposition 3.84. *The category of (left) modules over the Coxeter–Janus algebra* $\mathsf{WJ}[\mathcal{A}]$ *is equivalent to the category of Coxeter* $\mathcal{A}$*-bimonoids.*

PROOF. We construct functors between the two categories in either direction. It is straightforward to check that they define an equivalence.

Coxeter bimonoids to modules. Let h be a Coxeter $\mathcal{A}$-bimonoid. Then $\mathsf{h}[O]$ is a left $\mathsf{J}[\mathcal{A}]$-module with action defined by the two-sided characteristic operation

$$\mathtt{H}_{(F,F')} \cdot m := \mu_O^F \beta_{F,F'} \Delta_O^{F'}(m).$$

That is,

$$\mathtt{H}_{(G,G')} \cdot (\mathtt{H}_{(F,F')} \cdot m) = (\mathtt{H}_{(G,G')} \cdot \mathtt{H}_{(F,F')}) \cdot m \quad \text{and} \quad \mathtt{H}_{(O,O)} \cdot m = m.$$

For the first equality, we use the bimonoid axiom, support-naturality axiom, (co)associativity axiom, support-support axiom. The second equality follows from the support-unitality axiom and (co)unitality axiom. For more details, see [**11**, Lemma 10.32]. In addition, $\mathsf{h}[O]$ is a left W-module, that is, the diagrams

$$\begin{array}{ccc} & \mathsf{h}[O] & \\ {}^{u}\nearrow & & \searrow^{v} \\ \mathsf{h}[O] & \xrightarrow[vu]{} & \mathsf{h}[O] \end{array} \qquad\qquad (\mathsf{h}[O] \xrightarrow{e} \mathsf{h}[O]) = \mathrm{id}$$

commute. This follows by putting $F = F' = F'' = O$ in the type-type axiom and $F = O$ in the type-unitality axiom. Moreover, the actions by $\mathsf{J}[\mathcal{A}]$ and W satisfy the compatibility relation (1.164):

$$\begin{array}{ccccccc} \mathsf{h}[O] & \xrightarrow{\Delta_O^{F'}} & \mathsf{h}[F'] & \xrightarrow{\beta_{F,F'}} & \mathsf{h}[F] & \xrightarrow{\mu_O^F} & \mathsf{h}[O] \\ {\scriptstyle w}\downarrow & & {\scriptstyle w}\downarrow & & \downarrow{\scriptstyle w} & & \downarrow{\scriptstyle w} \\ \mathsf{h}[O] & \xrightarrow[\Delta_O^{wF'}]{} & \mathsf{h}[wF'] & \xrightarrow[\beta_{wF,wF'}]{} & \mathsf{h}[wF] & \xrightarrow[\mu_O^{wF}]{} & \mathsf{h}[O]. \end{array}$$

The first and last squares commute by the type-naturality axiom (of the coproduct and product, respectively), while the middle square commutes by the support-type axiom. Thus, $\mathsf{h}[O]$ is a left $\mathsf{WJ}[\mathcal{A}]$-module.

Further, if $f : \mathsf{h} \to \mathsf{k}$ is a morphism of Coxeter $\mathcal{A}$-bimonoids, then the component $f_O : \mathsf{h}[O] \to \mathsf{k}[O]$ is a map of left $\mathsf{WJ}[\mathcal{A}]$-modules as shown below.

$$\begin{array}{ccccccc} \mathsf{h}[O] & \xrightarrow{\Delta_O^{F'}} & \mathsf{h}[F'] & \xrightarrow{\beta_{F,F'}} & \mathsf{h}[F] & \xrightarrow{\mu_O^F} & \mathsf{h}[O] \\ {\scriptstyle f_O}\downarrow & & \downarrow{\scriptstyle f_{F'}} & & \downarrow{\scriptstyle f_F} & & \downarrow{\scriptstyle f_O} \\ \mathsf{k}[O] & \xrightarrow[\Delta_O^{F'}]{} & \mathsf{k}[F'] & \xrightarrow[\beta_{F,F'}]{} & \mathsf{k}[F] & \xrightarrow[\mu_O^F]{} & \mathsf{k}[O] \end{array} \qquad \begin{array}{ccc} \mathsf{h}[O] & \xrightarrow{w} & \mathsf{h}[O] \\ {\scriptstyle f_O}\downarrow & & \downarrow{\scriptstyle f_O} \\ \mathsf{k}[O] & \xrightarrow[w]{} & \mathsf{k}[O] \end{array}$$

The squares commute by (2.13), (2.29), (2.37).

Modules to Coxeter bimonoids. Let M be a left $\mathsf{WJ}[\mathcal{A}]$-module. We now construct a Coxeter $\mathcal{A}$-bimonoid h. Put

$$\mathsf{h}[F] := \mathtt{H}_{(F,F)} \cdot M.$$

In particular, $\mathsf{h}[O] = M$. Support and type morphisms are

$$
\begin{array}{ccc}
\mathsf{h}[F] & \xrightarrow{\beta_{G,F}} & \mathsf{h}[G] \\
\downarrow & & \downarrow \\
\mathsf{h}[O] & \xrightarrow[\mathtt{H}_{(G,F)}\cdot(-)]{} & \mathsf{h}[O]
\end{array}
\qquad
\begin{array}{ccc}
\mathsf{h}[F] & \xrightarrow{w} & \mathsf{h}[wF] \\
\downarrow & & \downarrow \\
\mathsf{h}[O] & \xrightarrow[w\cdot(-)]{} & \mathsf{h}[O],
\end{array}
$$

while the product and coproduct are

$$
\begin{array}{ccc}
\mathsf{h}[F] & \xrightarrow{\mu_A^F} & \mathsf{h}[A] \\
\downarrow & & \downarrow \\
\mathsf{h}[O] & = & \mathsf{h}[O]
\end{array}
\qquad
\begin{array}{ccc}
\mathsf{h}[A] & \xrightarrow{\Delta_A^F} & \mathsf{h}[F] \\
\downarrow & & \downarrow \\
\mathsf{h}[O] & \xrightarrow[\mathtt{H}_{(F,F)}\cdot(-)]{} & \mathsf{h}[O].
\end{array}
$$

These maps are well-defined and turn h into a Coxeter $\mathcal{A}$-bimonoid.

Further, if M and N are left $\mathsf{WJ}[\mathcal{A}]$-modules with h and k as the corresponding Coxeter $\mathcal{A}$-bimonoids, and $f : M \to N$ is a map of modules, then f restricts to linear maps

$$f_F : \mathsf{h}[F] \to \mathsf{k}[F],$$

one for each face F, and this family of maps constitutes a morphism $f : \mathsf{h} \to \mathsf{k}$ of Coxeter $\mathcal{A}$-bimonoids. □

Proposition 3.85. *For a scalar q, the category of (left) modules over the Coxeter-q-Janus algebra $\mathsf{WJ}_q[\mathcal{A}]$ is equivalent to the category of Coxeter $\mathcal{A}$-q-bimonoids.*

The same argument as in the proof of Proposition 3.84 extends.

3.13.2. Modules over the Coxeter–Tits algebra.

Proposition 3.86. *The category of left modules over the Coxeter–Tits algebra $\mathsf{W\Sigma}[\mathcal{A}]$ is equivalent to the category of cocommutative Coxeter $\mathcal{A}$-bimonoids. Dually, the category of right modules over the Coxeter–Tits algebra $\mathsf{W\Sigma}[\mathcal{A}]$ is equivalent to the category of commutative Coxeter $\mathcal{A}$-bimonoids.*

Proof. We explain the first statement. The argument is similar to that for Proposition 3.84. So we only provide an outline.

Cocommutative Coxeter bimonoids to modules. Let h be a cocommutative Coxeter $\mathcal{A}$-bimonoid. Then $\mathsf{h}[O]$ is a left $\Sigma[\mathcal{A}]$-module with action defined by the characteristic operation

$$\mathtt{H}_F \cdot m := \mu_O^F \Delta_O^F(m).$$

That is,

$$\mathtt{H}_G \cdot (\mathtt{H}_F \cdot m) = (\mathtt{H}_G \cdot \mathtt{H}_F) \cdot m \quad \text{and} \quad \mathtt{H}_O \cdot m = m.$$

In addition, $\mathsf{h}[O]$ is a left W-module in view of the type-type and type-unitality axioms. The two actions are compatible and yield a left $\mathsf{W\Sigma}[\mathcal{A}]$-module.

Modules to cocommutative Coxeter bimonoids. Let M be a left $\mathsf{W\Sigma}[\mathcal{A}]$-module. This yields a cocommutative Coxeter $\mathcal{A}$-bimonoid h by setting

$$\mathsf{h}[F] := \mathtt{H}_F \cdot M,$$

with support morphism $\beta_{G,F} : \mathsf{h}[F] \to \mathsf{h}[G]$ induced by action of $\mathtt{H}_G$, type morphism $w : \mathsf{h}[F] \to \mathsf{h}[wF]$ defined by restricting $w : \mathsf{h}[O] \to \mathsf{h}[O]$, product $\mu_A^F : \mathsf{h}[F] \to \mathsf{h}[A]$ defined by inclusion, and coproduct $\Delta_A^F : \mathsf{h}[A] \to \mathsf{h}[F]$ induced by action of $\mathtt{H}_F$. □

3.13.3. Modules over the Coxeter–Birkhoff algebra.

Proposition 3.87. *The category of (left) modules over the Coxeter–Birkhoff algebra* $\mathsf{W\Pi}[\mathcal{A}]$ *is equivalent to the category of bicommutative Coxeter* $\mathcal{A}$*-bimonoids.*

PROOF. The outline is as follows.

Bicommutative Coxeter bimonoids to modules. Let h be a bicommutative Coxeter $\mathcal{A}$-bimonoid. Then $\mathsf{h}[\perp]$ is a $\mathsf{\Pi}[\mathcal{A}]$-module with action defined by the commutative characteristic operation

$$\mathtt{H}_{\mathrm{X}} \cdot m := \mu_\perp^{\mathrm{X}} \Delta_\perp^{\mathrm{X}}(m).$$

In addition, $\mathsf{h}[\perp]$ is a left W-module in view of (2.15). The two actions are compatible and yield a left $\mathsf{W\Pi}[\mathcal{A}]$-module.

Modules to bicommutative Coxeter bimonoids. Let M be a left $\mathsf{W\Pi}[\mathcal{A}]$-module. This yields a bicommutative Coxeter $\mathcal{A}$-bimonoid h by setting

$$\mathsf{h}[\mathrm{X}] := \mathtt{H}_{\mathrm{X}} \cdot M,$$

with $w : \mathsf{h}[\mathrm{X}] \to \mathsf{h}[w\mathrm{X}]$ defined by restricting $w : \mathsf{h}[\perp] \to \mathsf{h}[\perp]$, product $\mu_{\mathrm{Z}}^{\mathrm{X}} : \mathsf{h}[\mathrm{X}] \to \mathsf{h}[\mathrm{Z}]$ defined by inclusion, and coproduct $\Delta_{\mathrm{Z}}^{\mathrm{X}} : \mathsf{h}[\mathrm{Z}] \to \mathsf{h}[\mathrm{X}]$ induced by action of $\mathtt{H}_{\mathrm{X}}$. □

3.13.4. Karoubi envelope of a semigroup. Let S be a semigroup. The *Karoubi envelope* $\mathsf{K}(S)$ of S is the category associated to S whose

- objects are idempotents of S, and
- morphisms $x \to y$ are elements $f \in S$ satisfying $fx = f = yf$.

The identity morphism at x is x, while composition is product in S. More precisely, composite of $x \xrightarrow{f} y$ and $y \xrightarrow{g} z$ is $x \xrightarrow{gf} z$.

The subcategory of $\mathsf{K}(S)$ consisting of all isomorphisms is the *Karoubi groupoid* of S.

Theorem 3.88. *For a monoid* S*, the categories* $[S, \mathsf{Vec}]$ *and* $[\mathsf{K}(S), \mathsf{Vec}]$ *are equivalent.*

Here S is viewed as a category with one object, with morphisms indexed by elements of S. The category S embeds inside $\mathsf{K}(S)$ and this induces the equivalence. Note: $[S, \mathsf{Vec}]$ is the same as the category of left S-modules.

Karoubi envelopes of monoids of interest to us are summarized in Table 3.5. The categories are as in Sections 2.1 and 2.9. Proposition 3.82 (except the q-case) follows by specializing Theorem 3.88 to these monoids and using Table 2.7.

The calculations build on those for the Tits, Birkhoff, Janus monoids done in [**11**, Section 11.8]. The summary in [**11**, Table 11.4] is reproduced in Table 3.6 for convenience. Also see Lemma 2.70 in this regard.

TABLE 3.5. Karoubi envelopes of monoids for reflection arrangements.

Monoid	Karoubi envelope	Karoubi groupoid
Coxeter–Tits monoid $\mathrm{W}\Sigma[\mathcal{A}]$	$\mathcal{A}$-$\mathtt{Cox}_{\mathrm{r}}^{\mathrm{d}}$	$\mathcal{A}$-$\mathtt{Cox}$
$\mathrm{W}\Sigma[\mathcal{A}]^{\mathrm{op}}$	$\mathcal{A}$-$\mathtt{Cox}_{\mathrm{c}}^{\mathrm{e}}$	$\mathcal{A}$-$\mathtt{Cox}$
Coxeter–Birkhoff monoid $\mathrm{W}\Pi[\mathcal{A}]$	$\mathcal{A}$-$\mathtt{Cox}_{\mathrm{r}}^{\mathrm{e}}$	$\mathcal{A}$-$\mathtt{Cox}'$ (equiv. to $\mathcal{A}$-$\mathtt{Cox}$)
Coxeter–Janus monoid $\mathrm{WJ}[\mathcal{A}]$	equiv. to $\mathcal{A}$-$\mathtt{Cox}_{\mathrm{c}}^{\mathrm{d}}$	equiv. to $\mathcal{A}$-$\mathtt{Cox}$

TABLE 3.6. Karoubi envelopes of monoids for hyperplane arrangements.

Monoid	Karoubi envelope	Karoubi groupoid
Tits monoid $\Sigma[\mathcal{A}]$	$\mathcal{A}$-$\mathtt{Hyp}_{\mathrm{r}}^{\mathrm{d}}$	$\mathcal{A}$-$\mathtt{Hyp}$
$\Sigma[\mathcal{A}]^{\mathrm{op}}$	$\mathcal{A}$-$\mathtt{Hyp}_{\mathrm{c}}^{\mathrm{e}}$	$\mathcal{A}$-$\mathtt{Hyp}$
Birkhoff monoid $\Pi[\mathcal{A}]$	$\mathcal{A}$-$\mathtt{Hyp}_{\mathrm{r}}^{\mathrm{e}}$	$\mathcal{A}$-$\mathtt{Hyp}'$ (equiv. to $\mathcal{A}$-$\mathtt{Hyp}$)
Janus monoid $\mathrm{J}[\mathcal{A}]$	equiv. to $\mathcal{A}$-$\mathtt{Hyp}_{\mathrm{c}}^{\mathrm{d}}$	equiv. to $\mathcal{A}$-$\mathtt{Hyp}$

Let us begin with the Coxeter–Tits monoid $\mathrm{W}\Sigma[\mathcal{A}]$.

Lemma 3.89. *An element (F, w) of $\mathrm{W}\Sigma[\mathcal{A}]$ is idempotent iff $w = e$. The element (H, w) is a morphism from (F, e) to (G, e) in $\mathsf{K}(\mathrm{W}\Sigma[\mathcal{A}])$ iff $Hw(F) = H = GH$. Further, it is an isomorphism iff $\mathrm{s}(w(F)) = \mathrm{s}(H)$ and $H = G$. In this situation, the inverse of (G, w) is (F, w^{-1}).*

PROOF. Suppose (F, w) is an idempotent. Then

$$(F, w) \cdot (F, w) = (Fw(F), w^2) = (F, w).$$

The condition $w^2 = w$ implies $w = e$. Conversely, (F, e) is clearly idempotent.

The element (H, w) is a morphism from (F, e) to (G, e) iff

$$(Hw(F), w) = (H, w) \cdot (F, e) = (H, w) = (G, e) \cdot (H, w) = (GH, w).$$

Now suppose (H, w) is an isomorphism with inverse (H', w'). Then

$$(H', w') \cdot (H, w) = (F, e) \quad \text{and} \quad (H, w) \cdot (H', w') = (G, e)$$

yield

$$w' = w^{-1}, \ H'w'(H) = F, \ Hw(H') = G.$$

Also,

$$Hw(F) = H = GH \quad \text{and} \quad H'w'(G) = H' = FH'.$$

The equation $H = GH$ implies $G \leq H$, while $Hw(H') = G$ implies $H \leq G$, so $H = G$. Similarly, $H' = F$. The remaining conditions can be rewritten as

$$Gw(F) = G \quad \text{and} \quad Fw^{-1}(G) = F.$$

Applying w to the second equation, we see that these two conditions are equivalent to $\mathrm{s}(w(F)) = \mathrm{s}(G)$. □

The set of idempotents of $\mathrm{W}\Sigma[\mathcal{A}]$ identifies with $\Sigma[\mathcal{A}]$ by writing F instead of (F,e). Using the normal form for morphisms in $\mathcal{A}\text{-}\mathtt{Cox}^{\mathtt{d}}_{\mathtt{r}}$ in Table 2.8 and morphisms in $\mathcal{A}\text{-}\mathtt{Cox}$ in Proposition 2.4, we see that the Karoubi envelope of $\mathrm{W}\Sigma[\mathcal{A}]$ is $\mathcal{A}\text{-}\mathtt{Cox}^{\mathtt{d}}_{\mathtt{r}}$, while the Karoubi groupoid is $\mathcal{A}\text{-}\mathtt{Cox}$.

Exercise 3.90. Consider the monoid $\mathrm{W}\Sigma[\mathcal{A}]^{\mathrm{op}}$ with product (1.158). It is isomorphic to the opposite of the Coxeter–Tits monoid. So its Karoubi envelope is equivalent to the category opposite to $\mathsf{K}(\mathrm{W}\Sigma[\mathcal{A}])$.

More precisely, check: An element (F,w) of $\mathrm{W}\Sigma[\mathcal{A}]^{\mathrm{op}}$ is idempotent iff $w=e$. The element (H,w) is a morphism from (F,e) to (G,e) in $\mathsf{K}(\mathrm{W}\Sigma[\mathcal{A}]^{\mathrm{op}})$ iff $w(F)H = H = HG$. Further, it is an isomorphism iff $w(F)=H$ and $\mathrm{s}(H)=\mathrm{s}(G)$.

Thus, the Karoubi envelope of $\mathrm{W}\Sigma[\mathcal{A}]^{\mathrm{op}}$ is $\mathcal{A}\text{-}\mathtt{Cox}^{\mathtt{e}}_{\mathtt{c}}$, while the Karoubi groupoid is $\mathcal{A}\text{-}\mathtt{Cox}$.

Exercise 3.91. Consider the Coxeter–Birkhoff monoid $\mathrm{W}\Pi[\mathcal{A}]$. Check: An element (X,w) of $\mathrm{W}\Pi[\mathcal{A}]$ is idempotent iff $w=e$. The element (Z,w) is a morphism from (X,e) to (Y,e) in $\mathsf{K}(\mathrm{W}\Pi[\mathcal{A}])$ iff $w(\mathrm{X}),\mathrm{Y}\le \mathrm{Z}$. Further, it is an isomorphism iff $w(\mathrm{X})=\mathrm{Z}=\mathrm{Y}$.

The set of idempotents of $\mathrm{W}\Pi[\mathcal{A}]$ identifies with $\Pi[\mathcal{A}]$ by writing X instead of (X,e). Use the normal form for morphisms in $\mathcal{A}\text{-}\mathtt{Cox}^{\mathtt{e}}_{\mathtt{r}}$ in Table 2.8 to deduce that the Karoubi envelope of $\mathrm{W}\Pi[\mathcal{A}]$ is $\mathcal{A}\text{-}\mathtt{Cox}^{\mathtt{e}}_{\mathtt{r}}$, while the Karoubi groupoid is $\mathcal{A}\text{-}\mathtt{Cox}'$.

Exercise 3.92. Consider the Coxeter–Janus monoid $\mathrm{WJ}[\mathcal{A}]$. Check: An element (F,F',w) of $\mathrm{WJ}[\mathcal{A}]$ is idempotent iff $w=e$. The element (H,H',w) is a morphism from (F,F',e) to (G,G',e) in $\mathsf{K}(\mathrm{WJ}[\mathcal{A}])$ iff $G\le H$ and $w(F')\le H'$. Further, it is an isomorphism iff $H=G$ and $H'=w(F')$.

The set of idempotents of $\mathrm{WJ}[\mathcal{A}]$ identifies with $\mathrm{J}[\mathcal{A}]$ by writing (F,F') instead of (F,F',e). Use the normal form for morphisms in $\mathcal{A}\text{-}\mathtt{Cox}^{\mathtt{d}}_{\mathtt{c}}$ in Table 2.8 and morphisms in $\mathcal{A}\text{-}\mathtt{Cox}$ in Proposition 2.4 to identify $\mathcal{A}\text{-}\mathtt{Cox}^{\mathtt{d}}_{\mathtt{c}}$ and $\mathcal{A}\text{-}\mathtt{Cox}$ with the full subcategories of the Karoubi envelope and Karoubi groupoid of $\mathrm{WJ}[\mathcal{A}]$ consisting of objects of the form (F,F,e). Moreover, these yield equivalences since objects of the form (F,F,e) cover all isomorphism classes in $\mathsf{K}(\mathrm{WJ}[\mathcal{A}])$. More precisely, $(F,F,e):(F,F,e)\to(F,F',e)$ is an isomorphism with inverse $(F,F',e):(F,F',e)\to(F,F,e)$.

3.14. Antipodes of Coxeter bimonoids

For a scalar q, we define the antipode of a Coxeter q-bimonoid using the Takeuchi formula, and study its basic properties. This section may be supplemented with the parallel more detailed discussion for antipodes of bimonoids in species in [**11**, Chapter 12].

3.14.1. Takeuchi formula. Let (h,μ,Δ) be a Coxeter q-bimonoid for a scalar q. Define its *antipode*

$$S:\mathsf{h}\to\mathsf{h}$$

with A-component given by the linear map

$$\big(S_A:\mathsf{h}[A]\to\mathsf{h}[A]\big) := \sum_{F:\,F\ge A} (-1)^{\dim(F)}\mu_A^F\Delta_A^F. \tag{3.159}$$

We refer to (3.159) as the *Takeuchi formula*. It may also be expressed as

$$S_A = (-1)^{\dim(A)}\,\mathbf{log}_0(\mathrm{id})_A, \tag{3.160}$$

where the rhs refers to the 0-logarithm of the identity map (3.114).

We write S_{h} instead of S if we need to show dependence on h.

Exercise 3.93 (Antipode in terms of flats and lunes). Check: The Z-component of the antipode S is given by the linear map

$$(S_{\mathrm{Z}} : \mathsf{h}[\mathrm{Z}] \to \mathsf{h}[\mathrm{Z}]) = \sum_{\mathrm{Y}:\, \mathrm{Y} \geq \mathrm{Z}} (-1)^{\dim(\mathrm{Y})} \sum_{\mathrm{L}:\, \mathrm{b}(\mathrm{L})=\mathrm{Z},\, \mathrm{c}(\mathrm{L})=\mathrm{Y}} \mu_{\mathrm{L}} \Delta_{\mathrm{L}}. \tag{3.161}$$

Exercise 3.94. Check: For a Coxeter q-bimonoid h, the antipode $S : \mathsf{h} \to \mathsf{h}$ is a map of Coxeter species. Also, a morphism of Coxeter q-bimonoids $\mathsf{h} \to \mathsf{h}'$ necessarily commutes with antipodes of h and h'.

Observation 3.95 (Signature functors). The antipode of a Coxeter q-bimonoid h commutes with the u-signature functor (2.70) and o-signature functor (2.73), that is,

$$S_{\mathsf{h}_{-}} = (S_{\mathsf{h}})_{-} \quad \text{and} \quad S_{\mathsf{h}^{-}} = (S_{\mathsf{h}})^{-}. \tag{3.162}$$

This is clear from (3.159) since product and coproduct components of h_{-} and h^{-} are the same as those of h.

3.14.2. Op and cop constructions.

Lemma 3.96. *For a commutative Coxeter bimonoid* h, *the antipode* S *is a morphism of Coxeter monoids. That is, for faces* $F \geq A$, *the first diagram below commutes.*

$$\begin{array}{ccc} \mathsf{h}[F] & \xrightarrow{S_F} & \mathsf{h}[F] \\ {\scriptstyle \mu_A^F}\downarrow & & \downarrow{\scriptstyle \mu_A^F} \\ \mathsf{h}[A] & \xrightarrow[S_A]{} & \mathsf{h}[A] \end{array} \qquad \begin{array}{ccc} \mathsf{h}[F] & \xrightarrow{S_F} & \mathsf{h}[F] \\ {\scriptstyle \Delta_A^F}\uparrow & & \uparrow{\scriptstyle \Delta_A^F} \\ \mathsf{h}[A] & \xrightarrow[S_A]{} & \mathsf{h}[A] \end{array} \tag{3.163}$$

Dually, for a cocommutative Coxeter bimonoid h, *the antipode* S *is a morphism of Coxeter comonoids. That is, for faces* $F \geq A$, *the second diagram above commutes.*

PROOF. This can be deduced from the corresponding result for species [**11**, Lemma 12.10]. □

This result can be reformulated using (3.118) and (3.120) as follows.

Lemma 3.97. *For a commutative Coxeter bimonoid* h, *the antipode* S *is a group-like series of* $\overline{\mathcal{C}}(\mathsf{h}, \mathsf{h})$. *For a cocommutative Coxeter bimonoid* h, *the antipode* S *is a group-like series of* $\mathcal{C}(\mathsf{h}, \mathsf{h})$.

A more general result is stated below. It follows from the corresponding result for species [**11**, Lemma 12.12]. It makes use of the op and cop constructions mentioned in Section 2.11.3.

Lemma 3.98. *For* $q \neq 0$, *let* h *be a Coxeter* q*-bimonoid with antipode* S. *Then:*

(1) *$S : {}^{\mathrm{op}}\mathsf{h} \to \mathsf{h}$ is a morphism of Coxeter monoids. Equivalently, for faces $F \geq A$, the diagram*

$$\begin{array}{ccccc} \mathsf{h}[F] & & \xrightarrow{\qquad \mu_A^F \qquad} & & \mathsf{h}[A] \\ {\scriptstyle S_F}\big\downarrow & & & & \big\downarrow{\scriptstyle S_A} \\ \mathsf{h}[F] & \xrightarrow[(\beta_q)_{A\overline{F},F}]{} & \mathsf{h}[A\overline{F}] & \xrightarrow[\mu_A^{A\overline{F}}]{} & \mathsf{h}[A] \end{array} \tag{3.164}$$

commutes.

(2) *$S : \mathsf{h} \to \mathsf{h}^{\mathrm{cop}}$ is a morphism of Coxeter comonoids. Equivalently, for faces $F \geq A$, the diagram*

$$\begin{array}{ccccc} \mathsf{h}[A] & & \xrightarrow{\qquad \Delta_A^F \qquad} & & \mathsf{h}[F] \\ {\scriptstyle S_A}\big\downarrow & & & & \big\downarrow{\scriptstyle S_F} \\ \mathsf{h}[A] & \xrightarrow[\Delta_A^{A\overline{F}}]{} & \mathsf{h}[A\overline{F}] & \xrightarrow[(\beta_{q^{-1}})_{F,A\overline{F}}]{} & \mathsf{h}[F] \end{array} \tag{3.165}$$

commutes.

For $q = 1$, (3.164) specializes to the first diagram in (3.163) when h is commutative, and (3.165) specializes to the second diagram in (3.163) when h is cocommutative.

Exercise 3.99. For $q \neq 0$, let h be a Coxeter q-bimonoid. Check:

(1) Antipodes of h, $\mathsf{h}^{\mathrm{op,cop}}$, ${}^{\mathrm{op,cop}}\mathsf{h}$ are all equal.

(2) The antipode of h is a Coxeter q-bimonoid morphism in two ways, namely, $\mathsf{h} \to \mathsf{h}^{\mathrm{op,cop}}$ and ${}^{\mathrm{op,cop}}\mathsf{h} \to \mathsf{h}$. (Use Lemma 3.98.)

Lemma 3.100. *For $q \neq 0$, for h a Coxeter q-bimonoid, S_{h} and $S_{{}^{\mathrm{op}}\mathsf{h}}$ (antipodes of h and ${}^{\mathrm{op}}\mathsf{h}$) are inverses under composition, that is, $S_{\mathsf{h}} S_{{}^{\mathrm{op}}\mathsf{h}} = \mathrm{id} = S_{{}^{\mathrm{op}}\mathsf{h}} S_{\mathsf{h}}$. In particular, S_{h} is invertible. The same result holds replacing ${}^{\mathrm{op}}\mathsf{h}$ with $\mathsf{h}^{\mathrm{cop}}$.*

PROOF. This can be deduced from the corresponding result for species [**11**, Lemma 12.16]. □

As a special case:

Lemma 3.101. *For a commutative or cocommutative Coxeter bimonoid, the antipode S is an involution under composition, that is, $SS = \mathrm{id}$.*

Recall: Op and cop constructions work in the more general setting of Coxeter b-species. The preceding results also work in this generality.

3.14.3. Logarithm of the antipode. There is a close connection between logarithm of the identity and logarithm of the antipode.

Lemma 3.102. *For a scalar q, for a Coxeter q-bimonoid h, for a face A,*

$$\mathbf{log}(S_{\mathsf{h}})_A = (-1)^{\dim(A)}\, \overline{\mathbf{log}}(\mathrm{id}_{\mathsf{h}})_A, \tag{3.166}$$

where $\overline{\mathbf{log}}$ refers to logarithm taken wrt $\overline{\boldsymbol{\mu}}$.

PROOF. This can be deduced from the corresponding result for species [**11**, Lemma 12.24]. The proof makes use of identity (1.119b) related to the noncommutative Zaslavsky formula. □

3.14.4. Antipodes of (co)free Coxeter bimonoids. We now write down cancellation-free formulas for antipodes of the universal constructions in Section 3.3. Interestingly, they involve the antipodal map on faces.

Theorem 3.103. *For a scalar q, for a Coxeter comonoid c, the antipode S of the Coxeter q-bimonoid $\mathcal{T}_q(\mathsf{c})$ on the A-component, on the F-summand, is*

$$S_A = q^{\mathrm{dist}(F,A\overline{F})} \sum_{G\colon A\overline{F}\leq G} (-1)^{\dim(G)} \Delta^G_{A\overline{F}} \beta_{A\overline{F},F}. \tag{3.167}$$

In particular, for a trivial Coxeter comonoid c, on the F-summand,

$$S_A = (-1)^{\dim(F)} q^{\mathrm{dist}(F,A\overline{F})}\, \beta_{A\overline{F},F}. \tag{3.168}$$

Dually:

Theorem 3.104. *For a scalar q, for a Coxeter monoid a, the antipode S of the Coxeter q-bimonoid $\mathcal{T}^\vee_q(\mathsf{a})$ on the A-component, on the G-summand, is*

$$S_A = (-1)^{\dim(G)} \sum_{F\colon A\leq F\leq A\overline{G}} q^{\mathrm{dist}(F,A\overline{F})}\, \beta_{F,A\overline{F}} \mu^G_{A\overline{F}}. \tag{3.169}$$

In particular, for a trivial Coxeter monoid a, on the G-summand,

$$S_A = (-1)^{\dim(G)} q^{\mathrm{dist}(G,A\overline{G})}\, \beta_{A\overline{G},G}. \tag{3.170}$$

These results follow from corresponding results for species [**11**, Theorems 12.52 and 12.53].

3.15. Notes

The theory developed in this chapter generalizes the classical theory of connected Joyal bimonoids. It builds on the theory of bimonoids in species developed in detail in [**11**, Part II]. The context for the latter is hyperplane arrangements (as opposed to reflection arrangements).

3.15.1. Cauchy powers. (Section 3.1.) The k-th Cauchy power of a Coxeter species (3.1) generalizes the k-th Cauchy power of a Joyal species. The latter is taken wrt the Cauchy product, denoted $\cdot$, which is a monoidal structure on Joyal species. Similarly, (3.2) and (3.3) generalize the iterated coproduct of a Joyal comonoid and iterated product of a Joyal monoid, respectively. See Table 3.7 for a comparison.

3.15.2. Primitive and decomposable filtrations. (Section 3.2.) The primitive filtration of a Coxeter comonoid generalizes the coradical filtration of a positive Joyal comonoid discussed in [**7**, Section 8.10]. Proposition 3.2 on injectivity generalizes [**7**, Proposition 8.46]. The primitive part of a connected Joyal comonoid and indecomposable part of a connected Joyal monoid were discussed in [**8**, Sections 5.5 and 5.6]. The indecomposable part functor was not explicitly considered in [**7**].

TABLE 3.7. Cauchy powers of Coxeter species and Joyal species.

Coxeter species p k-th Cauchy power p^k	Joyal species p k-th Cauchy power $\mathsf{p}^{\cdot k}$
Coxeter comonoid c $\Delta^{k-1} : \mathsf{c} \to \mathsf{c}^k$	Joyal comonoid c $\Delta^{k-1} : \mathsf{c} \to \mathsf{c}^{\cdot k}$
Coxeter monoid a $\mu^{k-1} : \mathsf{a}^k \to \mathsf{a}$	Joyal monoid a $\mu^{k-1} : \mathsf{a}^{\cdot k} \to \mathsf{a}$

3.15.3. Universal constructions. (Section 3.3.) The universal constructions for Coxeter species generalize corresponding constructions for Joyal species. Joyal described the free Joyal monoid and free commutative Joyal monoid in [**202**, Section 7, Examples 42 and 43]. These and related constructions such as the free Joyal bimonoid on a positive Joyal comonoid were discussed in our monograph [**7**, Chapter 11]. A more complete dictionary between objects in Table 3.2 and those in [**7**, Chapter 11] can be worked out using the information given on [**11**, Notes to Chapter 6, page 282].

The connection of classical (quasi)shuffles to the Tits product was explained in our monographs [**6**, page 125, Formula (6.33) and Remark], [**7**, Section 10.11.3], [**10**, Sections 6.5.4 and 6.5.5]. The significance of the q-distance function on faces for understanding q-deformations was brought forth in [**7**]; it is defined in [**7**, Section 10.5.3] and then widely used in the examples in [**7**, Chapter 12].

3.15.4. Abelianization. (Section 3.4.) Abelianization of Coxeter monoids generalizes a similar construction for Joyal monoids. The latter is mentioned in [**8**, Section 2.6]; it was not explicitly considered in [**7**]. Section 3.4.4 on (co)abelianization of universal constructions on Coxeter species generalizes a similar discussion for Joyal species in [**7**, Sections 11.6.2 and 11.7.4], [**8**, Section 7.6].

3.15.5. Hadamard product. (Sections 3.5, 3.6, 3.7.) The Hadamard product of Coxeter species (3.53) generalizes the Hadamard product of Joyal species. An early reference for the latter is by Smirnov [**326**, Section 1, page 577]. It was studied in our monograph [**7**, Section 8.13]; the definition is stated in [**7**, Formula (8.7)]. See also [**9**, Section 3.1], [**62**, Definition 6.3.1.3], [**234**, Section 5.1.12], [**317**, (5.1.4.1)]. For the set-theoretic context, see [**36**, Section 2.1, Formula (22)], [**259**, Section 3.2.2]. The classical analogue of Theorem 3.25 is [**7**, Propositions 8.58 and 9.5]. The analogue of Section 3.5.5 on the internal hom is [**7**, Section 8.13.3]. Sections 3.6 and 3.7 in the context of Joyal species were not considered in [**7**]. However, some ideas related to the copower $\mathsf{c} \triangleright \mathsf{a}$ are present in [**7**, Section 8.13.4]; one needs to lift the adjunction in [**7**, Proposition 8.67] to Joyal monoids.

The Hadamard product for species in hyperplane arrangements was studied in depth in [**11**, Chapters 8 and 15]. The operations (3.57) and (3.58) were defined in [**11**, (15.1) and (15.2)]; for Joyal species, the former was defined in [**9**, (3.1) and (3.2)]. Exercise 3.31 and Exercise 3.34 on q-deformations are however new even in the hyperplane setting.

For more detailed references on the Hadamard product, see [**11**, Notes to Chapters 8 and 15].

3.15.6. Exp-log correspondences. (Sections 3.8, 3.9, 3.10.) Series of Joyal species were treated in detail in our monograph [**8**, Section 12]. The bijections (3.125) and (3.126) for a Coxeter monoid and Coxeter bimonoid generalize similar bijections for a Joyal monoid and Joyal bimonoid [**8**, (188) and (189)]. Theorem 3.37 generalizes [**8**, Theorem 78]. Proposition 3.46 generalizes [**8**, Corollary 80]. In [**8**], we worked with the exponential and logarithmic power series; one may work more generally with zeta and Möbius series. Warning: Among exponential and equi-exponential series for Joyal monoids, only the latter were explicitly considered in [**8**, Section 12.2] (under the name exponential series). The condition in Exercise 3.68, item (2) generalizes [**8**, Condition (161)].

Exp-log correspondences for species in hyperplane arrangements were studied in depth in [**11**, Chapter 9]. Exercise 3.60 on q-deformations is however new. The exponential series considered in [**11**, Exercise 9.131] relate to the equi-exponential series in Section 3.10.8. Note very carefully: The exponential series in Section 3.10.8 are specific to Coxeter species since (3.131) invokes face-types.

For more detailed references on exp-log, see [**11**, Notes to Chapter 9].

3.15.7. Norm map. (Section 3.11.) The q-norm transformation (3.143) on Coxeter q-bimonoids generalizes the q-norm transformation on Joyal q-bimonoids [**7**, Sections 11.6.3 and 11.7.5], Lemma 3.78 generalizes [**7**, Theorem 11.35], diagrams (3.145) generalize [**7**, Diagrams (11.26) and (11.31)]. Sections 3.11.3 and 3.11.4 on the general q-norm map in the context of Joyal species were not considered in [**7**].

3.15.8. Characteristic operations. (Section 3.12.) Characteristic operations on Coxeter bimonoids generalize characteristic operations on Joyal bimonoids [**8**, Section 13]. Formulas (3.149) and (3.150) generalize [**8**, (207) and (208)]. In more detail: For a Joyal bimonoid h, for F a composition of I and $z \in \mathsf{h}[I]$, set

$$\mathtt{H}_F \cdot z := \mu^F \Delta^F(z), \tag{3.171}$$

where μ^F and Δ^F are iterated product and coproduct components of h. Two-sided characteristic operations on Joyal bimonoids were not considered in [**8**].

3.15.9. Antipode. (Section 3.14.) Formula (3.159) generalizes the Takeuchi formula for the antipode of a connected Joyal q-bimonoid [**7**, Proposition 8.13], [**8**, Section 5.3]. Formula (3.159) coincides with the one for bimonoids in $\mathcal{A}$-species in [**11**, Formula (12.1)]; formula (3.161) in terms of flats and lunes however is new. Theorems 3.103 and 3.104 generalize [**7**, Theorems 11.38 and 11.39].

Results on op and cop constructions in Section 3.14.2 work more generally for Coxeter b-species. Classically, this is related to the fact that such results work for Hopf monoids in any braided monoidal category (as opposed to a symmetric monoidal category), see for instance [**7**, Propositions 1.22 and 1.23, Corollary 1.24].

CHAPTER 4

Examples of Coxeter bimonoids

We discuss some important basic examples of Coxeter bimonoids. These include the exponential Coxeter bimonoid and Coxeter bimonoids constructed out of chambers, flats, faces of the arrangement. Their support and type morphisms, and products and coproducts have simple explicit formulas (involving the Tits product). These Coxeter bimonoids usually exhibit some freeness or cofreeness property which explains the abundance of morphisms from or into them. Some specific features are highlighted below.

- The exponential Coxeter bimonoid has two signed analogues which link to the u-signature and o-signature functors.
- Zeta and Möbius functions in their various avatars enter into the study of the Coxeter bimonoids of flats, faces, top-nested faces.
- Coxeter Lie species and Coxeter Zie species arise as primitive parts of the Coxeter comonoids of chambers and faces; this is how Lie theory enters into the picture.
- There is a decoration functor on Coxeter species constructed out of a module M over the Coxeter group W. We think of M as the space of decorations.

The above Coxeter bimonoids are Coxeter analogues of bimonoids in species studied in [**11**, Chapter 7]. Interestingly, support morphisms, products, coproducts (and hence antipodes) are defined by the same formulas as before. So most of the work is already done. The extra part is to understand type morphisms (connected to the Coxeter group W) and their interactions with the rest of the structure. For instance, in the decorated setting, one starts with a W-module M (as opposed to a vector space M).

4.1. Coxeter species characteristic of chambers

We begin with the Coxeter species x characteristic of chambers, and its decorated version x_M. The construction of x_M is functorial in M and embeds the category of left W-modules as a reflective and coreflective full subcategory of Coxeter species.

4.1.1. Coxeter species characteristic of chambers. Define the Coxeter species x by

$$\mathsf{x}[F] := \begin{cases} \Bbbk & \text{if } F \text{ is a chamber,} \\ 0 & \text{otherwise.} \end{cases} \tag{4.1}$$

Support and type morphisms are defined to be all identities, so diagrams (2.12a) and (2.12b) commute. (Note: On the zero space, the identity map is the same as the zero map.) This is the *Coxeter species characteristic of chambers*. It can be phrased in a similar manner in terms of flats; the relevant diagrams are (2.15).

4.1.2. Decorated version. More generally, for a left W-module M, define the Coxeter species x_M by

$$\mathsf{x}_M[F] := \begin{cases} M & \text{if } F \text{ is a chamber,} \\ 0 & \text{otherwise.} \end{cases} \tag{4.2}$$

Any support morphism β is the identity map, while any type morphism w is either zero or given by action of w on M. In other words,

$$\begin{array}{ccc} \mathsf{x}_M[C] & \xrightarrow{\beta_{D,C}} & \mathsf{x}_M[D] \\ \| & & \| \\ M & \xrightarrow[\text{id}]{} & M \end{array} \qquad \begin{array}{ccc} \mathsf{x}_M[C] & \xrightarrow{w(C)=D} & \mathsf{x}_M[D] \\ \| & & \| \\ M & \xrightarrow[w\cdot(-)]{} & M \end{array}$$

and the remaining maps are zero. In terms of flats,

$$\mathsf{x}_M[\mathrm{X}] := \begin{cases} M & \text{if } \mathrm{X} = \top, \\ 0 & \text{otherwise.} \end{cases} \tag{4.3}$$

When M is the trivial representation of W, we recover x. We view x_M as a decorated version of x, with M as the space of decorations.

Exercise 4.1. There is another way to define x_M on faces. To avoid confusion, we denote it by x'_M. We start with (4.2) as before, but support and type morphisms are

$$\begin{array}{ccc} \mathsf{x}'_M[C] & \xrightarrow{\beta_{D,C}} & \mathsf{x}'_M[D] \\ \| & & \| \\ M & \xrightarrow[d(D,C)\cdot(-)]{} & M \end{array} \qquad \begin{array}{ccc} \mathsf{x}'_M[C] & \xrightarrow{w(C)=D} & \mathsf{x}'_M[D] \\ \| & & \| \\ M & \xrightarrow[\text{id}]{} & M \end{array}$$

where d is the W-valued gallery distance in (1.12) (which depends on C_r). Check: x'_M is indeed a Coxeter species. Moreover,

$$\mathsf{x}'_M[C] \xrightarrow{u\cdot(-)} \mathsf{x}_M[C]$$

with $u(C_r) = C$ defines an isomorphism of Coxeter species from x'_M to x_M.

4.1.3. Categories of W-modules and Coxeter species. The construction of x_M is functorial in M. That is, a map $M \to M'$ of left W-modules gives rise to a map $\mathsf{x}_M \to \mathsf{x}_{M'}$ of Coxeter species. This yields a functor from the category of left W-modules to the category of Coxeter species

$$W\text{-}\mathsf{Mod} \to \mathcal{A}\text{-}\mathsf{CSp}, \qquad M \mapsto \mathsf{x}_M. \tag{4.4}$$

Moreover, this functor is full and faithful.

Exercise 4.2. Recall the functor (2.21) from Coxeter species to left W-modules which sends p to $\mathsf{p}[\top]$. Show: The functor (4.4) is both the left and right adjoint of (2.21). That is,

- a map of Coxeter species $\mathsf{x}_M \to \mathsf{p}$ is the same as a map of left W-modules $M \to \mathsf{p}[\top]$,
- a map of Coxeter species $\mathsf{p} \to \mathsf{x}_M$ is the same as a map of left W-modules $\mathsf{p}[\top] \to M$.

The functor (4.4) yields an isomorphism of the category W-Mod with the (reflective and coreflective) full subcategory of $\mathcal{A}$-CSp (formulated in terms of flats) consisting of those Coxeter species p for which $\mathsf{p}[\mathrm{X}] = 0$ whenever $\mathrm{X} \neq \top$.

Exercise 4.3. For a left W-module M, define a Coxeter species with $\bot$-component M and remaining components zero. This is similar to (4.3) with $\bot$ instead of $\top$. Check: This yields a functor from the category of left W-modules to the category of Coxeter species which is both the left and right adjoint of (2.20).

Observation 4.4. For a Coxeter comonoid c and Coxeter monoid a, any map of Coxeter species $f : \mathsf{x}_M \to \mathsf{c}$ is a coderivation, and any map of Coxeter species $g : \mathsf{a} \to \mathsf{x}_M$ is a derivation.

Exercise 4.5 (Duality). Check: $\mathsf{x}^* = \mathsf{x}$. More generally, for a left W-module M, we have $(\mathsf{x}_M)^* = \mathsf{x}_{M^*}$.

Exercise 4.6 (Signature functors). Recall the u-signature and o-signature functors on Coxeter species from Section 2.8. Check:

$$(\mathsf{x}_M)^- \cong \mathsf{x}_M \quad \text{and} \quad (\mathsf{x}_M)_- \cong \mathsf{x}_{M_-},$$

with both isomorphisms, on the C-component, given by scalar multiplication by $(-1)^{\mathrm{dist}(C_0,C)}$, where C_0 is a fixed chamber. For the second one, diagrams (2.13) take the form

$$\begin{array}{ccc} (\mathsf{x}_M)_-[C] & \xrightarrow{(-1)^{\mathrm{dist}(C_0,C)}} & \mathsf{x}_{M_-}[C] \\ \big\downarrow {\scriptstyle (-1)^{\mathrm{dist}(C,D)}} & & \big\downarrow {\scriptstyle \mathrm{id}} \\ (\mathsf{x}_M)_-[D] & \xrightarrow[(-1)^{\mathrm{dist}(C_0,D)}]{} & \mathsf{x}_{M_-}[D] \end{array} \qquad \begin{array}{ccc} (\mathsf{x}_M)_-[C] & \xrightarrow{(-1)^{\mathrm{dist}(C_0,C)}} & \mathsf{x}_{M_-}[C] \\ \big\downarrow {\scriptstyle w} & & \big\downarrow {\scriptstyle (-1)^{l(w)} w} \\ (\mathsf{x}_M)_-[wC] & \xrightarrow[(-1)^{\mathrm{dist}(C_0,wC)}]{} & \mathsf{x}_{M_-}[wC]. \end{array}$$

4.2. Exponential Coxeter bimonoid

We introduce the exponential Coxeter bimonoid E and its two signed analogues E^- and E_-. The former is bicommutative, the latter two are signed bicommutative. Further, the Hadamard product of E_- and E^- yields an unsigned analogue different from E, which we denote E_-^-. It is bicommutative. A summary of these objects is provided in Table 4.1; one can appreciate the subtle differences among them by comparing their minimum and maximum flat components.

TABLE 4.1. Exponential Coxeter bimonoid and its analogues.

E	Coxeter bimonoid	$\mathsf{E}[\bot]$ trivial rep.	$\mathsf{E}[\top]$ trivial rep.
E^-	signed Coxeter bimonoid	$\mathsf{E}^-[\bot]$ sign rep.	$\mathsf{E}^-[\top]$ trivial rep.
E_-	signed Coxeter bimonoid	$\mathsf{E}_-[\bot]$ trivial rep.	$\mathsf{E}_-[\top]$ sign rep.
E_-^-	Coxeter bimonoid	$\mathsf{E}_-^-[\bot]$ sign rep.	$\mathsf{E}_-^-[\top]$ sign rep.

Taking Hadamard product of a Coxeter species with E_- and E^- recovers the u-signature and o-signature functors defined in Section 2.8.

4.2.1. Exponential Coxeter species. We define the *exponential Coxeter species* E by setting $\mathsf{E}[A] := \Bbbk$ for any face A. Support and type morphisms

$$\beta_{G,F} : \mathsf{E}[F] \to \mathsf{E}[G] \quad \text{and} \quad w : \mathsf{E}[F] \to \mathsf{E}[wF] \tag{4.5}$$

are each defined to be the identity map $\Bbbk \to \Bbbk$. Note: For each A, $\mathsf{E}[A]$ is the trivial representation of the group W_A. For each $F \geq A$, define

$$\mu_A^F : \mathsf{E}[F] \to \mathsf{E}[A] \quad \text{and} \quad \Delta_A^F : \mathsf{E}[A] \to \mathsf{E}[F] \tag{4.6}$$

to be the identity map $\Bbbk \to \Bbbk$. This turns E into a Coxeter bimonoid. We call it the *exponential Coxeter bimonoid.*

Observe: E is bicommutative. So, in view of Proposition 2.43, one can express all the above in terms of flats instead of faces as follows. We set $\mathsf{E}[\mathrm{X}] := \Bbbk$ for any flat X, and

$$w : \mathsf{E}[\mathrm{X}] \to \mathsf{E}[\mathrm{X}'], \quad \mu_{\mathrm{Z}}^{\mathrm{X}} : \mathsf{E}[\mathrm{X}] \to \mathsf{E}[\mathrm{Z}], \quad \Delta_{\mathrm{Z}}^{\mathrm{X}} : \mathsf{E}[\mathrm{Z}] \to \mathsf{E}[\mathrm{X}] \tag{4.7}$$

to be the identity maps.

4.2.2. Signed exponential Coxeter species (over version). The exponential Coxeter bimonoid has a signed analogue, which we denote E^-. All its components are $\Bbbk$, support and type morphisms are

$$\begin{aligned} \beta_{G,F} : \mathsf{E}^-[F] &\to \mathsf{E}^-[G] & w : \mathsf{E}^-[F] &\to \mathsf{E}^-[wF] \\ 1 &\mapsto (-1)^{\mathrm{dist}(F,G)} & 1 &\mapsto (-1)^{l(w)}, \end{aligned} \tag{4.8}$$

and product and coproduct components are identities. This is a signed Coxeter bimonoid. We call it the *signed exponential Coxeter bimonoid.* It is signed bicommutative. Note: For each A, $\mathsf{E}^-[A]$ is the sign representation of the group W_A.

Another way to define E^- is via signature spaces as follows. The component $\mathsf{E}^-[A]$ is now the linear span of $\mathtt{H}_{C/A}$, with C a chamber greater than A, subject to the relations

$$\mathtt{H}_{C/A} = (-1)^{\mathrm{dist}(C,D)}\, \mathtt{H}_{D/A}$$

for $C, D \geq A$. It is one-dimensional. We write $\mathtt{H}_{[C/A]}$ for the class of $\mathtt{H}_{C/A}$. Any such element is a basis of $\mathsf{E}^-[A]$. Support and type morphisms are

$$\begin{aligned} \beta_{G,F} : \mathsf{E}^-[F] &\to \mathsf{E}^-[G] & w : \mathsf{E}^-[F] &\to \mathsf{E}^-[wF] \\ \mathtt{H}_{[C/F]} &\mapsto \mathtt{H}_{[GC/G]} & \mathtt{H}_{[C/F]} &\mapsto \mathtt{H}_{[wC/wF]}, \end{aligned} \tag{4.9}$$

and product and coproduct components are

$$\begin{aligned} \mu_A^F : \mathsf{E}^-[F] &\to \mathsf{E}^-[A] & \Delta_A^F : \mathsf{E}^-[A] &\to \mathsf{E}^-[F] \\ \mathtt{H}_{[C/F]} &\mapsto \mathtt{H}_{[C/A]} & \mathtt{H}_{[C/A]} &\mapsto (-1)^{\mathrm{dist}(C,FC)}\, \mathtt{H}_{[FC/F]}. \end{aligned} \tag{4.10}$$

In terms of flats, this amounts to setting

$$\mathsf{E}^-[\mathrm{X}] := \mathbf{E}^-[\mathrm{X}, \top], \tag{4.11}$$

the signature space over X as in Definition 2.48, and proceeding as on [**11**, page 287].

Exercise 4.7. To link the two definitions of E^-, fix a chamber C_0 and let

$$\mathsf{E}^-[F] \to \mathsf{E}^-[F], \qquad \mathtt{H}_{[C/F]} \mapsto (-1)^{\mathrm{dist}(C_0,C)}.$$

Verify that this is indeed an isomorphism of signed Coxeter bimonoids.

4.2.3. Signed exponential Coxeter species (under version). The exponential Coxeter bimonoid has another signed analogue, which we denote E_-. All its components are $\Bbbk$, support and type morphisms are

$$\begin{aligned} \beta_{G,F} : \mathsf{E}_-[F] &\to \mathsf{E}_-[G] & \qquad w : \mathsf{E}_-[F] &\to \mathsf{E}_-[wF] \\ 1 &\mapsto (-1)^{\mathrm{dist}(F,G)} & 1 &\mapsto 1, \end{aligned} \tag{4.12}$$

and product and coproduct components are identities. This is a signed Coxeter bimonoid which is signed bicommutative. One may also formulate E_- in terms of flats by setting

$$\mathsf{E}_-[\mathrm{X}] := \mathbf{E}^-[\bot, \mathrm{X}], \tag{4.13}$$

the signature space under X, and proceeding.

4.2.4. Exponential Coxeter species (under and over version). The exponential Coxeter bimonoid also has an unsigned analogue E_-^- as follows. All its components are $\Bbbk$, support and type morphisms are

$$\begin{aligned} \beta_{G,F} : \mathsf{E}_-^-[F] &\to \mathsf{E}_-^-[G] & \qquad w : \mathsf{E}_-^-[F] &\to \mathsf{E}_-^-[wF] \\ 1 &\mapsto 1 & 1 &\mapsto (-1)^{l(w)}, \end{aligned} \tag{4.14}$$

and product and coproduct components are identities. This is a bicommutative Coxeter bimonoid. In terms of flats, we set

$$\mathsf{E}_-^-[\mathrm{X}] := \mathbf{E}^-[\bot, \top]. \tag{4.15}$$

The two definitions of E_-^- can be related by fixing a chamber C_0 and setting

$$\beta_{\mathrm{X},F} : \mathsf{E}_-^-[F] \to \mathsf{E}_-^-[\mathrm{X}], \qquad 1 \mapsto \mathtt{H}_{[C_0/O]}.$$

Also possible is to set $\mathsf{E}_-^-[\mathrm{X}] := \Bbbk$ with type morphism w given by scalar multiplication by $(-1)^{l(w)}$ and product and coproduct components as identities.

4.2.5. Hadamard products. Recall from Section 3.5 the Hadamard product of Coxeter species. Observe:

$$\begin{gathered} \mathsf{E} \times \mathsf{E} = \mathsf{E}^- \times \mathsf{E}^- = \mathsf{E}_- \times \mathsf{E}_- = \mathsf{E}_-^- \times \mathsf{E}_-^- = \mathsf{E}, \\ \mathsf{E} \times \mathsf{E}^- = \mathsf{E}_- \times \mathsf{E}_-^- = \mathsf{E}^-, \qquad \mathsf{E} \times \mathsf{E}_- = \mathsf{E}^- \times \mathsf{E}_-^- = \mathsf{E}_-, \\ \mathsf{E} \times \mathsf{E}_-^- = \mathsf{E}_- \times \mathsf{E}^- = \mathsf{E}_-^-. \end{gathered} \tag{4.16}$$

(Two minus signs either in the subscript or in the superscript cancel each other out.) The last equality above, in terms of flats, on the X-component, using (4.11), (4.13), (4.15), amounts to an identification

$$\mathbf{E}^-[\bot, \mathrm{X}] \otimes \mathbf{E}^-[\mathrm{X}, \top] \xrightarrow{\cong} \mathbf{E}^-[\bot, \top].$$

For more context on this map, see (5.16).

Exercise 4.8. Refer to Exercise 2.19. Check entries in the last two columns in Table 4.1 pertaining to W-representations. Deduce: $\mathsf{E}, \mathsf{E}^-, \mathsf{E}_-, \mathsf{E}^-_-$ are not isomorphic as Coxeter species (assuming $|W| > 1$ and $\mathrm{char}(\Bbbk) \neq 2$).

4.2.6. Signature functors on Coxeter species. Recall the u-signature functor and o-signature functor on Coxeter species from Section 2.8.

Observation 4.9. For a Coxeter species p,

$$\mathsf{p} = \mathsf{p} \times \mathsf{E}, \quad \mathsf{p}_- = \mathsf{p} \times \mathsf{E}_-, \quad \mathsf{p}^- = \mathsf{p} \times \mathsf{E}^-, \quad \mathsf{p}^-_- = \mathsf{p} \times \mathsf{E}^-_-. \tag{4.17}$$

This also yields descriptions of $\mathsf{p}_-, \mathsf{p}^-, \mathsf{p}^-_-$ in terms of flats; they involve signature spaces under and over flats. Further, (4.17) extends to Coxeter (co, bi)monoids.

The fact that signature functors send Coxeter q-bimonoids to Coxeter $(-q)$-bimonoids can be formally understood as follows. Using (4.17) and the second part of Theorem 3.25, we deduce that the u-signature and o-signature functors are bilax. The latter fact is elaborated below.

Theorem 4.10. *For a scalar q, the o-signature functor*

$$(-)^- : (\mathcal{T}, \mathcal{T}^\vee, \lambda_q) \to (\mathcal{T}, \mathcal{T}^\vee, \lambda_{-q}),$$

with structure maps (4.18) *below, is a bilax isomorphism of bimonads. Moreover, the lax and colax structure maps are inverse isomorphisms (noting that $\mathcal{T}$ and $\mathcal{T}^\vee$ are identical as functors).*

The lax and colax structure maps of $(-)^-$, evaluated on a Coxeter species p on the A-component

$$\bigoplus_{F:\, F \geq A} \mathsf{p}[F] \otimes \mathsf{E}^-[F] \underset{\text{colax}}{\overset{\text{lax}}{\rightleftarrows}} \Big(\bigoplus_{F:\, F \geq A} \mathsf{p}[F] \Big) \otimes \mathsf{E}^-[A] \tag{4.18}$$

are the canonical identifications. In the signature space formulation of E^-, the maps on the F-summand are $\mathrm{id} \otimes \mu^F_A$ and $\mathrm{id} \otimes \Delta^F_A$, respectively, with μ^F_A and Δ^F_A as in (4.10). Since structure maps are isomorphisms, Exercise 3.17 is a formal consequence.

4.2.7. Freeness and cofreeness. Let x be the Coxeter species characteristic of chambers (4.1). Then

$$\begin{aligned} \mathsf{E} &= \mathcal{S}(\mathsf{x}) = \mathcal{S}^\vee(\mathsf{x}), & \mathsf{E}^- &= \mathcal{E}(\mathsf{x}) = \mathcal{E}^\vee(\mathsf{x}), \\ \mathsf{E}^-_- &= \mathcal{S}(\mathsf{x}_-) = \mathcal{S}^\vee(\mathsf{x}_-), & \mathsf{E}_- &= \mathcal{E}(\mathsf{x}_-) = \mathcal{E}^\vee(\mathsf{x}_-) \end{aligned} \tag{4.19}$$

as Coxeter bimonoids and signed Coxeter bimonoids, respectively. Similarly,

$$\mathcal{P}(\mathsf{E}) = \mathsf{x} \quad \text{and} \quad \mathcal{Q}(\mathsf{E}) = \mathsf{x}. \tag{4.20}$$

4.2.8. Primitive and group-like series. A series v of E is a family of scalars (v_{X}) indexed by flats X such that $v_{\mathrm{X}} = v_{w\mathrm{X}}$. Specializing (3.138): The space of series $\mathscr{S}(\mathsf{E})$ is a module over the invariant flat-incidence algebra. A series v of E is primitive iff $v_{\mathrm{X}} = 0$ for all $\mathrm{X} \neq \top$, and group-like iff $v_{\mathrm{X}} = v_{\mathrm{Y}}$ for all X and Y. The correspondence (3.141) between primitive and group-like series is evident. Moreover, for E, every primitive series is linear, and every group-like series is exponential. Compare with [**11**, Example 9.127], where E is viewed as a bimonoid in species.

4.2.9. Antipode. The antipode S of the Coxeter bimonoid E, on the A-component, is given by scalar multiplication by $(-1)^{\dim(\mathcal{A})}$ (which is independent of A). The same is true for antipodes of E^- and E_-, see (3.162) in this regard. The calculation makes use of the Euler characteristic of the sphere. For more detail, see [**11**, proof of formula (12.20)].

4.3. Coxeter bimonoid of chambers

For a scalar q, we introduce the Coxeter q-bimonoid of chambers, its dual, and the q-norm map linking the two. We also discuss primitive and group-like series, and antipode formulas.

4.3.1. Coxeter species of chambers. Define the Coxeter species Γ as follows. For a face A, let $\Gamma[A]$ be the linear span of chambers greater than A. We use the letter $\mathtt{H}$ for the canonical basis of $\Gamma[A]$. Support and type morphisms are

$$\begin{aligned} \beta_{G,F} : \Gamma[F] &\to \Gamma[G] & \qquad w : \Gamma[F] &\to \Gamma[wF] \\ \mathtt{H}_{C/F} &\mapsto \mathtt{H}_{GC/G} & \mathtt{H}_{C/F} &\mapsto \mathtt{H}_{wC/wF}. \end{aligned} \tag{4.21}$$

Note: For each A, $\Gamma[A]$ is the left regular representation of the group W_A. The verification of the support-type axiom boils down to the fact $w(G)w(C) = w(GC)$. We call Γ the *Coxeter species of chambers*.

4.3.2. Coxeter bimonoid of chambers. The Coxeter species of chambers Γ carries a Coxeter bimonoid structure with product and coproduct

$$\begin{aligned} \mu_A^F : \Gamma[F] &\to \Gamma[A] & \qquad \Delta_A^F : \Gamma[A] &\to \Gamma[F] \\ \mathtt{H}_{C/F} &\mapsto \mathtt{H}_{C/A} & \mathtt{H}_{C/A} &\mapsto \mathtt{H}_{FC/F}. \end{aligned} \tag{4.22}$$

We call it the *Coxeter bimonoid of chambers*. It is cocommutative but not commutative in general.

4.3.3. Coxeter q-bimonoid of chambers. More generally, for a scalar q, the Coxeter species of chambers carries a Coxeter q-bimonoid structure, denoted Γ_q, with product and coproduct

$$\begin{aligned} \mu_A^F : \Gamma_q[F] &\to \Gamma_q[A] & \qquad \Delta_A^F : \Gamma_q[A] &\to \Gamma_q[F] \\ \mathtt{H}_{C/F} &\mapsto \mathtt{H}_{C/A} & \mathtt{H}_{C/A} &\mapsto q^{\operatorname{dist}(C,FC)}\, \mathtt{H}_{FC/F}. \end{aligned} \tag{4.23}$$

We call it the *Coxeter q-bimonoid of chambers*. The product is the same as in (4.22), while the coproduct has been deformed using the q-distance function.

4.3.4. Coxeter bimonoid of chambers in terms of flats and lunes. Let us now describe Γ in terms of flats and lunes. For a flat Z, let $\Gamma[\mathrm{Z}]$ be the linear span of top-lunes with base Z. We use the letter $\mathtt{H}$ for the canonical basis of $\Gamma[\mathrm{Z}]$. Type morphisms are

$$w : \Gamma[\mathrm{X}] \to \Gamma[w\mathrm{X}], \qquad \mathtt{H}_{\mathrm{L}} \mapsto \mathtt{H}_{w\mathrm{L}}. \tag{4.24}$$

The product and coproduct are

$$\begin{aligned} \mu_{\mathrm{L}} : \Gamma[\mathrm{X}] &\to \Gamma[\mathrm{Z}] & \qquad \Delta_{\mathrm{Z}}^{\mathrm{X}} : \Gamma[\mathrm{Z}] &\to \Gamma[\mathrm{X}] \\ \mathtt{H}_{\mathrm{M}} &\mapsto \mathtt{H}_{\mathrm{L}\circ\mathrm{M}} & \mathtt{H}_{\mathrm{L}} &\mapsto \mathtt{H}_{\mathrm{X}\cdot\mathrm{L}}. \end{aligned} \tag{4.25}$$

The cocommutative bimonoid axiom (2.60) is checked below.

$$\begin{array}{ccc} \Gamma[\mathrm{X}] & \xrightarrow{\mu_{\mathrm{L}}} & \Gamma[\mathrm{Z}] \\ {\scriptstyle \Delta_{\mathrm{X}}^{\mathrm{X}\vee\mathrm{Y}}}\downarrow & & \downarrow{\scriptstyle \Delta_{\mathrm{Z}}^{\mathrm{Y}}} \\ \Gamma[\mathrm{X}\vee\mathrm{Y}] & \xrightarrow[\mu_{\mathrm{Y}\cdot\mathrm{L}}]{} & \Gamma[\mathrm{Y}] \end{array} \qquad \begin{array}{ccc} \mathtt{H}_{\mathrm{M}} & \longmapsto & \mathtt{H}_{\mathrm{L}\circ\mathrm{M}} \\ \downarrow & & \downarrow \\ \mathtt{H}_{\mathrm{Y}\cdot\mathrm{M}} & \longmapsto & \mathtt{H}_{(\mathrm{Y}\cdot\mathrm{L})\circ(\mathrm{Y}\cdot\mathrm{M})} = \mathtt{H}_{\mathrm{Y}\cdot(\mathrm{L}\circ\mathrm{M})} \end{array}$$

The last equality holds by (1.5).

4.3.5. Dual Coxeter q-bimonoid. For a scalar q, let Γ_q^* denote the Coxeter q-bimonoid dual to Γ_q. Let $\mathtt{M}$ denote the basis dual to the $\mathtt{H}$-basis. Dualizing formulas (4.23) yields the product and coproduct

$$\mu_A^F : \Gamma_q^*[F] \to \Gamma_q^*[A] \qquad\qquad \Delta_A^F : \Gamma_q^*[A] \to \Gamma_q^*[F]$$

$$(4.26)\qquad \mathtt{M}_{D/F} \mapsto \sum_{\substack{C:\, C \geq A \\ FC = D}} q^{\mathrm{dist}(FC,C)}\, \mathtt{M}_{C/A} \qquad\qquad \mathtt{M}_{C/A} \mapsto \begin{cases} \mathtt{M}_{C/F} & \text{if } F \leq C, \\ 0 & \text{otherwise.} \end{cases}$$

Exercise 4.11. Check: Γ and Γ^* are isomorphic as Coxeter species (under canonical identification of $\mathtt{H}$- and $\mathtt{M}$-bases). More generally, any finite-dimensional Coxeter species obtained by linearizing a Coxeter set-species is isomorphic to its dual.

4.3.6. Freeness and cofreeness. Let x be the Coxeter species characteristic of chambers (4.1). For a scalar q,

$$(4.27)\qquad \Gamma_q = \mathcal{T}_q(\mathsf{x}) \quad \text{and} \quad \Gamma_q^* = \mathcal{T}_q^\vee(\mathsf{x})$$

as Coxeter q-bimonoids. The (co)product of $\mathcal{T}_q(\mathsf{x})$ is given by (3.26) and of $\mathcal{T}_q^\vee(\mathsf{x})$ by (3.32) with $\mathsf{p} := \mathsf{x}$.

Exercise 4.12. Recall the o-signature functor (2.73). For a scalar q,

$$(\Gamma_q)^- = \Gamma_{-q} \quad \text{and} \quad (\Gamma_q^*)^- = \Gamma_{-q}^*$$

as Coxeter $(-q)$-bimonoids. Check this directly, and also deduce it from Exercise 3.17 using (4.27).

4.3.7. q-norm map on chambers. For a scalar q, the *q-norm map* on chambers

$$(4.28)\qquad \kappa_q : \Gamma_q \to \Gamma_q^*, \qquad \mathtt{H}_{C/A} \mapsto \sum_{D:\, D \geq A} q^{\mathrm{dist}(C,D)}\, \mathtt{M}_{D/A}$$

on the A-component, is a self-dual morphism of Coxeter q-bimonoids. This is the special case $\mathsf{p} := \mathsf{x}$ of (3.144). Hence by Lemma 3.78, the map (4.28) is an isomorphism when q is not a root of unity. This can also be checked directly.

Exercise 4.13. Let h be a Coxeter q-bimonoid for a scalar q. Show:

(1) If $f : \Gamma_q \to \mathsf{h}$ is a morphism of Coxeter monoids, then it is also a morphism of Coxeter q-bimonoids.
(2) If $g : \mathsf{h} \to \Gamma_q^*$ is a morphism of Coxeter comonoids, then it is also a morphism of Coxeter q-bimonoids.

(For item (1): Apply Theorems 3.7 and 3.11 for $\mathsf{p} := \mathsf{x}$ and use Observation 4.4 for $M := \Bbbk$. Alternatively, check directly by using formula (3.22). Dually, for item (2): Apply Theorems 3.12 and 3.16.)

Exercise 4.14. Check: The map

$$\mathsf{\Gamma} \twoheadrightarrow \mathsf{E}, \qquad \mathtt{H}_{C/A} \mapsto 1 \tag{4.29}$$

on the A-component, is a morphism of Coxeter bimonoids. This is an instance of the abelianization map, see (3.50) for $\mathsf{p} := \mathsf{x}$. For $\mathrm{char}(\Bbbk) \nmid |W|$, the map

$$\mathsf{E} \hookrightarrow \mathsf{\Gamma}, \qquad 1 \mapsto \frac{1}{|W_A|} \sum_{C:\, C \geq A} \mathtt{H}_{C/A} \tag{4.30}$$

on the A-component, is a Coxeter comonoid section to (4.29).

4.3.8. Primitive and group-like series. Specializing (3.122): The space of series $\mathscr{S}(\mathsf{\Gamma})$ is a left module over the invariant lune-incidence algebra. It coincides with the invariant lune-incidence module in [**10**, Sections 15.2.6 and 16.7.4]. The identification is done via

$$(f(A,C))_{A \leq C} \longleftrightarrow \sum_{C:\, C \geq A} f(A,C)\, \mathtt{H}_{C/A} \in \mathsf{\Gamma}[A] \quad \text{for each face } A. \tag{4.31}$$

In the right sum, all coefficients $f(A,C)$ are necessarily identical. Contrast with [**11**, Example 9.118], where $\mathsf{\Gamma}$ is viewed as a bimonoid in species.

Let $\mathrm{char}(\Bbbk) \nmid |W|$. Let x be the linear series of $\mathsf{\Gamma}$ defined by $x_C := \mathtt{H}_{C/C}$. The corresponding exponential series e of $\mathsf{\Gamma}$ under the bijection (3.134) is

$$e_A = \frac{1}{|W_A|} \sum_{C:\, C \geq A} \mathtt{H}_{C/A}. \tag{4.32}$$

This follows by specializing formula (3.135). Any other linear and exponential series of $\mathsf{\Gamma}$ are scalar multiples of x and e. Observe: $\mathsf{\Gamma}$ does not have any nonzero equi-exponential series (assuming $\mathrm{rk}(\mathcal{A}) \geq 1$); this is consistent with Exercise 3.64. Moreover, for $\mathsf{\Gamma}$, every primitive series is linear, and every group-like series is exponential. In other words, $\mathscr{X}(\mathsf{\Gamma}) = \mathscr{P}(\mathsf{\Gamma})$ and $\mathscr{E}(\mathsf{\Gamma}) = \mathscr{G}(\mathsf{\Gamma})$. Thus, for $\mathsf{\Gamma}$, the exp-log correspondence (3.126) coincides with (3.134). This bijection identifies with the one in [**10**, Theorem 16.40]; see also the discussion in Section 10.2.17.

Note: The exponential series (4.32) identifies with the Coxeter comonoid morphism (4.30) under (3.119).

Exercise 4.15. Let $\mathrm{char}(\Bbbk) \nmid |W|$. Show: For $\mathsf{\Gamma}^*$, the formula

$$e_A = \sum_{C:\, C \geq A} \mathtt{M}_{C/A} \tag{4.33}$$

defines an exponential series. Every other exponential series is a scalar multiple of this one. Also, every exponential series of $\mathsf{\Gamma}^*$ is equi-exponential, see Exercise 3.63.

4.3.9. Exponential series of a Coxeter monoid. Let $\mathrm{char}(\Bbbk) \nmid |W|$. For a Coxeter monoid a, we have the commutative diagram of identifications

$$\begin{array}{ccc} \mathsf{Mon}(\mathsf{CSp})(\mathsf{\Gamma}, \mathsf{a}) & \longleftrightarrow & \mathscr{E}(\mathsf{a}) \\ \updownarrow & & \updownarrow \\ \mathsf{CSp}(\mathsf{x}, \mathsf{a}) & \longleftrightarrow & \mathscr{X}(\mathsf{a}). \end{array} \tag{4.34}$$

The top-horizontal one is $f \mapsto e$ with

$$e_A = \frac{1}{|W_A|} \sum_{C:\, C \geq A} f_A(\mathtt{H}_{C/A}). \tag{4.35}$$

(This is the image of the exponential series (4.32) under f.) The bottom-horizontal one is (3.130). The left-vertical one is because Γ is the free monoid on x. The right-vertical one is (3.134).

For a Coxeter bimonoid h, we have

$$\begin{array}{ccccc} \mathsf{Mon}(\mathsf{CSp})(\mathsf{E},\mathsf{h}) & \hookrightarrow & \mathsf{Mon}(\mathsf{CSp})(\Gamma,\mathsf{h}) & \hookrightarrow & \mathsf{Comon}(\mathsf{CSp})(\mathsf{E},\mathsf{h}) \\ \| & & \| & & \| \\ \mathscr{E}^o(\mathsf{h}) & & \mathscr{E}(\mathsf{h}) & & \mathscr{G}(\mathsf{h}) \end{array}$$

with the inclusions induced from (4.29) and (4.30), respectively. The second map makes sense since by Exercise 4.13, item (1), any Coxeter monoid morphism $\Gamma \to \mathsf{h}$ is also a Coxeter comonoid morphism. This adds perspective to the inclusions $\mathscr{E}^o(\mathsf{h}) \subseteq \mathscr{E}(\mathsf{h}) \subseteq \mathscr{G}(\mathsf{h})$.

Exercise 4.16. This is a rephrasing of Exercise 3.69. Let $\mathrm{char}(\Bbbk) \nmid |W|$. For a Coxeter monoid a, we have the string of identifications

$$\mathscr{E}(\mathsf{a}) = \mathsf{Mon}(\mathsf{CSp})(\Gamma,\mathsf{a}) = \mathsf{CSp}(\mathsf{x},\mathsf{a}) = \mathsf{CSp}(\mathsf{x},\mathsf{a}_{ab}) = \mathsf{Mon}(\mathsf{CSp})(\mathsf{E},\mathsf{a}_{ab}) = \mathscr{E}(\mathsf{a}_{ab}).$$

More explicitly, we have the diagram of Coxeter monoids

$$\begin{array}{ccc} \Gamma & \longrightarrow & \mathsf{a} \\ \downarrow & \nearrow & \downarrow \\ \mathsf{E} & \longrightarrow & \mathsf{a}_{ab}, \end{array}$$

where the top- and bottom-horizontal maps correspond to elements of $\mathscr{E}(\mathsf{a})$ and $\mathscr{E}(\mathsf{a}_{ab})$, and the dotted arrow (if it exists) to an element of $\mathscr{E}^o(\mathsf{a})$.

4.3.10. Antipode. The antipode S of the Coxeter q-bimonoid Γ_q, on the A-component, is

$$S_A(\mathtt{H}_{C/A}) = (-1)^{\dim(\mathcal{A})} q^{\mathrm{dist}(C,A\overline{C})}\, \mathtt{H}_{A\overline{C}/A}. \tag{4.36}$$

Dually, the antipode S of Γ_q^*, on the A-component, is

$$S_A(\mathtt{M}_{D/A}) = (-1)^{\dim(\mathcal{A})} q^{\mathrm{dist}(D,A\overline{D})}\, \mathtt{M}_{A\overline{D}/A}. \tag{4.37}$$

See [**11**, Formulas (12.22) and (12.23)].

4.4. Coxeter bimonoid of flats

We introduce the Coxeter bimonoid of flats. It is bicommutative. We also discuss the norm map, primitive and group-like series, and antipode formulas.

4.4.1. Coxeter species of flats. Define the Coxeter species Π as follows. For a flat X, let $\Pi[\mathrm{X}]$ be the linear span of all flats greater than X. We use the letter $\mathtt{H}$ for the canonical basis of $\Pi[\mathrm{X}]$. Type morphisms are

$$w : \Pi[\mathrm{X}] \to \Pi[w\mathrm{X}], \qquad \mathtt{H}_{\mathrm{Y}/\mathrm{X}} \mapsto \mathtt{H}_{w\mathrm{Y}/w\mathrm{X}}. \tag{4.38}$$

This is the *Coxeter species of flats.*

4.4.2. Coxeter bimonoid of flats. The Coxeter species of flats Π carries a bicommutative Coxeter bimonoid structure with product and coproduct

$$\begin{aligned} \mu_{\mathrm{Z}}^{\mathrm{Y}} : \Pi[\mathrm{Y}] &\to \Pi[\mathrm{Z}] & \Delta_{\mathrm{Z}}^{\mathrm{Y}} : \Pi[\mathrm{Z}] &\to \Pi[\mathrm{Y}] \\ \mathtt{H}_{\mathrm{X/Y}} &\mapsto \mathtt{H}_{\mathrm{X/Z}} & \mathtt{H}_{\mathrm{X/Z}} &\mapsto \mathtt{H}_{\mathrm{X} \vee \mathrm{Y/Y}}. \end{aligned} \tag{4.39}$$

We call it the *Coxeter bimonoid of flats*.

4.4.3. $\mathtt{Q}$-basis. We now consider a second basis of the Coxeter bimonoid of flats. Define the $\mathtt{Q}$-basis of Π, on the Z-component, by either of the two equivalent formulas

$$\begin{aligned} \mathtt{H}_{\mathrm{X/Z}} &= \sum_{\mathrm{Y}:\, \mathrm{Y} \geq \mathrm{X}} \mathtt{Q}_{\mathrm{Y/Z}}, \\ \mathtt{Q}_{\mathrm{X/Z}} &= \sum_{\mathrm{Y}:\, \mathrm{Y} \geq \mathrm{X}} \mu(\mathrm{X}, \mathrm{Y})\, \mathtt{H}_{\mathrm{Y/Z}}. \end{aligned} \tag{4.40}$$

Type morphisms of Π in the $\mathtt{Q}$-basis are

$$w : \Pi[\mathrm{X}] \to \Pi[w\mathrm{X}], \qquad \mathtt{Q}_{\mathrm{Y/X}} \mapsto \mathtt{Q}_{w\mathrm{Y}/w\mathrm{X}}, \tag{4.41}$$

while the product and coproduct are

$$\begin{aligned} \mu_{\mathrm{Z}}^{\mathrm{Y}} : \Pi[\mathrm{Y}] &\to \Pi[\mathrm{Z}] & \Delta_{\mathrm{Z}}^{\mathrm{Y}} : \Pi[\mathrm{Z}] &\to \Pi[\mathrm{Y}] \\ \mathtt{Q}_{\mathrm{X/Y}} &\mapsto \mathtt{Q}_{\mathrm{X/Z}} & \mathtt{Q}_{\mathrm{X/Z}} &\mapsto \begin{cases} \mathtt{Q}_{\mathrm{X/Y}} & \text{if } \mathrm{X} \geq \mathrm{Y}, \\ 0 & \text{otherwise.} \end{cases} \end{aligned} \tag{4.42}$$

These (co)product formulas can be deduced from those for species [**11**, Proposition 7.6]. The calculations do not involve type morphisms.

4.4.4. Dual Coxeter bimonoid. Let Π^* denote the Coxeter bimonoid dual to Π. Let $\mathtt{M}$ denote the basis dual to the $\mathtt{H}$-basis. Similarly, let $\mathtt{P}$ denote the basis dual to the $\mathtt{Q}$-basis.

4.4.5. Freeness and cofreeness. We have

$$\Pi = \mathcal{S}(\mathsf{E}) \quad \text{and} \quad \Pi^* = \mathcal{S}^{\vee}(\mathsf{E}) \tag{4.43}$$

as Coxeter bimonoids, with E viewed as a Coxeter comonoid and Coxeter monoid, respectively, with identifications made using $\mathtt{H}$- and $\mathtt{M}$-bases. Also,

$$\Pi = \mathcal{S}(\mathsf{E}) \quad \text{and} \quad \Pi^* = \mathcal{S}^{\vee}(\mathsf{E}) \tag{4.44}$$

as Coxeter bimonoids, with E viewed as a Coxeter species, with identifications made using $\mathtt{Q}$- and $\mathtt{P}$-bases.

Corresponding results for species are spelt out in [**11**, Section 7.4.7].

4.4.6. Norm map on flats. Let (ξ_X) be an invariant family of scalars indexed by flats as in Definition 1.10. The *norm map* on flats

$$\kappa_\xi : \Pi \to \Pi^*, \qquad \mathtt{H}_{X/Z} \mapsto \sum_{Y:\, Y \geq Z} \xi_{X \vee Y}\, \mathtt{M}_{Y/Z} \tag{4.45}$$

on the Z-component, is a self-dual morphism of Coxeter bimonoids.

To see this: In the bijection (3.147), put $\mathsf{c} := \mathsf{E}$ and $\mathsf{a} := \mathsf{E}$, and take $f : \mathsf{c} \to \mathsf{a}$ on the Z-component to be scalar multiplication by ξ_Z. Then under the identification (4.43), the map $\widehat{f} : \Pi \to \Pi^*$ equals κ_ξ.

Theorem 4.17 (Isomorphism criterion). *The norm map on flats* (4.45) *is an isomorphism of Coxeter bimonoids iff*

$$\sum_{Y:\, Y \geq X} \mu(X, Y)\, \xi_Y \neq 0$$

for all flats X.

Proof. Recall the logarithm (3.92b). For the map $f : \mathsf{E} \to \mathsf{E}$ as above, $\log(f)_X$ is scalar multiplication by the lhs in the condition above. Now apply the isomorphism criterion in Theorem 3.77. □

The corresponding result for species is contained in [**11**, Theorem 7.10].

4.4.7. Primitive and indecomposable parts. We have

$$\mathcal{P}(\Pi) = \mathsf{E} \quad \text{and} \quad \mathcal{Q}(\Pi) = \mathsf{E}. \tag{4.46}$$

More precisely, $\mathcal{P}(\Pi)[Z]$ is one-dimensional and spanned by $\mathtt{Q}_{Z/Z}$. This follows from the coproduct formula in (4.42). Similarly, $\mathcal{Q}(\Pi)[Z]$ is spanned by the image of either $\mathtt{H}_{Z/Z}$ or $\mathtt{Q}_{Z/Z}$.

4.4.8. Primitive and group-like series. A series of Π is the same as a family of scalars $(f(Z,X))_{Z \leq X}$ such that $f(Z,X) = f(wZ, wX)$ for all $w \in W$. The identification is done via

$$(f(Z,X))_{Z \leq X} \longleftrightarrow \sum_{X:\, X \geq Z} f(Z,X)\, \mathtt{H}_{X/Z} \in \Pi[Z] \quad \text{for each flat Z.} \tag{4.47}$$

Specializing (3.138) and using the product formula in (4.39), we see that the module $\mathscr{S}(\Pi)$ is in fact the invariant flat-incidence algebra viewed as a left module over itself. Contrast with [**11**, Example 9.128], where Π is viewed as a bimonoid in species.

The unit element δ of the invariant flat-incidence algebra is a group-like series of Π. We call it the *universal series* of Π. The corresponding primitive series under the bijection (3.141) is μ. In terms of the identification (4.47),

$$\begin{aligned} \delta &\leftrightarrow \mathtt{H}_{Z/Z} \in \Pi[Z] \quad \text{for each Z,} \\ \mu &\leftrightarrow \mathtt{Q}_{Z/Z} \in \Pi[Z] \quad \text{for each Z,} \end{aligned} \tag{4.48}$$

where $\mathtt{Q}_{Z/Z}$ is the $\mathtt{Q}$-basis element defined in (4.40).

Let x be the linear series of Π defined by $x_\top := \mathtt{H}_{\top/\top}$. The corresponding exponential series e of Π under the bijection (3.142) is $e_X = \mathtt{H}_{\top/X}$. (It does not matter whether we write $\mathtt{H}$ or $\mathtt{Q}$ since they match on the maximum flat.) Any other linear and exponential series of $\sqcap$ are scalar multiples of x and e.

4.4.9. Antipode. The antipode S of the Coxeter bimonoid $\sqcap$, on the Z-component, in the H-basis, is

$$S_{\mathrm{Z}}(\mathtt{H}_{\mathrm{X}/\mathrm{Z}}) = \sum_{\mathrm{Y}:\, \mathrm{Y} \geq \mathrm{X}} (-1)^{\dim(\mathrm{Y})} c_{\mathrm{X}}^{\mathrm{Y}}\, \mathtt{H}_{\mathrm{Y}/\mathrm{Z}}, \tag{4.49}$$

and, in the Q-basis, is

$$S_{\mathrm{Z}}(\mathtt{Q}_{\mathrm{X}/\mathrm{Z}}) = (-1)^{\dim(\mathrm{X})} \mathtt{Q}_{\mathrm{X}/\mathrm{Z}}, \tag{4.50}$$

with $c_{\mathrm{X}}^{\mathrm{Y}}$ the number of chambers in $\mathcal{A}_{\mathrm{X}}^{\mathrm{Y}}$, the arrangement over X and under Y. See [**11**, Formulas (12.25) and (12.26)].

4.5. Coxeter bimonoid of faces

For a scalar q, we introduce the Coxeter q-bimonoid of faces, its dual, and q-norm maps linking the two (indexed by an invariant family of scalars). We also discuss primitive and group-like series, and antipode formulas.

4.5.1. Coxeter species of faces. Define the Coxeter species Σ as follows. For a face A, let $\Sigma[A]$ be the linear span of the set of faces greater than A. We use the letter $\mathtt{H}$ for the canonical basis of $\Sigma[A]$. Support and type morphisms are

$$\begin{aligned} \beta_{G,F} : \Sigma[F] &\to \Sigma[G] & \qquad w : \Sigma[F] &\to \Sigma[wF] \\ \mathtt{H}_{K/F} &\mapsto \mathtt{H}_{GK/G} & \mathtt{H}_{K/F} &\mapsto \mathtt{H}_{wK/wF}. \end{aligned} \tag{4.51}$$

We call Σ the *Coxeter species of faces*.

4.5.2. Coxeter bimonoid of faces. The Coxeter species of faces Σ carries a Coxeter bimonoid structure with product and coproduct

$$\begin{aligned} \mu_A^F : \Sigma[F] &\to \Sigma[A] & \qquad \Delta_A^F : \Sigma[A] &\to \Sigma[F] \\ \mathtt{H}_{K/F} &\mapsto \mathtt{H}_{K/A} & \mathtt{H}_{K/A} &\mapsto \mathtt{H}_{FK/F}. \end{aligned} \tag{4.52}$$

We call it the *Coxeter bimonoid of faces*. It is cocommutative but not commutative in general.

4.5.3. Coxeter q-bimonoid of faces. More generally, for a scalar q, the Coxeter species of faces carries a Coxeter q-bimonoid structure, denoted Σ_q, with product and coproduct

$$\begin{aligned} \mu_A^F : \Sigma_q[F] &\to \Sigma_q[A] & \qquad \Delta_A^F : \Sigma_q[A] &\to \Sigma_q[F] \\ \mathtt{H}_{K/F} &\mapsto \mathtt{H}_{K/A} & \mathtt{H}_{K/A} &\mapsto q^{\mathrm{dist}(K,F)}\, \mathtt{H}_{FK/F}. \end{aligned} \tag{4.53}$$

We call it the *Coxeter q-bimonoid of faces*. The product is the same as in (4.52), while the coproduct has been deformed using the q-distance function.

4.5.4. Q-basis. Let $\mathrm{char}(\Bbbk) \nmid |W|$. Fix an invariant noncommutative zeta function $\boldsymbol{\zeta}$ and its inverse Möbius function $\boldsymbol{\mu}$. Define a Q-basis of the Coxeter bimonoid Σ, on the A-component, by either of the two equivalent formulas

$$\mathtt{H}_{F/A} = \sum_{G:\, F \leq G} \boldsymbol{\zeta}(F, G)\, \mathtt{Q}_{G/A}, \qquad \mathtt{Q}_{F/A} = \sum_{G:\, F \leq G} \boldsymbol{\mu}(F, G)\, \mathtt{H}_{G/A}. \tag{4.54}$$

Support and type morphisms of Σ in the Q-basis are

$$\beta_{G,F} : \Sigma[F] \to \Sigma[G] \quad \mathtt{Q}_{K/F} \mapsto \mathtt{Q}_{GK/G} \qquad\qquad w : \Sigma[F] \to \Sigma[wF] \quad \mathtt{Q}_{K/F} \mapsto \mathtt{Q}_{wK/wF}, \tag{4.55}$$

while the product and coproduct are

$$\mu_A^F : \Sigma[F] \to \Sigma[A] \quad \mathtt{Q}_{K/F} \mapsto \mathtt{Q}_{K/A} \qquad\qquad \Delta_A^G : \Sigma[A] \to \Sigma[G] \quad \mathtt{Q}_{K/A} \mapsto \begin{cases} \mathtt{Q}_{GK/G} & \text{if } KG = K, \\ 0 & \text{otherwise.} \end{cases} \tag{4.56}$$

These (co)product formulas can be deduced from those for species [**11**, Proposition 7.22]. The calculations do not involve type morphisms.

4.5.5. Coxeter bimonoid of faces in terms of flats and lunes. The cocommutative Coxeter bimonoid of faces Σ can be described in terms of flats and lunes in the same way as Γ, with top-lunes replaced by arbitrary lunes. For instance, the product and coproduct are

$$\mu_{\mathrm{L}} : \Sigma[\mathrm{X}] \to \Sigma[\mathrm{Z}] \quad \mathtt{H}_{\mathrm{M}} \mapsto \mathtt{H}_{\mathrm{L} \circ \mathrm{M}} \qquad\qquad \Delta_{\mathrm{Z}}^{\mathrm{X}} : \Sigma[\mathrm{Z}] \to \Sigma[\mathrm{X}] \quad \mathtt{H}_{\mathrm{L}} \mapsto \mathtt{H}_{\mathrm{X} \cdot \mathrm{L}}. \tag{4.57}$$

4.5.6. Q-basis for q not a root of unity. Suppose q is not a root of unity. Recall the noncommutative q-zeta function $\boldsymbol{\zeta}_q$ and noncommutative q-Möbius function $\boldsymbol{\mu}_q$. Define the Q-basis of Σ_q, on the A-component, by either of the two equivalent formulas

$$\mathtt{H}_{F/A} = \sum_{G:\, F \leq G} \boldsymbol{\zeta}_q(F, G)\, \mathtt{Q}_{G/A}, \qquad \mathtt{Q}_{F/A} = \sum_{G:\, F \leq G} \boldsymbol{\mu}_q(F, G)\, \mathtt{H}_{G/A}. \tag{4.58}$$

These are the same as formulas (4.54) with $\boldsymbol{\zeta}$ and $\boldsymbol{\mu}$ replaced by $\boldsymbol{\zeta}_q$ and $\boldsymbol{\mu}_q$, respectively.

For q not a root of unity, support and type morphisms of Σ_q in the Q-basis are

$$\beta_{G,F} : \Sigma_q[F] \to \Sigma_q[G] \quad \mathtt{Q}_{K/F} \mapsto \mathtt{Q}_{GK/G} \qquad\qquad w : \Sigma_q[F] \to \Sigma_q[wF] \quad \mathtt{Q}_{K/F} \mapsto \mathtt{Q}_{wK/wF}, \tag{4.59}$$

while the product and coproduct are

$$\mu_A^F : \Sigma_q[F] \to \Sigma_q[A] \qquad \Delta_A^G : \Sigma_q[A] \to \Sigma_q[G]$$

$$(4.60) \qquad \mathtt{Q}_{K/F} \mapsto \mathtt{Q}_{K/A} \qquad \mathtt{Q}_{K/A} \mapsto \begin{cases} q^{\mathrm{dist}(K,G)}\, \mathtt{Q}_{GK/G} & \text{if } KG = K, \\ 0 & \text{otherwise.} \end{cases}$$

These (co)product formulas can be deduced from those for species [**11**, Proposition 7.24]. The gate property gets used in the calculation.

4.5.7. Eulerian idempotents. We return to the Coxeter bimonoid Σ. For a flat X/A of $\mathcal{A}_A$, put

$$(4.61) \qquad \mathtt{E}_{\mathrm{X}/A} = \sum_{\substack{F:\, F \geq A \\ \mathrm{s}(FA) = \mathrm{X}/A}} \zeta(A, F)\, \mathtt{Q}_{F/A}.$$

The $\mathtt{E}_{\mathrm{X}/A}$, as X/A varies, are eulerian idempotents of the Tits algebra of $\mathcal{A}_A$, see (1.134).

Exercise 4.18. Check: For any faces $A \leq G$,

$$(4.62) \qquad \Delta_A^G(\mathtt{E}_{\mathrm{X}/A}) = \begin{cases} \mathtt{E}_{\mathrm{X}/G} & \text{if } \mathrm{X} \geq \mathrm{s}(G), \\ 0 & \text{otherwise.} \end{cases}$$

(Use the coproduct formula in (4.56) and lune-additivity formula (1.79).)

4.5.8. Dual Coxeter q-bimonoid. For a scalar q, let Σ_q^* denote the Coxeter q-bimonoid dual to Σ_q. Let $\mathtt{M}$ denote the basis dual to the $\mathtt{H}$-basis. Similarly, let $\mathtt{P}$ denote the basis dual to the $\mathtt{Q}$-basis in both cases (4.54) and (4.58).

4.5.9. Freeness and cofreeness. For a scalar q,

$$(4.63) \qquad \Sigma_q = \mathcal{T}_q(\mathsf{E}) \quad \text{and} \quad \Sigma_q^* = \mathcal{T}_q^\vee(\mathsf{E})$$

as Coxeter q-bimonoids, with E viewed as a Coxeter comonoid and Coxeter monoid, respectively, with identifications made using $\mathtt{H}$- and $\mathtt{M}$-bases.

For $q = 1$, for $\mathrm{char}(\Bbbk) \nmid |W|$,

$$(4.64) \qquad \Sigma = \mathcal{T}(\mathsf{E}) \quad \text{and} \quad \Sigma^* = \mathcal{T}^\vee(\mathsf{E})$$

as Coxeter bimonoids, with E viewed as a Coxeter species, with identifications made using $\mathtt{Q}$- and $\mathtt{P}$-bases.

Similarly, for q not a root of unity,

$$(4.65) \qquad \Sigma_q = \mathcal{T}_q(\mathsf{E}) \quad \text{and} \quad \Sigma_q^* = \mathcal{T}_q^\vee(\mathsf{E})$$

as Coxeter q-bimonoids, with E viewed as a Coxeter species, with identifications made using $\mathtt{Q}$- and $\mathtt{P}$-bases.

Corresponding results for species are spelt out in more detail in [**11**, Section 7.6.11].

Exercise 4.19. For c a Coxeter comonoid and a a Coxeter monoid, recall the convolution Coxeter monoid $\mathrm{hom}^\times(\mathsf{c}, \mathsf{a})$ from Section 3.5.6. Check: Coxeter monoid morphisms from Σ to $\mathrm{hom}^\times(\mathsf{c}, \mathsf{a})$ are parametrized by Coxeter species maps from c to a. (Use Theorem 3.7, (4.63) with $q = 1$ and (3.62).)

Exercise 4.20. Let q be a scalar. For c a cocommutative Coxeter comonoid and k a Coxeter q-bimonoid, recall the Coxeter q-bimonoid $\mathcal{C}(\mathsf{c},\mathsf{k})_q$ from Exercise 3.31. Check: Coxeter q-bimonoid morphisms from Σ_q to $\mathcal{C}(\mathsf{c},\mathsf{k})_q$ are parametrized by Coxeter comonoid morphisms from c to k. (Use Theorem 3.9, (4.63) and (3.72).)

Dually: For h a Coxeter q-bimonoid and a a commutative Coxeter monoid, recall the Coxeter q-bimonoid $\overline{\mathcal{C}}(\mathsf{h},\mathsf{a})_q$ from Exercise 3.34. Check: Coxeter q-bimonoid morphisms from Σ_q to $\overline{\mathcal{C}}(\mathsf{h},\mathsf{a})_q$ are parametrized by Coxeter monoid morphisms from h to a. (Use Theorem 3.9, (4.63) and (3.78).)

Deduce as a consequence the existence of the Coxeter bimonoid morphisms $\Sigma \to \mathcal{C}(\mathsf{h},\mathsf{h})$ and $\Sigma \to \overline{\mathcal{C}}(\mathsf{h},\mathsf{h})$ in (3.152).

4.5.10. q-norm map on faces. Let (ξ_{X}) be an invariant family of scalars indexed by flats as in Definition 1.10. For a scalar q, the *q-norm map* on faces

$$\kappa_{q,\xi} : \Sigma_q \to \Sigma_q^*, \qquad \mathtt{H}_{F/A} \mapsto \sum_{G:\, G \geq A} \xi_{\mathrm{s}(FG)}\, q^{\mathrm{dist}(F,G)}\, \mathtt{M}_{G/A} \tag{4.66}$$

on the A-component, is a self-dual morphism of Coxeter q-bimonoids.

To see this: In the bijection (3.146), put $\mathsf{c} := \mathsf{E}$ and $\mathsf{a} := \mathsf{E}$, and take $f : \mathsf{c} \to \mathsf{a}$ on the A-component to be scalar multiplication by $\xi_{\mathrm{s}(A)}$. Then under the identification (4.63), the map $\widehat{f} : \Sigma_q \to \Sigma_q^*$ equals $\kappa_{q,\xi}$.

Theorem 4.21 (Isomorphism criterion). *For q not a root of unity, the q-norm map on faces (4.66) is an isomorphism of Coxeter q-bimonoids iff*

$$\sum_{\mathrm{Y}:\, \mathrm{Y} \geq \mathrm{X}} \mu_q(\mathrm{X},\mathrm{Y})\, \xi_{\mathrm{Y}} \neq 0$$

for all flats X.

Proof. Recall the q-logarithm (3.101d). For the map $f : \mathsf{E} \to \mathsf{E}$ as above, $\mathbf{log}_q(f)_{\mathrm{X}}$ is scalar multiplication by the lhs in the condition above. Now apply the isomorphism criterion in Theorem 3.79. ☐

The corresponding result for species is contained in [**11**, Theorem 7.38].

4.5.11. Primitive and group-like series. A series of Σ is the same as a family of scalars $(f(A,F))_{A \leq F}$ such that $f(A,F) = f(A',F')$ whenever (A,F) and (A',F') have either the same support or the same type. The identification is done via

$$(f(A,F))_{A \leq F} \longleftrightarrow \sum_{F:\, F \geq A} f(A,F)\, \mathtt{H}_{F/A} \in \Sigma[A] \quad \text{for each face } A. \tag{4.67}$$

Specializing (3.122) and using the product formula in (4.52), we see that the module $\mathscr{S}(\Sigma)$ is in fact the invariant lune-incidence algebra viewed as a left module over itself. Contrast with [**11**, Example 9.119], where Σ is viewed as a bimonoid in species.

Universal series. Let $\mathrm{char}(\Bbbk) \nmid |W|$. The unit element δ of the invariant lune-incidence algebra is a group-like series of Σ. We call it the *universal series* of Σ. The corresponding primitive series under the bijection (3.126) is $\boldsymbol{\mu}$. In terms of the identification (4.67),

$$\begin{aligned} \delta &\leftrightarrow \mathtt{H}_{A/A} \in \Sigma[A] \quad \text{for each } A, \\ \boldsymbol{\mu} &\leftrightarrow \mathtt{Q}_{A/A} \in \Sigma[A] \quad \text{for each } A, \end{aligned} \tag{4.68}$$

where $\mathtt{Q}_{A/A}$ is the $\mathtt{Q}$-basis element defined in (4.54). Similarly, for q not a root of unity, under the bijection (3.129), the corresponding primitive series is $\boldsymbol{\mu}_q$, or equivalently, the $\mathtt{Q}$-basis element $\mathtt{Q}_{A/A}$ defined in (4.58).

Takeuchi series. Let $\mathrm{char}(\Bbbk) \nmid |W|$. Define the *Takeuchi series*, denoted $\mathtt{Tak}$, of Σ by

$$\mathtt{Tak}_A := \sum_{F:\, F \geq A} (-1)^{\dim(F)}\, \mathtt{H}_{F/A}. \tag{4.69}$$

It is group-like. Its logarithm is the primitive series of Σ given by

$$\mathbf{log}(\mathtt{Tak})_A = (-1)^{\dim(A)}\, \overline{\mathtt{Q}}_{A/A}, \tag{4.70}$$

where the $\overline{\mathtt{Q}}$-basis is defined from the $\mathtt{H}$-basis via $\overline{\boldsymbol{\zeta}}$ and $\overline{\boldsymbol{\mu}}$. Equivalently,

$$\mathbf{log}(\mathtt{Tak})_A = (-1)^{\dim(A)}\, \overline{\mathbf{log}}(\delta)_A, \tag{4.71}$$

where δ denotes the universal series of Σ. See [**11**, Section 12.8.3] for corresponding results on species.

We mention that Exercises 4.19 and 4.20, in conjunction with the Takeuchi series, can be used to deduce certain antipode results in Section 3.14 such as Lemma 3.97 and Lemma 3.102.

Exponential series. Let $\mathrm{char}(\Bbbk) \nmid |W|$. Formula (4.32) defines an exponential series of Σ. Moreover, every exponential series of Σ is a scalar multiple of this one. One can check this directly or use (4.35) in view of (4.75) below.

4.5.12. Antipode. The antipode S of the Coxeter q-bimonoid Σ_q, on the A-component, in the $\mathtt{H}$-basis, is

$$S_A(\mathtt{H}_{F/A}) = q^{\mathrm{dist}(F, A\overline{F})} \sum_{G:\, A\overline{F} \leq G} (-1)^{\dim(G)}\, \mathtt{H}_{G/A}, \tag{4.72}$$

in the $\mathtt{Q}$-basis (4.54) for $q = 1$, is

$$S_A(\mathtt{Q}_{F/A}) = (-1)^{\dim(F)}\, \mathtt{Q}_{A\overline{F}/A}, \tag{4.73}$$

and, in the $\mathtt{Q}$-basis (4.58) for q not a root of unity, is

$$S_A(\mathtt{Q}_{F/A}) = (-1)^{\dim(F)} q^{\mathrm{dist}(F, A\overline{F})}\, \mathtt{Q}_{A\overline{F}/A}. \tag{4.74}$$

See [**11**, Formulas (12.32), (12.33), (12.34)].

4.5.13. Morphisms. Observe: The inclusion

$$\Gamma_q \hookrightarrow \Sigma_q \tag{4.75}$$

is a morphism of Coxeter q-bimonoids. Similarly, the support map

$$\pi : \Sigma \twoheadrightarrow \Pi, \qquad \mathtt{H}_{F/A} \mapsto \mathtt{H}_{\mathrm{s}(F/A)} \tag{4.76}$$

is a morphism of Coxeter bimonoids. The same formula holds on the $\mathtt{Q}$-basis as well.

4.6. Coxeter bimonoid of top-nested faces

For a scalar q, we introduce the Coxeter q-bimonoid of top-nested faces, its dual, and q-norm maps linking the two (indexed by an invariant family of scalars).

4.6.1. Coxeter species of top-nested faces. Define the Coxeter species $\widehat{\mathsf{Q}}$ as follows. For a face A, let $\widehat{\mathsf{Q}}[A]$ be the linear span of the set of top-nested faces with both coordinates greater than A. We use the letter $\mathtt{H}$ for the canonical basis of $\widehat{\mathsf{Q}}[A]$. Support and type morphisms are

$$\begin{aligned} \beta_{G,F} : \widehat{\mathsf{Q}}[F] &\to \widehat{\mathsf{Q}}[G] & \qquad w : \widehat{\mathsf{Q}}[F] &\to \widehat{\mathsf{Q}}[wF] \\ \mathtt{H}_{K/F,C/F} &\mapsto \mathtt{H}_{GK/G,GC/G} & \mathtt{H}_{K/F,C/F} &\mapsto \mathtt{H}_{wK/wF,wC/wF}. \end{aligned} \tag{4.77}$$

We call $\widehat{\mathsf{Q}}$ the *Coxeter species of top-nested faces*.

4.6.2. Coxeter q-bimonoid of top-nested faces. For a scalar q, the Coxeter species of top-nested faces carries a Coxeter q-bimonoid structure which we denote $\widehat{\mathsf{Q}}_q$. We call it the *Coxeter q-bimonoid of top-nested faces*. The product and coproduct are

$$\begin{aligned} \mu_A^K : \widehat{\mathsf{Q}}_q[K] &\to \widehat{\mathsf{Q}}_q[A] \\ \mathtt{H}_{F/K,C/K} &\mapsto \mathtt{H}_{F/A,C/A}, \end{aligned} \tag{4.78a}$$

$$\Delta_A^K : \widehat{\mathsf{Q}}_q[A] \to \widehat{\mathsf{Q}}_q[K]$$
$$\mathtt{H}_{F/A,C/A} \mapsto \begin{cases} q^{\mathrm{dist}(F,K)}\, \mathtt{H}_{KF/K,KC/K} & \text{if } FK \leq C, \\ 0 & \text{otherwise.} \end{cases} \tag{4.78b}$$

4.6.3. Dual Coxeter q-bimonoid. For a scalar q, let $\widehat{\mathsf{Q}}_q^*$ denote the Coxeter q-bimonoid dual to $\widehat{\mathsf{Q}}_q$. Let $\mathtt{M}$ denote the basis dual to the $\mathtt{H}$-basis.

4.6.4. Freeness and cofreeness. For a scalar q,

$$\widehat{\mathsf{Q}}_q = \mathcal{T}_q(\Gamma^*) \quad \text{and} \quad \widehat{\mathsf{Q}}_q^* = \mathcal{T}_q^\vee(\Gamma) \tag{4.79}$$

as Coxeter q-bimonoids, with Γ^* viewed as a Coxeter comonoid and Γ as a Coxeter monoid, with identifications made using $\mathtt{H}$- and $\mathtt{M}$-bases.

Corresponding results for species are explained in more detail in [**11**, Section 7.7.6].

4.6.5. q-norm map on top-nested faces. Let (ξ_X) be an invariant family of scalars indexed by flats as in Definition 1.10. For a scalar q, the *q-norm map* on top-nested faces

$$\kappa_{q,\xi} : \widehat{\mathsf{Q}}_q \to \widehat{\mathsf{Q}}_q^*, \quad \mathtt{H}_{F/A,C/A} \mapsto \sum_{\substack{(G,D):\, A \le G \le D \\ GC=D,\, FD=C}} \xi_{\mathrm{s}(FG)}\, q^{\mathrm{dist}(F,G)}\, \mathtt{M}_{G/A,D/A} \tag{4.80}$$

on the A-component, is a self-dual morphism of Coxeter q-bimonoids.

To see this: In the bijection (3.146), put $\mathsf{c} := \Gamma^*$ and $\mathsf{a} := \Gamma$, and take $f : \mathsf{c} \to \mathsf{a}$ to be the map which on the A-component sends $\mathtt{M}_{C/A}$ to $\xi_{\mathrm{s}(A)}\mathtt{H}_{C/A}$. Then under the identification (4.79), the map $\widehat{f} : \widehat{\mathsf{Q}}_q \to \widehat{\mathsf{Q}}_q^*$ equals $\kappa_{q,\xi}$.

Theorem 4.22 (Isomorphism criterion). *For q not a root of unity, the q-norm map on top-nested faces (4.80) is an isomorphism of Coxeter q-bimonoids iff*

$$\det\Bigg(\Big(\sum_{\substack{F':\, A \le F' \le C \\ F:A \le F \le D \\ \mathrm{s}(F)=\mathrm{s}(F'),\, FC=D}} \mu_q(A,F,F')\, \xi_{\mathrm{s}(F)}\Big)_{C,D \ge A}\Bigg) \neq 0$$

for all faces A. (In the above matrix, A is fixed, while C, D are the two varying indices. Each matrix entry is given by the sum over F, F'.)

PROOF. Recall the q-logarithm (3.99b). For the map $f : \Gamma^* \to \Gamma$ as above,

$$\mathbf{log}_q(f)_A(\mathtt{M}_{C/A}) = \sum_{D:\, D \ge A} \alpha_A^{C,D}\, \mathtt{H}_{D/A},$$

where $\alpha_A^{C,D}$ is the sum over F, F' in parenthesis above. Now apply the isomorphism criterion in Theorem 3.79. □

The corresponding result for species for $\xi \equiv 1$ is [**11**, Theorem 15.47].

Exercise 4.23. Let ξ be an invariant family of scalars. Define another invariant family of scalars $\mathrm{n}\,\xi$ indexed by faces by $(\mathrm{n}\,\xi)_A := |W_A|\, \xi_A$. Check: For a scalar q, we have a diagram of Coxeter q-bimonoids

$$\begin{array}{ccc} \widehat{\mathsf{Q}}_q & \xrightarrow{\kappa_{q,\xi}} & \widehat{\mathsf{Q}}_q^* \\ \uparrow & & \downarrow \\ \Sigma_q & \xrightarrow[\kappa_{q,\mathrm{n}\xi}]{} & \Sigma_q^*. \end{array} \tag{4.81}$$

The inclusion map, on the A-component, sends $\mathtt{H}_{F/A}$ to $\sum \mathtt{H}_{F/A,C/A}$ with the sum over all $C \ge F$. The surjective map is its dual.

4.6.6. Coxeter bimonoid of top-lunes. For any face A, let $\widehat{\Lambda}[A]$ be the linear span of the set of top-lunes with base greater than $\mathrm{s}(A)$. We use the letter $\mathtt{H}$ for its canonical basis. The map

$$\pi : \widehat{\mathsf{Q}} \twoheadrightarrow \widehat{\Lambda}, \qquad \mathtt{H}_{F/A,C/A} \mapsto \mathtt{H}_{\mathrm{s}(F/A,C/A)} \tag{4.82}$$

turns $\widehat{\Lambda}$ into a Coxeter bimonoid. We call it the *Coxeter bimonoid of top-lunes*.

4.7. Coxeter bimonoid of pairs of chambers

For a scalar q, we introduce the Coxeter q-bimonoid of pairs of chambers, its dual, and the self-duality map linking the two.

4.7.1. Coxeter species of pairs of chambers. Define the Coxeter species $\mathrm{I}\Gamma$ as follows. For a face A, let $\mathrm{I}\Gamma[A]$ be the linear span of the set of pairs of chambers both greater than A. We use the letter $\mathtt{K}$ for its canonical basis. Thus, a typical basis element is $\mathtt{K}_{C/A,D/A}$. Support and type morphisms are

$$\begin{aligned} &\beta_{G,F} : \mathrm{I}\Gamma[F] \to \mathrm{I}\Gamma[G] && w : \mathrm{I}\Gamma[F] \to \mathrm{I}\Gamma[wF] \\ &\mathtt{K}_{C/F,D/F} \mapsto \mathtt{K}_{GC/G,GD/G} && \mathtt{K}_{C/F,D/F} \mapsto \mathtt{K}_{wC/wF,wD/wF}. \end{aligned} \tag{4.83}$$

For the dual Coxeter species $\mathrm{I}\Gamma^*$, we use the letter $\mathtt{F}$ for the basis dual to $\mathtt{K}$. Define a second basis called the $\mathtt{M}$-basis related to the $\mathtt{F}$-basis by

$$\mathtt{F}_{E/A,D/A} = \sum_{\substack{C:\, C \geq A \\ C \text{ -- } E \text{ -- } D}} \mathtt{M}_{C/A,D/A}. \tag{4.84}$$

We use the letter $\mathtt{H}$ for the basis dual to $\mathtt{M}$. It is a second basis for $\mathrm{I}\Gamma$. The relation between the $\mathtt{K}$-basis and $\mathtt{H}$-basis is obtained by dualizing (4.84).

We refer to either $\mathrm{I}\Gamma$ or $\mathrm{I}\Gamma^*$ as the *Coxeter species of pairs of chambers*.

4.7.2. Coxeter q-bimonoid of pairs of chambers. For a scalar q, the Coxeter species $\mathrm{I}\Gamma^*$ carries a Coxeter q-bimonoid structure with product and coproduct

$$\begin{aligned} &\mu_A^K : \mathrm{I}\Gamma_q^*[K] \to \mathrm{I}\Gamma_q^*[A] \\ &\mathtt{F}_{C/K,D/K} \mapsto \sum_{\substack{D' :\, D' \geq A \\ KD' = D}} q^{\mathrm{dist}(D,D')}\, \mathtt{F}_{C/A,D'/A}, \end{aligned} \tag{4.85a}$$

$$\begin{aligned} &\Delta_A^G : \mathrm{I}\Gamma_q^*[A] \to \mathrm{I}\Gamma_q^*[G] \\ &\mathtt{F}_{C/A,D/A} \mapsto \begin{cases} \mathtt{F}_{GC/G,D/G} & \text{if } G \leq D, \\ 0 & \text{otherwise.} \end{cases} \end{aligned} \tag{4.85b}$$

To show dependence on q, we write $\mathrm{I}\Gamma_q^*$. We call it the *Coxeter q-bimonoid of pairs of chambers*. Its product and coproduct in the $\mathtt{M}$-basis are

$$\begin{aligned} &\mu_A^K : \mathrm{I}\Gamma_q^*[K] \to \mathrm{I}\Gamma_q^*[A] \\ &\mathtt{M}_{C/K,D/K} \mapsto \sum_{\substack{(C',D') :\, C' \geq A,\, D' \geq A \\ KC' = C,\, KD' = D \\ C' \text{ -- } KC' \text{ -- } KD' \text{ -- } D'}} q^{\mathrm{dist}(D,D')}\, M_{C'/A,D'/A}, \end{aligned} \tag{4.86a}$$

$$\begin{aligned} &\Delta_A^G : \mathrm{I}\Gamma_q^*[A] \to \mathrm{I}\Gamma_q^*[G] \\ &\mathtt{M}_{C/A,D/A} \mapsto \begin{cases} \mathtt{M}_{GC/G,D/G} & \text{if } G \leq D \text{ and } G \leq A\overline{C}, \\ 0 & \text{otherwise.} \end{cases} \end{aligned} \tag{4.86b}$$

These (co)product formulas can be deduced from those for species [**11**, Theorem 15.37].

4.7.3. Dual Coxeter q-bimonoid. For a scalar q, let $\mathbb{\Gamma}_q$ denote the Coxeter q-bimonoid dual to $\mathbb{\Gamma}^*_q$. (Co)product formulas in $\mathtt{K}$- and $\mathtt{H}$-bases can be obtained by dualizing the above formulas in $\mathtt{F}$- and $\mathtt{M}$-bases.

4.7.4. Self-duality. For $q \neq 0$, the map

$$s_q : \mathbb{\Gamma}_q \to \mathbb{\Gamma}^*_q, \qquad \mathtt{K}_{D/A,C/A} \mapsto q^{\operatorname{dist}(C,D)}\, \mathtt{F}_{C/A,D/A} \tag{4.87}$$

is an isomorphism of Coxeter q-bimonoids. In particular, $\mathbb{\Gamma}_q$ is self-dual if $q \neq 0$.

4.7.5. Primitive and indecomposable parts. Freeness and cofreeness. Using the coproduct formula of $\mathbb{\Gamma}^*$ in the $\mathtt{M}$-basis (4.86b), we see that

$$\{\mathtt{M}_{C/A,D/A} \mid D \wedge A\overline{C} = A\}$$

is a linear basis of $\mathcal{P}(\mathbb{\Gamma}^*)[A]$. Moreover, $\mathbb{\Gamma}^*$ is cofree on its primitive part. An explicit isomorphism of Coxeter comonoids

$$\mathbb{\Gamma}^* \xrightarrow{\cong} \mathcal{T}^\vee(\mathcal{P}(\mathbb{\Gamma}^*)), \tag{4.88}$$

on the A-component, sends $\mathtt{M}_{C/A,D/A}$ to $\mathtt{M}_{KC/K,D/K}$ with $K := D \wedge A\overline{C}$. The latter belongs to $\mathcal{P}(\mathbb{\Gamma}^*)[K]$. By duality, there is an isomorphism of Coxeter monoids

$$\mathbb{\Gamma} \cong \mathcal{T}(\mathcal{Q}(\mathbb{\Gamma})), \tag{4.89}$$

and $\mathbb{\Gamma}$ (and hence $\mathbb{\Gamma}^*$) is free as a Coxeter monoid on its indecomposable part.

4.7.6. Antipode. The antipode S of the Coxeter q-bimonoid $\mathbb{\Gamma}^*_q$, on the A-component, in the $\mathtt{F}$-basis, is

$$S_A(\mathtt{F}_{C/A,D/A}) = \sum (-1)^{\dim(C' \wedge D)} q^{\operatorname{dist}(D,D')}\, \mathtt{F}_{C'/A,D'/A}, \tag{4.90}$$

with sum over $(C'/A, D'/A)$ for which there is a unique face H greater than A with $HC = C'$ and $HD' = D$ (in which case $H = C' \wedge D$).

The antipode S of $\mathbb{\Gamma}^*_q$, on the A-component, in the $\mathtt{M}$-basis, is

$$S_A(\mathtt{M}_{C/A,D/A}) = (-1)^{\dim(A\overline{C} \wedge D)} \sum q^{\operatorname{dist}(D,D')}\, \mathtt{M}_{C'/A,D'/A}, \tag{4.91}$$

with sum over $(C'/A, D'/A)$ such that $C' - D - D'$ and there is a unique face H greater than A with $H\overline{C'} = A\overline{C}$ and $HD' = D$ (in which case $H = A\overline{C} \wedge D$).

See [**11**, Theorems 15.42 and 15.43].

4.7.7. Coxeter bimonoid morphisms. There is a commutative self-dual diagram of Coxeter bimonoids

$$\begin{array}{ccccc} \Sigma & \hookrightarrow & \mathbb{\Gamma} & & \\ \downarrow\!\!\!\!\downarrow & \searrow & & \searrow^{\cong} & \\ & & & & \mathbb{\Gamma}^* \\ \Pi & & & & \downarrow\!\!\!\!\downarrow \\ & \searrow & & \searrow & \\ & & \Pi^* & \hookrightarrow & \Sigma^*. \end{array} \tag{4.92}$$

The map $\mathsf{I\Gamma} \to \mathsf{I\Gamma}^*$ is (4.87) with $q = 1$. The top-horizontal map is

$$\Sigma \hookrightarrow \mathsf{I\Gamma}, \qquad \mathtt{H}_{F/A} \to \sum_{D:\, D \geq A} \mathtt{K}_{D/A, FD/A} \tag{4.93}$$

on the A-component, while $\mathsf{I\Gamma}^* \twoheadrightarrow \Sigma^*$ is its dual. One checks that the composite map $\Sigma \to \Sigma^*$ is the q-norm map on faces $\kappa_{q,\xi}$ in (4.66) with $q = 1$ and $\xi_F = |W_F|$, or equivalently, $\xi_{\mathrm{X}} = |W_{\mathrm{X}}|$. The map $\Sigma \twoheadrightarrow \Pi$ is as in (4.76) and $\Pi^* \hookrightarrow \Sigma^*$ is its dual. The map $\Pi \to \Pi^*$ is the norm map on flats κ_ξ in (4.45) with $\xi_{\mathrm{X}} = |W_{\mathrm{X}}|$. This map for species appears in [**11**, Theorem 7.11].

4.7.8. Coxeter bimonoid of chamber maps. We mention that $\mathsf{I\Gamma}$ identifies with $\mathrm{end}^\times(\Gamma)$, see [**11**, Map (15.39)]. Moreover, as an instance of the first diagram in (3.152), we have

$$\begin{array}{ccc} \Sigma & \hookrightarrow & \mathrm{end}^\times(\Gamma) = \mathsf{I\Gamma} \\ & \searrow \quad \nearrow & \\ & \mathcal{C}(\Gamma, \Gamma), & \end{array} \tag{4.94}$$

with the horizontal map same as in (4.93). We refer to $\mathcal{C}(\Gamma, \Gamma)$ as the *Coxeter bimonoid of chamber maps*. It is cocommutative.

4.8. Coxeter Lie species and Coxeter Zie species

We introduce Coxeter Lie species and Coxeter Zie species. They are the primitive parts of the Coxeter comonoids of chambers and faces, respectively.

4.8.1. Coxeter Lie species. Recall the Coxeter species of chambers Γ from Section 4.3. Let us write

$$z = \sum_{C:\, C \geq A} x^{C/A}\, \mathtt{H}_{C/A}$$

for a typical element of $\Gamma[A]$. The letter $\mathtt{H}$ denotes the $\mathtt{H}$-basis of Γ.

Let Lie be the Coxeter subspecies of Γ with A-component defined by

$$z \in \mathsf{Lie}[A] \iff \sum_{\substack{C:\, C \geq A \\ HC = D}} x^{C/A} = 0 \text{ for any } A < H \leq D. \tag{4.95}$$

The rhs is a linear system in variables $x^{C/A}$. One checks that support and type morphisms (4.21) of Γ indeed restrict to Lie.

We call Lie the *Coxeter Lie species*. It is the primitive part of the Coxeter comonoid of chambers Γ, that is,

$$\mathcal{P}(\Gamma) = \mathsf{Lie}. \tag{4.96}$$

The calculation is the same as for species [**11**, Lemma 7.64].

For q not a root of unity,

$$\mathcal{P}(\Gamma_q) = \mathsf{x}. \tag{4.97}$$

The calculation is the same as for species [**11**, Exercise 7.66].

4.8.2. Coxeter Zie species. Recall the Coxeter species of faces Σ from Section 4.5. Let us write

$$z = \sum_{F:\, F \geq A} x^{F/A}\, \mathtt{H}_{F/A}$$

for a typical element of $\Sigma[A]$. The letter $\mathtt{H}$ denotes the $\mathtt{H}$-basis of Σ.

Let Zie be the Coxeter subspecies of Σ with A-component defined by

$$z \in \mathsf{Zie}[A] \iff \sum_{\substack{F:\, F \geq A \\ HF = G}} x^{F/A} = 0 \text{ for any } A < H \leq G. \tag{4.98}$$

The rhs is a linear system in variables $x^{F/A}$. One checks that support and type morphisms (4.51) of Σ indeed restrict to Zie.

We call Zie the *Coxeter Zie species*. It is the primitive part of the Coxeter comonoid of faces Σ, that is,

$$\mathcal{P}(\Sigma) = \mathsf{Zie}. \tag{4.99}$$

The calculation is the same as for species [**11**, Lemma 7.69].

For q not a root of unity,

$$\mathcal{P}(\Sigma_q) = \mathsf{E}. \tag{4.100}$$

The calculation is the same as for species [**11**, Exercise 7.72].

4.9. Decoration functor on Coxeter species

Starting with a W-module M, we construct the bicommutative Coxeter bimonoid E_M. We call it the decorated exponential Coxeter bimonoid. When $M = \Bbbk$ with trivial W-action, E_M specializes to E. Related objects are summarized in Table 4.2 which generalizes Table 4.1.

TABLE 4.2. Decorated exponential Coxeter bimonoid and its analogues.

E_M	$\mathsf{E}_M \times \mathsf{E}$	Coxeter bimonoid
E^-_M	$\mathsf{E}_M \times \mathsf{E}^-$	signed Coxeter bimonoid
$\mathsf{E}^-_{M_-}$	$\mathsf{E}_M \times \mathsf{E}_-$	signed Coxeter bimonoid
E_{M_-}	$\mathsf{E}_M \times \mathsf{E}^-_-$	Coxeter bimonoid

Similarly, we construct the Coxeter bimonoid of decorated chambers Γ_M which generalizes Γ. In fact, every Coxeter species and Coxeter (co, bi)monoid has a decorated version obtained by taking Hadamard product with E_M. This yields a functor which we call the decoration functor. We study its interaction with duality, universal constructions, primitive and indecomposable parts, and antipodes.

4.9.1. Decorated exponential Coxeter species. To a left W-module M, we associate a bicommutative Coxeter bimonoid E_M as follows. All its components are M, support and type morphisms are

$$\begin{aligned} \beta_{G,F} : \mathsf{E}_M[F] &\to \mathsf{E}_M[G] & w : \mathsf{E}_M[F] &\to \mathsf{E}_M[wF] \\ m &\mapsto m & m &\mapsto w \cdot m, \end{aligned} \tag{4.101}$$

and product and coproduct components are identities. We refer to E_M as the *decorated exponential Coxeter bimonoid.* When M is the trivial and sign representation of W, we recover E and E^{-}_{-}, respectively, from Section 4.2. In terms of flats, $\mathsf{E}_M[\mathrm{X}] = M$ for all X, with type morphisms

$$w : \mathsf{E}_M[\mathrm{X}] \to \mathsf{E}_M[w\mathrm{X}], \qquad m \to w \cdot m,$$

and product and coproduct components $\mu^{\mathrm{X}}_{\mathrm{Z}}$ and $\Delta^{\mathrm{X}}_{\mathrm{Z}}$ all identities.

Observation 4.24. The construction E_M is functorial in M. That is, a map $M \to M'$ of left W-modules gives rise to a morphism $\mathsf{E}_M \to \mathsf{E}_{M'}$ of Coxeter bimonoids. This yields a functor

$$W\text{-}\mathsf{Mod} \to {}^{\mathrm{co}}\mathsf{Bimon}^{\mathrm{co}}(\mathcal{A}\text{-}\mathsf{CSp}), \qquad M \mapsto \mathsf{E}_M \tag{4.102}$$

from the category of left W-modules to the category of bicommutative Coxeter bimonoids. Moreover, it is full and faithful. Further, for left W-modules M and N,

$$\mathsf{E}_M \times \mathsf{E}_N = \mathsf{E}_{M \otimes N}.$$

Thus, the functor (4.102) is a strong monoidal functor wrt tensor product of W-modules and Hadamard product of Coxeter bimonoids.

Exercise 4.25. Check: For M a left W-module, $(\mathsf{E}_M)^* = \mathsf{E}_{M^*}$.

To a left W-module M, we also associate the signed Coxeter bimonoid

$$\mathsf{E}^{-}_M := \mathsf{E}_M \times \mathsf{E}^{-}. \tag{4.103}$$

All its components are M, support and type morphisms are

$$\begin{aligned} \beta_{G,F} : \mathsf{E}^{-}_M[F] &\to \mathsf{E}^{-}_M[G] & w : \mathsf{E}^{-}_M[F] &\to \mathsf{E}^{-}_M[wF] \\ m &\mapsto (-1)^{\mathrm{dist}(F,G)} m & m &\mapsto (-1)^{l(w)} w \cdot m, \end{aligned} \tag{4.104}$$

and product and coproduct components are identities. We refer to E^{-}_M as the *decorated signed exponential Coxeter bimonoid.* When M is the trivial and sign representation of W, we recover E^{-} and E_{-}, respectively, from Section 4.2. As before, we obtain a functor

$$W\text{-}\mathsf{Mod} \to (-1)\text{-}{}^{\mathrm{co}}\mathsf{Bimon}^{\mathrm{co}}(\mathcal{A}\text{-}\mathsf{CSp}), \qquad M \mapsto \mathsf{E}^{-}_M. \tag{4.105}$$

Apart from E_M and E^{-}_M, we can also consider E_{M_-} and $\mathsf{E}^{-}_{M_-}$ by replacing M by M_-. Their support and type morphisms can be written down explicitly. A summary of these objects is given in Table 4.2. When M is the trivial representation of W, it specializes to Table 4.1. In general, all four objects are different, even as Coxeter species. For the trivial representation, this fact is mentioned in Exercise 4.8.

Example 4.26. Recall the braid arrangement $\mathcal{B}^I$ on the set I. Its Coxeter group W is the group of permutations of I. For any vector space V, we have the left W-module $V^{\otimes I}$, with W acting by permuting the tensor factors. This yields the decorated exponential Coxeter bimonoid $\mathsf{E}_{V^{\otimes I}}$ and decorated signed exponential Coxeter bimonoid $\mathsf{E}^-_{V^{\otimes I}}$. In addition, we can consider both these objects with the W-action on $V^{\otimes I}$ twisted by the sign representation. □

Exercise 4.27. Recall from Section 1.14.2 the Coxeter–Birkhoff monoid $\mathrm{W\Pi}[\mathcal{A}]$. The map $\mathrm{W\Pi}[\mathcal{A}] \to W$ defined by $(\mathrm{X}, w) \mapsto w$ is a monoid morphism. Linearization yields an algebra morphism $\mathsf{W\Pi}[\mathcal{A}] \to \Bbbk W$. Check: The induced functor from the category of left W-modules to the category of left $\mathsf{W\Pi}[\mathcal{A}]$-modules, or equivalently by Proposition 3.87, to the category of bicommutative Coxeter $\mathcal{A}$-bimonoids coincides with (4.102).

4.9.2. Coxeter bimonoid of decorated chambers.

Let M be a left W-module. Define the Coxeter species Γ_M as follows. For a face A, let

$$\Gamma_M[A] := \bigoplus_{C:\, C \geq A} M.$$

For $m \in M$, let us write $m_{C/A} \in \Gamma_M[A]$ for the element m in the summand M indexed by $C \geq A$. Support and type morphisms are

$$\beta_{G,F} : \Gamma_M[F] \to \Gamma_M[G] \qquad\qquad w : \Gamma_M[F] \to \Gamma_M[wF] \tag{4.106}$$
$$m_{C/F} \mapsto m_{GC/G} \qquad\qquad m_{C/F} \mapsto (w \cdot m)_{wC/wF}.$$

The Coxeter species Γ_M carries a cocommutative Coxeter bimonoid structure with product and coproduct

$$\mu^F_A : \Gamma_M[F] \to \Gamma_M[A] \qquad\qquad \Delta^F_A : \Gamma_M[A] \to \Gamma_M[F] \tag{4.107}$$
$$m_{C/F} \mapsto m_{C/A} \qquad\qquad m_{C/A} \mapsto m_{FC/F}.$$

We call it the *Coxeter bimonoid of decorated chambers*. When M is the trivial representation of W, we recover Γ.

More generally, for a scalar q, define the Coxeter q-bimonoid of decorated chambers $\Gamma_{q,M}$ by deforming the coproduct component Δ^F_A by scalar multiplication by $q^{\mathrm{dist}(C,FC)}$. We recover Γ_q as a special case.

Exercise 4.28. Check: For a scalar q, for a left W-module M,

$$\Gamma_{q,M} = \Gamma_q \times \mathsf{E}_M$$

as Coxeter q-bimonoids. On the A-component, $m_{C/A}$ identifies with $\mathtt{H}_{C/A} \otimes m$.

For a scalar q, define the Coxeter q-bimonoid

$$\Gamma^*_{q,M} := \Gamma^*_q \times \mathsf{E}_M.$$

It can be made explicit in a manner similar to $\Gamma_{q,M}$ by using symbols $m_{C/A}$ which identify with $\mathtt{M}_{C/A} \otimes m$. For $q = 1$, we denote $\Gamma^* \times \mathsf{E}_M$ by Γ^*_M.

4.9.3. Freeness and cofreeness. Recall the Coxeter species x_M defined by (4.2). We have

$$\mathsf{E}_M = \mathcal{S}(\mathsf{x}_M) = \mathcal{S}^\vee(\mathsf{x}_M) \quad \text{and} \quad \mathsf{E}^-_M = \mathcal{E}(\mathsf{x}_M) = \mathcal{E}^\vee(\mathsf{x}_M) \tag{4.108}$$

as Coxeter bimonoids and signed Coxeter bimonoids, respectively. For a scalar q,

$$\Gamma_{q,M} = \mathcal{T}_q(\mathsf{x}_M) \quad \text{and} \quad \Gamma^*_{q,M} = \mathcal{T}^\vee_q(\mathsf{x}_M) \tag{4.109}$$

as Coxeter q-bimonoids. These generalize (4.19) and (4.27).

It follows from (3.50) that E_M is the abelianization of Γ_M and coabelianization of Γ^*_M. Similarly, from (3.52), E^-_M is the signed abelianization of $\Gamma_{-1,M}$ and signed coabelianization of $\Gamma^*_{-1,M}$.

Also note:

$$\mathcal{P}(\mathsf{E}_M) = \mathsf{x}_M \quad \text{and} \quad \mathcal{Q}(\mathsf{E}_M) = \mathsf{x}_M. \tag{4.110}$$

For q not a root of unity,

$$\mathcal{P}(\Gamma_{q,M}) = \mathsf{x}_M \quad \text{and} \quad \mathcal{Q}(\Gamma^*_{q,M}) = \mathsf{x}_M. \tag{4.111}$$

These generalize (4.20) and (4.97).

As a consequence of (4.108) and (4.110), we have commutative diagrams of functors

$$\begin{array}{ccc} \mathcal{A}\text{-}\mathsf{CSp} & \underset{\mathcal{P}}{\overset{\mathcal{S}}{\rightleftarrows}} & {}^{\mathrm{co}}\mathsf{Bimon}^{\mathrm{co}}(\mathcal{A}\text{-}\mathsf{CSp}) \\ \nwarrow & & \nearrow \\ & W\text{-}\mathsf{Mod} & \end{array} \qquad \begin{array}{ccc} {}^{\mathrm{co}}\mathsf{Bimon}^{\mathrm{co}}(\mathcal{A}\text{-}\mathsf{CSp}) & \underset{\mathcal{S}^\vee}{\overset{\mathcal{Q}}{\rightleftarrows}} & \mathcal{A}\text{-}\mathsf{CSp} \\ \nwarrow & & \nearrow \\ & W\text{-}\mathsf{Mod}, & \end{array}$$

the oblique arrows being (4.4) and (4.102). There are similar diagrams involving (4.105), with $\mathcal{S}$ and $\mathcal{S}^\vee$ replaced by $\mathcal{E}$ and $\mathcal{E}^\vee$.

4.9.4. q-norm map on decorated chambers. For a scalar q, the *q-norm map* on decorated chambers

$$\kappa_{q,M} : \Gamma_{q,M} \to \Gamma^*_{q,M}, \qquad m_{C/A} \mapsto \sum_{D:\, D \geq A} q^{\mathrm{dist}(C,D)}\, m_{D/A} \tag{4.112}$$

on the A-component, is a self-dual morphism of Coxeter q-bimonoids. This is the special case $\mathsf{p} := \mathsf{x}_M$ of (3.144). Hence by Lemma 3.78, the map (4.112) is an isomorphism when q is not a root of unity.

For $q = \pm 1$, we have commutative diagrams

$$\begin{array}{ccc} \Gamma_M & \xrightarrow{\kappa_{1,M}} & \Gamma^*_M \\ \downarrow & & \uparrow \\ \mathsf{E}_M & \xrightarrow[\cong]{} & \mathsf{E}_M \end{array} \qquad \begin{array}{ccc} \Gamma_{-1,M} & \xrightarrow{\kappa_{-1,M}} & \Gamma^*_{-1,M} \\ \downarrow & & \uparrow \\ \mathsf{E}^-_M & \xrightarrow[\cong]{} & \mathsf{E}^-_M \end{array} \tag{4.113}$$

of Coxeter bimonoids and of signed Coxeter bimonoids, respectively. This is the special case $\mathsf{p} := \mathsf{x}_M$ of (3.145).

Exercise 4.29. Check: The result of Exercise 4.13 holds more generally for $\Gamma_{q,M}$ instead of Γ_q, and $\Gamma^*_{q,M}$ instead of Γ^*_q.

4.9.5. Primitive and group-like series. Let $\mathrm{char}(\Bbbk) \nmid |W|$. Let us start with Γ_M. A series v of Γ_M is a family (v_F) indexed by faces F with $v_F \in \Gamma_M[F]$ such that $\beta_{G,F}(v_F) = v_G$ and $w(v_F) = v_{wF}$. Thus, $v_F = \sum_{w \in W_F} (w \cdot m)_{wC/F}$ for some $m \in M$ with $C \geq F$ fixed. Observe:

$$\mathscr{X}(\Gamma_M) = \mathscr{S}(\mathsf{x}_M) = M^W.$$

Explicitly, $m \in M^W$ yields the linear series x of Γ_M defined by $x_C := m_{C/C}$. The corresponding exponential series e of Γ_M under the bijection (3.134) is

$$e_A = \frac{1}{|W_A|} \sum_{C:\, C \geq A} m_{C/A}. \tag{4.114}$$

This follows by specializing formula (3.135). For $M = \Bbbk$, we recover (4.32) up to a scalar multiple.

Let us now look at E_M. We work with the flat formulation. A series v of E_M is a family (v_{X}) indexed by flats X with $v_{\mathrm{X}} \in M$ such that $w \cdot v_{\mathrm{X}} = v_{w\mathrm{X}}$. In particular, v_{X} is invariant under the group $\widehat{W}_{\mathrm{X}}$ for each X. A series v is primitive iff $v_{\mathrm{X}} = 0$ for all $\mathrm{X} \neq \top$. Thus,

$$\mathscr{P}(\mathsf{E}_M) = \mathscr{X}(\mathsf{E}_M) = \mathscr{S}(\mathsf{x}_M) = M^W.$$

A series v is group-like iff $v_{\mathrm{X}} = v_{\mathrm{Y}}$ for all X and Y. Observe: Every group-like series is exponential. Correspondences (3.141) and (3.142) between (linear) primitive series and (exponential) group-like series of E_M are evident.

Since $(\Gamma_M)_{ab} = \mathsf{E}_M$, the above observations are also consistent with Exercise 3.69.

4.9.6. Antipode. The antipode S of the Coxeter bimonoid E_M, on the A-component, is given by scalar multiplication by $(-1)^{\dim(\mathcal{A})}$ (which is independent of A).

The antipode S of the Coxeter q-bimonoid $\Gamma_{q,M}$ and of $\Gamma^*_{q,M}$, on the A-component, is

$$S_A(m_{C/A}) = (-1)^{\dim(\mathcal{A})} q^{\mathrm{dist}(C, A\overline{C})} m_{A\overline{C}/A}. \tag{4.115}$$

4.9.7. Decoration functor on Coxeter species. Fix a left W-module M. For a Coxeter species p, define the Coxeter species

$$\mathsf{p}_M := \mathsf{p} \times \mathsf{E}_M. \tag{4.116}$$

Thus, $\mathsf{p}_M[F] = \mathsf{p}[F] \otimes M$ for any face F. We refer to the functor

$$(-)_M : \mathcal{A}\text{-}\mathsf{CSp} \to \mathcal{A}\text{-}\mathsf{CSp}, \qquad \mathsf{p} \mapsto \mathsf{p}_M \tag{4.117}$$

as the *decoration functor* on Coxeter species. It preserves Coxeter monoids and Coxeter comonoids. More generally, for a scalar q, it preserves Coxeter q-bimonoids. Formally, by the second part of Theorem 3.25, the decoration functor

$$(-)_M : (\mathcal{T}, \mathcal{T}^\vee, \lambda_q) \to (\mathcal{T}, \mathcal{T}^\vee, \lambda_q)$$

is bilax. The lax and colax structure maps, evaluated on a Coxeter species p on the A-component

$$\bigoplus_{F:\,F\geq A} \mathsf{p}[F]\otimes M \underset{\text{colax}}{\overset{\text{lax}}{\rightleftarrows}} \Big(\bigoplus_{F:\,F\geq A} \mathsf{p}[F]\Big)\otimes M \tag{4.118}$$

are the canonical identifications, and inverses of each other. Since E_M is bicommutative, the decoration functor also preserves (co)commutative Coxeter (co)monoids.

Example 4.30. The Coxeter species x_M, as suggested by the notation, is the image of x under the decoration functor. Similarly, E_M, E^-_M, $\Gamma_{q,M}$, $\Gamma^*_{q,M}$ are images of E, E^-, Γ_q, Γ^*_q, respectively. □

For a Coxeter species p, we also define the Coxeter species

$$\mathsf{p}^-_M := \mathsf{p}\times\mathsf{E}^-_M. \tag{4.119}$$

This yields the *signed decoration functor*. In general, we have Coxeter species

$$\mathsf{p}_M = \mathsf{p}\times\mathsf{E}_M,\quad \mathsf{p}^-_M = \mathsf{p}^-\times\mathsf{E}_M,\quad \mathsf{p}^-_{M_-} = \mathsf{p}_-\times\mathsf{E}_M,\quad \mathsf{p}_{M_-} = \mathsf{p}^-_-\times\mathsf{E}_M.$$

For $\mathsf{p} := \mathsf{x}$, these four objects reduce to two, see Exercise 4.6.

Duality. We deduce from Exercise 4.25 that for p a Coxeter species, $(\mathsf{p}_M)^* = (\mathsf{p}^*)_{M^*}$ (assuming either p or M is finite-dimensional). For $\mathsf{p} := \mathsf{x}$, this recovers Exercise 4.5. Similarly, for h a Coxeter q-bimonoid, $(\mathsf{h}_M)^* = (\mathsf{h}^*)_{M^*}$ (assuming either h or M is finite-dimensional). In particular, $(\Gamma_{q,M})^* = \Gamma^*_{q,M^*}$.

Universal constructions. For a scalar q, a Coxeter comonoid c and Coxeter monoid a,

$$\mathcal{T}_q(\mathsf{c}\times\mathsf{E}_M) = \mathcal{T}_q(\mathsf{c})\times\mathsf{E}_M \quad\text{and}\quad \mathcal{T}^\vee_q(\mathsf{a}\times\mathsf{E}_M) = \mathcal{T}^\vee_q(\mathsf{a})\times\mathsf{E}_M \tag{4.120}$$

as Coxeter q-bimonoids. This is a formal consequence since structure maps (4.118) are isomorphisms.

Example 4.31. For $\mathsf{c} = \mathsf{a} := \mathsf{x}$, the above yields (4.109) from (4.27). Similarly, for $\mathsf{c} := \mathsf{E}$ and $\mathsf{a} := \mathsf{E}$, using (4.63), for a scalar q,

$$\Sigma_{q,M} := (\Sigma_q)_M = \mathcal{T}_q(\mathsf{E}_M) \quad\text{and}\quad \Sigma^*_{q,M} := (\Sigma^*_q)_M = \mathcal{T}^\vee_q(\mathsf{E}_M) \tag{4.121}$$

as Coxeter q-bimonoids, with identifications made using $\mathtt{H}$- and $\mathtt{M}$-bases. Similarly, using (4.64), for $\mathrm{char}(\Bbbk) \nmid |W|$,

$$\Sigma_M = \mathcal{T}(\mathsf{E}_M) \quad\text{and}\quad \Sigma^*_M = \mathcal{T}^\vee(\mathsf{E}_M) \tag{4.122}$$

as Coxeter bimonoids, with E_M viewed as a Coxeter species, with identifications made using $\mathtt{Q}$- and $\mathtt{P}$-bases. Similar results can be written using (4.65) for q not a root of unity. □

Similarly, for a Coxeter species p,

$$\mathcal{S}(\mathsf{p}\times\mathsf{E}_M) = \mathcal{S}(\mathsf{p})\times\mathsf{E}_M \quad\text{and}\quad \mathcal{E}(\mathsf{p}\times\mathsf{E}_M) = \mathcal{E}(\mathsf{p})\times\mathsf{E}_M \tag{4.123}$$

as Coxeter bimonoids and signed Coxeter bimonoids, respectively. For $\mathsf{p} := \mathsf{x}$, this recovers (4.108).

Primitive and indecomposable parts. For a Coxeter comonoid c and Coxeter monoid a,

$$\mathcal{P}(\mathsf{c}_M) = \mathcal{P}(\mathsf{c})_M \quad \text{and} \quad \mathcal{Q}(\mathsf{a}_M) = \mathcal{Q}(\mathsf{a})_M. \tag{4.124}$$

In particular,

$$\mathcal{P}(\mathsf{E}_M) = \mathsf{x}_M, \quad \mathcal{P}(\Gamma_M) = \mathsf{Lie}_M, \quad \mathcal{P}(\Sigma_M) = \mathsf{Zie}_M.$$

See also formula (6.14) in this regard. We use the terms decorated Lie elements and decorated Zie elements for elements in the components of Lie_M and Zie_M.

Abelianization. For a Coxeter monoid a and Coxeter comonoid c,

$$(\mathsf{a}_{ab})_M = (\mathsf{a}_M)_{ab} \quad \text{and} \quad (\mathsf{c}^{coab})_M = (\mathsf{c}_M)^{coab}. \tag{4.125}$$

Series. Recall: $\mathscr{X}(\mathsf{p}) = \mathsf{p}[\top]^W$. Hence, $\mathscr{X}(\mathsf{p}_M) = (\mathsf{p}[\top] \otimes M)^W$. In particular, $\mathscr{X}(\mathsf{E}_M) = \mathscr{X}(\Gamma_M) = \mathscr{X}(\Sigma_M) = M^W$ since the $\top$-component of all these Coxeter species is $\Bbbk$ with trivial W-action.

Antipode. For a scalar q, for a Coxeter q-bimonoid h,

$$S_{\mathsf{h} \times \mathsf{E}_M} = S_{\mathsf{h}} \times \mathrm{id}_{\mathsf{E}_M}. \tag{4.126}$$

For $\mathsf{h} := \Gamma_q$ and $\mathsf{h} := \Gamma_q^*$, this yields the antipode formula (4.115) from (4.36) and (4.37).

Functoriality. The decoration functor is natural in M. More precisely, for a scalar q, we have a functor

$$W\text{-}\mathsf{Mod} \times q\text{-}\mathsf{Bimon}(\mathcal{A}\text{-}\mathsf{CSp}) \to q\text{-}\mathsf{Bimon}(\mathcal{A}\text{-}\mathsf{CSp}), \qquad (M, \mathsf{h}) \mapsto \mathsf{h}_M \tag{4.127}$$

obtained by composing (4.102) with the Hadamard functor.

4.10. Notes

Examples in this chapter generalize well-known examples of Joyal bimonoids to the Coxeter setting. A comparison between the two is given in Table 4.3. Also see Table 1.9 in this regard.

TABLE 4.3. Relating Coxeter bimonoids and Joyal bimonoids.

Coxeter bimonoids	Joyal bimonoids
exponential Coxeter bimonoid	exponential Joyal bimonoid
Coxeter bimonoid of chambers	Joyal bimonoid of linear orders
Coxeter bimonoid of faces	Joyal bimonoid of set compositions
Coxeter bimonoid of top-nested faces	Joyal bimonoid of linear set compositions
Coxeter bimonoid of top-lunes	Joyal bimonoid of linear set partitions
Coxeter bimonoid of flats	Joyal bimonoid of set partitions
Coxeter bimonoid of pairs of chambers	Joyal bimonoid of pairs of linear orders

4.10.1. Joyal species. The Joyal bimonoids in Table 4.3 were studied in detail in our monograph [**7**, Chapter 12]. This work emphasized both the combinatorial and geometric viewpoints. For more specific pointers, see [**11**, page 331]. The Joyal monoid of linear orders first appeared in work of Barratt in a different guise and under the name "permutation bi-ring" [**30**, page 10]. For examples of Joyal set-species, see [**36**, Section 1.1]; for monoids in Joyal set-species, see [**259**].

Any module M over the symmetric group S_n gives rise to the Joyal species p with $\mathsf{p}[n] = M$, and $\mathsf{p}[j] = 0$ for $j \neq n$. This is the classical analogue of the construction in Exercise 4.3. A S_n-module of the form $M = V^{\otimes n}$ gives rise to the Joyal species p with $\mathsf{p}[1] = V$, and $\mathsf{p}[j] = 0$ for $j \neq 1$. This is the analogue of the construction (4.3). Note very carefully that it cannot handle arbitrary S_n-modules. This shows the flexibility of working with Coxeter species over Joyal species.

The decorated exponential Joyal species with I-component $V^{\otimes I}$ which appears implicitly in Example 4.26 is denoted E_V in [**7**, Example 8.18], see also [**290**, page 204], [**62**, Example 4.2.1.2]. Its signed analogue E_V^- is mentioned in [**7**, Section 19.7.1].

4.10.2. Species and Coxeter species. The Coxeter bimonoids in Table 4.3, in the less structured setting of hyperplane arrangements, were studied in detail in [**11**, Chapter 7 and Section 15.5]. To avoid repetition, our presentation of these examples here is brief. Formulas (4.25) and (4.57) are new. The results of Exercises 4.19 and 4.20 were not explicitly mentioned in [**11**]. The maps in diagram (4.92) first appeared in [**6**, Diagram (5.8)].

The material in Section 4.9 is built from a W-module M, so it is specific to Coxeter species. For the analogue in species (for hyperplane arrangements), one starts with a vector space M, and defines the decorated exponential species, and so on. This is briefly mentioned in [**11**, Section 7.2.6].

4.10.3. More examples of Coxeter bimonoids. There are many interesting examples of Coxeter bimonoids not discussed in this chapter. For instance, we have the Coxeter q-bimonoid of pairs of faces generalizing the Coxeter q-bimonoid of pairs of chambers; see [**11**, Table 15.1] in this regard. Other examples include the Coxeter bimonoids of charts and dicharts, Coxeter bimonoid of top-cones, Coxeter bimonoid of shards, and so on. These generalize the Joyal bimonoid of graphs [**7**, Section 13.2], [**8**, Section 9.4], the Joyal bimonoid of partial orders [**7**, Section 13.1], and so on. We plan to give more details in future work.

CHAPTER 5

Coxeter operads

Let $\mathcal{A}$ be a fixed reflection arrangement with Coxeter group W.

Coxeter dispecies and Coxeter operads. We introduce the category of Coxeter $\mathcal{A}$-dispecies. It carries a monoidal structure called the substitution product. Monoids wrt this product are Coxeter $\mathcal{A}$-operads, while comonoids are Coxeter $\mathcal{A}$-cooperads. Basic examples are the Coxeter commutative, associative, Lie operads which we denote **Com**, **As**, **Lie**, respectively. They have signed analogues, namely, $\mathbf{Com}^-$, $\mathbf{As}^-$, $\mathbf{Lie}^-$. (It turns out that $\mathbf{As}^-$ is isomorphic to **As**, so there is no need to consider it explicitly.) Duals of any of these Coxeter $\mathcal{A}$-operads provide examples of Coxeter $\mathcal{A}$-cooperads. They are denoted $\mathbf{Com}^*$, $\mathbf{As}^*$, $\mathbf{Lie}^*$, and so on.

We mention that the category of Coxeter $\mathcal{A}$-dispecies also carries another monoidal structure called the Hadamard product.

Coxeter operad monoids. We introduce the substitution operation of a Coxeter $\mathcal{A}$-species into a Coxeter $\mathcal{A}$-dispecies. This turns Coxeter $\mathcal{A}$-species into a left module category over the monoidal category of Coxeter $\mathcal{A}$-dispecies (under the substitution product). Thus, each Coxeter $\mathcal{A}$-operad gives rise to a monad on Coxeter $\mathcal{A}$-species. The Coxeter associative operad **As** yields the monad $\mathcal{T}$, while the Coxeter commutative operad **Com** yields the monad $\mathcal{S}$. Further, we have left modules over a Coxeter $\mathcal{A}$-operad which we call Coxeter operad monoids. They are the same as algebras over the corresponding monad. Thus, a Coxeter **As**-monoid is the same as a Coxeter monoid, while a Coxeter **Com**-monoid is the same as a commutative Coxeter monoid. Similarly, the Coxeter $\mathcal{A}$-operad $\mathbf{Com}^-$ yields the monad $\mathcal{E}$, and a Coxeter $\mathbf{Com}^-$-monoid is the same as a signed commutative Coxeter monoid. A Coxeter **Lie**-monoid is the same as a Coxeter Lie monoid; the latter is formally introduced in Chapter 6. Dual remarks apply to Coxeter $\mathcal{A}$-cooperads and comonads on Coxeter $\mathcal{A}$-species.

Operads. For any hyperplane arrangement $\mathcal{A}$, one can define notions of $\mathcal{A}$-dispecies and $\mathcal{A}$-operads. These were introduced in our previous monograph [**11**, Chapter 4]. When $\mathcal{A}$ is a reflection arrangement, one can impose additional structure on them by using the action of the Coxeter group. This results in notions of Coxeter $\mathcal{A}$-dispecies and Coxeter $\mathcal{A}$-operads.

Convention 5.1. Whenever convenient, we drop $\mathcal{A}$ from the notation, and write Coxeter dispecies and Coxeter operad instead of Coxeter $\mathcal{A}$-dispecies and Coxeter $\mathcal{A}$-operad. This is consistent with Convention 2.1 for Coxeter species.

Coxeter operads can be approached in many different ways. In this chapter, we approach them using flats. Other approaches involving top-lunes and face-types are developed later in Chapter 11.

5.1. Coxeter dispecies

We introduce Coxeter dispecies. For a reflection arrangement $\mathcal{A}$, the category of Coxeter $\mathcal{A}$-dispecies is the category of functors from $\mathcal{A}\text{-}\mathsf{dCox}$ to Vec, where $\mathcal{A}\text{-}\mathsf{dCox}$ is a certain finite groupoid whose objects are nested flats.

5.1.1. Category with nested flats as objects. Let $\mathcal{A}$ be a reflection arrangement with Coxeter group W. Recall: A nested flat is a pair of flats (X, Y) such that $\mathrm{X} \leq \mathrm{Y}$. Define the category D as follows. Objects are nested flats (X, Y) and morphisms are

$$w : (\mathrm{X}, \mathrm{Y}) \to (\mathrm{X}', \mathrm{Y}')$$

with $w \in W$ such that $w(\mathrm{X}) = \mathrm{X}'$ and $w(\mathrm{Y}) = \mathrm{Y}'$. A morphism u composed with a morphism v is the morphism vu (product in the group W). In other words, D is the groupoid of elements for the action of W on the set of nested flats.

Let $\mathcal{A}\text{-}\mathsf{dCox}$ denote the quotient category of D obtained by imposing the following relation. For any $w \in W$ which fixes a flat Y pointwise, the morphism $w : (\mathrm{X}, \mathrm{Y}) \to (\mathrm{X}, \mathrm{Y})$ equals the identity morphism. Equivalently, morphisms $w_1, w_2 : (\mathrm{X}, \mathrm{Y}) \to (\mathrm{X}', \mathrm{Y}')$ are equal whenever w_1 and w_2 agree on Y (that is, $w_1(F) = w_2(F)$ for any face F with support Y).

5.1.2. Coxeter dispecies. Define the category of Coxeter $\mathcal{A}$-dispecies to be the functor category

$$\mathcal{A}\text{-}\mathsf{CdSp} := [\mathcal{A}\text{-}\mathsf{dCox}, \mathsf{Vec}].$$

Explicitly: A Coxeter $\mathcal{A}$-dispecies $\mathbf{p}$ consists of a family of vector spaces $\mathbf{p}[\mathrm{X}, \mathrm{Y}]$, one for each $\mathrm{X} \leq \mathrm{Y}$, equipped with linear maps

$$w : \mathbf{p}[\mathrm{X}, \mathrm{Y}] \to \mathbf{p}[\mathrm{X}', \mathrm{Y}'],$$

one for each $w \in W$ satisfying $w(\mathrm{X}) = \mathrm{X}'$ and $w(\mathrm{Y}) = \mathrm{Y}'$, such that the diagrams

$$\begin{array}{ccc} & \mathbf{p}[\mathrm{X}', \mathrm{Y}'] & \\ {}^{u}\nearrow & & \searrow^{v} \\ \mathbf{p}[\mathrm{X}, \mathrm{Y}] & \xrightarrow[vu]{} & \mathbf{p}[\mathrm{X}'', \mathrm{Y}''] \end{array} \qquad (\mathbf{p}[\mathrm{X}, \mathrm{Y}] \xrightarrow{e} \mathbf{p}[\mathrm{X}, \mathrm{Y}]) = \mathrm{id} \tag{5.1}$$

commute. In addition,

$$(\mathbf{p}[\mathrm{X}, \mathrm{Y}] \xrightarrow{w_1} \mathbf{p}[\mathrm{X}', \mathrm{Y}']) = (\mathbf{p}[\mathrm{X}, \mathrm{Y}] \xrightarrow{w_2} \mathbf{p}[\mathrm{X}', \mathrm{Y}'])$$

whenever w_1 and w_2 agree on Y. Equivalently, it suffices to require that

$$(\mathbf{p}[\mathrm{X}, \mathrm{Y}] \xrightarrow{w} \mathbf{p}[\mathrm{X}, \mathrm{Y}]) = \mathrm{id}$$

whenever w fixes Y pointwise.

We refer to the linear maps w as structure maps of $\mathbf{p}$.

A map of Coxeter $\mathcal{A}$-dispecies $f : \mathbf{p} \to \mathbf{q}$ consists of a family of linear maps

$$f_{X,Y} : \mathbf{p}[X, Y] \to \mathbf{q}[X, Y],$$

one for each $X \leq Y$, such that the diagram

$$\begin{array}{ccc} \mathbf{p}[X, Y] & \xrightarrow{f_{X,Y}} & \mathbf{q}[X, Y] \\ {\scriptstyle w}\downarrow & & \downarrow{\scriptstyle w} \\ \mathbf{p}[X', Y'] & \xrightarrow[f_{X',Y'}]{} & \mathbf{q}[X', Y'] \end{array} \tag{5.2}$$

commutes for each morphism w.

A Coxeter $\mathcal{A}$-dispecies $\mathbf{p}$ is *finite-dimensional* if the vector spaces $\mathbf{p}[X, Y]$ have finite dimension for all $X \leq Y$.

5.1.3. Product and coproduct. The *zero Coxeter dispecies* is the Coxeter $\mathcal{A}$-dispecies with all components zero. It is the initial and terminal object in the category of Coxeter $\mathcal{A}$-dispecies.

For Coxeter $\mathcal{A}$-dispecies $\mathbf{p}$ and $\mathbf{q}$, define their *direct sum* $\mathbf{p} + \mathbf{q}$ by

$$(\mathbf{p} + \mathbf{q})[X, Y] := \mathbf{p}[X, Y] \oplus \mathbf{q}[X, Y]. \tag{5.3}$$

The structure maps are as follows. For a morphism $w : (X, Y) \to (X', Y')$, define the linear map

$$\mathbf{p}[X, Y] \oplus \mathbf{q}[X, Y] \xrightarrow{(w,w)} \mathbf{p}[X', Y'] \oplus \mathbf{q}[X', Y'].$$

This is the product and coproduct in the category of Coxeter $\mathcal{A}$-dispecies. It is clear that arbitrary products and coproducts also exist in this category.

5.1.4. Dispecies vs Coxeter dispecies. Let $\mathcal{A}$ be any hyperplane arrangement. Let $\mathcal{A}$-$\mathtt{dHyp}$ denote the discrete category whose objects are nested flats. (The only morphisms are identities.) The category of $\mathcal{A}$-dispecies is defined to be the functor category

$$\mathcal{A}\text{-}\mathsf{dSp} = [\mathcal{A}\text{-}\mathtt{dHyp}, \mathsf{Vec}].$$

For a reflection arrangement $\mathcal{A}$, the category $\mathcal{A}$-$\mathtt{dHyp}$ is a subcategory of $\mathcal{A}$-$\mathtt{dCox}$. Thus, every Coxeter $\mathcal{A}$-dispecies is also an $\mathcal{A}$-dispecies. More precisely, a Coxeter $\mathcal{A}$-dispecies is an $\mathcal{A}$-dispecies equipped with structure maps arising from the action of the Coxeter group on nested flats.

5.2. Coxeter operads

We introduce the substitution product on Coxeter dispecies, which turns the latter into a monoidal category. Monoids wrt this product are Coxeter operads, while comonoids wrt this product are Coxeter cooperads.

5.2.1. Substitution product of Coxeter dispecies. Let $\mathcal{A}$ be a reflection arrangement with Coxeter group W. Let $\mathbf{p}$ and $\mathbf{q}$ be two Coxeter $\mathcal{A}$-dispecies. Define the Coxeter $\mathcal{A}$-dispecies $\mathbf{p} \circ \mathbf{q}$ by

$$(\mathbf{p} \circ \mathbf{q})[\mathrm{X}, \mathrm{Z}] := \bigoplus_{\mathrm{Y}: \mathrm{X} \leq \mathrm{Y} \leq \mathrm{Z}} \mathbf{p}[\mathrm{X}, \mathrm{Y}] \otimes \mathbf{q}[\mathrm{Y}, \mathrm{Z}]. \tag{5.4}$$

The sum is over all flats Y which lie between X and Z. The structure maps are as follows. For a morphism $w : (\mathrm{X}, \mathrm{Z}) \to (\mathrm{X}', \mathrm{Z}')$, define the linear map

$$w : (\mathbf{p} \circ \mathbf{q})[\mathrm{X}, \mathrm{Z}] \to (\mathbf{p} \circ \mathbf{q})[\mathrm{X}', \mathrm{Z}']$$

by summing the linear maps

$$\mathbf{p}[\mathrm{X}, \mathrm{Y}] \otimes \mathbf{q}[\mathrm{Y}, \mathrm{Z}] \xrightarrow{w \otimes w} \mathbf{p}[\mathrm{X}', \mathrm{Y}'] \otimes \mathbf{q}[\mathrm{Y}', \mathrm{Z}'].$$

(A key observation is that when w_1 and w_2 agree on Z, they also agree on any $\mathrm{Y} \leq \mathrm{Z}$.) We refer to $\mathbf{p} \circ \mathbf{q}$ as the *substitution product* of $\mathbf{p}$ and $\mathbf{q}$. It defines a monoidal structure on the category of Coxeter $\mathcal{A}$-dispecies. The unit object is the Coxeter $\mathcal{A}$-dispecies $\mathbf{x}$ defined by

$$\mathbf{x}[\mathrm{X}, \mathrm{Y}] := \begin{cases} \Bbbk & \text{if } \mathrm{X} = \mathrm{Y}, \\ 0 & \text{otherwise,} \end{cases} \tag{5.5}$$

with structure maps defined to be all identities.

Exercise 5.2. Construct the internal hom for the substitution product of Coxeter dispecies.

5.2.2. Coxeter operads. A Coxeter $\mathcal{A}$-operad is a monoid in the monoidal category of Coxeter $\mathcal{A}$-dispecies $(\mathcal{A}\text{-}\mathtt{CdSp}, \circ, \mathbf{x})$. A morphism of Coxeter $\mathcal{A}$-operads is a morphism of monoids.

Explicitly, a Coxeter $\mathcal{A}$-operad is a Coxeter $\mathcal{A}$-dispecies $\mathbf{a}$ equipped with linear maps

$$\gamma : \mathbf{a}[\mathrm{X}, \mathrm{Y}] \otimes \mathbf{a}[\mathrm{Y}, \mathrm{Z}] \to \mathbf{a}[\mathrm{X}, \mathrm{Z}] \quad \text{and} \quad \eta : \Bbbk \to \mathbf{a}[\mathrm{X}, \mathrm{X}],$$

the former for each $\mathrm{X} \leq \mathrm{Y} \leq \mathrm{Z}$ and the latter for each X, subject to the axioms below.

Naturality. For any $\mathrm{X} \leq \mathrm{Y} \leq \mathrm{Z}$ and $w \in W$, the first diagram below commutes.

$$\begin{array}{ccc} \mathbf{a}[\mathrm{X}, \mathrm{Y}] \otimes \mathbf{a}[\mathrm{Y}, \mathrm{Z}] & \xrightarrow{\gamma} & \mathbf{a}[\mathrm{X}, \mathrm{Z}] \\ {\scriptstyle w \otimes w}\downarrow & & \downarrow{\scriptstyle w} \\ \mathbf{a}[\mathrm{X}', \mathrm{Y}'] \otimes \mathbf{a}[\mathrm{Y}', \mathrm{Z}'] & \xrightarrow[\gamma]{} & \mathbf{a}[\mathrm{X}', \mathrm{Z}'] \end{array} \qquad \begin{array}{ccc} \mathbf{a}[\mathrm{X}, \mathrm{X}] & \xrightarrow{w} & \mathbf{a}[\mathrm{X}', \mathrm{X}'] \\ {\scriptstyle \eta}\nwarrow & & \nearrow{\scriptstyle \eta} \\ & \Bbbk & \end{array} \tag{5.6a}$$

For any X and $w \in W$, the second diagram above commutes.

Associativity. For any $\mathrm{X} \leq \mathrm{Y} \leq \mathrm{Z} \leq \mathrm{W}$, the diagram

$$\begin{array}{ccc} \mathbf{a}[\mathrm{X}, \mathrm{Y}] \otimes \mathbf{a}[\mathrm{Y}, \mathrm{Z}] \otimes \mathbf{a}[\mathrm{Z}, \mathrm{W}] & \xrightarrow{\gamma \otimes \mathrm{id}} & \mathbf{a}[\mathrm{X}, \mathrm{Z}] \otimes \mathbf{a}[\mathrm{Z}, \mathrm{W}] \\ {\scriptstyle \mathrm{id} \otimes \gamma}\downarrow & & \downarrow{\scriptstyle \gamma} \\ \mathbf{a}[\mathrm{X}, \mathrm{Y}] \otimes \mathbf{a}[\mathrm{Y}, \mathrm{W}] & \xrightarrow[\gamma]{} & \mathbf{a}[\mathrm{X}, \mathrm{W}] \end{array} \tag{5.6b}$$

commutes.

Unitality. For any $\mathrm{X} \leq \mathrm{Y}$, the diagrams

$$\begin{array}{ccc} & \mathbf{a}[\mathrm{X},\mathrm{Y}] \otimes \mathbf{a}[\mathrm{Y},\mathrm{Y}] & \\ {}^{\mathrm{id}\otimes\eta}\nearrow & & \searrow^{\gamma} \\ \mathbf{a}[\mathrm{X},\mathrm{Y}] \otimes \Bbbk & \xrightarrow[\cong]{} & \mathbf{a}[\mathrm{X},\mathrm{Y}] \end{array} \qquad \begin{array}{ccc} & \mathbf{a}[\mathrm{X},\mathrm{X}] \otimes \mathbf{a}[\mathrm{X},\mathrm{Y}] & \\ {}^{\eta\otimes\mathrm{id}}\nearrow & & \searrow^{\gamma} \\ \Bbbk \otimes \mathbf{a}[\mathrm{X},\mathrm{Y}] & \xrightarrow[\cong]{} & \mathbf{a}[\mathrm{X},\mathrm{Y}] \end{array} \tag{5.6c}$$

commute.

We refer to γ as the *substitution map* of $\mathbf{a}$, and to η as the *unit map* of $\mathbf{a}$. We also refer to both of them as structure maps of $\mathbf{a}$.

A morphism of Coxeter $\mathcal{A}$-operads is a map $f : \mathbf{a} \to \mathbf{b}$ of Coxeter $\mathcal{A}$-dispecies such that the diagrams

$$\begin{array}{ccc} \mathbf{a}[\mathrm{X},\mathrm{Y}] \otimes \mathbf{a}[\mathrm{Y},\mathrm{Z}] & \xrightarrow{\gamma} & \mathbf{a}[\mathrm{X},\mathrm{Z}] \\ {\scriptstyle f_{\mathrm{X},\mathrm{Y}} \otimes f_{\mathrm{Y},\mathrm{Z}}}\downarrow & & \downarrow{\scriptstyle f_{\mathrm{X},\mathrm{Z}}} \\ \mathbf{b}[\mathrm{X},\mathrm{Y}] \otimes \mathbf{b}[\mathrm{Y},\mathrm{Z}] & \xrightarrow[\gamma]{} & \mathbf{b}[\mathrm{X},\mathrm{Z}] \end{array} \qquad \begin{array}{ccc} \mathbf{a}[\mathrm{X},\mathrm{X}] & \xrightarrow{f_{\mathrm{X},\mathrm{X}}} & \mathbf{b}[\mathrm{X},\mathrm{X}] \\ {}^{\eta}\nwarrow & & \nearrow^{\eta} \\ & \Bbbk & \end{array} \tag{5.7}$$

commute.

5.2.3. Connected Coxeter operads. A Coxeter $\mathcal{A}$-operad $\mathbf{a}$ is *connected* if the components $\mathbf{a}[\mathrm{X},\mathrm{X}]$ are one-dimensional for all X. In this case, the unit maps $\eta : \Bbbk \to \mathbf{a}[\mathrm{X},\mathrm{X}]$ are all linear isomorphisms. As a result, one can forget components of the form $\mathbf{a}[\mathrm{X},\mathrm{X}]$ without loss of information. This leads to an equivalent notion of a positive Coxeter operad. See [**11**, Section 4.4] for related ideas in the context of hyperplane arrangements.

5.2.4. Ideals and quotients. Define Coxeter subdispecies and Coxeter quotient dispecies in a manner similar to that for Coxeter species (Section 2.3.6). Similarly, for a map of Coxeter dispecies, we have notions of injectivity, surjectivity, kernel, cokernel, image, coimage.

An *ideal* of a Coxeter $\mathcal{A}$-operad $\mathbf{a}$ is a Coxeter $\mathcal{A}$-subdispecies $\mathbf{q}$ of $\mathbf{a}$ such that $\mathbf{a} \circ \mathbf{q} \to \mathbf{q}$ and $\mathbf{q} \circ \mathbf{a} \to \mathbf{q}$. For an ideal $\mathbf{q}$ of $\mathbf{a}$, the Coxeter quotient $\mathcal{A}$-dispecies $\mathbf{a}/\mathbf{q}$ is a Coxeter $\mathcal{A}$-operad, and the quotient map $\mathbf{a} \twoheadrightarrow \mathbf{a}/\mathbf{q}$ is a morphism of Coxeter $\mathcal{A}$-operads. For a Coxeter $\mathcal{A}$-subdispecies $\mathbf{p}$ of $\mathbf{a}$, there is a smallest ideal of $\mathbf{a}$ containing $\mathbf{p}$. We call it the ideal generated by $\mathbf{p}$.

5.2.5. Quadratic Coxeter operads. Recall: Direct sum is the coproduct in the category of Coxeter dispecies. Moreover, the substitution product of Coxeter dispecies distributes over arbitrary direct sums on either side. Hence, the *free Coxeter $\mathcal{A}$-operad* on a Coxeter $\mathcal{A}$-dispecies $\mathbf{e}$ is given by

$$\mathcal{F}_{\circ}(\mathbf{e}) := \underset{n \geq 0}{+} \mathbf{e}^{\circ n},$$

where $\mathbf{e}^{\circ n}$ is the n-fold substitution product of $\mathbf{e}$ with itself, and $+$ denotes direct sum. Explicitly,

$$\mathcal{F}_{\circ}(\mathbf{e})[\mathrm{X},\mathrm{Z}] = \bigoplus_{\mathrm{X} \leq \mathrm{Y}_1 \leq \cdots \leq \mathrm{Y}_k \leq \mathrm{Z}} \mathbf{e}[\mathrm{X},\mathrm{Y}_1] \otimes \mathbf{e}[\mathrm{Y}_1,\mathrm{Y}_2] \otimes \cdots \otimes \mathbf{e}[\mathrm{Y}_k,\mathrm{Z}],$$

with the sum over all multichains in the interval $[X, Z]$.

A Coxeter $\mathcal{A}$-operad $\mathbf{a}$ is *quadratic* if it can be written as a quotient of a free Coxeter $\mathcal{A}$-operad $\mathcal{F}_\circ(\mathbf{e})$ by an ideal generated by a Coxeter $\mathcal{A}$-subdispecies $\mathbf{r}$ of $\mathbf{e} \circ \mathbf{e}$. We denote such a presentation by $\mathbf{a} = \langle \mathbf{e} \,|\, \mathbf{r} \rangle$. We use the term *binary quadratic* if further $\mathbf{e}[X, Y] = 0$ unless Y covers X.

5.2.6. Coxeter cooperads. Dually, a Coxeter $\mathcal{A}$-cooperad is a comonoid in the monoidal category of Coxeter $\mathcal{A}$-dispecies $(\mathcal{A}\text{-}\mathtt{CdSp}, \circ, \mathbf{x})$. A morphism of Coxeter $\mathcal{A}$-cooperads is a morphism of comonoids.

Explicitly, a Coxeter $\mathcal{A}$-cooperad consists of a Coxeter $\mathcal{A}$-dispecies $\mathbf{c}$ equipped with linear maps

$$\mathbf{c}[X, Z] \to \mathbf{c}[X, Y] \otimes \mathbf{c}[Y, Z] \quad \text{and} \quad \mathbf{c}[X, X] \to \Bbbk$$

subject to naturality, coassociativity, counitality axioms. These are as in (5.6a)–(5.6c); we reverse appropriate arrows.

5.2.7. Duality functor on Coxeter dispecies. A duality functor can be defined on a functor category $[\mathsf{C}, \mathsf{Vec}]$ whenever C is a groupoid, see Remark 2.21. This applies to the category of Coxeter species with $\mathsf{C} = \mathcal{A}\text{-}\mathtt{Cox}$. Similarly, by taking $\mathsf{C} = \mathcal{A}\text{-}\mathtt{dCox}$, we obtain a *duality functor* on the category of Coxeter $\mathcal{A}$-dispecies. Explicitly, the dual of $\mathbf{p}$, denoted $\mathbf{p}^*$, is defined by

$$\mathbf{p}^*[X, Y] := \mathbf{p}[X, Y]^*.$$

The structure maps of $\mathbf{p}^*$ are obtained by dualizing those of $\mathbf{p}$.

The dual of a Coxeter $\mathcal{A}$-cooperad is a Coxeter $\mathcal{A}$-operad, while the dual of a finite-dimensional Coxeter $\mathcal{A}$-operad is a Coxeter $\mathcal{A}$-cooperad.

5.2.8. Coxeter set-operads. Let Set denote the category of sets. By replacing the codomain Vec by Set in the definition of Coxeter dispecies, one obtains the notion of Coxeter set-dispecies. The analogous notion of Coxeter set-species is given in Section 2.3.9.

One defines the substitution product of Coxeter set-dispecies as in (5.4) by replacing tensor product by cartesian product and direct sum by disjoint union. A Coxeter set-operad is a monoid in the monoidal category of Coxeter set-dispecies under substitution.

Linearization of a Coxeter set-dispecies (Coxeter set-operad) yields a Coxeter dispecies (Coxeter operad). Compare with the linearization functor from Coxeter set-species to Coxeter species in (2.27).

5.2.9. Operads vs Coxeter operads. Let $\mathcal{A}$ be any hyperplane arrangement. Formula (5.4) also defines a monoidal structure on the category of $\mathcal{A}$-dispecies. An $\mathcal{A}$-operad is a monoid in $(\mathcal{A}\text{-}\mathtt{dSp}, \circ, \mathbf{x})$. A morphism of $\mathcal{A}$-operads is a morphism of monoids.

For a reflection arrangement $\mathcal{A}$, the forgetful functor

$$(\mathcal{A}\text{-}\mathtt{CdSp}, \circ, \mathbf{x}) \to (\mathcal{A}\text{-}\mathtt{dSp}, \circ, \mathbf{x})$$

is a strong monoidal functor in the sense of [**7**, Section 3.6]; hence it induces a functor on the respective categories of monoids. Thus, every Coxeter $\mathcal{A}$-operad is also an $\mathcal{A}$-operad. More precisely, a Coxeter $\mathcal{A}$-operad is a Coxeter

$\mathcal{A}$-dispecies which carries an $\mathcal{A}$-operad structure and satisfies the naturality axiom (5.6a).

5.3. Hadamard product

Let $\mathcal{A}$ be a reflection arrangement with Coxeter group W. Let $\mathbf{p}$ and $\mathbf{q}$ be two Coxeter $\mathcal{A}$-dispecies. Define the Coxeter $\mathcal{A}$-dispecies $\mathbf{p} \times \mathbf{q}$ by

$$(\mathbf{p} \times \mathbf{q})[\mathrm{X}, \mathrm{Y}] := \mathbf{p}[\mathrm{X}, \mathrm{Y}] \otimes \mathbf{q}[\mathrm{X}, \mathrm{Y}]. \tag{5.8}$$

The structure maps are as follows. For a morphism $w : (\mathrm{X}, \mathrm{Y}) \to (\mathrm{X}', \mathrm{Y}')$, define the linear map

$$\mathbf{p}[\mathrm{X}, \mathrm{Y}] \otimes \mathbf{q}[\mathrm{X}, \mathrm{Y}] \xrightarrow{w \otimes w} \mathbf{p}[\mathrm{X}', \mathrm{Y}'] \otimes \mathbf{q}[\mathrm{X}', \mathrm{Y}'].$$

This is the *Hadamard product* of $\mathbf{p}$ and $\mathbf{q}$. It yields a monoidal structure on the category of Coxeter $\mathcal{A}$-dispecies. The unit object is the exponential Coxeter dispecies $\mathbf{E}$ defined later in Section 5.5.1. In fact, we have a symmetric monoidal category, with the braiding (symmetry) given by interchanging the tensor factors in (5.8).

If $\mathbf{a}$ and $\mathbf{b}$ are Coxeter $\mathcal{A}$-operads, then so is their Hadamard product $\mathbf{a} \times \mathbf{b}$. Moreover, the symmetry $\mathbf{a} \times \mathbf{b} \to \mathbf{b} \times \mathbf{a}$ is a morphism of Coxeter $\mathcal{A}$-operads. (The same is true for Coxeter $\mathcal{A}$-cooperads.) The structure maps of $\mathbf{a} \times \mathbf{b}$ are obtained by tensoring those of $\mathbf{a}$ and $\mathbf{b}$.

5.4. Coxeter operad monoids

We introduce the substitution operation of a Coxeter species into a Coxeter dispecies. This turns Coxeter species into a left module category over the monoidal category of Coxeter dispecies (under the substitution product). This allows us to define left modules over a Coxeter operad in the category of Coxeter species. These are Coxeter operad monoids. A Coxeter operad gives rise to a monad on Coxeter species, which results in an equivalence between Coxeter operad monoids and algebras over the corresponding monad. Dually, we have the notion of Coxeter cooperad comonoids, Coxeter cooperads giving rise to comonads on Coxeter species, and equivalences between Coxeter cooperad comonoids and coalgebras over corresponding comonads.

5.4.1. Coxeter species as a left module over Coxeter dispecies. Let $\mathcal{A}$ be a reflection arrangement with Coxeter group W. The category of Coxeter species is a left module category over the monoidal category of Coxeter dispecies as follows. For a Coxeter $\mathcal{A}$-dispecies $\mathbf{p}$ and a Coxeter $\mathcal{A}$-species m, define the Coxeter $\mathcal{A}$-species $\mathbf{p} \circ \mathsf{m}$ by

$$(\mathbf{p} \circ \mathsf{m})[\mathrm{X}] := \bigoplus_{\mathrm{Y}:\, \mathrm{Y} \geq \mathrm{X}} \mathbf{p}[\mathrm{X}, \mathrm{Y}] \otimes \mathsf{m}[\mathrm{Y}]. \tag{5.9}$$

The structure maps are as follows. For $w : \mathrm{X} \to \mathrm{X}'$, define the linear map

$$w : (\mathbf{p} \circ \mathsf{m})[\mathrm{X}] \to (\mathbf{p} \circ \mathsf{m})[\mathrm{X}']$$

by summing the linear maps

$$\mathbf{p}[\mathrm{X}, \mathrm{Y}] \otimes \mathsf{m}[\mathrm{Y}] \xrightarrow{w \otimes w} \mathbf{p}[\mathrm{X}', \mathrm{Y}'] \otimes \mathbf{m}[\mathrm{Y}'].$$

There are natural isomorphisms $(\mathbf{p} \circ \mathbf{q}) \circ \mathsf{m} \cong \mathbf{p} \circ (\mathbf{q} \circ \mathsf{m})$ and $\mathbf{x} \circ \mathsf{m} \cong \mathsf{m}$ for Coxeter $\mathcal{A}$-dispecies $\mathbf{p}$ and $\mathbf{q}$ and Coxeter $\mathcal{A}$-species m.

We call $\mathbf{p} \circ \mathsf{m}$ the substitution of m in $\mathbf{p}$.

5.4.2. Enriched hom and endomorphism Coxeter operad. We now describe the enriched hom for the left module category of Coxeter species. For background on enriched hom, see [**11**, Appendix B.2.4]. For Coxeter species m and n, $\mathrm{hom}^{\circ}(\mathsf{m}, \mathsf{n})$ is the Coxeter dispecies whose (X, Y)-component consists of all linear maps from $\mathsf{m}[\mathrm{Y}]$ to $\mathsf{n}[\mathrm{X}]$ invariant under conjugation by any $w \in W$ which fixes Y pointwise. The structure map

$$w : \mathrm{hom}^{\circ}(\mathsf{m}, \mathsf{n})[\mathrm{X}, \mathrm{Y}] \to \mathrm{hom}^{\circ}(\mathsf{m}, \mathsf{n})[\mathrm{X}', \mathrm{Y}']$$

is conjugation by w. By general principles [**11**, Proposition B.5], for a Coxeter species m,

$$\mathrm{end}^{\circ}(\mathsf{m}) := \mathrm{hom}^{\circ}(\mathsf{m}, \mathsf{m}) \tag{5.10}$$

is a Coxeter operad. This is the *endomorphism Coxeter operad* of m. In addition, following [**11**, Appendix B.3.3], the category of Coxeter species is enriched over the category of Coxeter dispecies, and is copowered by it by means of the left action (5.9).

Remark 5.3. Note very carefully: The components of $\mathrm{hom}^{\circ}(\mathsf{m}, \mathsf{n})$ for Coxeter species are different from those for the underlying species where we take all linear maps from $\mathsf{m}[\mathrm{Y}]$ to $\mathsf{n}[\mathrm{X}]$, see [**11**, Section 4.11.1].

5.4.3. Coxeter operad monoids. Recall: A Coxeter operad is a monoid wrt substitution. Thus, for a Coxeter operad $\mathbf{a}$, we have the category of left modules in Coxeter species. We call it the category of Coxeter $\mathbf{a}$-monoids and denote it by $\mathbf{a}\text{-}\mathsf{Mon}(\mathcal{A}\text{-}\mathsf{CSp})$. Let us spell this out.

A Coxeter $\mathbf{a}$-monoid is a Coxeter species m equipped with linear maps

$$\gamma : \mathbf{a}[\mathrm{X}, \mathrm{Y}] \otimes \mathsf{m}[\mathrm{Y}] \to \mathsf{m}[\mathrm{X}],$$

one for each $\mathrm{X} \leq \mathrm{Y}$, subject to the axioms below.

Naturality. For any $\mathrm{X} \leq \mathrm{Y}$ and $w \in W$, the diagram

$$\begin{array}{ccc} \mathbf{a}[\mathrm{X}, \mathrm{Y}] \otimes \mathsf{m}[\mathrm{Y}] & \xrightarrow{\gamma} & \mathsf{m}[\mathrm{X}] \\ {\scriptstyle w \otimes w}\downarrow & & \downarrow{\scriptstyle w} \\ \mathbf{a}[\mathrm{X}', \mathrm{Y}'] \otimes \mathsf{m}[\mathrm{Y}'] & \xrightarrow[\gamma]{} & \mathsf{m}[\mathrm{X}'] \end{array} \tag{5.11a}$$

commutes.

Associativity. For any $\mathrm{X} \leq \mathrm{Y} \leq \mathrm{Z}$, the diagram

$$\begin{array}{ccc} \mathbf{a}[\mathrm{X}, \mathrm{Y}] \otimes \mathbf{a}[\mathrm{Y}, \mathrm{Z}] \otimes \mathsf{m}[\mathrm{Z}] & \xrightarrow{\mathrm{id} \otimes \gamma} & \mathbf{a}[\mathrm{X}, \mathrm{Y}] \otimes \mathsf{m}[\mathrm{Y}] \\ {\scriptstyle \gamma \otimes \mathrm{id}}\downarrow & & \downarrow{\scriptstyle \gamma} \\ \mathbf{a}[\mathrm{X}, \mathrm{Z}] \otimes \mathsf{m}[\mathrm{Z}] & \xrightarrow[\gamma]{} & \mathsf{m}[\mathrm{X}] \end{array} \tag{5.11b}$$

commutes.

Unitality. For any X, the diagram

$$\begin{array}{ccc} & \mathbf{a}[X,X] \otimes \mathsf{m}[X] & \\ {}^{\eta\otimes \mathrm{id}}\nearrow & & \searrow^{\gamma} \\ \Bbbk \otimes \mathsf{m}[X] & \xrightarrow[\cong]{} & \mathsf{m}[X] \end{array} \tag{5.11c}$$

commutes.

We also refer to m as a Coxeter monoid over the Coxeter operad **a**. We refer to γ as the structure map of m.

A morphism of Coxeter **a**-monoids is a map $f : \mathsf{m} \to \mathsf{n}$ of Coxeter species such that for any $X \le Y$, the diagram

$$\begin{array}{ccc} \mathbf{a}[X,Y] \otimes \mathsf{m}[Y] & \xrightarrow{\gamma} & \mathsf{m}[X] \\ {\scriptstyle \mathrm{id}\otimes f_Y}\downarrow & & \downarrow{\scriptstyle f_X} \\ \mathbf{a}[X,Y] \otimes \mathsf{n}[Y] & \xrightarrow[\gamma]{} & \mathsf{n}[X] \end{array} \tag{5.12}$$

commutes.

5.4.4. Free Coxeter operad monoid. For a Coxeter operad **a**, the free Coxeter **a**-monoid over a Coxeter species p is given by $\mathbf{a} \circ \mathsf{p}$, with structure map defined by the composite

$$\mathbf{a} \circ (\mathbf{a} \circ \mathsf{p}) \xrightarrow{\cong} (\mathbf{a} \circ \mathbf{a}) \circ \mathsf{p} \xrightarrow{\gamma \circ \mathrm{id}} \mathbf{a} \circ \mathsf{p}.$$

The universal property of $\mathbf{a} \circ \mathsf{p}$ is stated below.

Theorem 5.4. *Let* **a** *be a Coxeter operad. Let* m *be a Coxeter* **a***-monoid,* p *a Coxeter species,* $f : \mathsf{p} \to \mathsf{m}$ *a map of Coxeter species. Then there exists a unique morphism of Coxeter* **a***-monoids* $\hat{f} : \mathbf{a} \circ \mathsf{p} \to \mathsf{m}$ *such that the diagram*

$$\begin{array}{ccc} \mathbf{a} \circ \mathsf{p} & \overset{\hat{f}}{\dashrightarrow} & \mathsf{m} \\ \uparrow & \nearrow_{f} & \\ \mathsf{p} & & \end{array}$$

commutes.

The vertical map is $\mathsf{p} \cong \mathbf{x} \circ \mathsf{p} \hookrightarrow \mathbf{a} \circ \mathsf{p}$. Explicitly, the map $\hat{f}$ is

$$\mathbf{a} \circ \mathsf{p} \xrightarrow{\mathrm{id} \circ f} \mathbf{a} \circ \mathsf{m} \xrightarrow{\gamma} \mathsf{m}. \tag{5.13}$$

5.4.5. Coxeter monoids over a quadratic Coxeter operad. For a Coxeter dispecies **e**, a Coxeter monoid over the free Coxeter operad $\mathcal{F}_{\circ}(\mathbf{e})$ is the same as a Coxeter species m equipped with a map $\mathbf{e} \circ \mathsf{m} \to \mathsf{m}$ of Coxeter species. More generally, for a quadratic Coxeter operad $\mathbf{a} = \langle \mathbf{e} \mid \mathbf{r} \rangle$, a Coxeter **a**-monoid is the same as a Coxeter species m equipped with a map $f : \mathbf{e} \circ \mathsf{m} \to \mathsf{m}$ of Coxeter species such that the composite map

$$\mathbf{r} \circ \mathsf{m} \to \mathbf{e} \circ \mathbf{e} \circ \mathsf{m} \xrightarrow{\mathrm{id} \circ f} \mathbf{e} \circ \mathsf{m} \xrightarrow{f} \mathsf{m}$$

is zero.

5.4.6. Coxeter operads and monads on Coxeter species. A Coxeter operad $\mathbf{a}$ gives rise to a monad $\mathcal{V}_{\mathbf{a}}$ on Coxeter species via $\mathcal{V}_{\mathbf{a}}(\mathsf{p}) := \mathbf{a} \circ \mathsf{p}$. Moreover, the category of Coxeter $\mathbf{a}$-monoids is isomorphic to the category of $\mathcal{V}_{\mathbf{a}}$-algebras.

5.4.7. Coxeter cooperad comonoids. Dually, for a Coxeter $\mathcal{A}$-cooperad $\mathbf{c}$, we have the category $\mathbf{c}\text{-}\mathsf{Comon}(\mathcal{A}\text{-}\mathsf{CSp})$ of Coxeter $\mathbf{c}$-comonoids. Explicitly, a Coxeter $\mathbf{c}$-comonoid is a Coxeter species m equipped with linear maps

$$\mathsf{m}[\mathrm{X}] \to \mathbf{c}[\mathrm{X}, \mathrm{Y}] \otimes \mathsf{m}[\mathrm{Y}],$$

one for each $\mathrm{X} \leq \mathrm{Y}$, such that the naturality, coassociativity, counitality axioms hold.

The cofree Coxeter $\mathbf{c}$-comonoid on a Coxeter species p is given by $\mathbf{c} \circ \mathsf{p}$. A Coxeter cooperad $\mathbf{c}$ gives rise to a comonad $\mathcal{U}_{\mathbf{c}}$ on Coxeter species via $\mathcal{U}_{\mathbf{c}}(\mathsf{p}) := \mathbf{c} \circ \mathsf{p}$. Moreover, the category of Coxeter $\mathbf{c}$-comonoids is isomorphic to the category of $\mathcal{U}_{\mathbf{c}}$-coalgebras.

5.5. Coxeter commutative, associative, Lie operads

We now define three important Coxeter operads, namely commutative, associative, Lie. We denote them by **Com**, **As**, **Lie**. These Coxeter operads are binary quadratic and in particular connected. Moreover, they have signed analogues, which we denote $\mathbf{Com}^-$, $\mathbf{As}^-$, $\mathbf{Lie}^-$. In fact, any Coxeter operad $\mathbf{a}$ has a signed partner $\mathbf{a}^-$. It turns out that **As** and $\mathbf{As}^-$ are canonically isomorphic, so there is no need to consider the latter explicitly.

Coxeter monoids over these Coxeter operads and their relation to algebras over corresponding monads on Coxeter species are summarized in Table 5.1. The free objects are listed in Table 5.2. Entries related to the Coxeter Lie operad are treated in Chapter 6.

TABLE 5.1. Coxeter operads and monads on Coxeter species.

Coxeter operad $\mathbf{a}$	monad $\mathcal{V}_{\mathbf{a}}$	Coxeter $\mathbf{a}$-monoid = $\mathcal{V}_{\mathbf{a}}$-algebra
As	$\mathcal{T}$	Coxeter monoid
Com	$\mathcal{S}$	commutative Coxeter monoid
$\mathbf{Com}^-$	$\mathcal{E}$	signed commutative Coxeter monoid
Lie	$\mathcal{PT}$	Coxeter Lie monoid
$\mathbf{Lie}^-$	$\mathcal{PT}_{-1}$	signed Coxeter Lie monoid

Dually, we have the Coxeter cooperads $\mathbf{Com}^*$, $\mathbf{As}^*$, $\mathbf{Lie}^*$ and their signed analogues. Coxeter comonoids over these Coxeter cooperads and their relation to coalgebras over corresponding comonads are summarized in Table 5.3. The cofree objects are listed in Table 5.4.

These ideas are developed further in Section 11.6.

TABLE 5.2. Free Coxeter operad monoids.

free Coxeter monoid	$\mathbf{As} \circ \mathsf{p} = \mathcal{T}(\mathsf{p})$
free commutative Coxeter monoid	$\mathbf{Com} \circ \mathsf{p} = \mathcal{S}(\mathsf{p})$
free signed commutative Coxeter monoid	$\mathbf{Com}^- \circ \mathsf{p} = \mathcal{E}(\mathsf{p})$
free Coxeter Lie monoid	$\mathbf{Lie} \circ \mathsf{p} = \mathcal{PT}(\mathsf{p})$
free signed Coxeter Lie monoid	$\mathbf{Lie}^- \circ \mathsf{p} = \mathcal{PT}_{-1}(\mathsf{p})$

TABLE 5.3. Coxeter cooperads and comonads on Coxeter species.

Coxeter cooperad $\mathbf{c}$	comonad $\mathcal{U}_{\mathbf{c}}$	Coxeter $\mathbf{c}$-comonoid = $\mathcal{U}_{\mathbf{c}}$-coalgebra
$\mathbf{As}^*$	$\mathcal{T}^\vee$	Coxeter comonoid
$\mathbf{Com}^*$	$\mathcal{S}^\vee$	cocommutative Coxeter comonoid
$(\mathbf{Com}^-)^*$	$\mathcal{E}^\vee$	signed cocommutative Coxeter comonoid
$\mathbf{Lie}^*$	$\mathcal{QT}^\vee$	Coxeter Lie comonoid
$(\mathbf{Lie}^-)^*$	$\mathcal{QT}^\vee_{-1}$	signed Coxeter Lie comonoid

TABLE 5.4. Cofree Coxeter cooperad comonoids.

cofree Coxeter comonoid	$\mathbf{As}^* \circ \mathsf{p} = \mathcal{T}^\vee(\mathsf{p})$
cofree cocommutative Coxeter comonoid	$\mathbf{Com}^* \circ \mathsf{p} = \mathcal{S}^\vee(\mathsf{p})$
cofree signed cocommutative Coxeter comonoid	$(\mathbf{Com}^-)^* \circ \mathsf{p} = \mathcal{E}^\vee(\mathsf{p})$
cofree Coxeter Lie comonoid	$\mathbf{Lie}^* \circ \mathsf{p} = \mathcal{QT}^\vee(\mathsf{p})$
cofree signed Coxeter Lie comonoid	$(\mathbf{Lie}^-)^* \circ \mathsf{p} = \mathcal{QT}^\vee_{-1}(\mathsf{p})$

5.5.1. Coxeter commutative operad. Define the *exponential Coxeter dispecies* $\mathbf{E}$ by

$$\mathbf{E}[\mathrm{X}, \mathrm{Y}] := \Bbbk.$$

The structure maps are defined to be all identities, that is,

$$w : \mathbf{E}[\mathrm{X}, \mathrm{Y}] \to \mathbf{E}[\mathrm{X}', \mathrm{Y}'], \qquad \Bbbk \xrightarrow{\cong} \Bbbk.$$

The exponential Coxeter dispecies carries a Coxeter operad structure with substitution and unit maps

$$\begin{array}{ccc} \mathbf{E}[\mathrm{X}, \mathrm{Y}] \otimes \mathbf{E}[\mathrm{Y}, \mathrm{Z}] \xrightarrow{\cong} \mathbf{E}[\mathrm{X}, \mathrm{Z}] & \qquad & \Bbbk \xrightarrow{\cong} \mathbf{E}[\mathrm{X}, \mathrm{X}] \\ 1 \otimes 1 \longmapsto 1 & & 1 \longmapsto 1. \end{array} \tag{5.14}$$

Axioms (5.6a)–(5.6c) are trivially satisfied. We call this the *Coxeter commutative operad* and denote it by $\mathbf{Com}$. It is the linearization of the Coxeter set-operad which is singleton in each component.

By reversing arrows in (5.14), we see that **E** also carries a Coxeter cooperad structure. We call this the *Coxeter commutative cooperad* and denote it by $\mathbf{Com}^*$. It is the Coxeter cooperad dual to **Com**.

(Co)commutative Coxeter (co)monoids. Observe: The category of Coxeter **Com**-monoids is the same as the category of commutative Coxeter monoids as described in Proposition 2.38 in terms of flats. To pass between a Coxeter **Com**-monoid a and a commutative Coxeter monoid a, we identify their structure maps

$$\mathbf{Com}[\mathrm{X},\mathrm{Y}] \otimes \mathsf{a}[\mathrm{Y}] \to \mathsf{a}[\mathrm{X}] \quad \text{and} \quad \mu_{\mathrm{X}}^{\mathrm{Y}} : \mathsf{a}[\mathrm{Y}] \to \mathsf{a}[\mathrm{X}].$$

Dually, the category of Coxeter $\mathbf{Com}^*$-comonoids is the same as the category of cocommutative Coxeter comonoids. Now we identify

$$\mathsf{c}[\mathrm{X}] \to \mathbf{Com}^*[\mathrm{X},\mathrm{Y}] \otimes \mathsf{c}[\mathrm{Y}] \quad \text{and} \quad \Delta_{\mathrm{X}}^{\mathrm{Y}} : \mathsf{c}[\mathrm{X}] \to \mathsf{c}[\mathrm{Y}].$$

For a Coxeter species p, using formulas (5.9) and (2.94),

$$(\mathbf{Com} \circ \mathsf{p})[\mathrm{X}] = \bigoplus_{\mathrm{Y} :\, \mathrm{Y} \geq \mathrm{X}} \mathbf{Com}[\mathrm{X},\mathrm{Y}] \otimes \mathsf{p}[\mathrm{Y}] = \bigoplus_{\mathrm{Y} :\, \mathrm{Y} \geq \mathrm{X}} \mathsf{p}[\mathrm{Y}] = \mathcal{S}(\mathsf{p})[\mathrm{X}].$$

We deduce from here that the monad on Coxeter species induced by **Com** is $\mathcal{S}$. Dually, the comonad on Coxeter species induced by $\mathbf{Com}^*$ is $\mathcal{S}^\vee$.

Binary quadratic structure. The Coxeter commutative operad **Com** is binary quadratic as follows. Define $\mathbf{e}$ by $\mathbf{e}[\mathrm{X},\mathrm{Y}] := \Bbbk$ when $\mathrm{X} \lessdot \mathrm{Y}$, and 0 otherwise. For $\mathrm{rk}(\mathrm{Z}/\mathrm{X}) = 2$, note that $(\mathbf{e} \circ \mathbf{e})[\mathrm{X},\mathrm{Z}]$ has a basis indexed by flats Y between X and Z. Define $\mathbf{r}[\mathrm{X},\mathrm{Z}]$ to be its subspace spanned by elements of the form $\mathrm{Y} - \mathrm{Y}'$. Then $\mathbf{Com} = \langle \mathbf{e} \mid \mathbf{r} \rangle$. This presentation yields the description of commutative Coxeter monoids in Proposition 2.41 in terms of the covering associativity axiom.

5.5.2. Coxeter associative operad. Define the *Coxeter dispecies of chambers* $\mathbf{\Gamma}$ by

$$\mathbf{\Gamma}[\mathrm{X},\mathrm{Y}] := \Gamma[\mathcal{A}_{\mathrm{X}}^{\mathrm{Y}}],$$

where the rhs is the vector space with basis consisting of chambers in the arrangement $\mathcal{A}_{\mathrm{X}}^{\mathrm{Y}}$. These are the same as lunes with base X and case Y. We denote basis elements either by $\mathtt{H}_{\mathrm{L}}$ or by $\mathtt{H}_{F/A}$, with the two related by $\mathrm{L} = \mathrm{s}(A,F)$. It is implicit that $\mathtt{H}_{F/A} = \mathtt{H}_{F'/A'}$ whenever $\mathrm{s}(A,F) = \mathrm{s}(A',F')$. The structure maps are defined by

$$w : \mathbf{\Gamma}[\mathrm{X},\mathrm{Y}] \to \mathbf{\Gamma}[\mathrm{X}',\mathrm{Y}'], \qquad \mathtt{H}_{\mathrm{L}} \mapsto \mathtt{H}_{w\mathrm{L}}, \quad \mathtt{H}_{F/A} \mapsto \mathtt{H}_{wF/wA}.$$

(Note: If w_1 and w_2 agree on $\mathrm{c}(\mathrm{L})$, then $w_1\mathrm{L} = w_2\mathrm{L}$.) The Coxeter dispecies of chambers carries a Coxeter operad structure with substitution and unit maps

$$\begin{array}{ccc} \mathbf{\Gamma}[\mathrm{X},\mathrm{Y}] \otimes \mathbf{\Gamma}[\mathrm{Y},\mathrm{Z}] \to \mathbf{\Gamma}[\mathrm{X},\mathrm{Z}] & & \Bbbk \xrightarrow{\cong} \mathbf{\Gamma}[\mathrm{X},\mathrm{X}] \\ \mathtt{H}_{\mathrm{L}} \otimes \mathtt{H}_{\mathrm{M}} \mapsto \mathtt{H}_{\mathrm{L}\circ\mathrm{M}} & & 1 \longmapsto \mathtt{H}_{\mathrm{X}} \\ \mathtt{H}_{F/A} \otimes \mathtt{H}_{G/F} \mapsto \mathtt{H}_{G/A} & & 1 \longmapsto \mathtt{H}_{A/A}. \end{array} \tag{5.15}$$

We call this the *Coxeter associative operad* and denote it by **As**. It is the linearization of a Coxeter set-operad. The naturality square in (5.6a) is checked below.

$$
\begin{array}{ccc}
\mathtt{H}_{\mathrm{L}} \otimes \mathtt{H}_{\mathrm{M}} & \longmapsto & \mathtt{H}_{\mathrm{L}\circ\mathrm{M}} \\
{\scriptstyle w\otimes w}\downarrow & & \downarrow{\scriptstyle w} \\
\mathtt{H}_{w\mathrm{L}} \otimes \mathtt{H}_{w\mathrm{M}} & \longmapsto & \mathtt{H}_{w\mathrm{L}\circ w\mathrm{M}} = \mathtt{H}_{w(\mathrm{L}\circ\mathrm{M})}
\end{array}
\qquad
\begin{array}{ccc}
\mathtt{H}_{F/A} \otimes \mathtt{H}_{G/F} & \longmapsto & \mathtt{H}_{G/A} \\
{\scriptstyle w\otimes w}\downarrow & & \downarrow{\scriptstyle w} \\
\mathtt{H}_{wF/wA} \otimes \mathtt{H}_{wG/wF} & \longmapsto & \mathtt{H}_{wG/wA}
\end{array}
$$

The associativity and unitality axioms can be checked similarly.

We let $\mathbf{As}^*$ denote the dual of **As** and call it the *Coxeter associative cooperad*. We use the letter $\mathtt{M}$ to denote the basis dual to $\mathtt{H}$.

Coxeter (co)monoids. The category of Coxeter **As**-monoids is the same as the category of Coxeter monoids as described in Proposition 2.24 in terms of flats and lunes. To pass between a Coxeter **As**-monoid a and a Coxeter monoid a, we identify their structure maps

$$\mathbf{As}[\mathrm{X},\mathrm{Y}] \otimes \mathsf{a}[\mathrm{Y}] \to \mathsf{a}[\mathrm{X}] \quad \text{and} \quad (\mu_{\mathrm{L}} : \mathsf{a}[\mathrm{Y}] \to \mathsf{a}[\mathrm{X}])_{\mathrm{L}}.$$

The latter is a family of linear maps indexed by lunes L with base X and case Y. More precisely, μ_{L} is the map arising from the basis element $\mathtt{H}_{\mathrm{L}}$. Dually, the category of Coxeter $\mathbf{As}^*$-comonoids is the same as the category of Coxeter comonoids.

For a Coxeter species p, using formulas (5.9) and (2.92),

$$(\mathbf{As} \circ \mathsf{p})[\mathrm{X}] = \bigoplus_{\mathrm{Y}:\, \mathrm{Y} \geq \mathrm{X}} \mathbf{As}[\mathrm{X},\mathrm{Y}] \otimes \mathsf{p}[\mathrm{Y}] = \bigoplus_{\mathrm{L}:\, \mathrm{b}(\mathrm{L}) = \mathrm{X}} \mathsf{p}[\mathrm{c}(\mathrm{L})] = \mathcal{T}(\mathsf{p})[\mathrm{X}].$$

We deduce from here that the monad on Coxeter species induced by **As** is $\mathcal{T}$. Dually, the comonad on Coxeter species induced by $\mathbf{As}^*$ is $\mathcal{T}^\vee$.

Binary quadratic structure. The Coxeter associative operad **As** is binary quadratic as follows. Define $\mathbf{e}$ by $\mathbf{e}[\mathrm{X},\mathrm{Y}] := \mathbf{\Gamma}[\mathrm{X},\mathrm{Y}]$ when $\mathrm{X} \lessdot \mathrm{Y}$, and 0 otherwise. For $\mathrm{rk}(\mathrm{Z}/\mathrm{X}) = 2$, note that $(\mathbf{e} \circ \mathbf{e})[\mathrm{X},\mathrm{Z}]$ has a basis indexed by pairs (L,M) of composable lunes each of slack 1 such that $\mathrm{b}(\mathrm{L}) = \mathrm{X}$ and $\mathrm{c}(\mathrm{M}) = \mathrm{Z}$. Define $\mathbf{r}[\mathrm{X},\mathrm{Z}]$ to be its subspace spanned by $(\mathrm{L},\mathrm{M}) - (\mathrm{L}',\mathrm{M}')$ whenever $\mathrm{L}\circ\mathrm{M} = \mathrm{L}'\circ\mathrm{M}'$. Then $\mathbf{As} = \langle \mathbf{e} \,|\, \mathbf{r} \rangle$. This presentation yields the description of Coxeter monoids in Proposition 2.25 in terms of the covering associativity axiom.

5.5.3. Coxeter Lie operad. Define the Coxeter dispecies **Lie** by

$$\mathbf{Lie}[\mathrm{X},\mathrm{Y}] := \mathsf{Lie}[\mathcal{A}_{\mathrm{X}}^{\mathrm{Y}}],$$

where the rhs is the space of Lie elements of $\mathcal{A}_{\mathrm{X}}^{\mathrm{Y}}$, see Section 1.13. It is a Coxeter subdispecies of $\mathbf{\Gamma}$. In fact, it is a Coxeter suboperad of **As**. We call this the *Coxeter Lie operad*. It carries a binary quadratic structure; the details are similar to the presentation of the Lie operad in [**11**, Example 4.12]. The core result here is the Joyal–Klyachko–Stanley theorem proved in [**10**, Section 14.5].

We let $\mathbf{Lie}^*$ denote the dual of **Lie** and call it the *Coxeter Lie cooperad*.

5.5.4. Signed Coxeter commutative operad. Define the Coxeter dispecies $\mathbf{E}^-$ by letting its components $\mathbf{E}^-[\mathrm{X},\mathrm{Y}]$ be the signature spaces as in Definition 2.48. The structure maps are defined by

$$w : \mathbf{E}^-[\mathrm{X},\mathrm{Y}] \to \mathbf{E}^-[\mathrm{X}',\mathrm{Y}'], \qquad \mathtt{H}_{[\mathrm{L}]} \mapsto \mathtt{H}_{[w\mathrm{L}]}, \quad \mathtt{H}_{[F/A]} \mapsto \mathtt{H}_{[wF/wA]}.$$

It carries a Coxeter operad structure with substitution and unit maps

$$\begin{array}{ccc} \mathbf{E}^-[\mathrm{X},\mathrm{Y}] \otimes \mathbf{E}^-[\mathrm{Y},\mathrm{Z}] \xrightarrow{\cong} \mathbf{E}^-[\mathrm{X},\mathrm{Z}] & & \Bbbk \xrightarrow{\cong} \mathbf{E}^-[\mathrm{X},\mathrm{X}] \\ \mathtt{H}_{[\mathrm{L}]} \otimes \mathtt{H}_{[\mathrm{M}]} \mapsto \mathtt{H}_{[\mathrm{L}\circ\mathrm{M}]} & & 1 \mapsto \mathtt{H}_{[\mathrm{X}]} \\ \mathtt{H}_{[F/A]} \otimes \mathtt{H}_{[G/F]} \mapsto \mathtt{H}_{[G/A]} & & 1 \mapsto \mathtt{H}_{[A/A]}. \end{array} \tag{5.16}$$

We call this the *signed Coxeter commutative operad* and denote it by $\mathbf{Com}^-$.

The category of Coxeter $\mathbf{Com}^-$-monoids is the same as the category of signed commutative Coxeter monoids. The monad on Coxeter species induced by $\mathbf{Com}^-$ is $\mathcal{E}$, see (2.98).

5.5.5. Signature functor on Coxeter dispecies. For a Coxeter dispecies $\mathbf{p}$, define its *signed partner* $\mathbf{p}^-$ by

$$\mathbf{p}^- := \mathbf{p} \times \mathbf{E}^-. \tag{5.17}$$

We refer to the assignment $\mathbf{p} \mapsto \mathbf{p}^-$ as the *signature functor* on Coxeter dispecies. Note: $\mathbf{E}$ and $\mathbf{E}^-$ are signed partners of each other. Further, for Coxeter dispecies $\mathbf{p}$ and $\mathbf{q}$, there are natural isomorphisms

$$(\mathbf{p} \circ \mathbf{q})^- \cong \mathbf{p}^- \circ \mathbf{q}^- \quad \text{and} \quad \mathbf{x}^- \cong \mathbf{x}. \tag{5.18}$$

Thus, the signature functor is strong, and it preserves Coxeter operads. Thus, if $\mathbf{a}$ is a Coxeter operad, then so is $\mathbf{a}^-$. It is the same as the Hadamard product of $\mathbf{a}$ with $\mathbf{Com}^-$.

Lemma 5.5. *Coxeter operads* $\mathbf{As}$ *and* $\mathbf{As}^-$ *are isomorphic. Explicitly, the morphism*

$$\mathbf{As}[\mathrm{X},\mathrm{Y}] \to \mathbf{As}^-[\mathrm{X},\mathrm{Y}], \qquad \mathtt{H}_{\mathrm{L}} \mapsto \mathtt{H}_{\mathrm{L}} \otimes \mathtt{H}_{[\mathrm{L}]}, \quad \mathtt{H}_{F/A} \mapsto \mathtt{H}_{F/A} \otimes \mathtt{H}_{[F/A]}$$

on the (X,Y)*-component is an isomorphism of Coxeter operads.*

PROOF. The check for diagram (5.2) is shown below.

$$\begin{array}{ccc} \mathtt{H}_{\mathrm{L}} & \longmapsto & \mathtt{H}_{\mathrm{L}} \otimes \mathtt{H}_{[\mathrm{L}]} \\ {\scriptstyle w}\downarrow & & \downarrow{\scriptstyle w\otimes w} \\ \mathtt{H}_{w\mathrm{L}} & \longmapsto & \mathtt{H}_{w\mathrm{L}} \otimes \mathtt{H}_{[w\mathrm{L}]} \end{array} \qquad \begin{array}{ccc} \mathtt{H}_{F/A} & \longmapsto & \mathtt{H}_{F/A} \otimes \mathtt{H}_{[F/A]} \\ {\scriptstyle w}\downarrow & & \downarrow{\scriptstyle w\otimes w} \\ \mathtt{H}_{wF/wA} & \longmapsto & \mathtt{H}_{wF/wA} \otimes \mathtt{H}_{[wF/wA]} \end{array}$$

Diagrams (5.7) can be checked using the substitution and unit maps (5.15) and (5.16). □

We refer to $\mathbf{Lie}^-$ as the *signed Coxeter Lie operad*, and to its dual as the *signed Coxeter Lie cooperad*. The latter is the same as $(\mathbf{Lie}^*)^-$ which is the signature functor applied to the Coxeter Lie cooperad.

5.5.6. Coxeter operad morphisms. There are morphisms

$$\mathbf{Lie} \hookrightarrow \mathbf{As} \twoheadrightarrow \mathbf{Com} \tag{5.19}$$

of Coxeter operads. The first map arises from the definition of the Coxeter Lie operad. The second map, on any (X, Y)-component, sends each basis element $\mathtt{H}_{\mathrm{L}}$ to $1 \in \Bbbk$. By dualizing (5.19), we obtain morphisms

$$\mathbf{Com}^* \hookrightarrow \mathbf{As}^* \twoheadrightarrow \mathbf{Lie}^* \tag{5.20}$$

of Coxeter cooperads.

5.5.7. Commutative, associative, Lie operads. For any hyperplane arrangement $\mathcal{A}$, one can define the commutative, associative, Lie operads, and their signed analogues. See [**11**, Sections 4.5 and 4.8]. They are defined as in the Coxeter case but by removing all references to the Coxeter group.

Exercise 5.6. Show: Coxeter operads **Com** and $\mathbf{Com}^-$ are not isomorphic (assuming $|W| > 1$ and $\mathrm{char}(\Bbbk) \neq 2$). However, their underlying operads are isomorphic.

5.5.8. Signed Coxeter operad monoids. Recall the u-signature and o-signature functors on Coxeter species from Section 2.8.

Exercise 5.7. Check: For $\mathbf{p}$ a Coxeter dispecies and m a Coxeter species,

$$\mathbf{p}^- \circ \mathsf{m}_- = (\mathbf{p} \circ \mathsf{m})_- \quad \text{and} \quad \mathbf{p}^- \circ \mathsf{m}^- = (\mathbf{p} \circ \mathsf{m})^-. \tag{5.21}$$

Deduce: For a Coxeter operad $\mathbf{a}$, if m is a Coxeter $\mathbf{a}$-monoid, then m_- and m^- are Coxeter $\mathbf{a}^-$-monoids. More formally, the u-signature and o-signature functors induce isomorphisms between categories of Coxeter $\mathbf{a}$-monoids and Coxeter $\mathbf{a}^-$-monoids.

We refer to a Coxeter $\mathbf{a}^-$-monoid as a *signed Coxeter* $\mathbf{a}$*-monoid.*

5.6. Notes

Coxeter operads for reflection arrangements are presented here for the first time, see Table II in the preface. This chapter follows the outline of [**11**, Chapter 4] which presented operads in the more general but less structured setting of hyperplane arrangements. A condensed summary of operads for arrangements was given earlier in [**10**, Section 15.9]. The precise connection between operads and Coxeter operads is mentioned in Section 5.2.9. Some concepts from [**11**, Chapter 4] whose Coxeter analogues are not discussed here include Hopf operads, quadratic duality, black and white circle products, bioperads, incidence algebra of an operad. We mention that Coxeter operads connect to invariant subalgebras of incidence algebras, and in particular, the Coxeter commutative operad to the invariant flat-incidence algebra, and the Coxeter associative operad to the invariant lune-incidence algebra.

Further theory of Coxeter operads is developed in Chapter 11. Coxeter operads generalize classical May operads; related references are given in Notes 11.9.

CHAPTER 6

Coxeter Lie monoids

Let $\mathcal{A}$ be a fixed reflection arrangement with Coxeter group W. All notions in this chapter depend on $\mathcal{A}$; however, for convenience, we usually do not show this dependence explicitly following Conventions 2.1 and 5.1.

Coxeter Lie monoids. We begin the study of Coxeter Lie monoids. These are natural companions of Coxeter monoids and commutative Coxeter monoids which we met in Chapter 2. The connection between the three is easiest to understand in terms of Coxeter operads. Recall: Commutative Coxeter monoids and Coxeter monoids are Coxeter **Com**-monoids and Coxeter **As**-monoids, respectively. Similarly, Coxeter Lie monoids are defined as Coxeter **Lie**-monoids. Also recall that Coxeter monoids can be formulated using the covering associativity axiom by using the presentation of **As**. Similarly, Coxeter Lie monoids can be formulated using antisymmetry and Jacobi identity by using the presentation of **Lie**. We also introduce the dual notion of Coxeter Lie comonoids. Both Coxeter Lie monoids and Coxeter Lie comonoids have signed analogues.

The toy example of rank one is developed in Section 7.8.

Interaction with Coxeter bimonoids. Coxeter Lie monoids interact fruitfully with Coxeter bimonoids. The primitive part $\mathcal{P}(\mathsf{h})$ of a Coxeter bimonoid h carries a Coxeter Lie monoid structure (via the commutator bracket). Dually, the indecomposable part $\mathcal{Q}(\mathsf{h})$ of a Coxeter bimonoid h carries a Coxeter Lie comonoid structure (via the cocommutator cobracket).

In the other direction, for a Coxeter Lie monoid g, we have the universal enveloping Coxeter monoid $\mathcal{U}(\mathsf{g})$ which carries a cocommutative Coxeter bimonoid structure. Dually, for a Coxeter Lie comonoid k, we have the universal coenveloping Coxeter comonoid $\mathcal{U}^\vee(\mathsf{k})$ which carries a commutative Coxeter bimonoid structure.

Lie monoids. For any hyperplane arrangement $\mathcal{A}$, one can define the notion of Lie monoids. These were introduced in our previous monograph [**11**, Chapter 16]. By imposing on them additional structure of type morphisms (which make sense only when $\mathcal{A}$ is a reflection arrangement), one obtains the notion of Coxeter Lie monoids.

Joyal Lie monoids. Connections to classical Lie theory. The notion of a Coxeter Lie monoid, when specialized to braid arrangements, relates to the classical notion of a Joyal Lie monoid. More precisely, the functor from positive Joyal species to Coxeter species (introduced in Chapter 2) induces a functor from positive Joyal Lie monoids to Coxeter Lie monoids.

6.1. Coxeter Lie monoids and Coxeter Lie comonoids

We introduce Coxeter Lie monoids. We give two definitions, one in terms of the Coxeter Lie operad, and the other in terms of a Lie bracket subject to antisymmetry and Jacobi identity. We also discuss the dual notion of Coxeter Lie comonoids. These notions have signed analogues linked to the signed Coxeter Lie operad.

6.1.1. Coxeter Lie monoids as Coxeter operad monoids. Recall the Coxeter Lie operad **Lie** from Section 5.5.3. Also recall Coxeter operad monoids from Section 5.4. A *Coxeter Lie monoid* is by definition a Coxeter **Lie**-monoid. Similarly, a morphism of Coxeter Lie monoids is a morphism of Coxeter **Lie**-monoids. We denote the category of Coxeter Lie monoids by $\mathsf{LieMon}(\mathcal{A}\text{-}\mathsf{CSp})$. We usually drop $\mathcal{A}$ from the notation for convenience.

Explicitly: A Coxeter Lie monoid is a Coxeter species g equipped with linear maps

$$\gamma_{\mathrm{X}}^{\mathrm{Y}} : \mathbf{Lie}[\mathrm{X}, \mathrm{Y}] \otimes \mathsf{g}[\mathrm{Y}] \to \mathsf{g}[\mathrm{X}],$$

one for each $\mathrm{X} \leq \mathrm{Y}$, subject to axioms (5.11a)–(5.11c) with $\mathbf{a} := \mathbf{Lie}$ and $\mathsf{m} := \mathsf{g}$.

A morphism $f : \mathsf{g} \to \mathsf{h}$ of Coxeter Lie monoids is a map of Coxeter species such that diagram (5.12) commutes with $\mathbf{a} := \mathbf{Lie}$, $\mathsf{m} := \mathsf{g}$, $\mathsf{n} := \mathsf{h}$.

6.1.2. Coxeter Lie monoids by generators and relations. We now come to the second definition. A *Coxeter Lie monoid* is a Coxeter species g equipped with linear maps

$$\nu_A^F : \mathsf{g}[F] \to \mathsf{g}[A],$$

one for each $A \lessdot F$, subject to the axioms below.

Naturality. For each support morphism $\beta_{B,A} : A \to B$, and $A \lessdot F$, the first diagram below commutes. For each type morphism $w : A \to wA$, and $A \lessdot F$, the second diagram below commutes.

$$\text{(6.1a)} \qquad \begin{array}{ccc} \mathsf{g}[F] & \xrightarrow{\beta_{BF,F}} & \mathsf{g}[BF] \\ {\scriptstyle \nu_A^F}\downarrow & & \downarrow{\scriptstyle \nu_B^{BF}} \\ \mathsf{g}[A] & \xrightarrow[\beta_{B,A}]{} & \mathsf{g}[B] \end{array} \qquad\qquad \begin{array}{ccc} \mathsf{g}[F] & \xrightarrow{w} & \mathsf{g}[wF] \\ {\scriptstyle \nu_A^F}\downarrow & & \downarrow{\scriptstyle \nu_{wA}^{wF}} \\ \mathsf{g}[A] & \xrightarrow[w]{} & \mathsf{g}[wA] \end{array}$$

Note: $A \lessdot F$ implies $B \lessdot BF$ and $wA \lessdot wF$.

Antisymmetry. For each $A \lessdot F$,

$$\text{(6.1b)} \qquad (\mathsf{g}[F] \xrightarrow{\nu_A^F} \mathsf{g}[A]) \;+\; (\mathsf{g}[F] \xrightarrow{\beta_{A\overline{F},F}} \mathsf{g}[A\overline{F}] \xrightarrow{\nu_A^{A\overline{F}}} \mathsf{g}[A]) \;= 0.$$

Note: $A \lessdot F$ implies $A \lessdot A\overline{F}$. For illustrations, see the first and third pictures in Figure 1.8.

Jacobi identity. For each face A of codimension two in a flat X,

$$\sum_{i=1}^{n} \left(\mathsf{g}[G_1] \xrightarrow{\beta_{G_i,G_1}} \mathsf{g}[G_i] \xrightarrow{\nu_{F_i}^{G_i}} \mathsf{g}[F_i] \xrightarrow{\nu_A^{F_i}} \mathsf{g}[A]\right) = 0, \tag{6.1c}$$

with integer n, and faces F_i, G_i as follows. Since X/A is a rank-two flat of $\mathcal{A}_A$, it supports $2n$ faces (all greater than A) for some $n \geq 2$. Let $G_1 \text{ -- } \dots \text{ -- } G_n$ be any minimal gallery consisting of half these faces. Let F_1 be the outer panel of G_1, and F_i be the panel between G_{i-1} and G_i for $2 \leq i \leq n$. See the illustration below with $A = O$.

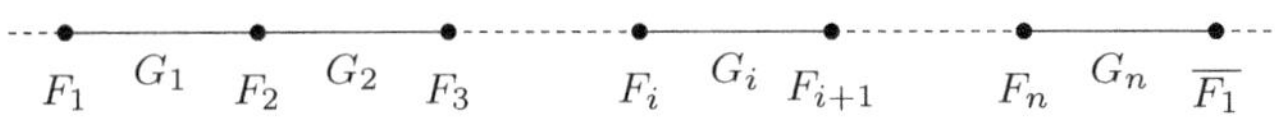

We denote a Coxeter Lie monoid by a pair (g, ν), or simply by g with ν understood. We refer to ν as the *Lie bracket*, and to the maps ν_A^F for $A \lessdot F$ as the Lie bracket components.

A morphism $f : \mathsf{g} \to \mathsf{h}$ of Coxeter Lie monoids is a map of Coxeter species such that for each $A \lessdot F$, the diagram

$$\begin{array}{ccc} \mathsf{g}[F] & \xrightarrow{f_F} & \mathsf{h}[F] \\ {\scriptstyle \nu_A^F}\big\downarrow & & \big\downarrow{\scriptstyle \nu_A^F} \\ \mathsf{g}[A] & \xrightarrow[f_A]{} & \mathsf{h}[A] \end{array} \tag{6.2}$$

commutes.

6.1.3. Coxeter Lie comonoids. Recall the Coxeter Lie cooperad $\mathbf{Lie}^*$. It is the dual of $\mathbf{Lie}$. A *Coxeter Lie comonoid* is by definition a Coxeter $\mathbf{Lie}^*$-comonoid. Similarly, a morphism of Coxeter Lie comonoids is a morphism of Coxeter $\mathbf{Lie}^*$-comonoids. We denote the category of Coxeter Lie comonoids by $\mathsf{LieComon}(\mathsf{CSp})$.

Explicitly: A Coxeter Lie comonoid is a Coxeter species k equipped with linear maps

$$\delta_{\mathrm{X}}^{\mathrm{Y}} : \mathsf{k}[\mathrm{X}] \to \mathbf{Lie}[\mathrm{X}, \mathrm{Y}]^* \otimes \mathsf{k}[\mathrm{Y}],$$

one for each $\mathrm{X} \leq \mathrm{Y}$, subject to the Coxeter cooperad comonoid axioms. Equivalently, it can be defined in terms of a *Lie cobracket* consisting of linear maps

$$\theta_A^F : \mathsf{k}[A] \to \mathsf{k}[F],$$

one for each $A \lessdot F$, subject to naturality, antisymmetry, Jacobi identity.

6.1.4. Signed analogue. Recall the signed Coxeter Lie operad $\mathbf{Lie}^-$. A *signed Coxeter Lie monoid* is by definition a Coxeter $\mathbf{Lie}^-$-monoid. A morphism of signed Coxeter Lie monoids is a morphism of Coxeter $\mathbf{Lie}^-$-monoids. We denote the category of signed Coxeter Lie monoids by $(-1)\text{-}\mathsf{LieMon}(\mathsf{CSp})$.

Explicitly: A signed Coxeter Lie monoid is a Coxeter species g equipped with linear maps

$$\gamma_{\mathrm{X}}^{\mathrm{Y}} : \mathbf{Lie}^-[\mathrm{X}, \mathrm{Y}] \otimes \mathsf{g}[\mathrm{Y}] \to \mathsf{g}[\mathrm{X}],$$

one for each $X \leq Y$, subject to axioms (5.11a)–(5.11c) with $\mathbf{a} := \mathbf{Lie}^-$ and $\mathsf{m} := \mathsf{g}$. Equivalently, it can be defined in terms of a Lie bracket, by inserting appropriate signs in (6.1b) and (6.1c).

Similarly, consider the signed Coxeter Lie cooperad $(\mathbf{Lie}^-)^*$. A *signed Coxeter Lie comonoid* is a Coxeter $(\mathbf{Lie}^-)^*$-comonoid. A morphism of signed Coxeter Lie comonoids is a morphism of Coxeter $(\mathbf{Lie}^-)^*$-comonoids. We denote the category of signed Coxeter Lie comonoids by $(-1)\text{-}\mathsf{LieComon}(\mathsf{CSp})$.

6.1.5. Signature functors. Recall the u-signature functor (2.70) on Coxeter species. If g is a Coxeter Lie monoid, then g_- is a signed Coxeter Lie monoid (with the same Lie bracket components ν_A^F), and vice versa. Thus, the categories of Coxeter Lie monoids and signed Coxeter Lie monoids are equivalent. This can also be deduced by employing the o-signature functor (2.73). (A more general result is given in Exercise 5.7.) Dual remarks apply to Coxeter Lie comonoids.

6.1.6. Lie monoids vs Coxeter Lie monoids. We build on the discussion in Section 2.7. Recall from [**11**, Chapter 16] that for any hyperplane arrangement $\mathcal{A}$, a Lie monoid is by definition a **Lie**-monoid (in the category of species). In particular, it is a species. Dually, a Lie comonoid is a $\mathbf{Lie}^*$-comonoid.

For a reflection arrangement $\mathcal{A}$, every Coxeter Lie monoid is also a Lie monoid. Conversely, one can view a Coxeter Lie monoid as a Lie monoid equipped with additional structure arising from the action of the Coxeter group. More precisely, a Coxeter Lie monoid is a Lie monoid g where in addition g carries a Coxeter species structure such that the second diagram in (6.1a) commutes. Similar remarks apply to Coxeter Lie comonoids.

6.1.7. Abelian Coxeter Lie monoids and Coxeter Lie comonoids. We say a Coxeter Lie monoid is *abelian* if $\gamma_X^Y = 0$ for all $X < Y$, or equivalently, $\nu_A^F = 0$ for all $A \lessdot F$. Every Coxeter species is an abelian Coxeter Lie monoid in a unique way. Thus, the category of Coxeter species is isomorphic to the full subcategory of abelian Coxeter Lie monoids.

Dually, a Coxeter Lie comonoid is *abelian* if $\delta_X^Y = 0$ for all $X < Y$, or equivalently, $\theta_A^F = 0$ for all $A \lessdot F$. Again every Coxeter species is an abelian Coxeter Lie comonoid in a unique way. Thus, the category of Coxeter species is also isomorphic to the full subcategory of abelian Coxeter Lie comonoids.

The same definitions and remarks apply to the signed setting.

6.2. (Co)commutator (co)bracket

Every Coxeter monoid carries a Coxeter Lie monoid structure via the commutator bracket. This arises by viewing the Coxeter Lie operad inside the Coxeter associative operad. Dually, every Coxeter comonoid carries a Coxeter Lie comonoid structure via the cocommutator cobracket. Both constructions have signed analogues.

6.2.1. Commutator bracket. The Coxeter Lie operad **Lie** is a Coxeter suboperad of the Coxeter associative operad **As**. Hence, there is an induced functor

$$\mathsf{Mon}(\mathsf{CSp}) \to \mathsf{LieMon}(\mathsf{CSp}). \tag{6.3}$$

We call this the *commutator functor*. In particular, every Coxeter monoid a carries a Coxeter Lie monoid structure via the composite map

$$\mathbf{Lie} \circ \mathsf{a} \hookrightarrow \mathbf{As} \circ \mathsf{a} \to \mathsf{a}.$$

For $A \lessdot F$, the Lie bracket component ν_A^F of a is

$$\nu_A^F = \mu_A^F - \mu_A^{A\overline{F}} \beta_{A\overline{F},F}. \tag{6.4}$$

We refer to ν as the *commutator bracket*.

Similarly, the signed Coxeter Lie operad $\mathbf{Lie}^-$ is a Coxeter suboperad of $\mathbf{As}^-$. By Lemma 5.5, the latter is isomorphic to **As**. Hence, every Coxeter monoid a also carries a signed Coxeter Lie monoid structure. For $A \lessdot F$, the Lie bracket component ν_A^F of a is

$$\nu_A^F = \mu_A^F - (-1)^{\mathrm{dist}(F,A\overline{F})} \mu_A^{A\overline{F}} \beta_{A\overline{F},F}. \tag{6.5}$$

We call ν the *signed commutator bracket*.

6.2.2. Cocommutator cobracket. Dually, the surjective morphism of Coxeter cooperads $\mathbf{As}^* \twoheadrightarrow \mathbf{Lie}^*$ induces a functor

$$\mathsf{Comon}(\mathsf{CSp}) \to \mathsf{LieComon}(\mathsf{CSp}). \tag{6.6}$$

We call this the *cocommutator functor*. In particular, every Coxeter comonoid c carries a Coxeter Lie comonoid structure via the composite map

$$\mathsf{c} \to \mathbf{As}^* \circ \mathsf{c} \twoheadrightarrow \mathbf{Lie}^* \circ \mathsf{c}.$$

For $A \lessdot F$, the Lie cobracket component θ_A^F is

$$\theta_A^F = \Delta_A^F - \beta_{F,A\overline{F}} \Delta_A^{A\overline{F}}. \tag{6.7}$$

We refer to θ as the *cocommutator cobracket*. This is the dual of (6.4).

Similarly, every Coxeter comonoid c carries a signed Coxeter Lie comonoid structure under the *signed cocommutator cobracket*

$$\theta_A^F = \Delta_A^F - (-1)^{\mathrm{dist}(F,A\overline{F})} \beta_{F,A\overline{F}} \Delta_A^{A\overline{F}}. \tag{6.8}$$

This is the dual of (6.5).

6.3. Primitive and indecomposable parts of Coxeter bimonoids

For any Coxeter bimonoid, the commutator bracket restricts to the primitive part of the Coxeter bimonoid. Dually, the cocommutator cobracket descends to the indecomposable part of the Coxeter bimonoid. These constructions have signed analogues.

6.3.1. Primitive part of a Coxeter bimonoid. A Coxeter bimonoid h is, in particular, a Coxeter monoid, and hence a Coxeter Lie monoid via the commutator functor (6.3), or equivalently, the commutator bracket (6.4). This Lie structure restricts to the primitive part $\mathcal{P}(\mathsf{h})$ as stated below.

Proposition 6.1. *If* h *is a Coxeter bimonoid, then* $\mathcal{P}(\mathsf{h})$ *is a Coxeter Lie submonoid of* h.

PROOF. This can be deduced from the corresponding result for bimonoids in species [**11**, Proposition 16.2]. □

This yields a functor

$$\mathcal{P} : \mathsf{Bimon}(\mathsf{CSp}) \to \mathsf{LieMon}(\mathsf{CSp}) \tag{6.9}$$

from the category of Coxeter bimonoids to the category of Coxeter Lie monoids.

Similarly, if h is a signed Coxeter bimonoid, then $\mathcal{P}(\mathsf{h})$ is a signed Coxeter Lie submonoid of h. This yields the signed analogue of (6.9).

6.3.2. Indecomposable part of a Coxeter bimonoid. Dually, a Coxeter bimonoid h is, in particular, a Coxeter comonoid, and hence a Coxeter Lie comonoid via the cocommutator functor (6.6), or equivalently, the cocommutator cobracket (6.7). This Lie structure descends to the indecomposable part $\mathcal{Q}(\mathsf{h})$ as stated below.

Proposition 6.2. *If* h *is a Coxeter bimonoid, then* $\mathcal{Q}(\mathsf{h})$ *is a Coxeter Lie quotient comonoid of* h.

PROOF. This can be deduced from the corresponding result for bimonoids in species [**11**, Section 16.8.4]. □

This is the dual of Proposition 6.1. It yields a functor

$$\mathcal{Q} : \mathsf{Bimon}(\mathsf{CSp}) \to \mathsf{LieComon}(\mathsf{CSp}) \tag{6.10}$$

from the category of Coxeter bimonoids to the category of Coxeter Lie comonoids.

Similarly, if h is a signed Coxeter bimonoid, then $\mathcal{Q}(\mathsf{h})$ is a signed Coxeter Lie quotient comonoid of h. This yields the signed analogue of (6.10).

6.4. (Co)free Coxeter Lie (co)monoids

Recall from Sections 3.3 and 5.5 that the free Coxeter monoid and free commutative Coxeter monoid on a Coxeter species p are given by $\mathbf{As} \circ \mathsf{p} = \mathcal{T}(\mathsf{p})$ and $\mathbf{Com} \circ \mathsf{p} = \mathcal{S}(\mathsf{p})$. We will now see that the free Coxeter Lie monoid on p is given by $\mathbf{Lie} \circ \mathsf{p} = \mathcal{P}\mathcal{T}(\mathsf{p})$. More formally, just as $\mathbf{As}$ and $\mathbf{Com}$ yield monads $\mathcal{T}$ and $\mathcal{S}$, $\mathbf{Lie}$ yields the monad $\mathcal{P}\mathcal{T}$ (arising from the adjunction between $\mathcal{T}$ and $\mathcal{P}$). See the Lie entries in Tables 5.1 and 5.2.

Dually, recall that the cofree Coxeter comonoid and cofree cocommutative Coxeter comonoid on p are given by $\mathbf{As}^* \circ \mathsf{p} = \mathcal{T}^\vee(\mathsf{p})$ and $\mathbf{Com}^* \circ \mathsf{p} = \mathcal{S}^\vee(\mathsf{p})$. Similarly, the cofree Coxeter Lie comonoid on p is given by $\mathbf{Lie}^* \circ \mathsf{p} = \mathcal{Q}\mathcal{T}^\vee(\mathsf{p})$.

This story also has signed analogues which we briefly mention.

6.4.1. Free Coxeter Lie monoid on a Coxeter species. The free Coxeter Lie monoid on a Coxeter species p is given by $\mathbf{Lie} \circ \mathsf{p}$. Thus,

$$(\mathbf{Lie} \circ \mathsf{p})[\mathrm{X}] = \bigoplus_{\mathrm{Y}:\, \mathrm{Y} \geq \mathrm{X}} \mathbf{Lie}[\mathrm{X}, \mathrm{Y}] \otimes \mathsf{p}[\mathrm{Y}]. \tag{6.11}$$

Its universal property is stated below. Let $\mathsf{p} \hookrightarrow \mathbf{Lie} \circ \mathsf{p}$ be the canonical inclusion.

Theorem 6.3. *Let* g *be a Coxeter Lie monoid,* p *a Coxeter species,* $f : \mathsf{p} \to \mathsf{g}$ *a map of Coxeter species. Then there exists a unique morphism of Coxeter Lie monoids* $\hat{f} : \mathbf{Lie} \circ \mathsf{p} \to \mathsf{g}$ *such that the diagram*

$$\begin{array}{ccc} \mathbf{Lie} \circ \mathsf{p} & \overset{\hat{f}}{\dashrightarrow} & \mathsf{g} \\ \uparrow & \nearrow_{f} & \\ \mathsf{p} & & \end{array}$$

commutes.

This specializes Theorem 5.4 to the Coxeter Lie operad. Explicitly, the map $\hat{f}$ is given by

$$\mathbf{Lie} \circ \mathsf{p} \xrightarrow{\mathrm{id} \circ f} \mathbf{Lie} \circ \mathsf{g} \xrightarrow{\gamma} \mathsf{g}.$$

Recall from Section 3.3.3 that functors $\mathcal{T}$ and $\mathcal{P}$ are adjoint between the category of Coxeter species and the category of Coxeter bimonoids. Thus, $\mathcal{PT}$ is a monad on Coxeter species. In fact, $\mathcal{PT}$-algebras are the same as Coxeter Lie monoids as stated below.

Proposition 6.4. *For a Coxeter species* p,

$$\mathcal{PT}(\mathsf{p}) = \mathbf{Lie} \circ \mathsf{p}.$$

The monad structure of $\mathcal{PT}$ *corresponds to the Coxeter operad structure of* **Lie**. *In particular, a* $\mathcal{PT}$*-algebra is the same as a Coxeter Lie monoid.*

PROOF. This can be deduced from the corresponding result for species [**11**, Proposition 16.6]. □

6.4.2. Cofree Coxeter Lie comonoid. The cofree Coxeter Lie comonoid on a Coxeter species p is given by $\mathbf{Lie}^* \circ \mathsf{p}$. Its universal property is stated below. Let $\mathbf{Lie}^* \circ \mathsf{p} \twoheadrightarrow \mathsf{p}$ be the canonical projection.

Theorem 6.5. *Let* k *be a Coxeter Lie comonoid,* p *a Coxeter species,* $f : \mathsf{k} \to \mathsf{p}$ *a map of Coxeter species. Then there exists a unique morphism of Coxeter Lie comonoids* $\hat{f} : \mathsf{k} \to \mathbf{Lie}^* \circ \mathsf{p}$ *such that the diagram*

$$\begin{array}{ccc} \mathsf{k} & \overset{\hat{f}}{\dashrightarrow} & \mathbf{Lie}^* \circ \mathsf{p} \\ & \searrow^{f} & \downarrow \\ & & \mathsf{p} \end{array}$$

commutes.

Recall from Section 3.3.6 that functors $\mathcal{T}^\vee$ and $\mathcal{Q}$ are adjoint between the category of Coxeter species and the category of Coxeter bimonoids. Thus, $\mathcal{Q}\mathcal{T}^\vee$ is a comonad on Coxeter species. In fact, $\mathcal{Q}\mathcal{T}^\vee$-coalgebras are the same as Coxeter Lie comonoids as stated below.

Proposition 6.6. *For a Coxeter species* p,

$$\mathcal{Q}\mathcal{T}^\vee(\mathsf{p}) = \mathbf{Lie}^* \circ \mathsf{p}.$$

The comonad structure of $\mathcal{Q}\mathcal{T}^\vee$ *corresponds to the Coxeter cooperad structure of* $\mathbf{Lie}^*$. *In particular, a* $\mathcal{Q}\mathcal{T}^\vee$*-coalgebra is the same as a Coxeter Lie comonoid.*

This is the dual of Proposition 6.4.

6.4.3. Signed analogue. Signed analogues of the preceding results can be deduced from the unsigned versions by conjugating by the u-signature functor (2.70) or the o-signature functor (2.73). They are recorded below.

The free signed Coxeter Lie monoid on a Coxeter species p is given by $\mathbf{Lie}^- \circ \mathsf{p}$. Thus,

$$(\mathbf{Lie}^- \circ \mathsf{p})[\mathrm{X}] = \bigoplus_{\mathrm{Y}:\, \mathrm{Y} \geq \mathrm{X}} \mathbf{Lie}^-[\mathrm{X}, \mathrm{Y}] \otimes \mathsf{p}[\mathrm{Y}].$$

Recall: Functors $\mathcal{T}_{-1}$ and $\mathcal{P}$ are adjoint between the category of Coxeter species and the category of signed Coxeter bimonoids. This yields the monad $\mathcal{P}\mathcal{T}_{-1}$ on Coxeter species whose algebras are signed Coxeter Lie monoids.

Dually, the cofree signed Coxeter Lie comonoid on a Coxeter species p is given by $(\mathbf{Lie}^-)^* \circ \mathsf{p}$. Coalgebras over the comonad $\mathcal{Q}\mathcal{T}^\vee_{-1}$ are signed Coxeter Lie comonoids.

6.4.4. Examples. We know from Proposition 6.4 that for a Coxeter species p, the primitive part of the Coxeter bimonoid $\mathcal{T}(\mathsf{p})$ is the free Coxeter Lie monoid on p. As special cases:

The Coxeter Lie species is the free Coxeter Lie monoid on x:

$$\mathsf{Lie} = \mathcal{P}(\Gamma) = \mathcal{P}(\mathcal{T}(\mathsf{x})) = \mathbf{Lie} \circ \mathsf{x}. \tag{6.12}$$

The first step used (4.96) and second step used (4.27) for $q = 1$.

Similarly, the Coxeter Zie species is the free Coxeter Lie monoid on the exponential Coxeter species E: For $\mathrm{char}(\Bbbk) \nmid |W|$,

$$\mathsf{Zie} = \mathcal{P}(\Sigma) = \mathcal{P}(\mathcal{T}(\mathsf{E})) = \mathbf{Lie} \circ \mathsf{E}. \tag{6.13}$$

The first step used (4.99) and second step used (4.64). Note very carefully: The $\mathtt{Q}$-basis of Σ enters here.

More generally, for a left W-module M,

$$\begin{aligned} \mathsf{Lie}_M &= \mathcal{P}(\Gamma_M) = \mathcal{P}(\mathcal{T}(\mathsf{x}_M)) = \mathbf{Lie} \circ \mathsf{x}_M, \\ \mathsf{Zie}_M &= \mathcal{P}(\Sigma_M) = \mathcal{P}(\mathcal{T}(\mathsf{E}_M)) = \mathbf{Lie} \circ \mathsf{E}_M, \end{aligned} \tag{6.14}$$

the latter if $\mathrm{char}(\Bbbk) \nmid |W|$. This used (4.109) for $q = 1$ and (4.122).

Dual remarks apply to the indecomposable parts of Γ^* and Σ^*, and more generally of Γ^*_M and Σ^*_M.

6.5. Universal (co)enveloping Coxeter (co)monoids

We construct the universal enveloping Coxeter monoid $\mathcal{U}(\mathsf{g})$ of a Coxeter Lie monoid g. It yields a functor $\mathcal{U}$ from Coxeter Lie monoids to Coxeter monoids which is left adjoint to the commutator functor. Moreover, $\mathcal{U}(\mathsf{g})$ carries a cocommutative Coxeter bimonoid structure. Thus, $\mathcal{U}$ also defines a functor from Coxeter Lie monoids to cocommutative Coxeter bimonoids. It is left adjoint to the primitive part functor $\mathcal{P}$.

Dually, we construct the universal coenveloping Coxeter comonoid $\mathcal{U}^{\vee}(\mathsf{k})$ of a Coxeter Lie comonoid k. It yields a functor $\mathcal{U}^{\vee}$ which is right adjoint to the cocommutator functor. Moreover, $\mathcal{U}^{\vee}(\mathsf{k})$ carries a commutative Coxeter bimonoid structure. Viewed in this manner, $\mathcal{U}^{\vee}$ is right adjoint to the indecomposable part functor $\mathcal{Q}$.

Similar considerations carry over to the signed setting.

6.5.1. Universal enveloping Coxeter monoid. Let g be a Coxeter Lie monoid. Define $\mathcal{U}(\mathsf{g})$ to be the largest Coxeter quotient monoid of $\mathcal{T}(\mathsf{g})$ such that the outside square in the diagram

$$\begin{array}{ccc} \mathbf{Lie} \circ \mathsf{g} & \xrightarrow{\gamma} & \mathsf{g} \\ \downarrow & & \downarrow \\ \mathbf{As} \circ \mathsf{g} & \xrightarrow{\cong} & \mathcal{T}(\mathsf{g}) \\ & & \downarrow \\ & & \mathcal{U}(\mathsf{g}) \end{array} \tag{6.15}$$

commutes. We call this the *universal enveloping Coxeter monoid* of g. It yields a functor

$$\mathcal{U} : \mathsf{LieMon}(\mathsf{CSp}) \to \mathsf{Mon}(\mathsf{CSp})$$

from the category of Coxeter Lie monoids to the category of Coxeter monoids. Explicitly, the kernel of $\mathcal{T}(\mathsf{g}) \twoheadrightarrow \mathcal{U}(\mathsf{g})$ has its A-component spanned by the H/A-relations defined in [**11**, Section 16.5] (where the context was species as opposed to Coxeter species). The key observation for this fact is that for any $w \in W$ and face A, the linear map

$$\mathcal{T}(\mathsf{g})[A] \xrightarrow{w} \mathcal{T}(\mathsf{g})[wA]$$

sends an H/A-relation to a wH/wA-relation.

Theorem 6.7. *The functor $\mathcal{U}$ is left adjoint to the commutator functor* (6.3). *That is, for a Coxeter Lie monoid g and Coxeter monoid a, there is a natural bijection*

$$\mathsf{Mon}(\mathsf{CSp})(\mathcal{U}(\mathsf{g}), \mathsf{a}) \xrightarrow{\cong} \mathsf{LieMon}(\mathsf{CSp})(\mathsf{g}, \mathsf{a}).$$

PROOF. This can be deduced from the corresponding result for species [**11**, Theorem 16.15] as follows. Let $f : \mathsf{g} \to \mathsf{a}$ and $\hat{f} : \mathcal{U}(\mathsf{g}) \to \mathsf{a}$ correspond to each other under the bijection in [**11**, Theorem 16.15]. Then f is a map of Coxeter species iff $\hat{f}$ is a map of Coxeter species. This fact can be deduced from a similar fact proved in Remark 3.8. Hence, the bijection restricts to the one in the above result. □

The unit of the adjunction is a morphism $i : \mathrm{g} \to \mathcal{U}(\mathrm{g})$ of Coxeter Lie monoids. It equals the composite map $\mathrm{g} \hookrightarrow \mathcal{T}(\mathrm{g}) \twoheadrightarrow \mathcal{U}(\mathrm{g})$.

So far $\mathcal{U}(\mathrm{g})$ has only been considered as a Coxeter monoid. In fact, the coproduct of $\mathcal{T}(\mathrm{g})$ descends to $\mathcal{U}(\mathrm{g})$ and turns the latter into a cocommutative Coxeter bimonoid. This follows from the corresponding fact for species [**11**, Proposition 16.18]. Thus, we also have a functor

$$\mathcal{U} : \mathsf{LieMon}(\mathsf{CSp}) \to {}^{\mathrm{co}}\mathsf{Bimon}(\mathsf{CSp})$$

from the category of Coxeter Lie monoids to the category of cocommutative Coxeter bimonoids.

Theorem 6.8. *The functor $\mathcal{U}$ is left adjoint to the primitive part functor $\mathcal{P}$. That is, for a Coxeter Lie monoid g and cocommutative Coxeter bimonoid h, there is a natural bijection*

$${}^{\mathrm{co}}\mathsf{Bimon}(\mathsf{CSp})(\mathcal{U}(\mathsf{g}), \mathsf{h}) \xrightarrow{\cong} \mathsf{LieMon}(\mathsf{CSp})(\mathsf{g}, \mathcal{P}(\mathsf{h})).$$

Proof. This can be deduced from the corresponding result for species [**11**, Theorem 16.25]. □

This can also be viewed as an adjunction between categories of Coxeter Lie monoids and Coxeter bimonoids.

Proposition 6.9. *For a Coxeter species* p,

$$\mathcal{U}(\mathbf{Lie} \circ \mathsf{p}) = \mathcal{T}(\mathsf{p}) \tag{6.16}$$

as Coxeter bimonoids.

Proof. This follows by composing adjunctions

$$\mathsf{CSp} \underset{\mathit{frg}}{\overset{\mathbf{Lie}\circ(-)}{\rightleftarrows}} \mathsf{LieMon}(\mathsf{CSp}) \underset{\mathcal{P}}{\overset{\mathcal{U}}{\rightleftarrows}} \mathsf{Bimon}(\mathsf{CSp}),$$

and comparing with the adjunction between $\mathcal{T}$ and $\mathcal{P}$ in Section 3.3.3. □

6.5.2. Universal coenveloping Coxeter comonoid. Let k be a Coxeter Lie comonoid. Define $\mathcal{U}^\vee(\mathsf{k})$ to be the largest Coxeter subcomonoid of $\mathcal{T}^\vee(\mathsf{k})$ such that the outside square in the diagram

$$\begin{array}{ccccc} \mathcal{U}^\vee(\mathsf{k}) & \hookrightarrow & \mathcal{T}^\vee(\mathsf{k}) & \xrightarrow{\cong} & \mathbf{As}^* \circ \mathsf{k} \\ & \searrow & \downarrow & & \downarrow \\ & & \mathsf{k} & \xrightarrow{\delta} & \mathbf{Lie}^* \circ \mathsf{k} \end{array} \tag{6.17}$$

commutes. We call $\mathcal{U}^\vee(\mathsf{k})$ the *universal coenveloping Coxeter comonoid* of k. It yields a functor

$$\mathcal{U}^\vee : \mathsf{LieComon}(\mathsf{CSp}) \to \mathsf{Comon}(\mathsf{CSp})$$

from the category of Coxeter Lie comonoids to the category of Coxeter comonoids.

Theorem 6.10. *The functor $\mathcal{U}^\vee$ is right adjoint to the cocommutator functor* (6.6). *That is, for a Coxeter Lie comonoid* k *and Coxeter comonoid* c, *there is a natural bijection*

$$\mathsf{LieComon}(\mathsf{CSp})(\mathsf{c},\mathsf{k}) \xrightarrow{\cong} \mathsf{Comon}(\mathsf{CSp})(\mathsf{c},\mathcal{U}^\vee(\mathsf{k})).$$

The counit of the adjunction is a morphism $p : \mathcal{U}^\vee(\mathsf{k}) \to \mathsf{k}$ of Coxeter Lie comonoids. It equals the composite map $\mathcal{U}^\vee(\mathsf{k}) \hookrightarrow \mathcal{T}^\vee(\mathsf{k}) \twoheadrightarrow \mathsf{k}$.

So far $\mathcal{U}^\vee(\mathsf{k})$ has only been considered as a Coxeter comonoid. In fact, it is a commutative Coxeter subbimonoid of $\mathcal{T}^\vee(\mathsf{k})$. This yields a functor

$$\mathcal{U}^\vee : \mathsf{LieComon}(\mathsf{CSp}) \to \mathsf{Bimon}^{\mathsf{co}}(\mathsf{CSp})$$

from the category of Coxeter Lie comonoids to the category of commutative Coxeter bimonoids.

Theorem 6.11. *The functor $\mathcal{U}^\vee$ is right adjoint to the indecomposable part functor $\mathcal{Q}$. That is, for a Coxeter Lie comonoid* k *and commutative Coxeter bimonoid* h, *there is a natural bijection*

$$\mathsf{LieComon}(\mathsf{CSp})(\mathcal{Q}(\mathsf{h}),\mathsf{k}) \xrightarrow{\cong} \mathsf{Bimon}^{\mathsf{co}}(\mathsf{CSp})(\mathsf{h},\mathcal{U}^\vee(\mathsf{k})).$$

Proposition 6.12. *For a Coxeter species* p,

$$\mathcal{U}^\vee(\mathbf{Lie}^* \circ \mathsf{p}) = \mathcal{T}^\vee(\mathsf{p})$$

as Coxeter bimonoids.

6.5.3. Signed analogue. For a signed Coxeter Lie monoid g, we have its universal enveloping Coxeter monoid $\mathcal{U}_{-1}(\mathsf{g})$. Dually, for a signed Coxeter Lie comonoid k, we have its universal coenveloping Coxeter comonoid $\mathcal{U}^\vee_{-1}(\mathsf{k})$. They can be defined as

$$\mathcal{U}_{-1}(\mathsf{g}) = \mathcal{U}(\mathsf{g}_-)_- \quad \text{and} \quad \mathcal{U}^\vee_{-1}(\mathsf{k}) = \mathcal{U}^\vee(\mathsf{k}_-)_-$$

by conjugating $\mathcal{U}$ and $\mathcal{U}^\vee$ by the u-signature functor (2.70). One can also employ the o-signature functor (2.73) instead. It follows that $\mathcal{U}_{-1}(\mathsf{g})$ is a signed cocommutative signed Coxeter bimonoid, and $\mathcal{U}^\vee_{-1}(\mathsf{k})$ is a signed commutative signed Coxeter bimonoid. The various adjunctions work the same way as in the unsigned case.

6.6. Joyal Lie monoids and Coxeter Lie monoids

Joyal Lie monoids were reviewed in [**11**, Section 17.7.1]. Recall from Section 2.12.2 that a Joyal species g gives rise to a Coxeter $\mathcal{B}^J$-species g^J by setting

$$\mathsf{g}^J[F] := \mathsf{g}[S_1] \otimes \cdots \otimes \mathsf{g}[S_k] \quad \text{for} \quad F = (S_1, \dots, S_k).$$

Now let g be a Joyal Lie monoid with Lie bracket components $[\,,\,]_{S,T}$. Then g^J is a Coxeter $\mathcal{B}^J$-Lie monoid with Lie bracket defined as follows. For $A \lessdot F$, note that F breaks some block of A into two blocks, say S and T. Define ν_A^F of g^J to be $[\,,\,]_{S,T}$ on the $\mathsf{g}[S] \otimes \mathsf{g}[T]$ factor of $\mathsf{g}^J[F]$, and identity on the other tensor factors. The discussion in [**11**, Section 17.7.3] shows that g^J is a

$\mathcal{B}^J$-Lie monoid. So the only additional check is the second diagram in (6.1a) which follows from [**11**, Axiom (17.40a)].

By a similar construction, a Joyal Lie comonoid k gives rise to a Coxeter $\mathcal{B}^J$-Lie comonoid k^J.

6.7. Notes

The notion of a Coxeter Lie monoid generalizes the classical notion of a positive Joyal Lie monoid as explained in Section 6.6. It builds on the less structured notion of a Lie monoid in species developed in detail in [**11**, Chapter 16]; for the precise connection, see Section 6.1.6.

6.7.1. Joyal Lie monoids. Joyal Lie monoids first appeared in work of Barratt [**30**, Definition 4] in the language of "twisted Lie algebras". They were considered later by Joyal [**203**], Goerss [**154**], Stover [**341**], Aubry [**23**], [**24**]; the adjunction between $\mathcal{P}$ and $\mathcal{U}$ is given in [**341**, Proposition 7.10]. A definition similar in spirit to the viewpoint of this book is given in our monographs [**8**, Section 2.9], [**11**, Section 17.7.1].

CHAPTER 7

Structure theory of Coxeter bimonoids

We now discuss structure theory of Coxeter bimonoids. These include the Leray–Samelson theorem, Loday–Ronco theorem and its q-generalization, and Cartier–Milnor–Moore theorem. They all specify adjoint equivalences between certain categories constructed out of Coxeter species. A summary is provided in Table 7.1. Related results include the Borel–Hopf theorem and Poincaré–Birkhoff–Witt theorem. We also discuss a family of results called the Hoffman–Newman–Radford theorems.

TABLE 7.1. Structure theorems for Coxeter bimonoids.

Theorem	Adjoint equivalence
Leray–Samelson	Coxeter species $\underset{\mathcal{P}}{\overset{\mathcal{S}}{\rightleftarrows}}$ bicom. Coxeter bimonoids
	bicom. Coxeter bimonoids $\underset{\mathcal{S}^{\vee}}{\overset{\mathcal{Q}}{\rightleftarrows}}$ Coxeter species
Loday–Ronco	Coxeter species $\underset{\mathcal{P}}{\overset{\mathcal{T}_0}{\rightleftarrows}}$ Coxeter 0-bimonoids
	Coxeter 0-bimonoids $\underset{\mathcal{T}_0^{\vee}}{\overset{\mathcal{Q}}{\rightleftarrows}}$ Coxeter species
q-generalization for q not a root of unity	Coxeter species $\underset{\mathcal{P}}{\overset{\mathcal{T}_q}{\rightleftarrows}}$ Coxeter q-bimonoids
	Coxeter q-bimonoids $\underset{\mathcal{T}_q^{\vee}}{\overset{\mathcal{Q}}{\rightleftarrows}}$ Coxeter species
Cartier–Milnor–Moore	Coxeter Lie monoids $\underset{\mathcal{P}}{\overset{\mathcal{U}}{\rightleftarrows}}$ cocom. Coxeter bimonoids
	com. Coxeter bimonoids $\underset{\mathcal{U}^{\vee}}{\overset{\mathcal{Q}}{\rightleftarrows}}$ Coxeter Lie comonoids

Species vs Coxeter species. The above results can be deduced from those for bimonoids in species proved in [**11**, Part III]. This is because the involved maps do not change. But they need to be maps of Coxeter species as opposed to species. Thus, whenever formulas involve noncommutative zeta and Möbius functions, the latter need to be invariant which requires $\mathrm{char}(\Bbbk) \nmid |W|$. This enters for instance in the Borel–Hopf and Poincaré–Birkhoff–Witt theorems. Apart from this issue, everything in [**11**, Part III] carries over to Coxeter species. To avoid repetition, our exposition will be brief.

Rank one. Let $\mathcal{A}$ be a rank-one arrangement. Its Coxeter group W has two elements. We discuss from first principles notions of Coxeter $\mathcal{A}$-species and Coxeter $\mathcal{A}$-(co, bi)monoids. We then develop the basic theory of these objects such as universal constructions, exp-log correspondences, and conclude with structure theorems. This provides a nice self-contained summary of the theory of Coxeter bimonoids in a toy example.

7.1. Borel–Hopf for (co)commutative Coxeter bimonoids

In this section, we assume $\mathrm{char}(\Bbbk) \nmid |W|$. The Borel–Hopf theorem says that any cocommutative Coxeter bimonoid is isomorphic as a Coxeter comonoid to the cofree cocommutative Coxeter comonoid on its primitive part. Dually, any commutative Coxeter bimonoid is isomorphic as a Coxeter monoid to the free commutative Coxeter monoid on its indecomposable part. These results have signed analogues.

7.1.1. Borel–Hopf for cocommutative Coxeter bimonoids (primitive part version). Recall from Section 3.3.10 the cofree cocommutative Coxeter comonoid $\mathcal{S}^\vee(\mathsf{p})$ on a Coxeter species p. Here we view $\mathcal{S}^\vee(\mathsf{p})$ inside $\mathcal{T}^\vee(\mathsf{p})$. Fix an invariant noncommutative zeta function $\boldsymbol{\zeta}$ and its inverse Möbius function $\boldsymbol{\mu}$. Let h be a cocommutative Coxeter bimonoid. Define a natural map

$$\mathcal{S}^\vee(\mathcal{P}(\mathsf{h})) \to \mathsf{h} \tag{7.1}$$

which on the A-component is

$$(x_F)_{F\geq A} \mapsto \sum_{F:\,F\geq A} \boldsymbol{\zeta}(A,F)\, \mu_A^F(x_F).$$

(In the element (x_F) above, by definition, $\beta_{G,F}(x_F) = x_G$ for any $F, G \geq A$.) Similarly, define a natural map

$$\mathsf{h} \to \mathcal{S}^\vee(\mathcal{P}(\mathsf{h})) \tag{7.2}$$

which on the A-component, into the K-summand, is

$$\mathbf{log}(\mathrm{id})_K \Delta_A^K = \sum_{G:\,G\geq K} \boldsymbol{\mu}(K,G)\, \mu_K^G \Delta_A^G.$$

It arises from cofreeness of $\mathcal{S}^\vee$. More precisely, apply Theorem 3.20 with $\mathsf{c} := \mathsf{h}$, $\mathsf{p} := \mathcal{P}(\mathsf{h})$, $f := \mathbf{log}(\mathrm{id}) : \mathsf{h} \twoheadrightarrow \mathcal{P}(\mathsf{h})$ as in Proposition 3.46, and use formula (3.108). This also shows that (7.2) is a Coxeter comonoid morphism.

Theorem 7.1. *Let* $\mathrm{char}(\Bbbk) \nmid |W|$. *For a cocommutative Coxeter bimonoid* h, *the maps* (7.1) *and* (7.2) *are inverse isomorphisms of Coxeter comonoids. In particular, as a Coxeter comonoid,* h *is the cofree cocommutative Coxeter comonoid on* $\mathcal{P}(\mathsf{h})$.

This is the *Borel–Hopf theorem* for cocommutative Coxeter bimonoids, and (7.1), (7.2) are the *Borel–Hopf isomorphisms* (primitive part version).

PROOF. This is a formal consequence of the corresponding result for species [**11**, Theorems 13.34 and 13.38]. The invariance of $\boldsymbol{\zeta}$ and $\boldsymbol{\mu}$ ensures that (7.1) and (7.2) are maps of Coxeter species. □

7.1.2. Borel–Hopf for commutative Coxeter bimonoids (indecomposable part version). Recall from Section 3.3.8 the free commutative Coxeter monoid $\mathcal{S}(\mathsf{p})$ on a Coxeter species p. Here we view $\mathcal{S}(\mathsf{p})$ as a quotient of $\mathcal{T}(\mathsf{p})$. Fix an invariant noncommutative zeta function $\boldsymbol{\zeta}$ and its inverse Möbius function $\boldsymbol{\mu}$. Let h be a commutative Coxeter bimonoid. Define a natural map

$$\mathsf{h} \to \mathcal{S}(\mathcal{Q}(\mathsf{h})) \tag{7.3}$$

which on the A-component sends $x \in \mathsf{h}[A]$ to the image of

$$\sum_{F:\, F \geq A} \boldsymbol{\zeta}(A, F)\, \Delta_A^F(x) \;\in\; \mathcal{T}(\mathsf{h})[A]$$

in the quotient $\mathcal{S}(\mathcal{Q}(\mathsf{h}))[A]$. Similarly, define a natural map

$$\mathcal{S}(\mathcal{Q}(\mathsf{h})) \to \mathsf{h} \tag{7.4}$$

which on the A-component sends a representative $x \in \mathsf{h}[K]$ to

$$\mu_A^K \, \mathbf{log}(\mathrm{id})_K = \sum_{G:\, G \geq K} \boldsymbol{\mu}(K, G)\, \mu_A^G \Delta_K^G(x).$$

It arises from freeness of $\mathcal{S}$. More precisely, apply Theorem 3.18 with $\mathsf{a} := \mathsf{h}$, $\mathsf{p} := \mathcal{Q}(\mathsf{h})$, $f := \mathbf{log}(\mathrm{id}) : \mathcal{Q}(\mathsf{h}) \hookrightarrow \mathsf{h}$ as in Proposition 3.46, and use formula (3.108). This also shows that (7.4) is a Coxeter monoid morphism.

Theorem 7.2. *Let* $\mathrm{char}(\Bbbk) \nmid |W|$. *For a commutative Coxeter bimonoid* h, *the maps* (7.3) *and* (7.4) *are inverse isomorphisms of Coxeter monoids. In particular, as a Coxeter monoid,* h *is the free commutative Coxeter monoid on* $\mathcal{Q}(\mathsf{h})$.

This is the *Borel–Hopf theorem* for commutative Coxeter bimonoids, and (7.3), (7.4) are the *Borel–Hopf isomorphisms* (indecomposable part version).

PROOF. This is a formal consequence of the corresponding result for species [**11**, Theorems 13.57 and 13.59]. □

Remark 7.3. Borel–Hopf isomorphisms for (co)commutative Coxeter bimonoids can also be written using the flat formulation of $\mathcal{S}(\mathsf{p}) = \mathcal{S}^\vee(\mathsf{p})$, see [**11**, Formulas (13.22), (13.24), (13.37), (13.40)]. This brings in the maps $\beta_{\mathrm{X},F}$ and $\beta_{F,\mathrm{X}}$.

7.1.3. Signed analogue. Borel–Hopf Theorems 7.1 and 7.2 have signed analogues. Corresponding theorems for species are stated on [**11**, pages 555 and 560].

7.2. Leray–Samelson for bicommutative Coxeter bimonoids

The Leray–Samelson theorem says that functors $\mathcal{S}$ and $\mathcal{P}$ determine an adjoint equivalence between categories of Coxeter species and bicommutative Coxeter bimonoids. Dually, functors $\mathcal{S}^\vee$ and $\mathcal{Q}$ determine an adjoint equivalence between the same two categories. These results have signed analogues.

7.2.1. Leray–Samelson (primitive part version). Recall from (3.38) the functor $\mathcal{S}$ from Coxeter species to bicommutative Coxeter bimonoids. It is left adjoint to the primitive part functor $\mathcal{P}$.

Theorem 7.4. *The adjunction*

$$\mathsf{CSp} \underset{\mathcal{P}}{\overset{\mathcal{S}}{\rightleftarrows}} {}^{\mathsf{co}}\mathsf{Bimon}^{\mathsf{co}}(\mathsf{CSp})$$

is an adjoint equivalence of categories.

This is the *Leray–Samelson theorem* for bicommutative Coxeter bimonoids. Thus, for p a Coxeter species, $\mathcal{P}(\mathcal{S}(\mathsf{p})) = \mathsf{p}$ as Coxeter species, and for h a bicommutative Coxeter bimonoid,

$$\mathcal{S}(\mathcal{P}(\mathsf{h})) \xrightarrow{\cong} \mathsf{h} \tag{7.5}$$

as Coxeter bimonoids. This isomorphism is the unit of the adjunction. It arises from the universal property of $\mathcal{S}$ in Theorem 3.19. Explicitly, the map, on the Z-component, on the X-summand, is $\mu^{\mathrm{X}}_{\mathrm{Z}}$. The inverse of the map (7.5), namely,

$$\mathsf{h} \xrightarrow{\cong} \mathcal{S}(\mathcal{P}(\mathsf{h})), \tag{7.6}$$

on the Z-component, into the X-summand, is

$$\log(\mathrm{id})_{\mathrm{X}} \Delta^{\mathrm{X}}_{\mathrm{Z}} = \sum_{\mathrm{Y}:\, \mathrm{Y} \geq \mathrm{X}} \mu(\mathrm{X}, \mathrm{Y})\, \mu^{\mathrm{Y}}_{\mathrm{X}} \Delta^{\mathrm{Y}}_{\mathrm{Z}} \;:\; \mathsf{h}[\mathrm{Z}] \to \mathcal{P}(\mathsf{h})[\mathrm{X}].$$

We refer to (7.5) and (7.6) as the *Leray–Samelson isomorphisms* (primitive part version).

PROOF. This is a formal consequence of the corresponding result for species [**11**, Propositions 13.9 and 13.14 and Theorem 13.11]. □

7.2.2. Leray–Samelson (indecomposable part version). Recall from (3.41) the functor $\mathcal{S}^\vee$ from Coxeter species to bicommutative Coxeter bimonoids. It is right adjoint to the indecomposable part functor $\mathcal{Q}$.

Theorem 7.5. *The adjunction*

$${}^{\mathsf{co}}\mathsf{Bimon}^{\mathsf{co}}(\mathsf{CSp}) \underset{\mathcal{S}^\vee}{\overset{\mathcal{Q}}{\rightleftarrows}} \mathsf{CSp}$$

is an adjoint equivalence of categories.

This is the *Leray–Samelson theorem* in dual form. Thus, for p a Coxeter species, $\mathcal{Q}(\mathcal{S}^\vee(\mathsf{p})) = \mathsf{p}$ as Coxeter species, and for h a bicommutative Coxeter bimonoid,

$$\mathsf{h} \xrightarrow{\cong} \mathcal{S}^\vee(\mathcal{Q}(\mathsf{h})) \tag{7.7}$$

as Coxeter bimonoids. This isomorphism is the counit of the adjunction. It arises from the universal property of $\mathcal{S}^\vee$ in Theorem 3.21. Explicitly, on the

Z-component, into the X-summand, it sends $x \in \mathsf{h}[\mathrm{Z}]$ to the image of $\Delta_{\mathrm{Z}}^{\mathrm{X}}(x)$ in the quotient $\mathcal{Q}(\mathsf{h})[\mathrm{X}]$. The inverse of the map (7.7), namely,

$$\mathcal{S}^{\vee}(\mathcal{Q}(\mathsf{h})) \xrightarrow{\cong} \mathsf{h}, \tag{7.8}$$

on the Z-component, on the X-summand, is

$$\mu_{\mathrm{Z}}^{\mathrm{X}} \log(\mathrm{id})_{\mathrm{X}} = \sum_{\mathrm{Y}:\, \mathrm{Y} \geq \mathrm{X}} \mu(\mathrm{X}, \mathrm{Y})\, \mu_{\mathrm{Z}}^{\mathrm{Y}} \Delta_{\mathrm{X}}^{\mathrm{Y}} \,:\, \mathcal{Q}(\mathsf{h})[\mathrm{X}] \to \mathsf{h}[\mathrm{Z}].$$

We refer to (7.7) and (7.8) as the *Leray–Samelson isomorphisms* (indecomposable part version).

PROOF. This is a formal consequence of the corresponding result for species [**11**, Propositions 13.22 and 13.24 and Theorem 13.21]. □

7.2.3. Connecting primitive and indecomposable part versions. Recall from Proposition 3.48 that for a bicommutative Coxeter bimonoid h, the maps pq and $\log(\mathrm{id})$ are inverses of each other. They allow us to connect the two versions of Leray–Samelson as follows.

For a bicommutative Coxeter bimonoid h, the diagram

$$\begin{array}{ccc} \mathcal{S}(\mathcal{P}(\mathsf{h})) = \mathcal{S}^{\vee}(\mathcal{P}(\mathsf{h})) & \underset{\mathcal{S}(\log(\mathrm{id}))}{\overset{\mathcal{S}(\mathrm{pq})}{\longleftrightarrow}} & \mathcal{S}(\mathcal{Q}(\mathsf{h})) = \mathcal{S}^{\vee}(\mathcal{Q}(\mathsf{h})) \\ & \nwarrow\!\searrow \quad \mathsf{h} \quad \swarrow\!\nearrow & \end{array} \tag{7.9}$$

commutes (with objects connected by pairs of inverse isomorphisms). The oblique maps are the four Leray–Samelson isomorphisms.

Exercise 7.6. Let $\mathrm{char}(\Bbbk) \nmid |W|$. Check: For a bicommutative Coxeter bimonoid h, the Borel–Hopf isomorphisms reduce to the Leray–Samelson isomorphisms. (Use Fact 1.80. Also see Remark 7.3.)

7.2.4. Signed analogue. Leray–Samelson Theorems 7.4 and 7.5 have signed analogues. Corresponding theorems for species are stated on [**11**, pages 545 and 546].

7.3. Rigidity of Coxeter q-bimonoids for q not a root of unity

The rigidity theorem of Coxeter q-bimonoids says that for q not a root of unity, functors $\mathcal{T}_q$ and $\mathcal{P}$ determine an adjoint equivalence between categories of Coxeter species and Coxeter q-bimonoids. Dually, functors $\mathcal{T}_q^{\vee}$ and $\mathcal{Q}$ determine an adjoint equivalence between the same two categories.

7.3.1. Rigidity of Coxeter q-bimonoids (primitive part version). Let q be a scalar which is not a root of unity. Recall the two-sided q-zeta function ζ_q and two-sided q-Möbius function $\boldsymbol{\mu}_q$. Recall from Section 3.3.6 the functor $\mathcal{T}_q^{\vee}$ from Coxeter species to Coxeter q-bimonoids. Let h be a Coxeter q-bimonoid. Define a natural map

$$\mathcal{T}_q^{\vee}(\mathcal{P}(\mathsf{h})) \to \mathsf{h}, \tag{7.10}$$

on the A-component, on the F'-summand, by

$$\sum_{\substack{F:\, F\geq A \\ \mathrm{s}(F)=\mathrm{s}(F')}} \zeta_q(A,F,F')\, \mu_A^F \beta_{F,F'}.$$

Similarly, define a natural map

$$\mathsf{h} \to \mathcal{T}_q^\vee(\mathcal{P}(\mathsf{h})), \tag{7.11}$$

on the A-component, into the F'-summand, by

$$\mathbf{log}_q(\mathrm{id})_{F'} \Delta_A^{F'} = \sum_{\substack{G,G'\geq F' \\ \mathrm{s}(G)=\mathrm{s}(G')}} \mu_q(F',G,G')\, \mu_{F'}^G \beta_{G,G'} \Delta_A^{G'}.$$

The sum is over both G and G'. This map arises from cofreeness of $\mathcal{T}_q^\vee$. More precisely, apply Theorem 3.16 with $\mathsf{p} := \mathcal{P}(\mathsf{h})$ and $f := \mathbf{log}_q(\mathrm{id}) : \mathcal{Q}(\mathsf{h}) \to \mathcal{P}(\mathsf{h})$ as in Proposition 3.50, and use formula (3.113). This also shows that (7.11) is a Coxeter q-bimonoid morphism.

Theorem 7.7. *For q not a root of unity, for a Coxeter q-bimonoid h, the maps* (7.10) *and* (7.11) *are inverse isomorphisms of Coxeter q-bimonoids.*

PROOF. This is a formal consequence of the corresponding result for species [**11**, Proposition 13.83]. □

Theorem 7.7 is the *rigidity theorem for Coxeter q-bimonoids*. It is reformulated below in Theorem 7.8 as an adjoint equivalence. Recall from Section 3.3.3 the functor $\mathcal{T}_q$ from Coxeter species to Coxeter q-bimonoids. It is left adjoint to the primitive part functor $\mathcal{P}$.

Theorem 7.8. *For q not a root of unity, the adjunction*

$$\mathsf{CSp} \underset{\mathcal{P}}{\overset{\mathcal{T}_q}{\rightleftarrows}} q\text{-}\mathsf{Bimon}(\mathsf{CSp})$$

is an adjoint equivalence of categories.

PROOF. This is a formal consequence of the corresponding result for species [**11**, Theorem 13.77]. □

Thus, for p a Coxeter species, $\mathcal{P}(\mathcal{T}_q(\mathsf{p})) = \mathsf{p}$ as Coxeter species, and for h a Coxeter q-bimonoid,

$$\mathcal{T}_q(\mathcal{P}(\mathsf{h})) \xrightarrow{\cong} \mathsf{h} \tag{7.12}$$

as Coxeter q-bimonoids. This isomorphism is the unit of the adjunction. It arises from the universal property of $\mathcal{T}_q$ in Theorem 3.11. Explicitly, on the A-component, it sends $x \in \mathcal{P}(\mathsf{h})[F]$ for $F \geq A$ to $\mu_A^F(x)$. Moreover, (7.12) equals the composite map

$$\mathcal{T}_q(\mathcal{P}(\mathsf{h})) \xrightarrow{\cong} \mathcal{T}_q^\vee(\mathcal{P}(\mathsf{h})) \xrightarrow{\cong} \mathsf{h},$$

where the first isomorphism is the q-norm map in (3.144) and second isomorphism is as in (7.10). A formula for the inverse of (7.12) is given in [**11**, Exercise 13.84].

7.3.2. Rigidity of Coxeter q-bimonoids (indecomposable part version). The functor $\mathcal{T}_q^\vee$ from Coxeter species to Coxeter q-bimonoids is right adjoint to the indecomposable part functor $\mathcal{Q}$. It yields the following dual version of Theorem 7.8.

Theorem 7.9. *For q not a root of unity, the adjunction*

$$q\text{-}\mathsf{Bimon}(\mathsf{CSp}) \underset{\mathcal{T}_q^\vee}{\overset{\mathcal{Q}}{\rightleftarrows}} \mathsf{CSp}$$

is an adjoint equivalence of categories.

PROOF. This is a formal consequence of the corresponding result for species [**11**, Theorem 13.93]. ☐

Formulas for the isomorphisms between h and $\mathcal{T}_q^\vee(\mathcal{Q}(\mathsf{h}))$ and between h and $\mathcal{T}_q(\mathcal{Q}(\mathsf{h}))$ are the same as those in [**11**, (13.52), (13.54), (13.55), (13.56)].

7.3.3. Connecting primitive and indecomposable part versions. Recall from Proposition 3.50 that for a Coxeter q-bimonoid h for q not a root of unity, the maps pq and $\mathbf{log}_q(\mathrm{id})$ are inverses of each other. They allow us to connect the two versions of the rigidity result as follows.

For a Coxeter q-bimonoid h for q not a root of unity, the diagram

(7.13)

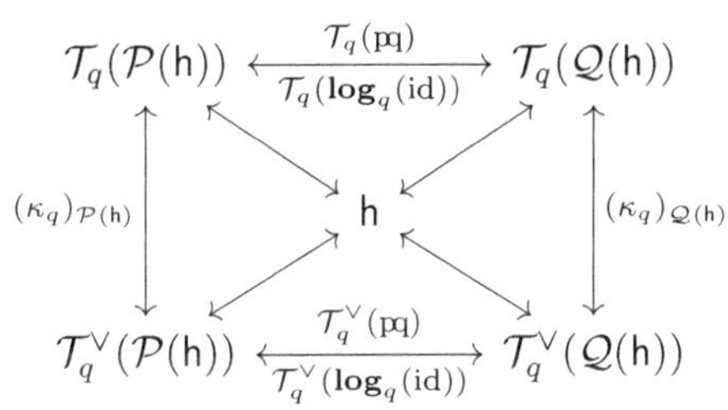

commutes (with objects connected by pairs of inverse isomorphisms).

7.4. Loday–Ronco for Coxeter 0-bimonoids

Let us now specialize the rigidity theorem of Coxeter q-bimonoids to $q = 0$. It says that the category of Coxeter 0-bimonoids is equivalent to the category of Coxeter species. We refer to this as the Loday–Ronco theorem.

7.4.1. Loday–Ronco. Theorem 7.8 specializes as follows.

Theorem 7.10. *The adjunction*

$$\mathsf{CSp} \underset{\mathcal{P}}{\overset{\mathcal{T}_0}{\rightleftarrows}} 0\text{-}\mathsf{Bimon}(\mathsf{CSp})$$

is an adjoint equivalence of categories.

This is the *Loday–Ronco theorem* for Coxeter 0-bimonoids. For h a Coxeter 0-bimonoid, the isomorphisms (7.10) and (7.12) coincide. The map

$$\mathcal{T}_0(\mathcal{P}(\mathsf{h})) = \mathcal{T}_0^\vee(\mathcal{P}(\mathsf{h})) \xrightarrow{\cong} \mathsf{h}, \tag{7.14}$$

on the A-component, on the F-summand, is μ_A^F. The inverse map

$$\mathsf{h} \xrightarrow{\cong} \mathcal{T}_0(\mathcal{P}(\mathsf{h})) = \mathcal{T}_0^\vee(\mathcal{P}(\mathsf{h})), \tag{7.15}$$

on the A-component, into the F-summand, is

$$\mathbf{log}_0(\mathrm{id})_F \Delta_A^F = \sum_{G\colon G \geq F} (-1)^{\mathrm{rk}(G/F)} \mu_F^G \Delta_A^G.$$

We refer to (7.14) and (7.15) as the *Loday–Ronco isomorphisms* (primitive part version).

The Loday–Ronco theorem in dual form is stated below. It is the specialization of Theorem 7.9.

Theorem 7.11. *The adjunction*

$$0\text{-}\mathsf{Bimon}(\mathsf{CSp}) \underset{\mathcal{T}_0^\vee}{\overset{\mathcal{Q}}{\rightleftarrows}} \mathsf{CSp}$$

is an adjoint equivalence of categories.

For supplementary reading, see the Loday–Ronco theorem for 0-bimonoids in species developed from first principles in [**11**, Section 13.1].

7.4.2. Hadamard product with one (co)free factor. We conclude with an interesting application of the Loday–Ronco theorem.

Theorem 7.12. *Let p and q be scalars. Let* h *be a Coxeter p-bimonoid,* k *a Coxeter q-bimonoid,* $\mathsf{h} \times \mathsf{k}$ *a Coxeter pq-bimonoid. Then:*

- *If* k *is free as a Coxeter monoid, then so is* $\mathsf{h} \times \mathsf{k}$.
- *If* k *is cofree as a Coxeter comonoid, then so is* $\mathsf{h} \times \mathsf{k}$.

PROOF. We explain the first part; the second is similar. Let k be free on the Coxeter species p. Then $\mathsf{h} \times \mathcal{T}_0(\mathsf{p})$ is a Coxeter 0-bimonoid, and hence free as a Coxeter monoid by Loday–Ronco Theorem 7.10. But as Coxeter monoids, $\mathsf{h} \times \mathcal{T}_0(\mathsf{p})$ is the same as $\mathsf{h} \times \mathsf{k}$, so the latter is free as well. □

Corollary 7.13. *Let p and q be scalars. Let* h *be a Coxeter p-bimonoid. Then the Coxeter pq-bimonoid*

- $\mathsf{h} \times \Gamma_q$ *is free as a Coxeter monoid,*
- $\mathsf{h} \times \Gamma_q^*$ *is cofree as a Coxeter comonoid.*

7.5. Hoffman–Newman–Radford

The Hoffman–Newman–Radford (HNR) theorems are a family of results that provide inverse isomorphisms between two universally constructed Coxeter bimonoids. These isomorphisms are constructed using zeta and Möbius functions. We call them HNR isomorphisms.

7.5.1. Shuffles and quasishuffles. Let $\mathrm{char}(\Bbbk) \nmid |W|$. Fix an invariant noncommutative zeta function $\boldsymbol{\zeta}$ and its inverse Möbius function $\boldsymbol{\mu}$.

Let c be a cocommutative Coxeter comonoid, and c_t its underlying Coxeter species with trivial coproduct. Recall the Coxeter bimonoid $\mathcal{T}(\mathsf{c})$ with concatenation product and dequasishuffle coproduct, and $\mathcal{T}(\mathsf{c}_t)$ with concatenation product and deshuffle coproduct. (Specialize the relevant entries in Table 3.2 to $q = 1$.) Define a natural map

$$\mathcal{T}(\mathsf{c}) \to \mathcal{T}(\mathsf{c}_t) \tag{7.16}$$

which on the A-component is

$$\bigoplus_{F:\, F \geq A} \mathsf{c}[F] \to \bigoplus_{H:\, H \geq A} \mathsf{c}[H], \qquad x \mapsto \sum_{H:\, H \geq F} \boldsymbol{\zeta}(F,H)\, \Delta^H_F(x)$$

for $x \in \mathsf{c}[F]$. Similarly, define a natural map

$$\mathcal{T}(\mathsf{c}_t) \to \mathcal{T}(\mathsf{c}) \tag{7.17}$$

which on the A-component is

$$\bigoplus_{F:\, F \geq A} \mathsf{c}[F] \to \bigoplus_{H:\, H \geq A} \mathsf{c}[H], \qquad x \mapsto \sum_{H:\, H \geq F} \boldsymbol{\mu}(F,H)\, \Delta^H_F(x)$$

for $x \in \mathsf{c}[F]$.

Theorem 7.14. *Let* $\mathrm{char}(\Bbbk) \nmid |W|$. *For a cocommutative Coxeter comonoid* c, *the maps* (7.16) *and* (7.17) *are inverse isomorphisms of cocommutative Coxeter bimonoids.*

PROOF. This is a formal consequence of the corresponding result for species [**11**, Propositions 14.40 and 14.42]. The invariance of $\boldsymbol{\zeta}$ and $\boldsymbol{\mu}$ ensures that (7.16) and (7.17) are maps of Coxeter species. □

Dually: Let a be a commutative Coxeter monoid, and a_t its underlying Coxeter species with trivial product. Recall the Coxeter bimonoid $\mathcal{T}^\vee(\mathsf{a})$ with quasishuffle product and deconcatenation coproduct, and $\mathcal{T}^\vee(\mathsf{a}_t)$ with shuffle product and deconcatenation coproduct. Define a natural map

$$\mathcal{T}^\vee(\mathsf{a}_t) \to \mathcal{T}^\vee(\mathsf{a}) \tag{7.18}$$

which on the A-component is

$$\bigoplus_{F:\, F \geq A} \mathsf{a}[F] \to \bigoplus_{G:\, G \geq A} \mathsf{a}[G], \qquad x \mapsto \sum_{G:\, A \leq G \leq F} \boldsymbol{\zeta}(G,F)\, \mu^F_G(x)$$

for $x \in \mathsf{a}[F]$. Similarly, define a natural map

$$\mathcal{T}^\vee(\mathsf{a}) \to \mathcal{T}^\vee(\mathsf{a}_t) \tag{7.19}$$

which on the A-component is

$$\bigoplus_{F:\, F \geq A} \mathsf{a}[F] \to \bigoplus_{G:\, G \geq A} \mathsf{a}[G], \qquad x \mapsto \sum_{G:\, A \leq G \leq F} \boldsymbol{\mu}(G,F)\, \mu^F_G(x)$$

for $x \in \mathsf{a}[F]$.

Theorem 7.15. *Let* $\mathrm{char}(\Bbbk) \nmid |W|$. *For a commutative Coxeter monoid* a, *the maps* (7.18) *and* (7.19) *are inverse isomorphisms of commutative Coxeter bimonoids.*

PROOF. This is a formal consequence of the corresponding result for species [**11**, Propositions 14.44 and 14.46]. □

We refer to (7.16) and (7.17) as *HNR isomorphisms* for cocommutative Coxeter bimonoids. Similarly, we refer to (7.18) and (7.19) as *HNR isomorphisms* for commutative Coxeter bimonoids. They depend on $\boldsymbol{\zeta}$ and $\boldsymbol{\mu}$.

7.5.2. Commutative aspects. For a cocommutative Coxeter comonoid c and commutative Coxeter monoid a, we have commutative diagrams

$$
(7.20)\qquad
\begin{array}{ccc}
\mathcal{T}(\mathsf{c}) & \underset{\boldsymbol{\mu}}{\overset{\boldsymbol{\zeta}}{\rightleftarrows}} & \mathcal{T}(\mathsf{c}_t) \\
{\scriptstyle \pi}\downarrow & & \downarrow{\scriptstyle \pi} \\
\mathcal{S}(\mathsf{c}) & \rightleftarrows & \mathcal{S}(\mathsf{c}_t)
\end{array}
\qquad\qquad
\begin{array}{ccc}
\mathcal{T}^\vee(\mathsf{a}_t) & \underset{\boldsymbol{\mu}}{\overset{\boldsymbol{\zeta}}{\rightleftarrows}} & \mathcal{T}^\vee(\mathsf{a}) \\
{\scriptstyle \pi^\vee}\uparrow & & \uparrow{\scriptstyle \pi^\vee} \\
\mathcal{S}^\vee(\mathsf{a}_t) & \rightleftarrows & \mathcal{S}^\vee(\mathsf{a}).
\end{array}
$$

The top-horizontal maps are the HNR isomorphisms for $\mathcal{T}$ and $\mathcal{T}^\vee$. The dotted maps are inverse isomorphisms which are induced from these via the vertical maps which are abelianization and coabelianization, see (3.49). Moreover, the dotted maps are unique, that is, they do not depend on $\boldsymbol{\zeta}$ and $\boldsymbol{\mu}$. They can be written in terms of ζ and μ using the flat formulation, and work in any field characteristic, see [**11**, Formulas (14.10), (14.12), (14.14), (14.16)].

Theorem 7.16 (General norm map). *For* c *a cocommutative Coxeter comonoid,* a *a commutative Coxeter monoid, let* f *and* $\widehat{f}$ *be as in* (3.147). *Then the composite map*

$$
(7.21)\qquad \mathcal{S}(\mathsf{c}_t) \xrightarrow{\cong} \mathcal{S}(\mathsf{c}) \xrightarrow{\widehat{f}} \mathcal{S}^\vee(\mathsf{a}) \xrightarrow{\cong} \mathcal{S}^\vee(\mathsf{a}_t),
$$

on the Z*-component, on the* X*-summand, is* $\log(f)_{\mathrm{X}}$. *Thus,* $\widehat{f}$ *is an isomorphism of Coxeter bimonoids iff* $\log(f)$ *is an isomorphism of Coxeter species.*

PROOF. This is a formal consequence of the corresponding result for species contained in [**11**, Exercises 14.29 and 14.30]. It subsumes Theorem 3.77. □

7.5.3. q**-shuffles and** q**-quasishuffles.** Let q be a scalar which is not a root of unity. Recall the two-sided q-zeta function ζ_q and two-sided q-Möbius function μ_q. Also recall the noncommutative q-zeta function $\boldsymbol{\zeta}_q$ and noncommutative q-Möbius function $\boldsymbol{\mu}_q$.

Let c be a Coxeter comonoid, and c_t its underlying Coxeter species with trivial coproduct. Recall from Table 3.2 the Coxeter q-bimonoid $\mathcal{T}_q(\mathsf{c})$ with concatenation product and q-dequasishuffle coproduct, and $\mathcal{T}_q(\mathsf{c}_t)$ with concatenation product and q-deshuffle coproduct. Define a natural map

$$
(7.22)\qquad \mathcal{T}_q(\mathsf{c}) \to \mathcal{T}_q(\mathsf{c}_t)
$$

which on the A-component is

$$\bigoplus_{F:\,F\geq A} \mathsf{c}[F] \to \bigoplus_{H:\,H\geq A} \mathsf{c}[H], \qquad x \mapsto \sum_{\substack{H,H'\geq F\\ \mathrm{s}(H)=\mathrm{s}(H')}} \zeta_q(F,H,H')\, \beta_{H,H'} \Delta_F^{H'}(x)$$

for $x \in \mathsf{c}[F]$. When c is cocommutative, this simplifies to

$$\bigoplus_{F:\,F\geq A} \mathsf{c}[F] \to \bigoplus_{H:\,H\geq A} \mathsf{c}[H], \qquad x \mapsto \sum_{H:\,H\geq F} \zeta_q(F,H)\, \Delta_F^{H}(x)$$

for $x \in \mathsf{c}[F]$. Similarly, define a natural map

$$\mathcal{T}_q(\mathsf{c}_t) \to \mathcal{T}_q(\mathsf{c}) \tag{7.23}$$

which on the A-component is

$$\bigoplus_{F:\,F\geq A} \mathsf{c}[F] \to \bigoplus_{H:\,H\geq A} \mathsf{c}[H], \qquad x \mapsto \sum_{\substack{H,H'\geq F\\ \mathrm{s}(H)=\mathrm{s}(H')}} \mu_q(F,H,H')\, \beta_{H,H'} \Delta_F^{H'}(x)$$

for $x \in \mathsf{c}[F]$. When c is cocommutative, this simplifies to

$$\bigoplus_{F:\,F\geq A} \mathsf{c}[F] \to \bigoplus_{H:\,H\geq A} \mathsf{c}[H], \qquad x \mapsto \sum_{H:\,H\geq F} \mu_q(F,H)\, \Delta_F^{H}(x)$$

for $x \in \mathsf{c}[F]$.

Theorem 7.17. *For q not a root of unity, for a Coxeter comonoid c, the maps (7.22) and (7.23) are inverse isomorphisms of Coxeter q-bimonoids.*

PROOF. This is a formal consequence of the corresponding result for species [**11**, Propositions 14.63, 14.65, 14.68, 14.70]. □

Dually: Let a be a Coxeter monoid, and a_t its underlying Coxeter species with trivial product. Recall the Coxeter q-bimonoid $\mathcal{T}_q^\vee(\mathsf{a})$ with q-quasishuffle product and deconcatenation coproduct, and $\mathcal{T}_q^\vee(\mathsf{a}_t)$ with q-shuffle product and deconcatenation coproduct. Define a natural map

$$\mathcal{T}_q^\vee(\mathsf{a}_t) \to \mathcal{T}_q^\vee(\mathsf{a}) \tag{7.24}$$

which on the A-component is

$$\bigoplus_{F:\,F\geq A} \mathsf{a}[F] \to \bigoplus_{G:\,G\geq A} \mathsf{a}[G], \qquad x \mapsto \sum_{\substack{G:\,A\leq G\leq F\\ F':\,F'\geq G,\,\mathrm{s}(F')=\mathrm{s}(F)}} \zeta_q(G,F',F)\, \mu_G^{F'} \beta_{F',F}(x)$$

for $x \in \mathsf{a}[F]$. When a is commutative, this simplifies to

$$\bigoplus_{F:\,F\geq A} \mathsf{a}[F] \to \bigoplus_{G:\,G\geq A} \mathsf{a}[G], \qquad x \mapsto \sum_{G:\,A\leq G\leq F} \zeta_q(G,F)\, \mu_G^{F}(x)$$

for $x \in \mathsf{a}[F]$. Similarly, define a natural map

$$\mathcal{T}_q^\vee(\mathsf{a}) \to \mathcal{T}_q^\vee(\mathsf{a}_t) \tag{7.25}$$

which on the A-component is

$$\bigoplus_{F:\,F\geq A} \mathsf{a}[F] \to \bigoplus_{G:\,G\geq A} \mathsf{a}[G], \qquad x \mapsto \sum_{\substack{G:\,A\leq G\leq F\\ F':\,F'\geq G,\,\mathrm{s}(F')=\mathrm{s}(F)}} \mu_q(G,F',F)\, \mu_G^{F'} \beta_{F',F}(x)$$

for $x \in \mathsf{a}[F]$. When a is commutative, this simplifies to

$$\bigoplus_{F\colon F \geq A} \mathsf{a}[F] \to \bigoplus_{G\colon G \geq A} \mathsf{a}[G], \qquad x \mapsto \sum_{G\colon A \leq G \leq F} \boldsymbol{\mu}_q(G,F)\, \mu_G^F(x)$$

for $x \in \mathsf{a}[F]$.

Theorem 7.18. *For q not a root of unity, for a Coxeter monoid a, the maps (7.24) and (7.25) are inverse isomorphisms of Coxeter q-bimonoids.*

Proof. This is a formal consequence of the corresponding result for species [**11**, Propositions 14.72, 14.73, 14.74, 14.75]. □

We collectively refer to (7.22), (7.23) and their duals (7.24), (7.25) as *HNR isomorphisms* for Coxeter q-bimonoids.

Theorem 7.19 (General q-norm map). *Suppose q is not a root of unity. For c a Coxeter comonoid, a a Coxeter monoid, let f and $\widehat{f}$ be as in (3.146). Then the composite*

$$\mathcal{T}_q(\mathsf{c}_t) \xrightarrow{\cong} \mathcal{T}_q(\mathsf{c}) \xrightarrow{\widehat{f}} \mathcal{T}_q^{\vee}(\mathsf{a}) \xrightarrow{\cong} \mathcal{T}_q^{\vee}(\mathsf{a}_t) \tag{7.26}$$

on the A-component, on the F-summand and into the F'-summand, is

$$\mathsf{c}[F] \to \mathsf{a}[F'], \qquad x \mapsto \begin{cases} (\beta_q)_{F',F}\, \mathbf{log}_q(f)_F(x) & \text{if } \mathrm{s}(F') = \mathrm{s}(F), \\ 0 & \text{otherwise.} \end{cases}$$

Thus, $\widehat{f}$ is an isomorphism of Coxeter q-bimonoids iff $\mathbf{log}_q(f)$ is an isomorphism of Coxeter species.

Proof. This result for species is not explicitly written in [**11**]; however, it follows immediately by applying [**11**, Exercise 9.93] to [**11**, Exercise 14.81]. It subsumes Theorem 3.79. □

7.6. Poincaré–Birkhoff–Witt

In this section, we assume $\mathrm{char}(\Bbbk) \nmid |W|$. The Poincaré–Birkhoff–Witt theorem says that for g a Coxeter Lie monoid, $\mathcal{U}(\mathsf{g})$ is isomorphic as a Coxeter comonoid to $\mathcal{S}(\mathsf{g})$. The isomorphism depends on the choice of an invariant noncommutative zeta function $\boldsymbol{\zeta}$. Equivalently, there exists an idempotent operator on $\mathcal{T}(\mathsf{g})$ of Coxeter comonoids with image $\mathcal{S}(\mathsf{g})$ and coimage $\mathcal{U}(\mathsf{g})$. This is the Solomon operator. Here $\mathcal{S}(\mathsf{g})$ is viewed inside $\mathcal{T}(\mathsf{g})$ via a Coxeter comonoid section to the abelianization map.

This result has dual and signed analogues.

7.6.1. PBW for Coxeter Lie monoids. For p a Coxeter species, the canonical surjection $\mathcal{T}(\mathsf{p}) \twoheadrightarrow \mathcal{S}(\mathsf{p})$ is an instance of the abelianization map, see (3.50). It is a Coxeter bimonoid morphism. We first construct Coxeter comonoid sections to this map.

Let $\mathrm{char}(\Bbbk) \nmid |W|$. Fix an invariant noncommutative zeta function $\boldsymbol{\zeta}$. For p a Coxeter species, define a natural map

$$\mathcal{S}(\mathsf{p}) \hookrightarrow \mathcal{T}(\mathsf{p}) \tag{7.27}$$

which on the A-component sends a representative $x \in \mathsf{p}[F]$ to

$$\sum_{\substack{F' : F' \geq A \\ \mathrm{s}(F') = \mathrm{s}(F)}} \zeta(A, F')\, \beta_{F',F}(x).$$

This map is an injective morphism of Coxeter comonoids which is a section to the abelianization map. The special case $\mathsf{p} := \mathsf{x}$ is given in (4.30).

For a Coxeter Lie monoid g, consider the diagram of Coxeter comonoids

$$\begin{array}{ccccc} & & \mathcal{T}(\mathsf{g}) & & \\ & \nearrow & & \searrow & \\ \mathcal{S}(\mathsf{g}) & & \dashrightarrow & & \mathcal{U}(\mathsf{g}). \end{array} \tag{7.28}$$

The inclusion is the map (7.27) for $\mathsf{p} := \mathsf{g}$. The surjective map is the canonical quotient. (It is in fact a Coxeter bimonoid morphism.)

Theorem 7.20. *Let* $\mathrm{char}(\mathbb{k}) \nmid |W|$. *For an invariant noncommutative zeta function* ζ *and Coxeter Lie monoid* g, *the dotted arrow in* (7.28) *is an isomorphism of Coxeter comonoids.*

This is the *Poincaré–Birkhoff–Witt theorem* for Coxeter Lie monoids, or PBW for short. The dotted arrow in (7.28) is a *Poincaré–Birkhoff–Witt isomorphism*, or PBW isomorphism for short.

PROOF. This can be deduced from the PBW theorem for Lie monoids in species [**11**, Theorem 17.9]. The key thing to note is that we need to start with an invariant noncommutative zeta function to ensure that the PBW isomorphism is a Coxeter species map (as opposed to just a species map). □

Corollary 7.21. *Let* $\mathrm{char}(\mathbb{k}) \nmid |W|$. *For a Coxeter Lie monoid* g, *the map* $i : \mathsf{g} \to \mathcal{U}(\mathsf{g})$ *is an injective morphism of Coxeter Lie monoids.*

For a Coxeter Lie monoid g, define the *Solomon operator*

$$\mathrm{Sol} : \mathcal{T}(\mathsf{g}) \to \mathcal{T}(\mathsf{g}) \tag{7.29}$$

to be the exponential (3.87a) of the composite map

$$\mathcal{T}(\mathsf{g}) \xrightarrow{\log(\mathrm{id})} \mathcal{PT}(\mathsf{g}) = \mathbf{Lie} \circ \mathsf{g} \xrightarrow{\gamma} \mathsf{g} \hookrightarrow \mathcal{T}(\mathsf{g}). \tag{7.30}$$

Theorem 7.22. *Let* $\mathrm{char}(\mathbb{k}) \nmid |W|$. *For an invariant noncommutative zeta function* ζ *and Coxeter Lie monoid* g, *the Solomon operator on* $\mathcal{T}(\mathsf{g})$ (7.29) *is an idempotent operator of Coxeter comonoids with image* $\mathcal{S}(\mathsf{g})$ *and coimage* $\mathcal{U}(\mathsf{g})$ *yielding the commutative diagram*

$$\begin{array}{ccc} \mathcal{T}(\mathsf{g}) & \xrightarrow{\mathrm{Sol}} & \mathcal{T}(\mathsf{g}) \\ \downarrow & & \uparrow \\ \mathcal{U}(\mathsf{g}) & \xrightarrow[\cong]{} & \mathcal{S}(\mathsf{g}). \end{array} \tag{7.31}$$

In particular, $\mathcal{S}(\mathsf{g})$ *and* $\mathcal{U}(\mathsf{g})$ *are isomorphic as Coxeter comonoids. Moreover, the inverse of* $\mathcal{U}(\mathsf{g}) \to \mathcal{S}(\mathsf{g})$ *is the PBW isomorphism in* (7.28).

Proof. This can be deduced from the corresponding theorem for Lie monoids in species [**11**, Theorem 17.32]. □

7.6.2. PBW for Coxeter Lie comonoids. We briefly discuss the dual situation. For a Coxeter species p, the canonical inclusion $\mathcal{S}^\vee(\mathsf{p}) \hookrightarrow \mathcal{T}^\vee(\mathsf{p})$ is an instance of the coabelianization map, see (3.50). It is a Coxeter bimonoid morphism. We now construct Coxeter monoid projections to this map. Fix an invariant noncommutative zeta function $\boldsymbol{\zeta}$, and define a natural map

$$\mathcal{T}^\vee(\mathsf{p}) \twoheadrightarrow \mathcal{S}^\vee(\mathsf{p}), \tag{7.32}$$

on the A-component, on the F-summand, by

$$\boldsymbol{\zeta}(A,F) \sum_{\substack{F' :\, F' \geq A \\ \mathrm{s}(F') = \mathrm{s}(F)}} \beta_{F',F}.$$

This map is a surjective morphism of Coxeter monoids which is a projection to the coabelianization map.

For a Coxeter Lie comonoid k, consider the diagram of Coxeter monoids

$$\begin{array}{ccc} & \mathcal{T}^\vee(\mathsf{k}) & \\ \nearrow & & \searrow \\ \mathcal{U}^\vee(\mathsf{k}) & \dashrightarrow & \mathcal{S}^\vee(\mathsf{k}). \end{array} \tag{7.33}$$

The surjective map is (7.32) for $\mathsf{p} := \mathsf{k}$. The injective map is the canonical inclusion. (It is in fact a Coxeter bimonoid morphism.)

Theorem 7.23. *Let* $\mathrm{char}(\mathbb{k}) \nmid |W|$. *For an invariant noncommutative zeta function* $\boldsymbol{\zeta}$ *and Coxeter Lie comonoid* k, *the dotted arrow in* (7.33) *is an isomorphism of Coxeter monoids.*

This is the *Poincaré–Birkhoff–Witt theorem* for Coxeter Lie comonoids. The dotted arrow in (7.33) is a *Poincaré–Birkhoff–Witt isomorphism*. The corresponding result for Lie comonoids in species is [**11**, Theorem 17.18].

Corollary 7.24. *Let* $\mathrm{char}(\mathbb{k}) \nmid |W|$. *For a Coxeter Lie comonoid* k, *the map* $p : \mathcal{U}^\vee(\mathsf{k}) \to \mathsf{k}$ *is a surjective morphism of Coxeter Lie comonoids.*

For a Coxeter Lie comonoid k, define the *Solomon operator*

$$\mathrm{Sol}^\vee : \mathcal{T}^\vee(\mathsf{k}) \to \mathcal{T}^\vee(\mathsf{k}) \tag{7.34}$$

to be the exponential of the composite map

$$\mathcal{T}^\vee(\mathsf{k}) \twoheadrightarrow \mathsf{k} \xrightarrow{\delta} \mathbf{Lie}^* \circ \mathsf{k} = \mathcal{Q}\mathcal{T}^\vee(\mathsf{k}) \xrightarrow{\mathbf{log}(\mathrm{id})} \mathcal{T}^\vee(\mathsf{k}).$$

Theorem 7.25. *Let* $\mathrm{char}(\mathbb{k}) \nmid |W|$. *For an invariant noncommutative zeta function* $\boldsymbol{\zeta}$ *and Coxeter Lie comonoid* k, *the Solomon operator on* $\mathcal{T}^\vee(\mathsf{k})$ (7.34) *is an idempotent operator of Coxeter monoids with image* $\mathcal{U}^\vee(\mathsf{k})$ *and coimage*

$\mathcal{S}^\vee(\mathsf{k})$ yielding the commutative diagram

$$\begin{array}{ccc} \mathcal{T}^\vee(\mathsf{k}) & \xrightarrow{\mathrm{Sol}^\vee} & \mathcal{T}^\vee(\mathsf{k}) \\ \big\downarrow & & \big\uparrow \\ \mathcal{S}^\vee(\mathsf{k}) & \xrightarrow[\cong]{} & \mathcal{U}^\vee(\mathsf{k}). \end{array} \tag{7.35}$$

In particular, $\mathcal{S}^\vee(\mathsf{k})$ and $\mathcal{U}^\vee(\mathsf{k})$ are isomorphic as Coxeter monoids. Moreover, the inverse of $\mathcal{S}^\vee(\mathsf{k}) \to \mathcal{U}^\vee(\mathsf{k})$ is the PBW isomorphism in (7.33).

7.6.3. Signed analogue. PBW Theorems 7.20 and 7.23 have signed analogues. Corresponding theorems for signed Lie (co)monoids in species are stated on [**11**, page 686].

7.7. Cartier–Milnor–Moore

The Cartier–Milnor–Moore theorem says that functors $\mathcal{U}$ and $\mathcal{P}$ determine an adjoint equivalence between categories of Coxeter Lie monoids and cocommutative Coxeter bimonoids. Dually, functors $\mathcal{U}^\vee$ and $\mathcal{Q}$ determine an adjoint equivalence between categories of Coxeter Lie comonoids and commutative Coxeter bimonoids. These results have signed analogues.

7.7.1. CMM for cocommutative Coxeter bimonoids. Recall the adjunction between $\mathcal{U}$ and $\mathcal{P}$ from Theorem 6.8.

Theorem 7.26. *The adjunction*

$$\mathsf{LieMon}(\mathsf{CSp}) \underset{\mathcal{P}}{\overset{\mathcal{U}}{\rightleftarrows}} {}^{\mathrm{co}}\mathsf{Bimon}(\mathsf{CSp})$$

is an adjoint equivalence of categories.

This is the *Cartier–Milnor–Moore theorem* for cocommutative Coxeter bimonoids, or CMM for short. It says that the counit and unit of the adjunction between $\mathcal{U}$ and $\mathcal{P}$ are natural isomorphisms:

$$\mathcal{U}(\mathcal{P}(\mathsf{h})) \xrightarrow{\cong} \mathsf{h} \quad \text{and} \quad \mathsf{g} \xrightarrow{\cong} \mathcal{P}(\mathcal{U}(\mathsf{g})).$$

PROOF. This can be deduced from the corresponding theorem for cocommutative bimonoids in species [**11**, Theorem 17.42]. That the unit and counit is an isomorphism can be deduced by forgetting the Coxeter structure, since none of the underlying objects change. Thus, no assumption on field characteristic is required. □

7.7.2. CMM for commutative Coxeter bimonoids. Recall the adjunction between $\mathcal{U}^\vee$ and $\mathcal{Q}$ from Theorem 6.11.

Theorem 7.27. *The adjunction*

$$\mathsf{Bimon}^{\mathrm{co}}(\mathsf{CSp}) \underset{\mathcal{U}^\vee}{\overset{\mathcal{Q}}{\rightleftarrows}} \mathsf{LieComon}(\mathsf{CSp})$$

is an adjoint equivalence of categories.

This is the *Cartier–Milnor–Moore theorem* for commutative Coxeter bimonoids. It says that the unit and counit of the adjunction between $\mathcal{U}^\vee$ and $\mathcal{Q}$ are natural isomorphisms:

$$\mathsf{h} \xrightarrow{\cong} \mathcal{U}^\vee(\mathcal{Q}(\mathsf{h})) \quad \text{and} \quad \mathcal{Q}(\mathcal{U}^\vee(\mathsf{k})) \xrightarrow{\cong} \mathsf{k}.$$

PROOF. The corresponding theorem for commutative bimonoids in species is [**11**, Theorem 17.53]. □

7.7.3. Signed analogue. CMM Theorems 7.26 and 7.27 have signed analogues as follows. Functors $\mathcal{U}_{-1}$ and $\mathcal{P}$ determine an adjoint equivalence between categories of signed Coxeter Lie monoids and signed cocommutative signed Coxeter bimonoids. Functors $\mathcal{U}^\vee_{-1}$ and $\mathcal{Q}$ determine an adjoint equivalence between categories of signed Coxeter Lie comonoids and signed commutative signed Coxeter bimonoids.

7.8. Coxeter bimonoids for a rank-one arrangement

Let $\mathcal{A}$ be a rank-one arrangement with chambers C and $\overline{C}$. Let w denote the nonidentity element of the Coxeter group. We now make explicit the notion of a Coxeter $\mathcal{A}$-bimonoid, and discuss its structure theory in a self-contained manner. We work with the formulation in terms of flats and lunes. So this discussion has a slightly different flavor from the general theory where we usually work with faces.

7.8.1. Coxeter bimonoids. We specialize objects in Sections 2.3, 2.4, 2.5, 2.6 to rank one.

Coxeter species. A Coxeter $\mathcal{A}$-species p consists of a pair of vector spaces denoted $\mathsf{p}[\bot]$ and $\mathsf{p}[\top]$, equipped with involutions

$$\tau : \mathsf{p}[\bot] \to \mathsf{p}[\bot] \quad \text{and} \quad \sigma : \mathsf{p}[\top] \to \mathsf{p}[\top].$$

We write this as $\mathsf{p} = (\mathsf{p}[\bot], \tau, \mathsf{p}[\top], \sigma)$, or simply as $\mathsf{p} = (\mathsf{p}[\bot], \mathsf{p}[\top])$ with σ and τ understood.

A map $f : \mathsf{p} \to \mathsf{q}$ of Coxeter $\mathcal{A}$-species consists of a pair of linear maps denoted $f_\bot : \mathsf{p}[\bot] \to \mathsf{q}[\bot]$ and $f_\top : \mathsf{p}[\top] \to \mathsf{q}[\top]$ such that $\tau f_\bot = f_\bot \tau$ and $\sigma f_\top = f_\top \sigma$. We write this as $f = (f_\bot, f_\top)$.

Coxeter monoids. A Coxeter $\mathcal{A}$-monoid is a Coxeter $\mathcal{A}$-species a equipped with two linear maps

$$\mu_C, \mu_{\overline{C}} : \mathsf{a}[\top] \to \mathsf{a}[\bot]$$

such that $\mu_{\overline{C}}\sigma = \tau\mu_C$, or equivalently, $\mu_C\sigma = \tau\mu_{\overline{C}}$. (The associativity axiom is vacuous in rank one.) A Coxeter $\mathcal{A}$-monoid is commutative if $\mu_C = \mu_{\overline{C}}$. It is signed commutative if $\mu_C = -\mu_{\overline{C}}$.

A morphism $f : \mathsf{a} \to \mathsf{b}$ of Coxeter $\mathcal{A}$-monoids is a map of Coxeter $\mathcal{A}$-species such that $\mu_C f_\top = f_\bot \mu_C$ and $\mu_{\overline{C}} f_\top = f_\bot \mu_{\overline{C}}$.

A Coxeter $\mathcal{A}$-q-bimonoid h consists of a pair of vector spaces $\mathsf{h}[\bot]$ and $\mathsf{h}[\top]$ equipped with linear maps

$$\mu_C, \mu_{\overline{C}} : \mathsf{h}[\top] \to \mathsf{h}[\bot], \ \sigma : \mathsf{h}[\top] \to \mathsf{h}[\top], \ \tau : \mathsf{h}[\bot] \to \mathsf{h}[\bot], \ \Delta_C, \Delta_{\overline{C}} : \mathsf{h}[\bot] \to \mathsf{h}[\top],$$

which satisfy $\sigma^2 = \tau^2 = \mathrm{id}$, $\mu_{\overline{C}}\sigma = \tau\mu_C$, $\Delta_{\overline{C}}\tau = \sigma\Delta_C$,
$\Delta_C\mu_C = \Delta_{\overline{C}}\mu_{\overline{C}} = \mathrm{id}$, $\Delta_{\overline{C}}\mu_C = \Delta_C\mu_{\overline{C}} = q\,\mathrm{id}$.

SUMMARY 7.1. Coxeter q-bimonoid in rank one.

A bicommutative Coxeter $\mathcal{A}$-bimonoid h consists of a pair of vector spaces $\mathsf{h}[\bot]$ and $\mathsf{h}[\top]$ equipped with linear maps

$$\mu : \mathsf{h}[\top] \to \mathsf{h}[\bot], \quad \sigma : \mathsf{h}[\top] \to \mathsf{h}[\top], \quad \tau : \mathsf{h}[\bot] \to \mathsf{h}[\bot], \quad \Delta : \mathsf{h}[\bot] \to \mathsf{h}[\top],$$

which satisfy $\sigma^2 = \tau^2 = \mathrm{id}$, $\mu\sigma = \tau\mu$, $\Delta\tau = \sigma\Delta$, $\Delta\mu = \mathrm{id}$.

SUMMARY 7.2. Bicommutative Coxeter bimonoid in rank one.

Coxeter comonoids. Dually, a Coxeter $\mathcal{A}$-comonoid is a Coxeter $\mathcal{A}$-species c equipped with two linear maps

$$\Delta_C, \Delta_{\overline{C}} : \mathsf{c}[\bot] \to \mathsf{c}[\top]$$

such that $\Delta_{\overline{C}}\tau = \sigma\Delta_C$, or equivalently, $\Delta_C\tau = \sigma\Delta_{\overline{C}}$. A Coxeter $\mathcal{A}$-comonoid is cocommutative if $\Delta_C = \Delta_{\overline{C}}$. It is signed cocommutative if $\Delta_C = -\Delta_{\overline{C}}$.

A morphism $f : \mathsf{c} \to \mathsf{d}$ of Coxeter $\mathcal{A}$-comonoids is a map of Coxeter $\mathcal{A}$-species such that $\Delta_C f_\bot = f_\top\Delta_C$ and $\Delta_{\overline{C}} f_\bot = f_\top\Delta_{\overline{C}}$.

Coxeter q-bimonoids. Combining the above notions: For a scalar q, a Coxeter $\mathcal{A}$-q-bimonoid is a tuple $(\mathsf{h}, \mu_C, \mu_{\overline{C}}, \Delta_C, \Delta_{\overline{C}})$, where h is a Coxeter $\mathcal{A}$-species, $(\mathsf{h}, \mu_C, \mu_{\overline{C}})$ is a Coxeter $\mathcal{A}$-monoid, $(\mathsf{h}, \Delta_C, \Delta_{\overline{C}})$ is a Coxeter $\mathcal{A}$-comonoid, and

$$\Delta_C\mu_C = \Delta_{\overline{C}}\mu_{\overline{C}} = \mathrm{id} \quad \text{and} \quad \Delta_{\overline{C}}\mu_C = \Delta_C\mu_{\overline{C}} = q\,\mathrm{id}\,.$$

These are different forms of the q-bimonoid axiom. The data present in a Coxeter $\mathcal{A}$-q-bimonoid is shown in Summary 7.1.

A morphism of Coxeter $\mathcal{A}$-q-bimonoids is a map of the underlying Coxeter $\mathcal{A}$-species which is a morphism of the underlying Coxeter $\mathcal{A}$-monoids and Coxeter $\mathcal{A}$-comonoids.

For $q = 1$, we use the term Coxeter $\mathcal{A}$-bimonoid, while for $q = -1$, we use the term signed Coxeter $\mathcal{A}$-bimonoid.

A Coxeter $\mathcal{A}$-bimonoid is bicommutative if it is both commutative and cocommutative. The data present in a bicommutative Coxeter $\mathcal{A}$-bimonoid is shown in Summary 7.2. We have written μ for $\mu_C = \mu_{\overline{C}}$ and Δ for $\Delta_C = \Delta_{\overline{C}}$.

7.8.2. Abelianization and coabelianization. We specialize the discussion in Section 3.4 to rank one. The abelianization of a Coxeter $\mathcal{A}$-monoid a, denoted a_{ab}, is

$$\mathsf{a}_{ab}[\top] = \mathsf{a}[\top] \quad \text{and} \quad \mathsf{a}_{ab}[\bot] = \operatorname{coker}(\mu_C - \mu_{\overline{C}})$$

with structure maps

$$\begin{array}{ccc} \mathsf{a}_{ab}[\top] & \overset{\sigma}{\dashrightarrow} & \mathsf{a}_{ab}[\top] \\ \| & & \| \\ \mathsf{a}[\top] & \xrightarrow[\sigma]{} & \mathsf{a}[\top] \end{array} \quad \begin{array}{ccc} \mathsf{a}_{ab}[\bot] & \overset{\tau}{\dashrightarrow} & \mathsf{a}_{ab}[\bot] \\ \twoheaduparrow & & \twoheaduparrow \\ \mathsf{a}[\bot] & \xrightarrow[\tau]{} & \mathsf{a}[\bot] \end{array} \quad \begin{array}{ccccc} \mathsf{a}_{ab}[\top] & & \overset{\mu_C = \mu_{\overline{C}}}{\dashrightarrow} & & \mathsf{a}_{ab}[\bot] \\ \| & & & & \| \\ \mathsf{a}[\top] & \xrightarrow[\mu_C]{} & \mathsf{a}[\bot] & \twoheadrightarrow & \operatorname{coker}(\mu_C - \mu_{\overline{C}}). \end{array}$$

Signed abelianization is defined similarly using $\operatorname{coker}(\mu_C + \mu_{\overline{C}})$.

The coabelianization of a Coxeter $\mathcal{A}$-comonoid c, denoted c^{coab}, is

$$\mathsf{c}^{coab}[\top] = \mathsf{c}[\top] \quad \text{and} \quad \mathsf{c}^{coab}[\bot] = \ker(\Delta_C - \Delta_{\overline{C}})$$

with structure maps

$$\begin{array}{ccc} \mathsf{c}^{coab}[\top] & \overset{\sigma}{\dashrightarrow} & \mathsf{c}^{coab}[\top] \\ \| & & \| \\ \mathsf{c}[\top] & \xrightarrow[\sigma]{} & \mathsf{c}[\top] \end{array} \quad \begin{array}{ccc} \mathsf{c}^{coab}[\bot] & \overset{\tau}{\dashrightarrow} & \mathsf{c}^{coab}[\bot] \\ \downarrow & & \downarrow \\ \mathsf{c}[\bot] & \xrightarrow[\tau]{} & \mathsf{c}[\bot] \end{array} \quad \begin{array}{ccccc} \mathsf{c}^{coab}[\bot] & & \overset{\Delta_C = \Delta_{\overline{C}}}{\dashrightarrow} & & \mathsf{c}^{coab}[\top] \\ \| & & & & \| \\ \ker(\Delta_C - \Delta_{\overline{C}}) & \hookrightarrow & \mathsf{c}[\bot] & \xrightarrow[\Delta_C]{} & \mathsf{c}[\top]. \end{array}$$

Signed coabelianization is defined similarly using $\ker(\Delta_C + \Delta_{\overline{C}})$.

Exercise 7.28. Check: The (co)abelianization of a Coxeter $\mathcal{A}$-bimonoid is again a Coxeter $\mathcal{A}$-bimonoid. (Use identities $\Delta_C(\mu_C - \mu_{\overline{C}}) = \Delta_{\overline{C}}(\mu_C - \mu_{\overline{C}}) = 0$ and $(\Delta_C - \Delta_{\overline{C}})\mu_C = (\Delta_C - \Delta_{\overline{C}})\mu_{\overline{C}} = 0$.)

7.8.3. Coxeter Lie monoids. We specialize objects in Section 6.1 to rank one.

Coxeter Lie monoids. A Coxeter $\mathcal{A}$-Lie monoid is a Coxeter $\mathcal{A}$-species g equipped with two linear maps

$$\nu_C, \nu_{\overline{C}} : \mathsf{g}[\top] \to \mathsf{g}[\bot]$$

such that $\nu_{\overline{C}}\sigma = \tau\nu_C$ and $\nu_C = -\nu_{\overline{C}}$. The latter axiom is antisymmetry. (Jacobi identity is vacuous in rank one.) We refer to both ν_C and $\nu_{\overline{C}}$ as the Lie bracket. A morphism $f : \mathsf{g} \to \mathsf{h}$ of Coxeter $\mathcal{A}$-Lie monoids is a map of Coxeter $\mathcal{A}$-species such that $\nu_C f_\top = f_\bot \, \nu_C$.

A signed Coxeter $\mathcal{A}$-Lie monoid is defined similarly using the signed antisymmetry axiom $\nu_C = \nu_{\overline{C}}$.

Coxeter Lie comonoids. Dually, a Coxeter $\mathcal{A}$-Lie comonoid is a Coxeter $\mathcal{A}$-species k equipped with two linear maps

$$\theta_C, \theta_{\overline{C}} : \mathsf{k}[\bot] \to \mathsf{k}[\top]$$

such that $\theta_{\overline{C}}\tau = \sigma\theta_C$ and $\theta_C = -\theta_{\overline{C}}$. The latter axiom is antisymmetry. We refer to both θ_C and $\theta_{\overline{C}}$ as the Lie cobracket. A morphism $f : \mathsf{k} \to \mathsf{h}$ of Coxeter $\mathcal{A}$-Lie comonoids is a map of Coxeter $\mathcal{A}$-species such that $f_\top \, \theta_C = \theta_C f_\bot$.

A signed Coxeter $\mathcal{A}$-Lie comonoid is defined similarly using the signed antisymmetry axiom $\theta_C = \theta_{\overline{C}}$.

7.8.4. Commutator bracket and cocommutator cobracket. Every Coxeter $\mathcal{A}$-monoid $(\mathsf{a}, \mu_C, \mu_{\overline{C}})$ carries a Coxeter $\mathcal{A}$-Lie monoid structure via the commutator bracket

$$\nu_C := \mu_C - \mu_{\overline{C}} \quad \text{and} \quad \nu_{\overline{C}} := \mu_{\overline{C}} - \mu_C.$$

Similarly, a is a signed Coxeter $\mathcal{A}$-Lie monoid via the signed commutator bracket $\nu_C = \nu_{\overline{C}} := \mu_C + \mu_{\overline{C}}$.

Dually, every Coxeter $\mathcal{A}$-comonoid $(\mathsf{c}, \Delta_C, \Delta_{\overline{C}})$ carries a Coxeter $\mathcal{A}$-Lie comonoid structure via the cocommutator cobracket

$$\theta_C := \Delta_C - \Delta_{\overline{C}} \quad \text{and} \quad \theta_{\overline{C}} := \Delta_{\overline{C}} - \Delta_C.$$

Similarly, c is a signed Coxeter $\mathcal{A}$-Lie comonoid via the signed cocommutator cobracket $\theta_C = \theta_{\overline{C}} := \Delta_C + \Delta_{\overline{C}}$.

7.8.5. Primitive and indecomposable parts. We specialize notions in Section 3.2 to rank one.

Primitive part. The primitive part $\mathcal{P}(\mathsf{c})$ of a Coxeter $\mathcal{A}$-comonoid c is

$$\mathcal{P}(\mathsf{c})[\top] = \mathsf{c}[\top] \quad \text{and} \quad \mathcal{P}(\mathsf{c})[\bot] = \ker \Delta_C \cap \ker \Delta_{\overline{C}}.$$

For h a Coxeter $\mathcal{A}$-bimonoid, $\mathcal{P}(\mathsf{h})$ is a Coxeter $\mathcal{A}$-Lie submonoid of h via the commutator bracket.

Indecomposable part. The indecomposable part $\mathcal{Q}(\mathsf{a})$ of a Coxeter $\mathcal{A}$-monoid a is

$$\mathcal{Q}(\mathsf{a})[\top] = \mathsf{a}[\top] \quad \text{and} \quad \mathcal{Q}(\mathsf{a})[\bot] = \mathsf{a}[\bot]/\operatorname{im} \mu_C + \operatorname{im} \mu_{\overline{C}}.$$

For h a Coxeter $\mathcal{A}$-bimonoid, $\mathcal{Q}(\mathsf{h})$ is a Coxeter $\mathcal{A}$-Lie quotient comonoid of h via the cocommutator cobracket.

Exercise 7.29. Verify Propositions 3.44 and 3.45 in rank one.

7.8.6. Universal constructions. We specialize the universal constructions in Section 3.3 to rank one, see Tables 7.2 and 7.3.

For a Coxeter $\mathcal{A}$-species p, the Coxeter $\mathcal{A}$-species $\mathcal{T}(\mathsf{p})$ is

$$\mathcal{T}(\mathsf{p})[\top] = \mathsf{p}[\top] \quad \text{and} \quad \mathcal{T}(\mathsf{p})[\bot] = \mathsf{p}[\bot] \oplus \mathsf{p}[\top] \oplus \mathsf{p}[\top],$$

with σ inherited from p, and τ given by $\tau(x, y, z) = (\tau(x), \sigma(z), \sigma(y))$. It is the free Coxeter $\mathcal{A}$-monoid on p with product $\mu_C(y) = (0, y, 0)$, $\mu_{\overline{C}}(z) = (0, 0, z)$. Moreover, it carries a Coxeter $\mathcal{A}$-q-bimonoid structure, see the first entry $\mathcal{T}_q(\mathsf{p})$ in Table 7.2. More generally, for a Coxeter $\mathcal{A}$-comonoid c, we have the Coxeter $\mathcal{A}$-q-bimonoid $\mathcal{T}_q(\mathsf{c})$.

Similarly, the Coxeter $\mathcal{A}$-species $\mathcal{S}(\mathsf{p})$ is

$$\mathcal{S}(\mathsf{p}) = (\mathsf{p}[\bot] \oplus \mathsf{p}[\top], (\tau, \sigma), \mathsf{p}[\top], \sigma).$$

This is the free commutative Coxeter $\mathcal{A}$-monoid on p with product $\mu_C(y) = \mu_{\overline{C}}(y) = (0, y)$, and so on.

Dually, the cofree Coxeter $\mathcal{A}$-comonoid on p is $\mathcal{T}^\vee(\mathsf{p})$ which as a Coxeter $\mathcal{A}$-species is the same as $\mathcal{T}(\mathsf{p})$ and its coproduct is $\Delta_C(x, y, z) = y$, $\Delta_{\overline{C}}(x, y, z) = z$. Moreover, it carries a Coxeter $\mathcal{A}$-q-bimonoid structure, see the first entry $\mathcal{T}_q^\vee(\mathsf{p})$ in Table 7.3. More generally, we have $\mathcal{T}_q^\vee(\mathsf{a})$, and so on.

TABLE 7.2. Free objects in rank one.

Coxeter q-bimonoid	Structure maps σ, τ, μ_C, $\mu_{\overline{C}}$, Δ_C, $\Delta_{\overline{C}}$
$\mathcal{T}_q(\mathsf{p}) = (\mathsf{p}[\bot] \oplus \mathsf{p}[\top] \oplus \mathsf{p}[\top], \mathsf{p}[\top])$	$\sigma(x) = \sigma(x)$, $\tau(x,y,z) = (\tau(x), \sigma(z), \sigma(y))$ $\mu_C(y) = (0,y,0)$, $\mu_{\overline{C}}(z) = (0,0,z)$ $\Delta_C(x,y,z) = y+qz$, $\Delta_{\overline{C}}(x,y,z) = qy+z$
$\mathcal{T}_q(\mathsf{c}) = (\mathsf{c}[\bot] \oplus \mathsf{c}[\top] \oplus \mathsf{c}[\top], \mathsf{c}[\top])$	σ, τ, μ_C, $\mu_{\overline{C}}$ as above $\Delta_C(x,y,z) = \Delta_C(x)+y+qz$ $\Delta_{\overline{C}}(x,y,z) = \Delta_{\overline{C}}(x)+qy+z$
$\mathcal{S}(\mathsf{p}) = (\mathsf{p}[\bot] \oplus \mathsf{p}[\top], \mathsf{p}[\top])$	$\sigma(x) = \sigma(x)$, $\tau(x,y) = (\tau(x), \sigma(y))$ $\mu_C(y) = \mu_{\overline{C}}(y) = (0,y)$ $\Delta_C(x,y) = \Delta_{\overline{C}}(x,y) = y$
$\mathcal{S}(\mathsf{c}) = (\mathsf{c}[\bot] \oplus \mathsf{c}[\top], \mathsf{c}[\top])$	σ, τ, μ_C, $\mu_{\overline{C}}$ as above $\Delta_C(x,y) = \Delta_C(x)+y$, $\Delta_{\overline{C}}(x,y) = \Delta_{\overline{C}}(x)+y$

TABLE 7.3. Cofree objects in rank one.

Coxeter q-bimonoid	Structure maps σ, τ, μ_C, $\mu_{\overline{C}}$, Δ_C, $\Delta_{\overline{C}}$
$\mathcal{T}_q^\vee(\mathsf{p}) = (\mathsf{p}[\bot] \oplus \mathsf{p}[\top] \oplus \mathsf{p}[\top], \mathsf{p}[\top])$	$\sigma(x) = \sigma(x)$, $\tau(x,y,z) = (\tau(x), \sigma(z), \sigma(y))$ $\Delta_C(x,y,z) = y$, $\Delta_{\overline{C}}(x,y,z) = z$ $\mu_C(y) = (0,y,qy)$, $\mu_{\overline{C}}(z) = (0,qz,z)$
$\mathcal{T}_q^\vee(\mathsf{a}) = (\mathsf{a}[\bot] \oplus \mathsf{a}[\top] \oplus \mathsf{a}[\top], \mathsf{a}[\top])$	σ, τ, Δ_C, $\Delta_{\overline{C}}$ as above $\mu_C(y) = (\mu_C(y), y, qy)$ $\mu_{\overline{C}}(z) = (\mu_{\overline{C}}(z), qz, z)$
$\mathcal{S}^\vee(\mathsf{p}) = (\mathsf{p}[\bot] \oplus \mathsf{p}[\top], \mathsf{p}[\top])$	$\sigma(x) = \sigma(x)$, $\tau(x,y) = (\tau(x), \sigma(y))$ $\Delta_C(x,y) = \Delta_{\overline{C}}(x,y) = y$ $\mu_C(y) = \mu_{\overline{C}}(y) = (0,y)$
$\mathcal{S}^\vee(\mathsf{a}) = (\mathsf{a}[\bot] \oplus \mathsf{a}[\top], \mathsf{a}[\top])$	σ, τ, Δ_C, $\Delta_{\overline{C}}$ as above $\mu_C(y) = (\mu_C(y), y)$, $\mu_{\overline{C}}(y) = (\mu_{\overline{C}}(y), y)$

We briefly mention the signed case. The Coxeter $\mathcal{A}$-species $\mathcal{E}(\mathsf{p})$ is

$$\mathcal{E}(\mathsf{p}) = (\mathsf{p}[\bot] \oplus \mathsf{p}[\top], (\tau, -\sigma), \mathsf{p}[\top], \sigma).$$

This is the free signed commutative Coxeter $\mathcal{A}$-monoid on p.

7.8.7. Signature functors. We specialize the u-signature functor and o-signature functor from Section 2.8 to rank one. For a Coxeter $\mathcal{A}$-species p, the Coxeter $\mathcal{A}$-species p_- and p^- are

$$\mathsf{p}_- = (\mathsf{p}[\bot], \tau, \mathsf{p}[\top], -\sigma) \quad \text{and} \quad \mathsf{p}^- = (\mathsf{p}[\bot], -\tau, \mathsf{p}[\top], \sigma).$$

Then,

$$\mathsf{p}_-^- := (\mathsf{p}_-)^- = (\mathsf{p}^-)_- = (\mathsf{p}[\bot], -\tau, \mathsf{p}[\top], -\sigma).$$

One may readily check that

$$\mathcal{T}(\mathsf{p}_-)_- \cong \mathcal{T}(\mathsf{p}) \cong \mathcal{T}(\mathsf{p}^-)^-, \quad \mathcal{S}(\mathsf{p}_-)_- = \mathcal{E}(\mathsf{p}) = \mathcal{S}(\mathsf{p}^-)^-.$$

This is a special case of Exercise 2.81.

7.8.8. Internal hom for Coxeter comonoids. We specialize the constructions from Section 3.6 to rank one. For Coxeter $\mathcal{A}$-comonoids c and d, the internal hom $\mathcal{C}(\mathsf{c},\mathsf{d})$ is as follows. An element of $\mathcal{C}(\mathsf{c},\mathsf{d})[\top]$ is a linear map $f_\top : \mathsf{c}[\top] \to \mathsf{d}[\top]$. An element of $\mathcal{C}(\mathsf{c},\mathsf{d})[\bot]$ consists of three linear maps $f_C, f_{\overline{C}} : \mathsf{c}[\top] \to \mathsf{d}[\top]$ and $f_\bot : \mathsf{c}[\bot] \to \mathsf{d}[\bot]$ satisfying $f_C \Delta_C = \Delta_C f_\bot$ and $f_{\overline{C}} \Delta_{\overline{C}} = \Delta_{\overline{C}} f_\bot$. The structure maps are

$$\sigma(f_\top) = \sigma f_\top \sigma, \quad \tau(f_\bot, f_C, f_{\overline{C}}) = (\tau f_\bot \tau, \sigma f_{\overline{C}} \sigma, \sigma f_C \sigma),$$
$$\Delta_C(f_\bot, f_C, f_{\overline{C}}) = f_C, \quad \Delta_{\overline{C}}(f_\bot, f_C, f_{\overline{C}}) = f_{\overline{C}}.$$

Now let c be a cocommutative Coxeter $\mathcal{A}$-comonoid and k a Coxeter $\mathcal{A}$-q-bimonoid. The Coxeter $\mathcal{A}$-q-bimonoid $\mathcal{C}(\mathsf{c},\mathsf{k})_q$ of Exercise 3.31 in rank one has product

$$\mu_C(f_\top) = (\mu_C f_\top \Delta_C, f_\top, q f_\top) \quad \text{and} \quad \mu_{\overline{C}}(f_\top) = (\mu_{\overline{C}} f_\top \Delta_{\overline{C}}, q f_\top, f_\top).$$

7.8.9. Universal measuring Coxeter comonoid. We specialize the constructions from Section 3.7 to rank one. For Coxeter $\mathcal{A}$-monoids a and b, the universal measuring Coxeter $\mathcal{A}$-comonoid $\overline{\mathcal{C}}(\mathsf{a},\mathsf{b})$ is as follows. An element of $\overline{\mathcal{C}}(\mathsf{a},\mathsf{b})[\top]$ is a linear map $f_\top : \mathsf{a}[\top] \to \mathsf{b}[\top]$. An element of $\overline{\mathcal{C}}(\mathsf{a},\mathsf{b})[\bot]$ consists of three linear maps $f_C, f_{\overline{C}} : \mathsf{a}[\top] \to \mathsf{b}[\top]$ and $f_\bot : \mathsf{a}[\bot] \to \mathsf{b}[\bot]$ satisfying $f_\bot \mu_C = \mu_C f_C$ and $f_\bot \mu_{\overline{C}} = \mu_{\overline{C}} f_{\overline{C}}$. The structure maps are defined similar to $\mathcal{C}(\mathsf{c},\mathsf{d})$ above.

Similarly, for h a Coxeter $\mathcal{A}$-q-bimonoid and a a commutative Coxeter $\mathcal{A}$-monoid, we have the Coxeter $\mathcal{A}$-q-bimonoid $\overline{\mathcal{C}}(\mathsf{h},\mathsf{a})_q$ of Exercise 3.34 in rank one with product defined similar to $\mathcal{C}(\mathsf{c},\mathsf{k})_q$ above.

7.8.10. Exp-log correspondence. We specialize the discussion in Sections 3.8 and 3.10 to rank one.

(Co)derivations and Coxeter (co)monoid morphisms. Let c be a Coxeter $\mathcal{A}$-comonoid and a a Coxeter $\mathcal{A}$-monoid. An element s of the invariant lune-incidence algebra acts on a map $f = (f_\bot, f_\top)$ from c to a by

$$s \circ f := (s(\bot) f_\bot + s(C)\, \mu_C f_\top \Delta_C + s(\overline{C})\, \mu_{\overline{C}} f_\top \Delta_{\overline{C}}, s(\top) f_\top).$$

This specializes (3.86). Let $\operatorname{char}(\Bbbk) \neq 2$. In rank one, there is a unique invariant $\boldsymbol{\zeta}$ and $\boldsymbol{\mu}$, namely,

$$\boldsymbol{\zeta}(\bot) = \boldsymbol{\zeta}(\top) = 1, \quad \boldsymbol{\zeta}(C) = \boldsymbol{\zeta}(\overline{C}) = 1/2,$$
$$\boldsymbol{\mu}(\bot) = \boldsymbol{\mu}(\top) = 1, \quad \boldsymbol{\mu}(C) = \boldsymbol{\mu}(\overline{C}) = -1/2.$$

This is equivalent to the rank-one data in Example 1.86. Hence,

$$\begin{aligned}\mathbf{exp}(f) &= (f_\perp + \tfrac{1}{2}\mu_C f_\top \Delta_C + \tfrac{1}{2}\mu_{\overline{C}} f_\top \Delta_{\overline{C}}, f_\top),\\ \mathbf{log}(f) &= (f_\perp - \tfrac{1}{2}\mu_C f_\top \Delta_C - \tfrac{1}{2}\mu_{\overline{C}} f_\top \Delta_{\overline{C}}, f_\top).\end{aligned}$$

Exercise 7.30. Verify Theorem 3.37 in rank one using the above formulas.

q-analogue. For c a Coxeter $\mathcal{A}$-comonoid and a a Coxeter $\mathcal{A}$-monoid, an element s of the invariant bilune-incidence algebra acts on a map $f = (f_\perp, f_\top)$ from c to a by

$$\begin{aligned}s \circ f := (s(\perp,\perp) f_\perp &+ s(C,C)\,\mu_C f_\top \Delta_C + s(\overline{C},\overline{C})\,\mu_{\overline{C}} f_\top \Delta_{\overline{C}}\\ &+ s(C,\overline{C})\,\mu_C f_\top \Delta_{\overline{C}} + s(\overline{C},C)\,\mu_{\overline{C}} f_\top \Delta_C, s(\top,\top) f_\top).\end{aligned}$$

This specializes (3.98). In particular,

$$\begin{aligned}\mathbf{exp}_q(f) &= (f_\perp + \tfrac{1}{1-q^2}(\mu_C f_\top \Delta_C + \mu_{\overline{C}} f_\top \Delta_{\overline{C}}) - \tfrac{q}{1-q^2}(\mu_C f_\top \Delta_{\overline{C}} + \mu_{\overline{C}} f_\top \Delta_C), f_\top),\\ \mathbf{log}_q(f) &= (f_\perp - \tfrac{1}{1-q^2}(\mu_C f_\top \Delta_C + \mu_{\overline{C}} f_\top \Delta_{\overline{C}}) + \tfrac{q}{1-q^2}(\mu_C f_\top \Delta_{\overline{C}} + \mu_{\overline{C}} f_\top \Delta_C), f_\top).\end{aligned}$$

Exercise 7.31. Verify Theorem 3.43 in rank one using the above formulas.

Primitive and group-like series. For p a Coxeter $\mathcal{A}$-species, a series v of p is a pair $(v_\perp, v_\top)$ with $v_\perp \in \mathsf{p}[\perp]$ and $v_\top \in \mathsf{p}[\top]$ such that $\tau(v_\perp) = v_\perp$ and $\sigma(v_\top) = v_\top$. For c a Coxeter $\mathcal{A}$-comonoid, we say a series v of c is primitive if $\Delta_C(v_\perp) = \Delta_{\overline{C}}(v_\perp) = 0$ and group-like if $\Delta_C(v_\perp) = \Delta_{\overline{C}}(v_\perp) = v_\top$. For a a Coxeter $\mathcal{A}$-monoid, an element s of the invariant lune-incidence algebra acts on a series $v = (v_\perp, v_\top)$ of a by

$$s \circ v := (s(\perp) v_\perp + s(C)\,\mu_C(v_\top) + s(\overline{C})\,\mu_{\overline{C}}(v_\top), s(\top) v_\top).$$

In particular, for $\mathrm{char}(\Bbbk) \neq 2$,

$$\begin{aligned}\mathbf{exp}(v) &= (v_\perp + \tfrac{1}{2}(\mu_C(v_\top) + \mu_{\overline{C}}(v_\top)), v_\top),\\ \mathbf{log}(v) &= (v_\perp - \tfrac{1}{2}(\mu_C(v_\top) + \mu_{\overline{C}}(v_\top)), v_\top).\end{aligned}$$

One may now verify Theorem 3.56 in rank one; it sets up a bijection between primitive and group-like series. Similarly, one may verify Theorem 3.59 in rank one by using

$$\begin{aligned}\mathbf{exp}_q(v) &= (v_\perp + \tfrac{1}{1+q}(\mu_C(v_\top) + \mu_{\overline{C}}(v_\top)), v_\top),\\ \mathbf{log}_q(v) &= (v_\perp - \tfrac{1}{1+q}(\mu_C(v_\top) + \mu_{\overline{C}}(v_\top)), v_\top).\end{aligned}$$

Linear and exponential series. For p a Coxeter $\mathcal{A}$-species, we say a series x of p is linear if $x_\perp = 0$. For a a Coxeter $\mathcal{A}$-monoid, we say a series e of a is exponential if $\mu_C(e_\top) + \mu_{\overline{C}}(e_\top) = 2\,e_\perp$, and equi-exponential if $\mu_C(e_\top) = \mu_{\overline{C}}(e_\top) = e_\perp$. Let $\mathrm{char}(\Bbbk) \neq 2$. It is clear that any equi-exponential series is exponential. The converse is true if a is commutative. Observe: A series e is exponential iff $\mathbf{log}(e)$ is linear. This verifies Lemma 3.66 in rank one. Similarly, observe: For a Coxeter $\mathcal{A}$-bimonoid h, any exponential series of h is group-like. This verifies Lemma 3.71 in rank one.

More generally, for a scalar q, we say a series e of a Coxeter $\mathcal{A}$-monoid a is q-exponential if $\mu_C(e_\top) + \mu_{\overline{C}}(e_\top) = (1+q)\,e_\perp$. For $q \neq \pm 1$, we have q-analogues of the preceding results.

7.8.11. q-norm map. We specialize the discussion in Section 3.11 to rank one. Let q be a scalar. For p a Coxeter $\mathcal{A}$-species, we have the q-norm map $(\kappa_q)_{\mathsf{p}} : \mathcal{T}_q(\mathsf{p}) \to \mathcal{T}_q^{\vee}(\mathsf{p})$ given by

$$((x, y, y'), z) \mapsto ((x, y + qy', qy + y'), z).$$

It is a morphism of Coxeter $\mathcal{A}$-q-bimonoids. This specializes (3.144). Observe: The above map is invertible for $q \neq \pm 1$. This verifies Lemma 3.78 in rank one.

More generally, for c a Coxeter $\mathcal{A}$-comonoid, a a Coxeter $\mathcal{A}$-monoid, and $f : \mathsf{c} \to \mathsf{a}$ a map of Coxeter $\mathcal{A}$-species, we have the morphism of Coxeter $\mathcal{A}$-q-bimonoids $\widehat{f} : \mathcal{T}_q(\mathsf{c}) \to \mathcal{T}_q^{\vee}(\mathsf{a})$ given by

$$\begin{aligned}((x, y, y'), z) \mapsto ((f_{\perp}(x) + \mu_C f_{\top}(y) + \mu_{\overline{C}} f_{\top}(y'), f_{\top}\Delta_C(x) + f_{\top}(y) + qf_{\top}(y'), \\ f_{\top}\Delta_{\overline{C}}(x) + qf_{\top}(y) + f_{\top}(y')), f_{\top}(z)).\end{aligned}$$

This specializes (3.146).

7.8.12. Universal enveloping Coxeter monoid. We specialize the construction in Section 6.5 to rank one. Let g be a Coxeter $\mathcal{A}$-Lie monoid. The universal enveloping Coxeter $\mathcal{A}$-monoid $\mathcal{U}(\mathsf{g})$ has components

$$\mathcal{U}(\mathsf{g})[\top] = \mathsf{g}[\top],$$
$$\mathcal{U}(\mathsf{g})[\perp] = \big(\mathsf{g}[\perp] \oplus \mathsf{g}[\top] \oplus \mathsf{g}[\top]\big)/(-\nu_C(y), y, -y) \text{ for } y \in \mathsf{g}[\top].$$

It is a cocommutative Coxeter $\mathcal{A}$-bimonoid and a quotient of $\mathcal{T}(\mathsf{g})$ (first entry in Table 7.2 for $q = 1$).

7.8.13. Structure theorems. We now discuss structure theorems for Coxeter bimonoids in rank one.

Borel–Hopf. Let $\operatorname{char}(\Bbbk) \neq 2$. Let h be a cocommutative Coxeter $\mathcal{A}$-bimonoid. Consider the maps (7.1) and (7.2). On the $\top$-component, they are both identity. On the $\perp$-component, they are

$$\mathcal{P}(\mathsf{h})[\perp] \oplus \mathsf{h}[\top] \to \mathsf{h}[\perp], \qquad (x, y) \mapsto x + \tfrac{1}{2}\mu_C(y) + \tfrac{1}{2}\mu_{\overline{C}}(y),$$
$$\mathsf{h}[\perp] \to \mathcal{P}(\mathsf{h})[\perp] \oplus \mathsf{h}[\top], \qquad x \mapsto (x - \tfrac{1}{2}\mu_C\Delta_C(x) - \tfrac{1}{2}\mu_{\overline{C}}\Delta_{\overline{C}}(x), \Delta_C(x)).$$

The same maps for $\mathcal{A}$-species are contained in [**11**, Example 13.39] as the special case $p = 1/2$.

Dually, let h be a commutative Coxeter $\mathcal{A}$-bimonoid. Consider the maps (7.3) and (7.4). On the $\top$-component, they are both identity. On the $\perp$-component, they are

$$\mathsf{h}[\perp] \to \mathcal{Q}(\mathsf{h})[\perp] \oplus \mathsf{h}[\top], \qquad x \mapsto (\bar{x}, \tfrac{1}{2}\Delta_C(x) + \tfrac{1}{2}\Delta_{\overline{C}}(x)),$$
$$\mathcal{Q}(\mathsf{h})[\perp] \oplus \mathsf{h}[\top] \to \mathsf{h}[\perp], \qquad (\bar{x}, y) \mapsto x - \tfrac{1}{2}\mu_C\Delta_C(x) - \tfrac{1}{2}\mu_{\overline{C}}\Delta_{\overline{C}}(x) + \mu_C(y).$$

One can directly check that each of the above pairs of maps consists of inverse isomorphisms. This verifies Theorems 7.1 and 7.2 in rank one.

Leray–Samelson. Let us now look at Theorems 7.4 and 7.5 in rank one. For p a Coxeter $\mathcal{A}$-species, $\mathcal{P}(\mathcal{S}(\mathsf{p})) = \mathcal{Q}(\mathcal{S}(\mathsf{p})) = \mathsf{p}$ using the third entry in Table 7.2. Now let h be a bicommutative Coxeter $\mathcal{A}$-bimonoid, see Summary 7.2. The involved inverse isomorphisms are

$$\mathcal{P}(\mathsf{h})[\bot] \oplus \mathsf{h}[\top] \to \mathsf{h}[\bot], \qquad (x, y) \mapsto x + \mu(y),$$
$$\mathsf{h}[\bot] \to \mathcal{P}(\mathsf{h})[\bot] \oplus \mathsf{h}[\top], \qquad x \mapsto (x - \mu\Delta(x), \Delta(x)),$$

and

$$\mathsf{h}[\bot] \to \mathcal{Q}(\mathsf{h})[\bot] \oplus \mathsf{h}[\top], \qquad x \mapsto (\bar{x}, \Delta(x)),$$
$$\mathcal{Q}(\mathsf{h})[\bot] \oplus \mathsf{h}[\top] \to \mathsf{h}[\bot], \qquad (\bar{x}, y) \mapsto x - \mu\Delta(x) + \mu(y).$$

The first pair of maps (involving the primitive part) for $\mathcal{A}$-species are given in [**11**, Example 13.17]. Note very carefully: In contrast to the Borel–Hopf isomorphisms, the above maps make sense in any field characteristic.

Loday–Ronco and its q-generalization. Let us now look at Theorem 7.8 in rank one. Let $q \neq \pm 1$. For p a Coxeter $\mathcal{A}$-species, $\mathcal{P}(\mathcal{T}_q(\mathsf{p})) = \mathsf{p}$ using the first entry in Table 7.2. Now let h be a Coxeter $\mathcal{A}$-q-bimonoid. The involved isomorphisms are

$$\mathcal{P}(\mathsf{h})[\bot] \oplus \mathsf{h}[\top] \oplus \mathsf{h}[\top] \to \mathsf{h}[\bot]$$
$$(x, y, z) \mapsto x + \tfrac{1}{1-q^2}(\mu_C(y) + \mu_{\overline{C}}(z)) - \tfrac{q}{1-q^2}(\mu_{\overline{C}}(y) + \mu_C(z)),$$

$$\mathsf{h}[\bot] \to \mathcal{P}(\mathsf{h})[\bot] \oplus \mathsf{h}[\top] \oplus \mathsf{h}[\top]$$
$$x \mapsto (x - \tfrac{1}{1-q^2}(\mu_C\Delta_C + \mu_{\overline{C}}\Delta_{\overline{C}})(x) + \tfrac{q}{1-q^2}(\mu_C\Delta_{\overline{C}} + \mu_{\overline{C}}\Delta_C)(x), \Delta_C(x), \Delta_{\overline{C}}(x)).$$

The same maps for $\mathcal{A}$-species are given in [**11**, Example 13.92]. Similarly, one can work out the dual Theorem 7.9 in rank one.

Setting $q = 0$, we recover Loday–Ronco Theorems 7.10 and 7.11 in rank one.

Hoffman–Newman–Radford. Let $\mathrm{char}(\Bbbk) \neq 2$. For c a cocommutative Coxeter $\mathcal{A}$-comonoid, let us specialize the HNR isomorphisms (7.16) and (7.17) to rank one. On the $\top$-component, the maps are identities, while on the $\bot$-component, they are

$$\mathcal{T}(\mathsf{c})[\bot] \to \mathcal{T}(\mathsf{c}_t)[\bot], \quad (x, y, z) \mapsto (x, \tfrac{1}{2}\Delta_C(x) + y, \tfrac{1}{2}\Delta_{\overline{C}}(x) + z)$$
$$\mathcal{T}(\mathsf{c}_t)[\bot] \to \mathcal{T}(\mathsf{c})[\bot], \quad (x, y, z) \mapsto (x, -\tfrac{1}{2}\Delta_C(x) + y, -\tfrac{1}{2}\Delta_{\overline{C}}(x) + z).$$

One may now directly verify Theorem 7.14 in rank one using formulas in Table 7.2.

Exercise 7.32. Let $q \neq \pm 1$. Make explicit the HNR isomorphisms (7.22) and (7.23) in rank one and verify Theorem 7.17 using formulas in Table 7.2.

Solomon operator and PBW. Let $\mathrm{char}(\Bbbk) \neq 2$. Let g be a Coxeter $\mathcal{A}$-Lie monoid. The Solomon operator $\mathrm{Sol} : \mathcal{T}(\mathsf{g}) \to \mathcal{T}(\mathsf{g})$ is as follows. On the $\top$-component, it is identity. On the $\bot$-component, it is

$$\mathrm{Sol}_\bot : \mathcal{T}(\mathsf{g})[\bot] \to \mathcal{T}(\mathsf{g})[\bot], \quad (x, y, z) \mapsto (x + \tfrac{1}{2}\nu_C(y - z), \tfrac{1}{2}(y + z), \tfrac{1}{2}(y + z)).$$

This specializes (7.29). See also [**11**, Exercise 17.63]. The map $\mathrm{Sol}_\perp$ has image $\mathcal{S}(\mathsf{g})[\perp]$ and coimage $\mathcal{U}(\mathsf{g})[\perp]$, with $\mathcal{S}(\mathsf{g})[\perp]$ viewed as a subspace of $\mathcal{T}(\mathsf{g})[\perp]$ via the section

$$(x, y) \mapsto (x, y/2, y/2).$$

This verifies Theorem 7.22, and hence also PBW Theorem 7.20 in rank one.

CMM. Let us look at Theorem 7.26 in rank one. The isomorphism between g and $\mathcal{PU}(\mathsf{g})$ says that the map

$$\mathsf{g}[\perp] \to \mathcal{U}(\mathsf{g})[\perp], \qquad x \mapsto \overline{(x, 0, 0)}$$

is injective, and its image is precisely $\mathcal{PU}(\mathsf{g})[\perp]$. Now let h be a cocommutative Coxeter $\mathcal{A}$-bimonoid. Using the commutator bracket on $\mathcal{P}(\mathsf{h})$,

$$\mathcal{UP}(\mathsf{h})[\perp] = \big(\mathcal{P}(\mathsf{h})[\perp] \oplus \mathsf{h}[\top] \oplus \mathsf{h}[\top]\big)/(-\mu_C(y) + \mu_{\overline{C}}(y), y, -y) \quad \text{for} \quad y \in \mathsf{h}[\top].$$

The isomorphism between $\mathcal{UP}(\mathsf{h})$ and h, on the $\perp$-component, is

$$\overline{(x, y, z)} \mapsto x + \mu_C(y) + \mu_{\overline{C}}(z) \quad \text{for} \quad x \in \mathcal{P}(\mathsf{h})[\perp] \text{ and } y, z \in \mathsf{h}[\top].$$

7.9. Notes

The results of this chapter generalize classical theorems for Joyal bimonoids. They can also be viewed as Coxeter analogues of corresponding results for bimonoids in species [**11**, Part III]. The context for the latter is hyperplane arrangements (as opposed to reflection arrangements).

7.9.1. Borel–Hopf. (Section 7.1.) Theorem 7.1 generalizes the classical result that in characteristic zero, any connected cocommutative Joyal bimonoid h is cofree cocommutative on $\mathcal{P}(\mathsf{h})$. The latter was proved in our monograph [**8**, Theorem 118] with the analogue of (7.1) defined using the uniform noncommutative zeta function. More generally, one may use any invariant noncommutative zeta function which arises from a zeta series (Section 1.16.5).

A result similar to [**8**, Theorem 118] was proved earlier by Stover [**341**, Proposition 12.2]. His exposition looks quite different since he worked with the formulation of Joyal species as a sequence of S_n-modules [**7**, page 237]. He considered the analogue of the map (7.1) (for any field characteristic) for a set-theoretic noncommutative zeta function and showed that it is an isomorphism of the underlying graded vector spaces. (He did not use the terminology of zeta functions which was introduced later in [**10**, Exercise 15.19].) On [**11**, page 715], we stated that Stover got an isomorphism of Joyal species but that was an inaccuracy.

Dually, Theorem 7.2 generalizes the result that in characteristic zero, any connected commutative Joyal bimonoid h is free commutative on $\mathcal{Q}(\mathsf{h})$.

7.9.2. Leray–Samelson. (Section 7.2.) Theorems 7.4 and 7.5 generalize the classical result that the category of connected bicommutative Joyal bimonoids is equivalent to the category of connected Joyal species.

7.9.3. Loday–Ronco. (Section 7.4.) Theorems 7.10 and 7.11 generalize the classical result that the category of connected Joyal 0-bimonoids is equivalent to the category of connected Joyal species. The latter was first proved in our monograph [**7**, Theorem 11.49], see also [**9**, Theorem 2.2]. In the involved isomorphisms, one uses the 0-zeta and 0-Möbius series.

7.9.4. Rigidity of Coxeter q-bimonoids for q not a root of unity. (Section 7.3.) The q-generalization of Loday–Ronco in Theorems 7.8 and 7.9 is new even for Joyal q-bimonoids. In the latter setting, in the involved isomorphisms, one uses the two-sided q-zeta and q-Möbius series.

7.9.5. Hoffman–Newman–Radford. (Section 7.5.) Theorem 7.14 generalizes the following classical result. In characteristic zero, for a positive cocommutative Joyal comonoid c, the Joyal bimonoid $\mathcal{T}(\mathsf{c})$ is isomorphic to $\mathcal{T}(\mathsf{c}_t)$, where c_t is the underlying Joyal species of c with trivial coproduct. In the analogues of (7.16) and (7.17), one may use any invariant noncommutative zeta and Möbius function which arises from a zeta and Möbius series.

Similar remarks apply to the other results in Section 7.5.

7.9.6. Poincaré–Birkhoff–Witt. (Section 7.6.) Theorem 7.20 generalizes the classical result that in characteristic zero, for g a positive Joyal Lie monoid, $\mathcal{U}(\mathsf{g})$ is isomorphic as a Joyal comonoid to $\mathcal{S}(\mathsf{g})$. The latter was proved in our monograph [**8**, Theorem 119] with the analogue of the PBW isomorphism in (7.28) defined using the uniform noncommutative zeta function. It may be replaced by any invariant noncommutative zeta function which arises from a zeta series. The Solomon operator for Joyal Lie monoids was defined in [**11**, Section 17.7.5].

PBW for Joyal Lie monoids (but without mention of Joyal comonoids) had been obtained earlier by Joyal [**203**, Section 4.2, Theorem 2]. Stover also proved a version of PBW in [**341**, Theorem 11.3]. The comments that we made above for his version of Borel–Hopf apply here as well.

Dually, Theorem 7.23 generalizes the result that in characteristic zero, for k a positive Joyal Lie comonoid, $\mathcal{U}^\vee(\mathsf{k})$ is isomorphic as a Joyal monoid to $\mathcal{S}^\vee(\mathsf{k})$.

7.9.7. Cartier–Milnor–Moore. (Section 7.7.) Theorem 7.26 generalizes the classical result that functors $\mathcal{U}$ and $\mathcal{P}$ determine an adjoint equivalence between categories of positive Joyal Lie monoids and connected cocommutative Joyal bimonoids. This result was first proved by Stover [**341**, Theorem 8.4]. It is also explained in our monograph [**8**, Theorem 120].

Dually, Theorem 7.27 generalizes the result that functors $\mathcal{U}^\vee$ and $\mathcal{Q}$ determine an adjoint equivalence between categories of positive Joyal Lie comonoids and connected commutative Joyal bimonoids.

Part II

Coxeter spaces

CHAPTER 8

Coxeter spaces and Coxeter bialgebras

We begin the study of Coxeter spaces. This notion can be seen in analogy with that of Coxeter species introduced in Chapter 2. Let $\mathcal{A}$ be a fixed reflection arrangement with Coxeter group W. We also fix a reference chamber C_r, so we have a pointed reflection arrangement $\boldsymbol{\alpha} = (\mathcal{A}, C_r)$. Top-nested faces, top-lunes, face-types of $\boldsymbol{\alpha}$ play a central role in our study. A useful summary is provided in Tables 8.1 and 8.2. Also, compare and contrast with Tables 2.1 and 2.2. Standard notations related to categories are given in the list of notations at the end of the book.

Coxeter spaces. We introduce the notion of a Coxeter $\mathcal{A}$-space. It can be formulated in terms of either top-nested faces or top-lunes of $\mathcal{A}$. In terms of top-nested faces: A Coxeter $\mathcal{A}$-space V attaches a vector space $\mathrm{V}[F, C]$ to every top-nested face (F, C) of $\mathcal{A}$ along with linear maps

$$\beta_{G,D,F,C} : \mathrm{V}[F, C] \to \mathrm{V}[G, D] \quad \text{and} \quad w : \mathrm{V}[F, C] \to \mathrm{V}[wF, wC],$$

the former whenever $\mathrm{s}(F, C) = \mathrm{s}(G, D)$, and the latter for each (F, C) and $w \in W$. These are support and type morphisms, respectively. They are subject to certain conditions, which we call support-support, support-type, type-type, support-unitality, type-unitality axioms.

In terms of top-lunes: A Coxeter $\mathcal{A}$-space V attaches a vector space $\mathrm{V}[\mathrm{L}]$ to every top-lune L of $\mathcal{A}$ along with linear maps

$$w : \mathrm{V}[\mathrm{L}] \to \mathrm{V}[w\mathrm{L}]$$

called type morphisms satisfying type-type and type-unitality axioms. The difference with the previous formulation is the absence of support morphisms and axioms involving them. This happens because top-nested faces are replaced by top-lunes. (Recall that the two are related by the support map.) A Coxeter $\mathcal{A}$-space can also be formulated using flats by fixing a chamber C and restricting to those top-lunes which contain C. Such top-lunes are in bijection with flats (via the base map).

We also introduce the equivalent notion of a Coxeter $\boldsymbol{\alpha}$-space in terms of face-types of $\boldsymbol{\alpha}$. A Coxeter $\boldsymbol{\alpha}$-space V attaches a vector space $\mathrm{V}[T]$ to every face-type T of $\boldsymbol{\alpha}$ along with an appropriate family of linear maps between $\mathrm{V}[T]$ and $\mathrm{V}[U]$ whenever T and U have the same support. Biface-types enter into this story.

TABLE 8.1. Axioms for Coxeter spaces and (co, bi)algebras.

Notion		Axioms
Coxeter space	top-nested faces	support-support, support-type, type-type, support-unitality, type-unitality axioms
	top-lunes	type-type, type-unitality axioms
Coxeter (co)algebra	top-nested faces	support-naturality, type-naturality, (co)associativity, (co)unitality axioms
	top-lunes	type-naturality, (co)associativity, (co)unitality axioms
Coxeter bialgebra	top-nested faces	bialgebra axiom
	top-lunes	bialgebra axiom
Coxeter q-bialgebra	top-nested faces	q-bialgebra axiom
	top-lunes	q-bialgebra axiom
(co)com. Coxeter (co)algebra	top-nested faces	(co)commutativity axiom
	top-lunes	(co)commutativity axiom

Coxeter algebras and coalgebras. Next, we introduce the notion of a Coxeter $\mathcal{A}$-algebra. This is a Coxeter $\mathcal{A}$-space A equipped with linear maps

$$\mu_{A,C}^{F,C} : \mathrm{A}[F, C] \to \mathrm{A}[A, C],$$

one for each $A \leq F \leq C$, subject to support-naturality, type-naturality, associativity, unitality axioms. Dually, a Coxeter $\mathcal{A}$-coalgebra is a Coxeter $\mathcal{A}$-space C equipped with linear maps

$$\Delta_{A,C}^{F,C} : \mathrm{C}[A, C] \to \mathrm{C}[F, C],$$

one for each $A \leq F \leq C$, subject to support-naturality, type-naturality, coassociativity, counitality axioms.

In terms of top-lunes: A Coxeter $\mathcal{A}$-algebra is a Coxeter $\mathcal{A}$-space A equipped with linear maps

$$\mu_{\mathrm{L}}^{\mathrm{M}} : \mathrm{A}[\mathrm{M}] \to \mathrm{A}[\mathrm{L}],$$

one for each $\mathrm{L} \leq' \mathrm{M}$, subject to type-naturality, associativity, unitality axioms. Note here the absence of the support-naturality axiom. Dually, a Coxeter $\mathcal{A}$-coalgebra C is defined using linear maps

$$\Delta_{\mathrm{L}}^{\mathrm{M}} : \mathrm{C}[\mathrm{L}] \to \mathrm{C}[\mathrm{M}]$$

for $\mathrm{L} \leq' \mathrm{M}$.

TABLE 8.2. Alternative axioms for Coxeter (co)algebras.

Notion	Alternative axioms
Coxeter (co)algebra	covering (co)associativity axiom
(co)commutative Coxeter (co)algebra	opposition (co)commutativity axiom
	covering (co)commutativity axiom

In terms of face-types: A Coxeter $\boldsymbol{\alpha}$-algebra is a Coxeter $\boldsymbol{\alpha}$-space A equipped with linear maps

$$\mu_Z^T : \mathrm{A}[T] \to \mathrm{A}[Z]$$

one for each $Z \le T$, subject to naturality, associativity, unitality axioms. Dually, a Coxeter $\boldsymbol{\alpha}$-coalgebra C is defined using linear maps

$$\Delta_Z^T : \mathrm{C}[Z] \to \mathrm{C}[T]$$

for $Z \le T$.

Coxeter bialgebras. By combining notions of Coxeter $\mathcal{A}$-algebra and Coxeter $\mathcal{A}$-coalgebra, we obtain the self-dual notion of a Coxeter $\mathcal{A}$-bialgebra. The product and coproduct satisfy a compatibility condition called the bialgebra axiom. We phrase it in terms of top-nested faces and top-lunes. By deforming this axiom by a scalar q, we obtain the more general notion of a Coxeter $\mathcal{A}$-q-bialgebra. For $q = -1$, we use the term signed Coxeter $\mathcal{A}$-bialgebra. We also pay attention to the case $q = 0$. There are equivalent notions with $\mathcal{A}$ replaced by $\boldsymbol{\alpha}$ which employ face-types.

Examples of Coxeter bialgebras are given in Chapter 10 and they could be read in parallel with the present chapter. The toy example of rank one is developed in Section 13.8.

(Co)commutative Coxeter (co)algebras. There is a notion of commutativity for a Coxeter $\mathcal{A}$-algebra, and cocommutativity for a Coxeter $\mathcal{A}$-coalgebra. There are equivalent notions with $\mathcal{A}$ replaced by $\boldsymbol{\alpha}$. A Coxeter $\mathcal{A}$- or $\boldsymbol{\alpha}$-bialgebra may be commutative, cocommutative, both or neither. Interestingly, face-type enumeration polynomials appear in identities involving products and coproducts when the Coxeter bialgebra is either commutative or cocommutative.

There is also a notion of signed commutativity for Coxeter $\mathcal{A}$-algebras and signed cocommutativity for Coxeter $\mathcal{A}$-coalgebras.

Signature functor. The categories of Coxeter bialgebras and signed Coxeter bialgebras are isomorphic. The same holds for the categories of commutative Coxeter algebras and signed commutative Coxeter algebras, and so on. We construct an explicit isomorphism called the u-signature functor by appropriately twisting support morphisms of the underlying Coxeter space.

Bimonads on Coxeter spaces. As for Coxeter species, the above notions are instances of general notions of category theory. We construct a bimonad on Coxeter $\mathcal{A}$-spaces which we denote $(\overline{\mathcal{T}}, \overline{\mathcal{T}}^{\vee}, \overline{\lambda})$. A $\overline{\mathcal{T}}$-algebra is the same as a Coxeter $\mathcal{A}$-algebra, a $\overline{\mathcal{T}}^{\vee}$-coalgebra is the same as a Coxeter $\mathcal{A}$-coalgebra, and a $(\overline{\mathcal{T}}, \overline{\mathcal{T}}^{\vee}, \overline{\lambda})$-bialgebra is the same as a Coxeter $\mathcal{A}$-bialgebra. Commutative aspects of the theory are handled in a similar manner.

Graded bialgebras. Connections to classical Hopf theory. The notion of a Coxeter $\boldsymbol{\alpha}$-bialgebra, when specialized to braid arrangements $\boldsymbol{\beta}^n$, relates to the classical notion of a graded bialgebra. More precisely, there is a functor from connected graded vector spaces to Coxeter $\boldsymbol{\beta}^n$-spaces which induces functors from connected graded (co, bi)algebras to Coxeter $\boldsymbol{\beta}^n$-(co, bi)algebras. Note: "connected graded" is equivalent to "positive-graded".

Convention 8.1. Whenever convenient, we drop $\mathcal{A}$ or $\boldsymbol{\alpha}$ from the notation, and write Coxeter space instead of Coxeter $\mathcal{A}$-space or Coxeter $\boldsymbol{\alpha}$-space, Coxeter algebra instead of Coxeter $\mathcal{A}$-algebra or Coxeter $\boldsymbol{\alpha}$-algebra, and so on. This is consistent with Convention 2.1 for Coxeter species.

8.1. Support and type morphisms

Starting with a reflection arrangement $\mathcal{A}$, we construct a finite category $\mathcal{A}\text{-}\overline{\mathtt{Cox}}$. It is a groupoid. Its objects are top-nested faces of $\mathcal{A}$. Its morphisms are generated by two kinds of morphisms called support and type morphisms. The former are between top-nested faces of the same support, the latter are between top-nested faces of the same type. They are subject to 5 relations called support-support, support-type, type-type, support-unitality, type-unitality relations. We obtain two normal forms for morphisms in $\mathcal{A}\text{-}\overline{\mathtt{Cox}}$. We also construct a category $\mathcal{A}\text{-}\overline{\mathtt{Cox}}'$ equivalent to $\mathcal{A}\text{-}\overline{\mathtt{Cox}}$ whose objects are top-lunes of $\mathcal{A}$ and whose morphisms are called type morphisms. These constructions are analogous to those of the categories $\mathcal{A}\text{-}\mathtt{Cox}$ and $\mathcal{A}\text{-}\mathtt{Cox}'$ from Section 2.1. We also briefly mention another equivalent category $\mathcal{A}\text{-}\overline{\mathtt{Cox}}''_C$ whose objects are flats. It is a subcategory of $\mathcal{A}\text{-}\mathtt{Cox}'$. We consider yet another category $\boldsymbol{\alpha}\text{-}\overline{\mathtt{Cox}}$ equivalent to $\mathcal{A}\text{-}\overline{\mathtt{Cox}}$ whose objects are face-types of the pointed reflection arrangement $\boldsymbol{\alpha}$. This category is isomorphic to the groupoid of biface-types introduced in Section 1.7.

The above categories are summarized in Table 8.3. Compare with Table 2.3.

TABLE 8.3. Base categories for Coxeter spaces.

Groupoid	Objects	Morphisms
$\mathcal{A}\text{-}\overline{\mathtt{Cox}}$	top-nested faces	support and type morphisms
$\mathcal{A}\text{-}\overline{\mathtt{Cox}}'$	top-lunes	type morphisms
$\mathcal{A}\text{-}\overline{\mathtt{Cox}}''_C$	flats	specific type morphisms
$\boldsymbol{\alpha}\text{-}\overline{\mathtt{Cox}}$	face-types	biface-types

8.1.1. Category with top-nested faces as objects. Fix a reflection arrangement $\mathcal{A}$ with Coxeter group W. We write e for the identity element of W.

Define the category $\mathcal{A}\text{-}\overline{\mathtt{Cox}}$ by a presentation. An object is a top-nested face (F, C) of $\mathcal{A}$. There are two different kinds of generating morphisms. There is a unique *support morphism*

$$\beta_{G,D,F,C} : (F, C) \to (G, D)$$

whenever $\mathrm{s}(F, C) = \mathrm{s}(G, D)$, and a unique *type morphism*

$$w : (F, C) \to (F', C')$$

whenever $\mathrm{t}(F) = \mathrm{t}(F')$, with $w \in W$ determined by the condition $C' = w(C)$ (which then implies $F' = w(F)$). These generating morphisms are subject to the 5 relations

$$\text{(8.1a)}\quad \begin{array}{ccc} (F,C) & \xrightarrow{\beta_{G,D,F,C}} & (G,D) \\ & {\scriptstyle \beta_{H,E,F,C}}\searrow \quad \swarrow{\scriptstyle \beta_{H,E,G,D}} & \\ & (H,E) & \end{array} \qquad \begin{array}{ccc} (F,C) & \xrightarrow{\beta_{G,D,F,C}} & (G,D) \\ {\scriptstyle w}\downarrow & & \downarrow{\scriptstyle w} \\ (F',C') & \xrightarrow[\beta_{G',D',F',C'}]{} & (G',D') \end{array} \qquad \begin{array}{ccc} (F,C) & \xrightarrow{u} & (G,D) \\ & {\scriptstyle vu}\searrow \quad \swarrow{\scriptstyle v} & \\ & (H,E) & \end{array}$$

$$\text{(8.1b)}\qquad \big((F,C) \xrightarrow{\beta_{F,C,F,C}} (F,C)\big) = \mathrm{id} \qquad\qquad \big((F,C) \xrightarrow{e} (F,C)\big) = \mathrm{id}\,.$$

We refer to diagrams (8.1a) as the *support-support relation*, *support-type relation*, *type-type relation*, respectively.

- In the support-support relation, $\mathrm{s}(F, C) = \mathrm{s}(G, D) = \mathrm{s}(H, E)$.
- In the support-type relation, $\mathrm{s}(F, C) = \mathrm{s}(G, D)$, and w takes F to F', C to C', G to G', D to D'.
- In the type-type relation, $\mathrm{t}(F) = \mathrm{t}(G) = \mathrm{t}(H)$, C is a chamber greater than F, and u takes C to D, v takes D to E, hence, vu takes C to E.

Similarly, we refer to conditions (8.1b) as the *support-unitality relation* and *type-unitality relation*, respectively. Here (F, C) is any top-nested face.

See the pictures in Figure 8.1. In relations involving type, vertices of the same type are shown with the same color.

Lemma 8.2. *The category $\mathcal{A}\text{-}\overline{\mathtt{Cox}}$ is a groupoid. The support morphisms*

$$\beta_{G,D,F,C} : (F, C) \to (G, D) \quad \textit{and} \quad \beta_{F,C,G,D} : (G, D) \to (F, C)$$

are inverses, while the type morphisms

$$w : (F, C) \to (F', C') \quad \textit{and} \quad w^{-1} : (F', C') \to (F, C)$$

are inverses.

PROOF. It suffices to establish the above claims about support and type morphisms. For the first, we use support-support and support-unitality relations, while for the second, we use type-type and type-unitality relations. $\square$

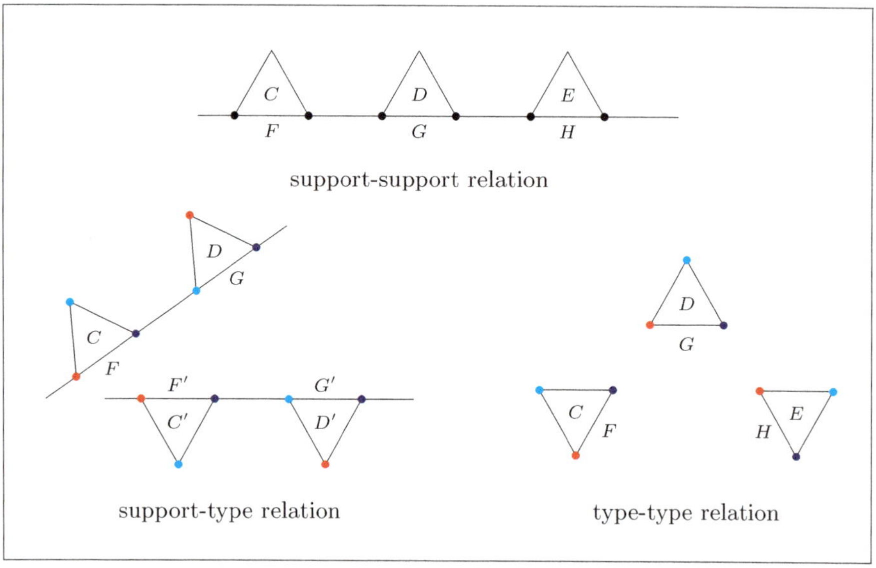

FIGURE 8.1. Support-support, support-type, type-type relations.

While writing support and type morphisms, we also employ the following alternative notations.

$$((F,C) \xrightarrow{\beta} (G,D)) = ((F,C) \xrightarrow{\beta_{G,D,F,C}} (G,D))$$

$$((F,C) \xrightarrow{w} (F',C')) = ((F,C) \xrightarrow{w(C)=C'} (F',C'))$$

The arrow label $w(C) = C'$ emphasizes that w is the (unique) Coxeter symmetry which sends C to C'.

Remark 8.3. Let us compare categories $\mathcal{A}\text{-}\overline{\mathtt{Cox}}$ and $\mathcal{A}\text{-}\mathtt{Cox}$. Both are generated by support and type morphisms. An important distinction is that type morphisms from a given object to another in $\mathcal{A}\text{-}\overline{\mathtt{Cox}}$ are unique (whenever they exist), while those in $\mathcal{A}\text{-}\mathtt{Cox}$ are not. The defining relations (which also bear the same names) can be compared using Figure 8.1 and Figure 2.1.

8.1.2. Normal forms for morphisms. We refer to an arbitrary morphism in $\mathcal{A}\text{-}\overline{\mathtt{Cox}}$ as a *support-type morphism*. It has two normal forms:

Lemma 8.4. *Every morphism in* $\mathcal{A}\text{-}\overline{\mathtt{Cox}}$ *can be uniquely expressed as*

- *a type morphism composed with a support morphism, and*
- *a support morphism composed with a type morphism.*

This is the analogue of Lemma 2.3. Since the proof is similar, we omit it. The two normal forms can be compared as follows. For a morphism σ in $\mathcal{A}\text{-}\overline{\mathtt{Cox}}$, in normal form,

$$\sigma = \beta w \iff \sigma = w\beta. \tag{8.2}$$

This follows from the support-type relation. Note very carefully: The same $w \in W$ occurs on both sides. Similarly, for composable morphisms σ and τ in $\mathcal{A}$-$\overline{\texttt{Cox}}$, in normal form,

$$\begin{aligned} \sigma = \beta u \text{ and } \tau = \beta v &\implies \tau\sigma = \beta v u, \\ \sigma = u\beta \text{ and } \tau = v\beta &\implies \tau\sigma = v u \beta. \end{aligned} \tag{8.3}$$

These normal forms yield two complementary descriptions of $\mathcal{A}$-$\overline{\texttt{Cox}}$. The analogue of Proposition 2.4 is as follows.

Proposition 8.5. *Objects of $\mathcal{A}$-$\overline{\texttt{Cox}}$ are top-nested faces of $\mathcal{A}$. A morphism $(F, C) \to (G, D)$ is an element $w \in W$ such that (wF, wC) and (G, D) have the same support. A morphism u composed with a morphism v is the morphism vu (product in the group W).*

A morphism can be written as

$$(F, C) \xrightarrow{\;w\;} (wF, wC) \xrightarrow{\;\beta\;} (G, D).$$

It may also be described as follows. Suppose H is a face of the same type as F and same support as G. Let w be the unique element of W which sends C to HD. Then w necessarily sends F to H since they have the same type. Thus,

$$\Big[(F, C) \xrightarrow{\;w\;} (wF, wC) \xrightarrow{\;\beta\;} (G, D)\Big] = \Big[(F, C) \xrightarrow{\;w\;} (H, HD) \xrightarrow{\;\beta\;} (G, D)\Big].$$

We refer to this as the morphism from (F, C) to (G, D) *cotransferred* by H. It may be visualized as follows.

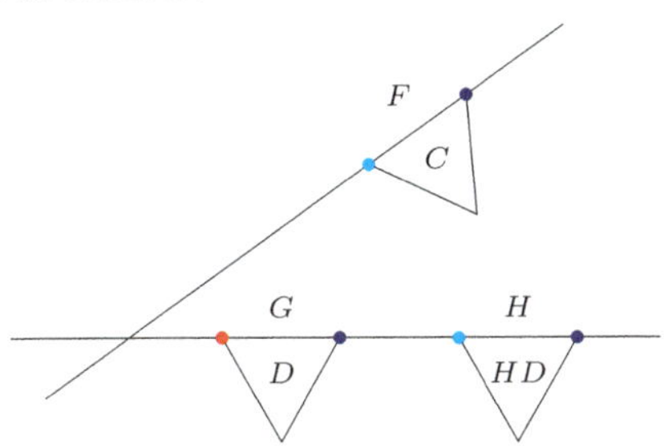

Thus, morphisms from (F, C) to (G, D) are indexed by faces H of the same type as F and same support as G.

The complementary description of $\mathcal{A}$-$\overline{\texttt{Cox}}$ is as follows. It is the analogue of Proposition 2.5.

Proposition 8.6. *Objects of $\mathcal{A}$-$\overline{\texttt{Cox}}$ are top-nested faces of $\mathcal{A}$. A morphism $(F, C) \to (G, D)$ is an element $w \in W$ such that (F, C) and $(w^{-1}G, w^{-1}D)$ have the same support. A morphism u composed with a morphism v is the morphism vu (product in the group W).*

A morphism can be written as

$$\begin{array}{ccc} & (w^{-1}G, w^{-1}D) & \\ {}^{\beta}\nearrow & & \searrow^{w} \\ (F,C) & & (G,D). \end{array}$$

Equivalently, let K be a face with the same support as F and the same type as G. Let $w \in W$ be such that $w(KC) = D$. Then

$$\begin{array}{ccc} & (w^{-1}G, w^{-1}D) & \\ {}^{\beta}\nearrow & & \searrow^{w} \\ (F,C) & & (G,D) \end{array} \quad = \quad \begin{array}{ccc} & (K, KC) & \\ {}^{\beta}\nearrow & & \searrow^{w} \\ (F,C) & & (G,D). \end{array}$$

We refer to this as the morphism from (F, C) to (G, D) *transferred* by K. Thus, morphisms from (F, C) to (G, D) are also indexed by faces K with the same support as F and same type as G.

Observation 8.7. A morphism $\sigma : (F, C) \to (G, D)$ is cotransferred by H and transferred by K iff $\sigma^{-1} : (G, D) \to (F, C)$ is cotransferred by K and transferred by H. This follows from Lemma 8.2 and the definition of (co)transfer.

In a similar manner:

Observation 8.8. For a morphism σ in $\mathcal{A}\text{-}\overline{\mathsf{Cox}}$, in normal form,

$$\sigma = \beta w = w\beta \iff \sigma^{-1} = \beta w^{-1} = w^{-1}\beta. \tag{8.4}$$

In long notation, for $\sigma : (F, C) \to (G, D)$, the normal forms are

$$\sigma = \beta_{G,D,wF,wC}\, w = w\, \beta_{w^{-1}G,w^{-1}D,F,C},$$
$$\sigma^{-1} = \beta_{F,C,w^{-1}G,w^{-1}D}\, w^{-1} = w^{-1}\, \beta_{wF,wC,G,D}.$$

Exercise 8.9. Show: There is a morphism from (F, C) to (G, D) in $\mathcal{A}\text{-}\overline{\mathsf{Cox}}$ iff F and G have the same support-type. In particular, existence of a morphism from (F, C) to (G, D) implies that F and G have the same rank.

Morphisms from (F, C) to itself in $\mathcal{A}\text{-}\overline{\mathsf{Cox}}$ are indexed by faces with the same support and same type as F. Deduce: The size of the automorphism group at (F, C) in $\mathcal{A}\text{-}\overline{\mathsf{Cox}}$ is $|\lambda|$, where λ is the support-type of F and $|\lambda|$ is as in Definition 1.8.

8.1.3. Category with top-lunes as objects.

Define the category $\mathcal{A}\text{-}\overline{\mathsf{Cox}}'$ as follows. Objects are top-lunes of $\mathcal{A}$. Morphisms are

$$w : \mathrm{L} \to \mathrm{L}'$$

with $w \in W$ such that $w(\mathrm{L}) = \mathrm{L}'$. We refer to these as *type morphisms*. A morphism u composed with a morphism v is the morphism vu (product in the group W). In other words,

$$\mathcal{A}\text{-}\overline{\mathsf{Cox}}' = \mathsf{el}_W(\widehat{\Lambda}[\mathcal{A}]), \tag{8.5}$$

the groupoid of elements for the action of W on the set of top-lunes $\widehat{\Lambda}[\mathcal{A}]$. For the general definition of the groupoid of elements, see Appendix A.2.3.

We now relate the category with top-nested faces as objects to the category with top-lunes as objects. Define a functor

$$\mathcal{A}\text{-}\overline{\mathtt{Cox}} \to \mathcal{A}\text{-}\overline{\mathtt{Cox}}' \tag{8.6}$$

which on objects sends (F, C) to its support $\mathrm{s}(F, C)$, and on morphisms sends $\beta : (F, C) \to (G, D)$ to $\mathrm{id} : \mathrm{s}(F, C) \to \mathrm{s}(G, D)$ and $w : (F, C) \to (F', C')$ to $w : \mathrm{s}(F, C) \to \mathrm{s}(F', C')$.

Proposition 8.10. *The functor* (8.6) *is an equivalence of categories.*

PROOF. We describe a functor in the other direction. For each top-lune L, pick a top-nested face sec(L) with support L. On objects, send L to sec(L), and on morphisms, send $w : \mathrm{L} \to \mathrm{L}'$ to $\beta w : \mathrm{sec(L)} \to \mathrm{sec(L')}$. That this assignment respects composition of morphisms follows from Proposition 8.5. It is straightforward to check that this functor yields an equivalence. □

The above result is the analogue of Proposition 2.8.

Exercise 8.11 (Skeletal category). Show: The category $\mathcal{A}\text{-}\overline{\mathtt{Cox}}$ is equivalent to the skeletal category obtained by taking disjoint union of the one-object categories associated to the subgroups W_ρ of W, one for each top-lune-type ρ. This is the analogue of Proposition 2.9.

8.1.4. Category with flats as objects.

For a chamber C, define the category $\mathcal{A}\text{-}\overline{\mathtt{Cox}}''_C$ as follows. Objects are flats of $\mathcal{A}$. Morphisms are

$$w : \mathrm{Y} \to \mathrm{Y}'$$

with $w \in W$ such that $w(\mathrm{Y}) = \mathrm{Y}'$, and w sends the top-lune containing C with base Y to the top-lune containing C with base Y′. A morphism u composed with a morphism v is the morphism vu (product in the group W). Observe: $\mathcal{A}\text{-}\overline{\mathtt{Cox}}''_C$ is a subcategory of the category $\mathcal{A}\text{-}\mathtt{Cox}'$ from (2.5).

Define $\mathcal{A}\text{-}\overline{\mathtt{Cox}}''$ to be the quotient category of $\mathcal{A}\text{-}\mathtt{Cox}'$ by imposing the following relation. For any $w \in W$ which fixes a flat Y pointwise, the morphism $w : \mathrm{Y} \to \mathrm{Y}$ equals the identity morphism.

Proposition 8.12. *For a chamber C, the composite functor*

$$\mathcal{A}\text{-}\overline{\mathtt{Cox}}''_C \hookrightarrow \mathcal{A}\text{-}\mathtt{Cox}' \twoheadrightarrow \mathcal{A}\text{-}\overline{\mathtt{Cox}}''$$

is an isomorphism of categories. (Both functors are identity on objects.)

For related ideas with nested flats instead of flats, see Proposition 11.1 and Exercise 11.2.

Exercise 8.13. Let us now relate the category with flats as objects to the category with top-lunes as objects. For a chamber C, define the functor

$$\mathcal{A}\text{-}\overline{\mathtt{Cox}}''_C \to \mathcal{A}\text{-}\overline{\mathtt{Cox}}' \tag{8.7}$$

which on objects sends Y to the unique top-lune containing C with base Y. Check: (8.7) is an equivalence of categories. For related ideas, see functor (11.2).

8.1.5. Category with face-types as objects. For a pointed reflection arrangement $\boldsymbol{\alpha} := (\mathcal{A}, C_r)$, define $\boldsymbol{\alpha}\text{-}\overline{\mathtt{Cox}}$ to be the full subcategory of $\mathcal{A}\text{-}\overline{\mathtt{Cox}}$ consisting of objects of the form (F, C_r). That is, we take top-nested faces with second coordinate C_r, and all morphisms between them. Such top-nested faces are in bijection with face-types. Thus, we may denote objects in $\boldsymbol{\alpha}\text{-}\overline{\mathtt{Cox}}$ by face-types. So, we write T instead of (F, C_r), with T being the type of F.

Lemma 8.14. *The inclusion functor*

$$\boldsymbol{\alpha}\text{-}\overline{\mathtt{Cox}} \to \mathcal{A}\text{-}\overline{\mathtt{Cox}}, \qquad T \mapsto (F, C_r), \tag{8.8}$$

where F is the face of C_r of type T, is an equivalence of categories.

PROOF. The functor is clearly full and faithful. It is also essentially surjective on objects: For any (G, D), there exists an isomorphism of the form $(G, D) \to (F, C_r)$. Take, for instance, F to be the face of C_r of the same type as G, and the Coxeter symmetry which sends D to C_r. Hence, the inclusion functor is an equivalence. □

Exercise 8.15. Check: An equivalence of categories in the direction opposite to (8.8) is given by the functor

$$\mathcal{A}\text{-}\overline{\mathtt{Cox}} \to \boldsymbol{\alpha}\text{-}\overline{\mathtt{Cox}}, \qquad (G, D) \mapsto \mathrm{t}(G). \tag{8.9}$$

On morphisms, $\sigma : (G, D) \to (H, E)$ maps to the composite

$$(\mathrm{t}(G), C_r) \xrightarrow{u} (G, D) \xrightarrow{\sigma} (H, E) \xrightarrow{v} (\mathrm{t}(H), C_r),$$

where u takes C_r to D, and v takes E to C_r.

Exercise 8.16. Check: $\boldsymbol{\alpha}\text{-}\overline{\mathtt{Cox}}$ is isomorphic to the full subcategory of $\mathcal{A}\text{-}\overline{\mathtt{Cox}}'$ consisting of objects of the form $\mathrm{s}(F, C_r)$. Deduce: For a face-type T, the group of morphisms from T to itself equals the stabilizer group W_{L} of some top-lune L.

To summarize, we have categorical equivalences

$$\begin{array}{ccccc} \mathcal{A}\text{-}\overline{\mathtt{Cox}} & \longrightarrow & \mathcal{A}\text{-}\overline{\mathtt{Cox}}' & \qquad & \text{top-nested faces} \xrightarrow{\ \mathrm{s}\ } \text{top-lunes} \\ & \nwarrow\!\!\searrow \quad \swarrow\!\!\nearrow & & & \nwarrow\!\!\searrow \quad \swarrow\!\!\nearrow \\ & \boldsymbol{\alpha}\text{-}\overline{\mathtt{Cox}} & & & \text{face-types.} \end{array} \tag{8.10}$$

The objects in the categories are shown on the right.

8.1.6. Connected components. We highlight two connected components of the category $\mathcal{A}\text{-}\overline{\mathtt{Cox}}$, and its equivalent categories $\mathcal{A}\text{-}\overline{\mathtt{Cox}}'$ and $\boldsymbol{\alpha}\text{-}\overline{\mathtt{Cox}}$.

Exercise 8.17 (Minimum, maximum top-lune-types). This is the analogue of Exercise 2.10. Check: The subcategories consisting of

- objects of the form (O, C) and morphisms between them in $\mathcal{A}\text{-}\overline{\mathtt{Cox}}$,
- top-lunes of the form $\mathrm{s}(O, C)$ and morphisms between them in $\mathcal{A}\text{-}\overline{\mathtt{Cox}}'$,
- the face-type $\emptyset$ and morphisms from $\emptyset$ to itself in $\boldsymbol{\alpha}\text{-}\overline{\mathtt{Cox}}$

are connected components which correspond to each other under equivalences (8.10). The last one is the one-arrow category. The first two are both isomorphic to $\mathsf{el}_W(\Gamma[\mathcal{A}])$, the groupoid of elements for the action of W on the set of chambers $\Gamma[\mathcal{A}]$. This is an indiscrete category, that is, there is a unique morphism from any C to any D. In particular, the only morphism from C to itself is the identity.

Similarly, the subcategories consisting of

- objects of the form (C, C) and morphisms between them in $\mathcal{A}\text{-}\overline{\mathtt{Cox}}$,
- the top-lune $\top$ and morphisms from $\top$ to itself in $\mathcal{A}\text{-}\overline{\mathtt{Cox}}'$,
- the face-type S and morphisms from S to itself in $\boldsymbol{\alpha}\text{-}\overline{\mathtt{Cox}}$

are connected components which correspond to each other under equivalences (8.10). The latter two are the same as C_W, the group W viewed as a one-object category.

In terms of Exercise 8.11, the above connected components correspond to top-lune-types $\mathrm{s}(\emptyset, S)$ and $\mathrm{s}(S, S)$, respectively.

8.1.7. Groupoid of biface-types. Recall the groupoid of biface-types from Section 1.7. It was also denoted $\boldsymbol{\alpha}\text{-}\overline{\mathtt{Cox}}$. The usage of the same notation is justified below.

Proposition 8.18. *The subcategory $\boldsymbol{\alpha}\text{-}\overline{\mathtt{Cox}}$ of $\mathcal{A}\text{-}\overline{\mathtt{Cox}}$ (as in this section) is isomorphic to the groupoid of biface-types $\boldsymbol{\alpha}\text{-}\overline{\mathtt{Cox}}$. Objects are face-types in both cases. Morphisms relate as follows.*

(1) *If $\mathrm{t}(F', F) = \sigma$ with $F \leq C_r$, then σ is transferred by F'.*
(2) *If $\mathrm{t}(G', G) = \sigma$ with $G' \leq C_r$, then σ is cotransferred by G.*

PROOF. The above sets up a bijection between morphisms in the two categories. This follows from Lemma 1.24. It remains to check that this bijection respects composition of morphisms which we leave as an exercise. □

Thus, there are two ways to think about $\boldsymbol{\alpha}\text{-}\overline{\mathtt{Cox}}$. Notions in one setting apply to the other using the above translation. For instance, biprojection by a biface-type in Section 1.7.4 applies to morphisms $(F, C_r) \to (G, C_r)$. Exercise 1.26 on existence of morphisms relates to Exercise 8.9. Exercise 1.29 on connected components relates to Exercise 8.17. Lemma 1.30 connects to normal forms, and Exercise 1.31 to (8.3) and (8.4).

We have introduced induced, local, opposite morphisms in the groupoid of biface-types. In Sections 8.2 and 8.3, we will develop these notions in detail in the category $\boldsymbol{\alpha}\text{-}\overline{\mathtt{Cox}}$ viewed as a subcategory of $\mathcal{A}\text{-}\overline{\mathtt{Cox}}$. In subsequent sections, we will treat the two versions of $\boldsymbol{\alpha}\text{-}\overline{\mathtt{Cox}}$ as virtually identical.

8.2. Induced and local morphisms

We define the notion of induced morphisms in the categories $\mathcal{A}\text{-}\overline{\mathtt{Cox}}$ and $\boldsymbol{\alpha}\text{-}\overline{\mathtt{Cox}}$. We also define classes of morphisms in $\mathcal{A}\text{-}\overline{\mathtt{Cox}}$ and $\boldsymbol{\alpha}\text{-}\overline{\mathtt{Cox}}$ local to a face and face-type, respectively. We introduce the notion of length of a morphism by using the gallery distance between chambers.

8.2.1. Induced morphisms. Let $\sigma : (A, C) \to (B, D)$ be a morphism in $\mathcal{A}\text{-}\overline{\mathtt{Cox}}$. For a face G between B and D, define $\sigma^*(G)$ to be the face between A and C inductively as follows.

- For $\sigma = \mathrm{id}$, put $\sigma^*(G) = G$.
- For $\sigma = \beta$, put $\sigma^*(G) = AG$, and for $\sigma = w$, put $\sigma^*(G) = w^{-1}(G)$.
- For composable morphisms σ and τ, put $(\tau\sigma)^*(G) = \sigma^*(\tau^*(G))$.

Further, there is an induced morphism $(\sigma^*(G), C) \to (G, D)$ which we continue to denote σ. It is id for id, β for β, w for w, and respects compositions.

Similarly, given $\sigma : (A, C) \to (B, D)$, for a face F between A and C, one defines the face $\sigma_*(F)$ between B and D and an induced morphism $\sigma : (F, C) \to (\sigma_*(F), D)$.

Induced morphisms can be visualized as follows.

$$
\begin{array}{ccc} (F,C) & \xrightarrow{\sigma} & (\sigma_*(F),D) \\ | & & | \\ (A,C) & \xrightarrow[\sigma]{} & (B,D) \end{array}
\qquad\qquad
\begin{array}{ccc} (\sigma^*(G),C) & \xrightarrow{\sigma} & (G,D) \\ | & & | \\ (A,C) & \xrightarrow[\sigma]{} & (B,D) \end{array}
\tag{8.11}
$$

The above discussion also applies to the subcategory $\boldsymbol{\alpha}$-$\overline{\mathtt{Cox}}$ as follows. Let $\sigma : Z \to Y$ be a morphism in $\boldsymbol{\alpha}$-$\overline{\mathtt{Cox}}$. For $Y \leq U$, there is a face-type $\sigma^*(U)$ greater than Z and an induced morphism $\sigma : \sigma^*(U) \to U$. Explicitly, $\sigma^*(U)$ is the type of $\sigma^*(G)$, where in the top-nested face notation $\sigma : (A, C_r) \to (B, C_r)$ and G is the face of C_r of type U. Similarly, for $Z \leq T$, there is a face-type $\sigma_*(T)$ greater than Y and an induced morphism $\sigma : T \to \sigma_*(T)$. Induced morphisms can now be visualized as follows.

$$
\begin{array}{ccc} T & \xrightarrow{\sigma} & \sigma_*(T) \\ | & & | \\ Z & \xrightarrow[\sigma]{} & Y \end{array}
\qquad\qquad
\begin{array}{ccc} \sigma^*(U) & \xrightarrow{\sigma} & U \\ | & & | \\ Z & \xrightarrow[\sigma]{} & Y \end{array}
\tag{8.12}
$$

This can be understood in the groupoid of biface-types $\boldsymbol{\alpha}$-$\overline{\mathtt{Cox}}$ as follows. Let (H', H) be a biface which represents $\sigma : Z \to Y$. For $Y \leq U$, let F' be a face greater than H' of type U. Then the type of HF' is $\sigma^*(U)$ and the biface (F', HF') represents $\sigma : \sigma^*(U) \to U$. (That σ does not depend on the specific choices made follows from Fact 1.4.) Similarly, for $Z \leq T$, let F be a face greater than H of type T. Then the type of $H'F$ is $\sigma_*(T)$ and the biface $(H'F, F)$ represents $\sigma : T \to \sigma_*(T)$. See also the explanation in Section 1.7.5.

Exercise 8.19. Check: For $\sigma : Z \to Y$ and $Z \leq T$, $Y \leq U$,

$$U = \sigma_*(T) \iff T = \sigma^*(U).$$

Moreover, in this situation, $(T \xrightarrow{\sigma} \sigma_*(T)) = (\sigma^*(U) \xrightarrow{\sigma} U)$. Compare with Exercise 2.11.

Exercise 8.20. Check: If $\sigma : Z \to Y$ in normal form is $\sigma = \beta w = w\beta$, then any induced $\sigma : T \to U$ (with $T = \sigma^*(U)$ and $U = \sigma_*(T)$) also has normal form $\sigma = \beta w = w\beta$ (with the same w). Compare with Exercise 2.12.

Exercise 8.21. Check: The following compatibility holds.

$$
\begin{array}{ccc} U & \xrightarrow{\sigma} & \sigma_*(U) \\ | & & | \\ T & \xrightarrow{\sigma} & \sigma_*(T) \\ | & & | \\ Z & \xrightarrow[\sigma]{} & Y \end{array}
\qquad\qquad
\begin{array}{ccc} \sigma^*(V) & \xrightarrow{\sigma} & V \\ | & & | \\ \sigma^*(U) & \xrightarrow{\sigma} & U \\ | & & | \\ Z & \xrightarrow[\sigma]{} & Y \end{array}
\tag{8.13}
$$

In the left diagram, we start with $\sigma : Z \to Y$ and $Z \leq T \leq U$, while in the right diagram, we start with $\sigma : Z \to Y$ and $Y \leq U \leq V$.

8.2.2. Morphisms local to a face, flat, face-type. A morphism $\sigma : (F,C) \to (G,D)$ in $\mathcal{A}\text{-}\overline{\texttt{Cox}}$ is *local* to a face A if $A \le F, G$ and σ is obtained by composing support and type morphisms involving only top-nested faces with first coordinate greater than A. Equivalently, $\sigma : (F,C) \to (G,D)$ is local to A if $A \le F, G$ and in normal form $\sigma = \beta w = w\beta$ with $w(A) = A$.

For each face A, we have the subcategory $\mathcal{A}_A\text{-}\overline{\texttt{Cox}}$ of $\mathcal{A}\text{-}\overline{\texttt{Cox}}$ whose objects are top-nested faces with first coordinate greater than A and morphisms are those local to A. As implied by the notation, this subcategory is an instance of $\mathcal{A}\text{-}\overline{\texttt{Cox}}$ with $\mathcal{A}$ replaced by $\mathcal{A}_A$.

Similarly, for a flat X, we have the subcategory $\mathcal{A}_{\mathrm{X}}\text{-}\overline{\texttt{Cox}}'$ of $\mathcal{A}\text{-}\overline{\texttt{Cox}}'$ whose objects are top-lunes with base greater than X, and morphisms are those which fix X pointwise. This subcategory is in fact an instance of $\mathcal{A}\text{-}\overline{\texttt{Cox}}'$ with $\mathcal{A}$ replaced by $\mathcal{A}_{\mathrm{X}}$.

Similarly, for a face-type Z, we have the category $\boldsymbol{\alpha}_Z\text{-}\overline{\texttt{Cox}}$. It is the full subcategory of $\mathcal{A}_A\text{-}\overline{\texttt{Cox}}$ consisting of top-nested faces with second coordinate C_r, and A is the face of C_r of type Z. In other words, objects are indexed by face-types greater than Z. We refer to a morphism $\sigma : T \to T'$ in $\boldsymbol{\alpha}_Z\text{-}\overline{\texttt{Cox}}$ as a morphism *local* to Z. Observe: $\boldsymbol{\alpha}_Z\text{-}\overline{\texttt{Cox}}$ is also a subcategory of $\boldsymbol{\alpha}\text{-}\overline{\texttt{Cox}}$. In fact, it is an instance of the latter with $\boldsymbol{\alpha}$ replaced by $\boldsymbol{\alpha}_Z$. See also Section 1.7.6.

To summarize, we have the commutative diagram

$$\begin{array}{ccccc}
\mathcal{A}_A\text{-}\overline{\texttt{Cox}} & & \hookrightarrow & & \mathcal{A}\text{-}\overline{\texttt{Cox}} & \\
\uparrow & \searrow & & & \uparrow & \searrow \\
 & & \mathcal{A}_{\mathrm{X}}\text{-}\overline{\texttt{Cox}}' & \hookrightarrow & & & \mathcal{A}\text{-}\overline{\texttt{Cox}}', \\
\cup & \nearrow & & & \cup & \nearrow \\
\boldsymbol{\alpha}_Z\text{-}\overline{\texttt{Cox}} & & \hookrightarrow & & \boldsymbol{\alpha}\text{-}\overline{\texttt{Cox}} &
\end{array}$$

where A is the face of C_r of type Z, and X is the support of A. The vertical and oblique inclusion functors are full. The left triangle is an instance of the right triangle.

Lemma 8.22. *A morphism $\sigma : Z \to Z'$ in $\boldsymbol{\alpha}\text{-}\overline{\texttt{Cox}}$ induces an isomorphism*

$$\boldsymbol{\alpha}_Z\text{-}\overline{\texttt{Cox}} \to \boldsymbol{\alpha}_{Z'}\text{-}\overline{\texttt{Cox}}, \qquad T \mapsto \sigma_*(T), \quad \tau \mapsto \sigma\tau\sigma^{-1} \tag{8.14}$$

of categories.

PROOF. This is the analogue of Lemma 2.14. We can apply a similar argument, and start by showing that $\mathcal{A}_A\text{-}\overline{\texttt{Cox}}$ and $\mathcal{A}_B\text{-}\overline{\texttt{Cox}}$ are isomorphic whenever A and B have either the same support or the same type. □

See also Exercise 1.21 in this regard.

8.2.3. Length of a morphism. Let $\sigma : (F,C) \to (G,D)$ be a morphism in the category $\mathcal{A}\text{-}\overline{\texttt{Cox}}$. Define the length of σ, denoted $l(\sigma)$, by

$$l(\sigma) = \operatorname{dist}(H,G) = \operatorname{dist}(HD,D), \tag{8.15a}$$

where H is the face with the same support as G which cotransfers σ, or equivalently, by

$$l(\sigma) = \operatorname{dist}(F,K) = \operatorname{dist}(C,KC), \tag{8.15b}$$

where K is the face with the same support as F which transfers σ.

This also defines length of a morphism $\sigma : T \to U$ in $\boldsymbol{\alpha}$-$\overline{\mathtt{Cox}}$ (since the latter is a subcategory of $\mathcal{A}$-$\overline{\mathtt{Cox}}$). Writing $\sigma = w\beta = \beta w$ in normal form, $l(\sigma) = l(w)$, the latter being the length of the Coxeter symmetry w wrt the reference chamber C_r, see (1.11). Alternatively, in the groupoid of biface-types $\boldsymbol{\alpha}$-$\overline{\mathtt{Cox}}$, we have $l(\sigma) = \mathrm{dist}(F, F')$, where (F', F) is any biface which represents σ, see (1.29).

Exercise 8.23. Show: In $\mathcal{A}$-$\overline{\mathtt{Cox}}$, and hence also in $\boldsymbol{\alpha}$-$\overline{\mathtt{Cox}}$:

(1) For σ and τ composable morphisms, $(-1)^{l(\sigma)}(-1)^{l(\tau)} = (-1)^{l(\tau\sigma)}$. Also, $l(\mathrm{id}) = 0$.
(2) The length of an induced morphism is the same as the length of the starting morphism, that is,

$$l((A, C) \xrightarrow{\sigma} (B, D)) = l((\sigma^*(G), C) \xrightarrow{\sigma} (G, D)) = l((F, C) \xrightarrow{\sigma} (\sigma_*(F), D)).$$

8.3. Opposite and adjacent support-type morphisms

We define two classes of morphisms in the category $\boldsymbol{\alpha}$-$\overline{\mathtt{Cox}}$, namely, opposite and adjacent support-type morphisms. The opposite support-type morphism starting from (F, C_r) is the morphism transferred by $\overline{F}$, the face opposite to F. Similarly, an adjacent support-type morphism starting from (F, C_r) is any morphism transferred by a face K adjacent to F. One can equivalently formulate these notions as morphisms going into (G, C_r) which are cotransferred by opposite or adjacent faces of G.

We also define the more general class of locally opposite support-type morphisms which includes both opposite and adjacent support-type morphisms.

8.3.1. Opposite support-type relation.

An important instance of the support-type relation in the category $\mathcal{A}$-$\overline{\mathtt{Cox}}$ is given below.

Lemma 8.24. *For F a face of a chamber C, and G the face of C of the same type as $\overline{F}$, the diagram*

$$\begin{array}{ccc} (F, C) & \xrightarrow{\beta_{\overline{F}, \overline{F}C, F, C}} & (\overline{F}, \overline{F}C) \\ {\scriptstyle w(C) = \overline{G}C}\Big\downarrow & & \Big\downarrow{\scriptstyle w(\overline{F}C) = C} \\ (\overline{G}, \overline{G}C) & \xrightarrow[\beta_{G, C, \overline{G}, \overline{G}C}]{} & (G, C) \end{array} \tag{8.16}$$

commutes.

PROOF. The point to note is that the Coxeter symmetry w, which sends $\overline{F}C$ to C, sends C to $\overline{G}C$. This is a special case of Fact 1.12. □

We refer to diagram (8.16) as an *opposite support-type relation.*

8.3.2. Opposite support-type morphisms. For a face-type T, define the morphism

$$\beta_{\overline{T},T} : T \to \overline{T} \tag{8.17}$$

in $\boldsymbol{\alpha}$-$\overline{\texttt{Cox}}$ to be the composite morphism $(F, C_r) \to (G, C_r)$ in diagram (8.16) with $C := C_r$, F of type T, G of type $\overline{T}$. It is the morphism transferred by $\overline{F}$, or equivalently, cotransferred by $\overline{G}$. We call this the *opposite support-type morphism* associated to T.

Lemma 8.25. *For a face-type T,*

$$\beta_{T,\overline{T}}\beta_{\overline{T},T} = \mathrm{id}\,. \tag{8.18}$$

FIRST PROOF. Pasting together the diagrams for $\beta_{\overline{T},T}$ and $\beta_{T,\overline{T}}$, we obtain

$$\begin{array}{ccccc}
(F, C_r) & \xrightarrow{\beta_{\overline{F},\overline{F}C_r,F,C_r}} & (\overline{F}, \overline{F}C_r) & & \\
\Big\downarrow {\scriptstyle w(C_r)=\overline{G}C_r} & & \Big\downarrow {\scriptstyle w(\overline{F}C_r)=C_r} & & \\
(\overline{G}, \overline{G}C_r) & \xrightarrow[\beta_{G,C_r,\overline{G},\overline{G}C_r}]{} & (G, C_r) & \xrightarrow{\beta_{\overline{G},\overline{G}C_r,G,C_r}} & (\overline{G}, \overline{G}C_r) \\
 & & \Big\downarrow {\scriptstyle w^{-1}(C_r)=\overline{F}C_r} & & \Big\downarrow {\scriptstyle w^{-1}(\overline{G}C_r)=C_r} \\
 & & (\overline{F}, \overline{F}C_r) & \xrightarrow[\beta_{F,C_r,\overline{F},\overline{F}C_r}]{} & (F, C_r).
\end{array}$$

We check that the composite map from (F, C_r) to itself is the identity.

Let us first go right-down-down-right. By the type-type relation, composite of the middle two arrows is identity, so they cancel off. Now by the support-support relation, composite of the first and last arrows is identity. This completes the check.

Alternatively, we can go down-right-right-down. First, by the support-support relation, composite of the middle two arrows is identity, and then by the type-type relation, composite of the first and last arrows is identity. □

SECOND PROOF. In the groupoid of biface-types $\boldsymbol{\alpha}$-$\overline{\texttt{Cox}}$, the morphisms $\beta_{\overline{T},T}$ are defined as in Section 1.7.7. Identity (8.18) is immediate from this point of view. □

8.3.3. Locally opposite support-type morphisms. We now formulate a more general notion of opposite support-type morphisms. For face-types $Z \leq T$, define the morphism

$$\beta_{Z\overline{T},T} : T \to Z\overline{T} \tag{8.19}$$

in $\boldsymbol{\alpha}$-$\overline{\texttt{Cox}}$ to be the morphism transferred by $A\overline{F}$ with A of type Z, F of type T, and $A \leq F \leq C_r$. This is the same as the morphism cotransferred by $A\overline{G}$ with A of type Z, G of type $Z\overline{T}$, and $A \leq G \leq C_r$. We call this a *locally opposite support-type morphism*. It is illustrated in Figure 8.2. The Coxeter element w is the one that appears in its normal form. The left picture is labeled with faces, while the right one is labeled with face-types.

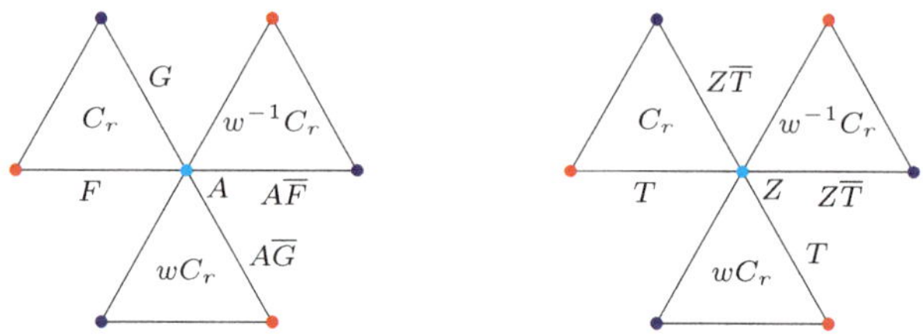

FIGURE 8.2. Locally opposite support-type morphism.

Observe: $\beta_{Z\overline{T},T}$ is local to Z and in fact an opposite support-type morphism in $\boldsymbol{\alpha}_Z\text{-}\overline{\mathtt{Cox}}$. When $Z = \emptyset$, it specializes to the morphism $\beta_{\overline{T},T}$ in (8.17).

Exercise 8.26. Check: The isomorphism (8.14) preserves opposite support-type morphisms. More precisely, for any $T \geq Z$, it sends $\beta_{Z\overline{T},T}$ to $\beta_{Z'\overline{T'},T'}$, where $T' = \sigma_*(T)$.

8.3.4. Adjacent support-type morphisms. We now discuss an important special case of locally opposite support-type morphisms. An *adjacent support-type morphism* is a locally opposite support-type morphism $\beta_{Z\overline{T},T}$ in which $Z \lessdot T$. This is indeed the case in the illustration in Figure 8.2. In this situation, F and $A\overline{F}$ (and G and $A\overline{G}$) are adjacent with A as their common panel. This explains our terminology.

Lemma 8.27. *The category $\boldsymbol{\alpha}\text{-}\overline{\mathtt{Cox}}$ is generated by adjacent support-type morphisms. In particular, it is also generated by locally opposite support-type morphisms.*

FIRST PROOF. Let $\sigma : (F, C_r) \to (G, C_r)$ be a morphism in $\boldsymbol{\alpha}\text{-}\overline{\mathtt{Cox}}$. Suppose it is transferred by H. Then, for any K with the same support as F (and H), we have the commutative diagram

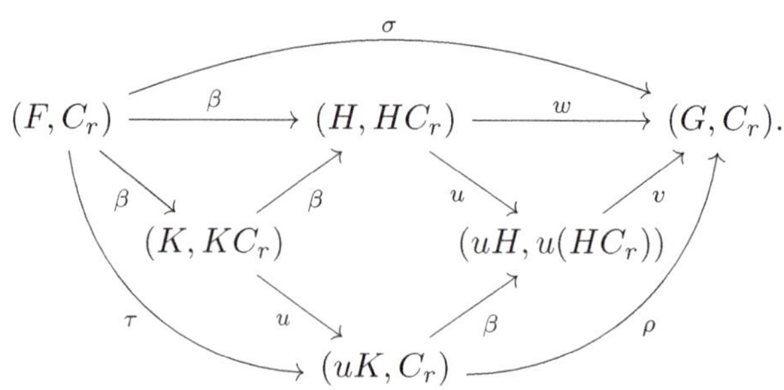

We have expressed σ as composite of two morphisms in $\boldsymbol{\alpha}\text{-}\overline{\mathtt{Cox}}$, namely, τ and ρ. Note: τ starts from (F, C_r) and is transferred by K, while ρ starts from (uK, C_r) and is transferred by uH, where u is the Coxeter symmetry which sends KC_r to C_r.

Now by choosing K to be any face adjacent to F with $F \text{--} K \text{--} H$, and applying induction on gallery distance, we deduce that σ can be expressed as a composite of morphisms each of which is transferred by a face adjacent to a face of C_r. These are precisely adjacent support-type morphisms. $\square$

SECOND PROOF. In the groupoid of biface-types $\boldsymbol{\alpha}\text{-}\overline{\mathsf{Cox}}$, the above argument can be seen more transparently as follows: Let the biface (H, F) represent σ. Then pick a minimal gallery from F to H, and write σ as a composite of morphisms represented by bifaces defined by adjacent faces in this gallery. $\square$

Exercise 8.28. Observe: For a face-type Z, any adjacent support-type morphism in $\boldsymbol{\alpha}_Z\text{-}\overline{\mathsf{Cox}}$ is given by $\beta_{Y\overline{T},T}$ for some $Z \leq Y \lessdot T$. Deduce: Any morphism in $\boldsymbol{\alpha}\text{-}\overline{\mathsf{Cox}}$ local to Z is generated by $\beta_{Y\overline{T},T}$ for $Z \leq Y \lessdot T$.

8.4. Coxeter spaces

We introduce Coxeter spaces. They are built out of the finite categories introduced in Section 8.1, see Table 8.3. For a reflection arrangement $\mathcal{A}$, the category of Coxeter $\mathcal{A}$-spaces is the category of functors from $\mathcal{A}\text{-}\overline{\mathsf{Cox}}$ to Vec. One may also employ instead the equivalent category $\mathcal{A}\text{-}\overline{\mathsf{Cox}}'$. Thus, a Coxeter space can be formulated using either top-nested faces or top-lunes, see the first entry in Table 8.1. There is yet another formulation using face-types obtained by using the groupoid of biface-types $\boldsymbol{\alpha}\text{-}\overline{\mathsf{Cox}}$.

The discussion here can be compared with the one for Coxeter species in Section 2.3.

8.4.1. Coxeter spaces in terms of top-nested faces.

Recall the category $\mathcal{A}\text{-}\overline{\mathsf{Cox}}$ from Section 8.1. A *Coxeter $\mathcal{A}$-space* is a functor

$$\mathrm{V} : \mathcal{A}\text{-}\overline{\mathsf{Cox}} \to \mathsf{Vec}.$$

A map of Coxeter $\mathcal{A}$-spaces $\mathrm{V} \to \mathrm{W}$ is a natural transformation. This defines the category of Coxeter $\mathcal{A}$-spaces which we denote $\mathcal{A}\text{-}\overline{\mathsf{C}}\mathsf{Sp}$. It is a functor category, and we also write

$$\mathcal{A}\text{-}\overline{\mathsf{C}}\mathsf{Sp} := [\mathcal{A}\text{-}\overline{\mathsf{Cox}}, \mathsf{Vec}].$$

For Coxeter spaces V and W, let $\mathcal{A}\text{-}\overline{\mathsf{C}}\mathsf{Sp}(\mathrm{V}, \mathrm{W})$ denote the space of all maps from V to W.

The value of a Coxeter space V on a top-nested face (F, C) is denoted $\mathrm{V}[F, C]$. This is the (F, C)-component of V. Using the presentation of $\mathcal{A}\text{-}\overline{\mathsf{Cox}}$, we obtain:

A Coxeter $\mathcal{A}$-space V consists of a family of vector spaces $\mathrm{V}[F, C]$, one for each top-nested face (F, C), equipped with linear maps

$$\beta_{G,D,F,C} : \mathrm{V}[F, C] \to \mathrm{V}[G, D] \qquad \text{and} \qquad w : \mathrm{V}[F, C] \to \mathrm{V}[F', C'].$$

These are support and type morphisms of V, respectively. The following diagrams must commute.

(8.20a)

$$\begin{array}{ccc} \mathrm{V}[F,C] & \xrightarrow{\beta_{G,D,F,C}} & \mathrm{V}[G,D] \\ & {\scriptstyle \beta_{H,E,F,C}} \searrow \quad \swarrow {\scriptstyle \beta_{H,E,G,D}} & \\ & \mathrm{V}[H,E] & \end{array} \qquad \begin{array}{ccc} \mathrm{V}[F,C] & \xrightarrow{\beta_{G,D,F,C}} & \mathrm{V}[G,D] \\ {\scriptstyle w}\downarrow & & \downarrow{\scriptstyle w} \\ \mathrm{V}[F',C'] & \xrightarrow[\beta_{G',D',F',C'}]{} & \mathrm{V}[G',D'] \end{array} \qquad \begin{array}{ccc} \mathrm{V}[F,C] & \xrightarrow{u} & \mathrm{V}[G,D] \\ & {\scriptstyle vu} \searrow \quad \swarrow {\scriptstyle v} & \\ & \mathrm{V}[H,E] & \end{array}$$

(8.20b) $$(\mathrm{V}[F,C] \xrightarrow{\beta_{F,C,F,C}} \mathrm{V}[F,C]) = \mathrm{id} \qquad (\mathrm{V}[F,C] \xrightarrow{e} \mathrm{V}[F,C]) = \mathrm{id}$$

The conventions are as in (8.1a)–(8.1b). We refer to diagrams (8.20a) as the *support-support axiom*, *support-type axiom*, *type-type axiom*, respectively. See the pictures in Figure 8.1. Similarly, we refer to conditions (8.20b) as the *support-unitality axiom* and *type-unitality axiom*, respectively.

A map of Coxeter $\mathcal{A}$-spaces $f : \mathrm{V} \to \mathrm{W}$ consists of a family of linear maps

$$f_{F,C} : \mathrm{V}[F,C] \to \mathrm{W}[F,C],$$

one for each top-nested face (F,C), such that the diagrams

(8.21) $$\begin{array}{ccc} \mathrm{V}[F,C] & \xrightarrow{f_{F,C}} & \mathrm{W}[F,C] \\ {\scriptstyle \beta_{G,D,F,C}}\downarrow & & \downarrow{\scriptstyle \beta_{G,D,F,C}} \\ \mathrm{V}[G,D] & \xrightarrow[f_{G,D}]{} & \mathrm{W}[G,D] \end{array} \qquad \begin{array}{ccc} \mathrm{V}[F,C] & \xrightarrow{f_{F,C}} & \mathrm{W}[F,C] \\ {\scriptstyle w}\downarrow & & \downarrow{\scriptstyle w} \\ \mathrm{V}[F',C'] & \xrightarrow[f_{F',C'}]{} & \mathrm{W}[F',C'] \end{array}$$

commute. Note: For any $\sigma : (F,C) \to (G,D)$, we get a naturality square. However, since σ is generated by support and type morphisms, it suffices to write naturality squares only for β and w.

Lemma 8.29. *Let* V *be a Coxeter* $\mathcal{A}$*-space. Then, for* F *a face of a chamber* C*, and* G *the face of* C *of the same type as* $\overline{F}$*,*

(8.22) $$\begin{array}{ccc} \mathrm{V}[F,C] & \xrightarrow{\beta_{\overline{F},\overline{F}C,F,C}} & \mathrm{V}[\overline{F},\overline{F}C] \\ {\scriptstyle w(C)=\overline{G}C}\downarrow & & \downarrow{\scriptstyle w(\overline{F}C)=C} \\ \mathrm{V}[\overline{G},\overline{G}C] & \xrightarrow[\beta_{G,C,\overline{G},\overline{G}C}]{} & \mathrm{V}[G,C]. \end{array}$$

PROOF. This follows from (8.16). □

A Coxeter $\mathcal{A}$-space V is *finite-dimensional* if its (F,C)-component $\mathrm{V}[F,C]$ has finite dimension for all top-nested faces (F,C).

Example 8.30 (Rank zero). Let $\mathcal{A}$ be a rank-zero reflection arrangement. A Coxeter $\mathcal{A}$-space V amounts to specifying a vector space $\mathrm{V}[O,O]$. (There is only one support morphism and one type morphism and both are identities.) Similarly, a map of Coxeter $\mathcal{A}$-spaces $\mathrm{V} \to \mathrm{W}$ amounts to specifying a linear map $\mathrm{V}[O,O] \to \mathrm{W}[O,O]$. Thus, the category of Coxeter $\mathcal{A}$-spaces is isomorphic to the category of vector spaces. □

Exercise 8.31 (Rank one). Let $\mathcal{A}$ be a rank-one reflection arrangement with chambers C and $\overline{C}$. Check: A Coxeter $\mathcal{A}$-space V amounts to specifying four vector spaces $\mathrm{V}[O,C]$, $\mathrm{V}[O,\overline{C}]$, $\mathrm{V}[C,C]$, $\mathrm{V}[\overline{C},\overline{C}]$ and three linear isomorphisms

$$\mathrm{V}[O,C] \xrightarrow{u} \mathrm{V}[O,\overline{C}], \qquad \mathrm{V}[C,C] \xrightarrow{\beta} \mathrm{V}[\overline{C},\overline{C}], \qquad \mathrm{V}[\overline{C},\overline{C}] \xrightarrow{v} \mathrm{V}[C,C]$$

such that the composite $v\beta$ is an involution. (The maps u, v are type morphisms, both corresponding to the nonidentity element of the Coxeter group W.) For related ideas, see Section 13.8.

Example 8.32 (Coxeter space of polynomials). Let $\mathcal{A}$ be any reflection arrangement. Define a Coxeter $\mathcal{A}$-space V as follows. For any top-nested face (F,C), define $\mathrm{V}[F,C] := \Bbbk$ and all support and type morphisms to be identities. Axioms (8.20a) and (8.20b) hold trivially. We call V the *Coxeter space of polynomials*. It is studied further in Section 10.2. □

8.4.2. Coxeter spaces in terms of top-lunes.

Proposition 8.10 yields the following alternative description of a Coxeter space. It is in terms of top-lunes instead of top-nested faces. It is the analogue of Proposition 2.17.

Proposition 8.33. *A Coxeter $\mathcal{A}$-space* V *is a family of vector spaces* $\mathrm{V}[\mathrm{L}]$, *one for each top-lune* L, *equipped with linear maps*

$$w : \mathrm{V}[\mathrm{L}] \to \mathrm{V}[\mathrm{L}']$$

whenever $w(\mathrm{L}) = \mathrm{L}'$, *such that the diagrams*

$$\begin{array}{ccc} & \mathrm{V}[\mathrm{L}'] & \\ {\scriptstyle u}\nearrow & & \searrow{\scriptstyle v} \\ \mathrm{V}[\mathrm{L}] & \xrightarrow[vu]{} & \mathrm{V}[\mathrm{L}''] \end{array} \qquad\qquad (\mathrm{V}[\mathrm{L}] \xrightarrow{e} \mathrm{V}[\mathrm{L}]) = \mathrm{id} \tag{8.23}$$

commute, the former whenever $u(\mathrm{L}) = \mathrm{L}'$ *and* $v(\mathrm{L}') = \mathrm{L}''$, *and the latter for every* L.

A map of Coxeter $\mathcal{A}$-spaces $f : \mathrm{V} \to \mathrm{W}$ *is a family of linear maps*

$$f_{\mathrm{L}} : \mathrm{V}[\mathrm{L}] \to \mathrm{W}[\mathrm{L}],$$

one for each top-lune L, *such that the diagram*

$$\begin{array}{ccc} \mathrm{V}[\mathrm{L}] & \xrightarrow{f_{\mathrm{L}}} & \mathrm{W}[\mathrm{L}] \\ {\scriptstyle w}\downarrow & & \downarrow{\scriptstyle w} \\ \mathrm{V}[\mathrm{L}'] & \xrightarrow[f_{\mathrm{L}'}]{} & \mathrm{W}[\mathrm{L}'] \end{array} \tag{8.24}$$

commutes whenever $w(\mathrm{L}) = \mathrm{L}'$.

We refer to the linear maps w as type morphisms of V, and to (8.23) as the *type-type axiom* and *type-unitality axiom*, respectively.

8.4.3. Relating top-nested faces and top-lunes. The connection between the two formulations of Coxeter spaces in terms of top-nested faces and top-lunes can be made explicit in a manner similar to the discussion in Section 2.3.3. Thus, for a top-nested face (F, C) with support L, we have inverse isomorphisms

$$\beta_{F,C,\mathrm{L}} : \mathrm{V}[\mathrm{L}] \to \mathrm{V}[F, C] \quad \text{and} \quad \beta_{\mathrm{L},F,C} : \mathrm{V}[F, C] \to \mathrm{V}[\mathrm{L}]. \tag{8.25}$$

Moreover, the diagrams

$$\begin{array}{ccc} \mathrm{V}[F,C] & \xrightarrow{\beta_{G,D,F,C}} & \mathrm{V}[G,D] \\ & {\scriptstyle\beta_{\mathrm{L},F,C}}\searrow \quad \swarrow{\scriptstyle\beta_{\mathrm{L},G,D}} & \\ & \mathrm{V}[\mathrm{L}] & \end{array} \qquad \begin{array}{ccc} \mathrm{V}[F,C] & \xrightarrow{w} & \mathrm{V}[wF,wC] \\ {\scriptstyle\beta_{\mathrm{L},F,C}}\downarrow & & \downarrow{\scriptstyle\beta_{w\mathrm{L},wF,wC}} \\ \mathrm{V}[\mathrm{L}] & \xrightarrow[w]{} & \mathrm{V}[w\mathrm{L}] \end{array} \tag{8.26}$$

commute. These two diagrams can be combined into one as follows. For a morphism $\sigma : (F, C) \to (G, D)$ in $\mathcal{A}$-$\overline{\mathsf{Cox}}$ with normal form $\sigma = \beta w = w\beta$, the diagram

$$\begin{array}{ccc} \mathrm{V}[F,C] & \xrightarrow{\sigma} & \mathrm{V}[G,D] \\ {\scriptstyle\beta_{\mathrm{L},F,C}}\downarrow & & \downarrow{\scriptstyle\beta_{w\mathrm{L},G,D}} \\ \mathrm{V}[\mathrm{L}] & \xrightarrow[w]{} & \mathrm{V}[w\mathrm{L}] \end{array} \tag{8.27}$$

commutes.

Exercise 8.34 (Rank one). This is a continuation of Exercise 8.31. Check: For a rank-one arrangement $\mathcal{A}$, a Coxeter $\mathcal{A}$-space V, in terms of top-lunes, amounts to specifying vector spaces $\mathrm{V}[C]$, $\mathrm{V}[\overline{C}]$, $\mathrm{V}[\top]$ and linear isomorphisms

$$\mathrm{V}[C] \xrightarrow{u} \mathrm{V}[\overline{C}] \qquad \text{and} \qquad \mathrm{V}[\top] \xrightarrow{w} \mathrm{V}[\top]$$

with $w^2 = \mathrm{id}$. Here C and $\overline{C}$ are chambers, and $\top$ is the maximum flat; they are all top-lunes. To connect to Exercise 8.31, take $\mathrm{V}[\top] = \mathrm{V}[C, C]$ with $\beta_{C,C,\top} = \mathrm{id}$, and $w = v\beta$.

Example 8.35 (Coxeter space of polynomials). This is a continuation of Example 8.32. In terms of top-lunes, the Coxeter space of polynomials can be formulated as follows. For any top-lune L, define $\mathrm{V}[\mathrm{L}] := \Bbbk$ and all type morphisms to be identities. Axioms (8.23) hold trivially. The maps $\beta_{F,C,\mathrm{L}}$ and $\beta_{\mathrm{L},F,C}$ are all identities; diagrams (8.26) commute trivially. □

8.4.4. Coxeter spaces in terms of top-lune-types. We may use Exercise 8.11 to give yet another formulation of Coxeter spaces as follows.

Proposition 8.36. *A Coxeter $\mathcal{A}$-space* V *is a family of W_ρ-modules, one for each top-lune-type ρ. A map of Coxeter $\mathcal{A}$-spaces is a family of W_ρ-module maps, one for each top-lune-type ρ.*

8.4.5. Coxeter spaces in terms of flats. We may use Exercise 8.13 to formulate Coxeter spaces in terms of flats. Fix a chamber C. A Coxeter $\mathcal{A}$-space V is a family of vector spaces $\mathrm{V}[\mathrm{Y}]$, one for each flat Y, equipped with linear maps

$$w : \mathrm{V}[\mathrm{Y}] \to \mathrm{V}[\mathrm{Y}']$$

whenever w sends the top-lune containing C with base Y to the top-lune containing C with base Y′. Moreover, these linear maps are compatible with the product in W.

Alternatively, following Proposition 8.12, one can use the quotient category $\mathcal{A}\text{-}\overline{\mathsf{Cox}}''$ to describe Coxeter spaces using flats which avoids mention of top-lunes. We now use linear maps, one for each $w \in W$, with the condition that $w : \mathrm{V}[\mathrm{Y}] \to \mathrm{V}[\mathrm{Y}]$ is identity if w fixes Y pointwise.

8.4.6. Coxeter spaces in terms of face-types. Recall the category $\boldsymbol{\alpha}\text{-}\overline{\mathsf{Cox}}$ from Section 8.1.5. It is isomorphic to the groupoid of biface-types from Section 1.7 as explained in Proposition 8.18. A *Coxeter $\boldsymbol{\alpha}$-space* is a functor

$$\mathrm{V} : \boldsymbol{\alpha}\text{-}\overline{\mathsf{Cox}} \to \mathsf{Vec}.$$

A map of Coxeter $\boldsymbol{\alpha}$-spaces V → W is a natural transformation. This defines the category of Coxeter $\boldsymbol{\alpha}$-spaces which we denote $\boldsymbol{\alpha}\text{-}\overline{\mathsf{C}}\mathsf{Sp}$. It is a functor category, and we also write

$$\boldsymbol{\alpha}\text{-}\overline{\mathsf{C}}\mathsf{Sp} := [\boldsymbol{\alpha}\text{-}\overline{\mathsf{Cox}}, \mathsf{Vec}].$$

The value of a Coxeter $\boldsymbol{\alpha}$-space V on a face-type T is denoted $\mathrm{V}[T]$.

Explicitly, a Coxeter $\boldsymbol{\alpha}$-space V consists of a family of vector spaces $\mathrm{V}[T]$, one for each face-type T, equipped with linear maps

$$\sigma : \mathrm{V}[T] \to \mathrm{V}[U],$$

one for each morphism $\sigma : T \to U$ in $\boldsymbol{\alpha}\text{-}\overline{\mathsf{Cox}}$, such that the diagrams

$$\begin{array}{ccc} & \mathrm{V}[U] & \\ {}^{\sigma}\nearrow & & \searrow^{\tau} \\ \mathrm{V}[T] & \xrightarrow[\tau\sigma]{} & \mathrm{V}[V] \end{array} \qquad\qquad (\mathrm{V}[T] \xrightarrow{\mathrm{id}} \mathrm{V}[T]) = \mathrm{id} \tag{8.28}$$

commute, the former for all $\sigma : T \to U$ and $\tau : U \to V$, and the latter for every T. Similarly, a map of Coxeter $\boldsymbol{\alpha}$-spaces $f : \mathrm{V} \to \mathrm{W}$ consists of a family of linear maps

$$f_T : \mathrm{V}[T] \to \mathrm{W}[T],$$

one for each face-type T, such that the diagram

$$\begin{array}{ccc} \mathrm{V}[T] & \xrightarrow{f_T} & \mathrm{W}[T] \\ {\scriptstyle\sigma}\downarrow & & \downarrow{\scriptstyle\sigma} \\ \mathrm{V}[U] & \xrightarrow[f_U]{} & \mathrm{W}[U] \end{array} \tag{8.29}$$

commutes for all $\sigma : T \to U$.

8.4.7. Relating top-nested faces and face-types. Recall: The categories $\mathcal{A}\text{-}\overline{\mathsf{Cox}}$ and $\boldsymbol{\alpha}\text{-}\overline{\mathsf{Cox}}$ are equivalent (Lemma 8.14 and Exercise 8.15). It follows:

Lemma 8.37. *The categories of Coxeter $\mathcal{A}$-spaces and Coxeter $\boldsymbol{\alpha}$-spaces are equivalent.*

Let us elaborate on this equivalence. Starting with a Coxeter $\mathcal{A}$-space V, set $\mathrm{V}[T] := \mathrm{V}[F, C_r]$, where F is the face of C_r of type T. For a morphism $\sigma : T \to U$ in $\boldsymbol{\alpha}$-$\overline{\mathsf{Cox}}$, define

$$\begin{array}{ccc} \mathrm{V}[T] & \overset{\sigma}{\dashrightarrow} & \mathrm{V}[U] \\ \| & & \| \\ \mathrm{V}[F, C_r] & \xrightarrow[\sigma]{} & \mathrm{V}[G, C_r], \end{array}$$

with F and G of types T and U, respectively. (The bottom arrow σ is a morphism in $\mathcal{A}$-$\overline{\mathsf{Cox}}$.) This yields a Coxeter $\boldsymbol{\alpha}$-space.

Conversely, starting with a Coxeter $\boldsymbol{\alpha}$-space V, set $\mathrm{V}[F, C] := \mathrm{V}[\mathrm{t}(F)]$. For a morphism $\sigma : (G, D) \to (H, E)$ in $\mathcal{A}$-$\overline{\mathsf{Cox}}$, define

$$\begin{array}{ccc} \mathrm{V}[G, D] & \overset{\sigma}{\dashrightarrow} & \mathrm{V}[H, E] \\ \| & & \| \\ \mathrm{V}[\mathrm{t}(G)] & \xrightarrow[v\sigma u]{} & \mathrm{V}[\mathrm{t}(H)], \end{array}$$

where u takes C_r to D, and v takes E to C_r. (The bottom arrow $v\sigma u$ is a morphism in $\boldsymbol{\alpha}$-$\overline{\mathsf{Cox}}$.) This yields a Coxeter $\mathcal{A}$-space.

Exercise 8.38 (W-modules). Check: For a Coxeter $\boldsymbol{\alpha}$-space V, the component $\mathrm{V}[S]$ is a left W-module. The W-action is defined by $w \cdot x := \sigma(x)$ where $\sigma : S \to S$ has normal form $\sigma = \beta w = w\beta$, or equivalently, $\sigma = \mathrm{t}(D, C)$ with $w = d(D, C)$. This yields a functor

$$\boldsymbol{\alpha}\text{-}\overline{\mathsf{C}}\mathsf{Sp} \to W\text{-}\mathsf{Mod}, \qquad \mathrm{V} \mapsto \mathrm{V}[S]. \tag{8.30}$$

(See Exercise 8.17 and Exercise 1.29 and the functor (1.26) in particular.)

TABLE 8.4. Coxeter spaces to W-modules.

Coxeter space	W-module	Coxeter space	W-module
$\mathrm{V}^1 = \mathrm{V}^{\underline{1}} = \mathrm{V}^{\overline{1}}$	$\mathrm{V}[S]$	V^k, $\mathrm{V}^{\underline{k}}$, $\mathrm{V}^{\overline{k}}$ for $k > 1$	0
$\overline{\mathcal{T}}(\mathrm{V})$	$\mathrm{V}[S]$	$\overline{\mathcal{S}}(\mathrm{V})$, $\overline{\mathcal{E}}(\mathrm{V})$	$\mathrm{V}[S]$
$\overline{\mathcal{P}}(\mathrm{C})$, C^{coab}	$\mathrm{C}[S]$	$\overline{\mathcal{Q}}(\mathrm{A})$, A_{ab}	$\mathrm{A}[S]$
$\mathrm{V} \times \mathrm{W}$	$\mathrm{V}[S] \otimes \mathrm{W}[S]$	$\hom^{\times}(\mathrm{V}, \mathrm{W})$	$\hom^{\otimes}(\mathrm{V}[S], \mathrm{W}[S])$
$\mathcal{C}(\mathrm{C}, \mathrm{D})$	$\hom^{\otimes}(\mathrm{C}[S], \mathrm{D}[S])$	$\overline{\mathcal{C}}(\mathrm{A}, \mathrm{B})$	$\hom^{\otimes}(\mathrm{A}[S], \mathrm{B}[S])$

The values of the functor (8.30) on certain Coxeter spaces which we will encounter in Chapter 9 are shown in Table 8.4. See Observation 9.4, Exercise 9.30, Exercise 9.32, Observation 9.33, Exercise 9.36 for more detail.

8.4.8. Product and coproduct. The *zero Coxeter space* is the Coxeter space with all components zero. It is both the initial and terminal object in the category of Coxeter spaces.

For Coxeter spaces V and W, define their *direct sum* $\mathrm{V}+\mathrm{W}$ by either of the equivalent formulas

$$(8.31)\qquad \begin{aligned}(\mathrm{V}+\mathrm{W})[A,C] &:= \mathrm{V}[A,C]\oplus \mathrm{W}[A,C],\\ (\mathrm{V}+\mathrm{W})[\mathrm{L}] &:= \mathrm{V}[\mathrm{L}]\oplus \mathrm{W}[\mathrm{L}],\\ (\mathrm{V}+\mathrm{W})[T] &:= \mathrm{V}[T]\oplus \mathrm{W}[T].\end{aligned}$$

The structure maps of $\mathrm{V}+\mathrm{W}$ are induced from those of V and W. Direct sum is both the product and coproduct in the category of Coxeter spaces.

8.4.9. Coxeter subspaces and Coxeter quotient spaces. The category of Coxeter spaces is $\Bbbk$-linear: If $f, g : \mathrm{V}\to \mathrm{W}$ are maps of Coxeter spaces, then so is $af+bg$ for any $a, b\in \Bbbk$ with

$$(8.32)\qquad \begin{aligned}(af+bg)_{A,C} &:= af_{A,C}+bg_{A,C},\\ (af+bg)_{\mathrm{L}} &:= af_{\mathrm{L}}+bg_{\mathrm{L}},\\ (af+bg)_{T} &:= af_{T}+bg_{T}.\end{aligned}$$

This can be seen from (8.21), (8.24), (8.29). This is the same as the situation for Coxeter species (Section 2.3.6). We can also talk of subspaces and quotient spaces of a Coxeter space, and of kernel, cokernel, image, coimage of a map between Coxeter spaces.

8.4.10. Duality functor on Coxeter spaces. For a Coxeter space V, define its dual Coxeter space V^* by dualizing each component. For instance, in terms of face-types,

$$\mathrm{V}^*[T] := \mathrm{V}[T]^*.$$

For any morphism $\sigma : T\to U$ in $\boldsymbol{\alpha}\text{-}\overline{\mathtt{Cox}}$, we have an invertible linear map $\sigma : \mathrm{V}[T]\to \mathrm{V}[U]$. Dualize it and take its inverse to get a linear map $\mathrm{V}^*[T]\to \mathrm{V}^*[U]$, and define this to be the structure map of V^* corresponding to σ.

Moreover, a map $\mathrm{V}\to \mathrm{W}$ of Coxeter spaces induces a map $\mathrm{W}^*\to \mathrm{V}^*$. This yields a functor from the category of Coxeter spaces to its opposite

$$(8.33)\qquad (-)^* : \boldsymbol{\alpha}\text{-}\overline{\mathtt{C}}\mathtt{Sp}\to \boldsymbol{\alpha}\text{-}\overline{\mathtt{C}}\mathtt{Sp}^{\mathrm{op}}.$$

This is the *duality functor* on Coxeter spaces. This is an instance of a general construction, see Remark 2.21.

8.4.11. Coxeter b-spaces and Coxeter lb-spaces. The notion of Coxeter spaces can be generalized by weakening the support-support axiom as follows. For a Coxeter lb-space, we only require the support-support axiom in (8.20a) whenever $F \text{--} G \text{--} H$, or equivalently, whenever $C \text{--} D \text{--} E$. For a Coxeter b-space, we require in addition that each support morphism $\beta_{G,D,F,C}$ is a linear isomorphism. Observe: A Coxeter b-space is a Coxeter space if in addition $\beta^{-1}_{G,D,F,C} = \beta_{F,C,G,D}$.

In terms of face-types: For a Coxeter lb-space, we require the first diagram in (8.28) whenever $l(\tau\sigma) = l(\tau)+l(\sigma)$. For a Coxeter b-space, we require in addition that each σ is a linear isomorphism.

8.5. Coxeter algebras and Coxeter coalgebras

In this section and the next two, we introduce Coxeter algebras, Coxeter coalgebras, Coxeter bialgebras. The latter has a q-analogue which we call Coxeter q-bialgebras. For $q = -1$, we use the term signed Coxeter bialgebras. We also pay attention to the case $q = 0$. A summary of the resulting categories along with their notations is given in Table 8.5. Here $\mathcal{A}$ denotes a fixed reflection arrangement. Alternatively, one can develop these notions relative to a pointed reflection arrangement $\boldsymbol{\alpha}$. In this case, we use the same notations for the categories with $\boldsymbol{\alpha}$ instead of $\mathcal{A}$. Many times, particularly in later chapters, we will drop $\mathcal{A}$ or $\boldsymbol{\alpha}$ from the notation, and write $\mathsf{Mon}(\overline{\mathsf{C}}\mathsf{Sp})$ and so on. This is consistent with Convention 8.1.

TABLE 8.5. Categories of Coxeter algebras, coalgebras, bialgebras.

Category	Description
$\mathsf{Mon}(\mathcal{A}\text{-}\overline{\mathsf{C}}\mathsf{Sp})$	Coxeter $\mathcal{A}$-algebras
$\mathsf{Comon}(\mathcal{A}\text{-}\overline{\mathsf{C}}\mathsf{Sp})$	Coxeter $\mathcal{A}$-coalgebras
$\mathsf{Bimon}(\mathcal{A}\text{-}\overline{\mathsf{C}}\mathsf{Sp})$	Coxeter $\mathcal{A}$-bialgebras
$q\text{-}\mathsf{Bimon}(\mathcal{A}\text{-}\overline{\mathsf{C}}\mathsf{Sp})$	Coxeter $\mathcal{A}$-q-bialgebras
$(-1)\text{-}\mathsf{Bimon}(\mathcal{A}\text{-}\overline{\mathsf{C}}\mathsf{Sp})$	signed Coxeter $\mathcal{A}$-bialgebras
$0\text{-}\mathsf{Bimon}(\mathcal{A}\text{-}\overline{\mathsf{C}}\mathsf{Sp})$	Coxeter $\mathcal{A}$-0-bialgebras

Compare the categories of Coxeter (co, bi)algebras in Table 8.5 with the categories of Coxeter (co, bi)monoids in Table 2.4. The two can be connected using Fock functors, see Section 15.1.

Coxeter (co, bi)algebras are structured objects in the category of Coxeter spaces. As with Coxeter spaces, we formulate them in terms of top-nested faces, top-lunes, face-types. They involve product and coproduct maps subject to a bunch of axioms, see the second, third, fourth entries in Table 8.1, and the first entry in Table 8.2.

8.5.1. Coxeter algebras in terms of top-nested faces.

Recall Coxeter $\mathcal{A}$-spaces from Section 8.4.1. In the present context, we use the letter A instead of V. A *Coxeter $\mathcal{A}$-algebra* is a Coxeter $\mathcal{A}$-space A equipped with linear maps

$$\mu_{A,C}^{F,C} : \mathrm{A}[F,C] \to \mathrm{A}[A,C],$$

one for each $A \leq F \leq C$, such that the diagrams

$$\text{(8.34a)}\qquad \begin{array}{ccc} \mathrm{A}[F,C] & \xrightarrow{\beta_{G,D,F,C}} & \mathrm{A}[G,D] \\ {\scriptstyle \mu_{A,C}^{F,C}}\downarrow & & \downarrow{\scriptstyle \mu_{B,D}^{G,D}} \\ \mathrm{A}[A,C] & \xrightarrow[\beta_{B,D,A,C}]{} & \mathrm{A}[B,D] \end{array} \qquad \begin{array}{ccc} \mathrm{A}[F,C] & \xrightarrow{w} & \mathrm{A}[F',C'] \\ {\scriptstyle \mu_{A,C}^{F,C}}\downarrow & & \downarrow{\scriptstyle \mu_{A',C'}^{F',C'}} \\ \mathrm{A}[A,C] & \xrightarrow[w]{} & \mathrm{A}[A',C'] \end{array}$$

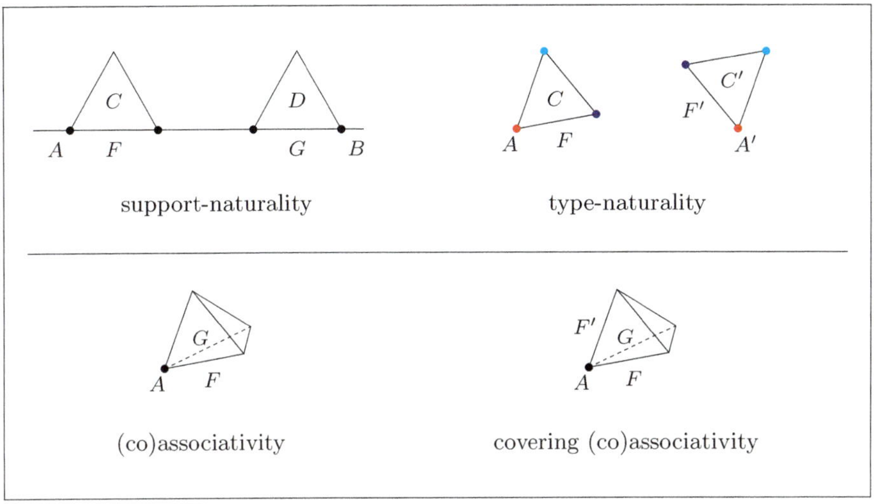

FIGURE 8.3. Naturality and (co)associativity axioms.

(8.34b)
$$\begin{array}{ccc} & \mathrm{A}[F,C] & \\ {}^{\mu^{G,C}_{F,C}}\nearrow & & \searrow^{\mu^{F,C}_{A,C}} \\ \mathrm{A}[G,C] & \xrightarrow[\mu^{G,C}_{A,C}]{} & \mathrm{A}[A,C] \end{array} \qquad (\mathrm{A}[A,C] \xrightarrow{\mu^{A,C}_{A,C}} \mathrm{A}[A,C]) = \mathrm{id}$$

commute.

- In the first diagram in (8.34a), $A \leq F \leq C$, $B \leq G \leq D$ and $\mathrm{s}(A, C) = \mathrm{s}(B, D)$, $\mathrm{s}(F, C) = \mathrm{s}(G, D)$, and in the second, $A \leq F \leq C$ and $w(A) = A'$, $w(F) = F'$, $w(C) = C'$.
- In the first diagram in (8.34b), $A \leq F \leq G \leq C$, and in the second, $A \leq C$.

We denote a Coxeter $\mathcal{A}$-algebra by a pair (A, μ), or simply by A with μ understood. The maps $\mu^{F,C}_{A,C}$ are product components of A. We refer to diagrams (8.34a) as the *support-naturality axiom* and *type-naturality axiom*, respectively, and to diagrams (8.34b) as the *associativity axiom* and *unitality axiom*, respectively. The first two pictures in Figure 8.3 illustrate naturality; the third picture illustrates associativity. In the first picture, A and B are to be pictured as opposite vertices.

Exercise 8.39. Check: In the support-naturality axiom in (8.34a), we necessarily have $F = AG$ and $G = BF$, and hence, $\mathrm{s}(A, F) = \mathrm{s}(B, G)$. (Use: Distinct faces of a chamber have distinct supports.)

Observation 8.40. Support-naturality and type-naturality axioms for a Coxeter $\mathcal{A}$-algebra A can be combined into one *naturality axiom* as follows. For a morphism

$\sigma : (A, C) \to (B, D)$ in $\mathcal{A}$-$\overline{\texttt{Cox}}$, the diagram

$$\begin{array}{ccc} \mathrm{A}[F,C] & \xrightarrow{\sigma} & \mathrm{A}[G,D] \\ {\scriptstyle \mu^{F,C}_{A,C}}\downarrow & & \downarrow{\scriptstyle \mu^{G,D}_{B,D}} \\ \mathrm{A}[A,C] & \xrightarrow[\sigma]{} & \mathrm{A}[B,D] \end{array} \tag{8.35}$$

commutes for $G = \sigma_*(F)$, or equivalently, $F = \sigma^*(G)$, with notation as in (8.11).

- Specializing σ to β and w, respectively, in (8.35) yields (8.34a).
- Pasting diagrams in (8.34a) for β and w given by $\sigma = \beta w = w\beta$ yields (8.35).

Compare with Observation 2.22.

A morphism $f : \mathrm{A} \to \mathrm{B}$ of Coxeter $\mathcal{A}$-algebras is a map of Coxeter $\mathcal{A}$-spaces such that for each $A \le F \le C$, the diagram

$$\begin{array}{ccc} \mathrm{A}[F,C] & \xrightarrow{f_{F,C}} & \mathrm{B}[F,C] \\ {\scriptstyle \mu^{F,C}_{A,C}}\downarrow & & \downarrow{\scriptstyle \mu^{F,C}_{A,C}} \\ \mathrm{A}[A,C] & \xrightarrow[f_{A,C}]{} & \mathrm{B}[A,C] \end{array} \tag{8.36}$$

commutes.

8.5.2. Coxeter algebras in terms of top-lunes. Recall the partial order $\le'$ on top-lunes (Section 1.2.4). The result below builds on Proposition 8.33. It is the analogue of Proposition 2.24 and has a similar proof.

Proposition 8.41. *A* Coxeter $\mathcal{A}$-algebra *is a Coxeter $\mathcal{A}$-space* A *equipped with linear maps*

$$\mu^{\mathrm{M}}_{\mathrm{L}} : \mathrm{A}[\mathrm{M}] \to \mathrm{A}[\mathrm{L}],$$

one for each $\mathrm{L} \le' \mathrm{M}$, *such that the diagrams*

$$\begin{array}{ccc} \mathrm{A}[\mathrm{M}] & \xrightarrow{w} & \mathrm{A}[\mathrm{M}'] \\ {\scriptstyle \mu^{\mathrm{M}}_{\mathrm{L}}}\downarrow & & \downarrow{\scriptstyle \mu^{\mathrm{M}'}_{\mathrm{L}'}} \\ \mathrm{A}[\mathrm{L}] & \xrightarrow[w]{} & \mathrm{A}[\mathrm{L}'] \end{array} \qquad \begin{array}{ccc} & \mathrm{A}[\mathrm{M}] & \\ {\scriptstyle \mu^{\mathrm{N}}_{\mathrm{M}}}\nearrow & & \searrow{\scriptstyle \mu^{\mathrm{M}}_{\mathrm{L}}} \\ \mathrm{A}[\mathrm{N}] & \xrightarrow[\mu^{\mathrm{N}}_{\mathrm{L}}]{} & \mathrm{A}[\mathrm{L}] \end{array} \qquad (\mathrm{A}[\mathrm{L}] \xrightarrow{\mu^{\mathrm{L}}_{\mathrm{L}}} \mathrm{A}[\mathrm{L}]) = \mathrm{id} \tag{8.37}$$

commute, the first diagram whenever $\mathrm{L} \le' \mathrm{M}$ *and* $w(\mathrm{L}) = \mathrm{L}'$, $w(\mathrm{M}) = \mathrm{M}'$, *the second for every* $\mathrm{L} \le' \mathrm{M} \le' \mathrm{N}$, *and the third for every* L.

A morphism $f : \mathrm{A} \to \mathrm{B}$ *of Coxeter $\mathcal{A}$-algebras is a map of Coxeter $\mathcal{A}$-spaces such that for each* $\mathrm{L} \le' \mathrm{M}$, *the diagram*

$$\begin{array}{ccc} \mathrm{A}[\mathrm{M}] & \xrightarrow{f_{\mathrm{M}}} & \mathrm{B}[\mathrm{M}] \\ {\scriptstyle \mu^{\mathrm{M}}_{\mathrm{L}}}\downarrow & & \downarrow{\scriptstyle \mu^{\mathrm{M}}_{\mathrm{L}}} \\ \mathrm{A}[\mathrm{L}] & \xrightarrow[f_{\mathrm{L}}]{} & \mathrm{B}[\mathrm{L}] \end{array} \tag{8.38}$$

commutes.

PROOF. Product components $\mu_{A,C}^{F,C}$ and $\mu_{\mathrm{L}}^{\mathrm{M}}$ in the two formulations of a Coxeter $\mathcal{A}$-algebra A are linked by the diagram

$$\begin{array}{ccc} \mathrm{A}[F,C] & \xrightarrow{\beta_{\mathrm{M},F,C}} & \mathrm{A}[\mathrm{M}] \\ {\scriptstyle \mu_{A,C}^{F,C}}\big\downarrow & & \big\downarrow{\scriptstyle \mu_{\mathrm{L}}^{\mathrm{M}}} \\ \mathrm{A}[A,C] & \xrightarrow[\beta_{\mathrm{L},A,C}]{} & \mathrm{A}[\mathrm{L}], \end{array} \tag{8.39}$$

where $\mathrm{M} = \mathrm{s}(F,C)$ and $\mathrm{L} = \mathrm{s}(A,C)$. The vertical maps define each other. If we start with the $\mu_{A,C}^{F,C}$, then the $\mu_{\mathrm{L}}^{\mathrm{M}}$ are well-defined due to the support-naturality axiom. Conversely, if we start with the $\mu_{\mathrm{L}}^{\mathrm{M}}$, then the $\mu_{A,C}^{F,C}$ satisfy the support-naturality axiom. Also, there is a clear correspondence between the remaining axioms. The checks make use of diagrams (8.26). □

We refer to diagrams (8.37) as the *type-naturality axiom*, *associativity axiom*, *unitality axiom*, respectively. Note very carefully how this formulation gets rid of the support-naturality axiom.

Remark 8.42. It is natural to ask what happens under the weaker hypothesis $\mathrm{L} \leq \mathrm{M}$ on top-lunes. In this situation, for each lune P with base $\mathrm{b}(\mathrm{L})$ and case $\mathrm{b}(\mathrm{M})$, there is a linear map $\mu_{\mathrm{L}}^{\mathrm{M}}(\mathrm{P})$ defined by the dotted arrow in the diagram

$$\begin{array}{ccc} \mathrm{A}[\mathrm{M}] & \xrightarrow{\mu_{\mathrm{P}\circ\mathrm{M}}^{\mathrm{M}}} & \mathrm{A}[\mathrm{P}\circ\mathrm{M}] \\ {\scriptstyle w}\big\downarrow & \searrow^{\mu_{\mathrm{L}}^{\mathrm{M}}(\mathrm{P})} & \big\downarrow{\scriptstyle w} \\ \mathrm{A}[\mathrm{M}'] & \xrightarrow[\mu_{\mathrm{L}}^{\mathrm{M}'}]{} & \mathrm{A}[\mathrm{L}], \end{array} \tag{8.40}$$

where w is the unique Coxeter symmetry which sends $\mathrm{P}\circ\mathrm{M}$ to L and fixes $\mathrm{b}(\mathrm{L})$. (Note: $\mathrm{P}\circ\mathrm{M} = \mathrm{PL}$, the Tits product of faces in $\mathcal{A}_{\mathrm{b}(\mathrm{L})}$.) To recover the usual product components, observe that for $\mathrm{L} \leq' \mathrm{M}$, we have $\mu_{\mathrm{L}}^{\mathrm{M}}(\mathrm{L}/\mathrm{M}) = \mu_{\mathrm{L}}^{\mathrm{M}}$. An illustration of how lunes P and L/M lie wrt the top-lunes L and M is shown below.

In the pictures, M is the visible hemisphere, L is the portion between two adjacent longitudes, P and L/M are the red longitudes.

8.5.3. Coxeter algebras in terms of face-types. Alternatively, one may work in the category of Coxeter $\boldsymbol{\alpha}$-spaces yielding the equivalent notion of a Coxeter $\boldsymbol{\alpha}$-algebra described below.

A *Coxeter* $\boldsymbol{\alpha}$*-algebra* is a Coxeter $\boldsymbol{\alpha}$-space A equipped with linear maps

$$\mu_Z^T : \mathrm{A}[T] \to \mathrm{A}[Z],$$

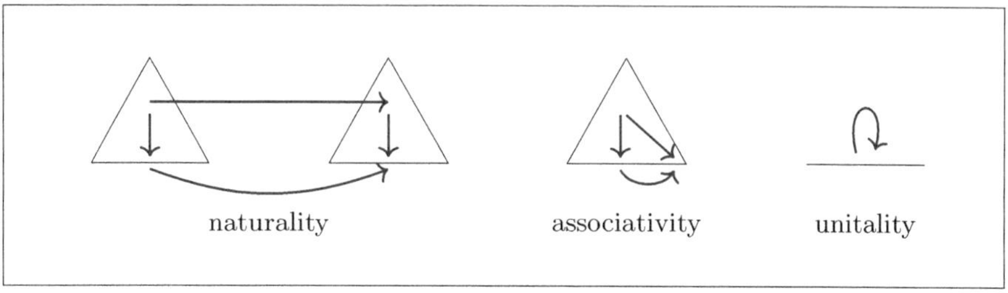

FIGURE 8.4. Coxeter algebra in terms of face-types.

one for each $Z \leq T$, such that the diagrams

$$
\begin{array}{ccc}
\mathrm{A}[T] & \xrightarrow{\sigma} & \mathrm{A}[T'] \\
{\scriptstyle \mu_Z^T}\downarrow & & \downarrow{\scriptstyle \mu_{Z'}^{T'}} \\
\mathrm{A}[Z] & \xrightarrow[\sigma]{} & \mathrm{A}[Z']
\end{array}
\qquad
\begin{array}{ccccc}
 & & \mathrm{A}[T] & & \\
 & {\scriptstyle \mu_T^U}\nearrow & & \searrow{\scriptstyle \mu_Z^T} & \\
\mathrm{A}[U] & & \xrightarrow[\mu_Z^U]{} & & \mathrm{A}[Z]
\end{array}
\qquad
(\mathrm{A}[Z] \xrightarrow{\mu_Z^Z} \mathrm{A}[Z]) = \mathrm{id} \tag{8.41}
$$

commute.

- In the first diagram, $\sigma : Z \to Z'$ is any morphism in $\boldsymbol{\alpha}$-$\overline{\texttt{Cox}}$, $Z \leq T$ and $T' = \sigma_*(T)$, or equivalently, $Z' \leq T'$ and $T = \sigma^*(T')$, with $\sigma : T \to T'$ the induced morphism as in (8.12) or (1.28).
- In the second diagram, $Z \leq T \leq U$.
- In the third diagram, Z is any face-type.

We refer to diagrams (8.41) as the *naturality axiom*, *associativity axiom*, *unitality axiom*, respectively. See Figure 8.4.

A morphism $f : \mathrm{A} \to \mathrm{B}$ of Coxeter $\boldsymbol{\alpha}$-algebras is a map of Coxeter $\boldsymbol{\alpha}$-spaces such that for each $Z \leq T$, the diagram

$$
\begin{array}{ccc}
\mathrm{A}[T] & \xrightarrow{f_T} & \mathrm{B}[T] \\
{\scriptstyle \mu_Z^T}\downarrow & & \downarrow{\scriptstyle \mu_Z^T} \\
\mathrm{A}[Z] & \xrightarrow[f_Z]{} & \mathrm{B}[Z]
\end{array} \tag{8.42}
$$

commutes.

Exercise 8.43. The categories of Coxeter $\mathcal{A}$-algebras and Coxeter $\boldsymbol{\alpha}$-algebras are equivalent. This builds on the equivalence in Lemma 8.37. Check: A Coxeter $\mathcal{A}$-algebra yields a Coxeter $\boldsymbol{\alpha}$-algebra, and vice versa using the recipes

$$
\begin{array}{ccc}
\mathrm{A}[T] & \overset{\mu_Z^T}{\dashrightarrow} & \mathrm{A}[Z] \\
\| & & \| \\
\mathrm{A}[F, C_r] & \xrightarrow[\mu_{A,C_r}^{F,C_r}]{} & \mathrm{A}[A, C_r],
\end{array}
\qquad
\begin{array}{ccc}
\mathrm{A}[F, C] & \overset{\mu_{A,C}^{F,C}}{\dashrightarrow} & \mathrm{A}[A, C] \\
\| & & \| \\
\mathrm{A}[\mathrm{t}(F)] & \xrightarrow[\mu_{\mathrm{t}(A)}^{\mathrm{t}(F)}]{} & \mathrm{A}[\mathrm{t}(A)],
\end{array}
$$

where in the first diagram, F and A are faces of C_r of types T and Z. (The naturality axiom in (8.41) corresponds to the naturality axiom (8.35) when $C = D = C_r$. There is no analogue of (8.34a) for face-types.)

8.5.4. Covering associativity axiom. In the definition of a Coxeter $\boldsymbol{\alpha}$-algebra, it is possible to restrict to those product components μ_Z^T, where T covers Z. The precise result is given below.

Proposition 8.44. *A Coxeter $\boldsymbol{\alpha}$-algebra is a Coxeter $\boldsymbol{\alpha}$-space* A *equipped with linear maps*

$$\mu_Z^T : \mathrm{A}[T] \to \mathrm{A}[Z]$$

whenever T covers Z, such that the diagrams

$$(8.43)\qquad \begin{array}{ccc} \mathrm{A}[T] & \xrightarrow{\sigma} & \mathrm{A}[T'] \\ {\scriptstyle \mu_Z^T}\downarrow & & \downarrow{\scriptstyle \mu_{Z'}^{T'}} \\ \mathrm{A}[Z] & \xrightarrow[\sigma]{} & \mathrm{A}[Z'] \end{array} \qquad\qquad \begin{array}{ccc} \mathrm{A}[U] & \xrightarrow{\mu_{T'}^U} & \mathrm{A}[T'] \\ {\scriptstyle \mu_T^U}\downarrow & & \downarrow{\scriptstyle \mu_Z^{T'}} \\ \mathrm{A}[T] & \xrightarrow[\mu_Z^T]{} & \mathrm{A}[Z] \end{array}$$

commute. In the first diagram, $\sigma : Z \to Z'$ is any morphism in $\boldsymbol{\alpha}$-$\overline{\mathsf{Cox}}$, T covers Z, $T' = \sigma_*(T)$ (and covers Z'). In the second diagram, U covers T and T' which in turn cover Z.

PROOF. The poset of face-types is a Boolean poset which is strongly connected, and the result follows as in the proof of Proposition 2.23. In this case, the argument is also easy to see directly. □

The first diagram in (8.43) is an instance of the naturality axiom. We refer to the second diagram in (8.43) as the *covering associativity axiom*. Note very carefully: It is a square instead of a triangle.

Exercise 8.45. Phrase Proposition 8.44 in terms of top-nested faces using product components $\mu_{A,C}^{F,C}$ only for $A \lessdot F \leq C$. See the last picture in Figure 8.3. Do the same in terms of top-lunes using $\mu_{\mathrm{L}}^{\mathrm{M}}$ only for $\mathrm{L} \lessdot' \mathrm{M}$.

8.5.5. Coxeter coalgebras in terms of top-nested faces. We now turn to the dual notion. A *Coxeter $\mathcal{A}$-coalgebra* is a Coxeter $\mathcal{A}$-space C equipped with linear maps

$$\Delta_{A,C}^{F,C} : \mathrm{C}[A, C] \to \mathrm{C}[F, C],$$

one for each $A \leq F \leq C$, such that the diagrams

$$(8.44a)\qquad \begin{array}{ccc} \mathrm{C}[F,C] & \xrightarrow{\beta_{G,D,F,C}} & \mathrm{C}[G,D] \\ {\scriptstyle \Delta_{A,C}^{F,C}}\uparrow & & \uparrow{\scriptstyle \Delta_{B,D}^{G,D}} \\ \mathrm{C}[A,C] & \xrightarrow[\beta_{B,D,A,C}]{} & \mathrm{C}[B,D] \end{array} \qquad \begin{array}{ccc} \mathrm{C}[F,C] & \xrightarrow{w} & \mathrm{C}[F',C'] \\ {\scriptstyle \Delta_{A,C}^{F,C}}\uparrow & & \uparrow{\scriptstyle \Delta_{A',C'}^{F',C'}} \\ \mathrm{C}[A,C] & \xrightarrow[w]{} & \mathrm{C}[A',C'] \end{array}$$

$$(8.44b)\qquad \begin{array}{ccc} & \mathrm{C}[F,C] & \\ {\scriptstyle \Delta_{A,C}^{F,C}}\nearrow & & \searrow{\scriptstyle \Delta_{F,C}^{G,C}} \\ \mathrm{C}[A,C] & \xrightarrow[\Delta_{A,C}^{G,C}]{} & \mathrm{C}[G,C] \end{array} \qquad\qquad (\mathrm{C}[A,C] \xrightarrow{\Delta_{A,C}^{A,C}} \mathrm{C}[A,C]) = \mathrm{id}$$

commute. (The conventions are as in the Coxeter algebra case.)

We denote a Coxeter $\mathcal{A}$-coalgebra by a pair (C, Δ), or simply by C with Δ understood. The maps $\Delta_{A,C}^{F,C}$ are coproduct components of C. We refer to diagrams (8.44a) as the *support-naturality axiom* and *type-naturality axiom*, respectively, and to diagrams (8.44b) as the *coassociativity axiom* and *counitality axiom*, respectively.

A morphism $f : \mathrm{C} \to \mathrm{D}$ of Coxeter $\mathcal{A}$-coalgebras is a map of Coxeter $\mathcal{A}$-spaces such that for each $A \leq F \leq C$, the diagram

$$
\begin{array}{ccc}
\mathrm{C}[F,C] & \xrightarrow{f_{F,C}} & \mathrm{D}[F,C] \\
{\scriptstyle \Delta_{A,C}^{F,C}}\uparrow & & \uparrow{\scriptstyle \Delta_{A,C}^{F,C}} \\
\mathrm{C}[A,C] & \xrightarrow[f_{A,C}]{} & \mathrm{D}[A,C]
\end{array}
\tag{8.45}
$$

commutes.

8.5.6. Coxeter coalgebras in terms of top-lunes. Equivalently, a Coxeter $\mathcal{A}$-coalgebra C can be formulated in terms of top-lunes by using linear maps

$$\Delta_{\mathrm{L}}^{\mathrm{M}} : \mathrm{C}[\mathrm{L}] \to \mathrm{C}[\mathrm{M}]$$

for $\mathrm{L} \leq' \mathrm{M}$. This would be the dual of Proposition 8.41. Similarly, for $\mathrm{L} \leq \mathrm{M}$ and lune P with base $\mathrm{b}(\mathrm{L})$ and case $\mathrm{b}(\mathrm{M})$, there is a linear map $\Delta_{\mathrm{L}}^{\mathrm{M}}(\mathrm{P})$ defined by the diagram dual to (8.40). For $\mathrm{L} \leq' \mathrm{M}$, we have $\Delta_{\mathrm{L}}^{\mathrm{M}}(\mathrm{L}/\mathrm{M}) = \Delta_{\mathrm{L}}^{\mathrm{M}}$.

8.5.7. Coxeter coalgebras in terms of face-types. Alternatively, working in the category of Coxeter $\boldsymbol{\alpha}$-spaces, we obtain the following.

A *Coxeter $\boldsymbol{\alpha}$-coalgebra* is a Coxeter $\boldsymbol{\alpha}$-space C equipped with linear maps

$$\Delta_{Z}^{T} : \mathrm{C}[Z] \to \mathrm{C}[T],$$

one for each $Z \leq T$, such that the diagrams

$$
\begin{array}{ccc}
\mathrm{C}[T] & \xrightarrow{\sigma} & \mathrm{C}[T'] \\
{\scriptstyle \Delta_{Z}^{T}}\uparrow & & \uparrow{\scriptstyle \Delta_{Z'}^{T'}} \\
\mathrm{C}[Z] & \xrightarrow[\sigma]{} & \mathrm{C}[Z']
\end{array}
\qquad
\begin{array}{ccc}
 & \mathrm{C}[T] & \\
{\scriptstyle \Delta_{Z}^{T}}\nearrow & & \searrow{\scriptstyle \Delta_{T}^{U}} \\
\mathrm{C}[Z] & \xrightarrow[\Delta_{Z}^{U}]{} & \mathrm{C}[U]
\end{array}
\qquad
(\mathrm{C}[Z] \xrightarrow{\Delta_{Z}^{Z}} \mathrm{C}[Z]) = \mathrm{id}
\tag{8.46}
$$

commute. (The conventions are as in the Coxeter algebra case.) We refer to diagrams (8.46) as the *naturality axiom*, *coassociativity axiom*, *counitality axiom*, respectively.

A morphism $f : \mathrm{C} \to \mathrm{D}$ of Coxeter $\boldsymbol{\alpha}$-coalgebras is a map of Coxeter $\boldsymbol{\alpha}$-spaces such that for each $Z \leq T$, the diagram

$$
\begin{array}{ccc}
\mathrm{C}[T] & \xrightarrow{f_{T}} & \mathrm{D}[T] \\
{\scriptstyle \Delta_{Z}^{T}}\uparrow & & \uparrow{\scriptstyle \Delta_{Z}^{T}} \\
\mathrm{C}[Z] & \xrightarrow[f_{Z}]{} & \mathrm{D}[Z]
\end{array}
\tag{8.47}
$$

commutes.

8.5.8. Covering coassociativity axiom. The dual of Proposition 8.44 is stated below.

Proposition 8.46. *A Coxeter $\boldsymbol{\alpha}$-coalgebra is a Coxeter $\boldsymbol{\alpha}$-space* C *equipped with linear maps*

$$\Delta_Z^T : \mathrm{C}[Z] \to \mathrm{C}[T]$$

whenever T covers Z, such that the diagrams

$$\begin{array}{ccc} \mathrm{C}[T] & \xrightarrow{\sigma} & \mathrm{C}[T'] \\ {\scriptstyle \Delta_Z^T}\uparrow & & \uparrow{\scriptstyle \Delta_{Z'}^{T'}} \\ \mathrm{C}[Z] & \xrightarrow[\sigma]{} & \mathrm{C}[Z'] \end{array} \qquad\qquad \begin{array}{ccc} \mathrm{C}[T'] & \xrightarrow{\Delta_{T'}^U} & \mathrm{C}[U] \\ {\scriptstyle \Delta_Z^{T'}}\uparrow & & \uparrow{\scriptstyle \Delta_T^U} \\ \mathrm{C}[Z] & \xrightarrow[\Delta_Z^T]{} & \mathrm{C}[T] \end{array} \tag{8.48}$$

commute. (The conventions are as in the Coxeter algebra case.)

The first diagram in (8.48) is an instance of the naturality axiom. We refer to the second diagram in (8.48) as the *covering coassociativity axiom*. It is a square instead of a triangle.

Exercise 8.47. Phrase Proposition 8.46 in terms of top-nested faces using $\Delta_{A,C}^{F,C}$ only for $A \lessdot F \leq C$. Do the same in terms of top-lunes using $\Delta_{\mathrm{L}}^{\mathrm{M}}$ only for $\mathrm{L} \lessdot' \mathrm{M}$.

8.5.9. Binary operations vs higher operations. Let A be a Coxeter $\boldsymbol{\alpha}$-algebra (formulated in terms of face-types). One may view its product components μ_Z^T as "higher operations" and μ_Z^T for $Z \lessdot T$ as "binary operations". In other words, a Coxeter $\boldsymbol{\alpha}$-algebra is defined using higher operations and Proposition 8.44 reformulates it using binary operations. A similar remark applies to a Coxeter $\boldsymbol{\alpha}$-coalgebra and its reformulation in Proposition 8.46. For Coxeter (co)algebras in terms of top-nested faces and top-lunes, see Exercises 8.45 and 8.47. For further results of this nature, see Exercises 8.64, 8.67, 8.74.

The binary operation viewpoint originates from the traditional approach to associative algebras and Hopf algebras; see the discussion in Section 8.14.12.

8.6. Coxeter bialgebras

We build on the discussion in Section 8.5 and define Coxeter bialgebras. These are analogues of Coxeter bimonoids from Section 2.5. However, in contrast to the bimonoid axiom which is set-theoretic, the bialgebra axiom is linear and involves a sum.

8.6.1. Coxeter bialgebras in terms of top-nested faces. Let V be a Coxeter $\mathcal{A}$-space, and suppose we are given faces

$$A \leq F_0 \leq G_0 \leq C \quad \text{and} \quad A \leq F_0' \leq G_0' \leq C.$$

Define a linear map

$$\overline{\lambda}_{F_0,G_0,F_0',G_0'} : \mathrm{V}[G_0, C] \to \mathrm{V}[G_0', C] \tag{8.49}$$

as follows. (The faces A and C are suppressed in the map notation.) See the first picture in Figure 8.5. Vertices with the same color have the same type.

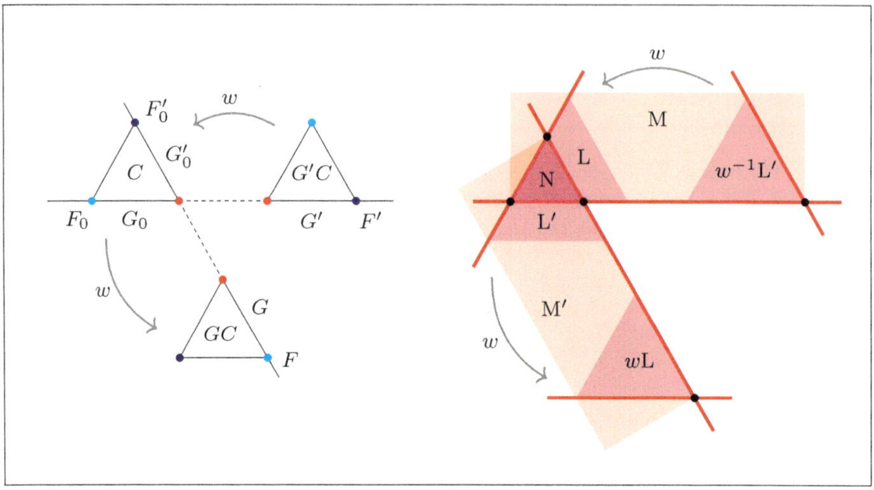

FIGURE 8.5. Bialgebra axiom.

Also, $A = O$ in the picture. The dotted lines are not edges; they indicate that G_0 and G', and G'_0 and G have the same support.

For each $A \leq F' \leq G'$ which satisfies

$$F_0F' = G_0,\ F'F_0 = G',\ \mathrm{t}(F') = \mathrm{t}(F'_0),\ \mathrm{t}(G') = \mathrm{t}(G'_0),$$

form the composite map

$$\mathrm{V}[G_0, C] \xrightarrow{\beta_{G',G'C,G_0,C}} \mathrm{V}[G', G'C] \xrightarrow{w(G'C)=C} \mathrm{V}[G'_0, C]. \tag{8.50a}$$

This is the map transferred by G'. The map (8.49) is obtained by summing over all allowed choices of (F', G'). It may happen that no choice works, in which case the map is 0.

Alternatively, for each $A \leq F \leq G$ which satisfies

$$F'_0F = G'_0,\ FF'_0 = G,\ \mathrm{t}(F) = \mathrm{t}(F_0),\ \mathrm{t}(G) = \mathrm{t}(G_0),$$

form the composite map

$$\mathrm{V}[G_0, C] \xrightarrow{w(C)=GC} \mathrm{V}[G, GC] \xrightarrow{\beta_{G'_0,C,G,GC}} \mathrm{V}[G'_0, C]. \tag{8.50b}$$

This is the map cotransferred by G. By the support-type axiom, (8.50b) equals (8.50a). So summing over all allowed choices of (F, G) also yields the map (8.49).

Example 8.48 (Rank three). The rank-three braid arrangement in Figure 1.5 is redrawn below.

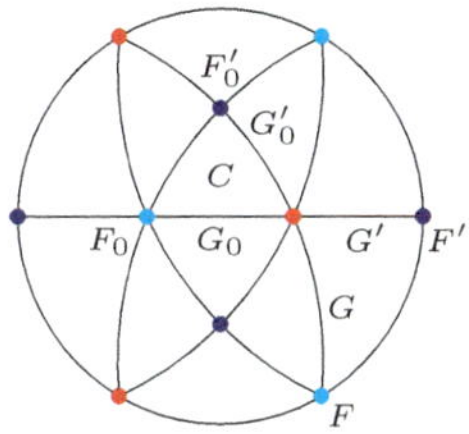

It shows an actual manifestation of the first picture in Figure 8.5. Interestingly, all three red vertices in that picture coincide in the picture above. Further, $\overline{\lambda}_{F_0,G_0,F'_0,G'_0}$ (with $A = O$) is the map transferred by G', or cotransferred by G. (There are no other choices for F, G, F', G' other than the ones shown above.)

Now consider the rank-three arrangement of type B. We work with the left front view in Figure 1.6.

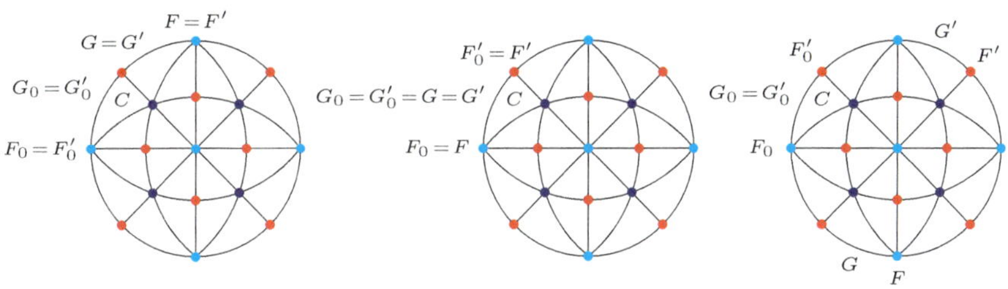

For F_0, G_0, F'_0, G'_0, C (with $A = O$) as in the first picture, there is only one choice of F, G, F', G' as shown. In the next two pictures, F'_0 is instead taken to be the red vertex of C. This yields two choices of F, G, F', G' which are, respectively, shown in these two pictures. □

Exercise 8.49. Check: For a Coxeter $\mathcal{A}$-space V, for faces F_0, G_0, F'_0, G'_0 all between A and C,

$$\left(\overline{\lambda}_{F_0,G_0,F'_0,G'_0} : \mathrm{V}[G_0, C] \to \mathrm{V}[G'_0, C]\right) = \sum \sigma.$$

The sum is over all morphisms $\sigma : (G_0, C) \to (G'_0, C)$ local to A (written in normal form as $\sigma = \beta w = w\beta$) satisfying either of the two equivalent sets of conditions

$$(8.51)\qquad \begin{aligned} (wF_0)F'_0 = wG_0 \quad &\text{and} \quad F'_0(wF_0) = G'_0, \\ F_0(w^{-1}F'_0) = G_0 \quad &\text{and} \quad (w^{-1}F'_0)F_0 = w^{-1}G'_0. \end{aligned}$$

Note: For any such σ, either of the two equivalent sets of conditions

$$\begin{aligned} (wG_0)C = wC \quad &\text{and} \quad G'_0(wC) = C, \\ G_0(w^{-1}C) = C \quad &\text{and} \quad (w^{-1}G'_0)C = w^{-1}C \end{aligned}$$

automatically holds.

Some special cases of (8.49) are highlighted in the exercises below.

Exercise 8.50. Check: For a Coxeter $\mathcal{A}$-space V, for faces F_0 and F'_0 both between A and C,

$$\left(\overline{\lambda}_{F_0,F_0,F'_0,F'_0} : \mathrm{V}[F_0, C] \to \mathrm{V}[F'_0, C]\right) = \sum \sigma,$$

with the sum over all morphisms $\sigma : (F_0, C) \to (F'_0, C)$ local to A.

Exercise 8.51. Check: For a Coxeter $\mathcal{A}$-space V, for faces G_0, F'_0, G'_0 all between A and C,

$$(\overline{\lambda}_{A,G_0,F'_0,G'_0} : \mathrm{V}[G_0, C] \to \mathrm{V}[G'_0, C]) = \begin{cases} \mathrm{id} & \text{if } F'_0 = G'_0 = G_0, \\ 0 & \text{otherwise.} \end{cases}$$

More precisely, in the first alternative, only $\sigma = \mathrm{id}$ contributes to $\overline{\lambda}$, while in the second alternative, no σ works. By symmetry, there is a similar statement for $\overline{\lambda}_{F_0,G_0,A,G'_0}$. Compare with Exercise 2.74.

Exercise 8.52. For a Coxeter $\mathcal{A}$-space V, for faces F_0 and F'_0 both between A and C, make explicit the maps

$$\overline{\lambda}_{C,C,F'_0,C} : \mathrm{V}[C, C] \to \mathrm{V}[C, C] \quad \text{and} \quad \overline{\lambda}_{F_0,C,C,C} : \mathrm{V}[C, C] \to \mathrm{V}[C, C].$$

(They involve the top-lunes $\mathrm{s}(F'_0/A, C/A)$ and $\mathrm{s}(F_0/A, C/A)$, respectively.) Obtain $\overline{\lambda}_{C,C,C,C}$ as a special case of either.

We are now ready to define a Coxeter bialgebra. It combines the notion of a Coxeter algebra and a Coxeter coalgebra.

A *Coxeter $\mathcal{A}$-bialgebra* is a triple $(\mathrm{H}, \mu, \Delta)$, where H is a Coxeter $\mathcal{A}$-space, (H, μ) is a Coxeter $\mathcal{A}$-algebra, (H, Δ) is a Coxeter $\mathcal{A}$-coalgebra, and for faces $A \leq F_0 \leq C$ and $A \leq F'_0 \leq C$, the diagram

(8.52)
$$\begin{CD}
\mathrm{H}[F_0, C] @>{\mu^{F_0,C}_{A,C}}>> \mathrm{H}[A, C] @>{\Delta^{F'_0,C}_{A,C}}>> \mathrm{H}[F'_0, C] \\
@V{(\Delta^{G_0,C}_{F_0,C})}VV @. @AA{(\mu^{G'_0,C}_{F'_0,C})}A \\
\displaystyle\bigoplus_{G_0 \colon F_0 \leq G_0 \leq C} \mathrm{H}[G_0, C] @>>{(\overline{\lambda}_{F_0,G_0,F'_0,G'_0})}> @. \displaystyle\bigoplus_{G'_0 \colon F'_0 \leq G'_0 \leq C} \mathrm{H}[G'_0, C]
\end{CD}$$

commutes, with $\overline{\lambda}$ as in (8.49). We refer to diagram (8.52) as the *bialgebra axiom*.

A morphism of Coxeter $\mathcal{A}$-bialgebras is a map of the underlying Coxeter $\mathcal{A}$-spaces which is a morphism of the underlying Coxeter $\mathcal{A}$-algebras and Coxeter $\mathcal{A}$-coalgebras.

The data present in a Coxeter $\mathcal{A}$-bialgebra is shown in Summary 8.1. Compare and contrast with Summary 2.1.

Lemma 8.53. *The bialgebra axiom* (8.52) *can be expressed as follows. For faces $A \leq F_0 \leq C$ and $A \leq F'_0 \leq C$, the diagram*

(8.53)
$$\begin{array}{ccc}
\mathrm{H}[F_0, C] & \xrightarrow{\sum_w \mu\beta\Delta w} & \mathrm{H}[F'_0, C] \\
& {\scriptstyle\mu} \searrow \quad \nearrow {\scriptstyle\Delta} & \\
& \mathrm{H}[A, C] &
\end{array}$$

commutes. The sum is over all $w \in W$ satisfying $w(A) = A$ and either of the equivalent conditions

(8.54)
$$\begin{aligned} (wF_0)C = wC \quad &\textit{and} \quad F'_0(wC) = C, \\ F_0(w^{-1}C) = C \quad &\textit{and} \quad (w^{-1}F'_0)C = w^{-1}C. \end{aligned}$$

A Coxeter $\mathcal{A}$-bialgebra H consists of linear maps

$$\beta_{G,D,F,C} : \mathrm{H}[F,C] \to \mathrm{H}[G,D], \qquad w : \mathrm{H}[F,C] \to \mathrm{H}[wF,wC],$$
$$\mu_{A,C}^{F,C} : \mathrm{H}[F,C] \to \mathrm{H}[A,C], \qquad \Delta_{A,C}^{F,C} : \mathrm{H}[A,C] \to \mathrm{H}[F,C],$$

which satisfy

- support-support, support-type, type-type, support-unitality, type-unitality axioms (8.20a) + (8.20b),
- support-naturality and type-naturality axioms (8.34a) + (8.44a),
- (co)associativity and (co)unitality axioms (8.34b) + (8.44b),
- bialgebra axiom (8.52).

There are a total of 14 axioms.

SUMMARY 8.1. Coxeter bialgebra in terms of top-nested faces.

The composite map $\mu\beta\Delta w$ *is*

$$\mathrm{H}[F_0,C] \xrightarrow{w} \mathrm{H}[wF_0,wC] \xrightarrow{\Delta} \mathrm{H}[(wF_0)F_0',wC] \xrightarrow{\beta} \mathrm{H}[F_0'(wF_0),C] \xrightarrow{\mu} \mathrm{H}[F_0',C].$$

This is the same as the composite map $w\mu\beta\Delta$*, namely,*

$$\mathrm{H}[F_0,C] \xrightarrow{\Delta} \mathrm{H}[F_0(w^{-1}F_0'),C] \xrightarrow{\beta} \mathrm{H}[(w^{-1}F_0')F_0,w^{-1}C]$$
$$\xrightarrow{\mu} \mathrm{H}[w^{-1}F_0',w^{-1}C] \xrightarrow{w} \mathrm{H}[F_0',C].$$

PROOF. Observe: The bialgebra axiom (8.52) can be rewritten as

$$\begin{array}{ccc} \mathrm{H}[F_0,C] & \xrightarrow{\sum \mu\sigma\Delta} & \mathrm{H}[F_0',C]. \\ & {\scriptstyle\mu}\searrow \quad \nearrow{\scriptstyle\Delta} & \\ & \mathrm{H}[A,C] & \end{array}$$

The sum is over all composable Δ, σ, μ, with σ satisfying the conditions in Exercise 8.49. Each summand can be understood using the diagram below.

$$\begin{array}{ccccccc} & & \mathrm{H}[w^{-1}G_0',w^{-1}C] & \xrightarrow{\mu} & \mathrm{H}[w^{-1}F_0',w^{-1}C] & & \\ & {\scriptstyle\beta}\nearrow & & \searrow{\scriptstyle w} & & \searrow{\scriptstyle w} & \\ \mathrm{H}[F_0,C] \xrightarrow{\Delta} & \mathrm{H}[G_0,C] & \xrightarrow{\sigma} & \mathrm{H}[G_0',C] & \xrightarrow{\mu} & \mathrm{H}[F_0',C] & \\ {\scriptstyle w}\searrow & & {\scriptstyle w}\searrow & \nearrow{\scriptstyle\beta} & & & \\ & \mathrm{H}[wF_0,wC] & \xrightarrow{\Delta} & \mathrm{H}[wG_0,wC] & & & \end{array}$$

The point is that given A, F_0, F_0', C, knowing $w \in W$ with $w(A) = A$ determines this entire diagram. One may then check that the conditions on w are captured by (8.54). The composite map $\mu\beta\Delta w$ is obtained by following the lower path in the above diagram. Similarly, following the upper path in the diagram yields the map $w\mu\beta\Delta$. $\square$

Equivalently:

Lemma 8.54. *For faces $A \leq F_0 \leq C$ and $A \leq F_0' \leq C$, the bialgebra axiom* (8.52) *can be expressed as the hexagon*

$$
\begin{array}{ccccc}
\mathrm{H}[F_0, C] & \xrightarrow{\ \mu\ } & \mathrm{H}[A, C] & \xrightarrow{\ \Delta\ } & \mathrm{H}[F_0', C] \\
\Big\downarrow{\scriptstyle w} & & & & \Big\uparrow{\scriptstyle \mu} \\
\mathrm{H}[wF_0, wC] & \xrightarrow[\Delta]{} & \mathrm{H}[(wF_0)F_0', wC] & \xrightarrow[\beta]{} & \mathrm{H}[F_0'(wF_0), C],
\end{array}
\tag{8.55a}
$$

or equivalently, as the hexagon

$$
\begin{array}{ccccc}
\mathrm{H}[F_0, C] & \xrightarrow{\ \mu\ } & \mathrm{H}[A, C] & \xrightarrow{\ \Delta\ } & \mathrm{H}[F_0', C] \\
\Big\downarrow{\scriptstyle \Delta} & & & & \Big\uparrow{\scriptstyle w} \\
\mathrm{H}[F_0(w^{-1}F_0'), C] & \xrightarrow[\beta]{} & \mathrm{H}[(w^{-1}F_0')F_0, w^{-1}C] & \xrightarrow[\mu]{} & \mathrm{H}[w^{-1}F_0', w^{-1}C],
\end{array}
\tag{8.55b}
$$

with the understanding that we sum over $w \in W$ satisfying $w(A) = A$ and conditions (8.54).

In the first hexagon, the morphism w appears on the left, while in the second hexagon, it appears on the right. To indicate that w is a variable, we have marked it in red. Note: The bialgebra axiom (8.52) can also be expressed as a hexagon in the forms

$$
\Delta\mu = \sum_{w} \mu\beta w\Delta \quad \text{and} \quad \Delta\mu = \sum_{w} \mu w\beta\Delta, \tag{8.55c}
$$

with w appearing in the second and third positions, respectively.

Lemma 8.55. *For defining a Coxeter $\mathcal{A}$-bialgebra, it suffices to require the bialgebra axiom* (8.52) *only for faces $A \lessdot F_0 \leq C$ and $A \lessdot F_0' \leq C$.*

Proof. The argument is similar to that for Lemma 2.27. Suppose F_0 and F_0' are faces between A and C, and $A \leq K \leq F_0'$. Let $u, v \in W$ be variables such that the two hexagons in the diagram below commute in the sense of (8.55a). In particular, $u(A) = A$ and $v(K) = K$.

(8.56)

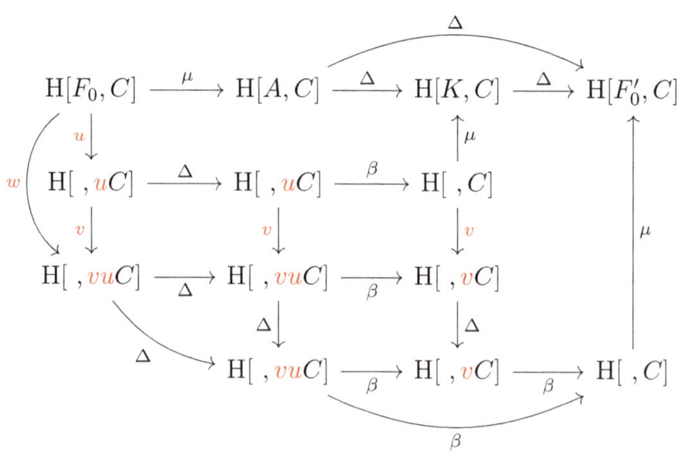

The first coordinates have been suppressed for convenience. By tagging on the three squares to the two hexagons, the outermost diagram becomes an instance of (8.55a) with $w = vu$. Conversely, given (8.55a) and a K between A and F_0', the diagram can be filled as above for a unique u and v. We omit the details. Compare with diagram (2.40). $\square$

The above result relates to Exercises 8.45 and 8.47 and the general discussion in Section 8.5.9.

8.6.2. Degeneracies of the bialgebra axiom. We now point out some special cases of the bialgebra axiom.

Exercise 8.56. Check: If either $F_0 = A$ or $F_0' = A$, then the bialgebra axiom (8.52) becomes a tautology, namely, $\Delta_{A,C}^{F_0',C} = \Delta_{A,C}^{F_0',C}$ or $\mu_{A,C}^{F_0,C} = \mu_{A,C}^{F_0,C}$. In more detail, for instance, when $F_0' = A$, the axiom becomes

$$\begin{array}{ccccc} \mathrm{H}[F_0, C] & \xrightarrow{\mu_{A,C}^{F_0,C}} & \mathrm{H}[A, C] & \xrightarrow{\mathrm{id}} & \mathrm{H}[A, C] \\ {\scriptstyle \mathrm{id}}\downarrow & & & & \uparrow{\scriptstyle \mu_{A,C}^{F_0,C}} \\ \mathrm{H}[F_0, C] & & \xrightarrow[\mathrm{id}]{\hspace{4em}} & & \mathrm{H}[F_0, C], \end{array}$$

with $G_0 = G_0' = F_0$. (Use Exercise 8.51.)

We now turn to the special case when either F_0 or F_0' equals C.

Exercise 8.57. Let H be a Coxeter $\mathcal{A}$-bialgebra. Check: For faces $A \le G \le C$, the first diagram below commutes, with the sum over $w \in W$ such that $w(A) = A$ and $G(wC) = C$. Similarly, for faces $A \le F \le C$, the second diagram below commutes, with the sum over $w \in W$ such that $w(A) = A$ and $F(w^{-1}C) = C$.

$$\begin{array}{ccccc} \mathrm{H}[C, C] & \xrightarrow{\mu} & \mathrm{H}[A, C] & \xrightarrow{\Delta} & \mathrm{H}[G, C] \\ {\scriptstyle w}\downarrow & & & & \uparrow{\scriptstyle \mu} \\ \mathrm{H}[wC, wC] & & \xrightarrow[\beta]{\hspace{4em}} & & \mathrm{H}[C, C] \end{array} \qquad \begin{array}{ccccc} \mathrm{H}[F, C] & \xrightarrow{\mu} & \mathrm{H}[A, C] & \xrightarrow{\Delta} & \mathrm{H}[C, C] \\ {\scriptstyle \Delta}\downarrow & & & & \uparrow{\scriptstyle w} \\ \mathrm{H}[C, C] & & \xrightarrow[\beta]{\hspace{4em}} & & \mathrm{H}[w^{-1}C, w^{-1}C] \end{array}$$

In particular, for any $A \le C$, the diagram below commutes, with the sum over $w \in W$ such that $w(A) = A$.

$$\begin{array}{ccccc} & & \mathrm{H}[w^{-1}C, w^{-1}C] & & \\ & \overset{\beta}{\nearrow} & & \overset{w}{\searrow} & \\ \mathrm{H}[C, C] & \xrightarrow{\mu} & \mathrm{H}[A, C] & \xrightarrow{\Delta} & \mathrm{H}[C, C] \\ & \underset{w}{\searrow} & & \underset{\beta}{\nearrow} & \\ & & \mathrm{H}[wC, wC] & & \end{array}$$

(See the related Exercise 8.52.)

8.6.3. Coxeter bialgebras in terms of top-lunes. We now formulate Coxeter bialgebras in terms of top-lunes by building on the description of Coxeter spaces in Proposition 8.33, Coxeter algebras in Proposition 8.41, and the dual description for Coxeter coalgebras. We employ notations of Section 1.2.4.

A Coxeter $\mathcal{A}$-bialgebra H consists of linear maps

$$\mu_{\mathrm{L}}^{\mathrm{M}} : \mathrm{H}[\mathrm{M}] \to \mathrm{H}[\mathrm{L}], \quad w : \mathrm{H}[\mathrm{L}] \to \mathrm{H}[w\mathrm{L}], \quad \Delta_{\mathrm{L}}^{\mathrm{M}} : \mathrm{H}[\mathrm{L}] \to \mathrm{H}[\mathrm{M}],$$

which satisfy

- type-type and type-unitality axioms (8.23),
- type-naturality, (co)associativity, (co)unitality axioms (8.37) + its dual,
- bialgebra axiom (8.59).

There are a total of 9 axioms.

SUMMARY 8.2. Coxeter bialgebra in terms of top-lunes.

Let V be a Coxeter $\mathcal{A}$-space, and suppose we are given top-lunes

$$\mathrm{N} \leq' \mathrm{L} \leq' \mathrm{M} \quad \text{and} \quad \mathrm{N} \leq' \mathrm{L}' \leq' \mathrm{M}'.$$

Define a linear map

$$(\overline{\lambda}_{\mathrm{L,M,L',M'}} : \mathrm{V}[\mathrm{M}] \to \mathrm{V}[\mathrm{M}']) := \sum w, \tag{8.57}$$

with the sum over $w \in W$ such that w fixes $\mathrm{b}(\mathrm{N})$ pointwise, $w\mathrm{M} = \mathrm{M}'$, and w satisfies either of the two equivalent sets of conditions

$$\begin{aligned} \mathrm{b}(w\mathrm{L}) \cdot \mathrm{N}/\mathrm{L}' = w(\mathrm{L}/\mathrm{M}) \quad &\text{and} \quad \mathrm{b}(\mathrm{L}') \cdot w(\mathrm{N}/\mathrm{L}) = \mathrm{L}'/\mathrm{M}', \\ \mathrm{b}(\mathrm{L}) \cdot w^{-1}(\mathrm{N}/\mathrm{L}') = \mathrm{L}/\mathrm{M} \quad &\text{and} \quad \mathrm{b}(w^{-1}\mathrm{L}') \cdot \mathrm{N}/\mathrm{L} = w^{-1}(\mathrm{L}'/\mathrm{M}'). \end{aligned} \tag{8.58}$$

(The top-lune N is suppressed in the map notation.) See the second picture in Figure 8.5. The top-lune N is shown as a chamber, so $\mathrm{b}(\mathrm{N}) = \perp$ in the picture.

Proposition 8.58. *A Coxeter $\mathcal{A}$-bialgebra is a triple* $(\mathrm{H}, \mu, \Delta)$, *where* H *is a Coxeter $\mathcal{A}$-space,* (H, μ) *is a Coxeter $\mathcal{A}$-algebra,* (H, Δ) *is a Coxeter $\mathcal{A}$-coalgebra, and for top-lunes* $\mathrm{N} \leq' \mathrm{L}$ *and* $\mathrm{N} \leq' \mathrm{L}'$, *the diagram*

$$\begin{array}{ccccc} \mathrm{H}[\mathrm{L}] & \xrightarrow{\mu_{\mathrm{N}}^{\mathrm{L}}} & \mathrm{H}[\mathrm{N}] & \xrightarrow{\Delta_{\mathrm{N}}^{\mathrm{L}'}} & \mathrm{H}[\mathrm{L}'] \\ {\scriptstyle (\Delta_{\mathrm{L}}^{\mathrm{M}})} \downarrow & & & & \uparrow {\scriptstyle (\mu_{\mathrm{L}'}^{\mathrm{M}'})} \\ \bigoplus_{\mathrm{M}:\, \mathrm{L} \leq' \mathrm{M}} \mathrm{H}[\mathrm{M}] & & \xrightarrow[(\overline{\lambda}_{\mathrm{L,M,L',M'}})]{} & & \bigoplus_{\mathrm{M}':\, \mathrm{L}' \leq' \mathrm{M}'} \mathrm{H}[\mathrm{M}'] \end{array} \tag{8.59}$$

commutes, with $\overline{\lambda}$ *as in* (8.57).

PROOF. We translate between diagrams (8.52) and (8.59) using diagrams (8.26), (8.39) and its dual. Conditions (8.51) translate to conditions (8.58). □

We refer to diagram (8.59) as the *bialgebra axiom.*

The data present in a Coxeter $\mathcal{A}$-bialgebra is shown in Summary 8.2. Compare and contrast with Summary 8.1 as well as with Summary 2.2.

Lemma 8.59. *For top-lunes* $\mathrm{N} \leq' \mathrm{L}$ *and* $\mathrm{N} \leq' \mathrm{L}'$, *the bialgebra axiom* (8.59) *can be expressed as*

$$\begin{array}{ccccc} \mathrm{H}[\mathrm{L}] & \xrightarrow{\ \mu\ } & \mathrm{H}[\mathrm{N}] & \xrightarrow{\ \Delta\ } & \mathrm{H}[\mathrm{L}'] \\ {\scriptstyle w}\big\downarrow & & & & \big\uparrow{\scriptstyle \mu} \\ \mathrm{H}[w\mathrm{L}] & \xrightarrow[\Delta]{} & \mathrm{H}[\mathrm{b}(w\mathrm{L}) \cdot \mathrm{L}'] & = & \mathrm{H}[\mathrm{b}(\mathrm{L}') \cdot w\mathrm{L}], \end{array} \tag{8.60a}$$

or equivalently, as

$$\begin{array}{ccccc} \mathrm{H}[\mathrm{L}] & \xrightarrow{\ \mu\ } & \mathrm{H}[\mathrm{N}] & \xrightarrow{\ \Delta\ } & \mathrm{H}[\mathrm{L}'] \\ {\scriptstyle \Delta}\big\downarrow & & & & \big\uparrow{\scriptstyle w} \\ \mathrm{H}[\mathrm{b}(\mathrm{L}) \cdot w^{-1}\mathrm{L}'] & = & \mathrm{H}[\mathrm{b}(w^{-1}\mathrm{L}') \cdot \mathrm{L}] & \xrightarrow[\mu]{} & \mathrm{H}[w^{-1}\mathrm{L}']. \end{array} \tag{8.60b}$$

The sum is over all $w \in W$ *fixing* $\mathrm{b}(\mathrm{N})$ *pointwise, and satisfying either of the two equivalent sets of conditions*

$$\begin{array}{ccc} \mathrm{b}(w\mathrm{L}) \cdot \mathrm{N} = w\mathrm{L} & \textit{and} & \mathrm{b}(\mathrm{L}') \cdot w\mathrm{N} = \mathrm{L}', \\ \mathrm{b}(\mathrm{L}) \cdot w^{-1}\mathrm{N} = \mathrm{L} & \textit{and} & \mathrm{b}(w^{-1}\mathrm{L}') \cdot \mathrm{N} = w^{-1}\mathrm{L}'. \end{array} \tag{8.61}$$

Note: For any such w, either of the two equivalent sets of conditions

$$w\mathrm{L} \leq' \mathrm{b}(w\mathrm{L}) \cdot \mathrm{L}', \quad \mathrm{b}(w\mathrm{L}) \cdot \mathrm{L}' = \mathrm{b}(\mathrm{L}') \cdot w\mathrm{L}, \quad \mathrm{L}' \leq' \mathrm{b}(\mathrm{L}') \cdot w\mathrm{L},$$

$$\mathrm{L} \leq' \mathrm{b}(\mathrm{L}) \cdot w^{-1}\mathrm{L}', \quad \mathrm{b}(\mathrm{L}) \cdot w^{-1}\mathrm{L}' = \mathrm{b}(w^{-1}\mathrm{L}') \cdot \mathrm{L}, \quad w^{-1}\mathrm{L}' \leq' \mathrm{b}(w^{-1}\mathrm{L}') \cdot \mathrm{L}$$

automatically holds.

PROOF. Translate from the hexagons (8.55a) and (8.55b). Conditions (8.54) translate to conditions (8.61). Note: Morphisms labeled β are now replaced by equalities. □

Exercise 8.60. Express the bialgebra axiom (8.59) in the form

$$\Delta\mu = \sum_{w} \mu w \Delta, \tag{8.62}$$

with w in the middle position. (Translate from either form in (8.55c).)

8.6.4. Coxeter bialgebras in terms of face-types. Let V be a Coxeter $\boldsymbol{\alpha}$-space, and suppose we are given face-types

$$Z \leq T \leq U \quad \text{and} \quad Z \leq T' \leq U'.$$

Define a linear map

$$\big(\overline{\lambda}_{TUT'U'} : \mathrm{V}[U] \to \mathrm{V}[U']\big) = \sum_{\substack{\sigma\colon U \to U' \\ \text{local to } Z \\ \text{biprojects } T \text{ and } T'}} \sigma. \tag{8.63}$$

(The face-type Z is suppressed in the map notation.) This map arises by taking $C := C_r$ in the discussion in Section 8.6.1 and replacing faces of C_r by their types. The sum is over all morphisms $\sigma : U \to U'$ local to Z (written in normal form as $\sigma = \beta w = w\beta$) satisfying conditions (8.51), where F_0, G_0, F_0', G_0' are faces of C_r of types T, U, T', U', respectively.

One can also understand (8.63) directly by working with the groupoid of biface-types, with biprojection as defined in Section 1.7.4.

Example 8.61 (Rank three). In the rank-three braid arrangement (Figure 1.5), take $Z = \emptyset$, $T =$ cyan, $U =$ cyan-red, $T' =$ blue, $U' =$ blue-red. To figure out $\overline{\lambda}_{TUT'U'}$, we focus on a rank-two flat as shown in Figure 1.7 and redrawn below.

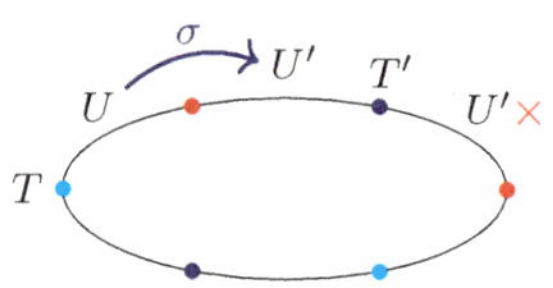

Fix an edge of type U and its vertex of type T. Now look for all edges of type U'. There are two such (with the vertex of type T' common to both), and only one of them satisfies the biprojection condition. Thus, $\overline{\lambda}_{TUT'U'} = \sigma$ as shown. This is equivalent to the type A calculation in Example 8.48.

Now, in the rank-three arrangement of type B (Figure 1.6), take $Z = \emptyset$, $T =$ cyan, $U = U' =$ cyan-red. See the illustrations below.

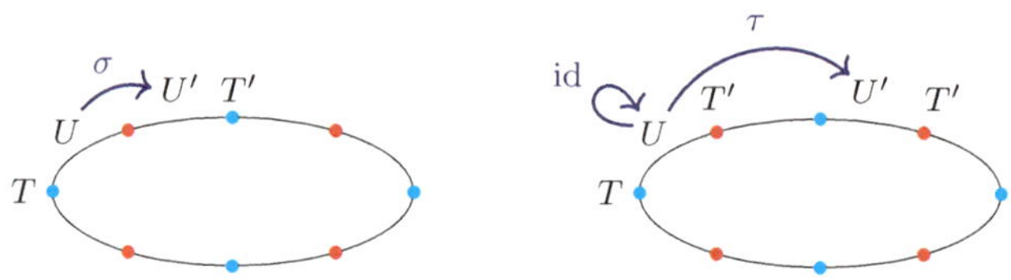

When $T' =$ cyan, $\overline{\lambda}_{TUT'U'} = \sigma$, with σ as in the left picture. When $T' =$ red, $\overline{\lambda}_{TUT'U'} = \mathrm{id} + \tau$, with id, τ as in the right picture. These are equivalent to the type B calculations in Example 8.48. They can also be interpreted as calculations for the dihedral arrangement of 4 lines with $U = U'$ as the chamber-type. □

A *Coxeter $\boldsymbol{\alpha}$-bialgebra* is a triple $(\mathrm{H}, \mu, \Delta)$, where H is a Coxeter $\boldsymbol{\alpha}$-space, (H, μ) is a Coxeter $\boldsymbol{\alpha}$-algebra, (H, Δ) is a Coxeter $\boldsymbol{\alpha}$-coalgebra, and for face-types $Z \leq T$ and $Z \leq T'$, the diagram

$$\begin{array}{ccccc} \mathrm{H}[T] & \xrightarrow{\mu_Z^T} & \mathrm{H}[Z] & \xrightarrow{\Delta_Z^{T'}} & \mathrm{H}[T'] \\ \Big\downarrow{\scriptstyle (\Delta_T^U)} & & & & \Big\uparrow{\scriptstyle (\mu_{T'}^{U'})} \\ \bigoplus_{U:\, T \leq U} \mathrm{H}[U] & & \xrightarrow[(\overline{\lambda}_{TUT'U'})]{} & & \bigoplus_{U':\, T' \leq U'} \mathrm{H}[U'] \end{array} \tag{8.64}$$

commutes, with $\overline{\lambda}$ as in (8.63). We refer to diagram (8.64) as the *bialgebra axiom*.

A morphism of Coxeter $\boldsymbol{\alpha}$-bialgebras is a map of the underlying Coxeter $\boldsymbol{\alpha}$-spaces which is a morphism of the underlying Coxeter $\boldsymbol{\alpha}$-algebras and Coxeter $\boldsymbol{\alpha}$-coalgebras.

The data present in a Coxeter $\boldsymbol{\alpha}$-bialgebra is shown in Summary 8.3.

Exercise 8.62. Check: The categories of Coxeter $\mathcal{A}$-bialgebras and Coxeter $\boldsymbol{\alpha}$-bialgebras are equivalent. This builds on the equivalence in Exercise 8.43.

A Coxeter $\boldsymbol{\alpha}$-bialgebra H consists of linear maps

$$\mu_Z^T : \mathrm{H}[T] \to \mathrm{H}[Z], \quad \sigma : \mathrm{H}[Z] \to \mathrm{H}[Z'], \quad \Delta_Z^T : \mathrm{H}[Z] \to \mathrm{H}[T],$$

which satisfy

- Coxeter $\boldsymbol{\alpha}$-space axioms (8.28),
- naturality, (co)associativity, (co)unitality axioms (8.41) + (8.46),
- bialgebra axiom (8.64).

There are a total of 9 axioms.

SUMMARY 8.3. Coxeter bialgebra in terms of face-types.

8.7. Coxeter q-bialgebras

We now turn to deformed Coxeter bialgebras. They are obtained by deforming the bialgebra axiom by the q-distance function for a scalar q. For $q = -1$, we use the term signed Coxeter $\mathcal{A}$-bialgebra. We also pay attention to the case $q = 0$.

8.7.1. Coxeter q-bialgebras in terms of top-nested faces. For a scalar q, for a Coxeter $\mathcal{A}$-space V, define the map

$$(\overline{\lambda}_q)_{F_0,G_0,F_0',G_0'} : \mathrm{V}[G_0, C] \to \mathrm{V}[G_0', C] \tag{8.65}$$

by replacing β by β_q in either (8.50a) or (8.50b). That is, we either multiply the map (8.50a) by $q^{\mathrm{dist}(C,G'C)}$ or (8.50b) by $q^{\mathrm{dist}(GC,C)}$. Both powers of q are the same. Equivalently, we multiply either map by $q^{l(w)}$, with $l(w)$ taken wrt the chamber C.

For a scalar q, a *Coxeter $\mathcal{A}$-q-bialgebra* is a triple $(\mathrm{H}, \mu, \Delta)$, where H is a Coxeter $\mathcal{A}$-space, (H, μ) is a Coxeter $\mathcal{A}$-algebra, (H, Δ) is a Coxeter $\mathcal{A}$-coalgebra, and for faces $A \leq F_0 \leq C$ and $A \leq F_0' \leq C$, the diagram

$$\begin{array}{ccccc}
\mathrm{H}[F_0, C] & \xrightarrow{\mu_{A,C}^{F_0,C}} & \mathrm{H}[A, C] & \xrightarrow{\Delta_{A,C}^{F_0',C}} & \mathrm{H}[F_0', C] \\
\Big\downarrow{\scriptstyle (\Delta_{F_0,C}^{G_0,C})} & & & & \Big\uparrow{\scriptstyle (\mu_{F_0',C}^{G_0',C})} \\
\displaystyle\bigoplus_{G_0 :\, F_0 \leq G_0 \leq C} \mathrm{H}[G_0, C] & & \xrightarrow[((\overline{\lambda}_q)_{F_0,G_0,F_0',G_0'})]{} & & \displaystyle\bigoplus_{G_0' :\, F_0' \leq G_0' \leq C} \mathrm{H}[G_0', C]
\end{array} \tag{8.66}$$

commutes, with $\overline{\lambda}_q$ as in (8.65). We refer to diagram (8.66) as the *q-bialgebra axiom*.

For $q = 1$, a Coxeter $\mathcal{A}$-q-bialgebra specializes to a Coxeter $\mathcal{A}$-bialgebra. For $q = -1$, we employ the term *signed Coxeter $\mathcal{A}$-bialgebra*. Similarly, for the (-1)-bialgebra axiom, we employ the term *signed bialgebra axiom*.

A morphism of Coxeter $\mathcal{A}$-q-bialgebras is a map of the underlying Coxeter $\mathcal{A}$-spaces which is a morphism of the underlying Coxeter $\mathcal{A}$-algebras and Coxeter $\mathcal{A}$-coalgebras. Note: q plays no role here.

Exercise 8.63. Check: Lemmas 8.53 and 8.54 generalize to Coxeter $\mathcal{A}$-q-bialgebras with β replaced by β_q, or equivalently, with w replaced by $q^{l(w)}w$ (where $l(w)$ is taken wrt the chamber C occurring in the diagrams). Similarly, (8.55c) can be written more generally as

$$\begin{aligned} \Delta\mu &= \sum_w \mu\beta_q w\Delta = \sum_w q^{l(w)}\mu\beta w\Delta, \\ \Delta\mu &= \sum_w \mu w\beta_q\Delta = \sum_w q^{l(w)}\mu w\beta\Delta. \end{aligned} \tag{8.67}$$

Observe: $w = \mathrm{id}$ satisfies conditions (8.54), so $\mu\Delta$ always appears as a term in the rhs of (8.67).

Exercise 8.64. Show: For defining a Coxeter $\mathcal{A}$-q-bialgebra, it suffices to require the q-bialgebra axiom (8.66) only for faces $A \lessdot F_0 \leq C$ and $A \lessdot F_0' \leq C$. This generalizes Lemma 8.55. (Repeatedly use diagram (8.56) and its dual with β_q instead of β. Compare with Exercise 2.30.)

Exercise 8.65. Check: Exercise 8.51 works for any $\overline{\lambda}_q$. Deduce: If either $F_0 = A$ or $F_0' = A$, then the q-bialgebra axiom (8.66) becomes a tautology. This generalizes Exercise 8.56.

Exercise 8.66. Check: Exercise 8.57 generalizes to Coxeter $\mathcal{A}$-q-bialgebras with β replaced by β_q.

8.7.2. Coxeter q-bialgebras in terms of top-lunes.

For a scalar q, for a Coxeter $\mathcal{A}$-space V, define the linear map

$$\big((\overline{\lambda}_q)_{\mathrm{L,M,L',M'}} : \mathrm{V[M]} \to \mathrm{V[M']}\big) := \sum q^{l(w)} w, \tag{8.68}$$

with the sum over the same $w \in W$ which appear in (8.57). Here $l(w)$ is the length of w taken wrt the top-lune N (which is a chamber in the arrangement over b(N)). Proposition 8.58 generalizes to Coxeter $\mathcal{A}$-q-bialgebras with diagram (8.59) replaced by

$$\begin{array}{ccccc} \mathrm{H[L]} & \xrightarrow{\mu^{\mathrm{L}}_{\mathrm{N}}} & \mathrm{H[N]} & \xrightarrow{\Delta^{\mathrm{L'}}_{\mathrm{N}}} & \mathrm{H[L']} \\ {\scriptstyle (\Delta^{\mathrm{M}}_{\mathrm{L}})}\big\downarrow & & & & \big\uparrow{\scriptstyle (\mu^{\mathrm{M'}}_{\mathrm{L'}})} \\ \bigoplus\limits_{\mathrm{M:\,L\leq' M}} \mathrm{H[M]} & & \xrightarrow[((\overline{\lambda}_q)_{\mathrm{L,M,L',M'}})]{} & & \bigoplus\limits_{\mathrm{M':\,L'\leq' M'}} \mathrm{H[M']} \end{array} \tag{8.69}$$

with $\overline{\lambda}_q$ as in (8.68). We refer to diagram (8.69) as the *q-bialgebra axiom*.

Exercise 8.67. Deduce from Exercise 8.64: For defining a Coxeter $\mathcal{A}$-q-bialgebra, it suffices to require the q-bialgebra axiom (8.69) only for top-lunes $\mathrm{N} \lessdot' \mathrm{L}$ and $\mathrm{N} \lessdot' \mathrm{L'}$.

Exercise 8.68. Check directly or deduce from Exercise 8.65: If either $\mathrm{L} = \mathrm{N}$ or $\mathrm{L'} = \mathrm{N}$, then the q-bialgebra axiom (8.69) becomes a tautology, namely, $\Delta^{\mathrm{L'}}_{\mathrm{N}} = \Delta^{\mathrm{L'}}_{\mathrm{N}}$ or $\mu^{\mathrm{L}}_{\mathrm{N}} = \mu^{\mathrm{L}}_{\mathrm{N}}$.

Exercise 8.69. Specialize the q-bialgebra axiom (8.69) to the case when either L or L′ equals ⊤. When L = L′ = ⊤, we get the following. For a Coxeter $\mathcal{A}$-bialgebra H, and top-lune N,

$$\Delta^{\top}_{\mathrm{N}}\mu^{\top}_{\mathrm{N}} = \sum q^{l(w)} w : \mathrm{H}[\top] \to \mathrm{H}[\top],$$

with the sum over $w \in W$ such that w fixes b(N) pointwise, and $l(w)$ is taken wrt N. (See Exercise 8.66.)

8.7.3. Coxeter q-bialgebras in terms of face-types.

For a scalar q, for a Coxeter $\boldsymbol{\alpha}$-space V, define the linear map

$$\big((\overline{\lambda}_q)_{TUT'U'} : \mathrm{V}[U] \to \mathrm{V}[U']\big) = \sum_{\substack{\sigma\colon U \to U' \\ \text{local to } Z \\ \text{biprojects } T \text{ and } T'}} q^{l(\sigma)}\sigma. \tag{8.70}$$

The sum is over the same morphisms σ which appear in (8.63). Recall: $l(\sigma)$ is the length of σ.

Example 8.70 (Rank three). Let us continue Example 8.61. For the rank-three braid arrangement, with Z, T, U, T', U' and σ as before, $(\overline{\lambda}_q)_{TUT'U'} = q^2\,\sigma$. Note very carefully: σ has length 2 and not 1. To figure out the length, look at the entire arrangement in Figure 1.5 and not just the rank-two flat in Figure 1.7.

For the rank-three arrangement of type B, with face-types as in Example 8.61, $(\overline{\lambda}_q)_{TUT'U'} = q\,\sigma$ when T' = cyan, and $(\overline{\lambda}_q)_{TUT'U'} = \mathrm{id} + q^4\,\tau$ when T' = red. For the dihedral arrangement of 4 lines, $(\overline{\lambda}_q)_{TUT'U'} = q\,\sigma$ and $(\overline{\lambda}_q)_{TUT'U'} = \mathrm{id} + q^2\,\tau$, respectively. Note very carefully: The coefficient of τ is different in the two cases. □

Observation 8.71. In particular,

$$\begin{aligned}\big((\overline{\lambda}_q)_{TTT'U'} : \mathrm{V}[T] \to \mathrm{V}[U']\big) &= \sum_{\substack{\sigma\colon T \to U' \\ \text{local to } Z \\ \text{projects } T \text{ on } T'}} q^{l(\sigma)}\sigma,\\ \big((\overline{\lambda}_q)_{TUT'T'} : \mathrm{V}[U] \to \mathrm{V}[T']\big) &= \sum_{\substack{\sigma\colon U \to T' \\ \text{local to } Z \\ \text{coprojects } T' \text{ on } T}} q^{l(\sigma)}\sigma.\end{aligned} \tag{8.71}$$

As a special case of either,

$$\big((\overline{\lambda}_q)_{TTT'T'} : \mathrm{V}[T] \to \mathrm{V}[T']\big) = \sum_{\substack{\sigma\colon T \to T' \\ \text{local to } Z}} q^{l(\sigma)}\,\sigma. \tag{8.72}$$

Exercise 8.72. Check: For a scalar q, for a Coxeter $\boldsymbol{\alpha}$-space V, for face-types $Z \leq U$ and $Z \leq T' \leq U'$,

$$\big((\overline{\lambda}_q)_{ZUT'U'} : \mathrm{V}[U] \to \mathrm{V}[U']\big) = \begin{cases}\mathrm{id} & \text{if } U = T' = U',\\ 0 & \text{otherwise.}\end{cases}$$

More precisely, in the first alternative, only $\sigma = \mathrm{id}$ contributes to $\overline{\lambda}_q$, while in the second alternative, no σ works. By symmetry, there is a similar statement for $(\overline{\lambda}_q)_{TUZU'}$.

Exercise 8.73 (Shuffles). Recall shuffles from Section 1.6.4. Use Exercise 1.35 to deduce: For a scalar q, for a Coxeter $\boldsymbol{\alpha}$-space V, for face-types $Z \leq T$ and $Z \leq T'$,

$$\big((\overline{\lambda}_q)_{SST'S} : \mathrm{V}[S] \to \mathrm{V}[S]\big) = \sum_{\zeta \in \mathrm{Sh}_{T'/Z}} q^{l(\zeta)}\,\zeta^{-1},$$

$$((\overline{\lambda}_q)_{TSSS} : \mathrm{V}[S] \to \mathrm{V}[S]) = \sum_{\zeta \in \mathrm{Sh}_{T/Z}} q^{l(\zeta)}\, \zeta.$$

As a special case of either,

$$((\overline{\lambda}_q)_{SSSS} : \mathrm{V}[S] \to \mathrm{V}[S]) = \sum_{w \in W_Z} q^{l(w)} w.$$

(Recall from Exercise 8.38 that $\mathrm{V}[S]$ is a left W-module.)

For a scalar q, a *Coxeter $\boldsymbol{\alpha}$-q-bialgebra* is a triple $(\mathrm{H}, \mu, \Delta)$, where H is a Coxeter $\boldsymbol{\alpha}$-space, (H, μ) is a Coxeter $\boldsymbol{\alpha}$-algebra, (H, Δ) is a Coxeter $\boldsymbol{\alpha}$-coalgebra, and for face-types $Z \leq T$ and $Z \leq T'$, the diagram

$$\begin{array}{ccccc} \mathrm{H}[T] & \xrightarrow{\mu_Z^T} & \mathrm{H}[Z] & \xrightarrow{\Delta_Z^{T'}} & \mathrm{H}[T'] \\ \Big\downarrow {\scriptstyle (\Delta_T^U)} & & & & \Big\uparrow {\scriptstyle (\mu_{T'}^{U'})} \\ \bigoplus_{U:\, T \leq U} \mathrm{H}[U] & & \xrightarrow[((\overline{\lambda}_q)_{TUT'U'})]{} & & \bigoplus_{U':\, T' \leq U'} \mathrm{H}[U'] \end{array} \tag{8.73}$$

commutes, with $\overline{\lambda}_q$ as in (8.70). We refer to diagram (8.73) as the *q-bialgebra axiom*.

For $q = 1$, a Coxeter $\boldsymbol{\alpha}$-q-bialgebra specializes to a Coxeter $\boldsymbol{\alpha}$-bialgebra. For $q = -1$, we employ the term *signed Coxeter $\boldsymbol{\alpha}$-bialgebra*. Similarly, for the (-1)-bialgebra axiom, we employ the term *signed bialgebra axiom*.

Analogues of Exercises 8.64, 8.65, 8.66 are given below.

Exercise 8.74. Show: For defining a Coxeter $\boldsymbol{\alpha}$-q-bialgebra, it suffices to require the q-bialgebra axiom (8.73) only for face-types $Z \lessdot T$ and $Z \lessdot T'$. See the discussion on covering face-types in Section 8.5.9 in this regard.

Exercise 8.75. Deduce from Exercise 8.72: If either $T = Z$ or $T' = Z$, then the q-bialgebra axiom (8.73) becomes a tautology, namely, $\Delta_Z^{T'} = \Delta_Z^{T'}$ or $\mu_Z^T = \mu_Z^T$.

Exercise 8.76. As special cases of the q-bialgebra axiom (8.73),

$$\Delta_Z^{T'} \mu_Z^S = \mu_{T'}^S (\overline{\lambda}_q)_{SST'S}, \quad \Delta_Z^S \mu_Z^T = (\overline{\lambda}_q)_{TSSS} \Delta_T^S, \quad \Delta_Z^S \mu_Z^S = (\overline{\lambda}_q)_{SSSS}.$$

Use formulas in Exercise 8.73 to make each rhs explicit. In particular: For a scalar q, for a Coxeter $\boldsymbol{\alpha}$-q-bialgebra H, for a face-type Z,

$$\Delta_Z^S \mu_Z^S = \sum_{w \in W_Z} q^{l(w)} w : \mathrm{H}[S] \to \mathrm{H}[S]. \tag{8.74}$$

Deduce using Fact 1.76 that for q not a root of unity, μ_Z^S is injective and Δ_Z^S is surjective. A more general result is given later in Exercise 13.32.

Example 8.77 (Rank zero). Let $\boldsymbol{\alpha}$ be a rank-zero pointed reflection arrangement. The category of Coxeter $\boldsymbol{\alpha}$-spaces is isomorphic to the category of vector spaces. Explicitly, a Coxeter $\boldsymbol{\alpha}$-space V corresponds to the vector space $\mathrm{V}[\emptyset]$. An equivalent result was given in Example 8.30. In fact, in a similar manner, the categories of Coxeter $\boldsymbol{\alpha}$-algebras, Coxeter $\boldsymbol{\alpha}$-coalgebras, Coxeter $\boldsymbol{\alpha}$-q-bialgebras are all isomorphic to the category of vector spaces. Compare with Example 2.31. □

Example 8.78 (Coordinate arrangement). The groupoid of biface-types of the coordinate arrangement was detailed in Section 1.15.2. For $\mathcal{C}^n$, the q-bialgebra axiom can be expressed as

$$\Delta_Z^{T'} \mu_Z^T = \sum q^{l(\sigma)} \mu_{T'}^{T \cup T'} \sigma \Delta_T^{T \cup T'}. \tag{8.75}$$

The sum is over all morphisms $\sigma : T \cup T' \to T \cup T'$ which preserve all elements of $Z \cup (T \cup T' \setminus T \cap T')$. Also, $l(\sigma)$ is the number of elements of $(T \cap T') \setminus Z$ which σ flips.

As the special case $T = T'$,

$$\Delta_Z^T \mu_Z^T = \sum_{\substack{\sigma : T \to T \\ \text{preserves all elements of } Z}} q^{l(\sigma)} \, \sigma.$$

Note very carefully: The rhs does not involve any face-types strictly greater than T. This is a special feature of the coordinate arrangement. □

The data present in a Coxeter q-bialgebra (in terms of top-nested faces, top-lunes, face-types) can be shown as in Summaries 8.1, 8.2, 8.3 with the bialgebra axiom replaced by the q-bialgebra axiom.

8.7.4. Coxeter 0-bialgebras in terms of top-nested faces.

For $q = 0$, the sum in (8.67) contains only one term, namely, the one for $w = \mathrm{id}$. So in this case, the q-bialgebra axiom (8.66) becomes set-theoretic and yields the following simpler description of Coxeter 0-bialgebras.

A Coxeter $\mathcal{A}$-0-bialgebra is a triple $(\mathrm{H}, \mu, \Delta)$, where H is a Coxeter $\mathcal{A}$-space, (H, μ) is a Coxeter $\mathcal{A}$-algebra, (H, Δ) is a Coxeter $\mathcal{A}$-coalgebra, and for faces $A \leq F \leq C$ and $A \leq F' \leq C$, the diagram

$$\begin{array}{ccc} \mathrm{H}[F, C] & \xrightarrow{\;\mu_{A,C}^{F,C}\;} & \mathrm{H}[A, C] \\ {\scriptstyle \Delta_{F,C}^{F \vee F', C}} \big\downarrow & & \big\downarrow {\scriptstyle \Delta_{A,C}^{F',C}} \\ \mathrm{H}[F \vee F', C] & \xrightarrow[\;\mu_{F',C}^{F \vee F', C}\;]{} & \mathrm{H}[F', C] \end{array} \tag{8.76}$$

commutes. Here $F \vee F'$ refers to the join of F and F' in the poset of faces. It exists since F and F' have a common upper bound, namely, C. We refer to diagram (8.76) as the 0-*bialgebra axiom*. Compare and contrast with the 0-bimonoid axiom (2.44) which involves an alternative.

Exercise 8.79. Deduce: For a Coxeter $\mathcal{A}$-0-bialgebra, for faces $A \leq F \leq C$, we have $\Delta_{A,C}^{F,C} \mu_{A,C}^{F,C} = \mathrm{id}$.

Exercise 8.80. Write down the 0-bialgebra axiom in terms of top-lunes by specializing the q-bialgebra axiom (8.69).

8.7.5. Coxeter 0-bialgebras in terms of face-types.

As a special case of (8.70),

$$\big((\overline{\lambda}_0)_{TUT'U'} : \mathrm{V}[U] \to \mathrm{V}[U']\big) = \begin{cases} \mathrm{id} & \text{if } T \cup T' = U = U', \\ 0 & \text{otherwise.} \end{cases} \tag{8.77}$$

Thus, for $q = 0$, the q-bialgebra axiom (8.73) simplifies to yield the following description of Coxeter 0-bialgebras.

A Coxeter $\boldsymbol{\alpha}$-0-bialgebra is a triple $(\mathrm{H}, \mu, \Delta)$, where H is a Coxeter $\boldsymbol{\alpha}$-space, (H, μ) is a Coxeter $\boldsymbol{\alpha}$-algebra, (H, Δ) is a Coxeter $\boldsymbol{\alpha}$-coalgebra, and for face-types $Z \leq T$ and $Z \leq T'$, the diagram

$$\begin{array}{ccc} \mathrm{H}[T] & \xrightarrow{\mu_Z^T} & \mathrm{H}[Z] \\ {\scriptstyle \Delta_T^{T\cup T'}}\big\downarrow & & \big\downarrow{\scriptstyle \Delta_Z^{T'}} \\ \mathrm{H}[T \cup T'] & \xrightarrow[\mu_{T'}^{T\cup T'}]{} & \mathrm{H}[T'] \end{array} \tag{8.78}$$

commutes. We refer to diagram (8.78) as the *0-bialgebra axiom*. It is set-theoretic. It can also be deduced from (8.76).

Exercise 8.81. Deduce: For a Coxeter $\boldsymbol{\alpha}$-0-bialgebra, for face-types $Z \leq T$, we have $\Delta_Z^T \mu_Z^T = \mathrm{id}$. This includes the case $q = 0$ of (8.74).

Observation 8.82. For a scalar q, the q-bialgebra axiom (8.73) has the form

$$\Delta_Z^{T'} \mu_Z^T = \mu_{T'}^{T\cup T'} \Delta_T^{T\cup T'} + \text{ terms involving nonzero powers of } q.$$

When $q = 0$, only the first term in the rhs survives and the axiom reduces to (8.78).

8.7.6. Coxeter υ-bialgebras. Coxeter q-bialgebras generalize further to Coxeter υ-bialgebras for any W-invariant distance function υ (Section 1.4.4) by replacing the q-bialgebra axiom by the *υ-bialgebra axiom*. The latter is defined using $\overline{\lambda}_\upsilon$ which arises by replacing β by β_υ in either (8.50a) or (8.50b). That is, we either multiply the map (8.50a) by $\upsilon_{C,G'C}$ or (8.50b) by $\upsilon_{GC,C}$.

8.7.7. Coxeter subalgebras and Coxeter quotient algebras. A Coxeter subalgebra of a Coxeter algebra A is a Coxeter subspace of A to which product components of A restrict. Similarly, one defines a Coxeter quotient algebra of A. We employ similar terminology for Coxeter coalgebras and Coxeter q-bialgebras.

The zero Coxeter algebra is the Coxeter algebra with all components zero. Its product components are all necessarily zero. Similarly, we have the zero Coxeter coalgebra and zero Coxeter q-bialgebra.

Exercise 8.83. Check: Direct sum (8.31) preserves Coxeter algebras, Coxeter coalgebras, and Coxeter q-bialgebras. Explicitly, if H and K are Coxeter q-bialgebras, then so is $\mathrm{H} + \mathrm{K}$ with product and coproduct induced from those of H and K.

Exercise 8.84. Check: If $f, g : \mathrm{A} \to \mathrm{B}$ are morphisms of Coxeter algebras, then so is $af + bg$ as defined in (8.32). (Use (8.36) or (8.38) or (8.42).) The same holds for morphisms of Coxeter coalgebras and Coxeter q-bialgebras. So these are all $\Bbbk$-linear categories. For related ideas, see Section 8.11.

8.7.8. Duality. Recall from Section 8.4.10 the duality functor on Coxeter spaces. The dual of a Coxeter algebra is a Coxeter coalgebra, and vice versa. Explicitly, for a Coxeter algebra A, coproduct components of A^* are obtained by dualizing product components of A, and for a Coxeter coalgebra C, product components of C^* are obtained by dualizing coproduct components of C. The dual of a Coxeter bialgebra is a Coxeter bialgebra. More generally, for a scalar q, the dual of a Coxeter q-bialgebra is a Coxeter q-bialgebra.

A Coxeter q-bialgebra H is *self-dual* if $\mathrm{H} \cong \mathrm{H}^*$. Similarly, a morphism of Coxeter q-bialgebras $f : \mathrm{H} \to \mathrm{H}^*$ is *self-dual* if $f = f^*$. (It is implicit here that H is finite-dimensional.)

8.8. (Co)commutative Coxeter (co)algebras

There is a notion of commutativity for Coxeter algebras and cocommutativity for Coxeter coalgebras. A Coxeter bialgebra may be commutative, cocommutative, both, or neither. A summary of the resulting categories along with their notations is given in Table 8.6. Here $\mathcal{A}$ denotes a fixed reflection arrangement. Alternatively, one may work relative to a pointed reflection arrangement $\boldsymbol{\alpha}$ in which case, we use the same notations with $\boldsymbol{\alpha}$ instead of $\mathcal{A}$. There is also a notion of signed commutativity for Coxeter algebras, signed cocommutativity for Coxeter coalgebras, and signed (co)commutativity for signed Coxeter bialgebras. To denote the resulting categories, we write a (-1) in front, see Table 8.7.

TABLE 8.6. Categories of (co)commutative Coxeter (co)algebras.

Category	Description
$\mathsf{Mon}^{\mathrm{co}}(\mathcal{A}\text{-}\bar{\mathsf{C}}\mathsf{Sp})$	com. Coxeter $\mathcal{A}$-algebras
${}^{\mathrm{co}}\mathsf{Comon}(\mathcal{A}\text{-}\bar{\mathsf{C}}\mathsf{Sp})$	cocom. Coxeter $\mathcal{A}$-coalgebras
$\mathsf{Bimon}^{\mathrm{co}}(\mathcal{A}\text{-}\bar{\mathsf{C}}\mathsf{Sp})$	com. Coxeter $\mathcal{A}$-bialgebras
${}^{\mathrm{co}}\mathsf{Bimon}(\mathcal{A}\text{-}\bar{\mathsf{C}}\mathsf{Sp})$	cocom. Coxeter $\mathcal{A}$-bialgebras
${}^{\mathrm{co}}\mathsf{Bimon}^{\mathrm{co}}(\mathcal{A}\text{-}\bar{\mathsf{C}}\mathsf{Sp})$	bicom. Coxeter $\mathcal{A}$-bialgebras

TABLE 8.7. Categories of signed (co)commutative Coxeter (co)algebras.

Category	Description
$(-1)\text{-}\mathsf{Mon}^{\mathrm{co}}(\mathcal{A}\text{-}\bar{\mathsf{C}}\mathsf{Sp})$	signed com. Coxeter $\mathcal{A}$-algebras
$(-1)\text{-}{}^{\mathrm{co}}\mathsf{Comon}(\mathcal{A}\text{-}\bar{\mathsf{C}}\mathsf{Sp})$	signed cocom. Coxeter $\mathcal{A}$-coalgebras
$(-1)\text{-}\mathsf{Bimon}^{\mathrm{co}}(\mathcal{A}\text{-}\bar{\mathsf{C}}\mathsf{Sp})$	signed com. signed Coxeter $\mathcal{A}$-bialgebras
$(-1)\text{-}{}^{\mathrm{co}}\mathsf{Bimon}(\mathcal{A}\text{-}\bar{\mathsf{C}}\mathsf{Sp})$	signed cocom. signed Coxeter $\mathcal{A}$-bialgebras
$(-1)\text{-}{}^{\mathrm{co}}\mathsf{Bimon}^{\mathrm{co}}(\mathcal{A}\text{-}\bar{\mathsf{C}}\mathsf{Sp})$	signed bicom. signed Coxeter $\mathcal{A}$-bialgebras

Compare with the categories of (co)commutative Coxeter (co)monoids in Table 2.5 and their signed analogues in Table 2.6. The two can be connected using Fock functors, see Section 15.2.

We define commutativity using the commutativity axiom. We also give variants called the opposition commutativity axiom and covering commutativity axiom. We formulate them in terms of top-nested faces, top-lunes, face-types. See the last entry in Table 8.1 and the second entry in Table 8.2.

8.8.1. Commutative Coxeter algebras in terms of top-nested faces. Recall Coxeter $\mathcal{A}$-algebras from Section 8.5.1. A Coxeter $\mathcal{A}$-algebra A is *commutative* if the diagram

$$\begin{array}{ccc} \mathrm{A}[F,C] & \xrightarrow{\ \sigma\ } & \mathrm{A}[F',C] \\ & {\scriptstyle \mu^{F,C}_{A,C}} \searrow \qquad \swarrow {\scriptstyle \mu^{F',C}_{A,C}} & \\ & \mathrm{A}[A,C] & \end{array} \tag{8.79}$$

commutes whenever $A \leq F \leq C$, $A \leq F' \leq C$ and $\sigma : (F,C) \to (F',C)$ is a morphism in $\mathcal{A}\text{-}\overline{\texttt{Cox}}$ local to A.

Lemma 8.85. *A Coxeter $\mathcal{A}$-algebra A is commutative iff the diagram*

$$\begin{array}{ccc} \mathrm{A}[F,C] & \xrightarrow{\ \beta_{G,D,F,C}\ } & \mathrm{A}[G,D] \\ {\scriptstyle \mu^{F,C}_{A,C}} \big\downarrow & & \big\downarrow {\scriptstyle \mu^{G,D}_{A,D}} \\ \mathrm{A}[A,C] & \xrightarrow[\ w(C)=D\ ]{} & \mathrm{A}[A,D] \end{array} \tag{8.80}$$

commutes whenever $A \leq F \leq C$, $A \leq G \leq D$, $\mathrm{s}(F,C) = \mathrm{s}(G,D)$, and w is the Coxeter symmetry which sends C to D (and hence A to itself).

PROOF. The argument is encapsulated in the diagram below.

$$\begin{array}{ccccc} \mathrm{A}[F,C] & \xrightarrow{\ \beta_{G,D,F,C}\ } & \mathrm{A}[G,D] & \xrightarrow{\ w^{-1}(D)=C\ } & \mathrm{A}[F',C] \\ {\scriptstyle \mu^{F,C}_{A,C}} \big\downarrow & & \big\downarrow {\scriptstyle \mu^{G,D}_{A,D}} & & \big\downarrow {\scriptstyle \mu^{F',C}_{A,C}} \\ \mathrm{A}[A,C] & \xrightarrow[\ w(C)=D\ ]{} & \mathrm{A}[A,D] & \xrightarrow[\ w^{-1}(D)=C\ ]{} & \mathrm{A}[A,C] \end{array}$$

(with a curved arrow $\sigma : \mathrm{A}[F,C] \to \mathrm{A}[F',C]$ above the top row and a curved arrow $\mathrm{id} : \mathrm{A}[A,C] \to \mathrm{A}[A,C]$ below the bottom row)

The second square commutes by the type-naturality axiom in (8.34a). The morphism σ is written in normal form $\sigma = w^{-1}\beta$. The lower triangle commutes by combining the type-type axiom in (8.20a) and type-unitality axiom in (8.20b). □

We refer to each of the diagrams (8.79) and (8.80) as the *commutativity axiom*. See the first two pictures in Figure 8.6. Note very carefully that this axiom involves both support and type morphisms. In contrast, the commutativity axiom (2.49) for Coxeter monoids involves only support morphisms.

8.8.2. Covering commutativity axiom in terms of top-nested faces.

Exercise 8.86. Show: A Coxeter $\mathcal{A}$-algebra A is commutative iff the diagram

$$\begin{array}{ccc} \mathrm{A}[F,C] & \xrightarrow{\ \beta_{A\overline{F},A\overline{F}C,F,C}\ } & \mathrm{A}[A\overline{F},A\overline{F}C] \\ {\scriptstyle \mu^{F,C}_{A,C}} \big\downarrow & & \big\downarrow {\scriptstyle \mu^{A\overline{F},A\overline{F}C}_{A,A\overline{F}C}} \\ \mathrm{A}[A,C] & \xrightarrow[\ w(C)=A\overline{F}C\ ]{} & \mathrm{A}[A,A\overline{F}C] \end{array} \tag{8.81}$$

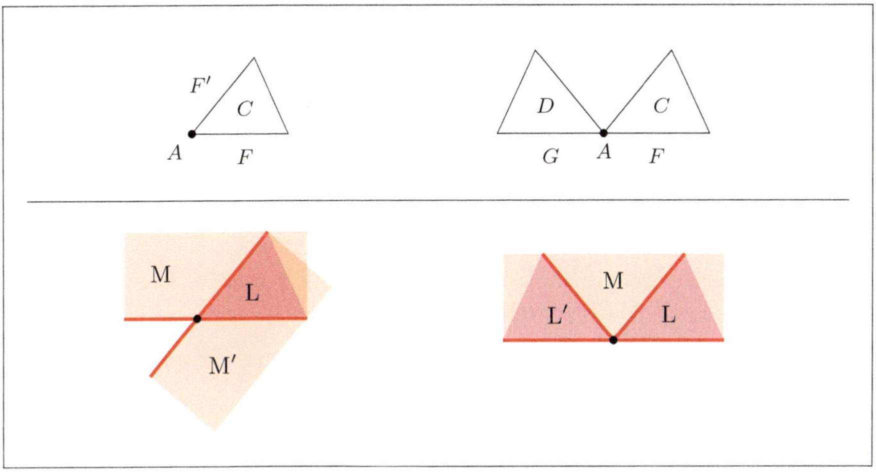

FIGURE 8.6. (Co)commutativity axiom for Coxeter (co)algebras.

commutes whenever $A \leq F \leq C$. We refer to diagram (8.81) as the *opposition commutativity axiom*. An illustration in rank three with $A = O$ is shown below.

commutativity axiom opposition commutativity axiom

A stronger result is given below.

Exercise 8.87. Show: For A a Coxeter $\mathcal{A}$-algebra, the following are equivalent.

- A is commutative.
- Commutativity axiom (8.79) holds whenever $A \lessdot F \leq C$ and $A \lessdot F' \leq C$.
- Commutativity axiom (8.80) holds whenever $A \lessdot F \leq C$ and $A \lessdot G \leq D$.
- Opposition commutativity axiom (8.81) holds whenever $A \lessdot F \leq C$.

We refer to either of these axioms as the *covering commutativity axiom*.

8.8.3. Commutative Coxeter algebras in terms of top-lunes. Commutativity can also be formulated in terms of top-lunes by using the description of Coxeter $\mathcal{A}$-algebras in Proposition 8.41.

Lemma 8.88. *A Coxeter $\mathcal{A}$-algebra* A *is commutative iff the diagram*

(8.82)
$$\begin{array}{ccc} \mathrm{A}[\mathrm{M}] & \xrightarrow{\ u\ } & \mathrm{A}[\mathrm{M}'] \\ & {\scriptstyle \mu^{\mathrm{M}}_{\mathrm{L}}} \searrow \quad \swarrow {\scriptstyle \mu^{\mathrm{M}'}_{\mathrm{L}}} & \\ & \mathrm{A}[\mathrm{L}] & \end{array}$$

commutes whenever $\mathrm{L} \leq' \mathrm{M}$, $\mathrm{L} \leq' \mathrm{M}'$ *and* $u : \mathrm{M} \to \mathrm{M}'$ *is such that it fixes* $\mathrm{b}(\mathrm{L})$ *pointwise.*

Proof. We translate between diagrams (8.79) and (8.82) using diagrams (8.27) and (8.39). □

Lemma 8.89. *A Coxeter $\mathcal{A}$-algebra* A *is commutative iff the diagram*

(8.83)

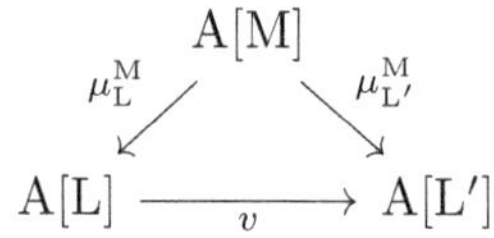

commutes whenever $\mathrm{L} \leq' \mathrm{M}$, $\mathrm{L}' \leq' \mathrm{M}$, $\mathrm{b(L)} = \mathrm{b(L')}$ *and* $v : \mathrm{L} \to \mathrm{L}'$ *is such that it fixes* $\mathrm{b(L)} = \mathrm{b(L')}$ *pointwise.*

Proof. This is similar to the previous proof. We now translate between diagrams (8.80) and (8.83). □

Exercise 8.90. Use diagrams below to deduce that (8.82) and (8.83) imply each other.

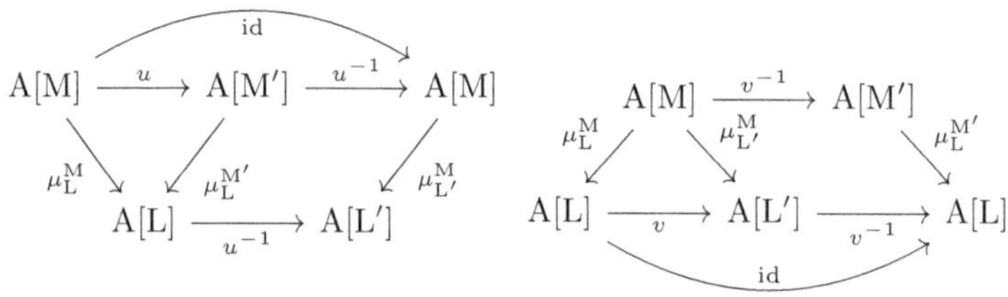

The parallelograms commute by the type-naturality axiom in (8.37). The above diagrams can be seen in analogy with the diagram in the proof of Lemma 8.85.

We refer to each of the diagrams (8.82) and (8.83) as the *commutativity axiom*. We sometimes refer to (8.83) as the *upside-down commutativity axiom*. See the last two pictures in Figure 8.6. Contrast these with the commutativity axiom (2.52) for Coxeter monoids.

Remark 8.91. We will see later in Section 11.6.2 that (8.82) and (8.83), respectively, arise from morphisms

$$(\mathrm{id}, u) : (\mathrm{L}, \mathrm{M}) \to (\mathrm{L}, \mathrm{M}') \quad \text{and} \quad (v, \mathrm{id}) : (\mathrm{L}, \mathrm{M}) \to (\mathrm{L}', \mathrm{M})$$

in a certain category whose objects are nested top-lunes.

Exercise 8.92. Show: A Coxeter $\mathcal{A}$-algebra A is commutative iff

$$(\mathrm{A}[\mathrm{M}] \xrightarrow{\mu_{\mathrm{L}}^{\mathrm{M}}(\mathrm{P})} \mathrm{A}[\mathrm{L}]) = (\mathrm{A}[\mathrm{M}] \xrightarrow{\mu_{\mathrm{L}}^{\mathrm{M}}(\mathrm{P}')} \mathrm{A}[\mathrm{L}]) \tag{8.84}$$

whenever $\mathrm{L} \leq \mathrm{M}$ and lunes P and P′ both have base b(L) and case b(M). The definition of $\mu_{\mathrm{L}}^{\mathrm{M}}(\mathrm{P})$ is as in (8.40).

8.8.4. Covering commutativity axiom in terms of top-lunes.

Exercise 8.93. Use Exercise 8.86 to deduce: A Coxeter $\mathcal{A}$-algebra A is commutative iff the diagram

(8.85)

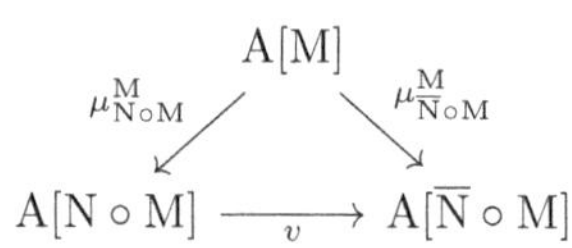

commutes whenever N and M are composable, and $v : \mathrm{N} \circ \mathrm{M} \to \overline{\mathrm{N}} \circ \mathrm{M}$ is such that it fixes b(N) pointwise. Here N is a lune, while M is a top-lune. We refer to diagram (8.85) as the *opposition commutativity axiom*. It is also upside-down. An illustration is shown below.

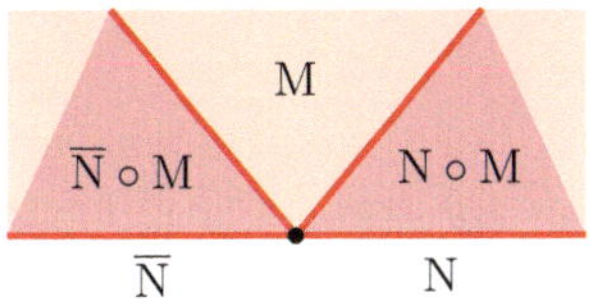

A stronger result is given below.

Exercise 8.94. Show: For A a Coxeter $\mathcal{A}$-algebra, the following are equivalent.

- A is commutative.
- Commutativity axiom (8.82) holds whenever $\mathrm{L} \lessdot \mathrm{M}$ and $\mathrm{L} \lessdot \mathrm{M}'$.
- Commutativity axiom (8.83) holds whenever $\mathrm{L} \lessdot \mathrm{M}$ and $\mathrm{L}' \lessdot \mathrm{M}$.
- Opposition commutativity axiom (8.85) holds whenever N has slack 1.

We refer to either of these axioms as the *covering commutativity axiom*. (Use Exercise 8.87.)

8.8.5. Commutative Coxeter algebras in terms of face-types.

Recall Coxeter $\boldsymbol{\alpha}$-algebras from Section 8.5.3. A Coxeter $\boldsymbol{\alpha}$-algebra A is *commutative* if the diagram

$$\begin{array}{ccc} \mathrm{A}[T] & \xrightarrow{\ \sigma\ } & \mathrm{A}[T'] \\ & {\scriptstyle \mu_Z^T}\searrow \quad \swarrow{\scriptstyle \mu_Z^{T'}} & \\ & \mathrm{A}[Z] & \end{array} \tag{8.86}$$

commutes whenever $Z \leq T$, $Z \leq T'$ and $\sigma : T \to T'$ is a morphism in $\boldsymbol{\alpha}\text{-}\overline{\mathsf{Cox}}$ local to Z. We refer to diagram (8.86) as the *commutativity axiom*.

This can be combined with the naturality axiom in (8.41) to obtain a more general diagram, see (11.53). It makes use of morphisms in a certain category whose objects are nested face-types.

Exercise 8.95. Check: Under the equivalence in Exercise 8.43, commutative Coxeter $\mathcal{A}$-algebras correspond to commutative Coxeter $\boldsymbol{\alpha}$-algebras. Compare axioms (8.79) and (8.86). There is no analogue of (8.80) for face-types.

8.8.6. Covering commutativity axiom in terms of face-types.

Recall the locally opposite support-type morphisms $\beta_{Z\overline{T},T}$ from (8.19).

Lemma 8.96. *A Coxeter $\boldsymbol{\alpha}$-algebra* A *is commutative iff the diagram*

$$\begin{array}{ccc} \mathrm{A}[T] & \xrightarrow{\ \beta_{Z\overline{T},T}\ } & \mathrm{A}[Z\overline{T}] \\ & {\scriptstyle \mu_Z^T}\searrow \quad \swarrow{\scriptstyle \mu_Z^{Z\overline{T}}} & \\ & \mathrm{A}[Z] & \end{array} \tag{8.87}$$

commutes whenever $Z \leq T$.

This corresponds to Exercises 8.86 and 8.93. We refer to diagram (8.87) as the *opposition commutativity axiom*. A stronger result is proved below. It corresponds to Exercises 8.87 and 8.94 (which can be proved similarly).

Lemma 8.97. *For* A *a Coxeter* $\boldsymbol{\alpha}$*-algebra, the following are equivalent.*

- A *is commutative.*
- *Commutativity axiom* (8.86) *holds whenever* $Z \lessdot T$ *and* $Z \lessdot T'$.
- *Opposition commutativity axiom* (8.87) *holds whenever* $Z \lessdot T$.

In these particular cases, the two axioms in fact coincide. We call this the *covering commutativity axiom*.

PROOF. It is clear that the first item implies the second, and the second implies the third. We now prove that the third item implies the first. In view of Exercise 8.28, it suffices to establish the commutativity axiom (8.86) for morphisms of the form $\sigma = \beta_{Y\overline{T},T}$ for some $Z \leq Y \lessdot T$. In this situation, diagram (8.86) can be filled in as follows.

$$\begin{array}{ccccc} \mathrm{A}[T] & & \xrightarrow{\ \beta_{Y\overline{T},T}\ } & & \mathrm{A}[Y\overline{T}] \\ & \searrow^{\mu^T_Y} & & {}^{\mu^{Y\overline{T}}_Y}\swarrow & \\ & & \mathrm{A}[Y] & & \\ {\scriptstyle\mu^T_Z}\searrow & & \downarrow{\scriptstyle\mu^Y_Z} & & \swarrow{\scriptstyle\mu^{Y\overline{T}}_Z} \\ & & \mathrm{A}[Z] & & \end{array}$$

The top triangle commutes by hypothesis. The two side triangles commute by the associativity axiom in (8.41). □

For Coxeter $\boldsymbol{\alpha}$-algebras as formulated in Proposition 8.44 (with product components involving cover relations), one can define commutativity using the covering commutativity axiom.

8.8.7. Cocommutative Coxeter coalgebras in terms of top-nested faces. We now briefly consider the dual notion. Recall Coxeter $\mathcal{A}$-coalgebras from Section 8.5.5. A Coxeter $\mathcal{A}$-coalgebra C is *cocommutative* if the diagram

$$\begin{array}{ccc} \mathrm{C}[F,C] & \xrightarrow{\ \sigma\ } & \mathrm{C}[F',C] \\ {\scriptstyle\Delta^{F,C}_{A,C}}\nwarrow & & \nearrow{\scriptstyle\Delta^{F',C}_{A,C}} \\ & \mathrm{C}[A,C] & \end{array} \tag{8.88}$$

commutes whenever $A \leq F \leq C$, $A \leq F' \leq C$ and $\sigma : (F,C) \to (F',C)$ is a morphism in $\mathcal{A}$-$\overline{\mathtt{Cox}}$ local to A.

Equivalently, a Coxeter $\mathcal{A}$-coalgebra C is cocommutative if the diagram

$$\begin{array}{ccc} \mathrm{C}[F,C] & \xrightarrow{\ \beta_{G,D,F,C}\ } & \mathrm{C}[G,D] \\ {\scriptstyle\Delta^{F,C}_{A,C}}\uparrow & & \uparrow{\scriptstyle\Delta^{G,D}_{A,D}} \\ \mathrm{C}[A,C] & \xrightarrow[\ w(C)=D\ ]{} & \mathrm{C}[A,D] \end{array} \tag{8.89}$$

commutes whenever $A \leq F \leq C$, $A \leq G \leq D$, $\mathrm{s}(F, C) = \mathrm{s}(G, D)$, and w is the Coxeter symmetry which sends C to D (and hence A to itself).

We refer to each of the diagrams (8.88) and (8.89) as the *cocommutativity axiom*. Contrast this with the cocommutativity axiom (2.57) for Coxeter comonoids which involves only support morphisms.

The duals to Exercises 8.86 and 8.87 hold. Thus, we have the *opposition cocommutativity axiom* and *covering cocommutativity axiom*.

8.8.8. Cocommutative Coxeter coalgebras in terms of top-lunes. Cocommutative Coxeter $\mathcal{A}$-coalgebras can also be formulated using top-lunes. This would be the dual of Lemmas 8.88 and 8.89. The *cocommutativity axiom* and *upside-down cocommutativity axiom*, respectively, are

$$(8.90) \qquad \begin{array}{ccc} \mathrm{C}[\mathrm{M}] & \xrightarrow{u} & \mathrm{C}[\mathrm{M}'] \\ {\scriptstyle \Delta^{\mathrm{M}}_{\mathrm{L}}} \nwarrow & & \nearrow {\scriptstyle \Delta^{\mathrm{M}'}_{\mathrm{L}}} \\ & \mathrm{C}[\mathrm{L}] & \end{array} \qquad\qquad \begin{array}{ccc} & \mathrm{C}[\mathrm{M}] & \\ {\scriptstyle \Delta^{\mathrm{M}}_{\mathrm{L}}} \nearrow & & \nwarrow {\scriptstyle \Delta^{\mathrm{M}}_{\mathrm{L}'}} \\ \mathrm{C}[\mathrm{L}] & \xrightarrow[v]{} & \mathrm{C}[\mathrm{L}'], \end{array}$$

with conventions as in the Coxeter algebra case. Similarly, the duals to Exercises 8.93 and 8.94 hold.

8.8.9. Cocommutative Coxeter coalgebras in terms of face-types. Recall Coxeter $\boldsymbol{\alpha}$-coalgebras from Section 8.5.7. A Coxeter $\boldsymbol{\alpha}$-coalgebra C is *cocommutative* if the diagram

$$(8.91) \qquad \begin{array}{ccc} \mathrm{C}[T] & \xrightarrow{\sigma} & \mathrm{C}[T'] \\ {\scriptstyle \Delta^{T}_{Z}} \nwarrow & & \nearrow {\scriptstyle \Delta^{T'}_{Z}} \\ & \mathrm{C}[Z] & \end{array}$$

commutes whenever $Z \leq T$, $Z \leq T'$ and $\sigma : T \to T'$ is a morphism in $\boldsymbol{\alpha}$-$\overline{\mathtt{Cox}}$ local to Z. We refer to diagram (8.91) as the *cocommutativity axiom*.

The duals to Lemmas 8.96 and 8.97 hold.

8.8.10. (Co, bi)commutative Coxeter bialgebras. The notion of commutativity also applies to Coxeter bialgebras. We say a Coxeter bialgebra is

- *commutative* if its underlying Coxeter algebra is commutative,
- *cocommutative* if its underlying Coxeter coalgebra is cocommutative,
- *bicommutative* if it is both commutative and cocommutative.

More generally, we apply these definitions to a Coxeter q-bialgebra.

8.8.11. Signed (co)commutativity. Recall from Section 8.2.3 the notion of length of a morphism in $\mathcal{A}$-$\overline{\mathtt{Cox}}$. The parity of this length plays a central role in signed aspects of commutativity.

In terms of face-types: A Coxeter $\boldsymbol{\alpha}$-algebra A is *signed commutative* if the diagram

$$\begin{array}{ccc} \mathrm{A}[T] & \xrightarrow{(-1)^{l(\sigma)}\sigma} & \mathrm{A}[T'] \\ & {\scriptstyle \mu^T_Z}\searrow \quad \swarrow{\scriptstyle \mu^{T'}_Z} & \\ & \mathrm{A}[Z] & \end{array} \tag{8.92}$$

commutes whenever $Z \leq T$, $Z \leq T'$ and $\sigma : T \to T'$ is a morphism in $\boldsymbol{\alpha}$-$\overline{\mathsf{Cox}}$ local to Z. We refer to diagram (8.92) as the *signed commutativity axiom.*

Dually, a Coxeter $\boldsymbol{\alpha}$-coalgebra C is *signed cocommutative* if the diagram

$$\begin{array}{ccc} \mathrm{C}[T] & \xrightarrow{(-1)^{l(\sigma)}\sigma} & \mathrm{C}[T'] \\ & {\scriptstyle \Delta^T_Z}\nwarrow \quad \nearrow{\scriptstyle \Delta^{T'}_Z} & \\ & \mathrm{C}[Z] & \end{array} \tag{8.93}$$

commutes whenever $Z \leq T$, $Z \leq T'$ and $\sigma : T \to T'$ is a morphism in $\boldsymbol{\alpha}$-$\overline{\mathsf{Cox}}$ local to Z. We refer to diagram (8.93) as the *signed cocommutativity axiom.*

In terms of top-nested faces, we replace σ by $(-1)^{l(\sigma)}\sigma$ in (8.79), and $\beta_{G,D,F,C}$ by $(-1)^{\mathrm{dist}(C,D)}\beta_{G,D,F,C}$ in (8.80). In terms of top-lunes, we replace u by $(-1)^{l(u)}u$ in (8.82), and v by $(-1)^{l(v)}v$ in (8.83).

We say a signed Coxeter bialgebra is *signed (co)commutative* if its underlying Coxeter (co)algebra is signed (co)commutative. Similarly, a signed Coxeter bialgebra is *signed bicommutative* if it is both signed commutative and signed cocommutative.

8.8.12. Coxeter double (co)algebras and Eckmann–Hilton. Define the category of Coxeter double algebras as follows. A *Coxeter $\boldsymbol{\alpha}$-double algebra* is a triple (A, μ, μ'), where A is a Coxeter $\boldsymbol{\alpha}$-space, (A, μ) and (A, μ') are Coxeter $\boldsymbol{\alpha}$-algebras, and for face-types $Z \leq T \leq U$ and $Z \leq T' \leq U'$, and morphism $\sigma : U \to U'$ local to Z which biprojects T and T', the diagram

$$\begin{array}{ccccc} \mathrm{A}[T] & \xrightarrow{\mu^T_Z} & \mathrm{A}[Z] & \xleftarrow{(\mu')^{T'}_Z} & \mathrm{A}[T'] \\ {\scriptstyle (\mu')^U_T}\uparrow & & & & \uparrow{\scriptstyle \mu^{U'}_{T'}} \\ \mathrm{A}[U] & & \xrightarrow[\sigma]{\qquad\qquad} & & \mathrm{A}[U'] \end{array} \tag{8.94}$$

commutes. We refer to diagram (8.94) as the *double algebra axiom.* A morphism of Coxeter $\boldsymbol{\alpha}$-double algebras is a map of the underlying Coxeter $\boldsymbol{\alpha}$-spaces which is a morphism of Coxeter $\boldsymbol{\alpha}$-algebras for both μ and μ'.

The category of Coxeter double coalgebras is defined dually.

Exercise 8.98. Show: The category of Coxeter double (co)algebras is equivalent to the category of (co)commutative Coxeter (co)algebras. This fact is an avatar of the Eckmann–Hilton argument.

8.9. Coxeter bialgebras and face-type enumeration

We now connect Coxeter bialgebras to face-type enumeration in Section 1.8. For this, we work with the formulation of Coxeter bialgebras in terms of face-types. Face-type enumeration polynomials appear in identities involving products and coproducts when the Coxeter bialgebra is either commutative or cocommutative; these identities will get used in some of the theory developed in Chapter 9. Moreover, these enumeration polynomials appear in (co)product formulas of certain Coxeter bialgebras studied in Chapter 10.

8.9.1. Bialgebra axiom and face-type enumeration. Note the striking similarity in the definitions of the map $(\overline{\lambda}_q)_{TUT'U'}$ in (8.70) and the polynomials $a_Z^{TUT'U'}(q)$ in (1.64a). As a special case, note the similarity between $(\overline{\lambda}_0)_{TUT'U'}$ in (8.77) and $a_Z^{TUT'U'}(0)$. The same remark applies to (8.71) and (1.64b), and to (8.72) and (1.64d). This is formalized below.

Observation 8.99. Let q be a scalar. Let $Z \leq T \leq U$ and $Z \leq T' \leq U'$.

(1) For $(\overline{\lambda}_q)_{TUT'U'}$, the sum of all coefficients is $a_Z^{TUT'U'}(q)$. Each coefficient is a power of q, so the number of terms that appear in the sum is $a_Z^{TUT'U'}$.
(2) For either $(\overline{\lambda}_q)_{TUT'T'}$ or $(\overline{\lambda}_q)_{T'T'TU}$ with $\mathrm{s}(U/Z) = \mathrm{s}(T'/Z)$, the sum of all coefficients is $a_Z^{TUT'}(q)$. The number of terms in the sum is $a_Z^{TUT'}$.
(3) For $(\overline{\lambda}_q)_{TTT'T'}$ with $\mathrm{s}(T/Z) = \mathrm{s}(T'/Z)$, the sum of all coefficients is $c_Z^{TT'}(q)$.

8.9.2. (Co)commutative Coxeter bialgebras. (Co)commutativity allows us to replace $\mu\sigma$ by μ, and $\sigma\Delta$ by Δ. That we can get rid of σ brings Observation 8.99 into play as illustrated below.

Lemma 8.100. *Let q be a scalar. Let $Z \leq T \leq U$ and $Z \leq T' \leq U'$. Then, for a commutative Coxeter $\boldsymbol{\alpha}$-algebra* A*, the diagram*

$$\begin{array}{ccc} \mathrm{A}[U] & \xrightarrow{(\overline{\lambda}_q)_{TUT'U'}} & \mathrm{A}[U'] \\ & {\scriptstyle a_Z^{TUT'U'}(q)\,\mu_Z^U} \searrow \quad \swarrow {\scriptstyle \mu_Z^{U'}} & \\ & \mathrm{A}[Z] & \end{array} \tag{8.95}$$

commutes. Dually, for a cocommutative Coxeter $\boldsymbol{\alpha}$-coalgebra C*, the diagram*

$$\begin{array}{ccc} \mathrm{C}[U] & \xrightarrow{(\overline{\lambda}_q)_{TUT'U'}} & \mathrm{C}[U'] \\ & {\scriptstyle \Delta_Z^U} \nwarrow \quad \nearrow {\scriptstyle a_Z^{TUT'U'}(q)\,\Delta_Z^{U'}} & \\ & \mathrm{C}[Z] & \end{array} \tag{8.96}$$

commutes.

PROOF. For diagram (8.95),

$$\mu_Z^{U'}\,(\overline{\lambda}_q)_{TUT'U'} = \sum q^{l(\sigma)} \mu_Z^{U'}\,\sigma = \Big(\sum q^{l(\sigma)}\Big)\,\mu_Z^U = a_Z^{TUT'U'}(q)\,\mu_Z^U.$$

We used the definition of $\overline{\lambda}_q$ in (8.70), the commutativity axiom (8.86), and Observation 8.99, item (1). The check for diagram (8.96) is similar. □

Exercise 8.101. Check: For a signed commutative Coxeter $\boldsymbol{\alpha}$-algebra A, diagram (8.95) commutes with exactly one of the two qs replaced by $-q$. The same statement holds for a signed cocommutative Coxeter $\boldsymbol{\alpha}$-coalgebra C and diagram (8.96).

Lemma 8.102. *Let q be a scalar. For a commutative Coxeter $\boldsymbol{\alpha}$-q-bialgebra, for face-types T and U greater than Z,*

$$\mu_Z^U \Delta_Z^U \mu_Z^T = \sum_{V:\, V \geq T} a_Z^{TVU}(q)\, \mu_Z^V \Delta_T^V. \tag{8.97}$$

Dually, for a cocommutative Coxeter $\boldsymbol{\alpha}$-q-bialgebra, for face-types T and U greater than Z,

$$\Delta_Z^T \mu_Z^U \Delta_Z^U = \sum_{V:\, V \geq T} a_Z^{TVU}(q)\, \mu_T^V \Delta_Z^V. \tag{8.98}$$

Proof. We prove the first statement. Let T and T' be face-types greater than Z. (The notation has been slightly changed for convenience.)

$$\begin{aligned}
\mu_Z^{T'} \Delta_Z^{T'} \mu_Z^T &= \sum_{U \geq T,\, U' \geq T'} \mu_Z^{T'} \mu_{T'}^{U'} (\overline{\lambda}_q)_{TUT'U'} \Delta_T^U \\
&= \sum_{U \geq T,\, U' \geq T'} \mu_Z^{U'} (\overline{\lambda}_q)_{TUT'U'} \Delta_T^U \\
&= \sum_{U \geq T,\, U' \geq T'} a_Z^{TUT'U'}(q)\, \mu_Z^U \Delta_T^U \\
&= \sum_{U \geq T} a_Z^{TUT'}(q)\, \mu_Z^U \Delta_T^U.
\end{aligned}$$

The sums are over both U and U'. The first step used the q-bialgebra axiom (8.73), the second step used the associativity axiom in (8.41), the next step used (8.95), the last step used the first identity in (1.44a) applied to $\boldsymbol{\alpha}_Z$. $\square$

8.10. Signature functor on Coxeter spaces

The unsigned and signed worlds are connected by a simple construction called the u-signature functor. It involves twisting support morphisms in a Coxeter space by the signed distance function. Compare and contrast with the discussion for Coxeter species in Section 2.8, where we have the u- and o-signature functors.

For a Coxeter $\mathcal{A}$-space V, define the Coxeter $\mathcal{A}$-space V_- with (F, C)-component $\mathrm{V}[F, C]$, and support and type morphisms

$$\begin{array}{ccc}
\mathrm{V}_-[F,C] & \xrightarrow{\ \beta_{G,D,F,C}\ } & \mathrm{V}_-[G,D] \\
\| & & \| \\
\mathrm{V}[F,C] & \xrightarrow[(-1)^{\mathrm{dist}(C,D)} \beta_{G,D,F,C}]{} & \mathrm{V}[G,D]
\end{array}
\qquad
\begin{array}{ccc}
\mathrm{V}_-[F,C] & \xrightarrow{\ w\ } & \mathrm{V}_-[F',C'] \\
\| & & \| \\
\mathrm{V}[F,C] & \xrightarrow[w]{} & \mathrm{V}[F',C'].
\end{array}$$

Further, a map $f : \mathrm{V} \to \mathrm{W}$ of Coxeter $\mathcal{A}$-spaces induces a map $f : \mathrm{V}_- \to \mathrm{W}_-$ of Coxeter $\mathcal{A}$-spaces. Thus, we have a functor

$$(-)_- : \mathcal{A}\text{-}\overline{\mathsf{C}}\mathsf{Sp} \to \mathcal{A}\text{-}\overline{\mathsf{C}}\mathsf{Sp}, \qquad \mathrm{V} \mapsto \mathrm{V}_-. \tag{8.99}$$

This is the *u-signature functor*. It is an isomorphism of categories (with its inverse being itself). It is also its own left and right adjoint.

In terms of face-types: For a Coxeter $\boldsymbol{\alpha}$-space V, define the Coxeter $\boldsymbol{\alpha}$-space V_- with T-component $\mathrm{V}[T]$ and for $\sigma : T \to U$ in $\boldsymbol{\alpha}$-$\overline{\mathsf{Cox}}$,

$$\begin{array}{ccc} \mathrm{V}_-[T] & \overset{\sigma}{\dashrightarrow} & \mathrm{V}_-[U] \\ \| & & \| \\ \mathrm{V}[T] & \xrightarrow[(-1)^{l(\sigma)}\sigma]{} & \mathrm{V}[U]. \end{array}$$

It follows from Exercise 8.23, item (1) that V_- is indeed a Coxeter $\boldsymbol{\alpha}$-space.

Exercise 8.103. Refer to Exercise 8.38. Check: For a Coxeter $\boldsymbol{\alpha}$-space V, the components $\mathrm{V}[S]$ and $\mathrm{V}_-[S]$ differ by the sign representation of W. Deduce: V and V_- are not isomorphic as Coxeter $\boldsymbol{\alpha}$-spaces in general (assuming $|W| > 1$ and $\mathrm{char}(\Bbbk) \neq 2$).

Observation 8.104. If A is a Coxeter $\mathcal{A}$-algebra, then so is A_- with product components $\mu_{A,C}^{F,C}$ the same as those of A. Further, if A is commutative, then A_- is signed commutative, and vice versa. Dually, if C is a Coxeter $\mathcal{A}$-coalgebra, then so is C_-. Further, if C is cocommutative, then C_- is signed cocommutative, and vice versa. Combining the two situations, if H is a Coxeter $\mathcal{A}$-q-bialgebra, then H_- is a Coxeter $\mathcal{A}$-$(-q)$-bialgebra. In particular, if H is a Coxeter $\mathcal{A}$-bialgebra, then H_- is a signed Coxeter $\mathcal{A}$-bialgebra, and vice versa.

This yields the following.

Lemma 8.105. *For a scalar q, there is an isomorphism of categories*

$$(-)_- : q\text{-}\mathsf{Bimon}(\mathcal{A}\text{-}\overline{\mathsf{C}}\mathsf{Sp}) \to (-q)\text{-}\mathsf{Bimon}(\mathcal{A}\text{-}\overline{\mathsf{C}}\mathsf{Sp}), \qquad \mathrm{H} \mapsto \mathrm{H}_- \tag{8.100}$$

whose left and right adjoint is again $(-)_-$. In particular, the categories of Coxeter $\mathcal{A}$-bialgebras and signed Coxeter $\mathcal{A}$-bialgebras are isomorphic.

Similarly, the categories of (co)commutative Coxeter $\mathcal{A}$-(co)algebras and signed (co)commutative Coxeter $\mathcal{A}$-(co)algebras are isomorphic.

Exercise 8.106. For V a Coxeter $\mathcal{A}$-space, define the Coxeter $\mathcal{A}$-space V' with (F, C)-component $\mathrm{V}[F, C]$, and support and type morphisms

$$\begin{array}{ccc} \mathrm{V}'[F,C] & \overset{\beta_{G,D,F,C}}{\dashrightarrow} & \mathrm{V}'[G,D] \\ \| & & \| \\ \mathrm{V}[F,C] & \xrightarrow[(-1)^{\mathrm{dist}(C,D)}\beta_{G,D,F,C}]{} & \mathrm{V}[G,D] \end{array} \qquad \begin{array}{ccc} \mathrm{V}'[F,C] & \overset{w}{\dashrightarrow} & \mathrm{V}'[F',C'] \\ \| & & \| \\ \mathrm{V}[F,C] & \xrightarrow[(-1)^{l(w)}w]{} & \mathrm{V}[F',C']. \end{array}$$

Show: $\mathrm{V}' \cong \mathrm{V}$. (Define the isomorphism using distance from a fixed chamber C_0.) These considerations extend to Coxeter algebras, Coxeter coalgebras, Coxeter q-bialgebras.

Deduce: $\mathrm{V}'_- \cong \mathrm{V}_-$. Support and type morphisms of V'_- are

$$\begin{array}{ccc} \mathrm{V}'_-[F,C] & \overset{\beta_{G,D,F,C}}{\dashrightarrow} & \mathrm{V}'_-[G,D] \\ \| & & \| \\ \mathrm{V}[F,C] & \xrightarrow[\beta_{G,D,F,C}]{} & \mathrm{V}[G,D] \end{array} \qquad \begin{array}{ccc} \mathrm{V}'_-[F,C] & \overset{w}{\dashrightarrow} & \mathrm{V}'_-[F',C'] \\ \| & & \| \\ \mathrm{V}[F,C] & \xrightarrow[(-1)^{l(w)}w]{} & \mathrm{V}[F',C']. \end{array}$$

Deduce: In terms of top-lunes, the Coxeter space V_- up to isomorphism has the same components as V with type morphisms

$$\begin{array}{ccc} \mathrm{V}_-[\mathrm{L}] & \overset{w}{\dashrightarrow} & \mathrm{V}_-[w\mathrm{L}] \\ \| & & \| \\ \mathrm{V}[\mathrm{L}] & \xrightarrow[(-1)^{l(w)}w]{} & \mathrm{V}[w\mathrm{L}]. \end{array}$$

8.11. Coxeter (co, bi)algebras as functor categories

Just like the category of Coxeter spaces, the categories of Coxeter algebras, Coxeter coalgebras, Coxeter bialgebras and their (co)commutative counterparts are also functor categories. In particular, they are all $\Bbbk$-linear categories. We now discuss the relevant base categories. The notations are summarized in Table 8.8. Similar to $\mathcal{A}$-$\overline{\mathsf{Cox}}$, these base categories are all defined by presentations. The objects are either top-nested faces, top-lunes or face-types, while the relations encode the various axioms listed in Table 8.1. In the case of face-types, we use the same notations for the categories with $\boldsymbol{\alpha}$ instead of $\mathcal{A}$. In the case of top-lunes, we use the $'$ symbol.

TABLE 8.8. Coxeter (co, bi)algebras as functor categories.

C	$[\mathsf{C}, \mathsf{Vec}]$	C	$[\mathsf{C}, \mathsf{Vec}]$
$\mathcal{A}$-$\overline{\mathsf{Cox}}^{\mathsf{d}}$	Coxeter $\mathcal{A}$-algebras	$\mathcal{A}$-$\overline{\mathsf{Cox}}^{\mathsf{d}}_{\mathsf{c}}$	Coxeter $\mathcal{A}$-bialgebras
$\mathcal{A}$-$\overline{\mathsf{Cox}}_{\mathsf{c}}$	Coxeter $\mathcal{A}$-coalgebras	$\mathcal{A}$-$\overline{\mathsf{Cox}}^{\mathsf{e}}_{\mathsf{c}}$	com. Cox. $\mathcal{A}$-bialgebras
$\mathcal{A}$-$\overline{\mathsf{Cox}}^{\mathsf{e}}$	com. Cox. $\mathcal{A}$-algebras	$\mathcal{A}$-$\overline{\mathsf{Cox}}^{\mathsf{d}}_{\mathsf{r}}$	cocom. Cox. $\mathcal{A}$-bialgebras
$\mathcal{A}$-$\overline{\mathsf{Cox}}_{\mathsf{r}}$	cocom. Cox. $\mathcal{A}$-coalgebras	$\mathcal{A}$-$\overline{\mathsf{Cox}}^{\mathsf{e}}_{\mathsf{r}}$	bicom. Cox. $\mathcal{A}$-bialgebras

The discussion parallels that of Coxeter (co, bi)monoids in Section 2.9.

8.11.1. Coxeter (co)algebras in terms of top-nested faces. Define the category $\mathcal{A}$-$\overline{\mathsf{Cox}}_{\mathsf{c}}$ by a presentation. Objects are top-nested faces of $\mathcal{A}$. Morphisms are generated by

$$(F, C) \xrightarrow{\Delta} (G, C), \quad (F, C) \xrightarrow{w} (F', C'), \quad (A, C) \xrightarrow{\beta} (B, D),$$

subject to the 9 relations

$$\text{(8.101a)} \qquad \begin{array}{ccc} & (G,D) & \\ {\scriptstyle\beta}\nearrow & & \searrow{\scriptstyle\beta} \\ (F,C) & \xrightarrow[\beta]{} & (H,E) \end{array} \qquad \begin{array}{ccc} & (F',C') & \\ {\scriptstyle u}\nearrow & & \searrow{\scriptstyle v} \\ (F,C) & \xrightarrow[vu]{} & (F'',C'') \end{array} \qquad \begin{array}{ccc} & (G,C) & \\ {\scriptstyle\Delta}\nearrow & & \searrow{\scriptstyle\Delta} \\ (F,C) & \xrightarrow[\Delta]{} & (H,C) \end{array}$$

$$\text{(8.101b)} \qquad \begin{array}{ccc} (F,C) & \xrightarrow{\beta} & (G,D) \\ {\scriptstyle w}\downarrow & & \downarrow{\scriptstyle w} \\ (F',C') & \xrightarrow[\beta]{} & (G',D') \end{array} \qquad \begin{array}{ccc} (F,C) & \xrightarrow{\beta} & (G,D) \\ {\scriptstyle\Delta}\uparrow & & \uparrow{\scriptstyle\Delta} \\ (A,C) & \xrightarrow[\beta]{} & (B,D) \end{array} \qquad \begin{array}{ccc} (F,C) & \xrightarrow{w} & (F',C') \\ {\scriptstyle\Delta}\uparrow & & \uparrow{\scriptstyle\Delta} \\ (A,C) & \xrightarrow[w]{} & (A',C') \end{array}$$

$$((F,C) \xrightarrow{\Delta} (F,C)) = \mathrm{id}$$
$$(8.101c) \qquad ((F,C) \xrightarrow{e} (F,C)) = \mathrm{id}$$
$$((F,C) \xrightarrow{\beta} (F,C)) = \mathrm{id}\,.$$

By convention, we write $\Delta : (F,C) \to (G,C)$ whenever $F \leq G \leq C$, $w : (F,C) \to (F',C')$ whenever $F' = w(F)$ and $C' = w(C)$, and $\beta : (A,C) \to (B,D)$ whenever $\mathrm{s}(A,C) - \mathrm{s}(B,D)$.

We can also give a presentation for $\mathcal{A}\text{-}\overline{\mathsf{Cox}}_{\mathsf{c}}$ by using a smaller set of generators. Namely, we allow $\Delta : (F,C) \to (G,C)$ only when G covers F, and modify the third relation in (8.101a) to

$$(8.102) \qquad \begin{array}{ccc} (F',C) & \xrightarrow{\Delta} & (G,C) \\ {\scriptstyle\Delta}\uparrow & & \uparrow{\scriptstyle\Delta} \\ (A,C) & \xrightarrow[\Delta]{} & (F,C) \end{array}$$

whenever G covers F and F', and they in turn cover A. It is also understood that in the second relation in (8.101b), G covers B and hence AG covers A, in the third relation in (8.101b), F covers A and hence F' covers A', and the first relation in (8.101c) is removed.

Exercise 8.107. Show: The category $\mathcal{A}\text{-}\overline{\mathsf{Cox}}_{\mathsf{c}}$ is generated by the category $\mathcal{A}\text{-}\overline{\mathsf{Cox}}$ and morphisms $\Delta : (F,C) \to (G,C)$ subject to the relations

$$(8.103) \qquad \begin{array}{ccc} (\sigma^*(G),C) & \xrightarrow{\sigma} & (G,D) \\ {\scriptstyle\Delta}\uparrow & & \uparrow{\scriptstyle\Delta} \\ (A,C) & \xrightarrow[\sigma]{} & (B,D) \end{array} \qquad \begin{array}{ccc} & (G,C) & \\ {\scriptstyle\Delta}\nearrow & & \searrow{\scriptstyle\Delta} \\ (F,C) & \xrightarrow[\Delta]{} & (H,C) \end{array} \qquad ((F,C) \xrightarrow{\Delta} (F,C)) = \mathrm{id}\,.$$

In the first relation, $\sigma : (A,C) \to (B,D)$ is any morphism in $\mathcal{A}\text{-}\overline{\mathsf{Cox}}$, with the remaining notation as in (8.11).

Alternatively, $\mathcal{A}\text{-}\overline{\mathsf{Cox}}_{\mathsf{c}}$ is generated by $\mathcal{A}\text{-}\mathsf{Cox}$ and $\Delta : (F,C) \to (G,C)$ whenever G covers F, subject to relation (8.102) and the first relation in (8.103) with G covering B.

Let $\mathcal{A}\text{-}\overline{\mathsf{Cox}}^{\mathsf{d}}$ denote the category obtained from $\mathcal{A}\text{-}\overline{\mathsf{Cox}}_{\mathsf{c}}$ by reversing arrows labeled Δ, and relabeling them μ. In other words, morphisms of $\mathcal{A}\text{-}\overline{\mathsf{Cox}}^{\mathsf{d}}$ are generated by

$$(G,C) \xrightarrow{\mu} (F,C), \quad (F,C) \xrightarrow{w} (F',C'), \quad (A,C) \xrightarrow{\beta} (B,D).$$

The relations are written down in a similar manner.

Proposition 8.108. *The category of Coxeter $\mathcal{A}$-coalgebras is equivalent to the functor category $[\mathcal{A}\text{-}\overline{\mathsf{Cox}}_{\mathsf{c}}, \mathsf{Vec}]$. Dually, the category of Coxeter $\mathcal{A}$-algebras is equivalent to $[\mathcal{A}\text{-}\overline{\mathsf{Cox}}^{\mathsf{d}}, \mathsf{Vec}]$.*

PROOF. The argument is straightforward and similar to that for Proposition 2.57. □

Remark 8.109. The relations used to present categories $\mathcal{A}\text{-}\overline{\mathsf{Cox}}_{\mathsf{c}}$ and $\mathcal{A}\text{-}\overline{\mathsf{Cox}}^{\mathsf{d}}$ can be named using the nomenclature in Table 8.1. For instance, diagrams (8.101a) are the *support-support relation*, *type-type relation*, *coassociativity relation*, respectively. The same remark applies to categories that appear in the discussion below.

8.11.2. Coxeter (co)algebras in terms of top-lunes. Define the category $\mathcal{A}\text{-}\overline{\mathsf{Cox}}'_{\mathsf{c}}$ by a presentation. Objects are top-lunes of $\mathcal{A}$. Morphisms are generated by

$$\mathrm{L} \xrightarrow{\Delta} \mathrm{M}, \qquad \mathrm{L} \xrightarrow{w} \mathrm{L}',$$

with $\mathrm{L} \leq' \mathrm{M}$ and $\mathrm{L}' = w\mathrm{L}$, subject to the relations

$$\begin{array}{ccc} & \mathrm{M} & \\ {}^{\Delta}\nearrow & & \searrow^{\Delta} \\ \mathrm{L} & \xrightarrow[\Delta]{} & \mathrm{N} \end{array} \qquad \begin{array}{ccc} \mathrm{M} & \xrightarrow{w} & \mathrm{M}' \\ {\scriptstyle\Delta}\uparrow & & \uparrow{\scriptstyle\Delta} \\ \mathrm{L} & \xrightarrow[w]{} & \mathrm{L}' \end{array} \qquad \begin{array}{ccc} & \mathrm{L}' & \\ {}^{u}\nearrow & & \searrow^{v} \\ \mathrm{L} & \xrightarrow[vu]{} & \mathrm{L}'' \end{array} \tag{8.104a}$$

$$(\mathrm{L} \xrightarrow{\Delta} \mathrm{L}) = \mathrm{id} \qquad\qquad (\mathrm{L} \xrightarrow{e} \mathrm{L}) = \mathrm{id}\,. \tag{8.104b}$$

Exercise 8.110. Show: The categories $\mathcal{A}\text{-}\overline{\mathsf{Cox}}_{\mathsf{c}}$ and $\mathcal{A}\text{-}\overline{\mathsf{Cox}}'_{\mathsf{c}}$ are equivalent. More precisely, the functor $\mathcal{A}\text{-}\overline{\mathsf{Cox}}_{\mathsf{c}} \to \mathcal{A}\text{-}\overline{\mathsf{Cox}}'_{\mathsf{c}}$ which on objects sends (F, C) to $\mathrm{s}(F, C)$, and on morphisms sends

$$\begin{gathered} ((F,C) \xrightarrow{\Delta} (G,C)) \longmapsto (\mathrm{s}(F,C) \xrightarrow{\Delta} \mathrm{s}(G,C)) \\ ((F,C) \xrightarrow{w} (F',C')) \longmapsto (\mathrm{s}(F,C) \xrightarrow{w} \mathrm{s}(F',C')) \\ ((A,C) \xrightarrow{\beta} (B,D)) \longmapsto (\mathrm{s}(A,C) \xrightarrow{\mathrm{id}} \mathrm{s}(B,D)) \end{gathered}$$

is an equivalence. Deduce: The category of Coxeter $\mathcal{A}$-coalgebras is equivalent to the functor category $[\mathcal{A}\text{-}\overline{\mathsf{Cox}}'_{\mathsf{c}}, \mathsf{Vec}]$. This corresponds to the description of Coxeter $\mathcal{A}$-coalgebras in terms of top-lunes mentioned at the end of Section 8.5.5.

Let $(\mathcal{A}\text{-}\overline{\mathsf{Cox}}^{\mathsf{d}})'$ denote the category obtained from $\mathcal{A}\text{-}\overline{\mathsf{Cox}}'_{\mathsf{c}}$ by reversing arrows labeled Δ, and relabeling them μ. It is equivalent to the category $\mathcal{A}\text{-}\overline{\mathsf{Cox}}^{\mathsf{d}}$. Hence the category of Coxeter $\mathcal{A}$-algebras is equivalent to the functor category $[(\mathcal{A}\text{-}\overline{\mathsf{Cox}}^{\mathsf{d}})', \mathsf{Vec}]$. This corresponds to the description of Coxeter $\mathcal{A}$-algebras in terms of top-lunes in Proposition 8.41.

8.11.3. Coxeter (co)algebras in terms of face-types. Similar to the construction of $\boldsymbol{\alpha}\text{-}\overline{\mathsf{Cox}}$ from $\mathcal{A}\text{-}\overline{\mathsf{Cox}}$ in Section 8.1.5, define $\boldsymbol{\alpha}\text{-}\overline{\mathsf{Cox}}_{\mathsf{c}}$ to be the full subcategory of $\mathcal{A}\text{-}\overline{\mathsf{Cox}}_{\mathsf{c}}$ consisting of objects of the form (F, C_r) (which identify with face-types).

Exercise 8.111. Show: The category $\boldsymbol{\alpha}\text{-}\overline{\mathsf{Cox}}_{\mathsf{c}}$ is generated by the category $\boldsymbol{\alpha}\text{-}\overline{\mathsf{Cox}}$ and morphisms $\Delta : T \to U$ whenever $T \leq U$, subject to the relations

$$\begin{array}{ccc} \sigma^*(V) & \xrightarrow{\sigma} & V \\ {\scriptstyle\Delta}\uparrow & & \uparrow{\scriptstyle\Delta} \\ T & \xrightarrow[\sigma]{} & U \end{array} \qquad \begin{array}{ccc} & U & \\ {}^{\Delta}\nearrow & & \searrow^{\Delta} \\ T & \xrightarrow[\Delta]{} & V \end{array} \qquad (T \xrightarrow{\Delta} T) = \mathrm{id}\,. \tag{8.105}$$

In the first relation, $\sigma : T \to U$ is any morphism in $\boldsymbol{\alpha}\text{-}\overline{\mathsf{Cox}}$, with the remaining notation as in (8.12). See also (1.28).

Alternatively: The category $\boldsymbol{\alpha}\text{-}\overline{\mathtt{Cox}}_{\mathsf{c}}$ is generated by the category $\boldsymbol{\alpha}\text{-}\overline{\mathtt{Cox}}$ and morphisms $\Delta : T \to U$ whenever U covers T, subject to the relations

$$\begin{array}{ccc} \sigma^*(V) & \xrightarrow{\sigma} & V \\ {\scriptstyle\Delta}\uparrow & & \uparrow{\scriptstyle\Delta} \\ T & \xrightarrow[\sigma]{} & U \end{array} \qquad\qquad \begin{array}{ccc} T' & \xrightarrow{\Delta} & U \\ {\scriptstyle\Delta}\uparrow & & \uparrow{\scriptstyle\Delta} \\ Z & \xrightarrow[\Delta]{} & T. \end{array} \tag{8.106}$$

In the first relation, $\sigma : T \to U$ is any morphism in $\boldsymbol{\alpha}\text{-}\overline{\mathtt{Cox}}$, V covers U (and hence $\sigma^*(V)$ covers T). In the second relation, U covers T and T' which in turn cover Z.

Similarly, define $\boldsymbol{\alpha}\text{-}\overline{\mathtt{Cox}}^{\mathsf{d}}$ to be the full subcategory of $\mathcal{A}\text{-}\overline{\mathtt{Cox}}^{\mathsf{d}}$ consisting of objects of the form (F, C_r). Alternatively, it can be obtained from $\boldsymbol{\alpha}\text{-}\overline{\mathtt{Cox}}_{\mathsf{c}}$ by reversing arrows labeled Δ, and relabeling them μ.

Proposition 8.108 holds with $\boldsymbol{\alpha}$ instead of $\mathcal{A}$, with Coxeter $\boldsymbol{\alpha}$-algebras and $\boldsymbol{\alpha}$-coalgebras as defined in Sections 8.5.3 and 8.5.7, respectively.

8.11.4. Coxeter bialgebras in terms of top-nested faces. A *$\mathbb{Z}$-linear category* is a category in which each hom-set carries an abelian group structure and composition of morphisms is bilinear. A functor between such categories is *$\mathbb{Z}$-linear* if maps on hom-sets are morphisms of abelian groups [**7**, Section 1.1.4].

Define the $\mathbb{Z}$-linear category $\mathcal{A}\text{-}\overline{\mathtt{Cox}}^{\mathsf{d}}_{\mathsf{c}}$ by a presentation. Objects are top-nested faces of $\mathcal{A}$. Morphisms are generated by

$$(F, C) \xrightarrow{\Delta} (G, C), \quad (G, C) \xrightarrow{\mu} (F, C), \quad (F, C) \xrightarrow{w} (F', C'), \quad (A, C) \xrightarrow{\beta} (B, D),$$

subject to the 14 relations

(8.107a)

$$\begin{array}{ccc} & (G,D) & \\ {\scriptstyle\beta}\nearrow & & \searrow{\scriptstyle\beta} \\ (F,C) & \xrightarrow[\beta]{} & (H,E) \end{array} \qquad \begin{array}{ccc} (F,C) & \xrightarrow{\beta} & (G,D) \\ {\scriptstyle w}\downarrow & & \downarrow{\scriptstyle w} \\ (F',C') & \xrightarrow[\beta]{} & (G',D') \end{array} \qquad \begin{array}{ccc} & (F',C') & \\ {\scriptstyle u}\nearrow & & \searrow{\scriptstyle v} \\ (F,C) & \xrightarrow[vu]{} & (F'',C'') \end{array}$$

$$\begin{array}{ccc} (F,C) & \xrightarrow{\beta} & (G,D) \\ {\scriptstyle\Delta}\uparrow & & \uparrow{\scriptstyle\Delta} \\ (A,C) & \xrightarrow[\beta]{} & (B,D) \end{array} \qquad \begin{array}{ccc} & (G,C) & \\ {\scriptstyle\Delta}\nearrow & & \searrow{\scriptstyle\Delta} \\ (F,C) & \xrightarrow[\Delta]{} & (H,C) \end{array} \qquad \begin{array}{ccc} (F,C) & \xrightarrow{w} & (F',C') \\ {\scriptstyle\Delta}\uparrow & & \uparrow{\scriptstyle\Delta} \\ (A,C) & \xrightarrow[w]{} & (A',C') \end{array} \tag{8.107b}$$

$$\begin{array}{ccc} (F,C) & \xrightarrow{\beta} & (G,D) \\ {\scriptstyle\mu}\downarrow & & \downarrow{\scriptstyle\mu} \\ (A,C) & \xrightarrow[\beta]{} & (B,D) \end{array} \qquad \begin{array}{ccc} & (G,C) & \\ {\scriptstyle\mu}\nearrow & & \searrow{\scriptstyle\mu} \\ (H,C) & \xrightarrow[\mu]{} & (F,C) \end{array} \qquad \begin{array}{ccc} (F,C) & \xrightarrow{w} & (F',C') \\ {\scriptstyle\mu}\downarrow & & \downarrow{\scriptstyle\mu} \\ (A,C) & \xrightarrow[w]{} & (A',C') \end{array} \tag{8.107c}$$

$$\begin{array}{cc} ((F,C) \xrightarrow{\Delta} (F,C)) = \mathrm{id} & ((F,C) \xrightarrow{e} (F,C)) = \mathrm{id} \\ ((F,C) \xrightarrow{\mu} (F,C)) = \mathrm{id} & ((F,C) \xrightarrow{\beta} (F,C)) = \mathrm{id} \end{array} \tag{8.107d}$$

$$
\text{(8.107e)} \qquad \begin{array}{ccccc}
(F, C) & \xrightarrow{\ \mu\ } & (A, C) & \xrightarrow{\ \Delta\ } & (F', C) \\
{\scriptstyle \Delta}\big\downarrow & & & & \big\uparrow{\scriptstyle \mu} \\
(F(w^{-1}F'), C) & \xrightarrow[w]{} & ((wF)F', wC) & \xrightarrow[\beta]{} & (F'(wF), C).
\end{array}
$$

The conventions on morphisms are the same as before. Diagram (8.107e) is the *bialgebra relation*. It is different from the rest, and requires further explanation. Here, w is a variable. More precisely, we sum over $w \in W$ satisfying the conditions $w(A) = A$, $(wF)C = wC$, $F'(wC) = C$. This is the reason why we need to go to a $\mathbb{Z}$-linear category.

Proposition 8.112. *The category of Coxeter $\mathcal{A}$-bialgebras is equivalent to the category of $\mathbb{Z}$-linear functors $[\mathcal{A}\text{-}\overline{\mathsf{Cox}}^{\mathsf{d}}_{\mathsf{c}}, \mathsf{Vec}]$.*

PROOF. This is similar to the proof of Proposition 8.108. The 14 relations listed above translate to the 14 axioms listed in Summary 8.1. The bialgebra relation (8.107e) translates to the bialgebra axiom (8.55c) (first form). □

We can give a smaller presentation for the category $\mathcal{A}\text{-}\overline{\mathsf{Cox}}^{\mathsf{d}}_{\mathsf{c}}$ by using only those generators where G covers F. The relations include (8.102) and its dual, and (8.107e) with F and F' covering A. For the latter, note that $(wF)F'$ may not cover wF, or equivalently, $F'(wF)$ may not cover F', so the Δ and μ linking these are obtained by composing the generators.

8.11.5. Coxeter bialgebras in terms of top-lunes. Define the $\mathbb{Z}$-linear category $(\mathcal{A}\text{-}\overline{\mathsf{Cox}}^{\mathsf{d}}_{\mathsf{c}})'$ by a presentation. Objects are top-lunes of $\mathcal{A}$. Morphisms are generated by

$$
\mathrm{L} \xrightarrow{\Delta} \mathrm{M}, \qquad \mathrm{L} \xrightarrow{w} \mathrm{L}', \qquad \mathrm{M} \xrightarrow{\mu} \mathrm{L}
$$

with $\mathrm{L} \leq' \mathrm{M}$ and $\mathrm{L}' = w\mathrm{L}$, subject to the 9 relations

$$
\text{(8.108a)} \qquad
\begin{array}{ccc}
& \mathrm{M} & \\
{\scriptstyle\Delta}\nearrow & & \searrow{\scriptstyle\Delta} \\
\mathrm{L} & \xrightarrow[\Delta]{} & \mathrm{N}
\end{array}
\qquad
\begin{array}{ccc}
& \mathrm{L}' & \\
{\scriptstyle u}\nearrow & & \searrow{\scriptstyle v} \\
\mathrm{L} & \xrightarrow[vu]{} & \mathrm{L}''
\end{array}
\qquad
\begin{array}{ccc}
& \mathrm{M} & \\
{\scriptstyle\mu}\nearrow & & \searrow{\scriptstyle\mu} \\
\mathrm{N} & \xrightarrow[\mu]{} & \mathrm{L}
\end{array}
$$

$$
\text{(8.108b)} \qquad
\begin{array}{ccc}
\mathrm{M} & \xrightarrow{w} & \mathrm{M}' \\
{\scriptstyle\Delta}\big\uparrow & & \big\uparrow{\scriptstyle\Delta} \\
\mathrm{L} & \xrightarrow[w]{} & \mathrm{L}'
\end{array}
\qquad
\begin{array}{ccccc}
\mathrm{L} & \xrightarrow{\mu} & \mathrm{N} & \xrightarrow{\Delta} & \mathrm{L}' \\
{\scriptstyle\Delta}\big\downarrow & & & & \big\uparrow{\scriptstyle\mu} \\
\mathrm{b}(\mathrm{L}) \cdot w^{-1}\mathrm{L}' & & \xrightarrow[w]{} & & \mathrm{b}(\mathrm{L}') \cdot w\mathrm{L}
\end{array}
\qquad
\begin{array}{ccc}
\mathrm{M} & \xrightarrow{w} & \mathrm{M}' \\
{\scriptstyle\mu}\big\downarrow & & \big\downarrow{\scriptstyle\mu} \\
\mathrm{L} & \xrightarrow[w]{} & \mathrm{L}'
\end{array}
$$

$$
\text{(8.108c)} \qquad (\mathrm{L} \xrightarrow{\Delta} \mathrm{L}) = \mathrm{id} \qquad (\mathrm{L} \xrightarrow{e} \mathrm{L}) = \mathrm{id} \qquad (\mathrm{L} \xrightarrow{\mu} \mathrm{L}) = \mathrm{id}\,.
$$

The middle diagram in (8.108b) is the *bialgebra relation*. Here, w is a variable. More precisely, we sum over $w \in W$ fixing $\mathrm{b}(\mathrm{N})$ pointwise and satisfying the conditions $\mathrm{b}(w\mathrm{L}) \cdot \mathrm{N} = w\mathrm{L}$ and $\mathrm{b}(\mathrm{L}') \cdot w\mathrm{N} = \mathrm{L}'$.

Exercise 8.113. Show: The $\mathbb{Z}$-linear categories $\mathcal{A}\text{-}\overline{\mathsf{Cox}}^{\mathsf{d}}_{\mathsf{c}}$ and $(\mathcal{A}\text{-}\overline{\mathsf{Cox}}^{\mathsf{d}}_{\mathsf{c}})'$ are equivalent. Deduce: The category of Coxeter $\mathcal{A}$-bialgebras is equivalent to the category of $\mathbb{Z}$-linear functors $[(\mathcal{A}\text{-}\overline{\mathsf{Cox}}^{\mathsf{d}}_{\mathsf{c}})', \mathsf{Vec}]$. This corresponds to the description of Coxeter $\mathcal{A}$-bialgebras in terms of top-lunes in Proposition 8.58. The 9 relations listed above translate to the 9 axioms listed in Summary 8.2. The bialgebra relation translates to the bialgebra axiom in the form (8.62).

8.11.6. (Co)commutativity in terms of top-nested faces. Define $\mathcal{A}\text{-}\overline{\mathsf{Cox}}_{\mathrm{r}}$ to be the category $\mathcal{A}\text{-}\overline{\mathsf{Cox}}_{\mathrm{c}}$ subject to the *cocommutativity relation*

$$\begin{array}{ccc} (F,C) & \xrightarrow{\ \sigma\ } & (F',C) \\ & {}_{\Delta}\nwarrow \quad \nearrow_{\Delta} & \\ & (A,C), & \end{array} \tag{8.109}$$

whenever $A \leq F \leq C$, $A \leq F' \leq C$ and $\sigma : (F,C) \to (F',C)$ is a morphism in $\mathcal{A}\text{-}\overline{\mathsf{Cox}}$ local to A.

Alternatively, $\mathcal{A}\text{-}\overline{\mathsf{Cox}}_{\mathrm{r}}$ is the category $\mathcal{A}\text{-}\overline{\mathsf{Cox}}_{\mathrm{c}}$ subject to the *cocommutativity relation*

$$\begin{array}{ccc} (F,C) & \xrightarrow{\ \beta\ } & (G,D) \\ {\scriptstyle\Delta}\uparrow & & \uparrow{\scriptstyle\Delta} \\ (A,C) & \xrightarrow[\ w\ ]{} & (A,D), \end{array} \tag{8.110}$$

whenever $A \leq F \leq C$, $A \leq G \leq D$, $\mathrm{s}(F,C) = \mathrm{s}(G,D)$, and w is the Coxeter symmetry which sends C to D (and hence A to itself).

In the definition of $\mathcal{A}\text{-}\overline{\mathsf{Cox}}_{\mathrm{r}}$, the cocommutativity relation (8.110) may be replaced by the *opposition cocommutativity relation*

$$\begin{array}{ccc} (F,C) & \xrightarrow{\ \beta\ } & (A\overline{F}, A\overline{F}C) \\ {\scriptstyle\Delta}\uparrow & & \uparrow{\scriptstyle\Delta} \\ (A,C) & \xrightarrow[\ w\ ]{} & (A, A\overline{F}C), \end{array} \tag{8.111}$$

whenever $A \leq F \leq C$. In fact, it suffices to require this relation whenever $A \lessdot F \leq C$. We refer to this as the *covering cocommutativity relation.*

Define $\mathcal{A}\text{-}\overline{\mathsf{Cox}}^{\mathrm{e}}$ to be the category $\mathcal{A}\text{-}\overline{\mathsf{Cox}}^{\mathrm{d}}$ subject to the relation dual to (8.109), or equivalently, dual to (8.110), or dual to (8.111). (That is, we reverse arrows labeled Δ and relabel them μ.)

Proposition 8.114. *The category of cocommutative Coxeter $\mathcal{A}$-coalgebras is equivalent to the functor category $[\mathcal{A}\text{-}\overline{\mathsf{Cox}}_{\mathrm{r}}, \mathsf{Vec}]$. Dually, the category of commutative Coxeter $\mathcal{A}$-algebras is equivalent to $[\mathcal{A}\text{-}\overline{\mathsf{Cox}}^{\mathrm{e}}, \mathsf{Vec}]$.*

The cocommutativity relation (8.109) translates to the cocommutativity axiom (8.88), while (8.110) translates to the cocommutativity axiom (8.89). Similarly, opposition and covering cocommutativity relations translate to opposition and covering cocommutativity axioms. The dual relations translate to commutativity axioms (8.79), (8.80), (8.81).

8.11.7. (Co)commutativity in terms of top-lunes. Define $\mathcal{A}\text{-}\overline{\mathtt{Cox}}'_{\mathrm{r}}$ to be the category $\mathcal{A}\text{-}\overline{\mathtt{Cox}}'_{\mathrm{c}}$ subject to either the *cocommutativity relation* or the equivalent *upside-down cocommutativity relation*

$$(8.112)\qquad \begin{array}{ccc} \mathrm{M} & \xrightarrow{\;u\;} & \mathrm{M}' \\ {\scriptstyle \Delta^{\mathrm{M}}_{\mathrm{L}}}\nwarrow & & \nearrow{\scriptstyle \Delta^{\mathrm{M}'}_{\mathrm{L}}} \\ & \mathrm{L} & \end{array} \qquad\qquad \begin{array}{ccc} & \mathrm{M} & \\ {\scriptstyle \Delta^{\mathrm{M}}_{\mathrm{L}}}\nearrow & & \nwarrow{\scriptstyle \Delta^{\mathrm{M}}_{\mathrm{L}'}} \\ \mathrm{L} & \xrightarrow[\;v\;]{} & \mathrm{L}'. \end{array}$$

In the first diagram, u is such that it fixes $\mathrm{b}(\mathrm{L})$ pointwise, while in the second diagram, $\mathrm{b}(\mathrm{L}) = \mathrm{b}(\mathrm{L}')$ and v is such that it fixes $\mathrm{b}(\mathrm{L}) = \mathrm{b}(\mathrm{L}')$ pointwise.

Building on Exercise 8.110, the categories $\mathcal{A}\text{-}\overline{\mathtt{Cox}}_{\mathrm{r}}$ and $\mathcal{A}\text{-}\overline{\mathtt{Cox}}'_{\mathrm{r}}$ are equivalent. Hence, the category of cocommutative Coxeter $\mathcal{A}$-coalgebras is equivalent to the functor category $[\mathcal{A}\text{-}\overline{\mathtt{Cox}}'_{\mathrm{r}}, \mathsf{Vec}]$. The cocommutativity relations (8.112) translate to the cocommutativity axioms (8.90).

Define $(\mathcal{A}\text{-}\overline{\mathtt{Cox}}^{\mathsf{e}})'$ to be the category $(\mathcal{A}\text{-}\overline{\mathtt{Cox}}^{\mathsf{d}})'$ subject to relations dual to (8.112). They translate to the commutativity axiom (8.82) and upside-down commutativity axiom (8.83).

8.11.8. (Co)commutativity in terms of face-types. Let $\boldsymbol{\alpha}\text{-}\overline{\mathtt{Cox}}_{\mathrm{r}}$ be the full subcategory of $\mathcal{A}\text{-}\overline{\mathtt{Cox}}_{\mathrm{r}}$ consisting of objects of the form (F, C_r). Then $\boldsymbol{\alpha}\text{-}\overline{\mathtt{Cox}}_{\mathrm{r}}$ is the same as the category $\boldsymbol{\alpha}\text{-}\overline{\mathtt{Cox}}_{\mathrm{c}}$ subject to the *cocommutativity relation*

$$(8.113)\qquad \begin{array}{ccc} T & \xrightarrow{\;\sigma\;} & T' \\ {\scriptstyle \Delta}\nwarrow & & \nearrow{\scriptstyle \Delta} \\ & Z & \end{array}$$

whenever $Z \leq T$, $Z \leq T'$ and σ is a morphism local to Z.

In this presentation, the cocommutativity relation may be replaced by the *opposition cocommutativity relation*

$$(8.114)\qquad \begin{array}{ccc} T & \xrightarrow{\;\beta_{Z\overline{T},T}\;} & Z\overline{T} \\ {\scriptstyle \Delta}\nwarrow & & \nearrow{\scriptstyle \Delta} \\ & Z, & \end{array}$$

whenever $Z \leq T$. In fact, it suffices to require this relation whenever $Z \lessdot T$. The latter is the same as the cocommutativity relation (8.113) for adjacent support-type morphisms σ. We refer to this as the *covering cocommutativity relation.*

Define $\boldsymbol{\alpha}\text{-}\overline{\mathtt{Cox}}^{\mathsf{e}}$ to be the subcategory of $\boldsymbol{\alpha}\text{-}\overline{\mathtt{Cox}}^{\mathsf{d}}$ subject to the relations dual to (8.113), or equivalently, dual to (8.114). (That is, we reverse arrows labeled Δ, and relabel them μ.)

Proposition 8.114 holds with $\boldsymbol{\alpha}$ in place of $\mathcal{A}$.

8.12. Bimonads on Coxeter spaces

We construct a monad $\overline{\mathcal{T}}$ on Coxeter spaces whose algebras are Coxeter algebras, a comonad $\overline{\mathcal{T}}^\vee$ whose coalgebras are Coxeter coalgebras, and a bimonad $(\overline{\mathcal{T}}, \overline{\mathcal{T}}^\vee, \overline{\lambda})$ whose bialgebras are Coxeter bialgebras. These also have commutative, signed commutative and q-analogues.

The discussion parallels that of Coxeter species in Section 2.10. For connections between the two using Fock functors, see Chapter 14.

8.12.1. Coxeter bialgebras in terms of top-nested faces. Let $\mathcal{A}$ be a reflection arrangement. The basic result is stated below.

Proposition 8.115. *The following are equivalences of categories.*

$$\begin{aligned} \overline{\mathcal{T}}\text{-algebras} &\cong \text{Coxeter } \mathcal{A}\text{-algebras} \\ \overline{\mathcal{T}}^\vee\text{-coalgebras} &\cong \text{Coxeter } \mathcal{A}\text{-coalgebras} \\ (\overline{\mathcal{T}}, \overline{\mathcal{T}}^\vee, \overline{\lambda})\text{-bialgebras} &\cong \text{Coxeter } \mathcal{A}\text{-bialgebras} \\ (\overline{\mathcal{T}}, \overline{\mathcal{T}}^\vee, \overline{\lambda}_q)\text{-bialgebras} &\cong \text{Coxeter } \mathcal{A}\text{-}q\text{-bialgebras} \end{aligned}$$

Let us first construct the functor $\overline{\mathcal{T}} : \mathcal{A}\text{-}\overline{\mathsf{C}}\mathsf{Sp} \to \mathcal{A}\text{-}\overline{\mathsf{C}}\mathsf{Sp}$. For a Coxeter $\mathcal{A}$-space V, define the Coxeter $\mathcal{A}$-space $\overline{\mathcal{T}}(\mathrm{V})$ on the (A, C)-component by

$$\overline{\mathcal{T}}(\mathrm{V})[A, C] := \bigoplus_{F:\, A \leq F \leq C} \mathrm{V}[F, C]. \tag{8.115}$$

Support and type morphisms of $\overline{\mathcal{T}}(\mathrm{V})$ are defined from those of V as follows.

- Suppose $\mathrm{s}(A, C) = \mathrm{s}(B, D)$. Then, for every F between A and C, there is a face $G := BF$ between B and D with $\mathrm{s}(G, D) = \mathrm{s}(F, C)$. The sum of the linear maps $\beta_{G,D,F,C} : \mathrm{V}[F, C] \to \mathrm{V}[G, D]$ yields the linear map $\beta_{B,D,A,C} : \overline{\mathcal{T}}(\mathrm{V})[A, C] \to \overline{\mathcal{T}}(\mathrm{V})[B, D]$.
- Suppose $w \in W$ and $w(A) = A'$, $w(C) = C'$. Then, for every F between A and C, there is a face $F' := w(F)$ between A' and C'. The sum of the linear maps $w : \mathrm{V}[F, C] \to \mathrm{V}[F', C']$ yields the linear map $w : \overline{\mathcal{T}}(\mathrm{V})[A, C] \to \overline{\mathcal{T}}(\mathrm{V})[A', C']$.

More abstractly, any morphism $\sigma : (A, C) \to (B, D)$ induces a morphism $\sigma : (F, C) \to (G, D)$, where $G = \sigma_*(F)$ as in (8.11). The sum of the linear maps $\sigma : \mathrm{V}[F, C] \to \mathrm{V}[G, D]$ yields the linear map

$$\sigma : \overline{\mathcal{T}}(\mathrm{V})[A, C] \to \overline{\mathcal{T}}(\mathrm{V})[B, D].$$

This turns $\overline{\mathcal{T}}(\mathrm{V})$ into a Coxeter $\mathcal{A}$-space. Further, a map $f : \mathrm{V} \to \mathrm{W}$ of Coxeter $\mathcal{A}$-spaces yields a map $\overline{\mathcal{T}}(f) : \overline{\mathcal{T}}(\mathrm{V}) \to \overline{\mathcal{T}}(\mathrm{W})$ of Coxeter $\mathcal{A}$-spaces whose (A, C)-component $\overline{\mathcal{T}}(f)_{A,C}$ is the sum of the linear maps $f_{F,C} : \mathrm{V}[F, C] \to \mathrm{W}[F, C]$ over all F between A and C. Thus $\overline{\mathcal{T}}$ is a functor. Note: It is $\Bbbk$-linear.

We now specify the monad structure on $\overline{\mathcal{T}}$. For that, first observe:

$$\overline{\mathcal{T}}\overline{\mathcal{T}}(\mathrm{V})[A, C] = \bigoplus_{(F, G):\, A \leq F \leq G \leq C} \mathrm{V}[G, C].$$

The sum is over both F and G. The structure map

$$\overline{\mathcal{T}}\,\overline{\mathcal{T}} \to \overline{\mathcal{T}}, \qquad \bigoplus_{(F,G)\colon A\leq F\leq G\leq C} \mathrm{V}[G,C] \to \bigoplus_{G\colon A\leq G\leq C} \mathrm{V}[G,C]$$

evaluated on V on the (A, C)-component sends each summand in the lhs identically to the matching summand in the rhs, while

$$\mathrm{id} \to \overline{\mathcal{T}}, \qquad \mathrm{V}[A,C] \to \bigoplus_{F\colon A\leq F\leq C} \mathrm{V}[F,C]$$

evaluated on V on the (A, C)-component is given by inclusion. One can check that $\overline{\mathcal{T}}$-algebras are the same as Coxeter $\mathcal{A}$-algebras (Section 8.5.1).

The comonad $\overline{\mathcal{T}}^\vee$ on Coxeter $\mathcal{A}$-spaces is constructed dually. As a functor, $\overline{\mathcal{T}}^\vee = \overline{\mathcal{T}}$. The comonad structure on $\overline{\mathcal{T}}^\vee$ is as follows. The map

$$\overline{\mathcal{T}}^\vee \to \overline{\mathcal{T}}^\vee\overline{\mathcal{T}}^\vee, \qquad \bigoplus_{G\colon A\leq G\leq C} \mathrm{V}[G,C] \to \bigoplus_{(F,G)\colon A\leq F\leq G\leq C} \mathrm{V}[G,C]$$

evaluated on V on the (A, C)-component sends each summand in the lhs identically to all matching summands in the rhs, while

$$\overline{\mathcal{T}}^\vee \to \mathrm{id}, \qquad \bigoplus_{F\colon A\leq F\leq C} \mathrm{V}[F,C] \to \mathrm{V}[A,C]$$

evaluated on V on the (A, C)-component sends $\mathrm{V}[A, C]$ to itself, and all other summands to zero. Coalgebras over $\overline{\mathcal{T}}^\vee$ are the same as Coxeter $\mathcal{A}$-coalgebras (Section 8.5.5).

Now define a mixed distributive law between $\overline{\mathcal{T}}$ and $\overline{\mathcal{T}}^\vee$ as follows. The map

$$(8.116) \qquad \overline{\lambda} : \overline{\mathcal{T}}\,\overline{\mathcal{T}}^\vee \to \overline{\mathcal{T}}^\vee\overline{\mathcal{T}}, \qquad \bigoplus_{\substack{(F_0,G_0):\\ A\leq F_0\leq G_0\leq C}} \mathrm{V}[G_0,C] \to \bigoplus_{\substack{(F_0',G_0'):\\ A\leq F_0'\leq G_0'\leq C}} \mathrm{V}[G_0',C]$$

evaluated on V on the (A, C)-component has matrix-components $\overline{\lambda}_{F_0,G_0,F_0',G_0'}$ defined by (8.49). (See Remark 2.72 for the meaning of matrix-component.)

Exercise 8.116. Check: (8.116) is a natural transformation. Thus, evaluated on V, it is a map of Coxeter $\mathcal{A}$-spaces, that is, diagrams (8.21) commute. (For the first diagram, use support-support and support-type axioms, while for the second diagram, use support-type and type-type axioms.)

Theorem 8.117. *The triple $(\overline{\mathcal{T}}, \overline{\mathcal{T}}^\vee, \overline{\lambda})$ is a bimonad on Coxeter $\mathcal{A}$-spaces. Its bialgebras are the same as Coxeter $\mathcal{A}$-bialgebras.*

PROOF. The argument is as in the proof of Theorem 2.75. Let us check the left diagram in (A.7a). Evaluated on V, on the (A, C)-component, it breaks as a sum of diagrams of the form shown below. The sum is indexed by appropriate choices of w, or equivalently, by appropriate choices of u and v.

The relationship between the two sets of choices is $w = vu$.

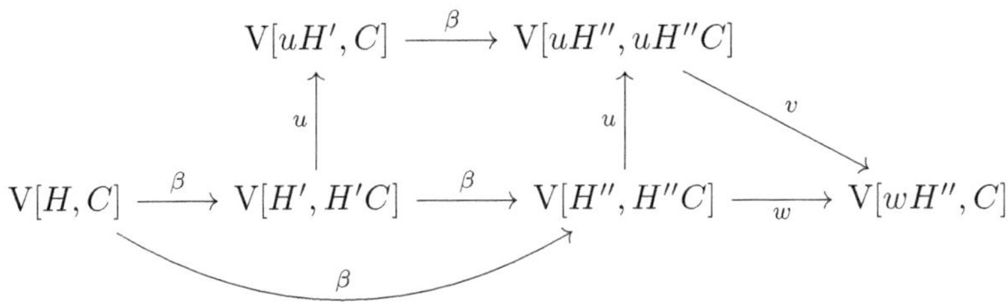

The notations are consistent with those in diagram (2.89). The starting summand is indexed by $A \leq F \leq H \leq C$, while the ending summand is indexed by $A \leq wF' \leq wG'' \leq wH'' \leq C$.

Suppose we choose an appropriate w. Then F', G'', H'' are determined, hence, as in the proof of Theorem 2.75, G, G', H' are also determined. As a result, u is determined by $u(H'C) = C$, and finally v is determined by $w = vu$. The triangles are support-support and type-type axioms, and the square is a support-type axiom. Observe: $(uH'')C = u(H''C)$, hence the notation $uH''C$ appearing in the diagram is unambiguous.

Similarly, suppose we choose an appropriate u and v. Then w is determined by $w = vu$.

The diagrams in (A.7b) can be checked using Exercise 8.51. □

More generally, for a scalar q, we define a mixed distributive law $\overline{\lambda}_q$ with matrix components (8.65). In fact, we define $\overline{\lambda}_v$ for any W-invariant distance function v (Section 8.7.6).

Theorem 8.118. *For a scalar q, the triple $(\overline{\mathcal{T}}, \overline{\mathcal{T}}^\vee, \overline{\lambda}_q)$ is a bimonad on Coxeter $\mathcal{A}$-spaces whose bialgebras are the same as Coxeter $\mathcal{A}$-q-bialgebras. More generally, for a W-invariant distance function v, $(\overline{\mathcal{T}}, \overline{\mathcal{T}}^\vee, \overline{\lambda}_v)$ is a bimonad whose bialgebras are the same as Coxeter $\mathcal{A}$-v-bialgebras.*

PROOF. We rerun the previous proof. For general q, β gets replaced by β_q, and the triangles and square continue to commute. This uses the gate property. The same remark applies to any W-invariant distance function v. Also Exercise 8.51 works in this generality. □

Exercise 8.119. Check: Theorems 8.117 and 8.118 extend to the categories of Coxeter lb-spaces and Coxeter b-spaces. Thus, we can talk of Coxeter algebras, Coxeter coalgebras, Coxeter q-bialgebras in these more general contexts. The explicit definitions stay the same as before.

8.12.2. Coxeter bialgebras in terms of top-lunes. The above discussion can be recast in terms of top-lunes. For a Coxeter $\mathcal{A}$-space V, the Coxeter $\mathcal{A}$-space $\overline{\mathcal{T}}(\mathrm{V})$ on the L-component is

$$\overline{\mathcal{T}}(\mathrm{V})[\mathrm{L}] := \bigoplus_{\mathrm{M}:\, \mathrm{L} \leq' \mathrm{M}} \mathrm{V}[\mathrm{M}]. \tag{8.117}$$

For $w \in W$, the type morphism

$$w : \overline{\mathcal{T}}(\mathrm{V})[\mathrm{L}] \to \overline{\mathcal{T}}(\mathrm{V})[w\mathrm{L}]$$

is the sum of the linear maps $w : \mathrm{V}[\mathrm{M}] \to \mathrm{V}[w\mathrm{M}]$ over all M greater than L wrt $\leq'$.

The monad structure on $\overline{\mathcal{T}}$ is as follows. The map

$$\overline{\mathcal{T}}\overline{\mathcal{T}} \to \overline{\mathcal{T}}, \qquad \bigoplus_{(\mathrm{L},\mathrm{M}):\, \mathrm{N} \leq' \mathrm{L} \leq' \mathrm{M}} \mathrm{V}[\mathrm{M}] \to \bigoplus_{\mathrm{M}:\, \mathrm{N} \leq' \mathrm{M}} \mathrm{V}[\mathrm{M}]$$

evaluated on V on the N-component sends each summand in the lhs identically to the matching summand in the rhs, while

$$\mathrm{id} \to \overline{\mathcal{T}}, \qquad \mathrm{V}[\mathrm{N}] \to \bigoplus_{\mathrm{L}:\, \mathrm{N} \leq' \mathrm{L}} \mathrm{V}[\mathrm{L}]$$

evaluated on V on the N-component is given by inclusion. Algebras over $\overline{\mathcal{T}}$ are the same as Coxeter $\mathcal{A}$-algebras (as formulated in Proposition 8.41).

The comonad $\overline{\mathcal{T}}^{\vee}$ is constructed dually. Its coalgebras are the same as Coxeter $\mathcal{A}$-coalgebras (as mentioned in Section 8.5.6).

The mixed distributive law

$$\overline{\lambda} : \overline{\mathcal{T}}\overline{\mathcal{T}}^{\vee} \to \overline{\mathcal{T}}^{\vee}\overline{\mathcal{T}}, \qquad \bigoplus_{\substack{(\mathrm{L},\mathrm{M}):\\ \mathrm{N} \leq' \mathrm{L} \leq' \mathrm{M}}} \mathrm{V}[\mathrm{M}] \to \bigoplus_{\substack{(\mathrm{L}',\mathrm{M}'):\\ \mathrm{N} \leq' \mathrm{L}' \leq' \mathrm{M}'}} \mathrm{V}[\mathrm{M}'] \tag{8.118}$$

evaluated on V on the N-component has matrix-components $\overline{\lambda}_{\mathrm{L},\mathrm{M},\mathrm{L}',\mathrm{M}'}$ defined by (8.57). More generally, for a scalar q, we have the mixed distributive law $\overline{\lambda}_q$ with matrix components (8.68). Bialgebras over the bimonad $(\overline{\mathcal{T}}, \overline{\mathcal{T}}^{\vee}, \overline{\lambda})$ are the same as Coxeter $\mathcal{A}$-bialgebras (as formulated in Proposition 8.58), while bialgebras over $(\overline{\mathcal{T}}, \overline{\mathcal{T}}^{\vee}, \overline{\lambda}_q)$ are the same as Coxeter $\mathcal{A}$-q-bialgebras (as in Section 8.7.2).

8.12.3. Coxeter bialgebras in terms of face-types. One may also view $\overline{\mathcal{T}}$ and $\overline{\mathcal{T}}^{\vee}$ as a monad and comonad on Coxeter $\boldsymbol{\alpha}$-spaces. In this case, we start with

$$\overline{\mathcal{T}}(\mathrm{V})[Z] = \overline{\mathcal{T}}^{\vee}(\mathrm{V})[Z] := \bigoplus_{T:\, Z \leq T} \mathrm{V}[T], \tag{8.119}$$

and proceed. A morphism $\sigma : Z \to Z'$ induces a morphism $\sigma : T \to T'$ with $T' = \sigma_*(T)$ and $T = \sigma^*(T')$ as in (8.12) or (1.28). The sum of the linear maps $\sigma : \mathrm{V}[T] \to \mathrm{V}[T']$ yields the linear map

$$\sigma : \overline{\mathcal{T}}(\mathrm{V})[Z] \to \overline{\mathcal{T}}(\mathrm{V})[Z'].$$

This is the Coxeter $\boldsymbol{\alpha}$-space structure of $\overline{\mathcal{T}}(\mathrm{V})$. The mixed distributive law

$$\overline{\lambda} : \overline{\mathcal{T}}\overline{\mathcal{T}}^{\vee} \to \overline{\mathcal{T}}^{\vee}\overline{\mathcal{T}}, \qquad \bigoplus_{\substack{(T,U):\\ Z \leq T \leq U}} \mathrm{V}[U] \to \bigoplus_{\substack{(T',U'):\\ Z \leq T' \leq U'}} \mathrm{V}[U'] \tag{8.120}$$

evaluated on V on the Z-component has matrix-components $\overline{\lambda}_{TUT'U'}$ defined by (8.63). More generally, for a scalar q, we have the mixed distributive law $\overline{\lambda}_q$ with matrix components (8.70).

Exercise 8.120. Verify Theorems 8.117 and 8.118 by working with the above definitions in terms of face-types. (Use the gate property for the groupoid of biface-types in Section 1.7.9.)

Algebras over $\overline{\mathcal{T}}$ and coalgebras over $\overline{\mathcal{T}}^{\vee}$ are the same as Coxeter $\boldsymbol{\alpha}$-algebras and Coxeter $\boldsymbol{\alpha}$-coalgebras in Sections 8.5.3 and 8.5.7, respectively. Similarly, bialgebras over $(\overline{\mathcal{T}}, \overline{\mathcal{T}}^{\vee}, \overline{\lambda})$ are the same as Coxeter $\boldsymbol{\alpha}$-bialgebras in Section 8.6.4, while bialgebras over $(\overline{\mathcal{T}}, \overline{\mathcal{T}}^{\vee}, \overline{\lambda}_q)$ are the same as Coxeter $\boldsymbol{\alpha}$-q-bialgebras in Section 8.7.3.

8.12.4. Bicommutative Coxeter bialgebras.

We now turn to commutative aspects. The basic result is stated below (in the setting of Coxeter $\boldsymbol{\alpha}$-spaces).

Proposition 8.121. *The following are equivalences of categories.*

$$\begin{aligned}
\overline{\mathcal{S}}\text{-algebras} &\cong \text{com. Coxeter } \boldsymbol{\alpha}\text{-algebras}\\
\overline{\mathcal{S}}^{\vee}\text{-coalgebras} &\cong \text{cocom. Coxeter } \boldsymbol{\alpha}\text{-coalgebras}\\
(\overline{\mathcal{S}}, \overline{\mathcal{T}}^{\vee}, \overline{\lambda})\text{-bialgebras} &\cong \text{com. Coxeter } \boldsymbol{\alpha}\text{-bialgebras}\\
(\overline{\mathcal{T}}, \overline{\mathcal{S}}^{\vee}, \overline{\lambda})\text{-bialgebras} &\cong \text{cocom. Coxeter } \boldsymbol{\alpha}\text{-bialgebras}\\
(\overline{\mathcal{S}}, \overline{\mathcal{S}}^{\vee}, \overline{\lambda})\text{-bialgebras} &\cong \text{bicom. Coxeter } \boldsymbol{\alpha}\text{-bialgebras}
\end{aligned}$$

We now construct the monad $\overline{\mathcal{S}} : \boldsymbol{\alpha}\text{-}\overline{\mathsf{C}}\mathsf{Sp} \to \boldsymbol{\alpha}\text{-}\overline{\mathsf{C}}\mathsf{Sp}$. For a Coxeter $\boldsymbol{\alpha}$-space V and face-type Z, define

$$\overline{\mathcal{S}}(\mathrm{V})[Z] := \operatorname{colim} \mathrm{V}_Z,$$

where $\mathrm{V}_Z : \boldsymbol{\alpha}_Z\text{-}\overline{\mathtt{Cox}} \to \mathsf{Vec}$ is the functor obtained by restricting V to the subcategory $\boldsymbol{\alpha}_Z\text{-}\overline{\mathtt{Cox}}$. Explicitly, it is the quotient space

$$\overline{\mathcal{S}}(\mathrm{V})[Z] = \Big(\bigoplus_{T:\, Z \leq T} \mathrm{V}[T] \Big)_{\mathrm{quo}}, \tag{8.121}$$

where $x \equiv \sigma(x)$ for any morphism $\sigma : T \to T'$ local to Z. Observe: $\mathrm{V}[Z]$ is a direct summand of $\overline{\mathcal{S}}(\mathrm{V})[Z]$ (since the only morphism from Z which is local to Z is the identity).

Suppose $\sigma : Z \to Z'$ is a morphism in $\boldsymbol{\alpha}\text{-}\overline{\mathtt{Cox}}$. It induces an isomorphism between functors V_Z and $\mathrm{V}_{Z'}$ (precomposed with the isomorphism (8.14)). Taking colimits yields a linear map

$$\overline{\mathcal{S}}(\mathrm{V})[Z] \to \overline{\mathcal{S}}(\mathrm{V})[Z'].$$

This turns $\overline{\mathcal{S}}(\mathrm{V})$ into a Coxeter space. Further, if $f : \mathrm{V} \to \mathrm{W}$ is a map of Coxeter spaces, then again taking a colimit yields a map $\overline{\mathcal{S}}(\mathrm{V}) \to \overline{\mathcal{S}}(\mathrm{W})$ of Coxeter spaces. Thus $\overline{\mathcal{S}}$ is a functor.

Observe:

$$\overline{\mathcal{S}}\overline{\mathcal{S}}(\mathrm{V})[Z] = \Big(\bigoplus_{(T,U):\, Z \le T \le U} \mathrm{V}[U] \Big)_{\mathrm{quo}}.$$

The sum is over both T and U, and $x \equiv \sigma(x)$ for any $\sigma : U \to U'$ local to T, or induced from a $\sigma : T \to T'$ local to Z. Sending the (T, U)-summand identically to the U-summand induces a linear map $\overline{\mathcal{S}}\overline{\mathcal{S}}(\mathrm{V})[Z] \to \overline{\mathcal{S}}(\mathrm{V})[Z]$. This yields a natural transformation $\overline{\mathcal{S}}\overline{\mathcal{S}} \to \overline{\mathcal{S}}$. There is also an obvious natural transformation $\mathrm{id} \to \overline{\mathcal{S}}$. These maps turn $\overline{\mathcal{S}}$ into a monad. Algebras over $\overline{\mathcal{S}}$ are the same as commutative Coxeter $\boldsymbol{\alpha}$-algebras (Section 8.8.5).

A comparison of (8.119) and (8.121) shows that there is a surjective natural transformation

$$\pi : \overline{\mathcal{T}} \to \overline{\mathcal{S}}. \tag{8.122}$$

We call this the *abelianization*. Formally, (id, π) is a lax functor of monads from $\overline{\mathcal{S}}$ to $\overline{\mathcal{T}}$ in the sense of (A.2).

Dually, define the comonad $\overline{\mathcal{S}}^\vee : \boldsymbol{\alpha}\text{-}\overline{\mathtt{C}}\mathtt{Sp} \to \boldsymbol{\alpha}\text{-}\overline{\mathtt{C}}\mathtt{Sp}$ by setting

$$\overline{\mathcal{S}}^\vee(\mathrm{V})[Z] := \lim \mathrm{V}_Z.$$

Explicitly, it is the subspace

$$\overline{\mathcal{S}}^\vee(\mathrm{V})[Z] = \Big(\bigoplus_{T:\, Z \le T} \mathrm{V}[T] \Big)^{\mathrm{sub}} \tag{8.123}$$

consisting of $(x_T)_{Z \le T}$ such that $\sigma(x_T) = x_{T'}$ for any morphism $\sigma : T \to T'$ local to Z. In particular, x_T is invariant under all morphisms from T to itself which are local to Z. Coalgebras over $\overline{\mathcal{S}}^\vee$ are the same as cocommutative Coxeter $\boldsymbol{\alpha}$-coalgebras (Section 8.8.9).

By construction, there is an injective natural transformation

$$\pi^\vee : \overline{\mathcal{S}}^\vee \to \overline{\mathcal{T}}^\vee. \tag{8.124}$$

We call this the *coabelianization*. Formally, $(\mathrm{id}, \pi^\vee)$ is a colax functor of comonads from $\overline{\mathcal{S}}^\vee$ to $\overline{\mathcal{T}}^\vee$ in the sense of (A.5).

The mixed distributive law $\overline{\lambda}$ between $\overline{\mathcal{T}}$ and $\overline{\mathcal{T}}^\vee$, via (co)abelianization, induces three more mixed distributive laws: one between $\overline{\mathcal{S}}$ and $\overline{\mathcal{T}}^\vee$, one between $\overline{\mathcal{T}}$ and $\overline{\mathcal{S}}^\vee$, one between $\overline{\mathcal{S}}$ and $\overline{\mathcal{S}}^\vee$. (This can be checked using Exercise 1.40.) We denote all of them by $\overline{\lambda}$. See the first diagram in (8.127) below. The last three entries in Proposition 8.121 may also now be verified.

Exercise 8.122. Reformulate (8.121) in terms of top-lunes as an appropriate quotient space

$$\overline{\mathcal{S}}(\mathrm{V})[\mathrm{L}] = \Big(\bigoplus_{\mathrm{M}:\, \mathrm{L} \le' \mathrm{M}} \mathrm{V}[\mathrm{M}] \Big)_{\mathrm{quo}}. \tag{8.125}$$

8.12.5. Signed bicommutative signed Coxeter bialgebras. The signed analogue of Proposition 8.121 is stated below.

Proposition 8.123. *The following are equivalences of categories.*

$$\begin{aligned}\overline{\mathcal{E}}\text{-algebras} &\cong \text{signed com. Coxeter } \boldsymbol{\alpha}\text{-algebras}\\ \overline{\mathcal{E}}^\vee\text{-coalgebras} &\cong \text{signed cocom. Coxeter } \boldsymbol{\alpha}\text{-coalgebras}\\ (\overline{\mathcal{E}},\overline{\mathcal{T}}^\vee,\overline{\lambda}_{-1})\text{-bialgebras} &\cong \text{signed com. signed Coxeter } \boldsymbol{\alpha}\text{-bialgebras}\\ (\overline{\mathcal{T}},\overline{\mathcal{E}}^\vee,\overline{\lambda}_{-1})\text{-bialgebras} &\cong \text{signed cocom. signed Coxeter } \boldsymbol{\alpha}\text{-bialgebras}\\ (\overline{\mathcal{E}},\overline{\mathcal{E}}^\vee,\overline{\lambda}_{-1})\text{-bialgebras} &\cong \text{signed bicom. signed Coxeter } \boldsymbol{\alpha}\text{-bialgebras}\end{aligned}$$

For the monad $\overline{\mathcal{E}}$, evaluated on a Coxeter $\boldsymbol{\alpha}$-space V, the component $\overline{\mathcal{E}}(\mathrm{V})[Z]$ is defined as in (8.121) as the quotient space with $x \equiv (-1)^{l(\sigma)}\sigma(x)$ for any morphism $\sigma : T \to T'$ local to Z. The structure maps of $\overline{\mathcal{E}}(\mathrm{V})$, however, do not involve any signs. Algebras over $\overline{\mathcal{E}}$ are the same as signed commutative Coxeter $\boldsymbol{\alpha}$-algebras.

Dually, for the comonad $\overline{\mathcal{E}}^\vee$, evaluated on a Coxeter $\boldsymbol{\alpha}$-space V, the component $\overline{\mathcal{E}}^\vee(\mathrm{V})[Z]$ is defined as in (8.123) as the subspace consisting of $(x_T)_{Z\leq T}$ such that $\sigma(x_T) = (-1)^{l(\sigma)}x_{T'}$ for any morphism $\sigma : T \to T'$ local to Z. Coalgebras over $\overline{\mathcal{E}}^\vee$ are the same as signed cocommutative Coxeter $\boldsymbol{\alpha}$-coalgebras.

By construction, there are surjective and injective natural transformations

$$\pi_{-1} : \overline{\mathcal{T}} \to \overline{\mathcal{E}} \qquad \text{and} \qquad \pi^\vee_{-1} : \overline{\mathcal{E}}^\vee \to \overline{\mathcal{T}}^\vee. \tag{8.126}$$

These are the *signed abelianization* and *signed coabelianization*, respectively. The mixed distributive law $\overline{\lambda}_{-1}$ between $\overline{\mathcal{T}}$ and $\overline{\mathcal{T}}^\vee$ induces three more mixed distributive laws via signed (co)abelianization. See the second diagram in (8.127) below. This leads to the last three entries in Proposition 8.123.

8.12.6. Bimonad summary. We have the following commutative diagrams of bilax functors between bimonads.

$$\begin{array}{ccccccc} (\overline{\mathcal{S}},\overline{\mathcal{S}}^\vee,\overline{\lambda}) & \longrightarrow & (\overline{\mathcal{T}},\overline{\mathcal{S}}^\vee,\overline{\lambda}) & \qquad & (\overline{\mathcal{E}},\overline{\mathcal{E}}^\vee,\overline{\lambda}_{-1}) & \longrightarrow & (\overline{\mathcal{T}},\overline{\mathcal{E}}^\vee,\overline{\lambda}_{-1})\\ \downarrow & & \downarrow & & \downarrow & & \downarrow\\ (\overline{\mathcal{S}},\overline{\mathcal{T}}^\vee,\overline{\lambda}) & \longrightarrow & (\overline{\mathcal{T}},\overline{\mathcal{T}}^\vee,\overline{\lambda}) & & (\overline{\mathcal{E}},\overline{\mathcal{T}}^\vee,\overline{\lambda}_{-1}) & \longrightarrow & (\overline{\mathcal{T}},\overline{\mathcal{T}}^\vee,\overline{\lambda}_{-1}) \end{array} \tag{8.127}$$

The functors are bilax in the sense of (A.8) with $\mathcal{F} :=$ id and the (co)lax structure given by (signed) (co)abelianization.

Exercise 8.124. Recall the u-signature functor (8.99). Check: For a Coxeter space V, we have natural identifications

$$\overline{\mathcal{T}}(\mathrm{V}_-)_- = \overline{\mathcal{T}}(\mathrm{V}), \qquad \overline{\mathcal{S}}(\mathrm{V}_-)_- = \overline{\mathcal{E}}(\mathrm{V}), \qquad \overline{\mathcal{E}}(\mathrm{V}_-)_- = \overline{\mathcal{S}}(\mathrm{V})$$

as Coxeter spaces. (Use Exercise 8.23, item (2).)

We mention that duality on Coxeter spaces is a bilax functor between $(\overline{\mathcal{T}},\overline{\mathcal{T}}^\vee,\overline{\lambda}_q)$ and $(\overline{\mathcal{T}}^\vee,\overline{\mathcal{T}},\overline{\lambda}_q)$, with the latter bimonad on the opposite category. This explains why duality preserves Coxeter q-bialgebras.

8.13. Opposite transformation

We introduce the opposite transformation for Coxeter spaces. It links a face-type to its locally opposite face-type. This construction yields op and cop constructions for Coxeter spaces.

The discussion parallels that of Coxeter species in Section 2.11.

8.13.1. Opposite transformation. Recall from (8.119) the functor $\overline{\mathcal{T}}$ on Coxeter spaces. There is a natural transformation $\bar{\tau} : \overline{\mathcal{T}} \to \overline{\mathcal{T}}$. On a Coxeter space V, the Z-component of $\bar{\tau}_{\mathrm{V}} : \overline{\mathcal{T}}(\mathrm{V}) \to \overline{\mathcal{T}}(\mathrm{V})$ is defined by summing the linear maps

$$\beta_{Z\overline{T},T} : \mathrm{V}[T] \to \mathrm{V}[Z\overline{T}]$$

over all $T \geq Z$. By Exercise 8.26, the isomorphism (8.14) preserves locally opposite support-type morphisms. So $\bar{\tau}_{\mathrm{V}}$ is a map of Coxeter spaces. Naturality in V is clear. Thus $\bar{\tau}$ is a natural transformation. We call it the *opposite transformation*. Since $\overline{\mathcal{T}} = \overline{\mathcal{T}}^{\vee}$ as functors, we can also write $\bar{\tau} : \overline{\mathcal{T}}^{\vee} \to \overline{\mathcal{T}}^{\vee}$.

Moreover, $(\mathrm{id}, \bar{\tau})$ is a lax functor of monads from $\overline{\mathcal{T}}$ to $\overline{\mathcal{T}}$, and a colax functor of comonads from $\overline{\mathcal{T}}^{\vee}$ to $\overline{\mathcal{T}}^{\vee}$. Going further, the functors $(\mathrm{id}, \tau, \mathrm{id})$ and $(\mathrm{id}, \mathrm{id}, \tau)$ both from $(\overline{\mathcal{T}}, \overline{\mathcal{T}}^{\vee}, \overline{\lambda})$ to itself are bilax.

8.13.2. Deformed opposite transformation. More generally, for a scalar q, define $\bar{\tau}_q : \overline{\mathcal{T}} \to \overline{\mathcal{T}}$ by employing the linear maps

$$(\beta_q)_{Z\overline{T},T} := q^{\mathrm{dist}(T, Z\overline{T})} \beta_{Z\overline{T},T}.$$

Note: The power of q is the length of the morphism $\beta_{Z\overline{T},T}$. For $q = 1$, we recover the opposite transformation. For $q = -1$, we use the term *signed opposite transformation*.

Proposition 8.125. *Let $q \neq 0$. The functors* $(\mathrm{id}, \bar{\tau}_{q^{-1}}, \mathrm{id})$ *and* $(\mathrm{id}, \mathrm{id}, \bar{\tau}_{q^{-1}})$ *both from*

$$(\overline{\mathcal{T}}, \overline{\mathcal{T}}^{\vee}, \overline{\lambda}_q) \to (\overline{\mathcal{T}}, \overline{\mathcal{T}}^{\vee}, \overline{\lambda}_{q^{-1}})$$

are bilax.

We leave this as an exercise. A relevant result is given in Exercise 1.41.

Exercise 8.126. Formulate $\bar{\tau}$, and more generally $\bar{\tau}_q$ in terms of top-nested faces starting with the functor $\overline{\mathcal{T}}$ as in (8.115). (Use the composite map $\mathrm{V}[F, C] \to \mathrm{V}[G, C]$ in (8.22), but local to A.)

8.13.3. (Co)abelianization and its signed analogue. Recall π and $\pi^{\vee}$ from (8.122) and (8.124). The diagrams

(8.128)
$$\begin{array}{ccc} \overline{\mathcal{T}} & \xrightarrow{\ \bar{\tau}\ } & \overline{\mathcal{T}} \\ & {\scriptstyle\pi}\searrow \quad \swarrow{\scriptstyle\pi} & \\ & \overline{\mathcal{S}} & \end{array} \qquad\qquad \begin{array}{ccc} \overline{\mathcal{T}}^{\vee} & \xrightarrow{\ \bar{\tau}\ } & \overline{\mathcal{T}}^{\vee} \\ & {\scriptstyle\pi^{\vee}}\nwarrow \quad \nearrow{\scriptstyle\pi^{\vee}} & \\ & \overline{\mathcal{S}}^{\vee} & \end{array}$$

commute. In fact, $(\overline{\mathcal{S}}, \pi)$ and $(\overline{\mathcal{S}}^{\vee}, \pi^{\vee})$ are, respectively, the equalizers of

$$\overline{\mathcal{T}} \underset{\mathrm{id}}{\overset{\bar{\tau}}{\rightrightarrows}} \overline{\mathcal{T}} \qquad \text{and} \qquad \overline{\mathcal{T}}^{\vee} \underset{\mathrm{id}}{\overset{\bar{\tau}}{\rightrightarrows}} \overline{\mathcal{T}}^{\vee}$$

in the category of monads and category of comonads on Coxeter spaces. (Note: The usage of equalizer is to match our convention of morphisms in the categories of (co)monads.) For closely related ideas, see Lemmas 8.96 and 8.97 and their proofs.

Now recall π_{-1} and $\pi^{\vee}_{-1}$ from (8.126). Then $(\bar{\mathcal{E}}, \pi_{-1})$ and $(\bar{\mathcal{E}}^{\vee}, \pi^{\vee}_{-1})$ are, respectively, the equalizers of

$$\overline{\mathcal{T}} \underset{\mathrm{id}}{\overset{\overline{\tau}_{-1}}{\rightrightarrows}} \overline{\mathcal{T}} \qquad \text{and} \qquad \overline{\mathcal{T}}^{\vee} \underset{\mathrm{id}}{\overset{\overline{\tau}_{-1}}{\rightrightarrows}} \overline{\mathcal{T}}^{\vee}$$

in the category of monads and category of comonads on Coxeter spaces.

8.13.4. Op and cop constructions. Fix $q \neq 0$. A Coxeter algebra (A, μ) has an *opposite Coxeter algebra* $\mathrm{A}^{\mathrm{op}} := (\mathrm{A}, \mu\beta_q)$ with the same underlying Coxeter space and with product $\mu\beta_q$ defined by

$$(\mu\beta_q)^T_Z : \mathrm{A}[T] \xrightarrow{(\beta_q)_{Z\overline{T},T}} \mathrm{A}[Z\overline{T}] \xrightarrow{\mu^{Z\overline{T}}_Z} \mathrm{A}[Z].$$

We write ${}^{\mathrm{op}}\mathrm{A} := (\mathrm{A}, \mu\beta_{q^{-1}})$. Dually, a Coxeter coalgebra (C, Δ) has an *opposite Coxeter coalgebra* ${}^{\mathrm{cop}}\mathrm{C} := (\mathrm{C}, \beta_q\Delta)$ with

$$(\beta_q\Delta)^T_Z : \mathrm{C}[Z] \xrightarrow{\Delta^{Z\overline{T}}_Z} \mathrm{C}[Z\overline{T}] \xrightarrow{(\beta_q)_{T,Z\overline{T}}} \mathrm{C}[T].$$

We write $\mathrm{C}^{\mathrm{cop}} := (\mathrm{C}, \beta_{q^{-1}}\Delta)$.

Proposition 8.127. *For $q \neq 0$, for $\mathrm{H} = (\mathrm{H}, \mu, \Delta)$ a Coxeter q-bialgebra,*

$$\mathrm{H}^{\mathrm{cop}} := (\mathrm{H}, \mu, \beta_{q^{-1}}\Delta) \quad \textit{and} \quad {}^{\mathrm{op}}\mathrm{H} := (\mathrm{H}, \mu\beta_{q^{-1}}, \Delta)$$

are Coxeter q^{-1}-bialgebras, and

$$\mathrm{H}^{\mathrm{op,cop}} := (\mathrm{H}, \mu\beta_q, \beta_{q^{-1}}\Delta) \quad \textit{and} \quad {}^{\mathrm{op,cop}}\mathrm{H} := (\mathrm{H}, \mu\beta_{q^{-1}}, \beta_q\Delta)$$

are Coxeter q-bialgebras. In fact,

$$\mathrm{H}^{\mathrm{op,cop}} = {}^{\mathrm{op}}(\mathrm{H}^{\mathrm{cop}}) \quad \textit{and} \quad {}^{\mathrm{op,cop}}\mathrm{H} = ({}^{\mathrm{op}}\mathrm{H})^{\mathrm{cop}}.$$

This can be either checked directly or formally deduced from Proposition 8.125.

The above considerations carry over to the more general setting of Coxeter b-spaces (Section 8.4.11). Here instead of $\beta_{q^{-1}}$ we employ the inverse of β_q.

8.14. Graded vector spaces and Coxeter spaces

We now specialize the reflection arrangement $\boldsymbol{\alpha} = (\mathcal{A}, C_r)$ to the braid arrangement $\boldsymbol{\beta}^n = (\mathcal{B}^{[n]}, 1| \dots |n)$ defined in Section 1.16.

We make explicit the groupoid of biface-types $\boldsymbol{\beta}^n$-$\overline{\mathtt{Cox}}$ using the combinatorics of integer compositions. We do the same for related concepts such as induced and local morphisms, opposite and adjacent support-type morphisms. We then formulate Coxeter $\boldsymbol{\beta}^n$-spaces and Coxeter $\boldsymbol{\beta}^n$-(co, bi)algebras. The q-bialgebra axiom takes an explicit form which involves filling a matrix with nonnegative integers with specified row and column sums.

Next, we relate classical Hopf theory to Coxeter $\boldsymbol{\beta}^n$-bialgebras. The key observation is that a positive-graded vector space yields a Coxeter $\boldsymbol{\beta}^n$-space whose value on a composition α is given by the tensor product of the graded components specified by the parts of α. Developing further, a positive-graded (co, bi)algebra yields a Coxeter $\boldsymbol{\beta}^n$-(co, bi)algebra.

8.14.1. Groupoid of biface-types. Fix a positive integer n. The *groupoid of biface-types* $\boldsymbol{\beta}^n$-$\overline{\texttt{Cox}}$ is as follows. There is one object for each composition α of n. A morphism $\alpha \to \beta$ is a bijection between the parts of α and parts of β which preserves the size of each part. For example,

$$(2, 3, 1, 2, 3, 2, 1) \to (1, 3, 2, 1, 3, 2, 2). \tag{8.129}$$

The bijection is specified by colors, the red 2 maps to the red 2, the blue 2 to the blue 2, and so on. Morphisms are composed by composing bijections.

There is a morphism between two compositions only if they have the same underlying partition. The connected components of $\boldsymbol{\beta}^n$-$\overline{\texttt{Cox}}$ are indexed by partitions of n. For compositions α and β with underlying partition $\lambda = (1^{m_1}, 2^{m_2}, \dots)$, the number of morphisms from α to β is $|\lambda| = m_1!m_2!\dots$.

8.14.2. (Co)projection by a morphism. Let $\sigma : \alpha \to \beta$ be a morphism.

- For $\gamma \leq \alpha$, we say σ coprojects β on γ if for each i, the parts of $(\alpha/\gamma)_i$ stay in the same order in β under σ.
- For $\gamma \leq \beta$, we say σ projects α on γ if σ^{-1} coprojects α on γ.

For example, the morphism $(1, 2, 2, 4, 3) \to (2, 3, 1, 2, 4)$ projects on $(5, 7)$.

8.14.3. Induced morphisms. Let $\sigma : \alpha \to \beta$ be a morphism. For $\alpha \leq \gamma$, the composition $\sigma_*(\gamma)$ is obtained by refining β using the bijection specified by σ, and $\gamma \to \sigma_*(\gamma)$ is the induced morphism (also denoted σ). Similarly, for $\beta \leq \gamma$, we have $\sigma^*(\gamma)$ refining α and the induced morphism $\sigma^*(\gamma) \to \gamma$.

Suppose $\alpha = (8, 2, 5)$ and $\beta = (5, 8, 2)$ and $\sigma : (8, 2, 5) \to (5, 8, 2)$ is the (unique) morphism between them. Then, for $\gamma = (3, 1, 4, 2, 3, 2)$ refining α,

$$\sigma_*(\gamma) = (3, 2, 3, 1, 4, 2) \quad \text{and} \quad \sigma : (3, 1, 4, 2, 3, 2) \to (3, 2, 3, 1, 4, 2).$$

Similarly, for $\gamma = (2, 3, 5, 3, 1, 1)$ refining β,

$$\sigma^*(\gamma) = (5, 3, 1, 1, 2, 3) \quad \text{and} \quad \sigma : (5, 3, 1, 1, 2, 3) \to (2, 3, 5, 3, 1, 1).$$

8.14.4. Local morphisms. Suppose α and β are compositions which both refine γ. Then a morphism $\sigma : \alpha \to \beta$ is local to γ if the bijection specified by σ operates within each part of γ. For instance, suppose $\gamma = (7, 1, 4)$, $\alpha = (3, 4, 1, 3, 1)$ and $\beta = (4, 3, 1, 1, 3)$. Consider the morphisms

$$(3, 4, 1, 3, 1) \to (4, 3, 1, 1, 3) \quad \text{and} \quad (3, 4, 1, 3, 1) \to (4, 3, 1, 1, 3)$$

both from α to β. The first one is local to γ, while the second one is not.

8.14.5. Length of a morphism. Let $\sigma : \alpha \to \beta$ be a morphism. Write $\alpha = (a_1, \dots, a_k)$. Define $\operatorname{Inv}(\alpha, \beta)$ to be the set of pairs (i, j) with $1 \leq i < j \leq k$ such that $\sigma(a_i)$ appears after $\sigma(a_j)$ in β. The length of σ is defined by

$$l(\sigma) = \sum_{(i,j)\in \operatorname{Inv}(\alpha,\beta)} a_i a_j. \tag{8.130}$$

For example, for $\sigma : (2, 1, 3, 1) \to (3, 2, 1, 1)$, we have $l(\sigma) = 6 + 3 + 1 = 10$. This specializes the definition given in (1.29) and in Section 8.2.3.

8.14.6. Opposite and adjacent support-type morphisms. Every composition α has an opposite $\overline{\alpha}$ and a morphism

$$\beta_{\overline{\alpha},\alpha} : \alpha \to \overline{\alpha} \tag{8.131}$$

which reverses the order of the parts of α. These specialize the opposite support-type morphisms in (8.17). (Warning: In the present discussion, β is also used to denote compositions.) A locally opposite support-type morphism

$$\beta_{\gamma\overline{\alpha},\alpha} : \alpha \to \gamma\overline{\alpha} \tag{8.132}$$

is obtained by starting with $\gamma \leq \alpha$ and for each part of γ reversing the portion of α which refines it. This specializes (8.19). For example,

$$(1, 2, 5, 3, 1, 4, 7, 6) \to (5, 2, 1, 3, 4, 1, 6, 7).$$

Here $\gamma = (8, 3, 5, 13)$, so $(1, 2, 5)$ has changed to $(5, 2, 1)$, (3) to (3), $(1, 4)$ to $(4, 1)$, $(7, 6)$ to $(6, 7)$. This morphism is local to γ. When γ is the composition with one part, we recover opposite support-type morphisms.

An adjacent support-type morphism is an adjacent transposition such as

$$(2, 1, 1, 3, 2) \to (2, 1, 1, 3, 2) \quad \text{or} \quad (2, 1, 1, 3, 2) \to (2, 1, 3, 1, 2).$$

The first one is local to $(2, 2, 3, 2)$, while the second one is local to $(2, 1, 4, 2)$. These are locally opposite support-type morphisms. The result of Lemma 8.27 in this case says that a permutation is generated by adjacent transpositions.

8.14.7. Coxeter β^n-spaces. A Coxeter β^n-space V is a family of vector spaces $\mathrm{V}[\alpha]$, one for each composition α of n, equipped with linear maps

$$\sigma : \mathrm{V}[\alpha] \to \mathrm{V}[\beta],$$

one for each morphism $\sigma : \alpha \to \beta$ in $\boldsymbol{\beta}^n\text{-}\overline{\mathtt{Cox}}$, such that the diagrams

$$\begin{array}{ccc} & \mathrm{V}[\beta] & \\ {}^{\sigma}\nearrow & & \searrow^{\tau} \\ \mathrm{V}[\alpha] & \xrightarrow[\tau\sigma]{} & \mathrm{V}[\gamma] \end{array} \qquad (\mathrm{V}[\alpha] \xrightarrow{\mathrm{id}_\alpha} \mathrm{V}[\alpha]) = \mathrm{id} \tag{8.133}$$

commute. These diagrams specialize (8.28).

8.14.8. Coxeter $\boldsymbol{\beta}^n$-algebras. A Coxeter $\boldsymbol{\beta}^n$-algebra is a Coxeter $\boldsymbol{\beta}^n$-space A equipped with linear maps

$$\mu_\alpha^\beta : \mathrm{A}[\beta] \to \mathrm{A}[\alpha],$$

one for each $\alpha \leq \beta$, subject to the *naturality*, *associativity*, *unitality axioms*

$$\begin{CD} \mathrm{A}[\beta] @>{\sigma}>> \mathrm{A}[\beta'] \\ @V{\mu_\alpha^\beta}VV @VV{\mu_{\alpha'}^{\beta'}}V \\ \mathrm{A}[\alpha] @>{\sigma}>> \mathrm{A}[\alpha'] \end{CD} \qquad \begin{array}{ccc} & \mathrm{A}[\beta] & \\ {\scriptstyle \mu_\beta^\gamma}\nearrow & & \searrow{\scriptstyle \mu_\alpha^\beta} \\ \mathrm{A}[\gamma] & \xrightarrow[\mu_\alpha^\gamma]{} & \mathrm{A}[\alpha] \end{array} \qquad (\mathrm{A}[\alpha] \xrightarrow{\mu_\alpha^\alpha} \mathrm{A}[\alpha]) = \mathrm{id}\,. \tag{8.134}$$

Here $\sigma : \alpha \to \alpha'$ is any morphism and $\beta' = \sigma_*(\beta)$, or equivalently, $\beta = \sigma^*(\beta')$. These diagrams specialize (8.41).

Proposition 8.128. *A Coxeter $\boldsymbol{\beta}^n$-algebra is a Coxeter $\boldsymbol{\beta}^n$-space* A *equipped with linear maps*

$$\mu_\alpha^\beta : \mathrm{A}[\beta] \to \mathrm{A}[\alpha],$$

one for each $\alpha \lessdot \beta$, such that the diagrams

$$\begin{CD} \mathrm{A}[\beta] @>{\sigma}>> \mathrm{A}[\beta'] \\ @V{\mu_\alpha^\beta}VV @VV{\mu_{\alpha'}^{\beta'}}V \\ \mathrm{A}[\alpha] @>{\sigma}>> \mathrm{A}[\alpha'] \end{CD} \qquad\qquad \begin{CD} \mathrm{A}[\gamma] @>{\mu_{\beta'}^\gamma}>> \mathrm{A}[\beta'] \\ @V{\mu_\beta^\gamma}VV @VV{\mu_\alpha^{\beta'}}V \\ \mathrm{A}[\beta] @>{\mu_\alpha^\beta}>> \mathrm{A}[\alpha] \end{CD} \tag{8.135}$$

commute. In the first diagram, β and β' cover α and α', respectively, with $\beta' = \sigma_*(\beta)$. In the second diagram, γ covers β and β' which in turn cover α.

This specializes Proposition 8.44 to $\boldsymbol{\beta}^n$. We refer to the second diagram in (8.135) as the *covering associativity axiom*. It is a square, not a triangle.

8.14.9. Coxeter $\boldsymbol{\beta}^n$-coalgebras. A Coxeter $\boldsymbol{\beta}^n$-coalgebra is a Coxeter $\boldsymbol{\beta}^n$-space C equipped with linear maps

$$\Delta_\alpha^\beta : \mathrm{C}[\alpha] \to \mathrm{C}[\beta],$$

one for each $\alpha \leq \beta$, satisfying diagrams dual to (8.134).

8.14.10. (Co)commutative Coxeter $\boldsymbol{\beta}^n$-(co)algebras. Let us now specialize (8.86). A Coxeter $\boldsymbol{\beta}^n$-algebra A is *commutative* if the diagram

$$\begin{array}{ccc} \mathrm{A}[\alpha] & \xrightarrow{\ \sigma\ } & \mathrm{A}[\alpha'] \\ {\scriptstyle \mu_\gamma^\alpha}\searrow & & \swarrow{\scriptstyle \mu_\gamma^{\alpha'}} \\ & \mathrm{A}[\gamma] & \end{array} \tag{8.136}$$

commutes for all morphisms $\sigma : \alpha \to \alpha'$ local to γ. We refer to (8.136) as the *commutativity axiom*. It suffices to restrict σ to opposite support-type morphisms local to γ or even further to adjacent support-type morphisms. This is the content of Lemmas 8.96 and 8.97 for $\boldsymbol{\beta}^n$.

Dually, specializing (8.91): A Coxeter $\boldsymbol{\beta}^n$-coalgebra C is *cocommutative* if the diagram

$$
\begin{array}{ccc}
\mathrm{C}[\alpha] & \xrightarrow{\ \sigma\ } & \mathrm{C}[\alpha'] \\
{\scriptstyle \Delta_\gamma^\alpha}\nwarrow & & \nearrow{\scriptstyle \Delta_\gamma^{\alpha'}} \\
& \mathrm{C}[\gamma] &
\end{array}
\tag{8.137}
$$

commutes for all morphisms $\sigma : \alpha \to \alpha'$ local to γ. We refer to (8.137) as the *cocommutativity axiom.*

8.14.11. Coxeter $\boldsymbol{\beta}^n$-bialgebras. For Coxeter $\boldsymbol{\beta}^n$-q-bialgebras, we need to understand how the q-bialgebra axiom works. Matrices naturally enter into its formulation; in rough terms, the interaction between the product and coproduct is captured by the interaction between rows and columns of a matrix. To make this precise, let us set up some notation.

Let $c = (c_{ij})$ be a $k \times l$ matrix of nonnegative integers with a specified block diagonal decomposition. (We also assume that each row and each column of c contains at least one nonzero entry.) Define

$$
\begin{aligned}
\beta_c &:= (c_{11}, \ldots, c_{1l}, \ldots, c_{k1}, \ldots, c_{kl})\widehat{\ },\\
\beta'_c &:= (c_{11}, \ldots, c_{k1}, \ldots, c_{1l}, \ldots, c_{kl})\widehat{\ },
\end{aligned}
$$

where hats indicate that zero entries are removed. Let $\sigma_c : \beta_c \to \beta'_c$ be the morphism which sends c_{ij} to c_{ij} (with zero entries removed). Observe using definition (8.130):

$$
l(\sigma_c) = \sum_{i<i',\, j>j'} c_{ij} c_{i'j'}.
$$

Define α_c by taking row sums of c, and α'_c by taking column sums of c. Finally, define γ_c by taking sums of diagonal block entries of c. Note: $\gamma_c \le \alpha_c \le \beta_c$ and $\gamma_c \le \alpha'_c \le \beta'_c$. Also, the morphism σ_c is local to γ_c.

For a scalar q, for a Coxeter $\boldsymbol{\beta}^n$-space V, for compositions γ, α, β, α', β' of n with $\gamma \le \alpha \le \beta$ and $\gamma \le \alpha' \le \beta'$, define the linear map

$$
\big((\overline{\lambda}_q)_{\alpha\beta\alpha'\beta'} : \mathrm{V}[\beta] \to \mathrm{V}[\beta']\big) = \sum_{\substack{c:\\ \gamma=\gamma_c \\ \alpha=\alpha_c,\, \alpha'=\alpha'_c \\ \beta=\beta_c,\, \beta'=\beta'_c}} q^{l(\sigma_c)} \sigma_c.
\tag{8.138}
$$

Let q be a scalar. A Coxeter $\boldsymbol{\beta}^n$-q-bialgebra is a Coxeter $\boldsymbol{\beta}^n$-space H equipped with structures of a Coxeter $\boldsymbol{\beta}^n$-algebra and a Coxeter $\boldsymbol{\beta}^n$-coalgebra such that for compositions α and α' which refine γ, the diagram

$$
\begin{array}{ccccc}
\mathrm{H}[\alpha] & \xrightarrow{\ \mu_\gamma^\alpha\ } & \mathrm{H}[\gamma] & \xrightarrow{\ \Delta_\gamma^{\alpha'}\ } & \mathrm{H}[\alpha'] \\
{\scriptstyle (\Delta_\alpha^\beta)}\Big\downarrow & & & & \Big\uparrow{\scriptstyle (\mu_{\alpha'}^{\beta'})} \\
\displaystyle\bigoplus_{\beta:\, \alpha\le\beta} \mathrm{H}[\beta] & & \xrightarrow[\ ((\overline{\lambda}_q)_{\alpha\beta\alpha'\beta'})\ ]{} & & \displaystyle\bigoplus_{\beta':\, \alpha'\le\beta'} \mathrm{H}[\beta']
\end{array}
\tag{8.139}
$$

commutes. We refer to (8.139) as the *q-bialgebra axiom*. Equivalently, for $\gamma \leq \alpha$ and $\gamma \leq \alpha'$,

$$\Delta_\gamma^{\alpha'} \mu_\gamma^\alpha = \sum_{\substack{c: \\ \gamma = \gamma_c \\ \alpha = \alpha_c,\, \alpha' = \alpha'_c}} q^{l(\sigma_c)}\, \mu_{\alpha'}^{\beta'_c}\, \sigma_c\, \Delta_\alpha^{\beta_c}. \tag{8.140}$$

8.14.12. Graded vector spaces. Let V be a positive-graded vector space, that is, $\mathrm{V} = (\mathrm{V}_i)_{i \geq 1}$, where each V_i is a vector space over $\Bbbk$. For a composition $\alpha = (a_1, \ldots, a_k)$ of n, put

$$\mathrm{V}^{[n]}[\alpha] := \mathrm{V}_{a_1} \otimes \cdots \otimes \mathrm{V}_{a_k}. \tag{8.141}$$

For a morphism $\sigma : \alpha \to \beta$, define the linear map $\sigma : \mathrm{V}^{[n]}[\alpha] \to \mathrm{V}^{[n]}[\beta]$ by rearranging the tensor factors according to the bijection specified by σ. For instance, for the morphism (8.129), the linear map is

$$\mathrm{V}_2 \otimes \mathrm{V}_3 \otimes \mathrm{V}_1 \otimes \mathrm{V}_2 \otimes \mathrm{V}_3 \otimes \mathrm{V}_2 \otimes \mathrm{V}_1 \to \mathrm{V}_1 \otimes \mathrm{V}_3 \otimes \mathrm{V}_2 \otimes \mathrm{V}_1 \otimes \mathrm{V}_3 \otimes \mathrm{V}_2 \otimes \mathrm{V}_2.$$

With these structure maps, $\mathrm{V}^{[n]}$ is a Coxeter $\boldsymbol{\beta}^n$-space.

Graded algebra. Now let A be a positive-graded algebra. (Since $\mathrm{A}_0 = 0$, it cannot have a unit.) Then for each composition γ of m, we have a linear map $\mathrm{A}^{[m]}[\gamma] \to \mathrm{A}_m$ obtained by iterating the product of A. For compositions $\alpha \leq \beta$ of n, define

$$\mu_\alpha^\beta : \mathrm{A}^{[n]}[\beta] \to \mathrm{A}^{[n]}[\alpha]$$

by tensoring the iterated product maps $\mathrm{A}^{[a_i]}[(\beta/\alpha)_i] \to \mathrm{A}_{a_i}$, where $\alpha = (a_1, \ldots, a_k)$, and $(\beta/\alpha)_i$ is the portion of β which refines a_i. For instance, for $\alpha = (3, 4)$ and $\beta = (2, 1, 2, 2)$,

$$\begin{array}{ccc} \mathrm{A}^{[7]}[(2,1,2,2)] & \overset{\mu_\alpha^\beta}{\dashrightarrow} & \mathrm{A}^{[7]}[(3,4)] \\ \| & & \| \\ \mathrm{A}_2 \otimes \mathrm{A}_1 \otimes \mathrm{A}_2 \otimes \mathrm{A}_2 & \xrightarrow[\mu^{2,1} \otimes \mu^{2,2}]{} & \mathrm{A}_3 \otimes \mathrm{A}_4. \end{array}$$

This turns $\mathrm{A}^{[n]}$ into a Coxeter $\boldsymbol{\beta}^n$-algebra for each $n \geq 1$. Further, if A is commutative, then so is $\mathrm{A}^{[n]}$. Note very carefully: The covering associativity axiom in (8.135) for $\mathrm{A}^{[n]}$ is in fact the usual associativity axiom of A (up to tensoring by identity maps). See the discussion in Section 8.5.9 in this regard.

Graded coalgebra. Dually, if C is a positive-graded coalgebra, then $\mathrm{C}^{[n]}$ is a Coxeter $\boldsymbol{\beta}^n$-coalgebra for each $n \geq 1$. Further, if C is cocommutative, then so is $\mathrm{C}^{[n]}$.

Graded q-bialgebra. Now let H be a positive-graded q-bialgebra for a scalar q. It satisfies the *q-bialgebra axiom*

$$\begin{aligned} \Delta(xy) = {} & \sum q^{|x_2||y_1|}\, x_1 y_1 \otimes x_2 y_2 \\ & + \sum q^{|x_2||y|}\, x_1 y \otimes x_2 + x_1 \otimes x_2 y + x y_1 \otimes y_2 + q^{|x||y_1|}\, y_1 \otimes x y_2 \\ & \qquad + x \otimes y + q^{|x||y|}\, y \otimes x, \end{aligned} \tag{8.142}$$

where $|\cdot|$ indicates degree and

$$\Delta(x) = \sum x_1 \otimes x_2 \quad \text{and} \quad \Delta(y) = \sum y_1 \otimes y_2.$$

The product and coproduct of $\mathrm{H}^{[n]}$ turn it into a Coxeter $\boldsymbol{\beta}^n$-q-bialgebra. This can be checked using Exercise 8.74. This yields a functor from positive-graded q-bialgebras to Coxeter $\boldsymbol{\beta}^n$-q-bialgebras for each $n \geq 1$.

Example 8.129 (Graded bialgebras to Coxeter bialgebras). Specific values of the above functor are mentioned later in Notes 10.7. In Table 10.8, the Hopf algebra of polynomials and its relatives are shown in the right column. Viewing them as positive-graded bialgebras by removing the component in degree 0, and applying the above functor yields the Coxeter $\boldsymbol{\beta}^n$-bialgebras in the left column. Tables 10.9 and 10.10 are to be interpreted in a similar manner. □

Exercise 8.130. Show: The q-bialgebra axiom (8.142) of a positive-graded q-bialgebra is equivalent to the *q-bialgebra axiom* of a connected graded q-bialgebra

$$\Delta(xy) = \sum q^{|x_2||y_1|}\, x_1 y_1 \otimes x_2 y_2. \tag{8.143}$$

This is the axiom more commonly considered in the literature.

Graded 0-bialgebra. We now specialize to $q = 0$. A positive-graded 0-bialgebra H satisfies the *0-bialgebra axiom*

$$\Delta(xy) = \Big(\sum x_1 \otimes x_2 y + x y_1 \otimes y_2\Big) + x \otimes y. \tag{8.144}$$

This is the case $q = 0$ of (8.142). Equivalently, for positive integers a, b, c, the diagrams

$$\begin{array}{ccc} \mathrm{H}_a \otimes \mathrm{H}_{b+c} & \xrightarrow{\mu^{a,b+c}} & \mathrm{H}_{a+b+c} \\ \big\downarrow{\scriptstyle \mathrm{id}\otimes\Delta^{b,c}} & & \big\downarrow{\scriptstyle \Delta^{a+b,c}} \\ \mathrm{H}_a \otimes \mathrm{H}_b \otimes \mathrm{H}_c & \xrightarrow[\mu^{a,b}\otimes\mathrm{id}]{} & \mathrm{H}_{a+b} \otimes \mathrm{H}_c \end{array} \qquad \begin{array}{ccc} \mathrm{H}_{a+b} \otimes \mathrm{H}_c & \xrightarrow{\mu^{a+b,c}} & \mathrm{H}_{a+b+c} \\ \big\downarrow{\scriptstyle \Delta^{a,b}\otimes\mathrm{id}} & & \big\downarrow{\scriptstyle \Delta^{a,b+c}} \\ \mathrm{H}_a \otimes \mathrm{H}_b \otimes \mathrm{H}_c & \xrightarrow[\mathrm{id}\otimes\mu^{b,c}]{} & \mathrm{H}_a \otimes \mathrm{H}_{b+c} \end{array}$$

$$\begin{array}{ccccc} & & \mathrm{H}_{a+b} & & \\ & \nearrow^{\mu^{a,b}} & & \searrow^{\Delta^{a,b}} & \\ \mathrm{H}_a \otimes \mathrm{H}_b & & \xrightarrow[\mathrm{id}]{} & & \mathrm{H}_a \otimes \mathrm{H}_b \end{array}$$

commute, with $\mu^{a,b}$ and $\Delta^{a,b}$ the product and coproduct components of H.

Exercise 8.131. Formulate the higher 0-bialgebra axiom for an iterated product followed by an iterated coproduct. Check directly that $\mathrm{H}^{[n]}$ is a Coxeter $\boldsymbol{\beta}^n$-0-bialgebra by comparing with the 0-bialgebra axiom (8.78).

Remark 8.132. One can attempt a similar analysis as done in this section for the arrangement of type B. Relevant combinatorial descriptions of faces and so on for type B can be found in [**10**, Section 6.7].

8.15. Notes

The material in this chapter is new. The historical beginnings of Coxeter bialgebras are briefly told in Notes 10.7.1.

8.15.1. Coxeter bialgebras vs Hopf algebras. The notion of a Coxeter bialgebra generalizes the classical notion of a connected graded Hopf algebra as explained in Section 8.14. A connected graded Hopf algebra can be formulated in terms of integer compositions which are face-types of the braid arrangement (Table 1.9). Thus, the connection between Hopf algebras and Coxeter bialgebras happens via the face-type formulation of the latter. It is intriguing that a Coxeter bialgebra can be formulated alternatively using top-nested faces, top-lunes, flats. Moreover, each formulation has a distinct flavor. A striking example is the upside-down commutativity axiom (8.83) in terms of top-lunes. It shows how rich the setting of reflection arrangements is, and the perspectives it brings to the classical theory. Exercises 8.83 and 8.84 are also very striking; absence of the tensor product makes this possible.

8.15.2. Hopf algebras. Some important sources in the classical literature are mentioned below. For more detailed references, see [**11**, Notes to Chapter 2].

Hopf theory originated in an influential paper of Hopf [**191**]. This paper is reprinted in his selected works [**192**, pages 119–151] and collected works [**193**, pages 686–746] (along with an english translation). Early references on the subject are by Samelson [**318**], Leray [**227**], Borel [**54**], Cartan and Eilenberg [**75**, Sections XI.4 and XI.8]. A systematic treatment was given by Cartier [**78**] and Milnor and Moore [**265**], [**266**], [**271**]. The first comprehensive work on Hopf algebras was carried out by Sweedler in his influential book [**345**]. Some later treatments are by Abe [**1**], Montgomery [**269**], Kassel [**209**, Chapter III], Majid [**247**, Chapter 1], Dăscălescu, Năstăsescu, Raianu [**93**], Hazewinkel, Gubareni, Kirichenko [**181**], Radford [**301**], Underwood [**355**], Skowroński and Yamagata [**325**, Part VI], May and Ponto [**255**, Part 5], Lorenz [**235**, Part IV], Cartier and Patras [**84**].

Associativity axiom. The associativity axiom for an algebra (A, μ) can be formulated in terms of binary operations or higher operations as shown below.

(8.145)
$$
\begin{array}{ccc}
A \otimes A \otimes A & \xrightarrow{\mu \otimes \mathrm{id}} & A \otimes A \\
{\scriptstyle \mathrm{id}\otimes\mu}\downarrow & & \downarrow{\scriptstyle \mu} \\
A \otimes A & \xrightarrow[\mu]{} & A
\end{array}
\qquad
\begin{array}{ccc}
A^{\otimes i_1} \otimes \cdots \otimes A^{\otimes i_k} & = & A^{\otimes n} \\
{\scriptstyle \mu^{i_1 - 1} \otimes \cdots \otimes \mu^{i_k - 1}}\downarrow & & \downarrow{\scriptstyle \mu^{n-1}} \\
A^{\otimes k} & \xrightarrow[\mu^{k-1}]{} & A
\end{array}
$$

In the second diagram, $(i_1, \ldots, i_k)$ is any composition of n, and $\mu^{n-1} : A^{\otimes n} \to A$ is the iterated product map. Note very carefully: The first diagram is a square, while the second diagram is a triangle. These correspond to the covering associativity axiom in (8.135) and associativity axiom in (8.134), respectively. See related remarks in Section 8.14.12.

Commutativity axiom. The commutativity axiom for a commutative algebra (A, μ) is formulated below in three different ways.

(8.146a)
$$
\begin{array}{ccc}
x_1 \otimes \cdots \otimes x_n & \longmapsto & x_{\sigma(1)} \otimes \cdots \otimes x_{\sigma(n)} \\[1ex]
A^{\otimes n} & \xrightarrow{\sigma} & A^{\otimes n} \\
& {\scriptstyle \mu^{n-1}}\searrow \quad \swarrow{\scriptstyle \mu^{n-1}} & \\
& A &
\end{array}
$$

Here σ is any permutation on n letters. This diagram corresponds to the commutativity axiom (8.136).

$$\begin{array}{ccc} x\otimes y & \longmapsto & y\otimes x \\ A\otimes A & \xrightarrow{\ \tau\ } & A\otimes A \\ & {\scriptstyle\mu}\searrow \quad \swarrow{\scriptstyle\mu} & \\ & A & \end{array} \qquad\qquad \begin{array}{ccc} x_1\otimes\cdots\otimes x_n & \longmapsto & x_n\otimes\cdots\otimes x_1 \\ A^{\otimes n} & \xrightarrow{\qquad\tau\qquad} & A^{\otimes n} \\ & {\scriptstyle\mu^{n-1}}\searrow \quad \swarrow{\scriptstyle\mu^{n-1}} & \\ & A & \end{array} \tag{8.146b}$$

Here τ reverses the order of the symbols as shown. These diagrams correspond, respectively, to the covering commutativity axiom in Lemma 8.97 and opposition commutativity axiom (8.87) in Lemma 8.96. See remarks after (8.136) in this regard.

Bialgebra axiom. The q-bialgebra axiom (8.73) generalizes the q-bialgebra axiom (8.142) of a positive-graded q-bialgebra. The latter is equivalent to the q-bialgebra axiom (8.143) of a connected graded q-bialgebra. Early references to (8.143) for $q = \pm 1$ are by Halpern [**172**, Axiom (1.1)], [**173**, Axiom (1.1)], Milnor and Moore [**265**, Definition 4.1, axiom (3)], [**266**, Definition 2.1, axiom (3)], Zisman [**367**, page 5].

An equivalent formulation of the q-bialgebra axiom (8.73) is given later in Exercise 9.29. In classical terms, for $q = \pm 1$, this says that product of a (signed) Hopf algebra is a morphism of coalgebras, and coproduct is a morphism of algebras. Early references for this observation are by Cartan and Eilenberg [**75**, Section XI.8, axiom (i)], Cartier [**78**, Section 2.1, first line of axiom (HA_2)], Milnor and Moore [**265**, page 227], [**266**, page 6].

Consideration of q-bialgebras for $q \neq \pm 1$ is less common in the literature. For an exposition, see [**7**, Section 2.3].

Braided bialgebras. A more flexible approach to classical bialgebras is to define the braiding "locally" using a Yang–Baxter operator. This idea is present in work of Gurevich [**168**, Definition in Section 6]. The term "braided bialgebra" is used by Takeuchi [**348**, Definition 5.1]. (The algebra part of this structure is a "braided algebra"; this notion is mentioned by Baez [**27**, page 886].) Some later references are [**7**, Section 19.9], [**211**, Chapter 6], [**20**, Definition 3.1]. In a way, braided bialgebras are closer to what we do since the underlying Coxeter space of a Coxeter bialgebra carries its own structure morphisms. In fact, this makes it possible to define the signature functor on Coxeter spaces in Section 8.10.

8.15.3. Presheaves on posets. Observe the formal similarity between the 0-bialgebra axiom (8.78) and bicommutative bimonoid axiom (2.59). The setting for the former is the Boolean poset of face-types, and for the latter is the poset of flats. See comments related to presheaves on posets on [**11**, page 133] in this regard.

8.15.4. Face-type enumeration. (Section 8.9.) The material here, and in particular, identities (8.97) and (8.98) look new even from a classical perspective. In the classical theory, the emphasis is mostly on binary operations as opposed to higher operations, see Section 8.5.9. Analogues of identities (8.97) and (8.98) for species were given in [**11**, (2.29) and (2.30)]; note very carefully that these do not require any enumeration polynomials. The same identities hold for Coxeter species as well.

CHAPTER 9

Basic theory of Coxeter bialgebras

In Chapter 8, we introduced Coxeter spaces, and then developed notions of Coxeter algebras, Coxeter coalgebras, Coxeter bialgebras. More generally, for a scalar q, we defined Coxeter q-bialgebras. The case $q = 1$ recovered Coxeter bialgebras; $q = -1$ yielded signed Coxeter bialgebras. We also saw commutative Coxeter algebras and cocommutative Coxeter coalgebras. We now proceed with the basic theory of these objects. We primarily work with the face-type formulation; hence, it suffices to be familiar with the groupoid of biface-types from Section 1.7 and how Coxeter bialgebras are formulated using face-types in Chapter 8. We discuss:

- Cauchy powers and commutative Cauchy powers of a Coxeter space,
- primitive part and primitive filtration of a Coxeter coalgebra, decomposable part and decomposable filtration of a Coxeter algebra; related adjunctions involving the trivial Coxeter (co)algebra functors; derivations and coderivations,
- universal constructions such as the (co)free Coxeter (co)algebra on a Coxeter space, the free Coxeter q-bialgebra on a Coxeter coalgebra and the cofree Coxeter q-bialgebra on a Coxeter algebra; similar constructions in the (signed) commutative setting; related adjunctions involving primitive and indecomposable part functors,
- (co)abelianization of a Coxeter (co)algebra and its universal property, showing that (co)commutative Coxeter (co)algebras is a (co)reflective subcategory of Coxeter (co)algebras,
- Hadamard product of Coxeter spaces, convolution Coxeter algebra, coconvolution Coxeter coalgebra; internal hom for Coxeter coalgebras, universal measuring Coxeter coalgebra, and related Coxeter bialgebras and their q-analogues, enrichment of Coxeter (co)algebras over Coxeter coalgebras and powers and copowers,
- exp-log correspondences between (co)derivations and (co)algebra morphisms arising from the action of invariant noncommutative zeta and Möbius functions (in the invariant lune-incidence algebra), logarithm of the identity morphism; exp-log correspondences between primitive and group-like series of a Coxeter bialgebra; their q-analogues for q not a root of unity,
- q-norm maps linking free and cofree objects arising from their universal properties, with connections to the q-logarithm for q not a root of unity,

- analysis of the bialgebra axiom on the primitive part of a Coxeter bialgebra, with $\Delta\mu = \sum \mu\sigma\Delta$ simplifying to $\Delta\mu = \sum \mu\sigma$,
- characteristic operations by face-types and invariant Tits algebra modules, commutative characteristic operations by flat-types and invariant Birkhoff algebra modules, two-sided characteristic operations by biface-types and invariant Janus algebra modules,
- antipode of a Coxeter bialgebra and its basic properties; computation of its logarithm using the noncommutative Zaslavsky formula.

Examples illustrating the above theory are provided in Chapter 10 and they could be read in parallel with the present chapter. The toy example of rank one is developed in Section 13.8.

The parallel discussion for Coxeter species is in Chapter 3. There are similarities as well as differences between the two situations.

9.1. Cauchy powers of a Coxeter space

We introduce Cauchy powers and commutative Cauchy powers of a Coxeter space. Coproduct components of a Coxeter coalgebra can be encapsulated as maps from the Coxeter coalgebra to its Cauchy powers. Dually, product components of a Coxeter algebra can be encapsulated as maps to the Coxeter algebra from its Cauchy powers.

The parallel discussion for Coxeter species is in Section 3.1.

9.1.1. Cauchy powers of a Coxeter space. Let V be a Coxeter space. For $k \geq 1$, define the Coxeter space V^k with Z-component given by

$$\mathrm{V}^k[Z] := \bigoplus_{\substack{T:\, T \geq Z \\ \mathrm{rk}(T/Z) = k-1}} \mathrm{V}[T]. \tag{9.1}$$

The structure maps of V^k are as follows. A morphism $\sigma : Z \to Z'$ induces a morphism $\sigma : T \to T'$ with $T' = \sigma_*(T)$ and $T = \sigma^*(T')$ as in (8.12) or (1.28). The sum of the linear maps $\sigma : \mathrm{V}[T] \to \mathrm{V}[T']$ yields the linear map

$$\sigma : \mathrm{V}^k[Z] \to \mathrm{V}^k[Z'].$$

We call V^k the k-th *Cauchy power* of the Coxeter space V. When $k = 1$, there is only one summand in (9.1) for $T = Z$, and so $\mathrm{V}^1 = \mathrm{V}$.

Example 9.1 (Rank three). Cauchy powers of V on a vertex-type in rank three are shown below.

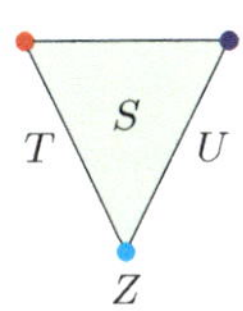

$$\mathrm{V}^4[Z] = \mathrm{V}^5[Z] = \cdots = 0$$
$$\mathrm{V}^3[Z] = \mathrm{V}[S]$$
$$\mathrm{V}^2[Z] = \mathrm{V}[T] \oplus \mathrm{V}[U]$$
$$\mathrm{V}^1[Z] = \mathrm{V}[Z]$$

Cauchy powers on other face-types can be written similarly. □

For a Coxeter coalgebra C, for $k \geq 1$, define

$$\Delta^{k-1} : \mathrm{C} \to \mathrm{C}^k, \qquad \Delta^{k-1} := (\Delta_Z^T)_{\mathrm{rk}(T/Z)=k-1}. \tag{9.2}$$

Note: $\Delta^0 = \mathrm{id}$ since $\Delta_Z^Z = \mathrm{id}$ by the counitality axiom in (8.46).

Dually, for a Coxeter algebra A, for $k \geq 1$, define

$$\mu^{k-1} : \mathrm{A}^k \to \mathrm{A}, \qquad \mu^{k-1} := (\mu_Z^T)_{\mathrm{rk}(T/Z)=k-1}. \tag{9.3}$$

Note: $\mu^0 = \mathrm{id}$ since $\mu_Z^Z = \mathrm{id}$ by the unitality axiom in (8.41).

Exercise 9.2. Check: Δ^{k-1} and μ^{k-1} are maps of Coxeter spaces. (Use the naturality axiom for the coproduct in (8.46) and for the product in (8.41).)

9.1.2. Commutative Cauchy powers of a Coxeter space.

Let V be a Coxeter space. For $k \geq 1$, define the Coxeter space $\mathrm{V}^{\underline{k}}$ with Z-component given by the quotient space

$$\mathrm{V}^{\underline{k}}[Z] := \Big(\bigoplus_{\substack{T:\, T \geq Z \\ \mathrm{rk}(T/Z)=k-1}} \mathrm{V}[T] \Big)_{\mathrm{quo}}, \tag{9.4}$$

where $x \equiv \sigma(x)$ for any morphism $\sigma : T \to T'$ local to Z, with T/Z and T'/Z' of rank $k-1$.

Similarly, for $k \geq 1$, define the Coxeter space $\mathrm{V}^{\overline{k}}$ with Z-component given by the subspace

$$\mathrm{V}^{\overline{k}}[Z] := \Big(\bigoplus_{\substack{T:\, T \geq Z \\ \mathrm{rk}(T/Z)=k-1}} \mathrm{V}[T] \Big)^{\mathrm{sub}} \tag{9.5}$$

consisting of elements $(x_T)_{\mathrm{rk}(T/Z)=k-1}$ such that $\sigma(x_T) = x_{T'}$ for any morphism $\sigma : T \to T'$ local to Z, with T/Z and T'/Z' of rank $k-1$.

We call either of $\mathrm{V}^{\underline{k}}$ and $\mathrm{V}^{\overline{k}}$ as the k-th *commutative Cauchy power* of the Coxeter space V. There are canonical maps

$$\mathrm{V}^{\overline{k}} \hookrightarrow \mathrm{V}^k \twoheadrightarrow \mathrm{V}^{\underline{k}} \tag{9.6}$$

of Coxeter spaces. For $k = 1$, the maps are identities with $\mathrm{V}^{\overline{1}} = \mathrm{V}^1 = \mathrm{V}^{\underline{1}} = \mathrm{V}$.

Exercise 9.3. Check: For a cocommutative Coxeter coalgebra C, the map Δ^{k-1} in (9.2) sends C into $\mathrm{C}^{\overline{k}}$. Dually, for a commutative Coxeter algebra A, the map μ^{k-1} in (9.3) to A factors through $\mathrm{A}^{\underline{k}}$. This is shown below.

$$\begin{array}{ccc} \mathrm{C} & \xrightarrow{\Delta^{k-1}} & \mathrm{C}^k \\ & \searrow & \uparrow \\ & & \mathrm{C}^{\overline{k}} \end{array} \qquad\qquad \begin{array}{ccc} \mathrm{A}^k & \xrightarrow{\mu^{k-1}} & \mathrm{A} \\ \downarrow & \nearrow & \\ \mathrm{A}^{\underline{k}} & & \end{array}$$

9.1.3. Examples.

Sums of (commutative) Cauchy powers of Coxeter spaces appear naturally in universal constructions which we will discuss in Section 9.3. Concrete examples of such sums occur in Chapter 10, see for instance (10.8), (10.35), (10.47), (10.61), (10.125).

Observation 9.4. For $k > 1$, for V a Coxeter space, $\mathrm{V}^k[S] = \mathrm{V}^{\underline{k}}[S] = \mathrm{V}^{\overline{k}}[S] = 0$. Hence, the functor (8.30) takes the Coxeter spaces V^k, $\mathrm{V}^{\underline{k}}$, $\mathrm{V}^{\overline{k}}$ to the zero W-module for $k > 1$. See Table 8.4.

9.2. Primitive and decomposable filtrations

Any Coxeter coalgebra carries a filtration called the primitive filtration. It arises by taking kernels of coproduct components from the Coxeter coalgebra to its Cauchy powers. The first term is the primitive part of the Coxeter coalgebra. Dually, any Coxeter algebra carries a filtration called the decomposable filtration. It arises by taking images of product components to the Coxeter algebra from its Cauchy powers. The first term is the decomposable part of the Coxeter algebra. Moding it out yields its indecomposable part. See Table 9.1.

TABLE 9.1. Primitive and decomposable filtrations.

Coxeter coalgebra C	primitive part $\overline{\mathcal{P}}(\mathrm{C})$ primitive filtration $\overline{\mathcal{P}}_1(\mathrm{C}) \subseteq \overline{\mathcal{P}}_2(\mathrm{C}) \subseteq \cdots \subseteq \mathrm{C}$
Coxeter algebra A	decomposable part $\overline{\mathcal{D}}(\mathrm{A})$ decomposable filtration $\mathrm{A} \supseteq \overline{\mathcal{D}}_1(\mathrm{A}) \supseteq \overline{\mathcal{D}}_2(\mathrm{A}) \supseteq \dots$ indecomposable part $\overline{\mathcal{Q}}(\mathrm{A})$

We formulate trivial Coxeter coalgebras (algebras) and relate them to the primitive (indecomposable) part functor. We also introduce the related notions of coderivations and derivations.

The parallel discussion for Coxeter species is in Section 3.2. Compare Table 9.1 with Table 3.1. For connections between the two using Fock functors, see Section 15.8.

9.2.1. Primitive filtration of a Coxeter coalgebra. Let C be a Coxeter coalgebra. Define the Coxeter subspace $\overline{\mathcal{P}}(\mathrm{C})$ by

$$\overline{\mathcal{P}}(\mathrm{C})[Z] := \bigcap_{T:\, T>Z} \ker(\Delta_Z^T : \mathrm{C}[Z] \to \mathrm{C}[T]). \tag{9.7}$$

For $Z = S$, the intersection is over the empty set and hence $\overline{\mathcal{P}}(\mathrm{C})[S] = \mathrm{C}[S]$.

Let $x \in \mathrm{C}[Z]$. If $\Delta_Z^T(x) = 0$ for all $T \gtrdot Z$, then by the coassociativity axiom in (8.46), $\Delta_Z^T(x) = 0$ for all $T > Z$. Thus,

$$\overline{\mathcal{P}}(\mathrm{C})[Z] = \bigcap_{T:\, T\gtrdot Z} \ker(\Delta_Z^T : \mathrm{C}[Z] \to \mathrm{C}[T]).$$

Equivalently, in terms of (9.2), $\overline{\mathcal{P}}(\mathrm{C}) = \ker \Delta^1$. We refer to $\overline{\mathcal{P}}(\mathrm{C})$ as the *primitive part* of C. This construction is functorial in C and yields the *primitive part functor* from the category of Coxeter coalgebras to the category of Coxeter spaces.

More generally, for $k \geq 1$, define the Coxeter subspace $\overline{\mathcal{P}}_k(\mathrm{C})$ by

$$\overline{\mathcal{P}}_k(\mathrm{C})[Z] := \bigcap_{\substack{T:\, T\geq Z \\ \mathrm{rk}(T/Z)\geq k}} \ker(\Delta_Z^T : \mathrm{C}[Z] \to \mathrm{C}[T]). \tag{9.8}$$

Again by the coassociativity axiom, it suffices to intersect over all $T \geq Z$ such that $\mathrm{rk}(T/Z) = k$. Thus, $\overline{\mathcal{P}}_k(\mathrm{C}) = \ker \Delta^k$. This is the k-th term of the *primitive filtration* of C. The first term is the primitive part of C, that is, $\overline{\mathcal{P}}_1 = \overline{\mathcal{P}}$. Observe:

$$\overline{\mathcal{P}}_1(\mathrm{C}) \subseteq \overline{\mathcal{P}}_2(\mathrm{C}) \subseteq \cdots \subseteq \mathrm{C} \quad \text{with} \quad \bigcup_{k \geq 1} \overline{\mathcal{P}}_k(\mathrm{C}) = \mathrm{C}. \tag{9.9}$$

More precisely, $\overline{\mathcal{P}}_k(\mathrm{C}) = \mathrm{C}$ as soon as k exceeds the rank of the arrangement.

Example 9.5 (Rank three). The primitive filtration of C on a vertex-type in rank three is spelt out below.

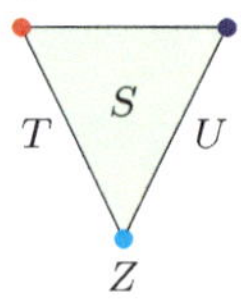

$$\overline{\mathcal{P}}_3(\mathrm{C})[Z] = \overline{\mathcal{P}}_4(\mathrm{C})[Z] = \cdots = \mathrm{C}[Z]$$
$$\overline{\mathcal{P}}_2(\mathrm{C})[Z] = \ker \Delta_Z^S$$
$$\overline{\mathcal{P}}(\mathrm{C})[Z] = \ker \Delta_Z^T \cap \ker \Delta_Z^U$$

Other face-types can be handled similarly. □

Proposition 9.6. *A morphism $f : \mathrm{C} \to \mathrm{D}$ of Coxeter coalgebras is injective iff the restriction $f : \overline{\mathcal{P}}(\mathrm{C}) \to \overline{\mathcal{P}}(\mathrm{D})$ is injective.*

PROOF. Forward implication is clear. The proof of the backward implication is formally the same as that of Proposition 3.2. For convenience, let us spell out the argument again. Suppose f is injective on the primitive part. We do a backward induction on the partial order on face-types. The component f_S is injective since $\mathrm{C}[S]$ belongs to the primitive part. By induction hypothesis, f_U is injective for all $U > T$. We want to show f_T is injective. Suppose $f_T(z) = 0$. Then, for any $U > T$, by (8.47), $f_U \Delta_T^U(z) = \Delta_T^U f_T(z) = 0$, and since f_U is injective, $\Delta_T^U(z) = 0$. So z is primitive, and hence $z = 0$. Thus f_T is injective which completes the induction step. □

9.2.2. Decomposable filtration of a Coxeter algebra. Let A be a Coxeter algebra. Define the Coxeter subspace $\overline{\mathcal{D}}(\mathrm{A})$ by

$$\overline{\mathcal{D}}(\mathrm{A})[Z] := \sum_{T:\, T > Z} \mathrm{im}(\mu_Z^T : \mathrm{A}[T] \to \mathrm{A}[Z]). \tag{9.10}$$

By the associativity axiom in (8.41), it suffices to sum only over face-types which cover Z. In terms of (9.3), $\overline{\mathcal{D}}(\mathrm{A}) = \mathrm{im}\, \mu^1$. We refer to $\overline{\mathcal{D}}(\mathrm{A})$ as the *decomposable part* of A.

More generally, for $k \geq 1$, define the Coxeter subspace $\overline{\mathcal{D}}_k(\mathrm{A})$ by

$$\overline{\mathcal{D}}_k(\mathrm{A})[Z] := \sum_{\substack{T:\, T \geq Z \\ \mathrm{rk}(T/Z) \geq k}} \mathrm{im}(\mu_Z^T : \mathrm{A}[T] \to \mathrm{A}[Z]). \tag{9.11}$$

Equivalently, $\overline{\mathcal{D}}_k(\mathrm{A}) = \mathrm{im}\, \mu^k$. This is the k-th term of the *decomposable filtration* of A. The first term is the decomposable part of A, that is, $\overline{\mathcal{D}}_1(\mathrm{A}) = \overline{\mathcal{D}}(\mathrm{A})$. Observe:

$$\mathrm{A} \supseteq \overline{\mathcal{D}}_1(\mathrm{A}) \supseteq \overline{\mathcal{D}}_2(\mathrm{A}) \supseteq \dots \quad \text{with} \quad \bigcap_{k \geq 1} \overline{\mathcal{D}}_k(\mathrm{A}) = 0. \tag{9.12}$$

More precisely, $\overline{\mathcal{D}}_k(\mathrm{A}) = 0$ as soon as k exceeds the rank of the arrangement.

9.2.3. Indecomposable part of a Coxeter algebra. Let A be a Coxeter algebra. The *indecomposable part* of A is the Coxeter quotient space

$$\overline{\mathcal{Q}}(\mathrm{A}) := \mathrm{A}/\overline{\mathcal{D}}(\mathrm{A}). \tag{9.13}$$

Equivalently, $\overline{\mathcal{Q}}(\mathrm{A}) = \operatorname{coker} \mu^1$. This construction is functorial in A and yields the *indecomposable part functor* from the category of Coxeter algebras to the category of Coxeter spaces.

Proposition 9.7. *A morphism $f : \mathrm{A} \to \mathrm{B}$ of Coxeter algebras is surjective iff the induced map $f : \overline{\mathcal{Q}}(\mathrm{A}) \to \overline{\mathcal{Q}}(\mathrm{B})$ is surjective.*

This result is the dual of Proposition 9.6 in view of (9.14) below and can be formally deduced from it.

9.2.4. Duality. For C a Coxeter coalgebra, $\overline{\mathcal{P}}_k(\mathrm{C})$ and $\overline{\mathcal{D}}_k(\mathrm{C}^*)$ are orthogonal complements of each other under the canonical pairing between C and C^*. Dually, for A a Coxeter algebra, $\overline{\mathcal{D}}_k(\mathrm{A})$ and $\overline{\mathcal{P}}_k(\mathrm{A}^*)$ are orthogonal complements of each other under the canonical pairing between A and A^*.

The case $k = 1$ can also be formulated as follows. For a Coxeter coalgebra C and Coxeter algebra A,

$$\overline{\mathcal{P}}(\mathrm{C})^* \cong \overline{\mathcal{Q}}(\mathrm{C}^*) \quad \text{and} \quad \overline{\mathcal{Q}}(\mathrm{A})^* \cong \overline{\mathcal{P}}(\mathrm{A}^*). \tag{9.14}$$

9.2.5. Trivial Coxeter (co)algebras. Every Coxeter space can be turned into a Coxeter algebra and a Coxeter coalgebra in a trivial way by setting

$$\mu_Z^T := \begin{cases} 0 & \text{if } T > Z, \\ \mathrm{id} & \text{if } T = Z, \end{cases} \qquad \Delta_Z^T := \begin{cases} 0 & \text{if } T > Z, \\ \mathrm{id} & \text{if } T = Z. \end{cases} \tag{9.15}$$

Axioms (8.41) and (8.46) clearly hold. We refer to this as a *trivial Coxeter algebra* and a *trivial Coxeter coalgebra*, respectively. This defines functors

$$trv : \overline{\mathsf{C}}\mathsf{Sp} \to \mathsf{Mon}(\overline{\mathsf{C}}\mathsf{Sp}) \quad \text{and} \quad trv : \overline{\mathsf{C}}\mathsf{Sp} \to \mathsf{Comon}(\overline{\mathsf{C}}\mathsf{Sp}). \tag{9.16}$$

These are the *trivial Coxeter algebra functor* and *trivial Coxeter coalgebra functor*, respectively. In fact, they map into the subcategories of commutative Coxeter algebras and cocommutative Coxeter coalgebras, respectively (since axioms (8.86) and (8.91) also clearly hold).

Lemma 9.8. *The trivial Coxeter coalgebra functor is left adjoint to the primitive part functor. Dually, the trivial Coxeter algebra functor is right adjoint to the indecomposable part functor.*

This is displayed below.

$$\overline{\mathsf{C}}\mathsf{Sp} \underset{\overline{\mathcal{P}}}{\overset{trv}{\rightleftarrows}} \mathsf{Comon}(\overline{\mathsf{C}}\mathsf{Sp}) \qquad \mathsf{Mon}(\overline{\mathsf{C}}\mathsf{Sp}) \underset{trv}{\overset{\overline{\mathcal{Q}}}{\rightleftarrows}} \overline{\mathsf{C}}\mathsf{Sp} \tag{9.17}$$

Explicitly:

- For C a trivial Coxeter coalgebra, a morphism of Coxeter coalgebras $\mathrm{C} \to \mathrm{D}$ is the same as a map of Coxeter spaces $\mathrm{C} \to \overline{\mathcal{P}}(\mathrm{D})$.

- For A a trivial Coxeter algebra, a morphism of Coxeter algebras $\mathrm{B} \to \mathrm{A}$ is the same as a map of Coxeter spaces $\overline{\mathcal{Q}}(\mathrm{B}) \to \mathrm{A}$.

9.2.6. Derivations and coderivations. We now define (co, bi)derivations.

- For a Coxeter space V and Coxeter coalgebra C, a map of Coxeter spaces $f : \mathrm{V} \to \mathrm{C}$ is a *coderivation* if it maps into $\overline{\mathcal{P}}(\mathrm{C})$.
- For a Coxeter algebra A and Coxeter space V, a map of Coxeter spaces $f : \mathrm{A} \to \mathrm{V}$ is a *derivation* if it factors through $\overline{\mathcal{Q}}(\mathrm{A})$.
- For a Coxeter algebra A and Coxeter coalgebra C, a map of Coxeter spaces $f : \mathrm{A} \to \mathrm{C}$ is a *biderivation* if it maps into $\overline{\mathcal{P}}(\mathrm{C})$ and factors through $\overline{\mathcal{Q}}(\mathrm{A})$.

The above definitions are illustrated below.

$$
\begin{array}{ccc}
\mathrm{V} & \xrightarrow{f} & \mathrm{C} \\
 & \searrow^{f} & \uparrow \\
 & & \overline{\mathcal{P}}(\mathrm{C})
\end{array}
\qquad
\begin{array}{ccc}
\mathrm{A} & \xrightarrow{f} & \mathrm{V} \\
\downarrow & \nearrow_{f} & \\
\overline{\mathcal{Q}}(\mathrm{A}) & &
\end{array}
\qquad
\begin{array}{ccc}
\mathrm{A} & \xrightarrow{f} & \mathrm{C} \\
\downarrow & & \uparrow \\
\overline{\mathcal{Q}}(\mathrm{A}) & \xrightarrow[f]{} & \overline{\mathcal{P}}(\mathrm{C})
\end{array}
$$

Note: A map is a biderivation iff it is both a derivation and a coderivation.

Observation 9.9. We have:

- A map f is a coderivation iff $\Delta^T_Z f_Z = 0$ for $T > Z$. Thus, a map of Coxeter spaces $f : \mathrm{V} \to \mathrm{C}$ is a coderivation iff f is a morphism of Coxeter coalgebras (with trivial coproduct on V).
- A map f is a derivation iff $f_Z \mu^T_Z = 0$ for $T > Z$. Thus, a map of Coxeter spaces $f : \mathrm{A} \to \mathrm{V}$ is a derivation iff f is a morphism of Coxeter algebras (with trivial product on V).

This can also be seen as a consequence of adjunctions (9.17).

Exercise 9.10. Formulate primitive part, decomposable part, and other notions above in terms of top-nested faces by replacing face-types Z and T, respectively, by top-nested faces (A, C) and (F, C), with $A \leq F \leq C$.

9.3. Universal constructions of Coxeter bialgebras

We construct the free Coxeter algebra $\overline{\mathcal{T}}(\mathrm{V})$ and cofree Coxeter coalgebra $\overline{\mathcal{T}}^{\vee}(\mathrm{V})$ on a Coxeter space V. In addition, for a scalar q, we construct the free Coxeter q-bialgebra $\overline{\mathcal{T}}_q(\mathrm{C})$ on a Coxeter coalgebra C, and the cofree Coxeter q-bialgebra $\overline{\mathcal{T}}_q{}^{\vee}(\mathrm{A})$ on a Coxeter algebra A. Every Coxeter space can be viewed as a Coxeter (co)algebra whose nontrivial (co)product components are zero. Thus, to every Coxeter space V, we have Coxeter q-bialgebras $\overline{\mathcal{T}}_q(\mathrm{V})$ and $\overline{\mathcal{T}}_q{}^{\vee}(\mathrm{V})$. These universal objects along with the terminology for their products and coproducts are summarized in Table 9.2.

We also discuss the free commutative Coxeter algebra $\overline{\mathcal{S}}(\mathrm{V})$ and cofree cocommutative Coxeter coalgebra $\overline{\mathcal{S}}^{\vee}(\mathrm{V})$ on a Coxeter space V. These have signed analogues which we denote $\overline{\mathcal{E}}(\mathrm{V})$ and $\overline{\mathcal{E}}^{\vee}(\mathrm{V})$.

We mention in passing that the existence of the above universal objects is a formal consequence of the existence of certain bimonads on Coxeter spaces which we constructed in Section 8.12.

TABLE 9.2. Universal Coxeter bialgebras.

Starting data	Coxeter q-bialgebra	Product	Coproduct
Coxeter coalgebra C	$\overline{\mathcal{T}}_q(\mathrm{C})$	concatenation	q-dequasishuffle
Coxeter algebra A	$\overline{\mathcal{T}}_q^{\vee}(\mathrm{A})$	q-quasishuffle	deconcatenation
Coxeter space V	$\overline{\mathcal{T}}_q(\mathrm{V})$	concatenation	q-deshuffle
Coxeter space V	$\overline{\mathcal{T}}_q^{\vee}(\mathrm{V})$	q-shuffle	deconcatenation
Coxeter space V	$\overline{\mathcal{T}}_0(\mathrm{V}) = \overline{\mathcal{T}}_0^{\vee}(\mathrm{V})$ for $q = 0$	concatenation = 0-shuffle	deconcatenation = 0-deshuffle

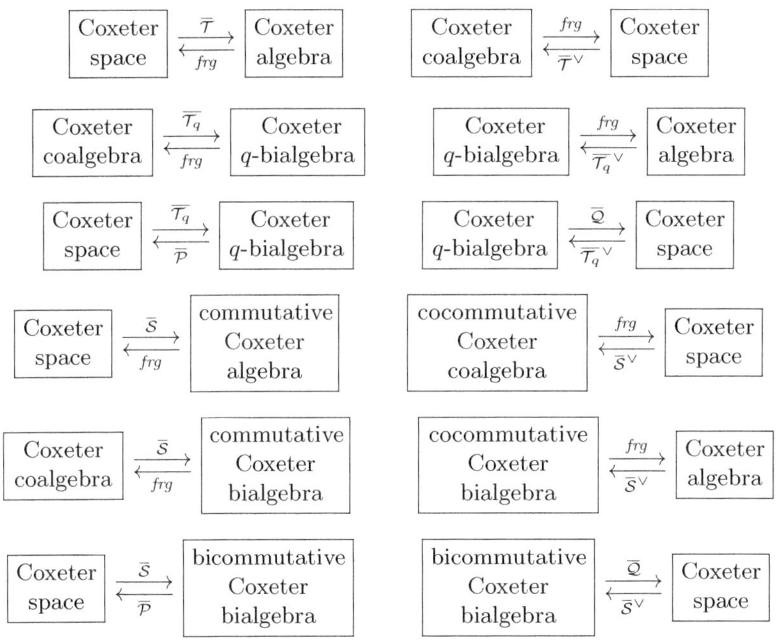

SUMMARY 9.1. Adjunctions involving Coxeter (co, bi)algebras.

Free and cofree objects are left and right adjoints, respectively, to forgetful functors which we denote *frg*. The main adjunctions are listed in Summary 9.1.

The above discussion parallels the one for Coxeter species in Section 3.3. Compare Table 9.2 with Table 3.2, and Summary 9.1 with Summary 3.1. For some connections between the two using Fock functors, see Section 15.4.

9.3.1. Free Coxeter algebra on a Coxeter space. For a Coxeter space V, recall from (8.119) the Coxeter space $\overline{\mathcal{T}}(\mathrm{V})$ defined by

$$\overline{\mathcal{T}}(\mathrm{V})[Z] := \bigoplus_{T\colon Z \leq T} \mathrm{V}[T]. \tag{9.18}$$

Observe:

$$\overline{\mathcal{T}}(\mathrm{V}) = \mathrm{V} + \mathrm{V}^2 + \mathrm{V}^3 + \dots, \tag{9.19}$$

the sum of Cauchy powers of V defined in (9.1). It carries a Coxeter algebra structure. The product component

$$\mu_Z^T : \overline{\mathcal{T}}(\mathrm{V})[T] \to \overline{\mathcal{T}}(\mathrm{V})[Z]$$

is the inclusion map which sends the summand $\mathrm{V}[U]$ (with $U \geq T$) to itself. We call μ the *concatenation product*. This yields a functor

$$\overline{\mathcal{T}} : \overline{\mathsf{C}}\mathsf{Sp} \to \mathsf{Mon}(\overline{\mathsf{C}}\mathsf{Sp}).$$

It is left adjoint to the forgetful functor. The unit of the adjunction is the inclusion $\mathrm{V} \hookrightarrow \overline{\mathcal{T}}(\mathrm{V})$ arising from (9.19). Thus, $\overline{\mathcal{T}}(\mathrm{V})$ is the *free Coxeter algebra* on the Coxeter space V. Its universal property is stated below.

Theorem 9.11. *Let* A *be a Coxeter algebra,* V *a Coxeter space,* $f : \mathrm{V} \to \mathrm{A}$ *a map of Coxeter spaces. Then there exists a unique morphism of Coxeter algebras* $\hat{f} : \overline{\mathcal{T}}(\mathrm{V}) \to \mathrm{A}$ *such that the diagram*

$$\begin{array}{ccc} \overline{\mathcal{T}}(\mathrm{V}) & \overset{\hat{f}}{\dashrightarrow} & \mathrm{A} \\ \uparrow & \nearrow_{f} & \\ \mathrm{V} & & \end{array}$$

commutes.

Explicitly, the map $\hat{f}$, on the Z-component, on the T-summand, is

$$\mathrm{V}[T] \xrightarrow{f_T} \mathrm{A}[T] \xrightarrow{\mu_Z^T} \mathrm{A}[Z]. \tag{9.20}$$

Observation 9.12. The indecomposable part of the free Coxeter algebra $\overline{\mathcal{T}}(\mathrm{V})$ is $\overline{\mathcal{Q}}(\overline{\mathcal{T}}(\mathrm{V})) = \mathrm{V}$. The V in the rhs is viewed as a quotient of $\overline{\mathcal{T}}(\mathrm{V})$ obtained by moding out the higher Cauchy powers of V which form the decomposable part of $\overline{\mathcal{T}}(\mathrm{V})$.

9.3.2. Free Coxeter q-bialgebra on a Coxeter coalgebra. Let q be a scalar. For a Coxeter coalgebra C, the free Coxeter algebra $\overline{\mathcal{T}}(\mathrm{C})$ carries a Coxeter q-bialgebra structure. To indicate dependence on q, we write $\overline{\mathcal{T}}_q(\mathrm{C})$. Its coproduct component $\Delta_Z^{T'}$ is defined by

$$\begin{array}{ccccc} \overline{\mathcal{T}}_q(\mathrm{C})[Z] & & \overset{\Delta_Z^{T'}}{\dashrightarrow} & & \overline{\mathcal{T}}_q(\mathrm{C})[T'] \\ \uparrow & & & & \uparrow \\ \mathrm{C}[T] & \xrightarrow[\Delta_T^{U}]{} & \mathrm{C}[U] & \xrightarrow[(\overline{\lambda}_q)_{TUT'U'}]{} & \mathrm{C}[U'] \end{array} \tag{9.21}$$

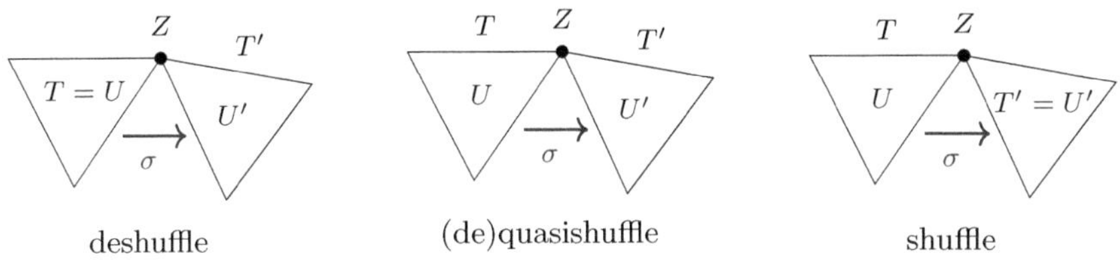

FIGURE 9.1. (De)shuffles and (de)quasishuffles.

for each $Z \leq T$. The map $\overline{\lambda}_q$ is as in (8.70). The bottom composite is not one map but a sum of maps over all choices of U and U'. To indicate that U and U' vary, we have marked them in red. Explicitly, $\Delta_Z^{T'}$ on the T-summand and into the U'-summand is

$$\sum_{U:\,U \geq T} \quad \sum_{\substack{\sigma:\,U \to U' \\ \text{local to } Z \\ \text{biprojects } T \text{ and } T'}} q^{l(\sigma)} \sigma \Delta_T^U. \tag{9.22}$$

See the middle picture in Figure 9.1. We call Δ the *q-dequasishuffle coproduct*. This yields a functor

$$\overline{\mathcal{T}}_q : \mathsf{Comon}(\overline{\mathsf{C}}\mathsf{Sp}) \to q\text{-}\mathsf{Bimon}(\overline{\mathsf{C}}\mathsf{Sp}). \tag{9.23}$$

It is left adjoint to the forgetful functor. Thus, $\overline{\mathcal{T}}_q(\mathrm{C})$ is the *free Coxeter q-bialgebra* on the Coxeter coalgebra C. Its universal property is stated below.

Theorem 9.13. *Let q be a scalar. Let* H *be a Coxeter q-bialgebra,* C *a Coxeter coalgebra, $f : \mathrm{C} \to \mathrm{H}$ a morphism of Coxeter coalgebras. Then there exists a unique morphism of Coxeter q-bialgebras $\hat{f} : \overline{\mathcal{T}}_q(\mathrm{C}) \to \mathrm{H}$ such that the diagram*

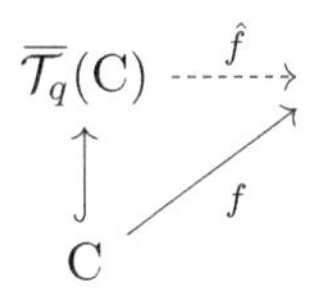

commutes.

Explicitly, the map $\hat{f}$ is given as before by formula (9.20).

Exercise 9.14. Let q be a scalar and C a Coxeter coalgebra. Check directly using coproduct formula (9.21) that the inclusion $\mathrm{C} \hookrightarrow \overline{\mathcal{T}}_q(\mathrm{C})$ is a morphism of Coxeter coalgebras. (Use Exercise 8.72.) Also check directly that the map $\hat{f} : \overline{\mathcal{T}}_q(\mathrm{C}) \to \mathrm{H}$ in Theorem 9.13 given by formula (9.20) is a morphism of Coxeter q-bialgebras. (For the check that $\hat{f}$ is first a map of Coxeter spaces and then a morphism of Coxeter algebras, we are in the setup of Theorem 9.11. For the Coxeter coalgebra part, use the q-bialgebra axiom (8.73) of H and connect it to (9.21).)

9.3.3. Left adjoint of the primitive part functor.

Precomposing (9.23) with the trivial Coxeter coalgebra functor (9.16) yields the functor

$$\overline{\mathcal{T}}_q : \overline{\mathsf{C}}\mathsf{Sp} \to q\text{-}\mathsf{Bimon}(\overline{\mathsf{C}}\mathsf{Sp}). \tag{9.24}$$

(We continue to denote it by $\overline{\mathcal{T}}_q$.) For a Coxeter space V, the product and coproduct of $\overline{\mathcal{T}}_q(\mathrm{V})$ are

$$\begin{array}{ccc} \overline{\mathcal{T}}_q(\mathrm{V})[T] & \xrightarrow{\mu_Z^T} & \overline{\mathcal{T}}_q(\mathrm{V})[Z] \\ \uparrow & & \uparrow \\ \mathrm{V}[U] & \xrightarrow[\mathrm{id}]{} & \mathrm{V}[U] \end{array} \qquad \begin{array}{ccc} \overline{\mathcal{T}}_q(\mathrm{V})[Z] & \xrightarrow{\Delta_Z^{T'}} & \overline{\mathcal{T}}_q(\mathrm{V})[T'] \\ \uparrow & & \uparrow \\ \mathrm{V}[T] & \xrightarrow[(\overline{\lambda}_q)_{TTT'U'}]{} & \mathrm{V}[U'], \end{array} \tag{9.25}$$

with $\overline{\lambda}_q$ as in (8.71). The product is concatenation as before, but note how the coproduct has simplified from (9.21). The sum is now only over U'. Explicitly, $\Delta_Z^{T'}$ on the T-summand and into the U'-summand is

$$\sum_{\substack{\sigma:\, T\to U' \\ \text{local to } Z \\ \text{projects } T \text{ on } T'}} q^{l(\sigma)}\sigma. \tag{9.26}$$

See the first picture in Figure 9.1. We call Δ the *q-deshuffle coproduct*.

Using the first adjunction in (9.17), we deduce that the functor $\overline{\mathcal{T}}_q$ is left adjoint to the primitive part functor (from Coxeter q-bialgebras to Coxeter spaces). This fact is phrased below as a universal property.

Theorem 9.15. *Let q be a scalar. Let* H *be a Coxeter q-bialgebra,* V *a Coxeter space, $f : \mathrm{V} \to \overline{\mathcal{P}}(\mathrm{H})$ a map of Coxeter spaces. Then there exists a unique morphism of Coxeter q-bialgebras $\hat{f} : \overline{\mathcal{T}}_q(\mathrm{V}) \to \mathrm{H}$ such that the diagram*

$$\begin{array}{ccc} \overline{\mathcal{T}}_q(\mathrm{V}) & \xrightarrow{\hat{f}} & \mathrm{H} \\ \uparrow & & \uparrow \\ \mathrm{V} & \xrightarrow[f]{} & \overline{\mathcal{P}}(\mathrm{H}) \end{array}$$

commutes.

Explicitly, the map $\hat{f}$, on the Z-component, on the T-summand, is

$$\mathrm{V}[T] \xrightarrow{f_T} \overline{\mathcal{P}}(\mathrm{H})[T] \hookrightarrow \mathrm{H}[T] \xrightarrow{\mu_Z^T} \mathrm{H}[Z]. \tag{9.27}$$

9.3.4. Cofree Coxeter coalgebra on a Coxeter space. Dually, one can construct the *cofree Coxeter coalgebra* on a Coxeter space V. It is denoted $\overline{\mathcal{T}}^\vee(\mathrm{V})$. As a Coxeter space, it equals $\overline{\mathcal{T}}(\mathrm{V})$. The coproduct component

$$\Delta_Z^T : \overline{\mathcal{T}}^\vee(\mathrm{V})[Z] \to \overline{\mathcal{T}}^\vee(\mathrm{V})[T]$$

is the projection which sends the summand $\mathrm{V}[U]$ (with $U \geq Z$) to itself if $U \geq T$, and to zero otherwise. We call Δ the *deconcatenation coproduct*. This yields a functor

$$\overline{\mathcal{T}}^\vee : \overline{\mathsf{C}}\mathsf{Sp} \to \mathsf{Comon}(\overline{\mathsf{C}}\mathsf{Sp}).$$

It is right adjoint to the forgetful functor. The counit of the adjunction is the projection $\overline{\mathcal{T}}^\vee(\mathrm{V}) \twoheadrightarrow \mathrm{V}$ arising from (9.19). The universal property of $\overline{\mathcal{T}}^\vee(\mathrm{V})$ can be phrased as follows.

Theorem 9.16. *Let* C *be a Coxeter coalgebra,* V *a Coxeter space,* $f : \mathrm{C} \to \mathrm{V}$ *a map of Coxeter spaces. Then there exists a unique morphism of Coxeter coalgebras* $\hat{f} : \mathrm{C} \to \overline{\mathcal{T}}^{\vee}(\mathrm{V})$ *such that the diagram*

$$\begin{array}{ccc} \mathrm{C} & \overset{\hat{f}}{\dashrightarrow} & \overline{\mathcal{T}}^{\vee}(\mathrm{V}) \\ & {\scriptstyle f}\searrow & \downarrow \\ & & \mathrm{V} \end{array}$$

commutes.

Explicitly, the map $\hat{f}$, on the Z-component, into the T-summand, is

$$\mathrm{C}[Z] \xrightarrow{\Delta_Z^T} \mathrm{C}[T] \xrightarrow{f_T} \mathrm{V}[T]. \tag{9.28}$$

Observation 9.17. The primitive part of the cofree Coxeter coalgebra $\overline{\mathcal{T}}^{\vee}(\mathrm{V})$ is $\overline{\mathcal{P}}(\overline{\mathcal{T}}^{\vee}(\mathrm{V})) = \mathrm{V}$. More generally, its primitive filtration is

$$\overline{\mathcal{P}}_k(\overline{\mathcal{T}}^{\vee}(\mathrm{V})) = \mathrm{V} + \mathrm{V}^2 + \cdots + \mathrm{V}^k, \qquad k \geq 1. \tag{9.29}$$

9.3.5. Cofree Coxeter q-bialgebra on a Coxeter algebra. Let q be a scalar. For a Coxeter algebra A, the cofree Coxeter coalgebra $\overline{\mathcal{T}}^{\vee}(\mathrm{A})$ carries a Coxeter q-bialgebra structure. To indicate dependence on q, we write $\overline{\mathcal{T}}_q^{\vee}(\mathrm{A})$. Its product component μ_Z^T is defined by

$$\begin{array}{ccccc} \overline{\mathcal{T}}_q^{\vee}(\mathrm{A})[T] & & \overset{\mu_Z^T}{\dashrightarrow} & & \overline{\mathcal{T}}_q^{\vee}(\mathrm{A})[Z] \\ \uparrow & & & & \uparrow \\ \mathrm{A}[U] & \xrightarrow[(\overline{\lambda}_q)_{TUT'U'}]{} & \mathrm{A}[U'] & \xrightarrow[\mu_{T'}^{U'}]{} & \mathrm{A}[T'] \end{array} \tag{9.30}$$

for each $T \leq U$. The map $\overline{\lambda}_q$ is as in (8.70). The bottom composite is a sum of maps over all choices of T' and U'. Explicitly, μ_Z^T on the U-summand and into the T'-summand is

$$\sum_{U' : U' \geq T'} \ \sum_{\substack{\sigma : U \to U' \\ \text{local to } Z \\ \text{biprojects } T \text{ and } T'}} q^{l(\sigma)} \mu_{T'}^{U'} \sigma. \tag{9.31}$$

See the middle picture in Figure 9.1. We call μ the *q-quasishuffle product*. This yields a functor

$$\overline{\mathcal{T}}_q^{\vee} : \mathsf{Mon}(\overline{\mathsf{C}}\mathsf{Sp}) \to q\text{-}\mathsf{Bimon}(\overline{\mathsf{C}}\mathsf{Sp}). \tag{9.32}$$

It is right adjoint to the forgetful functor. Thus, $\overline{\mathcal{T}}_q^{\vee}(\mathrm{A})$ is the *cofree Coxeter q-bialgebra* on the Coxeter algebra A. Its universal property is stated below.

Theorem 9.18. *Let q be a scalar. Let* H *be a Coxeter q-bialgebra,* A *a Coxeter algebra,* $f : \mathrm{H} \to \mathrm{A}$ *a morphism of Coxeter algebras. Then there* *exists a unique morphism of Coxeter q-bialgebras* $\hat{f} : \mathrm{H} \to \overline{\mathcal{T}}_q^{\vee}(\mathrm{A})$ *such that the diagram*

$$\begin{array}{ccc} \mathrm{H} & \overset{\hat{f}}{\dashrightarrow} & \overline{\mathcal{T}}_q^{\vee}(\mathrm{A}) \\ & \underset{f}{\searrow} & \downarrow \\ & & \mathrm{A} \end{array}$$

commutes.

Explicitly, the map $\hat{f}$ is given as before by formula (9.28).

9.3.6. Right adjoint of the indecomposable part functor. Precomposing (9.32) with the trivial Coxeter algebra functor (9.16) yields the functor

$$\overline{\mathcal{T}}_q^{\vee} : \overline{\mathsf{C}}\mathsf{Sp} \to q\text{-}\mathsf{Bimon}(\overline{\mathsf{C}}\mathsf{Sp}). \tag{9.33}$$

(We continue to denote it by $\overline{\mathcal{T}}_q^{\vee}$.) For a Coxeter space V, the product and coproduct of $\overline{\mathcal{T}}_q^{\vee}(\mathrm{V})$ are

$$\begin{array}{ccc} \overline{\mathcal{T}}_q^{\vee}(\mathrm{V})[T] & \overset{\mu_Z^T}{\dashrightarrow} & \overline{\mathcal{T}}_q^{\vee}(\mathrm{V})[Z] \\ \uparrow & & \uparrow \\ \mathrm{V}[U] & \xrightarrow[(\overline{\lambda}_q)_{TUT'T'}]{} & \mathrm{V}[T'] \end{array} \qquad \begin{array}{ccc} \overline{\mathcal{T}}_q^{\vee}(\mathrm{V})[Z] & \overset{\Delta_Z^T}{\dashrightarrow} & \overline{\mathcal{T}}_q^{\vee}(\mathrm{V})[T] \\ \uparrow & & \uparrow \\ \mathrm{V}[U] & \longrightarrow & \begin{cases} \mathrm{V}[U] & \text{if } U \geq T, \\ 0 & \text{otherwise,} \end{cases} \end{array} \tag{9.34}$$

with $\overline{\lambda}_q$ as in (8.71). The coproduct is deconcatenation as before, but the product has simplified from (9.30). The sum is now only over T'. Explicitly, μ_Z^T on the U-summand and into the T'-summand is

$$\sum_{\substack{\sigma\colon U \to T' \\ \text{local to } Z \\ \text{coprojects } T' \text{ on } T}} q^{l(\sigma)} \sigma. \tag{9.35}$$

See the last picture in Figure 9.1. In particular, μ_Z^T on the T-summand and into the T'-summand is

$$\sum_{\substack{\sigma\colon T \to T' \\ \text{local to } Z}} q^{l(\sigma)} \sigma. \tag{9.36}$$

We call μ the *q-shuffle product.*

Using the second adjunction in (9.17), we deduce that the functor $\overline{\mathcal{T}}_q^{\vee}$ is right adjoint to the indecomposable part functor (from Coxeter q-bialgebras to Coxeter spaces). This fact is phrased below as a universal property.

Theorem 9.19. *Let q be a scalar. Let* H *be a Coxeter q-bialgebra,* V *a Coxeter space,* $f : \overline{\mathcal{Q}}(\mathrm{H}) \to \mathrm{V}$ *a map of Coxeter spaces. Then there exists a unique morphism of Coxeter q-bialgebras* $\hat{f} : \mathrm{H} \to \overline{\mathcal{T}}_q^{\vee}(\mathrm{V})$ *such that the* *diagram*

$$\begin{array}{ccc} \mathrm{H} & \overset{\hat{f}}{\dashrightarrow} & \overline{\mathcal{T}}_q^{\vee}(\mathrm{V}) \\ \downarrow & & \downarrow \\ \overline{\mathcal{Q}}(\mathrm{H}) & \underset{f}{\longrightarrow} & \mathrm{V} \end{array}$$

commutes.

Explicitly, the map $\hat{f}$, on the Z-component, into the T-summand, is

$$\mathrm{H}[Z] \xrightarrow{\Delta_Z^T} \mathrm{H}[T] \twoheadrightarrow \overline{\mathcal{Q}}(\mathrm{H})[T] \xrightarrow{f_T} \mathrm{V}[T]. \tag{9.37}$$

9.3.7. Free and cofree Coxeter 0-bialgebra. Let us specialize to $q = 0$. For a Coxeter space V, the product and coproduct of $\overline{\mathcal{T}}_0(\mathrm{V})$ in (9.25) coincide with those of $\overline{\mathcal{T}}_0^{\vee}(\mathrm{V})$ in (9.34). They are given by concatenation (0-shuffle) and deconcatenation (0-deshuffle), respectively. They are rewritten below.

$$\begin{array}{ccc} \overline{\mathcal{T}}_0(\mathrm{V})[T] & \overset{\mu_Z^T}{\dashrightarrow} & \overline{\mathcal{T}}_0(\mathrm{V})[Z] \\ \uparrow & & \uparrow \\ \mathrm{V}[U] & \underset{\mathrm{id}}{\longrightarrow} & \mathrm{V}[U] \end{array} \qquad \begin{array}{ccc} \overline{\mathcal{T}}_0(\mathrm{V})[Z] & \overset{\Delta_Z^T}{\dashrightarrow} & \overline{\mathcal{T}}_0(\mathrm{V})[T] \\ \uparrow & & \uparrow \\ \mathrm{V}[U] & \longrightarrow & \begin{cases} \mathrm{V}[U] & \text{if } U \geq T, \\ 0 & \text{otherwise.} \end{cases} \end{array} \tag{9.38}$$

Thus, $\overline{\mathcal{T}}_0 = \overline{\mathcal{T}}_0^{\vee}$ as functors from the category of Coxeter spaces to the category of Coxeter 0-bialgebras.

Exercise 9.20. For a Coxeter space V, check that $\overline{\mathcal{T}}_0(\mathrm{V})$ is a Coxeter 0-bialgebra by verifying the 0-bialgebra axiom (8.78) using formulas (9.38). Note very carefully: This is the sense in which concatenation and deconcatenation are compatible.

Exercise 9.21 (0-dequasishuffle and 0-quasishuffle). We now specialize (9.21) and (9.30) to $q = 0$. Check: For a Coxeter coalgebra C, the coproduct of the Coxeter 0-bialgebra $\overline{\mathcal{T}}_0(\mathrm{C})$ is

$$\Delta_Z^{T'} : \bigoplus_{T:\, T \geq Z} \mathrm{C}[T] \to \bigoplus_{U':\, U' \geq T'} \mathrm{C}[U'], \qquad x \mapsto \Delta_T^{T' \cup T}(x)$$

for $x \in \mathrm{C}[T]$. Dually, for a Coxeter algebra A, the product of the Coxeter 0-bialgebra $\overline{\mathcal{T}}_0^{\vee}(\mathrm{A})$ is

$$\mu_Z^T : \bigoplus_{U:\, U \geq T} \mathrm{A}[U] \to \bigoplus_{T':\, T' \geq Z} \mathrm{A}[T'], \qquad x \mapsto \sum_{T':\, T' \cup T = U} \mu_{T'}^U(x)$$

for $x \in \mathrm{A}[U]$.

9.3.8. Free commutative Coxeter algebra. For a Coxeter space V, recall the Coxeter space $\overline{\mathcal{S}}(\mathrm{V})$ defined by (8.121). Observe:

$$\overline{\mathcal{S}}(\mathrm{V}) = \mathrm{V} + \mathrm{V}^{\underline{2}} + \mathrm{V}^{\underline{3}} + \dots, \tag{9.39}$$

the sum of commutative Cauchy powers of V defined in (9.4). It carries a commutative Coxeter algebra structure. The product component

$$\mu_Z^T : \overline{\mathcal{S}}(\mathrm{V})[T] \to \overline{\mathcal{S}}(\mathrm{V})[Z]$$

is induced from the inclusion map on summands, the point being that any morphism local to T is also local to Z, hence equivalent elements go to equivalent elements. Equivalently, $\overline{\mathcal{S}}(\mathrm{V})$ is the Coxeter quotient algebra of $\overline{\mathcal{T}}(\mathrm{V})$ under the canonical surjection $\overline{\mathcal{T}}(\mathrm{V}) \twoheadrightarrow \overline{\mathcal{S}}(\mathrm{V})$, see also (8.122). This is the *free commutative Coxeter algebra* on V. Its universal property is stated below. It is the commutative analogue of Theorem 9.11.

Theorem 9.22. *Let* A *be a commutative Coxeter algebra,* V *a Coxeter space,* $f : \mathrm{V} \to \mathrm{A}$ *a map of Coxeter spaces. Then there exists a unique morphism of Coxeter algebras* $\hat{f} : \overline{\mathcal{S}}(\mathrm{V}) \to \mathrm{A}$ *such that the diagram*

$$\begin{array}{ccc} \overline{\mathcal{S}}(\mathrm{V}) & \overset{\hat{f}}{\dashrightarrow} & \mathrm{A} \\ \uparrow & \nearrow_{f} & \\ \mathrm{V} & & \end{array}$$

commutes.

Explicitly, the map $\hat{f}$, on the Z-component, is induced from (9.20). It is well-defined in view of the commutative diagram

$$\begin{array}{ccccc} \mathrm{V}[T] & \xrightarrow{f_T} & \mathrm{A}[T] & & \\ \sigma\downarrow & & \sigma\downarrow & \searrow^{\mu_Z^T} & \mathrm{A}[Z] \\ \mathrm{V}[T'] & \xrightarrow[f_{T'}]{} & \mathrm{A}[T'] & \nearrow_{\mu_Z^{T'}} & \end{array}$$

for any $\sigma : T \to T'$ local to Z. Note use of the commutativity axiom (8.86).

Observation 9.23. The indecomposable part of the free commutative Coxeter algebra $\overline{\mathcal{S}}(\mathrm{V})$ is $\overline{\mathcal{Q}}(\overline{\mathcal{S}}(\mathrm{V})) = \mathrm{V}$. The V in the rhs is viewed as a quotient of $\overline{\mathcal{S}}(\mathrm{V})$ obtained by moding out the higher commutative Cauchy powers of V which form the decomposable part of $\overline{\mathcal{S}}(\mathrm{V})$.

9.3.9. Free commutative Coxeter bialgebra on a Coxeter coalgebra. For C a Coxeter coalgebra, $\overline{\mathcal{S}}(\mathrm{C})$ carries a Coxeter bialgebra structure as a quotient of $\overline{\mathcal{T}}(\mathrm{C})$. This is the free commutative Coxeter bialgebra on C. Further, if C is cocommutative, then $\overline{\mathcal{S}}(\mathrm{C})$ is bicommutative. Formally, we have functors

$$\overline{\mathcal{S}} : \mathsf{Comon}(\overline{\mathsf{C}}\mathsf{Sp}) \to \mathsf{Bimon}^{\mathsf{co}}(\overline{\mathsf{C}}\mathsf{Sp}) \quad \text{and} \quad \overline{\mathcal{S}} : {}^{\mathsf{co}}\mathsf{Comon}(\overline{\mathsf{C}}\mathsf{Sp}) \to {}^{\mathsf{co}}\mathsf{Bimon}^{\mathsf{co}}(\overline{\mathsf{C}}\mathsf{Sp})$$

which are left adjoints to the respective forgetful functors. By precomposing the latter with the trivial Coxeter coalgebra functor, we obtain the functor

$$\overline{\mathcal{S}} : \overline{\mathsf{C}}\mathsf{Sp} \to {}^{\mathsf{co}}\mathsf{Bimon}^{\mathsf{co}}(\overline{\mathsf{C}}\mathsf{Sp}). \tag{9.40}$$

It is left adjoint to the primitive part functor (from bicommutative Coxeter bialgebras to Coxeter spaces). This fact is phrased below as a universal property. It is the commutative analogue of Theorem 9.15.

Theorem 9.24. *Let* H *be a bicommutative Coxeter bialgebra,* V *a Coxeter space,* $f : \mathrm{V} \to \overline{\mathcal{P}}(\mathrm{H})$ *a map of Coxeter spaces. Then there exists a unique morphism of Coxeter bialgebras* $\hat{f} : \overline{\mathcal{S}}(\mathrm{V}) \to \mathrm{H}$ *such that the diagram*

$$\begin{array}{ccc} \overline{\mathcal{S}}(\mathrm{V}) & \xrightarrow{\hat{f}} & \mathrm{H} \\ \uparrow & & \uparrow \\ \mathrm{V} & \xrightarrow[f]{} & \overline{\mathcal{P}}(\mathrm{H}) \end{array}$$

commutes.

Explicitly, the map $\hat{f}$, on the Z-component, is induced from (9.20), as was the case in Theorem 9.22.

9.3.10. Cofree cocommutative Coxeter coalgebra. Dually, one can construct the *cofree cocommutative Coxeter coalgebra* on a Coxeter space V. It is denoted $\overline{\mathcal{S}}^{\vee}(\mathrm{V})$. As a Coxeter space, it is defined by (8.123). Observe:

$$\overline{\mathcal{S}}^{\vee}(\mathrm{V}) = \mathrm{V} + \mathrm{V}^{\overline{2}} + \mathrm{V}^{\overline{3}} + \dots, \tag{9.41}$$

the sum of commutative Cauchy powers of V defined in (9.5). The coproduct component

$$\Delta_Z^T : \overline{\mathcal{S}}^{\vee}(\mathrm{V})[Z] \to \overline{\mathcal{S}}^{\vee}(\mathrm{V})[T]$$

sends $(x_U)_{U \geq Z}$ to $(x_U)_{U \geq T}$. Equivalently, $\overline{\mathcal{S}}^{\vee}(\mathrm{V})$ is a Coxeter subcoalgebra of $\overline{\mathcal{T}}^{\vee}(\mathrm{V})$ under the canonical inclusion $\overline{\mathcal{S}}^{\vee}(\mathrm{V}) \hookrightarrow \overline{\mathcal{T}}^{\vee}(\mathrm{V})$, see also (8.124). The universal property of $\overline{\mathcal{S}}^{\vee}(\mathrm{V})$ can be phrased as follows. It is the commutative analogue of Theorem 9.16.

Theorem 9.25. *Let* C *be a cocommutative Coxeter coalgebra,* V *a Coxeter space,* $f : \mathrm{C} \to \mathrm{V}$ *a map of Coxeter spaces. Then there exists a unique morphism of Coxeter coalgebras* $\hat{f} : \mathrm{C} \to \overline{\mathcal{S}}^{\vee}(\mathrm{V})$ *such that the diagram*

$$\begin{array}{ccc} \mathrm{C} & \xrightarrow{\hat{f}} & \overline{\mathcal{S}}^{\vee}(\mathrm{V}) \\ & \searrow^{f} & \downarrow \\ & & \mathrm{V} \end{array}$$

commutes.

Explicitly, the map $\hat{f}$, on the Z-component, is induced from (9.28). It is well-defined in view of the commutative diagram

$$\begin{array}{cccc} & \mathrm{C}[T] & \xrightarrow{f_T} & \mathrm{V}[T] \\ {}^{\Delta_Z^T}\nearrow & \downarrow \sigma & & \downarrow \sigma \\ \mathrm{C}[Z] & & & \\ {}_{\Delta_Z^{T'}}\searrow & \mathrm{C}[T'] & \xrightarrow[f_{T'}]{} & \mathrm{V}[T'] \end{array}$$

for any $\sigma : T \to T'$ local to Z. Note use of the cocommutativity axiom (8.91).

Observation 9.26. The primitive part of the cofree cocommutative Coxeter coalgebra $\overline{\mathcal{S}}^{\vee}(\mathrm{V})$ is $\overline{\mathcal{P}}(\overline{\mathcal{S}}^{\vee}(\mathrm{V})) = \mathrm{V}$. More generally, its primitive filtration is

$$\overline{\mathcal{P}}_k(\overline{\mathcal{S}}^{\vee}(\mathrm{V})) = \mathrm{V} + \mathrm{V}^{\overline{2}} + \dots + \mathrm{V}^{\overline{k}}, \qquad k \geq 1. \tag{9.42}$$

9.3.11. Cofree cocommutative Coxeter bialgebra on a Coxeter algebra. Dually, for A a Coxeter algebra, $\overline{\mathcal{S}}^\vee(\mathrm{A})$ carries a Coxeter bialgebra structure. It sits inside $\overline{\mathcal{T}}^\vee(\mathrm{A})$. This is the cofree cocommutative Coxeter bialgebra on A. Further, if A is commutative, then $\overline{\mathcal{S}}^\vee(\mathrm{A})$ is bicommutative. Formally, we have functors

$$\overline{\mathcal{S}}^\vee : \mathsf{Mon}(\overline{\mathsf{C}}\mathsf{Sp}) \to {}^{\mathsf{co}}\mathsf{Bimon}(\overline{\mathsf{C}}\mathsf{Sp}) \quad \text{and} \quad \overline{\mathcal{S}}^\vee : \mathsf{Mon}^{\mathsf{co}}(\overline{\mathsf{C}}\mathsf{Sp}) \to {}^{\mathsf{co}}\mathsf{Bimon}^{\mathsf{co}}(\overline{\mathsf{C}}\mathsf{Sp})$$

which are right adjoints to the respective forgetful functors. By precomposing the latter with the trivial Coxeter algebra functor, we obtain the functor

$$\overline{\mathcal{S}}^\vee : \overline{\mathsf{C}}\mathsf{Sp} \to {}^{\mathsf{co}}\mathsf{Bimon}^{\mathsf{co}}(\overline{\mathsf{C}}\mathsf{Sp}). \tag{9.43}$$

It is right adjoint to the indecomposable part functor (from bicommutative Coxeter bialgebras to Coxeter spaces). This fact is phrased below as a universal property. It is the commutative analogue of Theorem 9.19.

Theorem 9.27. *Let* H *be a bicommutative Coxeter bialgebra,* V *a Coxeter space,* $f : \overline{\mathcal{Q}}(\mathrm{H}) \to \mathrm{V}$ *a map of Coxeter spaces. Then there exists a unique morphism of Coxeter bialgebras* $\hat{f} : \mathrm{H} \to \overline{\mathcal{S}}^\vee(\mathrm{V})$ *such that the diagram*

$$\begin{array}{ccc} \mathrm{H} & \overset{\hat{f}}{\dashrightarrow} & \overline{\mathcal{S}}^\vee(\mathrm{V}) \\ \downarrow & & \downarrow \\ \overline{\mathcal{Q}}(\mathrm{H}) & \underset{f}{\longrightarrow} & \mathrm{V} \end{array}$$

commutes.

Explicitly, the map $\hat{f}$, on the Z-component, is induced from (9.28), as was the case in Theorem 9.25.

Remark 9.28. We have defined two functors $\overline{\mathcal{S}}$ and $\overline{\mathcal{S}}^\vee$ from Coxeter spaces to bicommutative Coxeter bialgebras. It turns out that they are isomorphic via the norm map when $\mathrm{char}(\Bbbk) \nmid |W|$. See Lemma 9.94.

9.3.12. Signed analogue. For a Coxeter space V, we have the *free signed commutative Coxeter algebra* $\overline{\mathcal{E}}(\mathrm{V})$ and *cofree signed cocommutative Coxeter coalgebra* $\overline{\mathcal{E}}^\vee(\mathrm{V})$. They can be defined as

$$\overline{\mathcal{E}}(\mathrm{V}) := \overline{\mathcal{S}}(\mathrm{V}_-)_- \quad \text{and} \quad \overline{\mathcal{E}}^\vee(\mathrm{V}) := \overline{\mathcal{S}}^\vee(\mathrm{V}_-)_- \tag{9.44}$$

by conjugating $\overline{\mathcal{S}}$ and $\overline{\mathcal{S}}^\vee$ by the u-signature functor (8.99). We view the first as a Coxeter quotient algebra of $\overline{\mathcal{T}}_{-1}(\mathrm{V})$, and the second as a Coxeter subcoalgebra of $\overline{\mathcal{T}}^\vee_{-1}(\mathrm{V})$. Note:

$$\overline{\mathcal{T}}_{-1}(\mathrm{V}) = \overline{\mathcal{T}}(\mathrm{V}_-)_- \quad \text{and} \quad \overline{\mathcal{T}}^\vee_{-1}(\mathrm{V}) = \overline{\mathcal{T}}^\vee(\mathrm{V}_-)_-.$$

(More generally, for a scalar q, the conjugates of $\overline{\mathcal{T}}_q$ and $\overline{\mathcal{T}}_q^\vee$ are $\overline{\mathcal{T}}_{-q}$ and $\overline{\mathcal{T}}^\vee_{-q}$, respectively.)

For a Coxeter coalgebra C, we have the free signed commutative signed Coxeter bialgebra $\overline{\mathcal{E}}(\mathrm{C})$, and for a Coxeter algebra A, we have the cofree signed cocommutative signed Coxeter bialgebra $\overline{\mathcal{E}}^\vee(\mathrm{A})$. They can be defined as

$$\overline{\mathcal{E}}(\mathrm{C}) := \overline{\mathcal{S}}(\mathrm{C}_-)_- \quad \text{and} \quad \overline{\mathcal{E}}^\vee(\mathrm{A}) := \overline{\mathcal{S}}^\vee(\mathrm{A}_-)_-. \tag{9.45}$$

The first is a signed Coxeter quotient bialgebra of $\overline{\mathcal{T}}_{-1}(\mathrm{C})$, while the second is a signed Coxeter subbialgebra of $\overline{\mathcal{T}}^{\vee}_{-1}(\mathrm{A})$.

9.3.13. Alternative description of Coxeter bialgebras. For a Coxeter algebra (A, μ), putting together product components μ_Z^T yields a map of Coxeter spaces $\mu : \overline{\mathcal{T}}(\mathrm{A}) \to \mathrm{A}$. Diagram (8.29) follows from the naturality axiom in (8.41). Dually, for a Coxeter coalgebra (C, Δ), we have a map of Coxeter spaces $\Delta : \mathrm{C} \to \overline{\mathcal{T}}^{\vee}(\mathrm{C})$.

Exercise 9.29. Let q be a scalar. Let H be a Coxeter space equipped with maps μ and Δ such that (H, μ) is a Coxeter algebra and (H, Δ) is a Coxeter coalgebra. Show: The following are equivalent.

(1) $(\mathrm{H}, \mu, \Delta)$ is a Coxeter q-bialgebra.
(2) $\mu : \overline{\mathcal{T}}_q(\mathrm{H}) \to \mathrm{H}$ is a morphism of Coxeter coalgebras.
(3) $\Delta : \mathrm{H} \to \overline{\mathcal{T}}_q^{\vee}(\mathrm{H})$ is a morphism of Coxeter algebras.

The coproduct of $\overline{\mathcal{T}}_q(\mathrm{H})$ is given by q-dequasishuffle (9.21) with H viewed as a Coxeter coalgebra, while the product of $\overline{\mathcal{T}}_q^{\vee}(\mathrm{H})$ is given by q-quasishuffle (9.30) with H viewed as a Coxeter algebra.

In this situation, μ and Δ are in fact morphisms of Coxeter q-bialgebras. They are instances of the universal map $\hat{f}$ with $f = \mathrm{id}$ in Theorems 9.13 and 9.18.

Exercise 9.30. Check: The values of the functor (8.30) on the Coxeter spaces $\overline{\mathcal{T}}(\mathrm{V})$, $\overline{\mathcal{S}}(\mathrm{V})$, $\overline{\mathcal{S}}^{\vee}(\mathrm{V})$, $\overline{\mathcal{E}}(\mathrm{V})$, $\overline{\mathcal{E}}^{\vee}(\mathrm{V})$ all coincide and are equal to its value on V. (Use Observation 9.4 and the sums (9.19), (9.39), (9.41).) See also Table 8.4.

9.4. (Co)abelianizations of Coxeter (co)algebras

Abelianization turns a Coxeter algebra into a commutative Coxeter algebra. Dually, coabelianization turns a Coxeter coalgebra into a cocommutative Coxeter coalgebra. Both constructions apply to Coxeter bialgebras. The universality of these constructions is shown in Summary 9.2. They imply that (co)commutative Coxeter (co)algebras is a (co)reflective subcategory of Coxeter (co)algebras, and so on. We also have signed analogues.

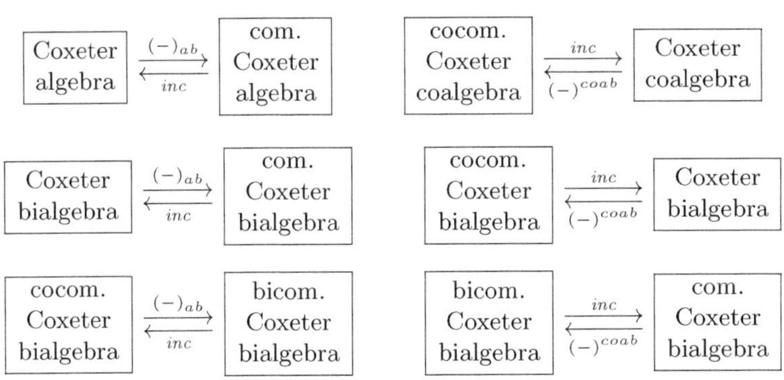

SUMMARY 9.2. Adjunctions involving (co)abelianization.

The parallel discussion for Coxeter species is in Section 3.4. Compare Summary 9.2 with Summary 3.2. For some connections between the two using Fock functors, see Section 16.5.

9.4.1. Abelianization of a Coxeter algebra. Let A be a Coxeter algebra. For each face-type Z, the linear span of the elements

$$\mu_Z^T(x) - \mu_Z^{T'}\sigma(x) \ \in \mathrm{A}[Z], \tag{9.46}$$

as x, T, T', σ vary, with $\sigma : T \to T'$ local to Z, yields a Coxeter subalgebra of A. That structure maps restrict can be deduced from Lemma 8.22, the first diagram in (8.28) and the naturality axiom in (8.41). That product components restrict can be deduced from the associativity axiom in (8.41). Taking quotient of A by this Coxeter subalgebra yields a commutative Coxeter algebra A_{ab} called the *abelianization* of A. The quotient map $\pi_{\mathrm{A}} : \mathrm{A} \twoheadrightarrow \mathrm{A}_{ab}$ is called the *abelianization map*.

A morphism of Coxeter algebras f from a Coxeter algebra A to a commutative Coxeter algebra B factors through π_{A} yielding a commutative diagram of Coxeter algebras

$$\begin{array}{ccc} \mathrm{A} & & \\ {\scriptstyle \pi_{\mathrm{A}}}\big\downarrow & \searrow^{f} & \\ \mathrm{A}_{ab} & \dashrightarrow & \mathrm{B}. \end{array}$$

Thus, A_{ab} is the largest commutative Coxeter quotient algebra of A. Equivalently, abelianization is left adjoint to the inclusion functor:

$$\mathsf{Mon}(\bar{\mathsf{C}}\mathsf{Sp}) \underset{inc}{\overset{(-)_{ab}}{\rightleftarrows}} \mathsf{Mon}^{\mathsf{co}}(\bar{\mathsf{C}}\mathsf{Sp}). \tag{9.47}$$

Hence, commutative Coxeter algebras is a reflective subcategory of Coxeter algebras via the inclusion functor (since this is full, faithful, and has a left adjoint).

Now let H be a Coxeter bialgebra. Then H_{ab} is a commutative Coxeter bialgebra via the quotient map π_{H}, see Exercise 9.31 below. Moreover, H_{ab} is bicommutative if H is cocommutative. Thus, (9.47) induces an adjunction between categories of Coxeter bialgebras and commutative Coxeter bialgebras

$$\mathsf{Bimon}(\bar{\mathsf{C}}\mathsf{Sp}) \underset{inc}{\overset{(-)_{ab}}{\rightleftarrows}} \mathsf{Bimon}^{\mathsf{co}}(\bar{\mathsf{C}}\mathsf{Sp}), \tag{9.48}$$

and also between categories of cocommutative Coxeter bialgebras and bicommutative Coxeter bialgebras.

Exercise 9.31. Check: For a Coxeter bialgebra H, the linear span of the elements (9.46) is closed under the coproduct of H.

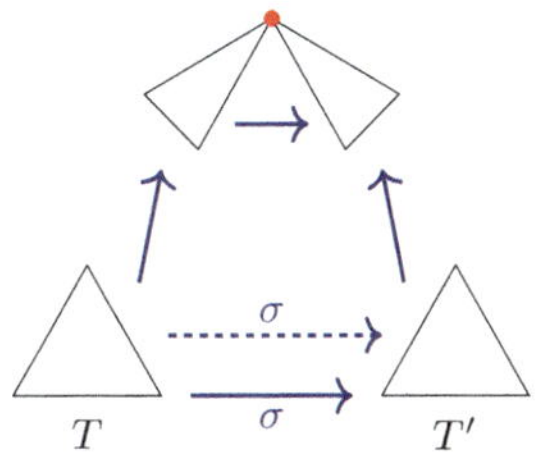

An illustration for $Z = \emptyset$ is shown above with coproduct taken on the red vertex-type. The oblique arrows indicate morphisms arising from the bialgebra axiom. (Use Exercise 1.40.)

9.4.2. Coabelianization of a Coxeter coalgebra. Dually, every Coxeter coalgebra C has a largest cocommutative Coxeter subcoalgebra C^{coab} called the *coabelianization*. The inclusion map $\pi_{\mathrm{C}}^{\vee} : \mathrm{C}^{coab} \hookrightarrow \mathrm{C}$ is called the *co-abelianization map*. Explicitly, $\mathrm{C}^{coab}[Z]$ consists of those $x \in \mathrm{C}[Z]$ for which

$$\sigma \Delta_Z^T(x) = \Delta_Z^{T'}(x) \tag{9.49}$$

for all $\sigma : T \to T'$ local to Z.

The image of a morphism of Coxeter coalgebras f from a cocommutative Coxeter coalgebra D to a Coxeter coalgebra C lies inside C^{coab} yielding a commutative diagram

$$\begin{array}{ccc} & & \mathrm{C} \\ & \overset{f}{\nearrow} & \uparrow \pi_{\mathrm{C}}^{\vee} \\ \mathrm{D} & \dashrightarrow & \mathrm{C}^{coab}. \end{array}$$

Thus, coabelianization is right adjoint to the inclusion functor:

$$ {}^{\mathrm{co}}\mathsf{Comon}(\overline{\mathsf{C}}\mathsf{Sp}) \underset{(-)^{coab}}{\overset{inc}{\rightleftarrows}} \mathsf{Comon}(\overline{\mathsf{C}}\mathsf{Sp}). \tag{9.50}$$

Hence, cocommutative Coxeter coalgebras is a coreflective subcategory of Coxeter coalgebras via the inclusion functor (since this is full, faithful, and has a right adjoint).

Now let H be a Coxeter bialgebra. Then H^{coab} is a cocommutative Coxeter bialgebra via the inclusion map $\pi_{\mathrm{H}}^{\vee}$. Moreover, H^{coab} is bicommutative if H is commutative. Thus, (9.50) induces an adjunction between categories of Coxeter bialgebras and cocommutative Coxeter bialgebras

$$ {}^{\mathrm{co}}\mathsf{Bimon}(\overline{\mathsf{C}}\mathsf{Sp}) \underset{(-)^{coab}}{\overset{inc}{\rightleftarrows}} \mathsf{Bimon}(\overline{\mathsf{C}}\mathsf{Sp}), \tag{9.51}$$

and also between categories of commutative Coxeter bialgebras and bicommutative Coxeter bialgebras.

9.4.3. Duality. For a Coxeter algebra A and Coxeter coalgebra C,

$$(\mathrm{A}_{ab})^* = (\mathrm{A}^*)^{coab} \quad \text{and} \quad (\mathrm{C}^{coab})^* = (\mathrm{C}^*)_{ab} \tag{9.52}$$

as Coxeter coalgebras and Coxeter algebras, respectively. Similarly, for a Coxeter bialgebra H,

$$(\mathrm{H}_{ab})^* = (\mathrm{H}^*)^{coab} \quad \text{and} \quad (\mathrm{H}^{coab})^* = (\mathrm{H}^*)_{ab} \tag{9.53}$$

as Coxeter bialgebras. Thus, abelianization and coabelianization are conjugates of each other wrt duality.

9.4.4. Signed (co)abelianization. By replacing σ by $(-1)^{l(\sigma)}\sigma$, the above discussion can be carried out in the signed setting. In this context, we use the terms *signed abelianization* and *signed coabelianization*. Thus, for the signed abelianization of A, denoted A_{-ab}, we take quotient by the elements

$$\mu_Z^T(x) - (-1)^{l(\sigma)}\mu_Z^{T'}\sigma(x) \ \in \mathrm{A}[Z],$$

while for the signed coabelianization of C, denoted C^{-coab}, we take the elements $x \in \mathrm{C}[Z]$ satisfying

$$(-1)^{l(\sigma)}\sigma\Delta_Z^T(x) = \Delta_Z^{T'}(x).$$

Formally,

$$(\mathrm{A}_-)_{ab} = (\mathrm{A}_{-ab})_- \quad \text{and} \quad (\mathrm{C}_-)^{coab} = (\mathrm{C}^{-coab})_-,$$

with $(-)_-$ being the u-signature functor (8.99).

9.4.5. Universal constructions. For a Coxeter coalgebra C and Coxeter algebra A,

$$\overline{\mathcal{S}}(\mathrm{C}) = \overline{\mathcal{T}}(\mathrm{C})_{ab} \quad \text{and} \quad \overline{\mathcal{S}}^\vee(\mathrm{A}) = \overline{\mathcal{T}}^\vee(\mathrm{A})^{coab} \tag{9.54}$$

as Coxeter bialgebras. In particular, this applies when C is a trivial Coxeter coalgebra, and when A is a trivial Coxeter algebra. In this case, we write: For a Coxeter space V,

$$\overline{\mathcal{S}}(\mathrm{V}) = \overline{\mathcal{T}}(\mathrm{V})_{ab} \quad \text{and} \quad \overline{\mathcal{S}}^\vee(\mathrm{V}) = \overline{\mathcal{T}}^\vee(\mathrm{V})^{coab} \tag{9.55}$$

as bicommutative Coxeter bialgebras. The abelianization map $\overline{\mathcal{T}}(\mathrm{V}) \twoheadrightarrow \overline{\mathcal{S}}(\mathrm{V})$ and coabelianization map $\overline{\mathcal{S}}^\vee(\mathrm{V}) \hookrightarrow \overline{\mathcal{T}}^\vee(\mathrm{V})$ are the canonical surjection and injection, respectively.

The proof of (9.54) is purely formal. By composing adjunctions, we see: The functor $\overline{\mathcal{T}}$ followed by abelianization is left adjoint to the forgetful functor from commutative Coxeter bialgebras to Coxeter coalgebras, and hence it coincides with $\overline{\mathcal{S}}$. This proves the first statement. The second is similar.

Similarly, for a Coxeter coalgebra C and Coxeter algebra A,

$$\overline{\mathcal{E}}(\mathrm{C}) = \overline{\mathcal{T}}_{-1}(\mathrm{C})_{-ab} \quad \text{and} \quad \overline{\mathcal{E}}^\vee(\mathrm{A}) = \overline{\mathcal{T}}^\vee_{-1}(\mathrm{A})^{-coab} \tag{9.56}$$

as signed Coxeter bialgebras. In particular, for a Coxeter space V,

$$\overline{\mathcal{E}}(\mathrm{V}) = \overline{\mathcal{T}}_{-1}(\mathrm{V})_{-ab} \quad \text{and} \quad \overline{\mathcal{E}}^\vee(\mathrm{V}) = \overline{\mathcal{T}}^\vee_{-1}(\mathrm{V})^{-coab} \tag{9.57}$$

as signed bicommutative signed Coxeter bialgebras. The signed abelianization map $\overline{\mathcal{T}}_{-1}(\mathrm{V}) \twoheadrightarrow \overline{\mathcal{E}}(\mathrm{V})$ and signed coabelianization map $\overline{\mathcal{E}}^\vee(\mathrm{V}) \hookrightarrow \overline{\mathcal{T}}^\vee_{-1}(\mathrm{V})$ are the canonical surjection and injection, respectively.

9.4.6. Primitive and indecomposable parts. For a Coxeter coalgebra C and Coxeter algebra A,

$$\overline{\mathcal{P}}(\mathrm{C}) = \overline{\mathcal{P}}(\mathrm{C}^{coab}) \quad \text{and} \quad \overline{\mathcal{Q}}(\mathrm{A}) = \overline{\mathcal{Q}}(\mathrm{A}_{ab}). \tag{9.58}$$

Exercise 9.32. Let C be a Coxeter coalgebra and A a Coxeter algebra. Check: The functor (8.30) takes both $\overline{\mathcal{P}}(\mathrm{C})$ and C^{coab} to $\mathrm{C}[S]$, and both $\overline{\mathcal{Q}}(\mathrm{A})$ and A_{ab} to $\mathrm{A}[S]$. See Table 8.4.

9.5. Hadamard product. Convolution Coxeter algebra

We introduce the Hadamard product of Coxeter spaces. It defines a symmetric monoidal structure on the category of Coxeter spaces, with the Coxeter space of polynomials as the unit object. Moreover, this structure extends to Coxeter algebras and coalgebras. We then construct the internal hom for the Hadamard product of Coxeter spaces, and discuss related constructions of the convolution Coxeter algebra and coconvolution Coxeter coalgebra. These ideas are continued in Sections 9.6 and 9.7. A part summary is provided in Table 9.3.

TABLE 9.3. Hadamard product of Coxeter spaces.

$\mathrm{V} \times \mathrm{W}$	Hadamard product of Coxeter spaces V and W
$\mathrm{hom}^{\times}(\mathrm{V}, \mathrm{W})$	internal hom of V and W
$\mathrm{C} \times \mathrm{D}$	Hadamard product of Coxeter coalgebras C and D
	copower of D by C
$\mathcal{C}(\mathrm{C}, \mathrm{D})$	enrichment of Coxeter coalgebras over itself internal hom of C and D
	power of D by C
$\overline{\mathcal{C}}(\mathrm{A}, \mathrm{B})$	enrichment of Coxeter algebras over Coxeter coalgebras universal measuring Coxeter coalgebra
$\mathrm{C} \triangleright \mathrm{A}$	copower of A by C
$\mathrm{hom}^{\times}(\mathrm{C}, \mathrm{A})$	convolution Coxeter algebra
	power of A by C

This parallels the discussion of the Hadamard product of Coxeter species in Sections 3.5, 3.6, 3.7. Compare Table 9.3 with Table 3.3. For connections between the two using Fock functors, see Section 17.3.

9.5.1. Coxeter space of polynomials. Let $\Bbbk[x]$ be the Coxeter space defined by $\Bbbk[x][T] = \Bbbk$ for all face-types T, with structure maps all identities. This is the *Coxeter space of polynomials*. Further, it carries a Coxeter algebra structure with product components also defined to be all identities. This is the *Coxeter algebra of polynomials*. Its dual is a Coxeter coalgebra which we denote $\Bbbk\{x\}$. We call it the *divided power Coxeter coalgebra*. Its coproduct components are all identities. In fact, $\Bbbk[x]$ and $\Bbbk\{x\}$ carry Coxeter bialgebra structures. They are studied in detail later in Section 10.2. See in particular (co)product formulas (10.9) and (10.12). Note: There is no distinction between $\Bbbk[x]$ and $\Bbbk\{x\}$ as Coxeter spaces.

9.5.2. Hadamard product. For Coxeter spaces V and W, define the Coxeter space $\mathrm{V} \times \mathrm{W}$ by

$$(\mathrm{V} \times \mathrm{W})[T] := \mathrm{V}[T] \otimes \mathrm{W}[T]. \tag{9.59}$$

For a morphism $\sigma : T \to U$, there is an induced linear map

$$\mathrm{V}[T] \otimes \mathrm{W}[T] \xrightarrow{\sigma \otimes \sigma} \mathrm{V}[U] \otimes \mathrm{W}[U],$$

which turns $\mathrm{V} \times \mathrm{W}$ into a Coxeter space. This is the *Hadamard product* of V and W. It defines a monoidal structure on the category of Coxeter spaces $\overline{\mathsf{C}}\mathsf{Sp}$. The unit object is the Coxeter space of polynomials. Interchanging the order of the tensor factors in (9.59) yields a braiding $\mathrm{V} \times \mathrm{W} \to \mathrm{W} \times \mathrm{V}$ (which is in fact a symmetry). Thus, $\overline{\mathsf{C}}\mathsf{Sp}$ is a symmetric monoidal category under the Hadamard product.

Observation 9.33. The functor (8.30) is a strong monoidal functor wrt Hadamard product of Coxeter spaces and tensor product of W-modules. That is, $(\mathrm{V} \times \mathrm{W})[S] = \mathrm{V}[S] \otimes \mathrm{W}[S]$ as W-modules. See also Table 8.4.

Observation 9.34. Hadamard product preserves Coxeter algebras. More precisely: If A and B are Coxeter algebras, then so is $\mathrm{A} \times \mathrm{B}$ with product components

$$\begin{array}{ccc} (\mathrm{A} \times \mathrm{B})[T] & \overset{\mu_Z^T}{\dashrightarrow} & (\mathrm{A} \times \mathrm{B})[Z] \\ \| & & \| \\ \mathrm{A}[T] \otimes \mathrm{B}[T] & \xrightarrow[\mu_Z^T \otimes \mu_Z^T]{} & \mathrm{A}[Z] \otimes \mathrm{B}[Z]. \end{array}$$

This turns the category of Coxeter algebras into a symmetric monoidal category. The unit object is the Coxeter algebra of polynomials $\Bbbk[x]$.

Dually, Hadamard product preserves Coxeter coalgebras. This turns the category of Coxeter coalgebras into a symmetric monoidal category. The unit object is the divided power Coxeter coalgebra $\Bbbk\{x\}$.

Hadamard product of (co)commutative Coxeter (co)algebras is again (co)commutative.

Remark 9.35. Hadamard product does not preserve Coxeter bialgebras in general. This is because the bialgebra axiom involves a sum (in contrast to the remaining axioms which are set-theoretic). This is in contrast to the situation for Coxeter bimonoids.

9.5.3. Internal hom. For Coxeter spaces V and W, let $\mathrm{hom}^{\times}(\mathrm{V}, \mathrm{W})$ denote the Coxeter space defined by

$$\mathrm{hom}^{\times}(\mathrm{V}, \mathrm{W})[Z] := \mathrm{Hom}_{\Bbbk}(\mathrm{V}[Z], \mathrm{W}[Z]). \tag{9.60}$$

The rhs is the space of all linear maps from $\mathrm{V}[Z]$ to $\mathrm{W}[Z]$. For a morphism $\sigma : Z \to Z'$, any linear map $f : \mathrm{V}[Z] \to \mathrm{W}[Z]$ induces a linear map

$$\begin{array}{ccc} \mathrm{V}[Z'] & \dashrightarrow & \mathrm{W}[Z'] \\ {\scriptstyle \sigma^{-1}}\downarrow & & \uparrow{\scriptstyle \sigma} \\ \mathrm{V}[Z] & \xrightarrow[f]{} & \mathrm{W}[Z]. \end{array} \tag{9.61}$$

With these structure maps, $\hom^{\times}(\mathrm{V}, \mathrm{W})$ is a Coxeter space. It is the internal hom for the Hadamard product on the category of Coxeter spaces. That is: For Coxeter spaces V, M, N, there is a natural bijection

$$\overline{\mathsf{C}}\mathsf{Sp}(\mathrm{V} \times \mathrm{M}, \mathrm{N}) \xrightarrow{\cong} \overline{\mathsf{C}}\mathsf{Sp}\big(\mathrm{V}, \hom^{\times}(\mathrm{M}, \mathrm{N})\big). \tag{9.62}$$

In particular, for Coxeter spaces M and N,

$$\overline{\mathsf{C}}\mathsf{Sp}(\mathrm{M}, \mathrm{N}) \cong \overline{\mathsf{C}}\mathsf{Sp}\big(\Bbbk[x], \hom^{\times}(\mathrm{M}, \mathrm{N})\big). \tag{9.63}$$

For a Coxeter space V, set

$$\mathrm{end}^{\times}(\mathrm{V}) := \hom^{\times}(\mathrm{V}, \mathrm{V}). \tag{9.64}$$

Exercise 9.36. Establish (9.62) using the natural bijection

$$\mathrm{Hom}_{\Bbbk}(\mathrm{V}[T] \otimes \mathrm{M}[T], \mathrm{N}[T]) \xrightarrow{\cong} \mathrm{Hom}_{\Bbbk}\big(\mathrm{V}[T], \mathrm{Hom}_{\Bbbk}(\mathrm{M}[T], \mathrm{N}[T])\big)$$

for each face-type T, which is the internal hom property for vector spaces. On the S-component, we also have

$$W\text{-}\mathsf{Mod}(\mathrm{V}[S] \otimes \mathrm{M}[S], \mathrm{N}[S]) \xrightarrow{\cong} W\text{-}\mathsf{Mod}(\mathrm{V}[S], \hom^{\otimes}(\mathrm{M}[S], \mathrm{N}[S]))$$

which is the internal hom property for W-modules [**11**, Proposition B.17].

9.5.4. Convolution Coxeter algebra.

For C a Coxeter coalgebra and A a Coxeter algebra, $\hom^{\times}(\mathrm{C}, \mathrm{A})$ carries a Coxeter algebra structure with product components

$$\mu_Z^T : \hom^{\times}(\mathrm{C}, \mathrm{A})[T] \to \hom^{\times}(\mathrm{C}, \mathrm{A})[Z], \qquad f \mapsto \mu_Z^T f \Delta_Z^T. \tag{9.65}$$

We refer to $\hom^{\times}(\mathrm{C}, \mathrm{A})$ as the *convolution Coxeter algebra*. If C is cocommutative and A is commutative, then $\hom^{\times}(\mathrm{C}, \mathrm{A})$ is commutative.

Dually, for A a Coxeter algebra and C a Coxeter coalgebra, $\hom^{\times}(\mathrm{A}, \mathrm{C})$ carries a Coxeter coalgebra structure with coproduct components

$$\Delta_Z^T : \hom^{\times}(\mathrm{A}, \mathrm{C})[Z] \to \hom^{\times}(\mathrm{A}, \mathrm{C})[T], \qquad f \mapsto \Delta_Z^T f \mu_Z^T. \tag{9.66}$$

We refer to $\hom^{\times}(\mathrm{A}, \mathrm{C})$ as the *coconvolution Coxeter coalgebra*. If A is commutative and C is cocommutative, then $\hom^{\times}(\mathrm{A}, \mathrm{C})$ is cocommutative.

Exercise 9.37. Check the assertions above. That is, (9.65) indeed yields a Coxeter algebra, and so on.

Special case. For a Coxeter coalgebra C and Coxeter algebra A,

$$\mathrm{C}^* \cong \hom^{\times}(\mathrm{C}, \Bbbk[x]) \quad \text{and} \quad \mathrm{A} \cong \hom^{\times}(\Bbbk\{x\}, \mathrm{A}) \tag{9.67}$$

as Coxeter algebras, and

$$\mathrm{A}^* \cong \hom^{\times}(\mathrm{A}, \Bbbk\{x\}) \quad \text{and} \quad \mathrm{C} \cong \hom^{\times}(\Bbbk[x], \mathrm{C}) \tag{9.68}$$

as Coxeter coalgebras.

Remark 9.38. For reasons similar to that in Remark 9.35, there is no biconvolution Coxeter bialgebra. This is again in contrast to the situation for Coxeter bimonoids.

Exercise 9.39. Formulate internal hom (9.60), convolution Coxeter algebra (9.65), coconvolution Coxeter coalgebra (9.66) in terms of top-nested faces by replacing face-types Z and T, respectively, by top-nested faces (A, C) and (F, C), with $A \leq F \leq C$.

9.6. Internal hom for Coxeter coalgebras

We construct the internal hom for Coxeter coalgebras wrt the Hadamard product. We write $\mathcal{C}(\mathrm{C}, \mathrm{D})$ for internal hom of C and D. For a cocommutative Coxeter coalgebra C and Coxeter bialgebra K, the Coxeter coalgebra $\mathcal{C}(\mathrm{C}, \mathrm{K})$ carries a Coxeter bialgebra structure. The existence of internal hom implies that the category of Coxeter coalgebras is enriched over itself, and it possesses powers and copowers. See Table 9.3.

9.6.1. Internal hom for Coxeter coalgebras. For Coxeter coalgebras C and D, we construct a Coxeter coalgebra $\mathcal{C}(\mathrm{C}, \mathrm{D})$ as follows. An element of $\mathcal{C}(\mathrm{C}, \mathrm{D})[Z]$ is a family of linear maps

$$f_{T/Z} : \mathrm{C}[T] \to \mathrm{D}[T],$$

one for each $Z \leq T$, such that the condition $f_{U/Z} \Delta_T^U = \Delta_T^U f_{T/Z}$ holds for each $Z \leq T \leq U$.

Example 9.40 (Rank three). The above conditions for a vertex-type in rank three are spelt out below.

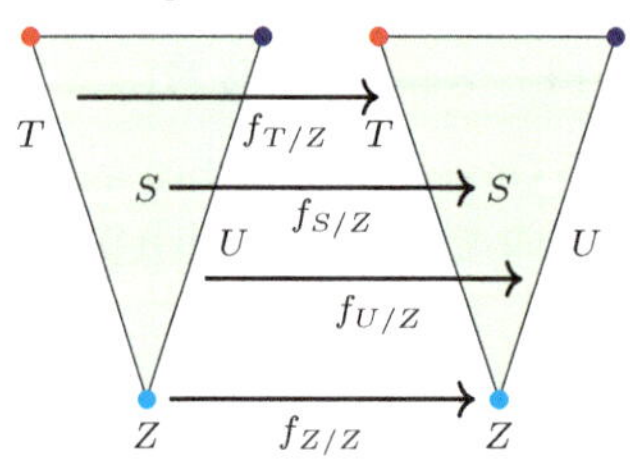

$$f_{T/Z}\Delta_Z^T = \Delta_Z^T f_{Z/Z} \qquad f_{S/Z}\Delta_T^S = \Delta_T^S f_{T/Z}$$
$$f_{U/Z}\Delta_Z^U = \Delta_Z^U f_{Z/Z} \qquad f_{S/Z}\Delta_U^S = \Delta_U^S f_{U/Z}$$
$$f_{S/Z}\Delta_Z^S = \Delta_Z^S f_{Z/Z}$$

Conditions for other face-types can be written similarly. □

We refer to $f_{Z/Z}$ as the base term, and to $f_{T/Z}$ for $T > Z$ as the higher terms. Now suppose we are given a morphism $\sigma : Z \to Z'$, and an element $(f_{T/Z})$ of $\mathcal{C}(\mathrm{C}, \mathrm{D})[Z]$. For each $T \geq Z$, consider the composite map

$$g_{T'/Z'} : \mathrm{C}[T'] \xrightarrow{\sigma^{-1}} \mathrm{C}[T] \xrightarrow{f_{T/Z}} \mathrm{D}[T] \xrightarrow{\sigma} \mathrm{D}[T']$$

with $T' := \sigma_*(T)$ and $\sigma : T \to T'$ the induced morphism as in (8.12) or (1.28). The linear maps

$$\sigma : \mathcal{C}(\mathrm{C}, \mathrm{D})[Z] \to \mathcal{C}(\mathrm{C}, \mathrm{D})[Z'], \qquad (f_{T/Z}) \mapsto (g_{T'/Z'}) \tag{9.69}$$

turn $\mathcal{C}(\mathrm{C}, \mathrm{D})$ into a Coxeter space. The coproduct component

$$\Delta_Z^T : \mathcal{C}(\mathrm{C}, \mathrm{D})[Z] \to \mathcal{C}(\mathrm{C}, \mathrm{D})[T], \qquad (f_{U/Z}) \mapsto (f_{U/T}), \tag{9.70}$$

is defined by restricting families to face-types greater than T, that is, $f_{U/T} := f_{U/Z}$ for $U \geq T$.

Exercise 9.41. Check: $\mathcal{C}(\mathrm{C}, \mathrm{D})$ is indeed a Coxeter coalgebra, that is, the naturality, coassociativity, counitality axioms in (8.46) hold.

This construction is the internal hom for the Hadamard product of Coxeter coalgebras. That is: For Coxeter coalgebras C, D, M, there is a natural bijection

$$\mathsf{Comon}(\overline{\mathsf{C}}\mathsf{Sp})(\mathrm{M} \times \mathrm{C}, \mathrm{D}) \xrightarrow{\cong} \mathsf{Comon}(\overline{\mathsf{C}}\mathsf{Sp})\big(\mathrm{M}, \mathcal{C}(\mathrm{C}, \mathrm{D})\big). \tag{9.71}$$

In particular, for Coxeter coalgebras C and D,

$$\mathsf{Comon}(\overline{\mathsf{C}}\mathsf{Sp})(\mathrm{C}, \mathrm{D}) \cong \mathsf{Comon}(\overline{\mathsf{C}}\mathsf{Sp})\big(\Bbbk\{x\}, \mathcal{C}(\mathrm{C}, \mathrm{D})\big). \tag{9.72}$$

The proof of (9.71) is sketched below.

Internal hom conditions. For Coxeter coalgebras C, D, M, a morphism of Coxeter coalgebras $\mathrm{M} \times \mathrm{C} \to \mathrm{D}$ is equivalent to a family of linear maps

$$h_Z : \mathrm{M}[Z] \otimes \mathrm{C}[Z] \to \mathrm{D}[Z],$$

one for each face-type Z, such that

$$h_{Z'}(\sigma(x) \otimes \sigma(y)) = \sigma h_Z(x \otimes y) \tag{9.73a}$$

for $\sigma : Z \to Z'$ and $x \in \mathrm{M}[Z]$ and $y \in \mathrm{C}[Z]$, and

$$h_T(\Delta_Z^T(x) \otimes \Delta_Z^T(y)) = \Delta_Z^T h_Z(x \otimes y) \tag{9.73b}$$

for $T \geq Z$ and $x \in \mathrm{M}[Z]$ and $y \in \mathrm{C}[Z]$. We refer to (9.73) as the *internal hom conditions* for Coxeter coalgebras.

Exercise 9.42. Check: The above data is the same as having a morphism of Coxeter coalgebras $\mathrm{M} \to \mathcal{C}(\mathrm{C}, \mathrm{D})$. On the Z-component, it sends $x \in \mathrm{M}[Z]$ to the family $(f_{T/Z}) \in \mathcal{C}(\mathrm{C}, \mathrm{D})[Z]$, with $f_{T/Z}$ the image of x under the composite

$$\mathrm{M}[Z] \xrightarrow{\Delta_Z^T} \mathrm{M}[T] \to \mathrm{Hom}_{\Bbbk}(\mathrm{C}[T], \mathrm{D}[T]), \qquad x \mapsto \Delta_Z^T(x) \mapsto f_{T/Z}.$$

The second map is induced by h_T as in Exercise 9.36. This establishes (9.71).

Exercise 9.43. Use the adjunction (9.50) to deduce: When C and D are cocommutative, $\mathcal{C}(\mathrm{C}, \mathrm{D})^{coab}$ is the internal hom for cocommutative Coxeter coalgebras. Explicitly, $\mathcal{C}(\mathrm{C}, \mathrm{D})^{coab}[Z]$ consists of those $(f_{T/Z}) \in \mathcal{C}(\mathrm{C}, \mathrm{D})[Z]$ satisfying $f_{T'/Z} = \sigma f_{T/Z} \sigma^{-1}$ for $\sigma : T \to T'$ local to Z.

Remark 9.44. Recall from Proposition 8.108 that the category of Coxeter coalgebras is a functor category. The above description of its internal hom can be deduced from a general result on functor categories, see Remark 3.29. The same remark applies to the category of cocommutative Coxeter coalgebras.

Exercise 9.45 (Primitive part). For Coxeter coalgebras C and D, define a map of Coxeter spaces

$$\mathcal{C}(\mathrm{C}, \mathrm{D}) \to \mathrm{hom}^{\times}(\mathrm{C}, \mathrm{D}), \qquad (f_{T/Z}) \mapsto f_{Z/Z} \tag{9.74}$$

on the Z-component. In other words, we map a family to its base term. Check: When restricted to the primitive part of $\mathcal{C}(\mathrm{C}, \mathrm{D})$, this map is injective, and in fact

$$\overline{\mathcal{P}}(\mathcal{C}(\mathrm{C}, \mathrm{D})) = \mathrm{hom}^{\times}(\mathrm{C}, \overline{\mathcal{P}}(\mathrm{D})). \tag{9.75}$$

Exercise 9.46. Formulate the Coxeter coalgebra $\mathcal{C}(\mathrm{C}, \mathrm{D})$ in terms of top-nested faces. (The (A, C)-component is a suitable family of linear maps indexed by faces F between A and C.) Compare with Exercise 9.39.

9.6.2. Coxeter bialgebra structure. We build on the preceding construction. For C a cocommutative Coxeter coalgebra and K a Coxeter bialgebra, $\mathcal{C}(\mathrm{C},\mathrm{K})$ carries a Coxeter bialgebra structure with product components

$$\mu_Z^{T'} : \mathcal{C}(\mathrm{C},\mathrm{K})[T'] \to \mathcal{C}(\mathrm{C},\mathrm{K})[Z], \qquad f \mapsto \tilde{f}, \tag{9.76}$$

where for $T \geq Z$, the map $\tilde{f}_{T/Z}$ is defined by summing the composite maps

$$\mathrm{C}[T] \xrightarrow{\Delta_T^U} \mathrm{C}[U] \xrightarrow{\sigma} \mathrm{C}[U'] \xrightarrow{f_{U'/T'}} \mathrm{K}[U'] \xrightarrow{\sigma^{-1}} \mathrm{K}[U] \xrightarrow{\mu_T^U} \mathrm{K}[T],$$

as U and U' vary, and $\sigma : U \to U'$ varies over all morphisms local to Z which biproject T and T'. Observe: $\tilde{f}_{Z/Z}$ is the composite map

$$\mathrm{C}[Z] \xrightarrow{\Delta_Z^{T'}} \mathrm{C}[T'] \xrightarrow{f_{T'/T'}} \mathrm{K}[T'] \xrightarrow{\mu_Z^{T'}} \mathrm{K}[Z].$$

In this case, there is only one choice for σ, namely, $\mathrm{id} : T' \to T'$. One needs to show that $\tilde{f}$ indeed belongs to $\mathcal{C}(\mathrm{C},\mathrm{K})[Z]$. The check $\tilde{f}_{V/Z}\Delta_T^V = \Delta_T^V \tilde{f}_{T/Z}$ is indicated in the diagram below.

$$\begin{array}{ccccccccccc}
\mathrm{C}[T] & \xrightarrow{\Delta} & \mathrm{C}[U] & \xrightarrow{\sigma} & \mathrm{C}[U'] & \xrightarrow{f} & \mathrm{K}[U'] & \xrightarrow{\sigma^{-1}} & \mathrm{K}[U] & \xrightarrow{\mu} & \mathrm{K}[T] \\
 & & {\scriptstyle\Delta}\downarrow & & & & & & \downarrow{\scriptstyle\Delta} & & \\
{\scriptstyle\Delta}\downarrow & & \mathrm{C}[W] & & \downarrow{\scriptstyle\Delta} & & \downarrow{\scriptstyle\Delta} & & \mathrm{K}[W] & & \downarrow{\scriptstyle\Delta} \\
 & & {\scriptstyle\tau}\downarrow & \searrow^{\sigma} & & & & \nearrow^{\sigma^{-1}} & \downarrow{\scriptstyle\tau} & & \\
\mathrm{C}[V] & \xrightarrow[\Delta]{} & \mathrm{C}[Y] & \xrightarrow[\pi]{} & \mathrm{C}[W'] & \xrightarrow[f]{} & \mathrm{K}[W'] & \xrightarrow[\pi^{-1}]{} & \mathrm{K}[Y] & \xrightarrow[\mu]{} & \mathrm{K}[V]
\end{array}$$

The varying face-types and morphisms are indicated in red. We use σ for the variable in $\tilde{f}_{T/Z}$, and π for the variable in $\tilde{f}_{V/Z}$. The morphism τ specifies how σ and π relate to each other. An illustration with $Z = \emptyset$ is shown below.

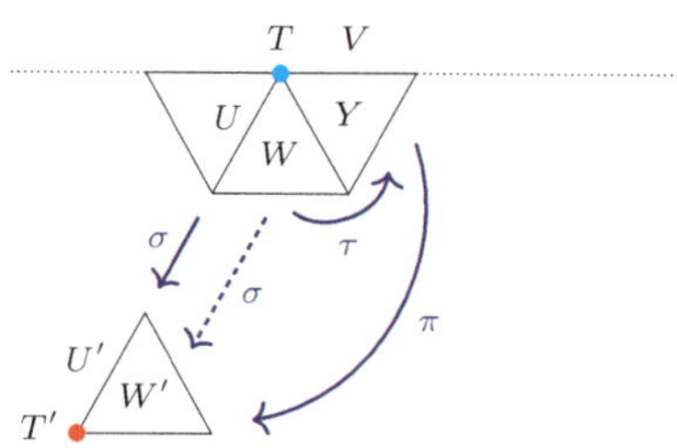

Note the similarity with Figure 1.19; the count for σ relates to $a^{TUT'}$, for τ to a_T^{VYU}, and for π to $a^{VYT'}$.

In the diagram, the left pentagon commutes by the cocommutativity axiom of C, while the right pentagon (after summing) commutes by the bialgebra axiom of K. The middle square commutes since $f \in \mathcal{C}(\mathrm{C},\mathrm{K})[T']$. For the remaining two squares, we use the naturality axiom for the coproduct of both C and K. For the triangles, we use (8.28).

Exercise 9.47. Check $\tilde{f}_{V/Z}\Delta_T^V = \Delta_T^V \tilde{f}_{T/Z}$ explicitly in the special case of the rank-three braid arrangement (Figure 1.5) with $Z = \emptyset$, $T' = $ red, $T = $ cyan, $V = $ cyan-blue. Note: $\tilde{f}$ depends on T'.

We point out that $\tilde{f}$ does not extend f. This is in contrast to the situation for Coxeter bimonoids (Section 3.6.2). We now briefly indicate the remaining checks. For the naturality axiom in (8.41), we use Lemma 8.22. For the unitality axiom, we use Exercise 8.72. For the associativity and bialgebra axioms, the idea of the proof of the gate identity (1.57) (for $q = 1$) enters.

If K is commutative, then so is $\mathcal{C}(\mathrm{C}, \mathrm{K})$.

Exercise 9.48. Let C be a cocommutative Coxeter coalgebra and K a Coxeter bialgebra. Check: The map (9.74) with $\mathrm{D} = \mathrm{K}$ is a morphism of Coxeter algebras, with convolution product on the latter. (Use the formula for $\tilde{f}_{Z/Z'}$.)

Exercise 9.49 (Deformation). Let q be a scalar. Use the gate property for the groupoid of biface-types to check: For C a cocommutative Coxeter coalgebra and K a Coxeter q-bialgebra, the Coxeter coalgebra $\mathcal{C}(\mathrm{C}, \mathrm{K})$ carries a Coxeter q-bialgebra structure with product component μ_Z^T defined as in (9.76) with each composite map multiplied by $q^{l(\sigma)}$. We denote it by $\mathcal{C}(\mathrm{C}, \mathrm{K})_q$. Setting $q = 1$ recovers $\mathcal{C}(\mathrm{C}, \mathrm{K})$. Exercise 9.48 extends to the q-setting.

9.6.3. Enrichment. The category of Coxeter coalgebras is a symmetric monoidal category under the Hadamard product. It is enriched over itself with hom-objects $\mathcal{C}(\mathrm{C}, \mathrm{D})$. That is, there are morphisms of Coxeter coalgebras

$$\mathcal{C}(\mathrm{N}, \mathrm{P}) \times \mathcal{C}(\mathrm{M}, \mathrm{N}) \to \mathcal{C}(\mathrm{M}, \mathrm{P}) \qquad \text{and} \qquad \Bbbk\{x\} \to \mathcal{C}(\mathrm{C}, \mathrm{C}),$$

the former for any Coxeter coalgebras M, N, P, and the latter for any Coxeter coalgebra C. The former is defined by composing families, and the latter by using families consisting of identity maps. The identification (9.72) says that the underlying category of this enriched category is indeed the category of Coxeter coalgebras.

This enriched category possesses powers and copowers. For Coxeter coalgebras C, D, M, the copower of C by M is $\mathrm{M} \times \mathrm{C}$, while the power of D by M is $\mathcal{C}(\mathrm{M}, \mathrm{D})$. That is:

$$\mathcal{C}(\mathrm{M} \times \mathrm{C}, \mathrm{D}) \cong \mathcal{C}(\mathrm{M}, \mathcal{C}(\mathrm{C}, \mathrm{D})) \cong \mathcal{C}(\mathrm{C}, \mathcal{C}(\mathrm{M}, \mathrm{D})). \tag{9.77}$$

Observe: The first isomorphism implies (9.71) by taking Coxeter coalgebra morphisms from $\Bbbk\{x\}$ and using (9.72).

9.7. Universal measuring Coxeter coalgebra

This section proceeds in parallel with Section 9.6. We construct the universal measuring Coxeter coalgebra $\overline{\mathcal{C}}(\mathrm{A}, \mathrm{B})$ from one Coxeter algebra A to another Coxeter algebra B. For a commutative Coxeter algebra A and Coxeter bialgebra H, the Coxeter coalgebra $\overline{\mathcal{C}}(\mathrm{H}, \mathrm{A})$ carries a Coxeter bialgebra structure.

The universal measuring Coxeter coalgebra can be used to enrich the category of Coxeter algebras over the category of Coxeter coalgebras. This enriched category possesses powers and copowers. See Table 9.3.

9.7.1. Universal measuring Coxeter coalgebra. For Coxeter algebras A and B, we construct a Coxeter coalgebra $\overline{\mathcal{C}}(\mathrm{A},\mathrm{B})$ as follows. An element of $\overline{\mathcal{C}}(\mathrm{A},\mathrm{B})[Z]$ is a family of linear maps

$$f_{T/Z} : \mathrm{A}[T] \to \mathrm{B}[T],$$

one for each $Z \leq T$, such that the condition $\mu_T^U f_{U/Z} = f_{T/Z}\mu_T^U$ holds for each $Z \leq T \leq U$. The structure morphisms are defined in a manner similar to the internal hom $\mathcal{C}(\mathrm{C},\mathrm{D})$, see (9.69) and (9.70). We call $\overline{\mathcal{C}}(\mathrm{A},\mathrm{B})$ the *universal measuring Coxeter coalgebra*. It satisfies the following universal property. For a Coxeter coalgebra C and Coxeter algebras A and B, there is a natural bijection

$$\mathsf{Mon}(\overline{\mathsf{C}}\mathsf{Sp})\big(\mathrm{A}, \mathrm{hom}^{\times}(\mathrm{C},\mathrm{B})\big) \xrightarrow{\cong} \mathsf{Comon}(\overline{\mathsf{C}}\mathsf{Sp})\big(\mathrm{C}, \overline{\mathcal{C}}(\mathrm{A},\mathrm{B})\big), \tag{9.78}$$

where $\mathrm{hom}^{\times}(\mathrm{C},\mathrm{B})$ is the convolution Coxeter algebra. In particular, for Coxeter algebras A and B,

$$\mathsf{Mon}(\overline{\mathsf{C}}\mathsf{Sp})(\mathrm{A},\mathrm{B}) \cong \mathsf{Comon}(\overline{\mathsf{C}}\mathsf{Sp})\big(\Bbbk\{x\}, \overline{\mathcal{C}}(\mathrm{A},\mathrm{B})\big). \tag{9.79}$$

The proof of (9.78) is sketched below.

Measuring conditions. For a Coxeter coalgebra C and Coxeter algebras A and B, a morphism of Coxeter algebras $\mathrm{A} \to \mathrm{hom}^{\times}(\mathrm{C},\mathrm{B})$ is equivalent to a family of linear maps

$$h_Z : \mathrm{C}[Z] \otimes \mathrm{A}[Z] \to \mathrm{B}[Z],$$

one for each face-type Z, such that

$$h_{Z'}(x \otimes \sigma(y)) = \sigma h_Z(\sigma^{-1}(x) \otimes y) \tag{9.80a}$$

for $\sigma : Z \to Z'$ and $x \in \mathrm{C}[Z']$ and $y \in \mathrm{A}[Z]$, and

$$h_Z(x \otimes \mu_Z^T(y)) = \mu_Z^T h_T(\Delta_Z^T(x) \otimes y) \tag{9.80b}$$

for $T \geq Z$ and $x \in \mathrm{C}[Z]$ and $y \in \mathrm{A}[T]$. We refer to (9.80) as the *measuring conditions*. Compare and contrast with the internal hom conditions (9.73).

Exercise 9.50. Check: The above data is the same as having a morphism of Coxeter coalgebras $\mathrm{C} \to \overline{\mathcal{C}}(\mathrm{A},\mathrm{B})$. On the Z-component, it sends $x \in \mathrm{C}[Z]$ to the family $(f_{T/Z}) \in \overline{\mathcal{C}}(\mathrm{A},\mathrm{B})[Z]$, with $f_{T/Z}$ the image of x under the composite

$$\mathrm{C}[Z] \xrightarrow{\Delta_Z^T} \mathrm{C}[T] \to \mathrm{Hom}_{\Bbbk}(\mathrm{A}[T],\mathrm{B}[T]), \qquad x \mapsto \Delta_Z^T(x) \mapsto f_{T/Z}.$$

The second map is induced by h_T as in Exercise 9.36. This establishes (9.78).

Exercise 9.51 (Primitive part). For Coxeter algebras A and B, define a map of Coxeter spaces

$$\overline{\mathcal{C}}(\mathrm{A},\mathrm{B}) \to \mathrm{hom}^{\times}(\mathrm{A},\mathrm{B}), \qquad (f_{T/Z}) \mapsto f_{Z/Z} \tag{9.81}$$

on the Z-component. Check: When restricted to the primitive part of $\overline{\mathcal{C}}(\mathrm{A},\mathrm{B})$, this map is injective, and in fact

$$\overline{\mathcal{P}}(\overline{\mathcal{C}}(\mathrm{A},\mathrm{B})) = \mathrm{hom}^{\times}(\overline{\mathcal{Q}}(\mathrm{A}),\mathrm{B}). \tag{9.82}$$

Compare and contrast with (9.75).

9.7.2. Coxeter bialgebra structure. We build on the preceding construction. For H a Coxeter bialgebra and A a commutative Coxeter algebra, $\overline{\mathcal{C}}(\mathrm{H},\mathrm{A})$ carries a Coxeter bialgebra structure with product components

$$\mu_Z^{T'} : \overline{\mathcal{C}}(\mathrm{H},\mathrm{A})[T'] \to \overline{\mathcal{C}}(\mathrm{H},\mathrm{A})[Z], \qquad f \mapsto \tilde{f}, \tag{9.83}$$

where for $T \geq Z$, the map $\tilde{f}_{T/Z}$ is defined by summing the composite maps

$$\mathrm{H}[T] \xrightarrow{\Delta_T^U} \mathrm{H}[U] \xrightarrow{\upsilon} \mathrm{H}[U'] \xrightarrow{f_{U'/T'}} \mathrm{A}[U'] \xrightarrow{\sigma^{-1}} \mathrm{A}[U] \xrightarrow{\mu_T^U} \mathrm{A}[T],$$

as U and U' vary, and $\sigma : U \to U'$ varies over all morphisms local to Z which biproject T and T'. The checks are similar to the one for $\mathcal{C}(\mathrm{C},\mathrm{K})$ in Section 9.6.2 and amount to reversing appropriate arrows. If H is cocommutative, then $\overline{\mathcal{C}}(\mathrm{H},\mathrm{A})$ is commutative.

Exercise 9.52. Let B be a commutative Coxeter algebra and A a Coxeter bialgebra. Check: The map (9.81) is a morphism of Coxeter algebras, with convolution product on the latter. Compare with Exercise 9.48.

Exercise 9.53 (Deformation). Let q be a scalar. Check: For H a Coxeter q-bialgebra and A a commutative Coxeter algebra, the Coxeter coalgebra $\overline{\mathcal{C}}(\mathrm{H},\mathrm{A})$ carries a Coxeter q-bialgebra structure with product component μ_Z^T defined as in (9.83) with each composite map multiplied by $q^{l(\sigma)}$. We denote it by $\overline{\mathcal{C}}(\mathrm{H},\mathrm{A})_q$. Setting $q = 1$ recovers $\overline{\mathcal{C}}(\mathrm{H},\mathrm{A})$. Exercise 9.52 extends to the q-setting. Compare with Exercise 9.49.

9.7.3. A quotient of the free Coxeter algebra. For a Coxeter coalgebra C and Coxeter algebra A, define $\mathrm{C} \triangleright \mathrm{A}$ to be the following Coxeter quotient algebra of $\overline{\mathcal{T}}(\mathrm{C}\times\mathrm{A})$. Its Z-component is the quotient of $\overline{\mathcal{T}}(\mathrm{C}\times\mathrm{A})[Z]$ by the linear span of elements of the form

$$x \otimes \mu_T^U(y) - \Delta_T^U(x) \otimes y, \tag{9.84}$$

for any $x \in \mathrm{C}[T]$ and $y \in \mathrm{A}[U]$ with $Z \leq T \leq U$. It satisfies the following universal property. For a Coxeter coalgebra C and Coxeter algebras A and B, there is a natural bijection

$$\mathsf{Mon}(\overline{\mathsf{C}}\mathsf{Sp})(\mathrm{C} \triangleright \mathrm{A}, \mathrm{B}) \xrightarrow{\cong} \mathsf{Comon}(\overline{\mathsf{C}}\mathsf{Sp})\big(\mathrm{C}, \overline{\mathcal{C}}(\mathrm{A},\mathrm{B})\big). \tag{9.85}$$

9.7.4. Enrichment. The category of Coxeter algebras is enriched over the symmetric monoidal category of Coxeter coalgebras with hom-objects $\overline{\mathcal{C}}(\mathrm{A},\mathrm{B})$. That is, there are morphisms of Coxeter coalgebras

$$\overline{\mathcal{C}}(\mathrm{N},\mathrm{P}) \times \overline{\mathcal{C}}(\mathrm{M},\mathrm{N}) \to \overline{\mathcal{C}}(\mathrm{M},\mathrm{P}) \qquad \text{and} \qquad \Bbbk\{x\} \to \overline{\mathcal{C}}(\mathrm{A},\mathrm{A}),$$

the former for any Coxeter algebras M, N, P, and the latter for any Coxeter algebra A. The former is defined by composing families, and the latter by using families consisting of identity maps. The identification (9.79) says that the underlying category of this enriched category is indeed the category of Coxeter algebras.

This enriched category possesses powers and copowers. For a Coxeter coalgebra C and Coxeter algebras A and B, the copower of A by C is $\mathrm{C} \triangleright \mathrm{A}$, while the power of B by C is the convolution Coxeter algebra $\hom^{\times}(\mathrm{C},\mathrm{B})$. That is:

$$\overline{\mathcal{C}}(\mathrm{C} \triangleright \mathrm{A}, \mathrm{B}) \cong \mathcal{C}(\mathrm{C}, \overline{\mathcal{C}}(\mathrm{A}, \mathrm{B})) \cong \overline{\mathcal{C}}(\mathrm{A}, \hom^{\times}(\mathrm{C}, \mathrm{B})). \tag{9.86}$$

Observe: The second isomorphism implies (9.78) by taking Coxeter coalgebra morphisms from $\Bbbk\{x\}$ and using (9.72) and (9.79). In a similar manner, the first isomorphism implies (9.85).

9.8. Exp-log between (co)derivations and (co)algebra morphisms

Recall the invariant lune-incidence algebra and invariant bilune-incidence algebra from Section 1.10. We define actions of these algebras on the space of all maps from a Coxeter coalgebra C to a Coxeter algebra A. See Table 9.4. We then use these to formulate various exp-log correspondences by employing zeta and Möbius functions. They provide bijections between spaces of (co, bi)derivations and spaces of Coxeter (co, bi)algebra morphisms. These ideas are continued in Sections 9.9 and 9.10.

TABLE 9.4. Exp-log on Coxeter spaces.

algebra action	exp-log and zeta-Möbius
$\mathrm{I}_{\mathrm{lune}}[\mathcal{A}]^W$ on $\overline{\mathsf{C}}\mathsf{Sp}(\mathrm{C},\mathrm{A})$	$\mathbf{exp}(f) = \boldsymbol{\zeta} \circ f$, $\mathbf{log}(f) = \boldsymbol{\mu} \circ f$
$\mathrm{I}_{\mathrm{bilune}}[\mathcal{A}]^W$ on $\overline{\mathsf{C}}\mathsf{Sp}(\mathrm{C},\mathrm{A})$	$\mathbf{exp}_q(f) = \zeta_q \circ f$, $\mathbf{log}_q(f) = \mu_q \circ f$
$\mathrm{I}_{\mathrm{lune}}[\mathcal{A}]^W$ on $\mathscr{S}(\mathrm{A})$	$\mathbf{exp}(v) = \boldsymbol{\zeta} \circ v$, $\mathbf{log}(v) = \boldsymbol{\mu} \circ v$ $\mathbf{exp}_q(v) = \boldsymbol{\zeta}_q \circ v$, $\mathbf{log}_q(v) = \boldsymbol{\mu}_q \circ v$

The parallel discussion for Coxeter species is in Section 3.8. Compare Table 9.4 with Table 3.4. For connections between the two using Fock functors, see Section 17.2.

9.8.1. Action of the invariant lune-incidence algebra.

Let C be a Coxeter coalgebra and A a Coxeter algebra. For $s \in \mathrm{I}_{\mathrm{lune}}[\mathcal{A}]^W$ and $f : \mathrm{C} \to \mathrm{A}$ a map of Coxeter spaces, define $s \circ f : \mathrm{C} \to \mathrm{A}$ by

$$(s \circ f)_Z := \sum_{T:\, T \geq Z} s(Z,T)\, \mu_Z^T f_T \Delta_Z^T. \tag{9.87}$$

This is a map of Coxeter spaces, that is, diagram (8.29) commutes. To see this, fix $\sigma : Z \to Z'$, and for any $T \geq Z$, consider the commutative diagram

$$\begin{array}{ccccccc} \mathrm{C}[Z] & \xrightarrow{\Delta_Z^T} & \mathrm{C}[T] & \xrightarrow{f_T} & \mathrm{A}[T] & \xrightarrow{\mu_Z^T} & \mathrm{A}[Z] \\ {\scriptstyle\sigma}\downarrow & & {\scriptstyle\sigma}\downarrow & & \downarrow{\scriptstyle\sigma} & & \downarrow{\scriptstyle\sigma} \\ \mathrm{C}[Z'] & \xrightarrow[\Delta_{Z'}^{T'}]{} & \mathrm{C}[T'] & \xrightarrow[f_{T'}]{} & \mathrm{A}[T'] & \xrightarrow[\mu_{Z'}^{T'}]{} & \mathrm{A}[Z']. \end{array}$$

The middle square is an instance of (8.29), while the side squares are instances of the naturality axioms in (8.41) and (8.46). Now multiply the above diagram by $s(Z,T) = s(Z',T')$ (which holds by Exercise 1.37) and sum over all $T \geq Z$.

The assignment $(s, f) \mapsto s \circ f$ defines a left action of $\mathrm{I}_{\mathrm{lune}}[\mathcal{A}]^W$ on $\overline{\mathtt{C}}\mathtt{Sp}(\mathrm{C}, \mathrm{A})$. The main check is shown below. For $s, t \in \mathrm{I}_{\mathrm{lune}}[\mathcal{A}]^W$,

$$\begin{aligned}(s \circ (t \circ f))_Z &= \sum_{T:\, T \geq Z} s(Z,T)\, \mu_Z^T (t \circ f)_T \Delta_Z^T \\ &= \sum_{T:\, T \geq Z} s(Z,T)\, \mu_Z^T \Big(\sum_{U:\, U \geq T} t(T,U)\, \mu_T^U f_U \Delta_T^U \Big) \Delta_Z^T \\ &= \sum_{U:\, U \geq Z} \Big(\sum_{T:\, U \geq T \geq Z} s(Z,T) t(T,U) \Big) \mu_Z^U f_U \Delta_Z^U \\ &= \sum_{U:\, U \geq Z} (st)(Z,U)\, \mu_Z^U f_U \Delta_Z^U \\ &= (st \circ f)_Z. \end{aligned}$$

The third step used the (co)associativity axioms in (8.41) and (8.46), and interchanged the order of the sums.

Exercise 9.54. Check: The assignment $(s, f) \mapsto s \circ f$ is linear in both s and f. (Use (8.32).)

Exercise 9.55. Check: The action (9.87) can be rewritten in terms of top-nested faces as

$$(s \circ f)_{A,C} = \sum_{F:\, A \leq F \leq C} s(A,F)\, \mu_{A,C}^{F,C} f_{F,C} \Delta_{A,C}^{F,C}, \tag{9.88}$$

and in terms of top-lunes as

$$(s \circ f)_{\mathrm{L}} = \sum_{\mathrm{M}:\, \mathrm{L} \leq' \mathrm{M}} s(\mathrm{L}/\mathrm{M})\, \mu_{\mathrm{L}}^{\mathrm{M}} f_{\mathrm{M}} \Delta_{\mathrm{L}}^{\mathrm{M}}, \tag{9.89}$$

where recall that L/M is the unique lune such that $\mathrm{L} = (\mathrm{L}/\mathrm{M}) \circ \mathrm{M}$.

9.8.2. Exp-log on Coxeter space maps.

Let $\mathrm{char}(\Bbbk) \nmid |W|$. Fix an invariant noncommutative zeta function $\boldsymbol{\zeta}$ and Möbius function $\boldsymbol{\mu}$ which are inverses of each other in $\mathrm{I}_{\mathrm{lune}}[\mathcal{A}]^W$. For a map of Coxeter spaces $f : \mathrm{C} \to \mathrm{A}$, we say $\boldsymbol{\zeta} \circ f$ is an *exponential* of f and $\boldsymbol{\mu} \circ f$ is a *logarithm* of f. We also use notations $\mathbf{exp}(f)$ and $\mathbf{log}(f)$. Using formula (9.87),

$$\mathbf{exp}(f)_Z = \sum_{T:\, T \geq Z} \boldsymbol{\zeta}(Z,T)\, \mu_Z^T f_T \Delta_Z^T, \tag{9.90a}$$

$$\mathbf{log}(f)_Z = \sum_{T:\, T \geq Z} \boldsymbol{\mu}(Z,T)\, \mu_Z^T f_T \Delta_Z^T. \tag{9.90b}$$

Since $\boldsymbol{\zeta}$ and $\boldsymbol{\mu}$ are inverses, we obtain inverse bijections

$$\overline{\mathtt{C}}\mathtt{Sp}(\mathrm{C}, \mathrm{A}) \underset{\mathbf{log}}{\overset{\mathbf{exp}}{\rightleftarrows}} \overline{\mathtt{C}}\mathtt{Sp}(\mathrm{C}, \mathrm{A}). \tag{9.91}$$

We refer to (9.91) as an *exp-log correspondence*. It depends on $\boldsymbol{\zeta}$ and $\boldsymbol{\mu}$.

9.8.3. Exp-log between (co)derivations and (co)algebra morphisms. Under additional hypotheses, the correspondence (9.91) restricts as follows.

Theorem 9.56. *Let* $\mathrm{char}(\Bbbk) \nmid |W|$. *For a cocommutative Coxeter coalgebra* C *and Coxeter bialgebra* K, *we have inverse bijections*

$$\overline{\mathsf{C}}\mathsf{Sp}\big(\mathrm{C}, \overline{\mathcal{P}}(\mathrm{K})\big) \underset{\mathbf{log}}{\overset{\mathbf{exp}}{\rightleftarrows}} \mathsf{Comon}(\overline{\mathsf{C}}\mathsf{Sp})(\mathrm{C}, \mathrm{K}). \tag{9.92}$$

Dually, for a Coxeter bialgebra H *and commutative Coxeter algebra* A, *we have inverse bijections*

$$\overline{\mathsf{C}}\mathsf{Sp}\big(\overline{\mathcal{Q}}(\mathrm{H}), \mathrm{A}\big) \underset{\mathbf{log}}{\overset{\mathbf{exp}}{\rightleftarrows}} \mathsf{Mon}(\overline{\mathsf{C}}\mathsf{Sp})(\mathrm{H}, \mathrm{A}). \tag{9.93}$$

The first statement says: For $f : \mathrm{C} \to \mathrm{K}$ a coderivation, $\mathbf{exp}(f) : \mathrm{C} \to \mathrm{K}$ is a Coxeter coalgebra morphism, and for $g : \mathrm{C} \to \mathrm{K}$ a Coxeter coalgebra morphism, $\mathbf{log}(g) : \mathrm{C} \to \mathrm{K}$ is a coderivation.

The second statement says: For $f : \mathrm{H} \to \mathrm{A}$ a derivation, $\mathbf{exp}(f) : \mathrm{H} \to \mathrm{A}$ is a Coxeter algebra morphism, and for $g : \mathrm{H} \to \mathrm{A}$ a Coxeter algebra morphism, $\mathbf{log}(g) : \mathrm{H} \to \mathrm{A}$ is a derivation.

PROOF. We explain the first statement. For f a coderivation, we check below that $\mathbf{exp}(f)$ is a Coxeter coalgebra morphism. For any $U \geq Z$,

$$\begin{aligned}
\Delta^U_Z(\boldsymbol{\zeta} \circ f)_Z &= \sum_{T:\, T \geq Z} \zeta(Z,T)\, \Delta^U_Z \mu^T_Z f_T \Delta^T_Z \\
&= \sum_{T:\, T \geq Z}\ \sum_{V:\, V \geq U} \zeta(Z,T)\, \mu^V_U \overline{\lambda}_{TTUV} f_T \Delta^T_Z \\
&= \sum_{T:\, T \geq Z}\ \sum_{V:\, V \geq U} \zeta(Z,T)\, \mu^V_U f_V \overline{\lambda}_{TTUV} \Delta^T_Z \\
&= \sum_{V:\, V \geq U} \Big(\sum_{T:\, T \geq Z} a^{TTUV}_Z \zeta(Z,T) \Big) \mu^V_U f_V \Delta^V_Z \\
&= \sum_{V:\, V \geq U} \Big(\sum_{\substack{T:\, T \geq Z \\ \mathrm{s}(T/Z) = \mathrm{s}(V/Z)}} a^{UVT}_Z \zeta(Z,T) \Big) \mu^V_U f_V \Delta^V_Z \\
&= \Big(\sum_{V:\, V \geq U} \zeta(U,V)\, \mu^V_U f_V \Delta^V_U \Big) \Delta^U_Z \\
&= (\boldsymbol{\zeta} \circ f)_U \Delta^U_Z .
\end{aligned}$$

The second step used the bialgebra axiom (8.64) and the hypothesis that f is a coderivation, that is, it maps into the primitive part. The third step interchanged f and $\overline{\lambda}$ using (8.29). The fourth step used (8.96) for $q = 1$ (which requires cocommutativity). The fifth step used (1.44b) applied to $\boldsymbol{\alpha}_Z$ for $q = 1$. The sixth step used the lune-additivity formula (1.82) and the coassociativity axiom in (8.46).

In the other direction, for g a Coxeter coalgebra morphism, we check below that $\mathbf{log}(g)$ is a coderivation. For any $U > Z$,

$$\begin{aligned}\Delta_Z^U(\boldsymbol{\mu} \circ g)_Z &= \sum_{T\colon T \geq Z} \boldsymbol{\mu}(Z,T)\, \Delta_Z^U \mu_Z^T g_T \Delta_Z^T \\ &= \sum_{T\colon T \geq Z} \boldsymbol{\mu}(Z,T)\, \Delta_Z^U \mu_Z^T \Delta_Z^T g_Z \\ &= \sum_{V\colon V \geq U} \Big(\sum_{T\colon T \geq Z} a_Z^{UVT} \boldsymbol{\mu}(Z,T) \Big) \mu_U^V \Delta_Z^V g_Z \\ &= 0.\end{aligned}$$

The second step used the hypothesis that g is a Coxeter coalgebra morphism. The third step used identity (8.98). The last step used the noncommutative Weisner formula (1.85). □

9.8.4. Opposite exponential and logarithm. For a map $f : \mathrm{C} \to \mathrm{A}$, let

$$\overline{\mathbf{exp}}(f)_Z = \sum_{T\colon T \geq Z} \boldsymbol{\zeta}(Z, Z\overline{T})\, \mu_Z^T f_T \Delta_Z^T, \tag{9.94a}$$

$$\overline{\mathbf{log}}(f)_Z = \sum_{T\colon T \geq Z} \boldsymbol{\mu}(Z, Z\overline{T})\, \mu_Z^T f_T \Delta_Z^T. \tag{9.94b}$$

In other words, $\overline{\mathbf{exp}}$ and $\overline{\mathbf{log}}$ are the exponential and logarithm associated to $\overline{\boldsymbol{\zeta}}$ and $\overline{\boldsymbol{\mu}}$, respectively.

9.8.5. Action of the invariant flat-incidence algebra and commutative exp-log. Let $\mathrm{char}(\Bbbk) \nmid |W|$. When C is cocommutative and A is commutative, the action (9.87) factors through the base-case map (1.78):

$$(s \circ f)_Z = \sum_{\lambda/Z} \frac{\mathrm{bc}(s)(\mathrm{X},\mathrm{Y})}{|\lambda/Z|}\, \mu_Z^T f_T \Delta_Z^T. \tag{9.95}$$

The sum is over all flat-types λ/Z of $\boldsymbol{\alpha}_Z$, and for each λ/Z, we choose a face-type T/Z of support λ/Z and a flat Y/X of type λ/Z. (Here X is a flat such that $\mathrm{t}(\mathrm{X}) = \mathrm{s}(Z)$.) The numbers $|\lambda/Z|$ are as in Definition 1.8.

In this situation, by Fact 1.80, all exp-log correspondences (9.91) reduce to a single correspondence. We call it the *commutative exp-log correspondence*. Further, we denote the exponential and logarithm of f by $\exp(f)$ and $\log(f)$, respectively. With notations as above, they are given by

$$\exp(f)_Z = \sum_{\lambda/Z} \frac{1}{|\lambda/Z|}\, \mu_Z^T f_T \Delta_Z^T, \tag{9.96a}$$

$$\log(f)_Z = \sum_{\lambda/Z} \frac{\mu(\mathrm{X},\mathrm{Y})}{|\lambda/Z|}\, \mu_Z^T f_T \Delta_Z^T. \tag{9.96b}$$

As a consequence of Theorem 9.56, we obtain:

Theorem 9.57. *Let* $\mathrm{char}(\Bbbk) \nmid |W|$. *For a cocommutative Coxeter bialgebra* H *and commutative Coxeter bialgebra* K, *we have inverse bijections*

$$\overline{\mathsf{C}}\mathsf{Sp}(\overline{\mathcal{Q}}(\mathrm{H}), \overline{\mathcal{P}}(\mathrm{K})) \underset{\log}{\overset{\exp}{\rightleftarrows}} \mathsf{Bimon}(\overline{\mathsf{C}}\mathsf{Sp})(\mathrm{H}, \mathrm{K}). \tag{9.97}$$

In other words: If $f : \mathrm{H} \to \mathrm{K}$ is a biderivation, then $\exp(f) : \mathrm{H} \to \mathrm{K}$ is a Coxeter bialgebra morphism. If $g : \mathrm{H} \to \mathrm{K}$ is a Coxeter bialgebra morphism, then $\log(g) : \mathrm{H} \to \mathrm{K}$ is a biderivation.

9.8.6. Action of the invariant bilune-incidence algebra. We now consider a more general setup. Let C be a Coxeter coalgebra and A a Coxeter algebra. For $s \in \mathrm{I}_{\mathrm{bilune}}[\mathcal{A}]^W$ and $f : \mathrm{C} \to \mathrm{A}$ a map of Coxeter spaces, define $s \circ f : \mathrm{C} \to \mathrm{A}$ by

$$\begin{aligned}(s \circ f)_Z &:= \sum_\sigma s(Z, T, \sigma, T')\, \mu_Z^T \sigma f_{T'} \Delta_Z^{T'} \\ &= \sum_\sigma s(Z, T, \sigma, T')\, \mu_Z^T f_T \sigma \Delta_Z^{T'}.\end{aligned} \tag{9.98}$$

The sums are over all $\sigma : T' \to T$ local to Z. Note: $\sigma f_{T'} = f_T \sigma$ by (8.29).

Exercise 9.58. Check: The map $s \circ f$ is a map of Coxeter spaces. (Use the diagram

$$\begin{array}{ccccccccc} \mathrm{C}[Z] & \xrightarrow{\Delta_Z^U} & \mathrm{C}[U] & \xrightarrow{f_U} & \mathrm{A}[U] & \xrightarrow{\sigma} & \mathrm{A}[T] & \xrightarrow{\mu_Z^T} & \mathrm{A}[Z] \\ \downarrow{\scriptstyle\rho} & & \downarrow{\scriptstyle\rho} & & \downarrow{\scriptstyle\rho} & & \downarrow{\scriptstyle\rho} & & \downarrow{\scriptstyle\rho} \\ \mathrm{C}[Z'] & \xrightarrow[\Delta_{Z'}^{U'}]{} & \mathrm{C}[U'] & \xrightarrow[f_{U'}]{} & \mathrm{A}[U'] & \xrightarrow[\rho\sigma\rho^{-1}]{} & \mathrm{A}[T'] & \xrightarrow[\mu_{Z'}^{T'}]{} & \mathrm{A}[Z'] \end{array}$$

starting with $\rho : Z \to Z'$. For the square involving σ, see Lemma 8.22. The side squares are the naturality axioms in (8.41) and (8.46).)

Moreover, check that the assignment $(s, f) \mapsto s \circ f$ defines a left action of $\mathrm{I}_{\mathrm{bilune}}[\mathcal{A}]^W$ on $\overline{\mathsf{C}}\mathsf{Sp}(\mathrm{C}, \mathrm{A})$.

Exercise 9.59 (Comparison of actions). Recall algebra morphisms p, q, i from (1.105). They can be used to relate the actions (9.87) and (9.98). Check:

- The action (9.98) restricts via the map i to the action (9.87).
- When C is cocommutative, the action (9.98) factors through the map p to induce the action (9.87).
- When A is commutative, the action (9.98) factors through the map q to induce the action (9.87).
- When C is cocommutative and A is commutative, the action factors further through the base-case map (1.78) to induce the action (9.95) (assuming $\mathrm{char}(\Bbbk) \nmid |W|$).

9.8.7. q-exp-log. Let q be a scalar which is not a root of unity. Recall the two-sided q-zeta function ζ_q and two-sided q-Möbius function μ_q. These are elements of the invariant bilune-incidence algebra. For a map of Coxeter spaces $f : \mathrm{C} \to \mathrm{A}$, we say $\mathbf{exp}_q(f) := \zeta_q \circ f$ and $\mathbf{log}_q(f) := \mu_q \circ f$ are the *q-exponential* and *q-logarithm* of f, respectively. Using formula (9.98),

$$\mathbf{exp}_q(f)_Z = \sum_\sigma \zeta_q(Z, T, \sigma, T')\, \mu_Z^T \sigma f_{T'} \Delta_Z^{T'}, \tag{9.99a}$$

$$\mathbf{log}_q(f)_Z = \sum_{\sigma} \mu_q(Z,T,\sigma,T')\, \mu_Z^T \sigma f_{T'} \Delta_Z^{T'}. \tag{9.99b}$$

Since ζ_q and μ_q are inverses, we obtain inverse bijections

$$\overline{\mathsf{C}}\mathsf{Sp}(\mathrm{C},\mathrm{A}) \underset{\mathbf{log}_q}{\overset{\mathbf{exp}_q}{\rightleftarrows}} \overline{\mathsf{C}}\mathsf{Sp}(\mathrm{C},\mathrm{A}). \tag{9.100}$$

This is the *q-exp-log correspondence*.

Observation 9.60. If either C is cocommutative or A is commutative, then

$$\mathbf{exp}_q(f)_Z = \sum_{T:\, T \geq Z} \zeta_q(Z,T)\, \mu_Z^T f_T \Delta_Z^T, \tag{9.101a}$$

$$\mathbf{log}_q(f)_Z = \sum_{T:\, T \geq Z} \mu_q(Z,T)\, \mu_Z^T f_T \Delta_Z^T. \tag{9.101b}$$

This follows from Exercise 9.59 and Fact 1.100.

Theorem 9.61. *Suppose q is not a root of unity. Then: for a Coxeter coalgebra* C *and Coxeter q-bialgebra* K*, we have inverse bijections*

$$\overline{\mathsf{C}}\mathsf{Sp}(\mathrm{C},\overline{\mathcal{P}}(\mathrm{K})) \underset{\mathbf{log}_q}{\overset{\mathbf{exp}_q}{\rightleftarrows}} \mathsf{Comon}(\overline{\mathsf{C}}\mathsf{Sp})(\mathrm{C},\mathrm{K}), \tag{9.102}$$

for a Coxeter q-bialgebra H *and Coxeter algebra* A*, we have inverse bijections*

$$\overline{\mathsf{C}}\mathsf{Sp}(\overline{\mathcal{Q}}(\mathrm{H}),\mathrm{A}) \underset{\mathbf{log}_q}{\overset{\mathbf{exp}_q}{\rightleftarrows}} \mathsf{Mon}(\overline{\mathsf{C}}\mathsf{Sp})(\mathrm{H},\mathrm{A}), \tag{9.103}$$

for Coxeter q-bialgebras H *and* K*, we have inverse bijections*

$$\overline{\mathsf{C}}\mathsf{Sp}(\overline{\mathcal{Q}}(\mathrm{H}),\overline{\mathcal{P}}(\mathrm{K})) \underset{\mathbf{log}_q}{\overset{\mathbf{exp}_q}{\rightleftarrows}} q\text{-}\mathsf{Bimon}(\overline{\mathsf{C}}\mathsf{Sp})(\mathrm{H},\mathrm{K}). \tag{9.104}$$

PROOF. The third bijection follows from the first two. To establish either bijection, use the two-sided q-lune-additivity formula (1.109) and two-sided q-Weisner formula (1.111), respectively. We leave the details as an exercise. □

Example 9.62. Suppose q is not a root of unity. Let C be a Coxeter coalgebra, A a Coxeter algebra, $f : \mathrm{C} \to \mathrm{A}$ a map of Coxeter spaces. Then the Coxeter algebra morphism $\hat{f} : \overline{\mathcal{T}}_q(\mathrm{C}) \to \mathrm{A}$ in Theorem 9.11 is the q-exponential (9.99a) of the derivation in the first diagram below.

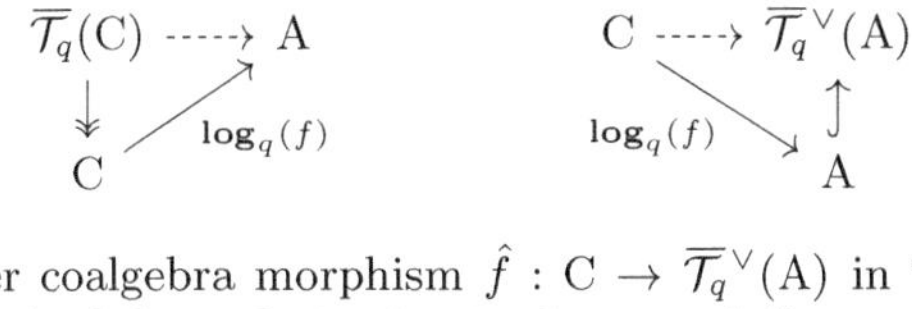

Dually, the Coxeter coalgebra morphism $\hat{f} : \mathrm{C} \to \overline{\mathcal{T}}_q^{\vee}(\mathrm{A})$ in Theorem 9.16 is the q-exponential (9.99a) of the coderivation in the second diagram above.

We explain the first part. By (9.103), $\mathbf{log}_q(\hat{f})$ factors through $\overline{\mathcal{Q}}(\overline{\mathcal{T}}_q(\mathrm{C})) = \mathrm{C}$, so it suffices to compute it on C. But by Exercise 9.14, C is a Coxeter subcoalgebra of $\overline{\mathcal{T}}_q(\mathrm{C})$, so we just get $\mathbf{log}_q(f)$ as required. □

9.8.8. 0-exp-log. The case $q = 0$ is of interest. Using formulas (1.118) for ζ_0 and μ_0, formulas (9.99a) and (9.99b) specialize to

$$\mathbf{exp}_0(f)_Z = \sum_{T:\, T \geq Z} \mu_Z^T f_T \Delta_Z^T, \tag{9.105a}$$

$$\mathbf{log}_0(f)_Z = \sum_{T:\, T \geq Z} (-1)^{\mathrm{rk}(T/Z)}\, \mu_Z^T f_T \Delta_Z^T. \tag{9.105b}$$

For $q = 0$, we refer to (9.100) as the 0-*exp-log correspondence*.

9.8.9. Naturality. The action (9.87) of the invariant lune-incidence algebra on $\overline{\mathsf{C}}\mathsf{Sp}(\mathrm{C}, \mathrm{A})$ is natural in C and A. That is, for $f : \mathrm{C}' \to \mathrm{C}$ a Coxeter coalgebra morphism and $g : \mathrm{A} \to \mathrm{A}'$ a Coxeter algebra morphism,

$$\overline{\mathsf{C}}\mathsf{Sp}(\mathrm{C}, \mathrm{A}) \to \overline{\mathsf{C}}\mathsf{Sp}(\mathrm{C}', \mathrm{A}'), \qquad h \mapsto ghf$$

is a map of left modules over the invariant lune-incidence algebra. (The map ghf is the composite f followed by h followed by g.) As a consequence, the exp-log correspondence (9.91) is natural in C and A. That is, for $f : \mathrm{C}' \to \mathrm{C}$ a Coxeter coalgebra morphism, $h : \mathrm{C} \to \mathrm{A}$ a Coxeter space map, $g : \mathrm{A} \to \mathrm{A}'$ a Coxeter algebra morphism,

$$\mathbf{exp}(ghf) = g\,\mathbf{exp}(h)f \quad \text{and} \quad \mathbf{log}(ghf) = g\,\mathbf{log}(h)f.$$

Similar remarks apply to the action (9.98) of the invariant bilune-incidence algebra, and to the q-exp-log correspondence (9.100).

Exercise 9.63. Let $\mathrm{char}(\Bbbk) \nmid |W|$. Let C be a Coxeter coalgebra, H a Coxeter bialgebra, A a Coxeter algebra. Suppose either C is cocommutative or A is commutative. Show:

(1) If $f : \mathrm{C} \to \mathrm{H}$ is a Coxeter coalgebra morphism, and $g : \mathrm{H} \to \mathrm{A}$ is a Coxeter algebra morphism, then $\mathbf{log}(gf) = \mathbf{log}(g)\,\mathbf{log}(f)$.
(2) If $f : \mathrm{C} \to \mathrm{H}$ is a coderivation, and $g : \mathrm{H} \to \mathrm{A}$ is a derivation, then $\mathbf{exp}(gf) = \mathbf{exp}(g)\,\mathbf{exp}(f)$.

Exercise 9.64. Suppose q is not a root of unity. Let C be a Coxeter coalgebra, H a Coxeter q-bialgebra, A a Coxeter algebra. Show:

(1) If $f : \mathrm{C} \to \mathrm{H}$ is a Coxeter coalgebra morphism, and $g : \mathrm{H} \to \mathrm{A}$ is a Coxeter algebra morphism, then $\mathbf{log}_q(gf) = \mathbf{log}_q(g)\,\mathbf{log}_q(f)$.
(2) If $f : \mathrm{C} \to \mathrm{H}$ is a coderivation, and $g : \mathrm{H} \to \mathrm{A}$ is a derivation, then $\mathbf{exp}_q(gf) = \mathbf{exp}_q(g)\,\mathbf{exp}_q(f)$.

9.9. Logarithm of the identity map

For any Coxeter bialgebra, there is a natural map from its primitive part to its indecomposable part. It is closely related to the logarithm of the identity map of a Coxeter bialgebra. These considerations extend to the q-setting.

The parallel discussion for Coxeter species is in Section 3.9.

9.9.1. Primitive part to indecomposable part. For a scalar q, for a Coxeter q-bialgebra H, we have the canonical map

$$\mathrm{pq}_{\mathrm{H}} : \overline{\mathcal{P}}(\mathrm{H}) \hookrightarrow \mathrm{H} \twoheadrightarrow \overline{\mathcal{Q}}(\mathrm{H}). \tag{9.106}$$

This map is natural in H, that is, for any morphism $f : \mathrm{H} \to \mathrm{K}$ of Coxeter q-bialgebras, the diagram

$$\begin{array}{ccc} \overline{\mathcal{P}}(\mathrm{H}) & \xrightarrow{f} & \overline{\mathcal{P}}(\mathrm{K}) \\ {\scriptstyle \mathrm{pq}_{\mathrm{H}}}\downarrow & & \downarrow{\scriptstyle \mathrm{pq}_{\mathrm{K}}} \\ \overline{\mathcal{Q}}(\mathrm{H}) & \xrightarrow[f]{} & \overline{\mathcal{Q}}(\mathrm{K}) \end{array} \tag{9.107}$$

commutes. This defines a natural transformation pq from $\overline{\mathcal{P}}$ to $\overline{\mathcal{Q}}$ (viewing both as functors from Coxeter q-bialgebras to Coxeter spaces).

Proposition 9.65. *Let* $\mathrm{char}(\Bbbk) \nmid |W|$. *Let* H *be a Coxeter bialgebra. Then:*

(1) H *is cocommutative iff the map* $\mathrm{pq}_{\mathrm{H}} : \overline{\mathcal{P}}(\mathrm{H}) \to \overline{\mathcal{Q}}(\mathrm{H})$ *is surjective.*
(2) H *is commutative iff the map* $\mathrm{pq}_{\mathrm{H}} : \overline{\mathcal{P}}(\mathrm{H}) \to \overline{\mathcal{Q}}(\mathrm{H})$ *is injective.*
(3) H *is bicommutative iff the map* $\mathrm{pq}_{\mathrm{H}} : \overline{\mathcal{P}}(\mathrm{H}) \to \overline{\mathcal{Q}}(\mathrm{H})$ *is bijective.*

PROOF. The proof is similar to the one for species [**11**, Proposition 5.56]; for Coxeter species, this result is given in Proposition 3.44. □

Proposition 9.66. *For a Coxeter q-bialgebra* H *for q not a root of unity, the map* $\mathrm{pq}_{\mathrm{H}} : \overline{\mathcal{P}}(\mathrm{H}) \to \overline{\mathcal{Q}}(\mathrm{H})$ *is bijective.*

PROOF. The proof is similar to the one for species [**11**, Proposition 5.58]; for Coxeter species, this result is given in Proposition 3.45. □

An improvement of the above result is given in Proposition 9.72 below.

9.9.2. Logarithm of the identity map. Let $\mathrm{char}(\Bbbk) \nmid |W|$. Fix an invariant noncommutative Möbius function $\boldsymbol{\mu}$. For a Coxeter q-bialgebra H, we have the operator $\mathbf{log}(\mathrm{id}) : \mathrm{H} \to \mathrm{H}$. This is a *logarithm of the identity map* on H. By formula (9.90b),

$$\mathbf{log}(\mathrm{id})_Z = \sum_{T:\, T \geq Z} \boldsymbol{\mu}(Z,T)\, \mu_Z^T \Delta_Z^T. \tag{9.108}$$

To emphasize dependence on H, we may also write $\mathbf{log}(\mathrm{id}_{\mathrm{H}})$. By (9.94b),

$$\overline{\mathbf{log}}(\mathrm{id})_Z = \sum_{T:\, T \geq Z} \boldsymbol{\mu}(Z,Z\overline{T})\, \mu_Z^T \Delta_Z^T. \tag{9.109}$$

Proposition 9.67. *Let* $\mathrm{char}(\Bbbk) \nmid |W|$. *Let* H *be a Coxeter bialgebra. For* H *cocommutative,* $\mathbf{log}(\mathrm{id})$ *is an idempotent operator on* H *with image* $\overline{\mathcal{P}}(\mathrm{H})$. *For* H *commutative,* $\mathbf{log}(\mathrm{id})$ *is an idempotent operator on* H *with coimage* $\overline{\mathcal{Q}}(\mathrm{H})$.

PROOF. We explain the first statement. Suppose H is cocommutative. That $\mathbf{log}(\mathrm{id})$ maps to $\overline{\mathcal{P}}(\mathrm{H})$ follows from (9.92). That it is identity on $\overline{\mathcal{P}}(\mathrm{H})$ follows from formula (9.108). □

The two situations when H is either cocommutative or commutative are, respectively, illustrated below.

$$
\begin{array}{ccc}
\mathrm{H} & \xrightarrow{\mathbf{log}(\mathrm{id})} & \mathrm{H} \\
\uparrow & \searrow & \uparrow \\
\overline{\mathcal{P}}(\mathrm{H}) & \xrightarrow[\mathbf{log}(\mathrm{id})=\mathrm{id}]{} & \overline{\mathcal{P}}(\mathrm{H})
\end{array}
\qquad\qquad
\begin{array}{ccc}
\mathrm{H} & \xrightarrow{\mathbf{log}(\mathrm{id})} & \mathrm{H} \\
\downarrow & \nearrow & \downarrow \\
\overline{\mathcal{Q}}(\mathrm{H}) & \xrightarrow[\mathbf{log}(\mathrm{id})=\mathrm{id}]{} & \overline{\mathcal{Q}}(\mathrm{H})
\end{array}
\tag{9.110}
$$

Exercise 9.68. For H a Coxeter bialgebra, for $U > Z$,

$$\sum_{T:\, T \geq Z} \mu(Z,T)\, \Delta^U_Z \mu^T_Z \Delta^T_Z = 0 \quad \text{and} \quad \sum_{T:\, T \geq Z} \mu(Z,T)\, \mu^T_Z \Delta^T_Z \mu^U_Z = 0,$$

the former if H is cocommutative, and the latter if H is commutative.

Deduce these formulas from Proposition 9.67. Alternatively, check them directly using identities (8.97), (8.98) and the noncommutative Weisner formula (1.85).

Exercise 9.69. Check: For H cocommutative, the first diagram below commutes. For H commutative, the second diagram below commutes.

$$
\begin{array}{ccc}
 & \overline{\mathcal{P}}(\mathrm{H}) & \\
{\scriptstyle \mathbf{log}(\mathrm{id}_{\mathrm{H}})} \nearrow & & \searrow {\scriptstyle \mathrm{pq}_{\mathrm{H}}} \\
\mathrm{H} & \longrightarrow & \overline{\mathcal{Q}}(\mathrm{H})
\end{array}
\qquad\qquad
\begin{array}{ccc}
 & \overline{\mathcal{Q}}(\mathrm{H}) & \\
{\scriptstyle \mathrm{pq}_{\mathrm{H}}} \nearrow & & \searrow {\scriptstyle \mathbf{log}(\mathrm{id}_{\mathrm{H}})} \\
\overline{\mathcal{P}}(\mathrm{H}) & \hookrightarrow & \mathrm{H}
\end{array}
\tag{9.111}
$$

The horizontal maps are the canonical maps.

Proposition 9.70. *Let* char($\Bbbk$) $\nmid |W|$. *For* H *a bicommutative Coxeter bialgebra,* log(id) *is an idempotent operator on* H *with image* $\overline{\mathcal{P}}$(H) *and coimage* $\overline{\mathcal{Q}}$(H) *yielding the commutative diagram of Coxeter spaces*

$$
\begin{array}{ccc}
\mathrm{H} & \xrightarrow{\log(\mathrm{id})} & \mathrm{H} \\
\downarrow & & \uparrow \\
\overline{\mathcal{Q}}(\mathrm{H}) & \xrightarrow[\cong]{} & \overline{\mathcal{P}}(\mathrm{H}).
\end{array}
\tag{9.112}
$$

In particular, $\overline{\mathcal{P}}$(H) *and* $\overline{\mathcal{Q}}$(H) *are isomorphic as Coxeter spaces. Moreover, the inverse of* $\overline{\mathcal{Q}}(\mathrm{H}) \to \overline{\mathcal{P}}(\mathrm{H})$ *is the map* pq_{H} *in* (9.106).

Proof. This can be deduced from Proposition 9.67. Alternatively, one may also employ (9.97). □

The above result adds to the content of Proposition 9.65, item (3).

Proposition 9.71. *Let* char($\Bbbk$) $\nmid |W|$. *For bicommutative Coxeter bialgebras* H *and* K, *the bijection* (9.97) *preserves isomorphisms.*

Proof. This can be deduced from Proposition 9.70 and Exercise 9.63. We leave the details as an exercise. They are formally the same as those involved in the proof of Proposition 3.49. □

Let q be a scalar which is not a root of unity. Let H be a Coxeter p-bialgebra for a scalar p possibly different from q. By formula (9.99b), the

q-logarithm of the identity map on H is

$$\mathbf{log}_q(\mathrm{id})_Z = \sum_{\sigma} \mu_q(Z,T,\sigma,T')\, \mu_Z^T \sigma \Delta_Z^{T'}, \tag{9.113}$$

with the sum over all $\sigma : T' \to T$ local to Z. In particular, the *0-logarithm of the identity map* on H is

$$\mathbf{log}_0(\mathrm{id})_Z = \sum_{T:\, T \geq Z} (-1)^{\mathrm{rk}(T/Z)}\, \mu_Z^T \Delta_Z^T. \tag{9.114}$$

This is a special case of (9.105b).

Proposition 9.72. *For* H *a Coxeter q-bialgebra for q not a root of unity,* $\mathbf{log}_q(\mathrm{id})$ *is an idempotent operator on* H *with image* $\overline{\mathcal{P}}(\mathrm{H})$ *and coimage* $\overline{\mathcal{Q}}(\mathrm{H})$ *yielding the commutative diagram of Coxeter spaces*

$$\begin{array}{ccc} \mathrm{H} & \xrightarrow{\mathbf{log}_q(\mathrm{id})} & \mathrm{H} \\ \downarrow & & \uparrow \\ \overline{\mathcal{Q}}(\mathrm{H}) & \xrightarrow[\cong]{} & \overline{\mathcal{P}}(\mathrm{H}). \end{array} \tag{9.115}$$

In particular, $\overline{\mathcal{P}}(\mathrm{H})$ *and* $\overline{\mathcal{Q}}(\mathrm{H})$ *are isomorphic as Coxeter spaces. Moreover, the inverse of* $\overline{\mathcal{Q}}(\mathrm{H}) \to \overline{\mathcal{P}}(\mathrm{H})$ *is the map* pq_{H} *in* (9.106).

PROOF. One may employ (9.104). □

The above result adds to the content of Proposition 9.66 (and also gives another proof for it).

Proposition 9.73. *For q not a root of unity, for Coxeter q-bialgebras* H *and* K*, the bijection* (9.104) *preserves isomorphisms.*

PROOF. This can be deduced from Proposition 9.72 and Exercise 9.64. We leave the details as an exercise. They are formally the same as those involved in the proof of Proposition 3.51. □

9.10. Exp-log between primitive and group-like series

We define the space of series of a Coxeter space. We also introduce primitive and group-like series of a Coxeter coalgebra. The invariant lune-incidence algebra acts on the space of series of a Coxeter algebra. See Table 9.4. This yields exp-log correspondences, one for each invariant noncommutative zeta or Möbius function. Each such correspondence restricts to a bijection between primitive series and group-like series of any Coxeter bialgebra. This story has a q-analogue for q not a root of unity. Moreover, exp-log correspondences in this section relate to those in Section 9.8 via the convolution Coxeter algebra, internal hom for Coxeter coalgebras, universal measuring Coxeter coalgebra. We also consider linear and exponential series of a Coxeter algebra; for a Coxeter bialgebra, a linear series is primitive while an exponential series is group-like.

In this section, we make small use of the Coxeter bialgebra of polynomials which was briefly mentioned in Section 9.5.1 and is fully developed later in Section 10.2.

The parallel discussion for Coxeter species is in Section 3.10. For connections between the two using Fock functors, see Section 17.1.

9.10.1. Series of a Coxeter space. Let V be a Coxeter space. A *series* of V is a family of elements $v_{F,C} \in \mathrm{V}[F, C]$, one for each top-nested face (F, C), such that

$$\beta_{G,D,F,C}(v_{F,C}) = v_{G,D} \quad \text{and} \quad w(v_{F,C}) = v_{wF,wC},$$

the former whenever $\mathrm{s}(F, C) = \mathrm{s}(G, D)$, and the latter for all $w \in W$. Equivalently, in terms of face-types, a series v of V is a family of elements $v_T \in \mathrm{V}[T]$, one for each face-type T, such that

$$\sigma(v_T) = v_U$$

for any morphism $\sigma : T \to U$. Let $\mathscr{S}(\mathrm{V})$ denote the space of series of V. Equivalently,

$$\mathscr{S}(\mathrm{V}) = \lim \mathrm{V} = \overline{\mathcal{S}}^{\vee}(\mathrm{V})[\emptyset].$$

This construction is functorial in V and defines a functor

$$\mathscr{S} : \overline{\mathsf{C}}\mathsf{Sp} \to \mathsf{Vec}$$

from the category of Coxeter spaces to the category of vector spaces.

Example 9.74 (Internal hom). For Coxeter spaces V and W, using (9.61):

$$\mathscr{S}(\hom^{\times}(\mathrm{V}, \mathrm{W})) = \overline{\mathsf{C}}\mathsf{Sp}(\mathrm{V}, \mathrm{W}). \tag{9.116}$$

As special cases, for a Coxeter space V,

$$\mathscr{S}(\mathrm{V}) = \overline{\mathsf{C}}\mathsf{Sp}(\Bbbk[x], \mathrm{V}) \quad \text{and} \quad \mathscr{S}(\mathrm{V}^*) = \overline{\mathsf{C}}\mathsf{Sp}(\mathrm{V}, \Bbbk[x]). \tag{9.117}$$

Interestingly, (9.116) can be formally deduced from the first statement in (9.117) by using (9.63). □

9.10.2. Primitive and group-like series of a Coxeter coalgebra. Let C be a Coxeter coalgebra. We define below two subspaces of $\mathscr{S}(\mathrm{C})$, namely, $\mathscr{P}(\mathrm{C})$ and $\mathscr{G}(\mathrm{C})$.

- A series v of C is *primitive* if $\Delta^T_Z(v_Z) = 0$ for all $T > Z$. Let $\mathscr{P}(\mathrm{C})$ denote the space of primitive series of C. Note: $\mathscr{P}(\mathrm{C}) = \mathscr{S}(\overline{\mathcal{P}}(\mathrm{C}))$.
- A series v of C is *group-like* if $\Delta^T_Z(v_Z) = v_T$ for all $T \geq Z$. Let $\mathscr{G}(\mathrm{C})$ denote the space of group-like series of C.

These constructions are functorial in C.

Example 9.75 (Enrichment). Recall from Section 9.6.1 the Coxeter coalgebra $\mathcal{C}(\mathrm{C}, \mathrm{D})$ associated to Coxeter coalgebras C and D. We have

$$\begin{aligned} \mathscr{P}(\mathcal{C}(\mathrm{C}, \mathrm{D})) &= \overline{\mathsf{C}}\mathsf{Sp}(\mathrm{C}, \overline{\mathcal{P}}(\mathrm{D})), \\ \mathscr{G}(\mathcal{C}(\mathrm{C}, \mathrm{D})) &= \mathsf{Comon}(\overline{\mathsf{C}}\mathsf{Sp})(\mathrm{C}, \mathrm{D}). \end{aligned} \tag{9.118}$$

As a special case, for a Coxeter coalgebra C,

$$\mathscr{P}(\mathrm{C}) = \overline{\mathsf{C}}\mathsf{Sp}(\Bbbk\{x\}, \overline{\mathcal{P}}(\mathrm{C})) \quad \text{and} \quad \mathscr{G}(\mathrm{C}) = \mathsf{Comon}(\overline{\mathsf{C}}\mathsf{Sp})(\Bbbk\{x\}, \mathrm{C}). \tag{9.119}$$

Interestingly, these two statements can be used to formally deduce (9.118) by using (9.63), (9.72), (9.75).

Recall from Section 9.7.1 the Coxeter coalgebra $\overline{\mathcal{C}}(\mathrm{A}, \mathrm{B})$ associated to Coxeter algebras A and B. We have

$$\begin{aligned} \mathscr{P}(\overline{\mathcal{C}}(\mathrm{A}, \mathrm{B})) &= \overline{\mathsf{C}}\mathsf{Sp}(\overline{\mathcal{Q}}(\mathrm{A}), \mathrm{B}), \\ \mathscr{G}(\overline{\mathcal{C}}(\mathrm{A}, \mathrm{B})) &= \mathsf{Mon}(\overline{\mathsf{C}}\mathsf{Sp})(\mathrm{A}, \mathrm{B}). \end{aligned} \tag{9.120}$$

As a special case, for a Coxeter algebra A,

$$\mathscr{P}(\mathrm{A}^*) = \overline{\mathsf{C}}\mathsf{Sp}(\overline{\mathcal{Q}}(\mathrm{A}), \Bbbk[x]) \quad \text{and} \quad \mathscr{G}(\mathrm{A}^*) = \mathsf{Mon}(\overline{\mathsf{C}}\mathsf{Sp})(\mathrm{A}, \Bbbk[x]). \tag{9.121}$$

This is how primitive series link to (co)derivations and group-like series to Coxeter (co)algebra morphisms. □

9.10.3. Action of the invariant lune-incidence algebra. Let A be a Coxeter algebra. For $s \in \mathrm{I}_{\mathrm{lune}}[\mathcal{A}]^W$ and $v \in \mathscr{S}(\mathrm{A})$, define $s \circ v \in \mathscr{S}(\mathrm{A})$ by

$$(s \circ v)_Z := \sum_{T:\, Z \leq T} s(Z,T)\, \mu_Z^T(v_T). \tag{9.122}$$

Let us check that $s \circ v$ is indeed a series. For $\sigma : Z \to Z'$,

$$\begin{aligned} \sigma((s \circ v)_Z) &= \sum_{T:\, T \geq Z} s(Z,T)\, \sigma \mu_Z^T(v_T) \\ &= \sum_{T:\, T \geq Z} s(Z,T)\, \mu_{Z'}^{T'} \sigma(v_T) \\ &= \sum_{T':\, T' \geq Z'} s(Z',T')\, \mu_{Z'}^{T'}(v_{T'}) \\ &= (s \circ v)_{Z'}. \end{aligned}$$

The second step used the naturality axiom in (8.41). The third step used that v is a series, and that $s(Z,T) = s(Z',T')$ which holds by Exercise 1.37.

The assignment $(s, v) \mapsto s \circ v$ defines a left action of $\mathrm{I}_{\mathrm{lune}}[\mathcal{A}]^W$ on $\mathscr{S}(\mathrm{A})$. The main check is shown below. For $s, t \in \mathrm{I}_{\mathrm{lune}}[\mathcal{A}]^W$,

$$\begin{aligned} (s \circ (t \circ v))_Z &= \sum_{T:\, Z \leq T} s(Z,T)\, \mu_Z^T((t \circ v)_T) \\ &= \sum_{T:\, Z \leq T} \sum_{U:\, T \leq U} s(Z,T) t(T,U)\, \mu_Z^T \mu_T^U(v_U) \\ &= \sum_{T:\, Z \leq T} \sum_{U:\, T \leq U} s(Z,T) t(T,U)\, \mu_Z^U(v_U) \\ &= \sum_{U:\, Z \leq U} \Big(\sum_{T:\, Z \leq T \leq U} s(Z,T) t(T,U) \Big) \mu_Z^U(v_U) \\ &= \sum_{U:\, Z \leq U} (st)(Z,U)\, \mu_Z^U(v_U) \\ &= ((st) \circ v)_Z \end{aligned}$$

Note use of the associativity axiom in (8.41).

9.10.4. Action of the invariant flat-incidence algebra. Let $\mathrm{char}(\Bbbk) \nmid |W|$. When A is commutative, the action (9.122) factors through the base-case map (1.78):

$$(s \circ v)_Z = \sum_{\lambda/Z} \frac{\mathrm{bc}(s)(\mathrm{X},\mathrm{Y})}{|\lambda/Z|} \, \mu_Z^T(v_T). \tag{9.123}$$

The sum is over all flat-types λ/Z of $\boldsymbol{\alpha}_Z$, and for each λ/Z, we choose a face-type T/Z of support λ/Z and a flat Y/X of type λ/Z. This is similar to formula (9.95).

9.10.5. Exp-log on series of a Coxeter algebra. Let $\mathrm{char}(\Bbbk) \nmid |W|$. Fix an invariant noncommutative zeta function $\boldsymbol{\zeta}$ and Möbius function $\boldsymbol{\mu}$ which are inverses of each other in $\mathrm{I}_{\mathrm{lune}}[\mathcal{A}]^W$. For $v \in \mathscr{S}(\mathrm{A})$, define

$$\mathbf{exp}(v)_Z := \sum_{T:\, T \geq Z} \boldsymbol{\zeta}(Z,T)\, \mu_Z^T(v_T), \tag{9.124a}$$

$$\mathbf{log}(v)_Z := \sum_{T:\, T \geq Z} \boldsymbol{\mu}(Z,T)\, \mu_Z^T(v_T). \tag{9.124b}$$

Since $\boldsymbol{\zeta}$ and $\boldsymbol{\mu}$ are inverses, we obtain inverse bijections

$$\mathscr{S}(\mathrm{A}) \underset{\mathbf{log}}{\overset{\mathbf{exp}}{\rightleftarrows}} \mathscr{S}(\mathrm{A}). \tag{9.125}$$

We refer to (9.125) as an *exp-log correspondence.*

9.10.6. Exp-log between primitive and group-like series. When A carries a Coxeter bialgebra structure, (9.125) restricts as follows.

Theorem 9.76. *Let* $\mathrm{char}(\Bbbk) \nmid |W|$. *For a Coxeter bialgebra* H, *we have inverse bijections*

$$\mathscr{P}(\mathrm{H}) \underset{\mathbf{log}}{\overset{\mathbf{exp}}{\rightleftarrows}} \mathscr{G}(\mathrm{H}). \tag{9.126}$$

These bijections depend on $\boldsymbol{\zeta}$ and $\boldsymbol{\mu}$.

PROOF. We need to show that **exp** and **log** map as stated. For v a primitive series of H, we check below that $\mathbf{exp}(v)$ is a group-like series. For $Z \leq T'$,

$$\begin{aligned}
\Delta_Z^{T'}(\mathbf{exp}(v)_Z) &= \sum_{T:\, T \geq Z} \boldsymbol{\zeta}(Z,T)\, \Delta_Z^{T'} \mu_Z^T(v_T) \\
&= \sum_{T:\, T \geq Z} \sum_{\substack{U:\, U \geq T \\ U':\, U' \geq T'}} \boldsymbol{\zeta}(Z,T)\, \mu_{T'}^{U'} \overline{\lambda}_{TUT'U'} \Delta_T^U(v_T) \\
&= \sum_{T:\, T \geq Z} \sum_{U':\, U' \geq T'} \boldsymbol{\zeta}(Z,T)\, \mu_{T'}^{U'} \overline{\lambda}_{TTT'U'}(v_T) \\
&= \sum_{T:\, T \geq Z} \sum_{U':\, U' \geq T'} a_Z^{TTT'U'} \boldsymbol{\zeta}(Z,T)\, \mu_{T'}^{U'}(v_{U'}) \\
&= \sum_{U':\, U' \geq T'} \Big(\sum_{T:\, T \geq Z} a_Z^{TTT'U'} \boldsymbol{\zeta}(Z,T) \Big) \mu_{T'}^{U'}(v_{U'})
\end{aligned}$$

$$
\begin{aligned}
&= \sum_{U' : U' \geq T'} \Big(\sum_{\substack{T : T \geq Z \\ \mathrm{s}(T/Z) = \mathrm{s}(U'/Z)}} a_Z^{T'U'T} \boldsymbol{\zeta}(Z,T) \Big) \mu_{T'}^{U'}(v_{U'}) \\
&= \sum_{U' : U' \geq T'} \boldsymbol{\zeta}(T',U') \, \mu_{T'}^{U'}(v_{U'}) \\
&= \mathbf{exp}(v)_{T'}.
\end{aligned}
$$

The first and last steps used definition (9.124a). The second step used the bialgebra axiom (8.64). Since v is a primitive series, $\Delta_T^U(v_T)$ is zero unless $U = T$. This was used in the third step. By Observation 8.99, item (1) for $q = 1$, $\overline{\lambda}_{TTT'U'}$ is the sum of $a_Z^{TTT'U'}$ morphisms, and since v is a series, each one takes v_T to $v_{U'}$. This was used in the fourth step. The sixth step used (1.42) and (1.44b) both for $q = 1$ applied to $\boldsymbol{\alpha}_Z$. The seventh step used the lune-additivity formula (1.82). The calculation leading up to the sixth step can be separated out; this is done later in (9.152).

Conversely, for v a group-like series of H, we check below that $\mathbf{log}(v)$ is a primitive series. For $Z < T'$,

$$
\begin{aligned}
\Delta_Z^{T'}(\mathbf{log}(v)_Z) &= \sum_{T : T \geq Z} \boldsymbol{\mu}(Z,T) \, \Delta_Z^{T'} \mu_Z^T(v_T) \\
&= \sum_{T : T \geq Z} \sum_{\substack{U : U \geq T \\ U' : U' \geq T'}} \boldsymbol{\mu}(Z,T) \, \mu_{T'}^{U'} \overline{\lambda}_{TUT'U'} \Delta_T^U(v_T) \\
&= \sum_{T : T \geq Z} \sum_{\substack{U : U \geq T \\ U' : U' \geq T'}} \boldsymbol{\mu}(Z,T) \, \mu_{T'}^{U'} \overline{\lambda}_{TUT'U'}(v_U) \\
&= \sum_{T : T \geq Z} \sum_{\substack{U : U \geq T \\ U' : U' \geq T'}} a_Z^{TUT'U'} \boldsymbol{\mu}(Z,T) \, \mu_{T'}^{U'}(v_{U'}) \\
&= \sum_{T : T \geq Z} \sum_{U' : U' \geq T'} a_Z^{T'U'T} \boldsymbol{\mu}(Z,T) \, \mu_{T'}^{U'}(v_{U'}) \\
&= \sum_{U' : U' \geq T'} \Big(\sum_{T : T \geq Z} a_Z^{T'U'T} \boldsymbol{\mu}(Z,T) \Big) \mu_{T'}^{U'}(v_{U'}) \\
&= 0.
\end{aligned}
$$

The first step used definition (9.124b). The second step used the bialgebra axiom (8.64). The third step used that v is group-like, and the next step that v is a series. The fifth step used the second identity in (1.44a) for $q = 1$ applied to $\boldsymbol{\alpha}_Z$. The last step used the noncommutative Weisner formula (1.85). □

Exercise 9.77 (Comparison of actions). Let us now compare the two actions (9.87) and (9.122) of the invariant lune-incidence algebra. In fact, either one can be seen as a special case of the other. Check:

(1) For C a Coxeter coalgebra and A a Coxeter algebra, the action (9.122) on the space of series of the convolution Coxeter algebra $\mathrm{hom}^{\times}(\mathrm{C}, \mathrm{A})$ coincides

with the action (9.87) via the identification (9.116) with $\mathrm{V} := \mathrm{C}$ and $\mathrm{W} := \mathrm{A}$. (Use product formula (9.65).)

(2) For A a Coxeter algebra, the action (9.87) for $\mathrm{C} := \Bbbk\{x\}$ coincides with the action (9.122) via the identification (9.117).

Deduce: The exp-log correspondences (9.91) and (9.125) imply each other.

Exercise 9.78 (Comparison of exp-logs). The exp-log correspondences (9.92) and (9.126) imply each other as follows.

(1) Apply (9.126) to the Coxeter bialgebra $\mathcal{C}(\mathrm{C}, \mathrm{K})$ from Section 9.6.2 to obtain (9.92) via the identification (9.118).
(2) Apply (9.92) with $\mathrm{C} := \Bbbk\{x\}$ to obtain (9.126) via the identification (9.119).

Similarly, apply (9.126) to the Coxeter bialgebra $\overline{\mathcal{C}}(\mathrm{H}, \mathrm{A})$ from Section 9.7.2 to obtain (9.93) via the identification (9.120).

Exercise 9.79. Check: The bijection (9.92) is the restriction of (9.125) applied to the convolution Coxeter algebra $\hom^{\times}(\mathrm{C}, \mathrm{K})$ via the diagram

$$\begin{array}{ccccc} \mathscr{P}(\mathcal{C}(\mathrm{C},\mathrm{K})) & \hookrightarrow & \mathscr{S}(\mathcal{C}(\mathrm{C},\mathrm{K})) & \hookleftarrow & \mathscr{G}(\mathcal{C}(\mathrm{C},\mathrm{K})) \\ \| & & \downarrow & & \| \\ \overline{\mathsf{C}}\mathsf{Sp}(\mathrm{C}, \overline{\mathcal{P}}(\mathrm{K})) & \hookrightarrow & \mathscr{S}(\hom^{\times}(\mathrm{C},\mathrm{K})) & \hookleftarrow & \mathsf{Comon}(\overline{\mathsf{C}}\mathsf{Sp})(\mathrm{C},\mathrm{K}). \end{array}$$

The vertical map in the middle is induced from (9.74) which is a morphism of Coxeter algebras by Exercise 9.48.

Similarly, the bijection (9.93) is the restriction of (9.125) applied to the convolution Coxeter algebra $\hom^{\times}(\mathrm{H}, \mathrm{A})$ via the diagram

$$\begin{array}{ccccc} \mathscr{P}(\overline{\mathcal{C}}(\mathrm{H},\mathrm{A})) & \hookrightarrow & \mathscr{S}(\overline{\mathcal{C}}(\mathrm{H},\mathrm{A})) & \hookleftarrow & \mathscr{G}(\overline{\mathcal{C}}(\mathrm{H},\mathrm{A})) \\ \| & & \downarrow & & \| \\ \overline{\mathsf{C}}\mathsf{Sp}(\overline{\mathcal{Q}}(\mathrm{H}), \mathrm{A}) & \hookrightarrow & \mathscr{S}(\hom^{\times}(\mathrm{H},\mathrm{A})) & \hookleftarrow & \mathsf{Mon}(\overline{\mathsf{C}}\mathsf{Sp})(\mathrm{H},\mathrm{A}). \end{array}$$

The vertical map in the middle is induced from (9.81) which is a morphism of Coxeter algebras by Exercise 9.52.

9.10.7. q-exp-log.

Let q be a scalar which is not a root of unity. Recall the noncommutative q-zeta function $\boldsymbol{\zeta}_q$ and noncommutative q-Möbius function $\boldsymbol{\mu}_q$. For $v \in \mathscr{S}(\mathrm{A})$, define

$$\mathbf{exp}_q(v)_Z := \sum_{T:\, T \geq Z} \boldsymbol{\zeta}_q(Z, T)\, \mu_Z^T(v_T), \tag{9.127a}$$

$$\mathbf{log}_q(v)_Z := \sum_{T:\, T \geq Z} \boldsymbol{\mu}_q(Z, T)\, \mu_Z^T(v_T). \tag{9.127b}$$

Since $\boldsymbol{\zeta}_q$ and $\boldsymbol{\mu}_q$ are inverses, we obtain inverse bijections

$$\mathscr{S}(\mathrm{A}) \underset{\mathbf{log}_q}{\overset{\mathbf{exp}_q}{\rightleftarrows}} \mathscr{S}(\mathrm{A}). \tag{9.128}$$

We refer to (9.128) as the *q-exp-log correspondence*.

Theorem 9.80. *For a Coxeter q-bialgebra* H *for q not a root of unity, we have inverse bijections*

$$\mathscr{P}(\mathrm{H}) \underset{\log_q}{\overset{\exp_q}{\rightleftarrows}} \mathscr{G}(\mathrm{H}). \tag{9.129}$$

PROOF. The calculations are similar to those in Theorem 9.76 replacing ζ by ζ_q, μ by μ_q, $\overline{\lambda}$ by $\overline{\lambda}_q$, $a_Z^{TUT'U'}$ by $a_Z^{TUT'U'}(q)$. We now employ the q-lune-additivity formula (1.93) and noncommutative q-Weisner formula (1.96). □

Exercise 9.81 (Comparison of q-exp-logs). Let us now consider the analogue of Exercise 9.78 for q not a root of unity.

(1) Apply (9.129) to the Coxeter q-bialgebra $\mathcal{C}(\mathrm{C},\mathrm{K})_q$ from Exercise 9.49 to obtain (9.102) (for C cocommutative) via the identification (9.118).
(2) Apply (9.102) with $\mathrm{C} := \Bbbk\{x\}$ to obtain (9.129) via the identification (9.119).

Similarly, apply (9.129) to the Coxeter q-bialgebra $\overline{\mathcal{C}}(\mathrm{H},\mathrm{A})_q$ from Exercise 9.53 to obtain (9.103) (for A commutative) via the identification (9.120).

Exercise 9.79 also extends to the q-setting.

9.10.8. Linear series of a Coxeter space. Let V be a Coxeter space. A series x of V is *linear* if $x_T = 0$ whenever $T \neq S$. Observe: A linear series x of V is the same as an element $x_S \in \mathrm{V}[S]^W$, the W-invariants of $\mathrm{V}[S]$. Let $\mathscr{X}(\mathrm{V})$ denote the space of linear series of V. We mention that

$$\mathscr{X}(\mathrm{V}) = \overline{\mathsf{C}}\mathsf{Sp}(\mathrm{X},\mathrm{V}), \tag{9.130}$$

where X is the Coxeter space defined later in Section 10.1.

Observe: For a Coxeter coalgebra C, any linear series of C is primitive, that is, $\mathscr{X}(\mathrm{C}) \subseteq \mathscr{P}(\mathrm{C})$.

Example 9.82. We build on Example 9.74. For Coxeter spaces V and W,

$$\mathscr{X}(\mathrm{hom}^{\times}(\mathrm{V},\mathrm{W})) = W\text{-}\mathsf{Mod}(\mathrm{V}[S],\mathrm{W}[S]).$$

In words, a linear series of $\mathrm{hom}^{\times}(\mathrm{V},\mathrm{W})$ is the same as a map $f : \mathrm{V} \to \mathrm{W}$ of Coxeter spaces such that $f_T = 0$ if $T \neq S$ which is the same as a map $f_S : \mathrm{V}[S] \to \mathrm{W}[S]$ of W-modules. See also Exercise 10.3. □

9.10.9. Exponential series of a Coxeter algebra. Let A be a Coxeter algebra. A series e of A is *exponential* if for all $T \geq Z$,

$$\mu_Z^T(e_T) = d_{T/Z}\, e_Z, \tag{9.131}$$

where $d_{T/Z}$ is the number of faces of type T/Z in $\boldsymbol{\alpha}_Z$ (Section 1.8.1). Let $\mathscr{E}(\mathrm{A})$ denote the space of exponential series of A. Observe:

$$\mathscr{E}(\mathrm{A}) = \mathsf{Mon}(\overline{\mathsf{C}}\mathsf{Sp})(\Bbbk\{x\},\mathrm{A}). \tag{9.132}$$

The point is that the same coefficient $d_{T/Z}$ appears in the product formula for $\Bbbk\{x\}$ in (10.12).

Exercise 9.83. Let $\mathrm{char}(\Bbbk) \nmid |W|$. Show: For A a Coxeter algebra, a series e of A is exponential iff $\mu_Z^S(e_S) = d_{S/Z}\, e_Z$ for all Z. (Use the multiplicative property (1.37) for $q = 1$ applied to $\boldsymbol{\alpha}_Z$.)

More generally, for a scalar q, a series e of A is *q-exponential* if

$$\mu_Z^T(e_T) = d_{T/Z}(q)\, e_Z \tag{9.133}$$

for all $T \geq Z$. Let $\mathscr{E}_q(\mathrm{A})$ denote the space of q-exponential series of A. Now

$$\mathscr{E}_q(\mathrm{A}) = \mathsf{Mon}(\overline{\mathsf{C}}\mathsf{Sp})(\Bbbk_q\{x\}, \mathrm{A}); \tag{9.134}$$

see the product formula in (10.13).

We also have a q-analogue of Exercise 9.83 for q not a root of unity with $d_{S/Z}$ replaced by $d_{S/Z}(q)$.

Exercise 9.84. Check: For q a scalar, $\mathscr{E}_q$ defines a functor from the category of Coxeter algebras to the category of vector spaces.

9.10.10. Exp-log between linear and exponential series.

Lemma 9.85. *Let* $\mathrm{char}(\Bbbk) \nmid |W|$. *For a Coxeter algebra* A, *any exp-log correspondence* (9.125) *restricts to inverse bijections*

$$\mathscr{X}(\mathrm{A}) \underset{\mathbf{log}}{\overset{\mathbf{exp}}{\rightleftarrows}} \mathscr{E}(\mathrm{A}). \tag{9.135}$$

Moreover, these bijections are independent of $\boldsymbol{\zeta}$ *and* $\boldsymbol{\mu}$.

Explicitly, for x a linear series and e an exponential series,

$$\mathbf{exp}(x)_Z = \frac{1}{d_{S/Z}}\, \mu_Z^S(x_S) \quad \text{and} \quad \mathbf{log}(e)_Z = \begin{cases} e_S & \text{if } Z = S, \\ 0 & \text{otherwise.} \end{cases} \tag{9.136}$$

Note: $e_S = x_S$ when x and e correspond to each other under (9.135).

PROOF. Using formula (1.84),

$$\mathbf{exp}(x)_Z = \sum_{T:\, T \geq Z} \boldsymbol{\zeta}(Z,T)\, \mu_Z^T(x_T) = \boldsymbol{\zeta}(Z,S)\, \mu_Z^S(x_S) = \frac{1}{d_{S/Z}}\, \mu_Z^S(x_S).$$

Using formula (1.89),

$$\mathbf{log}(e)_Z = \sum_{T:\, T \geq Z} \boldsymbol{\mu}(Z,T)\, \mu_Z^T(e_T) = \Big(\sum_{T:\, T \geq Z} \boldsymbol{\mu}(Z,T)\, d_{T/Z}\Big)\, e_Z$$
$$= \begin{cases} e_S & \text{if } Z = S, \\ 0 & \text{otherwise.} \end{cases}$$

This completes the checks. □

Corollary 9.86. *Let* $\mathrm{char}(\Bbbk) \nmid |W|$. *For a Coxeter algebra* A, *elements* $z \in \mathrm{A}[S]^W$ *parametrize exponential series* e *of* A *via* $e_T = \frac{1}{d_{S/T}}\, \mu_T^S(z)$.

Exercise 9.87. Let $\mathrm{char}(\Bbbk) \nmid |W|$. Check: For a Coxeter algebra A,

$$\mathscr{X}(\mathrm{A}) = \mathscr{X}(\mathrm{A}_{ab}) \quad \text{and} \quad \mathscr{E}(\mathrm{A}) = \mathscr{E}(\mathrm{A}_{ab})$$

induced from the abelianization map $\pi_{\mathrm{A}} : \mathrm{A} \twoheadrightarrow \mathrm{A}_{ab}$. Thus, the correspondence (9.135) only depends on the abelianization of A.

As a special case: For a Coxeter space V,

$$\mathscr{X}(\overline{\mathcal{T}}(\mathrm{V})) = \mathscr{X}(\overline{\mathcal{S}}(\mathrm{V})) = \mathscr{X}(\mathrm{V}) \quad \text{and} \quad \mathscr{E}(\overline{\mathcal{T}}(\mathrm{V})) = \mathscr{E}(\overline{\mathcal{S}}(\mathrm{V}))$$

induced from the abelianization map $\overline{\mathcal{T}}(\mathrm{V}) \twoheadrightarrow \overline{\mathcal{S}}(\mathrm{V})$.

For q not a root of unity, we have a q-analogue of Lemma 9.85 which gives inverse bijections

$$\mathscr{X}(\mathrm{A}) \underset{\mathbf{log}_q}{\overset{\mathbf{exp}_q}{\rightleftarrows}} \mathscr{E}_q(\mathrm{A}) \tag{9.137}$$

by restricting the q-exp-log correspondence (9.128). Relevant facts are formulas (1.95) and (1.99). In formula (9.136), we replace $d_{S/Z}$ by $d_{S/Z}(q)$. Thus, elements $z \in \mathrm{A}[S]^W$ parametrize q-exponential series e of A via $e_T = \frac{1}{d_{S/T}(q)}\, \mu_T^S(z)$.

Example 9.88 (Convolution Coxeter algebra). A linear series of $\hom^\times(\mathrm{C}, \mathrm{A})$ is the same as a map $f_S : \mathrm{C}[S] \to \mathrm{A}[S]$ of W-modules as in Example 9.82. By using (9.135), we deduce: An exponential series of $\hom^\times(\mathrm{C}, \mathrm{A})$ is a family of linear maps $g_T : \mathrm{C}[T] \to \mathrm{A}[T]$, one for each face-type T, such that g_S is a map of W-modules and $g_T = \frac{1}{d_{S/T}} \mu_T^S g_S \Delta_T^S$ for each T. See Exercise 9.77 in this regard.

Similarly, by using (9.137), we deduce: For q not a root of unity, a q-exponential series of $\hom^\times(\mathrm{C}, \mathrm{A})$ is as above with $d_{S/T}$ replaced by $d_{S/T}(q)$. □

A different way to arrive at the bijections (9.135) and (9.137) is suggested in Section 10.2.18.

Lemma 9.89. *Let* $\mathrm{char}(\Bbbk) \nmid |W|$. *For a Coxeter bialgebra* H, *any exponential series of* H *is group-like, that is,* $\mathscr{E}(\mathrm{H}) \subseteq \mathscr{G}(\mathrm{H})$. *Equivalently, a morphism* $\Bbbk\{x\} \to \mathrm{H}$ *of Coxeter algebras is necessarily a morphism of Coxeter coalgebras, and hence a morphism of Coxeter bialgebras.*

Direct proof. We do a backward induction on the partial order on face-types. For the induction step, we want to show $\Delta_Z^T(e_Z) = e_T$. For this, apply the bialgebra axiom (8.64) on $\Delta_Z^T \mu_Z^{T'}(e_{T'})$ for some fixed $T' > Z$. By the induction hypothesis along with Observation 8.99, item (1) for $q = 1$, (1.44a) for $q = 1$ for $\boldsymbol{\alpha}_Z$, the gate identity (1.59) for $q = 1$ for $\boldsymbol{\alpha}_Z$, we deduce that $\Delta_Z^T(d_{T'/Z} e_Z) = d_{T'/Z} e_T$. Now cancel off $d_{T'/Z}$ which is justified since $\mathrm{char}(\Bbbk) \nmid |W|$. This completes the induction step. In this argument, one can also for simplicity take $T' = S$ and use Exercise 8.76 which brings in T/Z-shuffles, and then use the multiplicative property (1.32) for $q = 1$. □

Proof using exp-log. By (9.135), $\mathbf{log}(e)$ is a linear series. Hence, it is primitive, which by (9.126) implies that e is group-like. □

To summarize:

Proposition 9.90. *For* $\mathrm{char}(\Bbbk) \nmid |W|$, *for a Coxeter bialgebra* H, *the diagram*

$$\begin{array}{ccccc} \mathscr{X}(\mathrm{H}) & \hookrightarrow & \mathscr{P}(\mathrm{H}) & \hookrightarrow & \mathscr{S}(\mathrm{H}) \\ \updownarrow & & \updownarrow & & \updownarrow \\ \mathscr{E}(\mathrm{H}) & \hookrightarrow & \mathscr{G}(\mathrm{H}) & \hookrightarrow & \mathscr{S}(\mathrm{H}) \end{array} \tag{9.138}$$

commutes. The vertical identifications are the exp-log correspondences in (9.125), (9.126), (9.135).

Proposition 9.91. *Let* $\operatorname{char}(\Bbbk) \nmid |W|$. *For a Coxeter bialgebra* H, *the following pieces of data are equivalent.*

(1) *An exponential series* e *of* H *(necessarily group-like).*
(2) *A linear series* x *of* H *(necessarily primitive).*
(3) *An element* $z \in \mathrm{H}[S]^W$, *the* W*-invariants of* $\mathrm{H}[S]$.
(4) *A Coxeter bialgebra morphism* $f : \Bbbk\{x\} \to \mathrm{H}$.

These pieces of data are linked by

$$\mathbf{log}(e) = x, \quad \mathbf{exp}(x) = e, \quad e_S = x_S = z, \quad f_T(x^{(T)}) = e_T$$

with formulas (9.136). See also Exercise 10.22 in this regard.

For q not a root of unity, we have q-analogues of Lemma 9.89 and Propositions 9.90 and 9.91 for a Coxeter q-bialgebra H with q-exponential series contained in group-like series, that is, $\mathscr{E}_q(\mathrm{H}) \subseteq \mathscr{G}(\mathrm{H})$.

Example 9.92. For C a cocommutative Coxeter coalgebra and K a Coxeter bialgebra, the Coxeter algebra morphism in Exercise 9.48 induces the diagram

$$\begin{array}{ccccc}
\mathscr{X}(\mathcal{C}(\mathrm{C},\mathrm{K})) & \hookrightarrow & \mathscr{S}(\mathcal{C}(\mathrm{C},\mathrm{K})) & \hookleftarrow & \mathscr{E}(\mathcal{C}(\mathrm{C},\mathrm{K})) \\
\| & & \downarrow & & \| \\
\mathscr{X}(\hom^{\times}(\mathrm{C},\mathrm{K})) & \hookrightarrow & \mathscr{S}(\hom^{\times}(\mathrm{C},\mathrm{K})) & \hookleftarrow & \mathscr{E}(\hom^{\times}(\mathrm{C},\mathrm{K})).
\end{array}$$

Linear and exponential series of $\hom^{\times}(\mathrm{C},\mathrm{K})$ were described in Example 9.88; the same descriptions apply to $\mathcal{C}(\mathrm{C},\mathrm{K})$. The above diagram can be combined with the first diagram in Exercise 9.79. The preceding also applies to $\mathcal{C}(\mathrm{C},\mathrm{K})_q$ for q not a root of unity with $\mathscr{E}_q$ instead of $\mathscr{E}$.

Similarly, for H a Coxeter bialgebra and A a commutative Coxeter algebra, we have the diagram

$$\begin{array}{ccccc}
\mathscr{X}(\overline{\mathcal{C}}(\mathrm{H},\mathrm{A})) & \hookrightarrow & \mathscr{S}(\overline{\mathcal{C}}(\mathrm{H},\mathrm{A})) & \hookleftarrow & \mathscr{E}(\overline{\mathcal{C}}(\mathrm{H},\mathrm{A})) \\
\| & & \downarrow & & \| \\
\mathscr{X}(\hom^{\times}(\mathrm{H},\mathrm{A}) & \hookrightarrow & \mathscr{S}(\hom^{\times}(\mathrm{H},\mathrm{A})) & \hookleftarrow & \mathscr{E}(\hom^{\times}(\mathrm{H},\mathrm{A})
\end{array}$$

induced by the Coxeter algebra morphism in Exercise 9.52. □

9.11. Norm transformation on Coxeter bialgebras

For a scalar q, the q-norm transformation $\overline{\kappa}_q$ is a natural transformation between functors $\overline{\mathcal{T}}_q$ and $\overline{\mathcal{T}}_q^{\vee}$ from the category of Coxeter spaces to the category of Coxeter q-bialgebras. Thus, for every Coxeter space V, we have the q-norm map $(\overline{\kappa}_q)_{\mathrm{V}}$ from $\overline{\mathcal{T}}_q(\mathrm{V})$ to $\overline{\mathcal{T}}_q^{\vee}(\mathrm{V})$ which is a morphism of Coxeter q-bialgebras. It arises from freeness of $\overline{\mathcal{T}}_q$ and cofreeness of $\overline{\mathcal{T}}_q^{\vee}$. It is an isomorphism when q is not a root of unity. For $q = 0$, $\overline{\kappa}_0$ is in fact the identity. It is of interest to understand the image of $\overline{\kappa}_q$ when q is a root of unity. For $q = 1$, the image of $\overline{\kappa}_1$ identifies with either $\overline{\mathcal{S}}$ or $\overline{\mathcal{S}}^{\vee}$ (if $\operatorname{char}(\Bbbk) \nmid |W|$), and in particular, yields an isomorphism between $\overline{\mathcal{S}}$ and $\overline{\mathcal{S}}^{\vee}$. Similarly, for $q = -1$, the image of $\overline{\kappa}_{-1}$ identifies with either $\overline{\mathcal{E}}$ or $\overline{\mathcal{E}}^{\vee}$.

More generally, we consider arbitrary Coxeter q-bialgebra morphisms of the form $\overline{\mathcal{T}}_q(\mathrm{C}) \to \overline{\mathcal{T}}_q^{\vee}(\mathrm{A})$ for C a Coxeter coalgebra and A a Coxeter algebra.

For q not a root of unity, we give a necessary and sufficient condition for any such map to be an isomorphism. Interestingly, it involves the q-logarithm.

The parallel discussion for Coxeter species is in Section 3.11. For connections between the two using Fock functors, see Sections 17.4 and 17.5.

9.11.1. q-norm map. Let q be a scalar. View $\overline{\mathcal{T}}_q$ and $\overline{\mathcal{T}}_q^\vee$ as functors from the category of Coxeter spaces to the category of Coxeter q-bialgebras as in (9.24) and (9.33). Define the *q-norm transformation*

$$\overline{\kappa}_q : \overline{\mathcal{T}}_q \to \overline{\mathcal{T}}_q^\vee \tag{9.139}$$

as follows. For a Coxeter space V, the map

$$(\overline{\kappa}_q)_{\mathrm{V}} : \overline{\mathcal{T}}_q(\mathrm{V}) \to \overline{\mathcal{T}}_q^\vee(\mathrm{V}), \tag{9.140}$$

on the Z-component, on the T-summand and into the T'-summand, is

$$\mathrm{V}[T] \to \mathrm{V}[T'], \qquad x \mapsto \sum_{\substack{\sigma\colon T \to T' \\ \text{local to } Z}} q^{l(\sigma)} \sigma(x).$$

This is zero if $\mathrm{s}(T/Z) \neq \mathrm{s}(T'/Z)$. We call (9.140) the *$q$-norm map*. It is a morphism of Coxeter q-bialgebras. It preserves Cauchy powers of V, and further, it is identity on the first Cauchy power $\mathrm{V}^1 = \mathrm{V}$.

One may visualize the q-norm map for $Z = \emptyset$ as shown below, with red and green shadings indicating faces of types T and T', respectively.

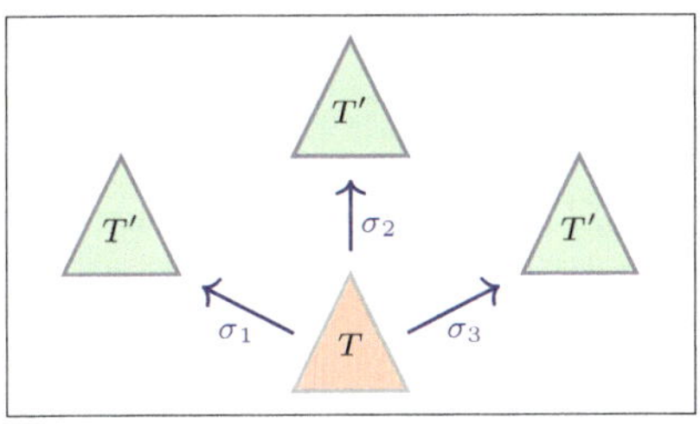

Exercise 9.93 (Universality). Check:

- In Theorem 9.15, for $\mathrm{H} := \overline{\mathcal{T}}_q^\vee(\mathrm{V})$ and $f := \mathrm{id}$, the map $\hat{f}$ equals (9.140).
- In Theorem 9.19, for $\mathrm{H} := \overline{\mathcal{T}}_q(\mathrm{V})$ and $f := \mathrm{id}$, the map $\hat{f}$ equals (9.140).

9.11.2. Norm map and signed norm map. Consider the cases $q = \pm 1$. We refer to $\overline{\kappa}_1$ as the norm transformation and to $\overline{\kappa}_{-1}$ as the signed norm transformation. They factor as follows.

$$\begin{array}{ccc} \overline{\mathcal{T}} & \xrightarrow{\overline{\kappa}_1} & \overline{\mathcal{T}}^\vee \\ \downarrow & & \uparrow \\ \overline{\mathcal{S}} & \longrightarrow & \overline{\mathcal{S}}^\vee \end{array} \qquad\qquad \begin{array}{ccc} \overline{\mathcal{T}}_{-1} & \xrightarrow{\overline{\kappa}_{-1}} & \overline{\mathcal{T}}_{-1}^\vee \\ \downarrow & & \uparrow \\ \overline{\mathcal{E}} & \longrightarrow & \overline{\mathcal{E}}^\vee \end{array} \tag{9.141}$$

The two diagrams are conjugates of each other wrt the u-signature functor (8.99). Let us focus on the first diagram. Functors $\overline{\mathcal{S}}$ and $\overline{\mathcal{S}}^\vee$ are as in (9.40) and (9.43), respectively. The vertical maps are abelianization and coabelianization, see (9.55). The bottom-horizontal map is an isomorphism when $\operatorname{char}(\Bbbk) \nmid |W|$. To see this: Since $\overline{\mathcal{S}}$ is defined as a colimit and $\overline{\mathcal{S}}^\vee$ as a limit and their domains are groupoids, the (co)limits reduce to (co)invariants. Now use the fact that for a finite group G and G-module M, with $\operatorname{char}(\Bbbk)$ not dividing $|G|$, the spaces of G-invariants and G-coinvariants of M are isomorphic via the norm map [**7**, Lemma 2.20]. Thus:

Lemma 9.94. *Let* $\operatorname{char}(\Bbbk) \nmid |W|$. *Functors* $\overline{\mathcal{S}}$ *and* $\overline{\mathcal{S}}^\vee$ *(from Coxeter spaces to bicommutative Coxeter bialgebras) are isomorphic. More precisely, for a Coxeter space* V, *there is a natural isomorphism of Coxeter bialgebras*

$$\overline{\mathcal{S}}(\mathrm{V}) \to \overline{\mathcal{S}}^\vee(\mathrm{V}) \tag{9.142}$$

which on the Z*-component sends a representative* $x \in \mathrm{V}[T]$ *for* $T \geq Z$ *to* $\sum_\sigma \sigma(x)$ *with the sum over all* $\sigma : T \to T'$ *local to* Z.

A similar statement holds for $\overline{\mathcal{E}}$ *and* $\overline{\mathcal{E}}^\vee$.

We continue to refer to (9.142) as the norm map.

Exercise 9.95. Let $\operatorname{char}(\Bbbk) \nmid |W|$. Fix an invariant noncommutative zeta function ζ. For a Coxeter space V, define $\overline{\mathcal{T}}^\vee(\mathrm{V}) \to \overline{\mathcal{T}}(\mathrm{V})$, on the Z-component, on the T-summand, by scalar multiplication by $\zeta(Z,T)$. Check: The composite map

$$\begin{array}{ccc} \overline{\mathcal{T}}^\vee(\mathrm{V}) & \longrightarrow & \overline{\mathcal{T}}(\mathrm{V}) \\ \uparrow & & \downarrow \\ \overline{\mathcal{S}}^\vee(\mathrm{V}) & \underset{\cong}{\dashrightarrow} & \overline{\mathcal{S}}(\mathrm{V}) \end{array} \tag{9.143}$$

is an isomorphism of Coxeter bialgebras. Moreover, it is the inverse of the norm map (9.142).

Exercise 9.96. Let $\operatorname{char}(\Bbbk) \nmid |W|$. For a Coxeter space V,

(i) use $\overline{\mathcal{P}}(\overline{\mathcal{S}}^\vee(\mathrm{V})) = \mathrm{V}$ to deduce $\overline{\mathcal{P}}(\overline{\mathcal{S}}(\mathrm{V})) = \mathrm{V}$,
(ii) use $\overline{\mathcal{Q}}(\overline{\mathcal{S}}(\mathrm{V})) = \mathrm{V}$ to deduce $\overline{\mathcal{Q}}(\overline{\mathcal{S}}^\vee(\mathrm{V})) = \mathrm{V}$.

Exercise 9.97. Let $\operatorname{char}(\Bbbk) \nmid |W|$. Check: The exponentials (9.96a) of the biderivations

$$\begin{array}{ccc} \overline{\mathcal{S}}(\mathrm{V}) & \dashrightarrow & \overline{\mathcal{S}}^\vee(\mathrm{V}) \\ \downarrow & & \uparrow \\ \mathrm{V} & \xrightarrow[\mathrm{id}]{} & \mathrm{V} \end{array} \qquad\qquad \begin{array}{ccc} \overline{\mathcal{S}}^\vee(\mathrm{V}) & \dashrightarrow & \overline{\mathcal{S}}(\mathrm{V}) \\ \downarrow & & \uparrow \\ \mathrm{V} & \xrightarrow[\mathrm{id}]{} & \mathrm{V} \end{array}$$

yield inverse isomorphisms of Coxeter bialgebras between $\overline{\mathcal{S}}(\mathrm{V})$ and $\overline{\mathcal{S}}^\vee(\mathrm{V})$ which are the norm map (9.142) and its inverse (9.143), respectively.

Exercise 9.98 (Universality). Check:

- In Theorem 9.24, for $\mathrm{H} := \overline{\mathcal{S}}^\vee(\mathrm{V})$ and $f := \mathrm{id}$, the map $\hat{f}$ equals (9.142).
- In Theorem 9.27, for $\mathrm{H} := \overline{\mathcal{S}}(\mathrm{V})$ and $f := \mathrm{id}$, the map $\hat{f}$ equals (9.142).

9.11.3. General q-norm map.

The q-norm map can be put in a more general setting as follows. It is the specialization of a more abstract discussion given in [**11**, Proposition C.33].

For a scalar q, for a Coxeter coalgebra C and Coxeter algebra A, there is a natural bijection

$$\overline{\mathsf{C}}\mathsf{Sp}(\mathrm{C},\mathrm{A}) \xrightarrow{\cong} q\text{-}\mathsf{Bimon}(\overline{\mathsf{C}}\mathsf{Sp})(\overline{\mathcal{T}_q}(\mathrm{C}), \overline{\mathcal{T}_q}^\vee(\mathrm{A})), \qquad f \mapsto \hat{f}. \tag{9.144}$$

For $f : \mathrm{C} \to \mathrm{A}$ a map of Coxeter spaces, the morphism $\widehat{f} : \overline{\mathcal{T}}_q(\mathrm{C}) \to \overline{\mathcal{T}}_q^{\vee}(\mathrm{A})$ of Coxeter q-bialgebras can be defined in two equivalent ways as follows.

- First lift f to a morphism $\mathrm{C} \to \overline{\mathcal{T}}_q^{\vee}(\mathrm{A})$ of Coxeter coalgebras, and then extend it to a morphism $\widehat{f}$.
- First extend f to a morphism $\overline{\mathcal{T}}_q(\mathrm{C}) \to \mathrm{A}$ of Coxeter algebras, and then lift it to a morphism $\widehat{f}$.

Explicitly, on the Z-component, on the T-summand and into the T'-summand, $\widehat{f}$ is

$$\begin{array}{ccccccc} & & & \mathrm{C}[U'] & & & \\ & & \overset{(\overline{\lambda}_q)_{TUT'U'}}{\nearrow} & & \overset{f_{U'}}{\searrow} & & \\ \mathrm{C}[T] & \xrightarrow[\Delta_T^U]{} & \mathrm{C}[U] & & & \mathrm{A}[U'] & \xrightarrow{\mu_{T'}^{U'}} \mathrm{A}[T'], \\ & & \underset{f_U}{\searrow} & & \underset{(\overline{\lambda}_q)_{TUT'U'}}{\nearrow} & & \\ & & & \mathrm{A}[U] & & & \end{array}$$

with $\overline{\lambda}_q$ as in (8.70). We sum over all possible choices of U and U'. These face-types are shown in red to indicate that they vary. The diamond commutes, so the two ways to go from $\mathrm{C}[T]$ to $\mathrm{A}[T']$ match. We call $\widehat{f}$ the general q-norm map associated to f.

Special case. Let V be a Coxeter space. View it as a trivial Coxeter coalgebra and trivial Coxeter algebra, and apply the bijection (9.144) to $f := \mathrm{id}$. The resulting morphism $\widehat{f} : \overline{\mathcal{T}}_q(\mathrm{V}) \to \overline{\mathcal{T}}_q^{\vee}(\mathrm{V})$ of Coxeter q-bialgebras on the Z-component, on the T-summand and into the T'-summand, is $(\overline{\lambda}_q)_{TTT'T'} : \mathrm{V}[T] \to \mathrm{V}[T']$. By formula (8.72), this coincides with the q-norm map (9.140).

More generally: Let $f : \mathrm{V} \to \mathrm{W}$ be a map of Coxeter spaces. Now apply (9.144) with V as a trivial Coxeter coalgebra and W as a trivial Coxeter algebra. Let us denote the resulting morphism $\widehat{f}$ by

$$\overline{\kappa}_{q,f} : \overline{\mathcal{T}}_q(\mathrm{V}) \to \overline{\mathcal{T}}_q^{\vee}(\mathrm{W}). \tag{9.145}$$

On the Z-component, on the T-summand and into the T'-summand, it is

$$\mathrm{V}[T] \to \mathrm{W}[T'], \qquad x \mapsto \sum_{\substack{\sigma:\, T \to T' \\ \text{local to } Z}} q^{l(\sigma)} \sigma f_T(x).$$

We recover the previous situation when $f = \mathrm{id}$. Any Coxeter q-bialgebra morphism $\overline{\mathcal{T}}_q(\mathrm{V}) \to \overline{\mathcal{T}}_q^{\vee}(\mathrm{W})$ is of the form $\overline{\kappa}_{q,f}$ for a unique $f : \mathrm{V} \to \mathrm{W}$.

The bijection (9.144) says that f can be recovered from $\widehat{f}$. In fact, as suggested by the notation, all components of f are present in components of $\widehat{f}$. More precisely, f_Z equals $\widehat{f}_Z$ on the Z-summand and into the Z-summand.

Exercise 9.99. Deduce: For a scalar q, Coxeter coalgebra C, Coxeter algebra A, let g and h be two Coxeter q-bialgebra morphisms from $\overline{\mathcal{T}}_q(\mathrm{C})$ to $\overline{\mathcal{T}}_q^{\vee}(\mathrm{A})$ such that for each face-type Z, the matrix-component of g_Z, on the Z-summand and into the Z-summand, agrees with that of h_Z. Then $g = h$.

9.11.4. General norm map. Let us now look at the case $q = 1$. For a cocommutative Coxeter coalgebra C and commutative Coxeter algebra A, there is a natural bijection

$$\overline{\mathsf{C}}\mathsf{Sp}(\mathrm{C}, \mathrm{A}) \xrightarrow{\cong} {}^{\mathrm{co}}\mathsf{Bimon}^{\mathrm{co}}(\overline{\mathsf{C}}\mathsf{Sp})(\overline{\mathcal{S}}(\mathrm{C}), \overline{\mathcal{S}}^{\vee}(\mathrm{A})), \qquad f \mapsto \widehat{f}, \tag{9.146}$$

with $\widehat{f}$ induced from the $\widehat{f}$ in (9.144) via

$$\begin{array}{ccc} \overline{\mathcal{T}}(\mathrm{C}) & \xrightarrow{\widehat{f}} & \overline{\mathcal{T}}^\vee(\mathrm{A}) \\ \downarrow & & \uparrow \\ \overline{\mathcal{S}}(\mathrm{C}) & \underset{\widehat{f}}{\dashrightarrow} & \overline{\mathcal{S}}^\vee(\mathrm{A}). \end{array} \tag{9.147}$$

We continue to refer to $\widehat{f} : \overline{\mathcal{S}}(\mathrm{C}) \to \overline{\mathcal{S}}^\vee(\mathrm{A})$ as the general norm map associated to f. Similarly, for $q = -1$, we have the general signed norm map from $\overline{\mathcal{E}}(\mathrm{C})$ to $\overline{\mathcal{E}}^\vee(\mathrm{A})$.

Theorem 9.100. *Let* $\mathrm{char}(\Bbbk) \nmid |W|$. *For* C *a cocommutative Coxeter coalgebra,* A *a commutative Coxeter algebra, let* f *and* $\widehat{f}$ *be as in* (9.146). *Then*

$$\widehat{f} : \overline{\mathcal{S}}(\mathrm{C}) \to \overline{\mathcal{S}}^\vee(\mathrm{A}) \textit{ is an isomorphism of Coxeter bialgebras}$$
$$\textit{iff}$$
$$\log(f) : \mathrm{C} \to \mathrm{A} \textit{ is an isomorphism of Coxeter spaces.}$$

In addition, if either C *or* A *is trivial, then* $\widehat{f}$ *is an isomorphism of Coxeter bialgebras iff* f *is an isomorphism of Coxeter spaces.*

PROOF. This is an instance of Proposition 9.71 with $\mathrm{H} := \overline{\mathcal{S}}(\mathrm{C})$ and $\mathrm{K} := \overline{\mathcal{S}}^\vee(\mathrm{A})$. We leave the details as an exercise. For a more constructive approach, see Theorem 13.42. □

We recover Lemma 9.94 from Theorem 9.100 by setting $\mathrm{C} = \mathrm{A} = \mathrm{V}$ with trivial Coxeter (co)algebra structures and $f = \mathrm{id}$.

9.11.5. General q-norm map for q not a root of unity.

Let us begin with the q-norm map.

Lemma 9.101. *Suppose* q *is not a root of unity. Then the* q*-norm transformation* $\overline{\kappa}_q$ *is an isomorphism. More precisely, for a Coxeter space* V*, the* q*-norm map* (9.140) *is an isomorphism of Coxeter* q*-bialgebras, which moreover induces an isomorphism of Coxeter spaces* $\mathrm{V}^k \to \mathrm{V}^k$ *for each* $k \geq 1$.

Explicitly, the inverse of (9.140), namely,

$$\overline{\mathcal{T}}_q^\vee(\mathrm{V}) \to \overline{\mathcal{T}}_q(\mathrm{V}), \tag{9.148}$$

on the Z-component, on the T'-summand and into the T-summand, is

$$\mathrm{V}[T'] \to \mathrm{V}[T], \qquad x \mapsto \sum_{\substack{\sigma\colon T' \to T \\ \text{local to } Z}} \zeta_q(Z, T, \sigma, T')\, \sigma(x).$$

Exercise 9.102. Check using the two-sided q-flat-additivity formula (1.110) that (9.140) and (9.148) are inverses for q not a root of unity.

Exercise 9.103 (Varchenko matrix). Recall the Coxeter space of polynomials $\Bbbk[x]$ from Section 9.5.1. Specialize Lemma 9.101 to $\mathrm{V} := \Bbbk[x]$ to deduce Theorem 1.56. The proof in Exercise 9.102 generalizes the one in Exercise 1.99.

Example 9.104 (Second Cauchy power). For q not a root of unity, and V a Coxeter space, let us make explicit the q-norm map and its inverse on the second Cauchy power of V (on the $\emptyset$-component). Accordingly, let T be any vertex-type. Put $\alpha := \operatorname{dist}(T, \overline{T})$.

- $\underline{T \neq \overline{T}}$. The q-norm map restricts to an invertible linear operator on the subspace $\mathrm{V}[T] \oplus \mathrm{V}[\overline{T}]$. This map and its inverse in matrix notation are

$$\begin{pmatrix} \mathrm{id} & q^\alpha\, \beta_{T,\overline{T}} \\ q^\alpha\, \beta_{\overline{T},T} & \mathrm{id} \end{pmatrix} \quad \text{and} \quad \frac{1}{1-q^{2\alpha}} \begin{pmatrix} \mathrm{id} & -q^\alpha\, \beta_{T,\overline{T}} \\ -q^\alpha\, \beta_{\overline{T},T} & \mathrm{id} \end{pmatrix}.$$

- $\underline{T = \overline{T}}$. The q-norm map restricts to an invertible linear operator on the subspace $\mathrm{V}[T]$. This map and its inverse are

$$\mathrm{id} + q^\alpha\, \beta_{\overline{T},T} \quad \text{and} \quad \tfrac{1}{1-q^{2\alpha}}(\mathrm{id} - q^\alpha\, \beta_{\overline{T},T}).$$

  In matrix notation, these matrices have size 1.

This verifies Lemma 9.101 directly for $k = 2$.

For the dihedral arrangement of d lines, $\alpha = d - 1$ (Example 1.20). By setting $\beta_{\overline{T},T}$ and $\beta_{T,\overline{T}}$ equal to 1, the above matrices on the left yield the Varchenko matrices of sizes 2 and 1 for rank-one flat-types mentioned in Example 1.57. □

Exercise 9.105 (Third Cauchy power). Work out the q-norm map on the third Cauchy power of a Coxeter space V (on the $\emptyset$-component) in the rank-three braid arrangement and arrangement of type B. Express it using a minimal set of generators for the groupoid of biface-types. Setting these generating morphisms equal to 1 will yield the Varchenko matrices in Example 1.58.

Exercise 9.106. Suppose q is not a root of unity, so the q-norm map (9.140) is an isomorphism. For a Coxeter space V,

(i) use $\overline{\mathcal{P}}(\overline{\mathcal{T}}_q^{\vee}(\mathrm{V})) = \mathrm{V}$ to deduce $\overline{\mathcal{P}}(\overline{\mathcal{T}}_q(\mathrm{V})) = \mathrm{V}$,
(ii) use $\overline{\mathcal{Q}}(\overline{\mathcal{T}}_q(\mathrm{V})) = \mathrm{V}$ to deduce $\overline{\mathcal{Q}}(\overline{\mathcal{T}}_q^{\vee}(\mathrm{V})) = \mathrm{V}$.

Exercise 9.107. Check: For q not a root of unity, the q-exponentials (9.99a) of the biderivations

$$\begin{array}{ccc} \overline{\mathcal{T}}_q(\mathrm{V}) & \dashrightarrow & \overline{\mathcal{T}}_q^{\vee}(\mathrm{V}) \\ \downarrow & & \uparrow \\ \mathrm{V} & \xrightarrow[\mathrm{id}]{} & \mathrm{V} \end{array} \qquad\qquad \begin{array}{ccc} \overline{\mathcal{T}}_q^{\vee}(\mathrm{V}) & \dashrightarrow & \overline{\mathcal{T}}_q(\mathrm{V}) \\ \downarrow & & \uparrow \\ \mathrm{V} & \xrightarrow[\mathrm{id}]{} & \mathrm{V} \end{array}$$

yield inverse isomorphisms of Coxeter q-bialgebras between $\overline{\mathcal{T}}_q(\mathrm{V})$ and $\overline{\mathcal{T}}_q^{\vee}(\mathrm{V})$ which are the q-norm map (9.140) and its inverse (9.148), respectively.

Exercise 9.108. Suppose q is not a root of unity. Show: The map $\overline{\kappa}_{q,f}$ in (9.145) is an isomorphism of Coxeter q-bialgebras iff f is an isomorphism of Coxeter spaces. Setting $f = \mathrm{id}$ recovers Lemma 9.101. (Express $\overline{\kappa}_{q,f}$ as the composite $(\overline{\kappa}_q)_{\mathrm{W}} \overline{\mathcal{T}}_q(f)$. Now use Lemma 9.101 and the fact that the functor $\overline{\mathcal{T}}_q$ reflects isomorphisms.)

Exercise 9.109. Suppose q is not a root of unity. Let $f : \mathrm{V} \to \mathrm{W}$ be a map of Coxeter spaces. Check: The q-exponential (9.99a) of the biderivation

$$\begin{array}{ccc} \overline{\mathcal{T}}_q(\mathrm{V}) & \dashrightarrow & \overline{\mathcal{T}}_q^{\vee}(\mathrm{W}) \\ \downarrow & & \uparrow \\ \mathrm{V} & \xrightarrow[f]{} & \mathrm{W} \end{array}$$

is the map $\bar{\kappa}_{q,f}$ in (9.145). (Use the two-sided q-flat-additivity formula (1.110).) Use Proposition 9.73 to deduce Exercise 9.108.

Let us now go to the general q-norm map.

Theorem 9.110. *Suppose q is not a root of unity. For* C *a Coxeter coalgebra,* A *a Coxeter algebra, let f and $\widehat{f}$ be as in* (9.144). *Then*

$$\widehat{f} : \overline{\mathcal{T}}_q(\mathrm{C}) \to \overline{\mathcal{T}}_q^{\vee}(\mathrm{A}) \text{ is an isomorphism of Coxeter } q\text{-bialgebras}$$
$$\textit{iff}$$
$$\mathbf{log}_q(f) : \mathrm{C} \to \mathrm{A} \text{ is an isomorphism of Coxeter spaces.}$$

In addition, if either C *or* A *is trivial, then $\widehat{f}$ is an isomorphism of Coxeter q-bialgebras iff f is an isomorphism of Coxeter spaces.*

PROOF. This follows from Proposition 9.73 and Exercise 9.111 below. For the last part, note that $\mathbf{log}_q(f) = f$ if either C or A is trivial. □

Exercise 9.111. Suppose q is not a root of unity. For C a Coxeter coalgebra, A a Coxeter algebra, let f and $\widehat{f}$ be as in (9.144). Deduce from Example 9.62 that $\widehat{f}$ is the q-exponential (9.99a) of the biderivation

$$\begin{array}{ccc} \overline{\mathcal{T}}_q(\mathrm{C}) & \dashrightarrow & \overline{\mathcal{T}}_q^{\vee}(\mathrm{A}) \\ \downarrow & & \uparrow \\ \mathrm{C} & \xrightarrow[\mathbf{log}_q(f)]{} & \mathrm{A}. \end{array}$$

In other words, (9.144) is composite of the bijections

$$\overline{\mathsf{C}}\mathsf{Sp}(\mathrm{C},\mathrm{A}) \xrightarrow{\mathbf{log}_q} \overline{\mathsf{C}}\mathsf{Sp}(\mathrm{C},\mathrm{A}) \xrightarrow{\mathbf{exp}_q} q\text{-}\mathsf{Bimon}(\overline{\mathsf{C}}\mathsf{Sp})(\overline{\mathcal{T}}_q(\mathrm{C}), \overline{\mathcal{T}}_q^{\vee}(\mathrm{A})).$$

The first is (9.100), while the second is an instance of (9.104).

We recover Lemma 9.101 from Theorem 9.110 by setting C = A = V with trivial Coxeter (co)algebra structures and $f = \mathrm{id}$. More generally, we recover Exercise 9.108 by setting C and A to be both trivial. In fact, this special case along with Exercise 9.111 can be used to give a more constructive proof of Theorem 9.110. This is explained later in Theorem 13.48.

9.11.6. 0-norm map. Observe: $\bar{\kappa}_0 = \mathrm{id}$, so Lemma 9.101 holds trivially for $q = 0$. More generally, for C a Coxeter coalgebra, A a Coxeter algebra, and $f : \mathrm{C} \to \mathrm{A}$, the map $\widehat{f} : \overline{\mathcal{T}}_0(\mathrm{C}) \to \overline{\mathcal{T}}_0^{\vee}(\mathrm{A})$, on the Z-component, on the T-summand and into the T'-summand, is the composite map

$$\mathrm{C}[T] \xrightarrow{\Delta_T^{T \cup T'}} \mathrm{C}[T \cup T'] \xrightarrow{f_{T \cup T'}} \mathrm{A}[T \cup T'] \xrightarrow{\mu_{T'}^{T \cup T'}} \mathrm{A}[T'].$$

By Theorem 9.110, $\widehat{f}$ is an isomorphism iff $\mathbf{log}_0(f)$ is an isomorphism, with $\mathbf{log}_0(f)$ as in (9.105b).

9.12. Bialgebra axiom on the primitive part

We now analyze how the bialgebra axiom works when restricted to the primitive part of a Coxeter bialgebra. The rhs of the bialgebra axiom $\Delta\mu = \sum \mu\sigma\Delta$ begins with a Δ whose nontrivial components are zero on the primitive part, so they drop out resulting in the simplified form $\Delta\mu = \sum \mu\sigma$. We also give some related results on the primitive filtration. These ideas are relevant to the study of structure results on Coxeter bialgebras in Chapter 13.

9.12.1. Coxeter q-bialgebras. The main observation is the following.

Lemma 9.112. *For a scalar q, let* H *be a Coxeter q-bialgebra. For face-types $Z \leq T$ and $Z \leq T'$ and $x \in \overline{\mathcal{P}}(\mathrm{H})[T]$,*

$$\Delta_Z^{T'} \mu_Z^T(x) = \sum_{\substack{U' : U' \geq T' \\ \mathrm{s}(U'/Z) = \mathrm{s}(T/Z)}} \; \sum_{\substack{\sigma : T \to U' \\ \text{local to } Z \\ \text{projects } T \text{ on } T'}} q^{l(\sigma)} \mu_{T'}^{U'} \sigma(x), \tag{9.149}$$

and as special cases,

$$\Delta_Z^{T'} \mu_Z^T(x) = \begin{cases} \sum_{\substack{\sigma : T \to T' \\ \text{local to } Z}} q^{l(\sigma)} \sigma(x) & \text{if } \mathrm{s}(T/Z) = \mathrm{s}(T'/Z), \\ 0 & \text{if } \mathrm{s}(T/Z) \not\geq \mathrm{s}(T'/Z). \end{cases} \tag{9.150}$$

PROOF. For (9.149): Apply the q-bialgebra axiom (8.73). The primitivity hypothesis forces $U = T$. Now use the first identity in (8.71).

For (9.150): Specialize (9.149). In the first alternative, $U' = T'$ is forced; the relevant identity is (8.72). In the second alternative, no U' works. □

Exercise 9.113. Let q be a scalar which is not a root of unity. For V a Coxeter space, for $k \geq 1$, define

$$(\sigma_q)^{k-1} : \mathrm{V}^k \to \mathrm{V}^k \quad \text{and} \quad (\sigma^q)^{k-1} : \mathrm{V}^k \to \mathrm{V}^k$$

as follows. The first map is the restriction of the q-norm map $(\overline{\kappa}_q)_{\mathrm{V}}$ in (9.140), and the second is its inverse which is the restriction of (9.148). The latter involves the two-sided q-zeta function.

Let H be a Coxeter q-bialgebra. Show: $\Delta^k \mu^k = (\sigma_q)^k$ on $\overline{\mathcal{P}}(\mathrm{H})^{k+1}$ for $k \geq 0$. (Use (9.150).) Moreover, for $k \geq 2$, for a face-type Z,

$$x \in \overline{\mathcal{P}}_k(\mathrm{H})[Z] \implies x - \mu_Z^{k-1} (\sigma^q)_Z^{k-1} \Delta_Z^{k-1}(x) \in \overline{\mathcal{P}}_{k-1}(\mathrm{H})[Z].$$

(Apply Δ_Z^{k-1} to the rhs and check that we get zero.)

9.12.2. Coxeter 0-bialgebras. We now specialize to $q = 0$.

Lemma 9.114. *Let* H *be a Coxeter* 0*-bialgebra. For face-types $Z \leq T$ and $Z \leq T'$ and $x \in \overline{\mathcal{P}}(\mathrm{H})[T]$,*

$$\Delta_Z^{T'} \mu_Z^T(x) = \begin{cases} \mu_{T'}^T(x) & \text{if } T \geq T', \\ 0 & \text{otherwise.} \end{cases} \tag{9.151}$$

PROOF. Apply the 0-bialgebra axiom (8.78). The primitivity hypothesis forces $T \cup T' = T$ which is the same as $T \geq T'$. Alternatively, put $q = 0$ in (9.149). □

Exercise 9.115. Let H be a Coxeter 0-bialgebra. Show: $\Delta^k \mu^k = \mathrm{id}$ on $\overline{\mathcal{P}}(\mathrm{H})^{k+1}$ for $k \geq 0$. (Use (9.151).) Moreover, for $k \geq 2$, for a face-type Z,

$$x \in \overline{\mathcal{P}}_k(\mathrm{H})[Z] \implies x - \mu_Z^{k-1} \Delta_Z^{k-1}(x) \in \overline{\mathcal{P}}_{k-1}(\mathrm{H})[Z].$$

This is the case $q = 0$ of Exercise 9.113.

9.12.3. Coxeter bialgebras.

Exercise 9.116. Let H be a Coxeter bialgebra. Let $T \geq Z$ and $x_T \in \overline{\mathcal{P}}(\mathrm{H})[T]$ with x_T invariant under all morphisms $T \to T$ local to Z. For any $Y \geq Z$ with $\mathrm{s}(Y/Z) = \mathrm{s}(T/Z)$, let x_Y denote the image of x_T under any morphism $T \to Y$ local to Z. Check: For any $T' \geq Z$,

$$\Delta_Z^{T'} \mu_Z^{T}(x_T) = \sum_{\substack{U' : U' \geq T' \\ \mathrm{s}(U'/Z) = \mathrm{s}(T/Z)}} a_Z^{T' U' T} \mu_{T'}^{U'}(x_{U'}). \tag{9.152}$$

(Use (9.149) and (1.64b) for $q = 1$. Also see the first calculation in the proof of Theorem 9.76.) As a special case: For any $T' \geq Z$ with $\mathrm{s}(T'/Z) = \mathrm{s}(T/Z)$,

$$\Delta_Z^{T'} \mu_Z^{T}(x_T) = |\mathrm{s}(T/Z)|\, x_{T'}, \tag{9.153}$$

with $|\mathrm{s}(T/Z)|$ as in Definition 1.8. (Use the first alternative in (9.150) and (1.64d) for $q = 1$.)

Exercise 9.117. Let H be a Coxeter bialgebra. Show: For face-types $Z \leq T, T', T''$ and $x \in \overline{\mathcal{P}}(\mathrm{H})[T]$ and $\sigma : T' \to T''$ local to Z,

$$\sigma \Delta_Z^{T'} \mu_Z^{T}(x) = \Delta_Z^{T''} \mu_Z^{T}(x).$$

(Use (9.149) for $q = 1$ and the special case in Exercise 1.40.)

Exercise 9.118. Let s be an element of the invariant lune-incidence algebra. For V a Coxeter space, for $k \geq 1$, define $s^{(k-1)} : \mathrm{V}^k \to \mathrm{V}^k$ on the Z-component, on the T-summand, to be scalar multiplication by $s(Z, T)$.

Let H be a cocommutative Coxeter bialgebra. Let $\boldsymbol{\zeta}$ be a noncommutative zeta function. Show: For $k \geq 2$, for a face-type Z,

$$x \in \overline{\mathcal{P}}_k(\mathrm{H})[Z] \implies x - \mu_Z^{k-1} \boldsymbol{\zeta}_Z^{(k-1)} \Delta_Z^{k-1}(x) \in \overline{\mathcal{P}}_{k-1}(\mathrm{H})[Z].$$

(Use the flat-additivity formula (1.83).)

9.12.4. Covering face-types. We now turn to the special case of covering face-types. For convenience, the argument is spelt out in detail.

Lemma 9.119. *Let* H *be a Coxeter bialgebra. Let* $Z \lessdot T$ *and* $x \in \overline{\mathcal{P}}(\mathrm{H})[T]$. *Then:*

$$\Delta_Z^{T'} \mu_Z^{T}(x) = 0 \text{ if } T' > Z,\ T' \neq T \text{ and } T' \neq Z\overline{T}. \tag{9.154a}$$

$$\Delta_Z^{T} \mu_Z^{T}(x) = \begin{cases} x & \text{if } Z\overline{T} \neq T, \\ x + \beta_{Z\overline{T},T}(x) & \text{if } Z\overline{T} = T. \end{cases} \tag{9.154b}$$

$$\Delta_Z^{Z\overline{T}} \mu_Z^{T}(x) = \begin{cases} \beta_{Z\overline{T},T}(x) & \text{if } Z\overline{T} \neq T, \\ x + \beta_{Z\overline{T},T}(x) & \text{if } Z\overline{T} = T. \end{cases} \tag{9.154c}$$

PROOF. In each case, we employ the bialgebra axiom (8.64). By the primitivity hypothesis, $\Delta_T^U(x) = 0$ for all $U > T$. Thus, we need to analyze the map $\overline{\lambda}_{TUT'U'}$ for $U = T$.

For (9.154a): Assume $T' > Z$, $T' \neq T$ and $T' \neq Z\overline{T}$. Then, for any face F/A of type T/Z and face F'/A of type T'/Z, FF' is strictly greater than F. Hence $\overline{\lambda}_{TUT'U'} = 0$ for $U = T$.

For (9.154b): First assume $Z\overline{T} \neq T$. Then, for $U = T$, $\overline{\lambda}_{TUTU'}$ is identity if $U' = T$ and zero otherwise. Next assume $Z\overline{T} = T$. Then, for $U = T$, $\overline{\lambda}_{TUTU'}$ is $\mathrm{id} + \beta_{Z\overline{T},T}$ if $U' = T$ and zero otherwise.

(9.154c) is similar. □

9.13. Characteristic operations by face-types

Recall the invariant Birkhoff, Tits, Janus algebras from Section 1.12. For any face-type Z, elements of the invariant Tits algebra of $\boldsymbol{\alpha}_Z$ determine characteristic operations on $\mathrm{H}[Z]$ for any Coxeter bialgebra H. These operations define a left action when H is cocommutative and a right action when H is commutative. When H is bicommutative, the action factors through the invariant support map to yield an action of the invariant Birkhoff algebra of $\boldsymbol{\alpha}_Z$. More generally, for a scalar q, elements of the invariant q-Janus algebra of $\boldsymbol{\alpha}_Z$ define a left action on $\mathrm{H}[Z]$ for any Coxeter q-bialgebra H via two-sided characteristic operations.

The parallel discussion for Coxeter species is in Sections 3.12 and 3.13.

9.13.1. Characteristic operations. For a face-type Z, let $\overline{\Sigma}[Z]$ denote the invariant Tits algebra of $\boldsymbol{\alpha}_Z$. Let H be a Coxeter bialgebra. For a face-type Z, for $z \in \overline{\Sigma}[Z]$ and $h \in \mathrm{H}[Z]$, define an element $z \cdot h \in \mathrm{H}[Z]$ by

$$z \cdot h := \sum_{T:\,T \geq Z} a^{T/Z}\, \mu_Z^T \Delta_Z^T(h) \quad \text{with} \quad z = \sum_{T:\,T \geq Z} a^{T/Z}\, \mathtt{H}_{T/Z}. \tag{9.155}$$

In particular,

$$\mathtt{H}_{T/Z} \cdot h := \mu_Z^T \Delta_Z^T(h). \tag{9.156}$$

We refer to this as a *characteristic operation*.

Lemma 9.120. *Let* H *be a Coxeter bialgebra. Then:*

- *For* $h \in \mathrm{H}[Z]$,

$$\mathtt{H}_{Z/Z} \cdot h = h. \tag{9.157}$$

- *If* H *is cocommutative, then for* $z, w \in \overline{\Sigma}[Z]$ *and* $h \in \mathrm{H}[Z]$,

$$(z \cdot w) \cdot h = z \cdot (w \cdot h). \tag{9.158}$$

- *If* H *is commutative, then for* $z, w \in \overline{\Sigma}[Z]$ *and* $h \in \mathrm{H}[Z]$,

$$(z \cdot w) \cdot h = w \cdot (z \cdot h). \tag{9.159}$$

In other words: When H is cocommutative, for any face-type Z, formula (9.155) defines a left action of the invariant Tits algebra $\overline{\Sigma}[Z]$ on the component $\mathrm{H}[Z]$. When H is commutative, there is a right action of $\overline{\Sigma}[Z]$ on $\mathrm{H}[Z]$ given by $h \cdot' z := z \cdot h$.

PROOF. Statement (9.157) follows from (co)unitality axioms in (8.41) and (8.46). The remaining statements are linear in z and w, so we may assume them to be basis elements. If H is cocommutative, then using identity (8.98) for $q = 1$,

$$\begin{aligned}
\mathtt{H}_{T/Z} \cdot (\mathtt{H}_{U/Z} \cdot h) &= \mu_Z^T \Delta_Z^T \mu_Z^U \Delta_Z^U (h) \\
&= \sum_{V:\, V \geq T} a_Z^{TVU} \mu_Z^T \mu_T^V \Delta_Z^V (h) \\
&= \sum_{V:\, V \geq T} a_Z^{TVU} \mu_Z^V \Delta_Z^V (h) \\
&= \sum_{V:\, V \geq T} a_Z^{TVU} \mathtt{H}_{V/Z} \cdot h \\
&= (\mathtt{H}_{T/Z} \cdot \mathtt{H}_{U/Z}) \cdot h.
\end{aligned}$$

The third step used the associativity axiom in (8.41); the last step used (1.156).

The case when H is commutative is proved similarly starting with identity (8.97) for $q = 1$. □

Example 9.121 (Invariant Tits algebra as a module over itself). The Coxeter bialgebra of face-types $\overline{\Sigma}$ is defined later in Section 10.4. Its Z-component $\overline{\Sigma}[Z]$ identifies with the invariant Tits algebra of $\boldsymbol{\alpha}_Z$. Since $\overline{\Sigma}$ is cocommutative, we may take $\mathrm{H} = \overline{\Sigma}$, resulting in a characteristic operation on each $\overline{\Sigma}[Z]$. This is the same as the product (1.156) in the invariant Tits algebra. Indeed, for face-types T and U greater than Z, using product and coproduct formulas (10.62),

$$\begin{aligned}
\mathtt{H}_{T/Z} \cdot \mathtt{H}_{U/Z} &= \mu_Z^T \Delta_Z^T (\mathtt{H}_{U/Z}) \\
&= \mu_Z^T \Big(\sum_{V:\, V \geq T} a_Z^{TVU} \mathtt{H}_{V/T} \Big) \\
&= \sum_{V:\, V \geq T} a_Z^{TVU} \mathtt{H}_{V/Z} \\
&= \mathtt{H}_{T/Z} \cdot \mathtt{H}_{U/Z}.
\end{aligned}$$

The starting term refers to a characteristic operation and ending term to the product in the invariant Tits algebra. □

Lemma 9.122. *Let* H *be a Coxeter bialgebra. Let* $Z \leq T$. *Then:*

- *For* $z \in \overline{\Sigma}[T]$ *and* $h \in \mathrm{H}[Z]$,

$$\mu_Z^T(z) \cdot h = \mu_Z^T\big(z \cdot \Delta_Z^T(h)\big). \tag{9.160}$$

- *If* H *is commutative, then for* $z \in \overline{\Sigma}[Z]$ *and* $h \in \mathrm{H}[T]$,

$$z \cdot \mu_Z^T(h) = \mu_Z^T\big(\Delta_Z^T(z) \cdot h\big). \tag{9.161}$$

- *If* H *is cocommutative, then for* $z \in \overline{\Sigma}[Z]$ *and* $h \in \mathrm{H}[Z]$,

$$\Delta_Z^T(z \cdot h) = \Delta_Z^T(z) \cdot \Delta_Z^T(h). \tag{9.162}$$

Here $\overline{\Sigma}$ refers to the Coxeter bialgebra of face-types $\overline{\Sigma}$ with product and coproduct (10.62).

PROOF. All statements are linear in z, so we may assume z is a basis element in each case.

- For (9.160): Take $z = \mathtt{H}_{U/T}$. Then

$$\mu_Z^T\big(z \cdot \Delta_Z^T(h)\big) = \mu_Z^T \mu_T^U \Delta_T^U \Delta_Z^T(h) = \mu_Z^U \Delta_Z^U(h) = \mathtt{H}_{U/Z} \cdot h = \mu_Z^T(z) \cdot h.$$

We used (co)associativity axioms of H in (8.41) and (8.46).

- For (9.161): Take $z = \mathtt{H}_{U/Z}$. Then

$$\begin{aligned}\mu_Z^T(\Delta_Z^T(z) \cdot h) &= \mu_Z^T\Big(\sum_{V:\, V \geq T} a_Z^{TVU}\, \mathtt{H}_{V/T} \cdot h\Big) = \mu_Z^T\Big(\sum_{V:\, V \geq T} a_Z^{TVU}\, \mu_T^V \Delta_T^V(h)\Big)\\ &= \sum_{V:\, V \geq T} a_Z^{TVU}\, \mu_Z^T \mu_T^V \Delta_T^V(h) = \sum_{V:\, V \geq T} a_Z^{TVU}\, \mu_Z^V \Delta_T^V(h)\\ &= \mu_Z^U \Delta_Z^U \mu_Z^T(h) = \mathtt{H}_{U/Z} \cdot \mu_Z^T(h) = z \cdot \mu_Z^T(h).\end{aligned}$$

We used the associativity axiom in (8.41) and identity (8.97).

- For (9.162): Take $z = \mathtt{H}_{U/Z}$. Then

$$\begin{aligned}\Delta_Z^T(z \cdot h) &= \Delta_Z^T \mu_Z^U \Delta_Z^U(h) = \sum_{V:\, V \geq T} a_Z^{TVU}\, \mu_T^V \Delta_Z^V(h)\\ &= \sum_{V:\, V \geq T} a_Z^{TVU}\, \mu_T^V \Delta_T^V \Delta_Z^T(h) = \Big(\sum_{V:\, V \geq T} a_Z^{TVU}\, \mathtt{H}_{V/T}\Big) \cdot \Delta_Z^T(h)\\ &= \Delta_Z^T(z) \cdot \Delta_Z^T(h).\end{aligned}$$

We used the coassociativity axiom in (8.46) and identity (8.98).

This completes the checks. □

9.13.2. Commutative characteristic operations. Let $\mathrm{char}(\Bbbk) \nmid |W|$. For a face-type Z, let $\overline{\Pi}[Z]$ denote the invariant Birkhoff algebra of $\boldsymbol{\alpha}_Z$. Let H be a bicommutative Coxeter bialgebra. Then the characteristic operation (9.156) factors through the invariant support map to yield an action of $\overline{\Pi}[Z]$ on $\mathrm{H}[Z]$:

$$\mathtt{H}_{\mathrm{s}(T/Z)} \cdot h := \mu_Z^T \Delta_Z^T(h). \tag{9.163}$$

This follows from Exercise 1.26 applied to $\boldsymbol{\alpha}_Z$ and (co)commutativity axioms (8.86) and (8.91). We refer to (9.163) as a *commutative characteristic operation*.

Example 9.123 (Invariant Birkhoff algebra as a module over itself). This is a continuation of Example 9.121. The Coxeter bialgebra of flat-types $\overline{\Pi}$ is defined later in Section 10.5. Its Z-component $\overline{\Pi}[Z]$ identifies with the invariant Birkhoff algebra of $\boldsymbol{\alpha}_Z$. Since $\overline{\Pi}$ is bicommutative, we may take $\mathrm{H} = \overline{\Pi}$, resulting in a commutative characteristic operation on each $\overline{\Pi}[Z]$. This is the same as the product in the invariant Birkhoff algebra. □

9.13.3. Two-sided characteristic operations. For a scalar q, for a face-type Z, let $\overline{\mathrm{J}}_q[Z]$ denote the invariant q-Janus algebra of $\boldsymbol{\alpha}_Z$. It has a basis of biface-types $\sigma : T' \to T$ local to Z. We denote a typical basis element by $\mathtt{H}_\sigma$ or for more clarity by $\mathtt{H}_{(T/Z,\sigma,T'/Z)}$.

For a scalar q, let H be a Coxeter q-bialgebra. For a face-type Z, for $z \in \overline{\mathrm{J}}_q[Z]$ and $h \in \mathrm{H}[Z]$, define an element $z \cdot h \in \mathrm{H}[Z]$ by

$$z \cdot h := \sum_{\substack{\sigma : T' \to T \\ \text{local to } Z}} a^\sigma \, \mu_Z^T \sigma \Delta_Z^{T'}(h) \quad \text{with} \quad z = \sum_\sigma a^\sigma \, \mathtt{H}_\sigma. \tag{9.164}$$

In particular,

$$\mathtt{H}_{(T/Z,\sigma,T'/Z)} \cdot h := \mu_Z^T \sigma \Delta_Z^{T'}(h). \tag{9.165}$$

We refer to this as a *two-sided characteristic operation*. Note very carefully: σ and not $q^{l(\sigma)}\sigma$ is employed in this definition.

Lemma 9.124. *For a scalar q, let* H *be a Coxeter q-bialgebra. Then for any $z, w \in \overline{\mathrm{J}}_q[Z]$ and $h \in \mathrm{H}[Z]$,*

$$(z \cdot w) \cdot h = z \cdot (w \cdot h) \quad \text{and} \quad \mathtt{H}_{(Z/Z,\mathrm{id},Z/Z)} \cdot h = h. \tag{9.166}$$

In other words, for a face-type Z, formula (9.164) defines a left action of the invariant q-Janus algebra $\overline{\mathrm{J}}_q[Z]$ on the component $\mathrm{H}[Z]$.

PROOF. The second statement follows from (co)unitality. For the first statement: For $z = \mathtt{H}_{(U/Z,\tau,U'/Z)}$ and $w = \mathtt{H}_{(T/Z,\sigma,T'/Z)}$,

$$\begin{aligned}
z \cdot (w \cdot h) &= \mu_Z^U \tau \Delta_Z^{U'} \, \mu_Z^T \sigma \Delta_Z^{T'}(h) \\
&= \sum_{\substack{\pi : V \to V' \\ \text{local to } Z \\ \text{biprojects } T \text{ and } U'}} q^{l(\pi)} \, \mu_Z^U \tau \mu_{U'}^{V'} \pi \Delta_T^V \sigma \Delta_Z^{T'}(h) \\
&= \sum_{\substack{\pi : V \to V' \\ \text{local to } Z \\ \text{biprojects } T \text{ and } U'}} q^{l(\pi)} \, \mu_Z^U \mu_U^{\tau_*(V')} \tau\pi\sigma \Delta_{T'}^{\sigma^*(V)} \Delta_Z^{T'}(h) \\
&= \sum_{\substack{\pi : V \to V' \\ \text{local to } Z \\ \text{biprojects } T \text{ and } U'}} q^{l(\pi)} \, \mu_Z^{\tau_*(V')} \tau\pi\sigma \Delta_Z^{\sigma^*(V)}(h) \\
&= \sum_{\rho \text{ local to } Z} a^{\tau\rho\sigma}(q) \, \mathtt{H}_\rho \cdot h \\
&= (z \cdot w) \cdot h.
\end{aligned}$$

We used the q-bialgebra axiom (8.73) and then naturality and (co)associativity axioms (8.41), (8.46). In the last two steps, we first put $\rho := \tau\pi\sigma$ and then used (1.146) and (1.147) applied to $\boldsymbol{\alpha}_Z$. □

9.13.4. Primitive and indecomposable parts. We now look at characteristic operations by invariant Zie elements from Section 1.13. We will see later in (10.100) that an invariant Zie element of $\overline{\Sigma}[Z]$ is the same as an element of $\overline{\mathcal{P}}(\overline{\Sigma})[Z]$. This fact is used below.

Proposition 9.125. *Let* H *be a cocommutative Coxeter bialgebra. Let Z be a face-type. Then the left action of any invariant Zie element of $\overline{\Sigma}[Z]$ sends* H$[Z]$ *to* $\overline{\mathcal{P}}(\mathrm{H})[Z]$*. Moreover, on* $\overline{\mathcal{P}}(\mathrm{H})[Z]$ *it acts by scalar multiplication by the coefficient of* $\mathtt{H}_{Z/Z}$ *in the invariant Zie element.*

In particular, any invariant special Zie element of $\overline{\Sigma}[Z]$ yields an idempotent operator on H$[Z]$ *with image* $\overline{\mathcal{P}}(\mathrm{H})[Z]$.

Proof. For $z \in \overline{\mathcal{P}}(\overline{\Sigma})[Z]$ and $h \in \mathrm{H}[Z]$, by (9.162),

$$\Delta_Z^T(z \cdot h) = \Delta_Z^T(z) \cdot \Delta_Z^T(h) = 0$$

for all $T > Z$. Thus $z \cdot h \in \overline{\mathcal{P}}(\mathrm{H})[Z]$. If h itself is primitive, then

$$z \cdot h = \Big(\sum_{T:\, T \geq Z} a^{T/Z} \mathtt{H}_{T/Z} \Big) \cdot h = \sum_{T:\, T \geq Z} a^{T/Z} \mathtt{H}_{T/Z} \cdot h = a^{Z/Z} h$$

as required. □

Dually:

Proposition 9.126. *Let* H *be a commutative Coxeter bialgebra. Let Z be a face-type. Then the right action of any invariant Zie element of $\overline{\Sigma}[Z]$ on* H$[Z]$ *factors through* $\overline{\mathcal{Q}}(\mathrm{H})[Z]$*. Moreover, on* $\overline{\mathcal{Q}}(\mathrm{H})[Z]$ *it acts by scalar multiplication by the coefficient of* $\mathtt{H}_{Z/Z}$ *in the invariant Zie element.*

In particular, any invariant special Zie element of $\overline{\Sigma}[Z]$ yields an idempotent operator on H$[Z]$ *with coimage* $\overline{\mathcal{Q}}(\mathrm{H})[Z]$.

Proof. This can be proved similar to the previous result using (9.161). □

Note the striking similarity of the above two results with Proposition 9.67. The connection is as follows. A typical invariant special Zie element of $\overline{\Sigma}[Z]$ is $\mathtt{Q}_{Z/Z}$ and its characteristic operation on H$[Z]$ is the same as $\mathbf{log}(\mathrm{id})_Z$. These ideas are developed further in Section 10.4.20, see diagrams (10.110) and (10.111) in particular.

For q not a root of unity, we now look at two-sided characteristic operation by the element

$$\mathtt{Q}_{(Z/Z,\mathrm{id},Z/Z)} = \mathtt{Q}^d_{(Z/Z,\mathrm{id},Z/Z)} = \sum_{\substack{\sigma:\, T' \to T \\ \text{local to } Z}} \mu_q(Z,T,\sigma,T')\, \mathtt{H}_{(T/Z,\sigma,T'/Z)}. \tag{9.167}$$

This formula arises by applying (1.144) to $\boldsymbol{\alpha}_Z$.

Proposition 9.127. *Let* H *be a Coxeter q-bialgebra for q not a root of unity. For Z any face-type,* $\mathtt{Q}_{(Z/Z,\mathrm{id},Z/Z)}$ *yields an idempotent operator on* H$[Z]$ *with image* $\overline{\mathcal{P}}(\mathrm{H})[Z]$ *and coimage* $\overline{\mathcal{Q}}(\mathrm{H})[Z]$.

This result is contained in Proposition 9.72. This is because two-sided characteristic operation by $\mathtt{Q}_{(Z/Z,\mathrm{id},Z/Z)}$ on H$[Z]$ is the same as $\mathbf{log}_q(\mathrm{id})_Z$.

9.13.5. Algebra modules and Coxeter bialgebras. We now formally relate the categories of modules over the invariant Janus, Tits, Birkhoff algebras to Coxeter bialgebras.

Invariant q-Janus algebra. For a scalar q, consider the invariant q-Janus algebra $\mathsf{J}_q[\mathcal{A}]^W$. In the notation of this section, it is $\overline{\mathrm{J}}_q[\emptyset]$. We saw in (9.166) that for H a Coxeter $\boldsymbol{\alpha}$-q-bialgebra, H$[\emptyset]$ is a left module over $\mathsf{J}_q[\mathcal{A}]^W$ via the two-sided characteristic operation

$$\mathtt{H}_{(T,\sigma,T')} \cdot h := \mu_\emptyset^T \sigma \Delta_\emptyset^{T'}(h). \tag{9.168}$$

Further, for $f : \mathrm{H} \to \mathrm{K}$ a morphism of Coxeter $\boldsymbol{\alpha}$-q-bialgebras, $f_\emptyset : \mathrm{H}[\emptyset] \to \mathrm{K}[\emptyset]$ is a map of $\mathsf{J}_q[\mathcal{A}]^W$-modules:

$$\begin{aligned}\mathtt{H}_{(T,\sigma,T')} \cdot f_\emptyset(h) &= \mu_\emptyset^T \sigma \Delta_\emptyset^{T'} f_\emptyset(h) = \mu_\emptyset^T \sigma f_{T'} \Delta_\emptyset^{T'}(h) \\ &= \mu_\emptyset^T f_T \sigma \Delta_\emptyset^{T'}(h) = f_\emptyset \mu_\emptyset^T \sigma \Delta_\emptyset^{T'}(h) = f_\emptyset(\mathtt{H}_{(T,\sigma,T')} \cdot h).\end{aligned}$$

Thus, we have a functor from the category of Coxeter $\boldsymbol{\alpha}$-q-bialgebras to the category of left $\mathsf{J}_q[\mathcal{A}]^W$-modules.

In particular, for $q = 1$, we have a functor from the category of Coxeter $\boldsymbol{\alpha}$-bialgebras to the category of left $\mathsf{J}[\mathcal{A}]^W$-modules.

Invariant Tits algebra. Consider the invariant Tits algebra $\Sigma[\mathcal{A}]^W$. In the notation of this section, it is $\overline{\Sigma}[\emptyset]$. We saw in (9.158) that for H a cocommutative Coxeter $\boldsymbol{\alpha}$-bialgebra, H$[\emptyset]$ is a left module over $\Sigma[\mathcal{A}]^W$ via the characteristic operation

$$\mathtt{H}_T \cdot h := \mu_\emptyset^T \Delta_\emptyset^T(h). \tag{9.169}$$

In fact, the two-sided characteristic operation reduces to this one when H is cocommutative. This yields a functor from the category of cocommutative Coxeter $\boldsymbol{\alpha}$-bialgebras to the category of left $\Sigma[\mathcal{A}]^W$-modules.

Similarly, there is a functor from the category of commutative Coxeter $\boldsymbol{\alpha}$-bialgebras to the category of right $\Sigma[\mathcal{A}]^W$-modules.

Invariant Birkhoff algebra. Let $\mathrm{char}(\Bbbk) \nmid |W|$. Consider the invariant Birkhoff algebra $\Pi[\mathcal{A}]^W$. In the notation of this section, it is $\overline{\Pi}[\emptyset]$. The commutative characteristic operations yield a functor from the category of bicommutative Coxeter $\boldsymbol{\alpha}$-bialgebras to the category of $\Pi[\mathcal{A}]^W$-modules.

Summary. We now summarize the preceding discussion. For an algebra A, let A-Mod denote the category of left A-modules. The following diagram of functors commutes.

(9.170)

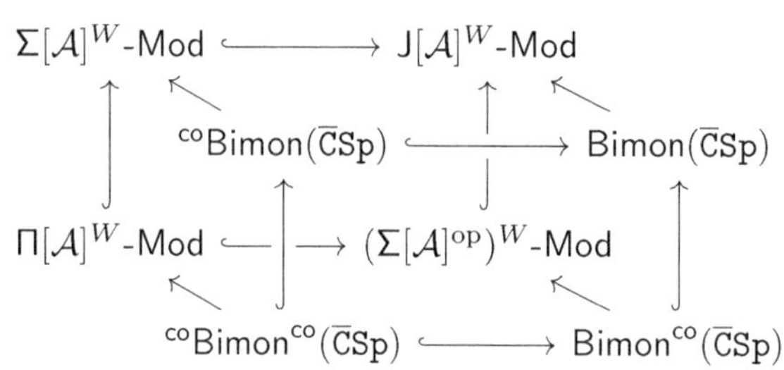

Consider the commutative diagram (1.151) of algebras. All morphisms are surjective. (For claims involving the invariant Birkhoff algebra, we assume $\mathrm{char}(\Bbbk) \nmid |W|$.) Hence, by passing to their module categories, we obtain the back face of the above cube. The front face simply says that every bicommutative Coxeter $\boldsymbol{\alpha}$-bialgebra is a (co)commutative Coxeter $\boldsymbol{\alpha}$-bialgebra which in turn is a Coxeter $\boldsymbol{\alpha}$-bialgebra. The front and back faces are linked by the functors discussed above.

9.14. Antipodes of Coxeter bialgebras

For a scalar q, we define the antipode of a Coxeter q-bialgebra using the Takeuchi formula, and study its basic properties. We compute its logarithm by using the noncommutative Zaslavsky formula. We also provide cancellation-free formulas for antipodes of (co)free Coxeter bialgebras.

The parallel discussion for Coxeter bimonoids is in Section 3.14. For connections between the two using Fock functors, see Section 17.6.

9.14.1. Takeuchi formula. Let $(\mathrm{H}, \mu, \Delta)$ be a Coxeter q-bialgebra for a scalar q. Define its *antipode*

$$S : \mathrm{H} \to \mathrm{H}$$

with Z-component given by the linear map

$$\big(S_Z : \mathrm{H}[Z] \to \mathrm{H}[Z]\big) := \sum_{T:\, T \geq Z} (-1)^{\dim(T)} \mu_Z^T \Delta_Z^T. \tag{9.171}$$

The sum is over all face-types T greater than Z, and $\dim(T)$ is the dimension of any face of type T.

We refer to (9.171) as the *Takeuchi formula*. It may also be expressed as

$$S_Z = (-1)^{\dim(Z)} \mathbf{log}_0(\mathrm{id})_Z, \tag{9.172}$$

where the rhs refers to the 0-logarithm of the identity map (9.114).

We write S_{H} instead of S if we need to show dependence on H.

Lemma 9.128. *For a scalar q, for a Coxeter q-bialgebra* H, *the antipode S is a map of Coxeter spaces. That is, the diagram*

$$\begin{array}{ccc} \mathrm{H}[Z] & \xrightarrow{S_Z} & \mathrm{H}[Z] \\ {\scriptstyle\sigma}\big\downarrow & & \big\downarrow{\scriptstyle\sigma} \\ \mathrm{H}[Z'] & \xrightarrow[S_{Z'}]{} & \mathrm{H}[Z'] \end{array} \tag{9.173}$$

commutes for all morphisms $\sigma : Z \to Z'$.

PROOF. Let $\sigma : Z \to Z'$ and $Z \leq T$. Then, by (8.12) or (1.28), we have the induced morphism $\sigma : T \to T'$, where $T' = \sigma_*(T)$. Employing the naturality axioms in (8.41) and (8.46), we see that the diagram

$$\begin{array}{ccccc} \mathrm{H}[Z] & \xrightarrow{\Delta_Z^T} & \mathrm{H}[T] & \xrightarrow{\mu_Z^T} & \mathrm{H}[Z] \\ {\scriptstyle\sigma}\big\downarrow & & \big\downarrow{\scriptstyle\sigma} & & \big\downarrow{\scriptstyle\sigma} \\ \mathrm{H}[Z'] & \xrightarrow[\Delta_{Z'}^{T'}]{} & \mathrm{H}[T'] & \xrightarrow[\mu_{Z'}^{T'}]{} & \mathrm{H}[Z'] \end{array}$$

commutes. Now multiply by $(-1)^{\dim(T)}$ and sum over all $T \geq Z$. □

9.14.2. Morphisms of Coxeter bialgebras.

Lemma 9.129. *Let q be a scalar. Let $(\mathrm{H}, \mu, \Delta)$ and $(\mathrm{H}', \mu', \Delta')$ be Coxeter q-bialgebras with antipodes S and S', respectively, and $f : \mathrm{H} \to \mathrm{H}'$ be a morphism of Coxeter q-bialgebras. Then the diagram*

$$\begin{array}{ccc} \mathrm{H}[Z] & \xrightarrow{S_Z} & \mathrm{H}[Z] \\ {\scriptstyle f_Z}\big\downarrow & & \big\downarrow{\scriptstyle f_Z} \\ \mathrm{H}'[Z] & \xrightarrow[S'_Z]{} & \mathrm{H}'[Z] \end{array} \tag{9.174}$$

commutes. In other words, a morphism of Coxeter q-bialgebras necessarily commutes with antipodes.

Proof. For any face-type $T \geq Z$, since f is a morphism of Coxeter algebras and Coxeter coalgebras, by (8.42) and (8.47), the diagram

$$\begin{array}{ccccc} \mathrm{H}[Z] & \xrightarrow{\Delta_Z^T} & \mathrm{H}[T] & \xrightarrow{\mu_Z^T} & \mathrm{H}[Z] \\ {\scriptstyle f_Z}\big\downarrow & & \big\downarrow{\scriptstyle f_T} & & \big\downarrow{\scriptstyle f_Z} \\ \mathrm{H}'[Z] & \xrightarrow[(\Delta')_Z^T]{} & \mathrm{H}'[T] & \xrightarrow[(\mu')_Z^T]{} & \mathrm{H}'[Z] \end{array}$$

commutes. Now multiply by $(-1)^{\dim(T)}$ and sum over all $T \geq Z$. □

9.14.3. Duality functor. For a scalar q, for a Coxeter q-bialgebra H,

$$S_{\mathrm{H}^*} = (S_{\mathrm{H}})^*. \tag{9.175}$$

In other words, the dual of the antipode of H is the antipode of H^*.

9.14.4. Signature functor. For a scalar q, the antipode of a Coxeter q-bialgebra H is compatible with the u-signature functor (8.99), that is,

$$S_{\mathrm{H}_-} = (S_{\mathrm{H}})_-. \tag{9.176}$$

9.14.5. Op and cop constructions. We begin by showing that the antipode of a (co)commutative Coxeter bialgebra is a morphism of Coxeter (co)algebras.

Lemma 9.130. *For a commutative Coxeter bialgebra* H, *the antipode* S *is a morphism of Coxeter algebras. That is, for face-types* $T \geq Z$, *the first diagram below commutes.*

$$\begin{array}{ccc} \mathrm{H}[T] & \xrightarrow{S_T} & \mathrm{H}[T] \\ {\scriptstyle \mu_Z^T}\downarrow & & \downarrow{\scriptstyle \mu_Z^T} \\ \mathrm{H}[Z] & \xrightarrow[S_Z]{} & \mathrm{H}[Z] \end{array} \qquad\qquad \begin{array}{ccc} \mathrm{H}[T] & \xrightarrow{S_T} & \mathrm{H}[T] \\ {\scriptstyle \Delta_Z^T}\uparrow & & \uparrow{\scriptstyle \Delta_Z^T} \\ \mathrm{H}[Z] & \xrightarrow[S_Z]{} & \mathrm{H}[Z] \end{array} \tag{9.177}$$

Dually, for a cocommutative Coxeter bialgebra H, *the antipode* S *is a morphism of Coxeter coalgebras. That is, for face-types* $T \geq Z$, *the second diagram above commutes.*

PROOF. We prove the first statement. The required calculation is done below.

$$\begin{aligned} S_Z \mu_Z^T &= \sum_{U:\, U \geq Z} (-1)^{\dim(U)} \mu_Z^U \Delta_Z^U \mu_Z^T \\ &= \sum_{U:\, U \geq Z} \sum_{V:\, V \geq T} (-1)^{\dim(U)} a_Z^{TVU} \mu_Z^V \Delta_T^V \\ &= \sum_{U:\, U \geq Z} \sum_{V:\, V \geq T} (-1)^{\dim(U)} a_Z^{TVU} \mu_Z^T \mu_T^V \Delta_T^V \\ &= \mu_Z^T \Big(\sum_{V:\, V \geq T} \Big(\sum_{U:\, U \geq Z} (-1)^{\dim(U)} a_Z^{TVU} \Big) \mu_T^V \Delta_T^V \Big) \\ &= \mu_Z^T \Big(\sum_{V:\, V \geq T} (-1)^{\dim(V)} \mu_T^V \Delta_T^V \Big) \\ &= \mu_Z^T S_T. \end{aligned}$$

The second equality used identity (8.97) for $q = 1$, the third equality used the associativity axiom in (8.41), the fifth equality used the alternating sign identity (1.63b) for $q = 1$ applied to $\boldsymbol{\alpha}_Z$. □

This result can be reformulated using (9.118) and (9.120) as follows.

Lemma 9.131. *For a commutative Coxeter bialgebra* H, *the antipode* S *is a group-like series of* $\overline{\mathcal{C}}(\mathrm{H}, \mathrm{H})$. *For a cocommutative Coxeter bialgebra* H, *the antipode* S *is a group-like series of* $\mathcal{C}(\mathrm{H}, \mathrm{H})$.

Another way to deduce this result using the group-like property of the Takeuchi series is given later in Exercise 10.104.

Exercise 9.132. Show: For a commutative Coxeter q-bialgebra H, for face-types $T \geq Z$,

$$S_Z \mu_Z^T = q^{\operatorname{dist}(T, Z\overline{T})} \mu_Z^T S_T. \tag{9.178}$$

(Replace a^{TVU} by $a^{TVU}(q)$ in the above calculation.)

We now generalize Lemma 9.130 to Coxeter q-bialgebras for $q \neq 0$. This makes use of the op and cop constructions in Proposition 8.127.

Lemma 9.133. *For $q \neq 0$, let* H *be a Coxeter q-bialgebra with antipode S. Then:*

(1) *$S : {}^{\mathrm{op}}\mathrm{H} \to \mathrm{H}$ is a morphism of Coxeter algebras. Equivalently, for face-types $T \geq Z$, the diagram*

$$
\begin{array}{ccccc}
\mathrm{H}[T] & & \xrightarrow{\quad \mu_Z^T \quad} & & \mathrm{H}[Z] \\
{\scriptstyle S_T}\downarrow & & & & \downarrow{\scriptstyle S_Z} \\
\mathrm{H}[T] & \xrightarrow[(\beta_q)_{Z\overline{T},T}]{} & \mathrm{H}[Z\overline{T}] & \xrightarrow[\mu_Z^{Z\overline{T}}]{} & \mathrm{H}[Z]
\end{array}
\tag{9.179}
$$

commutes.

(2) *$S : \mathrm{H} \to \mathrm{H}^{\mathrm{cop}}$ is a morphism of Coxeter coalgebras. Equivalently, for face-types $T \geq Z$, the diagram*

$$
\begin{array}{ccccc}
\mathrm{H}[Z] & & \xrightarrow{\quad \Delta_Z^T \quad} & & \mathrm{H}[T] \\
{\scriptstyle S_Z}\downarrow & & & & \downarrow{\scriptstyle S_T} \\
\mathrm{H}[Z] & \xrightarrow[\Delta_Z^{Z\overline{T}}]{} & \mathrm{H}[Z\overline{T}] & \xrightarrow[(\beta_{q^{-1}})_{T,Z\overline{T}}]{} & \mathrm{H}[T]
\end{array}
\tag{9.180}
$$

commutes.

For $q = 1$, (9.179) specializes to the first diagram in (9.177) when H is commutative, and (9.180) specializes to the second diagram in (9.177) when H is cocommutative.

PROOF. For diagram (9.179), a sketch of the calculation (with details omitted) is shown below.

$$
\begin{aligned}
S_Z \mu_Z^T &= \sum_{U:\, U \geq Z} (-1)^{\dim(U)} \mu_Z^U \Delta_Z^U \mu_Z^T \\
&= \sum_{V:\, V \geq T} (-1)^{\dim(V)} \mu_Z^{Z\overline{T}V} (\beta_q)_{Z\overline{T}V,V} \Delta_T^V \\
&= \sum_{V:\, V \geq T} (-1)^{\dim(V)} \mu_Z^{Z\overline{T}} \mu_{Z\overline{T}}^{Z\overline{T}V} (\beta_q)_{Z\overline{T}V,V} \Delta_T^V \\
&= \sum_{V:\, V \geq T} (-1)^{\dim(V)} \mu_Z^{Z\overline{T}} (\beta_q)_{Z\overline{T},T} \mu_T^V \Delta_T^V \\
&= \mu_Z^{Z\overline{T}} (\beta_q)_{Z\overline{T},T} S_T.
\end{aligned}
$$

Here $(\beta_q)_{Z\overline{T}V,V}$ denotes the morphism induced from $(\beta_q)_{Z\overline{T},T}$ using $T \leq V$. The second step is critical and can be established using the descent identity [**10**, (7.11a)], [**11**, (1.76a)]. Also relevant is Observation 1.34. Note very carefully: One cannot use identity (8.97) as in the proof of Lemma 9.130. In other words, we do not have a direct simplification of $\mu_Z^U \Delta_Z^U \mu_Z^T$. However, it is still possible to evaluate an alternating sum of such expressions by starting with the q-bialgebra axiom. □

Another approach to prove Lemma 9.133 is given in Exercise 9.140 below. It also builds from the descent identity.

Exercise 9.134. For $q \neq 0$, let H be a Coxeter q-bialgebra. Check:

(1) Antipodes of H, $\mathrm{H}^{\mathrm{op,cop}}$, ${}^{\mathrm{op,cop}}\mathrm{H}$ are all equal.
(2) The antipode of H is a Coxeter q-bialgebra morphism in two ways, namely, $\mathrm{H} \to \mathrm{H}^{\mathrm{op,cop}}$ and ${}^{\mathrm{op,cop}}\mathrm{H} \to \mathrm{H}$. (Use Lemma 9.133.)

Lemma 9.135. *For $q \neq 0$, for* H *a Coxeter q-bialgebra, S_{H} and $S_{{}^{\mathrm{op}}\mathrm{H}}$ (antipodes of* H *and* ${}^{\mathrm{op}}\mathrm{H}$*) are inverses under composition, that is, $S_{\mathrm{H}} S_{{}^{\mathrm{op}}\mathrm{H}} = \mathrm{id} = S_{{}^{\mathrm{op}}\mathrm{H}} S_{\mathrm{H}}$. In particular, S_{H} is invertible. The same result holds replacing* ${}^{\mathrm{op}}\mathrm{H}$ *with* $\mathrm{H}^{\mathrm{cop}}$.

PROOF. We check that $S_{\mathrm{H}} S_{{}^{\mathrm{op}}\mathrm{H}} = \mathrm{id}$; the rest is similar.

$$\begin{aligned}(S_{\mathrm{H}})_Z (S_{{}^{\mathrm{op}}\mathrm{H}})_Z &= \sum_{T:\, T \geq Z} (-1)^{\dim(T)} (S_{\mathrm{H}})_Z \mu_Z^{Z\overline{T}} (\beta_{q^{-1}})_{Z\overline{T},T} \Delta_Z^T \\ &= \sum_{T:\, T \geq Z} (-1)^{\dim(T)} \mu_Z^T (\beta_q)_{T,Z\overline{T}} (S_{\mathrm{H}})_{Z\overline{T}} (\beta_{q^{-1}})_{Z\overline{T},T} \Delta_Z^T \\ &= \sum_{T:\, T \geq Z} (-1)^{\dim(T)} \mu_Z^T (S_{\mathrm{H}})_T \Delta_Z^T \\ &= \sum_{T,U:\, U \geq T \geq Z} (-1)^{\dim(T)} (-1)^{\dim(U)} \mu_Z^U \Delta_Z^U \\ &= \mathrm{id}\,.\end{aligned}$$

In the first step, we used the Takeuchi formula for $(S_{{}^{\mathrm{op}}\mathrm{H}})_Z$. The second step used (9.179). The third step used (9.173) to cancel β_q and $\beta_{q^{-1}}$. The next step used the Takeuchi formula for $(S_{\mathrm{H}})_T$ and the (co)associativity axiom of H. In the last step, we first summed over T which forced $U = Z$. □

As a special case:

Lemma 9.136. *For a commutative or cocommutative Coxeter bialgebra, the antipode S is an involution under composition, that is, $SS = \mathrm{id}$.*

Recall: Op and cop constructions work in the more general setting of Coxeter b-spaces. The preceding results also work in this generality.

9.14.6. Logarithm of the antipode. There is a close connection between logarithm of the identity and logarithm of the antipode.

Lemma 9.137. *For a scalar q, for a Coxeter q-bialgebra* H*, for a face-type Z,*

$$\mathbf{log}(S_{\mathrm{H}})_Z = (-1)^{\dim(Z)} \overline{\mathbf{log}}(\mathrm{id}_{\mathrm{H}})_Z, \tag{9.181}$$

where $\overline{\mathbf{log}}$ refers to logarithm taken wrt $\overline{\boldsymbol{\mu}}$.

PROOF. We calculate:

$$\mathbf{log}(S_{\mathrm{H}})_Z = \sum_{T:\, T \geq Z} \boldsymbol{\mu}(Z,T)\, \mu_Z^T (S_{\mathrm{H}})_T \Delta_Z^T$$

$$= \sum_{U:\,U\geq Z} (-1)^{\dim(U)} \Big(\sum_{T:\,U\geq T\geq Z} \mu(Z,T) \Big) \mu_Z^U \Delta_Z^U$$
$$= (-1)^{\dim(Z)} \sum_{U:\,U\geq Z} \mu(Z, Z\overline{U})\, \mu_Z^U \Delta_Z^U$$
$$= (-1)^{\dim(Z)}\, \overline{\mathbf{log}}(\mathrm{id}_{\mathrm{H}})_Z.$$

The first and last steps used definitions (9.90b) and (9.109). The second step used the Takeuchi formula (9.171) and the (co)associativity axiom. The next step is critical and used identity (1.121b) related to the noncommutative Zaslavsky formula. □

Another way to deduce this result using logarithm of the Takeuchi series is given later in Exercise 10.105.

9.14.7. Antipodes of (co)free Coxeter bialgebras. We now write down cancellation-free formulas for antipodes of the universal constructions in Section 9.3. Interestingly, they involve the antipodal map on face-types.

Theorem 9.138. *For a scalar q, for a Coxeter coalgebra* C*, the antipode S of the Coxeter q-bialgebra $\overline{\mathcal{T}}_q(\mathrm{C})$ on the Z-component, on the T-summand, is*

$$S_Z = q^{\operatorname{dist}(T, Z\overline{T})} \sum_{U:\, Z\overline{T}\leq U} (-1)^{\dim(U)} \Delta^U_{Z\overline{T}} \beta_{Z\overline{T},T}. \tag{9.182}$$

In particular, for a trivial Coxeter coalgebra C*, on the T-summand,*

$$S_Z = (-1)^{\dim(T)} q^{\operatorname{dist}(T, Z\overline{T})}\, \beta_{Z\overline{T},T}. \tag{9.183}$$

Sketch of proof. We indicate the main steps. Start with the Takeuchi formula for $\overline{\mathcal{T}}_q(\mathrm{C})$. It involves (9.22) along with a sum over T' with a coefficient of $(-1)^{\dim(T')}$. Use Observation 1.34 and evaluate the sum using the descent identity [**10**, (7.11a)], [**11**, (1.76a)]. Only morphisms σ induced from $\beta_{Z\overline{T},T}$ survive. □

Dually:

Theorem 9.139. *For a scalar q, for a Coxeter algebra* A*, the antipode S of the Coxeter q-bialgebra $\overline{\mathcal{T}}_q^{\vee}(\mathrm{A})$ on the Z-component, on the U-summand, is*

$$S_Z = (-1)^{\dim(U)} \sum_{T:\, Z\leq T\leq Z\overline{U}} q^{\operatorname{dist}(T, Z\overline{T})}\, \beta_{T, Z\overline{T}} \mu^U_{Z\overline{T}}. \tag{9.184}$$

In particular, for a trivial Coxeter algebra A*, on the U-summand,*

$$S_Z = (-1)^{\dim(U)} q^{\operatorname{dist}(U, Z\overline{U})}\, \beta_{Z\overline{U},U}. \tag{9.185}$$

Exercise 9.140. For a scalar q, for a Coxeter q-bialgebra H, let $\mu : \overline{\mathcal{T}}_q(\mathrm{H}) \to \mathrm{H}$ and $\Delta : \mathrm{H} \to \overline{\mathcal{T}}_q^{\vee}(\mathrm{H})$ be the Coxeter q-bialgebra morphisms in Exercise 9.29. By Lemma 9.129, they commute with antipodes. Now use Theorems 9.138 and 9.139 to deduce Lemma 9.133.

Lemma 9.129 can also be applied to the morphisms $\overline{\mathcal{T}}(\mathrm{C}) \twoheadrightarrow \overline{\mathcal{S}}(\mathrm{C})$ and $\overline{\mathcal{S}}^{\vee}(\mathrm{A}) \hookrightarrow \overline{\mathcal{T}}^{\vee}(\mathrm{A})$. Thus, Theorems 9.138 and 9.139 for $q = 1$ also give information about antipodes of $\overline{\mathcal{S}}(\mathrm{C})$ and $\overline{\mathcal{S}}^{\vee}(\mathrm{A})$.

9.14.8. Antipode problem. Writing down a cancellation-free formula for the antipode of a specific Coxeter bialgebra in a given basis is a nontrivial problem. Some approaches are given below.

- The direct method is to substitute (co)product formulas of the given Coxeter bialgebra into the Takeuchi formula, and understand the resulting cancellations using for instance the alternating sign identities from Section 1.8.10.
- Once we know some antipode formulas, we can derive more by using compatibilities such as (9.174), (9.175), (9.176).
- If the given Coxeter bialgebra is free or cofree, then we can use Theorem 9.138 or 9.139 (which was established by the direct method).

Concrete examples of antipode formulas are given in Chapter 10; see in particular Sections 10.2.19, 10.3.20, 10.4.21, 10.5.9, 10.6.7.

9.15. Notes

The theory developed in this chapter is new. It generalizes the classical theory of connected graded Hopf algebras. We recover it by specializing to the braid arrangements (Section 8.14). Some important sources in the classical literature are mentioned below. More detailed references can be found in [**11**, Notes to chapters in Part II].

9.15.1. Cauchy powers and tensor powers. (Section 9.1.) For a vector space V, for $k \geq 1$, we have its k-th tensor power $V^{\otimes k}$. This is the classical analogue of the k-th Cauchy power of a Coxeter space (9.1). For a coalgebra C, we have iterated coproduct maps $\Delta^{k-1} : C \to C^{\otimes k}$, while for an algebra A, we have iterated product maps $\mu^{k-1} : A^{\otimes k} \to A$. These are classical analogues of maps (9.2) and (9.3). See also Notes 8.15.2 for a related discussion.

For a vector space V, the symmetric group S_k acts on $V^{\otimes k}$ by permuting the tensor factors yielding canonical linear maps

$$(V^{\otimes k})^{\mathrm{S}_k} \hookrightarrow V^{\otimes k} \twoheadrightarrow (V^{\otimes k})_{\mathrm{S}_k}.$$

This is the classical analogue of the maps in (9.6). Thus, invariants and coinvariants correspond to the two versions of commutative Cauchy powers of a Coxeter space V, namely, $\mathrm{V}^{\overline{k}}$ and $\mathrm{V}^{\underline{k}}$.

A comparison summary is provided in Table 9.5. For a more accurate comparison, we need to use graded vector spaces. Note in this regard that the tensor product of graded vector spaces is called the Cauchy product in [**7**, Section 2.1.1].

9.15.2. Primitive and decomposable filtrations. (Section 9.2.) Classical nondecomposable and primitive elements are present in work of Hopf [**191**, Section 14] and Leray [**227**, Sections 24 and 25] under the names "maximal" and "hypermaximal" elements, respectively. The term "primitive element" was used by Koszul [**214**, Section 10] and Borel [**54**, Formula (20.1)], [**55**, Formula (2.3)], [**56**, page 404].

The primitive filtration of a coalgebra appeared (in various levels of generality) in work of Cartier [**78**, Section 2.1, Formula (9)], [**79**, Section 2], Browder [**63**, page 155], Milnor and Moore [**265**, Definition 5.10], Sweedler [**345**, Section 9.1]. Early references for the decomposable filtration of an algebra (and its duality to the primitive filtration) are by Milnor and Moore [**265**, Definition 7.1], Larson [**220**, Definition 13.1], Browder [**63**, page 155], Cartier [**78**, bottom of page 2-05].

TABLE 9.5. Cauchy powers and tensor powers.

Coxeter space V	vector space V
k-th Cauchy power V^k	k-th tensor power $V^{\otimes k}$
$\mathrm{V}^{\overline{k}}$ and $\mathrm{V}^{\underline{k}}$	$(V^{\otimes k})^{\mathrm{S}_k}$ and $(V^{\otimes k})_{\mathrm{S}_k}$
Coxeter coalgebra C	coalgebra C
$\Delta^{k-1} : \mathrm{C} \to \mathrm{C}^k$	$\Delta^{k-1} : C \to C^{\otimes k}$
Coxeter algebra A	algebra A
$\mu^{k-1} : \mathrm{A}^k \to \mathrm{A}$	$\mu^{k-1} : A^{\otimes k} \to A$

For more detailed references on filtrations, see [**11**, Notes to Chapter 5].

9.15.3. Universal constructions. (Section 9.3.) The universal constructions on a Coxeter space V, Coxeter algebra A, Coxeter coalgebra C generalize corresponding constructions on a positive-graded vector space V, positive-graded algebra A, positive-graded coalgebra C. The latter are summarized in [**11**, Tables 6.2 and 6.3]. Our terminology of (de)concatenation and (de)quasishuffle for the (co)products (see Table 9.2) is carried over from the classical terminology. The special case when the Coxeter space V is concentrated in the S-component (where S is the chamber-type) is treated in Section 10.3. It generalizes the case when V is a positive-graded vector space concentrated in degree one. Related information on the tensor and shuffle Hopf algebras is given in Notes 10.7.4, see Table 10.9 in particular. References for constructions involving quasishuffles (which are more general) are given below.

(De)quasishuffle Hopf algebras. These Hopf algebras appeared in early work of Moore [**272**, pages 1–3] and Newman [**277**, Remarks 2.3 and 2.5], [**278**, Sections 1.13 and 1.14] (without any mention of quasishuffles). Newman and Radford described the quasishuffle product [**279**, Remark on page 1032]. The quasishuffle product had been considered earlier by Cartier in the context of Baxter algebras [**82**, Formulas (7), (8), (9)]. The quasishuffle Hopf algebra was studied later by Hoffman [**189**, Theorem 3.1]. See also the papers by Hazewinkel [**178**, Section 12] and Guo and Keigher [**167**, Theorem 3.5]. References for the example of the Hopf algebra of quasisymmetric functions are given in Notes 10.7.6, see also Table 10.10.

In the context of stochastic integrals, quasishuffles appeared in work of Gaines [**140**, Proposition 2.3], [**141**, Proposition 3], Li and Liu [**228**, Lemma 2.2], and in work of Hudson with Parthasarathy, Cohen, Eyre [**194**, Formula (4.3) and remarks after its proof], [**89**, Formula (1.3)], [**124**, Theorem 5.1].

For more detailed references on universal constructions of Hopf algebras, see [**11**, Notes to Chapter 6].

9.15.4. Abelianization. (Section 9.4.) Abelianization of an algebra to obtain a commutative algebra is a well-known classical construction. The dual construction of coabelianization of a coalgebra, or extensions of these to bialgebras are not easy to locate in the literature.

9.15.5. Hadamard product. (Sections 9.5, 9.6, 9.7.) The Hadamard product of Coxeter spaces (9.59) generalizes the Hadamard product of graded vector spaces. The latter is considered for instance in [**7**, Formula (2.3)], [**234**, Section 5.9.1], [**317**, Section 5.1.3]. Products with a similar flavor originated in work of Hadamard [**169**,

Formula (3)] and Schur [**322**, Section VII]. The internal hom (9.60) generalizes a similar formula given in [**7**, Section 2.1.5].

Convolution algebra. The convolution Coxeter algebra in Section 9.5.4 corresponds to the convolution algebra in classical Hopf theory. Early references for the latter are by Milnor and Moore [**265**, Proposition 8.2], [**266**, Proposition 6.3], Kostant [**213**, Section 1.1], Heyneman and Sweedler [**342**, page 14], [**345**, Section 4.0], [**182**, Section 1.5], Radford [**299**, Section 1.1], Hochschild [**187**, page 17], [**188**, page 5].

Internal hom for coalgebras. The Coxeter coalgebra $\mathcal{C}(\mathrm{C}, \mathrm{D})$ corresponds to the internal hom for classical coalgebras given for instance by Porst [**297**, Section 3.2]. The case of cocommutative coalgebras goes back to Barr [**28**, Theorem 5.3]. This is the classical analogue of Exercise 9.43.

Universal measuring coalgebra. The universal measuring Coxeter coalgebra $\overline{\mathcal{C}}(\mathrm{A}, \mathrm{B})$ corresponds to the universal measuring coalgebra introduced by Sweedler [**344**, Section 1], [**345**, Section 7.0]. Enrichment of algebras over coalgebras is implicit in Sweedler's work. It is explicitly mentioned by Barr [**28**, Sections 6 and 7] and Fox [**128**, Chapter III], [**129**, Sections 2 and 3]. The relations (9.84) that define $\mathrm{C} \triangleright \mathrm{A}$, the copower of A by C, correspond to the classical relations given by Fox [**130**, page 69] and much earlier by Wraith [**363**, Relations J.1 on page 154]. The Coxeter bialgebra $\overline{\mathcal{C}}(\mathrm{H}, \mathrm{A})$ in Section 9.7.2 corresponds to the bialgebra studied by Grünenfelder and Mastnak [**253**, Section 3], [**165**, Proposition 2.3], [**166**]. Exercise 9.53 about its q-deformation looks new even from the classical perspective.

For more detailed references on the Hadamard product and measurings, see [**11**, Notes to Chapters 8 and 15].

9.15.6. Exp-log correspondences. (Sections 9.8, 9.9, 9.10.) Invariant noncommutative zeta and Möbius functions generalize the classical exponential and logarithm functions. This is why exp-log correspondences are formulated in terms of zeta and Möbius functions.

(Co)derivations and (co)algebra morphisms. Theorem 9.56 relates to the classical fact that exponential (under the convolution product) of an ϵ-(co)derivation is a (co)algebra morphism. This fact is explained by Nichols [**280**, Chapter III, Theorem 6], [**281**, Lemma 7]. A treatment in the language of algebraic groups is given by Demazure and Gabriel [**96**, Section II.6, 3.7 and Section IV.2, Proposition 4.1].

Warning. The exponential of an endomorphism can also be calculated using composition powers rather than convolution powers. For the precise connection between the two points of view, see for instance [**188**, Chapter I, Proposition 2.1] or more generally [**212**, Proof of Proposition 2.1], [**269**, Lemma in Section 6.2].

Logarithm of the identity map. The map (9.108) generalizes the classical logarithm of the identity map which was studied by Nichols [**280**, page 51], [**281**, page 67]. Some later references are by Reutenauer [**304**, page 269], [**305**, pages 58 and 60], Garsia [**143**, page 362], Loday [**232**, Formula (4.5.2.5)], Patras [**287**, Theorem 3.5].

Zeta and Möbius series. Note very carefully: Theorems 9.56 and 9.76 yield a family of exp-log correspondences, one for each $\boldsymbol{\zeta}$ and $\boldsymbol{\mu}$. The same remark applies to logarithm of the identity map (9.108). This flexibility can be understood in the classical setting by employing notions of Section 1.16.5. Any zeta and Möbius series yields a classical exp-log correspondence (with (1.177) being the standard example). This fact (with emphasis on the Dynkin series (1.178)) was pointed out by Ebrahimi-Fard, Gracia-Bondía, Manchon, Patras [**108**, Theorem 4.1], [**250**, Theorem 6], [**111**,

Theorem 2.1]. They work with Lie idempotent series which are equivalent to Zie series; they, however, do not make contact with the notion of zeta or Möbius series.

Primitives and group-likes. Theorem 9.76 relates to the exp-log correspondence between primitives and group-likes given by Quillen [**298**, Appendix A, Proposition 2.6]. Quillen worked with the exponential and logarithmic power series, or equivalently, the zeta and Möbius series in (1.177). Related ideas are present in earlier work of Ree [**302**, (2.1.1) and (2.2.2)] and Hochschild [**186**, pages 109–110].

q-analogue. Theorem 9.61, q-logarithm of the identity (9.113), Theorem 9.80 look new even in the classical setting; we will give more detail in a separate exposition.

For additional references on exp-log, see [**11**, Notes to Chapter 9].

9.15.7. Norm map. (Section 9.11.) For a vector space V, the q-norm map $\kappa_q : \mathcal{T}_q(V) \to \mathcal{T}_q^\vee(V)$ is given in [**7**, Formula (2.67)]; for the case $q = 1$, see also [**33**, Sections 2.3.4 and 2.4.5]. This map is generalized by (9.140) when V is a positive-graded vector space. In a similar sense, diagrams (9.141) generalize [**7**, Diagrams (2.66) and (2.68)]. The fact that κ_q is an isomorphism when q is not a root of unity is shown by Duchamp, Klyachko, Krob, Thibon [**107**, Proposition 4.5], this is generalized by Lemma 9.101. For a concrete example of the norm map, see Notes 10.7.8. The material in Sections 9.11.3 and 9.11.4 looks new even in the classical setting.

9.15.8. Bialgebra axiom on the primitive part. (Section 9.12.) This material looks new even from a classical perspective, where the emphasis is on binary operations as opposed to higher operations, see Section 8.5.9. Lemma 9.119 is the only result related to binary operations. Results similar to those in this section for species were given in [**11**, Section 5.6]; they also hold for Coxeter species.

9.15.9. Characteristic operations. (Section 9.13.) Historically, characteristic operations appeared in a topological setting in work of Leray under the heading "puissances" [**227**, page 135].

Characteristic operations on Coxeter bialgebras in (9.156) generalize characteristic operations on graded Hopf algebras. The latter were considered by Patras [**288**, Definition II,1]. For a (co)commutative graded Hopf algebra H, for α a composition of n and $z \in \mathrm{H}_n$, set

$$\mathtt{H}_\alpha \cdot z := \mu^\alpha \Delta^\alpha(z), \tag{9.186}$$

where μ^α and Δ^α are iterated product and coproduct components of H. This defines an action on H_n of the Solomon descent algebra (isomorphic to the opposite of the invariant Tits algebra) for the braid arrangement on n letters; see [**288**, Theorem II,7 and subsequent remarks]. The special case of the tensor Hopf algebra $\mathcal{T}(V)$ is present in Reutenauer's book [**305**, Definition (9.1.2) and Theorem 9.2] with the general case hinted at in [**305**, Appendix 9.5.1]. The special case of the Hopf algebra of noncommutative symmetric functions NSym is mentioned in [**149**, Section 5.1], see also [**305**, Map (9.1.4)]. This corresponds to Example 9.121. Classical analogues of identities (9.160) and (9.162) for NSym are given in [**149**, Propositions 5.2 and 5.5].

The classical analogue of Proposition 9.125 for the tensor Hopf algebra appears in some form in [**305**, Lemma 8.15], [**149**, Corollary 5.17], [**215**, Theorem 3.1].

Characteristic operations also exist in the nongraded case, see [**151**, Section 1], [**287**, Definition 1.2]. They are called "power maps" in [**206**], [**229**, page 348], "Sweedler powers" in [**207**, Definition 2.1], "Hopf powers" in [**219**, Section 1], [**270**, Section 3], "Adams operators" in [**83**, page 578], [**5**, Section 1.2]. The classical analogue of the two-sided characteristic operations (9.165) for $q = 1$ was considered by Kashina, Sommerhäuser, Zhu [**208**, Section 1.3]. For additional information on characteristic operations, see [**11**, Notes to Chapter 10].

9.15.10. Antipode. (Section 9.14.) The notion of antipode of a Hopf algebra appeared in the book of Cartan and Eilenberg [**75**, Section XI.8]. It was later studied by Milnor and Moore [**271**, page 12, Theorem 5], [**265**, Proposition 8.2 and Definition 8.4], [**266**, Definition 6.9]. It is mentioned under different names in the early literature. The term "antipode" is now standard.

For a connected Hopf algebra, the antipode always exists. The Takeuchi formula for the antipode of a connected Hopf algebra appeared in work of Sweedler [**345**, Proof of Lemma 9.2.3] and Takeuchi [**347**, Proof of Lemma 14]. For later references, see for instance [**6**, Formula (3.2)], [**7**, Formula (2.55)], [**250**, Formula (18)], [**269**, Lemma 5.2.10], [**301**, Proposition 6.2.2 and Lemma 7.6.2], [**320**, page 314]. Compare with the formula (9.171).

For more detailed references on the antipode, see [**11**, Notes to Chapter 12].

CHAPTER 10

Examples of Coxeter bialgebras

We discuss some basic and important examples of Coxeter bialgebras. We develop them from first principles. They serve to understand the concepts introduced in Chapter 8. We also use them to illustrate the theory developed in Chapter 9. In particular, we discuss their self-duality, (co)freeness and universal properties, primitive and indecomposable parts, primitive and group-like series, and antipodes. We will primarily work with the face-type formulation, and also highlight connections to face-type enumeration.

Coxeter bialgebra of polynomials. We begin with the Coxeter bialgebra of polynomials $\Bbbk[x]$ and its dual which is the divided power Coxeter bialgebra $\Bbbk\{x\}$. Both are bicommutative. These Coxeter bialgebras admit q-deformations which we denote $\Bbbk_q[x]$ and $\Bbbk_q\{x\}$.

Tensor and shuffle Coxeter bialgebras. For a Coxeter group W, a left W-module M can be equivalently viewed as a Coxeter space X_M concentrated in the S-component. Starting with a left W-module M, we define the tensor Coxeter bialgebra $\overline{\mathcal{T}}(M)$ and shuffle Coxeter bialgebra $\overline{\mathcal{T}}^{\vee}(M)$, and their q-deformations $\overline{\mathcal{T}_q}(M)$ and $\overline{\mathcal{T}_q}^{\vee}(M)$. (We recover $\Bbbk_q[x]$ and $\Bbbk_q\{x\}$ when $M = \Bbbk$ with trivial W-action.) The tensor Coxeter bialgebra is cocommutative, while the shuffle Coxeter bialgebra is commutative. Related examples include the symmetric Coxeter bialgebras $\overline{\mathcal{S}}(M)$ and $\overline{\mathcal{S}}^{\vee}(M)$, and exterior signed Coxeter bialgebras $\overline{\mathcal{E}}(M)$ and $\overline{\mathcal{E}}^{\vee}(M)$.

Coxeter bialgebras of face-types. We introduce the cocommutative Coxeter bialgebra of face-types $\overline{\Sigma}$. It is the free Coxeter bialgebra on the divided power Coxeter coalgebra $\Bbbk\{x\}$. Further, it has close connections to the invariant Tits algebra. We also discuss the dual commutative Coxeter bialgebra of face-types $\overline{\Sigma}^*$. These Coxeter bialgebras admit q-deformations which we denote $\overline{\Sigma}_q$ and $\overline{\Sigma}_q^*$. For q not a root of unity, these two Coxeter q-bialgebras are isomorphic, and in fact we classify all isomorphisms between them.

Coxeter bialgebra of flat-types. We introduce the bicommutative Coxeter bialgebra of flat-types $\overline{\Pi}$. It is the abelianization of $\overline{\Sigma}$. Thus, it is the free commutative Coxeter bialgebra on the divided power Coxeter coalgebra $\Bbbk\{x\}$. It has close connections to the invariant Birkhoff algebra. Dually, we consider the bicommutative Coxeter bialgebra of flat-types $\overline{\Pi}^*$ which is the coabelianization of $\overline{\Sigma}^*$. The Coxeter bialgebra $\overline{\Pi}$ is self-dual, and in fact we classify all isomorphisms between $\overline{\Pi}$ and $\overline{\Pi}^*$.

Coxeter bialgebra of symmetries. We introduce the Coxeter bialgebra of symmetries $\overline{\mathrm{I\Gamma}}$. It is neither commutative nor cocommutative. It is self-dual, free as a Coxeter algebra and cofree as a Coxeter coalgebra. It admits a q-deformation which we denote $\overline{\mathrm{I\Gamma}}_q$. The Coxeter q-bialgebra of face-types $\overline{\Sigma}_q$ embeds inside $\overline{\mathrm{I\Gamma}}_q$, and dually, $\overline{\Sigma}_q^*$ is a quotient of $\overline{\mathrm{I\Gamma}}_q^*$.

The above examples of Coxeter bialgebras can be seen in analogy with the examples of Coxeter bimonoids in Chapter 4. The precise connections between the two using Fock functors are explained in Chapter 18.

10.1. Coxeter space characteristic of chambers

We begin with the Coxeter space X characteristic of chambers, and its decorated version X_M, where M is a left W-module. They play a basic role in the study of the Coxeter bialgebra of polynomials and the more general tensor Coxeter bialgebra to be discussed in Sections 10.2 and 10.3. The construction of X_M is functorial in M and embeds the category of left W-modules as a reflective and coreflective full subcategory of Coxeter spaces.

The parallel example of the Coxeter species x characteristic of chambers is given in Section 4.1. The connection between the two is made precise later in Section 18.1 using Fock functors.

10.1.1. Coxeter space characteristic of chambers. Define the Coxeter $\mathcal{A}$-space X by

$$\mathrm{X}[F,C] := \begin{cases} \Bbbk & \text{if } F = C, \\ 0 & \text{otherwise.} \end{cases} \tag{10.1}$$

Support and type morphisms are defined to be all identities, so diagrams (8.20a) and (8.20b) commute. (Note: On the zero space, the identity map is the same as the zero map.) This is the *Coxeter space characteristic of chambers*. It can be phrased in a similar manner in terms of top-lunes; the relevant diagrams are (8.23).

In terms of face-types, define the Coxeter $\boldsymbol{\alpha}$-space X by

$$\mathrm{X}[T] := \begin{cases} \Bbbk & \text{if } T = S, \\ 0 & \text{otherwise,} \end{cases} \tag{10.2}$$

with structure maps defined to be identities. The relevant diagrams are (8.28).

Exercise 10.1. Check: For a Coxeter space V,

$$\boldsymbol{\alpha}\text{-}\overline{\mathsf{C}}\mathsf{Sp}(\mathrm{X},\mathrm{V}) = \mathrm{V}[S]^W \quad \text{and} \quad \boldsymbol{\alpha}\text{-}\overline{\mathsf{C}}\mathsf{Sp}(\mathrm{V},\mathrm{X}) = (\mathrm{V}[S]_W)^*.$$

By Exercise 8.38, $\mathrm{V}[S]$ is a W-module, and $\mathrm{V}[S]^W$ and $\mathrm{V}[S]_W$ denote its W-invariants and W-coinvariants, respectively. In particular, Coxeter space maps from X to itself are scalar multiples of identity.

10.1.2. Decorated version. More generally, for a left W-module M, define the Coxeter $\mathcal{A}$-space X_M by

$$\mathrm{X}_M[F, C] := \begin{cases} M & \text{if } F = C, \\ 0 & \text{otherwise.} \end{cases} \tag{10.3}$$

Any support morphism β is the identity map, while any type morphism w is either zero or given by action of w on M. In other words,

$$\begin{array}{ccc} \mathrm{X}_M[C, C] & \xrightarrow{\beta_{D,D,C,C}} & \mathrm{X}_M[D, D] \\ \| & & \| \\ M & \xrightarrow[\mathrm{id}]{} & M \end{array} \qquad \begin{array}{ccc} \mathrm{X}_M[C, C] & \xrightarrow{w(C)=D} & \mathrm{X}_M[D, D] \\ \| & & \| \\ M & \xrightarrow[w\cdot(-)]{} & M \end{array}$$

and the remaining maps are zero.

Exercise 10.2. There is another way to define X_M on top-nested faces. To avoid confusion, we denote it by X'_M. We start with (10.3), but support and type morphisms are

$$\begin{array}{ccc} \mathrm{X}'_M[C, C] & \xrightarrow{\beta_{D,D,C,C}} & \mathrm{X}'_M[D, D] \\ \| & & \| \\ M & \xrightarrow[d(D,C)\cdot(-)]{} & M \end{array} \qquad \begin{array}{ccc} \mathrm{X}'_M[C, C] & \xrightarrow{w(C)=D} & \mathrm{X}'_M[D, D] \\ \| & & \| \\ M & \xrightarrow[\mathrm{id}]{} & M \end{array}$$

where d is the W-valued gallery distance in (1.12) (which depends on C_r). Check: X'_M is indeed a Coxeter space. Moreover,

$$\mathrm{X}'_M[C, C] \xrightarrow{u\cdot(-)} \mathrm{X}_M[C, C],$$

with $u(C_r) = C$, defines an isomorphism of Coxeter spaces from X'_M to X_M.

In terms of face-types, define the Coxeter $\boldsymbol{\alpha}$-space X_M by

$$\mathrm{X}_M[T] := \begin{cases} M & \text{if } T = S, \\ 0 & \text{otherwise.} \end{cases} \tag{10.4}$$

For a morphism $\sigma : S \to S$, the linear map $\sigma : \mathrm{X}_M[S] \to \mathrm{X}_M[S]$ is given by action of w on M, where $\sigma = w\beta = \beta w$ in normal form. (This w is the same as the Coxeter symmetry w_σ in Section 1.7.3.) The remaining maps are zero.

When M is the trivial representation of W, the Coxeter space X_M specializes to X. We view X_M as a decorated version of X, with M as the space of decorations.

10.1.3. Categories of W-modules and Coxeter spaces. The construction of X_M is functorial in M. That is, a map $M \to M'$ of left W-modules gives rise to a map $\mathrm{X}_M \to \mathrm{X}_{M'}$ of Coxeter $\boldsymbol{\alpha}$-spaces. This yields a functor from the category of left W-modules to the category of Coxeter $\boldsymbol{\alpha}$-spaces

$$W\text{-}\mathsf{Mod} \to \boldsymbol{\alpha}\text{-}\overline{\mathsf{C}}\mathsf{Sp}, \qquad M \mapsto \mathrm{X}_M. \tag{10.5}$$

Moreover, this functor is full and faithful. That is, a map $f : \mathrm{X}_M \to \mathrm{X}_{M'}$ of Coxeter $\boldsymbol{\alpha}$-spaces is the same as a map $f : M \to M'$ of left W-modules. See the diagrams below.

$$\begin{array}{ccc} \mathrm{X}_M[S] & \xrightarrow{f_S} & \mathrm{X}_{M'}[S] \\ {\scriptstyle\sigma}\downarrow & & \downarrow{\scriptstyle\sigma} \\ \mathrm{X}_M[S] & \xrightarrow[f_S]{} & \mathrm{X}_{M'}[S] \end{array} \qquad\qquad \begin{array}{ccc} M & \xrightarrow{f} & M' \\ {\scriptstyle w\cdot(\ )}\downarrow & & \downarrow{\scriptstyle w\cdot(\ )} \\ M & \xrightarrow[f]{} & M' \end{array}$$

Here σ and w are related by $\sigma = w\beta = \beta w$. Note: $f_T = 0$ for $T \neq S$.

Exercise 10.3. Recall the functor (8.30) from Coxeter $\boldsymbol{\alpha}$-spaces to left W-modules which sends V to $\mathrm{V}[S]$. It sends X_M to M. Show: The functor (10.5) is both the left and right adjoint of (8.30). That is,

- a map of Coxeter $\boldsymbol{\alpha}$-spaces $\mathrm{X}_M \to \mathrm{V}$ is the same as a map of left W-modules $M \to \mathrm{V}[S]$,
- a map of Coxeter $\boldsymbol{\alpha}$-spaces $\mathrm{V} \to \mathrm{X}_M$ is the same as a map of left W-modules $\mathrm{V}[S] \to M$.

For $M = \Bbbk$ with trivial W-action, we recover the result of Exercise 10.1.

The functor (10.5) yields an isomorphism of the category W-Mod with the (reflective and coreflective) full subcategory of $\boldsymbol{\alpha}$-$\overline{\mathsf{C}}\mathsf{Sp}$ consisting of those Coxeter $\boldsymbol{\alpha}$-spaces V for which $\mathrm{V}[T] = 0$ whenever $T \neq S$.

Observation 10.4. For a Coxeter coalgebra C and Coxeter algebra A, any map of Coxeter spaces $f : \mathrm{X}_M \to \mathrm{C}$ is a coderivation, and any map of Coxeter spaces $g : \mathrm{A} \to \mathrm{X}_M$ is a derivation.

Exercise 10.5. Check: $\mathrm{X}^* = \mathrm{X}$. More generally, for a left W-module M, we have $(\mathrm{X}_M)^* = \mathrm{X}_{M^*}$.

Exercise 10.6. Recall from Section 8.10 the u-signature functor $(-)_-$ on Coxeter $\boldsymbol{\alpha}$-spaces (in terms of face-types). Check: $(\mathrm{X}_M)_- = \mathrm{X}_{M_-}$. In terms of top-nested faces, we have an isomorphism $(\mathrm{X}_M)_- \cong \mathrm{X}_{M_-}$ which on the (C, C)-component is scalar multiplication by $(-1)^{\mathrm{dist}(C_0, C)}$ for a fixed chamber C_0.

10.2. Coxeter bialgebra of polynomials

We now introduce an important basic example of a Coxeter bialgebra. Each component of this Coxeter bialgebra equals $\Bbbk$. Any canonical basis element $1 \in \Bbbk$ plays the same role that a tensor product of monomials plays for the polynomial algebra, so we call this the Coxeter bialgebra of polynomials, and denote it by $\Bbbk[x]$. We also discuss some other close relatives; they are summarized in Table 10.1. Their basic properties are listed in Table 10.2. Here X refers to the Coxeter space characteristic of chambers from Section 10.1.

We formulate these Coxeter bialgebras in terms of face-types. The face-type enumerator and its q-analogue from Section 1.8 (which includes the Poincaré polynomial) enter into the (co)product formulas. Algebraic properties of the product and coproduct translate to properties of the Poincaré polynomial.

The Coxeter bialgebras $\Bbbk[x]$ and $\Bbbk\{x\}$ are duals of each other. More generally, for a scalar q, the Coxeter q-bialgebras $\Bbbk_q[x]$ and $\Bbbk_q\{x\}$ are duals of each other. The q-norm map $\overline{\kappa}_q$ which links them is self-dual, and hence its image, namely, $\Bbbk_q(x)$ is a self-dual Coxeter q-bialgebra. For q not a root

TABLE 10.1. Coxeter bialgebra of polynomials and its relatives.

1	$\Bbbk[x]$	Coxeter bialgebra of polynomials
	$\Bbbk\{x\}$	divided power Coxeter bialgebra
	$\Bbbk(x)$	image of the norm map $\bar{\kappa} : \Bbbk[x] \to \Bbbk\{x\}$
q	$\Bbbk_q[x]$	Coxeter q-bialgebra of polynomials
	$\Bbbk_q\{x\}$	divided power Coxeter q-bialgebra
	$\Bbbk_q(x)$	image of the q-norm map $\bar{\kappa}_q : \Bbbk_q[x] \to \Bbbk_q\{x\}$
-1	$\Bbbk_{-1}[x]$	signed Coxeter bialgebra of polynomials
	$\Bbbk_{-1}\{x\}$	divided power signed Coxeter bialgebra
	$\Bbbk[dx] = \Bbbk_{-1}(x)$	signed Coxeter bialgebra of dual numbers
0	$\Bbbk_0[x] = \Bbbk_0\{x\} = \Bbbk_0(x)$	Coxeter 0-bialgebra of polynomials

TABLE 10.2. Properties of the Coxeter bialgebra of polynomials.

Coxeter bialgebra	universal property	commutativity
$\Bbbk[x]$	$\overline{\mathcal{T}}(\mathrm{X}) = \overline{\mathcal{S}}(\mathrm{X})$	bicommutative
$\Bbbk\{x\}$	$\overline{\mathcal{T}}^\vee(\mathrm{X}) = \overline{\mathcal{S}}^\vee(\mathrm{X})$	bicommutative
$\Bbbk_q[x]$	$\overline{\mathcal{T}_q}(\mathrm{X})$	commutative
$\Bbbk_q\{x\}$	$\overline{\mathcal{T}_q}^\vee(\mathrm{X})$	cocommutative
$\Bbbk_{-1}[x]$	$\overline{\mathcal{T}}_{-1}(\mathrm{X})$	com. & signed cocom.
$\Bbbk_{-1}\{x\}$	$\overline{\mathcal{T}}^\vee_{-1}(\mathrm{X})$	signed com. & cocom.
$\Bbbk[dx]$	$\overline{\mathcal{E}}(\mathrm{X}) = \overline{\mathcal{E}}^\vee(\mathrm{X})$ if $\mathrm{char}(\Bbbk) \neq 2$	signed bicommutative
$\Bbbk_0[x] = \Bbbk_0\{x\}$	$\overline{\mathcal{T}_0}(\mathrm{X}) = \overline{\mathcal{T}_0}^\vee(\mathrm{X})$	bicommutative

of unity, $\bar{\kappa}_q$ is an isomorphism. This includes the case $q = 0$. For $q = 1$, $\bar{\kappa}$ is an isomorphism when $\mathrm{char}(\Bbbk) \nmid |W|$.

A series of $\Bbbk[x]$ is the same as an invariant family of scalars. For $\mathrm{char}(\Bbbk) \nmid |W|$, we refer to the exp-log correspondence on the space of series of $\Bbbk[x]$ as noncommutative Möbius inversion. Similarly, for q not a root of unity, for the q-exp-log correspondence, we use the term q-Möbius inversion.

The parallel examples of the exponential Coxeter bimonoid and Coxeter bimonoid of chambers are in Sections 4.2 and 4.3. Like the former, each component of the Coxeter bialgebra of polynomials is $\Bbbk$, while like the latter, it admits a q-deformation. These connections are made precise later in Section 18.2 using Fock functors.

10.2.1. Coxeter space of polynomials. Define the Coxeter $\boldsymbol{\alpha}$-space $\Bbbk[x]$ by setting

$$\Bbbk[x][T] := \Bbbk \tag{10.6}$$

for a face-type T. We write x^T for the basis element $1 \in \Bbbk[x][T]$ and call it a *Coxeter monomial*. For a morphism $\sigma : T \to U$, define

$$\sigma : \Bbbk[x][T] \to \Bbbk[x][U], \qquad \sigma(x^T) := x^U. \tag{10.7}$$

In other words, structure maps are all identities. So diagrams (8.28) commute, and we indeed have a Coxeter $\boldsymbol{\alpha}$-space. We call this the *Coxeter space of polynomials*.

Observation 10.7. As a Coxeter space,

$$\Bbbk[x] = \mathrm{X} + \mathrm{X}^2 + \mathrm{X}^3 + \dots, \tag{10.8}$$

the sum of Cauchy powers of X, with the latter as in (10.2).

Observation 10.8. The functor (8.30) takes the Coxeter space $\Bbbk[x]$ to the trivial representation of W. This is clear since all structure maps are identities. It also follows formally from (10.8) since X goes to the trivial representation of W under (8.30), while the higher Cauchy powers of X go to zero by Observation 9.4.

10.2.2. Coxeter bialgebra of polynomials. We now turn $\Bbbk[x]$ into a Coxeter $\boldsymbol{\alpha}$-bialgebra. The product and coproduct are

$$\begin{aligned} \mu_Z^T : \Bbbk[x][T] &\to \Bbbk[x][Z] & \qquad \Delta_Z^T : \Bbbk[x][Z] &\to \Bbbk[x][T] \\ x^T &\mapsto x^Z & x^Z &\mapsto d_{T/Z}\, x^T, \end{aligned} \tag{10.9}$$

where $d_{T/Z}$ is the number of faces of type T/Z in $\boldsymbol{\alpha}_Z$ (Section 1.8.1). We call this the *Coxeter bialgebra of polynomials*. It is bicommutative.

10.2.3. Coxeter q-bialgebra of polynomials. More generally, for a scalar q, the Coxeter $\boldsymbol{\alpha}$-space of polynomials carries a Coxeter $\boldsymbol{\alpha}$-q-bialgebra structure which we denote $\Bbbk_q[x]$. We call it the *Coxeter q-bialgebra of polynomials*. The product and coproduct are

$$\begin{aligned} \mu_Z^T : \Bbbk_q[x][T] &\to \Bbbk_q[x][Z] & \qquad \Delta_Z^T : \Bbbk_q[x][Z] &\to \Bbbk_q[x][T] \\ x^T &\mapsto x^Z & x^Z &\mapsto d_{T/Z}(q)\, x^T, \end{aligned} \tag{10.10}$$

where $d_{T/Z}(q)$ is the polynomial in q defined in Section 1.8.1. When $q = 1$, we recover the Coxeter bialgebra of polynomials. The product is the same as before, while the coproduct is deformed by the parameter q. Thus, $\Bbbk_q[x]$ is a commutative Coxeter q-bialgebra. Interestingly, $\Bbbk_q[x]$ continues to be cocommutative.

Observation 10.9. For the Coxeter q-bialgebra $\Bbbk_q[x]$, for a face-type T, using formulas (10.10),

$$\Delta_T^S \mu_T^S = d_{S/T}(q)\ \mathrm{id}.$$

This is indeed a special case of (8.74). We can see this from Observation 10.8 and formula (1.36).

10.2.4. Face-type enumeration identities. We check below that $\Bbbk_q[x]$ is indeed a bicommutative Coxeter q-bialgebra.

- For the product: The naturality, associativity, unitality axioms (8.41) are clear (since all maps are identities). Similarly, the commutativity axiom (8.86) is clear.
- For the coproduct: The naturality axiom follows from Exercise 1.67, the coassociativity axiom follows from the multiplicative property (1.32), the counitality axiom holds since $d_{Z/Z}(q) = 1$. The cocommutativity axiom (8.91) follows from Exercise 1.68. See the diagrams below.

$$\begin{array}{ccc} \Bbbk_q[x][T] & \xrightarrow{\text{id}} & \Bbbk_q[x][T'] \\ {\scriptstyle d_{T/Z}(q)\,\text{id}}\uparrow & & \uparrow{\scriptstyle d_{T'/Z'}(q)\,\text{id}} \\ \Bbbk_q[x][Z] & \xrightarrow[\text{id}]{} & \Bbbk_q[x][Z'] \end{array} \qquad \begin{array}{ccccc} & & \Bbbk_q[x][T] & & \\ & {\scriptstyle d_{T/Z}(q)\,\text{id}}\nearrow & & \searrow{\scriptstyle d_{U/T}(q)\,\text{id}} & \\ \Bbbk_q[x][Z] & & \xrightarrow[d_{U/Z}(q)\,\text{id}]{} & & \Bbbk_q[x][U] \end{array}$$

$$\left(\Bbbk_q[x][Z] \xrightarrow{d_{Z/Z}(q)\,\text{id}} \Bbbk_q[x][Z]\right) = \text{id} \qquad \begin{array}{ccccc} \Bbbk_q[x][T] & & \xrightarrow{\text{id}} & & \Bbbk_q[x][T'] \\ & {\scriptstyle d_{T/Z}(q)\,\text{id}}\nwarrow & & \nearrow{\scriptstyle d_{T'/Z}(q)\,\text{id}} & \\ & & \Bbbk_q[x][Z] & & \end{array}$$

- The q-bialgebra axiom (8.73) reduces to the gate identity (1.59) applied to $\boldsymbol{\alpha}_Z$. This can be seen using Observation 8.99, item (1) and the first identity in (1.44a) applied to $\boldsymbol{\alpha}_Z$. See the diagram below.

$$\begin{array}{ccccc} \Bbbk_q[x][T] & \xrightarrow{\text{id}} & \Bbbk_q[x][Z] & \xrightarrow{d_{T'/Z}(q)\,\text{id}} & \Bbbk_q[x][T'] \\ {\scriptstyle (d_{U/T}(q)\,\text{id})}\downarrow & & & & \uparrow{\scriptstyle (\text{id})} \\ \bigoplus_{U:\,T\leq U} \Bbbk_q[x][U] & & \xrightarrow[(a_Z^{TUT'U'}(q)\,\text{id})]{} & & \bigoplus_{U':\,T'\leq U'} \Bbbk_q[x][U'] \end{array}$$

  Alternatively: Since the product is commutative and each product component is an isomorphism, the q-bialgebra axiom (8.73) is in fact equivalent to (8.97), from where identity (1.59) applied to $\boldsymbol{\alpha}_Z$ pops out.

The existence of $\Bbbk_q[x]$ can also be asserted by viewing it as a universal object (Section 10.2.14). By reversing the logical flow, the above calculations then imply identities (1.32) and (1.59). See Section 18.2.3 for related comments.

10.2.5. Divided power Coxeter bialgebra. Let $\Bbbk\{x\}$ denote the Coxeter bialgebra dual to $\Bbbk[x]$. Thus,

$$\Bbbk\{x\}[T] := \Bbbk. \tag{10.11}$$

To distinguish from the previous notation, let us write $x^{(T)}$ for the canonical basis element $1 \in \Bbbk$ and call it a *divided power Coxeter monomial*. The product and coproduct are

$$
\begin{aligned}
\mu_Z^T : \Bbbk\{x\}[T] &\to \Bbbk\{x\}[Z] & \qquad \Delta_Z^T : \Bbbk\{x\}[Z] &\to \Bbbk\{x\}[T] \\
x^{(T)} &\mapsto d_{T/Z}\, x^{(Z)} & x^{(Z)} &\mapsto x^{(T)}.
\end{aligned}
\tag{10.12}
$$

We call this the *divided power Coxeter bialgebra*. It is bicommutative.

10.2.6. Divided power Coxeter q-bialgebra. More generally, for a scalar q, let $\Bbbk_q\{x\}$ denote the Coxeter q-bialgebra dual to $\Bbbk_q[x]$. The product and coproduct are

$$
\begin{aligned}
\mu_Z^T : \Bbbk_q\{x\}[T] &\to \Bbbk_q\{x\}[Z] & \qquad \Delta_Z^T : \Bbbk_q\{x\}[Z] &\to \Bbbk_q\{x\}[T] \\
x^{(T)} &\mapsto d_{T/Z}(q)\, x^{(Z)} & x^{(Z)} &\mapsto x^{(T)}.
\end{aligned}
\tag{10.13}
$$

We call this the *divided power Coxeter q-bialgebra*. When $q = 1$, we recover the divided power Coxeter bialgebra. The coproduct is the same as before, while the product is deformed by the parameter q. Thus, $\Bbbk_q\{x\}$ is a cocommutative Coxeter q-bialgebra. Interestingly, it continues to be commutative.

10.2.7. Duality. Let us formally record the duality mentioned above. For a scalar q, the Coxeter q-bialgebras $\Bbbk_q[x]$ and $\Bbbk_q\{x\}$ are duals of each other. That is,

$$\Bbbk_q[x]^* = \Bbbk_q\{x\} \quad \text{and} \quad \Bbbk_q\{x\}^* = \Bbbk_q[x]$$

as Coxeter q-bialgebras.

10.2.8. Norm map. The polynomial Coxeter bialgebra and divided power Coxeter bialgebra relate to each other by the *norm map* as follows.

Lemma 10.10. *The norm map*

$$\bar{\kappa} : \Bbbk[x] \to \Bbbk\{x\}, \qquad x^T \mapsto d_{S/T}\, x^{(T)} \tag{10.14}$$

on the T-component, is a self-dual morphism of Coxeter bialgebras. Moreover, it is an isomorphism iff $|W| \neq 0$ iff $\operatorname{char}(\Bbbk) \nmid |W|$.

PROOF. First we need to check that $\bar{\kappa}$ is a map of Coxeter spaces, that is, diagram (8.29) commutes. This check reduces to the identity $d_{S/T} = d_{S/U}$ for $\sigma : T \to U$. This is an instance of the result in Exercise 1.67. Next we need to check that $\bar{\kappa}$ is a morphism of Coxeter algebras and of Coxeter coalgebras, that is, diagrams (8.42) and (8.47) commute. See the diagrams below.

$$
\begin{array}{ccc}
\Bbbk[x][T] & \xrightarrow{d_{S/T}\,\mathrm{id}} & \Bbbk\{x\}[T] \\
{\scriptstyle \mathrm{id}}\downarrow & & \downarrow{\scriptstyle d_{T/Z}\,\mathrm{id}} \\
\Bbbk[x][Z] & \xrightarrow[d_{S/Z}\,\mathrm{id}]{} & \Bbbk\{x\}[Z]
\end{array}
\qquad\qquad
\begin{array}{ccc}
\Bbbk[x][T] & \xrightarrow{d_{S/T}\,\mathrm{id}} & \Bbbk\{x\}[T] \\
{\scriptstyle d_{T/Z}\,\mathrm{id}}\uparrow & & \uparrow{\scriptstyle \mathrm{id}} \\
\Bbbk[x][Z] & \xrightarrow[d_{S/Z}\,\mathrm{id}]{} & \Bbbk\{x\}[Z]
\end{array}
$$

Both reduce to the identity $d_{S/Z} = d_{S/T} d_{T/Z}$ for $Z \leq T \leq S$. This is an instance of the multiplicative property (1.32) for $q = 1$. (The two diagrams above are in fact duals of each other; hence, if one of them commutes, then so does the other.) This proves the first claim.

Observe: $\bar{\kappa}$ is an isomorphism iff $d_{S/T} \neq 0$ for all T. Now if $d_{S/T} \neq 0$ for some T, then $d_{S/U} \neq 0$ for all face-types $U \geq T$. In particular, $d_S \neq 0$ implies $d_{S/T} \neq 0$ for all T. Since $d_S = |W|$, the second claim follows. □

As a consequence:

Corollary 10.11. *When* $\mathrm{char}(\Bbbk) \nmid |W|$, *the Coxeter bialgebras* $\Bbbk[x]$ *and* $\Bbbk\{x\}$ *are isomorphic and hence self-dual.*

10.2.9. q-norm map. More generally, for a scalar q, the polynomial Coxeter q-bialgebra and divided power Coxeter q-bialgebra relate to each other by the *q-norm map* as follows.

Lemma 10.12. *For a scalar q, the q-norm map*

$$\bar{\kappa}_q : \Bbbk_q[x] \to \Bbbk_q\{x\}, \qquad x^T \mapsto d_{S/T}(q)\, x^{(T)} \tag{10.15}$$

on the T-component, is a self-dual morphism of Coxeter q-bialgebras. Moreover, it is an isomorphism iff $d_S(q) \neq 0$.

Proof. We generalize the proof of Lemma 10.10. First we need to check that $\bar{\kappa}_q$ is a map of Coxeter spaces. This reduces to the identity $d_{S/T}(q) = d_{S/U}(q)$ for $\sigma : T \to U$, which holds by Exercise 1.67. Next we need to check that $\bar{\kappa}_q$ is a morphism of Coxeter algebras and of Coxeter coalgebras. This reduces to the identity $d_{S/Z}(q) = d_{S/T}(q) d_{T/Z}(q)$ for $Z \leq T \leq S$, which is an instance of the multiplicative property (1.32). This proves the first claim.

Observe: $\bar{\kappa}_q$ is an isomorphism iff $d_{S/T}(q) \neq 0$ for all T. Now if $d_{S/T}(q) \neq 0$ for some T, then $d_{S/U}(q) \neq 0$ for all face-types $U \geq T$. In particular, $d_S(q) \neq 0$ implies $d_{S/T}(q) \neq 0$ for all T. This proves the second claim. □

Corollary 10.13. *For q not a root of unity, the q-norm map* (10.15) *is an isomorphism, and hence the Coxeter q-bialgebras $\Bbbk_q[x]$ and $\Bbbk_q\{x\}$ are isomorphic and self-dual.*

Proof. Define the i-th q-number to be $1 + q + \cdots + q^{i-1}$. We know from (1.34) that $d_S(q)$ can be written as a product of q-numbers. Since q is not a root of unity, the q-numbers are nonzero, and hence $d_S(q) \neq 0$. Now apply Lemma 10.12. □

The factorization (1.34) in fact yields the finer result given below.

Corollary 10.14. *Let $\boldsymbol{\alpha}$ be a pointed reflection arrangement with largest exponent $e \geq 1$. Let q be a primitive N-th root of unity for some $N > e + 1$. Then the q-norm map* (10.15) *is an isomorphism, and hence the Coxeter $\boldsymbol{\alpha}$-q-bialgebras $\Bbbk_q[x]$ and $\Bbbk_q\{x\}$ are isomorphic and self-dual.*

Example 10.15. Let q be a primitive N-th root of unity for some $N \geq 2$. Then:

- The q-norm map is an isomorphism for all braid arrangements with strictly less than N letters. For those with N or more letters, the q-norm is 0 on the S-component, and, in particular, is not an isomorphism. These facts follow from formula (1.167) for the Poincaré polynomial which is a q-factorial.

- For all $N > 30$, the q-norm map is an isomorphism for all exceptional arrangements. This follows from the fact that the largest exponent of an exceptional arrangement is 29, see [**46**, Appendix A1, Table I].

$\square$

For general context on the q-norm map, see Exercise 10.21.

10.2.10. Image of the q-norm map. For a scalar q, there is a self-dual Coxeter q-bialgebra, denoted $\Bbbk_q(x)$, fitting into a commutative diagram

$$\begin{array}{ccc} \Bbbk_q[x] & \xrightarrow{\overline{\kappa}_q} & \Bbbk_q\{x\} \\ & \searrow \quad \nearrow & \\ & \Bbbk_q(x) & \end{array} \tag{10.16}$$

of Coxeter q-bialgebras. Explicitly,

$$\Bbbk_q(x)[T] := \begin{cases} \Bbbk & \text{if } d_{S/T}(q) \neq 0, \\ 0 & \text{otherwise.} \end{cases} \tag{10.17}$$

The Coxeter q-bialgebra $\Bbbk_q(x)$ is called the image of the q-norm map.

10.2.11. Signed Coxeter bialgebras. Let us now look at the case $q = -1$. We have the signed Coxeter bialgebras $\Bbbk_{-1}[x]$ and $\Bbbk_{-1}\{x\}$ which are duals of each other. We call them the *signed Coxeter bialgebra of polynomials* and *divided power signed Coxeter bialgebra*, respectively.

Exercise 10.16. Check: $\Bbbk_{-1}[x]$ is commutative and signed cocommutative, while dually $\Bbbk_{-1}\{x\}$ is signed commutative and cocommutative. (For the signed part, use the identity in Exercise 1.68.)

The *signed norm map*

$$\overline{\kappa}_{-1} : \Bbbk_{-1}[x] \to \Bbbk_{-1}\{x\}, \qquad x^T \mapsto d_{S/T}(-1)\, x^{(T)} \tag{10.18}$$

on the T-component, is a self-dual morphism of signed Coxeter bialgebras. It is the case $q = -1$ of (10.15). Its image is discussed below.

10.2.12. Signed Coxeter bialgebra of dual numbers. Define the Coxeter space $\Bbbk[dx]$ by

$$\Bbbk[dx][T] := \begin{cases} \Bbbk & \text{if } T = S, \\ 0 & \text{otherwise.} \end{cases} \tag{10.19}$$

Let us write dx for the canonical basis element $1 \in \Bbbk$ in the S-component. The structure maps are all identities. In view of (1.35) and (10.17), observe:

$$\Bbbk[dx] = \Bbbk_{-1}(x).$$

This shows that $\Bbbk[dx]$ carries a signed Coxeter bialgebra structure. We call this the *signed Coxeter bialgebra of dual numbers*. The product and coproduct are defined in the only manner possible, namely, $\mu_S^S = \Delta_S^S = \mathrm{id}$, while all other components are zero.

Let us also verify the signed bialgebra axiom directly using these definitions. The relevant diagram to check is

$$\begin{array}{ccccc} \Bbbk[dx][S] & \xrightarrow{\mu_T^S} & \Bbbk[dx][T] & \xrightarrow{\Delta_T^S} & \Bbbk[dx][S] \\ {\scriptstyle \mathrm{id}}\downarrow & & & & \uparrow{\scriptstyle \mathrm{id}} \\ \Bbbk[dx][S] & & \xrightarrow[(\overline{\lambda}_{-1})_{SSSS}]{} & & \Bbbk[dx][S]. \end{array}$$

This is a special case of Observation 10.9. By Observation 8.99, item (1) and identities (1.44c) and (1.39) applied to $\boldsymbol{\alpha}_T$, we see that $(\overline{\lambda}_{-1})_{SSSS}$ is indeed scalar multiplication by $d_{S/T}(-1)$. By (1.35), this scalar is 1 if $S = T$ and 0 if $S > T$, as required.

10.2.13. Coxeter 0-bialgebra of polynomials. Let us now look at the case $q = 0$. Since $d_{T/Z}(0) = 1$, we obtain

$$\Bbbk_0[x] = \Bbbk_0\{x\} = \Bbbk_0(x),$$

with product and coproduct

$$\begin{aligned} \mu_Z^T : \Bbbk_0[x][T] &\to \Bbbk_0[x][Z] & \qquad \Delta_Z^T : \Bbbk_0[x][Z] &\to \Bbbk_0[x][T] \\ x^T &\mapsto x^Z & x^Z &\mapsto x^T. \end{aligned} \tag{10.20}$$

This is the Coxeter 0-bialgebra of polynomials as well as the divided power Coxeter 0-bialgebra. It is clearly self-dual. This is the case $q = 0$ of Corollary 10.13.

The 0-norm map $\overline{\kappa}_0$ is the identity, and, in particular, an isomorphism. It is also clear that formulas (10.20) satisfy the 0-bialgebra axiom (8.78). The result of Exercise 8.81 is also clear.

10.2.14. Freeness and cofreeness. Universal constructions were discussed in Section 9.3. We now discuss the (co)freeness properties of the Coxeter bialgebra of polynomials and its relatives. A summary is provided in Table 10.2.

Proposition 10.17. *Let* X *be the Coxeter space defined by* (10.2)*. Then, for a scalar* q*,*

$$\overline{\mathcal{T}_q}(\mathrm{X}) = \Bbbk_q[x] \quad \textit{and} \quad \overline{\mathcal{T}_q}^\vee(\mathrm{X}) = \Bbbk_q\{x\}$$

as Coxeter q*-bialgebras (under the canonical identifications).*

PROOF. For T a face-type, by (9.18), $\overline{\mathcal{T}_q}(\mathrm{X})[T] = \mathrm{X}[S] = \Bbbk$. Hence, all face-type components of $\overline{\mathcal{T}_q}(\mathrm{X})$ and $\Bbbk_q[x]$ equal $\Bbbk$, and we identify them canonically. Similarly, we identify $\overline{\mathcal{T}_q}^\vee(\mathrm{X})$ and $\Bbbk_q\{x\}$. To see that (co)products match, put $\mathrm{V} := \mathrm{X}$ in formulas (9.25) and (9.34), respectively, and use Observation 8.99, item (1) and identities (1.44c) and (1.39) applied to $\boldsymbol{\alpha}_Z$. □

For $q = 1$, the situation is

$$\overline{\mathcal{T}}(\mathrm{X}) = \overline{\mathcal{S}}(\mathrm{X}) = \Bbbk[x] \quad \text{and} \quad \overline{\mathcal{T}}^\vee(\mathrm{X}) = \overline{\mathcal{S}}^\vee(\mathrm{X}) = \Bbbk\{x\}.$$

In particular, $\Bbbk[x]$ is the free Coxeter algebra on X as well as the free commutative Coxeter algebra on X, while $\Bbbk\{x\}$ is the cofree Coxeter coalgebra on X as well as the cofree cocommutative Coxeter coalgebra on X.

For $q = -1$,

$$\overline{\mathcal{T}}_{-1}(\mathrm{X}) = \Bbbk_{-1}[x] \quad \text{and} \quad \overline{\mathcal{T}}^{\vee}_{-1}(\mathrm{X}) = \Bbbk_{-1}\{x\},$$

and assuming that $\mathrm{char}(\Bbbk) \neq 2$,

$$\overline{\mathcal{E}}(\mathrm{X}) = \overline{\mathcal{E}}^{\vee}(\mathrm{X}) = \Bbbk[dx].$$

We point out that as a Coxeter space, this is the same as X.

10.2.15. Classifying Coxeter bialgebra morphisms. We now use the above (co)freeness properties to classify Coxeter bialgebra morphisms from $\Bbbk[x]$ or into $\Bbbk\{x\}$.

Exercise 10.18 (Universal property). Let A be a Coxeter algebra. Check: Any $z \in \mathrm{A}[S]^W$ yields a Coxeter algebra morphism

$$\Bbbk[x] \to \mathrm{A}, \qquad x^T \mapsto \mu_T^S(z)$$

on the T-component. In particular, $x^S \mapsto z$. Conversely, every Coxeter algebra morphism $\Bbbk[x] \to \mathrm{A}$ arises in this manner from a unique z. (Proceed directly or use Theorem 9.11, formula (9.20) and Exercise 10.1.)

Dually, let C be a Coxeter coalgebra. Check: Any $f \in (\mathrm{C}[S]_W)^*$ yields a Coxeter coalgebra morphism

$$\mathrm{C} \to \Bbbk\{x\}, \qquad y \mapsto f(\Delta_T^S(y))\, x^{(T)}$$

on the T-component. Conversely, every Coxeter coalgebra morphism $\mathrm{C} \to \Bbbk\{x\}$ arises in this manner from a unique f.

Exercise 10.19. Let H be a Coxeter q-bialgebra for a scalar q. Show:

(1) If $f : \Bbbk_q[x] \to \mathrm{H}$ is a morphism of Coxeter algebras, then it is also a morphism of Coxeter q-bialgebras.
(2) If $g : \mathrm{H} \to \Bbbk_q\{x\}$ is a morphism of Coxeter coalgebras, then it is also a morphism of Coxeter q-bialgebras.

(For item (1): Apply Theorems 9.11 and 9.15 for $\mathrm{V} := \mathrm{X}$ and use Observation 10.4 for $M = \Bbbk$. Alternatively, check directly using the first formula in Exercise 10.18. Dually, for item (2): Apply Theorems 9.16 and 9.19 or use the second formula in Exercise 10.18.)

Exercise 10.20 (Universal property). Let H be a Coxeter q-bialgebra for a scalar q. Deduce:

(1) Coxeter q-bialgebra morphisms from $\Bbbk_q[x]$ to H are parametrized by elements of $\mathrm{H}[S]^W$.
(2) Coxeter q-bialgebra morphisms from H to $\Bbbk_q\{x\}$ are parametrized by elements of $(\mathrm{H}[S]_W)^*$.

Formulas for the morphisms are as in Exercise 10.18.

Exercise 10.21 (General q-norm map). For any scalar q, Coxeter q-bialgebra morphisms from $\Bbbk_q[x]$ to $\Bbbk_q\{x\}$ are scalar multiples of the q-norm map (10.15). (Recall from Exercise 8.84 that scalar multiplication preserves Coxeter q-bialgebra morphisms.)

(i) Check the above fact directly from definitions.
(ii) Deduce it from Exercise 10.20, item (1) or item (2).

(iii) Deduce it from the bijection (9.144) applied to C = A = X and the fact that Coxeter space maps from X to itself are scalar multiples of identity. This fits into the special case of (9.145). As a prelude: the q-norm map (10.15) is an instance of (9.140) for V := X, after the identification in Proposition 10.17. This follows from formula (1.36). Hence, Corollary 10.13 is an instance of Lemma 9.101. Similarly, Corollary 10.11 is an instance of Lemma 9.94.

Exercise 10.22. Let $\mathrm{char}(\Bbbk) \nmid |W|$. Deduce: For H a Coxeter bialgebra, elements $z \in \mathrm{H}[S]^W$ parametrize Coxeter bialgebra morphisms from $\Bbbk\{x\}$ to H via

$$\Bbbk\{x\}[T] \to \mathrm{H}[T], \qquad x^{(T)} \mapsto \tfrac{1}{d_{S/T}} \mu_T^S(z).$$

A similar statement holds for morphisms from $\Bbbk_q\{x\}$ to a Coxeter q-bialgebra H for q not a root of unity. (Use Exercise 10.20 and Lemmas 10.10 and 10.12.) A more substantial result is given in Proposition 9.91 and its q-analogue.

10.2.16. Primitive and indecomposable parts. For these notions, we refer to Section 9.2.

For the divided power Coxeter bialgebra,

$$\overline{\mathcal{P}}(\Bbbk\{x\}) = \mathrm{X}. \tag{10.21}$$

This follows from the coproduct formula in (10.12).

Similarly, for q not a root of unity,

$$\overline{\mathcal{P}}(\Bbbk_q[x]) = \overline{\mathcal{P}}(\Bbbk_q\{x\}) = \mathrm{X}. \tag{10.22}$$

The first equality follows from Corollary 10.13. The coproduct of $\Bbbk_q\{x\}$ does not depend on q, so the second equality follows from formula (10.21).

Exercise 10.23. Recall the Coxeter Lie species from Section 4.8.1. Show: For the Coxeter bialgebra of polynomials $\Bbbk[x]$, for any $A \leq C$,

$$\overline{\mathcal{P}}(\Bbbk[x])[A, C] = \mathsf{Lie}[A]^{W_A}, \tag{10.23}$$

where the rhs is the space of W_A-invariant Lie elements. This result has been formulated in terms of top-nested faces rather than face-types. (Use (1.31) for $q = 1$ applied to $\boldsymbol{\alpha}_Z$.) In addition, if $\mathrm{char}(\Bbbk) \nmid |W|$, then $\overline{\mathcal{P}}(\Bbbk[x]) = \mathrm{X}$. See also formula (12.30) in this regard.

Dually, for the Coxeter bialgebra of polynomials,

$$\overline{\mathcal{Q}}(\Bbbk[x]) = \mathrm{X}. \tag{10.24}$$

This follows from the product formula in (10.9).

Similarly, for q not a root of unity,

$$\overline{\mathcal{Q}}(\Bbbk_q\{x\}) = \overline{\mathcal{Q}}(\Bbbk_q[x]) = \mathrm{X}. \tag{10.25}$$

The first equality follows from Corollary 10.13. The product of $\Bbbk_q[x]$ does not depend on q, so the second equality follows from formula (10.24).

Exercise 10.24. Show: For the divided power Coxeter bialgebra $\Bbbk\{x\}$, for any $A \leq C$,

$$\overline{\mathcal{Q}}(\Bbbk\{x\})[A, C] = (\mathsf{Lie}[A]^*)_{W_A}, \tag{10.26}$$

where the rhs is the space of W_A-coinvariants of $\mathsf{Lie}[A]^*$. In addition, if $\mathrm{char}(\Bbbk) \nmid |W|$, then $\overline{\mathcal{Q}}(\Bbbk\{x\}) = \mathrm{X}$.

10.2.17. Primitive and group-like series. We now apply the discussion in Section 9.10 to the Coxeter bialgebra of polynomials $\Bbbk[x]$. A series of $\Bbbk[x]$ is a family of scalars $f(T)$, one for each face-type T, such that $f(T) = f(U)$ whenever $\mathrm{s}(T) = \mathrm{s}(U)$. This follows from Exercise 1.26. In other words, a series of $\Bbbk[x]$ is an invariant family of scalars in the sense of Definition 1.10. Thus, $\mathscr{S}(\Bbbk[x])$ has a basis indexed by flat-types. It is a left module over the invariant lune-incidence algebra via

$$(s \circ f)(Z) = \sum_{T:\, Z \leq T} s(Z,T)\, f(T) \tag{10.27}$$

for $s \in \mathrm{I}_{\mathrm{lune}}[\mathcal{A}]^W$ and series $f \in \mathscr{S}(\Bbbk[x])$. This action specializes (9.122). This follows from the product formula in (10.9). The module $\mathscr{S}(\Bbbk[x])$ is the same as the invariant lune-incidence module of [**10**, Section 16.7.4]. Since $\Bbbk[x]$ is commutative, (10.27) can be expressed in the form (9.123) as a module over the invariant flat-incidence algebra (assuming $\mathrm{char}(\Bbbk) \nmid |W|$).

Exp-log. Let $\mathrm{char}(\Bbbk) \nmid |W|$. The exp-log correspondence (9.125) takes the form

$$g(Z) = \sum_{T:\, Z \leq T} \boldsymbol{\zeta}(Z,T)\, f(T) \iff f(Z) = \sum_{T:\, Z \leq T} \boldsymbol{\mu}(Z,T)\, g(T). \tag{10.28}$$

We refer to this as *noncommutative Möbius inversion*. Note: It is independent of $\boldsymbol{\zeta}$ and $\boldsymbol{\mu}$.

Example 10.25 (Noncommutative Möbius inversion). Let $\mathrm{char}(\Bbbk) \nmid |W|$. For each face-type Z, put

$$f(Z) := \frac{|\mu(\mathrm{X}, \top)|}{|W_{\mathrm{X}}|} \quad \text{and} \quad g(Z) := 1,$$

where X is any flat such that $\mathrm{t}(\mathrm{X}) = \mathrm{s}(Z)$. The series f and g relate to each other by (10.28). The calculation in either direction is shown below.

$$\sum_{T:\, Z \leq T} \boldsymbol{\zeta}(Z,T)\, f(T) = \sum_{T:\, Z \leq T} (-1)^{\mathrm{rk}(S/T)} \boldsymbol{\zeta}(Z,T) \boldsymbol{\mu}(T,S) = 1 = g(Z),$$

$$\sum_{T:\, Z \leq T} \boldsymbol{\mu}(Z,T)\, g(T) = \sum_{T:\, Z \leq T} \boldsymbol{\mu}(Z,T) = (-1)^{\mathrm{rk}(S/Z)} \boldsymbol{\mu}(Z,S) = f(Z).$$

We used formula (1.86) and identities (1.121) related to the noncommutative Zaslavsky formula. □

Primitive and group-like series. Let $\mathrm{char}(\Bbbk) \nmid |W|$. We now turn to primitive and group-like series of $\Bbbk[x]$. A series f is primitive if $f(T) = 0$ for $T \neq S$. This follows from the last part of Exercise 10.23. A series g is group-like if it is a scalar multiple of (d_T). That (d_T) is group-like follows from the multiplicative property (1.32) for $q = 1$. Let us now directly verify Theorem 9.76 for $\Bbbk[x]$. For f primitive, using (1.84),

$$\mathbf{exp}(f)_Z = \boldsymbol{\zeta}(Z,S)\, f(S) = \tfrac{d_Z}{d_S}\, f(S),$$

while for g group-like, using (1.89),

$$\mathbf{log}(g)_Z = \sum_{T:\, Z \leq T} \boldsymbol{\mu}(Z,T)\, g(T) = \begin{cases} g(S) & \text{if } Z = S, \\ 0 & \text{otherwise.} \end{cases}$$

This bijection is the same as the one obtained in [**10**, Theorem 16.40].

For $\Bbbk[x]$, every group-like series is exponential and every primitive series is linear. So the above bijection is the same as the one in Lemma 9.85 and the above formulas are specializations of (9.136).

q-exp-log. A similar analysis can be carried out for the Coxeter q-bialgebra of polynomials $\Bbbk_q[x]$ for q not a root of unity. The q-exp-log correspondence (9.128) may be expressed as

$$g(Z) = \sum_{T:\, Z \leq T} \zeta_q(Z,T)\, f(T) \iff f(Z) = \sum_{T:\, Z\leq T} \boldsymbol{\mu}_q(Z,T)\, g(T). \tag{10.29}$$

We refer to this as *q-Möbius inversion.*

Example 10.26 (q-Möbius inversion). Suppose q is not a root of unity. For each face-type Z, put

$$f(Z) := (-1)^{\mathrm{rk}(S/Z)} \boldsymbol{\mu}_q(Z,S)\, q^{\mathrm{dist}(Z,\overline{Z})} \quad \text{and} \quad g(Z) := q^{\mathrm{dist}(Z,\overline{Z})}.$$

The series f and g relate to each other by (10.29). The calculation in either direction is shown below.

$$\begin{aligned}\sum_{T:\, Z\leq T} \zeta_q(Z,T)\, f(T) &= \sum_{T:\, Z\leq T} (-1)^{\mathrm{rk}(S/T)}\, \zeta_q(Z,T)\, \boldsymbol{\mu}_q(T,S)\, q^{\mathrm{dist}(T,\overline{T})} \\ &= q^{\mathrm{dist}(Z,\overline{Z})} = g(Z),\end{aligned}$$

$$\begin{aligned}\sum_{T:\, Z\leq T} \boldsymbol{\mu}_q(Z,T)\, g(T) &= \sum_{T:\, Z\leq T} \boldsymbol{\mu}_q(Z,T)\, q^{\mathrm{dist}(T,\overline{T})} \\ &= (-1)^{\mathrm{rk}(S/Z)}\, \boldsymbol{\mu}_q(Z,S)\, q^{\mathrm{dist}(Z,\overline{Z})} = f(Z).\end{aligned}$$

We used formula (1.15) and identities (1.124) related to the noncommutative q-Zaslavsky formula. For $q \neq 0$, we may normalize f and g as

$$\frac{1}{q^{l(w_\emptyset)}} f(Z) = \frac{(-1)^{\mathrm{rk}(S/Z)}}{q^{l(w_Z)}} \boldsymbol{\mu}_q(Z,S) \quad \text{and} \quad \frac{1}{q^{l(w_\emptyset)}} g(Z) = \frac{1}{q^{l(w_Z)}},$$

where w_Z is the longest element in W_Z. □

Exercise 10.27. Suppose q is not a root of unity. A series f of $\Bbbk_q[x]$ is primitive if $f(T) = 0$ for $T \neq S$. Group-like series of $\Bbbk_q[x]$ are scalar multiples of $(d_T(q))$. Use (1.95) and (1.99) to directly verify Theorem 9.80 for $\Bbbk_q[x]$. Every group-like series of $\Bbbk_q[x]$ is q-exponential and every primitive series is linear.

Exercise 10.28. Do the above analysis for the divided power Coxeter bialgebra $\Bbbk\{x\}$ instead of $\Bbbk[x]$ when $\mathrm{char}(\Bbbk) \nmid |W|$, and for $\Bbbk_q\{x\}$ instead of $\Bbbk_q[x]$ when q is not a root of unity. (We get nothing new in view of Lemmas 10.10 and 10.12.)

10.2.18. Exponential series of a Coxeter algebra.

Let $\mathrm{char}(\Bbbk) \nmid |W|$. For a Coxeter algebra A, the bijection (9.135) between linear and exponential series arises via the string of identifications

$$\begin{array}{ccccc}\overline{\mathsf{C}}\mathsf{Sp}(\mathrm{X},\mathrm{A}) & = & \mathsf{Mon}(\overline{\mathsf{C}}\mathsf{Sp})(\Bbbk[x],\mathrm{A}) & = & \mathsf{Mon}(\overline{\mathsf{C}}\mathsf{Sp})(\Bbbk\{x\},\mathrm{A}) \\ \| & & & & \| \\ \mathscr{X}(\mathrm{A}) & & & & \mathscr{E}(\mathrm{A}).\end{array}$$

The vertical equalities are (9.130) and (9.132). The first horizontal equality is because $\Bbbk[x]$ is free on X, while the second one is induced by the norm map (10.14).

For q not a root of unity, for a Coxeter algebra A, the bijection (9.137) between linear and q-exponential series arises by a similar string of identifications involving (9.130), (9.134), (10.15).

10.2.19. Antipode.

Lemma 10.29. *For a scalar q, the antipode S of the Coxeter q-bialgebra $\Bbbk_q[x]$, on the Z-component, is*

$$S_Z(x^Z) = (-1)^{\dim(\mathcal{A})} q^{l(w_Z)} x^Z, \tag{10.30}$$

where w_Z is the longest element in W_Z. In particular,

$$S_S(x^S) = (-1)^{\dim(\mathcal{A})} x^S.$$

PROOF. Substitute product and coproduct formulas (10.10) in the Takeuchi formula (9.171) and use the alternating sign identity (1.63a) applied to $\boldsymbol{\alpha}_Z$. Alternatively, start with the above formula for S_S which is clear, and apply formula (9.178) with $T := S$. □

Dually:

Lemma 10.30. *For a scalar q, the antipode S of the Coxeter q-bialgebra $\Bbbk_q\{x\}$, on the Z-component, is*

$$S_Z(x^{(Z)}) = (-1)^{\dim(\mathcal{A})} q^{l(w_Z)} x^{(Z)}, \tag{10.31}$$

where w_Z is the longest element in W_Z.

This formula is the same as (10.30) except for $x^{(Z)}$ in place of x^Z.

10.3. Tensor, shuffle, symmetric, exterior Coxeter bialgebras

Starting with a left W-module M, we construct the Coxeter bialgebra $\overline{\mathcal{T}}(M)$. We call it the tensor Coxeter bialgebra of M. We also discuss some other close relatives; they are summarized in Table 10.3. They include the shuffle Coxeter bialgebra $\overline{\mathcal{T}}^\vee(M)$, the symmetric Coxeter bialgebras (two versions), the exterior signed Coxeter bialgebras (two versions). In the symmetric and exterior cases, W-coinvariants of M appear in one version, while W-invariants of M appear in the other version. Basic properties of these Coxeter bialgebras are listed in Table 10.4. Here X_M refers to the decorated version of the Coxeter space characteristic of chambers from Section 10.1.

When $M = \Bbbk$ with trivial W-action, the tensor Coxeter bialgebra $\overline{\mathcal{T}}(M)$ specializes to the Coxeter bialgebra of polynomials $\Bbbk[x]$. In a similar manner, the remaining entries in Table 10.3 specialize to those in Table 10.1.

We formulate the tensor and shuffle Coxeter bialgebras in terms of face-types. Shuffles associated to face-types intervene in the (co)product formulas. The lengths of these shuffles appear in the q-deformations. Algebraic properties of the product and coproduct translate to enumeration identities in the group algebra of W from Section 1.9. We use the terminology of Table 10.5

TABLE 10.3. Tensor Coxeter bialgebra and its relatives.

1	$\overline{\mathcal{T}}(M)$	tensor Coxeter bialgebra
	$\overline{\mathcal{T}}^\vee(M)$	shuffle Coxeter bialgebra
	$\overline{\mathcal{S}}(M)$, $\overline{\mathcal{S}}^\vee(M)$	symmetric Coxeter bialgebras
q	$\overline{\mathcal{T}}_q(M)$	tensor Coxeter q-bialgebra
	$\overline{\mathcal{T}}_q^\vee(M)$	shuffle Coxeter q-bialgebra
-1	$\overline{\mathcal{T}}_{-1}(M)$	tensor signed Coxeter bialgebra
	$\overline{\mathcal{T}}_{-1}^\vee(M)$	shuffle signed Coxeter bialgebra
	$\overline{\mathcal{E}}(M)$, $\overline{\mathcal{E}}^\vee(M)$	exterior signed Coxeter bialgebras
0	$\overline{\mathcal{T}}_0(M) = \overline{\mathcal{T}}_0^\vee(M)$	tensor Coxeter 0-bialgebra

TABLE 10.4. Properties of the tensor Coxeter bialgebra.

Coxeter bialgebra	universal property	commutativity
$\overline{\mathcal{T}}_q(M)$	$\overline{\mathcal{T}}_q(\mathrm{X}_M)$	cocommutative for $q=1$ signed cocommutative for $q=-1$
$\overline{\mathcal{S}}(M)$	$\overline{\mathcal{S}}(\mathrm{X}_M)$	abelianization of $\overline{\mathcal{T}}(M)$, hence bicommutative
$\overline{\mathcal{E}}(M)$	$\overline{\mathcal{E}}(\mathrm{X}_M)$	signed abelianization of $\overline{\mathcal{T}}_{-1}(M)$, hence signed bicommutative
$\overline{\mathcal{T}}_q^\vee(M)$	$\overline{\mathcal{T}}_q^\vee(\mathrm{X}_M)$	commutative for $q=1$ signed commutative for $q=-1$
$\overline{\mathcal{S}}^\vee(M)$	$\overline{\mathcal{S}}^\vee(\mathrm{X}_M)$	coabelianization of $\overline{\mathcal{T}}^\vee(M)$, hence bicommutative
$\overline{\mathcal{E}}^\vee(M)$	$\overline{\mathcal{E}}^\vee(\mathrm{X}_M)$	signed coabelianization of $\overline{\mathcal{T}}_{-1}^\vee(M)$, hence signed bicommutative

for the products and coproducts. It is consistent with the terminology of Table 9.2.

We define the q-norm map $\overline{\kappa}_q$ from $\overline{\mathcal{T}}_q(M)$ to $\overline{\mathcal{T}}_q^\vee(M)$. It arises from freeness of $\overline{\mathcal{T}}_q(M)$ and cofreeness of $\overline{\mathcal{T}}_q^\vee(M)$. It is an isomorphism for q not a root of unity. For $q=1$, for $\mathrm{char}(\Bbbk) \nmid |W|$, its coimage and image are $\overline{\mathcal{S}}(M)$ and $\overline{\mathcal{S}}^\vee(M)$.

The parallel examples of the decorated exponential Coxeter bimonoid and Coxeter bimonoid of decorated chambers are in Section 4.9. Like the former, each component of the tensor Coxeter bialgebra is M, while like the latter, it admits a q-deformation. These connections are made precise later in Section 18.3 using Fock functors.

TABLE 10.5. Product and coproduct of the tensor Coxeter bialgebra.

Coxeter bialgebra	product	coproduct
$\overline{\mathcal{T}}_q(M)$	concatenation	q-deshuffle
$\overline{\mathcal{T}}_q^{\vee}(M)$	q-shuffle	deconcatenation
$\overline{\mathcal{T}}(M)$	concatenation	deshuffle
$\overline{\mathcal{T}}^{\vee}(M)$	shuffle	deconcatenation
$\overline{\mathcal{T}}_{-1}(M)$	concatenation	signed deshuffle
$\overline{\mathcal{T}}_{-1}^{\vee}(M)$	signed shuffle	deconcatenation
$\overline{\mathcal{T}}_0(M) = \overline{\mathcal{T}}_0^{\vee}(M)$	concatenation = 0-shuffle	deconcatenation = 0-deshuffle

10.3.1. Tensor Coxeter space. For a left W-module M, define the Coxeter space $\overline{\mathcal{T}}(M)$ by setting

$$\overline{\mathcal{T}}(M)[T] := M \tag{10.32}$$

for a face-type T. The structure maps are as follows. For a morphism $\sigma : T \to U$, define

$$\sigma : \overline{\mathcal{T}}(M)[T] \to \overline{\mathcal{T}}(M)[U], \qquad m \mapsto w \cdot m, \tag{10.33}$$

where $\sigma = w\beta = \beta w$ in normal form, and $w \cdot m$ is the action of w on $m \in M$. (This w is the same as the Coxeter symmetry w_σ in Section 1.7.3.) Observe from (8.3) or Exercise 1.31 that diagrams (8.28) commute. We call $\overline{\mathcal{T}}(M)$ the *tensor Coxeter space* of M.

Exercise 10.31. Check directly: The construction $\overline{\mathcal{T}}(M)$ is functorial in M, that is, there is a functor

$$W\text{-}\mathsf{Mod} \to \boldsymbol{\alpha}\text{-}\overline{\mathsf{C}}\mathsf{Sp}, \qquad M \mapsto \overline{\mathcal{T}}(M). \tag{10.34}$$

Alternatively and more formally, deduce it using the functor (1.27).

Observation 10.32. As a Coxeter space,

$$\overline{\mathcal{T}}(M) = \mathrm{X}_M + \mathrm{X}_M^2 + \mathrm{X}_M^3 + \dots, \tag{10.35}$$

the sum of Cauchy powers of X_M, with the latter as in (10.4). When $M = \Bbbk$ with trivial W-action, (10.35) specializes to (10.8).

Observation 10.33. The functor (8.30) takes the Coxeter space $\overline{\mathcal{T}}(M)$ to the W-module M. This is clear from definitions. It also follows formally from (10.35) since X_M goes to M under (8.30), while the higher Cauchy powers of X_M go to zero by Observation 9.4.

10.3.2. Tensor Coxeter bialgebra. We now turn $\overline{\mathcal{T}}(M)$ into a Coxeter bialgebra. The product and coproduct are

$$\begin{aligned} \mu_Z^T &: \overline{\mathcal{T}}(M)[T] \to \overline{\mathcal{T}}(M)[Z] & \Delta_Z^T &: \overline{\mathcal{T}}(M)[Z] \to \overline{\mathcal{T}}(M)[T] \\ m &\mapsto m & m &\mapsto \sum_{\zeta \in \mathrm{Sh}_{T/Z}} \zeta^{-1} \cdot m, \end{aligned} \tag{10.36}$$

with the sum over all T/Z-shuffles ζ in $\boldsymbol{\alpha}_Z$ (Section 1.6.4). We call this the *tensor Coxeter bialgebra* of the left W-module M. It is cocommutative. However, it is not commutative in general. We call μ the *concatenation product* and Δ the *deshuffle coproduct.*

10.3.3. Tensor Coxeter q-bialgebra. More generally, for a scalar q, the Coxeter space $\overline{\mathcal{T}}(M)$ carries a Coxeter q-bialgebra structure which we denote $\overline{\mathcal{T}}_q(M)$. We call it the *tensor Coxeter q-bialgebra* of M. The product and coproduct are

$$\begin{aligned} \mu_Z^T &: \overline{\mathcal{T}}_q(M)[T] \to \overline{\mathcal{T}}_q(M)[Z] & \Delta_Z^T &: \overline{\mathcal{T}}_q(M)[Z] \to \overline{\mathcal{T}}_q(M)[T] \\ m &\mapsto m & m &\mapsto \sum_{\zeta \in \mathrm{Sh}_{T/Z}} q^{l(\zeta)} \zeta^{-1} \cdot m. \end{aligned} \tag{10.37}$$

When $q = 1$, we recover the tensor Coxeter bialgebra. The product is the same as before, while the coproduct is deformed by multiplying each summand by a power of q. Recall: $l(\zeta)$ is the length of ζ. We call Δ the *q-deshuffle coproduct.*

Observation 10.34. For a scalar q, for the Coxeter q-bialgebra $\overline{\mathcal{T}}_q(M)$, for a face-type T, using formulas (10.37),

$$\Delta_T^S \mu_T^S(m) = \sum_{w \in W_T} q^{l(w)} w \cdot m$$

because $\mathrm{Sh}_{S/T} = W_T$. This is a special case of (8.74) by Observation 10.33. Further, if the action of W on M is trivial, then the rhs above simplifies to scalar multiplication by $d_{S/T}(q)$ by formula (1.36). When $M = \Bbbk$, this reduces to Observation 10.9.

10.3.4. Shuffle Coxeter bialgebra. Recall the Coxeter space $\overline{\mathcal{T}}(M)$. It also carries a dual Coxeter bialgebra structure. To distinguish from the previous notation, we write $\overline{\mathcal{T}}^\vee(M)$. The product and coproduct are

$$\begin{aligned} \mu_Z^T &: \overline{\mathcal{T}}^\vee(M)[T] \to \overline{\mathcal{T}}^\vee(M)[Z] & \Delta_Z^T &: \overline{\mathcal{T}}^\vee(M)[Z] \to \overline{\mathcal{T}}^\vee(M)[T] \\ m &\mapsto \sum_{\zeta \in \mathrm{Sh}_{T/Z}} \zeta \cdot m & m &\mapsto m. \end{aligned} \tag{10.38}$$

We call this the *shuffle Coxeter bialgebra* of the left W-module M. It is commutative. However, it is not cocommutative in general. We call μ the *shuffle product* and Δ the *deconcatenation coproduct.*

10.3.5. Shuffle Coxeter q-bialgebra. More generally, for a scalar q, the Coxeter space $\overline{\mathcal{T}}^\vee(M)$ carries a Coxeter q-bialgebra structure which we denote $\overline{\mathcal{T}}_q^\vee(M)$. We call it the *shuffle Coxeter q-bialgebra* of M. The product and coproduct are

$$\mu_Z^T : \overline{\mathcal{T}}_q^\vee(M)[T] \to \overline{\mathcal{T}}_q^\vee(M)[Z] \qquad \Delta_Z^T : \overline{\mathcal{T}}_q^\vee(M)[Z] \to \overline{\mathcal{T}}_q^\vee(M)[T]$$

$$m \mapsto \sum_{\zeta \in \mathrm{Sh}_{T/Z}} q^{l(\zeta)} \zeta \cdot m \qquad\qquad m \mapsto m. \tag{10.39}$$

When $q = 1$, we recover the shuffle Coxeter bialgebra. The coproduct is the same as before, while the product is deformed by multiplying each summand by a power of q. We call μ the *q-shuffle product*.

10.3.6. Enumeration identities in the Coxeter group algebra. Let us see why $\overline{\mathcal{T}}_q(M)$ and $\overline{\mathcal{T}}_q^\vee(M)$ are Coxeter q-bialgebras. These checks generalize the ones made for $\Bbbk_q[x]$ in Section 10.2.4.

- The associativity and unitality axioms in (8.41) for $\overline{\mathcal{T}}_q(M)$ and coassociativity and counitality axioms in (8.46) for $\overline{\mathcal{T}}_q^\vee(M)$ are clear. For naturality (in either case), we use Exercise 8.20.
- For the associativity axiom for $\overline{\mathcal{T}}_q^\vee(M)$ and coassociativity axiom for $\overline{\mathcal{T}}_q(M)$, we use the two identities in (1.66), respectively, which encode the multiplicative property of shuffles. For naturality, we use the two identities in (1.70).
- For the q-bialgebra axiom (8.73) for $\overline{\mathcal{T}}_q^\vee(M)$ and $\overline{\mathcal{T}}_q(M)$, we use the two identities in (1.69), respectively.
- For the commutativity axiom (8.86) for $\overline{\mathcal{T}}^\vee(M)$ and cocommutativity axiom (8.91) for $\overline{\mathcal{T}}(M)$, we use the two identities in (1.71), respectively.

Note very carefully: Unlike for $\Bbbk[x]$ the commutativity axiom does not work for $\overline{\mathcal{T}}(M)$ since $m \neq w \cdot m$ in general.

The existence of $\overline{\mathcal{T}}_q(M)$ and $\overline{\mathcal{T}}_q^\vee(M)$ can also be asserted by viewing them as universal objects (Section 10.3.15). By reversing the logical flow, the above calculations then imply the Coxeter group algebra identities. See Section 18.3.3 for related comments.

10.3.7. Norm map. The tensor Coxeter bialgebra and shuffle Coxeter bialgebra relate to each other by the *norm map* as follows.

Lemma 10.35. *The norm map*

$$\bar{\kappa}_M : \overline{\mathcal{T}}(M) \to \overline{\mathcal{T}}^\vee(M), \qquad m \mapsto \sum_{w \in W_T} w \cdot m \tag{10.40}$$

on the T-component, is a morphism of Coxeter bialgebras.

A more general result is given in Lemma 10.36 below.

10.3.8. q-norm map. For a scalar q, the tensor Coxeter q-bialgebra and shuffle Coxeter q-bialgebra relate to each other by the *q-norm map* as follows.

Lemma 10.36. *For a scalar q, the q-norm map*

$$\overline{\kappa}_{q,M} : \overline{\mathcal{T}}_q(M) \to \overline{\mathcal{T}}_q^{\vee}(M), \qquad m \mapsto \sum_{w \in W_T} q^{l(w)}\, w \cdot m \tag{10.41}$$

on the T-component, is a morphism of Coxeter q-bialgebras.

Proof. That $\overline{\kappa}_{q,M}$ is a map of Coxeter spaces follows from Exercise 1.21. The checks that $\overline{\kappa}_{q,M}$ is a morphism of Coxeter algebras and of Coxeter coalgebras, respectively, reduce to the two identities in (1.68) applied to $\boldsymbol{\alpha}_Z$. □

Corollary 10.37. *For q not a root of unity, the q-norm map (10.41) is an isomorphism, and hence the Coxeter q-bialgebras $\overline{\mathcal{T}}_q(M)$ and $\overline{\mathcal{T}}_q^{\vee}(M)$ are isomorphic.*

Proof. Apply Fact 1.76 to each W_T (which is a Coxeter group). □

For general context on the q-norm map, see Exercise 10.50.

Exercise 10.38. Deduce using Exercise 1.77 that for q not a root of unity, product and coproduct components of $\overline{\mathcal{T}}_q(M)$ and $\overline{\mathcal{T}}_q^{\vee}(M)$ are isomorphisms. This also follows from Corollary 10.37.

10.3.9. Signed Coxeter bialgebras. Let us now look at the case $q = -1$. We have the signed Coxeter bialgebras $\overline{\mathcal{T}}_{-1}(M)$ and $\overline{\mathcal{T}}_{-1}^{\vee}(M)$. We call them the *tensor signed Coxeter bialgebra* and *shuffle signed Coxeter bialgebra*, respectively. The former is signed cocommutative, the latter is signed commutative. We use the terms *signed deshuffle coproduct* and *signed shuffle product* for their coproduct and product, respectively.

The *signed norm map*

$$\overline{\kappa}_{-1,M} : \overline{\mathcal{T}}_{-1}(M) \to \overline{\mathcal{T}}_{-1}^{\vee}(M), \qquad m \mapsto \sum_{w \in W_T} (-1)^{l(w)}\, w \cdot m \tag{10.42}$$

on the T-component, is a morphism of signed Coxeter bialgebras. It is the case $q = -1$ of (10.41).

10.3.10. Tensor Coxeter 0-bialgebra. Let us now look at the case $q = 0$. Since $0^{l(\zeta)} = 0$ unless $\zeta = \mathrm{id}$, we obtain

$$\overline{\mathcal{T}}_0(M) = \overline{\mathcal{T}}_0^{\vee}(M),$$

with product and coproduct

$$\begin{array}{ll} \mu_Z^T : \overline{\mathcal{T}}_0(M)[T] \to \overline{\mathcal{T}}_0(M)[Z] & \qquad \Delta_Z^T : \overline{\mathcal{T}}_0(M)[Z] \to \overline{\mathcal{T}}_0(M)[T] \\ \qquad\quad m \mapsto m & \qquad\qquad m \mapsto m. \end{array} \tag{10.43}$$

Thus, the product is concatenation (0-shuffle), while the coproduct is deconcatenation (0-deshuffle). This is the tensor Coxeter 0-bialgebra as well as the shuffle Coxeter 0-bialgebra.

The 0-norm map $\overline{\kappa}_{0,M}$ is the identity, and, in particular, an isomorphism. It is also clear that formulas (10.43) satisfy the 0-bialgebra axiom (8.78). The result of Exercise 8.81 is also clear.

10.3.11. Symmetric Coxeter bialgebras. For a left W-module M, define Coxeter spaces $\overline{\mathcal{S}}(M)$ and $\overline{\mathcal{S}}^\vee(M)$ by setting

$$\overline{\mathcal{S}}(M)[T] := M_{W_T} \quad \text{and} \quad \overline{\mathcal{S}}^\vee(M)[T] := M^{W_T} \tag{10.44}$$

for a face-type T. The rhs refers, respectively, to coinvariants and invariants of M wrt the subgroup W_T. The Coxeter space structures of $\overline{\mathcal{S}}(M)$ and $\overline{\mathcal{S}}^\vee(M)$ are induced from those of $\overline{\mathcal{T}}(M)$ and $\overline{\mathcal{T}}^\vee(M)$ via the canonical maps

$$\overline{\mathcal{T}}(M) \twoheadrightarrow \overline{\mathcal{S}}(M) \quad \text{and} \quad \overline{\mathcal{S}}^\vee(M) \hookrightarrow \overline{\mathcal{T}}^\vee(M). \tag{10.45}$$

This follows from Exercise 1.21. We call the maps in (10.45) as the *abelianization map* and *coabelianization map*, respectively. In fact, these maps turn $\overline{\mathcal{S}}(M)$ and $\overline{\mathcal{S}}^\vee(M)$ into bicommutative Coxeter bialgebras. We refer to them together as the *symmetric Coxeter bialgebras*.

Exercise 10.39. Recall abelianization and coabelianization of a Coxeter bialgebra from Section 9.4. Check:

$$\overline{\mathcal{S}}(M) = \overline{\mathcal{T}}(M)_{ab} \quad \text{and} \quad \overline{\mathcal{S}}^\vee(M) = \overline{\mathcal{T}}^\vee(M)^{coab}. \tag{10.46}$$

This justifies the terminology used for the maps in (10.45).

Observation 10.40. This builds on Observation 10.32. As Coxeter spaces,

$$\begin{aligned} \overline{\mathcal{S}}(M) &= \mathrm{X}_M + \mathrm{X}_M^2 + \mathrm{X}_M^3 + \dots, \\ \overline{\mathcal{S}}^\vee(M) &= \mathrm{X}_M + \mathrm{X}_M^{\underline{2}} + \mathrm{X}_M^{\underline{3}} + \dots, \end{aligned} \tag{10.47}$$

the sums of commutative Cauchy powers of X_M. Recall from Section 9.1.2 that these come in two flavors, one of them goes with $\overline{\mathcal{S}}(M)$ and the other with $\overline{\mathcal{S}}^\vee(M)$.

Observation 10.41. This builds on Observation 10.33. The functor (8.30) takes both $\overline{\mathcal{S}}(M)$ and $\overline{\mathcal{S}}^\vee(M)$ to the W-module M. Note how the rhs of (10.35) and (10.47) all start with the term X_M.

10.3.12. Exterior signed Coxeter bialgebras. The symmetric Coxeter bialgebras have signed analogues defined as follows. For a left W-module M, recall the left W-module M_- obtained by tensoring M with the sign representation of W. Set

$$\overline{\mathcal{E}}(M)[T] := (M_-)_{W_T} \quad \text{and} \quad \overline{\mathcal{E}}^\vee(M)[T] := (M_-)^{W_T} \tag{10.48}$$

for a face-type T. In other words, we now take coinvariants and invariants wrt the action of W on M twisted by the sign representation. We refer to the canonical maps

$$\overline{\mathcal{T}}_{-1}(M) \twoheadrightarrow \overline{\mathcal{E}}(M) \quad \text{and} \quad \overline{\mathcal{E}}^\vee(M) \hookrightarrow \overline{\mathcal{T}}_{-1}^\vee(M) \tag{10.49}$$

as the *signed abelianization map* and *signed coabelianization map*, respectively. They turn $\overline{\mathcal{E}}(M)$ and $\overline{\mathcal{E}}^\vee(M)$ into signed bicommutative signed Coxeter bialgebras. We refer to them together as the *exterior signed Coxeter bialgebras*.

Exercise 10.42. Establish the signed analogue of formulas (10.46).

10.3.13. Image of the norm map and signed norm map. We have commutative diagrams

$$
\begin{array}{ccc}
\overline{\mathcal{T}}(M) & \xrightarrow{\overline{\kappa}_M} & \overline{\mathcal{T}}^{\vee}(M) \\
\downarrow & & \uparrow \\
\overline{\mathcal{S}}(M) & \longrightarrow & \overline{\mathcal{S}}^{\vee}(M)
\end{array}
\qquad\qquad
\begin{array}{ccc}
\overline{\mathcal{T}}_{-1}(M) & \xrightarrow{\overline{\kappa}_{-1,M}} & \overline{\mathcal{T}}^{\vee}_{-1}(M) \\
\downarrow & & \uparrow \\
\overline{\mathcal{E}}(M) & \longrightarrow & \overline{\mathcal{E}}^{\vee}(M)
\end{array}
\tag{10.50}
$$

of Coxeter bialgebras and of signed Coxeter bialgebras, respectively. The top-horizontal maps are the norm map and signed norm map, respectively. They induce the bottom-horizontal maps. We claim that they are isomorphisms when $\mathrm{char}(\Bbbk) \nmid |W|$. To see this, let us evaluate the first diagram above on the central face-type. This yields

$$
\begin{array}{ccc}
M & \longrightarrow & M \\
\downarrow & & \uparrow \\
M_W & \longrightarrow & M^W,
\end{array}
$$

where the top-horizontal map is given by the action of $\sum_{w \in W} w$. It is the norm map of the group W. It factors through W-coinvariants and maps into W-invariants and induces an isomorphism between them, see [**7**, Diagram (2.64) and Lemma 2.20]. On an arbitrary face-type T, we get the same diagram with W replaced by W_T.

To summarize:

Lemma 10.43. *Let* $\mathrm{char}(\Bbbk) \nmid |W|$. *The image and coimage of the norm map* $\overline{\kappa}_M$ *are the symmetric Coxeter bialgebras* $\overline{\mathcal{S}}^{\vee}(M)$ *and* $\overline{\mathcal{S}}(M)$ *isomorphic to each other, and of the signed norm map* $\overline{\kappa}_{-1,M}$ *are the exterior signed Coxeter bialgebras* $\overline{\mathcal{E}}^{\vee}(M)$ *and* $\overline{\mathcal{E}}(M)$ *isomorphic to each other.*

Exercise 10.44. Check: For a scalar q,

$$\overline{\mathcal{T}}_q(M_-)_- = \overline{\mathcal{T}}_{-q}(M)$$

as Coxeter $(-q)$-bialgebras. Similarly,

$$\overline{\mathcal{S}}(M_-)_- = \overline{\mathcal{E}}(M) \quad \text{and} \quad \overline{\mathcal{S}}^{\vee}(M_-)_- = \overline{\mathcal{E}}^{\vee}(M)$$

as signed Coxeter bialgebras. In fact, conjugating the left diagram in (10.50) by $(-)_-$ yields the right diagram, and vice versa.

10.3.14. Duality. Let M be a left W-module. Observe: For a scalar q,

$$\overline{\mathcal{T}}_q(M)^* = \overline{\mathcal{T}}_q^{\vee}(M^*) \quad \text{and} \quad \overline{\mathcal{T}}_q^{\vee}(M)^* = \overline{\mathcal{T}}_q(M^*)$$

as Coxeter q-bialgebras. Similarly,

$$\overline{\mathcal{S}}(M)^* = \overline{\mathcal{S}}^{\vee}(M^*) \quad \text{and} \quad \overline{\mathcal{S}}^{\vee}(M)^* = \overline{\mathcal{S}}(M^*)$$

as Coxeter bialgebras, and

$$\overline{\mathcal{E}}(M)^* = \overline{\mathcal{E}}^{\vee}(M^*) \quad \text{and} \quad \overline{\mathcal{E}}^{\vee}(M)^* = \overline{\mathcal{E}}(M^*)$$

as signed Coxeter bialgebras.

10.3.15. Freeness and cofreeness. We now discuss the (co)freeness properties of the tensor Coxeter bialgebra and its relatives. A summary is provided in Table 10.4.

Proposition 10.45. *Let* X_M *be the Coxeter space defined by* (10.4)*. Then, for a scalar* q*,*

$$\overline{\mathcal{T}}_q(M) = \overline{\mathcal{T}}_q(\mathrm{X}_M) \quad \text{and} \quad \overline{\mathcal{T}}_q^{\vee}(M) = \overline{\mathcal{T}}_q^{\vee}(\mathrm{X}_M)$$

as Coxeter q*-bialgebras (under the canonical identifications).*

Proof. The face-type components of the above Coxeter q-bialgebras are all M and we identify them canonically. (By (9.18), $\overline{\mathcal{T}}_q(\mathrm{X}_M)[T] = \mathrm{X}_M[S] = M$.) To see that (co)products match, put $\mathrm{V} := \mathrm{X}_M$ in formulas (9.25) and (9.34), respectively, and use Exercise 8.73. Observe how ζ^{-1} pops out in one formula, and ζ in the other. □

Similarly,

$$\begin{aligned} \overline{\mathcal{S}}(M) &= \overline{\mathcal{S}}(\mathrm{X}_M) \quad \text{and} \quad \overline{\mathcal{S}}^{\vee}(M) = \overline{\mathcal{S}}^{\vee}(\mathrm{X}_M), \\ \overline{\mathcal{E}}(M) &= \overline{\mathcal{E}}(\mathrm{X}_M) \quad \text{and} \quad \overline{\mathcal{E}}^{\vee}(M) = \overline{\mathcal{E}}^{\vee}(\mathrm{X}_M) \end{aligned}$$

as Coxeter bialgebras and signed Coxeter bialgebras, respectively. Formulas (10.47) are special cases of (9.39) and (9.41). Formulas (10.46) are a special case of (9.55).

In particular, $\overline{\mathcal{T}}(M)$ is the free Coxeter algebra on X_M, while $\overline{\mathcal{S}}(M)$ and $\overline{\mathcal{E}}(M)$ are the free commutative and free signed commutative Coxeter algebras on X_M, respectively. Dually, $\overline{\mathcal{T}}^{\vee}(M)$ is the cofree Coxeter coalgebra on X_M, while $\overline{\mathcal{S}}^{\vee}(M)$ and $\overline{\mathcal{E}}^{\vee}(M)$ are the cofree cocommutative and cofree signed cocommutative Coxeter coalgebras on X_M, respectively.

When $M = \Bbbk$ with trivial W-action, recall that $\mathrm{X}_M = \mathrm{X}$, and hence the above considerations specialize to those in Section 10.2.14.

Exercise 10.46. Use the functor (10.5) to deduce that the constructions $\overline{\mathcal{T}}_q(M)$, $\overline{\mathcal{T}}_q^{\vee}(M)$, $\overline{\mathcal{S}}(M)$, $\overline{\mathcal{S}}^{\vee}(M)$, $\overline{\mathcal{E}}(M)$, $\overline{\mathcal{E}}^{\vee}(M)$ are all functorial in M. In particular, this includes the result of Exercise 10.31.

10.3.16. Classifying Coxeter bialgebra morphisms. We now use the above (co)freeness properties to classify Coxeter bialgebra morphisms from $\overline{\mathcal{T}}(M)$ or into $\overline{\mathcal{T}}^{\vee}(M)$.

Exercise 10.47 (Universal property). Specialize Theorem 9.11 to deduce that the tensor Coxeter algebra satisfies the following universal property.

Let A be a Coxeter algebra, M a left W-module, $f : M \to \mathrm{A}[S]$ a map of left W-modules. Then there exists a unique morphism of Coxeter algebras $\hat{f} : \overline{\mathcal{T}}(M) \to \mathrm{A}$ such that the diagram

$$\begin{array}{ccc} \overline{\mathcal{T}}(M) & \xrightarrow{\hat{f}} & \mathrm{A} \\ \uparrow & \nearrow_{f} & \\ \mathrm{X}_M & & \end{array}$$

commutes. The inclusion $\mathrm{X}_M \hookrightarrow \overline{\mathcal{T}}(M)$ arises from (10.35). Explicitly, the map $\hat{f}$, on the Z-component, is

$$\overline{\mathcal{T}}(M)[Z] = M \xrightarrow{f} \mathrm{A}[S] \xrightarrow{\mu_Z^S} \mathrm{A}[Z].$$

(Recall from Exercise 10.3 that a left W-module map $M \to \mathrm{A}[S]$ is the same as a Coxeter space map $\mathrm{X}_M \to \mathrm{A}$.)

Similarly, specializing Theorem 9.16 yields a universal property for the Coxeter coalgebra $\overline{\mathcal{T}}^\vee(M)$.

The above considerations generalize Exercise 10.18.

Exercise 10.48. Check: The result of Exercise 10.19 holds more generally for $\overline{\mathcal{T}_q}(M)$ instead of $\Bbbk_q[x]$, and $\overline{\mathcal{T}_q}^\vee(M)$ instead of $\Bbbk_q\{x\}$.

Exercise 10.49 (Universal property). Let M be a left W-module. Let H be a Coxeter q-bialgebra for a scalar q. Deduce:

(1) Coxeter q-bialgebra morphisms from $\overline{\mathcal{T}_q}(M)$ to H are parametrized by left W-module maps $M \to \mathrm{H}[S]$.
(2) Coxeter q-bialgebra morphisms from H to $\overline{\mathcal{T}_q}^\vee(M)$ are parametrized by left W-module maps $\mathrm{H}[S] \to M$.

When $M = \Bbbk$ with trivial W-action, we recover the result of Exercise 10.20.

Exercise 10.50 (General q-norm map). Let M and M' be left W-modules. For a scalar q, Coxeter q-bialgebra morphisms $\overline{\mathcal{T}_q}(M) \to \overline{\mathcal{T}_q}^\vee(M')$ are parametrized by left W-module maps $M \to M'$. For $M = M'$ and $\mathrm{id} : M \to M'$, we recover the q-norm map (10.41).

(i) Deduce the above fact from Exercise 10.49, item (1) or item (2).
(ii) Deduce it from the bijection (9.144) applied to $\mathrm{C} = \mathrm{X}_M$ and $\mathrm{A} = \mathrm{X}_{M'}$ and the fact that Coxeter space maps $\mathrm{X}_M \to \mathrm{X}_{M'}$ are the same as W-module maps $M \to M'$. This fits into the special case of (9.145). As a prelude: The q-norm map (10.41) is an instance of (9.140) for $\mathrm{V} := \mathrm{X}_M$, after the identification in Proposition 10.45. Hence, Corollary 10.37 is an instance of Lemma 9.101. Further, diagrams (10.50) are obtained by evaluating diagrams (9.141) at X_M. Hence, Lemma 10.43 is an instance of Lemma 9.94.

When $M = M' = \Bbbk$ with trivial W-action, we recover the result of Exercise 10.21.

10.3.17. Primitive and indecomposable parts.

We have

$$\overline{\mathcal{P}}(\overline{\mathcal{T}}^\vee(M)) = \mathrm{X}_M \quad \text{and} \quad \overline{\mathcal{Q}}(\overline{\mathcal{T}}(M)) = \mathrm{X}_M. \tag{10.51}$$

This follows from the coproduct formula in (10.38) and product formula in (10.36), respectively.

Similarly, for q not a root of unity,

$$\begin{aligned} \overline{\mathcal{P}}(\overline{\mathcal{T}_q}(M)) &= \overline{\mathcal{P}}(\overline{\mathcal{T}_q}^\vee(M)) = \mathrm{X}_M, \\ \overline{\mathcal{Q}}(\overline{\mathcal{T}_q}^\vee(M)) &= \overline{\mathcal{Q}}(\overline{\mathcal{T}_q}(M)) = \mathrm{X}_M. \end{aligned} \tag{10.52}$$

We explain the first claim. The first equality follows from Corollary 10.37. The coproduct of $\overline{\mathcal{T}_q}^\vee(M)$ does not depend on q, so the second equality follows from the first formula in (10.51).

Exercise 10.51. The primitive part of the tensor Coxeter bialgebra $\overline{\mathcal{T}}(M)$ is related as follows to the primitive part of the Coxeter bimonoid of decorated chambers Γ_M defined in Section 4.9.2. Check: For $m \in M$,

$$m \in \overline{\mathcal{P}}(\overline{\mathcal{T}}(M))[Z] \iff \sum_{\zeta \in \mathrm{Sh}_{T/Z}} \zeta^{-1} \cdot m = 0 \text{ for all } T > Z$$
$$\iff (w \cdot m)_{w \in W_Z} \in \mathcal{P}(\Gamma_M)[A] \subseteq \bigoplus_{C:\, C \geq A} M,$$

where A is the face of C_r of type Z, and the element $w \cdot m$ belongs to the summand M indexed by $C := wC_r$. Hence,

$$\overline{\mathcal{P}}(\overline{\mathcal{T}}(M))[A, C] = (\mathsf{Lie}[A] \otimes M)^{W_A}. \tag{10.53}$$

These are invariant decorated Lie elements. See also formula (12.29) in this regard.

If $\mathrm{char}(\Bbbk) \nmid |W|$, then

$$\begin{aligned} \overline{\mathcal{P}}(\overline{\mathcal{S}}(M)) &= \overline{\mathcal{P}}(\overline{\mathcal{S}}^{\vee}(M)) = \mathrm{X}_M, \\ \overline{\mathcal{Q}}(\overline{\mathcal{S}}^{\vee}(M)) &= \overline{\mathcal{Q}}(\overline{\mathcal{S}}(M)) = \mathrm{X}_M. \end{aligned} \tag{10.54}$$

For the first equality in either claim, we use Lemma 10.43. The second equalities are clear from respective coproduct and product formulas (induced from those of $\overline{\mathcal{T}}^{\vee}(M)$ and $\overline{\mathcal{T}}(M)$).

10.3.18. Primitive and group-like series. We now apply the discussion in Section 9.10 to the tensor, shuffle, symmetric Coxeter bialgebras. A series v of the Coxeter space $\overline{\mathcal{T}}(M)$ consists of elements $v_T \in M$, one for each face-type T, such that $\sigma(v_T) = v_U$ for any morphism $\sigma : T \to U$. Observe: $v_\emptyset \in M$ is arbitrary, while $v_S \in M$ must be W-invariant, that is, $v_S \in M^W$. Thus, a linear series x of $\overline{\mathcal{T}}(M)$ is the same as an element $x_S \in M^W$.

Let $\mathrm{char}(\Bbbk) \nmid |W|$. For x a linear series (and hence also a primitive series) of $\overline{\mathcal{T}}(M)$,

$$\mathbf{exp}(x)_Z = \boldsymbol{\zeta}(Z, S)\, \mu_Z^S(x_S) = \boldsymbol{\zeta}(Z, S)\, x_S = \tfrac{1}{d_{S/Z}}\, x_S.$$

This is an exponential series (and hence also a group-like series) of $\overline{\mathcal{T}}(M)$. Applying $\mathbf{log}$ to it recovers x. The above formula is an instance of (9.136).

Let us now turn to $\overline{\mathcal{T}}^{\vee}(M)$. Every primitive series of $\overline{\mathcal{T}}^{\vee}(M)$ is linear. This follows from the first formula in (10.51). For such a series x,

$$\mathbf{exp}(x)_Z = \boldsymbol{\zeta}(Z, S)\, \mu_Z^S(x_S) = \boldsymbol{\zeta}(Z, S)\, d_{S/Z}\, x_S = x_S.$$

All group-like series of $\overline{\mathcal{T}}^{\vee}(M)$ are exponential and have the above form (with the same element $x_S \in M^W$ in each component).

For q not a root of unity, we can do the same analysis as above for $\overline{\mathcal{T}}_q(M)$ and $\overline{\mathcal{T}}_q{}^{\vee}(M)$ replacing $\mathbf{exp}$ by $\mathbf{exp}_q$, $\boldsymbol{\zeta}$ by $\boldsymbol{\zeta}_q$, $d_{S/Z}$ by $d_{S/Z}(q)$.

Let us now look at $\overline{\mathcal{S}}(M)$ and $\overline{\mathcal{S}}^{\vee}(M)$. So again let $\mathrm{char}(\Bbbk) \nmid |W|$. The primitive series of $\overline{\mathcal{S}}(M)$ and $\overline{\mathcal{S}}^{\vee}(M)$ are linear (and parametrized by $x_S \in M^W$). This follows from the first formula in (10.54). Hence every exponential series is group-like. For $\overline{\mathcal{S}}^{\vee}(M)$, it is of the form (x_S) (with the same element $x_S \in M^W$ in each component). For $\overline{\mathcal{S}}(M)$, it is of the form $(\frac{1}{d_{S/Z}}\, \bar{x}_S)$, where $\bar{x}_S$ denotes the image of x_S in the quotient M_{W_Z}.

When $M = \Bbbk$ with trivial W-action, $\overline{\mathcal{T}}(M) = \overline{\mathcal{S}}(M) = \Bbbk[x]$ and $\overline{\mathcal{T}}^{\vee}(M) = \overline{\mathcal{S}}^{\vee}(M) = \Bbbk\{x\}$, and the above considerations specialize to those in Section 10.2.17.

10.3.19. Characteristic operations.

Exercise 10.52. Apply (9.169) to $\mathrm{H} := \overline{\mathcal{T}}(M)$ to deduce: For a left W-module M, the invariant Tits algebra acts on M via

$$\mathtt{H}_T \cdot m := \sum_{\zeta \in \mathrm{Sh}_T} \zeta^{-1} \cdot m. \tag{10.55}$$

Use this to deduce the existence of the algebra morphism (1.152).

Similarly, apply (9.168) to $\mathrm{H} := \overline{\mathcal{T}}_q(M)$ to deduce: For a left W-module M, for a scalar q, the invariant q-Janus algebra acts on M via

$$\mathtt{H}_{(T,\sigma,T')} \cdot m := \sum_{\zeta \in \mathrm{Sh}_{T'}} q^{l(\zeta)} (w\zeta^{-1}) \cdot m, \tag{10.56}$$

with $\sigma = w\beta = \beta w$ in normal form. Use this to deduce the existence of the algebra morphism (1.153). To get (1.154), use $\mathrm{H} := \overline{\mathcal{T}}_q^{\vee}(M)$ instead.

10.3.20. Antipode.

Lemma 10.53. *For a scalar q, the antipode S of the Coxeter q-bialgebra $\overline{\mathcal{T}}_q(M)$, on the Z-component, is*

$$S_Z(m) = (-1)^{\dim(\mathcal{A})} q^{l(w_Z)} w_Z \cdot m, \tag{10.57}$$

where w_Z is the longest element in W_Z.

PROOF. Substitute product and coproduct formulas (10.37) in the Takeuchi formula (9.171) and use the second identity in (1.72) applied to $\boldsymbol{\alpha}_Z$. □

Lemma 10.54. *For a scalar q, the antipode S of the Coxeter q-bialgebra $\overline{\mathcal{T}}_q^{\vee}(M)$, on the Z-component, is*

$$S_Z(m) = (-1)^{\dim(\mathcal{A})} q^{l(w_Z)} w_Z \cdot m, \tag{10.58}$$

where w_Z is the longest element in W_Z.

PROOF. This formula is the same as (10.57). To prove it, we now use the first identity in (1.72) applied to $\boldsymbol{\alpha}_Z$. □

When $M = \Bbbk$ with trivial W-action, formulas (10.57) and (10.58) specialize to formulas (10.30) and (10.31), respectively.

Exercise 10.55. Show: The antipode S of both $\overline{\mathcal{S}}(M)$ and $\overline{\mathcal{S}}^{\vee}(M)$, on the Z-component, is

$$S_Z(m) = (-1)^{\dim(\mathcal{A})}\, m. \tag{10.59}$$

(Use (10.45) and formulas (10.57) and (10.58) for $q = 1$.)

Exercise 10.56. Deduce antipode formulas (10.57) and (10.58) as special cases of formulas (9.183) and (9.185) by taking the trivial Coxeter (co)algebra to be X_M. (Use Observation 1.39.)

10.4. Coxeter bialgebras of face-types

We introduce the Coxeter bialgebra of face-types $\overline{\Sigma}$. It is cocommutative but not commutative. It is free as a Coxeter algebra. In fact, it is the free Coxeter bialgebra on the divided power Coxeter coalgebra $\Bbbk\{x\}$. Thus, its T-component has a basis indexed by face-types greater than T. In addition to this canonical $\mathtt{H}$-basis, when $\operatorname{char}(\Bbbk) \nmid |W|$, we define a $\mathtt{Q}$-basis for $\overline{\Sigma}$. Invariant noncommutative zeta and Möbius functions intervene in the change of basis formulas. We also consider the dual Coxeter bialgebra of face-types $\overline{\Sigma}^*$. It is the cofree Coxeter bialgebra on the Coxeter algebra of polynomials $\Bbbk[x]$. A summary is given in Table 10.6. Face-type enumeration scalars enter into the coproduct formula of $\overline{\Sigma}$ and product formula of $\overline{\Sigma}^*$. These are structure constants of the invariant Tits algebra. The terminology for products and coproducts in Table 10.6 is motivated from classical Hopf algebras of noncommutative symmetric functions and quasisymmetric functions.

TABLE 10.6. Coxeter bialgebras of face-types.

Coxeter bialgebra		product	coproduct	universal property
$\overline{\Sigma}$	$\mathtt{H}$-basis for any $\Bbbk$	concatenation	dequasishuffle	$\overline{\mathcal{T}}(\Bbbk\{x\})$
	$\mathtt{Q}$-basis if $\operatorname{char}(\Bbbk) \nmid \lvert W\rvert$	concatenation	deshuffle	$\overline{\mathcal{T}}(\Bbbk\{x\}_t)$
$\overline{\Sigma}^*$	$\mathtt{M}$-basis for any $\Bbbk$	quasishuffle	deconcatenation	$\overline{\mathcal{T}}^\vee(\Bbbk[x])$
	$\mathtt{P}$-basis if $\operatorname{char}(\Bbbk) \nmid \lvert W\rvert$	shuffle	deconcatenation	$\overline{\mathcal{T}}^\vee(\Bbbk[x]_t)$

The Coxeter bialgebra $\overline{\Sigma}$ admits a deformation by a parameter q which we denote $\overline{\Sigma}_q$. It is a Coxeter q-bialgebra. This is done by replacing face-type enumeration scalars in the coproduct formula by face-type enumeration polynomials in q. These are q-analogues of the structure constants of the invariant Tits algebra. For q not a root of unity, in addition to the $\mathtt{H}$-basis, we define the $\mathtt{Q}$-basis for $\overline{\Sigma}_q$. Noncommutative q-zeta and q-Möbius functions intervene in the change of basis formulas. This includes the case $q = 0$. Similar considerations apply to $\overline{\Sigma}^*$. See Table 10.7. The terminology for the products and coproducts used here is consistent with that in Table 9.2.

For a scalar q, we define the q-norm map $\overline{\kappa}_{q,\xi}$ on face-types. It is a self-dual morphism of Coxeter q-bialgebras from $\overline{\Sigma}_q$ to $\overline{\Sigma}^*_q$. It depends on an invariant family of scalars ξ. For q not a root of unity, we give a necessary and sufficient condition for $\overline{\kappa}_{q,\xi}$ to be an isomorphism. It involves the invariant family of scalars η obtained from ξ by q-Möbius inversion. To show this, we express $\overline{\kappa}_{q,\xi}$ in the $\mathtt{Q}$-basis and its dual. In particular, this yields self-duality isomorphisms for the Coxeter q-bialgebra of face-types.

We introduce the universal and Takeuchi series as examples of group-like series of $\overline{\Sigma}$ and compute their logarithms. We connect $\overline{\Sigma}$ to the convolution Coxeter algebra, internal hom for Coxeter coalgebras, universal measuring

TABLE 10.7. Coxeter q-bialgebras of face-types.

Coxeter q-bialgebra		product	coproduct	universal property
$\overline{\Sigma}_q$	H-basis for any q	concatenation	q-dequasishuffle	$\overline{\mathcal{T}}_q(\Bbbk\{x\})$
	Q-basis for q not a root of unity	concatenation	q-deshuffle	$\overline{\mathcal{T}}_q(\Bbbk\{x\}_t)$
$\overline{\Sigma}^*_q$	M-basis for any q	q-quasishuffle	deconcatenation	$\overline{\mathcal{T}}_q^{\vee}(\Bbbk[x])$
	P-basis for q not a root of unity	q-shuffle	deconcatenation	$\overline{\mathcal{T}}_q^{\vee}(\Bbbk[x]_t)$

Coxeter coalgebra via morphisms which arise from freeness of $\overline{\Sigma}$. As an application, we use these morphisms to derive results about antipodes of Coxeter bialgebras by using the Takeuchi series.

The Coxeter bialgebras of face-types $\overline{\Sigma}$ and $\overline{\Sigma}^*$ play the same role for Coxeter bialgebras that the Coxeter bimonoids of faces $\mathbf{\Sigma}$ and $\mathbf{\Sigma}^*$ discussed in Section 4.5 play for Coxeter bimonoids. So they are of fundamental importance. The product of the Tits algebra, namely, $\mathtt{H}_F \cdot \mathtt{H}_G = \mathtt{H}_{FG}$, is set-theoretic, and this is what appears in the coproduct of $\mathbf{\Sigma}$. In contrast, the product of the invariant Tits algebra is not set-theoretic, that is, $\mathtt{H}_T \cdot \mathtt{H}_{T'}$ is a sum of H-basis elements, and it is their coefficients which appear in the coproduct of $\overline{\Sigma}$. This connection is made precise later in Section 18.4 using Fock functors.

10.4.1. Coxeter space of face-types. Define the Coxeter space $\overline{\Sigma}$ as follows. For a face-type T, let $\overline{\Sigma}[T]$ be the linear span of the set of face-types greater than T. This is the same as the face-types of $\boldsymbol{\alpha}_T$. We denote the canonical basis by H. The structure maps are as follows. For a morphism $\sigma : T \to U$,

$$\sigma : \overline{\Sigma}[T] \to \overline{\Sigma}[U], \qquad \mathtt{H}_{V/T} \mapsto \mathtt{H}_{\sigma_*(V)/U}, \tag{10.60}$$

with $\sigma_*(V)$ as defined in (8.12) and illustrated below. See also (1.28). In particular, $\mathtt{H}_{S/T} \mapsto \mathtt{H}_{S/U}$ since $\sigma_*(S) = S$.

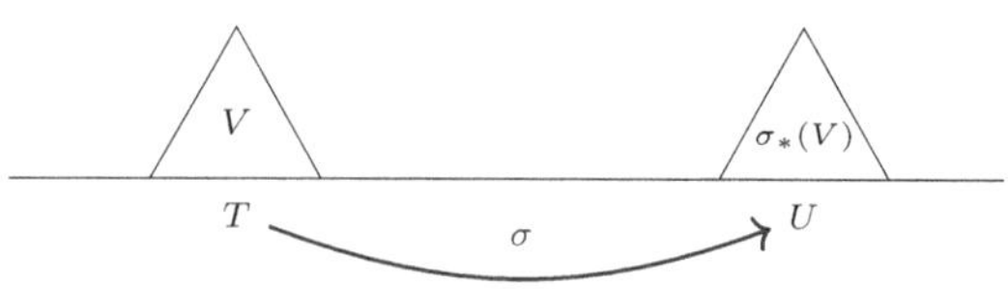

Axioms (8.28) express that $(\tau\sigma)_*(V) = \tau_*\sigma_*(V)$ and $\mathrm{id}_*(V) = V$. We call $\overline{\Sigma}$ the *Coxeter space of face-types*.

Observation 10.57. As a Coxeter space,

$$\overline{\Sigma} = \Bbbk[x] + \Bbbk[x]^2 + \Bbbk[x]^3 + \dots, \tag{10.61}$$

the sum of Cauchy powers of $\Bbbk[x]$. The Z-component of the rhs is the vector space $\bigoplus_{T \geq Z} \Bbbk[x][T]$. It has a basis indexed by face-types T greater than Z, and we identify this with the H-basis of $\overline{\Sigma}[Z]$.

Exercise 10.58. The functor (8.30) takes the Coxeter space $\overline{\Sigma}$ to the trivial representation of W, that is, $\overline{\Sigma}_q[S] = \Bbbk$ (spanned by $\mathtt{H}_{S/S}$) with trivial W-action. Check this directly or deduce it from (10.61) and Observation 10.8.

10.4.2. Coxeter bialgebra of face-types. We now turn $\overline{\Sigma}$ into a Coxeter bialgebra. The product and coproduct are

$$\begin{aligned} &\mu_Z^T : \overline{\Sigma}[T] \to \overline{\Sigma}[Z] && \Delta_Z^T : \overline{\Sigma}[Z] \to \overline{\Sigma}[T] \\ &\mathtt{H}_{U/T} \mapsto \mathtt{H}_{U/Z} && \mathtt{H}_{U/Z} \mapsto \sum_{V:\, V \geq T} a_Z^{TVU}\, \mathtt{H}_{V/T}, \end{aligned} \tag{10.62}$$

with the face-type enumeration scalars a_Z^{TVU} as defined in Section 1.8.3. (By formula (1.156), they are structure constants of the invariant Tits algebra of $\boldsymbol{\alpha}_Z$.) We refer to $\overline{\Sigma}$ as the *Coxeter bialgebra of face-types*. It is cocommutative. We call μ the *concatenation product* and Δ the *dequasishuffle coproduct.*

10.4.3. Coxeter q-bialgebra of face-types. More generally, for a scalar q, the Coxeter space $\overline{\Sigma}$ carries a Coxeter q-bialgebra structure which we denote $\overline{\Sigma}_q$. We call it the *Coxeter q-bialgebra of face-types*. The product and coproduct are

$$\begin{aligned} &\mu_Z^T : \overline{\Sigma}_q[T] \to \overline{\Sigma}_q[Z] && \Delta_Z^T : \overline{\Sigma}_q[Z] \to \overline{\Sigma}_q[T] \\ &\mathtt{H}_{U/T} \mapsto \mathtt{H}_{U/Z} && \mathtt{H}_{U/Z} \mapsto \sum_{V:\, V \geq T} a_Z^{TVU}(q)\, \mathtt{H}_{V/T}, \end{aligned} \tag{10.63}$$

with the face-type enumeration polynomials $a_Z^{TVU}(q)$ obtained by applying (1.38) to $\boldsymbol{\alpha}_Z$.

When $q = 1$, we recover the Coxeter bialgebra of face-types. The product is concatenation as before, while the coproduct is deformed by replacing the scalars a_Z^{TVU} by the polynomials $a_Z^{TVU}(q)$. We call Δ the *q-dequasishuffle coproduct*.

10.4.4. Face-type enumeration identities. The necessary checks that $\overline{\Sigma}_q$ is indeed a Coxeter q-bialgebra are elaborated below.

- For the product: The associativity and unitality axioms are clear. The naturality axiom follows from Exercise 8.21.
- For the coproduct: The naturality axiom follows from Exercise 1.70. The coassociativity axiom reduces to the gate identity (1.57) applied to $\boldsymbol{\alpha}_Z$ as shown below.

$$\begin{array}{ccc} \overline{\Sigma}_q[T] & & \\ \uparrow{\scriptstyle \Delta_Z^T} & \searrow^{\Delta_T^V} & \\ & & \overline{\Sigma}_q[V] \\ & \nearrow_{\Delta_Z^V} & \\ \overline{\Sigma}_q[Z] & & \end{array} \qquad \begin{array}{ccc} \sum_U a_Z^{TUT'}(q)\, \mathtt{H}_{U/T} & \longmapsto & \sum_Y \Big(\sum_U a_T^{VYU}(q)\, a_Z^{TUT'}(q) \Big)\, \mathtt{H}_{Y/V} \\ \uparrow & & \Vert \\ \mathtt{H}_{T'/Z} & \longmapsto & \sum_Y a_Z^{VYT'}(q)\, \mathtt{H}_{Y/V} \end{array}$$

  The counitality axiom follows from identity (1.40).
- The q-bialgebra axiom (8.73) involves a jungle of face-types, working through it yields the gate identity (1.62). We leave this as an exercise.

For $q = 1$, the cocommutativity axiom (8.91) follows from Exercise 1.69.

The existence of $\overline{\Sigma}_q$ can also be asserted by viewing it as a universal object (Section 10.4.13). By reversing the logical flow, the above calculations then imply identities (1.57) and (1.62). See Section 18.4.2 for related comments.

10.4.5. Embedding of the Coxeter bialgebra of polynomials. The Coxeter q-bialgebra of polynomials embeds inside the Coxeter q-bialgebra of face-types as follows.

Lemma 10.59. *For a scalar q, the map*

$$\mathbb{k}_q[x] \hookrightarrow \overline{\Sigma}_q, \qquad x^T \mapsto \mathtt{H}_{S/T} \tag{10.64}$$

on the T-component is a morphism of Coxeter q-bialgebras.

PROOF. Diagrams (8.29), (8.42), (8.47) are checked below.

$$\begin{array}{ccc} x^T & \longmapsto & \mathtt{H}_{S/T} \\ {\scriptstyle \sigma}\downarrow & & \downarrow{\scriptstyle \sigma} \\ x^U & \longmapsto & \mathtt{H}_{S/U} \end{array} \qquad \begin{array}{ccc} x^T & \longmapsto & \mathtt{H}_{S/T} \\ {\scriptstyle \mu^T_Z}\downarrow & & \downarrow{\scriptstyle \mu^T_Z} \\ x^Z & \longmapsto & \mathtt{H}_{S/Z} \end{array} \qquad \begin{array}{ccc} d_{T/Z}(q)\, x^T & \longmapsto & d_{T/Z}(q)\, \mathtt{H}_{S/T} \\ & & \| \\ {\scriptstyle \Delta^T_Z}\uparrow & & a^{TSS}_Z(q)\, \mathtt{H}_{S/T} \\ & & \uparrow{\scriptstyle \Delta^T_Z} \\ x^Z & \longmapsto & \mathtt{H}_{S/Z} \end{array}$$

The last diagram is interesting. The coproduct formula in (10.63) involves a sum, but $\Delta^T_Z(\mathtt{H}_{S/Z})$ has only one term. The equality in the diagram is identity (1.39) applied to $\boldsymbol{\alpha}_Z$. □

Exercise 10.60. Deduce Lemma 10.59 using the universal property of $\mathbb{k}_q[x]$. More precisely, apply Exercise 10.20, item (1) to the element $\mathtt{H}_{S/S} \in \overline{\Sigma}_q[S]$. In fact, any Coxeter q-bialgebra morphism $\mathbb{k}_q[x] \to \overline{\Sigma}_q$ is a scalar multiple of (10.64). (Use Exercise 10.58.) See Example 10.78 for a related discussion.

Exercise 10.61. Use Exercise 10.22 and Exercise 10.58 to deduce: If $\operatorname{char}(\mathbb{k}) \nmid |W|$, then any Coxeter bialgebra morphism $\mathbb{k}\{x\} \to \overline{\Sigma}$ is a scalar multiple of

$$\mathbb{k}\{x\} \hookrightarrow \overline{\Sigma}, \qquad x^{(T)} \mapsto \frac{1}{d_{S/T}} \mathtt{H}_{S/T}. \tag{10.65}$$

A similar statement holds for Coxeter q-bialgebra morphisms $\mathbb{k}_q\{x\} \to \overline{\Sigma}_q$ for q not a root of unity.

Exercise 10.62. For a scalar q, for the Coxeter q-bialgebra $\overline{\Sigma}_q$, for a face-type T,

$$\Delta^S_T \mu^S_T = d_{S/T}(q)\, \mathrm{id}\,.$$

Check this directly using formulas (10.63) or deduce it from Observation 10.9 and the fact that (10.64) is an isomorphism on the S-component. Also check that this identity is a special case of (8.74). (Use Exercise 10.58 and formula (1.36).)

10.4.6. Q-basis. Let $\mathrm{char}(\mathbb{k}) \nmid |W|$. Fix an invariant noncommutative zeta function $\boldsymbol{\zeta}$ and its inverse Möbius function $\boldsymbol{\mu}$. Define a Q-basis of $\overline{\Sigma}$, on the Z-component, by either of the two equivalent formulas

$$\mathtt{H}_{T/Z} = \sum_{U:\, T \leq U} \boldsymbol{\zeta}(T,U)\, \mathtt{Q}_{U/Z}, \qquad \mathtt{Q}_{T/Z} = \sum_{U:\, T \leq U} \boldsymbol{\mu}(T,U)\, \mathtt{H}_{U/Z}. \tag{10.66}$$

This is the same as (1.136) applied to $\boldsymbol{\alpha}_Z$ for each face-type Z. (Note: $\overline{\Sigma}[Z]$ identifies as a vector space with the invariant Tits algebra of $\boldsymbol{\alpha}_Z$.)

We now make the Coxeter bialgebra $\overline{\Sigma}$ explicit in the Q-basis. The upshot is that there is some simplification to be had.

Exercise 10.63. Let $\mathrm{char}(\mathbb{k}) \nmid |W|$. Check: The structure maps of the Coxeter space $\overline{\Sigma}$ in the Q-basis are as follows. For a morphism $\sigma : T \to U$,

$$\sigma : \overline{\Sigma}[T] \to \overline{\Sigma}[U], \qquad \mathtt{Q}_{V/T} \mapsto \mathtt{Q}_{\sigma_*(V)/U}. \tag{10.67}$$

The formula is the same as in the H-basis in (10.60). (Use Exercises 8.21 and 1.37.)

Proposition 10.64. *The product and coproduct of $\overline{\Sigma}$ in the Q-basis are*

$$\begin{aligned} &\mu_Z^T : \overline{\Sigma}[T] \to \overline{\Sigma}[Z] && \Delta_Z^T : \overline{\Sigma}[Z] \to \overline{\Sigma}[T] \\ &\mathtt{Q}_{U/T} \mapsto \mathtt{Q}_{U/Z} && \mathtt{Q}_{U/Z} \mapsto \sum_{\substack{V:\, V \geq T \\ \mathrm{s}(V/Z) = \mathrm{s}(U/Z)}} a_Z^{TVU}\, \mathtt{Q}_{V/T}, \end{aligned} \tag{10.68}$$

with scalars a_Z^{TVU} as defined in Section 1.8.3.

The product is concatenation as in the H-basis in (10.62). The coproduct simplifies with the "quasi" terms dropping out and we call it the *deshuffle coproduct*.

PROOF. We employ (10.62). The product calculation is easy.

$$\mu_Z^T(\mathtt{Q}_{Y/T}) = \mu_Z^T\Big(\sum_{U:\, Y \leq U} \boldsymbol{\mu}(Y,U)\, \mathtt{H}_{U/T} \Big) = \sum_{U:\, Y \leq U} \boldsymbol{\mu}(Y,U)\, \mathtt{H}_{U/Z} = \mathtt{Q}_{Y/Z}.$$

The coproduct calculation is more interesting.

$$\begin{aligned} \Delta_Z^T(\mathtt{Q}_{T'/Z}) &= \Delta_Z^T\Big(\sum_{U:\, U \geq T'} \boldsymbol{\mu}(T',U)\, \mathtt{H}_{U/Z} \Big) \\ &= \sum_{U:\, U \geq T'} \boldsymbol{\mu}(T',U) \sum_{V:\, V \geq T} a_Z^{TVU}\, \mathtt{H}_{V/T} \\ &= \sum_{V:\, V \geq T} \Big(\sum_{U:\, U \geq T'} a_Z^{TVU} \boldsymbol{\mu}(T',U) \Big) \mathtt{H}_{V/T} \\ &= \sum_{V:\, V \geq T} \Big(\sum_{\substack{Y:\, V \geq Y \geq T \\ \mathrm{s}(Y/Z) = \mathrm{s}(T'/Z)}} a_Z^{TYT'} \boldsymbol{\mu}(Y,V) \Big) \mathtt{H}_{V/T} \end{aligned}$$

$$= \sum_{\substack{Y:\,Y\geq T\\ \mathrm{s}(Y/Z)=\mathrm{s}(T'/Z)}} a_Z^{TYT'} \sum_{V:\,V\geq Y} \boldsymbol{\mu}(Y,V)\,\mathtt{H}_{V/T}$$

$$= \sum_{\substack{Y:\,Y\geq T\\ \mathrm{s}(Y/Z)=\mathrm{s}(T'/Z)}} a_Z^{TYT'}\,\mathtt{Q}_{Y/T}.$$

The fourth step is critical, and used identity (1.88) applied to $\boldsymbol{\alpha}_Z$.

Alternatively: By Example 9.121, the coproduct component Δ_Z^T is given by left multiplication by $\mathtt{H}_{T/Z}$ in $\overline{\Sigma}[Z]$ using formula (1.156) followed by viewing the result inside $\overline{\Sigma}[T]$ (under the canonical inclusion $\overline{\Sigma}[T] \hookrightarrow \overline{\Sigma}[Z]$ of vector spaces). So the coproduct formula in the Q-basis can be deduced from identity (1.140) applied to $\boldsymbol{\alpha}_Z$. This identity is equivalent to (1.88). □

10.4.7. Q-basis for q not a root of unity. Suppose q is not a root of unity. Recall the noncommutative q-zeta function $\boldsymbol{\zeta}_q$ and noncommutative q-Möbius function $\boldsymbol{\mu}_q$ from Section 1.10.6.

Define the Q-basis of $\overline{\Sigma}_q$, on the Z-component, by either of the two equivalent formulas

$$\mathtt{H}_{T/Z} = \sum_{U:\,T\leq U} \boldsymbol{\zeta}_q(T,U)\,\mathtt{Q}_{U/Z},$$
(10.69)
$$\mathtt{Q}_{T/Z} = \sum_{U:\,T\leq U} \boldsymbol{\mu}_q(T,U)\,\mathtt{H}_{U/Z}.$$

This is the same as formulas (10.66) with $\boldsymbol{\zeta}$ and $\boldsymbol{\mu}$ replaced by $\boldsymbol{\zeta}_q$ and $\boldsymbol{\mu}_q$, respectively.

Exercise 10.65. Check: For q not a root of unity, the structure maps of $\overline{\Sigma}_q$ in the Q-basis are as follows. For a morphism $\sigma : T \to U$,

$$\sigma : \overline{\Sigma}_q[T] \to \overline{\Sigma}_q[U], \qquad \mathtt{Q}_{V/T} \mapsto \mathtt{Q}_{\sigma_*(V)/U}. \tag{10.70}$$

(The formula and argument are the same as for the Q-basis of $\overline{\Sigma}$ in Exercise 10.63.)

Proposition 10.66. *For q not a root of unity, the product and coproduct of $\overline{\Sigma}_q$ in the Q-basis are*

$$\mu_Z^T : \overline{\Sigma}_q[T] \to \overline{\Sigma}_q[Z] \qquad\qquad \Delta_Z^T : \overline{\Sigma}_q[Z] \to \overline{\Sigma}_q[T]$$
(10.71)
$$\mathtt{Q}_{U/T} \mapsto \mathtt{Q}_{U/Z} \qquad\qquad \mathtt{Q}_{U/Z} \mapsto \sum_{\substack{V:\,V\geq T\\ \mathrm{s}(V/Z)=\mathrm{s}(U/Z)}} a_Z^{TVU}(q)\,\mathtt{Q}_{V/T},$$

with the polynomials $a_Z^{TVU}(q)$ obtained by applying (1.38) *to $\boldsymbol{\alpha}_Z$.*

We call Δ the *q-deshuffle coproduct*. It is simpler than the q-dequasishuffle coproduct in the H-basis in (10.63).

PROOF. The calculations are similar to those in the proof of Proposition 10.64. We now employ (10.63) and identity (1.98). □

Special case. For q not a root of unity, by formulas (10.63) and (10.71),

$$\begin{array}{ccccccc} \mathtt{H}_{S/T} & \overset{\mu_Z^T}{\longmapsto} & \mathtt{H}_{S/Z} & \qquad & \mathtt{H}_{S/Z} & \overset{\Delta_Z^T}{\longmapsto} & d_{T/Z}(q)\, \mathtt{H}_{S/T} \\ \| & & \| & & \| & & \| \\ \mathtt{Q}_{S/T} & \underset{\mu_Z^T}{\longmapsto} & \mathtt{Q}_{S/Z} & & \mathtt{Q}_{S/Z} & \underset{\Delta_Z^T}{\longmapsto} & d_{T/Z}(q)\, \mathtt{Q}_{S/T}. \end{array}$$

For the top arrow on the right, see the third diagram in the proof of Lemma 10.59.

Exercise 10.67 (Ranks one and two). In Examples 1.45 and 1.91, we computed $a_Z^{TVU}(q)$ and $\boldsymbol{\zeta}_q$ and $\boldsymbol{\mu}_q$ in ranks one and two.

In rank one, $\mu_\emptyset^S$ and $\Delta_\emptyset^S$ are the only nontrivial product and coproduct components. By formulas (10.63) and (10.71),

$$\begin{aligned} &\mu_\emptyset^S(\mathtt{H}_{S/S}) = \mathtt{H}_{S/\emptyset}, &&\Delta_\emptyset^S(\mathtt{H}_{\emptyset/\emptyset}) = \mathtt{H}_{S/S}, &&\Delta_\emptyset^S(\mathtt{H}_{S/\emptyset}) = (1+q)\, \mathtt{H}_{S/S}, \\ &\mu_\emptyset^S(\mathtt{Q}_{S/S}) = \mathtt{Q}_{S/\emptyset}, &&\Delta_\emptyset^S(\mathtt{Q}_{\emptyset/\emptyset}) = 0, &&\Delta_\emptyset^S(\mathtt{Q}_{S/\emptyset}) = (1+q)\, \mathtt{Q}_{S/S}. \end{aligned}$$

Check: These formulas transform into each other via change of basis (10.69), that is,

$$\mathtt{Q}_{S/S} = \mathtt{H}_{S/S}, \quad \mathtt{Q}_{S/\emptyset} = \mathtt{H}_{S/\emptyset}, \quad \mathtt{Q}_{\emptyset/\emptyset} = \mathtt{H}_{\emptyset/\emptyset} - \tfrac{1}{1+q}\, \mathtt{H}_{S/\emptyset}.$$

Write down (co)products in the H- and Q-bases for rank two arrangements, and check that they transform into each other via change of basis (10.69).

Note: For writing down (co)products in rank three, we need to compute face-type enumeration polynomials as indicated in Exercise 1.46.

10.4.8. Invariant Tits algebra and eulerian idempotents.

We return to the Coxeter bialgebra $\overline{\Sigma}$. Let $\mathrm{char}(\Bbbk) \nmid |W|$. For a face-type Z, we identify $\overline{\Sigma}[Z]$ with the invariant Tits algebra of $\boldsymbol{\alpha}_Z$ with basis elements $\mathtt{H}_{T/Z}$. For a flat-type λ/Z of $\boldsymbol{\alpha}_Z$, put

$$\mathtt{E}_{\lambda/Z} := \sum_{\substack{T:\, T \geq Z \\ \mathrm{s}(T/Z) = \lambda/Z}} \zeta(Z,T)\, \mathtt{Q}_{T/Z}. \tag{10.72}$$

In particular, for the minimum and maximum flat-types,

$$\mathtt{E}_{(\lambda/Z)_{\min}} = \mathtt{E}_{\mathrm{s}(Z/Z)} = \mathtt{Q}_{Z/Z} \quad \text{and} \quad \mathtt{E}_{(\lambda/Z)_{\max}} = \frac{1}{|W_Z|} \mathtt{Q}_{S/Z} = \frac{1}{|W_Z|} \mathtt{H}_{S/Z}.$$

The $\mathtt{E}_{\lambda/Z}$, as λ/Z varies, are the invariant eulerian idempotents of the invariant Tits algebra of $\boldsymbol{\alpha}_Z$, see (1.138).

Exercise 10.68. Check: For any face-types $Z \leq T$,

$$\Delta_Z^T(\mathtt{E}_{\mathrm{s}(T/Z)}) = \mathtt{E}_{\mathrm{s}(T/T)} = \mathtt{Q}_{T/T}. \tag{10.73}$$

(Use the coproduct formula in (10.68) and flat-additivity formula (1.83).)

For any flat-type λ/Z,

$$\sum_{T:\, \mathrm{s}(T/Z) = \lambda/Z} \zeta(Z,T)\, \mu_Z^T \Delta_Z^T(\mathtt{E}_{\lambda/Z}) = \mathtt{E}_{\lambda/Z}. \tag{10.74}$$

(Use identity (10.73) and product formula in (10.68).)

10.4.9. Dual Coxeter bialgebra of face-types. Let $\overline{\Sigma}^*$ be the Coxeter bialgebra dual to $\overline{\Sigma}$. We denote the basis dual to $\mathtt{H}$ by $\mathtt{M}$. The structure maps are as follows. For a morphism $\sigma : T \to U$,

$$\sigma : \overline{\Sigma}^*[T] \to \overline{\Sigma}^*[U], \qquad \mathtt{M}_{V/T} \mapsto \mathtt{M}_{\sigma_*(V)/U}. \tag{10.75}$$

The product and coproduct are

$$\begin{aligned} &\mu_Z^T : \overline{\Sigma}^*[T] \to \overline{\Sigma}^*[Z] && \Delta_Z^T : \overline{\Sigma}^*[Z] \to \overline{\Sigma}^*[T] \\ &\mathtt{M}_{V/T} \mapsto \sum_{U:\,U \geq Z} a_Z^{TVU}\, \mathtt{M}_{U/Z} && \mathtt{M}_{U/Z} \mapsto \begin{cases} \mathtt{M}_{U/T} & \text{if } U \geq T, \\ 0 & \text{otherwise.} \end{cases} \end{aligned} \tag{10.76}$$

The product is commutative. We call μ the *quasishuffle product* and Δ the *deconcatenation coproduct*.

Let $\mathrm{char}(\Bbbk) \nmid |W|$. Let $\mathtt{P}$ be the basis of $\overline{\Sigma}^*$ dual to $\mathtt{Q}$. Its product and coproduct formulas can be written by dualizing (10.68). The coproduct is deconcatenation as in the $\mathtt{M}$-basis. The product simplifies and we call it the *shuffle product*.

10.4.10. Dual Coxeter q-bialgebra of face-types. More generally, for a scalar q, let $\overline{\Sigma}_q^*$ denote the Coxeter q-bialgebra dual to $\overline{\Sigma}_q$. The product and coproduct are

$$\begin{aligned} &\mu_Z^T : \overline{\Sigma}_q^*[T] \to \overline{\Sigma}_q^*[Z] && \Delta_Z^T : \overline{\Sigma}_q^*[Z] \to \overline{\Sigma}_q^*[T] \\ &\mathtt{M}_{V/T} \mapsto \sum_{U:\,U \geq Z} a_Z^{TVU}(q)\, \mathtt{M}_{U/Z} && \mathtt{M}_{U/Z} \mapsto \begin{cases} \mathtt{M}_{U/T} & \text{if } U \geq T, \\ 0 & \text{otherwise.} \end{cases} \end{aligned} \tag{10.77}$$

The coproduct is deconcatenation as before, while the product is deformed by replacing the scalars a_Z^{TVU} by the polynomials $a_Z^{TVU}(q)$. We call it the *q-quasishuffle product*.

For q not a root of unity, let $\mathtt{P}$ be the basis of $\overline{\Sigma}_q^*$ dual to $\mathtt{Q}$. Its product and coproduct formulas can be written by dualizing (10.71). The coproduct is deconcatenation as in the $\mathtt{M}$-basis. The product simplifies and we call it the *q-shuffle product*.

10.4.11. Coxeter 0-bialgebras of face-types. Let us now look at the case $q = 0$. Specializing formulas (10.63) to $q = 0$ and using the first formula in (1.56): The product and coproduct of $\overline{\Sigma}_0$ in the $\mathtt{H}$-basis are

$$\begin{aligned} &\mu_Z^T : \overline{\Sigma}_0[T] \to \overline{\Sigma}_0[Z] && \Delta_Z^T : \overline{\Sigma}_0[Z] \to \overline{\Sigma}_0[T] \\ &\mathtt{H}_{U/T} \mapsto \mathtt{H}_{U/Z} && \mathtt{H}_{U/Z} \mapsto \mathtt{H}_{(T \cup U)/T} \end{aligned} \tag{10.78}$$

This is the *Coxeter 0-bialgebra of face-types*. The product is concatenation, while the coproduct is 0-dequasishuffle.

Specializing (10.69) to $q = 0$ and using formulas (1.115): The $\mathtt{Q}$-basis of $\overline{\Sigma}_0$, on the Z-component, relates to the $\mathtt{H}$-basis by

$$(10.79)\qquad \begin{aligned} \mathtt{H}_{T/Z} &= \sum_{U:\, T \leq U} \mathtt{Q}_{U/Z}, \\ \mathtt{Q}_{T/Z} &= \sum_{U:\, T \leq U} (-1)^{\mathrm{rk}(U/T)}\, \mathtt{H}_{U/Z}. \end{aligned}$$

Specializing formulas (10.71) to $q = 0$: The product and coproduct of $\overline{\Sigma}_0$ in the $\mathtt{Q}$-basis are

$$(10.80)\qquad \begin{aligned} &\mu_Z^T : \overline{\Sigma}_0[T] \to \overline{\Sigma}_0[Z] & &\Delta_Z^T : \overline{\Sigma}_0[Z] \to \overline{\Sigma}_0[T] \\ &\mathtt{Q}_{U/T} \mapsto \mathtt{Q}_{U/Z} & &\mathtt{Q}_{U/Z} \mapsto \begin{cases} \mathtt{Q}_{U/T} & \text{if } U \geq T, \\ 0 & \text{otherwise.} \end{cases} \end{aligned}$$

The product is concatenation as in the $\mathtt{H}$-basis, while the coproduct is deconcatenation (0-deshuffle).

Exercise 10.69. Check directly: Formulas (10.78) and (10.80) transform into each other via change of basis (10.79).

Now consider the dual Coxeter 0-bialgebra $\overline{\Sigma}_0^*$. Its product and coproduct in the $\mathtt{M}$-basis are

$$(10.81)\qquad \begin{aligned} &\mu_Z^T : \overline{\Sigma}_0^*[T] \to \overline{\Sigma}_0^*[Z] & &\Delta_Z^T : \overline{\Sigma}_0^*[Z] \to \overline{\Sigma}_0^*[T] \\ &\mathtt{M}_{V/T} \mapsto \sum_{\substack{U:\, U \geq Z \\ T \cup U = V}} \mathtt{M}_{U/Z} & &\mathtt{M}_{U/Z} \mapsto \begin{cases} \mathtt{M}_{U/T} & \text{if } T \leq U, \\ 0 & \text{otherwise.} \end{cases} \end{aligned}$$

The coproduct is deconcatenation, while the product is 0-quasishuffle. These formulas follow by specializing (10.77) to $q = 0$, or by dualizing (10.78). Similarly, the product and coproduct in the $\mathtt{P}$-basis can be written by dualizing (10.80). They are given by concatenation and deconcatenation, respectively.

Exercise 10.70. Check directly the 0-bialgebra axiom (8.78) for $\overline{\Sigma}_0$ in both $\mathtt{H}$- and $\mathtt{Q}$-bases and for $\overline{\Sigma}_0^*$ in both $\mathtt{M}$- and $\mathtt{P}$-bases.

10.4.12. $\mathtt{K}$-basis and $\mathtt{F}$-basis. Let q be a scalar. We briefly mention the $\mathtt{K}$-basis of $\overline{\Sigma}_q$ and the dual $\mathtt{F}$-basis of $\overline{\Sigma}_q^*$. They are, respectively, defined by

$$(10.82)\qquad \mathtt{H}_{U/Z} = \sum_{T:\, Z \leq T \leq U} \mathtt{K}_{T/Z} \qquad \text{and} \qquad \mathtt{F}_{T/Z} = \sum_{U:\, T \leq U} \mathtt{M}_{U/Z}$$

on the Z-component.

10.4.13. Freeness and cofreeness. We now discuss the (co)freeness properties of the Coxeter bialgebras of face-types. They are summarized in Tables 10.6 and 10.7.

The Coxeter q-bialgebra $\overline{\Sigma}_q$ is the free Coxeter q-bialgebra on $\Bbbk\{x\}$ viewed as a Coxeter coalgebra. Dually, $\overline{\Sigma}_q^*$ is the cofree Coxeter q-bialgebra on $\Bbbk[x]$ viewed as a Coxeter algebra. More precisely:

Proposition 10.71. *For a scalar q, there are isomorphisms of Coxeter q-bialgebras*

$$\overline{\Sigma}_q \xrightarrow{\cong} \overline{\mathcal{T}}_q(\Bbbk\{x\}) \quad \textit{and} \quad \overline{\Sigma}^*_q \xrightarrow{\cong} \overline{\mathcal{T}}_q^{\vee}(\Bbbk[x]).$$

On the Z-component, the first map sends $\mathtt{H}_{T/Z}$ to $x^{(T)}$ in the T-summand, while the second map sends $\mathtt{M}_{T/Z}$ to x^T in the T-summand.

The (co)product of $\overline{\Sigma}_q$ in the $\mathtt{H}$-basis is given by (10.63), while that of $\overline{\Sigma}^*_q$ in the $\mathtt{M}$-basis is given by (10.77).

Proof. For the first map: The nontrivial part is to check that the map is a Coxeter coalgebra morphism. Put $\mathrm{C} := \Bbbk\{x\}$ in (9.21). Since the Coxeter space structure maps and coproduct components of $\Bbbk\{x\}$ are identities, we first apply Observation 8.99, item (1) and then use the second identity in (1.44a) applied to $\boldsymbol{\alpha}_Z$.

For the second map: The nontrivial part now is to check that the map is a Coxeter algebra morphism. Put $\mathrm{A} := \Bbbk[x]$ in (9.30), and use Observation 8.99, item (1) and the first identity in (1.44a) applied to $\boldsymbol{\alpha}_Z$. □

Specializing Theorems 9.13 and 9.18, we see that $\overline{\Sigma}_q$ and $\overline{\Sigma}^*_q$ satisfy the following universal properties.

Theorem 10.72 (Universal property). *Let q be a scalar. For H a Coxeter q-bialgebra and $f : \Bbbk\{x\} \to \mathrm{H}$ a morphism of Coxeter coalgebras, there exists a unique morphism of Coxeter q-bialgebras $\hat{f} : \overline{\Sigma}_q \to \mathrm{H}$ such that the diagram*

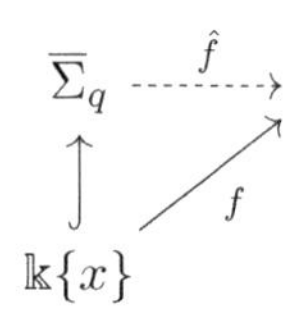

commutes, where $\Bbbk\{x\} \hookrightarrow \overline{\Sigma}_q$, on the Z-component, sends $x^{(Z)}$ to $\mathtt{H}_{Z/Z}$. The latter is a morphism of Coxeter coalgebras.

Explicitly, the map $\hat{f}$, on the Z-component, is

$$\hat{f}_Z(\mathtt{H}_{T/Z}) = \mu_Z^T(f_T(x^{(T)})) \tag{10.83}$$

for any $T \geq Z$.

Theorem 10.73 (Universal property). *Let q be a scalar. For H a Coxeter q-bialgebra and $f : \mathrm{H} \to \Bbbk[x]$ a morphism of Coxeter algebras, there exists a unique morphism of Coxeter q-bialgebras $\hat{f} : \mathrm{H} \to \overline{\Sigma}^*_q$ such that the diagram*

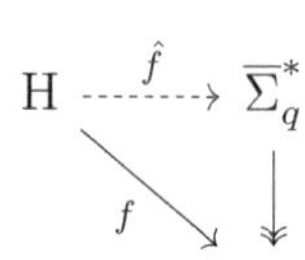

commutes, where $\overline{\Sigma}^*_q \twoheadrightarrow \Bbbk[x]$, on the Z-component, sends $\mathtt{M}_{T/Z}$ to x^Z if $T = Z$, and to 0 otherwise. The latter is a morphism of Coxeter algebras.

Explicitly, the map $\hat{f}$, on the Z-component, is

$$\hat{f}_Z(h) = \sum_{T:\, T \geq Z} \langle f_T \Delta^T_Z(h), x^T \rangle\, \mathtt{M}_{T/Z} \tag{10.84}$$

for any $h \in \mathrm{H}[Z]$. To clarify notation, the coefficient of $\mathtt{M}_{T/Z}$ is the coefficient of x^T in $f_T(\Delta^T_Z(h))$.

Let us now focus on $q = 1$. The Coxeter bialgebra $\overline{\Sigma}$ is also the free Coxeter bialgebra on $\Bbbk\{x\}$ viewed as a trivial Coxeter coalgebra. We denote the latter by $\Bbbk\{x\}_t$. Dually, $\overline{\Sigma}^*$ is the cofree Coxeter bialgebra on $\Bbbk[x]$ viewed as a trivial Coxeter algebra. We denote the latter by $\Bbbk[x]_t$. More precisely:

Proposition 10.74. *Let* $\mathrm{char}(\Bbbk) \nmid |W|$. *Then there are isomorphisms of Coxeter bialgebras*

$$\overline{\Sigma} \xrightarrow{\cong} \overline{\mathcal{T}}(\Bbbk\{x\}_t) \quad \textit{and} \quad \overline{\Sigma}^* \xrightarrow{\cong} \overline{\mathcal{T}}^{\vee}(\Bbbk[x]_t).$$

On the Z-component, the first map sends $\mathtt{Q}_{T/Z}$ *to* $x^{(T)}$ *in the T-summand, while the second map sends* $\mathtt{P}_{T/Z}$ *to* x^T *in the T-summand.*

Note very carefully: The $\mathtt{Q}$-basis of $\overline{\Sigma}$ and $\mathtt{P}$-basis of $\overline{\Sigma}^*$ are employed here. This explains the hypothesis on field characteristic. Relevant (co)product formulas are (10.68) and its dual.

Proof. This is similar to the proof of Proposition 10.71. To see that the first map is a Coxeter coalgebra morphism, put $q = 1$ and $\mathrm{V} := \Bbbk\{x\}$ in the second formula in (9.25), and then use Observation 8.99, item (2). Similarly, to see that the second map is a Coxeter algebra morphism, put $q = 1$ and $\mathrm{V} := \Bbbk[x]$ in the first formula in (9.34), and proceed. □

Similar considerations apply when q is not a root of unity. The Coxeter q-bialgebra $\overline{\Sigma}_q$ is the free Coxeter q-bialgebra on $\Bbbk\{x\}$ viewed as a trivial Coxeter coalgebra. Dually, $\overline{\Sigma}^*_q$ is the cofree Coxeter q-bialgebra on $\Bbbk[x]$ viewed as a trivial Coxeter algebra. More precisely:

Proposition 10.75. *For q not a root of unity, there are isomorphisms of Coxeter q-bialgebras*

$$\overline{\Sigma}_q \xrightarrow{\cong} \overline{\mathcal{T}}_q(\Bbbk\{x\}_t) \quad \textit{and} \quad \overline{\Sigma}^*_q \xrightarrow{\cong} \overline{\mathcal{T}}_q^{\vee}(\Bbbk[x]_t).$$

On the Z-component, the first map sends $\mathtt{Q}_{T/Z}$ *to* $x^{(T)}$ *in the T-summand, while the second map sends* $\mathtt{P}_{T/Z}$ *to* x^T *in the T-summand.*

The (co)product of $\overline{\Sigma}_q$ in the $\mathtt{Q}$-basis is given by (10.71). The argument is similar to the one above.

Exercise 10.76. Reformulate Propositions 10.74 and 10.75 as universal properties by specializing Theorems 9.15 and 9.19.

Remark 10.77 (Hoffman–Newman–Radford). Note very carefully how passage from the $\mathtt{H}$-basis to the $\mathtt{Q}$-basis of $\overline{\Sigma}$ (or of $\overline{\Sigma}_q$ when q is not a root of unity) trivialized the coproduct of $\Bbbk\{x\}$. Dually, passage from the $\mathtt{M}$-basis to the $\mathtt{P}$-basis of $\overline{\Sigma}^*$ trivialized the product of $\Bbbk[x]$. This is part of a general phenomenon studied in Section 13.5, see in particular Section 13.5.6.

Example 10.78. Consider the map $\mathrm{X} \hookrightarrow \Bbbk\{x\}$ which on the S-component is the canonical identification. It is a Coxeter coalgebra morphism with trivial coproduct on X. For a scalar q, applying the functor (9.23) yields a Coxeter q-bialgebra morphism $\overline{\mathcal{T}_q}(\mathrm{X}) \hookrightarrow \overline{\mathcal{T}_q}(\Bbbk\{x\})$ which identifies with (10.64) (using the $\mathtt{H}$-basis of $\overline{\Sigma}_q$).

For $q = 1$, we can work instead with the $\mathtt{Q}$-basis of $\overline{\Sigma}$. For this, view $\mathrm{X} \hookrightarrow \Bbbk\{x\}$ as a map of Coxeter spaces, and apply the functor (9.24). The resulting morphism takes x^T to $\mathtt{Q}_{S/T}$ which is the same as $\mathtt{H}_{S/T}$, so it is the same map as (10.64). □

10.4.14. q-norm map on face-types. Let (ξ_T) be an invariant family of scalars indexed by face-types as in Definition 1.10. For a scalar q, consider the map

$$\overline{\kappa}_{q,\xi} : \overline{\Sigma}_q \to \overline{\Sigma}_q^*, \qquad \mathtt{H}_{T/Z} \mapsto \sum_{T' : T' \geq Z} a_Z^{TT'}(q,\xi)\, \mathtt{M}_{T'/Z} \tag{10.85}$$

on the Z-component, with the polynomials $a_Z^{TT'}(q,\xi)$ obtained by applying (1.49) to $\boldsymbol{\alpha}_Z$. We call (10.85) the *q-norm map on face-types*. It depends on ξ.

Lemma 10.79. *The q-norm map on face-types* (10.85) *is a self-dual morphism of Coxeter q-bialgebras.*

Proof. We proceed here by direct calculation. A conceptual proof using freeness of $\overline{\Sigma}_q$ and cofreeness of $\overline{\Sigma}_q^*$ is given in Exercise 10.80 below.

Self-duality of $\overline{\kappa}_{q,\xi}$ requires $a_Z^{TT'}(q,\xi) = a_Z^{T'T}(q,\xi)$ which holds by (1.51) applied to $\boldsymbol{\alpha}_Z$. For $\overline{\kappa}_{q,\xi}$ to be a map of Coxeter spaces, we need the diagram

$$\begin{array}{ccc} \overline{\Sigma}_q[Z] & \xrightarrow{(\overline{\kappa}_{q,\xi})_Z} & \overline{\Sigma}_q^*[Z] \\ {\scriptstyle\sigma}\downarrow & & \downarrow{\scriptstyle\sigma} \\ \overline{\Sigma}_q[Z'] & \xrightarrow[(\overline{\kappa}_{q,\xi})_{Z'}]{} & \overline{\Sigma}_q^*[Z'] \end{array} \qquad \begin{array}{ccc} \mathtt{H}_{T/Z} & \longmapsto & \sum_{U : U \geq Z} a_Z^{TU}(q,\xi)\, \mathtt{M}_{U/Z} \\ \downarrow & & \downarrow \\ \mathtt{H}_{T'/Z'} & \longmapsto & \sum_{U' : U' \geq Z'} a_{Z'}^{T'U'}(q,\xi)\, \mathtt{M}_{U'/Z'} \end{array}$$

to commute for any $\sigma : Z \to Z'$. This requires $a_Z^{TU}(q,\xi) = a_{Z'}^{T'U'}(q,\xi)$ which holds by Exercise 1.70.

For $\overline{\kappa}_{q,\xi}$ to be a morphism of Coxeter algebras and of Coxeter coalgebras, we need the following diagrams, respectively, to commute.

$$\begin{array}{ccc} \overline{\Sigma}_q[Y] & \xrightarrow{(\overline{\kappa}_{q,\xi})_Y} & \overline{\Sigma}_q^*[Y] \\ {\scriptstyle\mu_Z^Y}\downarrow & & \downarrow{\scriptstyle\mu_Z^Y} \\ \overline{\Sigma}_q[Z] & \xrightarrow[(\overline{\kappa}_{q,\xi})_Z]{} & \overline{\Sigma}_q^*[Z] \end{array} \qquad \begin{array}{ccc} \overline{\Sigma}_q^*[Y] & \xleftarrow{(\overline{\kappa}_{q,\xi})_Y} & \overline{\Sigma}_q[Y] \\ {\scriptstyle\Delta_Z^Y}\uparrow & & \uparrow{\scriptstyle\Delta_Z^Y} \\ \overline{\Sigma}_q^*[Z] & \xleftarrow[(\overline{\kappa}_{q,\xi})_Z]{} & \overline{\Sigma}_q[Z] \end{array}$$

Since these are duals of each other, it suffices to check the first diagram. We calculate:

$$\begin{aligned} \mu_Z^Y(\overline{\kappa}_{q,\xi})_Y(\mathtt{H}_{T/Y}) &= \mu_Z^Y\Big(\sum_{U : U \geq Y} a_Y^{TU}(q,\xi)\, \mathtt{M}_{U/Y}\Big) \\ &= \sum_{U : U \geq Y} a_Y^{TU}(q,\xi) \sum_{V : V \geq Z} a_Z^{YUV}(q)\, \mathtt{M}_{V/Z} \end{aligned}$$

$$
\begin{aligned}
&= \sum_{V:\,V\geq Z} \Big(\sum_{U:\,U\geq Y} a_Y^{TU}(q,\xi)\, a_Z^{YUV}(q) \Big) \mathtt{M}_{V/Z} \\
&= \sum_{V:\,V\geq Z} a_Z^{TV}(q,\xi)\, \mathtt{M}_{V/Z} \\
&= (\overline{\kappa}_{q,\xi})_Z(\mathtt{H}_{T/Z}) \\
&= (\overline{\kappa}_{q,\xi})_Z \mu_Z^Y(\mathtt{H}_{T/Y}).
\end{aligned}
$$

The fourth step is critical, and used the gate identity (1.58) applied to $\boldsymbol{\alpha}_Z$. □

Exercise 10.80 (General q-norm map). The q-norm map on face-types (10.85) is an instance of the general q-norm map in Section 9.11.3. This is elaborated below. It yields another proof of Lemma 10.79.

Let (ξ_T) be an invariant family of scalars indexed by face-types as in Definition 1.10. Define the map $f_\xi : \Bbbk\{x\} \to \Bbbk[x]$ of Coxeter spaces by

$$
(f_\xi)_T : \Bbbk\{x\}[T] \to \Bbbk[x][T], \qquad x^{(T)} \mapsto \xi_T\, x^T.
$$

Use Exercise 1.26 to check that diagram (8.29) indeed commutes. In fact, every map $\Bbbk\{x\} \to \Bbbk[x]$ of Coxeter spaces is of the form f_ξ for a unique ξ. For instance, the identity map arises from $\xi \equiv 1$. (Note: The space of such maps coincides with $\mathscr{S}(\Bbbk[x])$ by (9.117). The latter is made explicit in Section 10.2.17.)

Now put $\mathrm{C} := \Bbbk\{x\}$ and $\mathrm{A} := \Bbbk[x]$ in the bijection (9.144) and apply it to f_ξ. Check that under the identifications in Proposition 10.71, the resulting morphism of Coxeter q-bialgebras $\widehat{f_\xi} : \overline{\Sigma}_q \to \overline{\Sigma}_q^*$ matches the q-norm map $\overline{\kappa}_{q,\xi}$ on face-types (10.85). Equivalently, obtain $\widehat{f_\xi}$ by appropriately employing the universal maps (10.83) and (10.84) one after the other in either order. See the diagram below.

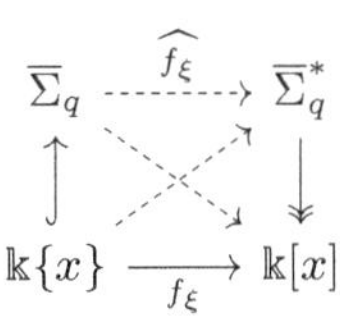

(Use identities (1.50) applied to $\boldsymbol{\alpha}_Z$.)

Exercise 10.81 (Self-dual morphisms). Deduce: For a scalar q, any morphism $\overline{\Sigma}_q \to \overline{\Sigma}_q^*$ of Coxeter q-bialgebras is necessarily of the form (10.85) for a unique invariant family of scalars ξ. In particular, it is always self-dual.

Remark 10.82. Exercise 10.80 establishes Lemma 10.79. By reversing the logical flow, the calculation in the proof of Lemma 10.79 implies the gate identity (1.58). See Section 10.4.4 for related comments.

Exercise 10.83 (Specialization of the invariant family of scalars). Check: The q-norm map on face-types (10.85), for $\xi \equiv 1$, is

$$
\overline{\kappa}_{q,\xi\equiv 1} : \overline{\Sigma}_q \to \overline{\Sigma}_q^*, \qquad \mathtt{H}_{T/Z} \mapsto \sum_{T':\,T'\geq Z} \Bigg(\sum_{\substack{w\in W_Z:\\ \mathrm{des}_Z(w)\leq T/Z \\ \mathrm{des}_Z(w^{-1})\leq T'/Z}} q^{l(w)} \Bigg) \mathtt{M}_{T'/Z}. \tag{10.86}
$$

(Apply formula (1.48) to $\boldsymbol{\alpha}_Z$.) The map (10.86), on the K-basis and into the F-basis, for $\xi \equiv 1$, is

$$(10.87)\qquad \overline{\kappa}_{q,\xi\equiv 1} : \overline{\Sigma}_q \to \overline{\Sigma}_q^*, \qquad \mathtt{K}_{T/Z} \mapsto \sum_{T':\,T' \geq Z} \Bigg(\sum_{\substack{w \in W_Z: \\ \mathrm{des}_Z(w) = T/Z \\ \mathrm{des}_Z(w^{-1}) = T'/Z}} q^{l(w)} \Bigg) \mathtt{F}_{T'/Z}.$$

(Use the change of basis formulas (10.82).)

Let (ξ_T) be an invariant family of scalars indexed by face-types as in Definition 1.10. For q not a root of unity, let (η_T) be the invariant family of scalars related to (ξ_T) by

$$(10.88)\qquad \xi_T = \sum_{U:\,U \geq T} \boldsymbol{\zeta}_q(T,U)\, \eta_U \quad \text{and} \quad \eta_T = \sum_{U:\,U \geq T} \boldsymbol{\mu}_q(T,U)\, \xi_U.$$

In other words, ξ and η relate to each other by q-Möbius inversion (10.29).

For q not a root of unity, we now give a necessary and sufficient numerical condition for the map (10.85) to be invertible. We do this via a change of basis. For that, recall from Section 1.8.7 the polynomials $c_Z^{TT'}(q)$ obtained by applying (1.52) to $\boldsymbol{\alpha}_Z$.

Theorem 10.84. *For q not a root of unity, the q-norm map on face-types* (10.85), *on the* Q*-basis and into the* P*-basis, is*

$$(10.89)\qquad \overline{\kappa}_{q,\xi} : \overline{\Sigma}_q \to \overline{\Sigma}_q^*, \qquad \mathtt{Q}_{T/Z} \mapsto \eta_T \sum_{\substack{T':\,T' \geq Z \\ \mathrm{s}(T'/Z) = \mathrm{s}(T/Z)}} c_Z^{TT'}(q)\, \mathtt{P}_{T'/Z}$$

on the Z-component. In particular, the two extreme cases are

$$\mathtt{Q}_{Z/Z} \mapsto \eta_Z\, \mathtt{P}_{Z/Z} \quad \textit{and} \quad \mathtt{Q}_{S/Z} \mapsto \eta_S\, d_{S/Z}(q)\, \mathtt{P}_{S/Z}.$$

This is a change of basis calculation. A nice way to proceed is given in the exercise below. Another proof is given later in Exercise 10.90.

Exercise 10.85. Show: Both maps (10.85) and (10.89), on the H-basis and into the P-basis, are

$$(10.90)\qquad \mathtt{H}_{T/Z} \mapsto \sum_{T':\,T' \geq Z} \eta_{T'}\, a_Z^{T'T'T}(q)\, \mathtt{P}_{T'/Z}$$

on the Z-component. (For the first calculation, use the gate identity (1.58) applied to $\boldsymbol{\alpha}_Z$ followed by the noncommutative q-Weisner formula (1.96). For the second calculation, use identity (1.97) applied to $\boldsymbol{\alpha}_Z$.)

Theorem 10.86 (Isomorphism criterion). *For q not a root of unity, the q-norm map on face-types* (10.85) *is an isomorphism of Coxeter q-bialgebras iff*

$$\eta_T = \sum_{U:\,U \geq T} \boldsymbol{\mu}_q(T,U)\, \xi_U \neq 0$$

for all face-types T.

Proof. We use Theorem 10.84. The map (10.89) splits as a direct sum over flat-types of $\boldsymbol{\alpha}_Z$. Now apply Theorem 1.56 (on invertibility of Varchenko matrices) to each summand. See Exercise 9.103 for a relevant perspective. □

In conjunction with Exercise 10.81, we classify all isomorphisms between $\overline{\Sigma}_q$ and $\overline{\Sigma}_q^*$ as follows.

Corollary 10.87 (Self-dual isomorphisms). *Suppose q is not a root of unity. Let (η_T) be an invariant family of nonzero scalars indexed by face-types. This yields an isomorphism (10.85) of Coxeter q-bialgebras, with (ξ_T) defined via (10.88). Conversely, every isomorphism $\overline{\Sigma}_q \to \overline{\Sigma}_q^*$ of Coxeter q-bialgebras arises in this manner from a unique nonzero η.*

Example 10.88 (Ranks one and two). Suppose q is not a root of unity and $\xi \equiv 1$. We make use of the calculations in Examples 1.50, 1.57, 1.91.

For a rank-one arrangement with face-types $\emptyset \lessdot S$, the map (10.85) on the $\emptyset$-component, on the H-basis and into the M-basis, is

$$\mathtt{H}_\emptyset \mapsto \mathtt{M}_\emptyset + \mathtt{M}_S, \qquad \mathtt{H}_S \mapsto \mathtt{M}_\emptyset + (1+q)\,\mathtt{M}_S,$$

and, on the Q-basis and into the P-basis, is

$$\mathtt{Q}_\emptyset \mapsto \tfrac{q}{1+q}\,\mathtt{P}_\emptyset, \qquad \mathtt{Q}_S \mapsto (1+q)\,\mathtt{P}_S.$$

In matrix form, the maps can be, respectively, written as

$$\begin{pmatrix} 1 & 1 \\ 1 & 1+q \end{pmatrix} \quad \text{and} \quad \begin{pmatrix} \frac{q}{1+q} & 0 \\ 0 & 1+q \end{pmatrix}.$$

The top-left entry in the latter is $\boldsymbol{\mu}_q(\emptyset,\emptyset) + \boldsymbol{\mu}_q(\emptyset, S)$. The matrices are invertible iff $q \neq 0$ (for q not a root of unity).

For the dihedral arrangement of d lines with face-types $\emptyset \lessdot T, U \lessdot S$, the map (10.85) on the $\emptyset$-component, on the H-basis and into the M-basis, in matrix form is

$$\begin{pmatrix} 1 & 1 & 1 & 1 \\ 1 & 1+q+q^3+\cdots+q^{d-2} & 1+q^2+q^4+\cdots+q^{d-1} & 1+q+\cdots+q^{d-1} \\ 1 & 1+q^2+q^4+\cdots+q^{d-1} & 1+q+q^3+\cdots+q^{d-2} & 1+q+\cdots+q^{d-1} \\ 1 & 1+q+\cdots+q^{d-1} & 1+q+\cdots+q^{d-1} & (1+q)(1+q+\cdots+q^{d-1}) \end{pmatrix}$$

for d odd, and

$$\begin{pmatrix} 1 & 1 & 1 & 1 \\ 1 & 1+q+q^3+\cdots+q^{d-1} & 1+q^2+q^4+\cdots+q^{d-2} & 1+q+\cdots+q^{d-1} \\ 1 & 1+q^2+q^4+\cdots+q^{d-2} & 1+q+q^3+\cdots+q^{d-1} & 1+q+\cdots+q^{d-1} \\ 1 & 1+q+\cdots+q^{d-1} & 1+q+\cdots+q^{d-1} & (1+q)(1+q+\cdots+q^{d-1}) \end{pmatrix}$$

for d even, and, on the Q-basis and into the P-basis, is

$$\begin{pmatrix} \frac{q^d(1-q^{d-1})}{(1-q^d)(1+q^{d-1})} & 0 & 0 & 0 \\ 0 & \frac{q}{1+q} & \frac{q^d}{1+q} & 0 \\ 0 & \frac{q^d}{1+q} & \frac{q}{1+q} & 0 \\ 0 & 0 & 0 & \frac{(1+q)(1-q^d)}{1-q} \end{pmatrix}$$

for d odd, and

$$\begin{pmatrix} \frac{q^d(1-q^{d-1})}{(1-q^d)(1+q^{d-1})} & 0 & 0 & 0 \\ 0 & \frac{q(1+q^{d-1})}{1+q} & 0 & 0 \\ 0 & 0 & \frac{q(1+q^{d-1})}{1+q} & 0 \\ 0 & 0 & 0 & \frac{(1+q)(1-q^d)}{1-q} \end{pmatrix}$$

for d even. Observe the block-diagonal form of the latter two matrices. The top-left entry is $\boldsymbol{\mu}_q(\emptyset,\emptyset) + \boldsymbol{\mu}_q(\emptyset,T) + \boldsymbol{\mu}_q(\emptyset,U) + \boldsymbol{\mu}_q(\emptyset,S)$. The matrices are invertible iff $q \neq 0$ (for q not a root of unity). □

Example 10.89 (Coordinate arrangement). Let $q \neq \pm 1$. Formulas (10.85) and (10.89) for the coordinate arrangement specialize to

$$\mathtt{H}_{T/Z} \mapsto \sum_{T' : T' \geq Z} (1+q)^{\mathrm{rk}((T\cap T')/Z)} \xi_{T\cup T'} \, \mathtt{M}_{T'/Z},$$

$$\mathtt{Q}_{T/Z} \mapsto \eta_T \, (1+q)^{\mathrm{rk}(T/Z)} \, \mathtt{P}_{T/Z}.$$

It is instructive to check directly that the first formula transforms into the second via the change of basis (10.69) (using the explicit formulas for $\boldsymbol{\zeta}_q$ and $\boldsymbol{\mu}_q$ for the coordinate arrangement). □

Exercise 10.90 (q-exponential and q-logarithm). We now give another proof of Theorems 10.84 and 10.86 below. It employs the q-exponential and q-logarithm.

Suppose q is not a root of unity. Let ξ and η be invariant families of scalars related as in (10.88) by q-Möbius inversion. Let f_ξ and $\widehat{f_\xi}$ be as in Exercise 10.80. Use formula (9.101b) to check that

$$\mathbf{log}_q(f_\xi)_T : \Bbbk\{x\}[T] \to \Bbbk[x][T], \qquad x^{(T)} \mapsto \eta_T \, x^T.$$

Use Theorem 9.110 to deduce: $\widehat{f_\xi}$ is an isomorphism iff $\eta_T \neq 0$ for all T. Moreover, by Exercise 9.111, $\widehat{f_\xi}$ is the q-exponential of $\mathbf{log}_q(f_\xi)$. Compute $\widehat{f_\xi}$ on the $\mathtt{Q}$-basis and $\mathtt{P}$-basis to derive the expression (10.89). (Use Exercise 1.99.) This is a special case of the calculation in Exercise 9.109. One may also compute it on the $\mathtt{H}$-basis and $\mathtt{M}$-basis to recover the expression (10.85), however, this calculation is much more intricate.

Following up on Remark 10.77, we mention that Theorems 10.84 and 10.86 (and the above proof) are generalized later in Theorem 13.48. The connection to this special case is made towards the end of Section 13.5.6.

Example 10.91 (0-norm map on face-types). Let us now look at the case $q = 0$. The 0-norm map on face-types, on the $\mathtt{H}$-basis and into the $\mathtt{M}$-basis, on the Z-component, is

$$\overline{\kappa}_{0,\xi} : \overline{\Sigma}_0 \to \overline{\Sigma}_0^*, \qquad \mathtt{H}_{T/Z} \mapsto \sum_{T' : T' \geq Z} \xi_{T\cup T'} \, \mathtt{M}_{T'/Z}, \tag{10.91}$$

and, on the $\mathtt{Q}$-basis and into the $\mathtt{P}$-basis, is

$$\overline{\kappa}_{0,\xi} : \overline{\Sigma}_0 \to \overline{\Sigma}_0^*, \qquad \mathtt{Q}_{T/Z} \mapsto \eta_T \, \mathtt{P}_{T/Z}. \tag{10.92}$$

It is an isomorphism iff $\eta_T \neq 0$ for all face-types T. We used the last two formulas in (1.55). □

Exercise 10.92 (Norm map on face-types). Let $\mathrm{char}(\Bbbk) \nmid |W|$. Fix mutually inverse invariant noncommutative zeta and Möbius functions $\boldsymbol{\zeta}$ and $\boldsymbol{\mu}$. Let (ξ_T) and (η_T) be invariant families of scalars such that

$$\xi_T = \sum_{U : U \geq T} \boldsymbol{\zeta}(T,U) \, \eta_U \quad \text{and} \quad \eta_T = \sum_{U : U \geq T} \boldsymbol{\mu}(T,U) \, \xi_U. \tag{10.93}$$

In other words, ξ and η relate to each other by noncommutative Möbius inversion (10.28).

Now consider the q-norm map on face-types (10.85) for $q = 1$, namely,

$$\overline{\kappa}_{1,\xi} : \overline{\Sigma} \to \overline{\Sigma}^*, \qquad \mathtt{H}_{T/Z} \mapsto \sum_{T' : T' \geq Z} a_Z^{TT'}(1, \xi)\, \mathtt{M}_{T'/Z} \tag{10.94}$$

on the Z-component. We call this the *norm map on face-types*. Show: This map, on the $\mathtt{Q}$-basis and into the $\mathtt{P}$-basis, is

$$\overline{\kappa}_{1,\xi} : \overline{\Sigma} \to \overline{\Sigma}^*, \qquad \mathtt{Q}_{T/Z} \mapsto \eta_T\, |\mathrm{s}(T/Z)| \sum_{\substack{T' : T' \geq Z \\ \mathrm{s}(T'/Z) = \mathrm{s}(T/Z)}} \mathtt{P}_{T'/Z} \tag{10.95}$$

on the Z-component, with $|\mathrm{s}(T/Z)|$ as in Definition 1.8. (Proceed as in Exercise 10.85 and show that the map is

$$\mathtt{H}_{T/Z} \mapsto \sum_{T' : T' \geq Z} \eta_{T'}\, a_Z^{T'T'T}\, \mathtt{P}_{T'/Z} \tag{10.96}$$

on the Z-component. Use the gate identity (1.58) with $q = 1$, noncommutative Weisner formula (1.85), identity (1.87).) For $\xi \equiv 1$, the norm map on face-types (10.94) specializes to

$$\overline{\kappa}_{1,\xi \equiv 1} : \overline{\Sigma} \to \overline{\Sigma}^*, \qquad \mathtt{H}_{T/Z} \mapsto \sum_{T' : T' \geq Z} a_Z^{TT'}\, \mathtt{M}_{T'/Z} \tag{10.97}$$

on the Z-component. The corresponding η has an explicit formula given by the series f in Example 10.25.

10.4.15. Primitive and indecomposable parts.

We have

$$\overline{\mathcal{P}}(\overline{\Sigma}^*) = \Bbbk[x] \quad \text{and} \quad \overline{\mathcal{Q}}(\overline{\Sigma}) = \Bbbk\{x\}. \tag{10.98}$$

More precisely, $\overline{\mathcal{P}}(\overline{\Sigma}^*)[Z]$ is the linear span of $\mathtt{M}_{Z/Z}$, while $\overline{\mathcal{Q}}(\overline{\Sigma})[Z]$ is the linear span of the image of $\mathtt{H}_{Z/Z}$. This follows from the coproduct formula in (10.76) and the product formula in (10.62), respectively.

Similarly, for q not a root of unity,

$$\begin{aligned} \overline{\mathcal{P}}(\overline{\Sigma}_q) &= \overline{\mathcal{P}}(\overline{\Sigma}_q^*) = \Bbbk[x], \\ \overline{\mathcal{Q}}(\overline{\Sigma}_q^*) &= \overline{\mathcal{Q}}(\overline{\Sigma}_q) = \Bbbk\{x\}. \end{aligned} \tag{10.99}$$

The second equality in either claim follows from (10.98). The first equality in either claim follows from Corollary 10.87. (Use for instance $\eta \equiv 1$.) More precisely, for q not a root of unity, $\overline{\mathcal{P}}(\overline{\Sigma}_q)[Z]$ is the linear span of $\mathtt{Q}_{Z/Z}$. That this is a primitive element is clear from the coproduct formula in (10.71). It is also instructive to check this using the coproduct formula in (10.63). The required calculation is shown below. For any $Z < T$,

$$\begin{aligned} \Delta_Z^T(\mathtt{Q}_{Z/Z}) &= \sum_{U : U \geq Z} \boldsymbol{\mu}_q(Z, U)\, \Delta_Z^T(\mathtt{H}_{U/Z}) \\ &= \sum_{V : V \geq T} \Big(\sum_{U : U \geq Z} a_Z^{TVU}(q) \boldsymbol{\mu}_q(Z, U) \Big) \mathtt{H}_{V/T} \\ &= 0. \end{aligned}$$

The sum in parenthesis above is zero by the noncommutative q-Weisner formula (1.96).

Exercise 10.93. Use the coproduct formula in (10.63) to check: For a scalar q, for a face-type Z,

$$\sum_{U:\, U \geq Z} x^{U/Z}\, \mathtt{H}_{U/Z} \in \overline{\mathcal{P}}(\overline{\Sigma}_q)[Z] \iff \sum_{U:\, U\geq Z} a_Z^{TVU}(q)\, x^{U/Z} = 0 \text{ for all } Z < T \leq V.$$

The rhs defines a linear system in variables $x^{U/Z}$.

For q not a root of unity, $x^{U/Z} := \boldsymbol{\mu}_q(Z,U)$ is a solution, and every other solution is a scalar multiple of this one.

The situation for $q = 1$ is elaborated below.

Exercise 10.94. Recall the Coxeter Zie species from Section 4.8.2. Show:

$$\overline{\mathcal{P}}(\overline{\Sigma})[A, C] = \mathsf{Zie}[A]^{W_A}, \tag{10.100}$$

where the rhs is the space of W_A-invariants of $\mathsf{Zie}[A]$. (These identify with the W_Z-invariant Zie elements of $\overline{\Sigma}[Z]$, where $Z = \mathrm{t}(A)$.) Compare with formula (10.23). See also formula (12.31) in this regard.

Exercise 10.95. Let $\mathrm{char}(\Bbbk) \nmid |W|$. Check using formula (9.108) that on the $\mathtt{Q}$-basis of $\overline{\Sigma}$,

$$\mathbf{log}(\mathrm{id})_Z(\mathtt{Q}_{U/Z}) = \sum_{\substack{V:\, V \geq Z \\ \mathrm{s}(V/Z) = \mathrm{s}(U/Z)}} \Bigg(\sum_{\substack{\alpha:\, U \to V \\ \text{local to } Z}} \mathrm{sln}_Z^{\alpha} \Bigg)\, \mathtt{Q}_{V/Z},$$

where sln_Z^{α} is the Solomon coefficient (1.126). (Use identity (1.127).) A similar formula can be written on the $\mathtt{H}$-basis. The above elements, as U varies, provide a spanning set for $\overline{\mathcal{P}}(\overline{\Sigma})[Z]$. For $U = Z$,

$$\mathbf{log}(\mathrm{id})_Z(\mathtt{H}_{Z/Z}) = \mathbf{log}(\mathrm{id})_Z(\mathtt{Q}_{Z/Z}) = \mathtt{Q}_{Z/Z}.$$

A general result is given later in Lemma 12.10.

10.4.16. Convolution Coxeter algebra.

Recall the convolution Coxeter algebra from Section 9.5.4.

Theorem 10.96. *For* C *a Coxeter coalgebra and* A *a Coxeter algebra, Coxeter algebra morphisms from* $\overline{\Sigma}$ *to* $\mathrm{hom}^{\times}(\mathrm{C}, \mathrm{A})$ *are parametrized by Coxeter space maps from* C *to* A.

More precisely, there is a bijection

$$\overline{\mathsf{C}}\mathsf{Sp}(\mathrm{C}, \mathrm{A}) \xrightarrow{\cong} \mathsf{Mon}(\overline{\mathsf{C}}\mathsf{Sp})(\overline{\Sigma}, \mathrm{hom}^{\times}(\mathrm{C}, \mathrm{A})), \quad f \mapsto \hat{f} \tag{10.101}$$

constructed as follows. A map $f : \mathrm{C} \to \mathrm{A}$ of Coxeter spaces, by (9.63), is the same as a Coxeter space map $f : \Bbbk\{x\} \to \mathrm{hom}^{\times}(\mathrm{C}, \mathrm{A})$ which by freeness of $\overline{\Sigma}$ extends uniquely to a Coxeter algebra morphism $\hat{f} : \overline{\Sigma} \to \mathrm{hom}^{\times}(\mathrm{C}, \mathrm{A})$. Explicitly, using formulas (10.83) and (9.65),

$$\hat{f}_Z(\mathtt{H}_{T/Z}) = \mu_Z^T f_T \Delta_Z^T.$$

10.4.17. Internal hom for Coxeter coalgebras. We now connect the Coxeter bialgebra of face-types $\overline{\Sigma}$ to the internal hom construction of Section 9.6. Recall the Coxeter q-bialgebra $\mathcal{C}(\mathrm{C},\mathrm{K})_q$ from Exercise 9.49.

Theorem 10.97. *Let q be a scalar. For* C *a cocommutative Coxeter coalgebra and* K *a Coxeter q-bialgebra, Coxeter q-bialgebra morphisms from $\overline{\Sigma}_q$ to $\mathcal{C}(\mathrm{C},\mathrm{K})_q$ are parametrized by Coxeter coalgebra morphisms from* C *to* K.

More precisely, there is a bijection

$$\mathsf{Comon}(\overline{\mathsf{C}}\mathsf{Sp})(\mathrm{C},\mathrm{K}) \xrightarrow{\cong} q\text{-}\mathsf{Bimon}(\overline{\mathsf{C}}\mathsf{Sp})(\overline{\Sigma}_q,\mathcal{C}(\mathrm{C},\mathrm{K})_q), \quad f \mapsto \hat{f} \tag{10.102}$$

as follows. A Coxeter coalgebra morphism $f:\mathrm{C}\to\mathrm{K}$, by (9.72), is the same as a Coxeter coalgebra morphism $f:\Bbbk\{x\}\to\mathcal{C}(\mathrm{C},\mathrm{K})$ which by Theorem 10.72 extends uniquely to a Coxeter q-bialgebra morphism $\hat{f}:\overline{\Sigma}_q\to\mathcal{C}(\mathrm{C},\mathrm{K})_q$. Explicitly, using formulas (10.83) and (9.76) (the latter with an additional factor of $q^{l(\sigma)}$), we obtain

$$\hat{f}_Z(\mathtt{H}_{T'/Z}) = (g_{T/Z}:\mathrm{C}[T]\to\mathrm{K}[T])_{T\geq Z},$$

with $g_{T/Z}$ given by summing

$$\mathrm{C}[T] \xrightarrow{\Delta_T^U} \mathrm{C}[U] \xrightarrow{q^{l(\sigma)}\sigma} \mathrm{C}[U'] \xrightarrow{f_{U'}} \mathrm{K}[U'] \xrightarrow{\sigma^{-1}} \mathrm{K}[U] \xrightarrow{\mu_T^U} \mathrm{K}[T],$$

as $\sigma: U\to U'$ varies over all morphisms local to Z which biproject T and T'. In particular, $g_{Z/Z}$ equals

$$\mathrm{C}[Z] \xrightarrow{\Delta_Z^{T'}} \mathrm{C}[T'] \xrightarrow{f_{T'}} \mathrm{K}[T'] \xrightarrow{\mu_Z^{T'}} \mathrm{K}[Z].$$

Note the similarity of this formula with the one for $\hat{f}_Z$ in (10.101).

10.4.18. Universal measuring Coxeter coalgebra. Dually, we now connect the Coxeter bialgebra of face-types $\overline{\Sigma}$ to the construction of Section 9.7. Recall the Coxeter q-bialgebra $\overline{\mathcal{C}}(\mathrm{H},\mathrm{A})_q$ from Exercise 9.53.

Theorem 10.98. *Let q be a scalar. For* H *a Coxeter q-bialgebra and* A *a commutative Coxeter algebra, Coxeter q-bialgebra morphisms from $\overline{\Sigma}_q$ to $\overline{\mathcal{C}}(\mathrm{H},\mathrm{A})_q$ are parametrized by Coxeter algebra morphisms from* H *to* A.

More precisely, there is a bijection

$$\mathsf{Mon}(\overline{\mathsf{C}}\mathsf{Sp})(\mathrm{H},\mathrm{A}) \xrightarrow{\cong} q\text{-}\mathsf{Bimon}(\overline{\mathsf{C}}\mathsf{Sp})(\overline{\Sigma}_q,\overline{\mathcal{C}}(\mathrm{H},\mathrm{A})_q), \quad f \mapsto \hat{f} \tag{10.103}$$

as follows. A Coxeter algebra morphism $f:\mathrm{H}\to\mathrm{A}$, by (9.79), is the same as a Coxeter coalgebra morphism $f:\Bbbk\{x\}\to\overline{\mathcal{C}}(\mathrm{H},\mathrm{A})$ which by Theorem 10.72 extends uniquely to a Coxeter q-bialgebra morphism $\hat{f}:\overline{\Sigma}_q\to\overline{\mathcal{C}}(\mathrm{H},\mathrm{A})_q$. This map can be made explicit using formulas (10.83) and (9.83) (the latter with an additional factor of $q^{l(\sigma)}$).

10.4.19. Universal maps from the Coxeter q-bialgebra of face-types.

Theorem 10.99. *Let q be a scalar. For a cocommutative Coxeter q-bialgebra* H*, there is a morphism of Coxeter q-bialgebras*

$$\overline{\Sigma}_q \to \mathcal{C}(\mathrm{H},\mathrm{H})_q, \qquad \mathtt{H}_{T'/Z} \mapsto (g_{T/Z})_{T\geq Z} \tag{10.104}$$

on the Z-component, with

$$g_{T'/Z} := \sum_{U:\, U\geq T} a_Z^{TUT'}(q)\, \mu_T^U \Delta_T^U. \tag{10.105}$$

In particular, $g_{Z/Z} = \mu_Z^{T'} \Delta_Z^{T'}$.

Similarly, for a commutative Coxeter q-bialgebra H*, there is a morphism of Coxeter q-bialgebras*

$$\overline{\Sigma}_q \to \overline{\mathcal{C}}(\mathrm{H},\mathrm{H})_q, \qquad \mathtt{H}_{T'/Z} \mapsto (g_{T/Z})_{T\geq Z} \tag{10.106}$$

on the Z-component, with $g_{T/Z}$ as in (10.105).

PROOF. Specialize (10.102) to $\mathrm{C} = \mathrm{K} := \mathrm{H}$ and $f := \mathrm{id}$, and observe that the Coxeter q-bialgebra morphism $\hat{f}$ specializes to (10.104). The map (10.106) is obtained similarly by specializing (10.103). □

Exercise 10.100. Let H be a Coxeter q-bialgebra for a scalar q. Check directly: For face-types $T' \geq Z$, the family $(g_{T/Z})$ defined by (10.105) is an element of $\mathcal{C}(\mathrm{H},\mathrm{H})_q[Z]$ when H is cocommutative, and an element of $\overline{\mathcal{C}}(\mathrm{H},\mathrm{H})_q[Z]$ when H is commutative. (Use identities (8.97) and (8.98) and gate identity (1.57) applied to $\boldsymbol{\alpha}_Z$.)

10.4.20. Primitive and group-like series.

We now apply the discussion in Section 9.10 to the Coxeter bialgebra of face-types $\overline{\Sigma}$. A series of $\overline{\Sigma}$ is the same as a family of scalars $(f(Z,T))_{Z\leq T}$ such that $f(Z,T) = f(Z',T')$ whenever (Z,T) and (Z',T') have the same support. The identification is done via

$$(f(Z,T))_{Z\leq T} \longleftrightarrow \sum_{T:\, T\geq Z} f(Z,T)\, \mathtt{H}_{T/Z} \in \overline{\Sigma}[Z] \quad \text{for each face-type } Z. \tag{10.107}$$

Specializing (9.122) and using the product formula in (10.62), we see that the module $\mathscr{S}(\overline{\Sigma})$ is in fact the invariant lune-incidence algebra viewed as a left module over itself.

Universal series. Let $\mathrm{char}(\Bbbk) \nmid |W|$. Consider the bijection (9.126) between primitive and group-like series of $\overline{\Sigma}$. An interesting example of this bijection is given by

$$\boldsymbol{\mu} \longleftrightarrow \delta, \tag{10.108}$$

where δ denotes the unit element of the invariant lune-incidence algebra. To see this, note: $\mathbf{exp}(\boldsymbol{\mu}) = \boldsymbol{\zeta} \circ \boldsymbol{\mu} = \delta$ and $\mathbf{log}(\delta) = \boldsymbol{\mu} \circ \delta = \boldsymbol{\mu}$. It is also clear that $\boldsymbol{\mu}$ is primitive and δ is group-like. In terms of the identification (10.107),

$$\begin{aligned} \delta &\leftrightarrow \mathtt{H}_{Z/Z} \in \overline{\Sigma}[Z] \quad \text{for each } Z,\\ \boldsymbol{\mu} &\leftrightarrow \mathtt{Q}_{Z/Z} \in \overline{\Sigma}[Z] \quad \text{for each } Z, \end{aligned} \tag{10.109}$$

where $\mathtt{Q}_{Z/Z}$ is the $\mathtt{Q}$-basis element defined in (10.66). We refer to δ as the *universal series* of $\overline{\Sigma}$.

Similarly, for q not a root of unity, we have the bijection (9.129) between primitive and group-like series of $\overline{\Sigma}_q$. Instead of (10.108), we now have

$$\boldsymbol{\mu}_q \longleftrightarrow \delta,$$

and $\boldsymbol{\mu}_q$ corresponds to the $\mathtt{Q}$-basis element $\mathtt{Q}_{Z/Z}$ defined in (10.69).

For a scalar q, specializing Exercise 9.14 to $\mathrm{C} := \mathbb{k}\{x\}$, we obtain a Coxeter coalgebra morphism

$$\mathbb{k}\{x\} \hookrightarrow \overline{\mathcal{T}}_q(\mathbb{k}\{x\}) = \overline{\Sigma}_q, \qquad x^{(Z)} \mapsto \mathtt{H}_{Z/Z}$$

on the Z-component; also see Theorem 10.72. By the second formula in (9.119), this yields a group-like series of $\overline{\Sigma}_q$. Observe: This group-like series is precisely δ.

Exercise 10.101. Let $f : \mathrm{C} \to \mathrm{K}$ and $\hat{f} : \overline{\Sigma} \to \mathcal{C}(\mathrm{C}, \mathrm{K})$ correspond to each other under the bijection (10.102) for $q = 1$. Since $\hat{f}$ is a Coxeter bialgebra morphism, it induces a map on the group-like (primitive) series of $\overline{\Sigma}$ and $\mathcal{C}(\mathrm{C}, \mathrm{K})$ which commutes with $\mathbf{log}$. Use identifications (9.118) to deduce the commutative diagram below on the left and example on the right.

$$(10.110)\qquad \begin{array}{ccc} \mathscr{G}(\overline{\Sigma}) & \xrightarrow{\mathscr{G}(\hat{f})} & \mathsf{Comon}(\overline{\mathsf{C}}\mathsf{Sp})(\mathrm{C}, \mathrm{K}) \\ {\scriptstyle \mathbf{log}}\downarrow & & \downarrow{\scriptstyle \mathbf{log}} \\ \mathscr{P}(\overline{\Sigma}) & \xrightarrow[\mathscr{P}(\hat{f})]{} & \overline{\mathsf{C}}\mathsf{Sp}(\mathrm{C}, \overline{\mathcal{P}}(\mathrm{K})) \end{array} \qquad \begin{array}{ccc} \delta & \longmapsto & f \\ \downarrow & & \downarrow \\ \boldsymbol{\mu} & \longmapsto & \mathbf{log}(f) \end{array}$$

As a special case, we may take $f = \mathrm{id}$ on a cocommutative Coxeter bialgebra.

There are similar diagrams for q not a root of unity which use instead $\overline{\Sigma}_q$, $\mathcal{C}(\mathrm{C}, \mathrm{K})_q$, $\mathbf{log}_q$, $\boldsymbol{\mu}_q$.

Exercise 10.102. Let $f : \mathrm{H} \to \mathrm{A}$ and $\hat{f} : \overline{\Sigma} \to \overline{\mathcal{C}}(\mathrm{H}, \mathrm{A})$ correspond to each other under the bijection (10.103) for $q = 1$. Use identifications (9.120) to deduce the commutative diagram below on the left and example on the right.

$$(10.111)\qquad \begin{array}{ccc} \mathscr{G}(\overline{\Sigma}) & \xrightarrow{\mathscr{G}(\hat{f})} & \mathsf{Mon}(\overline{\mathsf{C}}\mathsf{Sp})(\mathrm{H}, \mathrm{A}) \\ {\scriptstyle \mathbf{log}}\downarrow & & \downarrow{\scriptstyle \mathbf{log}} \\ \mathscr{P}(\overline{\Sigma}) & \xrightarrow[\mathscr{P}(\hat{f})]{} & \overline{\mathsf{C}}\mathsf{Sp}(\overline{\mathcal{Q}}(\mathrm{H}), \mathrm{A}) \end{array} \qquad \begin{array}{ccc} \delta & \longmapsto & f \\ \downarrow & & \downarrow \\ \boldsymbol{\mu} & \longmapsto & \mathbf{log}(f) \end{array}$$

As a special case, we may take $f = \mathrm{id}$ on a commutative Coxeter bialgebra.

There are similar diagrams for q not a root of unity which use instead $\overline{\Sigma}_q$, $\overline{\mathcal{C}}(\mathrm{H}, \mathrm{A})_q$, $\mathbf{log}_q$, $\boldsymbol{\mu}_q$.

Takeuchi series. Let $\mathrm{char}(\mathbb{k}) \nmid |W|$. Define the *Takeuchi series*, denoted $\mathtt{Tak}$, of $\overline{\Sigma}$ by

$$(10.112)\qquad \mathtt{Tak}_Z := \sum_{T:\, T \geq Z} (-1)^{\dim(T)}\, \mathtt{H}_{T/Z}.$$

It is group-like. The logarithm of $\mathtt{Tak}$ is the primitive series of $\overline{\Sigma}$ given by

$$(10.113)\qquad \mathbf{log}(\mathtt{Tak})_Z = (-1)^{\dim(Z)}\, \overline{\mathtt{Q}}_{Z/Z},$$

where the $\overline{\mathtt{Q}}$-basis is defined from the $\mathtt{H}$-basis via $\overline{\zeta}$ and $\overline{\mu}$. Equivalently,

$$\mathbf{log}(\mathtt{Tak})_Z = (-1)^{\dim(Z)}\,\overline{\mathbf{log}}(\delta)_Z, \tag{10.114}$$

where δ denotes the universal series of $\overline{\Sigma}$.

Exercise 10.103. Check that the Takeuchi series is group-like using the alternating sign identity (1.63b) for $q = 1$ applied to $\boldsymbol{\alpha}_Z$. Check formula (10.113) using (1.121b) related to the noncommutative Zaslavsky formula.

The above observations are intimately connected to antipode properties of Coxeter bialgebras; this is developed in the exercises below.

Exercise 10.104. Let $f : \mathrm{C} \to \mathrm{K}$ and $\hat{f} : \overline{\Sigma} \to \mathcal{C}(\mathrm{C}, \mathrm{K})$ be as in Exercise 10.101. Check: Another example of the left diagram in (10.110) is

$$\begin{array}{ccc} (\mathtt{Tak}_Z) & \longmapsto & ((-1)^{\dim(Z)}\,\mathbf{log}_0(f)_Z) \\ \downarrow & & \downarrow \\ ((-1)^{\dim(Z)}\,\overline{\mathbf{log}}(\delta)_Z) & \longmapsto & ((-1)^{\dim(Z)}\,\overline{\mathbf{log}}(f)_Z). \end{array} \tag{10.115}$$

We have a similar example identical in form for the left diagram in (10.111).

Deduce Lemma 9.131 as a consequence. (Take $f = \mathrm{id}$ and use (9.172). In other words, in the (co)commutative situation, the antipode can be obtained as the image of the Takeuchi series which is group-like.)

Exercise 10.105. Let $f : \mathrm{C} \to \mathrm{A}$ and $\hat{f} : \overline{\Sigma} \to \hom^{\times}(\mathrm{C}, \mathrm{A})$ correspond to each other under the bijection (10.101). Since $\hat{f}$ is a Coxeter algebra morphism, it induces a map from $\mathscr{S}(\overline{\Sigma})$ to $\mathscr{S}(\hom^{\times}(\mathrm{C}, \mathrm{A}))$ which commutes with $\mathbf{log}$. As an example, we have a diagram identical in form to (10.115) with the elements now viewed inside $\mathscr{S}(\overline{\Sigma})$ and $\overline{\mathtt{C}}\mathtt{Sp}(\mathrm{C}, \mathrm{A})$.

Deduce Lemma 9.137 as a consequence. (Take $f = \mathrm{id}$ and use (9.172). In other words, the logarithm of the antipode can be computed using the logarithm of the Takeuchi series.)

Exponential series. Let $\mathrm{char}(\Bbbk) \nmid |W|$. The series e of $\overline{\Sigma}$ defined by $e_T := \frac{1}{d_{S/T}}\mathtt{H}_{S/T}$ is exponential, and in particular group-like. This can be checked directly. It also follows by applying Proposition 9.91 to the morphism (10.65). The logarithm of this series is linear, that is, concentrated in the S-component where its value is $\mathtt{H}_{S/S}$. Also, any exponential series of $\overline{\Sigma}$ is a scalar multiple of e.

Similarly, for q not a root of unity, the series $e_T := \frac{1}{d_{S/T}(q)}\mathtt{H}_{S/T}$ of $\overline{\Sigma}_q$ is q-exponential, and in particular, group-like, and so on.

Example 10.106. This is a continuation of Example 9.88. The Coxeter algebra morphism $\hat{f} : \overline{\Sigma} \to \hom^{\times}(\mathrm{C}, \mathrm{A})$ in (10.101) sends the exponential series e of $\overline{\Sigma}$ to the exponential series $\hat{f}(e)$ of $\hom^{\times}(\mathrm{C}, \mathrm{A})$ given by $\hat{f}(e)_T = \frac{1}{d_{S/T}}\mu^S_T f_S \Delta^S_T$. There is a similar statement for q-exponential series. □

10.4.21. Antipode.

Lemma 10.107. *For a scalar q, the antipode S of the Coxeter q-bialgebra $\overline{\Sigma}_q$, on the Z-component, in the* H*-basis, is*

$$S_Z(\mathtt{H}_{U/Z}) = q^{\operatorname{dist}(U,Z\overline{U})} \sum_{V:\, Z\overline{U} \leq V} (-1)^{\dim(V)}\, \mathtt{H}_{V/Z}. \tag{10.116}$$

PROOF. Substitute (co)product formulas (10.63) in the Takeuchi formula (9.171) and use the alternating sign identity (1.63c) applied to $\boldsymbol{\alpha}_Z$. □

Lemma 10.108. *Let* $\operatorname{char}(\Bbbk) \nmid |W|$. *The antipode S of the Coxeter bialgebra $\overline{\Sigma}$, on the Z-component, in the* Q*-basis, is*

$$S_Z(\mathtt{Q}_{U/Z}) = (-1)^{\dim(U)}\, \mathtt{Q}_{Z\overline{U}/Z}. \tag{10.117}$$

PROOF. We proceed as in the proof of (10.116). Substitute (co)product formulas (10.68) in the Takeuchi formula (9.171) and use the alternating sign identity (1.63c) for $q = 1$ applied to $\boldsymbol{\alpha}_Z$. Now in addition to the condition $V \geq Z\overline{U}$, we have the condition $\mathrm{s}(V/Z) = \mathrm{s}(U/Z)$ which forces $V = Z\overline{U}$. □

Similarly:

Lemma 10.109. *For q not a root of unity, the antipode S of the Coxeter q-bialgebra $\overline{\Sigma}_q$, on the Z-component, in the* Q*-basis, is*

$$S_Z(\mathtt{Q}_{U/Z}) = (-1)^{\dim(U)} q^{\operatorname{dist}(U,Z\overline{U})}\, \mathtt{Q}_{Z\overline{U}/Z}. \tag{10.118}$$

Exercise 10.110. Dualize formula (10.116) to deduce: For a scalar q, the antipode S of the Coxeter q-bialgebra $\overline{\Sigma}^*_q$, on the Z-component, in the M-basis, is

$$S_Z(\mathtt{M}_{V/Z}) = (-1)^{\dim(V)} \sum_{U:\, Z\overline{U} \leq V} q^{\operatorname{dist}(U,Z\overline{U})}\, \mathtt{M}_{U/Z}. \tag{10.119}$$

Similarly, dualize formulas (10.117) and (10.118).

Exercise 10.111. Deduce antipode formulas (10.116) and (10.119) as special cases of formulas (9.182) and (9.184) by taking $\mathrm{C} := \Bbbk\{x\}$ and $\mathrm{A} := \Bbbk[x]$. Formulas (10.117), (10.118) and their duals can be deduced similarly from (9.183) and (9.185).

10.5. Coxeter bialgebra of flat-types

We introduce the Coxeter bialgebra of flat-types $\overline{\Pi}$. It is the free commutative Coxeter bialgebra on the divided power Coxeter coalgebra $\Bbbk\{x\}$. Thus, $\overline{\Pi}$ is the abelianization of the Coxeter bialgebra of face-types $\overline{\Sigma}$ (which is the free Coxeter bialgebra on $\Bbbk\{x\}$). In particular, it is bicommutative. In addition to the canonical H-basis, we also consider the Q-basis for $\overline{\Pi}$. We also consider the dual Coxeter bialgebra of flat-types $\overline{\Pi}^*$ with dual bases M and P. It is the cofree cocommutative Coxeter bialgebra on the Coxeter algebra of polynomials $\Bbbk[x]$. Thus, it is the coabelianization of $\overline{\Sigma}^*$ (which is the cofree Coxeter bialgebra on $\Bbbk[x]$).

The norm map on face-types from $\overline{\Sigma}$ to $\overline{\Sigma}^*$ (one for each invariant family of scalars ξ) factors through the abelianization and maps into the coabelianization to induce a self-dual morphism from $\overline{\Pi}$ to $\overline{\Pi}^*$ which we call the norm map on flat-types. We give a necessary and sufficient condition for this map to be an isomorphism. In particular, this yields self-duality isomorphisms for $\overline{\Pi}$.

The Coxeter bialgebras of flat-types $\overline{\Pi}$ and $\overline{\Pi}^*$ play the same role for Coxeter bialgebras that the Coxeter bimonoids of flats $\sqcap$ and $\sqcap^*$ discussed in Section 4.4 play for Coxeter bimonoids. So they are of fundamental importance. These connections are made precise later in Section 18.5 using Fock functors.

In this section, we assume $\mathrm{char}(\Bbbk) \nmid |W|$.

10.5.1. Flat-types. Let us first note down a few facts about flat-types. A flat-type of $\boldsymbol{\alpha}_T$ is generally denoted λ/T.

- For any $T \leq U$, a flat-type λ/U of $\boldsymbol{\alpha}_U$ yields a flat-type ρ/T of $\boldsymbol{\alpha}_T$: Let V/U be any face-type with support λ/U. Then ρ/T is the support of V/T. We write $\rho/T = i_T(\lambda/U)$. Observe: For $T \leq U \leq V$,
$$i_T(\lambda/V) = i_T i_U(\lambda/V).$$
Note very carefully: i_T viewed as a map from the set of flat-types of $\boldsymbol{\alpha}_U$ to the set of flat-types of $\boldsymbol{\alpha}_T$ is not injective in general. This is in contrast to what happens for faces, flats, face-types.
- For any morphism $\sigma : T \to U$, a flat-type λ/T of $\boldsymbol{\alpha}_T$ yields a flat-type μ/U of $\boldsymbol{\alpha}_U$: Let V/T be any face-type with support λ/T. Then μ/U is the support of $\sigma_*(V)/U$. We write $\mu/U = \sigma_*(\lambda/T)$. Further, if $\sigma : T \to U$ is a morphism local to Z, then
$$i_Z(\lambda/T) = i_Z(\sigma_*(\lambda/T)).$$

10.5.2. Coxeter bialgebra of flat-types. Define the Coxeter space $\overline{\Pi}$ as follows. For a face-type T, let $\overline{\Pi}[T]$ be the linear span of the set of flat-types of $\boldsymbol{\alpha}_T$. We denote the canonical basis by $\mathtt{H}$. Thus a basis element in the T-component is denoted $\mathtt{H}_{\lambda/T}$. The structure maps are as follows. For a morphism $\sigma : T \to U$,

$$\sigma : \overline{\Pi}[T] \to \overline{\Pi}[U], \qquad \mathtt{H}_{\lambda/T} \mapsto \mathtt{H}_{\sigma_*(\lambda/T)}. \tag{10.120}$$

We now turn $\overline{\Pi}$ into a Coxeter bialgebra. The product and coproduct are

$$\begin{array}{ll} \mu_Z^T : \overline{\Pi}[T] \to \overline{\Pi}[Z] & \Delta_Z^T : \overline{\Pi}[Z] \to \overline{\Pi}[T] \\ \mathtt{H}_{\lambda/T} \mapsto \mathtt{H}_{i_Z(\lambda/T)} & \mathtt{H}_{\mu/Z} \mapsto \displaystyle\sum_{\rho/T} \Bigg(\sum_{\substack{V:\, V \geq T \\ \mathrm{s}(V/T) = \rho/T}} a_Z^{TVU} \Bigg) \mathtt{H}_{\rho/T}, \end{array} \tag{10.121}$$

where U is any fixed face-type such that $\mathrm{s}(U/Z) = \mu/Z$, and a_Z^{TVU} are the face-type enumeration scalars defined in Section 1.8.3. We refer to $\overline{\Pi}$ as the *Coxeter bialgebra of flat-types*.

Recall the Coxeter bialgebra of face-types $\overline{\Sigma}$. It is clear from (co)product formulas (10.62) that

$$\pi : \overline{\Sigma} \twoheadrightarrow \overline{\Pi}, \qquad \mathtt{H}_{T/Z} \mapsto \mathtt{H}_{\mathrm{s}(T/Z)} \tag{10.122}$$

on the Z-component is a surjective morphism of Coxeter bialgebras. In fact, $\overline{\Pi}$ is the abelianization of $\overline{\Sigma}$. It follows that $\overline{\Pi}$ is bicommutative.

10.5.3. Q-basis. Define the Q-basis of $\overline{\Pi}$ by applying (1.131) to $\boldsymbol{\alpha}_Z$ for each face-type Z. (Note: $\overline{\Pi}[Z]$ identifies as a vector space with the invariant Birkhoff algebra of $\boldsymbol{\alpha}_Z$.) The map (10.122), on the Q-basis, is

$$\pi : \overline{\Sigma} \twoheadrightarrow \overline{\Pi}, \qquad \mathtt{Q}_{T/Z} \mapsto |\mathrm{s}(T/Z)|\, \mathtt{Q}_{\mathrm{s}(T/Z)} \tag{10.123}$$

on the Z-component, with $|\mathrm{s}(T/Z)|$ as in Definition 1.8.

Proposition 10.112. *The product and coproduct of* $\overline{\Pi}$ *in the* Q*-basis are*

$$\begin{aligned} &\mu_Z^T : \overline{\Pi}[T] \to \overline{\Pi}[Z] && \Delta_Z^T : \overline{\Pi}[Z] \to \overline{\Pi}[T] \\ &|\lambda/T|\, \mathtt{Q}_{\lambda/T} \mapsto |i_Z(\lambda/T)|\, \mathtt{Q}_{i_Z(\lambda/T)} && \mathtt{Q}_{\lambda/Z} \mapsto \sum_{\substack{\mu/T:\\ i_Z(\mu/T)=\lambda/Z}} \mathtt{Q}_{\mu/T}, \end{aligned} \tag{10.124}$$

PROOF. We may deduce these formulas from (co)product formulas (10.68) of $\overline{\Sigma}$ in the Q-basis by using the map (10.123). For the coproduct calculation, we use the gate identity (1.61) applied to $\boldsymbol{\alpha}_Z$. □

10.5.4. Dual Coxeter bialgebra of flat-types. Let $\overline{\Pi}^*$ denote the Coxeter bialgebra dual to $\overline{\Pi}$. Further, let M denote the basis dual to H, and P denote the basis dual to Q. One can explicitly write down (co)product formulas for $\overline{\Pi}^*$ in these bases by dualizing formulas for $\overline{\Pi}$. By general duality principles (9.53), $\overline{\Pi}^*$ is the coabelianization of $\overline{\Sigma}^*$.

10.5.5. Freeness and cofreeness. The Coxeter bialgebra $\overline{\Pi}$ is the free commutative Coxeter bialgebra on $\Bbbk\{x\}$ viewed as a Coxeter coalgebra. Dually, $\overline{\Pi}^*$ is the cofree cocommutative Coxeter bialgebra on $\Bbbk[x]$ viewed as a Coxeter algebra. More precisely:

Proposition 10.113. *There are isomorphisms of Coxeter bialgebras*

$$\overline{\Pi} \xrightarrow{\cong} \overline{\mathcal{S}}(\Bbbk\{x\}) \quad \textit{and} \quad \overline{\Pi}^* \xrightarrow{\cong} \overline{\mathcal{S}}^\vee(\Bbbk[x]).$$

On the Z*-component, they are, respectively, given by*

$$\mathtt{H}_{\mathrm{s}(T/Z)} \mapsto \overline{x^{(T)}} \quad \textit{and} \quad \mathtt{M}_{\lambda/Z} \to \sum_{\substack{T:\, T\geq Z\\ \mathrm{s}(T/Z)=\lambda/Z}} x^T,$$

where $\overline{x^{(T)}}$ *is the image of* $x^{(T)}$ *in the quotient map* $\overline{\mathcal{T}}(\Bbbk\{x\})[Z] \twoheadrightarrow \overline{\mathcal{S}}(\Bbbk\{x\})[Z]$.

PROOF. This can be formally deduced from Proposition 10.71 for $q = 1$ by using (9.54). □

Observation 10.114. As Coxeter spaces,

$$\begin{aligned} \overline{\Pi} &= \Bbbk[x] + \Bbbk[x]^{\underline{2}} + \Bbbk[x]^{\underline{3}} + \dots, \\ \overline{\Pi}^* &= \Bbbk[x] + \Bbbk[x]^{\overline{2}} + \Bbbk[x]^{\overline{3}} + \dots, \end{aligned} \tag{10.125}$$

the sums of commutative Cauchy powers of $\Bbbk[x]$. These are special cases of (9.39) and (9.41). In fact, $\overline{\Pi}$ and $\overline{\Pi}^*$ are canonically isomorphic as Coxeter spaces.

Observation 10.115. The functor (8.30) takes the Coxeter space $\overline{\Pi}$ to the trivial representation of W. This follows from (10.125) and Observation 10.8. Since $\overline{\Pi}$ is the abelianization of $\overline{\Sigma}$, we may also use Exercise 10.58 and Exercise 9.32.

Exercise 10.116. Reformulate Proposition 10.113 in terms of universal properties of $\overline{\Pi}$ and $\overline{\Pi}^*$. (For $\overline{\Sigma}_q$ and $\overline{\Sigma}_q^*$, this was done in Theorems 10.72 and 10.73.)

The Coxeter bialgebra $\overline{\Pi}$ is also the free commutative Coxeter bialgebra on $\Bbbk\{x\}$ viewed as a trivial Coxeter coalgebra. Dually, $\overline{\Pi}^*$ is the cofree cocommutative Coxeter bialgebra on $\Bbbk[x]$ viewed as a trivial Coxeter algebra. The identifications are made using the $\mathtt{Q}$-basis and $\mathtt{P}$-basis, respectively. (Note very carefully: These now also involve a coefficient.)

10.5.6. Primitive and indecomposable parts.

We have

$$\overline{\mathcal{P}}(\overline{\Pi}) = \Bbbk\{x\} \quad \text{and} \quad \overline{\mathcal{Q}}(\overline{\Pi}) = \Bbbk\{x\}. \tag{10.126}$$

More precisely, $\overline{\mathcal{P}}(\overline{\Pi})[Z]$ is one-dimensional and spanned by $\mathtt{Q}_{\mathrm{s}(Z/Z)}$. This follows from the coproduct formula in (10.124). Similarly, $\overline{\mathcal{Q}}(\overline{\Pi})[Z]$ is spanned by the image of either $\mathtt{H}_{\mathrm{s}(Z/Z)}$ or $\mathtt{Q}_{\mathrm{s}(Z/Z)}$.

10.5.7. Norm map on flat-types.

Recall the norm map on face-types $\overline{\kappa}_{1,\xi}$ from (10.94), with ξ an invariant family of scalars as in Definition 1.10.

Lemma 10.117. *Any invariant family of scalars ξ yields a morphism of Coxeter bialgebras*

$$\overline{\kappa}_\xi : \overline{\Pi} \to \overline{\Pi}^* \tag{10.127}$$

via the self-dual diagram

$$\begin{array}{ccc} \overline{\Sigma} & \xrightarrow{\overline{\kappa}_{1,\xi}} & \overline{\Sigma}^* \\ {\scriptstyle \pi}\big\downarrow & & \big\uparrow{\scriptstyle \pi^*} \\ \overline{\Pi} & \xrightarrow[\overline{\kappa}_\xi]{} & \overline{\Pi}^*, \end{array} \tag{10.128}$$

with $\overline{\kappa}_{1,\xi}$ factoring through the abelianization (10.122) *and mapping into the coabelianization. Moreover, every Coxeter bialgebra morphism $\overline{\Pi} \to \overline{\Pi}^*$ arises in this manner from a unique ξ.*

Proof. The existence of (10.128) follows from Exercise 1.55. That it yields all morphisms from $\overline{\Pi}$ to $\overline{\Pi}^*$ follows from Exercise 10.81 for $q = 1$.

For general considerations along the lines of Exercise 10.80, see (9.146). Diagram (10.128) is an instance of (9.147). Seen this way, by reversing the logical flow, we can deduce Exercise 1.55 as a consequence. □

We call the map $\overline{\kappa}_\xi$ in (10.127) the *norm map on flat-types.* We now give a necessary and sufficient numerical condition for $\overline{\kappa}_\xi$ to be invertible. We do this via a change of basis. Let η be the family defined from ξ by noncommutative Möbius inversion (10.93).

Theorem 10.118. *The norm map on flat-types* (10.127), *on the* Q*-basis and into the* P*-basis, is*

$$\bar{\kappa}_\xi : \overline{\Pi} \to \overline{\Pi}^*, \qquad |s(T/Z)|\, \mathtt{Q}_{s(T/Z)} \mapsto \eta_T\, \mathtt{P}_{s(T/Z)} \tag{10.129}$$

on the Z-component.

PROOF. This can be deduced from the formula for the norm map from $\overline{\Sigma}$ to $\overline{\Sigma}^*$, on the Q-basis and into the P-basis, given in (10.95). □

Theorem 10.119 (Isomorphism criterion). *The norm map on flat-types* (10.127) *is an isomorphism of Coxeter bialgebras iff*

$$\eta_T = \sum_{U:\, U \geq T} \boldsymbol{\mu}(T, U)\, \xi_U \neq 0$$

for each face-type T.

PROOF. This follows from Theorem 10.118. Alternatively, we can apply Theorem 9.100 (similar to what we did in Exercise 10.90). □

By taking $\eta \equiv 1$ for instance, we deduce that the Coxeter bialgebra $\overline{\Pi}$ is self-dual.

Example 10.120. Consider the norm map on face-types (10.97) for $\xi \equiv 1$. The corresponding η has an explicit formula given by the series f in Example 10.25. Substituting this into (10.129), we obtain: The norm map on flat-types (10.127), on the Q-basis and into the P-basis, for $\xi \equiv 1$, is

$$\bar{\kappa}_{\xi\equiv 1} : \overline{\Pi} \to \overline{\Pi}^*, \qquad |s(T/Z)|\, \mathtt{Q}_{s(T/Z)} \mapsto \frac{|\mu(\mathrm{X}, \top)|}{|W_{\mathrm{X}}|}\, \mathtt{P}_{s(T/Z)}, \tag{10.130}$$

where X is any flat such that $\mathrm{t}(\mathrm{X}) = \mathrm{s}(T)$. It is an isomorphism if $\mu(\mathrm{X}, \top) \neq 0$ for all flats X. This condition holds in characteristic zero [**10**, Proposition 1.80]. □

Compare and contrast Theorems 10.118 and 10.119 with Theorems 10.84 and 10.86.

10.5.8. Primitive and group-like series.

We now apply the discussion in Section 9.10 to the Coxeter bialgebra of flat-types $\overline{\Pi}$. The module $\mathscr{S}(\overline{\Pi})$ is the invariant flat-incidence algebra viewed as a left module over itself under the identification

$$(f(\mathrm{Z}, \mathrm{X}))_{\mathrm{Z} \leq \mathrm{X}} \longleftrightarrow \sum_{\lambda/T} \frac{g_{\lambda/T}}{|\lambda/T|}\, \mathtt{H}_{\lambda/T} \in \overline{\Pi}[T] \quad \text{for each face-type } T, \tag{10.131}$$

where $f(\mathrm{s}(F), \mathrm{X}) = g_{\mathrm{t}(\mathrm{X}/F)}$ for any flat X containing a face F of C_r, and $|\lambda/T|$ is as in Definition 1.8. We leave this as an exercise.

The unit element δ of the invariant flat-incidence algebra is a group-like series of $\overline{\Pi}$. We call it the *universal series* of $\overline{\Pi}$. The corresponding primitive series under the bijection (9.126) is μ. In terms of the identification (10.131),

$$\begin{aligned} \delta &\leftrightarrow \mathtt{H}_{s(T/T)} \in \overline{\Pi}[T] \quad \text{for each } T, \\ \mu &\leftrightarrow \mathtt{Q}_{s(T/T)} \in \overline{\Pi}[T] \quad \text{for each } T, \end{aligned} \tag{10.132}$$

where $\mathtt{Q}_{s(T/T)}$ is the Q-basis element of $\overline{\Pi}$.

Let x be the linear series of $\overline{\Pi}$ defined by $x_S := \mathtt{H}_{s(S/S)}$. The corresponding exponential series e of $\overline{\Pi}$ under the bijection (9.135) is $e_T = \frac{1}{d_{S/T}}\mathtt{H}_{s(S/T)}$. In the $\mathtt{Q}$-basis, $x_S = \mathtt{Q}_{s(S/S)}$ and $e_T = \mathtt{Q}_{s(S/T)}$. Any other linear and exponential series of $\overline{\Pi}$ are scalar multiples of x and e.

10.5.9. Antipode.

Lemma 10.121. *The antipode S of the Coxeter bialgebra $\overline{\Pi}$, on the Z-component, in the $\mathtt{H}$-basis, is*

$$S_Z(\mathtt{H}_{\lambda/Z}) = \sum_{\mu/Z\colon \mu/Z \geq \lambda/Z} (-1)^{\dim(\mu/Z)} c_{\lambda/Z,\mu/Z}\, \mathtt{H}_{\mu/Z}, \tag{10.133}$$

where $c_{\lambda/Z,\mu/Z}$ is the number of face-types of support μ/Z which are greater than a fixed face-type of support λ/Z.

Proof. This formula can be deduced from (10.116). Some points to note are as follows. The face-types U/Z and $Z\overline{U}/Z$ have the same support. The face-types V/Z and V have the same dimension. Similarly, μ/Z and $i_\emptyset(\mu/Z)$ have the same dimension. We prefer to use the former in the formula since it is less cumbersome. □

Lemma 10.122. *The antipode S of the Coxeter bialgebra $\overline{\Pi}$, on the Z-component, in the $\mathtt{Q}$-basis, is*

$$S_Z(\mathtt{Q}_{\lambda/Z}) = (-1)^{\dim(\lambda/Z)} \mathtt{Q}_{\lambda/Z}. \tag{10.134}$$

Proof. This formula can be deduced from (10.117). □

10.6. Coxeter bialgebra of symmetries

We build a Coxeter bialgebra $\overline{\mathrm{I\Gamma}}$ out of Coxeter symmetries of a reflection arrangement. We call it the Coxeter bialgebra of symmetries. It is neither commutative nor cocommutative. It is self-dual with the self-duality isomorphism defined by inverting the involved Coxeter symmetry. It is free as a Coxeter algebra and cofree as a Coxeter coalgebra. Moreover, it admits a q-deformation $\overline{\mathrm{I\Gamma}}_q$. The Coxeter q-bialgebra of face-types $\overline{\Sigma}_q$ embeds inside $\overline{\mathrm{I\Gamma}}_q$, and dually, $\overline{\Sigma}^*_q$ is a quotient of $\overline{\mathrm{I\Gamma}}^*_q$. Notions of descents, global descents and shuffles from Section 1.6 naturally enter into the discussion.

The Coxeter bialgebra of symmetries $\overline{\mathrm{I\Gamma}}$ plays the same role for Coxeter bialgebras that the Coxeter bimonoid of pairs of chambers $\mathrm{I\Gamma}$ discussed in Section 4.7 plays for Coxeter bimonoids. This connection is made precise later in Section 18.6 using Fock functors.

We do not supply any proofs in this section.

10.6.1. Coxeter space of symmetries. Define the Coxeter $\mathcal{A}$-space $\overline{\mathrm{I\Gamma}}$ as follows. For a top-nested face (A, C), let $\overline{\mathrm{I\Gamma}}[A, C]$ be the linear span of the set of chambers of $(\mathcal{A}_A, C/A)$. Equivalently, it is the linear span of the set of Coxeter symmetries of $(\mathcal{A}_A, C/A)$. We denote the canonical basis by $\mathtt{K}$, and index it either by chambers or by Coxeter symmetries depending on convenience. Thus, a typical basis element is denoted either $\mathtt{K}_{D/A}$ or $\mathtt{K}_u$ with

$u \in W_A$, the relation between the two being $u = d(C,D)$ and $D = uC$. Support and type morphisms are

$$\beta_{G,D,F,C} : \overline{\Gamma}[F,C] \to \overline{\Gamma}[G,D] \qquad w : \overline{\Gamma}[F,C] \to \overline{\Gamma}[wF,wC] \tag{10.135}$$
$$\mathtt{K}_{E/F} \mapsto \mathtt{K}_{GE/G} \qquad \mathtt{K}_{E/F} \mapsto \mathtt{K}_{wE/wF}.$$

We call $\overline{\Gamma}$ the *Coxeter space of symmetries*.

Alternatively, define the Coxeter $\boldsymbol{\alpha}$-space $\overline{\Gamma}$ as follows. For a face-type T, let $\overline{\Gamma}[T]$ be the linear span of the set of chambers (or Coxeter symmetries) of $\boldsymbol{\alpha}_T$. (A Coxeter symmetry of $\boldsymbol{\alpha}_T$ is the same as an element of W_T.) The structure maps are as follows. For a morphism $\sigma : T \to U$,

$$\sigma : \overline{\Gamma}[T] \to \overline{\Gamma}[U], \qquad \mathtt{K}_u \mapsto \mathtt{K}_{wuw^{-1}}, \tag{10.136}$$

where $u \in W_T$ and $\sigma = w\beta = \beta w$ in normal form. (See Exercise 1.21 in this regard.)

For the dual Coxeter space $\overline{\Gamma}^*$, we write $\mathtt{F}$ for the basis dual to $\mathtt{K}$. Now define the $\mathtt{H}$-basis of $\overline{\Gamma}$ and the dual $\mathtt{M}$-basis of $\overline{\Gamma}^*$ as follows. On the T-component,

$$\mathtt{H}_u = \sum_{\substack{v \in W_T: \\ v \leq_T u}} \mathtt{K}_v \qquad \text{and} \qquad \mathtt{F}_v = \sum_{\substack{u \in W_T: \\ v \leq_T u}} \mathtt{M}_u, \tag{10.137}$$

where $\leq_T$ is the weak left Bruhat order on W_T (Section 1.4.7).

10.6.2. Coxeter bialgebra of symmetries. We now turn $\overline{\Gamma}^*$ into a Coxeter bialgebra. The product and coproduct in the $\mathtt{F}$-basis are

$$\mu_A^F : \overline{\Gamma}^*[F,C] \to \overline{\Gamma}^*[A,C], \qquad \mathtt{F}_{D/F} \mapsto \sum_{D' :\, D' \geq A,\, FD' = D} \mathtt{F}_{D'/A}, \tag{10.138a}$$

$$\Delta_A^F : \overline{\Gamma}^*[A,C] \to \overline{\Gamma}^*[F,C], \qquad \mathtt{F}_{D/A} \mapsto \mathtt{F}_{wD/F}, \tag{10.138b}$$

where $w \in W_A$ is defined by $w(GC) = C$, and where G is the face of D of the same type as F.

We refer to $\overline{\Gamma}^*$ as the *Coxeter bialgebra of symmetries*. It is neither commutative nor cocommutative.

Proposition 10.123. *The product and coproduct of $\overline{\Gamma}^*$ in the* $\mathtt{M}$*-basis are*

$$\mu_A^F : \overline{\Gamma}^*[F,C] \to \overline{\Gamma}^*[A,C], \qquad \mathtt{M}_{D/F} \mapsto \sum_{\substack{(C',D'): \\ C' \geq A,\, D' \geq A \\ FD' = D,\, FC' = C \\ C' \text{ -- } C \text{ -- } D'}} \mathtt{M}_{d(C',D')C/A}, \tag{10.139a}$$

$$\Delta_A^F : \overline{\Gamma}^*[A,C] \to \overline{\Gamma}^*[F,C], \qquad \mathtt{M}_{D/A} \mapsto \begin{cases} \mathtt{M}_{wD/F} & \text{if } G \leq A\overline{C}, \\ 0 & \text{otherwise,} \end{cases} \tag{10.139b}$$

where $w \in W_A$ is defined by $w(GC) = C$, and where G is the face of D of the same type as F.

10.6.3. Dual Coxeter bialgebra of symmetries. Let $\overline{\Pi}$ denote the Coxeter bialgebra dual to $\overline{\Pi}^*$.

The product and coproduct of $\overline{\Pi}$ in the K-basis are

$$\mu_A^F : \overline{\Pi}[F,C] \to \overline{\Pi}[A,C], \qquad \mathtt{K}_{D/F} \mapsto \sum_{\zeta \in \mathrm{Sh}_{\mathrm{t}(F/A)}} \mathtt{K}_{\zeta D/A}, \tag{10.140a}$$

where ζ runs over all $\mathrm{t}(F/A)$-shuffles with reference chamber C/A, and

$$\Delta_A^F : \overline{\Pi}[A,C] \to \overline{\Pi}[F,C], \qquad \mathtt{K}_{D/A} \mapsto \mathtt{K}_{FD/F}. \tag{10.140b}$$

These are obtained by dualizing formulas (10.138a) and (10.138b).

The product and coproduct of $\overline{\Pi}$ in the H-basis are

$$\mu_A^F : \overline{\Pi}[F,C] \to \overline{\Pi}[A,C], \qquad \mathtt{H}_{D/F} \mapsto \mathtt{H}_{\zeta D/A}, \tag{10.141a}$$

where $\zeta = \theta(\mathrm{t}(F/A))$ is the longest $\mathrm{t}(F/A)$-shuffle with reference chamber C/A, and

$$\Delta_A^F : \overline{\Pi}[A,C] \to \overline{\Pi}[F,C], \qquad \mathtt{H}_{D/A} \mapsto \sum_{\substack{G:\, G \geq A \\ \mathrm{t}(G)=\mathrm{t}(F) \\ C \text{ -- } GC \text{ -- } GD \text{ -- } D}} \mathtt{H}_{d(GC,GD)C/F}. \tag{10.141b}$$

These are obtained by dualizing formulas (10.139a) and (10.139b).

10.6.4. Coxeter q-bialgebra of symmetries. Let q be a scalar. Consider the Coxeter bialgebra $\overline{\Pi}^*$. Its product in the F-basis (10.138a) can be deformed by inserting $q^{\mathrm{dist}(D,D')}$ inside the sum. This results in a q-analogue of $\overline{\Pi}^*$ which we denote $\overline{\Pi}_q^*$. It is a Coxeter q-bialgebra. Its product in the M-basis is given by inserting $q^{\mathrm{dist}(D,D')}$ inside the sum in formula (10.139a).

Let $\overline{\Pi}_q$ denote the Coxeter q-bialgebra dual to $\overline{\Pi}_q^*$. Its product is the same as that of $\overline{\Pi}$ while the coproduct is deformed by an appropriate power of q.

10.6.5. Self-duality. The Coxeter q-bialgebra $\overline{\Pi}_q$ is self-dual as follows.

Exercise 10.124. Let q be a nonzero scalar. Check: The map

$$\overline{\Pi}_q \to \overline{\Pi}_q^*, \qquad \mathtt{K}_w \mapsto q^{l(w)}\, \mathtt{F}_{w^{-1}} \tag{10.142}$$

on the T-component with $w \in W_T$ is a self-dual isomorphism of Coxeter q-bialgebras. (By self-duality, it suffices to either check that (10.142) is a morphism of Coxeter algebras or of Coxeter coalgebras. To match powers of q, use the gate property.)

10.6.6. Primitive and indecomposable parts. Freeness and cofreeness. For a face-type Z, for $w \in W_Z$, recall that $\mathrm{gdes}_Z(w)$ is the type of $A\overline{C} \wedge D/A$, where A is any face of type Z, and C and D are any chambers greater than A such that $d(C,D) = w$. This is the global descent of w.

Using the coproduct formula of $\overline{\Pi}^*$ in the M-basis (10.139b), one can deduce all of the following. The primitive part of $\overline{\Pi}^*$, on the Z-component, has linear basis

$$\{\mathtt{M}_w \mid w \in W_Z,\, \mathrm{gdes}_Z(w) = Z/Z\}.$$

More generally, the k-th term in the primitive filtration of $\overline{\Pi}^*$, on the Z-component, has linear basis

$$\{\mathtt{M}_w \mid w \in W_Z,\, \mathrm{rk}(\mathrm{gdes}_Z(w)) = k-1\}.$$

The $k = 1$ term recovers the primitive part.

As a Coxeter coalgebra, $\overline{\Pi}^*$ is cofree on its primitive part. More precisely, there is an isomorphism of Coxeter coalgebras

$$\overline{\Pi}^* \cong \overline{\mathcal{T}}^\vee(\overline{\mathcal{P}}(\overline{\Pi}^*)), \tag{10.143}$$

with $\overline{\mathcal{P}}(\overline{\Pi}^*)[T]$ (which is the T-summand of the Z-component of the rhs) spanned by $\mathtt{M}_w$ such that $\mathrm{gdes}_Z(w) = T/Z$. By duality, there is an isomorphism of Coxeter algebras

$$\overline{\Pi} \cong \overline{\mathcal{T}}(\overline{\mathcal{Q}}(\overline{\Pi})). \tag{10.144}$$

Thus, as a Coxeter algebra, $\overline{\Pi}$ (and hence $\overline{\Pi}^*$) is free on its indecomposable part.

10.6.7. Antipode. We begin with some preliminary definitions. For face-types $T \geq Z$ and $v \in W_Z$, define $v_T \in W_T$ by $v_T := d(HC, D)$, where C and D are any chambers such that $d(C, D) = v$, and H is the face of D of type T.

For a face-type Z and $v, w \in W_Z$, define the map

$$f_{v,w} : \{T \geq Z \mid w^{-1}v_T \text{ is a } T/Z\text{-shuffle }\} \longrightarrow W_Z, \qquad T \mapsto w^{-1}v_T.$$

Lemma 10.125. *The antipode S of the Coxeter bialgebra $\overline{\Pi}^*$, on the Z-component, in the* $\mathtt{F}$*-basis, is*

$$S_Z(\mathtt{F}_v) = \sum_{w \in W_Z} \Big(\sum (-1)^{\dim(T)}\Big)\mathtt{F}_w, \tag{10.145}$$

where the inside sum is over those face-types $T \geq Z$ for which the inverse image $f^{-1}_{v,w}(w^{-1}v_T)$ is the singleton set $\{T\}$.

Lemma 10.126. *The antipode S of the Coxeter bialgebra $\overline{\Pi}^*$, on the Z-component, in the* $\mathtt{M}$*-basis, is*

$$S_Z(\mathtt{M}_v) = (-1)^{\dim(\mathrm{gdes}_Z(v))} \sum_{w \in W_Z} c(v, w)\, \mathtt{M}_w, \tag{10.146}$$

where $T/Z = \mathrm{gdes}_Z(v)$ is the global descent of v, and $c(v, w)$ is the number of T/Z-shuffles ζ satisfying the conditions

(i) $v_T\zeta^{-1} \leq_Z w$,
(ii) *if* $v \leq_Z v'$ *and* $v'_T\zeta^{-1} \leq_Z w$, *then* $v = v'$,
(iii) *if* $\mathrm{des}_Z(\zeta) \leq R/Z \leq T/Z$ *and* $v_R\zeta^{-1} \leq_Z w$, *then* $R = T$,

where $\leq_Z$ denotes the weak left Bruhat order on W_Z.

In formulas (10.145) and (10.146), dimension of a face-type is defined to be the dimension of any face of that face-type.

10.6.8. Coxeter bialgebra morphisms. There is a commutative self-dual diagram of Coxeter bialgebras

(10.147)

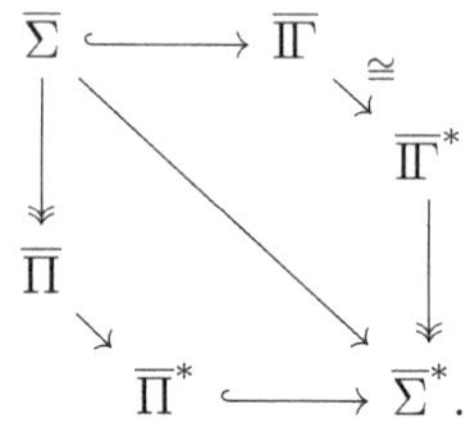

The morphism from $\overline{\Sigma}$ to $\overline{\Sigma}^*$ is the norm map on face-types (10.97) for $\xi \equiv 1$. Also see the expression for this map in (10.86) for $q = 1$, and the equivalent expression from the $\mathtt{K}$-basis to the $\mathtt{F}$-basis in (10.87) for $q = 1$. The part of the diagram below this map is an instance of the square (10.128) for $\xi \equiv 1$. In particular, the morphism from $\overline{\Pi}$ to $\overline{\Pi}^*$ is the one in (10.130).

The morphism from $\overline{\mathrm{I\Gamma}}$ to $\overline{\mathrm{I\Gamma}}^*$ is (10.142) for $q = 1$. Thus, on each component, it sends $\mathtt{K}_w$ to $\mathtt{F}_{w^{-1}}$. The injective morphism from $\overline{\Sigma}$ to $\overline{\mathrm{I\Gamma}}$, on the Z-component, is given by either of the three equivalent formulas

$$(10.148)\qquad \mathtt{K}_{T/Z} \mapsto \sum_{\substack{w \in W_Z:\\ \mathrm{des}_Z(w) = T/Z}} \mathtt{K}_w, \qquad \mathtt{H}_{T/Z} \mapsto \sum_{\substack{w \in W_Z:\\ \mathrm{des}_Z(w) \leq T/Z}} \mathtt{K}_w, \qquad \mathtt{H}_{T/Z} \mapsto \mathtt{H}_{\theta(T/Z)},$$

where recall that $\theta(T/Z)$ is the longest T/Z-shuffle. For the middle map, we may equivalently sum over $w \in \mathrm{Sh}_{T/Z}$. (See map (1.152) in this regard.) Dually, the surjective morphism from $\overline{\mathrm{I\Gamma}}^*$ to $\overline{\Sigma}^*$, on the Z-component, is given by either of the three equivalent formulas

(10.149)

$$\mathtt{F}_w \mapsto \mathtt{F}_{\mathrm{des}_Z(w)}, \qquad \mathtt{F}_w \mapsto \sum_{\substack{T:\, T \geq Z\\ \mathrm{des}_Z(w) \leq T/Z}} \mathtt{M}_{T/Z}, \qquad \mathtt{M}_w \mapsto \begin{cases} \mathtt{M}_{\mathrm{des}_Z(w)} & \text{if } w = \theta(T/Z),\\ 0 & \text{otherwise.} \end{cases}$$

Exercise 10.127. Let q be a scalar. The map $\Bbbk\{x\} \to \overline{\mathrm{I\Gamma}}_q$, which on the Z-component sends $x^{(Z)}$ to $\mathtt{K}_{\mathrm{id}}$, is a morphism of Coxeter coalgebras. By Theorem 10.72, this extends to a morphism of Coxeter q-bialgebras from $\overline{\Sigma}_q$ to $\overline{\mathrm{I\Gamma}}_q$. Use formula (10.83) to check that this universal map is given by (10.148) (with no dependence on q).

Dually, the map $\overline{\mathrm{I\Gamma}}^*_q \to \Bbbk[x]$, which on the Z-component sends $\mathtt{F}_w$ to x^Z if $w = \mathrm{id}$, and to 0 otherwise, is a morphism of Coxeter algebras. By Theorem 10.73, this lifts to a morphism of Coxeter q-bialgebras from $\overline{\mathrm{I\Gamma}}^*_q$ to $\overline{\Sigma}^*_q$. Use formula (10.84) to check that this universal map is given by (10.149).

Check: The composite morphism

$$\overline{\Sigma}_q \hookrightarrow \overline{\mathrm{I\Gamma}}_q \to \overline{\mathrm{I\Gamma}}^*_q \twoheadrightarrow \overline{\Sigma}^*_q$$

(with the middle map as in (10.142)) is the q-norm map (10.85) for $\xi \equiv 1$ which is the same as (10.87). (To get (10.85) for $\xi \equiv 1$, apply formula (1.48) to $\boldsymbol{\alpha}_Z$.)

Exercise 10.128 (Between face-types and Coxeter symmetries). There is a Coxeter bialgebra H which lies between $\overline{\Sigma}$ and $\overline{\mathrm{I\Gamma}}$ as follows. The component $\mathrm{H}[A, C]$

is the subspace of $\overline{\Pi}[A,C]$ consisting of those elements $\sum_{D\geq A} x^D\, \mathtt{K}_{D/A}$ which satisfy the condition below. Put $Z = \mathrm{t}(A)$.

For any face-type $T \geq Z$ and T/Z-shuffle ζ (with reference chamber C/A),

$$\sum_{D\colon D\geq A} x^D\, \mathtt{K}_{F\zeta^{-1}(D)/A} = \sum_{D\colon D\geq A} x^D\, \mathtt{K}_{FD/A}, \tag{10.150}$$

where F is the face of C of type T. In the lhs, the chamber $F\zeta^{-1}(D)$ can also be written as $\zeta^{-1}(GD)$ with $G := \zeta(F)$. In fact, one can formulate (10.150) alternatively by starting with $G \geq A$ of type T and defining ζ by $\zeta(C) := GC$.

Check: H is indeed a Coxeter subbialgebra of $\overline{\Pi}$. Moreover, the map $\overline{\Sigma} \hookrightarrow \overline{\Pi}$ given by (10.148) maps into H. Explicitly, $\mathtt{H}_{U/Z} \in \overline{\Sigma}[Z]$ yields the element

$$\sum_{H\colon H\geq A,\, \mathrm{t}(H/A)=U/Z} \mathtt{K}_{HC/A} \in \mathrm{H}[A,C].$$

One may check directly that it satisfies condition (10.150). Alternatively, using the universal property of $\overline{\Sigma}$ as in Exercise 10.127, it suffices to observe that $\mathtt{K}_{\mathrm{id}} = \mathtt{K}_{C/A}$ belongs to $\mathrm{H}[A,C]$. (This is the special case $U = Z$.)

Use coproduct formula (10.140b) and (4.95) to deduce:

$$\overline{\mathcal{P}}(\mathrm{H})[A,C] = \mathsf{Lie}[A]. \tag{10.151}$$

The K-basis in the lhs identifies with the H-basis in the rhs.

The connection of the Coxeter bialgebra H with the Coxeter bimonoid of chamber maps $\mathcal{C}(\Gamma,\Gamma)$ via the bosonic Fock functor is made in Section 18.6.5.

10.7. Notes

10.7.1. Beginnings of Coxeter bialgebras. Examples in this chapter are new. They generalize many well-known examples of connected graded Hopf algebras to the Coxeter setting, see Tables 10.8, 10.9, 10.10. These ideas were initiated in our first monograph [**6**] and fulfil the promise made on [**11**, page 333, last sentence]. The reason why this program could not be carried out in [**6**] is because the notion of Coxeter bialgebras did not exist at that time. In fact, that was precisely the notion that we were searching for. The search ended in the year 2011, when we first defined Coxeter bialgebras and started working out these examples beginning with the Coxeter bialgebra of polynomials. (See also the historical note in the preface.) For more examples beyond those in this chapter, see Notes 18.8.5.

Some important sources in the classical literature are mentioned below. More detailed information on Hopf algebras listed in the second column of Table 10.9 is given in [**11**, Notes to Chapter 6]. Similarly, for Hopf algebras listed in the second column of Table 10.10, see [**11**, Notes to Chapters 7 and 15].

10.7.2. Coxeter space characteristic of chambers. (Section 10.1.) The classical analogue of X_M is a graded vector space V concentrated in the degree one component. The latter gives rise to the S_n-module $V^{\otimes n}$ with action given by permutation of tensor factors. This is the analogue of the W-module M. This can be formally understood by the construction in Section 8.14.12.

10.7.3. Coxeter bialgebra of polynomials. (Section 10.2.) The Coxeter bialgebra of polynomials $\Bbbk[x]$ generalizes the classical Hopf algebra of polynomials

$\Bbbk[x]$ (in one variable x). To make the connection, we apply the construction in Section 8.14.12. For a composition $\alpha = (a_1, \dots, a_k)$ of n (which is a face-type of the braid arrangement on n letters), by (8.141),

$$\Bbbk[x]^{[n]}[\alpha] := \Bbbk[x]_{a_1} \otimes \cdots \otimes \Bbbk[x]_{a_k}. \tag{10.152}$$

This is the α-component of the Coxeter $\boldsymbol{\beta}^n$-bialgebra $\Bbbk[x]$ under the identification

$$x^{\alpha} := x^{a_1} \otimes \cdots \otimes x^{a_k}.$$

Similarly, the divided power Coxeter bialgebra $\Bbbk\{x\}$ generalizes the classical divided power Hopf algebra $\Bbbk\{x\}$. These considerations extend to the q-setting. A comparison summary is provided in Table 10.8. Also see Example 8.129 in this regard. The q-Chu–Vandermonde identity (1.170) is required for checking the q-bialgebra axiom for the q-Hopf algebra $\Bbbk_q[x]$. The more general q-multinomial identity (1.169) corresponds to the "higher" version of the classical q-bialgebra axiom for $\Bbbk_q[x]$. It is a special case of the gate identity (1.59) which we used for checking the q-bialgebra axiom for the Coxeter q-bialgebra $\Bbbk_q[x]$ in Section 10.2.4.

TABLE 10.8. Coxeter bialgebras and Hopf algebras of polynomials.

Coxeter bialgebras	Hopf algebras
Coxeter bialgebra of polynomials $\Bbbk[x]$	Hopf algebra of polynomials $\Bbbk[x]$
divided power Coxeter bialgebra $\Bbbk\{x\}$	divided power Hopf algebra $\Bbbk\{x\}$
Coxeter q-bialgebra of polynomials $\Bbbk_q[x]$	q-Hopf algebra of polynomials $\Bbbk_q[x]$
divided power Coxeter q-bialgebra $\Bbbk_q\{x\}$	divided power q-Hopf algebra $\Bbbk_q\{x\}$

A concise treatment of the Hopf algebras in Table 10.8 is given in [**7**, Examples 2.3 and 2.9]. The divided power Hopf algebra $\Bbbk\{x\}$ is also discussed in [**282**, Sections III and XI], [**1**, Example 2.6], [**269**, Example 5.6.8]. The q-Hopf algebra $\Bbbk_q[x]$ (as a cocommutative coalgebra) was introduced by Joni and Rota [**201**, Section VIII], see also [**346**]. For the dual $\Bbbk_q\{x\}$ as an algebra, see [**155**, Section 3, Formula (8)], [**105**, Example 4.9].

Primitive and group-like series. Noncommutative Möbius inversion (10.28) is explained in [**10**, Section 16.7.5]. However, Example 10.25 is new. Also, q-Möbius inversion (10.29) appears here for the first time.

For the positive bialgebra of polynomials $\Bbbk[x]_+$, primitive series are given by $(x, 0, 0, \dots)$ for $x \in \Bbbk$ with corresponding group-like series given by $(x, \frac{x^2}{2!}, \frac{x^3}{3!}, \dots)$. For q not a root of unity, for the positive q-bialgebra of polynomials $\Bbbk_q[x]_+$, primitive series are given by $(x, 0, 0, \dots)$ for $x \in \Bbbk$ with corresponding group-like series given by $(x, \frac{x^2}{(2)_q!}, \frac{x^3}{(3)_q!}, \dots)$. See the identity in [**209**, Proposition IV.2.4] in this regard. These considerations are generalized in Section 10.2.17 to the Coxeter bialgebra of polynomials.

10.7.4. Tensor and shuffle Coxeter bialgebras. Symmetric and exterior Coxeter bialgebras. (Section 10.3.) These Coxeter bialgebras generalize the tensor Hopf algebra, shuffle Hopf algebra, and so on. A comparison summary is provided in Table 10.9.

Tensor Hopf algebra. The tensor Coxeter bialgebra $\overline{\mathcal{T}}(M)$ associated to a W-module M generalizes the tensor Hopf algebra $\mathcal{T}(V)$ associated to a vector space V. The precise setting is to take the graded Hopf algebra $\mathcal{T}(V)$ with V in degree one. Thus, for each i, the component of $\mathcal{T}(V)$ of degree i is $V^{\otimes i}$. To make the connection to Coxeter bialgebras, for an integer composition $\alpha = (a_1, \dots, a_k)$ of n, by (8.141),

$$\mathcal{T}(V)^{[n]}[\alpha] = V^{\otimes a_1} \otimes \cdots \otimes V^{\otimes a_k} = V^{\otimes n}. \tag{10.153}$$

This is the α-component of the Coxeter $\boldsymbol{\beta}^n$-bialgebra $\overline{\mathcal{T}}(M)$, where M is the S_n-module $V^{\otimes n}$ with action given by permutation of tensor factors as in Notes 10.7.2. Also see Example 8.129 in this regard. Recall: In the universal property of $\mathcal{T}(V)$, we start with a linear map from V to an associative algebra A. This gives rise to a map of S_n-modules $V^{\otimes n} \to A^{\otimes n}$. This map is what we see in the universal property of $\overline{\mathcal{T}}(M)$ stated in Exercise 10.47.

Shuffle Hopf algebra. The shuffle Coxeter bialgebra $\overline{\mathcal{T}}^{\vee}(M)$ generalizes the shuffle Hopf algebra $\mathcal{T}^{\vee}(V)$. Shuffle algebras arose in work of Cartan [**74**, Section 8.4, page 8-07] and Ree [**302**, Section 1], and their Hopf structure was promoted by Sweedler [**345**, Chapter XII] and Reutenauer [**305**, Sections 1.4, 1.5 and Chapter 6]. Other early references are by Chen [**85**, Theorem 1.8] and André [**15**, page 25]. The q-shuffle product was considered by Duchamp, Klyachko, Krob, Thibon [**107**, Section 4.1].

TABLE 10.9. Tensor Coxeter bialgebras vs tensor Hopf algebras.

Coxeter bialgebras associated to a W-module M	Hopf algebras associated to a vector space V
tensor Coxeter bialgebra $\overline{\mathcal{T}}(M)$ shuffle Coxeter bialgebra $\overline{\mathcal{T}}^{\vee}(M)$	tensor Hopf algebra $\mathcal{T}(V)$ shuffle Hopf algebra $\mathcal{T}^{\vee}(V)$
tensor Coxeter q-bialgebra $\overline{\mathcal{T}_q}(M)$ shuffle Coxeter q-bialgebra $\overline{\mathcal{T}_q}^{\vee}(M)$	tensor q-Hopf algebra $\mathcal{T}_q(V)$ shuffle q-Hopf algebra $\mathcal{T}_q^{\vee}(V)$
tensor Coxeter 0-bialgebra $\overline{\mathcal{T}_0}(M) = \overline{\mathcal{T}_0}^{\vee}(M)$	tensor 0-Hopf algebra $\mathcal{T}_0(V) = \mathcal{T}_0^{\vee}(V)$
symmetric Coxeter bialgebras $\overline{\mathcal{S}}(M)$, $\overline{\mathcal{S}}^{\vee}(M)$	symmetric Hopf algebras $\mathcal{S}(V)$, $\mathcal{S}^{\vee}(V)$
exterior signed Coxeter bialgebras $\overline{\mathcal{E}}(M)$, $\overline{\mathcal{E}}^{\vee}(M)$	exterior signed Hopf algebras $\mathcal{E}(V)$, $\mathcal{E}^{\vee}(V)$

Symmetric Hopf algebra. The symmetric Coxeter bialgebra $\overline{\mathcal{S}}^{\vee}(M)$ generalizes the symmetric Hopf algebra $\mathcal{S}^{\vee}(V)$. As an algebra (along with its divided power structure), the latter is present in work of Cartan [**74**, Section 8.4, Proposition 4]. The additional bialgebra structure was considered by Cartier [**78**, Proposition 5 on page 3-09, Proposition 7 on page 3-14].

Norm map. The q-norm map (10.41) generalizes the map $\kappa_q : \mathcal{T}_q(V) \to \mathcal{T}_q^\vee(V)$ given in our monograph [**7**, Formula (2.67)]. The case $q = 1$ of the latter is present in earlier work of Benson [**33**, Sections 2.3.4 and 2.4.5]. That κ_q is an isomorphism when q is not a root of unity was shown by Duchamp, Klyachko, Krob, Thibon [**107**, Proposition 4.5]. This is generalized in Corollary 10.37.

The first diagram in (10.50) generalizes [**7**, Diagram (2.66)] linking the tensor algebra, shuffle algebra and the symmetric algebra. The second diagram in (10.50) generalizes [**7**, Diagram (2.68)] which now involves the exterior algebra.

Primitive and group-like series. For the positive symmetric bialgebra of polynomials $\Bbbk[x_1, \dots, x_n]_+$, primitive series are spanned by $(x_i, 0, 0, \dots)$ with corresponding group-like series given by $(x_i, \frac{x_i^2}{2!}, \frac{x_i^3}{3!}, \dots)$. The former is in fact a linear series and the latter an exponential series. Similar facts hold for the shuffle bialgebra of polynomials. These considerations are generalized in Section 10.3.18 to shuffle and symmetric Coxeter bialgebras.

Antipode formulas. Antipode formulas (10.57) and (10.58) of $\overline{\mathcal{T}_q}(M)$ and $\overline{\mathcal{T}_q}^\vee(M)$ generalize antipode formulas of the q-Hopf algebras $\mathcal{T}_q(V)$ and $\mathcal{T}_q^\vee(V)$. The latter for $q = 1$ are mentioned by Reutenauer [**305**, pages 19 and 35].

10.7.5. Coxeter bialgebra of flat-types. (Section 10.5.) Recall from Table 1.9: Flat-types of braid arrangements are the same as integer partitions.

Hopf algebra of symmetric functions. The Coxeter bialgebra of flat-types $\overline{\Pi}$ generalizes the Hopf algebra of symmetric functions Sym. The latter has a basis of integer partitions. It is often viewed as a subalgebra of the algebra of formal power series in countably many variables, see for instance the books by Fulton [**137**], Macdonald [**240**], Sagan [**314**], Stanley [**336**]. The Hopf algebra viewpoint on Sym originated in work of Geissinger [**146**] and Zelevinsky [**365**]. For later references, see the books by Blessenohl and Schocker [**50**, Chapter 3], Hazewinkel, Gubareni, Kirichenko [**181**, Chapter 4], Méliot [**258**, Chapter 2], Grinberg and Reiner [**159**, Chapter 2]. As an abstract Hopf algebra, it had appeared earlier in work of Moore on universal constructions [**272**, page 6, Definitions and Notations] and of Cartier on formal groups [**81**, Corollary 2], see also [**180**, Theorem 36.1.11].

Freeness. The universal property of $\overline{\Pi}^*$ mentioned in Exercise 10.116 generalizes the one for Sym [**4**, Theorem 4.3].

Antipode formulas. Antipode formulas (10.133) and (10.134) generalize classical antipode formulas of the Hopf algebra of symmetric functions, see for instance [**6**, page 35], [**258**, page 72], [**159**, Section 2.4].

10.7.6. Coxeter bialgebras of face-types. (Section 10.4.) Recall from Table 1.9: Face-types of braid arrangements are the same as integer compositions.

Hopf algebra of noncommutative symmetric functions. The Coxeter bialgebra of face-types $\overline{\Sigma}$ generalizes the Hopf algebra of noncommutative symmetric functions NSym. The latter has a basis of integer compositions. It was introduced by Gelfand, Krob, Lascoux, Leclerc, Retakh, Thibon [**149**] and studied further in a series of papers including [**215**], [**107**], [**216**], [**217**], [**106**]. This Hopf algebra had been considered earlier by Ditters in the context of formal groups [**99**, Section 1.2.7], [**100**, Section 1], [**101**, $Z(\infty, A)$ in Lemma 1.1], [**102**, Chapter I, Example 2.5, item (c)], see also [**103**, Sentence after Theorem 1.2] and [**180**, Section 36.3.8]. For recent references on NSym, see the books by Hazewinkel, Gubareni, Kirichenko [**181**, Chapter 6], Méliot [**258**, Chapter 6], Grinberg and Reiner [**159**, Chapter 5].

TABLE 10.10. Hopf algebra of symmetric functions and generalizations.

Coxeter bialgebras	Hopf algebras
Coxeter bialgebra of flat-types $\overline{\Pi}$, $\overline{\Pi}^*$	Hopf algebra of symmetric functions Sym
Coxeter bialgebra of face-types $\overline{\Sigma}$	Hopf algebra of noncommutative symmetric functions NSym
dual Coxeter bialgebra of face-types $\overline{\Sigma}^*$	Hopf algebra of quasisymmetric functions QSym
Coxeter bialgebra of symmetries $\overline{\Gamma}$, $\overline{\Gamma}^*$	Hopf algebra of permutations SSym

Hopf algebra of quasisymmetric functions. The dual Coxeter bialgebra of face-types $\overline{\Sigma}^*$ generalizes the Hopf algebra of quasisymmetric functions QSym. The latter has a basis of integer compositions. It was introduced by Gessel [**152**] as a subalgebra of the algebra of formal power series in countably many variables. It is discussed by Stanley [**336**, Section 7.19], Reutenauer [**305**, Section 9.4], Bertet, Krob, Morvan, Novelli, Phan, Thibon [**37**]. The Hopf algebra structure of QSym was introduced by Malvenuto [**248**, Section 4.1]. An explicit description of the quasishuffle product was given by Ehrenborg [**113**, Lemma 3.3].

Freeness. The universal property of $\overline{\Sigma}_q^*$ in Theorem 10.73 for $q = 1$ generalizes the universal property of QSym given by Aguiar, Bergeron, Sottile [**4**, Theorem 4.1]. The latter explains the abundance of morphisms to QSym. The result about $\overline{\Sigma}$ in Proposition 10.74 involving the $\mathtt{Q}$-basis generalizes the classical result that NSym is a tensor Hopf algebra. The latter is implicit in the discussion in [**149**, Section 3.1, pages 238–239]; it is explicitly stated in [**248**, Théorème 4.18].

Theorems 10.97, 10.98, 10.99 look new even in the classical setting.

$\mathtt{Q}$-basis. The $\mathtt{Q}$-basis of $\overline{\Sigma}$ defined in (10.66) (one for each invariant noncommutative zeta function) generalizes (up to normalization) the power-sum basis of the second kind of NSym [**149**, Definition 3.4]. The construction of the invariant eulerian idempotents in (10.72) generalizes the construction by Krob, Leclerc, Thibon [**215**, Section 3.4]. For more information, see [**10**, Section 16.11.6 and page 509]. Also see Section 1.16.5.

Primitive part. The primitive elements of NSym were described in [**149**, Proposition 3.10] and [**249**]. In the latter paper, the primitive elements were identified with a free Lie algebra. This is generalized in (10.100) and (12.31).

Primitive and group-like series. (Section 10.4.20.) Series of NSym, or at least particular examples of such, are considered in [**149**, Section 3.1] and in other literature on NSym such as [**350**, Section 2.1]. The Zie series in Section 1.16.5 are examples of primitive series of NSym. Group-like series of NSym are dual to characters on QSym; the latter are the main object of study in [**4**]. See the second formula in (9.121) in this regard.

Antipode formulas. The antipode formula (10.119) of $\overline{\Sigma}_q^*$ for $q = 1$ generalizes the antipode formula of QSym. The latter was obtained independently by Malvenuto and Reutenauer [**248**, Corollaire 4.20], [**249**, Corollary 2.3] and Ehrenborg [**113**, Proposition 3.4]. For some later references, see [**12**, Formula (1.7)], [**4**, Formula (4.7)], [**6**, page 36], [**7**, page 573], [**258**, Lemma 6.16], [**159**, Theorem 5.1.11]. The

q-analogue is given in our monograph [**7**, page 575]. Dually, the antipode formula (10.116) of $\overline{\Sigma}_q$ for $q = 1$ generalizes the antipode formula of NSym in [**149**, Proposition 3.9], see also [**181**, Formulas (3.4.9) and (6.1.5)], [**258**, Proposition 6.4].

10.7.7. Coxeter bialgebra of symmetries. (Section 10.6.) Recall: Coxeter symmetries of braid arrangements are the same as permutations.

Hopf algebra of permutations. The Coxeter bialgebra of symmetries $\overline{\Gamma}^*$ generalizes the Hopf algebra of permutations SSym. The latter (in the $\mathtt{F}$-basis) was introduced by Malvenuto in her thesis [**248**]. It was studied further by Aguiar and Sottile [**12**]; in particular, they introduced the $\mathtt{M}$-basis of SSym. The geometric perspective on SSym was developed in our monograph [**6**, Chapter 7]. Section 10.6 may be viewed as a natural continuation of the same set of ideas.

(Co)product formulas in the $\mathtt{M}$-basis of $\overline{\Gamma}^*$ in (10.139a) and (10.139b) and in the dual $\mathtt{H}$-basis of $\overline{\Gamma}$ in (10.141a) and (10.141b) generalize (co)product formulas of SSym obtained in [**12**, Theorems 3.1 and 4.1], [**6**, Theorems 7.3.3 and 7.3.6 and Theorems 7.5.2 and 7.5.5]. The self-duality of $\overline{\Gamma}$ obtained in Exercise 10.124 for $q = 1$ generalizes the self-duality of SSym which was obtained by Malvenuto in [**248**, Section 5.2] and in her paper with Reutenauer [**249**, Theorem 3.3], see also [**12**, Section 9], [**6**, Theorem 7.1.1].

The Coxeter q-bialgebra $\overline{\Gamma}_q$ generalizes the deformation of SSym considered by Foissy [**126**], see also [**7**, Section 17.2.7].

Book references on SSym are those by Blessenohl and Schocker [**50**, Chapter 5], Hazewinkel, Gubareni, Kirichenko [**181**, Chapter 7], Méliot [**258**, Section 12.1], Grinberg and Reiner [**159**, Chapter 8].

Freeness and primitive part. Section 10.6.6 on the primitive filtration of $\overline{\Gamma}^*$ generalizes a similar discussion on SSym in [**12**, Section 6], see also [**7**, Section 17.2.5]. Results related to freeness and the primitive part of SSym were obtained earlier by Poirier and Reutenauer [**296**, Section 2] and by Duchamp, Hivert, Thibon [**106**, Sections 3.3 and 3.4].

Antipode formulas. Antipode formulas (10.145) and (10.146) of the Coxeter bialgebra $\overline{\Gamma}^*$ generalize antipode formulas of SSym obtained by Aguiar and Sottile [**12**, Theorems 5.4 and 5.5], see also [**7**, Section 17.2.4].

10.7.8. Norm maps on face-types and flat-types and related Coxeter bialgebra morphisms.

Norm map. The norm map on face-types (10.97) for the invariant family of scalars $\xi \equiv 1$, on the $\mathtt{H}$-basis and into the $\mathtt{M}$-basis, (on the $\emptyset$-component) was given in [**6**, Definition 5.7.14 and Lemmas 2.6.1 and 2.6.2], but only as a linear map between vector spaces. The induced norm map on flat-types (10.130), on the $\mathtt{Q}$-basis and into the $\mathtt{P}$-basis, (on the $\emptyset$-component) was given in [**6**, Definition 5.7.15 and Lemma 5.7.1]. For braid arrangements, this yields a self-duality isomorphism for the Hopf algebra of symmetric functions. It corresponds to the well-known inner product on symmetric functions on the power-sum basis [**336**, Proposition 7.9.3]. This formula involves the numbers z_λ in [**6**, Example before Lemma 5.7.1], [**240**, Section I.2, page 24], [**258**, Proposition 2.2], [**336**, Formula (7.17)].

Morphisms of Hopf algebras. Diagram (10.147) linking the Coxeter bialgebras of face-types, flat-types, symmetries (as a diagram of vector spaces on the $\emptyset$-component) first appeared in [**6**, Diagram (5.17)]. The maps (10.148) and (10.149) (on the $\emptyset$-component) were given in [**6**, Definitions 5.7.10 and 5.7.11]. Diagram

(10.147) generalizes the diagram

$$\begin{array}{ccc} \mathrm{NSym} & \hookrightarrow & \mathrm{SSym} \\ \downarrow & \searrow & \downarrow \\ \mathrm{Sym} & \hookrightarrow & \mathrm{QSym} \end{array} \tag{10.154}$$

of Hopf algebras [**6**, Diagram (3.3)]. Formulas (10.86) and (10.87) for $q = 1$ generalize formulas for the above Hopf algebra morphism from NSym to QSym given in [**12**, Proposition 9.4].

q-norm map. Now consider the q-norm map on face-types (10.86) for the invariant family of scalars $\xi \equiv 1$, on the $\mathtt{H}$-basis and into the $\mathtt{M}$-basis (as a map of vector spaces on the $\emptyset$-component). It corresponds to the inner product on face-types on the $\mathtt{H}$-basis given in [**7**, Formula (10.138)]. The equivalent expression (10.87), on the $\mathtt{K}$-basis and into the $\mathtt{F}$-basis, (as a map of vector spaces on the $\emptyset$-component) corresponds to the inner product on face-types on the $\mathtt{K}$-basis given in [**7**, top of page 361]. The inner product on this basis for braid arrangements (symmetric groups) had appeared in work of Thibon and Ung [**351**, Formula (39)]. Specialized further to $q = 1$ recovers a result of Gessel [**152**, Theorem 5]; see also [**336**, Corollary 7.23.8].

Theorem 10.86, specialized to the invariant family of scalars $\xi \equiv 1$, gives a necessary and sufficient numerical condition for the map (10.86), or equivalently, (10.87) from $\overline{\Sigma}_q$ to $\overline{\Sigma}_q^*$ to be an isomorphism when q is not a root of unity. (Also see the rank-one and rank-two cases worked out in Example 10.88.) This provides a satisfactory answer to the question raised by Thibon and Ung [**351**, Section 3], see also [**179**, pages 80–81], [**7**, pages 361 and 575]. In fact, the answer that we have provided is in the greater generality of Coxeter bialgebras and for an arbitrary invariant family of scalars ξ. This fulfils the promise made on [**11**, page 636].

Hopf algebra between NSym and SSym. The Coxeter bialgebra H in Exercise 10.128 generalizes the Hopf algebra considered by Patras and Reutenauer [**289**, Theorems 9 and 11]. Condition (10.150), for T/Z a vertex-type, corresponds to the property in [**289**, Theorem 4]. This property is phrased in purely combinatorial terms with no reference to the Tits product. Formula (10.151) for the primitive part generalizes [**289**, Theorem 12].

CHAPTER 11

Coxeter operad algebras

Coxeter operads. We introduced Coxeter operads in Chapter 5, where we formulated them in terms of flats. We now revisit this approach and, in addition, develop two more approaches, one in terms of top-lunes and another in terms of face-types. A Coxeter operad is a Coxeter dispecies equipped with substitution and unit maps subject to naturality, associativity, unitality axioms. The substitution map in each of the three approaches is summarized in Table 11.1; partial orders on flats, top-lunes and face-types play a key role. Dual remarks apply to Coxeter cooperads.

TABLE 11.1. Substitution map of a Coxeter operad **a**.

flats	$\mathbf{a}[\mathrm{X},\mathrm{Y}] \otimes \mathbf{a}[\mathrm{Y},\mathrm{Z}] \to \mathbf{a}[\mathrm{X},\mathrm{Z}],\ \ \mathrm{X} \le \mathrm{Y} \le \mathrm{Z}$
top-lunes	$\mathbf{a}[\mathrm{L},\mathrm{M}] \otimes \mathbf{a}[\mathrm{M},\mathrm{N}] \to \mathbf{a}[\mathrm{L},\mathrm{N}],\ \ \mathrm{L} \le \mathrm{M} \le \mathrm{N}$
face-types	$\mathbf{a}[T,U] \otimes \mathbf{a}[U,V] \to \mathbf{a}[T,V],\ \ T \le U \le V$

In Chapter 5, we defined the Coxeter associative and commutative operads **As** and **Com** in terms of flats. We will now see how they work in the top-lunes and face-types formulation.

Coxeter operad algebras. Just like for Coxeter species, there is a substitution operation of a Coxeter space into a Coxeter dispecies. This turns Coxeter spaces into a left module category over the monoidal category of Coxeter dispecies. Thus, each Coxeter operad gives rise to a monad on Coxeter spaces. An algebra over such a monad is a Coxeter operad algebra. Its structure map is elaborated in Table 11.2.

TABLE 11.2. Structure map of a Coxeter **a**-algebra V.

top-nested faces	$\mathbf{a}[\mathrm{s}(A),\mathrm{s}(F)] \otimes \mathrm{V}[F,C] \to \mathrm{V}[A,C],\ \ A \le F \le C$
top-lunes	$\mathbf{a}[\mathrm{L},\mathrm{M}] \otimes \mathrm{V}[\mathrm{M}] \to \mathrm{V}[\mathrm{L}],\ \ \mathrm{L} \le \mathrm{M}$
face-types	$\mathbf{a}[T,U] \otimes \mathrm{V}[U] \to \mathrm{V}[T],\ \ T \le U$

Coxeter operads **As** and **Com** yield monads $\overline{\mathcal{T}}$ and $\overline{\mathcal{S}}$, respectively, which connects them to Coxeter algebras and commutative Coxeter algebras. The Coxeter operad $\mathbf{Com}^-$ yields the monad $\overline{\mathcal{E}}$ which connects it to signed commutative Coxeter algebras. Similarly, **Lie** connects to Coxeter Lie algebras;

these are introduced later in Chapter 12. Dual remarks apply to Coxeter cooperads and comonads on Coxeter spaces.

May operads. Connections to classical operad theory. The notion of a Coxeter operad, when specialized to braid arrangements, relates to the classical notion of a May operad. More precisely, there is a monoidal functor from Joyal species to Coxeter dispecies which is strong wrt the substitution products on either category. Hence, it carries May operads to Coxeter operads.

11.1. Nested flats, nested top-lunes, nested face-types

Fix a reflection arrangement $\mathcal{A}$ with Coxeter group W. We construct finite categories whose objects are nested flats, nested top-lunes, nested face-types, see Table 11.3. They are all groupoids and equivalent to one another.

TABLE 11.3. Base categories for Coxeter dispecies.

Groupoid	Objects
$\mathcal{A}$-$\mathtt{dCox}$, $\mathcal{A}$-$\mathtt{dCox}_C$	nested flats
$\mathcal{A}$-$\mathtt{dCox}'$	nested top-lunes
α-$\mathtt{dCox}$	nested face-types

11.1.1. Category with nested flats as objects.

11.1.1. Category with nested flats as objects. Recall from Section 5.1.1 the category D and its quotient category $\mathcal{A}$-$\mathtt{dCox}$. Objects in either category are nested flats. The category $\mathcal{A}$-$\mathtt{dCox}$ can be realized as a subcategory of D in many different ways. This is explained below.

For a chamber C, let $\mathcal{A}$-$\mathtt{dCox}_C$ denote the category whose objects are nested flats (X, Y) and morphisms are

$$w : (\mathrm{X}, \mathrm{Y}) \to (\mathrm{X}', \mathrm{Y}')$$

with $w \in W$ such that $w(\mathrm{X}) = \mathrm{X}'$, $w(\mathrm{Y}) = \mathrm{Y}'$, and w sends the top-lune containing C with base Y to the top-lune containing C with base Y'. A morphism u composed with a morphism v is the morphism vu (product in the group W). Thus, $\mathcal{A}$-$\mathtt{dCox}_C$ is a subcategory of D with the same set of objects, but with a restricted set of morphisms.

Proposition 11.1. *For a chamber C, the composite functor*

$$\mathcal{A}\text{-}\mathtt{dCox}_C \hookrightarrow \mathsf{D} \twoheadrightarrow \mathcal{A}\text{-}\mathtt{dCox}$$

is an isomorphism of categories. (Both functors are identity on objects.)

PROOF. Let C be any chamber. Suppose Y and Y' are flats and $w \in W$ is such that $w(\mathrm{Y}) = \mathrm{Y}'$. Then w can be uniquely expressed in the normal form $w = vu$, with u and v subject to the following conditions: u sends the top-lune containing C with base Y to the top-lune containing C with base Y', while v fixes Y' pointwise. This is related to and can be deduced from (1.9). Note:

The normal form depends on the choice of C. One can deduce from here that the maps

$$\mathcal{A}\text{-}\mathtt{dCox}_C((\mathrm{X},\mathrm{Y}),(\mathrm{X}',\mathrm{Y}')) \to \mathcal{A}\text{-}\mathtt{dCox}((\mathrm{X},\mathrm{Y}),(\mathrm{X}',\mathrm{Y}'))$$

are bijections, as required. □

Exercise 11.2. For a chamber C, consider the normal form defined in the above proof. Show:

- Morphisms w_1 and w_2 in D both from (X,Y) to $(\mathrm{X}',\mathrm{Y}')$ are equal in the quotient category $\mathcal{A}\text{-}\mathtt{dCox}$ iff there is a u such that $w_1 = v_1 u$ and $w_2 = v_2 u$ in normal form.
- If w and w' are composable morphisms in D with normal forms vu and $v'u'$, then the normal form of $w'w$ is $(v'u'vu'^{-1})(u'u)$.

Exercise 11.3 (Semidirect product). Show: For any chamber C,

$$\mathsf{D} = \mathcal{F} \rtimes \mathcal{A}\text{-}\mathtt{dCox}_C. \tag{11.1}$$

This is an example of the Grothendieck construction (Appendix A.2.1) where $\mathcal{F}$ is a functor from $\mathcal{A}\text{-}\mathtt{dCox}_C$ to the category of groups. It sends the object (X,Y) to the group W_{Y}, and a morphism w in $\mathcal{A}\text{-}\mathtt{dCox}_C$ to the group morphism $W_{\mathrm{Y}} \to W_{\mathrm{Y}'}$ given by conjugation by w. The semidirect product (11.1) may be viewed as a many-object generalization of (1.9).

Exercise 11.4. Check: $\mathcal{A}\text{-}\mathtt{dCox}_C$ is isomorphic to the category whose objects are nested flats (X,Y) and morphisms are

$$(u,w) : (\mathrm{X},\mathrm{Y}) \to (\mathrm{X}',\mathrm{Y}')$$

with $u, w \in W$ such that

- $u(\mathrm{X}) = \mathrm{X}'$ and $w(\mathrm{Y}) = \mathrm{Y}'$,
- u (resp. w) sends the top-lune containing C with base X (resp. Y) to the top-lune containing C with base X' (resp. Y'),
- u and w agree on X (that is, $u(F) = w(F)$ whenever $\mathrm{s}(F) = \mathrm{X}$).

Morphisms are composed coordinatewise, that is, $(u',w')(u,w) = (u'u, w'w)$. (Use Exercise 1.18.)

Observation 11.5. Restricted to objects of the form $(\mathrm{X},\top)$, the three categories in Proposition 11.1 are identical. The automorphism group at $(\mathrm{X},\top)$ is $\widehat{W_{\mathrm{X}}}$. The subgroup W_{X} consists of morphisms of the form (id, w).

11.1.2. Category with nested top-lunes as objects.

Recall from Section 1.2.4 the partial orders $\leq$ and $\leq'$ on top-lunes. A nested top-lune is a pair of top-lunes (L,M) such that $\mathrm{L} \leq \mathrm{M}$. Define the category $\mathcal{A}\text{-}\mathtt{dCox}'$ as follows. Objects are nested top-lunes (L,M). Morphisms are

$$(u,w) : (\mathrm{L},\mathrm{M}) \to (\mathrm{L}',\mathrm{M}')$$

with $u, w \in W$ such that $u(\mathrm{L}) = \mathrm{L}'$, $w(\mathrm{M}) = \mathrm{M}'$, and u and w agree on the base of L (that is, $u(F) = w(F)$ whenever $\mathrm{s}(F) = \mathrm{b}(\mathrm{L})$). An illustration is shown in Figure 11.1. Here M and M' are the visible hemispheres, and L and L' are portions between two adjacent longitudes. Morphisms are composed coordinatewise, that is, $(u',w')(u,w) = (u'u, w'w)$. The category $\mathcal{A}\text{-}\mathtt{dCox}'$ is a groupoid. The inverse of (u,w) is (u^{-1}, w^{-1}).

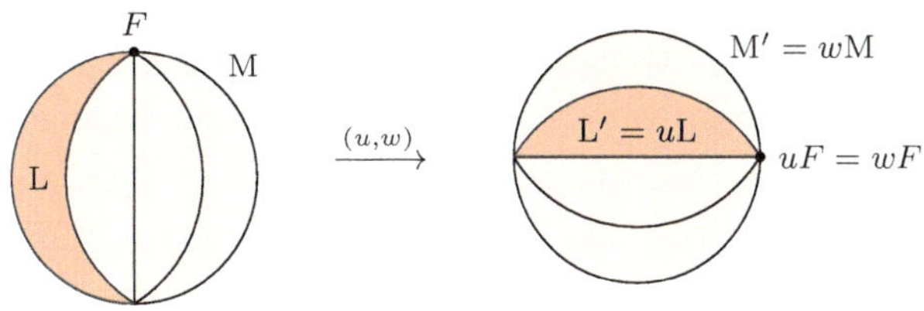

FIGURE 11.1. Morphisms between nested top-lunes.

In a morphism (u, w) as above, note very carefully that $w(\mathrm{L}) \neq \mathrm{L}'$ in general. In fact:

Lemma 11.6. *Let* (L, M) *and* $(\mathrm{L}', \mathrm{M}')$ *be nested top-lunes. Let* $w \in W$ *be such that* $w(\mathrm{M}) = \mathrm{M}'$ *and* $w(\mathrm{b}(\mathrm{L})) = \mathrm{b}(\mathrm{L}')$. *Then there exists a unique* $u \in W$ *such that* $(u, w) : (\mathrm{L}, \mathrm{M}) \to (\mathrm{L}', \mathrm{M}')$.

PROOF. This follows from Exercise 1.18. More precisely, u is the unique Coxeter symmetry which agrees with w on $\mathrm{b}(\mathrm{L})$ and satisfies $u(\mathrm{L}) = \mathrm{L}'$. □

Corollary 11.7. *If* $(u_1, w), (u_2, w) : (\mathrm{L}, \mathrm{M}) \to (\mathrm{L}', \mathrm{M}')$, *then* $u_1 = u_2$.

PROOF. This is a restatement of the uniqueness assertion in Lemma 11.6. □

Corollary 11.8. *Suppose* $\mathrm{L}_1, \mathrm{L}_2 \leq \mathrm{M}$ *and* $\mathrm{b}(\mathrm{L}_1) = \mathrm{b}(\mathrm{L}_2)$. *Then there exists a unique* $u \in W$ *such that* $(u, \mathrm{id}) : (\mathrm{L}_1, \mathrm{M}) \to (\mathrm{L}_2, \mathrm{M})$.

PROOF. This is the special case $w = \mathrm{id}$ of Lemma 11.6. □

An illustration is shown below.

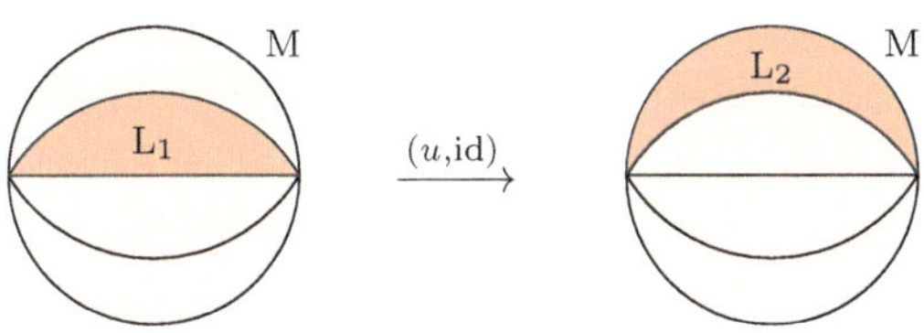

Lemma 11.9. *A morphism* $(u, w) : (\mathrm{L}, \mathrm{M}) \to (\mathrm{L}', \mathrm{M}')$ *factors as*

$$\begin{array}{ccc} (\mathrm{L}, \mathrm{M}) & \xrightarrow{(u,w)} & (\mathrm{L}', \mathrm{M}') \\ & {\scriptstyle (u,u)} \searrow \quad \nearrow {\scriptstyle (\mathrm{id}, wu^{-1})} & \\ & (\mathrm{L}', u\mathrm{M}) & \end{array} \qquad \textit{and} \qquad \begin{array}{ccc} (\mathrm{L}, \mathrm{M}) & \xrightarrow{(u,w)} & (\mathrm{L}', \mathrm{M}') \\ & {\scriptstyle (w,w)} \searrow \quad \nearrow {\scriptstyle (uw^{-1}, \mathrm{id})} & \\ & (w\mathrm{L}, \mathrm{M}'). & \end{array}$$

PROOF. We explain the first factorization; the second is similar. It is clear that $(u, u) : (\mathrm{L}, \mathrm{M}) \to (u\mathrm{L}, u\mathrm{M})$ is a morphism. Since we are in a groupoid, $(\mathrm{id}, wu^{-1}) = (u, w)(u^{-1}, u^{-1})$ is uniquely determined. □

Observation 11.10. For $\mathrm{L} \leq \mathrm{M} \leq \mathrm{N}$ and $\mathrm{L}' \leq \mathrm{M}' \leq \mathrm{N}'$, and morphisms

$$(u, v) : (\mathrm{L}, \mathrm{M}) \to (\mathrm{L}', \mathrm{M}') \quad \text{and} \quad (v, w) : (\mathrm{M}, \mathrm{N}) \to (\mathrm{M}', \mathrm{N}'),$$

we have the morphism $(u, w) : (\mathrm{L}, \mathrm{N}) \to (\mathrm{L}', \mathrm{N}')$.

Conversely: For a morphism $(u, w) : (\mathrm{L}, \mathrm{N}) \to (\mathrm{L}', \mathrm{N}')$ and M between L and N, there exists a unique M′ between L′ and N′ and a unique $v \in W$ such that there are morphisms (u, v) and (v, w) as above.

(The first statement is clear. The converse follows from Fact 1.13 and Exercise 1.18.)

Exercise 11.11. Check: The full subcategory of $\mathcal{A}$-$\mathtt{dCox}'$ consisting of objects (L, M) with $\mathrm{L} \leq' \mathrm{M}$ is equivalent to $\mathcal{A}$-$\mathtt{dCox}'$. (Use Exercise 1.16 and Corollary 11.8.) The factorizations in Lemma 11.9 work in this subcategory as well. What about the converse in Observation 11.10?

Exercise 11.12. Show: For $\mathrm{L} \leq \mathrm{M}$, there exists M′ and $w \in W$ such that $\mathrm{L} \leq' \mathrm{M}'$ and $(\mathrm{id}, w) : (\mathrm{L}, \mathrm{M}) \to (\mathrm{L}, \mathrm{M}')$.

11.1.3. Relating nested flats and nested top-lunes. For a chamber C, consider the functor

$$\mathcal{A}\text{-}\mathtt{dCox}_C \to \mathcal{A}\text{-}\mathtt{dCox}', \qquad (\mathrm{X}, \mathrm{Y}) \mapsto (\mathrm{L}, \mathrm{M}), \tag{11.2}$$

where L (resp. M) is the unique top-lune containing C with base X (resp. Y). On morphisms, it sends

$$w : (\mathrm{X}, \mathrm{Y}) \to (\mathrm{X}', \mathrm{Y}') \;\text{ to }\; (u, w) : (\mathrm{L}, \mathrm{M}) \to (\mathrm{L}', \mathrm{M}'),$$

with u uniquely determined by w by Lemma 11.6. Alternatively, in the formulation of $\mathcal{A}$-$\mathtt{dCox}_C$ in Exercise 11.4, it sends (u, w) to (u, w). This functor is full and faithful and injective on the set of objects. As a consequence:

Lemma 11.13. *For a chamber C, the category $\mathcal{A}$-$\mathtt{dCox}_C$ is isomorphic to the full subcategory of $\mathcal{A}$-$\mathtt{dCox}'$ consisting of those objects (L, M) for which* L *(and hence* M*) contains C.*

Let us temporarily denote the above full subcategory by E. In fact, it is equivalent to $\mathcal{A}$-$\mathtt{dCox}'$. This is because for any object (L, M) in the latter, there is an isomorphic object in E. For concreteness, we take this to be $(\mathrm{L}_0, \mathrm{M}_0)$, where L_0 (resp. M_0) is the unique top-lune containing C with base $\mathrm{b}(\mathrm{L})$ (resp. $\mathrm{b}(\mathrm{M})$). Also, we fix the isomorphism to be

$$(u_0, w_0) : (\mathrm{L}, \mathrm{M}) \to (\mathrm{L}_0, \mathrm{M}_0),$$

where u_0 (resp. w_0) sends L to L_0 (resp. M to M_0) fixing their common base pointwise. This specifies an explicit equivalence between E and $\mathcal{A}$-$\mathtt{dCox}'$. Combining with Lemma 11.13, we obtain:

Proposition 11.14. *For a chamber C, the categories $\mathcal{A}$-$\mathtt{dCox}_C$ and $\mathcal{A}$-$\mathtt{dCox}'$ are equivalent.*

The functor in one direction is defined above. In the other direction, it is

$$\mathcal{A}\text{-}\mathtt{dCox}' \to \mathcal{A}\text{-}\mathtt{dCox}_C, \qquad (\mathrm{L}, \mathrm{M}) \mapsto (\mathrm{b}(\mathrm{L}), \mathrm{b}(\mathrm{M})).$$

On morphisms, it sends

$$(u, w) : (\mathrm{L}, \mathrm{M}) \to (\mathrm{L}', \mathrm{M}') \;\text{ to }\; w_0' w w_0^{-1} : (\mathrm{b}(\mathrm{L}), \mathrm{b}(\mathrm{M})) \to (\mathrm{b}(\mathrm{L}'), \mathrm{b}(\mathrm{M}'))$$

which arises from the composite

$$(\mathrm{L}_0, \mathrm{M}_0) \xrightarrow{(u_0^{-1}, w_0^{-1})} (\mathrm{L}, \mathrm{M}) \xrightarrow{(u, w)} (\mathrm{L}', \mathrm{M}') \xrightarrow{(u_0', w_0')} (\mathrm{L}_0', \mathrm{M}_0').$$

In other words, we conjugate (u, w) by our chosen isomorphisms.

Observation 11.15. Following Observation 11.5, the automorphism group at the object $(\mathrm{L}, \top)$ in $\mathcal{A}$-$\mathtt{dCox}'$ is $\widehat{W}_{\mathrm{b(L)}}$. More precisely, it consists of morphisms (u, w) with $w \in \widehat{W}_{\mathrm{b(L)}}$ and $u \in W_{\mathrm{L}}$ obtained from w via the semidirect decomposition (1.9). The subgroup $W_{\mathrm{b(L)}}$ corresponds to morphisms (id, w).

11.1.4. Category with nested face-types as objects. Let $\boldsymbol{\alpha} = (\mathcal{A}, C_r)$ be a pointed reflection arrangement. Recall: A nested face-type is a pair of face-types (T, U) such that $T \leq U$. Also recall from Section 8.1.5 the category $\boldsymbol{\alpha}$-$\overline{\mathtt{Cox}}$ with face-types as objects. Now define the category $\boldsymbol{\alpha}$-$\mathtt{dCox}$ as follows. Objects are nested face-types (T, U). Morphisms are

$$(\sigma, \tau) : (T, U) \to (T', U'),$$

where $\sigma : T \to T'$ and $\tau : U \to U'$ are morphisms in $\boldsymbol{\alpha}$-$\overline{\mathtt{Cox}}$ subject to the condition that $H \leq K$ where σ is transferred by H, and τ by K. In this situation, we say (σ, τ) is transferred by the nested face (H, K). Morphisms are composed coordinatewise, that is, $(\sigma', \tau')(\sigma, \tau) = (\sigma'\sigma, \tau'\tau)$. This requires a check which is stated below.

Exercise 11.16. Check: If (σ, τ) and (σ', τ') are transferred by nested faces, then so is $(\sigma'\sigma, \tau'\tau)$.

Alternatively, one can use the groupoid of biface-types $\boldsymbol{\alpha}$-$\overline{\mathtt{Cox}}$ defined in Section 1.7 to describe morphisms in $\boldsymbol{\alpha}$-$\mathtt{dCox}$. A pair (σ, τ) is a morphism if there exist bifaces (F', F) and (G', G) which represent σ and τ, respectively, such that $F \leq G$ and $F' \leq G'$. This can be visualized as in Figure 11.2.

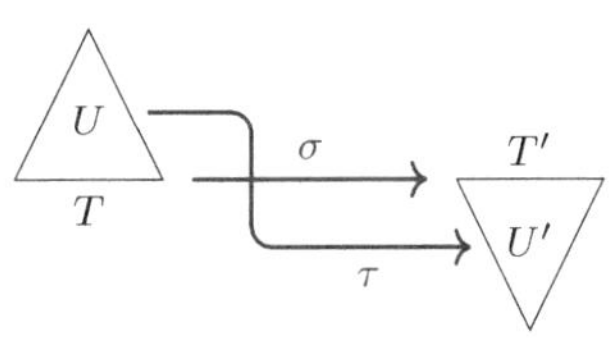

FIGURE 11.2. Morphisms between nested face-types.

The category $\boldsymbol{\alpha}$-$\mathtt{dCox}$ is a groupoid. The inverse of (σ, τ) is (σ^{-1}, τ^{-1}).

Lemma 11.17. *A morphism $(\sigma, \tau) : (T, U) \to (T', U')$ factors as*

$$\begin{array}{ccc} (T, U) & \xrightarrow{(\sigma,\tau)} & (T', U') \\ & {\scriptstyle (\sigma,\sigma)} \searrow \quad \nearrow {\scriptstyle (\mathrm{id},\tau')} & \\ & (T', \sigma_*(U)), & \end{array}$$

with $\tau' : \sigma_*(U) \to U'$ a uniquely determined morphism local to T'. (Here $\sigma : U \to \sigma_*(U)$ is the morphism induced from $\sigma : T \to T'$.)

Observe: In the above diagram, $\tau = \tau'\sigma$ is the gate property factorization of the morphism τ wrt the face-type T'.

Compare and contrast with Lemma 11.9. We have the analogue of the first factorization, but not of the second. For the analogue of Observation 11.10: If $(\sigma, \tau) : (T, U) \to (T', U')$ and $(\tau, \rho) : (U, V) \to (U', V')$ are morphisms, then so is $(\sigma, \rho) : (T, V) \to (T', V')$. However, the converse does not hold in $\boldsymbol{\alpha}$-dCox. The reason is that for faces G and K of the same support and $F \leq G$, in general, there does *not* exist a face $H \leq K$ which has the same support as F, see Remark 1.2. This is of significance.

11.1.5. Relating nested flats and nested face-types. Consider the functor

$$\boldsymbol{\alpha}\text{-dCox} \to \mathcal{A}\text{-dCox}_{C_r}, \qquad (T, U) \mapsto (\mathrm{s}(F), \mathrm{s}(G)), \tag{11.3}$$

where F and G are faces of C_r of types T and U, respectively. On morphisms it works as follows.

Let $F \leq G \leq C_r$ be such that F has support X and type T, and G has support Y and type U. Similarly, let $F' \leq G' \leq C_r$ be such that F' has support X$'$ and type T', and G' has support Y$'$ and type U'. The morphism

$$(\sigma, \tau) : (T, U) \to (T', U') \;\text{ goes to }\; w : (\mathrm{X}, \mathrm{Y}) \to (\mathrm{X}', \mathrm{Y}'),$$

where $\tau = w\beta = \beta w$ in normal form (Lemma 8.4). Explicitly, if (σ, τ) is transferred by (H, K), then w is defined by $w(KC_r) = C_r$. Note: $wH = F'$ and $wK = G'$. Thus, $w(\mathrm{X}) = \mathrm{X}'$, $w(\mathrm{Y}) = \mathrm{Y}'$, and w sends the top-lune containing C_r with base Y to the top-lune containing C_r with base Y$'$, as required.

This functor is full and faithful. To see this, we describe the inverse map on morphisms. With notations as above,

$$w : (\mathrm{X}, \mathrm{Y}) \to (\mathrm{X}', \mathrm{Y}') \;\text{ goes to }\; (\sigma, \tau) : (T, U) \to (T', U'),$$

the morphism transferred by (H, K), with H and K defined by $wH = F'$ and $wK = G'$.

Moreover, since distinct faces of C_r have distinct supports, the functor is injective on the set of objects. As a consequence:

Lemma 11.18. *The category $\boldsymbol{\alpha}$-dCox is isomorphic to the full subcategory of $\mathcal{A}$-dCox$_{C_r}$ whose objects are of the form $(\mathrm{s}(F), \mathrm{s}(G))$ with $F \leq G \leq C_r$.*

The functor is also essentially surjective on objects: For an object (X, Y) in $\mathcal{A}$-dCox$_{C_r}$, there exists an isomorphism of the form

$$w : (\mathrm{X}, \mathrm{Y}) \to (\mathrm{s}(F), \mathrm{s}(G)).$$

For instance, take any nested face (H, K) with $\mathrm{s}(H) = \mathrm{X}$ and $\mathrm{s}(K) = \mathrm{Y}$. Let F and G be the faces of C_r of the same type as H and K, and define w by $w(KC_r) = C_r$. Note: $wH = F$ and $wK = G$.

Thus:

Proposition 11.19. *The categories $\boldsymbol{\alpha}$-dCox and $\mathcal{A}$-dCox$_{C_r}$ are equivalent.*

Exercise 11.20. Fix an object (X, Y) in $\mathcal{A}$-dCox$_{C_r}$. Show: Morphisms from (X, Y) that land on some object in the subcategory $\boldsymbol{\alpha}$-dCox are indexed by nested faces (H, K) such that $\mathrm{s}(H) = \mathrm{X}$ and $\mathrm{s}(K) = \mathrm{Y}$.

By composing (11.2) and (11.3), we obtain the functor

$$\alpha\text{-}\mathtt{dCox} \to \mathcal{A}\text{-}\mathtt{dCox}', \qquad (T, U) \mapsto (\mathrm{s}(F, C_r), \mathrm{s}(G, C_r)), \tag{11.4}$$

where F and G are faces of C_r of types T and U, respectively. In fact, it maps into the subcategory defined in Exercise 11.11. A more precise result is given below.

Exercise 11.21. Deduce: α-$\mathtt{dCox}$ is isomorphic to the full subcategory of $\mathcal{A}$-$\mathtt{dCox}'$ consisting of objects of the form $(\mathrm{s}(F, C_r), \mathrm{s}(G, C_r))$. (Combine Lemma 11.13 and Lemma 11.18.)

Observation 11.22. Following Observation 11.15, the automorphism group at the object (T, S) in α-$\mathtt{dCox}$ is $\widehat{W}_{\mathrm{X}}$, where X supports the face of C_r of type T. The subgroup $W_{\mathrm{X}} = W_T$ corresponds to morphisms of the form (id, σ).

11.2. Coxeter dispecies

Coxeter dispecies were introduced in Section 5.1 as functors from $\mathcal{A}$-$\mathtt{dCox}$ to Vec. We now discuss some alternative approaches to Coxeter dispecies by replacing $\mathcal{A}$-$\mathtt{dCox}$ with the other equivalent categories in Table 11.3.

11.2.1. Coxeter dispecies in terms of flats. Let us formulate Coxeter dispecies using $\mathcal{A}$-$\mathtt{dCox}_C$ as the base category instead of $\mathcal{A}$-$\mathtt{dCox}$. In other words,

$$\mathcal{A}\text{-}\mathtt{CdSp} := [\mathcal{A}\text{-}\mathtt{dCox}_C, \mathsf{Vec}].$$

The formulation again involves nested flats.

Explicitly: Fix a chamber C. A Coxeter $\mathcal{A}$-dispecies $\mathbf{p}$ consists of a family of vector spaces $\mathbf{p}[\mathrm{X}, \mathrm{Y}]$, one for each $\mathrm{X} \leq \mathrm{Y}$, equipped with linear maps

$$w : \mathbf{p}[\mathrm{X}, \mathrm{Y}] \to \mathbf{p}[\mathrm{X}', \mathrm{Y}'],$$

one for each $w \in W$ such that $w(\mathrm{X}) = \mathrm{X}'$, $w(\mathrm{Y}) = \mathrm{Y}'$, and w sends the top-lune containing C with base Y to the top-lune containing C with base Y', and such that diagrams (5.1) commute. Following Exercise 11.4, we can also work with linear maps

$$(u, w) : \mathbf{p}[\mathrm{X}, \mathrm{Y}] \to \mathbf{p}[\mathrm{X}', \mathrm{Y}'].$$

Note very carefully the difference between this formulation and the previous one. We can pass between the two by using the isomorphism of categories in Proposition 11.1.

11.2.2. Coxeter dispecies in terms of top-lunes. Alternatively, one can employ the category with nested top-lunes as objects, and define

$$\mathcal{A}\text{-}\mathtt{CdSp} := [\mathcal{A}\text{-}\mathtt{dCox}', \mathsf{Vec}].$$

In this formulation: A Coxeter $\mathcal{A}$-dispecies $\mathbf{p}$ consists of a family of vector spaces $\mathbf{p}[\mathrm{L}, \mathrm{M}]$, one for each $\mathrm{L} \leq \mathrm{M}$, equipped with linear maps

$$(u, w) : \mathbf{p}[\mathrm{L}, \mathrm{M}] \to \mathbf{p}[\mathrm{L}', \mathrm{M}'],$$

one for each morphism (u, w), such that the diagrams

$$\begin{array}{ccc} & \mathbf{p}[\mathrm{L}', \mathrm{M}'] & \\ {}^{(u,w)}\nearrow & & \searrow^{(u',w')} \\ \mathbf{p}[\mathrm{L}, \mathrm{M}] & \xrightarrow[(u'u,w'w)]{} & \mathbf{p}[\mathrm{L}'', \mathrm{M}''] \end{array} \qquad (\mathbf{p}[\mathrm{L}, \mathrm{M}] \xrightarrow{(e,e)} \mathbf{p}[\mathrm{L}, \mathrm{M}]) = \mathrm{id} \tag{11.5}$$

commute.

A map of Coxeter $\mathcal{A}$-dispecies $f : \mathbf{p} \to \mathbf{q}$ consists of a family of linear maps

$$f_{\mathrm{L},\mathrm{M}} : \mathbf{p}[\mathrm{L}, \mathrm{M}] \to \mathbf{q}[\mathrm{L}, \mathrm{M}],$$

one for each $\mathrm{L} \leq \mathrm{M}$, such that the diagram

$$\begin{array}{ccc} \mathbf{p}[\mathrm{L}, \mathrm{M}] & \xrightarrow{f_{\mathrm{L},\mathrm{M}}} & \mathbf{q}[\mathrm{L}, \mathrm{M}] \\ {\scriptstyle (u,w)}\downarrow & & \downarrow{\scriptstyle (u,w)} \\ \mathbf{p}[\mathrm{L}', \mathrm{M}'] & \xrightarrow[f_{\mathrm{L}',\mathrm{M}'}]{} & \mathbf{q}[\mathrm{L}', \mathrm{M}'] \end{array} \tag{11.6}$$

commutes for all (u, w).

Instead of $\mathcal{A}$-dCox', one may also employ its equivalent full subcategory defined in Exercise 11.11 to formulate $\mathbf{p}$. In this case, the components $\mathbf{p}[\mathrm{L}, \mathrm{M}]$ are indexed by $\mathrm{L} \leq' \mathrm{M}$.

11.2.3. Coxeter dispecies in terms of face-types. Define the category of Coxeter $\boldsymbol{\alpha}$-dispecies to be the functor category

$$\boldsymbol{\alpha}\text{-}\mathsf{CdSp} := [\boldsymbol{\alpha}\text{-}\mathsf{dCox}, \mathsf{Vec}].$$

Explicitly: A Coxeter $\boldsymbol{\alpha}$-dispecies $\mathbf{p}$ consists of a family of vector spaces $\mathbf{p}[T, U]$, one for each $T \leq U$, equipped with linear maps

$$(\sigma, \tau) : \mathbf{p}[T, U] \to \mathbf{p}[T', U'],$$

one for each morphism (σ, τ), such that the diagrams

$$\begin{array}{ccc} & \mathbf{p}[T', U'] & \\ {}^{(\sigma,\tau)}\nearrow & & \searrow^{(\sigma',\tau')} \\ \mathbf{p}[T, U] & \xrightarrow[(\sigma'\sigma,\tau'\tau)]{} & \mathbf{p}[T'', U''] \end{array} \qquad (\mathbf{p}[T, U] \xrightarrow{(\mathrm{id},\mathrm{id})} \mathbf{p}[T, U]) = \mathrm{id} \tag{11.7}$$

commute.

A map of Coxeter $\boldsymbol{\alpha}$-dispecies $f : \mathbf{p} \to \mathbf{q}$ consists of a family of linear maps

$$f_{T,U} : \mathbf{p}[T, U] \to \mathbf{q}[T, U],$$

one for each $T \leq U$, such that the diagram

$$\begin{array}{ccc} \mathbf{p}[T, U] & \xrightarrow{f_{T,U}} & \mathbf{q}[T, U] \\ {\scriptstyle (\sigma,\tau)}\downarrow & & \downarrow{\scriptstyle (\sigma,\tau)} \\ \mathbf{p}[T', U'] & \xrightarrow[f_{T',U'}]{} & \mathbf{q}[T', U'] \end{array} \tag{11.8}$$

commutes for all (σ, τ).

11.2.4. Relating nested flats and nested face-types. The categories of Coxeter $\mathcal{A}$-dispecies and Coxeter $\boldsymbol{\alpha}$-dispecies are equivalent. This follows from Proposition 11.19. Let us make this explicit. Clearly, a Coxeter $\mathcal{A}$-dispecies $\mathbf{p}$ yields a Coxeter $\boldsymbol{\alpha}$-dispecies by restricting the components of $\mathbf{p}$ using Lemma 11.18. In other words,

$$\mathbf{p}[T, U] := \mathbf{p}[\mathrm{s}(F), \mathrm{s}(G)],$$

where F and G are faces of C_r of types T and U, respectively. Conversely, starting with a Coxeter $\boldsymbol{\alpha}$-dispecies $\mathbf{p}$, we construct a Coxeter $\mathcal{A}$-dispecies as follows (by using Exercise 11.20). For a nested flat (X, Y), let $\mathsf{C}_{\mathrm{X},\mathrm{Y}}$ be the indiscrete category whose objects are nested faces (H, K) such that $\mathrm{s}(H) = \mathrm{X}$ and $\mathrm{s}(K) = \mathrm{Y}$. Consider the functor

$$\mathsf{C}_{\mathrm{X},\mathrm{Y}} \to \mathsf{Vec}, \qquad (H, K) \mapsto \mathbf{p}[\mathrm{t}(H), \mathrm{t}(K)].$$

On morphisms, it sends $(H, K) \to (H', K')$ to the linear map

$$\mathbf{p}[\mathrm{t}(H), \mathrm{t}(K)] \xrightarrow{(\sigma,\tau)} \mathbf{p}[\mathrm{t}(H'), \mathrm{t}(K')],$$

where (σ, τ) is transferred by (wH', wK'), where w is defined by $w(KC_r) = C_r$, or equivalently, σ is represented by (H', H) and τ by (K', K). Define $\mathbf{p}[\mathrm{X}, \mathrm{Y}]$ to be the colimit of this functor. Explicitly,

$$\mathbf{p}[\mathrm{X}, \mathrm{Y}] = \Big(\bigoplus_{\substack{(H,K)\colon H \leq K \\ \mathrm{s}(H)=\mathrm{X},\, \mathrm{s}(K)=\mathrm{Y}}} \mathbf{p}[\mathrm{t}(H), \mathrm{t}(K)] \Big)_{\mathrm{quo}}, \tag{11.9}$$

where a (H, K)-summand is identified with a (H', K')-summand via the linear isomorphism (σ, τ) above. (The quotient space is isomorphic to any of its summands.) Further, for any $w : (\mathrm{X}, \mathrm{Y}) \to (\mathrm{X}', \mathrm{Y}')$, there is a linear map

$$w : \mathbf{p}[\mathrm{X}, \mathrm{Y}] \to \mathbf{p}[\mathrm{X}', \mathrm{Y}']$$

obtained by identifying the (H, K)-summand with the (wH, wK)-summand. (Recall: A Coxeter symmetry is type-preserving.) This turns $\mathbf{p}$ into a Coxeter $\mathcal{A}$-dispecies.

The above constructions also work more or less the same way for the formulation of Coxeter $\mathcal{A}$-dispecies from Section 5.1.2.

11.3. Coxeter operads

Coxeter operads were introduced in Section 5.2 where they were formulated in terms of flats. We revisit this approach and, in addition, develop two more approaches, one in terms of top-lunes and another in terms of face-types. This builds on discussion in Section 11.2.

11.3.1. Substitution product in terms of flats. Recall from (5.4) the substitution product on Coxeter dispecies. Let us see how it works in the formulation of Section 11.2.1. Accordingly, fix a chamber C. The formula is the same as before, namely,

$$(\mathbf{p} \circ \mathbf{q})[\mathrm{X}, \mathrm{Z}] := \bigoplus_{\mathrm{Y}\colon \mathrm{X} \leq \mathrm{Y} \leq \mathrm{Z}} \mathbf{p}[\mathrm{X}, \mathrm{Y}] \otimes \mathbf{q}[\mathrm{Y}, \mathrm{Z}]. \tag{11.10}$$

The structure maps, however, work slightly differently. With notation as before, we now sum over

$$\mathbf{p}[\mathrm{X},\mathrm{Y}] \otimes \mathbf{q}[\mathrm{Y},\mathrm{Z}] \xrightarrow{v\otimes w} \mathbf{p}[\mathrm{X}',\mathrm{Y}'] \otimes \mathbf{q}[\mathrm{Y}',\mathrm{Z}'],$$

where v is the unique element in W which agrees with w on Y, and sends the top-lune containing C with base Y to the top-lune containing C with base Y′.

Coxeter operads work as before, with a minor change in the naturality square in (5.6a), where $w \otimes w$ is replaced by $v \otimes w$, with v as above.

Exercise 11.23. Check: The above structure maps, in terms of the morphisms in Exercise 11.4, work as follows. For a morphism $(u, w) : (\mathrm{X}, \mathrm{Z}) \to (\mathrm{X}', \mathrm{Z}')$, define the linear map

$$(u, w) : (\mathbf{p} \circ \mathbf{q})[\mathrm{X},\mathrm{Z}] \to (\mathbf{p} \circ \mathbf{q})[\mathrm{X}',\mathrm{Z}']$$

by summing the linear maps

$$\mathbf{p}[\mathrm{X},\mathrm{Y}] \otimes \mathbf{q}[\mathrm{Y},\mathrm{Z}] \xrightarrow{(u,v)\otimes(v,w)} \mathbf{p}[\mathrm{X}',\mathrm{Y}'] \otimes \mathbf{q}[\mathrm{Y}',\mathrm{Z}'],$$

with v uniquely determined by Y (and defined as above).

Express the naturality square in (5.6a) in these terms.

11.3.2. Substitution product in terms of top-lunes. Recall: Coxeter $\mathcal{A}$-dispecies can be phrased using top-lunes. The substitution product in this formulation is defined as

$$(\mathbf{p} \circ \mathbf{q})[\mathrm{L},\mathrm{N}] := \bigoplus_{\mathrm{M}:\, \mathrm{L} \leq \mathrm{M} \leq \mathrm{N}} \mathbf{p}[\mathrm{L},\mathrm{M}] \otimes \mathbf{q}[\mathrm{M},\mathrm{N}]. \tag{11.11}$$

The sum is over all top-lunes M which lie between L and N. By Fact 1.13, this corresponds to flats Y which lie between b(L) and b(N). This is the main observation to relate (11.11) and (11.10).

The structure maps are as follows. For a morphism $(u, w) : (\mathrm{L}, \mathrm{N}) \to (\mathrm{L}', \mathrm{N}')$, define the linear map

$$(u, w) : (\mathbf{p} \circ \mathbf{q})[\mathrm{L},\mathrm{N}] \to (\mathbf{p} \circ \mathbf{q})[\mathrm{L}',\mathrm{N}']$$

by summing the linear maps

$$\mathbf{p}[\mathrm{L},\mathrm{M}] \otimes \mathbf{q}[\mathrm{M},\mathrm{N}] \xrightarrow{(u,v)\otimes(v,w)} \mathbf{p}[\mathrm{L}',\mathrm{M}'] \otimes \mathbf{q}[\mathrm{M}',\mathrm{N}'],$$

with M′ and v uniquely determined by M using the converse in Observation 11.10. This ties with the description in Exercise 11.23.

The unit object for (11.11) is the Coxeter $\mathcal{A}$-dispecies $\mathbf{x}$ defined by

$$\mathbf{x}[\mathrm{L},\mathrm{M}] := \begin{cases} \Bbbk & \text{if } \mathrm{L} = \mathrm{M}, \\ 0 & \text{otherwise}, \end{cases} \tag{11.12}$$

with structure maps defined to be all identities. Compare with (5.5).

11.3.3. Coxeter operads in terms of top-lunes. Let us now formulate Coxeter operads in the language of top-lunes. A Coxeter $\mathcal{A}$-operad is a Coxeter $\mathcal{A}$-dispecies $\mathbf{a}$ equipped with linear maps

$$\gamma : \mathbf{a}[\mathrm{L}, \mathrm{M}] \otimes \mathbf{a}[\mathrm{M}, \mathrm{N}] \to \mathbf{a}[\mathrm{L}, \mathrm{N}] \quad \text{and} \quad \eta : \Bbbk \to \mathbf{a}[\mathrm{L}, \mathrm{L}],$$

the former for each $\mathrm{L} \leq \mathrm{M} \leq \mathrm{N}$ and the latter for each L, subject to the axioms below.

Naturality. For any $(u, v) : (\mathrm{L}, \mathrm{M}) \to (\mathrm{L}', \mathrm{M}')$ and $(v, w) : (\mathrm{M}, \mathrm{N}) \to (\mathrm{M}', \mathrm{N}')$, the first diagram below commutes.

$$(11.13a) \qquad \begin{array}{ccc} \mathbf{a}[\mathrm{L}, \mathrm{M}] \otimes \mathbf{a}[\mathrm{M}, \mathrm{N}] & \xrightarrow{\ \gamma\ } & \mathbf{a}[\mathrm{L}, \mathrm{N}] \\ {\scriptstyle (u,v)\otimes(v,w)}\downarrow & & \downarrow{\scriptstyle (u,w)} \\ \mathbf{a}[\mathrm{L}', \mathrm{M}'] \otimes \mathbf{a}[\mathrm{M}', \mathrm{N}'] & \xrightarrow[\gamma]{} & \mathbf{a}[\mathrm{L}', \mathrm{N}'] \end{array} \qquad \begin{array}{ccc} \mathbf{a}[\mathrm{L}, \mathrm{L}] & \xrightarrow{(u,u)} & \mathbf{a}[\mathrm{L}', \mathrm{L}'] \\ {\scriptstyle \eta}\nwarrow & & \nearrow{\scriptstyle \eta} \\ & \Bbbk & \end{array}$$

For any $(u, u) : (\mathrm{L}, \mathrm{L}) \to (\mathrm{L}', \mathrm{L}')$, the second diagram above commutes.

Associativity. For any $\mathrm{L} \leq \mathrm{M} \leq \mathrm{N} \leq \mathrm{P}$, the diagram

$$(11.13b) \qquad \begin{array}{ccc} \mathbf{a}[\mathrm{L}, \mathrm{M}] \otimes \mathbf{a}[\mathrm{M}, \mathrm{N}] \otimes \mathbf{a}[\mathrm{N}, \mathrm{P}] & \xrightarrow{\gamma\otimes\mathrm{id}} & \mathbf{a}[\mathrm{L}, \mathrm{N}] \otimes \mathbf{a}[\mathrm{N}, \mathrm{P}] \\ {\scriptstyle \mathrm{id}\otimes\gamma}\downarrow & & \downarrow{\scriptstyle \gamma} \\ \mathbf{a}[\mathrm{L}, \mathrm{M}] \otimes \mathbf{a}[\mathrm{M}, \mathrm{P}] & \xrightarrow[\gamma]{} & \mathbf{a}[\mathrm{L}, \mathrm{P}] \end{array}$$

commutes.

Unitality. For any $\mathrm{L} \leq \mathrm{M}$, the diagrams

$$(11.13c) \qquad \begin{array}{ccc} & \mathbf{a}[\mathrm{L}, \mathrm{M}] \otimes \mathbf{a}[\mathrm{M}, \mathrm{M}] & \\ {\scriptstyle \mathrm{id}\otimes\eta}\nearrow & & \searrow{\scriptstyle \gamma} \\ \mathbf{a}[\mathrm{L}, \mathrm{M}] \otimes \Bbbk & \xrightarrow[\cong]{} & \mathbf{a}[\mathrm{L}, \mathrm{M}] \end{array} \qquad \begin{array}{ccc} & \mathbf{a}[\mathrm{L}, \mathrm{L}] \otimes \mathbf{a}[\mathrm{L}, \mathrm{M}] & \\ {\scriptstyle \eta\otimes\mathrm{id}}\nearrow & & \searrow{\scriptstyle \gamma} \\ \Bbbk \otimes \mathbf{a}[\mathrm{L}, \mathrm{M}] & \xrightarrow[\cong]{} & \mathbf{a}[\mathrm{L}, \mathrm{M}] \end{array}$$

commute.

A morphism of Coxeter $\mathcal{A}$-operads is a map $f : \mathbf{a} \to \mathbf{b}$ of Coxeter $\mathcal{A}$-dispecies such that the diagrams

$$(11.14) \qquad \begin{array}{ccc} \mathbf{a}[\mathrm{L}, \mathrm{M}] \otimes \mathbf{a}[\mathrm{M}, \mathrm{N}] & \xrightarrow{\ \gamma\ } & \mathbf{a}[\mathrm{L}, \mathrm{N}] \\ {\scriptstyle f_{\mathrm{L},\mathrm{M}}\otimes f_{\mathrm{M},\mathrm{N}}}\downarrow & & \downarrow{\scriptstyle f_{\mathrm{L},\mathrm{N}}} \\ \mathbf{b}[\mathrm{L}, \mathrm{M}] \otimes \mathbf{b}[\mathrm{M}, \mathrm{N}] & \xrightarrow[\gamma]{} & \mathbf{b}[\mathrm{L}, \mathrm{N}] \end{array} \qquad \begin{array}{ccc} \mathbf{a}[\mathrm{L}, \mathrm{L}] & \xrightarrow{f_{\mathrm{L},\mathrm{L}}} & \mathbf{b}[\mathrm{L}, \mathrm{L}] \\ {\scriptstyle \eta}\nwarrow & & \nearrow{\scriptstyle \eta} \\ & \Bbbk & \end{array}$$

commute.

Remark 11.24. The substitution product (11.11) is less convenient to phrase in terms of $\leq'$, see Exercise 1.17 in this regard. Nonetheless, a Coxeter operad can be equivalently formulated as above by replacing all occurrences of $\leq$ with $\leq'$. The point is that the additional substitution maps are uniquely determined by the naturality square in (11.13a).

11.3.4. Substitution product in terms of face-types. Let us now translate the substitution product to Coxeter $\boldsymbol{\alpha}$-dispecies. It is given by

$$(\mathbf{p} \circ \mathbf{q})[T, V] = \bigoplus_{\mathrm{Y}:\, \mathrm{s}(F) \leq \mathrm{Y} \leq \mathrm{s}(G)} \mathbf{p}[\mathrm{s}(F), \mathrm{Y}] \otimes \mathbf{q}[\mathrm{Y}, \mathrm{s}(G)], \tag{11.15}$$

where F and G are faces of C_r of types T and V, respectively. The sum is over all flats Y between $\mathrm{s}(F)$ and $\mathrm{s}(G)$. The tensor factors in the summands in the rhs are defined as in (11.9).

Observation 11.25. For any $F \leq K \leq G$, the flat $\mathrm{Y} = \mathrm{s}(K)$ occurs as a summand in (11.15). In this manner,

$$\bigoplus_{U:\, T \leq U \leq V} \mathbf{p}[T, U] \otimes \mathbf{q}[U, V] \subseteq (\mathbf{p} \circ \mathbf{q})[T, V], \tag{11.16}$$

with the lhs a direct summand of the rhs (after identifying $\mathbf{p}[T, U]$ with $\mathbf{p}[\mathrm{s}(F), \mathrm{Y}]$, and $\mathbf{q}[U, V]$ with $\mathbf{q}[\mathrm{Y}, \mathrm{s}(G)]$). In general, this inclusion is proper since not all flats Y have the special form where they support a face of C_r, see the illustration below.

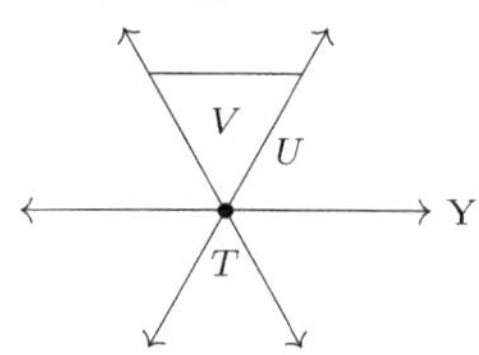

Special case. For the pair of minimum and maximum face-types,

$$(\mathbf{p} \circ \mathbf{q})[\emptyset, S] = \bigoplus_{\mathrm{Y}} \mathbf{p}[\bot, \mathrm{Y}] \otimes \mathbf{q}[\mathrm{Y}, \top]. \tag{11.17}$$

The sum is over all flats Y. Those that support a face of C_r contribute summands of the form $\mathbf{p}[\emptyset, U] \otimes \mathbf{q}[U, S]$.

Naturality for the substitution product (11.15) works as follows. For $(\sigma, \tau) : (T, V) \to (T', V')$, define the linear map

$$(\mathbf{p} \circ \mathbf{q})[T, V] \to (\mathbf{p} \circ \mathbf{q})[T', V']$$

by applying $w \otimes w$ on each summand, where $\tau = w\beta = \beta w$ in normal form. Note very carefully: This map does not preserve the direct summand in (11.16), so we do indeed need the extra summands in (11.15).

The unit object for (11.15) is the Coxeter $\boldsymbol{\alpha}$-dispecies $\mathbf{x}$ defined by

$$\mathbf{x}[T, U] := \begin{cases} \mathbb{k} & \text{if } T = U, \\ 0 & \text{otherwise}, \end{cases} \tag{11.18}$$

with structure maps all identities.

11.3.5. Internal hom for the substitution product. We begin with an interesting observation.

Observation 11.26. Let $\mathbf{p}$, $\mathbf{q}$, $\mathbf{r}$ be Coxeter $\boldsymbol{\alpha}$-dispecies. A map from $\mathbf{p} \circ \mathbf{q}$ to $\mathbf{r}$ is the same as the following. For each $T \leq U \leq V$, there is a linear map

$$\gamma : \mathbf{p}[T, U] \otimes \mathbf{q}[U, V] \to \mathbf{r}[T, V]$$

such that for each $(\sigma,\tau):(T,U)\to(T',U')$ and $(\tau,\rho):(U,V)\to(U',V')$, the diagram

$$\begin{array}{ccc} \mathbf{p}[T,U]\otimes\mathbf{q}[U,V] & \xrightarrow{\gamma} & \mathbf{r}[T,V] \\ {\scriptstyle(\sigma,\tau)\otimes(\tau,\rho)}\big\downarrow & & \big\downarrow{\scriptstyle(\sigma,\rho)} \\ \mathbf{p}[T',U']\otimes\mathbf{q}[U',V'] & \xrightarrow[\gamma]{} & \mathbf{r}[T',V'] \end{array} \tag{11.19}$$

commutes.

(Note very carefully: This description does not involve flats; it is purely in terms of face-types. The key point is that the map on a Y-summand, when Y does not support a face of C_r, is determined by naturality: Represent such a Y-summand by $\mathbf{p}[\mathrm{t}(H),\mathrm{t}(K)]\otimes\mathbf{q}[\mathrm{t}(K),\mathrm{t}(N)]$ for $H\le K\le N$. The morphism $(\sigma,\rho):(T,V)\to(T',V')$, with (σ,ρ) transferred by (H,N), takes this summand to a summand of the required form.)

We use this to describe the internal hom for the substitution product. For Coxeter $\boldsymbol{\alpha}$-dispecies $\mathbf{q}$ and $\mathbf{r}$, define $\mathrm{hom}^\circ(\mathbf{q},\mathbf{r})$ as follows. An element of $\mathrm{hom}^\circ(\mathbf{q},\mathbf{r})[T,U]$ is a family of linear maps $f_V:\mathbf{q}[U,V]\to\mathbf{r}[T,V]$, one for each $V\ge U$, such that the diagram

$$\begin{array}{ccc} \mathbf{q}[U,V] & \xrightarrow{f_V} & \mathbf{r}[T,V] \\ {\scriptstyle(\mathrm{id},\rho)}\big\downarrow & & \big\downarrow{\scriptstyle(\mathrm{id},\rho)} \\ \mathbf{q}[U,V'] & \xrightarrow[f_{V'}]{} & \mathbf{r}[T,V'] \end{array}$$

commutes for each $\rho:V\to V'$ local to U. The structure map

$$(\sigma,\tau):\mathrm{hom}^\circ(\mathbf{q},\mathbf{r})[T,U]\to\mathrm{hom}^\circ(\mathbf{q},\mathbf{r})[T',U']$$

sends a family f to the family whose V'-component, for $V'\ge U'$, is

$$\mathbf{q}[U',V']\xrightarrow{(\tau^{-1},\tau^{-1})}\mathbf{q}[U,V]\xrightarrow{f_V}\mathbf{r}[T,V]\xrightarrow{(\sigma,\tau)}\mathbf{r}[T',V'],$$

with $\tau:V\to V'$ induced from $\tau:U\to U'$.

The internal hom can also be described in terms of flats or top-lunes. For the former, see Exercise 5.2.

Exercise 11.27. Let $\mathbf{p}$, $\mathbf{q}$, $\mathbf{r}$ be Coxeter $\boldsymbol{\alpha}$-dispecies. Describe a map of Coxeter $\boldsymbol{\alpha}$-dispecies from $\mathbf{r}$ to $\mathbf{p}\circ\mathbf{q}$ as above by reversing the arrows involving γ.

11.3.6. Coxeter operads in terms of face-types. A Coxeter $\boldsymbol{\alpha}$-operad is a monoid in $(\boldsymbol{\alpha}\text{-}\mathtt{CdSp},\circ)$. A morphism of Coxeter $\boldsymbol{\alpha}$-operads is a morphism of monoids.

Explicitly: A Coxeter $\boldsymbol{\alpha}$-operad is a Coxeter $\boldsymbol{\alpha}$-dispecies $\mathbf{a}$ equipped with linear maps

$$\gamma:\mathbf{a}[T,U]\otimes\mathbf{a}[U,V]\to\mathbf{a}[T,V]\quad\text{and}\quad\eta:\Bbbk\to\mathbf{a}[T,T],$$

the former for each $T\le U\le V$ and the latter for each T, subject to the axioms below.

Naturality. For any $(\sigma, \tau) : (T, U) \to (T', U')$ and $(\tau, \rho) : (U, V) \to (U', V')$, the first diagram below commutes.

$$
\begin{array}{ccc}
\mathbf{a}[T,U] \otimes \mathbf{a}[U,V] & \xrightarrow{\gamma} & \mathbf{a}[T,V] \\
{\scriptstyle (\sigma,\tau)\otimes(\tau,\rho)}\big\downarrow & & \big\downarrow{\scriptstyle (\sigma,\rho)} \\
\mathbf{a}[T',U'] \otimes \mathbf{a}[U',V'] & \xrightarrow[\gamma]{} & \mathbf{a}[T',V']
\end{array}
\qquad
\begin{array}{ccc}
\mathbf{a}[T,T] & \xrightarrow{(\sigma,\sigma)} & \mathbf{a}[T',T'] \\
{\scriptstyle \eta}\nwarrow & & \nearrow{\scriptstyle \eta} \\
& \Bbbk &
\end{array}
\tag{11.20a}
$$

For any $(\sigma, \sigma) : (T, T) \to (T', T')$, the second diagram above commutes.

Associativity. For any $T \leq U \leq V \leq Y$, the diagram

$$
\begin{array}{ccc}
\mathbf{a}[T,U] \otimes \mathbf{a}[U,V] \otimes \mathbf{a}[V,Y] & \xrightarrow{\gamma\otimes \mathrm{id}} & \mathbf{a}[T,V] \otimes \mathbf{a}[V,Y] \\
{\scriptstyle \mathrm{id}\otimes\gamma}\big\downarrow & & \big\downarrow{\scriptstyle \gamma} \\
\mathbf{a}[T,U] \otimes \mathbf{a}[U,Y] & \xrightarrow[\gamma]{} & \mathbf{a}[T,Y]
\end{array}
\tag{11.20b}
$$

commutes.

Unitality. For any $T \leq U$, the diagrams

$$
\begin{array}{ccc}
& \mathbf{a}[T,U] \otimes \mathbf{a}[U,U] & \\
{\scriptstyle \mathrm{id}\otimes\eta}\nearrow & & \searrow{\scriptstyle \gamma} \\
\mathbf{a}[T,U] \otimes \Bbbk & \xrightarrow[\cong]{} & \mathbf{a}[T,U]
\end{array}
\qquad
\begin{array}{ccc}
& \mathbf{a}[T,T] \otimes \mathbf{a}[T,U] & \\
{\scriptstyle \eta\otimes\mathrm{id}}\nearrow & & \searrow{\scriptstyle \gamma} \\
\Bbbk \otimes \mathbf{a}[T,U] & \xrightarrow[\cong]{} & \mathbf{a}[T,U]
\end{array}
\tag{11.20c}
$$

commute.

A morphism of Coxeter $\boldsymbol{\alpha}$-operads is a map $f : \mathbf{a} \to \mathbf{b}$ of Coxeter $\boldsymbol{\alpha}$-dispecies such that the diagrams

$$
\begin{array}{ccc}
\mathbf{a}[T,U] \otimes \mathbf{a}[U,V] & \xrightarrow{\gamma} & \mathbf{a}[T,V] \\
{\scriptstyle f_{T,U}\otimes f_{U,V}}\big\downarrow & & \big\downarrow{\scriptstyle f_{T,V}} \\
\mathbf{b}[T,U] \otimes \mathbf{b}[U,V] & \xrightarrow[\gamma]{} & \mathbf{b}[T,V]
\end{array}
\qquad
\begin{array}{ccc}
\mathbf{a}[T,T] & \xrightarrow{f_{T,T}} & \mathbf{b}[T,T] \\
{\scriptstyle \eta}\nwarrow & & \nearrow{\scriptstyle \eta} \\
& \Bbbk &
\end{array}
\tag{11.21}
$$

commute.

Observation 11.28. Using the factorization of (σ, τ) in Lemma 11.17, the naturality square in (11.20a) breaks into three separate diagrams

$$
\begin{array}{ccc}
\mathbf{a}[T,U,V] & \xrightarrow{\gamma} & \mathbf{a}[T,V] \\
{\scriptstyle (\sigma,\sigma,\sigma)}\big\downarrow & & \big\downarrow{\scriptstyle (\sigma,\sigma)} \\
\mathbf{a}[T',U',V'] & \xrightarrow[\gamma]{} & \mathbf{a}[T',V']
\end{array}
\tag{11.22a}
$$

$$
\begin{array}{ccc}
\mathbf{a}[T,U,V] & \xrightarrow{\gamma} & \mathbf{a}[T,V] \\
{\scriptstyle (\mathrm{id},\tau,\tau)}\big\downarrow & & \big\downarrow{\scriptstyle (\mathrm{id},\tau)} \\
\mathbf{a}[T,U',V'] & \xrightarrow[\gamma]{} & \mathbf{a}[T,V']
\end{array}
\qquad
\begin{array}{ccc}
\mathbf{a}[T,U,V] & \xrightarrow{\gamma} & \mathbf{a}[T,V] \\
{\scriptstyle (\mathrm{id},\mathrm{id},\rho)}\big\downarrow & & \big\downarrow{\scriptstyle (\mathrm{id},\rho)} \\
\mathbf{a}[T,U,V'] & \xrightarrow[\gamma]{} & \mathbf{a}[T,V'],
\end{array}
\tag{11.22b}
$$

with shorthand notations

$$
\mathbf{a}[T,U,V] := \mathbf{a}[T,U] \otimes \mathbf{a}[U,V] \quad \text{and} \quad (\sigma,\tau,\rho) := (\sigma,\tau) \otimes (\tau,\rho).
$$

Morphisms involved in these diagrams are illustrated below.

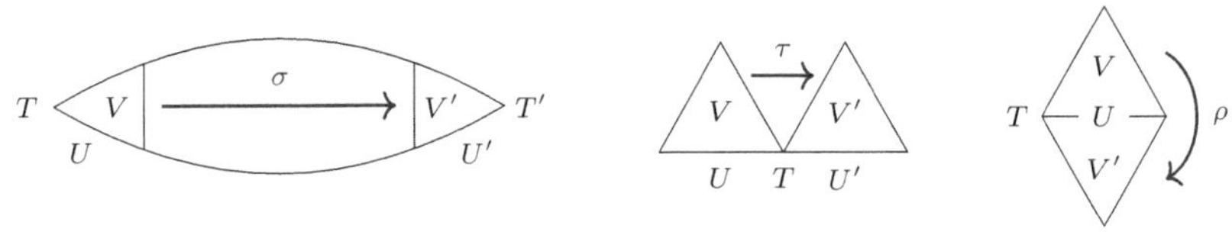

Let us consider the special case when $T = \emptyset$ and $V = S$. The naturality square is given by

$$(11.23)\qquad \begin{array}{ccc} \mathbf{a}[\emptyset, U, S] & \xrightarrow{\gamma} & \mathbf{a}[\emptyset, S] \\ {\scriptstyle (\mathrm{id},\tau,\rho)}\downarrow & & \downarrow{\scriptstyle (\mathrm{id},\rho)} \\ \mathbf{a}[\emptyset, U', S] & \xrightarrow[\gamma]{} & \mathbf{a}[\emptyset, S], \end{array}$$

which breaks into the two diagrams

$$(11.24)\qquad \begin{array}{ccc} \mathbf{a}[\emptyset, U, S] & \xrightarrow{\gamma} & \mathbf{a}[\emptyset, S] \\ {\scriptstyle (\mathrm{id},\tau,\tau)}\downarrow & & \downarrow{\scriptstyle (\mathrm{id},\tau)} \\ \mathbf{a}[\emptyset, U', S] & \xrightarrow[\gamma]{} & \mathbf{a}[\emptyset, S] \end{array} \qquad \begin{array}{ccc} \mathbf{a}[\emptyset, U, S] & \xrightarrow{\gamma} & \mathbf{a}[\emptyset, S] \\ {\scriptstyle (\mathrm{id},\mathrm{id},\rho)}\downarrow & & \downarrow{\scriptstyle (\mathrm{id},\rho)} \\ \mathbf{a}[\emptyset, U, S] & \xrightarrow[\gamma]{} & \mathbf{a}[\emptyset, S]. \end{array}$$

11.3.7. Connected Coxeter operads. Recall the notion of a connected Coxeter operad from Section 5.2.3. In terms of top-lunes, a Coxeter operad $\mathbf{a}$ is *connected* if the unit maps $\eta : \Bbbk \to \mathbf{a}[\mathrm{L}, \mathrm{L}]$ are all linear isomorphisms. In terms of face-types, a Coxeter operad $\mathbf{a}$ is *connected* if $\eta : \Bbbk \to \mathbf{a}[T, T]$ are all linear isomorphisms.

11.3.8. Coxeter cooperads. Let us briefly revisit Coxeter cooperads from Section 5.2.6. In terms of flats as in Section 11.2.1, they work as before, with a minor change in the naturality axiom.

In terms of top-lunes, a Coxeter cooperad $\mathbf{c}$ consists of linear maps

$$\mathbf{c}[\mathrm{L}, \mathrm{N}] \to \mathbf{c}[\mathrm{L}, \mathrm{M}] \otimes \mathbf{c}[\mathrm{M}, \mathrm{N}] \quad \text{and} \quad \mathbf{c}[\mathrm{L}, \mathrm{L}] \to \Bbbk$$

subject to naturality, coassociativity, counitality axioms. These are as in (11.13a)–(11.13c); we reverse appropriate arrows.

In terms of face-types, a Coxeter cooperad $\mathbf{c}$ consists of linear maps

$$\mathbf{c}[T, V] \to \mathbf{c}[T, U] \otimes \mathbf{c}[U, V] \quad \text{and} \quad \mathbf{c}[T, T] \to \Bbbk$$

subject to axioms (11.20a)–(11.20c) with appropriate arrows reversed.

11.4. Coxeter spaces as a module over Coxeter dispecies

We introduce the substitution operation of a Coxeter space into a Coxeter dispecies. This turns Coxeter spaces into a left module category over the monoidal category of Coxeter dispecies (wrt the substitution product). We present three versions of the substitution operation, one with top-lunes, one with top-nested faces, one with face-types.

11.4.1. Substitution operation in terms of top-lunes. Recall the formulation of Coxeter dispecies and of Coxeter spaces in terms of top-lunes (Section 11.2.2 and Section 8.4.2).

Let $\mathbf{p}$ be a Coxeter $\mathcal{A}$-dispecies and V a Coxeter $\mathcal{A}$-space. We now proceed to define the Coxeter $\mathcal{A}$-space $\mathbf{p} \circ \mathrm{V}$. For a top-lune L, let $\mathcal{A}_{\mathrm{L}}\text{-}\overline{\mathsf{Cox}}$ denote the category whose objects are top-lunes M greater than L and morphisms are $w : \mathrm{M} \to \mathrm{M}'$ with $w \in W$ such that $w(\mathrm{M}) = \mathrm{M}'$ and w fixes the base of L pointwise. We refer to these morphisms as morphisms local to L. Equivalently, $\mathcal{A}_{\mathrm{L}}\text{-}\overline{\mathsf{Cox}}$ is the subcategory of $\mathcal{A}\text{-}\mathsf{dCox}'$ consisting of objects of the form (L, M) (with first coordinate L) and morphisms of the form $(\mathrm{id}, w) : (\mathrm{L}, \mathrm{M}) \to (\mathrm{L}, \mathrm{M}')$. Consider the functor

$$\mathcal{A}_{\mathrm{L}}\text{-}\overline{\mathsf{Cox}} \longrightarrow \mathsf{Vec}, \qquad \mathrm{M} \longmapsto \mathbf{p}[\mathrm{L}, \mathrm{M}] \otimes \mathrm{V}[\mathrm{M}].$$

It sends a morphism $w : \mathrm{M} \to \mathrm{M}'$ local to L to the linear map

$$\mathbf{p}[\mathrm{L}, \mathrm{M}] \otimes \mathrm{V}[\mathrm{M}] \xrightarrow{(\mathrm{id},w)\otimes w} \mathbf{p}[\mathrm{L}, \mathrm{M}'] \otimes \mathrm{V}[\mathrm{M}'].$$

Define $(\mathbf{p} \circ \mathrm{V})[\mathrm{L}]$ to be the colimit of this functor. Explicitly,

$$(\mathbf{p} \circ \mathrm{V})[\mathrm{L}] = \Big(\bigoplus_{\mathrm{M}:\, \mathrm{L} \leq \mathrm{M}} \mathbf{p}[\mathrm{L}, \mathrm{M}] \otimes \mathrm{V}[\mathrm{M}] \Big)_{\mathrm{quo}}, \tag{11.25}$$

where, for any $w : \mathrm{M} \to \mathrm{M}'$ local to L, an element in the M-summand identifies with its image in the M′-summand under $(\mathrm{id}, w) \otimes w$. Further, for a morphism $w : \mathrm{L} \to \mathrm{L}'$, the linear map

$$(\mathbf{p} \circ \mathrm{V})[\mathrm{L}] \to (\mathbf{p} \circ \mathrm{V})[\mathrm{L}']$$

is induced from the linear maps

$$\mathbf{p}[\mathrm{L}, \mathrm{M}] \otimes \mathrm{V}[\mathrm{M}] \xrightarrow{(w,w)\otimes w} \mathbf{p}[\mathrm{L}', \mathrm{M}'] \otimes \mathrm{V}[\mathrm{M}'].$$

This completes the construction of the Coxeter $\mathcal{A}$-space $\mathbf{p} \circ \mathrm{V}$. We call it the substitution of V in $\mathbf{p}$.

Proposition 11.29. *The functor*

$$\mathsf{CdSp} \times \overline{\mathsf{C}}\mathsf{Sp} \to \overline{\mathsf{C}}\mathsf{Sp}, \qquad (\mathbf{p}, \mathrm{V}) \mapsto \mathbf{p} \circ \mathrm{V}$$

turns $\overline{\mathsf{C}}\mathsf{Sp}$ *into a left module category over the monoidal category* $(\mathsf{CdSp}, \circ)$.

That is, for Coxeter dispecies $\mathbf{p}$ and $\mathbf{q}$, and a Coxeter space V, there are natural isomorphisms

$$\mathbf{p} \circ (\mathbf{q} \circ \mathrm{V}) \xrightarrow{\cong} (\mathbf{p} \circ \mathbf{q}) \circ \mathrm{V} \qquad \text{and} \qquad \mathbf{x} \circ \mathrm{V} \xrightarrow{\cong} \mathrm{V}. \tag{11.26}$$

PROOF. We need to check (11.26). The second isomorphism is clear. For the first one, we evaluate both sides on a top-lune L. The rhs yields

$$((\mathbf{p} \circ \mathbf{q}) \circ \mathrm{V})[\mathrm{L}] = \operatorname*{colim}_{\mathrm{L} \leq \mathrm{N}} \bigoplus_{\mathrm{L} \leq \mathrm{M} \leq \mathrm{N}} \mathbf{p}[\mathrm{L}, \mathrm{M}] \otimes \mathbf{q}[\mathrm{M}, \mathrm{N}] \otimes \mathrm{V}[\mathrm{N}].$$

The direct sum is over M, while the colimit is over N. For a morphism $w : \mathrm{N} \to \mathrm{N}'$ local to L, the induced linear map on the M-summand is

$$\mathbf{p}[\mathrm{L}, \mathrm{M}] \otimes \mathbf{q}[\mathrm{M}, \mathrm{N}] \otimes \mathrm{V}[\mathrm{N}] \xrightarrow{(\mathrm{id},v)\otimes(v,w)\otimes w} \mathbf{p}[\mathrm{L}, \mathrm{M}'] \otimes \mathbf{q}[\mathrm{M}', \mathrm{N}'] \otimes \mathrm{V}[\mathrm{N}'],$$

with M′ and v uniquely determined by M using the converse in Observation 11.10.

Similarly, the lhs yields

$$(\mathbf{p} \circ (\mathbf{q} \circ \mathrm{V}))[\mathrm{L}] = \operatorname*{colim}_{\mathrm{L} \leq \mathrm{M}} \big(\mathbf{p}[\mathrm{L},\mathrm{M}] \otimes \operatorname*{colim}_{\mathrm{M} \leq \mathrm{N}} \mathbf{q}[\mathrm{M},\mathrm{N}] \otimes \mathrm{V}[\mathrm{N}]\big)$$
$$= \operatorname*{colim}_{\mathrm{L} \leq \mathrm{M} \leq \mathrm{N}} \mathbf{p}[\mathrm{L},\mathrm{M}] \otimes \mathbf{q}[\mathrm{M},\mathrm{N}] \otimes \mathrm{V}[\mathrm{N}].$$

Using the first factorization in Lemma 11.9, the two colimits have been combined into one by considering the category whose objects are (M, N) with $\mathrm{L} \leq \mathrm{M} \leq \mathrm{N}$, and morphisms are morphisms $(v, w) : (\mathrm{M}, \mathrm{N}) \to (\mathrm{M}', \mathrm{N}')$ of nested top-lunes such that v (and hence w) is local to L. For such a morphism, the induced linear map is

$$\mathbf{p}[\mathrm{L},\mathrm{M}] \otimes \mathbf{q}[\mathrm{M},\mathrm{N}] \otimes \mathrm{V}[\mathrm{N}] \xrightarrow{(\mathrm{id},v)\otimes(v,w)\otimes w} \mathbf{p}[\mathrm{L},\mathrm{M}'] \otimes \mathbf{q}[\mathrm{M}',\mathrm{N}'] \otimes \mathrm{V}[\mathrm{N}'].$$

Comparing the two expressions, we see that property (11.26) holds. □

Remark 11.30. In formula (11.25), we may use the partial order $\leq'$ instead of $\leq$. This follows from Exercise 11.12.

11.4.2. Substitution operation in terms of flats. We briefly mention how the substitution operation works in terms of flats. Let $\mathbf{p}$ be a Coxeter $\mathcal{A}$-dispecies and V a Coxeter $\mathcal{A}$-space. Fix a chamber C. For a flat X,

$$(\mathbf{p} \circ \mathrm{V})[\mathrm{X}] = \Big(\bigoplus_{\mathrm{Y}:\, \mathrm{X} \leq \mathrm{Y}} \mathbf{p}[\mathrm{X},\mathrm{Y}] \otimes \mathrm{V}[\mathrm{Y}]\Big)_{\mathrm{quo}}. \tag{11.27}$$

The quotient identifies summands via the morphisms $(\mathrm{id}, w) \otimes w$; the notation is as in Exercise 11.4.

11.4.3. Substitution operation in terms of top-nested faces. Let $\mathbf{p}$ be a Coxeter $\mathcal{A}$-dispecies and V a Coxeter $\mathcal{A}$-space. We now formulate $\mathbf{p} \circ \mathrm{V}$ using nested flats for $\mathbf{p}$ and top-nested faces for V. Recall the category $\mathcal{A}_A\text{-}\overline{\mathsf{Cox}}$ consisting of morphisms between top-nested faces which are local to A. For a top-nested face (A, C), let $\mathcal{A}_{A,C}\text{-}\overline{\mathsf{Cox}}$ denote the full subcategory of $\mathcal{A}_A\text{-}\overline{\mathsf{Cox}}$ consisting of top-nested faces with second coordinate C. Consider the functor

$$\mathcal{A}_{A,C}\text{-}\overline{\mathsf{Cox}} \longrightarrow \mathsf{Vec}, \qquad (F, C) \longmapsto \mathbf{p}[\mathrm{s}(A), \mathrm{s}(F)] \otimes \mathrm{V}[F, C].$$

It sends a morphism $\tau : (F, C) \to (G, C)$ local to A to the linear map

$$\mathbf{p}[\mathrm{s}(A), \mathrm{s}(F)] \otimes \mathrm{V}[F, C] \xrightarrow{w \otimes \tau} \mathbf{p}[\mathrm{s}(A), \mathrm{s}(G)] \otimes \mathrm{V}[G, C],$$

where $\tau = w\beta = \beta w$ in normal form. Define $(\mathbf{p} \circ \mathrm{V})[A, C]$ to be the colimit of this functor. Explicitly,

$$(\mathbf{p} \circ \mathrm{V})[A, C] = \Big(\bigoplus_{F:\, A \leq F \leq C} \mathbf{p}[\mathrm{s}(A), \mathrm{s}(F)] \otimes \mathrm{V}[F, C]\Big)_{\mathrm{quo}}, \tag{11.28}$$

where, for any $\tau : (F, C) \to (G, C)$ local to A, an element in the F-summand identifies with its image in the G-summand under $w \otimes \tau$. Further, for a morphism $\sigma : (A, C) \to (B, D)$, the linear map

$$(\mathbf{p} \circ \mathrm{V})[A, C] \to (\mathbf{p} \circ \mathrm{V})[B, D]$$

is induced from the linear maps

$$\mathbf{p}[\mathrm{s}(A), \mathrm{s}(F)] \otimes \mathrm{V}[F, C] \xrightarrow{w \otimes \sigma} \mathbf{p}[\mathrm{s}(B), \mathrm{s}(\sigma_*(F))] \otimes \mathrm{V}[\sigma_*(F), D],$$

where $\sigma = w\beta = \beta w$ in normal form.

11.4.4. Substitution operation in terms of face-types. We now formulate the substitution operation using face-types. Accordingly, let $\mathbf{p}$ be a Coxeter $\boldsymbol{\alpha}$-dispecies and V a Coxeter $\boldsymbol{\alpha}$-space. We employ the category $\boldsymbol{\alpha}_T\text{-}\overline{\mathsf{Cox}}$ whose objects are face-types greater than T and morphisms are those local to T. Consider the functor

$$\boldsymbol{\alpha}_T\text{-}\overline{\mathsf{Cox}} \longrightarrow \mathsf{Vec}, \qquad U \longmapsto \mathbf{p}[T, U] \otimes \mathrm{V}[U].$$

It sends a morphism $\tau : U \to U'$ local to T to the linear map

$$\mathbf{p}[T, U] \otimes \mathrm{V}[U] \xrightarrow{(\mathrm{id}, \tau) \otimes \tau} \mathbf{p}[T, U'] \otimes \mathrm{V}[U'].$$

Define $(\mathbf{p} \circ \mathrm{V})[T]$ to be the colimit of this functor. Explicitly,

$$(\mathbf{p} \circ \mathrm{V})[T] = \Big(\bigoplus_{U:\, T \leq U} \mathbf{p}[T, U] \otimes \mathrm{V}[U] \Big)_{\mathrm{quo}}, \tag{11.29}$$

where, for any $\tau : U \to U'$ local to T, an element in the U-summand identifies with its image in the U'-summand under $(\mathrm{id}, \tau) \otimes \tau$. Further, for a morphism $\sigma : T \to T'$, the linear map

$$(\mathbf{p} \circ \mathrm{V})[T] \to (\mathbf{p} \circ \mathrm{V})[T']$$

is induced from the linear maps

$$\mathbf{p}[T, U] \otimes \mathrm{V}[U] \xrightarrow{(\sigma, \sigma) \otimes \sigma} \mathbf{p}[T', \sigma_*(U)] \otimes \mathrm{V}[\sigma_*(U)].$$

Exercise 11.31. Show: For Coxeter $\boldsymbol{\alpha}$-dispecies $\mathbf{p}$ and $\mathbf{q}$ and Coxeter $\boldsymbol{\alpha}$-space V,

$$(\mathbf{p} \circ (\mathbf{q} \circ \mathrm{V}))[T] = \Big(\bigoplus_{U, Y:\, T \leq U \leq Y} \mathbf{p}[T, U] \otimes \mathbf{q}[U, Y] \otimes \mathrm{V}[Y] \Big)_{\mathrm{quo}},$$

with summands identified via the morphisms $(\mathrm{id}, \sigma) \otimes (\sigma, \tau) \otimes \tau$. Identify the above expression with $((\mathbf{p} \circ \mathbf{q}) \circ \mathrm{V})[T]$. (The summands in $\mathbf{p} \circ \mathbf{q}$ for flats Y not supporting a face of C_T can be removed without loss of information since we are taking a colimit.) Use this to deduce property (11.26).

This way of establishing property (11.26) is different from the one using (11.25). Note very carefully the distinction between (11.15) and (11.16). Also compare and contrast with (11.11).

Exercise 11.32. Check:

$$(\mathbf{p}_1 + \mathbf{p}_2) \circ \mathrm{V} \cong \mathbf{p}_1 \circ \mathrm{V} + \mathbf{p}_2 \circ \mathrm{V} \quad \text{and} \quad \mathbf{p} \circ (\mathrm{V}_1 + \mathrm{V}_2) \cong \mathbf{p} \circ \mathrm{V}_1 + \mathbf{p} \circ \mathrm{V}_2.$$

11.4.5. Enriched hom and endomorphism Coxeter operad. For definition of enriched hom, we refer to [**11**, Appendix B.2.4]. We begin with a basic observation.

Observation 11.33. Let $\mathbf{p}$ be a Coxeter $\boldsymbol{\alpha}$-dispecies, and V and W be Coxeter $\boldsymbol{\alpha}$-spaces. A map from $\mathbf{p} \circ \mathrm{V}$ to W of Coxeter $\boldsymbol{\alpha}$-spaces is the same as the following. For each $T \leq U$, there is a linear map

$$\gamma : \mathbf{p}[T, U] \otimes \mathrm{V}[U] \to \mathrm{W}[T]$$

such that for each $(\sigma, \tau) : (T, U) \to (T', U')$, the diagram

$$\begin{array}{ccc} \mathbf{p}[T,U] \otimes \mathrm{V}[U] & \xrightarrow{\gamma} & \mathrm{W}[T] \\ {\scriptstyle (\sigma,\tau)\otimes\tau}\downarrow & & \downarrow{\scriptstyle \sigma} \\ \mathbf{p}[T',U'] \otimes \mathrm{V}[U'] & \xrightarrow[\gamma]{} & \mathrm{W}[T'] \end{array} \tag{11.30}$$

commutes.

(This can be deduced from Lemma 11.17: Using the definition of $\mathbf{p} \circ \mathrm{V}$, we get two diagrams, one for (σ, σ) and one for (id, τ), which can be combined into a single diagram for (σ, τ).)

We use this to describe the enriched hom for the left module category of Coxeter spaces. For Coxeter $\boldsymbol{\alpha}$-spaces V and W, $\mathrm{hom}^\circ(\mathrm{V}, \mathrm{W})$ is the Coxeter $\boldsymbol{\alpha}$-dispecies whose (T, U)-component consists of all linear maps from $\mathrm{V}[U]$ to $\mathrm{W}[T]$. The structure map

$$(\sigma, \tau) : \mathrm{hom}^\circ(\mathrm{V}, \mathrm{W})[T, U] \to \mathrm{hom}^\circ(\mathrm{V}, \mathrm{W})[T', U']$$

sends f to the composite

$$\mathrm{V}[U'] \xrightarrow{\tau^{-1}} \mathrm{V}[U] \xrightarrow{f} \mathrm{W}[T] \xrightarrow{\sigma} \mathrm{W}[T'].$$

By general principles [**11**, Proposition B.5], for a Coxeter $\boldsymbol{\alpha}$-space V,

$$\mathrm{end}^\circ(\mathrm{V}) := \mathrm{hom}^\circ(\mathrm{V}, \mathrm{V}) \tag{11.31}$$

is a Coxeter $\boldsymbol{\alpha}$-operad. This is the *endomorphism Coxeter operad* of V. Explicitly, γ is given by composition of maps, while η sends $1 \in \Bbbk$ to the identity map. In addition, following [**11**, Appendix B.3.3], the category of Coxeter $\boldsymbol{\alpha}$-spaces is enriched over the category of Coxeter $\boldsymbol{\alpha}$-dispecies, and is copowered by it by means of the left action (11.29).

Exercise 11.34. Check directly: For any Coxeter $\boldsymbol{\alpha}$-space V, $\mathrm{end}^\circ(\mathrm{V})$ as defined above is a Coxeter $\boldsymbol{\alpha}$-operad.

Exercise 11.35. Formulate the enriched hom and endomorphism Coxeter operad in terms of top-lunes.

11.4.6. Substitution operation (limits version). Let $\mathbf{p}$ be a Coxeter dispecies and V a Coxeter space. By replacing colimit by limit in the definition of $\mathbf{p} \circ \mathrm{V}$, we obtain another substitution operation which we denote $\mathbf{p} \circ^\vee \mathrm{V}$. It can be described using top-lunes as

$$(\mathbf{p} \circ^\vee \mathrm{V})[\mathrm{L}] = \Big(\bigoplus_{\mathrm{M}:\, \mathrm{L} \leq \mathrm{M}} \mathbf{p}[\mathrm{L}, \mathrm{M}] \otimes \mathrm{V}[\mathrm{M}] \Big)^{\mathrm{sub}} \tag{11.32}$$

consisting of elements invariant under $w : \mathrm{M} \to \mathrm{M}'$ local to L.

In terms of top-nested faces,

$$(11.33)\qquad (\mathbf{p} \circ^\vee \mathrm{V})[A, C] = \Big(\bigoplus_{F:\, A \leq F \leq C} \mathbf{p}[\mathrm{s}(A), \mathrm{s}(F)] \otimes \mathrm{V}[F, C] \Big)^{\mathrm{sub}},$$

consisting of elements invariant under $\tau : (F, C) \to (G, C)$ local to A.

In terms of face-types,

$$(11.34)\qquad (\mathbf{p} \circ^\vee \mathrm{V})[T] = \Big(\bigoplus_{U:\, T \leq U} \mathbf{p}[T, U] \otimes \mathrm{V}[U] \Big)^{\mathrm{sub}},$$

consisting of elements invariant under $\tau : U \to U'$ local to T.

As indicated by the notation, $\circ$ and $\circ^\vee$ are conjugates of each other wrt duality (in finite-dimensional setting).

Exercise 11.36. Check: $\mathbf{p} \circ \mathrm{V} \cong \mathbf{p} \circ^\vee \mathrm{V}$ if $\mathrm{char}(\Bbbk) \nmid |W|$.

Exercise 11.37. Let $\mathrm{char}(\Bbbk) \nmid |W|$. If $f : \mathbf{p} \hookrightarrow \mathbf{q}$ is an injective map of Coxeter dispecies, then $f \circ \mathrm{id} : \mathbf{p} \circ \mathrm{V} \hookrightarrow \mathbf{q} \circ \mathrm{V}$ is an injective map of Coxeter spaces. For the dual operation $\circ^\vee$, this works in any field characteristic. There is a similar statement for a surjective map $\mathbf{p} \twoheadrightarrow \mathbf{q}$.

11.5. Coxeter operad algebras

In Section 11.4, we saw the substitution operation of a Coxeter space into a Coxeter dispecies. This allows us to define Coxeter operad algebras. Dually, one defines Coxeter cooperad coalgebras.

The parallel discussion of Coxeter (co)operad (co)monoids arising from the substitution operation of a Coxeter species is in Section 5.4. For connections between the two using Fock functors, see Section 15.6.

11.5.1. Coxeter operad algebras. We saw in Proposition 11.29 that Coxeter spaces is a left module category over the monoidal category of Coxeter dispecies wrt the substitution product. Thus, for a Coxeter operad $\mathbf{a}$, which is a monoid wrt substitution, one can define the category of left $\mathbf{a}$-modules in Coxeter spaces. We call it the category of Coxeter $\mathbf{a}$-algebras and denote it by $\mathbf{a}\text{-}\mathsf{Mon}(\overline{\mathsf{C}}\mathsf{Sp})$. By definition: A *Coxeter* $\mathbf{a}$-*algebra* is a Coxeter space V equipped with a map of Coxeter spaces $\mathbf{a} \circ \mathrm{V} \to \mathrm{V}$ which is associative and unital. A morphism of Coxeter $\mathbf{a}$-algebras is a map of Coxeter spaces $\mathrm{V} \to \mathrm{W}$ which commutes with the respective structure maps.

Let us now make this explicit. We present three versions, one with top-lunes, one with top-nested faces, one with face-types.

11.5.2. Coxeter operad algebras in terms of top-lunes. A Coxeter $\mathbf{a}$-algebra is a Coxeter space V equipped with linear maps

$$\gamma : \mathbf{a}[\mathrm{L}, \mathrm{M}] \otimes \mathrm{V}[\mathrm{M}] \to \mathrm{V}[\mathrm{L}],$$

one for each $\mathrm{L} \leq \mathrm{M}$, subject to the axioms below. We denote these maps $\gamma_{\mathrm{L}}^{\mathrm{M}}$ if we need to make the dependence on L and M explicit.

Naturality. For each morphism $(u, w) : (\mathrm{L}, \mathrm{M}) \to (\mathrm{L}', \mathrm{M}')$, the diagram

$$\begin{array}{ccc} \mathbf{a}[\mathrm{L}, \mathrm{M}] \otimes \mathrm{V}[\mathrm{M}] & \xrightarrow{\gamma} & \mathrm{V}[\mathrm{L}] \\ {\scriptstyle (u,w)\otimes w}\downarrow & & \downarrow{\scriptstyle u} \\ \mathbf{a}[\mathrm{L}', \mathrm{M}'] \otimes \mathrm{V}[\mathrm{M}'] & \xrightarrow[\gamma]{} & \mathrm{V}[\mathrm{L}'] \end{array} \tag{11.35a}$$

commutes. Using the first factorization of (u, w) in Lemma 11.9, this diagram breaks into two separate diagrams

$$\begin{array}{ccc} \mathbf{a}[\mathrm{L}, \mathrm{M}] \otimes \mathrm{V}[\mathrm{M}] & \xrightarrow{\gamma} & \mathrm{V}[\mathrm{L}] \\ {\scriptstyle (u,u)\otimes u}\downarrow & & \downarrow{\scriptstyle u} \\ \mathbf{a}[\mathrm{L}', \mathrm{M}'] \otimes \mathrm{V}[\mathrm{M}'] & \xrightarrow[\gamma]{} & \mathrm{V}[\mathrm{L}'] \end{array} \qquad \begin{array}{ccc} \mathbf{a}[\mathrm{L}, \mathrm{M}] \otimes \mathrm{V}[\mathrm{M}] & \searrow^{\gamma} & \\ {\scriptstyle (\mathrm{id},w)\otimes w}\downarrow & & \mathrm{V}[\mathrm{L}]. \\ \mathbf{a}[\mathrm{L}, \mathrm{M}'] \otimes \mathrm{V}[\mathrm{M}'] & \nearrow_{\gamma} & \end{array} \tag{11.35b}$$

Associativity. For each $\mathrm{L} \leq \mathrm{M} \leq \mathrm{N}$, the diagram

$$\begin{array}{ccc} \mathbf{a}[\mathrm{L}, \mathrm{M}] \otimes \mathbf{a}[\mathrm{M}, \mathrm{N}] \otimes \mathrm{V}[\mathrm{N}] & \xrightarrow{\mathrm{id}\otimes\gamma} & \mathbf{a}[\mathrm{L}, \mathrm{M}] \otimes \mathrm{V}[\mathrm{M}] \\ {\scriptstyle \gamma\otimes\mathrm{id}}\downarrow & & \downarrow{\scriptstyle \gamma} \\ \mathbf{a}[\mathrm{L}, \mathrm{N}] \otimes \mathrm{V}[\mathrm{N}] & \xrightarrow[\gamma]{} & \mathrm{V}[\mathrm{L}] \end{array} \tag{11.35c}$$

commutes.

Unitality. For each L, the diagram

$$\begin{array}{ccc} & \mathbf{a}[\mathrm{L}, \mathrm{L}] \otimes \mathrm{V}[\mathrm{L}] & \\ {\scriptstyle \eta\otimes\mathrm{id}}\nearrow & & \searrow{\scriptstyle \gamma} \\ \Bbbk \otimes \mathrm{V}[\mathrm{L}] & \xrightarrow[\cong]{} & \mathrm{V}[\mathrm{L}] \end{array} \tag{11.35d}$$

commutes.

We also refer to V as a Coxeter algebra over the Coxeter operad **a**. We refer to γ as the structure map of V.

A morphism of Coxeter **a**-algebras is a map $f : \mathrm{V} \to \mathrm{W}$ of Coxeter spaces such that the diagram

$$\begin{array}{ccc} \mathbf{a}[\mathrm{L}, \mathrm{M}] \otimes \mathrm{V}[\mathrm{M}] & \xrightarrow{\gamma} & \mathrm{V}[\mathrm{L}] \\ {\scriptstyle \mathrm{id}\otimes f_{\mathrm{M}}}\downarrow & & \downarrow{\scriptstyle f_{\mathrm{L}}} \\ \mathbf{a}[\mathrm{L}, \mathrm{M}] \otimes \mathrm{W}[\mathrm{M}] & \xrightarrow[\gamma]{} & \mathrm{W}[\mathrm{L}] \end{array} \tag{11.36}$$

commutes.

Observation 11.38. Using the second factorization of (u, w) in Lemma 11.9, diagram (11.35a) breaks into two separate diagrams. The first of these is equivalent to the first diagram in (11.35b). The second of these is shown below.

For each morphism $(u, \mathrm{id}) : (\mathrm{L}, \mathrm{M}) \to (\mathrm{L}', \mathrm{M})$, the diagram

$$\begin{array}{ccc} \mathbf{a}[\mathrm{L}, \mathrm{M}] \otimes \mathrm{V}[\mathrm{M}] & \xrightarrow{\gamma} & \mathrm{V}[\mathrm{L}] \\ {\scriptstyle (u,\mathrm{id})\otimes\mathrm{id}}\downarrow & & \downarrow{\scriptstyle u} \\ \mathbf{a}[\mathrm{L}', \mathrm{M}] \otimes \mathrm{V}[\mathrm{M}] & \xrightarrow[\gamma]{} & \mathrm{V}[\mathrm{L}'] \end{array} \tag{11.37}$$

commutes.

Remark 11.39. A Coxeter operad algebra can be equivalently formulated as above by replacing all occurrences of $\leq$ with $\leq'$. The point is that the additional structure maps are uniquely determined by the naturality square in (11.35a). See Remarks 11.24 and 11.30 in this regard.

11.5.3. Coxeter operad algebras in terms of top-nested faces.

A Coxeter **a**-algebra is a Coxeter space V equipped with linear maps

$$\gamma : \mathbf{a}[\mathrm{s}(A), \mathrm{s}(F)] \otimes \mathrm{V}[F, C] \to \mathrm{V}[A, C],$$

one for each $A \leq F \leq C$, subject to the axioms below. We denote these maps $\gamma_{A,C}^{F,C}$ if we need to make the dependence on A, F, C explicit.

Naturality. For each morphism $\sigma : (A, C) \to (B, D)$ and F between A and C, the diagram

$$\begin{array}{ccc} \mathbf{a}[\mathrm{s}(A), \mathrm{s}(F)] \otimes \mathrm{V}[F, C] & \xrightarrow{\gamma} & \mathrm{V}[A, C] \\ {\scriptstyle u \otimes \sigma}\downarrow & & \downarrow{\scriptstyle \sigma} \\ \mathbf{a}[\mathrm{s}(B), \mathrm{s}(G)] \otimes \mathrm{V}[G, D] & \xrightarrow[\gamma]{} & \mathrm{V}[B, D] \end{array} \tag{11.38a}$$

commutes, where $G = \sigma_*(F)$ and $\sigma = u\beta = \beta u$ in normal form. In addition, for any morphism $\tau : (F, C) \to (G, C)$ local to A, the diagram

$$\begin{array}{ccc} \mathbf{a}[\mathrm{s}(A), \mathrm{s}(F)] \otimes \mathrm{V}[F, C] & & \\ & \searrow^{\gamma} & \\ {\scriptstyle v \otimes \tau}\downarrow & & \mathrm{V}[A, C] \\ & \nearrow_{\gamma} & \\ \mathbf{a}[\mathrm{s}(A), \mathrm{s}(G)] \otimes \mathrm{V}[G, C] & & \end{array} \tag{11.38b}$$

commutes, with $\tau = v\beta = \beta v$ in normal form.

Associativity. For each $A \leq F \leq G \leq C$, the diagram

$$\begin{array}{ccc} \mathbf{a}[\mathrm{s}(A), \mathrm{s}(F)] \otimes \mathbf{a}[\mathrm{s}(F), \mathrm{s}(G)] \otimes \mathrm{V}[G, C] & \xrightarrow{\mathrm{id} \otimes \gamma} & \mathbf{a}[\mathrm{s}(A), \mathrm{s}(F)] \otimes \mathrm{V}[F, C] \\ {\scriptstyle \gamma \otimes \mathrm{id}}\downarrow & & \downarrow{\scriptstyle \gamma} \\ \mathbf{a}[\mathrm{s}(A), \mathrm{s}(G)] \otimes \mathrm{V}[G, C] & \xrightarrow[\gamma]{} & \mathrm{V}[A, C] \end{array} \tag{11.38c}$$

commutes.

Unitality. For each $A \leq C$, the diagram

$$\begin{array}{ccc} & \mathbf{a}[\mathrm{s}(A), \mathrm{s}(A)] \otimes \mathrm{V}[A, C] & \\ {\scriptstyle \eta \otimes \mathrm{id}}\nearrow & & \searrow{\scriptstyle \gamma} \\ \Bbbk \otimes \mathrm{V}[A, C] & \xrightarrow[\cong]{} & \mathrm{V}[A, C] \end{array} \tag{11.38d}$$

commutes.

A morphism of Coxeter **a**-algebras is a map $f : \mathrm{V} \to \mathrm{W}$ of Coxeter spaces such that the diagram

$$\begin{array}{ccc} \mathbf{a}[\mathrm{s}(A), \mathrm{s}(F)] \otimes \mathrm{V}[F, C] & \xrightarrow{\gamma} & \mathrm{V}[A, C] \\ {\scriptstyle \mathrm{id} \otimes f_{F,C}} \downarrow & & \downarrow {\scriptstyle f_{A,C}} \\ \mathbf{a}[\mathrm{s}(A), \mathrm{s}(F)] \otimes \mathrm{W}[F, C] & \xrightarrow[\gamma]{} & \mathrm{W}[A, C] \end{array} \tag{11.39}$$

commutes.

11.5.4. Coxeter operad algebras in terms of face-types. A Coxeter **a**-algebra is a Coxeter space V equipped with linear maps

$$\gamma : \mathbf{a}[T, U] \otimes \mathrm{V}[U] \to \mathrm{V}[T],$$

one for each $T \leq U$, subject to the axioms below. We denote these maps γ_T^U if we need to make the dependence on T and U explicit.

Naturality. For each morphism $(\sigma, \tau) : (T, U) \to (T', U')$, the diagram

$$\begin{array}{ccc} \mathbf{a}[T, U] \otimes \mathrm{V}[U] & \xrightarrow{\gamma} & \mathrm{V}[T] \\ {\scriptstyle (\sigma,\tau)\otimes\tau} \downarrow & & \downarrow {\scriptstyle \sigma} \\ \mathbf{a}[T', U'] \otimes \mathrm{V}[U'] & \xrightarrow[\gamma]{} & \mathrm{V}[T'] \end{array} \tag{11.40a}$$

commutes. Using the factorization in Lemma 11.17, this diagram breaks into two separate diagrams

$$\begin{array}{ccc} \mathbf{a}[T, U] \otimes \mathrm{V}[U] & \xrightarrow{\gamma} & \mathrm{V}[T] \\ {\scriptstyle (\sigma,\sigma)\otimes\sigma} \downarrow & & \downarrow {\scriptstyle \sigma} \\ \mathbf{a}[T', U'] \otimes \mathrm{V}[U'] & \xrightarrow[\gamma]{} & \mathrm{V}[T'] \end{array} \qquad \begin{array}{ccc} \mathbf{a}[T, U] \otimes \mathrm{V}[U] & \searrow^{\gamma} & \\ {\scriptstyle (\mathrm{id},\tau)\otimes\tau} \downarrow & & \mathrm{V}[T]. \\ \mathbf{a}[T, U'] \otimes \mathrm{V}[U'] & \nearrow_{\gamma} & \end{array} \tag{11.40b}$$

Associativity. For each $T \leq U \leq Y$, the diagram

$$\begin{array}{ccc} \mathbf{a}[T, U] \otimes \mathbf{a}[U, Y] \otimes \mathrm{V}[Y] & \xrightarrow{\mathrm{id}\otimes\gamma} & \mathbf{a}[T, U] \otimes \mathrm{V}[U] \\ {\scriptstyle \gamma\otimes\mathrm{id}} \downarrow & & \downarrow {\scriptstyle \gamma} \\ \mathbf{a}[T, Y] \otimes \mathrm{V}[Y] & \xrightarrow[\gamma]{} & \mathrm{V}[T] \end{array} \tag{11.40c}$$

commutes.

Unitality. For each T, the diagram

$$\begin{array}{ccc} & \mathbf{a}[T, T] \otimes \mathrm{V}[T] & \\ {\scriptstyle \eta\otimes\mathrm{id}} \nearrow & & \searrow {\scriptstyle \gamma} \\ \Bbbk \otimes \mathrm{V}[T] & \xrightarrow[\cong]{} & \mathrm{V}[T] \end{array} \tag{11.40d}$$

commutes.

A morphism of Coxeter $\mathbf{a}$-algebras is a map $f : \mathrm{V} \to \mathrm{W}$ of Coxeter spaces such that the diagram

$$\begin{array}{ccc} \mathbf{a}[T,U] \otimes \mathrm{V}[U] & \xrightarrow{\gamma} & \mathrm{V}[T] \\ {\scriptstyle \mathrm{id} \otimes f_U}\big\downarrow & & \big\downarrow{\scriptstyle f_T} \\ \mathbf{a}[T,U] \otimes \mathrm{W}[U] & \xrightarrow[\gamma]{} & \mathrm{W}[T] \end{array} \tag{11.41}$$

commutes.

11.5.5. Free Coxeter operad algebra. For a Coxeter operad $\mathbf{a}$, the free Coxeter $\mathbf{a}$-algebra on a Coxeter space V is given by $\mathbf{a} \circ \mathrm{V}$, with structure map defined by the composite

$$\mathbf{a} \circ (\mathbf{a} \circ \mathrm{V}) \xrightarrow{\cong} (\mathbf{a} \circ \mathbf{a}) \circ \mathrm{V} \xrightarrow{\gamma \circ \mathrm{id}} \mathbf{a} \circ \mathrm{V}.$$

The universal property of $\mathbf{a} \circ \mathrm{V}$ is stated below.

Theorem 11.40. *Let* $\mathbf{a}$ *be a Coxeter operad. Let* M *be a Coxeter* $\mathbf{a}$*-algebra,* V *a Coxeter space,* $f : \mathrm{V} \to \mathrm{M}$ *a map of Coxeter spaces. Then there exists a unique morphism of Coxeter* $\mathbf{a}$*-algebras* $\hat{f} : \mathbf{a} \circ \mathrm{V} \to \mathrm{M}$ *such that the diagram*

$$\begin{array}{ccc} \mathbf{a} \circ \mathrm{V} & \overset{\hat{f}}{\dashrightarrow} & \mathrm{M} \\ \big\uparrow & \nearrow_{f} & \\ \mathrm{V} & & \end{array}$$

commutes.

The vertical map is $\mathrm{V} \cong \mathbf{x} \circ \mathrm{V} \hookrightarrow \mathbf{a} \circ \mathrm{V}$. Explicitly, the map $\hat{f}$ is

$$\mathbf{a} \circ \mathrm{V} \xrightarrow{\mathrm{id} \circ f} \mathbf{a} \circ \mathrm{M} \xrightarrow{\gamma} \mathrm{M}. \tag{11.42}$$

Example 11.41 (Free Coxeter operad algebra on one generator). Recall the Coxeter space characteristic of chambers X from Section 10.1. For a Coxeter dispecies $\mathbf{p}$, by definitions (11.28), (11.25), (11.29),

$$\begin{aligned} (\mathbf{p} \circ \mathrm{X})[A, C] &= \mathbf{p}[\mathrm{s}(A), \top]_{W_A}, \\ (\mathbf{p} \circ \mathrm{X})[\mathrm{L}] &= \mathbf{p}[\mathrm{L}, \top]_{W_{\mathrm{b(L)}}}, \\ (\mathbf{p} \circ \mathrm{X})[T] &= \mathbf{p}[T, S]_{W_T}. \end{aligned} \tag{11.43}$$

In the rhs, we take coinvariants wrt the groups W_A, $W_{\mathrm{b(L)}}$, W_T. The group actions arise from Observations 11.5, 11.15, 11.22. For $\mathbf{a}$ a Coxeter operad, $\mathbf{a} \circ \mathrm{X}$ is the free Coxeter $\mathbf{a}$-algebra on X. We call it the free Coxeter $\mathbf{a}$-algebra on one generator.

More generally, for a left W-module M,

$$\begin{aligned} (\mathbf{p} \circ \mathrm{X}_M)[A, C] &= (\mathbf{p}[\mathrm{s}(A), \top] \otimes M)_{W_A}, \\ (\mathbf{p} \circ \mathrm{X}_M)[\mathrm{L}] &= (\mathbf{p}[\mathrm{L}, \top] \otimes M)_{W_{\mathrm{b(L)}}}, \\ (\mathbf{p} \circ \mathrm{X}_M)[T] &= (\mathbf{p}[T, S] \otimes M)_{W_T}. \end{aligned} \tag{11.44}$$

The coinvariants are taken wrt the diagonal action. When $M = \Bbbk$ with trivial W-action, we recover (11.43).

For the dual operation $\circ^{\vee}$, we have formulas as above with coinvariants replaced by invariants. The relevant definitions are (11.33), (11.32), (11.34). □

11.5.6. Coxeter algebras over a quadratic Coxeter operad. For a Coxeter dispecies $\mathbf{e}$, a Coxeter algebra over the free Coxeter operad $\mathcal{F}_\circ(\mathbf{e})$ is the same as a Coxeter space V equipped with a map $\mathbf{e} \circ \mathrm{V} \to \mathrm{V}$ of Coxeter spaces. More generally, for a quadratic Coxeter operad $\mathbf{a} = \langle \mathbf{e} \,|\, \mathbf{r} \rangle$, a Coxeter $\mathbf{a}$-algebra is the same as a Coxeter space V equipped with a map $f : \mathbf{e} \circ \mathrm{V} \to \mathrm{V}$ of Coxeter spaces such that the composite map

$$\mathbf{r} \circ \mathrm{V} \to \mathbf{e} \circ \mathbf{e} \circ \mathrm{V} \xrightarrow{\mathrm{id} \circ f} \mathbf{e} \circ \mathrm{V} \xrightarrow{f} \mathrm{V}$$

is zero.

11.5.7. Coxeter operads and monads on Coxeter spaces. A Coxeter operad $\mathbf{a}$ gives rise to a monad $\overline{\mathcal{V}}_{\mathbf{a}}$ on Coxeter spaces via $\overline{\mathcal{V}}_{\mathbf{a}}(\mathrm{V}) := \mathbf{a} \circ \mathrm{V}$. Moreover, the category of Coxeter $\mathbf{a}$-algebras is isomorphic to the category of $\overline{\mathcal{V}}_{\mathbf{a}}$-algebras.

11.5.8. Coxeter cooperad coalgebras. Dually, for a Coxeter cooperad $\mathbf{c}$, we have the category $\mathbf{c}$-$\mathsf{Comon}(\overline{\mathsf{C}}\mathsf{Sp})$ of Coxeter $\mathbf{c}$-coalgebras. The latter are left $\mathbf{c}$-comodules in Coxeter spaces wrt the dual substitution operation $\circ^\vee$. In terms of top-lunes, a Coxeter $\mathbf{c}$-coalgebra V consists of linear maps

$$\mathrm{V}[\mathrm{L}] \to \mathbf{c}[\mathrm{L}, \mathrm{M}] \otimes \mathrm{V}[\mathrm{M}]$$

subject to naturality, coassociativity, counitality axioms. These are as in (11.35a)–(11.35d); we reverse appropriate arrows.

In terms of top-nested faces, a Coxeter $\mathbf{c}$-coalgebra V consists of linear maps

$$\mathrm{V}[A, C] \to \mathbf{c}[\mathrm{s}(A), \mathrm{s}(F)] \otimes \mathrm{V}[F, C]$$

subject to axioms (11.38a)–(11.38d) with appropriate arrows reversed.

In terms of face-types, a Coxeter $\mathbf{c}$-coalgebra V consists of linear maps

$$\mathrm{V}[T] \to \mathbf{c}[T, U] \otimes \mathrm{V}[U]$$

subject to axioms (11.40a)–(11.40d) with appropriate arrows reversed.

The cofree Coxeter $\mathbf{c}$-coalgebra on a Coxeter space V is given by $\mathbf{c} \circ^\vee \mathrm{V}$. A Coxeter cooperad $\mathbf{c}$ gives rise to a comonad $\overline{\mathcal{U}}_{\mathbf{c}}$ on Coxeter spaces via $\overline{\mathcal{U}}_{\mathbf{c}}(\mathrm{V}) := \mathbf{c} \circ^\vee \mathrm{V}$. Moreover, the category of Coxeter $\mathbf{c}$-coalgebras is isomorphic to the category of $\overline{\mathcal{U}}_{\mathbf{c}}$-coalgebras.

11.6. Coxeter commutative, associative, Lie operads revisited

Recall from Section 5.5 the Coxeter commutative, associative, Lie operads, namely, **Com**, **As**, **Lie**. We now reformulate them in terms of top-lunes and face-types. Recall from Section 11.5 that for any Coxeter operad, we have the notion of a Coxeter operad algebra. For **As** and **Com**, this recovers Coxeter algebras and commutative Coxeter algebras, respectively. See Table 11.4. As a consequence, the free Coxeter algebra and free commutative Coxeter algebra are given by $\mathbf{As} \circ \mathrm{V}$ and $\mathbf{Com} \circ \mathrm{V}$, respectively. See Table 11.5. Entries related to **Lie** are treated later in Section 12.4. They require an assumption on the field characteristic.

TABLE 11.4. Coxeter operads and monads on Coxeter spaces.

Coxeter operad $\mathbf{a}$	monad $\bar{\mathcal{V}}_{\mathbf{a}}$	Coxeter $\mathbf{a}$-algebra = $\bar{\mathcal{V}}_{\mathbf{a}}$-algebra
As	$\bar{\mathcal{T}}$	Coxeter algebra
Com	$\bar{\mathcal{S}}$	commutative Coxeter algebra
$\mathbf{Com}^-$	$\bar{\mathcal{E}}$	signed commutative Coxeter algebra
Lie	$\bar{\mathcal{P}}\bar{\mathcal{T}}$	Coxeter Lie algebra
$\mathbf{Lie}^-$	$\bar{\mathcal{P}}\bar{\mathcal{T}}_{-1}$	signed Coxeter Lie algebra

TABLE 11.5. Free Coxeter operad algebras.

free Coxeter algebra	$\mathbf{As} \circ \mathrm{V} = \bar{\mathcal{T}}(\mathrm{V})$
free commutative Coxeter algebra	$\mathbf{Com} \circ \mathrm{V} = \bar{\mathcal{S}}(\mathrm{V})$
free signed commutative Coxeter algebra	$\mathbf{Com}^- \circ \mathrm{V} = \bar{\mathcal{E}}(\mathrm{V})$
free Coxeter Lie algebra	$\mathbf{Lie} \circ \mathrm{V} = \bar{\mathcal{P}}\bar{\mathcal{T}}(\mathrm{V})$
free signed Coxeter Lie algebra	$\mathbf{Lie}^- \circ \mathrm{V} = \bar{\mathcal{P}}\bar{\mathcal{T}}_{-1}(\mathrm{V})$

Dually, for a Coxeter cooperad, we have the notion of a Coxeter cooperad coalgebra. For $\mathbf{As}^*$ and $\mathbf{Com}^*$, this recovers Coxeter coalgebras and cocommutative Coxeter coalgebras, respectively. See Table 11.6. As a consequence, the cofree Coxeter coalgebra and cofree cocommutative Coxeter coalgebra are given by $\mathbf{As}^* \circ^\vee \mathrm{V}$ and $\mathbf{Com}^* \circ^\vee \mathrm{V}$, respectively. See Table 11.7.

TABLE 11.6. Coxeter cooperads and comonads on Coxeter spaces.

Coxeter cooperad $\mathbf{c}$	comonad $\bar{\mathcal{U}}_{\mathbf{c}}$	Coxeter $\mathbf{c}$-coalgebra = $\bar{\mathcal{U}}_{\mathbf{c}}$-coalgebra
$\mathbf{As}^*$	$\bar{\mathcal{T}}^\vee$	Coxeter coalgebra
$\mathbf{Com}^*$	$\bar{\mathcal{S}}^\vee$	cocommutative Coxeter coalgebra
$(\mathbf{Com}^-)^*$	$\bar{\mathcal{E}}^\vee$	signed cocommutative Coxeter coalgebra
$\mathbf{Lie}^*$	$\bar{\mathcal{Q}}\bar{\mathcal{T}}^\vee$	Coxeter Lie coalgebra
$(\mathbf{Lie}^-)^*$	$\bar{\mathcal{Q}}\bar{\mathcal{T}}^\vee_{-1}$	signed Coxeter Lie coalgebra

11.6.1. Coxeter algebras. Recall the Coxeter associative operad **As** from Section 5.5.2 where it was formulated using flats.

Coxeter associative operad in terms of top-lunes. Let us now reformulate **As** in terms of top-lunes. For $\mathrm{L} \leq \mathrm{M}$, define $\mathbf{As}[\mathrm{L}, \mathrm{M}]$ to be the vector space with basis consisting of lunes P with base b(L) and case b(M). We use the letter

TABLE 11.7. Cofree Coxeter cooperad coalgebras.

cofree Coxeter coalgebra	$\mathbf{As}^* \circ^\vee \mathrm{V} = \overline{\mathcal{T}}^\vee(\mathrm{V})$
cofree cocommutative Coxeter coalgebra	$\mathbf{Com}^* \circ^\vee \mathrm{V} = \overline{\mathcal{S}}^\vee(\mathrm{V})$
cofree signed cocommutative Coxeter coalgebra	$(\mathbf{Com}^-)^* \circ^\vee \mathrm{V} = \overline{\mathcal{E}}^\vee(\mathrm{V})$
cofree Coxeter Lie coalgebra	$\mathbf{Lie}^* \circ^\vee \mathrm{V} = \overline{\mathcal{Q}\mathcal{T}}^\vee(\mathrm{V})$
cofree signed Coxeter Lie coalgebra	$(\mathbf{Lie}^-)^* \circ^\vee \mathrm{V} = \overline{\mathcal{Q}\mathcal{T}}^\vee_{-1}(\mathrm{V})$

$\mathtt{H}$ to denote basis elements. The structure maps are

$$(u,w) : \mathbf{As}[\mathrm{L},\mathrm{M}] \to \mathbf{As}[\mathrm{L}',\mathrm{M}'], \qquad \mathtt{H}_{\mathrm{P}} \mapsto \mathtt{H}_{w\mathrm{P}}. \tag{11.45}$$

The substitution and unit maps are

$$\begin{aligned} \mathbf{As}[\mathrm{L},\mathrm{M}] \otimes \mathbf{As}[\mathrm{M},\mathrm{N}] &\to \mathbf{As}[\mathrm{L},\mathrm{N}] & \qquad \Bbbk &\xrightarrow{\cong} \mathbf{As}[\mathrm{L},\mathrm{L}] \\ \mathtt{H}_{\mathrm{P}} \otimes \mathtt{H}_{\mathrm{Q}} &\mapsto \mathtt{H}_{\mathrm{P}\circ\mathrm{Q}} & 1 &\longmapsto \mathtt{H}_{\mathrm{L}/\mathrm{L}}, \end{aligned} \tag{11.46}$$

where L/L is the unique lune with base and case both equal b(L). The naturality square in (11.13a) is checked below.

$$\begin{array}{ccc} \mathtt{H}_{\mathrm{P}} \otimes \mathtt{H}_{\mathrm{Q}} & \longmapsto & \mathtt{H}_{\mathrm{P}\circ\mathrm{Q}} \\ {\scriptstyle (u,v)\otimes(v,w)} \downarrow & & \downarrow {\scriptstyle (u,w)} \\ \mathtt{H}_{v\mathrm{P}} \otimes \mathtt{H}_{w\mathrm{Q}} & \longmapsto & \mathtt{H}_{v\mathrm{P}\circ w\mathrm{Q}} = \mathtt{H}_{w(\mathrm{P}\circ\mathrm{Q})} \end{array}$$

Note very carefully: $v\mathrm{P} = w\mathrm{P}$ since v and w agree on the case of P.

In view of Remark 11.24, we may restrict to those components $\mathbf{As}[\mathrm{L},\mathrm{M}]$ for which $\mathrm{L} \leq' \mathrm{M}$.

Coxeter associative operad in terms of face-types. Let us now reformulate **As** in terms of face-types. For $T \leq U$, define $\mathbf{As}[T,U]$ to be the vector space with basis consisting of morphisms $\alpha : U \to U'$ local to T, as U' varies over face-types greater than T. The structure maps are

$$(\sigma,\tau) : \mathbf{As}[T,U] \to \mathbf{As}[T',U'], \qquad \mathtt{H}_\alpha \mapsto \mathtt{H}_{\sigma\alpha\tau^{-1}}. \tag{11.47}$$

Note very carefully: The σ in $\sigma\alpha\tau^{-1}$ refers to the morphism induced from the starting σ, see the illustration below.

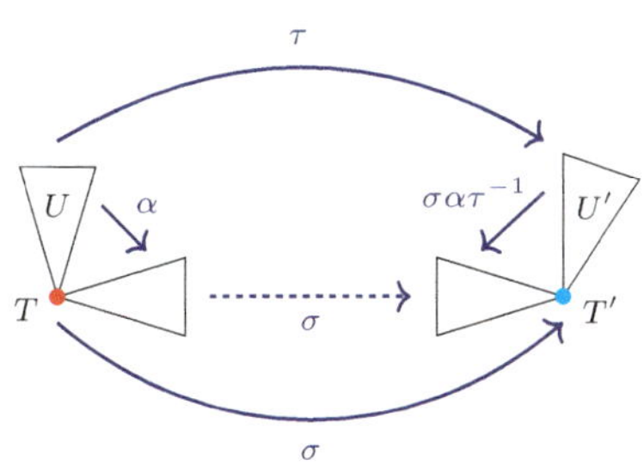

(For induced morphisms, see the pictures in Figure 1.16.) The substitution and unit maps are

$$(11.48) \qquad \begin{array}{c} \mathbf{As}[T,U] \otimes \mathbf{As}[U,V] \to \mathbf{As}[T,V] \\ \mathtt{H}_\alpha \otimes \mathtt{H}_\beta \mapsto \mathtt{H}_{\alpha\beta} \end{array} \qquad \begin{array}{c} \Bbbk \xrightarrow{\cong} \mathbf{As}[T,T] \\ 1 \longmapsto \mathtt{H}_{\mathrm{id}}. \end{array}$$

The α in $\alpha\beta$ refers to the morphism induced from the starting α, see the illustration below with $T = \emptyset$.

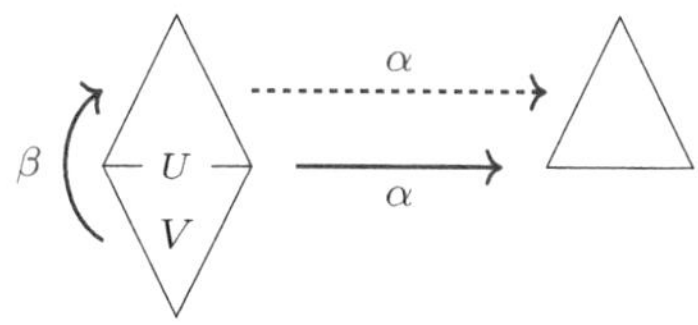

The naturality square in (11.20a) is checked below.

$$\begin{array}{ccc} \mathtt{H}_\alpha \otimes \mathtt{H}_\beta & \longmapsto & \mathtt{H}_{\alpha\beta} \\ \downarrow & & \downarrow \\ \mathtt{H}_{\sigma\alpha\tau^{-1}} \otimes \mathtt{H}_{\tau\beta\rho^{-1}} & \longmapsto & \mathtt{H}_{\sigma\alpha\beta\rho^{-1}} \end{array}$$

Coxeter algebras in terms of top-lunes. The category of Coxeter **As**-algebras as formulated in Section 11.5.2 is the same as the category of Coxeter algebras as described in Proposition 8.41 in terms of top-lunes. This can be understood as follows. Let A be a Coxeter **As**-algebra. For each $\mathrm{L} \le \mathrm{M}$, this entails a family of linear maps

$$\mu_{\mathrm{P}} : \mathrm{A}[\mathrm{M}] \to \mathrm{A}[\mathrm{L}]$$

indexed by lunes P with base b(L) and case b(M). In view of Remark 11.39, we may restrict to $\mathrm{L} \le' \mathrm{M}$. Using the naturality triangle in (11.35b), we may restrict further to the maps

$$\mu_{\mathrm{L}/\mathrm{M}} : \mathrm{A}[\mathrm{M}] \to \mathrm{A}[\mathrm{L}].$$

We identify these with the maps $\mu_{\mathrm{L}}^{\mathrm{M}}$ in Proposition 8.41. The naturality square in (11.35b) translates to the naturality axiom in (8.37). Similarly, axioms (11.35c) and (11.35d) translate to the associativity and unitality axioms in (8.37). The maps μ_{P} identify with the maps $\mu_{\mathrm{L}}^{\mathrm{M}}(\mathrm{P})$ defined by diagram (8.40).

Similarly, for a Coxeter space V, using formulas (11.25) and (8.117),

$$(11.49) \qquad (\mathbf{As} \circ \mathrm{V})[\mathrm{L}] \cong \bigoplus_{\mathrm{M}:\, \mathrm{L} \le' \mathrm{M}} \mathrm{V}[\mathrm{M}] = \overline{\mathcal{T}}(\mathrm{V})[\mathrm{L}].$$

Let us spell out the first isomorphism. Without loss of generality, suppose L is a chamber. Call it C. Then

$$\bigoplus_{\mathrm{M}:\, \mathrm{L} \le \mathrm{M}} \mathbf{As}[\mathrm{L},\mathrm{M}] \otimes \mathrm{V}[\mathrm{M}] = \bigoplus_{\substack{\mathrm{M}:\, \mathrm{L} \le \mathrm{M} \\ F:\, \mathrm{s}(F) = \mathrm{b}(\mathrm{M})}} \mathrm{V}[\mathrm{M}].$$

Now consider the quotient space in (11.25). A summand V[M] indexed by F identifies with a summand V[M′] indexed by G iff $\mathrm{t}(F) = \mathrm{t}(G)$. Moreover, the identification is unique and given by the morphism $(\mathrm{id}, w) \otimes w$, where w sends FC to GC, and, in particular, F to G. See the illustration below.

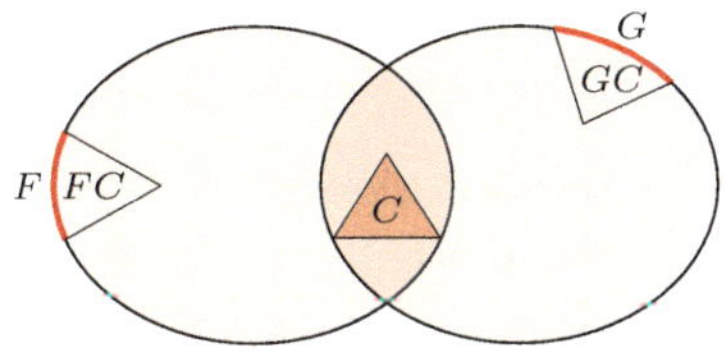

It follows that every summand uniquely identifies with a summand of the form $\mathrm{V}[\mathrm{M}]$ with $\mathrm{L} \leq' \mathrm{M}$ indexed by $F \leq C$.

We deduce from here that the monad on Coxeter spaces induced by **As** is $\overline{\mathcal{T}}$. Thus, Theorem 9.11 is a special case of Theorem 11.40 with $\mathbf{a} := \mathbf{As}$.

Coxeter algebras in terms of face-types. The category of Coxeter **As**-algebras as formulated in Section 11.5.4 is the same as the category of Coxeter algebras as described in Section 8.5.3 in terms of face-types. This can be seen as follows. Let A be a Coxeter **As**-algebra. For each $T \leq U$, this entails a family of linear maps

$$\mu_\alpha : \mathrm{A}[U] \to \mathrm{A}[T]$$

indexed by morphisms $\alpha : U \to U'$ local to T. Using the naturality triangle in (11.40b), we may restrict to the maps

$$\mu_{\mathrm{id}} : \mathrm{A}[U] \to \mathrm{A}[T]$$

which we then identify with μ_T^U. In other words,

$$\mathbf{As}[T, U] \otimes \mathrm{A}[U] \to \mathrm{A}[T], \qquad \mathtt{H}_\alpha \otimes v \mapsto \mu_T^{U'} \alpha(v) \tag{11.50}$$

for $\alpha : U \to U'$ local to T. The naturality square in (11.40b) translates to the naturality axiom in (8.41). Similarly, axioms (11.40c) and (11.40d) translate to the associativity and unitality axioms in (8.41).

Similarly, for a Coxeter space V, one checks that

$$(\mathbf{As} \circ \mathrm{V})[T] \cong \bigoplus_{U : T \leq U} \mathrm{V}[U] = \overline{\mathcal{T}}(\mathrm{V})[T]. \tag{11.51}$$

The first isomorphism is induced from the map

$$\mathbf{As}[T, U] \otimes \mathrm{V}[U] \to \bigoplus_{U' : T \leq U'} \mathrm{V}[U'], \qquad \mathtt{H}_\alpha \otimes v \mapsto \alpha(v).$$

Exercise 11.42. Show: A Coxeter **As**-algebra formulated in terms of top-nested faces in Section 11.5.3 is the same as a Coxeter algebra as formulated in Section 8.5.1.

The binary quadratic structure of **As** yields the descriptions of Coxeter algebras in Section 8.5.4 in terms of the covering associativity axiom.

Exercise 11.43. Show: $\mathbf{As} \circ \mathrm{V} \cong \mathbf{As} \circ^\vee \mathrm{V}$ as Coxeter spaces in any field characteristic. (Use our argument for computing $(\mathbf{As} \circ \mathrm{V})[\mathrm{L}]$ in (11.49).) This is an improvement on Exercise 11.36 for $\mathbf{p} := \mathbf{As}$.

Coxeter coalgebras. Recall the Coxeter associative cooperad $\mathbf{As}^*$. It is the dual of $\mathbf{As}$. The category of Coxeter $\mathbf{As}^*$-coalgebras is the same as the category of Coxeter coalgebras. The comonad on Coxeter spaces induced by $\mathbf{As}^*$ is $\overline{\mathcal{T}}^\vee$.

11.6.2. Commutative Coxeter algebras. Recall the Coxeter commutative operad $\mathbf{Com}$ from Section 5.5.1.

Coxeter commutative operad in terms of top-lunes. In terms of top-lunes, $\mathbf{Com}[\mathrm{L}, \mathrm{M}] = \Bbbk$ for all $\mathrm{L} \leq \mathrm{M}$ with structure maps all identities. In view of Remark 11.24, we may restrict to those components for which $\mathrm{L} \leq' \mathrm{M}$.

Coxeter commutative operad in terms of face-types. In terms of face-types, $\mathbf{Com}[T, U] = \Bbbk$ for all $T \leq U$ with structure maps all identities.

Commutative Coxeter algebras in terms of top-lunes. The category of Coxeter $\mathbf{Com}$-algebras as formulated in Section 11.5.2 is the same as the category of commutative Coxeter algebras as described in Section 8.8.3 in terms of top-lunes. To pass between a Coxeter $\mathbf{Com}$-algebra A and a commutative Coxeter algebra A, we identify their structure maps

$$\mathbf{Com}[\mathrm{L}, \mathrm{M}] \otimes \mathrm{A}[\mathrm{M}] \to \mathrm{A}[\mathrm{L}] \quad \text{and} \quad \mu_{\mathrm{L}}^{\mathrm{M}} : \mathrm{A}[\mathrm{M}] \to \mathrm{A}[\mathrm{L}].$$

The naturality triangle in (11.35b) corresponds to the commutativity axiom (8.82). Interestingly, diagram (11.37) reduces to a triangle which corresponds to the upside-down commutativity axiom (8.83).

Similarly, for a Coxeter space V, using formulas (11.25) and (8.125),

$$(\mathbf{Com} \circ \mathrm{V})[\mathrm{L}] = \Big(\bigoplus_{\mathrm{M}:\, \mathrm{L} \leq' \mathrm{M}} \mathrm{V}[\mathrm{M}] \Big)_{\mathrm{quo}} = \overline{\mathcal{S}}(\mathrm{V})[\mathrm{L}]. \tag{11.52}$$

We deduce from here that the monad on Coxeter spaces induced by $\mathbf{Com}$ is $\overline{\mathcal{S}}$. Thus, Theorem 9.22 is a special case of Theorem 11.40 with $\mathbf{a} := \mathbf{Com}$.

Commutative Coxeter algebras in terms of face-types. The category of Coxeter $\mathbf{Com}$-algebras as formulated in Section 11.5.4 is the same as the category of commutative Coxeter algebras as described in Section 8.8.5 in terms of face-types. The translation is

$$\mathbf{Com}[T, U] \otimes \mathrm{A}[U] \to \mathrm{A}[T] \quad \text{and} \quad \mu_T^U : \mathrm{A}[U] \to \mathrm{A}[T].$$

The naturality triangle in (11.40b) corresponds to the commutativity axiom (8.86). The naturality square in (11.40a) translates to the diagram

$$\begin{array}{ccc} \mathrm{A}[U] & \xrightarrow{\tau} & \mathrm{A}[U'] \\ {\scriptstyle \mu_T^U}\downarrow & & \downarrow{\scriptstyle \mu_{T'}^{U'}} \\ \mathrm{A}[T] & \xrightarrow[\sigma]{} & \mathrm{A}[T'] \end{array} \tag{11.53}$$

for any $(\sigma, \tau) : (T, U) \to (T', U')$. This is the analogue of diagram (2.50) for commutative Coxeter monoids.

Similarly, for a Coxeter space V, using formulas (11.29) and (8.121),

$$(\mathbf{Com} \circ \mathrm{V})[T] = \Big(\bigoplus_{U:\, T \le U} \mathrm{V}[U] \Big)_{\mathrm{quo}} = \bar{\mathcal{S}}(\mathrm{V})[T], \tag{11.54}$$

with quotient taken wrt morphisms local to T.

Exercise 11.44. Do Exercise 11.42 for a Coxeter **Com**-algebra and tie up with the formulations of the commutativity axioms in (8.79) and (8.80).

The binary quadratic structure of **Com** yields the descriptions of commutative Coxeter algebras in Sections 8.8.2, 8.8.4, 8.8.6 in terms of the covering commutativity axiom.

Exercise 11.45. Repeat the above analysis for the signed Coxeter commutative operad $\mathbf{Com}^-$ from Section 5.5.4. In particular, show that the category of Coxeter $\mathbf{Com}^-$-algebras is the same as the category of signed commutative Coxeter algebras. Also see Exercise 11.54.

Exercise 11.46. Check: For $\mathrm{V} := \Bbbk[x]$, the endomorphism Coxeter operad $\mathrm{end}^\circ(\mathrm{V})$ in (11.31) is **Com**. More generally, describe $\mathrm{end}^\circ(\mathrm{V})$ for $\mathrm{V} := \overline{\mathcal{T}}(M)$.

Cocommutative Coxeter coalgebras. Recall the Coxeter commutative cooperad $\mathbf{Com}^*$. It is the dual of **Com**. The category of Coxeter $\mathbf{Com}^*$-coalgebras is the same as the category of cocommutative Coxeter coalgebras. The comonad on Coxeter spaces induced by $\mathbf{Com}^*$ is $\bar{\mathcal{S}}^\vee$.

11.6.3. Coxeter Lie algebras.

Recall the Coxeter Lie operad **Lie** from Section 5.5.3. It is a Coxeter suboperad of **As**.

Coxeter Lie operad in terms of top-lunes. In terms of top-lunes, $\mathbf{Lie}[\mathrm{L}, \mathrm{M}]$ is the subspace of $\mathbf{As}[\mathrm{L}, \mathrm{M}]$ consisting of those

$$\sum_{\substack{\mathrm{P}:\, \mathrm{b(P)=b(L)} \\ \mathrm{c(P)=b(M)}}} x^{\mathrm{P}}\, \mathtt{H}_{\mathrm{P}}$$

such that for any lune Q with $\mathrm{b(L)} < \mathrm{b(Q)}$ and $\mathrm{c(Q)} = \mathrm{b(M)}$,

$$\sum_{\mathrm{P}:\, \mathrm{P} \le \mathrm{Q}} x^{\mathrm{P}} = 0. \tag{11.55}$$

The sum is over those lunes P which appear in the lune decomposition of Q over the flat b(L), see [**10**, Proposition 3.23]. We refer to (11.55) as the *lune criterion* for Lie elements.

Coxeter Lie operad in terms of face-types. In terms of face-types, $\mathbf{Lie}[T, U]$ is the subspace of $\mathbf{As}[T, U]$ consisting of those

$$\sum_{U':\, U' \ge T} \ \sum_{\substack{\alpha:\, U \to U' \\ \text{local to } T}} x^{\alpha}\, \mathtt{H}_{\alpha}$$

such that for any $T < T' \le V$ and $\beta : U \to V$ local to T,

$$\sum_{\alpha:\, \beta\alpha^{-1} \text{projects } U' \text{ on } T'} x^{\alpha} = 0. \tag{11.56}$$

Here U' denotes the codomain of α. We refer to (11.56) as the *lune criterion* for Lie elements (face-type version). An illustration is shown in Figure 11.3 with $T = \emptyset$ and T' a vertex-type.

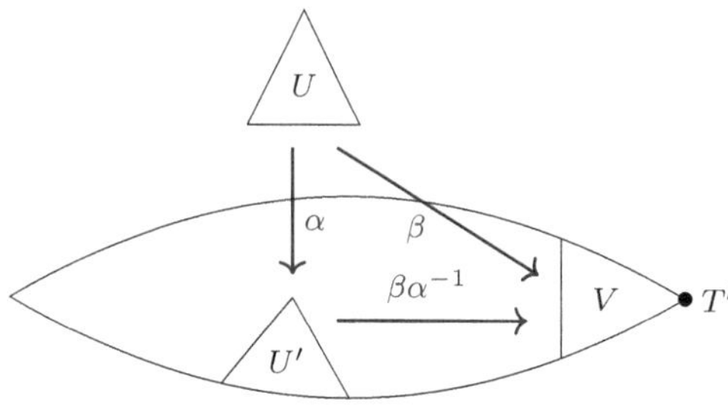

FIGURE 11.3. Lune criterion for Lie elements.

Exercise 11.47. Let $\mathrm{char}(\Bbbk) \nmid |W|$. Show: For $Z \leq T$,

$$\sum_{U:\, U \geq Z} \sum_{\substack{\alpha:\, T \to U \\ \text{local to } Z}} \mathrm{sln}^{\alpha}_{Z}\, \mathtt{H}_{\alpha} \in \mathbf{Lie}[Z, T],$$

where $\mathrm{sln}^{\alpha}_{Z}$ is the Solomon coefficient (1.126). (Use noncommutative Weisner formula (1.85).) A similar result which implies this one was given in [**11**, Lemma 17.24].

Coxeter Lie algebras which arise out of the Coxeter Lie operad are studied in Chapter 12.

11.6.4. Free Coxeter algebras. We saw in Section 10.2.14 that $\Bbbk[x]$ is the free Coxeter algebra as well as the free commutative Coxeter algebra on X. From the viewpoint of Coxeter operads, this can be written as

$$\Bbbk[x] = \mathbf{As} \circ \mathrm{X} = \mathbf{Com} \circ \mathrm{X}.$$

For $\mathbf{p}$ equal to either $\mathbf{As}$ or $\mathbf{Com}$, observe that the components in (11.43) equal $\Bbbk$, as expected.

More generally from Section 10.3.15, for a left W-module M,

$$\overline{\mathcal{T}}(M) = \mathbf{As} \circ \mathrm{X}_M \quad \text{and} \quad \overline{\mathcal{S}}(M) = \mathbf{Com} \circ \mathrm{X}_M.$$

These specialize (11.44).

Dually, the cofree Coxeter coalgebra as well as the cofree cocommutative Coxeter coalgebra on X is

$$\Bbbk\{x\} = \mathbf{As}^* \circ^{\vee} \mathrm{X} = \mathbf{Com}^* \circ^{\vee} \mathrm{X},$$

and more generally, for a left W-module M,

$$\overline{\mathcal{T}}^{\vee}(M) = \mathbf{As}^* \circ^{\vee} \mathrm{X}_M \quad \text{and} \quad \overline{\mathcal{S}}^{\vee}(M) = \mathbf{Com}^* \circ^{\vee} \mathrm{X}_M.$$

11.7. Signature functor on Coxeter dispecies

Recall the signature functor on Coxeter dispecies from Section 5.5.5. In this section, we develop another equivalent approach. We continue to employ the same terminology. Notation though is different, with $(-)_-$ instead of $(-)^-$.

11.7.1. Signature functor. For a Coxeter dispecies $\mathbf{p}$, define its *signed partner* $\mathbf{p}_-$ with (X, Y)-component $\mathbf{p}[\mathrm{X},\mathrm{Y}]$ and structure morphisms

$$\begin{array}{ccc} \mathbf{p}_-[\mathrm{X},\mathrm{Y}] & \xrightarrow{(u,w)} & \mathbf{p}_-[\mathrm{X}',\mathrm{Y}'] \\ \| & & \| \\ \mathbf{p}[\mathrm{X},\mathrm{Y}] & \xrightarrow[(-1)^{l(w)-l(u)}(u,w)]{} & \mathbf{p}[\mathrm{X}',\mathrm{Y}'], \end{array}$$

with (u,w) as in Exercise 11.4. Note very carefully: This definition only involves parities of $l(u)$ and $l(w)$, so these lengths can be taken wrt any reference chamber. We refer to the assignment $\mathbf{p} \mapsto \mathbf{p}_-$ as the *signature functor* on Coxeter dispecies.

A similar definition expresses $\mathbf{p}_-$ in terms of top-lunes. In terms of face-types, $\mathbf{p}_-$ has (T,U)-component $\mathbf{p}[T,U]$ with structure morphisms

$$\begin{array}{ccc} \mathbf{p}_-[T,U] & \xrightarrow{(\sigma,\tau)} & \mathbf{p}_-[T',U'] \\ \| & & \| \\ \mathbf{p}[T,U] & \xrightarrow[(-1)^{l(\tau)-l(\sigma)}(\sigma,\tau)]{} & \mathbf{p}[T',U']. \end{array}$$

Exercise 11.48. Check: For Coxeter dispecies $\mathbf{p}$ and $\mathbf{q}$,

$$\mathbf{p}_- \circ \mathbf{q}_- = (\mathbf{p} \circ \mathbf{q})_- \quad \text{and} \quad \mathbf{x}_- = \mathbf{x}. \tag{11.57}$$

(Use Exercise 11.23.) Thus, the signature functor is a strong monoidal functor wrt substitution. Deduce: If $\mathbf{a}$ is a Coxeter operad, then so is $\mathbf{a}_-$. The substitution and unit maps of $\mathbf{a}_-$ are the same as that of $\mathbf{a}$.

11.7.2. Signed Coxeter commutative operad. Recall from Section 5.5.1 the exponential Coxeter dispecies $\mathbf{E}$. Applying the signature functor yields $\mathbf{E}_-$ whose (X, Y)-component is $\Bbbk$ and structure morphism (u,w) is scalar multiplication by $(-1)^{l(w)-l(u)}$. The description in terms of top-lunes is similar. In terms of face-types, $\mathbf{E}_-$ has (T,U)-component $\Bbbk$ with structure morphism (σ,τ) given by scalar multiplication by $(-1)^{l(\tau)-l(\sigma)}$.

The Coxeter dispecies $\mathbf{E}_-$ carries a Coxeter operad structure which we denote $\mathbf{Com}_-$. We call it the *signed Coxeter commutative operad.* It is obtained by applying the signature functor to $\mathbf{Com}$. The substitution map of $\mathbf{Com}_-$ consists of the identifications $\Bbbk \otimes \Bbbk \cong \Bbbk$ just as for $\mathbf{Com}$.

Exercise 11.49. Recall the Hadamard product of Coxeter dispecies from Section 5.3. Check: For $\mathbf{p}$ a Coxeter dispecies, $\mathbf{p}_- = \mathbf{p} \times \mathbf{E}_-$, and for $\mathbf{a}$ a Coxeter operad, $\mathbf{a}_- = \mathbf{a} \times \mathbf{Com}_-$.

11.7.3. Comparison between signature functors. Recall the Coxeter operad $\mathbf{Com}^-$ and its underlying Coxeter dispecies $\mathbf{E}^-$ from Section 5.5.4.

Lemma 11.50. *The Coxeter dispecies* $\mathbf{E}^-$ *and* $\mathbf{E}_-$ *are isomorphic. Explicitly, the map*

$$\mathbf{E}^-[\mathrm{X},\mathrm{Y}] \to \mathbf{E}_-[\mathrm{X},\mathrm{Y}], \qquad \mathtt{H}_{[F/A]} \mapsto (-1)^{\mathrm{dist}(AC,FC)}$$

on the (X, Y)*-component is an isomorphism of Coxeter dispecies. Moreover, it induces an isomorphism of Coxeter operads from* $\mathbf{Com}^-$ *to* $\mathbf{Com}_-$*.*

Here C is the chamber fixed in the nested-flat formulation of Coxeter dispecies (Section 11.2.1).

Proof. We indicate the required checks. The map is:

- well-defined since for $\mathrm{s}(F/A) = \mathrm{s}(F'/A') = \mathrm{Y}/\mathrm{X}$,

$$\begin{array}{ccc} \mathtt{H}_{[F/A]} & \longmapsto & (-1)^{\mathrm{dist}(AC,FC)} \\ \| & & \| \\ (-1)^{\mathrm{dist}(A'F,F')}\,\mathtt{H}_{[F'/A']} & \longmapsto & (-1)^{\mathrm{dist}(A'F,F')}(-1)^{\mathrm{dist}(A'C,F'C)}, \end{array}$$

- a map of Coxeter dispecies since

$$\begin{array}{ccc} \mathtt{H}_{[F/A]} & \longmapsto & (-1)^{\mathrm{dist}(AC,FC)} \\ {\scriptstyle (u,w)}\downarrow & & \downarrow{\scriptstyle (u,w)} \\ \mathtt{H}_{[wF/wA]} & \longmapsto & (-1)^{\mathrm{dist}((wA)C,(wF)C)} = (-1)^{\mathrm{dist}(AC,FC)}(-1)^{l(w)-l(u)}, \end{array}$$

- a morphism of Coxeter operads since for $A \leq F \leq G$,

$$\begin{array}{ccc} \mathtt{H}_{[F/A]} \otimes \mathtt{H}_{[G/F]} & \longmapsto & (-1)^{\mathrm{dist}(AC,FC)}(-1)^{\mathrm{dist}(FC,GC)} 1 \otimes 1 \\ \downarrow & & \downarrow \\ \mathtt{H}_{[G/A]} & \longmapsto & (-1)^{\mathrm{dist}(AC,GC)}. \end{array}$$

We used familiar properties of the distance function between faces. □

Lemma 11.51. *For a Coxeter dispecies* $\mathbf{p}$*, the Coxeter dispecies* $\mathbf{p}^-$ *and* $\mathbf{p}_-$ *are isomorphic. Similarly, for a Coxeter operad* $\mathbf{a}$*, the Coxeter operads* $\mathbf{a}^-$ *and* $\mathbf{a}_-$ *are isomorphic.*

Proof. Recall $\mathbf{p}^- = \mathbf{p} \times \mathbf{E}^-$ and $\mathbf{p}_- = \mathbf{p} \times \mathbf{E}_-$, and use Lemma 11.50. The same argument applies to $\mathbf{a}^-$ and $\mathbf{a}_-$. □

Exercise 11.52. Check: Coxeter operads $\mathbf{As}$ and $\mathbf{As}_-$ are isomorphic via the map

$$\mathbf{As}[T,U] \to \mathbf{As}_-[T,U], \qquad \mathtt{H}_\alpha \mapsto (-1)^{l(\alpha)}\,\mathtt{H}_\alpha$$

on the (T,U)-component. (Use Exercise 8.23.) This result is equivalent to the one obtained in Lemma 5.5 in view of Lemma 11.51.

We refer to $\mathbf{Lie}_-$ as the *signed Coxeter Lie operad*. This is the same terminology that we used for $\mathbf{Lie}^-$.

11.7.4. Signed Coxeter operad algebras.

Recall the u-signature functor (8.99) on Coxeter spaces.

Exercise 11.53. Check: For $\mathbf{p}$ a Coxeter dispecies and V a Coxeter space,

$$\mathbf{p}_- \circ \mathrm{V}_- = (\mathbf{p} \circ \mathrm{V})_-. \tag{11.58}$$

Deduce: For a Coxeter operad $\mathbf{a}$, if V is a Coxeter $\mathbf{a}$-algebra, then V_- is a Coxeter $\mathbf{a}_-$-algebra. More formally, the u-signature functor (8.99) induces an isomorphism between categories of Coxeter $\mathbf{a}$-algebras and Coxeter $\mathbf{a}_-$-algebras.

We refer to a Coxeter $\mathbf{a}_-$-algebra as a *signed Coxeter* $\mathbf{a}$*-algebra*.

Exercise 11.54. Check: A Coxeter $\mathbf{Com}_-$-algebra is the same as a signed commutative Coxeter algebra. (Identify structure maps

$$\mathbf{Com}_-[T,U] \otimes \mathrm{A}[U] \to \mathrm{A}[T] \quad \text{and} \quad \mu_T^U : \mathrm{A}[U] \to \mathrm{A}[T].$$

The naturality triangle in (11.40b) corresponds to the signed commutativity axiom (8.92).)

11.8. May operads and Coxeter operads

We recall May operads and explain how they give rise to a Coxeter operad for each braid arrangement. This is done via a monoidal functor from Joyal species to Coxeter dispecies which is strong wrt the substitution product. We present two versions, one using flats and the other using face-types.

Joyal species were reviewed in Section 2.12. For background information about the braid arrangement, see Section 1.16.

11.8.1. May operads (First version). A positive Joyal species associates a vector space to each nonempty finite set, natural wrt bijections.

For positive Joyal species p and q, define the positive Joyal species $\mathsf{p} \circ \mathsf{q}$ by

$$(\mathsf{p} \circ \mathsf{q})[I] := \bigoplus_{\mathrm{X} \vdash I} \mathsf{p}[\mathrm{X}] \otimes \Big(\bigotimes_{S \in \mathrm{X}} \mathsf{q}[S] \Big), \tag{11.59}$$

where the direct sum is over all partitions X of the set I. This defines a monoidal structure on the category of positive Joyal species. The unit object is the positive Joyal species x characteristic of singletons, namely,

$$\mathsf{x}[I] := \begin{cases} \Bbbk & \text{if } I \text{ is a singleton,} \\ 0 & \text{otherwise.} \end{cases} \tag{11.60}$$

A *May operad* is a monoid in this monoidal category. In particular, a May operad p entails maps

$$\mathsf{p}[\mathrm{X}] \otimes \Big(\bigotimes_{S \in \mathrm{X}} \mathsf{p}[S] \Big) \to \mathsf{p}[I], \tag{11.61}$$

one for each partition X of I. An illustration with $I = \{a, b, c, d, e\}$ and $\mathrm{X} = \{bd, ae, c\}$ is shown below.

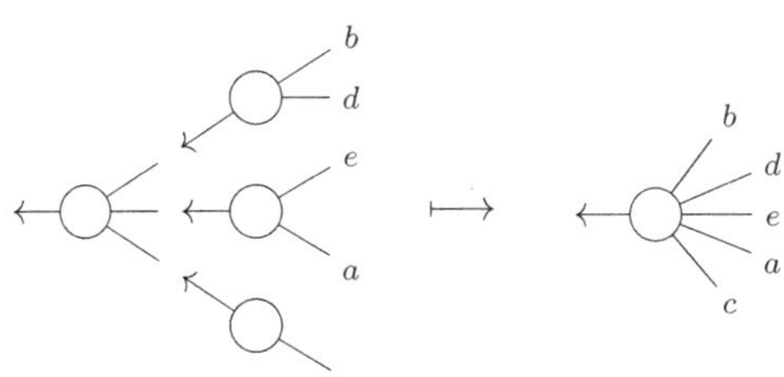

Dually, a *May cooperad* is a comonoid in the above monoidal category.

11.8.2. May operads to Coxeter operads (First version). Fix a finite set I. Recall the braid arrangement $\mathcal{B}^I$ on I. To a positive Joyal species p, we associate a Coxeter $\mathcal{B}^I$-dispecies $\mathbf{p}$ with (X, Y)-component defined by

$$\mathbf{p}[\mathrm{X}, \mathrm{Y}] := \bigotimes_{S \in \mathrm{X}} \mathsf{p}[\mathrm{Y}_S], \tag{11.62}$$

where Y_S is the set whose elements are the blocks of Y which refine the block S of X.

The Coxeter group of $\mathcal{B}^I$ is the group of bijections from I to itself. For a bijection $w : I \to I$, there is an induced linear map

$$w : \mathbf{p}[\mathrm{X}, \mathrm{Y}] \to \mathbf{p}[w\mathrm{X}, w\mathrm{Y}],$$

where wX (resp. wY) denotes the partition whose blocks are obtained by applying w to the blocks of X (resp. Y). Axioms (5.1) clearly hold. The condition w fixes Y pointwise means that w preserves each block of Y. In this case, for $\mathrm{X} \leq \mathrm{Y}$, w also preserves each block of X, and further the induced map $\mathrm{Y}_S \to \mathrm{Y}_S$ is identity for each $S \in \mathrm{X}$. As a result, $w : \mathbf{p}[\mathrm{X}, \mathrm{Y}] \to \mathbf{p}[\mathrm{X}, \mathrm{Y}]$ equals the identity map. Thus, $\mathbf{p}$ is a Coxeter $\mathcal{B}^I$-dispecies.

This construction yields a functor from the category of positive Joyal species to the category of Coxeter $\mathcal{B}^I$-dispecies. Let us temporarily denote it by $\mathcal{F}$. For positive Joyal species p and q, there are natural isomorphisms

$$\mathcal{F}(\mathsf{p}) \circ \mathcal{F}(\mathsf{q}) \xrightarrow{\cong} \mathcal{F}(\mathsf{p} \circ \mathsf{q}) \quad \text{and} \quad \mathbf{x} \xrightarrow{\cong} \mathcal{F}(\mathsf{x}),$$

with products as in (5.4) and (11.59). The second isomorphism is clear. For the first one, we evaluate both sides on say the component (X, Z). In both cases, the sum splits over partitions Y that lie between X and Z, and corresponding summands identify by rearranging tensor factors. We omit the details. Thus, $\mathcal{F}$ is a strong monoidal functor. Since strong functors preserve (co)monoids, $\mathcal{F}$ sends a May (co)operad to a Coxeter $\mathcal{B}^I$-(co)operad.

Remark 11.55. We do not discuss classical associative, commutative, Lie operads here, see for instance [**7**, Appendix B.1.4]. But we mention that the Coxeter operads **As**, **Com**, **Lie** defined in Section 5.5, when specialized to the braid arrangement, arise from their classical counterparts via the above construction.

The internal hom for (11.59) described in [**7**, Proposition B.26] can be compared with the enriched hom construction in Section 5.4.2.

11.8.3. May operads (Second version). There is another formulation of Joyal species which we mention. A positive Joyal species p is a sequence of vector spaces $\mathsf{p}[n]$, for $n \geq 1$, with a S_n-module structure on each $\mathsf{p}[n]$, where S_n is the symmetric group on n letters. In this formulation, (11.59) can be expressed as

$$(\mathsf{p} \circ \mathsf{q})[n] := \bigoplus_{k \geq 1} \big(\mathsf{p}[k] \otimes \mathsf{q}^{\cdot k}[n]\big)_{\mathrm{S}_k}. \tag{11.63}$$

Here $\mathsf{q}^{\cdot k}$ is the positive Joyal species defined by

$$\mathsf{q}^{\cdot k}[n] := \bigoplus_{(m_1, \dots, m_k) \vDash n} \mathrm{Ind}_{\mathrm{S}_{m_1} \times \cdots \times \mathrm{S}_{m_k}}^{\mathrm{S}_n} \mathsf{q}[m_1] \otimes \cdots \otimes \mathsf{q}[m_k].$$

The sum is over all compositions of n with k parts. In each summand, $\mathsf{q}[m_1] \otimes \cdots \otimes \mathsf{q}[m_k]$ is a module over $\mathrm{S}_{m_1} \times \cdots \times \mathrm{S}_{m_k}$ which is induced up to a module over S_n. Note: S_k acts on $\mathsf{q}^{\cdot k}[n]$ by permuting the tensor factors. The coinvariants in (11.63) are taken wrt the diagonal action of S_k.

In this formulation, a May operad p entails maps

$$\mathsf{p}[k] \otimes \big(\mathsf{p}[m_1] \otimes \cdots \otimes \mathsf{p}[m_k]\big) \to \mathsf{p}[n], \tag{11.64}$$

one for each composition $(m_1, \ldots, m_k)$ of n.

11.8.4. Category with nested face-types as objects. Fix a positive integer n. Recall the braid arrangement $\boldsymbol{\beta}^n$ with reference chamber $C_r = 1|\ldots|n$. The category $\boldsymbol{\beta}^n$-dCox is as follows. An object is a pair of compositions (α, β) of n with $\alpha \leq \beta$. A morphism

$$(\sigma, \tau) : (\alpha, \beta) \to (\alpha', \beta')$$

consists of a pair of morphisms $\sigma : \alpha \to \alpha'$ and $\tau : \beta \to \beta'$ in the category $\boldsymbol{\beta}^n$-$\overline{\mathsf{Cox}}$ as illustrated in (8.129) with the following additional condition: if σ maps the i-th part of α to the j-th part of α', then τ must map a part of β which refines the i-th part of α to a part of β' which refines the j-th part of α'. For example,

$$((5,2,3),(1,3,1,2,1,2)) \to ((2,3,5),(2,2,1,3,1,1)).$$

11.8.5. May operads to Coxeter operads (Second version). Let us understand the passage from May operads to Coxeter operads using the second formulation of Joyal species. Here it is more natural to work with Coxeter $\boldsymbol{\beta}^n$-dispecies. Thus, to a positive Joyal species p, we now associate a Coxeter $\boldsymbol{\beta}^n$-dispecies $\mathbf{p}$. The (α, β)-component of $\mathbf{p}$ is defined by

$$\mathbf{p}[\alpha, \beta] := \bigotimes_{i=1}^{k} \mathsf{p}[\deg(\beta/\alpha)_i], \tag{11.65}$$

where k is the number of parts of α, and $\deg(\beta/\alpha)_i$ is the number of parts of β which refine the i-th part of α. For example,

$$\mathbf{p}[(5,2,3),(1,3,1,2,1,2)] = \mathsf{p}[3] \otimes \mathsf{p}[1] \otimes \mathsf{p}[2].$$

Here $(1,3,1)$ refines (5), (2) refines (2), $(1,2)$ refines (3), and they have 3, 1, 2 parts, respectively.

For $(\sigma, \tau) : (\alpha, \beta) \to (\alpha', \beta')$, the linear map $\mathbf{p}[\alpha, \beta] \to \mathbf{p}[\alpha', \beta']$ is defined as follows. The morphism $\sigma : \alpha \to \alpha'$ specifies a bijection between the tensor factors of $\mathbf{p}[\alpha, \beta]$ and $\mathbf{p}[\alpha', \beta']$, while the morphism $\tau : \beta \to \beta'$ maps each tensor factor of $\mathbf{p}[\alpha, \beta]$ to the corresponding tensor factor of $\mathbf{p}[\alpha', \beta']$ by acting by an appropriate element of that symmetric group.

This yields a strong monoidal functor from positive Joyal species to Coxeter $\boldsymbol{\beta}^n$-dispecies. To verify this, we need to compare the products (11.15) and (11.63). This is more intricate than the check in the earlier formulation.

11.9. Notes

The material in this chapter is new. It builds on the material in Chapter 5 where Coxeter operads were introduced. The connection between Coxeter operads and May operads explained in Section 11.8 partly fulfils the promise made in the last paragraph on [**11**, page 204]. We first formulated Coxeter operad algebras in the year 2011.

11.9.1. May operads. The concept of a May operad matured under different names in work of several authors such as Boardman and Vogt [**51**], Lazard [**223**], [**224**], Lambek [**218**], Stasheff [**339**]. Its first formal definition is due to May [**254**, Definition 1.1]. An early reference is the paper by Kelly [**210**, Section 4] from 1971. Book references are those by Smirnov [**327**], Markl, Shnider, Stasheff [**252**], Leinster [**225**], Loday and Vallette [**234**], Bremner and Dotsenko [**62**], Fresse [**133**], [**134**], [**135**]. Mac Lane [**238**, Section VI.4] discusses how an associative binary operation is equivalent to a family of higher operations with suitable compatibilities.

When a May operad is defined using (11.64), one needs to impose two S_n-equivariance conditions which correspond to the two diagrams in (11.24). In [**252**, Definition 1.4], the first diagram is given but not the second which is a mistake. In [**234**, Proposition 5.3.1], the second condition is given but the first condition is not properly formulated. Both conditions are correctly stated in Smirnov's paper [**326**, Section 1, page 577] and in his book [**327**, Section 1.2]. Alternatively, one can combine the two conditions into one as in (11.23). This is done by Leinster [**225**, (2.7) and Fig. 2-H].

11.9.2. Substitution operation. Coxeter spaces generalize graded vector spaces via the construction of Section 8.14. The substitution operation $\mathbf{p} \circ \mathrm{V}$ defined in (11.29) corresponds to the well-known formula for the free May operad algebra on a vector space V, namely,

$$\mathbf{p} \circ V = \bigoplus_{n \geq 1} \left(\mathbf{p}[n] \otimes V^{\otimes n}\right)_{\mathrm{S}_n}, \tag{11.66}$$

see [**153**, Formula (1.3.5)], [**327**, Section 5.2, page 96], [**252**, Definition 1.24], [**234**, Section 5.1.2]. Property (11.26) corresponds to [**234**, Proposition 5.1.3].

The substitution operation $\mathbf{p} \circ^{\vee} \mathrm{V}$ defined in (11.34) corresponds to the formula

$$\mathbf{p} \circ^{\vee} V = \bigoplus_{n \geq 1} \left(\mathbf{p}[n] \otimes V^{\otimes n}\right)^{\mathrm{S}_n}. \tag{11.67}$$

This is considered by Fresse [**131**, Section 1.1.1], [**132**, Sections 1.2.12]. Another alternative is to use direct product instead of direct sum in the above formula, see [**7**, Formula (B.17)] in this regard.

11.9.3. Future work. We plan to discuss Koszul theory of Coxeter operads. For **As**, **Com**, **Lie**, this generalizes the Hochschild, Harrison, Chevalley–Eilenberg chain complexes to (reflection) arrangements. Another important example is the Coxeter Poisson operad. Its modules in Coxeter species and in Coxeter spaces yield, respectively, notions of Coxeter Poisson monoids and Coxeter Poisson algebras.

CHAPTER 12

Coxeter Lie algebras

Coxeter Lie algebras. We begin the study of Coxeter Lie algebras. They can be seen in analogy with Coxeter Lie monoids introduced in Chapter 6, and similar general remarks apply. A Coxeter Lie algebra is defined as a Coxeter algebra over the Coxeter Lie operad. We formulate it in terms of top-nested faces, top-lunes, face-types. Using the presentation of the Coxeter Lie operad, we equivalently define it in terms of a Lie bracket subject to antisymmetry and Jacobi identity. This parallels the classical definition of a Lie algebra. We also introduce the dual notion of a Coxeter Lie coalgebra. These notions also have signed analogues.

The toy example of rank one is developed in Section 13.8.

Primitive and indecomposable parts. Coxeter Lie algebras interact fruitfully with Coxeter bialgebras introduced in Chapter 8. The primitive part $\overline{\mathcal{P}}(\mathrm{H})$ of a Coxeter bialgebra H is a Coxeter Lie algebra via the commutator bracket. Dually, the indecomposable part $\overline{\mathcal{Q}}(\mathrm{H})$ of a Coxeter bialgebra H is a Coxeter Lie coalgebra via the cocommutator cobracket. This yields functors from Coxeter bialgebras to Coxeter Lie algebras and to Coxeter Lie coalgebras, respectively.

(Co)free Coxeter Lie (co)algebras. Recall from Chapters 9 and 11 that the free Coxeter algebra on a Coxeter space V is given by $\mathbf{As} \circ \mathrm{V} = \overline{\mathcal{T}}(\mathrm{V})$ and the free commutative Coxeter algebra by $\mathbf{Com} \circ \mathrm{V} = \overline{\mathcal{S}}(\mathrm{V})$. Interestingly, the free Coxeter Lie algebra on V is given by $\mathbf{Lie} \circ \mathrm{V} = \overline{\mathcal{P}}\overline{\mathcal{T}}(\mathrm{V})$ which brings into play the deshuffle coproduct of $\overline{\mathcal{T}}(\mathrm{V})$. Dually, the cofree Coxeter Lie coalgebra on V is given by $\mathbf{Lie}^* \circ^\vee \mathrm{V} = \overline{\mathcal{Q}}\overline{\mathcal{T}}^\vee(\mathrm{V})$ involving the shuffle product of $\overline{\mathcal{T}}^\vee(\mathrm{V})$.

Universal (co)enveloping Coxeter (co)algebras. To a Coxeter Lie algebra G, we associate the universal enveloping Coxeter algebra $\overline{\mathcal{U}}(\mathrm{G})$. It carries a cocommutative Coxeter bialgebra structure. This yields a functor from Coxeter Lie algebras to Coxeter bialgebras which is left adjoint to the primitive part functor.

Dually, to a Coxeter Lie coalgebra K, we associate the universal coenveloping Coxeter coalgebra $\overline{\mathcal{U}}^\vee(\mathrm{K})$. This yields a functor from Coxeter Lie coalgebras to Coxeter bialgebras which is right adjoint to the indecomposable part functor.

Graded Lie algebras. Connections to classical Lie theory. The notion of a Coxeter Lie algebra, when specialized to braid arrangements, relates to the classical notion of a graded Lie algebra. More precisely, the functor from positive-graded vector spaces to Coxeter spaces (introduced in Chapter 8) induces a functor from positive-graded Lie algebras to Coxeter Lie algebras.

12.1. Coxeter Lie algebras and Coxeter Lie coalgebras

We introduce Coxeter Lie algebras. We give two definitions, one in terms of the Coxeter Lie operad, and the other in terms of a Lie bracket subject to antisymmetry and Jacobi identity. Sometimes we may use the terms Coxeter $\mathcal{A}$-Lie algebra or Coxeter $\boldsymbol{\alpha}$-Lie algebra for emphasis or clarity. We also discuss the dual notion of Coxeter Lie coalgebras. These notions have signed analogues linked to the signed Coxeter Lie operad.

The discussion here parallels that of Coxeter Lie monoids and Coxeter Lie comonoids in Section 6.1. For connections between the two using Fock functors, see Section 15.7.

12.1.1. Coxeter Lie algebras as Coxeter operad algebras. Recall the Coxeter Lie operad **Lie** from Sections 5.5.3 and 11.6.3. Also recall Coxeter operad algebras from Section 11.5. A *Coxeter Lie algebra* is by definition a Coxeter **Lie**-algebra. Similarly, a morphism of Coxeter Lie algebras is a morphism of Coxeter **Lie**-algebras. This is made more explicit below. We denote the category of Coxeter Lie algebras by $\mathsf{LieMon}(\mathcal{A}\text{-}\overline{\mathsf{C}}\mathsf{Sp})$ or $\mathsf{LieMon}(\boldsymbol{\alpha}\text{-}\overline{\mathsf{C}}\mathsf{Sp})$ depending on whether we work with top-nested faces/top-lunes or face-types. We will usually drop $\mathcal{A}$ or $\boldsymbol{\alpha}$ from the notation for convenience.

Top-nested faces. A Coxeter $\mathcal{A}$-Lie algebra is a Coxeter $\mathcal{A}$-space G equipped with linear maps

$$\gamma_{A,C}^{F,C} : \mathbf{Lie}[\mathrm{s}(A), \mathrm{s}(F)] \otimes \mathrm{G}[F, C] \to \mathrm{G}[A, C],$$

one for each $A \leq F \leq C$, subject to axioms (11.38a)–(11.38d) with $\mathbf{a} := \mathbf{Lie}$ and $\mathrm{V} := \mathrm{G}$.

A morphism $f : \mathrm{G} \to \mathrm{H}$ of Coxeter $\mathcal{A}$-Lie algebras is a map of Coxeter $\mathcal{A}$-spaces such that diagram (11.39) commutes with $\mathbf{a} := \mathbf{Lie}$, $\mathrm{V} := \mathrm{G}$, $\mathrm{W} := \mathrm{H}$.

Top-lunes. A Coxeter $\mathcal{A}$-Lie algebra is a Coxeter $\mathcal{A}$-space G equipped with linear maps

$$\gamma_{\mathrm{L}}^{\mathrm{M}} : \mathbf{Lie}[\mathrm{L}, \mathrm{M}] \otimes \mathrm{G}[\mathrm{M}] \to \mathrm{G}[\mathrm{L}],$$

one for each $\mathrm{L} \leq \mathrm{M}$, subject to axioms (11.35a)–(11.35d) with $\mathbf{a} := \mathbf{Lie}$ and $\mathrm{V} := \mathrm{G}$.

A morphism $f : \mathrm{G} \to \mathrm{H}$ of Coxeter $\mathcal{A}$-Lie algebras is a map of Coxeter $\mathcal{A}$-spaces such that diagram (11.36) commutes with $\mathbf{a} := \mathbf{Lie}$, $\mathrm{V} := \mathrm{G}$, $\mathrm{W} := \mathrm{H}$.

Face-types. A Coxeter $\boldsymbol{\alpha}$-Lie algebra is a Coxeter $\boldsymbol{\alpha}$-space G equipped with linear maps

$$\gamma_{T}^{U} : \mathbf{Lie}[T, U] \otimes \mathrm{G}[U] \to \mathrm{G}[T],$$

one for each $T \leq U$, subject to axioms (11.40a)–(11.40d) with $\mathbf{a} := \mathbf{Lie}$ and $\mathrm{V} := \mathrm{G}$.

A morphism $f : \mathrm{G} \to \mathrm{H}$ of Coxeter $\boldsymbol{\alpha}$-Lie algebras is a map of Coxeter $\boldsymbol{\alpha}$-spaces such that diagram (11.41) commutes with $\mathbf{a} := \mathbf{Lie}$, $\mathrm{V} := \mathrm{G}$, $\mathrm{W} := \mathrm{H}$.

12.1.2. Coxeter Lie algebras by generators and relations. We now come to the second definition. It makes use of the presentation of the Coxeter Lie operad. A Coxeter $\mathcal{A}$-Lie algebra is a Coxeter $\mathcal{A}$-space G equipped with linear maps

$$\nu_{A,C}^{F,C} : \mathrm{G}[F, C] \to \mathrm{G}[A, C],$$

one for each $A \lessdot F \leq C$, subject to the axioms below.

Naturality. For each morphism $\sigma : (A, C) \to (B, D)$ and $A \lessdot F \leq C$, the diagram

$$\begin{array}{ccc} \mathrm{G}[F,C] & \xrightarrow{\ \sigma\ } & \mathrm{G}[G,D] \\ {\scriptstyle \nu_{A,C}^{F,C}}\downarrow & & \downarrow{\scriptstyle \nu_{B,D}^{G,D}} \\ \mathrm{G}[A,C] & \xrightarrow[\ \sigma\ ]{} & \mathrm{G}[B,D] \end{array} \tag{12.1a}$$

commutes, with $G = \sigma_*(F)$ and $\sigma : (F, C) \to (G, D)$ the induced morphism as in (8.11).

Antisymmetry. For each $A \lessdot F \leq C$,

$$(\mathrm{G}[F,C] \xrightarrow{\nu_{A,C}^{F,C}} \mathrm{G}[A,C]) \ + \ (\mathrm{G}[F,C] \xrightarrow{\beta_{A\overline{F},A\overline{F}C,F,C}} \mathrm{G}[A\overline{F}, A\overline{F}C] \xrightarrow{w(A\overline{F}C)=C} \mathrm{G}[G,C] \xrightarrow{\nu_{A,C}^{G,C}} \mathrm{G}[A,C]) \ = 0. \tag{12.1b}$$

Jacobi identity. For each $A \lessdot F \lessdot G \leq C$,

$$\sum_{i=1}^{n} (\mathrm{G}[G,C] \xrightarrow{\beta_{G_i,G_iC,G,C}} \mathrm{G}[G_i, G_iC] \xrightarrow{\nu_{F_i,G_iC}^{G_i,G_iC}} \mathrm{G}[F_i, G_iC] \xrightarrow{\nu_{A,G_iC}^{F_i,G_iC}} \mathrm{G}[A, G_iC] \xrightarrow{w(G_iC)=C} \mathrm{G}[A,C]) \ = 0. \tag{12.1c}$$

Here $F_1 := F$, $G_1 := G$, and the remaining F_i and G_i are defined as in (6.1c). See the illustration below with $A = O$.

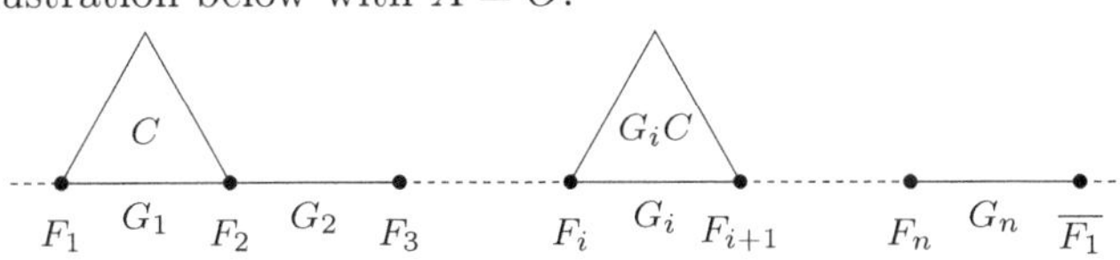

A morphism $f : \mathrm{G} \to \mathrm{H}$ of Coxeter $\mathcal{A}$-Lie algebras is a map of Coxeter $\mathcal{A}$-spaces such that for each $A \lessdot F \leq C$, the diagram

$$\begin{array}{ccc} \mathrm{G}[F,C] & \xrightarrow{f_{F,C}} & \mathrm{H}[F,C] \\ {\scriptstyle \nu_{A,C}^{F,C}}\downarrow & & \downarrow{\scriptstyle \nu_{A,C}^{F,C}} \\ \mathrm{G}[A,C] & \xrightarrow[f_{A,C}]{} & \mathrm{H}[A,C] \end{array} \tag{12.2}$$

commutes.

Exercise 12.1. Formulate the above definition of a Coxeter Lie algebra in terms of top-lunes.

A Coxeter $\boldsymbol{\alpha}$-Lie algebra is a Coxeter $\boldsymbol{\alpha}$-space G equipped with linear maps

$$\nu_Z^T : \mathrm{G}[T] \to \mathrm{G}[Z],$$

one for each $Z \lessdot T$, subject to the axioms below.

Naturality. For each morphism $\sigma : Z \to Z'$ and $Z \lessdot T$, the diagram

$$\begin{array}{ccc} \mathrm{G}[T] & \xrightarrow{\ \sigma\ } & \mathrm{G}[T'] \\ {\scriptstyle \nu_Z^T}\downarrow & & \downarrow{\scriptstyle \nu_{Z'}^{T'}} \\ \mathrm{G}[Z] & \xrightarrow[\sigma]{} & \mathrm{G}[Z'] \end{array} \tag{12.3a}$$

commutes, with $T' = \sigma_*(T)$ and $\sigma : T \to T'$ the induced morphism as in (8.12) or (1.28).

Antisymmetry. For each $Z \lessdot T$,

$$(\mathrm{G}[T] \xrightarrow{\nu_Z^T} \mathrm{G}[Z]) \; + \; (\mathrm{G}[T] \xrightarrow{\beta_{Z\overline{T},T}} \mathrm{G}[Z\overline{T}] \xrightarrow{\nu_Z^{Z\overline{T}}} \mathrm{G}[Z]) \; = 0. \tag{12.3b}$$

For illustrations of $\beta_{Z\overline{T},T}$, see the first and third pictures in Figure 1.17.

Jacobi identity. For each $Z \lessdot T \lessdot U$,

$$\sum_{i=1}^{n} (\mathrm{G}[U] \xrightarrow{\sigma_i} \mathrm{G}[U_i] \xrightarrow{\nu_{T_i}^{U_i}} \mathrm{G}[T_i] \xrightarrow{\nu_Z^{T_i}} \mathrm{G}[Z]) \; = 0. \tag{12.3c}$$

Let F_1 and G_1 be faces of C_r of types T and U, respectively. This then defines faces F_i and G_i as in (12.1c). Let T_i denote the type of F_i, and U_i the type of G_i. The linear map $\sigma_i : \mathrm{G}[U] \to \mathrm{G}[U_i]$ is the one transferred by G_i or represented by the biface (G_i, G_1).

A morphism $f : \mathrm{G} \to \mathrm{H}$ of Coxeter $\boldsymbol{\alpha}$-Lie algebras is a map of Coxeter $\boldsymbol{\alpha}$-spaces such that for each $Z \lessdot T$, the diagram

$$\begin{array}{ccc} \mathrm{G}[T] & \xrightarrow{\ f_T\ } & \mathrm{H}[T] \\ {\scriptstyle \nu_Z^T}\downarrow & & \downarrow{\scriptstyle \nu_Z^T} \\ \mathrm{G}[Z] & \xrightarrow[f_Z]{} & \mathrm{H}[Z] \end{array} \tag{12.4}$$

commutes.

For a Coxeter Lie algebra (G, ν), we refer to ν as the *Lie bracket*, and to the maps $\nu_{A,C}^{F,C}$ and ν_Z^T as the Lie bracket components.

12.1.3. Coxeter Lie coalgebras. Recall the Coxeter Lie cooperad $\mathbf{Lie}^*$. It is the dual of $\mathbf{Lie}$. A *Coxeter Lie coalgebra* is by definition a Coxeter $\mathbf{Lie}^*$-coalgebra. Similarly, a morphism of Coxeter Lie coalgebras is a morphism of Coxeter $\mathbf{Lie}^*$-coalgebras. We denote the category of Coxeter Lie coalgebras by $\mathsf{LieComon}(\overline{\mathsf{C}}\mathsf{Sp})$.

Explicitly: A Coxeter $\boldsymbol{\alpha}$-Lie coalgebra is a Coxeter $\boldsymbol{\alpha}$-space K equipped with linear maps

$$\delta_T^U : \mathrm{K}[T] \to \mathbf{Lie}[T, U]^* \otimes \mathrm{K}[U],$$

one for each $T \leq U$, subject to the Coxeter cooperad coalgebra axioms. In terms of generators and relations, a Coxeter $\boldsymbol{\alpha}$-Lie coalgebra is a Coxeter $\boldsymbol{\alpha}$-space K equipped with linear maps

$$\theta_Z^T : \mathrm{K}[Z] \to \mathrm{K}[T],$$

one for each $Z \lessdot T$, subject to the axioms below.

Naturality. For each morphism $\sigma : Z \to Z'$ and $Z \lessdot T$, the diagram

$$\text{(12.5a)} \qquad \begin{array}{ccc} \mathrm{K}[T] & \xrightarrow{\ \sigma\ } & \mathrm{K}[T'] \\ {\scriptstyle \theta_Z^T}\uparrow & & \uparrow{\scriptstyle \theta_{Z'}^{T'}} \\ \mathrm{K}[Z] & \xrightarrow[\sigma]{} & \mathrm{K}[Z'] \end{array}$$

commutes, with $T' = \sigma_*(T)$ and $\sigma : T \to T'$ the induced morphism.

Antisymmetry. For each $Z \lessdot T$,

$$\text{(12.5b)} \qquad (\mathrm{K}[Z] \xrightarrow{\theta_Z^T} \mathrm{K}[T]) \ + \ (\mathrm{K}[Z] \xrightarrow{\theta_Z^{Z\overline{T}}} \mathrm{K}[Z\overline{T}] \xrightarrow{\beta_{T,Z\overline{T}}} \mathrm{K}[T]) \ = 0.$$

Jacobi identity. For each $Z \lessdot T \lessdot U$,

$$\text{(12.5c)} \qquad \sum_{i=1}^{n} (\mathrm{K}[Z] \xrightarrow{\theta_Z^{T_i}} \mathrm{K}[T_i] \xrightarrow{\theta_{T_i}^{U_i}} \mathrm{K}[U_i] \xrightarrow{\sigma_i} \mathrm{K}[U]) \ = 0,$$

with notation similar to (12.3c).

A morphism $f : \mathrm{K} \to \mathrm{H}$ of Coxeter $\boldsymbol{\alpha}$-Lie coalgebras is a map of Coxeter $\boldsymbol{\alpha}$-spaces such that for each $Z \lessdot T$, the diagram

$$\text{(12.6)} \qquad \begin{array}{ccc} \mathrm{K}[T] & \xrightarrow{\ f_T\ } & \mathrm{H}[T] \\ {\scriptstyle \theta_Z^T}\uparrow & & \uparrow{\scriptstyle \theta_Z^T} \\ \mathrm{K}[Z] & \xrightarrow[f_Z]{} & \mathrm{H}[Z] \end{array}$$

commutes.

The above may also be formulated using top-nested faces or top-lunes instead of face-types. For a Coxeter Lie coalgebra (G, θ), we refer to θ as the *Lie cobracket*.

12.1.4. Signed analogue. The preceding notions have signed analogues where we pay attention to parity of the lengths of the morphisms involved.

Recall: There are two equivalent definitions of a signed Coxeter Lie operad, namely, $\mathbf{Lie}^-$ and $\mathbf{Lie}_-$. A *signed Coxeter Lie algebra* is by definition a Coxeter $\mathbf{Lie}^-$-algebra. A morphism of signed Coxeter Lie algebras is a morphism of Coxeter $\mathbf{Lie}^-$-algebras. We denote the category of signed Coxeter Lie algebras by (-1)-$\mathsf{LieMon}(\overline{\mathsf{C}}\mathsf{Sp})$.

In terms of generators and relations, a signed Coxeter Lie algebra is a Coxeter $\mathcal{A}$-space equipped with a Lie bracket ν subject to axioms (12.1a), (12.1b), (12.1c), with appropriate sign changes in the latter two. In (12.1b), we multiply the second term by $(-1)^{\mathrm{dist}(C, A\overline{F}C)}$, and in (12.1c), we multiply each summand by $(-1)^{\mathrm{dist}(C,G_iC)}$. Equivalently, a signed Coxeter Lie algebra is a Coxeter $\boldsymbol{\alpha}$-space equipped with a Lie bracket ν subject to the axioms below.

Naturality. For each morphism $\sigma : Z \to Z'$ and $Z \lessdot T$, the diagram

$$\begin{array}{ccc} \mathrm{G}[T] & \xrightarrow{\sigma} & \mathrm{G}[T'] \\ {\scriptstyle \nu_Z^T}\big\downarrow & & \big\downarrow{\scriptstyle \nu_{Z'}^{T'}} \\ \mathrm{G}[Z] & \xrightarrow[\sigma]{} & \mathrm{G}[Z'] \end{array} \tag{12.7a}$$

commutes, with $T' = \sigma_*(T)$ and $\sigma : T \to T'$ the induced morphism.

Signed antisymmetry. For each $Z \lessdot T$,

$$(\mathrm{G}[T] \xrightarrow{\nu_Z^T} \mathrm{G}[Z]) + (-1)^{\mathrm{dist}(T,Z\overline{T})}(\mathrm{G}[T] \xrightarrow{\beta_{Z\overline{T},T}} \mathrm{G}[Z\overline{T}] \xrightarrow{\nu_Z^{Z\overline{T}}} \mathrm{G}[Z]) = 0. \tag{12.7b}$$

Note: $\mathrm{dist}(T, Z\overline{T})$ is the length of the morphism $\beta_{Z\overline{T},T}$.

Signed Jacobi identity. For each $Z \lessdot T \lessdot U$,

$$\sum_{i=1}^{n} (-1)^{l(\sigma_i)}(\mathrm{G}[U] \xrightarrow{\sigma_i} \mathrm{G}[U_i] \xrightarrow{\nu_{T_i}^{U_i}} \mathrm{G}[T_i] \xrightarrow{\nu_Z^{T_i}} \mathrm{G}[Z]) = 0, \tag{12.7c}$$

with notation as in (12.3c).

Similarly, consider the signed Coxeter Lie cooperad $(\mathbf{Lie}^-)^*$. A *signed Coxeter Lie coalgebra* is by definition a Coxeter $(\mathbf{Lie}^-)^*$-coalgebra. A morphism of signed Coxeter Lie coalgebras is a morphism of Coxeter $(\mathbf{Lie}^-)^*$-coalgebras. We denote the category of signed Coxeter Lie coalgebras by (-1)-$\mathsf{LieComon}(\bar{\mathsf{C}}\mathsf{Sp})$. In terms of generators and relations, a signed Coxeter Lie coalgebra can be defined using a Lie cobracket, with signed antisymmetry and signed Jacobi identity dual to (12.7b) and (12.7c).

12.1.5. Signature functor. Recall the u-signature functor (8.99) on Coxeter spaces. If G is a Coxeter Lie algebra, then G_- is a signed Coxeter Lie algebra (with the same Lie bracket components ν_Z^T), and vice versa. Thus, the categories of Coxeter Lie algebras and signed Coxeter Lie algebras are isomorphic. (A more general result is given in Exercise 11.53.) Dual remarks apply to Coxeter Lie coalgebras.

12.1.6. Abelian Coxeter Lie algebras and Coxeter Lie coalgebras. We say a Coxeter Lie algebra is *abelian* if the structure maps $\gamma_{A,C}^{F,C}$ and γ_T^U are zero for all $A < F \leq C$ and $T < U$. Equivalently, a Coxeter Lie algebra is abelian if the Lie brackets $\nu_{A,C}^{F,C}$ and ν_Z^T are zero for all $A \lessdot F \leq C$ and $Z \lessdot T$. Observe: Every Coxeter space is an abelian Coxeter Lie algebra in a unique way. Thus, the category of Coxeter spaces is isomorphic to the full subcategory of abelian Coxeter Lie algebras.

Dually, a Coxeter Lie coalgebra is *abelian* if the structure maps δ_T^U are zero for all $T < U$, or equivalently, the Lie cobrackets θ_Z^T are zero for all $Z \lessdot T$. Again every Coxeter space is an abelian Coxeter Lie coalgebra in a unique way. Thus, the category of Coxeter spaces is also isomorphic to the full subcategory of abelian Coxeter Lie coalgebras.

The same definitions and remarks apply to the signed setting.

12.2. (Co)commutator (co)bracket

Every Coxeter algebra carries a Coxeter Lie algebra structure via the commutator bracket. This arises by viewing the Coxeter Lie operad inside the Coxeter associative operad. Dually, every Coxeter coalgebra carries a Coxeter Lie coalgebra structure via the cocommutator cobracket. Both constructions have signed analogues.

The parallel discussion for Coxeter monoids is in Section 6.2.

12.2.1. Commutator bracket. Since **Lie** is a Coxeter suboperad of **As**, there is an induced functor

$$\mathsf{Mon}(\overline{\mathsf{C}}\mathsf{Sp}) \to \mathsf{LieMon}(\overline{\mathsf{C}}\mathsf{Sp}). \tag{12.8}$$

We call this the *commutator functor*. Thus, every Coxeter algebra (A, μ) carries a Coxeter Lie algebra structure via the composite map

$$\mathbf{Lie} \circ \mathrm{A} \to \mathbf{As} \circ \mathrm{A} \to \mathrm{A}.$$

Explicitly, using (11.50),

$$\begin{aligned} \gamma_T^U : \mathbf{Lie}[T, U] \otimes \mathrm{A}[U] &\longrightarrow \mathrm{A}[T] \\ \sum_{\alpha} x^{\alpha}\, \mathtt{H}_{\alpha} \otimes v &\longmapsto \sum_{\alpha} x^{\alpha} \mu_T^{U'} \alpha(v), \end{aligned} \tag{12.9}$$

with the sums over $\alpha : U \to U'$ local to T, as U' varies. Equivalently, one can work with a Coxeter algebra A as formulated in Proposition 8.44 which only involves product components μ_Z^T for $Z \lessdot T$. In this situation, the induced Lie bracket component ν_Z^T of A is

$$\nu_Z^T := \mu_Z^T - \mu_Z^{Z\overline{T}} \beta_{Z\overline{T},T}. \tag{12.10}$$

We refer to ν as the *commutator bracket.*

If the Coxeter algebra A is commutative, then $\nu_Z^T = 0$ for all $Z \lessdot T$, and the underlying Lie structure is abelian.

Exercise 12.2. Check directly: The commutator bracket (12.10) satisfies antisymmetry (12.3b) and Jacobi identity (12.3c).

12.2.2. Signed commutator bracket. Since **Lie** is a Coxeter suboperad of **As**, the signed Coxeter Lie operad $\mathbf{Lie}^-$ is a Coxeter suboperad of $\mathbf{As}^-$. The latter is isomorphic to **As** by Lemma 5.5. Hence, there is an induced functor

$$\mathsf{Mon}(\overline{\mathsf{C}}\mathsf{Sp}) \to (-1)\text{-}\mathsf{LieMon}(\overline{\mathsf{C}}\mathsf{Sp}). \tag{12.11}$$

We call this the *signed commutator functor*. Equivalently, one can view $\mathbf{Lie}_-$ as a Coxeter suboperad of $\mathbf{As}_-$ and use Exercise 11.52. Thus, every Coxeter algebra (A, μ) carries a signed Coxeter Lie algebra structure via

$$\begin{aligned} \gamma_T^U : \mathbf{Lie}_-[T, U] \otimes \mathrm{A}[U] &\longrightarrow \mathrm{A}[T] \\ \sum_\alpha x^\alpha \, \mathtt{H}_\alpha \otimes v &\longmapsto \sum_\alpha (-1)^{l(\alpha)} x^\alpha \mu_T^{U'} \alpha(v). \end{aligned} \tag{12.12}$$

Equivalently, for $Z \lessdot T$, the Lie bracket component ν_Z^T of A is

$$\nu_Z^T := \mu_Z^T - (-1)^{\mathrm{dist}(T, Z\overline{T})} \mu_Z^{Z\overline{T}} \beta_{Z\overline{T}, T}. \tag{12.13}$$

We refer to ν as the *signed commutator bracket*. Compare and contrast with (12.10).

12.2.3. Cocommutator cobracket. Dually, the surjective morphism of Coxeter cooperads $\mathbf{As}^* \twoheadrightarrow \mathbf{Lie}^*$ induces a functor

$$\mathsf{Comon}(\overline{\mathsf{C}}\mathsf{Sp}) \to \mathsf{LieComon}(\overline{\mathsf{C}}\mathsf{Sp}). \tag{12.14}$$

We call this the *cocommutator functor*. Thus, every Coxeter coalgebra C carries a Coxeter Lie coalgebra structure via the composite map

$$\mathrm{C} \to \mathbf{As}^* \circ^\vee \mathrm{C} \to \mathbf{Lie}^* \circ^\vee \mathrm{C}.$$

For C formulated as in Proposition 8.46 using coproduct components Δ_Z^T for $Z \lessdot T$, the Lie cobracket component θ_Z^T is

$$\theta_Z^T := \Delta_Z^T - \beta_{T, Z\overline{T}} \Delta_Z^{Z\overline{T}}. \tag{12.15}$$

We refer to θ as the *cocommutator cobracket*. This is the dual of (12.10).

12.2.4. Signed cocommutator cobracket. Similarly, we have a functor

$$\mathsf{Comon}(\overline{\mathsf{C}}\mathsf{Sp}) \to (-1)\text{-}\mathsf{LieComon}(\overline{\mathsf{C}}\mathsf{Sp}). \tag{12.16}$$

We call this the *signed cocommutator functor*. Every Coxeter coalgebra C carries a signed Coxeter Lie coalgebra structure under the *signed cocommutator cobracket*

$$\theta_Z^T := \Delta_Z^T - (-1)^{\mathrm{dist}(T, Z\overline{T})} \beta_{T, Z\overline{T}} \Delta_Z^{Z\overline{T}}. \tag{12.17}$$

This is the dual of (12.13).

12.3. Primitive and indecomposable parts of Coxeter bialgebras

For any Coxeter bialgebra, the commutator bracket restricts to the primitive part of the Coxeter bialgebra. This yields a functor from the category of Coxeter bialgebras to the category of Coxeter Lie algebras which we continue to call the primitive part functor. Dually, the cocommutator cobracket descends to the indecomposable part of the Coxeter bialgebra. This yields a functor from the category of Coxeter bialgebras to the category of Coxeter Lie coalgebras which we continue to call the indecomposable part functor. These constructions have signed analogues.

The parallel discussion for Coxeter bimonoids is in Section 6.3.

12.3.1. Primitive part of a Coxeter bialgebra. A Coxeter bialgebra H is, in particular, a Coxeter algebra, and hence a Coxeter Lie algebra via the commutator functor (12.8), or equivalently, the commutator bracket (12.10). We now prove that this Lie structure restricts to the primitive part $\overline{\mathcal{P}}(\mathrm{H})$.

Proposition 12.3. *If* H *is a Coxeter bialgebra, then* $\overline{\mathcal{P}}(\mathrm{H})$ *is a Coxeter Lie subalgebra of* H.

We have given two definitions of a Coxeter Lie algebra. We explain below how the proof works with either definition.

FIRST PROOF. For any $T \leq U$, we need to show that the dotted arrow in the diagram

$$\begin{array}{ccc} \mathbf{Lie}[T,U] \otimes \overline{\mathcal{P}}(\mathrm{H})[U] & \hookrightarrow & \mathbf{As}[T,U] \otimes \mathrm{H}[U] \\ \vdots & & \downarrow \\ \overline{\mathcal{P}}(\mathrm{H})[T] & \hookrightarrow & \mathrm{H}[T] \end{array}$$

exists. We continue the notation in (12.9). Let $v \in \overline{\mathcal{P}}(\mathrm{H})[U]$. For $T' > T$,

$$\begin{aligned} \Delta_T^{T'}\Big(\sum_\alpha x^\alpha \mu_T^{U'} \alpha(v)\Big) &= \sum_\alpha x^\alpha \Delta_T^{T'} \mu_T^{U'} \alpha(v) \\ &= \sum_{\substack{\beta\colon U \to V \\ \text{local to } T}} \Big(\sum_{\alpha\colon \beta\alpha^{-1} \text{projects } U' \text{ on } T'} x^\alpha\Big) \mu_{T'}^{V} \beta(v) \\ &= 0, \end{aligned}$$

as required. The second step used (9.149) with $q = 1$ which amounts to using the bialgebra axiom (8.64) and the primitivity hypothesis. The sum in parenthesis in zero by the lune criterion (11.56). See the picture in Figure 11.3. It is implicit that the V occurring above varies over face-types greater than T' such that $\mathrm{s}(V/T) = \mathrm{s}(U/T)$. □

SECOND PROOF. For any $Z \lessdot T$, let ν_Z^T be the commutator bracket on H defined by (12.10). We need to check that it restricts to $\overline{\mathcal{P}}(\mathrm{H})$, that is, the dotted arrow in the diagram

$$\begin{array}{ccc} \overline{\mathcal{P}}(\mathrm{H})[T] & \hookrightarrow & \mathrm{H}[T] \\ \vdots & & \downarrow{\scriptstyle \nu_Z^T} \\ \overline{\mathcal{P}}(\mathrm{H})[Z] & \hookrightarrow & \mathrm{H}[Z] \end{array}$$

exists. Accordingly, let $v \in \overline{\mathcal{P}}(\mathrm{H})[T]$. We want to show that $\nu_Z^T(v) \in \overline{\mathcal{P}}(\mathrm{H})[Z]$, that is, $\Delta_Z^{T'}(\nu_Z^T(v)) = 0$ for any $T' \gtrdot Z$. There are three cases. We employ Lemma 9.119.

- $T' \neq T$ and $T' \neq Z\overline{T}$. In this case, by (9.154a),

$$\Delta_Z^{T'}(\mu_Z^T(v)) = 0 = \Delta_Z^{T'}(\mu_Z^{Z\overline{T}} \beta_{Z\overline{T},T}(v)).$$

  The second equality holds since $\beta_{Z\overline{T},T}(v) \in \overline{\mathcal{P}}(\mathrm{H})[Z\overline{T}]$.

- $\underline{T' = T}$. In this case, reversing the roles of T and $Z\overline{T}$ in (9.154c) and applying it to $\beta_{Z\overline{T},T}(v)$,

$$\Delta_Z^T \mu_Z^{Z\overline{T}} \beta_{Z\overline{T},T}(v) = \begin{cases} v & \text{if } Z\overline{T} \neq T, \\ v + \beta_{Z\overline{T},T}(v) & \text{if } Z\overline{T} = T. \end{cases}$$

  We used (8.18). Now by (9.154b), the above is equal to $\Delta_Z^T(\mu_Z^T(v))$.
- $\underline{T' = Z\overline{T}}$. We can deduce from the previous case that

$$\Delta_Z^{Z\overline{T}}(\mu_Z^T(v)) = \Delta_Z^{Z\overline{T}}(\mu_Z^{Z\overline{T}} \beta_{Z\overline{T},T}(v)).$$

Thus, in all three cases,

$$\Delta_Z^{T'}(\nu_Z^T(v)) = \Delta_Z^{T'}(\mu_Z^T(v)) - \Delta_Z^{T'}(\mu_Z^{Z\overline{T}} \beta_{Z\overline{T},T}(v)) = 0$$

as required. □

This yields a functor

$$\overline{\mathcal{P}} : \mathsf{Bimon}(\overline{\mathsf{C}}\mathsf{Sp}) \to \mathsf{LieMon}(\overline{\mathsf{C}}\mathsf{Sp}) \tag{12.18}$$

from the category of Coxeter bialgebras to the category of Coxeter Lie algebras.

12.3.2. Indecomposable part of a Coxeter bialgebra. Dually, a Coxeter bialgebra H is, in particular, a Coxeter coalgebra, and hence a Coxeter Lie coalgebra via the cocommutator functor (12.14), or equivalently, the cocommutator cobracket (12.15). This Lie structure descends to the indecomposable part $\overline{\mathcal{Q}}(\mathrm{H})$ as stated below.

Proposition 12.4. *If* H *is a Coxeter bialgebra, then* $\overline{\mathcal{Q}}(\mathrm{H})$ *is a Coxeter Lie quotient coalgebra of* H.

This is the dual of Proposition 12.3. It yields a functor

$$\overline{\mathcal{Q}} : \mathsf{Bimon}(\overline{\mathsf{C}}\mathsf{Sp}) \to \mathsf{LieComon}(\overline{\mathsf{C}}\mathsf{Sp}) \tag{12.19}$$

from the category of Coxeter bialgebras to the category of Coxeter Lie coalgebras.

12.3.3. Signed analogue. The preceding results have signed analogues. They can be deduced from the unsigned versions by conjugating by the u-signature functor (8.99). They are stated below.

We have a functor

$$\overline{\mathcal{P}} : (-1)\text{-}\mathsf{Bimon}(\overline{\mathsf{C}}\mathsf{Sp}) \to (-1)\text{-}\mathsf{LieMon}(\overline{\mathsf{C}}\mathsf{Sp}). \tag{12.20}$$

That is, the primitive part of a signed Coxeter bialgebra is a signed Coxeter Lie algebra.

Exercise 12.5. Verify the above assertion directly by making suitable sign changes in either of the two proofs of Proposition 12.3.

Dually, we have a functor

$$\overline{\mathcal{Q}} : (-1)\text{-}\mathsf{Bimon}(\overline{\mathsf{C}}\mathsf{Sp}) \to (-1)\text{-}\mathsf{LieComon}(\overline{\mathsf{C}}\mathsf{Sp}). \tag{12.21}$$

That is, the indecomposable part of a signed Coxeter bialgebra is a signed Coxeter Lie coalgebra.

12.3.4. Examples. For a cocommutative Coxeter coalgebra C and Coxeter bialgebra K, recall from Section 9.6.2 the Coxeter bialgebra $\mathcal{C}(\mathrm{C},\mathrm{K})$. By formula (9.75), its primitive part is $\mathrm{hom}^{\times}(\mathrm{C},\overline{\mathcal{P}}(\mathrm{K}))$ which by Proposition 12.3 is a Coxeter Lie algebra. Explicitly, the structure map

$$\gamma_T^U : \mathbf{Lie}[T,U] \otimes \mathrm{hom}^{\times}(\mathrm{C},\overline{\mathcal{P}}(\mathrm{K}))[U] \to \mathrm{hom}^{\times}(\mathrm{C},\overline{\mathcal{P}}(\mathrm{K}))[T]$$

is equivalent to the composite

$$\begin{aligned}\mathbf{Lie}[T,U] &\otimes \mathrm{hom}^{\times}(\mathrm{C},\overline{\mathcal{P}}(\mathrm{K}))[U] \otimes \mathrm{C}[T] \\ &\xrightarrow{\mathrm{id}\otimes\mathrm{id}\otimes\Delta_T^U} \mathbf{Lie}[T,U] \otimes \mathrm{hom}^{\times}(\mathrm{C},\overline{\mathcal{P}}(\mathrm{K}))[U] \otimes \mathrm{C}[U] \\ &\qquad\qquad \to \mathbf{Lie}[T,U] \otimes \overline{\mathcal{P}}(\mathrm{K})[U] \xrightarrow{\gamma_T^U} \overline{\mathcal{P}}(\mathrm{K})[T],\end{aligned}$$

where the last map is the structure map of $\overline{\mathcal{P}}(\mathrm{K})$. Equivalently, the Lie bracket

$$\nu_Z^T : \mathrm{hom}^{\times}(\mathrm{C},\overline{\mathcal{P}}(\mathrm{K}))[T] \to \mathrm{hom}^{\times}(\mathrm{C},\overline{\mathcal{P}}(\mathrm{K}))[Z]$$

sends the map f to the composite

$$\mathrm{C}[Z] \xrightarrow{\Delta_Z^T} \mathrm{C}[T] \xrightarrow{f} \overline{\mathcal{P}}(\mathrm{K})[T] \xrightarrow{\nu_Z^T} \overline{\mathcal{P}}(\mathrm{K})[Z],$$

where the last map is the Lie bracket of $\overline{\mathcal{P}}(\mathrm{K})$.

Similar considerations apply to the Coxeter bialgebra $\overline{\mathcal{C}}(\mathrm{H},\mathrm{A})$ from Section 9.7.2. By formula (9.82), its primitive part is $\mathrm{hom}^{\times}(\overline{\mathcal{Q}}(\mathrm{H}),\mathrm{A})$. This carries a Coxeter Lie algebra structure which involves structure of $\overline{\mathcal{Q}}(\mathrm{H})$ as a Coxeter Lie coalgebra. Here we apply Proposition 12.4.

12.4. (Co)free Coxeter Lie (co)algebras

Recall from Section 11.6 that the free Coxeter algebra and free commutative Coxeter algebra on a Coxeter space V are given by $\mathbf{As} \circ \mathrm{V}$ and $\mathbf{Com} \circ \mathrm{V}$. Alternatively, they can be described as $\overline{\mathcal{T}}(\mathrm{V})$ and $\overline{\mathcal{S}}(\mathrm{V})$ as in Section 9.3 (without making any direct reference to Coxeter operads). We now turn to the Lie side of this story, and explain the Lie entries in Tables 11.4 and 11.5. The free Coxeter Lie algebra on V is given by $\mathbf{Lie} \circ \mathrm{V}$. Alternatively, it can be described as $\overline{\mathcal{P}}\overline{\mathcal{T}}(\mathrm{V})$, the primitive part of $\overline{\mathcal{T}}(\mathrm{V})$. More formally, just as $\mathbf{As}$ and $\mathbf{Com}$ yield monads $\overline{\mathcal{T}}$ and $\overline{\mathcal{S}}$, $\mathbf{Lie}$ yields the monad $\overline{\mathcal{P}}\overline{\mathcal{T}}$ (arising from the adjunction between $\overline{\mathcal{T}}$ and $\overline{\mathcal{P}}$).

Dually, recall that the cofree Coxeter coalgebra and cofree cocommutative Coxeter coalgebra on V are given by $\mathbf{As}^* \circ^{\vee} \mathrm{V} = \overline{\mathcal{T}}^{\vee}(\mathrm{V})$ and $\mathbf{Com}^* \circ^{\vee} \mathrm{V} = \overline{\mathcal{S}}^{\vee}(\mathrm{V})$. Similarly, we will see that the cofree Coxeter Lie coalgebra on V is given by $\mathbf{Lie}^* \circ^{\vee} \mathrm{V} = \overline{\mathcal{Q}}\overline{\mathcal{T}}^{\vee}(\mathrm{V})$, the indecomposable part of $\overline{\mathcal{T}}^{\vee}(\mathrm{V})$.

This story also has signed analogues which we briefly mention.

12.4.1. Free Coxeter Lie algebra on a Coxeter space. Let us specialize the discussion in Section 11.5.5 to $\mathbf{a} := \mathbf{Lie}$.

The free Coxeter Lie algebra on a Coxeter space V is given by $\mathbf{Lie} \circ \mathrm{V}$. Its T-component is

$$(\mathbf{Lie} \circ \mathrm{V})[T] = \Big(\bigoplus_{U:\, T \leq U} \mathbf{Lie}[T, U] \otimes \mathrm{V}[U] \Big)_{\mathrm{quo}}, \tag{12.22}$$

with quotient taken wrt morphisms local to T as in (11.29).

Example 12.6. For the Coxeter space $\mathrm{V} := \mathrm{X}_M$,

$$(\mathbf{Lie} \circ \mathrm{X}_M)[\emptyset] = (\mathbf{Lie}[\emptyset, S] \otimes M)_W, \tag{12.23}$$

where the W-coinvariants are taken wrt the diagonal action of W. This specializes (11.44) to $\mathbf{p} := \mathbf{Lie}$ and $T := \emptyset$. □

The universal property of $\mathbf{Lie} \circ \mathrm{V}$ is stated below. Let $\mathrm{V} \hookrightarrow \mathbf{Lie} \circ \mathrm{V}$ be the canonical inclusion.

Theorem 12.7. *Let* G *be a Coxeter Lie algebra,* V *a Coxeter space,* $f : \mathrm{V} \to \mathrm{G}$ *a map of Coxeter spaces. Then there exists a unique morphism of Coxeter Lie algebras* $\hat{f} : \mathbf{Lie} \circ \mathrm{V} \to \mathrm{G}$ *such that the diagram*

$$\begin{array}{ccc} \mathbf{Lie} \circ \mathrm{V} & \overset{\hat{f}}{\dashrightarrow} & \mathrm{G} \\ \uparrow & \nearrow_{f} & \\ \mathrm{V} & & \end{array}$$

commutes.

Explicitly, the map $\hat{f}$ is

$$\mathbf{Lie} \circ \mathrm{V} \xrightarrow{\mathrm{id} \circ f} \mathbf{Lie} \circ \mathrm{G} \xrightarrow{\gamma} \mathrm{G}, \tag{12.24}$$

where γ is the structure map of G.

This specializes Theorem 11.40 and formula (11.42) to $\mathbf{a} := \mathbf{Lie}$.

12.4.2. Monad for Coxeter Lie algebras. Recall from Section 9.3.3 that functors $\overline{\mathcal{T}}$ and $\overline{\mathcal{P}}$ are adjoint between the category of Coxeter spaces and the category of Coxeter bialgebras. Thus, $\overline{\mathcal{P}}\overline{\mathcal{T}}$ is a monad on Coxeter spaces. There is an injective natural transformation

$$\iota : \overline{\mathcal{P}}\overline{\mathcal{T}} \to \overline{\mathcal{T}}$$

between functors from the category of Coxeter spaces to itself. Note very carefully: The two $\overline{\mathcal{T}}$s stand for different functors. The left $\overline{\mathcal{T}}$ goes from Coxeter spaces to Coxeter bialgebras, while the right $\overline{\mathcal{T}}$ goes from Coxeter spaces to itself by forgetting the bialgebra structure. The latter is a monad as discussed in Section 8.12. It is straightforward to check that (id, ι) is a lax functor of monads from $\overline{\mathcal{T}}$ to $\overline{\mathcal{P}}\overline{\mathcal{T}}$ in the sense of (A.2). Combining with the abelianization (8.122), we have morphisms of monads

$$\overline{\mathcal{P}}\overline{\mathcal{T}} \xrightarrow{\iota} \overline{\mathcal{T}} \xrightarrow{\pi} \overline{\mathcal{S}}.$$

Recall: $\overline{\mathcal{T}}$-algebras are Coxeter algebras, and $\overline{\mathcal{S}}$-algebras are commutative Coxeter algebras. We show below that $\overline{\mathcal{P}}\overline{\mathcal{T}}$-algebras are the same as Coxeter Lie algebras.

Lemma 12.8. *For a Coxeter space* V *and face-type Z, the primitive part of the Coxeter bialgebra $\overline{\mathcal{T}}(\mathrm{V})$ on the Z-component, namely, $\overline{\mathcal{P}}\overline{\mathcal{T}}(\mathrm{V})[Z]$ consists of elements $(v_U)_{U \geq Z}$ such that*

$$\sum_{\substack{\sigma\colon U \to V \\ \text{local to } Z \\ \text{projects } U \text{ on } T'}} \sigma(v_U) = 0 \tag{12.25}$$

for any face-types $Z < T' \leq V$.

This follows from the coproduct formula (9.26) of $\overline{\mathcal{T}_q}(\mathrm{V})$ for $q = 1$. An illustration for $Z = \emptyset$ is shown below.

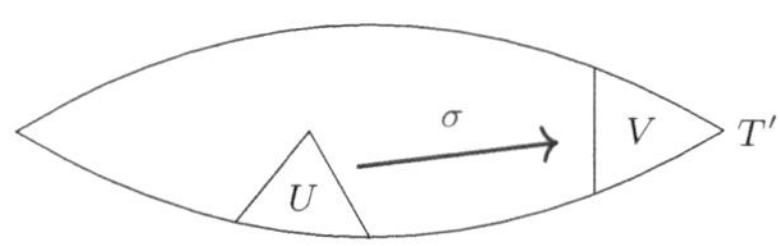

Proposition 12.9. *Let* $\mathrm{char}(\Bbbk) \nmid |W|$. *For a Coxeter space* V,

$$\overline{\mathcal{P}}\overline{\mathcal{T}}(\mathrm{V}) = \mathbf{Lie} \circ \mathrm{V}.$$

The monad structure on $\overline{\mathcal{P}}\overline{\mathcal{T}}$ corresponds to the Coxeter operad structure of **Lie**. *In particular, a $\overline{\mathcal{P}}\overline{\mathcal{T}}$-algebra is the same as a Coxeter Lie algebra.*

Recall from (11.51) and (11.54) that the monad $\overline{\mathcal{T}}$ corresponds to the Coxeter operad **As**, and $\overline{\mathcal{S}}$ to **Com**. The correspondence between $\overline{\mathcal{P}}\overline{\mathcal{T}}$ and **Lie** comes from the diagram

$$\begin{array}{ccccc} \mathbf{Lie} \circ \mathrm{V} & \hookrightarrow & \mathbf{As} \circ \mathrm{V} & \twoheadrightarrow & \mathbf{Com} \circ \mathrm{V} \\ \cong \downarrow & & \cong \downarrow & & \cong \downarrow \\ \overline{\mathcal{P}}\overline{\mathcal{T}}(\mathrm{V}) & \hookrightarrow & \overline{\mathcal{T}}(\mathrm{V}) & \twoheadrightarrow & \overline{\mathcal{S}}(\mathrm{V}). \end{array} \tag{12.26}$$

Since $\mathbf{Lie} \hookrightarrow \mathbf{As}$, the map $\mathbf{Lie} \circ \mathrm{V} \hookrightarrow \mathbf{As} \circ \mathrm{V}$ is injective. (See Exercise 11.37.)

PROOF. For the existence of the dotted arrow, we need to show that the image of $\mathbf{Lie} \circ \mathrm{V}$ in $\overline{\mathcal{T}}(\mathrm{V})$ lands inside $\overline{\mathcal{P}}\overline{\mathcal{T}}(\mathrm{V})$. For that, consider for $Z \leq T$,

$$\mathbf{Lie}[Z,T] \otimes \mathrm{V}[T] \hookrightarrow \mathbf{As}[Z,T] \otimes \mathrm{V}[T] \to \bigoplus_{\substack{U\colon U \geq Z \\ \mathrm{s}(U/Z) = \mathrm{s}(T/Z)}} \mathrm{V}[U]$$

$$\Big(\sum_{U\colon U \geq Z} \sum_{\substack{\alpha\colon T \to U \\ \text{local to } Z}} x^{\alpha}\, \mathtt{H}_{\alpha} \Big) \otimes v \longmapsto \sum_{U\colon U \geq Z} \sum_{\substack{\alpha\colon T \to U \\ \text{local to } Z}} x^{\alpha}\, \alpha(v),$$

where the term in parenthesis is a Lie element. We check below that the element in the rhs belongs to $\overline{\mathcal{P}}\overline{\mathcal{T}}(\mathrm{V})[Z]$, that is, it satisfies condition (12.25).

Accordingly, for any $Z < T' \leq V$,

$$\sum_{\substack{\sigma: U \to V \\ \text{local to } Z \\ \text{projects } U \text{ on } T'}} \sum_{\substack{\alpha: T \to U \\ \text{local to } Z}} x^{\alpha}\, \sigma\alpha(v) = \sum_{\substack{\beta: T \to V \\ \text{local to } Z}} \Bigg(\sum_{\substack{\alpha:\, \beta\alpha^{-1} \text{local to } Z \\ \text{projects } U \text{ on } T'}} x^{\alpha} \Bigg) \beta(v) = 0,$$

as required. The term in parenthesis is zero by the lune criterion (11.56) for Lie elements. See the picture in Figure 11.3.

Thus, the dotted arrow exists. It remains to show that it is surjective. For this, we make use of the idempotent operator $\mathbf{log}(\mathrm{id})$ on $\overline{\mathcal{T}}(\mathrm{V})$ whose image is $\overline{\mathcal{P}}\overline{\mathcal{T}}(\mathrm{V})$, see Proposition 9.67. By Lemma 12.10 below, we see that the element $\mathbf{log}(\mathrm{id})_Z(v)$ for $v \in \mathrm{V}[T]$ arises from the Lie element in Exercise 11.47. This proves surjectivity of the dotted arrow. □

12.4.3. The free Coxeter algebra projects onto the free Coxeter Lie algebra. Consider the operator $\mathbf{log}(\mathrm{id})$ given by formula (9.108) acting on the cocommutative Coxeter bialgebra $\overline{\mathcal{T}}(\mathrm{V})$.

Lemma 12.10. *Let* $\mathrm{char}(\Bbbk) \nmid |W|$. *The map* $\mathbf{log}(\mathrm{id})_Z$ *on* $\overline{\mathcal{T}}(\mathrm{V})[Z]$, *on the* T*-summand, is*

$$\sum_{V:\, V \geq Z} \sum_{\substack{\alpha: T \to V \\ \text{local to } Z}} \mathrm{sln}_Z^{\alpha}\, \alpha,$$

where sln_Z^{α} *is the Solomon coefficient* (1.126). *Only those summands* $\mathrm{V}[V]$ *for* $V \geq Z$ *show up for which* $\mathrm{s}(V/Z) = \mathrm{s}(T/Z)$.

PROOF. We employ the coproduct formula (9.26) of $\overline{\mathcal{T}}_q(\mathrm{V})$ for $q = 1$. Thus, for $v \in \mathrm{V}[T]$,

$$\begin{aligned}
\mathbf{log}(\mathrm{id})_Z(v) &= \sum_{T':\, T' \geq Z} \boldsymbol{\mu}(Z, T')\, \mu_Z^{T'} \Delta_Z^{T'}(v) \\
&= \sum_{\substack{T', V: \\ V \geq T' \geq Z}} \boldsymbol{\mu}(Z, T') \sum_{\substack{\alpha: T \to V \\ \text{local to } Z \\ \text{projects } T \text{ on } T'}} \alpha(v), \\
&= \sum_{V:\, V \geq Z} \sum_{\substack{\alpha: T \to V \\ \text{local to } Z}} \Bigg(\sum_{\substack{T':\, Z \leq T' \leq V \\ \alpha \text{ projects } T \text{ on } T'}} \boldsymbol{\mu}(Z, T') \Bigg) \alpha(v) \\
&= \sum_{V:\, V \geq Z} \sum_{\substack{\alpha: T \to V \\ \text{local to } Z}} \mathrm{sln}_Z^{\alpha}\, \alpha(v),
\end{aligned}$$

as required. The last claim used Exercise 1.26. □

The above result was used in the proof of Proposition 12.9. Now as a consequence, we see that $\mathbf{log}(\mathrm{id})$ projects the free Coxeter algebra onto the free Coxeter Lie algebra. This is formally stated below.

Corollary 12.11. *Let* $\mathrm{char}(\Bbbk) \nmid |W|$. *For* V *a Coxeter space,* $\mathbf{log}(\mathrm{id})$ *is an idempotent operator on* $\overline{\mathcal{T}}(\mathrm{V}) = \mathbf{As} \circ \mathrm{V}$ *with image* $\overline{\mathcal{P}}\overline{\mathcal{T}}(\mathrm{V}) = \mathbf{Lie} \circ \mathrm{V}$.

12.4.4. Cofree Coxeter Lie coalgebra. Dually, the cofree Coxeter Lie coalgebra on a Coxeter space V is given by $\mathbf{Lie}^* \circ^\vee \mathrm{V}$. Its universal property is stated below. Let $\mathbf{Lie}^* \circ^\vee \mathrm{V} \twoheadrightarrow \mathrm{V}$ be the canonical projection.

Theorem 12.12. *Let* K *be a Coxeter Lie coalgebra,* V *a Coxeter space,* $f : \mathrm{K} \to \mathrm{V}$ *a map of Coxeter spaces. Then there exists a unique morphism of Coxeter Lie coalgebras* $\hat{f} : \mathrm{K} \to \mathbf{Lie}^* \circ^\vee \mathrm{V}$ *such that the diagram*

$$\begin{array}{ccc} \mathrm{K} & \overset{\hat{f}}{\dashrightarrow} & \mathbf{Lie}^* \circ^\vee \mathrm{V} \\ & \searrow^{f} & \downarrow \\ & & \mathrm{V} \end{array}$$

commutes.

Explicitly, the map $\hat{f}$ is

$$\mathrm{K} \xrightarrow{\delta} \mathbf{Lie}^* \circ^\vee \mathrm{K} \xrightarrow{\mathrm{id} \circ^\vee f} \mathbf{Lie}^* \circ^\vee \mathrm{V}, \tag{12.27}$$

where δ is the structure map of K.

Recall from Section 9.3.6 that functors $\overline{\mathcal{T}}^\vee$ and $\overline{\mathcal{Q}}$ are adjoint between the category of Coxeter spaces and the category of Coxeter bialgebras. Thus, $\overline{\mathcal{Q}}\overline{\mathcal{T}}^\vee$ is a comonad on Coxeter spaces.

Proposition 12.13. *Let* $\mathrm{char}(\Bbbk) \nmid |W|$. *For a Coxeter space* V,

$$\overline{\mathcal{Q}}\overline{\mathcal{T}}^\vee(\mathrm{V}) = \mathbf{Lie}^* \circ^\vee \mathrm{V}.$$

The comonad structure of $\overline{\mathcal{Q}}\overline{\mathcal{T}}^\vee$ *corresponds to the Coxeter cooperad structure of* $\mathbf{Lie}^*$. *In particular, a* $\overline{\mathcal{Q}}\overline{\mathcal{T}}^\vee$*-coalgebra is the same as a Coxeter Lie coalgebra.*

This is the dual of Proposition 12.9. The correspondence between $\overline{\mathcal{Q}}\overline{\mathcal{T}}^\vee$ and $\mathbf{Lie}^*$ comes from the diagram

$$\begin{array}{ccccc} \mathbf{Com}^* \circ^\vee \mathrm{V} & \hookrightarrow & \mathbf{As}^* \circ^\vee \mathrm{V} & \twoheadrightarrow & \mathbf{Lie}^* \circ^\vee \mathrm{V} \\ \cong \downarrow & & \cong \downarrow & & \cong \downarrow \\ \overline{\mathcal{S}}^\vee(\mathrm{V}) & \hookrightarrow & \overline{\mathcal{T}}^\vee(\mathrm{V}) & \twoheadrightarrow & \overline{\mathcal{Q}}\overline{\mathcal{T}}^\vee(\mathrm{V}). \end{array} \tag{12.28}$$

This is dual to diagram (12.26).

As a consequence, we see that $\mathbf{log}(\mathrm{id})$ injects the cofree Coxeter Lie coalgebra into the cofree Coxeter coalgebra. This is formally stated below.

Corollary 12.14. *Let* $\mathrm{char}(\Bbbk) \nmid |W|$. *For* V *a Coxeter space,* $\mathbf{log}(\mathrm{id})$ *is an idempotent operator on* $\overline{\mathcal{T}}^\vee(\mathrm{V}) = \mathbf{As}^* \circ^\vee \mathrm{V}$ *with coimage* $\overline{\mathcal{Q}}\overline{\mathcal{T}}^\vee(\mathrm{V}) = \mathbf{Lie}^* \circ^\vee \mathrm{V}$.

This is the dual of Corollary 12.11.

12.4.5. Signed analogue. The preceding results have signed analogues. They can be deduced from the unsigned versions by conjugating by the u-signature functor (8.99). They are recorded below.

The free signed Coxeter Lie algebra on a Coxeter space V is given by $\mathbf{Lie}^- \circ \mathrm{V}$. (This also follows by specializing the discussion in Section 11.5.5 to $\mathbf{a} := \mathbf{Lie}^-$.) It satisfies a universal property as in Theorem 12.7 with $\mathbf{Lie}^-$ instead of **Lie**. Recall: Functors $\overline{\mathcal{T}}_{-1}$ and $\overline{\mathcal{P}}$ are adjoint between the category of Coxeter spaces and the category of signed Coxeter bialgebras. This yields the monad $\overline{\mathcal{P}}\overline{\mathcal{T}}_{-1}$ on Coxeter spaces whose algebras are signed Coxeter Lie algebras.

Dually, the cofree signed Coxeter Lie coalgebra on a Coxeter space V is given by $(\mathbf{Lie}^-)^* \circ^\vee \mathrm{V}$. Coalgebras over the comonad $\overline{\mathcal{Q}}\overline{\mathcal{T}}^\vee_{-1}$ are signed Coxeter Lie coalgebras.

12.4.6. Examples. Let $\mathrm{char}(\Bbbk) \nmid |W|$. We know from Proposition 12.9 that for a Coxeter space V, the primitive part of the free Coxeter bialgebra $\overline{\mathcal{T}}(\mathrm{V})$ is the free Coxeter Lie algebra on V. As special cases:

For a left W-module M, the primitive part of the tensor Coxeter bialgebra $\overline{\mathcal{T}}(M)$ is the free Coxeter Lie algebra on X_M:

$$\overline{\mathcal{P}}(\overline{\mathcal{T}}(M)) = \overline{\mathcal{P}}(\overline{\mathcal{T}}(\mathrm{X}_M)) = \mathbf{Lie} \circ \mathrm{X}_M. \tag{12.29}$$

The first step used that $\overline{\mathcal{T}}(M)$ is free on X_M. In particular, when M is the trivial representation of W,

$$\overline{\mathcal{P}}(\Bbbk[x]) = \overline{\mathcal{P}}(\overline{\mathcal{T}}(\mathrm{X})) = \mathbf{Lie} \circ \mathrm{X}. \tag{12.30}$$

Compare with formulas (10.23) and (10.53).

The primitive part of the Coxeter bialgebra of face-types $\overline{\Sigma}$ is the free Coxeter Lie algebra on $\Bbbk\{x\}$:

$$\overline{\mathcal{P}}(\overline{\Sigma}) = \overline{\mathcal{P}}(\overline{\mathcal{T}}(\Bbbk\{x\}_t)) = \mathbf{Lie} \circ \Bbbk\{x\}. \tag{12.31}$$

The first step used Proposition 10.74 which says that $\overline{\Sigma}$ is free on $\Bbbk\{x\}_t$. This involves the $\mathtt{Q}$-basis of $\overline{\Sigma}$. Compare with formula (10.100).

Dual remarks apply to the indecomposable parts of $\overline{\mathcal{T}}^\vee(M)$ and $\overline{\Sigma}^*$.

12.5. Universal (co)enveloping Coxeter (co)algebras

We construct the universal enveloping Coxeter algebra $\overline{\mathcal{U}}(\mathrm{G})$ of a Coxeter Lie algebra G. It yields a functor $\overline{\mathcal{U}}$ from Coxeter Lie algebras to Coxeter algebras which is left adjoint to the commutator functor. Moreover, $\overline{\mathcal{U}}(\mathrm{G})$ carries a cocommutative Coxeter bialgebra structure. Thus, $\overline{\mathcal{U}}$ also defines a functor from Coxeter Lie algebras to cocommutative Coxeter bialgebras. It is left adjoint to the primitive part functor $\overline{\mathcal{P}}$.

Dually, we construct the universal coenveloping Coxeter coalgebra $\overline{\mathcal{U}}^\vee(\mathrm{K})$ of a Coxeter Lie coalgebra K. It yields a functor $\overline{\mathcal{U}}^\vee$ which is right adjoint to the cocommutator functor. Moreover, $\overline{\mathcal{U}}^\vee(\mathrm{K})$ carries a commutative Coxeter bialgebra structure. Viewed in this manner, $\overline{\mathcal{U}}^\vee$ is right adjoint to the indecomposable part functor $\overline{\mathcal{Q}}$.

Similar considerations carry over to the signed setting.

12.5.1. Universal enveloping Coxeter algebra. Let G be a Coxeter Lie algebra. Define $\overline{\mathcal{U}}(\mathrm{G})$ to be the largest Coxeter quotient algebra of $\overline{\mathcal{T}}(\mathrm{G})$ such that the outside square in the diagram

$$\begin{array}{ccccc} \mathbf{Lie} \circ \mathrm{G} & \xrightarrow{\gamma} & \mathrm{G} & & \\ \downarrow & & \downarrow & \searrow & \\ \mathbf{As} \circ \mathrm{G} & \xrightarrow{\cong} & \overline{\mathcal{T}}(\mathrm{G}) & \twoheadrightarrow & \overline{\mathcal{U}}(\mathrm{G}) \\ & \searrow & & & \uparrow \\ & & \longrightarrow & & \overline{\mathcal{U}}(\mathrm{G}) \end{array} \tag{12.32}$$

commutes. We call this the *universal enveloping Coxeter algebra* of G. Explicitly, for face-types $Z \leq T \leq U$, the diagram

$$\begin{array}{ccccc} \mathbf{Lie}[T,U] \otimes \mathrm{G}[U] & \xrightarrow{\gamma_T^U} & \mathrm{G}[T] & \searrow & \\ \Big\downarrow & \textit{relation} & & & \displaystyle\bigoplus_{Y:\, Y \geq Z} \mathrm{G}[Y] = \overline{\mathcal{T}}(\mathrm{G})[Z] \\ \mathbf{As}[T,U] \otimes \mathrm{G}[U] & \longrightarrow & \displaystyle\bigoplus_{\substack{V:\, V \geq T \\ \mathrm{s}(V/T) = \mathrm{s}(U/T)}} \mathrm{G}[V] & \nearrow & \end{array} \tag{12.33}$$

commutes after passing to the quotient $\overline{\mathcal{U}}(\mathrm{G})[Z]$. The bottom-horizontal map sends $\mathtt{H}_\alpha \otimes v$ to $\alpha(v)$ as stated after (11.51).

The above diagram is specifying relations that we need to impose on $\overline{\mathcal{T}}(\mathrm{G})$ to obtain $\overline{\mathcal{U}}(\mathrm{G})$. In the terminology of [**11**, Section 16.5], it specifies a T/Z-relation arising from the face-type U/T.

Exercise 12.15. Check directly: The relations (12.33) are preserved under $\sigma : Z \to Z'$, namely, a T/Z-relation arising from U/T maps to a T'/Z'-relation arising from U'/T', where $T' = \sigma_*(T)$ and $U' = \sigma_*(U)$.

Lemma 12.16. *For* (G, ν) *a Coxeter Lie algebra,* $\overline{\mathcal{U}}(\mathrm{G})$ *is the quotient of* $\overline{\mathcal{T}}(\mathrm{G})$ *by the Coxeter subalgebra generated by the relations*

$$v - \beta_{Z\overline{T},T}(v) - \nu_Z^T(v), \quad v \in \mathrm{G}[T] \tag{12.34}$$

for $Z \lessdot T$.

Proof. Observe: (12.34) is a Z/Z-relation arising from the face-type T/Z. The key fact now is that this relation generates all "higher" relations. This is because **Lie** is a binary quadratic Coxeter operad and, in particular, it is generated in rank one. The formal details are similar to the proof of [**11**, Lemma 16.12]. □

The construction of $\overline{\mathcal{U}}(\mathrm{G})$ is functorial in G and yields a functor

$$\overline{\mathcal{U}} : \mathsf{LieMon}(\overline{\mathsf{C}}\mathsf{Sp}) \to \mathsf{Mon}(\overline{\mathsf{C}}\mathsf{Sp})$$

from the category of Coxeter Lie algebras to the category of Coxeter algebras.

Theorem 12.17. *The functor $\overline{\mathcal{U}}$ is left adjoint to the commutator functor* (12.8). *That is, for a Coxeter Lie algebra* G *and Coxeter algebra* A, *there is a natural bijection*

$$\mathsf{Mon}(\overline{\mathsf{C}}\mathsf{Sp})(\overline{\mathcal{U}}(\mathrm{G}), \mathrm{A}) \xrightarrow{\cong} \mathsf{LieMon}(\overline{\mathsf{C}}\mathsf{Sp})(\mathrm{G}, \mathrm{A}).$$

First proof. We indicate how the bijection works. The necessary checks are formally similar to those in the first proof of [**11**, Theorem 16.15] and we omit them. Starting with a morphism $\mathrm{G} \to \mathrm{A}$ of Coxeter Lie algebras, extend it to a morphism $\overline{\mathcal{T}}(\mathrm{G}) \to \mathrm{A}$ of Coxeter algebras, and then factor it through $\overline{\mathcal{T}}(\mathrm{G}) \twoheadrightarrow \overline{\mathcal{U}}(\mathrm{G})$ to obtain $\overline{\mathcal{U}}(\mathrm{G}) \to \mathrm{A}$. Conversely, starting with a morphism $\overline{\mathcal{U}}(\mathrm{G}) \to \mathrm{A}$ of Coxeter algebras, precompose it with $\mathrm{G} \hookrightarrow \overline{\mathcal{T}}(\mathrm{G}) \twoheadrightarrow \overline{\mathcal{U}}(\mathrm{G})$ to obtain a morphism $\mathrm{G} \to \mathrm{A}$ of Coxeter Lie algebras. □

Second proof. Let (G, ν) be a Coxeter Lie algebra and (A, μ) a Coxeter algebra. Suppose $f : \mathrm{G} \to \mathrm{A}$ is a map of Coxeter spaces. Then f is a morphism of Coxeter Lie algebras iff

$$f_Z \nu_Z^T = (\mu_Z^T - \mu_Z^{Z\overline{T}} \beta_{Z\overline{T},T}) f_T \quad \text{for} \quad T \gtrdot Z.$$

This used (12.4) and (12.10). The map f extends to a map of Coxeter algebras $\overline{\mathcal{T}}(\mathrm{G}) \to \mathrm{A}$ which on the Z-component, on the T-summand, is

$$\mathrm{G}[T] \xrightarrow{f_T} \mathrm{A}[T] \xrightarrow{\mu_Z^T} \mathrm{A}[Z].$$

By Lemma 12.16, this map factors through $\overline{\mathcal{U}}(\mathrm{G})$ iff the elements (12.34) map to zero. Writing this out yields the exact same condition written above. □

The adjunction in Theorem 12.17 is reformulated below as a universal property. The unit of the adjunction is a morphism $i : \mathrm{G} \to \overline{\mathcal{U}}(\mathrm{G})$ of Coxeter Lie algebras. It equals the composite map $\mathrm{G} \hookrightarrow \overline{\mathcal{T}}(\mathrm{G}) \twoheadrightarrow \overline{\mathcal{U}}(\mathrm{G})$.

Theorem 12.18. *Let* A *be a Coxeter algebra,* G *a Coxeter Lie algebra,* $f : \mathrm{G} \to \mathrm{A}$ *a morphism of Coxeter Lie algebras. Then there exists a unique morphism of Coxeter algebras* $\hat{f} : \overline{\mathcal{U}}(\mathrm{G}) \to \mathrm{A}$ *such that the diagram*

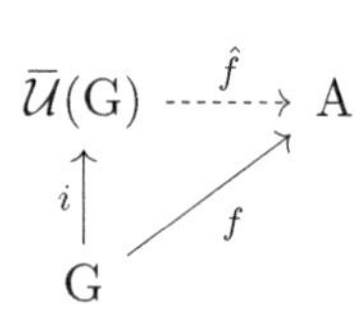

commutes.

So far $\overline{\mathcal{U}}(\mathrm{G})$ has only been considered as a Coxeter algebra. In fact, the deshuffle coproduct of $\overline{\mathcal{T}}(\mathrm{G})$ descends to $\overline{\mathcal{U}}(\mathrm{G})$ and turns the latter into a cocommutative Coxeter bialgebra. This can be checked directly. An alternative way to arrive at this fact is explained below.

Lemma 12.19. *For a Coxeter algebra* A, *the inclusion map* $\mathrm{A} \hookrightarrow \overline{\mathcal{T}}^{\vee}(\mathrm{A})$ *is a morphism of Coxeter Lie algebras.*

Here both A and $\overline{\mathcal{T}}^\vee(\mathrm{A})$ are Coxeter algebras, the latter with the quasishuffle product (9.31) with $q = 1$. They are then viewed as Coxeter Lie algebras via the commutator bracket.

Proof. We check diagram (12.4) which is

$$\begin{array}{ccc} \mathrm{A}[T] & \hookrightarrow & \overline{\mathcal{T}}^\vee(\mathrm{A})[T] \\ {\scriptstyle \nu_Z^T}\downarrow & & \downarrow{\scriptstyle \nu_Z^T} \\ \mathrm{A}[Z] & \hookrightarrow & \overline{\mathcal{T}}^\vee(\mathrm{A})[Z]. \end{array}$$

Both ν_Z^T are components of commutator brackets. We need to evaluate the second ν_Z^T only on the summand $\mathrm{A}[T]$. It has the form

$$\mathrm{A}[T] \to \mathrm{A}[Z] \oplus \mathrm{A}[T] \oplus \mathrm{A}[Z\overline{T}].$$

The first component $\mathrm{A}[T] \to \mathrm{A}[Z]$ of this map agrees with the commutator bracket component ν_Z^T of A. The remaining two components are

$$\mathrm{A}[T] \xrightarrow{\mathrm{id} - \beta_{T,Z\overline{T}}\beta_{Z\overline{T},T}} \mathrm{A}[T] \quad \text{and} \quad \mathrm{A}[T] \xrightarrow{\beta_{Z\overline{T},T} - \beta_{Z\overline{T},T}} \mathrm{A}[Z\overline{T}]$$

which are both zero. □

Exercise 12.20. For A a Coxeter algebra, consider the cocommutative Coxeter bialgebra $\overline{\mathcal{S}}^\vee(\mathrm{A})$. Check: $\overline{\mathcal{P}}(\overline{\mathcal{S}}^\vee(\mathrm{A})) = \mathrm{A}$ as Coxeter Lie algebras, with the commutator bracket on the latter. (Use Lemma 12.19.)

As a consequence of Lemma 12.19, we deduce that the composite map

$$\mathrm{G} \xrightarrow{i} \overline{\mathcal{U}}(\mathrm{G}) \hookrightarrow \overline{\mathcal{T}}^\vee(\overline{\mathcal{U}}(\mathrm{G}))$$

is a morphism of Coxeter Lie algebras. Now apply Theorem 12.18 to obtain a morphism $\overline{\mathcal{U}}(\mathrm{G}) \to \overline{\mathcal{T}}^\vee(\overline{\mathcal{U}}(\mathrm{G}))$ of Coxeter algebras. By Exercise 9.29, this yields a Coxeter bialgebra structure on $\overline{\mathcal{U}}(\mathrm{G})$. By construction, the image of G is primitive in $\overline{\mathcal{U}}(\mathrm{G})$, so Theorem 9.15 (with $q = 1$) yields a map of Coxeter bialgebras $\overline{\mathcal{T}}(\mathrm{G}) \to \overline{\mathcal{U}}(\mathrm{G})$. This map is clearly the usual one. Thus, we also have a functor

$$\overline{\mathcal{U}} : \mathsf{LieMon}(\overline{\mathsf{C}}\mathsf{Sp}) \to {}^{\mathsf{co}}\mathsf{Bimon}(\overline{\mathsf{C}}\mathsf{Sp})$$

from the category of Coxeter Lie algebras to the category of cocommutative Coxeter bialgebras.

Theorem 12.21. *The functor $\overline{\mathcal{U}}$ is left adjoint to the primitive part functor $\overline{\mathcal{P}}$. That is, for a Coxeter Lie algebra* G *and cocommutative Coxeter bialgebra* H, *there is a natural bijection*

$${}^{\mathsf{co}}\mathsf{Bimon}(\overline{\mathsf{C}}\mathsf{Sp})(\overline{\mathcal{U}}(\mathrm{G}), \mathrm{H}) \xrightarrow{\cong} \mathsf{LieMon}(\overline{\mathsf{C}}\mathsf{Sp})(\mathrm{G}, \overline{\mathcal{P}}(\mathrm{H})).$$

Proof. The proof is formally the same as that of [**11**, Theorem 16.25]. □

This can also be viewed as an adjunction between categories of Coxeter Lie algebras and Coxeter bialgebras. It is phrased below as a universal property.

Theorem 12.22. *Let* H *be a Coxeter bialgebra,* G *a Coxeter Lie algebra,* $f : \mathrm{G} \to \overline{\mathcal{P}}(\mathrm{H})$ *a morphism of Coxeter Lie algebras. Then there exists a unique morphism of Coxeter bialgebras* $\hat{f} : \overline{\mathcal{U}}(\mathrm{G}) \to \mathrm{H}$ *such that the diagram*

$$\begin{array}{ccc} \overline{\mathcal{U}}(\mathrm{G}) & \overset{\hat{f}}{\dashrightarrow} & \mathrm{H} \\ {\scriptstyle i}\uparrow & & \uparrow \\ \mathrm{G} & \underset{f}{\longrightarrow} & \overline{\mathcal{P}}(\mathrm{H}) \end{array}$$

commutes.

Proposition 12.23. *For a Coxeter space* V*,*

$$\overline{\mathcal{U}}(\mathbf{Lie} \circ \mathrm{V}) = \overline{\mathcal{T}}(\mathrm{V})$$

as Coxeter bialgebras.

Proof. This follows by composing adjunctions

$$\overline{\mathsf{C}}\mathsf{Sp} \underset{frg}{\overset{\mathbf{Lie}\circ(-)}{\rightleftarrows}} \mathsf{LieMon}(\overline{\mathsf{C}}\mathsf{Sp}) \underset{\overline{\mathcal{P}}}{\overset{\overline{\mathcal{U}}}{\rightleftarrows}} \mathsf{Bimon}(\overline{\mathsf{C}}\mathsf{Sp}),$$

and comparing with the adjunction between $\overline{\mathcal{T}}$ and $\overline{\mathcal{P}}$ in Section 9.3.3. □

12.5.2. Universal coenveloping Coxeter coalgebra. Let K be a Coxeter Lie coalgebra. Define $\overline{\mathcal{U}}^{\vee}(\mathrm{K})$ to be the largest Coxeter subcoalgebra of $\overline{\mathcal{T}}^{\vee}(\mathrm{K})$ such that the outside square in the diagram

(12.35)
$$\begin{array}{ccccc} \overline{\mathcal{U}}^{\vee}(\mathrm{K}) & \hookrightarrow & \overline{\mathcal{T}}^{\vee}(\mathrm{K}) & \underset{\cong}{\longrightarrow} & \mathbf{As}^* \circ^{\vee} \mathrm{K} \\ & \searrow & \downarrow\!\!\downarrow & & \downarrow \\ & & \mathrm{K} & \underset{\delta}{\longrightarrow} & \mathbf{Lie}^* \circ^{\vee} \mathrm{K} \end{array}$$

commutes. We call $\overline{\mathcal{U}}^{\vee}(\mathrm{K})$ the *universal coenveloping Coxeter coalgebra* of K. It yields a functor

$$\overline{\mathcal{U}}^{\vee} : \mathsf{LieComon}(\overline{\mathsf{C}}\mathsf{Sp}) \to \mathsf{Comon}(\overline{\mathsf{C}}\mathsf{Sp})$$

from the category of Coxeter Lie coalgebras to the category of Coxeter coalgebras.

Theorem 12.24. *The functor* $\overline{\mathcal{U}}^{\vee}$ *is right adjoint to the cocommutator functor* (12.14)*. That is, for a Coxeter Lie coalgebra* K *and Coxeter coalgebra* C*, there is a natural bijection*

$$\mathsf{LieComon}(\overline{\mathsf{C}}\mathsf{Sp})(\mathrm{C},\mathrm{K}) \xrightarrow{\cong} \mathsf{Comon}(\overline{\mathsf{C}}\mathsf{Sp})(\mathrm{C},\overline{\mathcal{U}}^{\vee}(\mathrm{K})).$$

The adjunction in Theorem 12.24 is reformulated below as a universal property. The counit of the adjunction is a morphism $p : \overline{\mathcal{U}}^{\vee}(\mathrm{K}) \to \mathrm{K}$ of Coxeter Lie coalgebras. It equals the composite map $\overline{\mathcal{U}}^{\vee}(\mathrm{K}) \hookrightarrow \overline{\mathcal{T}}^{\vee}(\mathrm{K}) \twoheadrightarrow \mathrm{K}$.

Theorem 12.25. *Let* C *be a Coxeter coalgebra,* K *a Coxeter Lie coalgebra,* $f : \mathrm{C} \to \mathrm{K}$ *a morphism of Coxeter Lie coalgebras. Then there exists a unique morphism of Coxeter coalgebras* $\hat{f} : \mathrm{C} \to \overline{\mathcal{U}}^\vee(\mathrm{K})$ *such that the diagram*

$$\begin{array}{ccc} \mathrm{C} & \overset{\hat{f}}{\dashrightarrow} & \overline{\mathcal{U}}^\vee(\mathrm{K}) \\ & \underset{f}{\searrow} & \downarrow{\scriptstyle p} \\ & & \mathrm{K} \end{array}$$

commutes.

So far $\overline{\mathcal{U}}^\vee(\mathrm{K})$ has only been considered as a Coxeter coalgebra. In fact, it is a commutative Coxeter subbialgebra of $\overline{\mathcal{T}}^\vee(\mathrm{K})$. This yields a functor

$$\overline{\mathcal{U}}^\vee : \mathsf{LieComon}(\overline{\mathsf{C}}\mathsf{Sp}) \to \mathsf{Bimon}^{\mathsf{co}}(\overline{\mathsf{C}}\mathsf{Sp})$$

from the category of Coxeter Lie coalgebras to the category of commutative Coxeter bialgebras.

Theorem 12.26. *The functor* $\overline{\mathcal{U}}^\vee$ *is right adjoint to the indecomposable part functor* $\overline{\mathcal{Q}}$*. That is, for a Coxeter Lie coalgebra* K *and commutative Coxeter bialgebra* H*, there is a natural bijection*

$$\mathsf{LieComon}(\overline{\mathsf{C}}\mathsf{Sp})(\overline{\mathcal{Q}}(\mathrm{H}), \mathrm{K}) \xrightarrow{\cong} \mathsf{Bimon}^{\mathsf{co}}(\overline{\mathsf{C}}\mathsf{Sp})(\mathrm{H}, \overline{\mathcal{U}}^\vee(\mathrm{K})).$$

This can also be viewed as an adjunction between categories of Coxeter bialgebras and Coxeter Lie coalgebras. It is phrased below as a universal property.

Theorem 12.27. *Let* H *be a Coxeter bialgebra,* K *a Coxeter Lie coalgebra,* $f : \overline{\mathcal{Q}}(\mathrm{H}) \to \mathrm{K}$ *a morphism of Coxeter Lie coalgebras. Then there exists a unique morphism of Coxeter bialgebras* $\hat{f} : \mathrm{H} \to \overline{\mathcal{U}}^\vee(\mathrm{K})$ *such that the diagram*

$$\begin{array}{ccc} \mathrm{H} & \overset{\hat{f}}{\dashrightarrow} & \overline{\mathcal{U}}^\vee(\mathrm{K}) \\ \downarrow & & \downarrow{\scriptstyle p} \\ \overline{\mathcal{Q}}(\mathrm{H}) & \underset{f}{\longrightarrow} & \mathrm{K} \end{array}$$

commutes.

Proposition 12.28. *For a Coxeter space* V*,*

$$\overline{\mathcal{U}}^\vee(\mathbf{Lie}^* \circ^\vee \mathrm{V}) = \overline{\mathcal{T}}^\vee(\mathrm{V})$$

as Coxeter bialgebras.

PROOF. This follows by composing adjunctions

$$\mathsf{Bimon}(\overline{\mathsf{C}}\mathsf{Sp}) \underset{\overline{\mathcal{U}}^\vee}{\overset{\overline{\mathcal{Q}}}{\rightleftarrows}} \mathsf{LieComon}(\overline{\mathsf{C}}\mathsf{Sp}) \underset{\mathbf{Lie}^* \circ^\vee (-)}{\overset{frg}{\rightleftarrows}} \overline{\mathsf{C}}\mathsf{Sp}$$

and comparing with the adjunction between $\overline{\mathcal{T}}^\vee$ and $\overline{\mathcal{Q}}$ in Section 9.3.6. □

12.5.3. Signed analogue. The preceding constructions and results have signed analogues. For a signed Coxeter Lie algebra G, we have its universal enveloping Coxeter algebra $\overline{\mathcal{U}}_{-1}(\mathrm{G})$. Dually, for a signed Coxeter Lie coalgebra K, we have its universal coenveloping Coxeter coalgebra $\overline{\mathcal{U}}^{\vee}_{-1}(\mathrm{K})$. They can be defined as

$$\overline{\mathcal{U}}_{-1}(\mathrm{G}) = \overline{\mathcal{U}}(\mathrm{G}_{-})_{-} \quad \text{and} \quad \overline{\mathcal{U}}^{\vee}_{-1}(\mathrm{K}) = \overline{\mathcal{U}}^{\vee}(\mathrm{K}_{-})_{-}$$

by conjugating $\overline{\mathcal{U}}$ and $\overline{\mathcal{U}}^{\vee}$ by the u-signature functor (8.99). It follows that $\overline{\mathcal{U}}_{-1}(\mathrm{G})$ is a signed cocommutative signed Coxeter bialgebra, and $\overline{\mathcal{U}}^{\vee}_{-1}(\mathrm{K})$ is a signed commutative signed Coxeter bialgebra. The signed analogue of relation (12.34) is

$$v - (-1)^{\mathrm{dist}(T, Z\overline{T})} \beta_{Z\overline{T},T}(v) - \nu_Z^T(v), \quad v \in \mathrm{G}[T] \tag{12.36}$$

for $Z \lessdot T$. The various adjunctions work the same way as in the unsigned case.

Exercise 12.29. Check: For G an abelian Coxeter Lie algebra, $\overline{\mathcal{U}}(\mathrm{G}) \cong \overline{\mathcal{S}}(\mathrm{G})$ with the quotient map $\overline{\mathcal{T}}(\mathrm{G}) \twoheadrightarrow \overline{\mathcal{U}}(\mathrm{G})$ specializing to the abelianization map. A similar remark applies to the dual and signed analogues.

12.6. Graded Lie algebras and Coxeter Lie algebras

We build on the discussion in Section 8.14. Recall: $\boldsymbol{\beta}^n$ is the braid arrangement on n letters with reference chamber $C_r = 1|\dots|n$. We make explicit the notion of Coxeter $\boldsymbol{\beta}^n$-Lie algebras using the combinatorics of integer compositions. Next, we relate classical Lie theory to this notion by showing that any positive-graded Lie algebra yields a Coxeter $\boldsymbol{\beta}^n$-Lie algebra.

12.6.1. Coxeter $\boldsymbol{\beta}^n$-Lie algebras. Coxeter $\boldsymbol{\beta}^n$-spaces were made explicit in Section 8.14.7. A Coxeter $\boldsymbol{\beta}^n$-Lie algebra is a Coxeter $\boldsymbol{\beta}^n$-space G equipped with linear maps

$$\nu_\alpha^\beta : \mathrm{G}[\beta] \to \mathrm{G}[\alpha],$$

one for each $\alpha \lessdot \beta$, subject to the axioms below.

Naturality. For each morphism $\sigma : \alpha \to \alpha'$ and $\alpha \lessdot \beta$, the diagram

$$\begin{array}{ccc} \mathrm{G}[\beta] & \xrightarrow{\sigma} & \mathrm{G}[\beta'] \\ {\scriptstyle \nu_\alpha^\beta}\downarrow & & \downarrow{\scriptstyle \nu_{\alpha'}^{\beta'}} \\ \mathrm{G}[\alpha] & \xrightarrow[\sigma]{} & \mathrm{G}[\alpha'] \end{array} \tag{12.37a}$$

commutes, where $\beta' = \sigma_*(\beta)$ refines α' using the bijection specified by σ.

Antisymmetry. For each $\alpha \lessdot \beta$,

$$(\mathrm{G}[\beta] \xrightarrow{\nu_\alpha^\beta} \mathrm{G}[\alpha]) \; + \; (\mathrm{G}[\beta] \xrightarrow{\beta_{\alpha\overline{\beta},\beta}} \mathrm{G}[\alpha\overline{\beta}] \xrightarrow{\nu_\alpha^{\alpha\overline{\beta}}} \mathrm{G}[\alpha]) \; = 0, \tag{12.37b}$$

with the morphism $\beta_{\alpha\overline{\beta},\beta}$ as in (8.132). Since β covers α, it is obtained by breaking a part of α into two. Reversing the order of these two parts yields $\alpha\overline{\beta}$, with $\beta_{\alpha\overline{\beta},\beta}$ being the corresponding transposition. It is an adjacent support-type morphism.

Jacobi identity. There are two cases.

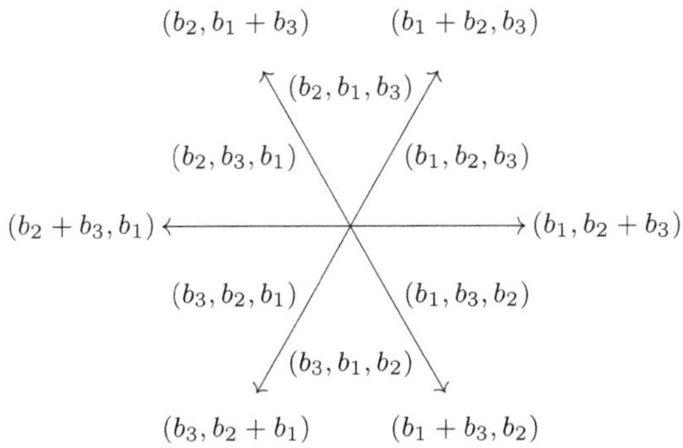

First case. Suppose

$$\alpha = (x, b_1 + b_2 + b_3, y), \quad \beta = (x, b_1, b_2 + b_3, y), \quad \gamma = (x, b_1, b_2, b_3, y)$$

for some compositions x and y and positive integers b_1, b_2, b_3. Then

(12.37c)
$$\begin{aligned}
&\big(\mathrm{G}[(x, b_1, b_2, b_3, y)] \xrightarrow{\nu} \mathrm{G}[(x, b_1, b_2 + b_3, y)] \xrightarrow{\nu} \mathrm{G}[(x, b_1 + b_2 + b_3, y)]\big) \\
&\quad + \big(\mathrm{G}[(x, b_1, b_2, b_3, y)] \xrightarrow{\sigma} \mathrm{G}[(x, b_2, b_1, b_3, y)] \\
&\quad\quad \xrightarrow{\nu} \mathrm{G}[(x, b_1 + b_2, b_3, y)] \xrightarrow{\nu} \mathrm{G}[(x, b_1 + b_2 + b_3, y)]\big) \\
&\quad\quad + \big(\mathrm{G}[(x, b_1, b_2, b_3, y)] \xrightarrow{\sigma} \mathrm{G}[(x, b_2, b_3, b_1, y)] \\
&\quad\quad\quad \xrightarrow{\nu} \mathrm{G}[(x, b_2, b_1 + b_3, y)] \xrightarrow{\nu} \mathrm{G}[(x, b_1 + b_2 + b_3, y)]\big) \;=\; 0.
\end{aligned}$$

The lhs is a sum of three maps all from $\mathrm{G}[\gamma]$ to $\mathrm{G}[\alpha]$ with the first one being $\nu_\alpha^\beta \nu_\beta^\gamma$.

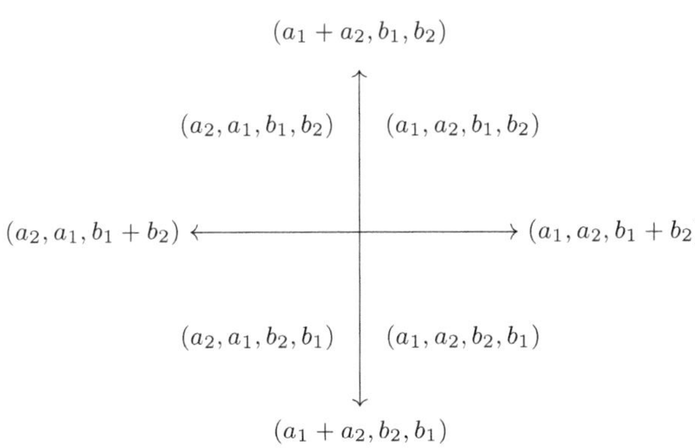

Second case. Suppose

$$\alpha = (x, a_1 + a_2, b_1 + b_2, y), \quad \beta = (x, a_1, a_2, b_1 + b_2, y), \quad \gamma = (x, a_1, a_2, b_1, b_2, y)$$

for some compositions x and y and positive integers a_1, a_2, b_1, b_2. Then

(12.37d)

$$\begin{aligned}\big(\mathrm{G}[(x,a_1,a_2,b_1,b_2,y)] &\xrightarrow{\nu} \mathrm{G}[(x,a_1,a_2,b_1{+}b_2,y)] \xrightarrow{\nu} \mathrm{G}[(x,a_1{+}a_2,b_1{+}b_2,y)]\big)\\ &+ \big(\mathrm{G}[(x,a_1,a_2,b_1,b_2,y)] \xrightarrow{\sigma} \mathrm{G}[(x,a_2,a_1,b_1,b_2,y)]\\ &\xrightarrow{\nu} \mathrm{G}[(x,a_1+a_2,b_1,b_2,y)] \xrightarrow{\nu} \mathrm{G}[(x,a_1+a_2,b_1+b_2,y)]\big) = 0.\end{aligned}$$

The lhs is a sum of two maps both from $\mathrm{G}[\gamma]$ to $\mathrm{G}[\alpha]$ with the first one being $\nu_\alpha^\beta \nu_\beta^\gamma$.

The above axioms specialize (12.3). Note very carefully that there are two cases for the Jacobi identity: When $\mathrm{rk}(\gamma/\alpha) = 2$, either γ breaks a part of α into three subparts, or breaks two parts of α into two subparts each.

Exercise 12.30. In the definition of a Coxeter Lie algebra, there is a Jacobi identity (12.3c) for each $\alpha \lessdot \beta \lessdot \gamma$. Show: Any such Jacobi identity, for instance, for

$$\alpha = (x, b_1 + b_2 + b_3, y), \quad \beta = (x, b_1 + b_2, b_3, y), \quad \gamma = (x, b_1, b_2, b_3, y)$$

can be deduced from axioms (12.37a), (12.37b), (12.37c), (12.37d).

Exercise 12.31. Use naturality (12.37a) and antisymmetry (12.37b) to show

$$\begin{aligned}\big(\mathrm{G}[(x,b_1,b_2,b_3,y)] &\xrightarrow{\nu} \mathrm{G}[(x,b_1,b_2+b_3,y)] \xrightarrow{\nu} \mathrm{G}[(x,b_1+b_2+b_3,y)]\big)\\ &+ \big(\mathrm{G}[(x,b_1,b_2,b_3,y)] \xrightarrow{\sigma} \mathrm{G}[(x,b_2,b_3,b_1,y)]\\ &\xrightarrow{\nu} \mathrm{G}[(x,b_2+b_3,b_1,y)] \xrightarrow{\nu} \mathrm{G}[(x,b_1+b_2+b_3,y)]\big) = 0.\end{aligned}$$

12.6.2. Graded Lie algebras. Now let G be a positive-graded Lie algebra. Let us recall the definition. First of all $\mathrm{G} = (\mathrm{G}_i)_{i \geq 1}$ is a positive-graded vector space. For any positive integers b_1 and b_2, there is a linear map

$$\mathrm{G}_{b_1} \otimes \mathrm{G}_{b_2} \to \mathrm{G}_{b_1+b_2}, \qquad x \otimes y \mapsto [x,y]$$

called the *Lie bracket* which satisfies two axioms:

Antisymmetry says

$$(\mathrm{G}_{b_1} \otimes \mathrm{G}_{b_2} \to \mathrm{G}_{b_1+b_2}) + (\mathrm{G}_{b_1} \otimes \mathrm{G}_{b_2} \to \mathrm{G}_{b_2} \otimes \mathrm{G}_{b_1} \to \mathrm{G}_{b_1+b_2}) = 0,$$

or equivalently,

$$[x,y] + [y,x] = 0.$$

Jacobi identity says the sum of the three maps

$$\begin{gathered}\mathrm{G}_{b_1} \otimes \mathrm{G}_{b_2} \otimes \mathrm{G}_{b_3} \to \mathrm{G}_{b_1} \otimes \mathrm{G}_{b_2+b_3} \to \mathrm{G}_{b_1+b_2+b_3},\\ \mathrm{G}_{b_1} \otimes \mathrm{G}_{b_2} \otimes \mathrm{G}_{b_3} \to \mathrm{G}_{b_2} \otimes \mathrm{G}_{b_1} \otimes \mathrm{G}_{b_3} \to \mathrm{G}_{b_1+b_2} \otimes \mathrm{G}_{b_3} \to \mathrm{G}_{b_1+b_2+b_3},\\ \mathrm{G}_{b_1} \otimes \mathrm{G}_{b_2} \otimes \mathrm{G}_{b_3} \to \mathrm{G}_{b_2} \otimes \mathrm{G}_{b_3} \otimes \mathrm{G}_{b_1} \to \mathrm{G}_{b_2} \otimes \mathrm{G}_{b_1+b_3} \to \mathrm{G}_{b_1+b_2+b_3}\end{gathered}$$

is zero, or equivalently,

$$[x,[y,z]] + [[y,x],z] + [y,[z,x]] = 0.$$

For each n, we have the Coxeter $\boldsymbol{\beta}^n$-space $\mathrm{G}^{[n]}$ whose components are defined as in (8.141). We now turn $\mathrm{G}^{[n]}$ into a Coxeter $\boldsymbol{\beta}^n$-Lie algebra. For compositions $\alpha \lessdot \beta$ of n, define

$$\nu_\alpha^\beta : \mathrm{G}^{[n]}[\beta] \to \mathrm{G}^{[n]}[\alpha]$$

by tensoring the Lie bracket map with identity maps on the left and right. The integers b_1 and b_2 are chosen to be the parts of β which refine the part $b_1 + b_2$ of α. Then antisymmetry implies (12.37b), while Jacobi identity implies (12.37c). The other axioms (12.37a) and (12.37d) follow directly from the definitions.

12.7. Notes

The material in this chapter is new. The notion of a Coxeter Lie algebra generalizes the classical notion of a positive-graded Lie algebra as explained in Section 12.6. We give some pointers to the classical literature below. More detailed references can be found in [**11**, Notes to Chapters 16 and 17].

12.7.1. Lie algebras. The abstract definition of a Lie algebra appeared in a paper by Jacobson [**197**, Section 1]. Other early references are by Birkhoff [**44**, Section 1] and Chevalley [**86**, Section IV.II, Definition 2]. Lie algebras were called infinitesimal groups in the older literature.

Lie coalgebras appeared independently in works of André [**14**, page 359], [**15**, Section 3] and Michaelis [**260**], [**261**, Section 1], [**262**, Definition 1.4], [**263**, Remark 3.52]. Some early references are by Nichols [**280**, Chapter III, page 63], [**281**, Section 3], Neisendorfer [**276**, page 432].

12.7.2. Universal enveloping algebras. The universal enveloping algebra of a Lie algebra appeared implicitly in work of Poincaré [**294**, page 1066], [**295**, page 225]. Later references are by Birkhoff [**44**], Witt [**361**], Harish-Chandra [**174**, page 900], [**175**, page 50], Gelfand [**147**], [**148**]. Its structure as a cocommutative Hopf algebra appeared implicitly in the book by Cartan and Eilenberg [**75**, Section XIII.5]. Later references are by Cartier [**78**, beginning of Section 3.4] and Milnor and Moore [**265**, Definition 5.7], [**266**, Proposition 5.7].

12.7.3. Free Lie algebras. Free Lie algebras originated in work of Hall [**170**], Magnus [**244**], [**245**], Witt [**361**]. Other early references are by Birkhoff [**44**, Theorem 3], Cartan and Eilenberg [**75**, Chapter XIII, Exercises 5, 6, 7, 8]. A comprehensive treatment is given by Reutenauer [**305**], [**306**]. Formula (12.23) corresponds to the well-known formula for the free Lie algebra on a vector space V, namely,

$$\mathbf{Lie} \circ \mathrm{V} = \bigoplus_{n \geq 1} \big(\mathbf{Lie}[n] \otimes V^{\otimes n}\big)_{\mathrm{S}_n}. \tag{12.38}$$

This is a special case of (11.66).

12.7.4. Signed analogue. The notion of a signed Coxeter Lie algebra generalizes the classical notion of a signed positive-graded Lie algebra. Early references for the latter are by Cartier [**77**, Section 1], Gerstenhaber [**150**, Section 2], Milnor and Moore [**265**, Proposition 5.2], [**266**, Lemma 5.3]. A closely related notion is that of Lie superalgebras. Early references are by Nijenhuis and Richardson [**283**, Section 1], Ross [**311**], [**312**, Section 2], Berezin and Kac [**34**, Formula (3.6)]. The first book on the subject is by Scheunert [**319**]. A more recent reference is the book by Musson [**275**].

CHAPTER 13

Structure theory of Coxeter bialgebras

The structure theory of Coxeter bialgebras parallels that of Coxeter bimonoids in Chapter 7. The Leray–Samelson theorem, Loday–Ronco theorem and its q-generalization, and Cartier–Milnor–Moore theorem all specify adjoint equivalences between certain categories constructed out of Coxeter spaces. A summary is provided in Table 13.1. Compare with Table 7.1. Related results include the Borel–Hopf theorem and Poincaré–Birkhoff–Witt theorem. We also discuss a family of results called the Hoffman–Newman–Radford theorems.

A comparison of structure theories of Coxeter bimonoids and Coxeter bialgebras using Fock functors is made in Sections 17.7 and 17.8.

TABLE 13.1. Structure theorems for Coxeter bialgebras.

Theorem	Adjoint equivalence
Leray–Samelson	Coxeter spaces $\underset{\overline{\mathcal{P}}}{\overset{\overline{\mathcal{S}}}{\rightleftarrows}}$ bicom. Coxeter bialgebras
	bicom. Coxeter bialgebras $\underset{\overline{\mathcal{S}}^\vee}{\overset{\overline{\mathcal{Q}}}{\rightleftarrows}}$ Coxeter spaces
Loday–Ronco	Coxeter spaces $\underset{\overline{\mathcal{P}}}{\overset{\overline{\mathcal{T}}_0}{\rightleftarrows}}$ Coxeter 0-bialgebras
	Coxeter 0-bialgebras $\underset{\overline{\mathcal{T}}_0^\vee}{\overset{\overline{\mathcal{Q}}}{\rightleftarrows}}$ Coxeter spaces
q-generalization for q not a root of unity	Coxeter spaces $\underset{\overline{\mathcal{P}}}{\overset{\overline{\mathcal{T}}_q}{\rightleftarrows}}$ Coxeter q-bialgebras
	Coxeter q-bialgebras $\underset{\overline{\mathcal{T}}_q^\vee}{\overset{\overline{\mathcal{Q}}}{\rightleftarrows}}$ Coxeter spaces
Cartier–Milnor–Moore	Coxeter Lie algebras $\underset{\overline{\mathcal{P}}}{\overset{\overline{\mathcal{U}}}{\rightleftarrows}}$ cocom. Coxeter bialgebras
	com. Coxeter bialgebras $\underset{\overline{\mathcal{U}}^\vee}{\overset{\overline{\mathcal{Q}}}{\rightleftarrows}}$ Coxeter Lie coalgebras

Borel–Hopf and Leray–Samelson. Let $\mathrm{char}(\Bbbk) \nmid |W|$. The Borel–Hopf theorem says that for H a cocommutative Coxeter bialgebra, $\mathrm{H} \cong \overline{\mathcal{S}}^\vee(\overline{\mathcal{P}}(\mathrm{H}))$ as Coxeter coalgebras, and dually, for H a commutative Coxeter bialgebra, $\mathrm{H} \cong \overline{\mathcal{S}}(\overline{\mathcal{Q}}(\mathrm{H}))$ as Coxeter algebras. These isomorphisms depend on the choice of an invariant noncommutative zeta function $\boldsymbol{\zeta}$ or Möbius function $\boldsymbol{\mu}$. When H is bicommutative, dependence on $\boldsymbol{\zeta}$ or $\boldsymbol{\mu}$ disappears, and moreover, we obtain isomorphisms

$$\mathrm{H} \cong \overline{\mathcal{S}}^{\vee}(\overline{\mathcal{P}}(\mathrm{H})) \cong \overline{\mathcal{S}}(\overline{\mathcal{P}}(\mathrm{H})) \cong \overline{\mathcal{S}}(\overline{\mathcal{Q}}(\mathrm{H})) \cong \overline{\mathcal{S}}^{\vee}(\overline{\mathcal{Q}}(\mathrm{H}))$$

of Coxeter bialgebras. This is the Leray–Samelson theorem. It shows that the category of bicommutative Coxeter bialgebras is equivalent to the category of Coxeter spaces. To understand the multiple isomorphisms above, recall that $\overline{\mathcal{S}}(\mathrm{V}) \cong \overline{\mathcal{S}}^{\vee}(\mathrm{V})$ as bicommutative Coxeter bialgebras and $\overline{\mathcal{P}}(\mathrm{H}) \cong \overline{\mathcal{Q}}(\mathrm{H})$ as Coxeter spaces.

These theorems have signed analogues for signed (co, bi)commutative signed Coxeter bialgebras, with $\overline{\mathcal{S}}$ and $\overline{\mathcal{S}}^{\vee}$ replaced by $\overline{\mathcal{E}}$ and $\overline{\mathcal{E}}^{\vee}$, respectively.

Loday–Ronco and its q-generalization. For H a Coxeter q-bialgebra for q not a root of unity, we have isomorphisms

$$\mathrm{H} \cong \overline{\mathcal{T}_q}^{\vee}(\overline{\mathcal{P}}(\mathrm{H})) \cong \overline{\mathcal{T}_q}(\overline{\mathcal{P}}(\mathrm{H})) \cong \overline{\mathcal{T}_q}(\overline{\mathcal{Q}}(\mathrm{H})) \cong \overline{\mathcal{T}_q}^{\vee}(\overline{\mathcal{Q}}(\mathrm{H}))$$

of Coxeter q-bialgebras. They involve the two-sided q-zeta and q-Möbius functions $\boldsymbol{\zeta}_q$ and $\boldsymbol{\mu}_q$. This is the rigidity theorem for Coxeter q-bialgebras. It shows that the category of Coxeter q-bialgebras for q not a root of unity is equivalent to the category of Coxeter spaces. To understand the multiple isomorphisms above, recall that $\overline{\mathcal{T}_q}(\mathrm{V}) \cong \overline{\mathcal{T}_q}^{\vee}(\mathrm{V})$ as Coxeter q-bialgebras and $\overline{\mathcal{P}}(\mathrm{H}) \cong \overline{\mathcal{Q}}(\mathrm{H})$ as Coxeter spaces. This is similar to the Leray–Samelson setup.

In particular, the category of Coxeter 0-bialgebras is equivalent to the category of Coxeter spaces. This special case is the Loday–Ronco theorem.

Hoffman–Newman–Radford. The Hoffman–Newman–Radford theorems are a family of results that provide inverse isomorphisms between two universally constructed Coxeter bialgebras. The two Coxeter bialgebras have the same underlying Coxeter space and also share either the same product or coproduct. If they have the same coproduct, then their products differ, with one defined using shuffles and the other using quasishuffles. Similarly, if they have the same product, then their coproducts differ, with one defined using deshuffles and the other using dequasishuffles. The HNR theorems use zeta and Möbius functions to do a change of basis to pass from one (co)product to the other. This can be pictured as

$$\text{dequasishuffle} \underset{\boldsymbol{\mu}}{\overset{\boldsymbol{\zeta}}{\rightleftarrows}} \text{deshuffle} \qquad \text{and} \qquad \text{shuffle} \underset{\boldsymbol{\mu}}{\overset{\boldsymbol{\zeta}}{\rightleftarrows}} \text{quasishuffle}.$$

Poincaré–Birkhoff–Witt. Let $\mathrm{char}(\Bbbk) \nmid |W|$. The Poincaré–Birkhoff–Witt theorem says that for G a Coxeter Lie algebra, $\overline{\mathcal{U}}(\mathrm{G}) \cong \overline{\mathcal{S}}(\mathrm{G})$ as Coxeter coalgebras, and dually, for K a Coxeter Lie coalgebra, $\overline{\mathcal{U}}^{\vee}(\mathrm{K}) \cong \overline{\mathcal{S}}^{\vee}(\mathrm{K})$ as Coxeter algebras. These isomorphisms depend on the choice of an invariant noncommutative zeta function $\boldsymbol{\zeta}$ or Möbius function $\boldsymbol{\mu}$. (Note the similarity with the Borel–Hopf theorem.) Equivalently, PBW says that there is an idempotent operator on $\overline{\mathcal{T}}(\mathrm{G})$ of Coxeter coalgebras with image $\overline{\mathcal{S}}(\mathrm{G})$ and coimage $\overline{\mathcal{U}}(\mathrm{G})$, and dually, an idempotent operator on $\overline{\mathcal{T}}^{\vee}(\mathrm{K})$ of Coxeter algebras with image $\overline{\mathcal{U}}^{\vee}(\mathrm{K})$ and coimage $\overline{\mathcal{S}}^{\vee}(\mathrm{K})$. These are the Solomon operators (involving the Solomon coefficients). The PBW theorem has a signed analogue for signed Coxeter Lie (co)algebras which makes use of $\overline{\mathcal{T}}_{-1}$, $\overline{\mathcal{U}}_{-1}$, $\overline{\mathcal{E}}$, and so on.

Cartier–Milnor–Moore. Let $\mathrm{char}(\Bbbk) \nmid |W|$. The Cartier–Milnor–Moore theorem says that functors $\overline{\mathcal{U}}$ and $\overline{\mathcal{P}}$ determine an adjoint equivalence between categories of Coxeter Lie algebras and cocommutative Coxeter bialgebras, and dually, functors $\overline{\mathcal{U}}^{\vee}$ and $\overline{\mathcal{Q}}$ determine an adjoint equivalence between categories of Coxeter Lie coalgebras and commutative Coxeter bialgebras. This theorem has a signed analogue.

The Leray–Samelson theorem for bicommutative Coxeter bialgebras is a restriction of the adjoint equivalence in CMM. This is illustrated below.

$$
\begin{array}{ccc}
\text{Coxeter Lie algebra} & \xleftrightarrow{\text{CMM}} & \text{cocommutative Coxeter bialgebra} \\
\uparrow & & \uparrow \\
\text{Coxeter space} & \xleftrightarrow{\text{LS}} & \text{bicommutative Coxeter bialgebra} \\
\downarrow & & \downarrow \\
\text{Coxeter Lie coalgebra} & \xleftrightarrow[\text{CMM}]{} & \text{commutative Coxeter bialgebra}
\end{array}
$$

(The category of Coxeter spaces is isomorphic to the category of abelian Coxeter Lie algebras and also to the category of abelian Coxeter Lie coalgebras.)

Rank one. Let $\mathcal{A}$ be a rank-one arrangement. Its Coxeter group W has two elements. We discuss from first principles notions of Coxeter $\mathcal{A}$-space and Coxeter $\mathcal{A}$-(co, bi)algebras. We then develop the basic theory of these objects such as universal constructions, exp-log correspondences, and conclude with structure theorems. This provides a nice self-contained summary of the theory of Coxeter bialgebras in a toy example.

13.1. Borel–Hopf for (co)commutative Coxeter bialgebras

In this section, we assume $\mathrm{char}(\Bbbk) \nmid |W|$. Any cocommutative Coxeter bialgebra is isomorphic as a Coxeter coalgebra to the cofree cocommutative Coxeter coalgebra on its primitive part. In fact, we construct a family of isomorphisms, one for each invariant noncommutative zeta function $\boldsymbol{\zeta}$. The inverse isomorphisms can be described using invariant noncommutative Möbius functions $\boldsymbol{\mu}$. This is the Borel–Hopf theorem for cocommutative Coxeter bialgebras. We also interpret the involved Coxeter coalgebra isomorphisms as exponentials of coderivations. Finer information on the isomorphisms can be obtained by using invariant eulerian families corresponding to $\boldsymbol{\zeta}$ and $\boldsymbol{\mu}$.

Dually, any commutative Coxeter bialgebra is isomorphic as a Coxeter algebra to the free commutative Coxeter algebra on its indecomposable part. This is the Borel–Hopf theorem for commutative Coxeter bialgebras.

The above statements have signed analogues. These are the Borel–Hopf theorems for signed (co)commutative signed Coxeter bialgebras.

The parallel discussion for Coxeter species is in Section 7.1. There we work with faces, while here we work with face-types.

13.1.1. Borel–Hopf for cocommutative Coxeter bialgebras (primitive part version). Recall from Section 9.3.10 the cofree cocommutative Coxeter coalgebra $\overline{\mathcal{S}}^\vee(\mathrm{V})$ on a Coxeter space V. Fix an invariant noncommutative zeta function $\boldsymbol{\zeta}$ and its inverse Möbius function $\boldsymbol{\mu}$. Let H be a cocommutative Coxeter bialgebra. Define a natural map

$$\overline{\mathcal{S}}^\vee(\overline{\mathcal{P}}(\mathrm{H})) \to \mathrm{H} \tag{13.1}$$

which on the Z-component is

$$(x_T)_{T \geq Z} \mapsto \sum_{T:\, T \geq Z} \boldsymbol{\zeta}(Z,T)\, \mu_Z^T(x_T).$$

(In the element (x_T) above, by definition, $\sigma(x_T) = x_{T'}$ for any $\sigma : T \to T'$ local to Z.) Similarly, define a natural map

$$\mathrm{H} \to \overline{\mathcal{S}}^\vee(\overline{\mathcal{P}}(\mathrm{H})) \tag{13.2}$$

which on the Z-component, into the T-summand, is

$$\mathbf{log}(\mathrm{id})_T \Delta_Z^T = \sum_{U:\, U \geq T} \boldsymbol{\mu}(T,U)\, \mu_T^U \Delta_Z^U.$$

It arises from cofreeness of $\overline{\mathcal{S}}^\vee$. More precisely, apply Theorem 9.25 with $\mathrm{C} := \mathrm{H}$, $\mathrm{V} := \overline{\mathcal{P}}(\mathrm{H})$, $f := \mathbf{log}(\mathrm{id}) : \mathrm{H} \twoheadrightarrow \overline{\mathcal{P}}(\mathrm{H})$ as in Proposition 9.67, and use formula (9.108). This also shows that (13.2) is a Coxeter coalgebra morphism.

Theorem 13.1. *Let* $\mathrm{char}(\Bbbk) \nmid |W|$. *For a cocommutative Coxeter bialgebra* H, *the maps* (13.1) *and* (13.2) *are inverse isomorphisms of Coxeter coalgebras. In particular, as a Coxeter coalgebra,* H *is the cofree cocommutative Coxeter coalgebra on* $\overline{\mathcal{P}}(\mathrm{H})$.

This is the *Borel–Hopf theorem* for cocommutative Coxeter bialgebras, and (13.1), (13.2) are the *Borel–Hopf isomorphisms* (primitive part version).

PROOF. We proceed directly. For any face-type Z, the composite linear map

$$\mathrm{H}[Z] \to \overline{\mathcal{S}}^\vee(\overline{\mathcal{P}}(\mathrm{H}))[Z] \to \mathrm{H}[Z]$$

is

$$\begin{aligned}
&\sum_{T:\, T \geq Z} \boldsymbol{\zeta}(Z,T)\, \mu_Z^T \Big(\sum_{U:\, U \geq T} \boldsymbol{\mu}(T,U)\, \mu_T^U \Delta_Z^U \Big) \\
&\quad = \sum_{T:\, T \geq Z}\ \sum_{U:\, U \geq T} \boldsymbol{\zeta}(Z,T)\, \boldsymbol{\mu}(T,U)\, \mu_Z^T \mu_T^U \Delta_Z^U \\
&\quad = \sum_{U:\, U \geq Z} \Big(\sum_{T:\, U \geq T \geq Z} \boldsymbol{\zeta}(Z,T)\, \boldsymbol{\mu}(T,U) \Big) \mu_Z^U \Delta_Z^U \\
&\quad = \mu_Z^Z \Delta_Z^Z \\
&\quad = \mathrm{id},
\end{aligned}$$

as required. We used the associativity axiom in (8.41) and that $\boldsymbol{\zeta}\boldsymbol{\mu} = \mathrm{id}$.

For any face-type Z, the composite linear map

$$\overline{\mathcal{S}}^\vee(\overline{\mathcal{P}}(\mathrm{H}))[Z] \to \mathrm{H}[Z] \to \overline{\mathcal{S}}^\vee(\overline{\mathcal{P}}(\mathrm{H}))[Z]$$

on the element $(x_T)_{T\geq Z}$, into the T'-component, is

$$
\begin{aligned}
&\sum_{U:\,U\geq T'} \mu(T',U)\,\mu^U_{T'}\Delta^U_Z\Big(\sum_{T:\,T\geq Z}\zeta(Z,T)\,\mu^T_Z(x_T)\Big)\\
&\quad= \sum_{U:\,U\geq T'}\ \sum_{T:\,T\geq Z} \mu(T',U)\,\zeta(Z,T)\,\mu^U_{T'}\Delta^U_Z\mu^T_Z(x_T)\\
&\quad= \sum_{U:\,U\geq T'}\ \sum_{T:\,T\geq Z}\ \sum_{\substack{Y:\,Y\geq U\\ \mathrm{s}(Y/Z)=\mathrm{s}(T/Z)}} \mu(T',U)\,\zeta(Z,T)\,\mu^U_{T'}\mu^Y_U\overline{\lambda}_{TTUY}(x_T)\\
&\quad= \sum_{Y:\,Y\geq T'}\ \sum_{U:\,Y\geq U\geq T'} \mu(T',U)\Big(\sum_{\substack{T:\,T\geq Z\\ \mathrm{s}(T/Z)=\mathrm{s}(Y/Z)}}\zeta(Z,T)\,a^{UYT}_Z\Big)\mu^Y_{T'}(x_Y)\\
&\quad= \sum_{Y:\,Y\geq T'}\Big(\sum_{U:\,Y\geq U\geq T'} \mu(T',U)\,\zeta(U,Y)\Big)\mu^Y_{T'}(x_Y)\\
&\quad= \mu^{T'}_{T'}(x_{T'})\\
&\quad= x_{T'},
\end{aligned}
$$

as required. We used the primitivity hypothesis on x_T, the bialgebra axiom (8.64), the associativity axiom in (8.41), the lune-additivity formula (1.82) and that $\boldsymbol{\mu\zeta} = \mathrm{id}$. Also there are a^{UYT}_Z morphisms in $\overline{\lambda}_{TTUY}$ by Observation 8.99, item (2) and each one of them takes x_T to x_Y. In the second step, we can also directly use (9.152).

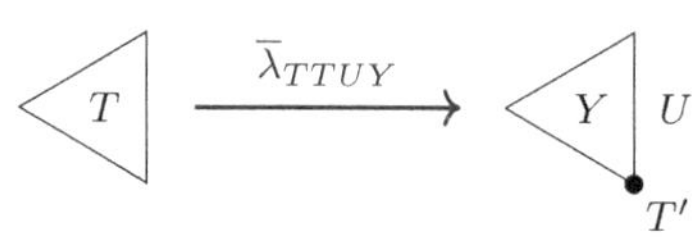

A helpful picture with $Z = \emptyset$ is shown above. □

Remark 13.2 (Proof by induction). It is possible to establish that (13.1) is an isomorphism without making any reference to its inverse (13.2). For injectivity, we use Proposition 9.6 and for surjectivity, we induct on the primitive filtration of H and use Exercise 9.118. For a similar argument for $\mathcal{A}$-species, see [**11**, proof of Theorem 13.34].

13.1.2. Exponentials of coderivations. Borel–Hopf isomorphisms can be interpreted as exponentials of coderivations as follows.

Exercise 13.3. For a cocommutative Coxeter bialgebra H, consider the following coderivations f and g.

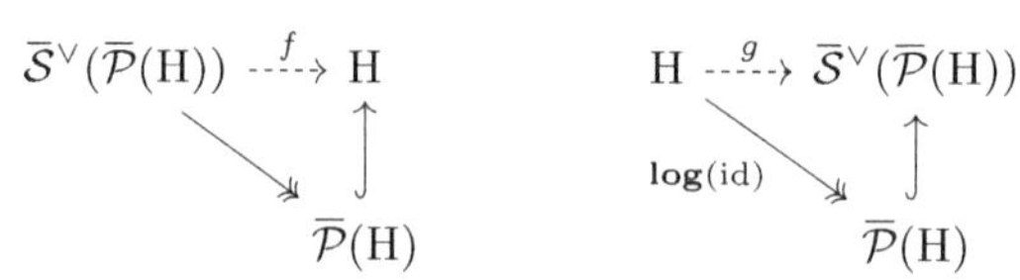

Check: The maps (13.1) and (13.2) are exponentials (9.90a) of f and g, respectively. Deduce from Theorem 9.56 that **exp**(f) and **exp**(g) are Coxeter coalgebra morphisms. Moreover, check that they are inverses of each other by using the inverse relationship between **exp** and **log**. This gives another proof of Theorem 13.1 which avoids the explicit calculations made above.

13.1.3. Characteristic operations. We now obtain finer information on the Borel–Hopf isomorphisms by using characteristic operations by invariant eulerian idempotents in the invariant Tits algebra. This proceeds independently of the preceding discussion.

Lemma 13.4. *Let* $\operatorname{char}(\Bbbk) \nmid |W|$. *For a cocommutative Coxeter bialgebra* H *and invariant eulerian family* E, *for a face-type* Z,

$$\mathrm{H}[Z] = \bigoplus_{\lambda/Z} \mathtt{E}_{\lambda/Z} \cdot \mathrm{H}[Z],$$

with the sum over all flat-types λ/Z *of* $\boldsymbol{\alpha}_Z$.

Proof. Recall from Section 9.13.1 that the invariant Tits algebra $\overline{\Sigma}[Z]$ acts on $\mathrm{H}[Z]$. Now use that the elements $\mathtt{E}_{\lambda/Z}$ in (10.72) determine a complete system of primitive orthogonal idempotents of $\overline{\Sigma}[Z]$. □

For a Coxeter space V and flat-type λ/Z of $\boldsymbol{\alpha}_Z$, define

$$\overline{\mathcal{S}}^{\vee}_{\lambda/Z}(\mathrm{V})[Z] := \Big(\bigoplus_{T \geq Z:\, \mathrm{s}(T/Z) = \lambda/Z} \mathrm{V}[T] \Big)^{\mathrm{sub}} \tag{13.3}$$

consisting of (x_T) such that any morphism $T \to T'$ local to Z sends x_T to $x_{T'}$. These are finer pieces of $\overline{\mathcal{S}}^{\vee}(\mathrm{V})$ than the commutative Cauchy powers of V in (9.5) where flat-types of the same rank are lumped together.

Proposition 13.5. *Let* $\operatorname{char}(\Bbbk) \nmid |W|$. *For* H *a cocommutative Coxeter bialgebra, for a flat-type* λ/Z *of* $\boldsymbol{\alpha}_Z$, *there are inverse linear isomorphisms*

$$\overline{\mathcal{S}}^{\vee}_{\lambda/Z}(\overline{\mathcal{P}}(\mathrm{H}))[Z] \underset{(\Delta_Z^T)}{\overset{(\zeta(Z,T)\mu_Z^T)}{\rightleftarrows}} \mathtt{E}_{\lambda/Z} \cdot \mathrm{H}[Z] \tag{13.4}$$

as T *varies over* $\mathrm{s}(T/Z) = \lambda/Z$. *In particular, for the minimum flat-type,*

$$\overline{\mathcal{P}}(\mathrm{H})[Z] = \mathtt{E}_{\mathrm{s}(Z/Z)} \cdot \mathrm{H}[Z] \tag{13.5}$$

and for the maximum flat-type, there are inverse linear isomorphisms

$$\mathrm{H}[S]^{W_Z} \underset{\Delta_Z^S}{\overset{\frac{1}{|W_Z|}\mu_Z^S}{\rightleftarrows}} \frac{1}{|W_Z|}\mathtt{H}_{S/Z} \cdot \mathrm{H}[Z]. \tag{13.6}$$

Proof. We first check that the lhs of (13.4) maps into the rhs. Accordingly, let (x_T) belong to the lhs. Since $\mathtt{E}_{\lambda/Z}$ is an idempotent, we need to show that

$$\mathtt{E}_{\lambda/Z} \cdot \big(\sum \zeta(Z,T)\, \mu_Z^T(x_T) \big) = \sum \zeta(Z,T)\, \mu_Z^T(x_T),$$

with the sums over face-types T/Z of support λ/Z. We calculate:

$$\begin{aligned}
&\mathtt{E}_{\lambda/Z} \cdot \big(\sum \zeta(Z,T)\, \mu_Z^T(x_T)\big)\\
&\quad = \big(\sum \zeta(Z,T')\, \mathtt{Q}_{T'/Z}\big) \cdot \big(\sum \zeta(Z,T)\, \mu_Z^T(x_T)\big)\\
&\quad = \big(\sum \zeta(Z,T')\, \mathtt{H}_{T'/Z} + \text{higher terms}\big) \cdot \big(\sum \zeta(Z,T)\, \mu_Z^T(x_T)\big)\\
&\quad = \sum \zeta(Z,T')\, \mu_Z^{T'} \big(\sum \zeta(Z,T)\, \Delta_Z^{T'} \mu_Z^T(x_T)\big)\\
&\quad = \sum \zeta(Z,T')\, \mu_Z^{T'}(x_{T'}).
\end{aligned}$$

The first step used (10.72); we used a different variable T' to avoid confusion with T. The second step used (10.66) in crude form for each $\mathtt{Q}_{T'/Z}$. In the third step, for base terms, we used definition of the action (9.156). Higher terms do not contribute by the second alternative in (9.150). The last step used (9.153) and the flat-additivity formula (1.83).

We now check that the rhs of (13.4) maps into the lhs. Since $\mathtt{E}_{\mathrm{s}(Z/Z)}$ is an invariant special Zie element of $\overline{\Sigma}[Z]$, the special case (13.5) follows from Proposition 9.125. By the cocommutativity axiom (8.91), we see that (Δ_Z^T) maps $\mathrm{H}[Z]$ into $\overline{\mathcal{S}}^{\vee}_{\lambda/Z}(\mathrm{H})[Z]$. Further, for $\mathrm{s}(T/Z) = \lambda/Z$,

$$\Delta_Z^T(\mathtt{E}_{\lambda/Z} \cdot \mathrm{H}[Z]) = \Delta_Z^T(\mathtt{E}_{\lambda/Z}) \cdot \Delta_Z^T(\mathrm{H}[Z]) \subseteq \mathtt{E}_{\mathrm{s}(T/T)} \cdot \mathrm{H}[T] = \overline{\mathcal{P}}(\mathrm{H})[T].$$

We used (9.162), (10.73), (13.5). It follows that (Δ_Z^T) maps $\mathtt{E}_{\lambda/Z} \cdot \mathrm{H}[Z]$ to

$$\overline{\mathcal{S}}^{\vee}_{\lambda/Z}(\overline{\mathcal{P}}(\mathrm{H}))[Z] = \overline{\mathcal{S}}^{\vee}_{\lambda/Z}(\mathrm{H})[Z] \cap \Big(\bigoplus_{T \geq Z:\, \mathrm{s}(T/Z) = \lambda/Z} \overline{\mathcal{P}}(\mathrm{H})[T] \Big).$$

Finally, we need to check that composites of the two maps in (13.4) in either order are identities. In one order,

$$\begin{aligned}
\sum \zeta(Z,T)\, \mu_Z^T \Delta_Z^T(\mathtt{E}_{\lambda/Z} \cdot x) &= \sum \zeta(Z,T)\, \mathtt{H}_{T/Z} \cdot (\mathtt{E}_{\lambda/Z} \cdot x)\\
&= \sum \zeta(Z,T)\, (\mathtt{H}_{T/Z} \cdot \mathtt{E}_{\lambda/Z}) \cdot x\\
&= \sum \zeta(Z,T)\, \mu_Z^T \Delta_Z^T(\mathtt{E}_{\lambda/Z}) \cdot x\\
&= \mathtt{E}_{\lambda/Z} \cdot x.
\end{aligned}$$

The first step is definition of the action (9.156), the second step used (9.158), the third step used Example 9.121, and the last step used (10.74).

In the other order, we check that the composite is identity by using (9.153) and (1.83) (as in the last step of the first calculation above). □

By summing isomorphisms (13.4) over all flat-types λ/Z of $\boldsymbol{\alpha}_Z$ and using Lemma 13.4, we obtain Z-components of isomorphisms (13.1) and (13.2). This gives another proof of Theorem 13.1. To claim an independent proof, we do need to check that these isomorphisms are Coxeter coalgebra morphisms. We leave this as an exercise.

13.1.4. Primitive filtration of a cocommutative Coxeter bialgebra. The primitive filtration of a cofree cocommutative Coxeter coalgebra is given by (9.42). Using Borel–Hopf Theorem 13.1 and Proposition 13.5, we get:

Proposition 13.6. *Let* $\mathrm{char}(\Bbbk) \nmid |W|$*. The primitive filtration of a cocommutative Coxeter bialgebra* H *is*

$$\overline{\mathcal{P}}_k(\mathrm{H})[Z] = \bigoplus_{\lambda/Z:\, \mathrm{rk}(\lambda/Z)<k} \mathtt{E}_{\lambda/Z} \cdot \mathrm{H}[Z].$$

The sum is over all flat-types of $\boldsymbol{\alpha}_Z$ *of rank strictly less than* k*, and* $\mathtt{E}$ *is any invariant eulerian family.*

13.1.5. Primitively generated Coxeter bialgebras. Let A be a Coxeter algebra and V a Coxeter subspace of A. The Coxeter subalgebra of A generated by V, denoted $\langle \mathrm{V} \rangle$, is the smallest Coxeter subalgebra of A containing V. Explicitly, for a face-type Z,

$$\langle \mathrm{V} \rangle[Z] = \sum_{T:\, T \geq Z} \mu_Z^T(\mathrm{V}[T]). \tag{13.7}$$

We say V *generates* A if $\langle \mathrm{V} \rangle = \mathrm{A}$. This happens iff the morphism $\overline{\mathcal{T}}(\mathrm{V}) \to \mathrm{A}$ of Coxeter algebras in Theorem 9.11 is surjective.

Exercise 13.7. Let H be a Coxeter bialgebra. We say H is *primitively generated* if H is generated as a Coxeter algebra by its primitive part, that is, if $\mathrm{H} = \langle \overline{\mathcal{P}}(\mathrm{H}) \rangle$. Let $\mathrm{char}(\Bbbk) \nmid |W|$. Show: H is primitively generated iff H is cocommutative. (For the forward implication, use Exercise 9.117. For the backward implication, use the Borel–Hopf theorem.)

13.1.6. Borel–Hopf for commutative Coxeter bialgebras (indecomposable part version). Recall from Section 9.3.8 the free commutative Coxeter algebra $\overline{\mathcal{S}}(\mathrm{V})$ on a Coxeter space V. Fix an invariant noncommutative zeta function $\boldsymbol{\zeta}$ and its inverse Möbius function $\boldsymbol{\mu}$. Let H be a commutative Coxeter bialgebra. Define a natural map

$$\mathrm{H} \to \overline{\mathcal{S}}(\overline{\mathcal{Q}}(\mathrm{H})) \tag{13.8}$$

which on the Z-component sends $x \in \mathrm{H}[Z]$ to the image of

$$\sum_{T:\, T \geq Z} \boldsymbol{\zeta}(Z,T)\, \Delta_Z^T(x) \;\in\; \overline{\mathcal{T}}(\mathrm{H})[Z]$$

in the quotient $\overline{\mathcal{S}}(\overline{\mathcal{Q}}(\mathrm{H}))[Z]$. Similarly, define a natural map

$$\overline{\mathcal{S}}(\overline{\mathcal{Q}}(\mathrm{H})) \to \mathrm{H} \tag{13.9}$$

which on the Z-component sends a representative $x \in \mathrm{H}[T]$ to

$$\mu_Z^T \,\mathbf{log}(\mathrm{id})_T = \sum_{U:\, U \geq T} \boldsymbol{\mu}(T,U)\, \mu_Z^U \Delta_T^U(x).$$

It arises from freeness of $\overline{\mathcal{S}}$. More precisely, apply Theorem 9.22 with $\mathrm{A} := \mathrm{H}$, $\mathrm{V} := \overline{\mathcal{Q}}(\mathrm{H})$, $f := \mathbf{log}(\mathrm{id}) : \overline{\mathcal{Q}}(\mathrm{H}) \hookrightarrow \mathrm{H}$ as in Proposition 9.67, and use formula (9.108). This also shows that (13.9) is a Coxeter algebra morphism.

Theorem 13.8. *Let* $\operatorname{char}(\Bbbk) \nmid |W|$. *For a commutative Coxeter bialgebra* H, *the maps* (13.8) *and* (13.9) *are inverse isomorphisms of Coxeter algebras. In particular, as a Coxeter algebra,* H *is the free commutative Coxeter algebra on* $\overline{\mathcal{Q}}(\mathrm{H})$.

This is the *Borel–Hopf theorem* for commutative Coxeter bialgebras, and (13.8), (13.9) are the *Borel–Hopf isomorphisms* (indecomposable part version). They can be interpreted as exponentials of derivations as follows.

Exercise 13.9. For a commutative Coxeter bialgebra H, consider the following derivations f and g.

$$\begin{array}{ccc} \mathrm{H} & \overset{f}{\dashrightarrow} & \overline{\mathcal{S}}(\overline{\mathcal{Q}}(\mathrm{H})) \\ \downarrow & \nearrow & \\ \overline{\mathcal{Q}}(\mathrm{H}) & & \end{array} \qquad\qquad \begin{array}{ccc} \overline{\mathcal{S}}(\overline{\mathcal{Q}}(\mathrm{H})) & \overset{g}{\dashrightarrow} & \mathrm{H} \\ \downarrow & \nearrow_{\mathbf{log}(\mathrm{id})} & \\ \overline{\mathcal{Q}}(\mathrm{H}) & & \end{array}$$

Check: The maps (13.8) and (13.9) are exponentials (9.90a) of f and g, respectively. Deduce from Theorem 9.56 that $\mathbf{exp}(f)$ and $\mathbf{exp}(g)$ are Coxeter algebra morphisms.

13.1.7. Examples. Recall from Section 10.3 the tensor and shuffle Coxeter bialgebras. The former is cocommutative, the latter is commutative. Hence, by Borel–Hopf Theorems 13.1 and 13.8:

Proposition 13.10. *Let* $\operatorname{char}(\Bbbk) \nmid |W|$. *The tensor Coxeter bialgebra* $\overline{\mathcal{T}}(M)$ *is cofree cocommutative on its primitive part, while the shuffle Coxeter bialgebra* $\overline{\mathcal{T}}^{\vee}(M)$ *is free commutative on its indecomposable part. That is,*

$$\overline{\mathcal{T}}(M) \cong \overline{\mathcal{S}}^{\vee}(\overline{\mathcal{P}}(\overline{\mathcal{T}}(M))) \quad \textit{and} \quad \overline{\mathcal{T}}^{\vee}(M) \cong \overline{\mathcal{S}}(\overline{\mathcal{Q}}(\overline{\mathcal{T}}^{\vee}(M)))$$

as Coxeter coalgebras and Coxeter algebras, respectively.

The primitive part $\overline{\mathcal{P}}(\overline{\mathcal{T}}(M))$ was described in (10.53) and involves invariant decorated Lie elements.

Recall from Section 10.4 the Coxeter bialgebra $\overline{\Sigma}$ and its dual $\overline{\Sigma}^{*}$. The former is cocommutative, the latter is commutative. Hence:

Proposition 13.11. *Let* $\operatorname{char}(\Bbbk) \nmid |W|$. *The Coxeter bialgebra of face-types* $\overline{\Sigma}$ *is cofree cocommutative on its primitive part, while* $\overline{\Sigma}^{*}$ *is free commutative on its indecomposable part. That is,*

$$\overline{\Sigma} \cong \overline{\mathcal{S}}^{\vee}(\overline{\mathcal{P}}(\overline{\Sigma})) \quad \textit{and} \quad \overline{\Sigma}^{*} \cong \overline{\mathcal{S}}(\overline{\mathcal{Q}}(\overline{\Sigma}^{*}))$$

as Coxeter coalgebras and Coxeter algebras, respectively.

The primitive part $\overline{\mathcal{P}}(\overline{\Sigma})$ was described in (10.100) and involves invariant Zie elements.

13.1.8. Signed analogue. Borel–Hopf Theorems 13.1 and 13.8 have signed analogues which we briefly mention.

For a signed cocommutative signed Coxeter bialgebra H, define maps

$$\overline{\mathcal{E}}^{\vee}(\overline{\mathcal{P}}(\mathrm{H})) \to \mathrm{H} \quad \text{and} \quad \mathrm{H} \to \overline{\mathcal{E}}^{\vee}(\overline{\mathcal{P}}(\mathrm{H})) \tag{13.10}$$

by the same formulas as in (13.1) and (13.2). In the first map, when we write (x_T), we now mean $(-1)^{l(\sigma)}\sigma(x_T) = x_{T'}$ for $\sigma : T \to T'$ local to Z. Similarly, the second map lands inside such "signed invariant" elements.

Theorem 13.12. *Let* $\mathrm{char}(\Bbbk) \nmid |W|$. *For a signed cocommutative signed Coxeter bialgebra* H, *the maps* (13.10) *are inverse isomorphisms of Coxeter coalgebras. In particular, as a Coxeter coalgebra,* H *is the cofree signed cocommutative Coxeter coalgebra on* $\overline{\mathcal{P}}(\mathrm{H})$.

PROOF. This can be deduced from Theorem 13.1 by conjugating by the u-signature functor (8.99). Alternatively, one can rerun the two calculations in its proof. The only change is in one step of the second calculation where we now write $(\overline{\lambda}_{-1})_{TTUY}(x_T)$ and equate it to x_Y. □

This is the *Borel–Hopf theorem* for signed cocommutative signed Coxeter bialgebras.

Dually, for a signed commutative signed Coxeter bialgebra H, define maps

$$\mathrm{H} \to \overline{\mathcal{E}}(\overline{\mathcal{Q}}(\mathrm{H})) \quad \text{and} \quad \overline{\mathcal{E}}(\overline{\mathcal{Q}}(\mathrm{H})) \to \mathrm{H} \tag{13.11}$$

by the same formulas as in (13.8) and (13.9).

Theorem 13.13. *Let* $\mathrm{char}(\Bbbk) \nmid |W|$. *For a signed commutative signed Coxeter bialgebra* H, *the maps* (13.11) *are inverse isomorphisms of Coxeter algebras. In particular, as a Coxeter algebra,* H *is the free signed commutative Coxeter algebra on* $\overline{\mathcal{Q}}(\mathrm{H})$.

This is the *Borel–Hopf theorem* for signed commutative signed Coxeter bialgebras.

13.2. Leray–Samelson for bicommutative Coxeter bialgebras

In this section, we assume $\mathrm{char}(\Bbbk) \nmid |W|$. Any bicommutative Coxeter bialgebra is free as a commutative Coxeter algebra on its indecomposable part and cofree as a cocommutative Coxeter coalgebra on its primitive part. This follows from the Borel–Hopf theorems in Section 13.1. In this situation, there is in fact a categorical equivalence between Coxeter spaces and bicommutative Coxeter bialgebras. This is the Leray–Samelson theorem. It comes in two flavors, one involving the primitive part functor and the other involving the indecomposable part functor. We write down formulas for the involved Coxeter bialgebra isomorphisms and interpret them as exponentials of biderivations. The logarithm of the identity enters into the story.

The Leray–Samelson theorem also has a signed analogue which gives a categorical equivalence between Coxeter spaces and signed bicommutative signed Coxeter bialgebras.

Corresponding results for Coxeter species are given in Section 7.2. There we work with flats, while here we continue to work with face-types.

13.2.1. Leray–Samelson (primitive part version). Recall from (9.40) the functor $\overline{\mathcal{S}}$ from Coxeter spaces to bicommutative Coxeter bialgebras. It is left adjoint to the primitive part functor $\overline{\mathcal{P}}$.

Theorem 13.14. *Let* $\operatorname{char}(\Bbbk) \nmid |W|$. *The adjunction*

$$\overline{\mathsf{C}}\mathsf{Sp} \underset{\overline{\mathcal{P}}}{\overset{\overline{\mathcal{S}}}{\rightleftarrows}} {}^{\mathrm{co}}\mathsf{Bimon}^{\mathrm{co}}(\overline{\mathsf{C}}\mathsf{Sp})$$

is an adjoint equivalence of categories.

This is the *Leray–Samelson theorem* for bicommutative Coxeter bialgebras. Thus, for V a Coxeter space, $\overline{\mathcal{P}}(\overline{\mathcal{S}}(\mathrm{V})) = \mathrm{V}$ as Coxeter spaces, and for H a bicommutative Coxeter bialgebra,

$$\overline{\mathcal{S}}(\overline{\mathcal{P}}(\mathrm{H})) \xrightarrow{\cong} \mathrm{H} \tag{13.12}$$

as Coxeter bialgebras. This isomorphism is the unit of the adjunction. It arises from the universal property of $\overline{\mathcal{S}}$ in Theorem 9.24. Explicitly, on the Z-component, it sends a representative $x \in \overline{\mathcal{P}}(\mathrm{H})[T]$ for $T \geq Z$ to $\mu_Z^T(x)$. Moreover, (13.12) equals the composite map

$$\overline{\mathcal{S}}(\overline{\mathcal{P}}(\mathrm{H})) \xrightarrow{\cong} \overline{\mathcal{S}}^{\vee}(\overline{\mathcal{P}}(\mathrm{H})) \xrightarrow{\cong} \mathrm{H},$$

where the first isomorphism is the norm map in (9.142) and second isomorphism is as in (13.1).

Proof. The claim $\overline{\mathcal{P}}(\overline{\mathcal{S}}(\mathrm{V})) = \mathrm{V}$ is item (i) of Exercise 9.96. For the claim that (13.12) is an isomorphism, we check that it equals the above composite map. This composite sends the representative $x \in \overline{\mathcal{P}}(\mathrm{H})[T]$ for $T \geq Z$ to

$$\begin{aligned}\sum_{\substack{T' \geq Z:\\ \mathrm{s}(T'/Z)=\mathrm{s}(T/Z)}} \; \sum_{\substack{\sigma: T \to T'\\ \text{local to } Z}} \zeta(Z,T')\, \mu_Z^{T'} \sigma(x) &= \Big(|\mathrm{s}(T/Z)| \sum_{\substack{T' \geq Z:\\ \mathrm{s}(T'/Z)=\mathrm{s}(T/Z)}} \zeta(Z,T') \Big) \mu_Z^T(x)\\ &= \mu_Z^T(x),\end{aligned}$$

as required. The first step used the commutativity axiom (8.86). The second step used the flat-additivity formula (1.83). Note very carefully how $\boldsymbol{\zeta}$ drops out of the calculation. □

Exercise 13.15. Show: The inverse of the map (13.12), namely,

$$\mathrm{H} \xrightarrow{\cong} \overline{\mathcal{S}}(\overline{\mathcal{P}}(\mathrm{H})), \tag{13.13}$$

on the Z-component, sends $x \in \mathrm{H}[Z]$ to the image of

$$\sum_{U \geq T \geq Z} \boldsymbol{\zeta}(Z,T)\, \boldsymbol{\mu}(T,U)\, \mu_T^U \Delta_Z^U(x) \;\in\; \overline{\mathcal{T}}(\overline{\mathcal{P}}(\mathrm{H}))[Z]$$

in the quotient $\overline{\mathcal{S}}(\overline{\mathcal{P}}(\mathrm{H}))[Z]$. The sum is over both T and U. In this formula, $\boldsymbol{\zeta}$ is any noncommutative zeta function and $\boldsymbol{\mu}$ any noncommutative Möbius function (not necessarily inverse to $\boldsymbol{\zeta}$). Moreover, (13.13) equals the composite map

$$\mathrm{H} \to \overline{\mathcal{S}}^{\vee}(\overline{\mathcal{P}}(\mathrm{H})) \to \overline{\mathcal{S}}(\overline{\mathcal{P}}(\mathrm{H})),$$

with the first isomorphism as in (13.2) and second as in (9.143).

We refer to (13.12) and (13.13) as the *Leray–Samelson isomorphisms* (primitive part version). They are uniquely defined in contrast to the Borel–Hopf isomorphisms (13.1) and (13.2) which depend on $\boldsymbol{\zeta}$ and $\boldsymbol{\mu}$.

13.2.2. Exponentials of biderivations. When H is bicommutative, the maps (13.1) and (13.2) are isomorphisms of Coxeter bialgebras. We deduce from Exercise 13.3 that they are exponentials of the biderivations

$$\begin{array}{ccc} \overline{\mathcal{S}}^{\vee}(\overline{\mathcal{P}}(\mathrm{H})) & \dashrightarrow & \mathrm{H} \\ \big\downarrow & & \big\uparrow \\ \overline{\mathcal{P}}(\mathrm{H}) & \xrightarrow[\mathrm{id}]{} & \overline{\mathcal{P}}(\mathrm{H}) \end{array} \qquad \begin{array}{ccc} \mathrm{H} & \dashrightarrow & \overline{\mathcal{S}}^{\vee}(\overline{\mathcal{P}}(\mathrm{H})) \\ \big\downarrow & & \big\uparrow \\ \overline{\mathcal{Q}}(\mathrm{H}) & \xrightarrow[\log(\mathrm{id})]{} & \overline{\mathcal{P}}(\mathrm{H}). \end{array}$$

Similarly, one checks that the maps (13.12) and (13.13) are exponentials of the biderivations

$$\begin{array}{ccc} \overline{\mathcal{S}}(\overline{\mathcal{P}}(\mathrm{H})) & \dashrightarrow & \mathrm{H} \\ \big\downarrow & & \big\uparrow \\ \overline{\mathcal{P}}(\mathrm{H}) & \xrightarrow[\mathrm{id}]{} & \overline{\mathcal{P}}(\mathrm{H}) \end{array} \qquad \begin{array}{ccc} \mathrm{H} & \dashrightarrow & \overline{\mathcal{S}}(\overline{\mathcal{P}}(\mathrm{H})) \\ \big\downarrow & & \big\uparrow \\ \overline{\mathcal{Q}}(\mathrm{H}) & \xrightarrow[\log(\mathrm{id})]{} & \overline{\mathcal{P}}(\mathrm{H}). \end{array}$$

Also see Exercise 9.97 in this regard.

Exercise 13.16. Use the above interpretation of (13.12) and (13.13) as exponentials to formally check that they are inverses of each other.

13.2.3. Commutative characteristic operations. Regarding finer information on the isomorphisms, observe that Lemma 13.4 and Proposition 13.5 also apply to a bicommutative Coxeter bialgebra. In this situation, we can replace $\mathsf{E}_{\lambda/Z}$ by $\mathsf{Q}_{\lambda/Z}$ which is an idempotent element of the invariant Birkhoff algebra of $\boldsymbol{\alpha}_Z$.

13.2.4. Leray–Samelson (indecomposable part version). Recall from (9.43) the functor $\overline{\mathcal{S}}^{\vee}$ from Coxeter spaces to bicommutative Coxeter bialgebras. It is right adjoint to the indecomposable part functor $\overline{\mathcal{Q}}$.

Theorem 13.17. *Let* $\mathrm{char}(\Bbbk) \nmid |W|$. *The adjunction*

$${}^{\mathrm{co}}\mathsf{Bimon}^{\mathrm{co}}(\overline{\mathsf{C}}\mathsf{Sp}) \underset{\overline{\mathcal{S}}^{\vee}}{\overset{\overline{\mathcal{Q}}}{\rightleftarrows}} \overline{\mathsf{C}}\mathsf{Sp}$$

is an adjoint equivalence of categories.

This is the *Leray–Samelson theorem* in dual form. Thus, for V a Coxeter space, $\overline{\mathcal{Q}}(\overline{\mathcal{S}}^{\vee}(\mathrm{V})) = \mathrm{V}$ as Coxeter spaces, and for H a bicommutative Coxeter bialgebra,

$$\mathrm{H} \xrightarrow{\cong} \overline{\mathcal{S}}^{\vee}(\overline{\mathcal{Q}}(\mathrm{H})) \tag{13.14}$$

as Coxeter bialgebras. This isomorphism is the counit of the adjunction. It arises from the universal property of $\overline{\mathcal{S}}^{\vee}$ in Theorem 9.27. Explicitly, on the

Z-component, into the T-summand, it sends $x \in \mathrm{H}[Z]$ to the image of $\Delta_Z^T(x)$ in the quotient $\overline{\mathcal{Q}}(\mathrm{H})[T]$. Moreover, (13.14) equals the composite map

$$\mathrm{H} \xrightarrow{\cong} \overline{\mathcal{S}}(\overline{\mathcal{Q}}(\mathrm{H})) \xrightarrow{\cong} \overline{\mathcal{S}}^{\vee}(\overline{\mathcal{Q}}(\mathrm{H}))$$

with the first isomorphism as in (13.8) and second as in (9.142).

Proof. The claim $\overline{\mathcal{Q}}(\overline{\mathcal{S}}^{\vee}(\mathrm{V})) = \mathrm{V}$ is item (ii) of Exercise 9.96. For the claim that (13.14) is an isomorphism, we use the flat-additivity formula (1.83) to check that it equals the above composite map (as in the proof of Theorem 13.14). □

The inverse of the map (13.14), namely,

$$\overline{\mathcal{S}}^{\vee}(\overline{\mathcal{Q}}(\mathrm{H})) \xrightarrow{\cong} \mathrm{H}, \tag{13.15}$$

on the Z-component, sends the representative $(x_T)_{T \geq Z} \in \overline{\mathcal{S}}^{\vee}(\mathrm{H})[Z]$ to

$$\sum_{U \geq T \geq Z} \zeta(Z,T)\, \boldsymbol{\mu}(T,U)\, \mu_Z^U \Delta_T^U(x_T) \ \in \ \mathrm{H}[Z].$$

The sum is over both T and U. Moreover, (13.15) equals the composite map

$$\overline{\mathcal{S}}^{\vee}(\overline{\mathcal{Q}}(\mathrm{H})) \to \overline{\mathcal{S}}(\overline{\mathcal{Q}}(\mathrm{H})) \to \mathrm{H},$$

with the first isomorphism as in (9.143) and second as in (13.9).

We refer to (13.14) and (13.15) as the *Leray–Samelson isomorphisms* (indecomposable part version).

13.2.5. Exponentials of biderivations. When H is bicommutative, the maps (13.8) and (13.9) are isomorphisms of Coxeter bialgebras. We deduce from Exercise 13.9 that they are exponentials of the biderivations

$$\begin{array}{ccc} \mathrm{H} & \dashrightarrow & \overline{\mathcal{S}}(\overline{\mathcal{Q}}(\mathrm{H})) \\ \downarrow & & \uparrow \\ \overline{\mathcal{Q}}(\mathrm{H}) & \xrightarrow[\mathrm{id}]{} & \overline{\mathcal{Q}}(\mathrm{H}) \end{array} \qquad \begin{array}{ccc} \overline{\mathcal{S}}(\overline{\mathcal{Q}}(\mathrm{H})) & \dashrightarrow & \mathrm{H} \\ \downarrow & & \uparrow \\ \overline{\mathcal{Q}}(\mathrm{H}) & \xrightarrow[\log(\mathrm{id})]{} & \overline{\mathcal{P}}(\mathrm{H}). \end{array}$$

Similarly, one checks that the maps (13.14) and (13.15) are exponentials of the biderivations

$$\begin{array}{ccc} \mathrm{H} & \dashrightarrow & \overline{\mathcal{S}}^{\vee}(\overline{\mathcal{Q}}(\mathrm{H})) \\ \downarrow & & \uparrow \\ \overline{\mathcal{Q}}(\mathrm{H}) & \xrightarrow[\mathrm{id}]{} & \overline{\mathcal{Q}}(\mathrm{H}) \end{array} \qquad \begin{array}{ccc} \overline{\mathcal{S}}^{\vee}(\overline{\mathcal{Q}}(\mathrm{H})) & \dashrightarrow & \mathrm{H} \\ \downarrow & & \uparrow \\ \overline{\mathcal{Q}}(\mathrm{H}) & \xrightarrow[\log(\mathrm{id})]{} & \overline{\mathcal{P}}(\mathrm{H}). \end{array}$$

13.2.6. Connecting primitive and indecomposable part versions. Recall from Proposition 9.70 that for a bicommutative Coxeter bialgebra H, the maps pq and $\log(\mathrm{id})$ are inverses of each other. They allow us to connect the two versions of Leray–Samelson as follows.

For a bicommutative Coxeter bialgebra H, the diagram

$$
\begin{array}{ccccc}
\overline{\mathcal{S}}(\overline{\mathcal{P}}(\mathrm{H})) & \xleftrightarrow[\overline{\mathcal{S}}(\log(\mathrm{id}))]{\overline{\mathcal{S}}(\mathrm{pq})} & & & \overline{\mathcal{S}}(\overline{\mathcal{Q}}(\mathrm{H})) \\
\uparrow & \nwarrow & & \nearrow & \uparrow \\
& \mathrm{LS} \searrow & & \swarrow \mathrm{BH} & \\
\mathrm{norm} & & \mathrm{H} & & \mathrm{norm} \\
& \mathrm{BH} \nearrow & & \nwarrow \mathrm{LS} & \\
\downarrow & \swarrow & & \searrow & \downarrow \\
\overline{\mathcal{S}}^{\vee}(\overline{\mathcal{P}}(\mathrm{H})) & \xleftrightarrow[\overline{\mathcal{S}}^{\vee}(\log(\mathrm{id}))]{\overline{\mathcal{S}}^{\vee}(\mathrm{pq})} & & & \overline{\mathcal{S}}^{\vee}(\overline{\mathcal{Q}}(\mathrm{H}))
\end{array}
\tag{13.16}
$$

commutes (with objects connected by pairs of inverse isomorphisms). The oblique maps in the top triangle are (13.12), (13.13), (13.8), (13.9), while those in the bottom triangle are (13.1), (13.2), (13.14), (13.15). The vertical isomorphisms are instances of the norm map (9.142) and its inverse (9.143).

Exercise 13.18. A more general relation between the Borel–Hopf and Leray–Samelson isomorphisms is as follows.

Check: For a commutative Coxeter bialgebra H, the first diagram below commutes. The horizontal maps are injective.

$$
\begin{array}{ccc}
\overline{\mathcal{S}}(\overline{\mathcal{P}}(\mathrm{H})) & \xhookrightarrow{\overline{\mathcal{S}}(\mathrm{pq})} & \overline{\mathcal{S}}(\overline{\mathcal{Q}}(\mathrm{H})) \\
\mathrm{LS}\,\big\uparrow\!\!\downarrow & & \big\downarrow\!\!\uparrow\,\mathrm{BH} \\
\mathrm{H}^{coab} & \xhookrightarrow[\pi_{\mathrm{H}}^{\vee}]{} & \mathrm{H}
\end{array}
\qquad\qquad
\begin{array}{ccc}
\overline{\mathcal{S}}^{\vee}(\overline{\mathcal{P}}(\mathrm{H})) & \xrightarrow{\overline{\mathcal{S}}^{\vee}(\mathrm{pq})} & \overline{\mathcal{S}}^{\vee}(\overline{\mathcal{Q}}(\mathrm{H})) \\
\mathrm{BH}\,\big\uparrow\!\!\downarrow & & \big\downarrow\!\!\uparrow\,\mathrm{LS} \\
\mathrm{H} & \xrightarrow[\pi_{\mathrm{H}}]{} & \mathrm{H}_{ab}
\end{array}
$$

For a cocommutative Coxeter bialgebra H, the second diagram above commutes. The horizontal maps are surjective.

Note: In order to employ the LS isomorphisms, we first use (9.58).

Example 13.19 (Norm map). For a Coxeter space V, the norm map $\overline{\mathcal{S}}(\mathrm{V}) \to \overline{\mathcal{S}}^{\vee}(\mathrm{V})$ in (9.142) is an example of a Leray–Samelson isomorphism in two ways. It is an instance of (13.12) with $\mathrm{H} = \overline{\mathcal{S}}^{\vee}(\mathrm{V})$ and an instance of (13.14) with $\mathrm{H} = \overline{\mathcal{S}}(\mathrm{V})$. This follows from Exercise 9.98. It also then follows that the inverse isomorphism $\overline{\mathcal{S}}^{\vee}(\mathrm{V}) \to \overline{\mathcal{S}}(\mathrm{V})$ in (9.143) is an instance of (13.13) and (13.15). □

Example 13.20 (Coxeter bialgebra of flat-types). Recall from Section 10.5 the Coxeter bialgebra of flat-types $\overline{\Pi}$. It is bicommutative. Hence, by Leray–Samelson Theorems 13.14 and 13.17,

$$\overline{\Pi} \cong \overline{\mathcal{S}}(\overline{\mathcal{P}}(\overline{\Pi})) \cong \overline{\mathcal{S}}^{\vee}(\overline{\mathcal{Q}}(\overline{\Pi}))$$

as Coxeter bialgebras. This is consistent with observations about (co)freeness, primitive and indecomposable parts, and self-duality of $\overline{\Pi}$ from Section 10.5. □

13.2.7. Signed analogue. Leray–Samelson Theorems 13.14 and 13.17 have signed analogues which are stated below. They can be deduced from the unsigned versions by conjugating by the u-signature functor (8.99).

Theorem 13.21. *Let* $\mathrm{char}(\Bbbk) \nmid |W|$. *The adjunction*

$$\overline{\mathsf{C}}\mathsf{Sp} \underset{\overline{\mathcal{P}}}{\overset{\overline{\mathcal{E}}}{\rightleftarrows}} (-1)\text{-}^{\mathsf{co}}\mathsf{Bimon}^{\mathsf{co}}(\overline{\mathsf{C}}\mathsf{Sp})$$

is an adjoint equivalence of categories.

Theorem 13.22. *Let* $\mathrm{char}(\Bbbk) \nmid |W|$. *The adjunction*

$$(-1)\text{-}^{\mathsf{co}}\mathsf{Bimon}^{\mathsf{co}}(\overline{\mathsf{C}}\mathsf{Sp}) \underset{\overline{\mathcal{E}}^{\vee}}{\overset{\overline{\mathcal{Q}}}{\rightleftarrows}} \overline{\mathsf{C}}\mathsf{Sp}$$

is an adjoint equivalence of categories.

These are the *Leray–Samelson theorems* for signed bicommutative signed Coxeter bialgebras.

13.3. Rigidity of Coxeter q-bialgebras for q not a root of unity

For q not a root of unity, the category of Coxeter q-bialgebras is equivalent to the category of Coxeter spaces. We call this the rigidity theorem of Coxeter q-bialgebras. It comes in two flavors, one involving the primitive part functor and the other involving the indecomposable part functor. Moreover, within each flavor, there are two subflavors depending on whether we use the functor $\overline{\mathcal{T}}_q$ or $\overline{\mathcal{T}}_q^{\vee}$. (Recall: The two are isomorphic via the q-norm map.) We write down formulas for the involved Coxeter q-bialgebra isomorphisms in terms of the two-sided q-zeta and q-Möbius functions ζ_q and $\boldsymbol{\mu}_q$, and interpret them as q-exponentials of biderivations. The q-logarithm of the identity enters into the story. The exposition here follows the combined pattern of Sections 13.1 and 13.2. We bring out this similarity towards the end of the section.

The parallel discussion for Coxeter species is in Section 7.3. There we work with local bifaces, while here we work with local biface-types.

13.3.1. Rigidity of Coxeter q-bialgebras (primitive part version). Let q be a scalar which is not a root of unity. Recall the two-sided q-zeta function ζ_q and q-Möbius function $\boldsymbol{\mu}_q$. Recall from Section 9.3.6 the functor $\overline{\mathcal{T}}_q^{\vee}$ from Coxeter spaces to Coxeter q-bialgebras. Let H be a Coxeter q-bialgebra. Define a natural map

$$\overline{\mathcal{T}}_q^{\vee}(\overline{\mathcal{P}}(\mathrm{H})) \to \mathrm{H}, \tag{13.17}$$

on the Z-component, on the T'-summand, by

$$\sum_{T\colon T \geq Z} \; \sum_{\substack{\sigma\colon T' \to T \\ \text{local to } Z}} \zeta_q(Z, T, \sigma, T')\, \mu_Z^T\, \sigma.$$

Similarly, define a natural map

$$\mathrm{H} \to \overline{\mathcal{T}}_q^{\vee}(\overline{\mathcal{P}}(\mathrm{H})), \tag{13.18}$$

on the Z-component, into the T'-summand, by

$$\mathbf{log}_q(\mathrm{id})_{T'} \Delta_Z^{T'} = \sum_{\substack{U, U'\colon \\ U, U' \geq T'}} \; \sum_{\substack{\sigma\colon U' \to U \\ \text{local to } T'}} \boldsymbol{\mu}_q(T', U, \sigma, U')\, \mu_{T'}^U \sigma \Delta_Z^{U'}.$$

This map arises from cofreeness of $\overline{\mathcal{T}}_q^{\vee}$. More precisely, apply Theorem 9.19 with $\mathrm{V} := \overline{\mathcal{P}}(\mathrm{H})$ and $f := \mathbf{log}_q(\mathrm{id}) : \overline{\mathcal{Q}}(\mathrm{H}) \to \overline{\mathcal{P}}(\mathrm{H})$ as in Proposition 9.72, and use formula (9.113). This also shows that (13.18) is a Coxeter q-bialgebra morphism.

Theorem 13.23. *For q not a root of unity, for a Coxeter q-bialgebra* H, *the maps* (13.17) *and* (13.18) *are inverse isomorphisms of Coxeter q-bialgebras.*

PROOF. We can proceed directly as in the proof of Theorem 13.1. We leave these calculations as an exercise. They are slightly more involved now due to the two-sided q-zeta and q-Möbius functions. Alternatively, one may use Exercise 13.24 below. □

Exercise 13.24. For a Coxeter q-bialgebra H for q not a root of unity, consider the following biderivations f and g.

$$
\begin{array}{ccc}
\overline{\mathcal{T}}_q^{\vee}(\overline{\mathcal{P}}(\mathrm{H})) & \xrightarrow{\ f\ } & \mathrm{H} \\
\big\downarrow & & \big\uparrow \\
\overline{\mathcal{P}}(\mathrm{H}) & \xrightarrow[\mathrm{id}]{} & \overline{\mathcal{P}}(\mathrm{H})
\end{array}
\qquad\qquad
\begin{array}{ccc}
\mathrm{H} & \xrightarrow{\ g\ } & \overline{\mathcal{T}}_q^{\vee}(\overline{\mathcal{P}}(\mathrm{H})) \\
\big\downarrow & & \big\uparrow \\
\overline{\mathcal{Q}}(\mathrm{H}) & \xrightarrow[\mathbf{log}_q(\mathrm{id})]{} & \overline{\mathcal{P}}(\mathrm{H})
\end{array}
$$

Check: The maps (13.17) and (13.18) are q-exponentials (9.99a) of f and g, respectively. (For the second calculation, use (9.36) and the two-sided q-flat-additivity formula (1.110).) Now deduce from Theorem 9.61 that $\mathbf{exp}_q(f)$ and $\mathbf{exp}_q(g)$ are Coxeter q-bialgebra morphisms. Moreover, deduce that they are inverses of each other by justifying the steps below.

$$
\begin{aligned}
\mathbf{exp}_q(f)\,\mathbf{exp}_q(g) &= \mathbf{exp}_q(fg) = \mathbf{exp}_q(\mathbf{log}_q(\mathrm{id})) = \mathrm{id}, \\
\mathbf{exp}_q(g)\,\mathbf{exp}_q(f) &= \mathbf{exp}_q(gf) = \mathbf{exp}_q(\mathbf{log}_q(\mathrm{id})) = \mathrm{id}\,.
\end{aligned}
$$

(Use Exercise 9.64, item (2).)

Theorem 13.23 is the *rigidity theorem for Coxeter q-bialgebras*. It is reformulated below in Theorem 13.25 as an adjoint equivalence. Recall from Section 9.3.3 the functor $\overline{\mathcal{T}}_q$ from Coxeter spaces to Coxeter q-bialgebras. It is left adjoint to the primitive part functor $\overline{\mathcal{P}}$.

Theorem 13.25. *For q not a root of unity, the adjunction*

$$
\overline{\mathsf{C}}\mathsf{Sp} \underset{\overline{\mathcal{P}}}{\overset{\overline{\mathcal{T}}_q}{\rightleftarrows}} q\text{-}\mathsf{Bimon}(\overline{\mathsf{C}}\mathsf{Sp})
$$

is an adjoint equivalence of categories.

Thus, for V a Coxeter space, $\overline{\mathcal{P}}(\overline{\mathcal{T}}_q(\mathrm{V})) = \mathrm{V}$ as Coxeter spaces, and for H a Coxeter q-bialgebra,

$$
\overline{\mathcal{T}}_q(\overline{\mathcal{P}}(\mathrm{H})) \xrightarrow{\cong} \mathrm{H} \tag{13.19}
$$

as Coxeter q-bialgebras. This isomorphism is the unit of the adjunction. It arises from the universal property of $\overline{\mathcal{T}}_q$ in Theorem 9.15. Explicitly, on the Z-component, it sends $x \in \overline{\mathcal{P}}(\mathrm{H})[T]$ for $T \geq Z$ to $\mu_Z^T(x)$. Moreover, (13.19) equals the composite map

$$
\overline{\mathcal{T}}_q(\overline{\mathcal{P}}(\mathrm{H})) \xrightarrow{\cong} \overline{\mathcal{T}}_q^{\vee}(\overline{\mathcal{P}}(\mathrm{H})) \xrightarrow{\cong} \mathrm{H},
$$

where the first isomorphism is the q-norm map in (9.140) and second isomorphism is as in (13.17).

PROOF. The claim $\overline{\mathcal{P}}(\overline{\mathcal{T}_q}(\mathrm{V})) = \mathrm{V}$ is item (i) of Exercise 9.106. For the claim that (13.19) is an isomorphism, we check that it equals the above composite map. This composite sends $x \in \overline{\mathcal{P}}(\mathrm{H})[T]$ for $T \geq Z$ to

$$\sum_{\substack{T',T'':\\ T',T''\geq Z}} \sum_{\substack{\sigma: T\to T'\\ \tau: T'\to T''}} \zeta_q(Z,T'',\tau,T')\, q^{l(\sigma)}\, \mu_Z^{T''}\, \tau\sigma(x)$$
$$= \sum_{\substack{T'':T''\geq Z\\ \rho: T\to T''}} \Bigg(\sum_{\substack{T':T'\geq Z\\ \sigma: T\to T'\\ \tau: T'\to T''\\ \rho=\tau\sigma}} \zeta_q(Z,T'',\tau,T')\, q^{l(\sigma)} \Bigg) \mu_Z^{T''}\, \rho(x)$$
$$= \mu_Z^{T}(x),$$

as required. In the calculation above, it is implicit that σ, τ, ρ are local to Z. The second step used the two-sided q-flat-additivity formula (1.110). Note very carefully how ζ_q drops out of the calculation. Compare with the calculation in the proof of Theorem 13.14. □

Exercise 13.26. Show: The inverse of the map (13.19), namely,

$$\mathrm{H} \xrightarrow{\cong} \overline{\mathcal{T}_q}(\overline{\mathcal{P}}(\mathrm{H})), \tag{13.20}$$

on the Z-component, into the T-summand, is

$$\sum_{T':T'\geq Z} \sum_{\substack{\sigma: T'\to T\\ \text{local to } Z}} \zeta_q(Z,T,\sigma,T')\, \sigma\, \mathbf{log}_q(\mathrm{id})_{T'}\, \Delta_Z^{T'}.$$

Moreover, (13.20) equals the composite map

$$\mathrm{H} \to \overline{\mathcal{T}_q}^{\vee}(\overline{\mathcal{P}}(\mathrm{H})) \to \overline{\mathcal{T}_q}(\overline{\mathcal{P}}(\mathrm{H})),$$

with the first isomorphism as in (13.18) and second as in (9.148).

Exercise 13.27. For a Coxeter q-bialgebra H for q not a root of unity, consider the following biderivations f and g.

$$\begin{array}{ccc} \overline{\mathcal{T}_q}(\overline{\mathcal{P}}(\mathrm{H})) & \xrightarrow{f} & \mathrm{H} \\ \downarrow & & \uparrow \\ \overline{\mathcal{P}}(\mathrm{H}) & \xrightarrow[\mathrm{id}]{} & \overline{\mathcal{P}}(\mathrm{H}) \end{array} \qquad \begin{array}{ccc} \mathrm{H} & \xrightarrow{g} & \overline{\mathcal{T}_q}(\overline{\mathcal{P}}(\mathrm{H})) \\ \downarrow & & \uparrow \\ \overline{\mathcal{Q}}(\mathrm{H}) & \xrightarrow[\mathbf{log}_q(\mathrm{id})]{} & \overline{\mathcal{P}}(\mathrm{H}) \end{array}$$

Check: The maps (13.19) and (13.20) are q-exponentials (9.99a) of f and g, respectively. Compare with Exercise 13.24 using Exercise 9.107. One can apply Exercise 9.64 to pass from one situation to the other.

13.3.2. Two-sided characteristic operations.

We now obtain finer information on the isomorphisms in the rigidity theorem by using two-sided characteristic operations by certain $\mathtt{Q}$-basis and $\mathtt{Q}^d$-basis elements in the invariant q-Janus algebra.

Lemma 13.28. *Let* H *be a Coxeter q-bialgebra for q not a root of unity. Then, for any face-type Z,*

$$\mathrm{H}[Z] = \bigoplus_{T:\, T \geq Z} \mathtt{Q}_{(T/Z,\mathrm{id},T/Z)} \cdot \mathrm{H}[Z] = \bigoplus_{T':\, T' \geq Z} \mathtt{Q}^{d}_{(T'/Z,\mathrm{id},T'/Z)} \cdot \mathrm{H}[Z].$$

PROOF. Recall from Section 9.13.3 that the invariant q-Janus algebra $\overline{\mathrm{J}}_q[Z]$ acts on $\mathrm{H}[Z]$. Now use that the $\mathtt{Q}_{(T/Z,\mathrm{id},T/Z)}$, as T varies, are mutually orthogonal idempotents and add up to the unit element, and similarly, for the $\mathtt{Q}^{d}_{(T'/Z,\mathrm{id},T'/Z)}$ as T' varies. This is a part of Theorem 1.108 applied to $\boldsymbol{\alpha}_Z$. □

Exercise 13.29. Let H be a Coxeter q-bialgebra for q not a root of unity. Check: For face-types $Z \leq T$, the map

$$\mu_Z^T : \overline{\mathcal{P}}(\mathrm{H})[T] \xrightarrow{\cong} \mathtt{Q}_{(T/Z,\mathrm{id},T/Z)} \cdot \mathrm{H}[Z]$$

is a linear isomorphism, and summing over T yields (13.19). Similarly, for face-types $Z \leq T'$, the map

$$\sum_{\substack{\sigma:\, T' \to T \\ \text{local to } Z}} \zeta_q(Z,T,\sigma,T')\, \mu_Z^T\, \sigma : \overline{\mathcal{P}}(\mathrm{H})[T'] \xrightarrow{\cong} \mathtt{Q}^{d}_{(T'/Z,\mathrm{id},T'/Z)} \cdot \mathrm{H}[Z]$$

is a linear isomorphism, and summing over T' yields (13.17). (Use formulas (1.145) applied to $\boldsymbol{\alpha}_Z$ and (9.150) and the two-sided q-flat-additivity formula (1.110).)

In particular,

$$\overline{\mathcal{P}}(\mathrm{H})[Z] = \mathtt{Q}_{(Z/Z,\mathrm{id},Z/Z)} \cdot \mathrm{H}[Z].$$

Compare with Proposition 9.127.

13.3.3. Rigidity of Coxeter q-bialgebras (indecomposable part version).

The functor $\overline{\mathcal{T}}_q^{\vee}$ from Coxeter spaces to Coxeter q-bialgebras is right adjoint to the indecomposable part functor $\overline{\mathcal{Q}}$. It yields the following dual version of Theorem 13.25.

Theorem 13.30. *For q not a root of unity, the adjunction*

$$q\text{-}\mathsf{Bimon}(\overline{\mathsf{C}}\mathsf{Sp}) \underset{\overline{\mathcal{T}}_q^{\vee}}{\overset{\overline{\mathcal{Q}}}{\rightleftarrows}} \overline{\mathsf{C}}\mathsf{Sp}$$

is an adjoint equivalence of categories.

The four Coxeter q-bialgebra isomorphisms involved in the dual story are highlighted below. Define a natural map

$$\mathrm{H} \xrightarrow{\cong} \overline{\mathcal{T}}_q(\overline{\mathcal{Q}}(\mathrm{H})), \tag{13.21}$$

on the Z-component, into the T'-summand, by

$$\sum_{T:\, T \geq Z} \sum_{\substack{\sigma:\, T \to T' \\ \text{local to } Z}} \zeta_q(Z,T',\sigma,T)\, \sigma\, \Delta_Z^T.$$

Its inverse

$$\overline{\mathcal{T}}_q(\overline{\mathcal{Q}}(\mathrm{H})) \xrightarrow{\cong} \mathrm{H}, \tag{13.22}$$

on the Z-component, on the T'-summand, is

$$\mu_Z^{T'} \, \mathbf{log}_q(\mathrm{id})_{T'} = \sum_{\substack{U,U':\\ U,U' \geq T'}} \; \sum_{\substack{\sigma: U' \to U \\ \text{local to } T'}} \mu_q(T', U, \sigma, U') \, \mu_Z^U \sigma \Delta_{T'}^{U'}.$$

This map arises from freeness of $\overline{\mathcal{T}}_q$ by applying Theorem 9.15 with $\mathrm{V} := \overline{\mathcal{Q}}(\mathrm{H})$ and $f := \mathbf{log}_q(\mathrm{id})$. The maps (13.21) and (13.22) are q-exponentials of the biderivations

$$\begin{array}{ccc} \mathrm{H} & \dashrightarrow & \overline{\mathcal{T}}_q(\overline{\mathcal{Q}}(\mathrm{H})) \\ \downarrow & & \uparrow \\ \overline{\mathcal{Q}}(\mathrm{H}) & \xrightarrow[\mathrm{id}]{} & \overline{\mathcal{Q}}(\mathrm{H}) \end{array} \qquad \begin{array}{ccc} \overline{\mathcal{T}}_q(\overline{\mathcal{Q}}(\mathrm{H})) & \dashrightarrow & \mathrm{H} \\ \downarrow & & \uparrow \\ \overline{\mathcal{Q}}(\mathrm{H}) & \xrightarrow[\mathbf{log}_q(\mathrm{id})]{} & \overline{\mathcal{P}}(\mathrm{H}). \end{array}$$

Similarly, define a natural map

$$\mathrm{H} \xrightarrow{\cong} \overline{\mathcal{T}}_q^{\vee}(\overline{\mathcal{Q}}(\mathrm{H})), \tag{13.23}$$

on the Z-component, into the T-summand, by Δ_Z^T followed by projection to $\overline{\mathcal{Q}}(\mathrm{H})[T]$. This isomorphism is the counit of the adjunction in Theorem 13.30. It arises from the universal property of $\overline{\mathcal{T}}_q^{\vee}$ in Theorem 9.19. Its inverse

$$\overline{\mathcal{T}}_q^{\vee}(\overline{\mathcal{Q}}(\mathrm{H})) \xrightarrow{\cong} \mathrm{H}, \tag{13.24}$$

on the Z-component, on the T-summand, is

$$\sum_{T': T' \geq Z} \; \sum_{\substack{\sigma: T \to T' \\ \text{local to } Z}} \zeta_q(Z, T', \sigma, T) \, \mu_Z^{T'} \, \sigma \, \mathbf{log}_q(\mathrm{id})_T.$$

The maps (13.23) and (13.24) are q-exponentials of the biderivations

$$\begin{array}{ccc} \mathrm{H} & \dashrightarrow & \overline{\mathcal{T}}_q^{\vee}(\overline{\mathcal{Q}}(\mathrm{H})) \\ \downarrow & & \uparrow \\ \overline{\mathcal{Q}}(\mathrm{H}) & \xrightarrow[\mathrm{id}]{} & \overline{\mathcal{Q}}(\mathrm{H}) \end{array} \qquad \begin{array}{ccc} \overline{\mathcal{T}}_q^{\vee}(\overline{\mathcal{Q}}(\mathrm{H})) & \dashrightarrow & \mathrm{H} \\ \downarrow & & \uparrow \\ \overline{\mathcal{Q}}(\mathrm{H}) & \xrightarrow[\mathbf{log}_q(\mathrm{id})]{} & \overline{\mathcal{P}}(\mathrm{H}). \end{array}$$

They relate to maps (13.21) and (13.22) via the q-norm map (9.140) and its inverse (9.148).

13.3.4. Connecting primitive and indecomposable part versions. Recall from Proposition 9.72 that for a Coxeter q-bialgebra H for q not a root of unity, the maps $\mathfrak{pq}$ and $\mathbf{log}_q(\mathrm{id})$ are inverses of each other. They allow us to connect the two versions of the rigidity result as follows.

For a Coxeter q-bialgebra H for q not a root of unity, the diagram

$$
\begin{array}{ccccc}
\overline{\mathcal{T}}_q(\overline{\mathcal{P}}(\mathrm{H})) & \xrightleftharpoons[\overline{\mathcal{T}}_q(\mathbf{log}_q(\mathrm{id}))]{\overline{\mathcal{T}}_q(\mathrm{pq})} & & & \overline{\mathcal{T}}_q(\overline{\mathcal{Q}}(\mathrm{H})) \\
& \nwarrow\searrow & & \swarrow\nearrow & \\
(\overline{\kappa}_q)_{\overline{\mathcal{P}}(\mathrm{H})} \updownarrow & & \mathrm{H} & & \updownarrow (\overline{\kappa}_q)_{\overline{\mathcal{Q}}(\mathrm{H})} \\
& \swarrow\nearrow & & \nwarrow\searrow & \\
\overline{\mathcal{T}}_q^{\vee}(\overline{\mathcal{P}}(\mathrm{H})) & \xrightleftharpoons[\overline{\mathcal{T}}_q^{\vee}(\mathbf{log}_q(\mathrm{id}))]{\overline{\mathcal{T}}_q^{\vee}(\mathrm{pq})} & & & \overline{\mathcal{T}}_q^{\vee}(\overline{\mathcal{Q}}(\mathrm{H}))
\end{array}
\tag{13.25}
$$

commutes (with objects connected by pairs of inverse isomorphisms). The oblique maps in the top triangle are (13.19), (13.20), (13.21), (13.22), while those in the bottom triangle are (13.17), (13.18), (13.23), (13.24). The vertical maps are instances of the q-norm map (9.140) and its inverse (9.148).

Example 13.31 (**q-norm map**). For q not a root of unity, for a Coxeter space V, the q-norm map $\overline{\mathcal{T}}_q(\mathrm{V}) \to \overline{\mathcal{T}}_q^{\vee}(\mathrm{V})$ in (9.140) is an instance of (13.19) with $\mathrm{H} = \overline{\mathcal{T}}_q^{\vee}(\mathrm{V})$ and an instance of (13.23) with $\mathrm{H} = \overline{\mathcal{T}}_q(\mathrm{V})$. This follows from Exercise 9.93. It also then follows that the inverse map $\overline{\mathcal{T}}_q^{\vee}(\mathrm{V}) \to \overline{\mathcal{T}}_q(\mathrm{V})$ in (9.148) is an instance of (13.20) and (13.24). □

Exercise 13.32. Let H be a Coxeter q-bialgebra for q not a root of unity. Then H is free as a Coxeter algebra and cofree as a Coxeter coalgebra. Deduce: For H, all product components μ_Z^T are injective, while all coproduct components Δ_Z^T are surjective. The special case $T = S$ is mentioned in Exercise 8.76. See also Exercise 10.38.

13.3.5. Similarity with Borel–Hopf and Leray–Samelson.

There are similarities between the Coxeter q-bialgebra rigidity theorem and the Borel–Hopf and Leray–Samelson theorems. They are brought out in Table 13.2.

TABLE 13.2. Comparing structure theorems of Coxeter bialgebras.

Primitive part functor		Indecomposable part functor	
$\overline{\mathcal{T}}_q^{\vee}$ (13.17), (13.18)	$\overline{\mathcal{S}}^{\vee}$ (13.1), (13.2)	$\overline{\mathcal{T}}_q$ (13.21), (13.22)	$\overline{\mathcal{S}}$ (13.8), (13.9)
$\overline{\mathcal{T}}_q$ (13.19), (13.20)	$\overline{\mathcal{S}}$ (13.12), (13.13)	$\overline{\mathcal{T}}_q^{\vee}$ (13.23), (13.24)	$\overline{\mathcal{S}}^{\vee}$ (13.14), (13.15)
Theorem 13.25	Theorem 13.14	Theorem 13.30	Theorem 13.17

Also compare Exercises 13.24 and 13.27 with Section 13.2.2, where the involved isomorphisms (primitive part version) are interpreted as exponentials of biderivations. One can do a similar comparison for the indecomposable part version. Also compare diagrams (13.16) and (13.25).

Remark 13.33. Recall: The unsigned and signed versions of Borel–Hopf relate to each other by conjugation by the u-signature functor (8.99). The same remark applies to Leray–Samelson. For the rigidity theorem for Coxeter q-bialgebras, conjugation by (8.99) has the effect of changing q to $-q$.

Remark 13.34 (Proof by induction). Alternatively, one can prove the rigidity theorem for Coxeter q-bialgebras by induction on the primitive filtration by employing Exercise 9.113. The details are similar to the argument for $\mathcal{A}$-species spelt out in [**11**, proof of Proposition 13.76]. Compare with Remark 13.2 for Borel–Hopf.

13.4. Loday–Ronco for Coxeter 0-bialgebras

Let us now specialize the rigidity theorem of Coxeter q-bialgebras to $q = 0$. It says that the category of Coxeter 0-bialgebras is equivalent to the category of Coxeter spaces. We refer to this as the Loday–Ronco theorem. It is much easier to handle than the general case due to the simple explicit formulas for the two-sided 0-zeta and 0-Möbius functions ζ_0 and μ_0, and hence for the involved isomorphisms.

13.4.1. Loday–Ronco (primitive part version). Recall formulas (1.118) for ζ_0 and μ_0. Theorem 13.25 specializes as follows.

Theorem 13.35. *The adjunction*

$$\overline{\mathsf{C}}\mathsf{Sp} \underset{\overline{\mathcal{P}}}{\overset{\overline{\mathcal{T}_0}}{\rightleftarrows}} 0\text{-}\mathsf{Bimon}(\overline{\mathsf{C}}\mathsf{Sp})$$

is an adjoint equivalence of categories.

This is the *Loday–Ronco theorem* for Coxeter 0-bialgebras. Let us make the involved Coxeter 0-bialgebra isomorphisms explicit. Recall: For V a Coxeter space, $\overline{\mathcal{T}_0}(\mathrm{V}) = \overline{\mathcal{T}_0}^{\vee}(\mathrm{V})$ as Coxeter 0-bialgebras with product and coproduct (9.38). Hence, for H a Coxeter 0-bialgebra, the isomorphisms (13.17) and (13.19) coincide, and so do their inverses (13.18) and (13.20). The map

$$\overline{\mathcal{T}_0}(\overline{\mathcal{P}}(\mathrm{H})) = \overline{\mathcal{T}_0}^{\vee}(\overline{\mathcal{P}}(\mathrm{H})) \xrightarrow{\cong} \mathrm{H}, \tag{13.26}$$

on the Z-component, on the T-summand, is μ^T_Z. The inverse map

$$\mathrm{H} \xrightarrow{\cong} \overline{\mathcal{T}_0}(\overline{\mathcal{P}}(\mathrm{H})) = \overline{\mathcal{T}_0}^{\vee}(\overline{\mathcal{P}}(\mathrm{H})), \tag{13.27}$$

on the Z-component, into the T-summand, is

$$\mathbf{log}_0(\mathrm{id})_T \Delta^T_Z = \sum_{U:\, U \geq T} (-1)^{\mathrm{rk}(U/T)} \mu^U_T \Delta^U_Z.$$

We refer to (13.26) and (13.27) as the *Loday–Ronco isomorphisms* (primitive part version).

Exercise 13.36. Check directly: (13.26) and (13.27) are inverses of each other. (Use identity (9.151).) This yields a proof of Theorem 13.35.

Alternatively, following Remark 13.34, one can prove the Loday–Ronco theorem by induction on the primitive filtration by using Exercise 9.115. The details for $\mathcal{A}$-species are spelt out in [**11**, proof of Proposition 13.1].

13.4.2. Loday–Ronco (indecomposable part version). Loday–Ronco in dual form is stated below. It is the specialization of Theorem 13.30.

Theorem 13.37. *The adjunction*

$$\text{0-Bimon}(\overline{\mathsf{C}}\mathsf{Sp}) \underset{\overline{\mathcal{T}_0}^{\vee}}{\overset{\overline{\mathcal{Q}}}{\rightleftarrows}} \overline{\mathsf{C}}\mathsf{Sp}$$

is an adjoint equivalence of categories.

The involved isomorphisms are as follows. For H a Coxeter 0-bialgebra, the isomorphisms (13.21) and (13.23) coincide, and so do their inverses (13.22) and (13.24). The map

$$\text{H} \xrightarrow{\cong} \overline{\mathcal{T}_0}(\overline{\mathcal{Q}}(\text{H})) = \overline{\mathcal{T}_0}^{\vee}(\overline{\mathcal{Q}}(\text{H})), \tag{13.28}$$

on the Z-component, into the T-summand, is Δ_Z^T followed by projection to $\overline{\mathcal{Q}}(\text{H})[T]$. The inverse map

$$\overline{\mathcal{T}_0}(\overline{\mathcal{Q}}(\text{H})) = \overline{\mathcal{T}_0}^{\vee}(\overline{\mathcal{Q}}(\text{H})) \xrightarrow{\cong} \text{H}, \tag{13.29}$$

on the Z-component, on the T-summand, is

$$\mu_Z^T \, \mathbf{log}_0(\text{id})_T = \sum_{U:\, U \geq T} (-1)^{\text{rk}(U/T)} \mu_Z^U \Delta_T^U.$$

The map as written is from H$[T]$ to H$[Z]$, the point being that it factors through $\overline{\mathcal{Q}}(\text{H})[T]$.

We refer to (13.28) and (13.29) as the *Loday–Ronco isomorphisms* (indecomposable part version).

13.5. Hoffman–Newman–Radford

The Hoffman–Newman–Radford (HNR) theorems are a family of results that provide inverse isomorphisms between two universally constructed Coxeter bialgebras. These isomorphisms are constructed using zeta and Möbius functions. We call them HNR isomorphisms. As an illustration, we explain how these ideas relate to the study of the Coxeter bialgebras of face-types in Section 10.4.

The parallel discussion for Coxeter species is in Section 7.5. There we work with faces, while here we work with face-types.

13.5.1. Shuffles and quasishuffles. Let $\text{char}(\Bbbk) \nmid |W|$. Fix an invariant noncommutative zeta function $\boldsymbol{\zeta}$ and its inverse Möbius function $\boldsymbol{\mu}$.

Let C be a cocommutative Coxeter coalgebra, and C_t its underlying Coxeter space with trivial coproduct. Recall the Coxeter bialgebra $\overline{\mathcal{T}}(\text{C})$ with concatenation product and dequasishuffle coproduct, and $\overline{\mathcal{T}}(\text{C}_t)$ with concatenation product and deshuffle coproduct. (Specialize the relevant entries in Table 9.2 to $q = 1$.) Define a natural map

$$\overline{\mathcal{T}}(\text{C}) \to \overline{\mathcal{T}}(\text{C}_t) \tag{13.30}$$

which on the Z-component is

$$\bigoplus_{T:\,T\geq Z} \mathrm{C}[T] \to \bigoplus_{U:\,U\geq Z} \mathrm{C}[U], \qquad x \mapsto \sum_{U:\,U\geq T} \boldsymbol{\zeta}(T,U)\, \Delta_T^U(x)$$

for $x \in \mathrm{C}[T]$. Similarly, define a natural map

$$\overline{\mathcal{T}}(\mathrm{C}_t) \to \overline{\mathcal{T}}(\mathrm{C}) \tag{13.31}$$

which on the Z-component is

$$\bigoplus_{T:\,T\geq Z} \mathrm{C}[T] \to \bigoplus_{U:\,U\geq Z} \mathrm{C}[U], \qquad x \mapsto \sum_{U:\,U\geq T} \boldsymbol{\mu}(T,U)\, \Delta_T^U(x)$$

for $x \in \mathrm{C}[T]$.

Exercise 13.38. Let $\mathrm{char}(\Bbbk) \nmid |W|$. For a cocommutative Coxeter coalgebra C, check:

(1) The composite map $\mathrm{C} \hookrightarrow \overline{\mathcal{T}}(\mathrm{C}) \to \overline{\mathcal{T}}(\mathrm{C}_t)$ is a Coxeter coalgebra morphism. It is the exponential (9.90a) of the coderivation $\mathrm{C} \hookrightarrow \overline{\mathcal{T}}(\mathrm{C}_t)$. Deduce from Theorem 9.13 for $q = 1$ that (13.30) is a Coxeter bialgebra morphism.

(2) The composite map $\mathrm{C} \hookrightarrow \overline{\mathcal{T}}(\mathrm{C}_t) \to \overline{\mathcal{T}}(\mathrm{C})$ is a coderivation. It is the logarithm (9.90b) of the Coxeter coalgebra morphism $\mathrm{C} \hookrightarrow \overline{\mathcal{T}}(\mathrm{C})$. Deduce from Theorem 9.15 for $q = 1$ that (13.31) is a Coxeter bialgebra morphism.

Theorem 13.39. *Let* $\mathrm{char}(\Bbbk) \nmid |W|$. *For a cocommutative Coxeter coalgebra* C, *the maps* (13.30) *and* (13.31) *are inverse isomorphisms of cocommutative Coxeter bialgebras.*

PROOF. By Exercise 13.38, the maps are Coxeter bialgebra morphisms. Moreover, they are inverse to each other (since $\boldsymbol{\zeta}$ and $\boldsymbol{\mu}$ are inverses). This is a straightforward calculation using the product (1.76). Also observe that the matrix form of both maps is unitriangular. □

We refer to (13.30) and (13.31) as *HNR isomorphisms* for cocommutative Coxeter bialgebras. They depend on $\boldsymbol{\zeta}$ and $\boldsymbol{\mu}$.

Let us now look at the dual situation. Let A be a commutative Coxeter algebra, and A_t its underlying Coxeter space with trivial product. Recall the Coxeter bialgebra $\overline{\mathcal{T}}^\vee(\mathrm{A})$ with quasishuffle product and deconcatenation coproduct, and $\overline{\mathcal{T}}^\vee(\mathrm{A}_t)$ with shuffle product and deconcatenation coproduct. Define a natural map

$$\overline{\mathcal{T}}^\vee(\mathrm{A}_t) \to \overline{\mathcal{T}}^\vee(\mathrm{A}) \tag{13.32}$$

which on the Z-component is

$$\bigoplus_{T:\,T\geq Z} \mathrm{A}[T] \to \bigoplus_{V:\,V\geq Z} \mathrm{A}[V], \qquad x \mapsto \sum_{V:\,Z\leq V\leq T} \boldsymbol{\zeta}(V,T)\, \mu_V^T(x)$$

for $x \in \mathrm{A}[T]$. Similarly, define a natural map

$$\overline{\mathcal{T}}^\vee(\mathrm{A}) \to \overline{\mathcal{T}}^\vee(\mathrm{A}_t) \tag{13.33}$$

which on the Z-component is

$$\bigoplus_{T:\,T\geq Z} \mathrm{A}[T] \to \bigoplus_{V:\,V\geq Z} \mathrm{A}[V], \qquad x \mapsto \sum_{V:\,Z\leq V\leq T} \boldsymbol{\mu}(V,T)\, \mu_V^T(x)$$

for $x \in \mathrm{A}[T]$.

The dual of Exercise 13.38 holds. It yields the following result which is the dual of Theorem 13.39.

Theorem 13.40. *Let* $\operatorname{char}(\Bbbk) \nmid |W|$. *For a commutative Coxeter algebra* A, *the maps* (13.32) *and* (13.33) *are inverse isomorphisms of commutative Coxeter bialgebras.*

We refer to (13.32) and (13.33) as *HNR isomorphisms* for commutative Coxeter bialgebras.

Observation 13.41. For a cocommutative Coxeter coalgebra C and commutative Coxeter algebra A, the diagrams

$$\begin{array}{ccc} \overline{\mathcal{T}}(\mathrm{C}) & \underset{\mu}{\overset{\zeta}{\rightleftarrows}} & \overline{\mathcal{T}}(\mathrm{C}_t) \\ & \searrow \quad \swarrow & \\ & \mathrm{C} & \end{array} \qquad\qquad \begin{array}{ccc} \overline{\mathcal{T}}^\vee(\mathrm{A}_t) & \underset{\mu}{\overset{\zeta}{\rightleftarrows}} & \overline{\mathcal{T}}^\vee(\mathrm{A}) \\ & \nwarrow \quad \nearrow & \\ & \mathrm{A} & \end{array}$$

commute. The horizontal maps are the HNR isomorphisms. The oblique maps are the canonical surjections and injections.

13.5.2. Commutative aspects. For a cocommutative Coxeter coalgebra C and commutative Coxeter algebra A, we have commutative diagrams

$$\begin{array}{ccc} \overline{\mathcal{T}}(\mathrm{C}) & \underset{\mu}{\overset{\zeta}{\rightleftarrows}} & \overline{\mathcal{T}}(\mathrm{C}_t) \\ {\scriptstyle\pi}\downarrow & & \downarrow{\scriptstyle\pi} \\ \overline{\mathcal{S}}(\mathrm{C}) & \rightleftarrows & \overline{\mathcal{S}}(\mathrm{C}_t) \end{array} \qquad\qquad \begin{array}{ccc} \overline{\mathcal{T}}^\vee(\mathrm{A}_t) & \underset{\mu}{\overset{\zeta}{\rightleftarrows}} & \overline{\mathcal{T}}^\vee(\mathrm{A}) \\ {\scriptstyle\pi^\vee}\uparrow & & \uparrow{\scriptstyle\pi^\vee} \\ \overline{\mathcal{S}}^\vee(\mathrm{A}_t) & \rightleftarrows & \overline{\mathcal{S}}^\vee(\mathrm{A}). \end{array} \tag{13.34}$$

The top-horizontal maps are the HNR isomorphisms for $\overline{\mathcal{T}}$ and $\overline{\mathcal{T}}^\vee$. The dotted maps are inverse isomorphisms which are induced from these via the vertical maps which are abelianization and coabelianization, see (9.54). Moreover, the dotted maps are unique, that is, they do not depend on $\boldsymbol{\zeta}$ and $\boldsymbol{\mu}$.

Theorem 13.42 (General norm map). *Let* $\operatorname{char}(\Bbbk) \nmid |W|$. *For* C *a cocommutative Coxeter coalgebra,* A *a commutative Coxeter algebra, let f and $\widehat{f}$ be as in* (9.146). *Then the composite map*

$$\overline{\mathcal{S}}(\mathrm{C}_t) \xrightarrow{\cong} \overline{\mathcal{S}}(\mathrm{C}) \xrightarrow{\widehat{f}} \overline{\mathcal{S}}^\vee(\mathrm{A}) \xrightarrow{\cong} \overline{\mathcal{S}}^\vee(\mathrm{A}_t), \tag{13.35}$$

on the Z-component, sends a representative $x \in \mathrm{C}[T]$ to $\sum_\sigma \sigma \log(f)_T(x)$, with the sum over all $\sigma : T \to T'$ local to Z. Thus, $\widehat{f}$ is an isomorphism of Coxeter bialgebras iff $\log(f)$ *is an isomorphism of Coxeter spaces.*

The proof is similar to that of Theorem 13.48 below, so we omit it.

13.5.3. q-shuffles and q-quasishuffles. Let q be a scalar which is not a root of unity. Recall the two-sided q-zeta function ζ_q and q-Möbius function μ_q, and the noncommutative q-zeta function $\boldsymbol{\zeta}_q$ and q-Möbius function $\boldsymbol{\mu}_q$.

Let C be a Coxeter coalgebra, and C_t its underlying Coxeter space with trivial coproduct. Recall from Table 9.2 the Coxeter q-bialgebra $\overline{\mathcal{T}}_q(\mathrm{C})$ with concatenation product and q-dequasishuffle coproduct, and $\overline{\mathcal{T}}_q(\mathrm{C}_t)$ with concatenation product and q-deshuffle coproduct. Define a natural map

$$\overline{\mathcal{T}}_q(\mathrm{C}) \to \overline{\mathcal{T}}_q(\mathrm{C}_t) \tag{13.36}$$

which on the Z-component is

$$\bigoplus_{T:\,T\geq Z} \mathrm{C}[T] \to \bigoplus_{U:\,U\geq Z} \mathrm{C}[U], \qquad x \mapsto \sum_{\substack{\sigma:\,U'\to U\\ \text{local to } T}} \zeta_q(T,U,\sigma,U')\,\sigma\Delta_T^{U'}(x)$$

for $x \in \mathrm{C}[T]$. When C is cocommutative, this simplifies to

$$\bigoplus_{T:\,T\geq Z} \mathrm{C}[T] \to \bigoplus_{U:\,U\geq Z} \mathrm{C}[U], \qquad x \mapsto \sum_{U:\,U\geq T} \zeta_q(T,U)\,\Delta_T^U(x)$$

for $x \in \mathrm{C}[T]$. This used the cocommutativity axiom (8.91) and Fact 1.100. Similarly, define a natural map

$$\overline{\mathcal{T}}_q(\mathrm{C}_t) \to \overline{\mathcal{T}}_q(\mathrm{C}) \tag{13.37}$$

which on the Z-component is

$$\bigoplus_{T:\,T\geq Z} \mathrm{C}[T] \to \bigoplus_{U:\,U\geq Z} \mathrm{C}[U], \qquad x \mapsto \sum_{\substack{\sigma:\,U'\to U\\ \text{local to } T}} \mu_q(T,U,\sigma,U')\,\sigma\Delta_T^{U'}(x)$$

for $x \in \mathrm{C}[T]$. When C is cocommutative, this simplifies to

$$\bigoplus_{T:\,T\geq Z} \mathrm{C}[T] \to \bigoplus_{U:\,U\geq Z} \mathrm{C}[U], \qquad x \mapsto \sum_{U:\,U\geq T} \mu_q(T,U)\,\Delta_T^U(x)$$

for $x \in \mathrm{C}[T]$.

Exercise 13.43. For q not a root of unity, for a Coxeter coalgebra C, check:

(1) The composite map $\mathrm{C} \hookrightarrow \overline{\mathcal{T}}_q(\mathrm{C}) \to \overline{\mathcal{T}}_q(\mathrm{C}_t)$ is a Coxeter coalgebra morphism. It is the q-exponential (9.99a) of the coderivation $\mathrm{C} \hookrightarrow \overline{\mathcal{T}}_q(\mathrm{C}_t)$. Deduce from Theorem 9.13 that (13.36) is a Coxeter q-bialgebra morphism.

(2) The composite map $\mathrm{C} \hookrightarrow \overline{\mathcal{T}}_q(\mathrm{C}_t) \to \overline{\mathcal{T}}_q(\mathrm{C})$ is a coderivation. It is the q-logarithm (9.99b) of the Coxeter coalgebra morphism $\mathrm{C} \hookrightarrow \overline{\mathcal{T}}_q(\mathrm{C})$. Deduce from Theorem 9.15 that (13.37) is a Coxeter q-bialgebra morphism.

When C is cocommutative, we can use the simpler formulas (9.101a) and (9.101b) for the q-exponential and q-logarithm.

Theorem 13.44. *For q not a root of unity, for a Coxeter coalgebra* C*, the maps* (13.36) *and* (13.37) *are inverse isomorphisms of Coxeter q-bialgebras.*

Proof. By Exercise 13.43, the maps are Coxeter q-bialgebra morphisms. Moreover, they are inverse to each other (since ζ_q and μ_q are inverses). This is a straightforward calculation using the alternative to (1.102) mentioned in Remark 1.97. □

Exercise 13.45. Check either directly or using Example 9.62: The maps (13.36) and (13.37) are q-exponentials (9.99a) of the biderivations

$$\begin{array}{ccc} \overline{\mathcal{T}}_q(\mathrm{C}) & \dashrightarrow & \overline{\mathcal{T}}_q(\mathrm{C}_t) \\ \downarrow & & \uparrow \\ \mathrm{C} & \xrightarrow[\mathrm{id}]{} & \mathrm{C} \end{array} \qquad\qquad \begin{array}{ccc} \overline{\mathcal{T}}_q(\mathrm{C}_t) & \dashrightarrow & \overline{\mathcal{T}}_q(\mathrm{C}) \\ \downarrow & & \uparrow \\ \mathrm{C} & \xrightarrow[\cong]{} & \mathcal{P}\overline{\mathcal{T}}_q(\mathrm{C}). \end{array}$$

The bottom-horizontal map in the second diagram arises from the coderivation in Exercise 13.43, item (2).

Dually: Let A be a Coxeter algebra, and A_t its underlying Coxeter space with trivial product. Recall the Coxeter q-bialgebra $\overline{\mathcal{T}}_q^{\vee}(\mathrm{A})$ with q-quasishuffle product and deconcatenation coproduct, and $\overline{\mathcal{T}}_q^{\vee}(\mathrm{A}_t)$ with q-shuffle product and deconcatenation coproduct. Define a natural map

$$\overline{\mathcal{T}}_q^{\vee}(\mathrm{A}_t) \to \overline{\mathcal{T}}_q^{\vee}(\mathrm{A}) \tag{13.38}$$

which on the Z-component is

$$\bigoplus_{T:\, T \geq Z} \mathrm{A}[T] \to \bigoplus_{V:\, V \geq Z} \mathrm{A}[V], \qquad x \mapsto \sum_{\substack{V:\, Z \leq V \leq T \\ \sigma:\, T \to T' \text{ local to } V}} \zeta_q(V, T', \sigma, T)\, \mu_V^{T'} \sigma(x)$$

for $x \in \mathrm{A}[T]$. When A is commutative, this simplifies to

$$\bigoplus_{T:\, T \geq Z} \mathrm{A}[T] \to \bigoplus_{V:\, V \geq Z} \mathrm{A}[V], \qquad x \mapsto \sum_{V:\, Z \leq V \leq T} \zeta_q(V, T)\, \mu_V^{T}(x)$$

for $x \in \mathrm{A}[T]$. Similarly, define a natural map

$$\overline{\mathcal{T}}_q^{\vee}(\mathrm{A}) \to \overline{\mathcal{T}}_q^{\vee}(\mathrm{A}_t) \tag{13.39}$$

which on the Z-component is

$$\bigoplus_{T:\, T \geq Z} \mathrm{A}[T] \to \bigoplus_{V:\, V \geq Z} \mathrm{A}[V], \qquad x \mapsto \sum_{\substack{V:\, Z \leq V \leq T \\ \sigma:\, T \to T' \text{ local to } V}} \mu_q(V, T', \sigma, T)\, \mu_V^{T'} \sigma(x)$$

for $x \in \mathrm{A}[T]$. When A is commutative, this simplifies to

$$\bigoplus_{T:\, T \geq Z} \mathrm{A}[T] \to \bigoplus_{V:\, V \geq Z} \mathrm{A}[V], \qquad x \mapsto \sum_{V:\, Z \leq V \leq T} \boldsymbol{\mu}_q(V, T)\, \mu_V^{T}(x)$$

for $x \in \mathrm{A}[T]$.

The dual of Exercise 13.43 holds. It yields the following result which is the dual of Theorem 13.44. Similarly, the dual of Exercise 13.45 holds.

Theorem 13.46. *For q not a root of unity, for a Coxeter algebra* A, *the maps* (13.38) *and* (13.39) *are inverse isomorphisms of Coxeter q-bialgebras.*

We collectively refer to (13.36), (13.37) and their duals (13.38), (13.39) as *HNR isomorphisms* for Coxeter q-bialgebras.

Observation 13.47. For q not a root of unity, for a Coxeter coalgebra C and Coxeter algebra A, the diagrams

$$
\begin{array}{ccccc}
\overline{\mathcal{T}}_q(\mathrm{C}) & \underset{\mu_q}{\overset{\varsigma_q}{\rightleftarrows}} & \overline{\mathcal{T}}_q(\mathrm{C}_t) \\
& \searrow \quad \swarrow & \\
& \mathrm{C} &
\end{array}
\qquad\qquad
\begin{array}{ccc}
\overline{\mathcal{T}}_q^{\vee}(\mathrm{A}_t) & \underset{\mu_q}{\overset{\varsigma_q}{\rightleftarrows}} & \overline{\mathcal{T}}_q^{\vee}(\mathrm{A}) \\
& \nwarrow \quad \nearrow & \\
& \mathrm{A} &
\end{array}
\tag{13.40}
$$

commute. The horizontal maps are the HNR isomorphisms. The oblique maps are the canonical surjections and injections.

13.5.4. Conjugation by HNR isomorphisms.

For q not a root of unity, for a Coxeter coalgebra C and Coxeter algebra A, we have bijections

$$q\text{-}\mathsf{Bimon}(\bar{\mathsf{C}}\mathsf{Sp})(\overline{\mathcal{T}}_q(\mathrm{C}), \overline{\mathcal{T}}_q^{\vee}(\mathrm{A})) \rightleftarrows q\text{-}\mathsf{Bimon}(\bar{\mathsf{C}}\mathsf{Sp})(\overline{\mathcal{T}}_q(\mathrm{C}_t), \overline{\mathcal{T}}_q^{\vee}(\mathrm{A}_t))$$

given by

$$
\begin{aligned}
(\overline{\mathcal{T}}_q(\mathrm{C}) \xrightarrow{g} \overline{\mathcal{T}}_q^{\vee}(\mathrm{A})) &\longmapsto (\overline{\mathcal{T}}_q(\mathrm{C}_t) \xrightarrow{\cong} \overline{\mathcal{T}}_q(\mathrm{C}) \xrightarrow{g} \overline{\mathcal{T}}_q^{\vee}(\mathrm{A}) \xrightarrow{\cong} \overline{\mathcal{T}}_q^{\vee}(\mathrm{A}_t)) \\
(\overline{\mathcal{T}}_q(\mathrm{C}_t) \xrightarrow{h} \overline{\mathcal{T}}_q^{\vee}(\mathrm{A}_t)) &\longmapsto (\overline{\mathcal{T}}_q(\mathrm{C}) \xrightarrow{\cong} \overline{\mathcal{T}}_q(\mathrm{C}_t) \xrightarrow{h} \overline{\mathcal{T}}_q^{\vee}(\mathrm{A}_t) \xrightarrow{\cong} \overline{\mathcal{T}}_q^{\vee}(\mathrm{A})).
\end{aligned}
$$

The isomorphisms are the HNR isomorphisms. Moreover, these bijections fit into the commutative triangle

$$
\begin{array}{ccc}
q\text{-}\mathsf{Bimon}(\bar{\mathsf{C}}\mathsf{Sp})(\overline{\mathcal{T}}_q(\mathrm{C}), \overline{\mathcal{T}}_q^{\vee}(\mathrm{A})) & \rightleftarrows & q\text{-}\mathsf{Bimon}(\bar{\mathsf{C}}\mathsf{Sp})(\overline{\mathcal{T}}_q(\mathrm{C}_t), \overline{\mathcal{T}}_q^{\vee}(\mathrm{A}_t)) \\
\mathbf{exp}_q \searrow \nwarrow \mathbf{log}_q & & \mathbf{log}_q \nearrow \swarrow \mathbf{exp}_q \\
& \bar{\mathsf{C}}\mathsf{Sp}(\mathrm{C}, \mathrm{A}). &
\end{array}
$$

The q-exp-log correspondences are instances of (9.104). To obtain this triangle, we can use naturality of the q-exp-log correspondence (Section 9.8.9) along with diagrams (13.40). Alternatively, we may also employ Exercise 13.45, its dual and Exercise 9.64.

As an application, we give a constructive proof of Theorem 9.110 below.

Theorem 13.48 (General q-norm map). *Suppose q is not a root of unity. For C a Coxeter coalgebra, A a Coxeter algebra, let f and $\widehat{f}$ be as in (9.144). Then the composite map*

$$\overline{\mathcal{T}}_q(\mathrm{C}_t) \xrightarrow{\cong} \overline{\mathcal{T}}_q(\mathrm{C}) \xrightarrow{\widehat{f}} \overline{\mathcal{T}}_q^{\vee}(\mathrm{A}) \xrightarrow{\cong} \overline{\mathcal{T}}_q^{\vee}(\mathrm{A}_t), \tag{13.41}$$

on the Z-component, on the T-summand and into the T'-summand, is

$$\mathrm{C}[T] \to \mathrm{A}[T'], \qquad x \mapsto \sum_{\substack{\sigma: T \to T' \\ \text{local to } Z}} q^{l(\sigma)} \sigma\, \mathbf{log}_q(f)_T(x).$$

Thus, $\widehat{f}$ is an isomorphism of Coxeter q-bialgebras iff $\mathbf{log}_q(f)$ is an isomorphism of Coxeter spaces.

PROOF. By Exercise 9.111, $\widehat{f}$ is the q-exponential of $\mathbf{log}_q(f)$. Hence, by the commutative triangle above, the conjugate of $\widehat{f}$ by the HNR isomorphisms is the q-exponential of

$$\begin{array}{ccc} \overline{\mathcal{T}}_q(\mathrm{C}_t) & \dashrightarrow & \overline{\mathcal{T}}_q^{\vee}(\mathrm{A}_t) \\ \downarrow & & \uparrow \\ \mathrm{C} & \xrightarrow[\mathbf{log}_q(f)]{} & \mathrm{A}. \end{array}$$

Now apply Exercise 9.109 (with f replaced by $\mathbf{log}_q(f)$) to obtain the formula in the theorem. The last claim follows from Exercise 9.108. □

13.5.5. 0-shuffles and 0-quasishuffles. Let us now specialize the preceding discussion to $q = 0$. Recall formulas (1.118) for $\boldsymbol{\zeta}_0$ and $\boldsymbol{\mu}_0$.

For a Coxeter coalgebra C, the natural maps

$$\begin{aligned} \overline{\mathcal{T}}_0(\mathrm{C}) &\to \overline{\mathcal{T}}_0(\mathrm{C}_t) & \qquad \overline{\mathcal{T}}_0(\mathrm{C}_t) &\to \overline{\mathcal{T}}_0(\mathrm{C}) \\ x &\mapsto \sum_{U:\, U \geq T} \Delta_T^U(x) & \qquad x &\mapsto \sum_{U:\, U \geq T} (-1)^{\mathrm{rk}(U/T)} \Delta_T^U(x) \end{aligned} \tag{13.42}$$

on the Z-component, for $x \in \mathrm{C}[T]$ with $T \geq Z$, are inverse isomorphisms of Coxeter 0-bialgebras.

Dually, for a Coxeter algebra A, the natural maps

$$\begin{aligned} \overline{\mathcal{T}}_0^{\vee}(\mathrm{A}_t) &\to \overline{\mathcal{T}}_0^{\vee}(\mathrm{A}) & \qquad \overline{\mathcal{T}}_0^{\vee}(\mathrm{A}) &\to \overline{\mathcal{T}}_0^{\vee}(\mathrm{A}_t) \\ x &\mapsto \sum_{V:\, Z \leq V \leq T} \mu_V^T(x) & \qquad x &\mapsto \sum_{V:\, Z \leq V \leq T} (-1)^{\mathrm{rk}(T/V)} \mu_V^T(x) \end{aligned} \tag{13.43}$$

on the Z-component, for $x \in \mathrm{A}[T]$ with $T \geq Z$, are inverse isomorphisms of Coxeter 0-bialgebras.

Exercise 13.49. Check directly: (13.42) and (13.43) are morphisms of Coxeter 0-bialgebras. (Use formulas in Exercise 9.21.)

Let us now specialize Theorem 13.48 to $q = 0$. Let $\widehat{f}$ be as in Section 9.11.6. Then the composite map

$$\overline{\mathcal{T}}_0(\mathrm{C}_t) \xrightarrow{\cong} \overline{\mathcal{T}}_0(\mathrm{C}) \xrightarrow{\widehat{f}} \overline{\mathcal{T}}_0^{\vee}(\mathrm{A}) \xrightarrow{\cong} \overline{\mathcal{T}}_0^{\vee}(\mathrm{A}_t),$$

on the Z-component, on the T-summand and into the T'-summand, is

$$\mathrm{C}[T] \to \mathrm{A}[T'], \qquad x \mapsto \begin{cases} \mathbf{log}_0(f)_T(x) & \text{if } T = T', \\ 0 & \text{otherwise.} \end{cases}$$

13.5.6. Coxeter bialgebras of face-types. Recall the Coxeter bialgebras of face-types $\overline{\Sigma}$ and $\overline{\Sigma}^*$ from Section 10.4. The relation between the H-basis and Q-basis (10.66) and the dual relation between the M-basis and P-basis can be understood in the HNR context as explained below.

Let $\operatorname{char}(\Bbbk) \nmid |W|$. The following are commutative diagrams of Coxeter bialgebras.

$$
(13.44) \qquad
\begin{array}{ccc}
\overline{\mathcal{T}}(\Bbbk\{x\}) & \underset{\boldsymbol{\mu}}{\overset{\boldsymbol{\zeta}}{\rightleftarrows}} & \overline{\mathcal{T}}(\Bbbk\{x\}_t) \\
 & {\scriptstyle\cong}\nwarrow \quad \nearrow{\scriptstyle\cong} & \\
 & \overline{\Sigma} &
\end{array}
\qquad\qquad
\begin{array}{ccc}
\overline{\mathcal{T}}^{\vee}(\Bbbk[x]_t) & \underset{\boldsymbol{\mu}}{\overset{\boldsymbol{\zeta}}{\rightleftarrows}} & \overline{\mathcal{T}}^{\vee}(\Bbbk[x]) \\
 & {\scriptstyle\cong}\nwarrow \quad \nearrow{\scriptstyle\cong} & \\
 & \overline{\Sigma}^{*} &
\end{array}
$$

The horizontal maps in the two diagrams are the HNR isomorphisms specialized to $\mathrm{C} := \Bbbk\{x\}$ and $\mathrm{A} := \Bbbk[x]$, respectively. The oblique maps are the isomorphisms in Proposition 10.71 for $q = 1$ and in Proposition 10.74.

Similarly, for q not a root of unity, the relation between the $\mathtt{H}$-basis and $\mathtt{Q}$-basis (10.69) and its dual relation can be understood as follows. For q not a root of unity, the following are commutative diagrams of Coxeter q-bialgebras.

$$
(13.45) \qquad
\begin{array}{ccc}
\overline{\mathcal{T}}_q(\Bbbk\{x\}) & \underset{\boldsymbol{\mu}_q}{\overset{\boldsymbol{\zeta}_q}{\rightleftarrows}} & \overline{\mathcal{T}}_q(\Bbbk\{x\}_t) \\
 & {\scriptstyle\cong}\nwarrow \quad \nearrow{\scriptstyle\cong} & \\
 & \overline{\Sigma}_q &
\end{array}
\qquad\qquad
\begin{array}{ccc}
\overline{\mathcal{T}}_q^{\vee}(\Bbbk[x]_t) & \underset{\boldsymbol{\mu}_q}{\overset{\boldsymbol{\zeta}_q}{\rightleftarrows}} & \overline{\mathcal{T}}_q^{\vee}(\Bbbk[x]) \\
 & {\scriptstyle\cong}\nwarrow \quad \nearrow{\scriptstyle\cong} & \\
 & \overline{\Sigma}_q^{*} &
\end{array}
$$

The horizontal maps are the HNR isomorphisms specialized to $\mathrm{C} := \Bbbk\{x\}$ and $\mathrm{A} := \Bbbk[x]$. Note: The simpler formulas involving $\boldsymbol{\zeta}_q$ and $\boldsymbol{\mu}_q$ come into play here since $\Bbbk\{x\}$ is cocommutative and $\Bbbk[x]$ is commutative. The oblique maps are the isomorphisms in Propositions 10.71 and 10.75.

Now recall the q-norm map on face-types (10.85). It is defined from the $\mathtt{H}$-basis to the $\mathtt{M}$-basis. For q not a root of unity, (10.89) expresses it from the $\mathtt{Q}$-basis to the $\mathtt{P}$-basis. This is an instance of formula (13.41) in Theorem 13.48. Moreover, the proof in Exercise 10.90 specializes the proof of Theorem 13.48.

13.6. Poincaré–Birkhoff–Witt

In this section, we assume $\operatorname{char}(\Bbbk) \nmid |W|$. Recall: For any Coxeter Lie algebra G, its universal enveloping Coxeter algebra $\overline{\mathcal{U}}(\mathrm{G})$ carries a cocommutative Coxeter bialgebra structure. The Poincaré–Birkhoff–Witt theorem (PBW) says that $\overline{\mathcal{U}}(\mathrm{G})$ is isomorphic as a Coxeter coalgebra to $\overline{\mathcal{S}}(\mathrm{G})$. The isomorphism depends on the choice of an invariant noncommutative zeta function $\boldsymbol{\zeta}$. Equivalently, PBW says that there exists an idempotent operator on $\overline{\mathcal{T}}(\mathrm{G})$ of Coxeter coalgebras with image $\overline{\mathcal{S}}(\mathrm{G})$ and coimage $\overline{\mathcal{U}}(\mathrm{G})$. This is the Solomon operator. Here $\overline{\mathcal{S}}(\mathrm{G})$ is viewed inside $\overline{\mathcal{T}}(\mathrm{G})$ via a Coxeter coalgebra section to the abelianization map.

Dually, for any Coxeter Lie coalgebra K, its universal coenveloping Coxeter coalgebra $\overline{\mathcal{U}}^{\vee}(\mathrm{K})$ is isomorphic as a Coxeter algebra to $\overline{\mathcal{S}}^{\vee}(\mathrm{K})$. Now there is an idempotent operator on $\overline{\mathcal{T}}^{\vee}(\mathrm{K})$ of Coxeter algebras with image $\overline{\mathcal{U}}^{\vee}(\mathrm{K})$ and coimage $\overline{\mathcal{S}}^{\vee}(\mathrm{K})$. Here $\overline{\mathcal{S}}^{\vee}(\mathrm{K})$ is viewed as a quotient of $\overline{\mathcal{T}}^{\vee}(\mathrm{K})$ via a Coxeter algebra projection to the coabelianization map.

These results have signed analogues.

The parallel discussion for Coxeter species is in Section 7.6. There we work with faces, while here we work with face-types.

13.6.1. Coxeter coalgebra sections to the abelianization map. For V a Coxeter space, the canonical surjection $\overline{\mathcal{T}}(\mathrm{V}) \twoheadrightarrow \overline{\mathcal{S}}(\mathrm{V})$ is an instance of the abelianization map, see (9.55). It is a Coxeter bialgebra morphism. We now construct Coxeter coalgebra sections to this map.

Let $\mathrm{char}(\Bbbk) \nmid |W|$. Fix an invariant noncommutative zeta function $\boldsymbol{\zeta}$. For V a Coxeter space, define a natural map

$$\overline{\mathcal{S}}(\mathrm{V}) \hookrightarrow \overline{\mathcal{T}}(\mathrm{V}) \tag{13.46}$$

which on the Z-component sends a representative $x \in \mathrm{V}[T]$ to

$$\sum_{T' :\, T' \geq Z} \zeta(Z, T') \sum_{\substack{\sigma :\, T \to T' \\ \text{local to } Z}} \sigma(x).$$

Exercise 13.50. Check: The map (13.46) is well-defined, that is, it does not depend on the chosen representative.

Exercise 13.51. Check: The map (13.46) is an exponential (9.90a) of the coderivation $\overline{\mathcal{S}}(\mathrm{V}) \twoheadrightarrow \mathrm{V} \hookrightarrow \overline{\mathcal{T}}(\mathrm{V})$. The first map is the canonical projection, while the second is the canonical inclusion.

Lemma 13.52. *The map* (13.46) *is an injective morphism of Coxeter coalgebras which is a section to the abelianization map.*

Proof. The first statement in Theorem 9.56 applied to Exercise 13.51 shows that (13.46) is a Coxeter coalgebra morphism. Moreover, the sum of the coefficients in its expression is 1 by the flat-additivity formula (1.83). Hence, it is a section to the abelianization map. □

13.6.2. PBW for Coxeter Lie algebras. Let $\mathrm{char}(\Bbbk) \nmid |W|$. Fix an invariant noncommutative zeta function $\boldsymbol{\zeta}$. For a Coxeter Lie algebra G, consider the diagram of Coxeter coalgebras

$$\begin{array}{ccc} & \overline{\mathcal{T}}(\mathrm{G}) & \\ \nearrow & & \searrow \\ \overline{\mathcal{S}}(\mathrm{G}) & \dashrightarrow & \overline{\mathcal{U}}(\mathrm{G}). \end{array} \tag{13.47}$$

The inclusion is the map (13.46) for $\mathrm{V} := \mathrm{G}$. The surjective map is the canonical quotient. (It is in fact a Coxeter bialgebra morphism.)

Theorem 13.53. *Let* $\mathrm{char}(\Bbbk) \nmid |W|$. *For an invariant noncommutative zeta function* ζ *and Coxeter Lie algebra* G, *the dotted arrow in* (13.47) *is an isomorphism of Coxeter coalgebras.*

This is the *Poincaré–Birkhoff–Witt theorem* for Coxeter Lie algebras, or PBW for short. The dotted arrow in (13.47) is a *Poincaré–Birkhoff–Witt isomorphism*, or PBW isomorphism for short. To prove this theorem, we construct an idempotent operator on $\overline{\mathcal{T}}(\mathrm{G})$ with image $\overline{\mathcal{S}}(\mathrm{G})$ and coimage $\overline{\mathcal{U}}(\mathrm{G})$, see Theorem 13.55 below.

Corollary 13.54. *Let* $\mathrm{char}(\Bbbk) \nmid |W|$. *For a Coxeter Lie algebra* G, *the map* $i : \mathrm{G} \to \overline{\mathcal{U}}(\mathrm{G})$ *is an injective morphism of Coxeter Lie algebras.*

PROOF. Since the map (13.46) is identity on V, we see that i is the restriction of the dotted arrow in (13.47) to G, and hence injective. □

This result justifies the usage of the term "enveloping" for $\overline{\mathcal{U}}(\mathrm{G})$.

13.6.3. Solomon operator. Let $\mathrm{char}(\Bbbk) \nmid |W|$. Fix an invariant noncommutative zeta function $\boldsymbol{\zeta}$. For a Coxeter Lie algebra G, consider the composite map

$$\overline{\mathcal{T}}(\mathrm{G}) \xrightarrow{\mathbf{log}(\mathrm{id})} \overline{\mathcal{P}}\overline{\mathcal{T}}(\mathrm{G}) = \mathbf{Lie} \circ \mathrm{G} \xrightarrow{\gamma} \mathrm{G} \hookrightarrow \overline{\mathcal{T}}(\mathrm{G}). \tag{13.48}$$

The first map is as in Lemma 12.10 for V := G and involves the Solomon coefficients. The equality is by Proposition 12.9 for V := G. The map γ is the Lie structure map of G. The last map is the canonical inclusion. Since G is contained in $\overline{\mathcal{P}}\overline{\mathcal{T}}(\mathrm{G})$, (13.48) is a coderivation. Moreover, since $\overline{\mathcal{T}}(\mathrm{G})$ is cocommutative, its exponential is a Coxeter coalgebra morphism. We call this the *Solomon operator* and denote it by

$$\mathrm{Sol} : \overline{\mathcal{T}}(\mathrm{G}) \to \overline{\mathcal{T}}(\mathrm{G}). \tag{13.49}$$

Explicitly, on the Z-component, on the T-summand and into the T'-summand, it is

$$\boldsymbol{\zeta}(Z, T')\, \gamma_{T'}\, \mathbf{log}(\mathrm{id})_{T'} \Big(\sum_{V:\, V \geq T'} \ \sum_{\substack{\sigma:\, T \to V \\ \text{local to } Z \\ \text{projects } T \text{ on } T'}} \sigma \Big).$$

Theorem 13.55. *Let* $\mathrm{char}(\Bbbk) \nmid |W|$. *For an invariant noncommutative zeta function* $\boldsymbol{\zeta}$ *and Coxeter Lie algebra* G, *the Solomon operator on* $\overline{\mathcal{T}}(\mathrm{G})$ (13.49) *is an idempotent operator of Coxeter coalgebras with image* $\overline{\mathcal{S}}(\mathrm{G})$ *and coimage* $\overline{\mathcal{U}}(\mathrm{G})$ *yielding the commutative diagram*

$$\begin{array}{ccc} \overline{\mathcal{T}}(\mathrm{G}) & \xrightarrow{\mathrm{Sol}} & \overline{\mathcal{T}}(\mathrm{G}) \\ \downarrow & & \uparrow \\ \overline{\mathcal{U}}(\mathrm{G}) & \xrightarrow[\cong]{} & \overline{\mathcal{S}}(\mathrm{G}). \end{array} \tag{13.50}$$

In particular, $\overline{\mathcal{S}}(\mathrm{G})$ *and* $\overline{\mathcal{U}}(\mathrm{G})$ *are isomorphic as Coxeter coalgebras. Moreover, the inverse of* $\overline{\mathcal{U}}(\mathrm{G}) \to \overline{\mathcal{S}}(\mathrm{G})$ *is the PBW isomorphism in* (13.47).

This result can be proved by following the outline in [**11**, Section 17.4.3]. Two important lemmas which get used in the proof are recorded below.

Lemma 13.56. *For the Coxeter bialgebra* $\overline{\mathcal{T}}(\mathrm{V})$, *for* $T \geq Z$ *and* $z \in \mathrm{V}[T]$,

$$z = \sum_{T':\, T' \geq Z} \boldsymbol{\zeta}(Z, T')\, \mathbf{log}(\mathrm{id})_{T'} \Big(\sum_{\substack{\sigma:\, T \to V \\ \textit{local to } Z \\ \textit{projects } T \textit{ on } T'}} \sigma(z) \Big). \tag{13.51}$$

PROOF. The rhs equals $\mathbf{exp}(\mathbf{log}(\mathrm{id}))_Z(z)$. □

Lemma 13.57. *For the Coxeter bialgebra $\overline{\mathcal{T}}(\mathrm{V})$, for $T \geq Z$ and $z \in \mathrm{V}[T]$,*

$$\sum_{T' : T' \geq Z} \sum_{\substack{\sigma : T \to T' \\ \text{local to } Z}} \zeta(Z, T') \, \mathbf{log}(\mathrm{id})_Z \, \sigma(z) = \begin{cases} z & \text{if } T = Z, \\ 0 & \text{if } T > Z. \end{cases} \tag{13.52}$$

PROOF. Let us denote the map (13.46) by f. Let $\bar{z}$ denote the image of z in $\overline{\mathcal{S}}(\mathrm{V})[Z]$. Then the lhs equals $\mathbf{log}(\mathrm{id})_Z f_Z(\bar{z}) = \mathbf{log}(f)_Z(\bar{z})$ which equals the rhs by using Exercise 13.51. □

13.6.4. Abelian Coxeter Lie algebras. Let G be an abelian Coxeter Lie algebra. The map (13.48) specializes to $\overline{\mathcal{T}}(\mathrm{G}) \twoheadrightarrow \mathrm{G} \hookrightarrow \overline{\mathcal{T}}(\mathrm{G})$, with the first map the canonical surjection and second the canonical inclusion. Its exponential, on the Z-component, on the T-summand and into the T'-summand, is

$$\zeta(Z, T') \sum_{\substack{\sigma : T \to T' \\ \text{local to } Z}} \sigma. \tag{13.53}$$

This is the Solomon operator (13.49) when G is abelian. Diagram (13.50) specializes to

$$\begin{array}{ccc} \overline{\mathcal{T}}(\mathrm{G}) & \xrightarrow{\mathrm{Sol}} & \overline{\mathcal{T}}(\mathrm{G}) \\ \downarrow & & \uparrow \\ \overline{\mathcal{S}}(\mathrm{G}) & \xrightarrow[\mathrm{id}]{} & \overline{\mathcal{S}}(\mathrm{G}). \end{array}$$

In other words, $\overline{\mathcal{S}}(\mathrm{G})$ is a retract of the idempotent operator Sol. For definition of retract, see for instance [**11**, Section A.5.1]. In this case, the PBW isomorphism is identity.

13.6.5. PBW for Coxeter Lie coalgebras. The preceding discussion has a dual version which we briefly go over. For a Coxeter space V, the canonical inclusion $\overline{\mathcal{S}}^{\vee}(\mathrm{V}) \hookrightarrow \overline{\mathcal{T}}^{\vee}(\mathrm{V})$ is an instance of the coabelianization map, see (9.55). It is a Coxeter bialgebra morphism. We now construct Coxeter algebra projections to this map. For that, fix an invariant noncommutative zeta function $\boldsymbol{\zeta}$, and define a natural map

$$\overline{\mathcal{T}}^{\vee}(\mathrm{V}) \twoheadrightarrow \overline{\mathcal{S}}^{\vee}(\mathrm{V}), \tag{13.54}$$

on the Z-component, on the T-summand, by

$$\zeta(Z, T) \sum_{T' : T' \geq Z} \sum_{\substack{\sigma : T \to T' \\ \text{local to } Z}} \sigma.$$

It is an exponential (9.90a) of the derivation $\overline{\mathcal{T}}^{\vee}(\mathrm{V}) \twoheadrightarrow \mathrm{V} \hookrightarrow \overline{\mathcal{S}}^{\vee}(\mathrm{V})$. The first map is the canonical projection, while the second is the canonical inclusion.

Lemma 13.58. *The map* (13.54) *is a surjective morphism of Coxeter algebras which is a projection to the coabelianization map.*

For a Coxeter Lie coalgebra K, consider the diagram of Coxeter algebras

$$\begin{array}{ccc} & \overline{\mathcal{T}}^{\vee}(\mathrm{K}) & \\ \nearrow & & \searrow \\ \overline{\mathcal{U}}^{\vee}(\mathrm{K}) & \dashrightarrow & \overline{\mathcal{S}}^{\vee}(\mathrm{K}). \end{array} \tag{13.55}$$

The surjective map is (13.54) for $\mathrm{V} := \mathrm{K}$. The injective map is the canonical inclusion. (It is in fact a Coxeter bialgebra morphism.)

Theorem 13.59. *Let* $\mathrm{char}(\Bbbk) \nmid |W|$. *For an invariant noncommutative zeta function* ζ *and Coxeter Lie coalgebra* K, *the dotted arrow in* (13.55) *is an isomorphism of Coxeter algebras.*

This is the *Poincaré–Birkhoff–Witt theorem* for Coxeter Lie coalgebras. The dotted arrow in (13.55) is a *Poincaré–Birkhoff–Witt isomorphism.*

Corollary 13.60. *Let* $\mathrm{char}(\Bbbk) \nmid |W|$. *For a Coxeter Lie coalgebra* K, *the map* $p : \overline{\mathcal{U}}^{\vee}(\mathrm{K}) \to \mathrm{K}$ *is a surjective morphism of Coxeter Lie coalgebras.*

Let K be a Coxeter Lie coalgebra. Consider the composite map

$$\overline{\mathcal{T}}^{\vee}(\mathrm{K}) \twoheadrightarrow \mathrm{K} \xrightarrow{\delta} \mathbf{Lie}^{*} \circ^{\vee} \mathrm{K} = \overline{\mathcal{Q}}\overline{\mathcal{T}}^{\vee}(\mathrm{K}) \xrightarrow{\mathbf{log}(\mathrm{id})} \overline{\mathcal{T}}^{\vee}(\mathrm{K}). \tag{13.56}$$

This is a derivation. Since $\overline{\mathcal{T}}^{\vee}(\mathrm{K})$ is commutative, its exponential is a Coxeter algebra morphism. We call this the *Solomon operator* and denote it by

$$\mathrm{Sol}^{\vee} : \overline{\mathcal{T}}^{\vee}(\mathrm{K}) \to \overline{\mathcal{T}}^{\vee}(\mathrm{K}). \tag{13.57}$$

Theorem 13.61. *Let* $\mathrm{char}(\Bbbk) \nmid |W|$. *For an invariant noncommutative zeta function* ζ *and Coxeter Lie coalgebra* K, *the Solomon operator on* $\overline{\mathcal{T}}^{\vee}(\mathrm{K})$ (13.57) *is an idempotent operator of Coxeter algebras with image* $\overline{\mathcal{U}}^{\vee}(\mathrm{K})$ *and coimage* $\overline{\mathcal{S}}^{\vee}(\mathrm{K})$ *yielding the commutative diagram*

$$\begin{array}{ccc} \overline{\mathcal{T}}^{\vee}(\mathrm{K}) & \xrightarrow{\mathrm{Sol}^{\vee}} & \overline{\mathcal{T}}^{\vee}(\mathrm{K}) \\ \big\downarrow & & \big\uparrow \\ \overline{\mathcal{S}}^{\vee}(\mathrm{K}) & \xrightarrow[\cong]{} & \overline{\mathcal{U}}^{\vee}(\mathrm{K}). \end{array} \tag{13.58}$$

In particular, $\overline{\mathcal{S}}^{\vee}(\mathrm{K})$ *and* $\overline{\mathcal{U}}^{\vee}(\mathrm{K})$ *are isomorphic as Coxeter algebras. Moreover, the inverse of* $\overline{\mathcal{S}}^{\vee}(\mathrm{K}) \to \overline{\mathcal{U}}^{\vee}(\mathrm{K})$ *is the PBW isomorphism in* (13.55).

Let K be an abelian Coxeter Lie coalgebra. The map (13.56) specializes to $\overline{\mathcal{T}}^{\vee}(\mathrm{K}) \twoheadrightarrow \mathrm{K} \hookrightarrow \overline{\mathcal{T}}^{\vee}(\mathrm{K})$, with the first map the canonical surjection and second the canonical inclusion. Its exponential, on the Z-component, on the T-summand and into the T'-summand, is

$$\zeta(Z, T) \sum_{\substack{\sigma : T \to T' \\ \text{local to } Z}} \sigma. \tag{13.59}$$

This is the Solomon operator (13.57) when K is abelian. Diagram (13.58) specializes to

$$\begin{array}{ccc} \overline{\mathcal{T}}^{\vee}(\mathrm{K}) & \xrightarrow{\mathrm{Sol}^{\vee}} & \overline{\mathcal{T}}^{\vee}(\mathrm{K}) \\ \big\downarrow & & \big\uparrow \\ \overline{\mathcal{S}}^{\vee}(\mathrm{K}) & \xrightarrow[\mathrm{id}]{} & \overline{\mathcal{S}}^{\vee}(\mathrm{K}). \end{array}$$

In this case, the PBW isomorphism is identity.

13.6.6. Signed analogue. PBW Theorems 13.53 and 13.59 have signed analogues which are stated below. They can be deduced from the unsigned versions by conjugating by the u-signature functor (8.99).

For V a Coxeter space, define a natural map

$$\overline{\mathcal{E}}(\mathrm{V}) \hookrightarrow \overline{\mathcal{T}}_{-1}(\mathrm{V}) \tag{13.60}$$

which on the Z-component sends a representative $x \in \mathrm{V}[T]$ to

$$\sum_{T' : T' \geq Z} \zeta(Z, T') \sum_{\substack{\sigma : T \to T' \\ \text{local to } Z}} (-1)^{l(\sigma)} \sigma(x).$$

Compare and contrast with (13.46). The above is a Coxeter coalgebra section to the signed abelianization map $\overline{\mathcal{T}}_{-1}(\mathrm{V}) \twoheadrightarrow \overline{\mathcal{E}}(\mathrm{V})$, see (9.57).

For a signed Coxeter Lie algebra G, consider the diagram of Coxeter coalgebras

$$\begin{array}{ccc} & \overline{\mathcal{T}}_{-1}(\mathrm{G}) & \\ \nearrow & & \searrow \\ \overline{\mathcal{E}}(\mathrm{G}) & \dashrightarrow & \overline{\mathcal{U}}_{-1}(\mathrm{G}). \end{array} \tag{13.61}$$

The inclusion is the map (13.60) for V := G. The surjective map is the canonical quotient. (It is in fact a morphism of signed Coxeter bialgebras.)

Theorem 13.62. *Let* $\mathrm{char}(\Bbbk) \nmid |W|$. *For an invariant noncommutative zeta function* ζ *and signed Coxeter Lie algebra* G, *the dotted arrow in* (13.61) *is an isomorphism of Coxeter coalgebras.*

PROOF. This can be deduced from Theorem 13.53 by conjugation by the u-signature functor (8.99). □

This is PBW for signed Coxeter Lie algebras.

Dually, for a signed Coxeter Lie coalgebra K, we have a diagram of Coxeter algebras

$$\begin{array}{ccc} & \overline{\mathcal{T}}^{\vee}_{-1}(\mathrm{K}) & \\ \nearrow & & \searrow \\ \overline{\mathcal{U}}^{\vee}_{-1}(\mathrm{K}) & \dashrightarrow & \overline{\mathcal{E}}^{\vee}(\mathrm{K}). \end{array} \tag{13.62}$$

The surjective map is a Coxeter algebra projection to the signed coabelianization map.

Theorem 13.63. *Let* $\mathrm{char}(\Bbbk) \nmid |W|$. *For an invariant noncommutative zeta function* ζ *and signed Coxeter Lie coalgebra* K, *the dotted arrow in* (13.62) *is an isomorphism of Coxeter algebras.*

This is PBW for signed Coxeter Lie coalgebras.

13.7. Cartier–Milnor–Moore

In this section, we assume $\mathrm{char}(\Bbbk) \nmid |W|$. The Cartier–Milnor–Moore theorem (CMM) says that functors $\overline{\mathcal{U}}$ and $\overline{\mathcal{P}}$ determine an adjoint equivalence between categories of Coxeter Lie algebras and cocommutative Coxeter bialgebras. Dually, functors $\overline{\mathcal{U}}^{\vee}$ and $\overline{\mathcal{Q}}$ determine an adjoint equivalence between categories of Coxeter Lie coalgebras and commutative Coxeter bialgebras. These results have signed analogues. CMM is an extension of Leray–Samelson from the bicommutative to the (co)commutative setting.

The parallel result for Coxeter species is given in Section 7.7.

13.7.1. CMM for cocommutative Coxeter bialgebras. The following result connects Borel–Hopf and PBW.

Lemma 13.64. *Let* $\mathrm{char}(\Bbbk) \nmid |W|$. *For a cocommutative Coxeter bialgebra* H, *the diagram*

$$\begin{array}{ccc} \overline{\mathcal{S}}(\overline{\mathcal{P}}(\mathrm{H})) & \xrightarrow{\cong} & \overline{\mathcal{U}}(\overline{\mathcal{P}}(\mathrm{H})) \\ {\scriptstyle\cong}\downarrow & & \downarrow{\scriptstyle\cong} \\ \overline{\mathcal{S}}^{\vee}(\overline{\mathcal{P}}(\mathrm{H})) & \xrightarrow[\cong]{} & \mathrm{H} \end{array} \tag{13.63}$$

commutes.

The vertical maps are Coxeter bialgebra isomorphisms, while the horizontal maps are Coxeter coalgebra isomorphisms. The left-vertical map is the norm map (9.142) for $\mathrm{V} := \overline{\mathcal{P}}(\mathrm{H})$. The top-horizontal map is the PBW isomorphism in (13.47) for $\mathrm{G} := \overline{\mathcal{P}}(\mathrm{H})$. The bottom-horizontal map is the Borel–Hopf isomorphism (13.1). The right-vertical map is the counit of the adjunction between $\overline{\mathcal{U}}$ and $\overline{\mathcal{P}}$ in Theorem 12.21. It is an isomorphism because the remaining three maps are. The horizontal maps depend on ζ, but the vertical maps do not.

PROOF. The composite in either direction, on the Z-component, on a representative $x \in \overline{\mathcal{P}}(\mathrm{H})[T]$, is

$$\sum_{T' : T' \geq Z} \zeta(Z, T') \sum_{\substack{\sigma : T \to T' \\ \text{local to } Z}} \mu_Z^{T'} \sigma(x).$$

(This formula appeared in the proof of Theorem 13.14 where it simplified to $\mu_Z^T(x)$ since H was bicommutative there.) Note: Going across and down is the same as the composite

$$\overline{\mathcal{S}}(\overline{\mathcal{P}}(\mathrm{H})) \to \overline{\mathcal{T}}(\overline{\mathcal{P}}(\mathrm{H})) \to \mathrm{H},$$

with the first map as in (13.46) with $\mathrm{V} := \overline{\mathcal{P}}(\mathrm{H})$, and the second map the counit of the adjunction between $\overline{\mathcal{T}}$ and $\overline{\mathcal{P}}$. Equivalently, the latter arises from Theorem 9.15 for $q = 1$ applied to $\mathrm{V} := \overline{\mathcal{P}}(\mathrm{H})$ and $f := \mathrm{id}$. □

Theorem 13.65. *Let* $\mathrm{char}(\Bbbk) \nmid |W|$. *The adjunction*

$$\mathsf{LieMon}(\overline{\mathsf{C}}\mathsf{Sp}) \underset{\overline{\mathcal{P}}}{\overset{\overline{\mathcal{U}}}{\rightleftarrows}} {}^{\mathsf{co}}\mathsf{Bimon}(\overline{\mathsf{C}}\mathsf{Sp})$$

is an adjoint equivalence of categories.

This is the *Cartier–Milnor–Moore theorem* for cocommutative Coxeter bialgebras, or CMM for short. It says that the counit and unit of the adjunction between $\overline{\mathcal{U}}$ and $\overline{\mathcal{P}}$ are natural isomorphisms:

$$\overline{\mathcal{U}}(\overline{\mathcal{P}}(\mathrm{H})) \xrightarrow{\cong} \mathrm{H} \quad \text{and} \quad \mathrm{G} \xrightarrow{\cong} \overline{\mathcal{P}}(\overline{\mathcal{U}}(\mathrm{G})).$$

PROOF. The first isomorphism is a part of diagram (13.63). The second follows from the PBW isomorphism $\overline{\mathcal{S}}(\mathrm{G}) \cong \overline{\mathcal{U}}(\mathrm{G})$ and $\overline{\mathcal{P}}(\overline{\mathcal{S}}(\mathrm{G})) = \mathrm{G}$ which holds by Exercise 9.96, item (i). □

Exercise 13.66. Let $\mathrm{char}(\Bbbk) \nmid |W|$. Deduce from Exercise 12.20: For A a Coxeter algebra with the commutator bracket, $\overline{\mathcal{U}}(\mathrm{A}) = \overline{\mathcal{S}}^{\vee}(\mathrm{A})$ as Coxeter bialgebras.

13.7.2. CMM for commutative Coxeter bialgebras.

The preceding discussion has a dual version which we state below.

Lemma 13.67. *Let* $\mathrm{char}(\Bbbk) \nmid |W|$. *For a commutative Coxeter bialgebra* H, *the diagram*

$$\begin{array}{ccc} \mathrm{H} & \xrightarrow{\quad\cong\quad} & \overline{\mathcal{S}}(\overline{\mathcal{Q}}(\mathrm{H})) \\ {\scriptstyle\cong}\big\downarrow & & \big\downarrow{\scriptstyle\cong} \\ \overline{\mathcal{U}}^{\vee}(\overline{\mathcal{Q}}(\mathrm{H})) & \xrightarrow[\cong]{} & \overline{\mathcal{S}}^{\vee}(\overline{\mathcal{Q}}(\mathrm{H})) \end{array} \tag{13.64}$$

commutes.

The vertical maps are Coxeter bialgebra isomorphisms, while the horizontal maps are Coxeter algebra isomorphisms. The right-vertical map is the norm map (9.142) for $\mathrm{V} := \overline{\mathcal{Q}}(\mathrm{H})$. The bottom-horizontal map is the PBW isomorphism in (13.55) for $\mathrm{K} := \overline{\mathcal{Q}}(\mathrm{H})$. The top-horizontal map is the Borel–Hopf isomorphism (13.8). The left-vertical map is the unit of the adjunction between $\overline{\mathcal{U}}^{\vee}$ and $\overline{\mathcal{Q}}$ in Theorem 12.26. It is an isomorphism because the remaining three maps are.

Theorem 13.68. *Let* $\mathrm{char}(\Bbbk) \nmid |W|$. *The adjunction*

$$\mathsf{Bimon}^{\mathsf{co}}(\overline{\mathsf{C}}\mathsf{Sp}) \underset{\overline{\mathcal{U}}^{\vee}}{\overset{\overline{\mathcal{Q}}}{\rightleftarrows}} \mathsf{LieComon}(\overline{\mathsf{C}}\mathsf{Sp})$$

is an adjoint equivalence of categories.

This is the *Cartier–Milnor–Moore theorem* for commutative Coxeter bialgebras. It says that the unit and counit of the adjunction between $\overline{\mathcal{U}}^\vee$ and $\overline{\mathcal{Q}}$ are natural isomorphisms:

$$\mathrm{H} \xrightarrow{\cong} \overline{\mathcal{U}}^\vee(\overline{\mathcal{Q}}(\mathrm{H})) \quad \text{and} \quad \overline{\mathcal{Q}}(\overline{\mathcal{U}}^\vee(\mathrm{K})) \xrightarrow{\cong} \mathrm{K}.$$

The first isomorphism is a part of diagram (13.64). The second follows from the PBW isomorphism $\overline{\mathcal{U}}^\vee(\mathrm{K}) \cong \overline{\mathcal{S}}^\vee(\mathrm{K})$ and $\overline{\mathcal{Q}}(\overline{\mathcal{S}}^\vee(\mathrm{K})) = \mathrm{K}$.

13.7.3. Examples. This is a continuation of Section 13.1.7. We upgrade Propositions 13.10 and 13.11 by applying CMM Theorems 13.65 and 13.68.

For the tensor and shuffle Coxeter bialgebras,

$$\overline{\mathcal{T}}(M) \cong \overline{\mathcal{U}}(\overline{\mathcal{P}}(\overline{\mathcal{T}}(M))) \quad \text{and} \quad \overline{\mathcal{T}}^\vee(M) \cong \overline{\mathcal{U}}^\vee(\overline{\mathcal{Q}}(\overline{\mathcal{T}}^\vee(M))) \tag{13.65}$$

as Coxeter bialgebras. Thus, $\overline{\mathcal{T}}(M)$ is the universal enveloping Coxeter algebra of $\overline{\mathcal{P}}(\overline{\mathcal{T}}(M))$ which is a free Coxeter Lie algebra by (12.29). Dually, $\overline{\mathcal{T}}^\vee(M)$ is the universal coenveloping Coxeter coalgebra of $\overline{\mathcal{Q}}(\overline{\mathcal{T}}^\vee(M))$ which is a cofree Coxeter Lie coalgebra.

Similarly, for the Coxeter bialgebras of face-types,

$$\overline{\Sigma} \cong \overline{\mathcal{U}}(\overline{\mathcal{P}}(\overline{\Sigma})) \quad \text{and} \quad \overline{\Sigma}^* \cong \overline{\mathcal{U}}^\vee(\overline{\mathcal{Q}}(\overline{\Sigma}^*)) \tag{13.66}$$

as Coxeter bialgebras. Thus, $\overline{\Sigma}$ is the universal enveloping Coxeter algebra of $\overline{\mathcal{P}}(\overline{\Sigma})$ which is a free Coxeter Lie algebra by (12.31). Dually, $\overline{\Sigma}^*$ is the universal coenveloping Coxeter coalgebra of $\overline{\mathcal{Q}}(\overline{\Sigma}^*)$ which is a cofree Coxeter Lie coalgebra.

13.7.4. Signed analogue. CMM Theorems 13.65 and 13.68 have signed analogues which are stated below. They can be deduced from the unsigned versions by conjugating by the u-signature functor (8.99).

Theorem 13.69. *Let* $\mathrm{char}(\mathbb{k}) \nmid |W|$. *The adjunction*

$$(-1)\text{-}\mathsf{LieMon}(\overline{\mathsf{C}}\mathsf{Sp}) \underset{\overline{\mathcal{P}}}{\overset{\overline{\mathcal{U}}_{-1}}{\rightleftarrows}} (-1)\text{-}^{\mathsf{co}}\mathsf{Bimon}(\overline{\mathsf{C}}\mathsf{Sp})$$

is an adjoint equivalence of categories.

This is CMM for signed cocommutative signed Coxeter bialgebras.

Theorem 13.70. *Let* $\mathrm{char}(\mathbb{k}) \nmid |W|$. *The adjunction*

$$(-1)\text{-}\mathsf{Bimon}^{\mathsf{co}}(\overline{\mathsf{C}}\mathsf{Sp}) \underset{\overline{\mathcal{U}}^\vee_{-1}}{\overset{\overline{\mathcal{Q}}}{\rightleftarrows}} (-1)\text{-}\mathsf{LieComon}(\overline{\mathsf{C}}\mathsf{Sp})$$

is an adjoint equivalence of categories.

This is CMM for signed commutative signed Coxeter bialgebras.

13.7.5. Specialization to Leray–Samelson. The adjoint equivalence in CMM restricts to the adjoint equivalence in Leray–Samelson as follows.

$$\begin{array}{ccc} \mathsf{LieMon}(\overline{\mathsf{C}}\mathsf{Sp}) & \underset{\overline{\mathcal{P}}}{\overset{\overline{\mathcal{U}}}{\rightleftarrows}} & {}^{\mathrm{co}}\mathsf{Bimon}(\overline{\mathsf{C}}\mathsf{Sp}) \\ \uparrow & & \uparrow \\ \overline{\mathsf{C}}\mathsf{Sp} & \underset{\overline{\mathcal{P}}}{\overset{\overline{\mathcal{S}}}{\rightleftarrows}} & {}^{\mathrm{co}}\mathsf{Bimon}^{\mathrm{co}}(\overline{\mathsf{C}}\mathsf{Sp}) \end{array}$$

In particular, for H a bicommutative Coxeter bialgebra, the map $\overline{\mathcal{U}}\overline{\mathcal{P}}(\mathrm{H}) \to \mathrm{H}$ specializes to $\overline{\mathcal{S}}\overline{\mathcal{P}}(\mathrm{H}) \to \mathrm{H}$ in (13.12).

Similar diagrams hold in the dual and signed settings.

13.8. Coxeter bialgebras for a rank-one arrangement

Let $\mathcal{A}$ be a rank-one arrangement with chambers C and $\overline{C}$. Let w denote the nonidentity element of the Coxeter group W. Let $\boldsymbol{\alpha}$ be the pointed reflection arrangement with C as the reference chamber. There are two face-types, namely, $\emptyset$ and S. We now make explicit the notion of a Coxeter $\boldsymbol{\alpha}$-bialgebra, and discuss its structure theory in a self-contained manner.

It is insightful to compare with the parallel discussion for Coxeter bimonoids in Section 7.8. In contrast to that setting, (co)invariants wrt the two-element group $\mathbb{Z}_2$ appear in the present setting, thus forcing us to stay away from field characteristic 2 more often.

13.8.1. Coxeter bialgebras. We specialize objects in Sections 8.4, 8.5, 8.6, 8.7, 8.8 to rank one.

Groupoid of biface-types. The *groupoid of biface-types* $\boldsymbol{\alpha}$-$\overline{\mathsf{Cox}}$ consists of two objects $\emptyset$ and S. Apart from the two identity morphisms, there is a morphism $\sigma : S \to S$ with $\sigma^2 = \mathrm{id}$. It has length 1. See the picture below.

$$\mathrm{id} \circlearrowright \emptyset \qquad\qquad \mathrm{id} \circlearrowright S \circlearrowleft \sigma$$

Coxeter spaces. A Coxeter $\boldsymbol{\alpha}$-space V consists of a pair of vector spaces denoted $\mathrm{V}[\emptyset]$ and $\mathrm{V}[S]$, with the latter equipped with an involution

$$\sigma : \mathrm{V}[S] \to \mathrm{V}[S].$$

We write this as $\mathrm{V} = (\mathrm{V}[\emptyset], \mathrm{V}[S], \sigma)$, or simply as $\mathrm{V} = (\mathrm{V}[\emptyset], \mathrm{V}[S])$ with σ understood.

A map $f : \mathrm{V} \to \mathrm{W}$ of Coxeter $\boldsymbol{\alpha}$-spaces consists of a pair of linear maps denoted $f_\emptyset : \mathrm{V}[\emptyset] \to \mathrm{W}[\emptyset]$ and $f_S : \mathrm{V}[S] \to \mathrm{W}[S]$ such that $\sigma f_S = f_S\, \sigma$. We write this as $f = (f_\emptyset, f_S)$.

Coxeter algebras. A Coxeter $\boldsymbol{\alpha}$-algebra is a Coxeter $\boldsymbol{\alpha}$-space A equipped with a linear map

$$\mu : \mathrm{A}[S] \to \mathrm{A}[\emptyset].$$

(The associativity axiom is vacuous in rank one.) A Coxeter $\boldsymbol{\alpha}$-algebra is commutative if $\mu\sigma = \mu$. It is signed commutative if $\mu\sigma + \mu = 0$. We call μ the product map. If $\mu = 0$, then we say the product is trivial.

A Coxeter $\boldsymbol{\alpha}$-q-bialgebra H consists of a pair of vector spaces $\mathrm{H}[\emptyset]$ and $\mathrm{H}[S]$ equipped with linear maps

$$\mu : \mathrm{H}[S] \to \mathrm{H}[\emptyset], \quad \sigma : \mathrm{H}[S] \to \mathrm{H}[S], \quad \Delta : \mathrm{H}[\emptyset] \to \mathrm{H}[S],$$

which satisfy $\sigma^2 = \mathrm{id}$ and $\Delta\mu = \mathrm{id} + q\,\sigma$.

SUMMARY 13.1. Coxeter q-bialgebra in rank one.

A morphism $f : \mathrm{A} \to \mathrm{B}$ of Coxeter $\boldsymbol{\alpha}$-algebras is a map of Coxeter $\boldsymbol{\alpha}$-spaces such that $\mu f_S = f_\emptyset\, \mu$.

Coxeter coalgebras. Dually, a Coxeter $\boldsymbol{\alpha}$-coalgebra is a Coxeter $\boldsymbol{\alpha}$-space C equipped with a linear map

$$\Delta : \mathrm{C}[\emptyset] \to \mathrm{C}[S].$$

It is cocommutative if $\sigma\Delta = \Delta$. It is signed cocommutative if $\sigma\Delta + \Delta = 0$. We call Δ the coproduct map. If $\Delta = 0$, then we say the coproduct is trivial.

A morphism $f : \mathrm{C} \to \mathrm{D}$ of Coxeter $\boldsymbol{\alpha}$-coalgebras is a map of Coxeter $\boldsymbol{\alpha}$-spaces such that $f_S\, \Delta = \Delta f_\emptyset$.

Coxeter bialgebras. A Coxeter $\boldsymbol{\alpha}$-bialgebra is a triple $(\mathrm{H}, \mu, \Delta)$, where H is a Coxeter $\boldsymbol{\alpha}$-space, (H, μ) is a Coxeter $\boldsymbol{\alpha}$-algebra, (H, Δ) is a Coxeter $\boldsymbol{\alpha}$-coalgebra, and

$$\Delta\mu = \mathrm{id} + \sigma$$

as maps from $\mathrm{H}[S]$ to itself. This equation is the bialgebra axiom.

A morphism of Coxeter $\boldsymbol{\alpha}$-bialgebras is a map of the underlying Coxeter $\boldsymbol{\alpha}$-spaces which is a morphism of the underlying Coxeter $\boldsymbol{\alpha}$-algebras and Coxeter $\boldsymbol{\alpha}$-coalgebras.

Example 13.71. Take $\mathrm{H}[\emptyset] = 0$ and $\mathrm{H}[S]$ arbitrary with $\sigma = -\mathrm{id}$. Then H is a Coxeter $\boldsymbol{\alpha}$-bialgebra with trivial product and coproduct, that is, μ and Δ are the zero maps. □

Coxeter q-bialgebras. More generally, for a scalar q, a Coxeter $\boldsymbol{\alpha}$-q-bialgebra is a triple $(\mathrm{H}, \mu, \Delta)$, where H is a Coxeter $\boldsymbol{\alpha}$-space, (H, μ) is a Coxeter $\boldsymbol{\alpha}$-algebra, (H, Δ) is a Coxeter $\boldsymbol{\alpha}$-coalgebra, and

$$\Delta\mu = \mathrm{id} + q\,\sigma$$

as maps from $\mathrm{H}[S]$ to itself. This equation is the q-bialgebra axiom. It is an instance of (8.74). It also specializes (8.75) to $n = 1$. The data present in a Coxeter $\boldsymbol{\alpha}$-q-bialgebra is shown in Summary 13.1.

A morphism of Coxeter $\boldsymbol{\alpha}$-q-bialgebras is a map of the underlying Coxeter $\boldsymbol{\alpha}$-spaces which is a morphism of the underlying Coxeter $\boldsymbol{\alpha}$-algebras and Coxeter $\boldsymbol{\alpha}$-coalgebras.

The signed bialgebra axiom says that $\Delta\mu = \mathrm{id} - \sigma$. Similarly, the 0-bialgebra axiom says that $\Delta\mu = \mathrm{id}$. It is set-theoretic. See Exercise 8.81.

13.8.2. Abelianization and coabelianization. We specialize the discussion in Section 9.4 to rank one. The abelianization of a Coxeter $\boldsymbol{\alpha}$-algebra A, denoted A_{ab}, is

$$\mathrm{A}_{ab}[S] = \mathrm{A}[S] \quad \text{and} \quad \mathrm{A}_{ab}[\emptyset] = \mathrm{coker}(\mu\sigma - \mu)$$

with structure maps

$$\begin{array}{ccc} \mathrm{A}_{ab}[S] & \overset{\sigma}{\dashrightarrow} & \mathrm{A}_{ab}[S] \\ \| & & \| \\ \mathrm{A}[S] & \xrightarrow[\sigma]{} & \mathrm{A}[S] \end{array} \qquad \begin{array}{ccccc} \mathrm{A}_{ab}[S] & & \overset{\mu}{\dashrightarrow} & & \mathrm{A}_{ab}[\emptyset] \\ \| & & & & \| \\ \mathrm{A}[S] & \xrightarrow[\mu]{} & \mathrm{A}[\emptyset] & \twoheadrightarrow & \mathrm{coker}(\mu\sigma - \mu). \end{array}$$

Signed abelianization is defined similarly using $\mathrm{coker}(\mu\sigma + \mu)$.

The coabelianization of a Coxeter $\boldsymbol{\alpha}$-coalgebra C, denoted C^{coab}, is

$$\mathrm{C}^{coab}[S] = \mathrm{C}[S] \quad \text{and} \quad \mathrm{C}^{coab}[\emptyset] = \ker(\sigma\Delta - \Delta)$$

with structure maps

$$\begin{array}{ccc} \mathrm{C}^{coab}[S] & \overset{\sigma}{\dashrightarrow} & \mathrm{C}^{coab}[S] \\ \| & & \| \\ \mathrm{C}[S] & \xrightarrow[\sigma]{} & \mathrm{C}[S] \end{array} \qquad \begin{array}{ccccc} \mathrm{C}^{coab}[\emptyset] & & \overset{\Delta}{\dashrightarrow} & & \mathrm{C}^{coab}[S] \\ \| & & & & \| \\ \ker(\sigma\Delta - \Delta) & \hookrightarrow & \mathrm{C}[\emptyset] & \xrightarrow[\Delta]{} & \mathrm{C}[S]. \end{array}$$

Signed coabelianization is defined similarly using $\ker(\sigma\Delta + \Delta)$.

Exercise 13.72. Check: The (co)abelianization of a Coxeter $\boldsymbol{\alpha}$-bialgebra is again a Coxeter $\boldsymbol{\alpha}$-bialgebra. (Use identities $\Delta(\mu\sigma - \mu) = 0 = (\sigma\Delta - \Delta)\mu$.) This specializes Exercise 9.31 to rank one.

Exercise 13.73. Check: For a Coxeter $\boldsymbol{\alpha}$-bialgebra H, the composite map

$$\mathrm{H}[S] \xrightarrow{\mu\sigma - \mu} \mathrm{H}[\emptyset] \xrightarrow{\sigma\Delta - \Delta} \mathrm{H}[S]$$

is zero. Moreover, the homology group $\ker(\sigma\Delta - \Delta)/\operatorname{im}(\mu\sigma - \mu)$ identifies with $(\mathrm{H}^{coab})_{ab}[\emptyset] = (\mathrm{H}_{ab})^{coab}[\emptyset]$.

13.8.3. Tensor Coxeter bialgebra and other examples. In Chapter 10, we discussed many examples of Coxeter bialgebras. The tensor Coxeter bialgebra and its relatives from Sections 10.2 and 10.3, when specialized to rank one, take the form shown in Table 13.3. Here M is a W-module, which is the same as a vector space M equipped with an involution $w : M \to M$. Also, $\bar{y}$ denotes the image of $y \in M$ in M_W.

Similarly, the Coxeter bialgebras of face-types and their relatives from Sections 10.4, 10.5, 10.6, when specialized to rank one, take the form shown in Table 13.4. Formulas are written in the H-basis for $\overline{\Sigma}_q$, M-basis for $\overline{\Sigma}^*_q$, K-basis for $\overline{\Pi\Gamma}_q$, F-basis for $\overline{\Pi\Gamma}^*_q$. In rank one, there is no distinction between face-types and flat-types, hence $\overline{\Pi} = \overline{\Sigma}$ and $\overline{\Pi}^* = \overline{\Sigma}^*$.

For a scalar q and a pair of scalars $(\xi_\emptyset, \xi_S)$, the map

$$\overline{\kappa}_{q,\xi} : \overline{\Sigma}_q \to \overline{\Sigma}^*_q, \quad ((x, y), z) \mapsto ((\xi_\emptyset x + \xi_S y, \xi_S x + (1+q)\,\xi_S y), \xi_S z)$$

TABLE 13.3. Tensor Coxeter bialgebra in rank one.

Coxeter q-bialgebra	involution σ	product μ	coproduct Δ
$\Bbbk_q[x] = (\Bbbk, \Bbbk)$	$\sigma(y) = y$	$\mu(y) = y$	$\Delta(x) = (1+q)\,x$
$\Bbbk_q\{x\} = (\Bbbk, \Bbbk)$	$\sigma(y) = y$	$\mu(y) = (1+q)\,y$	$\Delta(x) = x$
$\overline{\mathcal{T}}_q(M) = (M, M)$	$\sigma(y) = w \cdot y$	$\mu(y) = y$	$\Delta(x) = x + q\,(w \cdot x)$
$\overline{\mathcal{T}}_q^{\vee}(M) = (M, M)$	$\sigma(y) = w \cdot y$	$\mu(y) = y + q\,(w \cdot y)$	$\Delta(x) = x$
$\overline{\mathcal{T}}_0(M) = \overline{\mathcal{T}}_0^{\vee}(M)$	$\sigma(y) = w \cdot y$	$\mu(y) = y$	$\Delta(x) = x$
$\overline{\mathcal{S}}(M) = (M_W, M)$	$\sigma(y) = w \cdot y$	$\mu(y) = \bar{y}$	$\Delta(\bar{x}) = x + w \cdot x$
$\overline{\mathcal{S}}^{\vee}(M) = (M^W, M)$	$\sigma(y) = w \cdot y$	$\mu(y) = y + w \cdot y$	$\Delta(x) = x$

TABLE 13.4. Coxeter bialgebras of face-types in rank one.

Coxeter q-bialgebra	involution σ	product μ	coproduct Δ
$\overline{\Sigma}_q = (\Bbbk \oplus \Bbbk, \Bbbk)$	$\sigma(z) = z$	$\mu(z) = (0, z)$	$\Delta(x, y) = x + (1+q)\,y$
$\overline{\Sigma}_q^* = (\Bbbk \oplus \Bbbk, \Bbbk)$	$\sigma(z) = z$	$\mu(z) = (z, (1+q)\,z)$	$\Delta(x, y) = y$
$\overline{\Pi} = \overline{\Sigma} = (\Bbbk \oplus \Bbbk, \Bbbk)$	$\sigma(z) = z$	$\mu(z) = (0, z)$	$\Delta(x, y) = x + 2y$
$\overline{\Pi}^* = \overline{\Sigma}^* = (\Bbbk \oplus \Bbbk, \Bbbk)$	$\sigma(z) = z$	$\mu(z) = (z, 2z)$	$\Delta(x, y) = y$
$\overline{\Gamma}_q = (\Bbbk \oplus \Bbbk, \Bbbk)$	$\sigma(z) = z$	$\mu(z) = (z, z)$	$\Delta(x, y) = x + q\,y$
$\overline{\Gamma}_q^* = (\Bbbk \oplus \Bbbk, \Bbbk)$	$\sigma(z) = z$	$\mu(z) = (z, q\,z)$	$\Delta(x, y) = x + y$

is a morphism of Coxeter $\boldsymbol{\alpha}$-q-bialgebras. This can be checked directly. This map specializes (10.85). Setting $q = 1$, we obtain

$$\overline{\Pi} \to \overline{\Pi}^*, \quad ((x, y), z) \mapsto ((\xi_\emptyset x + \xi_S y, \xi_S x + 2\,\xi_S y), \xi_S z).$$

This is an isomorphism of Coxeter $\boldsymbol{\alpha}$-bialgebras iff $\xi_S \neq 0$ and $2\,\xi_\emptyset \neq \xi_S$ (assuming $\mathrm{char}(\Bbbk) \neq 2$). This is an instance of the isomorphism criterion in Theorem 10.119.

For a scalar q, the map

$$\overline{\Sigma}_q \to \overline{\Gamma}_q, \quad ((x, y), z) \mapsto ((x + y, y), z)$$

is an isomorphism of Coxeter $\boldsymbol{\alpha}$-q-bialgebras. This map specializes the middle map in (10.148). Similarly, the map

$$\overline{\Gamma}_q \to \overline{\Gamma}_q^*, \quad ((x, y), z) \mapsto ((x, q\,y), z)$$

is an isomorphism of Coxeter $\boldsymbol{\alpha}$-q-bialgebras. This map specializes (10.142).

It also follows that the Coxeter $\boldsymbol{\alpha}$-bialgebra H in Exercise 10.128 equals $\overline{\Gamma}$. In fact, one can directly check that conditions (10.150) hold in rank one.

13.8.4. Modules over the invariant Tits algebra. We specialize ideas from Section 9.13 to rank one. Consider the invariant Tits algebra $\Sigma[\mathcal{A}]^W$ in rank one from Example 1.105. It is isomorphic to the group algebra of W by Example 1.113. Thus a $\Sigma[\mathcal{A}]^W$-module is the same as a W-module. (We do not need to specify left or right module, since the algebras are commutative.)

For a cocommutative Coxeter $\boldsymbol{\alpha}$-bialgebra H, the characteristic operation

$$\mathtt{H}_S \cdot z = \mu\Delta(z) \quad \text{and} \quad w \cdot z := \mu\Delta(z) - z$$

defines, equivalently, an action of $\Sigma[\mathcal{A}]^W$ and of W on $\mathrm{H}[\emptyset]$. This is a special case of (9.169). This yields a functor from the category of cocommutative Coxeter $\boldsymbol{\alpha}$-bialgebras to the category of $\Sigma[\mathcal{A}]^W$-modules, or equivalently, of W-modules. Let us call it $\mathcal{F}$. It is essentially surjective on objects: For any W-module M, applying $\mathcal{F}$ to the Coxeter $\boldsymbol{\alpha}$-bialgebra $\overline{\mathcal{T}}(M)$ recovers M. However, $\overline{\mathcal{T}}(M)$ may not be the only Coxeter $\boldsymbol{\alpha}$-bialgebra (up to isomorphism) which yields M. For instance, consider $M = \Bbbk \oplus \Bbbk$ with action

$$\mathtt{H}_S \cdot (x, y) := (0, x + 2y) \quad \text{and} \quad w \cdot (x, y) := (-x, x + y).$$

Observe: This is the action of $\Sigma[\mathcal{A}]^W$ on itself. Hence, the Coxeter $\boldsymbol{\alpha}$-bialgebra $\overline{\Sigma}$ also yields M by Example 9.121. Moreover, it is not isomorphic to $\overline{\mathcal{T}}(M)$ (for instance, the dimensions of their S-components are different). This shows that $\mathcal{F}$ is not an equivalence.

Exercise 13.74. Let $M = \Bbbk \oplus \Bbbk$ be the $\Sigma[\mathcal{A}]^W$-module as above. Check: Module maps $f : M \to M$ are characterized by $f(\mathtt{H}_\emptyset) = a\,\mathtt{H}_\emptyset + b\,\mathtt{H}_S$ for scalars a and b. Coxeter $\boldsymbol{\alpha}$-bialgebra morphisms from $\overline{\Sigma}$ to $\overline{\mathcal{T}}(M)$ are also similarly characterized. A morphism has two components, namely,

$$\overline{\Sigma}[\emptyset] \to \overline{\mathcal{T}}(M)[\emptyset] \quad \text{and} \quad \overline{\Sigma}[S] \to \overline{\mathcal{T}}(M)[S].$$

The first map sends $\mathtt{H}_\emptyset$ to $a\,\mathtt{H}_\emptyset + b\,\mathtt{H}_S$ and $\mathtt{H}_S$ to $(a + 2b)\,\mathtt{H}_S$. The second map sends $\mathtt{H}_S$ to $(a + 2b)\,\mathtt{H}_S$. However, Coxeter $\boldsymbol{\alpha}$-bialgebra morphisms from $\overline{\mathcal{T}}(M)$ to $\overline{\Sigma}$ are characterized by a single scalar c. The components

$$\overline{\mathcal{T}}(M)[\emptyset] \to \overline{\Sigma}[\emptyset] \quad \text{and} \quad \overline{\mathcal{T}}(M)[S] \to \overline{\Sigma}[S]$$

both send $\mathtt{H}_\emptyset$ to $c\,\mathtt{H}_S$ and $\mathtt{H}_S$ to $2c\,\mathtt{H}_S$. This shows that $\mathcal{F}$ is not full.

Exercise 13.75. For a scalar q, consider the invariant q-Janus algebra $\mathsf{J}_q[\mathcal{A}]^W$ in rank one from Example 1.109. Check directly: For a Coxeter $\boldsymbol{\alpha}$-q-bialgebra H, the two-sided characteristic operations

$$\mathtt{H}_{(S,\mathrm{id},S)} \cdot z = \mu\Delta(z) \quad \text{and} \quad \mathtt{H}_{(S,\sigma,S)} \cdot z = \mu\sigma\Delta(z)$$

define an action of $\mathsf{J}_q[\mathcal{A}]^W$ on $\mathrm{H}[\emptyset]$. This is a special case of (9.168). This yields a functor from the category of Coxeter $\boldsymbol{\alpha}$-q-bialgebras to the category of $\mathsf{J}_q[\mathcal{A}]^W$-modules. Similarly, for $q \neq \pm 1$, write down the action of the $\mathtt{Q}$-basis elements.

13.8.5. Coxeter Lie algebras.

We specialize objects in Section 12.1 to rank one.

Coxeter Lie algebras. A Coxeter $\boldsymbol{\alpha}$-Lie algebra is a Coxeter $\boldsymbol{\alpha}$-space G equipped with a linear map

$$\nu : \mathrm{G}[S] \to \mathrm{G}[\emptyset],$$

called the Lie bracket, which satisfies $\nu\sigma + \nu = 0$. This is antisymmetry. (Jacobi identity is vacuous in rank one.) A morphism $f : \mathrm{G} \to \mathrm{H}$ of Coxeter $\boldsymbol{\alpha}$-Lie algebras is a map of Coxeter $\boldsymbol{\alpha}$-spaces such that $\nu f_S = f_\emptyset\,\nu$.

A signed Coxeter $\boldsymbol{\alpha}$-Lie algebra is defined similarly using the signed antisymmetry axiom $\nu\sigma = \nu$.

Coxeter Lie coalgebras. Dually, a Coxeter $\boldsymbol{\alpha}$-Lie coalgebra is a Coxeter $\boldsymbol{\alpha}$-space K equipped with a linear map

$$\theta : \mathrm{K}[\emptyset] \to \mathrm{K}[S],$$

called the Lie cobracket, which satisfies $\sigma\theta + \theta = 0$. This is antisymmetry. A morphism $f : \mathrm{K} \to \mathrm{H}$ of Coxeter $\boldsymbol{\alpha}$-Lie coalgebras is a map of Coxeter $\boldsymbol{\alpha}$-spaces such that $f_S\,\theta = \theta f_\emptyset$.

A signed Coxeter $\boldsymbol{\alpha}$-Lie coalgebra is defined similarly using the signed antisymmetry axiom $\sigma\theta = \theta$.

13.8.6. Commutator bracket and cocommutator cobracket. Every Coxeter $\boldsymbol{\alpha}$-algebra (A, μ) carries a Coxeter $\boldsymbol{\alpha}$-Lie algebra structure via the commutator bracket

$$\nu := \mu - \mu\sigma.$$

Let us check antisymmetry: $\nu\sigma + \nu = (\mu - \mu\sigma)\sigma + (\mu - \mu\sigma) = 0$. Similarly, A is a signed Coxeter $\boldsymbol{\alpha}$-Lie algebra via the signed commutator bracket $\nu := \mu + \mu\sigma$.

Dually, every Coxeter $\boldsymbol{\alpha}$-coalgebra (C, Δ) carries a Coxeter $\boldsymbol{\alpha}$-Lie coalgebra structure via the cocommutator cobracket

$$\theta := \Delta - \sigma\Delta.$$

One may directly check antisymmetry. Similarly, C is a signed Coxeter $\boldsymbol{\alpha}$-Lie coalgebra via the signed cocommutator cobracket $\theta := \Delta + \sigma\Delta$.

13.8.7. Substitution of a Coxeter space. Recall from Section 11.1.4 the category $\boldsymbol{\alpha}$-dCox. In rank one, it is given by

$$(\mathrm{id},\mathrm{id}) \circlearrowright (\emptyset, \emptyset) \qquad (\mathrm{id},\mathrm{id}) \circlearrowright (\emptyset, S) \circlearrowleft (\mathrm{id},\sigma) \qquad (\mathrm{id},\mathrm{id}) \circlearrowright (S, S) \circlearrowleft (\sigma,\sigma)$$

with $(\mathrm{id}, \sigma)^2 = (\mathrm{id}, \mathrm{id})$ and $(\sigma, \sigma)^2 = (\mathrm{id}, \mathrm{id})$. A Coxeter $\boldsymbol{\alpha}$-dispecies $\mathbf{p}$ is a functor from $\boldsymbol{\alpha}$-dCox to Vec. So it consists of vector spaces $\mathbf{p}[\emptyset, \emptyset]$, $\mathbf{p}[\emptyset, S]$, $\mathbf{p}[S, S]$ equipped with involutions

$$(\mathrm{id}, \sigma) : \mathbf{p}[\emptyset, S] \to \mathbf{p}[\emptyset, S] \quad \text{and} \quad (\sigma, \sigma) : \mathbf{p}[S, S] \to \mathbf{p}[S, S].$$

Recall from (11.29) the substitution operation $\mathbf{p} \circ \mathrm{V}$ of a Coxeter $\boldsymbol{\alpha}$-space V in a Coxeter $\boldsymbol{\alpha}$-dispecies $\mathbf{p}$. In rank one,

$$\begin{aligned}(\mathbf{p} \circ \mathrm{V})[\emptyset] &= (\mathbf{p}[\emptyset, \emptyset] \otimes \mathrm{V}[\emptyset]) \oplus (\mathbf{p}[\emptyset, S] \otimes \mathrm{V}[S])_\sigma, \\ (\mathbf{p} \circ \mathrm{V})[S] &= \mathbf{p}[S, S] \otimes \mathrm{V}[S].\end{aligned}$$

In the first formula, in the second summand, we take coinvariants of the diagonal action of σ, where the action of σ on $\mathbf{p}[\emptyset, S]$ is by (id, σ). In the second formula, the action of σ on $(\mathbf{p} \circ \mathrm{V})[S]$ is by $(\sigma, \sigma) \otimes \sigma$.

Let us now specialize Section 11.5.4 to rank one. For a connected Coxeter operad $\mathbf{a}$, a Coxeter $\mathbf{a}$-algebra is a Coxeter space V equipped with a linear map

$$(\mathbf{a}[\emptyset, S] \otimes \mathrm{V}[S])_\sigma \to \mathrm{V}[\emptyset].$$

For the Coxeter associative, commutative, Lie operads, this reduces, respectively, to

$$\mathbf{As}\quad \mathrm{V}[S] \to \mathrm{V}[\emptyset], \qquad \mathbf{Com}\quad \mathrm{V}[S]_{\sigma} \to \mathrm{V}[\emptyset], \qquad \mathbf{Lie}\quad \mathrm{V}[S]_{-\sigma} \to \mathrm{V}[\emptyset].$$

This is consistent with the definitions given earlier for Coxeter $\boldsymbol{\alpha}$-algebras, commutative Coxeter $\boldsymbol{\alpha}$-algebras, Coxeter $\boldsymbol{\alpha}$-Lie algebras. The morphisms (5.19) yield maps of Coxeter $\boldsymbol{\alpha}$-spaces

$$\mathbf{Lie} \circ \mathrm{V} \hookrightarrow \mathbf{As} \circ \mathrm{V} \twoheadrightarrow \mathbf{Com} \circ \mathrm{V}.$$

Explicitly, on the $\emptyset$-component,

$$\mathrm{V}[\emptyset] \oplus \mathrm{V}[S]_{-\sigma} \hookrightarrow \mathrm{V}[\emptyset] \oplus \mathrm{V}[S] \twoheadrightarrow \mathrm{V}[\emptyset] \oplus \mathrm{V}[S]_{\sigma}.$$

The first map sends $(x, \bar{y})$ to $(x, y - \sigma(y))$, while the second map sends (x, y) to $(x, \bar{y})$. (For injectivity of the first map, we assume $\mathrm{char}(\Bbbk) \neq 2$.)

13.8.8. Primitive and indecomposable parts. We specialize notions in Section 9.2 to rank one.

Primitive part. Define the primitive part $\overline{\mathcal{P}}(\mathrm{C})$ of a Coxeter $\boldsymbol{\alpha}$-coalgebra C by

$$\overline{\mathcal{P}}(\mathrm{C})[S] = \mathrm{C}[S] \quad \text{and} \quad \overline{\mathcal{P}}(\mathrm{C})[\emptyset] = \ker \Delta.$$

For H a Coxeter $\boldsymbol{\alpha}$-bialgebra, $\overline{\mathcal{P}}(\mathrm{H})$ is a Coxeter $\boldsymbol{\alpha}$-Lie subalgebra of H via the commutator bracket $\nu = \mu - \mu\sigma$. This is equivalent to saying that $\Delta\nu = 0$. Let us check this directly: $\Delta\nu = \Delta(\mu - \mu\sigma) = (\mathrm{id} + \sigma) - (\mathrm{id} + \sigma)\sigma = 0$.

Indecomposable part. Define the indecomposable part $\overline{\mathcal{Q}}(\mathrm{A})$ of a Coxeter $\boldsymbol{\alpha}$-algebra A by

$$\overline{\mathcal{Q}}(\mathrm{A})[S] = \mathrm{A}[S] \quad \text{and} \quad \overline{\mathcal{Q}}(\mathrm{A})[\emptyset] = \mathrm{coker}\, \mu.$$

For H a Coxeter $\boldsymbol{\alpha}$-bialgebra, $\overline{\mathcal{Q}}(\mathrm{H})$ is a Coxeter $\boldsymbol{\alpha}$-Lie quotient coalgebra of H via the cocommutator cobracket $\theta = \Delta - \sigma\Delta$. This is equivalent to saying that $\theta\mu = 0$.

Primitive part to indecomposable part. For a scalar q, for a Coxeter $\boldsymbol{\alpha}$-q-bialgebra H, we have the canonical map

$$\mathrm{pq}_{\mathrm{H}} : \ker \Delta \hookrightarrow \mathrm{H}[\emptyset] \twoheadrightarrow \mathrm{coker}\, \mu.$$

This specializes (9.106).

We record below two elementary but interesting results. Item (2) in both results specializes Propositions 9.65 and 9.66.

Lemma 13.76. *Let* $\mathrm{char}(\Bbbk) \neq 2$.

(1) *For a Coxeter* $\boldsymbol{\alpha}$*-space* V*, the linear operators* $\mathrm{id} + \sigma$ *and* $\mathrm{id} - \sigma$ *on* $\mathrm{V}[S]$ *are quasi-idempotent and mutually orthogonal.*
(2) *Let* $(\mathrm{H}, \mu, \Delta)$ *be a Coxeter* $\boldsymbol{\alpha}$*-bialgebra. Then* H *is*
   - *commutative iff* $\ker \Delta$ *and* $\mathrm{im}\, \mu$ *intersect trivially,*
   - *cocommutative iff* $\ker \Delta$ *and* $\mathrm{im}\, \mu$ *span* $\mathrm{H}[\emptyset]$*,*
   - *bicommutative iff* $\mathrm{H}[\emptyset] = \ker \Delta \oplus \mathrm{im}\, \mu$.

Explicitly, when H is bicommutative, the decomposition of $y \in \mathrm{H}[\emptyset]$ is

$$y = \big(y - \tfrac{1}{2}\,\mu\Delta(y)\big) + \tfrac{1}{2}\,\mu\Delta(y),$$

with the first term in the kernel of Δ and second term in the image of μ. This can be equivalently written as $y = \mathtt{E}_{\lambda_{\min}} \cdot y + \mathtt{E}_{\lambda_{\max}} \cdot y$ using characteristic operations by the invariant eulerian idempotents from Example 1.105.

Lemma 13.77. *Let $q \neq \pm 1$.*

(1) *For a Coxeter $\boldsymbol{\alpha}$-space* V, *the linear operator* $\mathrm{id} + q\,\sigma$ *on* $\mathrm{V}[S]$ *is invertible with inverse* $\frac{1}{1-q^2}(\mathrm{id} - q\,\sigma)$.

(2) *For* $(\mathrm{H}, \mu, \Delta)$ *a Coxeter $\boldsymbol{\alpha}$-q-bialgebra, μ is injective, Δ surjective, and*

$$\mathrm{H}[\emptyset] = \ker \Delta \oplus \operatorname{im} \mu.$$

Explicitly, the decomposition of $y \in \mathrm{H}[\emptyset]$ is

$$y = \big(y - \tfrac{1}{1-q^2}\,\mu(\mathrm{id} - q\,\sigma)\Delta(y)\big) + \big(\tfrac{1}{1-q^2}\,\mu(\mathrm{id} - q\,\sigma)\Delta(y)\big),$$

with the first term in the kernel of Δ and second term in the image of μ. This can be equivalently written as $y = \mathtt{Q}_{(\emptyset,\mathrm{id},\emptyset)} \cdot y + \mathtt{Q}_{(S,\mathrm{id},S)} \cdot y$ using two-sided characteristic operations by the $\mathtt{Q}$-basis elements from Example 1.109.

13.8.9. Universal constructions of Coxeter bialgebras. In Section 9.3, we discussed many universal constructions. We now specialize them to rank one, see Tables 13.5 and 13.6. The spaces $\mathrm{V}[S]_\sigma$ and $\mathrm{V}[S]^\sigma$ are coinvariants and invariants, respectively, of $\mathrm{V}[S]$ for the group $\{\mathrm{id}, \sigma\}$.

TABLE 13.5. Free objects in rank one.

Coxeter q-bialgebra	product μ	coproduct Δ
$\overline{\mathcal{T}}_q(\mathrm{V}) = (\mathrm{V}[\emptyset] \oplus \mathrm{V}[S], \mathrm{V}[S], \sigma)$	$\mu(y) = (0, y)$	$\Delta(x, y) = y + q\,\sigma(y)$
$\overline{\mathcal{T}}_q(\mathrm{C}) = (\mathrm{C}[\emptyset] \oplus \mathrm{C}[S], \mathrm{C}[S], \sigma)$	$\mu(y) = (0, y)$	$\Delta(x, y) = \Delta(x) + y + q\,\sigma(y)$
$\overline{\mathcal{S}}(\mathrm{V}) = (\mathrm{V}[\emptyset] \oplus \mathrm{V}[S]_\sigma, \mathrm{V}[S], \sigma)$	$\mu(y) = (0, \bar{y})$	$\Delta(x, \bar{y}) = y + \sigma(y)$
$\overline{\mathcal{S}}(\mathrm{C}) = (\mathrm{C}[\emptyset] \oplus \mathrm{C}[S]_\sigma, \mathrm{C}[S], \sigma)$	$\mu(y) = (0, \bar{y})$	$\Delta(x, \bar{y}) = \Delta(x) + y + \sigma(y)$

For a Coxeter $\boldsymbol{\alpha}$-space V, the free Coxeter $\boldsymbol{\alpha}$-algebra $\overline{\mathcal{T}}(\mathrm{V})$ is

$$\overline{\mathcal{T}}(\mathrm{V})[S] = \mathrm{V}[S] \quad \text{and} \quad \overline{\mathcal{T}}(\mathrm{V})[\emptyset] = \mathrm{V}[\emptyset] \oplus \mathrm{V}[S],$$

with σ inherited from V and $\mu(y) = (0, y)$. It carries a Coxeter $\boldsymbol{\alpha}$-q-bialgebra structure which we denote $\overline{\mathcal{T}}_q(\mathrm{V})$, see the first entry in Table 13.5. More generally, for a Coxeter $\boldsymbol{\alpha}$-coalgebra (C, Δ), we have the Coxeter $\boldsymbol{\alpha}$-q-bialgebra $\overline{\mathcal{T}}_q(\mathrm{C})$. When $\Delta = 0$, the coproduct formula $\Delta(x) + y + q\,\sigma(y)$ simplifies to $y + q\,\sigma(y)$, and we recover $\overline{\mathcal{T}}_q(\mathrm{V})$. Now consider the case $q = 1$. The Coxeter bialgebras $\overline{\mathcal{S}}(\mathrm{V})$ and $\overline{\mathcal{S}}(\mathrm{C})$ are abelianizations of $\overline{\mathcal{T}}(\mathrm{V})$ and $\overline{\mathcal{T}}(\mathrm{C})$, respectively; in particular, they are commutative.

Exercise 13.78. Check directly the universal properties of $\overline{\mathcal{T}}(\mathrm{V})$ and $\overline{\mathcal{T}}_q(\mathrm{C})$ in Theorems 9.11 and 9.13 in rank one.

Exercise 13.79. Let $\mathrm{char}(\Bbbk) \neq 2$. Check: The identification $\mathbf{As} \circ \mathrm{V} \cong \overline{\mathcal{T}}(\mathrm{V})$ restricts to $\mathbf{Lie} \circ \mathrm{V} \cong \overline{\mathcal{P}\mathcal{T}}(\mathrm{V})$. Note: $\overline{\mathcal{P}\mathcal{T}}(\mathrm{V})[\emptyset]$ consists of (x, y) such that $y + \sigma(y) = 0$. This verifies Proposition 12.9 in rank one.

TABLE 13.6. Cofree objects in rank one.

Coxeter q-bialgebra	product μ	coproduct Δ
$\overline{\mathcal{T}_q}^\vee(\mathrm{V}) = (\mathrm{V}[\emptyset] \oplus \mathrm{V}[S], \mathrm{V}[S], \sigma)$	$\mu(y) = (0, y + q\,\sigma(y))$	$\Delta(x, y) = y$
$\overline{\mathcal{T}_q}^\vee(\mathrm{A}) = (\mathrm{A}[\emptyset] \oplus \mathrm{A}[S], \mathrm{A}[S], \sigma)$	$\mu(y) = (\mu(y), y + q\,\sigma(y))$	$\Delta(x, y) = y$
$\overline{\mathcal{S}}^\vee(\mathrm{V}) = (\mathrm{V}[\emptyset] \oplus \mathrm{V}[S]^\sigma, \mathrm{V}[S], \sigma)$	$\mu(y) = (0, y + \sigma(y))$	$\Delta(x, y) = y$
$\overline{\mathcal{S}}^\vee(\mathrm{A}) = (\mathrm{A}[\emptyset] \oplus \mathrm{A}[S]^\sigma, \mathrm{A}[S], \sigma)$	$\mu(y) = (\mu(y), y + \sigma(y))$	$\Delta(x, y) = y$

Dually, for a Coxeter $\boldsymbol{\alpha}$-space V, the cofree Coxeter $\boldsymbol{\alpha}$-coalgebra $\overline{\mathcal{T}}^\vee(\mathrm{V})$ is

$$\overline{\mathcal{T}}^\vee(\mathrm{V})[S] = \mathrm{V}[S] \quad \text{and} \quad \overline{\mathcal{T}}^\vee(\mathrm{V})[\emptyset] = \mathrm{V}[\emptyset] \oplus \mathrm{V}[S],$$

with σ inherited from V and $\Delta(x, y) = y$. It carries a Coxeter $\boldsymbol{\alpha}$-q-bialgebra structure which we denote $\overline{\mathcal{T}_q}^\vee(\mathrm{V})$, see the first entry in Table 13.6. More generally, we have $\overline{\mathcal{T}_q}^\vee(\mathrm{A})$, and finally, $\overline{\mathcal{S}}^\vee(\mathrm{V})$ and $\overline{\mathcal{S}}^\vee(\mathrm{A})$ as coabelianizations of $\overline{\mathcal{T}}^\vee(\mathrm{V})$ and $\overline{\mathcal{T}}^\vee(\mathrm{A})$, respectively.

We briefly mention the signed case. For a Coxeter $\boldsymbol{\alpha}$-space V, the free signed commutative Coxeter $\boldsymbol{\alpha}$-algebra $\overline{\mathcal{E}}(\mathrm{V})$ is

$$\overline{\mathcal{E}}(\mathrm{V})[S] = \mathrm{V}[S] \quad \text{and} \quad \overline{\mathcal{E}}(\mathrm{V})[\emptyset] = \mathrm{V}[\emptyset] \oplus \mathrm{V}[S]_{-\sigma},$$

while the cofree signed cocommutative Coxeter $\boldsymbol{\alpha}$-coalgebra $\overline{\mathcal{E}}^\vee(\mathrm{V})$ is

$$\overline{\mathcal{E}}^\vee(\mathrm{V})[S] = \mathrm{V}[S] \quad \text{and} \quad \overline{\mathcal{E}}^\vee(\mathrm{V})[\emptyset] = \mathrm{V}[\emptyset] \oplus \mathrm{V}[S]^{-\sigma}.$$

Exercise 13.80. This is specific to rank one. Show:

- A Coxeter $\boldsymbol{\alpha}$-algebra (A, μ) is free iff μ is injective.
- A Coxeter $\boldsymbol{\alpha}$-coalgebra (C, Δ) is cofree iff Δ is surjective.

13.8.10. Signature functor.

We specialize the u-signature functor in Section 8.10 to rank one. For a Coxeter $\boldsymbol{\alpha}$-space $\mathrm{V} = (\mathrm{V}[\emptyset], \mathrm{V}[S], \sigma)$, the Coxeter $\boldsymbol{\alpha}$-space V_- is

$$\mathrm{V}_- = (\mathrm{V}[\emptyset], \mathrm{V}[S], -\sigma).$$

One may readily check that

$$\overline{\mathcal{T}}(\mathrm{V}_-)_- = \overline{\mathcal{T}}(\mathrm{V}), \qquad \overline{\mathcal{S}}(\mathrm{V}_-)_- = \overline{\mathcal{E}}(\mathrm{V}), \qquad \overline{\mathcal{E}}(\mathrm{V}_-)_- = \overline{\mathcal{S}}(\mathrm{V})$$

as Coxeter $\boldsymbol{\alpha}$-spaces. This is a special case of Exercise 8.124.

If A is a Coxeter $\boldsymbol{\alpha}$-algebra, then so is A_- (with the same μ). Further, if A is commutative, then A_- is signed commutative, and vice versa. Dual remarks apply to a Coxeter $\boldsymbol{\alpha}$-coalgebra. Combining the two situations, if H is a Coxeter $\boldsymbol{\alpha}$-q-bialgebra, then H_- is a Coxeter $\boldsymbol{\alpha}$-$(-q)$-bialgebra. Similarly, if G is a Coxeter $\boldsymbol{\alpha}$-Lie algebra, then G_- is a signed Coxeter $\boldsymbol{\alpha}$-Lie algebra (with the same ν). Dual remark applies to a Coxeter $\boldsymbol{\alpha}$-Lie coalgebra.

13.8.11. Internal hom for Coxeter coalgebras. We specialize the constructions from Section 9.6 to rank one. For Coxeter $\boldsymbol{\alpha}$-coalgebras C and D, the internal hom $\mathcal{C}(\mathrm{C},\mathrm{D})$ is as follows. An element of $\mathcal{C}(\mathrm{C},\mathrm{D})[S]$ is a linear map $f_{S/S} : \mathrm{C}[S] \to \mathrm{D}[S]$. An element of $\mathcal{C}(\mathrm{C},\mathrm{D})[\emptyset]$ is a pair of linear maps $f_{S/\emptyset} : \mathrm{C}[S] \to \mathrm{D}[S]$ and $f_{\emptyset/\emptyset} : \mathrm{C}[\emptyset] \to \mathrm{D}[\emptyset]$ satisfying $f_{S/\emptyset}\Delta = \Delta f_{\emptyset/\emptyset}$. The involution σ is

$$\sigma : \mathcal{C}(\mathrm{C},\mathrm{D})[S] \to \mathcal{C}(\mathrm{C},\mathrm{D})[S], \qquad f_{S/S} \mapsto \sigma f_{S/S}\, \sigma,$$

with the two σs around $f_{S/S}$ being the involutions of $\mathrm{C}[S]$ and $\mathrm{D}[S]$. The coproduct Δ of $\mathcal{C}(\mathrm{C},\mathrm{D})$ is

$$\Delta : \mathcal{C}(\mathrm{C},\mathrm{D})[\emptyset] \to \mathcal{C}(\mathrm{C},\mathrm{D})[S], \qquad (f_{\emptyset/\emptyset}, f_{S/\emptyset}) \mapsto f_{S/\emptyset}.$$

Now let C be a cocommutative Coxeter $\boldsymbol{\alpha}$-coalgebra and K a Coxeter $\boldsymbol{\alpha}$-q-bialgebra. We work out some details of the Coxeter $\boldsymbol{\alpha}$-q-bialgebra $\mathcal{C}(\mathrm{C},\mathrm{K})_q$ of Exercise 9.49 in rank one. The product $\mu : \mathcal{C}(\mathrm{C},\mathrm{K})_q[S] \to \mathcal{C}(\mathrm{C},\mathrm{K})_q[\emptyset]$ is $f \mapsto \tilde{f}$, with $\tilde{f}_{\emptyset/\emptyset} = \mu f_{S/S}\Delta$ and $\tilde{f}_{S/\emptyset} = f_{S/S} + q\,\sigma f_{S/S}\,\sigma$. We first check that $\tilde{f}_{S/\emptyset}\Delta = \Delta\tilde{f}_{\emptyset/\emptyset}$. This is done below.

$$\begin{aligned}\Delta\tilde{f}_{\emptyset/\emptyset} = \Delta\mu f_{S/S}\Delta = (\mathrm{id} + q\,\sigma) f_{S/S}\Delta &= f_{S/S}\Delta + q\,\sigma f_{S/S}\,\sigma\Delta \\ &= (f_{S/S} + q\,\sigma f_{S/S}\,\sigma)\Delta = \tilde{f}_{S/\emptyset}\Delta.\end{aligned}$$

Note how the q-bialgebra axiom of K and cocommutativity of C got used. Finally, the q-bialgebra axiom of $\mathcal{C}(\mathrm{C},\mathrm{K})_q$ is verified below.

$$\Delta\mu(f_{S/S}) = \tilde{f}_{S/\emptyset} = f_{S/S} + q\,\sigma f_{S/S}\,\sigma = (\mathrm{id} + q\,\sigma)(f_{S/S}).$$

13.8.12. Universal measuring Coxeter coalgebra. We specialize the constructions from Section 9.7 to rank one. For Coxeter $\boldsymbol{\alpha}$-algebras A and B, the universal measuring Coxeter $\boldsymbol{\alpha}$-coalgebra $\overline{\mathcal{C}}(\mathrm{A},\mathrm{B})$ is as follows. An element of $\overline{\mathcal{C}}(\mathrm{A},\mathrm{B})[S]$ is a linear map $f_{S/S} : \mathrm{A}[S] \to \mathrm{B}[S]$. An element of $\overline{\mathcal{C}}(\mathrm{A},\mathrm{B})[\emptyset]$ is a pair of linear maps $f_{S/\emptyset} : \mathrm{A}[S] \to \mathrm{B}[S]$ and $f_{\emptyset/\emptyset} : \mathrm{A}[\emptyset] \to \mathrm{B}[\emptyset]$ satisfying $f_{\emptyset/\emptyset}\mu = \mu f_{S/\emptyset}$. The involution σ and coproduct Δ are defined similar to $\mathcal{C}(\mathrm{C},\mathrm{D})$ above.

Similarly, for H a Coxeter $\boldsymbol{\alpha}$-q-bialgebra and A a commutative Coxeter $\boldsymbol{\alpha}$-algebra, we have the Coxeter $\boldsymbol{\alpha}$-q-bialgebra $\overline{\mathcal{C}}(\mathrm{H},\mathrm{A})_q$ of Exercise 9.53 in rank one whose product μ is defined similar to $\mathcal{C}(\mathrm{C},\mathrm{K})_q$ above.

13.8.13. Exp-log correspondence. We specialize the discussion in Sections 9.8 and 9.10 to rank one.

(Co)derivations and Coxeter (co)algebra morphisms. For C a Coxeter $\boldsymbol{\alpha}$-coalgebra and A a Coxeter $\boldsymbol{\alpha}$-algebra, an element s of the invariant lune-incidence algebra acts on a map $f = (f_\emptyset, f_S)$ from C to A by

$$s \circ f := (s(\emptyset,\emptyset) f_\emptyset + s(\emptyset, S)\, \mu f_S \Delta, s(S,S) f_S).$$

This specializes (9.87). Thus, for $\mathrm{char}(\Bbbk) \neq 2$, specializing (9.90a) and (9.90b) and using the rank-one data in Example 1.86,

$$\mathbf{exp}(f) = (f_\emptyset + \tfrac{1}{2}\mu f_S\Delta, f_S) \quad \text{and} \quad \mathbf{log}(f) = (f_\emptyset - \tfrac{1}{2}\mu f_S\Delta, f_S).$$

Let us now directly verify the first statement in Theorem 9.56 in rank one; it sets up a bijection between coderivations and Coxeter coalgebra morphisms. For $f = (f_\emptyset, f_S)$ from C to K such that $\Delta f_\emptyset = 0$,

$$\Delta\, \mathbf{exp}(f)_\emptyset = \Delta(f_\emptyset + \tfrac{1}{2}\mu f_S \Delta) = \tfrac{1}{2}\Delta\mu f_S\Delta = \tfrac{1}{2}(\mathrm{id} + \sigma) f_S \Delta$$
$$= \tfrac{1}{2} f_S(\mathrm{id} + \sigma)\Delta = f_S\Delta = \mathbf{exp}(f)_S\Delta,$$

as required. Similarly, for $g = (g_\emptyset, g_S)$ from C to K such that $\Delta g_\emptyset = g_S\Delta$,

$$\Delta\, \mathbf{log}(g)_\emptyset = \Delta(g_\emptyset - \tfrac{1}{2}\mu g_S\Delta) = \Delta g_\emptyset - \tfrac{1}{2}\Delta\mu g_S\Delta = \Delta g_\emptyset - \tfrac{1}{2}(\mathrm{id} + \sigma) g_S\Delta$$
$$= \Delta g_\emptyset - \tfrac{1}{2} g_S(\mathrm{id} + \sigma)\Delta = \Delta g_\emptyset - g_S\Delta = 0,$$

as required.

q-analogue. For C a Coxeter $\boldsymbol{\alpha}$-coalgebra and A a Coxeter $\boldsymbol{\alpha}$-algebra, an element s of the invariant bilune-incidence algebra acts on a map $f = (f_\emptyset, f_S)$ from C to A by

$$s \circ f := (s(\emptyset, \emptyset, \mathrm{id}, \emptyset) f_\emptyset + s(\emptyset, S, \mathrm{id}, S)\, \mu f_S\Delta + s(\emptyset, S, \sigma, S)\, \mu\sigma f_S\Delta,$$
$$s(S, S, \mathrm{id}, S) f_S).$$

This specializes (9.98). Thus, for $q \neq \pm 1$, for $f : \mathrm{C} \to \mathrm{A}$, specializing (9.99a) and (9.99b), and using Examples 1.91 and 1.98,

$$\mathbf{exp}_q(f) = (f_\emptyset + \tfrac{1}{1-q^2}\, \mu f_S\Delta - \tfrac{q}{1-q^2}\, \mu\sigma f_S\Delta, f_S),$$
$$\mathbf{log}_q(f) = (f_\emptyset - \tfrac{1}{1-q^2}\, \mu f_S\Delta + \tfrac{q}{1-q^2}\, \mu\sigma f_S\Delta, f_S).$$

If either C is cocommutative or A is commutative, then

$$\mathbf{exp}_q(f) = (f_\emptyset + \tfrac{1}{1+q}\, \mu f_S\Delta, f_S) \quad \text{and} \quad \mathbf{log}_q(f) = (f_\emptyset - \tfrac{1}{1+q}\, \mu f_S\Delta, f_S).$$

Exercise 13.81. Verify Theorem 9.61 in rank one using the above formulas.

Primitive and group-like series. For V a Coxeter $\boldsymbol{\alpha}$-space, a series v of V is a pair $(v_\emptyset, v_S)$ with $v_\emptyset \in \mathrm{V}[\emptyset]$ and $v_S \in \mathrm{V}[S]$ such that $\sigma(v_S) = v_S$. For C a Coxeter $\boldsymbol{\alpha}$-coalgebra, we say a series v of C is primitive if $\Delta(v_\emptyset) = 0$ and group-like if $\Delta(v_\emptyset) = v_S$. For A a Coxeter $\boldsymbol{\alpha}$-algebra, an element s of the invariant lune-incidence algebra acts on a series $v = (v_\emptyset, v_S)$ of A by

$$s \circ v := (s(\emptyset, \emptyset) v_\emptyset + s(\emptyset, S)\, \mu(v_S), s(S, S) v_S).$$

This specializes (9.122). Thus, for $\mathrm{char}(\Bbbk) \neq 2$, specializing (9.124a) and (9.124b),

$$\mathbf{exp}(v) = (v_\emptyset + \tfrac{1}{2}\mu(v_S), v_S) \quad \text{and} \quad \mathbf{log}(v) = (v_\emptyset - \tfrac{1}{2}\mu(v_S), v_S).$$

One can now directly verify Theorem 9.76 in rank one; it sets up a bijection between primitive and group-like series.

Similarly, for $q \neq \pm 1$, specializing (9.127a) and (9.127b),

$$\mathbf{exp}_q(v) = (v_\emptyset + \tfrac{1}{1+q}\, \mu(v_S), v_S) \quad \text{and} \quad \mathbf{log}_q(v) = (v_\emptyset - \tfrac{1}{1+q}\, \mu(v_S), v_S).$$

One can now directly verify Theorem 9.80 in rank one.

Linear and exponential series. For V a Coxeter $\boldsymbol{\alpha}$-space, we say a series x of V is linear if $x_\emptyset = 0$. For A a Coxeter $\boldsymbol{\alpha}$-algebra, we say a series e of A is exponential if $\mu(e_S) = 2\, e_\emptyset$. Let $\mathrm{char}(\Bbbk) \neq 2$. Observe: A series e is exponential iff $\mathbf{log}(e)$ is linear. This verifies Lemma 9.85 in rank one. Similarly, observe: For a Coxeter $\boldsymbol{\alpha}$-bialgebra H, any exponential series of H is group-like. This verifies Lemma 9.89 in rank one.

More generally, for a scalar q, we say a series e of a Coxeter $\boldsymbol{\alpha}$-algebra A is q-exponential if $\mu(e_S) = (1+q)\, e_\emptyset$. For $q \neq \pm 1$, we have q-analogues of the preceding results.

13.8.14. q-norm map. We specialize the discussion in Section 9.11 to rank one. Let q be a scalar. For V a Coxeter $\boldsymbol{\alpha}$-space, we have the q-norm map $(\bar{\kappa}_q)_{\mathrm{V}} : \overline{\mathcal{T}}_q(\mathrm{V}) \to \overline{\mathcal{T}}_q^{\vee}(\mathrm{V})$ given by

$$((x,y),z) \mapsto ((x, y + q\,\sigma(y)), z).$$

It is a morphism of Coxeter $\boldsymbol{\alpha}$-q-bialgebras. This specializes (9.140). By Lemma 13.77, item (1), the above map is invertible for $q \neq \pm 1$. This verifies Lemma 9.101 in rank one.

More generally, for C a Coxeter $\boldsymbol{\alpha}$-coalgebra, A a Coxeter $\boldsymbol{\alpha}$-algebra, and $f : \mathrm{C} \to \mathrm{A}$ a map of Coxeter $\boldsymbol{\alpha}$-spaces, we have the morphism of Coxeter $\boldsymbol{\alpha}$-q-bialgebras $\widehat{f} : \overline{\mathcal{T}}_q(\mathrm{C}) \to \overline{\mathcal{T}}_q^{\vee}(\mathrm{A})$ given by

$$((x,y),z) \mapsto ((f_\emptyset(x) + \mu f_S(y), f_S\Delta(x) + f_S(y) + q\,\sigma f_S(y)), f_S(z)).$$

This specializes (9.144). By Theorem 9.110, $\widehat{f}$ is an isomorphism iff $\mathbf{log}_q(f)$ is an isomorphism. This can be checked directly by manipulating the above formula for $\widehat{f}$. A conceptual proof is given later in Exercise 13.83.

13.8.15. Universal enveloping Coxeter algebra. We specialize the construction in Section 12.5 to rank one. Let (G, ν) be a Coxeter $\boldsymbol{\alpha}$-Lie algebra. The Lie bracket $\nu : \mathrm{G}[S] \to \mathrm{G}[\emptyset]$ is a linear map which satisfies $\nu\sigma + \nu = 0$. The universal enveloping Coxeter $\boldsymbol{\alpha}$-algebra $\overline{\mathcal{U}}(\mathrm{G})$ has components

$$\overline{\mathcal{U}}(\mathrm{G})[S] = \mathrm{G}[S], \qquad \overline{\mathcal{U}}(\mathrm{G})[\emptyset] = \big(\mathrm{G}[\emptyset] \oplus \mathrm{G}[S]\big)_{\mathrm{quo}},$$

where $\nu(y) \equiv y - \sigma(y)$ for $y \in \mathrm{G}[S]$. The product and coproduct are

$$\begin{aligned} \mu : \overline{\mathcal{U}}(\mathrm{G})[S] &\to \overline{\mathcal{U}}(\mathrm{G})[\emptyset] & \qquad \Delta : \overline{\mathcal{U}}(\mathrm{G})[\emptyset] &\to \overline{\mathcal{U}}(\mathrm{G})[S] \\ y &\mapsto \overline{(0,y)} & \overline{(x,y)} &\mapsto y + \sigma(y), \end{aligned}$$

with bar indicating the image in the quotient. This turns $\overline{\mathcal{U}}(\mathrm{G})$ into a cocommutative Coxeter $\boldsymbol{\alpha}$-bialgebra. It is a quotient of $\overline{\mathcal{T}}(\mathrm{G})$ (first entry in Table 13.5 for $q = 1$).

Dually, let (K, θ) be a Coxeter $\boldsymbol{\alpha}$-Lie coalgebra. The universal coenveloping Coxeter $\boldsymbol{\alpha}$-coalgebra $\overline{\mathcal{U}}^{\vee}(\mathrm{K})$ has components

$$\overline{\mathcal{U}}^{\vee}(\mathrm{K})[S] = \mathrm{K}[S], \qquad \overline{\mathcal{U}}^{\vee}(\mathrm{K})[\emptyset] = \big(\mathrm{K}[\emptyset] \oplus \mathrm{K}[S]\big)^{\mathrm{sub}}$$

consisting of (x, y) such that $\theta(x) = y - \sigma(y)$. The product and coproduct are

$$\mu : \overline{\mathcal{U}}^\vee(\mathrm{K})[S] \to \overline{\mathcal{U}}^\vee(\mathrm{K})[\emptyset] \qquad \Delta : \overline{\mathcal{U}}^\vee(\mathrm{K})[\emptyset] \to \overline{\mathcal{U}}^\vee(\mathrm{K})[S]$$
$$y \mapsto (0, y + \sigma(y)) \qquad (x, y) \mapsto y.$$

This is a commutative Coxeter $\boldsymbol{\alpha}$-subbialgebra of $\overline{\mathcal{T}}^\vee(\mathrm{K})$ (first entry in Table 13.6 for $q = 1$).

Exercise 13.82. Let $\mathrm{char}(\Bbbk) \neq 2$. Check directly: The map $i : \mathrm{G} \to \overline{\mathcal{U}}(\mathrm{G})$ is injective and $p : \overline{\mathcal{U}}^\vee(\mathrm{K}) \to \mathrm{K}$ is surjective. This proves Corollaries 13.54 and 13.60 in rank one. Injectivity or surjectivity may fail without the characteristic assumption.

13.8.16. Structure theorems. We now look at structure theorems for Coxeter bialgebras in rank one.

Borel–Hopf. Let us verify Theorems 13.1 and 13.8 in rank one. Let $\mathrm{char}(\Bbbk) \neq 2$. Let H be a cocommutative Coxeter $\boldsymbol{\alpha}$-bialgebra. Consider the maps (13.1) and (13.2). On the S-component, they are both identity. On the $\emptyset$-component, they are

$$\overline{\mathcal{P}}(\mathrm{H})[\emptyset] \oplus \mathrm{H}[S]^\sigma \to \mathrm{H}[\emptyset], \qquad (x, y) \mapsto x + \tfrac{1}{2}\mu(y),$$
$$\mathrm{H}[\emptyset] \to \overline{\mathcal{P}}(\mathrm{H})[\emptyset] \oplus \mathrm{H}[S]^\sigma, \qquad x \mapsto (x - \tfrac{1}{2}\mu\Delta(x), \Delta(x)).$$

One can directly check that these are inverses of each other. Also, the second map is well-defined, that is, the first coordinate lies in the primitive part and the second in the invariant part.

Dually, let H be a commutative Coxeter $\boldsymbol{\alpha}$-bialgebra. Consider the maps (13.8) and (13.9). On the S-component, they are both identity. On the $\emptyset$-component, they are

$$\mathrm{H}[\emptyset] \to \overline{\mathcal{Q}}(\mathrm{H})[\emptyset] \oplus \mathrm{H}[S]_\sigma, \qquad x \mapsto (\bar{x}, \tfrac{1}{2}\overline{\Delta(x)}),$$
$$\overline{\mathcal{Q}}(\mathrm{H})[\emptyset] \oplus \mathrm{H}[S]_\sigma \to \mathrm{H}[\emptyset], \qquad (\bar{x}, \bar{y}) \mapsto x - \tfrac{1}{2}\mu\Delta(x) + \mu(y).$$

Again, one can directly check that these are inverses of each other, and that the second map is well-defined.

Leray–Samelson. Let us verify Theorem 13.14 in rank one. Let $\mathrm{char}(\Bbbk) \neq 2$. Let V be a Coxeter $\boldsymbol{\alpha}$-space. Recall from Table 13.5 the bicommutative Coxeter $\boldsymbol{\alpha}$-bialgebra $\overline{\mathcal{S}}(\mathrm{V})$. Its components are

$$\overline{\mathcal{S}}(\mathrm{V})[S] = \mathrm{V}[S] \quad \text{and} \quad \overline{\mathcal{S}}(\mathrm{V})[\emptyset] = \mathrm{V}[\emptyset] \oplus \mathrm{V}[S]_\sigma,$$

with product $\mu(y) = (0, \bar{y})$ and coproduct $\Delta(x, \bar{y}) = y + \sigma(y)$. The statement $\overline{\mathcal{P}}(\overline{\mathcal{S}}(\mathrm{V})) = \mathrm{V}$ is equivalent to saying that for $y \in \mathrm{V}[S]$,

$$y + \sigma(y) = 0 \iff y = z - \sigma(z) \text{ for some } z \in \mathrm{V}[S],$$

that is, $\ker(\mathrm{id} + \sigma) = \mathrm{im}(\mathrm{id} - \sigma)$. This follows from item (1) of Lemma 13.76.

Now let H be a bicommutative Coxeter $\boldsymbol{\alpha}$-bialgebra. To see that $\overline{\mathcal{S}}(\overline{\mathcal{P}}(\mathrm{H}))$ is isomorphic to H, we look at the $\emptyset$-component of the map (13.12), namely,

$$\overline{\mathcal{P}}(\mathrm{H})[\emptyset] \oplus \mathrm{H}[S]_\sigma \to \mathrm{H}[\emptyset], \qquad (x, \bar{y}) \mapsto x + \mu(y).$$

Now we use Lemma 13.76, item (2), along with the observation that $\mu : \mathrm{H}[S]_\sigma \to \mathrm{im}(\mu)$ is an isomorphism, with inverse $y \mapsto \frac{1}{2}\overline{\Delta(y)}$. This is consistent with the expression for the map (13.13).

Similarly, one can verify the dual Theorem 13.17 in rank one.

Loday–Ronco and its q-generalization. Let us verify Theorem 13.25 in rank one. Let $q \neq \pm 1$. Let V be a Coxeter $\boldsymbol{\alpha}$-space. Recall from Table 13.5 the Coxeter $\boldsymbol{\alpha}$-q-bialgebra $\overline{\mathcal{T}_q}(\mathrm{V})$. Its components are

$$\overline{\mathcal{T}_q}(\mathrm{V})[S] = \mathrm{V}[S] \quad \text{and} \quad \overline{\mathcal{T}_q}(\mathrm{V})[\emptyset] = \mathrm{V}[\emptyset] \oplus \mathrm{V}[S],$$

with product $\mu(y) = (0, y)$ and coproduct $\Delta(x, y) = y + q\,\sigma(y)$. The statement $\overline{\mathcal{P}}(\overline{\mathcal{T}_q}(\mathrm{V})) = \mathrm{V}$ is equivalent to saying that for $y \in \mathrm{V}[S]$,

$$y + q\,\sigma(y) = 0 \iff y = 0.$$

This follows from item (1) of Lemma 13.77.

Now let H be a Coxeter $\boldsymbol{\alpha}$-q-bialgebra. The statement that $\overline{\mathcal{T}_q}(\overline{\mathcal{P}}(\mathrm{H}))$ is isomorphic to H can be deduced from item (2) of Lemma 13.77. This is because, on the $\emptyset$-component, the map (13.19) is

$$\overline{\mathcal{P}}(\mathrm{H})[\emptyset] \oplus \mathrm{H}[S] \to \mathrm{H}[\emptyset], \qquad (x, y) \mapsto x + \mu(y).$$

Similarly, one can verify the dual Theorem 13.30 in rank one.

Setting $q = 0$, we recover Loday–Ronco Theorems 13.35 and 13.37.

Hoffman–Newman–Radford. Let $\mathrm{char}(\Bbbk) \neq 2$. For C a cocommutative Coxeter $\boldsymbol{\alpha}$-coalgebra, let us specialize the HNR isomorphisms (13.30) and (13.31) to rank one. On the S-component, the maps are identities, while on the $\emptyset$-component, they are

$$\begin{array}{ll} \overline{\mathcal{T}}(\mathrm{C})[\emptyset] \to \overline{\mathcal{T}}(\mathrm{C}_t)[\emptyset] & \overline{\mathcal{T}}(\mathrm{C}_t)[\emptyset] \to \overline{\mathcal{T}}(\mathrm{C})[\emptyset] \\ \quad (x, y) \mapsto (x, \tfrac{1}{2}\Delta(x) + y) & \quad (x, y) \mapsto (x, -\tfrac{1}{2}\Delta(x) + y). \end{array}$$

One may now directly verify Theorem 13.39 in rank one using formulas in Table 13.5.

Let $q \neq \pm 1$. For C a Coxeter $\boldsymbol{\alpha}$-coalgebra, let us specialize the HNR isomorphisms (13.36) and (13.37) to rank one. On the S-component, the maps are identities, while on the $\emptyset$-component, they are

$$\overline{\mathcal{T}_q}(\mathrm{C})[\emptyset] \to \overline{\mathcal{T}_q}(\mathrm{C}_t)[\emptyset], \quad (x, y) \mapsto (x, \tfrac{1}{1-q^2}\Delta(x) - \tfrac{q}{1-q^2}\sigma\Delta(x) + y),$$
$$\overline{\mathcal{T}_q}(\mathrm{C}_t)[\emptyset] \to \overline{\mathcal{T}_q}(\mathrm{C})[\emptyset], \quad (x, y) \mapsto (x, -\tfrac{1}{1-q^2}\Delta(x) + \tfrac{q}{1-q^2}\sigma\Delta(x) + y).$$

When C is cocommutative, they specialize to

$$\begin{array}{ll} \overline{\mathcal{T}_q}(\mathrm{C})[\emptyset] \to \overline{\mathcal{T}_q}(\mathrm{C}_t)[\emptyset] & \overline{\mathcal{T}_q}(\mathrm{C}_t)[\emptyset] \to \overline{\mathcal{T}_q}(\mathrm{C})[\emptyset] \\ \quad (x, y) \mapsto (x, \tfrac{1}{1+q}\Delta(x) + y) & \quad (x, y) \mapsto (x, -\tfrac{1}{1+q}\Delta(x) + y). \end{array}$$

One may now directly verify Theorem 13.44 in rank one using formulas in Table 13.5.

Similarly, one may verify the dual Theorems 13.40 and 13.46 in rank one by specializing (13.32), (13.33), (13.38), (13.39).

Exercise 13.83. Let f and $\widehat{f}$ be as in Section 13.8.14. Check: The composite map

$$\overline{\mathcal{T}}_q(\mathrm{C}_t) \xrightarrow{\cong} \overline{\mathcal{T}}_q(\mathrm{C}) \xrightarrow{\widehat{f}} \overline{\mathcal{T}}_q^{\vee}(\mathrm{A}) \xrightarrow{\cong} \overline{\mathcal{T}}_q^{\vee}(\mathrm{A}_t)$$

is

$$((x,y),z) \mapsto ((f_\emptyset(x) - \tfrac{1}{1-q^2}\mu f_S\Delta(x) + \tfrac{q}{1-q^2}\mu\sigma f_S\Delta(x), f_S(y) + q\,\sigma f_S(y)), f_S(z)).$$

This verifies Theorem 13.48, and hence also Theorem 9.110 in rank one.

Solomon operator and PBW. Let $\mathrm{char}(\Bbbk) \neq 2$. Let G be a Coxeter $\boldsymbol{\alpha}$-Lie algebra. The Solomon operator $\mathrm{Sol} : \overline{\mathcal{T}}(\mathrm{G}) \to \overline{\mathcal{T}}(\mathrm{G})$ is as follows. On the S-component, it is identity. On the $\emptyset$-component, it is

$$\mathrm{Sol}_\emptyset : \overline{\mathcal{T}}(\mathrm{G})[\emptyset] \to \overline{\mathcal{T}}(\mathrm{G})[\emptyset], \quad (x,y) \mapsto (x + \tfrac{1}{2}\nu(y), \tfrac{1}{2}(y + \sigma(y))).$$

This specializes (13.49). One may directly check that it has image $\overline{\mathcal{S}}(\mathrm{G})[\emptyset]$ and coimage $\overline{\mathcal{U}}(\mathrm{G})[\emptyset]$. Here $\overline{\mathcal{S}}(\mathrm{G})[\emptyset]$ is viewed as a subspace of $\overline{\mathcal{T}}(\mathrm{G})[\emptyset]$ via the section

$$(x, \bar{y}) \mapsto (x, \tfrac{1}{2}(y + \sigma(y))).$$

This verifies Theorem 13.55, and hence also PBW Theorem 13.53 in rank one.

CMM. Let $\mathrm{char}(\Bbbk) \neq 2$. Let us verify Theorem 13.65 in rank one. The isomorphism between G and $\overline{\mathcal{P}}\overline{\mathcal{U}}(\mathrm{G})$ says that the map

$$\mathrm{G}[\emptyset] \to \overline{\mathcal{U}}(\mathrm{G})[\emptyset], \qquad x \mapsto \overline{(x,0)}$$

is injective, and its image is precisely $\overline{\mathcal{P}}\overline{\mathcal{U}}(\mathrm{G})[\emptyset]$. For the first part, suppose $\overline{(x,0)} = 0$, that is, $(x,0) = (-\nu(y), y - \sigma(y))$ for some y. Then $y = \sigma(y)$. Hence, $0 = \nu\sigma(y) + \nu(y) = 2\nu(y)$ from which we conclude $x = 0$. For the second part, note that a primitive element has the form $\overline{(x,y)}$ with $y + \sigma(y) = 0$. In this case, $\overline{(0,2y)} = \overline{(0, y - \sigma(y))} = \overline{(\nu(y), 0)}$. Therefore, $\overline{(x,y)} = \overline{(x + \frac{1}{2}\nu(y), 0)}$.

Now let H be a cocommutative Coxeter $\boldsymbol{\alpha}$-bialgebra. Using the commutator bracket on $\overline{\mathcal{P}}(\mathrm{H})$,

$$\overline{\mathcal{U}}\overline{\mathcal{P}}(\mathrm{H})[\emptyset] = \big(\overline{\mathcal{P}}(\mathrm{H})[\emptyset] \oplus \mathrm{H}[S]\big)/(-\mu(y) + \mu\sigma(y), y - \sigma(y)) \;\text{ for }\; y \in \mathrm{H}[S].$$

Let us check that $\overline{\mathcal{U}}\overline{\mathcal{P}}(\mathrm{H})$ is isomorphic to H. On the $\emptyset$-component, the map is

$$\overline{(x,y)} \mapsto x + \mu(y) \;\text{ for }\; x \in \overline{\mathcal{P}}(\mathrm{H})[\emptyset] \text{ and } y \in \mathrm{H}[S].$$

For injectivity, suppose $x + \mu(y) = 0$. Then $0 = \Delta\mu(y) = (\mathrm{id} + \sigma)(y)$ and hence $(\mathrm{id} - \sigma)(y) = 2y$. Thus, $(-\mu(y), y)$ is a relation and $\overline{(x,y)} = 0$. For surjectivity, note that $z \in \mathrm{H}[\emptyset]$ can be expressed as

$$z = x + \mu(y) \;\text{ for }\; y = \tfrac{1}{2}\Delta(z) \;\text{ and }\; x = z - \tfrac{1}{2}\mu\Delta(z).$$

This assertion is contained in item (2) of Lemma 13.76.

Exercise 13.84. Specialize the discussion in Sections 8.14 and 12.6 to the braid arrangement on 2 letters. The latter is a rank-one arrangement. Check: The description of a Coxeter $\boldsymbol{\beta}^2$-space, of a Coxeter $\boldsymbol{\beta}^2$-algebra, and so on given there coincide with those given in this section.

13.9. Notes

The results of this chapter generalize classical theorems from Hopf theory (in the category of connected graded vector spaces). One recovers these classical theorems by specializing to the braid arrangements (Sections 8.14 and 12.6); we will give more details in a separate work. Classically, we assume that the field characteristic is zero; this is the analogue of our assumption $\mathrm{char}(\Bbbk) \nmid |W|$.

Some important sources in the classical literature are mentioned below. More detailed references can be found in [**11**, Notes to Chapters 13, 14, 17].

13.9.1. Borel–Hopf. (Section 13.1.) Theorem 13.8 generalizes the classical result that a connected graded commutative Hopf algebra is free on its indecomposable part. Dually, Theorem 13.1 generalizes the result that a connected graded cocommutative Hopf algebra is cofree on its primitive part. These results have signed analogues which are generalized by Theorems 13.13 and 13.12.

The first result originated in work of Hopf [**190**], [**191**, Satz I] in the form of an example. Hopf worked in the signed setting. Borel provided a purely algebraic formulation of Hopf's result [**54**, Theorem 6.1]. For the second result, early references are by Heyneman and Sweedler [**343**, Theorem 4 and second paragraph on page 516], [**345**, page 275], [**183**, Corollary 4.2.7]. It is developed further by Grünenfelder [**164**, Theorem on page 575]. In the context of formal groups, it is given by Dieudonné [**98**, Section II.2.4, Proposition 1 and Section II.2.5].

Exponential and logarithm. Both results are nicely treated by Nichols [**280**, Chapter III, Theorem 7], [**281**, Theorem 8 and remark on page 71]. He worked with the exponential and logarithmic power series, see (1.177). In this classical setting, the Borel–Hopf isomorphisms (13.1), (13.2), (13.8), (13.9) can be defined more generally using any zeta and Möbius series.

Characteristic operations. For the proof in Section 13.1.3 using invariant eulerian idempotents, in the classical setting, one can use the Krob–Leclerc–Thibon idempotents. The more standard choice is to use the Garsia–Reutenauer idempotents. An approach of this nature is given by Patras and Cartier [**286**, Lemma I,5,4 and Corollary I,5,5], [**288**, Proposition III,5], [**83**, proof of Theorem 3.8.1], [**84**, proof of Theorem 4.3.1] and Fresse [**134**, Theorem 7.2.16]. They work with coarser idempotents that arise from the diagonalization of the Adams elements. For additional remarks, see [**11**, page 573].

Freeness of shuffle algebras and QSym. Proposition 13.10 generalizes the classical result that the shuffle algebra is a free commutative algebra. This was analyzed by Radford [**300**, Theorem 3.1.1, item (e)]. An early treatment is given by Melançon and Reutenauer [**257**, Theorem on page 589], [**305**, Corollary 5.5 and Theorem 6.1], [**306**, Corollary 5]. For more references, see [**11**, page 676].

Proposition 13.11 generalizes the classical result that NSym is cofree cocommutative on its primitive part, and QSym is free commutative on its indecomposable part. For the latter, see for instance [**248**, Corollaire 4.19], [**189**, Section 6, Example 1].

The above results are consequences of the classical Borel–Hopf theorem for connected graded (co)commutative Hopf algebras.

13.9.2. Leray–Samelson. (Section 13.2.) Theorems 13.14 and 13.17 generalize the classical result that the category of connected graded bicommutative Hopf algebras is equivalent to the category of positive-graded vector spaces. Theorems 13.21 and 13.22 generalize its signed analogue.

This result originated in a conjecture of Hopf [**191**, Section 37] which was proved by Samelson [**318**, Chapter I, Section 3, Satz I and Satz I'] and later improved by Leray [**227**, Section 25, pages 132–135]. An algebraic treatment of this result was given by Milnor and Moore [**265**, Theorems 7.16 and 7.20 and Proposition 7.21], [**266**, Theorem 4.10 and Proposition 4.25], [**271**, Theorems 6 and 7 on pages 27 and 28], [**273**, Theorem 1].

13.9.3. Loday–Ronco. (Section 13.4.) Theorems 13.35 and 13.37 generalize the classical result that the category of connected graded 0-Hopf algebras is equivalent to the category of positive-graded vector spaces. The latter is due to Loday and Ronco [**233**, Theorem 2.6]. It is also formulated in [**7**, Theorem 2.13], [**230**, Theorem 4.2.1]. In this classical setting, the Loday–Ronco isomorphisms (13.26), (13.27), (13.28), (13.29) involve the 0-zeta and 0-Möbius series.

13.9.4. Rigidity of Coxeter q-bialgebras for q not a root of unity. (Section 13.3.) The q-generalization of Loday–Ronco in Theorems 13.25 and 13.30 looks new even from a classical perspective. Classically, the isomorphisms involve the two-sided q-zeta and q-Möbius series; we will elaborate on this separately.

13.9.5. Hoffman–Newman–Radford. (Section 13.5.) Theorem 13.40 generalizes the classical result that shuffle Hopf algebras and quasishuffle Hopf algebras (which are both commutative) are isomorphic. Dually, Theorem 13.39 generalizes the result that deshuffle Hopf algebras and dequasishuffle Hopf algebras (which are both cocommutative) are isomorphic.

The first result appeared in work of Newman and Radford [**279**, Theorem 1.12]. The dual result had appeared earlier (somewhat implicitly) in Newman's thesis [**277**, Theorem 2.18], [**278**, Theorem 2.16]. Both results were later discovered independently by Hoffman [**189**, Theorems 2.5, 3.3, 4.1]. For later references, see [**251**, Theorem 3], [**109**, Definition 3.14], [**127**, Lemma 5.2], [**70**, Theorem 1], [**26**, Proposition 5.1], [**267**, Proposition 2], [**72**, Theorem 2, item (2)]. Hoffman made explicit use of the exponential and logarithmic power series, see (1.177). More generally, one can use any zeta and Möbius series.

The special case in the first diagram in (13.44) generalizes a result about the Hopf algebra NSym which goes back to Ditters [**99**, Theorem 3.5.2, item (a)], [**101**, Theorem 3.3]. Recall from Table 10.10 that the Coxeter bialgebra of face-types $\overline{\Sigma}$ generalizes NSym. Ditters considered NSym in the context of formal groups. NSym as a noncommutative generalization of the Hopf algebra of symmetric functions was defined and studied later in [**149**].

Theorems 13.44 and 13.46 look new even in the classical setting. Classically, one uses the two-sided q-zeta and q-Möbius series; we will elaborate on this separately.

13.9.6. Poincaré–Birkhoff–Witt. (Section 13.6.) The Poincaré–Birkhoff–Witt theorem originated in work of Poincaré [**294**, page 1066, second-to-last paragraph], [**295**, page 232, last sentence of Section III]. It was rediscovered later independently by Birkhoff [**44**] and Witt [**361**]. The precise classical analogue of Theorem 13.53 was given later by Smoke [**328**, pages 11–12], [**329**, page 467] and Quillen [**298**, Appendix B, Theorem 2.3]. (The latter worked in the differential graded setting.)

In the classical setting, the map (13.46) and PBW isomorphism in (13.47) are usually defined using the uniform noncommutative zeta function which arises from the exponential power series. They can be defined more generally using any invariant noncommutative zeta function which arises from a zeta series.

Early references to the Poincaré–Birkhoff–Witt theorem for Lie coalgebras are by André [**15**, Theorem 11] and Michaelis [**260**], [**261**, Theorem on page 52], [**262**, Theorem 3.8]. This is the analogue of Theorem 13.59.

Solomon operator. The map (13.48) and operator (13.49) generalize the map and operator defined by Solomon in the classical setting [**332**, Formulas (1.2) and (1.3)]. He implicitly worked with the uniform noncommutative zeta function, where there is an explicit formula for the Solomon coefficients [**11**, Formula (17.41)]. Solomon's formulas are mentioned in Dixmier's book [**104**, Section 2.8.12, items (a), (b)].

13.9.7. Cartier–Milnor–Moore. (Section 13.7.) Theorem 13.65 generalizes the classical Cartier–Milnor–Moore theorem for connected graded cocommutative Hopf algebras. It first appeared in work of Cartier [**78**, Section 3.5, Propositions 6 and 8], [**79**, Theorem 1] and Milnor and Moore [**265**, Theorem 5.18], [**266**, Theorem 5.22]. The latter worked in the signed setting which is generalized in Theorem 13.69. Other early references are by Quillen [**298**, Appendix B, Theorem 4.5], Sweedler [**342**, Theorem 4 on page 103], [**345**, Theorem 13.0.1 and Proposition 13.0.2], Grünenfelder [**162**, Theorems I.3.15 and III.3.3], [**163**, Theorems 1.3 and 3.2]. In the language of formal groups, it is given by Cartier [**80**, Theorem 3], Gabriel [**138**, Exposé VIIB, Corollary 3.3.2], Serre [**323**, Part II, Section V.6, Theorem 3], Fröhlich [**136**, Section II.2, Theorem 1], Demazure [**95**, page 42], Ditters [**102**, Theorem on page 28], Hazewinkel [**180**, Theorem 14.2.3].

Theorem 13.68 generalizes the classical Cartier–Milnor–Moore theorem for connected graded commutative Hopf algebras. Early references for the latter are by André [**15**, Theorems 17 and 20], [**16**, Theorems 1 and 2], Sjödin [**324**, Theorem 2], Nichols [**280**, Chapter III, Theorems 12 and 14], [**281**, Theorems 11 and 14]. In the language of algebraic groups, it is given by Demazure and Gabriel [**96**, Section IV.2, Corollary 4.5] and Hochschild [**188**, Chapter XVI, Theorem 4.2].

13.9.8. Rank-one arrangement. (Section 13.8.) Recall that a rank-one arrangement is the same as the braid arrangement on 2 letters. In this sense, the rank-one example is present in classical Hopf theory. However, unlike the classical theory, our theory is local, so the rank-one example is complete by itself. Also, our presentation of it is geometric (and not combinatorial) giving insights into how to generalize it to reflection arrangements.

Part III

Fock functors

CHAPTER 14

Fock functors

We studied Coxeter species in Part I and Coxeter spaces in Part II relative to a reflection arrangement $\mathcal{A}$ with Coxeter group W. We now introduce Fock functors. They go from the category of Coxeter species to the category of Coxeter spaces. They are summarized in Table 14.1 and elaborated below.

TABLE 14.1. Fock functors.

$\mathcal{K}$, $\mathcal{K}^{\vee}$	full Fock functors
$\mathcal{K}_{-1}$, $\mathcal{K}^{\vee}_{-1}$	signed full Fock functors
$\mathcal{K}_{q}$, $\mathcal{K}^{\vee}_{q}$	q-full Fock functors
$\overline{\mathcal{K}}$, $\overline{\mathcal{K}}^{\vee}$	bosonic Fock functors
$\overline{\mathcal{K}}_{-1}$, $\overline{\mathcal{K}}^{\vee}_{-1}$	fermionic Fock functors
$\Im_q$	anyonic Fock functor
$\mathcal{K}_0 = \mathcal{K}^{\vee}_0 = \Im_0$	free Fock functor

Bilax functors between bimonads play a key role in our discussion. They are reviewed in Appendix A.1. The main point is that Fock functors are bilax, and interactions between them preserve these bilax structures.

Full Fock functors. Recall the bimonad $(\mathcal{T}, \mathcal{T}^{\vee}, \lambda)$ on Coxeter species and bimonad $(\overline{\mathcal{T}}, \overline{\mathcal{T}}^{\vee}, \overline{\lambda})$ on Coxeter spaces. The full Fock functors $\mathcal{K}$ and $\mathcal{K}^{\vee}$ carry bilax structures wrt these bimonads. They are conjugates of each other wrt duality. Moreover, $\mathcal{K}$ carries an induced colax structure wrt comonads $\mathcal{S}^{\vee}$ and $\overline{\mathcal{S}}^{\vee}$, while $\mathcal{K}^{\vee}$ carries an induced lax structure wrt monads $\mathcal{S}$ and $\overline{\mathcal{S}}$.

Signed full Fock functors. We also have signed full Fock functors $\mathcal{K}_{-1}$ and $\mathcal{K}^{\vee}_{-1}$ obtained by precomposing full Fock functors $\mathcal{K}$ and $\mathcal{K}^{\vee}$ with the o-signature functor on Coxeter species. Since the latter is bilax, so are the signed full Fock functors. The signed distance function on chambers intervenes in the colax structure of $\mathcal{K}_{-1}$ and dually in the lax structure of $\mathcal{K}^{\vee}_{-1}$.

Bosonic Fock functors. The bosonic Fock functors $\overline{\mathcal{K}}$ and $\overline{\mathcal{K}}^{\vee}$ are constructed from full Fock functors $\mathcal{K}$ and $\mathcal{K}^{\vee}$ by, respectively, taking coinvariants and invariants wrt the family of subgroups W_T (indexed by face-types T) of the Coxeter group W. The two bosonic Fock functors are isomorphic when $\mathrm{char}(\Bbbk) \nmid |W|$. They inherit a bilax structure from full Fock functors wrt bimonads $(\mathcal{T}, \mathcal{T}^\vee, \lambda)$ and $(\overline{\mathcal{T}}, \overline{\mathcal{T}}^\vee, \overline{\lambda})$. Interestingly, bilax structure maps are now isomorphisms. Moreover, they also induce bilax structures wrt bimonads $(\mathcal{S}, \mathcal{S}^\vee, \lambda)$ and $(\overline{\mathcal{S}}, \overline{\mathcal{S}}^\vee, \overline{\lambda})$. The full Fock functors can be reconstructed from bosonic Fock functors using the Coxeter bimonoid of chambers Γ.

Fermionic Fock functors. The fermionic Fock functors $\overline{\mathcal{K}}_{-1}$ and $\overline{\mathcal{K}}^\vee_{-1}$ are constructed similarly from signed full Fock functors $\mathcal{K}_{-1}$ and $\mathcal{K}^\vee_{-1}$ by taking (co)invariants with the W-action twisted by the sign representation. They also inherit bilax structures. Alternatively, they can be obtained from bosonic Fock functors by precomposing with the o-signature functor on Coxeter species.

Deformed full Fock functors. The bilax structure of full Fock functors $\mathcal{K}$ and $\mathcal{K}^\vee$ can be deformed by a scalar q by using the q-distance function on chambers. We indicate this by writing $\mathcal{K}_q$ and $\mathcal{K}^\vee_q$ and call them q-full Fock functors. The values $q = \pm 1$ recover full Fock functors and signed full Fock functors, respectively. Precomposing $\mathcal{K}_q$ and $\mathcal{K}^\vee_q$ with the o-signature functor yields $\mathcal{K}_{-q}$ and $\mathcal{K}^\vee_{-q}$, respectively. Interestingly, one can bring another scalar p into the story by deforming the mixed distributive laws so that $\mathcal{K}_q$ and $\mathcal{K}^\vee_q$ are bilax wrt bimonads $(\mathcal{T}, \mathcal{T}^\vee, \lambda_p)$ and $(\overline{\mathcal{T}}, \overline{\mathcal{T}}^\vee, \overline{\lambda}_{pq})$. (This also applies when $q = 1$.) The q-full Fock functors can also be reconstructed from bosonic Fock functors using the Coxeter q-bimonoid of chambers Γ_q.

Anyonic Fock functor. For a scalar q, there is a norm transformation κ_q which links $\mathcal{K}_q$ and $\mathcal{K}^\vee_q$. It is self-dual. Its image is the anyonic Fock functor $\mathfrak{I}_q$. For q not a root of unity, κ_q is an isomorphism, and so $\mathfrak{I}_q$ is isomorphic to both q-full Fock functors $\mathcal{K}_q$ and $\mathcal{K}^\vee_q$. The values $q = \pm 1$ of $\mathfrak{I}_q$ specialize to bosonic and fermionic Fock functors, respectively (assuming $\mathrm{char}(\Bbbk) \nmid |W|$).

Free Fock functor. For $q = 0$, the q-full Fock functors $\mathcal{K}_q$ and $\mathcal{K}^\vee_q$ coincide as bilax functors (with the norm transformation κ_0 specializing to the identity). We refer to $\mathcal{K}_0 = \mathcal{K}^\vee_0 = \mathfrak{I}_0$ as the free Fock functor.

Decorated Fock functors. Let M be a left W-module. Fock functors have decorated versions which we denote $\mathcal{K}_M$, $\mathcal{K}^\vee_M$, and so on. They are constructed by precomposing Fock functors with the decoration functor $(-)_M$ on Coxeter species. Since the latter is bilax, decorated Fock functors are bilax as well. We also have notions such as the decorated norm transformation.

Fock functors and Coxeter operads. From an operadic point of view, (co)lax properties of Fock functors mentioned above pertain to Coxeter associative and commutative (co)operads. More generally, we also consider interactions of Fock functors with (co)monads arising from any Coxeter (co)operad. In particular, this brings in the Coxeter Lie (co)operad into the picture.

14.1. Full Fock functors

We construct full Fock functors from the category of Coxeter species to the category of Coxeter spaces. There are two of them, namely, $\mathcal{K}$ and $\mathcal{K}^\vee$. They are both bilax wrt bimonads $(\mathcal{T}, \mathcal{T}^\vee, \lambda)$ and $(\overline{\mathcal{T}}, \overline{\mathcal{T}}^\vee, \overline{\lambda})$. As suggested by the notation, the bilax structures of $\mathcal{K}$ and $\mathcal{K}^\vee$ are related through duality. The lax structure of $\mathcal{K}$ is the canonical inclusion, while the colax structure is a surjection defined by employing type morphisms. For $\mathcal{K}^\vee$, the situation is the other way round, where the colax structure is the canonical projection while the lax structure is an injection defined using type morphisms.

14.1.1. Full Fock functor. Coxeter species to Coxeter spaces. Recall from Sections 2.1 and 8.1 the categories $\mathcal{A}\text{-}\mathsf{Cox}$ and $\mathcal{A}\text{-}\overline{\mathsf{Cox}}$ whose objects are faces and top-nested faces, respectively. These serve as base categories for Coxeter species and Coxeter spaces, respectively. The starting observation is that there is a functor

$$\mathcal{A}\text{-}\overline{\mathsf{Cox}} \to \mathcal{A}\text{-}\mathsf{Cox}, \qquad (A, C) \mapsto A, \tag{14.1}$$

which on objects forgets the second coordinate. Support and type morphisms in $\mathcal{A}\text{-}\overline{\mathsf{Cox}}$ go, respectively, to support and type morphisms in $\mathcal{A}\text{-}\mathsf{Cox}$. A comparison of (8.1a)–(8.1b) with (2.1a)–(2.1b) shows that relations map to relations. So the functor is well-defined on morphisms. Since

$$\mathcal{A}\text{-}\mathsf{CSp} = [\mathcal{A}\text{-}\mathsf{Cox}, \mathsf{Vec}] \qquad \text{and} \qquad \mathcal{A}\text{-}\overline{\mathsf{C}}\mathsf{Sp} = [\mathcal{A}\text{-}\overline{\mathsf{Cox}}, \mathsf{Vec}],$$

we obtain an induced functor

$$\mathcal{K} : \mathcal{A}\text{-}\mathsf{CSp} \to \mathcal{A}\text{-}\overline{\mathsf{C}}\mathsf{Sp}, \qquad \mathcal{K}(\mathsf{p})[A, C] := \mathsf{p}[A] \tag{14.2}$$

from the category of Coxeter species to the category of Coxeter spaces. Here p is a Coxeter species, and $\mathcal{K}(\mathsf{p})$ is the Coxeter space it maps to. Support and type morphisms of $\mathcal{K}(\mathsf{p})$ in terms of those of p are

$$\begin{array}{ccccccc} \mathcal{K}(\mathsf{p})[F,C] & \xrightarrow{\beta_{G,D,F,C}} & \mathcal{K}(\mathsf{p})[G,D] & \qquad & \mathcal{K}(\mathsf{p})[F,C] & \xrightarrow{w} & \mathcal{K}(\mathsf{p})[F',C'] \\ \| & & \| & & \| & & \| \\ \mathsf{p}[F] & \xrightarrow[\beta_{G,F}]{} & \mathsf{p}[G] & & \mathsf{p}[F] & \xrightarrow[w]{} & \mathsf{p}[F']. \end{array} \tag{14.3}$$

They satisfy (8.20a) and (8.20b) since those of p satisfy (2.12a) and (2.12b).

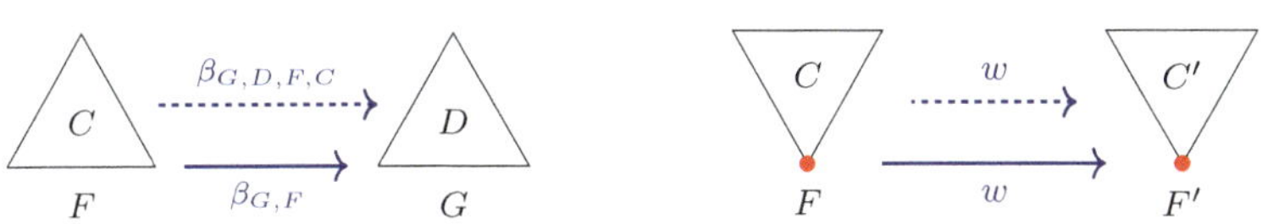

A map $f : \mathsf{p} \to \mathsf{q}$ of Coxeter species induces the map $\mathcal{K}(f) : \mathcal{K}(\mathsf{p}) \to \mathcal{K}(\mathsf{q})$ of Coxeter spaces with components

$$\begin{array}{ccc} \mathcal{K}(\mathsf{p})[F,C] & \xrightarrow{\mathcal{K}(f)_{F,C}} & \mathcal{K}(\mathsf{q})[F,C] \\ \| & & \| \\ \mathsf{p}[F] & \xrightarrow[f_F]{} & \mathsf{q}[F]. \end{array} \tag{14.4}$$

We refer to $\mathcal{K}$ as the *full Fock functor*. It is $\Bbbk$-linear; this can either be seen directly or deduced from the existence of the functor (14.1).

14.1.2. Bilax structure. Recall from Sections 2.10 and 8.12 the bimonads $(\mathcal{T}, \mathcal{T}^\vee, \lambda)$ and $(\overline{\mathcal{T}}, \overline{\mathcal{T}}^\vee, \overline{\lambda})$ on the categories $\mathcal{A}$-CSp and $\mathcal{A}$-$\overline{\mathsf{C}}\mathsf{Sp}$, respectively. See formulas (2.87) and (8.115) in particular. We now equip $\mathcal{K}$ with a bilax structure wrt these bimonads.

Define a natural transformation $\varphi : \overline{\mathcal{T}}\mathcal{K} \to \mathcal{K}\mathcal{T}$ as follows. For a Coxeter species p, consider φ_{p} on the (A, C)-component:

$$
\begin{array}{ccc}
\overline{\mathcal{T}}\mathcal{K}(\mathsf{p})[A, C] & \hookrightarrow & \mathcal{K}\mathcal{T}(\mathsf{p})[A, C] \\
\| & & \| \\
\displaystyle\bigoplus_{F_0 \colon A \leq F_0 \leq C} \mathsf{p}[F_0] & & \displaystyle\bigoplus_{F \colon A \leq F} \mathsf{p}[F].
\end{array}
\tag{14.5}
$$

Both sums are over faces, but there are less summands in the lhs than in the rhs since in the former F_0 must be a face of C. The map (14.5) is defined to be the canonical inclusion. It is easy to see that φ_{p} is a map of Coxeter spaces, and it is natural in p.

Define a natural transformation $\psi : \mathcal{K}\mathcal{T}^\vee \to \overline{\mathcal{T}}^\vee\mathcal{K}$ as follows. For a Coxeter species p, consider ψ_{p} on the (A, C)-component:

$$
\begin{array}{ccc}
\mathcal{K}\mathcal{T}^\vee(\mathsf{p})[A, C] & \twoheadrightarrow & \overline{\mathcal{T}}^\vee\mathcal{K}(\mathsf{p})[A, C] \\
\| & & \| \\
\displaystyle\bigoplus_{F \colon A \leq F} \mathsf{p}[F] & & \displaystyle\bigoplus_{F_0 \colon A \leq F_0 \leq C} \mathsf{p}[F_0].
\end{array}
\tag{14.6}
$$

Now the situation has reversed; there are more summands in the lhs than in the rhs. We proceed as follows. For each F, choose F_0 to be that face of C which has the same type as F. Let $w \in W$ be the unique Coxeter symmetry which sends FC to C (and hence F to F_0 since the action is type-preserving).

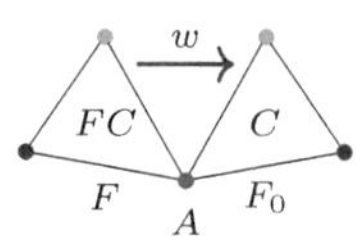

The map (14.6) is the sum of the linear maps $\mathsf{p}[w] : \mathsf{p}[F] \to \mathsf{p}[F_0]$. As a shorthand, we write it as

$$
\bigoplus_{F \colon A \leq F} \mathsf{p}[F] \xrightarrow[w(FC) = C]{\mathsf{p}[w]} \bigoplus_{F_0 \colon A \leq F_0 \leq C} \mathsf{p}[F_0].
\tag{14.7}
$$

It is convenient sometimes to drop $\mathsf{p}[w]$ and just label the map by $w(FC) = C$.

Exercise 14.1. Use support-type and type-type axioms in (2.12a) to check that ψ_{p} is indeed a map of Coxeter spaces, that is, diagrams (8.21) commute.

Theorem 14.2. *The full Fock functor*

$$
(\mathcal{K}, \varphi, \psi) : (\mathcal{T}, \mathcal{T}^\vee, \lambda) \to (\overline{\mathcal{T}}, \overline{\mathcal{T}}^\vee, \overline{\lambda})
$$

is bilax, with φ as the lax and ψ as the colax structure. Also, $\psi\varphi = \mathrm{id}$ *(noting that $\mathcal{T}$ and $\mathcal{T}^\vee$, and $\overline{\mathcal{T}}$ and $\overline{\mathcal{T}}^\vee$ are identical as functors).*

PROOF. The last part is clear. The remaining checks are done below.

ψ is colax. In (A.5), the right diagram clearly commutes. The left diagram is as follows.

$$\begin{array}{ccccc} \mathcal{K}\mathcal{T}^\vee\mathcal{T}^\vee & \xrightarrow{\psi\mathcal{T}^\vee} & \overline{\mathcal{T}}^\vee\mathcal{K}\mathcal{T}^\vee & \xrightarrow{\overline{\mathcal{T}}^\vee\psi} & \overline{\mathcal{T}}^\vee\overline{\mathcal{T}}^\vee\mathcal{K} \\ \uparrow & & & & \uparrow \\ \mathcal{K}\mathcal{T}^\vee & & \xrightarrow[\psi]{\qquad\qquad} & & \overline{\mathcal{T}}^\vee\mathcal{K} \end{array}$$

Evaluating on a Coxeter species p on the (A, C)-component:

$$\begin{array}{ccccc} \bigoplus\limits_{A\le F\le G} \mathsf{p}[G] & \xrightarrow[u(FC)=C]{\mathsf{p}[u]} & \bigoplus\limits_{\substack{A\le F_0\le C\\ A\le F_0\le G_1}} \mathsf{p}[G_1] & \xrightarrow[v(G_1C)=C]{\mathsf{p}[v]} & \bigoplus\limits_{A\le F_0\le G_0\le C} \mathsf{p}[G_0] \\ \uparrow & & & & \uparrow \\ \bigoplus\limits_{A\le G} \mathsf{p}[G] & & \xrightarrow[w(GC)=C]{\mathsf{p}[w]} & & \bigoplus\limits_{A\le G_0\le C} \mathsf{p}[G_0]. \end{array} \tag{14.8}$$

Here A and C are fixed; the sums are over the remaining faces. An illustration with $A = O$ is shown in Figure 14.1. The element u is the Coxeter symmetry which moves the two left triangles onto the two right triangles.

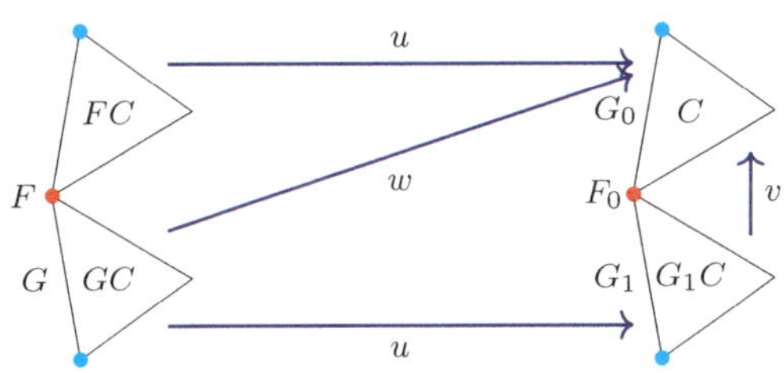

FIGURE 14.1. Full Fock functor $\mathcal{K}$ is colax.

We have $u(G) = G_1$, $u(FC) = C$, $v(G_1C) = C$. Multiplying the first two yields $u(GC) = G_1C$. Now applying v yields $vu(GC) = C$. Since $w(GC) = C$, we conclude that $vu = w$. Hence $\mathsf{p}[w] = \mathsf{p}[v]\mathsf{p}[u]$ by the type-type axiom in (2.12a), and commutativity of diagram (14.8) follows.

φ is lax. The right diagram in (A.2) clearly commutes. The check for the left diagram is similar to (14.8) above, the arrows are reversed, and the horizontal maps are inclusions; so it commutes.

φ and ψ are compatible. Diagram (A.8) is as follows. It connects the mixed distributive laws λ and $\overline{\lambda}$.

$$\begin{array}{ccccc} \overline{\mathcal{T}}\mathcal{K}\mathcal{T}^\vee & \xhookrightarrow{\varphi\mathcal{T}^\vee} & \mathcal{K}\mathcal{T}\mathcal{T}^\vee & \xrightarrow{\mathcal{K}\lambda} & \mathcal{K}\mathcal{T}^\vee\mathcal{T} \\ {\scriptstyle \overline{\mathcal{T}}\psi}\downarrow & & & & \downarrow{\scriptstyle \psi\mathcal{T}} \\ \overline{\mathcal{T}}\overline{\mathcal{T}}^\vee\mathcal{K} & \xrightarrow[\overline{\lambda}\mathcal{K}]{} & \overline{\mathcal{T}}^\vee\overline{\mathcal{T}}\mathcal{K} & \xhookrightarrow[\overline{\mathcal{T}}^\vee\varphi]{} & \overline{\mathcal{T}}^\vee\mathcal{K}\mathcal{T} \end{array}$$

Evaluating on a Coxeter species p on the (A, C)-component, we obtain the following. The two inclusion maps have been suppressed, and $\overline{\lambda}$ appears in the bottom-horizontal composite.

$$\begin{array}{ccccc}
\displaystyle\bigoplus_{\substack{A\le F_0\le G\\ A\le F_0\le C}} \mathsf{p}[G] & \xrightarrow{(\beta_{G',G})} & \displaystyle\bigoplus_{A\le F'\le G'} \mathsf{p}[G'] & & \\
{\scriptstyle \mathsf{p}[u]}\Big\downarrow{\scriptstyle u(GC)=C} & & {\scriptstyle u(F'C)=F_1C}\Big\downarrow{\scriptstyle \mathsf{p}[u]} & \overset{w(F'C)=C}{\underset{\mathsf{p}[w]}{\searrow}} & \\
\displaystyle\bigoplus_{A\le F_0\le G_0\le C} \mathsf{p}[G_0] & \xrightarrow[(\beta_{G_1,G_0})]{} & \displaystyle\bigoplus_{A\le F_1\le G_1} \mathsf{p}[G_1] & \xrightarrow[v(F_1C)=C]{\mathsf{p}[v]} & \displaystyle\bigoplus_{A\le F_0'\le G_0'\le C} \mathsf{p}[G_0']
\end{array} \tag{14.9}$$

Here A and C are fixed; the sums are over the remaining faces. An illustration with $A = O$ is shown in Figure 14.2. The faces F_0' and G_0' of C have not been shown for simplicity.

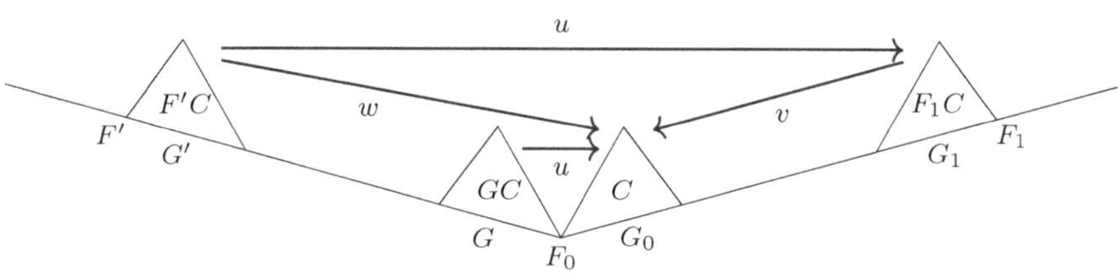

FIGURE 14.2. Lax-colax compatibility for $\mathcal{K}$.

The square and triangle in (14.9) commute by support-type and type-type axioms in (2.12a) (after noting $w = vu$). A key point is that $G'C = F'F_0C = F'C$, and similarly, $G_1C = F_1C$. □

Exercise 14.3. Work out diagram (14.9) in the special case $F_0 = A$.

14.1.3. Conjugate full Fock functor. By general duality principles, the functor $\mathcal{K}$ has a conjugate denoted $\mathcal{K}^\vee$. We refer to $\mathcal{K}^\vee$ also as the *full Fock functor*. We describe it explicitly. Define

$$\mathcal{K}^\vee : \mathcal{A}\text{-}\mathsf{CSp} \to \mathcal{A}\text{-}\overline{\mathsf{C}}\mathsf{Sp}, \qquad \mathcal{K}^\vee(\mathsf{p})[A, C] := \mathsf{p}[A]. \tag{14.10}$$

As a functor, $\mathcal{K}^\vee$ is the same as $\mathcal{K}$.

Define a natural transformation $\psi^\vee : \overline{\mathcal{T}}\mathcal{K}^\vee \to \mathcal{K}^\vee\mathcal{T}$ as follows. For a Coxeter species p, consider $\psi^\vee_\mathsf{p}$ on the (A, C)-component:

$$\begin{array}{ccc}
\overline{\mathcal{T}}\mathcal{K}^\vee(\mathsf{p})[A, C] & \hookrightarrow & \mathcal{K}^\vee\mathcal{T}(\mathsf{p})[A, C] \\
\| & & \| \\
\displaystyle\bigoplus_{F_0 \colon A\le F_0\le C} \mathsf{p}[F_0] & & \displaystyle\bigoplus_{F \colon A\le F} \mathsf{p}[F].
\end{array} \tag{14.11}$$

We define this summand-wise, so fix F_0. Let F be any face greater than A which has the same type as F_0. Let $w \in W$ be the unique Coxeter symmetry which sends C to FC (and hence F_0 to F since the action is type-preserving).

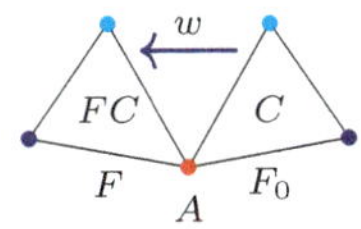

The map (14.11) is the sum of the linear maps $\mathsf{p}[w] : \mathsf{p}[F_0] \to \mathsf{p}[F]$ over all choices of F. As a shorthand, we write it as

$$\bigoplus_{F_0 :\, A \leq F_0 \leq C} \mathsf{p}[F_0] \xrightarrow[w(C)=FC]{\mathsf{p}[w]} \bigoplus_{F :\, A \leq F} \mathsf{p}[F]. \tag{14.12}$$

Define a natural transformation $\varphi^\vee : \mathcal{K}^\vee \mathcal{T}^\vee \to \overline{\mathcal{T}}^\vee \mathcal{K}^\vee$ as follows. For a Coxeter species p, consider $\varphi^\vee_{\mathsf{p}}$ on the (A, C)-component:

$$\begin{array}{ccc} \mathcal{K}^\vee \mathcal{T}^\vee(\mathsf{p})[A, C] & \longrightarrow\!\!\!\!\to & \overline{\mathcal{T}}^\vee \mathcal{K}^\vee(\mathsf{p})[A, C] \\ \| & & \| \\ \bigoplus_{F :\, A \leq F} \mathsf{p}[F] & & \bigoplus_{F_0 :\, A \leq F_0 \leq C} \mathsf{p}[F_0]. \end{array} \tag{14.13}$$

This is defined to be the canonical projection: If F is a face of C, then $\mathsf{p}[F]$ maps identically to the same summand occurring in the rhs. If F is not a face of C, then $\mathsf{p}[F]$ maps to zero.

Theorem 14.4. *The full Fock functor*

$$(\mathcal{K}^\vee, \psi^\vee, \varphi^\vee) : (\mathcal{T}, \mathcal{T}^\vee, \lambda) \to (\overline{\mathcal{T}}, \overline{\mathcal{T}}^\vee, \overline{\lambda})$$

is bilax, with $\psi^\vee$ as the lax and $\varphi^\vee$ as the colax structure. Also, $\varphi^\vee \psi^\vee = \mathrm{id}$.

Remark 14.5 (Deformed bilax structures). Functors $\mathcal{K}$ and $\mathcal{K}^\vee$ continue to be bilax even when the laws λ and $\overline{\lambda}$ are deformed to λ_p and $\overline{\lambda}_p$ for a scalar p. This point is addressed more fully in Section 14.2.

14.1.4. Conjugation by duality functors. We elaborate on conjugation. Here we work with the full subcategories of finite-dimensional Coxeter species and Coxeter spaces.

Consider the composite of bilax functors

$$(\mathcal{T}^\vee, \mathcal{T}, \lambda) \xrightarrow{(-)^*} (\mathcal{T}, \mathcal{T}^\vee, \lambda) \xrightarrow{\mathcal{K}} (\overline{\mathcal{T}}, \overline{\mathcal{T}}^\vee, \overline{\lambda}) \xrightarrow{(-)^*} (\overline{\mathcal{T}}^\vee, \overline{\mathcal{T}}, \overline{\lambda})$$

obtained by conjugating $\mathcal{K}$ by the duality functors in either category. This is a bilax functor from $\mathcal{A}$-$\mathsf{CSp}^{\mathrm{op}}$ to $\mathcal{A}$-$\overline{\mathsf{C}}\mathsf{Sp}^{\mathrm{op}}$. Now pass to the opposite categories. The resulting bilax functor is precisely $\mathcal{K}^\vee$. Explicitly, for a finite-dimensional Coxeter species p,

$$\varphi^\vee_{\mathsf{p}} = (\varphi_{\mathsf{p}^*})^* \quad \text{and} \quad \psi^\vee_{\mathsf{p}} = (\psi_{\mathsf{p}^*})^*$$

after using the identification $(\mathsf{p}^*)^* = \mathsf{p}$.

If we start with $\mathcal{K}^\vee$ instead, then the resulting bilax functor is $\mathcal{K}$. This is expected since duality is an involution. Hence we say $\mathcal{K}$ and $\mathcal{K}^\vee$ are conjugates of each other wrt duality (in finite-dimensional setting).

In general, what we can say is the following. For any Coxeter species p (not necessarily finite-dimensional),

$$(\varphi_{\mathsf{p}})^* = \varphi^{\vee}_{\mathsf{p}^*}, \quad (\varphi^{\vee}_{\mathsf{p}})^* = \varphi_{\mathsf{p}^*}, \quad (\psi_{\mathsf{p}})^* = \psi^{\vee}_{\mathsf{p}^*}, \quad (\psi^{\vee}_{\mathsf{p}})^* = \psi_{\mathsf{p}^*}.$$

14.1.5. Full Fock functors in terms of face-types. Using the inclusion of the category $\boldsymbol{\alpha}\text{-}\overline{\mathsf{Cox}}$ in $\mathcal{A}\text{-}\overline{\mathsf{Cox}}$, we get

$$\mathcal{K}(\mathsf{p})[T] = \mathcal{K}^{\vee}(\mathsf{p})[T] = \mathsf{p}[F], \tag{14.14}$$

where F is the face of C_r of type T. Similarly, bilax structures of $\mathcal{K}$ and $\mathcal{K}^{\vee}$ can be expressed in terms of face-types by putting $C := C_r$ and replacing (A, C_r) by the type of A.

14.1.6. Full Fock functors in terms of flats and lunes. Recall categories $\mathcal{A}\text{-}\mathsf{Cox}'$ and $\mathcal{A}\text{-}\overline{\mathsf{Cox}}'$ from (2.5) and (8.5) whose objects are flats and top-lunes, respectively. Similar to $\mathcal{A}\text{-}\mathsf{Cox}$ and $\mathcal{A}\text{-}\overline{\mathsf{Cox}}$, they also serve as base categories for Coxeter species and Coxeter spaces, respectively. This allows us to formulate full Fock functors in terms of flats and lunes as follows. There is a functor

$$\mathcal{A}\text{-}\overline{\mathsf{Cox}}' \to \mathcal{A}\text{-}\mathsf{Cox}', \qquad \mathrm{L} \mapsto \mathrm{b}(\mathrm{L}), \tag{14.15}$$

that is, on objects, it sends a top-lune to its base. (Recall: Any map of W-sets $X \to Y$ yields a functor $\mathsf{el}_W(X) \to \mathsf{el}_W(Y)$ between corresponding groupoids of elements.) This induces the functor

$$\mathcal{K} : \mathcal{A}\text{-}\mathsf{CSp} \to \mathcal{A}\text{-}\overline{\mathsf{C}}\mathsf{Sp}, \qquad \mathcal{K}(\mathsf{p})[\mathrm{L}] := \mathsf{p}[\mathrm{b}(\mathrm{L})]. \tag{14.16}$$

Type morphisms of $\mathcal{K}(\mathsf{p})$ in terms of those of p are

$$\begin{array}{ccc} \mathcal{K}(\mathsf{p})[\mathrm{L}] & \overset{w}{\dashrightarrow} & \mathcal{K}(\mathsf{p})[\mathrm{L}'] \\ \| & & \| \\ \mathsf{p}[\mathrm{b}(\mathrm{L})] & \underset{w}{\longrightarrow} & \mathsf{p}[\mathrm{b}(\mathrm{L}')]. \end{array} \tag{14.17}$$

The bilax structure of $\mathcal{K}$ is as follows. For a Coxeter species p and top-lune L, using (2.92) and (8.117),

$$\begin{array}{ccccc} \mathcal{K}\mathcal{T}(\mathsf{p})[\mathrm{L}] & = & \displaystyle\bigoplus_{\mathrm{P}:\, \mathrm{b}(\mathrm{P})=\mathrm{b}(\mathrm{L})} \mathsf{p}[\mathrm{c}(\mathrm{P})] & = & \mathcal{K}\mathcal{T}^{\vee}(\mathsf{p})[\mathrm{L}] \\ \uparrow{\scriptstyle \varphi_{\mathsf{p}}} & & & & \downarrow{\scriptstyle \psi_{\mathsf{p}}} \\ \overline{\mathcal{T}}\mathcal{K}(\mathsf{p})[\mathrm{L}] & = & \displaystyle\bigoplus_{\mathrm{M}:\, \mathrm{L} \leq' \mathrm{M}} \mathsf{p}[\mathrm{b}(\mathrm{M})] & = & \overline{\mathcal{T}}^{\vee}\mathcal{K}(\mathsf{p})[\mathrm{L}]. \end{array} \tag{14.18}$$

The smaller sum is over top-lunes M which are greater than L wrt $\leq'$, while the larger sum is over lunes P with the same base as L.

The map φ_{p} is inclusion; it sends the M-summand to the L/M-summand with $\mathrm{id} : \mathsf{p}[\mathrm{b}(\mathrm{M})] \to \mathsf{p}[\mathrm{c}(\mathrm{L}/\mathrm{M})]$ (recalling that $\mathrm{b}(\mathrm{L}/\mathrm{M}) = \mathrm{b}(\mathrm{L})$ and $\mathrm{c}(\mathrm{L}/\mathrm{M}) = \mathrm{b}(\mathrm{M})$). The map ψ_{p} sends the P-summand to the unique M-summand such that $\mathrm{t}(\mathrm{L}/\mathrm{M}) = \mathrm{t}(\mathrm{P})$ wrt the Coxeter group $W_{\mathrm{b}(\mathrm{L})}$ of $\mathcal{A}_{\mathrm{b}(\mathrm{L})}$. The linear map

$\mathsf{p}[\mathrm{c(P)}] \to \mathsf{p}[\mathrm{b(M)}]$ is induced by $w \in W_{\mathrm{b(L)}}$ which sends PL to L (and P to L/M) where PL is the Tits product of faces in $\mathcal{A}_{\mathrm{b(L)}}$. See the pictures below.

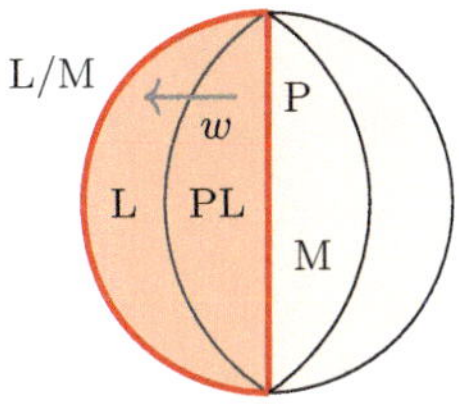

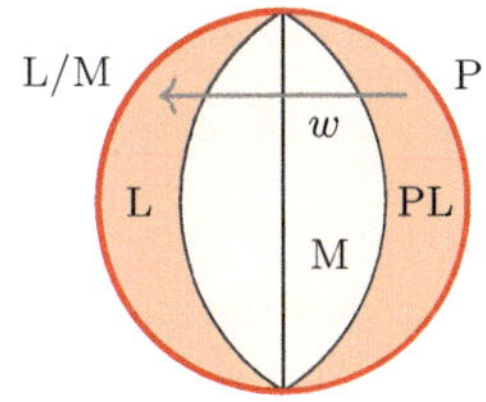

In these pictures, M is the visible hemisphere, L and PL are portions between two adjacent longitudes, P and L/M are the red longitudes, w takes PL to L. As a shorthand, we write ψ_{p} as

$$\bigoplus_{\mathrm{P}:\, \mathrm{b(P)}=\mathrm{b(L)}} \mathsf{p}[\mathrm{c(P)}] \xrightarrow[w(\mathrm{PL})=\mathrm{L}]{\mathsf{p}[w]} \bigoplus_{\mathrm{M}:\, \mathrm{L} \leq' \mathrm{M}} \mathsf{p}[\mathrm{b(M)}]. \tag{14.19}$$

The conjugate functor $\mathcal{K}^\vee$ in terms of flats and lunes is defined dually.

Remark 14.6. Alternatively, instead of $\mathcal{A}\text{-}\overline{\mathsf{Cox}}'$, one can use the base category $\mathcal{A}\text{-}\overline{\mathsf{Cox}}''_C$ (see Exercise 8.13). Recall: There is a functor

$$\mathcal{A}\text{-}\overline{\mathsf{Cox}}''_C \hookrightarrow \mathcal{A}\text{-}\mathsf{Cox}', \qquad \mathrm{X} \mapsto \mathrm{X}. \tag{14.20}$$

Both categories have flats as objects, but the former has fewer morphisms than the latter. In this setting, we write $\mathcal{K}(\mathsf{p})[\mathrm{X}] := \mathsf{p}[\mathrm{X}]$ and proceed. There is also a second option when using flats for Coxeter spaces: Instead of $\mathcal{A}\text{-}\overline{\mathsf{Cox}}''_C$, one can use the isomorphic category $\mathcal{A}\text{-}\overline{\mathsf{Cox}}''$ following Proposition 8.12.

14.2. Deformed full Fock functors

We build on the discussion in Section 14.1. For a scalar q, we deform the colax structure of $\mathcal{K}$ (and dually the lax structure of $\mathcal{K}^\vee$) by multiplying the intervening type morphisms by a power of q involving the distance function on chambers. This gives rise to q-full Fock functors $\mathcal{K}_q$ and $\mathcal{K}^\vee_q$. They are bilax. (The gate property of distance functions plays a critical role here.) Setting $q = 1$ recovers full Fock functors. Something special happens for $q = 0$: $\mathcal{K}_0$ and $\mathcal{K}^\vee_0$ coincide. We call this the free Fock functor. The case $q = -1$ is also of interest; we refer to $\mathcal{K}_{-1}$ and $\mathcal{K}^\vee_{-1}$ as signed full Fock functors.

14.2.1. Deformed full Fock functors. For a scalar q, recall from Theorems 2.77 and 8.118 that the mixed distributive laws λ and $\overline{\lambda}$ can be deformed to λ_q and $\overline{\lambda}_q$ with $q = 1$ recovering the previous case.

For scalars p and q, we now deform $(\mathcal{K}, \varphi, \psi)$ to $(\mathcal{K}, \varphi, \psi_q)$ so that the latter is bilax wrt the laws λ_p and $\overline{\lambda}_{pq}$. The functor $\mathcal{K}$ and its lax structure φ remain as before in (14.2) and (14.5). The colax structure ψ given by (14.7) is deformed to ψ_q by

$$\bigoplus_{F:\, A \leq F} \mathsf{p}[F] \xrightarrow[w(FC)=C]{q^{\mathrm{dist}(FC,C)}\mathsf{p}[w]} \bigoplus_{F_0:\, A \leq F_0 \leq C} \mathsf{p}[F_0]. \tag{14.21}$$

In other words, we multiply each nonzero component of ψ with an appropriate power of q.

Theorem 14.7. *For scalars p and q, the q-full Fock functor*

$$(\mathcal{K}, \varphi, \psi_q) : (\mathcal{T}, \mathcal{T}^\vee, \lambda_p) \to (\overline{\mathcal{T}}, \overline{\mathcal{T}}^\vee, \overline{\lambda}_{pq})$$

is bilax.

Proof. We extend the proof of Theorem 14.2.

ψ_q is colax. This reduces to diagram (14.8) (which we have checked) and the identity

$$q^{\operatorname{dist}(GC,C)} = q^{\operatorname{dist}(G_1C,C)} q^{\operatorname{dist}(FC,C)}.$$

See Figure 14.1. Since u takes FC to C and GC to G_1C and is distance-preserving, we have $\operatorname{dist}(G_1C, C) = \operatorname{dist}(GC, FC)$. By the gate property, $C \text{ -- } FC \text{ -- } GC$, or equivalently, $\operatorname{dist}(GC, C) = \operatorname{dist}(GC, FC) + \operatorname{dist}(FC, C)$. Combining the two, we see that powers of q on both sides match.

φ is lax. The same as before.

φ and ψ_q are compatible. This reduces to diagram (14.9) (which we have checked) and the identity

$$p^{\operatorname{dist}(G,G')} q^{\operatorname{dist}(F'C,C)} = q^{\operatorname{dist}(GC,C)} (pq)^{\operatorname{dist}(G_0,G_1)}.$$

See Figure 14.2. Since u takes G to G_0 and G' to G_1, we have $\operatorname{dist}(G, G') = \operatorname{dist}(G_0, G_1)$. Thus powers of p on both sides match. By the gate property, $F'C \text{ -- } GC \text{ -- } C$, or equivalently, $\operatorname{dist}(F'C, C) = \operatorname{dist}(F'C, GC) + \operatorname{dist}(GC, C)$. Further, $\operatorname{dist}(G, G') = \operatorname{dist}(F'C, GC)$. Hence powers of q also match. ☐

Dually, we can deform $(\mathcal{K}^\vee, \psi^\vee, \varphi^\vee)$ to $(\mathcal{K}^\vee, \psi_q^\vee, \varphi^\vee)$. The functor $\mathcal{K}^\vee$ and its colax structure $\varphi^\vee$ remain as before in (14.10) and (14.13). The lax structure $\psi^\vee$ given by (14.12) is deformed to $\psi_q^\vee$ by

$$\bigoplus_{F_0\colon A \le F_0 \le C} \mathsf{p}[F_0] \xrightarrow[w(C)=FC]{q^{\operatorname{dist}(C,FC)} \mathsf{p}[w]} \bigoplus_{F\colon A \le F} \mathsf{p}[F]. \tag{14.22}$$

In other words, we multiply each nonzero component of $\psi^\vee$ with an appropriate power of q.

Theorem 14.8. *For scalars p and q, the q-full Fock functor*

$$(\mathcal{K}^\vee, \psi_q^\vee, \varphi^\vee) : (\mathcal{T}, \mathcal{T}^\vee, \lambda_p) \to (\overline{\mathcal{T}}, \overline{\mathcal{T}}^\vee, \overline{\lambda}_{pq})$$

is bilax.

It is convenient to write $\mathcal{K}_q := (\mathcal{K}, \varphi, \psi_q)$ and $\mathcal{K}_q^\vee := (\mathcal{K}^\vee, \psi_q^\vee, \varphi^\vee)$. Note very carefully: Theorems 14.7 and 14.8, when specialized to $q = 1$, are more general than Theorems 14.2 and 14.4 because of the additional scalar p, see Remark 14.5. Note also that the definitions of $\mathcal{K}_q$ and $\mathcal{K}_q^\vee$ only depend on q and not on p.

The q-full Fock functors $\mathcal{K}_q$ and $\mathcal{K}_q^\vee$ are conjugates of each other wrt duality (in finite-dimensional setting). This generalizes considerations in Section 14.1.4.

14.2.2. Deformed full Fock functors in terms of flats and lunes. The colax structure ψ_q of $\mathcal{K}_q$ in terms of flats and lunes is given by

$$\bigoplus_{\mathrm{P}:\, \mathrm{b(P)=b(L)}} \mathsf{p}[\mathrm{c(P)}] \xrightarrow[w(\mathrm{PL})=\mathrm{L}]{q^{\mathrm{dist(L,PL)}}\mathsf{p}[w]} \bigoplus_{\mathrm{M}:\, \mathrm{L}\leq'\mathrm{M}} \mathsf{p}[\mathrm{b(M)}]. \tag{14.23}$$

This deforms (14.19). The lax structure $\psi_q^\vee$ of $\mathcal{K}_q^\vee$ in terms of flats and lunes is similar.

14.2.3. Free Fock functor. Let us specialize to $q = 0$. For a scalar p,

$$\mathcal{K}_0 = \mathcal{K}_0^\vee \tag{14.24}$$

as bilax functors from $(\mathcal{T}, \mathcal{T}^\vee, \lambda_p)$ to $(\overline{\mathcal{T}}, \overline{\mathcal{T}}^\vee, \overline{\lambda}_0)$. In other words,

$$\varphi = \psi_0^\vee \quad \text{and} \quad \psi_0 = \varphi^\vee,$$

given by inclusion and projection, respectively, as in (14.5) and (14.13). (The dependence on type morphisms disappears.) We call (14.24) the *free Fock functor*. It is self-dual.

14.2.4. Signed full Fock functors. Also of interest is the case $q = -1$. We refer to $\mathcal{K}_{-1}$ and $\mathcal{K}_{-1}^\vee$ as *signed full Fock functors*. The colax structure of $\mathcal{K}_{-1}$ and dually the lax structure of $\mathcal{K}_{-1}^\vee$ involve the signed distance function on chambers.

Exercise 14.9 (Distance functions). Deformed Fock functors may be generalized even further as follows. Let υ and υ' be W-invariant distance functions as in Section 1.4.4. Define their Hadamard product $\upsilon' \times \upsilon$ by $(\upsilon' \times \upsilon)_{C,D} := \upsilon'_{C,D}\, \upsilon_{C,D}$. Show: There is a bilax functor

$$\mathcal{K}_\upsilon : (\mathcal{T}, \mathcal{T}^\vee, \lambda_{\upsilon'}) \to (\overline{\mathcal{T}}, \overline{\mathcal{T}}^\vee, \overline{\lambda}_{\upsilon' \times \upsilon})$$

with the colax structure of $\mathcal{K}_\upsilon$ defined by replacing $q^{\mathrm{dist}(FC,C)}$ in (14.21) by $\upsilon_{FC,C}$. Specializing to $\upsilon := \upsilon_q$ and $\upsilon' := \upsilon_p$ recovers $\mathcal{K}_q$.

Exercise 14.10 (Lax braidings). Check: $\mathcal{K}$ and $\mathcal{K}^\vee$ can be viewed as bilax functors from the category of Coxeter lb-species to the category of Coxeter lb-spaces. That is, Theorems 14.7 and 14.8 hold in this generality.

14.3. Full Fock functors and signature functors

For a scalar q, precomposing by the o-signature functor on Coxeter species transforms q-full Fock functors $\mathcal{K}_q$ and $\mathcal{K}_q^\vee$ to $\mathcal{K}_{-q}$ and $\mathcal{K}_{-q}^\vee$, respectively. In particular, this provides a way to link full Fock functors and signed full Fock functors. In contrast, the u-signature functors on Coxeter species and on Coxeter spaces commute with both $\mathcal{K}_q$ and $\mathcal{K}_q^\vee$.

14.3.1. Over version. Recall from (2.73) the o-signature functor $(-)^-$ on Coxeter species. Here we view it as the Hadamard product with E^- as in Section 4.2.6, with E^- defined by (4.9) and (4.10). The o-signature functor is bilax by Theorem 4.10. We start with a basic observation.

Lemma 14.11. *There is a commutative diagram of functors*

$$\begin{array}{ccc} \mathtt{CSp} & \xrightarrow{(-)^-} & \mathtt{CSp} \\ & {\scriptstyle\mathcal{K}}\searrow \quad \swarrow{\scriptstyle\mathcal{K}} & \\ & \overline{\mathtt{CSp}}. & \end{array}$$

Thus, for p a Coxeter species, $\mathcal{K}(\mathsf{p}^-)$ and $\mathcal{K}(\mathsf{p})$ are naturally isomorphic as Coxeter spaces. On the (A, C)-component, the map is defined by

$$\begin{array}{ccc} \mathcal{K}(\mathsf{p}^-)[A,C] & \xrightarrow{\cong} & \mathcal{K}(\mathsf{p})[A,C] \\ \| & & \| \\ \mathsf{p}[A] \otimes \mathsf{E}^-[A] & & \mathsf{p}[A] \end{array}, \qquad x \otimes \mathtt{H}_{[C/A]} \mapsto x. \tag{14.25}$$

Note very carefully the choice of the basis element $\mathtt{H}_{[C/A]} \in \mathsf{E}^-[A]$.

PROOF. We need to check diagrams (8.21) with $\mathrm{V} := \mathcal{K}(\mathsf{p}^-)$ and $\mathrm{W} := \mathcal{K}(\mathsf{p})$. This can be done using (14.3) as follows. For $x \in \mathsf{p}[F]$,

$$\begin{array}{ccc} \mathcal{K}(\mathsf{p}^-)[F,C] & \xrightarrow{\cong} & \mathcal{K}(\mathsf{p})[F,C] \\ {\scriptstyle\beta_{G,D,F,C}}\downarrow & & \downarrow{\scriptstyle\beta_{G,D,F,C}} \\ \mathcal{K}(\mathsf{p}^-)[G,D] & \xrightarrow[\cong]{} & \mathcal{K}(\mathsf{p})[G,D] \end{array} \qquad \begin{array}{ccc} x \otimes \mathtt{H}_{[C/F]} & \longmapsto & x \\ \downarrow & & \downarrow \\ \beta_{G,F}(x) \otimes \mathtt{H}_{[D/G]} & \longmapsto & \beta_{G,F}(x) \end{array}$$

$$\begin{array}{ccc} \mathcal{K}(\mathsf{p}^-)[F,C] & \xrightarrow{\cong} & \mathcal{K}(\mathsf{p})[F,C] \\ {\scriptstyle w}\downarrow & & \downarrow{\scriptstyle w} \\ \mathcal{K}(\mathsf{p}^-)[F',C'] & \xrightarrow[\cong]{} & \mathcal{K}(\mathsf{p})[F',C'] \end{array} \qquad \begin{array}{ccc} x \otimes \mathtt{H}_{[C/F]} & \longmapsto & x \\ \downarrow & & \downarrow \\ w(x) \otimes \mathtt{H}_{[C'/F']} & \longmapsto & w(x). \end{array}$$

Naturality in p is easy to see; we use (14.4). □

We now upgrade Lemma 14.11 as follows.

Proposition 14.12. *For scalars p and q, the diagram*

$$\begin{array}{ccc} (\mathcal{T}, \mathcal{T}^\vee, \lambda_p) & \xrightarrow{(-)^-} & (\mathcal{T}, \mathcal{T}^\vee, \lambda_{-p}) \\ & {\scriptstyle\mathcal{K}_{-q}}\searrow \quad \swarrow{\scriptstyle\mathcal{K}_q} & \\ & (\overline{\mathcal{T}}, \overline{\mathcal{T}}^\vee, \overline{\lambda}_{-pq}) & \end{array}$$

commutes (up to isomorphism) of bilax functors. The same result holds with $\mathcal{K}_q$ replaced by $\mathcal{K}_q^\vee$.

We may write this in short as

$$\mathcal{K}_q((-)^-) \xrightarrow{\cong} \mathcal{K}_{-q}(-). \tag{14.26}$$

PROOF. We need to check that this isomorphism respects lax and colax structures, see diagrams (A.3) and (A.6). The lax part is straightforward (with q

playing no role). For the colax part, we need to check that the diagram

$$\begin{array}{ccc} \mathcal{K}_q(\mathcal{T}^\vee(\mathsf{p})^-) & \xrightarrow{\text{colax}} & \overline{\mathcal{T}}^\vee(\mathcal{K}_q(\mathsf{p}^-)) \\ \cong\downarrow & & \downarrow\cong \\ \mathcal{K}_{-q}(\mathcal{T}^\vee(\mathsf{p})) & \xrightarrow[\text{colax}]{} & \overline{\mathcal{T}}^\vee(\mathcal{K}_{-q}(\mathsf{p})) \end{array}$$

commutes. On the (A, C)-component, we have

$$\begin{array}{ccccc} \bigoplus_{A \le F} \mathsf{p}[F] \otimes \mathsf{E}^-[A] & \xrightarrow{\text{colax}} & \bigoplus_{A \le F} \mathsf{p}[F] \otimes \mathsf{E}^-[F] & \xrightarrow{\text{colax}} & \bigoplus_{A \le F_0 \le C} \mathsf{p}[F_0] \otimes \mathsf{E}^-[F_0] \\ \cong\downarrow & & & & \downarrow\cong \\ \bigoplus_{A \le F} \mathsf{p}[F] & & \xrightarrow[\text{colax}]{\hspace{8em}} & & \bigoplus_{A \le F_0 \le C} \mathsf{p}[F_0]. \end{array}$$

Here A and C are fixed; the sums are over the remaining faces. The first top-horizontal map is the colax structure of the o-signature functor in (4.18) (in the signature space formulation), while the second top-horizontal map is the colax structure of $\mathcal{K}_q$ in (14.21). The commutativity is checked below.

$$\begin{array}{ccccc} x \otimes \mathtt{H}_{[C/A]} & \longmapsto & (-1)^{\mathrm{dist}(C,FC)} x \otimes \mathtt{H}_{[FC/F]} & \longmapsto & (-q)^{\mathrm{dist}(C,FC)} x_0 \otimes \mathtt{H}_{[C/F_0]} \\ \big\downarrow & & & & \big\downarrow \\ x & & \longmapsto & & (-q)^{\mathrm{dist}(C,FC)}\, x_0, \end{array}$$

where $x \in \mathsf{p}[F]$ and $x_0 := \mathsf{p}[u](x)$ with $u \in W$ defined by $u(FC) = C$. This proves (14.26).

The check for $\mathcal{K}_q^\vee$ is similar. In finite-dimensional setting, the statements for $\mathcal{K}_q$ and $\mathcal{K}_q^\vee$ imply each other by taking conjugates wrt duality, and using self-duality of the o-signature functor. □

As the special case $q = 1$ of Proposition 14.12, full Fock functors and signed full Fock functors are obtained from each other by precomposing with the o-signature functor. Another way of proving Proposition 14.12 is given later in Exercise 14.25, item (1).

14.3.2. Under version. Recall the u-signature functors (2.70) and (8.99) on Coxeter species and on Coxeter spaces, both denoted $(-)_-$. They carry bilax structures via canonical identifications $\mathcal{T}(\mathsf{p}_-) = \mathcal{T}(\mathsf{p})_-$ and $\overline{\mathcal{T}}(\mathrm{V}_-) = \overline{\mathcal{T}}(\mathrm{V})_-$.

Exercise 14.13. Check: For scalars p and q, the diagram

$$\begin{array}{ccc} (\mathcal{T}, \mathcal{T}^\vee, \lambda_p) & \xrightarrow{(-)_-} & (\mathcal{T}, \mathcal{T}^\vee, \lambda_{-p}) \\ \mathcal{K}_q\downarrow & & \downarrow\mathcal{K}_q \\ (\overline{\mathcal{T}}, \overline{\mathcal{T}}^\vee, \overline{\lambda}_{pq}) & \xrightarrow[(-)_-]{} & (\overline{\mathcal{T}}, \overline{\mathcal{T}}^\vee, \overline{\lambda}_{-pq}) \end{array}$$

commutes (up to isomorphism) of bilax functors. The same result holds with $\mathcal{K}_q$ replaced by $\mathcal{K}_q^\vee$.

A related result is given later in Exercise 14.18. In fact, this result can be used to derive Exercise 14.13, see Exercise 14.25, item (2).

14.4. Bosonic and fermionic Fock functors

We now introduce bosonic Fock functors and fermionic Fock functors. They are bilax, and moreover, their structure maps are isomorphisms. They are built out of full Fock functors and signed full Fock functors, respectively. They also come in two flavors each. The bosonic ones are $\overline{\mathcal{K}}$ and $\overline{\mathcal{K}}^\vee$, while fermionic ones are $\overline{\mathcal{K}}_{-1}$ and $\overline{\mathcal{K}}^\vee_{-1}$. The functor $\overline{\mathcal{K}}$ is constructed from $\mathcal{K}$ by taking coinvariants wrt the subgroups W_A of W. Dually, $\overline{\mathcal{K}}^\vee$ is constructed from $\mathcal{K}^\vee$ by taking invariants instead of coinvariants. In a similar manner, $\overline{\mathcal{K}}_{-1}$ and $\overline{\mathcal{K}}^\vee_{-1}$ are constructed from $\mathcal{K}_{-1}$ and $\mathcal{K}^\vee_{-1}$, the difference is that (co)invariants are taken by twisting the action of the subgroup W_A by its sign representation. Bosonic and fermionic Fock functors transform into each other by precomposing by the o-signature functor on Coxeter species.

14.4.1. Bosonic Fock functor (coinvariant version). Define

$$\overline{\mathcal{K}} : \mathcal{A}\text{-}\mathsf{CSp} \to \mathcal{A}\text{-}\overline{\mathsf{CSp}}, \qquad \overline{\mathcal{K}}(\mathsf{p})[A, C] := \mathsf{p}[A]_{W_A} \tag{14.27}$$

for p a Coxeter species. (Recall from Section 2.3.7 that the subgroup W_A acts on $\mathsf{p}[A]$ via type morphisms.) Suppose $\sigma : (A, C) \to (B, D)$ is a morphism in $\mathcal{A}\text{-}\overline{\mathsf{Cox}}$. The functor (14.1) induces a morphism $A \to B$ which we may again denote σ. By (2.24), σ induces a linear map

$$\mathsf{p}[A]_{W_A} \to \mathsf{p}[B]_{W_B}.$$

This turns $\overline{\mathcal{K}}(\mathsf{p})$ into a Coxeter space. Now suppose $f : \mathsf{p} \to \mathsf{q}$ is a map of Coxeter species. Then, for a face A, by (2.25), the component f_A induces a linear map

$$\mathsf{p}[A]_{W_A} \to \mathsf{q}[A]_{W_A}.$$

This yields a map $\overline{\mathcal{K}}(\mathsf{p}) \to \overline{\mathcal{K}}(\mathsf{q})$ of Coxeter spaces. Thus $\overline{\mathcal{K}}$ is a functor. It is $\Bbbk$-linear. We refer to $\overline{\mathcal{K}}$ as the *bosonic Fock functor* (coinvariant version). It also follows that the quotient maps

$$\mathcal{K}(\mathsf{p}) \twoheadrightarrow \overline{\mathcal{K}}(\mathsf{p}), \qquad \mathsf{p}[A] \twoheadrightarrow \mathsf{p}[A]_{W_A}$$

on the A-component, define a natural transformation $\mathcal{K} \twoheadrightarrow \overline{\mathcal{K}}$.

Recall from (14.5) and (14.6) that $\mathcal{K}$ is equipped with lax and colax structures φ and ψ. Define similar structure maps $\overline{\varphi}$ and $\overline{\psi}$ for $\overline{\mathcal{K}}$ by commutativity of the diagrams below.

$$\begin{array}{ccc} \overline{\mathcal{T}}\mathcal{K} & \xhookrightarrow{\varphi} & \mathcal{K}\mathcal{T} \\ \big\downarrow & & \big\downarrow \\ \overline{\mathcal{T}}\overline{\mathcal{K}} & \xrightarrow[\overline{\varphi}]{} & \overline{\mathcal{K}}\mathcal{T} \end{array} \qquad \begin{array}{ccc} \displaystyle\bigoplus_{F_0 \colon A \leq F_0 \leq C} \mathsf{p}[F_0] & \hookrightarrow & \displaystyle\bigoplus_{F \colon A \leq F} \mathsf{p}[F] \\ \big\downarrow & & \big\downarrow \\ \displaystyle\bigoplus_{F_0 \colon A \leq F_0 \leq C} \mathsf{p}[F_0]_{W_{F_0}} & \dashrightarrow & \Big(\displaystyle\bigoplus_{F \colon A \leq F} \mathsf{p}[F]\Big)_{W_A} \end{array} \tag{14.28}$$

$$
(14.29) \qquad
\begin{array}{ccc}
\mathcal{K}\mathcal{T}^\vee & \xrightarrow{\psi} & \mathcal{T}^\vee\mathcal{K} \\
\downarrow & & \downarrow \\
\overline{\mathcal{K}}\mathcal{T}^\vee & \xrightarrow[\overline{\psi}]{} & \overline{\mathcal{T}}^\vee\overline{\mathcal{K}}
\end{array}
\qquad
\begin{array}{ccc}
\bigoplus_{F\colon A\leq F} \mathsf{p}[F] & \longrightarrow & \bigoplus_{F_0\colon A\leq F_0\leq C} \mathsf{p}[F_0] \\
\downarrow & & \downarrow \\
\Big(\bigoplus_{F\colon A\leq F} \mathsf{p}[F]\Big)_{W_A} & \dashrightarrow & \bigoplus_{F_0\colon A\leq F_0\leq C} \mathsf{p}[F_0]_{W_{F_0}}
\end{array}
$$

On the right, we evaluated on a Coxeter species p on the (A, C)-component. The vertical maps are all surjections. First note that $\overline{\varphi}$ and $\overline{\psi}$ are well-defined: This is obvious for $\overline{\varphi}$ since W_{F_0} is a subgroup of W_A. For $\overline{\psi}$, the essential observation is that for any $w \in W_A$ which takes F to say G, there is a $w_0 \in W_{F_0}$ such that the diagram

$$
\begin{array}{ccc}
F & \xrightarrow{w} & G \\
{\scriptstyle u(FC)=C}\downarrow & & \downarrow{\scriptstyle v(GC)=C} \\
F_0 & \xrightarrow[w_0]{} & F_0
\end{array}
$$

commutes, namely, $w_0 = vwu^{-1}$.

Theorem 14.14. *For a scalar p, the bosonic Fock functor*

$$
(\overline{\mathcal{K}}, \overline{\varphi}, \overline{\psi}) : (\mathcal{T}, \mathcal{T}^\vee, \lambda_p) \to (\overline{\mathcal{T}}, \overline{\mathcal{T}}^\vee, \overline{\lambda}_p)
$$

is bilax. Further, $\mathcal{K} \twoheadrightarrow \overline{\mathcal{K}}$ is a morphism of bilax functors. Moreover, bilax structure maps $\overline{\varphi}$ and $\overline{\psi}$ of $\overline{\mathcal{K}}$ are isomorphisms. In fact, they are inverses of each other (noting that $\mathcal{T}$ and $\mathcal{T}^\vee$, and $\overline{\mathcal{T}}$ and $\overline{\mathcal{T}}^\vee$ are identical as functors).

Proof. The first part follows by construction (since $\mathcal{K}$ is bilax). For the second part: Since $\psi\varphi = \mathrm{id}$, we have $\overline{\psi}\overline{\varphi} = \mathrm{id}$. Following previous notations, the composite in the other direction is

$$
\varphi\psi : \bigoplus_{F\colon A\leq F} \mathsf{p}[F] \to \bigoplus_{F\colon A\leq F} \mathsf{p}[F], \qquad \mathsf{p}[F] \xrightarrow{w(FC)=C} \mathsf{p}[F_0],
$$

which is clearly identity on W_A-coinvariants. ☐

14.4.2. Bosonic Fock functor (invariant version). Define the functor

$$
(14.30) \qquad \overline{\mathcal{K}}^\vee : \mathcal{A}\text{-}\mathsf{CSp} \to \mathcal{A}\text{-}\overline{\mathsf{C}}\mathsf{Sp}, \qquad \overline{\mathcal{K}}^\vee(\mathsf{p})[A, C] := \mathsf{p}[A]^{W_A}
$$

for p a Coxeter species. It is $\Bbbk$-linear. We refer to $\overline{\mathcal{K}}^\vee$ as the *bosonic Fock functor* (invariant version). The inclusion maps

$$
\overline{\mathcal{K}}^\vee(\mathsf{p}) \hookrightarrow \mathcal{K}^\vee(\mathsf{p}), \qquad \mathsf{p}[A]^{W_A} \hookrightarrow \mathsf{p}[A]
$$

on the A-component, define a natural transformation $\overline{\mathcal{K}}^\vee \hookrightarrow \mathcal{K}^\vee$. The lax and colax structure maps $\psi^\vee$ and $\varphi^\vee$ of $\mathcal{K}^\vee$ from (14.11) and (14.13) restrict to maps $\overline{\psi}^\vee$ and $\overline{\varphi}^\vee$ for $\overline{\mathcal{K}}^\vee$ via diagrams shown below.

$$\begin{array}{ccc} \overline{\mathcal{T}}\mathcal{K}^\vee & \overset{\psi^\vee}{\hookrightarrow} & \mathcal{K}^\vee\mathcal{T} \\ \uparrow & & \uparrow \\ \overline{\mathcal{T}}\overline{\mathcal{K}}^\vee & \underset{\overline{\psi}^\vee}{\dashrightarrow} & \overline{\mathcal{K}}^\vee\mathcal{T} \end{array} \qquad \begin{array}{ccc} \bigoplus_{F_0 \colon A \le F_0 \le C} \mathsf{p}[F_0] & \hookrightarrow & \bigoplus_{F \colon A \le F} \mathsf{p}[F] \\ \uparrow & & \uparrow \\ \bigoplus_{F_0 \colon A \le F_0 \le C} \mathsf{p}[F_0]^{W_{F_0}} & \dashrightarrow & \Big(\bigoplus_{F \colon A \le F} \mathsf{p}[F]\Big)^{W_A} \end{array} \tag{14.31}$$

$$\begin{array}{ccc} \mathcal{K}^\vee\mathcal{T}^\vee & \overset{\varphi^\vee}{\twoheadrightarrow} & \overline{\mathcal{T}}^\vee\mathcal{K}^\vee \\ \uparrow & & \uparrow \\ \overline{\mathcal{K}}^\vee\mathcal{T}^\vee & \underset{\overline{\varphi}^\vee}{\dashrightarrow} & \overline{\mathcal{T}}^\vee\overline{\mathcal{K}}^\vee \end{array} \qquad \begin{array}{ccc} \bigoplus_{F \colon A \le F} \mathsf{p}[F] & \twoheadrightarrow & \bigoplus_{F_0 \colon A \le F_0 \le C} \mathsf{p}[F_0] \\ \uparrow & & \uparrow \\ \Big(\bigoplus_{F \colon A \le F} \mathsf{p}[F]\Big)^{W_A} & \dashrightarrow & \bigoplus_{F_0 \colon A \le F_0 \le C} \mathsf{p}[F_0]^{W_{F_0}} \end{array} \tag{14.32}$$

The vertical maps are all inclusions. To check that $\overline{\varphi}^\vee$ and $\overline{\psi}^\vee$ are well-defined: For $\overline{\varphi}^\vee$, if $(x_F)_{F \ge A}$ is W_A-invariant, then x_{F_0} for any $A \le F_0 \le C$ is W_{F_0}-invariant since W_{F_0} is a subgroup of W_A. For $\overline{\psi}^\vee$, suppose x_{F_0} is W_{F_0}-invariant. For any F greater than A of the same type as F_0, put $x_F := u(x_{F_0})$ with $u(C) = FC$. We need to check that $w(x_F) = x_G$ for any $w(F) = G$ (with F and G greater than A and of the same type as F_0). This follows from the diagram

$$\begin{array}{ccc} F & \xrightarrow{w} & G \\ {\scriptstyle u(C)=FC}\uparrow & & \uparrow{\scriptstyle v(C)=GC} \\ F_0 & \xrightarrow[w_0]{} & F_0, \end{array}$$

where $w_0 = v^{-1}wu$, and the fact that $w_0(x_{F_0}) = x_{F_0}$.

Theorem 14.15. *For a scalar p, the bosonic Fock functor*

$$(\overline{\mathcal{K}}^\vee, \overline{\psi}^\vee, \overline{\varphi}^\vee) : (\mathcal{T}, \mathcal{T}^\vee, \lambda_p) \to (\overline{\mathcal{T}}, \overline{\mathcal{T}}^\vee, \overline{\lambda}_p)$$

is bilax. Further, $\overline{\mathcal{K}}^\vee \hookrightarrow \mathcal{K}^\vee$ is a morphism of bilax functors. Moreover, bilax structure maps $\overline{\psi}^\vee$ and $\overline{\varphi}^\vee$ of $\overline{\mathcal{K}}^\vee$ are isomorphisms, and inverses of each other.

14.4.3. Conjugation by duality functors. Similar to the situation for $\mathcal{K}$ and $\mathcal{K}^\vee$ in Section 14.1.4, functors $\overline{\mathcal{K}}$ and $\overline{\mathcal{K}}^\vee$ are conjugates of each other wrt duality (in finite-dimensional setting). Explicitly, for a finite-dimensional Coxeter species p,

$$\overline{\varphi}^\vee_{\mathsf{p}} = (\overline{\varphi}_{\mathsf{p}^*})^* \quad \text{and} \quad \overline{\psi}^\vee_{\mathsf{p}} = (\overline{\psi}_{\mathsf{p}^*})^*$$

after using the identification $(\mathsf{p}^*)^* = \mathsf{p}$. In general, for any Coxeter species p,

$$(\overline{\varphi}_{\mathsf{p}})^* = \overline{\varphi}^\vee_{\mathsf{p}^*} \quad \text{and} \quad (\overline{\psi}_{\mathsf{p}})^* = \overline{\psi}^\vee_{\mathsf{p}^*}.$$

14.4.4. Bosonic Fock functors in terms of face-types. Building on the description in (14.14),

$$\overline{\mathcal{K}}(\mathsf{p})[T] = \mathsf{p}[F]_{W_F} \quad \text{and} \quad \overline{\mathcal{K}}^{\vee}(\mathsf{p})[T] = \mathsf{p}[F]^{W_F}, \tag{14.33}$$

where F is the face of C_r of type T. Similarly, bilax structures of $\overline{\mathcal{K}}$ and $\overline{\mathcal{K}}^{\vee}$ can be expressed in terms of face-types.

14.4.5. Fermionic Fock functors. The *fermionic Fock functors* denoted $\overline{\mathcal{K}}_{-1}$ and $\overline{\mathcal{K}}^{\vee}_{-1}$ are obtained by precomposing bosonic Fock functors $\overline{\mathcal{K}}$ and $\overline{\mathcal{K}}^{\vee}$ with the o-signature functor on Coxeter species:

$$\overline{\mathcal{K}}_{-1}(-) := \overline{\mathcal{K}}((-)^{-}) \qquad \text{and} \qquad \overline{\mathcal{K}}^{\vee}_{-1}(-) := \overline{\mathcal{K}}^{\vee}((-)^{-}). \tag{14.34}$$

They are $\Bbbk$-linear. It follows from (14.26) that there are natural transformations $\mathcal{K}_{-1} \twoheadrightarrow \overline{\mathcal{K}}_{-1}$ and $\overline{\mathcal{K}}^{\vee}_{-1} \hookrightarrow \mathcal{K}^{\vee}_{-1}$. Explicitly,

$$\overline{\mathcal{K}}_{-1}(\mathsf{p})[A, C] := (\mathsf{p}[A] \otimes \mathsf{E}^{-}[A])_{W_A} \text{ and } \overline{\mathcal{K}}^{\vee}_{-1}(\mathsf{p})[A, C] := (\mathsf{p}[A] \otimes \mathsf{E}^{-}[A])^{W_A}.$$

Recall: $\mathsf{E}^{-}[A]$ is the sign representation of W_A. In contrast to the situation for bosonic Fock functors, we now take W_A invariants and coinvariants of $\mathsf{p}[A]$ wrt the usual action twisted by the sign representation of W_A.

Theorem 14.16. *For a scalar p, the fermionic Fock functors*

$$\overline{\mathcal{K}}_{-1}, \overline{\mathcal{K}}^{\vee}_{-1} : (\mathcal{T}, \mathcal{T}^{\vee}, \lambda_p) \to (\overline{\mathcal{T}}, \overline{\mathcal{T}}^{\vee}, \overline{\lambda}_{-p})$$

are bilax. Further, $\mathcal{K}_{-1} \twoheadrightarrow \overline{\mathcal{K}}_{-1}$ and $\overline{\mathcal{K}}^{\vee}_{-1} \hookrightarrow \mathcal{K}^{\vee}_{-1}$ are morphisms of bilax functors. Moreover, bilax structure maps of $\overline{\mathcal{K}}_{-1}$ and $\overline{\mathcal{K}}^{\vee}_{-1}$ are isomorphisms.

PROOF. This follows from Theorems 14.14 and 14.15 by precomposing $\overline{\mathcal{K}}$ and $\overline{\mathcal{K}}^{\vee}$ with the o-signature functor which is bilax by Theorem 4.10. For instance, the lax structure of $\overline{\mathcal{K}}_{-1}$ is

$$\overline{\mathcal{T}}\overline{\mathcal{K}}_{-1}(\mathsf{p}) = \overline{\mathcal{T}}\overline{\mathcal{K}}(\mathsf{p}^{-}) \xrightarrow{\cong} \overline{\mathcal{K}}\mathcal{T}(\mathsf{p}^{-}) \xrightarrow{\cong} \overline{\mathcal{K}}(\mathcal{T}(\mathsf{p})^{-}) = \overline{\mathcal{K}}_{-1}\mathcal{T}(\mathsf{p}).$$

It is an isomorphism since lax structure maps of $\overline{\mathcal{K}}$ and $(-)^{-}$ are both isomorphisms. □

Remark 14.17. Instead of (14.34), one can also approach fermionic Fock functors directly similar to bosonic Fock functors, with (co)invariants replaced by signed (co)invariants. In diagrams (14.28), (14.29), (14.31), (14.32), we replace full Fock functors by signed full Fock functors, and bosonic Fock functors by fermionic Fock functors. The signed distance function on chambers enters while checking diagram commutativity.

Bosonic and fermionic Fock functors transform into each other via the o-signature functor. Their behavior wrt u-signature functors is given below.

Exercise 14.18. Check: For a scalar p, the diagrams

$$\begin{array}{ccc} (\mathcal{T}, \mathcal{T}^{\vee}, \lambda_p) & \xrightarrow{(-)_{-}} & (\mathcal{T}, \mathcal{T}^{\vee}, \lambda_{-p}) \\ {\scriptstyle \overline{\mathcal{K}}}\downarrow & & \downarrow{\scriptstyle \overline{\mathcal{K}}} \\ (\overline{\mathcal{T}}, \overline{\mathcal{T}}^{\vee}, \overline{\lambda}_p) & \xrightarrow[(-)_{-}]{} & (\overline{\mathcal{T}}, \overline{\mathcal{T}}^{\vee}, \overline{\lambda}_{-p}) \end{array} \qquad \begin{array}{ccc} (\mathcal{T}, \mathcal{T}^{\vee}, \lambda_p) & \xrightarrow{(-)_{-}} & (\mathcal{T}, \mathcal{T}^{\vee}, \lambda_{-p}) \\ {\scriptstyle \overline{\mathcal{K}}_{-1}}\downarrow & & \downarrow{\scriptstyle \overline{\mathcal{K}}_{-1}} \\ (\overline{\mathcal{T}}, \overline{\mathcal{T}}^{\vee}, \overline{\lambda}_{-p}) & \xrightarrow[(-)_{-}]{} & (\overline{\mathcal{T}}, \overline{\mathcal{T}}^{\vee}, \overline{\lambda}_p) \end{array}$$

commute (up to isomorphism) of bilax functors. The same result holds with $\overline{\mathcal{K}}$ and $\overline{\mathcal{K}}_{-1}$ replaced by $\overline{\mathcal{K}}^{\vee}$ and $\overline{\mathcal{K}}^{\vee}_{-1}$.

14.4.6. Lax structures wrt comonads and vice versa. Recall from Theorems 14.14 and 14.15 that bilax structure maps of bosonic Fock functors are isomorphisms. This leads to the following.

Proposition 14.19. *We have colax functors of monads*

$$(\overline{\mathcal{K}}, \overline{\psi}) : \mathcal{T} \to \overline{\mathcal{T}} \quad \textit{and} \quad (\overline{\mathcal{K}}^{\vee}, \overline{\varphi}^{\vee}) : \mathcal{T} \to \overline{\mathcal{T}},$$

and lax functors of comonads

$$(\overline{\mathcal{K}}, \overline{\varphi}) : \mathcal{T}^{\vee} \to \overline{\mathcal{T}}^{\vee} \quad \textit{and} \quad (\overline{\mathcal{K}}^{\vee}, \overline{\psi}^{\vee}) : \mathcal{T}^{\vee} \to \overline{\mathcal{T}}^{\vee}.$$

Proof. We explain the claim about $(\overline{\mathcal{K}}, \overline{\psi})$; the rest are similar. Start with the lax structure $\overline{\varphi}$ of $\overline{\mathcal{K}}$ wrt $\mathcal{T}$ and $\overline{\mathcal{T}}$ from (14.28). Its inverse $\overline{\psi}$ is then a colax structure wrt $\mathcal{T}$ and $\overline{\mathcal{T}}$. In other words, we have inverse isomorphisms

$$\overline{\varphi} : \overline{\mathcal{T}}\overline{\mathcal{K}} \xrightarrow{\cong} \overline{\mathcal{K}}\mathcal{T} \quad \text{and} \quad \overline{\psi} : \overline{\mathcal{K}}\mathcal{T} \xrightarrow{\cong} \overline{\mathcal{T}}\overline{\mathcal{K}}.$$

(As a map, $\overline{\psi}$ is the same as the colax structure wrt $\mathcal{T}^{\vee}$ and $\overline{\mathcal{T}}^{\vee}$ from (14.29), however, comonads are not relevant here.) □

We have a similar result for fermionic Fock functors (since their bilax structure maps are also isomorphisms by Theorem 14.16).

14.4.7. Compatibility of (co)lax structures with norm maps. For a scalar q, recall q-norm transformations $\kappa_q : \mathcal{T} \to \mathcal{T}^{\vee}$ and $\overline{\kappa}_q : \overline{\mathcal{T}} \to \overline{\mathcal{T}}^{\vee}$ from (3.143) and (9.139), respectively. We do not write $\mathcal{T}_q$ and so on since here we view them as functors from Coxeter species (or from Coxeter spaces) to itself.

Lemma 14.20. *For a scalar q, the diagrams*

$$\begin{array}{ccc} \mathcal{K}\mathcal{T} & \xrightarrow{\mathcal{K}\kappa_q} & \mathcal{K}\mathcal{T}^{\vee} \\ {\scriptstyle \varphi}\uparrow & & \downarrow{\scriptstyle \psi} \\ \overline{\mathcal{T}}\mathcal{K} & \xrightarrow[\overline{\kappa}_q\mathcal{K}]{} & \overline{\mathcal{T}}^{\vee}\mathcal{K} \end{array} \qquad\qquad \begin{array}{ccc} \mathcal{K}^{\vee}\mathcal{T} & \xrightarrow{\mathcal{K}^{\vee}\kappa_q} & \mathcal{K}^{\vee}\mathcal{T}^{\vee} \\ {\scriptstyle \psi^{\vee}}\uparrow & & \downarrow{\scriptstyle \varphi^{\vee}} \\ \overline{\mathcal{T}}\mathcal{K}^{\vee} & \xrightarrow[\overline{\kappa}_q\mathcal{K}^{\vee}]{} & \overline{\mathcal{T}}^{\vee}\mathcal{K}^{\vee} \end{array} \tag{14.35}$$

commute. (The two diagrams are conjugates of each other wrt duality.) Similarly, the diagrams

$$\begin{array}{ccc} \overline{\mathcal{K}}\mathcal{T} & \xrightarrow{\overline{\mathcal{K}}\kappa_q} & \overline{\mathcal{K}}\mathcal{T}^{\vee} \\ {\scriptstyle \overline{\varphi}}\uparrow{\scriptstyle \cong} & & {\scriptstyle \cong}\downarrow{\scriptstyle \overline{\psi}} \\ \overline{\mathcal{T}}\overline{\mathcal{K}} & \xrightarrow[\overline{\kappa}_q\overline{\mathcal{K}}]{} & \overline{\mathcal{T}}^{\vee}\overline{\mathcal{K}} \end{array} \qquad\qquad \begin{array}{ccc} \overline{\mathcal{K}}^{\vee}\mathcal{T} & \xrightarrow{\overline{\mathcal{K}}^{\vee}\kappa_q} & \overline{\mathcal{K}}^{\vee}\mathcal{T}^{\vee} \\ {\scriptstyle \overline{\psi}^{\vee}}\uparrow{\scriptstyle \cong} & & {\scriptstyle \cong}\downarrow{\scriptstyle \overline{\varphi}^{\vee}} \\ \overline{\mathcal{T}}\overline{\mathcal{K}}^{\vee} & \xrightarrow[\overline{\kappa}_q\overline{\mathcal{K}}^{\vee}]{} & \overline{\mathcal{T}}^{\vee}\overline{\mathcal{K}}^{\vee} \end{array} \tag{14.36}$$

commute.

Proof. Let us understand the first diagram in (14.35). Recall: $\overline{\kappa}_q$ involves σ, κ_q involves β, and ψ involves w. The commutativity of the diagram amounts to using the normal form $\sigma = w\beta$ (or a transfer). Similarly, for the second diagram in (14.35), we use the normal form $\sigma = \beta w$ (or a cotransfer).

Diagrams (14.36) follow from diagrams (14.35) by replacing $\mathcal{K}$ and $\mathcal{K}^\vee$ with $\overline{\mathcal{K}}$ and $\overline{\mathcal{K}}^\vee$. $\square$

Exercise 14.21. Check: Diagrams (14.35) commute more generally with ψ and $\psi^\vee$ replaced by ψ_p and $\psi_p^\vee$, respectively, and $\bar{\kappa}_q$ replaced by $\bar{\kappa}_{pq}$, where p is another scalar. By setting $p = -1$, deduce analogues of diagrams (14.36) for fermionic Fock functors by replacing $\overline{\mathcal{K}}$ and $\overline{\mathcal{K}}^\vee$ with $\overline{\mathcal{K}}_{-1}$ and $\overline{\mathcal{K}}_{-1}^\vee$, and $\bar{\kappa}_q$ with $\bar{\kappa}_{-q}$.

14.5. Full Fock functors from bosonic Fock functors

Interestingly, it is possible to express full Fock functors in terms of bosonic or fermionic Fock functors by using the Coxeter bimonoid of chambers from Section 4.3. Precise results are given below. For that recall from Theorem 3.25 that taking Hadamard product with a Coxeter bimonoid is a bilax functor.

Proposition 14.22. *For scalars p and q, there are isomorphisms of bilax functors*

$$\mathcal{K}_q(-) \cong \overline{\mathcal{K}}(\mathsf{\Gamma}_q \times (-)) \qquad \textit{and} \qquad \mathcal{K}_q^\vee(-) \cong \overline{\mathcal{K}}^\vee(\mathsf{\Gamma}_q^* \times (-))$$

from $(\mathcal{T}, \mathcal{T}^\vee, \lambda_p)$ to $(\overline{\mathcal{T}}, \overline{\mathcal{T}}^\vee, \overline{\lambda}_{pq})$.

PROOF. We explain the first isomorphism. For a Coxeter species p, define a map of Coxeter spaces $\overline{\mathcal{K}}(\mathsf{\Gamma}_q \times \mathsf{p}) \to \mathcal{K}_q(\mathsf{p})$ with (A, C)-component

$$(\mathsf{\Gamma}_q[A] \otimes \mathsf{p}[A])_{W_A} \to \mathsf{p}[A], \qquad \overline{\mathtt{H}_{C/A} \otimes x} \mapsto x.$$

The bar indicates the coinvariant class. Since $\mathsf{\Gamma}_q[A]$ is the left regular representation of W_A, this map is an isomorphism. It remains to check that the map respects lax and colax structures, see diagrams (A.3) and (A.6). For the lax part, the necessary diagram is shown below.

$$\begin{array}{ccccc}
\overline{\mathcal{T}}\overline{\mathcal{K}}(\mathsf{\Gamma}_q \times \mathsf{p})[A,C] & \longrightarrow & \overline{\mathcal{K}}\mathcal{T}(\mathsf{\Gamma}_q \times \mathsf{p})[A,C] & \longrightarrow & \overline{\mathcal{K}}(\mathsf{\Gamma}_q \times \mathcal{T}(\mathsf{p}))[A,C] \\
\| & & \| & & \| \\
\bigoplus_{A \le F_0 \le C} (\mathsf{\Gamma}_q[F_0] \otimes \mathsf{p}[F_0])_{W_{F_0}} & \to & \big(\bigoplus_{A \le F} \mathsf{\Gamma}_q[F] \otimes \mathsf{p}[F]\big)_{W_A} & \to & \big(\mathsf{\Gamma}_q[A] \otimes \bigoplus_{A \le F} \mathsf{p}[F]\big)_{W_A} \\
\downarrow & & & & \downarrow \\
\bigoplus_{A \le F_0 \le C} \mathsf{p}[F_0] & & \longrightarrow & & \bigoplus_{A \le F} \mathsf{p}[F] \\
\| & & & & \| \\
\overline{\mathcal{T}}\mathcal{K}_q(\mathsf{p})[A,C] & & \longrightarrow & & \mathcal{K}_q\mathcal{T}(\mathsf{p})[A,C]
\end{array}$$

Here A and C are fixed; the sums are over the remaining faces. Its commutativity is checked below. For $x \in \mathsf{p}[F_0]$,

$$\begin{array}{ccccc}
\overline{\mathtt{H}_{C/F_0} \otimes x} & \longmapsto & \overline{\mathtt{H}_{C/F_0} \otimes x} & \longmapsto & \overline{\mathtt{H}_{C/A} \otimes x} \\
\downarrow & & & & \downarrow \\
x & & \longmapsto & & x.
\end{array}$$

For the colax part, there is a similar diagram with arrows reversed, with $\mathcal{T}$ and $\overline{\mathcal{T}}$ replaced by $\mathcal{T}^\vee$ and $\overline{\mathcal{T}}^\vee$. Its commutativity is checked below. For $x \in \mathsf{p}[F]$,

$$\begin{array}{ccccc} q^{\operatorname{dist}(C,FC)}\overline{\mathtt{H}_{C/F_0} \otimes w(x)} & \longleftarrow & q^{\operatorname{dist}(C,FC)}\overline{\mathtt{H}_{FC/F} \otimes x} & \longleftarrow & \overline{\mathtt{H}_{C/A} \otimes x} \\ \downarrow & & & & \downarrow \\ q^{\operatorname{dist}(C,FC)} w(x) & & \longleftarrow & & x. \end{array}$$

Here w is the Coxeter symmetry which sends FC to C. □

Exercise 14.23. This pertains to the second isomorphism in Proposition 14.22. Check: For a Coxeter species p, the map of Coxeter spaces $\mathcal{K}_q^\vee(\mathsf{p}) \to \overline{\mathcal{K}}^\vee(\Gamma_q^* \times \mathsf{p})$ with (A, C)-component

$$\mathsf{p}[A] \to (\Gamma_q[A]^* \otimes \mathsf{p}[A])^{W_A}, \qquad x \mapsto \sum_{w \in W_A} \mathtt{M}_{wC/A} \otimes w \cdot x$$

is an isomorphism, and respects lax and colax structures.

Proposition 14.24. *For scalars p and q, there are isomorphisms of bilax functors*

$$\mathcal{K}_q(-) \cong \overline{\mathcal{K}}_{-1}(\Gamma_{-q} \times (-)) \quad \textit{and} \quad \mathcal{K}_q^\vee(-) \cong \overline{\mathcal{K}}_{-1}^\vee(\Gamma_{-q}^* \times (-))$$

from $(\mathcal{T}, \mathcal{T}^\vee, \lambda_p)$ to $(\overline{\mathcal{T}}, \overline{\mathcal{T}}^\vee, \overline{\lambda}_{pq})$.

PROOF. This follows from Proposition 14.22 by employing the o-signature functor and using (14.34) and Exercise 4.12. □

Exercise 14.25. Proposition 14.22 can be used to deduce facts about q-full Fock functors from those about bosonic Fock functors and the Coxeter q-bimonoid of chambers. For instance:

(1) Use Exercise 4.12 to deduce Proposition 14.12.
(2) Use the first diagram in Exercise 14.18 to deduce Exercise 14.13.

14.6. Fock functors and commutativity

We continue the study of bilax properties of Fock functors, now focussing on commutative counterparts of monads $\mathcal{T}$ and $\overline{\mathcal{T}}$ and their duals, namely, $\mathcal{S}$ and $\overline{\mathcal{S}}$ and their duals. For this, we employ the opposite transformations on Coxeter species and on Coxeter spaces. Among the two full Fock functors, $\mathcal{K}$ carries an induced colax structure wrt $\mathcal{S}^\vee$ and $\overline{\mathcal{S}}^\vee$, while $\mathcal{K}^\vee$ carries an induced lax structure wrt $\mathcal{S}$ and $\overline{\mathcal{S}}$. In contrast, both bosonic Fock functors $\overline{\mathcal{K}}$ and $\overline{\mathcal{K}}^\vee$ carry induced bilax structures. Similar facts hold wrt $\mathcal{E}$ and $\overline{\mathcal{E}}$ and their duals. There are also similar results for signed full Fock functors and fermionic Fock functors; in this case, $\mathcal{S}$ and $\overline{\mathcal{E}}$ go together, and so on. These results can be understood uniformly by employing the deformed opposite transformations.

We also introduce the half-twist transformation θ on the full Fock functor $\mathcal{K}$. Unlike $\mathcal{K}^\vee$, the lax structure on $\mathcal{K}$ does not commute with the opposite transformations. The point is that one can still formulate a weaker form of commutativity by using θ.

14.6.1. Opposite transformations. We briefly recall the opposite transformations τ and $\bar{\tau}$ discussed in Sections 2.11 and 8.13.

For a Coxeter species p and faces $F \geq A$, consider the isomorphism

$$\beta_{A\overline{F},F} : \mathsf{p}[F] \to \mathsf{p}[A\overline{F}].$$

Summing over F with A fixed yields a linear map $\mathcal{T}(\mathsf{p})[A] \to \mathcal{T}(\mathsf{p})[A]$. This is natural in p, so we obtain a natural transformation $\mathcal{T} \to \mathcal{T}$ which is τ. Since $\mathcal{T} = \mathcal{T}^\vee$ as functors, we may also write $\mathcal{T}^\vee$ instead of $\mathcal{T}$. More generally, for a scalar q, replacing β by β_q yields $\tau_q : \mathcal{T} \to \mathcal{T}$. For $q \neq 0$, τ_q is invertible, and $(\tau_q)^{-1} = \tau_{q^{-1}}$. In particular, $\tau^{-1} = \tau$.

For a Coxeter space V and face-types $T \geq Z$, consider the isomorphism

$$\beta_{Z\overline{T},T} : \mathrm{V}[T] \to \mathrm{V}[Z\overline{T}].$$

Summing over T with Z fixed yields a linear map $\overline{\mathcal{T}}(\mathrm{V})[Z] \to \overline{\mathcal{T}}(\mathrm{V})[Z]$. (For the top-nested faces formulation, see Exercise 8.126.) This is natural in V, so we obtain a natural transformation $\overline{\mathcal{T}} \to \overline{\mathcal{T}}$ which is $\bar{\tau}$. Since $\overline{\mathcal{T}} = \overline{\mathcal{T}}^\vee$ as functors, we may also write $\overline{\mathcal{T}}^\vee$ instead of $\overline{\mathcal{T}}$. More generally, for a scalar q, replacing β by β_q yields $\bar{\tau}_q : \overline{\mathcal{T}} \to \overline{\mathcal{T}}$. For $q \neq 0$, $\bar{\tau}_q$ is invertible, and $(\bar{\tau}_q)^{-1} = \bar{\tau}_{q^{-1}}$. In particular, $\bar{\tau}^{-1} = \bar{\tau}$.

14.6.2. Full Fock functors and commutativity. Recall: The monad $\mathcal{S}$ can be constructed from the monad $\mathcal{T}$, and the comonad $\mathcal{S}^\vee$ from the comonad $\mathcal{T}^\vee$ by equalizing τ with the identity. The same statements hold for $\overline{\mathcal{S}}$ and $\overline{\mathcal{S}}^\vee$ wrt $\overline{\mathcal{T}}$, $\overline{\mathcal{T}}^\vee$ and $\bar{\tau}$. We have seen how Fock functors interact with the $\mathcal{T}$s. So to understand their interaction with the $\mathcal{S}$s, we need to understand how they interact with the τs.

Proposition 14.26. *The colax structure ψ of $\mathcal{K}$ in* (14.6) *commutes with τ and $\bar{\tau}$. That is, the diagram*

$$\begin{array}{ccc} \mathcal{K}\mathcal{T}^\vee & \xrightarrow{\mathcal{K}\tau} & \mathcal{K}\mathcal{T}^\vee \\ {\scriptstyle\psi}\downarrow & & \downarrow{\scriptstyle\psi} \\ \overline{\mathcal{T}}^\vee\mathcal{K} & \xrightarrow[\bar{\tau}\mathcal{K}]{} & \overline{\mathcal{T}}^\vee\mathcal{K} \end{array} \tag{14.37}$$

commutes.

PROOF. Evaluate diagram (14.37) on a Coxeter species p on the (A, C)-component. All morphisms carry a summand isomorphically to exactly one summand. Restricting to the relevant summands, the diagram reduces to

$$\begin{array}{ccccc} \mathsf{p}[F] & & \xrightarrow{\beta_{A\overline{F},F}} & & \mathsf{p}[A\overline{F}] \\ {\scriptstyle u(FC)=C}\downarrow & & \swarrow{\scriptstyle u(A\overline{F}C)=A\overline{F_0}C} & & \downarrow{\scriptstyle w(A\overline{F}C)=C} \\ \mathsf{p}[F_0] & \xrightarrow[\beta_{A\overline{F_0},F_0}]{} & \mathsf{p}[A\overline{F_0}] & \xrightarrow[v(A\overline{F_0}C)=C]{} & \mathsf{p}[F_0']. \end{array}$$

Here F is a face greater than A, F_0 is the face of C of the same type as F, and F_0' is the face of C of the same type as $A\overline{F}$ or $A\overline{F_0}$. The square commutes by the support-type axiom (2.14). The triangle commutes by the type-type axiom in (2.12a) since $w = vu$ with both taking $A\overline{F}C$ to C. So the above diagram commutes as required. □

This yields the commutative diagram

$$\begin{array}{ccccc} (\mathcal{A}\text{-}\mathtt{CSp}, \mathcal{S}^\vee) & \longrightarrow & (\mathcal{A}\text{-}\mathtt{CSp}, \mathcal{T}^\vee) & \overset{(\mathrm{id},\tau)}{\underset{(\mathrm{id},\mathrm{id})}{\rightrightarrows}} & (\mathcal{A}\text{-}\mathtt{CSp}, \mathcal{T}^\vee) \\ {\scriptstyle(\mathcal{K},\psi)}\downarrow & & {\scriptstyle(\mathcal{K},\psi)}\downarrow & & \downarrow{\scriptstyle(\mathcal{K},\psi)} \\ (\mathcal{A}\text{-}\overline{\mathtt{C}}\mathtt{Sp}, \overline{\mathcal{S}}^\vee) & \longrightarrow & (\mathcal{A}\text{-}\overline{\mathtt{C}}\mathtt{Sp}, \overline{\mathcal{T}}^\vee) & \overset{(\mathrm{id},\overline{\tau})}{\underset{(\mathrm{id},\mathrm{id})}{\rightrightarrows}} & (\mathcal{A}\text{-}\overline{\mathtt{C}}\mathtt{Sp}, \overline{\mathcal{T}}^\vee) \end{array}$$

in the category of comonads. In each arrow label, the first entry is the functor and second entry is its colax structure. We have an induced colax structure $\psi : \mathcal{K}\mathcal{S}^\vee \to \overline{\mathcal{S}}^\vee \mathcal{K}$ fitting into the commutative diagram

$$\begin{array}{ccc} \mathcal{K}\mathcal{T}^\vee & \xrightarrow{\psi} & \overline{\mathcal{T}}^\vee \mathcal{K} \\ \uparrow & & \uparrow \\ \mathcal{K}\mathcal{S}^\vee & \underset{\psi}{\dashrightarrow} & \overline{\mathcal{S}}^\vee \mathcal{K}. \end{array} \tag{14.38}$$

It also follows that $(\mathcal{K}, \varphi, \psi)$ is bilax from $(\mathcal{T}, \mathcal{S}^\vee, \lambda)$ to $(\overline{\mathcal{T}}, \overline{\mathcal{S}}^\vee, \overline{\lambda})$ fitting into the commutative diagram

$$\begin{array}{ccccc} (\mathcal{T}, \mathcal{S}^\vee, \lambda) & \longrightarrow & (\mathcal{T}, \mathcal{T}^\vee, \lambda) & \overset{(\mathrm{id},\mathrm{id},\tau)}{\underset{(\mathrm{id},\mathrm{id},\mathrm{id})}{\rightrightarrows}} & (\mathcal{T}, \mathcal{T}^\vee, \lambda) \\ {\scriptstyle(\mathcal{K},\varphi,\psi)}\downarrow & & {\scriptstyle(\mathcal{K},\varphi,\psi)}\downarrow & & \downarrow{\scriptstyle(\mathcal{K},\varphi,\psi)} \\ (\overline{\mathcal{T}}, \overline{\mathcal{S}}^\vee, \overline{\lambda}) & \longrightarrow & (\overline{\mathcal{T}}, \overline{\mathcal{T}}^\vee, \overline{\lambda}) & \overset{(\mathrm{id},\mathrm{id},\overline{\tau})}{\underset{(\mathrm{id},\mathrm{id},\mathrm{id})}{\rightrightarrows}} & (\overline{\mathcal{T}}, \overline{\mathcal{T}}^\vee, \overline{\lambda}) \end{array}$$

in the category of bimonads. See Propositions 2.82 and 8.125 in this regard.

Exercise 14.27. Check the existence of diagram (14.38) directly working with (2.96) and (8.124).

Proposition 14.28. *The lax structure $\psi^\vee$ of $\mathcal{K}^\vee$ in* (14.11) *commutes with τ and $\overline{\tau}$. That is, the diagram*

$$\begin{array}{ccc} \overline{\mathcal{T}}\mathcal{K}^\vee & \xrightarrow{\overline{\tau}\mathcal{K}^\vee} & \overline{\mathcal{T}}\mathcal{K}^\vee \\ {\scriptstyle\psi^\vee}\downarrow & & \downarrow{\scriptstyle\psi^\vee} \\ \mathcal{K}^\vee\mathcal{T} & \underset{\mathcal{K}^\vee\tau}{\longrightarrow} & \mathcal{K}^\vee\mathcal{T} \end{array} \tag{14.39}$$

commutes.

This is the dual of Proposition 14.26. It implies the existence of a lax structure $\psi^\vee : \overline{\mathcal{S}}\mathcal{K}^\vee \to \mathcal{K}^\vee\mathcal{S}$ fitting into the commutative diagram

$$\begin{array}{ccc} \overline{\mathcal{T}}\mathcal{K}^\vee & \xrightarrow{\ \psi^\vee\ } & \mathcal{K}^\vee\mathcal{T} \\ \big\downarrow & & \big\downarrow \\ \overline{\mathcal{S}}\mathcal{K}^\vee & \underset{\psi^\vee}{\dashrightarrow} & \mathcal{K}^\vee\mathcal{S}. \end{array} \tag{14.40}$$

It also follows that $(\mathcal{K}^\vee, \psi^\vee, \varphi^\vee)$ is bilax from $(\mathcal{S}, \mathcal{T}^\vee, \lambda)$ to $(\overline{\mathcal{S}}, \overline{\mathcal{T}}^\vee, \overline{\lambda})$.

14.6.3. Bosonic Fock functors and commutativity. In contrast to diagram (14.39), the diagram

$$\begin{array}{ccc} \overline{\mathcal{T}}\mathcal{K} & \xrightarrow{\ \overline{\tau}\mathcal{K}\ } & \overline{\mathcal{T}}\mathcal{K} \\ {\scriptstyle\varphi}\big\downarrow & & \big\downarrow{\scriptstyle\varphi} \\ \mathcal{K}\mathcal{T} & \xrightarrow[\ \mathcal{K}\tau\ ]{} & \mathcal{K}\mathcal{T} \end{array} \tag{14.41}$$

does *not* commute: Evaluating the above diagram on a Coxeter species p on the (A, C)-component, the summand $\mathsf{p}[F_0]$ with $A \leq F_0 \leq C$, going in one direction maps to $\mathsf{p}[F_0']$ for some $A \leq F_0' \leq C$, while in the other direction it maps to $\mathsf{p}[A\overline{F_0}]$. Note: $A\overline{F_0}$ is a face of $A\overline{C}$. So the diagram cannot possibly commute in general. However, this problem disappears if we pass to coinvariants leading to the following result.

Proposition 14.29. *The lax and colax structures of* $\overline{\mathcal{K}}$ *in* (14.28) *and* (14.29) *commute with* τ *and* $\overline{\tau}$. *That is, the diagrams*

$$\begin{array}{ccc} \overline{\mathcal{T}}\overline{\mathcal{K}} & \xrightarrow{\ \overline{\tau}\overline{\mathcal{K}}\ } & \overline{\mathcal{T}}\overline{\mathcal{K}} \\ {\scriptstyle\overline{\varphi}}\big\downarrow & & \big\downarrow{\scriptstyle\overline{\varphi}} \\ \overline{\mathcal{K}}\mathcal{T} & \xrightarrow[\ \overline{\mathcal{K}}\tau\ ]{} & \overline{\mathcal{K}}\mathcal{T} \end{array} \qquad\qquad \begin{array}{ccc} \overline{\mathcal{K}}\mathcal{T}^\vee & \xrightarrow{\ \overline{\mathcal{K}}\tau\ } & \overline{\mathcal{K}}\mathcal{T}^\vee \\ {\scriptstyle\overline{\psi}}\big\downarrow & & \big\downarrow{\scriptstyle\overline{\psi}} \\ \overline{\mathcal{T}}^\vee\overline{\mathcal{K}} & \xrightarrow[\ \overline{\tau}\overline{\mathcal{K}}\ ]{} & \overline{\mathcal{T}}^\vee\overline{\mathcal{K}} \end{array} \tag{14.42}$$

commute. Similarly, the lax and colax structures of $\overline{\mathcal{K}}^\vee$ *in* (14.31) *and* (14.32) *commute with* τ *and* $\overline{\tau}$.

PROOF. The first diagram is induced from (14.41) and second diagram from (14.37). □

Thus, we have an induced lax structure $\overline{\varphi} : \overline{\mathcal{S}}\overline{\mathcal{K}} \to \overline{\mathcal{K}}\mathcal{S}$ and colax structure $\overline{\psi} : \overline{\mathcal{K}}\mathcal{S}^\vee \to \overline{\mathcal{S}}^\vee\overline{\mathcal{K}}$ fitting into commutative diagrams

$$\begin{array}{ccc} \overline{\mathcal{T}}\overline{\mathcal{K}} & \xrightarrow{\ \overline{\varphi}\ } & \overline{\mathcal{K}}\mathcal{T} \\ \big\downarrow & & \big\downarrow \\ \overline{\mathcal{S}}\overline{\mathcal{K}} & \underset{\overline{\varphi}}{\dashrightarrow} & \overline{\mathcal{K}}\mathcal{S} \end{array} \qquad\qquad \begin{array}{ccc} \overline{\mathcal{K}}\mathcal{T}^\vee & \xrightarrow{\ \overline{\psi}\ } & \overline{\mathcal{T}}^\vee\overline{\mathcal{K}} \\ \big\uparrow & & \big\uparrow \\ \overline{\mathcal{K}}\mathcal{S}^\vee & \underset{\overline{\psi}}{\dashrightarrow} & \overline{\mathcal{S}}^\vee\overline{\mathcal{K}}. \end{array} \tag{14.43}$$

There are similar diagrams for $\overline{\mathcal{K}}^\vee$, namely,

$$(14.44)\qquad \begin{array}{ccc} \overline{\mathcal{T}}\overline{\mathcal{K}}^\vee & \xrightarrow{\overline{\psi}^\vee} & \overline{\mathcal{K}}^\vee\mathcal{T} \\ \downarrow & & \downarrow \\ \overline{\mathcal{S}}\overline{\mathcal{K}}^\vee & \underset{\overline{\psi}^\vee}{\dashrightarrow} & \overline{\mathcal{K}}^\vee\mathcal{S} \end{array} \qquad\qquad \begin{array}{ccc} \overline{\mathcal{K}}^\vee\mathcal{T}^\vee & \xrightarrow{\overline{\varphi}^\vee} & \overline{\mathcal{T}}^\vee\overline{\mathcal{K}}^\vee \\ \uparrow & & \uparrow \\ \overline{\mathcal{K}}^\vee\mathcal{S}^\vee & \underset{\overline{\varphi}^\vee}{\dashrightarrow} & \overline{\mathcal{S}}^\vee\overline{\mathcal{K}}^\vee. \end{array}$$

Theorem 14.30. *Bosonic Fock functors $\overline{\mathcal{K}}$ and $\overline{\mathcal{K}}^\vee$ are bilax wrt bimonads*

$$(\mathcal{T},\mathcal{T}^\vee,\lambda) \to (\overline{\mathcal{T}},\overline{\mathcal{T}}^\vee,\overline{\lambda}), \qquad (\mathcal{T},\mathcal{S}^\vee,\lambda) \to (\overline{\mathcal{T}},\overline{\mathcal{S}}^\vee,\overline{\lambda}),$$
$$(\mathcal{S},\mathcal{T}^\vee,\lambda) \to (\overline{\mathcal{S}},\overline{\mathcal{T}}^\vee,\overline{\lambda}), \qquad (\mathcal{S},\mathcal{S}^\vee,\lambda) \to (\overline{\mathcal{S}},\overline{\mathcal{S}}^\vee,\overline{\lambda}).$$

They connect the first diagram in (2.100) *to the first diagram in* (8.127).

Moreover, structure maps $\overline{\varphi} : \overline{\mathcal{S}}\overline{\mathcal{K}} \to \overline{\mathcal{K}}\mathcal{S}$ and $\overline{\varphi}^\vee : \overline{\mathcal{K}}^\vee\mathcal{S}^\vee \to \overline{\mathcal{S}}^\vee\overline{\mathcal{K}}^\vee$ are isomorphisms. If $\mathrm{char}(\Bbbk) \nmid |W|$, *then $\overline{\psi} : \overline{\mathcal{K}}\mathcal{S}^\vee \to \overline{\mathcal{S}}^\vee\overline{\mathcal{K}}$ and $\overline{\psi}^\vee : \overline{\mathcal{S}}\overline{\mathcal{K}}^\vee \to \overline{\mathcal{K}}^\vee\mathcal{S}$ are also isomorphisms.*

PROOF. The first part follows from the preceding discussion. For the second part, a direct way to proceed is to write down the inverse isomorphisms. We leave the details as an exercise. For $\overline{\varphi}$ and $\overline{\varphi}^\vee$, an alternative way to proceed is given later in Exercises 16.25 and 16.33. □

Exercise 14.31. Improve the second part of Theorem 14.30 as follows. If $\mathrm{char}(\Bbbk) \nmid |W|$, then $\overline{\varphi}$ and $\overline{\psi}$ are inverses of each other when we identify $\mathcal{S}$ and $\mathcal{S}^\vee$, and $\overline{\mathcal{S}}$ and $\overline{\mathcal{S}}^\vee$ via the norm transformations κ_1 and $\overline{\kappa}_1$. The same remark applies to $\overline{\varphi}^\vee$ and $\overline{\psi}^\vee$. (Put $q = 1$ in diagram (14.36) and use the left diagrams in (3.145) and (9.141).)

14.6.4. Deformed commutativity. So far we have worked with opposite transformations. We can work more generally with deformed opposite transformations. Full Fock functors interact with them as follows.

Proposition 14.32. *For nonzero scalars p and q, the diagrams*

$$(14.45)\qquad \begin{array}{ccc} \mathcal{K}\mathcal{T}^\vee & \xrightarrow{\mathcal{K}\tau_p} & \mathcal{K}\mathcal{T}^\vee \\ {\scriptstyle \psi_{q^{-1}}}\downarrow & & \downarrow{\scriptstyle \psi_q} \\ \overline{\mathcal{T}}^\vee\mathcal{K} & \underset{\overline{\tau}_{pq}\mathcal{K}}{\longrightarrow} & \overline{\mathcal{T}}^\vee\mathcal{K} \end{array} \qquad\qquad \begin{array}{ccc} \overline{\mathcal{T}}\mathcal{K}^\vee & \xrightarrow{\overline{\tau}_{pq}\mathcal{K}^\vee} & \overline{\mathcal{T}}\mathcal{K}^\vee \\ {\scriptstyle \psi_q^\vee}\downarrow & & \downarrow{\scriptstyle \psi_{q^{-1}}^\vee} \\ \mathcal{K}^\vee\mathcal{T} & \underset{\mathcal{K}^\vee\tau_p}{\longrightarrow} & \mathcal{K}^\vee\mathcal{T} \end{array}$$

commute.

PROOF. We explain the first diagram. Continuing the proof of Proposition 14.26, we need to check in addition that

$$p^{\mathrm{dist}(F,A\overline{F})} q^{\mathrm{dist}(A\overline{F}C,C)} = q^{-\mathrm{dist}(FC,C)} (pq)^{\mathrm{dist}(F_0,A\overline{F_0})}.$$

Since F and F_0 have the same type, $\mathrm{dist}(F, A\overline{F}) = \mathrm{dist}(F_0, A\overline{F_0})$, so powers of p are equal. For powers of q, we use the identity

$$\mathrm{dist}(F, A\overline{F}) = \mathrm{dist}(FC, C) + \mathrm{dist}(A\overline{F}C, C).$$

This holds because $FC \,\text{--}\, C \,\text{--}\, A\overline{F}C$, see [**11**, (1.19)]. □

Special case. We have:

- For $p = q = 1$, the above result recovers Propositions 14.26 and 14.28.
- For $p = -1$ and $q = 1$, diagrams (14.45) imply the existence of a colax structure $\psi : \mathcal{K}\mathcal{E}^\vee \to \overline{\mathcal{E}}^\vee\mathcal{K}$ and a lax structure $\psi^\vee : \overline{\mathcal{E}}\mathcal{K}^\vee \to \mathcal{K}^\vee\mathcal{E}$.
- For $q = -1$, we have similar implications involving the signed full Fock functors $\mathcal{K}_{-1}$ and $\mathcal{K}^\vee_{-1}$.

Proposition 14.32 can be upgraded as follows.

Proposition 14.33. *For nonzero scalars p and q, the diagrams*

$$\begin{array}{ccc} (\mathcal{T},\mathcal{T}^\vee,\lambda_{p^{-1}}) & \xrightarrow{(\mathrm{id},\mathrm{id},\tau_p)} & (\mathcal{T},\mathcal{T}^\vee,\lambda_p) \\ {\scriptstyle \mathcal{K}_{q^{-1}}}\big\downarrow & & \big\downarrow{\scriptstyle \mathcal{K}_q} \\ (\overline{\mathcal{T}},\overline{\mathcal{T}}^\vee,\overline{\lambda}_{(pq)^{-1}}) & \xleftarrow[(\mathrm{id},\mathrm{id},\overline{\tau}_{(pq)^{-1}})]{} & (\overline{\mathcal{T}},\overline{\mathcal{T}}^\vee,\overline{\lambda}_{pq}) \end{array}$$

$$\begin{array}{ccc} (\mathcal{T},\mathcal{T}^\vee,\lambda_{p^{-1}}) & \xrightarrow{(\mathrm{id},\tau_p,\mathrm{id})} & (\mathcal{T},\mathcal{T}^\vee,\lambda_p) \\ {\scriptstyle \mathcal{K}^\vee_{q^{-1}}}\big\downarrow & & \big\downarrow{\scriptstyle \mathcal{K}^\vee_q} \\ (\overline{\mathcal{T}},\overline{\mathcal{T}}^\vee,\overline{\lambda}_{(pq)^{-1}}) & \xleftarrow[(\mathrm{id},\overline{\tau}_{(pq)^{-1}},\mathrm{id})]{} & (\overline{\mathcal{T}},\overline{\mathcal{T}}^\vee,\overline{\lambda}_{pq}) \end{array}$$

commute as bilax functors.

PROOF. Functors involving opposite transformations in the above diagrams are bilax by Propositions 2.82 and 8.125. To check the diagrams, we need to compose bilax structures. We elaborate on the first diagram. The lax structures of $\mathcal{K}_q$ and $\mathcal{K}_{q^{-1}}$ are identical, and the other two lax structures are identities, so there is nothing to check. Composing colax structures yields a diagram equivalent to the first diagram in (14.45). □

Exercise 14.34. Check: Theorem 14.30 also holds with $\mathcal{S}$, $\overline{\mathcal{S}}$, $\mathcal{S}^\vee$, $\overline{\mathcal{S}}^\vee$ replaced by $\mathcal{E}$, $\overline{\mathcal{E}}$, $\mathcal{E}^\vee$, $\overline{\mathcal{E}}^\vee$, and λ, $\overline{\lambda}$ replaced by λ_{-1}, $\overline{\lambda}_{-1}$. The bosonic Fock functors now connect the second diagram in (2.100) to the second diagram in (8.127).

Similarly, fermionic Fock functors carry bilax structures wrt these various bimonads. They connect the first (second) diagram in (2.100) to the second (first) diagram in (8.127).

14.6.5. Half-twist transformation.

Define the *half-twist transformation*

(14.46) $$\theta : \mathcal{K} \to \mathcal{K}$$

as follows. For p a Coxeter species, θ_{p} on the (A, C)-component is

(14.47)
$$\begin{array}{ccc} \mathcal{K}(\mathsf{p})[A,C] & \xrightarrow[w(C)=A\overline{C}]{\mathsf{p}[w]} & \mathcal{K}(\mathsf{p})[A,C] \\ \| & & \| \\ \mathsf{p}[A] & & \mathsf{p}[A]. \end{array}$$

(Since C and $A\overline{C}$ are both greater than A, $w(C) = A\overline{C}$ forces $w(A) = A$.) Let us check that $\theta_{\mathsf{p}} : \mathcal{K}(\mathsf{p}) \to \mathcal{K}(\mathsf{p})$ is a map of Coxeter spaces. Diagrams

(8.21) take the form:

$$\begin{array}{ccc} \mathsf{p}[A] & \xrightarrow{w(C)=A\overline{C}} & \mathsf{p}[A] \\ {\scriptstyle \beta_{B,A}}\downarrow & & \downarrow{\scriptstyle \beta_{B,A}} \\ \mathsf{p}[B] & \xrightarrow[w(BC)=B\overline{C}]{} & \mathsf{p}[B] \end{array} \qquad \begin{array}{ccc} \mathsf{p}[A] & \xrightarrow{w(C)=A\overline{C}} & \mathsf{p}[A] \\ {\scriptstyle u}\downarrow & & \downarrow{\scriptstyle u} \\ \mathsf{p}[A'] & \xrightarrow[w'(C')=A'\overline{C'}]{} & \mathsf{p}[A'] \end{array}$$

In the first diagram, A and B have the same support. So $w(B) = B$, and hence $w(BC) = w(B)w(C) = BA\overline{C} = B\overline{C}$. This justifies the label on the bottom-horizontal arrow, and the diagram commutes by the support-type axiom in (2.12a). In the second diagram, u sends A to A' and C to C'. Hence by Fact 1.11, it sends $A\overline{C}$ to $A'\overline{C'}$ (which are opposite to C and C' in the stars of A and A'). Thus $w'u = uw$ since both send C to $A'\overline{C'}$, and the diagram commutes by the type-type axiom in (2.12a).

Exercise 14.35. Check: θ_{p} is natural in p.

Observe: $\theta^2 = \mathrm{id}$ since if $w(C) = A\overline{C}$, then by Fact 1.11, $w(A\overline{C}) = C$ and hence $w^2 = \mathrm{id}$. Thus θ is invertible with its inverse being itself.

Proposition 14.36. *The half-twist transformation is an isomorphism*

$$\theta : (\mathcal{K}, \varphi^b, \psi) \to (\mathcal{K}, \varphi, \psi)$$

of bilax functors, where $(\mathcal{K}, \varphi^b, \psi)$ *refers to the composite*

$$(\mathcal{T}, \mathcal{T}^\vee, \lambda) \xrightarrow{(\mathrm{id},\tau,\mathrm{id})} (\mathcal{T}, \mathcal{T}^\vee, \lambda) \xrightarrow{\mathcal{K}} (\overline{\mathcal{T}}, \overline{\mathcal{T}}^\vee, \overline{\lambda}) \xrightarrow{(\mathrm{id},\overline{\tau},\mathrm{id})} (\overline{\mathcal{T}}, \overline{\mathcal{T}}^\vee, \overline{\lambda}).$$

The colax structure of $(\mathcal{K}, \varphi^b, \psi)$ continues to be ψ, while the lax structure φ^b is obtained from φ by conjugating by the opposite transformations. Since diagram (14.41) does not commute, φ^b differs from φ.

We have seen why θ is an isomorphism. So the additional claim is that θ is a morphism of lax functors and of colax functors:

(14.48a)
$$\begin{array}{ccccccc} \overline{\mathcal{T}}\mathcal{K} & \xrightarrow{\overline{\tau}\mathcal{K}} & \overline{\mathcal{T}}\mathcal{K} & \xrightarrow{\varphi} & \mathcal{K}\mathcal{T} & \xrightarrow{\mathcal{K}\tau} & \mathcal{K}\mathcal{T} \\ {\scriptstyle \overline{\mathcal{T}}\theta}\downarrow & & & & & & \downarrow{\scriptstyle \theta\mathcal{T}} \\ \overline{\mathcal{T}}\mathcal{K} & & & \xrightarrow[\varphi]{} & & & \mathcal{K}\mathcal{T} \end{array}$$

(14.48b)
$$\begin{array}{ccc} \mathcal{K}\mathcal{T}^\vee & \xrightarrow{\psi} & \overline{\mathcal{T}}^\vee\mathcal{K} \\ {\scriptstyle \theta\mathcal{T}^\vee}\downarrow & & \downarrow{\scriptstyle \overline{\mathcal{T}}^\vee\theta} \\ \mathcal{K}\mathcal{T}^\vee & \xrightarrow[\psi]{} & \overline{\mathcal{T}}^\vee\mathcal{K}. \end{array}$$

PROOF. Let us first check diagram (14.48b). Evaluate it on a Coxeter species p on the (A, C)-component. All morphisms carry a summand isomorphically to exactly one summand. Restricting to the relevant summands, the diagram reduces to

$$\begin{array}{ccc} \mathsf{p}[F] & \xrightarrow{u(FC)=C} & \mathsf{p}[F_0] \\ {\scriptstyle w(C)=A\overline{C}}\Big\downarrow & & \Big\downarrow{\scriptstyle v(C)=F_0\overline{C}} \\ \mathsf{p}[G] & \xrightarrow[u'(GC)=C]{} & \mathsf{p}[F_0]. \end{array}$$

Here F is any face greater than A, F_0 is the face of C of the same type as F, and $G = w(F)$ where w takes C to $A\overline{C}$. First note that $w(FC) = w(F)w(C) = GA\overline{C} = G\overline{C}$. Further, u' takes GC to C, so it takes the opposite of GC in the star of G to the opposite of C in the star of F_0. In other words, $u'(G\overline{C}) = F_0\overline{C}$. Thus, $u'w = vu$ since both take FC to $F_0\overline{C}$, and the above diagram commutes by the type-type axiom in (2.12a).

Let us now check diagram (14.48a). Proceeding as above, we reduce to

$$\begin{array}{ccccc} & & \mathsf{p}[A\overline{F_0}] & \xrightarrow{v(A\overline{F_0}C)=C} & \mathsf{p}[G_0] \\ & \nearrow^{\beta_{A\overline{F_0},F_0}} & \Big\downarrow{\scriptstyle \beta_{F_0,A\overline{F_0}}} & & \Big\downarrow{\scriptstyle \beta_{A\overline{G_0},G_0}} \\ \mathsf{p}[F_0] & \xrightarrow[\mathrm{id}]{} & \mathsf{p}[F_0] & \xrightarrow[v(C)=A\overline{G_0}C]{} & \mathsf{p}[A\overline{G_0}] \\ {\scriptstyle u(C)=F_0\overline{C}}\Big\downarrow & & & & {\scriptstyle w(C)=A\overline{C}}\Big\downarrow{\scriptstyle w(A\overline{G_0}C)=F_0\overline{C}} \\ \mathsf{p}[F_0] & & \xrightarrow[\mathrm{id}]{\qquad\qquad} & & \mathsf{p}[F_0]. \end{array}$$

Here F_0 and G_0 are faces of C greater than A, and G_0 and $A\overline{F_0}$ have the same type, and hence so do F_0 and $A\overline{G_0}$. The vertical maps of the pentagon come from θ. The bottom id comes from the bottom φ, while the id of the top φ has been suppressed.

The triangle clearly commutes. The square is an instance of the support-type axiom (2.14) with all arrows reversed and $F := A\overline{G_0}$; the point is that $v(A\overline{F_0}C) = C$ and $v(C) = A\overline{G_0}C$ are consistent. Let us check that $w(C) = A\overline{C}$ and $w(A\overline{G_0}C) = F_0\overline{C}$ are consistent: $w(C) = A\overline{C}$ implies $w(A\overline{C}) = C$ and hence $w(A\overline{G_0}) = F_0$. Thus $w(A\overline{G_0}C) = w(A\overline{G_0})w(C) = F_0A\overline{C} = F_0\overline{C}$. Now $u = wv$ since both send C to $F_0\overline{C}$, so the pentagon commutes by the type-type axiom in (2.12a). ☐

More generally, for a scalar q, define the natural transformation

$$\theta_q : \mathcal{K} \to \mathcal{K} \tag{14.49}$$

by deforming the map (14.47) to

$$\begin{array}{ccc} \mathcal{K}(\mathsf{p})[A,C] & \xrightarrow[w(C)=A\overline{C}]{q^{\mathrm{dist}(C,A\overline{C})}\mathsf{p}[w]} & \mathcal{K}(\mathsf{p})[A,C] \\ \| & & \| \\ \mathsf{p}[A] & & \mathsf{p}[A]. \end{array} \tag{14.50}$$

We call θ_q the *q-half-twist transformation*.

Proposition 14.37. *Let p and q be nonzero scalars. The q-half-twist transformation is an isomorphism*

$$\theta_q : (\mathcal{K}, \varphi', \psi'_q) \to (\mathcal{K}, \varphi, \psi_q)$$

of bilax functors, where $(\mathcal{K}, \varphi', \psi'_q)$ refers to the composite

$$\begin{array}{ccccc} (\mathcal{T}, \mathcal{T}^\vee, \lambda_p) & \xrightarrow{(\mathrm{id}, \tau_p, \tau_{p^{-1}})} & (\mathcal{T}, \mathcal{T}^\vee, \lambda_p) & & \\ & & \Big\downarrow \mathcal{K}_q & & \\ & & (\overline{\mathcal{T}}, \overline{\mathcal{T}}^\vee, \overline{\lambda}_{pq}) & \xrightarrow{(\mathrm{id}, \overline{\tau}_{(pq)^{-1}}, \overline{\tau}_{pq})} & (\overline{\mathcal{T}}, \overline{\mathcal{T}}^\vee, \overline{\lambda}_{pq}). \end{array}$$

Explicitly: The diagrams

$$\begin{array}{ccccccc} \overline{\mathcal{T}}\mathcal{K} & \xrightarrow{\overline{\tau}_{(pq)^{-1}}\mathcal{K}} & \overline{\mathcal{T}}\mathcal{K} & \xrightarrow{\varphi} & \mathcal{K}\mathcal{T} & \xrightarrow{\mathcal{K}\tau_p} & \mathcal{K}\mathcal{T} \\ {\scriptstyle \overline{\mathcal{T}}\theta_q}\Big\downarrow & & & & & & \Big\downarrow{\scriptstyle \theta_q\mathcal{T}} \\ \overline{\mathcal{T}}\mathcal{K} & & & \xrightarrow[\varphi]{\hspace{6em}} & & & \mathcal{K}\mathcal{T} \end{array} \tag{14.51a}$$

$$\begin{array}{ccccccc} \mathcal{K}\mathcal{T}^\vee & \xrightarrow{\mathcal{K}\tau_{p^{-1}}} & \mathcal{K}\mathcal{T}^\vee & \xrightarrow{\psi_q} & \overline{\mathcal{T}}^\vee\mathcal{K} & \xrightarrow{\overline{\tau}_{pq}\mathcal{K}} & \overline{\mathcal{T}}^\vee\mathcal{K} \\ {\scriptstyle \theta_q\mathcal{T}^\vee}\Big\downarrow & & & & & & \Big\downarrow{\scriptstyle \overline{\mathcal{T}}^\vee\theta_q} \\ \mathcal{K}\mathcal{T}^\vee & & & \xrightarrow[\psi_q]{\hspace{6em}} & & & \overline{\mathcal{T}}^\vee\mathcal{K} \end{array} \tag{14.51b}$$

commute.

Proof. Let us first check diagram (14.51b). For $p = q = 1$, by (14.37), the top-horizontal composite is ψ, so the diagram reduces to (14.48b), and hence commutes. Going to general p and q, with notations as usual, the relevant diagram is

$$\begin{array}{ccccccccc} \mathsf{p}[F] & \xrightarrow{p^{-\operatorname{dist}(F, A\overline{F})}} & \mathsf{p}[A\overline{F}] & \xrightarrow{q^{\operatorname{dist}(A\overline{F}C, C)}} & \mathsf{p}[F_0] & \xrightarrow{(pq)^{\operatorname{dist}(F_0, A\overline{F_0})}} & \mathsf{p}[A\overline{F_0}] & \longrightarrow & \mathsf{p}[G_0] \\ \Big\downarrow{\scriptstyle q^{\operatorname{dist}(C, A\overline{C})}} & & & & & & & & \Big\downarrow{\scriptstyle q^{\operatorname{dist}(C, G_0\overline{C})}} \\ \mathsf{p}[G] & & & & \xrightarrow[q^{\operatorname{dist}(GC, C)}]{\hspace{8em}} & & & & \mathsf{p}[G_0]. \end{array}$$

The powers of p cancel each other out. To see equality of powers of q, the relevant identities are

$$\begin{aligned} \operatorname{dist}(C, A\overline{C}) &= \operatorname{dist}(C, G_0\overline{C}) + \operatorname{dist}(G_0\overline{C}, A\overline{C}), \\ \operatorname{dist}(G_0\overline{C}, A\overline{C}) &= \operatorname{dist}(G_0, A\overline{G_0}) = \operatorname{dist}(F_0, A\overline{F_0}), \\ \operatorname{dist}(A\overline{F}C, C) &= \operatorname{dist}(F\overline{C}, A\overline{C}) = \operatorname{dist}(GC, C). \end{aligned}$$

For the first identity, note that $C \,\text{--}\, G_0\overline{C} \,\text{--}\, A\overline{C}$. The w which sends C to $A\overline{C}$, sends $A\overline{C}$ to C, F to G, F_0 to $A\overline{G_0}$, $A\overline{F_0}$ to G_0. This justifies the last step in the next two identities. The first step in the last identity follows by taking opposites in the star of A.

Now we look at diagram (14.51a). The case $p = q = 1$ was done in diagram (14.48a). Going to the case of general p and q, four of the six arrows in the outer hexagon (of the diagram in that proof) get multiplied by scalars. This is shown below.

$$\begin{array}{ccccccc}
\mathsf{p}[F_0] & \xrightarrow{(pq)^{-\operatorname{dist}(A\overline{F_0},F_0)}} & \mathsf{p}[A\overline{F_0}] & \longrightarrow & \mathsf{p}[G_0] & \xrightarrow{p^{\operatorname{dist}(A\overline{G_0},G_0)}} & \mathsf{p}[A\overline{G_0}] \\
\Big\downarrow q^{\operatorname{dist}(C,F_0\overline{C})} & & & & & & \Big\downarrow q^{\operatorname{dist}(C,A\overline{C})} \\
\mathsf{p}[F_0] & & & \longrightarrow & & & \mathsf{p}[F_0].
\end{array}$$

Powers of p cancel each other out, while powers of q match. □

14.7. Norm transformation between Fock functors

There is a morphism of bilax functors which relates full Fock functors $\mathcal{K}$ and $\mathcal{K}^\vee$. This is the norm transformation κ. The construction is similar to that of the norm map associated to a module over a finite group. In our case, the finite groups are the subgroups W_T of W, and modules are components of a Coxeter species. More generally, for a scalar q, one can deform κ using powers of q, the exponents being gallery distances between appropriate chambers. The resulting q-norm transformation κ_q is a morphism of bilax functors from $\mathcal{K}_q$ to $\mathcal{K}_q^\vee$.

The norm transformation κ induces an isomorphism between bosonic Fock functors $\overline{\mathcal{K}}$ and $\overline{\mathcal{K}}^\vee$, while κ_{-1} induces an isomorphism between fermionic Fock functors $\overline{\mathcal{K}}_{-1}$ and $\overline{\mathcal{K}}^\vee_{-1}$ (when $\operatorname{char}(\Bbbk) \nmid |W|$). For q not a root of unity, κ_q is an isomorphism between $\mathcal{K}_q$ and $\mathcal{K}_q^\vee$.

Note very carefully: Apart from q, there is another scalar p involved in the discussion (present in the mixed distributive laws) which we may not always make explicit.

14.7.1. Norm transformation between full Fock functors. Define the *norm transformation*

(14.52)
$$\kappa : \mathcal{K} \to \mathcal{K}^\vee$$

as follows. For p a Coxeter species, κ_{p} on the (A, C)-component is

(14.53)
$$\begin{array}{ccc}
\mathcal{K}(\mathsf{p})[A,C] & \longrightarrow & \mathcal{K}^\vee(\mathsf{p})[A,C] \\
\| & & \| \\
\mathsf{p}[A] & & \mathsf{p}[A]
\end{array}, \qquad \kappa_{\mathsf{p}}(x) = \sum_{w \in W_A} \mathsf{p}[w](x).$$

Exercise 14.38. Check: κ_{p} is a map of Coxeter spaces, and it is natural in p.

Proposition 14.39. *The norm transformation $\kappa : \mathcal{K} \to \mathcal{K}^\vee$ is a morphism of bilax functors.*

PROOF. We first check that κ is a morphism of lax functors. Diagram (A.3) is as follows.

$$
\begin{array}{ccc}
\overline{\mathcal{T}}\mathcal{K} & \xhookrightarrow{\varphi} & \mathcal{K}\mathcal{T} \\
{\scriptstyle \overline{\mathcal{T}}\kappa}\downarrow & & \downarrow{\scriptstyle \kappa\mathcal{T}} \\
\overline{\mathcal{T}}\mathcal{K}^\vee & \xhookrightarrow[\psi^\vee]{} & \mathcal{K}^\vee\mathcal{T}
\end{array}
\qquad\qquad
\begin{array}{ccc}
\bigoplus_{F_0\colon A\le F_0\le C} \mathsf{p}[F_0] & \hookrightarrow & \bigoplus_{F\colon A\le F} \mathsf{p}[F] \\
{\scriptstyle \sum \mathsf{p}[u]}\downarrow & & \downarrow{\scriptstyle \sum \mathsf{p}[w]} \\
\bigoplus_{F_0\colon A\le F_0\le C} \mathsf{p}[F_0] & \xhookrightarrow[\sum \mathsf{p}[v]]{} & \bigoplus_{F\colon A\le F} \mathsf{p}[F]
\end{array}
$$

On the right, we evaluated on a Coxeter species p on the (A, C)-component. Since all the action is happening in the star of A, we may assume without loss of generality that $A = O$. Now for every face F_0 of C, letting $T = \mathrm{t}(F_0)$, u varies over W_T, v varies over Sh_T, and w varies over all elements of W. The diagram then commutes by the bijection in (1.67) and the fact that $\mathsf{p}[w] = \mathsf{p}[v]\mathsf{p}[u]$ by the type-type axiom in (2.12a).

Similarly, one checks diagram (A.6) which says that κ is a morphism of colax functors. In finite-dimensional setting, this can be deduced from the lax case by using self-duality of κ. (The self-duality of κ is addressed in Section 14.7.3 below.) □

14.7.2. q-norm transformation between q-full Fock functors. For a scalar q, the norm transformation κ can be deformed to

$$\kappa_q : \mathcal{K}_q \to \mathcal{K}_q^\vee \tag{14.54}$$

as follows. For p a Coxeter species, $(\kappa_q)_{\mathsf{p}}$ on the (A, C)-component is

$$
\begin{array}{ccc}
\mathcal{K}_q(\mathsf{p})[A, C] & \longrightarrow & \mathcal{K}_q^\vee(\mathsf{p})[A, C] \\
\| & & \| \\
\mathsf{p}[A] & & \mathsf{p}[A]
\end{array}
\quad , \qquad (\kappa_q)_{\mathsf{p}}(x) = \sum_{w \in W_A} q^{l(w)} \mathsf{p}[w](x). \tag{14.55}
$$

In other words, the linear map $\mathsf{p}[A] \to \mathsf{p}[A]$ is given by action of the element

$$\sum_{w \in W_A} q^{l(w)} w,$$

where we view $\mathsf{p}[A]$ as a $\Bbbk W_A$-module. (We need to linearize W_A since we want to take a sum.) The length of w is wrt the chamber C under consideration, that is, $l(w) = \mathrm{dist}(C, wC)$.

We call κ_q the *q-norm transformation*. Setting $q = 1$ recovers κ. We refer to κ_{-1} as the *signed norm transformation*. For $q = 0$, functors $\mathcal{K}_0$ and $\mathcal{K}_0^\vee$ coincide as noted in (14.24), and the 0-norm transformation κ_0 is the identity.

Proposition 14.40. *For scalars p and q, the q-norm transformation $\kappa_q : \mathcal{K}_q \to \mathcal{K}_q^\vee$ is a morphism of bilax functors. The bilax functors are from $(\mathcal{T}, \mathcal{T}^\vee, \lambda_p)$ to $(\overline{\mathcal{T}}, \overline{\mathcal{T}}^\vee, \overline{\lambda}_{pq})$.*

PROOF. This generalizes Proposition 14.39. We follow the same proof and use in addition that $q^{l(w)} = q^{l(u)} q^{l(v)}$ (which holds since the bijection in (1.67) is length-preserving). The gate property played a critical role here. □

14.7.3. Self-duality. Assume finite-dimensionality. Just as for Fock functors, one can conjugate the norm transformation wrt duality. More precisely, the conjugate of κ, denoted $\kappa^\vee$, is obtained by taking the composite

$$(\mathcal{T}^\vee,\mathcal{T},\lambda) \xrightarrow{(-)^*} (\mathcal{T},\mathcal{T}^\vee,\lambda) \underset{\mathcal{K}^\vee}{\overset{\mathcal{K}}{\rightrightarrows}}\ \Downarrow\kappa\ (\overline{\mathcal{T}},\overline{\mathcal{T}}^\vee,\overline{\lambda}) \xrightarrow{(-)^*} (\overline{\mathcal{T}}^\vee,\overline{\mathcal{T}},\overline{\lambda})$$

and passing to the opposite categories. The last step reverses the direction of morphisms. So $\kappa^\vee$ again goes from $(\mathcal{K}^\vee)^\vee$ (which is the same as $\mathcal{K}$) to $\mathcal{K}^\vee$.

Exercise 14.41. Check: κ is self-dual, that is, $\kappa^\vee = \kappa$. More generally, for q a scalar, κ_q is self-dual, that is, $\kappa_q^\vee = \kappa_q$.

14.7.4. Interaction with signature functors. The following result builds on Proposition 14.12.

Proposition 14.42. *For a scalar q, precomposing $\kappa_q : \mathcal{K}_q \to \mathcal{K}_q^\vee$ with the o-signature functor on Coxeter species yields $\kappa_{-q} : \mathcal{K}_{-q} \to \mathcal{K}_{-q}^\vee$.*

PROOF. We calculate the component $\kappa_q(\mathsf{p}^-)[A,C]$.

$$\begin{array}{ccc} \mathcal{K}_q(\mathsf{p}^-)[A,C] & \longrightarrow & \mathcal{K}_q^\vee(\mathsf{p}^-)[A,C] \\ \| & & \| \\ \mathsf{p}^-[A] & & \mathsf{p}^-[A] \end{array}, \quad x \otimes \mathtt{H}_{[C/A]} \mapsto \sum_{w \in W_A} q^{l(w)} \mathsf{p}^-[w](x \otimes \mathtt{H}_{[C/A]}).$$

Now

$$\mathsf{p}^-[w](x \otimes \mathtt{H}_{[C/A]}) = \mathsf{p}[w](x) \otimes \mathtt{H}_{[wC/A]} = (-1)^{l(w)} \mathsf{p}[w](x) \otimes \mathtt{H}_{[C/A]}.$$

Substituting this in the formula and using isomorphism (14.25), we see that the component map equals $\kappa_{-q}(\mathsf{p})[A,C]$. □

Exercise 14.43. We build on Exercise 14.13. Check: For q a scalar, $\kappa_q : \mathcal{K}_q \to \mathcal{K}_q^\vee$ commutes with the u-signature functors on Coxeter species and on Coxeter spaces.

14.7.5. Norm transformation between bosonic Fock functors. The map κ_{p} on the (A,C)-component (14.53) is an instance of the norm map in group theory, where the group is W_A and left W_A-module is $\mathsf{p}[A]$. See [**7**, Section 2.5.2]. It maps into W_A-invariants and factors through W_A-coinvariants:

$$\begin{array}{ccc} \mathsf{p}[A] & \longrightarrow & \mathsf{p}[A] \\ \downarrow & & \uparrow \\ \mathsf{p}[A]_{W_A} & \dashrightarrow & \mathsf{p}[A]^{W_A}. \end{array}$$

Further, by standard group theory [**7**, Lemma 2.20], the induced bottom-horizontal map is an isomorphism when $\operatorname{char}(\Bbbk) \nmid |W|$. We denote the resulting natural transformation $\overline{\mathcal{K}} \to \overline{\mathcal{K}}^\vee$ by $\overline{\kappa}$.

Proposition 14.44. *The diagram of morphisms of bilax functors*

$$\begin{array}{ccc} \mathcal{K} & \xrightarrow{\kappa} & \mathcal{K}^\vee \\ \big\downarrow & & \big\uparrow \\ \overline{\mathcal{K}} & \xrightarrow[\overline{\kappa}]{} & \overline{\mathcal{K}}^\vee \end{array} \tag{14.56}$$

commutes. If $\mathrm{char}(\Bbbk) \nmid |W|$, *then* $\overline{\kappa}$ *is an isomorphism and* $\overline{\mathcal{K}}$ *and* $\overline{\mathcal{K}}^\vee$ *are isomorphic as bilax functors.*

Exercise 14.45. Consider the setting of Theorem 14.30. Check: If $\mathrm{char}(\Bbbk) \nmid |W|$, then the norm transformation $\overline{\kappa} : \overline{\mathcal{K}} \to \overline{\mathcal{K}}^\vee$ is an isomorphism of bilax functors from $(\mathcal{S}, \mathcal{S}^\vee, \lambda)$ to $(\overline{\mathcal{S}}, \overline{\mathcal{S}}^\vee, \overline{\lambda})$.

14.7.6. Signed norm transformation between fermionic Fock functors. Let us now look at $q = -1$. Define $\overline{\kappa}_{-1}$ to be $\overline{\kappa}$ precomposed with the o-signature functor.

Proposition 14.46. *The diagram of morphisms of bilax functors*

$$\begin{array}{ccc} \mathcal{K}_{-1} & \xrightarrow{\kappa_{-1}} & \mathcal{K}^\vee_{-1} \\ \big\downarrow & & \big\uparrow \\ \overline{\mathcal{K}}_{-1} & \xrightarrow[\overline{\kappa}_{-1}]{} & \overline{\mathcal{K}}^\vee_{-1} \end{array} \tag{14.57}$$

commutes. If $\mathrm{char}(\Bbbk) \nmid |W|$, *then* $\overline{\kappa}_{-1}$ *is an isomorphism and* $\overline{\mathcal{K}}_{-1}$ *and* $\overline{\mathcal{K}}^\vee_{-1}$ *are isomorphic as bilax functors.*

PROOF. Precompose diagram (14.56) with the o-signature functor and use (14.34) and Proposition 14.42 with $q = 1$. □

Exercise 14.47. We build on Exercise 14.18. Check: $\overline{\kappa} : \overline{\mathcal{K}} \to \overline{\mathcal{K}}^\vee$ commutes with the u-signature functors on Coxeter species and on Coxeter spaces. The same result holds for $\overline{\kappa}_{-1} : \overline{\mathcal{K}}_{-1} \to \overline{\mathcal{K}}^\vee_{-1}$.

14.7.7. q-norm for q not a root of unity. The behavior of κ_q for q not a root of unity is substantially different than for $q = \pm 1$.

Proposition 14.48. *For q not a root of unity, the q-norm transformation $\kappa_q : \mathcal{K}_q \to \mathcal{K}^\vee_q$ is an isomorphism of bilax functors.*

PROOF. In view of Proposition 14.40, we only need to show that for a Coxeter species p, the (A, C)-component $\mathsf{p}[A] \to \mathsf{p}[A]$ of $(\kappa_q)_{\mathsf{p}}$ is a linear isomorphism. This follows by applying Fact 1.76 to each W_A. □

14.8. Anyonic Fock functor

For a scalar q, recall the q-norm transformation κ_q from Section 14.7. It is a morphism of bilax functors from $\mathcal{K}_q$ to $\mathcal{K}^\vee_q$. So the image of κ_q is a bilax functor. This is the anyonic Fock functor $\Im_q$. For q not a root of unity, it is isomorphic to the q-full Fock functors. In particular, for $q = 0$, it coincides with the free Fock functor. For $q = \pm 1$, $\Im_q$ is isomorphic to bosonic and fermionic Fock functors, respectively (when $\mathrm{char}(\Bbbk) \nmid |W|$).

14.8.1. Anyonic Fock functor. For a scalar q, let $\Im_q$ denote the image of the q-norm transformation $\kappa_q : \mathcal{K}_q \to \mathcal{K}_q^\vee$. We call $\Im_q$ the *anyonic Fock functor*. Explicitly, evaluated on a Coxeter species p, the component $\Im_q(\mathsf{p})[A, C]$ is the subspace of $\mathcal{K}_q^\vee(\mathsf{p})[A, C]$ consisting of elements of the form

$$\sum_{w \in W_A} q^{l(w)} \mathsf{p}[w](x),$$

as x varies over elements of $\mathsf{p}[A]$. The length of w is wrt the chamber C.

Proposition 14.49. *For scalars p and q, the anyonic Fock functor*

$$\Im_q : (\mathcal{T}, \mathcal{T}^\vee, \lambda_p) \to (\overline{\mathcal{T}}, \overline{\mathcal{T}}^\vee, \overline{\lambda}_{pq})$$

carries a bilax structure and fits into the commutative diagram of morphisms of bilax functors

(14.58)
$$\begin{array}{ccc} \mathcal{K}_q & \xrightarrow{\kappa_q} & \mathcal{K}_q^\vee \\ & \searrow \quad \nearrow & \\ & \Im_q. & \end{array}$$

PROOF. This builds on Proposition 14.40 and follows from general principles of (co)images of morphisms: The dotted arrows in the diagrams

$$\begin{array}{ccccc} \overline{\mathcal{T}}\mathcal{K}_q & \longrightarrow & \overline{\mathcal{T}}\Im_q & \longrightarrow & \overline{\mathcal{T}}\mathcal{K}_q^\vee \\ {\scriptstyle \varphi}\downarrow & & \vdots & & \downarrow{\scriptstyle \psi_q^\vee} \\ \mathcal{K}_q\mathcal{T} & \longrightarrow & \Im_q\mathcal{T} & \longrightarrow & \mathcal{K}_q^\vee\mathcal{T} \end{array} \qquad \begin{array}{ccccc} \mathcal{K}_q\mathcal{T}^\vee & \longrightarrow & \Im_q\mathcal{T}^\vee & \longrightarrow & \mathcal{K}_q^\vee\mathcal{T}^\vee \\ {\scriptstyle \psi_q}\downarrow & & \vdots & & \downarrow{\scriptstyle \varphi^\vee} \\ \overline{\mathcal{T}}^\vee\mathcal{K}_q & \longrightarrow & \overline{\mathcal{T}}^\vee\Im_q & \longrightarrow & \overline{\mathcal{T}}^\vee\mathcal{K}_q^\vee \end{array}$$

exist. This is because the functor $\overline{\mathcal{T}} = \overline{\mathcal{T}}^\vee$ on the $\mathbb{k}$-linear category of Coxeter spaces is exact, and hence preserves (co)images. The diagrams imply that the lax and colax structures of $\mathcal{K}_q^\vee$ restrict to $\Im_q$. The bilax axioms for $\Im_q$ then follow from those for $\mathcal{K}_q^\vee$ by restriction. This completes the argument.

More concretely, one directly checks that the lax and colax structures of $\mathcal{K}_q^\vee$ restrict to $\Im_q$. For the lax structure $\psi_q^\vee$ in (14.22): For $x \in \mathsf{p}[F_0]$,

$$\begin{array}{ccc} \displaystyle\bigoplus_{F_0 :\, A \leq F_0 \leq C} \mathsf{p}[F_0] & \xrightarrow[v(C)=FC]{q^{\operatorname{dist}(C,FC)}\mathsf{p}[v]} & \displaystyle\bigoplus_{F:\, A \leq F} \mathsf{p}[F] \\ \displaystyle\sum_{u \in W_{F_0}} q^{l(u)} \mathsf{p}[u](x) & \longmapsto & \displaystyle\sum_{w \in W_A} q^{l(w)} \mathsf{p}[w](x). \end{array}$$

This calculation used the bijection in (1.67). For the colax structure $\varphi^\vee$ in (14.13): For $x \in \mathsf{p}[F]$,

$$\begin{array}{ccc} \displaystyle\bigoplus_{F:\, A \leq F} \mathsf{p}[F] & \longrightarrow & \displaystyle\bigoplus_{F_0 :\, A \leq F_0 \leq C} \mathsf{p}[F_0] \\ \displaystyle\sum_{w \in W_A} q^{l(w)} \mathsf{p}[w](x) & \longmapsto & \displaystyle\sum_{u \in W_{F_0}} q^{l(u)} \mathsf{p}[u]\big(q^{l(v)} \mathsf{p}[v](x)\big), \end{array}$$

where υ is the Coxeter symmetry defined by $\upsilon(FC) = C$, and F_0 is the face of C of the same type as F. $\square$

Special case. Let us now look at some specific values of q.

- For $q = \pm 1$, we have commutative diagrams of morphisms of bilax functors

$$
\begin{array}{ccc}
\mathcal{K} & \xrightarrow{\quad\kappa\quad} & \mathcal{K}^\vee \\
\big\downarrow & & \big\uparrow \\
\overline{\mathcal{K}} \twoheadrightarrow & \Im & \hookrightarrow \overline{\mathcal{K}}^\vee
\end{array}
\qquad\qquad
\begin{array}{ccc}
\mathcal{K}_{-1} & \xrightarrow{\quad\kappa_{-1}\quad} & \mathcal{K}^\vee_{-1} \\
\big\downarrow & & \big\uparrow \\
\overline{\mathcal{K}}_{-1} \twoheadrightarrow & \Im_{-1} & \hookrightarrow \overline{\mathcal{K}}^\vee_{-1}.
\end{array}
$$

  (We write $\Im$ for $\Im_1$.) If $\operatorname{char}(\Bbbk) \nmid |W|$, then the anyonic Fock functor $\Im_q$ is isomorphic to bosonic Fock functors for $q = 1$, and to fermionic Fock functors for $q = -1$; that is, $\overline{\mathcal{K}} \cong \Im \cong \overline{\mathcal{K}}^\vee$ and $\overline{\mathcal{K}}_{-1} \cong \Im_{-1} \cong \overline{\mathcal{K}}^\vee_{-1}$. This follows from Propositions 14.44 and 14.46.
- For q not a root of unity, by Proposition 14.48, the anyonic Fock functor $\Im_q$ is isomorphic to q-full Fock functors $\mathcal{K}_q$ and $\mathcal{K}^\vee_q$.
- For $q = 0$, functors $\mathcal{K}_0$ and $\mathcal{K}^\vee_0$ coincide, and the 0-norm transformation is the identity. Thus, $\Im_0$ is the free Fock functor.

14.8.2. Self-duality. Assume finite-dimensionality. Recall: The conjugate of $\mathcal{K}_q$ is $\mathcal{K}^\vee_q$, and κ_q is self-dual. Building on this fact, diagram (14.58) is self-dual. In particular, the anyonic Fock functor is self-dual.

14.9. Fock functors and Coxeter operads

Recall that monads $\mathcal{T}$ and $\mathcal{S}$ on Coxeter species and monads $\overline{\mathcal{T}}$ and $\overline{\mathcal{S}}$ on Coxeter spaces arise from the Coxeter associative and commutative operads, respectively. So far, we have studied interactions of Fock functors with these monads. Now we look at some aspects of this theory for monads arising from any Coxeter operad. We construct, in this generality, a lax structure on the full Fock functor $\mathcal{K}^\vee$, and an induced lax structure on the bosonic Fock functor $\overline{\mathcal{K}}^\vee$. There are similar results for the other bosonic Fock functor $\overline{\mathcal{K}}$. We also briefly look at dual results for Coxeter cooperads.

14.9.1. Monads arising from a Coxeter operad. Fix a Coxeter operad $\mathbf{a}$. Recall from Sections 5.4.6 and 11.5.7 the endofunctors

$$\mathcal{V}_{\mathbf{a}} : \mathcal{A}\text{-}\mathsf{CSp} \to \mathcal{A}\text{-}\mathsf{CSp}, \qquad \mathsf{m} \mapsto \mathbf{a} \circ \mathsf{m}$$

and

$$\overline{\mathcal{V}}_{\mathbf{a}} : \mathcal{A}\text{-}\overline{\mathsf{C}}\mathsf{Sp} \to \mathcal{A}\text{-}\overline{\mathsf{C}}\mathsf{Sp}, \qquad \mathrm{V} \mapsto \mathbf{a} \circ \mathrm{V}$$

on Coxeter species and Coxeter spaces, respectively. Since $\mathbf{a}$ is a Coxeter operad, these are monads. For the Coxeter associative operad $\mathbf{As}$ and Coxeter commutative operad $\mathbf{Com}$,

$$\mathcal{V}_{\mathbf{As}} = \mathcal{T}, \quad \overline{\mathcal{V}}_{\mathbf{As}} = \overline{\mathcal{T}}, \quad \mathcal{V}_{\mathbf{Com}} = \mathcal{S}, \quad \overline{\mathcal{V}}_{\mathbf{Com}} = \overline{\mathcal{S}}.$$

See Tables 5.1 and 11.4.

14.9.2. Lax structure on a full Fock functor. Let $\mathcal{K}^\vee$ be the full Fock functor in (14.10). We proceed to define a lax structure $\psi^\vee : \overline{\mathcal{V}}_{\mathbf{a}}\mathcal{K}^\vee \to \mathcal{K}^\vee\mathcal{V}_{\mathbf{a}}$ generalizing (14.11). We work with the flat formulation of Coxeter operads. Let us first evaluate both sides on a Coxeter species m on the (A, C)-component. Accordingly, using definitions (5.9) and (11.28),

$$\overline{\mathcal{V}}_{\mathbf{a}}\mathcal{K}^\vee(\mathsf{m})[A,C] = (\mathbf{a} \circ \mathcal{K}^\vee(\mathsf{m}))[A,C] = \Big(\bigoplus_{A \leq F_0 \leq C} \mathbf{a}[\mathrm{s}(A), \mathrm{s}(F_0)] \otimes \mathsf{m}[F_0] \Big)_{\mathrm{quo}},$$

where, for any morphism $(F_0, C) \to (G_0, C)$ local to A, an element in the F_0-summand identifies with its image in the G_0-summand, and

$$\mathcal{K}^\vee\mathcal{V}_{\mathbf{a}}(\mathsf{m})[A,C] = \mathcal{K}^\vee(\mathbf{a} \circ \mathsf{m})[A,C] = (\mathbf{a} \circ \mathsf{m})[A] = \bigoplus_{\mathrm{s}(A) \leq \mathrm{X}} \mathbf{a}[\mathrm{s}(A), \mathrm{X}] \otimes \mathsf{m}[\mathrm{X}].$$

Define the linear map

$$\psi^\vee_{\mathsf{m}} : \overline{\mathcal{V}}_{\mathbf{a}}\mathcal{K}^\vee(\mathsf{m})[A,C] \to \mathcal{K}^\vee\mathcal{V}_{\mathbf{a}}(\mathsf{m})[A,C] \tag{14.59}$$

as follows. First fix a face F_0 between A and C. Let F be any face greater than A which has the same type as F_0. Let $w \in W$ be the unique Coxeter symmetry which sends C to FC (and hence F_0 to F since the action is type-preserving). This induces a linear map

$$\mathbf{a}[\mathrm{s}(A), \mathrm{s}(F_0)] \otimes \mathsf{m}[F_0] \xrightarrow{\mathbf{a}[w] \otimes \mathsf{m}[w]} \mathbf{a}[\mathrm{s}(A), \mathrm{s}(F)] \otimes \mathsf{m}[F].$$

Summing over all choices of F (and employing $\beta_{\mathrm{X},F}$) yields a linear map

$$\mathbf{a}[\mathrm{s}(A), \mathrm{s}(F_0)] \otimes \mathsf{m}[F_0] \to \bigoplus_{\mathrm{X}:\, \mathrm{s}(A) \leq \mathrm{X}} \mathbf{a}[\mathrm{s}(A), \mathrm{X}] \otimes \mathsf{m}[\mathrm{X}]. \tag{14.60}$$

Now take direct sum over all F_0 between A and C.

Fact 14.50. The resulting linear map factors through morphisms $(F_0, C) \to (G_0, C)$ local to A to yield

$$\Big(\bigoplus_{A \leq F_0 \leq C} \mathbf{a}[\mathrm{s}(A), \mathrm{s}(F_0)] \otimes \mathsf{m}[F_0] \Big)_{\mathrm{quo}} \longrightarrow \bigoplus_{\mathrm{s}(A) \leq \mathrm{X}} \mathbf{a}[\mathrm{s}(A), \mathrm{X}] \otimes \mathsf{m}[\mathrm{X}]. \tag{14.61}$$

PROOF. Let us first assume $\mathbf{a} = \mathbf{Com}$. Consider the morphism $(F_0, C) \to (G_0, C)$ local to A transferred by K (which is greater than A). Thus K has the same support as F_0 and same type as G_0. For any face F of the same type as F_0 which is greater than A, the diagram

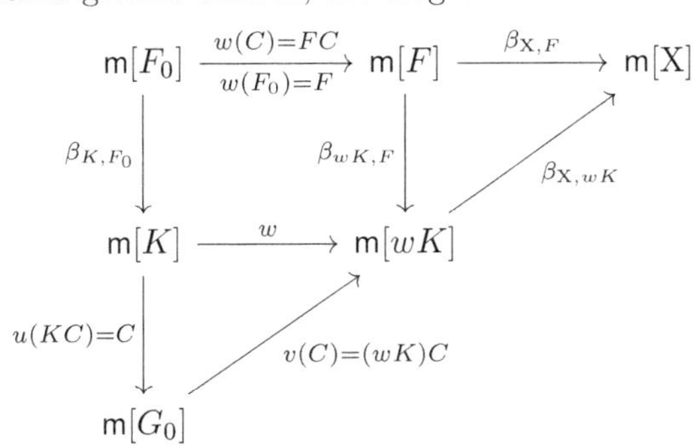

commutes. For the bottom triangle, note that $w(KC) = (wK)(wC) = (wK)FC = (wK)C$ (since F has the same support as wK). Now, summing over all choices of F, composite of the top-horizontal maps equals (14.60), while composite of the diagonal maps equals (14.60) with G_0 instead of F_0. This completes the argument. For general $\mathbf{a}$, we need to take tensor product of the above diagram with appropriate components of $\mathbf{a}$. The details are straightforward. □

This completes the definition of the lax structure (14.59).

Remark 14.51. It is clear that $\mathbf{a} = \mathbf{As}$ recovers the lax structure (14.11). The argument for $\mathbf{a} = \mathbf{Com}$ in the above proof yields a natural transformation $\psi^\vee : \overline{\mathcal{S}}\mathcal{K}^\vee \to \mathcal{K}^\vee\mathcal{S}$. This is the same as the one obtained in (14.40). There it was obtained from $\psi^\vee : \overline{\mathcal{T}}\mathcal{K}^\vee \to \mathcal{K}^\vee\mathcal{T}$, while here we are working directly with **Com** without any reference to **As**.

Theorem 14.52. *For a Coxeter operad* $\mathbf{a}$*, the full Fock functor*

$$(\mathcal{K}^\vee, \psi^\vee) : \mathcal{V}_{\mathbf{a}} \to \overline{\mathcal{V}}_{\mathbf{a}}$$

is lax.

PROOF. The right diagram in (A.2) clearly commutes. The left diagram is

$$\begin{array}{ccccc} \overline{\mathcal{V}}_{\mathbf{a}}\overline{\mathcal{V}}_{\mathbf{a}}\mathcal{K}^\vee & \longrightarrow & \overline{\mathcal{V}}_{\mathbf{a}}\mathcal{K}^\vee\mathcal{V}_{\mathbf{a}} & \longrightarrow & \mathcal{K}^\vee\mathcal{V}_{\mathbf{a}}\mathcal{V}_{\mathbf{a}} \\ \downarrow & & & & \downarrow \\ \overline{\mathcal{V}}_{\mathbf{a}}\mathcal{K}^\vee & & \longrightarrow & & \mathcal{K}^\vee\mathcal{V}_{\mathbf{a}}. \end{array}$$

This reduces to the diagram below, with $A \leq C$ fixed. (For convenience, we have flipped the diagram, and also put $\mathrm{s}(F_0) = \mathrm{X}_0$ and $\mathrm{s}(G_0) = \mathrm{Y}_0$.)

$$\begin{array}{ccc} \displaystyle\bigoplus_{A \leq F_0 \leq G_0 \leq C} \mathbf{a}[\mathrm{s}(A), \mathrm{X}_0] \otimes \mathbf{a}[\mathrm{X}_0, \mathrm{Y}_0] \otimes \mathsf{m}[G_0] & \longrightarrow & \displaystyle\bigoplus_{A \leq G_0 \leq C} \mathbf{a}[\mathrm{s}(A), \mathrm{Y}_0] \otimes \mathsf{m}[G_0] \\ \downarrow & & \\ \displaystyle\bigoplus_{\substack{A \leq F_0 \leq C \\ \mathrm{X}_0 \leq \mathrm{Y}'}} \mathbf{a}[\mathrm{s}(A), \mathrm{X}_0] \otimes \mathbf{a}[\mathrm{X}_0, \mathrm{Y}'] \otimes \mathsf{m}[\mathrm{Y}'] & & \downarrow \\ \downarrow & & \\ \displaystyle\bigoplus_{\mathrm{s}(A) \leq \mathrm{X} \leq \mathrm{Y}} \mathbf{a}[\mathrm{s}(A), \mathrm{X}] \otimes \mathbf{a}[\mathrm{X}, \mathrm{Y}] \otimes \mathsf{m}[\mathrm{Y}] & \longrightarrow & \displaystyle\bigoplus_{\mathrm{s}(A) \leq \mathrm{Y}} \mathbf{a}[\mathrm{s}(A), \mathrm{Y}] \otimes \mathsf{m}[\mathrm{Y}] \end{array}$$

Let us start with an element in the (F_0, G_0)-summand. Going across and down: first use substitution of $\mathbf{a}$, and then apply (14.60) (with G_0 instead of F_0). There is one term for each face of type $\mathrm{t}(G_0)$ which is greater than A. Going down and across: first apply (14.60) (with G_0 and F_0 instead of F_0 and A), so there is one term for each face of type $\mathrm{t}(G_0)$ which is greater than F_0. Then again apply (14.60), so there is one term for each face of type $\mathrm{t}(F_0)$ which is greater than A. Finally use substitution of $\mathbf{a}$. We know that faces of type $\mathrm{t}(G_0)$ (greater than A) can be written as a disjoint union indexed by the face of type $\mathrm{t}(F_0)$ which they contain. Then, following the two directions, one checks that the terms match. □

14.9.3. Lax structure on a bosonic Fock functor (invariant version). The lax structure $\psi^\vee$ of the full Fock functor $\mathcal{K}^\vee$ in Theorem 14.52 induces a lax structure $\overline{\psi}^\vee$ on the bosonic Fock functor $\overline{\mathcal{K}}^\vee$ via the diagram

$$\begin{array}{ccc} \overline{\mathcal{V}}_{\mathbf{a}}\mathcal{K}^\vee & \xrightarrow{\psi^\vee} & \mathcal{K}^\vee\mathcal{V}_{\mathbf{a}} \\ \uparrow & & \uparrow \\ \overline{\mathcal{V}}_{\mathbf{a}}\overline{\mathcal{K}}^\vee & \underset{\overline{\psi}^\vee}{\dashrightarrow} & \overline{\mathcal{K}}^\vee\mathcal{V}_{\mathbf{a}}. \end{array} \tag{14.62}$$

Explicitly, (14.61) induces a linear map

$$\Big(\bigoplus_{A\le F_0\le C} \mathbf{a}[\mathrm{s}(A),\mathrm{s}(F_0)]\otimes \mathsf{m}[F_0]^{W_{F_0}}\Big)_{\text{quo}} \longrightarrow \Big(\bigoplus_{\mathrm{s}(A)\le \mathrm{X}} \mathbf{a}[\mathrm{s}(A),\mathrm{X}]\otimes \mathsf{m}[\mathrm{X}]\Big)^{W_A}. \tag{14.63}$$

Note: This linear map splits as a direct sum over flat-types (local to A).

Theorem 14.53. *For a Coxeter operad* $\mathbf{a}$, *the bosonic Fock functor*

$$(\overline{\mathcal{K}}^\vee, \overline{\psi}^\vee) : \mathcal{V}_{\mathbf{a}} \to \overline{\mathcal{V}}_{\mathbf{a}}$$

is lax. Further, $\overline{\mathcal{K}}^\vee \hookrightarrow \mathcal{K}^\vee$ *is a morphism of lax functors. Moreover, if* $\mathrm{char}(\Bbbk) \nmid |W|$, *then the map* $\overline{\psi}^\vee$ *is an isomorphism.*

Proof. The first part follows by construction. For the second part, we define the inverse of (14.63). Call it f. Fix a flat-type λ (local to A). Let $x = (x_{\mathrm{X}})$ be a W_A-invariant for the λ-summand, that is, the x_{X} are nonzero only if X has type λ. Pick a face F_0 between A and C whose support is one such X. Since x is W_A-invariant, the element $x_{\mathrm{s}(F_0)}$ (after using $\beta_{F_0,\mathrm{s}(F_0)}$) can be viewed as an element of $\mathbf{a}[\mathrm{s}(A),\mathrm{s}(F_0)] \otimes \mathsf{m}[F_0]^{W_{F_0}}$. Define $f(x)$ to be $1/|\lambda|$ times the equivalence class of $x_{\mathrm{s}(F_0)}$, where $|\lambda|$ is the number of faces of a given type in $\mathrm{s}(F_0)$ (local to A) as in Definition 1.8. (This is where field characteristic hypothesis is used.) It is straightforward to check that f is well-defined and inverse to $\overline{\psi}^\vee$. □

14.9.4. Lax structure on a bosonic Fock functor (coinvariant version). Similar to $\overline{\mathcal{K}}^\vee$, for a Coxeter operad $\mathbf{a}$, we also have a lax structure on the bosonic Fock functor $\overline{\mathcal{K}}$, which we write as $\overline{\varphi} : \overline{\mathcal{V}}_{\mathbf{a}}\overline{\mathcal{K}} \to \overline{\mathcal{K}}\mathcal{V}_{\mathbf{a}}$. Evaluated on a Coxeter species m on the (A, C)-component, the linear map

$$\Big(\bigoplus_{A\le F_0\le C} \mathbf{a}[\mathrm{s}(A),\mathrm{s}(F_0)]\otimes \mathsf{m}[F_0]_{W_{F_0}}\Big)_{\text{quo}} \longrightarrow \Big(\bigoplus_{\mathrm{s}(A)\le \mathrm{X}} \mathbf{a}[\mathrm{s}(A),\mathrm{X}]\otimes \mathsf{m}[\mathrm{X}]\Big)_{W_A} \tag{14.64}$$

is induced from

$$\mathbf{a}[\mathrm{s}(A),\mathrm{s}(F_0)] \otimes \mathsf{m}[F_0] \xrightarrow{\mathrm{id}\otimes\beta_{\mathrm{s}(F_0),F_0}} \mathbf{a}[\mathrm{s}(A),\mathrm{s}(F_0)] \otimes \mathsf{m}[\mathrm{s}(F_0)].$$

(Comparing with (14.63), invariants are replaced by coinvariants. An important distinction is that the definition of the linear map now does not involve a sum.)

Theorem 14.54. *For a Coxeter operad* $\mathbf{a}$*, the bosonic Fock functor*

$$(\overline{\mathcal{K}}, \overline{\varphi}) : \mathcal{V}_{\mathbf{a}} \to \overline{\mathcal{V}}_{\mathbf{a}}$$

is lax. Moreover, $\overline{\varphi}$ *is an isomorphism (in any field characteristic).*

PROOF. The special case $\mathbf{a} := \mathbf{Com}$ recovers the isomorphism $\overline{\mathcal{S}}\overline{\mathcal{K}} \to \overline{\mathcal{K}}\mathcal{S}$ in Theorem 14.30. The argument in the general and special cases are more or less identical. □

Remark 14.55. Note very carefully: We have not claimed any lax structure on the full Fock functor $\mathcal{K}$ (in the generality of Coxeter operads). Theorem 14.2 says that $\mathcal{K}$ carries a lax structure wrt $\mathbf{a} := \mathbf{As}$. Importantly, this does not induce a lax structure wrt $\mathbf{Com}$. This is the same as the observation that diagram (14.41) does not commute.

14.9.5. Naturality of lax structures.

We now consider what happens when the Coxeter operad $\mathbf{a}$ varies.

Proposition 14.56. *Let* $\mathbf{a} \to \mathbf{b}$ *be a morphism of Coxeter operads. Then the identity functors on* $\mathcal{A}$-CSp *and* $\mathcal{A}$-$\overline{\mathsf{CSp}}$

$$\mathrm{id} : \mathcal{V}_{\mathbf{b}} \to \mathcal{V}_{\mathbf{a}} \quad \textit{and} \quad \mathrm{id} : \overline{\mathcal{V}}_{\mathbf{b}} \to \overline{\mathcal{V}}_{\mathbf{a}}$$

are lax. Further, we have the commutative diagram of lax functors

$$(14.65) \qquad \begin{array}{ccc} \mathcal{V}_{\mathbf{b}} & \xrightarrow{\mathrm{id}} & \mathcal{V}_{\mathbf{a}} \\ {\scriptstyle \mathcal{K}^{\vee}}\big\downarrow & & \big\downarrow{\scriptstyle \mathcal{K}^{\vee}} \\ \overline{\mathcal{V}}_{\mathbf{b}} & \xrightarrow[\mathrm{id}]{} & \overline{\mathcal{V}}_{\mathbf{a}}. \end{array}$$

In other words, the lax structure $\psi^{\vee}$ *of* $\mathcal{K}^{\vee}$ *is natural in* $\mathbf{a}$.

PROOF. Since $\mathbf{a} \to \mathbf{b}$ is a morphism of Coxeter operads, the identity functors are lax. The check for (14.65) boils down to the diagram below.

$$(14.66) \qquad \begin{array}{ccc} \mathbf{a}[\mathrm{s}(A), \mathrm{s}(F_0)] \otimes \mathcal{K}^{\vee}(\mathsf{m})[F_0, C] & \longrightarrow & \mathbf{b}[\mathrm{s}(A), \mathrm{s}(F_0)] \otimes \mathcal{K}^{\vee}(\mathsf{m})[F_0, C] \\ \Big\| & & \Big\| \\ \mathbf{a}[\mathrm{s}(A), \mathrm{s}(F_0)] \otimes \mathsf{m}[F_0] & \longrightarrow & \mathbf{b}[\mathrm{s}(A), \mathrm{s}(F_0)] \otimes \mathsf{m}[F_0] \\ \big\downarrow & & \big\downarrow \\ \bigoplus_{F} \mathbf{a}[\mathrm{s}(A), \mathrm{s}(F)] \otimes \mathsf{m}[F] & \longrightarrow & \bigoplus_{F} \mathbf{b}[\mathrm{s}(A), \mathrm{s}(F)] \otimes \mathsf{m}[F] \\ \big\downarrow & & \big\downarrow \\ \bigoplus_{\mathrm{s}(A) \leq \mathrm{X}} \mathbf{a}[\mathrm{s}(A), \mathrm{X}] \otimes \mathsf{m}[\mathrm{X}] & \longrightarrow & \bigoplus_{\mathrm{s}(A) \leq \mathrm{X}} \mathbf{b}[\mathrm{s}(A), \mathrm{X}] \otimes \mathsf{m}[\mathrm{X}] \end{array}$$

The horizontal maps are induced by $\mathbf{a} \to \mathbf{b}$. □

Diagram (14.65) also commutes with $\overline{\mathcal{K}}^{\vee}$ instead of $\mathcal{K}^{\vee}$, and further the two diagrams are compatible with $\overline{\mathcal{K}}^{\vee} \hookrightarrow \mathcal{K}^{\vee}$. This is shown below.

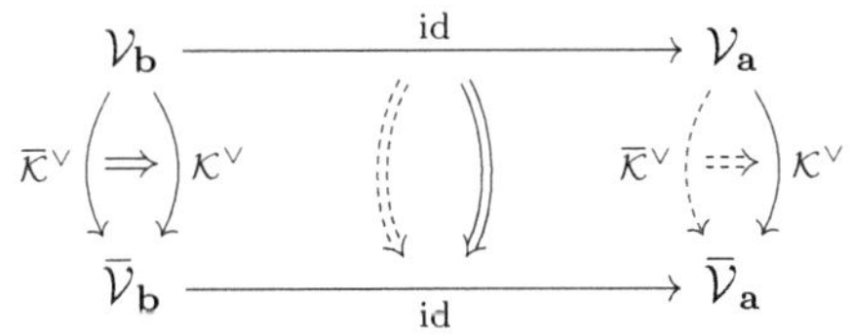

This is a commutative diagram of 2-cells in the 2-category of monads. The natural transformations on the curved surfaces of the cylinders are the canonical identifications.

Diagram (14.65) also commutes with $\overline{\mathcal{K}}$ instead of $\mathcal{K}^{\vee}$.

14.9.6. Lax structures on signed full Fock and fermionic Fock functors. There are lax structures on the signed full Fock functor $\mathcal{K}^{\vee}_{-1}$ and on fermionic Fock functors $\overline{\mathcal{K}}^{\vee}_{-1}$ and $\overline{\mathcal{K}}_{-1}$. They are obtained from those of $\mathcal{K}^{\vee}$, $\overline{\mathcal{K}}^{\vee}$, $\overline{\mathcal{K}}$ by precomposing with the lax structure of the o-signature functor on Coxeter species. Note very carefully that the latter connects $\mathcal{V}_{\mathbf{a}}$ and $\mathcal{V}_{\mathbf{a}^-}$.

14.9.7. Dual results for a Coxeter cooperad. Fix a Coxeter cooperad $\mathbf{c}$. This yields comonads

$$\mathcal{U}_{\mathbf{c}} : \mathcal{A}\text{-}\mathsf{CSp} \to \mathcal{A}\text{-}\mathsf{CSp}, \qquad \mathsf{m} \mapsto \mathbf{c} \circ \mathsf{m}$$

and

$$\overline{\mathcal{U}}_{\mathbf{c}} : \mathcal{A}\text{-}\overline{\mathsf{C}}\mathsf{Sp} \to \mathcal{A}\text{-}\overline{\mathsf{C}}\mathsf{Sp}, \qquad \mathrm{V} \mapsto \mathbf{c} \circ^{\vee} \mathrm{V}$$

on Coxeter species and Coxeter spaces, respectively.

The full Fock functor $\mathcal{K}$ carries a colax structure $\psi : \mathcal{K}\mathcal{U}_{\mathbf{c}} \to \overline{\mathcal{U}}_{\mathbf{c}}\mathcal{K}$. Evaluated on a Coxeter species m on the (A, C)-component,

$$\psi_{\mathsf{m}} : \mathcal{K}\mathcal{U}_{\mathbf{c}}(\mathsf{m})[A, C] \to \overline{\mathcal{U}}_{\mathbf{c}}\mathcal{K}(\mathsf{m})[A, C] \tag{14.67}$$

is defined in a manner dual to (14.59). Note:

$$\mathcal{K}\mathcal{U}_{\mathbf{c}}(\mathsf{m})[A, C] = \mathcal{K}(\mathbf{c} \circ \mathsf{m})[A, C] = (\mathbf{c} \circ \mathsf{m})[A] = \bigoplus_{\mathrm{s}(A) \leq \mathrm{X}} \mathbf{c}[\mathrm{s}(A), \mathrm{X}] \otimes \mathsf{m}[\mathrm{X}],$$

$$\overline{\mathcal{U}}_{\mathbf{c}}\mathcal{K}(\mathsf{m})[A, C] = (\mathbf{c} \circ^{\vee} \mathcal{K}(\mathsf{m}))[A, C] = \Big(\bigoplus_{A \leq F_0 \leq C} \mathbf{c}[\mathrm{s}(A), \mathrm{s}(F_0)] \otimes \mathsf{m}[F_0] \Big)^{\mathrm{sub}}$$

consisting of elements (x_{F_0}) such that any $(F_0, C) \to (G_0, C)$ local to A carries x_{F_0} to x_{G_0}. The map (14.67) on the X-summand is defined by summing over all faces F greater than A of support X.

Further, the colax structure ψ of $\mathcal{K}$ induces a colax structure $\overline{\psi}$ on $\overline{\mathcal{K}}$ via the diagram

$$\begin{array}{ccc} \mathcal{K}\mathcal{U}_{\mathbf{c}} & \xrightarrow{\psi} & \overline{\mathcal{U}}_{\mathbf{c}}\mathcal{K} \\ \downarrow & & \downarrow \\ \overline{\mathcal{K}}\mathcal{U}_{\mathbf{c}} & \xrightarrow[\overline{\psi}]{} & \overline{\mathcal{U}}_{\mathbf{c}}\overline{\mathcal{K}}. \end{array} \tag{14.68}$$

To summarize:

Theorem 14.57. *For a Coxeter cooperad* $\mathbf{c}$*, the Fock functors*

$$(\mathcal{K}, \psi) : \mathcal{U}_{\mathbf{c}} \to \overline{\mathcal{U}}_{\mathbf{c}} \quad \text{and} \quad (\overline{\mathcal{K}}, \overline{\psi}) : \mathcal{U}_{\mathbf{c}} \to \overline{\mathcal{U}}_{\mathbf{c}}$$

are colax. Further, $\mathcal{K} \twoheadrightarrow \overline{\mathcal{K}}$ *is a morphism of colax functors. Moreover, if* $\operatorname{char}(\Bbbk) \nmid |W|$*, then the map* $\overline{\psi}$ *is an isomorphism.*

This is dual to Theorems 14.52 and 14.53. The dual of Proposition 14.56 is stated below.

Proposition 14.58. *Let* $\mathbf{c} \to \mathbf{d}$ *be a morphism of Coxeter cooperads. Then*

$$\mathrm{id} : \mathcal{U}_{\mathbf{c}} \to \mathcal{U}_{\mathbf{d}} \quad \text{and} \quad \mathrm{id} : \overline{\mathcal{U}}_{\mathbf{c}} \to \overline{\mathcal{U}}_{\mathbf{d}}$$

are colax. Further, we have the commutative diagram of colax functors

$$\begin{array}{ccc} \mathcal{U}_{\mathbf{c}} & \xrightarrow{\mathrm{id}} & \mathcal{U}_{\mathbf{d}} \\ {\scriptstyle \mathcal{K}}\downarrow & & \downarrow{\scriptstyle \mathcal{K}} \\ \overline{\mathcal{U}}_{\mathbf{c}} & \xrightarrow[\mathrm{id}]{} & \overline{\mathcal{U}}_{\mathbf{d}}. \end{array} \tag{14.69}$$

This diagram also commutes with $\overline{\mathcal{K}}$ *instead of* $\mathcal{K}$.

There is also a colax structure on $\overline{\mathcal{K}}^{\vee}$, namely,

$$\Big(\bigoplus_{\mathrm{s}(A) \leq \mathrm{X}} \mathbf{c}[\mathrm{s}(A), \mathrm{X}] \otimes \mathsf{m}[\mathrm{X}] \Big)^{W_A} \longrightarrow \Big(\bigoplus_{A \leq F_0 \leq C} \mathbf{c}[\mathrm{s}(A), \mathrm{s}(F_0)] \otimes \mathsf{m}[F_0]^{W_{F_0}} \Big)^{\mathrm{sub}} \tag{14.70}$$

which is dual to (14.64). It is an isomorphism. Also, diagram (14.69) commutes with $\overline{\mathcal{K}}^{\vee}$ instead of $\mathcal{K}$.

14.9.8. Norm transformation. Recall the norm transformation $\kappa : \mathcal{K} \to \mathcal{K}^{\vee}$ in (14.52) and the induced $\overline{\kappa} : \overline{\mathcal{K}} \to \overline{\mathcal{K}}^{\vee}$ in (14.56).

Proposition 14.59. *For* $\mathbf{a}$ *a Coxeter operad,* $\overline{\kappa} : \overline{\mathcal{K}} \to \overline{\mathcal{K}}^{\vee}$ *is a morphism of lax functors from* $\mathcal{V}_{\mathbf{a}}$ *to* $\overline{\mathcal{V}}_{\mathbf{a}}$*. Dually, for* $\mathbf{c}$ *a Coxeter cooperad,* $\overline{\kappa} : \overline{\mathcal{K}} \to \overline{\mathcal{K}}^{\vee}$ *is a morphism of colax functors from* $\mathcal{U}_{\mathbf{c}}$ *to* $\overline{\mathcal{U}}_{\mathbf{c}}$*. Moreover,* $\overline{\kappa}$ *is an isomorphism if* $\operatorname{char}(\Bbbk) \nmid |W|$.

PROOF. Similar to the argument in the proof of Proposition 14.39. □

14.10. Decorated Fock functors

Fock functors can be generalized by using a module M over the Coxeter group W. We refer to them collectively as decorated Fock functors. A summary is given in Table 14.2. When $M = \Bbbk$ with trivial W-action, we recover the Fock functors in Table 14.1.

TABLE 14.2. Decorated Fock functors.

$\mathcal{K}_M$, $\mathcal{K}^\vee_M$	decorated full Fock functors
$\mathcal{K}_{-1,M}$, $\mathcal{K}^\vee_{-1,M}$	decorated signed full Fock functors
$\mathcal{K}_{q,M}$, $\mathcal{K}^\vee_{q,M}$	decorated q-full Fock functors
$\overline{\mathcal{K}}_M$, $\overline{\mathcal{K}}^\vee_M$	decorated bosonic Fock functors
$\overline{\mathcal{K}}_{-1,M}$, $\overline{\mathcal{K}}^\vee_{-1,M}$	decorated fermionic Fock functors
$\Im_{q,M}$	decorated anyonic Fock functor
$\mathcal{K}_{0,M} = \mathcal{K}^\vee_{0,M} = \Im_{0,M}$	decorated free Fock functor

14.10.1. Decorated full Fock functors. Fix a left W-module M. Recall the decoration functor on Coxeter species $(-)_M$ from Section 4.9.7. It is defined by taking Hadamard product with E_M. Define the decorated full Fock functor $\mathcal{K}_M$ from Coxeter species to Coxeter spaces by

$$\mathcal{K}_M := \mathcal{K}((-)_M). \tag{14.71}$$

Explicitly, evaluated on a Coxeter species p, the (A, C)-component is

$$\mathcal{K}_M(\mathsf{p})[A, C] = \mathsf{p}[A] \otimes M. \tag{14.72}$$

This generalizes (14.2). Support and type morphisms of $\mathcal{K}_M(\mathsf{p})$ are

$$\begin{array}{ccccccc} \mathcal{K}_M(\mathsf{p})[F,C] & \xrightarrow{\beta_{G,D,F,C}} & \mathcal{K}_M(\mathsf{p})[G,D] & & \mathcal{K}_M(\mathsf{p})[F,C] & \xrightarrow{w} & \mathcal{K}_M(\mathsf{p})[F',C'] \\ \| & & \| & & \| & & \| \\ \mathsf{p}[F] \otimes M & \xrightarrow[\beta_{G,F}\otimes \mathrm{id}]{} & \mathsf{p}[G] \otimes M & & \mathsf{p}[F] \otimes M & \xrightarrow[w\otimes w]{} & \mathsf{p}[F'] \otimes M. \end{array} \tag{14.73}$$

These generalize (14.3). Here we used support and type morphisms of E_M in (4.101).

Moreover, $\mathcal{K}_M$ is bilax since it is composite of $\mathcal{K}$ and $(-)_M$ both of which are bilax. Explicitly, its lax and colax structure maps

$$\bigoplus_{F_0 \colon A \leq F_0 \leq C} \mathsf{p}[F_0] \otimes M \;\underset{\text{colax}}{\overset{\text{lax}}{\rightleftarrows}}\; \bigoplus_{F \colon A \leq F} \mathsf{p}[F] \otimes M \tag{14.74}$$

are, respectively, given by inclusion, and by sending a F-summand to the F_0-summand via $w \otimes w$ (where w is defined by $w(FC) = C$ and F_0 is the face of C of the same type as F). These generalize (14.5) and (14.6).

Similarly, one defines the decorated full Fock functor $\mathcal{K}^\vee_M$. More generally, for a scalar q, define $\mathcal{K}_{q,M}$ and $\mathcal{K}^\vee_{q,M}$ by precomposing q-full Fock functors $\mathcal{K}_q$ and $\mathcal{K}^\vee_q$ by $(-)_M$. These are the decorated q-full Fock functors. They are bilax from $(\mathcal{T}, \mathcal{T}^\vee, \lambda_p)$ to $(\overline{\mathcal{T}}, \overline{\mathcal{T}}^\vee, \overline{\lambda}_{pq})$. The case $q = -1$ yields the decorated signed full Fock functors, while $q = 0$ yields the decorated free Fock functor.

14.10.2. Decorated bosonic and fermionic Fock functors. Decorated bosonic Fock functors $\overline{\mathcal{K}}_M$ and $\overline{\mathcal{K}}^\vee_M$ are defined by precomposing bosonic Fock functors $\overline{\mathcal{K}}$ and $\overline{\mathcal{K}}^\vee$ by $(-)_M$. Thus,

$$\overline{\mathcal{K}}_M(\mathsf{p})[A,C] = (\mathsf{p}[A] \otimes M)_{W_A} \quad \text{and} \quad \overline{\mathcal{K}}^\vee_M(\mathsf{p})[A,C] = (\mathsf{p}[A] \otimes M)^{W_A}, \tag{14.75}$$

with (co)invariants taken wrt the diagonal action of W_A on $\mathsf{p}[A] \otimes M$. These generalize (14.27) and (14.30). The bilax structure maps of $\overline{\mathcal{K}}_M$ and $\overline{\mathcal{K}}^\vee_M$ are isomorphisms. Also, if $\mathrm{char}(\Bbbk) \nmid |W|$, then the two are isomorphic to each other.

Similarly, decorated fermionic Fock functors are obtained by precomposing fermionic Fock functors by the decoration functor, or equivalently, bosonic Fock functors by the signed decoration functor, see (4.119).

Exercise 14.60. Use Proposition 14.22 to deduce: For scalars p and q, for a left W-module M, there are isomorphisms of bilax functors

$$\mathcal{K}_{q,M}(-) \cong \overline{\mathcal{K}}_M(\mathsf{\Gamma}_q \times (-)) \qquad \text{and} \qquad \mathcal{K}^\vee_{q,M}(-) \cong \overline{\mathcal{K}}^\vee_M(\mathsf{\Gamma}^*_q \times (-))$$

from $(\mathcal{T}, \mathcal{T}^\vee, \lambda_p)$ to $(\overline{\mathcal{T}}, \overline{\mathcal{T}}^\vee, \overline{\lambda}_{pq})$.

14.10.3. Decorated norm transformation.

For a scalar q, precomposing the q-norm transformation (14.54) with the decoration functor yields the morphism of bilax functors

$$\kappa_{q,M} : \mathcal{K}_{q,M} \to \mathcal{K}^\vee_{q,M}. \tag{14.76}$$

We call this the *decorated q-norm transformation*. For $q = 1$, we write κ_M and refer to it as the *decorated norm transformation*. Precomposing diagram (14.56) with the decoration functor yields

$$\begin{array}{ccc} \mathcal{K}_M & \xrightarrow{\kappa_M} & \mathcal{K}^\vee_M \\ \big\downarrow & & \big\uparrow \\ \overline{\mathcal{K}}_M & \xrightarrow[\overline{\kappa}_M]{} & \overline{\mathcal{K}}^\vee_M. \end{array} \tag{14.77}$$

It is a commutative diagram of morphisms of bilax functors. This generalizes (14.56). There is a similar diagram which generalizes (14.57).

For a scalar q, let $\mathfrak{I}_{q,M}$ denote the image of $\kappa_{q,M}$. We call it the *decorated anyonic Fock functor*. It fits into the commutative diagram

$$\begin{array}{ccc} \mathcal{K}_{q,M} & \xrightarrow{\kappa_{q,M}} & \mathcal{K}^\vee_{q,M} \\ & \searrow \quad \nearrow & \\ & \mathfrak{I}_{q,M}. & \end{array} \tag{14.78}$$

This generalizes (14.58).

14.11. Notes

The material in this chapter is new. It was originally discovered in the year 2011. (See also the historical note in the preface.) This chapter fulfils the promise made on [**11**, page 762]. The theory presented here generalizes the classical theory of Fock functors from the category of Joyal species to the category of graded vector spaces developed in [**7**, Part III]. The generalization can be understood using the constructions of Sections 2.12.2 and 8.14.12. Note very carefully: In this book, bilax functors are between bimonads, while in [**7**], bilax functors are between braided monoidal categories. A precise dictionary with the classical theory is given below.

14.11.1. Fock functors. (Sections 14.1–14.5.) Theorems 14.2 and 14.4 generalize similar results about full Fock functors $\mathcal{K}$ and $\mathcal{K}^\vee$ in [**7**, Theorems 15.3 and 15.6]. The lax and colax structures of $\mathcal{K}$, namely, (14.5) and (14.6) (defined in geometric terms) generalize the combinatorial procedures of shifting and standardization [**7**, Notation 2.5 and page 520]. Theorems 14.7 and 14.8 about q-full Fock functors $\mathcal{K}_q$ and $\mathcal{K}_q^\vee$ generalize [**7**, Theorems 16.1 and 16.2], with gallery distance between chambers replacing the Schubert statistic; the connection between the two concepts is already mentioned in [**7**, Section 10.13.1]. The fact in Theorems 14.14 and 14.15 that bilax structure maps of bosonic Fock functors $\overline{\mathcal{K}}$ and $\overline{\mathcal{K}}^\vee$ are inverse isomorphisms is the bistrong property in [**7**, Theorems 15.3 and 15.6]. Proposition 14.12 involving the o-signature functor generalizes [**7**, Proposition 16.5]; gallery distance between chambers and the Schubert statistic appear in the respective proofs. The relation between bosonic and fermionic Fock functors in (14.34) generalizes [**7**, (16.7)]. Proposition 14.22 generalizes [**7**, Proposition 16.6], while Proposition 14.24 generalizes [**7**, Proposition 16.22]. The classical analogue of Lemma 14.20 was not mentioned in [**7**].

14.11.2. Fock functors and commutativity. (Section 14.6.) Results concerning $\mathcal{S}$, $\bar{\mathcal{S}}$ and their duals generalize classical results as shown in Table 14.3. The latter were given in [**7**, Propositions 15.26 and 15.28]. In fact, these properties, namely, $\mathcal{K}^\vee$ is braided lax, $\mathcal{K}$ is braided colax, $\overline{\mathcal{K}}^\vee$ and $\overline{\mathcal{K}}$ are braided bilax also underlie the more general results of Section 14.9 such as Theorems 14.52, 14.53, 14.57.

TABLE 14.3. Commutativity under Fock functors.

Reflection arrangement	Braid arrangement
$(\mathcal{K},\psi) : \mathcal{S}^\vee \to \bar{\mathcal{S}}^\vee$ colax	$(\mathcal{K},\psi)$ braided colax
$(\mathcal{K}^\vee,\psi^\vee) : \mathcal{S} \to \bar{\mathcal{S}}$ lax	$(\mathcal{K}^\vee,\psi^\vee)$ braided lax
$(\overline{\mathcal{K}},\bar{\varphi},\bar{\psi}) : (\mathcal{S},\mathcal{S}^\vee) \to (\bar{\mathcal{S}},\bar{\mathcal{S}}^\vee)$ bilax	$(\overline{\mathcal{K}},\bar{\varphi},\bar{\psi})$ braided bilax
$(\overline{\mathcal{K}}^\vee,\bar{\psi}^\vee,\bar{\varphi}^\vee) : (\mathcal{S},\mathcal{S}^\vee) \to (\bar{\mathcal{S}},\bar{\mathcal{S}}^\vee)$ bilax	$(\overline{\mathcal{K}}^\vee,\bar{\psi}^\vee,\bar{\varphi}^\vee)$ braided bilax

Proposition 14.33 generalizes [**7**, Proposition 16.26]. Propositions 14.36 and 14.37 on the half-twist transformation generalize [**7**, Propositions 15.30 and 16.27].

14.11.3. Norm transformation. Anyonic Fock functor. (Sections 14.7–14.8.) Proposition 14.39 generalizes the norm transformation between full Fock functors in [**7**, Proposition 15.20], while Proposition 14.40 generalizes the q-norm transformation between q-full Fock functors in [**7**, Proposition 16.15]. Shuffles in their geometric and combinatorial avatars appear in the respective proofs. Proposition 14.48 generalizes [**7**, Theorem 16.18]; the proof using Fact 1.76 that we have given here is conceptually clearer. Proposition 14.42 involving the o-signature functor generalizes [**7**, Proposition 16.16]. Diagrams (14.56) and (14.57) generalize [**7**, Diagrams (15.15) and (16.8)].

The anyonic Fock functor $\mathfrak{I}_q$ in Section 14.8 generalizes the anyonic Fock functor $\mathfrak{I}_q$ in [**7**, Section 16.3.5]. The general fact about (co)images of morphisms of bilax functors mentioned in the proof of Proposition 14.49 is the analogue of [**7**, Theorem 3.116].

14.11.4. Decorated Fock functors. (Section 14.10.) Decorated Fock functors from Coxeter species to Coxeter spaces generalize decorated Fock functors from Joyal species to graded vector spaces defined in [**7**, Part III, Chapter 19]. The former are associated to a W-module M, and the latter to a vector space V. The connection is that V yields a S_n-module $V^{\otimes n}$ for each n, see also Example 4.26. Definitions (14.72) and (14.75) correspond to [**7**, Definition 19.1], see also formulas (11.66) and (11.67) in this regard. Definition (14.71) corresponds to [**7**, (19.2)]. Diagram (14.77) on the decorated norm transformation corresponds to [**7**, Diagram (19.3)].

CHAPTER 15

Coxeter bimonoids and Coxeter bialgebras

Fock functors were introduced in Chapter 14. They come in various flavors, see the summary in Table 14.1. We now begin the study of their interactions with Coxeter bimonoids and Coxeter bialgebras. Examples are given in Chapter 18, and these could be read in parallel with the present chapter. The toy example of rank one is developed in Section 18.7.

Coxeter bimonoids to Coxeter bialgebras. Recall that full Fock functors $\mathcal{K}$ and $\mathcal{K}^\vee$ carry bilax structures wrt bimonads $(\mathcal{T}, \mathcal{T}^\vee, \lambda)$ and $(\overline{\mathcal{T}}, \overline{\mathcal{T}}^\vee, \overline{\lambda})$ on Coxeter species and Coxeter spaces, respectively. As a consequence: If h is a Coxeter bimonoid, then $\mathcal{K}(\mathsf{h})$ and $\mathcal{K}^\vee(\mathsf{h})$ are Coxeter bialgebras. The (co)products of these Coxeter bialgebras can be expressed in terms of the (co)product of h and (co)lax structures of $\mathcal{K}$ and $\mathcal{K}^\vee$. Since bosonic Fock functors inherit bilax structures from full Fock functors, they also send Coxeter bimonoids to Coxeter bialgebras. Thus, for a Coxeter bimonoid h, we also have Coxeter bialgebras $\overline{\mathcal{K}}(\mathsf{h})$ and $\overline{\mathcal{K}}^\vee(\mathsf{h})$.

More generally, for scalars p and q, the q-full Fock functors $\mathcal{K}_q$ and $\mathcal{K}_q^\vee$ send a Coxeter p-bimonoid to a Coxeter pq-bialgebra. In particular, signed full Fock functors $\mathcal{K}_{-1}$ and $\mathcal{K}_{-1}^\vee$ send Coxeter bimonoids to signed Coxeter bialgebras. The same holds for fermionic Fock functors.

All of the above constructions are functorial. Moreover, these functors are $\Bbbk$-linear. They are compactly shown in full generality in Summary 15.1.

q-full Fock functors	Coxeter p-bimonoid	$\xrightarrow{\mathcal{K}_q,\, \mathcal{K}_q^\vee}$	Coxeter pq-bialgebra
bosonic Fock functors	Coxeter p-bimonoid	$\xrightarrow{\overline{\mathcal{K}},\, \overline{\mathcal{K}}^\vee}$	Coxeter p-bialgebra
fermionic Fock functors	Coxeter p-bimonoid	$\xrightarrow{\overline{\mathcal{K}}_{-1},\, \overline{\mathcal{K}}_{-1}^\vee}$	Coxeter $(-p)$-bialgebra

SUMMARY 15.1. Coxeter bimonoids to Coxeter bialgebras.

Since $\mathcal{K}$ and $\mathcal{K}^\vee$ are conjugates wrt duality, for h any Coxeter bimonoid, $\mathcal{K}(\mathsf{h}^*) = \mathcal{K}^\vee(\mathsf{h})^*$ as Coxeter bialgebras. Similar statements hold for $\overline{\mathcal{K}}$ and $\overline{\mathcal{K}}^\vee$, and so on.

Commutative Coxeter monoids to commutative Coxeter algebras. Since Fock functors are bilax, they are, in particular, lax and colax. Hence, they send Coxeter (co)monoids to Coxeter (co)algebras. The full Fock functor $\mathcal{K}$ preserves (signed) cocommutativity, $\mathcal{K}^\vee$ preserves (signed) commutativity, while bosonic Fock functors preserve both. Similarly, the signed full Fock functor $\mathcal{K}_{-1}$ interchanges cocommutativity and signed cocommutativity, $\mathcal{K}^\vee_{-1}$ interchanges commutativity and signed commutativity, while fermionic Fock functors interchange both.

A part summary is shown below.

$$\boxed{\begin{array}{c}\text{com.}\\ \text{Coxeter}\\ \text{monoid}\end{array}} \xrightarrow[\overline{\mathcal{K}},\, \overline{\mathcal{K}}^\vee]{\mathcal{K}^\vee} \boxed{\begin{array}{c}\text{com.}\\ \text{Coxeter}\\ \text{algebra}\end{array}} \qquad \boxed{\begin{array}{c}\text{cocom.}\\ \text{Coxeter}\\ \text{comonoid}\end{array}} \xrightarrow[\overline{\mathcal{K}},\, \overline{\mathcal{K}}^\vee]{\mathcal{K}} \boxed{\begin{array}{c}\text{cocom.}\\ \text{Coxeter}\\ \text{coalgebra}\end{array}}$$

Coxeter operad monoids to Coxeter operad algebras. The above ideas extend to the operadic setting. For a Coxeter operad $\mathbf{a}$, functors $\mathcal{K}^\vee$, $\overline{\mathcal{K}}^\vee$, $\overline{\mathcal{K}}$ send a Coxeter $\mathbf{a}$-monoid to a Coxeter $\mathbf{a}$-algebra. In particular, they send a Coxeter Lie monoid to a Coxeter Lie algebra. Dually, for a Coxeter cooperad $\mathbf{c}$, functors $\mathcal{K}$, $\overline{\mathcal{K}}$, $\overline{\mathcal{K}}^\vee$ send a Coxeter $\mathbf{c}$-comonoid to a Coxeter $\mathbf{c}$-coalgebra. In particular, they send a Coxeter Lie comonoid to a Coxeter Lie coalgebra. This is summarized below.

$$\boxed{\begin{array}{c}\text{Coxeter}\\ \text{operad}\\ \text{monoid}\end{array}} \xrightarrow[\overline{\mathcal{K}},\, \overline{\mathcal{K}}^\vee]{\mathcal{K}^\vee} \boxed{\begin{array}{c}\text{Coxeter}\\ \text{operad}\\ \text{algebra}\end{array}} \qquad \boxed{\begin{array}{c}\text{Coxeter}\\ \text{cooperad}\\ \text{comonoid}\end{array}} \xrightarrow[\overline{\mathcal{K}},\, \overline{\mathcal{K}}^\vee]{\mathcal{K}} \boxed{\begin{array}{c}\text{Coxeter}\\ \text{cooperad}\\ \text{coalgebra}\end{array}}$$

$$\boxed{\begin{array}{c}\text{Coxeter Lie}\\ \text{monoid}\end{array}} \xrightarrow[\overline{\mathcal{K}},\, \overline{\mathcal{K}}^\vee]{\mathcal{K}^\vee} \boxed{\begin{array}{c}\text{Coxeter Lie}\\ \text{algebra}\end{array}} \qquad \boxed{\begin{array}{c}\text{Coxeter Lie}\\ \text{comonoid}\end{array}} \xrightarrow[\overline{\mathcal{K}},\, \overline{\mathcal{K}}^\vee]{\mathcal{K}} \boxed{\begin{array}{c}\text{Coxeter Lie}\\ \text{coalgebra}\end{array}}$$

(Co)freeness properties. Bosonic and fermionic Fock functors preserve both freeness and cofreeness. Thus, a bosonic Fock functor sends a free (commutative) Coxeter monoid to a free (commutative) Coxeter algebra, and so on. This fact generalizes to Coxeter (co)operads. In particular, a bosonic Fock functor sends a free Coxeter Lie monoid to a free Coxeter Lie algebra.

Primitive and indecomposable parts. The functor $\overline{\mathcal{K}}^\vee$ commutes with the primitive part functor which goes from Coxeter bimonoids (Coxeter bialgebras) to Coxeter Lie monoids (Coxeter Lie algebras). Dually, $\overline{\mathcal{K}}$ commutes with the indecomposable part functor which goes from Coxeter bimonoids (Coxeter bialgebras) to Coxeter Lie comonoids (Coxeter Lie coalgebras).

Universal enveloping functors. We also study how Fock functors connect the universal enveloping Coxeter monoid of a Coxeter Lie monoid to the universal enveloping Coxeter algebra of a Coxeter Lie algebra. This has connections to commutativity properties of Fock functors: recall that the universal enveloping functor specializes to the free commutative functor when the Lie structure is trivial.

15.1. Coxeter bimonoids to Coxeter bialgebras

Fock functors possess bilax structures, and hence send Coxeter bimonoids to Coxeter bialgebras, see Summary 15.1. In addition, there are relations among Fock functors compatible with their bilax structures. Thus, starting with a single Coxeter bimonoid, one obtains a variety of Coxeter bialgebras related to one another by morphisms of Coxeter bialgebras.

15.1.1. Coxeter (co, bi)monoids to Coxeter (co, bi)algebras. The main result is as follows.

Theorem 15.1. *Let p and q be scalars. Let h be a Coxeter p-bimonoid. Then:*

- $\mathcal{K}_q(\mathsf{h})$ *and* $\mathcal{K}_q^\vee(\mathsf{h})$ *are Coxeter pq-bialgebras.*
- *In particular,* $\mathcal{K}(\mathsf{h})$ *and* $\mathcal{K}^\vee(\mathsf{h})$ *are Coxeter p-bialgebras, and* $\mathcal{K}_{-1}(\mathsf{h})$ *and* $\mathcal{K}_{-1}^\vee(\mathsf{h})$ *are Coxeter $(-p)$-bialgebras.*
- $\overline{\mathcal{K}}(\mathsf{h})$ *and* $\overline{\mathcal{K}}^\vee(\mathsf{h})$ *are Coxeter p-bialgebras, and* $\overline{\mathcal{K}}_{-1}(\mathsf{h})$ *and* $\overline{\mathcal{K}}_{-1}^\vee(\mathsf{h})$ *are Coxeter $(-p)$-bialgebras.*
- $\Im_q(\mathsf{h})$ *is a Coxeter pq-bialgebra.*

Moreover, these constructions are functorial in h.

PROOF. Recall: A bilax functor of bimonads preserves bialgebras. Theorems 14.7 and 14.8 say that $\mathcal{K}_q$ and $\mathcal{K}_q^\vee$ are bilax. Further, a bialgebra wrt $(\mathcal{T}, \mathcal{T}^\vee, \lambda_p)$ is a Coxeter p-bimonoid, while a bialgebra wrt $(\overline{\mathcal{T}}, \overline{\mathcal{T}}^\vee, \overline{\lambda}_{pq})$ is a Coxeter pq-bialgebra (Theorems 2.77 and 8.118). Putting all facts together, the first item follows. The last two items are similar, we use Theorems 14.14, 14.15, 14.16 and Proposition 14.49. □

In particular, full Fock functors $\mathcal{K}$ and $\mathcal{K}^\vee$ send (signed) Coxeter bimonoids to (signed) Coxeter bialgebras, while signed full Fock functors $\mathcal{K}_{-1}$ and $\mathcal{K}_{-1}^\vee$ send Coxeter bimonoids to signed Coxeter bialgebras, and signed Coxeter bimonoids to Coxeter bialgebras.

Theorem 15.2. *For a scalar p, let h be a Coxeter p-bimonoid. Then there are morphisms of Coxeter p-bialgebras*

$$\mathcal{K}(\mathsf{h}) \twoheadrightarrow \overline{\mathcal{K}}(\mathsf{h}) \quad \textit{and} \quad \overline{\mathcal{K}}^\vee(\mathsf{h}) \hookrightarrow \mathcal{K}^\vee(\mathsf{h}),$$

and of Coxeter $(-p)$-bialgebras

$$\mathcal{K}_{-1}(\mathsf{h}) \twoheadrightarrow \overline{\mathcal{K}}_{-1}(\mathsf{h}) \quad \textit{and} \quad \overline{\mathcal{K}}_{-1}^\vee(\mathsf{h}) \hookrightarrow \mathcal{K}_{-1}^\vee(\mathsf{h}).$$

PROOF. We use morphisms $\mathcal{K} \twoheadrightarrow \overline{\mathcal{K}}$ and $\overline{\mathcal{K}}^\vee \hookrightarrow \mathcal{K}^\vee$ of bilax functors in Theorems 14.14 and 14.15, and their signed analogues in Theorem 14.16. □

Remark 15.3. The following is implicit in Theorem 15.1 but worth pointing out separately. Since Fock functors are bilax, they are, in particular, lax and colax. Hence, they send Coxeter (co)monoids to Coxeter (co)algebras. A similar remark applies to many other results in this chapter; for convenience, we state them for Coxeter bimonoids and Coxeter bialgebras, but they are also valid more simply for Coxeter (co)monoids and Coxeter (co)algebras.

Exercise 15.4. We saw in Chapter 14 that Fock functors are $\Bbbk$-linear from the category of Coxeter species to the category of Coxeter spaces. Deduce: They continue to be $\Bbbk$-linear from the category of Coxeter (co, bi)monoids to the category of Coxeter (co, bi)algebras.

15.1.2. Fock spaces. We use the term *Fock space* for the value of a Fock functor. More precisely, $\mathcal{K}(\mathsf{h})$ and $\mathcal{K}^\vee(\mathsf{h})$ are full Fock spaces, $\mathcal{K}_{-1}(\mathsf{h})$ and $\mathcal{K}^\vee_{-1}(\mathsf{h})$ are signed full Fock spaces, $\overline{\mathcal{K}}(\mathsf{h})$ and $\overline{\mathcal{K}}^\vee(\mathsf{h})$ are bosonic Fock spaces, $\overline{\mathcal{K}}_{-1}(\mathsf{h})$ and $\overline{\mathcal{K}}^\vee_{-1}(\mathsf{h})$ are fermionic Fock spaces, and so on.

15.1.3. General (co)product formulas. Let $(\mathcal{F}, \varphi, \psi)$ be any Fock functor. If (a, μ) is a Coxeter monoid, then $\mathcal{F}(\mathsf{a})$ is a Coxeter algebra with product given by the composite

$$\overline{\mathcal{T}}\mathcal{F}(\mathsf{a}) \xrightarrow{\varphi_{\mathsf{a}}} \mathcal{F}\mathcal{T}(\mathsf{a}) \xrightarrow{\mathcal{F}\mu} \mathcal{F}(\mathsf{a})$$

(which involves the lax structure of $\mathcal{F}$ and product of a). If (c, Δ) is a Coxeter comonoid, then $\mathcal{F}(\mathsf{c})$ is a Coxeter coalgebra with coproduct given by the composite

$$\mathcal{F}(\mathsf{c}) \xrightarrow{\mathcal{F}\Delta} \mathcal{F}\mathcal{T}^\vee(\mathsf{c}) \xrightarrow{\psi_{\mathsf{c}}} \overline{\mathcal{T}}^\vee\mathcal{F}(\mathsf{c})$$

(which involves the colax structure of $\mathcal{F}$ and coproduct of c). Here μ and Δ are interpreted as in Sections 3.3.13 and 9.3.13. If $(\mathsf{h}, \mu, \Delta)$ is a Coxeter p-bimonoid, then $\mathcal{F}(\mathsf{h})$ carries both a product and a coproduct.

The above formulas come from the abstract theory of bimonads. We make them more explicit in the discussion below.

15.1.4. Deformed full Fock spaces. Suppose $(\mathsf{h}, \mu, \Delta)$ is a Coxeter p-bimonoid for a scalar p. We now describe the Coxeter pq-bialgebras $\mathcal{K}_q(\mathsf{h})$ and $\mathcal{K}^\vee_q(\mathsf{h})$. We work with the formulation in terms of faces and top-nested faces. Firstly, recall from (14.2) and (14.10) that

$$\mathcal{K}_q(\mathsf{h})[A, C] = \mathcal{K}^\vee_q(\mathsf{h})[A, C] := \mathsf{h}[A]. \tag{15.1}$$

(The underlying functors do not depend on q.) Using (14.5) and (14.21), for $A \leq F \leq C$, the product and coproduct of $\mathcal{K}_q(\mathsf{h})$ are

$$\begin{array}{ccccccc} \mathcal{K}_q(\mathsf{h})[F, C] & \xrightarrow{\mu^{F,C}_{A,C}} & \mathcal{K}_q(\mathsf{h})[A, C] & \qquad & \mathcal{K}_q(\mathsf{h})[A, C] & \xrightarrow{\Delta^{F,C}_{A,C}} & \mathcal{K}_q(\mathsf{h})[F, C] \\ \| & & \| & & \| & & \| \\ \mathsf{h}[F] & \xrightarrow[\mu^F_A]{} & \mathsf{h}[A] & & \mathsf{h}[A] & \longrightarrow & \mathsf{h}[F], \end{array} \tag{15.2}$$

where the map $\mathsf{h}[A] \to \mathsf{h}[F]$ in the coproduct formula is

$$x \longmapsto \sum_{\substack{G:\, G \geq A \\ \mathrm{t}(G) = \mathrm{t}(F)}} q^{\operatorname{dist}(GC, C)}\, w \cdot \Delta^G_A(x),$$

with w the Coxeter symmetry which sends GC to C. By type-naturality axiom of Δ in (2.36a), the formula can be rewritten as

$$x \longmapsto \Delta_A^F\Big(\sum_{\substack{G:\, G \geq A \\ \mathrm{t}(G)=\mathrm{t}(F)}} q^{\mathrm{dist}(GC,C)}\, w \cdot x \Big).$$

Dually, using (14.22) and (14.13), for $A \leq F \leq C$, the product and coproduct of $\mathcal{K}_q^\vee(\mathsf{h})$ are

$$(15.3)\qquad \begin{array}{ccc} \mathcal{K}_q^\vee(\mathsf{h})[F,C] & \overset{\mu_{A,C}^{F,C}}{\dashrightarrow} & \mathcal{K}_q^\vee(\mathsf{h})[A,C] \\ \| & & \| \\ \mathsf{h}[F] & \longrightarrow & \mathsf{h}[A] \end{array} \qquad \begin{array}{ccc} \mathcal{K}_q^\vee(\mathsf{h})[A,C] & \overset{\Delta_{A,C}^{F,C}}{\dashrightarrow} & \mathcal{K}_q^\vee(\mathsf{h})[F,C] \\ \| & & \| \\ \mathsf{h}[A] & \underset{\Delta_A^F}{\longrightarrow} & \mathsf{h}[F], \end{array}$$

where the map $\mathsf{h}[F] \to \mathsf{h}[A]$ in the product formula is

$$x \mapsto \sum_{\substack{G:\, G \geq A \\ \mathrm{t}(G)=\mathrm{t}(F)}} q^{\mathrm{dist}(GC,C)}\, \mu_A^G(w \cdot x),$$

with w the Coxeter symmetry which sends C to GC. By type-naturality axiom of μ in (2.28a), the formula can be rewritten as

$$x \longmapsto \sum_{\substack{G:\, G \geq A \\ \mathrm{t}(G)=\mathrm{t}(F)}} q^{\mathrm{dist}(GC,C)}\, w \cdot \mu_A^F(x).$$

Exercise 15.5. Check directly: If a is a Coxeter monoid, then $\mathcal{K}_q^\vee(\mathsf{a})$ is a Coxeter algebra with product given by the first formula in (15.3). That is, if a satisfies axioms (2.28a) and (2.28b), then $\mathcal{K}_q^\vee(\mathsf{a})$ satisfies axioms (8.34a) and (8.34b). Dually, if c is a Coxeter comonoid, then $\mathcal{K}_q(\mathsf{c})$ is a Coxeter coalgebra with coproduct given by the second formula in (15.2).

Exercise 15.6. Check: Full Fock functors preserve categorical (co)products. That is, for scalars p and q, for Coxeter p-bimonoids h and k,

$$\mathcal{K}_q(\mathsf{h}+\mathsf{k}) = \mathcal{K}_q(\mathsf{h}) + \mathcal{K}_q(\mathsf{k}) \quad \text{and} \quad \mathcal{K}_q^\vee(\mathsf{h}+\mathsf{k}) = \mathcal{K}_q^\vee(\mathsf{h}) + \mathcal{K}_q^\vee(\mathsf{k})$$

as Coxeter pq-bialgebras. See Exercises 2.32 and 8.83 in this regard.

15.1.5. Free Fock spaces. Suppose $(\mathsf{h}, \mu, \Delta)$ is a Coxeter p-bimonoid for a scalar p. Then

$$(15.4)\qquad \mathcal{K}_0(\mathsf{h}) = \mathcal{K}_0^\vee(\mathsf{h})$$

as Coxeter 0-bialgebras. This follows from (14.24). For $A \leq F \leq C$, the product and coproduct of $\mathcal{K}_0(\mathsf{h})$ are

$$(15.5)\qquad \begin{array}{ccc} \mathcal{K}_0(\mathsf{h})[F,C] & \overset{\mu_{A,C}^{F,C}}{\dashrightarrow} & \mathcal{K}_0(\mathsf{h})[A,C] \\ \| & & \| \\ \mathsf{h}[F] & \underset{\mu_A^F}{\longrightarrow} & \mathsf{h}[A] \end{array} \qquad \begin{array}{ccc} \mathcal{K}_0(\mathsf{h})[A,C] & \overset{\Delta_{A,C}^{F,C}}{\dashrightarrow} & \mathcal{K}_0(\mathsf{h})[F,C] \\ \| & & \| \\ \mathsf{h}[A] & \underset{\Delta_A^F}{\longrightarrow} & \mathsf{h}[F]. \end{array}$$

Exercise 15.7. Check directly: Formulas (15.5) satisfy the 0-bialgebra axiom (8.76). (Use the q-bimonoid axiom (2.42) for $q := p$ and $FG = GF$.)

15.1.6. Bosonic and fermionic Fock spaces. Suppose $(\mathsf{h}, \mu, \Delta)$ is a Coxeter p-bimonoid for a scalar p. Recall from (14.27) and (14.30) that

$$\overline{\mathcal{K}}(\mathsf{h})[A, C] = \mathsf{h}[A]_{W_A} \quad \text{and} \quad \overline{\mathcal{K}}^{\vee}(\mathsf{h})[A, C] = \mathsf{h}[A]^{W_A}. \tag{15.6}$$

The product and coproduct of the Coxeter p-bialgebra $\overline{\mathcal{K}}(\mathsf{h})$ are induced from formulas (15.2) (for $q = 1$) of $\mathcal{K}(\mathsf{h})$ via the quotient maps $\mathsf{h}[A] \twoheadrightarrow \mathsf{h}[A]_{W_A}$. Similarly, the product and coproduct of $\overline{\mathcal{K}}^{\vee}(\mathsf{h})$ are induced from formulas (15.3) (for $q = 1$) of $\mathcal{K}^{\vee}(\mathsf{h})$ via the inclusion maps $\mathsf{h}[A]^{W_A} \hookrightarrow \mathsf{h}[A]$.

Explicitly, for product of $\overline{\mathcal{K}}(\mathsf{h})$ and coproduct of $\overline{\mathcal{K}}^{\vee}(\mathsf{h})$, we have commutative diagrams

$$\begin{array}{ccc} \mathsf{h}[F] & \xrightarrow{\mu_A^F} & \mathsf{h}[A] \\ \downarrow & & \downarrow \\ \mathsf{h}[F]_{W_F} & \dashrightarrow & \mathsf{h}[A]_{W_A} \end{array} \qquad\qquad \begin{array}{ccc} \mathsf{h}[A] & \xrightarrow{\Delta_A^F} & \mathsf{h}[F] \\ \uparrow & & \uparrow \\ \mathsf{h}[A]^{W_A} & \dashrightarrow & \mathsf{h}[F]^{W_F}. \end{array} \tag{15.7}$$

These are easy to see directly from the type-naturality axioms for μ and Δ, and the fact that W_F is a subgroup of W_A for $F \geq A$.

Similarly, (co)products of the Coxeter $(-p)$-bialgebras $\overline{\mathcal{K}}_{-1}(\mathsf{h})$ and $\overline{\mathcal{K}}_{-1}^{\vee}(\mathsf{h})$ are induced from formulas (15.2) and (15.3) (for $q = -1$) by using signed (co)invariants.

15.1.7. Fock spaces in terms of face-types. Let us reformulate full Fock spaces in terms of face-types. Let h be a Coxeter p-bimonoid for a scalar p. Recall from (14.14) that

$$\mathcal{K}_q(\mathsf{h})[T] = \mathcal{K}_q^{\vee}(\mathsf{h})[T] := \mathsf{h}[F], \tag{15.8}$$

where F is the face of C_r of type T. For $Z \leq T$, the product and coproduct of $\mathcal{K}_q(\mathsf{h})$ are

$$\begin{array}{ccc} \mathcal{K}_q(\mathsf{h})[T] & \overset{\mu_Z^T}{\dashrightarrow} & \mathcal{K}_q(\mathsf{h})[Z] \\ \| & & \| \\ \mathsf{h}[F] & \xrightarrow[\mu_A^F]{} & \mathsf{h}[A] \end{array} \qquad\qquad \begin{array}{ccc} \mathcal{K}_q(\mathsf{h})[Z] & \overset{\Delta_Z^T}{\dashrightarrow} & \mathcal{K}_q(\mathsf{h})[T] \\ \| & & \| \\ \mathsf{h}[A] & \longrightarrow & \mathsf{h}[F], \end{array} \tag{15.9}$$

where A and F are faces of C_r of types Z and T, respectively, and the map $\mathsf{h}[A] \to \mathsf{h}[F]$ is as in (15.2). Dually, the product and coproduct of $\mathcal{K}_q^{\vee}(\mathsf{h})$ are

$$\begin{array}{ccc} \mathcal{K}_q^{\vee}(\mathsf{h})[T] & \overset{\mu_Z^T}{\dashrightarrow} & \mathcal{K}_q^{\vee}(\mathsf{h})[Z] \\ \| & & \| \\ \mathsf{h}[F] & \longrightarrow & \mathsf{h}[A] \end{array} \qquad\qquad \begin{array}{ccc} \mathcal{K}_q^{\vee}(\mathsf{h})[Z] & \overset{\Delta_Z^T}{\dashrightarrow} & \mathcal{K}_q^{\vee}(\mathsf{h})[T] \\ \| & & \| \\ \mathsf{h}[A] & \xrightarrow[\Delta_A^F]{} & \mathsf{h}[F], \end{array} \tag{15.10}$$

with notations as above, and the map $\mathsf{h}[F] \to \mathsf{h}[A]$ as in (15.3).

Bosonic Fock spaces in terms of face-types can be described similarly starting with (14.33).

15.1.8. Fock spaces in terms of flats and lunes. For h a Coxeter p-bimonoid, we now describe the Coxeter pq-bialgebra $\mathcal{K}_q(\mathsf{h})$ in terms of flats and lunes. Firstly, for a top-lune L,

$$\mathcal{K}_q(\mathsf{h})[\mathrm{L}] = \mathsf{h}[\mathrm{b}(\mathrm{L})]. \tag{15.11}$$

For top-lunes $\mathrm{L} \leq' \mathrm{M}$, the product and coproduct of $\mathcal{K}_q(\mathsf{h})$ are given by

$$\begin{array}{ccc} \mathcal{K}_q(\mathsf{h})[\mathrm{M}] & \xrightarrow{\mu^{\mathrm{M}}_{\mathrm{L}}} & \mathcal{K}_q(\mathsf{h})[\mathrm{L}] \\ \| & & \| \\ \mathsf{h}[\mathrm{b}(\mathrm{M})] & \xrightarrow[\mu_{\mathrm{L}/\mathrm{M}}]{} & \mathsf{h}[\mathrm{b}(\mathrm{L})] \end{array} \qquad \begin{array}{ccc} \mathcal{K}_q(\mathsf{h})[\mathrm{L}] & \xrightarrow{\Delta^{\mathrm{M}}_{\mathrm{L}}} & \mathcal{K}_q(\mathsf{h})[\mathrm{M}] \\ \| & & \| \\ \mathsf{h}[\mathrm{b}(\mathrm{L})] & \longrightarrow & \mathsf{h}[\mathrm{b}(\mathrm{M})], \end{array} \tag{15.12}$$

where the map $\mathsf{h}[\mathrm{b}(\mathrm{L})] \to \mathsf{h}[\mathrm{b}(\mathrm{M})]$ in the coproduct formula is

$$x \longmapsto \sum_{\substack{\mathrm{P}:\, \mathrm{b}(\mathrm{P})=\mathrm{b}(\mathrm{L}) \\ \mathrm{t}(\mathrm{P})=\mathrm{t}(\mathrm{L}/\mathrm{M})}} q^{\mathrm{dist}(\mathrm{PL},\mathrm{L})}\, w \cdot (\Delta_{\mathrm{P}}(x)),$$

with $w(\mathrm{PL}) = \mathrm{L}$ (and hence $w(\mathrm{P}) = \mathrm{L}/\mathrm{M}$).

The Coxeter pq-bialgebra $\mathcal{K}^{\vee}_q(\mathsf{h})$ can be described dually. For $q = 0$, we have $\mathcal{K}_0(\mathsf{h}) = \mathcal{K}^{\vee}_0(\mathsf{h})$ with product and coproduct

$$\begin{array}{ccc} \mathcal{K}_0(\mathsf{h})[\mathrm{M}] & \xrightarrow{\mu^{\mathrm{M}}_{\mathrm{L}}} & \mathcal{K}_0(\mathsf{h})[\mathrm{L}] \\ \| & & \| \\ \mathsf{h}[\mathrm{b}(\mathrm{M})] & \xrightarrow[\mu_{\mathrm{L}/\mathrm{M}}]{} & \mathsf{h}[\mathrm{b}(\mathrm{L})] \end{array} \qquad \begin{array}{ccc} \mathcal{K}_0(\mathsf{h})[\mathrm{L}] & \xrightarrow{\Delta^{\mathrm{M}}_{\mathrm{L}}} & \mathcal{K}_0(\mathsf{h})[\mathrm{M}] \\ \| & & \| \\ \mathsf{h}[\mathrm{b}(\mathrm{L})] & \xrightarrow[\Delta_{\mathrm{L}/\mathrm{M}}]{} & \mathsf{h}[\mathrm{b}(\mathrm{M})]. \end{array} \tag{15.13}$$

15.1.9. Relations among Fock spaces. The Coxeter q-bimonoid of chambers from Section 4.3 can be used to relate q-full Fock spaces with bosonic and fermionic Fock spaces.

Proposition 15.8. *For scalars p and q, for a Coxeter p-bimonoid* h, *there are isomorphisms of Coxeter pq-bialgebras*

$$\mathcal{K}_q(\mathsf{h}) \cong \overline{\mathcal{K}}(\Gamma_q \times \mathsf{h}), \quad \mathcal{K}^{\vee}_q(\mathsf{h}) \cong \overline{\mathcal{K}}^{\vee}(\Gamma^*_q \times \mathsf{h}),$$
$$\mathcal{K}_q(\mathsf{h}) \cong \overline{\mathcal{K}}_{-1}(\Gamma_{-q} \times \mathsf{h}), \quad \mathcal{K}^{\vee}_q(\mathsf{h}) \cong \overline{\mathcal{K}}^{\vee}_{-1}(\Gamma^*_{-q} \times \mathsf{h}).$$

PROOF. This follows from Propositions 14.22 and 14.24. □

15.1.10. Interaction with signature functors. Recall the o-signature functor $(-)^{-}$ on Coxeter species, and the u-signature functors $(-)_{-}$ on Coxeter species and on Coxeter spaces.

Observation 15.9. For scalars p and q, for a Coxeter p-bimonoid h,

$$\mathcal{K}_q(\mathsf{h}^{-}) = \mathcal{K}_{-q}(\mathsf{h}) \quad \text{and} \quad \mathcal{K}_q(\mathsf{h}_{-}) = \mathcal{K}_q(\mathsf{h})_{-} \tag{15.14}$$

as Coxeter $(-pq)$-bialgebras. The same identities hold with $\mathcal{K}_q$ replaced by $\mathcal{K}^{\vee}_q$. This follows from Proposition 14.12 and Exercise 14.13.

Similarly, for a scalar p, for a Coxeter p-bimonoid h,

$$\begin{aligned} \overline{\mathcal{K}}(\mathsf{h}^{-}) &= \overline{\mathcal{K}}_{-1}(\mathsf{h}), & \overline{\mathcal{K}}_{-1}(\mathsf{h}^{-}) &= \overline{\mathcal{K}}(\mathsf{h}), \\ \overline{\mathcal{K}}(\mathsf{h}_{-}) &= \overline{\mathcal{K}}(\mathsf{h})_{-}, & \overline{\mathcal{K}}_{-1}(\mathsf{h}_{-}) &= \overline{\mathcal{K}}_{-1}(\mathsf{h})_{-} \end{aligned} \tag{15.15}$$

as Coxeter $(-p)$-bialgebras and Coxeter p-bialgebras, respectively. The same identities hold for $\overline{\mathcal{K}}^\vee$ and $\overline{\mathcal{K}}^\vee_{-1}$. This follows from (14.34) and Exercise 14.18.

Exercise 15.10. Deduce identities (15.14) from (15.15) using Proposition 15.8.

15.1.11. Duality. Observe from formulas (15.2) and (15.3): For scalars p and q, for h a Coxeter p-bimonoid,

$$\mathcal{K}_q(\mathsf{h}^*) = \mathcal{K}^\vee_q(\mathsf{h})^* \quad \text{and} \quad \mathcal{K}^\vee_q(\mathsf{h}^*) = \mathcal{K}_q(\mathsf{h})^* \tag{15.16}$$

as Coxeter pq-bialgebras.

Similarly, for a scalar p, for h a Coxeter p-bimonoid,

$$\overline{\mathcal{K}}(\mathsf{h}^*) = \overline{\mathcal{K}}^\vee(\mathsf{h})^* \quad \text{and} \quad \overline{\mathcal{K}}^\vee(\mathsf{h}^*) = \overline{\mathcal{K}}(\mathsf{h})^* \tag{15.17}$$

as Coxeter p-bialgebras (the former if h is finite-dimensional).

15.1.12. Trivial product and coproduct. Suppose a is a Coxeter monoid with trivial product, that is, $\mu_A^F = 0$ for all $F > A$. Then the Coxeter algebras $\mathcal{K}(\mathsf{a})$, $\mathcal{K}^\vee(\mathsf{a})$, $\overline{\mathcal{K}}(\mathsf{a})$, $\overline{\mathcal{K}}^\vee(\mathsf{a})$ all have trivial products, that is, $\mu_{A,C}^{F,C} = 0$ for all $F > A$. A similar remark applies to trivial coproducts.

15.2. Commutativity under Fock functors

We saw that Fock functors send Coxeter bimonoids to Coxeter bialgebras. Let us now consider the commutativity issue. Here the core question is of the following nature. For $\mathcal{F}$ a Fock functor, if a is a commutative Coxeter monoid, then is $\mathcal{F}(\mathsf{a})$ a commutative Coxeter algebra, and similarly, if c is a cocommutative Coxeter comonoid, then is $\mathcal{F}(\mathsf{c})$ a cocommutative Coxeter coalgebra? It turns out that of the two full Fock functors, one preserves commutativity, while the other preserves cocommutativity. In contrast, the bosonic (and hence fermionic) Fock functors preserve both.

15.2.1. Commutativity under Fock functors. To answer the commutativity question, one needs to understand how $\mathcal{F}$ interacts with monads $\mathcal{S}$ and $\mathcal{S}^\vee$. This was essentially done in Section 14.6 in Propositions 14.26, 14.28, 14.29, 14.32, where we studied interactions of Fock functors with opposite transformations. As a consequence:

Theorem 15.11. *For full Fock functors and bosonic Fock functors,*

1. *$\mathcal{K}$ preserves (signed) cocommutativity, that is, if a Coxeter comonoid* c *is (signed) cocommutative, then so is the Coxeter coalgebra* $\mathcal{K}(\mathsf{c})$,
2. *$\mathcal{K}^\vee$ preserves (signed) commutativity, that is, if a Coxeter monoid* a *is (signed) commutative, then so is the Coxeter algebra* $\mathcal{K}^\vee(\mathsf{a})$,
3. *$\overline{\mathcal{K}}$ and $\overline{\mathcal{K}}^\vee$ preserve both (signed) commutativity and (signed) cocommutativity.*

Theorem 15.12. *For signed full Fock functors and fermionic Fock functors,*

1. *$\mathcal{K}_{-1}$ interchanges cocommutativity and signed cocommutativity, that is, if* c *is cocommutative, then* $\mathcal{K}_{-1}(\mathsf{c})$ *is signed cocommutative, and vice versa,*

(2) $\mathcal{K}^{\vee}_{-1}$ *interchanges commutativity and signed commutativity, that is, if* a *is commutative, then* $\mathcal{K}^{\vee}_{-1}(\mathsf{a})$ *is signed commutative, and vice versa,*

(3) $\overline{\mathcal{K}}_{-1}$ *and* $\overline{\mathcal{K}}^{\vee}_{-1}$ *interchange commutativity and signed commutativity, and also cocommutativity and signed cocommutativity.*

Exercise 15.13. Verify Theorem 15.11, item (2) directly. That is, if a satisfies the commutativity axiom (2.49), then $\mathcal{K}^{\vee}(\mathsf{a})$ satisfies the commutativity axiom (8.79). The signed case is similar. The abstract argument for proving item (2) of Theorem 15.11 using opposite transformations, in direct terms, translates to the opposition commutativity axioms (2.51) and (8.81). We emphasize that Lemma 2.16 plays a key role here.

Dual remarks apply to Theorem 15.11, item (1).

15.2.2. Op and cop constructions.

We now look at some more general considerations. Recall the op and cop constructions on Coxeter (co)monoids from Section 2.11.3 and on Coxeter (co)algebras from Section 8.13.4.

Exercise 15.14. Let p and q be nonzero scalars. Use Proposition 14.33 to deduce: For a Coxeter comonoid c and Coxeter monoid a,

$$\mathcal{K}_{q^{-1}}(\mathsf{c}^{\mathrm{cop}}) = \mathcal{K}_q(\mathsf{c})^{\mathrm{cop}} \quad \text{and} \quad \mathcal{K}^{\vee}_{q^{-1}}({}^{\mathrm{op}}\mathsf{a}) = {}^{\mathrm{op}}\mathcal{K}^{\vee}_q(\mathsf{a}), \tag{15.18}$$

where cop on c and op on a are taken wrt the scalar p, while cop on $\mathcal{K}_q(\mathsf{c})$ and op on $\mathcal{K}^{\vee}_q(\mathsf{a})$ are taken wrt the scalar pq. Further, for a Coxeter p-bimonoid h,

$$\mathcal{K}_{q^{-1}}(\mathsf{h}^{\mathrm{cop}}) = \mathcal{K}_q(\mathsf{h})^{\mathrm{cop}} \quad \text{and} \quad \mathcal{K}^{\vee}_{q^{-1}}({}^{\mathrm{op}}\mathsf{h}) = {}^{\mathrm{op}}\mathcal{K}^{\vee}_q(\mathsf{h}) \tag{15.19}$$

as Coxeter $(pq)^{-1}$-bialgebras.

Deduce Theorem 15.11, items (1), (2) as a consequence: $p = 1$, $q = 1$ deal with commutativity and $p = -1$, $q = 1$ with signed commutativity.

Check: (15.18) and (15.19) also hold for $q = 1$, with $\mathcal{K}$ and $\mathcal{K}^{\vee}$ replaced by $\overline{\mathcal{K}}$ and $\overline{\mathcal{K}}^{\vee}$. This relates to Theorem 15.11, item (3).

Exercise 15.15 (Half-twist transformation). Use Proposition 14.36 to deduce: For a Coxeter monoid a, the map

$$\mathcal{K}(\mathsf{a}^{\mathrm{op}}) \xrightarrow{\cong} \mathcal{K}(\mathsf{a})^{\mathrm{op}} \tag{15.20}$$

whose (A, C)-component is (14.47) with $\mathsf{p} := \mathsf{a}$ is an isomorphism of Coxeter algebras. Here the op on a and on $\mathcal{K}(\mathsf{a})$ are taken wrt the scalar 1.

Let p and q be nonzero scalars. Use Proposition 14.37 to deduce: For a Coxeter p-bimonoid h, the map

$$\mathcal{K}_q(\mathsf{h}^{\mathrm{op,cop}}) \xrightarrow{\cong} \mathcal{K}_q(\mathsf{h})^{\mathrm{op,cop}} \tag{15.21}$$

whose (A, C)-component is (14.50) with $\mathsf{p} := \mathsf{h}$ is an isomorphism of Coxeter pq-bialgebras.

15.3. Norm map between Fock spaces

Recall the q-norm transformation κ_q from Section 14.7. Its bilax property allows us to relate the two q-full Fock spaces with each other.

Proposition 15.16. *For scalars p and q, for a Coxeter p-bimonoid* h, *we have a morphism of Coxeter pq-bialgebras*

$$(\kappa_q)_{\mathsf{h}} : \mathcal{K}_q(\mathsf{h}) \to \mathcal{K}^{\vee}_q(\mathsf{h}). \tag{15.22}$$

Explicitly, $(\kappa_q)_{\mathsf{h}}$ on the (A, C)-component is

$$\begin{array}{ccc} \mathcal{K}_q(\mathsf{h})[A,C] & \longrightarrow & \mathcal{K}_q^{\vee}(\mathsf{h})[A,C] \\ \| & & \| \\ \mathsf{h}[A] & & \mathsf{h}[A] \end{array}, \qquad (\kappa_q)_{\mathsf{h}}(x) = \sum_{w \in W_A} q^{l(w)} \mathsf{h}[w](x).$$

It is an isomorphism if q is not a root of unity.

PROOF. The first statement follows from Proposition 14.40 and the second from Proposition 14.48. □

Proposition 15.17. *For p a scalar, and h a Coxeter p-bimonoid, there are commutative diagrams*

$$(15.23) \qquad \begin{array}{ccc} \mathcal{K}(\mathsf{h}) & \xrightarrow{\kappa_{\mathsf{h}}} & \mathcal{K}^{\vee}(\mathsf{h}) \\ \downarrow & & \uparrow \\ \overline{\mathcal{K}}(\mathsf{h}) & \xrightarrow[\overline{\kappa}_{\mathsf{h}}]{} & \overline{\mathcal{K}}^{\vee}(\mathsf{h}) \end{array} \qquad \begin{array}{ccc} \mathcal{K}_{-1}(\mathsf{h}) & \xrightarrow{(\kappa_{-1})_{\mathsf{h}}} & \mathcal{K}_{-1}^{\vee}(\mathsf{h}) \\ \downarrow & & \uparrow \\ \overline{\mathcal{K}}_{-1}(\mathsf{h}) & \xrightarrow[(\overline{\kappa}_{-1})_{\mathsf{h}}]{} & \overline{\mathcal{K}}_{-1}^{\vee}(\mathsf{h}), \end{array}$$

the former of Coxeter p-bialgebras, and the latter of Coxeter $(-p)$-bialgebras. In addition, $\overline{\kappa}_{\mathsf{h}}$ and $(\overline{\kappa}_{-1})_{\mathsf{h}}$ are isomorphisms if $\mathrm{char}(\Bbbk) \nmid |W|$.

PROOF. This follows from Propositions 14.44 and 14.46. □

Example 15.18. Recall the Coxeter species of chambers Γ from Section 4.3. The map $\kappa_\Gamma : \mathcal{K}(\Gamma) \to \mathcal{K}^{\vee}(\Gamma)$ on the (A, C)-component is

$$\Gamma[A] \to \Gamma[A], \qquad \mathtt{H}_{D/A} \mapsto \sum_{w \in W_A} \mathtt{H}_{wD/A} = \sum_{D' :\, D' \geq A} \mathtt{H}_{D'/A}.$$

The image of this linear map is one-dimensional, so κ_Γ is far from being an isomorphism. In contrast: The map $(\kappa_q)_\Gamma : \mathcal{K}_q(\Gamma) \to \mathcal{K}_q^{\vee}(\Gamma)$ on the (A, C)-component is

$$\Gamma[A] \to \Gamma[A], \qquad \mathtt{H}_{D/A} \mapsto \sum_{w \in W_A} q^{l(w)} \mathtt{H}_{wD/A},$$

with $l(w)$ taken wrt C. This map is an isomorphism when q is not a root of unity. See [**10**, page 232] in this regard. □

Recall the anyonic Fock functor $\mathfrak{S}_q$ from Section 14.8.

Proposition 15.19. *For scalars p and q, for a Coxeter p-bimonoid h, we have the commutative diagram*

$$(15.24) \qquad \begin{array}{ccc} \mathcal{K}_q(\mathsf{h}) & \xrightarrow{(\kappa_q)_{\mathsf{h}}} & \mathcal{K}_q^{\vee}(\mathsf{h}) \\ & \searrow \quad \nearrow & \\ & \mathfrak{S}_q(\mathsf{h}) & \end{array}$$

of Coxeter pq-bialgebras.

PROOF. This follows from Proposition 14.49. □

Observation 15.20. For scalars p and q, for a Coxeter p-bimonoid h,

$$(\kappa_q)_{\mathsf{h}^-} = (\kappa_{-q})_{\mathsf{h}} \quad \text{and} \quad (\kappa_q)_{\mathsf{h}_-} = ((\kappa_q)_{\mathsf{h}})_- \tag{15.25}$$

as morphisms of Coxeter $(-pq)$-bialgebras. This builds on (15.14), and follows from Proposition 14.42 and Exercise 14.43.

Similarly, for a scalar p, for a Coxeter p-bimonoid h,

$$\begin{aligned} \overline{\kappa}_{\mathsf{h}^-} &= (\overline{\kappa}_{-1})_{\mathsf{h}}, \qquad & (\overline{\kappa}_{-1})_{\mathsf{h}^-} &= \overline{\kappa}_{\mathsf{h}}, \\ \overline{\kappa}_{\mathsf{h}_-} &= (\overline{\kappa}_{\mathsf{h}})_-, \qquad & (\overline{\kappa}_{-1})_{\mathsf{h}_-} &= ((\overline{\kappa}_{-1})_{\mathsf{h}})_- \end{aligned} \tag{15.26}$$

as morphisms of Coxeter $(-p)$-bialgebras and Coxeter p-bialgebras, respectively. See (15.15) and Exercise 14.47 in this regard.

15.4. (Co)freeness under Fock functors

Recall the universal constructions on Coxeter species from Section 3.3 and on Coxeter spaces from Section 9.3. These can be linked together using Fock functors by employing their lax and colax structures. For bosonic and fermionic Fock functors one can say more, namely, that they preserve both freeness and cofreeness. This is because their lax and colax structures are isomorphisms.

15.4.1. (Co)freeness under full Fock functors. Recall the free Coxeter monoid $\mathcal{T}(\mathsf{p})$ and Coxeter q-bimonoid $\mathcal{T}_q(\mathsf{c})$ from Sections 3.3.1 and 3.3.2, and the free Coxeter algebra $\overline{\mathcal{T}}(\mathrm{V})$ and Coxeter q-bialgebra $\overline{\mathcal{T}_q}(\mathrm{C})$ from Sections 9.3.1 and 9.3.2.

Proposition 15.21. *For a Coxeter species* p, *there are morphisms of Coxeter algebras*

$$\varphi_{\mathsf{p}} : \overline{\mathcal{T}}(\mathcal{K}(\mathsf{p})) \hookrightarrow \mathcal{K}(\mathcal{T}(\mathsf{p})) \quad \textit{and} \quad \psi_{\mathsf{p}}^{\vee} : \overline{\mathcal{T}}(\mathcal{K}^{\vee}(\mathsf{p})) \hookrightarrow \mathcal{K}^{\vee}(\mathcal{T}(\mathsf{p}))$$

given by lax structures of $\mathcal{K}$ *and* $\mathcal{K}^{\vee}$ *in* (14.5) *and* (14.11), *respectively. More generally, for a scalar* q, *for a Coxeter comonoid* c, *there are morphisms of Coxeter* q*-bialgebras*

$$\varphi_{\mathsf{c}} : \overline{\mathcal{T}_q}(\mathcal{K}(\mathsf{c})) \hookrightarrow \mathcal{K}(\mathcal{T}_q(\mathsf{c})) \quad \textit{and} \quad \psi_{\mathsf{c}}^{\vee} : \overline{\mathcal{T}_q}(\mathcal{K}^{\vee}(\mathsf{c})) \hookrightarrow \mathcal{K}^{\vee}(\mathcal{T}_q(\mathsf{c})).$$

PROOF. The second part follows from Theorems 14.2 and 14.4 and [**11**, Proposition C.36]. For the first part, we may either directly use [**11**, Proposition C.35], or deduce it as a special case of the second part by viewing p as a trivial Coxeter comonoid. □

Dually, recall the cofree Coxeter comonoid $\mathcal{T}^{\vee}(\mathsf{p})$ and Coxeter q-bimonoid $\mathcal{T}_q^{\vee}(\mathsf{a})$ from Sections 3.3.4 and 3.3.5, and the cofree Coxeter coalgebra $\overline{\mathcal{T}}^{\vee}(\mathrm{V})$ and Coxeter q-bialgebra $\overline{\mathcal{T}_q}^{\vee}(\mathrm{A})$ from Sections 9.3.4 and 9.3.5.

Proposition 15.22. *For a Coxeter species* p, *there are morphisms of Coxeter coalgebras*

$$\psi_{\mathsf{p}} : \mathcal{K}(\mathcal{T}^{\vee}(\mathsf{p})) \twoheadrightarrow \overline{\mathcal{T}}^{\vee}(\mathcal{K}(\mathsf{p})) \quad \textit{and} \quad \varphi_{\mathsf{p}}^{\vee} : \mathcal{K}^{\vee}(\mathcal{T}^{\vee}(\mathsf{p})) \twoheadrightarrow \overline{\mathcal{T}}^{\vee}(\mathcal{K}^{\vee}(\mathsf{p}))$$

given by colax structures of $\mathcal{K}$ *and* $\mathcal{K}^\vee$ *in* (14.6) *and* (14.13)*, respectively. More generally, for a scalar* q*, for a Coxeter monoid* a*, there are morphisms of Coxeter* q*-bialgebras*

$$\psi_{\mathsf{a}} : \mathcal{K}(\mathcal{T}_q^\vee(\mathsf{a})) \twoheadrightarrow \overline{\mathcal{T}}_q^\vee(\mathcal{K}(\mathsf{a})) \quad \text{and} \quad \varphi_{\mathsf{a}}^\vee : \mathcal{K}^\vee(\mathcal{T}_q^\vee(\mathsf{a})) \twoheadrightarrow \overline{\mathcal{T}}_q^\vee(\mathcal{K}^\vee(\mathsf{a})).$$

Exercise 15.23. Check: In Proposition 15.21, the map $\varphi_{\mathsf{p}} : \overline{\mathcal{T}}(\mathcal{K}(\mathsf{p})) \to \mathcal{K}(\mathcal{T}(\mathsf{p}))$ arises from the universal property of $\overline{\mathcal{T}}$ in Theorem 9.11 applied to the map $\mathcal{K}(\mathsf{p}) \to \mathcal{K}(\mathcal{T}(\mathsf{p}))$ arising from the inclusion $\mathsf{p} \hookrightarrow \mathcal{T}(\mathsf{p})$. A similar remark applies to the map $\psi_{\mathsf{p}}^\vee$.

Dually, in Proposition 15.22, the map $\psi_{\mathsf{p}} : \mathcal{K}(\mathcal{T}^\vee(\mathsf{p})) \to \overline{\mathcal{T}}^\vee(\mathcal{K}(\mathsf{p}))$ arises from the universal property of $\overline{\mathcal{T}}^\vee$ in Theorem 9.16 applied to the map $\mathcal{K}(\mathcal{T}^\vee(\mathsf{p})) \to \mathcal{K}(\mathsf{p})$ arising from the projection $\mathcal{T}^\vee(\mathsf{p}) \twoheadrightarrow \mathsf{p}$. A similar remark applies to the map $\varphi_{\mathsf{p}}^\vee$.

These results also follow formally from the general theory of bimonads.

15.4.2. Bosonic Fock functors preserve (co)freeness.

The bilax structure maps of bosonic Fock functors $\overline{\mathcal{K}}$ and $\overline{\mathcal{K}}^\vee$ are isomorphisms, in contrast to those of full Fock functors $\mathcal{K}$ and $\mathcal{K}^\vee$. See the last part of Theorems 14.14 and 14.15. As a result, bosonic Fock functors preserve freeness and cofreeness. More precisely:

Proposition 15.24. *For a Coxeter species* p*, there are isomorphisms of Coxeter algebras*

$$\overline{\varphi}_{\mathsf{p}} : \overline{\mathcal{T}}(\overline{\mathcal{K}}(\mathsf{p})) \xrightarrow{\cong} \overline{\mathcal{K}}(\mathcal{T}(\mathsf{p})) \quad \text{and} \quad \overline{\psi}_{\mathsf{p}}^\vee : \overline{\mathcal{T}}(\overline{\mathcal{K}}^\vee(\mathsf{p})) \xrightarrow{\cong} \overline{\mathcal{K}}^\vee(\mathcal{T}(\mathsf{p}))$$

given by lax structures of $\overline{\mathcal{K}}$ *and* $\overline{\mathcal{K}}^\vee$ *in* (14.28) *and* (14.31)*, respectively. More generally, for a scalar* q*, for a Coxeter comonoid* c*, there are isomorphisms of Coxeter* q*-bialgebras*

$$\overline{\varphi}_{\mathsf{c}} : \overline{\mathcal{T}}_q(\overline{\mathcal{K}}(\mathsf{c})) \xrightarrow{\cong} \overline{\mathcal{K}}(\mathcal{T}_q(\mathsf{c})) \quad \text{and} \quad \overline{\psi}_{\mathsf{c}}^\vee : \overline{\mathcal{T}}_q(\overline{\mathcal{K}}^\vee(\mathsf{c})) \xrightarrow{\cong} \overline{\mathcal{K}}^\vee(\mathcal{T}_q(\mathsf{c})).$$

Dually:

Proposition 15.25. *For a Coxeter species* p*, there are isomorphisms of Coxeter coalgebras*

$$\overline{\psi}_{\mathsf{p}} : \overline{\mathcal{K}}(\mathcal{T}^\vee(\mathsf{p})) \xrightarrow{\cong} \overline{\mathcal{T}}^\vee(\overline{\mathcal{K}}(\mathsf{p})) \quad \text{and} \quad \overline{\varphi}_{\mathsf{p}}^\vee : \overline{\mathcal{K}}^\vee(\mathcal{T}^\vee(\mathsf{p})) \xrightarrow{\cong} \overline{\mathcal{T}}^\vee(\overline{\mathcal{K}}^\vee(\mathsf{p}))$$

given by colax structures of $\overline{\mathcal{K}}$ *and* $\overline{\mathcal{K}}^\vee$ *in* (14.29) *and* (14.32)*, respectively. More generally, for a scalar* q*, for a Coxeter monoid* a*, there are isomorphisms of Coxeter* q*-bialgebras*

$$\overline{\psi}_{\mathsf{a}} : \overline{\mathcal{K}}(\mathcal{T}_q^\vee(\mathsf{a})) \xrightarrow{\cong} \overline{\mathcal{T}}_q^\vee(\overline{\mathcal{K}}(\mathsf{a})) \quad \text{and} \quad \overline{\varphi}_{\mathsf{a}}^\vee : \overline{\mathcal{K}}^\vee(\mathcal{T}_q^\vee(\mathsf{a})) \xrightarrow{\cong} \overline{\mathcal{T}}_q^\vee(\overline{\mathcal{K}}^\vee(\mathsf{a})).$$

The maps in Propositions 15.24 and 15.25 arise from universal properties as in Exercise 15.23.

Exercise 15.26. By the preceding results:

- If a is a free Coxeter monoid, then $\overline{\mathcal{K}}(\mathsf{a})$ is a free Coxeter algebra.
- If c is a cofree Coxeter comonoid, then $\overline{\mathcal{K}}^\vee(\mathsf{c})$ is a cofree Coxeter coalgebra.

Show: The converses are false in general.

15.4.3. Fermionic Fock functors preserve (co)freeness. Similarly, using Theorem 14.16, we deduce that fermionic Fock functors preserve freeness and cofreeness. This is summarized below.

Proposition 15.27. *Let q be a scalar. For a Coxeter comonoid c, there are isomorphisms of Coxeter $(-q)$-bialgebras*

$$\overline{\mathcal{T}}_{-q}(\overline{\mathcal{K}}_{-1}(\mathsf{c})) \cong \overline{\mathcal{K}}_{-1}(\mathcal{T}_q(\mathsf{c})) \quad \textit{and} \quad \overline{\mathcal{T}}_{-q}(\overline{\mathcal{K}}^{\vee}_{-1}(\mathsf{c})) \cong \overline{\mathcal{K}}^{\vee}_{-1}(\mathcal{T}_q(\mathsf{c})).$$

For a Coxeter monoid a, there are isomorphisms of Coxeter $(-q)$-bialgebras

$$\overline{\mathcal{K}}_{-1}(\mathcal{T}^{\vee}_q(\mathsf{a})) \cong \overline{\mathcal{T}}^{\vee}_{-q}(\overline{\mathcal{K}}_{-1}(\mathsf{a})) \quad \textit{and} \quad \overline{\mathcal{K}}^{\vee}_{-1}(\mathcal{T}^{\vee}_q(\mathsf{a})) \cong \overline{\mathcal{T}}^{\vee}_{-q}(\overline{\mathcal{K}}^{\vee}_{-1}(\mathsf{a})).$$

This result can also be formally deduced from Propositions 15.24 and 15.25. For instance, the first isomorphism is obtained as the composite

$$\overline{\mathcal{T}}_{-q}\overline{\mathcal{K}}_{-1}(\mathsf{c}) \cong \overline{\mathcal{T}}_{-q}\overline{\mathcal{K}}(\mathsf{c}^-) \cong \overline{\mathcal{K}}\mathcal{T}_{-q}(\mathsf{c}^-) \cong \overline{\mathcal{K}}(\mathcal{T}_q(\mathsf{c})^-) \cong \overline{\mathcal{K}}_{-1}\mathcal{T}_q(\mathsf{c}).$$

This used Exercise 3.17.

15.4.4. Commutative aspects. As a consequence of Propositions 14.26 and 14.28 and subsequent discussion, we have the following commutative analogue for full Fock functors.

Proposition 15.28. *For a Coxeter comonoid c, there is a morphism of commutative Coxeter bialgebras*

$$\psi^{\vee}_{\mathsf{c}} : \overline{\mathcal{S}}(\mathcal{K}^{\vee}(\mathsf{c})) \to \mathcal{K}^{\vee}(\mathcal{S}(\mathsf{c})).$$

For a Coxeter monoid a, there is a morphism of cocommutative Coxeter bialgebras

$$\psi_{\mathsf{a}} : \mathcal{K}(\mathcal{S}^{\vee}(\mathsf{a})) \to \overline{\mathcal{S}}^{\vee}(\mathcal{K}(\mathsf{a})).$$

Similarly, using Theorem 14.30, we have the following commutative analogue for bosonic Fock functors.

Proposition 15.29. *For a Coxeter comonoid c, there are isomorphisms of commutative Coxeter bialgebras*

$$\overline{\varphi}_{\mathsf{c}} : \overline{\mathcal{S}}(\overline{\mathcal{K}}(\mathsf{c})) \xrightarrow{\cong} \overline{\mathcal{K}}(\mathcal{S}(\mathsf{c})) \quad \textit{and} \quad \overline{\psi}^{\vee}_{\mathsf{c}} : \overline{\mathcal{S}}(\overline{\mathcal{K}}^{\vee}(\mathsf{c})) \xrightarrow{\cong} \overline{\mathcal{K}}^{\vee}(\mathcal{S}(\mathsf{c})),$$

the latter if $\mathrm{char}(\mathbb{k}) \nmid |W|$. *For a Coxeter monoid a, there are isomorphisms of cocommutative Coxeter bialgebras*

$$\overline{\psi}_{\mathsf{a}} : \overline{\mathcal{K}}(\mathcal{S}^{\vee}(\mathsf{a})) \xrightarrow{\cong} \overline{\mathcal{S}}^{\vee}(\overline{\mathcal{K}}(\mathsf{a})) \quad \textit{and} \quad \overline{\varphi}^{\vee}_{\mathsf{a}} : \overline{\mathcal{K}}^{\vee}(\mathcal{S}^{\vee}(\mathsf{a})) \xrightarrow{\cong} \overline{\mathcal{S}}^{\vee}(\overline{\mathcal{K}}^{\vee}(\mathsf{a})),$$

the former if $\mathrm{char}(\mathbb{k}) \nmid |W|$.

There are similar isomorphisms with $\mathcal{S}$ replaced by $\mathcal{E}$, and for fermionic Fock functors. See Exercise 14.34.

15.4.5. Abelianization map and its section.

Exercise 15.30. Let $\operatorname{char}(\Bbbk) \nmid |W|$. Fix an invariant noncommutative zeta function $\boldsymbol{\zeta}$. Check: For a Coxeter species p, the diagrams

$$\begin{array}{ccc} \overline{\mathcal{T}}\mathcal{K}^\vee(\mathsf{p}) & \xrightarrow{\psi_{\mathsf{p}}^\vee} & \mathcal{K}^\vee\mathcal{T}(\mathsf{p}) \\ \downarrow & & \downarrow \\ \overline{\mathcal{S}}\mathcal{K}^\vee(\mathsf{p}) & \xrightarrow[\psi_{\mathsf{p}}^\vee]{} & \mathcal{K}^\vee\mathcal{S}(\mathsf{p}) \end{array} \qquad \begin{array}{ccc} \overline{\mathcal{T}}\mathcal{K}^\vee(\mathsf{p}) & \xrightarrow{\psi_{\mathsf{p}}^\vee} & \mathcal{K}^\vee\mathcal{T}(\mathsf{p}) \\ \uparrow & & \uparrow \\ \overline{\mathcal{S}}\mathcal{K}^\vee(\mathsf{p}) & \xrightarrow[\psi_{\mathsf{p}}^\vee]{} & \mathcal{K}^\vee\mathcal{S}(\mathsf{p}) \end{array} \tag{15.27}$$

$$\begin{array}{ccc} \overline{\mathcal{T}}\overline{\mathcal{K}}^\vee(\mathsf{p}) & \xrightarrow{\overline{\psi}_{\mathsf{p}}^\vee} & \overline{\mathcal{K}}^\vee\mathcal{T}(\mathsf{p}) \\ \downarrow & & \downarrow \\ \overline{\mathcal{S}}\overline{\mathcal{K}}^\vee(\mathsf{p}) & \xrightarrow[\overline{\psi}_{\mathsf{p}}^\vee]{} & \overline{\mathcal{K}}^\vee\mathcal{S}(\mathsf{p}) \end{array} \qquad \begin{array}{ccc} \overline{\mathcal{T}}\overline{\mathcal{K}}^\vee(\mathsf{p}) & \xrightarrow{\overline{\psi}_{\mathsf{p}}^\vee} & \overline{\mathcal{K}}^\vee\mathcal{T}(\mathsf{p}) \\ \uparrow & & \uparrow \\ \overline{\mathcal{S}}\overline{\mathcal{K}}^\vee(\mathsf{p}) & \xrightarrow[\overline{\psi}_{\mathsf{p}}^\vee]{} & \overline{\mathcal{K}}^\vee\mathcal{S}(\mathsf{p}) \end{array} \tag{15.28}$$

commute, the left ones as Coxeter bialgebras and right ones as Coxeter coalgebras. In the left diagrams, the left-vertical maps are instances of (9.55), while the right-vertical maps are induced from (3.50). In the right diagrams, the left-vertical maps are instances of (13.46), while the right-vertical maps are induced from (7.27). These depend on $\boldsymbol{\zeta}$. There are similar diagrams for $\overline{\mathcal{K}}$.

Dually: For a Coxeter species p, the diagrams

$$\begin{array}{ccc} \mathcal{K}\mathcal{T}^\vee(\mathsf{p}) & \xrightarrow{\psi_{\mathsf{p}}} & \overline{\mathcal{T}}^\vee\mathcal{K}(\mathsf{p}) \\ \uparrow & & \uparrow \\ \mathcal{K}\mathcal{S}^\vee(\mathsf{p}) & \xrightarrow[\psi_{\mathsf{p}}]{} & \overline{\mathcal{S}}^\vee\mathcal{K}(\mathsf{p}) \end{array} \qquad \begin{array}{ccc} \mathcal{K}\mathcal{T}^\vee(\mathsf{p}) & \xrightarrow{\psi_{\mathsf{p}}} & \overline{\mathcal{T}}^\vee\mathcal{K}(\mathsf{p}) \\ \downarrow & & \downarrow \\ \mathcal{K}\mathcal{S}^\vee(\mathsf{p}) & \xrightarrow[\psi_{\mathsf{p}}]{} & \overline{\mathcal{S}}^\vee\mathcal{K}(\mathsf{p}) \end{array} \tag{15.29}$$

$$\begin{array}{ccc} \overline{\mathcal{K}}\mathcal{T}^\vee(\mathsf{p}) & \xrightarrow{\overline{\psi}_{\mathsf{p}}} & \overline{\mathcal{T}}^\vee\overline{\mathcal{K}}(\mathsf{p}) \\ \uparrow & & \uparrow \\ \overline{\mathcal{K}}\mathcal{S}^\vee(\mathsf{p}) & \xrightarrow[\overline{\psi}_{\mathsf{p}}]{} & \overline{\mathcal{S}}^\vee\overline{\mathcal{K}}(\mathsf{p}) \end{array} \qquad \begin{array}{ccc} \overline{\mathcal{K}}\mathcal{T}^\vee(\mathsf{p}) & \xrightarrow{\overline{\psi}_{\mathsf{p}}} & \overline{\mathcal{T}}^\vee\overline{\mathcal{K}}(\mathsf{p}) \\ \downarrow & & \downarrow \\ \overline{\mathcal{K}}\mathcal{S}^\vee(\mathsf{p}) & \xrightarrow[\overline{\psi}_{\mathsf{p}}]{} & \overline{\mathcal{S}}^\vee\overline{\mathcal{K}}(\mathsf{p}) \end{array} \tag{15.30}$$

commute, the left ones as Coxeter bialgebras and right ones as Coxeter algebras. The vertical maps arise from the coabelianization maps and their projections (7.32) and (13.54). The latter depend on $\boldsymbol{\zeta}$. There are similar diagrams for $\overline{\mathcal{K}}^\vee$.

15.4.6. (Co)freeness of full Fock spaces. An interesting application is given below. It combines the above ideas with properties of the Hadamard product of Coxeter species.

Theorem 15.31. *For scalars p and q, for a Coxeter p-bimonoid h, the Coxeter pq-bialgebra*

- *$\mathcal{K}_q(\mathsf{h})$ is free as a Coxeter algebra,*
- *$\mathcal{K}_q^\vee(\mathsf{h})$ is cofree as a Coxeter coalgebra.*

FIRST PROOF. For the first part: Using Corollary 7.13, write $\mathsf{h} \times \Gamma_q \cong \mathcal{T}(\mathsf{p})$ for some Coxeter species p. This is an isomorphism of Coxeter monoids. Applying the functor $\overline{\mathcal{K}}$, we obtain isomorphisms of Coxeter algebras

$$\mathcal{K}_q(\mathsf{h}) \cong \overline{\mathcal{K}}(\mathsf{h} \times \Gamma_q) \cong \overline{\mathcal{K}}(\mathcal{T}(\mathsf{p})) \cong \overline{\mathcal{T}}(\overline{\mathcal{K}}(\mathsf{p})).$$

The first isomorphism is given in Proposition 15.8, while the last isomorphism is given in Proposition 15.24.

The second part is similar, with isomorphisms of Coxeter coalgebras

$$\mathcal{K}_q^\vee(\mathsf{h}) \cong \overline{\mathcal{K}}^\vee(\mathsf{h} \times \Gamma_q^*) \cong \overline{\mathcal{K}}^\vee(\mathcal{T}^\vee(\mathsf{p})) \cong \overline{\mathcal{T}}^\vee(\overline{\mathcal{K}}^\vee(\mathsf{p}))$$

using Propositions 15.8 and 15.25. □

SECOND PROOF. In the previous proof, Loday–Ronco Theorem 7.10 got used. Alternatively, we can prove the previous result using Loday–Ronco Theorem 13.35 as follows. Observe: $\mathcal{K}_q(\mathsf{h}) = \mathcal{K}_0(\mathsf{h})$ as Coxeter algebras (since the product does not depend on q). But the latter is a Coxeter 0-bialgebra, and hence free. The argument for $\mathcal{K}_q^\vee(\mathsf{h})$ is similar. □

We also point out that for h a Coxeter 0-bimonoid, $\mathcal{K}_q(\mathsf{h})$ and $\mathcal{K}_q^\vee(\mathsf{h})$ are Coxeter 0-bialgebras, and hence both free and cofree.

15.5. Nonnegativity of the Boolean transform

We introduce the notion of type series of a reflection arrangement, and define its Boolean transform. We define the Hilbert type series of a Coxeter bimonoid and show that its Boolean transform is nonnegative. The proof uses universal constructions and Fock functors, but the statement of the result does not require any familiarity with them. All objects in this section are assumed to be finite-dimensional.

15.5.1. Boolean transform. Let $\mathcal{A}$ be a reflection arrangement. A *type series* of $\mathcal{A}$ is a family (a_T) of scalars, one for each face-type T of $\mathcal{A}$. The *Boolean transform* of (a_T) is the type series (b_T) defined by either of the two equivalent formulas

$$a_T = \sum_{U:\, U \geq T} b_U, \qquad b_T = \sum_{U:\, U \geq T} (-1)^{\operatorname{rk}(U/T)} a_U. \tag{15.31}$$

Since the poset of face-types is Boolean, this is an instance of Möbius inversion.

15.5.2. Hilbert type series of Coxeter species and Coxeter spaces. For a Coxeter space V, we have the type series $a_T := \dim \mathrm{V}[T]$. We call this the *Hilbert type series* of V.

Proposition 15.32. *Let* $\mathrm{A} = \overline{\mathcal{T}}(\mathrm{V})$ *be a free Coxeter algebra. Then the Boolean transform of the Hilbert type series of* A *agrees with the Hilbert type series of* V*, and in particular, is nonnegative.*

PROOF. This follows from construction of $\overline{\mathcal{T}}$, see formula (8.119). □

Now suppose p is a Coxeter species. Then in view of the isomorphisms $\mathsf{p}[F] \to \mathsf{p}[wF]$, we deduce that the dimension of $\mathsf{p}[F]$ only depends on the type of F. This allows us to define $a_T := \dim \mathsf{p}[F]$, where F is any face of type T. We call this the *Hilbert type series* of p.

Observe: The Coxeter species p and Coxeter space $\mathcal{K}(\mathsf{p})$ have the same Hilbert type series.

15.5.3. Hilbert type series of a Coxeter bimonoid.

Proposition 15.33. *For a scalar p, for a Coxeter p-bimonoid h, the Boolean transform of the Hilbert type series of h is nonnegative.*

PROOF. Since h and $\mathcal{K}(\mathsf{h})$ have the same Hilbert type series, and the latter is a free Coxeter algebra by Theorem 15.31, the result follows from Proposition 15.32. □

The above result gives a numerical necessary condition for the existence of a Coxeter bimonoid. For instance: A Coxeter species p on a rank-one arrangement (with chambers C and $\overline{C}$) for which $\dim \mathsf{p}[O] < \dim \mathsf{p}[C]$ cannot carry a Coxeter bimonoid structure. We know this to be true since Δ_O^C is surjective and μ_O^C is injective.

15.5.4. Hilbert type series of a free Coxeter monoid. The Boolean transform of the Hilbert type series of a free Coxeter monoid has a nice explicit formula as follows.

Proposition 15.34. *For a Coxeter species p, the Boolean transform of the Hilbert type series of the free Coxeter monoid $\mathcal{T}(\mathsf{p})$ is*

$$b_Z = \sum_{F:\, F \wedge C = A} \dim \mathsf{p}[F],$$

where C is a fixed chamber, and A is its face of type Z.

PROOF. This can be directly checked using (15.31). However, let us proceed systematically. Since $\Gamma = \mathcal{T}(\mathsf{x})$, we deduce from Proposition 3.26 that $\mathcal{T}(\mathsf{p}) \times \Gamma = \mathcal{T}(\mathsf{p} \star \mathsf{x})$ as Coxeter monoids. Applying the functor $\overline{\mathcal{K}}$, we obtain isomorphisms of Coxeter algebras

$$\mathcal{K}(\mathcal{T}(\mathsf{p})) \cong \overline{\mathcal{K}}(\mathcal{T}(\mathsf{p}) \times \Gamma) \cong \overline{\mathcal{K}}(\mathcal{T}(\mathsf{p} \star \mathsf{x})) \cong \overline{\mathcal{T}}(\overline{\mathcal{K}}(\mathsf{p} \star \mathsf{x}))$$

as in the proof of Theorem 15.31. Thus, the required Boolean transform is the Hilbert type series of $\overline{\mathcal{K}}(\mathsf{p} \star \mathsf{x})$. Making this explicit using formula (3.57) yields the above formula. □

15.6. Coxeter operad monoids to Coxeter operad algebras

It follows from results in Section 14.9 that Fock functors $\mathcal{K}^\vee$ and $\overline{\mathcal{K}}^\vee$ send Coxeter operad monoids to Coxeter operad algebras. There are similar results for the bosonic Fock functor $\overline{\mathcal{K}}$. Dually, $\mathcal{K}$, $\overline{\mathcal{K}}$, $\overline{\mathcal{K}}^\vee$ send Coxeter cooperad comonoids to Coxeter cooperad coalgebras. Details follow.

15.6.1. Coxeter operad algebras.

We now record the consequences of the existence of lax functors between monads, and morphisms between them.

Theorem 15.35. *For* $\mathbf{a}$ *a Coxeter operad and* m *a Coxeter* $\mathbf{a}$*-monoid,* $\mathcal{K}^\vee(\mathsf{m})$ *and* $\overline{\mathcal{K}}^\vee(\mathsf{m})$ *are Coxeter* $\mathbf{a}$*-algebras, and* $\overline{\mathcal{K}}^\vee(\mathsf{m}) \hookrightarrow \mathcal{K}^\vee(\mathsf{m})$ *is a morphism of Coxeter* $\mathbf{a}$*-algebras.*

PROOF. This follows from Theorems 14.52 and 14.53. □

Explicitly, suppose for each $\mathrm{X} \leq \mathrm{Y}$,

$$\gamma_{\mathrm{X}}^{\mathrm{Y}} : \mathbf{a}[\mathrm{X}, \mathrm{Y}] \otimes \mathsf{m}[\mathrm{Y}] \to \mathsf{m}[\mathrm{X}]$$

are structure maps of m. Then, for each face F_0 between A and C, the structure map of $\mathcal{K}^\vee(\mathsf{m})$, namely,

$$\gamma_{A,C}^{F_0,C} : \mathbf{a}[\mathrm{s}(A), \mathrm{s}(F_0)] \otimes \mathcal{K}^\vee(\mathsf{m})[F_0, C] \to \mathcal{K}^\vee(\mathsf{m})[A, C],$$

or equivalently,

$$\gamma_{A,C}^{F_0,C} : \mathbf{a}[\mathrm{s}(A), \mathrm{s}(F_0)] \otimes \mathsf{m}[F_0] \to \mathsf{m}[A]$$

is defined by summing the composite maps

$$\mathbf{a}[\mathrm{s}(A), \mathrm{s}(F_0)] \otimes \mathsf{m}[F_0] \xrightarrow{\mathbf{a}[w] \otimes \mathsf{m}[w]} \mathbf{a}[\mathrm{s}(A), \mathrm{s}(F)] \otimes \mathsf{m}[F]$$

$$\xrightarrow{\mathrm{id} \otimes \beta_{\mathrm{s}(F),F}} \mathbf{a}[\mathrm{s}(A), \mathrm{s}(F)] \otimes \mathsf{m}[\mathrm{s}(F)] \xrightarrow{\gamma_{\mathrm{s}(A)}^{\mathrm{s}(F)}} \mathsf{m}[\mathrm{s}(A)] \xrightarrow{\beta_{A,\mathrm{s}(A)}} \mathsf{m}[A]$$

over all F which are greater than A and have the same type as F_0, and with $w \in W$ defined by $w(C) = FC$.

Proposition 15.36. *Let* $\mathbf{a} \to \mathbf{b}$ *be a morphism of Coxeter operads, and* m *be a Coxeter* $\mathbf{b}$*-monoid. Then* $\mathcal{K}^\vee(\mathsf{m})$ *is a Coxeter* $\mathbf{a}$*-algebra in two different ways:*

- *First using* $\mathbf{a} \to \mathbf{b}$*,* m *is a Coxeter* $\mathbf{a}$*-monoid, and hence* $\mathcal{K}^\vee(\mathsf{m})$ *is a Coxeter* $\mathbf{a}$*-algebra.*
- *First* $\mathcal{K}^\vee(\mathsf{m})$ *is a Coxeter* $\mathbf{b}$*-algebra, and then using* $\mathbf{a} \to \mathbf{b}$*, it is also a Coxeter* $\mathbf{a}$*-algebra.*

The two Coxeter $\mathbf{a}$*-algebra structures on* $\mathcal{K}^\vee(\mathsf{m})$ *are equal.*

The same result holds with $\overline{\mathcal{K}}^\vee$ *instead of* $\mathcal{K}^\vee$*.*

PROOF. This follows from diagram (14.65). More specifically, in diagram (14.66), apply $\sum_{\mathrm{X} \geq \mathrm{s}(A)} \gamma_{\mathrm{s}(A)}^{\mathrm{X}}$ to the final term to end in $\mathsf{m}[\mathrm{s}(A)]$. Going down and across is the first structure, and going across and down is the second structure of a Coxeter $\mathbf{a}$-algebra on $\mathcal{K}^\vee(\mathsf{m})$. □

Proposition 15.37. *Let* $\mathrm{char}(\Bbbk) \nmid |W|$*. For* $\mathbf{a}$ *a Coxeter operad and* m *a Coxeter species, there is an isomorphism of Coxeter* $\mathbf{a}$*-algebras*

$$\overline{\mathcal{K}}^\vee(\mathbf{a} \circ \mathsf{m}) \cong \mathbf{a} \circ \overline{\mathcal{K}}^\vee(\mathsf{m}).$$

In particular, $\overline{\mathcal{K}}^\vee$ *of a free Coxeter* $\mathbf{a}$*-monoid is a free Coxeter* $\mathbf{a}$*-algebra.*

PROOF. This follows from Theorem 14.53. □

The above results are also valid for the functor $\overline{\mathcal{K}}$ instead of $\overline{\mathcal{K}}^\vee$. We elaborate briefly. The relevant result is Theorem 14.54.

Theorem 15.38. *For* $\mathbf{a}$ *a Coxeter operad and* m *a Coxeter* $\mathbf{a}$*-monoid,* $\overline{\mathcal{K}}(\mathsf{m})$ *is a Coxeter* $\mathbf{a}$*-algebra.*

Explicitly, the structure map of $\overline{\mathcal{K}}(\mathsf{m})$, namely,

$$\gamma_{A,C}^{F_0,C} : \mathbf{a}[\mathrm{s}(A), \mathrm{s}(F_0)] \otimes \overline{\mathcal{K}}(\mathsf{m})[F_0, C] \to \overline{\mathcal{K}}(\mathsf{m})[A, C],$$

or equivalently,

$$\gamma_{A,C}^{F_0,C} : \mathbf{a}[\mathrm{s}(A), \mathrm{s}(F_0)] \otimes \mathsf{m}[F_0]_{W_{F_0}} \to \mathsf{m}[A]_{W_A}$$

is induced from the composite map

$$\mathbf{a}[\mathrm{s}(A), \mathrm{s}(F_0)] \otimes \mathsf{m}[F_0] \xrightarrow{\mathrm{id} \otimes \beta_{\mathrm{s}(F_0), F_0}} \mathbf{a}[\mathrm{s}(A), \mathrm{s}(F_0)] \otimes \mathsf{m}[\mathrm{s}(F_0)] \xrightarrow{\gamma_{\mathrm{s}(A)}^{\mathrm{s}(F_0)}} \mathsf{m}[\mathrm{s}(A)] \xrightarrow{\beta_{A,\mathrm{s}(A)}} \mathsf{m}[A].$$

Proposition 15.39. *For* $\mathbf{a}$ *a Coxeter operad and* m *a Coxeter* $\mathbf{a}$*-species,*

$$\overline{\mathcal{K}}(\mathbf{a} \circ \mathsf{m}) \cong \mathbf{a} \circ \overline{\mathcal{K}}(\mathsf{m})$$

as Coxeter $\mathbf{a}$*-algebras (in any field characteristic).*

We also have signed analogues of the preceding results by replacing $\mathcal{K}^\vee$, $\overline{\mathcal{K}}^\vee$, $\overline{\mathcal{K}}$ by $\mathcal{K}_{-1}^\vee$, $\overline{\mathcal{K}}_{-1}^\vee$, $\overline{\mathcal{K}}_{-1}$, respectively.

15.6.2. Coxeter cooperad coalgebras. Let us now record the dual results.

Theorem 15.40. *For* $\mathbf{c}$ *a Coxeter cooperad and* m *a Coxeter* $\mathbf{c}$*-comonoid,* $\mathcal{K}(\mathsf{m})$ *and* $\overline{\mathcal{K}}(\mathsf{m})$ *are Coxeter* $\mathbf{c}$*-coalgebras, and* $\mathcal{K}(\mathsf{m}) \twoheadrightarrow \overline{\mathcal{K}}(\mathsf{m})$ *is a morphism of Coxeter* $\mathbf{c}$*-coalgebras.*

PROOF. This follows from Theorem 14.57. □

Proposition 15.41. *Let* $\mathrm{char}(\Bbbk) \nmid |W|$*. For* $\mathbf{c}$ *a Coxeter cooperad and* m *a Coxeter species, there is an isomorphism of Coxeter* $\mathbf{c}$*-coalgebras*

$$\overline{\mathcal{K}}(\mathbf{c} \circ \mathsf{m}) \cong \mathbf{c} \circ^\vee \overline{\mathcal{K}}(\mathsf{m}).$$

In particular, $\overline{\mathcal{K}}$ *of a cofree Coxeter* $\mathbf{c}$*-comonoid is a cofree Coxeter* $\mathbf{c}$*-coalgebra.*

PROOF. This follows from the last statement in Theorem 14.57. □

Proposition 15.42. *Let* $\mathbf{c} \to \mathbf{d}$ *be a morphism of Coxeter cooperads, and* m *be a Coxeter* $\mathbf{c}$*-comonoid. Then* $\mathcal{K}(\mathsf{m})$ *is a Coxeter* $\mathbf{d}$*-coalgebra in two different ways:*

- *First using* $\mathbf{c} \to \mathbf{d}$*,* m *is a Coxeter* $\mathbf{d}$*-comonoid, and hence* $\mathcal{K}(\mathsf{m})$ *is a Coxeter* $\mathbf{d}$*-coalgebra.*
- *First* $\mathcal{K}(\mathsf{m})$ *is a Coxeter* $\mathbf{c}$*-coalgebra, and then using* $\mathbf{c} \to \mathbf{d}$*, it is also a Coxeter* $\mathbf{d}$*-coalgebra.*

The two Coxeter $\mathbf{d}$*-coalgebra structures on* $\mathcal{K}(\mathsf{m})$ *are equal.*

The same result holds with $\overline{\mathcal{K}}$ *instead of* $\mathcal{K}$*.*

PROOF. This follows from diagram (14.69). □

The above results are also valid for $\overline{\mathcal{K}}^\vee$ instead of $\overline{\mathcal{K}}$. In particular, $\overline{\mathcal{K}}^\vee$ of a cofree Coxeter **c**-comonoid is a cofree Coxeter **c**-coalgebra (in any field characteristic).

15.6.3. Norm map.

Exercise 15.43. Let $\mathrm{char}(\Bbbk) \nmid |W|$. Deduce from Proposition 14.59: For **a** a Coxeter operad and m a Coxeter **a**-monoid, $\overline{\mathcal{K}}(\mathsf{m}) \cong \overline{\mathcal{K}}^\vee(\mathsf{m})$ as Coxeter **a**-algebras. Dually, for **c** a Coxeter cooperad and m a Coxeter **c**-comonoid, $\overline{\mathcal{K}}(\mathsf{m}) \cong \overline{\mathcal{K}}^\vee(\mathsf{m})$ as Coxeter **c**-coalgebras.

15.6.4. Specializations. For Coxeter operads $\mathbf{a} = \mathbf{As}, \mathbf{Com}, \mathbf{Com}^-$, Theorem 15.35 recovers that $\mathcal{K}^\vee$ and $\overline{\mathcal{K}}^\vee$ send Coxeter monoids to Coxeter algebras, and preserve (signed) commutativity. Proposition 15.36 clearly holds for the morphism $\mathbf{As} \to \mathbf{Com}$. Propositions 15.37 and 15.39 recover that bosonic Fock functors send free (commutative) Coxeter monoids to free (commutative) Coxeter algebras.

Dually, Theorem 15.40 recovers that $\mathcal{K}$ and $\overline{\mathcal{K}}$ send Coxeter comonoids to Coxeter coalgebras, and preserve (signed) cocommutativity.

We will look at $\mathbf{a} = \mathbf{Lie}$ and $\mathbf{c} = \mathbf{Lie}^*$ in Section 15.7.

15.7. Coxeter Lie monoids to Coxeter Lie algebras

We record how Fock functors preserve Lie structures. These are special cases of the generalities in Section 15.6. Alternatively, it is possible to approach them directly using Lie (co)brackets.

15.7.1. Coxeter Lie monoids to Coxeter Lie algebras. For g a Coxeter Lie monoid, $\mathcal{K}^\vee(\mathsf{g})$ and $\overline{\mathcal{K}}^\vee(\mathsf{g})$ are Coxeter Lie algebras. This is a special case of Theorem 15.35 for $\mathbf{a} := \mathbf{Lie}$.

Explicitly, for $A \leq F_0 \leq C$, the structure map of $\mathcal{K}^\vee(\mathsf{g})$, namely,

$$\gamma_{A,C}^{F_0,C} : \mathbf{Lie}[\mathrm{s}(A), \mathrm{s}(F_0)] \otimes \mathcal{K}^\vee(\mathsf{g})[F_0, C] \to \mathcal{K}^\vee(\mathsf{g})[A, C]$$

is defined by summing the composite maps

$$\mathbf{Lie}[\mathrm{s}(A), \mathrm{s}(F_0)] \otimes \mathsf{g}[F_0] \xrightarrow{w \otimes w} \mathbf{Lie}[\mathrm{s}(A), \mathrm{s}(F)] \otimes \mathsf{g}[F]$$
$$\xrightarrow{\mathrm{id} \otimes \beta_{\mathrm{s}(F),F}} \mathbf{Lie}[\mathrm{s}(A), \mathrm{s}(F)] \otimes \mathsf{g}[\mathrm{s}(F)] \xrightarrow{\gamma_{\mathrm{s}(A)}^{\mathrm{s}(F)}} \mathsf{g}[\mathrm{s}(A)] \xrightarrow{\beta_{A,\mathrm{s}(A)}} \mathsf{g}[A]$$

over all $F \geq A$ with $\mathrm{t}(F/A) = \mathrm{t}(F_0/A)$ and $w \in W$ defined by $w(C) = FC$. Here $\gamma_{\mathrm{s}(A)}^{\mathrm{s}(F)}$ refers to the structure map of g. This formula comes from the recipe provided after Theorem 15.35 which in turn arises from the lax structure of $\mathcal{K}^\vee$. In terms of Lie brackets: For $A \lessdot F_0 \leq C$,

$$\begin{array}{ccc} \mathcal{K}^\vee(\mathsf{g})[F_0, C] & \xrightarrow{\nu_{A,C}^{F_0,C}} & \mathcal{K}^\vee(\mathsf{g})[A, C] \\ \| & & \| \\ \mathsf{g}[F_0] & \longrightarrow & \mathsf{g}[A] \end{array} \tag{15.32}$$

is the sum of the composite maps

$$\mathsf{g}[F_0] \xrightarrow{w(C)=FC} \mathsf{g}[F] \xrightarrow{\nu_A^F} \mathsf{g}[A]$$

as F varies over all faces which cover A and have the same type as F_0. The Lie bracket (15.32) restricts to $\overline{\mathcal{K}}^\vee(\mathsf{g})$.

Exercise 15.44. Check directly: $\mathcal{K}^\vee(\mathsf{g})$ with Lie bracket (15.32) is indeed a Coxeter Lie algebra. (The check for the antisymmetry axiom (12.1b) is similar to the check for the commutativity axiom in Exercise 15.13 with Lemma 2.16 playing a key role. The check for the Jacobi identity (12.1c) has similarities to the check for the associativity axiom in Exercise 15.5.)

Similarly, for g a Coxeter Lie monoid, $\overline{\mathcal{K}}(\mathsf{g})$ is a Coxeter Lie algebra. This is a special case of Theorem 15.38 for $\mathbf{a} = \mathbf{Lie}$. In terms of Lie brackets: For $A \lessdot F \leq C$,

$$\begin{array}{ccc} \overline{\mathcal{K}}(\mathsf{g})[F,C] & \overset{\nu_{A,C}^{F,C}}{\dashrightarrow} & \overline{\mathcal{K}}(\mathsf{g})[A,C] \\ \| & & \| \\ \mathsf{g}[F]_{W_F} & \longrightarrow & \mathsf{g}[A]_{W_A} \end{array} \tag{15.33}$$

is induced from $\nu_A^F : \mathsf{g}[F] \to \mathsf{g}[A]$. It is easy to see directly that $\overline{\mathcal{K}}(\mathsf{g})$ is a Coxeter Lie algebra (in contrast to the more complicated check in Exercise 15.44).

Note: If the Lie structure on g is trivial, then so are the Lie structures on $\mathcal{K}^\vee(\mathsf{g})$, $\overline{\mathcal{K}}^\vee(\mathsf{g})$, $\overline{\mathcal{K}}(\mathsf{g})$.

If g is a free Coxeter Lie monoid, then $\overline{\mathcal{K}}(\mathsf{g})$ and $\overline{\mathcal{K}}^\vee(\mathsf{g})$ are free Coxeter Lie algebras (the latter if $\mathrm{char}(\Bbbk) \nmid |W|$). This is a special case of Propositions 15.37 and 15.39.

15.7.2. Coxeter monoids to Coxeter Lie algebras.

For a a Coxeter monoid, the Coxeter space $\mathcal{K}^\vee(\mathsf{a})$ can be viewed as a Coxeter Lie algebra in two different ways, namely:

- view it as the image under $\mathcal{K}^\vee$ of the Coxeter Lie monoid a, or
- view the Coxeter algebra $\mathcal{K}^\vee(\mathsf{a})$ as a Coxeter Lie algebra.

The two Lie structures coincide. This is a special case of Proposition 15.36 with $\mathbf{a} := \mathbf{Lie}$ and $\mathbf{b} := \mathbf{As}$. Alternatively, this fact can be checked directly by working with Lie brackets. Explicitly, the Lie bracket of $\mathcal{K}^\vee(\mathsf{a})$, namely,

$$\nu_{A,C}^{F_0,C} : \mathcal{K}^\vee(\mathsf{a})[F_0,C] \to \mathcal{K}^\vee(\mathsf{a})[A,C]$$

is obtained by summing

$$\mathsf{a}[F_0] \xrightarrow{w(C)=FC} \mathsf{a}[F] \xrightarrow{\mu_A^F - \mu_A^{A\overline{F}} \beta_{A\overline{F},F}} \mathsf{a}[A]$$

as F varies over all faces which cover A and have the same type as F_0.

Similar statements can be made for $\overline{\mathcal{K}}^\vee(\mathsf{a})$ and $\overline{\mathcal{K}}(\mathsf{a})$.

15.7.3. Coxeter Lie comonoids to Coxeter Lie coalgebras. Dually, for k a Coxeter Lie comonoid, $\mathcal{K}(\mathsf{k})$ and $\overline{\mathcal{K}}(\mathsf{k})$ are Coxeter Lie coalgebras. This is a special case of Theorem 15.40 for $\mathbf{c} := \mathbf{Lie}^*$. In terms of Lie cobrackets: For $A \lessdot F_0 \leq C$,

$$\begin{array}{ccc} \mathcal{K}(\mathsf{k})[A,C] & \xrightarrow{\theta^{F_0,C}_{A,C}} & \mathcal{K}(\mathsf{k})[F_0,C] \\ \| & & \| \\ \mathsf{k}[A] & \longrightarrow & \mathsf{k}[F_0] \end{array} \tag{15.34}$$

is the sum of the composite maps

$$\mathsf{k}[A] \xrightarrow{\theta^F_A} \mathsf{k}[F] \xrightarrow{w(FC)=C} \mathsf{k}[F_0]$$

as F varies over all faces which cover A and have the same type as F_0. The Lie cobracket of $\overline{\mathcal{K}}(\mathsf{k})$ is induced from (15.34). Similarly, $\overline{\mathcal{K}}^\vee(\mathsf{k})$ is a Coxeter Lie coalgebra with Lie cobracket, for $A \lessdot F \leq C$,

$$\begin{array}{ccc} \overline{\mathcal{K}}^\vee(\mathsf{k})[A,C] & \xrightarrow{\theta^{F,C}_{A,C}} & \overline{\mathcal{K}}^\vee(\mathsf{k})[F,C] \\ \| & & \| \\ \mathsf{k}[A]^{W_A} & \longrightarrow & \mathsf{k}[F]^{W_F} \end{array} \tag{15.35}$$

induced from $\theta^F_A : \mathsf{k}[A] \to \mathsf{k}[F]$.

If k is a cofree Coxeter Lie comonoid, then $\overline{\mathcal{K}}^\vee(\mathsf{k})$ and $\overline{\mathcal{K}}(\mathsf{k})$ are cofree Coxeter Lie coalgebras (the latter if $\operatorname{char}(\mathbb{k}) \nmid |W|$).

15.7.4. Coxeter comonoids to Coxeter Lie coalgebras. For c a Coxeter comonoid, the Coxeter space $\mathcal{K}(\mathsf{c})$ can be viewed as a Coxeter Lie coalgebra in two different ways, namely:

- view it as the image under $\mathcal{K}$ of the Coxeter Lie comonoid c, or
- view the Coxeter coalgebra $\mathcal{K}(\mathsf{c})$ as a Coxeter Lie coalgebra.

The two Lie structures coincide. This is a special case of Proposition 15.42 with $\mathbf{c} := \mathbf{As}^*$ and $\mathbf{d} := \mathbf{Lie}^*$. Explicitly, the Lie cobracket of $\mathcal{K}(\mathsf{c})$, namely,

$$\theta^{F_0,C}_{A,C} : \mathcal{K}(\mathsf{c})[A,C] \to \mathcal{K}(\mathsf{c})[F_0,C]$$

is obtained by summing

$$\mathsf{c}[A] \xrightarrow{\Delta^F_A - \beta_{F,A\overline{F}}\Delta^{A\overline{F}}_A} \mathsf{c}[F] \xrightarrow{w(FC)=C} \mathsf{c}[F_0]$$

as F varies over all faces which cover A and have the same type as F_0.

Similar remarks apply to $\overline{\mathcal{K}}(\mathsf{c})$ and $\overline{\mathcal{K}}^\vee(\mathsf{c})$.

15.7.5. Norm map.

Exercise 15.45. Let $\operatorname{char}(\mathbb{k}) \nmid |W|$. Check directly: For g a Coxeter Lie monoid, the norm map yields an isomorphism $\overline{\mathcal{K}}(\mathsf{g}) \cong \overline{\mathcal{K}}^\vee(\mathsf{g})$ of Coxeter Lie algebras, and for k a Coxeter Lie comonoid, an isomorphism $\overline{\mathcal{K}}(\mathsf{k}) \cong \overline{\mathcal{K}}^\vee(\mathsf{k})$ of Coxeter Lie coalgebras. These are the special cases $\mathbf{a} := \mathbf{Lie}$ and $\mathbf{c} := \mathbf{Lie}^*$ of Exercise 15.43.

15.7.6. Signed analogue. All of the preceding discussion also works in the signed setting. For g a signed Coxeter Lie monoid, $\mathcal{K}^\vee(\mathsf{g})$, $\overline{\mathcal{K}}^\vee(\mathsf{g})$, $\overline{\mathcal{K}}(\mathsf{g})$ are signed Coxeter Lie algebras. The relevant Coxeter operad is $\mathbf{a} = \mathbf{Lie}^-$. Similarly, for a a Coxeter monoid, the two signed Coxeter Lie algebra structures on each of $\mathcal{K}^\vee(\mathsf{a})$, $\overline{\mathcal{K}}^\vee(\mathsf{a})$, $\overline{\mathcal{K}}(\mathsf{a})$ match.

Dual statements hold for signed Coxeter Lie comonoids.

15.8. Primitive and indecomposable parts under Fock functors

Recall the primitive and indecomposable parts of Coxeter bimonoids from Section 6.3 and of Coxeter bialgebras from Section 12.3. They carry Coxeter Lie (co)monoid and Coxeter Lie (co)algebra structures, respectively. We now study how Fock functors interact with these constructions.

15.8.1. Primitive part. Recall from Section 15.7.2 that for a a Coxeter monoid, the two (signed) Coxeter Lie algebra structures on each of the Coxeter spaces $\mathcal{K}^\vee(\mathsf{a})$, $\overline{\mathcal{K}}^\vee(\mathsf{a})$, $\overline{\mathcal{K}}(\mathsf{a})$ match.

Proposition 15.46. *For a (signed) Coxeter bimonoid* h*, the diagram*

$$\begin{array}{ccccc} \mathcal{K}^\vee\big(\mathcal{P}(\mathsf{h})\big) & \hookrightarrow & \overline{\mathcal{P}}\big(\mathcal{K}^\vee(\mathsf{h})\big) & \hookrightarrow & \mathcal{K}^\vee(\mathsf{h}) \\ \uparrow & & \uparrow & & \uparrow \\ \overline{\mathcal{K}}^\vee\big(\mathcal{P}(\mathsf{h})\big) & = & \overline{\mathcal{P}}\big(\overline{\mathcal{K}}^\vee(\mathsf{h})\big) & \hookrightarrow & \overline{\mathcal{K}}^\vee(\mathsf{h}) \end{array}$$

commutes as (signed) Coxeter Lie algebras.

Proof. The inclusions in the second square are obvious. We check below the horizontal inclusion and equality in the first square (as Coxeter spaces). Since all spaces involved are (signed) Coxeter Lie subalgebras of $\mathcal{K}^\vee(\mathsf{h})$, the claim regarding the "Lie" part is automatic.

For any top-nested face (A, C), consider the commutative diagram

$$\begin{array}{ccccc} \mathsf{h}[A] & \xrightarrow{(\Delta_A^F)} & \bigoplus_{F:\, A<F} \mathsf{h}[F] & \xrightarrow{\varphi_{\mathsf{h}}^\vee} & \bigoplus_{F_0:\, A<F_0\leq C} \mathsf{h}[F_0] \\ \uparrow & & \uparrow & & \uparrow \\ \mathsf{h}[A]^{W_A} & \longrightarrow & \Big(\bigoplus_{F:\, A<F} \mathsf{h}[F]\Big)^{W_A} & \xrightarrow[\overline{\varphi}_{\mathsf{h}}^\vee]{\cong} & \bigoplus_{F_0:\, A<F_0\leq C} \mathsf{h}[F_0]^{W_{F_0}}. \end{array}$$

For the second square, see (14.32). By definition (3.7), $\mathcal{K}^\vee\big(\mathcal{P}(\mathsf{h})\big)[A, C]$ is the kernel of (Δ_A^F), while by definition (9.7), $\overline{\mathcal{P}}\big(\mathcal{K}^\vee(\mathsf{h})\big)[A, C]$ is the kernel of $(\Delta_{A,C}^{F_0,C}) = \varphi_{\mathsf{h}}^\vee \circ (\Delta_A^F)$. Hence,

$$\mathcal{K}^\vee\big(\mathcal{P}(\mathsf{h})\big)[A, C] \subseteq \overline{\mathcal{P}}\big(\mathcal{K}^\vee(\mathsf{h})\big)[A, C].$$

More directly, an element $x \in \mathsf{h}[A]$ belongs to the first space if $\Delta_A^F(x) = 0$ for every F strictly greater than A, while it belongs to the second space if $\Delta_A^{F_0}(x) = 0$ for every F_0 strictly greater than A but less than C.

The same argument on the bottom-horizontal maps shows that

$$\overline{\mathcal{K}}^{\vee}\big(\mathcal{P}(\mathsf{h})\big)[A,C] \subseteq \overline{\mathcal{P}}\big(\overline{\mathcal{K}}^{\vee}(\mathsf{h})\big)[A,C].$$

(Note: In the lhs we need to take W_A-invariants of the kernel of (Δ_A^F), but this is the same as taking kernel of the W_A-invariants. More formally, taking invariants is left exact.) Further, since $\overline{\varphi}_{\mathsf{h}}^{\vee}$ is an isomorphism, the above inclusion is an equality. □

Proposition 15.47. *For a (signed) Coxeter bimonoid* h*, the diagram*

$$\begin{array}{ccccc}
\mathcal{K}\big(\mathcal{P}(\mathsf{h})\big) & \hookrightarrow & \overline{\mathcal{P}}\big(\mathcal{K}(\mathsf{h})\big) & \hookrightarrow & \mathcal{K}(\mathsf{h}) \\
\downarrow & & \downarrow & & \downarrow \\
\overline{\mathcal{K}}\big(\mathcal{P}(\mathsf{h})\big) & \longrightarrow & \overline{\mathcal{P}}\big(\overline{\mathcal{K}}(\mathsf{h})\big) & \hookrightarrow & \overline{\mathcal{K}}(\mathsf{h})
\end{array}$$

commutes as Coxeter spaces. The bottom-horizontal maps are morphisms of (signed) Coxeter Lie algebras. Moreover, if $\mathrm{char}(\Bbbk) \nmid |W|$*, then*

$$\overline{\mathcal{K}}\big(\mathcal{P}(\mathsf{h})\big) \xrightarrow{\cong} \overline{\mathcal{P}}\big(\overline{\mathcal{K}}(\mathsf{h})\big). \tag{15.36}$$

PROOF. This result is similar to Proposition 15.46 and can be proved along the same lines. We note some differences. The functor $\mathcal{K}$ does not preserve Lie monoids in general, so the top-horizontal row is only an inclusion of Coxeter spaces. Since the definition of $\overline{\mathcal{K}}$ is in terms of coinvariants, this functor is right exact. If $\mathrm{char}(\Bbbk) \nmid |W|$, then it is also left exact, and (15.36) follows. □

15.8.2. Indecomposable part. Recall from Section 15.7.4 that for c a Coxeter comonoid, the two (signed) Coxeter Lie coalgebra structures on each of the Coxeter spaces $\mathcal{K}(\mathsf{c})$, $\overline{\mathcal{K}}(\mathsf{c})$, $\overline{\mathcal{K}}^{\vee}(\mathsf{c})$ match.

Proposition 15.48. *For a (signed) Coxeter bimonoid* h*, the diagram*

$$\begin{array}{ccccc}
\mathcal{K}(\mathsf{h}) & \twoheadrightarrow & \overline{\mathcal{Q}}\big(\mathcal{K}(\mathsf{h})\big) & \twoheadrightarrow & \mathcal{K}\big(\mathcal{Q}(\mathsf{h})\big) \\
\downarrow & & \downarrow & & \downarrow \\
\overline{\mathcal{K}}(\mathsf{h}) & \twoheadrightarrow & \overline{\mathcal{Q}}\big(\overline{\mathcal{K}}(\mathsf{h})\big) & = & \overline{\mathcal{K}}\big(\mathcal{Q}(\mathsf{h})\big)
\end{array}$$

commutes as (signed) Coxeter Lie coalgebras.

PROOF. This is similar to the proof of Proposition 15.46. The relevant commutative diagram for a top-nested face (A, C) is given below.

$$\begin{array}{ccccc}
\displaystyle\bigoplus_{F_0\colon A<F_0\leq C} \mathsf{h}[F_0] & \xrightarrow{\varphi_{\mathsf{h}}} & \displaystyle\bigoplus_{F\colon A<F} \mathsf{h}[F] & \xrightarrow{(\mu_A^F)} & \mathsf{h}[A] \\
\downarrow & & \downarrow & & \downarrow \\
\displaystyle\bigoplus_{F_0\colon A<F_0\leq C} \mathsf{h}[F_0]_{W_{F_0}} & \xrightarrow[\overline{\varphi}_{\mathsf{h}}]{\cong} & \Big(\displaystyle\bigoplus_{F\colon A<F} \mathsf{h}[F]\Big)_{W_A} & \longrightarrow & \mathsf{h}[A]_{W_A}
\end{array}$$

For the first square, see diagram (14.28). □

Proposition 15.49. *For a (signed) Coxeter bimonoid* h*, the diagram*

$$\begin{array}{ccccc}
\mathcal{K}^\vee(\mathsf{h}) & \twoheadrightarrow & \overline{\mathcal{Q}}\big(\mathcal{K}^\vee(\mathsf{h})\big) & \twoheadrightarrow & \mathcal{K}^\vee\big(\mathcal{Q}(\mathsf{h})\big) \\
\uparrow & & \uparrow & & \uparrow \\
\overline{\mathcal{K}}^\vee(\mathsf{h}) & \twoheadrightarrow & \overline{\mathcal{Q}}\big(\overline{\mathcal{K}}^\vee(\mathsf{h})\big) & \longrightarrow & \overline{\mathcal{K}}^\vee\big(\mathcal{Q}(\mathsf{h})\big)
\end{array}$$

commutes as Coxeter spaces. The bottom-horizontal maps are morphisms of (signed) Coxeter Lie coalgebras. Moreover, if $\mathrm{char}(\Bbbk) \nmid |W|$*, then*

$$\overline{\mathcal{Q}}\big(\overline{\mathcal{K}}^\vee(\mathsf{h})\big) \xrightarrow{\cong} \overline{\mathcal{K}}^\vee\big(\mathcal{Q}(\mathsf{h})\big). \tag{15.37}$$

This is the dual of Proposition 15.47.

Remark 15.50. For a Coxeter comonoid c, we have diagrams of Coxeter spaces as in Propositions 15.46 and 15.47. The point is that we are now starting with a less structured object (Coxeter comonoid rather than Coxeter bimonoid), so the output is also less structured (Coxeter space rather than Coxeter Lie algebra). Dually, for a Coxeter monoid a, we have diagrams of Coxeter spaces as in Propositions 15.48 and 15.49.

Exercise 15.51. Check: For a Coxeter species p, the diagrams

$$\begin{array}{ccc}
\overline{\mathcal{T}}\mathcal{K}^\vee(\mathsf{p}) & \xrightarrow{\psi^\vee_{\mathsf{p}}} & \mathcal{K}^\vee\mathcal{T}(\mathsf{p}) \\
\downarrow & & \uparrow \\
\overline{\mathcal{T}}\mathcal{K}^\vee\mathcal{P}\mathcal{T}(\mathsf{p}) & \longrightarrow & \overline{\mathcal{T}}\overline{\mathcal{P}}\mathcal{K}^\vee\mathcal{T}(\mathsf{p})
\end{array}
\qquad
\begin{array}{ccc}
\mathcal{K}\mathcal{T}^\vee(\mathsf{p}) & \xrightarrow{\psi_{\mathsf{p}}} & \overline{\mathcal{T}}^\vee\mathcal{K}(\mathsf{p}) \\
\downarrow & & \uparrow \\
\overline{\mathcal{T}}^\vee\overline{\mathcal{Q}}\mathcal{K}\mathcal{T}^\vee(\mathsf{p}) & \longrightarrow & \overline{\mathcal{T}}^\vee\mathcal{K}\mathcal{Q}\mathcal{T}^\vee(\mathsf{p})
\end{array} \tag{15.38}$$

commute as Coxeter bialgebras. The vertical maps are induced from units and counits of adjunctions between $\mathcal{T}$ and $\mathcal{P}$, and $\overline{\mathcal{T}}$ and $\overline{\mathcal{P}}$, and $\mathcal{Q}$ and $\mathcal{T}^\vee$, and $\overline{\mathcal{Q}}$ and $\overline{\mathcal{T}}^\vee$ (Theorems 3.11, 3.16, 9.15, 9.19 with $q = 1$).

Either of the diagrams in (15.38) also holds for the other full Fock functor as well as for both bosonic Fock functors.

More generally, for a scalar q and Coxeter species p, diagrams (15.38) commute as Coxeter q-bialgebras, with $\mathcal{T}$ replaced by $\mathcal{T}_q$, and so on.

15.9. Universal enveloping constructions under Fock functors

Recall universal enveloping Coxeter monoids from Section 6.5 and universal enveloping Coxeter algebras from Section 12.5. We now study how Fock functors interact with these constructions. This has connections to considerations in Section 14.6.

15.9.1. Full Fock functors.

Proposition 15.52. *For a Coxeter Lie monoid* g*, the lax structure* $\psi^\vee_{\mathsf{g}}$ *of* $\mathcal{K}^\vee$ *induces the left diagram below.*

$$\begin{array}{ccc}
\overline{\mathcal{T}}\mathcal{K}^\vee(\mathsf{g}) & \xrightarrow{\psi^\vee_{\mathsf{g}}} & \mathcal{K}^\vee\mathcal{T}(\mathsf{g}) \\
\downarrow & & \downarrow \\
\overline{\mathcal{U}}\mathcal{K}^\vee(\mathsf{g}) & \underset{\psi^\vee_{\mathsf{g}}}{\dashrightarrow} & \mathcal{K}^\vee\mathcal{U}(\mathsf{g})
\end{array}
\qquad
\begin{array}{ccc}
\mathcal{K}\mathcal{T}^\vee(\mathsf{k}) & \xrightarrow{\psi_{\mathsf{k}}} & \overline{\mathcal{T}}^\vee\mathcal{K}(\mathsf{k}) \\
\uparrow & & \uparrow \\
\mathcal{K}\mathcal{U}^\vee(\mathsf{k}) & \underset{\psi_{\mathsf{k}}}{\dashrightarrow} & \overline{\mathcal{U}}^\vee\mathcal{K}(\mathsf{k})
\end{array} \tag{15.39}$$

For a Coxeter Lie comonoid k, *the colax structure* ψ_{k} *of* $\mathcal{K}$ *induces the right diagram above.*

PROOF. This can be checked directly. For instance, for the first diagram, we check that the relation in (12.33) with $\mathrm{G} := \mathcal{K}^{\vee}(\mathsf{g})$ maps to a sum of relations under $\psi^{\vee}_{\mathsf{g}}$. We leave the details as an exercise. Alternatively, this diagram can be expressed as the composite

$$\begin{array}{ccccccc} \overline{\mathcal{T}}\mathcal{K}^{\vee}(\mathsf{g}) & \longrightarrow & \overline{\mathcal{T}}\mathcal{K}^{\vee}\mathcal{P}\mathcal{T}(\mathsf{g}) & \longrightarrow & \overline{\mathcal{T}}\overline{\mathcal{P}}\mathcal{K}^{\vee}\mathcal{T}(\mathsf{g}) & \longrightarrow & \mathcal{K}^{\vee}\mathcal{T}(\mathsf{g}) \\ \downarrow & & \downarrow & & \downarrow & & \downarrow \\ \overline{\mathcal{U}}\mathcal{K}^{\vee}(\mathsf{g}) & \longrightarrow & \overline{\mathcal{U}}\mathcal{K}^{\vee}\mathcal{P}\mathcal{U}(\mathsf{g}) & \longrightarrow & \overline{\mathcal{U}}\overline{\mathcal{P}}\mathcal{K}^{\vee}\mathcal{U}(\mathsf{g}) & \longrightarrow & \mathcal{K}^{\vee}\mathcal{U}(\mathsf{g}). \end{array}$$

For the top line, we use the first diagram in (15.38). □

When Lie structures on g and k are trivial, the diagrams in (15.39) reduce to (14.40) and (14.38), respectively.

15.9.2. Bosonic Fock functors. Bosonic Fock functors commute with universal (co)enveloping Coxeter (co)monoid and (co)algebra functors:

Proposition 15.53. *For a Coxeter Lie monoid* g, *the lax structure* $\overline{\psi}^{\vee}_{\mathsf{g}}$ *of* $\overline{\mathcal{K}}^{\vee}$ *induces the left diagram below.*

$$(15.40)\qquad \begin{array}{ccc} \overline{\mathcal{T}}\overline{\mathcal{K}}^{\vee}(\mathsf{g}) & \xrightarrow{\overline{\psi}^{\vee}_{\mathsf{g}}} & \overline{\mathcal{K}}^{\vee}\mathcal{T}(\mathsf{g}) \\ \downarrow & & \downarrow \\ \overline{\mathcal{U}}\overline{\mathcal{K}}^{\vee}(\mathsf{g}) & \underset{\overline{\psi}^{\vee}_{\mathsf{g}}}{\dashrightarrow} & \overline{\mathcal{K}}^{\vee}\mathcal{U}(\mathsf{g}) \end{array} \qquad \begin{array}{ccc} \overline{\mathcal{K}}\mathcal{T}^{\vee}(\mathsf{k}) & \xrightarrow{\overline{\psi}_{\mathsf{k}}} & \overline{\mathcal{T}}^{\vee}\overline{\mathcal{K}}(\mathsf{k}) \\ \uparrow & & \uparrow \\ \overline{\mathcal{K}}\mathcal{U}^{\vee}(\mathsf{k}) & \underset{\overline{\psi}_{\mathsf{k}}}{\dashrightarrow} & \overline{\mathcal{U}}^{\vee}\overline{\mathcal{K}}(\mathsf{k}) \end{array}$$

For a Coxeter Lie comonoid k, *the colax structure* $\overline{\psi}_{\mathsf{k}}$ *of* $\overline{\mathcal{K}}$ *induces the right diagram above. Moreover, if* $\mathrm{char}(\Bbbk) \nmid |W|$, *then the dotted arrows are isomorphisms.*

PROOF. Diagrams (15.40) can be deduced from diagrams (15.39). This proves the first claim. For the second claim, we can either proceed directly or use CMM as follows.

For the first dotted arrow,

$$\overline{\mathcal{U}}\overline{\mathcal{K}}^{\vee}(\mathsf{g}) = \overline{\mathcal{U}}\overline{\mathcal{K}}^{\vee}\mathcal{P}\mathcal{U}(\mathsf{g}) = \overline{\mathcal{U}}\overline{\mathcal{P}}\overline{\mathcal{K}}^{\vee}\mathcal{U}(\mathsf{g}) = \overline{\mathcal{K}}^{\vee}\mathcal{U}(\mathsf{g}).$$

The first step used CMM Theorem 7.26, while the last step used CMM Theorem 13.65. The second step used Proposition 15.46.

Dually, for the second dotted arrow,

$$\overline{\mathcal{K}}\mathcal{U}^{\vee}(\mathsf{k}) = \overline{\mathcal{U}}^{\vee}\overline{\mathcal{Q}}\overline{\mathcal{K}}\mathcal{U}^{\vee}(\mathsf{k}) = \overline{\mathcal{U}}^{\vee}\overline{\mathcal{K}}\mathcal{Q}\mathcal{U}^{\vee}(\mathsf{k}) = \overline{\mathcal{U}}^{\vee}\overline{\mathcal{K}}(\mathsf{k}).$$

The first and last steps used dual CMM Theorems 13.68 and 7.27. The second step used Proposition 15.48.

To reconcile the above strings of identifications with diagrams (15.40), we can use diagrams (15.38) with $\overline{\mathcal{K}}^{\vee}$ and $\overline{\mathcal{K}}$ instead of $\mathcal{K}^{\vee}$ and $\mathcal{K}$ (as in the proof of Proposition 15.52). □

Proposition 15.54. *For a Coxeter Lie monoid* g*, the lax structure* $\overline{\varphi}_{\mathsf{g}}$ *of* $\overline{\mathcal{K}}$ *induces the left diagram below.*

$$
\begin{array}{ccc}
\overline{\mathcal{T}}\overline{\mathcal{K}}(\mathsf{g}) & \xrightarrow{\overline{\varphi}_{\mathsf{g}}} & \overline{\mathcal{K}}\mathcal{T}(\mathsf{g}) \\
\downarrow & & \downarrow \\
\overline{\mathcal{U}}\overline{\mathcal{K}}(\mathsf{g}) & \underset{\overline{\varphi}_{\mathsf{g}}}{\dashrightarrow} & \overline{\mathcal{K}}\mathcal{U}(\mathsf{g})
\end{array}
\qquad
\begin{array}{ccc}
\overline{\mathcal{K}}^{\vee}\mathcal{T}^{\vee}(\mathsf{k}) & \xrightarrow{\overline{\varphi}_{\mathsf{k}}^{\vee}} & \overline{\mathcal{T}}^{\vee}\overline{\mathcal{K}}^{\vee}(\mathsf{k}) \\
\uparrow & & \uparrow \\
\overline{\mathcal{K}}^{\vee}\mathcal{U}^{\vee}(\mathsf{k}) & \underset{\overline{\varphi}_{\mathsf{k}}^{\vee}}{\dashrightarrow} & \overline{\mathcal{U}}^{\vee}\overline{\mathcal{K}}^{\vee}(\mathsf{k})
\end{array}
\tag{15.41}
$$

For a Coxeter Lie comonoid k*, the colax structure* $\overline{\varphi}_{\mathsf{k}}^{\vee}$ *of* $\overline{\mathcal{K}}^{\vee}$ *induces the right diagram above. Moreover, the dotted arrows are isomorphisms (in any field characteristic).*

Sketch of proof. We can proceed directly as in the proof of Proposition 15.52. For instance, in the first diagram, a relation (which defines $\overline{\mathcal{U}}$ as a quotient of $\overline{\mathcal{T}}$) maps to a relation under $\overline{\varphi}_{\mathsf{g}}$ (which defines $\mathcal{U}$ as a quotient of $\mathcal{T}$). To see this, we need to rearrange terms using the W-action; this is possible since we are working in the quotient space of W-coinvariants. □

When Lie structures on g and k are trivial, diagrams (15.40) and (15.41) reduce to (14.43) and (14.44).

15.10. Notes

The material in this chapter is new.

15.10.1. Stover constructions. Section 15.1 generalizes constructions of (signed) graded bialgebras from Joyal bimonoids. More precisely, a Joyal bimonoid h yields graded bialgebras $\mathcal{K}(\mathsf{h})$, $\mathcal{K}^{\vee}(\mathsf{h})$, $\overline{\mathcal{K}}(\mathsf{h})$, $\overline{\mathcal{K}}^{\vee}(\mathsf{h})$ and their signed versions $\mathcal{K}_{-1}(\mathsf{h})$, $\mathcal{K}^{\vee}_{-1}(\mathsf{h})$, $\overline{\mathcal{K}}_{-1}(\mathsf{h})$, $\overline{\mathcal{K}}^{\vee}_{-1}(\mathsf{h})$. These constructions first appeared in work of Stover [**341**, Proposition 14.6 and comments in 14.7]; for later references, see for instance [**290**], [**291**], [**292**] or the more recent [**84**, Section 9.6]. Stover formulated these constructions in the language of twisted bialgebras. Interestingly, he focussed on the signed case, mentioning at the end that everything could be done without signs. Stover did not consider graded q-bialgebras.

15.10.2. Bilax structure of Fock functors. Joyal bimonoids were first formulated in our monograph [**7**, Sections 8.2 and 8.3]. Stover constructions in this language and with emphasis on the bilax structure of Fock functors were explained in [**7**, Section 15.2]. In fact, one central principle of [**7**] was to explain how the abstract study of Fock functors has powerful implications for Hopf algebras. The present chapter is built on the same principle. The precise dictionary of some of the results in this chapter with those in [**7**] is as follows.

- Theorem 15.1 on the passage from Coxeter bimonoids to Coxeter bialgebras generalizes [**7**, Theorems 15.12 and 16.7]. Theorem 15.2 and Proposition 15.8 generalize [**7**, Theorem 15.13].
- For Theorem 15.11 on commutativity, see [**7**, Corollary 15.27 and comments after Proposition 15.28]. Exercise 15.14 generalizes [**7**, Corollary 16.28]. Exercise 15.15 generalizes [**7**, Corollary 16.29 and discussion on page 537 in Section 15.5.2].

- Proposition 15.16 on the q-norm map generalizes [**7**, Theorem 16.19]. Diagram (15.23) generalizes [**7**, Diagrams (15.16) and (16.9)]. Example 15.18 generalizes [**7**, Example 16.17].
- Theorem 15.31 on (co)freeness of full Fock spaces generalizes [**7**, Proposition 16.11], [**230**, Theorem 4.2.2]; see also [**9**, Section 3.3].
- Classical analogues of Propositions 15.21, 15.22, 15.28 on full Fock functors were not mentioned in [**7**]; in the present setup, they arise very naturally. Propositions 15.24 and 15.25 on bosonic Fock functors can also be approached through the theory of adjoints; see Exercises 16.25, 16.33, 16.53 and related references in Notes 16.12.
- The results in Section 15.4.5 look new even classically.
- Propositions 15.46 and 15.47 on the primitive part generalize [**7**, Propositions 15.35 and 15.36]. Classical analogues of Propositions 15.48 and 15.49 on the indecomposable part were not explicitly given in [**7**]. The material in Section 15.9 appears to be new.

The results of Section 15.5 on the Boolean transform generalize those in [**9**, Section 4.4]. Theorem 15.35 relates to [**7**, Corollary 4.37] which says that a braided lax monoidal functor preserves operad monoids. (Recall from the second column in Table 14.3 that $\mathcal{K}^\vee$ and $\overline{\mathcal{K}}^\vee$ are braided lax.)

CHAPTER 16

Adjoints of Fock functors

Fock functors were introduced and their basic properties studied in Chapters 14 and 15. Their main feature is that they send Coxeter (co, bi)monoids to Coxeter (co, bi)algebras. At a more basic level, they send Coxeter species to Coxeter spaces. We now construct adjoints of these functors.

Adjoints of bosonic Fock functors. Trivialization functor and its lifts. We introduce the trivialization functor $\mathsf{t}_{(-)}$ from Coxeter spaces to Coxeter species. The construction builds on the process of turning a vector space into a W-module with trivial W-action. This motivates our terminology. The left and right adjoints of $\mathsf{t}_{(-)}$ are the two bosonic Fock functors $\overline{\mathcal{K}}$ and $\overline{\mathcal{K}}^\vee$ (which we recall are constructed using W-coinvariants and W-invariants).

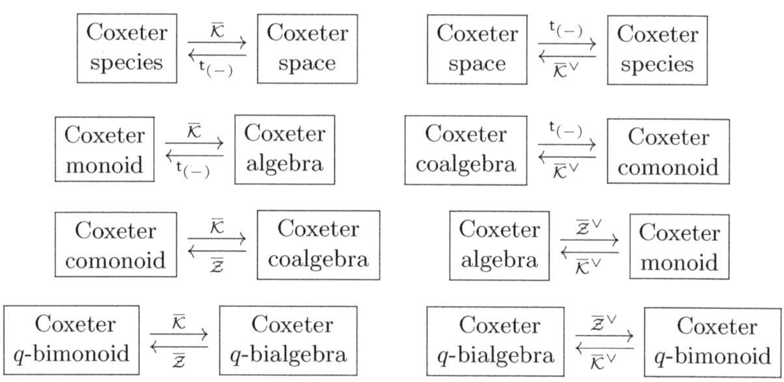

SUMMARY 16.1. Adjunctions involving bosonic Fock functors.

We lift these two adjunctions to the categories of Coxeter (co, bi)monoids and Coxeter (co, bi)algebras, see Summary 16.1. Functors $\overline{\mathcal{Z}}$ and $\overline{\mathcal{Z}}^\vee$ are built out of $\mathsf{t}_{(-)}$. The functor $\mathsf{t}_{(-)}$ is self-dual, while $\overline{\mathcal{Z}}$ and $\overline{\mathcal{Z}}^\vee$ are conjugates of each other wrt duality. Thus, in Summary 16.1, adjunctions on the left are conjugates of adjunctions on the right. We also recall that $\overline{\mathcal{K}}$ and $\overline{\mathcal{K}}^\vee$ are isomorphic functors via the norm map (when $\mathrm{char}(\Bbbk) \nmid |W|$). Interestingly, the adjoint functors $\mathsf{t}_{(-)}$ and $\overline{\mathcal{Z}}$ are sections of $\overline{\mathcal{K}}$, while $\mathsf{t}_{(-)}$ and $\overline{\mathcal{Z}}^\vee$ are sections of $\overline{\mathcal{K}}^\vee$ (with claims about $\overline{\mathcal{Z}}$ and $\overline{\mathcal{Z}}^\vee$ assuming $\mathrm{char}(\Bbbk) \nmid |W|$). Thus, they are full and faithful. Also, Coxeter q-bialgebras can be viewed both as a reflective and coreflective subcategory of Coxeter q-bimonoids.

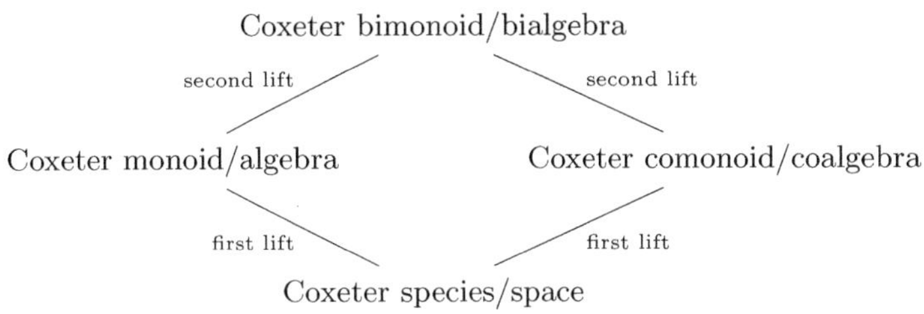

FIGURE 16.1. The adjunction lifting process.

The above adjunction lifts can be understood abstractly using notions of lax-lax, colax-colax, colax-lax adjunctions and the adjoint lifting theorem from Appendix A.1.3. The process is summarized in Figure 16.1. Starting with $(\overline{\mathcal{K}}, \mathsf{t}_{(-)})$ yields the left column in Summary 16.1, while $(\mathsf{t}_{(-)}, \overline{\mathcal{K}}^{\vee})$ yields the right column in Summary 16.1.

Adjoints of bosonic Fock functors preserving commutativity. We also consider commutative aspects. For instance, we lift the adjunction between $\mathsf{t}_{(-)}$ and $\overline{\mathcal{K}}^{\vee}$ to the categories of commutative Coxeter monoids and commutative Coxeter algebras. Here, $\mathsf{t}_{(-)}$ lifts to $\overline{\mathcal{Z}}^{\vee}_{ab}$ which is $\overline{\mathcal{Z}}^{\vee}$ followed by the abelianization functor. Dually, for lifting the adjunction between $\overline{\mathcal{K}}$ and $\mathsf{t}_{(-)}$, we consider $\overline{\mathcal{Z}}^{coab}$ which is $\overline{\mathcal{Z}}$ followed by the coabelianization functor. All such adjunctions are listed in Summary 16.2.

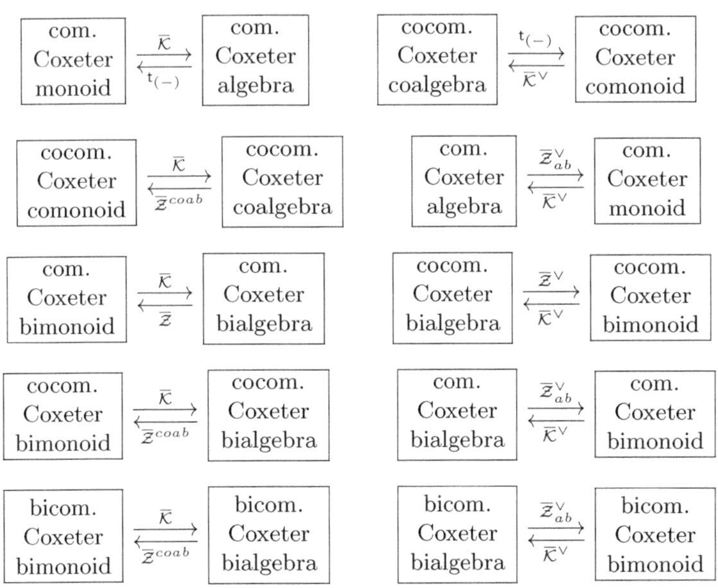

SUMMARY 16.2. Adjunctions preserving commutativity.

There are similar adjunctions preserving signed (co)commutativity. We use notations $\overline{\mathcal{Z}}^{\vee}_{-ab}$ and $\overline{\mathcal{Z}}^{-coab}$ in this setting.

Adjoints of fermionic Fock functors. We obtain adjoints of fermionic Fock functors from those of bosonic Fock functors in Summaries 16.1 and 16.2 by employing the o-signature functor on Coxeter species. For instance, we have adjunctions

$$\boxed{\begin{matrix}\text{Coxeter}\\ \text{species}\end{matrix}} \underset{(\mathsf{t}_{(-)})^{-}}{\overset{\overline{\mathcal{K}}_{-1}}{\rightleftarrows}} \boxed{\begin{matrix}\text{Coxeter}\\ \text{space}\end{matrix}} \qquad \boxed{\begin{matrix}\text{Coxeter}\\ \text{space}\end{matrix}} \underset{\overline{\mathcal{K}}^{\vee}_{-1}}{\overset{(\mathsf{t}_{(-)})^{-}}{\rightleftarrows}} \boxed{\begin{matrix}\text{Coxeter}\\ \text{species}\end{matrix}}$$

and so on. We refer to $(\mathsf{t}_{(-)})^{-}$ as the signed trivialization functor.

Adjoints of full Fock functors. We construct adjoints of full Fock functors as shown in Summary 16.3. The functor $\mathcal{L}$ is constructed by taking Hadamard product of the trivialization functor $\mathsf{t}_{(-)}$ with the Coxeter species of chambers Γ. Functors $\mathcal{Z}$ and $\mathcal{Z}^{\vee}$ are similarly constructed from $\overline{\mathcal{Z}}$ and $\overline{\mathcal{Z}}^{\vee}$ with Γ used more intricately. Moreover, $\mathcal{Z}$ and $\mathcal{Z}^{\vee}$ can be deformed to $\mathcal{Z}_q$ and $\mathcal{Z}^{\vee}_q$ by replacing Γ by Γ_q. We also consider $\mathcal{Z}^{\vee}_{ab}$ and $\mathcal{Z}^{coab}$; recall here that $\mathcal{K}$ preserves cocommutativity, while $\mathcal{K}^{\vee}$ preserves commutativity. Functors $\mathcal{Z}^{\vee}_{ab}$ and $\mathcal{Z}^{coab}$ have signed analogues $\mathcal{Z}^{\vee}_{-ab}$ and $\mathcal{Z}^{-coab}$.

$$\boxed{\begin{matrix}\text{Coxeter}\\ \text{species}\end{matrix}} \underset{\mathcal{L}^{\vee}}{\overset{\mathcal{K}=\mathcal{K}^{\vee}}{\rightleftarrows}} \boxed{\begin{matrix}\text{Coxeter}\\ \text{space}\end{matrix}} \qquad \boxed{\begin{matrix}\text{Coxeter}\\ \text{space}\end{matrix}} \underset{\mathcal{K}=\mathcal{K}^{\vee}}{\overset{\mathcal{L}}{\rightleftarrows}} \boxed{\begin{matrix}\text{Coxeter}\\ \text{species}\end{matrix}}$$

$$\boxed{\begin{matrix}\text{Coxeter}\\ \text{comonoid}\end{matrix}} \underset{\mathcal{L}^{\vee}}{\overset{\mathcal{K}^{\vee}}{\rightleftarrows}} \boxed{\begin{matrix}\text{Coxeter}\\ \text{coalgebra}\end{matrix}} \qquad \boxed{\begin{matrix}\text{Coxeter}\\ \text{algebra}\end{matrix}} \underset{\mathcal{K}}{\overset{\mathcal{L}}{\rightleftarrows}} \boxed{\begin{matrix}\text{Coxeter}\\ \text{monoid}\end{matrix}}$$

$$\boxed{\begin{matrix}\text{Coxeter}\\ \text{comonoid}\end{matrix}} \underset{\mathcal{Z}_q}{\overset{\mathcal{K}_q}{\rightleftarrows}} \boxed{\begin{matrix}\text{Coxeter}\\ \text{coalgebra}\end{matrix}} \qquad \boxed{\begin{matrix}\text{Coxeter}\\ \text{algebra}\end{matrix}} \underset{\mathcal{K}^{\vee}_q}{\overset{\mathcal{Z}^{\vee}_q}{\rightleftarrows}} \boxed{\begin{matrix}\text{Coxeter}\\ \text{monoid}\end{matrix}}$$

$$\boxed{\begin{matrix}\text{cocom.}\\ \text{Coxeter}\\ \text{comonoid}\end{matrix}} \underset{\mathcal{Z}^{coab}}{\overset{\mathcal{K}}{\rightleftarrows}} \boxed{\begin{matrix}\text{cocom.}\\ \text{Coxeter}\\ \text{coalgebra}\end{matrix}} \qquad \boxed{\begin{matrix}\text{com.}\\ \text{Coxeter}\\ \text{algebra}\end{matrix}} \underset{\mathcal{K}^{\vee}}{\overset{\mathcal{Z}^{\vee}_{ab}}{\rightleftarrows}} \boxed{\begin{matrix}\text{com.}\\ \text{Coxeter}\\ \text{monoid}\end{matrix}}$$

SUMMARY 16.3. Adjunctions involving full Fock functors.

Coxeter operad algebras to Coxeter operad monoids. We briefly initiate how adjoints of Fock functors work in the generality of Coxeter operads. See Summary 16.4; here **a** denotes a Coxeter operad and **c** a Coxeter cooperad.

For the Coxeter associative and commutative operads, we recover as special cases the adjunctions $(\overline{\mathcal{K}}, \mathsf{t}_{(-)})$ and $(\overline{\mathcal{Z}}^{\vee}, \overline{\mathcal{K}}^{\vee})$ between Coxeter monoids and Coxeter algebras in Summary 16.1, their commutative versions $(\overline{\mathcal{K}}, \mathsf{t}_{(-)})$ and $(\overline{\mathcal{Z}}^{\vee}_{ab}, \overline{\mathcal{K}}^{\vee})$ in Summary 16.2, as well as conjugates of these adjunctions between (cocommutative) Coxeter comonoids and (cocommutative) Coxeter

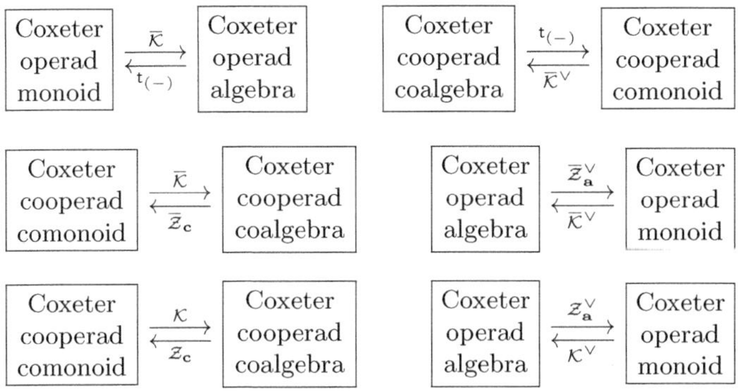

SUMMARY 16.4. Coxeter operad monoids and algebras.

coalgebras. It also includes the adjunctions $(\mathcal{Z}^\vee, \mathcal{K}^\vee)$ and $(\mathcal{Z}^\vee_{ab}, \mathcal{K}^\vee)$ and their conjugates in Summary 16.3.

Coxeter Lie algebras to Coxeter Lie monoids. The adjunctions in Summary 16.4 specialize to adjunctions between Coxeter Lie (co)monoids and Coxeter Lie (co)algebras; see Summary 16.5. Moreover, these interact with adjunctions between (co)commutative Coxeter bimonoids and (co)commutative Coxeter bialgebras via the universal (co)enveloping and primitive and indecomposable part functors. This is specific to the Lie case. Also functors $\mathsf{t}_{(-)}$ and $\overline{\mathcal{Z}}_{\mathbf{Lie}^*}$ are sections of $\overline{\mathcal{K}}$, while $\mathsf{t}_{(-)}$ and $\overline{\mathcal{Z}}^\vee_{\mathbf{Lie}}$ are sections of $\overline{\mathcal{K}}^\vee$.

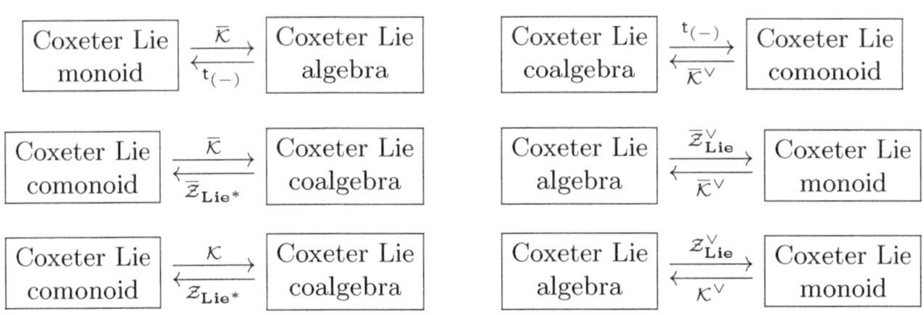

SUMMARY 16.5. Coxeter Lie monoids and Coxeter Lie algebras.

Observation 16.1. Since Fock functors are $\Bbbk$-linear, so are all their adjoints. The $\Bbbk$-linearity of these adjoint functors such as $\mathsf{t}_{(-)}$, $\mathcal{L}$, $\overline{\mathcal{Z}}$, $\overline{\mathcal{Z}}^\vee$ can also be checked directly using their explicit constructions.

16.1. Trivialization functor

We introduce the trivialization functor $\mathsf{t}_{(-)}$ from Coxeter spaces to Coxeter species. In rough terms, it associates to a Coxeter space V the Coxeter species t_V with components matching those of V and equipped with the trivial

W-action. There are canonical (co)lax structures on the trivialization functor which induce functors from Coxeter (co)algebras to Coxeter (co)monoids. The relevant monads and comonads on Coxeter species and Coxeter spaces such as $\mathcal{T}$ and $\overline{\mathcal{T}}$ were constructed in Sections 2.10 and 8.12.

16.1.1. Trivialization functor. Coxeter spaces to Coxeter species. Fix a pointed reflection arrangement $\boldsymbol{\alpha} = (\mathcal{A}, C_r)$. Define a functor

$$\mathsf{t}_{(-)} : \boldsymbol{\alpha}\text{-}\overline{\mathtt{C}}\mathtt{Sp} \to \mathcal{A}\text{-}\mathtt{CSp}, \qquad \mathsf{t}_{\mathrm{V}}[F] := \mathrm{V}[\mathrm{t}(F)], \tag{16.1}$$

where V is a Coxeter space, F is any face, and $\mathrm{t}(F)$ is its type. The structure maps of t_{V} are as follows.

$$\begin{array}{ccc} \mathsf{t}_{\mathrm{V}}[F] & \overset{\beta}{\dashrightarrow} & \mathsf{t}_{\mathrm{V}}[G] \\ \| & & \| \\ \mathrm{V}[\mathrm{t}(F)] & \underset{\sigma}{\longrightarrow} & \mathrm{V}[\mathrm{t}(G)] \end{array} \qquad \begin{array}{ccc} \mathsf{t}_{\mathrm{V}}[F] & \overset{w}{\dashrightarrow} & \mathsf{t}_{\mathrm{V}}[F'] \\ \| & & \| \\ \mathrm{V}[\mathrm{t}(F)] & \underset{\mathrm{id}}{\longrightarrow} & \mathrm{V}[\mathrm{t}(F')] \end{array} \tag{16.2}$$

- For a support morphism $\beta : F \to G$, define $\mathsf{t}_{\mathrm{V}}[F] \to \mathsf{t}_{\mathrm{V}}[G]$ to be the morphism σ transferred by uG, where u is any element of W which takes F to a face of C_r, or equivalently, cotransferred by vF, where v is any element of W which takes G to a face of C_r. Alternatively, σ is the biface-type represented by the biface (G, F), see Proposition 8.18.
- For a type morphism $w : F \to F'$, define $\mathsf{t}_{\mathrm{V}}[F] \to \mathsf{t}_{\mathrm{V}}[F']$ to be the identity map. (This makes sense since $\mathrm{t}(F) = \mathrm{t}(F')$.)

This turns t_{V} into a Coxeter species. A map $f : \mathrm{V} \to \mathrm{W}$ of Coxeter spaces induces a map of Coxeter species $\mathsf{t}_f : \mathsf{t}_{\mathrm{V}} \to \mathsf{t}_{\mathrm{W}}$ via

$$\begin{array}{ccc} \mathsf{t}_{\mathrm{V}}[F] & \overset{(\mathsf{t}_f)_F}{\dashrightarrow} & \mathsf{t}_{\mathrm{W}}[F] \\ \| & & \| \\ \mathrm{V}[\mathrm{t}(F)] & \underset{f_{\mathrm{t}(F)}}{\longrightarrow} & \mathrm{W}[\mathrm{t}(F)]. \end{array} \tag{16.3}$$

Hence $\mathsf{t}_{(-)}$ is a functor, which we refer to as the *trivialization functor*. It is $\Bbbk$-linear; this can either be seen directly or deduced from the existence of the functor (16.4) below.

Observation 16.2. The two diagrams in (16.2) can be combined into one as

$$\begin{array}{ccc} \mathsf{t}_{\mathrm{V}}[F] & \overset{\sigma'}{\dashrightarrow} & \mathsf{t}_{\mathrm{V}}[G] \\ \| & & \| \\ \mathrm{V}[\mathrm{t}(F)] & \underset{\sigma}{\longrightarrow} & \mathrm{V}[\mathrm{t}(G)], \end{array}$$

where $\sigma' = \beta w = w\beta$ in normal form, and σ corresponds to β as in the first diagram in (16.2). Note very carefully the difference between σ' and σ.

Exercise 16.3. Check: There is a functor

$$\mathcal{A}\text{-}\mathtt{Cox} \to \boldsymbol{\alpha}\text{-}\overline{\mathtt{Cox}}, \qquad F \mapsto \mathrm{t}(F), \quad (F \xrightarrow{\sigma'} G) \mapsto (\mathrm{t}(F) \xrightarrow{\sigma} \mathrm{t}(G)), \tag{16.4}$$

with notation as above. In other words, $\sigma' \mapsto \sigma$ respects composition of morphisms. By passing to functor categories, (16.4) yields (16.1).

Remark 16.4. One can also formulate the trivialization functor in terms of top-nested faces instead of face-types. Define

$$\mathtt{t}_{(-)} : \mathcal{A}\text{-}\overline{\mathtt{C}}\mathtt{Sp} \to \mathcal{A}\text{-}\mathtt{CSp}, \qquad \mathtt{t}_{\mathrm{V}}[F] := \mathrm{V}[F, FC_r], \tag{16.5}$$

with structure maps

$$\begin{array}{ccc} \mathtt{t}_{\mathrm{V}}[F] & \overset{\beta}{\dashrightarrow} & \mathtt{t}_{\mathrm{V}}[G] \\ \| & & \| \\ \mathrm{V}[F, FC_r] & \xrightarrow[\beta]{} & \mathrm{V}[G, GC_r] \end{array} \qquad \begin{array}{ccc} \mathtt{t}_{\mathrm{V}}[F] & \overset{w}{\dashrightarrow} & \mathtt{t}_{\mathrm{V}}[F'] \\ \| & & \| \\ \mathrm{V}[F, FC_r] & \xrightarrow[u(FC_r)=F'C_r]{} & \mathrm{V}[F', F'C_r], \end{array}$$

and so on. Alternatively, one can define $\mathtt{t}_{\mathrm{V}}[F] := \mathrm{V}[F_0, C_r]$, where F_0 is the face of C_r of the same type as F, and proceed. The two definitions of $\mathtt{t}_{\mathrm{V}}$ are not the same but they yield isomorphic functors.

16.1.2. Coxeter (co)algebras to Coxeter (co)monoids.

For a Coxeter algebra (A, μ), the Coxeter species $\mathtt{t}_{\mathrm{A}}$ is a Coxeter monoid with product defined by the first diagram below.

$$\begin{array}{ccc} \mathtt{t}_{\mathrm{A}}[F] & \overset{\mu_A^F}{\dashrightarrow} & \mathtt{t}_{\mathrm{A}}[A] \\ \| & & \| \\ \mathrm{A}[\mathrm{t}(F)] & \xrightarrow[\mu_{\mathrm{t}(A)}^{\mathrm{t}(F)}]{} & \mathrm{A}[\mathrm{t}(A)] \end{array} \qquad \begin{array}{ccc} \mathtt{t}_{\mathrm{C}}[A] & \overset{\Delta_A^F}{\dashrightarrow} & \mathtt{t}_{\mathrm{C}}[F] \\ \| & & \| \\ \mathrm{C}[\mathrm{t}(A)] & \xrightarrow[\Delta_{\mathrm{t}(A)}^{\mathrm{t}(F)}]{} & \mathrm{C}[\mathrm{t}(F)] \end{array} \tag{16.6}$$

Similarly, by reversing arrows as in the second diagram above, we see that $\mathtt{t}_{\mathrm{C}}$ is a Coxeter comonoid for any Coxeter coalgebra C.

Note: If the product on A is trivial, then so is the product on $\mathtt{t}_{\mathrm{A}}$. Similarly, if the coproduct on C is trivial, then so is the coproduct on $\mathtt{t}_{\mathrm{C}}$.

Exercise 16.5. Check:

- If a Coxeter algebra A is (signed) commutative, then so is the Coxeter monoid $\mathtt{t}_{\mathrm{A}}$. (Use commutativity axioms (2.49) and (8.86) and their signed analogues (2.61) and (8.92).)
- If a Coxeter coalgebra C is (signed) cocommutative, then so is the Coxeter comonoid $\mathtt{t}_{\mathrm{C}}$. (Use cocommutativity axioms (2.57) and (8.91) and their signed analogues (2.62) and (8.93).)

16.1.3. Lax and colax structures.

More formally, $\mathtt{t}_{(-)}$ carries a lax structure wrt monads $\mathcal{T}$ and $\overline{\mathcal{T}}$ and colax structure wrt comonads $\mathcal{T}^{\vee}$ and $\overline{\mathcal{T}}^{\vee}$ such that it sends Coxeter (co)algebras to Coxeter (co)monoids in the manner described by formulas (16.6). Evaluated on a Coxeter space V on the A-component, the lax and colax structures are

$$\bigoplus_{F:\, F \geq A} \mathrm{V}[\mathrm{t}(F)] \underset{\text{colax}}{\overset{\text{lax}}{\rightleftarrows}} \bigoplus_{T:\, T \geq \mathrm{t}(A)} \mathrm{V}[T], \tag{16.7}$$

with a summand on one side identified with all matching summands on the other side. The first sum is over faces, while the second is over face-types.

Exercise 16.6. Check: (16.7) indeed defines lax and colax structures on $\mathtt{t}_{(-)}$. Further, $\mathtt{t}_{(-)}$ carries a lax structure wrt monads $\mathcal{S}$ and $\overline{\mathcal{S}}$ and colax structure wrt comonads $\mathcal{S}^\vee$ and $\overline{\mathcal{S}}^\vee$ which are induced from (16.7). In other words, we have commutative diagrams

$$
\begin{array}{ccc}
\mathcal{T}\mathsf{t}_{(-)} & \longrightarrow & \mathsf{t}_{(-)}\overline{\mathcal{T}} \\
\big\downarrow & & \big\downarrow \\
\mathcal{S}\mathsf{t}_{(-)} & \longrightarrow & \mathsf{t}_{(-)}\overline{\mathcal{S}}
\end{array}
\qquad\qquad
\begin{array}{ccc}
\mathcal{T}^\vee\mathsf{t}_{(-)} & \longleftarrow & \mathsf{t}_{(-)}\overline{\mathcal{T}}^\vee \\
\big\uparrow & & \big\uparrow \\
\mathcal{S}^\vee\mathsf{t}_{(-)} & \longleftarrow & \mathsf{t}_{(-)}\overline{\mathcal{S}}^\vee.
\end{array}
\tag{16.8}
$$

The same remark applies to monads $\mathcal{E}$ and $\overline{\mathcal{E}}$ and comonads $\mathcal{E}^\vee$ and $\overline{\mathcal{E}}^\vee$.

Exercise 16.7. Check: $\mathsf{t}_{(-)}$ carries a colax structure wrt monads $\mathcal{T}$ and $\overline{\mathcal{T}}$ and lax structure wrt comonads $\mathcal{T}^\vee$ and $\overline{\mathcal{T}}^\vee$ given by the same formulas (16.7). Moreover, there is an induced colax structure wrt monads $\mathcal{S}$ and $\overline{\mathcal{S}}$ and lax structure wrt comonads $\mathcal{S}^\vee$ and $\overline{\mathcal{S}}^\vee$. In other words, we have commutative diagrams

$$
\begin{array}{ccc}
\mathcal{T}\mathsf{t}_{(-)} & \longleftarrow & \mathsf{t}_{(-)}\overline{\mathcal{T}} \\
\big\downarrow & & \big\downarrow \\
\mathcal{S}\mathsf{t}_{(-)} & \longleftarrow & \mathsf{t}_{(-)}\overline{\mathcal{S}}
\end{array}
\qquad\qquad
\begin{array}{ccc}
\mathcal{T}^\vee\mathsf{t}_{(-)} & \longrightarrow & \mathsf{t}_{(-)}\overline{\mathcal{T}}^\vee \\
\big\uparrow & & \big\uparrow \\
\mathcal{S}^\vee\mathsf{t}_{(-)} & \longrightarrow & \mathsf{t}_{(-)}\overline{\mathcal{S}}^\vee.
\end{array}
\tag{16.9}
$$

(Compare and contrast with diagrams (16.8). Note the reversal of the horizontal arrows.) The same remark applies to monads $\mathcal{E}$ and $\overline{\mathcal{E}}$ and comonads $\mathcal{E}^\vee$ and $\overline{\mathcal{E}}^\vee$.

Remark 16.8. The lax and colax structures of $\mathsf{t}_{(-)}$ wrt either monads or comonads arise as mates of colax and lax structures, respectively, of bosonic Fock functors. This fact is elaborated in Section 16.4.

16.1.4. Self-duality. For V a Coxeter space, $\mathsf{t}_{\mathrm{V}^*} = \mathsf{t}_{\mathrm{V}}^*$ as Coxeter species, and this identification is natural in V. In finite-dimensional setting, the functor $\mathsf{t}_{(-)}$ from Coxeter spaces to Coxeter species is self-dual.

For A a Coxeter algebra and C a Coxeter coalgebra,

$$\mathsf{t}_{\mathrm{A}^*} = \mathsf{t}_{\mathrm{A}}^* \quad \text{and} \quad \mathsf{t}_{\mathrm{C}^*} = \mathsf{t}_{\mathrm{C}}^*$$

as Coxeter comonoids and Coxeter monoids, respectively. A similar statement holds for lax and colax structures (16.7) of $\mathsf{t}_{(-)}$.

16.1.5. Trivialization functor in terms of flats and lunes. One can formulate the trivialization functor in terms of flats and lunes by defining

$$\mathsf{t}_{(-)} : \mathcal{A}\text{-}\overline{\mathsf{C}}\mathsf{Sp} \to \mathcal{A}\text{-}\mathsf{CSp}, \qquad \mathsf{t}_{\mathrm{V}}[\mathrm{X}] := \mathrm{V}[\mathrm{L}], \tag{16.10}$$

where L is the top-lune with base X containing C_r. The structure maps are

$$
\begin{array}{ccc}
\mathsf{t}_{\mathrm{V}}[\mathrm{X}] & \overset{w}{\dashrightarrow} & \mathsf{t}_{\mathrm{V}}[\mathrm{X}'] \\
\| & & \| \\
\mathrm{V}[\mathrm{L}] & \underset{u}{\longrightarrow} & \mathrm{V}[\mathrm{L}'],
\end{array}
$$

where u is the unique Coxeter symmetry which agrees with w on X and takes L to L′. This description can be deduced from Remark 16.4.

For A a Coxeter algebra, t_{A} is a Coxeter monoid and for C a Coxeter coalgebra, t_{C} is a Coxeter comonoid. Their product and coproduct are as follows. For a lune P with base Z and case X,

$$
\begin{array}{ccc}
\mathtt{t}_{\mathrm{A}}[\mathrm{X}] & \xrightarrow{\mu_{\mathrm{P}}} & \mathtt{t}_{\mathrm{A}}[\mathrm{Z}] \\
\| & & \| \\
\mathrm{A}[\mathrm{M}] & \xrightarrow[\mu_{\mathrm{L}}^{\mathrm{M}}(\mathrm{P})]{} & \mathrm{A}[\mathrm{L}]
\end{array}
\qquad
\begin{array}{ccc}
\mathtt{t}_{\mathrm{C}}[\mathrm{Z}] & \xrightarrow{\Delta_{\mathrm{P}}} & \mathtt{t}_{\mathrm{C}}[\mathrm{X}] \\
\| & & \| \\
\mathrm{C}[\mathrm{L}] & \xrightarrow[\Delta_{\mathrm{L}}^{\mathrm{M}}(\mathrm{P})]{} & \mathrm{C}[\mathrm{M}],
\end{array}
\tag{16.11}
$$

where $\mu_{\mathrm{L}}^{\mathrm{M}}(\mathrm{P})$ is as in (8.40) and $\Delta_{\mathrm{L}}^{\mathrm{M}}(\mathrm{P})$ is defined dually.

The lax and colax structures of $\mathtt{t}_{(-)}$ on a Coxeter space V on the Z-component, namely,

$$
\bigoplus_{\mathrm{P}:\, \mathrm{b}(\mathrm{P})=\mathrm{Z}} \mathrm{V}[\mathrm{M}] \underset{\text{colax}}{\overset{\text{lax}}{\rightleftarrows}} \bigoplus_{\mathrm{M}':\, \mathrm{L} \leq' \mathrm{M}'} \mathrm{V}[\mathrm{M}']
\tag{16.12}
$$

are as follows. Here L is the top-lune with base Z containing C_r. In the lhs, M is determined by P as the top-lune with base c(P) containing C_r. There are more summands in the lhs than in the rhs. Each summand V[M] in the lhs identifies with a summand V[M′] in the rhs via $w : \mathrm{M} \to \mathrm{M}'$ where w fixes Z and $w(\mathrm{PL}) = \mathrm{L}$. This yields the lax structure. For the colax structure, V[M′] in the rhs identifies with all matching summands V[M] in the lhs as above.

Exercise 16.9. Repeat Exercise 16.5 in terms of flats and lunes. (Use commutativity axioms (2.52) and (8.84) and their duals.)

We mention that one can also work with the flat formulation of Coxeter spaces and write $\mathtt{t}_{\mathrm{V}}[\mathrm{X}] := \mathrm{V}[\mathrm{X}]$, and so on.

16.2. Hadamard product with chambers

We now consider functors built out of the trivialization functor in Section 16.1 by taking Hadamard product with chambers.

16.2.1. Hadamard product with chambers. Recall the Hadamard product of Coxeter species from Section 3.5. Also recall the Coxeter bimonoid of chambers Γ from Section 4.3. Put

$$
\mathcal{L} := \Gamma \times \mathtt{t}_{(-)} \quad \text{and} \quad \mathcal{L}^{\vee} := \Gamma^{*} \times \mathtt{t}_{(-)}.
\tag{16.13}
$$

These are functors from Coxeter spaces to Coxeter species. They are $\Bbbk$-linear.

Functors $\mathcal{L}$ and $\mathcal{L}^{\vee}$ are isomorphic in view of Exercise 4.11. However, they differ when we start going to Coxeter algebras and Coxeter monoids; so we keep them notationally separate.

16.2.2. Coxeter (co)algebras to Coxeter (co)monoids. We know that Hadamard product preserves Coxeter (co)monoids. Thus, for A a Coxeter algebra, $\mathcal{L}(\mathrm{A})$ is a Coxeter monoid with product

$$
\begin{aligned}
\mu_A^F : \Gamma[F] \otimes \mathtt{t}_{\mathrm{A}}[F] &\longrightarrow \Gamma[A] \otimes \mathtt{t}_{\mathrm{A}}[A] \\
\mathtt{H}_{C/F} \otimes x &\longmapsto \mathtt{H}_{C/A} \otimes \mu_{\mathrm{t}(A)}^{\mathrm{t}(F)}(x).
\end{aligned}
\tag{16.14}
$$

This used the product of Γ in (4.22).

Similarly, for C a Coxeter coalgebra, $\mathcal{L}(C)$ is a Coxeter comonoid with coproduct

$$\begin{aligned} \Delta_A^F : \Gamma[A] \otimes \mathrm{t}_{\mathrm{C}}[A] &\longrightarrow \Gamma[F] \otimes \mathrm{t}_{\mathrm{C}}[F] \\ \mathtt{H}_{C/A} \otimes x &\longmapsto \mathtt{H}_{FC/F} \otimes \Delta_{\mathrm{t}(A)}^{\mathrm{t}(F)}(x). \end{aligned} \tag{16.15}$$

This used the coproduct of Γ in (4.22).

Similar considerations apply to the functor $\mathcal{L}^\vee$. Now we use the product and coproduct of Γ^*.

16.2.3. Lax and colax structures. More formally: The functor $\mathcal{L}$ carries a lax structure induced from the lax structure of $\mathrm{t}_{(-)}$ in (16.7) and the product of Γ in (4.22) and using Theorem 3.25. Evaluated on a Coxeter space V on the A-component, the lax structure is

$$\begin{aligned} \bigoplus_{F:\, F \geq A} \Gamma[F] \otimes \mathrm{V}[\mathrm{t}(F)] &\longrightarrow \Gamma[A] \otimes \bigoplus_{T:\, T \geq \mathrm{t}(A)} \mathrm{V}[T] \\ \mathtt{H}_{C/F} \otimes x &\longmapsto \mathtt{H}_{C/A} \otimes x \end{aligned} \tag{16.16}$$

for $x \in \mathrm{V}[\mathrm{t}(F)]$.

Exercise 16.10. Check: The lax structure (16.16) of $\mathcal{L}$ is an isomorphism. (Use Observation 1.5 local to A.)

Similarly, the functor $\mathcal{L}$ carries a colax structure induced from the colax structure of $\mathrm{t}_{(-)}$ in (16.7) and the coproduct of Γ in (4.22). Evaluated on a Coxeter space V on the A-component, the colax structure is

$$\begin{aligned} \Gamma[A] \otimes \bigoplus_{T:\, T \geq \mathrm{t}(A)} \mathrm{V}[T] &\longrightarrow \bigoplus_{F:\, F \geq A} \Gamma[F] \otimes \mathrm{V}[\mathrm{t}(F)] \\ \mathtt{H}_{C/A} \otimes x &\longmapsto \sum_{\substack{F:\, F \geq A \\ \mathrm{t}(F) = T}} \mathtt{H}_{FC/F} \otimes x \end{aligned} \tag{16.17}$$

for $x \in \mathrm{V}[T]$.

Exercise 16.11. In a similar manner, deduce that the functor $\mathcal{L}^\vee$ also carries lax and colax structures. It is now the colax structure which is an isomorphism. Check that it is given by

$$\mathtt{M}_{C/A} \otimes x \mapsto \mathtt{M}_{C/F} \otimes x, \tag{16.18}$$

where $x \in \mathrm{V}[T]$ and F is the face of C of type T. This is the inverse of (16.16) if we identify $\mathtt{M}_{C/A}$ with $\mathtt{H}_{C/A}$.

16.3. Adjoints of bosonic Fock functors

We now show that the left and right adjoints of the trivialization functor $\mathrm{t}_{(-)}$ are the two bosonic Fock functors $\overline{\mathcal{K}}$ and $\overline{\mathcal{K}}^\vee$. Moreover, $\mathrm{t}_{(-)}$ is a section of both of them.

16.3.1. Adjoints of bosonic Fock functors. Recall the bosonic Fock functors $\overline{\mathcal{K}}$ and $\overline{\mathcal{K}}^\vee$ from (14.27) and (14.30). We now establish the adjunctions

$$\mathcal{A}\text{-}\mathsf{CSp} \underset{\mathsf{t}_{(-)}}{\overset{\overline{\mathcal{K}}}{\rightleftarrows}} \boldsymbol{\alpha}\text{-}\overline{\mathsf{C}}\mathsf{Sp} \qquad \boldsymbol{\alpha}\text{-}\overline{\mathsf{C}}\mathsf{Sp} \underset{\overline{\mathcal{K}}^\vee}{\overset{\mathsf{t}_{(-)}}{\rightleftarrows}} \mathcal{A}\text{-}\mathsf{CSp}. \tag{16.19}$$

Let us begin with the second one.

Proposition 16.12. *The functor* $\mathsf{t}_{(-)}$ *is left adjoint to* $\overline{\mathcal{K}}^\vee$ *between categories of Coxeter species and Coxeter spaces. That is, for a Coxeter space* V *and Coxeter species* p, *there is a natural bijection*

$$\mathcal{A}\text{-}\mathsf{CSp}(\mathsf{t}_\mathrm{V}, \mathsf{p}) \xrightarrow{\cong} \boldsymbol{\alpha}\text{-}\overline{\mathsf{C}}\mathsf{Sp}(\mathrm{V}, \overline{\mathcal{K}}^\vee(\mathsf{p})).$$

PROOF. For a group G, recall the following adjunction between vector spaces and G-modules. A vector space V can be turned into a G-module via the trivial action. This functor is left adjoint to the functor which sends a G-module M to its space of invariants M^G. Our situation is very similar. Let F_0 be the face of C_r of type T. Then having a linear map

$$f_T : \mathrm{V}[T] \to \mathsf{p}[F_0]^{W_{F_0}}$$

is equivalent to having a map

$$f_{F_0} : \mathsf{t}_\mathrm{V}[F_0] \to \mathsf{p}[F_0]$$

of W_{F_0}-modules. Now a map $f : \mathrm{V} \to \overline{\mathcal{K}}^\vee(\mathsf{p})$ of Coxeter spaces consists of a family of linear maps f_T subject to conditions in the first diagram below. Similarly, a map $f : \mathsf{t}_\mathrm{V} \to \mathsf{p}$ of Coxeter species consists of a family of linear maps f_F subject to conditions in the second diagram below.

$$\begin{array}{ccc} \mathrm{V}[T] & \xrightarrow{f_T} & \mathsf{p}[F_0]^{W_{F_0}} \\ {\scriptstyle\sigma}\downarrow & & \downarrow{\scriptstyle\sigma} \\ \mathrm{V}[U] & \xrightarrow[f_U]{} & \mathsf{p}[G_0]^{W_{G_0}} \end{array} \qquad\qquad \begin{array}{ccc} \mathsf{t}_\mathrm{V}[F] & \xrightarrow{f_F} & \mathsf{p}[F] \\ {\scriptstyle\sigma}\downarrow & & \downarrow{\scriptstyle\sigma} \\ \mathsf{t}_\mathrm{V}[G] & \xrightarrow[f_G]{} & \mathsf{p}[G] \end{array}$$

There are more diagrams on the right than on the left since F is not required to be a face of C_r. However, the maps f_F are uniquely determined by

$$\begin{array}{ccc} \mathsf{t}_\mathrm{V}[F] & \xrightarrow{f_F} & \mathsf{p}[F] \\ \| & & \uparrow{\scriptstyle u} \\ \mathsf{t}_\mathrm{V}[F_0] & \xrightarrow[f_{F_0}]{} & \mathsf{p}[F_0], \end{array}$$

where u is any element of W which takes F_0 to F. This does not depend on the specific choice of u since f_{F_0} maps into W_{F_0}-invariants. Finally, one checks that the left diagrams imply the extra right diagrams. For $\sigma = w : F \to F'$, we have

$$\begin{array}{ccccccc} & & \mathsf{t}_\mathrm{V}[F] & \xrightarrow{\mathrm{id}} & \mathsf{t}_\mathrm{V}[F'] & & \\ & {\scriptstyle f_{F_0}}\swarrow & \downarrow{\scriptstyle f_F} & & {\scriptstyle f_{F'}}\downarrow & \searrow{\scriptstyle f_{F_0}} & \\ \mathsf{p}[F_0] & \xrightarrow[u]{} & \mathsf{p}[F] & \xrightarrow[w]{} & \mathsf{p}[F'] & \xleftarrow[v]{} & \mathsf{p}[F_0]. \end{array}$$

The check for $\sigma = \beta$ is similar. □

Proposition 16.13. *The functor* $\mathsf{t}_{(-)}$ *is right adjoint to* $\overline{\mathcal{K}}$ *between categories of Coxeter species and Coxeter spaces. That is, for a Coxeter space* V *and Coxeter species* p*, there is a natural bijection*

$$\boldsymbol{\alpha}\text{-}\overline{\mathsf{C}}\mathsf{Sp}(\overline{\mathcal{K}}(\mathsf{p}), \mathrm{V}) \xrightarrow{\cong} \mathcal{A}\text{-}\mathsf{CSp}(\mathsf{p}, \mathsf{t}_{\mathrm{V}}).$$

PROOF. This is the dual of Proposition 16.12. We now use that for a group G, the functor given by the trivial action is right adjoint to the functor which sends a G-module M to its space of coinvariants M_G. The remaining details are similar. □

In finite-dimensional setting, Propositions 16.12 and 16.13 imply each other. (Recall: $\overline{\mathcal{K}}$ and $\overline{\mathcal{K}}^{\vee}$ are conjugates wrt duality and $\mathsf{t}_{(-)}$ is self-dual.)

16.3.2. Sections of bosonic Fock functors. The trivialization functor is a section of both bosonic Fock functors. This is formalized below.

Proposition 16.14. *The unit of the adjunction* $(\mathsf{t}_{(-)}, \overline{\mathcal{K}}^{\vee})$ *is an isomorphism. That is, for a Coxeter space* V*, the map* $\mathrm{V} \to \overline{\mathcal{K}}^{\vee}(\mathsf{t}_{\mathrm{V}})$ *is an isomorphism of Coxeter spaces.*

PROOF. We have $\overline{\mathcal{K}}^{\vee}(\mathsf{t}_{\mathrm{V}})[T] = \mathsf{t}_{\mathrm{V}}[F_0]^{W_{F_0}} = \mathrm{V}[T]$, where F_0 is the face of C_r of type T. The second step used that the action of W_{F_0} on $\mathsf{t}_{\mathrm{V}}[F_0]$ is trivial. □

Dually:

Proposition 16.15. *The counit of the adjunction* $(\overline{\mathcal{K}}, \mathsf{t}_{(-)})$ *is an isomorphism. That is, for a Coxeter space* V*, the map* $\overline{\mathcal{K}}(\mathsf{t}_{\mathrm{V}}) \to \mathrm{V}$ *is an isomorphism of Coxeter spaces.*

PROOF. We have $\overline{\mathcal{K}}(\mathsf{t}_{\mathrm{V}})[T] = \mathsf{t}_{\mathrm{V}}[F_0]_{W_{F_0}} = \mathrm{V}[T]$, as in the previous proof. □

Exercise 16.16. Check: For a Coxeter species p, the counit $\mathsf{t}_{\overline{\mathcal{K}}^{\vee}(\mathsf{p})} \to \mathsf{p}$ and unit $\mathsf{p} \to \mathsf{t}_{\overline{\mathcal{K}}(\mathsf{p})}$, on a face F, are

$$\begin{array}{ccc} \mathsf{t}_{\overline{\mathcal{K}}^{\vee}(\mathsf{p})}[F] & \dashrightarrow & \mathsf{p}[F] \\ \| & & \uparrow u \\ \mathsf{p}[F_0]^{W_{F_0}} & \hookrightarrow & \mathsf{p}[F_0] \end{array} \qquad \begin{array}{ccc} \mathsf{p}[F] & \dashrightarrow & \mathsf{t}_{\overline{\mathcal{K}}(\mathsf{p})}[F] \\ u^{-1}\downarrow & & \| \\ \mathsf{p}[F_0] & \twoheadrightarrow & \mathsf{p}[F_0]_{W_{F_0}}, \end{array} \tag{16.20}$$

where F_0 is the face of C_r of the same type as F, and u is any element of W which takes F_0 to F.

16.3.3. Reflective and coreflective subcategories. As a formal consequence of Propositions 16.14 and 16.15: The trivialization functor is full and faithful. Further:

Proposition 16.17. *The category of Coxeter spaces can be viewed as a reflective and coreflective subcategory of the category of Coxeter species via the trivialization functor.*

Explicitly, the above full subcategory consists of those Coxeter species p for which the component $\mathsf{p}[F]$ only depends on the type of F and all type morphisms are identities.

Exercise 16.18. Check directly: The above subcategory of Coxeter species identifies with the category of Coxeter spaces in terms of face-types defined in Section 8.4.6.

Exercise 16.19. Check: The trivialization functor in (16.5) induces an isomorphism between the full subcategories of

- Coxeter $\mathcal{A}$-spaces consisting of V for which $\mathrm{V}[F, C] = 0$ when $F \neq C$, and
- Coxeter $\mathcal{A}$-species consisting of p for which $\mathsf{p}[F] = 0$ when F is not a chamber.

Moreover, both subcategories are equivalent to the category of left W-modules, see Exercises 4.2 and 10.3.

16.4. Lifts of adjunctions involving bosonic Fock functors

Recall from Section 16.3 the adjunctions between Coxeter species and Coxeter spaces involving the trivialization functor and bosonic Fock functors. We now lift them to Coxeter (co, bi)monoids and Coxeter (co, bi)algebras as shown in Summary 16.1 following the process outlined in Figure 16.1.

16.4.1. Coxeter (co)algebras and (co)monoids (First lift, part 1).

We first lift the adjunctions in (16.19) to

$$\mathsf{Mon}(\mathsf{CSp}) \underset{\mathsf{t}_{(-)}}{\overset{\overline{\mathcal{K}}}{\rightleftarrows}} \mathsf{Mon}(\overline{\mathsf{C}}\mathsf{Sp}) \qquad \mathsf{Comon}(\overline{\mathsf{C}}\mathsf{Sp}) \underset{\overline{\mathcal{K}}^{\vee}}{\overset{\mathsf{t}_{(-)}}{\rightleftarrows}} \mathsf{Comon}(\mathsf{CSp}). \tag{16.21}$$

Let us begin with the second one.

Proposition 16.20. *The functor* $\mathsf{t}_{(-)}$ *is left adjoint to* $\overline{\mathcal{K}}^{\vee}$ *between categories of Coxeter comonoids and Coxeter coalgebras. That is, for a Coxeter coalgebra* C *and Coxeter comonoid* c*, there is a natural bijection*

$$\mathsf{Comon}(\mathsf{CSp})(\mathsf{t}_{\mathrm{C}}, \mathsf{c}) \xrightarrow{\cong} \mathsf{Comon}(\overline{\mathsf{C}}\mathsf{Sp})(\mathrm{C}, \overline{\mathcal{K}}^{\vee}(\mathsf{c})).$$

Moreover, the adjunction restricts to the full subcategories of (signed) cocommutative Coxeter comonoids and (signed) cocommutative Coxeter coalgebras.

PROOF. The bijection is defined as in Proposition 16.12. We need to verify that under this bijection, morphisms of Coxeter comonoids correspond to morphisms of Coxeter coalgebras. The key observation is that for a Coxeter comonoid c, the coproduct of $\overline{\mathcal{K}}^{\vee}(\mathsf{c})$ is defined using

$$\begin{array}{ccc} \mathsf{c}[A] & \xrightarrow{\Delta_A^F} & \mathsf{c}[F] \\ \uparrow & & \uparrow \\ \mathsf{c}[A]^{W_A} & \dashrightarrow & \mathsf{c}[F]^{W_F}, \end{array}$$

see the second diagram in (15.7).

The last claim follows from the fact that $\mathsf{t}_{(-)}$ and $\overline{\mathcal{K}}^{\vee}$ preserve (signed) cocommutativity, see Exercise 16.5 and Theorem 15.11. □

A more formal way to proceed is given below; see Remark 16.24 for the connection.

Proposition 16.21. *Wrt comonads $\mathcal{T}^\vee$ and $\overline{\mathcal{T}}^\vee$ on Coxeter species and Coxeter spaces,*

(1) $(\mathsf{t}_{(-)}, \overline{\mathcal{K}}^\vee)$ *is a colax-colax adjunction,*
(2) $(\mathsf{t}_{(-)}, \overline{\mathcal{K}}^\vee)$ *is a colax-lax adjunction.*

The same holds wrt comonads $\mathcal{S}^\vee$ and $\overline{\mathcal{S}}^\vee$, and comonads $\mathcal{E}^\vee$ and $\overline{\mathcal{E}}^\vee$.

PROOF. Let us first understand item (2). This is related to Proposition 14.19; the relevant ideas are repeated here for convenience. Start with the colax structure $\overline{\varphi}^\vee$ on $\overline{\mathcal{K}}^\vee$ from (14.32). Its inverse $\overline{\psi}^\vee$ is a lax structure wrt comonads $\mathcal{T}^\vee$ and $\overline{\mathcal{T}}^\vee$. In other words, we have inverse isomorphisms

$$\overline{\varphi}^\vee : \overline{\mathcal{K}}^\vee \mathcal{T}^\vee \xrightarrow{\cong} \overline{\mathcal{T}}^\vee \overline{\mathcal{K}}^\vee \quad \text{and} \quad \overline{\psi}^\vee : \overline{\mathcal{T}}^\vee \overline{\mathcal{K}}^\vee \xrightarrow{\cong} \overline{\mathcal{K}}^\vee \mathcal{T}^\vee.$$

(As a map, $\overline{\psi}^\vee$ is the same as the lax structure wrt monads $\mathcal{T}$ and $\overline{\mathcal{T}}$ from (14.31), however, monads are not relevant here). So by (A.18), there is a unique colax structure on $\mathsf{t}_{(-)}$ given by the composite

$$\mathsf{t}_{\overline{\mathcal{T}}^\vee(\mathrm{V})} \to \mathsf{t}_{\overline{\mathcal{T}}^\vee \overline{\mathcal{K}}^\vee \mathsf{t}_\mathrm{V}} \to \mathsf{t}_{\overline{\mathcal{K}}^\vee \mathcal{T}^\vee \mathsf{t}_\mathrm{V}} \to \mathcal{T}^\vee \mathsf{t}_\mathrm{V} \tag{16.22}$$

such that $(\mathsf{t}_{(-)}, \overline{\mathcal{K}}^\vee)$ is a colax-lax adjunction. The above composite coincides with the colax structure defined by formula (16.7), see Exercise 16.22. This proves item (2), and item (1) also then follows.

For comonads $\mathcal{S}^\vee$ and $\overline{\mathcal{S}}^\vee$, one can proceed similarly. By Theorem 14.30, the colax structure $\overline{\varphi}^\vee$ on $\overline{\mathcal{K}}^\vee$ wrt these comonads is an isomorphism. See also Exercise 16.25 below. So we can take inverse of $\overline{\varphi}^\vee$ which then induces a colax structure on $\mathsf{t}_{(-)}$ which is the bottom-horizontal map in the second diagram in (16.8). The comonads $\mathcal{E}^\vee$ and $\overline{\mathcal{E}}^\vee$ can be handled similarly. □

Exercise 16.22. Check: The composite (16.22) on the A-component equals the backward map in (16.7). (The crux of the calculation is that when we use $\overline{\psi}^\vee$, we go from a face of C_r to all faces of the same type via type morphisms, but here type morphisms are identities, so we just end up identifying a summand with all matching summands.)

Exercise 16.23. Check item (1) of Proposition 16.21 directly. That is, for a Coxeter space V and Coxeter species p, the diagrams

$$\begin{array}{ccc} \overline{\mathcal{T}}^\vee(\mathrm{V}) & \longrightarrow & \overline{\mathcal{K}}^\vee \mathsf{t}_{\overline{\mathcal{T}}^\vee(\mathrm{V})} \\ \downarrow & & \downarrow \\ \overline{\mathcal{T}}^\vee \overline{\mathcal{K}}^\vee \mathsf{t}_\mathrm{V} & \longleftarrow & \overline{\mathcal{K}}^\vee \mathcal{T}^\vee \mathsf{t}_\mathrm{V} \end{array} \qquad \begin{array}{ccc} \mathsf{t}_{\overline{\mathcal{T}}^\vee \overline{\mathcal{K}}^\vee(\mathsf{p})} & \longrightarrow & \mathcal{T}^\vee \mathsf{t}_{\overline{\mathcal{K}}^\vee(\mathsf{p})} \\ \uparrow & & \downarrow \\ \mathsf{t}_{\overline{\mathcal{K}}^\vee \mathcal{T}^\vee(\mathsf{p})} & \longrightarrow & \mathcal{T}^\vee(\mathsf{p}) \end{array}$$

commute. These specialize diagrams (A.16).

Remark 16.24. Proposition 16.20 follows from item (1) of Proposition 16.21 by applying Lemma A.2. It can also be deduced from item (2) of Proposition 16.21 by applying the adjoint lifting theorem (comonad version of Theorem A.3). Since the lax structure of $\overline{\mathcal{K}}^\vee$ is an isomorphism, the lift of $\overline{\mathcal{K}}^\vee$ coincides with itself.

Exercise 16.25. Deduce: The diagrams

$$\begin{array}{ccc} \mathsf{Comon}(\overline{\mathsf{C}}\mathsf{Sp}) & \xrightarrow{frg} & \overline{\mathsf{C}}\mathsf{Sp} \\ {\scriptstyle \mathsf{t}_{(-)}}\big\downarrow & & \big\downarrow{\scriptstyle \mathsf{t}_{(-)}} \\ \mathsf{Comon}(\mathsf{CSp}) & \xrightarrow[frg]{} & \mathsf{CSp} \end{array} \qquad \begin{array}{ccc} \mathsf{Comon}(\overline{\mathsf{C}}\mathsf{Sp}) & \xleftarrow{\overline{\mathcal{T}}^\vee} & \overline{\mathsf{C}}\mathsf{Sp} \\ {\scriptstyle \overline{\mathcal{K}}^\vee}\big\uparrow & & \big\uparrow{\scriptstyle \overline{\mathcal{K}}^\vee} \\ \mathsf{Comon}(\mathsf{CSp}) & \xleftarrow[\mathcal{T}^\vee]{} & \mathsf{CSp} \end{array}$$

commute. Functors in the left diagram are left adjoints of corresponding functors in the right diagram. In more detail: For a Coxeter species p, the colax structure $\overline{\varphi}_{\mathsf{p}}^\vee : \overline{\mathcal{K}}^\vee \mathcal{T}^\vee(\mathsf{p}) \to \overline{\mathcal{T}}^\vee \overline{\mathcal{K}}^\vee(\mathsf{p})$ is an isomorphism of Coxeter coalgebras. This result was also obtained in the first part of Proposition 15.25.

Similarly, the diagrams

$$\begin{array}{ccc} {}^{\mathrm{co}}\mathsf{Comon}(\overline{\mathsf{C}}\mathsf{Sp}) & \xrightarrow{frg} & \overline{\mathsf{C}}\mathsf{Sp} \\ {\scriptstyle \mathsf{t}_{(-)}}\big\downarrow & & \big\downarrow{\scriptstyle \mathsf{t}_{(-)}} \\ {}^{\mathrm{co}}\mathsf{Comon}(\mathsf{CSp}) & \xrightarrow[frg]{} & \mathsf{CSp} \end{array} \qquad \begin{array}{ccc} {}^{\mathrm{co}}\mathsf{Comon}(\overline{\mathsf{C}}\mathsf{Sp}) & \xleftarrow{\overline{\mathcal{S}}^\vee} & \overline{\mathsf{C}}\mathsf{Sp} \\ {\scriptstyle \overline{\mathcal{K}}^\vee}\big\uparrow & & \big\uparrow{\scriptstyle \overline{\mathcal{K}}^\vee} \\ {}^{\mathrm{co}}\mathsf{Comon}(\mathsf{CSp}) & \xleftarrow[\mathcal{S}^\vee]{} & \mathsf{CSp} \end{array}$$

commute. For a Coxeter species p, the colax structure $\overline{\varphi}_{\mathsf{p}}^\vee : \overline{\mathcal{K}}^\vee \mathcal{S}^\vee(\mathsf{p}) \to \overline{\mathcal{S}}^\vee \overline{\mathcal{K}}^\vee(\mathsf{p})$ is an isomorphism of cocommutative Coxeter coalgebras. This result was also obtained as a part of Proposition 15.29. We have a similar diagram with $\mathcal{S}^\vee$ and $\overline{\mathcal{S}}^\vee$ replaced by $\mathcal{E}^\vee$ and $\overline{\mathcal{E}}^\vee$.

Convention 16.26. In this chapter, we will encounter many diagrams like the one in Exercise 16.25. By convention, we put categories related to Coxeter spaces in the top line and those related to Coxeter species in the bottom line. As a result, Fock functors point up while their adjoints point down. Also, when two such diagrams are next to each other, functors in the left diagram will be left adjoints of corresponding functors in the right diagram.

Proposition 16.27. *The unit of the adjunction* $(\mathsf{t}_{(-)}, \overline{\mathcal{K}}^\vee)$ *is an isomorphism. That is, for a (cocommutative) Coxeter coalgebra* C*, the map* $\mathrm{C} \to \overline{\mathcal{K}}^\vee(\mathsf{t}_{\mathrm{C}})$ *is an isomorphism of (cocommutative) Coxeter coalgebras.*

Proof. This follows formally from Proposition 16.14 using that a morphism of Coxeter coalgebras is an isomorphism if it is an isomorphism of the underlying Coxeter spaces. □

Exercise 16.28. Recall the functors *trv* in (3.18) and (9.16). Use the first diagram below to deduce the second diagram by taking right adjoints.

$$\begin{array}{ccc} \overline{\mathsf{C}}\mathsf{Sp} & \xrightarrow{trv} & \mathsf{Comon}(\overline{\mathsf{C}}\mathsf{Sp}) \\ {\scriptstyle \mathsf{t}_{(-)}}\big\downarrow & & \big\downarrow{\scriptstyle \mathsf{t}_{(-)}} \\ \mathsf{CSp} & \xrightarrow[trv]{} & \mathsf{Comon}(\mathsf{CSp}) \end{array} \qquad \begin{array}{ccc} \overline{\mathsf{C}}\mathsf{Sp} & \xleftarrow{\overline{\mathcal{P}}} & \mathsf{Comon}(\overline{\mathsf{C}}\mathsf{Sp}) \\ {\scriptstyle \overline{\mathcal{K}}^\vee}\big\uparrow & & \big\uparrow{\scriptstyle \overline{\mathcal{K}}^\vee} \\ \mathsf{CSp} & \xleftarrow[\mathcal{P}]{} & \mathsf{Comon}(\mathsf{CSp}) \end{array}$$

This result is implicitly contained in Proposition 15.46 (see Remark 15.50).

Let us now turn to the dual picture.

Proposition 16.29. *The functor* $\mathsf{t}_{(-)}$ *is right adjoint to* $\overline{\mathcal{K}}$ *between categories of Coxeter monoids and Coxeter algebras. That is, for a Coxeter algebra* A *and Coxeter monoid* a*, there is a natural bijection*

$$\mathsf{Mon}(\overline{\mathsf{C}}\mathsf{Sp})(\overline{\mathcal{K}}(\mathsf{a}), \mathrm{A}) \xrightarrow{\cong} \mathsf{Mon}(\mathsf{CSp})(\mathsf{a}, \mathsf{t}_{\mathrm{A}}).$$

Moreover, the adjunction restricts to the full subcategories of (signed) commutative Coxeter monoids and (signed) commutative Coxeter algebras.

PROOF. This is the dual of Proposition 16.20. It can be seen directly or formally deduced as mentioned in Remark 16.32 below. □

Proposition 16.30. *Wrt monads $\mathcal{T}$ and $\overline{\mathcal{T}}$ on Coxeter species and Coxeter spaces,*

(1) $(\overline{\mathcal{K}}, \mathsf{t}_{(-)})$ *is a lax-lax adjunction,*
(2) $(\overline{\mathcal{K}}, \mathsf{t}_{(-)})$ *is a colax-lax adjunction.*

The same holds wrt monads $\mathcal{S}$ and $\overline{\mathcal{S}}$, and monads $\mathcal{E}$ and $\overline{\mathcal{E}}$.

PROOF. This is the dual of Proposition 16.21. We can prove it in a similar manner by starting with the lax structure $\overline{\varphi}$ of $\overline{\mathcal{K}}$ which is an isomorphism and taking its inverse. By (A.17), there is an induced lax structure on $\mathsf{t}_{(-)}$ given by the composite

$$\mathcal{T}\mathsf{t}_{\mathrm{V}} \to \mathsf{t}_{\overline{\mathcal{K}}\mathcal{T}\mathsf{t}_{\mathrm{V}}} \to \mathsf{t}_{\overline{\mathcal{T}}\overline{\mathcal{K}}\mathsf{t}_{\mathrm{V}}} \to \mathsf{t}_{\overline{\mathcal{T}}(\mathrm{V})} \tag{16.23}$$

which coincides with the one defined by formula (16.7), see Exercise 16.31. This proves item (2), and item (1) also follows.

The lax structure $\overline{\varphi}$ of $\overline{\mathcal{K}}$ is also an isomorphism wrt $\mathcal{S}$ and $\overline{\mathcal{S}}$ by Theorem 14.30. See also Exercise 16.33 below. The induced lax structure of $\mathsf{t}_{(-)}$ wrt $\mathcal{S}$ and $\overline{\mathcal{S}}$ is the bottom-horizontal map in the first diagram in (16.8). □

Exercise 16.31. Check: The composite (16.23) on the A-component equals the forward map in (16.7).

Remark 16.32. Proposition 16.29 follows from item (1) of Proposition 16.30 by applying Lemma A.2. It can also be deduced from item (2) of Proposition 16.30 by applying the adjoint lifting theorem (Theorem A.3). Since the colax structure of $\overline{\mathcal{K}}$ is an isomorphism, the lift of $\overline{\mathcal{K}}$ coincides with itself.

Exercise 16.33. This is the dual of Exercise 16.25. Deduce: The diagrams

$$\begin{array}{ccc} \overline{\mathsf{CSp}} & \xrightarrow{\overline{\mathcal{T}}} & \mathsf{Mon}(\overline{\mathsf{CSp}}) \\ {\scriptstyle \overline{\mathcal{K}}}\uparrow & & \uparrow{\scriptstyle \overline{\mathcal{K}}} \\ \mathsf{CSp} & \xrightarrow[\mathcal{T}]{} & \mathsf{Mon}(\mathsf{CSp}) \end{array} \qquad \begin{array}{ccc} \overline{\mathsf{CSp}} & \xleftarrow{frg} & \mathsf{Mon}(\overline{\mathsf{CSp}}) \\ {\scriptstyle \mathsf{t}_{(-)}}\downarrow & & \downarrow{\scriptstyle \mathsf{t}_{(-)}} \\ \mathsf{CSp} & \xleftarrow[frg]{} & \mathsf{Mon}(\mathsf{CSp}) \end{array}$$

$$\begin{array}{ccc} \overline{\mathsf{CSp}} & \xrightarrow{\overline{\mathcal{S}}} & \mathsf{Mon}^{\mathrm{co}}(\overline{\mathsf{CSp}}) \\ {\scriptstyle \overline{\mathcal{K}}}\uparrow & & \uparrow{\scriptstyle \overline{\mathcal{K}}} \\ \mathsf{CSp} & \xrightarrow[\mathcal{S}]{} & \mathsf{Mon}^{\mathrm{co}}(\mathsf{CSp}) \end{array} \qquad \begin{array}{ccc} \overline{\mathsf{CSp}} & \xleftarrow{frg} & \mathsf{Mon}^{\mathrm{co}}(\overline{\mathsf{CSp}}) \\ {\scriptstyle \mathsf{t}_{(-)}}\downarrow & & \downarrow{\scriptstyle \mathsf{t}_{(-)}} \\ \mathsf{CSp} & \xleftarrow[frg]{} & \mathsf{Mon}^{\mathrm{co}}(\mathsf{CSp}) \end{array}$$

commute. In more detail: For a Coxeter species p, the lax structure $\overline{\varphi}_{\mathsf{p}} : \overline{\mathcal{T}}\overline{\mathcal{K}}(\mathsf{p}) \to \overline{\mathcal{K}}\mathcal{T}(\mathsf{p})$ is an isomorphism of Coxeter algebras, and $\overline{\varphi}_{\mathsf{p}} : \overline{\mathcal{S}}\overline{\mathcal{K}}(\mathsf{p}) \to \overline{\mathcal{K}}\mathcal{S}(\mathsf{p})$ is an isomorphism of commutative Coxeter algebras. These results were also obtained in Propositions 15.24 and 15.29.

Proposition 16.34. *The counit of the adjunction $(\overline{\mathcal{K}}, \mathsf{t}_{(-)})$ is an isomorphism. That is, for a (commutative) Coxeter algebra A, the map $\overline{\mathcal{K}}(\mathsf{t}_{\mathrm{A}}) \to \mathrm{A}$ is an isomorphism of (commutative) Coxeter algebras.*

PROOF. This is the dual of Proposition 16.27 and follows formally from Proposition 16.15. ☐

16.4.2. Coxeter (co)algebras and (co)monoids (First lift, part 2). We now lift the adjunctions in (16.19) to

$$\mathsf{Comon}(\mathsf{CSp}) \underset{\overline{\mathcal{Z}}}{\overset{\overline{\mathcal{K}}}{\rightleftarrows}} \mathsf{Comon}(\overline{\mathsf{C}}\mathsf{Sp}) \qquad \mathsf{Mon}(\overline{\mathsf{C}}\mathsf{Sp}) \underset{\overline{\mathcal{K}}^\vee}{\overset{\overline{\mathcal{Z}}^\vee}{\rightleftarrows}} \mathsf{Mon}(\mathsf{CSp}). \tag{16.24}$$

Compare and contrast with the situation in Section 16.4.1. Since the (co)lax structures of $\mathsf{t}_{(-)}$ in (16.7) are not isomorphisms, its lifts are not itself, thus leading to new functors $\overline{\mathcal{Z}}$ and $\overline{\mathcal{Z}}^\vee$. Contrast with Remarks 16.24 and 16.32.

We begin with the second adjunction. Let (A, μ) be a Coxeter algebra. Consider the Coxeter species t_A and the free Coxeter monoid $\mathcal{T}(\mathsf{t}_\mathrm{A})$ on it:

$$\mathcal{T}(\mathsf{t}_\mathrm{A})[A] = \bigoplus_{F:\, A \le F} \mathrm{A}[\mathrm{t}(F)],$$

with structure maps induced from (16.2) and product given by inclusion. (The Coxeter algebra structure of A has played no role so far.) We now define a Coxeter submonoid $\mathcal{I}_\mathrm{A}$ of $\mathcal{T}(\mathsf{t}_\mathrm{A})$. Its A-component is linearly spanned by elements (16.25) given below.

For each face K greater than A, and $x \in \mathrm{A}[U]$ with $U > \mathrm{t}(K)$, consider

$$\mu_T^U(x) - \sum x \quad \in \mathrm{A}[\mathrm{t}(K)] \oplus \bigoplus_{\substack{G:\, K \le G \\ \mathrm{t}(G) = U}} \mathrm{A}[U], \tag{16.25}$$

where $T = \mathrm{t}(K)$. The direct sum is over all faces G which are greater than K and of type U. The same element x is chosen from each of the summands $\mathrm{A}[\mathrm{t}(G)] = \mathrm{A}[U]$. Observe how the Coxeter algebra structure of A enters here via product components μ_T^U. A helpful illustration is shown below. (Also see the picture in Observation 1.5.)

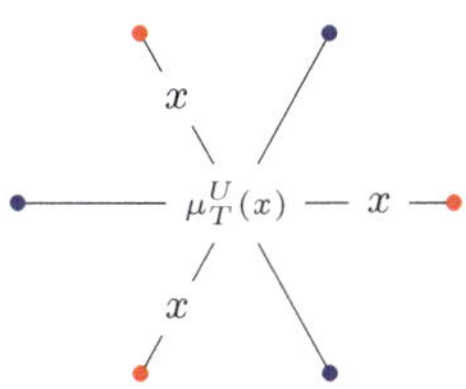

We need to check that diagrams (2.13) commute with $\mathsf{p} := \mathcal{I}_\mathrm{A}$, $\mathsf{q} := \mathcal{T}(\mathsf{t}_\mathrm{A})$ and f the inclusion. The second diagram is clear, while the first follows from the naturality axiom of μ in (8.41). It is also clear that diagram (2.29) commutes with $\mathsf{a} := \mathcal{I}_\mathrm{A}$, $\mathsf{b} := \mathcal{T}(\mathsf{t}_\mathrm{A})$ and f the inclusion. Thus, $\mathcal{I}_\mathrm{A}$ is indeed a Coxeter submonoid of $\mathcal{T}(\mathsf{t}_\mathrm{A})$. Denote the Coxeter quotient monoid by $\overline{\mathcal{Z}}^\vee(\mathrm{A})$. A morphism $\mathrm{A} \to \mathrm{B}$ of Coxeter algebras induces the diagram

$$\begin{array}{ccc} \mathcal{T}(\mathsf{t}_\mathrm{A}) & \longrightarrow & \mathcal{T}(\mathsf{t}_\mathrm{B}) \\ \downarrow & & \downarrow \\ \overline{\mathcal{Z}}^\vee(\mathrm{A}) & \dashrightarrow & \overline{\mathcal{Z}}^\vee(\mathrm{B}) \end{array}$$

of Coxeter monoids. This yields a functor $\overline{\mathcal{Z}}^{\vee}$ from the category of Coxeter algebras to the category of Coxeter monoids along with a natural transformation $\mathcal{T}\mathsf{t}_{(-)} \to \overline{\mathcal{Z}}^{\vee}$. The functor $\overline{\mathcal{Z}}^{\vee}$ is $\Bbbk$-linear.

Proposition 16.35. *The functor $\overline{\mathcal{Z}}^{\vee}$ is left adjoint to $\overline{\mathcal{K}}^{\vee}$ between categories of Coxeter monoids and Coxeter algebras. That is, for a Coxeter algebra* A *and Coxeter monoid* a, *there is a natural bijection*

$$\mathsf{Mon}(\mathsf{CSp})(\overline{\mathcal{Z}}^{\vee}(\mathrm{A}), \mathsf{a}) \xrightarrow{\cong} \mathsf{Mon}(\overline{\mathsf{C}}\mathsf{Sp})(\mathrm{A}, \overline{\mathcal{K}}^{\vee}(\mathsf{a})).$$

PROOF. Suppose we are given a morphism $f : \mathrm{A} \to \overline{\mathcal{K}}^{\vee}(\mathsf{a})$ of Coxeter algebras. In particular, the diagram

$$\begin{array}{ccccc} \mathrm{A}[T] & \xrightarrow{f_T} & \mathsf{a}[F_0]^{W_{F_0}} & \xrightarrow{w} & \mathsf{a}[F]^{W_F} \\ {\scriptstyle \mu^T_Z}\big\downarrow & & & & \big\downarrow{\scriptstyle \mu^F_{A_0}} \\ \mathrm{A}[Z] & & \xrightarrow[f_Z]{\qquad\qquad\qquad} & & \mathsf{a}[A_0]^{W_{A_0}} \end{array}$$

commutes for any $Z \leq T$. Here A_0 and F_0 are faces of C_r of types Z and T, respectively. The red mark indicates that we are summing over all F greater than A_0 which are of type T. For each such F, w is any group element which sends F_0 to F.

By Proposition 16.12, we obtain a map $\mathsf{t}_{\mathrm{A}} \to \mathsf{a}$ of Coxeter species, and hence by Theorem 3.7, a morphism $\mathcal{T}(\mathsf{t}_{\mathrm{A}}) \to \mathsf{a}$ of Coxeter monoids. Explicitly: for faces $A \leq F$, consider the composite linear map

$$\mathrm{A}[\mathsf{t}(F)] \xrightarrow{f_{\mathsf{t}(F)}} \mathsf{a}[F_0] \xrightarrow{w} \mathsf{a}[F] \xrightarrow{\mu^F_A} \mathsf{a}[A].$$

Here F_0 is the face of C_r of the same type as F. For the second map, we choose any w which sends F_0 to F. The particular choice does not matter since the first map lands inside W_{F_0}-invariants. Summing over F for a fixed A defines the linear map $\mathcal{T}(\mathsf{t}_{\mathrm{A}})[A] \to \mathsf{a}[A]$. Further, it sends element (16.25) to zero. This is elaborated in the diagram below.

$$\begin{array}{ccccccc} \mathrm{A}[U] & \xrightarrow{f_U} & \mathsf{a}[G_0] & \xrightarrow[u^{-1}w]{} & \mathsf{a}[G'] & \xrightarrow[u]{} & \mathsf{a}[G] & & \\ & & & \multicolumn{3}{c}{\curvearrowright^{\,w}\ (\mathsf{a}[G_0] \to \mathsf{a}[G])} & & \searrow{\scriptstyle \mu^G_A} & \\ {\scriptstyle \mu^U_T}\big\downarrow & & & & {\scriptstyle \mu^{G'}_{K_0}}\big\downarrow & & {\scriptstyle \mu^G_K}\big\downarrow & & \\ \mathrm{A}[T] & & \xrightarrow[f_T]{\qquad\qquad} & & \mathsf{a}[K_0] & \xrightarrow[u]{} & \mathsf{a}[K] & \xrightarrow[\mu^K_A]{} & \mathsf{a}[A] \end{array}$$

Here K_0 and G_0 are faces of C_r of types T and U, respectively, K is a fixed face of type T, and u is a fixed group element which takes K_0 to K. Further, we are summing over all G greater than K of type U, and over all G' greater than K_0 of type U. Thus, we get the diagram

$$\begin{array}{ccc} \mathcal{T}(\mathsf{t}_{\mathrm{A}}) & & \\ \big\downarrow & \searrow & \\ \overline{\mathcal{Z}}^{\vee}(\mathrm{A}) & \dashrightarrow & \mathsf{a} \end{array}$$

of Coxeter monoids, which yields the required morphism $g : \overline{\mathcal{Z}}^{\vee}(\mathrm{A}) \to \mathsf{a}$.

Conversely, suppose we are given a morphism $g : \overline{\mathcal{Z}}^{\vee}(\mathrm{A}) \to \mathsf{a}$ of Coxeter monoids. Reversing the above steps yields a map $f : \mathrm{A} \to \overline{\mathcal{K}}^{\vee}(\mathsf{a})$ of Coxeter spaces. Explicitly: for each $A \leq F$, we have a linear map $g_{F/A} : \mathrm{A}[\mathsf{t}(F)] \to \mathsf{a}[A]$. Since g is a morphism of Coxeter monoids, $g_{F/A}$ is forced to be the composite

$$\mathrm{A}[\mathsf{t}(F)] \xrightarrow{g_{F/F}} \mathsf{a}[F] \xrightarrow{\mu_A^F} \mathsf{a}[A].$$

Further, $g_{F/F}$ maps into the space of W_F-invariants. Since, by hypothesis, elements (16.25) maps to zero under the $g_{F/A}$, we conclude that f is a morphism of Coxeter algebras. □

More formally, Proposition 16.35 can be deduced by proceeding as follows.

Proposition 16.36. *Wrt monads $\mathcal{T}$ and $\overline{\mathcal{T}}$ on Coxeter species and Coxeter spaces, $(\mathsf{t}_{(-)}, \overline{\mathcal{K}}^{\vee})$ is a colax-lax adjunction. The same holds wrt monads $\mathcal{S}$ and $\overline{\mathcal{S}}$, and monads $\mathcal{E}$ and $\overline{\mathcal{E}}$.*

PROOF. Start with the lax structure $\overline{\psi}^{\vee}$ on $\overline{\mathcal{K}}^{\vee}$ from (14.31). So by (A.18), there is a unique colax structure on $\mathsf{t}_{(-)}$ such that $(\mathsf{t}_{(-)}, \overline{\mathcal{K}}^{\vee})$ is a colax-lax adjunction. One checks that this coincides with the colax structure defined by formula (16.7) and mentioned in Exercise 16.7. The calculation is identical to the one in Exercise 16.22.

Wrt monads $\mathcal{S}$ and $\overline{\mathcal{S}}$, we proceed similarly. In this case, the colax structure on $\mathsf{t}_{(-)}$ is the bottom-horizontal map in the first diagram in (16.9). □

Exercise 16.37. Apply the adjoint lifting theorem (Theorem A.3) to the first statement in Proposition 16.36. To construct the lift of $\mathsf{t}_{(-)}$, coequalize the maps (A.20) which specialize to

$$\mathcal{T}\mathsf{t}_{\overline{\mathcal{T}}(\mathrm{A})} \to \mathcal{T}\mathcal{T}\mathsf{t}_{\mathrm{A}} \to \mathcal{T}\mathsf{t}_{\mathrm{A}} \quad \text{and} \quad \mathcal{T}\mathsf{t}_{\overline{\mathcal{T}}(\mathrm{A})} \to \mathcal{T}\mathsf{t}_{\mathrm{A}}$$

yielding relations (16.25). Thus the lift of $\mathsf{t}_{(-)}$ is $\overline{\mathcal{Z}}^{\vee}$ and Proposition 16.35 follows.

We deduce that the diagrams

$$\begin{array}{ccc} \overline{\mathsf{CSp}} & \xrightarrow{\overline{\mathcal{T}}} & \mathsf{Mon}(\overline{\mathsf{CSp}}) \\ {\scriptstyle \mathsf{t}_{(-)}}\big\downarrow & & \big\downarrow{\scriptstyle \overline{\mathcal{Z}}^{\vee}} \\ \mathsf{CSp} & \xrightarrow[\mathcal{T}]{} & \mathsf{Mon}(\mathsf{CSp}) \end{array} \qquad \begin{array}{ccc} \overline{\mathsf{CSp}} & \xleftarrow{frg} & \mathsf{Mon}(\overline{\mathsf{CSp}}) \\ {\scriptstyle \overline{\mathcal{K}}^{\vee}}\big\uparrow & & \big\uparrow{\scriptstyle \overline{\mathcal{K}}^{\vee}} \\ \mathsf{CSp} & \xleftarrow[frg]{} & \mathsf{Mon}(\mathsf{CSp}) \end{array} \tag{16.26}$$

commute.

Exercise 16.38 (Unit and counit). Check: The unit and counit of $(\overline{\mathcal{Z}}^{\vee}, \overline{\mathcal{K}}^{\vee})$ can be expressed in terms of those of $(\mathsf{t}_{(-)}, \overline{\mathcal{K}}^{\vee})$ as follows. For a Coxeter algebra A and Coxeter monoid a, the diagrams

$$\begin{array}{ccc} \overline{\mathcal{K}}^{\vee}(\mathsf{t}_{\mathrm{A}}) & \longrightarrow & \overline{\mathcal{K}}^{\vee}\mathcal{T}(\mathsf{t}_{\mathrm{A}}) \\ \big\uparrow & & \big\downarrow \\ \mathrm{A} & \longrightarrow & \overline{\mathcal{K}}^{\vee}\overline{\mathcal{Z}}^{\vee}(\mathrm{A}) \end{array} \qquad \begin{array}{ccc} \mathcal{T}\mathsf{t}_{\overline{\mathcal{K}}^{\vee}(\mathsf{a})} & \longrightarrow & \mathcal{T}(\mathsf{a}) \\ \big\downarrow & & \big\downarrow \\ \overline{\mathcal{Z}}^{\vee}\overline{\mathcal{K}}^{\vee}(\mathsf{a}) & \longrightarrow & \mathsf{a} \end{array} \tag{16.27}$$

commute. The map $\mathsf{t}_{\mathrm{A}} \to \mathcal{T}(\mathsf{t}_{\mathrm{A}})$ is the inclusion, while $\mathcal{T}(\mathsf{a}) \to \mathsf{a}$ is the product of a, with a viewed as an algebra over the monad $\mathcal{T}$.

Exercise 16.39. Let A be a Coxeter algebra. Fix a face-type Z. Suppose we are given an element $x_{U/T} \in \mathrm{A}[U]$ for each $Z \le T < U$ and an element $y_U \in \mathrm{A}[U]$ for each $Z < U$ such that

$$\sum_{T:\, Z \le T < U} x_{U/T} - \sum_{V:\, U < V} \mu_U^V(x_{V/U}) = y_U$$

for each $U > Z$. Show:

$$\sum_{V:\, Z<V} \mu_Z^V(x_{V/Z}) = \sum_{V:\, Z<V} \mu_Z^V(y_V).$$

In particular, if $y_U = 0$ for all $U > Z$, then

$$\sum_{V:\, Z<V} \mu_Z^V(x_{V/Z}) = 0.$$

(For each $U > Z$, apply μ_Z^U to the given equation, and then add all of them to obtain the required identity.)

Proposition 16.40. *Let* $\mathrm{char}(\Bbbk) \nmid |W|$*. The unit of the adjunction* $(\overline{\mathcal{Z}}^\vee, \overline{\mathcal{K}}^\vee)$ *is an isomorphism. That is, for a Coxeter algebra* A*, the map* $\mathrm{A} \to \overline{\mathcal{K}}^\vee \overline{\mathcal{Z}}^\vee(\mathrm{A})$ *is an isomorphism of Coxeter algebras.*

Explicitly, the isomorphism is given by the composite

$$\mathrm{A} \to \overline{\mathcal{K}}^\vee(\mathrm{t_A}) \to \overline{\mathcal{K}}^\vee \mathcal{T}(\mathrm{t_A}) \to \overline{\mathcal{K}}^\vee(\overline{\mathcal{Z}}^\vee(\mathrm{A}))$$

as in the first diagram in (16.27).

Proof. Let us denote the above map of Coxeter algebras by f. For any face-type Z, we need to show that f_Z is a linear isomorphism. Accordingly, fix a face-type Z and let A_0 be the face of C_r of type Z. Consider the exact sequence of vector spaces

$$0 \to \mathcal{I}_\mathrm{A}[A_0]^{W_{A_0}} \hookrightarrow \mathcal{T}(\mathrm{t_A})[A_0]^{W_{A_0}} \twoheadrightarrow \overline{\mathcal{Z}}^\vee(\mathrm{A})[A_0]^{W_{A_0}} \to 0.$$

The middle space is linearly spanned by the elements

$$\sum x \quad \in \bigoplus_{\substack{F:\, A_0 \le F \\ \mathrm{t}(F)=T}} \mathrm{A}[T]$$

for $x \in \mathrm{A}[T]$. In view of (16.25), this element maps to $\mu_Z^T(x)$ which is in $\mathrm{A}[Z]$. This shows that f_Z is surjective. For injectivity, we need to show that $\mathcal{I}_\mathrm{A}[A_0]^{W_{A_0}} \cap \mathrm{A}[Z] = 0$. Observe: Any element of $\mathcal{I}_\mathrm{A}[A_0]^{W_{A_0}}$ can be written in the form

$$\sum_{\substack{(T,U):\\ Z \le T < U}} \sum_{\substack{K:\, A_0 \le K \\ \mathrm{t}(K)=T}} \Big(\mu_T^U(x_{U/T}) - \sum_{\substack{G:\, K \le G \\ \mathrm{t}(G)=U}} x_{U/T} \Big)$$

for suitable $x_{U/T} \in \mathrm{A}[U]$. Now suppose this element belongs to $\mathrm{A}[Z]$. This imposes a condition, one for each $U > Z$, as written in Exercise 16.39 with $y_U = 0$. The conclusion of Exercise 16.39 then implies that this element is 0 as required. □

Exercise 16.41 (Free Coxeter algebra). Verify the first diagram in (16.26) by directly computing $\overline{\mathcal{Z}}^{\vee}$ of the free Coxeter algebra $\mathrm{A} := \overline{\mathcal{T}}(\mathrm{V})$ to be $\mathcal{T}(\mathrm{t_V})$ using relations (16.25). Moreover, show by direct calculation that $\overline{\mathcal{K}}^{\vee}(\overline{\mathcal{Z}}^{\vee}(\mathrm{A})) = \overline{\mathcal{K}}^{\vee}(\mathcal{T}\mathrm{t_V}) = \overline{\mathcal{T}}(\mathrm{V}) = \mathrm{A}$ (in any field characteristic).

Let us now turn to the dual picture. Let (C, Δ) be a Coxeter coalgebra. Consider the Coxeter species $\mathrm{t_C}$ and the cofree Coxeter comonoid $\mathcal{T}^{\vee}(\mathrm{t_C})$ on it:

$$\mathcal{T}^{\vee}(\mathrm{t_C})[A] = \bigoplus_{F\colon A \leq F} \mathrm{C}[\mathrm{t}(F)],$$

with structure maps induced from (16.2) and coproduct given by projection. Let $\overline{\mathcal{Z}}(\mathrm{C})$ denote the Coxeter subcomonoid of $\mathcal{T}^{\vee}(\mathrm{t_C})$ whose A-component consists of elements $(x_F)_{F \geq A}$ such that for each face K greater than A, and $U > \mathrm{t}(K)$,

$$\Delta^U_T(x_K) = \sum_{\substack{G\colon K \leq G \\ \mathrm{t}(G) = U}} x_G \quad \in \mathrm{C}[U], \tag{16.28}$$

where $T = \mathrm{t}(K)$. This yields a functor $\overline{\mathcal{Z}}$ from the category of Coxeter coalgebras to the category of Coxeter comonoids along with a natural transformation $\overline{\mathcal{Z}} \to \mathcal{T}^{\vee}\mathrm{t}_{(-)}$. The functor $\overline{\mathcal{Z}}$ is $\Bbbk$-linear.

Proposition 16.42. *The functor $\overline{\mathcal{Z}}$ is right adjoint to $\overline{\mathcal{K}}$ between categories of Coxeter comonoids and Coxeter coalgebras. That is, for a Coxeter coalgebra C and Coxeter comonoid c, there is a natural bijection*

$$\mathsf{Comon}(\overline{\mathsf{C}}\mathsf{Sp})(\overline{\mathcal{K}}(\mathsf{c}), \mathrm{C}) \xrightarrow{\cong} \mathsf{Comon}(\mathsf{CSp})(\mathsf{c}, \overline{\mathcal{Z}}(\mathrm{C})).$$

Sketch of proof. This is the dual of Proposition 16.35 and we proceed similarly. A map $f : \overline{\mathcal{K}}(\mathsf{c}) \to \mathrm{C}$ of Coxeter spaces is the same as a map $g : \mathsf{c} \to \mathrm{t_C}$ of Coxeter species by Proposition 16.13. The map g uniquely lifts to a morphism $\hat{g} : \mathsf{c} \to \mathcal{T}^{\vee}(\mathrm{t_C})$ of Coxeter comonoids by Theorem 3.12. The key observation is that f is a morphism of Coxeter coalgebras iff $\hat{g}$ maps into the Coxeter subcomonoid $\overline{\mathcal{Z}}(\mathrm{C})$. □

More formally, Proposition 16.42 can be deduced by proceeding as follows.

Proposition 16.43. *Wrt comonads $\mathcal{T}^{\vee}$ and $\overline{\mathcal{T}}^{\vee}$ on Coxeter species and Coxeter spaces, $(\overline{\mathcal{K}}, \mathrm{t}_{(-)})$ is a colax-lax adjunction. The same holds wrt comonads $\mathcal{S}^{\vee}$ and $\overline{\mathcal{S}}^{\vee}$, and comonads $\mathcal{E}^{\vee}$ and $\overline{\mathcal{E}}^{\vee}$.*

Proof. The colax structure $\overline{\psi}$ on $\overline{\mathcal{K}}$ from (14.29) induces a unique lax structure on $\mathrm{t}_{(-)}$ which coincides with the one mentioned in Exercise 16.7. The calculation is identical to the one in Exercise 16.31.

Wrt comonads $\mathcal{S}^{\vee}$ and $\overline{\mathcal{S}}^{\vee}$, the lax structure on $\mathrm{t}_{(-)}$ is the bottom-horizontal map in the second diagram in (16.9). □

Exercise 16.44. Check: The lift of $\mathrm{t}_{(-)}$ wrt $\mathcal{T}^{\vee}$ and $\overline{\mathcal{T}}^{\vee}$ is indeed $\overline{\mathcal{Z}}$, that is, obtain conditions (16.28) by equalizing the maps

$$\mathcal{T}^{\vee}\mathrm{t_C} \to \mathcal{T}^{\vee}\mathcal{T}^{\vee}\mathrm{t_C} \to \mathcal{T}^{\vee}\mathrm{t}_{\overline{\mathcal{T}}^{\vee}(\mathrm{C})} \quad \text{and} \quad \mathcal{T}^{\vee}\mathrm{t_C} \to \mathcal{T}^{\vee}\mathrm{t}_{\overline{\mathcal{T}}^{\vee}(\mathrm{C})}.$$

This is an instance of the comonad version of the adjoint lifting theorem. This proves Proposition 16.42.

We deduce that the diagrams

$$
\begin{array}{ccc}
\overline{\mathsf{C}}\mathsf{Sp} & \xleftarrow{frg} & \mathsf{Comon}(\overline{\mathsf{C}}\mathsf{Sp}) \\
\overline{\mathcal{K}}\uparrow & & \uparrow\overline{\mathcal{K}} \\
\mathsf{CSp} & \xleftarrow[frg]{} & \mathsf{Comon}(\mathsf{CSp})
\end{array}
\qquad
\begin{array}{ccc}
\overline{\mathsf{C}}\mathsf{Sp} & \xrightarrow{\overline{\mathcal{T}}^\vee} & \mathsf{Comon}(\overline{\mathsf{C}}\mathsf{Sp}) \\
\mathrm{t}_{(-)}\downarrow & & \downarrow\overline{\mathcal{Z}} \\
\mathsf{CSp} & \xrightarrow[\mathcal{T}^\vee]{} & \mathsf{Comon}(\mathsf{CSp})
\end{array}
\tag{16.29}
$$

commute.

The dual of Proposition 16.40 is stated below.

Proposition 16.45. *Let* $\mathrm{char}(\Bbbk) \nmid |W|$. *The counit of the adjunction* $(\overline{\mathcal{K}}, \overline{\mathcal{Z}})$ *is an isomorphism. That is, for a Coxeter coalgebra* C, *the map* $\overline{\mathcal{K}}\overline{\mathcal{Z}}(\mathrm{C}) \to \mathrm{C}$ *is an isomorphism of Coxeter coalgebras.*

Explicitly, the isomorphism is given by the composite

$$\overline{\mathcal{K}}(\overline{\mathcal{Z}}(\mathrm{C})) \to \overline{\mathcal{K}}\mathcal{T}^\vee(\mathrm{t_C}) \to \overline{\mathcal{K}}(\mathrm{t_C}) \to \mathrm{C}.$$

On the Z-component,

$$\overline{\mathcal{K}}(\overline{\mathcal{Z}}(\mathrm{C}))[Z] = \overline{\mathcal{Z}}(\mathrm{C})[A_0]_{W_{A_0}} \xrightarrow{\cong} \mathrm{C}[Z],$$

where A_0 is the face of C_r of type Z.

Exercise 16.46. Check the surjectivity of the above isomorphism as follows. An element $x \in \mathrm{C}[Z]$ arises from the representative $(x_F)_{F \geq A_0}$ in $\overline{\mathcal{Z}}(\mathrm{C})[A_0]$ defined by

$$x_F := \tfrac{1}{d_{T/Z}} \Delta_Z^T(x)$$

with $T = \mathrm{t}(F)$. Use the multiplicative property (1.32) for $q = 1$ to check that (x_F) indeed satisfies conditions (16.28).

16.4.3. Coxeter q-bialgebras and q-bimonoids (Second lift). We now lift the adjunctions in (16.21) and (16.24) to

$$q\text{-}\mathsf{Bimon}(\mathsf{CSp}) \underset{\overline{\mathcal{Z}}}{\overset{\overline{\mathcal{K}}}{\rightleftarrows}} q\text{-}\mathsf{Bimon}(\overline{\mathsf{C}}\mathsf{Sp}) \qquad q\text{-}\mathsf{Bimon}(\overline{\mathsf{C}}\mathsf{Sp}) \underset{\overline{\mathcal{K}}^\vee}{\overset{\overline{\mathcal{Z}}^\vee}{\rightleftarrows}} q\text{-}\mathsf{Bimon}(\mathsf{CSp}).$$

To get to either adjunction, there are two ways to do the second lift as indicated in Figure 16.1.

We begin with the second adjunction. Let q be a scalar. Let H be a Coxeter q-bialgebra. Then $\mathrm{t_H}$ is a Coxeter comonoid. Consider the free Coxeter q-bimonoid $\mathcal{T}_q(\mathrm{t_H})$ on $\mathrm{t_H}$ as in Section 3.3.2. Thus,

$$\mathcal{T}_q(\mathrm{t_H})[A] = \bigoplus_{F:\, A \leq F} \mathrm{H}[\mathrm{t}(F)]$$

with product components given by inclusion, and coproduct components by

$$\Delta_A^G : \mathrm{H}[\mathrm{t}(F)] \xrightarrow[\Delta_F^{FG}]{\Delta_{\mathrm{t}(F)}^{\mathrm{t}(FG)}} \mathrm{H}[\mathrm{t}(FG)] \xrightarrow[(\beta_q)_{GF,FG}]{q^{l(\mathrm{t}(GF,FG))}\, \mathrm{t}(GF,FG)} \mathrm{H}[\mathrm{t}(GF)]$$

for $A \leq F$. The labels below the arrows refer to the general definition, while those above the arrows show how they specialize to the present example. The coproduct of H enters via the second diagram in (16.6), while β is defined by the first diagram in (16.2).

Lemma 16.47. *For a scalar q, for a Coxeter q-bialgebra H, the quotient map $\mathcal{T}_q(\mathrm{t_H}) \twoheadrightarrow \overline{\mathcal{Z}}^\vee(\mathrm{H})$ induces a Coxeter q-bimonoid structure on $\overline{\mathcal{Z}}^\vee(\mathrm{H})$.*

PROOF. We know $\overline{\mathcal{Z}}^\vee(\mathrm{H})$ is a Coxeter quotient monoid of $\mathcal{T}_q(\mathrm{t_H})$. It remains to check that the coproduct of $\mathcal{T}_q(\mathrm{t_H})$ preserves relations (16.25). In fact, one only needs to consider relations in which $K = A$:

$$\mu_Z^T(x) - \sum_{F:\, F\geq A,\, \mathrm{t}(F)=T} x, \qquad x \in \mathrm{H}[T],\ \mathrm{t}(A) = Z.$$

(These generate the remaining relations via the product of $\mathcal{T}_q(\mathrm{t_H})$ which is inclusion). Using the q-bialgebra axiom (8.73) of H, we show below that the coproduct Δ_A^G takes the above relation to a sum of such relations. Putting $\mathrm{t}(G) = T'$,

$$\begin{aligned}
&\Delta_Z^{T'}\mu_Z^T(x) - \sum_{F:\, F\geq A,\, \mathrm{t}(F)=T} q^{l(\mathrm{t}(GF,FG))}\, \mathrm{t}(GF,FG)\Delta_T^{\mathrm{t}(FG)}(x) \\
&= \Delta_Z^{T'}\mu_Z^T(x) - \sum_{H:\, H\geq G}\ \sum_{\substack{F:\, F\geq A \\ \mathrm{t}(F)=T,\, GF=H}} q^{l(\mathrm{t}(H,FG))}\, \mathrm{t}(H,FG)\Delta_T^{\mathrm{t}(FG)}(x) \\
&= \Delta_Z^{T'}\mu_Z^T(x) - \sum_{U':\, U'\geq T'}\ \sum_{\substack{H:\, H\geq G \\ \mathrm{t}(H)=U'}}\ \sum_{\substack{\sigma:\, U\to U' \text{ local to } Z \\ \text{biprojects } T \text{ and } T'}} q^{l(\sigma)}\sigma\Delta_T^U(x) \\
&= \sum_{U':\, U'\geq T'} \Big(\mu_{T'}^{U'}(x_{U'}) - \sum_{\substack{H:\, H\geq G \\ \mathrm{t}(H)=U'}} x_{U'} \Big)
\end{aligned}$$

where

$$x_{U'} := \sum_{U:\, U\geq T}\ \sum_{\substack{\sigma:\, U\to U' \text{ local to } Z \\ \text{biprojects } T \text{ and } T'}} q^{l(\sigma)}\sigma\Delta_T^U(x) \qquad \text{for each } U' \geq T'.$$

The term in parenthesis is 0 for $U' = T'$; it is a relation for $U' > T'$.

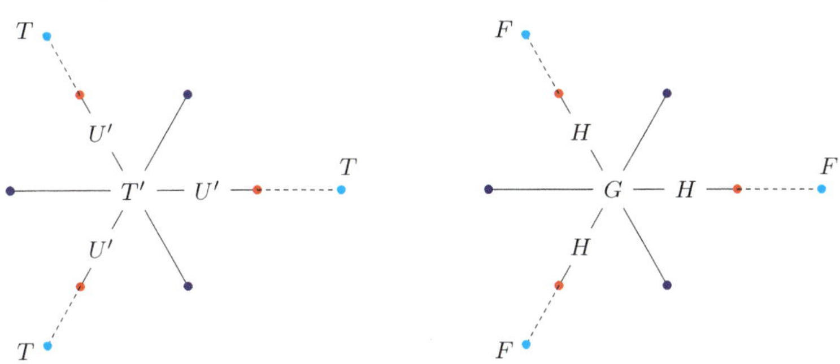

A helpful illustration with $Z = \emptyset$ and $A = O$ is shown above. The left picture uses face-types, while the right picture uses faces. The dotted lines are not edges; they are portions of rank-two flats. □

Proposition 16.48. *Let q be a scalar. The functor $\overline{\mathcal{Z}}^\vee$ is left adjoint to $\overline{\mathcal{K}}^\vee$ between categories of Coxeter q-bimonoids and Coxeter q-bialgebras. That is, for a Coxeter q-bialgebra* H *and Coxeter q-bimonoid* h, *there is a natural bijection*

$$q\text{-}\mathsf{Bimon}(\mathsf{CSp})(\overline{\mathcal{Z}}^\vee(\mathrm{H}), \mathsf{h}) \xrightarrow{\cong} q\text{-}\mathsf{Bimon}(\overline{\mathsf{C}}\mathsf{Sp})(\mathrm{H}, \overline{\mathcal{K}}^\vee(\mathsf{h})).$$

PROOF. The bijection is defined as in Proposition 16.35. We need to verify that under this bijection, morphisms of Coxeter comonoids correspond to morphisms of Coxeter coalgebras. This follows from Proposition 16.20 and Lemma 16.47: Start with a morphism $\mathrm{H} \to \overline{\mathcal{K}}^\vee(\mathsf{h})$ of Coxeter q-bialgebras. This yields a morphism $\mathsf{t}_\mathrm{H} \to \mathsf{h}$ of Coxeter comonoids. By Theorem 3.9, this is equivalent to a morphism $\mathcal{T}_q(\mathsf{t}_\mathrm{H}) \to \mathsf{h}$ of Coxeter q-bimonoids, which induces a morphism $\overline{\mathcal{Z}}^\vee(\mathrm{H}) \to \mathsf{h}$. These steps are reversible. □

More formally, Proposition 16.48 can be deduced from item (1) of Proposition 16.49 below (which builds on Proposition 16.35) by applying Lemma A.2. Further, by working things out, one can also derive Lemma 16.47 as explained in Exercise 16.50 below. This is one way to do the second lift.

Proposition 16.49. *For a scalar q, wrt comonads $\mathcal{T}^\vee$ and $\overline{\mathcal{T}}^\vee$ on Coxeter monoids and Coxeter algebras,*

(1) $(\overline{\mathcal{Z}}^\vee, \overline{\mathcal{K}}^\vee)$ *is a colax-colax adjunction,*
(2) $(\overline{\mathcal{Z}}^\vee, \overline{\mathcal{K}}^\vee)$ *is a colax-lax adjunction.*

Note very carefully: The comonads $\mathcal{T}^\vee$ and $\overline{\mathcal{T}}^\vee$ above are induced from the ones on Coxeter species and Coxeter spaces via the mixed distributive laws λ_q and $\overline{\lambda}_q$. See Remark A.1.

PROOF. The formal argument is exactly as in the proof of Proposition 16.21, so we do not repeat it. The colax structure of $\overline{\mathcal{Z}}^\vee$ is given by the composite

$$\overline{\mathcal{Z}}^\vee\overline{\mathcal{T}}^\vee(\mathrm{A}) \to \overline{\mathcal{Z}}^\vee\overline{\mathcal{T}}^\vee\overline{\mathcal{K}}^\vee\overline{\mathcal{Z}}^\vee(\mathrm{A}) \to \overline{\mathcal{Z}}^\vee\overline{\mathcal{K}}^\vee\mathcal{T}^\vee\overline{\mathcal{Z}}^\vee(\mathrm{A}) \to \mathcal{T}^\vee\overline{\mathcal{Z}}^\vee(\mathrm{A}),$$

with the unit and counit of $(\overline{\mathcal{Z}}^\vee, \overline{\mathcal{K}}^\vee)$ defined by diagrams (16.27). This expression is simplified in the exercise below. □

Exercise 16.50 (Morphism of colax functors). Check: For a scalar q, for a Coxeter algebra A, the diagram of Coxeter monoids

(16.30)
$$\begin{array}{ccccc} \mathcal{T}\mathsf{t}_{\overline{\mathcal{T}}^\vee(\mathrm{A})} & \longrightarrow & \mathcal{T}\mathcal{T}^\vee\mathsf{t}_\mathrm{A} & \longrightarrow & \mathcal{T}^\vee\mathcal{T}\mathsf{t}_\mathrm{A} \\ \downarrow & & & & \downarrow \\ \overline{\mathcal{Z}}^\vee\overline{\mathcal{T}}^\vee(\mathrm{A}) & & \longrightarrow & & \mathcal{T}^\vee\overline{\mathcal{Z}}^\vee(\mathrm{A}) \end{array}$$

commutes. It connects the colax structures of $\mathcal{T}\mathsf{t}_{(-)}$ and $\overline{\mathcal{Z}}^\vee$. The former is obtained by composing the colax functors

$$\mathsf{Mon}(\overline{\mathsf{C}}\mathsf{Sp}) \xrightarrow{frg} \overline{\mathsf{C}}\mathsf{Sp} \xrightarrow{\mathsf{t}_{(-)}} \mathsf{CSp} \xrightarrow{\mathcal{T}} \mathsf{Mon}(\mathsf{CSp}).$$

(Note how the mixed distributive law λ_q enters the picture via the last map.) In other words, the natural transformation $\mathcal{T}\mathsf{t}_{(-)} \to \overline{\mathcal{Z}}^\vee$ is a morphism of colax functors from Coxeter algebras to Coxeter monoids. Deduce Lemma 16.47 as a consequence.

Theorem 16.51. *Let* $\operatorname{char}(\Bbbk) \nmid |W|$. *Let* q *be a scalar. The unit of the adjunction* $(\overline{\mathcal{Z}}^\vee, \overline{\mathcal{K}}^\vee)$ *is an isomorphism. That is, for a Coxeter* q*-bialgebra* H*, the map* $\mathrm{H} \to \overline{\mathcal{K}}^\vee \overline{\mathcal{Z}}^\vee(\mathrm{H})$ *is an isomorphism of Coxeter* q*-bialgebras.*

Explicitly, the map is defined as the composite

$$\mathrm{H} \to \overline{\mathcal{K}}^\vee(\mathsf{t}_\mathrm{H}) \to \overline{\mathcal{K}}^\vee \mathcal{T}_q(\mathsf{t}_\mathrm{H}) \to \overline{\mathcal{K}}^\vee(\overline{\mathcal{Z}}^\vee(\mathrm{H})).$$

PROOF. This follows formally from Proposition 16.40. □

Exercise 16.52. Deduce Proposition 16.48 by applying the adjoint lifting theorem to the following result which builds on Proposition 16.20 and Proposition 16.36.

For a scalar q, wrt monads $\mathcal{T}$ and $\overline{\mathcal{T}}$ on Coxeter comonoids and Coxeter coalgebras, $(\mathsf{t}_{(-)}, \overline{\mathcal{K}}^\vee)$ is a colax-lax adjunction.

This is the other way to do the second lift.

Exercise 16.53. Check: For a scalar q, the diagrams

$$\begin{array}{ccc} q\text{-}\mathsf{Bimon}(\overline{\mathsf{C}}\mathsf{Sp}) & \xrightarrow{frg} & \mathsf{Mon}(\overline{\mathsf{C}}\mathsf{Sp}) \\ {\scriptstyle \overline{\mathcal{Z}}^\vee}\downarrow & & \downarrow{\scriptstyle \overline{\mathcal{Z}}^\vee} \\ q\text{-}\mathsf{Bimon}(\mathsf{CSp}) & \xrightarrow[frg]{} & \mathsf{Mon}(\mathsf{CSp}) \end{array} \qquad \begin{array}{ccc} q\text{-}\mathsf{Bimon}(\overline{\mathsf{C}}\mathsf{Sp}) & \xleftarrow{\overline{\mathcal{T}}_q^{\vee}} & \mathsf{Mon}(\overline{\mathsf{C}}\mathsf{Sp}) \\ {\scriptstyle \overline{\mathcal{K}}^\vee}\uparrow & & \uparrow{\scriptstyle \overline{\mathcal{K}}^\vee} \\ q\text{-}\mathsf{Bimon}(\mathsf{CSp}) & \xleftarrow[\mathcal{T}_q^\vee]{} & \mathsf{Mon}(\mathsf{CSp}) \end{array}$$

commute. In more detail: For a Coxeter monoid a, the colax structure $\overline{\varphi}_\mathsf{a}^\vee$: $\overline{\mathcal{K}}^\vee \mathcal{T}_q^\vee(\mathsf{a}) \to \overline{\mathcal{T}}_q{}^\vee \overline{\mathcal{K}}^\vee(\mathsf{a})$ is an isomorphism of Coxeter q-bialgebras. This result was also obtained in the second part of Proposition 15.25.

In the first part of Exercise 16.25, corresponding functors in the two diagrams form colax-lax adjunctions wrt monads $\mathcal{T}$ and $\overline{\mathcal{T}}$. Applying the adjoint lifting theorem yields the two diagrams above. This extends considerations in Exercise 16.52.

Exercise 16.54. Check: For a scalar q, the diagrams

$$\begin{array}{ccc} \mathsf{Comon}(\overline{\mathsf{C}}\mathsf{Sp}) & \xrightarrow{\overline{\mathcal{T}}_q} & q\text{-}\mathsf{Bimon}(\overline{\mathsf{C}}\mathsf{Sp}) \\ {\scriptstyle \mathsf{t}_{(-)}}\downarrow & & \downarrow{\scriptstyle \overline{\mathcal{Z}}^\vee} \\ \mathsf{Comon}(\mathsf{CSp}) & \xrightarrow[\mathcal{T}_q]{} & q\text{-}\mathsf{Bimon}(\mathsf{CSp}) \end{array} \qquad \begin{array}{ccc} \mathsf{Comon}(\overline{\mathsf{C}}\mathsf{Sp}) & \xleftarrow{frg} & q\text{-}\mathsf{Bimon}(\overline{\mathsf{C}}\mathsf{Sp}) \\ {\scriptstyle \overline{\mathcal{K}}^\vee}\uparrow & & \uparrow{\scriptstyle \overline{\mathcal{K}}^\vee} \\ \mathsf{Comon}(\mathsf{CSp}) & \xleftarrow[frg]{} & q\text{-}\mathsf{Bimon}(\mathsf{CSp}) \end{array}$$

commute. Corresponding functors in the two diagrams in (16.26) form colax-lax adjunctions wrt comonads $\mathcal{T}^\vee$ and $\overline{\mathcal{T}}^\vee$. Applying the adjoint lifting theorem yields the two diagrams above. This extends considerations in Proposition 16.49.

Exercise 16.55. Deduce from Exercises 16.28 and 16.54: For a scalar q, the diagrams

$$\begin{array}{ccc} \overline{\mathsf{C}}\mathsf{Sp} & \xrightarrow{\overline{\mathcal{T}}_q} & q\text{-}\mathsf{Bimon}(\overline{\mathsf{C}}\mathsf{Sp}) \\ {\scriptstyle \mathsf{t}_{(-)}}\downarrow & & \downarrow{\scriptstyle \overline{\mathcal{Z}}^\vee} \\ \mathsf{CSp} & \xrightarrow[\mathcal{T}_q]{} & q\text{-}\mathsf{Bimon}(\mathsf{CSp}) \end{array} \qquad \begin{array}{ccc} \overline{\mathsf{C}}\mathsf{Sp} & \xleftarrow{\overline{\mathcal{P}}} & q\text{-}\mathsf{Bimon}(\overline{\mathsf{C}}\mathsf{Sp}) \\ {\scriptstyle \overline{\mathcal{K}}^\vee}\uparrow & & \uparrow{\scriptstyle \overline{\mathcal{K}}^\vee} \\ \mathsf{CSp} & \xleftarrow[\mathcal{P}]{} & q\text{-}\mathsf{Bimon}(\mathsf{CSp}) \end{array}$$

commute.

Use Theorems 7.8 and 13.25 to deduce: For q not a root of unity, as functors between categories of Coxeter q-bimonoids and Coxeter q-bialgebras,

$$\overline{\mathcal{Z}}^\vee = \mathcal{T}_q \mathsf{t}_{(-)} \overline{\mathcal{P}} \quad \text{and} \quad \overline{\mathcal{K}}^\vee = \overline{\mathcal{T}}_q \overline{\mathcal{K}}^\vee \mathcal{P}, \tag{16.31}$$

with $\overline{\mathcal{Z}}^{\vee}$ a section of $\overline{\mathcal{K}}^{\vee}$ (in any field characteristic). This improves upon Theorem 16.51 when q is not a root of unity. Since Coxeter q-bialgebras are free when q is not a root of unity, this improvement can also be deduced from Exercise 16.41.

Observation 16.56. For H a cocommutative Coxeter bialgebra, t_{H} is a cocommutative Coxeter comonoid, so $\mathcal{T}(\mathsf{t}_{\mathrm{H}})$ is a cocommutative Coxeter bimonoid, and hence so is its quotient $\overline{\mathcal{Z}}^{\vee}(\mathrm{H})$. Thus, the adjunction in Proposition 16.48 for $q = 1$ restricts to the full subcategories of cocommutative Coxeter bimonoids and cocommutative Coxeter bialgebras.

Let us now briefly look at the dual picture. For a scalar q, for a Coxeter q-bialgebra H, the inclusion $\overline{\mathcal{Z}}(\mathrm{H}) \hookrightarrow \mathcal{T}_q^{\vee}(\mathsf{t}_{\mathrm{H}})$ induces a Coxeter q-bimonoid structure on $\overline{\mathcal{Z}}(\mathrm{H})$. Here $\mathcal{T}_q^{\vee}(\mathsf{t}_{\mathrm{H}})$ is the cofree Coxeter q-bimonoid on t_{H} viewed as a Coxeter monoid.

Proposition 16.57. *Let q be a scalar. The functor $\overline{\mathcal{Z}}$ is right adjoint to $\overline{\mathcal{K}}$ between categories of Coxeter q-bimonoids and Coxeter q-bialgebras. That is, for a Coxeter q-bialgebra* H *and Coxeter q-bimonoid* h, *there is a natural bijection*

$$q\text{-}\mathsf{Bimon}(\overline{\mathsf{C}}\mathsf{Sp})(\overline{\mathcal{K}}(\mathsf{h}), \mathrm{H}) \xrightarrow{\cong} q\text{-}\mathsf{Bimon}(\mathsf{CSp})(\mathsf{h}, \overline{\mathcal{Z}}(\mathrm{H})).$$

This is dual to Proposition 16.48 and can be proved in a similar manner.

Theorem 16.58. *Let* $\mathrm{char}(\Bbbk) \nmid |W|$. *Let q be a scalar. The counit of the adjunction $(\overline{\mathcal{K}}, \overline{\mathcal{Z}})$ is an isomorphism. That is, for a Coxeter q-bialgebra* H, *the map* $\overline{\mathcal{K}}\overline{\mathcal{Z}}(\mathrm{H}) \to \mathrm{H}$ *is an isomorphism of Coxeter q-bialgebras.*

PROOF. This is dual to Theorem 16.51 and follows formally from Proposition 16.45. □

Exercise 16.59. This is the dual of Exercise 16.53. Check: For a scalar q, the diagrams

$$\begin{array}{ccc} \mathsf{Comon}(\overline{\mathsf{C}}\mathsf{Sp}) & \xrightarrow{\overline{\mathcal{T}}_q} & q\text{-}\mathsf{Bimon}(\overline{\mathsf{C}}\mathsf{Sp}) \\ {\scriptstyle \overline{\mathcal{K}}}\uparrow & & \uparrow{\scriptstyle \overline{\mathcal{K}}} \\ \mathsf{Comon}(\mathsf{CSp}) & \xrightarrow[\mathcal{T}_q]{} & q\text{-}\mathsf{Bimon}(\mathsf{CSp}) \end{array} \qquad \begin{array}{ccc} \mathsf{Comon}(\overline{\mathsf{C}}\mathsf{Sp}) & \xleftarrow{frg} & q\text{-}\mathsf{Bimon}(\overline{\mathsf{C}}\mathsf{Sp}) \\ {\scriptstyle \overline{\mathcal{Z}}}\downarrow & & \downarrow{\scriptstyle \overline{\mathcal{Z}}} \\ \mathsf{Comon}(\mathsf{CSp}) & \xleftarrow[frg]{} & q\text{-}\mathsf{Bimon}(\mathsf{CSp}) \end{array}$$

commute. These are lifts of the first two diagrams in Exercise 16.33 whose corresponding functors form colax-lax adjunctions wrt comonads $\mathcal{T}^{\vee}$ and $\overline{\mathcal{T}}^{\vee}$.

Observation 16.60. For H a commutative Coxeter bialgebra, t_{H} is commutative, hence so is $\mathcal{T}^{\vee}(\mathsf{t}_{\mathrm{H}})$, and hence so is $\overline{\mathcal{Z}}(\mathrm{H})$. Thus, the adjunction in Proposition 16.57 for $q = 1$ restricts to the full subcategories of commutative Coxeter bimonoids and commutative Coxeter bialgebras.

16.4.4. Duality.

Functors $\overline{\mathcal{Z}}$ and $\overline{\mathcal{Z}}^{\vee}$ are conjugates of each other wrt duality. We elaborate below.

Exercise 16.61. Show: For a Coxeter algebra A and Coxeter q-bialgebra H,

$$\overline{\mathcal{Z}}^{\vee}(\mathrm{A})^* = \overline{\mathcal{Z}}(\mathrm{A}^*) \quad \text{and} \quad \overline{\mathcal{Z}}^{\vee}(\mathrm{H})^* = \overline{\mathcal{Z}}(\mathrm{H}^*) \tag{16.32}$$

as Coxeter comonoids and Coxeter q-bimonoids, respectively. Similarly, for a finite-dimensional Coxeter coalgebra C and Coxeter q-bialgebra H,

$$\overline{\mathcal{Z}}^{\vee}(\mathrm{C}^*) = \overline{\mathcal{Z}}(\mathrm{C})^* \quad \text{and} \quad \overline{\mathcal{Z}}^{\vee}(\mathrm{H}^*) = \overline{\mathcal{Z}}(\mathrm{H})^* \tag{16.33}$$

as Coxeter monoids and Coxeter q-bimonoids, respectively. (Use similar results on bosonic Fock functors in Section 15.1.11 and that the duality functor is self-adjoint between categories of Coxeter monoids (algebras) and Coxeter comonoids (coalgebras).)

16.4.5. Reflective and coreflective subcategories. This builds on ideas in Section 16.3.3. Functors $\mathsf{t}_{(-)}$, $\overline{\mathcal{Z}}$, $\overline{\mathcal{Z}}^{\vee}$ are full and faithful (with claims about $\overline{\mathcal{Z}}$ and $\overline{\mathcal{Z}}^{\vee}$ assuming $\mathrm{char}(\mathbb{k}) \nmid |W|$). Further:

Proposition 16.62. *The category of Coxeter coalgebras can be viewed as a coreflective and reflective subcategory of the category of Coxeter comonoids via functors* $\mathsf{t}_{(-)}$ *and* $\overline{\mathcal{Z}}$, *respectively (the latter when* $\mathrm{char}(\mathbb{k}) \nmid |W|$*).*

PROOF. This is a formal consequence of Propositions 16.27 and 16.45. □

Proposition 16.63. *The category of Coxeter algebras can be viewed as a reflective and coreflective subcategory of the category of Coxeter monoids via functors* $\mathsf{t}_{(-)}$ *and* $\overline{\mathcal{Z}}^{\vee}$, *respectively (the latter when* $\mathrm{char}(\mathbb{k}) \nmid |W|$*).*

PROOF. This is a formal consequence of Propositions 16.34 and 16.40. □

Theorem 16.64. *Let* $\mathrm{char}(\mathbb{k}) \nmid |W|$. *For a scalar q, the category of Coxeter q-bialgebras can be viewed as a reflective and coreflective subcategory of the category of Coxeter q-bimonoids via functors* $\overline{\mathcal{Z}}$ *and* $\overline{\mathcal{Z}}^{\vee}$, *respectively.*

PROOF. This is a formal consequence of Theorems 16.51 and 16.58. □

We point out that field characteristic assumption is not required when q is not a root of unity.

16.5. Bosonic Fock functors and abelianization

Recall (co)abelianization from Sections 3.4 and 9.4. Bosonic Fock functors commute with (co)abelianization. We elaborate below.

For a Coxeter monoid a and Coxeter comonoid c,

$$\overline{\mathcal{K}}(\mathsf{a})_{ab} = \overline{\mathcal{K}}(\mathsf{a}_{ab}) \quad \text{and} \quad \overline{\mathcal{K}}^{\vee}(\mathsf{c}^{coab}) = \overline{\mathcal{K}}^{\vee}(\mathsf{c})^{coab} \tag{16.34}$$

as commutative Coxeter algebras and cocommutative Coxeter coalgebras, respectively. In more detail: The map $\mathsf{a} \twoheadrightarrow \mathsf{a}_{ab}$ induces $\overline{\mathcal{K}}(\mathsf{a}) \to \overline{\mathcal{K}}(\mathsf{a}_{ab})$ which factors through the abelianization to yield $\overline{\mathcal{K}}(\mathsf{a})_{ab} \to \overline{\mathcal{K}}(\mathsf{a}_{ab})$. That this is an isomorphism can be checked directly. Another way is shown in Exercise 16.65 below. Dually, the map $\mathsf{c}^{coab} \hookrightarrow \mathsf{c}$ induces $\overline{\mathcal{K}}^{\vee}(\mathsf{c}^{coab}) \to \overline{\mathcal{K}}^{\vee}(\mathsf{c})$ which maps into the coabelianization to yield an isomorphism $\overline{\mathcal{K}}^{\vee}(\mathsf{c}^{coab}) \to \overline{\mathcal{K}}^{\vee}(\mathsf{c})^{coab}$.

Similarly, for a Coxeter bimonoid h,

$$\overline{\mathcal{K}}(\mathsf{h})_{ab} = \overline{\mathcal{K}}(\mathsf{h}_{ab}) \quad \text{and} \quad \overline{\mathcal{K}}^{\vee}(\mathsf{h}^{coab}) = \overline{\mathcal{K}}^{\vee}(\mathsf{h})^{coab} \tag{16.35}$$

as Coxeter bialgebras. This can be deduced from (16.34).

Exercise 16.65. Use the second diagram below to deduce the first diagram.

$$\begin{array}{ccc} \mathsf{Mon}(\overline{\mathsf{C}}\mathsf{Sp}) & \xrightarrow{(-)_{ab}} & \mathsf{Mon}^{\mathsf{co}}(\overline{\mathsf{C}}\mathsf{Sp}) \\ {\scriptstyle \overline{\mathcal{K}}}\uparrow & & \uparrow{\scriptstyle \overline{\mathcal{K}}} \\ \mathsf{Mon}(\mathsf{CSp}) & \xrightarrow[(-)_{ab}]{} & \mathsf{Mon}^{\mathsf{co}}(\mathsf{CSp}) \end{array} \qquad \begin{array}{ccc} \mathsf{Mon}(\overline{\mathsf{C}}\mathsf{Sp}) & \hookleftarrow & \mathsf{Mon}^{\mathsf{co}}(\overline{\mathsf{C}}\mathsf{Sp}) \\ {\scriptstyle \mathrm{t}_{(-)}}\downarrow & & \downarrow{\scriptstyle \mathrm{t}_{(-)}} \\ \mathsf{Mon}(\mathsf{CSp}) & \hookleftarrow & \mathsf{Mon}^{\mathsf{co}}(\mathsf{CSp}) \end{array}$$

This proves the first identity in (16.34). Similarly, prove the second identity in (16.34). For identities (16.35), employ functors $\overline{\mathcal{Z}}$ and $\overline{\mathcal{Z}}^{\vee}$ instead of $\mathrm{t}_{(-)}$.

Exercise 16.66. Show: If $\mathrm{char}(\mathbb{k}) \nmid |W|$, then (16.34) and (16.35) also hold with roles of $\overline{\mathcal{K}}$ and $\overline{\mathcal{K}}^{\vee}$ interchanged. (Use that $\overline{\mathcal{K}}$ and $\overline{\mathcal{K}}^{\vee}$ are naturally isomorphic in this situation.)

16.6. Adjoints preserving commutativity

In Section 16.4 we saw some adjunctions between (co)commutative Coxeter (co)monoids and (co)commutative Coxeter (co)algebras involving the trivialization functor; see the last part of Propositions 16.20 and 16.29. Other similar results were mentioned in Observations 16.56 and 16.60. We now look at more results of this nature. A complete list is given in Summary 16.2.

16.6.1. Commutative Coxeter algebras and monoids.

For A a Coxeter algebra, recall the Coxeter monoid $\overline{\mathcal{Z}}^{\vee}(\mathrm{A})$ from Section 16.4.2.

Observation 16.67. The abelianization $\overline{\mathcal{Z}}^{\vee}(\mathrm{A})_{ab}$, in explicit terms, is the quotient of $\mathcal{T}(\mathrm{t}_{\mathrm{A}})$ by the Coxeter submonoid whose A-component is linearly spanned by elements (16.25) and

$$x - \beta_{G,F}(x), \quad \text{or equivalently,} \quad x - \mathrm{t}(G,F)(x), \qquad x \in \mathrm{A}[\mathrm{t}(F)] \tag{16.36}$$

for G and F both greater than A and with the same support. The second expression is obtained using definition of β in the first diagram in (16.2).

Define the functor $\overline{\mathcal{Z}}^{\vee}_{ab}$ to be the composite

$$\mathsf{Mon}^{\mathsf{co}}(\overline{\mathsf{C}}\mathsf{Sp}) \to \mathsf{Mon}(\overline{\mathsf{C}}\mathsf{Sp}) \xrightarrow{\overline{\mathcal{Z}}^{\vee}} \mathsf{Mon}(\mathsf{CSp}) \xrightarrow{(-)_{ab}} \mathsf{Mon}^{\mathsf{co}}(\mathsf{CSp}).$$

Thus, $\overline{\mathcal{Z}}^{\vee}_{ab}(\mathrm{A}) = \overline{\mathcal{Z}}^{\vee}(\mathrm{A})_{ab}$ for a commutative Coxeter algebra A.

Proposition 16.68. *The functor $\overline{\mathcal{Z}}^{\vee}_{ab}$ is left adjoint to $\overline{\mathcal{K}}^{\vee}$ between categories of commutative Coxeter monoids and commutative Coxeter algebras. That is, for a commutative Coxeter algebra* A *and commutative Coxeter monoid* a, *there is a natural bijection*

$$\mathsf{Mon}^{\mathsf{co}}(\mathsf{CSp})(\overline{\mathcal{Z}}^{\vee}_{ab}(\mathrm{A}), \mathsf{a}) \xrightarrow{\cong} \mathsf{Mon}^{\mathsf{co}}(\overline{\mathsf{C}}\mathsf{Sp})(\mathrm{A}, \overline{\mathcal{K}}^{\vee}(\mathsf{a})).$$

PROOF. Composing the adjunction (3.46) with $(\overline{\mathcal{Z}}^{\vee}, \overline{\mathcal{K}}^{\vee})$ yields an adjunction between categories of Coxeter algebras and commutative Coxeter monoids. Since commutative Coxeter algebras form a full subcategory of the former and $\overline{\mathcal{K}}^{\vee}$ preserves commutativity, the result follows. □

More formally, Proposition 16.68 arises by applying the adjoint lifting theorem (Theorem A.3) to the colax-lax adjunction $(\mathtt{t}_{(-)}, \overline{\mathcal{K}}^\vee)$ wrt monads $\mathcal{S}$ and $\overline{\mathcal{S}}$ in Proposition 16.36. In particular, there is a natural transformation $\mathcal{S}\mathtt{t}_{(-)} \to \overline{\mathcal{Z}}^\vee_{ab}$ between categories of commutative Coxeter algebras and commutative Coxeter monoids. It is the same as the one obtained by applying abelianization to $\mathcal{T}\mathtt{t}_{(-)} \to \overline{\mathcal{Z}}^\vee$.

Similar to (16.26), we deduce that the diagram

$$\begin{array}{ccc} \overline{\mathsf{C}}\mathsf{Sp} & \xrightarrow{\overline{\mathcal{S}}} & \mathsf{Mon}^{\mathsf{co}}(\overline{\mathsf{C}}\mathsf{Sp}) \\ {\scriptstyle \mathtt{t}_{(-)}}\downarrow & & \downarrow{\scriptstyle \overline{\mathcal{Z}}^\vee_{ab}} \\ \mathsf{CSp} & \xrightarrow[\mathcal{S}]{} & \mathsf{Mon}^{\mathsf{co}}(\mathsf{CSp}) \end{array} \tag{16.37}$$

commutes.

Proposition 16.69. *Let* $\mathrm{char}(\Bbbk) \nmid |W|$. *The unit of the adjunction* $(\overline{\mathcal{Z}}^\vee_{ab}, \overline{\mathcal{K}}^\vee)$ *is an isomorphism. That is, for a commutative Coxeter algebra* A, *the map* $\mathrm{A} \to \overline{\mathcal{K}}^\vee \overline{\mathcal{Z}}^\vee_{ab}(\mathrm{A})$ *is an isomorphism of commutative Coxeter algebras.*

This map is the same as the composite

$$\mathrm{A} \to \overline{\mathcal{K}}^\vee(\overline{\mathcal{Z}}^\vee(\mathrm{A})) \to \overline{\mathcal{K}}^\vee(\overline{\mathcal{Z}}^\vee_{ab}(\mathrm{A})).$$

PROOF. Surjectivity follows from Proposition 16.40. For injectivity, we proceed as in the proof there. We again apply Exercise 16.39. This time the y_V come into play, however, commutativity of A implies $\sum_{V:\, Z<V} \mu^V_Z(y_V) = 0$, so we have the same conclusion.

Alternatively, in view of Proposition 16.40, it suffices to show that the second map in the above composite is an isomorphism. This can be deduced from Exercise 16.66. □

For H a Coxeter bialgebra, $\overline{\mathcal{Z}}^\vee(\mathrm{H})$ is a Coxeter bimonoid, and hence its abelianization $\overline{\mathcal{Z}}^\vee(\mathrm{H})_{ab}$ is a commutative Coxeter bimonoid. Proceeding as before, we get a functor $\overline{\mathcal{Z}}^\vee_{ab}$ from commutative Coxeter bialgebras to commutative Coxeter bimonoids. From Proposition 16.48 for $q = 1$, we deduce:

Proposition 16.70. *The functor* $\overline{\mathcal{Z}}^\vee_{ab}$ *is left adjoint to* $\overline{\mathcal{K}}^\vee$ *between categories of commutative Coxeter bimonoids and commutative Coxeter bialgebras. That is, for a commutative Coxeter bialgebra* H *and commutative Coxeter bimonoid* $\mathtt{h}$, *there is a natural bijection*

$$\mathsf{Bimon}^{\mathsf{co}}(\mathsf{CSp})(\overline{\mathcal{Z}}^\vee_{ab}(\mathrm{H}), \mathtt{h}) \xrightarrow{\cong} \mathsf{Bimon}^{\mathsf{co}}(\overline{\mathsf{C}}\mathsf{Sp})(\mathrm{H}, \overline{\mathcal{K}}^\vee(\mathtt{h})).$$

In conjunction with Proposition 16.69, we obtain:

Theorem 16.71. *Let* $\mathrm{char}(\Bbbk) \nmid |W|$. *For a commutative Coxeter bialgebra* H, *the unit of the adjunction* $\mathrm{H} \to \overline{\mathcal{K}}^\vee \overline{\mathcal{Z}}^\vee_{ab}(\mathrm{H})$ *is an isomorphism of commutative Coxeter bialgebras.*

Similar to (16.26), we deduce that the diagram

$$\begin{array}{ccc} \mathsf{Comon}(\overline{\mathsf{C}}\mathsf{Sp}) & \xrightarrow{\overline{\mathcal{S}}} & \mathsf{Bimon}^{\mathsf{co}}(\overline{\mathsf{C}}\mathsf{Sp}) \\ {\scriptstyle \mathsf{t}_{(-)}}\downarrow & & \downarrow{\scriptstyle \overline{\mathcal{Z}}^{\vee}_{ab}} \\ \mathsf{Comon}(\mathsf{CSp}) & \xrightarrow[\mathcal{S}]{} & \mathsf{Bimon}^{\mathsf{co}}(\mathsf{CSp}) \end{array} \tag{16.38}$$

commutes.

16.6.2. Cocommutative Coxeter coalgebras and comonoids. For C a Coxeter coalgebra, recall the Coxeter comonoid $\overline{\mathcal{Z}}(\mathrm{C})$.

Observation 16.72. The coabelianization $\overline{\mathcal{Z}}(\mathrm{C})^{coab}$, in explicit terms, is the Coxeter subcomonoid of $\mathcal{T}^{\vee}(\mathsf{t}_{\mathrm{C}})$ whose A-component consists of elements $(x_F)_{F\geq A}$ which satisfy conditions (16.28) and

$$x_G = \beta_{G,F}(x_F), \quad \text{or equivalently,} \quad x_G = \mathrm{t}(G,F)(x_F) \tag{16.39}$$

for G and F both greater than A and with the same support.

Define the functor $\overline{\mathcal{Z}}^{coab}$ to be the composite

$$^{\mathsf{co}}\mathsf{Comon}(\overline{\mathsf{C}}\mathsf{Sp}) \to \mathsf{Comon}(\overline{\mathsf{C}}\mathsf{Sp}) \xrightarrow{\overline{\mathcal{Z}}} \mathsf{Comon}(\mathsf{CSp}) \xrightarrow{(-)^{coab}} {}^{\mathsf{co}}\mathsf{Comon}(\mathsf{CSp}).$$

Thus, $\overline{\mathcal{Z}}^{coab}(\mathrm{C}) = \overline{\mathcal{Z}}(\mathrm{C})^{coab}$ for a cocommutative Coxeter coalgebra C.

Proposition 16.73. *The functor $\overline{\mathcal{Z}}^{coab}$ is right adjoint to $\overline{\mathcal{K}}$ between categories of cocommutative Coxeter comonoids and cocommutative Coxeter coalgebras. That is, for a cocommutative Coxeter coalgebra* C *and cocommutative Coxeter comonoid* c, *there is a natural bijection*

$$^{\mathsf{co}}\mathsf{Comon}(\overline{\mathsf{C}}\mathsf{Sp})(\overline{\mathcal{K}}(\mathsf{c}), \mathrm{C}) \xrightarrow{\cong} {}^{\mathsf{co}}\mathsf{Comon}(\mathsf{CSp})(\mathsf{c}, \overline{\mathcal{Z}}^{coab}(\mathrm{C})).$$

This is dual to Proposition 16.68 and can be proved in the same manner. Now, there is a natural transformation $\overline{\mathcal{Z}}^{coab} \to \mathcal{S}^{\vee}\mathsf{t}_{(-)}$ obtained by applying coabelianization to $\overline{\mathcal{Z}} \to \mathcal{T}^{\vee}\mathsf{t}_{(-)}$. Subsequent results also have dual statements. For instance:

Theorem 16.74. *Let* $\mathrm{char}(\Bbbk) \nmid |W|$. *For a cocommutative Coxeter bialgebra* H, *the counit of the adjunction* $\overline{\mathcal{K}}\overline{\mathcal{Z}}^{coab}(\mathrm{H}) \to \mathrm{H}$ *is an isomorphism of cocommutative Coxeter bialgebras.*

Similarly, the diagram

$$\begin{array}{ccc} \mathsf{Mon}(\overline{\mathsf{C}}\mathsf{Sp}) & \xrightarrow{\overline{\mathcal{S}}^{\vee}} & {}^{\mathsf{co}}\mathsf{Bimon}(\overline{\mathsf{C}}\mathsf{Sp}) \\ {\scriptstyle \mathsf{t}_{(-)}}\downarrow & & \downarrow{\scriptstyle \overline{\mathcal{Z}}^{coab}} \\ \mathsf{Mon}(\mathsf{CSp}) & \xrightarrow[\mathcal{S}^{\vee}]{} & {}^{\mathsf{co}}\mathsf{Bimon}(\mathsf{CSp}) \end{array} \tag{16.40}$$

commutes.

Exercise 16.75. Check: The element of $\overline{\mathcal{Z}}(\mathrm{C})[A_0]$ defined in Exercise 16.46 belongs to $\overline{\mathcal{Z}}(\mathrm{C})^{coab}[A_0]$ when C is cocommutative. (Use Exercise 1.68.)

16.6.3. Bicommutative Coxeter bialgebras and bimonoids. We build on Observation 16.56.

Observation 16.76. If H is bicommutative, then so is $\overline{\mathcal{Z}}^{\vee}_{ab}(\mathrm{H})$. Thus, the adjunction in Proposition 16.70 restricts to the full subcategories of bicommutative Coxeter bimonoids and bicommutative Coxeter bialgebras.

A similar remark applies to the dual of Proposition 16.70 building on Observation 16.60.

16.6.4. Reflective and coreflective subcategories. We have commutative analogues of Propositions 16.62 and 16.63 and Theorem 16.64 for $q = 1$, with $\overline{\mathcal{Z}}$ and $\overline{\mathcal{Z}}^{\vee}$ replaced by $\overline{\mathcal{Z}}^{coab}$ and $\overline{\mathcal{Z}}^{\vee}_{ab}$, respectively.

16.6.5. Signed analogue. We briefly comment on the signed commutative case. Define the functor $\overline{\mathcal{Z}}^{\vee}_{-ab}$ to be the composite

$$(-1)\text{-}\mathsf{Mon}^{\mathsf{co}}(\overline{\mathsf{C}}\mathsf{Sp}) \xrightarrow{\overline{\mathcal{Z}}^{\vee}} \mathsf{Mon}(\mathsf{CSp}) \xrightarrow{(-)_{-ab}} (-1)\text{-}\mathsf{Mon}^{\mathsf{co}}(\mathsf{CSp}),$$

where $(-)_{-ab}$ sends a Coxeter monoid to its signed abelianization.

Proposition 16.77. *The functor $\overline{\mathcal{Z}}^{\vee}_{-ab}$ is left adjoint to $\overline{\mathcal{K}}^{\vee}$ between categories of signed commutative Coxeter monoids and signed commutative Coxeter algebras.*

This is the signed analogue of Proposition 16.68. Further: If H is a signed commutative signed Coxeter bialgebra, then $\overline{\mathcal{Z}}^{\vee}_{-ab}(\mathrm{H})$ is a signed commutative signed Coxeter bimonoid, and this construction is left adjoint to $\overline{\mathcal{K}}^{\vee}$ between categories of signed commutative signed Coxeter bimonoids and signed commutative signed Coxeter bialgebras. This is the signed analogue of Proposition 16.70. Here we made use of Proposition 16.48 for $q = -1$.

We deduce that the diagrams

$$\begin{array}{ccc} \overline{\mathsf{C}}\mathsf{Sp} & \xrightarrow{\overline{\mathcal{E}}} & (-1)\text{-}\mathsf{Mon}^{\mathsf{co}}(\overline{\mathsf{C}}\mathsf{Sp}) \\ {\scriptstyle \mathsf{t}_{(-)}}\downarrow & & \downarrow{\scriptstyle \overline{\mathcal{Z}}^{\vee}_{-ab}} \\ \mathsf{CSp} & \xrightarrow[\mathcal{E}]{} & (-1)\text{-}\mathsf{Mon}^{\mathsf{co}}(\mathsf{CSp}) \end{array} \tag{16.41}$$

$$\begin{array}{ccc} \mathsf{Comon}(\overline{\mathsf{C}}\mathsf{Sp}) & \xrightarrow{\overline{\mathcal{E}}} & (-1)\text{-}\mathsf{Bimon}^{\mathsf{co}}(\overline{\mathsf{C}}\mathsf{Sp}) \\ {\scriptstyle \mathsf{t}_{(-)}}\downarrow & & \downarrow{\scriptstyle \overline{\mathcal{Z}}^{\vee}_{-ab}} \\ \mathsf{Comon}(\mathsf{CSp}) & \xrightarrow[\mathcal{E}]{} & (-1)\text{-}\mathsf{Bimon}^{\mathsf{co}}(\mathsf{CSp}) \end{array} \tag{16.42}$$

commute.

Dually, define the functor $\overline{\mathcal{Z}}^{-coab}$ to be the composite

$$(-1)\text{-}^{\mathsf{co}}\mathsf{Comon}(\overline{\mathsf{C}}\mathsf{Sp}) \xrightarrow{\overline{\mathcal{Z}}} \mathsf{Comon}(\mathsf{CSp}) \xrightarrow{(-)^{-coab}} (-1)\text{-}^{\mathsf{co}}\mathsf{Comon}(\mathsf{CSp}),$$

where $(-)^{-coab}$ sends a Coxeter comonoid to its signed coabelianization. It is right adjoint to $\overline{\mathcal{K}}$ between categories of signed cocommutative Coxeter comonoids and signed cocommutative Coxeter coalgebras. This is the signed analogue of Proposition 16.73.

16.6.6. Coxeter algebras to commutative Coxeter monoids.

Lemma 16.78. *For* A *a Coxeter algebra and* C *a Coxeter coalgebra,*

$$\overline{\mathcal{Z}}^{\vee}(\mathrm{A})_{ab} = \overline{\mathcal{Z}}^{\vee}_{ab}(\mathrm{A}_{ab}) \quad \textit{and} \quad \overline{\mathcal{Z}}(\mathrm{C})^{coab} = \overline{\mathcal{Z}}^{coab}(\mathrm{C}^{coab}) \tag{16.43}$$

as Coxeter monoids and Coxeter comonoids, respectively. Similarly, for H *a Coxeter bialgebra,*

$$\overline{\mathcal{Z}}^{\vee}(\mathrm{H})_{ab} = \overline{\mathcal{Z}}^{\vee}_{ab}(\mathrm{H}_{ab}) \quad \textit{and} \quad \overline{\mathcal{Z}}(\mathrm{H})^{coab} = \overline{\mathcal{Z}}^{coab}(\mathrm{H}^{coab}) \tag{16.44}$$

as Coxeter bimonoids.

Similar statements hold for signed (co)abelianization.

Proof. We explain the second identity in (16.43). We need to show that for A a face, $\overline{\mathcal{Z}}(\mathrm{C})^{coab}[A] \subseteq \overline{\mathcal{Z}}^{coab}(\mathrm{C}^{coab})[A]$. (The reverse inclusion is clear.) Thus, we need to check that

$$(x_F)_{F \geq A} \in \overline{\mathcal{Z}}(\mathrm{C})^{coab}[A] \implies x_K \in \mathrm{C}^{coab}[\mathrm{t}(K)] \text{ for all } K \geq A.$$

For that, let $\sigma : U \to U'$ be any morphism local to T, where $T = \mathrm{t}(K)$. Then

$$\sigma \Delta^U_T(x_K) = \sigma\Big(\sum_{\substack{G:\, K \leq G \\ \mathrm{t}(G) = U}} x_G\Big) = \sum_{\substack{G':\, K \leq G' \\ \mathrm{t}(G') = U'}} x_{G'} = \Delta^{U'}_T(x_K)$$

as required. The first and last steps used (16.28). The middle step used (16.39): $\sigma(x_G) = x_{G'}$ whenever $\mathrm{t}(G', G) = \sigma$. □

A formal way to prove Lemma 16.78 is given below.

Exercise 16.79. Use the first diagram below to deduce the second diagram.

$$\begin{array}{ccc} \mathsf{Comon}(\overline{\mathsf{C}}\mathsf{Sp}) & \hookleftarrow & {}^{\mathrm{co}}\mathsf{Comon}(\overline{\mathsf{C}}\mathsf{Sp}) \\ {\scriptstyle \overline{\mathcal{K}}}\uparrow & & \uparrow{\scriptstyle \overline{\mathcal{K}}} \\ \mathsf{Comon}(\mathsf{CSp}) & \hookleftarrow & {}^{\mathrm{co}}\mathsf{Comon}(\mathsf{CSp}) \end{array} \qquad \begin{array}{ccc} \mathsf{Comon}(\overline{\mathsf{C}}\mathsf{Sp}) & \xrightarrow{(-)^{coab}} & {}^{\mathrm{co}}\mathsf{Comon}(\overline{\mathsf{C}}\mathsf{Sp}) \\ {\scriptstyle \overline{\mathcal{Z}}}\downarrow & & \downarrow{\scriptstyle \overline{\mathcal{Z}}^{coab}} \\ \mathsf{Comon}(\mathsf{CSp}) & \xrightarrow[(-)^{coab}]{} & {}^{\mathrm{co}}\mathsf{Comon}(\mathsf{CSp}) \end{array}$$

This proves the second identity in (16.43). Similarly, prove the first identity in (16.43), and identities (16.44).

Example 16.80. For a Coxeter space V,

$$\overline{\mathcal{Z}}^{\vee}(\overline{\mathcal{T}}(\mathrm{V}))_{ab} = \mathcal{S}(\mathrm{t}_{\mathrm{V}}) \quad \text{and} \quad \overline{\mathcal{Z}}(\overline{\mathcal{T}}^{\vee}(\mathrm{V}))^{coab} = \mathcal{S}^{\vee}(\mathrm{t}_{\mathrm{V}})$$

as Coxeter monoids and Coxeter comonoids, respectively. Similarly, for A a Coxeter algebra and C a Coxeter coalgebra,

$$\overline{\mathcal{Z}}^{\vee}(\overline{\mathcal{T}}(\mathrm{C}))_{ab} = \mathcal{S}(\mathrm{t}_{\mathrm{C}}) \quad \text{and} \quad \overline{\mathcal{Z}}(\overline{\mathcal{T}}^{\vee}(\mathrm{A}))^{coab} = \mathcal{S}^{\vee}(\mathrm{t}_{\mathrm{A}})$$

as Coxeter bimonoids. This can be checked using (3.49) and Exercise 16.54 and its dual with $q = 1$. Alternatively, one can first use Lemma 16.78 and then proceed similarly using (9.54) and (16.38) and its dual. □

16.7. Adjoints of fermionic Fock functors

Recall from Section 2.8 that the o-signature functor is an adjoint equivalence between the category of Coxeter (co)monoids and itself, and between categories of Coxeter bimonoids and signed Coxeter bimonoids, and so on. Also recall that bosonic and fermionic Fock functors relate to each other by the o-signature functor. Thus, adjoints of the former can be used to deduce adjoints of the latter. We briefly elaborate below.

16.7.1. Signed trivialization functor and its lifts. Recall left adjoints of the bosonic Fock functor $\overline{\mathcal{K}}^{\vee}$ from Propositions 16.12, 16.20, 16.35, 16.48. Composing with the o-signature functor and using (14.34) yields the following left adjoints of the fermionic Fock functor $\overline{\mathcal{K}}^{\vee}_{-1}$.

$$\overline{\mathsf{C}}\mathsf{Sp} \underset{\overline{\mathcal{K}}^{\vee}_{-1}}{\overset{(\mathsf{t}_{(-)})^{-}}{\rightleftarrows}} \mathsf{CSp} \tag{16.45}$$

$$\mathsf{Comon}(\overline{\mathsf{C}}\mathsf{Sp}) \underset{\overline{\mathcal{K}}^{\vee}_{-1}}{\overset{(\mathsf{t}_{(-)})^{-}}{\rightleftarrows}} \mathsf{Comon}(\mathsf{CSp}) \qquad \mathsf{Mon}(\overline{\mathsf{C}}\mathsf{Sp}) \underset{\overline{\mathcal{K}}^{\vee}_{-1}}{\overset{(\overline{\mathcal{Z}}^{\vee})^{-}}{\rightleftarrows}} \mathsf{Mon}(\mathsf{CSp}) \tag{16.46}$$

$$q\text{-}\mathsf{Bimon}(\overline{\mathsf{C}}\mathsf{Sp}) \underset{\overline{\mathcal{K}}^{\vee}_{-1}}{\overset{(\overline{\mathcal{Z}}^{\vee})^{-}}{\rightleftarrows}} (-q)\text{-}\mathsf{Bimon}(\mathsf{CSp}). \tag{16.47}$$

We refer to $(\mathsf{t}_{(-)})^{-}$ as the *signed trivialization functor*. It identifies Coxeter spaces with the full subcategory of those Coxeter species p for which the component $\mathsf{p}[F]$ only depends on the type of F, and a type morphism w is scalar multiplication by $(-1)^{l(w)}$.

In a similar manner, one obtains right adjoints of the fermionic Fock functor $\overline{\mathcal{K}}_{-1}$.

16.7.2. Commutative aspects. Also from Propositions 16.68 and 16.77, we deduce: $(\overline{\mathcal{Z}}^{\vee}_{ab})^{-}$ and $\overline{\mathcal{K}}^{\vee}_{-1}$ are adjoints between categories of commutative Coxeter algebras and signed commutative Coxeter monoids, while $(\overline{\mathcal{Z}}^{\vee}_{-ab})^{-}$ and $\overline{\mathcal{K}}^{\vee}_{-1}$ are adjoints between categories of signed commutative Coxeter algebras and commutative Coxeter monoids.

16.8. Adjoints of full Fock functors

We now turn to adjoints of full Fock functors as shown in Summary 16.3. Both $\mathcal{K}$ and $\mathcal{K}^{\vee}$ have left adjoints between categories of Coxeter monoids and Coxeter algebras, and right adjoints between categories of Coxeter comonoids and Coxeter coalgebras. These considerations extend to q-full Fock functors. We proceed as follows. Recall from Section 14.5 that full Fock functors can be constructed from bosonic Fock functors via the Coxeter bimonoid of chambers Γ and its dual Γ^*. Thus adjoints of full Fock functors can be deduced from those of bosonic Fock functors (constructed in preceding sections).

16.8.1. Adjoints of full Fock functors. Recall the functor $\mathcal{L}$ from Section 16.2.

Proposition 16.81. *The functor $\mathcal{L}$ is left adjoint to $\mathcal{K} = \mathcal{K}^\vee$ between categories of Coxeter species and Coxeter spaces. That is, for a Coxeter space* V *and Coxeter species* p, *there is a natural bijection*

$$\mathsf{CSp}(\mathcal{L}(\mathrm{V}), \mathsf{p}) \xrightarrow{\cong} \overline{\mathsf{C}}\mathsf{Sp}(\mathrm{V}, \mathcal{K}(\mathsf{p})).$$

PROOF. Recall from Proposition 14.22 for $q = 1$ that $\mathcal{K}^\vee \cong \overline{\mathcal{K}}^\vee(\mathsf{\Gamma}^* \times (-))$ and from (16.13) that $\mathcal{L} = \mathsf{\Gamma} \times \mathsf{t}_{(-)}$. The result now follows by composing adjunctions

$$\overline{\mathsf{C}}\mathsf{Sp} \underset{\overline{\mathcal{K}}^\vee}{\overset{\mathsf{t}_{(-)}}{\rightleftarrows}} \mathsf{CSp} \underset{\mathsf{\Gamma}^* \times (-)}{\overset{\mathsf{\Gamma} \times (-)}{\rightleftarrows}} \mathsf{CSp}.$$

The first adjunction is as in Proposition 16.12; the second is the specialization of (3.61) with $\mathsf{m} := \mathsf{\Gamma}$ noting that $\mathsf{\Gamma}^* \times (-)$ is the same as $\hom^\times(\mathsf{\Gamma}, -)$. □

Exercise 16.82. Compose units and counits of adjunctions, respectively, in the proof of Proposition 16.81 to check: The unit and counit of the adjunction $(\mathcal{L}, \mathcal{K})$ are

$$\begin{array}{ccc} \mathrm{V} \to \mathcal{K}(\mathsf{\Gamma} \times \mathsf{t}_\mathrm{V}) & \qquad & \mathsf{\Gamma} \times \mathsf{t}_{\mathcal{K}(\mathsf{p})} \to \mathsf{p} \\ \mathrm{V}[T] \to \mathsf{\Gamma}[F_0] \otimes \mathrm{V}[T] & & \mathsf{\Gamma}[A] \otimes \mathsf{p}[A_0] \to \mathsf{p}[A] \\ x \mapsto \mathtt{H}_{C_r/F_0} \otimes x & & \mathtt{H}_{C/A} \otimes x \mapsto w \cdot x. \end{array}$$

For the unit, V is a Coxeter space, T is a face-type, F_0 is the face of C_r of type T. For the counit, p is a Coxeter species, A is a face, A_0 is the face of C_r of the same type as A, w is the Coxeter symmetry defined by $w(C_r) = C$. (Use the isomorphism of $\mathcal{K}(\mathsf{p}) = \mathcal{K}^\vee(\mathsf{p})$ with $\overline{\mathcal{K}}^\vee(\mathsf{\Gamma}^* \times \mathsf{p})$ in Exercise 14.23, along with the canonical maps $\mathsf{\Gamma} \times \mathsf{\Gamma}^* \to \mathsf{E}$ and $\mathsf{E} \to \mathsf{\Gamma} \times \mathsf{\Gamma}^*$.)

Proposition 16.83. *The functor $\mathcal{L}^\vee$ is right adjoint to $\mathcal{K} = \mathcal{K}^\vee$ between categories of Coxeter species and Coxeter spaces. That is, for a Coxeter space* V *and Coxeter species* p, *there is a natural bijection*

$$\overline{\mathsf{C}}\mathsf{Sp}(\mathcal{K}^\vee(\mathsf{p}), \mathrm{V}) \xrightarrow{\cong} \mathsf{CSp}(\mathsf{p}, \mathcal{L}^\vee(\mathrm{V})).$$

PROOF. This is similar to the previous proof. Now we use $\mathcal{K} \cong \overline{\mathcal{K}}(\mathsf{\Gamma} \times (-))$ and $\mathcal{L}^\vee = \mathsf{\Gamma}^* \times \mathsf{t}_{(-)}$. We compose adjunctions

$$\mathsf{CSp} \underset{\mathsf{\Gamma}^* \times (-)}{\overset{\mathsf{\Gamma} \times (-)}{\rightleftarrows}} \mathsf{CSp} \underset{\mathsf{t}_{(-)}}{\overset{\overline{\mathcal{K}}}{\rightleftarrows}} \overline{\mathsf{C}}\mathsf{Sp},$$

with the second one as in Proposition 16.13. □

16.8.2. Coxeter (co)algebras and (co)monoids (Part 1).

Proposition 16.84. *The functor $\mathcal{L}$ is left adjoint to $\mathcal{K}$ between categories of Coxeter monoids and Coxeter algebras. That is, for a Coxeter algebra* A *and Coxeter monoid* a, *there is a natural bijection*

$$\mathsf{Mon}(\mathsf{CSp})(\mathcal{L}(\mathrm{A}), \mathsf{a}) \xrightarrow{\cong} \mathsf{Mon}(\overline{\mathsf{C}}\mathsf{Sp})(\mathrm{A}, \mathcal{K}(\mathsf{a})).$$

PROOF. We build on Proposition 16.81. Start with the lax structure φ of $\mathcal{K}$. So by (A.18), there is a unique colax structure on $\mathcal{L}$. In fact, it is invertible and its inverse coincides with the lax structure in (16.16), see Exercise 16.85 below. This then yields a lax-lax adjunction, and the result follows by applying Lemma A.2. □

Exercise 16.85. Use Exercise 16.82 to check: The composite

$$\mathcal{L}\overline{\mathcal{T}} \to \mathcal{L}\overline{\mathcal{T}}\mathcal{K}\mathcal{L} \to \mathcal{L}\mathcal{K}\mathcal{T}\mathcal{L} \to \mathcal{T}\mathcal{L},$$

evaluated on V, on the A-component, equals the inverse of (16.16). In contrast, the above composite with $\mathcal{K}^\vee$ instead of $\mathcal{K}$ (with the middle arrow using the lax structure of $\mathcal{K}^\vee$) yields (16.17).

We deduce that the diagrams

$$\begin{array}{ccc} \overline{\mathsf{CSp}} & \xrightarrow{\overline{\mathcal{T}}} & \mathsf{Mon}(\overline{\mathsf{CSp}}) \\ {\scriptstyle\mathcal{L}}\downarrow & & \downarrow{\scriptstyle\mathcal{L}} \\ \mathsf{CSp} & \xrightarrow[\mathcal{T}]{} & \mathsf{Mon}(\mathsf{CSp}) \end{array} \qquad \begin{array}{ccc} \overline{\mathsf{CSp}} & \xleftarrow{frg} & \mathsf{Mon}(\overline{\mathsf{CSp}}) \\ {\scriptstyle\mathcal{K}}\uparrow & & \uparrow{\scriptstyle\mathcal{K}} \\ \mathsf{CSp} & \xleftarrow[frg]{} & \mathsf{Mon}(\mathsf{CSp}) \end{array} \tag{16.48}$$

commute.

Dually:

Proposition 16.86. *The functor $\mathcal{L}^\vee$ is right adjoint to $\mathcal{K}^\vee$ between categories of Coxeter comonoids and Coxeter coalgebras. That is, for a Coxeter coalgebra* C *and Coxeter comonoid* c, *there is a natural bijection*

$$\mathsf{Comon}(\overline{\mathsf{CSp}})(\mathcal{K}^\vee(\mathsf{c}), \mathrm{C}) \xrightarrow{\cong} \mathsf{Comon}(\mathsf{CSp})(\mathsf{c}, \mathcal{L}^\vee(\mathrm{C})).$$

PROOF. This builds on Proposition 16.83. The argument is similar to the proof of Proposition 16.84. We start with the colax structure $\varphi^\vee$ of $\mathcal{K}^\vee$ and proceed. □

We deduce that the diagrams

$$\begin{array}{ccc} \overline{\mathsf{CSp}} & \xleftarrow{frg} & \mathsf{Comon}(\overline{\mathsf{CSp}}) \\ {\scriptstyle\mathcal{K}^\vee}\uparrow & & \uparrow{\scriptstyle\mathcal{K}^\vee} \\ \mathsf{CSp} & \xleftarrow[frg]{} & \mathsf{Comon}(\mathsf{CSp}) \end{array} \qquad \begin{array}{ccc} \overline{\mathsf{CSp}} & \xrightarrow{\overline{\mathcal{T}}^\vee} & \mathsf{Comon}(\overline{\mathsf{CSp}}) \\ {\scriptstyle\mathcal{L}^\vee}\downarrow & & \downarrow{\scriptstyle\mathcal{L}^\vee} \\ \mathsf{CSp} & \xrightarrow[\mathcal{T}^\vee]{} & \mathsf{Comon}(\mathsf{CSp}) \end{array} \tag{16.49}$$

commute.

16.8.3. Coxeter (co)algebras and (co)monoids (Part 2).

Recall from Section 16.4.2 the functor $\overline{\mathcal{Z}}^\vee$ from Coxeter algebras to Coxeter monoids. For a Coxeter algebra A, define

$$\mathcal{Z}^\vee(\mathrm{A}) := \Gamma \triangleright \overline{\mathcal{Z}}^\vee(\mathrm{A}), \tag{16.50}$$

with the operation $\triangleright$ as in Section 3.7.3 and Γ viewed as a Coxeter comonoid. This yields another functor $\mathcal{Z}^\vee$ from Coxeter algebras to Coxeter monoids.

Proposition 16.87. *The functor $\mathcal{Z}^\vee$ is left adjoint to $\mathcal{K}^\vee$ between categories of Coxeter monoids and Coxeter algebras.*

PROOF. This follows by composing adjunctions

$$\mathsf{Mon}(\overline{\mathsf{C}}\mathsf{Sp}) \underset{\overline{\mathcal{K}}^\vee}{\overset{\overline{\mathcal{Z}}^\vee}{\rightleftarrows}} \mathsf{Mon}(\mathsf{CSp}) \underset{\Gamma^* \times (-)}{\overset{\Gamma \triangleright (-)}{\rightleftarrows}} \mathsf{Mon}(\mathsf{CSp}).$$

The first adjunction is as in Proposition 16.35, while the second follows by combining (3.77) and (3.83) with $\mathsf{c} := \Gamma$.

Alternatively, we compose colax-lax adjunctions wrt monads $\mathcal{T}$ and $\overline{\mathcal{T}}$ in the proof of Proposition 16.81. The resulting lax structure on $\mathcal{K}^\vee$ is the usual one, while the colax structure on $\mathcal{L}$ is the one in (16.17) (also mentioned in the second part of Exercise 16.85). Now lift $\mathcal{L}$ using the adjoint lifting theorem (Theorem A.3) by coequalizing the maps

$$\mathcal{T}\mathcal{L}\overline{\mathcal{T}}(\mathrm{A}) \to \mathcal{T}\mathcal{T}\mathcal{L}(\mathrm{A}) \to \mathcal{T}\mathcal{L}(\mathrm{A}) \quad \text{and} \quad \mathcal{T}\mathcal{L}\overline{\mathcal{T}}(\mathrm{A}) \to \mathcal{T}\mathcal{L}(\mathrm{A})$$

to obtain $\mathcal{Z}^\vee$; see exercise below. This is a specialization of (A.20). □

Exercise 16.88. Show: For A a Coxeter algebra, $\mathcal{Z}^\vee(\mathrm{A})$ is the Coxeter quotient monoid of $\mathcal{T}\mathcal{L}(\mathrm{A}) = \mathcal{T}(\Gamma \times \mathsf{t}_{\mathrm{A}})$ whose A-component is the quotient of $\mathcal{T}(\Gamma \times \mathsf{t}_{\mathrm{A}})[A]$ by the linear span of the elements

$$\mathtt{H}_{D/K} \otimes \mu_T^U(x) - \sum \mathtt{H}_{GD/G} \otimes x \ \in \ (\Gamma[K] \otimes \mathrm{A}[T]) \oplus \bigoplus_{\substack{G:\, K \leq G \\ \mathrm{t}(G) = U}} (\Gamma[G] \otimes \mathrm{A}[U]), \tag{16.51}$$

where $A \leq K \leq D$, $x \in \mathrm{A}[U]$ with $U > \mathrm{t}(K)$, and $T = \mathrm{t}(K)$.

We deduce that the diagrams

$$\begin{array}{ccc} \overline{\mathsf{C}}\mathsf{Sp} & \xrightarrow{\overline{\mathcal{T}}} & \mathsf{Mon}(\overline{\mathsf{C}}\mathsf{Sp}) \\ {\scriptstyle \mathcal{L}}\downarrow & & \downarrow{\scriptstyle \mathcal{Z}^\vee} \\ \mathsf{CSp} & \xrightarrow[\mathcal{T}]{} & \mathsf{Mon}(\mathsf{CSp}) \end{array} \qquad \begin{array}{ccc} \overline{\mathsf{C}}\mathsf{Sp} & \xleftarrow{frg} & \mathsf{Mon}(\overline{\mathsf{C}}\mathsf{Sp}) \\ {\scriptstyle \mathcal{K}^\vee}\uparrow & & \uparrow{\scriptstyle \mathcal{K}^\vee} \\ \mathsf{CSp} & \xleftarrow[frg]{} & \mathsf{Mon}(\mathsf{CSp}) \end{array} \tag{16.52}$$

commute. Compare and contrast with diagrams (16.48).

Dually: Recall also from Section 16.4.2 the functor $\overline{\mathcal{Z}}$ from Coxeter coalgebras to Coxeter comonoids. For a Coxeter coalgebra C, define

$$\mathcal{Z}(\mathrm{C}) := \mathcal{C}(\Gamma, \overline{\mathcal{Z}}(\mathrm{C})), \tag{16.53}$$

with $\mathcal{C}(\mathsf{c}, \mathsf{d})$ the internal hom for Coxeter comonoids in Section 3.6. This yields another functor $\mathcal{Z}$ from Coxeter coalgebras to Coxeter comonoids.

Proposition 16.89. *The functor $\mathcal{Z}$ is right adjoint to $\mathcal{K}$ between categories of Coxeter comonoids and Coxeter coalgebras.*

PROOF. This follows by composing adjunctions

$$\mathsf{Comon}(\mathsf{CSp}) \underset{\mathcal{C}(\Gamma, -)}{\overset{\Gamma \times (-)}{\rightleftarrows}} \mathsf{Comon}(\mathsf{CSp}) \underset{\overline{\mathcal{Z}}}{\overset{\overline{\mathcal{K}}}{\rightleftarrows}} \mathsf{Comon}(\overline{\mathsf{C}}\mathsf{Sp}).$$

The second adjunction is as in Proposition 16.42, while the first is the specialization of (3.71) with $\mathsf{c} := \Gamma$.

Alternatively, we compose colax-lax adjunctions wrt comonads $\mathcal{T}^\vee$ and $\overline{\mathcal{T}}^\vee$ in the proof of Proposition 16.83. The resulting colax structure on $\mathcal{K}$ is the usual one. Now lift $\mathcal{L}^\vee$ using the comonad version of the adjoint lifting theorem by equalizing the maps

$$\mathcal{T}^\vee\mathcal{L}^\vee(\mathrm{C}) \to \mathcal{T}^\vee\mathcal{T}^\vee\mathcal{L}^\vee(\mathrm{C}) \to \mathcal{T}^\vee\mathcal{L}^\vee\overline{\mathcal{T}}^\vee(\mathrm{C}) \quad \text{and} \quad \mathcal{T}^\vee\mathcal{L}^\vee(\mathrm{C}) \to \mathcal{T}^\vee\mathcal{L}^\vee\overline{\mathcal{T}}^\vee(\mathrm{C})$$

to obtain $\mathcal{Z}$; see exercise below. □

Exercise 16.90. Show: For C a Coxeter coalgebra, $\mathcal{Z}(\mathrm{C})$ is the Coxeter subcomonoid of $\mathcal{T}^\vee\mathcal{L}^\vee(\mathrm{C}) = \mathcal{T}^\vee(\Gamma^* \times \mathrm{t_C})$ whose A-component consists of families of elements $x_{D/F} \in \mathrm{C}[\mathrm{t}(F)]$ indexed by $A \leq F \leq D$ such that for each $A \leq K \leq D$ and $U > \mathrm{t}(K)$,

$$\Delta_T^U(x_{D/K}) = \sum_{\substack{G:\, K \leq G \\ \mathrm{t}(G)=U}} x_{GD/G} \quad \in \mathrm{C}[U], \tag{16.54}$$

where $T = \mathrm{t}(K)$. Note: $x_{D/F}$ corresponds to $\mathtt{M}_{D/F} \otimes x_{D/F} \in (\Gamma^* \times \mathrm{t_C})[F]$.

We deduce that the diagrams

$$\begin{array}{ccc} \overline{\mathsf{C}}\mathsf{Sp} & \xleftarrow{\ \mathit{frg}\ } & \mathsf{Comon}(\overline{\mathsf{C}}\mathsf{Sp}) \\ {\scriptstyle\mathcal{K}}\uparrow & & \uparrow{\scriptstyle\mathcal{K}} \\ \mathsf{CSp} & \xleftarrow[\ \mathit{frg}\ ]{} & \mathsf{Comon}(\mathsf{CSp}) \end{array} \qquad \begin{array}{ccc} \overline{\mathsf{C}}\mathsf{Sp} & \xrightarrow{\ \overline{\mathcal{T}}^\vee\ } & \mathsf{Comon}(\overline{\mathsf{C}}\mathsf{Sp}) \\ {\scriptstyle\mathcal{L}^\vee}\downarrow & & \downarrow{\scriptstyle\mathcal{Z}} \\ \mathsf{CSp} & \xrightarrow[\ \mathcal{T}^\vee\ ]{} & \mathsf{Comon}(\mathsf{CSp}) \end{array} \tag{16.55}$$

commute. Compare and contrast with diagrams (16.49).

16.8.4. Deformed full Fock functors. Let us now take a look at q-full Fock functors $\mathcal{K}_q$ and $\mathcal{K}_q^\vee$. Propositions 16.81 and 16.83 work the same way as before since the scalar q plays no role at the level of Coxeter species and Coxeter spaces. The same remark applies to Propositions 16.84 and 16.86 since the product of $\mathcal{K}_q$ and coproduct of $\mathcal{K}_q^\vee$ do not depend on q.

Now we come to Propositions 16.87 and 16.89. Here we have an honest generalization. For any scalar q, define

$$\mathcal{Z}_q^\vee(\mathrm{A}) := \Gamma_q \rhd \overline{\mathcal{Z}}^\vee(\mathrm{A}) \quad \text{and} \quad \mathcal{Z}_q(\mathrm{C}) := \mathcal{C}(\Gamma_q, \overline{\mathcal{Z}}(\mathrm{C})). \tag{16.56}$$

In other words, replace Γ by Γ_q in definitions (16.50) and (16.53). This has the effect of putting in the factor $q^{\mathrm{dist}(D,GD)}$ inside the sum in relation (16.51) and in condition (16.54). By the same arguments, we see:

Proposition 16.91. *For a scalar q, the functor $\mathcal{Z}_q^\vee$ is left adjoint to $\mathcal{K}_q^\vee$ between categories of Coxeter monoids and Coxeter algebras, while $\mathcal{Z}_q$ is right adjoint to $\mathcal{K}_q$ between categories of Coxeter comonoids and Coxeter coalgebras.*

16.8.5. Commutative aspects. Define the functor $\mathcal{Z}_{ab}^\vee$ to be the composite

$$\mathsf{Mon}^{\mathsf{co}}(\overline{\mathsf{C}}\mathsf{Sp}) \to \mathsf{Mon}(\overline{\mathsf{C}}\mathsf{Sp}) \xrightarrow{\mathcal{Z}^\vee} \mathsf{Mon}(\mathsf{CSp}) \xrightarrow{(-)_{ab}} \mathsf{Mon}^{\mathsf{co}}(\mathsf{CSp})$$

and $\mathcal{Z}^{coab}$ to be the composite

$$^{\mathsf{co}}\mathsf{Comon}(\overline{\mathsf{C}}\mathsf{Sp}) \to \mathsf{Comon}(\overline{\mathsf{C}}\mathsf{Sp}) \xrightarrow{\mathcal{Z}} \mathsf{Comon}(\mathsf{CSp}) \xrightarrow{(-)^{coab}} {}^{\mathsf{co}}\mathsf{Comon}(\mathsf{CSp}).$$

Proposition 16.92. *The functor $\mathcal{Z}^{\vee}_{ab}$ is left adjoint to $\mathcal{K}^{\vee}$ between categories of commutative Coxeter monoids and commutative Coxeter algebras.*

The functor $\mathcal{Z}^{coab}$ is right adjoint to $\mathcal{K}$ between categories of cocommutative Coxeter comonoids and cocommutative Coxeter coalgebras.

PROOF. The argument is formally the same as for Propositions 16.68 and 16.73. □

The functors $\mathcal{Z}^{\vee}_{ab}$ and $\mathcal{Z}^{coab}$ have signed analogues, namely, $\mathcal{Z}^{\vee}_{-ab}$ and $\mathcal{Z}^{-coab}$ which are defined by using signed (co)abelianization. This yields the signed analogue of Proposition 16.92.

16.9. Coxeter operad algebras to Coxeter operad monoids

We briefly initiate how adjoints of Fock functors work in the generality of Coxeter operads, see Summary 16.4. For the Coxeter associative and commutative operads, we recover the setups in previous sections.

16.9.1. Trivialization functor. For a Coxeter operad $\mathbf{a}$, if V is a Coxeter $\mathbf{a}$-algebra, then $\mathrm{t_V}$ is a Coxeter $\mathbf{a}$-monoid. Explicitly, for $\mathrm{Z} \leq \mathrm{X}$, the product component of $\mathrm{t_V}$ is given by the composite map

$$\mathbf{a}[\mathrm{Z},\mathrm{X}] \otimes \mathrm{t_V}[\mathrm{X}] \xrightarrow{\mathrm{id}\otimes\beta_{F,\mathrm{X}}} \mathbf{a}[\mathrm{s}(A),\mathrm{s}(F)] \otimes \mathrm{t_V}[F]$$
$$\xrightarrow{w^{-1}\otimes\mathrm{id}} \mathbf{a}[\mathrm{s}(A_0),\mathrm{s}(F_0)] \otimes \mathrm{t_V}[F_0] \xrightarrow{\gamma} \mathrm{t_V}[A_0] = \mathrm{t_V}[A] \xrightarrow{\beta_{\mathrm{Z},A}} \mathrm{t_V}[\mathrm{Z}].$$

Here $A \leq F$ are any faces with $\mathrm{s}(F/A) = \mathrm{X}/\mathrm{Z}$, faces $A_0 \leq F_0 \leq C_r$ are defined by $\mathrm{t}(F_0/A_0) = \mathrm{t}(F/A)$, and w is any element of W such that $w(F_0/A_0) = F/A$. Here γ refers to the structure map of V.

Dually, for a Coxeter cooperad $\mathbf{c}$, if V is a Coxeter $\mathbf{c}$-coalgebra, then $\mathrm{t_V}$ is a Coxeter $\mathbf{c}$-comonoid. Explicitly, for $\mathrm{Z} \leq \mathrm{X}$, the coproduct component of $\mathrm{t_V}$ is given by the composite map

$$\mathrm{t_V}[\mathrm{Z}] \xrightarrow{\beta_{A,\mathrm{Z}}} \mathrm{t_V}[A] = \mathrm{t_V}[A_0] \to \mathbf{c}[\mathrm{s}(A_0),\mathrm{s}(F_0)] \otimes \mathrm{t_V}[F_0]$$
$$\xrightarrow{w\otimes\mathrm{id}} \mathbf{c}[\mathrm{s}(A),\mathrm{s}(F)] \otimes \mathrm{t_V}[F] \xrightarrow{\mathrm{id}\otimes\beta_{\mathrm{X},F}} \mathbf{c}[\mathrm{Z},\mathrm{X}] \otimes \mathrm{t_V}[\mathrm{X}]$$

with A_0, F_0, A, F, w defined as in the previous paragraph.

Observe: For Coxeter operads $\mathbf{a} = \mathbf{As}, \mathbf{Com}, \mathbf{Com}^-$ and Coxeter cooperads $\mathbf{c} = \mathbf{As}^*, \mathbf{Com}^*, (\mathbf{Com}^-)^*$, we recover the results in Section 16.1.2.

Exercise 16.93. Let $\mathbf{a} \to \mathbf{b}$ be a morphism of Coxeter operads, and V be a Coxeter $\mathbf{b}$-algebra. Check: The two ways in which $\mathrm{t_V}$ can be viewed as a Coxeter $\mathbf{a}$-monoid are equal. There is a dual statement for Coxeter cooperads. Similar results for Fock functors were given in Propositions 15.36 and 15.42.

16.9.2. Lax and colax structures. More formally, $\mathrm{t}_{(-)}$ carries a lax structure wrt monads $\mathcal{V}_{\mathbf{a}}$ and $\overline{\mathcal{V}}_{\mathbf{a}}$ and colax structure wrt comonads $\mathcal{U}_{\mathbf{c}}$ and $\overline{\mathcal{U}}_{\mathbf{c}}$ such that it sends Coxeter $\mathbf{a}$-algebras to Coxeter $\mathbf{a}$-monoids and Coxeter $\mathbf{c}$-coalgebras to Coxeter $\mathbf{c}$-comonoids in the manner described above.

For the lax structure $\mathcal{V}_{\mathbf{a}}\mathrm{t}_{(-)} \to \mathrm{t}_{(-)}\overline{\mathcal{V}}_{\mathbf{a}}$, the map

$$\mathcal{V}_{\mathbf{a}}\mathrm{t}_{\mathrm{V}}[A] = \bigoplus_{\mathrm{X}:\, \mathrm{X} \geq \mathrm{s}(A)} \mathbf{a}[\mathrm{s}(A), \mathrm{X}] \otimes \mathrm{t}_{\mathrm{V}}[\mathrm{X}]$$
$$\longrightarrow \mathrm{t}_{\overline{\mathcal{V}}_{\mathbf{a}}(\mathrm{V})}[A] = \Big(\bigoplus_{F_0:\, A_0 \leq F_0 \leq C_r} \mathbf{a}[\mathrm{s}(A_0), \mathrm{s}(F_0)] \otimes \mathrm{V}[F_0, C_r] \Big)_{\mathrm{quo}}$$

(with A_0 the face of C_r of the same type as A) is induced from

$$\mathbf{a}[\mathrm{s}(A), \mathrm{X}] \otimes \mathrm{t}_{\mathrm{V}}[\mathrm{X}] \xrightarrow{\mathrm{id} \otimes \beta_{F,\mathrm{X}}} \mathbf{a}[\mathrm{s}(A), \mathrm{X}] \otimes \mathrm{t}_{\mathrm{V}}[F] \xrightarrow{w^{-1} \otimes \mathrm{id}} \mathbf{a}[\mathrm{s}(A_0), \mathrm{s}(F_0)] \otimes \mathrm{V}[F_0, C_r].$$

Here F is any face greater than A of support X, faces $A_0 \leq F_0 \leq C_r$ are defined by $\mathrm{t}(F_0/A_0) = \mathrm{t}(F/A)$, and $w \in W$ is such that $w(F_0/A_0) = F/A$.

Dually, for the colax structure $\mathrm{t}_{(-)}\overline{\mathcal{U}}_{\mathbf{c}} \to \mathcal{U}_{\mathbf{c}}\mathrm{t}_{(-)}$, the map

$$\mathrm{t}_{\overline{\mathcal{U}}_{\mathbf{c}}(\mathrm{V})}[A] = \Big(\bigoplus_{F_0:\, A_0 \leq F_0 \leq C_r} \mathbf{c}[\mathrm{s}(A_0), \mathrm{s}(F_0)] \otimes \mathrm{V}[F_0, C_r] \Big)^{\mathrm{sub}}$$
$$\longrightarrow \mathcal{U}_{\mathbf{c}}\mathrm{t}_{\mathrm{V}}[A] = \bigoplus_{\mathrm{X}:\, \mathrm{X} \geq \mathrm{s}(A)} \mathbf{c}[\mathrm{s}(A), \mathrm{X}] \otimes \mathrm{t}_{\mathrm{V}}[\mathrm{X}]$$

into the X-summand is induced from

$$\mathbf{c}[\mathrm{s}(A_0), \mathrm{s}(F_0)] \otimes \mathrm{V}[F_0, C_r] \xrightarrow{w \otimes \mathrm{id}} \mathbf{c}[\mathrm{s}(A), \mathrm{X}] \otimes \mathrm{t}_{\mathrm{V}}[F] \xrightarrow{\mathrm{id} \otimes \beta_{\mathrm{X},F}} \mathbf{c}[\mathrm{s}(A), \mathrm{X}] \otimes \mathrm{t}_{\mathrm{V}}[\mathrm{X}],$$

with notations as in the previous paragraph.

16.9.3. Coxeter (co)operad (co)algebras and (co)monoids (Part 1).

Proposition 16.94. *For a Coxeter operad* $\mathbf{a}$, *the functor* $\mathrm{t}_{(-)}$ *is right adjoint to* $\overline{\mathcal{K}}$ *between categories of Coxeter* $\mathbf{a}$*-monoids and Coxeter* $\mathbf{a}$*-algebras.*

For a Coxeter cooperad $\mathbf{c}$, *the functor* $\mathrm{t}_{(-)}$ *is left adjoint to* $\overline{\mathcal{K}}^{\vee}$ *between categories of Coxeter* $\mathbf{c}$*-comonoids and Coxeter* $\mathbf{c}$*-coalgebras.*

PROOF. We explain the first statement. The formal way to proceed is as follows. The lax structure $\overline{\varphi}$ of $\overline{\mathcal{K}}$ in Theorem 14.54 is an isomorphism, so its inverse is a colax structure. By (A.17), there is an induced lax structure on $\mathrm{t}_{(-)}$ given by the composite

$$\mathcal{V}_{\mathbf{a}}\mathrm{t}_{\mathrm{V}} \to \mathrm{t}_{\overline{\mathcal{K}}\mathcal{V}_{\mathbf{a}}\mathrm{t}_{\mathrm{V}}} \to \mathrm{t}_{\overline{\mathcal{V}}_{\mathbf{a}}\overline{\mathcal{K}}\mathrm{t}_{\mathrm{V}}} \to \mathrm{t}_{\overline{\mathcal{V}}_{\mathbf{a}}(\mathrm{V})}.$$

One checks that this coincides with the lax structure defined in Section 16.9.2. Now apply Lemma A.2. □

Proposition 16.95. *For a Coxeter operad* $\mathbf{a}$, *for a Coxeter* $\mathbf{a}$*-algebra* V, *the counit of the adjunction* $\overline{\mathcal{K}}(\mathrm{t}_{\mathrm{V}}) \to \mathrm{V}$ *is an isomorphism of Coxeter* $\mathbf{a}$*-algebras.*

For a Coxeter cooperad $\mathbf{c}$, *for a Coxeter* $\mathbf{c}$*-coalgebra* V, *the unit of the adjunction* $\mathrm{V} \to \overline{\mathcal{K}}^{\vee}(\mathrm{t}_{\mathrm{V}})$ *is an isomorphism of Coxeter* $\mathbf{c}$*-coalgebras.*

PROOF. This follows formally from Propositions 16.14 and 16.15 using the fact that a morphism of Coxeter $\mathbf{a}$-algebras or of Coxeter $\mathbf{c}$-coalgebras is an isomorphism if it is an isomorphism of the underlying Coxeter spaces. □

The first statement in Proposition 16.94 specializes to Proposition 16.29 when $\mathbf{a} = \mathbf{As}, \mathbf{Com}, \mathbf{Com}^-$, while the second statement specializes to Proposition 16.20 when $\mathbf{c} = \mathbf{As}^*, \mathbf{Com}^*, (\mathbf{Com}^-)^*$. In a similar manner, Proposition 16.95 specializes to Propositions 16.27 and 16.34. Interestingly, the proofs of the above two results are formally the same as in the special cases.

Exercise 16.96. Deduce: For a Coxeter operad $\mathbf{a}$ and Coxeter cooperad $\mathbf{c}$, the diagrams

$$\begin{array}{ccc} \overline{\mathsf{C}}\mathsf{Sp} & \xrightarrow{\overline{\mathcal{V}}_{\mathbf{a}}} & \mathbf{a}\text{-}\mathsf{Mon}(\overline{\mathsf{C}}\mathsf{Sp}) \\ {\scriptstyle \overline{\mathcal{K}}}\uparrow & & \uparrow{\scriptstyle \overline{\mathcal{K}}} \\ \mathsf{CSp} & \xrightarrow[\mathcal{V}_{\mathbf{a}}]{} & \mathbf{a}\text{-}\mathsf{Mon}(\mathsf{CSp}) \end{array} \qquad \begin{array}{ccc} \overline{\mathsf{C}}\mathsf{Sp} & \xleftarrow{\mathit{frg}} & \mathbf{a}\text{-}\mathsf{Mon}(\overline{\mathsf{C}}\mathsf{Sp}) \\ {\scriptstyle \mathsf{t}_{(-)}}\downarrow & & \downarrow{\scriptstyle \mathsf{t}_{(-)}} \\ \mathsf{CSp} & \xleftarrow[\mathit{frg}]{} & \mathbf{a}\text{-}\mathsf{Mon}(\mathsf{CSp}) \end{array}$$

$$\begin{array}{ccc} \mathbf{c}\text{-}\mathsf{Comon}(\overline{\mathsf{C}}\mathsf{Sp}) & \xrightarrow{\mathit{frg}} & \overline{\mathsf{C}}\mathsf{Sp} \\ {\scriptstyle \mathsf{t}_{(-)}}\downarrow & & \downarrow{\scriptstyle \mathsf{t}_{(-)}} \\ \mathbf{c}\text{-}\mathsf{Comon}(\mathsf{CSp}) & \xrightarrow[\mathit{frg}]{} & \mathsf{CSp} \end{array} \qquad \begin{array}{ccc} \mathbf{c}\text{-}\mathsf{Comon}(\overline{\mathsf{C}}\mathsf{Sp}) & \xleftarrow{\overline{\mathcal{U}}_{\mathbf{c}}} & \overline{\mathsf{C}}\mathsf{Sp} \\ {\scriptstyle \overline{\mathcal{K}}^{\vee}}\uparrow & & \uparrow{\scriptstyle \overline{\mathcal{K}}^{\vee}} \\ \mathbf{c}\text{-}\mathsf{Comon}(\mathsf{CSp}) & \xleftarrow[\mathcal{U}_{\mathbf{c}}]{} & \mathsf{CSp} \end{array}$$

commute. This generalizes Exercises 16.25 and 16.33.

16.9.4. Coxeter (co)operad (co)algebras and (co)monoids (Part 2). For a Coxeter operad $\mathbf{a}$, one can lift the trivialization functor $\mathsf{t}_{(-)}$ to $\overline{\mathcal{Z}}^{\vee}_{\mathbf{a}}$ to obtain the left adjoint of $\overline{\mathcal{K}}^{\vee}$ between categories of Coxeter $\mathbf{a}$-monoids and Coxeter $\mathbf{a}$-algebras. We omit the details. Dually, for a Coxeter cooperad $\mathbf{c}$, one can lift $\mathsf{t}_{(-)}$ to $\overline{\mathcal{Z}}_{\mathbf{c}}$ to obtain the right adjoint of $\overline{\mathcal{K}}$ between categories of Coxeter $\mathbf{c}$-comonoids and Coxeter $\mathbf{c}$-coalgebras. One can do the same for full Fock functors $\mathcal{K}^{\vee}$ and $\mathcal{K}$ by starting with $\mathcal{L}$ instead of $\mathsf{t}_{(-)}$ to obtain functors $\mathcal{Z}^{\vee}_{\mathbf{a}}$ and $\mathcal{Z}_{\mathbf{c}}$.

TABLE 16.1. Coxeter operad algebras to Coxeter operad monoids.

$\mathbf{a}$	$\mathbf{As}$	$\mathbf{Com}$	$\mathbf{Com}^-$	$\mathbf{c}$	$\mathbf{As}^*$	$\mathbf{Com}^*$	$(\mathbf{Com}^-)^*$
$\overline{\mathcal{Z}}^{\vee}_{\mathbf{a}}$	$\overline{\mathcal{Z}}^{\vee}$	$\overline{\mathcal{Z}}^{\vee}_{ab}$	$\overline{\mathcal{Z}}^{\vee}_{-ab}$	$\overline{\mathcal{Z}}_{\mathbf{c}}$	$\overline{\mathcal{Z}}$	$\overline{\mathcal{Z}}^{coab}$	$\overline{\mathcal{Z}}^{-coab}$
$\mathcal{Z}^{\vee}_{\mathbf{a}}$	$\mathcal{Z}^{\vee}$	$\mathcal{Z}^{\vee}_{ab}$	$\mathcal{Z}^{\vee}_{-ab}$	$\mathcal{Z}_{\mathbf{c}}$	$\mathcal{Z}$	$\mathcal{Z}^{coab}$	$\mathcal{Z}^{-coab}$

Special cases discussed in preceding sections are shown in Table 16.1. We will look at $\mathbf{a} = \mathbf{Lie}$ and $\mathbf{c} = \mathbf{Lie}^*$ in Section 16.10.

Exercise 16.97. Deduce: For a Coxeter operad $\mathbf{a}$ and Coxeter cooperad $\mathbf{c}$, the diagrams

$$\begin{array}{ccc} \overline{\mathsf{C}}\mathsf{Sp} & \xrightarrow{\overline{\mathcal{V}}_{\mathbf{a}}} & \mathbf{a}\text{-}\mathsf{Mon}(\overline{\mathsf{C}}\mathsf{Sp}) \\ {\scriptstyle \mathsf{t}_{(-)}}\downarrow & & \downarrow{\scriptstyle \overline{\mathcal{Z}}^{\vee}_{\mathbf{a}}} \\ \mathsf{CSp} & \xrightarrow[\mathcal{V}_{\mathbf{a}}]{} & \mathbf{a}\text{-}\mathsf{Mon}(\mathsf{CSp}) \end{array} \qquad \begin{array}{ccc} \overline{\mathsf{C}}\mathsf{Sp} & \xleftarrow{\mathit{frg}} & \mathbf{a}\text{-}\mathsf{Mon}(\overline{\mathsf{C}}\mathsf{Sp}) \\ {\scriptstyle \overline{\mathcal{K}}^{\vee}}\uparrow & & \uparrow{\scriptstyle \overline{\mathcal{K}}^{\vee}} \\ \mathsf{CSp} & \xleftarrow[\mathit{frg}]{} & \mathbf{a}\text{-}\mathsf{Mon}(\mathsf{CSp}) \end{array}$$

$$\begin{array}{ccc} \overline{\mathsf{C}}\mathsf{Sp} & \xleftarrow{frg} & \mathsf{c\text{-}Comon}(\overline{\mathsf{C}}\mathsf{Sp}) \\ {\scriptstyle \overline{\mathcal{K}}}\uparrow & & \uparrow{\scriptstyle \overline{\mathcal{K}}} \\ \mathsf{CSp} & \xleftarrow[frg]{} & \mathsf{c\text{-}Comon}(\mathsf{CSp}) \end{array} \qquad \begin{array}{ccc} \overline{\mathsf{C}}\mathsf{Sp} & \xrightarrow{\overline{\mathcal{U}}_{\mathbf{c}}} & \mathsf{c\text{-}Comon}(\overline{\mathsf{C}}\mathsf{Sp}) \\ {\scriptstyle \mathsf{t}_{(-)}}\downarrow & & \downarrow{\scriptstyle \overline{\mathcal{Z}}_{\mathbf{c}}} \\ \mathsf{CSp} & \xrightarrow[\mathcal{U}_{\mathbf{c}}]{} & \mathsf{c\text{-}Comon}(\mathsf{CSp}) \end{array}$$

commute. This generalizes diagrams (16.26) and (16.29).

16.10. Coxeter Lie algebras to Coxeter Lie monoids

We now look at adjoints of Fock functors in the context of Lie theory, see Summary 16.5. These specialize the generalities in Section 16.9, see Summary 16.4. Alternatively, it is possible to partly avoid the abstraction of Coxeter operads and approach these adjunctions directly using Lie (co)brackets. We also explain how they interact with adjunctions between (co)commutative Coxeter bimonoids and (co)commutative Coxeter bialgebras via the universal (co)enveloping and primitive and indecomposable part functors.

16.10.1. Trivialization functor. For G a Coxeter Lie algebra, t_{G} is a Coxeter Lie monoid and for K a Coxeter Lie coalgebra, t_{K} is a Coxeter Lie comonoid, with Lie bracket and Lie cobracket

$$(16.57)\qquad \begin{array}{ccc} \mathsf{t}_{\mathrm{G}}[F] & \xrightarrow{\nu_A^F} & \mathsf{t}_{\mathrm{G}}[A] \\ \| & & \| \\ \mathrm{G}[\mathsf{t}(F)] & \xrightarrow[\nu_{\mathsf{t}(A)}^{\mathsf{t}(F)}]{} & \mathrm{G}[\mathsf{t}(A)] \end{array} \qquad \begin{array}{ccc} \mathsf{t}_{\mathrm{K}}[A] & \xrightarrow{\theta_A^F} & \mathsf{t}_{\mathrm{K}}[F] \\ \| & & \| \\ \mathrm{K}[\mathsf{t}(A)] & \xrightarrow[\theta_{\mathsf{t}(A)}^{\mathsf{t}(F)}]{} & \mathrm{K}[\mathsf{t}(F)] \end{array}$$

for $A \lessdot F$.

This can be checked directly. Alternatively, it is the special case $\mathbf{a} = \mathbf{Lie}$ and $\mathbf{c} = \mathbf{Lie}^*$ in Section 16.9.1.

Exercise 16.98. For A a Coxeter algebra, t_{A} can be viewed as a Coxeter Lie monoid in two different ways using commutator brackets (6.4) and (12.10), respectively. Check: The two structures coincide. There is a dual statement for a Coxeter coalgebra C. These can also be seen as special cases of Exercise 16.93.

16.10.2. Coxeter Lie (co)algebras and (co)monoids (Part 1). The adjunctions in Section 16.3 lift as follows.

Proposition 16.99. *The functor $\mathsf{t}_{(-)}$ is right adjoint to $\overline{\mathcal{K}}$ between categories of Coxeter Lie monoids and Coxeter Lie algebras.*

The functor $\mathsf{t}_{(-)}$ is left adjoint to $\overline{\mathcal{K}}^{\vee}$ between categories of Coxeter Lie comonoids and Coxeter Lie coalgebras.

PROOF. This can be checked directly using the explicit description of these functors in terms of Lie (co)brackets: For $\overline{\mathcal{K}}$ in (15.33), for $\overline{\mathcal{K}}^{\vee}$ in (15.35) and for $\mathsf{t}_{(-)}$ in (16.57). Alternatively, it is the specialization of Proposition 16.94 for $\mathbf{a} = \mathbf{Lie}$ and $\mathbf{c} = \mathbf{Lie}^*$. □

Proposition 16.100. *For a Coxeter Lie algebra G, the counit of the adjunction $\overline{\mathcal{K}}(\mathsf{t}_{\mathrm{G}}) \to \mathrm{G}$ is an isomorphism of Coxeter Lie algebras.*

For a Coxeter Lie coalgebra K, the unit of the adjunction $\mathrm{K} \to \overline{\mathcal{K}}^{\vee}(\mathsf{t}_{\mathrm{K}})$ is an isomorphism of Coxeter Lie coalgebras.

PROOF. This follows formally from Propositions 16.14 and 16.15. Alternatively, it is the specialization of Proposition 16.95 for $\mathbf{a} = \mathbf{Lie}$ and $\mathbf{c} = \mathbf{Lie}^*$. □

As a formal consequence, we have diagrams as in Exercise 16.96 with $\mathbf{a} = \mathbf{Lie}$ and $\mathbf{c} = \mathbf{Lie}^*$.

Exercise 16.101. Adjunctions between Coxeter Lie (co)monoids and Coxeter Lie (co)algebras interact with adjunctions between (co)commutative Coxeter bimonoids and (co)commutative Coxeter bialgebras. We elaborate below.

The diagrams

$$\begin{array}{ccc} \mathsf{LieMon}(\overline{\mathsf{CSp}}) & \xrightarrow{\overline{\mathcal{U}}} & {}^{\mathrm{co}}\mathsf{Bimon}(\overline{\mathsf{CSp}}) \\ {\scriptstyle \overline{\mathcal{K}}}\uparrow & & \uparrow{\scriptstyle \overline{\mathcal{K}}} \\ \mathsf{LieMon}(\mathsf{CSp}) & \xrightarrow[\mathcal{U}]{} & {}^{\mathrm{co}}\mathsf{Bimon}(\mathsf{CSp}) \end{array} \qquad \begin{array}{ccc} \mathsf{LieMon}(\overline{\mathsf{CSp}}) & \xleftarrow{\overline{\mathcal{P}}} & {}^{\mathrm{co}}\mathsf{Bimon}(\overline{\mathsf{CSp}}) \\ {\scriptstyle \mathrm{t}_{(-)}}\downarrow & & \downarrow{\scriptstyle \overline{\mathcal{Z}}^{coab}} \\ \mathsf{LieMon}(\mathsf{CSp}) & \xleftarrow[\mathcal{P}]{} & {}^{\mathrm{co}}\mathsf{Bimon}(\mathsf{CSp}) \end{array}$$

commute. The left one arises from the dotted isomorphism in the first diagram in (15.41), and the right one is then obtained by taking right adjoints.

Use CMM Theorem 7.26 to deduce:

$$\overline{\mathcal{K}} = \overline{\mathcal{U}}\mathcal{K}\mathcal{P} \quad \text{and} \quad \overline{\mathcal{Z}}^{coab} = \mathcal{U}\mathrm{t}_{(-)}\overline{\mathcal{P}}$$

as functors between categories of cocommutative Coxeter bimonoids and cocommutative Coxeter bialgebras. Use CMM Theorem 13.65 and Proposition 16.100 to recover the result of Theorem 16.74.

All of the above have dual statements involving functors $\mathcal{U}^\vee$, $\overline{\mathcal{U}}^\vee$, $\mathcal{Q}$, $\overline{\mathcal{Q}}$, $\overline{\mathcal{K}}^\vee$, $\overline{\mathcal{Z}}^\vee_{ab}$ which lead up to Theorem 16.71.

Exercise 16.102. Check directly the second diagram in Exercise 16.101 starting with the cocommutative Coxeter bialgebra $\overline{\mathcal{S}}^\vee(\mathrm{A})$ for A a Coxeter algebra. The end result is the Coxeter Lie monoid t_A. (Use Exercise 12.20, its Coxeter species analogue, diagram (16.40) and finally Exercise 16.98.)

Exercise 16.103. Deduce: The diagrams

$$\begin{array}{ccc} \overline{\mathsf{CSp}} & \xrightarrow{\overline{\mathcal{S}}} & {}^{\mathrm{co}}\mathsf{Bimon}^{\mathrm{co}}(\overline{\mathsf{CSp}}) \\ {\scriptstyle \overline{\mathcal{K}}}\uparrow & & \uparrow{\scriptstyle \overline{\mathcal{K}}} \\ \mathsf{CSp} & \xrightarrow[\mathcal{S}]{} & {}^{\mathrm{co}}\mathsf{Bimon}^{\mathrm{co}}(\mathsf{CSp}) \end{array} \qquad \begin{array}{ccc} \overline{\mathsf{CSp}} & \xleftarrow{\overline{\mathcal{P}}} & {}^{\mathrm{co}}\mathsf{Bimon}^{\mathrm{co}}(\overline{\mathsf{CSp}}) \\ {\scriptstyle \mathrm{t}_{(-)}}\downarrow & & \downarrow{\scriptstyle \overline{\mathcal{Z}}^{coab}} \\ \mathsf{CSp} & \xleftarrow[\mathcal{P}]{} & {}^{\mathrm{co}}\mathsf{Bimon}^{\mathrm{co}}(\mathsf{CSp}) \end{array}$$

commute. Use Leray–Samelson Theorem 7.4, to deduce:

$$\overline{\mathcal{K}} = \overline{\mathcal{S}}\mathcal{K}\mathcal{P} \quad \text{and} \quad \overline{\mathcal{Z}}^{coab} = \mathcal{S}\mathrm{t}_{(-)}\overline{\mathcal{P}}$$

as functors between categories of bicommutative Coxeter bimonoids and bicommutative Coxeter bialgebras.

There is also a dual pair of diagrams involving $\mathcal{S}^\vee$, $\overline{\mathcal{S}}^\vee$, $\mathcal{Q}$, $\overline{\mathcal{Q}}$.

16.10.3. Coxeter Lie (co)algebras and (co)monoids (Part 2). As a companion to the preceding discussion: One can lift the trivialization functor $\mathrm{t}_{(-)}$ (from Coxeter spaces to Coxeter species) to $\overline{\mathcal{Z}}^\vee_{\mathbf{Lie}}$ to obtain the left adjoint of $\overline{\mathcal{K}}^\vee$ between categories of Coxeter Lie monoids and Coxeter Lie algebras. Explicitly, $\overline{\mathcal{Z}}^\vee_{\mathbf{Lie}}(\mathrm{G})$ is the quotient of the free Coxeter Lie monoid $\mathbf{Lie} \circ \mathrm{t}_\mathrm{G}$ obtained by coequalizing the maps

$$\mathbf{Lie} \circ \mathrm{t}_{\mathbf{Lie}\circ\mathrm{G}} \to \mathbf{Lie} \circ \mathbf{Lie} \circ \mathrm{t}_\mathrm{G} \to \mathbf{Lie} \circ \mathrm{t}_\mathrm{G} \quad \text{and} \quad \mathbf{Lie} \circ \mathrm{t}_{\mathbf{Lie}\circ\mathrm{G}} \to \mathbf{Lie} \circ \mathrm{t}_\mathrm{G}.$$

The first composite uses the structure map of the Coxeter Lie operad and the colax structure of $\mathsf{t}_{(-)}$ (obtained as the mate of the lax structure of $\overline{\mathcal{K}}^\vee$ wrt $\mathbf{Lie}$). The second map uses the structure map of G.

Dually, one can lift $\mathsf{t}_{(-)}$ to $\overline{\mathcal{Z}}_{\mathbf{Lie}^*}$ to obtain the right adjoint of $\overline{\mathcal{K}}$ between categories of Coxeter Lie comonoids and Coxeter Lie coalgebras.

As a formal consequence, we have diagrams as in Exercise 16.97 with $\mathbf{a} = \mathbf{Lie}$ and $\mathbf{c} = \mathbf{Lie}^*$. We now pursue ideas similar to those in Exercise 16.101.

Proposition 16.104. *The diagrams*

$$
\begin{array}{ccc}
\mathsf{LieMon}(\overline{\mathsf{C}}\mathsf{Sp}) & \xrightarrow{\overline{\mathcal{U}}} & \mathsf{Bimon}(\overline{\mathsf{C}}\mathsf{Sp}) \\
{\scriptstyle \overline{\mathcal{Z}}^\vee_{\mathbf{Lie}}}\downarrow & & \downarrow{\scriptstyle \overline{\mathcal{Z}}^\vee} \\
\mathsf{LieMon}(\mathsf{CSp}) & \xrightarrow[\mathcal{U}]{} & \mathsf{Bimon}(\mathsf{CSp})
\end{array}
\qquad
\begin{array}{ccc}
\mathsf{LieMon}(\overline{\mathsf{C}}\mathsf{Sp}) & \xleftarrow{\overline{\mathcal{P}}} & \mathsf{Bimon}(\overline{\mathsf{C}}\mathsf{Sp}) \\
{\scriptstyle \overline{\mathcal{K}}^\vee}\uparrow & & \uparrow{\scriptstyle \overline{\mathcal{K}}^\vee} \\
\mathsf{LieMon}(\mathsf{CSp}) & \xleftarrow[\mathcal{P}]{} & \mathsf{Bimon}(\mathsf{CSp})
\end{array}
$$

$$
\begin{array}{ccc}
\mathsf{LieComon}(\overline{\mathsf{C}}\mathsf{Sp}) & \xleftarrow{\overline{\mathcal{Q}}} & \mathsf{Bimon}(\overline{\mathsf{C}}\mathsf{Sp}) \\
{\scriptstyle \overline{\mathcal{K}}}\uparrow & & \uparrow{\scriptstyle \overline{\mathcal{K}}} \\
\mathsf{LieComon}(\mathsf{CSp}) & \xleftarrow[\mathcal{Q}]{} & \mathsf{Bimon}(\mathsf{CSp})
\end{array}
\qquad
\begin{array}{ccc}
\mathsf{LieComon}(\overline{\mathsf{C}}\mathsf{Sp}) & \xrightarrow{\overline{\mathcal{U}}^\vee} & \mathsf{Bimon}(\overline{\mathsf{C}}\mathsf{Sp}) \\
{\scriptstyle \overline{\mathcal{Z}}_{\mathbf{Lie}^*}}\downarrow & & \downarrow{\scriptstyle \overline{\mathcal{Z}}} \\
\mathsf{LieComon}(\mathsf{CSp}) & \xrightarrow[\mathcal{U}^\vee]{} & \mathsf{Bimon}(\mathsf{CSp})
\end{array}
$$

commute.

In the first pair of diagrams, we can also work with cocommutative Coxeter bimonoids and bialgebras, while in the second pair of diagrams, we can work with commutative Coxeter bimonoids and bialgebras.

PROOF. The two diagrams involving $\overline{\mathcal{K}}$ and $\overline{\mathcal{K}}^\vee$ commute by the equalities in Propositions 15.46 and 15.48. The other two diagrams then follow by taking adjoints. ☐

Proposition 16.105. *As functors between categories of Coxeter Lie monoids and Coxeter Lie algebras,*

$$\overline{\mathcal{Z}}^\vee_{\mathbf{Lie}} = \mathcal{P}\overline{\mathcal{Z}}^\vee\overline{\mathcal{U}} \quad \textit{and} \quad \overline{\mathcal{K}}^\vee = \overline{\mathcal{P}}\overline{\mathcal{K}}^\vee\mathcal{U}, \tag{16.58}$$

and as functors between categories of Coxeter Lie comonoids and Coxeter Lie coalgebras,

$$\overline{\mathcal{Z}}_{\mathbf{Lie}^*} = \mathcal{Q}\overline{\mathcal{Z}}\overline{\mathcal{U}}^\vee \quad \textit{and} \quad \overline{\mathcal{K}} = \overline{\mathcal{Q}}\overline{\mathcal{K}}\mathcal{U}^\vee. \tag{16.59}$$

PROOF. For the first claim,

$$\mathcal{P}\overline{\mathcal{Z}}^\vee\overline{\mathcal{U}} = \mathcal{P}\mathcal{U}\overline{\mathcal{Z}}^\vee_{\mathbf{Lie}} = \overline{\mathcal{Z}}^\vee_{\mathbf{Lie}} \quad \text{and} \quad \overline{\mathcal{P}}\overline{\mathcal{K}}^\vee\mathcal{U} = \overline{\mathcal{K}}^\vee\mathcal{P}\mathcal{U} = \overline{\mathcal{K}}^\vee.$$

This used Proposition 16.104 and CMM Theorem 7.26. For the second claim, we use the dual CMM Theorem 7.27. ☐

Remark 16.106. That $\overline{\mathcal{Z}}^\vee_{\mathbf{Lie}}$ is left adjoint to $\overline{\mathcal{K}}^\vee$ can also be seen formally from the above result. For this, we view $\mathcal{P}$ as the left adjoint and $\mathcal{U}$ as the right adjoint between categories of Coxeter Lie monoids and cocommutative Coxeter Lie bimonoids. A similar remark applies to $\overline{\mathcal{Z}}_{\mathbf{Lie}^*}$ and $\overline{\mathcal{K}}$.

Proposition 16.107. *Let* $\mathrm{char}(\Bbbk) \nmid |W|$. *For a Coxeter Lie algebra* G, *the unit of the adjunction* $\mathrm{G} \to \overline{\mathcal{K}}^{\vee}\overline{\mathcal{Z}}^{\vee}_{\mathbf{Lie}}(\mathrm{G})$ *is an isomorphism of Coxeter Lie algebras.*

For a Coxeter Lie coalgebra K, *the counit of the adjunction* $\overline{\mathcal{K}}\overline{\mathcal{Z}}_{\mathbf{Lie}^*}(\mathrm{K}) \to \mathrm{K}$ *is an isomorphism of Coxeter Lie coalgebras.*

Proof. Building on the preceding remark: The adjunction $(\overline{\mathcal{Z}}^{\vee}_{\mathbf{Lie}}, \overline{\mathcal{K}}^{\vee})$ can be viewed as a composite of three adjunctions. The units of all of them are isomorphisms by CMM Theorem 7.26, Observation 16.56 and Theorem 16.51 for $q = 1$, and CMM Theorem 13.65. The latter two require the characteristic assumption. The first claim follows. The second claim is similar.

Alternatively:

$$\overline{\mathcal{K}}^{\vee}\overline{\mathcal{Z}}^{\vee}_{\mathbf{Lie}}(\mathrm{G}) = \overline{\mathcal{K}}^{\vee}\mathcal{P}\overline{\mathcal{Z}}^{\vee}\overline{\mathcal{U}}(\mathrm{G}) = \overline{\mathcal{P}}\mathcal{K}^{\vee}\overline{\mathcal{Z}}^{\vee}\overline{\mathcal{U}}(\mathrm{G}) = \overline{\mathcal{P}}\overline{\mathcal{U}}(\mathrm{G}) = \mathrm{G},$$
$$\overline{\mathcal{K}}\overline{\mathcal{Z}}_{\mathbf{Lie}^*}(\mathrm{K}) = \overline{\mathcal{K}}\mathcal{Q}\overline{\mathcal{Z}}\overline{\mathcal{U}}^{\vee}(\mathrm{K}) = \overline{\mathcal{Q}}\mathcal{K}\overline{\mathcal{Z}}\overline{\mathcal{U}}^{\vee}(\mathrm{K}) = \overline{\mathcal{Q}}\overline{\mathcal{U}}^{\vee}(\mathrm{K}) = \mathrm{K}.$$

In the first calculation, we used Proposition 15.46 and Theorem 16.51 for $q = 1$ and CMM. In the second, we used Proposition 15.48 and Theorem 16.58 for $q = 1$ and dual CMM. □

Exercise 16.108. Deduce: The diagrams

$$\begin{array}{ccc} \overline{\mathsf{C}}\mathsf{Sp} & \xrightarrow{\overline{\mathcal{S}}} & {}^{\mathrm{co}}\mathsf{Bimon}^{\mathrm{co}}(\overline{\mathsf{C}}\mathsf{Sp}) \\ {\scriptstyle \mathsf{t}_{(-)}}\downarrow & & \downarrow{\scriptstyle \overline{\mathcal{Z}}^{\vee}_{ab}} \\ \mathsf{CSp} & \xrightarrow[\mathcal{S}]{} & {}^{\mathrm{co}}\mathsf{Bimon}^{\mathrm{co}}(\mathsf{CSp}) \end{array} \qquad \begin{array}{ccc} \overline{\mathsf{C}}\mathsf{Sp} & \xleftarrow{\overline{\mathcal{P}}} & {}^{\mathrm{co}}\mathsf{Bimon}^{\mathrm{co}}(\overline{\mathsf{C}}\mathsf{Sp}) \\ {\scriptstyle \overline{\mathcal{K}}^{\vee}}\uparrow & & \uparrow{\scriptstyle \overline{\mathcal{K}}^{\vee}} \\ \mathsf{CSp} & \xleftarrow[\mathcal{P}]{} & {}^{\mathrm{co}}\mathsf{Bimon}^{\mathrm{co}}(\mathsf{CSp}) \end{array}$$

commute. (See diagram (16.38) in this regard.) Use Leray–Samelson Theorem 13.14 to deduce: If $\mathrm{char}(\Bbbk) \nmid |W|$, then

$$\overline{\mathcal{Z}}^{\vee}_{ab} = \mathcal{S}\mathsf{t}_{(-)}\overline{\mathcal{P}} \quad \text{and} \quad \overline{\mathcal{K}}^{\vee} = \overline{\mathcal{S}}\mathcal{K}^{\vee}\mathcal{P}$$

as functors between categories of bicommutative Coxeter bimonoids and bicommutative Coxeter bialgebras.

There is also a dual pair of diagrams involving $\mathcal{S}^{\vee}$, $\overline{\mathcal{S}}^{\vee}$, $\mathcal{Q}$, $\overline{\mathcal{Q}}$.

Remark 16.109. One can also lift the functor $\mathcal{L}$ instead of $\mathsf{t}_{(-)}$ to obtain functors $\mathcal{Z}^{\vee}_{\mathbf{Lie}}$ and $\mathcal{Z}_{\mathbf{Lie}^*}$, the former as left adjoint to $\mathcal{K}^{\vee}$ between categories of Coxeter Lie monoids and Coxeter Lie algebras, and the latter as right adjoint to $\mathcal{K}$ between categories of Coxeter Lie comonoids and Coxeter Lie coalgebras.

16.11. (Co)freeness under Fock functor adjoints

We build on the discussion in Section 15.4.

Proposition 16.110. *Let* $\mathrm{char}(\Bbbk) \nmid |W|$. *For* A *a Coxeter algebra,*

- A *is a free Coxeter algebra iff* $\overline{\mathcal{Z}}^{\vee}(\mathrm{A})$ *is a free Coxeter monoid,*
- A *is a free commutative Coxeter algebra iff* $\overline{\mathcal{Z}}^{\vee}_{\mathrm{ab}}(\mathrm{A})$ *is a free commutative Coxeter monoid.*

Proof. Let us prove the first statement. Suppose A is free, thus $\mathrm{A} = \overline{\mathcal{T}}(\mathrm{V})$ for some Coxeter space V. Applying $\overline{\mathcal{Z}}^{\vee}$ to both sides, and using the first diagram in (16.26), we obtain $\overline{\mathcal{Z}}^{\vee}(\mathrm{A}) = \mathcal{T}(\mathsf{t}_{\mathrm{V}})$ which shows that $\overline{\mathcal{Z}}^{\vee}(\mathrm{A})$ is free.

Conversely: Suppose $\overline{\mathcal{Z}}^{\vee}(\mathrm{A}) = \mathcal{T}(\mathsf{p})$ for some Coxeter species p. Applying $\overline{\mathcal{K}}^{\vee}$ to both sides, and using Propositions 15.24 and 16.40, we obtain $\mathrm{A} = \overline{\mathcal{T}}(\overline{\mathcal{K}}^{\vee}(\mathsf{p}))$. This used the hypothesis on field characteristic.

The second statement can be proved similarly using diagram (16.37) and Propositions 15.29 and 16.69. □

Dually:

Proposition 16.111. *Let* $\mathrm{char}(\Bbbk) \nmid |W|$. *For* C *a Coxeter coalgebra,*

- C *is a cofree Coxeter coalgebra iff* $\overline{\mathcal{Z}}(\mathrm{C})$ *is a cofree Coxeter comonoid,*
- C *is a cofree cocommutative Coxeter coalgebra iff* $\overline{\mathcal{Z}}^{coab}(\mathrm{C})$ *is a cofree cocommutative Coxeter comonoid.*

Contrast with the result in Exercise 15.26.

Exercise 16.112. Let $\mathrm{char}(\Bbbk) \nmid |W|$. Show: A Coxeter Lie algebra G is free iff $\overline{\mathcal{Z}}^{\vee}_{\mathbf{Lie}}(\mathrm{G})$ is a free Coxeter Lie monoid. Dually, a Coxeter Lie coalgebra K is cofree iff $\overline{\mathcal{Z}}_{\mathbf{Lie}^*}(\mathrm{K})$ is a cofree Coxeter Lie comonoid.

16.12. Notes

The material in this chapter is new. It generalizes and develops further the theory of adjoints of Fock functors from the category of Joyal species to the category of graded vector spaces developed in [**7**, Chapter 18]. Compare Summaries 16.1 and 16.3 with [**7**, Tables 18.1 and 18.2]. We point out that adjunctions between Joyal bimonoids and graded bialgebras were not considered in [**7**]; thus, adjunctions between Coxeter bimonoids and Coxeter bialgebras considered in Section 16.4.3 are new even classically. Lax-lax and colax-colax structures were used in [**7**] to lift adjunctions, but the technique of lifting colax-lax structures using the adjoint lifting theorem was not employed in [**7**]. Note very carefully: In this book, for a colax-lax adjunction, we need to specify whether it is between monads or comonads, while in [**7**] it is always between monoidal categories.

Warning. In [**7**], the letter $\mathcal{Q}$ was used instead of $\mathcal{Z}$ for lifts of the trivialization functor. We do not use $\mathcal{Q}$ here to avoid conflict with the indecomposable part functor for which we are using $\mathcal{Q}$.

16.12.1. Trivialization functor. The trivialization functor $\mathsf{t}_{(-)}$ between Coxeter spaces and Coxeter species in Section 16.1 generalizes the trivialization functor $\mathsf{t}_{(-)}$ between graded vector spaces and Joyal species in [**7**, Section 18.1.1]. The adjunction in Proposition 16.13 generalizes the one in [**7**, Proposition 18.3]. The lax and colax structures of $\mathsf{t}_{(-)}$ in (16.7) generalize those in [**7**, (18.1) and paragraph after Proposition 18.1]. Proposition 16.29 and the first part of Proposition 16.99 generalize [**7**, Proposition 18.5]. Proposition 16.30 on lax-lax and colax-lax adjunctions generalizes [**7**, Proposition 18.4]. Diagrams in Exercise 16.33 generalize those in [**7**, Proposition 18.9]; the first diagram in Exercise 16.59 for $q = 1$ generalizes the first diagram in [**7**, (18.10)].

The classical analogue of the signed trivialization functor briefly considered in Section 16.7 (in the context of fermionic Fock functors) was considered by Stover [**341**, last paragraph in Section 14.1].

16.12.2. Lifts of the trivialization functor. (Sections 16.4, 16.6, 16.10, 16.11.) The adjunction in Proposition 16.35 between $\overline{\mathcal{Z}}^\vee$ and $\overline{\mathcal{K}}^\vee$ generalizes the one in [**7**, Proposition 18.23]; relations (16.25) generalize [**7**, Relations (18.23)]. Similarly, Proposition 16.68 involving $\overline{\mathcal{Z}}^\vee_{ab}$ generalizes [**7**, Proposition 18.25]. Analogues of the conjugate functors $\overline{\mathcal{Z}}$ and $\overline{\mathcal{Z}}^{coab}$ were mentioned on [**7**, page 591]; however, conditions (16.28) and (16.39) were not explicitly stated. The analogue of Proposition 16.48 involving Coxeter bimonoids and Coxeter bialgebras was not considered in [**7**].

Analogues of $\overline{\mathcal{Z}}^\vee_{\mathbf{Lie}}$ and $\overline{\mathcal{Z}}_{\mathbf{Lie}^*}$ were not considered in [**7**]. Exercise 16.101 and Propositions 16.104 and 16.105 are also new in this sense.

Propositions 16.110 and 16.111 on (co)freeness under Fock functor adjoints as well as Lemma 16.78 look new even from a classical perspective.

16.12.3. Hadamard product with chambers. (Sections 16.2 and 16.8.) The adjunction in Proposition 16.81 between $\mathcal{L}$ and $\mathcal{K}$ generalizes [**7**, Proposition 18.17]. Unit and counit formulas in Exercise 16.82 generalize those in [**7**, Proof of Proposition 18.17]. Proposition 16.83 generalizes [**7**, Proposition 18.12]. The adjunction in the former is colax-lax wrt monads $\mathcal{T}$ and $\overline{\mathcal{T}}$. However, since the colax structure of $\mathcal{K}$ is not an isomorphism, it does not yield a lax-lax adjunction; compare with the first paragraph of [**7**, Remark 18.13]. The adjunction is also colax-lax wrt comonads $\mathcal{T}^\vee$ and $\overline{\mathcal{T}}^\vee$ as used in the proof of Proposition 16.89. This relates to the second paragraph of [**7**, Remark 18.13]. The lax structure of $\mathcal{L}$ in (16.16) generalizes the one in [**7**, (18.11)]; its inverse (which has the same form as (16.18)) generalizes [**7**, (18.12)]. Proposition 16.84 generalizes [**7**, Proposition 18.19]. Diagrams (16.48) generalize [**7**, Diagrams (18.16)].

The adjunction in Proposition 16.87 between $\mathcal{Z}^\vee$ and $\mathcal{K}^\vee$ generalizes the one in [**7**, Proposition 18.22]. The description of $\mathcal{Z}^\vee$ in Exercise 16.88 generalizes the one in [**7**, Section 18.5.2]. Similarly, the part of Proposition 16.92 involving $\mathcal{Z}^\vee_{ab}$ generalizes [**7**, Proposition 18.24]. Analogues of the conjugate functors $\mathcal{Z}$ and $\mathcal{Z}^{coab}$ were mentioned on [**7**, page 591] but not made explicit. Analogues of functors $\mathcal{Z}_q$ and $\mathcal{Z}^\vee_q$ and hence of Proposition 16.91 were not considered in [**7**].

16.12.4. Sections of Fock functors. Reflective and coreflective subcategories. Classical analogues of Propositions 16.14, 16.15, 16.27, 16.34, 16.100 involving sections of Fock functors are not mentioned in [**7**]. Similarly, analogues of Propositions 16.40, 16.45, 16.69, 16.107 and Theorems 16.51, 16.58, 16.71, 16.74 (which are less trivial) are also not in [**7**]. This is tied to the discussion of reflective and coreflective subcategories in Sections 16.3.3, 16.4.5, 16.6.4.

16.12.5. Bosonic Fock functors and abelianization. (Section 16.5.) Classical analogue of this material was not considered in [**7**].

CHAPTER 17

Structure theory under Fock functors

Fock functors connect the theory of Coxeter bimonoids to the theory of Coxeter bialgebras. We saw such connections in Chapter 15 between universal constructions of Coxeter bimonoids and Coxeter bialgebras. These arose naturally via lax and colax structures of Fock functors. We saw as a consequence that bosonic and fermionic Fock functors preserve (co)freeness. We also saw connections between primitive and indecomposable parts of Coxeter bimonoids and Coxeter bialgebras, and between universal enveloping functors of Coxeter Lie monoids and Coxeter Lie algebras.

Taking these ideas forward, we now see how Fock functors and their adjoints connect

- series of Coxeter species and series of Coxeter spaces,
- action of the invariant bilune-incidence algebra on Coxeter species maps and its action on Coxeter space maps,
- the convolution Coxeter monoid and convolution Coxeter algebra,
- the internal hom for Coxeter comonoids and internal hom for Coxeter coalgebras wrt Hadamard products,
- the universal measuring Coxeter comonoid and universal measuring Coxeter coalgebra,
- exp-log correspondences for Coxeter bimonoids and exp-log correspondences for Coxeter bialgebras,
- the norm map for Coxeter species and norm map for Coxeter spaces,
- the antipode of a Coxeter bimonoid and antipode of a Coxeter bialgebra,
- structure theorems for Coxeter bimonoids and structure theorems for Coxeter bialgebras.

For a comparison summary, see Tables VI and IX in the main introduction.

Another aspect of these connections is that one can prove results about Coxeter bialgebras using corresponding results about Coxeter bimonoids. This makes use of the key observation from Chapter 16 that adjoint functors $\mathsf{t}_{(-)}$ and $\overline{\mathcal{Z}}$ are sections of $\overline{\mathcal{K}}$, while $\mathsf{t}_{(-)}$ and $\overline{\mathcal{Z}}^{\vee}$ are sections of $\overline{\mathcal{K}}^{\vee}$ (enabling us to view Coxeter spaces and Coxeter bialgebras as (co)reflective subcategories of Coxeter species and Coxeter bimonoids). We illustrate this technique on several results including structure theorems of Coxeter bialgebras.

17.1. Series of Coxeter species and Coxeter spaces

Recall series of Coxeter species from Section 3.10 and series of Coxeter spaces from Section 9.10. We now connect the two situations using the bosonic

Fock functor (invariant version). As an application, we explain how exp-log correspondences between primitive and group-like series for Coxeter bialgebras can be derived from those for Coxeter bimonoids.

17.1.1. Series of a Coxeter species. Let p be a Coxeter species. Applying the bosonic Fock functor $\overline{\mathcal{K}}^\vee$ to it yields the Coxeter space $\overline{\mathcal{K}}^\vee(\mathsf{p})$. In terms of faces-types, it is defined by (14.33).

Lemma 17.1. *For a Coxeter species* p*, there is a canonical identification*

$$\mathscr{S}(\mathsf{p}) \xrightarrow{\cong} \mathscr{S}(\overline{\mathcal{K}}^\vee(\mathsf{p})), \qquad (v_F) \mapsto (v_T). \tag{17.1}$$

The lhs is the space of series of p*, while the rhs is the space of series of the Coxeter space* $\overline{\mathcal{K}}^\vee(\mathsf{p})$*.*

Let us elaborate on the notation. Start with a series $v = (v_F)$ of p. It is a family indexed by faces F. For a face-type U, define $v_U := v_G$, where G is the face of type U of the reference chamber C_r. By construction, v_U belongs to $\mathsf{p}[G]^{W_G}$. The family (v_T), indexed by face-types T, defines a series of $\overline{\mathcal{K}}^\vee(\mathsf{p})$. It makes sense to continue to denote it by v.

Exercise 17.2. Prove Lemma 17.1. That is, check that (v_T) is indeed a series, and (17.1) is a linear isomorphism.

17.1.2. Actions of the invariant lune-incidence algebra. Recall: The invariant lune-incidence algebra $\mathrm{I}_{\mathrm{lune}}[\mathcal{A}]^W$ acts on the space of series of a Coxeter monoid as well as on the space of series of a Coxeter algebra. The module structures are (3.122) and (9.122), respectively.

Proposition 17.3. *For a Coxeter monoid* a *and the Coxeter algebra* $\overline{\mathcal{K}}^\vee(\mathsf{a})$*, the identification*

$$\mathscr{S}(\mathsf{a}) \xrightarrow{\cong} \mathscr{S}(\overline{\mathcal{K}}^\vee(\mathsf{a})) \tag{17.2}$$

is an isomorphism of left modules over $\mathrm{I}_{\mathrm{lune}}[\mathcal{A}]^W$*.*

PROOF. Let $s \in \mathrm{I}_{\mathrm{lune}}[\mathcal{A}]^W$. Let v denote a series of a as well as the corresponding series of $\overline{\mathcal{K}}^\vee(\mathsf{a})$ under (17.1). For a face-type T, let F be the face of C_r of type T. We need to check that $(s \circ v)_F = (s \circ v)_T$. This is done below.

$$\begin{aligned}(s \circ v)_F &= \sum_{G:\, F \leq G} s(F,G)\, \mu_F^G(v_G)\\ &= \sum_{U:\, T \leq U} s(T,U) \sum_{G:\, F \leq G,\, \mathrm{t}(G)=U} \mu_F^G(v_G)\\ &= \sum_{U:\, T \leq U} s(T,U)\, \mu_T^U(v_U)\\ &= (s \circ v)_T.\end{aligned}$$

The first and last steps used definitions of the module structures. The third step used the relation between the products of a and $\overline{\mathcal{K}}^\vee(\mathsf{a})$. See the formula after (15.3) for $q = 1$. In the present case, $w \cdot v_U = v_G$, since v is a series. □

Exercise 17.4 (Fermionic Fock functor). Use (15.15) for $\overline{\mathcal{K}}^\vee$ and $\overline{\mathcal{K}}^\vee_{-1}$ to deduce: For a a Coxeter monoid, $\mathscr{S}(\mathsf{a}^-) \cong \mathscr{S}(\overline{\mathcal{K}}^\vee_{-1}(\mathsf{a}))$ as left modules over $\mathrm{I}_{\mathrm{lune}}[\mathcal{A}]^W$.

17.1.3. Linear and exponential series. Recall: For a Coxeter species and a Coxeter space, we can talk of linear series.

Observation 17.5. For a Coxeter species p, the identification (17.1) restricts to an isomorphism

$$\mathscr{X}(\mathsf{p}) \xrightarrow{\cong} \mathscr{X}(\overline{\mathcal{K}}^\vee(\mathsf{p})). \tag{17.3}$$

Recall: For a Coxeter monoid and a Coxeter algebra, we can talk of exponential series.

Proposition 17.6. *For a Coxeter monoid* a *and the Coxeter algebra* $\overline{\mathcal{K}}^\vee(\mathsf{a})$*, the identification* (17.2) *restricts to an isomorphism*

$$\mathscr{E}(\mathsf{a}) \xrightarrow{\cong} \mathscr{E}(\overline{\mathcal{K}}^\vee(\mathsf{a})). \tag{17.4}$$

PROOF. This can be seen directly. Alternatively, when $\operatorname{char}(\Bbbk) \nmid |W|$, (17.3) and (17.4) imply each other in view of Proposition 17.3 and bijections (3.134) and (9.135). □

Thus:

Proposition 17.7. *Let* $\operatorname{char}(\Bbbk) \nmid |W|$*. For a Coxeter monoid* a*, there are commutative diagrams*

$$\begin{array}{ccc} \mathscr{X}(\mathsf{a}) & \xrightarrow{\ \mathbf{exp}\ } & \mathscr{E}(\mathsf{a}) \\ {\scriptstyle\cong}\downarrow & & \downarrow{\scriptstyle\cong} \\ \mathscr{X}(\overline{\mathcal{K}}^\vee(\mathsf{a})) & \xrightarrow[\mathbf{exp}]{} & \mathscr{E}(\overline{\mathcal{K}}^\vee(\mathsf{a})) \end{array} \qquad \begin{array}{ccc} \mathscr{X}(\mathsf{a}) & \xleftarrow{\ \mathbf{log}\ } & \mathscr{E}(\mathsf{a}) \\ {\scriptstyle\cong}\downarrow & & \downarrow{\scriptstyle\cong} \\ \mathscr{X}(\overline{\mathcal{K}}^\vee(\mathsf{a})) & \xleftarrow[\mathbf{log}]{} & \mathscr{E}(\overline{\mathcal{K}}^\vee(\mathsf{a})). \end{array}$$

17.1.4. Primitive and group-like series. Recall: For a Coxeter comonoid and a Coxeter coalgebra, we can talk of primitive series and group-like series.

Proposition 17.8. *For a Coxeter comonoid* c *and the Coxeter coalgebra* $\overline{\mathcal{K}}^\vee(\mathsf{c})$*, the identification*

$$\mathscr{S}(\mathsf{c}) \xrightarrow{\cong} \mathscr{S}(\overline{\mathcal{K}}^\vee(\mathsf{c}))$$

restricts to isomorphisms

$$\mathscr{P}(\mathsf{c}) \xrightarrow{\cong} \mathscr{P}(\overline{\mathcal{K}}^\vee(\mathsf{c})) \quad \text{and} \quad \mathscr{G}(\mathsf{c}) \xrightarrow{\cong} \mathscr{G}(\overline{\mathcal{K}}^\vee(\mathsf{c})). \tag{17.5}$$

PROOF. We compare the coproducts of c and $\overline{\mathcal{K}}^\vee(\mathsf{c})$ using (15.7). First, let us consider the case of group-like series. We follow the notation of (17.1). If the series (v_F) is group-like, then so is (v_T) since

$$\Delta_T^U(v_T) = \Delta_{F_0}^{G_0}(v_{F_0}) = v_{G_0} = v_U,$$

where F_0 and G_0 are faces of C_r of types T and U. Conversely, suppose (v_T) is a group-like series. Consider any $F \leq G$. Say F has type T and G has type U. Let F_0 and G_0 be faces of C_r of types T and U. Further, choose a Coxeter symmetry w such that $wF_0 = F$ and $wG_0 = G$. Then

$$\begin{aligned} \Delta_F^G(v_F) = \Delta_F^G(w \cdot v_{F_0}) &= w \cdot \Delta_{F_0}^{G_0}(v_{F_0}) \\ &= w \cdot \Delta_T^U(v_T) = w \cdot v_U = w \cdot v_{G_0} = v_G. \end{aligned}$$

The second step used the type-naturality axiom (2.36a). This shows that (v_F) is a group-like series.

The case of primitive series is similar. Alternatively, one can argue formally as follows.

$$\mathscr{P}(\mathsf{c}) = \mathscr{S}(\mathcal{P}(\mathsf{c})) = \mathscr{S}(\overline{\mathcal{K}}^\vee(\mathcal{P}(\mathsf{c}))) = \mathscr{S}(\overline{\mathcal{P}}(\overline{\mathcal{K}}^\vee(\mathsf{c}))) = \mathscr{P}(\overline{\mathcal{K}}^\vee(\mathsf{c})).$$

The third step used Proposition 15.46. □

17.1.5. Exp-log correspondences. Recall from Theorems 3.56 and 9.76 the exp-log correspondences between primitive series and group-like series of a Coxeter bimonoid and a Coxeter bialgebra, respectively. The two situations can be compared as follows.

Theorem 17.9. *Let* $\mathrm{char}(\Bbbk) \nmid |W|$. *For a Coxeter bimonoid* h, *there are commutative diagrams*

$$\begin{array}{ccc} \mathscr{P}(\mathsf{h}) & \xrightarrow{\mathbf{exp}} & \mathscr{G}(\mathsf{h}) \\ \cong\downarrow & & \downarrow\cong \\ \mathscr{P}(\overline{\mathcal{K}}^\vee(\mathsf{h})) & \xrightarrow[\mathbf{exp}]{} & \mathscr{G}(\overline{\mathcal{K}}^\vee(\mathsf{h})) \end{array} \qquad \begin{array}{ccc} \mathscr{P}(\mathsf{h}) & \xleftarrow{\mathbf{log}} & \mathscr{G}(\mathsf{h}) \\ \cong\downarrow & & \downarrow\cong \\ \mathscr{P}(\overline{\mathcal{K}}^\vee(\mathsf{h})) & \xleftarrow[\mathbf{log}]{} & \mathscr{G}(\overline{\mathcal{K}}^\vee(\mathsf{h})). \end{array}$$

PROOF. This follows by combining Propositions 17.3 and 17.8. □

Shifting our viewpoint, Propositions 17.3 and 17.8 along with Theorem 3.56 can also be used to derive Theorem 9.76 as follows.

PROOF OF THEOREM 9.76 VIA FOCK FUNCTORS. The vertical maps in the above diagrams are isomorphisms. So existence of the top-horizontal maps implies existence of the bottom-horizontal maps. This yields the correspondence (9.126) for $\mathrm{H} = \overline{\mathcal{K}}^\vee(\mathsf{h})$. Further, by Theorem 16.51, every Coxeter bialgebra necessarily has this form. This completes the argument. □

Similarly, the q-exp-log correspondences in Theorems 3.59 and 9.80 can be compared. For a Coxeter q-bimonoid h for q not a root of unity, we have commutative diagrams as above with $\mathbf{exp}_q$ and $\mathbf{log}_q$.

Remark 17.10. Preceding results show that when looking at series, the Coxeter species world does not contain any more information than the Coxeter space world. However, the fact remains that Coxeter species (due to their additional structure) are easier to handle than Coxeter spaces. We will look at more general considerations in Sections 17.2 and 17.3.

17.2. Actions of the invariant bilune-incidence algebra

Recall from Sections 3.8 and 9.8 that the invariant bilune-incidence algebra $\mathrm{I}_{\mathrm{bilune}}[\mathcal{A}]^W$ acts on $\mathsf{CSp}(\mathsf{c},\mathsf{a})$ and on $\overline{\mathsf{C}}\mathsf{Sp}(\mathrm{C},\mathrm{A})$. We now relate the two actions in three different ways:

(1) via bosonic Fock functors,
(2) via adjunctions between the trivialization and bosonic Fock functors,
(3) via the trivialization functor by employing a number operator.

As an application of item (2), we deduce exp-log correspondences between (co)derivations and (co)algebra morphisms for Coxeter space maps from those for Coxeter species maps.

17.2.1. Bosonic Fock functors (First way). Recall from Section 14.4 the two bosonic Fock functors $\overline{\mathcal{K}}$ and $\overline{\mathcal{K}}^{\vee}$.

Lemma 17.11. *For a Coxeter comonoid* c *and Coxeter monoid* a,

$$\mathsf{CSp}(\mathsf{c},\mathsf{a}) \to \overline{\mathsf{C}}\mathsf{Sp}(\overline{\mathcal{K}}(\mathsf{c}),\overline{\mathcal{K}}(\mathsf{a})) \quad \textit{and} \quad \mathsf{CSp}(\mathsf{c},\mathsf{a}) \to \overline{\mathsf{C}}\mathsf{Sp}(\overline{\mathcal{K}}^{\vee}(\mathsf{c}),\overline{\mathcal{K}}^{\vee}(\mathsf{a})) \tag{17.6}$$

are maps of left modules over $\mathrm{I}_{\mathrm{bilune}}[\mathcal{A}]^W$.

That is: The maps (17.6) are $\Bbbk$-linear, and for $s \in \mathrm{I}_{\mathrm{bilune}}[\mathcal{A}]^W$ and $f : \mathsf{c} \to \mathsf{a}$ a map of Coxeter species,

$$s \circ \overline{\mathcal{K}}(f) = \overline{\mathcal{K}}(s \circ f) \quad \text{and} \quad s \circ \overline{\mathcal{K}}^{\vee}(f) = \overline{\mathcal{K}}^{\vee}(s \circ f), \tag{17.7}$$

with actions given by (3.97) and (9.98).

PROOF. Linearity is clear since the functors $\overline{\mathcal{K}}$ and $\overline{\mathcal{K}}^{\vee}$ are $\Bbbk$-linear. We now prove the first identity in (17.7). Let us first compute $s \circ \mathcal{K}(f)$ employing the coproduct of $\mathcal{K}(\mathsf{c})$ and product of $\mathcal{K}(\mathsf{a})$ given by formulas (15.2) with $q = 1$ and $C := C_r$. This will also show why identities (17.7) do not work for full Fock functors in general. A helpful illustration to understand the calculation is shown below.

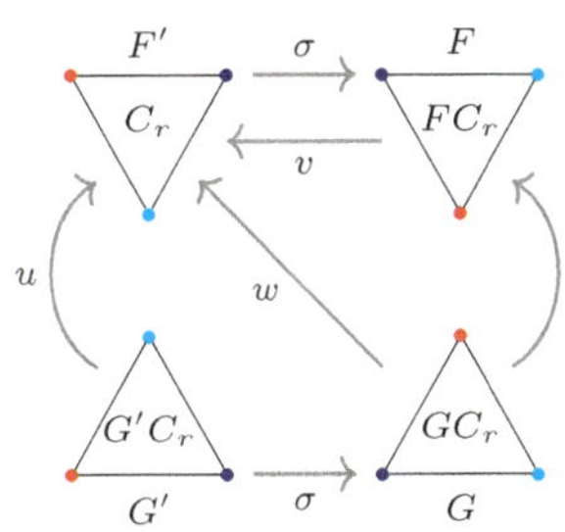

On the Z-component,

$$\begin{aligned}
(s \circ \mathcal{K}(f))_Z &= \sum_{\substack{\sigma: T' \to T \\ \text{local to } Z}} s(Z, T, \sigma, T')\, \mu_Z^T \mathcal{K}(f)_T \sigma \Delta_Z^{T'} \\
&= \sum_{\substack{\sigma: T' \to T \\ \text{local to } Z}} \sum_{\substack{G' : G' \geq A \\ \mathrm{t}(G') = T'}} s(Z, T, \sigma, T')\, \mu_Z^T \mathcal{K}(f)_T \sigma u \Delta_A^{G'} \\
&= \sum_{T, T' \geq Z} \sum_{\substack{F : F \geq A \\ \mathrm{t}(F) = T \\ \mathrm{s}(F) = \mathrm{s}(F')}} \sum_{\substack{G' : G' \geq A \\ \mathrm{t}(G') = T'}} s(A, F, F')\, \mu_A^{F_0} f_{F_0} v \beta_{F,F'} u \Delta_A^{G'} \\
&= \sum_{\substack{G, G' \geq A \\ \mathrm{s}(G) = \mathrm{s}(G')}} s(A, G, G')\, \mu_A^{wG} f_{wG} w \beta_{G,G'} \Delta_A^{G'}
\end{aligned}$$

$$= \sum_{\substack{G,G' \geq A \\ \mathrm{s}(G)=\mathrm{s}(G')}} s(A,G,G')\, w\mu_A^G f_G \beta_{G,G'} \Delta_A^{G'}.$$

Observe how the face-type notation gradually evolved into the face notation. Here A, F_0, F' are faces of C_r of types Z, T, T', respectively. Coxeter symmetries u, v, w are defined by $u(G'C_r) = C_r$, $v(FC_r) = C_r$, $w(GC_r) = C_r$. Moreover, $u(GC_r) = FC_r$, $w = vu$, $u(G) = F$, $u(G') = F'$, $v(F) = F_0$, $w(G) = F_0$. Also, $\sigma = \mathrm{t}(F,F') = \mathrm{t}(G,G')$. In the calculation, we gradually moved Coxeter symmetries from right to left. Note very carefully: The end result is not $\mathcal{K}(s \circ f)_Z$ because of the extra w at the left end. This problem disappears if we replace $\mathcal{K}$ by $\overline{\mathcal{K}}$; now we land in the space of coinvariants, so we can get rid of the w.

For the second identity, the calculation is similar. We use formulas (15.3). Now we move Coxeter symmetries from left to right; we can get rid of the w at the right end since we start from the space of invariants. □

Exercise 17.12. Deduce: For a Coxeter q-bimonoid h for q not a root of unity,

$$\mathbf{log}_q(\mathrm{id}_{\overline{\mathcal{K}}(\mathsf{h})}) = \overline{\mathcal{K}}(\mathbf{log}_q(\mathrm{id}_{\mathsf{h}})) \quad \text{and} \quad \mathbf{log}_q(\mathrm{id}_{\overline{\mathcal{K}}^{\vee}(\mathsf{h})}) = \overline{\mathcal{K}}^{\vee}(\mathbf{log}_q(\mathrm{id}_{\mathsf{h}})). \tag{17.8}$$

(Put $f = \mathrm{id}$ and $s = \boldsymbol{\mu}_q$ in (17.7).)

Exercise 17.13. Use the algebra morphism i in (1.105) to deduce: (17.6) are also maps of left modules over $\mathrm{I}_{\mathrm{lune}}[\mathcal{A}]^W$. Thus, identities (17.7) hold for $s \in \mathrm{I}_{\mathrm{lune}}[\mathcal{A}]^W$ with actions given by (3.85) and (9.87). Check these identities directly by simplifying the calculations in the proof of Lemma 17.11. A different way to derive this result is given in Exercise 17.23.

Deduce: For a Coxeter bimonoid h,

$$\mathbf{log}(\mathrm{id}_{\overline{\mathcal{K}}(\mathsf{h})}) = \overline{\mathcal{K}}(\mathbf{log}(\mathrm{id}_{\mathsf{h}})) \quad \text{and} \quad \mathbf{log}(\mathrm{id}_{\overline{\mathcal{K}}^{\vee}(\mathsf{h})}) = \overline{\mathcal{K}}^{\vee}(\mathbf{log}(\mathrm{id}_{\mathsf{h}})). \tag{17.9}$$

(Put $f = \mathrm{id}$ and $s = \boldsymbol{\mu}$.)

Exercise 17.14 (Fermionic Fock functors). For $s \in \mathrm{I}_{\mathrm{bilune}}[\mathcal{A}]^W$ and $f : \mathsf{c} \to \mathsf{a}$,

$$s_- \circ \overline{\mathcal{K}}_{-1}(f) = \overline{\mathcal{K}}_{-1}(s \circ f) \quad \text{and} \quad s_- \circ \overline{\mathcal{K}}^{\vee}_{-1}(f) = \overline{\mathcal{K}}^{\vee}_{-1}(s \circ f), \tag{17.10}$$

with $s_-(A,F,F') := (-1)^{\mathrm{dist}(F,F')} s(A,F,F')$. Deduce these identities from (17.7) and also check them directly. (For the first identity, start the calculation with the signed full Fock functor $\mathcal{K}_{-1}$ and take signed coinvariants at the very end.)

17.2.2. Adjunctions and exp-log correspondences (Second way).

Recall from Section 16.1 the trivialization functor $\mathsf{t}_{(-)}$. Its left and right adjoints are the two bosonic Fock functors.

Exercise 17.15. Check: For a Coxeter coalgebra C and Coxeter monoid a, there is an isomorphism

$$\mathsf{CSp}(\mathsf{t}_{\mathrm{C}}, \mathsf{a}) \xrightarrow{\cong} \overline{\mathsf{C}}\mathsf{Sp}(\mathrm{C}, \overline{\mathcal{K}}^{\vee}(\mathsf{a})) \tag{17.11}$$

of left modules over $\mathrm{I}_{\mathrm{bilune}}[\mathcal{A}]^W$. The module structures are (3.97) and (9.98), respectively, and the map is the natural bijection of the adjunction $(\mathsf{t}_{(-)}, \overline{\mathcal{K}}^{\vee})$ in Proposition 16.12. (Put $\mathsf{c} := \mathsf{t}_{\mathrm{C}}$ in the second map in (17.6) and use the fact that $\mathrm{C} \cong \overline{\mathcal{K}}^{\vee}\mathsf{t}_{\mathrm{C}}$ as Coxeter coalgebras.)

Similarly, for a Coxeter comonoid $\mathtt{c}$ and Coxeter algebra A, there is an isomorphism

$$\overline{\mathtt{C}}\mathtt{Sp}(\overline{\mathcal{K}}(\mathtt{c}), \mathrm{A}) \xrightarrow{\cong} \mathtt{CSp}(\mathtt{c}, \mathtt{t}_{\mathrm{A}}) \tag{17.12}$$

of left modules over $\mathrm{I}_{\mathrm{bilune}}[\mathcal{A}]^W$, the map being the natural bijection of the adjunction $(\overline{\mathcal{K}}, \mathtt{t}_{(-)})$ in Proposition 16.13. (Put $\mathtt{a} := \mathtt{t}_{\mathrm{A}}$ in the first map in (17.6) and use the fact that $\overline{\mathcal{K}}\mathtt{t}_{\mathrm{A}} \cong \mathrm{A}$ as Coxeter algebras.)

A different way to derive a slightly weaker result is given in Exercise 17.22.

Exercise 17.16. Deduce the exp-log correspondences in Theorem 9.56 from those in Theorem 3.37. (Use Propositions 15.46 and 15.48 and Propositions 16.20 and 16.29.) Similarly, deduce the q-exp-log correspondences in Theorem 9.61 from those in Theorem 3.43.

Exercise 17.17. Put $\mathrm{C} := \Bbbk\{x\}$ in (17.11) to deduce Proposition 17.3. (Use (3.117) and (9.117).)

17.2.3. Number operator and trivialization functor (Third way).

For a Coxeter space V, define a map of Coxeter species

$$\mathrm{n}_{\mathrm{V}} : \mathtt{t}_{\mathrm{V}} \to \mathtt{t}_{\mathrm{V}} \tag{17.13}$$

which on the A-component is scalar multiplication by $|W_A|$. Diagrams (2.13) commute because $|W_F| = |W_G|$ when $\mathrm{s}(F) = \mathrm{s}(G)$, and $|W_F| = |W_{F'}|$ when $w(F) = F'$. We refer to n_{V} as a *number operator*. It is an isomorphism of Coxeter species when $\mathrm{char}(\Bbbk) \nmid |W|$.

Lemma 17.18. *For a Coxeter coalgebra* C *and Coxeter algebra* A,

$$\overline{\mathtt{C}}\mathtt{Sp}(\mathrm{C}, \mathrm{A}) \to \mathtt{CSp}(\mathtt{t}_{\mathrm{C}}, \mathtt{t}_{\mathrm{A}}), \qquad f \mapsto \mathtt{t}_f \mathrm{n}_{\mathrm{C}} \tag{17.14}$$

is a map of left modules over $\mathrm{I}_{\mathrm{bilune}}[\mathcal{A}]^W$. *If* $\mathrm{char}(\Bbbk) \nmid |W|$, *then* (17.14) *is an isomorphism of modules; moreover,* $f : \mathrm{C} \to \mathrm{A}$ *is an isomorphism iff* $\mathtt{t}_f \mathrm{n}_{\mathrm{C}} : \mathtt{t}_{\mathrm{C}} \to \mathtt{t}_{\mathrm{A}}$ *is an isomorphism.*

That is: The map (17.14) is $\Bbbk$-linear, and for $s \in \mathrm{I}_{\mathrm{bilune}}[\mathcal{A}]^W$, $f : \mathrm{C} \to \mathrm{A}$ a map of Coxeter spaces, and face A,

$$(s \circ \mathtt{t}_f \mathrm{n}_{\mathrm{C}})_A = |W_A| \, (s \circ f)_{\mathrm{t}(A)} \tag{17.15}$$

with lhs as in (3.97) and rhs as in (9.98). Since $\mathtt{t}_f \mathrm{n}_{\mathrm{C}} = \mathrm{n}_{\mathrm{A}} \mathtt{t}_f$ as maps of Coxeter species from $\mathtt{t}_{\mathrm{C}}$ to $\mathtt{t}_{\mathrm{A}}$, we may also write $\mathrm{n}_{\mathrm{A}} \mathtt{t}_f$ instead of $\mathtt{t}_f \mathrm{n}_{\mathrm{C}}$ in the above formula.

Direct proof. Linearity is clear since the functor $\mathtt{t}_{(-)}$ is $\Bbbk$-linear. We now prove (17.15). Using (16.6) and the first diagram in (16.2),

$$\begin{aligned}(s \circ \mathtt{t}_f \mathrm{n}_{\mathrm{C}})_A &= \sum_{\substack{G, G' \geq A \\ \mathrm{s}(G) = \mathrm{s}(G')}} s(A, G, G') \, \mu_A^G (\mathtt{t}_f \mathrm{n}_{\mathrm{C}})_G \beta_{G,G'} \Delta_A^{G'} \\ &= \sum_{\substack{G, G' \geq A \\ \mathrm{s}(G) = \mathrm{s}(G')}} s(A, G, G') \, |W_G| \, \mu_{\mathrm{t}(A)}^{\mathrm{t}(G)} f_{\mathrm{t}(G)} \, \mathrm{t}(G, G') \Delta_{\mathrm{t}(A)}^{\mathrm{t}(G')}\end{aligned}$$

$$
\begin{aligned}
&= \sum_{\substack{\sigma:T'\to T\\ \text{local to } \mathrm{t}(A)}} s(\mathrm{t}(A),T,\sigma,T')\,|W_T| \sum_{\substack{G,G'\geq A\\ \mathrm{t}(G,G')=\sigma}} \mu^{T}_{\mathrm{t}(A)} f_T \sigma \Delta^{T'}_{\mathrm{t}(A)}\\
&= \sum_{\substack{\sigma:T'\to T\\ \text{local to } \mathrm{t}(A)}} s(\mathrm{t}(A),T,\sigma,T')\,|W_A|\, \mu^{T}_{\mathrm{t}(A)} f_T \sigma \Delta^{T'}_{\mathrm{t}(A)}\\
&= |W_A|\,(s\circ f)_{\mathrm{t}(A)}.
\end{aligned}
$$

In the third step, we can also pull $\mu^{T}_{\mathrm{t}(A)} f_T \sigma \Delta^{T'}_{\mathrm{t}(A)}$ outside the inner sum, so we are left to count the number of bifaces local to A of type σ. By Lemma 1.24, this equals the number of faces greater than A of type T. This number multiplied by $|W_T|$ equals $|W_A|$ which is independent of T.

For the last part of the lemma, note that the functor $\mathtt{t}_{(-)}$ is full and faithful, and hence reflects isomorphisms. □

Second proof using identities (17.7). Identity (17.15) is a lot simpler to check than either identity in (17.7). In fact, either of the latter can be used to deduce the former as follows. We start with $f : \mathrm{C} \to \mathrm{A}$. We put $\mathtt{t}_f\mathrm{n}_{\mathrm{C}}$ for f in say the identity involving $\overline{\mathcal{K}}$. The rhs equals $(s\circ \mathtt{t}_f\mathrm{n}_{\mathrm{C}})_A$. In the lhs, we express $\overline{\mathcal{K}}(\mathtt{t}_f\mathrm{n}_{\mathrm{C}})$ as the composite $\overline{\mathcal{K}}(\mathtt{t}_{\mathrm{C}}) \to \overline{\mathcal{K}}^{\vee}(\mathtt{t}_{\mathrm{C}}) \to \overline{\mathcal{K}}(\mathtt{t}_{\mathrm{A}})$, where the first map is the norm map between bosonic Fock functors (which in this case is scalar multiplication by $|W_A|$) and second map is f. Now we use naturality to pull the first map out of $s\circ\overline{\mathcal{K}}(\mathtt{t}_f\mathrm{n}_{\mathrm{C}})$ (since it is a morphism of Coxeter coalgebras), so we are left with $s\circ f$ with $|W_A|$ outside. □

Observation 17.19. The ideas in the proof above are encapsulated in the diagram below. For $\mathrm{char}(\Bbbk) \nmid |W|$, the composite

$$
\begin{array}{ccccc}
\overline{\mathtt{C}}\mathtt{Sp}(\mathrm{C},\mathrm{A}) & \xrightarrow{\cong} & \mathtt{CSp}(\mathtt{t}_{\mathrm{C}},\mathtt{t}_{\mathrm{A}}) & \xrightarrow{\cong} & \overline{\mathtt{C}}\mathtt{Sp}(\overline{\mathcal{K}}^{\vee}\mathtt{t}_{\mathrm{C}},\overline{\mathcal{K}}^{\vee}\mathtt{t}_{\mathrm{A}})\\
& & \downarrow\cong & & \downarrow\cong\\
& & \overline{\mathtt{C}}\mathtt{Sp}(\overline{\mathcal{K}}\mathtt{t}_{\mathrm{C}},\overline{\mathcal{K}}\mathtt{t}_{\mathrm{A}}) & \xrightarrow[\cong]{} & \overline{\mathtt{C}}\mathtt{Sp}(\overline{\mathcal{K}}^{\vee}\mathtt{t}_{\mathrm{C}},\overline{\mathcal{K}}\mathtt{t}_{\mathrm{A}}) \xrightarrow[\cong]{} \overline{\mathtt{C}}\mathtt{Sp}(\mathrm{C},\mathrm{A})
\end{array}
$$

is identity, with each map a module isomorphism. The first map involves multiplication by a number. The square commutes by naturality of $\overline{\mathcal{K}}^{\vee} \xrightarrow{\cong} \overline{\mathcal{K}}$ which involves division by a number.

Exercise 17.20. Deduce: For q not a root of unity, $f : \mathrm{C} \to \mathrm{A}$ a map of Coxeter spaces, and face A,

$$
\mathbf{log}_q(\mathtt{t}_f\mathrm{n}_{\mathrm{C}})_A = |W_A|\,\mathbf{log}_q(f)_{\mathrm{t}(A)} \tag{17.16}
$$

with lhs as in (3.99b) and rhs as in (9.99b).

17.3. Internal hom for Coxeter species and Coxeter spaces

We now understand how the trivialization functor and bosonic Fock functors connect the convolution Coxeter monoid and convolution Coxeter algebra, the internal hom for Coxeter comonoids and internal hom for Coxeter coalgebras, the universal measuring Coxeter comonoid and universal measuring Coxeter coalgebra.

17.3.1. Convolution Coxeter monoid and algebra. Recall the convolution Coxeter monoid $\hom^\times(\mathsf{c},\mathsf{a})$ from Section 3.5.6 and convolution Coxeter algebra $\hom^\times(\mathrm{C},\mathrm{A})$ from Section 9.5.4.

Lemma 17.21. *For a Coxeter coalgebra* C *and Coxeter monoid* a,

$$\overline{\mathcal{K}}^\vee(\hom^\times(\mathsf{t}_\mathrm{C},\mathsf{a})) = \hom^\times(\mathrm{C},\overline{\mathcal{K}}^\vee(\mathsf{a})) \tag{17.17}$$

as Coxeter algebras.

Similarly, for a Coxeter comonoid c *and Coxeter algebra* A,

$$\overline{\mathcal{K}}^\vee(\hom^\times(\mathsf{c},\mathsf{t}_\mathrm{A})) = \hom^\times(\overline{\mathcal{K}}(\mathsf{c}),\mathrm{A}) \tag{17.18}$$

as Coxeter algebras.

PROOF. Let c be a Coxeter comonoid and a a Coxeter monoid. An element of $\hom^\times(\mathsf{c},\mathsf{a})[A]^{W_A}$ is the same as a W_A-module map $\mathsf{c}[A]\to\mathsf{a}[A]$. For $A\le C_r$, this is an element of $\overline{\mathcal{K}}^\vee(\hom^\times(\mathsf{c},\mathsf{a}))[Z]$, where Z is the type of A.

- For $\mathsf{c} := \mathsf{t}_\mathrm{C}$, this is the same as a linear map $\mathrm{C}[Z]\to\mathsf{a}[A]^{W_A}$ which yields the identification (17.17).
- For $\mathsf{a} := \mathsf{t}_\mathrm{A}$, this is the same as a linear map $\mathsf{c}[A]_{W_A}\to\mathrm{A}[Z]$ which yields the identification (17.18).

One checks that these are indeed identifications of Coxeter algebras. A more general check is required for the maps (17.19) in Exercise 17.23 below. □

Special case. We have:

- For $\mathrm{C} := \Bbbk\{x\}$, identity (17.17) reduces to $\overline{\mathcal{K}}^\vee(\mathsf{a}) = \overline{\mathcal{K}}^\vee(\mathsf{a})$.
- For $\mathrm{A} := \Bbbk[x]$, identity (17.18) reduces to $\overline{\mathcal{K}}^\vee(\mathsf{c}^*) = \overline{\mathcal{K}}(\mathsf{c})^*$.

This follows from identities (3.67) and (9.67) and the fact that $\mathsf{t}_{\Bbbk\{x\}} = \mathsf{E}$ as Coxeter comonoids and $\mathsf{t}_{\Bbbk[x]} = \mathsf{E}$ as Coxeter monoids (stated formally later in (18.15)).

Exercise 17.22 (Series). Apply $\mathscr{S}$ to identity (17.17) and use Proposition 17.3 to obtain an isomorphism

$$\begin{array}{ccc}\mathscr{S}(\hom^\times(\mathsf{t}_\mathrm{C},\mathsf{a})) & \xrightarrow{\cong} & \mathscr{S}(\hom^\times(\mathrm{C},\overline{\mathcal{K}}^\vee(\mathsf{a}))) \\ \| & & \| \\ \mathsf{CSp}(\mathsf{t}_\mathrm{C},\mathsf{a}) & & \overline{\mathsf{C}}\mathsf{Sp}(\mathrm{C},\overline{\mathcal{K}}^\vee(\mathsf{a}))\end{array}$$

of left modules over $\mathrm{I}_{\mathrm{lune}}[\mathcal{A}]^W$. The module structures are (3.85) and (9.87), respectively. (Use item (1) of Exercises 3.57 and 9.77.) This map is the same as in (17.11). However, note very carefully that here we got an isomorphism only over the invariant lune-incidence algebra. Similarly, applying $\mathscr{S}$ to identity (17.18) yields the map in (17.12).

Exercise 17.23. Check: For a Coxeter comonoid c and Coxeter monoid a, there are Coxeter algebra morphisms

$$\begin{gathered}\overline{\mathcal{K}}^\vee(\hom^\times(\mathsf{c},\mathsf{a}))\to\hom^\times(\overline{\mathcal{K}}(\mathsf{c}),\overline{\mathcal{K}}(\mathsf{a})),\\ \overline{\mathcal{K}}^\vee(\hom^\times(\mathsf{c},\mathsf{a}))\to\hom^\times(\overline{\mathcal{K}}^\vee(\mathsf{c}),\overline{\mathcal{K}}^\vee(\mathsf{a})).\end{gathered} \tag{17.19}$$

On the Z-component (with A the face of C_r of type Z), they send a W_A-module map $\mathsf{c}[A]\to\mathsf{a}[A]$ to the induced maps $\mathsf{c}[A]_{W_A}\to\mathsf{a}[A]_{W_A}$ and $\mathsf{c}[A]^{W_A}\to\mathsf{a}[A]^{W_A}$, respectively.

Put $\mathsf{c} := \mathsf{t}_{\mathrm{C}}$ in the second map in (17.19) to deduce (17.17). Similarly, put $\mathsf{a} := \mathsf{t}_{\mathrm{A}}$ in the first map in (17.19) to deduce (17.18).

Apply $\mathscr{S}$ to maps (17.19) to recover the result of Exercise 17.13.

17.3.2. Internal hom for Coxeter comonoids and coalgebras. Recall the internal hom for Coxeter comonoids $\mathcal{C}(\mathsf{c},\mathsf{d})$ from Section 3.6 and for Coxeter coalgebras $\mathcal{C}(\mathrm{C},\mathrm{D})$ from Section 9.6.

Lemma 17.24. *For a Coxeter coalgebra* C *and Coxeter comonoid* d,

$$\overline{\mathcal{K}}^{\vee}(\mathcal{C}(\mathsf{t}_{\mathrm{C}},\mathsf{d})) = \mathcal{C}(\mathrm{C},\overline{\mathcal{K}}^{\vee}(\mathsf{d})) \tag{17.20}$$

as Coxeter coalgebras.

PROOF. Let c and d be Coxeter comonoids. An element of $\mathcal{C}(\mathsf{c},\mathsf{d})[A]^{W_A}$ is an element $(f_{F/A})$ of $\mathcal{C}(\mathsf{c},\mathsf{d})[A]$ which is W_A-invariant, that is, which satisfies the additional condition $wf_{F/A} = f_{wF/A}w$ for all $F \geq A$ and $w \in W_A$. (Recall: $f_{F/A} : \mathsf{c}[F] \to \mathsf{d}[F]$.) For $A \leq C_r$, this is an element of $\overline{\mathcal{K}}^{\vee}(\mathcal{C}(\mathsf{c},\mathsf{d}))[Z]$, where Z is the type of A. It is determined and can be characterized by the subfamily of maps $f_{F/A}$ with $A \leq F \leq C_r$. The equivalent conditions are that each $f_{F/A}$ is a map of W_F-modules and $f_{G/A}\Delta^G_F = \Delta^G_F f_{F/A}$ holds for each $A \leq F \leq G \leq C_r$.

Now suppose $\mathsf{c} := \mathsf{t}_{\mathrm{C}}$. In this case, the invariance condition reduces to having maps $\mathsf{t}_{\mathrm{C}}[F] \to \mathsf{d}[F]^{W_F}$ which are the same as $\mathrm{C}[T] \to \overline{\mathcal{K}}^{\vee}(\mathsf{d})[T]$, where T is the type of F. This family determines an element of $\mathcal{C}(\mathrm{C},\overline{\mathcal{K}}^{\vee}(\mathsf{d}))[Z]$, where Z is the type of A. Thus, we obtain the required identification (17.20) on the Z-component. It is straightforward to check that this is an identification of Coxeter coalgebras. □

Special case. For $\mathrm{C} := \Bbbk\{x\}$, identity (17.20) reduces to the tautology $\overline{\mathcal{K}}^{\vee}(\mathsf{d}) = \overline{\mathcal{K}}^{\vee}(\mathsf{d})$. This is because $\mathcal{C}(\mathsf{E},\mathsf{d}) = \mathsf{d}$ and $\mathcal{C}(\Bbbk\{x\},\mathrm{D}) = \mathrm{D}$.

Exercise 17.25 (Primitive and group-like series). Apply $\mathscr{G}$ to identity (17.20) and use Proposition 17.8 to obtain

$$\begin{array}{ccc} \mathscr{G}(\mathcal{C}(\mathsf{t}_{\mathrm{C}},\mathsf{d})) & \xrightarrow{\cong} & \mathscr{G}(\mathcal{C}(\mathrm{C},\overline{\mathcal{K}}^{\vee}(\mathsf{d}))) \\ \| & & \| \\ \mathsf{Comon}(\mathsf{CSp})(\mathsf{t}_{\mathrm{C}},\mathsf{d}) & & \mathsf{Comon}(\overline{\mathsf{C}}\mathsf{Sp})(\mathrm{C},\overline{\mathcal{K}}^{\vee}(\mathsf{d})). \end{array}$$

This is the bijection in Proposition 16.20. Similarly, apply $\mathscr{P}$ to obtain

$$\begin{array}{ccc} \mathscr{P}(\mathcal{C}(\mathsf{t}_{\mathrm{C}},\mathsf{d})) & \xrightarrow{\cong} & \mathscr{P}(\mathcal{C}(\mathrm{C},\overline{\mathcal{K}}^{\vee}(\mathsf{d}))) \\ \| & & \| \\ \mathsf{CSp}(\mathsf{t}_{\mathrm{C}},\mathcal{P}(\mathsf{d})) & & \overline{\mathsf{C}}\mathsf{Sp}(\mathrm{C},\overline{\mathcal{P}}\overline{\mathcal{K}}^{\vee}(\mathsf{d})). \end{array}$$

This is the bijection in Proposition 16.12 in view of Proposition 15.46. (For the vertical equalities, use identities (3.118) and (9.118).)

Exercise 17.26. This builds on Lemma 17.24. Check: For a scalar q, for a cocommutative Coxeter coalgebra C and Coxeter q-bimonoid k,

$$\overline{\mathcal{K}}^{\vee}(\mathcal{C}(\mathsf{t}_{\mathrm{C}},\mathsf{k})_q) = \mathcal{C}(\mathrm{C},\overline{\mathcal{K}}^{\vee}(\mathsf{k}))_q \tag{17.21}$$

as Coxeter q-bialgebras. Thus, this is also an identification of Coxeter algebras. A more general check is required for the map (17.23) in Exercise 17.27 below.

Exercise 17.27. Check: For Coxeter comonoids c and d, there is a Coxeter coalgebra morphism

$$\overline{\mathcal{K}}^{\vee}(\mathcal{C}(\mathsf{c},\mathsf{d})) \to \mathcal{C}(\overline{\mathcal{K}}^{\vee}(\mathsf{c}), \overline{\mathcal{K}}^{\vee}(\mathsf{d})). \tag{17.22}$$

On the Z-component (with A the face of C_r of type Z), it sends the family $(f_{F/A})$ to the family $(f_{T/Z})$, where $f_{T/Z} : \mathsf{c}[F]^{W_F} \to \mathsf{d}[F]^{W_F}$ with $A \leq F \leq C_r$ and $\mathrm{t}(F) = T$ is induced from $f_{F/A} : \mathsf{c}[F] \to \mathsf{d}[F]$.

In view of the bijection (9.71), the map (17.22) corresponds to the Coxeter coalgebra morphism

$$\overline{\mathcal{K}}^{\vee}(\mathcal{C}(\mathsf{c},\mathsf{d})) \times \overline{\mathcal{K}}^{\vee}(\mathsf{c}) \to \overline{\mathcal{K}}^{\vee}(\mathsf{d}),$$

which on the Z-component (with A the face of C_r of type Z) is

$$\mathcal{C}(\mathsf{c},\mathsf{d})[A]^{W_A} \otimes \mathsf{c}[A]^{W_A} \to \mathsf{d}[A]^{W_A}, \qquad (f_{F/A}) \otimes x \mapsto f_{A/A}(x).$$

For a scalar q, for a cocommutative Coxeter comonoid c and Coxeter q-bimonoid k, there is a Coxeter q-bialgebra morphism

$$\overline{\mathcal{K}}^{\vee}(\mathcal{C}(\mathsf{c},\mathsf{k})_q) \to \mathcal{C}(\overline{\mathcal{K}}^{\vee}(\mathsf{c}), \overline{\mathcal{K}}^{\vee}(\mathsf{k}))_q \tag{17.23}$$

given by the same formula as (17.22).

Put $\mathsf{c} := \mathsf{t}_{\mathrm{C}}$ in (17.22) and (17.23) to deduce (17.20) and (17.21).

17.3.3. Universal measuring Coxeter comonoid and coalgebra.

Recall the universal measuring Coxeter comonoid $\overline{\mathcal{C}}(\mathsf{a},\mathsf{b})$ from Section 3.7 and universal measuring Coxeter coalgebra $\overline{\mathcal{C}}(\mathrm{A},\mathrm{B})$ from Section 9.7.

Lemma 17.28. *For a Coxeter monoid* a *and Coxeter algebra* B,

$$\overline{\mathcal{K}}^{\vee}(\overline{\mathcal{C}}(\mathsf{a},\mathsf{t}_{\mathrm{B}})) = \overline{\mathcal{C}}(\overline{\mathcal{K}}(\mathsf{a}),\mathrm{B}) \tag{17.24}$$

as Coxeter coalgebras.

PROOF. Similar to that of Lemma 17.24. □

Exercise 17.29. Formulate and prove analogues of Exercises 17.25, 17.26, 17.27.

17.4. Norm maps on Coxeter species and Coxeter spaces

Recall the norm map on Coxeter species from Section 3.11 and the norm map on Coxeter spaces from Section 9.11. In this section and the next, we relate the two norm maps in three different ways:

(1) via full Fock and bosonic Fock functors,
(2) via adjunctions between the trivialization and bosonic Fock functors,
(3) via the trivialization functor and its lifts.

Note the similarity with the philosophy of Section 17.2. As a part of item (1), we show that bosonic Fock functors when applied to any general q-norm map of Coxeter q-bimonoids yield a general q-norm map of Coxeter q-bialgebras.

17.4.1. Norm maps linked by Fock functors. We begin by linking the q-norm maps κ_q and $\overline{\kappa}_q$ in (3.144) and (9.140) using full Fock and bosonic Fock functors.

Lemma 17.30. *For a scalar q and Coxeter species p, the diagrams*

$$
\begin{array}{ccc}
\mathcal{K}\mathcal{T}_q(\mathsf{p}) & \xrightarrow{\mathcal{K}(\kappa_q)_{\mathsf{p}}} & \mathcal{K}\mathcal{T}_q^{\vee}(\mathsf{p}) \\
{\scriptstyle \varphi_{\mathsf{p}}}\uparrow & & \downarrow{\scriptstyle \psi_{\mathsf{p}}} \\
\overline{\mathcal{T}}_q\mathcal{K}(\mathsf{p}) & \xrightarrow[(\overline{\kappa}_q)_{\mathcal{K}(\mathsf{p})}]{} & \overline{\mathcal{T}}_q^{\vee}\mathcal{K}(\mathsf{p})
\end{array}
\qquad
\begin{array}{ccc}
\mathcal{K}^{\vee}\mathcal{T}_q(\mathsf{p}) & \xrightarrow{\mathcal{K}^{\vee}(\kappa_q)_{\mathsf{p}}} & \mathcal{K}^{\vee}\mathcal{T}_q^{\vee}(\mathsf{p}) \\
{\scriptstyle \psi_{\mathsf{p}}^{\vee}}\uparrow & & \downarrow{\scriptstyle \varphi_{\mathsf{p}}^{\vee}} \\
\overline{\mathcal{T}}_q\mathcal{K}^{\vee}(\mathsf{p}) & \xrightarrow[(\overline{\kappa}_q)_{\mathcal{K}^{\vee}(\mathsf{p})}]{} & \overline{\mathcal{T}}_q^{\vee}\mathcal{K}^{\vee}(\mathsf{p})
\end{array}
\tag{17.25}
$$

$$
\begin{array}{ccc}
\overline{\mathcal{K}}\mathcal{T}_q(\mathsf{p}) & \xrightarrow{\overline{\mathcal{K}}(\kappa_q)_{\mathsf{p}}} & \overline{\mathcal{K}}\mathcal{T}_q^{\vee}(\mathsf{p}) \\
{\scriptstyle \overline{\varphi}_{\mathsf{p}}}\uparrow{\scriptstyle \cong} & & {\scriptstyle \cong}\downarrow{\scriptstyle \overline{\psi}_{\mathsf{p}}} \\
\overline{\mathcal{T}}_q\overline{\mathcal{K}}(\mathsf{p}) & \xrightarrow[(\overline{\kappa}_q)_{\overline{\mathcal{K}}(\mathsf{p})}]{} & \overline{\mathcal{T}}_q^{\vee}\overline{\mathcal{K}}(\mathsf{p})
\end{array}
\qquad
\begin{array}{ccc}
\overline{\mathcal{K}}^{\vee}\mathcal{T}_q(\mathsf{p}) & \xrightarrow{\overline{\mathcal{K}}^{\vee}(\kappa_q)_{\mathsf{p}}} & \overline{\mathcal{K}}^{\vee}\mathcal{T}_q^{\vee}(\mathsf{p}) \\
{\scriptstyle \overline{\psi}_{\mathsf{p}}^{\vee}}\uparrow{\scriptstyle \cong} & & {\scriptstyle \cong}\downarrow{\scriptstyle \overline{\varphi}_{\mathsf{p}}^{\vee}} \\
\overline{\mathcal{T}}_q\overline{\mathcal{K}}^{\vee}(\mathsf{p}) & \xrightarrow[(\overline{\kappa}_q)_{\overline{\mathcal{K}}^{\vee}(\mathsf{p})}]{} & \overline{\mathcal{T}}_q^{\vee}\overline{\mathcal{K}}^{\vee}(\mathsf{p})
\end{array}
\tag{17.26}
$$

commute as Coxeter q-bialgebras.

Proof. This follows from Lemma 14.20. The additional observation is that all maps involved are morphisms of Coxeter q-bialgebras. □

17.4.2. q-norm maps are isomorphisms. For q not a root of unity, the q-norm maps κ_q and $\overline{\kappa}_q$ are isomorphisms, see Lemma 3.78 and Lemma 9.101. We now explain how the latter can be deduced from the former using Proposition 16.14.

Proof of Lemma 9.101 via Fock functors. Let q be a scalar which is not a root of unity. Given a Coxeter space V, choose a Coxeter species p such that $\overline{\mathcal{K}}^{\vee}(\mathsf{p}) = \mathrm{V}$. For instance, we may take $\mathsf{p} := \mathsf{t}_{\mathrm{V}}$ by using Proposition 16.14. Now consider the second diagram in (17.26). The vertical maps are the lax and colax structures of $\overline{\mathcal{K}}^{\vee}$. These are isomorphisms by Theorem 14.15. By Lemma 3.78, the q-norm map on Coxeter species κ_q is an isomorphism. Hence, the top-horizontal map is an isomorphism, and the result follows. Instead of $\overline{\mathcal{K}}^{\vee}$, one can also employ $\overline{\mathcal{K}}$ and the first diagram in (17.26). □

17.4.3. General norm maps linked by Fock functors. We now turn to the general q-norm maps $\widehat{f}$ in (3.146) and (9.144). Recall: κ_q and $\overline{\kappa}_q$ are special cases of these.

Lemma 17.31. *Let $f : \mathsf{c} \to \mathsf{a}$ be a map of Coxeter species from a Coxeter comonoid to a Coxeter monoid. For a scalar q, the diagrams*

$$
\begin{array}{ccc}
\mathcal{K}\mathcal{T}_q(\mathsf{c}) & \xrightarrow{\mathcal{K}(\widehat{f})} & \mathcal{K}\mathcal{T}_q^{\vee}(\mathsf{a}) \\
{\scriptstyle \varphi_{\mathsf{c}}}\uparrow & & \downarrow{\scriptstyle \psi_{\mathsf{a}}} \\
\overline{\mathcal{T}}_q\mathcal{K}(\mathsf{c}) & \xrightarrow[\widehat{\mathcal{K}(f)}]{} & \overline{\mathcal{T}}_q^{\vee}\mathcal{K}(\mathsf{a})
\end{array}
\qquad
\begin{array}{ccc}
\mathcal{K}^{\vee}\mathcal{T}_q(\mathsf{c}) & \xrightarrow{\mathcal{K}^{\vee}(\widehat{f})} & \mathcal{K}^{\vee}\mathcal{T}_q^{\vee}(\mathsf{a}) \\
{\scriptstyle \psi_{\mathsf{c}}^{\vee}}\uparrow & & \downarrow{\scriptstyle \varphi_{\mathsf{a}}^{\vee}} \\
\overline{\mathcal{T}}_q\mathcal{K}^{\vee}(\mathsf{c}) & \xrightarrow[\widehat{\mathcal{K}^{\vee}(f)}]{} & \overline{\mathcal{T}}_q^{\vee}\mathcal{K}^{\vee}(\mathsf{a})
\end{array}
\tag{17.27}
$$

$$
\begin{CD}
\overline{\mathcal{K}}\mathcal{T}_q(\mathsf{c}) @>{\overline{\mathcal{K}}(\widehat{f})}>> \overline{\mathcal{K}}\mathcal{T}_q^{\vee}(\mathsf{a}) \\
@A{\overline{\varphi}_{\mathsf{c}}}A{\cong}A @V{\cong}V{\overline{\psi}_{\mathsf{a}}}V \\
\overline{\mathcal{T}}_q\overline{\mathcal{K}}(\mathsf{c}) @>>{\widehat{\overline{\mathcal{K}}(f)}}> \overline{\mathcal{T}}_q^{\vee}\overline{\mathcal{K}}(\mathsf{a})
\end{CD}
\qquad
\begin{CD}
\overline{\mathcal{K}}^{\vee}\mathcal{T}_q(\mathsf{c}) @>{\overline{\mathcal{K}}^{\vee}(\widehat{f})}>> \overline{\mathcal{K}}^{\vee}\mathcal{T}_q^{\vee}(\mathsf{a}) \\
@A{\overline{\psi}_{\mathsf{c}}^{\vee}}A{\cong}A @V{\cong}V{\overline{\varphi}_{\mathsf{a}}^{\vee}}V \\
\overline{\mathcal{T}}_q\overline{\mathcal{K}}^{\vee}(\mathsf{c}) @>>{\widehat{\overline{\mathcal{K}}^{\vee}(f)}}> \overline{\mathcal{T}}_q^{\vee}\overline{\mathcal{K}}^{\vee}(\mathsf{a})
\end{CD}
\tag{17.28}
$$

commute as Coxeter q-bialgebras.

Here $\mathcal{K}(f) : \mathcal{K}(\mathsf{c}) \to \mathcal{K}(\mathsf{a})$ and so on are the induced maps of Coxeter spaces from a Coxeter coalgebra to a Coxeter algebra.

The above result is more general than Lemma 17.30. Also the proof given below is different; it emphasizes the role of universal properties of $\mathcal{T}_q$ and $\mathcal{T}_q^{\vee}$.

PROOF. Let us look at the first diagram in (17.27), the rest are similar. The composite φ_{c} followed by $\mathcal{K}(\widehat{f})$ followed by ψ_{a} is a morphism of Coxeter q-bialgebras. To check that it agrees with $\widehat{\mathcal{K}(f)}$, it suffices to check that on the Z-component, on the Z-summand and into the Z-summand, it equals $\mathcal{K}(f)_Z$ (see Exercise 9.99). But this is clear from definitions. In the special situation of Lemma 17.30, it is even simpler to see. □

Exercise 17.32. Diagrams (17.27) generalize further to deformed full Fock functors. Show: For scalars p and q, the diagrams

$$
\begin{CD}
\mathcal{K}_p\mathcal{T}_q(\mathsf{c}) @>{\mathcal{K}_p(\widehat{f})}>> \mathcal{K}_p\mathcal{T}_q^{\vee}(\mathsf{a}) \\
@A{\varphi_{\mathsf{c}}}AA @VV{(\psi_p)_{\mathsf{a}}}V \\
\overline{\mathcal{T}}_{pq}\mathcal{K}_p(\mathsf{c}) @>>{\widehat{\mathcal{K}_p(f)}}> \overline{\mathcal{T}}_{pq}^{\vee}\mathcal{K}_p(\mathsf{a})
\end{CD}
\qquad
\begin{CD}
\mathcal{K}_p^{\vee}\mathcal{T}_q(\mathsf{c}) @>{\mathcal{K}_p^{\vee}(\widehat{f})}>> \mathcal{K}_p^{\vee}\mathcal{T}_q^{\vee}(\mathsf{a}) \\
@A{(\psi_p^{\vee})_{\mathsf{c}}}AA @VV{\varphi_{\mathsf{a}}^{\vee}}V \\
\overline{\mathcal{T}}_{pq}\mathcal{K}_p^{\vee}(\mathsf{c}) @>>{\widehat{\mathcal{K}_p^{\vee}(f)}}> \overline{\mathcal{T}}_{pq}^{\vee}\mathcal{K}_p^{\vee}(\mathsf{a})
\end{CD}
\tag{17.29}
$$

commute as Coxeter pq-bialgebras.

As a special case, diagrams (17.25) also generalize to deformed full Fock functors: replace $\mathcal{K}$ and $\mathcal{K}^{\vee}$ with $\mathcal{K}_p$ and $\mathcal{K}_p^{\vee}$ and so on. This can also be deduced from Exercise 14.21.

17.4.4. General norm maps linked by adjunctions.

Recall from Section 16.4 the functors $\overline{\mathcal{Z}}^{\vee}$ and $\overline{\mathcal{Z}}$ which lift $\mathsf{t}_{(-)}$.

Exercise 17.33. Let q be a scalar. Check: For a Coxeter coalgebra C and Coxeter monoid a, the diagram

$$
\begin{CD}
\mathsf{CSp}(\mathsf{t}_{\mathrm{C}}, \mathsf{a}) @>{\cong}>> q\text{-}\mathsf{Bimon}(\mathsf{CSp})(\overline{\mathcal{Z}}^{\vee}\overline{\mathcal{T}}_q(\mathrm{C}), \mathcal{T}_q^{\vee}(\mathsf{a})) \\
@V{\cong}VV @VV{\cong}V \\
\overline{\mathsf{CSp}}(\mathrm{C}, \overline{\mathcal{K}}^{\vee}(\mathsf{a})) @>>{\cong}> q\text{-}\mathsf{Bimon}(\overline{\mathsf{CSp}})(\overline{\mathcal{T}}_q(\mathrm{C}), \overline{\mathcal{K}}^{\vee}\mathcal{T}_q^{\vee}(\mathsf{a}))
\end{CD}
\tag{17.30}
$$

commutes. The horizontal maps send a map to its associated general q-norm map as in (3.146) and (9.144), noting that $\overline{\mathcal{Z}}^{\vee}\overline{\mathcal{T}}_q(\mathrm{C}) = \mathcal{T}_q(\mathsf{t}_{\mathrm{C}})$ and $\overline{\mathcal{K}}^{\vee}\mathcal{T}_q^{\vee}(\mathsf{a}) = \overline{\mathcal{T}}_q^{\vee}\overline{\mathcal{K}}^{\vee}(\mathsf{a})$. The vertical maps are natural bijections of adjunctions $(\mathsf{t}_{(-)}, \overline{\mathcal{K}}^{\vee})$ and $(\overline{\mathcal{Z}}^{\vee}, \overline{\mathcal{K}}^{\vee})$ in Propositions 16.12 and 16.48. (Use the second diagram in (17.28) along with the first diagram in (16.27). Alternatively, proceed directly using the strategy in the proof of Lemma 17.31.)

Similarly, for a Coxeter comonoid c and Coxeter algebra A, the diagram

$$
(17.31)\qquad
\begin{array}{ccc}
\overline{\mathsf{C}}\mathsf{Sp}(\overline{\mathcal{K}}(\mathsf{c}), \mathrm{A}) & \xrightarrow{\cong} & q\text{-}\mathsf{Bimon}(\overline{\mathsf{C}}\mathsf{Sp})(\overline{\mathcal{K}}\mathcal{T}_q(\mathsf{c}), \overline{\mathcal{T}_q}^{\vee}(\mathrm{A})) \\
{\scriptstyle\cong}\downarrow & & \downarrow{\scriptstyle\cong} \\
\mathsf{CSp}(\mathsf{c}, \mathsf{t}_{\mathrm{A}}) & \xrightarrow[\cong]{} & q\text{-}\mathsf{Bimon}(\mathsf{CSp})(\mathcal{T}_q(\mathsf{c}), \overline{\mathcal{Z}}\overline{\mathcal{T}_q}^{\vee}(\mathrm{A}))
\end{array}
$$

commutes. The vertical maps are those in Propositions 16.13 and 16.57.

17.4.5. Isomorphism criteria for general q-norm maps. We develop further the discussion in Section 17.4.2.

We connect the isomorphism criteria in Theorems 3.79 and 9.110 when $\mathrm{C} = \overline{\mathcal{K}}(\mathsf{c})$ and $\mathrm{A} = \overline{\mathcal{K}}(\mathsf{a})$.

Observation 17.34. For $f : \mathsf{c} \to \mathsf{a}$ a Coxeter species map and q not a root of unity:

- If $\widehat{f} : \mathcal{T}_q(\mathsf{c}) \to \mathcal{T}_q^{\vee}(\mathsf{a})$ is an isomorphism, then so is $\overline{\mathcal{K}}(\widehat{f}) = \widehat{\overline{\mathcal{K}}(f)}$. The equality holds by the first diagram in (17.28).
- If $\mathbf{log}_q(f) : \mathsf{c} \to \mathsf{a}$ is an isomorphism, then so is $\overline{\mathcal{K}}(\mathbf{log}_q(f)) = \mathbf{log}_q(\overline{\mathcal{K}}(f))$. The equality holds by the first identity in (17.7) applied to $s = \boldsymbol{\mu}_q$.

The same remarks apply to $\overline{\mathcal{K}}^{\vee}$ instead of $\overline{\mathcal{K}}$.

Remark 17.35. We can carry out a similar analysis when $\mathsf{c} = \mathsf{t}_{\mathrm{C}}$ and $\mathrm{A} = \overline{\mathcal{K}}^{\vee}(\mathsf{a})$. This pertains to diagram (17.30). Note: The vertical maps in this diagram preserve isomorphisms. Also relevant is module map (17.11). A similar remark applies when $\mathrm{C} = \overline{\mathcal{K}}(\mathsf{c})$ and $\mathsf{a} = \mathsf{t}_{\mathrm{A}}$. Now we employ diagram (17.31) and module map (17.12).

This discussion is continued in Section 17.5.4.

17.5. Transformation between bosonic Fock functor adjoints

Let q be a scalar. Recall from Section 16.4.3 the functors $\overline{\mathcal{Z}}$ and $\overline{\mathcal{Z}}^{\vee}$ from the category of Coxeter q-bialgebras to the category of Coxeter q-bimonoids. We define a natural transformation z from $\overline{\mathcal{Z}}^{\vee}$ to $\overline{\mathcal{Z}}$. It is induced from the general q-norm map associated to a number operator. It is an isomorphism when q is not a root of unity and $\mathrm{char}(\Bbbk) \nmid |W|$. Moreover, when applied to any general q-norm map of Coxeter q-bialgebras, it yields a general q-norm map of Coxeter q-bimonoids. We also briefly consider commutative aspects.

17.5.1. General q-norm map of a number operator. Let H be a Coxeter q-bialgebra for a scalar q. Recall the number operator from (17.13). View $\mathrm{n_H} : \mathsf{t}_{\mathrm{H}} \to \mathsf{t}_{\mathrm{H}}$ as a map from a Coxeter comonoid to a Coxeter monoid. Applying (3.146) yields a morphism of Coxeter q-bimonoids

$$
(17.32)\qquad \widehat{\mathrm{n_H}} : \mathcal{T}_q(\mathsf{t}_{\mathrm{H}}) \to \mathcal{T}_q^{\vee}(\mathsf{t}_{\mathrm{H}}).
$$

On the A-component, on the F-summand and into the G-summand, it is

$$
\mathrm{H}[\mathrm{t}(F)] \xrightarrow{\Delta^{\mathrm{t}(FG)}_{\mathrm{t}(F)}} \mathrm{H}[\mathrm{t}(FG)] \xrightarrow{|W_{FG}|q^{l(\sigma)}\sigma} \mathrm{H}[\mathrm{t}(GF)] \xrightarrow{\mu^{\mathrm{t}(GF)}_{\mathrm{t}(G)}} \mathrm{H}[\mathrm{t}(G)]
$$

with $\sigma = \mathrm{t}(GF, FG)$. The middle map is obtained using definition of β in the first diagram in (16.2).

As a special case of (17.15) with $f := \mathrm{id}_{\mathrm{H}}$, we obtain: For $s \in \mathrm{I}_{\mathrm{bilune}}[\mathcal{A}]^W$ and face A,

$$(s \circ \mathrm{n}_{\mathrm{H}})_A = |W_A|\,(s \circ \mathrm{id}_{\mathrm{H}})_{\mathrm{t}(A)} \tag{17.33}$$

with lhs as in (3.97) and rhs as in (9.98). In particular, for q not a root of unity,

$$\mathbf{log}_q(\mathrm{n}_{\mathrm{H}})_A = |W_A|\;\mathbf{log}_q(\mathrm{id}_{\mathrm{H}})_{\mathrm{t}(A)} \tag{17.34}$$

with lhs as in (3.99b) and rhs as in (9.99b). The map in the rhs is far from being an isomorphism; in fact, it factors as in diagram (9.115). Thus, in view of Theorem 3.79, the map $\widehat{\mathrm{n}_{\mathrm{H}}}$ also cannot be an isomorphism; we will look at this in detail below for arbitrary q.

17.5.2. Transformation between bosonic Fock functor adjoints.

Proposition 17.36. *For a scalar q and Coxeter q-bialgebra* H, *there is a morphism of Coxeter q-bimonoids*

$$\mathrm{z}_{\mathrm{H}} : \overline{\mathcal{Z}}^{\vee}(\mathrm{H}) \to \overline{\mathcal{Z}}(\mathrm{H}) \tag{17.35}$$

such that the diagram of Coxeter q-bimonoids

$$\begin{array}{ccc} \mathcal{T}_q(\mathrm{t}_{\mathrm{H}}) & \xrightarrow{\widehat{\mathrm{n}_{\mathrm{H}}}} & \mathcal{T}_q^{\vee}(\mathrm{t}_{\mathrm{H}}) \\ \downarrow & & \uparrow \\ \overline{\mathcal{Z}}^{\vee}(\mathrm{H}) & \xrightarrow[\mathrm{z}_{\mathrm{H}}]{} & \overline{\mathcal{Z}}(\mathrm{H}) \end{array} \tag{17.36}$$

commutes.

When H is finite-dimensional, the dual of (17.36) is the same diagram with H* instead of H.

PROOF. We need to check that $\widehat{\mathrm{n}_{\mathrm{H}}}$ takes relations (16.25) to zero, so it factors through $\overline{\mathcal{Z}}^{\vee}(\mathrm{H})$ and elements in the image of $\widehat{\mathrm{n}_{\mathrm{H}}}$ satisfy conditions (16.28), so it maps into $\overline{\mathcal{Z}}(\mathrm{H})$. In either case, it suffices to assume $K = A$; this is because $\widehat{\mathrm{n}_{\mathrm{H}}}$ is a morphism of Coxeter monoids and of Coxeter comonoids. Either check can be done using the q-bialgebra axiom (8.73) for H. The scalars $|W_{FG}|$ play a critical role and justify why the number operator is defined the way it is. We leave these checks as an exercise. Note: When H is finite-dimensional, one check implies the other by duality. □

The map z_{H} is natural in H, yielding a natural transformation $\mathrm{z} : \overline{\mathcal{Z}}^{\vee} \to \overline{\mathcal{Z}}$.

Theorem 17.37. *For a scalar q and Coxeter q-bialgebra* H, *the map* (17.35) *is an isomorphism when q is not a root of unity and* $\mathrm{char}(\Bbbk) \nmid |W|$.

PROOF. Suppose q is not a root of unity and $\mathrm{char}(\Bbbk) \nmid |W|$. Consider the diagram of Coxeter q-bimonoids

$$
\begin{array}{ccccc}
 & & \overset{\cong}{\curvearrowright} & & \\
\mathcal{T}_q(\mathrm{t}_{\overline{\mathcal{P}}(\mathrm{H})}) & \hookrightarrow & \mathcal{T}_q(\mathrm{t}_{\mathrm{H}}) & \twoheadrightarrow & \overline{\mathcal{Z}}^{\vee}(\mathrm{H}) \\
{\scriptstyle\cong}\downarrow & & \downarrow{\scriptstyle\widehat{\mathrm{n}_{\mathrm{H}}}} & & {\scriptstyle \mathrm{z}_{\mathrm{H}}}\downarrow{\scriptstyle\cong} \\
\mathcal{T}_q^{\vee}(\mathrm{t}_{\overline{\mathcal{Q}}(\mathrm{H})}) & \twoheadleftarrow & \mathcal{T}_q^{\vee}(\mathrm{t}_{\mathrm{H}}) & \hookleftarrow & \overline{\mathcal{Z}}(\mathrm{H}). \\
 & & \underset{\cong}{\curvearrowleft} & &
\end{array}
$$

We need to show that z_{H} is an isomorphism. This will follow once we establish the remaining three isomorphisms. The composite of the top-horizontal maps is an isomorphism since it is an isomorphism when $\mathrm{H} = \overline{\mathcal{T}}_q(\mathrm{V})$ (with both ends being $\mathcal{T}_q(\mathrm{t}_{\mathrm{V}})$) and every Coxeter q-bialgebra H is of this form by the rigidity theorem. See also (16.31) in this regard. Similarly, the composite of the bottom-horizontal maps is an isomorphism. The left-vertical map (defined as the composite of three maps) is the general q-norm map associated to the map $\mathrm{t}_{\overline{\mathcal{P}}(\mathrm{H})} \to \mathrm{t}_{\overline{\mathcal{Q}}(\mathrm{H})}$ whose A-component is $\mathrm{pq}_{\mathrm{H}} : \overline{\mathcal{P}}(\mathrm{H})[\mathrm{t}(A)] \to \overline{\mathcal{Q}}(\mathrm{H})[\mathrm{t}(A)]$ scalar multiplied by $|W_A|$. The latter is an isomorphism by Proposition 9.66, and hence so is the former by the last part of Theorem 3.79. □

Exercise 17.38. Let $\mathrm{char}(\Bbbk) \nmid |W|$. Thus $\overline{\mathcal{K}} \cong \overline{\mathcal{K}}^{\vee}$. Let H be a Coxeter q-bialgebra for a scalar q. Define a morphism $\overline{\mathcal{Z}}^{\vee}(\mathrm{H}) \to \overline{\mathcal{Z}}(\mathrm{H})$ of Coxeter q-bimonoids in two different ways as follows.

- Since $\overline{\mathcal{Z}}$ is a section of $\overline{\mathcal{K}}$, we have $\mathrm{H} \xrightarrow{\cong} \overline{\mathcal{K}}\overline{\mathcal{Z}}(\mathrm{H}) \xrightarrow{\cong} \overline{\mathcal{K}}^{\vee}\overline{\mathcal{Z}}(\mathrm{H})$. Now use that $\overline{\mathcal{Z}}^{\vee}$ is left adjoint to $\overline{\mathcal{K}}^{\vee}$.
- Since $\overline{\mathcal{Z}}^{\vee}$ is a section of $\overline{\mathcal{K}}^{\vee}$, we have $\overline{\mathcal{K}}\overline{\mathcal{Z}}^{\vee}(\mathrm{H}) \xrightarrow{\cong} \overline{\mathcal{K}}^{\vee}\overline{\mathcal{Z}}^{\vee}(\mathrm{H}) \xrightarrow{\cong} \mathrm{H}$. Now use that $\overline{\mathcal{Z}}$ is right adjoint to $\overline{\mathcal{K}}$.

Check: The two ways yield the same morphism which in fact coincides with the morphism z_{H} in (17.35). Moreover, the composite

$$
\begin{array}{ccccccc}
\mathrm{H} & \xrightarrow{\cong} & \overline{\mathcal{K}}^{\vee}\overline{\mathcal{Z}}^{\vee}(\mathrm{H}) & \xrightarrow[\cong]{\overline{\mathcal{K}}^{\vee}(\mathrm{z}_{\mathrm{H}})} & \overline{\mathcal{K}}^{\vee}\overline{\mathcal{Z}}(\mathrm{H}) & & \\
 & & {\scriptstyle\cong}\downarrow & & \downarrow{\scriptstyle\cong} & & \\
 & & \overline{\mathcal{K}}\overline{\mathcal{Z}}^{\vee}(\mathrm{H}) & \xrightarrow[\overline{\mathcal{K}}(\mathrm{z}_{\mathrm{H}})]{\cong} & \overline{\mathcal{K}}\overline{\mathcal{Z}}(\mathrm{H}) & \xrightarrow[\cong]{} & \mathrm{H}
\end{array}
$$

is identity, with each map an isomorphism of Coxeter q-bialgebras.

17.5.3. General norm maps linked by the trivialization functor and its lifts.

We continue the ideas in Lemma 17.31 and Exercise 17.33. The natural transformation $\mathrm{z} : \overline{\mathcal{Z}}^{\vee} \to \overline{\mathcal{Z}}$ when applied to any general q-norm map of Coxeter q-bialgebras yields a general q-norm map of Coxeter q-bimonoids. More precisely:

Lemma 17.39. *For q a scalar and $f : \mathrm{C} \to \mathrm{A}$ a map of Coxeter spaces,*

$$\begin{array}{ccc} \overline{\mathcal{Z}}^{\vee}(\overline{\mathcal{T}}_q(\mathrm{C})) & \xrightarrow{\overline{\mathcal{Z}}(\widehat{f})\,\mathrm{z}_{\overline{\mathcal{T}}_q(\mathrm{C})} \,=\, \mathrm{z}_{\overline{\mathcal{T}}_q{}^{\vee}(\mathrm{A})}\overline{\mathcal{Z}}^{\vee}(\widehat{f})} & \overline{\mathcal{Z}}(\overline{\mathcal{T}}_q{}^{\vee}(\mathrm{A})) \\ \| & & \| \\ \mathcal{T}_q(\mathtt{t}_{\mathrm{C}}) & \xrightarrow[\widehat{\mathtt{t}_f\mathrm{n}_{\mathrm{C}}} \,=\, \widehat{\mathrm{n}_{\mathrm{A}}\mathtt{t}_f}]{} & \mathcal{T}_q^{\vee}(\mathtt{t}_{\mathrm{A}}) \end{array} \tag{17.37}$$

as Coxeter q-bimonoids, with hats as in (3.146) *and* (9.144).

The equality in the top-horizontal arrow is by naturality of z wrt $\widehat{f}$.

PROOF. One way to expand the above diagram is as follows.

$$\begin{array}{ccccc} \mathcal{T}_q(\mathtt{t}_{\mathrm{C}}) & & \xrightarrow{\quad\widehat{\mathtt{t}_f\mathrm{n}_{\mathrm{C}}}\quad} & & \mathcal{T}_q^{\vee}(\mathtt{t}_{\mathrm{A}}) \\ \downarrow & & & & \uparrow \\ \mathcal{T}_q\mathtt{t}_{\overline{\mathcal{T}}_q(\mathrm{C})} & \xrightarrow{\widehat{\mathrm{n}_{\overline{\mathcal{T}}_q(\mathrm{C})}}} & \mathcal{T}_q^{\vee}\mathtt{t}_{\overline{\mathcal{T}}_q(\mathrm{C})} & \xrightarrow{\mathcal{T}_q^{\vee}\mathtt{t}_{\widehat{f}}} & \mathcal{T}_q^{\vee}\mathtt{t}_{\overline{\mathcal{T}}_q{}^{\vee}(\mathrm{A})} \\ \downarrow & & \uparrow & & \uparrow \\ \overline{\mathcal{Z}}^{\vee}(\overline{\mathcal{T}}_q(\mathrm{C})) & \xrightarrow[\mathrm{z}_{\overline{\mathcal{T}}_q(\mathrm{C})}]{} & \overline{\mathcal{Z}}(\overline{\mathcal{T}}_q(\mathrm{C})) & \xrightarrow[\overline{\mathcal{Z}}(\widehat{f})]{} & \overline{\mathcal{Z}}(\overline{\mathcal{T}}_q{}^{\vee}(\mathrm{A})) \end{array}$$

(with curved isomorphisms $\cong$ from $\mathcal{T}_q(\mathtt{t}_{\mathrm{C}})$ to $\overline{\mathcal{Z}}^{\vee}(\overline{\mathcal{T}}_q(\mathrm{C}))$ on the left and from $\overline{\mathcal{Z}}(\overline{\mathcal{T}}_q{}^{\vee}(\mathrm{A}))$ to $\mathcal{T}_q^{\vee}(\mathtt{t}_{\mathrm{A}})$ on the right)

The first square commutes by definition of z. The second square commutes by naturality of $\overline{\mathcal{Z}} \to \mathcal{T}_q^{\vee}\mathtt{t}_{(-)}$ wrt $\widehat{f}$. For the pentagon, we observe that the composite obtained by going down, across, across, up, when restricted to $\mathtt{t}_{\mathrm{C}}$ and projected on $\mathtt{t}_{\mathrm{A}}$, yields the map $\mathtt{t}_f\mathrm{n}_{\mathrm{C}}$, so it must equal $\widehat{\mathtt{t}_f\mathrm{n}_{\mathrm{C}}}$. □

Exercise 17.40. Check the pentagon in the above diagram directly by evaluating on the A-component, on the F-summand and into the G-summand. (In the argument above, we looked at the case $F = G = A$.)

17.5.4. Isomorphism criteria for general q-norm maps.

This is a continuation of ideas in Section 17.4.5.

Theorem 17.41. *Suppose q is not a root of unity and* $\mathrm{char}(\Bbbk) \nmid |W|$. *Then for $f : \mathrm{C} \to \mathrm{A}$ a map of Coxeter spaces,*

$$\widehat{f} : \overline{\mathcal{T}}_q(\mathrm{C}) \to \overline{\mathcal{T}}_q{}^{\vee}(\mathrm{A}) \textit{ is an isomorphism of Coxeter } q\textit{-bialgebras}$$

iff

$$\widehat{\mathtt{t}_f\mathrm{n}_{\mathrm{C}}} : \mathcal{T}_q(\mathtt{t}_{\mathrm{C}}) \to \mathcal{T}_q^{\vee}(\mathtt{t}_{\mathrm{A}}) \textit{ is an isomorphism of Coxeter } q\textit{-bimonoids.}$$

PROOF. We combine Theorems 3.79 and 9.110 with identity (17.16) which implies that $\mathbf{log}_q(\mathtt{t}_f\mathrm{n}_{\mathrm{C}})$ is an isomorphism iff $\mathbf{log}_q(f)$ is an isomorphism. This used the hypothesis on field characteristic.

Alternatively: We use Theorem 17.37 and diagram (17.37). Also, functors $\overline{\mathcal{Z}}$ and $\overline{\mathcal{Z}}^{\vee}$ are full and faithful, and hence reflect isomorphisms. □

17.5.5. Commutative aspects. Recall functors $\overline{\mathcal{Z}}^{\vee}_{ab}$ and $\overline{\mathcal{Z}}^{coab}$ from Section 16.6. For a Coxeter bialgebra H, the map $\mathrm{z_H}$ in (17.35) factors as

$$\begin{array}{ccc} \overline{\mathcal{Z}}^{\vee}(\mathrm{H}) & \xrightarrow{\mathrm{z_H}} & \overline{\mathcal{Z}}(\mathrm{H}) \\ \downarrow & \nearrow_{\mathrm{z_H}} & \\ \overline{\mathcal{Z}}^{\vee}_{ab}(\mathrm{H}) & & \end{array} \qquad \begin{array}{ccc} \overline{\mathcal{Z}}^{\vee}(\mathrm{H}) & \xrightarrow{\mathrm{z_H}} & \overline{\mathcal{Z}}(\mathrm{H}) \\ \downarrow & & \uparrow \\ \overline{\mathcal{Z}}^{\vee}_{ab}(\mathrm{H}) & \xrightarrow[\mathrm{z_H}]{} & \overline{\mathcal{Z}}^{coab}(\mathrm{H}) \end{array} \qquad \begin{array}{ccc} \overline{\mathcal{Z}}^{\vee}(\mathrm{H}) & \xrightarrow{\mathrm{z_H}} & \overline{\mathcal{Z}}(\mathrm{H}) \\ & \searrow^{\mathrm{z_H}} & \uparrow \\ & & \overline{\mathcal{Z}}^{coab}(\mathrm{H}) \end{array} \tag{17.38}$$

when H is, respectively, commutative, bicommutative, cocommutative. This is because $\overline{\mathcal{Z}}(\mathrm{H})$ is commutative when H is, and $\overline{\mathcal{Z}}^{\vee}(\mathrm{H})$ is cocommutative when H is. We continue to denote the induced maps by $\mathrm{z_H}$. Exercise 17.38 works for each one of them.

Theorem 17.42. *Let* $\mathrm{char}(\mathbb{k}) \nmid |W|$. *For a bicommutative Coxeter bialgebra* H, *the map* $\mathrm{z_H} : \overline{\mathcal{Z}}^{\vee}_{ab}(\mathrm{H}) \to \overline{\mathcal{Z}}^{coab}(\mathrm{H})$ *in* (17.38) *is an isomorphism.*

PROOF. We proceed as in the proof of Theorem 17.37 employing instead the diagram

$$\begin{array}{ccccc} & & \overset{\cong}{\frown\!\!\!\longrightarrow} & & \\ \mathcal{S}(\mathrm{t}_{\overline{\mathcal{P}}(\mathrm{H})}) & \hookrightarrow & \mathcal{S}(\mathrm{t_H}) & \twoheadrightarrow & \overline{\mathcal{Z}}^{\vee}_{ab}(\mathrm{H}) \\ {\scriptstyle\cong}\,\vdots & & \downarrow{\scriptstyle \widehat{\mathrm{n}_\mathrm{H}}} & & {\scriptstyle \mathrm{z_H}}\downarrow{\scriptstyle\cong} \\ \mathcal{S}^{\vee}(\mathrm{t}_{\overline{\mathcal{Q}}(\mathrm{H})}) & \twoheadleftarrow & \mathcal{S}^{\vee}(\mathrm{t_H}) & \hookleftarrow & \overline{\mathcal{Z}}^{coab}(\mathrm{H}). \\ & & \underset{\cong}{\longleftarrow\!\!\!\smile} & & \end{array}$$

The middle-vertical map is induced from (17.32) and we have denoted it again by $\widehat{\mathrm{n}_\mathrm{H}}$. We leave further details as an exercise. Relevant results are Proposition 9.65, item (3), Theorem 3.77, and the Leray–Samelson theorem for bicommutative Coxeter bialgebras. □

Exercise 17.43. Formulate and prove commutative analogues of diagram (17.37) and Theorem 17.41.

17.6. Antipodes under Fock functors

Recall the antipode of a Coxeter bimonoid from Section 3.14 and of a Coxeter bialgebra from Section 9.14. We now relate the two via bosonic Fock functors.

17.6.1. Antipodes under Fock functors. For a Coxeter q-bimonoid h, let S_{h} denote its antipode. It is a map $\mathsf{h} \to \mathsf{h}$ of Coxeter species defined by the Takeuchi formula (3.159). Let $\mathcal{F}$ be any Fock functor. Applying $\mathcal{F}$ yields a map $\mathcal{F}(\mathsf{h}) \to \mathcal{F}(\mathsf{h})$ of Coxeter spaces. It is natural to ask whether this coincides with $S_{\mathcal{F}(\mathsf{h})}$, the antipode of $\mathcal{F}(\mathsf{h})$. (The antipode for Coxeter q-bialgebras is defined by the Takeuchi formula (9.171).) In words, do Fock functors preserve antipodes? We address this below.

Proposition 17.44. *For a scalar q, for a Coxeter q-bimonoid* h,

$$\overline{\mathcal{K}}(S_{\mathsf{h}}) = S_{\overline{\mathcal{K}}(\mathsf{h})}, \qquad \overline{\mathcal{K}}^{\vee}(S_{\mathsf{h}}) = S_{\overline{\mathcal{K}}^{\vee}(\mathsf{h})},$$
$$\overline{\mathcal{K}}_{-1}(S_{\mathsf{h}}) = S_{\overline{\mathcal{K}}_{-1}(\mathsf{h})}, \qquad \overline{\mathcal{K}}^{\vee}_{-1}(S_{\mathsf{h}}) = S_{\overline{\mathcal{K}}^{\vee}_{-1}(\mathsf{h})}.$$

In other words, bosonic and fermionic Fock functors preserve antipodes.

PROOF. We first explain the argument for $\overline{\mathcal{K}}$. Fix $A \leq C$. Then, for any face G greater than A, consider the commutative diagram

$$\begin{array}{ccccccc}
\mathsf{h}[A] & \xrightarrow{\Delta_A^G} & \mathsf{h}[G] & \xrightarrow{\mu_A^G} & \mathsf{h}[A] & & \\
 & & {\scriptstyle w}\downarrow & & {\scriptstyle w}\downarrow & \searrow & \\
 & & \mathsf{h}[F] & \xrightarrow[\mu_A^F]{} & \mathsf{h}[A] & \twoheadrightarrow & \mathsf{h}[A]_{W_A},
\end{array}$$

where F is the face between A and C of the same type as G, and w is the Coxeter symmetry which sends GC to C. Now multiply by $(-1)^{\dim(G)}$ and sum over all G greater than A. This map factors through $\mathsf{h}[A]_{W_A}$. Going along the top yields $\overline{\mathcal{K}}(S_{\mathsf{h}})$, while going along the bottom yields $S_{\overline{\mathcal{K}}(\mathsf{h})}$ (in view of formulas (15.2) for $q = 1$). This completes the argument.

For $\overline{\mathcal{K}}^{\vee}$, we employ the dual diagram

$$\begin{array}{ccccccc}
 & & \mathsf{h}[A] & \xrightarrow{\Delta_A^G} & \mathsf{h}[G] & \xrightarrow{\mu_A^G} & \mathsf{h}[A] \\
 & \nearrow & \uparrow{\scriptstyle w} & & \uparrow{\scriptstyle w} & & \\
\mathsf{h}[A]^{W_A} & \hookrightarrow & \mathsf{h}[A] & \xrightarrow[\Delta_A^F]{} & \mathsf{h}[F] & &
\end{array}$$

and formulas (15.3) for $q = 1$.

We know from (3.162) that the o-signature functor preserves antipodes. Hence claims for $\overline{\mathcal{K}}_{-1}$ and $\overline{\mathcal{K}}^{\vee}_{-1}$ also follow. □

Remark 17.45. Full Fock functors $\mathcal{K}$ and $\mathcal{K}^{\vee}$ do not preserve antipodes in general. Going back to the above proof, we no longer have the triangle at the right end of the first diagram for $\mathcal{K}$, and dually, the triangle at the left end of the second diagram for $\mathcal{K}^{\vee}$. So the arguments do not go through for $\mathcal{K}$ and $\mathcal{K}^{\vee}$.

17.6.2. Logarithm of antipodes. Logarithm of the antipode of a Coxeter q-bimonoid was computed in (3.166), and of a Coxeter q-bialgebra in (9.181). We now show how to deduce the latter from the former.

PROOF OF FORMULA (9.181) VIA FOCK FUNCTORS. For a scalar q, for a Coxeter q-bimonoid h, for a face-type Z,

$$\begin{aligned}
\mathbf{log}(S_{\overline{\mathcal{K}}(\mathsf{h})})_Z &= \mathbf{log}(\overline{\mathcal{K}}(S_{\mathsf{h}}))_Z = \overline{\mathcal{K}}(\mathbf{log}(S_{\mathsf{h}}))_Z \\
&= (-1)^{\dim(Z)}\, \overline{\mathcal{K}}(\overline{\mathbf{log}}(\mathrm{id}_{\mathsf{h}}))_Z = (-1)^{\dim(Z)}\, \overline{\mathbf{log}}(\mathrm{id}_{\overline{\mathcal{K}}(\mathsf{h})})_Z.
\end{aligned}$$

We used Proposition 17.44, Exercise 17.13 and formula (3.166). Now use that every Coxeter q-bialgebra is of the form $\overline{\mathcal{K}}(\mathsf{h})$ for some Coxeter q-bimonoid h (assuming $\mathrm{char}(\Bbbk) \nmid |W|$).

For the above argument, we can also employ $\overline{\mathcal{K}}^{\vee}$ instead of $\overline{\mathcal{K}}$. □

17.6.3. Op and cop constructions. We now relate commutative aspects of antipodes of Coxeter bimonoids and Coxeter bialgebras.

PROOF OF LEMMA 9.130 VIA FOCK FUNCTORS. Let h be a commutative Coxeter bimonoid. By Lemma 3.96, S_{h} is a morphism of Coxeter monoids. Hence, $\overline{\mathcal{K}}(S_{\mathsf{h}}) = S_{\overline{\mathcal{K}}(\mathsf{h})}$ is a morphism of Coxeter algebras. Thus, Lemma 9.130 holds for $\mathrm{H} := \overline{\mathcal{K}}(\mathsf{h})$. This implies the general case since every commutative Coxeter bialgebra is of this form (assuming $\mathrm{char}(\mathbb{k}) \nmid |W|$). The cocommutative case is similar. □

We can handle the equivalent Lemma 9.131 as follows.

PROOF OF LEMMA 9.131 VIA FOCK FUNCTORS. Let h be a cocommutative Coxeter bimonoid. By Lemma 3.97, S_{h} is a group-like series of $\mathcal{C}(\mathsf{h}, \mathsf{h})$. Hence, it is a group-like series of $\overline{\mathcal{K}}^{\vee}(\mathcal{C}(\mathsf{h}, \mathsf{h}))$ by Proposition 17.8. By (17.22), it maps to the group-like series $\overline{\mathcal{K}}^{\vee}(S_{\mathsf{h}}) = S_{\overline{\mathcal{K}}^{\vee}(\mathsf{h})}$ of $\mathcal{C}(\overline{\mathcal{K}}^{\vee}(\mathsf{h}), \overline{\mathcal{K}}^{\vee}(\mathsf{h}))$. Thus, Lemma 9.131 holds for $\mathrm{H} := \overline{\mathcal{K}}^{\vee}(\mathsf{h})$, and the general case follows (assuming $\mathrm{char}(\mathbb{k}) \nmid |W|$). The commutative case is similar. □

More generally:

PROOF OF LEMMA 9.133 VIA FOCK FUNCTORS. We explain item (1). Let h be a Coxeter q-bimonoid. By item (1) of Lemma 3.98, $S_{\mathsf{h}} : {}^{\mathrm{op}}\mathsf{h} \to \mathsf{h}$ is a morphism of Coxeter monoids. Hence, $\overline{\mathcal{K}}^{\vee}(S_{\mathsf{h}}) : \overline{\mathcal{K}}^{\vee}({}^{\mathrm{op}}\mathsf{h}) \to \overline{\mathcal{K}}^{\vee}(\mathsf{h})$ is a morphism of Coxeter algebras. But $\overline{\mathcal{K}}^{\vee}(S_{\mathsf{h}}) = S_{\overline{\mathcal{K}}^{\vee}(\mathsf{h})}$ and $\overline{\mathcal{K}}^{\vee}({}^{\mathrm{op}}\mathsf{h}) = {}^{\mathrm{op}}\overline{\mathcal{K}}^{\vee}(\mathsf{h})$, the latter by Exercise 15.14. Thus, item (1) of Lemma 9.133 holds for $\mathrm{H} := \overline{\mathcal{K}}^{\vee}(\mathsf{h})$ which then implies the general case (assuming $\mathrm{char}(\mathbb{k}) \nmid |W|$). □

17.6.4. Antipodes of (co)free Coxeter bimonoids and bialgebras.

Exercise 17.46. Deduce antipode formulas in Theorems 9.138 and 9.139 from antipode formulas in Theorems 3.103 and 3.104 by tracing through the identifications

$$\overline{\mathcal{K}}^{\vee}(S_{\mathcal{T}_q(\mathrm{t_C})}) = S_{\overline{\mathcal{K}}^{\vee}\mathcal{T}_q(\mathrm{t_C})} = S_{\overline{\mathcal{T}_q}\overline{\mathcal{K}}^{\vee}(\mathrm{t_C})} = S_{\overline{\mathcal{T}_q}(\mathrm{C})},$$
$$\overline{\mathcal{K}}(S_{\mathcal{T}_q^{\vee}(\mathrm{t_A})}) = S_{\overline{\mathcal{K}}\mathcal{T}_q^{\vee}(\mathrm{t_A})} = S_{\overline{\mathcal{T}_q}^{\vee}\overline{\mathcal{K}}(\mathrm{t_A})} = S_{\overline{\mathcal{T}_q}^{\vee}(\mathrm{A})}.$$

The middle identifications arise from the (co)lax structures of $\overline{\mathcal{K}}$ and $\overline{\mathcal{K}}^{\vee}$ being isomorphisms.

17.7. Structure theorems under Fock functors

We discussed the structure theory of Coxeter bimonoids in Chapter 7 and structure theory of Coxeter bialgebras in Chapter 13. There is a striking parallel between the two, both in terms of results and proofs. Since Coxeter species are easier to handle than Coxeter spaces, it is natural to ask whether one can derive structure theorems for Coxeter bialgebras from corresponding theorems for Coxeter bimonoids. This is possible and a way to go is to use bosonic Fock functors and their adjoints to go back and forth between the two worlds. We have seen this technique in preceding sections as well.

17.7.1. Borel–Hopf. Borel–Hopf for cocommutative Coxeter bimonoids is stated in Theorem 7.1, and for cocommutative Coxeter bialgebras is stated in Theorem 13.1. Dual results for commutative Coxeter bimonoids and commutative Coxeter bialgebras are stated in Theorems 7.2 and 13.8.

Lemma 17.47 (Borel–Hopf isomorphisms). *Let* $\mathrm{char}(\Bbbk) \nmid |W|$. *For* h *a cocommutative Coxeter bimonoid and* $\mathrm{H} = \overline{\mathcal{K}}^\vee(\mathsf{h})$, *the diagrams*

$$
\begin{array}{ccc}
\overline{\mathcal{K}}^\vee(\mathcal{S}^\vee(\mathcal{P}(\mathsf{h}))) & \xrightarrow{\cong} & \overline{\mathcal{K}}^\vee(\mathsf{h}) \\
\| & & \| \\
\overline{\mathcal{S}}^\vee(\overline{\mathcal{P}}(\mathrm{H})) & \xrightarrow[\cong]{} & \mathrm{H}
\end{array}
\qquad
\begin{array}{ccc}
\overline{\mathcal{K}}^\vee(\mathsf{h}) & \xrightarrow{\cong} & \mathcal{K}^\vee(\mathcal{S}^\vee(\mathcal{P}(\mathsf{h}))) \\
\| & & \| \\
\mathrm{H} & \xrightarrow[\cong]{} & \overline{\mathcal{S}}^\vee(\overline{\mathcal{P}}(\mathrm{H}))
\end{array}
\tag{17.39}
$$

commute. The top-horizontal maps are induced from (7.1) *and* (7.2), *the bottom-horizontal maps are* (13.1) *and* (13.2).

For h *a commutative Coxeter bimonoid and* $\mathrm{H} = \overline{\mathcal{K}}(\mathsf{h})$, *the diagrams*

$$
\begin{array}{ccc}
\overline{\mathcal{K}}(\mathsf{h}) & \xrightarrow{\cong} & \overline{\mathcal{K}}(\mathcal{S}(\mathcal{Q}(\mathsf{h}))) \\
\| & & \| \\
\mathrm{H} & \xrightarrow[\cong]{} & \overline{\mathcal{S}}(\overline{\mathcal{Q}}(\mathrm{H}))
\end{array}
\qquad
\begin{array}{ccc}
\overline{\mathcal{K}}(\mathcal{S}(\mathcal{Q}(\mathsf{h}))) & \xrightarrow{\cong} & \overline{\mathcal{K}}(\mathsf{h}) \\
\| & & \| \\
\overline{\mathcal{S}}(\overline{\mathcal{Q}}(\mathrm{H})) & \xrightarrow[\cong]{} & \mathrm{H}
\end{array}
\tag{17.40}
$$

commute. The top-horizontal maps are induced from (7.3) *and* (7.4), *the bottom-horizontal maps are* (13.8) *and* (13.9).

PROOF. We explain (17.39). The second diagram follows from (17.9). For the first diagram, let us compute the bottom-horizontal map by going up, across, down. In short, we will use (7.1) to derive (13.1). The composite on the Z-component (with A the face of C_r of type Z), evaluated on $(x_T)_{T \geq Z}$, is

$$
\begin{aligned}
\sum_{T:\, T \geq Z} \sum_{\substack{F:\, F \geq A \\ \mathrm{t}(F) = T}} \zeta(A, F)\, \mu_A^F w_F(x_T) &= \sum_{T:\, T \geq Z} \zeta(Z, T) \sum_{\substack{F:\, F \geq A \\ \mathrm{t}(F) = T}} \mu_A^F w_F(x_T) \\
&= \sum_{T:\, T \geq Z} \zeta(Z, T)\, \mu_Z^T(x_T),
\end{aligned}
$$

where w_F is the Coxeter symmetry defined by $w_F(C_r) = FC_r$. Note very carefully: (13.1) has fewer terms than (7.1); the compensation comes from μ_Z^T which involves a sum of the μ_A^F (using (15.10) with $q = 1$). □

We now see how Borel–Hopf for Coxeter spaces can be deduced from Borel–Hopf for Coxeter species.

PROOF OF THEOREM 13.8 VIA FOCK FUNCTORS. Every commutative Coxeter bialgebra H is of the form $\overline{\mathcal{K}}(\mathsf{h})$ for some commutative Coxeter bimonoid h. For instance, we may take $\mathsf{h} := \overline{\mathcal{Z}}(\mathrm{H})$ following Observation 16.60. Alternatively, we may take $\mathsf{h} := \overline{\mathcal{Z}}_{ab}^\vee(\mathrm{H})$ following Theorem 16.71 and the fact that $\overline{\mathcal{K}}$ and $\overline{\mathcal{K}}^\vee$ are isomorphic (Proposition 14.44). Now use commutativity of diagrams (17.40) and Theorem 7.2 to deduce that the bottom-horizontal maps in (17.40) are inverse isomorphisms. □

Dually, one can use diagrams (17.39), Observation 16.56 and Theorem 7.1 to deduce Theorem 13.1.

17.7.2. Leray–Samelson. Leray–Samelson (primitive part version) for bicommutative Coxeter bimonoids is stated in Theorem 7.4, and for bicommutative Coxeter bialgebras is stated in Theorem 13.14. We now explain how the latter can be deduced from the former.

PROOF OF THEOREM 13.14 VIA FOCK FUNCTORS. For a Coxeter space V,

$$\overline{\mathcal{P}}\overline{\mathcal{S}}(\mathrm{V}) = \overline{\mathcal{P}}\overline{\mathcal{K}}^\vee \overline{\mathcal{Z}}^\vee_{ab}\overline{\mathcal{S}}(\mathrm{V}) = \overline{\mathcal{K}}^\vee \mathcal{P}\mathcal{S}(\mathrm{t_V}) = \overline{\mathcal{K}}^\vee(\mathrm{t_V}) = \mathrm{V}.$$

The first step used Theorem 16.71. In the second step, we used Proposition 15.46 to move $\overline{\mathcal{P}}$ past $\overline{\mathcal{K}}^\vee$, and the first diagram in Exercise 16.108 to move $\overline{\mathcal{S}}$ past $\overline{\mathcal{Z}}^\vee_{ab}$. The third step used Theorem 7.4. The last step used Proposition 16.14.

For a bicommutative Coxeter bialgebra H,

$$\overline{\mathcal{S}}\overline{\mathcal{P}}(\mathrm{H}) = \overline{\mathcal{S}}\overline{\mathcal{P}}\overline{\mathcal{K}}^\vee \overline{\mathcal{Z}}^\vee_{ab}(\mathrm{H}) = \overline{\mathcal{K}}^\vee \mathcal{S}\mathcal{P}\overline{\mathcal{Z}}^\vee_{ab}(\mathrm{H}) = \overline{\mathcal{K}}^\vee \overline{\mathcal{Z}}^\vee_{ab}(\mathrm{H}) = \mathrm{H}.$$

The first and last steps used Theorem 16.71. In the second step, we first used Proposition 15.46 to move $\overline{\mathcal{P}}$ past $\overline{\mathcal{K}}^\vee$ and then Proposition 15.29 to move $\overline{\mathcal{S}}$ past $\overline{\mathcal{K}}^\vee$. The third step used Theorem 7.4. Note very carefully: $\overline{\mathcal{Z}}^\vee_{ab}(\mathrm{H})$ is bicommutative by Observation 16.76.

Tracing through the above equalities, one checks that they agree with the unit and counit of the adjunction between $\overline{\mathcal{S}}$ and $\overline{\mathcal{P}}$. □

Dually, for the indecomposable part versions, one can start with Theorem 7.5 to deduce Theorem 13.17.

PROOF OF THEOREM 13.17 VIA FOCK FUNCTORS. The calculations proceed as follows.

$$\overline{\mathcal{Q}}\overline{\mathcal{S}}^\vee(\mathrm{V}) = \overline{\mathcal{Q}}\overline{\mathcal{K}}\overline{\mathcal{Z}}^{coab}\overline{\mathcal{S}}^\vee(\mathrm{V}) = \overline{\mathcal{K}}\mathcal{Q}\mathcal{S}^\vee(\mathrm{t_V}) = \overline{\mathcal{K}}(\mathrm{t_V}) = \mathrm{V},$$

$$\overline{\mathcal{S}}^\vee\overline{\mathcal{Q}}(\mathrm{H}) = \overline{\mathcal{S}}^\vee\overline{\mathcal{Q}}\overline{\mathcal{K}}\overline{\mathcal{Z}}^{coab}(\mathrm{H}) = \overline{\mathcal{K}}\mathcal{S}^\vee\mathcal{Q}\overline{\mathcal{Z}}^{coab}(\mathrm{H}) = \overline{\mathcal{K}}\overline{\mathcal{Z}}^{coab}(\mathrm{H}) = \mathrm{H}.$$

We used Proposition 15.48. □

17.7.3. Rigidity of Coxeter q-bialgebras for q not a root of unity. The rigidity theorem (primitive part version) for Coxeter q-bimonoids is stated in Theorem 7.8, and for Coxeter q-bialgebras is stated in Theorem 13.25. We now explain how the latter can be deduced from the former (under the additional assumption $\mathrm{char}(\Bbbk) \nmid |W|$). The argument is formally identical to the one above for Leray–Samelson.

PROOF OF THEOREM 13.25 VIA FOCK FUNCTORS. Let q be a scalar which is not a root of unity. For a Coxeter space V,

$$\overline{\mathcal{P}}\overline{\mathcal{T}}_q(\mathrm{V}) = \overline{\mathcal{P}}\overline{\mathcal{K}}^\vee \overline{\mathcal{Z}}^\vee\overline{\mathcal{T}}_q(\mathrm{V}) = \overline{\mathcal{K}}^\vee \mathcal{P}\mathcal{T}_q(\mathrm{t_V}) = \overline{\mathcal{K}}^\vee(\mathrm{t_V}) = \mathrm{V}.$$

The first step used Theorem 16.51. In the second step, we used Proposition 15.46 to move $\overline{\mathcal{P}}$ past $\overline{\mathcal{K}}^\vee$, and Exercise 16.54 to move $\overline{\mathcal{T}}_q$ past $\overline{\mathcal{Z}}^\vee$. The third step used Theorem 7.8. The last step used Proposition 16.14.

Similarly, for a Coxeter q-bialgebra H,

$$\overline{\mathcal{T}_q}\overline{\mathcal{P}}(\mathrm{H}) = \overline{\mathcal{T}_q}\overline{\mathcal{P}}\overline{\mathcal{K}}^\vee\overline{\mathcal{Z}}^\vee(\mathrm{H}) = \overline{\mathcal{K}}^\vee\mathcal{T}_q\mathcal{P}\overline{\mathcal{Z}}^\vee(\mathrm{H}) = \overline{\mathcal{K}}^\vee\overline{\mathcal{Z}}^\vee(\mathrm{H}) = \mathrm{H}.$$

The first and last steps used Theorem 16.51. In the second step, we first used Proposition 15.46 to move $\overline{\mathcal{P}}$ past $\overline{\mathcal{K}}^\vee$, and then Proposition 15.24 to move $\overline{\mathcal{T}_q}$ past $\overline{\mathcal{K}}^\vee$. The third step used Theorem 7.8.

Tracing through the above equalities, one checks that they agree with the unit and counit of the adjunction between $\overline{\mathcal{T}_q}$ and $\overline{\mathcal{P}}$. □

Dually, for the indecomposable part versions, one can start with Theorem 7.9 to deduce Theorem 13.30.

Exercise 17.48. Check directly: For q not a root of unity, for h a Coxeter q-bimonoid and $\mathrm{H} = \overline{\mathcal{K}}^\vee(\mathsf{h})$, the diagrams

$$\begin{array}{ccc} \overline{\mathcal{K}}^\vee(\mathcal{T}_q^\vee(\mathcal{P}(\mathsf{h}))) & \xrightarrow{\cong} & \overline{\mathcal{K}}^\vee(\mathsf{h}) \\ \| & & \| \\ \overline{\mathcal{T}_q}^\vee(\overline{\mathcal{P}}(\mathrm{H})) & \xrightarrow[\cong]{} & \mathrm{H} \end{array} \qquad \begin{array}{ccc} \overline{\mathcal{K}}^\vee(\mathsf{h}) & \xrightarrow{\cong} & \overline{\mathcal{K}}^\vee(\mathcal{T}_q^\vee(\mathcal{P}(\mathsf{h}))) \\ \| & & \| \\ \mathrm{H} & \xrightarrow[\cong]{} & \overline{\mathcal{T}_q}^\vee(\overline{\mathcal{P}}(\mathrm{H})) \end{array} \tag{17.41}$$

commute. The top-horizontal maps are induced from (7.10) and (7.11), the bottom-horizontal maps are (13.17) and (13.18). (For the second diagram, use (17.8). For the first diagram, proceed as in the proof of Lemma 17.47.)

We also have diagrams (17.41) with $\mathcal{T}_q$ and $\overline{\mathcal{T}_q}$ instead of $\mathcal{T}_q^\vee$ and $\overline{\mathcal{T}_q}^\vee$. In fact the first of these is implicitly mentioned in the last line of the above proof (and is a lot easier to check than the first diagram in (17.41)). One can pass between the two sets of diagrams using the second diagram in (17.26) involving q-norm maps.

There are similar diagrams involving the indecomposable part instead of the primitive part.

17.7.4. Hoffman–Newman–Radford. Recall the HNR results for Coxeter bimonoids from Section 7.5 and for Coxeter bialgebras from Section 13.5. Theorem 7.14 says that for a cocommutative Coxeter comonoid c, the Coxeter bimonoids $\mathcal{T}(\mathsf{c})$ and $\mathcal{T}(\mathsf{c}_t)$ are isomorphic. The analogous result for a cocommutative Coxeter coalgebra is given in Theorem 13.39. We now show how the latter can be deduced from the former.

Proof of Theorem 13.39 via Fock functors. Let C be a cocommutative Coxeter coalgebra. Put $\mathsf{c} := \mathsf{t}_{\mathrm{C}}$. This is a cocommutative Coxeter comonoid. Moreover, $\overline{\mathcal{K}}^\vee(\mathsf{c}) = \mathrm{C}$ by Proposition 16.27. It follows that $\overline{\mathcal{K}}^\vee(\mathsf{c}_t) = \mathrm{C}_t$. Then

$$\overline{\mathcal{T}}(\mathrm{C}) = \overline{\mathcal{T}}\overline{\mathcal{K}}^\vee(\mathsf{c}) = \overline{\mathcal{K}}^\vee\mathcal{T}(\mathsf{c}) \xrightarrow{\cong} \overline{\mathcal{K}}^\vee\mathcal{T}(\mathsf{c}_t) = \overline{\mathcal{T}}\overline{\mathcal{K}}^\vee(\mathsf{c}_t) = \overline{\mathcal{T}}(\mathrm{C}_t)$$

as cocommutative Coxeter bialgebras. The middle arrow is induced from the HNR isomorphism (7.16). We explained the first and last equalities. The remaining two equalities hold by Proposition 15.24 for $q = 1$. One checks that the above composite is precisely the HNR isomorphism (13.30). Similarly, by reversing the middle arrow above, we use (7.17) to obtain (13.31). □

Similarly, Theorem 7.17 says that for q not a root of unity, for a Coxeter comonoid c, the Coxeter q-bimonoids $\mathcal{T}_q(\mathsf{c})$ and $\mathcal{T}_q(\mathsf{c}_t)$ are isomorphic. The analogous result for a Coxeter coalgebra is given in Theorem 13.44. We now show how the latter can be deduced from the former. The argument is formally identical to the one above.

PROOF OF THEOREM 13.44 VIA FOCK FUNCTORS. Let q be a scalar which is not a root of unity. Let C be a Coxeter coalgebra. Then $\mathsf{c} := \mathsf{t}_{\mathrm{C}}$ is a Coxeter comonoid such that $\overline{\mathcal{K}}^{\vee}(\mathsf{c}) = \mathrm{C}$. Thus

$$\overline{\mathcal{T}}_q(\mathrm{C}) = \overline{\mathcal{T}}_q\overline{\mathcal{K}}^{\vee}(\mathsf{c}) = \overline{\mathcal{K}}^{\vee}\mathcal{T}_q(\mathsf{c}) \xrightarrow{\cong} \overline{\mathcal{K}}^{\vee}\mathcal{T}_q(\mathsf{c}_t) = \overline{\mathcal{T}}_q\overline{\mathcal{K}}^{\vee}(\mathsf{c}_t) = \overline{\mathcal{T}}_q(\mathrm{C}_t)$$

as Coxeter q-bialgebras. The middle arrow is induced from the HNR isomorphism (7.22). One checks that the above composite is precisely the HNR isomorphism (13.36). Similarly, by reversing the middle arrow above, we use (7.23) to obtain (13.37). □

Dually, in the Coxeter monoid and Coxeter algebra setting, one can start with Theorem 7.15 to deduce Theorem 13.40. Similarly, one can start with Theorem 7.18 to deduce Theorem 13.46.

17.8. Lie theory under Fock functors

Coxeter Lie monoids were introduced in Chapter 6 and Coxeter Lie algebras in Chapter 12. They were linked by Fock functors and their adjoints in Sections 15.7 and 16.10 with related material in Sections 15.8 and 15.9. We now carry these ideas forward. In particular, continuing in the spirit of Section 17.7, we show how to deduce structure theorems such as PBW and CMM for Coxeter Lie algebras from those for Coxeter Lie monoids.

17.8.1. Monad for Coxeter Lie algebras. Recall from Proposition 6.4 that $\mathbf{Lie} \circ \mathsf{p} = \mathcal{PT}(\mathsf{p})$ for a Coxeter species p. The same statement for a Coxeter space V is given in Proposition 12.9. We now show how the latter can be deduced from the former.

PROOF OF PROPOSITION 12.9 VIA FOCK FUNCTORS. For a Coxeter space V, pick a Coxeter species p such that $\overline{\mathcal{K}}^{\vee}(\mathsf{p}) = \mathrm{V}$. For instance, we may take $\mathsf{p} := \mathsf{t}_{\mathrm{V}}$ by using Proposition 16.14. Then

$$\mathbf{Lie} \circ \overline{\mathcal{K}}^{\vee}(\mathsf{p}) = \overline{\mathcal{K}}^{\vee}(\mathbf{Lie} \circ \mathsf{p}) = \overline{\mathcal{K}}^{\vee}\mathcal{PT}(\mathsf{p}) = \overline{\mathcal{P}}\overline{\mathcal{K}}^{\vee}\mathcal{T}(\mathsf{p}) = \overline{\mathcal{P}}\overline{\mathcal{T}}\overline{\mathcal{K}}^{\vee}(\mathsf{p}).$$

The first and last steps used Proposition 15.37 for $\mathbf{a}$ equal to $\mathbf{Lie}$ and $\mathbf{As}$. For the last step, a more direct way is to use Proposition 15.24. The middle steps used Propositions 6.4 and 15.46. □

Dually, one can use Proposition 6.6 to deduce Proposition 12.13.

17.8.2. PBW isomorphisms. We now link the PBW isomorphisms for Coxeter Lie monoids and Coxeter Lie algebras using bosonic Fock functors.

Proposition 17.49. *Let* $\mathrm{char}(\Bbbk) \nmid |W|$. *For an invariant noncommutative zeta function* ζ *and Coxeter Lie monoid* g, *the diagram of Coxeter coalgebras*

$$
\begin{array}{ccccc}
\overline{\mathcal{S}}\overline{\mathcal{K}}^{\vee}(\mathrm{g}) & & \xrightarrow{\cong} & & \overline{\mathcal{K}}^{\vee}\mathcal{S}(\mathrm{g}) & \\
 & \searrow & & & \downarrow\cong & \searrow \\
\cong\downarrow & & \overline{\mathcal{T}}\overline{\mathcal{K}}^{\vee}(\mathrm{g}) & \xrightarrow{\cong} & & & \overline{\mathcal{K}}^{\vee}\mathcal{T}(\mathrm{g}) \\
 & \swarrow & & & \downarrow & \swarrow \\
\overline{\mathcal{U}}\overline{\mathcal{K}}^{\vee}(\mathrm{g}) & & \xrightarrow[\cong]{} & & \overline{\mathcal{K}}^{\vee}\mathcal{U}(\mathrm{g}) &
\end{array}
\tag{17.42}
$$

commutes. There is a similar diagram for $\overline{\mathcal{K}}$.

The left triangle is (13.47) for $\mathrm{G} := \overline{\mathcal{K}}^{\vee}(\mathrm{g})$, the vertical map being the PBW isomorphism. The right triangle is $\overline{\mathcal{K}}^{\vee}$ applied to (7.28), the vertical map being $\overline{\mathcal{K}}^{\vee}$ applied to the PBW isomorphism. The oblique faces are as in (15.28) and (15.40).

Proof. The only additional claim is commutativity of the back face. This is a formal consequence of commutativity of the remaining faces. □

17.8.3. PBW. We now explain how to deduce PBW for Coxeter Lie algebras using PBW for Coxeter Lie monoids.

Proof of PBW Theorem 13.53 via Fock functors. Let us go back to the prism (17.42). That the first vertical map is an isomorphism is implied by the remaining isomorphisms. The same remark applies to the prism involving $\overline{\mathcal{K}}$. This proves PBW Theorem 13.53 for $\mathrm{G} = \overline{\mathcal{K}}^{\vee}(\mathrm{g})$ and for $\mathrm{G} = \overline{\mathcal{K}}(\mathrm{g})$. But by Proposition 16.100, every Coxeter Lie algebra is of the form $\overline{\mathcal{K}}(\mathrm{g})$. This then proves Theorem 13.53 in general, as required. Independently, by Proposition 16.107, every Coxeter Lie algebra is also of the form $\overline{\mathcal{K}}^{\vee}(\mathrm{g})$, but we point out that our proof of this fact uses CMM Theorem 13.65 which is nontrivial. Also recall from Exercise 15.45 that $\overline{\mathcal{K}}(\mathrm{g}) \cong \overline{\mathcal{K}}^{\vee}(\mathrm{g})$ which is an elementary fact. □

Similar remarks apply to PBW Theorem 7.23 for Coxeter Lie comonoids and PBW Theorem 13.59 for Coxeter Lie coalgebras.

17.8.4. Solomon operator. Recall Solomon operator (7.29) for a Coxeter Lie monoid and Solomon operator (13.49) for a Coxeter Lie algebra. The two can be linked using bosonic Fock functors as follows.

Proposition 17.50. *Let* $\mathrm{char}(\Bbbk) \nmid |W|$. *For an invariant noncommutative zeta function* ζ *and Coxeter Lie monoid* g, *the diagrams of Coxeter coalgebras*

$$
\begin{array}{ccc}
\overline{\mathcal{T}}\overline{\mathcal{K}}^{\vee}(\mathrm{g}) & \xrightarrow{\mathrm{Sol}} & \overline{\mathcal{T}}\overline{\mathcal{K}}^{\vee}(\mathrm{g}) \\
\cong\downarrow & & \downarrow\cong \\
\overline{\mathcal{K}}^{\vee}\mathcal{T}(\mathrm{g}) & \xrightarrow[\mathrm{Sol}]{} & \overline{\mathcal{K}}^{\vee}\mathcal{T}(\mathrm{g})
\end{array}
\qquad
\begin{array}{ccc}
\overline{\mathcal{T}}\overline{\mathcal{K}}(\mathrm{g}) & \xrightarrow{\mathrm{Sol}} & \overline{\mathcal{T}}\overline{\mathcal{K}}(\mathrm{g}) \\
\cong\downarrow & & \downarrow\cong \\
\overline{\mathcal{K}}\mathcal{T}(\mathrm{g}) & \xrightarrow[\mathrm{Sol}]{} & \overline{\mathcal{K}}\mathcal{T}(\mathrm{g})
\end{array}
\tag{17.43}
$$

commute.

We elaborate on the first diagram. The top-horizontal map is Solomon operator (13.49) for $\mathrm{G} := \overline{\mathcal{K}}^\vee(\mathrm{g})$; the bottom-horizontal map is $\overline{\mathcal{K}}^\vee$ applied to Solomon operator (7.29). They make use of the Lie structure of g. The two vertical maps are the same and equal the isomorphism of Coxeter bialgebras $\overline{\psi}^\vee_{\mathsf{c}}$ in Proposition 15.24 with $\mathsf{c} := \mathrm{g}$ with trivial Coxeter coalgebra structure. It is the same as the lax structure of $\overline{\mathcal{K}}^\vee$ wrt the Coxeter associative operad **As**. The Lie structure of g is not relevant here.

PROOF. We explain the first diagram. For that, first consider the commutative diagram

$$\begin{array}{ccccccccc}
\overline{\mathcal{T}}\overline{\mathcal{K}}^\vee(\mathrm{g}) & \longrightarrow & \overline{\mathcal{P}}\overline{\mathcal{T}}\overline{\mathcal{K}}^\vee(\mathrm{g}) = \mathbf{Lie} \circ \overline{\mathcal{K}}^\vee(\mathrm{g}) & \longrightarrow & \overline{\mathcal{K}}^\vee(\mathrm{g}) & \hookrightarrow & \overline{\mathcal{T}}\overline{\mathcal{K}}^\vee(\mathrm{g}) \\
\cong\downarrow & & \cong\downarrow \qquad\qquad \cong\downarrow & & \| & & \downarrow\cong \\
\overline{\mathcal{K}}^\vee\mathcal{T}(\mathrm{g}) & \longrightarrow & \overline{\mathcal{K}}^\vee\mathcal{P}\mathcal{T}(\mathrm{g}) = \overline{\mathcal{K}}^\vee(\mathbf{Lie} \circ \mathrm{g}) & \longrightarrow & \overline{\mathcal{K}}^\vee(\mathrm{g}) & \hookrightarrow & \overline{\mathcal{K}}^\vee\mathcal{T}(\mathrm{g}).
\end{array}$$

The top line is (13.48) for $\mathrm{G} := \overline{\mathcal{K}}^\vee(\mathrm{g})$; the bottom line is $\overline{\mathcal{K}}^\vee$ applied to (7.30). For the first square, we use (17.9) with $\mathsf{h} := \mathcal{T}(\mathrm{g})$ and naturality of $\mathbf{log}(\mathrm{id})$. The middle vertical map is the lax structure of $\overline{\mathcal{K}}^\vee$ wrt the Coxeter operad **Lie**. This explains why the square to its right commutes: The top-horizontal map is the Lie structure of $\overline{\mathcal{K}}^\vee(\mathrm{g})$ and it is written as a composite using the lax structure of $\overline{\mathcal{K}}^\vee$ and the Lie structure of g. Since $\overline{\mathcal{K}}^\vee$ is lax, the last square commutes; it is an instance of the right diagram in (A.2).

Now take exponential of both the top and bottom lines in the above diagram and again use Exercise 17.13. □

We now extend the above result to the factorizations of the two Solomon operators.

Proposition 17.51. *Let* $\mathrm{char}(\Bbbk) \nmid |W|$. *For an invariant noncommutative zeta function* ζ *and Coxeter Lie monoid* g, *the diagram of Coxeter coalgebras*

(17.44)

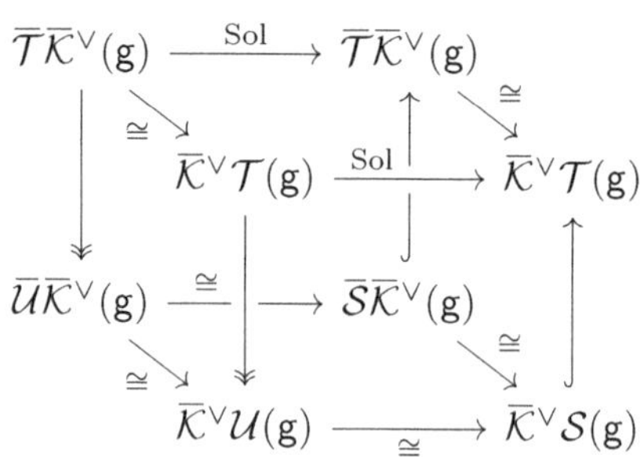

commutes. The top face is the first diagram in (17.43). *The back face is diagram* (13.50) *for* $\mathrm{G} := \overline{\mathcal{K}}^\vee(\mathrm{g})$; *the front face is* $\overline{\mathcal{K}}^\vee$ *applied to diagram* (7.31). *The side faces are as in* (15.40) *and* (15.28).

There is a similar diagram for $\overline{\mathcal{K}}$.

PROOF. The only additional claim is commutativity of the bottom face. This is a formal consequence of commutativity of the remaining faces. □

We now explain how the factorization of the Solomon operator for Coxeter Lie algebras can be obtained from the one for Coxeter Lie monoids.

PROOF OF THEOREM 13.55 VIA FOCK FUNCTORS. Let us revisit the cube (17.44). Suppose we do not assume any knowledge of diagram (13.50). Then we do not have the map $\overline{\mathcal{U}}\overline{\mathcal{K}}^{\vee}(\mathrm{g}) \to \overline{\mathcal{S}}\overline{\mathcal{K}}^{\vee}(\mathrm{g})$ to start with. However, using the rest of the cube, we can deduce the existence of this map along with commutativity of the back and bottom faces. The same remark applies to the cube involving $\overline{\mathcal{K}}$. This yields diagram (13.50) for $\mathrm{G} = \overline{\mathcal{K}}^{\vee}(\mathrm{g})$ and for $\mathrm{G} = \overline{\mathcal{K}}(\mathrm{g})$ which takes care of the general case.

It remains to check that Solomon operator (13.49) is idempotent. Using that Solomon operator (7.29) is idempotent and by diagrams (17.43), we deduce that (13.49) is idempotent for $\mathrm{G} = \overline{\mathcal{K}}^{\vee}(\mathrm{g})$ and for $\mathrm{G} = \overline{\mathcal{K}}(\mathrm{g})$ which again takes care of the general case. □

Similar remarks apply to Solomon operator (7.34) for a Coxeter Lie comonoid and Solomon operator (13.57) for a Coxeter Lie coalgebra.

17.8.5. CMM. Recall CMM Theorem 7.26 for Coxeter Lie monoids and CMM Theorem 13.65 for Coxeter Lie algebras. We now show how the latter can be deduced from the former.

PROOF OF THEOREM 13.65 VIA FOCK FUNCTORS. To show $\overline{\mathcal{P}}\overline{\mathcal{U}}(\mathrm{G}) = \mathrm{G}$, we write $\mathrm{G} = \overline{\mathcal{K}}(\mathrm{g})$ and use compatibility of the universal enveloping and primitive part functors with $\overline{\mathcal{K}}$:

$$\overline{\mathcal{P}}\overline{\mathcal{U}}(\mathrm{G}) = \overline{\mathcal{P}}\overline{\mathcal{U}}\overline{\mathcal{K}}(\mathrm{g}) = \overline{\mathcal{P}}\overline{\mathcal{K}}\mathcal{U}(\mathrm{g}) = \overline{\mathcal{K}}\mathcal{P}\mathcal{U}(\mathrm{g}) = \overline{\mathcal{K}}(\mathrm{g}) = \mathrm{G}.$$

Similarly, to show $\overline{\mathcal{U}}\overline{\mathcal{P}}(\mathrm{H}) = \mathrm{H}$, we write $\mathrm{H} = \overline{\mathcal{K}}(\mathsf{h})$ and proceed:

$$\overline{\mathcal{U}}\overline{\mathcal{P}}(\mathrm{H}) = \overline{\mathcal{U}}\overline{\mathcal{P}}\overline{\mathcal{K}}(\mathsf{h}) = \overline{\mathcal{U}}\overline{\mathcal{K}}\mathcal{P}(\mathsf{h}) = \overline{\mathcal{K}}\mathcal{U}\mathcal{P}(\mathsf{h}) = \overline{\mathcal{K}}(\mathsf{h}) = \mathrm{H}.$$

In both cases, the key step is to use Theorem 7.26. □

Similar remarks apply to CMM Theorem 7.27 for Coxeter Lie comonoids and CMM Theorem 13.68 for Coxeter Lie coalgebras.

17.9. Notes

17.9.1. Series. (Sections 17.1, 17.2, 17.3.) Comparison of series of Coxeter species and Coxeter spaces looks new even from a classical perspective; see the brief comments in [**8**, Section 12.5] in this regard. Similarly, comparison of actions of the invariant bilune-incidence algebra, comparison of internal homs, comparison of universal measurings are new in the classical setting.

17.9.2. General norm maps. (Sections 17.4 and 17.5.) Classical analogues of Lemma 17.30, Theorem 17.41 on comparison of isomorphism criteria, and other results are not in [**7**]. In fact, general norm maps were not considered in [**7**].

17.9.3. Antipode. (Section 17.6.) For Proposition 17.44, see [**7**, Paragraph at the end of Section 15.2.1]. Exercise 17.46 is new from a classical perspective.

17.9.4. Structure results. (Sections 17.7 and 17.8.) This material is new from a classical perspective.

CHAPTER 18

Examples of Fock spaces

We saw examples of Coxeter bimonoids in Chapter 4 and of Coxeter bialgebras in Chapter 10. We now link them using Fock functors. See Table 18.1. An entry in the left column relates to the corresponding entry in the right column via a Fock functor. For instance, the Coxeter bimonoids $\mathsf{\Gamma}$, $\mathsf{\Gamma}^*$, E give rise to the Coxeter bialgebras $\Bbbk[x]$, $\Bbbk\{x\}$, and so on. As a consequence, one can assert the existence of and derive properties of the latter (such as commutativity and freeness) from those of the former. One can also compute the primitive and indecomposable parts and antipode formulas of the latter from those of the former. In addition, interrelationships among Fock functors allow us to relate Coxeter bialgebras arising from the same Coxeter bimonoid. For instance, for a W-module M, the Coxeter bialgebras $\overline{\mathcal{T}}(M)$, $\overline{\mathcal{T}}^\vee(M)$, $\overline{\mathcal{S}}(M)$, $\overline{\mathcal{S}}^\vee(M)$ can be constructed and related to each other starting with the single Coxeter bimonoid E_M.

TABLE 18.1. Relating Coxeter bimonoids and Coxeter bialgebras.

Coxeter bimonoids		Coxeter bialgebras
$\mathsf{\Gamma}$, $\mathsf{\Gamma}^*$ (chambers) E		$\Bbbk[x]$, $\Bbbk\{x\}$
$\mathsf{\Gamma}_M$, $\mathsf{\Gamma}^*_M$ (decorated chambers) E_M	Fock $\longrightarrow$ functor	$\overline{\mathcal{T}}(M)$, $\overline{\mathcal{T}}^\vee(M)$ $\overline{\mathcal{S}}(M)$, $\overline{\mathcal{S}}^\vee(M)$
Σ, Σ^* (faces) $\widehat{\mathsf{Q}}$, $\widehat{\mathsf{Q}}^*$ (top-nested faces)		$\overline{\Sigma}$, $\overline{\Sigma}^*$ (face-types)
Π (flats)		$\overline{\Pi}$ (flat-types)
$\mathsf{I\Gamma}$ (pairs of chambers)		$\overline{\mathsf{I\Gamma}}$ (Coxeter symmetries)

The above ideas extend to the q-setting. For instance, the Coxeter q-bialgebras $\Bbbk_q[x]$ and $\Bbbk_q\{x\}$ can be obtained either by applying q-full Fock functors to E or bosonic Fock functors to $\mathsf{\Gamma}_q$ and $\mathsf{\Gamma}^*_q$. Similar considerations apply to $\overline{\mathcal{T}_q}(M)$ and $\overline{\mathcal{T}_q}^\vee(M)$, and so on.

Thus, Fock functors provide an alternative way to deduce many of the results in Chapter 10. Interestingly, they also give an alternative way to deduce many of the enumeration identities in Sections 1.8 and 1.9.

Adjoints. The trivialization functor sends Coxeter spaces to Coxeter species. It is adjoint to the bosonic Fock functors. Its lifts send Coxeter (co, bi)algebras to Coxeter (co, bi)monoids. We compute the values of these functors on some of the Coxeter bialgebras in the right column in Table 18.1.

Rank one. Let $\mathcal{A}$ be a rank-one arrangement. We know from general theory that Fock functors carry Coxeter $\mathcal{A}$-bimonoids to Coxeter $\mathcal{A}$-bialgebras. Their adjoints go in the other direction. We make these constructions explicit in this toy example.

18.1. Coxeter space characteristic of chambers as a Fock space

Recall the Coxeter species characteristic of chambers x and its decorated version x_M from Section 4.1, and the Coxeter space characteristic of chambers X and its decorated version X_M from Section 10.1. We now link them using Fock functors.

Lemma 18.1. *We have*

$$\begin{gathered}\mathcal{K}(\mathsf{x}) = \overline{\mathcal{K}}(\mathsf{x}) = \mathrm{X} \quad \text{and} \quad \mathcal{K}^\vee(\mathsf{x}) = \overline{\mathcal{K}}^\vee(\mathsf{x}) = \mathrm{X}, \\ \overline{\mathcal{K}}_{-1}(\mathsf{x}) = \overline{\mathcal{K}}^\vee_{-1}(\mathsf{x}) = \mathrm{X}\end{gathered} \tag{18.1}$$

as Coxeter spaces. More generally, for a left W-module M,

$$\begin{gathered}\mathcal{K}(\mathsf{x}_M) = \overline{\mathcal{K}}(\mathsf{x}_M) = \mathrm{X}_M \quad \text{and} \quad \mathcal{K}^\vee(\mathsf{x}_M) = \overline{\mathcal{K}}^\vee(\mathsf{x}_M) = \mathrm{X}_M, \\ \overline{\mathcal{K}}_{-1}(\mathsf{x}_M) = \overline{\mathcal{K}}^\vee_{-1}(\mathsf{x}_M) = \mathrm{X}_M\end{gathered} \tag{18.2}$$

as Coxeter spaces.

PROOF. We elaborate on one identity in (18.2). For a top-nested face (A, C),

$$\overline{\mathcal{K}}^\vee(\mathsf{x}_M)[A, C] = \mathsf{x}_M[A]^{W_A} = \begin{cases} M & \text{if } A = C \\ 0 & \text{otherwise} \end{cases} = \mathrm{X}_M[A, C].$$

Also, support and type morphisms of $\overline{\mathcal{K}}^\vee(\mathsf{x}_M)$ agree with those of X_M. So they are equal as Coxeter spaces. The other calculations are identical: Whether we write (signed) invariants or coinvariants or neither does not matter since W_C is the trivial group for any C. □

Exercise 18.2 (Trivialization functor). Check:

$$\mathsf{t}\mathrm{X} = \mathsf{x}. \tag{18.3}$$

More generally, for a left W-module M,

$$\mathsf{t}\mathrm{X}_M = \mathsf{x}'_M \quad \text{and} \quad \mathsf{t}\mathrm{X}_M = \mathsf{x}_M,\ \mathsf{t}\mathrm{X}'_M = \mathsf{x}'_M, \tag{18.4}$$

the former when working with definitions (10.4) and (16.1), and the latter when working with definitions (10.3) and (16.5). Recall that x_M and x'_M are isomorphic, and so are X_M and X'_M.

18.2. Coxeter bialgebra of polynomials as a Fock space

Recall the exponential Coxeter bimonoid E from Section 4.2 and Coxeter bimonoid of chambers Γ from Section 4.3. Similarly, recall the Coxeter bialgebra of polynomials $\Bbbk[x]$ and its relatives from Section 10.2. We now link them as well as their basic properties using Fock functors and their adjoints. As an application, we deduce many of the enumeration identities in Section 1.8.

18.2.1. Exponential Coxeter bimonoid. Let us compute values of Fock functors on the exponential Coxeter bimonoid E. We start with the q-full Fock functors.

Lemma 18.3. *For a scalar q,*

$$\mathcal{K}_q(\mathsf{E}) = \Bbbk_q[x] \quad \text{and} \quad \mathcal{K}_q^\vee(\mathsf{E}) = \Bbbk_q\{x\} \tag{18.5}$$

as Coxeter q-bialgebras.

Proof. We begin with the first claim. For T a face-type, $\mathcal{K}_q(\mathsf{E})$ and $\Bbbk_q[x]$ on the T-component both equal $\Bbbk$. Moreover, structure maps are all identities. So they are isomorphic as Coxeter spaces. We now verify that under this identification, the product and coproduct of $\mathcal{K}_q(\mathsf{E})$ given by formulas (15.9) match those of $\Bbbk_q[x]$ in (10.10). The product part is clear, while the coproduct part works as follows.

$$\Delta_Z^T(x^Z) = \sum_{G:\, G \geq A,\, \mathrm{t}(G)=T} q^{\mathrm{dist}(GC_r, C_r)}\, x^T = d_{T/Z}(q)\, x^T,$$

where A is the face of C_r of type Z. The last step used definition of $d_{T/Z}(q)$ which is to apply (1.30) to $\boldsymbol{\alpha}_Z$.

For the second claim, we proceed similarly using formulas (15.10) and match them with (10.13). Alternatively, we can deduce the second claim from the first by duality. In more detail, we use that E is self-dual, $\Bbbk_q[x]$ and $\Bbbk_q\{x\}$ are duals of each other, and $\mathcal{K}_q$ and $\mathcal{K}_q^\vee$ are conjugates of each other wrt duality which then implies that $\mathcal{K}_q(\mathsf{h})$ and $\mathcal{K}_q^\vee(\mathsf{h}^*)$ are duals of each other (see Section 15.1.11). □

Now let us go to bosonic and fermionic Fock functors.

Lemma 18.4. *We have*

$$\overline{\mathcal{K}}(\mathsf{E}) = \Bbbk[x] \quad \text{and} \quad \overline{\mathcal{K}}^\vee(\mathsf{E}) = \Bbbk\{x\} \tag{18.6}$$

as Coxeter bialgebras, and if $\mathrm{char}(\Bbbk) \neq 2$, then

$$\overline{\mathcal{K}}_{-1}(\mathsf{E}) = \overline{\mathcal{K}}_{-1}^\vee(\mathsf{E}) = \Bbbk[dx] \tag{18.7}$$

as signed Coxeter bialgebras. The latter is the signed Coxeter bialgebra of dual numbers.

Proof. We have $\mathcal{K}(\mathsf{E}) = \overline{\mathcal{K}}(\mathsf{E})$ and $\mathcal{K}^\vee(\mathsf{E}) = \overline{\mathcal{K}}^\vee(\mathsf{E})$ since $\mathsf{E}[A]$ is the trivial representation of W_A for each A. So (18.6) follows from (18.5) for $q = 1$. The claims in (18.7) can be checked directly. □

Exercise 18.5. For the signed exponential Coxeter bimonoid E^{-},

$$\mathcal{K}_q(\mathsf{E}^-) = \Bbbk_{-q}[x] \quad \text{and} \quad \mathcal{K}_q^\vee(\mathsf{E}^-) = \Bbbk_{-q}\{x\} \tag{18.8}$$

as Coxeter $(-q)$-bialgebras. Deduce this from (15.14) and also check directly. Similarly,

$$\mathcal{K}_q(\mathsf{E}_-) = \Bbbk_q[x]_- \quad \text{and} \quad \mathcal{K}_q^\vee(\mathsf{E}_-) = \Bbbk_q\{x\}_- \tag{18.9}$$

as Coxeter $(-q)$-bialgebras.

18.2.2. Coxeter bimonoid of chambers. The Coxeter bialgebra of polynomials can also be obtained from the Coxeter bimonoid of chambers:

Lemma 18.6. *For a scalar q,*

$$\begin{aligned} \overline{\mathcal{K}}(\Gamma_q) = \overline{\mathcal{K}}^\vee(\Gamma_q) &= \Bbbk_q[x], \\ \overline{\mathcal{K}}(\Gamma_q^*) = \overline{\mathcal{K}}^\vee(\Gamma_q^*) &= \Bbbk_q\{x\} \end{aligned} \tag{18.10}$$

as Coxeter q-bialgebras, and

$$\begin{aligned} \overline{\mathcal{K}}_{-1}(\Gamma_q) = \overline{\mathcal{K}}_{-1}^\vee(\Gamma_q) &= \Bbbk_{-q}[x], \\ \overline{\mathcal{K}}_{-1}(\Gamma_q^*) = \overline{\mathcal{K}}_{-1}^\vee(\Gamma_q^*) &= \Bbbk_{-q}\{x\} \end{aligned} \tag{18.11}$$

as Coxeter $(-q)$-bialgebras.

PROOF. The first claim in (18.10) can be handled directly as in the proof of Lemma 18.3. The key observation is that $\Gamma[A]$ is the left regular representation of W_A, hence its space of W_A-invariants and of W_A-coinvariants both equal $\Bbbk$. Part of the first claim also follows from (18.5) and Proposition 15.8 with $\mathsf{h} := \mathsf{E}$. The second claim in (18.10) follows from the first by duality. Finally, (18.11) can be deduced from (18.10) by employing (15.15) and Exercise 4.12. □

Remark 18.7. We can also consider values of full Fock functors on the Coxeter q-bimonoid of chambers and its dual. This yields Coxeter q-bialgebras $\mathcal{K}(\Gamma_q)$, $\mathcal{K}^\vee(\Gamma_q)$, and so on. The Coxeter q-bialgebra of symmetries arises in this manner, see Exercise 18.49.

18.2.3. Face-type enumeration identities. It is easy to see that E is a Coxeter bimonoid. Hence, by general theory, $\mathcal{K}_q(\mathsf{E})$ is a Coxeter q-bialgebra. It then follows from Lemma 18.3 that $\Bbbk_q[x]$ with product and coproduct (10.10) is a Coxeter q-bialgebra. In particular, the coproduct is coassociative and the q-bialgebra axiom holds. As a consequence, by reversing the logical flow in Section 10.2.4, we deduce identities (1.32) and (1.59).

One can also deduce these enumeration identities by starting with the Coxeter q-bimonoid Γ_q and using Lemma 18.6.

18.2.4. q-norm map. Let us now connect the q-norm transformation κ_q between q-full Fock functors to the q-norm map $\overline{\kappa}_q$ between the Coxeter q-bialgebras of polynomials and divided powers.

Lemma 18.8. *For a scalar q, we have the diagram of Coxeter q-bialgebras*

$$\begin{array}{ccc} \mathcal{K}_q(\mathsf{E}) & \xrightarrow{(\kappa_q)_{\mathsf{E}}} & \mathcal{K}_q^\vee(\mathsf{E}) \\ \| & & \| \\ \Bbbk_q[x] & \xrightarrow[\bar{\kappa}_q]{} & \Bbbk_q\{x\}. \end{array} \tag{18.12}$$

The top-horizontal map is as in (15.22)*; the bottom-horizontal map is* (10.15).

PROOF. This follows from (1.36) and the fact that $\mathsf{E}[A]$ is the trivial representation of W_A for each face A. □

Proposition 15.16 with $\mathsf{h} := \mathsf{E}$ (which says something about $(\kappa_q)_{\mathsf{E}}$) implies Lemma 10.12 and Corollary 10.13 (which says something about $\bar{\kappa}_q$), thus providing an alternative way to prove these two results. Yet another way is given in Exercise 18.9 below. Also, by definitions,

$$\Im_q(\mathsf{E}) = \Bbbk_q(x), \tag{18.13}$$

and diagram (15.24) with $\mathsf{h} := \mathsf{E}$ coincides with diagram (10.16).

Exercise 18.9. Check: The q-norm map on polynomials can be obtained from the q-norm map on chambers as follows. For a scalar q, we have diagrams of Coxeter q-bialgebras

$$\begin{array}{ccc} \overline{\mathcal{K}}(\Gamma_q) & \xrightarrow{\overline{\mathcal{K}}(\kappa_q)} & \overline{\mathcal{K}}(\Gamma_q^*) \\ \| & & \| \\ \Bbbk_q[x] & \xrightarrow[\bar{\kappa}_q]{} & \Bbbk_q\{x\} \end{array} \qquad \begin{array}{ccc} \overline{\mathcal{K}}^\vee(\Gamma_q) & \xrightarrow{\overline{\mathcal{K}}^\vee(\kappa_q)} & \overline{\mathcal{K}}^\vee(\Gamma_q^*) \\ \| & & \| \\ \Bbbk_q[x] & \xrightarrow[\bar{\kappa}_q]{} & \Bbbk_q\{x\}. \end{array} \tag{18.14}$$

The top-horizontal maps are induced from (4.28); the bottom-horizontal maps are as in (10.15). This provides another derivation of Lemma 10.12 and Corollary 10.13.

Diagrams (18.14) can be viewed as the special case $\mathsf{p} := \mathsf{x}$ of diagrams (17.26). See the discussion in Section 18.2.5 in this regard.

18.2.5. Freeness and cofreeness. Since the Coxeter bialgebra of polynomials and its relatives arise from the Coxeter bimonoids Γ, Γ^*, E, (co)freeness properties of the former can be deduced from those of the latter. This gives an alternative way of deducing results in Section 10.2.14.

For instance, for a scalar q,

$$\Bbbk_q[x] = \overline{\mathcal{K}}(\Gamma_q) = \overline{\mathcal{K}}\mathcal{T}_q(\mathsf{x}) = \overline{\mathcal{T}}_q\overline{\mathcal{K}}(\mathsf{x}) = \overline{\mathcal{T}}_q(\mathrm{X})$$

as Coxeter q-bialgebras,

$$\Bbbk[x] = \overline{\mathcal{K}}(\mathsf{E}) = \overline{\mathcal{K}}\mathcal{S}(\mathsf{x}) = \overline{\mathcal{S}}\overline{\mathcal{K}}(\mathsf{x}) = \overline{\mathcal{S}}(\mathrm{X})$$

as Coxeter bialgebras, and for $\mathrm{char}(\Bbbk) \neq 2$,

$$\Bbbk[dx] = \overline{\mathcal{K}}_{-1}(\mathsf{E}) = \overline{\mathcal{K}}(\mathsf{E}^-) = \overline{\mathcal{K}}\mathcal{E}(\mathsf{x}) = \overline{\mathcal{E}}\overline{\mathcal{K}}(\mathsf{x}) = \overline{\mathcal{E}}(\mathrm{X})$$

as signed Coxeter bialgebras. In the last case, another way to proceed is

$$\Bbbk[dx] = \overline{\mathcal{K}}_{-1}(\mathsf{E}) = \overline{\mathcal{K}}_{-1}\mathcal{S}(\mathsf{x}) = \overline{\mathcal{E}}\overline{\mathcal{K}}_{-1}(\mathsf{x}) = \overline{\mathcal{E}}(\mathrm{X}).$$

In these calculations, we used Propositions 15.24 and 15.29 which say that bosonic Fock functors preserve freeness. We also used (4.19), (4.27), (15.15), (18.1).

Dually, for a scalar q,

$$\Bbbk_q\{x\} = \overline{\mathcal{K}}^\vee(\Gamma_q^*) = \overline{\mathcal{K}}^\vee\mathcal{T}_q^\vee(\mathsf{x}) = \overline{\mathcal{T}_q}^\vee\overline{\mathcal{K}}^\vee(\mathsf{x}) = \overline{\mathcal{T}_q}^\vee(\mathrm{X})$$

as Coxeter q-bialgebras, and so on.

Alternatively, changing perspective, assuming (co)freeness properties, one can deduce

$$\overline{\mathcal{K}}(\Gamma_q) = \overline{\mathcal{K}}\mathcal{T}_q(\mathsf{x}) = \overline{\mathcal{T}_q}\overline{\mathcal{K}}(\mathsf{x}) = \overline{\mathcal{T}_q}(\mathrm{X}) = \Bbbk_q[x]$$

which is part of the claim in (18.10), and so on.

18.2.6. Primitive and indecomposable parts. In a similar manner, one can deduce results of Section 10.2.16.

To deduce formulas (10.21) and (10.24):

$$\overline{\mathcal{P}}(\Bbbk\{x\}) = \overline{\mathcal{P}}\overline{\mathcal{K}}^\vee(\mathsf{E}) = \overline{\mathcal{K}}^\vee\mathcal{P}(\mathsf{E}) = \overline{\mathcal{K}}^\vee(\mathsf{x}) = \mathrm{X},$$
$$\overline{\mathcal{Q}}(\Bbbk[x]) = \overline{\mathcal{Q}}\overline{\mathcal{K}}(\mathsf{E}) = \overline{\mathcal{K}}\mathcal{Q}(\mathsf{E}) = \overline{\mathcal{K}}(\mathsf{x}) = \mathrm{X}.$$

We used Propositions 15.46 and 15.48, and also (4.20) and (18.1). Alternatively, instead of E, one can also use Γ^* and Γ in the two calculations, respectively.

To deduce formulas (10.23) and (10.26):

$$\overline{\mathcal{P}}(\Bbbk[x]) = \overline{\mathcal{P}}\overline{\mathcal{K}}^\vee(\Gamma) = \overline{\mathcal{K}}^\vee\mathcal{P}(\Gamma) = \overline{\mathcal{K}}^\vee(\mathsf{Lie}),$$
$$\overline{\mathcal{Q}}(\Bbbk\{x\}) = \overline{\mathcal{Q}}\overline{\mathcal{K}}(\Gamma^*) = \overline{\mathcal{K}}\mathcal{Q}(\Gamma^*) = \overline{\mathcal{K}}(\mathsf{Lie}^*).$$

The last steps used (4.96) and its dual. Explicitly, the (A, C)-component of $\overline{\mathcal{P}}(\Bbbk[x])$ is the space of W_A-invariants of $\mathsf{Lie}[A]$, and of $\overline{\mathcal{Q}}(\Bbbk\{x\})$ is the space of W_A-coinvariants of $\mathsf{Lie}[A]^*$.

For the results in (10.22) and (10.25): For q not a root of unity,

$$\overline{\mathcal{P}}(\Bbbk_q[x]) = \overline{\mathcal{P}}\overline{\mathcal{K}}^\vee(\Gamma_q) = \overline{\mathcal{K}}^\vee\mathcal{P}(\Gamma_q) = \overline{\mathcal{K}}^\vee(\mathsf{x}) = \mathrm{X},$$
$$\overline{\mathcal{Q}}(\Bbbk_q\{x\}) = \overline{\mathcal{Q}}\overline{\mathcal{K}}(\Gamma_q^*) = \overline{\mathcal{K}}\mathcal{Q}(\Gamma_q^*) = \overline{\mathcal{K}}(\mathsf{x}) = \mathrm{X}.$$

We used (4.97) and its dual. Compare and contrast with the previous calculations for $q = 1$.

18.2.7. Exp-log correspondences. The exp-log correspondence for $\Bbbk[x]$ between primitive and group-like series was discussed in Section 10.2.17. It is strikingly similar to the exp-log correspondence for Γ outlined in Section 4.3.8. We now have an explanation for this similarity, namely, put $\mathsf{h} = \Gamma$ in Theorem 17.9 and use $\overline{\mathcal{K}}^\vee(\Gamma) = \Bbbk[x]$.

Exercise 18.10. Relate the space of series and exp-log correspondence of E in Section 4.2.8 with that of $\overline{\mathcal{K}}^\vee(\mathsf{E}) = \Bbbk\{x\}$.

18.2.8. Fock functor adjoints. Let us now compute values of the adjoints of Fock functors on $\Bbbk[x]$ and its relatives. This is independent of the preceding discussion.

Lemma 18.11. *For the trivialization functor* $\mathsf{t}_{(-)}$,

$$\mathsf{t}_{\Bbbk[x]} = \mathsf{E} \text{ as Coxeter monoids,} \qquad (18.15)$$
$$\mathsf{t}_{\Bbbk\{x\}} = \mathsf{E} \text{ as Coxeter comonoids.}$$

Similarly, for functors $\mathcal{L}$ *and* $\mathcal{L}^\vee$,

$$\mathcal{L}(\Bbbk[x]) = \Gamma \text{ as Coxeter monoids,} \qquad (18.16)$$
$$\mathcal{L}^\vee(\Bbbk\{x\}) = \Gamma^* \text{ as Coxeter comonoids.}$$

PROOF. The first claim follows from formulas (16.6). The second follows from formula (16.14) and its dual. □

Recall: There are also functors such as $\overline{\mathcal{Z}}$ and $\overline{\mathcal{Z}}^\vee$ which produce Coxeter bimonoids out of Coxeter bialgebras. The functor $\overline{\mathcal{Z}}^\vee$ preserves freeness while $\overline{\mathcal{Z}}$ preserves cofreeness, so their values on the Coxeter bialgebra of polynomials and divided power Coxeter bialgebra can be computed as follows.

Lemma 18.12. *For a scalar* q,

$$\overline{\mathcal{Z}}^\vee(\Bbbk_q[x]) = \Gamma_q \quad \textit{and} \quad \overline{\mathcal{Z}}(\Bbbk_q\{x\}) = \Gamma_q^*$$

as Coxeter q*-bimonoids.*

PROOF. For the first identity,

$$\overline{\mathcal{Z}}^\vee(\Bbbk_q[x]) = \overline{\mathcal{Z}}^\vee\overline{\mathcal{T}_q}(\mathrm{X}) = \mathcal{T}_q(\mathsf{t}_\mathrm{X}) = \mathcal{T}_q(\mathsf{x}) = \Gamma_q.$$

The second step used the left diagram in Exercise 16.55. The third step used (18.3). Dually,

$$\overline{\mathcal{Z}}(\Bbbk_q\{x\}) = \overline{\mathcal{Z}}\overline{\mathcal{T}_q}^\vee(\mathrm{X}) = \mathcal{T}_q^\vee(\mathsf{t}_\mathrm{X}) = \mathcal{T}_q^\vee(\mathsf{x}) = \Gamma_q^*$$

as required. □

Exercise 18.13. Make the identifications in Lemma 18.12 explicit as follows. For a scalar q, the map

$$\overline{\mathcal{Z}}(\Bbbk_q\{x\}) \xrightarrow{\cong} \Gamma_q^*, \qquad (x_F)_{F \geq A} \mapsto \sum_{C:\, C \geq A} x_C \, \mathtt{M}_{C/A} \qquad (18.17)$$

on the A-component, is an isomorphism of Coxeter q-bimonoids. (The key observation is that the scalars x_C determine the remaining scalars by $x_F := \sum_{C \geq F} x_C$ and the resulting element (x_F) indeed satisfies conditions (16.28).) Similarly, we have the isomorphism

$$\overline{\mathcal{Z}}^\vee(\Bbbk_q[x]) \xrightarrow{\cong} \Gamma_q, \qquad \overline{x}_F \mapsto x_F \sum_{C:\, C \geq F} \mathtt{H}_{C/A} \qquad (18.18)$$

on the A-component with $F \geq A$. The bar on top of x indicates an element in the quotient space.

Exercise 18.14. We now supplement Exercise 18.9. Specialize diagram (17.37) to $\mathrm{id} : \mathrm{X} \to \mathrm{X}$ to obtain: For a scalar q,

$$\begin{array}{ccc} \overline{\mathcal{Z}}^{\vee}(\Bbbk_q[x]) & \xrightarrow[\mathrm{z}_{\Bbbk_q\{x\}}\overline{\mathcal{Z}}^{\vee}(\bar{\kappa}_q)]{\overline{\mathcal{Z}}(\bar{\kappa}_q)\,\mathrm{z}_{\Bbbk_q[x]}} & \overline{\mathcal{Z}}(\Bbbk_q\{x\}) \\ \| & & \| \\ \Gamma_q & \xrightarrow[\kappa_q]{} & \Gamma_q^* \end{array} \tag{18.19}$$

as Coxeter q-bimonoids, with κ_q and $\bar{\kappa}_q$ the q-norm maps in (4.28) and (10.15). (The number operator is not manifest since X is characteristic of chambers.) In this way, interestingly, we can recover the q-norm map on chambers from the q-norm map on polynomials.

For this example, Theorem 17.41 says that when q is not a root of unity, κ_q is an isomorphism iff $\bar{\kappa}_q$ is an isomorphism. In fact, both statements are always true. (The field characteristic assumption is not required here since the number operator is not playing any role.)

Lemma 18.15. *We have*

$$\overline{\mathcal{Z}}^{\vee}_{ab}(\Bbbk[x]) = \overline{\mathcal{Z}}^{coab}(\Bbbk\{x\}) = \mathsf{E}$$

as Coxeter bimonoids, and for $\mathrm{char}(\Bbbk) \neq 2$,

$$\overline{\mathcal{Z}}^{\vee}_{-ab}(\Bbbk[dx]) = \overline{\mathcal{Z}}^{-coab}(\Bbbk[dx]) = \mathsf{E}^-$$

as signed Coxeter bimonoids.

PROOF. By employing diagrams (16.38) and (16.42), we obtain

$$\overline{\mathcal{Z}}^{\vee}_{ab}(\Bbbk[x]) = \overline{\mathcal{Z}}^{\vee}_{ab}\overline{\mathcal{S}}(\mathrm{X}) = \mathcal{S}(\mathsf{t}_{\mathrm{X}}) = \mathcal{S}(\mathsf{x}) = \mathsf{E},$$
$$\overline{\mathcal{Z}}^{\vee}_{-ab}(\Bbbk[dx]) = \overline{\mathcal{Z}}^{\vee}_{-ab}\overline{\mathcal{E}}(\mathrm{X}) = \mathcal{E}(\mathsf{t}_{\mathrm{X}}) = \mathcal{E}(\mathsf{x}) = \mathsf{E}^-.$$

The dual calculations are similar.

Alternatively, to deduce the first statement in this result, one can employ Lemma 18.12 with $q = 1$ along with $\Gamma_{ab} = (\Gamma^*)^{coab} = \mathsf{E}$. □

Exercise 18.16. Check: The identification $\overline{\mathcal{Z}}^{coab}(\Bbbk\{x\}) = \mathsf{E}$ arises from the co-abelianization of (18.17) with $q = 1$. Now the scalars x_C indexed by chambers are all equal. This condition arises from (16.39).

Remark 18.17. Recall: $\overline{\mathcal{Z}}^{\vee}$ and $\overline{\mathcal{Z}}^{\vee}_{ab}$ are sections of $\overline{\mathcal{K}}^{\vee}$, and $\overline{\mathcal{Z}}$ and $\overline{\mathcal{Z}}^{coab}$ are sections of $\overline{\mathcal{K}}$ (under field characteristic assumption). Thus, Lemmas 18.12 and 18.15 recover parts of Lemmas 18.4 and 18.6.

18.2.9. Antipode formulas.

Exercise 18.18. Use Lemma 18.6 and Proposition 17.44 to deduce antipode formulas of Coxeter q-bialgebras $\Bbbk_q[x]$ and $\Bbbk_q\{x\}$ in (10.30) and (10.31) from those of Coxeter q-bimonoids Γ_q and Γ_q^* in (4.36) and (4.37). Deduce the alternating sign identity (1.63a) as a consequence.

For $q = 1$, alternatively use Lemma 18.4 and the antipode formula of E to deduce antipode formulas of $\Bbbk[x]$ and $\Bbbk\{x\}$.

18.3. Tensor Coxeter bialgebra as a Fock space

Recall the decorated exponential Coxeter bimonoid E_M and Coxeter bimonoid of decorated chambers Γ_M from Section 4.9. Similarly, recall the tensor Coxeter bialgebra $\overline{\mathcal{T}}(M)$ and its relatives from Section 10.3. We now link them using Fock functors and their adjoints. As an application, we deduce the Coxeter group algebra identities in Section 1.9. This generalizes considerations in Section 18.2.

18.3.1. Decorated exponential Coxeter bimonoid. Let us compute values of Fock functors on the decorated exponential Coxeter bimonoid. This is equivalent to evaluating decorated Fock functors on the exponential Coxeter bimonoid.

Lemma 18.19. *For a scalar q, for a left W-module M,*

$$\mathcal{K}_q(\mathsf{E}_M) = \overline{\mathcal{T}}_q(M) \quad \text{and} \quad \mathcal{K}_q^\vee(\mathsf{E}_M) = \overline{\mathcal{T}}_q^\vee(M) \tag{18.20}$$

as Coxeter q-bialgebras.

PROOF. The proof of Lemma 18.3 generalizes in a straightforward manner. This also explains why we use ζ^{-1} in the coproduct formula (10.37) of $\overline{\mathcal{T}}_q(M)$ and ζ in the product formula (10.39) of $\overline{\mathcal{T}}_q^\vee(M)$. □

The two identities in (18.20) can be linked via duality by using Exercise 4.25 and the discussion in Section 10.3.14.

Observation 18.20. Recall: E_M is bicommutative. For $q = 1$, applying Theorem 15.11, we deduce that $\overline{\mathcal{T}}(M)$ is cocommutative and $\overline{\mathcal{T}}^\vee(M)$ is commutative. Similarly, for $q = -1$, applying Theorem 15.12, we deduce that $\overline{\mathcal{T}}_{-1}(M)$ is signed cocommutative and $\overline{\mathcal{T}}_{-1}^\vee(M)$ is signed commutative.

Lemma 18.21. *For a left W-module M,*

$$\overline{\mathcal{K}}(\mathsf{E}_M) = \overline{\mathcal{S}}(M) \quad \text{and} \quad \overline{\mathcal{K}}^\vee(\mathsf{E}_M) = \overline{\mathcal{S}}^\vee(M) \tag{18.21}$$

as Coxeter bialgebras, and

$$\overline{\mathcal{K}}_{-1}(\mathsf{E}_M) = \overline{\mathcal{E}}(M) \quad \text{and} \quad \overline{\mathcal{K}}_{-1}^\vee(\mathsf{E}_M) = \overline{\mathcal{E}}^\vee(M) \tag{18.22}$$

as signed Coxeter bialgebras.

PROOF. We explain the first claim in (18.21); the rest are similar. For T a face-type, $\overline{\mathcal{K}}(\mathsf{E}_M)$ and $\overline{\mathcal{S}}(M)$ on the T-component both equal M_{W_T}. Further, their structure maps are induced from those of $\mathcal{K}(\mathsf{E}_M)$ and $\overline{\mathcal{T}}(M)$, respectively, under the canonical quotient maps $M \twoheadrightarrow M_{W_T}$. So the result follows from the first claim in (18.20) with $q = 1$. □

18.3.2. Coxeter bimonoid of decorated chambers. The tensor Coxeter bialgebra can also be obtained from the Coxeter bimonoid of decorated chambers as follows.

Lemma 18.22. *For a scalar q, for a left W-module M,*

$$\begin{aligned} \overline{\mathcal{K}}(\Gamma_{q,M}) &= \overline{\mathcal{K}}^\vee(\Gamma_{q,M}) = \overline{\mathcal{T}}_q(M), \\ \overline{\mathcal{K}}(\Gamma^*_{q,M}) &= \overline{\mathcal{K}}^\vee(\Gamma^*_{q,M}) = \overline{\mathcal{T}}_q^\vee(M) \end{aligned} \tag{18.23}$$

as Coxeter q-bialgebras, and

$$
\begin{aligned}
\overline{\mathcal{K}}_{-1}(\Gamma_{q,M}) = \overline{\mathcal{K}}^{\vee}_{-1}(\Gamma_{q,M}) = \overline{\mathcal{T}}_{-q}(M), \\
\overline{\mathcal{K}}_{-1}(\Gamma^{*}_{q,M}) = \overline{\mathcal{K}}^{\vee}_{-1}(\Gamma^{*}_{q,M}) = \overline{\mathcal{T}}^{\vee}_{-q}(M)
\end{aligned}
\tag{18.24}
$$

as Coxeter $(-q)$-bialgebras.

PROOF. This generalizes Lemma 18.6. For the proof, we can generalize the argument given there. The key observation now is that for $\Gamma[A]\otimes M$, the space of W_A-invariants and of W_A-coinvariants both equal M. Further calculation uses the bijection in (1.67). □

Lemma 18.22 can also be rephrased as evaluations of decorated Fock functors on Γ_q and Γ^*_q. For instance,

$$
\overline{\mathcal{K}}_M(\Gamma_q) = \overline{\mathcal{K}}^{\vee}_M(\Gamma_q) = \overline{\mathcal{T}}_q(M),
\tag{18.25}
$$

and so on. This is purely formal:

$$
\overline{\mathcal{K}}(\Gamma_{q,M}) = \overline{\mathcal{K}}(\Gamma_q \times \mathsf{E}_M) = \overline{\mathcal{K}}((\Gamma_q)_M) = \overline{\mathcal{K}}_M(\Gamma_q),
$$

and so on.

Exercise 18.23. Use Lemma 18.22 for $q = \pm 1$ to deduce the commutativity results in Observation 18.20.

18.3.3. Enumeration identities in the Coxeter group algebra.

It is easy to see that E_M is a Coxeter bimonoid. It then follows from Lemma 18.19 that $\overline{\mathcal{T}}_q(M)$ with product and coproduct (10.37) is a Coxeter q-bialgebra. Similarly, by Observation 18.20, $\overline{\mathcal{T}}(M)$ is cocommutative. Dual statements hold for $\overline{\mathcal{T}}_q^{\vee}(M)$. As a consequence, by reversing the logical flow in Section 10.3.6, we deduce the Coxeter group algebra identities in Section 1.9. The case $M = \Bbbk$ is already interesting; see Section 18.2.3.

Instead of E_M, one can also start with $\Gamma_{q,M}$ and use Lemma 18.22 to deduce these Coxeter group algebra identities.

18.3.4. q-norm map.

Let us now connect the q-norm transformation κ_q between q-full Fock functors to the q-norm map $\overline{\kappa}_{q,M}$ between the tensor and shuffle Coxeter q-bialgebras.

Lemma 18.24. *For a scalar q, for a left W-module M, we have the diagram of Coxeter q-bialgebras*

$$
\begin{array}{ccc}
\mathcal{K}_q(\mathsf{E}_M) & \xrightarrow{(\kappa_q)_{\mathsf{E}_M}} & \mathcal{K}^{\vee}_q(\mathsf{E}_M) \\
\| & & \| \\
\overline{\mathcal{T}}_q(M) & \xrightarrow[\overline{\kappa}_{q,M}]{} & \overline{\mathcal{T}}_q^{\vee}(M).
\end{array}
\tag{18.26}
$$

The top-horizontal map is as in (15.22)*; the bottom-horizontal map is* (10.41).

This generalizes Lemma 18.8 and the proof is even more transparent in this case. It provides an alternative way to obtain Lemma 10.36 and Corollary 10.37. Yet another way is given in Exercise 18.25 below.

Specializing diagram (15.23) to $\mathsf{h} := \mathsf{E}_M$ yields commutative diagrams

$$\begin{array}{ccc} \mathcal{K}(\mathsf{E}_M) & \xrightarrow{\kappa_{\mathsf{E}_M}} & \mathcal{K}^\vee(\mathsf{E}_M) \\ \downarrow & & \uparrow \\ \overline{\mathcal{K}}(\mathsf{E}_M) & \xrightarrow[\overline{\kappa}_{\mathsf{E}_M}]{} & \overline{\mathcal{K}}^\vee(\mathsf{E}_M) \end{array} \qquad \begin{array}{ccc} \mathcal{K}_{-1}(\mathsf{E}_M) & \xrightarrow{(\kappa_{-1})_{\mathsf{E}_M}} & \mathcal{K}^\vee_{-1}(\mathsf{E}_M) \\ \downarrow & & \uparrow \\ \overline{\mathcal{K}}_{-1}(\mathsf{E}_M) & \xrightarrow[(\overline{\kappa}_{-1})_{\mathsf{E}_M}]{} & \overline{\mathcal{K}}^\vee_{-1}(\mathsf{E}_M) \end{array} \tag{18.27}$$

of Coxeter bialgebras and of signed Coxeter bialgebras, respectively. Alternatively, these diagrams can also be obtained by evaluating diagram (14.77) (involving decorated Fock functors) on E and E^-, respectively. Now by using the identifications in Lemmas 18.19, 18.21, 18.24, we recover diagrams (10.50). The result in Lemma 10.43 also follows as a consequence.

Exercise 18.25. This generalizes Exercise 18.9. Check: The q-norm map between tensors and shuffles can be obtained from the q-norm map on decorated chambers as follows. For a scalar q, for a left W-module M, we have diagrams of Coxeter q-bialgebras

$$\begin{array}{ccc} \overline{\mathcal{K}}(\mathsf{\Gamma}_{q,M}) & \xrightarrow{\overline{\mathcal{K}}(\kappa_{q,M})} & \overline{\mathcal{K}}(\mathsf{\Gamma}^*_{q,M}) \\ \| & & \| \\ \overline{\mathcal{T}_q}(M) & \xrightarrow[\overline{\kappa}_{q,M}]{} & \overline{\mathcal{T}_q}^\vee(M) \end{array} \qquad \begin{array}{ccc} \overline{\mathcal{K}}^\vee(\mathsf{\Gamma}_{q,M}) & \xrightarrow{\overline{\mathcal{K}}^\vee(\kappa_{q,M})} & \overline{\mathcal{K}}^\vee(\mathsf{\Gamma}^*_{q,M}) \\ \| & & \| \\ \overline{\mathcal{T}_q}(M) & \xrightarrow[\overline{\kappa}_{q,M}]{} & \overline{\mathcal{T}_q}^\vee(M). \end{array} \tag{18.28}$$

The top-horizontal maps are induced from (4.112); the bottom-horizontal maps are as in (10.41). This provides another derivation of Lemma 10.36 and Corollary 10.37.

Diagrams (18.28) can be viewed as the special case $\mathsf{p} := \mathsf{x}_M$ of diagrams (17.26). See the discussion in Section 18.3.5 in this regard.

18.3.5. Freeness and cofreeness. Since the tensor Coxeter bialgebra and its relatives arise from the Coxeter bimonoids $\mathsf{\Gamma}_M$, $\mathsf{\Gamma}^*_M$, E_M, (co)freeness properties of the former can be deduced from those of the latter. This gives an alternative way of deducing results in Section 10.3.15.

For a scalar q,

$$\overline{\mathcal{T}_q}(M) = \overline{\mathcal{K}}(\mathsf{\Gamma}_{q,M}) = \overline{\mathcal{K}}\mathcal{T}_q(\mathsf{x}_M) = \overline{\mathcal{T}_q}\overline{\mathcal{K}}(\mathsf{x}_M) = \overline{\mathcal{T}_q}(\mathrm{X}_M)$$

as Coxeter q-bialgebras,

$$\overline{\mathcal{S}}(M) = \overline{\mathcal{K}}(\mathsf{E}_M) = \overline{\mathcal{K}}\mathcal{S}(\mathsf{x}_M) = \overline{\mathcal{S}}\overline{\mathcal{K}}(\mathsf{x}_M) = \overline{\mathcal{S}}(\mathrm{X}_M)$$

as Coxeter bialgebras,

$$\overline{\mathcal{E}}(M) = \overline{\mathcal{K}}_{-1}(\mathsf{E}_M) = \overline{\mathcal{K}}_{-1}\mathcal{S}(\mathsf{x}_M) = \overline{\mathcal{E}}\overline{\mathcal{K}}_{-1}(\mathsf{x}_M) = \overline{\mathcal{E}}(\mathrm{X}_M)$$

as signed Coxeter bialgebras. In the last case, alternatively,

$$\overline{\mathcal{E}}(M) = \overline{\mathcal{K}}_{-1}(\mathsf{E}_M) = \overline{\mathcal{K}}(\mathsf{E}^-_M) = \overline{\mathcal{K}}\mathcal{E}(\mathsf{x}_M) = \overline{\mathcal{E}}\overline{\mathcal{K}}(\mathsf{x}_M) = \overline{\mathcal{E}}(\mathrm{X}_M).$$

Dually,

$$\overline{\mathcal{T}_q}^\vee(M) = \overline{\mathcal{K}}^\vee(\mathsf{\Gamma}^*_{q,M}) = \overline{\mathcal{K}}^\vee\mathcal{T}_q^\vee(\mathsf{x}_M) = \overline{\mathcal{T}_q}^\vee\overline{\mathcal{K}}^\vee(\mathsf{x}_M) = \overline{\mathcal{T}_q}^\vee(\mathrm{X}_M),$$

and so on.

We used Propositions 15.24, 15.25, 15.29 as well as (4.108), (4.109), (15.15), (18.2).

18.3.6. Primitive and indecomposable parts. In a similar manner, one can deduce results of Section 10.3.17 and more.

To deduce formulas (10.51):

$$\begin{aligned}\overline{\mathcal{P}}(\overline{\mathcal{T}}^\vee(M)) &= \overline{\mathcal{P}}\overline{\mathcal{K}}^\vee(\Gamma^*_M) = \overline{\mathcal{K}}^\vee\mathcal{P}(\Gamma^*_M) = \overline{\mathcal{K}}^\vee(\mathsf{x}_M) = \mathrm{X}_M,\\ \overline{\mathcal{Q}}(\overline{\mathcal{T}}(M)) &= \overline{\mathcal{Q}}\overline{\mathcal{K}}(\Gamma_M) = \overline{\mathcal{K}}\mathcal{Q}(\Gamma_M) = \overline{\mathcal{K}}(\mathsf{x}_M) = \mathrm{X}_M.\end{aligned}$$

We used Propositions 15.46 and 15.48.

Formula (12.29) can be derived as follows.

$$\begin{aligned}\overline{\mathcal{P}}(\overline{\mathcal{T}}(M)) &= \overline{\mathcal{P}}\overline{\mathcal{T}}(\mathrm{X}_M) = \overline{\mathcal{P}}\overline{\mathcal{T}}\overline{\mathcal{K}}^\vee(\mathsf{x}_M) = \overline{\mathcal{P}}\overline{\mathcal{K}}^\vee\mathcal{T}(\mathsf{x}_M) = \overline{\mathcal{K}}^\vee\mathcal{P}\mathcal{T}(\mathsf{x}_M)\\ &= \overline{\mathcal{K}}^\vee\mathcal{P}(\Gamma_M) = \overline{\mathcal{K}}^\vee(\mathbf{Lie}\circ\mathsf{x}_M) = \mathbf{Lie}\circ\overline{\mathcal{K}}^\vee(\mathsf{x}_M) = \mathbf{Lie}\circ\mathrm{X}_M\end{aligned}$$

as Coxeter Lie algebras. Lie entered into the calculation via formula (6.14) or more generally via Proposition 6.4. The equality in red used Proposition 15.37 which assumes $\mathrm{char}(\Bbbk) \nmid |W|$. Without this assumption, one can stop at the previous step and obtain

$$\overline{\mathcal{P}}(\overline{\mathcal{T}}(M))[A,C] = (\mathbf{Lie}\circ\mathsf{x}_M)[A]^{W_A} = (\mathbf{Lie}[\mathrm{s}(A),\top]\otimes M)^{W_A} = (\mathsf{Lie}[A]\otimes M)^{W_A}.$$

This recovers formula (10.53).

Dually,

$$\begin{aligned}\overline{\mathcal{Q}}(\overline{\mathcal{T}}^\vee(M)) &= \overline{\mathcal{Q}}\overline{\mathcal{T}}^\vee(\mathrm{X}_M) = \overline{\mathcal{Q}}\overline{\mathcal{T}}^\vee\overline{\mathcal{K}}(\mathsf{x}_M) = \overline{\mathcal{Q}}\overline{\mathcal{K}}\mathcal{T}^\vee(\mathsf{x}_M) = \overline{\mathcal{K}}\mathcal{Q}\mathcal{T}^\vee(\mathsf{x}_M)\\ &= \overline{\mathcal{K}}\mathcal{Q}(\Gamma^*_M) = \overline{\mathcal{K}}(\mathbf{Lie}^*\circ\mathsf{x}_M) = \mathbf{Lie}^*\circ^\vee\overline{\mathcal{K}}(\mathsf{x}_M) = \mathbf{Lie}^*\circ^\vee\mathrm{X}_M\end{aligned}$$

as Coxeter Lie coalgebras. The equality in red used Proposition 15.41.

For the results in (10.52): For q not a root of unity,

$$\begin{aligned}\overline{\mathcal{P}}(\overline{\mathcal{T}}_q(M)) &= \overline{\mathcal{P}}\overline{\mathcal{K}}^\vee(\Gamma_{q,M}) = \overline{\mathcal{K}}^\vee\mathcal{P}(\Gamma_{q,M}) = \overline{\mathcal{K}}^\vee(\mathsf{x}_M) = \mathrm{X}_M,\\ \overline{\mathcal{Q}}(\overline{\mathcal{T}}_q{}^\vee(M)) &= \overline{\mathcal{Q}}\overline{\mathcal{K}}(\Gamma^*_{q,M}) = \overline{\mathcal{K}}\mathcal{Q}(\Gamma^*_{q,M}) = \overline{\mathcal{K}}(\mathsf{x}_M) = \mathrm{X}_M.\end{aligned}$$

We used (4.111). Compare and contrast with the previous calculations for $q = 1$.

For formulas (10.54):

$$\begin{aligned}\overline{\mathcal{P}}(\overline{\mathcal{S}}^\vee(M)) &= \overline{\mathcal{P}}\overline{\mathcal{K}}^\vee(\mathsf{E}_M) = \overline{\mathcal{K}}^\vee\mathcal{P}(\mathsf{E}_M) = \overline{\mathcal{K}}^\vee(\mathsf{x}_M) = \mathrm{X}_M,\\ \overline{\mathcal{Q}}(\overline{\mathcal{S}}(M)) &= \overline{\mathcal{Q}}\overline{\mathcal{K}}(\mathsf{E}_M) = \overline{\mathcal{K}}\mathcal{Q}(\mathsf{E}_M) = \overline{\mathcal{K}}(\mathsf{x}_M) = \mathrm{X}_M.\end{aligned}$$

We used (4.110).

18.3.7. Exp-log correspondences. Since $\overline{\mathcal{K}}^\vee(\Gamma_M) = \overline{\mathcal{T}}(M)$, by Proposition 17.3, the space of series of Γ_M and $\overline{\mathcal{T}}(M)$ match. Next, by Theorem 17.9, the exp-log correspondences of Γ_M and $\overline{\mathcal{T}}(M)$ match. In particular, by Proposition 17.7, the exponential series of Γ_M and $\overline{\mathcal{T}}(M)$ match. This can all be checked directly using the descriptions given in Sections 4.9.5 and 10.3.18.

Since $\overline{\mathcal{K}}^\vee(\mathsf{E}_M) = \overline{\mathcal{S}}^\vee(M)$, by Theorem 17.9, the exp-log correspondences of E_M and $\overline{\mathcal{S}}^\vee(M)$ match. This can be seen directly by comparing the analysis for E_M and $\overline{\mathcal{S}}^\vee(M)$.

18.3.8. Fock functor adjoints. We generalize Lemmas 18.12 and 18.15.

Lemma 18.26. *For a scalar q, for a left W-module M,*

$$\overline{\mathcal{Z}}^\vee(\overline{\mathcal{T}_q}(M)) = \Gamma_{q,M} \quad \text{and} \quad \overline{\mathcal{Z}}(\overline{\mathcal{T}_q}^\vee(M)) = \Gamma^*_{q,M}$$

as Coxeter q-bimonoids.

For $q = 1$, this also illustrates the general fact that $\overline{\mathcal{Z}}^\vee$ preserves cocommutativity while $\overline{\mathcal{Z}}$ preserves commutativity.

PROOF. The calculations go as follows.

$$\overline{\mathcal{Z}}^\vee(\overline{\mathcal{T}_q}(M)) = \overline{\mathcal{Z}}^\vee\overline{\mathcal{T}_q}(\mathrm{X}_M) = \mathcal{T}_q(\mathtt{t}\mathrm{x}_M) = \mathcal{T}_q(\mathtt{x}'_M) \cong \mathcal{T}_q(\mathtt{x}_M) = \Gamma_{q,M},$$

$$\overline{\mathcal{Z}}(\overline{\mathcal{T}_q}^\vee(M)) = \overline{\mathcal{Z}}\overline{\mathcal{T}_q}^\vee(\mathrm{X}_M) = \mathcal{T}_q^\vee(\mathtt{t}\mathrm{x}_M) = \mathcal{T}_q^\vee(\mathtt{x}'_M) \cong \mathcal{T}_q^\vee(\mathtt{x}_M) = \Gamma^*_{q,M}.$$

We used the first identity in (18.4) followed by Exercise 4.1. □

Exercise 18.27. Check: The second identification in Lemma 18.26 works as follows. For a scalar q, the map

$$\overline{\mathcal{Z}}(\overline{\mathcal{T}_q}^\vee(M)) \xrightarrow{\cong} \Gamma^*_{q,M}, \qquad (x_F)_{F\geq A} \mapsto \sum_{C:\, C\geq A} \mathtt{M}_{C/A} \otimes u \cdot x_C \tag{18.29}$$

on the A-component, is an isomorphism of Coxeter q-bimonoids. Here $u \in W$ is defined by $u(C_r) = C$; it depends on the choice of C. This map generalizes (18.17): instead of a scalar, each x_F is now an element of M.

Lemma 18.28. *For a left W-module M,*

$$\overline{\mathcal{Z}}^\vee_{ab}(\overline{\mathcal{S}}(M)) = \overline{\mathcal{Z}}^{coab}(\overline{\mathcal{S}}^\vee(M)) = \mathsf{E}_M$$

as Coxeter bimonoids, and

$$\overline{\mathcal{Z}}^\vee_{-ab}(\overline{\mathcal{E}}(M)) = \overline{\mathcal{Z}}^{-coab}(\overline{\mathcal{E}}^\vee(M)) = \mathsf{E}^-_M$$

as signed Coxeter bimonoids.

The calculations are similar; we leave them as exercises. Alternatively, we can deduce Lemma 18.28 from Lemma 18.26 by using Lemma 16.78.

18.3.9. Antipode formulas. We generalize Exercise 18.18.

Exercise 18.29. Use Lemma 18.22 and Proposition 17.44 to deduce antipode formulas of Coxeter q-bialgebras $\overline{\mathcal{T}_q}(M)$ and $\overline{\mathcal{T}_q}^\vee(M)$ in (10.57) and (10.58) from those of Coxeter q-bimonoids $\Gamma_{q,M}$ and $\Gamma^*_{q,M}$ in (4.115). Deduce the alternating sign identities in (1.72) as a consequence.

Similarly, use Lemma 18.21 and the antipode formula of E_M to deduce antipode formulas of $\overline{\mathcal{S}}(M)$ and $\overline{\mathcal{S}}^\vee(M)$ in (10.59).

18.4. Coxeter bialgebra of face-types as a bosonic Fock space

Recall the Coxeter bimonoid of faces Σ from Section 4.5, and the Coxeter bialgebra of face-types $\overline{\Sigma}$ from Section 10.4. We now link them using the bosonic Fock functor. Put more simply, $\overline{\Sigma}$ relates to the invariant Tits algebra in the same manner that the Tits algebra relates to Σ. This relationship can be used to establish many properties of $\overline{\Sigma}$ from those of Σ. We also connect the Coxeter bimonoid of top-nested faces $\widehat{\mathsf{Q}}$ from Section 4.6 to $\overline{\Sigma}$.

18.4.1. Coxeter bimonoid of faces. Let q be a scalar. Applying full Fock functors and bosonic Fock functors to the Coxeter q-bimonoid of faces Σ_q yields Coxeter q-bialgebras. Let us make $\mathcal{K}^\vee(\Sigma_q)$ and $\overline{\mathcal{K}}^\vee(\Sigma_q)$ explicit. By (15.1),

$$\mathcal{K}^\vee(\Sigma_q)[F,C] := \Sigma_q[F].$$

Its support and type morphisms are induced from those of Σ_q. Its product and coproduct are as follows. For $A \leq F \leq C$,

$$\begin{array}{ccc} \mathcal{K}^\vee(\Sigma_q)[F,C] & \overset{\mu^{F,C}_{A,C}}{\dashrightarrow} & \mathcal{K}^\vee(\Sigma_q)[A,C] \\ \| & & \| \\ \Sigma_q[F] & \longrightarrow & \Sigma_q[A] \end{array} \qquad \begin{array}{ccc} \mathcal{K}^\vee(\Sigma_q)[A,C] & \overset{\Delta^{F,C}_{A,C}}{\dashrightarrow} & \mathcal{K}^\vee(\Sigma_q)[F,C] \\ \| & & \| \\ \Sigma_q[A] & \underset{\Delta^F_A}{\longrightarrow} & \Sigma_q[F], \end{array}$$

with the map $\Sigma_q[F] \to \Sigma_q[A]$ given by

$$x \longmapsto \sum_{\substack{G:\, \mathtt{t}(G)=\mathtt{t}(F)\\ w(C)=GC}} \mu^G_A(w \cdot x).$$

These specialize formulas (15.3) with $q = 1$. (This q is not to be confused with the q in the discussion.) This defines the Coxeter q-bialgebra $\mathcal{K}^\vee(\Sigma_q)$. For $q = 1$, it is neither commutative nor cocommutative.

Let us now consider $\overline{\mathcal{K}}^\vee(\Sigma_q)$. Its (F,C)-component is the subspace of $\Sigma_q[F]$ invariant under the subgroup W_F. Since support and type morphisms preserve invariants, $\overline{\mathcal{K}}^\vee(\Sigma_q)$ is a Coxeter subspace of $\mathcal{K}^\vee(\Sigma_q)$. In fact, by general theory, it is a Coxeter subbialgebra of $\mathcal{K}^\vee(\Sigma_q)$. Projecting a W_A-invariant element (in the star of A) on F produces an invariant element in the star of F. This is how the coproduct works. Conversely, given an invariant element in the star of F, by taking a copy of it in the star of G, as G varies over faces greater than A of the same type as F, and summing all such copies produces an invariant element in the star of A. This is how the product works.

Lemma 18.30. *For a scalar q, under the identification* (1.155),

$$\overline{\mathcal{K}}^\vee(\Sigma_q) \xrightarrow{\cong} \overline{\Sigma}_q \tag{18.30}$$

as Coxeter q-bialgebras.

PROOF. We need to check that the above descriptions of the product and coproduct (translated to face-types by putting $C := C_r$) match those given in (10.63). Let $U \geq T \geq Z$. For a face F of type T, let $\mathtt{H}_{U/F}$ denote the sum of all faces of type U greater than F. The product formula expresses the fact that $\mathtt{H}_{U/Z} = \sum_F \mathtt{H}_{U/F}$, as F varies over all faces greater than Z which are of type T. See Observation 1.5 in this regard. The coproduct formula can be deduced similarly using (1.38). □

Exercise 18.31 ($\mathtt{Q}$-bases). Deduce (co)product formulas (10.68) for the Coxeter bialgebra $\overline{\Sigma}$ from (co)product formulas (4.56) for the Coxeter bimonoid Σ. Similarly, for q not a root of unity, deduce formulas (10.71) for $\overline{\Sigma}_q$ from formulas (4.60) for Σ_q.

Exercise 18.32 (Invariant eulerian idempotents). Deduce coproduct formula (10.73) from coproduct formula (4.62). (Use (1.137) applied to $\boldsymbol{\alpha}_Z$.)

Exercise 18.33. Check: For a scalar q, we have the diagram of Coxeter q-bialgebras

$$\begin{array}{ccc} \overline{\mathcal{K}}^{\vee}(\Gamma_q) & \longrightarrow & \overline{\mathcal{K}}^{\vee}(\Sigma_q) \\ \| & & \| \\ \Bbbk_q[x] & \longrightarrow & \overline{\Sigma}_q, \end{array} \tag{18.31}$$

with the top-horizontal map induced from (4.75) and bottom-horizontal map (10.64).

Exercise 18.34. Specializing the first diagram in (15.23): For a scalar q and $\operatorname{char}(\Bbbk) \nmid |W|$, we have the diagram of Coxeter q-bialgebras

$$\begin{array}{ccc} \mathcal{K}(\Sigma_q) & \xrightarrow{\kappa_{\Sigma_q}} & \mathcal{K}^{\vee}(\Sigma_q) \\ \downarrow & & \uparrow \\ \overline{\mathcal{K}}(\Sigma_q) & \xrightarrow[\overline{\kappa}_{\Sigma_q}]{\cong} & \overline{\mathcal{K}}^{\vee}(\Sigma_q) \end{array} \qquad \begin{array}{ccc} \Sigma_q[A] & \longrightarrow & \Sigma_q[A] \\ \downarrow & & \uparrow \\ \Sigma_q[A]_{W_A} & \xrightarrow{\cong} & \Sigma_q[A]^{W_A} \end{array} \tag{18.32}$$

on the (A,C)-component. The top-horizontal map is the action by $\sum_{w \in W_A} w$, so it sends $\mathtt{H}_{F/A}$ to $|W_F| \sum \mathtt{H}_{G/A}$, with the sum over all G greater than A of the same type as F.

Deduce: For a scalar q and $\operatorname{char}(\Bbbk) \nmid |W|$,

$$\overline{\mathcal{K}}(\Sigma_q) \xrightarrow{\cong} \overline{\Sigma}_q, \qquad \overline{\mathtt{H}}_{F/A} \leftrightarrow |W_F|\, \mathtt{H}_{\mathrm{t}(F/A)}, \tag{18.33}$$

where bar on top of $\mathtt{H}$ indicates image in the quotient space. Also check this fact directly.

By dualizing (18.30) and (18.33), we obtain: For a scalar q,

$$\overline{\mathcal{K}}(\Sigma_q^*) \xrightarrow{\cong} \overline{\Sigma}_q^* \quad \text{and} \quad \overline{\mathcal{K}}^{\vee}(\Sigma_q^*) \xrightarrow{\cong} \overline{\Sigma}_q^*, \tag{18.34}$$

the latter if $\operatorname{char}(\Bbbk) \nmid |W|$.

18.4.2. Face-type enumeration identities. We start with Σ_q which is a Coxeter q-bimonoid. It follows from (18.30) that $\overline{\Sigma}_q$ with product and coproduct (10.63) is a Coxeter q-bialgebra. In particular, the coproduct is coassociative and the q-bialgebra axiom holds. As a consequence, by reversing the logical flow in Section 10.4.4, we deduce identities (1.57) and (1.62).

Further, for $q = 1$, Σ is cocommutative, and hence by Theorem 15.11, so is $\overline{\Sigma}$. Again by reversing the logical flow, we recover Exercise 1.69.

18.4.3. Freeness and cofreeness. Recall: For q a scalar, Σ_q is the free Coxeter q-bimonoid on the exponential Coxeter comonoid E, while $\overline{\Sigma}_q$ is the free Coxeter q-bialgebra on the Coxeter coalgebra $\Bbbk\{x\}$. Similarly, there are dual cofreeness results. The identifications are done using $\mathtt{H}$- and $\mathtt{M}$-bases. See (4.63) and Proposition 10.71. We now show how the former can be used to deduce the latter:

$$\begin{aligned} \overline{\Sigma}_q &= \overline{\mathcal{K}}^{\vee}(\Sigma_q) = \overline{\mathcal{K}}^{\vee}\mathcal{T}_q(\mathsf{E}) = \overline{\mathcal{T}}_q\overline{\mathcal{K}}^{\vee}(\mathsf{E}) = \overline{\mathcal{T}}_q(\Bbbk\{x\}), \\ \overline{\Sigma}_q^* &= \overline{\mathcal{K}}(\Sigma_q^*) = \overline{\mathcal{K}}\mathcal{T}_q^{\vee}(\mathsf{E}) = \overline{\mathcal{T}}_q^{\vee}\overline{\mathcal{K}}(\mathsf{E}) = \overline{\mathcal{T}}_q^{\vee}(\Bbbk[x]). \end{aligned}$$

In these calculations, we also used Propositions 15.24 and 15.25 and (18.6).

Similarly, the (co)freeness results in Propositions 10.74 and 10.75 can be deduced from those in (4.64) and (4.65). The calculations work exactly as above; however, note very carefully that the identifications now involve Q- and P-bases.

Exercise 18.35. Check: For a Coxeter comonoid c and Coxeter monoid a, the diagram

$$\begin{array}{ccc} \mathsf{CSp}(\mathsf{c},\mathsf{a}) & \xrightarrow{\cong} & \mathsf{Mon}(\mathsf{CSp})(\mathbf{\Sigma}, \hom^{\times}(\mathsf{c},\mathsf{a})) \\ \downarrow & & \downarrow \\ \overline{\mathsf{C}}\mathsf{Sp}(\overline{\mathcal{K}}(\mathsf{c}),\overline{\mathcal{K}}(\mathsf{a})) & \xrightarrow[\cong]{} & \mathsf{Mon}(\overline{\mathsf{C}}\mathsf{Sp})(\overline{\Sigma}, \hom^{\times}(\overline{\mathcal{K}}(\mathsf{c}),\overline{\mathcal{K}}(\mathsf{a}))) \end{array}$$

commutes. The horizontal bijections are as in Exercise 4.19 and Theorem 10.96. The left-vertical map is obtained by applying $\overline{\mathcal{K}}$. The right-vertical map is obtained by applying $\overline{\mathcal{K}}^{\vee}$, followed by the first map in (17.19). There is a similar diagram with $\overline{\mathcal{K}}^{\vee}$ instead of $\overline{\mathcal{K}}$.

Formulate similar diagrams using Exercise 4.20 and Theorems 10.97 and 10.98.

18.4.4. Primitive and indecomposable parts. In a similar manner, one can deduce results of Section 10.4.15. Some illustrations are given below.

To deduce formula (10.100):

$$\overline{\mathcal{P}}(\overline{\Sigma}) = \overline{\mathcal{P}}\overline{\mathcal{K}}^{\vee}(\Sigma) = \overline{\mathcal{K}}^{\vee}\mathcal{P}(\Sigma) = \overline{\mathcal{K}}^{\vee}(\mathsf{Zie}).$$

We used Proposition 15.46 and also (4.99). If $\mathrm{char}(\Bbbk) \nmid |W|$, then this calculation can be continued further as follows to deduce (12.31).

$$\overline{\mathcal{P}}(\overline{\Sigma}) = \overline{\mathcal{K}}^{\vee}(\mathsf{Zie}) = \overline{\mathcal{K}}^{\vee}(\mathbf{Lie} \circ \mathsf{E}) = \mathbf{Lie} \circ \overline{\mathcal{K}}^{\vee}(\mathsf{E}) = \mathbf{Lie} \circ \Bbbk\{x\}.$$

We used (6.13) and Proposition 15.37.

Dually,

$$\begin{aligned} \overline{\mathcal{Q}}(\overline{\Sigma}^{*}) = \overline{\mathcal{Q}}\overline{\mathcal{K}}(\Sigma^{*}) = \overline{\mathcal{K}}\mathcal{Q}(\Sigma^{*}) = \overline{\mathcal{K}}(\mathsf{Zie}^{*}) \\ = \overline{\mathcal{K}}(\mathbf{Lie}^{*} \circ \mathsf{E}) = \mathbf{Lie}^{*} \circ^{\vee} \overline{\mathcal{K}}(\mathsf{E}) = \mathbf{Lie}^{*} \circ^{\vee} \Bbbk[x]. \end{aligned}$$

We used the dual of (6.13) and Proposition 15.41.

For the results in (10.99): For q not a root of unity,

$$\begin{aligned} \overline{\mathcal{P}}(\overline{\Sigma}_q) = \overline{\mathcal{P}}\overline{\mathcal{K}}^{\vee}(\Sigma_q) = \overline{\mathcal{K}}^{\vee}\mathcal{P}(\Sigma_q) = \overline{\mathcal{K}}^{\vee}(\mathsf{E}) = \Bbbk\{x\}, \\ \overline{\mathcal{Q}}(\overline{\Sigma}_q^{*}) = \overline{\mathcal{Q}}\overline{\mathcal{K}}(\Sigma_q^{*}) = \overline{\mathcal{K}}\mathcal{Q}(\Sigma_q^{*}) = \overline{\mathcal{K}}(\mathsf{E}) = \Bbbk[x]. \end{aligned}$$

We used (4.100) and its dual. The above are equalities of Coxeter spaces, so it does not matter whether we write $\Bbbk\{x\}$ or $\Bbbk[x]$.

18.4.5. Exp-log correspondences. The exp-log correspondence for $\overline{\Sigma}$ between primitive and group-like series was discussed in Section 10.4.20. It is strikingly similar to the exp-log correspondence for Σ in Section 4.5.11. We now have an explanation for this similarity, namely, put $\mathsf{h} = \Sigma$ in Theorem 17.9 and use $\overline{\mathcal{K}}^{\vee}(\Sigma) = \overline{\Sigma}$. One can check directly that the universal series, Takeuchi series, exponential series of Σ and $\overline{\Sigma}$ match.

18.4.6. Coxeter bimonoid of top-nested faces. Recall the Coxeter q-bimonoid of top-nested faces $\widehat{\mathtt{Q}}_q$ from Section 4.6.

Lemma 18.36. *For a scalar q,*

$$\overline{\mathcal{K}}(\widehat{\mathtt{Q}}_q) = \overline{\Sigma}_q \quad \textit{and} \quad \overline{\mathcal{K}}(\widehat{\mathtt{Q}}^*_q) = \overline{\Sigma}^*_q \tag{18.35}$$

as Coxeter q-bialgebras. Explicitly,

$$\begin{array}{ccc} \overline{\mathcal{K}}(\widehat{\mathtt{Q}}_q)[A, C_r] \xrightarrow{\cong} \overline{\Sigma}_q[A, C_r] & \qquad & \overline{\mathcal{K}}(\widehat{\mathtt{Q}}^*_q)[A, C_r] \xrightarrow{\cong} \overline{\Sigma}^*_q[A, C_r] \\ \overline{\mathtt{H}}_{F/A, C/A} \longmapsto \mathtt{H}_{\mathrm{t}(F)/\,\mathrm{t}(A)} & & \overline{\mathtt{M}}_{G/A, D/A} \longmapsto \mathtt{M}_{\mathrm{t}(G)/\,\mathrm{t}(A)}. \end{array}$$

Here $\overline{\mathtt{H}}_{F/A,C/A}$ refers to the class of $\mathtt{H}_{F/A,C/A}$ under the action of W_A. The same remark applies to $\overline{\mathtt{M}}_{G/A,D/A}$.

PROOF. For the first claim: The product part is clear. For the coproduct part, we use (1.45) applied to $\boldsymbol{\alpha}_{\mathrm{t}(A)}$. Similarly, for the second claim: The coproduct part is clear. For the product part, we use (1.38) applied to $\boldsymbol{\alpha}_{\mathrm{t}(A)}$. □

18.4.7. q-norm map on faces, top-nested faces, face-types.

Exercise 18.37. Check directly: The q-norm map on face-types can be obtained from the q-norm map on top-nested faces as follows. For a scalar q, and an invariant family of scalars ξ as in Definition 1.10, we have a diagram of Coxeter q-bialgebras

$$\begin{array}{ccc} \overline{\mathcal{K}}(\widehat{\mathtt{Q}}_q) & \xrightarrow{\overline{\mathcal{K}}(\kappa_{q,\xi})} & \overline{\mathcal{K}}(\widehat{\mathtt{Q}}^*_q) \\ \| & & \| \\ \overline{\Sigma}_q & \xrightarrow[\overline{\kappa}_{q,\xi}]{} & \overline{\Sigma}^*_q. \end{array} \tag{18.36}$$

The top-horizontal map is induced from (4.80); the bottom-horizontal map is (10.85). This provides another derivation of Lemma 10.79.

Alternatively, starting with the diagram of Coxeter spaces

$$\begin{array}{ccc} \overline{\mathcal{K}}(\Gamma^*) & \xrightarrow{\overline{\mathcal{K}}(h_\xi)} & \overline{\mathcal{K}}(\Gamma) \\ \| & & \| \\ \Bbbk\{x\} & \xrightarrow[f_\xi]{} & \Bbbk[x] \end{array} \qquad \begin{array}{c} h_\xi(\mathtt{M}_{C/A}) := \xi_{\mathrm{s}(A)}\mathtt{H}_{C/A} \\ \\ f_\xi(x^{(T)}) := \xi_T x^T \end{array} \tag{18.37}$$

from a Coxeter coalgebra to a Coxeter algebra, deduce diagram (18.36) by using $\overline{\mathcal{K}}(\widehat{h_\xi}) = \widehat{\overline{\mathcal{K}}(h_\xi)} = \widehat{f_\xi}$. The first equality is as in the first diagram in (17.28).

Let us now look at the isomorphism problem for general q-norm maps. If $\kappa_{q,\xi} : \widehat{\mathtt{Q}}_q \to \widehat{\mathtt{Q}}^*_q$ is an isomorphism, then so is $\overline{\kappa}_{q,\xi} : \overline{\Sigma}_q \to \overline{\Sigma}^*_q$ in view of (18.36). Thus, the isomorphism criterion in Theorem 4.22 implies the isomorphism criterion in Theorem 10.86. This can also be seen directly as follows. In the first criterion, summing the matrix entries with C fixed and D varying yields the same value independent of C. Since the determinant is nonzero, this common value must also be nonzero, and this is precisely the second criterion. (To match notations, take $T = \mathrm{t}(A)$ and $U = \mathrm{t}(F')$.) Put differently, in the notation of diagram (18.37), if $\mathbf{log}_q(h_\xi)$ is an isomorphism, then so is $\mathbf{log}_q(f_\xi)$. See also the generalities in Section 17.4.5.

Exercise 18.38. We now relate the q-norm map on faces and the q-norm map on face-types. Let ξ be an invariant family of scalars as in Definition 1.10. Define another invariant family of scalars $\mathrm{n}\,\xi$ indexed by faces by $(\mathrm{n}\,\xi)_A := |W_A|\,\xi_A$. Deduce from diagrams (4.81) and (18.36): For a scalar q and $\mathrm{char}(\Bbbk) \nmid |W|$, we have a diagram of Coxeter q-bialgebras

$$\begin{array}{ccc} \overline{\mathcal{K}}(\boldsymbol{\Sigma}_q) & \xrightarrow{\overline{\mathcal{K}}(\kappa_{q,\mathrm{n}\xi})} & \overline{\mathcal{K}}(\boldsymbol{\Sigma}^*_q) \\ \| & & \| \\ \overline{\Sigma}_q & \xrightarrow[\overline{\kappa}_{q,\xi}]{} & \overline{\Sigma}^*_q. \end{array} \tag{18.38}$$

The top-horizontal map is induced from (4.66); the bottom-horizontal map is (10.85). Note very carefully: In the former we use $\mathrm{n}\,\xi$ instead of ξ.

Let us now specialize Theorem 17.41 to the map $f_\xi : \Bbbk\{x\} \to \Bbbk[x]$ in (18.37) which on the T-component is scalar multiplication by ξ_T. Firstly, using (18.15), we see that $\mathsf{t}_{f_\xi}\,\mathrm{n}_{\Bbbk\{x\}} : \mathsf{E} \to \mathsf{E}$ on the A-component is scalar multiplication by $(\mathrm{n}\,\xi)_A$. Next,

$$\widehat{f_\xi} = \overline{\kappa}_{q,\xi} : \overline{\Sigma}_q \to \overline{\Sigma}^*_q \quad \text{and} \quad \widehat{\mathsf{t}_{f_\xi}\,\mathrm{n}_{\Bbbk\{x\}}} = \kappa_{q,\mathrm{n}\xi} : \boldsymbol{\Sigma}_q \to \boldsymbol{\Sigma}^*_q.$$

The first fact is mentioned in Exercise 10.80. Deduce: For q not a root of unity and $\mathrm{char}(\Bbbk) \nmid |W|$,

$$\overline{\kappa}_{q,\xi} \text{ is an isomorphism} \iff \kappa_{q,\mathrm{n}\xi} \text{ is an isomorphism.}$$

Thus, the isomorphism criteria in Theorem 10.86 and Theorem 4.21 are equivalent, with the latter for $\mathrm{n}\,\xi$ (assuming $\mathrm{char}(\Bbbk) \nmid |W|$). Check this directly. This is a manifestation of identity (17.16).

18.4.8. Fock functor adjoints.

Lemma 18.39. *For a scalar q,*

$$\overline{\mathcal{Z}}^\vee(\overline{\Sigma}_q) = \boldsymbol{\Sigma}_q \quad \textit{and} \quad \overline{\mathcal{Z}}(\overline{\Sigma}^*_q) = \boldsymbol{\Sigma}^*_q$$

as Coxeter q-bimonoids.

For $q = 1$, this also illustrates the general fact that $\overline{\mathcal{Z}}^\vee$ preserves cocommutativity while $\overline{\mathcal{Z}}$ preserves commutativity.

Proof. This is similar to Lemma 18.12. The two calculations are

$$\overline{\mathcal{Z}}^\vee(\overline{\Sigma}_q) = \overline{\mathcal{Z}}^\vee\overline{\mathcal{T}}_q(\Bbbk\{x\}) = \mathcal{T}_q(\mathsf{t}_{\Bbbk\{x\}}) = \mathcal{T}_q(\mathsf{E}) = \boldsymbol{\Sigma}_q,$$
$$\overline{\mathcal{Z}}(\overline{\Sigma}^*_q) = \overline{\mathcal{Z}}\overline{\mathcal{T}}_q^\vee(\Bbbk[x]) = \mathcal{T}_q^\vee(\mathsf{t}_{\Bbbk[x]}) = \mathcal{T}_q^\vee(\mathsf{E}) = \boldsymbol{\Sigma}^*_q.$$

They are dual to each other. In the first calculation, E is viewed as a Coxeter comonoid, while in the second, it is viewed as a Coxeter monoid. □

Exercise 18.40. Check: The identifications in Lemma 18.39 work as follows. For a scalar q, the map

$$\overline{\mathcal{Z}}(\overline{\Sigma}^*_q) \xrightarrow{\cong} \boldsymbol{\Sigma}^*_q, \qquad (x_F)_{F \geq A} \mapsto \sum_{F:\, F \geq A} \langle x_F, \mathtt{M}_{\mathrm{t}(F)/\,\mathrm{t}(F)} \rangle\, \mathtt{M}_{F/A} \tag{18.39}$$

on the A-component, is an isomorphism of Coxeter q-bimonoids. Here x_F is a linear combination of face-types greater than $\mathrm{t}(F)$ (written in the $\mathtt{M}$-basis) and

$\langle x_F, \mathtt{M}_{\mathrm{t}(F)/\,\mathrm{t}(F)} \rangle$ picks the coefficient of the smallest face-type, namely, $\mathrm{t}(F)$. Similarly, we have the isomorphism

$$\overline{\mathcal{Z}}^\vee(\overline{\Sigma}_q) \xrightarrow{\cong} \Sigma_q, \qquad \overline{x}_F \mapsto \sum_{G:\, G \geq F} \langle x_F, \mathtt{H}_{\mathrm{t}(G)/\,\mathrm{t}(F)} \rangle\, \mathtt{H}_{G/A} \tag{18.40}$$

on the A-component with $F \geq A$ and bar on x indicating the image in the quotient.

Exercise 18.41. We now supplement Exercise 18.38. Specialize diagram (17.37) to the map f_ξ in (18.37) to obtain: For a scalar q,

$$\begin{array}{ccc} \overline{\mathcal{Z}}^\vee(\overline{\Sigma}_q) & \xrightarrow[\mathrm{z}_{\overline{\Sigma}^*_q}\, \overline{\mathcal{Z}}^\vee(\overline{\kappa}_{q,\xi})]{\overline{\mathcal{Z}}(\overline{\kappa}_{q,\xi})\, \mathrm{z}_{\overline{\Sigma}_q}} & \overline{\mathcal{Z}}(\overline{\Sigma}^*_q) \\ \| & & \| \\ \Sigma_q & \xrightarrow[\kappa_{q,\mathrm{n}\xi}]{} & \Sigma^*_q \end{array} \tag{18.41}$$

as Coxeter q-bimonoids.

18.4.9. Antipode formulas.

Exercise 18.42. Use Proposition 17.44 to deduce antipode formulas of the Coxeter q-bialgebra $\overline{\Sigma}_q$ in (10.116), (10.117), (10.118) from antipode formulas of the Coxeter q-bimonoid Σ_q in (4.72), (4.73), (4.74). Deduce the alternating sign identity (1.63c) as a consequence.

18.5. Coxeter bialgebra of flat-types as a bosonic Fock space

In this section, we assume $\mathrm{char}(\Bbbk) \nmid |W|$. Recall the Coxeter bimonoid of flats $\sqcap$ from Section 4.4, and the Coxeter bialgebra of flat-types $\overline{\Pi}$ from Section 10.5. Both are bicommutative. We now link them using bosonic Fock functors.

18.5.1. Coxeter bimonoid of flats. Applying the bosonic Fock functor to the Coxeter bimonoid of flats $\sqcap$ yields the Coxeter bialgebra of flat-types $\overline{\Pi}$:

Lemma 18.43. *Under the identification* (1.130) *applied to each* $\boldsymbol{\alpha}_Z$,

$$\overline{\mathcal{K}}^\vee(\sqcap) \xrightarrow{\cong} \overline{\Pi} \tag{18.42}$$

as Coxeter bialgebras. Moreover, we have the diagram of Coxeter bialgebras

$$\begin{array}{ccc} \overline{\mathcal{K}}^\vee(\Sigma) & \xrightarrow{\overline{\mathcal{K}}^\vee(\pi)} & \overline{\mathcal{K}}^\vee(\sqcap) \\ \| & & \| \\ \overline{\Sigma} & \xrightarrow[\pi]{} & \overline{\Pi}, \end{array} \tag{18.43}$$

with the top-horizontal map induced from (4.76) *and bottom-horizontal map* (10.122).

PROOF. Identity (18.42) can be checked directly. Alternatively, it can be deduced using diagram (18.43). Note very carefully: (1.130) specifies the map (18.42) on both the $\mathtt{H}$-basis and $\mathtt{Q}$-basis. □

In view of (18.42), formulas (10.124) can be deduced from formulas (4.42). Diagram (18.43) can be viewed as an example of the fact that bosonic Fock functors preserve abelianization (Exercise 16.66).

Exercise 18.44. Deduce using (18.33):

$$\overline{\mathcal{K}}(\mathsf{\Pi}) \xrightarrow{\cong} \overline{\Pi}, \qquad \overline{\mathtt{H}}_{\mathrm{X/Z}} \leftrightarrow |W_{\mathrm{X}}|\, \mathtt{H}_{\mathrm{t(X/Z)}}, \tag{18.44}$$

where bar on $\mathtt{H}$ indicates image in the quotient space.

By duality,

$$\overline{\mathcal{K}}(\mathsf{\Pi}^*) \xrightarrow{\cong} \overline{\Pi}^* \quad \text{and} \quad \overline{\mathcal{K}}^{\vee}(\mathsf{\Pi}^*) \xrightarrow{\cong} \overline{\Pi}^*. \tag{18.45}$$

18.5.2. Freeness and cofreeness. Recall: $\mathsf{\Pi}$ is the free commutative Coxeter bimonoid on E viewed either as a Coxeter species or a Coxeter comonoid, while $\overline{\Pi}$ is the free commutative Coxeter bialgebra on $\Bbbk\{x\}$ viewed either as a Coxeter space or a Coxeter coalgebra. Similarly, there are dual cofreeness results. See (4.43) and (4.44), and Proposition 10.113 and subsequent discussion. The latter set of results can be deduced from the former as follows.

$$\begin{aligned} \overline{\Pi} &= \overline{\mathcal{K}}^{\vee}(\mathsf{\Pi}) = \overline{\mathcal{K}}^{\vee}\mathcal{S}(\mathsf{E}) = \overline{\mathcal{S}}\overline{\mathcal{K}}^{\vee}(\mathsf{E}) = \overline{\mathcal{S}}(\Bbbk\{x\}), \\ \overline{\Pi}^* &= \overline{\mathcal{K}}(\mathsf{\Pi}^*) = \overline{\mathcal{K}}\mathcal{S}^{\vee}(\mathsf{E}) = \overline{\mathcal{S}}^{\vee}\overline{\mathcal{K}}(\mathsf{E}) = \overline{\mathcal{S}}^{\vee}(\Bbbk[x]). \end{aligned}$$

This used Proposition 15.29 and also (18.6).

18.5.3. Primitive and indecomposable parts. Similarly, the primitive and indecomposable parts of $\overline{\Pi}$ in (10.126) can be computed from those of $\mathsf{\Pi}$ in (4.46):

$$\begin{aligned} \overline{\mathcal{P}}(\overline{\Pi}) &= \overline{\mathcal{P}}\overline{\mathcal{K}}^{\vee}(\mathsf{\Pi}) = \overline{\mathcal{K}}^{\vee}\mathcal{P}(\mathsf{\Pi}) = \overline{\mathcal{K}}^{\vee}(\mathsf{E}) = \Bbbk\{x\}, \\ \overline{\mathcal{Q}}(\overline{\Pi}) &= \overline{\mathcal{Q}}\overline{\mathcal{K}}^{\vee}(\mathsf{\Pi}) = \overline{\mathcal{K}}^{\vee}\mathcal{Q}(\mathsf{\Pi}) = \overline{\mathcal{K}}^{\vee}(\mathsf{E}) = \Bbbk\{x\}. \end{aligned}$$

Note: Since we are assuming $\mathrm{char}(\Bbbk) \nmid |W|$, it does not matter whether we write $\overline{\mathcal{K}}$ or $\overline{\mathcal{K}}^{\vee}$, or $\Bbbk\{x\}$ or $\Bbbk[x]$.

18.5.4. Exp-log correspondences. The exp-log correspondence for $\overline{\Pi}$ between primitive and group-like series was discussed in Section 10.5.8 and for $\mathsf{\Pi}$ in Section 4.4.8. The similarity between the two is explained by $\overline{\mathcal{K}}^{\vee}(\mathsf{\Pi}) = \overline{\Pi}$ and Theorem 17.9. One can check directly that the universal series and exponential series of $\mathsf{\Pi}$ and $\overline{\Pi}$ match.

18.5.5. Norm map on flats and flat-types.

Exercise 18.45. This is the commutative analogue of Exercise 18.38. Following the same notations, check: We have a diagram of Coxeter bialgebras

$$\begin{array}{ccc} \overline{\mathcal{K}}(\mathsf{\Pi}) & \xrightarrow{\overline{\mathcal{K}}(\kappa_{\mathrm{n}\xi})} & \overline{\mathcal{K}}(\mathsf{\Pi}^*) \\ \| & & \| \\ \overline{\Pi} & \xrightarrow[\overline{\kappa}_{\xi}]{} & \overline{\Pi}^*. \end{array} \tag{18.46}$$

The top-horizontal map is induced from (4.45); the bottom-horizontal map is (10.127). Also check that the isomorphism criteria in Theorem 4.17 and Theorem 10.119 are consistent.

18.5.6. Fock functor adjoints.

Lemma 18.46. *We have*

$$\overline{\mathcal{Z}}^{\vee}_{ab}(\overline{\Pi}) = \Pi \quad \textit{and} \quad \overline{\mathcal{Z}}^{coab}(\overline{\Pi}^*) = \Pi^*$$

as Coxeter bimonoids.

PROOF. The two calculations (which are dual to each other) are

$$\overline{\mathcal{Z}}^{\vee}_{ab}(\overline{\Pi}) = \overline{\mathcal{Z}}^{\vee}_{ab}\overline{\mathcal{S}}(\mathbb{k}\{x\}) = \mathcal{S}(\mathsf{t}_{\mathbb{k}\{x\}}) = \mathcal{S}(\mathsf{E}) = \Pi,$$
$$\overline{\mathcal{Z}}^{coab}(\overline{\Pi}^*) = \overline{\mathcal{Z}}^{coab}\overline{\mathcal{S}}^{\vee}(\mathbb{k}[x]) = \mathcal{S}^{\vee}(\mathsf{t}_{\mathbb{k}[x]}) = \mathcal{S}^{\vee}(\mathsf{E}) = \Pi^*.$$

The second steps used diagrams (16.38) and (16.40). □

18.5.7. Antipode formulas.

Exercise 18.47. Use Proposition 17.44 to deduce antipode formulas (10.133) and (10.134) for $\overline{\Pi}$ from antipode formulas (4.49) and (4.50) for Π.

18.6. Coxeter bialgebra of symmetries as a bosonic Fock space

Recall the Coxeter bimonoid of pairs of chambers $\mathsf{I\Gamma}$ from Section 4.7, and the Coxeter bialgebra of symmetries $\overline{\mathrm{I\Gamma}}$ from Section 10.6. We now link them using bosonic Fock functors.

18.6.1. Coxeter bimonoid of pairs of chambers. Let q be a scalar. Applying the bosonic Fock functor to the Coxeter q-bimonoid of pairs of chambers yields the Coxeter q-bialgebra of symmetries:

$$\overline{\mathcal{K}}^{\vee}(\mathsf{I\Gamma}^*_q) = \overline{\mathrm{I\Gamma}}^*_q. \tag{18.47}$$

On the (A, C)-component, the identification is

$$\mathtt{F}_w = \sum_{d(D,D')=w} \mathtt{F}_{D/A,D'/A}, \quad \text{or equivalently,} \quad \mathtt{M}_w = \sum_{d(D,D')=w} \mathtt{M}_{D/A,D'/A},$$

where $w \in W_A$. Similarly, we also have

$$\overline{\mathcal{K}}(\mathsf{I\Gamma}^*_q) = \overline{\mathrm{I\Gamma}}^*_q. \tag{18.48}$$

On the (A, C)-component, the identification is

$$\overline{\mathtt{F}}_{D/A,D'/A} = \mathtt{F}_{d(D,D')}, \quad \text{or equivalently,} \quad \overline{\mathtt{M}}_{D/A,D'/A} = \mathtt{M}_{d(D,D')},$$

where bar on top of $\mathtt{F}$ and $\mathtt{M}$ refers to the coinvariant class.

Dually,

$$\overline{\mathcal{K}}^{\vee}(\mathsf{I\Gamma}_q) = \overline{\mathrm{I\Gamma}}_q \quad \text{and} \quad \overline{\mathcal{K}}(\mathsf{I\Gamma}_q) = \overline{\mathrm{I\Gamma}}_q. \tag{18.49}$$

Exercise 18.48 ($\mathtt{M}$-bases). Deduce (co)product formulas (10.139) of $\overline{\mathrm{I\Gamma}}^*_q$ from (co)product formulas (4.86) of $\mathsf{I\Gamma}^*_q$.

Exercise 18.49. Use Proposition 15.8 and the Coxeter analogue of [**11**, (15.37)] to deduce: $\mathcal{K}^{\vee}_q(\Gamma) = \overline{\mathrm{I\Gamma}}^*_q$ and $\mathcal{K}_q(\Gamma^*) = \overline{\mathrm{I\Gamma}}_q$. See Remark 18.7 in this regard.

18.6.2. Self-duality. The self-duality of the Coxeter q-bimonoid $\mathbb{\Gamma}^*_q$ implies the self-duality of the Coxeter q-bialgebra $\overline{\mathbb{\Gamma}}^*_q$. More explicitly, one can deduce the existence of the map (10.142) from the map (4.87).

Exercise 18.50. Use diagram (4.92) of Coxeter bimonoids to deduce diagram (10.147) of Coxeter bialgebras.

18.6.3. Freeness and cofreeness. Facts about primitive part and cofreeness of $\overline{\mathbb{\Gamma}}^*$ in (10.143) can also be deduced from corresponding facts for $\mathbb{\Gamma}^*$ in (4.88) via

$$\overline{\mathcal{P}}(\overline{\mathbb{\Gamma}}^*) = \overline{\mathcal{P}}\overline{\mathcal{K}}^\vee(\mathbb{\Gamma}^*) = \overline{\mathcal{K}}^\vee\mathcal{P}(\mathbb{\Gamma}^*)$$

and

$$\overline{\mathbb{\Gamma}}^* = \overline{\mathcal{K}}^\vee(\mathbb{\Gamma}^*) = \overline{\mathcal{K}}^\vee\mathcal{T}^\vee\mathcal{P}(\mathbb{\Gamma}^*) = \overline{\mathcal{T}}^\vee\overline{\mathcal{K}}^\vee\mathcal{P}(\mathbb{\Gamma}^*) = \overline{\mathcal{T}}^\vee\overline{\mathcal{P}}(\overline{\mathbb{\Gamma}}^*).$$

A similar remark applies to indecomposable part and freeness.

18.6.4. Antipode formulas.

Exercise 18.51. Use Proposition 17.44 and antipode formulas (4.90) and (4.91) of $\mathbb{\Gamma}^*_q$ to deduce antipode formulas (10.145) and (10.146) of $\overline{\mathbb{\Gamma}}^*_q$. (Some detail on how the translation works can be found on [**7**, page 570].)

18.6.5. Coxeter bimonoid of chamber maps. Recall the Coxeter bimonoid of chamber maps $\mathcal{C}(\Gamma, \Gamma)$ from Section 4.7.8. Applying $\overline{\mathcal{K}}^\vee$ to diagram (4.94) and using (18.30) and (18.49) with $q = 1$ yields the diagram

$$\begin{array}{ccc} \overline{\Sigma} & \hookrightarrow & \overline{\mathbb{\Gamma}} \\ & \searrow \quad \nearrow & \\ & \overline{\mathcal{K}}^\vee(\mathcal{C}(\Gamma, \Gamma)) & \end{array} \tag{18.50}$$

of Coxeter bialgebras.

Exercise 18.52. Check: $\overline{\mathcal{K}}^\vee(\mathcal{C}(\Gamma, \Gamma))$ equals the Coxeter bialgebra H considered in Exercise 10.128.

Employing (3.73) and (4.96), we obtain

$$\overline{\mathcal{P}}\overline{\mathcal{K}}^\vee(\mathcal{C}(\Gamma, \Gamma)) = \overline{\mathcal{K}}^\vee\mathcal{P}(\mathcal{C}(\Gamma, \Gamma)) = \overline{\mathcal{K}}^\vee(\mathrm{hom}^\times(\Gamma, \mathcal{P}(\Gamma))) = \overline{\mathcal{K}}^\vee(\mathrm{hom}^\times(\Gamma, \mathsf{Lie})).$$

The A-component of $\mathrm{hom}^\times(\Gamma, \mathsf{Lie})$ consists of linear maps from $\Gamma[A]$ to $\mathsf{Lie}[A]$. The group W_A acts on this space by conjugation, and the space of invariants consists of W_A-module maps which identifies with $\mathsf{Lie}[A]$. Thus,

$$\overline{\mathcal{P}}\overline{\mathcal{K}}^\vee(\mathcal{C}(\Gamma, \Gamma))[A, C] = \mathsf{Lie}[A]. \tag{18.51}$$

This recovers formula (10.151).

18.7. Fock functors for a rank-one arrangement

Let $\mathcal{A}$ be a rank-one arrangement with chambers C and $\overline{C}$. Let w denote the nonidentity element of the Coxeter group W. Let $\boldsymbol{\alpha}$ be the pointed rank-one reflection arrangement with C as the reference chamber. We now see how Fock functors and their adjoints work for this example. Coxeter $\boldsymbol{\alpha}$-bialgebras are discussed in Section 13.8. Coxeter $\mathcal{A}$-bimonoids are in Section 7.8; however, there we worked with flats and lunes while here we work with faces.

18.7.1. Fock spaces. We specialize constructions of Fock spaces in Section 15.1 to rank one.

Let p and q be scalars. Let h be a Coxeter $\mathcal{A}$-p-bimonoid. Applying the q-full Fock functor $\mathcal{K}_q$ yields the Coxeter $\boldsymbol{\alpha}$-pq-bialgebra $\mathcal{K}_q(\mathsf{h})$. As a Coxeter $\boldsymbol{\alpha}$-space, its components are

$$\mathcal{K}_q(\mathsf{h})[\emptyset] := \mathsf{h}[O] \quad \text{and} \quad \mathcal{K}_q(\mathsf{h})[S] := \mathsf{h}[C],$$

with involution σ on $\mathcal{K}_q(\mathsf{h})[S]$ defined by $w\beta_{\overline{C},C} = \beta_{C,\overline{C}}\, w$ (where β and w refer to support and type morphisms of h). The product and coproduct of $\mathcal{K}_q(\mathsf{h})$ are

$$\begin{array}{ccc} \mathcal{K}_q(\mathsf{h})[S] & \overset{\mu}{\dashrightarrow} & \mathcal{K}_q(\mathsf{h})[\emptyset] \\ \| & & \| \\ \mathsf{h}[C] & \xrightarrow[\mu_O^C]{} & \mathsf{h}[O] \end{array} \qquad \begin{array}{ccc} \mathcal{K}_q(\mathsf{h})[\emptyset] & \overset{\Delta}{\dashrightarrow} & \mathcal{K}_q(\mathsf{h})[S] \\ \| & & \| \\ \mathsf{h}[O] & \xrightarrow[\Delta_O^C + q\, w \Delta_O^{\overline{C}}]{} & \mathsf{h}[C]. \end{array}$$

These specialize formulas (15.9) to rank one. The pq-bialgebra axiom for $\mathcal{K}_q(\mathsf{h})$ can be directly checked using the p-bimonoid axiom for h:

$$\Delta\mu = (\Delta_O^C + q\, w\Delta_O^{\overline{C}})\mu_O^C = \mathrm{id} + pq\, w\beta_{\overline{C},C} = \mathrm{id} + pq\, \sigma.$$

Dually, applying the conjugate q-full Fock functor $\mathcal{K}_q^\vee$ to h yields the Coxeter $\boldsymbol{\alpha}$-pq-bialgebra $\mathcal{K}_q^\vee(\mathsf{h})$. It is identical to $\mathcal{K}_q(\mathsf{h})$ as a Coxeter $\boldsymbol{\alpha}$-space; its product and coproduct are

$$\begin{array}{ccc} \mathcal{K}_q^\vee(\mathsf{h})[S] & \overset{\mu}{\dashrightarrow} & \mathcal{K}_q^\vee(\mathsf{h})[\emptyset] \\ \| & & \| \\ \mathsf{h}[C] & \xrightarrow[\mu_O^C + q\, \mu_O^{\overline{C}} w]{} & \mathsf{h}[O] \end{array} \qquad \begin{array}{ccc} \mathcal{K}_q^\vee(\mathsf{h})[\emptyset] & \overset{\Delta}{\dashrightarrow} & \mathcal{K}_q^\vee(\mathsf{h})[S] \\ \| & & \| \\ \mathsf{h}[O] & \xrightarrow[\Delta_O^C]{} & \mathsf{h}[C]. \end{array}$$

These specialize formulas (15.10) to rank one. Observe: $\mathcal{K}_0(\mathsf{h}) = \mathcal{K}_0^\vee(\mathsf{h})$ as Coxeter $\boldsymbol{\alpha}$-0-bialgebras, with μ and Δ given by μ_O^C and Δ_O^C, respectively. This is a special case of (15.4).

The bosonic Fock functors $\overline{\mathcal{K}}$ and $\overline{\mathcal{K}}^\vee$ applied to h yield the Coxeter $\boldsymbol{\alpha}$-p-bialgebras $\overline{\mathcal{K}}(\mathsf{h})$ and $\overline{\mathcal{K}}^\vee(\mathsf{h})$ with components

$$\overline{\mathcal{K}}(\mathsf{h})[\emptyset] := \mathsf{h}[O]_W, \quad \overline{\mathcal{K}}(\mathsf{h})[S] := \mathsf{h}[C],$$
$$\overline{\mathcal{K}}^\vee(\mathsf{h})[\emptyset] := \mathsf{h}[O]^W, \quad \overline{\mathcal{K}}^\vee(\mathsf{h})[S] := \mathsf{h}[C].$$

The former is a quotient of $\mathcal{K}(\mathsf{h})$, while the latter is contained in $\mathcal{K}^\vee(\mathsf{h})$. Similarly, for fermionic Fock functors $\overline{\mathcal{K}}_{-1}$ and $\overline{\mathcal{K}}_{-1}^\vee$, taking signed (co)invariants, we obtain Coxeter $\boldsymbol{\alpha}$-$(-p)$-bialgebras $\overline{\mathcal{K}}_{-1}(\mathsf{h})$ and $\overline{\mathcal{K}}_{-1}^\vee(\mathsf{h})$. The former is a quotient of $\mathcal{K}_{-1}(\mathsf{h})$, while the latter is contained in $\mathcal{K}_{-1}^\vee(\mathsf{h})$.

Exercise 18.53. Verify directly the commutativity properties of Fock functors in Theorems 15.11 and 15.12 in rank one.

Exercise 18.54 (q-norm map). Verify directly Proposition 15.16 in rank one. In other words, $(\kappa_q)_{\mathsf{h}} : \mathcal{K}_q(\mathsf{h}) \to \mathcal{K}_q^\vee(\mathsf{h})$ given by $\mathrm{id} + qw$ on the $\emptyset$-component, and id on the S-component, is a morphism of Coxeter $\boldsymbol{\alpha}$-pq-bialgebras. It is an isomorphism when $q \neq \pm 1$.

Exercise 18.55 (Freeness). Use Exercise 13.80 to deduce Theorem 15.31 in rank one.

18.7.2. Trivialization functor. We specialize constructions in Section 16.1 to rank one.

For a Coxeter $\boldsymbol{\alpha}$-space V, define the Coxeter $\mathcal{A}$-species t_{V} by

$$\mathsf{t}_{\mathrm{V}}[O] := \mathrm{V}[\emptyset] \quad \text{and} \quad \mathsf{t}_{\mathrm{V}}[C] = \mathsf{t}_{\mathrm{V}}[\overline{C}] := \mathrm{V}[S]$$

with support morphisms $\beta_{\overline{C},C} = \beta_{C,\overline{C}} := \sigma$ and type morphisms all identities. (Here σ refers to the involution on $\mathrm{V}[S]$.) Observe:

$$\overline{\mathcal{K}}(\mathsf{t}_{\mathrm{V}}) = \overline{\mathcal{K}}^{\vee}(\mathsf{t}_{\mathrm{V}}) = \mathrm{V}$$

which verifies Propositions 16.14 and 16.15 in rank one.

For a Coxeter $\boldsymbol{\alpha}$-algebra (A, μ), the Coxeter $\mathcal{A}$-species t_{A} is a Coxeter $\mathcal{A}$-monoid with product components $\mu_O^C = \mu_O^{\overline{C}} := \mu$. If A is commutative, then so is t_{A}. Dually, for a Coxeter $\boldsymbol{\alpha}$-coalgebra (C, Δ), the Coxeter $\mathcal{A}$-species t_{C} is a Coxeter $\mathcal{A}$-comonoid with coproduct components $\Delta_O^C = \Delta_O^{\overline{C}} := \Delta$. If C is cocommutative, then so is t_{C}. Observe:

$$\overline{\mathcal{K}}(\mathsf{t}_{\mathrm{A}}) = \mathrm{A} \quad \text{and} \quad \overline{\mathcal{K}}^{\vee}(\mathsf{t}_{\mathrm{C}}) = \mathrm{C}$$

which verifies Propositions 16.27 and 16.34 in rank one.

18.7.3. Lifts of the trivialization functor. We specialize constructions in Section 16.4 to rank one.

Let (A, μ) be a Coxeter $\boldsymbol{\alpha}$-algebra. Consider the Coxeter $\mathcal{A}$-species t_{A} and the free Coxeter $\mathcal{A}$-monoid $\mathcal{T}(\mathsf{t}_{\mathrm{A}})$ on it. Its components are

$$\mathcal{T}(\mathsf{t}_{\mathrm{A}})[O] := \mathrm{A}[\emptyset] \oplus \mathrm{A}[S] \oplus \mathrm{A}[S] \quad \text{and} \quad \mathcal{T}(\mathsf{t}_{\mathrm{A}})[C] = \mathcal{T}(\mathsf{t}_{\mathrm{A}})[\overline{C}] := \mathrm{A}[S].$$

The two copies of $\mathrm{A}[S]$ are indexed by chambers, the first by C and second by $\overline{C}$. The support morphisms are $\beta_{\overline{C},C} = \beta_{C,\overline{C}} := \sigma$. One type morphism is

$$w : \mathcal{T}(\mathsf{t}_{\mathrm{A}})[O] \to \mathcal{T}(\mathsf{t}_{\mathrm{A}})[O], \qquad (x, y, z) \mapsto (x, z, y)$$

and remaining type morphisms are identities. Product components are

$$\mu_O^C : \mathcal{T}(\mathsf{t}_{\mathrm{A}})[C] \to \mathcal{T}(\mathsf{t}_{\mathrm{A}})[O], \qquad x \mapsto (0, x, 0),$$
$$\mu_O^{\overline{C}} : \mathcal{T}(\mathsf{t}_{\mathrm{A}})[\overline{C}] \to \mathcal{T}(\mathsf{t}_{\mathrm{A}})[O], \qquad x \mapsto (0, 0, x).$$

Let us now look at the Coxeter $\mathcal{A}$-monoid $\overline{\mathcal{Z}}^{\vee}(\mathrm{A})$. It is a quotient of $\mathcal{T}(\mathsf{t}_{\mathrm{A}})$. The C- and $\overline{C}$-components are $\mathrm{A}[S]$ which are the same as before, while

$$\overline{\mathcal{Z}}^{\vee}(\mathrm{A})[O] := \big(\mathrm{A}[\emptyset] \oplus \mathrm{A}[S] \oplus \mathrm{A}[S]\big)/(\mu(x), -x, -x), \quad x \in \mathrm{A}[S].$$

The relations $(\mu(x), -x, -x)$ specialize (16.25).

For a scalar q, let $(\mathrm{H}, \mu, \Delta)$ be a Coxeter $\boldsymbol{\alpha}$-q-bialgebra. Then $\mathcal{T}_q(\mathsf{t}_{\mathrm{H}})$ is a Coxeter $\mathcal{A}$-q-bimonoid. Product components are as above, while coproduct components are

$$\Delta_O^C : \mathcal{T}_q(\mathsf{t}_{\mathrm{H}})[O] \to \mathcal{T}_q(\mathsf{t}_{\mathrm{H}})[C], \qquad (x, y, z) \mapsto \Delta(x) + y + q\,\sigma(z),$$
$$\Delta_O^{\overline{C}} : \mathcal{T}_q(\mathsf{t}_{\mathrm{H}})[O] \to \mathcal{T}_q(\mathsf{t}_{\mathrm{H}})[\overline{C}], \qquad (x, y, z) \mapsto \Delta(x) + q\,\sigma(y) + z.$$

The quotient map $\mathcal{T}_q(\mathrm{t_H}) \twoheadrightarrow \overline{\mathcal{Z}}^\vee(\mathrm{H})$ induces a Coxeter $\mathcal{A}$-q-bimonoid structure on $\overline{\mathcal{Z}}^\vee(\mathrm{H})$. This is because $\Delta\mu(x) - x - q\,\sigma(x) = 0$ by the q-bialgebra axiom. This verifies Lemma 16.47 in rank one. Observe: For $q = 1$, if H is cocommutative, then so is $\overline{\mathcal{Z}}^\vee(\mathrm{H})$. This is the content of Observation 16.56 in rank one.

Let us now look at the dual picture. Let (C, Δ) be a Coxeter $\boldsymbol{\alpha}$-coalgebra. The cofree Coxeter $\mathcal{A}$-comonoid $\mathcal{T}^\vee(\mathrm{t_C})$ has components

$$\mathcal{T}^\vee(\mathrm{t_C})[O] := \mathrm{C}[\emptyset] \oplus \mathrm{C}[S] \oplus \mathrm{C}[S] \quad \text{and} \quad \mathcal{T}^\vee(\mathrm{t_C})[C] = \mathcal{T}^\vee(\mathrm{t_C})[\overline{C}] := \mathrm{C}[S].$$

Its coproduct components are

$$\Delta_O^C : \mathcal{T}^\vee(\mathrm{t_C})[O] \to \mathcal{T}^\vee(\mathrm{t_C})[C], \qquad (x, y, z) \mapsto y,$$
$$\Delta_O^{\overline{C}} : \mathcal{T}^\vee(\mathrm{t_C})[O] \to \mathcal{T}^\vee(\mathrm{t_C})[\overline{C}], \qquad (x, y, z) \mapsto z.$$

The Coxeter $\mathcal{A}$-subcomonoid $\overline{\mathcal{Z}}(\mathrm{C})$ of $\mathcal{T}^\vee(\mathrm{t_C})$ has C- and $\overline{C}$-components $\mathrm{C}[S]$ and O-component consisting of (x, y, z) which satisfy $\Delta(x) = y+z$. The latter specializes conditions (16.28) to rank one.

For a scalar q, for H a Coxeter $\boldsymbol{\alpha}$-q-bialgebra, $\mathcal{T}_q^\vee(\mathrm{t_H})$ is a Coxeter $\mathcal{A}$-q-bimonoid with coproduct components as above and product components

$$\mu_O^C : \mathcal{T}_q^\vee(\mathrm{t_H})[C] \to \mathcal{T}_q^\vee(\mathrm{t_H})[O], \qquad y \mapsto (\mu(y), y, q\,\sigma(y)),$$
$$\mu_O^{\overline{C}} : \mathcal{T}_q^\vee(\mathrm{t_H})[\overline{C}] \to \mathcal{T}_q^\vee(\mathrm{t_H})[O], \qquad z \mapsto (\mu(z), q\,\sigma(z), z).$$

Observe: $\overline{\mathcal{Z}}(\mathrm{H})$ is indeed a Coxeter $\mathcal{A}$-q-subbimonoid of $\mathcal{T}_q^\vee(\mathrm{t_H})$. Also, for $q = 1$, if H is commutative, then so is $\overline{\mathcal{Z}}(\mathrm{H})$.

Example 18.56. Let H be the Coxeter $\boldsymbol{\alpha}$-bialgebra in Example 13.71. Then

$$\overline{\mathcal{Z}}^\vee(\mathrm{H})[O] = \big(\mathrm{H}[S] \oplus \mathrm{H}[S]\big)/(x, x), \quad x \in \mathrm{H}[S].$$

Product components μ_O^C and $\mu_O^{\overline{C}}$ map into the two coordinates, respectively. The coproduct component Δ_O^C is the first coordinate minus the second, while $\Delta_O^{\overline{C}}$ is the second coordinate minus the first. Dually,

$$\overline{\mathcal{Z}}(\mathrm{H})[O] = \{(x, -x)\} \subseteq \mathrm{H}[S] \oplus \mathrm{H}[S], \quad x \in \mathrm{H}[S]$$

with coproduct and product

$$\Delta_O^C(x, -x) = x, \quad \Delta_O^{\overline{C}}(x, -x) = -x, \quad \mu_O^C(x) = (x, -x), \quad \mu_O^{\overline{C}}(x) = (-x, x).$$

□

Exercise 18.57. Check directly Theorems 16.51 and 16.58 in rank one: For a scalar q, for H a Coxeter $\boldsymbol{\alpha}$-q-bialgebra, $\overline{\mathcal{K}}^\vee\overline{\mathcal{Z}}^\vee(\mathrm{H})$ and $\overline{\mathcal{K}}\overline{\mathcal{Z}}(\mathrm{H})$ are isomorphic to H when $\mathrm{char}(\mathbb{k}) \neq 2$. The characteristic assumption is not necessary when $q \neq \pm 1$.

18.7.4. Adjoints preserving commutativity. We specialize constructions in Section 16.6 to rank one. For A a commutative Coxeter $\boldsymbol{\alpha}$-algebra, $\overline{\mathcal{Z}}_{ab}^\vee(\mathrm{A})$ is the abelianization of $\overline{\mathcal{Z}}^\vee(\mathrm{A})$. Its components are

$$\overline{\mathcal{Z}}_{ab}^\vee(\mathrm{A})[C] = \overline{\mathcal{Z}}_{ab}^\vee(\mathrm{A})[\overline{C}] = \mathrm{A}[S],$$
$$\overline{\mathcal{Z}}_{ab}^\vee(\mathrm{A})[O] = \big(\mathrm{A}[\emptyset] \oplus \mathrm{A}[S] \oplus \mathrm{A}[S]\big)/(\mu(x), -x, -x), (0, x, -\sigma(x)).$$

Dually, for C a cocommutative Coxeter $\boldsymbol{\alpha}$-coalgebra, $\overline{\mathcal{Z}}^{coab}(\mathrm{C})$ is the co-abelianization of $\overline{\mathcal{Z}}(\mathrm{C})$. Its components are

$$\overline{\mathcal{Z}}^{coab}(\mathrm{C})[C] = \overline{\mathcal{Z}}^{coab}(\mathrm{C})[\overline{C}] = \mathrm{C}[S],$$

$$\overline{\mathcal{Z}}^{coab}(\mathrm{C})[O] = \{(x, y, z) \mid \Delta(x) = y + z,\ z = \sigma(y)\} \subseteq \mathrm{C}[\emptyset] \oplus \mathrm{C}[S] \oplus \mathrm{C}[S].$$

Exercise 18.58. Check directly Proposition 16.69 in rank one: For A a commutative Coxeter $\boldsymbol{\alpha}$-algebra, $\overline{\mathcal{K}}^{\vee}\overline{\mathcal{Z}}^{\vee}_{ab}(\mathrm{A})$ is isomorphic to A when $\mathrm{char}(\Bbbk) \neq 2$. Dually, for C a cocommutative Coxeter $\boldsymbol{\alpha}$-coalgebra, $\overline{\mathcal{K}}\overline{\mathcal{Z}}^{coab}(\mathrm{C})$ is isomorphic to C when $\mathrm{char}(\Bbbk) \neq 2$.

Exercise 18.59. Check directly Lemma 16.78 in rank one.

18.7.5. Hadamard product with chambers. We specialize constructions in Section 16.2 to rank one. For a Coxeter $\boldsymbol{\alpha}$-algebra (A, μ), define the Coxeter $\mathcal{A}$-monoid $\mathcal{L}(\mathrm{A})$ by

$$\mathcal{L}(\mathrm{A})[O] := \mathrm{A}[\emptyset] \oplus \mathrm{A}[\emptyset] \quad \text{and} \quad \mathcal{L}(\mathrm{A})[C] = \mathcal{L}(\mathrm{A})[\overline{C}] := \mathrm{A}[S]$$

with support morphisms $\beta_{\overline{C},C} = \beta_{C,\overline{C}} := \sigma$ and type morphism on the O-component interchanging the two summands, and type morphisms relating the C- and $\overline{C}$-components being identities. Product components are $\mu_O^C(x) := (\mu(x), 0)$ and $\mu_O^{\overline{C}}(x) := (0, \mu(x))$.

Dually: For a Coxeter $\boldsymbol{\alpha}$-coalgebra (C, Δ), the Coxeter $\mathcal{A}$-comonoid $\mathcal{L}^{\vee}(\mathrm{C})$ is given by

$$\mathcal{L}^{\vee}(\mathrm{C})[O] := \mathrm{C}[\emptyset] \oplus \mathrm{C}[\emptyset] \quad \text{and} \quad \mathcal{L}^{\vee}(\mathrm{C})[C] = \mathcal{L}^{\vee}(\mathrm{C})[\overline{C}] := \mathrm{C}[S]$$

with coproduct components $\Delta_O^C(x, y) := \Delta(x)$ and $\Delta_O^{\overline{C}}(x, y) := \Delta(y)$.

We specialize constructions in Section 16.8 to rank one. Let q be a scalar. For a Coxeter $\boldsymbol{\alpha}$-algebra (A, μ), define the Coxeter $\mathcal{A}$-monoid $\mathcal{Z}_q^{\vee}(\mathrm{A})$ whose C- and $\overline{C}$-components are $\mathrm{A}[S]$, while O-component is

$$\big(\mathrm{A}[\emptyset] \oplus \mathrm{A}[\emptyset] \oplus \mathrm{A}[S] \oplus \mathrm{A}[S]\big)/(\mu(x), 0, -x, -qx), (0, \mu(x), -qx, -x)$$

on which the type morphism interchanges the first two summands and last two summands. Product components μ_O^C and $\mu_O^{\overline{C}}$ send $\mathrm{A}[S]$ to its image in the third and last summands, respectively.

Dually, for a Coxeter $\boldsymbol{\alpha}$-coalgebra (C, Δ), define the Coxeter $\mathcal{A}$-comonoid $\mathcal{Z}_q(\mathrm{C})$ whose C- and $\overline{C}$-components are $\mathrm{C}[S]$, while O-component consists of

$$(x, x', y, y') \in \mathrm{C}[\emptyset] \oplus \mathrm{C}[\emptyset] \oplus \mathrm{C}[S] \oplus \mathrm{C}[S]$$

such that $\Delta(x) = y + qy'$ and $\Delta(x') = qy + y'$. Coproduct components Δ_O^C and $\Delta_O^{\overline{C}}$ project on the third and last coordinates, respectively.

18.7.6. Lie theory. We now specialize the constructions in Section 15.7 to rank one.

For g a Coxeter $\mathcal{A}$-Lie monoid, $\mathcal{K}^\vee(\mathsf{g})$, $\overline{\mathcal{K}}^\vee(\mathsf{g})$, $\overline{\mathcal{K}}(\mathsf{g})$ are Coxeter $\boldsymbol{\alpha}$-Lie algebras with Lie brackets

$$\begin{array}{ccc}\mathcal{K}^\vee(\mathsf{g})[S] & \xrightarrow{\nu} & \mathcal{K}^\vee(\mathsf{g})[\emptyset] \\ \| & & \| \\ \mathsf{g}[C] & \xrightarrow[\nu_O^C+\nu_O^{\overline{C}}w]{} & \mathsf{g}[O]\end{array} \qquad \begin{array}{ccc}\overline{\mathcal{K}}^\vee(\mathsf{g})[S] & \xrightarrow{\nu} & \overline{\mathcal{K}}^\vee(\mathsf{g})[\emptyset] \\ \| & & \| \\ \mathsf{g}[C] & \xrightarrow[\nu_O^C+\nu_O^{\overline{C}}w]{} & \mathsf{g}[O]^W\end{array} \qquad \begin{array}{ccc}\overline{\mathcal{K}}(\mathsf{g})[S] & \xrightarrow{\nu} & \overline{\mathcal{K}}(\mathsf{g})[\emptyset] \\ \| & & \| \\ \mathsf{g}[C] & \xrightarrow[\nu_O^C]{} & \mathsf{g}[O]_W.\end{array}$$

If we try to define a Lie bracket for $\mathcal{K}(\mathsf{g})$ with the same formula as for $\overline{\mathcal{K}}(\mathsf{g})$, then antisymmetry fails; thus, it is important that we go into the quotient space of W-coinvariants.

Dually, for k a Coxeter $\mathcal{A}$-Lie comonoid, $\mathcal{K}(\mathsf{k})$, $\overline{\mathcal{K}}(\mathsf{k})$, $\overline{\mathcal{K}}^\vee(\mathsf{k})$ are Coxeter $\boldsymbol{\alpha}$-Lie coalgebras with Lie cobrackets

$$\begin{array}{ccc}\mathcal{K}(\mathsf{k})[\emptyset] & \xrightarrow{\theta} & \mathcal{K}(\mathsf{k})[S] \\ \| & & \| \\ \mathsf{k}[O] & \xrightarrow[\theta_O^C+w\theta_O^{\overline{C}}]{} & \mathsf{k}[C]\end{array} \qquad \begin{array}{ccc}\overline{\mathcal{K}}(\mathsf{k})[\emptyset] & \xrightarrow{\theta} & \overline{\mathcal{K}}(\mathsf{k})[S] \\ \| & & \| \\ \mathsf{k}[O]_W & \xrightarrow[\theta_O^C+w\theta_O^{\overline{C}}]{} & \mathsf{k}[C]\end{array} \qquad \begin{array}{ccc}\overline{\mathcal{K}}^\vee(\mathsf{k})[\emptyset] & \xrightarrow{\theta} & \overline{\mathcal{K}}^\vee(\mathsf{k})[S] \\ \| & & \| \\ \mathsf{k}[O]^W & \xrightarrow[\theta_O^C]{} & \mathsf{k}[C].\end{array}$$

Universal enveloping functors. Let us now look at the results in Section 15.9 in rank one. To verify the first diagram in (15.39): The top-horizontal map is

$$\mathsf{g}[O] \oplus \mathsf{g}[C] \to \mathsf{g}[O] \oplus \mathsf{g}[C] \oplus \mathsf{g}[\overline{C}], \qquad (x, y) \mapsto (x, y, w(y)).$$

This is the lax structure $\psi_{\mathsf{g}}^\vee$ of $\mathcal{K}^\vee$. It sends the relation

$$(-\nu(y), y - \sigma(y)) = (-\nu_O^C(y) - \nu_O^{\overline{C}}w(y), y - \beta_{C,\overline{C}}w(y))$$

(which defines $\overline{\mathcal{U}}$ as a quotient of $\overline{\mathcal{T}}$) to the sum of the two relations

$$(-\nu_O^C(y), y, -\beta_{\overline{C},C}(y)) \quad \text{and} \quad (-\nu_O^{\overline{C}}w(y), -\beta_{C,\overline{C}}w(y), w(y))$$

(which define $\mathcal{U}$ as a quotient of $\mathcal{T}$).

Similarly, for the first diagram in (15.41): The top-horizontal map is

$$\mathsf{g}[O]_W \oplus \mathsf{g}[C] \to (\mathsf{g}[O] \oplus \mathsf{g}[C] \oplus \mathsf{g}[\overline{C}])_W, \qquad (x, y) \mapsto (x, y, 0).$$

This is the lax structure $\overline{\varphi}_{\mathsf{g}}$ of $\overline{\mathcal{K}}$. It sends the relation

$$(-\nu(y), y - \sigma(y)) = (-\nu_O^C(y), y - w\beta_{\overline{C},C}(y))$$

to the relation

$$(-\nu_O^C(y), y - w\beta_{\overline{C},C}(y), 0) = (-\nu_O^C(y), y, -\beta_{\overline{C},C}(y)).$$

The equality holds since we are in the space of W-coinvariants. One can also see that the induced map is an isomorphism.

Trivialization functor and its lifts. We now specialize constructions in Section 16.10 to rank one. For a Coxeter $\boldsymbol{\alpha}$-Lie algebra (G, ν), the Coxeter $\mathcal{A}$-species t_{G} is a Coxeter $\mathcal{A}$-Lie monoid with Lie bracket $\nu_O^C = \nu_O^{\overline{C}} := \nu$. Dually, for a Coxeter $\boldsymbol{\alpha}$-Lie coalgebra (K, θ), the Coxeter $\mathcal{A}$-species t_{K} is a Coxeter $\mathcal{A}$-Lie comonoid with Lie cobracket $\theta_O^C = \theta_O^{\overline{C}} := \theta$. Observe:

$$\overline{\mathcal{K}}(\mathsf{t}_{\mathrm{G}}) = \mathrm{G} \quad \text{and} \quad \overline{\mathcal{K}}^{\vee}(\mathsf{t}_{\mathrm{K}}) = \mathrm{K}$$

which verifies Proposition 16.100 in rank one.

For (G, ν) a Coxeter $\boldsymbol{\alpha}$-Lie algebra, the Coxeter $\mathcal{A}$-Lie monoid $\overline{\mathcal{Z}}^{\vee}_{\mathbf{Lie}}(\mathrm{G})$ has components

$$\overline{\mathcal{Z}}^{\vee}_{\mathbf{Lie}}(\mathrm{G})[C] = \overline{\mathcal{Z}}^{\vee}_{\mathbf{Lie}}(\mathrm{G})[\overline{C}] = \mathrm{G}[S],$$
$$\overline{\mathcal{Z}}^{\vee}_{\mathbf{Lie}}(\mathrm{G})[O] = \big(\mathrm{G}[\emptyset] \oplus \mathrm{G}[S]\big)/(-\nu(y), y - \sigma(y)) \text{ for } y \in \mathrm{G}[S]$$

with support morphisms $\beta_{\overline{C},C} = \beta_{C,\overline{C}} := \sigma$ and type morphism

$$w : \overline{\mathcal{Z}}^{\vee}_{\mathbf{Lie}}(\mathrm{G})[O] \to \overline{\mathcal{Z}}^{\vee}_{\mathbf{Lie}}(\mathrm{G})[O], \qquad \overline{(x,y)} \mapsto \overline{(x, -\sigma(y))}.$$

The remaining type morphisms relating the C- and $\overline{C}$-components are identities. The Lie brackets are

$$\nu_O^C(x) = \overline{(0,x)} \quad \text{and} \quad \nu_O^{\overline{C}}(x) = \overline{(0, -\sigma(x))}.$$

One can readily check that $\overline{\mathcal{K}}^{\vee}\overline{\mathcal{Z}}^{\vee}_{\mathbf{Lie}}(\mathrm{G}) = \mathrm{G}$ when $\mathrm{char}(\Bbbk)$ is not 2. This is the first part of Proposition 16.107.

Exercise 18.60. Verify that $\overline{\mathcal{Z}}^{\vee}_{\mathbf{Lie}}(\mathrm{G}) = \mathcal{P}\overline{\mathcal{Z}}^{\vee}\overline{\mathcal{U}}(\mathrm{G})$ in rank one. This is part of the claim in Proposition 16.105.

18.7.7. Comparison of series. A series v of a Coxeter $\mathcal{A}$-species p is a triple $(v_O, v_C, v_{\overline{C}})$ such that $v_O \in \mathsf{p}[O]$, $v_C \in \mathsf{p}[C]$, $v_{\overline{C}} \in \mathsf{p}[\overline{C}]$ with $w(v_O) = v_O$, $\beta_{\overline{C},C}(v_C) = v_{\overline{C}}$, $w(v_C) = v_{\overline{C}}$. A series v of the Coxeter $\boldsymbol{\alpha}$-space $\overline{\mathcal{K}}^{\vee}(\mathsf{p})$ is a pair $(v_\emptyset, v_S)$ such that $v_\emptyset \in \mathsf{p}[O]^W$ and $v_S \in \mathsf{p}[C]$ with $\sigma(v_S) = v_S$. Since $\sigma = w\beta_{\overline{C},C}$, it is clear that $(v_O, v_C, v_{\overline{C}}) \mapsto (v_O, v_C)$ specifies an isomorphism $\mathscr{S}(\mathsf{p}) \to \mathscr{S}(\overline{\mathcal{K}}^{\vee}(\mathsf{p}))$. This verifies Lemma 17.1 in rank one.

Exercise 18.61. Verify directly the remaining results in Section 17.1 in rank one by taking forward the above analysis.

Let us now directly verify the first identity in (17.7) in rank one. The calculation below specializes the one given in its proof.

$$\begin{aligned}
(s \circ \mathcal{K}(f))_\emptyset &= s(\emptyset, \emptyset, \mathrm{id}, \emptyset) f_\emptyset + s(\emptyset, S, \mathrm{id}, S)\, \mu f_S \Delta + s(\emptyset, S, \sigma, S)\, \mu f_S \sigma \Delta \\
&= s(O,O,O) f_O + s(O,C,C)\, \mu_O^C f_C (\Delta_O^C + w\Delta_O^{\overline{C}}) + s(O,C,\overline{C})\, \mu_O^C f_C w \beta_{\overline{C},C} (\Delta_O^C + w\Delta_O^{\overline{C}}) \\
&= s(O,O,O) f_O + s(O,C,C)\, [\mu_O^C f_C \Delta_O^C + w\mu_O^{\overline{C}} f_{\overline{C}} \Delta_O^{\overline{C}}] \\
&\qquad + s(O,C,\overline{C})\, [w\mu_O^{\overline{C}} f_{\overline{C}} \beta_{\overline{C},C} \Delta_O^C + \mu_O^C f_C \beta_{C,\overline{C}} \Delta_O^{\overline{C}}].
\end{aligned}$$

This is $\mathcal{K}(s \circ f)_\emptyset$, except for the presence of w. This problem disappears when we replace $\mathcal{K}$ by $\overline{\mathcal{K}}$.

Exercise 18.62. Similarly, verify directly the remaining results in Section 17.2 and also the results in Section 17.3 in rank one.

18.7.8. Transformation between bosonic Fock functor adjoints. We now specialize the discussion in Sections 17.4 and 17.5 to rank one.

For a scalar q, for H a Coxeter $\boldsymbol{\alpha}$-q-bialgebra, consider the morphism $\widehat{\mathrm{n}_{\mathrm{H}}} : \mathcal{T}_q(\mathrm{t_H}) \to \mathcal{T}_q^{\vee}(\mathrm{t_H})$ of Coxeter $\mathcal{A}$-q-bimonoids in (17.32). On the O-component, it is a linear map from $\mathrm{H}[\emptyset] \oplus \mathrm{H}[S] \oplus \mathrm{H}[S]$ to itself which in matrix notation can be written as

$$\begin{pmatrix} 2 & \mu & \mu \\ \Delta & \mathrm{id} & q\,\sigma \\ \Delta & q\,\sigma & \mathrm{id} \end{pmatrix}.$$

Right multiplying by the column vector $(\mu(x), -x, -x)$ yields zero, so $\widehat{\mathrm{n}_{\mathrm{H}}}$ factors through $\overline{\mathcal{Z}}^{\vee}(\mathrm{H})$. Similarly, multiplying by (x, y, z) yields $(2x + \mu(y) + \mu(z), \Delta(x) + y + q\sigma(z), \Delta(x) + q\sigma(y) + z)$ and applying Δ to the first coordinate yields the sum of the remaining two coordinates, so $\widehat{\mathrm{n}_{\mathrm{H}}}$ goes into $\overline{\mathcal{Z}}(\mathrm{H})$. This yields the morphism $\mathrm{z_H} : \overline{\mathcal{Z}}^{\vee}(\mathrm{H}) \to \overline{\mathcal{Z}}(\mathrm{H})$ of Coxeter $\mathcal{A}$-q-bimonoids in (17.35) and verifies Proposition 17.36 in rank one.

Exercise 18.63. Verify directly Theorem 17.37 in rank one. More precisely, the map $\mathrm{z_H}$ above is an isomorphism when $q \neq \pm 1$ and $\mathrm{char}(\Bbbk) \neq 2$.

Exercise 18.64. Theorem 17.41 in rank one reduces to the following statement. Let $q \neq \pm 1$ and $\mathrm{char}(\Bbbk) \neq 2$. For $f : \mathrm{C} \to \mathrm{A}$ a map of $\boldsymbol{\alpha}$-Coxeter spaces, consider the matrices

$$\begin{pmatrix} f_\emptyset & \mu f_S \\ f_S\Delta & (\mathrm{id} + q\,\sigma) f_S \end{pmatrix} \quad \text{and} \quad \begin{pmatrix} 2\,f_\emptyset & \mu f_S & \mu f_S \\ f_S\Delta & f_S & q\,\sigma f_S \\ f_S\Delta & q\,\sigma f_S & f_S \end{pmatrix},$$

the first specifying a linear map from $\mathrm{C}[\emptyset] \oplus \mathrm{C}[S]$ to $\mathrm{A}[\emptyset] \oplus \mathrm{A}[S]$ and the second from $\mathrm{C}[\emptyset] \oplus \mathrm{C}[S] \oplus \mathrm{C}[S]$ to $\mathrm{A}[\emptyset] \oplus \mathrm{A}[S] \oplus \mathrm{A}[S]$. Show: The first matrix is invertible iff the second matrix is invertible.

Exercise 18.65. Check directly Lemma 17.30 in rank one. Explicitly, in the first diagram in (17.25), the bottom-horizontal map, on the $\emptyset$-component, is

$$\mathsf{p}[O] \oplus \mathsf{p}[C] \to \mathsf{p}[O] \oplus \mathsf{p}[C], \qquad (x, y) \mapsto (x, y + q\, w \beta_{\overline{C}, C}(y)).$$

More generally, check directly Lemma 17.31 and Exercise 17.32 in rank one.

18.7.9. Antipodes under Fock functors. Let us look at Proposition 17.44 in rank one. We focus on the first identity. For h a Coxeter $\mathcal{A}$-q-bimonoid,

$$(S_{\mathsf{h}})_O = (-1)^{\dim(O)}(\mathrm{id} - \mu_O^C \Delta_O^C - \mu_O^{\overline{C}} \Delta_O^{\overline{C}}),$$

$$\begin{aligned}(S_{\mathcal{K}(\mathsf{h})})_\emptyset = (-1)^{\dim(O)}(\mathrm{id} - \mu\Delta) &= (-1)^{\dim(O)}(\mathrm{id} - \mu_O^C(\Delta_O^C + w\Delta_O^{\overline{C}})) \\ &= (-1)^{\dim(O)}(\mathrm{id} - \mu_O^C \Delta_O^C - w\mu_O^{\overline{C}} \Delta_O^{\overline{C}}).\end{aligned}$$

If we replace $\mathcal{K}$ by $\overline{\mathcal{K}}$, then we can remove the w in the last expression since we are in the space of W-coinvariants, and we see that $\overline{\mathcal{K}}(S_{\mathsf{h}}) = S_{\overline{\mathcal{K}}(\mathsf{h})}$.

18.8. Notes

In [**7**, Part III], we showed how several well-known Hopf algebras arise as values of Fock functors on Joyal bimonoids. The material in this chapter generalizes this to the setting of Coxeter bialgebras and Coxeter bimonoids. Some helpful pointers are provided below.

18.8.1. Coxeter bialgebra of polynomials. (Section 18.2.) Lemmas 18.3, 18.4, 18.8 generalize [**7**, Examples 15.14 and 16.30]. Lemma 18.6 generalizes [**7**, first paragraph in Example 15.17 and last paragraph in 16.32]. The observations in (18.15) and (18.16) generalize those in [**7**, (18.2) and (18.13)].

18.8.2. Tensor Coxeter bialgebra. (Section 18.3.) Lemmas 18.19, 18.21, 18.24 generalize [**7**, Examples 15.16, 16.31, 19.4]. For (18.25) with $q = 1$, see [**7**, Example 19.31].

18.8.3. Coxeter bialgebra of face-types and its relatives. (Sections 18.4, 18.5, 18.6.) Section 18.4 may be compared with [**7**, Section 17.3], Section 18.5 with [**7**, Section 17.4], Section 18.6 with [**7**, Section 17.2]. Section 18.6.5 is however new.

In general, it is an interesting problem to compute values of Fock functor adjoints on Coxeter bialgebras. For instance, we saw $\overline{\mathscr{Z}}^{\vee}_{ab}(\overline{\Pi}) = \sqcap$; however we did not compute $\overline{\mathscr{Z}}(\overline{\Pi})$. The latter is a Coxeter bimonoid trapped between Σ and $\sqcap$. We plan to address such problems in future work.

18.8.4. Enumeration identities. In this chapter, we used existence of certain Coxeter bialgebras to derive several enumeration identities; see Sections 18.2.3, 18.3.3, 18.4.2. Similarly, antipode considerations allowed us to derive alternating sign identities, see Exercises 18.18, 18.29, 18.42. This fulfils the promise made on [**10**, page 508] where Coxeter bialgebras were explicitly mentioned for the first time.

18.8.5. More examples of Coxeter bialgebras. As a general principle, Coxeter bimonoids are easier to construct than Coxeter bialgebras. We saw many examples of Coxeter bialgebras in Chapter 10, and saw in this chapter how they arise by applying Fock functors to the Coxeter bimonoids in Chapter 4. In a similar way, the Coxeter bimonoids mentioned in Notes 4.10.3 yield many more examples of Coxeter bialgebras. The latter generalize the Hopf algebras of graphs [**320**, Sections 12 and 13], [**7**, Section 17.5.3], and so on.

APPENDIX A

Category theory

A.1. Bimonads and bilax functors

We briefly review monads, comonads, bimonads and their algebras, coalgebras, bialgebras.

A.1.1. Monads, comonads, bimonads. A *monad* on a category C is a functor $\mathcal{V} : \mathsf{C} \to \mathsf{C}$ equipped with natural transformations $\mu : \mathcal{V}\mathcal{V} \to \mathcal{V}$ and $\iota : \mathrm{id} \to \mathcal{V}$ such that the diagrams

$$(\text{A.1}) \qquad \begin{CD} \mathcal{V}\mathcal{V}\mathcal{V} @>{\mathcal{V}\mu}>> \mathcal{V}\mathcal{V} \\ @V{\mu\mathcal{V}}VV @VV{\mu}V \\ \mathcal{V}\mathcal{V} @>>{\mu}> \mathcal{V} \end{CD} \qquad \begin{array}{ccc} & \mathcal{V}\mathcal{V} & \\ {\scriptstyle \iota\mathcal{V}}\nearrow & & \searrow{\scriptstyle \mu} \\ \mathcal{V} & = \!= \!= \!= & \mathcal{V} \end{array} \qquad \begin{array}{ccc} & \mathcal{V}\mathcal{V} & \\ {\scriptstyle \mathcal{V}\iota}\nearrow & & \searrow{\scriptstyle \mu} \\ \mathcal{V} & = \!= \!= \!= & \mathcal{V} \end{array}$$

commute.

Let $\mathcal{V}$ be a monad on C, and $\mathcal{V}'$ a monad on C'. A *lax functor* from $\mathcal{V}$ to $\mathcal{V}'$ is a functor $\mathcal{F} : \mathsf{C} \to \mathsf{C}'$ equipped with a natural transformation $\varphi : \mathcal{V}'\mathcal{F} \to \mathcal{F}\mathcal{V}$ such that the diagrams

$$(\text{A.2}) \qquad \begin{CD} \mathcal{V}'\mathcal{V}'\mathcal{F} @>{\mathcal{V}'\varphi}>> \mathcal{V}'\mathcal{F}\mathcal{V} @>{\varphi\mathcal{V}}>> \mathcal{F}\mathcal{V}\mathcal{V} \\ @V{\mu'\mathcal{F}}VV @. @VV{\mathcal{F}\mu}V \\ \mathcal{V}'\mathcal{F} @>>{\varphi}> @>>> \mathcal{F}\mathcal{V} \end{CD} \qquad \begin{array}{ccc} \mathcal{V}'\mathcal{F} & \xrightarrow{\ \varphi\ } & \mathcal{F}\mathcal{V} \\ {\scriptstyle \iota'\mathcal{F}}\nwarrow & & \nearrow{\scriptstyle \mathcal{F}\iota} \\ & \mathcal{F} & \end{array}$$

commute. We use the notation $(\mathcal{F}, \varphi) : \mathcal{V} \to \mathcal{V}'$.

Let $(\mathcal{F}, \varphi)$ and $(\tilde{\mathcal{F}}, \tilde{\varphi})$ be lax functors from $\mathcal{V}$ to $\mathcal{V}'$. Then a morphism $(\mathcal{F}, \varphi) \to (\tilde{\mathcal{F}}, \tilde{\varphi})$ is a natural transformation $\theta : \mathcal{F} \to \tilde{\mathcal{F}}$ such that the diagram

$$(\text{A.3}) \qquad \begin{CD} \mathcal{V}'\mathcal{F} @>{\varphi}>> \mathcal{F}\mathcal{V} \\ @V{\mathcal{V}'\theta}VV @VV{\theta\mathcal{V}}V \\ \mathcal{V}'\tilde{\mathcal{F}} @>>{\tilde{\varphi}}> \tilde{\mathcal{F}}\mathcal{V} \end{CD}$$

commutes.

This defines the 2-category $m(\mathsf{Cat})$ whose 0-cells are monads, 1-cells are lax functors of monads, 2-cells are morphisms of lax functors.

A *comonad* on a category C is a functor $\mathcal{U} : \mathsf{C} \to \mathsf{C}$ equipped with natural transformations $\Delta : \mathcal{U} \to \mathcal{U}\mathcal{U}$ and $\epsilon : \mathcal{U} \to \mathrm{id}$ such that the diagrams

$$\text{(A.4)}\qquad \begin{array}{ccc} \mathcal{UUU} & \xleftarrow{\mathcal{U}\Delta} & \mathcal{UU} \\ {\scriptstyle \Delta\mathcal{U}}\uparrow & & \uparrow{\scriptstyle \Delta} \\ \mathcal{UU} & \xleftarrow[\Delta]{} & \mathcal{U} \end{array} \qquad \begin{array}{ccc} & \mathcal{UU} & \\ {\scriptstyle \epsilon\mathcal{U}}\swarrow & & \nwarrow{\scriptstyle \Delta} \\ \mathcal{U} & = & \mathcal{U} \end{array} \qquad \begin{array}{ccc} & \mathcal{UU} & \\ {\scriptstyle \mathcal{U}\epsilon}\swarrow & & \nwarrow{\scriptstyle \Delta} \\ \mathcal{U} & = & \mathcal{U} \end{array}$$

commute.

Let $\mathcal{U}$ be a comonad on C, and $\mathcal{U}'$ a comonad on C'. A *colax functor* from $\mathcal{U}$ to $\mathcal{U}'$ is a functor $\mathcal{F} : \mathsf{C} \to \mathsf{C}'$ equipped with a natural transformation $\psi : \mathcal{F}\mathcal{U} \to \mathcal{U}'\mathcal{F}$ such that the diagrams

$$\text{(A.5)}\qquad \begin{array}{ccccc} \mathcal{FUU} & \xrightarrow{\psi\mathcal{U}} & \mathcal{U}'\mathcal{FU} & \xrightarrow{\mathcal{U}'\psi} & \mathcal{U}'\mathcal{U}'\mathcal{F} \\ {\scriptstyle \mathcal{F}\Delta}\uparrow & & & & \uparrow{\scriptstyle \Delta'\mathcal{F}} \\ \mathcal{FU} & & \xrightarrow[\psi]{} & & \mathcal{U}'\mathcal{F} \end{array} \qquad \begin{array}{ccc} \mathcal{FU} & \xrightarrow{\psi} & \mathcal{U}'\mathcal{F} \\ {\scriptstyle \mathcal{F}\epsilon}\searrow & & \swarrow{\scriptstyle \epsilon'\mathcal{F}} \\ & \mathcal{F} & \end{array}$$

commute. We use the notation $(\mathcal{F}, \psi) : \mathcal{U} \to \mathcal{U}'$.

Let $(\mathcal{F}, \varphi)$ and $(\tilde{\mathcal{F}}, \tilde{\varphi})$ be colax functors from $\mathcal{U}$ to $\mathcal{U}'$. Then a morphism $(\mathcal{F}, \varphi) \to (\tilde{\mathcal{F}}, \tilde{\varphi})$ is a natural transformation $\theta : \mathcal{F} \to \tilde{\mathcal{F}}$ such that the diagram

$$\text{(A.6)}\qquad \begin{array}{ccc} \mathcal{FU} & \xrightarrow{\psi} & \mathcal{U}'\mathcal{F} \\ {\scriptstyle \theta\mathcal{U}}\downarrow & & \downarrow{\scriptstyle \mathcal{U}'\theta} \\ \tilde{\mathcal{F}}\mathcal{U} & \xrightarrow[\tilde{\psi}]{} & \mathcal{U}'\tilde{\mathcal{F}} \end{array}$$

commutes.

This defines the 2-category $c(\mathsf{Cat})$ whose 0-cells are comonads, 1-cells are colax functors of comonads, 2-cells are morphisms of colax functors.

A *bimonad* on a category C is a triple $(\mathcal{V}, \mathcal{U}, \lambda)$, where $\mathcal{V}$ is a monad on C, $\mathcal{U}$ is a comonad on C, and $\lambda : \mathcal{VU} \to \mathcal{UV}$ is a natural transformation such that the diagrams

$$\text{(A.7a)}\qquad \begin{array}{ccccc} \mathcal{VUU} & \xrightarrow{\lambda\mathcal{U}} & \mathcal{UVU} & \xrightarrow{\mathcal{U}\lambda} & \mathcal{UUV} \\ {\scriptstyle \mathcal{V}\Delta}\uparrow & & & & \uparrow{\scriptstyle \Delta\mathcal{V}} \\ \mathcal{VU} & & \xrightarrow[\lambda]{} & & \mathcal{UV} \end{array} \qquad \begin{array}{ccccc} \mathcal{VVU} & \xrightarrow{\mathcal{V}\lambda} & \mathcal{VUV} & \xrightarrow{\lambda\mathcal{V}} & \mathcal{UVV} \\ {\scriptstyle \mu\mathcal{U}}\downarrow & & & & \downarrow{\scriptstyle \mathcal{U}\mu} \\ \mathcal{VU} & & \xrightarrow[\lambda]{} & & \mathcal{UV} \end{array}$$

$$\text{(A.7b)}\qquad \begin{array}{ccc} & \mathcal{V} & \\ {\scriptstyle \mathcal{V}\epsilon}\nearrow & & \nwarrow{\scriptstyle \epsilon\mathcal{V}} \\ \mathcal{VU} & \xrightarrow[\lambda]{} & \mathcal{UV} \end{array} \qquad \begin{array}{ccc} & \mathcal{U} & \\ {\scriptstyle \iota\mathcal{U}}\swarrow & & \searrow{\scriptstyle \mathcal{U}\iota} \\ \mathcal{VU} & \xrightarrow[\lambda]{} & \mathcal{UV} \end{array}$$

commute. We refer to λ as a *mixed distributive law* linking $\mathcal{V}$ and $\mathcal{U}$.

Let $(\mathcal{V}, \mathcal{U}, \lambda)$ be a bimonad on C, and $(\mathcal{V}', \mathcal{U}', \lambda')$ a bimonad on C'. A *bilax functor* $(\mathcal{V}, \mathcal{U}, \lambda) \to (\mathcal{V}', \mathcal{U}', \lambda')$ is a triple $(\mathcal{F}, \varphi, \psi)$ such that $\mathcal{F} : \mathsf{C} \to \mathsf{C}'$ is a functor, $(\mathcal{F}, \varphi) : \mathcal{V} \to \mathcal{V}'$ is a lax functor of monads, $(\mathcal{F}, \psi) : \mathcal{U} \to \mathcal{U}'$ is a colax functor of comonads, and the diagram

(A.8)
$$\begin{array}{ccccc}
\mathcal{V}'\mathcal{F}\mathcal{U} & \xrightarrow{\varphi\mathcal{U}} & \mathcal{F}\mathcal{V}\mathcal{U} & \xrightarrow{\mathcal{F}\lambda} & \mathcal{F}\mathcal{U}\mathcal{V} \\
{\scriptstyle \mathcal{V}'\psi}\downarrow & & & & \downarrow{\scriptstyle \psi\mathcal{V}} \\
\mathcal{V}'\mathcal{U}'\mathcal{F} & \xrightarrow[\lambda'\mathcal{F}]{} & \mathcal{U}'\mathcal{V}'\mathcal{F} & \xrightarrow[\mathcal{U}'\varphi]{} & \mathcal{U}'\mathcal{F}\mathcal{V}
\end{array}$$

commutes.

Let $(\mathcal{F}, \varphi, \psi)$ and $(\tilde{\mathcal{F}}, \tilde{\varphi}, \tilde{\psi})$ be bilax functors from $(\mathcal{V}, \mathcal{U}, \lambda)$ to $(\mathcal{V}', \mathcal{U}', \lambda')$. A morphism $(\mathcal{F}, \varphi, \psi) \to (\tilde{\mathcal{F}}, \tilde{\varphi}, \tilde{\psi})$ is a natural transformation $\theta : \mathcal{F} \to \tilde{\mathcal{F}}$ such that diagrams (A.3) and (A.6) commute.

This defines the 2-category $b(\mathsf{Cat})$ whose 0-cells are bimonads, 1-cells are bilax functors, 2-cells are morphisms of bilax functors.

By reversing appropriate arrows, one defines a colax functor of monads, and a lax functor of comonads. Explicitly: A *colax functor* of monads $\mathcal{V} \to \mathcal{V}'$ is a functor $\mathcal{F} : \mathsf{C} \to \mathsf{C}'$ equipped with $\psi : \mathcal{F}\mathcal{V} \to \mathcal{V}'\mathcal{F}$ such that the diagrams

(A.9)
$$\begin{array}{ccccc}
\mathcal{V}'\mathcal{V}'\mathcal{F} & \xleftarrow{\mathcal{V}'\psi} & \mathcal{V}'\mathcal{F}\mathcal{V} & \xleftarrow{\psi\mathcal{V}} & \mathcal{F}\mathcal{V}\mathcal{V} \\
{\scriptstyle \mu'\mathcal{F}}\downarrow & & & & \downarrow{\scriptstyle \mathcal{F}\mu} \\
\mathcal{V}'\mathcal{F} & & \xleftarrow[\psi]{} & & \mathcal{F}\mathcal{V}
\end{array}
\qquad
\begin{array}{ccc}
\mathcal{V}'\mathcal{F} & \xleftarrow{\psi} & \mathcal{F}\mathcal{V} \\
{\scriptstyle \iota'\mathcal{F}}\nwarrow & & \nearrow{\scriptstyle \mathcal{F}\iota} \\
& \mathcal{F} &
\end{array}$$

commute. A *lax functor* from $\mathcal{U}$ to $\mathcal{U}'$ is a functor $\mathcal{F} : \mathsf{C} \to \mathsf{C}'$ equipped with $\varphi : \mathcal{U}'\mathcal{F} \to \mathcal{F}\mathcal{U}$ such that the diagrams

(A.10)
$$\begin{array}{ccccc}
\mathcal{F}\mathcal{U}\mathcal{U} & \xleftarrow{\varphi\mathcal{U}} & \mathcal{U}'\mathcal{F}\mathcal{U} & \xleftarrow{\mathcal{U}'\varphi} & \mathcal{U}'\mathcal{U}'\mathcal{F} \\
{\scriptstyle \mathcal{F}\Delta}\uparrow & & & & \uparrow{\scriptstyle \Delta'\mathcal{F}} \\
\mathcal{F}\mathcal{U} & & \xleftarrow[\varphi]{} & & \mathcal{U}'\mathcal{F}
\end{array}
\qquad
\begin{array}{ccc}
\mathcal{F}\mathcal{U} & \xleftarrow{\varphi} & \mathcal{U}'\mathcal{F} \\
{\scriptstyle \mathcal{F}\epsilon}\searrow & & \swarrow{\scriptstyle \epsilon'\mathcal{F}} \\
& \mathcal{F} &
\end{array}$$

commute.

A.1.2. Monad algebras, comonad coalgebras, bimonad bialgebras. Let $\mathcal{V}$ be a monad. A $\mathcal{V}$*-algebra* is an object A equipped with a morphism $\alpha : \mathcal{V}A \to A$ such that the diagrams

(A.11)
$$\begin{array}{ccc}
\mathcal{V}\mathcal{V}A & \xrightarrow{\mu A} & \mathcal{V}A \\
{\scriptstyle \mathcal{V}\alpha}\downarrow & & \downarrow{\scriptstyle \alpha} \\
\mathcal{V}A & \xrightarrow[\alpha]{} & A
\end{array}
\qquad
\begin{array}{ccc}
& \mathcal{V}A & \\
{\scriptstyle \iota A}\nearrow & & \searrow{\scriptstyle \alpha} \\
A & = & A
\end{array}$$

commute. A morphism $f : A \to B$ of $\mathcal{V}$-algebras is a map $A \to B$ such that the diagram

(A.12)
$$\begin{array}{ccc}
\mathcal{V}A & \xrightarrow{\mathcal{V}f} & \mathcal{V}B \\
\downarrow & & \downarrow \\
A & \xrightarrow[f]{} & B
\end{array}$$

commutes. Let $\mathcal{V}$-Alg denote the category of $\mathcal{V}$-algebras. A lax functor of monads $(\mathcal{F}, \varphi) : \mathcal{V} \to \mathcal{V}'$ induces a functor $\mathcal{V}$-$\mathsf{Alg} \to \mathcal{V}'$-$\mathsf{Alg}$, and a morphism of lax functors induces a natural transformation.

Let $\mathcal{U}$ be a comonad. A *$\mathcal{U}$-coalgebra* is an object C equipped with a morphism $C \to \mathcal{U}C$ such that the diagrams

$$\text{(A.13)} \qquad \begin{array}{ccc} \mathcal{U}\mathcal{U}C & \longleftarrow & \mathcal{U}C \\ \uparrow & & \uparrow \\ \mathcal{U}C & \longleftarrow & C \end{array} \qquad\qquad \begin{array}{ccc} & \mathcal{U}C & \\ \swarrow & & \nwarrow \\ C & \Longleftarrow\!\!\Longrightarrow & C \end{array}$$

commute. A morphism of $\mathcal{U}$-coalgebras is defined similarly. Let $\mathcal{U}$-Coalg denote the category of $\mathcal{U}$-coalgebras. A colax functor of comonads $(\mathcal{F}, \psi) : \mathcal{U} \to \mathcal{U}'$ induces a functor $\mathcal{U}$-$\mathsf{Coalg} \to \mathcal{U}'$-$\mathsf{Coalg}$, and a morphism of colax functors induces a natural transformation.

A *bialgebra over a bimonad* $(\mathcal{V}, \mathcal{U}, \lambda)$ is a triple (H, μ, Δ) such that (H, μ) is an algebra over $\mathcal{V}$, (H, Δ) is a coalgebra over $\mathcal{U}$, and the diagram

$$\text{(A.14)} \qquad \begin{array}{ccccc} \mathcal{V}H & \longrightarrow & H & \longrightarrow & \mathcal{U}H \\ \downarrow & & & & \uparrow \\ \mathcal{V}\mathcal{U}H & & \longrightarrow & & \mathcal{U}\mathcal{V}H \end{array}$$

commutes. A morphism of bialgebras is a map $H \to H'$ which is a morphism of the underlying algebras and coalgebras. Let $(\mathcal{V}, \mathcal{U}, \lambda)$-$\mathsf{Bialg}$ denote the category of $(\mathcal{V}, \mathcal{U}, \lambda)$-bialgebras. A bilax functor of bimonads $(\mathcal{F}, \varphi, \psi) : (\mathcal{V}, \mathcal{U}, \lambda) \to (\mathcal{V}', \mathcal{U}', \lambda')$ induces a functor $(\mathcal{V}, \mathcal{U}, \lambda)$-$\mathsf{Bialg} \to (\mathcal{V}', \mathcal{U}', \lambda')$-$\mathsf{Bialg}$, and a morphism of bilax functors induces a natural transformation.

Remark A.1. A bialgebra over a bimonad can be viewed as a coalgebra in a category of algebras, and vice versa. In more detail: Let $(\mathcal{V}, \mathcal{U}, \lambda)$ be a bimonad. Then $\mathcal{U}$ induces a comonad on $\mathcal{V}$-Alg (using the mixed distributive law λ), and the resulting $\mathcal{U}$-coalgebras are the same as $(\mathcal{V}, \mathcal{U}, \lambda)$-bialgebras. Dually, $\mathcal{V}$ induces a monad on $\mathcal{U}$-Coalg, and the resulting $\mathcal{V}$-algebras are the same as $(\mathcal{V}, \mathcal{U}, \lambda)$-bialgebras.

A.1.3. Lifting adjunctions. Suppose $(\mathcal{F}, \mathcal{G})$ is an adjunction between categories C and C'. Let $\mathcal{V}$ and $\mathcal{V}'$ be monads on C and C' and $\mathcal{F}$ and $\mathcal{G}$ be lax wrt them. The adjunction $(\mathcal{F}, \mathcal{G})$ is *lax-lax* if the diagrams

$$\text{(A.15)} \qquad \begin{array}{ccc} \mathcal{V} & \longrightarrow & \mathcal{G}\mathcal{F}\mathcal{V} \\ \downarrow & & \uparrow \\ \mathcal{V}\mathcal{G}\mathcal{F} & \longrightarrow & \mathcal{G}\mathcal{V}'\mathcal{F} \end{array} \qquad\qquad \begin{array}{ccc} \mathcal{F}\mathcal{V}\mathcal{G} & \longleftarrow & \mathcal{V}'\mathcal{F}\mathcal{G} \\ \downarrow & & \downarrow \\ \mathcal{F}\mathcal{G}\mathcal{V}' & \longrightarrow & \mathcal{V}' \end{array}$$

commute. Dually, one defines colax-colax adjunctions wrt comonads $\mathcal{U}$ and $\mathcal{U}'$ using diagrams

$$\text{(A.16)} \qquad \begin{array}{ccc} \mathcal{U} & \longrightarrow & \mathcal{G}\mathcal{F}\mathcal{U} \\ \downarrow & & \downarrow \\ \mathcal{U}\mathcal{G}\mathcal{F} & \longleftarrow & \mathcal{G}\mathcal{U}'\mathcal{F} \end{array} \qquad\qquad \begin{array}{ccc} \mathcal{F}\mathcal{U}\mathcal{G} & \longrightarrow & \mathcal{U}'\mathcal{F}\mathcal{G} \\ \uparrow & & \downarrow \\ \mathcal{F}\mathcal{G}\mathcal{U}' & \longrightarrow & \mathcal{U}'. \end{array}$$

Lemma A.2. *A lax-lax adjunction between* C *and* C' *wrt monads* $\mathcal{V}$ *and* $\mathcal{V}'$ *lifts to an adjunction between* $\mathcal{V}$-Alg *and* $\mathcal{V}'$-Alg. *Dually, a colax-colax adjunction between* C *and* C' *wrt comonads* $\mathcal{U}$ *and* $\mathcal{U}'$ *lifts to an adjunction between* $\mathcal{U}$-Coalg *and* $\mathcal{U}'$-Coalg.

Let C (resp. C') be a category equipped with a monad $\mathcal{V}$ (resp. $\mathcal{V}'$). Then colax structures on $\mathcal{F}$ correspond to lax structures on $\mathcal{G}$. Explicitly: For a colax structure $\psi : \mathcal{F}\mathcal{V} \to \mathcal{V}'\mathcal{F}$, one defines the lax structure

$$\varphi : \mathcal{V}\mathcal{G} \to \mathcal{G}\mathcal{F}\mathcal{V}\mathcal{G} \to \mathcal{G}\mathcal{V}'\mathcal{F}\mathcal{G} \to \mathcal{G}\mathcal{V}'. \tag{A.17}$$

Conversely, for a lax structure $\varphi : \mathcal{V}\mathcal{G} \to \mathcal{G}\mathcal{V}'$, one defines the colax structure

$$\psi : \mathcal{F}\mathcal{V} \to \mathcal{F}\mathcal{V}\mathcal{G}\mathcal{F} \to \mathcal{F}\mathcal{G}\mathcal{V}'\mathcal{F} \to \mathcal{V}'\mathcal{F}. \tag{A.18}$$

The structures φ and ψ are called *mates* of each other. In this situation, we say $(\mathcal{F}, \psi)$ and $(\mathcal{G}, \varphi)$ is a *colax-lax adjunction*. Equivalently, an adjunction is colax-lax iff either of the two diagrams

$$\begin{array}{ccc} \mathcal{V} & \longrightarrow & \mathcal{G}\mathcal{F}\mathcal{V} \\ \downarrow & & \downarrow \\ \mathcal{V}\mathcal{G}\mathcal{F} & \longrightarrow & \mathcal{G}\mathcal{V}'\mathcal{F} \end{array} \qquad\qquad \begin{array}{ccc} \mathcal{F}\mathcal{V}\mathcal{G} & \longrightarrow & \mathcal{V}'\mathcal{F}\mathcal{G} \\ \downarrow & & \downarrow \\ \mathcal{F}\mathcal{G}\mathcal{V}' & \longrightarrow & \mathcal{V}' \end{array} \tag{A.19}$$

commute.

Suppose $(\mathcal{F}, \psi)$ is a colax functor from $\mathcal{V}$ to $\mathcal{V}'$. Given a $\mathcal{V}$-algebra A with structure map $\alpha : \mathcal{V}A \to A$, consider the maps

$$\begin{gathered} \mathcal{V}'\mathcal{F}\mathcal{V}(A) \xrightarrow{\mathcal{V}'\psi} \mathcal{V}'\mathcal{V}'\mathcal{F}(A) \xrightarrow{\mu'\mathcal{F}} \mathcal{V}'\mathcal{F}(A), \\ \mathcal{V}'\mathcal{F}\mathcal{V}(A) \xrightarrow{\mathcal{V}'\mathcal{F}(\alpha)} \mathcal{V}'\mathcal{F}(A). \end{gathered} \tag{A.20}$$

These maps form a reflexive pair in the category $\mathcal{V}'$-Alg. The joint splitting is $\mathcal{V}'\mathcal{F}(\iota)$. Assume that the coequalizer of such a pair exists, and call it $\mathcal{F}'(A)$. This then defines a functor

$$\mathcal{F}' : \mathcal{V}\text{-}\mathsf{Alg} \to \mathcal{V}'\text{-}\mathsf{Alg}.$$

Moreover, functors $\mathcal{F}$ and $\mathcal{F}'$ intertwine the free algebra functors, that is, $\mathcal{F}'(\mathcal{V}(X)) = \mathcal{V}'(\mathcal{F}(X))$ for any object X of C. Explicitly, the coequalizer arrow to $\mathcal{F}'(\mathcal{V}(X))$ is given by the composite

$$\mathcal{V}'\mathcal{F}(\mathcal{V}(X)) \xrightarrow{\mathcal{V}'\psi} \mathcal{V}'\mathcal{V}'\mathcal{F}(X) \xrightarrow{\mu'\mathcal{F}} \mathcal{V}'(\mathcal{F}(X)).$$

It is split by the morphism of $\mathcal{V}'$-algebras $\mathcal{V}'\mathcal{F}(\iota) : \mathcal{V}'\mathcal{F}(X) \to \mathcal{V}'\mathcal{F}(\mathcal{V}(X))$.

Theorem A.3. *Let* $(\mathcal{F}, \psi)$ *and* $(\mathcal{G}, \varphi)$ *be a colax-lax adjunction between* C *and* C' *wrt monads* $\mathcal{V}$ *and* $\mathcal{V}'$. *Suppose the coequalizer of* (A.20) *exists yielding the functor* $\mathcal{F}'$ *as above. Then* $(\mathcal{F}, \mathcal{G})$ *lifts to an adjunction* $(\mathcal{F}', \mathcal{G})$ *between* $\mathcal{V}$-Alg *and* $\mathcal{V}'$-Alg.

This is the *adjoint lifting theorem* (monad version). We mention that colax-lax adjunctions and the adjoint lifting theorem can also be formulated for comonads instead of monads.

Special case. In the above setup, suppose in addition that the colax structure ψ of $\mathcal{F}$ is an isomorphism. Then in fact the coequalizer of (A.20) exists, and equals $\mathcal{F}(A)$. In other words, $\mathcal{F}' = \mathcal{F}$. The coequalizer arrow to $\mathcal{F}(A)$ is

$$\mathcal{V}'\mathcal{F}(A) \xrightarrow{\psi^{-1}} \mathcal{F}\mathcal{V}(A) \xrightarrow{\mathcal{F}(\alpha)} \mathcal{F}(A).$$

The intertwining of $\mathcal{F}$ with the free algebra functors is $\psi : \mathcal{F}(\mathcal{V}(X)) \xrightarrow{\cong} \mathcal{V}'(\mathcal{F}(X))$. By Theorem A.3, $(\mathcal{F}, \mathcal{G})$ lifts to an adjunction $(\mathcal{F}, \mathcal{G})$ between $\mathcal{V}$-Alg and $\mathcal{V}'$-Alg. In this situation, $(\mathcal{F}, \mathcal{G})$ is also a lax-lax adjunction, with ψ^{-1} as the lax structure on $\mathcal{F}$; so we can also obtain the adjunction-lift by an application of Lemma A.2.

A.1.4. Examples. The main examples of bimonads in this book are on the categories of Coxeter species and Coxeter spaces, see Sections 2.10 and 8.12. Bialgebras over them yield Coxeter bimonoids and Coxeter bialgebras. The main examples of bilax functors are summarized in Table A.1.

TABLE A.1. Examples of bilax functors.

Hadamard functor	Theorem 3.25
u-signature and o-signature functors	Theorem 4.10
decoration functor	Section 4.9.7
q-full Fock functors	Theorems 14.2, 14.4, 14.7, 14.8
bosonic/fermionic Fock functors	Theorems 14.14, 14.15, 14.16, 14.30
anyonic Fock functor	Proposition 14.49
decorated Fock functors	Section 14.10

Examples of adjunction-lifting in this book occur in Chapter 16. Lax-lax, colax-colax, colax-lax adjunctions appear for instance in Propositions 16.21, 16.30, 16.36, 16.43, 16.49. The special case mentioned above of the adjoint lifting theorem arises when $\mathcal{F}$ is a bosonic Fock functor (whose (co)lax structures are isomorphisms), see Remark 16.32 for instance.

A.2. Semidirect products

We discuss the semidirect product of a monoid by a category on which it acts. We present it as a special case of the Grothendieck construction for a category equipped with a functor to Cat.

A.2.1. Grothendieck construction. Let C be a category and $\mathcal{F} : \mathsf{C} \to \mathsf{Cat}$ a functor. The *Grothendieck construction* on $\mathcal{F}$ is the category $\mathcal{F} \rtimes \mathsf{C}$ defined as follows. An object is a pair (x, a) with a an object of C and x an object of $\mathcal{F}(a)$. A morphism $(x, a) \to (y, b)$ is a pair (γ, g) with $g : a \to b$ a morphism in C and $\gamma : \mathcal{F}(g)(x) \to y$ a morphism in the category $\mathcal{F}(b)$.

C	Cat	$\mathcal{F}(b)$	$\mathcal{F} \rtimes \mathsf{C}$
$a \xrightarrow{g} b$	$\mathcal{F}(a) \xrightarrow{\mathcal{F}(g)} \mathcal{F}(b)$	$\mathcal{F}(g)(x) \xrightarrow{\gamma} y$	$(x, a) \xrightarrow{(\gamma, g)} (y, b)$

Composition and identities are as indicated below.

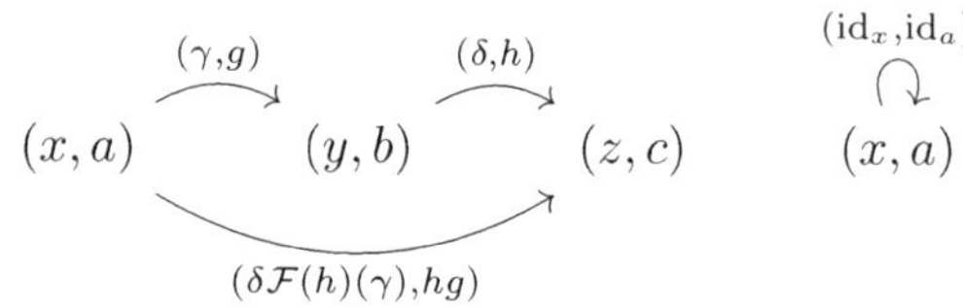

Note: In addition to (γ, g) as above, we are given

$$\mathcal{F}(b) \xrightarrow{\mathcal{F}(h)} \mathcal{F}(c) \quad \text{and} \quad \mathcal{F}(h)(y) \xrightarrow{\delta} z$$

(a functor and a morphism in $\mathcal{F}(c)$, respectively). Therefore the composite

$$\mathcal{F}(h)\mathcal{F}(g)(x) \xrightarrow{\mathcal{F}(h)(\gamma)} \mathcal{F}(h)(y) \xrightarrow{\delta} z$$

is a morphism in $\mathcal{F}(c)$ from $\mathcal{F}(hg)(x)$ to z. Associativity and unitality in $\mathcal{F} \rtimes \mathsf{C}$ follow from functoriality of $\mathcal{F}$.

There is a functor from $\mathcal{F} \rtimes \mathsf{C}$ to C that sends (x, a) to a and (γ, g) to g.

Special case. If C is a groupoid and $\mathcal{F}$ factors through the category of groupoids, then $\mathcal{F} \rtimes \mathsf{C}$ is a groupoid. The inverse of (γ, g) is $(\mathcal{F}(g^{-1})(\gamma^{-1}), g^{-1})$.

Special case (Category of elements). Suppose $\mathcal{F}$ factors through the category of sets:

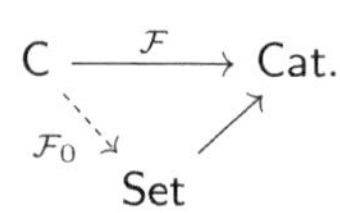

Here $\mathsf{Set} \to \mathsf{Cat}$ is the functor that sends a set X to the discrete category with object set X. In this case, a morphism $(x, a) \to (y, b)$ in $\mathcal{F} \rtimes \mathsf{C}$ is a morphism $g : a \to b$ in C such that $\mathcal{F}_0(g)(x) = y$. The category $\mathcal{F} \rtimes \mathsf{C}$ is known as the *category of elements* of $\mathcal{F}_0$. We denote it $\mathsf{el}_{\mathsf{C}}(\mathcal{F}_0)$; another commonly used notation is $\int_{\mathsf{C}} \mathcal{F}_0$. If in addition C is a groupoid, then $\mathcal{F} \rtimes \mathsf{C}$ is a groupoid (by the previous special case); this is the *groupoid of elements* of $\mathcal{F}_0$.

Special case. Suppose $\mathcal{F}$ factors through the category of monoids:

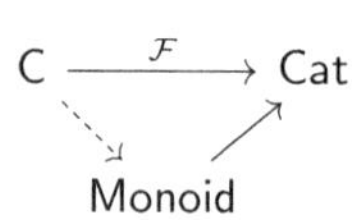

Here $\mathsf{Monoid} \to \mathsf{Cat}$ is the functor that sends a monoid M to the category with one object $*$ and endomorphism monoid M. In this case, $\mathcal{F} \rtimes \mathsf{C}$ has the same objects as C, and a morphism $a \to b$ in $\mathcal{F} \rtimes \mathsf{C}$ is a pair (γ, g) with $g : a \to b$ a morphism in C and γ an element in the monoid $\mathcal{F}(b)$. There is a faithful functor $\mathsf{C} \to \mathcal{F} \rtimes \mathsf{C}$ that sends a to $(*, a)$ and $g : a \to b$ to $(1_b, g)$, where 1_b is the unit element in $\mathcal{F}(b)$. If in addition C is a groupoid and $\mathcal{F}$ factors through the category of groups, then $\mathcal{F} \rtimes \mathsf{C}$ is a groupoid. An example of such a groupoid appears in this book in (11.1).

A.2.2. Generators and relations. Let us return to the general situation. We are given a functor $\mathcal{F} : \mathsf{C} \to \mathsf{Cat}$. Let $\mathcal{F}^0 : \mathsf{C} \to \mathsf{Set}$ denote the composite

$$\begin{array}{ccc} \mathsf{C} & \xrightarrow{\mathcal{F}} & \mathsf{Cat} \\ & {\scriptstyle \mathcal{F}^0} \searrow \quad \swarrow & \\ & \mathsf{Set}, & \end{array}$$

where $\mathsf{Cat} \to \mathsf{Set}$ is the functor that sends a category to its object set. Consider the category of elements $\mathsf{el}_{\mathsf{C}}(\mathcal{F}^0)$. It has the same objects as $\mathcal{F} \rtimes \mathsf{C}$. There is a faithful functor $\mathsf{el}_{\mathsf{C}}(\mathcal{F}^0) \to \mathcal{F} \rtimes \mathsf{C}$ that is identity on objects and sends a morphism $g : (x, a) \to (y, b)$ in $\mathsf{el}_{\mathsf{C}}(\mathcal{F}^0)$ to the morphism $(\mathrm{id}_y, g) : (x, a) \to (y, b)$ in $\mathcal{F} \rtimes \mathsf{C}$.

Fix an object a of C. There is a faithful functor $\mathcal{F}(a) \to \mathcal{F} \rtimes \mathsf{C}$ that sends an object x of $\mathcal{F}(a)$ to the object (x, a) of $\mathcal{F} \rtimes \mathsf{C}$, and a morphism $\gamma : x \to y$ in $\mathcal{F}(a)$ to the morphism $(\gamma, \mathrm{id}_a) : (x, a) \to (y, a)$ in $\mathcal{F} \rtimes \mathsf{C}$.

In this manner, we may view each category $\mathcal{F}(a)$, as a runs over objects of C, as well as the category $\mathsf{el}_{\mathsf{C}}(\mathcal{F}^0)$, as subcategories of $\mathcal{F} \rtimes \mathsf{C}$. Taken all together, they generate the category $\mathcal{F} \rtimes \mathsf{C}$, and the relation

$$\text{(A.21)} \qquad \begin{array}{ccc} (x, a) & \xrightarrow{(\gamma, \mathrm{id}_a)} & (y, a) \\ {\scriptstyle (\mathrm{id}_{x'}, g)} \downarrow & & \downarrow {\scriptstyle (\mathrm{id}_{y'}, g)} \\ (x', b) & \xrightarrow[(\mathcal{F}(g)(\gamma), \mathrm{id}_b)]{} & (y', b) \end{array}$$

holds, where $\gamma : x \to y$ is a morphism in $\mathcal{F}(a)$, $x' = \mathcal{F}(g)(x)$, and $y' = \mathcal{F}(g)(y)$. Indeed, both composites in $\mathcal{F} \rtimes \mathsf{C}$ equal $(\mathcal{F}(g)(\gamma), g)$.

A.2.3. Semidirect product of a monoid by a category. Given a monoid G, let C denote the category with one object $*$ and endomorphism monoid $\mathsf{C}(*, *) = G$. A functor $\mathcal{F} : \mathsf{C} \to \mathsf{Cat}$ consists of a category $\mathsf{F} := \mathcal{F}(*)$ together with a monoid morphism from G to the set of endofunctors of F (a monoid under composition). In this case, we say G acts on the category F. Explicitly, such an action consists of a functor $\mathcal{F}(g) : \mathsf{F} \to \mathsf{F}$ for each $g \in G$ such that

$$\mathcal{F}(1) = \mathrm{id}_{\mathsf{F}} \quad \text{and} \quad \mathcal{F}(g)\mathcal{F}(h) = \mathcal{F}(gh)$$

for all $g, h \in G$. Let X be the object set of F. It follows that G acts on X by

$$g \cdot x := \mathcal{F}(g)(x).$$

Given $\gamma : x \to y$ in F, let us write

$$g \cdot \gamma := \mathcal{F}(g)(\gamma).$$

This is a morphism $g \cdot x \to g \cdot y$ in F.

The semidirect product of G by F, denoted $\mathsf{F} \rtimes G$, is the Grothendieck construction on the functor $\mathcal{F}$. We call it a *semidirect product category*. Explicitly, the object set is X and the set of morphisms from x to y is

$$(\mathsf{F} \rtimes G)(x, y) := \{(\gamma, g) \mid g \in G, \ \gamma \in \mathsf{F}(g \cdot x, y)\}.$$

Composition and identities are as indicated below.

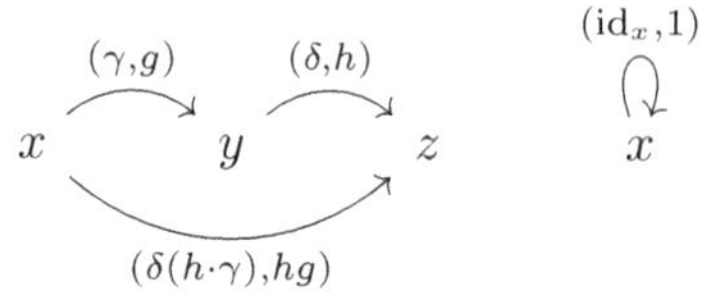

The discussion in Appendix A.2.2 specializes as follows. Consider the category of elements $\mathsf{el}_G(X)$. (As mentioned below, this itself is a semidirect product category.) There is a faithful functor $\mathsf{el}_G(X) \to \mathsf{F} \rtimes G$ that is the identity on objects and sends a morphism $(x, g) : x \to y$ in $\mathsf{el}_G(X)$ to the morphism $(\mathrm{id}_y, g) : x \to y$ in $\mathsf{F} \rtimes G$. In addition, there is a faithful functor $\mathsf{F} \to \mathsf{F} \rtimes G$ that is the identity on objects. It sends a morphism $\gamma : x \to y$ in F to the morphism $(\gamma, 1) : x \to y$ in $\mathsf{F} \rtimes G$. In this manner, we may view both F and $\mathsf{el}_G(X)$ as subcategories of $\mathsf{F} \rtimes G$. Together, they generate the category $\mathsf{F} \rtimes G$, and as a special case of (A.21), the relation

$$\begin{array}{ccc} x & \xrightarrow{(\gamma,1)} & y \\ {\scriptstyle (\mathrm{id}_{x'},g)}\big\downarrow & & \big\downarrow{\scriptstyle (\mathrm{id}_{y'},g)} \\ x' & \xrightarrow[(g\cdot\gamma,1)]{} & y' \end{array} \tag{A.22}$$

holds, where $\gamma \in \mathsf{F}(x, y)$, $x' = g \cdot x$, and $y' = g \cdot y$. Indeed, both composites in $\mathsf{F} \rtimes G$ equal $(g \cdot \gamma, g)$.

Special case (Semidirect product groupoid). If G is a group and F is a groupoid, then $\mathsf{F} \rtimes G$ is a groupoid. We call it a *semidirect product groupoid*. The inverse of (γ, g) is $(g^{-1} \cdot \gamma^{-1}, g^{-1})$. Examples of such groupoids in this book appear in Lemma 2.70. In this result, G is the Coxeter group, and F is constructed out of faces and flats of the corresponding reflection arrangement.

Special case (Category of elements for a monoid action on a set). Let G be a monoid. If F is the discrete category on the set X, an action of G on F boils down to an action of G on the set X. A morphism from x to y in $\mathsf{F} \rtimes G$ is a pair (x, g) such that $g \cdot x = y$ and $\mathsf{F} \rtimes G = \mathsf{el}_G(X)$. When G is a group, $\mathsf{el}_G(X)$ is a groupoid (by the previous special case); this is the groupoid of elements of the G-action. Many examples of such groupoids appear in this book with G as the Coxeter group; see for instance (2.5), (8.5).

Special case (Semidirect product monoid). Let G and H be monoids. If F is the one-object category with endomorphism monoid H, an action of G on F boils down to an action by endomorphisms of the monoid G on the monoid H. The category $\mathsf{F} \rtimes G$ has one object and a morphism therein is a pair (h, g) in $H \times G$. It is the one-object category associated to the semidirect product monoid $H \rtimes G$. Examples of semidirect product monoids in this book appear in Section 1.14, with G as the Coxeter group and H as the Birkhoff, Tits, Janus monoids. An example of a semidirect product group where both G and H are groups is (1.9).

A.3. Notes

A.3.1. Bimonads and bilax functors. (Appendix A.1.) This is a brief summary of [**11**, Appendix C] on higher monads, where more details can be found. Historical information related to monads, distributive laws and the adjoint lifting theorem is given in [**11**, Notes to Appendix C].

There is a parallel between higher monads and higher monoidal categories, and in particular, between bimonads and braided monoidal categories [**11**, page 762]. The Hopf theory in [**7**] was formulated in terms of braided monoidal categories, while the Hopf theory in [**11**] and the present book is formulated in terms of bimonads.

A.3.2. Semidirect products. (Appendix A.2.) The Grothendieck construction goes back to Grothendieck [**160**, Section 9]. It is discussed for instance by Barr and Wells [**29**, Section 12.2]. In fact, Grothendieck formulated his construction more generally for a category equipped with a 2-functor to Cat [**160**, Section 8]. The construction in this generality is discussed by Borceux [**53**, proof of Theorem 8.3.1], Jacobs [**196**, Section 1.10], Johnson and Yau [**199**, Section 10.1], Johnstone [**200**, Section B1.3, Definition 1.3.1].

The special case of the category of elements goes back to Ehresmann [**114**, pages 52 and 53], [**116**, Section II.1.B]. It is discussed for instance in our monograph [**7**, Example A.23] and by Barr and Wells [**29**, Section 12.2.3], Borceux [**52**, Definition 1.6.4], Johnstone [**200**, Section A1.1, Proposition 1.1.7], Mac Lane and Moerdijk [**239**, Section VII.6, page 386], Riehl [**307**, Section 2.4], Roman [**309**, Example 22], Spivak [**334**, Section 6.2.2]. For the special case of a monoid action on a set, see [**7**, Example A.14], [**307**, Examples 1.5.18 and 2.4.10]. For the special case of a semidirect product monoid, see [**29**, Sections 12.2.5, 12.2.6, 12.2.7].

Groupoids. The Grothendieck construction when C is a groupoid and $\mathcal{F} : \mathsf{C} \to \mathsf{Cat}$ factors through the category of groupoids is present in Brown's work [**67**, Section 2]. He credits Frölich for this construction.

The special case of the groupoid of elements is explicitly mentioned by Ehresmann in his paper cited above [**114**, page 53]. This case is also discussed by Gabriel and Zisman [**139**, Appendix I, Section 1, top of page 140], Brown, Higgins, Sivera [**69**, Section B.7, page 591]. The special case of the semidirect product groupoid is discussed by Brown [**68**, Section 11.4]. The special case of a group action on a set goes back to Reidemeister [**303**, First chapter, Section 15]; it also appeared in Mackey's work on ergodic theory [**242**, Section 2 and Section 7, Remark (c)], [**243**, Section 11, page 204]. Finally, the special case of the semidirect product group is the most common and well-known (and also probably the oldest) of all semidirect products; it is the one taught in introductory algebra classes.

References

This is a list of all sources that are referred to in the book. The numbers at the end of each entry indicate the pages on which the item is cited.

1. E. Abe, *Hopf algebras*, Cambridge Tracts in Mathematics, vol. 74, Cambridge University Press, Cambridge-New York, 1980, Translated from the Japanese by Hisae Kinoshita and Hiroko Tanaka. 380, 516
2. P. Abramenko and K. S. Brown, *Buildings*, Graduate Texts in Mathematics, vol. 248, Springer, New York, 2008, Theory and applications. 97, 98
3. J. Adámek, *Theory of mathematical structures*, D. Reidel Publishing Co., Dordrecht, 1983. 156
4. M. Aguiar, N. Bergeron, F. Sottile, *Combinatorial Hopf algebras and generalized Dehn-Sommerville relations*, Compos. Math. **142** (2006), no. 1, 1–30. 518, 519
5. M. Aguiar and A. Lauve, *The characteristic polynomial of the Adams operators on graded connected Hopf algebras*, Algebra Number Theory **9** (2015), no. 3, 547–583. 454
6. M. Aguiar and S. Mahajan, *Coxeter groups and Hopf algebras*, Fields Institute Monographs, vol. 23, American Mathematical Society, Providence, RI, 2006, With a foreword by Nantel Bergeron. xi, xii, 40, 92, 95, 96, 97, 156, 214, 245, 455, 515, 518, 519, 520, 521
7. ———, *Monoidal functors, species and Hopf algebras*, CRM Monograph Series, vol. 29, American Mathematical Society, Providence, RI, 2010, With forewords by Kenneth Brown and Stephen Chase and André Joyal. xi, xii, xiii, xv, 38, 43, 95, 96, 97, 115, 156, 157, 213, 214, 215, 245, 251, 297, 361, 381, 432, 451, 452, 453, 454, 455, 478, 516, 518, 519, 520, 521, 558, 560, 639, 673, 684, 685, 686, 712, 713, 757, 758, 785, 807, 815, 825
8. ———, *Hopf monoids in the category of species*, Hopf algebras and tensor categories, Contemp. Math., vol. 585, Amer. Math. Soc., Providence, RI, 2013, pp. 17–124. xi, xii, xv, 156, 213, 214, 215, 245, 272, 297, 298, 785
9. ———, *On the Hadamard product of Hopf monoids*, Canad. J. Math. **66** (2014), no. 3, 481–504. 214, 297, 713
10. ———, *Topics in hyperplane arrangements*, Mathematical Surveys and Monographs, vol. 226, American Mathematical Society, Providence, RI, 2017. xii, xv, xvi, xix, 10, 20, 21, 23, 25, 27, 29, 31, 33, 34, 36, 37, 48, 49, 57, 61, 62, 63, 65, 66, 78, 81, 83, 84, 87, 90, 95, 96, 97, 98, 121, 150, 158, 214, 224, 258, 260, 297, 379, 448, 450, 469, 470, 509, 516, 519, 553, 696, 815
11. ———, *Bimonoids for hyperplane arrangements*, Encyclopedia of Mathematics and its Applications, vol. 173, Cambridge University Press, Cambridge, 2020. xi, xii, xiv, xv, xvi, xvii, xix, 6, 9, 10, 11, 20, 21, 36, 48, 55, 63, 65, 66, 68, 72, 74, 76, 78, 81, 82, 84, 93, 95, 96, 97, 98, 103, 115, 116, 119, 131, 135, 136, 137, 139, 146, 147, 148, 149, 150, 153, 154, 156, 157, 158, 159, 161, 162, 165, 166, 167, 168, 169, 174, 176, 178, 180, 181, 183, 185, 187, 188, 190, 191, 192, 193, 194, 195, 196, 200, 202, 203, 205, 206, 208, 210, 211, 212, 213, 214, 215, 216, 220, 221, 222, 224, 225, 226, 227, 228, 229, 230, 231, 232, 233, 234, 235, 236, 237, 238, 245, 246, 250, 253, 258, 260, 261, 264, 266, 267, 269, 270, 271, 272, 273, 274, 275, 276, 277, 278, 279, 280, 281, 282, 283, 284, 285, 286, 287, 288, 295, 296, 297, 298, 380, 381, 405, 419, 432, 448, 450, 451, 452, 453, 454, 455, 515,

521, 541, 554, 560, 577, 578, 579, 585, 590, 606, 616, 617, 638, 640, 666, 684, 697, 806, 824, 825, 847
12. M. Aguiar and F. Sottile, *Structure of the Malvenuto-Reutenauer Hopf algebra of permutations*, Adv. Math. **191** (2005), no. 2, 225–275. 96, 519, 520, 521
13. M. Aigner, *A course in enumeration*, Graduate Texts in Mathematics, vol. 238, Springer, Berlin, 2007. 97
14. M. André, *L'algèbre de Lie d'un anneau local*, Symposia Mathematica, Vol. IV (INDAM, Rome, 1968/69), Academic Press, London, 1970, pp. 337–375. 585
15. ———, *Hopf algebras with divided powers*, J. Algebra **18** (1971), 19–50. 517, 585, 640
16. ———, *Hopf and Eilenberg-MacLane algebras*, Reports of the Midwest Category Seminar, V (Zürich, 1970), Lecture Notes in Mathematics, Vol. 195, Springer, Berlin, 1971, pp. 1–28. 640
17. G. E. Andrews, R. Askey, R. Roy, *Special functions*, Encyclopedia of Mathematics and its Applications, vol. 71, Cambridge University Press, Cambridge, 1999. 97
18. G. E. Andrews and K. Eriksson, *Integer partitions*, Cambridge University Press, Cambridge, 2004. 97
19. M. A. Arbib and E. G. Manes, *Arrows, structures, and functors*, Academic Press, New York-London, 1975, The categorical imperative. 156
20. A. Ardizzoni and C. Menini, *Milnor-Moore categories and monadic decomposition*, J. Algebra **448** (2016), 488–563. 381
21. W. A. Arvola, *Complexified real arrangements of hyperplanes*, Manuscripta Math. **71** (1991), no. 3, 295–306. 96
22. M. D. Atkinson, *Solomon's descent algebra revisited*, Bull. London Math. Soc. **24** (1992), no. 6, 545–551. 97
23. M. Aubry, *Twisted Lie algebras and idempotent of Dynkin*, Sém. Lothar. Combin. **62** (2009/10), Art. B62b, 22. 272
24. ———, *Hall basis of twisted Lie algebras*, J. Algebraic Combin. **32** (2010), no. 2, 267–286. 272
25. S. Awodey, *Category theory*, second ed., Oxford Logic Guides, vol. 52, Oxford University Press, Oxford, 2010. xiv
26. H. Bachmann, *The algebra of bi-brackets and regularized multiple Eisenstein series*, J. Number Theory **200** (2019), 260–294. 639
27. J. C. Baez, *Hochschild homology in a braided tensor category*, Trans. Amer. Math. Soc. **344** (1994), no. 2, 885–906. 381
28. M. Barr, *Coalgebras over a commutative ring*, J. Algebra **32** (1974), no. 3, 600–610. 453
29. M. Barr and C. Wells, *Category theory for computing science*, Repr. Theory Appl. Categ. (2012), no. 22, xviii+538. 825
30. M. G. Barratt, *Twisted Lie algebras*, Geometric applications of homotopy theory (Proc. Conf., Evanston, Ill., 1977), II, Lecture Notes in Math., vol. 658, Springer, Berlin, 1978, pp. 9–15. 157, 245, 272
31. H. Belbachir, *A combinatorial contribution to the multinomial Chu-Vandermonde convolution*, Les Annales RECITS **1** (2014), 27–32. 97
32. J. Bénabou, *Structures algébriques dans les catégories*, Cahiers Topologie Géom. Différentielle **10** (1968), 1–126. 156
33. D. B. Benson, *Bialgebras: some foundations for distributed and concurrent computation*, Fund. Inform. **12** (1989), no. 4, 427–486. 454, 518
34. F. A. Berezin and G. I. Kac, *Lie groups with commuting and anticommuting parameters*, Mat. Sb. (N.S.) **82 (124)** (1970), 343–359. 585
35. F. Bergeron, N. Bergeron, R. B. Howlett, D. E. Taylor, *A decomposition of the descent algebra of a finite Coxeter group*, J. Algebraic Combin. **1** (1992), no. 1, 23–44. 95, 97
36. F. Bergeron, G. Labelle, P. Leroux, *Combinatorial species and tree-like structures*, Cambridge University Press, Cambridge, 1998, Translated from the 1994 French original by Margaret Readdy, With a foreword by Gian-Carlo Rota. 156, 214, 245

37. K. Bertet, D. Krob, M. Morvan, J.-C. Novelli, T. H. D. Phan, J.-Y. Thibon, *An overview of Λ-type operations on quasi-symmetric functions*, Comm. Algebra **29** (2001), no. 9, 4277–4303, Special issue dedicated to Alexei Ivanovich Kostrikin. 519
38. Bhaskaracharya, *Siddhantashiromani*, 1150. 96
39. I. Bialynicki-Birula, B. Mielnik, J. Plebański, *Explicit solution of the continuous Baker-Campbell-Hausdorff problem and a new expression for the phase operator*, Ann. of Phys. **51** (1969), no. 1, 187–200. 98
40. T. P. Bidigare, *Hyperplane arrangement face algebras and their associated Markov chains*, Ph.D. thesis, University of Michigan, 1997. 95
41. T. P. Bidigare, P. Hanlon, D. Rockmore, *A combinatorial description of the spectrum for the Tsetlin library and its generalization to hyperplane arrangements*, Duke Math. J. **99** (1999), no. 1, 135–174. 95, 96
42. L. J. Billera, K. S. Brown, P. Diaconis, *Random walks and plane arrangements in three dimensions*, Amer. Math. Monthly **106** (1999), no. 6, 502–524. 96
43. G. Birkhoff, *On the structure of abstract algebras*, Mathematical Proceedings of the Cambridge Philosophical Society **31** (1935), no. 4, 433–454. 95
44. ———, *Representability of Lie algebras and Lie groups by matrices*, Ann. of Math. (2) **38** (1937), no. 2, 526–532. 585, 639
45. ———, *Lattice theory*, third ed., American Mathematical Society Colloquium Publications, vol. 25, American Mathematical Society, Providence, RI, 1979. 95
46. A. Björner and F. Brenti, *Combinatorics of Coxeter groups*, Graduate Texts in Mathematics, vol. 231, Springer, New York, 2005. 96, 97, 98, 465
47. A. Björner, M. Las Vergnas, B. Sturmfels, N. White, G. M. Ziegler, *Oriented matroids*, Encyclopedia of Mathematics and its Applications, vol. 46, Cambridge University Press, Cambridge, 1993. 96
48. R. G. Bland, *Complementary orthogonal subspaces of R^n and orientability of matroids*, Ph.D. thesis, Cornell University, 1974. 95
49. D. Blessenohl and H. Laue, *Algebraic combinatorics related to the free Lie algebra*, Séminaire Lotharingien de Combinatoire (Thurnau, 1992), Prépubl. Inst. Rech. Math. Av., vol. 1993/33, Univ. Louis Pasteur, Strasbourg, 1993, pp. 1–21. 97
50. D. Blessenohl and M. Schocker, *Noncommutative character theory of the symmetric group*, Imperial College Press, London, 2005. 518, 520
51. J. M. Boardman and R. M. Vogt, *Homotopy invariant algebraic structures on topological spaces*, Springer-Verlag, Berlin, 1973, Lecture Notes in Mathematics, Vol. 347. 560
52. F. Borceux, *Handbook of categorical algebra. 1*, Encyclopedia of Mathematics and its Applications, vol. 50, Cambridge University Press, Cambridge, 1994, Basic category theory. 825
53. ———, *Handbook of categorical algebra. 2*, Encyclopedia of Mathematics and its Applications, vol. 51, Cambridge University Press, Cambridge, 1994, Categories and structures. 825
54. A. Borel, *Sur la cohomologie des espaces fibrés principaux et des espaces homogènes de groupes de Lie compacts*, Ann. of Math. (2) **57** (1953), 115–207. 380, 451, 638
55. ———, *Sur l'homologie et la cohomologie des groupes de Lie compacts connexes*, Amer. J. Math. **76** (1954), 273–342. 451
56. ———, *Topology of Lie groups and characteristic classes*, Bull. Amer. Math. Soc. **61** (1955), 397–432. 451
57. A. V. Borovik and A. Borovik, *Mirrors and reflections*, Universitext, Springer, New York, 2010, The geometry of finite reflection groups. 98
58. A. V. Borovik, I. M. Gelfand, N. White, *Coxeter matroids*, Progress in Mathematics, vol. 216, Birkhäuser Boston, Inc., Boston, MA, 2003. 98
59. N. Bourbaki, *Éléments de mathématique. 22. Première partie: Les structures fondamentales de l'analyse. Livre 1: Théorie des ensembles. Chapitre 4: Structures*, Actualités Sci. Ind. no. 1258, Hermann, Paris, 1957. 156

60. ______, *Elements of mathematics. Theory of sets*, Translated from the French, Hermann, Publishers in Arts and Science, Paris; Addison-Wesley Publishing Co., Reading, Mass.-London-Don Mills, Ont., 1968. 156
61. ______, *Éléments de mathématique*, Masson, Paris, 1981, Groupes et algèbres de Lie. Chapitres 4, 5 et 6. 98
62. M. R. Bremner and V. Dotsenko, *Algebraic operads*, CRC Press, Boca Raton, FL, 2016, An algorithmic companion. 214, 245, 560
63. W. Browder, *On differential Hopf algebras*, Trans. Amer. Math. Soc. **107** (1963), 153–176. 451
64. K. S. Brown, *Semigroups, rings, and Markov chains*, J. Theoret. Probab. **13** (2000), no. 3, 871–938. 95, 96
65. ______, *Semigroup and ring theoretical methods in probability*, Representations of finite dimensional algebras and related topics in Lie theory and geometry, Fields Inst. Commun., vol. 40, Amer. Math. Soc., Providence, RI, 2004, pp. 3–26. 95
66. K. S. Brown and P. Diaconis, *Random walks and hyperplane arrangements*, Ann. Probab. **26** (1998), no. 4, 1813–1854. 95
67. R. Brown, *Groupoids as coefficients*, Proc. London Math. Soc. (3) **25** (1972), 413–426. 825
68. ______, *Topology and groupoids*, BookSurge, LLC, Charleston, SC, 2006. 825
69. R. Brown, P. J. Higgins, R. Sivera, *Nonabelian algebraic topology*, EMS Tracts in Mathematics, vol. 15, European Mathematical Society (EMS), Zürich, 2011, Filtered spaces, crossed complexes, cubical homotopy groupoids. 157, 825
70. Y. Bruned, C. Curry, K. Ebrahimi-Fard, *Quasi-shuffle algebras and renormalisation of rough differential equations*, Bull. Lond. Math. Soc. **52** (2020), no. 1, 43–63. 639
71. I. Bucur and A. Deleanu, *Introduction to the theory of categories and functors*, With the collaboration of Peter J. Hilton and Nicolae Popescu. Pure and Applied Mathematics, Vol. XIX, Interscience Publication John Wiley & Sons, Ltd., London-New York-Sydney, 1968. 156
72. V. C. Bui, G. H. E. Duchamp, H. N. Minh, Q. H. Ngô, C. Tollu, *(Pure) transcendence bases in φ-deformed shuffle bialgebras*, Sém. Lothar. Combin. **74** ([2015-2018]), Art. B74f, 22. 639
73. L. Carlitz, *Sums of products of multinomial coefficients*, Elem. Math. **18** (1963), 37–39. 97
74. H. Cartan, *Séminaire Henri Cartan de l'Ecole Normale Supérieure, 1954/1955. Algèbres d'Eilenberg-MacLane et homotopie*, Secrétariat mathématique, 11 rue Pierre Curie, Paris, 1956, 2ème éd. 517
75. H. Cartan and S. Eilenberg, *Homological algebra*, Princeton University Press, Princeton, N. J., 1956. 380, 381, 455, 585
76. R. W. Carter, *Simple groups of Lie type*, Wiley Classics Library, John Wiley & Sons, Inc., New York, 1989, Reprint of the 1972 original, A Wiley-Interscience Publication. 98
77. P. Cartier, *Effacement dans la cohomologie des algèbres de Lie*, Séminaire N. Bourbaki; Volume 3. Exposé no. 116, 1954–1956, pp. 161–167. 585
78. ______, *Hyperalgèbres et groupes de Lie formels*, Séminaire Sophus Lie; Volume 2, Secrétariat mathématique, Paris, 1955–56. 380, 381, 451, 517, 585, 640
79. ______, *Théorie différentielle des groupes algébriques*, C. R. Acad. Sci. Paris **244** (1957), 540–542. 451, 640
80. ______, *Groupes algébriques et groupes formels*, Colloq. Théorie des Groupes Algébriques (Bruxelles, 1962), Librairie Universitaire, Louvain; GauthierVillars, Paris, 1962, pp. 87–111. 640
81. ______, *Groupes formels associés aux anneaux de Witt généralisés*, C. R. Acad. Sci. Paris Sér. A-B **265** (1967), A49–A52. 518
82. ______, *On the structure of free Baxter algebras*, Adv. Math. **9** (1972), 253–265. 452
83. ______, *A primer of Hopf algebras*, Frontiers in number theory, physics, and geometry. II, Springer, Berlin, 2007, pp. 537–615. 454, 638

84. P. Cartier and F. Patras, *Classical Hopf algebras and their applications*, Algebra and Applications, vol. 29, Springer, Cham, 2021. 157, 380, 638, 712
85. K.-t. Chen, *Algebraic paths*, J. Algebra **10** (1968), 8–36. 517
86. C. Chevalley, *Theory of Lie Groups. I*, Princeton Mathematical Series, vol. 8, Princeton University Press, Princeton, N. J., 1946. 585
87. ———, *Sur certains groupes simples*, Tôhoku Math. J. (2) **7** (1955), 14–66. 96
88. S.-C. Chu, *Ssu yüan yü chien (Precious mirror of the four elements)*, 1303. 97
89. P. B. Cohen, T. M. W. Eyre, R. L. Hudson, *Higher order Itô product formula and generators of evolutions and flows*, Proceedings of the International Quantum Structures Association, Quantum Structures '94 (Prague, 1994), vol. 34, 1995, pp. 1481–1486. 452
90. C. Contou-Carrère, *Buildings and Schubert schemes*, CRC Press, Boca Raton, FL, 2017. 98
91. H. S. M. Coxeter, *Discrete groups generated by reflections*, Ann. of Math. (2) **35** (1934), no. 3, 588–621. 95
92. ———, *Regular polytopes*, third ed., Dover Publications, Inc., New York, 1973. 98
93. S. Dăscălescu, C. Năstăsescu, Ş. Raianu, *Hopf algebras*, Monographs and Textbooks in Pure and Applied Mathematics, vol. 235, Marcel Dekker, Inc., New York, 2001, An introduction. 380
94. M. W. Davis, *The geometry and topology of Coxeter groups*, London Mathematical Society Monographs Series, vol. 32, Princeton University Press, Princeton, NJ, 2008. 96, 98
95. M. Demazure, *Lectures on p-divisible groups*, Lecture Notes in Mathematics, Vol. 302, Springer-Verlag, Berlin-New York, 1972. 640
96. M. Demazure and P. Gabriel, *Groupes algébriques. Tome I: Géométrie algébrique, généralités, groupes commutatifs*, Masson & Cie, Éditeur, Paris; North-Holland Publishing Co., Amsterdam, 1970. 453, 640
97. P. Deshpande, *Arrangements of spheres and projective spaces*, Rocky Mountain J. Math. **46** (2016), no. 5, 1447–1487. 96
98. J. Dieudonné, *Introduction to the theory of formal groups*, Marcel Dekker, Inc., New York, 1973, Pure and Applied Mathematics, 20. 638
99. E. J. Ditters, *Curves and exponential series in the theory of noncommutative formal groups*, Ph.D. thesis, University of Nijmegen, 1969. 518, 639
100. ———, *Sur une série exponentielle non commutative définie sur les corps de caractéristique p*, C. R. Acad. Sci. Paris Sér. A-B **268** (1969), A580–A582. 518
101. ———, *Curves and formal (co)groups*, Invent. Math. **17** (1972), 1–20. 518, 639
102. ———, *Groupes formels*, U. E. R. Mathématique, Université de Paris XI, Orsay, 1975, Cours de 3e cycle 1973–1974, Publications Mathématiques d'Orsay, No. 149 75.42. 518, 640
103. E. J. Ditters and A. C. J. Scholtens, *Free polynomial generators for the Hopf algebra QSym of quasisymmetric functions*, J. Pure Appl. Algebra **144** (1999), no. 3, 213–227. 518
104. J. Dixmier, *Enveloping algebras*, Graduate Studies in Mathematics, vol. 11, American Mathematical Society, Providence, RI, 1996, Revised reprint of the 1977 translation. 640
105. P. Doubilet, G.-C. Rota, R. P. Stanley, *On the foundations of combinatorial theory. VI. The idea of generating function*, Proceedings of the Sixth Berkeley Symposium on Mathematical Statistics and Probability (Univ. California, Berkeley, Calif., 1970/1971), Vol. II: Probability theory, 1972, pp. 267–318. 516
106. G. H. E. Duchamp, F. Hivert, J.-Y. Thibon, *Noncommutative symmetric functions. VI. Free quasi-symmetric functions and related algebras*, Internat. J. Algebra Comput. **12** (2002), no. 5, 671–717. 518, 520
107. G. H. E. Duchamp, A. Klyachko, D. Krob, J.-Y. Thibon, *Noncommutative symmetric functions. III. Deformations of Cauchy and convolution algebras*, Discrete Math. Theor. Comput. Sci. **1** (1997), no. 1, 159–216, Lie computations (Marseille, 1994).

454, 517, 518

108. K. Ebrahimi-Fard, J. M. Gracia-Bondía, F. Patras, *A Lie theoretic approach to renormalization*, Comm. Math. Phys. **276** (2007), no. 2, 519–549. 98, 453
109. K. Ebrahimi-Fard, S. J. A. Malham, F. Patras, A. Wiese, *The exponential Lie series for continuous semimartingales*, Proc. A. **471** (2015), no. 2184, 20150429, 19. 96, 98, 639
110. ———, *Flows and stochastic Taylor series in Itô calculus*, J. Phys. A **48** (2015), no. 49, 495202, 17. 98
111. K. Ebrahimi-Fard, D. Manchon, F. Patras, *A noncommutative Bohnenblust-Spitzer identity for Rota-Baxter algebras solves Bogoliubov's recursion*, J. Noncommut. Geom. **3** (2009), no. 2, 181–222. 98, 453
112. B. Eckmann and P. J. Hilton, *Group-like structures in general categories. I. Multiplications and comultiplications*, Math. Ann. **145** (1961/1962), 227–255. 157
113. R. Ehrenborg, *On posets and Hopf algebras*, Adv. Math. **119** (1996), no. 1, 1–25. 519
114. C. Ehresmann, *Gattungen von lokalen Strukturen*, Jber. Deutsch. Math.-Verein. **60** (1957), 49–77. 825
115. ———, *Catégories structurées*, Ann. Sci. École Norm. Sup. (3) **80** (1963), 349–426. 157
116. ———, *Catégories et structures*, Dunod, Paris, 1965. 156, 157, 825
117. S. Eilenberg and S. Mac Lane, *On the groups of $H(\Pi, n)$. I*, Ann. of Math. (2) **58** (1953), 55–106. 96
118. H. Epstein, *Trees*, Nuclear Phys. B **912** (2016), 151–171. 95
119. H. Epstein, V. J. Glaser, R. Stora, *Geometry of the n point p space function of quantum field theory*, Hyperfunctions and theoretical physics (Rencontre, Nice, 1973; dédié à la mémoire de A. Martineau), 1975, pp. 143–162. Lecture Notes in Math., Vol. 449. 95
120. ———, *General properties of the n-point functions in local quantum field theory*, Structural analysis of collision amplitudes, North-Holland Publishing Co., Amsterdam-New York-Oxford, 1976, Lectures delivered at Les Houches, École d'Été de Physique Théorique, 2-27 June 1975, pp. 7–94. 95
121. D. C. Ernst, *Cell complexes for arrangements with group actions*, Master's thesis, Northern Arizona University, 2000. 96
122. T. Ernst, *A comprehensive treatment of q-calculus*, Birkhäuser/Springer Basel AG, Basel, 2012. 97
123. G. Etienne, *Linear extensions of finite posets and a conjecture of G. Kreweras on permutations*, Discrete Math. **52** (1984), no. 1, 107–111. 95
124. T. M. W. Eyre, *Quantum stochastic calculus and representations of Lie superalgebras*, Lecture Notes in Mathematics, vol. 1692, Springer-Verlag, Berlin, 1998. 452
125. Z. Fiedorowicz and J.-L. Loday, *Crossed simplicial groups and their associated homology*, Trans. Amer. Math. Soc. **326** (1991), no. 1, 57–87. 157
126. L. Foissy, *Quantification de l'algèbre de Hopf de Malvenuto et Reutenauer*, preprint 2005. 520
127. L. Foissy, F. Patras, J.-Y. Thibon, *Deformations of shuffles and quasi-shuffles*, Ann. Inst. Fourier (Grenoble) **66** (2016), no. 1, 209–237. 639
128. T. F. Fox, *Universal coalgebras*, Ph.D. thesis, McGill University (Canada), 1976. 453
129. ———, *The coalgebra enrichment of algebraic categories*, Comm. Algebra **9** (1981), no. 3, 223–234. 453
130. ———, *The tensor product of Hopf algebras*, Rend. Istit. Mat. Univ. Trieste **24** (1992), no. 1-2, 65–71 (1994). 453
131. B. Fresse, *On the homotopy of simplicial algebras over an operad*, Trans. Amer. Math. Soc. **352** (2000), no. 9, 4113–4141. 157, 560
132. ———, *Koszul duality of operads and homology of partition posets*, Homotopy theory: relations with algebraic geometry, group cohomology, and algebraic K-theory, Contemp. Math., vol. 346, Amer. Math. Soc., Providence, RI, 2004, pp. 115–215. 560

133. ———, *Modules over operads and functors*, Lecture Notes in Mathematics, vol. 1967, Springer-Verlag, Berlin, 2009. 560
134. ———, *Homotopy of operads and Grothendieck-Teichmüller groups. Part 1*, Mathematical Surveys and Monographs, vol. 217, American Mathematical Society, Providence, RI, 2017, The algebraic theory and its topological background. 560, 638
135. ———, *Homotopy of operads and Grothendieck-Teichmüller groups. Part 2*, Mathematical Surveys and Monographs, vol. 217, American Mathematical Society, Providence, RI, 2017, The applications of (rational) homotopy theory methods. 560
136. A. Fröhlich, *Formal groups*, Lecture Notes in Mathematics, No. 74, Springer-Verlag, Berlin-New York, 1968. 640
137. W. Fulton, *Young tableaux*, London Mathematical Society Student Texts, vol. 35, Cambridge University Press, Cambridge, 1997, With applications to representation theory and geometry. 518
138. P. Gabriel, *Exposé VII, Etude infinitesimale des schemas en groupes*, Schémas en groupes. I: Propriétés générales des schémas en groupes, Séminaire de Géométrie Algébrique du Bois Marie 1962/64 (SGA 3). Dirigé par M. Demazure et A. Grothendieck. Lecture Notes in Mathematics, Vol. 151, Springer-Verlag, Berlin-New York, 1970, pp. xv+564. 640
139. P. Gabriel and M. Zisman, *Calculus of fractions and homotopy theory*, Ergebnisse der Mathematik und ihrer Grenzgebiete, Band 35, Springer-Verlag New York, Inc., New York, 1967. 825
140. J. G. Gaines, *The algebra of iterated stochastic integrals*, Stochastics Stochastics Rep. **49** (1994), no. 3-4, 169–179. 452
141. ———, *A basis for iterated stochastic integrals*, Math. Comput. Simulation **38** (1995), no. 1-3, 7–11, Probabilités numériques (Paris, 1992). 452
142. P. Garrett, *Buildings and classical groups*, Chapman & Hall, London, 1997. 98
143. A. M. Garsia, *Combinatorics of the free Lie algebra and the symmetric group*, Analysis, et cetera, Academic Press, Boston, MA, 1990, pp. 309–382. 453
144. A. M. Garsia and C. Reutenauer, *A decomposition of Solomon's descent algebra*, Adv. Math. **77** (1989), no. 2, 189–262. 95, 97, 98
145. C. F. Gauss, *Summatio quarumdam serierum singularium*, (1808). 97
146. L. Geissinger, *Hopf algebras of symmetric functions and class functions*, Combinatoire et représentation du groupe symétrique (Actes Table Ronde C.N.R.S., Univ. Louis-Pasteur Strasbourg, Strasbourg, 1976), Springer, Berlin, 1977, pp. 168–181. Lecture Notes in Math., Vol. 579. 518
147. I. M. Gelfand, *The center of an infinitesimal group ring*, Mat. Sbornik N.S. **26(68)** (1950), 103–112. 585
148. ———, *Center of the infinitesimal group ring*, Collected papers. Vol. II, Edited by S. G. Gindikin, V. W. Guillemin, A. A. Kirillov, B. Kostant and S. Sternberg, Springer-Verlag, Berlin, 1988, pp. 22–30. 585
149. I. M. Gelfand, D. Krob, A. Lascoux, B. Leclerc, V. S. Retakh, J.-Y. Thibon, *Noncommutative symmetric functions*, Adv. Math. **112** (1995), no. 2, 218–348. 98, 454, 518, 519, 520, 639
150. M. Gerstenhaber, *The cohomology structure of an associative ring*, Ann. of Math. (2) **78** (1963), 267–288. 585
151. M. Gerstenhaber and S. D. Schack, *The shuffle bialgebra and the cohomology of commutative algebras*, J. Pure Appl. Algebra **70** (1991), no. 3, 263–272. 454
152. I. M. Gessel, *Multipartite P-partitions and inner products of skew Schur functions*, Combinatorics and algebra (Boulder, Colo., 1983), Amer. Math. Soc., Providence, RI, 1984, pp. 289–317. 95, 519, 521
153. V. Ginzburg and M. Kapranov, *Koszul duality for operads*, Duke Math. J. **76** (1994), no. 1, 203–272. 560
154. P. G. Goerss, *Barratt's desuspension spectral sequence and the Lie ring analyzer*, Quart. J. Math. Oxford Ser. (2) **44** (1993), no. 173, 43–85. 272

155. J. Goldman and G.-C. Rota, *On the foundations of combinatorial theory. IV. Finite vector spaces and Eulerian generating functions*, Studies in Appl. Math. **49** (1970), 239–258. 516
156. H. W. Gould, *The q-series generalization of a formula of Sparre Andersen*, Math. Scand. **9** (1961), 90–94. 97
157. M. Grandis, *Higher dimensional categories*, World Scientific Publishing Co. Pte. Ltd., Hackensack, NJ, 2020, From double to multiple categories. 157
158. C. Greene, *On the Möbius algebra of a partially ordered set*, Adv. Math. **10** (1973), 177–187. 95
159. D. Grinberg and V. Reiner, *Hopf algebras in combinatorics*, available at arXiv:1409.8356. 518, 519, 520
160. A. Grothendieck, *Exposé VI, Catégories fibrées et descente*, Séminaire de Géométrie Algébrique du Bois Marie 1960–1961 (SGA 1), Lecture Notes in Mathematics, Vol. 224, Springer-Verlag, Berlin-New York, 1971. 825
161. L. C. Grove and C. T. Benson, *Finite reflection groups*, second ed., Graduate Texts in Mathematics, vol. 99, Springer-Verlag, New York, 1985. 98
162. L. A. Grünenfelder, *Über die struktur von Hopf-algebren*, Ph.D. thesis, ETH-Zurich, 1969. 640
163. ———, *Hopf-algebren und coradikal*, Math. Z. **116** (1970), 166–182. 640
164. ———, *(Co-)homology of commutative coalgebras*, Comm. Algebra **16** (1988), no. 3, 541–576. 638
165. L. A. Grünenfelder and M. Mastnak, *On bimeasurings*, J. Pure Appl. Algebra **204** (2006), no. 2, 258–269. 453
166. ———, *On bimeasurings. II*, J. Pure Appl. Algebra **209** (2007), no. 3, 823–832. 453
167. L. Guo and W. Keigher, *Baxter algebras and shuffle products*, Adv. Math. **150** (2000), no. 1, 117–149. 452
168. D. I. Gurevich, *Algebraic aspects of the quantum Yang-Baxter equation*, Algebra i Analiz **2** (1990), no. 4, 119–148. 381
169. J. S. Hadamard, *Théorème sur les séries entières*, Acta Math. **22** (1899), no. 1, 55–63. 452
170. P. Hall, *A Contribution to the Theory of Groups of Prime-Power Order*, Proc. London Math. Soc. (2) **36** (1934), 29–95. 585
171. ———, *The Eulerian functions of a group*, Quarterly J. Math. **7** (1936), no. 1, 134–151. 97
172. E. Halpern, *On the structure of hyperalgebras. Class 1 Hopf algebras*, Portugal. Math. **17** (1958), 127–147. 381
173. ———, *Twisted polynomial hyperalgebras*, Mem. Amer. Math. Soc. no. **29** (1958), 61 pp. (1958). 381
174. Harish-Chandra, *On representations of Lie algebras*, Ann. of Math. (2) **50** (1949), 900–915. 585
175. ———, *On some applications of the universal enveloping algebra of a semisimple Lie algebra*, Trans. Amer. Math. Soc. **70** (1951), 28–96. 585
176. M. Hartl, T. Pirashvili, C. Vespa, *Polynomial functors from algebras over a set-operad and nonlinear Mackey functors*, Int. Math. Res. Not. IMRN (2015), no. 6, 1461–1554. 158
177. M. Hasse and L. Michler, *Theorie der Kategorien*, Mathematische Monographien, Band 7, VEB Deutscher Verlag der Wissenschaften, Berlin, 1966. 156
178. M. Hazewinkel, *Generalized overlapping shuffle algebras*, J. Math. Sci. (New York) **106** (2001), no. 4, 3168–3186, Pontryagin Conference, 8, Algebra (Moscow, 1998). 452
179. ———, *Symmetric functions, noncommutative symmetric functions, and quasisymmetric functions*, Acta Appl. Math. **75** (2003), no. 1-3, 55–83, Monodromy and differential equations (Moscow, 2001). 521
180. ———, *Formal groups and applications*, AMS Chelsea Publishing, Providence, RI, 2012, Corrected reprint of the 1978 original. 518, 640

181. M. Hazewinkel, N. Gubareni, V. V. Kirichenko, *Algebras, rings and modules*, Mathematical Surveys and Monographs, vol. 168, American Mathematical Society, Providence, RI, 2010, Lie algebras and Hopf algebras. 380, 518, 520
182. R. G. Heyneman and M. E. Sweedler, *Affine Hopf algebras. I*, J. Algebra **13** (1969), 192–241. 453
183. ______, *Affine Hopf algebras. II*, J. Algebra **16** (1970), 271–297. 638
184. H. Hiller, *Geometry of Coxeter groups*, Research Notes in Mathematics, vol. 54, Pitman (Advanced Publishing Program), Boston, Mass.-London, 1982. 98
185. P. J. Hilton, *Homotopy theory and duality*, Gordon and Breach Science Publishers, New York-London-Paris, 1965. 157
186. G. P. Hochschild, *The structure of Lie groups*, Holden-Day, Inc., San Francisco-London-Amsterdam, 1965. 454
187. ______, *Introduction to affine algebraic groups*, Holden-Day, Inc., San Francisco, Calif.-Cambridge-Amsterdam, 1971. 453
188. ______, *Basic theory of algebraic groups and Lie algebras*, Graduate Texts in Mathematics, vol. 75, Springer-Verlag, New York-Berlin, 1981. 453, 640
189. M. E. Hoffman, *Quasi-shuffle products*, J. Algebraic Combin. **11** (2000), no. 1, 49–68. 452, 638, 639
190. H. Hopf, *Sur la topologie des groupes clos de Lie et de leurs généralisations*, C. R. Acad. Sci. Paris **208** (1939), 1266–1267. 638
191. ______, *Über die Topologie der Gruppen-Mannigfaltigkeiten und ihre Verallgemeinerungen*, Ann. of Math. (2) **42** (1941), 22–52. 380, 451, 638, 639
192. ______, *Selecta Heinz Hopf*, Herausgegeben zu seinem 70. Geburtstag von der Eidgenössischen Technischen Hochschule Zürich, Springer-Verlag, Berlin-New York, 1964. 380
193. ______, *Collected papers/Gesammelte Abhandlungen*, Springer Collected Works in Mathematics, Springer, Heidelberg, 2013, Edited and with a preface and foreword by Beno Eckmann, With appendices by Peter J. Hilton, P. Alexandroff, Eckmann and Hopf. Reprint of the 2001 edition. 380
194. R. L. Hudson and K. R. Parthasarathy, *The Casimir chaos map for* $\mathrm{U}(N)$, Tatra Mt. Math. Publ. **3** (1993), 81–88, Measure theory (Liptovský Ján, 1993). 452
195. J. E. Humphreys, *Reflection groups and Coxeter groups*, Cambridge Studies in Advanced Mathematics, vol. 29, Cambridge University Press, Cambridge, 1990. 96, 97, 98
196. B. Jacobs, *Categorical logic and type theory*, Studies in Logic and the Foundations of Mathematics, vol. 141, North-Holland Publishing Co., Amsterdam, 1999. 825
197. N. Jacobson, *Rational methods in the theory of Lie algebras*, Ann. of Math. (2) **36** (1935), no. 4, 875–881. 585
198. J. L. W. V. Jensen, *Studier over en afhandling af Gauss*, Nyt tidsskrift for matematik **29** (1918), 29–36. 97
199. N. Johnson and D. Yau, *2-dimensional categories*, Oxford University Press, Oxford, 2021. 157, 825
200. P. T. Johnstone, *Sketches of an elephant: a topos theory compendium. Vol. 1*, Oxford Logic Guides, vol. 43, The Clarendon Press, Oxford University Press, New York, 2002. 825
201. S. A. Joni and G.-C. Rota, *Coalgebras and bialgebras in combinatorics*, Umbral calculus and Hopf algebras (Norman, Okla., 1978), Contemp. Math., vol. 6, Amer. Math. Soc., Providence, RI, 1982, pp. 1–47. 516
202. A. Joyal, *Une théorie combinatoire des séries formelles*, Adv. Math. **42** (1981), no. 1, 1–82. 156, 214
203. ______, *Foncteurs analytiques et espèces de structures*, Combinatoire énumérative (Montreal, Que., 1985/Quebec, Que., 1985), Lecture Notes in Math., vol. 1234, Springer, Berlin, 1986, pp. 126–159. 156, 272, 298
204. V. G. Kac and P. Cheung, *Quantum calculus*, Universitext, Springer-Verlag, New York, 2002. 97

205. R. Kane, *Reflection groups and invariant theory*, CMS Books in Mathematics, vol. 5, Springer-Verlag, New York, 2001. 98
206. Y. Kashina, *A generalized power map for Hopf algebras*, Hopf algebras and quantum groups (Brussels, 1998), Lecture Notes in Pure and Appl. Math., vol. 209, Dekker, New York, 2000, pp. 159–175. 454
207. Y. Kashina, S. Montgomery, S.-H. Ng, *On the trace of the antipode and higher indicators*, Israel J. Math. **188** (2012), 57–89. 454
208. Y. Kashina, Y. Sommerhäuser, Y. Zhu, *On higher Frobenius-Schur indicators*, Mem. Amer. Math. Soc. **181** (2006), no. 855, viii+65. 455
209. C. Kassel, *Quantum groups*, Graduate Texts in Mathematics, vol. 155, Springer-Verlag, New York, 1995. 97, 380, 516
210. G. M. Kelly, *On the operads of J. P. May*, Repr. Theory Appl. Categ. (2005), no. 13, 1–13. 157, 560
211. V. K. Kharchenko, *Quantum Lie theory*, Lecture Notes in Mathematics, vol. 2150, Springer, Cham, 2015, A multilinear approach. 381
212. M. Koppinen, *A Skolem-Noether theorem for coalgebra measurings*, Arch. Math. (Basel) **57** (1991), no. 1, 34–40. 453
213. B. Kostant, *Groups over Z*, Algebraic Groups and Discontinuous Subgroups (Proc. Sympos. Pure Math., Boulder, Colo., 1965), Amer. Math. Soc., Providence, RI, 1966, pp. 90–98. 453
214. J.-L. Koszul, *Homologie et cohomologie des algèbres de Lie*, Bull. Soc. Math. France **78** (1950), 65–127. 451
215. D. Krob, B. Leclerc, J.-Y. Thibon, *Noncommutative symmetric functions. II. Transformations of alphabets*, Internat. J. Algebra Comput. **7** (1997), no. 2, 181–264. 98, 454, 518, 519
216. D. Krob and J.-Y. Thibon, *Noncommutative symmetric functions. IV. Quantum linear groups and Hecke algebras at* $q = 0$, J. Algebraic Combin. **6** (1997), no. 4, 339–376. 518
217. ———, *Noncommutative symmetric functions. V. A degenerate version of* $U_q(\mathrm{gl}_N)$, Internat. J. Algebra Comput. **9** (1999), no. 3-4, 405–430, Dedicated to the memory of Marcel-Paul Schützenberger. 518
218. J. Lambek, *Deductive systems and categories. II. Standard constructions and closed categories*, Category Theory, Homology Theory and their Applications, I (Battelle Institute Conference, Seattle, Wash., 1968, Vol. One), Springer, Berlin, 1969, pp. 76–122. 560
219. R. Landers, S. Montgomery, P. Schauenburg, *Hopf powers and orders for some bismash products*, J. Pure Appl. Algebra **205** (2006), no. 1, 156–188. 454
220. R. G. Larson, *Hopf algebras and group algebras*, Ph.D. thesis, The University of Chicago, 1965. 451
221. M. Las Vergnas, *Matroïdes orientables*, C. R. Acad. Sci. Paris Sér. A-B **280** (1975), Ai, A61–A64. 98
222. ———, *Convexity in oriented matroids*, J. Combin. Theory Ser. B **29** (1980), no. 2, 231–243. 98
223. M. Lazard, *Lois de groupes et analyseurs*, Ann. Sci. Ecole Norm. Sup. (3) **72** (1955), 299–400. 560
224. ———, *Lois de groupes et analyseurs*, Séminaire Bourbaki, Vol. 3, Soc. Math. France, Paris, 1995, pp. Exp. No. 109, 77–91. 560
225. T. Leinster, *Higher operads, higher categories*, London Mathematical Society Lecture Note Series, vol. 298, Cambridge University Press, Cambridge, 2004. 560
226. ———, *Basic category theory*, Cambridge Studies in Advanced Mathematics, vol. 143, Cambridge University Press, Cambridge, 2014. xiv
227. J. Leray, *Sur la forme des espaces topologiques et sur les points fixes des représentations*, J. Math. Pures Appl. (9) **24** (1945), 95–167. 380, 451, 454, 639
228. C. W. Li and X. Q. Liu, *Algebraic structure of multiple stochastic integrals with respect to Brownian motions and Poisson processes*, Stochastics Stochastics Rep. **61** (1997),

no. 1-2, 107–120. 452

229. V. V. Linchenko and S. Montgomery, *A Frobenius-Schur theorem for Hopf algebras*, Algebr. Represent. Theory **3** (2000), no. 4, 347–355, Special issue dedicated to Klaus Roggenkamp on the occasion of his 60th birthday. 454
230. M. Livernet, *From left modules to algebras over an operad: application to combinatorial Hopf algebras*, Ann. Math. Blaise Pascal **17** (2010), no. 1, 47–96. 639, 713
231. M. Livernet and F. Patras, *Lie theory for Hopf operads*, J. Algebra **319** (2008), no. 12, 4899–4920. 157
232. J.-L. Loday, *Cyclic homology*, second ed., Grundlehren der Mathematischen Wissenschaften, vol. 301, Springer-Verlag, Berlin, 1998. 453
233. J.-L. Loday and M. O. Ronco, *On the structure of cofree Hopf algebras*, J. Reine Angew. Math. **592** (2006), 123–155. 639
234. J.-L. Loday and B. Vallette, *Algebraic operads*, Grundlehren der Mathematischen Wissenschaften, vol. 346, Springer, Heidelberg, 2012. 214, 452, 560
235. M. Lorenz, *A tour of representation theory*, Graduate Studies in Mathematics, vol. 193, American Mathematical Society, Providence, RI, 2018. 380
236. S. Mac Lane, *Cohomology theory of Abelian groups*, Proceedings of the International Congress of Mathematicians, Cambridge, Mass., 1950, vol. 2, Amer. Math. Soc., Providence, R. I., 1952, pp. 8–14. 96
237. ———, *Homology*, Classics in Mathematics, Springer-Verlag, Berlin, 1995, Reprint of the 1975 edition. 96
238. ———, *Categories for the working mathematician*, second ed., Graduate Texts in Mathematics, vol. 5, Springer-Verlag, New York, 1998. 108, 560
239. S. Mac Lane and I. Moerdijk, *Sheaves in geometry and logic*, Universitext, Springer-Verlag, New York, 1994. 825
240. I. G. Macdonald, *Symmetric functions and Hall polynomials*, second ed., The Clarendon Press Oxford University Press, New York, 1995, With contributions by A. Zelevinsky, Oxford Science Publications. 518, 520
241. K. C. H. Mackenzie, *Double Lie algebroids and second-order geometry. I*, Adv. Math. **94** (1992), no. 2, 180–239. 157
242. G. W. Mackey, *Ergodic theory, group theory, and differential geometry*, Proc. Nat. Acad. Sci. U.S.A. **50** (1963), 1184–1191. 825
243. ———, *Ergodic theory and virtual groups*, Math. Ann. **166** (1966), 187–207. 825
244. W. Magnus, *Beziehungen zwischen Gruppen und Idealen in einem speziellen Ring*, Math. Ann. **111** (1935), no. 1, 259–280. 585
245. ———, *Über Beziehungen zwischen höheren Kommutatoren*, J. Reine Angew. Math. **177** (1937), 105–115. 585
246. S. Mahajan, *Shuffles, shellings and projections*, Ph.D. thesis, Cornell University, 2002. 96
247. S. Majid, *Foundations of quantum group theory*, Cambridge University Press, Cambridge, 1995. 380
248. C. Malvenuto, *Produits et coproduits des fonctions quasi-symétriques et de l'algèbre des descents*, Ph.D. thesis, Laboratoire de combinatoire et d'informatique mathématique (LACIM), Univ. du Québec à Montréal, 1994. 519, 520, 638
249. C. Malvenuto and C. Reutenauer, *Duality between quasi-symmetric functions and the Solomon descent algebra*, J. Algebra **177** (1995), no. 3, 967–982. 519, 520
250. D. Manchon, *Hopf algebras in renormalisation*, Handbook of algebra. Vol. 5, Elsevier/North-Holland, Amsterdam, 2008, pp. 365–427. 98, 453, 455
251. D. Manchon and S. Paycha, *Nested sums of symbols and renormalized multiple zeta values*, Int. Math. Res. Not. IMRN (2010), no. 24, 4628–4697. 639
252. M. Markl, S. Shnider, J. D. Stasheff, *Operads in algebra, topology and physics*, Mathematical Surveys and Monographs, vol. 96, American Mathematical Society, Providence, RI, 2002. 560
253. M. Mastnak, *On the cohomology of a smash product of Hopf algebras*, available at arXiv:math/0210123. 453

254. J. P. May, *The geometry of iterated loop spaces*, Springer-Verlag, Berlin, 1972, Lectures Notes in Mathematics, Vol. 271. 560
255. J. P. May and K. Ponto, *More concise algebraic topology*, Chicago Lectures in Mathematics, University of Chicago Press, Chicago, IL, 2012, Localization, completion, and model categories. 380
256. P. McMullen and E. Schulte, *Abstract regular polytopes*, Encyclopedia of Mathematics and its Applications, vol. 92, Cambridge University Press, Cambridge, 2002. 98
257. G. Melançon and C. Reutenauer, *Lyndon words, free algebras and shuffles*, Canad. J. Math. **41** (1989), no. 4, 577–591. 638
258. P.-L. Méliot, *Representation theory of symmetric groups*, Discrete Mathematics and its Applications, CRC Press, Boca Raton, FL, 2017. 518, 519, 520
259. M. A. Méndez, *Set operads in combinatorics and computer science*, SpringerBriefs in Mathematics, Springer, Cham, 2015. 156, 214, 245
260. W. Michaelis, *Lie coalgebras (with a proof of an analogue of the Poincaré-Birkhoff-Witt theorem)*, Ph.D. thesis, University of Washington, 1974. 585, 640
261. ———, *Lie coalgebras*, Adv. Math. **38** (1980), no. 1, 1–54. 585, 640
262. ———, *The dual Poincaré-Birkhoff-Witt theorem*, Adv. Math. **57** (1985), no. 2, 93–162. 585, 640
263. ———, *Coassociative coalgebras*, Handbook of algebra, Vol. 3, Elsevier/North-Holland, Amsterdam, 2003, pp. 587–788. 585
264. B. Mielnik and J. Plebański, *Combinatorial approach to Baker-Campbell-Hausdorff exponents*, Ann. Inst. H. Poincaré Sect. A (N.S.) **12** (1970), 215–254. 98
265. J. W. Milnor and J. C. Moore, *On the structure of Hopf algebras*, Ann. of Math. (2) **81** (1965), 211–264. 380, 381, 451, 453, 455, 585, 639, 640
266. ———, *On the structure of Hopf algebras (preprint, 1959)*, Collected papers of John Milnor. V. Algebra, Edited by Hyman Bass and T. Y. Lam, American Mathematical Society, Providence, RI, 2010, pp. 7–36. 380, 381, 453, 455, 585, 639, 640
267. H. N. Minh, *Structure of polyzetas and Lyndon words*, Vietnam J. Math. **41** (2013), no. 4, 409–450. 639
268. A. F. Möbius, *Über eine besondere Art von Umkehrung der Reihen*, J. Reine Angew. Math. **9** (1832), 105–123. 97
269. S. Montgomery, *Hopf algebras and their actions on rings*, CBMS Regional Conference Series in Mathematics, vol. 82, American Mathematical Society, Providence, RI, 1993. 380, 453, 455, 516
270. S. Montgomery, M. D. Vega, S. Witherspoon, *Hopf automorphisms and twisted extensions*, J. Algebra Appl. **15** (2016), no. 6, 1650103, 14. 454
271. J. C. Moore, *Mimeographed notes of the algebraic topology seminar, Princeton*, 1957–1958. 380, 455, 639
272. ———, *Algèbres de Hopf universelles*, Séminaire Henri Cartan; Volume 12. Exposé no. 10, Secrétariat mathématique, Paris, 1959–1960, pp. 1–11. 452, 518
273. ———, *Compléments sur les algèbres de Hopf*, Séminaire Henri Cartan; Volume 12. Exposé no. 4, Secrétariat mathématique, Paris, 1959–1960, pp. 1–12. 639
274. P. Moszkowski, *Puissance d'un triplet de formes*, C. R. Acad. Sci. Paris Sér. I Math. **299** (1984), no. 4, 93–95. 95
275. I. M. Musson, *Lie superalgebras and enveloping algebras*, Graduate Studies in Mathematics, vol. 131, American Mathematical Society, Providence, RI, 2012. 585
276. J. Neisendorfer, *Lie algebras, coalgebras and rational homotopy theory for nilpotent spaces*, Pacific J. Math. **74** (1978), no. 2, 429–460. 585
277. K. W. Newman, *Topics in the theory of irreducible Hopf algebras*, Ph.D. thesis, Cornell University, 1970. 452, 639
278. ———, *The structure of free irreducible, cocommutative Hopf algebras*, J. Algebra **29** (1974), 1–26. 452, 639
279. K. W. Newman and D. E. Radford, *The cofree irreducible Hopf algebra on an algebra*, Amer. J. Math. **101** (1979), no. 5, 1025–1045. 452, 639

280. W. D. Nichols, *Bialgebras*, Ph.D. thesis, The University of Chicago, 1975. 453, 585, 638, 640
281. ———, *Pointed irreducible bialgebras*, J. Algebra **57** (1979), no. 1, 64–76. 453, 585, 638, 640
282. W. D. Nichols and M. E. Sweedler, *Hopf algebras and combinatorics*, Umbral calculus and Hopf algebras (Norman, Okla., 1978), Contemp. Math., vol. 6, Amer. Math. Soc., Providence, RI, 1982, pp. 49–84. 516
283. A. Nijenhuis and R. W. Richardson, Jr., *Cohomology and deformations of algebraic structures*, Bull. Amer. Math. Soc. **70** (1964), 406–411. 585
284. J.-C. Novelli, F. Patras, J.-Y. Thibon, *Natural endomorphisms of quasi-shuffle Hopf algebras*, Bull. Soc. Math. France **141** (2013), no. 1, 107–130. 98
285. P. Orlik and H. Terao, *Arrangements of hyperplanes*, Grundlehren der Mathematischen Wissenschaften, vol. 300, Springer-Verlag, Berlin, 1992. 96
286. F. Patras, *Homothéties simpliciales*, Ph.D. thesis, Université Paris 7, 1992. 638
287. ———, *La décomposition en poids des algèbres de Hopf*, Ann. Inst. Fourier (Grenoble) **43** (1993), no. 4, 1067–1087. 453, 454
288. ———, *L'algèbre des descentes d'une bigèbre graduée*, J. Algebra **170** (1994), no. 2, 547–566. 454, 638
289. F. Patras and C. Reutenauer, *Lie representations and an algebra containing Solomon's*, J. Algebraic Combin. **16** (2002), no. 3, 301–314 (2003). 521
290. ———, *On descent algebras and twisted bialgebras*, Mosc. Math. J. **4** (2004), no. 1, 199–216, 311. 157, 245, 712
291. F. Patras and M. Schocker, *Twisted descent algebras and the Solomon-Tits algebra*, Adv. Math. **199** (2006), no. 1, 151–184. 157, 712
292. ———, *Trees, set compositions and the twisted descent algebra*, J. Algebraic Combin. **28** (2008), no. 1, 3–23. 157, 712
293. T. K. Petersen, *Eulerian numbers*, Birkhäuser Advanced Texts: Basler Lehrbücher, Birkhäuser/Springer, New York, 2015, With a foreword by Richard Stanley. 98
294. H. Poincaré, *Sur les groupes continus*, C. R. Acad. Sci. Paris **128** (1899), 1065–1069. 585, 639
295. ———, *Sur les groupes continus*, Trans. Cambr. Philos. Soc. **18** (1900), 220–255. 585, 639
296. S. Poirier and C. Reutenauer, *Algèbres de Hopf de tableaux*, Ann. Sci. Math. Québec **19** (1995), no. 1, 79–90. 520
297. H.-E. Porst, *On categories of monoids, comonoids, and bimonoids*, Quaest. Math. **31** (2008), no. 2, 127–139. 453
298. D. Quillen, *Rational homotopy theory*, Ann. of Math. (2) **90** (1969), 205–295. 454, 639, 640
299. D. E. Radford, *Rationality and the theory of coalgebras*, Ph.D. thesis, The University of North Carolina at Chapel Hill, 1970. 453
300. ———, *A natural ring basis for the shuffle algebra and an application to group schemes*, J. Algebra **58** (1979), no. 2, 432–454. 638
301. ———, *Hopf algebras*, Series on Knots and Everything, vol. 49, World Scientific Publishing Co. Pte. Ltd., Hackensack, NJ, 2012. 380, 455
302. R. Ree, *Lie elements and an algebra associated with shuffles*, Ann. of Math. (2) **68** (1958), 210–220. 454, 517
303. K. Reidemeister, *Einführung in die kombinatorische Topologie*, Chelsea Publishing Co., New York, 1950, Reprint of the 1932 original. 825
304. C. Reutenauer, *Theorem of Poincaré-Birkhoff-Witt, logarithm and symmetric group representations of degrees equal to Stirling numbers*, Combinatoire énumérative (Montreal, Que., 1985/Quebec, Que., 1985), Lecture Notes in Math., vol. 1234, Springer, Berlin, 1986, pp. 267–284. 453
305. ———, *Free Lie algebras*, London Mathematical Society Monographs. New Series, vol. 7, The Clarendon Press, Oxford University Press, New York, 1993, Oxford Science Publications. 98, 453, 454, 517, 518, 519, 585, 638

306. ______, *Free Lie algebras*, Handbook of algebra, Vol. 3, Elsevier/North-Holland, Amsterdam, 2003, pp. 887–903. 585, 638
307. E. Riehl, *Category theory in context*, Aurora, Dover Publications, 2016. xiv, 825
308. J. Riordan, *Combinatorial identities*, Robert E. Krieger Publishing Co., Huntington, N.Y., 1979, Reprint of the 1968 original. 97
309. S. Roman, *An introduction to the language of category theory*, Compact Textbooks in Mathematics, Birkhäuser/Springer, Cham, 2017. xiv, 825
310. M. Ronan, *Lectures on buildings*, University of Chicago Press, Chicago, IL, 2009, Updated and revised. 98
311. L. E. Ross, *On representations and cohomology of graded Lie algebras*, Ph.D. thesis, University of California, Berkeley, 1964. 585
312. ______, *Representations of graded Lie algebras*, Trans. Amer. Math. Soc. **120** (1965), 17–23. 585
313. G.-C. Rota, *On the foundations of combinatorial theory. I. Theory of Möbius functions*, Z. Wahrscheinlichkeitstheorie und Verw. Gebiete **2** (1964), 340–368 (1964). 97
314. B. E. Sagan, *The symmetric group*, second ed., Graduate Texts in Mathematics, vol. 203, Springer-Verlag, New York, 2001, Representations, combinatorial algorithms, and symmetric functions. 518
315. F. V. Saliola, *On the quiver of the descent algebra*, J. Algebra **320** (2008), no. 11, 3866–3894. 95
316. ______, *The face semigroup algebra of a hyperplane arrangement*, Canad. J. Math. **61** (2009), no. 4, 904–929. 95
317. S. V. Sam and A. Snowden, *Introduction to twisted commutative algebras*, available at arXiv:1209.5122. 214, 452
318. H. Samelson, *Beiträge zur Topologie der Gruppen-Mannigfaltigkeiten*, Ann. of Math. (2) **42** (1941), 1091–1137. 380, 639
319. M. Scheunert, *The theory of Lie superalgebras*, Lecture Notes in Mathematics, vol. 716, Springer, Berlin, 1979, An introduction. 585
320. W. R. Schmitt, *Incidence Hopf algebras*, J. Pure Appl. Algebra **96** (1994), no. 3, 299–330. 455, 815
321. M. Schocker, *The module structure of the Solomon-Tits algebra of the symmetric group*, J. Algebra **301** (2006), no. 2, 554–586. 95, 98
322. J. Schur, *Bemerkungen zur Theorie der beschränkten Bilinearformen mit unendlich vielen Veränderlichen*, J. Reine Angew. Math. **140** (1911), 1–28. 453
323. J.-P. Serre, *Lie algebras and Lie groups*, Lecture Notes in Mathematics, vol. 1500, Springer-Verlag, Berlin, 2006, 1964 lectures given at Harvard University, Corrected fifth printing of the second (1992) edition. 640
324. G. Sjödin, *Hopf algebras and derivations*, J. Algebra **64** (1980), no. 1, 218–229. 640
325. A. Skowroński and K. Yamagata, *Frobenius algebras. I*, EMS Textbooks in Mathematics, European Mathematical Society (EMS), Zürich, 2011, Basic representation theory. 380
326. V. A. Smirnov, *On the cochain complex of topological spaces*, Mat. Sb. (N.S.) **115(157)** (1981), no. 1, 146–158, 160. 214, 560
327. ______, *Simplicial and operad methods in algebraic topology*, Translations of Mathematical Monographs, vol. 198, American Mathematical Society, Providence, RI, 2001, Translated from the Russian manuscript by G. L. Rybnikov. 560
328. W. H. Smoke, *Differential operators on homogeneous spaces*, Ph.D. thesis, University of California, Berkeley, 1965. 639
329. ______, *Invariant differential operators*, Trans. Amer. Math. Soc. **127** (1967), 460–494. 639
330. L. Solomon, *The orders of the finite Chevalley groups*, J. Algebra **3** (1966), 376–393. 96, 97
331. ______, *The Burnside algebra of a finite group*, J. Combinatorial Theory **2** (1967), 603–615. 95

332. ———, *On the Poincaré-Birkhoff-Witt theorem*, J. Combinatorial Theory **4** (1968), 363–375. 98, 640
333. ———, *A Mackey formula in the group ring of a Coxeter group*, J. Algebra **41** (1976), no. 2, 255–264. 95
334. D. I. Spivak, *Category theory for the sciences*, MIT Press, Cambridge, MA, 2014. 825
335. R. P. Stanley, *Hyperplane arrangements, parking functions and tree inversions*, Mathematical essays in honor of Gian-Carlo Rota (Cambridge, MA, 1996), Progr. Math., vol. 161, Birkhäuser Boston, Boston, MA, 1998, pp. 359–375. 97
336. ———, *Enumerative combinatorics. Vol. 2*, Cambridge Studies in Advanced Mathematics, vol. 62, Cambridge University Press, Cambridge, 1999, With a foreword by Gian-Carlo Rota. 518, 519, 520, 521
337. ———, *An introduction to hyperplane arrangements*, Geometric combinatorics, IAS/Park City Math. Ser., vol. 13, Amer. Math. Soc., Providence, RI, 2007, pp. 389–496. 97
338. ———, *Enumerative combinatorics. Volume 1*, second ed., Cambridge Studies in Advanced Mathematics, vol. 49, Cambridge University Press, Cambridge, 2012. 97
339. J. D. Stasheff, *Homotopy associativity of H-spaces. I*, Trans. Amer. Math. Soc. **108** (1963), 275–292; II, 293–312. 560
340. R. Steinberg, *Endomorphisms of linear algebraic groups*, Memoirs of the American Mathematical Society, No. 80, American Mathematical Society, Providence, R.I., 1968. 97
341. C. R. Stover, *The equivalence of certain categories of twisted Lie and Hopf algebras over a commutative ring*, J. Pure Appl. Algebra **86** (1993), no. 3, 289–326. 157, 272, 297, 298, 712, 757
342. M. E. Sweedler, *Cocommutative Hopf algebras with antipode*, Ph.D. thesis, Massachusetts Institute of Technology, 1966. 453, 640
343. ———, *Hopf algebras with one grouplike element*, Trans. Amer. Math. Soc. **127** (1967), 515–526. 638
344. ———, *The Hopf algebra of an algebra applied to field theory*, J. Algebra **8** (1968), 262–276. 453
345. ———, *Hopf algebras*, Mathematics Lecture Note Series, W. A. Benjamin, Inc., New York, 1969. xv, 380, 451, 453, 455, 517, 638, 640
346. E. J. Taft, *Combinatorial sequences as sequences of divided powers*, Proceedings of the thirteenth Southeastern conference on combinatorics, graph theory and computing (Boca Raton, Fla., 1982), vol. 36, 1982, pp. 23–26. 516
347. M. Takeuchi, *Free Hopf algebras generated by coalgebras*, J. Math. Soc. Japan **23** (1971), 561–582. 455
348. ———, *Survey of braided Hopf algebras*, New trends in Hopf algebra theory (La Falda, 1999), Contemp. Math., vol. 267, Amer. Math. Soc., Providence, RI, 2000, pp. 301–323. 381
349. S. Tauber, *On multinomial coefficients*, Amer. Math. Monthly **70** (1963), 1058–1063. 97
350. J.-Y. Thibon, *Lectures on noncommutative symmetric functions*, Interaction of combinatorics and representation theory, MSJ Mem., vol. 11, Math. Soc. Japan, Tokyo, 2001, pp. 39–94. 519
351. J.-Y. Thibon and B.-C.-V. Ung, *Quantum quasi-symmetric functions and Hecke algebras*, J. Phys. A **29** (1996), no. 22, 7337–7348. 521
352. A. Thomas, *Geometric and topological aspects of Coxeter groups and buildings*, Zurich Lectures in Advanced Mathematics, European Mathematical Society (EMS), Zürich, 2018. 98
353. J. Tits, *Buildings of spherical type and finite BN-pairs*, Lecture Notes in Mathematics, Vol. 386, Springer-Verlag, Berlin-New York, 1974. 95
354. ———, *Two properties of Coxeter complexes*, J. Algebra **41** (1976), no. 2, 265–268, Appendix to "A Mackey formula in the group ring of a Coxeter group" (J. Algebra **41** (1976), no. 2, 255–264) by Louis Solomon. 95

355. R. G. Underwood, *Fundamentals of Hopf algebras*, Universitext, Springer, Cham, 2015. 380

356. A.-T. Vandermonde, *Mémoire sur des irrationnelles de différens ordres avec une application au cercle*, Mem. Acad. Roy. Sci. Paris (1772), 489–498. 97

357. A. Varchenko, *Bilinear form of real configuration of hyperplanes*, Adv. Math. **97** (1993), no. 1, 110–144. 96

358. M. Ward, *The algebra of lattice functions*, Duke Math. J. **5** (1939), no. 2, 357–371. 97

359. L. Weisner, *Abstract theory of inversion of finite series*, Trans. Amer. Math. Soc. **38** (1935), no. 3, 474–484. 97

360. R. M. Weiss, *The structure of spherical buildings*, Princeton University Press, Princeton, NJ, 2003. 98

361. E. Witt, *Treue Darstellung Liescher Ringe*, J. Reine Angew. Math. **177** (1937), 152–160. 585, 639

362. ———, *Spiegelungsgruppen und Aufzählung halbeinfacher Liescher Ringe*, Abh. Math. Sem. Hansischen Univ. **14** (1941), 289–322. 95, 97

363. G. C. Wraith, *Hopf algebras over Hopf algebras*, Ann. Mat. Pura Appl. (4) **76** (1967), 149–163. 453

364. T. Zaslavsky, *Facing up to arrangements: face-count formulas for partitions of space by hyperplanes*, Mem. Amer. Math. Soc. **1** (1975), no. issue 1, 154, vii+102. 98

365. A. V. Zelevinsky, *Representations of finite classical groups*, Lecture Notes in Mathematics, vol. 869, Springer-Verlag, Berlin, 1981, A Hopf algebra approach. 518

366. G. M. Ziegler, *Algebraic combinatorics of hyperplane arrangements*, Ph.D. thesis, Massachusetts Institute of Technology, 1987. 95, 96

367. M. Zisman, *Espaces de Hopf, algèbres de Hopf*, Séminaire Henri Cartan; Volume 12. Exposé no. 2, Secrétariat mathématique, Paris, 1959–1960, pp. 1–17. 381

List of Notations

Important notations used in the book are listed below, grouped according to topic. We begin with standard notations related to number systems, posets, vector spaces, categories. These are followed by notations more specific to this book. We also indicate chapters, sections, equation numbers where these notations can be located. In this book, examples, exercises, observations, remarks, tables, figures and end-of-chapter notes are in small font. Examples and proofs are terminated by the □ symbol.

Abbreviations

iff	if and only if
lhs, rhs	left-hand side, right-hand side
wrt	with respect to

Number systems

$[n]$	set $\{1, 2, \dots, n\}$
$\mathbb{N}$	set of nonnegative integers $\{0, 1, 2, \dots\}$
$\mathbb{Z}$, $\mathbb{Z}_2$	set of integers, set of integers modulo 2
$\mathbb{Q}$	set of rational numbers
$\mathbb{R}$	set of real numbers
$\Bbbk$	field
$\mathbb{Z}[q]$	polynomial ring over $\mathbb{Z}$ in the variable q

Elements of a field $\Bbbk$ are called scalars. The symbol $\nmid$ stands for "does not divide". The condition $\operatorname{char}(\Bbbk) \nmid |W|$ appears at several places in the book. It means that the characteristic of the field $\Bbbk$ does not divide the order of the Coxeter group W. This condition replaces the characteristic zero assumption made in classical Hopf theory while proving structure theorems.

Posets

$x \leq y$	x is smaller than y, or, y is greater than x
$x < y$	x is strictly smaller than y, or, y is strictly greater than x
$x \lessdot y$	x is covered by y, or, y covers x
$\operatorname{rk}(P)$	rank of the graded poset P
$\operatorname{rk}(x)$	rank of the element x in a graded poset
$\wedge$, $\vee$	meet, join (whenever they exist)

Most sets associated to a reflection arrangement that we consider such as the set of faces, flats, face-types, flat-types, and so on carry additional structure of a graded poset (to which the above notations apply).

Vector spaces

We write U, V, W for vector spaces; they are always over a fixed field $\Bbbk$.

V^*	dual of the vector space V
V^{sub}, V_{quo}	a subspace of V, a quotient space of V
$\mathrm{Hom}_{\Bbbk}(V, W)$	space of $\Bbbk$-linear maps from V to W
$\mathrm{End}_{\Bbbk}(V)$	space of $\Bbbk$-linear endomorphisms of V
$V \otimes W$	tensor product of V and W over $\Bbbk$
$\ker(f)$, $\mathrm{coker}(f)$	kernel, cokernel of a linear map f
$\mathrm{im}(f)$, $\mathrm{coim}(f)$	image, coimage of a linear map f

The notations V^{sub} and V_{quo} are not standard; they are convenient to describe limits and colimits of functors into the category of vector spaces; see for instance (8.121), (8.123), (11.25), (11.32).

Categories

Notations for some standard categories are as follows.

Set	category of sets
Vec	category of vector spaces over a fixed field $\Bbbk$
Monoid	category of monoids
Group	category of groups
Cat	category of small categories and functors between them
W-Mod	category of left W-modules for the Coxeter group W

We write C, D for categories, and $\mathcal{F}$, $\mathcal{G}$ for functors. For objects a and b in C, we write $\mathsf{C}(a, b)$ for the set of morphisms from a to b.

C^{op}	category opposite to C
$[\mathsf{C}, \mathsf{D}]$	category of functors from C to D
inc, *frg*	inclusion functor, forgetful functor
$\mathsf{C} \underset{\mathcal{G}}{\overset{\mathcal{F}}{\rightleftarrows}} \mathsf{D}$	adjunction with left adjoint $\mathcal{F}$ and right adjoint $\mathcal{G}$
C_W	one-object category of the Coxeter group W

The notation C^{op} mainly occurs in discussions related to duality. The categories of Coxeter species, Coxeter dispecies, Coxeter spaces are functor categories of the form $[\mathsf{C}, \mathsf{Vec}]$. More examples of this nature are given in Section 2.9 and Section 8.11. The main examples of adjunctions in this book are shown in Summaries 3.1, 9.1 and in Summaries 16.1, 16.2, 16.3, 16.4, 16.5.

$\mathcal{F} \rtimes \mathsf{C}$	Grothendieck construction on $\mathcal{F} : \mathsf{C} \to \mathsf{Cat}$
$\mathsf{el}_{\mathsf{C}}(\mathcal{F})$, $\int_{\mathsf{C}} \mathcal{F}$	category of elements for $\mathcal{F} : \mathsf{C} \to \mathsf{Set}$
$\mathsf{F} \rtimes G$	semidirect product of the monoid G by the category F
$\mathsf{el}_G(X)$	category of elements for the action of the monoid G on the set X

The above notations appear in Appendix A.2.

Monoids and algebras

The book focuses on particular monoids such as the Tits monoid, and on particular algebras such as the lune-incidence algebra. We have specific notations to denote these important objects. If required, we denote a general monoid by M or G and an algebra by A. All our algebras are over a fixed field $\Bbbk$.

Coxeter groups and reflection arrangements

The notations below all appear in Chapter 1.

Reflection arrangements.

$\mathcal{A}$	hyperplane arrangement or reflection arrangement
$\mathcal{A}^{\mathrm{X}}$, $\mathcal{A}_{\mathrm{X}}$, $\mathcal{A}_{\mathrm{Y}}^{\mathrm{X}}$	arrangements under and over flats of $\mathcal{A}$
$\boldsymbol{\alpha} = (\mathcal{A}, C_r)$	pointed reflection arrangement with reference chamber C_r
$\mathrm{rk}(\mathcal{A})$	rank of the arrangement $\mathcal{A}$
$\mathcal{C}^n$	coordinate arrangement of rank n

Coxeter groups.

W, (W, S)	Coxeter group, Coxeter system
W_F, W_T, W_{X}, W_{L}	stabilizer subgroups of W
$\Bbbk W$	group algebra of W over the field $\Bbbk$
$\Bbbk W \to \Bbbk$	augmentation map (1.65)
M, N	left W-modules
$w \cdot m$ or $w(m)$	action of $w \in W$ on $m \in M$
M^*	dual of the W-module M (Section 1.5)
M_-	W-module M tensored with the sign representation of W
M_W and M^W	W-coinvariants and W-invariants of M

We often refer to the trivial and sign representations of W without employing any specific notation for them.

Faces, flats, lunes. Basic notations related to faces, flats, lunes, and so on are given in Tables 1.1 and 1.2. Additional related notations from Sections 1.1 and 1.2 are given below.

O	central face
$\overline{F}$, $\overline{\mathrm{L}}$	face opposite to F, lune opposite to L
$\overline{T}$, $Z\overline{T}$	face-type opposite to T, face-type opposite to T local to Z
K/F and $F\overline{K}/F$	pair of opposite faces in the star of F
$F \wedge G$, FG	meet of F and G, Tits product of F and G
$\mathrm{X} \wedge \mathrm{Y}$, $\mathrm{X} \vee \mathrm{Y}$	meet of X and Y, join of X and Y
$\bot$, $\top$	minimum flat, maximum flat
$\lambda_{\min}$, $\lambda_{\max}$	minimum flat-type, maximum flat-type
$\lvert\lambda\rvert$	number associated to the flat-type λ (Definition 1.8)
ξ	invariant family of scalars (Definition 1.10)
$\mathrm{s}(F)$, $\mathrm{s}(H, G)$	support of the face F, support of the nested face (H, G)
$\mathrm{b}(\mathrm{L})$, $\mathrm{c}(\mathrm{L})$, $\mathrm{sk}(\mathrm{L})$	base, case, slack of the lune L
$\mathrm{L} \circ \mathrm{M}$	composite of morphisms L and M in the category of lunes
$\mathrm{X} \cdot \mathrm{L}$	action of the flat X on the lune L
$\mathrm{b}(\mathrm{L}, \mathrm{L}')$, $\mathrm{c}(\mathrm{L}, \mathrm{L}')$	base, case of the bilune $(\mathrm{L}, \mathrm{L}')$
$(\mathrm{L}, \mathrm{L}') \circ (\mathrm{M}, \mathrm{M}')$	composite of morphisms in the category of bilunes
s, t	support map, type map
b, c	base map, case map

The support map s and type map t are defined in a variety of related contexts, see (1.4), (1.6), (1.7), (1.8).

Gallery distance. These notations are from Section 1.4.

C -- D -- E	minimal gallery from C to E passing through D
F -- G -- H	minimal gallery between faces F, G, H of the same support
$\text{dist}(C, D)$	gallery distance between chambers C and D
$\text{dist}(F, G)$	distance between faces F and G
$\text{dist}(\mathrm{L}, \mathrm{M})$	distance between lunes L and M with the same base
$d(C, D)$	W-valued gallery distance between C and D, (1.12)
$l(w)$	length of the element $w \in W$, (1.11)
$w_{\emptyset}$, w_T	longest element in W, longest element in the subgroup W_T
$\text{dist}(T, \overline{T})$	distance between a face-type T and its opposite $\overline{T}$, (1.14)
$u \leq v$	weak left Bruhat order on W, (1.16)
v	abstract distance function on faces
v_q	q-distance function on faces, (1.10)

Descents and shuffles. These notations are from Section 1.6.

$\text{des}(C, D)$	descent of D wrt C, (1.17)
$\text{des}(w)$	descent of $w \in W$
$\text{gdes}(C, D)$	global descent of D wrt C, (1.18)
$\text{gdes}(w)$	global descent of $w \in W$
$\text{des}(F, G)$, $\overline{\text{des}}(F, G)$	two notions of descent of G wrt F, (1.20)
Sh_T, Sh_T^*	set of T-shuffles, (1.21), (1.22), set of T-deshuffles, (1.23)
$\theta(T)$	longest T-shuffle for a face-type T

Groupoid of biface-types. The groupoid of biface-types $\boldsymbol{\alpha}$-$\overline{\mathtt{Cox}}$ is introduced in Section 1.7 with an alternative viewpoint given in Section 8.1.

σ, τ, ρ	morphisms in $\boldsymbol{\alpha}$-$\overline{\mathtt{Cox}}$
$l(\sigma)$	length of σ, (1.29)
$\beta_{\overline{T},T}$	a specific morphism from a face-type T to its opposite $\overline{T}$
$\beta_{Z\overline{T},T}$	local version of $\beta_{\overline{T},T}$

Face-type enumeration. Face-type enumeration is discussed in Section 1.8. Polynomials occurring in face-type enumeration such as $d_T(q)$, its local version $d_{T/Z}(q)$, and their generalizations $a^{TUT'}(q)$ and $a^{TUT'U'}(q)$ and their local versions $a_Z^{TUT'}(q)$ and $a_Z^{TUT'U'}(q)$ are summarized in Table 1.4. We also mention the Varchenko matrices A_λ, one for each flat-type λ, constructed out of the polynomials $c^{TT'}(q)$.

Incidence algebras. Incidence algebras, zeta and Möbius functions and related notions are introduced in Sections 1.10 and 1.11. The main notations are summarized in Table 1.5. The algebra morphisms bc, p, q between incidence algebras are shown in Table 1.6.
The Solomon coefficients $\text{sln}_A^{G,H}$ and sln_Z^{α} are introduced in Section 1.11.3. They depend on the choice of a noncommutative Möbius function.

Birkhoff algebra, Tits algebra, Janus algebra. The Birkhoff algebra, Tits algebra, Janus algebra, and their Coxeter analogues are introduced in Section 1.12 and Section 1.14. The main notations are summarized in Table 1.7 and Table 1.8. The element $\mathtt{E}_{\mathrm{X}}$ is an eulerian idempotent of the Tits algebra, with $\mathtt{E}_{\perp}$ the first eulerian idempotent.

Braid arrangements. Braid arrangements are a family of reflection arrangements indexed by finite sets. They are discussed in Section 1.16. See also Table 1.9.

$\mathcal{B}^J$	braid arrangement on the finite set J

$F \vDash J$, $\mathrm{X} \vdash J$	F is a composition of J, X is a partition of J
$\boldsymbol{\beta}^n$	braid arrangement on the set $[n]$ with reference chamber $1\|\ldots\|n$
S_n	symmetric group on n letters
$\alpha \vDash n$, $\lambda \vdash n$	α is a composition of n, λ is a partition of n
$(\boldsymbol{\zeta}_\gamma)$, $(\boldsymbol{\mu}_\gamma)$	zeta series, Möbius series
$(\mathtt{Q}_{(n)})$	Zie series
$(n)_q!$	q-factorial of the positive integer n

Coxeter species and Coxeter spaces

Coxeter species and Coxeter spaces are studied in Parts I and II. Some basic notations are listed in Table III in the main introduction.

Coxeter species and Coxeter spaces. Coxeter species are introduced in Section 2.3, and Coxeter (co, bi)monoids in Sections 2.4 and 2.5. We write β for support morphisms, w for type morphisms, μ for product, Δ for coproduct.

$\mathsf{p}+\mathsf{q}$	direct sum of Coxeter species p and q, (2.22)
p^*	dual of the Coxeter species p
p^k	k-th Cauchy power of the Coxeter species p (Section 3.1)
$\mathsf{p}^{\overline{k}}$	k-th commutative Cauchy power of the Coxeter species p
a_t, c_t	a as a trivial Coxeter monoid, c as a trivial Coxeter comonoid
a_{ab}, c^{coab}	abelianization of a, coabelianization of c (Section 3.4)
a_{-ab}, c^{-coab}	signed abelianization of a, signed coabelianization of c
$\mathsf{h}^{\mathrm{cop}}$, ${}^{\mathrm{op}}\mathsf{h}$	op and cop constructions on h (Section 2.11.3)
S_{h}	antipode of the Coxeter bimonoid h (Section 3.14)

Coxeter spaces are introduced in Section 8.4, and Coxeter (co, bi)algebras in Sections 8.5, 8.6, 8.7. In this context, we use capital letters instead of small as is done for vector spaces and algebras.

$\mathrm{V}+\mathrm{W}$	direct sum of Coxeter spaces V and W, (8.31)
V^*	dual of the Coxeter space V
V^k	k-th Cauchy power of the Coxeter space V (Section 9.1)
$\mathrm{V}^{\underline{k}}$, $\mathrm{V}^{\overline{k}}$	k-th commutative Cauchy powers of the Coxeter space V
A_t, C_t	A as a trivial Coxeter algebra, C as a trivial Coxeter coalgebra
A_{ab}, C^{coab}	abelianization of A, coabelianization of C (Section 9.4)
A_{-ab}, C^{-coab}	signed abelianization of A, signed coabelianization of C
$\mathrm{H}^{\mathrm{cop}}$, ${}^{\mathrm{op}}\mathrm{H}$	op and cop constructions on H (Section 8.13.4)
S_{H}	antipode of the Coxeter bialgebra H (Section 9.14)

Categories related to Coxeter species and Coxeter spaces. Notations for the base categories for Coxeter species, and for Coxeter monoids and so on are summarized in Tables 2.3 and 2.7. Similarly, notations for the base categories for Coxeter spaces, and for Coxeter algebras and so on are summarized in Tables 8.3 and 8.8. The general convention is to write $\mathtt{Cox}$ for Coxeter species and $\overline{\mathtt{Cox}}$ for Coxeter spaces.

The category of Coxeter species is denoted $\mathcal{A}$-$\mathtt{CSp}$. Notations for the categories of Coxeter monoids, comonoids, and so on are summarized in Tables 2.4, 2.5, 2.6. Signed analogues follow the same notations with an additional (-1) in front. The category of Coxeter spaces is denoted either $\mathcal{A}$-$\overline{\mathtt{C}}\mathtt{Sp}$ or $\boldsymbol{\alpha}$-$\overline{\mathtt{C}}\mathtt{Sp}$. Notations for the categories of Coxeter algebras, coalgebras, and so on are summarized in Tables 8.5, 8.6, 8.7. The general convention is to write $\mathtt{CSp}$ for Coxeter species and $\overline{\mathtt{C}}\mathtt{Sp}$ for

Coxeter spaces. In [**11**], we used Sp for species and Hyp for its base category; these notations appear briefly in this book when we compare species and Coxeter species.

Some important functors are listed below.

$\Bbbk(-)$	linearization functor from Coxeter set-species to Coxeter species, (2.27)
trv	trivial Coxeter (co)monoid and (co)algebra functors, (3.18) and (9.16)
$(-)_{-}$	u-signature functors on Coxeter species and spaces, (2.70) and (8.99)
$(-)^{-}$	o-signature functor on Coxeter species, (2.73)
$(-)_M$	decoration functor on Coxeter species, (4.117)

Primitive and decomposable filtrations. Primitive and decomposable filtrations of Coxeter comonoids and Coxeter monoids are introduced in Section 3.2. See Table 3.1.

$\mathcal{P}(\mathsf{c})$	primitive part of the Coxeter comonoid c
$\mathcal{P}_k(\mathsf{c})$	k-th term of the primitive filtration of the Coxeter comonoid c
$\mathcal{D}(\mathsf{a})$	decomposable part of the Coxeter monoid a
$\mathcal{D}_k(\mathsf{a})$	k-th term of the decomposable filtration of the Coxeter monoid a
$\mathcal{Q}(\mathsf{a})$	indecomposable part of the Coxeter monoid a
pq_{h}	map from $\mathcal{P}(\mathsf{h})$ to $\mathcal{Q}(\mathsf{h})$ of a Coxeter q-bimonoid h, (3.106)

Primitive and decomposable filtrations of Coxeter coalgebras and Coxeter algebras are introduced in Section 9.2. See Table 9.1. In this context, we write $\overline{\mathcal{P}}$, $\overline{\mathcal{D}}$, $\overline{\mathcal{Q}}$ instead of $\mathcal{P}$, $\mathcal{D}$, $\mathcal{Q}$.

$\overline{\mathcal{P}}(\mathrm{C})$	primitive part of the Coxeter coalgebra C
$\overline{\mathcal{P}}_k(\mathrm{C})$	k-th term of the primitive filtration of the Coxeter coalgebra C
$\overline{\mathcal{D}}(\mathrm{A})$	decomposable part of the Coxeter algebra A
$\mathcal{D}_k(\mathrm{A})$	k-th term of the decomposable filtration of the Coxeter algebra A
$\overline{\mathcal{Q}}(\mathrm{A})$	indecomposable part of the Coxeter algebra A
pq_{H}	map from $\overline{\mathcal{P}}(\mathrm{H})$ to $\overline{\mathcal{Q}}(\mathrm{H})$ of a Coxeter q-bialgebra H, (9.106)

Universal constructions. Universal constructions on Coxeter species are introduced in Section 3.3, see Table 3.2 for constructions of Coxeter q-bimonoids and Summary 3.1 for a complete list. Some of these constructions are highlighted below.

$\mathcal{T}(\mathsf{p})$	free Coxeter monoid on the Coxeter species p
$\mathcal{T}_q(\mathsf{c})$	free Coxeter q-bimonoid on the Coxeter comonoid c
$\mathcal{T}^{\vee}(\mathsf{p})$	cofree Coxeter comonoid on the Coxeter species p
$\mathcal{T}_q^{\vee}(\mathsf{a})$	cofree Coxeter q-bimonoid on the Coxeter monoid a
$\mathcal{S}(\mathsf{p})$	free commutative Coxeter monoid on the Coxeter species p
$\mathcal{S}^{\vee}(\mathsf{p})$	cofree cocommutative Coxeter comonoid on the Coxeter species p

Universal constructions on Coxeter spaces are introduced in Section 9.3, see Table 9.2 and Summary 9.1. The same constructions as above, but for Coxeter spaces, are written below. In this context, we write $\overline{\mathcal{T}}$ and $\overline{\mathcal{S}}$ instead of $\mathcal{T}$ and $\mathcal{S}$.

$\overline{\mathcal{T}}(\mathrm{V})$	free Coxeter algebra on the Coxeter space V
$\overline{\mathcal{T}}_q(\mathrm{C})$	free Coxeter q-bialgebra on the Coxeter coalgebra C
$\overline{\mathcal{T}}^{\vee}(\mathrm{V})$	cofree Coxeter coalgebra on the Coxeter space V
$\overline{\mathcal{T}}_q^{\vee}(\mathrm{A})$	cofree Coxeter q-bialgebra on the Coxeter algebra A
$\overline{\mathcal{S}}(\mathrm{V})$	free commutative Coxeter algebra on the Coxeter space V
$\overline{\mathcal{S}}^{\vee}(\mathrm{V})$	cofree cocommutative Coxeter coalgebra on the Coxeter space V

Hadamard product. The Hadamard product of Coxeter species (and all notations below) are introduced in Sections 3.5, 3.6, 3.7. Also see Table 3.3.

$\mathsf{p} \times \mathsf{q}$	Hadamard product of Coxeter species p and q, (3.53), (3.55)
$\mathsf{p} \star \mathsf{q}$, $\mathsf{p} \diamond \mathsf{q}$	operations on Coxeter species p and q, (3.57), (3.58)
$\mathrm{hom}^{\times}(\mathsf{p}, \mathsf{q})$	internal hom for the Hadamard product of Coxeter species, (3.59)
$\mathrm{hom}^{\times}(\mathsf{c}, \mathsf{a})$	convolution Coxeter monoid, (3.64), (3.65)
$\mathrm{hom}^{\times}(\mathsf{a}, \mathsf{c})$	coconvolution Coxeter comonoid, (3.66)
$\mathrm{hom}^{\times}(\mathsf{h}, \mathsf{k})$	biconvolution Coxeter bimonoid
$\mathrm{end}^{\times}(\mathsf{h})$	specialization of the biconvolution Coxeter bimonoid
$\mathcal{C}(\mathsf{c}, \mathsf{d})$	internal hom for the Hadamard product of Coxeter comonoids
$\overline{\mathcal{C}}(\mathsf{a}, \mathsf{b})$	universal measuring Coxeter comonoid

The Hadamard product of Coxeter spaces are introduced in Sections 9.5, 9.6, 9.7. Also see Table 9.3. The notations are exactly the same as above. We point out that there is no biconvolution Coxeter bialgebra.

$\mathrm{V} \times \mathrm{W}$	Hadamard product of Coxeter spaces V and W, (9.59)
$\mathrm{hom}^{\times}(\mathrm{V}, \mathrm{W})$	internal hom for the Hadamard product of Coxeter spaces, (9.60)
$\mathrm{hom}^{\times}(\mathrm{C}, \mathrm{A})$	convolution Coxeter algebra, (9.65)
$\mathrm{hom}^{\times}(\mathrm{A}, \mathrm{C})$	coconvolution Coxeter coalgebra, (9.66)
$\mathcal{C}(\mathrm{C}, \mathrm{D})$	internal hom for the Hadamard product of Coxeter coalgebras
$\overline{\mathcal{C}}(\mathrm{A}, \mathrm{B})$	universal measuring Coxeter coalgebra

Exponential and logarithm. Actions of incidence algebras on maps of Coxeter species from a Coxeter comonoid to a Coxeter monoid are introduced in Section 3.8. Similar ideas for series of a Coxeter species are introduced in Section 3.10. This leads to various avatars of the exponential and logarithm. See Table 3.4.

$\mathbf{exp}(f)$, $\mathbf{log}(f)$	exponential, logarithm of a map of Coxeter species $f : \mathsf{c} \to \mathsf{a}$
$\mathbf{log}(\mathrm{id})$	logarithm of the identity map on a Coxeter bimonoid h
$\overline{\mathbf{exp}}$, $\overline{\mathbf{log}}$	opposite of **exp**, opposite of **log**
$\mathbf{exp}_q(f)$, $\mathbf{log}_q(f)$	q-exponential, q-logarithm of f for q not a root of unity
$\mathbf{log}_q(\mathrm{id})$	q-logarithm of the identity map on a Coxeter q-bimonoid h

$\mathscr{S}(\mathsf{p})$	series of the Coxeter species p
$\mathscr{E}^o(\mathsf{a})$	equi-exponential series of the Coxeter monoid a
$\mathscr{X}(\mathsf{a})$, $\mathscr{E}(\mathsf{a})$	linear, exponential series of the Coxeter monoid a
$\mathscr{P}(\mathsf{c})$, $\mathscr{G}(\mathsf{c})$	primitive, group-like series of the Coxeter comonoid c

The same notations as above are used for Coxeter spaces, so we do not repeat them. The relevant material is introduced in Sections 9.8 and 9.10, see Table 9.4. We point out that there is no notion of equi-exponential series for Coxeter algebras.

Examples. Examples of Coxeter species below are introduced in Chapter 4. Many of them carry a Coxeter bimonoid structure and also have q-analogues. See Tables 4.1 and 4.2 for entries related to E.

x	Coxeter species characteristic of chambers
E	exponential Coxeter species
E^-, E_-	signed analogues of E (over and under versions)
E_M	exponential Coxeter species decorated by the W-module M
E^-_M	decorated signed exponential Coxeter species
Γ, Π, Σ	Coxeter species of chambers, flats, faces
Lie, Zie	Coxeter Lie species, Coxeter Zie species
$\widehat{\mathsf{Q}}$, $\widehat{\Lambda}$	Coxeter species of top-nested faces, top-lunes

$\mathbb{I}\Gamma$	Coxeter species of pairs of chambers
$\mathcal{C}(\Gamma, \Gamma)$	Coxeter species of chamber maps

Examples of Coxeter spaces below are introduced in Chapter 10. Many of them carry a Coxeter bialgebra structure and also have q-analogues. See Tables 10.1, 10.2, 10.3, 10.4, 10.5, 10.6, 10.7 for detailed summaries of some of these.

X	Coxeter space characteristic of chambers
$\mathbb{k}[x]$, $\mathbb{k}\{x\}$	Coxeter bialgebras of polynomials and divided powers
$\mathbb{k}[dx]$	signed Coxeter bialgebra of dual numbers
$\overline{\mathcal{T}}(M)$, $\overline{\mathcal{T}}^{\vee}(M)$	tensor and shuffle Coxeter bialgebras
$\overline{\mathcal{S}}(M)$, $\overline{\mathcal{S}}^{\vee}(M)$	symmetric Coxeter bialgebras
$\overline{\mathcal{E}}(M)$, $\overline{\mathcal{E}}^{\vee}(M)$	exterior signed Coxeter bialgebras
$\overline{\Sigma}$, $\overline{\Sigma}^*$	Coxeter bialgebras of face-types dual to each other
$\overline{\Pi}$, $\overline{\Pi}^*$	self-dual Coxeter bialgebra of flat-types
$\overline{\mathbb{I}\Gamma}$, $\overline{\mathbb{I}\Gamma}^*$	self-dual Coxeter bialgebra of symmetries

Important examples of Joyal bimonoids such as the exponential Joyal bimonoid and Joyal bimonoids of linear orders, set compositions (along with their generalizations to Coxeter bimonoids) are shown in Table 4.3. Similarly, examples of graded Hopf algebras such as the Hopf algebra of polynomials $\mathbb{k}[x]$, tensor Hopf algebra $\mathcal{T}(V)$, Hopf algebra of quasisymmetric functions QSym (along with their generalizations to Coxeter bialgebras) are shown in Tables 10.8, 10.9, 10.10.

Monads on the category of Coxeter species. The (co, bi)monads on the category of Coxeter species listed below are introduced in Section 2.10.

$\mathcal{T}$	monad for Coxeter monoids
$\mathcal{S}$	monad for commutative Coxeter monoids
$\mathcal{E}$	monad for signed commutative Coxeter monoids
$\mathcal{T}^{\vee}$	comonad for Coxeter comonoids
$\mathcal{S}^{\vee}$	comonad for cocommutative Coxeter comonoids
$\mathcal{E}^{\vee}$	comonad for signed cocommutative Coxeter comonoids
$(\mathcal{T}, \mathcal{T}^{\vee}, \lambda)$	bimonad for Coxeter bimonoids
$\tau : \mathcal{T} \to \mathcal{T}$	opposite transformation

Analogous (co, bi)monads for Coxeter spaces are introduced in Section 8.12. The general convention is to put a bar of top of the letters in this setting.

$\overline{\mathcal{T}}$	monad for Coxeter algebras
$\overline{\mathcal{S}}$	monad for commutative Coxeter algebras
$\overline{\mathcal{E}}$	monad for signed commutative Coxeter algebras
$\overline{\mathcal{T}}^{\vee}$	comonad for Coxeter coalgebras
$\overline{\mathcal{S}}^{\vee}$	comonad for cocommutative Coxeter coalgebras
$\overline{\mathcal{E}}^{\vee}$	comonad for signed cocommutative Coxeter coalgebras
$(\overline{\mathcal{T}}, \overline{\mathcal{T}}^{\vee}, \overline{\lambda})$	bimonad for Coxeter bialgebras
$\overline{\tau} : \overline{\mathcal{T}} \to \overline{\mathcal{T}}$	opposite transformation

Coxeter Lie monoids and Coxeter Lie algebras

Coxeter Lie monoids and Coxeter Lie comonoids are introduced in Chapter 6. We denote the category of Coxeter Lie monoids by $\mathsf{LieMon}(\mathsf{CSp})$, and of Coxeter Lie comonoids by $\mathsf{LieComon}(\mathsf{CSp})$.

g	Coxeter Lie monoid

k	Coxeter Lie comonoid
ν_A^F for $A \lessdot F$	Lie bracket component of a Coxeter Lie monoid
θ_A^F for $A \lessdot F$	Lie cobracket component of a Coxeter Lie comonoid
$\mathbf{Lie} \circ \mathsf{p}$	free Coxeter Lie monoid on the Coxeter species p, (6.11)
$\mathcal{PT}$, $\mathcal{PT}_{-1}$	monad for Coxeter Lie monoids, for signed Coxeter Lie monoids
$\mathcal{U}(\mathsf{g})$	universal enveloping Coxeter monoid of g
$\mathbf{Lie}^* \circ \mathsf{p}$	cofree Coxeter Lie comonoid on the Coxeter species p
$\mathcal{U}^\vee(\mathsf{k})$	universal coenveloping Coxeter comonoid of k
Sol, $\mathrm{Sol}^\vee$	Solomon operators

Coxeter Lie algebras and Coxeter Lie coalgebras are introduced in Chapter 12. We denote the category of Coxeter Lie algebras by $\mathsf{LieMon}(\overline{\mathsf{C}}\mathsf{Sp})$, and of Coxeter Lie coalgebras by $\mathsf{LieComon}(\overline{\mathsf{C}}\mathsf{Sp})$. The same notations as above, but for Coxeter Lie (co)algebras are written below. In this context, we write $\overline{\mathcal{U}}$ instead of $\mathcal{U}$.

G	Coxeter Lie algebra
K	Coxeter Lie coalgebra
ν_Z^T for $Z \lessdot T$	Lie bracket component of a Coxeter Lie algebra
θ_Z^T for $Z \lessdot T$	Lie cobracket component of a Coxeter Lie coalgebra
$\mathbf{Lie} \circ \mathrm{V}$	free Coxeter Lie algebra on the Coxeter space V, (12.22)
$\overline{\mathcal{PT}}$, $\overline{\mathcal{PT}}_{-1}$	monad for Coxeter Lie algebras, for signed Coxeter Lie algebras
$\overline{\mathcal{U}}(\mathrm{G})$	universal enveloping Coxeter algebra of G
$\mathbf{Lie}^* \circ^\vee \mathrm{V}$	cofree Coxeter Lie coalgebra on the Coxeter space V
$\overline{\mathcal{U}}^\vee(\mathrm{K})$	universal coenveloping Coxeter coalgebra of K
Sol, $\mathrm{Sol}^\vee$	Solomon operators

Coxeter dispecies and Coxeter operads

Coxeter dispecies and Coxeter operads (and all notations below) are introduced in Chapters 5 and 11. We denote the category of Coxeter dispecies by $\mathcal{A}$-$\mathtt{CdSp}$.

$\mathbf{p}$, $\mathbf{q}$, $\mathbf{r}$	Coxeter dispecies
$\mathbf{p} + \mathbf{q}$	direct sum of Coxeter dispecies $\mathbf{p}$ and $\mathbf{q}$, (5.3)
$\mathbf{p} \circ \mathbf{q}$	substitution product of Coxeter dispecies $\mathbf{p}$ and $\mathbf{q}$, (5.4)
$\mathbf{x}$	unit object for the substitution product, (5.5)
$\mathbf{p} \times \mathbf{q}$	Hadamard product of Coxeter dispecies $\mathbf{p}$ and $\mathbf{q}$, (5.8)
$\mathbf{p}^-$	signed partner of the Coxeter dispecies $\mathbf{p}$
$\mathbf{p} \circ \mathsf{m}$	substitution of the Coxeter species m in the Coxeter dispecies $\mathbf{p}$, (5.9)
$\mathbf{p} \circ \mathrm{V}$	substitution of the Coxeter space V in the Coxeter dispecies $\mathbf{p}$, (11.29)

$\mathbf{a}$, $\mathbf{b}$	Coxeter operads
$\mathbf{c}$, $\mathbf{d}$	Coxeter cooperads
$\mathcal{F}_\circ(\mathbf{e})$	free Coxeter operad on the Coxeter dispecies $\mathbf{e}$
$\langle \mathbf{e} \mid \mathbf{r} \rangle$	presentation of a quadratic Coxeter operad
$\mathcal{V}_\mathbf{a}$	monad on Coxeter species induced by the Coxeter operad $\mathbf{a}$
$\mathcal{U}_\mathbf{c}$	comonad on Coxeter species induced by the Coxeter cooperad $\mathbf{c}$
$\overline{\mathcal{V}}_\mathbf{a}$	monad on Coxeter spaces induced by the Coxeter operad $\mathbf{a}$
$\overline{\mathcal{U}}_\mathbf{c}$	comonad on Coxeter spaces induced by the Coxeter cooperad $\mathbf{c}$

Examples. Examples below are present in Sections 5.5 and 11.6.

$\mathbf{E}$	exponential Coxeter dispecies
$\mathbf{Com}$, $\mathbf{Com}^*$	Coxeter commutative operad, Coxeter commutative cooperad
$\mathbf{\Gamma}$	Coxeter dispecies of chambers

As, **As***	Coxeter associative operad, Coxeter associative cooperad
Lie, **Lie***	Coxeter Lie operad, Coxeter Lie cooperad
Com$^-$	signed Coxeter commutative operad

Fock functors

Fock functors are studied in Part III. We decorate the letter $\mathcal{K}$ is various ways to denote Fock functors, see Table 14.1.

$\mathcal{K}$, $\mathcal{K}^\vee$	full Fock functors
$\mathcal{K}_q$, $\mathcal{K}_q^\vee$	q-full Fock functors
$\overline{\mathcal{K}}$, $\overline{\mathcal{K}}^\vee$	bosonic Fock functors
$\overline{\mathcal{K}}_{-1}$, $\overline{\mathcal{K}}_{-1}^\vee$	fermionic Fock functors
$\Im_q$	anyonic Fock functor
$\mathcal{K}_0 = \mathcal{K}_0^\vee = \Im_0$	free Fock functor
$\mathcal{K}_M$	decorated full Fock functor for the left W-module M

$\kappa : \mathcal{K} \to \mathcal{K}^\vee$	norm transformation
$\kappa_q : \mathcal{K}_q \to \mathcal{K}_q^\vee$	q-norm transformation
κ_M	decorated norm transformation
$\theta : \mathcal{K} \to \mathcal{K}$	half-twist transformation
$\theta_q : \mathcal{K} \to \mathcal{K}$	q-half-twist transformation

Adjoints of Fock functors. These are studied in Chapter 16. The main notations for Fock functor adjoints are clear from Summaries 16.1, 16.2, 16.3, 16.4, 16.5. Some important ones are listed below.

$\mathsf{t}_{(-)}$	trivialization functor
$(\mathsf{t}_{(-)})^-$	signed trivialization functor
$\overline{\mathcal{Z}}^\vee$	left adjoint of $\overline{\mathcal{K}}^\vee$ for Coxeter monoids/algebras
$\overline{\mathcal{Z}}_{ab}^\vee$	left adjoint of $\overline{\mathcal{K}}^\vee$ for commutative Coxeter monoids/algebras
$\overline{\mathcal{Z}}$	right adjoint of $\overline{\mathcal{K}}$ for Coxeter comonoids/coalgebras
$\overline{\mathcal{Z}}^{coab}$	right adjoint of $\overline{\mathcal{K}}$ for cocommutative Coxeter comonoids/coalgebras
$\mathcal{L}$, $\mathcal{L}^\vee$	adjoints of the full Fock functors, (16.13)

List of Tables

The tables listed below summarize important objects, constructions, examples and results relevant to the material presented in the book.

I	Coxeter bimonoids and Joyal bimonoids.	xi
II	Coxeter operads and May operads.	xii
III	Coxeter species and Coxeter spaces.	2
IV	Coxeter bialgebras and face-type enumeration.	5
V	Coxeter bialgebras and shuffles.	6
VI	Hopf theory for Coxeter species and Coxeter spaces.	7
VII	Lie theory for Coxeter species and Coxeter spaces.	10
VIII	Relations in Coxeter operad presentations.	11
IX	Comparison of structure theorems.	11
X	Structure theorems via zeta and Möbius functions.	12
1.1	Objects associated to a reflection arrangement $\mathcal{A}$.	19
1.2	Object-types associated to a reflection arrangement $\mathcal{A}$.	20
1.3	Examples of reflection arrangements.	21
1.4	Face-type enumeration polynomials.	49
1.5	Incidence algebras.	63
1.6	Algebra morphisms between incidence algebras.	63
1.7	Birkhoff, Tits, Janus algebras.	78
1.8	Monoids and their linearized algebras.	86
1.9	Geometry and combinatorics.	90
2.1	Axioms for Coxeter species and (co, bi)monoids.	102
2.2	Alternative axioms for Coxeter (co)monoids.	103
2.3	Base categories for Coxeter species.	104
2.4	Categories of Coxeter monoids, comonoids, bimonoids.	117
2.5	Categories of (co)commutative Coxeter (co)monoids.	128
2.6	Categories of signed (co)commutative Coxeter (co)monoids.	128
2.7	Coxeter (co, bi)monoids as functor categories.	139
2.8	Normal forms for morphisms.	146
3.1	Primitive and decomposable filtrations.	161
3.2	Universal Coxeter bimonoids.	164
3.3	Hadamard product of Coxeter species.	176
3.4	Exp-log on Coxeter species.	185
3.5	Karoubi envelopes of monoids for reflection arrangements.	209
3.6	Karoubi envelopes of monoids for hyperplane arrangements.	209

3.7 Cauchy powers of Coxeter species and Joyal species. 214

4.1 Exponential Coxeter bimonoid and its analogues. 218
4.2 Decorated exponential Coxeter bimonoid and its analogues. 238
4.3 Relating Coxeter bimonoids and Joyal bimonoids. 244

5.1 Coxeter operads and monads on Coxeter species. 255
5.2 Free Coxeter operad monoids. 256
5.3 Coxeter cooperads and comonads on Coxeter species. 256
5.4 Cofree Coxeter cooperad comonoids. 256

7.1 Structure theorems for Coxeter bimonoids. 273
7.2 Free objects in rank one. 292
7.3 Cofree objects in rank one. 292

8.1 Axioms for Coxeter spaces and (co, bi)algebras. 302
8.2 Alternative axioms for Coxeter (co)algebras. 303
8.3 Base categories for Coxeter spaces. 304
8.4 Coxeter spaces to W-modules. 322
8.5 Categories of Coxeter algebras, coalgebras, bialgebras. 324
8.6 Categories of (co)commutative Coxeter (co)algebras. 347
8.7 Categories of signed (co)commutative Coxeter (co)algebras. 347
8.8 Coxeter (co, bi)algebras as functor categories. 358

9.1 Primitive and decomposable filtrations. 385
9.2 Universal Coxeter bialgebras. 389
9.3 Hadamard product of Coxeter spaces. 403
9.4 Exp-log on Coxeter spaces. 412
9.5 Cauchy powers and tensor powers. 452

10.1 Coxeter bialgebra of polynomials and its relatives. 460
10.2 Properties of the Coxeter bialgebra of polynomials. 460
10.3 Tensor Coxeter bialgebra and its relatives. 472
10.4 Properties of the tensor Coxeter bialgebra. 472
10.5 Product and coproduct of the tensor Coxeter bialgebra. 473
10.6 Coxeter bialgebras of face-types. 483
10.7 Coxeter q-bialgebras of face-types. 484
10.8 Coxeter bialgebras and Hopf algebras of polynomials. 516
10.9 Tensor Coxeter bialgebras vs tensor Hopf algebras. 517
10.10 Hopf algebra of symmetric functions and generalizations. 519

11.1 Substitution map of a Coxeter operad **a**. 522
11.2 Structure map of a Coxeter **a**-algebra V. 522
11.3 Base categories for Coxeter dispecies. 523
11.4 Coxeter operads and monads on Coxeter spaces. 548
11.5 Free Coxeter operad algebras. 548
11.6 Coxeter cooperads and comonads on Coxeter spaces. 548
11.7 Cofree Coxeter cooperad coalgebras. 549

13.1 Structure theorems for Coxeter bialgebras. 586
13.2 Comparing structure theorems of Coxeter bialgebras. 605

13.3 Tensor Coxeter bialgebra in rank one. 626
13.4 Coxeter bialgebras of face-types in rank one. 626
13.5 Free objects in rank one. 630
13.6 Cofree objects in rank one. 631

14.1 Fock functors. 643
14.2 Decorated Fock functors. 683
14.3 Commutativity under Fock functors. 685

16.1 Coxeter operad algebras to Coxeter operad monoids. 752

18.1 Relating Coxeter bimonoids and Coxeter bialgebras. 786

A.1 Examples of bilax functors. 821

List of Figures

Most figures below are schematic pictures of reflection arrangements that serve to bring out the geometry present in Hopf theory.

I	Chart of interdependence of chapters.	xvi
II	Coxeter species vs Coxeter spaces.	3
1.1	Tits product of faces.	22
1.2	Visualizing a biface.	23
1.3	Rank-two dihedral arrangements.	24
1.4	Rank-three coordinate arrangement.	24
1.5	Rank-three braid arrangement (Type A).	25
1.6	Rank-three arrangement of type B.	25
1.7	Rank-two flat in the rank-three braid arrangement.	27
1.8	Opposite faces in stars.	28
1.9	Lune in a rank-four arrangement.	30
1.10	Partial orders on top-lunes.	31
1.11	Gate property in terms of a minimal gallery.	35
1.12	Two dual ways to picture shuffles.	40
1.13	Morphisms in the groupoid of biface-types.	41
1.14	Coxeter symmetry w_σ of a biface-type σ.	44
1.15	Face-types biprojected by a biface-type.	45
1.16	Induced morphisms in the groupoid of biface-types.	46
1.17	Opposite morphisms in the groupoid of biface-types.	47
1.18	Gate property in terms of biface-types.	47
1.19	Gate identity for face-type enumeration.	57
1.20	Product in the invariant biface-incidence algebra.	71
1.21	Two-sided q-lune-additivity formula.	73
1.22	Two-sided q-Weisner formula.	74
1.23	Structure constants of the invariant q-Janus algebra.	82
2.1	Support-support, support-type, type-type relations.	105
2.2	Naturality and (co)associativity axioms.	118
2.3	Bimonoid axiom.	123
2.4	(Co)commutativity axiom for Coxeter (co)monoids.	129
2.5	Covering (co)commutativity axiom for Coxeter (co)monoids.	130
8.1	Support-support, support-type, type-type relations.	306
8.2	Locally opposite support-type morphism.	316
8.3	Naturality and (co)associativity axioms.	325

8.4	Coxeter algebra in terms of face-types.	328
8.5	Bialgebra axiom.	332
8.6	(Co)commutativity axiom for Coxeter (co)algebras.	349
9.1	(De)shuffles and (de)quasishuffles.	391
11.1	Morphisms between nested top-lunes.	525
11.2	Morphisms between nested face-types.	527
11.3	Lune criterion for Lie elements.	554
14.1	Full Fock functor $\mathcal{K}$ is colax.	647
14.2	Lax-colax compatibility for $\mathcal{K}$.	648
16.1	The adjunction lifting process.	715

List of Summaries

The summaries listed below compile together all the different equivalent definitions of Coxeter bimonoids and Coxeter bialgebras as well as all important adjunctions considered in the book.

2.1	Coxeter bimonoid in terms of faces.	123
2.2	Coxeter bimonoid in terms of flats and lunes.	124
2.3	Bicommutative Coxeter bimonoid in terms of flats.	133
3.1	Adjunctions involving Coxeter (co, bi)monoids.	165
3.2	Adjunctions involving (co)abelianization.	174
7.1	Coxeter q-bimonoid in rank one.	289
7.2	Bicommutative Coxeter bimonoid in rank one.	289
8.1	Coxeter bialgebra in terms of top-nested faces.	335
8.2	Coxeter bialgebra in terms of top-lunes.	338
8.3	Coxeter bialgebra in terms of face-types.	341
9.1	Adjunctions involving Coxeter (co, bi)algebras.	389
9.2	Adjunctions involving (co)abelianization.	399
13.1	Coxeter q-bialgebra in rank one.	624
15.1	Coxeter bimonoids to Coxeter bialgebras.	687
16.1	Adjunctions involving bosonic Fock functors.	714
16.2	Adjunctions preserving commutativity.	715
16.3	Adjunctions involving full Fock functors.	716
16.4	Coxeter operad monoids and algebras.	717
16.5	Coxeter Lie monoids and Coxeter Lie algebras.	717

Author Index

The page numbers indicate the places where the author's work is being cited.

A

Abe, Eiichi, 380, 516
Abramenko, Peter, 97, 98
Adámek, Jiří, 156
Aguiar, Marcelo, xi–xvii, xix, 6, 9–11, 20, 21, 23, 25, 27, 29, 31, 33, 34, 36–38, 40, 43, 48, 49, 55, 57, 61–63, 65, 66, 68, 72, 74, 76, 78, 81–85, 87, 90, 92, 93, 95–98, 103, 115, 116, 119, 121, 131, 135–137, 139, 146–150, 153, 154, 156–159, 161, 162, 165–169, 174, 176, 178, 180, 181, 183, 185, 187, 188, 190–196, 200, 202, 203, 205, 206, 208, 210–216, 220–222, 224–238, 245, 246, 250, 251, 253, 258, 260, 261, 264, 266, 267, 269–288, 295–298, 361, 379–381, 405, 419, 432, 448, 450–455, 469, 470, 478, 509, 515, 516, 518–521, 541, 553, 554, 558, 560, 577–579, 585, 590, 606, 616, 617, 638–640, 666, 673, 684–686, 696, 697, 712, 713, 757, 758, 785, 806, 807, 815, 824, 825, 847
Aigner, Martin, 97
André, Michel, 517, 585, 640
Andrews, George E., 97
Arbib, Michael A., 156
Ardizzoni, Alessandro, 381
Arvola, William Arthur, 96
Askey, Richard, 97
Atkinson, Michael D., 97
Aubry, Marc, 272
Awodey, Steve, xiv

B

Bachmann, Henrik, 639
Baez, John C., 381
Barr, Michael, 453, 825
Barratt, Michael G., 157, 245, 272
Belbachir, Hacéne, 97
Bénabou, Jean, 156
Benson, Clark T., 98
Benson, David B., 454, 518
Berezin, Felix Alexandrovich, 585
Bergeron, François, 95, 97, 156, 214, 245
Bergeron, Nantel, 95, 97, 518, 519
Bertet, Karell, 519
Bhaskaracharya, 96
Bialynicki-Birula, I., 98
Bidigare, Thomas Patrick, 95, 96
Billera, Louis J., 96
Birkhoff, Garrett, 95, 585, 639
Björner, Anders, 96–98, 465
Bland, Robert Gary, 95
Blessenohl, Dieter, 97, 518, 520
Boardman, J. Michael, 560
Borceux, Francis, 825
Borel, Armand, 380, 451, 638
Borovik, Alexandre V., 98
Borovik, Anna, 98
Bourbaki, Nicolas, 98, 156
Bremner, Murray R., 214, 245, 560
Brenti, Francesco, 96–98, 465
Browder, William, 451
Brown, Kenneth Stephen, 95–98
Brown, Ronald, 157, 825
Bruned, Yvain, 639
Bucur, Ion, 156
Bui, Van Chien, 639

C

Carlitz, Leonard, 97
Cartan, Henri, 380, 381, 455, 517, 585
Carter, Roger William, 98
Cartier, Pierre, 157, 380, 381, 451, 452, 454, 517, 518, 585, 638, 640, 712

Chen, Kuo-tsai, 517
Cheung, Pokman, 97
Chevalley, Claude, 96, 585
Chu, Shih-Chieh, 97
Cohen, P. Beazley, 452
Contou-Carrère, Carlos, 98
Coxeter, H. S. M., 95, 98
Curry, Charles, 639

D
Dăscălescu, Sorin, 380
Davis, Michael W., 96, 98
Deleanu, Aristide, 156
Demazure, Michel, 453, 640
Deshpande, Priyavrat, 96
Diaconis, Persi, 95, 96
Dieudonné, Jean, 638
Ditters, E. J., 518, 639, 640
Dixmier, Jacques, 640
Dotsenko, Vladimir, 214, 245, 560
Doubilet, Peter, 516
Duchamp, Gérard H. E., 454, 517, 518, 520, 639

E
Ebrahimi-Fard, Kurusch, 96, 98, 453, 454, 639
Eckmann, Beno, 157
Ehrenborg, Richard, 519
Ehresmann, Charles, 156, 157, 825
Eilenberg, Samuel, 96, 380, 381, 455, 585
Epstein, Henri, 95
Eriksson, Kimmo, 97
Ernst, Dana C., 96
Ernst, Thomas, 97
Etienne, Gwihen, 95
Eyre, Timothy M. W., 452

F
Fiedorowicz, Zbigniew, 157
Foissy, Loïc, 520, 639
Fox, Thomas F., 453
Fresse, Benoit, 157, 560, 638
Fröhlich, Albrecht, 640
Fulton, William, 518

G
Gabriel, Pierre, 453, 640, 825
Gaines, Jessica G., 452
Garrett, Paul, 98
Garsia, Adriano M., 95, 97, 98, 453
Gauss, Carl Friedrich, 97
Geissinger, Ladnor, 518
Gelfand, Israel M., 98, 454, 518–520, 585, 639
Gerstenhaber, Murray, 454, 585
Gessel, Ira M., 95, 519, 521
Ginzburg, Victor, 560
Glaser, Vladimir Jurko, 95
Goerss, Paul G., 272
Goldman, Jay, 516
Gould, Henry Wadsworth, 97
Gracia-Bondía, José M., 98, 453
Grandis, Marco, 157
Greene, Curtis, 95
Grinberg, Darij, 518–520
Grothendieck, Alexandre, 825
Grove, Larry C., 98
Grünenfelder, Luzius A., 453, 638, 640
Gubareni, Nadiya, 380, 518, 520
Guo, Li, 452
Gurevich, Dmitriĭ I., 381

H
Hadamard, Jacques Salomon, 453
Hall, Philip, 97, 585
Halpern, Edward, 381
Hanlon, Phil, 95, 96
Harish-Chandra, 585
Hartl, Manfred, 158
Hasse, Maria, 156
Hazewinkel, Michiel, 380, 452, 518, 520, 521, 640
Heyneman, Robert G., 453, 638
Higgins, Philip J., 157, 825
Hiller, Howard, 98
Hilton, Peter J., 157
Hivert, Florent, 518, 520
Hochschild, Gerhard P., 453, 454, 640
Hoffman, Michael E., 452, 638, 639
Hopf, Heinz, 380, 451, 638, 639
Howlett, Robert B., 95, 97
Hudson, Robin Lyth, 452
Humphreys, James E., 97, 98

J
Jacobs, Bart, 825
Jacobson, Nathan, 585
Jensen, J. L. W. V., 97
Johnson, Niles, 157, 825
Johnstone, Peter T., 825
Joni, S. A., 516
Joyal, André, 156, 214, 272, 298

K
Kac, George I., 585
Kac, Victor G., 97
Kane, Richard, 98
Kapranov, Mikhail, 560
Kashina, Yevgenia, 454, 455

Kassel, Christian, 97, 380, 516
Keigher, William, 452
Kelly, Gregory Maxwell, 157, 560
Kharchenko, Vladislav Kirillovich, 381
Kirichenko, Vladimir V., 380, 518, 520
Klyachko, Alexander, 454, 517, 518
Koppinen, Markku, 453
Kostant, Bertram, 453
Koszul, Jean-Louis, 451
Krob, Daniel, 98, 454, 517–520, 639

L
Labelle, Gilbert, 156, 214, 245
Lambek, Joachim, 560
Landers, Rachel, 454
Larson, Richard Gustavus, 451
Las Vergnas, Michel, 96, 98
Lascoux, Alain, 98, 454, 518–520, 639
Laue, Hartmut, 97
Lauve, Aaron, 454
Lazard, Michel, 560
Leclerc, Bernard, 98, 454, 518–520, 639
Leinster, Tom, xiv, 560
Leray, Jean, 380, 451, 454, 639
Leroux, Pierre, 156, 214, 245
Li, Chun Wah, 452
Linchenko, Vitali Vladim, 454
Liu, Xiao Qing, 452
Livernet, Muriel, 157, 639, 713
Loday, Jean-Louis, 157, 214, 452, 453, 560, 639
Lorenz, Martin, 380

M
Mac Lane, Saunders, 96, 108, 560, 825
Macdonald, Ian G., 518, 520
Mackenzie, Kirill Charles Howard, 157
Mackey, George Whitelaw, 825
Magnus, Wilhelm, 585
Mahajan, Swapneel, xi–xvii, xix, 6, 9–11, 20, 21, 23, 25, 27, 29, 31, 33, 34, 36–38, 40, 43, 48, 49, 55, 57, 61–63, 65, 66, 68, 72, 74, 76, 78, 81–85, 87, 90, 92, 93, 95–98, 103, 115, 116, 119, 121, 131, 135–137, 139, 146–150, 153, 154, 156–159, 161, 162, 165–169, 174, 176, 178, 180, 181, 183, 185, 187, 188, 190–196, 200, 202, 203, 205, 206, 208, 210–216, 220–222, 224–238, 245, 246, 250, 251, 253, 258, 260, 261, 264, 266, 267, 269–288, 295–298, 361, 379–381, 405, 419, 432, 448, 450–455, 469, 470, 478, 509, 515, 516, 518–521, 541, 553, 554, 558, 560, 577–579, 585, 590, 606, 616, 617, 638–640, 666, 673, 684–686, 696, 697, 712, 713, 757, 758, 785, 806, 807, 815, 824, 825, 847
Majid, Shahn, 380
Malham, Simon J. A., 96, 98, 639
Malvenuto, Claudia, 519, 520, 638
Manchon, Dominique, 98, 453–455, 639
Manes, Ernest G., 156
Markl, Martin, 560
Mastnak, Mitja, 453
May, Jon Peter, 380, 560
McMullen, Peter, 98
Melançon, Guy, 638
Méliot, Pierre-Loïc, 518–520
Méndez, Miguel A., 156, 214, 245
Menini, Claudia, 381
Michaelis, Walter, 585, 640
Michler, Lothar, 156
Mielnik, Bogdan, 98
Milnor, John Willard, 380, 381, 451, 453, 455, 585, 639, 640
Minh, Hoang Ngoc, 639
Möbius, August Ferdinand, 97
Moerdijk, Ieke, 825
Montgomery, Susan, 380, 453–455, 516
Moore, John Coleman, 380, 381, 451–453, 455, 518, 585, 639, 640
Morvan, Michel, 519
Moszkowski, Paul, 95
Musson, Ian M., 585

N
Năstăsescu, Constantin, 380
Neisendorfer, Joseph, 585
Newman, Kenneth W., 452, 639
Ng, Siu-Hung, 454
Ngô, Quoc Hoan, 639
Nichols, Warren Douglas, 453, 516, 585, 638, 640
Nijenhuis, Albert, 585
Novelli, Jean-Christophe, 98, 519

O
Orlik, Peter, 96

P
Parthasarathy, Kalyanapuram Rangachari, 452

Patras, Frédéric, 96, 98, 157, 245, 380, 453, 454, 521, 638, 639, 712
Paycha, Sylvie, 639
Petersen, T. Kyle, 98
Phan, Thi Ha Duong, 519
Pirashvili, Teimuraz, 158
Plebański, Jerzy, 98
Poincaré, Henri, 585, 639
Poirier, Stéphane, 520
Ponto, Kathleen, 380
Porst, Hans-E., 453

Q

Quillen, Daniel, 454, 639, 640

R

Radford, David E., 380, 452, 453, 455, 638, 639
Raianu, Şerban, 380
Ree, Rimhak, 454, 517
Reidemeister, Kurt, 825
Reiner, Victor, 518–520
Retakh, Vladimir S., 98, 454, 518–520, 639
Reutenauer, Christophe, 95, 97, 98, 157, 245, 453, 454, 517–521, 585, 638, 712
Richardson, Roger Wolcott, Jr., 585
Riehl, Emily, xiv, 825
Riordan, John, 97
Rockmore, Dan, 95, 96
Roman, Steven, xiv, 825
Ronan, Mark, 98
Ronco, María Ofelia, 639
Ross, Leonard Erwin, 585
Rota, Gian-Carlo, 97, 516
Roy, Ranjan, 97

S

Sagan, Bruce E., 518
Saliola, Franco V., 95
Sam, Steven V., 214, 452
Samelson, Hans, 380, 639
Schack, Samuel Donald, 454
Schauenburg, Peter, 454
Scheunert, Manfred, 585
Schmitt, William Russell, 455, 815
Schocker, Manfred, 95, 98, 157, 518, 520, 712
Scholtens, A. C. J., 518
Schulte, Egon, 98
Schur, J., 453
Serre, Jean-Pierre, 640
Shnider, Steven, 560
Sivera, Rafael, 157, 825
Sjödin, Gunnar, 640
Skowroński, Andrzej, 380
Smirnov, Vladimir Alekseevich, 214, 560
Smoke, William Henry, 639
Snowden, Andrew, 214, 452
Solomon, Louis, 95–98, 640
Sommerhäuser, Yorck, 455
Sottile, Frank, 96, 518–521
Spivak, David I., 825
Stanley, Richard Peter, 97, 516, 518–521
Stasheff, James Dillon, 560
Steinberg, Robert, 97
Stora, Raymond, 95
Stover, Christopher R., 157, 272, 297, 298, 712, 757
Sturmfels, Bernd, 96
Sweedler, Moss Eisenberg, xv, 380, 451, 453, 455, 516, 517, 638, 640

T

Taft, Earl J., 516
Takeuchi, Mitsuhiro, 381, 455
Tauber, Selmo, 97
Taylor, Donald E., 95, 97
Terao, Hiroaki, 96
Thibon, Jean-Yves, 98, 454, 517–521, 639
Thomas, Anne, 98
Tits, Jacques, 95
Tollu, Christophe, 639

U

Underwood, Robert G., 380
Ung, B.-C.-V., 521

V

Vallette, Bruno, 214, 452, 560
Vandermonde, Alexandre-Théophile, 97
Varchenko, Alexandre, 96
Vega, Maria D., 454
Vespa, Christine, 158
Vogt, Rainer M., 560

W

Ward, Morgan, 97
Weisner, Louis, 97
Weiss, Richard M., 98
Wells, Charles, 825
White, Neil, 96, 98
Wiese, Anke, 96, 98, 639
Witherspoon, Sarah, 454
Witt, Ernst, 95, 97, 585, 639
Wraith, Gavin C., 453

Y

Yamagata, Kunio, 380

Yau, Donald, 157, 825

Z

Zaslavsky, Thomas, 98
Zelevinsky, Andrey V., 518
Zhu, Yongchang, 455
Ziegler, Günter Matthias, 95, 96
Zisman, Michel, 381, 825

Subject Index

The page numbers shown in bold indicate the place where the corresponding item is defined. Multiple bold numbers for the same item mean that the term is used in multiple related senses.

Alphanumeric
0-Möbius function, **75**
0-Möbius series, 297, 639
0-bialgebra axiom
    classical, **379**
    in rank one, **624**
    in terms of face-types, **346**, 395, 466, 476
    in terms of top-nested faces, **345**
0-bimonoid axiom
    in terms of faces, **126**
    in terms of flats and lunes, **126**
0-deshuffle coproduct, 164, 389, 473
0-exp-log correspondence, **190**, **418**
0-logarithm of the identity map, **192**, 211, **421**, 445
0-norm map
    between tensors and shuffles, **476**
    on face-types, **498**
    on polynomials, **466**
0-shuffle product, 164, 389, 473
0-zeta function, **75**
0-zeta series, 297, 639
W-invariant distance function, 36, 126, 149, 346, 367, 653
W-valued gallery distance, **36**, 39, 43, 217, 458
$\Bbbk$-linear category, **115**, 127, 323, 346
$\Bbbk$-linear functor, **115**, 147, 365, 645, 656, 657, 659, 690, 717, 718, 721, 763, 765
$\mathbb{Z}$-linear category, **361**, 362
$\mathbb{Z}$-linear functor, **361**
v-bialgebra axiom, **346**
v-bimonoid axiom, **127**
q-Chu–Vandermonde identity, **92**, 97, 516
q-Janus algebra, **80**, 95, 204
q-Möbius function, **70**
q-Möbius inversion, **470**, 496
q-Möbius series, **94**
q-Zaslavsky formula, **76**
q-bialgebra axiom
    classical, **378, 379**
    for the coordinate arrangement, **345**
    in rank one, **624**
    in terms of compositions, **378**
    in terms of face-types, **344**
    in terms of top-lunes, **342**
    in terms of top-nested faces, **341**
q-bimonoid axiom
    in rank one, **289**
    in terms of faces, **125**
    in terms of flats and lunes, **126**
q-binomial coefficient, **92**, 97
q-dequasishuffle coproduct, 164, **167**, 389, **391**, 484, **485**
q-deshuffle coproduct, 164, **167**, 389, **392**, 473, **474**, 484, **488**
q-distance function, **36**
q-exp-log correspondence
    for Coxeter spaces, **417**, **426**
    for Coxeter species, **189**, **196**
q-exponential
    of a map of Coxeter spaces, **416**, 601, 602
    of a map of Coxeter species, **189**
q-exponential series
    in rank one, **294**, **634**
    of a Coxeter algebra, **428**
    of a Coxeter monoid, **198**
q-factorial, **91**, 464
q-flat-additivity formula, **68**, 69
    for type A, **94**
q-full Fock functors, **651**

q-half-twist transformation, **669**
q-logarithm
  of a map of Coxeter spaces, **416**
  of a map of Coxeter species, **189**
  of the identity map, **192**, **421**
q-lune-additivity formula, **68**, 68, 196, 427
  for type A, **94**
q-multinomial coefficient, **92**
q-norm map
  between tensors and shuffles, **476**, 796
  for the coordinate arrangement, **498**
  on chambers, **223**, 790, 793
  on Coxeter q-bialgebras, **431**
  on Coxeter q-bimonoids, **201**
  on decorated chambers, **241**, 796
  on face-types, **494**, 614, 802, 803
  on faces, **231**, 803
  on polynomials, **464**, 790, 793
  on top-nested faces, **234**, 802
q-norm transformation
  between q-full Fock functors, **672**
  for Coxeter spaces, **431**
  for Coxeter species, **201**
q-number, **91**, 97, 464
q-quasishuffle product, 164, **169**, 389, **393**, 484, **490**
q-shuffle product, 164, **169**, 389, **394**, 473, **475**, 484, **490**
q-zeta function, **70**
q-zeta series, **94**

A
abelian Coxeter Lie algebra, **566**, 617
abelian Coxeter Lie coalgebra, **566**, 618
abelian Coxeter Lie comonoid, **264**
abelian Coxeter Lie monoid, **264**
abelianization, **151**, **174**, **370**, **400**
  in rank one, **289**, **625**
abelianization map, **174**, 198, 282, 285, **400**, 428, **477**, 609, 615
Adams elements, 638
Adams operator, 454
adjacent support-type morphism, **316**, 364, 582
  for type A, **375**
adjoint lifting theorem, 726, 728, 731, 734, 737, 741, 748, 749, **820**
algebra over a monad, 147, 365, **818**
algebraic group, 453, 640
alternating sign identity, **58**, 62, 97, 447, 471, 504, 505, 793
ambient space of an arrangement, **20**
antipode
  formulas for Coxeter bialgebras, 471, 482, 504, 510, 513
  formulas for Coxeter bimonoids, 222, 225, 228, 232, 236, 242
  of a (co)free Coxeter bialgebra, 450
  of a (co)free Coxeter bimonoid, 213
  of a Coxeter q-bialgebra, **445**
  of a Coxeter q-bimonoid, **210**
  of a Hopf algebra, 455
  of a Joyal bimonoid, 215
antisymmetry, **262**, **563–565**, **582**
anyonic Fock functor, **675**, 696
arrangement of type B, **21**
arrangement of type D, **21**
associativity axiom
  for commutative Coxeter monoids, **130**
  for Coxeter algebras, **325**, **327**, **328**, 550, 551
  for Coxeter monoids, **118**, **120**
  for type A, **376**
augmentation map, **60**

B
base of a lune, **30**
base-case map, **65**, 187, 199, 415, 424
Baxter algebra, 452
bialgebra axiom
  in rank one, **624**
  in terms of face-types, **340**
  in terms of top-lunes, **338**
  in terms of top-nested faces, **334**
bialgebra over a bimonad, 147, 365, **819**
bialgebra relation
  in terms of top-lunes, **362**
  in terms of top-nested faces, **362**
bicommutative bimonoid axiom, **132**
bicommutative bimonoid relation, **145**
bicommutative Coxeter bialgebra, **353**
bicommutative Coxeter bimonoid, **132**
  in rank one, **289**
biconvolution Coxeter bimonoid, **179**
biderivation
  for Coxeter spaces, **388**, 416, 432, 435, 597, 598, 601, 602
  for Coxeter species, **163**, 188
biface, **22**
  for type A, **91**
biface-incidence algebra, **70**
biface-type, **28**
  for type A, **91**
bilax functor between bimonads, 153, 177, 221, 243, 371, 372, 646,

649, 652, 657–659, 666, 675, 683, **817**
bilune, **32**
bilune-incidence algebra, **71**
bilune-type, **33**
bimonad
on a category, **817**
on Coxeter spaces, **365**
on Coxeter species, **147**
bimonoid axiom
in terms of faces, **122**
in terms of flats and lunes, **124**
bimonoid relation
in terms of faces, **142**
in terms of flats and lunes, **143**
bioperad, 260
Birkhoff algebra, **78**, 95
Birkhoff monoid, **22**, 95
bistrong property, 685
black and white circle products, 260
Boolean transform, **701**
Borel–Hopf isomorphisms, **274, 275**, **589**, **594**
classical, 638
Borel–Hopf theorem
classical, 297, 638
for cocommutative Coxeter bialgebras, **589**
for cocommutative Coxeter bimonoids, **274**
for commutative Coxeter bialgebras, **594**
for commutative Coxeter bimonoids, **275**
for signed cocommutative signed Coxeter bialgebras, **595**
for signed commutative signed Coxeter bialgebras, **595**
in rank one, **295**, **635**
via Fock functors, 779
bosonic Fock functor, **656, 657**
braid arrangement, **21**, **90**, 154, 240, 373, 557, 582
braided algebra, 381
braided bialgebra, 381

C

Cartier–Milnor–Moore theorem
classical, 298, 640
for cocommutative Coxeter bialgebras, **621**
for cocommutative Coxeter bimonoids, **287**
for commutative Coxeter bialgebras, **622**
for commutative Coxeter bimonoids, **288**
for signed cocommutative signed Coxeter bialgebras, **622**
for signed commutative signed Coxeter bialgebras, **622**
in rank one, **297**, **637**
case of a lune, **30**
category of bilunes, **32**
category of elements, **822**, 825
category of lunes, **30**, 119, 158
Cauchy power
of a Coxeter space, **383**, 390, 435, 451, 461, 473
of a Coxeter species, **160**, 165, 213
Cauchy product
of graded vector spaces, 451
of Joyal species, 157, 213
central face, **21**
central face-type, **24**
chamber, **21**
characteristic operation
by a face, **203**, 207
by a face-type, **439**, 591
by a set composition, **215**
by an integer composition, **454**
coabelianization, **151**, **175**, **370**, **401**
in rank one, **290**, **625**
coabelianization map, **175**, 282, 286, **401**, **477**, 609, 617
coalgebra over a comonad, 147, 365, **819**
coassociativity axiom
for Coxeter coalgebras, **330**
for Coxeter comonoids, **121**
coassociativity relation, **141**, **360**
cocommutative bimonoid axiom, **133**, 223
cocommutative Coxeter bialgebra, **353**
cocommutative Coxeter bimonoid, **132**
cocommutative Coxeter coalgebra, **352**
cocommutative Coxeter comonoid, **131**
cocommutativity axiom
for Coxeter coalgebras, **353**
for Coxeter comonoids, **131, 132**
in terms of compositions, **377**
cocommutativity relation, **143**, **363, 364**
cocommutator cobracket, **265**, **568**
in rank one, **291**, **628**
cocommutator functor, **265**, **568**
coconvolution Coxeter coalgebra, **405**
coconvolution Coxeter comonoid, **179**
coderivation
for Coxeter spaces, **388**, 414, 590
for Coxeter species, **163**, 187

cofree cocommutative Coxeter coalgebra, **397**
cofree cocommutative Coxeter comonoid, **172**
cofree Coxeter q-bialgebra, **393**
cofree Coxeter q-bimonoid, **169**
cofree Coxeter coalgebra, **392**
cofree Coxeter comonoid, **168**
cofree Coxeter Lie coalgebra, **575**
cofree Coxeter Lie comonoid, **267**
cofree signed cocommutative Coxeter coalgebra, **398**
cofree signed cocommutative Coxeter comonoid, **173**
colax functor
  between comonads, 152, 370, **817**
  between monads, 660, **818**
colax-lax adjunction, **820**
commutative bimonoid axiom, **133**
commutative Cauchy power
  of a Coxeter space, **384**, 395, 397, 477, 507
  of a Coxeter species, **160**, 170
commutative characteristic operation
  by a flat, **204**, 208
  by a flat-type, **441**
commutative Coxeter algebra, **348**
commutative Coxeter bialgebra, **353**
commutative Coxeter bimonoid, **132**
commutative Coxeter monoid, **127**
commutative exp-log correspondence, **188**, **200**, **415**
commutativity axiom
  for Coxeter algebras, **348**, **350**, **351**, 552
  for Coxeter monoids, **128**, **129**
  in terms of compositions, **376**
commutativity relation, **144**
commutator bracket, **265**, **567**
  in rank one, **291**, **628**
commutator functor, **265**, 269, **567**, 578
comonad on a category, **817**
composition, **91**, 93, 374
concatenation product, 164, **166**, 389, **390**, 473, **474**, 483, 484, **485**
connected Coxeter operad, **250**, **537**
convolution algebra, 453
convolution Coxeter algebra, **405**, 425, 429
convolution Coxeter monoid, **179**, 195, 199
coordinate arrangement, **21**, 87
coreflective subcategory, 175, 401, 724, 739, 743, 758
cotransferred morphism, **307**
counitality axiom
  for Coxeter coalgebras, **330**
  for Coxeter comonoids, **121**
covering associativity axiom
  for commutative Coxeter monoids, **131**, 257
  for Coxeter algebras, **329**, 551
  for Coxeter monoids, **119**, **121**, 258
  for type A, **376**, 378
covering coassociativity axiom
  for Coxeter coalgebras, **331**
  for Coxeter comonoids, **122**
covering cocommutativity axiom
  for Coxeter coalgebras, **353**
  for Coxeter comonoids, **132**
covering cocommutativity relation, **363**, **364**
covering commutativity axiom
  for Coxeter algebras, **349**, **351**, **352**, 553
  for Coxeter monoids, **129**, **131**
Coxeter 0-bialgebra
  in terms of face-types, **345**
  in terms of top-nested faces, **345**
  of face-types, **490**
  of polynomials, **466**
Coxeter 0-bimonoid
  in terms of faces, **125**
  in terms of flats and lunes, **126**
Coxeter **a**-algebra, **542**
Coxeter v-bimonoid, **126**
Coxeter v-bialgebra, **346**
Coxeter q-bialgebra
  for type A, **377**
  in rank one, **624**
  in terms of face-types, **344**
  in terms of top-lunes, **342**
  in terms of top-nested faces, **341**
  of face-types, **485**
  of polynomials, **461**
  of symmetries, **512**
Coxeter q-bimonoid
  in rank one, **289**
  in terms of faces, **124**
  in terms of flats and lunes, **126**
  of chambers, **222**
  of faces, **228**
  of pairs of chambers, **235**
  of pairs of faces, 245
  of top-nested faces, **233**
Coxeter algebra
  for type A, **376**
  in rank one, **623**
  in terms of face-types, **327**

in terms of top-lunes, **326**
in terms of top-nested faces, **324**
of polynomials, **403**
Coxeter associative cooperad, **258**, 552
Coxeter associative operad, **258**, 548
Coxeter b-space, **323**, 373, 449
Coxeter b-species, **116**, 154, 212, 215
Coxeter bialgebra
in rank one, **624**
in terms of face-types, **340**
in terms of top-lunes, **337**
in terms of top-nested faces, **334**
of face-types, 440, **485**, 576, 594, 622, 798
of flat-types, 441, **506**, 599, 804
of polynomials, **461**, 788
of symmetries, **511**, 806
Coxeter bimonoid
in rank one, **289**
in terms of faces, **122**
in terms of flats and lunes, **124**
of chamber maps, **237**, 807
of chambers, **222**
of charts, 245
of decorated chambers, **240**, 481
of dicharts, 245
of faces, **228**
of flats, **226**
of shards, 245
of top-cones, 245
of top-lunes, **234**
Coxeter coalgebra
for type A, **376**
in rank one, **624**
in terms of face-types, **330**
in terms of top-lunes, **330**
in terms of top-nested faces, **329**
Coxeter commutative cooperad, **257**, 553
Coxeter commutative operad, **256**, 552
Coxeter comonoid
in rank one, **289**
in terms of faces, **121**
in terms of flats and lunes, **122**
Coxeter complex, **21**
Coxeter cooperad, **251**
Coxeter cooperad coalgebra, **547**
Coxeter cooperad comonoid, **255**
Coxeter dispecies
in terms of face-types, **530**
in terms of flats, **247**, 529
in terms of top-lunes, **529**
of chambers, **257**
Coxeter double algebra, **354**
Coxeter double coalgebra, **354**
Coxeter double comonoid, **135**
Coxeter double monoid, **135**
Coxeter group, **20**
Coxeter lb-space, **323**
Coxeter lb-species, **116**
Coxeter Lie algebra, **562**
in rank one, **627**
Coxeter Lie coalgebra, **564**
in rank one, **628**
Coxeter Lie comonoid, **263**
in rank one, **290**
Coxeter Lie cooperad, **258**, 263, 564
Coxeter Lie monoid, **262**
in rank one, **290**
Coxeter Lie operad, **258**, 262, 553, 562
Coxeter Lie species, **237**, 268, 468
Coxeter monoid
in rank one, **288**
in terms of faces, **117**
in terms of flats and lunes, **119**
Coxeter monomial, **461**
Coxeter operad
in terms of face-types, **535**
in terms of nested flats, **249**
in terms of top-lunes, **533**
Coxeter operad algebra, **542**
Coxeter operad monoid, **253**
Coxeter Poisson operad, 560
Coxeter quotient dispecies, 250
Coxeter quotient species, 115
Coxeter set-dispecies, **251**
Coxeter set-operad, **251**
Coxeter set-species, **116**
Coxeter space
characteristic of chambers, **457**, 546, 787
for type A, **375**
in rank one, **623**
in terms of face-types, **321**
in terms of top-lune-types, **320**
in terms of top-lunes, **319**
in terms of top-nested faces, **317**
of face-types, **484**
of polynomials, **319, 320**, **403**, 434, **461**
of symmetries, **511**
Coxeter species
characteristic of chambers, **217**, 221, 223, 787
in rank one, **288**
in terms of faces, **111**
in terms of flat-types, **114**
in terms of flats, **113**
of chambers, **222**
of faces, **228**

of flats, **225**
of pairs of chambers, **235**
of top-nested faces, **233**
Coxeter subalgebra, 346
Coxeter subdispecies, 250
Coxeter submonoid, 127
Coxeter subspecies, 115
Coxeter system, **23**
Coxeter Zie species, **238**, 268, 500
Coxeter-q-Janus algebra, **86**, 207
Coxeter–Birkhoff algebra, **86**, 208
Coxeter–Birkhoff monoid, **85**, 240
Coxeter–Janus algebra, **86**, 205
Coxeter–Janus monoid, **86**
Coxeter–Tits algebra, **86**, 207
Coxeter–Tits monoid, **85**

D
decomposable filtration
of a Coxeter algebra, **386**
of a Coxeter monoid, **162**
decomposable part
of a Coxeter algebra, **386**
of a Coxeter monoid, **162**
deconcatenation coproduct, 164, **168**, 389, **392**, 473, **474**, 483, 484, **490**
decorated q-norm transformation, **684**
decorated anyonic Fock functor, **684**
decorated exponential Coxeter bimonoid, **239**
decorated Fock functors, **682**, 794, 795
decorated Lie element, **244**
decorated norm transformation, **684**
decorated signed exponential Coxeter bimonoid, **239**
decorated Zie element, **244**
decoration functor on Coxeter species, **242**, 683
dequasishuffle coproduct, 483, **485**
derivation
for Coxeter spaces, **388**, 414, 594
for Coxeter species, **163**, 187
descent, **38, 39**, 96
descent identity, 97, 448, 450
deshuffle coproduct, 473, **474**, 483, **487**
diagonal category, 158
diagonal groupoid, 157
dihedral arrangement, **21**
direct sum
of Coxeter (co, bi)algebras, **346**
of Coxeter (co, bi)monoids, **127**
of Coxeter dispecies, **248**
of Coxeter spaces, **323**
of Coxeter species, **114**
divided power Coxeter 0-bialgebra, **466**
divided power Coxeter q-bialgebra, **463**
divided power Coxeter bialgebra, **463**
divided power Coxeter coalgebra, **403**
divided power Coxeter monomial, **462**
divided power signed Coxeter bialgebra, **465**
double algebra axiom, **354**
double category, 157
double monoid axiom, **135**
duality functor
on Coxeter dispecies, **251**
on Coxeter spaces, **323**
on Coxeter species, **116**
Dynkin series, **94**, 98, 453

E
Eckmann–Hilton argument, 135, 157, 354
edge, **21**
edge-type, **24**
elementary algebra, 83
endomorphism Coxeter operad, **253**, **541**, 553
enriched hom, 253, 541
equi-exponential series
in rank one, **294**
of a Coxeter monoid, **197**
eulerian idempotent, **79**, 230
exp-log correspondence
for Coxeter spaces, **413**, **424**
for Coxeter species, **186**, **195**
in rank one, **293**, **632**
exponential
of a map of Coxeter spaces, **413**
of a map of Coxeter species, **186, 187**
exponential Coxeter bimonoid, **176**, **219**
exponential Coxeter dispecies, **256**
exponential Coxeter species, **176**, **219**
exponential Joyal bimonoid, 244
exponential power series, 93, 215, 638, 639
exponential series
in rank one, **294**, **634**
of a commutative Coxeter monoid, 197
of a Coxeter algebra, **427**, 470, 761
of a Coxeter monoid, **197**, 224, 761
of the Coxeter bialgebra of face-types, **504**, 801
of the Coxeter bialgebra of flat-types, **510**, 805
of the Coxeter bimonoid of faces, **232**, 801

of the Coxeter bimonoid of flats, **227**, 805
exponents of a Coxeter system, **50**
exterior signed Coxeter bialgebras, **477**
exterior signed Hopf algebra, 517

F
face, **21**
for type A, **90**
face-incidence algebra, **64**
face-type, **24**
for type A, **91**
face-type enumerator, **49**
fermionic Fock functors, **659**
finite-dimensional Coxeter dispecies, **248**
finite-dimensional Coxeter space, **318**
finite-dimensional Coxeter species, **112**
first eulerian idempotent, **79**
flat, **22**
for type A, **91**
flat-additivity formula, **65**, 66, 489, 592, 596, 598
for type A, **93**
flat-incidence algebra, **63**
flat-type, **26**
for type A, **91**
Fock space, **690**
forgetful functor, 843
formal group, 518, 638, 640
free commutative Coxeter algebra, **396**
free commutative Coxeter monoid, **171**
free Coxeter q-bialgebra, **391**
free Coxeter q-bimonoid, **167**
free Coxeter algebra, **390**
free Coxeter Lie algebra, **572**
free Coxeter Lie monoid, **267**
free Coxeter monoid, **166**
free Coxeter operad, **250**
free Fock functor, **653**
free signed commutative Coxeter algebra, **398**
free signed commutative Coxeter monoid, **173**
free signed Coxeter Lie algebra, **576**
free signed Coxeter Lie monoid, **268**
full Fock functor, **645**, **648**
functor category, 111, 136, 139, 247, 248, 317, 321

G
gallery distance, **35**
Garsia–Reutenauer idempotents, **94**, 98, 638
gate identity, **57**, 92, 97, 409, 462, 485, 495, 507, 516
gate property
for the groupoid of biface-types, **47**, 71, 88, 369, 409, 527
in terms of a minimal gallery, **35**, 57, 95, 97, 125, 150, 182, 230, 367, 512, 652
global descent, **38, 39**, 96, 512
good reflection arrangement, 23
graded vector space, 373
Grothendieck construction, 524, **821**, 825
group coinvariants, 38, 115, 432, 451, 457, 477, 559, 656, 673, 724
group invariants, 38, 115, 432, 451, 457, 477, 657, 673, 723
group-like series
in rank one, **294**, **633**
of a cocommutative Coxeter comonoid, 194
of a Coxeter coalgebra, **422**, 761
of a Coxeter comonoid, **194**, 761
groupoid of biface-types, **41**, 59, 70, 81, 311, 312, 527
for type A, **374**
in rank one, **623**
groupoid of elements, 108, 157, 247, 308, 310, **822**, **824**, 825

H
Hadamard functor, **177**
Hadamard product
of Coxeter dispecies, **252**, 555
of Coxeter spaces, **404**
of Coxeter species, **177**
of graded vector spaces, 452
of Joyal species, 214
half-twist transformation, **667**, 685, 695
Hilbert type series, **701**
HNR isomorphisms
for cocommutative Coxeter bialgebras, **608**
for cocommutative Coxeter bimonoids, **282**
for commutative Coxeter bialgebras, **609**
for commutative Coxeter bimonoids, **282**
for Coxeter q-bialgebras, **611**
for Coxeter q-bimonoids, **284**
Hopf algebra
of graphs, 815
of noncommutative symmetric functions, 454, 518

of permutations, 520
of polynomials, 515
of quasisymmetric functions, 519
of symmetric functions, 518
Hopf operad, 260
Hopf power, 454

I
ideal of a Coxeter operad, **250**
incidence algebra of an operad, 260
inclusion functor, 843
indecomposable part
in rank one, **291**, **629**
of a Coxeter algebra, **387**
of a Coxeter monoid, **162**
indecomposable part functor, **162**, **387**
infinitesimal group, 585
internal hom conditions, **407**
internal hom for the Hadamard product
classical, 453
in rank one, **293**, **632**
of Coxeter coalgebras, **406**, 768
of Coxeter comonoids, **180**, 768
of Coxeter spaces, **404**
of Coxeter species, **178**
invariant q-Janus algebra, **80**, 95, 442, 444, 482, 603, 627
invariant biface-incidence algebra, **70**
invariant bilune-incidence algebra, **71**, 188, 416
invariant Birkhoff algebra, **78**, 95, 441, 444, 507
invariant decorated Lie element, 481, 594
invariant eulerian family, **79**, 591
invariant eulerian idempotent, **80**, 489
invariant family of scalars, **27**, 54, 227, 231, 234, 494–496, 508, 520, 802, 803
invariant flat-incidence algebra, **64**, 187, 199, 227, 469
invariant Janus algebra, **80**
invariant Lie element, 84, 468
invariant lune-incidence algebra, **64**, 186, 194, 231, 412, 423, 469, 502
invariant lune-incidence module, 469
invariant noncommutative Möbius function, **66**, 79, 186, 191, 195, 229, 274, 275, 281, 413, 419, 424, 487, 589, 593, 607
invariant noncommutative zeta function, **66**, 79, 186, 195, 229, 274, 275, 281, 286, 413, 424, 487, 589, 593, 607, 617
invariant support map, **83**, 441
invariant Tits algebra, **79**, 95, 439, 444, 454, 482, 485, 489, 591, 626
invariant Zie element, 84, 443, 500, 594

J
Jacobi identity, **263**, **563–565**, **583**
Janus algebra, **80**, 95
Janus monoid, **23**, 95, 158
Joyal bimonoid
of graphs, 245
of linear orders, 244
of linear set compositions, 244
of linear set partitions, 244
of pairs of linear orders, 244
of partial orders, 245
of set compositions, 244
of set partitions, 244
Joyal set-species, 156
Joyal species, 154
Joyal–Klyachko–Stanley theorem, 258

K
Karoubi envelope, **208**
Karoubi groupoid, **208**
Koszul theory, 560
Krob–Leclerc–Thibon idempotents, **94**, 98, 638

L
lax braiding, 116
lax functor
between comonads, 660, **818**
between monads, 151, 370, **816**
left regular band, **22**
length
of a biface-type, **47**
of a biface-type for type A, **375**
of a Coxeter symmetry, **36**
of a quasishuffle, **94**
Leray–Samelson isomorphisms, **276, 277**, **597, 598**
Leray–Samelson theorem
classical, 297, 638
for bicommutative Coxeter bialgebras, **596, 597**
for bicommutative Coxeter bimonoids, **276**
for signed bicommutative signed Coxeter bialgebras, **600**
in rank one, **296**, **635**
via Fock functors, 780
Lie algebra, 585
Lie bracket
classical, **584**

in rank one, **290**, **627**
of a Coxeter Lie algebra, **564**
of a Coxeter Lie monoid, **263**
Lie coalgebra, 585
Lie cobracket
in rank one, **290**, **628**
of a Coxeter Lie coalgebra, **565**
of a Coxeter Lie comonoid, **263**
Lie element, 84, 258
Lie idempotent, 98
Lie idempotent series, 98, 454
Lie superalgebra, 585
linear hyperplane arrangement, **20**
linear reflection arrangement, **20**
linear series
in rank one, **294**, **634**
of a Coxeter space, **427**, 761
of a Coxeter species, **196**, 761
linearization functor, **116**, 251
local biface, **32**
local biface-type, **33**
locally opposite support-type morphism, **315**, 351
for type A, **375**
Loday–Ronco isomorphisms, **280**, **606**, **607**
classical, 639
Loday–Ronco theorem
classical, 297, 639
for 0-bimonoids, 280
for Coxeter 0-bialgebras, **606**
for Coxeter 0-bimonoids, **279**
in rank one, **296**, **636**
logarithm
of a map of Coxeter spaces, **413**
of a map of Coxeter species, **186**, **187**
of the identity map, **191**, **192**, **419**, 453
logarithmic power series, 93, 215, 638, 639
longest element in a Coxeter group, **36**, 470, 471, 482
longest shuffle, **39**
lune, **29**
for type A, **91**
lune criterion, **553**, **554**, 569, 574
lune decomposition, 553
lune identity, 97
lune of Hippocrates, **96**
lune-additivity formula, **65**, 66, 187, 195, 414
for type A, **93**
lune-incidence algebra, **64**
lune-type, **32**
for type A, **91**

M
Möbius function, **64**, 187, 200
Möbius series, **93**, 98, 215, 298, 453, 638, 639
Mackey functor, 158
mates, **820**
matrix-component, **148**, 366
May cooperad, **557**
May operad, **557**, 560
measuring conditions, **410**
mixed distributive law, **817**
monad on a category, **816**
morphism of bilax functors, 657, 658, 671, 672, 684
multiplicative property
of face-type enumerators, **50**, 427, 462, 464, 469, 734
of shuffles, **60**, 475

N
naturality axiom
for Coxeter algebras, **325**, **328**
for Coxeter coalgebras, **330**
for Coxeter monoids, **118**
for type A, **376**
nested face, **29**, 64
nested face-type, **32**, 52, 64, 527
nested flat, **63**, 247, 523
nested top-lune, 524
noncommutative 0-Möbius function, **75**
noncommutative 0-zeta function, **75**
noncommutative q-Möbius function, **68**, 196, 229, 282, 426, 488, 609
noncommutative q-Weisner formula, **68**, 69, 196, 427, 496, 499
for type A, **94**
noncommutative q-Zaslavsky formula, **76**, 98, 470
noncommutative q-zeta function, **68**, 196, 229, 282, 426, 488, 609
noncommutative Möbius function, **65**
noncommutative Möbius inversion, **469**, 498, 508
noncommutative Weisner formula, **65**, 67, 187, 195, 415, 499
for type A, **93**
noncommutative Zaslavsky formula, **76**, 98, 213, 450, 469, 504
noncommutative zeta function, **65**
norm map
between tensors and shuffles, **475**
on face-types, **499**, 520
on flat-types, **508**, 520, 805

on flats, **227**, 805
on polynomials, **463**
norm transformation, **671**
number operator, **765**

O
o-signature functor
in rank one, **292**
on Coxeter species, **138**, 175, 211, 221, 264, 271, 653, 659, 673, 745
on species, 157
one-object category, 40
op and cop constructions, **154**, 211, 215, **373**, 446, 695, 778
operad, **251**
opposite Coxeter algebra, **373**
opposite Coxeter coalgebra, **373**
opposite Coxeter comonoid, **154**
opposite Coxeter monoid, **154**
opposite face, **28**
opposite lune, **30**
opposite support-type morphism, **315**
for type A, **375**
opposite support-type relation, **314**
opposite transformation
for Coxeter spaces, **372**, 663
for Coxeter species, **153**, 663
opposition cocommutativity axiom
for Coxeter coalgebras, **353**
for Coxeter comonoids, **132**
opposition cocommutativity relation, **363, 364**
opposition commutativity axiom
for Coxeter algebras, **349**, **351, 352**
for Coxeter monoids, **129**, **131**

P
panel, **21**
partition, **91**, 374
permutation bi-ring, 245
Poincaré polynomial, **50**, 88, 92, 96, 198, 199, 464
Poincaré–Birkhoff–Witt isomorphism, **285, 286**, **615**, **618**
classical, 639
Poincaré–Birkhoff–Witt theorem
classical, 298, 639
for Coxeter Lie algebras, **615**
for Coxeter Lie coalgebras, **618**
for Coxeter Lie comonoids, **286**
for Coxeter Lie monoids, **285**
for signed Coxeter Lie algebras, **619**
for signed Coxeter Lie coalgebras, **620**
in rank one, **296**, **637**
pointed reflection arrangement, **21**
positive Coxeter operad, 250
positive Joyal species, 557
positive-graded q-bialgebra, **378**
positive-graded algebra, **378**
positive-graded coalgebra, **378**
positive-graded Lie algebra, **584**
positive-graded vector space, **378**
power map, 454
presheaves on posets, 381
primitive filtration
of a Coxeter coalgebra, **386**
of a Coxeter comonoid, **161**
primitive part
in rank one, **291**, **629**
of a Coxeter coalgebra, **385**
of a Coxeter comonoid, **161**
primitive part functor, **161**, 163, 270, **385**
primitive series
in rank one, **294**, **633**
of a cocommutative Coxeter comonoid, 194
of a Coxeter coalgebra, **422**, 761
of a Coxeter comonoid, **194**, 761
primitively generated Coxeter bialgebra, **593**
projective invariant lune-incidence algebra, **70**
projective lune-incidence algebra, **70**

Q
quadratic Coxeter operad, **251**
quadratic duality, 260
quasishuffle for type A, 92, 452
quasishuffle product, 483, **490**
for type A, 452, 519

R
random walk, 97
reference chamber, **21**
reflective subcategory, 174, 400, 724, 739, 743, 758
regular band, **23**
retract, 617
rigidity theorem
for Coxeter q-bialgebras, **601**
for Coxeter q-bimonoids, **278**

S
Schubert statistic, 685
self-dual

Coxeter q-bialgebra, **347**
Coxeter q-bimonoid, **127**
morphism of Coxeter q-bialgebras, **347**
morphism of Coxeter q-bimonoids, **127**
semidirect product category, **823**
semidirect product group, 34, **824**, 825
semidirect product groupoid, 136, 146, **824**, 825
semidirect product monoid, 85, **824**, 825
series
of a Coxeter space, **422**
of a Coxeter species, **193**
of a species, 194
set composition, **90**
set partition, **91**
set-theoretic noncommutative zeta function, 297
shuffle, 39, 46, 60, 83, 96, 343, 685
for the coordinate arrangement, **88**
for type A, 92, 96, 214
shuffle Coxeter 0-bialgebra, **476**
shuffle Coxeter q-bialgebra, **475**
shuffle Coxeter bialgebra, **474**, 594, 622
shuffle Hopf algebra, 517
shuffle product, 473, **474**, 483, **490**
shuffle signed Coxeter bialgebra, **476**
sign representation, **37**, 219, 477, 659
signature functor on Coxeter dispecies, **259**, **555**
signature space, **134**, 152, 220, 259
signed abelianization, **152**, **175**, **371**, **402**
signed abelianization map, **477**, 619
signed antisymmetry, **566**
signed bialgebra axiom, **341**, **344**, 466
in rank one, **624**
signed bicommutative signed Coxeter bialgebra, **354**
signed bicommutative signed Coxeter bimonoid, **134**
signed bimonoid axiom, **125**
signed coabelianization, **152**, **175**, **371**, **402**
signed coabelianization map, **477**, 619
signed cocommutative Coxeter coalgebra, **354**
signed cocommutative Coxeter comonoid, **134**
signed cocommutativity axiom, **134**, **354**
signed cocommutator cobracket, **265**, **568**
signed cocommutator functor, **568**
signed commutative Coxeter algebra, **354**, 553, 557
signed commutative Coxeter monoid, **133**
signed commutativity axiom, **133**, **354**, 557
signed commutator bracket, **265**, **568**
signed commutator functor, **567**
signed Coxeter $\boldsymbol{\alpha}$-bialgebra, **344**
signed Coxeter $\mathcal{A}$-bialgebra, **341**
signed Coxeter bialgebra of dual numbers, **465**, 788
signed Coxeter bialgebra of polynomials, **465**
signed Coxeter bimonoid, **125**
in rank one, **289**
signed Coxeter commutative operad, **259**, 553, **555**
signed Coxeter Lie algebra, **565**
signed Coxeter Lie coalgebra, **566**
signed Coxeter Lie comonoid, **264**
signed Coxeter Lie cooperad, **259**, 264, 566
signed Coxeter Lie monoid, **263**
signed Coxeter Lie operad, **259**, 263, **556**, 565
signed decoration functor on Coxeter species, **243**, 684
signed deshuffle coproduct, 473, **476**
signed distance function, **36**
signed exponential Coxeter bimonoid, **219**
signed exponential Coxeter species, 219
signed full Fock functors, **653**
signed Jacobi identity, **566**
signed norm map, **465**, **476**
signed norm transformation, **672**
signed opposite transformation
for Coxeter spaces, **372**
for Coxeter species, **153**
signed partner, **259**, **555**
signed positive-graded Lie algebra, 585
signed shuffle product, 473, **476**
signed trivialization functor, **745**, 757
skeletal category, 43, 108, 309
skeleton of a groupoid, 43, 108
slack of a lune, **30**
Solomon coefficient, **77**, 98, 500, 554, 574, 616, 640
Solomon descent algebra, **83**, 92, 95, 454
Solomon operator
classical, 298, 640
Coxeter Lie algebra version, **616**
Coxeter Lie coalgebra version, **618**
Coxeter Lie comonoid version, **286**

Coxeter Lie monoid version, **285**
in rank one, **296**, **637**
special Zie element, **84**, 98
species, **135**
split-semisimple algebra, 78, 79
star of a face, **23**
stationary distribution, 97
stochastic integral, 452
strong monoidal functor, 239, 251, 404, 558, 559
structure constants
of the invariant q-Janus algebra, 82
of the invariant Tits algebra, 80
substitution map of a Coxeter operad, **250**, 522
substitution product
of Coxeter dispecies, **249**, 531
of Coxeter set-dispecies, 251
support
of a face, **22**
of a face-type, **26**
of a local biface, **32**
of a local biface-type, **33**
of a nested face, **29**
of a nested face-type, **32**
support map, **83**, 233
support morphism, **104**, **305**
support-naturality axiom
for Coxeter algebras, **325**
for Coxeter coalgebras, **330**
for Coxeter comonoids, **121**
for Coxeter monoids, **118**
support-support axiom, **112**, **318**
support-support relation, **105**, **141**, **305**, **360**
support-type axiom, **112**, **318**
support-type morphism, **106**, **306**
support-type of a face, **27**
support-type relation, **105**, **305**
support-unitality axiom, **112**, **318**
support-unitality relation, **105**, **305**
Sweedler power, 454
symmetric Coxeter bialgebras, **477**
symmetric group, 90, 156, 245, 451, 558
symmetric Hopf algebra, 517

T
Takeuchi formula
for connected Hopf algebras, 455
for connected Joyal bimonoids, 215
for Coxeter bialgebras, **445**
for Coxeter bimonoids, **210**
Takeuchi series
of the Coxeter bialgebra of face-types, **503**, 801
of the Coxeter bimonoid of faces, **232**, 801
tensor q-Hopf algebra, 517
tensor Coxeter 0-bialgebra, **476**
tensor Coxeter q-bialgebra, **474**
tensor Coxeter bialgebra, **474**, 576, 594, 622, 794
tensor Coxeter space, **473**
tensor Hopf algebra, 454, 517
tensor signed Coxeter bialgebra, **476**
Tits algebra, **79**, 95
Tits monoid, **21**, 95
Tits product for type A, **91**
top-lune, **29**
top-lune-type, 33, 309
top-nested face, **29**
top-nested face-type, 33
top-star of a face, **23**
transferred morphism, **308**
trivial Coxeter algebra, **387**
trivial Coxeter algebra functor, **387**, 394, 398
trivial Coxeter coalgebra, **387**
trivial Coxeter coalgebra functor, **387**, 391, 396
trivial Coxeter comonoid, **163**
trivial Coxeter comonoid functor, **163**, 167, 171
trivial Coxeter monoid, **163**
trivial Coxeter monoid functor, **163**, 169, 172
trivial representation, **37**, 461, 485, 508
trivialization functor, **718**
twisted algebra, 157
twisted Hopf algebra, 157
twisted Lie algebra, 272
two-sided 0-Möbius function, **75**
two-sided 0-zeta function, **75**
two-sided q-flat-additivity formula, **72**, 203, 434, 601–603
two-sided q-lune-additivity formula, **72**, 190, 417
two-sided q-Möbius function, **72**, 189, 277, 282, 416, 600, 609
two-sided q-Möbius series, **94**, 298, 639
two-sided q-Weisner formula, **72**, 190, 417
two-sided q-zeta function, **72**, 189, 277, 282, 416, 600, 609
two-sided q-zeta series, **94**, 298, 639
two-sided characteristic operation
by a biface, **205**, 206
by a biface-type, **442**, 602
type
of a biface, **28**

of a bilune, **33**
of a face, **24**
of a flat, **26**
of a local biface, **33**
of a lune, **32**
of a nested face, **32**
type map, **24**
type morphism, **104**, **108**, **305**, **308**
type series, **701**
type-naturality axiom
for commutative Coxeter monoids, **130**
for Coxeter algebras, **325**, **327**
for Coxeter coalgebras, **330**
for Coxeter comonoids, **121**
for Coxeter monoids, **118**, **120**
type-type axiom, **112, 113**, **318, 319**
type-type relation, **105**, **141**, **305**, **360**
type-unitality axiom, **112, 113**, **318, 319**
type-unitality relation, **105**, **305**

U
u-signature functor
in rank one, **292**, **631**
on Coxeter spaces, **357**, 398, 402, 431, 556, 566, 582
on Coxeter species, **138**, 173, 175, 201, 211, 221, 264, 271
on species, 157
uniform noncommutative zeta function, 297, 298, 639
unitality axiom
for commutative Coxeter monoids, **130**
for Coxeter algebras, **325**, **327, 328**, 550, 551
for Coxeter monoids, **118**, **120**
for type A, **376**
universal coenveloping Coxeter coalgebra, **580**
universal coenveloping Coxeter comonoid, **270**
universal enveloping algebra, 585
universal enveloping Coxeter algebra, **577**
in rank one, **634**
universal enveloping Coxeter monoid, **269**
in rank one, **295**
universal measuring coalgebra, 453
universal measuring Coxeter coalgebra, **410**, 769
in rank one, **632**
universal measuring Coxeter comonoid, **183**, 769
in rank one, **293**
universal series
of the Coxeter bialgebra of face-types, **503**, 801
of the Coxeter bialgebra of flat-types, **509**, 805
of the Coxeter bimonoid of faces, **232**, 801
of the Coxeter bimonoid of flats, **227**, 805
upside-down cocommutativity axiom, **353**
upside-down commutativity axiom, **350**, 380, 552

V
vapraka, **96**
Varchenko factorization theorem, 62
Varchenko matrix, **55**, 74, 89, 97, 434, 496
vertex, **21**
vertex-type, **24**

W
weak left Bruhat order, **37**, 511
Weisner formula, **64**, 188, 200
Witt identity, 97, 98

Y
Yang–Baxter operator, 381

Z
Zaslavsky formula, **76**, 98
zero Coxeter (co, bi)algebra, **346**
zero Coxeter (co, bi)monoid, **127**
zero Coxeter dispecies, **248**
zero Coxeter space, **322**
zero Coxeter species, **114**
zeta function, **64**, 187, 200
zeta series, **93**, 98, 215, 297, 298, 453, 638, 639
Zie element, 84
Zie series, **94**, 98, 454, 519